OXFORD MEDICAL PUBLICATIONS

OXFORD TEXTBOOK OF SPORTS MEDICINE

EDITORS

MARK HARRIES

Consultant Physician,
Northwick Park Hospital,
and Clinical Director,
British Olympic Medical Centre

CLYDE WILLIAMS

Professor of Sports Science,
Loughborough University

WILLIAM D. STANISH

Professor of Surgery,
Dalhousie University, Halifax,
Nova Scotia, Canada

LYLE J. MICHELI

Director, Divison of Sports Medicine,
Children's Hospital, Boston, MA, USA

OXFORD TEXTBOOK OF SPORTS MEDICINE

Edited by

MARK HARRIES, CLYDE WILLIAMS, WILLIAM D. STANISH, LYLE J. MICHELI

New York   Oxford   Tokyo
OXFORD UNIVERSITY PRESS
1996

Oxford University Press, Walton Street, Oxford OX2 6DP

Oxford   New York   Toronto
Delhi   Bombay   Calcutta   Madras   Karachi
Kuala Lumpur   Singapore   Hong Kong   Tokyo
Nairobi   Dar es Salaam   Cape Town
Melbourne   Auckland   Madrid

and associated companies in
Berlin   Ibadan

Oxford is a trade mark of Oxford University Press

Published in the United States
by Oxford University Press Inc., New York

Reprinted 1995
New as Paperback 1995

A catalogue record for this book is available from the British Library

Library of Congress Cataloging in Publication Data
Oxford textbook of sports medicine / edited by
Mark Harries ... [et al.].
p.   cm. — (Oxford Medical publications)
Includes bibliographical references.
1. Sports medicine.   I. Harries, Mark.   II. Series.
[DNLM: 1. Sports Medicine.   QT 260 098 1994]
RC1210.096   1994   617.1'027—dc20   93–23591
ISBN 0 19 262009 6 Hbk
ISBN 0 19 262010 X Pbk

Typeset by Create, Bath
Printed in Great Britain by
Butler & Tanner Ltd, Frome

Foreword

There are several reasons why the standard of sports medicine has lagged behind that of general medicine. First, when clinics are overcrowded, neither the consultant nor other patients may relish the notion of an athlete, injured, as they might see it, as a result of voluntary activity, being granted priority for urgent treatment. Second, since sports medicine cannot be easily defined or circumscribed, confusion arises which lowers standards of care. Sports medicine encompasses cardiology, respiratory medicine, orthopaedic surgery, traumatology and many other specialties. For years the existence of clinics for sports medicine in Britain has depended too much on the individual enthusiasm of dedicated specialists, who were prepared to create, often in their own time, clinics adjusted to suit the needs of athletes. I take the view that general medicine and surgery have a duty to learn from study of the diverse range of acute sports injury, so that the lessons can be applied to the management of traumatic injuries which also occur in other circumstances. Moreover, sport usually provides for the sports physician an enthusiastic subject, eager to regain his capacity to compete as soon as possible, which is not always true of other patients. Sports medicine specialists can also advise governing bodies of sport on the best ways to organize complex competitive rules in order to reduce the possibility of injuries.

For all these reasons the editors are to be praised for having compiled this much-needed comprehensive textbook of sports medicine. I hope it will also encourage examining authorities in medicine and surgery to include questions related to sports medicine in finals papers and higher examinations as a spur to raising standards of diagnosis and treatment. With some 25 per cent of the population, including children and students, now regularly taking part in sport, the problem of treatment of injuries can no longer be regarded as a 'Cinderella' area of medicine.

Sir Roger Bannister

Preface

We are all encouraged to take exercise as one contribution to a healthy life-style. However, participation in exercise and sport carries with it the risk of injury. Injuries occurring in the course of recreational activity are treated in exactly the same way as those sustained in the work-place or during the daily round of domestic activities. Why then sports medicine?

Sports medicine has evolved over the last 50 years from a core activity of treating injuries to one which now uses a multidisciplinary approach to the care of those injured whilst participating in sport. The rationale for this approach (and the legitimate claim on the title 'sports medicine') is that those who look after the injured are professionally obliged to offer advice on how injuries can be both treated and avoided.

Practitioners in all branches of sports medicine must now be well informed about those activities which have the potential to lead to injury. Understanding the physicial and physiological demands that heavy and sustained activity places on participants in sport, whatever their age, requires a special knowledge of the adaptive responses to exercise.

In writing the *Oxford Textbook of Sports Medicine* we have included sections on environmental sciences as well as aspects of basic physiology, psychology, and biomechanics. While coverage is comprehensive, we have not included highly specialized subjects such as the fine details of surgical procedures. Each chapter offers the experience and perspective of its author in a way which informs; often the approach challenges the basis on which cherished beliefs have been built. We have tried above all to produce a concise reference work that will support all those involved in the broad spectrum of sports medicine

To our authors we offer our thanks and trust that despite all, we have managed to remain friends. Our thanks also go to the staff of Oxford University Press for their help and guidance.

Contents

Contributors

JOHN ANNETT
  Professor of Psychology, University of Warwick, Coventry, UK
PER-OLOF ÅSTRAND
  Professor Emeritus, Physiology, Karolinska Institute, Stockholm, Sweden
DARREN W. BOOTH
  Consultant Physiotherapist, Department of Athletics, Acadia University, Wolfville, Nova Scotia, Canada
MARK K. BOWEN
  Assistant Professor of Orthopedic Surgery, Northwestern University Medical School, Chicago, Illinois, USA
OWEN H. BRADY
  Orthopaedic Registrar, St Mary's Orthopaedic Hospital, Cappagh, Dublin, Ireland
JEAN-LOUIS BRIARD
  Chirurgie Orthopédique et Réparatrice, Traumatologie Clinique du Cèdre, Bois Guillaume, France
ROBERT M. BROCK
  Consultant Orthopaedic Surgeon, North York General Hospital, Willowdale, Ontario, Canada
GUGLIELMO CERULLO
  Orthopaedic Surgeon, Clinica Valle Giulia, Rome, Italy
PAUL S. COOPER
  Instructor in Orthopedic Surgery, Pennsylvania State University, University Park, USA
DAVID A. COWAN
  Director, Drug Control Centre, King's College London, University of London, UK
JAY S. COX
  Professor of Orthopedic Surgery, Pennsylvania State University, University Park, USA
SANDRA L. CURWIN
  Assistant Professor, School of Physiotherapy, Dalhousie University, Halifax, Nova Scotia, Canada
FOSCO DE PAULIS
  Chief, Computed Tomography Service, Santa Maria di Collemaggio Hospital and Professor of Anthropology and Anthropometry, Superior Institute of Sports Education, L'Aquila, Italy
JADE E. DILLON
  Associate Chief of Ambulatory Care Services, Canadian Forces Base, Halifax, Nova Scotia, Canada
DAVID H. ELLIOTT
  Shell Professor in Occupational Health, The Robens Institute of Health and Safety, University of Surrey, UK
EJNAR ERIKSSON
  Professor of Sports Medicine; Head, Department of Sports Orthopaedic Surgery, Karolinska Hospital, Stockholm, Sweden
MALCOLM A. FERGUSON-SMITH
  Professor of Pathology, University of Cambridge and Director, East Anglian Regional Genetics Service, Addenbrooke's Hospital, Cambridge, UK
PETER A. FRICKER
  Director of Medical Services, Australian Institute of Sport, Canberra, Australia
J. ROBERT GIFFIN
  Research Fellow, Orthopaedic and Sport Medicine Clinic of Nova Scotia, Dalhousie University, Halifax, Canada

FRANK GOLDEN
  Surgeon Rear Admiral, Royal Navy, RN Hospital, Gosport, Hampshire, UK
SALEEM K. GOOLAMALI
  Consultant Dermatologist, Northwick Park Hospital, Harrow, Middlesex, UK
G. F. GOUBRAN
  Consultant Oral and Maxillo-Facial Surgeon, Central Middlesex and Ealing Hospital, London, and Clementine Churchill Hospital, Harrow, Middlesex, UK
ANN C. GRANDJEAN
  Director, International Center for Sports Nutrition, Omaha, Nebraska, USA
GARRY GREENFIELD
  Fellow in Orthopaedic Surgery, Victoria General Hospital, Halifax, Nova Scotia, Canada
ASHLEY GROSSMAN
  Reader in Neuroendocrinology, St Bartholomew's Hospital, London
R. DONALD HAGAN
  Senior Scientist, Naval Health Research Center, San Diego, California, USA
LEW HARDY
  Senior Lecturer in Sport Psychology, University of Wales in Bangor, UK
MARK HARRIES
  Consultant Physician, Northwick Park Hospital; Clinical Director (Hon.), British Olympic Medical Centre, Northwick Park Hospital, Harrow, Middlesex, UK
JAN HENRIKSSON
  Professor of Physiology, Karolinska Institute, Stockholm, Sweden
ROBERT C. HICKNER
  Karolinska Institute, Stockholm, Sweden
TREVOR A. HOWLETT
  Consultant Endocrinologist, Leicester Royal Infirmary, UK
BRIAN J. HURSON
  Consultant Orthopaedic Surgeon, St Vincent's Hospital, Dublin, Ireland
J. C. HYNDMAN
  Head, Department of Orthopaedic Surgery, Izaak Walton Killam Hospital for Children, Halifax, Nova Scotia, Canada
ROBERT J. JOHNSON
  Professor of Orthopedic Surgery and Head of Division of Sports Medicine, Department of Orthopedics and Rehabilitation, University of Vermont College of Medicine, Burlington, USA
PETER R. M. JONES
  Professor of Human Functional Anatomy, Department of Human Sciences, University of Loughborough, Leicestershire, UK
GRAHAM JONES
  Senior Lecturer in Sport Psychology, Loughborough University, Leicestershire, UK
HENRYK K. A. LAKOMY
  Lecturer, Loughborough University, Leicestershire, UK
ANGUS M. McBRYDE Jr
  Professor and Chairman, Department of Orthopedic Surgery, University of South Alabama, Mobile, USA

GARY R. McGILLIVARY
Lecturer, Division of Orthopaedic Surgery, Department of Surgery, Dalhousie University, Halifax, Nova Scotia, Canada

TERRY R. MALONE
Director and Associate Professor of Physical Therapy, University of Kentucky, Lexington, USA

R. J. MAUGHAN
Senior Lecturer, University Medical School, Aberdeen, UK

DARRELL MENARD
Health Promotion Coordinator, Directorate of Health Protection and Promotion, Ottawa, Canada

LYLE J. MICHELI
Director, Division of Sports Medicine, Children's Hospital, Boston, Massachusetts, USA

JAMES S. MILLEDGE
Consultant Physician, Northwick Park Hospital, Harrow, Middlesex, UK

JEFFREY MINKOFF
Clinical Professor of Orthopedics, New York University Medical Center; Attending Orthopedic Surgeon, Lenox Hill Hospital and the Hospital for Joint Disease Orthopedic Institute, New York, USA

MICHAEL F. MURPHY
Director of Emergency Medicine, Isaak Walton Killam Children's Hospital; Assistant Professor of Anaesthesia, Dalhousie University, Halifax, Nova Scotia, Canada

BENNO M. NIGG
Professor of Biomechanics, University of Calgary, Alberta, Canada

N. G. NORGAN
Senior lecturer in Applied Human Physiology, Department of Human Sciences, University of Loughborough, Leicestershire, UK

ROBIN J. NORTHCOTE
Consultant Cardiologist, Victoria Infirmary, Glasgow, UK

PATRICK O'GRADY
General Practitioner, Parrsboro, Nova Scotia, Canada

BARRY W. OAKES
Senior Lecturer, Department of Anatomy, Monash University; Director, The Sports Medicine Centres of Victoria, Melbourne, Australia

GEOFFREY PASVOL
Professor of Infectious Diseases and Tropical Medicine, St Mary's Hospital Medical School, Imperial College, London

GIANCARLO PUDDU
Orthopaedic Surgeon, Clinical Valle Giulia, Rome, Italy

PETER B. RAVEN
Professor of Physiology, Texas College of Osteopathic Medicine, Fort Worth, USA

JONATHAN REEVE
MRC External Scientific Staff, Department of Medicine, Addenbrooke's Hospital, Cambridge, UK

DAVID C. REID
Professor of Orthopaedic Surgery, Adjunct Professor of Rehabilitation Medicine, Honorary Professor of Physical Education, University of Alberta, Edmonton, Canada

THOMAS REILLY
Professor of Sports Science, School of Human Sciences, Liverpool John Moore's University, Liverpool, UK

PER A. F. H. RENSTRÖM
Professor of Sports Medicine, Department of Orthopedics and Rehabilitation, University of Vermont, Burlington, USA

R. MITCHELL RUBINOVICH
Associate Professor, Department of Surgery, McGill University, Montreal, Quebec, Canada

JAIME S. RUUD
Sports Nutrition Consultant, International Center for Sports Nutrition, Omaha, Nebraska, USA

KENT SAHLIN
Associate Professor, Department of Physiology III, Karolinska Institute and Department of Sport and Health Sciences, University of Sports, Stockholm, Sweden

MICHAEL L. SCHWARTZ
Associate Professor of Surgery, University of Toronto; Staff Neurosurgeon and Director of the Neurotrauma Program, Sunnybrook Health Science Centre, Toronto, Ontario, Canada

ALBERTO SELVANETTI
Sports Medicine, Clinica Valle Giulia, Rome, Italy

LEONARD M. SHAPIRO
Consultant Cardiologist, Papworth and Addenbrooke's Hospitals, Cambridge, UK

BARRY G. SIMONSON
Orthopedic Attending Surgeon, South Nassau Communities Hospital, Oceanside, New York, USA

BERTIL SJÖDIN
Senior Research Officer, The National Research Establishment, Stockholm, Sweden

WILLIAM D. STANISH
Professor of Surgery, Dalhousie University, Halifax, Nova Scotia, Canada

JOHN R. SUTTON
Professor of Medicine and Head, Department of Biological Science, Faculty of Health Science, University of Sydney, Australia

JAN SVEDENHAG
Consultant Physician, Department of Clinical Physiology, Huddinge University Hospital, Stockholm, Sweden

CHARLES H. TATOR
Professor and Chairman, Division of Neurosurgery, University of Toronto; Program Director, Toronto Hospital Neurological Centre; Director, Canadian Paraplegic Association Spinal Cord Injury Research Laboratory, Toronto, Ontario, Canada

JACK TAUNTON
Professor and Codirector, Allan McGavin Sports Medicine Centre, University of British Columbia, Vancouver, Canada

MICHAEL J. TIPTON
Senior Research Officer, Robens Institute, University of Surrey, Guildford, UK

NANCY E. VINCENT
Orthopaedic Surgery and Sports Medicine Consultant, Lethbridge, Alberta, Canada

SUSAN A. WARD
Professor of Anesthesiology and Physiology, UCLA School of Medicine, Los Angeles, California, USA

RUSSELL F. WARREN
Professor of Orthopedic Surgery, Cornell Medical College; Director, Sports Medicine/Shoulder Service, Hospital for Special Surgery, New York, USA

BRIAN J. WHIPP
Professor of Physiology, St George's Hospital Medical School, London

CLYDE WILLIAMS
  Professor of Sports Science; Head of Department of Physical Education, Sports Science, and Recreation Management, Loughborough University, Leicestershire, UK

ROGER L. WOLMAN
  Senior Registrar, Royal National Orthopaedic Hospital, Stanmore, Middlesex, UK

ROBERT M. WOOD
  Resident in Orthopaedic Surgery, University of Alberta, Edmonton, Canada

ROBERT A. YANCEY
  Consultant Physician, Mary Bridge Children's Hospital, Tacoma, Washington, USA

Introduction—Man as an athlete*

PER-OLOF ÅSTRAND

The purpose of this introductory chapter on man as an athlete is to put the textbook in a broad historical perspective for readers.

A brief sketch of our evolutionary history is presented to remind us that it has taken a long time for us to become the way we are. The history of competitive sports is summarized, with special emphasis on the ancient and modern Olympic Games, and the reasons for the improvement in world records during this century are discussed with examples mainly from track and field events.

OUR BIOLOGICAL HERITAGE

According to the natural sciences it is believed that our solar system was created some 4600 million years ago.[1] Evidently, the atmosphere surrounding our planet at that time did not contain oxygen. This was a prerequisite for the evolution of life from non-living organic matter. Without atmospheric oxygen there was no high altitude ozone, and hence ultraviolet radiation from the Sun reached the surface of the Earth. This radiation then provided the energy for the photosynthesis of organic compounds from such molecules as water, carbon dioxide, and ammonia. The process that enabled living organisms to capture solar energy for the synthesis of organic molecules (e.g. glucose) can be clearly traced in fossils that are about 3500 million years old. Similarly, the familiar anaerobic fermentation (i.e. glycolysis) is probably the oldest energy-extracting pathway found in life on Earth.

The ancient organisms split the water molecule by photosynthesis, gradually releasing free oxygen into the atmosphere. It may have taken some 2000 million years to create an atmosphere in which one out of every five molecules was oxygen. As oxygen became toxic for many of the original oxygen producers, new metabolic patterns (i.e. aerobic energy yield) were developed that utilized oxygen as a hydrogen acceptor.

A new milestone in the biological evolution was reached approximately 1500 million years ago when a unicellular organism with a nucleus, the eukaryote, developed.[2] The energy-absorbing and energy-yielding processes typical of our cell activities today, such as the ATP–ADP system, are merely copies of events that occurred thousands of millions of years ago. ATP is the principal medium for the storage and exchange of energy in almost all living organisms. However, the store of ATP is very limited because it is a heavy fuel. Within a period of 24 h, an individual will use energy equivalent to the energy stored in ATP weighing 50–100 per cent more than his or her own body weight, depending on how physically active he or she is. Therefore, very rapid resynthesis of ATP is necessary, and the anaerobic processes, which are several thousand million years old, are supplemented by the aerobic energy yield taking place inside the mitochondria.

Thus, over thousands of millions of years of evolution, a unicellular living organism was created. By some sort of trial and error, the fundamental biological principles for maintaining life were developed and they are still in efficient operation. A comprehensive textbook of biochemistry written some 1500 million years ago would no doubt still be up to date in its treatment of the functions of the cell.

Evolution was now ready for the next major step—the creation of larger animals. That stage probably began 700 million years ago.[3] In this evolution of larger animals, the individual cell retained its original size (i.e. the same size as the unicellular organism living more than 1000 million years ago) but more cells were grouped together to increase the size of the organism.

As an inevitable consequence of grouping thousands of millions of some 200 different types of cells together in one organism (the human being), the individual cell lost its intimate contact with the external environment. This problem was solved by bathing each cell in water (i.e. the interstitial fluid). Like the amoeba, each cell in our body (with some exceptions) is surrounded by fluid, the composition of which is basically very similar to that of the ancient oceans. The organism brought the sea water with it, so to speak, in a bag made of skin.

During the diversification of the multicellular organisms, which occurred over the last 700 million years, new types of organisms appeared and dispersions took place within groups which were already established. It should be noted that the history of mammals covers the last 220 million years, if not longer. The first primates (the order including man) can be traced back some 60 to 70 million years to a period when the dinosaurs still dominated the scene. With the extinction of the dinosaurs, there was a mammalian dispersion into vacant niches. Another evolutionary explosion occurred, with a dispersion of flowering plants, birds, and mammals.

What then are the mechanisms that underlie the origin of species and the evolutionary relationships among them (i.e. Darwinism)? Lewin[4] has summarized the current views held by different researchers in this field. According to modern ideas, evolution is a consequence of the gradual accumulation of genetic differences due to point mutations and rearrangements in the chromosomes. The direction of an evolutionary change is determined by natural selection, promoting the variants that are best fitted to their environment. However, the fact remains that fossils do not generally document a smooth transition from old morphologies to new ones. This was also discussed by Darwin. For millions of years species remain unchanged in the fossil record, suddenly to be replaced by something that is substantially different but clearly related.[4]

Because the accumulation of small genetic changes cannot exclusively explain the development of new species, a new theory called punctuated equilibrium has been advanced. According to this theory, individual species may remain virtually unchanged for long periods of time. Then they are suddenly punctuated by abrupt events in the environment and a new species arises from the original stock. It is conceivable, however, that future fossil records may fill many of the gaps and provide some of the missing links. It may have been only 5 million or as many as 20 million years ago that the family tree of primates developed a branch, the hominids, which finally resulted in *Homo sapiens sapiens*, the only surviving hominid. Not until about 4 million years ago do the African fossils

*Parts of this chapter have been published earlier (see Åstrand P-O, Rodahl K. *Textbook of Work Physiology*. 3rd edn. New York: McGraw-Hill, 1986; ch. 1).

reveal the presence of the hominid genus *Australopithecus*. The pelvis permitted an upright posture and bipedal gait with the arms free. There are archaeological records of tools, pebble choppers, and small stones that are probably more than 3 million years old.[5] Tool-making was thus established before there was a marked brain expansion in the hominid stock. Although a few varieties have been identified, the *Australopithecus* was a relatively homogeneous genus that survived for more than 2 million years. The next well-identified member of our family tree may have been the first true man. *Homo habilis* existed from 2.3 to 1.5 million years ago. He was replaced by *Homo erectus*, who had a modern pelvis and moved with a striding gait. *Homo erectus* lived as hunters and food gatherers and had a wide geographical range. Their body height was probably 150 to 160 cm. They made use of fire, as evidenced by a hominid occupation site 1.4 million years old.

The general public is probably most familiar with Neanderthal man (*Homo sapiens neanderthalensis*) who, from archaeological findings, appears to have been well established some 200 000 years ago.[6] Neanderthals were skilled hunters of large and small game, forming bands similar to those of more recent hunting people, and were probably linked into tribal groupings, or at least groups with a common language. They formed a human population complex extending from Gibraltar across Europe into East Asia. The Neanderthal population was as homogeneous as the human population of today. On average, the brain encased in the Neanderthal skull was slightly larger than the brain of modern man. Although the Neanderthals had the same postural abilities, manual dexterity, and range and character of movement as modern man, they had more massive limb bones and a larger muscular mass and power. The departure of the Neanderthals occurred some 35 000 years ago. When they disappeared from the scene anatomically, modern man, *Homo sapiens sapiens*, was already in existence.

There are different opinions concerning '. . . the latest phase in human origins—the emergence of people like you and me, our species *Homo sapiens*, with its widespread varieties of physique and color'.[6] One hypothesis is that modern humans evolved in Africa and then spread throughout the world, developing racial features in the process. Modern humans and Neanderthals could be distinct lines that diverged from a common ancestor more than 200 000 years ago in Africa and Europe respectively. At a later stage they spread and in some parts of the world they shared the environment.

An alternative hypothesis is a 'gene-flow' model with a genetic contribution varying from region to region, with the rate of intermixture gradually increasing as modern man evolved. Stringer[6] points out that in the gene-flow model racial features preceded the appearance of modern man, whereas the African model reverses the order. He supports the African model with dispersal of early modern humans from Africa within the past 100 000 years. However, the dating of our origin as modern man is controversial.

A human being living 50 000 years ago probably had the same potential for physical and intellectual performance, playing a piano or constructing a computer, as anyone living today. Therefore, from all indications, *Homo sapiens sapiens* has remained biologically unchanged for at least 50 000 years. By 30 000 years ago, modern man had spread to nearly all parts of the world. It was not

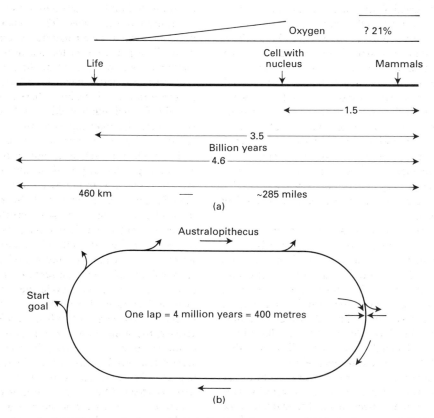

Fig. 1 (a) The history (approximately 4.6 billion years) of our planet is illustrated as proportional to a journey of 460 km (285 miles). After some 300 km (186 miles) we find a eukaryote—a cell with a nucleus. A textbook dealing with the structure and metabolism of that cell probably would not need much updating to describe the basic histochemistry of our cells. (b) The history of the hominids, starting with *Australopithecus*, covers the last 400 m (see text). Farming and agriculture started after 399 m (10 000 years ago). *Homo sapiens neanderthaliensis* died out at 3.5 m (35 000 years ago). The 20th century covers the last 10 mm.

until some 10 000 years ago that the transition from a roaming hunter and food gatherer to a stationary farmer began.

To illustrate the evolutionary time-scale, let us compare the 4600 million years that our planet has existed with a 460-km journey (Fig. 1). Life began after the first 100 km of the trip had been covered. Another 200 km was required before the unicellular organism with a nucleus was born. Multicellular animals were living when we arrived at the 400-km mark. Evolutionary radiation of the mammalian stock began somewhere around the 453-km mark. The first hominid appeared approximately 6 km later. *Australopithecus* joined the journey 200 to 450 m from the end, and the Neanderthals disappeared about 3.5 m from the finish line. The cultivation of land and keeping of livestock occurred 1 m from our present position. A 100-year-old person today has covered a distance of just 0.01 m or 10 mm of the 460-km journey.

The purpose of this brief summary has been to present an outline of our genetic background. Many structures and functions are common to different species in the animal kingdom. For instance, there appears to be no fundamental difference in structure, chemistry, or function between the neurones and synapses in man and those in a squid, a snail, or a leech.[7] Therefore we can learn a great deal from studying different species. It is remarkable that all living organisms have a genetic code based on the same principles. For instance, data indicate that man and the chimpanzee share more than 99 per cent of their genetic material.[8] However, minimal genetic changes can affect major morphological modifications. Consequently, we should be careful when extrapolating findings from one species to another, because over millions of years many species undergo minor or major modifications in their physical and other characteristics. In general, evolution is a very conservative process. For instance, all vertebrates, including the hominids, have backbones of a similar complicated design. This supports the hypothesis that backbones have evolved only once, i.e. all vertebrates share a common ancestor with a backbone.

In general, vertebrate locomotion is genetically programmed. Fish can swim and birds can walk as soon as they hatch. Many species of mammals are well developed at birth. Thus some are able to walk or run as soon as they are born, and some of these are able to attain a speed of 35 km/h when they are only a few days old. Evidently, survival may depend on their ability to escape. In the case of humans, who are utterly helpless at birth and entirely dependent on parental care, it may be advantageous not to be able to move very far from their parents until a reasonable level of maturity has been attained.

The evolutionary process continues, and mammalian history has been a wave of extinctions, which have been particularly severe for large mammals including the hominids. Extinctions are a measure of the success of evolution in adapting organisms, because particular adaptations provide entry into a relatively empty niche. In the balance between existence and extinction, the odds are not favourable. It has been estimated that 2000 million species have appeared on the Earth during the last 700 million years, but the number of multicellular species now living is only a few million (i.e. only 0.1–0.2 per cent have survived).

The cortex of the human brain mirrors man's evolutionary success. Just as the proportions of the human hand, with its large opposable and muscular thumb, reflect successful adaptation for life in trees and later for the use of tools, so does the anatomy of the human brain reflect a successful adaptation for manual and intellectual skills.

In the same way as upright walking and tool-making were unique adaptations of the earlier phases of human evolution, the physiological capacity for speech was the biological basis for the later stages. Indeed, it is by language that human social systems are mediated. Speech is the form of behaviour that differentiates man from other animals. The use of language to transfer knowledge and experience from one generation to the next has enabled man, biologically unchanged for tens of thousands of years, to accelerate progress. In addition, language has enabled humans to apply their endowed intellectual resources in a technical revolution leading to entirely new and complex tools, weapons, shelters, boats, wheeled locomotion, exploratory voyages, and the attainment of the impossible—space travel. Nevertheless, in the midst of these splendid achievements, there are those who wonder whether the evolution of the human brain has gone too far. Although its ability to conceive, invent, create, and construct is astonishing, it remains to be seen whether or not it has retained or developed equally well its capacity for ethical conduct or responsible application of its endowed potential. When our ancestors roamed around in small bands, any destructive consequence of their activity was quite limited. However, because of social developments and technical innovations, basically the same brain is now capable of self-destruction.

Like all higher animals, man is basically designed for mobility. Consequently, our locomotive apparatus and service organs constitute the main part of our total body mass. The shape and dimensions of the human skeleton and musculature are such that the human body cannot compete with a gazelle in speed or an elephant in sturdiness, but it is indeed outstanding in diversity. The basic instrument of mobility is the muscle. It is a very old tissue. As already mentioned, the earliest animal fossils were burrowers living some 700 million years ago. Using muscle force, these animals could dig into the seabed. They retained the metabolic pathways developed when the air had no oxygen (i.e. the anaerobic energy yield). The pyruvic acid formed in our muscles under anaerobic conditions is now removed by the formation of lactate. One old-fashioned alternative could have been the transformation of the pyruvate into ethyl alcohol. There may be those who now regret that the skeletal muscles did not select this alternative route. Had this occurred, producing pyruvate by exercising to exhaustion or running uphill might have been a very popular endeavour!

The skeletal muscle is unique in that it can vary its metabolic rate to a greater degree than any other tissue. In fact, active skeletal muscles may increase their oxidative processes to more than 50 times the resting level. Such an enormous variation in metabolic rate must necessarily create serious problems for the muscle cell because, although the consumption of fuel and oxygen increases 50-fold, the rate of removal of heat, carbon dioxide, water, and waste products must also be increased. To maintain the chemical and physical equilibrium of the cell, there must be an enormous increase in the exchange of molecules between intracellular and extracellular fluid (i.e. fresh fluid must continuously flush the exercising cell). When muscles are thrown into vigorous activity, the ability to maintain the internal equilibria necessary to continue the exercise is entirely dependent on those organs that service the muscle's circulation. Food intake, digestion and handling of substrates, kidney function, and water balance are also strongly affected by variation in metabolic rate.

Almost 100 per cent of the biological existence of our species has been dominated by outdoor activity. Hunting and foraging for food and other necessities have been conditions of human life for millions of years. We are adapted to that style of life. This applies to our emotional, social, and intellectual skills. After a brief spell in an agrarian culture, we have ended up in an urbanized highly technological society. There is obviously no way to revert to our natural way of life, which was not without its problems. With insight into our biological heritage, however, we may yet be able to modify our current life-style. Understanding of the function of the body at rest, as well as during exercise under various conditions, is important as a basis for optimizing our existence.

Children are definitely spontaneously physically active. Unfortunately, in our modern society, we discourage this activity by furnishing houses and apartments to fit parents' needs, keeping children indoors in schools and doing homework for many hours, creating heavily overpopulated 'concrete deserts', and producing television programmes to capture their attention. Children should keep quiet and stay clean and neat! Vigorous physical activity is antisocial behaviour in too many circumstances. From the time of puberty, human nature has an inclination towards physical laziness. There is no appetite centre for physical activity.

Some years ago, the author visited the Bushmen of the Kalahari Desert, probably the last remaining Stone Age people. They followed the life-style of true hunters and food gatherers. Gathering sufficient food meant trudging long distances, for the men in their hunting efforts, and for the women and children in their collection of berries, melons, roots, and various plants. The sequence of walking, stopping, squatting to dig, and walking again is physically demanding. When the women gather enough and return home, they still have to collect and carry firewood for the cooking and the night fire. For most of the year, the game and food plants are not found in any abundance. To gather enough to eat, the Bushmen have to exercise for hours almost every day. The driving factor for the habitual physical activity is hunger and thirst, and not a particular love for exercise. The author never saw an adult Bushman out jogging, but the walking was fast! The Bushmen are well trained with emphasis on endurance.

HISTORY OF SPORTS

For obvious reasons we do not know anything about athletic activities during the Stone Age. People probably liked games and plays, and they sang and danced. There is definite evidence in sculptures, reliefs, and paintings some 5000 years old that Egyptians exercised. The hieroglyphic sign for swimming dates from the same period.

It is not until the Olympic Games began that the history of organized athletic activities can be revealed. While the origin of the Olympic Games is not known exactly, there is an historical record of the ancient games beginning in Olympia in the western Peloponnisos, Greece, in 776 BC. Thereafter they were held at 4-year intervals, until AD 394 when they were abolished. There are various traditional explanations of the origin of the games. One attributes the festival to Heracles, the most famous Greek hero. In art and literature he is represented as an enormously strong man of moderate height, a huge eater and drinker, very amorous, and generally kind but with occasional outbursts of brutal rage. This great fighter and hunter is famous for the Twelve Labours, or *Dodekathlos*, which include the capture of the lion of Nemea, the cleansing of the stables of Augeias in Elis, the capture of the Cretan Bull, and seizing the cattle of Geryon. Another myth tells us that there was a chariot race between Pelops and king Oenomaus, who used to challenge the suitors of his daughter Hippodameia. Pelops successfully persuaded Oenomaus' servant to remove the wheel spindle pins; the chariot crashed and the king was killed. Unfair play is not a modern phenomenon in sport! Pelops married Hippodameia and became King of Pisa. He con-

quered Olympia where, for the glory of Zeus, he arranged competitions.

The earlier Olympic programmes consisted almost exclusively of exercises of the Spartan type, testing endurance and strength with a special view to war. Later, more and more events were added: chariot races and horse races, wrestling and boxing, and the pentathlon including long jumping, quoit (discus) throwing, javelin throwing, running, and wrestling. Winners became national heroes: musicians sang their praises and sculptures preserved their strength and beauty in marble.

Olympia became an expression of the Greek ideas that the body of man as well as his intellect and spirit has a glory, that the body and mind should alike be disciplined, and that it is by the harmonious discipline of both that men best honour Zeus. It should be noted that women were not allowed as competitors or, except for the priestesses of Demeter, as spectators.[9]

Baron Pierre de Coubertin took the initiative of reviving the Olympic Games for athletes of all countries of the world, regardless of national rivalries, jealousies, and differences of all kinds, and with all considerations of politics, race, religion, wealth, and social status eliminated. At an enthusiastic conference at the Sorbonne, Paris, in 1894, it was decided that the games of the First Olympiad of the modern cycle should take place in Athens in 1896. However, we know too well that at times the high ideals expressed by Pierre de Coubertin have been neglected.

MODERN ATHLETICS

Many modern sports have their origins in games played with and without balls and equipment enjoyed during the Middle Ages and later. Preparation for hunting, war, and defence made training in archery, javelin throwing, fencing, shooting, boxing, and wrestling necessary for survival. During the Middle Ages the knights' tournaments were frequently a life-and-death struggle.

Historians claim that Great Britain was the cradle of modern organized sports and competitions. At the beginning of the nineteenth century sports were introduced in schools, and in the middle of that century championships were arranged at colleges and universities and competitions took place between them, for example the famous Boat Race between Oxford and Cambridge. In 1880 the British Amateur Athletic Association was founded. As mentioned above, a few years later Baron Pierre de Coubertin launched the flagship of sport, the Olympic Games.

WHY ARE SPORTS RECORDS IMPROVING?

Several factors, which vary in importance depending on the characteristics of the sport, must be considered. The following factors will be discussed in detail:

    selection from a larger and healthier population;
    better training methods and preparation;
    improved techniques;
    improved materials;
    psychological aspects;
    scientific support;
    doping;
    physiological aspects.

Examples chosen mainly from track and field events will be used in this discussion. Figures 2 to 7 illustrate the development of world records in several sports from the beginning of this century, when systematic documentation of the world's best performances started. For obvious reasons, one can trace the effects of two world wars in the statistics: there is a hiatus in progress during, and for some years after, the wars.

Selection from a larger and healthier population

More and more individuals, particularly women, are attracted by sports activities. Increasing numbers of nations are represented in the sports arena. As preventive and curative health measures become more successful throughout the Third World, millions of teenagers should have a chance to enjoy sports. These factors make it more likely that individuals with talent for a particular sport will be noticed by the experts.

Better training methods and preparation

Training volume has increased and training methods have improved dramatically. Today, top athletes are not 'true amateurs' as in the days when the Olympic oath included the statement that athletes did not compete for improvement of their economic status. Certainly, top athletes have always managed to make money from their sport, but today this is permissible and can involve large sums. In other words, athletes can now devote more time to training and can train year round in an optimal climate.

Improved techniques

In some events, changes in the rules have made developments in techniques possible. In the early rules for the high jump it was stated that when passing the cross-bar (1) the jumper's buttocks should be on a lower level than his or her head and (2) the feet should precede the head. In 1936 the rules were changed and the only restriction was that the take-off should be accomplished from one foot. Until then the scissors style had dominated. However, the 'western roll', originally introduced by Horine in 1910, had given H.M. Osborne the gold medal in the 1924 Olympic Games although the judges had great problems in deciding whether or not his jumping style conformed to the rules. Probably this and similar incidents made it necessary to change the rules. Dick Fosbury's victory with his 'flop style' in the 1968 Olympic Games provided a spectacular introduction of today's dominant technique.

Covering the circle for the shot put, discus, and hammer throw with rubber material or concrete facilitated the development of new techniques.

When the breaststroke was modified to increase speed, a new event was born—the butterfly. Both styles are energetically expensive, and so it is realistic that the longest distance swum in competition is 200 m. Some techniques have been prohibited for safety reasons, namely climbing on the pole in the pole vault and turning a somersault in the long jump. In Spain the technique in a traditional sport was adopted for the javelin throw. The thrower initiates the throw with fast rotations, and the back part of the javelin is prepared with soap to reduce the friction against the palm when the javelin glides out of the hand. For obvious reasons, this 'soap style' is forbidden.

Improved materials

Technological innovations have played an important role in the performance explosion in many events. The introduction of artificial surface on tracks improved conditions and, most importantly, maintained consistent lane quality throughout a competition. Previously, the inner lane of the track often deteriorated as it became worn by large numbers of feet. The Pan American Games in 1967 were the first major event to be held on an artificial surface. The modern materials on the thrower's circle also provided long-lasting and equal conditions for all competitors. When starting blocks were permitted, the start in sprint events became faster. The introduction of the fibreglass pole improved world records in the pole vault in the 1960s. This is well illustrated in Fig. 2. Such a pole was first introduced during the latter part of the 1950s, and an American, Alburley Dooley, was a pioneer in the development of a technique that efficiently utilized its elastic properties.

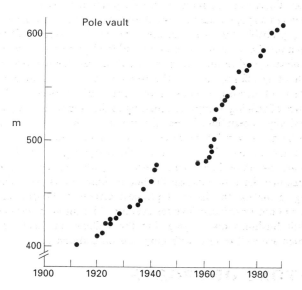

Fig. 2 In 1960, Don Bragg broke the world record by jumping 480 cm with his steel pole. In 1961, the first record using a fibreglass pole was set (483 cm), and 2 years later John Pennel jumped 520 cm. The fibreglass pole effectively stores some of the athlete's energy developed during the run and, with good technique, that energy can be utilized at exactly the right moment. During the last 20 years, this catapult pole has changed very little in quality. Therefore the continuous improvement in records must be due to better skill and power of the record breakers. Without the development of foam rubber mats on which to land, jumping with fibreglass poles would be dangerous. (Reproduced from ref. 15, with permission.)

Without a foam rubber mat, the landing after the high jump and pole vault would be hazardous. In fact, without this equipment the 'flop style' and the fibreglass pole would be too dangerous to use.

The aerodynamic properties of the javelin and discus have been improved. The long throws, as illustrated in Fig. 3, can be a threat to spectators. Also, it is often difficult to judge whether or not the javelin hits the ground with its point first. In 1986 a rule was passed which introduced a new javelin model with different aerodynamic characteristics from those of the traditional model. It is estimated that Hohn's world record (104.8 m) is equivalent to a throw of approximately 85 m with the new model. The length-reducing effect is less pronounced with shorter throws. An 'old' 90-m mark would now be comparable to approximately 78 m but

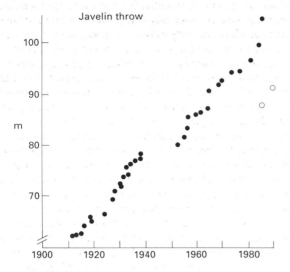

Fig. 3 In 1932 Matti Järvinen of Finland, holder of the world record in javelin throwing, predicted that one day someone would throw the javelin farther than 100 m. His record at that time did not exceed the 75 m mark. In 1984 Uwe Hohn (DDR) achieved a throw of 104.80 m. There has been a dramatic development of the aerodynamic characteristics of the javelin over the years, with Frank Held (United States) as a pioneer. He broke the world record in 1953 and again in 1955. Held threw a specially designed javelin some 4 m farther than was possible with traditional equipment, but this design was not approved by the authorities. One problem is to combine the javelin's aerodynamic ability to 'float' on the air with a landing on its tip in accordance with the rules. A javelin throw is more effective if the body is bent into an extreme 'bow' before the throw. In fact, Järvinen had already applied this 'bow'. Stretched muscles can develop more force than muscles at a shorter initial length. The javelin throw is anatomically very demanding, and most throwers suffer from orthopaedic problems, particularly in the elbow, at least once during their careers. The champion discus thrower Adolfo Consolini (Italy) is reported to have achieved a throw of 114 m using the forbidden 'soap style' (see text). (Reproduced from ref. 15, with permission.)

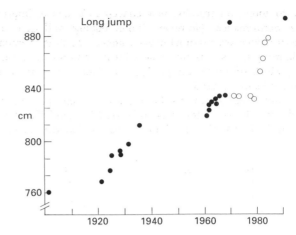

Fig. 4 Bob Beamon's meteoric jump, sending him 8.9 m from the take-off point at the 1968 Olympic Games in Mexico City, was spectacular. 'No other world track and field record excels the previous best performance by a comparable margin ... Beamon's feat outshines all others. It is unlikely that the 8.90 meter record will ever be beaten' (Ernst Jokl). The open circles denote the best results achieved during the years following 1968. Carl Lewis gradually came closer. However, it was Mike Powell who broke the record with a jump of 8.95 m in 1991. (Reproduced from ref. 15, with permission.)

60 m is comparable to 59 m. There is some sort of badminton ball effect. In addition, there is a much better chance of the javelin landing correctly on its point. The introduction of rules and equipment which reduce the performance as evaluated in metres is unique to javelin throwing.

Psychological aspects

Good performance depends on expectation, which in part determines tactics. With a world record of 3 minutes 30 seconds in the 1500-m run, the goal may be to run in 3 minutes 29 seconds but not in 3 minutes 20 seconds. In some events there are no barriers, i.e. the limits are unknown. A perfect example is Bob Beamon's 890-cm aerial trip in the long jump in 1968. That jump improved the world record by 55 cm (Fig. 4)! At that time the world high jump record was 228 cm. Who would have dreamed of putting the cross-bar up to 240 cm, which, if successfully passed, would have resulted in a similar improvement in the record?

From a psychological viewpoint, it is often a handicap to run at the front of the pack for most of a 1500-m race. This drawback has, at least partly, a physiological background. Even if there is no wind, the speed of the runner causes significant air resistance.

Running behind another competitor in a 'shielded' position can save 4 to 6 per cent of the energy cost.[10] From both a psychological and a physiological aspect, a steady speed throughout the race up to the final spurt usually gives the best time.

Scientific support

It is difficult to prove to what extent medical science has helped athletes in their pursuit of new records. Often the athletes have been one step ahead, applying trial and error methods, followed by the physiologists whose studies have revealed mechanisms that can explain why a particular regimen can enhance performance. However, basic research from the 1930s and 1940s, confirmed in more recent studies, has proved that diet and fluid balance can affect physical performance decisively. As early as 1939, Christensen and Hansen[11] reported that a carbohydrate-rich diet improved endurance in heavy exercise and that training could have a glycogen-saving effect that enhanced aerobic capacity. Scientific data supported the belief in the beneficial effects of a warm-up before high-intensity exercise.

Unfortunately, many athletes are injured during training and competition. Physicians and physical therapists try, often successfully, to enable the athlete to return to the arena as quickly as possible. Methods for treatment and rehabilitation are examined critically. If successful, they are then available to everyone.

Doping

It is a tragedy that so many athletes, coaches, and physicians break the rules in their ambitions to win and break world records. It is in only a few disciplines that pills can improve performance beyond what the athlete can achieve through will power and stimulation from a cheering crowd. Actually, he or she usually performs better without doping. However, many coaches and athletes appear to be willing to adopt, uncritically, new concepts claimed to improve athletic performance. However, 'blood doping' will definitely significantly increase maximal oxygen uptake which is of importance in many sports. It is now believed that the hormone erythropoietin, which stimulates the production of red cells in the bone marrow, is being misused.

It is interesting that the world record in the shot put only increased from 21.78 m in the mid-1960s to 22.02 m in 1982 (it now stands at 23.12 m) (Fig. 5). During that period the intake of

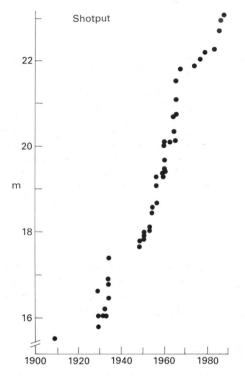

Fig. 5 In the early days of the shot put event, it was predominantly a one-arm affair. Jack Torrance put the shot up on the finger tips and could thereby add extra power from the wrist. He reached a 'phenomenal' 17.40 m in 1934, a result that survived attacks until 1948 (with the Second World War as a restriction in between). Parry O'Brien, in particular, introduced a new technique, starting in a low position with the back facing the direction of throw. The strength of leg and trunk muscles became more important, and it was logical that more strength training was included in preparation for the competitive season. O'Brien dominated the scene from 1953 until his last record noted in 1959. At the beginning of the 1950s the cinders in the ring were replaced by concrete, which facilitated the introduction of new techniques. In the mid-1970s some shot putters launched a rotation to initiate the put. As mentioned in the text, up to the mid-1980s there was a trend towards stagnation in the breaking of records despite the growing popularity of the use of anabolic steroids. (Reproduced from ref. 15, with permission.)

anabolic steroids increased dramatically and probably involved many world-class shot putters. It is tempting to conclude that the intake or injection of such hormones has not contributed significantly to the world record statistics in the shot put.

Physiological aspects

Are the athletes of today superior in their physiological potential compared with their predecessors? A high maximal oxygen up-

take, or aerobic power, is essential for success in sports utilizing large muscle groups in all-out efforts for several minutes or longer. In activities in which the body weight is carried, this aerobic power is related to body weight (oxygen uptake in ml/kg.min). However, in exercises such as rowing and swimming, oxygen uptake is given in litres per minute. In 1937, Robinson et al.[12] reported that Don Lash, who held the world record for the 2 mile race, attained a maximal oxygen uptake of 81.5 ml/kg.min when running on a treadmill. Today, competitors record similar values but they run faster. (Lash's record was 8 minutes 58.4 seconds; today the best time recorded is 8 minutes 13.8 seconds). This fact is intriguing. Evidently, modern training principles allow the athlete to exercise at or closer to maximal aerobic power for longer periods of time. Another reason for improved performance is a higher power and capacity of the anaerobic metabolic pathways (see below).

Better shoes and tracks cannot explain the superiority of modern athletes. Ronald Clark's time of 27 minutes 39.4 seconds for 10 000 m on a cinder track in 1965 is not far from Henry Rono's time of 27 minutes 22.5 seconds on an artificial surface in 1978. For the 100-m distance, James Hines' time of 10.03 seconds on a cinder lane at sea level is not significantly slower than his time of 9.95 seconds on an artificial surface in the same year but at high altitude, which favours the sprinter because air resistance is reduced. The better results noted on distances demanding a high maximal oxygen uptake cannot be explained by better tracks and lanes.

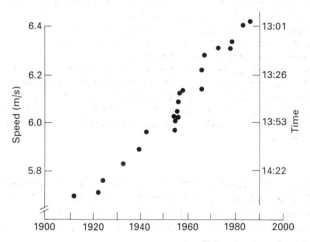

Fig. 6 The world records for the 5000 m track event follow a relatively straight line from 1920 onwards. The introduction of an artificial track surface did not noticeably improve the records. Extrapolation to the world record for the year 2000 is tempting. (Reproduced from ref. 15, with permission.)

There is a personal limit for maximal oxygen uptake. For example, a Swedish cross-country skier who in 1955 had just qualified for the national team had at that time a maximal oxygen uptake of 5.48 l/min. In 1963 it was about the same (5.60 l/min), but he had trained almost daily during the intervening 8 years and had successfully participated in two Olympic games and two world championships, winning several gold medals. In repeated tests during 1955 another skier never exceeded 5.88 l/min in maximal oxygen uptake. He trained intensively and competed successfully until 1964, winning a gold medal in the 50-km race in that year's

Olympic Games. There are few data from longitudinal studies of top athletes in running disciplines. There are indications that running times improve despite a stagnation in the maximal aerobic power. Training can produce a slight improvement in running economy and the ability to run faster before a continuous accumulation of lactate sets in.[10]

No methods are available for an exact measurement of an individual's maximal anaerobic power and capacity. Therefore we do not know whether today's athletes have a better anaerobic metabolism to support the contractile machinery of exercising muscles than did earlier generations. An increase in blood lactate concentration reflects a breakdown of glycogen in muscle. However, one cannot calculate from blood concentration how much lactate is produced. It is interesting to note that the lactate concentration is usually higher if measured after an important competition than after an all-out test in the laboratory. Apparently, the tolerance for high lactate values and low pH can be modified by psychological factors. It is remarkable that the pH in the arterial blood can fall below 7.0 after repeated 1 minute maximal runs. A comparison of peak blood lactate concentrations after maximal physical performance in the laboratory or in connection with competition in top athletes does not indicate any differences over the last 30 years.

There is a continuous increase in body height H in most developed countries. If a proportional increase in all dimensions is assumed, maximal strength, related to the surface area of the muscles, should be proportional to H, and maximal work and torque should be proportional to H^3. The mean height of the participants in the decathlon in the 1960 Olympic Games in Rome was 184 cm. Approximately 30 years earlier the average height was 176 cm. These heights compare as 1.045:1, their muscle strength as 1.09:1, and their work or torque as 1.14:1. Therefore, owing to different dimensions, the decathlete taller by 4.5 per cent can be expected to be 9 per cent stronger and to have the capacity to do 14 per cent more work than the shorter competitor. These advantages are particularly evident in such events as throwing the javelin and discus and putting the shot. If these data are extrapolated to the development of world records in the shot put, simi-lar changes in body dimensions alone could explain a gain from 16 m in 1930 to approximately 18 m in 1960. But in that year the shot was put 20 m. No doubt, in sports in which body size influences the results, from a democratic point of view the competitors should be classified according to weight as in boxing, wrestling, and weight lifting. However, such classifications would be unrealistic in track and field events.

Sexual dimorphism

World records for women and men are compared in Table 1. In swimming, the highest speeds attained by women are, on average, 91.4 per cent of those reached by men. In track events women are relatively slower, with top speeds at the 90.6 per cent level. In speed skating, women reach 93.2 per cent of men's world record speeds. In cycling the percentage is 87.1 per cent. Perhaps a number of women with a talent for cycling have not yet discovered this discipline.

The largest difference between women and men in world records in track and field events is noticed in the high jump, with the women's cross-bar reaching 85.7 per cent of the men's 244 cm, and the long jump, in which the best woman jumped 84.0 per cent of Mike Powell's 895 cm. We have no explanation of why women are particularly 'inferior' in jumping.

In sports testing strength (bench press, squat, deadlift) the weight handled by women is on average 60.7 per cent of the men's world record (range 55.9–68.1 per cent in weight classes 52–82.5 kg).

It is interesting to follow the development of results for women and men over the years. In 1950 the highest speed during the 100-m run for women was 88.7 per cent of the best male performance (94.6 per cent today). In 1950 in the 800-m race the highest speed for women was 80.2 per cent of that for men. Few women competed over longer distances, even though the 800-m race appeared in the Olympic Games as early as 1928. In the high jump, the women's record was 81 per cent of the men's best result in 1950, 83 per cent in 1960, 84 per cent in 1970, and, as mentioned, close to 86 per cent in 1990. In the long jump the figures are 77 per

Table 1 Women's world records compared with those of men

Track and field		Swimming		Speed skating		Cycling	
Race	Women	Race	Women	Race	Women	Race	Women
Running		*Freestyle*		500 m	93.2%	1 km	83.6%
100 m	94.6%	50 m	87.3%	1000 m	93.5%	10 km	91.7%
200 m	92.4%	100 m	88.5%	1500 m	93.9%	100 km	85.9%
400 m	90.9%	200 m	90.8%	5000 m	92.3%		
800 m	89.8%	400 m	93.1%				
1500 m	90.1%	800 m	94.9%				
3000 m	89.4%	1500 m	94.0%				
5000 m	88.7%	*Breaststroke*					
10 000 m	89.8%	100 m	90.6%				
Marathon	89.9%	200 m	89.6%				
4 × 100 m	91.4%	*Butterfly*					
4 × 400 m	90.3%	100 m	91.2%				
High jump	85.7%	200 m	92.3%				
Long jump	84.0%	*Backstroke*					
		100 m	90.0%				
		200 m	91.9%				

The data are taken from a selection of records valid at the end of 1991. The women's times (speeds) are given as percentage of the men's records (= 100 per cent). A similar calculation is performed for the jumping events.

cent, 78 per cent, 77 per cent, and 84 per cent respectively. The women's records have gradually crept closer to the men's levels. There are speculations, which have gained world-wide interest in the mass media, that women will eventually catch up with the men's world records, sooner in the marathon than in other events.[13] Whipp and Ward[13] extrapolate from world record progression, expressed as mean running velocity versus historical time, for women and men respectively and conclude that women may overtake men in the marathon by the year 2000 and in the 200-m sprint by the year 2050. If this were to happen it would be tempting for a female marathon runner to masquerade as a man and race in the men's event in the next century. Whipp and Ward ignore the fact that there are basic genetically fixed differences that are decisive for physical performance demanding muscular strength and high aerobic power even when related to body mass. Women cannot compensate by 'natural methods' for their lower blood haemoglobin concentration during maximal aerobic effort. There are no physiological data indicating that women have a particularly high potential for long-distance events such as the marathon.

World records in swimming compared with running

Swimming appears to be the discipline that has the world record for breaking world records. During the 1972 Olympic games in Munich the swimming competitors produced 30 world records in 29 events. There are claims that swimming is still a developing discipline. Figure 7 illustrates the advances in 200-m freestyle swimming and 800-m running since the turn of the century. The reason for selecting these events is that the present world records are not far from each other with regard to the times involved. Records in running can survive for years, but in swimming the records have been broken frequently since 1950. One factor to

consider is that the percentage increase in the number of pools over the years exceeds the increase in the number of new running tracks. Specific training of muscular strength and flexibility may improve swimming performance more than running ability. Many countries are handicapped because swimming cannot be recommended as a recreational sport where waters are polluted and can cause serious diseases (e.g. bilharzia (schistosomiasis) in tropical areas).

Altitude training

Is training at high altitude beneficial to the oxygen transport system? The reason for raising this question here is that a study of the improvement of world records can enlighten the discussion. Acclimatization is essential in the preparation for optimal performance in events demanding high aerobic power if the competition takes place at high altitude. The 1968 Olympic Games in Mexico City were not the first challenge forcing athletes to face new environmental conditions. In the 1960 Winter Olympic games in Squaw Valley the athletes had to gasp at an altitude of approximately 2000 m (6600 feet).

It is a common belief that training at high altitude will also enhance performance at lower altitudes. The history of world records does not support this hypothesis. In the years 1966 and 1967, and particularly in 1968, the cream of the world's athletes spent long periods of time in Mexico City or at similar altitudes. Many scientific studies were conducted on these athletes. However, few world records in middle- and long-distance running were broken in those years. Nor were there any spectacular new records in swimming. New records would be expected at sea level if a sojourn at high altitude elicited an additional improvement in maximal aerobic power and endurance. In fact, in 1968, when all Olympic candidates were extremely well prepared, there were no new world records in middle- and long-distance running.

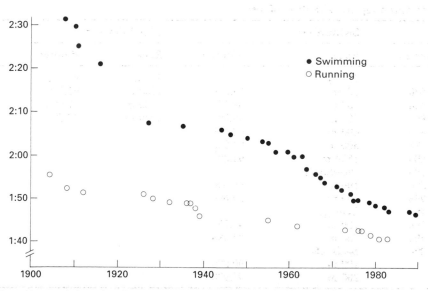

Fig. 7 The development of world records in swimming and running is shown. Swimming speed for the 200-m freestyle has improved by 15.9 per cent since the 1920s, whereas the speed for running 800 m has improved by only 8.8 per cent. (Reproduced from ref. 15, with permission.)

The record

Eric Segal[14] writes:

'But without any question, the most famous barrier in sports history was the four minute mile. And yet I seriously wonder if it would have acquired such mystique if the first man to break it had not been that eloquent obsessive, Roger Bannister. Anyone who reads his autobiography cannot help but sense what an all-consuming fixation it was for him to run merely two seconds faster than he ever had before. After all, Glenn Cunningham, that great miler of the 1930s, claimed that he had often run better than four minutes in practice sessions, but it had never seemed important to do so officially in a race. He had just wanted to win. Not Bannister. To him, breaking four minutes for the mile would be extending the 'ultimate' in human capability. Indeed, as a doctor Bannister believed that the effort required would be so enormous that the runner would expend his entire oxygen reserve a few yards before the finish and have to complete the race as a semiconscious reflex action. He planned his race according to his own theories. The rest is history. The photos of Bannister after his epoch-making run on May 6, 1954, show a totally exhausted man who by hint of courage and scientific preparation systematically depleted himself of all his resources to surpass all previous limits. His face shows that his body had not an ounce in reserve. He could not have done an instant better. The ultimate time for the mile had to be 3:59.4. This was, I repeat, May 6, 1954. And yet on August 7 of the same year, Bannister ran 3:58.8 without collapsing. Paradoxically, running a mile even faster proved to be less exhausting. Because once he had surpassed the four-minute 'limit', there was no magic in 3:58. Nor did there appear to be in 3:50. Values are what they are, Hamlet tells us, 'because thinking makes it so'. On July 18, 1979, the London Daily Express wrote the following: 'Coe came up to the final straight looking almost relaxed, hardly gasping . . . what a fantastic contrast to the complete exhaustion of . . . Sir Roger Bannister'. The limits are, of course, purely mental. Sebastian Coe was so relaxed in Oslo because he did not consider 3:49 any kind of ultimate. And when he finally begins to think in these terms, some mad idealist will appear in track shoes and prove him wrong. There are simply no absolute limits.'

Swedes like to remember the comments of two runners who broke many world records in middle-distance running in the 1940s, Gunder Hägg and Arne Andersson: first of all, a mile is not a Swedish distance; secondly, the goal for them was to win, not to make spectacular times.

CONCLUSION

Many factors have contributed to improvements in sports world records. The complexities of the disciplines are decisive for the quantitative impact of the various factors. Probably the basic endowment of the human being has been the most stable factor.

Changes in training methods, tactics, techniques, rules, equipment, material, and economy, and an increase in the number of people engaged in sports have all contributed to the improvements. It will probably be many years before we can write the final history of all world records—records that will never be surpassed.

A retardation in the curves can be seen in some events but not in others. As mentioned previously, in the javelin event it will be very difficult to beat Hohn's record owing to equipment modifications. In certain disciplines not discussed here one can, with great confidence, say that 'this record can never be improved'. One such discipline is shooting. With all bullets awarded 10 points, present rules do not permit any improvement.

It should be emphasized that sports that cannot be evaluated by world records are also very popular with participants and spectators: racket sports, American football, boxing, cricket, soccer, gymnastics, skiing, and golf, to mention a few. Beating a world record is not an essential stimulus for action.

Parts of the text in the section entitled 'Why are sports records improving?' have been published elsewhere,[15] and are reproduced here with permission from the copyright holder.

REFERENCES

1. Dickerson RE. Chemical evolution and the origin of life. *Scientific American* 1978; **239**(3): 62–78.
2. Vidal G. The oldest eukaryotic cells. *Scientific American* 1984; **250**(2): 32–41.
3. Valentine JW. The evolution of multicellular plants and animals. *Scientific American* 1978; **239**(3): 104–17.
4. Lewin R. Evolutionary theory under fire. *Science* 1980; **210**: 883–7.
5. Lewin R. Ethiopian stone tools are world's oldest. *Science* 1981; **211**: 806–7.
6. Stringer CB. The emergence of modern humans. *Scientific American* 1990; **264**: 68–74.
7. Kandel ER. Small systems of neurons. *Scientific American* 1979; **239**(3): 60–70.
8. Washburn SL. The evolution of man. *Scientific American* 1978; **239**(3): 146–54.
9. *Encyclopaedia Britannica*. London: William Benton, 1963.
10. Åstrand P-O, Rodahl K. *Textbook of work physiology*. 3rd edn. New York: McGraw-Hill, 1986: Chapters 7, 10.
11. Christensen EH, Hansen O. Arbeitsfähigkeit und Ehrnärung. *Skandinavischen Archiv für Physiologie* 1939; **81**: 160–71.
12. Robinson S, Edwards HT, Dill DB. New records in human power. *Science* 1937; **85**: 409–10.
13. Whipp BJ, Ward SA. Will women soon outrun men? *Nature, London* 1992; **355**(6355): 25.
14. Segal E. Reflections on the right to one'.s own limit. In: Pabst U, ed. *Baden–Baden report: the limits of sports. 11th Olympic Congress 1980*. Munich: Nationales Olympisches Komitee für Deutschland, 1981.
15. Åstrand P-O, Borgström A. Why are sports records improving? In: Strauss R, ed. *Drugs and performance in sports*. Philadelphia: WB Saunders, 1987; 147–63.

Sports science 1

1.1.1 Respiratory responses of athletes to exercise

BRIAN J. WHIPP AND SUSAN A. WARD

INTRODUCTION

The appropriateness of the ventilatory response to exercise is best considered with respect not to the actual level of ventilation achieved, but to the degree of arterial blood-gas homeostasis for moderate exercise and by the degree of compensatory hyperventilation at work rates that engender a metabolic acidosis. However, these pulmonary responses should not be considered in isolation; they are only one part of an integrated system which is comprised of external pulmonary and internal tissue gas exchange components linked by the circulation. Therefore the systems operate as a coupled unit during exercise: mechanically-coupled anatomically and control-coupled physiologically.

In highly trained athletes, high intensity exercise imposes such pulmonary demands that the effective limits of the convective fluid flow—both air through the airways and blood through the pulmonary capillary bed—can be approached or even exceeded. Furthermore, the large ventilatory demands can require such a high level of respiratory muscle power that a large fraction of the total increase of cardiac output can be required by the respiratory muscles. This, in addition to diverting blood flow from the muscles generating the external work, can also exceed the ability of the respiratory muscles to provide ventilation wholly aerobically with the potential consequence being respiratory muscle fatigue.

Thus the assessment of the athlete's respiratory system performance during exercise is not straightforward. However, it should address three inter-related issues.

(i) To what extent are the 'requirements' met? The major requirements are for arterial P_{O_2}, P_{CO_2} and pH regulation. These variables need to be determined directly or, when this is not possible, reliable and valid estimators should be used.

(ii) What is the 'cost' of meeting these 'requirements'? This necessitates an assessment of (a) how much ventilation (and its pattern) is utilized to meet the requirements and (b) the amount of respiratory muscle work involved, together with its oxygen and blood flow cost.

(iii) To what extent is the system 'constrained' or 'limited'? This requires estimation of whether the effective limits of the system are achieved or approached for example with respect to limiting air flow, volume change, and gas exchange efficiency. Flow-volume and tidal-volume inspiratory capacity considerations, the maximum voluntary ventilation, and the maximum sustained ventilatory capacity provide useful frames of reference for mechanical reserve, and the alveolar–arterial P_{O_2} difference provides an important index of gas exchange efficiency.

VENTILATORY REQUIREMENTS DURING EXERCISE

Ventilatory response characteristics

The pattern of homeostatic regulation is not uniform during exercise; it varies with the 'intensity' of exercise. Consequently, it is important to determine the intensity domain of a particular work rate in order to establish whether the respiratory response is appropriate. Although there is no generally agreed procedure for normalizing work intensity, we believe that two widely used procedures, the 'met' increment and the percentage of maximal oxygen uptake $\dot{V}_{O_2}$max fail to meet the demands of critical scrutiny in this regard, at least with respect to the respiratory system.

The onset of the metabolic (lactic) acidaemia of exercise does not occur at a common 'met' increment in different individuals. Consequently, different subjects with the same 'met' level can have markedly different degrees of metabolic acidaemia. Similarly, there is a wide variation in the percentage $\dot{V}_{O_2}$max at which metabolic acidosis becomes evident. In normal individuals, this can range from 40 to 80 per cent of $\dot{V}_{O_2}$max$_2$ tending to be greater the fitter the subject but with a large variability at any particular level.

For these reasons, it is preferable to utilize the measured or estimated degree of metabolic acidaemia as the index of exercise intensity. The range of work rates within which there is not a sustained metabolic acidaemia can be considered to be of moderate intensity: in the athlete, this range is substantial. Work rates at which blood lactate concentration and pH are elevated, but eventually stabilize or even decrease as the exercise continues, are of heavy intensity. Those higher work rates at which blood lactate concentrations and [H⁺] are not only elevated but increase inexorably throughout the test are of severe intensity.

For moderate constant-load exercise, oxygen uptake $\dot{V}_{O_2}$ increases monoexponentially with a time constant τ that typically ranges from 20 to 40 s, such that a steady state is attained within about 3 min.[2-6] However, as $\tau(\dot{V}_{O_2})$ appears to be shorter in athletes,[7] the time to steady state is also reduced. The CO_2 output $\dot{V}_{CO_2}$ takes longer to reach a steady state, as it has a longer τ.[2-6] The time constant for ventilation $\dot{V}_E$ is longer still, and consequently a steady state is normally not attained for 4 to 5 min.[2-6,8] The lack of metabolic acidosis and the attainment of steady states in $\dot{V}_{O_2}$, $\dot{V}_{CO_2}$, and $\dot{V}_E$ allows exercise in this intensity domain to be sustained for long periods with relative ease.

During heavy exercise, $\dot{V}_E$ is further stressed not only by metabolic acidosis per se but also by the gas exchange consequences of the acidosis. In response to constant-load exercise within this domain, blood lactate concentration and arterial pH (pH$_a$) eventually stabilize,[9-11] as do the circulating levels of adrenaline and noradrenaline.[11] $\dot{V}_{O_2}$, $\dot{V}_{CO_2}$, and $\dot{V}_E$ also attain new steady state levels. However, relative to moderate exercise, the $\dot{V}_{O_2}$ response develops far more slowly, such that its time constant is prolonged, thus delaying the attainment of a steady state. Furthermore, at these work rates a further acidosis-related component of $\dot{V}_{O_2}$ is superimposed upon this initial response, adding to the steady state oxygen cost, and hence ventilatory cost, of the exercise.[12,13]

In contrast, in the severe intensity domain, both $\dot{V}_{O_2}$ and lactate concentration continue to increase throughout the work until $\dot{V}_{O_2}$max and the limit of exercise tolerance are attained.[10,11] Likewise, pH$_a$ continues to fall,[10,11] but at a rate that depends on the degree of respiratory compensation. In this domain, the duration for which a given work rate can be sustained is hyperbolically related to the work rate.[11,14,15]

As there is no systematic metabolic (or respiratory) acid–base derangement in the steady state of moderate exercise, pH_a is regulated as a result of ventilation changing in proportion to $\dot{V}CO_2$, i.e. through regulation of PCO_2a. This is implicit in the Henderson–Hasselbalch equation:

$$pH_a = pK' + \log\left(\frac{[HCO_3^-]_a}{\alpha PCO_2}\right) \qquad (1)$$

where α is the solubility coefficient for CO_2 in plasma and is equal to 0.03 mmol/mmHg at 37°C; pK', which is related to the first 'apparent' ionization constant of carbonic acid, is equal to 6.1; $[HCO_3^-]_a$ is the arterial bicarbonate concentration; and PCO_2a is the arterial PCO_2. The influence of $\dot{V}CO_2$ and ventilation on the regulated level of PCO_2a can be incorporated into this consideration by means of the 'alveolar air' equation for the 'ideal' lung*[16–18]

$$\dot{V}_A(BTPS) = \frac{863\,\dot{V}CO_2\,(STPD)}{PCO_2a} \qquad (2)$$

where $\dot{V}CO_2$ (the CO_2 production rate) is expressed at standard temperature and pressure, dry (STPD), and the alveolar ventilation $\dot{V}_A$ is expressed at body temperature, ambient pressure, saturated with water vapour (BTPS). Substituting for PCO_2a in equation (1) yields

$$pH_a = pK' + \log\left(\frac{[HCO_3^-]_a}{25.6}\right)\frac{\dot{V}_A}{\dot{V}CO_2} \qquad (3)$$

where the constant 25.6 is the product of the conversion constant 863 and α. It should be noted that as long as $[HCO_3^-]_a$ is unaltered, pH_a can only be regulated if $\dot{V}_A/\dot{V}CO_2$ is maintained constant i.e. there is a proportional increase of $\dot{V}_A$ with $\dot{V}CO_2$ during the work.

However, it is the total ventilation $\dot{V}_E$ that is controlled by the respiratory system, and the extent to which this is translated into alveolar ventilation is dictated by the physiological dead space fraction of the breath V_D/V_T:

$$\dot{V}_A = \dot{V}_E(1 - V_D/V_T) \qquad (4)$$

Therefore, substituting for $\dot{V}_A$ in equation (3), we obtain

$$pH_a = pK' + \log\left[\frac{[HCO_3^-]_a}{25.6}\frac{\dot{V}_E}{\dot{V}CO_2}(1 - V_D/V_T)\right] \qquad (5)$$

This integration of the Henderson–Hasslebalch and alveolar air equations allows pH regulation relative to the demand for CO_2 clearance to be considered in terms of three distinct components:

(1) the acid–base set-point component $[HCO_3^-]_a$;
(2) the respiratory control component $\dot{V}_E/\dot{V}CO_2$;
(3) the ventilatory efficiency component $(1 - V_D/V_T)$.

The metabolic acidosis of heavy and severe exercise leads to additional drive to $\dot{V}_E$. In normal subjects, this is thought to derive predominantly from metabolic acidaemia;[19,20] other influences also contribute, however, such as increased levels of circulating catecholamines, further increased levels of circulating potassium ions, sufficiently high body temperature and in some subjects anxiety, pain, and apprehension.[4,21–23] In the extremely fit subjects who develop arterial hypoxaemia at these works rates,[24] there is

presumably an additional drive via peripheral chemoreception. As a result $\dot{V}_E$ increases at a greater rate than $\dot{V}CO_2$, causing PCO_2a to fall. It is the magnitude of this compensatory reduction in PCO_2a that constrains the fall of pH_a.

These considerations emphasize the importance of establishing the transitional work rate between the moderate and heavy intensities of exercise. This has been termed the lactate threshold θ_L, and represents the highest work rate that can be sustained without sustained lactic acidosis.[25–28] Likewise, it is also important to identify the work rate that separates the heavy and severe intensity domains. This has been termed the fatigue threshold* θ_F[11] or critical power; it appears to correspond to the highest supra-θ_L work rate that can be sustained without blood lactate and $[H^+]$ concentrations, and $\dot{V}O_2$ continuing to increase throughout the work bout.[10,11]

However, an important caveat to this consideration of the ventilatory determinants of PCO_2a regulation is the recognition that the control studies under laboratory conditions may not be appropriate for consideration of actual athletic performance.

Estimation of functional parameters

We consider here the technique that identify the parameters which partition the intensity domains of exercise.

Lactate threshold

A wide range of techniques have been advocated for estimation of θ_L, and include both direct and indirect estimators (Table 1). Various exercise regimens have been used, but the most useful has proved to be the rapid-incremental test, in which the work rate is increased in a continuous ramp profile or by a constant small increments (e.g. 15–25 W for healthy subjects) at a regular interval (typically each minute or less) until the subject's limit of tolerance is reached.

θ_L is highly task specific. It occurs at an appreciably lower $\dot{V}O_2$ for arm exercise than for leg exercise, and is typically lower for cycle ergometry than for treadmill exercise.

The response profile of blood lactate concentration is often not the most sensitive estimator for θ_L, as there may be no clear breakpoint in the profile that can be identified with sufficient confidence, at least for some investigators. As a component of the blood lactate concentration increase reflects pyruvate-dependent increases, a clearer estimate is provided by the ratio of lactate to pyruvate concentrations.[30] Another approach has been to transform the lactate concentration (or $[HCO_3^-]_a$) response to a logarithmic function to linearize its rising phase.[31] The intersection of this phase with the earlier region of shallower slope (i.e. the moderate exercise region) coincides with non-invasive estimators of θ_L (described in detail below).

However, one can forego the necessity for serial blood sampling and even in many cases enhance the discriminability of θ_L by utilizing a particular cluster of pulmonary gas exchange responses. The compensatory hyperventilation for the metabolic acidosis of heavy and severe exercise coincides with the increase in lactate concentrations and decrease in $[HCO_3^-]$ in tests for which the work rate is incremented in a quasi-steady-state manner (i.e. increment durations of more than 4 min), i.e. $\dot{V}_E$ begins to increase more rapidly then $\dot{V}CO_2$ (and $\dot{V}O_2$) and PCO_2a is reduced.[26,32] In

---

* The 'ideal' lung is characterized by a diffusion equilibrium between alveolar gas and pulmonary end-capillary blood, regional matching of ventilation to perfusion, and the absence of a right-to-left vascular shunt. Therefore alveolar PO_2 and PO_2 equal arterial PO_2 and PCO_2.

---

* The parameter θ_F has been termed the fatigue threshold[11] not in the sense that there is no fatigue at lower work rates but that the fatigue will be of a different character (e.g. no metabolic acidosis) from that which results in the attainment of $\dot{V}O_2max$.

Table 1 Markers of the lactate threshold

Marker	Comments
Incremental work tests	
1. Blood and muscle $[L^-]/[P^-]$	Best single estimate that $[L^-]$ increase is probably consequent to anaerobiosis (i.e. corrects for pyruvate-dependent $[L^-]$ changes). Arterialized capillary blood or arterialized venous blood from high flow to low metabolic rate region (e.g. dorsum of hand) is acceptable. Direct venous samples are inadequate
2. Blood $[L^-]$ and $[HCO_3^-]$	As $[L^-]$ and $[HCO_3^-]$ changes are often curvilinear above θ_L, a plot of log $[L^-]$ or log $[HCO_3^-]$ vs. $\dot{V}_{O_2}$ or, even better, log $[L^-]$ or log $[HCO_3^-]$ vs. log $\dot{V}_{O_2}$ gives linear components intersecting precisely at θ_L
3. Ventilatory response variables	In subjects with normal chemosensitivity *and* normal pulmonary mechanics, the beginning of a systematic increase in $\dot{V}_E/\dot{V}_{O_2}$ and Po_2et without P_{CO_2et} decreasing provides a valid index of θ_L: R beginning to increase more rapidly and $\dot{V}_E/\dot{V}_{CO_2}$ becoming relatively constant (usually after having decreased) often, but not always, gives added support; breathing frequency also often begins to increase more rapidly
4. V-slope method	When the early kinetic non-linearity and the hyperventilatory region are discarded, a plot of $\dot{V}_{O_2}$ vs. $\dot{V}_{O_2}$ gives two linear components, the intersection of which occurs at θ_L (validated, to date, only for 1 min incremental tests). As this method only relies on the accelerated $\dot{V}_{CO_2}$ above θ_L, it is expected to be valid even when $\dot{V}_E$ does not respond as required in method 3
Constant-load tests	
5. Blood $[L^-]$ and $[HCO_3^-]$	Sustained change is required to know if exercise is supra-θ_L. Several studies are needed to determine the location of θ_L
6. $\Delta\dot{V}_{O_{2(6-3)}}$	For sub-θ_L exercise $\Delta\dot{V}_{O_{2(6-3)}}$ is zero (i.e. steady state as attained after 3 min). When $\Delta\dot{V}_{O_{2(6-3)}}$ is positive, a slow $\dot{V}_{O_2}$ kinetic phase is evident, the magnitude of which correlates well with the increase in blood $[L^-]$. Although several tests are needed to determine the location of θ_L, this index is useful for confirming that a particular work rate is or is not greater than θ_L

$[L^-]$, lactate concentration; $[P^-]$, pyruvate concentration
Reproduced from ref. 9, with permission.

contrast, however, when the work-rate incrementation rate is more rapid, the compensatory hyperventilation is strikingly attenuated (Fig. 1). In this situation there is a range of work rates immediately above θ_L within which $\dot{V}_E$ increases with $\dot{V}_{CO_2}$ in approximately the same proportionality as for moderate exercise. Under these conditions, $\dot{V}_{CO_2}$ has contributions from both metabolic sources and HCO_3^- buffering reactions. Therefore, as $P_{CO_2}a$ does not fall in this region, it has been termed the range of isocapnic buffering.[33] Respiratory compensation for lactic acidosis (i.e. decreasing $P_{CO_2}a$) only begins for rapid-incremental tests (Fig. 1) at a work rate which is typically about midway between θ_L and $\dot{V}_{O_2}$max.[34]

The phenomenon of isocapnic buffering forms the basis for probably the most widely used techniques for θ_L estimation.[33] Thus, for a rapid-incremental protocol, θ_L is taken as the $\dot{V}_{O_2}$ (not the work rate) at which the alveolar (end-tidal) oxygen pressure P_{CO_2}et and the ventilatory equivalent for oxygen ($\dot{V}_E/\dot{V}_{O_2}$) start to rise systematically without a simultaneous fall in the end-tidal CO_2 pressure P_{CO_2}et.[1,29]

An additional approach, known as the V-slope technique,[35] has arisen out of the recognition that in some individuals such as those with insensitive peripheral chemoreceptors, the respiratory system may be compromised in its ability to generate the required increase in $\dot{V}_E$ at higher work rates. As a result, if there is no additional ventilatory response, or it is poor, it proves difficult or impossible to discriminate θ_L according to these criteria.[34]

However, the work rate at which metabolic acidaemia first becomes evident is also associated with an accelerated rate of CO_2 production.[36] This reflects the additional contribution from HCO_3^--mediated buffering of lactic acid.[37] For example, in order to sustain a given ATP production rate under anaerobic conditions relative to aerobic conditions, the glycogen utilization rate must

increase more than 12-fold—a major concern, of course, for the marathon runner who runs too fast early in the race. The formation of two lactate molecules from one glucosyl unit of glycogen yields only three ATP molecules, compared with the 37 that result from its complete aerobic catabolism to CO_2 and water. Therefore the ATP production rate can only be sustained if the glycolytic flux increases by a factor $37/3 = 12.3$. The resulting 24.6 mEq lactic acid yield will be predominantly buffered by the HCO_3^- system;[9,38] $[HCO_3^-]_a$ will therefore decrease by about 22 mEq (i.e. some 90 per cent of 24.6). This yields an increase of about 22 mmol in CO_2 production. However 6 mmol of this replaces the CO_2 that would have been produced aerobically for this rate of ATP formation (i.e. as no oxygen is used in the glycogen-to-lactate catabolism, no CO_2 is directly released). The net increase in the CO_2 production rate is therefore 22 mmol anaerobic $CO_2 - 6$ mmol aerobic $CO_2 = 16$ mmol net CO_2 yield. This results in an increase in $\dot{V}_{CO_2}$ for these reactions during supra-θ_L exercise by a factor of about 2.5 relative to purely aerobic conditions.

It should be noted that this treatment considers the volume released from the blood compartment with no arterial hypocapnia. It also does not consider the changes within the exercising muscle compartment, which has a resting intracellular $[HCO_3^-]$ less than half that of blood, but which interacts with the blood compartment through ion-exchange mechanisms. Interested readers are referred to articles by Wasserman *et al.*[1] and Jones[39] for further consideration of this topic.

Beaver *et al.*[35] proposed that θ_L could be identified from the relationship between $\dot{V}_{CO_2}$ and $\dot{V}_{O_2}$. This relationship, which is characterized by a relatively linear relationship during moderate exercise, evidences an increased, but still essentially linear, slope within the isocapnic buffering phase. The intersection of these two linear phases has been shown to agree closely with the beginning of

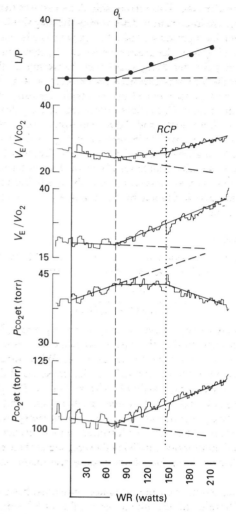

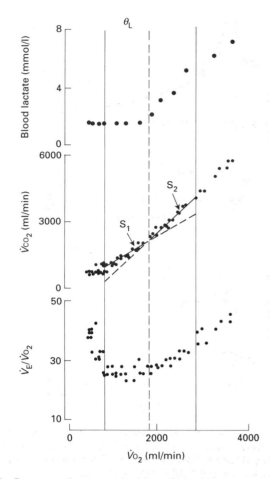

Fig. 1 Responses in a normal subject of the ratio of blood lactate to pyruvate concentrations (L/P), ventilatory equivalents for CO_2 and O_2 ($\dot{V}_E/\dot{V}CO_2$, $\dot{V}_E/\dot{V}O_2$), and end-tidal gas tensions (PCO_2et PO_2et), to an exhausting incremental exercise test (15 W/min) as a function of $\dot{V}O_2$. The full vertical line indicates start of the test, the broken vertical line indicates the lactate threshold θ_L, and the dotted vertical line indicates the onset of the respiratory compensation phase (RCP) for lactic acidosis. It should be noted that θ_L occurs at the work rate where both $\dot{V}_E/\dot{V}O_2$ and PO_2et start to increase, but PCO_2et does not yet fall. Typically, the respiratory compensation phase occurs approximately midway between θ_L and $\dot{V}O_2$max.

the increase in blood lactate concentration and the ratio of lactate to pyruvate concentrations and with the decrease in $[HCO_3^-]$[35] (Fig. 2).

It is important to point out that the resolving power of these non-invasive approaches depends on the rate at which the work rate is incremented. The more slowly the work rate is incremented the more slowly blood $[HCO_3^-]$ will fall, and therefore the smaller will be the contribution to $\dot{V}CO_2$ from the blood-buffering reactions.[40,41] Therefore rapid-incremental work rates are preferable. However, it should be recognized that the total extra volume yield of CO_2 accruing above θ_L is considerably greater for slow-incremental than for fast-incremental tests owing to the reduction of PCO_2a and the tissue CO_2 stores with which it equilibrates. This is because of the greater compensatory hyperventilation in the slow tests.[42]

Fig. 2 Responses in a normal subject of blood lactate concentration, $\dot{V}CO_2$, and the ventilatory equivalent for oxygen ($\dot{V}_E/\dot{V}O_2$) to an exhausting incremental test (20 W/min) as a function of $\dot{V}O_2$. The broken vertical line indicates the lactate threshold θ_L, and the left- and right-hand solid vertical lines indicate the region of interest for the V-slope analysis.[35] The V-slope parameters, S_1 and S_2 are the slopes of the regressions of the sub-θ_L and supra-θ_L regions respectively of the $\dot{V}CO_2$–$\dot{V}O_2$ relationship. It should be noted that the intersection of these two regressions coincides with θ_L. (Modified from ref. 13.)

Fatigue threshold

The tolerable duration for high intensity exercise decreases hyperbolically as a function of the absolute work rate[11,14,15] (Fig. 3)

$$(P - \theta_F)\, t = W' \tag{6}$$

where θ_F corresponds to the lower limit or asymptotic power. Interestingly, W' has the units of work and hence represents a constant amount of work that can be performed above θ_F, regardless of the rate at which it is performed. Therefore it can be regarded as an energy store comprised of oxygen stores, a phosphagen pool, and a source related to anaerobic glycolysis and the consequent production and accumulation of lactate.[11,14,15]

The asymptote θ_F (or critical power[14]) has been shown to represent the highest work rate for which a steady state can be attained in pulmonary gas exchange, blood acid–base status, and blood lactate concentration, given sufficient time.[10,11] θ_F can therefore be regarded as reflecting a rate of energy pool reconstitution which dictates the maximum power that can be sustained without a continued and progressive anaerobic contri-

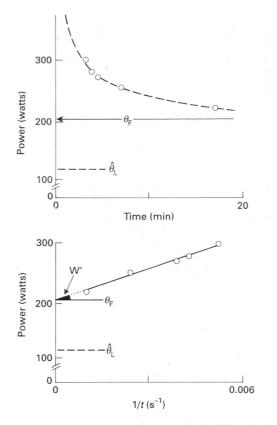

Fig. 3 The upper panel shows the power–duration P–t relationship for high intensity exercise. The lower panel shows the determination of parameters W' and θ_F from the linear P–$1/t$ formulation. θ_L is the estimated lactate threshold. (Modified from ref. 43.)

bution. Like θ_L, θ_F is high in trained individuals and can be increased by training.[43]

In many subjects, this higher sustainable lactate level occurs at about 4 to 5 mEq/l. This may be the reason that the work rate yielding a 4 mEq/l blood lactate concentration has proved such a useful index of endurance performance.

θ_F can be readily estimated by transforming the P–t relationship (equation (6)) into its linear formulation

$$P = W'/t + \theta_F \tag{7}$$

where W' and θ_F are the slope and intercept respectively (Fig. 3). The parameters W' and θ_F are determined from the linear regression of P versus $1/t$ from a series of discrete bouts of exhausting supra-θ_L constant-load exercise, preferably performed on different days.[11] Four or five tests typically provide good definition of the relationship, as long as the tolerable duration is spaced relatively evenly on the $1/t$ axis (i.e. not evenly with respect to t).

It is important to emphasize that this relationship is unlikely to provide a precise representation of the actual physiological behaviour at the very extremes of performance because of distorting factors such as (a) limitations of muscular (mechanical) force generation for the very highest power requirements and (b) constraints resulting from substrate provision and thermoregulatory or body fluid requirements for markedly prolonged exercise.[11]

Determinants of ventilatory requirements
Alveolar ventilation

The alveolar air equation described earlier (equation (2)) provides the basic frame of reference for predicting the ventilatory require-

ment of a particular task. This relationship demonstrates that both $\dot{V}_{CO_2}$ and $P_{CO_2}a$ are important in determining the magnitude of the $\dot{V}_A$ requirement for a particular task. The greater the demand for metabolic CO_2 clearance, the greater will be the $\dot{V}_A$ requirement at a particular set point for $P_{CO_2 a}$ regulation. Lowering of the set point, for example for a sea-level native sojourning at high altitude or with hyperventilation induced by metabolic acidosis, will require a larger increase in $\dot{V}_A$ to effect a given rate of CO_2 clearance.

Furthermore, both $\dot{V}_{CO_2}$ and $P_{CO_2}a$ determine the compensatory potential of a particular increment of $\dot{V}_A$. As a result, the greatest stress to potential encroachment on the functional ventilatory limits for the elite athlete results from the combination of high levels of $\dot{V}_{CO_2}$ and low levels of $P_{CO_2}a$; the less fit individual is protected from such limitation by the relatively low achievable levels of metabolic rate.

Total ventilation

Substituting for $\dot{V}_A$ in equation (2) from equation (4) provides the total ventilatory requirement for a given work rate:

$$\dot{V}_E = \frac{863\,\dot{V}_{CO_2}}{P_{CO_2}a\,(1 - V_D/V_T)} \tag{8}$$

It should be noted that, on cursory inspection, equation (8) does not appear to characterize the linear isocapnic $\dot{V}_E$–$\dot{V}_{CO_2}$ relationship for steady state moderate exercise. What makes $\dot{V}_E$ linearly related to $\dot{V}_{CO_2}$ isocapnically under these conditions is that V_D/V_T decreases hyperbolically with respect to $\dot{V}_{CO_2}$.[11]

Equation (8) indicates that the three variables $\dot{V}_{CO_2}$, $P_{CO_2}a$ and V_D/V_T collectively determine the overall ventilatory requirements for exercise. This can be illustrated with reference to an elite athlete and an untrained individual exercising at their respective θ_L and $\dot{V}_{O_2}max$ levels based upon some reasonable assumptions.

For example, assume that the untrained individual has a steady state $\dot{V}_{O_2}$ of 2 l/min and the athlete has a value of 5 l/min at θ_L. With the further assumption that carbohydrate is the substrate undergoing oxidation (i.e. $RQ = 1.0$), the corresponding requirements for metabolic CO_2 clearance (i.e. $\dot{V}_{CO_2}$) will also be 2.0 l/min for the untrained subject and 5.0 l/min for the athlete. Assume that both have a normal set-point $P_{CO_2}a$ of 40 mmHg. The $\dot{V}_A$ requirements will be 43 l/min for the untrained subject and 108 l/min for the athlete (equation (2)). However, V_D/V_T is likely to be somewhat lower for the athlete owing to the higher work rate at θ_L. Therefore the untrained individual with V_D/V_T 0.15 at θ_L has a $\dot{V}_E$ requirement of some 50 l/min; for the athlete, with $V_D/V_T = 0.1$, this requirement is 120 l/min (equation (4)).

A similar analysis can be applied at maximum exercise, with the assumption that $\dot{V}_{O_2}max$ is 3 l/min for the untrained subject and 7 l/min for the elite athlete. However, while carbohydrate is still likely to be the dominant metabolic substrate ($RQ = 1.0$), R is now dissociated from RQ owing to the additional CO_2 clearance coming from HCO_3^- buffering of lactic acid and the reduction of CO_2 stores resulting from the compensatory hyperventilation. With the reasonable assumption that R at maximum is 1.2 in both subjects,[1] $\dot{V}_{CO_2}max$ will be 3.6 l/min for the untrained subject and 8.4 l/min for the athlete. Allowing for a 10 mmHg reduction of $P_{CO_2}a$ at maximum (i.e. $P_{CO_2}a = 30$ mmHg), the $\dot{V}_A$ requirements will be 103 l/min and 241 l/min (equation (2)) for the untrained individual and the athlete respectively. However, both subjects are now likely to have a similar V_D/V_T at a maximum of 0.1, which yields a $\dot{V}_E$ requirement of 114 l/min for the untrained subject and 267 l/min for the athlete (equation (4)).

This is an impossible value, even for a large athlete. Consequently, if the athlete's $\dot{V}_E$ is mechanically limited to about $200\,l/$min, equations (2) and (4) indicate that PCO_2a would not be reduced below 40 mmHg, given an unchanged $\dot{V}CO_2$max of $8.4\,l/$min. The fall of arterial and muscle pH would therefore not be constrained; rather, rapid fatigue would be the consequence. It should also be noted that arterial hypoxaemia would also result, as evidenced by the lower $\dot{V}_E/\dot{V}O_2$.

Some investigators[45,46] have suggested that athletes tend to have low peripheral chemosensitivity to hypoxia (as do their non-athletic relatives). As both hypoxia and H^+ are sensed by the carotid bodies (and even single afferents), it is tempting to speculate that they are also insensitive to the exercise-induced metabolic acidaemia. Whether this reduced peripheral chemosensitivity also ameliorates the sensation of shortness of breath in the athlete remains to be determined.

THE COSTS OF MEETING THE VENTILATORY REQUIREMENTS OF EXERCISE

The increased ventilatory requirements for arterial blood-gas and acid–base homeostasis during exercise exact increased mechanical cost, with respect to both work and power, from both the respiratory and cardiac pumps.

Ventilatory costs

The provision of energy at a sufficient rate to support the increased costs incurred by the respiratory muscles during exercise requires increased local vascular perfusion, as well as adequate levels of stored substrates that can readily be mobilized. Without this, respiratory muscle fatigue ensues.

Mechanical costs

The respiratory power, i.e. work rate, generated during the breathing cycle is manifested as the pressure changes which distend the lungs via the chest wall (the elastic component or elastance) and generate air flow through the conducting airways (the resistive component or resistance). The third component of the total pulmonary impedance, the accelerative component (or inertance) is often disregarded as it is usually such a small component of the total. This is because, although the chest wall has a large mass, its acceleration during breathing is low, and although the acceleration of the inspired gas can be high, its mass is low.

During exercise, the respiratory power increases curvilinearly with respect to $\dot{V}_E$ such that a greater increment in power is required to establish a given increase in $\dot{V}_E$ as the work rate increases.[47–52] In moderately fit subjects exercising near maximum, respiratory muscle power has been estimated to be only about 30 per cent of that achieved on a maximal volitional test (i.e. the maximum voluntary ventilation*(MVV)).[53] The average in the more recent work of Klas and Dempsey[54] was 40 per cent.

Similarly, the maximum exercise $\dot{V}_E$ is appreciably less than the MVV in such subjects. This difference has been termed the 'breathing reserve' (Fig. 4).[1] As the MVV is largely independent of fitness and training status, a highly fit athlete who is capable of achieving high levels of $\dot{V}_E$ during exercise has a lower breathing reserve than does the less fit subject.

---

* The $\dot{V}_E$ that may be volitionally attained for periods as short as 15 s.

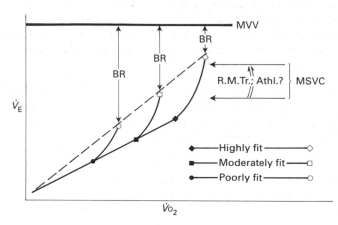

Fig. 4 Schematic representation of the $\dot{V}_E$ response to exhausting incremental exercise as a function of $\dot{V}O_2$ for poorly fit (circles), normal (squares), and highly fit (diamonds) subjects. The $\dot{V}_E$ responses are similar below the lactate threshold (full symbols) regardless of fitness. However, $\dot{V}_E$ at maximum exercise $\dot{V}_E$max is progressively higher the greater the maximum $\dot{V}O_2$ (open symbols), and the breathing reserve (BR) therefore becomes reduced (BR = MVV − $\dot{V}_E$max). The MVV is largely independent of fitness and training. However, the maximum sustained ventilatory capacity (MSVC) can be increased by endurance training of the respiratory muscles (R.M.Tr) and is considered to be high in highly fit athletes (Athl.?). (Modified from ref. 55.)

The greater mechanical cost of ventilation at high work rates is predictable on the basis of the increased contributions from turbulence (and even inertia) when air flow is high.[56,57] In addition, there may also be an increased elastic work of breathing owing to a decreased compliance of the lungs in the tidal range during exercise;[58–61] this effect has been ascribed to the increase in pulmonary blood volume.[61] Other studies, in contrast, have reported no change in compliance[62,63], or even an increase.[47] However, the progressive increase of end-inspiratory lung volume that occurs as V_T increases during exercise will tend to encroach on the upper poorly compliant region of the compliance curve,[64] although this effect will be ameliorated by the decrease in end-expiratory lung volume.

It is important to recognize that estimates of the components of respiratory muscle work often fail to take into account several factors that may be important at maximal work rates in elite athletes, i.e. with extremely high levels of $\dot{V}_E$ and high breathing frequencies. For example, there may be considerable distortion of the chest wall at these high work rates,[65,66] and its inertial contributions to the total impedance may no longer be neglected.[67,68]

Attempts have been made to partition respiratory muscle power to individual muscles of respiration.[69] However, those estimates that are presently available have been obtained during maximal volitional ventilatory manoeuvres; we do not believe that this condition is useful as a mechanical or metabolic analogue of ventilation during muscular exercise.

Metabolic costs

The mechanical costs of ventilation appear to be satisfied, in large part, by the aerobic energy-yielding process in the respiratory muscles. Thus, during exercise, the oxygen consumption of the respiratory muscles $\dot{V}O_2$rm increases progressively relative to $\dot{V}_E$.

However, the relationship is concave upwards, i.e. a greater increment in $\dot{V}_{O_2}$rm is required to establish a given increase in $\dot{V}_E$ as work rate increases.[49,67,70-77] It is important to point out that estimation of $\dot{V}_{O_2}$rm in humans is technically quite difficult owing to its small size relative to whole-body $\dot{V}_{O_2}$.

A further confounding influence is the fact that if P_{CO_2}a is allowed to fall in tests which attempt to mimic the exercise hyperpnoea at rest, then the whole-body $\dot{V}_{O_2}$ is increased by the alkalosis by about 10 per cent per 10 mmHg reduction in P_{CO_2}a.[78] This is large compared with the actual $\dot{V}_{O_2}$ cost of exercise ventilation.

However, while $\dot{V}_{O_2}$rm is small when $\dot{V}_E$ is at resting levels, it can be a significant factor during near-maximal exercise, particularly in highly fit athletes who attain extremely high levels of $\dot{V}_E$. Normal subjects have been reported to attain a $\dot{V}_{O_2}$rm of some 0.5 l/min for a $\dot{V}_E$ in excess of 120 l/min, i.e. about 14 per cent of the total $\dot{V}_{O_2}$ for a subject with a $\dot{V}_{O_2}$max of 4 l/min. Shephard[76] has reported that, in the $\dot{V}_E$ range of 90–130 l/min, $\dot{V}_{O_2}$rm increases by about 4.4 ml/min for each V_E increase of 1 l/min. In athletes who attain much higher levels of $\dot{V}_E$ at maximum, $\dot{V}_{O_2}$rm is presumably even greater.

The vascular perfusion of the respiratory muscles $\dot{Q}_{rm}$ plays an important supportive role in the generation of $\dot{V}_{O_2}$rm. As the arterial oxygen content C_{O_2}a remains essentially stable over the entire work-rate range (however, P_{O_2}a can decrease at high work rates in highly trained athletes,[74] as described below), the increased oxygen delivery to the respiratory muscles necessary to sustain increased levels of $\dot{V}_E$ depends on an appropriate increase in $\dot{Q}_{rm}$.

What is less certain in humans is the exact magnitude of the $\dot{Q}_{rm}$ response during exercise, and how it is partitioned between the contributing muscle groups. Whipp and Pardy[55] have provided an estimate of the $\dot{Q}_{rm}$ response to maximum exercise in moderately fit subjects using the $\dot{V}_{O_2}$rm data of Shephard[76] together with the assumption that the arteriovenous oxygen content difference across the respiratory muscles increases to 15 ml/100 ml at maximum. This yielded a value for $\dot{Q}_{rm}$ at maximum exercise of 3.8 l/min (i.e. $\dot{V}_{O_2}$rm = 570 ml/min and $\dot{V}_E$ = 130 l/min) or about 15 per cent of a maximal cardiac output of 25 l/min. However, extrapolating to the human respiratory musculature from blood flow measurements in the diaphragm of the dog made by Robertson et al.,[79] Bye et al.[80] have suggested that $\dot{Q}_{rm}$ may be as high as 8 l/min and $\dot{V}_{O_2}$rm in excess of 1 l/min at maximum exercise. Johnson[81] has even argued that the upper limit for sustained operation of the respiratory muscles may be set by the cardiovascular system through the vascular pressure-flow properties of the respiratory muscles.

These considerations raise the important issue of whether the demands for oxygen utilization by the respiratory muscles during exercise outstrip the ability of vascular supply mechanisms to deliver oxygen at the appropriate rate. Do the respiratory muscles resemble other skeletal muscles in having a demonstrable lactate threshold?

It is known that the duration for which a particular level of $\dot{V}_E$ can be sustained bears an inverse curvilinear relationship to $\dot{V}_E$ during voluntary hyperpnoea. We can infer that such a relationship also holds for the hyperpnoea of muscular exercise (Fig. 5(b)).[77,82] However, while this relationship may be qualitatively similar, there may be significant quantitative differences, both with regard to the actual mechanical cost of achieving a given level of $\dot{V}_E$ (i.e. the respiratory muscle recruitment pattern and mechanical efficiencies are likely to be different) and the metabolic cost of the $\dot{V}_E$ which is attained.

The actual response resembles the hyperbolic whole-body power–duration relationship described earlier (Fig. 3) (equations (6) and (7)). Whether the $\dot{V}_E$–duration relationship is also well described by a hyperbolic function and whether it provides a reasonably accurate representation of the relationship that actually obtains during high intensity exercise has not yet been established.

In the context of the $\dot{V}_E$–duration relationship, the MVV constitutes an upper limit for exercise ventilation.* An index of respiratory muscle endurance is provided by the horizontal $\dot{V}_E$asymptote—the maximum sustained ventilatory capacity (MSVC),[77,83] which may be usefully regarded as the highest level of ventilation which can be sustained for 'long periods'. As a corollary, respiratory muscle fatigue is considered to develop when $\dot{V}_E$ exceeds MSVC.[84,85] In moderately fit individuals, the MSVC lies in the range of 55 to 80 per cent of the MVV;[77,82,86-89] however, in highly fit athletes, it may be as much as 90 per cent of MVV.[90]

Of course the actual $\dot{V}_E$ attained during maximum exercise is appreciably greater than the MSVC and apparently lies even 'beyond' the $\dot{V}_E$ duration curve itself. The explanation for this is given in Fig. 5. The $\dot{V}_E$ response to exhausting incremental exercise and the subject's sustainable $\dot{V}_E$ versus time plot are presented in Figs. 5(a) and 5(b). In Fig. 5(c) both relationships share a common axis; however, the fatiguing component of the respiratory muscles does not begin until $\dot{V}_E$ exceeds MSVC (i.e. at MSVC (t_0)). When the actual $\dot{V}_E$ reaches the MSVC curve (point (i)), the subject is capable of sustaining that level of $\dot{V}_E$ for time $t_{(i)}$, i.e. the area $x+y$ determines the end-point of the respiratory muscle fatigue. The subject's actual $\dot{V}_E$–t integral at this time is 'only' area x. Consequently, the subject can continue until the attainment of a $\dot{V}_E$ at which area z is equal to area y which is at a much higher $\dot{V}_E$ (i.e. equivalent to sustaining the constant-level $\dot{V}_E$ shown at point (ii) for the time $t_{(ii)}$). Whether certain categories of athlete ever attain this limiting fatigue value before metabolic factors in the task-subserving muscles lead to exercise limitation remains to be conclusively demonstrated.

It is tempting to speculate that the MSVC represents the highest $\dot{V}_E$ that can be maintained without a sustained and progressive metabolic acidosis of the respiratory muscles i.e. analogous to the fatigue threshold θ_F or critical power for whole-body exercise. Furthermore, the actual characterization of the $\dot{V}_E$–duration relationship has not been well described in elite athletes, who are likely to have highly trained respiratory musculature. There is particular paucity of information on athletes such as swimmers and rowers, in whom the demands of performance impose a mechanical imperative on the pattern of breathing.

Both $\dot{V}_{O_2}$rm and respiratory muscle endurance increase in normal individuals after specific respiratory muscle endurance training.[77,79] Whether this increased $\dot{V}_{O_2}$rm is the result of an increased $\dot{Q}_{rm}$ an improved oxygen extraction, or both is not known. Also, it is not known whether the regional 'matching' of $\dot{Q}_{rm}$ to local metabolic rate is optimized to any extent by training strategies.

Perfusion costs

In contrast with the respiratory muscles, cardiac power and cardiac output appear to be essentially linearly related throughout the entire work-rate range.[81]

---

* This is based on the assumption that the respiratory muscles operate during the MVV manoeuvre in the same manner as for the spontaneous breathing exercise (i.e. with respect to recruitment pattern, mechanical cost, and efficiency of contraction) and that there is no exercise-induced bronchodilatation or bronchoconstriction.

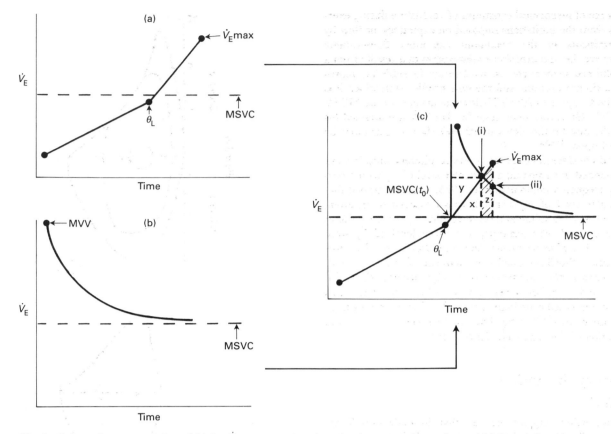

Fig. 5 Schematic representation of (a) the $\dot{V}_E$ response to exhausting incremental exercise and (b) the relationship between sustainable $\dot{V}_E$ and exercise duration in a normal subject. These relationships are superimposed in (c). Respiratory muscle fatigue starts to develop at MSVC (t_0 i.e. when $\dot{V}_E$ = MSVC. At point (i), where $\dot{V}_E$ crosses the MSVC curve, that level of $\dot{V}_E$ can be sustained for the entire period $t_{(i)}$. However, it should be noted that a much higher level of $\dot{V}_E$ can actually be attained during exercise $\dot{V}_E$max. See text for discussion.

Cardiac muscle power has been estimated to be about 30 times greater than the respiratory muscle power in normal subjects at rest.[81] However, at maximum exercise, the respiratory muscle power may be as much as three times greater than the cardiac power.[81] In fact, the actual difference at maximum exercise may be more marked than this, as no respiratory compensation for the metabolic acidosis was included in the $\dot{V}_E$ response profiles. Therefore respiratory muscle power may have been underestimated both absolutely and relative to the cardiac power at the highest work rates attainable in the athlete (see above).

These observations on the relative costs of ventilatory and cardiac pump function support the contention that although the myocardial oxygen consumption increases with increases in cardiac output, it is likely to be a trivially small component of the total[81,91] even at the high maximal cardiac outputs that can be achieved by some highly fit athletes.

CONSTRAINTS AND LIMITATIONS ON RESPIRATORY SYSTEM PERFORMANCE

The mechanical and metabolic costs of ventilation during muscular exercise are themselves sources of constraint and potential limitation.[55]

In the context of the respiratory system, constraints refers to a condition in which the ventilatory response achieved is lower than the required response owing to the influence of an opposing mechanism, despite the system's not being limited from further increases in $\dot{V}_E$ (for example the presence of applied restrictive loads).

In contrast, limitation refers to situations in which the variable is actually prevented from increasing, despite an increased ventilatory requirement. For example, reduced lung recoil and/or increased airways resistance limits the maximum expiratory air flow attainable during exercise (at a particular lung volume) despite further increases in ventilatory drive; increased elastance (i.e. reduced compliance) in patients with diffuse interstitial fibrosis limits the achievable tidal volume during exercise.

Ventilatory constraints

The operating limit for tidal volume V_T in humans ranges from zero to vital capacity. The breathing frequency can range from zero to about 5 to 7 Hz,[92] with the upper limit being set by the rate at which the neuromuscular apparatus can generate rapid alternating movements.[67,92] In moderately fit individuals, however, the ventilatory pump during maximum exercise typically operates at a frequency of no more than 1 Hz and a V_T of only about 50 to 60 per cent of the vital capacity. As the maximum V_T attained during exercise does not increase appreciably as a result of physical training or increased fitness, further increases in $\dot{V}_E$ must be accomplished through breathing frequency.

One source of mechanical constraint of ventilation during exercise arises from the limitations imposed on expiratory air flow by the determinants of the maximum expiratory flow–volume (MEFV) curve, i.e. the expiratory relationship that results from a forced volitional manoeuvre. In moderately fit subjects, ventilation at maximum exercise, and the flow profiles with which it is accomplished, appear to fall well below the maxima of the MEFV curve.[60,61,93,94] However, when high levels of $\dot{V}_E$ are attained (as would be the case for the highly fit athlete), these maxima may be encroached upon.[54,94,95]

A second consideration relates to the respiratory muscle work that is expended in sustaining a particular level of $\dot{V}_E$ with a particular V_T–frequency combination. In 1925, Rohrer[56] argued that any particular level of $\dot{V}_E$ would have an associated breathing frequency at which a minimal level of respiratory work would, in theory, be exacted. For example, if a given level of $\dot{V}_E$ were generated with a high frequency (and therefore a low V_T), this would increase the flow-resistive component of the respiratory work. In contrast, when the same level of $\dot{V}_E$ is accomplished with a large V_T and therefore a very low breathing frequency, this results in an increased contribution from the elastic component of the respiratory work[75,96] if the lung volume encroaches on the flatter position of the lung compliance curve.

Ventilatory limitations

Mechanical

Several observations support the view that, in moderately fit individuals, ventilation is neither mechanically limited nor associated with respiratory muscle fatigue during maximal exercise:

(a) such subjects can increase $\dot{V}_E$ substantially above the spontaneous maximal value at maximum exercise by volitional means;

(b) inspiratory muscle fatigue cannot be demonstrated in such subjects at high work rates;[97]

(c) the ratio of maximum exercise $\dot{V}_E$ to MVV is relatively low, i.e. about 60 to 70 per cent;[95,98–100]

(d) the spontaneous expiratory flow–volume curve at high work rates does not encroach on the boundaries of the MEFV curve.[60,61,93]

In contrast, evidence of both ventilatory–mechanical limitation and inspiratory muscle fatigue may emerge in fitter subjects. For example, in subjects with $\dot{V}_{O_2}$max of about 5 to 6 l/min and a maximum exercise $\dot{V}_E$ of about 110 to 160 l/min (i.e. some 80 per cent of MVV), Hesser et al.[53] reported that the spontaneous expiratory flow–volume curve impacted on the outer envelope of the MEFV curve during maximal exercise. Similar findings have been reported by Olafsson and Hyatt[60] Grimby et al.[95], Johnson,[81] and Klas and Dempsey[54] (Fig. 6). However, it appears that such subjects only generate sufficient pleural pressures to establish the maximum flow.[54,59,95] In contrast, subjects generate appreciably greater (and wasteful) pressures during MVV manoeuvres.[54] Furthermore, Bye et al.[80] have demonstrated evidence of diaphragmatic fatigue in terms of both electromyographic criteria and as a result of reduced maximum transdiaphragmatic pressures after exhausting exercise.

However, one should be cautious about using the resting MEFV curve, and even the resting MVV curve as the frame of reference for deciding whether there is air flow limitation during exercise. The reasons for this are as follows.

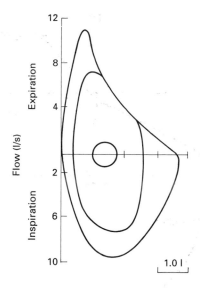

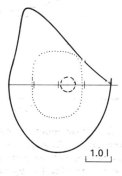

Fig. 6 Spontaneous flow–volume curves generated at rest (inner loop) and maximal exercise (middle loop) in a fit subject (left-hand panel) and a normal untrained subject (right-hand panel) compared with their maximum flow–volume curves (outer loop). See text for discussion. (Modified from ref. 54 and from ref. 93.)

(a) Locating the spontaneous expiratory flow–volume curve on the MEFV curve is crucial. It is often assumed that the subject's total lung capacity does not change. It is better to 'trap' a flow–volume display oscillographically on a particular breath and then have an MEFV manoeuvre performed immediately thereafter, ensuring that the subject really goes to the extremes of thoracic excursion on the manoeuvre.

(b) Some bronchodilatation can result during exercise from the increased circulating levels of catecholamines—this presumably accounts for the increased MVV that has been reported during exercise.

(c) In many subjects, the maximum expiratory effort flow–volume manoeuvre does not yield that optimum maximum expiratory flow–volume curve, although this is more likely to be a factor in older athletes with diminished lung recoil and airways function than in healthy young athletes.

There is also now a consistent body of evidence which demonstrates that, when helium is used to replace nitrogen in the inspirate at high work rates, $\dot{V}_E$ abruptly increases with consequent hypocapnia,[101–103] even when care is taken to mask the sudden sensation of cold in the airways associated with helium breathing.[102] As helium is about 39 per cent less dense than nitrogen, it is thought that the turbulent component of air-flow resistance is reduced when helium is breathed. Therefore, it appears that there is normally a component of air-flow constraint at high work rates. This will presumably be appreciable in elite athletes who attain high levels of $\dot{V}_E$. However, as airway dimensions play an important role in the onset and magnitude of the air-flow turbulence, lung structure will also be important.

It should be noted that reduction of the vital capacity and respiratory muscle strength and endurance have only been demonstrated consistently in athletes following prolonged exercise.[100,104,105]

Consequently, at high work rates an athlete is confronted with a control dilemma: if $\dot{V}_E$ is increased further to provide respiratory compensation for the metabolic acidosis (i.e. to constrain the pH fall in blood and exercising muscle), the additional $\dot{V}_E$ necessary can be so large (as described above) that it is likely to induce ventilatory mechanical limitation and possibly ventilatory muscle fatigue.[106] However, the absence of this additional $\dot{V}_E$, although tending to protect the subject from pulmonary mechanical limitation, predisposes to both arterial hypoxaemia and a more precipitous fall in blood and muscle pH impairing contractile responses[107] and exacerbating intramuscular fatigue.

Metabolic

The respiratory muscles may also reach their metabolic limits during maximum exercise in highly fit athletes. This reflects not only that $\dot{V}_E$ at maximum exercise is high in such subjects, but also that the relationship between $\dot{V}O_2$rm and $\dot{V}_E$ becomes steeper in this $\dot{V}_E$ range.[49,67,70,71–75,83,108] Thus a greater component of the systemic oxygen supply is diverted away from the exercising limb musculature at high levels of $\dot{V}_E$.

This raises the notion of a theoretical limiting (or maximal) level of $\dot{V}_E$ above which the oxygen requirement of the respiratory muscles becomes sufficiently large that it requires the entire further increment in whole-body $\dot{V}O_2$ to be satisfied. This has been estimated by several investigators to lie in the range 120 to 160 litres/min,[48,67,76] i.e. levels of $\dot{V}_E$ which are not uncommon in elite athletes. If the energetic requirements of the exercising limb musculature are preferentially met, then the muscles of respiration would be predisposed towards fatigue. Conversely, if the requirements of the respiratory muscles are to be met, the limb muscles would be predisposed to premature fatigue.

Perfusion-related limitation

It is important to understand not only whey some athletic subjects do not become hypoxaemic at high work rates, but also why there is not a systematic hyperoxaemia. End-tidal, mean alveolar, and 'ideal' alveolar PO_2 increase systematically above θ_L; arterial PO_2 does not! Consequently, the difference between alveolar (ideal) and arterial PO_2 ($[A-a]O_2$) increase further, reaching 20 to 30 mmHg at high work rates.

At sea level, moderately fit individuals maintain arterial PO_2 and oxygen saturation at, or close to, resting values even during exhausting work rates.[109–111] As the topographic distribution of $\dot{V}_A$ and $\dot{Q}$ has been shown to improve during upright exercise,[112–114] it has been assumed that the distribution of alveolar ventilation to perfusion $\dot{V}_A/\dot{Q}$ within the lung is also improved and hence that the lung exchanges gas more efficiently during exercise than at rest.

However, this notion has been brought into question as a result of recent investigations utilizing the multiple inert gas technique

of Wagner et al.[115] For example, Gledhill et al.[111] have demonstrated that the mean $\dot{V}_A/\dot{Q}$ for the entire lung increased during steady state exercise. This must occur because

$$\dot{V}_A(FO_2I - FO_2A) \Leftarrow \dot{V}_{O_2} \Rightarrow \dot{Q}(CO_2a - CO_2\bar{v})$$

and therefore

$$\frac{\dot{V}_A}{\dot{Q}} = \frac{CO_2a - CO_2\bar{v}}{FO_2I - FO_2A}$$

For simplicity, we have disregarded the effect of the different inspired and expired tidal volumes which occurs when $R \neq 1$ as it does not materially affect the argument. That is, $CO_2\bar{v}$ decreases systematically whereas FO_2A remains relatively constant as work rate increases in this range; therefore $\dot{V}_A/\dot{Q}$ increases. In contrast, however, the $\dot{V}_A/\dot{Q}$ dispersion within lung was reported to be increased and to constitute a major influence on the widening of the $[A-a]O_2$ that is normally seen during moderate and heavy exercise. Likewise, Hammond et al.[116] demonstrated increased $\dot{V}_A/\dot{Q}$ inequality at high work rates (i.e. $\dot{V}O_2 > 3$ l/min). However, they argued that this could not entirely account for the widened $[A-a]O_2$. The mechanisms of the gas exchange impairment implicit in the wider dispersion of $\dot{V}_A/\dot{Q}$ at high work rates remain to be elucidated.

In contrast, at more moderate work rates ($\dot{V}O_2 < 2$ l/min), Derkes[117] reported the $\dot{Q}$ distribution to be unchanged (or even narrowed). Furthermore, neither Hammond et al.[116] nor Gale et al.[118] were able to detect statistically significant changes in $\dot{V}_A/\dot{Q}$ dispersion.

The recent work of Hammond et al.[116] and Dempsey et al.,[24] coupled with the earlier work of Johnson,[119] implicates diffusion impairment as a contributor to the widened $[A-a]O_2$, at least for very high work rates. On the basis of the demonstration that PO_2a fell at such work rates in highly fit subjects, Dempsey et al.[24] have suggested that a diffusion impairment could arise from a pulmonary capillary transit time that was too short (owing to the high pulmonary blood flow) to allow diffusion equilibrium to be attained throughout the lung. The observation of Hammond et al.[116] agree with this view. Neither the increased dispersion of $\dot{V}_A/\dot{Q}$ nor the increased postpulmonary shunt that these investigators demonstrated in highly fit subjects (exercising in the steady state at 300 W) were sufficiently large to account entirely for the widening of $[A-a]O_2$ at this high work rate.

Whether there is equilibrium of pulmonary end-capillary PO_2 with the alveolar PO_2 depends both on the rate at which PO_2 rises in the capillary bed (i.e. its time constant) and the time available for exchange. As the diffusing capacity DO_2L for oxygen is a crucial determinant of the time constant, a large DO_2L in addition to a long residence time in the pulmonary capillary bed, is vital in preventing arterial hypoxaemia. The reader is referred to refs. 119 and 120 for a more comprehensive discussion of limitations of diffusive gas exchange during exercise.

Normally, approximately 0.3 s is required for the PO_2 (and hence haemoglobin saturation) of the blood traversing the pulmonary capillary bed to increase to its equilibrium with the alveolar gas; this will be longer during exercise, however, particularly in trained athletes.[24,81,120] Therefore, if the flow through the capillary bed is sufficiently great, the blood will leave the gas-exchanging region and pass into the pulmonary veins before equilibrium occurs; hypoxaemia is the inevitable consequence.

Whether or not there is sufficient residence time in the

pulmonary capillary bed depends on both the capillary volume V_c and the pulmonary blood flow $\dot{Q}$, i.e.

$$T_{tr} = \frac{V_c \; (ml)}{\dot{Q} \; (ml/s)}$$

where T_{tr} is the mean transit time.

Therefore, to prevent critical reductions of the pulmonary capillary residence time, the athlete must recruit capillary volume as the capillary flow increases. However, there is a morphological limit to the capillary dimension.[121]

The relationship between T_{tr} and V_c is presented in Fig. 7. Note that, if there were to be no increase in the resting V_c of about 80 ml, then the critical transit time of 0.3 s would be achieved at a $\dot{V}_{O_2}$ of only 2 l/min. Doubling the capillary volume to 160 ml would lead to the critical transit time occurring at a $\dot{V}_{O_2}$ of 'only' 5 l/min. At a $\dot{V}_{O_2}$ of 7 l/min, this occurs even if the capillary bed has expanded to 200 ml. The broken line in the left-hand panel reflects the increase in the time necessary for equilibrium as work rate increases.[24,81,120] However, this relationship is presently uncertain at high work rates and hence this line should only be considered an approximation, but one that is notionally important for the mechanism.

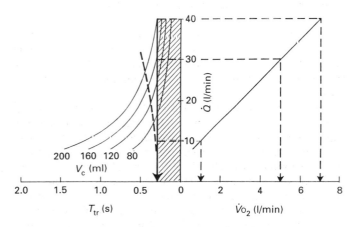

Fig. 7 Schematic representations of the relationship between cardiac output $\dot{Q}$ and $\dot{V}_{O_2}$ (right-hand panel) and of the relationship between cardiac output $\dot{Q}$ and the mean pulmonary capillary transit time T_{tr}, with superimposed isopleths for pulmonary capillary blood volume V_c (left-hand panel). The full vertical line reflects the increase in the time necessary for diffusion equilibrium at higher work rates. See text for discussion.

It should further be recognized that there is a distribution of both pulmonary capillary length and diameter, and therefore probably a similar distribution of transit time. As transit times longer than the critical do not further increase oxygenation but the shorter ones reduce it further, then the actual hypoxaemia-inducing effect will begin to be manifest at work rates less than suggested by the mean transit time consideration.

A useful rule of thumb to determine the minimum pulmonary capillary volume (ml) that will maintain the mean transit time at 0.3 s is to multiply the cardiac output (l/min) by 5 (for a mean transit time of 0.6 s, the multiplier is 10), i.e.

$$V_c(T_{tr} = 0.3) = 5\dot{Q} \quad \text{or} \quad V_c(T_{tr} = 0.6) = 10\dot{Q}$$

The ability of the elite athlete to perform intense exercise without developing arterial hypoxaemia depends critically on such factors in order to minimize or obviate the development of diffusion impairment[120] which provides additional stress on already taxed mechanisms of tissue energy provision and ventilatory control and which can therefore limit exercise performance. This is supported by the fact that when the hypoxaemia was prevented by increasing the inhaled oxygen fraction to maintain resting levels of $P_{O_2}a$, exercise tolerance increased significantly.[122]

Consideration of Fig. 7 raises the question of why all subjects exercising at high levels of pulmonary blood flow do not develop arterial hypoxaemia. The answer is likely to reside in the fact that the genetic make-up of some athletes leads to development of a large-volume pulmonary capillary bed. Consequently, it is likely that for future record performances in athletic events, an athlete will have to be 'elite' not only in the metabolic and cardiovascular capabilities, but also in terms of pulmonary capabilities.

This is likely to be dominated by hereditary factors, as physical training does not systematically improve these indices of pulmonary function. However, it should be pointed out that this concept is based on studies in adults i.e. when the lung is mature. Whether these adult dimensions can be improved by training during the period of lung growth remains to be determined.

A further potential concern for the elite athlete is that the high pulmonary vascular pressures during exercise[123] predispose to increased pulmonary interstitial oedema. Whether or not there is a net fluid flux J across the alveolar capillary membrane during exercise will depend upon the difference between the hydrostatic pressure tending to 'push' water out of the vascular bed and the osmotic pressure tending to 'pull' it back. These forces can be quantified using the Starling equation, i.e.

$$J = K(P_c - P_i) - K\sigma(\Pi_c - \Pi_i) \tag{9}$$

where K is the filtration coefficient, P_c and Π_c are respectively the vascular hydrostatic and colloid osmotic pressures in the capillary, and P_i and Π_i are these pressures in the pulmonary interstitial (i.e. perimicrovascular) space. The term σ is the 'reflection coefficient', which is an index of the degree of the membrane impermeability to protein.

Not surprisingly, perhaps, the behaviour of many of the variables in the Starling equation during exercise is poorly understood. However, while transpulmonary microvascular pressure and the area of the perfused surface both increase during exercise, there is no evidence of impaired capillary permeability.

O'Brodovich and Coates[124] cite three safety mechanisms that prevent pulmonary oedema: (a) increased lung lymph flow, (b) fluid movement into the interstitial space which will dilute its protein concentration and hence lower its osmotic pressure, and (c) the low compliance of the interstitial space which will lead to a large increase in perimicrovascular pressure for a small change in volume. Mechanisms (a) and (b) occur even with mild exercise, and (c) becomes operative if oedema occurs. Based upon a careful analysis of the available evidence, O'Brodovich and Coates[124], and also Brower and Permutt[125], have concluded that these safety mechanisms are adequate to prevent the development of pulmonary oedema in normal humans. Furthermore, the evidence suggesting the development of interstitial oedema during severe exercise is not only indirect, but inconclusive. However, it is important to establish conclusively whether or not this occurs in highly trained athletes, because the presence of such oedema during severe exercise would predispose the athlete to (a) reduced gas exchange efficiency and hypoxemia, and (b) pulmonary J-receptor stimulation with its attendant tachypnoea and possibly increased respiratory sensation and even decreased exercise tolerance through inhibition of spinal motor neurones.[126]

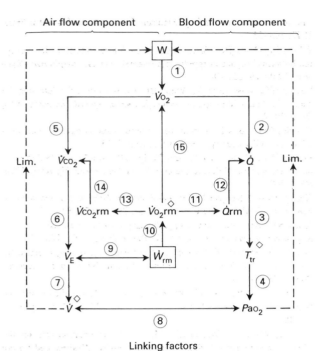

Linking factors

1. Work efficiency
2. Tissue O_2 extraction
3. Pulmonary capillary blood volume
4. Pulmonary capillary transit time dispersion
5. Respiratory exchange ratio (metabolic and buffering components)
6. Arterial p_{CO_2}
7. Pulmonary mechanics; breathing pattern
8. Carotid chemoreceptor stimulation
9. Respiratory muscle mechanical efficiency
10. Respiratory muscle work efficiency
11. Respiratory muscle O_2 extraction
12. Respiratory muscle fractional utilization of $\dot{Q}$
13. Respiratory muscle respiratory quotient
14. Respiratory muscle fractional contribution to $\dot{V}_{CO_2}$
15. Respiratory muscle fractional utilization of $\dot{V}_{O_2}$

Fig. 8 Schematic representation of the stress cascade predisposing to pulmonary limitation (i.e. air-flow limitation, respiratory muscle fatigue, and arterial hypoxaemia) during exercise. The full arrows connect the links in the cascade and the broken lines indicate limiting (Lim.) feedback influences on work rate. Full diamonds indicate points at which a large-value potential for the particular variable is crucial for minimizing the likelihood of limitation. See text for discussion.

CONCLUSION

As data on the performance of the pulmonary system during competition, or even under mock competitive conditions, are sparse, this chapter has focused on the pulmonary consequences of high and sustained metabolic rate. Our considerations are therefore based almost entirely on experiments performed in the laboratory. Space constraints also precluded consideration of the young and maturing athlete, and the effects of ageing or environmental stressors on pulmonary performance limits.

The influence of preparatory breathing responses for certain events (e.g. the deep inspiration followed by explosive expiration used prior to high power events such as shot-putting, sprinting, and weight-lifting) on the subsequent pattern of response is largely unknown, as is the influence of deliberately chosen modifications of spontaneous breathing control. An example of this is the

'breath-play' method[127] which has been suggested for athletes. This proposes what is termed 'upside-down' breathing, which demands active expiration with (attempts at) passive inhalation. Clearly, the control of blood-gas and acid–base status during actual competition, especially when it involves such peculiar modifications of normal control, awaits exploration.

What is clear is that pulmonary function is likely to be an important determinant of success in a wide range of athletic performance. Genetics will naturally play a predominant role in determining the extent to which the pulmonary system capacities will be sufficient to allow the athlete's full metabolic potential to be expressed i.e. as shown schematically in Fig. 8. In this figure we show potential pulmonary limitations to sustaining the required work rate. The full lines connect the links in the stress cascade which can lead to air-flow limitation, respiratory muscle fatigue, and also arterial hypoxaemia. The broken lines demonstrate the limiting feedback on the work rate. The open diamonds represent points at which a large-value potential for the particular variable is crucial for minimizing the likelihood of limitation. It should be noted that, in our terminology, the maximum metabolic rate of the task-performing muscles would be constrained by the pulmonary limitations, i.e. the systemic $\dot{V}_{O_2}$ has the potential to increase further under these conditions.

To what extent training focused on the pulmonary system, particularly during the period of lung growth, will be capable of extending the pulmonary limits, or even decreasing the rate of their deterioration with age, will probably prove to be a topic of not inconsiderable interest for future record-breaking performances.

The well-known assertion of Karpovich[125] that 'A large vital capacity does not make a champion' is not only true but remains the case, we believe, even if 'vital capacity' is replaced by 'pulmonary vascular dimensions and volume-specific air flow'. However, we appear to be entering an era in which these factors are necessary to allow an athlete to become a champion.

REFERENCES

1. Wassermann K, Hansen JE, Sue DY, Whipp BJ. *Principles of exercise testing and interpretation*. Philadelphia: Lea and Febiger, 1987.
2. Linnarsson D. Dynamics of pulmonary gas exchange and heart rate changes at start and end of exercise. *Acta Physiologica Scandinavica* 1974; **415** (Suppl.): 1–68.
3. Casaburi R, Whipp BJ, Wasserman K, Beaver WL, Koyal SN. Ventilatory and gas exchange dynamics in response to sinusoidal work. *Journal of Applied Physiology* 1977; **42**: 300–11.
4. Whipp BJ. The control of the exercise hyperpnea. In: Hornbein T, Ed. *The regulation of breathing*. New York: Marcel Dekker 1981; 1069–1139.
5. Hughson RL, Morrissey M. Delayed kinetics of respiratory gas exchange in the transition from prior exercise. *Journal of Applied Physiology* 1982; **52**: 921–9.
6. Griffiths TG, Henson LC, Whipp BJ. Influence of inspired oxygen concentration on the dynamics of the exercise hyperpnea. *Journal of Physiology (London)* 1986; **380**: 387–407.
7. Hagberg JM, Hickson RC, Ehsani AA, Holloszy JO. Faster adjustment to and from recovery from submaximal exercise in the trained state. *Journal of Applied Physiology* 1980; **48**: 218–24.
8. Herxheimer H, Kost R. Das verhaltnis von sauerstoffaufnahme and kohlensaurausscheidung zur ventilation bei harter muskelarbeit. *Zeitschrift für Klinische Medizin* 1932; **108**: 240–7.
9. Wasserman K, Van Kessel AL, Burton CC. Interaction of physiological mechanisms during exercise. *Journal of Applied Physiology* 1967; **22**: 71–85.
10. Roston WL, Whipp BJ, Davis JA, Effros RM, Wasserman K. Oxygen

uptake kinetics and lactate concentration during exercise in man. *American Review of Respiratory Disease* 1987; **135**: 1080–4.

11. Poole DC, Ward SA, Gardner GW, Whipp BJ. Metabolic and respiratory profile of the upper limit for prolonged exercise in man. *Ergonomics* 1988; **31**: 1265–79.

12. Whipp BJ. Dynamics of pulmonary gas exchange. *Circulation* 1987; **76**; vi–18–28.

13. Henson LC, Poole DC, Whipp BJ. Fitness as a determinant of oxygen uptake response to constant-load exercise. *European Journal of Applied Physiology* 1989; **59**: 21–8.

14. Moritani T, Nagata A, de Vries HA, Munro M. Critical power as a measure of physical work capacity and anaerobic threshold. *Ergonomics* 1981; **24**: 339–50.

15. Hughson RL, Orok CJ, Staudt LE. A high velocity treadmill running test to assess endurance running potential. *International Journal of Sports Medicine* 1984; **5**: 23–5.

16. Rahn H. A concept of mean alveolar air and the ventilation–bloodflow relationships during pulmonary gas exchange. *American Journal of Physiology* 1949; **158**: 21–30.

17. Riley RL, Cournand A. 'Ideal' alveolar air and the analysis of ventilation–perfusion relationships in the lung. *Journal of Applied Physiology* 1949; **1**: 825–47.

18. Rahn H, Fenn WO. *A graphical analysis of the respiratory gas exchange.* The O_2–CO_2 diagram. Washington, DC: American Physiology Society, 1955.

19. Sutton J, Jones NL, Towes CJ. Growth hormone secretion in acid–base alterations at rest and during exercise. *Clinical Science and Molecular Medicine* 1976; **50**: 241–7.

20. Wasserman K, Whipp BJ, Koyal SN, Cleary MG. Effect of carotid body resection on ventilatory and acid–base control during exercise. *Journal of Applied Physiology* 1975; **39**: 354–58.

21. Dejours P. Control of respiration in muscular exercise. In: Fenn W, Rahn H, eds. *Handbook of physiology, Vol. 1, Respiration.* Washington DC: American Physiological Society 1964: 631–48.

22. Cunningham DJC. Integrative aspects of the regulation of breathing; a person view. In: Widdcombe JG, ed. *MTP International Reviews of Science*, Series 1, *Physiology* Vol. 2, *Respiration*. Baltimore: University Park Press 1974: 303–69.

23. Linton RAF, Band DM. The effect of potassium on carotid chemoreceptor activity and ventilation in the cat. *Respiration Physiology* 1985; **59**: 65–70.

24. Dempsey JA, Hanson P, Henderson K. Exercise-induced alveolar hypoxemia in healthy human subjects at sea level. *Journal of Physiology (London)* 1984; **355**: 161–75.

25. Wasserman K, McIlroy MB. Detecting the threshold of anaerobic metabolism. *American Journal of Cardiology* 1964; **14**: 844–52.

26. Wasserman K, Whipp BJ, Koyal SN, Beaver WL. Anaerobic threshold and respiratory gas exchange during exercise. *Journal of Applied Physiology* 1973; **35**: 236–43.

27. Reinhard U, Muller PH, Schmullingm R-M. Determination of anaerobic threshold by the ventilation equivalent in normal individuals. *Respiration* 1979; **38**: 36–42.

28. Stegmann H, Kindermann W, Schnabel A. Lactate kinetics and individual anaerobic threshold. *International Journal of Sports Medicine* 1981; **2**: 160–5.

29. Whipp BJ, Ward SA, Wasserman K. Respiratory markers of the anaerobic threshold. *Advances in Cardiology* 1986; **35**: 47–64.

30. Wasserman K, Beaver WL, Davis JA, Pu JZ, Heber D, Whipp BJ. Lactate, pyruvate and lactate-to-pyruvate ratio during exercise and recovery. *Journal of Applied Physiology* 1985; **59**: 935–40.

31. Beaver WL, Wasserman K, Whipp BJ. Improved detection of the lactate threshold during exercise using a log–log transformation. *Journal of Applied Physiology* 1986; **59**: 1936–40.

32. Wasserman K, Whipp BJ. Exercise physiology in health and disease. *American Review of Respiratory Disease* 1975; **112**: 219–49.

33. Wasserman K, Whipp BJ, Casaburi R, Beaver WL, Brown HV. CO_2 flow to the lungs and ventilatory control. In: Dempsey JA, Reed CE, eds.

Muscular exercise and the lung. Madison: University of Wisconsin Press, 1977: 103–35.

34. Whipp BJ, Davis JA, Wasserman K. Ventilatory control of the 'isocapnic buffering' region in rapidly-incremental exercise. *Respiration Physiology* 1989; **76**: 357–68.

35. Beaver WL, Wasserman K, Whipp BJ. A new method for detecting the anaerobic threshold by gas exchange. *Journal of Applied Physiology* 1986; **60**: 2020–7.

36. Douglas CG. Co-ordination of the respiration and circulation with variations in bodily activity. *Lancet* 1927; **ii**: 213–18.

37. Beaver WL, Wasserman K, and Whipp BJ. Bicarbonate buffering of lactic acid generated during exercise. *Journal of Applied Physiology* 1986; **60**: 472–8.

38. Keul J, Doll E, Keppler D. In: Jokl E, ed. *Oxidative energy metabolism of human muscle.* Baltimore: University Park Press, 1972: 52–202.

39. Jones NL. Acid–base physiology. In: Crystal RG, West JB, eds. *The lung: scientific foundations.* New York: Raven Press, 1991: 1251–65.

40. Whipp BJ, Mahler M. Dynamics of gas exchange during exercise In: West JB, ed. *Pulmonary gas exchange*, Vol. II. New York: Academic Press, 1980: 33–96.

41. Wasserman K, Beaver WL, Whipp BJ. Gas exchange theory and the lactic acidosis (anaerobic) threshold. *Circulation* 1991; **81**: (Suppl.): II–14–30.

42. Shea K, Ward SA, Whipp BJ. Effect of incrementation rate on the 'V-slope' components during progressive exercise. *Medicine and Science in Sports and Exercise* 1991; **23**:(4): S1.

43. Poole DP, Ward SA, Whipp BJ. Effect of training on the metabolic and respiratory profile of heavy and severe exercise. *European Journal of Applied Physiology* 1990; **59**: 421–9.

44. Whipp BJ, Ward SA. Ventilatory control dynamics during muscular exercise in man. *International Journal of Sports Medicine* 1980; **1**: 146–59.

45. Weil JV, Swanson GD. Peripheral chemoreceptors and the control of breathing. In: Whipp BJ, Wasserman K, eds. *Lung biology in health and disease*, Vol. 52 *Exercise: pulmonary physiology and pathophysiology.* New York: Dekker, 1991: 371–403.

46. Ohyabu Y, Honda Y. Exercise and ventilatory chemosensitivities. *Annals of Physiology and Anthropometry* 1990; **9**: 117–21.

47. McIlroy MR, Marshall R, Christie RV. The work of breathing in normal subjects. *Clinical Science* 1954; **13**: 127–36.

48. Margaria R, Milic-Emili G, Petit JM, Cavagna G. Mechanical work of breathing during muscular exercise. *Journal of Applied Physiology* 1960; **15**: 354–8.

49. McGregor M, Becklake MR. The relationship of oxygen cost of breathing to respiratory mechanical work and respiratory force. *Journal of Clinical Investigation* 1961; **40**: 971–80.

50. Milic-Emili G, Petit JM, Deroanne R. Mechanical work of breathing during exercise in trained and untrained subjects. *Journal of Applied Physiology* 1962; **17**: 43–6.

51. Thoden JS, Dempsey JA, Reddan WG, Birnbaum ML, Forster HV, Grover RF, Rankin J. Ventilatory work during steady-state response to exercise. *Federation Proceedings* 1969; **28**: 1316–21.

52. Holmgren A, Herzog P, Astrom H. Work of breathing during exercise in healthy young men and women. *Scandinavian Journal of Clinical Laboratory Investigation* 1973; **31**: 165–74.

53. Hesser CM, Linnarsson D, Fagraeus L. Pulmonary mechanics and work of breathing at maximal ventilation and raised air pressure. *Journal of Applied Physiology* 1981; **50**: 747–53.

54. Klas JV, Dempsey JA. Voluntary versus reflex regulation of maximal exercise flow: volume loops. *American Review of Respiratory Disease* 1989; **139**: 150–6.

55. Whipp BJ, Pardy R. Breathing during exercise. In: Macklem PT, Mead J, eds. *Handbook of physiology, respiration (pulmonary mechanics).* Washington, DC: American Physiological Society, 1986: 605–29.

56. Rohrer F. Physiologie der Atembewegung. In: Bethe ATJ, von Bergmann G, Embden G, Ellinger A, eds. *Handbuch der normalen und pathogischen Physiologie.* Berlin: Springer, 1925: 70–127.

57. Otis AB, Fenn WO, Rahn H. Mechanics of breathing in man. *Journal of Applied Physiology* 1950; **2**: 592–607.

58. Hanson JS, Tabakin BS, Levy AM, Falsetti HL. Alterations in pulmonary mechanics with airway obstruction during rest and exercise. *Journal of Applied Physiology* 1965; **20**: 664–8.

59. Gilbert R, Auchincloss JH. Mechanics of breathing in normal subjects during brief, severe exercise. *Journal of Laboratory and Clinical Medicine* 1969; **73**: 439–50.

60. Olafsson S, Hyatt RE. Ventilatory mechanics and expiratory flow limitation during exercise in normal subjects. *Journal of Clinical Investigation* 1969; **48**: 564–73.

61. Stubbing DG, Pengelly LD, Morse JLC, Jones NL. Pulmonary mechanics during exercise in normal males. *Journal of Applied Physiology* 1980; **49**: 506–10.

62. Granath A, Horie E, Linderholm H. Compliance and resistance of the lungs in the sitting and supine positions at rest and during exercise. *Scandinavian Journal of Clinical Laboratory Investigation* 1959; **11**: 226–34.

63. Chiang ST, Steigbigel NH, Lyons HA. Pulmonary compliance and non-elastic resistance during treadmill exercise. *Journal of Applied Physiology* 1965; **20**: 1194–8.

64. Jones NL, Killian KJ, Stubbings DG. The thorax in exercise. In: Roussos Ch, Macklem PT, eds. *Lung biology in health and disease*, Vol. 29, *The Thorax*. New York: Dekker, 1988: 627–62.

65. Goldman MD, Grimby G, Mead J. Mechanical work of breathing derived from rib cage and abdominal V–P partitioning. *Journal of Applied Physiology* 1976; **41**: 752–63.

66. Grimby G, Goldman M, Mead J. Respiratory muscle action inferred from rib cage and abdominal V–P partitioning. *Journal of Applied Physiology* 1976; **41**: 739–51.

67. Otis AB. The work of breathing. *Physiological Reviews* 1954; **34**: 449–58.

68. Otis AB. The work of breathing. In: Fenn WO, Rahn H, eds. *Handbook of physiology, Section 3, Vol. I Respiration*, Chapter 17. Washington, DC: American Physiological Society, 1964: 463–76.

69. Rochester DF, Farkas GA, Lu JY. Contractility of the in situ human diaphragm: Assessment based on dimensional analysis. In: Sieck GC, ed. *Respiratory muscles and their neuromotor control*. New York: Liss, 1987.

70. McKerrow CB, Otis AB. Oxygen cost of hyperventilation. *Journal of Applied Physiology* 1956; **9**: 375–9.

71. Campbell EJM, Westlake EK, Cherniack RM. Simple methods of estimating oxygen consumption and efficiency of the muscles of breathing. *Journal of Applied Physiology* 1957; **11**: 303–8.

72. Bartlett RG Jr, Brubach HF, Specht H. Oxygen cost of breathing. *Journal of Applied Physiology* 1958; **12**: 413–24.

73. Cherniack RM. The oxygen consumption and efficiency of respiratory muscles in health and emphysema. *Journal of Clinical Investigation* 1959; **38**: 494–9.

74. Fritts HN, Filler J, Fishman AB, Cournand A. The efficiency of ventilation during voluntary hyperpnea. *Journal of Clinical Investigation* 1959; **38**: 1339–48.

75. Milic-Emili G, Petit JM. Mechanical efficiency of breathing. *Journal of Applied Physiology* 1960; **15**: 359–62.

76. Shephard RJ. The oxygen cost of breathing during vigorous exercise. *Quarterly Journal of Experimental Physiology* 1966; **51**: 336–50.

77. Leith DE, Bradley M. Ventilatory muscle strength and endurance training. *Journal of Applied Physiology* 1976; **41**: 508–16.

78. Karetsky MS, Cain SM. Factors controlling O_2 uptake. *Chest* 1972; **61** (Suppl.): 48S–9S.

79. Robertson CH Jr, Foster GH, Johnson RL Jr. The relationship of respiratory failure to the oxygen consumption of, lactate production by, and distribution of blood flow among respiratory muscles during increasing inspiratory resistance. *Journal of Clinical Investigation* 1977; **59**: 31–42.

80. Bye PTP, Esau SA, Walley KR, Macklem PT, Pardy RL. Ventilatory muscles during exercise in air and oxygen in normal men. *Journal of Applied Physiology* 1984; **56**: 464–71.

81. Johnson RL Jr. Heart–lung interactions in the transport of oxygen. In: Scharf SM, Cassidy SS, eds. *Lung biology in health and disease*, Vol. 42 *Heart–Lung interactions in health and disease*. New York: Dekker, 1989: 5–41.

82. Tenney SM, Reese RE. The ability to sustain great breathing efforts. *Respiration Physiology* 1968; **5**: 187–201.

83. Bradley ME, Leith DE. Ventilatory muscle training and the oxygen cost of sustained hyperpnea. *Journal of Applied Physiology* 1978; **45**: 885–92.

84. Belman MJ, Mittman C. Ventilatory muscle training improves exercise capacity in chronic obstructive pulmonary disease patients. *American Review of Respiratory Disease* 1980; **121**: 273–280.

85. Bai TR, Rabinovitch BJ, Pardy RL. Near-maximal voluntary hyperpnea and ventilatory muscle function. *Journal of Applied Physiology* 1984; **57**: 1742–8.

86. Shephard RJ. The maximum sustained voluntary ventilation in exercise. *Clinical Science* 1967; **32**: 167–76.

87. Freedman S. Sustained maximum voluntary ventilation. *Respiration Physiology* 1970; **8**: 230–44.

88. Keens TG, Krastings IRB, Wannamaker EM, Levison H, Crozier DN, Bryan A. Ventilatory muscle endurance training in normal subjects and patients with cystic fibrosis. *American Review of Respiratory Disease* 1977; **116**: 853–60.

89. Peress L, McLean P, Woolf CR, Zamel N. Ventilatory muscle training in obstructive lung disease. *Bulletin Euorpéen de Physiopathologie Respiratoire* 1979; **15**: 91–2.

90. Folinsbee LJ, Wallace ES, Bedi JF, Horvath SM. Respiratory patterns and control during unrestrained human running. In: Whipp BJ, Wiberg DM, eds. *Modelling and control of breathing*. New York: Elsevier, 1983: 205–12.

91. Dill DB, Edwards HT, Bauer PS, Levenson EJ. Physical performance in relation to external temperature. *Arbeitsphysiologie* 1931; **4**: 508–18.

92. Otis AB, Guyatt AR. The maximal frequency of breathing in man at various tidal volumes. *Respiration Physiology* 1968; **5**: 118–29.

93. Leaver DJ, Pride NB. Flow–volume curves and expiratory pressures during exercise in patients with chronic airway obstruction. *Scandinavian Journal of Respiratory Disease* 1971; **77** (Suppl.): 23–7.

94. Jensen JE, Lyager S, Pederson OF. The relationship between maximum ventilation, breathing patterns and mechanical limitation of ventilation. *Journal of Physiology* (London) 1980; **309**: 521–32.

95. Grimby G, Saltin B, Wilhelmsen L. Pulmonary flow–volume and pressure–volume relationship during submaximal and maximal exercise in young well-trained men. *Bulletin Européen de Physiopathologie Respiratoire* 1971; **7**: 157–72.

96. Milic-Emili G, Petit JM, Deroanne R. The effects of respiratory rate on the mechanical work of breathing during muscular exercise. *Internationale Zeitschrift für Angewandte Physiologie Einschliesslich Arbeitsphysiologie* 1960; **18**: 330–40.

97. Macklem PT. Discussion of 'Inspiratory muscle fatigue as a factor limiting exercise' by A. Grassino, D. Gross, P. T. Macklem, C. Roussos, and G. Zagelbaum. *Bulletin Européen de Physiopathologie Respiratoire* 1979; **15**: 111–15.

98. Pierce AK, Luterman D, Loudermilk J, Blomqvist G, Johnson RL Jr. Exercise ventilatory patterns in normal subjects and patients with airway obstruction. *Journal of Applied Physiology* 1968; **25**: 249–54.

99. Brown HV, Wasserman K, Whipp BJ. Strategies of exercise testing in chronic lung disease. *Bulletin Européen de Physiopathologie Respiratoire* 1977; **13**: 409–23.

100. Bye PTP, Farkas GA, Roussos C. Respiratory factors limiting exercise. *Annual Review of Physiology* 1983; **45**: 439–51.

101. Nattie EE, Tenny SM. The ventilatory response to resistance unloading during muscular exercise. *Respiratory Physiology* 1970; **10**: 249–62.

102. Ward SA, Whipp BJ, Poon CS. Density-dependent air flow and ventilatory control in exercise. *Respiration Physiology* 1982; **49**: 267–77.

103. Hussain SNA, Pardy RL, Dempsey JA. Mechanical impedance as determinant of inspiratory neural drive during exercise. *Journal of Applied Physiology* 1985; **59**: 365–73.

104. Gordon B, Levine SA, Wilmaers A. Observations on a group of marathon runners. *Archives of Internal Medicine* 1924; **33**: 425.

105. Warren GL, Cureton KJ, Sparling PB. Does lung function limit performance in a 24–hour ultramarathon? *Physiology* 1989; **78**: 253–64.

106. Whipp BJ, Davis JA. Does ventilatory mechanics limit maximum exercise? In: Russo P, Gass G, eds. *Exercise, a workshop*. Sydney: Cumberland College, 1979; 20–6.

107. Edwards RHT. Human muscle function and fatigue. In: Porter R, Whelan J, eds. *Human muscle fatigue*. London: Pitman, 1981: 1–34.

108. Pardy RL, Hussain SNA, Macklem PT. The ventilatory pump in exercise. *Clinics in Chest Medicine* 1984; **5**: 35–49.

109. Hansen JE, Stelter GP, Vogel JA. Arterial pyruvate, lactate, pH, and P_{O_2} during work at sea level and high altitude. *Journal of Applied Physiology* 1967; **23**: 523–30.

110. Jones NL. Exercise testing in pulmonary evaluation: Rationale, methods, and the normal respiratory response to exercise. *New England Journal of Medicine* 1975; **293**: 541–4.

111. Gledhill N, Froese AB, Dempsey JA. Ventilation to perfusion distribution during exercise in health. In: Dempsey J, Reed C, eds. *Muscular exercise and the lung*. Madison: University of Wisconsin Press, 1977: 325–43.

112. West BJ, Dollery CT. Distribution of blood flow and ventilation–perfusion ratio in the lung measured with radioactive CO_2. *Journal of Applied Physiology* 1960; **15**: 405–10.

113. Bryan AC, Bentivoglio LG, Beerel F, MacLeish H, Zidulka A, Bates DV. Factors affecting regional distribution of ventilation and perfusion in the lung. *Journal of Applied Physiology* 1964; **19**: 395–402.

114. Bake B, Bjure J, Widimsky J. The effect of sitting and graded exercise on the distribution of pulmonary blood flow in healthy subjects studied with the 133–xenon technique. *Scandinavian Journal of Clinical Laboratory Investigation* 1968; **22**: 99–106.

115. Wagner PD, Saltzman HA, West JB. Measurement of continuous distributions of ventilation-perfusion ratios: theory. *Journal of Applied Physiology* 1974; **36**: 588–99.

116. Hammond MD, Gale GE, Kapitan KS, Ries A, Wagner PD. Pulmonary gas exchange in humans during exercise at sea level. *Journal of Applied Physiology* 1986; **60**: 1590–8.

117. Derks CM. Ventilation–perfusion distribution in young and old volunteers during mild exercise. *Bulletin Européen de Physiopathologie Respiratoire* 1980; **16**: 145–54.

118. Gale GE, Jorre-Bueno J, Moon R, Saltzman HA, Wagner PD. V_A/Q inequality in normal man during exercise at sea level and simulated altitude. *Journal of Applied Physiology* 1985; **58**: 978–88.

119. Johnson RL Jr, Taylor HF, DeGraff AC Jr. Functional significance of a low diffusing capacity for carbon monoxide. *Journal of Clinical Investigation* 1965; **44**: 789–800.

120. Hughes JMB. Diffusive gas exchange. In: Whipp BJ, Wasserman K, eds. *Lung Biology in Health and Disease*, Vol. 52, *Exercise: pulmonary physiology and pathophysiology*. New York: Dekker, 1991: 143–71.

121. Weibel ER. *The pathway for oxygen*. Harvard University Press, 1984.

122. Powers SK, Lawler J, Dempsey JA, Dodd S, Landry G. Effects of incomplete pulmonary gas exchange on $V_{O_2 max}$. *Journal of Applied Physiology* 1989; **66**: 2491–5.

123. Reeves JT, Dempsey JA, Grover RF. Pulmonary circulation during exercise. In: Weir EK, Reeves JT, eds. *Lung biology in health and disease*, Vol. 38, *Pulmonary vascular physiology and pathophysiology*. New York: Dekker, 1989: 107–33.

124. O'Brodovich H, Coates G. Lung water and solute movement during exercise. In: Whipp BJ, Wasserman K, eds. *Lung biology in health and disease*, Vol. 52, *Exercise: pulmonary physiology and pathophysiology*. New York, Dekker, 1991: 253–70.

125. Brower R, Permutt S. Exercise and the pulmonary circulation. In: Whipp BJ, Wasserman K, eds. *Lung biology in health and disease*, Vol. 52, *Exercise pulmonary physiology and pathophysiology*. New York: Dekker, 1991: 201–20.

126. Paintal AS. The mechanism of excitation of type-J receptors and the J-reflex. In: Porter R, ed. *Breathing: Hering–Breuer Centenary Symposium*. London: Churchill, 1970: 59.

127. Jackson I. *The breath play approach to whole life fitness*. New York: Doubleday, 1986.

128. Karpovich PV. *Physiology of muscular activity*. Philadelphia: Saunders, 1962: 133.

1.1.2 Training-induced adaptations in skeletal muscle

JAN HENRIKSSON* AND ROBERT C. HICKNER

INTRODUCTION

Skeletal muscle cells possess a quite remarkable capacity for adaptation to changes in metabolic demand. Endurance training, for instance, induces marked adaptive changes in several structural components and metabolic variables in the engaged skeletal muscles. Among the observed changes with different training regimens are those involving the muscle's content of metabolic enzymes, the sensitivity to hormones, and the composition of the contracting filaments. Other adaptations affect membrane transport processes and the muscular capillary network. The adaptive changes in metabolic enzymes and capillaries are the best described consequences of endurance training, and both these factors are likely to be important determinants for an individual's physical working capacity. The enhanced muscle glucose transport and insulin sensitivity represents another major effect of exercise and training on muscle metabolism. This chapter is devoted to a closer look at the cellular adaptation to endurance training, particularly with regard to changes in the referred variables.

ESTIMATION OF THE METABOLIC CAPACITY OF SKELETAL MUSCLE

The muscle biopsy procedure,[1] whereby small (10–100 mg) muscle pieces can be sampled, can be combined with sensitive biochemical techniques to permit estimation of the capacity of different metabolic pathways in human muscle. Today, biochemical measurements can be done even at the single-fibre level.[2] The most important metabolic pathways for energy delivery in exercising muscle are glycolysis/glycogenolysis, fatty acid oxidation, the citric acid cycle, and the respiratory chain. The capacity of these pathways is limited mainly by the amount of pathway

*JH was supported by grants from the Swedish Medical Research Council, the Karolinska Institute, the Research Council of the Swedish Sports Federation, and the National Institutes of Health, USA.

enzymes contained in the cell. In this context, some enzymes, the rate-limiting or flux-generating ones, are more important than others. Such enzymes have low activity and constitute bottlenecks in the pathways. Theoretically, therefore, an increased concentration of the rate-limiting enzyme with unchanged concentrations of the other pathway enzymes would be sufficient to increase the capacity of the entire metabolic pathway.[3] Generally, however, with a change in the capacity of a metabolic pathway, for example due to training or inactivity, the content of all enzymes, whether rate-limiting or not, changes in the same direction. This can be illustrated by the changes recorded in the anterior tibial muscle of the rabbit in response to chronic electrical stimulation. In this situation there is an almost identical decrease in all glycolytic enzymes, although only one, phosphofructokinase, is considered to be rate limiting (Figs. 1 and 2). The reason for this is not entirely clear, but it may be that relatively constant proportions of the different enzymes in a certain metabolic pathway are necessary in order to maintain the metabolic equilibrium of a cell. Therefore it is possible to obtain a good estimation of the cellular capacity of a specific metabolic pathway just by measuring the content (maximal activity) of any one of its enzymes. Thus the choice of enzymes for analyses is largely dependent on the simplicity and speed of the available analytical methods. A list of enzymes, the levels of which are commonly used as a measure of the capacity of their respective metabolic pathways, is given below.

Glycolysis: phosphofructokinase, lactate dehydrogenase (LDH)
Fatty acid oxidation: 3-hydroxyacylCoA dehydrogenase
Citric acid cycle: citrate synthase, succinate dehydrogenase
Respiratory chain: cytochrome c oxidase.

THE MAXIMAL ADAPTABILITY OF SKELETAL MUSCLE: RESULTS OF EXPERIMENTS INVOLVING CHRONIC ELECTRICAL STIMULATION

It is of considerable theoretical interest to know how skeletal muscle adapts to a maximal training stimulus. This knowledge can be used as a frame of reference against which the effects of, for instance, different endurance training regimens can be compared and evaluated. One way of obtaining such information has been to subject a normally rather inactive muscle, such as the rabbit anterior tibial muscle, to chronic electrical stimulation. This can be done in a way that is essentially painless to the animal. During anaesthesia, a stimulator (of fingertip size) is implanted under aseptic conditions so that, when activated, it subjects the anterior tibial muscle to chronic stimulation via the common peroneal nerve. The stimulator is activated non-invasively by means of an electronic flash-gun after the rabbit has been allowed to recover from the operation. When stimulating with a continuous train of pulses at a frequency of 10 Hz, as in most studies, there is a very small-amplitude oscillation of the hindpaw but without any observable effect on the use of the limb in posture control and locomotion or on the general well-being of the animal. These investigations have revealed a quite remarkable capacity of skeletal muscle for adaptation to the extreme metabolic demand imposed by the chronic stimulation. Therefore this response will be described in some detail in the present chapter before turning to a discussion of the effects of more physiological endurance training regimens.

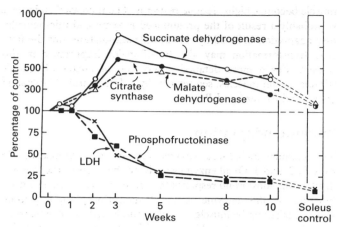

Fig. 1 Enzyme changes induced by chronic electrical muscle stimulation. The rabbit anterior tibial muscle was stimulated at 10 impulses/second, 24 h a day, for 3 days to 10 weeks. The figure depicts changes in three oxidative and two glycolytic enzymes: succinate dehydrogenase, citrate synthase, and malate dehydrogenase are enzymes in the citric acid cycle, LDH and 6-phosphofructokinase are involved in glycolysis. The value for unstimulated control muscles has been set at 100 per cent. To the right, enzyme levels are given for the slow-twitch soleus muscle in unstimulated control rabbits, thus illustrating that, as a result of the chronic stimulation, the originally fast-twitch anterior tibial muscle acquires a clearly higher oxidative enzyme content than a normal slow-twitch muscle. (Reproduced from ref. 4 with permission.)

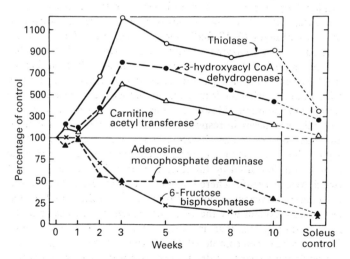

Fig. 2 Enzyme changes induced by chronic electrical muscle stimulation. Thiolase and 3-hydroxyacyl CoA dehydrogenase represent the fat degradation (fatty acid oxidation) pathway and carnitine acetyl transferase is one of the enzymes required for transfer of fatty acids from the cytosol into the mitochondria. Adenosine monophosphate (AMP) deaminase is related to the enzymes involved in high-energy phosphate transfer, whereas Fructose bisphosphatase catalyses the reversal of the 6-phosphofructokinase reaction (Fig. 1) and is therefore required for the regeneration of glycogen from lactate. For further explanations, see the caption to Fig. 1. (Reproduced from ref. 4 with permission.)

The rabbit anterior tibial muscle is a predominantly fast muscle, containing not more than 6 per cent slow-twitch fibres. However, the chronic stimulation programme results in a striking fibre type transformation so that, after stimulation durations of 5 to 6 weeks or more the anterior tibial muscle contains slow-twitch fibres only. Simultaneously, the normally very fatiguable anterior tibial muscle becomes highly fatigue resistant. The increased endurance is probably a result of the pronounced enzyme and microcirculatory adaptation induced by the chronic stimulation, but the fibre type transformation may also be of major importance in this respect. For a more detailed description of effects of chronic stimulation on muscles, see reviews by Salmons and Henriksson,[5] Jolesz and Sreter,[6] and Pette.[7]

Enzyme adaptation

The stimulation-induced enzyme changes are summarized in Figs. 1, 2, and 3. Of all enzymes analysed, hexokinase (not shown) displays the most rapid response to chronic stimulation. The main function of this enzyme is to channel (by phosphorylation) glucose, taken up by muscle, into the muscle cell's glycolytic pathway. An increase in hexokinase (which upon chronic stimulation was demonstrable within 1 day, more than doubled in 3 days, and increased 11-fold at its peak) is not normally seen in endurance training, but the observed response illustrates the capacity of skeletal muscle to adapt rapidly to the use of blood glucose as the preferred energy substrate. The absence of an increase in hexokinase in response to endurance training makes sense physiologically, since, with a large trained muscle mass, a large consumption of blood-derived glucose during exercise would rapidly override the capacity of the liver to replenish the consumed blood glucose.

The enzymes of the citric acid cycle and fatty acid β-oxidation are commonly referred to as oxidative enzymes. These enzymes display large increases on stimulation, but the increases occur somewhat later than that of hexokinase. In our investigations, maximal changes (6–12-fold) were reached after 2 to 5 weeks. Thereafter, the enzyme concentrations decreased somewhat before stabilizing at a lower level (Figs. 1 and 2). It is not known whether this two-phase pattern of change is specific for chronic stimulation or whether it may also occur with certain endurance training programmes. However, to date there has been no report that this may occur in response to endurance training in man or any other species. It may therefore be speculated that the secondary decline in oxidative enzymes, which is observed with chronic stimulation, is caused by the fibre-type transformation from fast twitch to slow twitch. There is evidence that the latter fibre type is more energy efficient at slow contractions,[9] and this could then at least partly explain why the levels of oxidative enzymes could be kept at slightly less than maximal values after fibre-type transformation had been completed. Contrary to what was observed for the oxidative enzymes, a drastic decrease was noted with chronic stimulation with respect to the glycolytic enzymes as well as creatine kinase (i.e. enzymes supplying energy to the muscle during short-term intense exercise). Following 2 months of continuous stimulation, only a fifth of the initial glycolytic enzyme content remained in the anterior tibial muscle. On discontinuation of stimulation, the enzymes whose activity had increased again declined towards the initial level, at first rapidly and later more slowly, and had returned to normal after 5 to 6 weeks. This was also true of the glycolytic enzymes, which increased when stimulation was stopped, but normalization occurred more rectilinearly with these enzymes.[10] It may be noted that the normally large variation in enzyme content among different fibres in the same muscle, and even between those of the same type, is reduced with chronic stimulation (Fig. 3). However, no corresponding information is available with regard to the effect of endurance training, although some information can be found in Chi et al.[11]

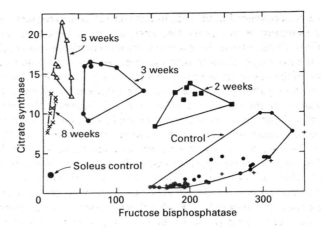

Fig. 3 Enzyme changes in single skeletal muscle fibres induced by chronic electrical muscle stimulation. The anterior tibial muscle of the rabbit was stimulated as described in the text to Fig. 1. Single fibres were isolated by microdissection from muscles stimulated for different periods of time (2, 3, 5 and 8 weeks respectively) as well as from unstimulated control muscles. Citrate synthase is a member of the citric acid cycle, and was therefore used as a measure of the fibre's oxidative capacity, while fructose bisphosphatase catalyses the reversal of the phosphofructokinase reaction in glycolysis and was used as a measure of glycolytic capacity. The average value for fibres in the control slow-twitch soleus muscle is included for reference (for explanation see the text to Fig 1) All fibres in a normal unstimulated (control) muscle have a high content of glycolytic enzymes, whereas the content of oxidative enzymes varies 10-fold. The chronic stimulation induces a high oxidative capacity in all fibres, whereas the glycolytic capacity decreases to low levels. Figures are in moles (citrate synthase) or millimoles (fructose bisphosphatase) per kilogram dry weight per hour at 20°C. Each symbol (except soleus control) denotes one individual fibre. (Reproduced from ref. 8 with permission.)

Other adaptations induced by chronic stimulation

There is a doubling of the number of blood capillaries per unit muscle cross-sectional area, thus greatly improving the muscle's blood supply.[12] The time course of this change has not been studied in detail, but preliminary data indicate that it is roughly similar to that of the oxidative enzymes. Concomitant with these changes there is, as mentioned above, a dramatic improvement in the muscle's endurance. In our investigations the endurance of the muscle has been measured as an index: the remaining muscle force following a 5-min period of intense muscle stimulation divided by the muscle force exerted during the first few contractions. This index increases from a normal value of 0.5 to 1.0 in muscle that has been continuously stimulated for 6 weeks. Following discontinuation of the chronic stimulation, the fatiguability again increases with the time needed for normalization (5–6 weeks) being similar to that of the metabolic enzymes and the capillary supply. An interesting general observation from these chronic stimulation experiments is that the different biochemical and morphological adaptations to chronic stimulation fit into a 'first in, last out' pattern for the response to stimulation and recovery. This means that the earlier the stage at which a parameter changes during the course of stimulation, the later the stage at which it returns to control levels during recovery. (For further information see Brown et al.[10])

This summary of what is likely to be the maximal activity-induced adaptability of skeletal muscle might serve as a background to a description of the effect of endurance training on skeletal muscle characteristics. It may be argued that chronic stimulation in the rabbit is quite different from endurance training in man. However, when we sought to reconcile observations from chronic stimulation and endurance training,[5] it was found that results from the two experimental approaches differ only in degree. The properties that change in response to exercise are also those that change at an early stage of stimulation; the properties that are resistant to change under exercise conditions change only after prolonged stimulation. Therefore there is a hierarchy of stability in the properties of skeletal muscle, which is also revealed by the rate at which a parameter returns to control values following cessation of stimulation (see above).

HUMAN SKELETAL MUSCLE: EFFECTS OF ENDURANCE TRAINING

The first observations of the effects of endurance training on metabolic enzymes in skeletal muscle (rat) were made by Russian investigators in the 1950s, but a detailed investigation of these changes was first performed by Holloszy and coworkers.[13] Petrén et al.[14] had already shown in 1937 that the capillary network of rat skeletal muscle was influenced by training. The first human studies on muscle metabolic enzymes were published around 1970.[15,16] During the 1970s and 1980s improved methodology allowed more detailed studies of both humans and other species. For the human studies small muscle biopsy specimens (20–100 mg) were obtained, usually from the thigh muscle, but also from other muscles like the gastrocnemius, deltoid, and triceps. In these studies different groups of individuals were compared, for example untrained persons versus athletes in different sports, or alternatively a group of previously untrained individuals was studied repeatedly with muscle biopsies taken during a training period. Despite the fact that different parts of the same muscle may often differ with regard to fibre-type composition, capillary density, and enzyme content, the muscle biopsy technique has proved to be surprisingly useful for these studies. With this technique, relatively small changes can be detected, such as a change in enzyme content or capillary density of 15 to 20 per cent, when a group of five or six subjects is studied with single biopsies before and after training. Generally, however, the biopsy technique is not sensitive enough to allow conclusions to be drawn from the analysis of a single sample. With several samples from the same muscle, the methodological error is markedly reduced.

Enzyme changes

An illustration of the effects of endurance training on human muscle is the observed difference between endurance athletes and untrained individuals (Fig. 4). With regard to oxidative enzymes (i.e. enzymes of fatty acid oxidation, the citric acid cycle, and the respiratory chain), the values are approximately threefold higher in the trained thigh muscle of the athletes than in the thigh muscle of untrained individuals. With total inactivity, such as in muscle encased in plaster after an injury, the content of oxidative enzymes decreases to 70 to 75 per cent of the 'untrained level'. It can be speculated that lower levels than this would not be compatible with survival of the muscle cell. The maximal range of oxidative enzyme content in the human thigh muscle is therefore approxi-

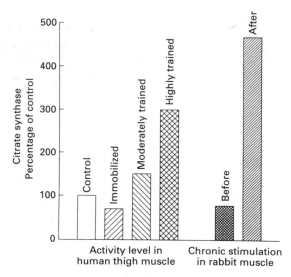

Fig. 4. The influence of the physical fitness level on skeletal muscle oxidative capacity measured as the content of citrate synthase. Muscle tissue from normal sedentary individuals (controls) has been compared with muscle subjected to encasement in plaster after injury (immobilized) or to 2 to 3 months of moderate endurance training as well as with values recorded in top-class cyclists and long-distance runners (highly trained). As a further comparison, the corresponding values from the rabbit anterior tibial muscle before and after 3 to 5 weeks of chronic electrical stimulation are indicated on the right. (The human data have generously been placed at our disposal by Dr Eva Jansson, Department of Clinical Physiology, Karolinska Hospital, Stockholm; the results regarding chronic stimulation are from ref. 4.)

mately fourfold, but it can be assumed that a very long training time would be required for an individual to cover this whole range. A comparison with chronically stimulated rabbit muscle (Fig. 4) reveals that muscles of endurance athletes have approximately 40 per cent lower levels of oxidative enzymes than these chronically stimulated muscles. The difference with respect to fat oxidation enzymes is somewhat greater. If possible differences between the rabbit and man are ignored, this result can be taken to indicate that the trained muscles of our best endurance athletes have an oxidative capacity that is one-half to two-thirds of the theoretically attainable maximal level.

Another important question concerns the magnitude of the enzyme changes that can be attained with a few weeks or months of more moderate endurance training regimens. Here, information is available from a large number of investigations in which different research groups have studied the effects of 2 to 3 months of training on the oxidative enzyme content of leg or arm muscles. These studies have usually involved bouts of 30 to 60 min of exercise at intensities corresponding to 70 to 80 per cent of V_{O_2}-max three to five times per week. With a group of previously untrained individuals, the general finding is an approximately 40 to 50 per cent increase in the content of oxidative enzymes in the trained muscle (Fig. 5). This increase occurs gradually over 6 to 8 weeks of training, with the most rapid change taking place during the first 3 weeks.[18]

Exercise intensity and duration of training

A question of practical importance is the intensity and duration of training needed to obtain optimal results with respect to enzyme adaptation. For an untrained person, some increase in the muscle

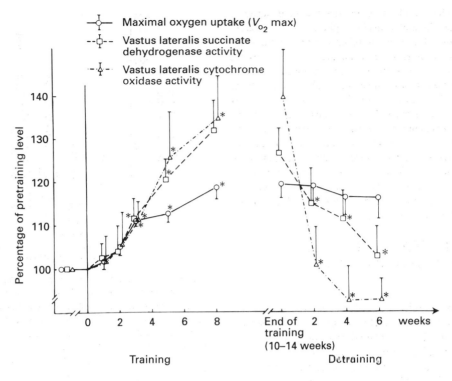

Fig. 5 The effect of endurance training on the content of oxidative enzymes in human skeletal muscle. A group of previously untrained subjects trained for 10 to 14 weeks on bicycle ergometers (40 min/day, 4 days/week; the rate of work corresponded to 80 per cent of the maximal oxygen uptake) and was subsequently studied for 6 weeks after cessation of training. Thigh muscle biopsy samples were analysed for the oxidative enzymes succinate dehydrogenase (of the citric acid cycle) and cytochrome c oxidase (the last enzyme of the respiratory chain). In addition, the maximal oxygen uptake during cycling was determined using the Douglas bag technique. It is noteworthy that, in the post-training period, the whole-body $V_{O_2}max$ is maintained significantly longer than the muscle oxidative enzyme content. (Modified from ref. 17.)

content of oxidative enzymes can be obtained by fairly light running (jogging), but the enzyme adaptation becomes much more marked if the training intensity is increased to work rates demanding 70 to 80 per cent of the individual's $V_{O_2}max$. In theory, even higher training intensities would result in a further enhancement of the muscle's oxidative capacity, but in practical terms this may not be so since another important factor is the duration of the training bouts. There are no conclusive human data illustrating this interdependence of exercise intensity and duration, but interesting information can be obtained from a detailed study on the rat by Dudley et al.[19] These workers subjected rats to training in the form of treadmill running 5 days per week for 2 months at varying speeds and daily training durations (Fig. 6). The rats trained at six different running speeds, demanding approximately 60, 70, 80, 95, 105, and 115 per cent of their maximal oxygen uptake. At the two highest speeds, the exercise was performed intermittently. For each speed, the muscle enzyme adaptation increased with the duration of the daily exercise, but no additional training effect was noted when the daily duration exceeded 45 to 60 min. At the two highest speeds, the rats could only tolerate exercise for 30 and 15 min daily, but this was sufficient for marked training effects to occur. As might be expected, the initial period of the daily exercise bout gave the highest training effect per unit of training time, with

successively smaller effects for the following periods. In a previous study by Fitts et al.[20] rats trained for 10, 30, 60, or 120 min/day on a motor-driven treadmill and displayed progressively larger increases in the gastrocnemius muscle oxidative capacity with increasing exercise duration. Beyond 120 min there were no further increases. The available data thus indicate that beyond a certain time the law of diminishing returns applies, i.e. less adaptations occur per unit increase in training duration.

Fibre type differences

An interesting observation in the animal study by Dudley et al.[19] was that the training response differed markedly between the fibre types. For a training effect to occur in the fast glycolytic fibres (type IIb), the exercise intensity had to require at least 80 per cent of the rat's $V_{O_2}max$ (Fig. 6). Higher running speeds gave successively better training effects in this fibre type. In contrast, for the fast oxidative glycolytic fibre type (IIa), the training effect increased with increasing speeds up to an intensity (speed) demanding 80 per cent of $V_{O_2}max$. Higher running speeds did not result in an enhanced training effect. The slow-twitch fibre type (type I) responded in yet another fashion. In this fibre type, the training effect increased up to a running speed demanding 80 per cent of

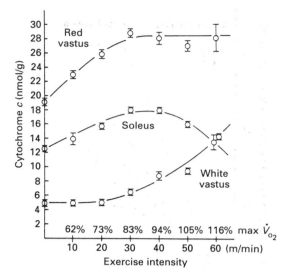

Fig. 6 The effects of endurance training on the oxidative enzyme content of different muscles in the rat. The rats were trained on a treadmill 5 days per week for 2 months at varying running speeds (usually 45–60 min daily). The muscle content of cytochrome c, a respiratory chain component, was measured as an indicator of muscle oxidative capacity. The three muscles, representing different fibre types, were red vastus (fast-twitch oxidative glycolytic, type 2a), soleus (slow-twitch, type 1), and white vastus (fast-twitch glycolytic, type 2b). The approximate percentage of the rat's maximal oxygen uptake demanded at each speed is indicated. (Reproduced from ref. 19 with permission.)

the V_{O_2}max. Paradoxically, higher speeds than this resulted in a decreased training effect. These fibre-type specific training effects are likely to be explained by the recruitment pattern of different muscles during running. The deep part of the thigh muscle, which was the source of the fast oxidative glycolytic fibres in the cited study, is fully activated during running at 30 m/min; consequently, this training speed results in a maximal training effect. The superficial part of the thigh muscle, which was the source of the fast glycolytic fibres, has a higher activation threshold. No training effect is therefore seen at low running speeds, but above the activation threshold there is a linear relationship between running speed and training effect. The soleus muscle, which contains almost exclusively slow-twitch (type I) fibres, is fully activated at 30 m/min. Therefore, higher running speeds than this would not be expected to increase the training effect. However, there is no obvious explanation of the finding in the study by Dudley *et al.*[19] that higher running speeds than 30 m/min resulted in a successively diminished training effect in this muscle.

Fibre-type recruitment in man

The results of the training studies in laboratory rats clearly illustrate that knowledge of the fibre-type recruitment pattern during exercise is essential when trying to predict the effects of different training regimens. For humans, quite detailed information is available on cycle ergometer exercise at different rates of work. It has been shown that, as in the rat, the slow-twitch (type I) muscle fibres are the first to be activated and are kept activated even at high exercise intensities. With increasing rates of work there is a recruitment of the fast-twitch motor units, with type IIa followed by type IIb. It is believed that fibres (motor units) of all types are

recruited at exercise intensities demanding more than 80 to 85 per cent of V_{O_2}max. However, very strenuous exercise is probably required to activate maximally all the type IIb motor units in a given muscle (for references, see Saltin and Gollnick).[18]

This fibre-type recruitment pattern probably also applies to other activities, such as running. However, it is possible that the recruitment of high-threshold IIb units is less marked in running than in cycling, especially at high exercise intensities when cycling may involve quite forceful pedalling. Training at an exercise intensity slightly above that resulting in a marked increase in the blood lactate concentration is generally sufficient for a maximal recruitment of the muscle's slow-twitch (type I) fibres. However, there is no evidence that more intense exercise would decrease the training effect in this fibre type as it did in the rat studies referred to above. A maximal training effect on the muscle's IIa and IIb fibres demands higher rates of work—how high may depend on the percentage fibre-type composition of the particular muscle. Available evidence indicates that with long exercise durations and the resulting glycogen depletion, there is a time-dependent increase in the recruitment of higher threshold motor units; therefore long-duration exercise would be expected to result in an increased training effect.[18,21,22] It can be concluded that, for a large training effect per unit of training time, it is advisable to use high training intensities. With very heavy exercise, however, the duration of the exercise bouts may be insufficient for an optimal training effect. The rat study by Dudley and colleagues[19] gives some hints about the optimal balance between the intensity and duration of training, but there are still insufficient human data available.

Effects of endurance training on glycolytic enzymes, capillarization, and fibre types

The muscle cell's content of glycolytic enzymes is not, or only marginally, affected by endurance training programmes of 2 to 6 months' duration. The content of glycolytic enzymes is normally low in the skeletal muscles of endurance athletes, but this finding is entirely explained by the large percentage of slow-twitch fibres in their muscles. The content of glycolytic enzymes in this fibre type is normally only half that in fast-twitch fibres. The mean glycolytic enzyme level of slow-twitch or fast-twitch muscle fibres in athletes has been found to be normal, or even slightly enhanced.[11,23] This finding is in accord with what has been observed during chronic stimulation (see above), when there is a complete fibre-type transformation from fast-twitch glycolytic (type IIb) to slow-twitch (type I) fibres. In this situation the glycolytic enzyme content of the muscle is decreased to 20 per cent of the initial level (Figs. 1–3), a decrease which reflects the large difference in glycolytic potential between fast-twitch glycolytic and slow-twitch fibres in the rabbit.[8] It therefore seems as if the type of muscle fibre, based on its composition of myofibrillar proteins, is a strong determinant of its glycolytic enzyme content. The same is not true of most of the oxidative enzymes, which change with training and inactivity completely independently of the specific myofibrillar protein isoforms of the fibre.

Skeletal muscle capillarization in man is rapidly enhanced with endurance training. Two months of training at high submaximal exercise intensities is sufficient to increase the total number of muscle capillaries by 50 per cent (Fig. 7).[24–26] The difference between endurance athletes and untrained individuals with respect to the capillary count per muscle fibre (leg muscles) has been found to be two- to threefold.[18] There is a lack of information

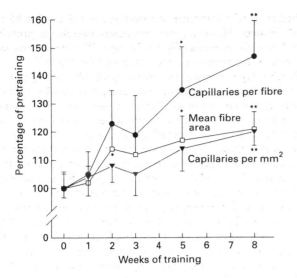

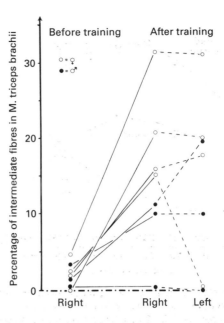

Fig. 7 The effect of 2 months of endurance training (identical with that shown in Fig. 5) on capillary formation in the human thigh muscle. Muscle biopsies were histochemically stained for capillaries, which appear as dark spots in a muscle cross-section (amylase-periodic acid—Schiff reagent method). The number of capillaries per fibre was calculated in each muscle cross-section as the number of cross-sected capillaries divided by the number of cross-sected muscle fibres. * ($p < 0.05$) and ** ($p < 0.01$) denote significant differences from pretraining values. (Modified from ref. 24.)

Fig. 8 Increase in myofibrillar ATPase intermediate human skeletal muscle fibres in response to extensive endurance training. Muscle biopsies were obtained from the triceps brachii muscle (upper arm) in seven individuals before and after training and were investigated by means of histochemical fibre typing (myofibrillar ATPase). The training consisted of skiing with sledges for 500 miles over a period of 36 days. It is believed that the emergence of intermediate fibres reflects on-going fibre-type transformation from fast to slow twitch. (Reproduced from ref. 30 with permission.)

about the extent to which capillary neoformation is dependent upon training intensity and duration. However, it is known that less intense training regimens often result in oxidative enzyme increases without any change in capillarization.

When stains for myofibrillar ATPase have been used as the basis for fibre-type classification, most longitudinal studies in man have failed to demonstrate an interconversion of fibre types (i.e. fast-twitch to slow-twitch) in response to endurance training. The stable nature of a muscle's fibre-type composition is further illustrated by the results of chronic stimulation studies in rabbits. Although in this situation there is a gradual and complete replacement of fast-twitch by slow-twitch fibres, quite long periods of chronic stimulation are required. The fibre-type changes are also the first to revert to normal when stimulation is discontinued.[10] On the basis of these findings, the high percentage of slow-twitch (type I) fibres in endurance athletes and the opposite finding in sprinters have therefore been ascribed to genetic factors.[27] However, endurance training is known to lead to a complete type transformation within the fast-twitch (type II) fibres from type IIb to type IIa.[28,29]

The concept that endurance training does not change the relative occurrence of fast- and slow-twitch fibres has been challenged in recent years. Thus it has been shown, for example, that (1) endurance training of long duration leads to the appearance of fibres intermediate between fast and slow-twitch (Fig. 8), (2) the muscles of the dominant leg in different types of athletes, such as badminton players, contain a significantly increased percentage of slow-twitch fibres, and (3) in several studies of detraining the percentage of fast-twitch muscle fibres increases. In accord with these results are analyses of the myofibrillar protein isoform pattern within single muscle fibres. Bauman *et al.*[31] demonstrated the appearance with training of fast-twitch fibres containing a mixed pattern of fast and slow myofibrillar protein isoforms. Therefore it

is reasonable to conclude that extensive endurance training will result in an enhanced percentage of slow-twitch fibres. The extent to which this might occur still remains to be demonstrated. The probable reasons that fibre type transformation was not seen in the early studies are that (1) these studies were too short and (2) the muscles investigated were postural muscles and therefore relatively trained even in the pretraining state. (For a detailed discussion, see Schantz.[32])

RATE OF LOSS OF THE TRAINING-INDUCED ADAPTATION IN SKELETAL MUSCLE OXIDATIVE CAPACITY AND CAPILLARIZATION FOLLOWING CESSATION OF TRAINING

An increase in the oxidative capacity of a muscle, induced by 2 months of endurance training, is lost in 4 to 6 weeks if the training is stopped (Fig. 5). This loss of muscle oxidative enzymes occurs faster than the decrease in muscle capillarization[33] and in the whole-body maximal oxygen uptake that can be attained during cycling (Fig. 5). The time course of the decrease in muscle oxidative enzyme content following cessation of training agrees well with that observed following cessation of chronic stimulation in the rabbit (see above). In the latter case, however, the return to pretraining levels of both muscle capillarization and oxidative enzyme content occur simultaneously. There has been only one detailed investigation of the enzyme changes that take place during the detraining of individuals who have done endurance training for several years (very well trained, although not top, athletes).[11] It was found that the oxidative capacity of the slow-twitch fibres rapidly decreased with detraining to the level found in untrained control subjects. Interestingly, however, the oxidative capacity of the fast-twitch fibres, despite the detraining decrease, maintained an elevated level throughout the 12-week period of detraining studied. One theory put forward was that, because of prolonged endurance training, changes had occurred in the normal impulse pattern of fast motor neurones in the spinal cord. In the latter study, as in Fig. 5, the relative decrease in oxidative enzymes occurred faster than the decline in Vo_2max. In the referred study, however, there was also a significant decrease in Vo_2max after 12 days of detraining (these data are reported in ref. 34). It is possible that these individuals were kept more inactive during the de-training phase compared with the study depicted in Fig. 5.

POSSIBLE INTRACELLULAR SIGNALS MEDIATING THE ENZYME ADAPTATION TO TRAINING

The enzymes, as well as other protein molecules, have a limited life-span. They are built up and degraded in a continuous cycle in which the biological half-life of many of the mitochondrial enzymes is about a week and that of the glycolytic enzymes is one to a few days. Accordingly, the cellular content of a certain enzyme is the result of this balance between synthesis and degradation. It has been shown that a change in the rate of synthesis of enzyme proteins is the most important factor in explaining the enzyme changes resulting from chronic stimulation or training.[7,35,36] An interesting area of research at present is to explore the biochemical mechanisms underlying the altered rate of enzyme synthesis, i.e., how the information that there is a need for an increased amount of oxidative enzymes in the muscle cell is transferred to the genes. Suggested mediators include the following: decreases in the concentration of ATP or other high energy phosphate compounds; a decreased oxygen tension; an increased sympathoadrenal stimulation of the muscle cell; substances released from the motor nerve; calcium-induced diacylglycerol release with subsequent activation of protein kinase C. The availability of advanced genetic techniques as well as improved cell culture systems has led to a renewed interest in this area of research, and this will probably lead to a better understanding of the mechanisms whereby the skeletal muscle cell adapts to different normal and pathological states.

TRAINING-INDUCED CHANGES IN MUSCLE GLUCOSE UPTAKE AND INSULIN ACTION

Another major effect of exercise and physical training on muscle metabolism is that of an increased glucose uptake. Glucose uptake into the cell occurs by markedly different processes in different tissues. In skeletal muscle and adipocytes, as well as fibroblasts, cellular glucose transport occurs by facilitated diffusion. This process is passive in the sense that glucose moves from higher concentrations to lower concentrations without requiring energy. The transport involves a mobile carrier molecule which facilitates hexose transport through the membrane. This mechanism results in saturation kinetics of the glucose transport process. The Michaelis constant K_m for the transport of glucose from the outer to the inner surface of the cell membrane (i.e. the glucose concentration at which the transport rate is half-maximal) is approximately 5 to 10 mmol in several tissues including skeletal muscle.[37] Thus the transport rate responds to fluctuations in the blood glucose level, but, in addition, it is regulated by hormones and other extra- and intracellular factors. Specifically, contractile activity and insulin are the two most potent stimulators of skeletal muscle glucose transport.

Following an acute bout of exercise, glucose uptake into skeletal muscle is stimulated. This is partly a direct effect, occurring independently of insulin.[38,39] Recent evidence indicates that the effect is initiated by the increase in cytoplasmic calcium.[40] In addition, however, the sensitivity of the muscle glucose transport process to insulin is also increased.[41–43] Both the direct and insulin-mediated effects of contraction are sustained into the post-exercise period. The direct effect seems to be reversed within a few hours,[44] whereas the enhanced insulin sensitivity as a result of an acute exercise bout (which is not detectable until the direct effect has been partially reversed) lasts longer (usually 1–2 days).[39] The rate of reversal varies and seems to depend upon the refilling of glycogen stores (see the following discussion under insulin action and glycogen storage). Recently, Cartee *et al.*[45] obtained evidence that the enhanced insulin sensitivity is indicative of a process which is not limited to insulin, but which may be thought of as a non-specific increase in the susceptibility of glucose transport to stimulation by a variety of agents. There is evidence that the increased insulin action following an acute bout of exercise is to a large extent a local effect restricted to the muscle groups that were recruited during the exercise session.[46]

It is believed that muscle contraction and insulin stimulate glucose uptake through two separate pathways. The reasons for this belief are (1) that the effect of contractile activity and a maximal dose of insulin are additive *in vitro*,[38,39,47,48] and possibly even synergistic *in vivo*,[48] and (2) that the time courses of the effects are different.[37] However, a detailed description of how these two pathways differ cannot be given at present.

GLUCOSE TRANSPORT FOLLOWING A SINGLE BOUT OF EXERCISE

Glucose is transported into mammalian skeletal muscle by a glucose transporter. The most abundant glucose transporter in mammalian skeletal muscle is the insulin-regulatable glucose transporter GLUT4.[49] A second isoform of the glucose transporter (GLUT1) has also been isolated from skeletal muscle, although the role of this transporter is likely to be restricted to basal conditions. Under basal conditions the major portion of the cellular GLUT4 content is present in an intracellular pool, but upon stimulation (through contraction or insulin) the transporters are believed to be translocated to the sarcolemma where they facilitate glucose transport. In order to increase glucose uptake, either the number of transporters in the plasma membrane or the activity of the available transporters must increase. Glucose uptake following a single bout of exercise appears to be enhanced by an increase in both the number of transporters and transporter activity.[50–52] With respect to insulin-mediated effects, these increases result in enhanced insulin action at physiological insulin concentrations[46,53,54] and probably also at maximally stimulating insulin concentrations.[39,48,55] The term 'insulin sensitivity' is defined as the concentration of insulin required for half-maximal activation of glucose transport; however, it is not always possible to measure glucose transport at maximal insulin concentrations. In such cases, insulin sensitivity is often incorrectly said to be improved if there is an increased glucose transport activity at a submaximal insulin concentration. In this case, it would be more correct to state that there is an enhancement of the insulin-mediated glucose uptake at the given submaximal insulin concentration. Insulin responsiveness is said to be improved when an enhanced effect of a maximal insulin stimulus is registered.

GLUCOSE TRANSPORT IN RESPONSE TO ENDURANCE TRAINING

Although, as indicated in the previous discussion, it is a well-established fact that an acute exercise bout leads to an increase in muscle glucose uptake, the question of whether training produces an additional effect has been debated.

Contraction-stimulated glucose transport

There is very little information available regarding possible training effects, but a recent investigation by Coggan et al.[56] suggested that endurance training may in fact reduce glucose transport during submaximal exercise in man. Idström et al.[57] and Ivy et al.[58] failed to demonstrate any chronic training effect in the rat on contraction-induced glucose transport in the perfused hindquarter, but these studies were complicated by the fact that insulin was present in the systems. This problem was avoided in a recent study using the perfused hindquarter by Plough et al.[55] who found an increased contraction-stimulated glucose-transport rate with training, but only in slow-twitch fibres. This effect, which could not be ascribed to an influence of the last training session, was paralleled by signs of an increased abundance of glucose transporter proteins.

Methods of determining insulin-stimulated glucose transport

Glucose tolerance, or the ability of the body to process an ingested glucose load, can easily be measured using the oral glucose tolerance test. In this test, subjects ingest a drink containing 75 to 100 g of glucose. Blood samples are obtained during the following 3 h for the analysis of plasma insulin and glucose. The data are then plotted, with the glucose tolerance being inversely related to the areas under the glucose and insulin curves.

Whole-body insulin sensitivity and responsiveness can be determined using the hyperinsulinaemic euglycaemic clamp procedure developed by DeFronzo et al.[59] During this procedure, insulin is infused via the brachial vein at a low dose (less than 100 mU/m^2.min), followed by a high dose (more than 2000 mU/m^2.min) during two consecutive 2-h periods. Blood samples are taken every 5 min for the monitoring of blood glucose levels, and glucose is infused via the brachial vein at rates necessary to keep the blood glucose constant. Insulin sensitivity and responsiveness are then determined from the amount of glucose infused via the brachial vein during the low and high-dose stages. As skeletal muscle is the primary site of glucose disposal under these conditions,[60,61] it is generally presumed that skeletal muscle insulin sensitivity or responsiveness is being measured by this technique. Furthermore,

skeletal muscle appears to be the major site of the increased peripheral insulin sensitivity associated with physical training.[62,63]

Results with respect to insulin-mediated glucose disposal and glucose tolerance

The fact that training can increase insulin-mediated glucose disposal has been demonstrated in a number of studies,[54,55,64,65] but most of this effect is likely to be mediated by short-term effects of the last exercise bouts, which are lost within a few days following the cessation of training (see below). In healthy individuals these changes in insulin action are not usually accompanied by similar changes in glucose tolerance[66–68] since the plasma insulin level during a glucose tolerance test changes in a reciprocal manner relative to the changes in insulin action (Fig. 9).

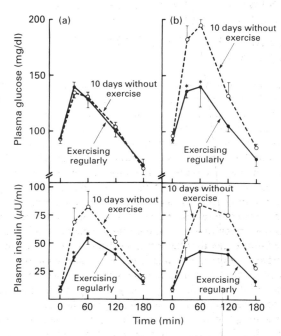

Fig. 9 Plasma glucose and plasma insulin concentrations during an oral glucose tolerance test where 100 g of glucose was ingested at time zero.

(a) Ten master athletes who did not show deterioration in glucose tolerance with inactivity. However, the increase in plasma insulin concentration during the glucose tolerance test indicates that insulin sensitivity was impaired with inactivity.

(b) Four master athletes who showed deterioration in glucose tolerance in response to 10 days without exercise compared with 16 to 18 h after a usual training session when they were exercising regularly.* Exercising regularly vs. 10 days without exercise, $p < 0.05$ (mean±SEM). (Reproduced from ref. 69 with permission.)

In accordance with the notion that the effect of the last exercise bout is the important factor, the enhanced insulin-mediated glucose disposal in trained individuals has been found to be rapidly lost after the cessation of exercise training in a number of studies.[69–72] King et al.[70] studied nine endurance-trained male and female subjects in the trained state (within 24 h of the last exercise bout) and again after 10 days of physical inactivity using the hyperinsulinaemic euglycaemic clamp procedure. When the plasma insulin concentration was maintained at approximately 80

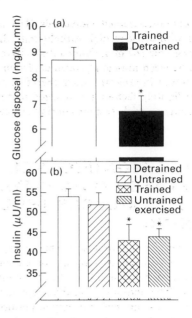

Fig. 10 Effects of exercise and inactivity on insulin sensitivity.
(a) Nine well-trained individuals were studied using the hyperinsulinaemic euglycaemic clamp procedure in the trained exercising state (16 h after the last exercise bout) and again after 10 days of physical inactivity. Insulin sensitivity is estimated from the whole-body glucose disposal rate at a submaximal plasma insulin concentration (78 μU/ml) (mean ± SEM). (From ref. 70.)
(b) Insulin sensitivity (the insulin concentration eliciting 50 per cent of maximal insulin-mediated glucose disposal) was measured in seven trained individuals in the habitual state 15 h after the last training bout and 5 days after the last training session (detrained). In addition, seven untrained subjects were studied at rest and after 60 min of bicycle exercise at 150 W (mean±SEM). (Hyperinsulinaemic euglycaemic clamp experiments from refs. 54 and 71.)

μU/ml (high physiological level), the glucose disposal rate averaged 8.7±0.5 mg/kg.min before and 6.7±0.6 mg/kg.min after 10 days of inactivity ($p < 0.001$) (Fig. 10 (a)). These results were supported by those of Mikines and coworkers[54,71] (Fig. 10 (b)), who observed a reduction of insulin sensitivity in physically trained men 5 days after the cessation of training to levels identical with those observed in untrained individuals, who in turn could increase their insulin sensitivity to the same level as that seen in well-trained individuals by performing just one 60-min exercise bout. Nagasawa et al.[72] studied insulin-mediated glucose disposal in rats after 5 weeks of wheel running, as well as 1, 2, 3, 7, and 14 days after the last exercise session, using the euglycaemic insulin clamp technique. They found insulin action to be enhanced in the trained rats compared with sedentary controls; however, the enhanced insulin-mediated glucose disposal was lost 3 days after the cessation of training. Thus most investigations performed to date indicate that the enhanced glucose uptake at submaximal insulin concentrations noted in trained individuals is not a true training-induced adaptation but merely an effect of the last exercise bout.

Results with respect to insulin responsiveness

There are recent reports that endurance training leads to an increased number of glucose transporters in muscle owing to an increased synthesis of transporter protein.[55,73] Fibre-type differences may play a role in this process, since Rodnick et al.[73] found an increased transporter number in fast-twitch muscle only, with no change in the slow-twitch soleus muscle. It is possible that such fibre-type-related differences may explain the variable results which have been noted with respect to the change in insulin responsiveness in connection with endurance training. Several studies have shown that endurance training results in an increased insulin responsiveness in skeletal muscle,[54,65,74,75] which seems consistent with the increased muscle content of glucose transporters as indicated above. Mikines et al.[71] claim that this adaptation is not due to residual effects of the last bouts of exercise. The reason is that they found insulin responsiveness to be clearly higher (40 per cent) in endurance-trained than in untrained subjects, and that these results were not influenced if the trained subjects had (a) performed one training session or (b) stopped training for 5 days. In untrained subjects, a 1-h exercise session resulted in a 15 per cent increase in insulin responsiveness, but not to the level found in trained subjects. However, in the light of the conflicting data reported by King et al.,[64,70] the question of whether endurance training actually results in a long-term adaptation with respect to insulin responsiveness must still be left open. These authors found no difference in insulin responsiveness with respect to glucose uptake between 11 endurance-trained individuals in the habitual state 16 h after a normal exercise session and a group of 11 untrained controls. In conformity with these results, there was no change in the insulin responsiveness of the trained subjects following 10 days without exercise.

It can be concluded that endurance training leads to an increased insulin sensitivity and possibly also responsiveness in the engaged skeletal muscles. Whether this represents a long-term adaptation or is just the result of consistently performing a single bout of exercise cannot yet be determined. This question may be purely academic, however, and the results demonstrate the importance of regularly performed exercise to protect against the development of insulin resistance, for example with ageing, or to improve insulin action in such pathological states as obesity and type II diabetes. It may also be noted that not all types of exercise are associated with increased insulin action, as one bout of eccentric exercise has been found to be associated with insulin resistance (see below for discussion).[55,76]

METABOLIC SIGNIFICANCE OF THE TRAINING-INDUCED ADAPTATION IN INSULIN ACTION

Glycogen storage

The main routes of disposal of the glucose taken up by the muscle cell in response to insulin are (1) glycogen formation, (2) oxidation, and (3) lactate formation. Although oxidation of glucose increases in response to insulin infusion,[54,71] this increased oxidation does not account for the entire glucose uptake seen in response to insulin. It should also be noted that much of the increase in oxygen consumption in response to insulin is due to increased glycogen synthesis.[46] Lactate formation in response to insulin has been estimated to account for only 10 per cent of the glucose uptake in human skeletal muscle.[46] Therefore it can be concluded that a significant portion of the glucose uptake in skeletal muscle is directed towards glycogen synthesis.[46,77,78]

Thus a major change associated with the increased insulin action in endurance-trained skeletal muscle is likely to be an increased glycogen storage capacity which allows faster replenishment of muscle glycogen stores following exercise bouts. However, the increased rate of insulin-stimulated glycogen synthesis observed in trained individuals is not great enough to account for the entire increase in glucose uptake which is due to exercise training.[54,78] It has been shown that the concentration of muscle glycogen is higher in trained than in untrained individuals. The resting glycogen concentration in untrained human muscle has been found to range between 70 and 110 mmol/kg wet weight of muscle,[18,79,80] while that in endurance-trained muscle may range from 140 to over 230 mmol/kg wet weight.[80,81] Particularly convincing evidence that local and not just dietary factors are responsible for this difference in storage capacity between trained and untrained muscle comes from studies of one-leg training.[82,83] In these studies it was found that the trained leg possessed from 6 to 60 mmol glucose units/kg wet weight more glycogen than the untrained leg. However, this increased level of glycogen is reduced to that of untrained muscle upon detraining or immobilization.[84] Muscle glycogen synthase activity has also been found to be higher in trained than in untrained individuals,[71,85,86] and this enzyme can be activated by insulin.[87,88]

The replenishment of glycogen stores following a bout of exercise is associated with a reversal of the enhanced insulin sensitivity.[89] Gulve et al.[89] found that this reversal was related to glycogen concentration and not to the increased glucose transport into the muscle, but could say nothing about the effectors beyond the level of glucose transport. However, it is unlikely that the reversal of insulin action is solely linked to glycogen replenishment per se, since an enhanced glycogen synthesis was found in response to insulin even when the glycogen concentration in the muscle had returned to control levels.[46,77] With regard to muscle insulin responsiveness, Cartree et al.[41] found a complete reversal following an exercise bout even when rats were fed a fat diet after exercise in order to keep the muscle glycogen concentration well below that measured in fasted sedentary rats. The fact that a corresponding reversal was not observed with respect to insulin sensitivity may indicate different controlling mechanisms for insulin sensitivity and responsiveness.

As stated above, the enhanced insulin action is likely to facilitate a rapid resynthesis of muscle glycogen after each exercise bout in endurance-trained individuals. This is particularly crucial during times of frequent training when muscle glycogen stores can be reduced daily. As the enhanced insulin action is reversed upon refilling the glycogen stores, usually to concentrations above those measured in the sedentary state before exercise (supercompensation)[41], the differences found between trained and untrained individuals with regard to muscle glycogen levels may be merely an effect of the last exercise bout.

Muscle glycogen concentration and performance

The initial muscle glycogen concentration is important for sustained exercise (longer than 1 hour) at work rates above 60 to 80 per cent of V_{O_2}max. Bergström et al.[90] found that subjects exercising at 75 per cent of the V_{O_2}max could sustain exercise for 114 min when muscle glycogen stores were at concentrations comparable with those found in resting untrained muscle (97 mmol/kg wet weight), but were able to exercise for only 57 min at this intensity when muscle glycogen concentrations were 35 mmol/kg wet weight. When the muscle glycogen level was comparable with

that found in endurance-trained individuals (184 mmol/kg wet weight (attained with a 3-day high carbohydrate diet), subjects were able to exercise for 167 min. Other authors have confirmed a close association between muscle glycogen depletion and fatigue-.[91,92] Glycogen storage in skeletal muscle therefore appears to be one of the major limiting factors in prolonged performance, and accordingly preservation of these stores during exercise is of great importance.

METABOLIC SIGNIFICANCE OF THE INCREASED OXIDATIVE CAPACITY IN MUSCLE INDUCED BY TRAINING

After the cessation of chronic stimulation (see above), the restoration of a normal muscle oxidative enzyme content and capillarization follows a time course similar to that of the normalization of muscle endurance. This indicates that the adaptations described are of importance for the muscle's capacity to perform prolonged exercise. This can be further illustrated by an investigation in which a group of subjects underwent one-leg endurance training on a cycle ergometer during a period of 6 weeks. With one well-trained leg (the level of the oxidative enzyme succinate dehydrogenase was 25 per cent higher than in the untrained leg), the subjects then performed a two-leg endurance exercise at 70 per cent of V_{O_2}max, in which both legs performed identically. The energy metabolism of the two legs could be analysed and compared by means of arterial and venous catheterization and muscle biopsy analysis.

As illustrated in Fig. 11, there was a significantly smaller release of lactate from the trained leg than from the untrained one, and a significantly larger percentage of the energy output in the trained leg stemmed from fat combustion. In the following, a review will be given of a large number of studies showing that at the same absolute exercise intensity (and possibly even at the same relative intensity (percentage V_{O_2}max)), trained individuals rely more than untrained ones on fat as an energy substrate. This is the case despite the fact that, at a given rate of work, the plasma level of free fatty acids is either similar or lower in endurance-trained subjects.

Endurance training and glycogen depletion rates

Training can result in reduced muscle glycogen utilization during exercise in several ways. The amount of muscle glycogen utilization and lactate formation in the rat plantaris muscle subjected to 3 min of electrical stimulation has been found to be reduced after training.[93] The cause of this decreased muscle glycogen utilization upon initiation of exercise was not clear, but it was accompanied by smaller increases in inorganic phosphate, AMP, and estimated free ADP concentrations in the trained than in the untrained muscles during the electrically induced contractions. Furthermore, the trained muscles displayed smaller decreases in ATP and phosphocreatine concentrations. The authors concluded that the observed changes were consequences of the adaptive increase in muscle mitochondria (see below and Fig. 12 for hypothetical mechanisms). Furthermore they suggested that reduced levels of inorganic phosphate in trained muscle might have played a role in the decreased initial burst of glycolysis. The validity of these

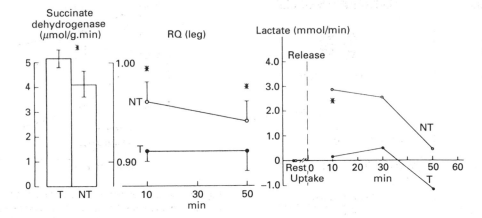

Fig. 11 The metabolic significance of the training-induced adaptation of human skeletal muscle. A group of subjects underwent one-leg endurance training on the bicycle ergometer during a period of 6 weeks. With one well-trained leg (T) (the level of succinate dehydrogenase is 25 per cent higher than in the non-trained leg (NT) (left-hand figure)), the subjects performed a two-leg bicycle ergometer exercise at 70 per cent of V_{O_2}max, in which both legs performed identically. Arterial and venous catheterization made possible measurements of the oxygen uptake V_{O_2} and the carbon dioxide production V_{CO_2} of both legs separately. The V_{CO_2}/V_{O_2} ratio, known as the respiratory quotient RQ, indicates the relative contributions of carbohydrate and fat to the oxidative metabolism; an RQ of 1.0 indicates oxidation of carbohydrate only and an RQ of 0.7 indicates oxidation of fat only. Thus the middle figure indicates that fat is a more important energy source for the trained leg than for the untrained one. Accompanying the greater use of carbohydrates in the untrained leg, there is a larger formation and release of lactate (right-hand figure). In the trained leg, the lactate release is low, and towards the end of the exercise bout there is even a tendency towards an uptake of lactate from the blood ($n=6$, mean ± SEM). (Modified from ref. 83.)

results was supported by Jansson and Kaijser[94] and Green et al.,[95] who found that training exerts its greatest effect in reducing glycogen degradation early in exercise (Fig. 13). Green et al.[95] were also able to show that this effect occurs in both fast- and slow-twitch fibres. Glycogen depletion has been found to be reduced during prolonged exercise in trained individuals compared with untrained individuals working at the same absolute rate (the same rate of oxygen consumption),[20] although the rate of glycogen depletion is likely to be similar if the subjects are exercising at the same relative exercise intensity (same percentage of V_{O_2}max).[96] Saltin and Karlsson[97] found the latter to be true when the same subjects were studied before and after training; however, Jansson and Kaijser[94] have recently demonstrated reduced glycogen utilization in the muscles of endurance-trained subjects when exercise was performed at the same relative intensity (65 per cent of V_{O_2}max) (Fig. 13).

Lactate formation

It is well documented that endurance-trained individuals have lower blood lactate levels than untrained individuals during exercise at both the same absolute exercise intensity[98–101] and the same relative exercise intensity (percentage of V_{O_2}max).[94,96,102] This is probably attributable to a decreased rate of lactate formation in trained muscles, as indicated by several studies of the arteriovenous difference for lactate. Jansson and Kaijser[94] studied trained and untrained individuals exercising at an identical relative intensity and found in this case that the arteriofemoral vein difference for lactate tended to be higher in the trained than in the untrained

individuals at both 15 min and 60 min of exercise. This is in accord with the one-leg training study shown in Fig. 11. The fact that glycogen depletion occurs at lower rates in trained than in untrained muscle supports the notion of a deceased rate of lactate formation in trained muscle (see the preceding text for references). The enzyme isoforms of LDH are also known to shift to favour the LDH_{1-2} (heart) isoforms over the LDH_{4-5} (muscle) isoforms following endurance training.[103] This adaptation in favour of the heart isoform increases lactate conversion to pyruvate in aerobically trained skeletal muscle and decreases the conversion of pyruvate to lactate. This may well be a major cause of the reduced glycogen utilization in trained compared with untrained individuals. The reduction of lactate production with training is also supported by the findings of Favier et al.,[104] who observed that after 3 min of electrical stimulation of the gastrocnemius–plantaris–soleus muscle group, the lactate concentration in all muscle fibre types was considerably lower in trained than in untrained animals. Thus the decreased lactate production with training occurs in all fibre types, although lactate production is related to the type of fibre,[105] with slow-twitch fibres producing less lactate. In addition to the proposed mechanism described in Fig. 12 and the shift in the LDH isoenzyme pattern, the increased capacity of the malate–aspartate shuttle system for the transport of NADH electrons from the cytosol into the mitochondria probably also contributes to the reduced lactate production in response to training.[106] However, these changes, which result in lower blood lactate concentration after training, are lost upon cessation of training. Costill[107] found that postexercise blood lactate levels increased gradually towards pretraining

TRAINED MUSCLE   UNTRAINED MUSCLE

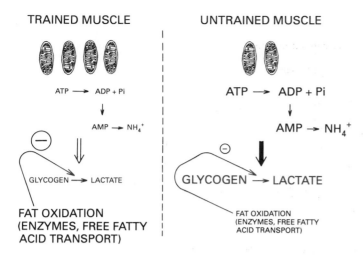

FAT OXIDATION
(ENZYMES, FREE FATTY
ACID TRANSPORT)

Fig. 12 A hypothetical biochemical mechanism whereby a large concentration of oxidative enzymes (i.e. citric acid cycle and fat oxidation enzymes and respiratory chain components) in trained muscle would lead to a greater reliance on fat metabolism, a lower rate of lactate formation, and sparing of muscle glycogen during exercise. The increased content of oxidative enzymes in trained skeletal muscle is explained to a large extent by a larger mitochondrial volume (volume fraction), indicated schematically with mitochondrial symbols. Suppose that, in the untrained muscle, there are only half as many enzyme molecules of the citric acid cycle and half as many components of the respiratory chain than in the trained muscle (which is a reasonable assumption (see Fig. 4)). Owing to the lower enzymatic capacity and mitochondrial volume fraction of the untrained muscle, it follows that, at a given rate of work, i.e. at a given rate of oxygen uptake, each mitochondrial unit has to be activated twice as much in the untrained as in the trained muscle. An important component of this activation is the increased level of the degradation products of ATP (e.g. ADP), which are the result of muscle contractions. These substances must thus be stabilized at a higher concentration in the untrained than in the trained muscle. However, these ATP degradation products are also powerful stimulators of the glycolytic pathway, thus leading to a higher glycolytic rate in untrained muscle, resulting in a greater lactate release and carbohydrate oxidation. The fat oxidation rate is higher in trained muscle mainly due to the higher content of enzymes of fatty acid transport and oxidation. This leads to a more pronounced inhibition of glycolysis in the trained than in the untrained muscle, where the rate of fat oxidation is lower (the glucose–fatty acid cycle[113]). Together, these factors lead to sparing of glycogen during exercise in trained skeletal muscle. (Reproduced from ref. 13 with permission.)

levels during 4 weeks of detraining. The data given in this section exclude the possibility that the decreased blood lactate concentration during exercise in the trained state is simply due to an increased clearance of lactate as has been suggested previously.[108,109] However, an increased lactate clearance in endurance-trained individuals cannot be ruled out, but further studies are needed to evaluate this possibility.

Utilization of blood-derived glucose

In addition to a decreased utilization of muscle glycogen, the carbohydrate-sparing effect of training also involves a decreased utilization of blood-derived glucose. In the study by Coggan et al.[56] this could account for approximately half the total decrease in

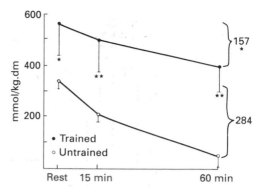

Fig. 13 Muscle glycogen concentration in the quadriceps femoris muscle, vastus lateralis, at rest and during bicycle exercise in untrained (open circles) and trained (full circles) subjects. All subjects exercised at an identical relative intensity (65 per cent of Vo_2max). Glycogen values are expressed as mmol glucose units/kg dry weight ($n=5$, mean ± SEM). Statistical comparisons are between untrained and trained subjects: $*p<0.05$, $**p<0.01$. (Reproduced from ref. 94 with permission.)

carbohydrate oxidation following training during the final 30 min of a 2-h cycle ergometer exercise session at 60 per cent of the pretraining Vo_2max. Similar results were reported in the study by Jansson and Kaijser.[94] They found no change with training in blood glucose extraction by the legs after 15 min of bicycle exercise at 65 per cent of Vo_2max, but after 60 min of exercise the blood glucose extraction was considerably lower in the trained subjects, corresponding to only 5 per cent of the total oxidative metabolism versus 23 per cent for the untrained subjects. This low blood glucose utilization may explain why the trained, unlike the untrained, subjects in this study were able to maintain or even increase their blood glucose concentration throughout the exercise. A lower utilization of blood glucose during exercise in the trained state could explain the reduced liver glycogen depletion during exercise reported earlier in rats after training.[20,110] Jansson and Kaijser[94] suggested that the increased blood glucose extraction by the legs of the untrained subjects was secondary to their low muscle glycogen concentration during the later stages of the exercise session. This would be in accord with Essén et al.[111] and Gollnick et al.,[112] who demonstrated an inverse relationship between blood glucose extraction and muscle glycogen concentration.

POSSIBLE MECHANISM BEHIND THE GLYCOGEN-SPARING EFFECT

Taken together, the available evidence, including the results of the one-leg training study (Fig. 11), indicates that the lower reliance on carbohydrates in endurance-trained individuals may be explained to a large extent by local factors within the trained muscle. Such factors may be a larger utilization of intracellular or extracellular adipose tissue stores (see below), but the high content of mitochondrial oxidative enzymes is also likely to be important in this respect. This assumption is supported by the results of a study on the rat in which it could be shown that the amount of glycogen (muscle plus liver) remaining after an endurance exercise test on a rodent treadmill was directly proportional to the muscle's content of oxidative enzymes.[20] Figure 12 shows a possible biochemical mechanism whereby a large concentration of oxidative enzymes (i.e. citric acid cycle and fat oxidation enzymes and respiratory chain components) leads to a situation in which a major portion of the energy supply is derived from fat metabolism, with a lower rate of lactate formation and sparing of muscle glycogen during exercise. The training-induced enhancement of muscle capillarization probably contributes to the metabolic adaptation seen in trained muscle. A conceivable mechanism for this effect might involve an augmented muscle supply of oxygen and fatty acids. An additional factor that many contribute to the lower carbohydrate utilization during prolonged exercise in the trained state may be increased levels of citrate, which has been implicated as a major inhibitor of phosphofructokinase and therefore glycolysis.[113] This effect may, in combination with other factors, such as inhibition of pyruvate dehydrogenase activity or of the cellular glucose uptake, be important in leading to reduced carbohydrate utilization in times of high fatty acid oxidation. This metabolic link between fat and carbohydrate metabolism has been termed the glucose-fatty acid cycle (or Randle cycle[113]). As pointed out by Green et al.,[114] there may in addition be other adaptive changes, particularly during the early part of training, which lead to a reduction in anaerobic glycolysis and carbohydrate utilization during exercise. These may include the distribution of the workload to more muscles and muscle fibres, thereby lessening the work done by a single muscle fibre.

UTILIZATION OF FREE FATTY ACIDS

As is evident from the previous discussion, it is known from a large number of studies that, at the same absolute exercise intensity, trained individuals rely more than untrained individuals on fat as an energy substrate. The source of this increased fat supply has been debated, however, since the plasma levels of free fatty acids are often lower in endurance-trained subjects.[115] Hurley et al.[101] studied nine male subjects before and after a 12-week programme of endurance training. When exercising at the same absolute intensity (64 per cent of the pretraining V_{O_2}max before and after training, plasma free fatty acid and glycerol concentrations were found to be lower in the trained state than in the untrained state. Despite this, the respiratory exchange ratio was reduced in the trained state, indicating a greater reliance on fat oxidation. Muscle triglyceride utilization was found to be twice as great (12.8 ± 5.5 mmol/kg dry weight compared with 26.1 ± 9.3 mmol/kg dry weight) (Fig. 14) and muscle glycogen utilization to be 41 per cent lower in the trained, as opposed to the untrained state. The increased muscular capacity for free fatty acid oxidation was reflected in the 90 per cent increase in the level of β-hydroxyacyl-CoA dehydrogenase, which was used as a marker for the mitochondrial enzymes involved in free fatty acid oxidation. It was concluded that the greater utilization of fat in the trained compared with the untrained state was fuelled by increased lipolysis of intramuscular triglycerides. This conclusion was supported by Jansson and Kaijser,[94] who concluded that the reduced reliance on carbohydrate metabolism in trained compared with untrained individuals exercising at the same relative exercise intensity would have been covered by intramuscular triglycerides. They based this conclusion on the finding of no difference between trained and untrained individuals in the ratio of plasma free fatty acid extraction to oxygen extraction by the working legs.

Therefore, the notion that the muscle oxidation rate of plasma-

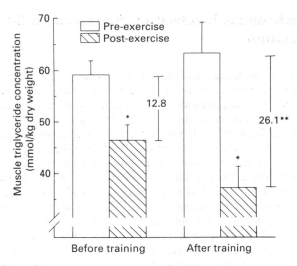

Fig. 14 Utilization of muscle triglyceride during exercise in the untrained and the trained state. Nine previously untrained subjects performed a prolonged bout of bicycle ergometer exercise of the same absolute intensity before and after adapting to a strenuous 12-week endurance training programme. The exercise test required 64 per cent of the pretraining V_{O_2}max. Muscle biopsies were obtained from the quadriceps femoris muscle, vastus lateralis. *Pre-exercise versus post-exercise, $p < 0.05$; **before training versus after training, $p < 0.01$. (Modified from ref. 101.)

derived free fatty acids depends only on the plasma free fatty acid concentration and blood flow[116,117] still seems undisputed. However, there is still a limited amount of information available in this area. Kiens et al.[118] found no decrease in intramuscular triglycerides during a 2-h bout of one-leg knee extension exercise in either the trained or the untrained leg of subjects trained unilaterally on the bicycle ergometer. The authors speculated that this could be attributable to the small increase in sympathoadrenal activity during one-leg, as opposed to two-leg exercise. This sympathoadrenal activation may have been too low to induce intramuscular lipolysis. In this case it is possible that the observed increased reliance on fat metabolism observed after training was due to an increased uptake of free fatty acids from the blood.

Intramuscular triglyceride concentration

As intramuscular triglycerides appear to be an important factor in the sparing of muscle glycogen, one might expect an increased resting level of intramuscular triglycerides in trained muscle. However, this has not been definitively demonstrated in human muscle. Three studies in which the same subjects were studied before and after training demonstrated an increase in muscle triglycerides of approximately 50 per cent with training.[119–121] Howald et al.[122] reported a significant increase in the volume density of intracellular lipid in fast-twitch, but not slow-twitch, fibres in an electron micrography study of endurance training. However, Hurley et al.[101] found no significant difference between pre- and post-training muscle triglyceride concentrations in subjects who underwent an intensive 12-week running and cycling training programme. Conflicting results have been found in studies on the rat, where both a decrease[123,124] and no change[125] of muscle triglyceride concentration in response to training have been noted.

Mechanisms involved in the increased fat oxidation

If neither the utilization of plasma free fatty acids nor the concentration of intramuscular triglycerides is clearly affected by training, one must conclude that changes responsible for increased fatty acid utilization in the trained state would be related to (1) increased lipolysis of existing intramuscular or plasma triglycerides, (2) increased transport of fatty acids into the mitochondria, and/or (3) the increased number of mitochondria within the muscle. The increased mitochondrial density and the increased content of mitochondrial enzymes in aerobically trained muscle are accompanied by increases in the enzymes involved in activation, transfer into the mitochondria and β-oxidation of fatty acids.[13,18,126] Paulussen and Veerkamp[127] have presented evidence that this adaptation may include increases in the low molecular weight cytosolic fatty acid binding proteins, which may play an important role in the intracellular transport and targeting of fatty acids. A change in the activity of regulatory molecules may also be included in this response, as a 36 per cent decrease in malonyl-CoA (an inhibitor of carnitine acyltransferase I activity) has been noted during 30 min of treadmill exercise in rats.[128] This would lead to an increased oxidation of fatty acids. Whether malonyl-CoA is reduced in trained individuals is not known at present, however. At a given exercise intensity, these adaptations in skeletal muscle would permit the rate of fatty acid oxidation to be higher in the trained than in the untrained muscle, even in the presence of a lower intracellular fatty acid concentration in the trained state. The latter could be secondary to the lower sympathoadrenal activation in the trained state[129] which, unopposed would lead to decreased lipolysis, not only of adipose tissue but also of intramuscular triglycerides. Since β-receptor mechanisms regulate skeletal muscle triglyceride hydrolysis,[130] it is possible that an increased β-receptor density may at least partially oppose the lower sympathoadrenal activation in the trained state. However, to date, increased density of β-receptors has been found in response to training only in the rat.[131,132] Martin et al.[133] found no increase in β-receptor density in human subjects following 12 weeks of endurance training. Another potentially more important factor regulating intramuscular triglyceride utilization may be a training-induced increase in hormone-sensitive lipase, the enzyme which hydrolyses intracellular triglycerides into fatty acids.[134] However, on this issue, no information is currently available.

Another potential source of free fatty acids for skeletal muscle is the hydrolysis of intravascular triglycerides, catalysed by the enzyme lipoprotein lipase, which is located on the intraluminal surface of the capillaries. There is still conflicting evidence, however, as to whether[135,136] or not[137] this enzyme is increased by endurance training. Lipoprotein lipase is known to be activated following a single bout of exercise[138] and it is possible that the resulting variability may mask possible effects of training. Yki-Järvinen et al.[139] recently presented evidence of a feedback mechanism which serves to maintain a certain rate of cellular fatty acid oxidation under conditions of changing inflow of plasma free fatty acids. This mechanism is supposed to involve stimulation of lipoprotein lipase at times of lowered intracellular free fatty acid concentration resulting from either insufficient hormone-sensitive lipase activity or lowered plasma free fatty acid concentrations. Whether such mechanisms are important in explaining the increased fat reliance in endurance-trained muscle remains to be demonstrated.

AMINO-ACID CONCENTRATIONS IN MUSCLE AND PLASMA

In the postabsorptive state, skeletal muscle constitutes the major source of circulating amino acids in the body. Despite this, very little information is available concerning the influence of endurance training on muscle production and release of individual amino acids. We have found some evidence that trained individuals may have higher basal amino-acid concentrations in skeletal muscle and plasma than untrained individuals.[140] In this study, we compared seven endurance-trained individuals (V_{O_2}max $= 59.4 \pm 2.5$ ml/kg.min) and eight sedentary controls (V_{O_2}max $= 38.7 \pm 1.4$ ml/kg.min). It was found that the average amino-acid concentrations in skeletal muscle were higher in the trained group for eight of eleven measured amino acids, but there was a statistically significant difference only for glutamate (+39 per cent, $p < 0.05$) and taurine (+36 per cent, $p < 0.05$). With respect to the basal plasma amino-acid concentrations, 10 out of the 11 amino acids (all except serine) displayed higher levels in the trained subjects. However, the differences were quite small (16.1 ± 5.5 per cent, mean$\pm$SD) and reached statistical significance only in the case of plasma phenylalanine (15 per cent higher in the trained group, $p < 0.05$). There is only one reference in the literature to a possible difference in this direction.[141] Twelve members of a major university track team who ran an average of 110 km per week were compared with 13 controls who ran less than 5 km per week and it was found that the trained subjects had significantly higher plasma concentrations of leucine (41 per cent), isoleucine (27 per cent), and tyrosine (23 per cent). Evidently, more research is needed to confirm these preliminary findings and, if they are confirmed, to determine the underlying mechanisms responsible for these muscular adaptations.

EXERCISE-INDUCED MUSCLE DAMAGE

Although concentric contractions dominate in most activities involved in exercise and training, the majority of activities also include movements requiring eccentric muscle contractions. It is known that an acute bout of eccentric exercise may result in negative effects on skeletal muscle. It is also known that muscle soreness, lasting from a few days to a week, almost invariably follows eccentric exercise in unaccustomed individuals.[142] Such negative effects of eccentric exercise may include a reduction in maximal voluntary force, as well as in maximal twitch and tetanic tension, leading to greater fatiguability.[143,144] Oedema, increased plasma creatine kinase levels,[145] and ultrastructural damage, such as myofibrillar disorganization and Z-band disruption, are also apparent.[146,147] Excretion of 3-methyl histidine, an indicator of protein degradation within skeletal muscle, has also been found to increase following eccentric exercise. Incidentally, it has been suggested that this apparent increase in protein degradation may be a prerequisite for the exercise-induced hypertrophy of skeletal muscle.[145]

In addition to these effects of eccentric exercise, a failure to replenish glycogen stores following such exercise has been noted in a number of studies.[76,148] It has recently been suggested that this defect may be due to the development of insulin resistance following eccentric exercise. Using the hyperinsulinaemic euglycaemic clamp technique, Kirwan et al.[76] found that insulin-mediated glucose disposal was reduced by about 40 per cent in subjects 2 days after a 30-min bout of downhill running. Ploug et al.[55] also observed the development of insulin resistance in the red gastrocnemius and soleus muscles of untrained rats subjected to a single 6-h swimming bout. However, these effects are likely to disappear when the eccentric exercise is performed regularly. Thus prior exposure to the specific eccentric exercise has been found to induce protection against several of these negative effects.[146,149,150] The mechanism by which this adaptation occurs is not known, but it may be an acute response to eccentric exercise as indicated in the study by Byrnes et al.[151] These authors found that a single 30-min bout of downhill running on a 10 per cent decline reduced or eliminated the delayed onset of muscle soreness and the increased circulating creatine kinase levels normally associated with eccentric exercise following subsequent downhill running up to 6 weeks later.

SUMMARY

A proposal suggesting how the increased mitochondrial density may mechanistically influence endurance performance has been put forward.[152,153] It was reasoned that ADP and/or ATP concentrations may be changed from their homeostatic levels only half as much in muscles which contain twice the mitochondrial density, a hypothesis which is supported by the data of Saltin and Karlsson,[102] Constable et al.[93] and Dudley et al.[154] This increased proportion of ATP/(ADP plus inorganic phosphate) during exercise in trained muscle would inhibit phosphofructokinase and result in less stimulation of glycogen phosphorylase, thereby slowing glycolysis and the accompanying glycogen depletion and lactate production. In addition, the increased fat oxidation in trained skeletal muscle is likely to have a major role in the sparing of muscle glycogen.[155] Furthermore, Nosek et al.[156] found a strong correlation between inorganic phosphate, $H_2PO_4^-$, and reduction in contraction force, indicating that a lower increase in inorganic phosphate, and thus $H_2PO_4^-$, with training would result in a smaller decrease in force production of the muscle at the same exercise intensity after aerobic training. The adaptations mentioned above are accompanied by the increased capillarization in trained skeletal muscle, which provides a longer mean transit time for the oxygen and substrate exchange between blood and tissue, as well as by a possible increase in Na^+–K^+ pump activity with training, which would enhance re-uptake of K^+ and thereby delay fatigue of the contraction process.[157] After several years of endurance training, fibre-type transformations from fast twitch to slow twitch,[32] which would serve to increase the overall muscle oxidative capacity and possibly also reduce energy expenditure,[9] would be expected to have a further positive influence on the endurance capacity of the muscle. The increased insulin action in the skeletal muscle of individuals regularly involved in endurance training demonstrates the importance of consistent exercise if insulin action is to be improved in pathological states such as obesity and type II diabetes as well as to protect against the development of insulin resistance with ageing. However, the described training-induced adaptations in skeletal muscle must be considered in the light of adaptations in all other organs and organ systems of the body in order to obtain an accurate picture of how endurance training may influence the metabolic homeostasis of the organism as a whole.

REFERENCES

1. Bergström J. Muscle electrolytes in man. *Scandinavian Journal of Clinical and Laboratory Investigation* 1962; **68** (Suppl.):1–110.

2. Lowry OH, Passonneau JV. *A flexible system of enzymatic analysis*, New York: Academic Press, 1972.

3. Newsholme EA, Leech, AR. The thermodynamic structure of a metabolic pathway. In: Newsholme EA, Leech AR, eds. *Biochemistry for the medical sciences*. Chichester: Wiley, 1983: 38–42.

4. Henriksson J, et al. Chronic stimulation of mammalian muscle: changes in enzymes of six metabolic pathways. *American Journal of Physiology* 1986; **251**: C614–32.

5. Salmons S, Henriksson J. The adaptive response of skeletal muscle to increased use. *Muscle and Nerve* 1981; **4**: 94–105.

6. Jolesz, F, Sreter FA. Development, innervation, and activity-pattern induced changes in skeletal muscle. *Annual Review of Physiology* 1981; **43**: 531–52.

7. Pette D. Activity-induced fast to slow transitions in mammalian muscle. *Medicine and Science in Sports and Exercise* 1984; **16**: 517–28.

8. Chi MM-Y et al. Chronic stimulation of mammalian muscle: enzyme changes in individual fibres. *American Journal of Physiology* 1986; **251**:-C633–42.

9. Crow M, Kushmerick MJ. Chemical energetics of slow- and fast-twitch muscles of the mouse. *Journal of General Physiology* 1982; **79**: 147–166.

10. Brown JMC, Henriksson, J, Salmons S. Restoration of fast muscle characteristics following cessation of chronic stimulation: physiological, histochemical and metabolic changes during slow-to-fast transformation. *Proceedings of the Royal Society of London Series B*, 1989; **235**:321–46.

11. Chi MM-Y et al. Effects of detraining on enzymes of energy metabolism in individual human muscle fibres. *American Journal of Physiology* 1983; **244**:C276–87.

12. Brown MD, Cotter MA, Hudlicka O, Vrbová G. The effects of different patterns of muscle activity on capillary density, mechanical properties and structure of slow and fast rabbit muscles. *Pflügers Archiv* 1976; **361**:241–50.

13. Holloszy JO, Booth FW. Biochemical adaptations to endurance exercise in muscle. *Annual Reviews of Physiology* 1976; **38**: 273–91.

14. Petrén T, Sjöstrand, T, Sylvén B. Der Einfluss des Trainings auf die Häufigkeit der Capillaren in Herz- und Skeletmuskulatur. *Arbeitsphysiologie* 1937; **9**: 376–86.

15. Varnauskas E, Björntorp P, Fahlén M, Prerovsky I, Stenberg J. Effects of physical training on exercise blood flow and enzymatic activity in skeletal muscle. *Cardiovascular Research* 1970; **4**: 418–22.

16. Morgan TE, Cobb LA, Short FA, Ross, R, Gunn DR. Effects of long-term exercise on human muscle mitochondria. In: Pernow B, Saltin B, eds. *Muscle metabolism during exercise*. New York: Plenum Press, 1971: 87–95.

17. Henriksson J, Reitman JS. Time course of changes in human skeletal muscle succinate dehydrogenase and cytochrome oxidase activities and maximal oxygen uptake with physical activity and inactivity. *Acta Physiologica Scandinavica* 1977; **99**: 91–7.

18. Saltin B, Gollnick PD. Skeletal muscle adaptability: significance for metabolism and performance. In: Peachey LD, Adrian RH, Geiger SR. eds. *Handbook of physiology*, Section 10, *Skeletal muscle*. Bethesda MD: American Physiological Society, 1983: 555–631.

19. Dudley GA, Abraham WM, Terjung RL. Influence of exercise intensity and duration on biochemical adaptations in skeletal muscle. *Journal of Applied Physiology* 1982; **53**(4): 844–50.

20. Fitts RH, Booth FW, Winder WW, Holloszy JO. Skeletal muscle respiratory capacity, endurance, and glycogen utilization. *American Journal of Physiology* 1975; **228**(4): 1029–33.

21. Völlestad NK, Vaage O, Hermansen L. Muscle glycogen depletion patterns in type I and subgroups of type II fibres during prolonged severe exercise in man. *Acta Physiologica Scandinavica* 1984; **122**: 433–41.

22. Ball-Burnett M, Green HJ, Houston ME. Energy metabolism in human slow and fast twitch fibres during prolonged cycle exercise. *Journal of Physiology (London)* 1991; **437**: 257–67.

23. Essén-Gustavsson B, Henriksson J. Enzyme levels in pools of microdissected human muscle fibres of identified type. Adaptive response to exercise. *Acta Physiologica Scandinavica* 1984; **120**: 505–15.

24. Andersen P, Henriksson J. Capillary supply of the quadriceps femoris muscle of man: adaptive response to exercise. *Journal of Physiology (London)* 1977; **270**: 677–90.

25. Brodal P, Ingjer F, Hermansen L. Capillary supply of skeletal muscle fibres in untrained and endurance-trained men. *American Journal of Physiology* 1977; **232**: H705–12.

26. Ingjer F. Effects of endurance training on muscle fibre ATP-ase activity, capillary supply and mitochondrial content in man. *Journal of Physiology (London)* 1979; **294**: 419–22.

27. Komi PV, Viitasalo JT, Havu M, Thorstensson A, Karlsson J. Physiological and structural performance capacity: effect of heredity. In: Komi PV, ed. *International Series of Biomechanics*. Baltimore: University Park Press, 1976: 118–23.

28. Andersen P, Henriksson J. Training induced changes in the subgroups of human type II skeletal muscle fibres. *Acta Physiologica Scandinavica* 1977; **99**:123–5.

29. Jansson E, Kaijser L. Muscle adaptation to extreme endurance training in man. *Acta Physiologica Scandinavica* 1977; **100**: 315–24.

30. Schantz P, Henriksson J. Increases in myofibrillar ATPase intermediate human skeletal muscle fibres in response to endurance training. *Muscle and Nerve* 1983; **6**: 553–6.

31. Baumann H, Jäggi M, Soland F, Howald H, Schaub MC. Exercise training induces transitions of myosin isoform subunits within histochemically typed human muscle fibres. *Pflügers Archiv* 1987; **409**: 349–60.

32. Schantz P. Plasticity of human skeletal muscle. *Acta Physiologica Scandinavica* 1986; **128** (Suppl 558): 1–62.

33. Schantz P, Henriksson J, Jansson E. Adaptation of human skeletal muscle to endurance training of long duration. *Clinical Physiology* 1983; **3**: 141–51.

34. Coyle EF, Martin III WH, Sinacore DR, Joyner MJ, Hagberg JM, Holloszy JO. Time course of loss of adaptations after stopping prolonged intense endurance training. *Journal of Applied Physiology* 1984; **57**(6): 1857–64.

35. Booth FW, Holloszy JO. Cytochrome c turnover in rat skeletal muscles. *Journal of Biological Chemistry* 1977; **252**: 416–19.

36. Williams RS, Salmons S, Newsholme EA, Kaufman RE, Mellor J. Regulation of nuclear and mitochondrial expression by contractile activity in skeletal muscle. *Journal of Biological Chemistry* 1986; **261**: 376–80.

37. Wallberg-Henriksson H. Glucose transport into skeletal muscle. Influence of contractile activity, insulin, catecholamines and diabetes mellitus. *Acta Physiologica Scandinavica* 1987; **131** (Suppl 564): 1–80.

38. Nesher R, Karl IE, Kipnis DM. Dissociation of effects of insulin and contraction on glucose transport in rat epitrochlearis muscle. *American Journal of Physiology* 1985; **249**: C226–32.

39. Wallberg-Henriksson H, Constable SH, Young DA, Holloszy JO. Glucose transport into rat skeletal muscle: interaction between exercise and insulin. *Journal of Applied Physiology* 1988; **65**(2): 909–13.

40. Youn JH, Gulve EA, Holloszy JO. Calcium stimulates glucose transport in skeletal muscle by a pathway independent of contraction. *American Journal of Physiology* 1991; **260**: C555–61.

41. Cartee GD, Young DA, Sleeper MD, Zierath J, Wallberg-Henriksson H, Holloszy JO. Prolonged increase in insulin-stimulated glucose transport in muscle after exercise. *American Journal of Physiology* 1989; **256**: E494–9.

42. Richter EA, Garetto LP, Goodman MN, Ruderman NB. Muscle glucose metabolism following exercise in the rat. Increased sensitivity to insulin. *Journal of Clinical Investigation* 1982; **69**: 785–93.

43. Zorzano A, Balon TW, Goodman MN, Ruderman NB. Additive effects of prior exercise and insulin on glucose and AIB uptake by rat muscle. *American Journal of Physiology* 1986; **251**: E21–6.

44. Young DA, Wallberg-Henriksson H, Sleeper MD, Holloszy JO. Reversal of the exercise-induced increase in muscle permeability to glucose. *American Journal of Physiology* 1987; **253**: E331–5.

45. Cartee GD, Holloszy JO. Exercise increases susceptibility of muscle glucose transport to activation by various stimuli. *American Journal of Physiology* 1990; **258**: E390–3.

46. Richter EA, Mikines KJ, Galbo H, Kiens B. Effect of exercise on insulin action in human skeletal muscle. *Journal of Applied Physiology* 1989; **66**(2): 876–85.

47. Garetto LP, Richter EA, Goodman MN, Ruderman NB. Enhanced muscle glucose metabolism after exercise in the rat: the two phases. *American Journal of Physiology* 1984; **246**: E471–5.

48. Wasserman DH *et al.* Interaction of exercise and insulin action in humans. *American Journal of Physiology* 1991; **260**: E37–45

49. Klip A, Paquet MR. Glucose transport and glucose transporters in muscle and their metabolic regulation. *Diabetes Care* 1990; **13**: 228–43.

50. Douen AG, Ramlal T, Klip A, Young DA, Cartee GD, Holloszy JO. Exercise-induced increase in glucose transporters in plasma membranes of rat skeletal muscle. *Endocrinology* 1989; **124**: 449–54.

51. Sternlicht E, Barnard RJ, Grimditch GK. Exercise and insulin stimulate skeletal muscle glucose transport through different mechanisms. *American Journal of Physiology* 1989; **256**: E227–30.

52. Goodyear LJ, Hirshman MF, King PA, Horton ED, Thompson CM, Horton ES. Skeletal muscle plasma membrane glucose transport and glucose transporters after exercise. *Journal of Applied Physiology* 1990;**68**(1): 193–8.

53. Ivy JL, Holloszy JO. Persistent increase in glucose uptake by rat skeletal muscle following exercise. *American Journal of Physiology* 1981; **241**:C200–3.

54. Mikines KJ, Sonne B, Farrell PA, Tronier B, Galbo H. Effect of training on the dose-response relationship for insulin action in men. *Journal of Applied Physiology* 1989; **66**(2): 695–703.

55. Ploug T *et al.* Effect of endurance training on glucose transport capacity and glucose transporter expression in rat skeletal muscle. *American Journal of Physiology* 1990; **259**: E778–86.

56. Coggan AR, Kohrt WM, Spina RJ, Bier DM, Holloszy JO. Endurance training decreases plasma glucose turnover and oxidation during moderate-intensity exercise in men. *Journal of Applied Physiology* 1990; **68**(3): 990–6.

57. Idström J-P, Elander A, Soussi B, Scherstén T, Bylund-Fellenius AC. Influence of endurance training on glucose transport and uptake in rat skeletal muscle. *American Journal of Physiology* 1986; **251**: H903–7.

58. Ivy JL, Young JC, McLane, JA, Fell RD, Holloszy JO. Exercise training and glucose uptake by skeletal muscle in rats. *Journal of Applied Physiology* 1983; **55**(5): 1393–6.

59. DeFronzo RA, Tobin JD, Andres R. Glucose clamp technique: a method for quantifying insulin secretion and resistance. *American Journal of Physiology* 1979; **237**: E214–23.

60. DeFronzo RA, Jacot E, Jequier E, Maeder E, Wahren J, Felber JP. The effect of insulin on the disposal of intravenous glucose. *Diabetes* 1981; **30**: 1000–7.

61. Katz LD, Glickman MG, Rapoport S, Ferrannini E, DeFronzo RA. Splanchnic and peripheral disposal of oral glucose in man. *Diabetes* 1983; **32**: 675–9.

62. Horton ES. Role and management of exercise in diabetes mellitus. *Diabetes Care* 1988; **11**: 201–11.

63. Koivisto VA, Yki-Järvinen H, Defronzo RA. Physical training and insulin sensitivity. *Diabetes Metabolism Reviews* 1986; **1**: 445–81.

64. King DS, Dalsky GP, Staten MA, Clutter WE, Van Houten DR, Holloszy JO. Insulin action and secretion in endurance-trained and untrained humans. *Journal of Applied Physiology* 1987; **63**(6): 2247–52.

65. Rodnick KJ, Reaven GM, Azhar S, Goodman MN, Mondon CE. Effects of insulin on carbohydrate and protein metabolism in voluntary running rats. *American Journal of Physiology* 1990; **259**: E706–14.

66. Heath GW, Gavin III JR, Hinderliter JM, Hagberg JM, Bloomfield SA, Holloszy JO. Effects of exercise and lack of exercise on glucose tolerance and insulin sensitivity. *Journal of Applied Physiology* 1983; **55**(2): 512–17.

67. LeBlanc, J, Nadeau A, Richard D, Tremblay A. Studies on the sparing effect of exercise on insulin requirements in human subjects. *Metabolism* 1981; **30**: 1119–24.

68. Seals DR *et al.* Glucose tolerance in young and older athletes and sedentary men. *Journal of Applied Physiology* 1984; **56**(6): 1521–5.

69. Rogers MA, King DS, Hagberg JM, Ehsani AA, Holloszy JO. Effect of 10 days of physical inactivity on glucose tolerance in master athletes. *Journal of Applied Physiology* 1990; **68**(5): 1833–7.

70. King DS *et al.* Effects of exercise and lack of exercise on insulin sensitivity and responsiveness. *Journal of Applied Physiology* 1988; **64**(5): 1942–6.

71. Mikines KJ, Sonne B, Tronier B, Galbo H. Effects of acute exercise and detraining on insulin action in trained men. *Journal of Applied Physiology* 1989; **66**(2): 704–11.

72. Nagasawa J, Sato Y, Ishiko T. Effect of training and detraining on in vivo insulin sensitivity. *International Journal of Sports Medicine* 1990; **11**: 107–110.

73. Rodnick KJ, Holloszy JO, Mondon CE, James DE. Effect of exercise training on insulin-regulatable glucose-transporter protein levels in rat skeletal muscle. *Diabetes* 1990; **39**: 1425–9.

74. James DE, Kraegen EW, Chisholm DJ. Effect of exercise training on in vivo insulin-action in individual tissues of the rat. *Journal of Clinical Investigation* 1985; **76**: 657–66.

75. Davis TA, Klahr S, Tegtmeyer ED, Osborne DF, Howard TL, Karl IE. Glucose metabolism in epitrochlearis muscle of acutely exercised and trained rats. *American Journal of Physiology* 1986; **250**: E137–43.

76. Kirwan JP, Hickner RC, Yarasheski KE, Kohrt WM, Wiethop BV, Holloszy JO. Eccentric exercise induces transient insulin resistance in healthy individuals. *Journal of Applied Physiology* 1992; **72**(6): 2197–202.

77. Langfort J, Budohoski L, Newsholme EA. Effect of various types of acute exercise and exercise training on the insulin sensitivity of rat soleus muscle measured in vitro. *Pflügers Archiv* 1988; **412**: 101–5.

78. Kern M, Tapscott EB, Downes DL, Frisell WR, Dohm GL. Insulin resistance induced by high-fat feeding is only partially reversed by exercise training. *Pflügers Archiv* 1990; **417**: 79–83.

79. Gollnick PD, Armstrong RB, Saubert CW, Piehl K, Saltin B. Enzyme activity and fibre composition in skeletal muscle of untrained and trained men. *Journal of Applied Physiology* 1972; **33**(3): 312–19.

80. Hultman E, Nilsson LH. Liver glycogen in man: effect of different diets and muscular exercise. In: Pernow B, Saltin B, eds. *Muscle metabolism during exercise*. New York: Plenum, 1971: 143–51.

81. Costill DL. Carbohydrates for exercise: dietary demands for optimal performance. *International Journal of Sports Medicine* 1988; **9**: 1–18.

82. Saltin B *et al.* The nature of the training response; peripheral and central adaptations to one-legged exercise. *Acta Physiologica Scandinavica* 1976; **96**: 289–305.

83. Henriksson J. Training induced adaptation of skeletal muscle and metabolism during submaximal exercise. *Journal of Physiology (London)* 1977; **270**: 661–75.

84. Häggmark, T. A study of morphologic and enzymatic properties of the skeletal muscles after injuries and immobilization in man. Unpublished thesis, Karolinska Institutet, 1978.

85. Piehl K, Adolfsson S, Nazar K. Glycogen storage and glycogen synthetase activity in trained and untrained muscle of man. *Acta Physiologica Scandinavica* 1974; **90**: 779–88.

86. Tesch P, Piehl K, Wilson G, Karlsson J. Physiological investigations of Swedish elite canoe competitors. *Medicine and Science in Sports and Exercise* 1976; **8**(4): 214–18.

87. Bogardus C, Ravussin E, Robbins DC, Wolfe RR, Horton ES, Sims EAH. Effects of physical training and diet therapy on carbohydrate metabolism in patients with glucose intolerance and non-insulin-dependent diabetes mellitus. *Diabetes* 1984; **33**: 311–18.

88. Devlin JT, Horton ES. Effects of prior high-intensity exercise on glucose metabolism in normal and insulin-resistant men. *Diabetes* 1985; **34**: 973–79.

89. Gulve EA, Cartee GD, Zierath JR, Corpus VM, Holloszy JO. Reversal of enhanced muscle glucose transport after exercise: roles of insulin and glucose. *American Journal of Physiology* 1990; **259**: E685–91.

90. Bergström J, Hermansen L, Hultman E, Saltin B. Diet, muscle glycogen and physical performance. *Acta Physiologica Scandinavica* 1967; **71**: 140–50.

91. Karlsson J, Saltin B. Diet, muscle glycogen, and endurance performance. *Journal of Applied Physiology* 1971; **31**(2): 203–6.

92. Sherman WM, Costill DL. The marathon: dietary manipulation to optimize performance. *American Journal of Sports Medicine* 1984; **12**(1): 44–51.

93. Constable SH, Favier RJ, McLane JA, Fell RD, Chen M, Holloszy JO. Energy metabolism in contracting rat skeletal muscle: adaptation to exercise training. *American Journal of Physiology* 1987; **253**: C316–22.

94. Jansson E, Kaijser L. Substrate utilization and enzymes in skeletal muscle of extremely endurance-trained men. *Journal of Applied Physiology* 1987; **62**(3): 999–1005.

95. Green HJ, Smith D, Murphy P, Fraser I. Training-induced alterations in muscle glycogen utilization in fibre-specific types during prolonged exercise. *Canadian Journal of Physiology and Pharmacology* 1990; **68**: 1372–6.

96. Hermansen L, Hultman E., Saltin B. Muscle glycogen during prolonged severe exercise. *Acta Physiologica Scandinavica* 1967; **71**: 129–39.

97. Saltin B, Karlsson J. Muscle glycogen utilization during work of different intensities. In: Pernow B, Saltin B, eds. *Muscle metabolism during exercise*. New York: Plenum, 1971: 289–99.

98. Bang O. The lactate content of the blood during and after muscular exercise in man. *Skandinavisches Archiv für Physiologie* 1936; **74** (Suppl. 10): 51–82.

99. Cobb LA, Johnson WP. Hemodynamic relationships of anaerobic metabolism and plasma free fatty acids during prolonged, strenuous exercise in trained and untrained subjects. *Journal of Clinical Investigation* 1963; **42**: 800–10.

100. Ekblom B, Åstrand P-O, Saltin B, Stenberg J, Wallström B. Effect of training on circulatory response to exercise. *Journal of Applied Physiology* 1968; **24**: 518–28.

101. Hurley BF, Nemeth PM, Martin III WH, Hagberg JM, Dalsky GP, Holloszy JO. Muscle triglyceride utilization during exercise: effect of training. *Journal of Applied Physiology* 1986; **60**(2): 562–7.

102. Saltin B, Karlsson J. Muscle ATP, CP, and lactate during exercise after physical conditioning. In: Pernow B, Saltin B, eds. *Muscle metabolism during exercise*. New York: Plenum, 1971: 395–9.

103. Karlsson J, Sjödin B, Thorstensson A, Hultén B, Frith K. LDH isozymes in skeletal muscles of endurance and strength trained athletes. *Acta Physiologica Scandinavica* 1975; **93**: 150–6.

104. Favier RJ, Constable SH, Chen M, Holloszy JO. Endurance exercise training reduces lactate production. *Journal of Applied Physiology* 1986; **61**(3): 885–9.

105. Stanley WC, Gertz EW, Wisneski JA, Neese RA, Morris DL, Brooks GA. Lactate extraction during net lactate release in legs of humans during exercise. *Journal of Applied Physiology* 1986; **60**(4): 1116–20.

106. Schantz PG, Sjöberg B, Svedenhag J. Malate–aspartate and alpha-glycerophosphate shuttle enzyme levels in human skeletal muscle: methodological considerations and effect of endurance training. *Acta Physiologica Scandinavica* 1986; **128**: 397–407.

107. Costill DL, Fink WJ, Hargreaves M, King DS, Thomas R, Fielding R. Metabolic characteristics of skeletal muscle during detraining from competitive swimming. *Medicine and Science in Sports and Exercise* 1985; **17**: 339–43.

108. Donovan CM, Pagliassotti MJ. Enhanced efficiency of lactate removal after endurance training. *Journal of Applied Physiology* 1990; **68**(3): 1053–8.

109. Donovan CM, Brooks GA. Endurance training affects lactate clearance, not lactate production. *American Journal of Physiology* 1983; **244**: E83–92.

110. Baldwin KM, Fitts RH, Booth FW, Winder WW, Holloszy JO. Depletion of muscle and liver glycogen during exercise. Protective effect of training. *Pflügers Archiv* 1975; **354**: 203–12.

111. Essén B, Hagenfeldt L, Kaijser L. Utilization of blood-borne and intramuscular substrates during continuous and intermittent exercise in man. *Journal of Physiology (London)* 1977; **265**: 489–506.

112. Gollnick PD, Pernow, B, Essén B, Jansson E, Saltin B. Availability of glycogen and plasma FFA for substrate utilization in leg muscle of man during exercise *Clinical Physiology* 1981; **1**: 27–42.

113. Randle PJ, Kerbey AL, Espinal J. Mechanisms decreasing glucose oxidation in diabetes and starvation: role of lipid fuels and hormones. *Diabetes/Metabolism Reviews* 1988; **4**: 623–38.

114. Green HJ, Jones S, Ball-Burnett ME, Smith D, Livesey J, Farrance BW. Early muscular and metabolic adaptations to prolonged exercise training in humans. *Journal of Applied Physiology* 1991; **70**(5): 2032–8.

115. Holloszy JO. Metabolic consequences of endurance exercise training. In: Horton ES, Terjung RL eds. *Exercise, nutrition and energy metabolism.* New York: Macmillan, 1988: 116–131.

116. Hagenfeldt L. Turnover of individual free fatty acids in man. *Federation Proceedings* 1975; **34**: 2236–40.

117. Groop LC et al. Glucose and free fatty acid metabolism in non-insulin dependent diabetes mellitus: evidence for multiple sites of insulin resistance. *Journal of Clinical Investigation* 1989; **84**: 205–13.

118. Kiens B, Saltin B, Christensen NJ, Essén-Gustavsson B. Private communication, November 1991.

119. Morgan TE, Short FA, Cobb LA. Effect of long-term exercise on skeletal muscle lipid composition. *American Journal of Physiology* 1969; **216**(1): 82–6.

120. Bylund-Fellenius AC et al. Physical training in man. Skeletal muscle metabolism in relation to muscle morphology and running ability. *European Journal of Applied Physiology and Occupational Physiology* 1977; **36**: 151–69.

121. Staron RS, Hikida RS, Hagerman FC, Dudley GA, Murray TF. Human skeletal muscle fibre type adaptability to various workloads. *Journal of Histochemistry and Cytochemistry* 1984; **32**: 146–52.

122. Howald H, Hoppeler H, Claasen H, Mathieu O, Straub R. Influences of endurance training on the ultrastructural composition of the different muscle fibre types in humans. *Pflügers Archiv* 1985; **403**: 369–76.

123. Fröberg SO. Effects of training and of acute exercise in trained rats. *Metabolism* 1971; **20**: 1044–51.

124. Fröberg SO, Östman I, Sjöstrand O. Effect of training on esterified fatty acids and carnitine in muscle and on lipolysis in adipose tissue in vitro. *Acta Physiologica Scandinavica* 1972; **86**: 166–74.

125. Górski J, Kiryluk T. The post-exercise recovery of triglycerides in rat tissues. *European Journal of Applied Physiology and Occupational Physiology* 1980; **45**: 33–41.

126. Molé PA, Oscai LB, Holloszy JO. Adaptation of muscle to exercise. Increase in levels of palmityl COA synthetase, carnitine palmityltransferase, and palmityl CoA dehydrogenase, and in the capacity to oxidize fatty acids. *Journal of Clinical Investigation* 1971; **50**: 2323–30.

127. Paulussen RJA, Veerkamp JH. Intracellular fatty-acid binding proteins. Characteristics and function. In: Hilderson HJ, ed. *Subcellular biochemistry: intracellular transfer of lipid molecules.* New York: Plenum Press, 1990.

128. Winder WW, Arogyasami J, Barton RJ, Elayan IM, Vehrs PR. Muscle malonyl-CoA decreases during exercise. *Journal of Applied Physiology* 1989; **67**(6): 2230–3.

129. Winder WW, Hickson RC, Hagberg JM, Ehsani AA, McLane JA. Training-induced changes in hormonal and metabolic responses to submaximal exercise. *Journal of Applied Physiology* 1979; **46**: 766–71.

130. Stankiewicz-Choroszucha B, Górski J. Effect of beta-adrenergic blockade on intramuscular triglyceride mobilization during exercise. *Experientia* 1978; **34**: 357–8.

131. Williams RS, Caron MG, Daniel K. Skeletal muscle β-adrenergic receptors: variations due to fibre type and training. *American Journal of Physiology* 1984; **246**: E160–7.

132. Buckenmeyer PJ, Goldfarb AH, Partilla JS, Pineyro MA, Dax EM. Endurance training, not acute exercise, differentially alters β-receptors and cyclase in skeletal fibre types. *American Journal of Physiology* 1990; **258**: E71–7.

133. Martin WH, Coggan AR, Spina RJ, Saffitz JE. Effects of fibre type and training an β-adrenoceptor density in human skeletal muscle. *American Journal of Physiology* 1989; **257**: E736–42.

134. Oscai LB, Essig DA, Palmer WK. Lipase regulation of muscle triglyceride hydrolysis. *Journal of Applied Physiology* 1990; **69**(5): 1571–7.

135. Svedenhag J, Lithell H, Juhlin-Dannfelt A, Henriksson J. Increase in skeletal muscle lipoprotein lipase following endurance training in man. *Atherosclerosis* 1983; **49**: 203–7.

136. Nikkilä EA, Taskinen M-R, Rehunen S, Härkönen M. Lipoprotein lipase activity in adipose tissue and skeletal muscle of runners—relation to serum lipoproteins. *Metabolism* 1978; **27**: 1661–9.

137. Stubbe I, Hansson P, Gustafson A, Nilsson-Ehle P. Plasma lipoproteins and lipolytic enzyme activities during endurance training in sedentary men: changes in high-density lipoprotein subfractions and composition. *Metabolism* 1983; **32**: 1120–8.

138. Lithell H, Cedermark M, Fröberg J, Tesch P, Karlsson J. Increase of lipoprotein-lipase activity in skeletal muscle during heavy exercise-relation to epinephrine excretion. *Metabolism* 1981; **30**: 1130–8.

139. Yki-Järvinen H, Puhakainen I, Saloranta C, Groop L, Taskinen M-R. Demonstration of a novel feedback mechanism between FFA oxidation from intracellular and intravascular sources. *American Journal of Physiology* 1991; **260**: E680–9.

140. Henriksson J. Effect of exercise on amino acid concentrations in skeletal muscle and plasma. *Journal of Experimental Biology* 1991; **160**: 149–65.

141. Einspahr KJ, Tharp G. Influence of endurance training on plasma amino acid concentrations in humans at rest and after intense exercise. *International Journal of Sports Medicine* 1989; **10**: 233–6.

142. Ebbeling CB, Clarkson PM. Exercise-induced muscle damage and adaptation. *Sports Medicine* 1989; **7**: 207–34.

143. Davies CTM, White MJ. Muscle weakness following eccentric work in man. *Pflügers Archiv* 1981; **392**: 168—71.

144. Newham DJ, Jones DA, Edwards RHT. Large delayed plasma creatine kinase changes after stepping exercise. *Muscle and Nerve* 1983; **6**: 380–5.

145. Evans WJ, Cannon JG. The metabolic effects of exercise induced muscle damage. In: Holloszy JO, ed. *Exercise and sports sciences reviews.* Baltimore: Williams & Wilkins, 1991: 99–125.

146. Fridén J, Seger J, Sjöström M, Ekblom B. Adaptive response in human skeletal muscle subjected to prolonged eccentric training. *International Journal of Sports Medicine* 1983; **4**: 177–83.

147. Warhol MJ, Siegel AJ, Evans WJ, Silverman LM. Skeletal muscle injury and repair in marathon runners after competition. *American Journal of Pathology* 1985; **118**: 331–9.

148. Costill DL, Pascoe DD, Fink WJ, Robergs RA, Barr SI, Pearson D. Impaired muscle glycogen resynthesis after eccentric exercise. *Journal of Applied Physiology* 1990; **69**(1): 46–50.

149. Hunter JB, Critz JB. Effect of training on plasma enzyme levels in man. *Journal of Applied Physiology* 1971; **31**: 20–3.

150. Schwane JA, Armstrong RB. Effect of training on skeletal muscle injury from downhill running in rats. *Journal of Applied Physiology* 1983; **55**: 969–75.

151. Byrnes WC, Clarkson PM, White JS, Hsieh SS, Frykman PN, Maughan RJ. Delayed onset muscle soreness following repeated bouts of downhill running. *Journal of Applied Physiology* 1985; **59**: 710–5.

152. Holloszy JO. Biochemical adaptations to exercise: aerobic metabolism. In: Wilmore JH, ed. *Exercise and sports sciences reviews.* New York: Academic Press, 1973:45–71.

153. Booth FW, Thomason DB. Molecular and cellular adaptation of muscle in response to exercise: perspectives of various models. *Physiological Reviews* 1991; **71**: 541–85.

154. Dudley GA, Tullson PC, Terjung RL. Influence of mitochondrial content on the sensitivity of respiratory control. *Journal of Biological Chemistry* 1987; **262**(19): 9109–14.

155. Holloszy JO, Coyle EF. Adaptations of skeletal muscle to endurance exercise and their metabolic consequences. *Journal of Applied Physiology* 1984; **56**: 831–8.

156. Nosek TM, Fender KY, Godt RE. It is diprotonated inorganic phosphate that depresses force in skinned skeletal muscle fibres. *Science* 1987; **236**: 191–3.

157. Kjeldsen K, Nörgaard A, Hau C. Exercise-induced hyperkalaemia can be reduced in human subjects by moderate training without change in skeletal muscle Na, K-ATPase concentration. *European Journal of Clinical Investigation* 1990; **20**: 642–7.

1.1.3 Acid–base balance during high intensity exercise

KENT SAHLIN

INTRODUCTION

The hydrogen ion is unique in its high reactivity and its abundance in biological fluids. Most of the hydrogen ions are bound and the cellular concentration of the free form is only about 10^{-7} mol/l, corresponding to a pH of 7. Despite its low value, changes in H^+ concentration will have a profound effect on various biochemical and physiological processes.

Generation and removal of H^+ at the cellular level, and in the whole body are controlled by a series of mechanisms by which disturbances of the acid–base homeostasis are minimized or counteracted. At the whole-body level acid–base balance is maintained by renal extrusion of acids/bases and by variation in respiratory carbon dioxide elimination. Cellular acid–base balance is maintained by (1) intracellular buffering processes, (2) feedback control of metabolic pathways and physiological processes, and (3) transmembrane transport of acids and bases.

This chapter will discuss the influence of exercise on the cellular acid–base balance and the role of acidosis in physical performance.

EXERCISE-INDUCED CHANGES IN ACID–BASE BALANCE

Exercise results in increased energy demand and thus increased rates of the energetic processes. Since all the energetic processes influence the acid–base balance, exercise is a potential problem for cellular homeostasis.

Increased carbon dioxide production

The end-product of oxidative metabolism is carbon dioxide which, after hydration to carbonic acid, can dissociate to protons and bicarbonate:

$$CO_2 \rightleftarrows H_2CO_3 \rightleftarrows H^+ + HCO_3^- \qquad (1)$$

Exercise can increase CO_2 production and thus the proton load by a factor of about 20. However, normal circulation and respiration will balance the increased CO_2 production through increased CO_2 elimination. Therefore the hydrogen ion concentration can remain normal despite the increased proton load. In fact the ventilation of CO_2 during high intensity exercise will be in excess of the CO_2 formation and constitutes one mechanism counteracting the acidosis induced by lactic acid accumulation.

Hydrolysis of ATP

The immediate energy source for practically all energy-requiring processes is the hydrolysis of ATP:

$$ATP \rightarrow ADP + Pi + n_1 H^+ \qquad (2)$$

where Pi denotes inorganic phosphate and n_1 is between zero and unity. The rate of ATP hydrolysis is very high in a contracting muscle, but the acid–base disturbance is negligible since the reverse reaction (ATP formation) proceeds at an almost identical rate.

Breakdown of phosphocreatine

The breakdown of phosphocreatine (PCr) is the energetic process that can rephosphorylate ADP at the highest rate and is an important energy source during short bursts of high intensity exercise:

$$PCr + ADP + n_2 H^+ \rightarrow ATP + creatine \qquad (3)$$

Combination of equations (2) and (3) gives

$$PCr + (n_2 - n_1)H^+ \rightarrow creatine + Pi \qquad (4)$$

Breakdown of phosphocreatine is therefore associated with a removal of protons, with the stoichiometry being dependent on the cellular pH.[1] At rest, when intracellular pH is close to 7.0, about 0.4 mol H^+ are consumed per mole of phosphocreatine utilized, giving a theoretical maximal uptake of protons of about 32 mmol/kg dry weight (PCr = 80 mmol/kg dry weight, pH 7.0). During the initial state of a contraction when the removal of H^+ (due to phosphocreatine breakdown) exceeds the release of H^+ (due to lactic acid formation) one would expect an increase in pH. An alkaline shift of about 0.2 pH units has also been demonstrated during the first few seconds of contraction in frog skeletal muscle[2] and in myophosphorylase-deficient patients.[3,4] Conversely, when phosphocreatine is resynthesized during the recovery period following exercise, reaction (4) is reversed and H^+ is released. This post-exercise proton load explains why the nadir of muscle pH is not reached at the end of exercise, at the peak of muscle lactate, but after recovery for about a minute.[5]

Formation of lactic acid

Exercise-induced acid–base disturbances are normally associated with lactic acidosis. Oxidation of carbohydrates to CO_2 and water in the mitochondria results in complete utilization of the stored energy. Incomplete degradation of carbohydrates to lactate occurs when the energy demand exceeds the oxidative capacity or during conditions of cellular hypoxia. Formation of lactate exploits only 8 per cent of the stored energy in carbohydrates and is therefore an inefficient method of fuel utilization. However, owing to the requirements of a high rate of ADP rephosphorylation and the shortage of oxygen, lactate formation is an important energy source during high intensity exercise.

Since lactic acid has a pK_a of about 3.7 it is almost completely dissociated under physiological conditions. Lactate formation is therefore associated with a stoichiometric release of protons:

$$\frac{1}{2}\ glucose\ (glycogen) \rightarrow La + H^+ \qquad (5)$$

$$ADP \frown ATP$$

where La denotes lactic acid. If the uptake of protons due to ADP rephosphorylation is included in the balance, the picture becomes more complex.[6] However, since there are no major changes in the

ATP or ADP concentrations during exercise, the changes in proton concentration due to ADP rephosphorylation and ATP hydrolysis can be neglected. Metabolism of lactate, either through a reversal of reaction (5) (which normally occurs in the liver) or through transformation to pyruvate and subsequent oxidation in the mitochondria, will result in an equivalent uptake of protons.

Lactate has a key position in exercise physiology, and the changes in blood and muscle lactate have been thoroughly investigated under a variety of conditions. This interest stems from the close connection between lactate accumulation and muscle fatigue, which was observed by Berzelius in 1877 and later substantiated by Fletcher and Hopkins[7] in 1907 and in numerous other studies since then.

Lactate formation can be related to fatigue either through an increased breakdown of intramuscular glycogen, resulting in a shortage of fuel, or through lactate-induced acidosis. The first mechanism will be important during prolonged exercise and may explain the close relation between performance during long-distance running and the running speed corresponding to the lactate threshold.[8] During high intensity exercise lactate accumulation in muscle and blood will be excessive and the resulting acidosis will influence both biochemical and physiological processes.

Breakdown of the whole muscle glycogen store (400 mmol glucosyl units/kg dry weight) to lactate would release 800 mmol of protons and result in a decrease in muscle pH to less than 3 which would cause irreversible cellular damage. Therefore powerful protective mechanisms are required to prevent exessive acidosis. These mechanisms form the basis for the acidotic impairment of the contraction process.

INFLUENCE OF TRANSMEMBRANE LACTATE FLUX ON ACID–BASE BALANCE

Only a limited amount of lactic acid can accumulate in the muscle cell before the contraction process is impaired. Therefore extrusion of lactic acid to the extracellular fluid will increase anaerobic energy utilization and prolong exercise duration. The release of lactate from the working muscle increases when the intensity increases and is related to muscle lactate (Fig. 1). Previous studies of lactate release during cycling suggested that there was an upper limit of lactate release of about 4 to 5 mmol/min.[9] However, later studies with the one-leg knee-extension model have shown that lactate release can increase by up to 20 mmol/min.[12] It was suggested that the relatively low lactate release during cycling was due to a limitation in the perfusion rather than in the sarcolemma transport.[12]

The proportion of the lactate that can be extruded from the muscle is dependent on the intensity, duration, and nature of the exercise. More than 30 per cent of the formed lactate was released from the muscle during maximal one-leg knee extension,[12] whereas during maximal cycling at 100 per cent V_{O_2}max about 17 per cent was released.[10] Thus it is clear that, despite large increases in blood lactate, most of the lactate produced during high intensity exercise remains within the muscle. However, during prolonged exercise there is only a moderate increase of muscle lactate but a continuous release of lactate, which in this situation will correspond to a major proportion of the lactate formed.

The importance of lactate translocation for acid–base balance is related to whether hydrogen ions are translocated at the same rate

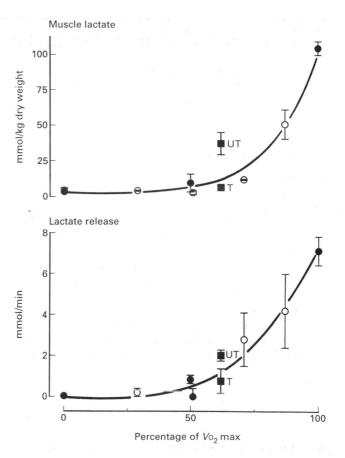

Fig. 1 Muscle lactate and lactate release after cycling for 4 to 15 min at different intensities. Symbols are means ± SE from four to eight subjects in three different studies: ○ ref. 9; ● ref. 10; ■ ref. 11.

or at a different rate from lactate. Some studies suggest a more rapid efflux of protons than of lactate in contracting frog and rat muscle[13] and in humans during the early recovery period following exhaustive exercise.[14] However, later studies have shown that the venous–arterial difference for base deficit (corresponding to the release of H^+) is equal to that of lactate during both exercise and the early recovery period[15] and suggest that the transport of lactate corresponds to an equal translocation of H^+. The extent of the translocation occurring through a carrier mechanism or through passive diffusion of undissociated lactate is at present uncertain.

During prolonged exercise at a moderate intensity lactate is continuously released from the working muscles. Despite this large release of lactate from the working muscle, blood lactate is maintained fairly constant which demonstrates that the formation of lactate is balanced by an equal rate of uptake by the liver and non-active muscle tissues. The continuous release of lactate from the working muscle during a long period should correspond to a large flux of protons to the blood. However, the perturbation of extracellular pH is small and suggests that the removal of lactate corresponds to a similar removal of protons.

CHANGES IN MUSCLE AND BLOOD pH DURING EXERCISE

Arterial blood pH at rest is 7.4, whereas intramuscular pH is close to 7.0 (Table 1). Because of the negative potential of the cell one

Table 1 Skeletal muscle pH at rest and after exercise to fatigue in man ($\bar{x} \pm$SE)

Method	Muscle pH at rest	Type of exercise	Duration of exercise (min)	Muscle pH after exercise	Reference
DMO	6.92 ± 0.04 (n=6)	—	—	—	16
Muscle homogenate	7.02 ± 0.01	Maximal running	0.5	6.78 ± 0.03	17
Muscle homogenate	6.92 ± 0.03 (n=11)	Running or cycling to fatigue	2	6.41 ± 0.04 (n=8)	18
CO$_2$	7.00 ± 0.02 (n = 13)	Cycling to fatigue at about 100% of V_{O_2}max	5–6	6.4 ± 0.05 (n = 6)	19
Muscle homogenate	7.08 ± 0.01 (n = 12)	Isometric contraction to fatigue at 66% of MVC	0.8	6.56 ± 0.02 (n = 8)	20
NMR	7.01 ± 0.00 (n = 17)	Repeated maximal contractions with the forearm until tension decreased to 73% of initial	4	6.24 ± 0.09 (n = 17)	21
NMR	7.03 ± 0.01	Repeated maximal contraction with the first dorsal interosseous muscle during ischaemic conditions; force declined to 64% of MVC	0.8	6.51 ± 0.02	22

DMO, 5,5-dimethyl-2,5-oxazolidinedione. The principle of the method is based on the distribution of a weak acid (DMO) over the cell membrane.
CO$_2$ method is based on estimates of intracellular HCO$_3^-$.
NMR method is based on estimation of the dissociation of H$_2$PO$_4^-$ by ^{31}P-NMR.
MVC, maximal voluntary contraction/force.

would expect a much higher concentration of H$^+$ (corresponding to pH 6.0) if the hydrogen ions were to be in electrochemical equilibrium over the cell membrane. The low transmembrane permeability of H$^+$ and HCO$_3^-$ and the presence of energy-consuming transport systems ensure that cellular pH can be maintained far above the value corresponding to the electrochemical equilibrium for H$^+$.

During high intensity exercise lactate accumulates in both muscle and blood, and the acid–base homeostasis is disturbed. The decrease in pH is attenuated by the buffering processes (see below) and by hyperventilation by which arterial $P\mathrm{CO_2}$ is decreased. Capillary blood pH can decrease at fatigue to 7.1–7.2, or even below 7.0 during extreme conditions.[18]

Muscle pH decreases from about 7.1 at rest to about 6.4–6.8 at fatigue when measured with the homogenate technique (Table 1). The homogenate technique will also include interstitial fluid and blood and in some cases the values obtained will underestimate the true changes. With the CO$_2$ technique, where intracellular pH is calculated using the Hendersson–Hasselbach equation, pH at rest in human muscle is 7.0 and the value after cycling to fatigue is 6.4. With the nuclear magnetic resonance (NMR) technique intracellular pH can be measured non-invasively by monitoring the position of the ^{31}P peak of inorganic phosphate. Intracellular pH at rest obtained using this technique is similar to the others (Table 1), but pH at fatigue is generally lower and values below pH 6 have been reported in the contracting forearm muscle[21] and in the flexor digitorum flexiliralis muscle.[5]

BUFFER CAPACITY

During high intensity exercises the release of H$^+$ is mainly due to accumulation of lactic acid, which can amount to more than 30 mmol/l muscle water. The change in pH is determined both by the amount of acid or base added and by the cellular buffering

capacity β. If 30 mmol/l of H$^+$ ions were added to an unbuffered solution, pH would decrease to less than 2. However, because of the buffering processes more than 99 per cent of the released H$^+$ ions are bound within the cell and the changes in the free H$^+$ concentration and pH are therefore diminished. The decrease in muscle pH is linearly related to accumulation of lactate (Fig. 2) and the slope of the line is related to the buffer capacity.

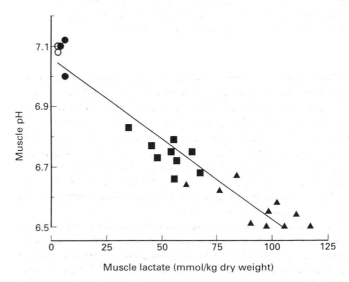

Fig. 2 Relationship between muscle pH (determined by the homogenate technique) and muscle lactate content. Muscle samples were taken from the quadriceps femoris muscle in man at rest (○), after 10 min circulatory occlusion (●), and after isometric contraction at 68 per cent of the maximum voluntary contraction force (MVC) sustained for 25 s (■) or to fatigue (45 s) (▲). $y=-0.00537x + 7.06$; $r=0.96$. (Data from ref. 20.)

From the exercise-induced changes in lactate, pyruvate, and pH in the quadriceps femoris muscle of man β during dynamic exercise has been estimated to be 73 mmol/l pH.[23] Similar values have been obtained after *in-vitro* titration of homogenates quadriceps femoris muscle,[17] or of gastrocnemius and triceps brachii muscle.[24]

The magnitude and the composition of β in skeletal muscle is different from that in plasma and erythrocytes (Fig. 3). Phosphate compounds are of major importance for β in skeletal muscle but have a negligible influence on β in erythrocytes and plasma.

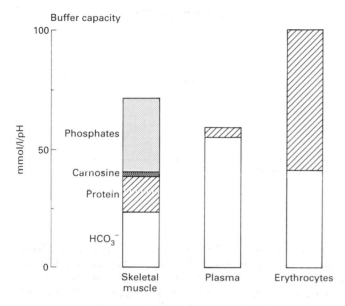

Fig. 3 Buffer capacity of skeletal muscle, plasma, and erythrocytes. The buffering of HCO_3^-–CO_2 was calculated from 2.3 (HCO_3^-) according to Woodbury[25] and the assumption of an open system. In a closed system where CO_2 is trapped (e.g. ischaemia) the buffer capacity of CO_2–HCO_3^- is negligible. The buffering by proteins in erythrocytes and plasma is from Woodbury.[25] The composition of buffer capacity in skeletal muscle is from Sahlin.[23]

The CO_2 system is less important for β in skeletal muscle than in the other tissues and is explained by the lower HCO_3^- concentration. The change in pH after a change in P_{CO_2} will be smaller if the concentration of HCO_3^- is low and the non-bicarbonate buffering is high. Consequently, changes in P_{CO_2} will have much less influence on intracellular pH than on extracellular pH. It can be speculated that one of the physiological advantages in maintaining the H^+ concentration displaced from its electrochemical equilibrium is to maintain a potential respiratory influence on intracellular acid–base homeostasis.

During high intensity exercise high values of β may improve the exercise capacity since the decrease in pH for a certain increase in lactate will be diminished. After anaerobic training for 8 weeks the buffer capacity increased by 37 per cent (range 12–50 per cent, $n=7$).[26] The increased buffer capacity enabled the subjects to accumulate more lactic acid and the capacity to perform exercise also increased accordingly. Two weeks of high altitude training increased both the buffer capacity in muscle (by 6 per cent) and the capacity for short-term running.[24] In a cross-sectional study buffering capacity was found to be higher in well-trained subjects involved in sports with a high degree of anaerobic energy utiliz-

ation (ice hockey, football) compared with a control group of untrained subjects.[27]

In contrast, no change in β (measured by *in-vitro* titration of muscle) was found after 8 weeks of spring training.[17] Despite the unchanged β the subjects increased their sprinting capacity and were able to accumulate more lactate in the muscle. The reason for the divergence between studies is unclear but may be related to methodological differences.

ACID–BASE CHANGES AND EXERCISE PERFORMANCE

During high intensity exercise large changes occur in intracellular pH (Table 1). Several studies have shown a parallel decrease in pH and in maximal force, and increases in the intracellular concentration of H^+ are often implicated as a cause of fatigue. This idea is supported by the finding that blood pH continues to fall during intermittent exercise, whereas muscle pH decreases to about the same level after several exercise bouts, indicating a limiting muscle pH.[18]

A series of experiments have been performed in order to investigate the influence of altered acid–base status on exercise performance. Alkalosis has been induced by administration of $NaHCO_3$ and acidosis by administration of NH_4Cl. During high intensity exercise of short duration the exercise performance is generally unaffected by treatment with HCO_3^- or NH_4Cl.[78,79] However, during exercise periods of longer duration induced alkalosis has been shown to improve performance and result in a higher muscle and blood lactate content at fatigue.[30,31] Improvements in exercise performance during alkalotic conditions have also been observed during intermittent exercise.[32]

The divergent results may be related to the observation that intracellular pH was affected in some studies but not in others. Changes in extracellular pH are unlikely to result in similar changes in intracellular pH. In fact, under certain conditions such as administration of ammonium chloride, lactate, or bicarbonate one may even expect intracellular pH to change transiently in the opposite direction to that of extracellular pH. Extracellular pH is thus a poor indicator of both the magnitude and the direction of the change in intracellular pH.

It is possible that the improved performance which was observed in some studies during alkalotic conditions was at least partly related to an increased efflux of lactic acid since alkalosis has been shown to increase post-exercise blood lactate.[29,30,31]

Fatigue during high intensity exercise is associated with lactic acid and acidosis in many cases but it is clear that under other conditions fatigue can occur when muscle pH is unchanged or even increased. Thus patients with glycogen phosphorylase deficiency (McArdle's disease) and phosphofructokinase deficiency have a low exercise tolerance although their muscle pH at fatigue is increased.[4] Similarly, fatigue during prolonged submaximal exercise coincides with glycogen depletion and not with lactate accumulation.

ACIDOSIS AND FATIGUE: POSSIBLE MECHANISMS

Acidosis could impair the contraction process through different mechanisms (Fig. 4). Increased hydrogen ion concentration could interfere with the energy supply, which could secondarily affect one or several steps in the contraction process. Alternatively, an

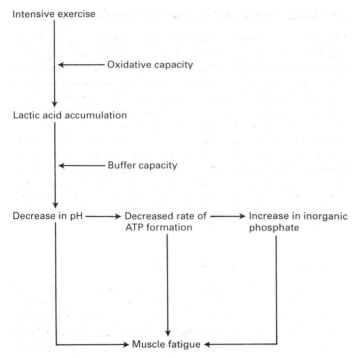

Fig. 4 Possible links between acidosis and fatigue. Increased oxidative capacity and buffer capacity will attenuate the decrease in pH for a given exercise. Diprotonated inorganic phosphate ($H_2PO_4^-$) increases when phosphocreatine and pH decrease.

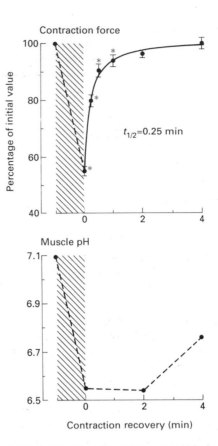

Fig. 5 Maximum force and estimated muscle pH before and after a sustained isometric contraction to fatigue in man. Muscle pH before and immediately after contraction was calculated from lactate content and the relation between pH and lactate shown in Fig. 1. Muscle pH during recovery was calculated from the decrease in H^+ (due to lactate removal), the increase in H^+ (due to phosphocreatine resynthesis), and the buffer capacity. (Results reproduced with permission from ref. 35.)

increased concentration of $H_2PO_4^-$ occurring secondary to the increase in H^+, could be the fatiguing factor.

A third possibility is that increased H^+ ion concentration interferes directly with the contractile machinery. This last hypothesis is supported by findings from *in-vitro* studies with skinned muscle fibres where the cellular membrane is removed and the intracellular chemical composition is altered.[33,34]

The time courses of metabolic recovery and force recovery after a fatiguing contraction have been studied in man. Maximal force reverted to the precontraction value with a half-time of about 15 s, and after 2 min of recovery the force was not statistically different from that before contraction (Fig. 5). In contrast, recovery in lactate and muscle pH occurred with a slower time course and after 2 min recovery the calculated muscle pH was similar to the immediate postcontraction value. A similar time course of pH recovery has been observed previously with the ^{31}P NMR technique.[5] These data for humans demonstrate that the maximal contraction force is not necessarily reduced during acidotic conditions. Consequently, the depression of force at low pH observed *in vitro* with the skinned fibre preparation does not apply to humans *in vivo*.

Studies in animals have shown that decreases in tension are linearly related to increases in $H_2PO_4^-$.[35] Since the pK_a of $H_2PO_4^-$ is about 6.8, a decrease in pH in the physiological range will transform HPO_4^{2-} to $H_2PO_4^-$. Thus acidosis will both increase the total inorganic phosphate concentration in muscle, through a displacement of the creatine kinase equilibrium towards phosphocreatine depletion (see below) and increase the proportion of inorganic phosphate present as $H_2PO_4^-$.

In-vitro experiments with the skinned fibre preparation have shown that increases in inorganic phosphate and H^+ impair the contraction process.[34] However, the relevance of these *in-vitro*

experiments to *in-vivo* conditions is unclear. In contracting human muscle it was shown that decline in force was related to the increase in $H_2PO_4^-$,[37] whereas in another study of a myophosphorylase-deficient patient no relation was observed between $H_2PO_4^-$ and decline in tension.[22]

ENERGETIC DEFICIENCY AND FATIGUE

Acidosis and glycolysis

It has been known for a long time that glycolysis is enhanced during alkalosis and inhibited during acidosis. The main regulatory step in glycolysis is phosphofructokinase, whose activity shows a marked pH dependence.[38] The rate of glycolysis (measured as the rate of lactate accumulation) has been determined during isometric contraction sustained to fatigue (Fig. 6). Despite a pronounced decrease in muscle pH, the glycolytic rate is maintained. Similar data are available from animal studies.[39] Thus it is clear that a low pH does not necessarily limit glycolysis. Reversal of the pH inhibition of phosphofructokinase is probably achieved through increases in inorganic phosphate, adenosine monophosphate (AMP), ADP, and hexose phosphates[39] which are allosteric activators of phosphofructokinase.

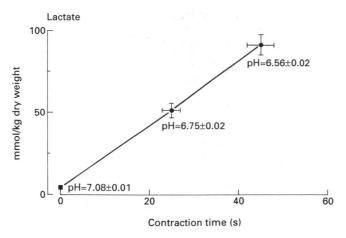

Fig. 6 Muscle lactate during isometric contraction at 68 per cent of maximal voluntary contraction to fatigue. Corresponding values of muscle pH are shown. Results are means ± SE from ref. 20.

Acidotic inhibition of glycolysis can also occur at the level of glycogen phosphorylase.[40] However, increases in AMP and inosine monophosphate, which are allosteric activators of phosphorylase b, may counteract the acidotic inhibition of glycogenolysis and restore the glycogenolytic rate.

In conclusion, many studies have demonstrated that the glycolytic chain is sensitive to acidosis. However, the pH inhibition of phosphofructokinase and phosphorylase can be reversed and the glycolytic rate maintained through counter regulatory mechanisms. However, the price of maintaining a high rate of glycolysis at a low pH is increased levels of ADP and AMP. It is possible that these changes could be involved in the impairment of the contraction process.

Acidosis and phosphocreatine

Regeneration of ATP can occur at a high rate by the creatine kinase reaction at the expense of phosphocreatine. A decrease of muscle content of phosphocreatine occurs already at a low work load,[41] and when exercise intensity increases, phosphocreatine decreases progressively. The decrease in phosphocreatine has been shown to be curvilinearly related to muscle lactate content.[42] The relationship seems to be independent of type, intensity, and duration of the exercise, and has been suggested to reflect a steady state condition in the muscle influenced by the increased H^+ ion concentration.[42] The involvement of H^+ in the creatine kinase reaction and the increase in ADP will displace the equilibrium towards phosphocreatine breakdown:

$$H^+ + ADP + PCr \rightarrow ATP + Cr$$

where phosphocreatine denotes phosphocreatine and Cr denotes creatine.

Breakdown of phosphocreatine is the energetic process which can rephosphorylate ADP at the highest rate. Therefore depletion of phosphocreatine will severely decrease the maximal rate of ATP formation with a potential impairment of the contraction process.

During aerobic recovery phosphocreatine is rapidly resynthesized despite high lactate levels and the relationship between phosphocreatine and lactate disappears. The rapid recovery of phosphocreatine during aerobic conditions $(t_{1/2} \approx 30 \text{ s})$[43] will rapidly restore the maximal rate of ADP rephosphorylation and may be related to the rapid recovery in maximal force.[35]

Acidosis and aerobic energy production

Changes of extramitochondrial pH in the range 6.5 to 7.0 have only a minor influence on mitochondrial respiration,[44,45] whereas severe inhibition is noted at pH 6.0.[44] The absence of physiological CO_2 tensions in these experiments makes a translation to an *invivo* situation difficult since the presence of the CO_2 system may be necessary to transfer an extramitochondrial pH change to the mitochondria. When the pH gradient between mitochondria and the surrounding medium was abolished by addition of ionophores, maximal mitochondrial respiration showed a marked pH dependence at pH 6.4, where it was about 50 per cent of the peak value at pH 7.1.[46] Therefore it is possible that aerobic energy transduction will be affected by the exercise-induced acidosis. However, the high oxygen uptake maintained during high intensity exercise to fatigue does conflict with this contention.

CONCLUDING REMARKS

The increased energy demand during exercise is associated with large increases in the rates of the energetic processes and therefore also in the flux of protons. Both the aerobic and the anaerobic processes cause an increased proton flux. Formation of CO_2 is balanced by increased removal of CO_2 through respiration and therefore acid–base balance is not adversely affected by the oxidative processes.

The increased proton formation due to lactate accumulation causes a decrease in tissue pH, the extent of which depends on the buffer capacity. The decrease in extracellular pH will facilitate tissue oxygen uptake through a decreased binding of oxygen to haemoglobin and through local vasodilation and therefore will have a positive influence on the energy transduction. However, a decrease in intracellular pH will interfere with the intracellular energetic processes and decrease the maximal rate of ADP rephosphorylation.

The inhomogeneity of muscle, the difficulties in arresting the metabolism rapidly, and the high turnover of ADP and AMP limit the ability to assess the true concentration of ADP and AMP in a contracting muscle. However, the activity of adenylate kinase and AMP deaminase in skeletal muscle is very high, and therefore increases in ADP and AMP, which are potent activators of AMP deaminase, may be reflected by increases in inosine monophosphate. Since the further metabolism of inosine monophosphate is a rather slow process, one can use inosine monophosphate formation as an indirect measure of energetic deficiency. Recent studies have shown that formation of inosine monophosphate is increased during exercise to fatigue under a variety of conditions[47] and support the hypothesis that in many cases fatigue is related to a decreased capacity to generate ATP.

It is concluded that muscle fatigue during high intensity exercise is generally associated with acidosis. Evidence suggests that the failure of the contraction process is caused by energetic deficiency mediated by the decrease in pH.

REFERENCES

1. Hultman E, Sahlin K. Acid–base balance during exercise. In: Hutton RS, Miller DI, eds. *Exercise and sport sciences reviews*. Philadelphia: Franklin Institute Press, 1981: 41–128.
2. Dawson JM, Gadian DG, Wilkie DR. Contraction and recovery of living muscles studied by ^{31}P nuclear magnetic resonance. *Journal of Physiology* 1977; **267**: 703–35.
3. Lewis SF, Haller RG, Cook JD, Nunnally RL. Muscle fatigue in

McArdle's disease studied by 31P-NMR: effect of glucose infusion. *Journal of Clinical Investigation* 1985; **76**: 556–60.

4. Ross BD, Radda GK, Gadian DG, Rocker G, Esiri M, Falconer-Smith J. Examination of a case of suspected McArdle's syndrome by 31P nuclear magnetic resonance. *New England Journal of Medicine* 1981; **304**: 1338–42.

5. Taylor DJ, Bore PJ, Styles P, Gadian DG, Radda GK. Bioenergetics of intact human muscle a 31P nuclear magnetic resonance study. *Molecular Biology and Medicine* 1983; **1**: 77–94.

6. Hochachka PW, Mommsen TP. Protons and anaerobiosis. *Science* 1983, **219**: 1392–7.

7. Fletcher WW, Hopkins FG. Lactic acid in mammalian muscle. *Journal of Physiology* (London) 1907; **35**: 247–303.

8. Sjödin B, Svedenhag J. Applied physiology of marathon running. *Sports Medicine* 1985; **2**: 83–99.

9. Jorfeldt L, Juhlin-Dannfelt A, Karlsson J. Lactate release in relation to tissue lactate in human skeletal muscle during exercise. *Journal of Applied Physiology* 1978; **44**: 350–2.

10. Katz A, Broberg S, Sahlin K, Wahren J. Muscle ammonia and amino, acid metabolism during dynamic exercise in man. *Clinical Physiology* 1986; **6**: 365–79.

11. Jansson E, Kaijser L. Substrate utilization and enzymes in skeletal muscle of extremely endurance-trained men. *Journal of Applied Physiology* 1987; **62**: 999–1005.

12. Saltin B. Anaerobic capacity: past, present and prospective. In: Taylor AW, Gollnick PD, eds. *Biochemistry of exercise* VII Champaign, IL: Human Kinetics, 1990: 387–412.

13. Benade AJS, Heisler N. Comparison of efflux rates of hydrogen and lactate ions from isolated muscles *in vitro*. *Respiration Physiology* 1978; **32**: 369–80.

14. Sahlin K, Alvestrand A, Brandt R, Hultman E. Acid–base balance in blood during exhaustive bicycle exercise and the following recovery period. *Acta Physiologia Scandinavica* 1978; **104**: 370–2.

15. Katz A, Sahlin K, Juhlin-Dannfelt A. Effect of β-adrenoceptor blockade on H$^+$ and K$^+$ flux in exercising man. *Journal of Applied Physiology* 1986; **250**: C834–40.

16. Maschio G, Bazzato G, Bertaglia E, Sardini D, Mioni G, D'Angelo A, Marzo A. Intracellular pH and electrolyte content of skeletal muscle in patients with chronic renal acidosis. *Nephron* 1970; **7**: 481–7.

17. Nevill ME, Boobis LH, Brooks S, Williams C. Effect of training on muscle metabolism during treadmill sprinting. *Journal of Applied Physiology* 1989; **67**: 2376–82.

18. Hermansen L, Osnes JB. Blood and muscle pH after maximal exercise in man. *Journal of Applied Physiology* 1972; **32**: 304–8.

19. Sahlin K, Alvestrand A, Brandt R, Hultman E. Intracellular pH and bicarbonate concentration in human muscle during recovery from exercise. *Journal of Applied Physiology* 1978; **45**: 474–80.

20. Sahlin K, Harris RC, Hultman E. Creatine kinase equilibrium and lactate content compared with muscle pH in tissue samples obtained after isometric exercise. *Biochemical Journal* 1975; **152**: 173–80.

21. Wilson JR, McCully KK, Mancini DM, Boden B, Chance B. Relationship of muscular fatigue to pH and diprotonated Pi in humans: a ^{31}P-NMR study. *Journal of Applied Physiology* 1988; **64**: 33–9.

22. Cady EB, Jones DA, Lynn J, Newham DJ. Changes in force and intracellular metabolites during fatigue of fatigued human skeletal muscle. *Journal of Physiology* 1989; **418**: 311–25.

23. Sahlin K. Intracellular pH and energy metabolism in skeletal muscle of man with special reference to exercise. Thesis, Stockholm, *Acta Physiologica Scandinavica* 1978, Suppl. 455.

24. Mizuno M, Juel C, Bro-Rasmussen T, Mygind E. Schibye B, Rasmussen B, Saltin B. Limb skeletal muscle adaptation in athletes after training at altitude. *Journal of Applied Physiology* 1990; **68**: 496–502.

25. Woodbury W. Regulation of pH. In: Ruch TC, Patton HD, eds. *Physiology and biophysics*. London: Saunders, 1965: 899–934.

26. Sharp RL, Costill DL, Fink WJ, King DS. The effects of eight weeks of bicycle ergometer spring training on buffer capacity. *International Journal of Sports Medicine* 1983; **7**: 13–7.

27. Sahlin K, Henriksson J. Buffer capacity and lactate accumulation in skeletal muscle of trained and untrained men. *Acta Physiologica Scandinavica* 1984; **122**: 331–9.

28. Kindermann W, Keul J, Huber G. Physical exercise after induced alkalosis (bicarbonate or tris-buffer). *European Journal of Applied Physiology* 1977; **37**: 197–204.

29. Katz A, Costill DL, King DS, Hargreaves M, Fink WJ. Maximal exercise tolerance after induced alkalosis. *International Journal of Sports Medicine* 1984; **5**: 107–10.

30. Jones NL, Sutton JR, Taylor R, Toews CJ. Effect of pH on cardiorespiratory and metabolic responses to exercise. *Journal of Applied Physiology* 1977; **43**: 959–64.

31. Sutton JR, Jones NL, Toews CJ. Effect of pH on muscle glycolysis during exercise. *Clinical Science* 1981; **61**: 331–8.

32. Costill DL, Verstappen F, Kuipers H, Janssen E, Fink W. Acid–base balance during repeated bouts of exercise. Influence of NaHCO$_3$. *International Journal of Sports Medicine* 1984, **5**: 228–31.

33. Donaldsson SKB, Hermansen L, Bolles L. Differential, direct effects of H$^+$ on Ca^{2+}—activated force of skinned fibers from the soleus, cardiac and adductor magnus muscles of rabbits. *Pflügers Archiv* 1978; **376**: 55–65.

34. Cooke R, Franks K, Luciani GB, Pate E. The inhibition of rabbit skeletal muscle contraction by hydrogen ions and phosphate. *Journal of Physiology* (London) 1988; **395**: 77–97.

35. Sahlin K, Ren J. Relationship of contraction capacity to metabolic changes during recovery from a fatiguing contraction. *Journal of Applied Physiology* 1989; **67**: 648–54.

36. Wilkie Dr. Muscular fatigue: effects of hydrogen ions and inorganic phosphate. *Federation Proceedings* 1986; **45**: 2921–3.

37. Blum H, Schnall MD, Chance B, Buzby GP. Intracellular sodium flux and high-energy phosphorus metabolites in ischemic skeletal muscle. *American Journal of Physiology* 1988; **255**: C377–84.

38. Trivedi B, Danforth WH. Effect of pH on the kinetics of frog muscle phosphofructokinase. *Journal of Biological Chemistry* 1966; **241**: 4110–14.

39. Dobson GP, Yamamoto E, Hochachka PW. Phosphofructokinase control in muscle: nature and reversal of pH dependent ATP inhibition. *American Journal of Physiology* 1986; **250**: R71–6.

40. Chasiotis D, Hultman E, Sahlin K. Acidotic depression of cyclic AMP accumulation and phosphorylase *b* to a transformation in skeletal muscle of man. *Journal of Physiology* (London) 1983; **335**: 197–204.

41. Hultman E, Bergström J, Mclennan Anderson N. Breakdown and resynthesis of phosphorylcreatine and adenosine triphosphate in connection with muscular work in man. *Scandinavian Journal of Clinical Laboratory Investigation* 1967; **19**: 56–66.

42. Harris RC, Sahlin K, Hultman E. Phosphagen and lactate contents of m. quadriceps femoris of man after exercise. *Journal of Applied Physiology* 1977; **43**: 852–7.

43. Harris RC, Edwards RHT, Hultman E, Nordesjö LO, Nylind B, Sahlin K. The time course of phosphorylcreatine resynthesis during recovery of the quadriceps muscle in man. *Pflugers Archiv* 1981; **389**: 277–82.

44. Mitchelson KR, Hird FJR. Effect of pH and halothane on muscle and liver mitochondria. *American Journal of Physiology* 1973; **225**: 1393–8.

45. Tobin RB, Macherer CR, Mehlman MA. pH effects on oxidative phosphorylation of rat liver mitochondria. *American* Journal of Physiology 1972; **223**: 83–8.

46. Hansford RG. Some properties of pyruvate and 2–oxoglutarate oxidation by blowfly flight-muscle mitochondria. *Biochemical Journal* 1972; **127**: 271–83.

47. Sahlin K, Broberg S. Adenine nucleotide depletion in human muscle during exercise: causality and significance of AMP deamination. *International Journal of Sports Medicine* 1990; **11**: S62–7.

1.2.1 Energy intake of athletes

ANN C. GRANDJEAN AND JAIME S. RUUD

INTRODUCTION

The energy requirement of an individual has been defined as 'the level of energy intake from food that will balance energy expenditure when the individual has a body size and composition, and level of physical activity, consistent with long-term good health; and that will allow for the maintenance of economically necessary and socially desirable physical activity'.[1]

For the serious competitive athlete, concerns about energy go beyond health or socially desirable physical activity. It is imperative that energy intake supports the training and competitive schedule which will allow the athlete to achieve his or her personal best. Maintaining adequate energy levels, weight loss, and weight gain can have profound consequence and significance. Coaches, athletes, and sports scientists are all interested in the energy requirements of athletes.

Basal metabolic requirement is the major factor affecting total energy needs. The second-largest component is the energy expended in physical activity. Athletes as a subpopulation have tremendous variability in the amount of energy expended for physical activity owing to the wide variance in demands of the various sports as well as differences in training programmes.

When reviewing the literature on energy intakes of athletes, the assessment methods utilized, factors influencing energy expenditure, and the intra- and interindividual variations must all be considered.

DIETARY ASSESSMENT: METHODS AND LIMITATIONS

Measuring energy intake in humans is a difficult task. Even under the best of circumstances, a 10 per cent error in precision is expected.[2] Several methods have been utilized to assess the energy intakes of given populations including 24-hour recall, diet history, food frequency questionnaires, and food records. Most of these methods have some degree of validity when measured against one another.[3] The method of choice depends on the purpose of the survey, the sample size, and the funds and personnel available.[4] The 24-hour recall and the 3-day food record are the tools most frequently used to collect dietary data on athletes.

The 24-hour recall

The 24-hour recall is commonly used to measure group means and trends in food and nutrient intakes.[5,6] It is not an appropriate tool for assessing the usual diet of individuals.[3,6-9] The major difficulty in analysing data derived from 24-hour recalls or records is the considerable intraindividual variation in intake. Use of the 24-hour recall also tends to underestimate portion sizes and/or completely omit some food items.[10] Karvetti and Knuts[6] found that common food items like bread, potatoes, and coffee were easy to recall, while less frequently consumed food items, such as cooked vegetables, were often forgotten. Furthermore, the 24-hour recall is prone to over-reporting low intakes and under-reporting high intakes.[6,8,11]

Food records

A number of investigators have used food records to determine the dietary intake of subjects. The number of days for which food records are obtained affects the appropriate use of the data, with greater restrictions placed on the interpretation of information for 1 day than for 3 to 7 days.

Van Erp-Baart et al.,[12] in a nationwide survey of the nutritional habits of elite athletes, found that, with a small number of subjects, recording for 4 days including the weekend gives better information than a 24-hour recall. Gersovitz et al.[8] compared the validity of a 24-hour recall and a 7-day written food record and found that both methods provided equally accurate estimates of mean nutrient intake. They also collected intake data on a second group of subjects for comparison with the 7-day records. Regression analysis revealed that the percentage of usable records declined from 85 per cent after day 2 to 60 per cent by day 7, with accuracy of the records declining on days 5, 6, and 7. Additionally, analysis revealed that while age, sex, and education did not affect the accuracy of the records submitted, the subjects returning food records for the full 7 days were predominantly the more highly educated.

Food frequency and diet history methods

Food frequency questionnaires and diet histories require that subjects recall their usual dietary intake over a period of time. These methods are employed frequently to estimate the intake of foods by rank or category according to frequency of consumption rather than to provide a quantitative measure of actual intake. The food frequency questionnaire is probably the least precise method of assessing energy intake because it is difficult to obtain a true picture of the usual diet.[13] Bergman et al.[14] found that, compared with the 3-day food record, the food frequency questionnaire resulted in significantly higher estimates for energy, carbohydrate, protein, and several vitamins and minerals.

Duplicate food collections

Some investigators believe that collecting a duplicate sample of foods and beverages consumed and analysing it in a laboratory is the most accurate method of measuring energy intake.[4,15] This technique, which is expensive and time-consuming, also requires a high degree of co-operation from the subjects.

During a 1-year study in which food intakes were recorded daily, subjects were asked to make duplicate food collections during four 1-week periods.[16] Mean food energy intakes calculated from diet records for the group were compared with the mean values calculated from intakes during the food collection periods. Mean energy intake of the subjects was 12.9 per cent less during food collection periods than the yearly mean. One possible explanation may be that subjects alter food selection or decrease food intake during food collection periods.

Covert observation has been used effectively to assess food intake of individuals.[17,18] While this method may be practicable for use in metabolic units, because of the need for trained observers and the requirement that subjects eat in a predetermined setting, this technique has limited use. However, it may be applicable if athletes are eating in an athletic dining hall, but utilization is limited when studying large numbers of subjects.

To summarize, several methods are used to assess food intake. The 24-h recall, although reliable for measuring group means, is not a practical tool for assessing the usual diet of individuals. Neither are food frequency questionnaires. The majority of surveys reviewed for this chapter used 3-, 4-, and 7-day diet records. Although accuracy may decline as the length of the recording period increases,[8] athletes are generally accustomed to keeping records and have a positive attitude towards nutrition.[19,20] Therefore it can be assumed that dietary intake data collected on athletes are probably more reliable than those for other sections of the general population.

Double-labelled water technique

The double-labelled water technique has been validated for measuring energy expenditure in free-living subjects and, as a result, may serve as a reference for validating the accuracy of energy intake.[21] In athletes, comparisons between reported energy intake and energy expenditure using the double-labelled water technique have been made,[22,23] and large discrepancies have been reported. Researchers have suggested that the differences between reported energy intake and expenditure are probably due to restricted eating during the recording period[22] or under-reporting by subjects.[24] While the double-labelled water method offers potential for studying subjects in 'real world' conditions, it is a relatively new and costly technique and may be impractical for use in large studies. More research is needed before conclusions can be drawn. While studies using double-labelled water suggest, for the most part, that individuals tend to underestimate energy intakes,[21] the consistency of underestimated energy intakes may be reason for suspicion. Repeatability does not equate to validity. While some studies on athletes using double-labelled water have estimated expenditures and intakes for periods of up to 8 days, caution is indicated. It is recognized that while energy intake and expenditure may balance over a period of time, this is not the case on a daily basis and may be particularly relevant in athletes whose training intensity varies over time. Improved methods of assessing habitual food intake and a better understanding of how to analyse and interpret intake and expenditure data are needed.

Special considerations for females

It is important that studies assessing the food intake of women consider the different stages of the menstrual cycle. Changes in energy intake during the menstrual cycle have been reported. In a study by Gong et al.,[25] the voluntary energy intakes of seven women were evaluated by the weighed intake method over an entire menstrual cycle. Average energy intake was higher during the luteal phase (8541 kJ/day) (2040 kcal/day) than during the follicular phase (7674 kJ/day) (1833 kcal/day). Tarasuk and Beaton[26] also found higher energy and fat intakes during the pre-menstruum. They concluded that both physiological and behavioural factors may influence energy and macronutrient intake during the menstrual cycle.

Tai et al.[27] have investigated whether the phase of the menstrual cycle affects energy expenditure. The resting metabolic rate was examined in six healthy young women during the various stages of

a single menstrual cycle, and no differences were found. Therefore, while intake appears to vary, expenditure may not.

STANDARDS FOR EVALUATING DIETARY INTAKE

The quality of a diet is determined by the nutrient content. Several methods have been used to evaluate nutritional quality, ranging from a checklist that grossly determines variety to more sophisticated computer dietary analysis programmes. Recommended dietary allowances are amounts of specific nutrients that are recommended for populations grouped by age and sex.[28] Recommendations are based on intakes greater than those needed to prevent disease and/or deficiencies.

Energy

With the exception of energy, the recommended dietary allowances are set 'high enough to meet an upper level of requirement variability among individuals within the groups'. The recommended dietary allowances for energy reflect the average requirements for individuals.

The United States recommended energy allowances are presented in Table 1. These energy values are based on individuals with a light to moderate activity level. They were calculated using the World Health Organization (WHO) equations for resting energy expenditure and then multiplying by an activity factor.[28] The average energy intake is 12 142 kJ (2900 kcal) or 167 kJ/kg/day (40 kcal/kg.day) for the reference 19–24 year old male (72 kg) and 9211 kJ (2200 kcal) or 159 kJ/kg.day (38 kcal/kg.day) for the reference female of the same age (58 kg). For people with greater physical activity or of a larger or smaller body size, further adjustments must be made. Table 2 shows the activity factors associated with various levels of physical activity.

While the recommended dietary allowance is widely used for interpreting food consumption data, one should be aware of its limitations. As stated by the authors:[28]

Recommended dietary allowances are neither minimal requirements nor necessarily optimal levels of intake ... Rather, recommended dietary allowances are safe and adequate levels reflecting the state of knowledge concerning a nutrient, its bioavailability and variation among the U.S. population.

International nutrition standards for athletes do not exist. The standards utilized by scientists evaluating dietary intake of athletes include guidelines from their respective countries, WHO recommendations, and in many cases the recommended dietary allowances established by the National Research Council of the United States. The data considered in this chapter were collected on subjects from a variety of countries. Although based on the United States population, the United States recommended dietary allowance was selected for reference and comparison because it was the standard used by 75 per cent of the authors whose studies have been cited in this chapter.

Energy distribution

While recommended energy allowances and actual energy intake are based on years of surveys and research, the interest in energy distribution has a much shorter history and recommendations are

Table 1 Median heights and weights and recommended energy intake

Category	Age (years) or condition	Weight (kg (lb))	Height (cm (in))	REE[a] (kJ/day (kcal/day))	Average energy allowance[b] (kJ (kcal))		
					× REE	Per kg	Per day[c]
Infants	0.0–0.5	6 (13)	60 (24)	1340 (320)		452 (108)	2721 (650)
	0.5–1.0	9 (20)	71 (28)	2093 (500)		410 (98)	3559 (850)
Children	1–3	13 (29)	90 (35)	3098 (740)		427 (102)	5443 (1300)
	4–6	20 (44)	112 (44)	3977 (950)		377 (90)	7536 (1800)
	7–10	28 (62)	132 (52)	4731 (1130)		293 (70)	8374 (2000)
Males	11–14	45 (99)	157 (62)	6029 (1440)	1.70	230 (55)	10 467 (2500)
	15–18	66 (145)	176 (69)	7369 (1760)	1.67	188 (45)	12 560 (3000)
	19–24	72 (160)	177 (70)	7453 (1780)	1.67	167 (40)	12 142 (2900)
	25–50	79 (174)	176 (70)	7536 (1800)	1.60	155 (37)	12 142 (2900)
	51+	77 (170)	173 (68)	6406 (1530)	1.50	126 (30)	9630 (2300)
Females	11–14	46 (101)	157 (62)	5485 (1310)	1.67	197 (47)	9211 (2200)
	15–18	55 (120)	163 (64)	5736 (1370)	1.60	167 (40)	9211 (2200)
	19–24	58 (128)	164 (65)	5652 (1350)	1.60	159 (38)	9211 (2200)
	25–50	63 (138)	163 (64)	5777 (1380)	1.55	151 (36)	9211 (2200)
	51+	65 (143)	160 (63)	5359 (1280)	1.50	126 (30)	7955 (1900)
Pregnant	1st trimester						+0 (+0)
	2nd trimester						+1256 (+300)
	3rd trimester						+1256 (+300)
Lactating	1st 6 months						+2093 (+500)
	2nd 6 months						+2093 (+500)

Kilojules calculated from kilocalories as appear in reference. REE, recommended energy intake.
[a]Calculation based on FAO equations and then rounded.
[b]In the range of light to moderate activity, the coefficient of variations is ±20 per cent.
[c]Figure is rounded.
Adapted from ref. 28.

more arbitrarily based. Although the relationships of carbohydrate and protein to athletic performance have been extensively researched, the collective results do not allow conclusive recommendations for all athletes regarding the ideal distribution of energy. Many experts contend that the same basic dietary principles that promote good health for the general public will maximize performance for most athletes. However, there is evidence that some forms of heavy training increase the requirements for certain nutrients. Research is not adequate to allow recommendations on an individual or by-sport basis. Therefore general dietary recommendations have been made for all athletes.[29–33] It is recommended that protein contribute 10 to 15 per cent of the total energy intake and carbohydrate contribute 50 to 60 per cent. Research assessing the risk of heart disease and cancer has led to the recommendation that fats should be limited to 30 per cent of total calories. For athletes involved in prolonged strenuous endurance training, the relative contribution of specific nutrients may change, with carbohydrates providing 60 to 70 per cent of total energy intake.

While agreement exists regarding the value of carbohydrate intake for endurance activity (competition or training), there is a lack of agreement of the carbohydrate needs for short-term or intermittent high intensity exercise performance. Essen and co-workers[34,35] evaluated the contribution from carbohydrate and lipids to oxidative metabolism during intermittent intense exercise compared with 60 min of continuous exercise. The data show that less glycogen and more lipids were used when exercise was performed intermittently.

More recently, Symons and Jacobs[36] evaluated the effects of

Table 2 Factors for estimating daily allowances at various levels of physical activity for men and women (ages 19—50)

Level of activity	Activity factor[a] × REE	Energy expenditure[b] (kJ/kg.day (kcal/kg.day))
Very light		
Men	1.3	130 (31)
Women	1.3	126 (30)
Light		
Men	1.6	159 (38)
Women	1.5	147 (35)
Moderate		
Men	1.7	172 (41)
Women	1.6	155 (37)
Heavy		
Men	2.1	209 (50)
Women	1.9	184 (44)
Exceptional		
Men	2.4	243 (58)
Women	2.2	214 (51)

Kilojules calculated from kilocalories as appear in reference.
REE, recommended energy intake.
[a]Based on examples presented by WHO.[1]
[b]REE was computed from formulae in Table 3–1 of ref. 28 and is the average of values for median weights of persons aged 19–24 and 25–74 years: males, 100 kJ/kg (24.0 kcal/kg); females, 97 kJ/kg (23.2 kcal/kg).

glycogen availability on short-term high intensity exercise performance. Subjects performed a variety of exercise tasks after consuming a mixed diet and then again on a low carbohydrate diet. The authors concluded that short-term high intensity exercise performance is not adversely affected when intramuscular glycogen concentrations are low.

When the results of such research are coupled with dietary consumption data on elite athletes, one must ask if the 60 to 70 per cent recommendation for dietary carbohydrate is necessary for all athletes. Reasonable thought would indicate that neither an absolute recommendation for grams of carbohydrate nor a recommendation based on percentage energy intake is appropriate. If the primary purpose of pre-exercise and post-exercise carbohydrate intake is to replenish glycogen stores, then it seems reasonable that the amount of carbohydrate needed should be proportional to the amount of muscle mass, with consideration given to the length and intensity of effort. A more operative and more logical recommendation would be one based on body weight or lean body mass with a factor for activity intensity and duration. However, until more definitive research is conducted, the only standard available is the percentage of energy discussed above.

FACTORS INFLUENCING ENERGY EXPENDITURE

Total energy expenditure is influenced by three major factors: resting metabolic rate, thermogenesis, and physical activity.[37] Each of these factors is affected directly or indirectly by age, sex, body size, and climate.[38]

Resting metabolic rate

Resting metabolic rate is usually the largest contributor to total daily energy expenditure, accounting for approximately 60 to 75 per cent of daily energy needs.[39] It is a measure of the energy required for maintenance of normal body functions and homeostasis, including a small component related to the sympathetic nervous system.[40] Resting metabolic rate is measured several hours after the last meal or any physical activity with the individual lying down or sitting quietly. Basal metabolic rate is defined as the energy expended in the morning upon awakening before any physical activity and at least 12 to 14 h after the last meal. Although the basal metabolic rate may be slightly lower than the resting metabolic rate, the difference is less than 10 per cent and thus the terms are often used interchangeably.[28]

Several factors are known to affect resting metabolic rate, including body composition, age, sex, body temperature, state of nutrition, menstruation, and endocrine function. In persons of similar age, sex, height, and weight, differences in fat-free mass account for approximately 80 per cent of the variance in resting metabolic rate.[41]

Thermic effect of food

The thermic effect of food, also known as diet-induced thermogenesis, is the increase in energy expenditure above the resting metabolic rate that can be measured for several hours following a meal.[40] The thermic effect of food includes the energy costs of digestion, absorption, metabolism, and storage of food in the body.[37] According to Garrow,[42] the thermic effect reaches a maximum 1 h after a meal and can be measured for up to 4 h and sometimes even longer. However, it is relatively small, accounting

for approximately 7 to 10 per cent of the total energy expenditure.[43,44]

Physical activity

The third major factor directly influencing energy expenditure is physical activity. Of all the components, it is the most variable among individuals.[40] The amount of energy expended by an athlete depends on several factors including age, sex, height, weight, body composition, type of sport, physical conditioning, clothing worn, playing surface, environment in which the activity takes place, and frequency, intensity, and duration of the event or training session. Body size affects energy expenditure more than any other single factor.

Physical activity may also affect total energy requirement indirectly through its effect on resting metabolic rate and thermic effect of food.

The effects of exercise on resting metabolic rate

Several studies have found differences in the resting metabolic rate between trained and untrained subjects.[45–47] Poehlman et al.[46] reported a higher resting metabolic rate in endurance-trained men compared with moderately trained and untrained subjects. Similar results were also reported by Tremblay and coworkers[48,49] and LeBlanc et al.[50] Owen et al.[51] determined the energy expenditure of 44 healthy lean and obese women, including eight trained female athletes, by direct calorimetry. The female athletes showed greater increases in resting metabolic rate per gain in body weight than non-athletic women. Other studies, however, have found no significant difference between trained and untrained individuals.[52–54] A better understanding will only follow improved methodology and consistency in the equipment and techniques used to measure the effect of exercise on resting metabolic rate. Until such time, conclusions cannot be drawn.

The effects of exercise on thermic effect of food

Studies of the effects of exercise on the thermic effect of food have reported inconsistent findings. Some data show that the thermic effect of a meal increases with improved fitness[55,56] whereas other studies found no such relationship.[50,54,57] Davis et al.[55] reported that the thermic effect of food increased in response to training and that the increase correlated with the change in Vo_2. Poehlman et al.[46] examined the thermic effect of food in 28 men with a wide range of fitness levels and maximal aerobic capacities. The thermal effect of food was higher in the moderately trained as compared with untrained and highly trained subjects. However, LeBlanc et al.[50] concluded that exercise training reduces the thermic effect of food, indicating a possible increase in food efficiency and a sparing effect on glucose utilization in highly trained subjects.

ENERGY INTAKE OF ATHLETES REPORTED IN THE LITERATURE

When reviewing survey data, it is difficult to make comparisons of athletes' energy intakes due to differences in research design, sample size, body size, gender, level of competition (elite, amateur, collegiate), and the manner in which data are reported.

Athletes' energy intakes also vary due to the stage of training. Hickson et al.[58] reported that the mean energy intakes of intercollegiate male soccer players were higher during preseason conditioning (18 807 kJ/day) (4492 kcal/day) than during the competitive season (14 009 kJ/day) (3346 kcal/day). Results of a study by Ellsworth et al.[59] showed that the mean energy intakes of both male and female Nordic skiers were higher during training than during the competitive season.

One would assume that energy requirements would increase during the competitive season and periods of heavy training. However, data do not always support this theory. Nutter[60] found that the mean energy intake of 24 female athletes was less than that recommended for their mean body weight and age during their competitive season and after their season ended. The authors concluded that the dietary practices of females may be influenced more by their desire to be thin than changes in exercise training.

Athletes' energy intakes reported in the literature are summarized by sex and sport in Tables 3 to 10. As can be seen, there is a wide range both between and within sport groups. Male swimmers, cyclists, triathletes, and basketball players had among the highest mean daily energy intakes, ranging from 14 790 to 24 700 kJ/day (3533–5900 kcal/day). The lowest mean energy intakes were noted for female gymnasts, dancers, and figure skaters (4915–8328 kJ/day (1174–1989 kcal/day)), and for male gymnasts (8709 kJ/day (2080 kcal/day)).

Reporting on a body weight basis adjusts for some of the factors affecting daily energy intake. Therefore, where possible, comparisons are made as such or are otherwise indicated.

On a kilojoules per kilogram body weight (BW) basis, the adolescent swimmers studied by Berning et al.[61] had the highest mean energy intakes for male and female athletes (283 kJ/kg BW (67.6 kcal/kg BW) and 257.1 kJ/kg BW (61.4 kcal/kg BW) respectively) (Table 4). The wrestlers studied by Grandjean[62] had the lowest energy intake at 118.1 kJ/kg BW (28.2 kcal/kg BW) (Table 10). Data on most of the wrestlers in Grandjean's study were collected during the competitive season, thus reflecting the weight control practices of the sport.

Energy distribution

The studies summarized in this chapter revealed a great variance in the distribution of energy from protein, carbohydrate, and fat, as shown in Tables 3 to 10. Overall, mean intakes of protein ranged from 10 to 22 per cent, carbohydrate from 38 to 67 per cent, and fat from 20 to 42 per cent. The cyclists and triathletes generally consumed a higher percentage of energy from carbohydrate (50–67 per cent) than other sport groups.

It is of interest that the range of fat intakes was less for weight-lifters, wrestlers, and football and basketball players than for runners, swimmers, cyclists, and triathletes. Overall, data from the studies reported herein indicate that the diets of many of the athletes studied are lower in carbohydrate and higher in fat than recommended. However, mean intakes alone should be viewed with caution in the absence of more descriptive data such as standard deviation, ranges, or frequency distributions.

INTAKES BY SPORT AND SEX

Runners

The mean energy intakes of male and female runners are presented in Table 3. Total energy intakes and kilojoules per kilogram body weight were fairly consistent for all studies reviewed, with the exception of the male and female marathon runners studied by Nieman et al.[66] who had the lowest energy intakes of all groups.

Energy intakes of some distance runners are lower than the predicted requirement.[64,70] Fogelholm[64] reported that the mean energy intake of 59 male Finnish endurance athletes was 12 800 kJ/day (190 kJ/kg) (3160 kcal/day (45 kcal/kg)), while energy expenditure was estimated at 15 491 kJ/day (3700 kcal/day). Similarly, Deuster et al.[70] reported a mean daily energy intake of 10 036 kJ (2397 kcal) for 51 highly trained female runners, while energy expenditure was estimated at 10 886 kJ/day (2600 kcal/day). Thus an obvious question is: are these athletes continually consuming less energy than they actually expend? If this were the case, the deficit would eventually present itself as weight loss.

Table 3 Energy intakes and calorie distribution for runners

Reference	N	Age	Energy intake		Carbohydrate (%)	Protein (%)	Fat (%)
			kJ (kcal)	kJ/kg (kcal/kg)			
Males							
63	15	36	13 783 (3292)	—	49.0	10.0	26.0
12[a]	56	30	—	193.0 (46.1)	50.0	—	33.0
62	10	26[b]	12 703 (3034)	193.0 (46.1)[b]	49.0	17.0	34.0
64	59	22	12 800 (3160)	190.0 (45.4)	53.0	15.0	31.0
65	34	—	12 388 (2959)	175.8 (42.0)	39.8	13.8	40.8
66	291	40	10 576 (2526)	146.5 (35.0)	51.8	16.6	30.9
67	35	21–22	12 644 (3020)	—	—	—	—
68	10	—	17 254 (4121)	—	49.0	14.0	36.0
Females							
69	8	27	10 421 (2489)	202.2 (48.3)	53.7	15.9	30.5
70	51	29	10 036 (2397)	193.8 (46.3)	55.0	13.0	32.0
65	27	—	9990 (2386)	175.8 (42.0)	39.5	14.2	41.1
12[a]	18	31	—	168.0 (40.1)	55.0	—	28.0
66	56	38	7821 (1868)	143.6 (34.3)	52.7	15.8	32.0
67	17	21–22	8482 (2026)	—	—	—	—
71	10	34	9512 (2272)	—	48.0	14.0	38.0

[a]Calorie distribution data were estimated from information presented in a different format by the original authors.
[b]Unpublished data.

Table 4 Energy intakes and calorie distribution for swimmers

Reference	N	Age	Energy intake		Carbohydrate (%)	Protein (%)	Fat (%)
			kJ (kcal)	kJ/kg (kcal/kg)			
Males							
61	22	16	21 863 (5222)	283.0 (67.6)	45.6	12.6	42.8
12[a]	20	18	—	221.0 (52.8)	48.0	—	37.0
62	15	19[b]	16 823 (4018)	217.3 (51.9)[b]	51.0	14.0	35.0
74	10	16	14 790 (3533)	209.0 (49.9)	51.2	15.9	33.5
75	13	17–23	20 231 (4832)	—	—	—	—
Females							
61	21	15	14 959 (3573)	257.1 (61.4)	47.9	12.0	41.4
76	6	22	10 350 (2472)	165.8 (39.6)	53.0	15.0	32.0
74	10	16	8636 (2063)	146.0 (34.9)	54.0	17.0	29.9
77	19	19	10 438 (2493)	—	54.1	12.7	33.1
68	7	—	9412 (2248)	—	44.0	17.0	36.0
78	9	—	10 333 (2468)	—	48.9	16.2	35.8

[a]Calorie distribution data were estimated from information presented in a different format by the original authors.
[b]Unpublished data.

Error in measurement and calculation could be a factor. Another feasible explanation is the increased efficiency in trained athletes.[72,73] Factors such as non-sport-related daily activity, clothing, and climate should also be considered.

The runners' diets varied considerably with respect to the distribution of energy, particularly for carbohydrate and fat. Carbohydrate intake ranged from 39 to 55 per cent and fat from 26 to 41 per cent, while protein accounted for 10 to 17 per cent of total energy. None of the groups studied had mean carbohydrate intakes equal to the 60 to 70 per cent frequently recommended for athletes who compete in endurance events.[30,33]

Swimmers

The energy intakes of male and female swimmers are shown in Table 4. Mean intakes ranged from 209 to 283 kJ/kg (49.9–67.6 kcal/kg) for males and from 146.0 to 257.1 kJ/kg (34.9–61.4 kcal/kg) for females.

The lowest mean energy intakes for female swimmers were reported by Barr.[74] Barr believes that the low intakes reflect the swimmers' attempts to achieve a lower than average weight for height.

In a study of Vallieres et al.,[76] energy balance was evaluated over a 30-day training cycle in six elite female swimmers. Estimated mean energy expenditure (11 204 kJ/day) (2676 kcal/day) did not differ significantly from mean energy intake (10 350 kJ/day (2472 kcal/day) ($p > 0.05$). The results indicate that the female swimmers compensated for the extra energy expenditure associated with their training regimen.[76]

For both male and female swimmers, the percentage of energy provided by carbohydrate ranged from 48 to 54 per cent, with 12 to 17 per cent coming from protein and 29.9 to 42.8 per cent from fat. Like the data on runners, many swimmers consume diets lower in carbohydrate and higher in fat than often recommended.

Cyclists

The mean energy intakes of cyclists reported in the literature are presented in Table 5. 'Trained' female cyclists studied by Keith et al.[82] had an average caloric intake of 7457 kJ/day (1781 kcal/day), which is less than the United States recommended dietary allowance of 9211 kJ/day (2200 kcal/day) established for non-athletic women. Grandjean[62] reported the highest energy intake for female cyclists. It should be noted that Keith and colleagues defined

Table 5 Energy intakes and calorie distribution for cyclists

Reference	N	Age	Energy intake		Carbohydrate (%)	Protein (%)	Fat (%)
			kJ (kcal)	kJ/kg (kcal/kg)			
Males							
12[a]	14	20	—	253.0 (60.4)	58.0	—	30.0
62	18	21[b]	17 350 (4144)	242.0 (57.8)[b]	46.0	15.0	40.0
79	6	21	16 303 (3894)	—	53.8	13.5	32.6
80	4	—	24 700 (5900)	—	62.0	15.0	23.0
81	13	20	—	—	62.5	12.5	—
Females							
62	12	26[b]	12 682 (3029)	213.5 (51.0)[b]	51.0	13.0	36.0
12[a]	21	23	—	164.0 (39.2)	52.0	—	33.0
82	8	22	7457 (1781)	—	60.0	14.0	26.0

[a]Calorie distribution data were estimated from information presented in a different format by the original authors.
[b]Unpublished data.

Table 6 Energy intakes and calorie distribution for triathletes

Reference	N	Age	Energy intake		Carbohydrate (%)	Protein (%)	Fat (%)
			kJ (kcal)	kJ/kg (kcal/kg)			
Males							
12[a]	33	26	—	272.0 (65.0)	50.0	—	36.0
84	25	27	17 145 (4095)	247.0 (59.0)	59.5	13.0	27.0
85	48	—	21 897 (5230)	—	67.6	10.6	20.1
86	19	43.6	15 169 (3623)	—	—	—	—
Females							
85	34	—	17 371 (4149)	—	66.2	11.6	21.2
86	10	38.6	10 358 (2474)	—	—	—	—

[a]Calorie distribution data were estimated from information presented in a different format by the original authors.

trained as 'cycling for a minimum of 5 days per week, 1 hour per day'. The training mileage of elite female cyclists studied by Grandjean ranged from 200 to 400 miles per week.

Of the males, the highest mean energy intake (24 700 kJ/day) (5900 kcal/day) was reported by Saris et al.[80] who studied cyclists taking part in the Tour de France, which is considered to be one of the most strenuous of all endurance competitions. The estimated mean energy expenditure for the cyclists was 25 400 kJ/day (6067 kcal/day), very close to their actual intake. The highest mean daily value for energy was during one of the mountain stages of the race when intakes averaged 32 400 kJ/day (7739 kcal/day).

Compared with data on runners and swimmers, the cyclists' relative contribution of carbohydrate was higher, ranging from 46 to 62 per cent of total energy. It should be noted that the relatively high carbohydrate intake in the study by Saris et al.[80] reflects the use of carbohydrate beverages during the Tour de France and therefore may not be representative of normal daily intake.

Triathletes

Few studies have been conducted on triathletes, which, in part, is a reflection of the short history of the sport. As noted by Applegate,[83] the energy intakes of triathletes can be staggering or rather modest depending on the type and intensity of the training programme.

As shown in Table 6, the mean energy intakes of triathletes reported in the literature range from 15 169 to 21 897 kJ/day (3623–5230 kcal/day) for males, and from 10 358 to 17 371 kJ/day (2474–4149 kcal/day) for females. The female athletes studied by Green et al.[85] consumed almost twice as much energy as those studied by Khoo et al.[86] Unfortunately, the authors did not report data as kilocalories per kilogram body weight, which would allow for a more accurate comparison.

Because of their intensive training, triathletes have increased energy, carbohydrate, and possibly protein requirements. During a bout of aerobic exercise, leucine oxidation can increase several-fold in proportion to exercise intensity,[87] there is a net release of amino acids and ammonia from muscles,[88] and urinary nitrogen increases immediately after exercise.[87] These observations have suggested that endurance athletes may have increased need for protein or individual amino acids.[89] However, it appears that a diet providing 12 to 15 per cent of the total energy as protein (1.0–1.5g/kg BW) is sufficient to support any increased requirement as long as energy needs are met.[31]

Protein provided 10 to 13 per cent of the triathletes' total energy, while carbohydrate contributed from 50 to 67 per cent (Table 6). Fat intake tended to meet the criteria for a prudent diet, with all but one group consuming 30 per cent or less of their calories as fat.

Gymnasts

Reported energy intakes for female gymnasts range from 6498 to 8101 kJ/day (1552–1935 kcal/day) (Table 7), with protein intakes

Table 7 Energy intakes and calorie distribution for gymnasts

Reference	N	Age	Energy intake		Carbohydrate (%)	Protein (%)	Fat (%)
			kJ (kcal)	kJ/kg (kcal/kg)			
Males							
68	10	—	8709 (2080)	—	44.0	15.0	39.0
Females							
62	10	16[b]	8101 (1935)	182.1 (43.5)[b]	49.0	15.0	36.0
90	26	12	6498 (1552)	179.2 (42.8)	47.7	15.3	36.0
12[a]	11	15	—	158.0 (37.7)	52.0	—	32.0
91	97	13	7695 (1838)	—	49.0	15.0	36.0
92	9	19	7649 (1827)	134.0 (32.0)	—	—	—
93	13	15	8051 (1923)	—	46.1	15.4	38.3
94	22	11–14	7143 (1706)	—	52.7	15.0	32.5
	29	7–10	6912 (1651)	—	52.3	15.9	32.1

[a]Calorie distribution data were estimated from information presented in a different format by the original authors.
[b]Unpublished data.

Table 8 Energy intakes and calorie distribution for dancers and figure skaters

Reference	N	Age	Energy intake		Carbohydrate (%)	Protein (%)	Fat (%)
			kJ (kcal)	kJ/kg (kcal/kg)			
Dancers							
Males							
101	10	26	12 422 (2967)	—	38.4	16.5	42.1
Females							
102	14	24	8328 (1989)	142.4 (34.0)	48.4	15.1	35.9
95	34	22	5686 (1358)	—	—	—	—
103	21	21	7432 (1775)	—	—	—	—
68	9	—	7993 (1909)	—	52.0	17.0	31.0
101	12	24	7005 (1673)	—	50.1	14.2	38.0
Figure skaters							
Males							
62	15	18[a]	11 137 (2660)	184.2 (44.0)[a]	47.0	17.0	33.4
104	17	21	12 129 (2897)	—	—	—	—
Females							
62	29	15[a]	7574 (1809)	167.1 (39.9)[a]	52.0	15.0	33.2
104	23	18	4915 (1174)	—	—	—	—

[a]Unpublished data.

remaining similar (15 per cent) for all groups of gymnasts. All groups obtained less than 55 per cent of their energy from carbohydrate, and fat intakes ranged from 32 to 39 per cent.

A number of factors can affect the energy intake of young athletes. According to Calabrese et al.[95] females are often concerned about their appearance and reduce food intake in an effort to reduce body fat. Severely restricting energy intake can result in failure to grow, impaired maturation, glycogen depletion, and fatigue.[96,97] As energy intake decreases, so does nutrient intake.[19,98,99]

The methods used to assess dietary intake may underestimate actual energy values.[4,15,100] Nevertheless, even though a wide range of energy intakes may exist within a group, it is striking that the energy intakes of female gymnasts are consistently low, particularly when considered as total daily intake. Considering data as kilocalories per kilogram body weight reveals a more positive representation.

Dancers and figure skaters

Dietary surveys of female dancers and figure skaters generally indicate low energy intakes. As shown in Table 8, reported energy intakes ranged from 4915 kJ/day to 8328 kJ/day (1174–1989 kcal/day). As with female gymnasts, intakes of female dancers and figure skaters are less than the United States recommended dietary allowance of 9211 kJ (2200 kcal). However, on a kilojoules per kilogram body weight basis, the energy intake of elite female figure skaters studied by Grandjean[62] exceeded the 167 kJ/kg BW (40 kcal/kg BW) average energy allowance cited by the National Research Council for the 15-year-old reference female weighing 55 kg (Table 1). The figure skaters in Grandjean's study averaged 15 years and 46.8 kg.

Dahlstrom et al.[102] reported the highest mean energy intake for female dancers (8328 kJ) (1989 kcal). However, on a kilojoules per kilogram body weight basis, the energy intake of these Swedish dancers (142 kJ/kg) (34 kcal/kg) was lower than their predicted energy requirement. Dahlstrom and coworkers estimated the energy requirement to be 10 287 kJ/day (2457 kcal/day or 176

kJ/kg (42 kcal/kg), using two different methods of calculation (Appendix).

Some researchers[95,101] have raised the question as to why dancers are not consistently losing weight if reported energy intakes truly reflect day-to-day eating habits. One explanation may be the high incidence of binge eating.[95] Other contributory factors include the decrease in resting metabolic rate that reportedly occurs in response to energy restriction[105] and, as mentioned previously, possible errors in methods available for calculating energy expenditure and/or intake.[102]

Limited data are available on the energy distribution for figure skaters and dancers. On the data reported, the most remarkable finding is the relatively high fat intake of male dancers in the study by Cohen et al.[101] However, their mean intake of 42 per cent of energy from fat is not unlike the intakes of several groups of athletes reported herein. The female dancers studied by Short and Short[68] had the highest mean protein intake (17 per cent of total energy) and the lowest fat intake (31 per cent) compared with other female dancers and figure skaters.

American football and basketball players

The average energy intake and percentage of energy from protein, carbohydrate, and fat for American football players and male and female basketball players are presented in Table 9. The energy intakes for football players ranged from 15 043 to 20 319 kJ/day (3593–4853 kcal/day). The protein intakes, as a percentage of energy, were almost identical (16 per cent) for all groups of football players with the exception of the linemen studied by Hickson et al.[106] who averaged 22 per cent of energy from protein. Meat was a major component in the diets of Hickson's football players, providing 33 per cent of the total energy, 63 per cent of the protein, and 45 per cent of the total fat intake.

Of the studies on basketball players, both the males and females surveyed by Short and Short[68] had the highest mean energy intakes reported with 23 237 and 13 565 kJ/day (5550 and 3240 kcal/day) respectively. The mean energy intakes of female basketball players reported by Nowak et al.[107] and Hickson et al.[92] were 7243 kJ/day

Table 9 Energy intakes and calorie distribution for American football and basketball players

Reference	N	Age	Energy intake		Carbohydrate (%)	Protein (%)	Fat (%)
			kJ (kcal)	kJ/kg (kcal/kg)			
American football (males)							
Linemen							
62	25	19[a]	16 584 (3961)	162.0 (38.7)[a]	48.0	16.3	36.0
106	11	19	15 043 (3593)	—	39.2	22.0	39.0
68	23	—	20 193 (4823)	—	43.0	16.0	41.0
Non-linemen							
62	30	19[a]	16 019 (3826)	194.3 (46.4)[a]	45.0	16.3	39.0
68	33	—	20 319 (4853)	—	44.0	16.0	38.0
Basketball							
Males							
62	11	21[a]	17 065 (4076)	188.4 (45.0)[a]	44.0	15.4	41.0
107	16	19	14 897 (3558)	—	48.0	17.0	34.0
68	13	—	23 237 (5550)	—	42.0	15.0	41.0
Females							
92	13	19	8 353 (1995)	125.6 (30.0)	—	—	—
107	10	19	7 243 (1730)	—	52.0	16.0	32.0
68	9	—	13 565 (3240)	—	46.0	14.0	40.0

[a]Unpublished data.

and 8353 kJ/day (1730 kcal/day and 1995 kcal/day) respectively, significantly less than subjects in the study by Short and Short.[68] The basketball players in the study by Hickson *et al.* averaged heights of 175 cm and weights of 68.3 kg, and data were collected during the competitive season with regular training, although dietary data were not collected on the days before, of, or following competition. In the study by Nowak *et al.* the female basketball players averaged heights of 178 cm and weights of 71.7 kg, and their training schedule was limited at the time of data collection.

Weight-lifters

The energy and macronutrient intakes of weight-lifters are shown in Table 10. The elite weight-lifters studied by Chen *et al.*[108] had the highest mean energy intake, averaging 238.6 kJ/kg BW (57 kcal/kg BW). Their percentages of energy from protein, carbohydrate, and fat were 22 per cent, 38 per cent, and 40 per cent respectively.

Limited data are available regarding the eating habits of elite weight-lifters. The perception is that weight-lifters consume large amounts of protein. The data reviewed here do not consistently support that impression. Protein intakes for these athletes ranged from 18 to 22 per cent of total energy, similar to the intakes of American football players.

The protein requirements of athletes in general, and particularly for athletes in sports such as weight-lifting, are controversial. The protein intake recommended by the International Weight-lifting Federation is 2.0 g/kg BW. Many nutritionists feel that this amount is excessive because (1) high protein intakes increase the 'work load' of the kidney,[109] (2) diets containing animal products are often associated with higher fat intakes, and (3) high protein diets may increase calcium loss.[110] However, there are not enough data on athletes to support these viewpoints, although many nutritionists agree that an intake greater than the recommended dietary allowance is appropriate for athletes consuming plant-based diets, endurance athletes, and athletes during periods of heavy weight training.[32,62,81,89]

Recent data suggest that the protein requirements of strength-trained subjects may be slightly higher than those of sedentary individuals.[111] Strength training involves powerful intermittent concentric contractions which could increase protein breakdown if prolonged.[112] Because strength-trained athletes generally con-

Table 10 Energy intakes and calorie distribution for weight-lifters and wrestlers

Reference	N	Age	Energy intake		Carbohydrate (%)	Protein (%)	Fat (%)
			kJ (kcal)	kJ/kg (kcal/kg)			
Weight-lifters (males)							
108[a]	10	21	19 247 (4597)	238.6 (57.0)	38.0	22.0	40.0
62	28	23[b]	15 253 (3643)	178.4 (42.6)[b]	43.0	18.0	39.0
12[c]	7	27	—	167.0 (39.9)	40.0	—	38.0
Wrestlers (males)							
62	10	21[b]	9018 (2154)	118.1 (28.2)[b]	54.0	12.0	34.0
68	7	—	16 098 (3845)	—	53.0	14.0	29.0

[a]Elite weight-lifters
[b]Unpublished data.
[c]Calorie distribution data were estimated from information presented in a different format by the original authors.

sume adequate energy, the increased need for protein is usually covered by a normal diet. However, further studies are necessary to determine the protein needs of weight-lifters who are attempting to lose weight, increase weight, do not consume adequate energy, and/or consume a plant-based diet.

Wrestlers

The minimum amount of energy required per day for wrestlers depends on body weight, surface area, and activity level.[113] As shown in Table 10, the mean energy consumption for wrestlers in Grandjean's study[62] was 118.1 kJ/kg BW (28.2 kcal/kg BW) or 9018 kJ/day (2154 kcal/day). As was indicated earlier, data for 90 per cent of the wrestlers were collected during the competitive season and therefore reflect the dietary weight control practices of the subjects. In contrast, Short and Short[68] reported a higher mean energy intake during the competitive season (16 098 kJ/day (3845 kcal/day). However, the 1-day intakes of wrestlers in the study by Short and Short varied from 327 (kJ/day (78 kcal/day) to more than 46 055 kJ/day (11 000 kcal/day) for the same athlete. Although Short and Short collected their data during the competitive season and also included wrestlers who were making weight, the data may have been biased in view of the fact that, prior to the subjects recording their diets, 'the senior author spoke to the whole team at practice session concerning good nutritional practices in relation to physical performance and answered questions about weight control and nutritional supplements'.

Recently, the metabolic effects of repeated weight loss and regain have been studied in adolescent wrestlers.[114] Twenty-seven wrestlers were classified as weight-cyclers (reduced weight 10 times or more during the season) or non-weight-cyclers. Resting metabolic rate was measured by indirect calorimetry, and body composition was evaluated using skin folds. The weight-cyclers had a significantly lower mean resting metabolic rate than the non-weight-cyclers (154.6 kJ/m².h versus 177.2 kJ/m².h ($p < 0.05$). However, consideration must be given to the limitation of a single resting metabolic rate measurement. Interindividual variation, or what is 'normal' for a given individual, can be determined only by serial measurements.

It has been shown that the reduction in resting metabolic rate and total energy requirements appears to be temporary and returns to normal with feeding and weight regain.[115] Therefore, when considering the energy intake of wrestlers, data would be more meaningful if they included weight loss practices at the time of and preceding data collection, the number of weeks that the subjects have been hyper-, hypo-, or isocaloric, the amount of weight lost as a percentage of body weight, and information regarding the competitive season (preseason, during season, post-season).

CONCLUSION

Under the best of circumstances, there is inherent error in assessing energy intake with one point of possible error being the data collection phase. However, many athletes have been found to be impeccable record-keepers, probably for a variety of reasons, including their high level of self-discipline, their familiarity in record-keeping (such as training logs), and their substantial interest in knowing the quality of their diet.

Even if data collection is error free, comparisons of the reported energy intakes of athletes is difficult owing to differences in research design, sample size, gender, body size, level of competition, and manner in which the data are reported. Even with these limitations, some generalization can and should be made in an effort to add to the growing body of knowledge regarding the dietary habits of athletes. In making comparisons, however, it is important to remember that athletes' energy intakes also vary with the stage of training, intensity of effort, and weight control practices, in addition to numerous other factors.

On the basis of studies comparing estimated energy expenditure with reported intakes, it has been suggested that athletes are under-reporting intakes. If this is true, then it must be kept in mind that the calculated distribution of energy has also been skewed. The studies summarized in this chapter revealed a great variance in the mean percentage of energy for protein, carbohydrate, and fat. Existing data indicate that the diets of some athletes approximate the recommendations made for the promotion of health while others do not.

The ideal diet for an individual athlete, or even for a given sport, still eludes us. Broad recommendations for protein, carbohydrate, and fat are made for all athletes. However, the data that serve as the basis for these recommendations have been, for the most part, conducted on subjects performing endurance-type exercise. For example, it has been recommended that to maintain glycogen stores during repeated days of training, athletes should consume at least 8 to 10 g carbohydrate/kg BW.day.[116] Under conditions of extreme endurance exercise, it is suggested that 12 to 13 carbohydrate g/kg BW/day be consumed.[80] However, it is not known whether athletes consuming 5 to 8 g of dietary carbohydrate will have impaired performance, and dietary recommendations for short-term muscular activity have not been established.

A protein intake of 1.0 to 1.5 g/kg.day is recommended for most athletes. This level can be provided by a mixed diet with 12 to 15 per cent of its energy contributed by protein, assuming that adequate energy is consumed. Energy intake and biological value of the protein are primary factors influencing protein requirements. Weight-conscious athletes (e.g. wrestlers, figure skaters, gymnasts) and vegetarians may require a more protein-dense diet to cover needs.

Information acquired from dietary surveys should allow researchers to draw conclusions about the eating habits of athletes and the effect of diet on sports performance. However, this is not possible if the methodologies used to measure these factors are not valid and consistent. Some of the inherent problems can be overcome by (1) using larger sample sizes,[117] (2) collecting data at more frequent intervals over longer time periods,[60] and (3) considering the level of athlete, the type of sport, and the training season. Studies assessing the food intake of female athletes should account for menstrual cycle variations. Weight control practices should also be considered, and, more importantly, researchers need to collect, analyse, and report data in a consistent manner. Only when a substantial pool of such data accumulates will more accurate profiles emerge and allow for more specific recommendations to be made.

APPENDIX ESTIMATION OF THE DAILY ENERGY REQUIREMENT IN FEMALE DANCERS

Method 1 The energy requirement of the dancers was estimated using the formula recommended by FAO/WHO/UNU:[1]

$$E_{req} = (0.0621 \times \text{body weight} + 2.0357) \times 1.82$$

where the factor 1.82 is a standard factor for subjects with high habitual physical activity.

Method 2 The energy requirement of the dancers was also estimated from indirect energy expenditure measurement during training sessions ($E_{\text{exp train}}$), which was added to the energy requirement for the remainder of the 24-h measuring period

$$E_{\text{req}} = \frac{(0.0621 \times \text{body weight} + 2.0357) \times 1.56 \times 21.35}{24} + E_{\text{exp train}}$$

where the factor 1.56 is a standard factor for subjects with 'normal' habitual physical activity.

Adapted from Dahlstrom *et al.*[102]

REFERENCES

1. World Health Organization. *Energy and protein requirements*. Report of Joint FAO/WHO/UNU Expert Consultation. Technical Report Series 724. Geneva: World Health Organization, 1985.
2. Stern, JS, Grivetti L, Castonguay TW. Energy intake: uses and misuses. *International Journal of Obesity* 1984; 8: 535–41.
3. Block G. A review of validations of dietary assessment methods. *American Journal of Epidemiology* 1982; 115: 492–505.
4. Pekkarinen M. Methodology in the collection of food consumption data. In: Bourne GH, ed. *World review of nutrition and dietetics*, Vol. 12. Basel: S. Karger, 1970, pp 145–71.
5. Fanelli MT, Stevenhagen KJ. Consistency of energy and nutrient intakes of older adults: 24-hour recall vs. 1-day food record. *Journal of the American Dietetic Association* 1986; 86: 665–7.
6. Karvetti RL, Knuts LR. Validity of the 24-hour dietary recall. *Journal of the American Dietetic Association* 1985; 85: 1437–42.
7. Beaton GH, Milner J, McGuire V, Feather TE, Little JA. Source of variance in 24-hour dietary recall data: implications for nutrition study design and interpretation. Carbohydrate sources, vitamins, and minerals. *American Journal of Clinical Nutrition* 1983; 37: 986–95.
8. Gersovitz M, Madden JP, Smiciklas-Wright H. Validity of the 24-hr. dietary recall and seven-day record for group comparisons. *Journal of the American Dietetic Association* 1978; 73: 48–55.
9. Todd KS, Hudes M, Calloway DH. Food intake measurement: problems and approaches. *American Journal of Clinical Nutrition* 1983; 37: 139–46.
10. Acheson KJ, Campbell IT, Edholm OG, Miller DS, Stock MJ. The measurement of food and energy intake in man—an evaluation of some techniques. *American Journal of Clinical Nutrition* 1980; 33: 1147–54.
11. Carter RL, Sharbaugh CO, Stapell CA. Reliability and validity of the 24-hour recall. *Journal of the American Dietetic Association* 1981; 79: 542–7.
12. van Erp-Baart AMJ, Saris WHM, Binkhorst RA, Vos JA, Elvers JWH. Nationwide survey on nutritional habits in elite athletes. Part 1. Energy, carbohydrate, protein, and fat intake. *International Journal of Sports Medicine* 1989; 10: S3–10.
13. Abramson JH, Slome C, Kosovsky C. Food frequency interview as an epidemiological tool. *American Journal of Public Health* 1963; 53: 1093–101.
14. Bergman EA, Boyungs JC, Erickson ML. Comparison of a food frequency questionnaire and a 3-day diet record. *Journal of the American Dietetic Association* 1990; 90: 1431–3.
15. Marr JW. Individual dietary surveys: purposes and methods. In: Bourne GH, ed. *World review of nutrition and dietetics*, vol. 13. Basel: Karger, 1971: 105–64.
16. Kim WW, Mertz W, Judd JT, Marshall MW, Kelsay JL, Prather ES. Effect of making duplicate food collections on nutrient intakes calculated from diet records. *American Journal of Clinical Nutrition* 1984; 40: 1333–7.
17. Woo R, Garrow JS, Pi-Sunyer FX. Effect of exercise on spontaneous calorie intake in obesity. *American Journal of Clinical Nutrition* 1982; 36: 470–7.
18. Woo R, Garrow JS, Pi-Sunyer FX. Voluntary food intake during prolonged exercise in obese women. *American Journal of Clinical Nutrition* 1982; 36: 478–84.
19. Perron M, Endres J. Knowledge, attitudes, and dietary practices of female athletes. *Journal of the American Dietetic Association* 1985; 85: 573–6.
20. Werblow JA, Fox HM, Henneman A. Nutritional knowledge, attitudes, and food patterns of women athletes. *Journal of the American Dietetic Association* 1978; 73: 242–5.
21. Schoeller DA. How accurate is self-reported dietary energy intake? *Nutrition Reviews* 1990; 48: 373–9.
22. Schulz LO, Alger S, Harper I, Wilmore JH, Ravussin E. Energy expenditure of elite female runners measured by respiratory chamber and doubly labeled water. *Journal of Applied Physiology* 1992; 72: 23–8.
23. Westerterp KR, Saris WHM, van Es M, ten Hoor F. Use of the doubly labeled water technique in humans during heavy sustained exercise. *Journal of Applied Physiology* 1986; 61: 2162–7.
24. Wilmore JH *et al*. Is there energy conservation in amenorrheic compared with eumenorrheic distance runners? *Journal of Applied Physiology* 1992; 72: 15–22.
25. Gong EJ, Garrel D, Calloway DH. Menstrual cycle and voluntary food intake. *American Journal of Clinical Nutrition* 1989; 49: 252–8.
26. Tarasuk V, Beaton GH. Menstrual-cycle patterns in energy and macronutrient intake. *American Journal of Clinical Nutrition* 1991; 53: 442–7.
27. Tai M, Castillo P, Young L, Pi-Sunyer FX. Resting metabolic rate during four phases of the menstrual cycle. *American Journal of Clinical Nutrition* 1992; 56: 771 (Abstract).
28. National Research Council. *Recommended Dietary Allowances*, 10th edn. Subcommittee on the Tenth Edition of the recommended dietary allowances, Food and Nutrition Board, Commission on Life Sciences. Washington, DC: National Academy of Sciences, 1989.
29. Åstrand PO, Rodahl K., eds. *Textbook of work physiology*. New York: McGraw Hill, 1970.
30. Costill DL. Carbohydrate nutrition before, during, and after exercise. *Federation Proceedings* 1985; 44: 364–8.
31. Leaf A, Frisa KB. Eating for health or for athletic performance? *American Journal of Clinical Nutrition* 1989; 49: 1066–9.
32. Lemon PWR, Yarasheski KE, Dolny DG. The importance of protein for athletes. *Sports Medicine* 1984; 1: 474–84.
33. Sherman WM. Carbohydrates, muscle glycogen and muscle glycogen supercompensation. In: Williams MH, ed. *Ergogenic aids in sport*. Champaign, IL: Human Kinetics, 1983; 3–26.
34. Essen B. Glycogen depletion of different fibre types in human skeletal muscle during intermittent and continuous exercise. *Acta Physiologica Scandinavica* 1978; 103: 446–55.
35. Essen B, Hagenfeldt L,. Kaijser L. Utilization of blood-borne and intramuscular substrates during continuous and intermittent exercise in man. *Journal of Physiology* 1977; 265: 489–506.
36. Symons JD, Jacobs I. High-intensity exercise performance is not impaired by low intramuscular glycogen. *Medicine and Science in Sports and Exercise* 1989; 21: 550–7.
37. Poehlman ET, Horton ES. The impact of food intake and exercise on energy expenditure. *Nutrition Reviews* 1989; 47: 129–37.
38. Pellett PL. Food energy requirements in humans. *American Journal of Clinical Nutrition* 1990; 51: 711–22.
39. Devlin JT, Horton ES. Energy requirements. In: Brown ML, ed. *Present knowledge in nutrition*, 6th edn. Washington, DC: The Nutrition Foundation, 1990: 1–6.
40. Horton ES. Introduction: an overview of the assessment and regulation of energy balance in humans. *American Journal of Clinical Nutrition* 1983; 38: 972–7.
41. Bogardus C *et al*. Familial dependence of the resting metabolic rate. *New England Journal of Medicine* 1986; 315: 96–100.
42. Garrow JS. *Energy balance and obesity in man*. Amsterdam: North-Holland, 1974: 136.
43. Jequier E, Schutz Y. Long-term measurements of energy expenditure in humans using a respiration chamber. *American Journal of Clinical Nutrition* 1983; 38: 989–98.

44. Ravussin E, Lillioja S, Anderson TE, Christin L, Bogardus C. Determinants of 24-hour energy expenditure in man. *Journal of Clinical Investigation* 1986; **78**: 1568–78.

45. Lennon D, Nagle F, Stratman F, Shrago E, Dennis S. Diet and exercise training effects on resting metabolic rate. *International Journal of Obesity* 1985; **9**: 39–47.

46. Poehlman ET, Melby CL, Badylak SF, Calles J. Aerobic fitness and resting energy expenditure in young adult males. *Metabolism* 1989; **38**: 85–90.

47. Poehlman ET, Melby CL, Badylak SF. Resting metabolic rate and postprandial thermogenesis in highly trained and untrained males. *American Journal of Clinical Nutrition* 1988; **47**: 793–8.

48. Tremblay A, Fontaine E, Poehlman ET, Mitchell D, Perron L, Bouchard C. The effect of exercise-training on resting metabolic rate in lean and moderately obese individuals. *International Journal of Obesity* 1986; **10**: 511–17.

49. Tremblay A, Fontaine E, Nadeau A. Contribution of postexercise increment in glucose storage to variations in glucose-induced thermogenesis in endurance athletes. *Canadian Journal of Physiology and Pharmacology* 1985; **63**: 1165–9.

50. LeBlanc J, Diamond P, Cote, J, Labrie, A. Hormonal factors in reduced postprandial heat production of exercise-trained subjects. *Journal of Applied Physiology* 1984; **56**: 772–6.

51. Owen OE et al. A reappraisal of caloric requirements in healthy women. *American Journal of Clinical Nutrition* 1986; **44**: 1–19.

52. Freedman-Akabas S, Colt E, Kissileff HR, Pi-Sunyer FX. Lack of sustained increase in V_{O_2} following exercise in fit and unfit subjects. *American Journal of Clinical Nutrition* 1985; **41**: 545–9.

53. Tremblay A, Nadeau A, Despres JP, St-Jean L, Theriault G, Bouchard C. Long-term exercise training with constant energy intake. 2: Effect on glucose metabolism and resting energy expenditure. *International Journal of Obesity* 1990; **14**: 75–84.

54. Tremblay A, Cote J, LeBlanc J. Diminished dietary thermogenesis in exercise-trained human subjects. *European Journal of Applied Physiology* 1983; **52**: 1–4.

55. Davis JR, Tagliaferro AR, Kertzer R, Gerardo T, Nichols J, Wheeler J. Variations in dietary-induced thermogenesis and body fatness with aerobic capacity. *European Journal of Applied Physiology* 1983; **50**: 319–29.

56. Hill JO, Heymsfield SB, McManus C, DiGirolamo M. Meal size and thermic response to food in male subjects as a function of maximum aerobic capacity. *Metabolism* 1984; **33**: 743–9.

57. Poehlman ET, Despres JP, Bessette H, Fontaine E, Tremblay A, Bouchard C. Influence of caffeine on the resting metabolic rate of exercise-trained and inactive subjects. *Medicine and Science in Sports and Exercise* 1985; **17**: 689–94.

58. Hickson JF, Schrader JW, Pivarnik JM, Stockton JE. Nutritional intake from food sources of soccer athletes during two stages of training. *Nutrition Reports International* 1986; **34**: 85–91.

59. Ellsworth NM, Hewitt BF, Haskell WL. Nutrient intake of elite male and female Nordic skiers. *Physician and Sportsmedicine* 1985; **13**: 78–92.

60. Nutter J. Seasonal changes in female athletes' diets. *International Journal of Sport Nutrition* 1991; **1**: 395–407.

61. Berning JR, Troup JP, VanHandel PJ, Daniels J, Daniels N. The nutritional habits of young adolescent swimmers. *International Journal of Sport Nutrition* 1991; **1**: 240–8.

62. Grandjean AC. Macronutrient intake of US athletes compared with the general population and recommendations made for athletes. *American Journal of Clinical Nutrition* 1989; **49**: 1070–6.

63. Peters AJ, Dressendorfer RH, Rimar J, Keen CL. Diets of endurance runners competing in a 20-day road race. *Physician and Sportsmedicine* 1986; **14**: 63–70.

64. Fogelholm M. Estimated energy expenditure, diet and iron status of male Finnish endurance athletes: a cross-sectional study. *Scandinavian Journal of Sports Science* 1989; **11**: 59–63.

65. Blair SN, Ellsworth NM, Haskell WL, Stern MP, Farquhar JW, Wood PD. Comparison of nutrient intake in middle-aged men and women runners and controls. *Medicine and Science in Sports Exercise* 1981; **13**: 310–15.

66. Nieman DC, Butler JV, Pollett LM, Dietrich SJ, Lutz RD. Nutrient intake of marathon runners. *Journal of the American Dietetic Association* 1989; **89**: 1273–8.

67. Clement DB, Asmundson RC. Nutritional intake and hematological parameters in endurance runners. *Physician and Sportsmedicine* 1982; **10**: 37–43.

68. Short SH, Short WR. Four-year study of university athletes' dietary intake. *Journal of the American Dietetic Association* 1983; **82**: 632–45.

69. Grandjean AC. Unpublished data.

70. Deuster PA, Kyle SB, Moser PB, Vigersky RA, Singh A, Schoomaker EB. Nutritional survey of highly trained women runners. *American Journal of Clinical Nutrition* 1986; **44**: 954–62.

71. Manore, MM, Besenfelder PD, Wells CL, Carroll SS, Hooker SP. Nutrient intakes and iron status in female long-distance runners during training. *Journal of American Dietetic Association* 1989; **89**: 257–9.

72. Barr SI. Women, nutrition and exercise: a review of athletes' intakes and a discussion of energy balance in active women. *Progress in Food and Nutrition Science* 1987; **11**: 307–61.

73. Janssen GME, Graef CJJ, Saris WHM. Food intake and body composition in novice athletes during a training period to run a marathon. *International Journal of Sports Medicine* 1989; **10**: S17–21.

74. Barr SI. Energy and nutrient intakes of elite adolescent swimmers. *Journal of the Canadian Dietetic Association* 1989; **50**: 20–4.

75. Porcello LA. Dietary intakes of competitive college athletes. In: *Ross Symposium on Nutrient Utilization During Exercise*. Columbus, OH: Ross Laboratories, 1983: 125–30.

76. Vallieres F, Tremblay A, St-Jean L. Study of the energy balance and the nutritional status of highly trained female swimmers. *Nutrition Research* 1989; **9**: 699–708.

77. Tilgner SA, Schiller MR. Dietary intakes of female college athletes: the need for nutrition education. *Journal of the American Dietetic Association* 1989; **89**: 967–9.

78. Smith MP, Mendez J, Druckenmiller M, Kris-Etherton PM. Exercise intensity, dietary intake, and high-density lipoprotein cholesterol in young female competitive swimmers. *American Journal of Clinical Nutrition* 1982; **36**: 251–5.

79. Johnson A et al. Psychological, nutritional and physical status of Olympic road cyclists. *British Journal of Sports Medicine* 1985; **19**: 11–14.

80. Saris WHM, van Erp-Baart MA, Brouns F, Westerterp KR, ten Hoor F. Study on food intake and energy expenditure during extreme sustained exercise: the Tour de France. *International Journal of Sports Medicine* 1989; **10**: S26–31.

81. Brouns F et al. The effect of diet manipulation and repeated sustained exercise on nitrogen balance, a controlled Tour de France simulation study, Part 3. In: Brouns F, ed. *Food and fluid related aspects in highly trained athletes*. Haarlem: De Vrieseborch, 1988: 73–9.

82. Keith RE, O'Keeffe KA, Alt LA, Young KL. Dietary status of trained female cyclists. *Journal of the American Dietetic Association* 1989; **89**: 1620–3.

83. Applegate E. Nutritional concerns of the ultraendurance triathlete. *Medicine and Science in Sports and Exercise* 1989; **21**: S205–8.

84. Burke LM, Read RSD. Diet patterns of elite Australian male triathletes. *Physician and Sportsmedicine* 1987; **15**: 140–55.

85. Green DR, Gibbons C, O'Toole M, Hiller WBO. An evaluation of dietary intakes of triathletes: are recommended dietary allowances being met? *Journal of the American Dietetic Association* 1989; **89**: 1653–4.

86. Khoo CS, Rawson NE, Robinson ML, Stevenson RJ. Nutrient intake and eating habits of triathletes. *Annals of Sports Medicine* 1987; **3**: 144–50.

87. Millward DJ, Davies CTM, Halliday D, Wolman SL, Matthews D, Rennie M. Effect of exercise on protein metabolism in humans as explored with stable isotopes. *Federation Proceedings* 1982; **41**: 2686–91.

88. Felig P, Wahren J. Fuel homeostasis in exercise. *New England Journal of Medicine* 1975; **293**: 1078–84.

89. Evans WJ, Fisher EC, Hoerr RA, Young VR. Protein metabolism and endurance exercise. *Physician and Sportsmedicine* 1983; **11**: 63–72.

90. Reggiani E, Arras GB, Trabacca S, Senarega D, Chiodini G. Nutrition status and body composition of adolescent female gymnasts. *Journal of Sports Medicine and Physical Fitness* 1989; **29**: 285–8.

91. Loosli AR, Benson J, Gillien DM, Bourdet K. Nutrition habits and knowledge in competitive adolescent female gymnasts. *Physician and Sportsmedicine* 1986; **14**: 118–30.

92. Hickson JF, Schrader J, Trischler LC. Dietary intakes of female basketball and gymnastics athletes. *Journal of the American Dietetic Association* 1986; **86**: 251–3.

93. Moffatt RJ. Dietary status of elite female high school gymnasts: inadequacy of vitamin and mineral intake. *Journal of the American Dietetic Association* 1984; **84**: 1361–3.

94. Benardot D, Schwarz M, Heller DW. Nutrient intake in young, highly competitive gymnasts. *Journal of the American Dietetic Association* 1989; **89**: 401–3.

95. Calabrese LH *et al*. Menstrual abnormalities, nutritional patterns, and body composition in female classical ballet dancers. *Physician and Sportsmedicine* 1983; **11**: 86–98.

96. Askew EW. Role of fat metabolism in exercise. In: Hecker AL, ed. *Clinics in sports medicine*, Vol. 3. Philadelphia: WB Saunders, 1984: 605–21.

97. Pugliese MT, Lifshitz F, Grad G, Fort P, Marks-Katz M. Fear of obesity: a cause of short stature and delayed puberty. *New England Journal of Medicine* 1983; **309**: 513–17.

98. Grandjean AC. Profile of nutritional beliefs and practices of the elite athlete. In: Butts NK, Gushiken TT, Zarins B, eds. *The elite athlete*. Jamaica, NY: Spectrum, 1985: 239–47.

99. Welch PK, Zager KA, Endres J, Poon SW. Nutrition education, body composition, and dietary intake of female college athletes. *Physician and Sportsmedicine* 1987; **15**: 63–74.

100. Beaton GH *et al*. Sources of variance in 24 hour dietary recall data: implications for nutrition study design and interpretation. *American Journal of Clinical Nutrition* 1979; **32**: 2456–59.

101. Cohen JL, Potosnak L, Frank O, Baker H. A nutritional and hematologic assessment of elite ballet dancers. *Physician and Sportsmedicine* 1985; **13**: 43–54.

102. Dahlstrom M, Jansson E, Nordevang E, Kaijser L. Discrepancy between estimated energy intake and requirement in female dancers. *Clinical Physiology* 1990; **10**: 11–25.

103. Evers CL. Dietary intake and symptoms of anorexia nervosa in female university dancers. *Journal of the American Dietetic Association* 1987; **87**: 66–8.

104. Rucinski A. Relationship of body image and dietary intake of competitive ice skaters. *Journal of the American Dietetic Association* 1989; **89**: 98–100.

105. Bray GA. Effect of caloric restriction on energy expenditure in obese patients. *Lancet* 1969; August 23: 397–8.

106. Hickson JF, Wolinsky I, Pivarnik JM, Neuman EA, Itak JF, Stockton JE. Nutritional profile of football athletes eating from a training table. *Nutrition Research* 1987; 7: 27–34.

107. Nowak RK, Knudsen KS, Schulz LO. Body composition and nutrient intakes of college men and women basketball players. *Journal of the American Dietetic Association* 1988; **88**: 575–8.

108. Chen JD *et al*. Nutritional problems and measures in elite and amateur athletes. *American Journal of Clinical Nutrition* 1989; **49**:1084–9.

109. Brenner BM, Meyer TW, Hostetter TH. Dietary protein intake and the progressive nature of kidney disease: the role of hemodynamically mediated glomerular injury in the pathogenesis of progressive glomerular sclerosis in aging, renal ablation, and intrinsic renal disease. *New England Journal of Medicine* 1982; **307**: 652–9.

110. Zemel MB, Schuette SA, Hegsted M, Linkswiler HM. Role of the sulfur-containing amino acids in protein-induced hypercalciuria in men. *Journal of Nutrition* 1981; **111**: 545–52.

111. Tarnopolsky MA, MacDougall JD, Atkinson SA. Influence of protein intake and training status on nitrogen balance and lean body mass. *Journal of Applied Physiology* 1988; **64**: 187–93.

112. Meredith CN. Protein needs and protein supplements in strength-trained men. In: *Ross Symposium on Muscle Development: Nutritional Alternatives to Anabolic Steroids*. Columbus, OH: Ross Laboratories, 1988: 68–72.

113. Tipton CM. Making and maintaining weight for interscholastic wrestling. *Gatorade Sports Science Institute Sports Science Exchange* 1990; 2(22).

114. Steen SN, Oppliger RA, Brownell KD. Metabolic effects of repeated weight loss and regain in adolescent wrestlers. *Journal of the American Medical Association* 1988; **260**: 47–50.

115. Melby CL, Schmidt WD, Corrigan D. Resting metabolic rate in weight-cycling collegiate wrestlers compared with physically active, noncycling control subjects. *American Journal of Clinical Nutrition* 1990; **52**: 409–14.

116. Sherman WM, Wimer GS. Insufficient dietary carbohydrate during training: does it impair athletic performance? *International Journal of Sport Nutrition* 1991; 1: 28–44.

117. Errors in reporting habitual energy intake. *Nutrition Reviews* 1991; **49**: 215–17.

1.2.2 Diet and sports performance

CLYDE WILLIAMS

INTRODUCTION

Interest in the influences of food on the capacity for physical activity is as old as mankind. From earliest times, certain foods were regarded as essential preparation for physical confrontation, whether this was on the battlefields of history or in the stadia of ancient Greece.[1] More recently, this interest has extended to understanding the mechanisms by which the consumption of certain foods influences exercise capacity. These studies draw upon skills from a wide range of disciplines within the biological sciences. This scientific interest in the nutritional influences on sports performance is collectively described as 'sports nutrition'. From the practical point of view, sports nutrition can be regarded as the study of how to use commonly available foods to support sportspersons during their preparation for participation in and recovery from sport and exercise.

The contributions of sports nutrition are not limited to those on improving the performances of sportspersons. Nutritional studies of human exercise capacity also provide us with a better understanding of the mechanisms underpinning the digestion, absorption, and metabolism of foods. For example, the detrimental influence of dehydration on physical performance has prompted studies of the factors which play a significant role in determining the rate of gastric emptying and intestinal absorption of water.[2] As a result of the research interest in this topic we now have a much better understanding of how the composition of a fluid can influence its uptake.[3] This information not only helps sportspersons but also contributes to the parent discipline of gastroenterology.[4]

Although the aim of sports nutrition is to understand the influence of diet on sports performance, there are, unfortunately, only a few studies on the links between food intake and sports performance during actual competition. Therefore our knowledge about sports nutrition has as its basis the results of well-controlled laboratory studies of food intake and exercise performance. These laboratory studies can be divided into two broad categories: those that have assessed nutritional influences on endurance capacity, and those that have reported nutritional influences on endurance performance. Endurance capacity is described in terms of the time to exhaustion during exercise of constant intensity, whereas endurance performance is the time taken to complete a prescribed exercise task such as running an agreed distance.

Other than for distance running, there is not as yet an extensive sport-specific nutrition literature. Therefore the recommended nutritional strategies for athletes preparing for or recovering from sport and exercise are mainly derived from the results of laboratory studies. Before any recommendations can be made, however, it is necessary to have some idea about the relative energy demands of different sports. To this end, we can divide sports into two simple categories based on the pattern of energy expenditure of their participants. There are the endurance sports, such as long-distance running, swimming, cycling, walking, and cross-country skiing, and the multiple-sprint sports, such as football, hockey, tennis, basketball, and soccer. The multiple-sprint sports involve a mixture of brief periods of exercise of maximum intensity followed by recovery periods of rest or light activity.

Many sportspersons do not regard themselves as athletes either because, in Europe the term has a specific meaning and is therefore inappropriately applied or because they believe that they do not train frequently or hard enough to be regarded as athletes. Nevertheless, for the purpose of this chapter sportspersons will be referred to collectively as athletes. This chapter will deal with the influences of nutrition on exercise performance as a background to understanding the links between nutrition and sports performance.

DIET: SOME GENERAL CONSIDERATIONS

Before examining the evidence for the links between diet and exercise capacity it is worth making several general, though nevertheless important, points about the diet of athletes. The overall assumption is that athletes eat a well-balanced healthy diet which provides them with sufficient energy to cover their daily energy expenditure. Health professionals argue that a healthy diet is one which provides us with at least 50 per cent of our daily energy intake in the form of carbohydrates, 35 per cent or less from fats, and the remainder from proteins. The common message is that we should move from high fat meat-based diets to those that are made up of more carbohydrates and fresh fruits and vegetables.[5]

However, athletes should have diets which include more carbohydrate-containing foods than recommended by the health professionals. Their diets should be such that about 60 per cent of their daily energy intake is obtained from carbohydrates, 30 per cent or less from fat, and 10 to 15 per cent from proteins.[6]

Bread, potatoes, pasta, rice, and vegetables are often called complex carbohydrates, whereas foods that have a low fibre content and contain a significant proportion of simple sugars, such as glucose, sucrose, and fructose, are referred to as simple carbohydrates. The common assumption is that the simple carbohydrates increase blood glucose concentration rapidly after they are eaten, whereas the complex carbohydrates do not produce such a rapid rise. However, this is not the case with all simple carbohydrates, or indeed with all complex carbohydrates. A functional definition of carbohydrate-containing foods is one that describes their influence on blood glucose concentrations. The glycaemic index of food is one way of describing the extent to which blood glucose concentration changes after the food has been eaten.[7] The reference value of 100 is assigned to white bread and not, as might be expected, to the simple sugars. Carbohydrates which have a high glycaemic index are white bread, glucose, rice, sweet corn, and potatoes, whereas carbohydrates which have low glycaemic indices are apples, dates, peaches, fructose, and milk ice cream. The simple sugars glucose, sucrose, and fructose are not all high glycaemic carbohydrates; glucose has a value close to 100, whereas sucrose is about 86 and fructose is less than 60. The glycaemic index provides important information about a carbohydrate, which is particularly useful when designing diets to deliver glucose to working muscles rapidly.[8]

Although we are encouraged to eat less fat and more carbohydrates, we should not overlook the fact that fat plays several important roles in our diets. Fat is the body's most efficient fuel. It contributes 37 kJ (9 kcal) per gram to energy metabolism, whereas the equivalent value for carbohydrate is 17 kJ (4 kcal) and protein is 17 kJ (4 kcal). Fat is also a carrier of the fat-soluble vitamins A, D, E, and K, and provides the essential fatty acids linolenic acid and linoleic acid. These are essential because they form integral parts of cell membranes, particularly the membranes of nerve cells.

Many athletes share the same beliefs as the first Olympians about the beneficial values of high protein diets. There is a long held, but misguided, belief that athletes who want to develop strength must consume large quantities of meat. The World Health Organization (WHO)[9] recommends a daily protein intake of 1 g/kg body weight (BW). However, the protein intakes of even female endurance athletes, who are not normally preoccupied with gaining strength, are about 1.5 g/kg BW/day, and so they are well above the WHO recommendations. However, these recommendations are too low for athletes involved in prolonged heavy training. These athletes should eat the equivalent of 1.2 to 1.7 g protein/kg BW/day.[10]

Dietary protein provides us with 20 amino acids from which all the cells in the body and their structural contents are created. They are the body's building blocks which contribute to the repair and reproduction of tissues; they also contribute to physiological regulation by the formation of hormones and enzymes. Of the 20 amino acids, eight must be provided by our diet because they cannot be manufactured by our own metabolism. High protein foods such as meat, eggs, and fish contain all the essential amino acids and are therefore regarded as being of high biological value. Cereals and vegetables contain only some of the essential amino acids and so are of low biological value. Combining several of the low biological foods to make a meal is a strategy for overcoming the limited number of essential amino acids available from vegetables and cereals. This mutual supplementation of dietary protein is an effective way of obtaining a full complement of essential amino acids without contributions from meat and meat products. A well-balanced vegetarian diet does not impair endurance capacity in particular and physical fitness in general. Kiens[11] has recently shown that the endurance capacity of non-vegetarians was no different after they had consumed a lacto-ovovegetarian diet for 6 weeks than before changing their diet. Although there were no signs of iron deficiency after 6 weeks on a vegetarian diet, it is worth remembering that vegetable non-haem iron is less readily absorbed than animal haem iron. The large fibre intake which is characteristic of the diets of vegetarians may influence the absorption of some essential nutrients such as iron and calcium. Therefore the iron and calcium status of vegetarians, particularly women, should be checked during routine health screening.

In sports which have well-defined weight categories, athletes often fall into the trap of trying to lose weight by reducing their food intake while at the same time trying to train hard.[12] The usual consequence of this approach to 'making weight' is a greater loss of exercise capacity than of body weight. A more successful approach is to maintain a constant food intake and gradually increase energy expenditure. A weekly weight loss of about 1 kg requires either an

Table 1 Carbohydrate exchanges

Amount (g)	Food description	Energy (kcal)	CHO (g)
High carbohydrate exchanges*			
20 (3 tbs)	Cornflakes	74	17
20 (5 tbs)	Rice Krispies	75	18
25 (2)	Ryvita/crispbread	80	18
100 (2 egg sized)	Boiled potatoes	80	20
100 (1 small)	Jacket potatoes	85	20
60 (heaped tbs)	Boiled rice	74	18
65	Boiled spaghetti	76	17
130 (⅓ tin)	Spaghetti in tomato sauce	77	16
100 (3 tbs)	Sweetcorn	76	16
130	Jelly (made-up)	77	18
25 (1 slice)	Swiss roll	76	16
24 (2)	Jaffa Cakes	88	16
Medium carbohydrate exchanges†			
20 (3 tbs)	Fruit 'n' Fibre	70	15
22 (1)	Shredded Wheat	72	15
20 (3 tbs)	Shreddies	66	15
30 (5 tbs)	Bran cereal	82	13
20 (heaped tbs)	Porridge oats (raw weight)	80	15
20 (heaped tbs)	Muesli	74	13
20 (3)	Cream crackers	88	14
20 (1)	Weetabix	68	14
20 (⅔ slice)	White bread	70	15
30 (1 slice)	Wholemeal bread	65	13
15 (1)	Sweet biscuits (e.g. digestives)	71	10
70 (large scoop)	Mashed potatoes	83	13
50 (1 small)	Roast potatoes	79	14
30 (3)	Chips	76	11
100 (3 tbs)	Processed peas	80	14
125 (3 tbs)	Baked beans	80	13
80 (2 tbs)	Cooked pulses (e.g. lentils)	76	14
75 (½ pot)	Fruit yoghurt (low fat)	72	14
80 (⅕ tin)	Canned rice pudding	73	12
35 (tbs)	Fruit crumble	73	13
20 (½ individual)	Fruit pie (individual)	74	11
20 (¼ slice)	Plain fruit cake	71	12
20 (1)	Jam tart (individual)	77	13
20 (3 heaped tsp)	Drinking chocolate/Horlicks	76	15
100 (3 tbs)	Frozen peas (boiled)	69	10

* The specific weights of all of these food items contain one high carbohydrate (CHO) exchange and are interchangeable.
† The specific weights of all of these food items contain one medium carbohydrate (CHO) exchange and are interchangeable.

Table 2 Meat exchanges

Amount (g)	Food description	Energy (kcal)	CHO (g)
50 (thick slice)	Lean beef (cooked weight)	78	0
35 (tbs)	Minced beef	80	0
35 (1 cube)	Stewed beef	78	0
30 (1 small)	Beefburgers	79	2
35 (thin slice)	Corned beef	76	0
60 (2 tbs)	Bolognese sauce	83	2
40 (thick slice)	Lean roast lamb	76	0
60 (cutlet)	Lean lamb chop	74	0
60 (small chop)	Lean pork chop	80	0
25 (1 thin)	Grilled pork sausage	80	3
60 (2 thin slices)	Lean ham	72	0
30 (1 rasher)	Grilled bacon (lean)	88	0
25 (2 tsps)	Liver pâte/sausage	78	1
35 (small slice)	Fried liver	81	1
25 (⅛ small tin)	Tuna fish in oil (drained)	73	0
67 (⅓ small tin)	Tuna fish in brine (drained)	80	0
80 (medium fillet)	Grilled white fish	76	0
50 (2)	Sardines in tomato sauce	86	0
20 (small chunk)	Cheddar cheese	81	0
75 (2 tbsp)	Cottage cheese	81	0
25 (small chunk)	Edam/Gouda cheese	85	0
30 (½)	Fried egg	70	0
60 (1)	Boiled egg	88	0
35 (½)	Scrambled egg	86	0
40 (⅔ egg)	Omelette	76	0
50 (1)	Poached egg	78	0
55 (large slice)	Roast chicken	83	0
55 (large slice)	Roast turkey	77	0

The specific weights of all of these food items contain one meat exchange and are interchangeable.

increase in daily energy expenditure of 4200 kJ (1000 kcal) or a reduction in energy intake by the same amount. Of course, a slower weight loss is the healthier option. To maintain a high training volume while trying to reduce body fat content, it is essential (a) to maintain a high carbohydrate intake and (b) to reduce the fat content of the diet but (c) without reducing the overall energy intake. This dietary manipulation can be achieved effectively by using the food exchange system. This concept is based on the allocation of food to six groups in which servings of different foods with the same energy value are listed (Tables 1–6). Those wishing to decrease their fat intake and increase their carbohydrate intake, for example, are directed to decrease the meat intake by cutting out some meat exchanges and replacing them with the equivalent number of carbohydrate exchanges. For example, the daily energy and nutrient intake of a young female distance runner whose diet has been altered to increase her carbohydrate intake is shown in Table 7. She was able to increase her carbohydrate intake, without changing her daily energy intake, simply by choosing different yet familiar foods with similar energy values. Whenever possible, dietary manipulation of this kind should be undertaken with the guidance of a sports dietitian, particularly if athletes are trying to lose weight in preparation for competition. There are also helpful books which help translate sports nutrition principles into meals.[13, 14]

Athletes who eat a wide variety of foods in sufficient quantity to cover their daily energy expenditures do not require vitamin and

Table 3 Fat exchanges

Amount (g)	Food description	Energy (kcal)	CHO (g)
10 (pat)	Margarine/butter	73	0
20 (2 pats)	Low fat spread (e.g. Gold®)	73	0
30 (3 pats)	Very low fat spread	81	0
20 (small chunk)	Cheddar cheese	81	0
20 (2 tsp)	Cream cheese	88	0
30 (large triangle)	Cheese spread	85	0
17 (small chunk)	Stilton cheese	79	0
35 (2 tbsp)	Single cream	74	1
12 (med)	French dressing	79	0
10 (heaped tsp)	Mayonnaise	72	0
15 (15)	Peanuts/other nuts	86	1
35 (¼)	Avocado pear	78	1

The specific weights of all the food items contain one fat exchange and are interchangeable.

mineral supplements. There is no good evidence to suggest that vitamin[15] and mineral[16] supplementation of a well-balanced diet improves performance. However, the diets of people who are trying to lose weight are, of necessity, low in energy. Low energy diets, even if they contain the recommended proportions of carbohydrate, fat, and protein, can lead to vitamin and mineral deficiency. Therefore those people who are physically active and on low energy diets are most at risk and should seek the advice of a sports dietitian. Of particular concern are sportswomen who reduce their energy intake and so run the risk of not consuming enough iron- and calcium-containing foods.

DIET AND ENDURANCE CAPACITY

Christensen and Hansen[17] were the first to explore the link between diet and exercise capacity systematically. In these studies the endurance capacity of a group of subjects was examined, on a cycle ergometer, after 3 to 4 days on a normal diet, a diet of fat and protein, and then a diet rich in carbohydrate. After the period on a high carbohydrate diet their endurance capacity doubled in comparison with the exercise times they achieved after consuming their normal mixed diets. In contrast, the fat and protein diet reduced their exercise capacity to almost half that achieved after their normal mixed diets. Thus these studies clearly showed the benefits of eating a high carbohydrate diet before prolonged exercise and were the first to establish the importance of the carbohydrate content in the diets of athletes preparing for competition.

Table 4 Milk exchanges

Amount (g)	Food description	Energy (kcal)	CHO (g)
150 (small glass)	Whole milk	98	7
200 (⅓ pint)	Semi-skimmed milk	98	10
250 (large glass)	Skimmed milk	83	13
75 (¼ carton)	Custard	88	13
50 (small scoop)	Ice cream	83	10
150 (small bowl)	'Cream of' … soup	80	9
150 (pot)	Yoghurt (natural)	84	11

The specific weights of all the food items contain one milk exchange and are interchangeable.

Table 5 Fruit exchanges

Amount (g)	Food description	Energy (kcal)	CHO (g)
125 (medium)	Apple (with core)	44	12
50 (½)	Banana (flesh only)	47	12
80 (16)	Grapes	48	12
140 (medium)	Orange (no skin)	49	12
200 (small)	Grapefruit (no skin)	44	11
200 (slice)	Honeydew melon	42	10
75 (2 tbsp)	Stewed fruit with sugar	50	13
150 (4 tbsp)	Stewed fruit, no sugar	48	12
50 (⅛ tin)	Tinned fruit in syrup	49	13
120 (wine glass)	Unsweetened orange juice	46	11
40 (medium)	Orange squash	43	11
70 (⅕ can)	Lucozade®	48	13
200 (medium glass)	Lemonade	42	11
100 (⅓ can)	Coca cola	39	11
20 (⅓ measure)	Ribena (concentrate)	46	12
15 (heaped tsp)	Marmalade or jam	39	10
15 (heaped tsp)	Honey	43	12
10 (2 tsp)	Sugar	40	11
20 (2 tbsp)	Raisins	50	12

The specific weights of all the food items contain one fruit exchange and are interchangeable.

A high carbohydrate diet increases the stores of liver and muscle glycogen. Glycogen is a polymer of glucose which is described in terms of the amount of glucose released during hydrolysis, i.e. as millimoles of glucosyl units per kilogram of muscle. For example, human skeletal muscle has a glycogen concentration in the range 60 to 150 mmol glucosyl units/kg wet weight (WW) or 258 to 645 mmol glucosyl units/kg dry weight (DW). Many studies report glycogen concentrations and those of the glycolytic intermediates as dry weight values because they freeze dry their muscle samples before analysis. Freeze-dried muscle samples are easier to handle and analyse than fresh muscle. On the basis that the water content of human skeletal muscle is approximately 77 per cent, wet weight concentrations can be converted to dry weight values by using a conversion factor of 4.3.

Table 6 Miscellaneous exchanges

Amount (g)	Food description	Exchanges
70 (1 mini)	Cheese and tomato pizza	1 medium CHO + 1 fat
25 (¼ slice)	Cheesecake	0.5 medium CHO + 1 fat
45 (½)	Doughnut	1 medium CHO + 1 fat
100 (large portion)	White sauce	1 medium CHO + 1 fat
50 (1 small)	Pancakes	1 medium CHO + 1 fat
75 (small slice)	Meat pie	1 medium CHO + 1 meat + 1 fat
75 (½ small)	Cornish pasty	1 medium CHO + 1 meat + 1 fat
25 (bag)	Potato crisps	1 medium CHO + 1 fat
20 (⅓ bar)	Mars® bar, Snickers®, etc.	1 medium CHO + 1 fat
32 (⅔ bar)	Chocolate bar	1 medium CHO + 1 fat

These foods contain more than one exchange.

Table 7 An example of changing the carbohydrate intake of a female athlete using a food exchanges system

	Normal diet	Recommended diet
Energy (kcal)	2130	2120
Protein (g (%))	96 (18)	94 (18)
Fat (g (%))	95 (40)	56 (24)
CHO (g (%))	241 (42)	(331 (58)
Alc (g (%))	0 (0)	0 (0)
Exchanges		
High CHO	7	8
Med CHO	5	5
Meat	6	6
Fruit	2	9
Fat	5	0
Milk	3	3

Foods removed from diet	
20 g (2 pats) butter	3 Fat exchanges
10 g (heaped tsp) mayonnaise	1 Fat exchange
60 g (3 small chunks) Cheddar	2 Fat + 1 meat exchange

Foods added to diet	
30 g (2 heaped tsp) jam	2 Fruit exchanges
24 g (2) jaffa cakes	1 High CHO exchange
75 g (2 heaped tbsp) cottage cheese	1 Meat exchange
120 g (wine glass) orange juice	1 Fruit exchange
200 (average glass) lemonade	1 Fruit exchange
150 g (2 small) bananas	3 Fruit exchanges

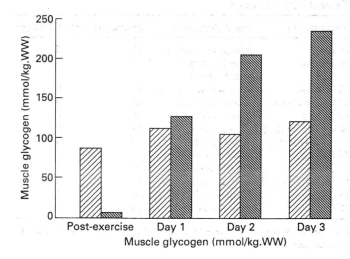

Fig. 1 Human glycogen concentrations before and after prolonged heavy single leg exercise in the active leg (dark shading) and the resting leg (light shading). After ref. 20.

The size of the liver glycogen store depends on whether the person is fed or fasted. When fed, an adult man, with a liver weighing about 1.8 kg, has a liver glycogen concentration of approximately 550 mmol glucosyl units WW, whereas after an overnight fast the concentration of glycogen falls to about 200 mmol WW. After a number of days on a high carbohydrate diet the liver glycogen concentration can increase to as much as 1000 mmol WW.[18] Interestingly, however, an overnight fast does not appear to lower human muscle glycogen concentration as is the case with liver glycogen.[19]

Bergstrom and Hultman[20] were first to show that eating a high carbohydrate diet in the days after exercise increased muscle glycogen stores. They used a modified form of the original Duchenne biopsy needle to obtain repeated samples of muscle to follow the changes in glycogen resynthesis after prolonged exercise. The experimental design was novel because it involved single leg exercise with the contralateral leg acting as a resting control. Bergstrom and Hultman were their own subjects, and used one leg for exercise while the other acted as a non-exercised control. They exercised to exhaustion by placing a cycle ergometer between them. Muscle samples were taken from the vastus laterali of the exercise and the control legs before exercise, immediately after exercise, and at 24-h intervals for 3 days. During the 3 days following the experiment, the two subjects consumed a carbohydrate-rich diet which amounted to about 550 to 650 g of carbohydrate a day. The muscle glycogen concentration in the exercised legs increased to values twice as high as those in the control legs during the 3-day recovery period (Fig. 1). This 'local phenomenon' of increased muscle glycogen concentration is described as 'glycogen supercompensation'.

Ahlborg et al.[21] demonstrated the link between endurance performance during cycle ergometry and pre-exercise muscle glycogen concentration. They found a correlation of 0.87 between initial muscle glycogen concentrations and endurance times for

nine subjects. The importance of muscle glycogen during prolonged exercise was also confirmed in subsequent studies which showed that fatigue occurs when muscle glycogen concentrations are reduced to low values.[22–25] Therefore it is not surprising that attempts were made to find methods of increasing muscle glycogen stores in preparation for prolonged exercise. To this end Bergstrom et al.[26] also examined the influence of different nutritional states on the resynthesis of glycogen during recovery from prolonged exhaustive exercise. In their study they found that a diet low in carbohydrate and high in fat and protein for 2 to 3 days after prolonged submaximal exercise produced a delayed muscle glycogen resynthesis, but when this was followed by a high carbohydrate diet for the same period of time glycogen supercompensation occurred (Fig. 2). This dietary manipulation not only increased the pre-exercise muscle glycogen concentration but also resulted in a significant improvement in endurance

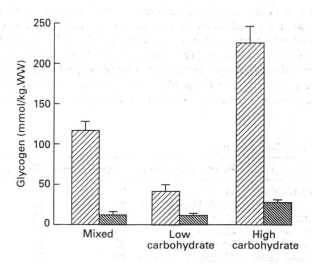

Fig. 2 Human muscle glycogen concentration before and immediately after prolonged submaximal cycling to exhaustion after a mixed diet, after 3 days on a diet low in carbohydrate and high in protein and fat, and after 3 days on a high carbohydrate diet (mean ± SD). After ref. 26.

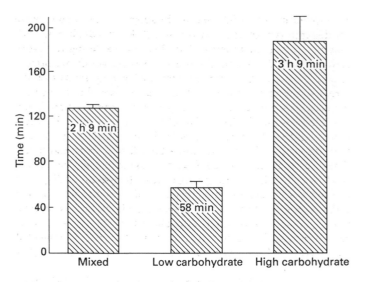

Fig. 3 Exercise time to exhaustion after a mixed diet, after 3 days on a diet low in carbohydrate and high in protein and fat, and after 3 days on a high carbohydrate diet (mean ± SD). After ref. 26.

capacity (Fig. 3). Although this original method of carbohydrate loading was recommended to athletes preparing for endurance activities,[27] the dietary procedure involving low carbohydrate with high fat and protein is an unpleasant experience which undermines the confidence of athletes before competition. Therefore Sherman et al.[28] explored alternative ways of increasing pre-exercise glycogen stores without including a period on a diet high in fat and protein. They found that a carbohydrate-rich diet consumed for 3 days prior to competition, accompanied by a decrease in training intensity, results in increased muscle glycogen concentrations of the same size as those achieved with the traditional carbohydrate-loading procedure.

The benefits of carbohydrate loading before prolonged submaximal exercise have been shown mainly during cycling. There are only a few studies of the influence of carbohydrate loading on running performance, and not all the available evidence supports the recommended dietary practice. Therefore it is worth considering the results of both the studies which have found that carbohydrate loading enhances performance and those that have not in order to establish whether there are good grounds for recommending this dietary practice to runners.

In one study, Goforth et al.[29] showed that the endurance time of nine well-trained runners increased from 120 to 130 min at treadmill speeds equivalent to 80 per cent V_{O_2}max. The dietary manipulation during the carbohydrate-loading procedure was well controlled. Isocaloricity of each phase of the carbohydrate loading was achieved using a combination of liquid and solid foods. The carbohydrate intake of the control diet was quite high and so this, together with the high relative exercise intensity (per cent V_{O_2}max, may explain why there was only a 9 per cent increase in endurance capacity. This increase in performance was quite modest in comparison with the improvements of 57 per cent reported in the cycle ergometer studies of the late 1960s.[30]

A clear benefit from increasing the carbohydrate content of runners' diets is achieved when the exercise intensity is less intense. Brewer et al.[31] modified the normal diets of 30 runners during the 3 days following a treadmill run to exhaustion at 70 per cent V_{O_2}max. They modified the runners' normal mixed diets by providing additional protein, complex carbohydrates, or simple carbohydrates. The complex carbohydrate group supplemented their normal mixed diet with bread, potatoes, rice, or pasta. The simple carbohydrate group ate their normal mixed diet but increased their carbohydrate intake with chocolates. After recovery for 3 days the runners completed a second treadmill time trial to exhaustion. Running times increased after both high carbohydrate diets. The complex carbohydrate group improved their running times by 26 per cent, and the simple carbohydrate group improved by 23 per cent. However, there was no improvement in the performance times of the protein group, confirming that the carbohydrate content of the diet is the important nutrient and that the changes were not simply the consequence of a greater energy intake.

Not all studies have shown this clear benefit in endurance capacity after dietary carbohydrate loading. Madsen et al.[32] reported that carbohydrate loading increased muscle glycogen concentrations in the gastrocnemius muscles of their subjects by 25 per cent. Their subjects ran to exhaustion on a level treadmill at speeds equivalent to 75 to 80 per cent V_{O_2}max on two occasions. The time to exhaustion on the first trial, after the subjects consumed their normal mixed diet, was 70 min, whereas after carbohydrate loading, using the method described by Sherman et al.,[28] they recorded 77 min. However, this 10 per cent improvement in endurance capacity was not statistically significant. Furthermore, these running times are considerably less than would be expected from well-trained runners. The amount of muscle glycogen used by the gastrocnemius muscles was the same in both trials. Furthermore, the average glycogen concentrations at exhaustion, after the control and carbohydrate trials, were almost the same (553 and 434 mmol glucosyl units/kg DW respectively). These values do not reach the very low levels normally associated with fatigue during cycling. Histochemical examination of samples from the gastrocnemius muscles of the runners did not show selective glycogen depletion in any of the type I or type II fibres. Therefore selective glycogen depletion in one or other of the fibre populations in the gastrocnemius muscles was probably not the cause of fatigue. Therefore it is difficult to offer reasons why these runners did not continue unless, of course, glycogen depletion occurred in the other muscle groups which also play a central role in running, such as the soleus and the quadriceps. The evidence that glycogen depletion in the gastrocnemius muscles is the main limitation to endurance running is not particularly strong. The original studies designed to answer the question about the involvement of different muscles in fatigue during running suffered from the small number of subjects in the original studies so that glycogen utilization could not be examined statistically.[33] Nevertheless, the running times of the well-trained subjects in the study by Madsen et al.[32] were too short to be the result of glycogen depletion. Perhaps they found the invasive nature of the study an obstacle to their attempts to run to exhaustion.

DIET AND ENDURANCE PERFORMANCE

Endurance capacity studies provide opportunities to examine the influences of dietary changes on fuel utilization, thermoregulation, and exercise tolerance without the complication of changes in exercise intensity. However, competitors in endurance races do not maintain a constant pace throughout.

The choice of pace is influenced by the maximum oxygen uptake of the athlete, his or her training status, the duration of the race, and the general feelings of fatigue. Therefore studies designed to

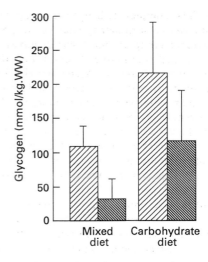

Fig. 4 Human muscle glycogen concentrations before (light striped shaded bars) and immediately after (dark bars) a 30-km cross-country race before which runners consumed either their normal mixed diets or undertook dietary carbohydrate loading (mean ± SD). After ref. 25.

examine the influence of diet on race performance are limited to providing information about performance times and the changes in the physiological conditions of the competitors. Comparisons of different dietary conditions are difficult because competitors have freedom to dictate their own pace in order to achieve the best possible performance times.

One of the best examples of the influence of dietary manipulation on endurance performance was reported by Karlsson and Saltin.[25] They addressed themselves to the question of whether or not an increased pre-exercise glycogen concentration improves running speed as well as endurance capacity during a 30-km cross-country race. Two groups of runners were used: one group undertook carbohydrate loading, by the original form of this procedure, prior to the race, while the other group remained on their normal mixed diet. In the second part of the study, the race conditions were recreated for the 10 runners 3 weeks later and the dietary preparation of the two was reversed. The time to complete the 30-km race was improved by 8 min (135.0 min compared with 143.0 min) when the subjects increased their pre-race muscle glycogen concentrations by carbohydrate loading. Their running speeds were not improved during the early part of the race as a consequence of carbohydrate loading, but they were able to sustain their optimum pace for longer. Interestingly, post-exercise glycogen concentrations after carbohydrate loading were similar to the pre-exercise values of the runners when they were eating their normal mixed diets (Fig. 4). Therefore the dietary preparation for the race provided glycogen stores which were more than adequate for muscle metabolism under these conditions. Thus it is obvious that their performances were not limited by the availability of their glycogen stores.

In a more recent study, the influence of carbohydrate loading on running performance during a simulated 30-km race was conducted using a laboratory treadmill.[34] One of the aims of this study was to determine at what point during the race runners began to show signs of fatigue and how this was modified by dietary manipulation. The treadmill was instrumented so that the subject controlled their own speeds using a light-weight hand-held switch.[35] Changes in speed, time, and distance elapsed were all displayed on

a computer screen in full view of the subjects. The runners were divided into two groups after the first 30-km treadmill time trial. One group increased their carbohydrate intake during the 7-day recovery period, whereas the other group ate additional protein and fat in order to match the increased energy intakes of the carbohydrate group. Although there was no overall improvement in performance times for the two groups, the carbohydrate group ran faster during the last 10 km of the simulated race. Furthermore, eight of the nine runners in the carbohydrate group had faster times for 30 km than during their first attempt and better times than the control group. Even though the carbohydrate group ran faster than the control group, after carbohydrate loading they had lower adrenaline concentrations. This was attributed to the carbohydrate loading and subsequent maintenance of normal blood glucose concentrations throughout the race. Noradrenaline concentrations increased, as expected, during the simulated 30-km races following normal dietary conditions and after carbohydrate loading.

Not all studies have reported clear benefits of carbohydrate loading on running performances. Sherman et al.[28] found no differences in performance when a group of six well-trained endurance athletes completed three races over 20.9 km on an indoor 200-m track. Three different dietary procedures were used to prepare for the races: a low carbohydrate diet followed by 3 to 4 days on a high carbohydrate diet (low/high) (104 g versus 542 g of carbohydrate), a normal mixed diet followed by the same period on a high carbohydrate diet (mixed/high) (352g versus 542 g of carbohydrate), and a normal mixed diet for the whole of the preparatory period before the race (mixed/mixed) (353 g of carbohydrate). The running times for the three races were 83.5 min (low/mixed), 83.63 min (mixed/high), and 82.95 min (mixed/mixed). Both the carbohydrate loading procedures (low/high and mixed/high) increased muscle glycogen concentrations in the gastrocnemius muscles of the runners prior to the races. The subjects who consumed their normal mixed diet throughout the week before the race also showed increased muscle glycogen concentrations, but not to the same extent as in the carbohydrate-loading trials (Fig. 5).

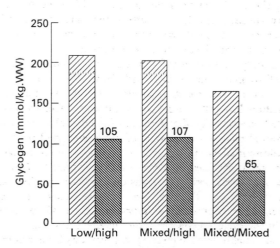

Fig. 5 Human muscle glycogen concentrations of a group of well-trained runners before and after three 20.9-km races on an indoor 200 m track following three dietary methods of increasing the prerace glycogen concentrations. The numbers refer to the amount of glycogen used in each race. After ref. 28.

The fastest time was recorded when the runners simply tapered their training and consumed their normal mixed diets. Under these conditions the runners had lower pre-race muscle glycogen concentrations and used less glycogen during the race (Fig. 5). It is worth noting that the pre-race muscle glycogen concentrations, even without carbohydrate loading, were high, and were also higher than the values reported by Karlsson and Saltin[25] for their subjects before carbohydrate loading. Therefore the muscle glycogen concentrations of the subjects in the study reported by Sherman *et al.*[28] were more than sufficient to meet the demands imposed upon them by races over 20.9 km. It is clear from this study, however, that well-trained runners need only taper their training in preparation for races over the half-marathon distance. They do not need to undertake any dietary manipulation in preparation for races over this or shorter distances.

The study by Sherman *et al.*[28] confirmed three important observations which have been reported separately in other studies. First, high glycogen concentrations can be achieved without undergoing the low carbohydrate phase of the traditional carbohydrate-loading procedure;[27] second, carbohydrate loading after running to exhaustion does produced glycogen supercompensation as effectively as has been repeatedly shown in cycling studies; third more glycogen is used by glycogen-loaded muscle with its normal complement of carbohydrate.

DIET AND HIGH INTENSITY EXERCISE

Only a limited amount of information is available on the influence of diet on exercise of maximal intensity and brief duration. One of the reasons for the limited amount of research on this topic has been the lack of suitable laboratory methods for studying the metabolic and physiological responses to maximal exercise. Inexpensive microcomputers are now widely available, and recently they have been used to record the rapid changes in power outputs during maximal exercise of short duration. As a result of the introduction of this technology into laboratories considerable advances have been made in our knowledge of fatigue during maximal exercise.[36] Using cycle ergometers and treadmills linked

to microcomputers in conjunction with the needle biopsy procedure has provided opportunities to describe the relationship between metabolic events in skeletal muscles and maximal power output.

It is now clear that glycogen depletion is not a limiting factor during maximal exercise of short duration. Some evidence for this conclusion is illustrated in Fig. 6, which is a summary of a series of studies on muscle glycogen changes during exercise of maximal intensity and different durations. It shows the muscle glycogen concentrations before and immediately after 6 s and 30 s of maximal exercise during cycling, using a modified Wingate protocol,[37] glycogen concentrations before and after 30 s of sprint running on a non-motorized treadmill, and glycogen concentrations before and after 120 s of running on a motorized treadmill at speeds equivalent to 110 per cent $\dot{V}O_2$max.[38, 39] During maximal exercise lasting no more than 30 seconds the loss of glycogen is more rapid in type II fibres than in type I fibres.[40] Nevertheless, carbohydrate loading does not influence the power output during repeated 30 second periods of maximal exercise separated by 15 minutes of recovery (Fig. 7).[41]

Carbohydrate loading has been reported to improve performance during high, but not maximal, intensity exercise. Maughan and Poole[42] reported a 36 per cent improvement in cycling time to exhaustion (6.65 min versus 4.87 min), at 105 per cent $\dot{V}O_2$max, after undergoing the traditional carbohydrate-loading procedure. In subsequent studies this group concluded that the improvement in exercise tolerance was a consequence of an increased buffering capacity rather than the additional glycogen stores.[43]

During 30 s of maximal exercise, peak power output is achieved within 6 s. Phosphocreatine and anaerobic glycogenolysis contributed equally to the high rate of ATP production during this early period of maximal exercise.[44] As the subjects continued to exercise for the remaining 24 s their power output decreases along with the concentration of phosphocreatine and so ATP resynthesis occurs almost entirely at the expense of anaerobic glycogenolysis. Nevertheless, without sufficient phosphocreatine to resynthesize ATP

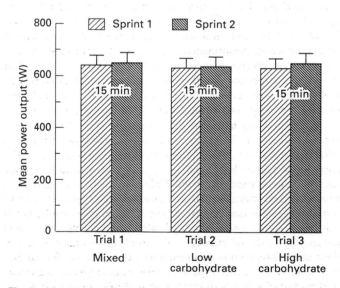

Fig. 7 Mean power outputs during two 30-s periods of maximal exercise (sprints) on a cycle ergometer (Wingate protocol) separate by 15 min after a normal mixed diet, a diet low in carbohydrate, and a high carbohydrate diet. The two sprints were completed after each phase of the traditional method of carbohydrate loading (mean ± SD). After ref. 41.

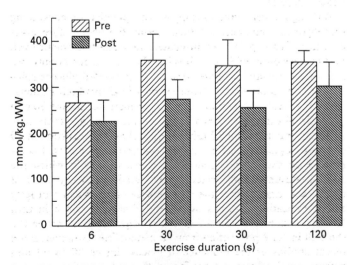

Fig. 6 Human muscle glycogen concentrations before and after maximal exercise of different durations: (from left to right) 6 s and 30 s of cycling, 30 s of sprinting on a non-motorized treadmill and 120 s of running on a motorized treadmill (mean ± SEM).

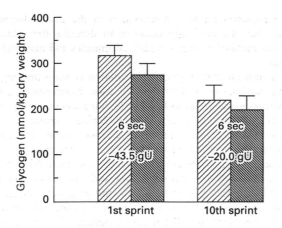

Fig. 8 Human muscle glycogen concentrations before and after the first and the tenth 6-s period of maximal exercise (sprint) on a cycle ergometer. Each sprint was followed by a resting recovery of 30 s. Glycogen utilization in the first and tenth sprints are shown as mmol glucosyl units/kg DW (mean ± SD).

rapidly from ADP, peak power output cannot be sustained even in the presence of adequate glycogen stores.

Even though there is rapid utilization of muscle glycogen during several brief periods of maximal exercise the rate of glycogenolysis decreases as exercise continues. For example, in a series of 10 maximal sprints of 6 seconds duration and 30 s recovery on a cycle ergometer, glycogen degradation was reduced by half during the last sprint (Fig. 8).[45] This glycogen sparing is probably the consequence of an increase in the aerobic metabolism of glycogen and free fatty acids.[46]

Performance during sports which involve several brief sprints may not be improved by carbohydrate loading. Sports which demand that their participants perform a combination of submaximal running and brief periods of sprinting, such as soccer, reduces muscle glycogen concentrations to critically low values.[47] Performance is impaired when this occurs, and so carbohydrate loading would probably be of benefit to participants in multiple-sprint sports.[48] Bangsbo et al.[49] showed that an increased carbohydrate intake, prior to a field test which simulates activities common to soccer, improves performance. Their subjects were seven professional soccer players who performed two tests, one after their normal mixed diet and the other after 2 days on a high carbohydrate diet (65 per cent carbohydrate, i.e. 600g/day).[50] The field test consisted of two parts. During the first part, the subjects performed 46 min of walking and running, simulating the activity pattern common to soccer. The second part was a performance run on a laboratory treadmill. The subjects ran at high (15 s) and low (10 s) speeds on a treadmill until exhaustion. After the high carbohydrate diet the subjects were able to run 0.9 km further than when the test was performed after a normal mixed diet. Muscle glycogen concentrations were not determined during this particular study, but it confirms the recommendations of Saltin[47] and Jacobs et al.[48] that soccer players should increase their carbohydrate intake as part of their match preparation. Unfortunately, there are too few studies on the influence of carbohydrate intake, muscle glycogen concentration, and physical performance on multiple-sprint sports.

PRECOMPETITION MEALS

In most studies of the influence of diet on exercise capacity, subjects fasted overnight before undertaking tests of exercise tolerance. However, there is considerable interest in the optimum pre-competition meal by athletes. The pragmatic advice offered to athletes is that they should eat an easily digestible high carbohydrate meal no later than 3 to 4 h before competition. There is some evidence to support this recommendation, but only a few studies have used solid rather than liquid meals.

In one such study, subjects ate a breakfast consisting of bread, cereal, milk, and fruit juice (200 g of carbohydrate) 4 h before exercise to exhaustion. In addition, a chocolate bar was eaten 5 min before an hour of cycling.[51] This meal increased pre-exercise muscle glycogen concentration (15 per cent) and improved work capacity after 45 min of cycling at 80 per cent V_{O_2}max when compared with exercise following no food intake. The carbohydrate content of pre-exercise meals is also important. Sherman et al.[52] showed that although small amounts of carbohydrate (46 g and 156 g) consumed 4 h before intermittent cycling exercise improved performance, the consumption of a larger amount of carbohydrate (312 g) was even more effective in improving cycling performance. The more effective meal, which contained 312 g of carbohydrate, is larger than the meals normally consumed by athletes before exercise. Athletes' breakfasts usually contain about 100 to 120 g of carbohydrate.[53] However, a compromise must be reached between eating sufficient food to provide enough carbohydrate but not so much as to cause gastrointestinal disturbances during subsequent exercise. Sherman et al.[52] avoided these potential problems by liquidizing the pre-exercise meals that they gave their subjects. Furthermore, they used cycling rather than running, and so the potential for gastrointestinal disturbances was further reduced. Although these studies confirm the principle that a large carbohydrate intake 3 to 4 h before exercise improves endurance performance compared with pre-exercise fasting, care has to be taken when translating this principle into practice. High carbohydrate foods in large amounts may cause gastrointestinal disturbances, and so one way around this potential problem is to supplement pre-competition meals with carbohydrate solutions. This nutritional strategy even allows athletes to take in carbohydrate to good effect as late as 1 h before exercise. Sherman et al.[54] showed that performance is improved during submaximal cycling lasting longer than 90 min when a solution containing the equivalent of 1.1 to 2.2 g/kg BW of carbohydrate is consumed 1 h before exercise.

Although recent studies support the practice of consuming carbohydrate-containing solutions during the hour before exercise, there is a popular, yet misguided, view that this practice will lead to early onset of fatigue. It is based on the results of only two studies. In one study runners drank a concentrated glucose solution (25 per cent) 45 min before submaximal treadmill running. The rate of utilization of muscle glycogen increased during 30 min of treadmill running.[55] When this pre-exercise carbohydrate consumption was repeated 30 min before cycling to exhaustion at 80 per cent V_{O_2}max, exercise time was reduced by 19 per cent compared with the values obtained in a control trial in which the cyclists consumed only water.[56] This paradoxical early onset of fatigue was explained as the consequence of a reduction in fatty acid mobilization, induced by hyperinsulinaemia, which in turn led to an increased rate of glycogen degradation to cover the shortfall in ATP production from fat metabolism. However, more recent studies of the influence of glucose intake 30 to 60 min before exercise have not confirmed that this practice causes a decrease in endurance capacity during either cycling[57, 58] or running.[59] Even consuming carbohydrate in the form of chocolate bar 30 min, or indeed 5 min, before prolonged submaximal exercise has no detrimental affect on endurance capacity[60] and may even improve performance.[51]

Consuming carbohydrate foods or fluids immediately, rather than within the hour before prolonged submaximal exercise, does not cause hyperinsulinaemia or a reduction in endurance performance.[35,51] The increase in noradrenaline at the start of exercise probably inhibits the release of insulin from the islets of Langerhans.[61] Therefore an increase in the blood glucose concentration of athletes during exercise does not produce the same rise in plasma insulin concentration as when they are at rest. Thus, there are advantages of not only increasing carbohydrate intake during the 3 to 4 days before prolonged exercise but also consuming a carbohydrate meal 3 to 4 h before exercise. If eating a carbohydrate meal 3 to 4 h before exercise causes gastrointestinal discomfort during exercise, then an alternative is to drink a carbohydrate solution before exercise and every 20 min during exercise.[62] Evidence in support of this recommendation as an effective means of not only increasing the carbohydrate intake but also improving sports performance is provided by a study of the diet and goal-scoring ability of a soccer team. Muckle[63] reported an increase in the number of goals scored when players supplemented their normal diets with a glucose syrup solution (46 per cent) during the day before the game and when they took this supplement again 30 min before competition. The glucose syrup supplementation was provided for a team of professional players for 20 matches while the remaining 20 matches of the season were used as the control. The number of goals scored in the second half of each game increased and there was a reduction in the number of goals conceded. Drinking a glucose solution also improves players' work rates during a soccer match. Kirkendall et al.[64] filmed a game in which a group of players consumed a concentrated carbohydrate solution (ca. 15.5 per cent) before the game and a half-time. The players who consumed the carbohydrate solution covered more ground during the second half of the game than those players who consumed a sweet placebo solution. The results of this study confirm the benefits of consuming additional carbohydrate before a game reported by Muckle[63] 15 years earlier.

CARBOHYDRATE INTAKE DURING EXERCISE

Drinking carbohydrate-containing solutions immediately before and during exercise improves endurance capacity and performance.[65-71] However, the amounts which appear to be effective in improving exercise capacity vary widely between studies. For example, Bjorkman et al.[66] reported an improvement of 18 per cent in endurance capacity (137 min vs. 116 min) for their active but untrained subjects when they drank a 7 per cent glucose solution (250 ml) every 20 min, equivalent to 52.5 g/h, during cycling to exhaustion at an intensity of 68 per cent $\dot{V}O_2$max. In a similar study, Coyle et al.[67] showed that well-trained cyclists improved their exercise time to exhaustion by 33 per cent (3 h vs. 4 h) at a similar exercise intensity (70 per cent $\dot{V}O_2$max ingesting a carbohydrate solution every 20 min at a rate which was twice as great (100 g/h) as that reported by Bjorkman et al.[66] In contrast, Maughan et al.[70] showed that cycling time to exhaustion improved by 29 per cent (90.8 min versus 70.2 min) when their subjects drank a dilute carbohydrate–electrolyte solution (4 g per cent); however, when they drank a concentrated glucose solution (36 per cent), their performance was no different from their endurance capacity during a water trial. It is difficult to explain why such a range of concentrations and dosages of carbohydrate solutions are

all so effective in improving endurance performance. There may be an effective threshold concentration, above which the additional carbohydrate is stored in the liver or indeed contributes to glycogen synthesis in non-active muscle fibres.[72] Of course, to be on the safe side athletes may adopt the approach that if a recommended amount is effective, then twice that amount may be even more effective. However, during cycling or any sport where body weight is supported, gastrointestinal discomfort does not occur to the same extent after drinking large volumes of fluids as it does during running sports. Therefore the question about the optimum volume and concentration of a carbohydrate solution is very real for sports involving running or walking where gastrointestinal disturbances have a greater potential to impair performance than during cycling.

Another interesting aspect of these studies is that Bjorkman et al.[66] found that their subjects used less glycogen as a result of their carbohydrate intake during exercise when they drank only water, whereas Coyle et al.[67] showed quite clearly that the rate and amount of glycogen used in working skeletal muscles was the same during the carbohydrate and the water placebo trials. However, during the carbohydrate trial the muscle glycogen concentration at exhaustion was the same after 4 h as it was after 3 h. This suggests that carbohydrate intake during the first 3 h of submaximal cycling did not lead to glycogen sparing, but the additional hour of exercise was at the expense of the exogenous carbohydrate supply. Coggan and Coyle[73] conclude from a review of the studies on carbohydrate ingestion, muscle glycogen, and endurance performance that the weight of available evidence does not support the proposal that glycogen sparing is the reason for improvement exercise capacity. However, this conclusion was based on a review of cycling studies; the same conclusion may not be true for carbohydrate ingestion and improvements in endurance performance during running. Glycogen sparing may occur during running, especially in the quadriceps muscles of runners when they drink a carbohydrate solution throughout exercise.[74] This would explain the improvement in endurance capacity [71] and endurance performance[35] of runners who drink a dilute carbohydrate solution throughout the run. It appears that performance towards the end of exercise benefits from ingesting carbohydrate solutions (Fig. 9). This is because the exogenous glucose supply is used by working muscles mainly during the later stages of prolonged exercise when muscle glycogen concentration is low.[69, 75, 76] However, further studies on carbohydrate ingestion, muscle glycogen utilization, and running performance are needed to confirm or deny the presence of glycogen sparing.

It appears that drinking a concentrated glucose solution towards the end of prolonged cycling also delays fatigue and so improves endurance capacity. Coggan and Coyle[69] showed that cyclists who ingested a carbohydrate solution (3 g/kg BW) after 135 min of submaximal cycling maintained their normal blood glucose concentrations and improved their endurance capacity by 36 min. This study supports the proposal that supplying carbohydrate late in exercise provides substrate for muscle metabolism when glycogen stores are low.

DIET AND RECOVERY

Recovery from exercise is not a passive process. Tissues undergo repair and reproduction, fluid balance is restored, and energy stores are replaced. Carbohydrate replacement is one of the most important events during recovery. When several days separate periods of exercise or participation in sport, a normal mixed diet

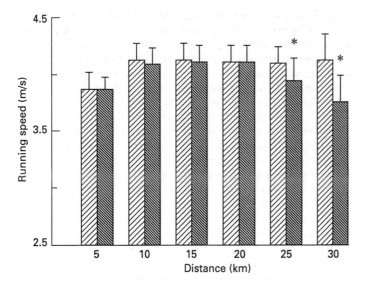

Fig. 9 Running speeds during two simulated 30-km races on a motorized treadmill. On one occasion the runners drank a 5 per cent carbohydrate–electrolyte solution (light striped bars) throughout the race, whereas on the other occasion they drank only water. Runners were required to complete the first 5 km at the same pace during two races; thereafter they were free to select their own speeds in order to complete the distance as fast as possible. Running speeds were faster (*) during the last 5 km of the simulated races when runners drank the carbohydrate–electrolyte solution (mean ± SD). After ref. 35.

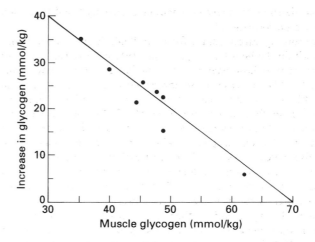

Fig. 10 Relationship between the post-match muscle glycogen concentrations of professional soccer players and the increase in muscle glycogen concentrations after 24 h recovery during which these players ate high carbohydrate diets (glycogen concentrations are expressed as mmol/kg WW). After ref. 48.

containing about 4 to 5 g/kg BW of carbohydrate is sufficient to replace muscle glycogen stores. However, daily training or competition makes considerable demands on the body's carbohydrate stores. Therefore the normally high carbohydrate intake of athletes may not be enough to prevent a gradual reduction in this important fuel store. For example, even when the daily carbohydrate intake is 5 g/kg BW cycling or running for an hour each day, gradually delays the daily restoration of muscle glycogen stores.[77] Increasing the carbohydrate intake to 8 g/kg BW per day may not be enough to prevent a significant reduction in muscle glycogen concentrations after five successive days of hard training.[78] These studies underline the importance of prescribing adequate amounts of carbohydrate for athletes in training and justifies the need for more frequent recovery days between periods of intense training.

The muscle cell membrane is more permeable to glucose immediately after exercise than before exercise. It appears that exercise changes the characteristics of the sarcolemma such that glucose permeability is improved and muscle has an increased insulin sensitivity. The two effects appear to be additive.[79] In addition, glycogen synthase, the enzyme complex responsible for glycogen synthesis, is also in its most active form immediately after exercise. There is an inverse relationship between muscle glycogen concentration and the concentration of glycogen synthase in the active form.[80,81] The specificity of this relationship has been highlighted in a study which examined, both biochemically and histochemically, the activity of this enzyme in samples of different populations of muscle fibres obtained from subjects who had exercised to exhaustion. The activity of glycogen synthase was greatest in the type I fibres, i.e. the slow contracting oxidative fibres.[81] From more recent work it can be concluded that this fibre

population is most involved during submaximal cycling and running to exhaustion.[71, 82] From the foregoing, it is not surprising that there is also an inverse relationship between the concentration of glycogen in skeletal muscle immediately after exercise and the amount resynthesized during 24 h of recovery (Fig. 10).[48] Muscle biopsy samples taken from a group of professional football players immediately after a soccer match showed a range of glycogen concentrations. Those players with very low post-exercise glycogen concentrations had the greatest glycogen concentrations a day later, whereas those players with a moderate level of muscle glycogen at the end of the game replaced only a relatively small amount during the recovery period. Therefore, even though there is evidence to suggest that the duration of exercise plays a significant part in activating glycogen synthase after exercise,[83] the circumstantial evidence suggests that the post-exercise muscle glycogen concentration has a more influential role.[84]

The highest glycogen resynthesis rate occurs during the first few hours of recovery.[85] Drinking a carbohydrate solution, which provided the equivalent of 2 g/kg BW, immediately after prolonged heavy exercise produced a muscle glycogen resynthesis rate of about 5 to 6 mmol/kg.h.[86] This rate of glycogen resynthesis (about 5 per cent per hour) is about 300 per cent greater than under conditions where no carbohydrate is ingested immediately after exercise. When carbohydrate intake was delayed for 2 h, the rate of glycogen resynthesis was 47 per cent slower than when carbohydrate was provided immediately after exercise. Blom *et al.*[87] found that the maximum rate of glycogen resynthesis occurred when their subjects consumed the equivalent of only 0.7 g/kg BW of carbohydrate every 2 h during the first 6 h of recovery. This amounts to about 50 g of carbohydrate every 2 h for a person weighing 70 kg. At present, the available evidence suggests that the optimum amount of carbohydrate needed for the rapid post-exercise glycogen resynthesis is about 1 g/kg BW which should be consumed immediately after exercise and at 2-h intervals until meal time.

The type of carbohydrate eaten during the recovery period is also an important consideration for those who want to replace their muscle glycogen stores quickly. Kiens[11] compared the muscle glycogen resynthesis rates when her subjects were fed either high or low glycaemic foods. Seven well-trained athletes were fed a diet in which 70 per cent of their energy was obtained from either high or low glycaemic foods for 2 days after their muscle glycogen stores were reduced by prolonged cycling. The post-exercise diet of high glycaemic foods produced a glycogen resynthesis rate which restored the carbohydrate store to 70 per cent of pre-exercise concentrations within 6 h, where the low gly-caemic high carbohydrate diet restored muscle glycogen stores to only 34 per cent of their pre-exercise values. This study con-tributes to the evidence which supports the recommendation that glycogen resynthesis after exercise is most rapid when athletes consume high glycaemic carbohydrates.[88]

Adding some protein to the carbohydrate solution increases the rate of post-exercise glycogen synthesis to a greater extent than can be achieved with a carbohydrate solution alone.[89] The addition of protein increases the concentrations of plasma insulin to a greater extent than when only a carbohydrate solution is con-sumed after exercise. An increased insulin concentration will not only increase the transport of glucose into muscle cells but will also help restore the potassium balance across muscle membranes. However, the optimum concentrations of protein and carbo-hydrate for rapid glycogen synthesis have yet to be established. Prolonged heavy exercise has an anorexic effect, particularly in hot weather. Therefore drinking after exercise is generally more attractive than eating carbohydrate-containing foods. But drink-ing what type of fluid? Carbohydrate–electrolyte solution such as 'sports drinks' not only provide glucose for rapid glycogen re-synthesis but also help to complete the process of rehydration during recovery.[90] The sports drinks appear to be more effective as rehydration agents than either water or some of the most popular soft drinks.[91] However, dehydration does not appear to prevent glycogen resynthesis during recovery from exercise.[92]

There are conditions, other than low carbohydrate diets, which can interfere with glycogen resynthesis. Hard training or pro-longed competition causes residual fatigue which is often ac-companied by muscle soreness and lethargy. The extreme stress of exercise and competition may lower the plasma concentrations of the body's natural anabolic hormones.[93] Testosterone has been shown to play a role in converting inactive glycogen synthase into its active form,[80] and so the depression of this hormone may delay the replenishment of the muscle's carbohydrate store.[93] Extreme muscle soreness is the result of damaged muscle fibres and the inflammatory response which is associated with the healing pro-cess. Glycogen resynthesis is delayed in skeletal muscle fibres that have experienced microtrauma as a result of too much eccentric activity.[94]

DIET, RECOVERY, AND PERFORMANCE

The full recovery of muscle glycogen stores can take as long as 2 days,[95] although shorter recovery times are possible when initial glycogen stores are not high.[48, 84] However, the relevant informa-tion for athletes is not the rate of glycogen resynthesis but the rate of recovery of their fitness. Keizer et al.[96] tried to answer this important question. They reported that muscle glycogen stores can be replenished in only 22 h by administering either liquid or solid carbohydrates during the first 5 h of recovery. After the first 5 h their subjects consumed foods which were consistent in com-position and quantity with their normal diets. The amounts of carbohydrate ingested during the 22-h recovery period were 580 g by the liquid group and 602 g by the solid group. Before and after

the recovery period an incremental cycle ergometer test was used to assess the maximum physical work capacity of the subjects. Their maximal physical work capacity was 7.3 per cent lower after 24 h of recovery, even though their glycogen stores had returned to pre-exercise values. Therefore it appears that the replenish-ment of muscle glycogen concentration alone is not sufficient to restore maximal work capacity. One additional practical observa-tion from this study was that when the subjects were allowed to eat *ad libitum* after exercise their muscle glycogen concentrations were significantly lower than when their food was prescribed and prepared for them.

Nevill et al.[97] came to a similar conclusion to Keizer et al.[96] after studying the influence of a high carbohydrate recovery diet on the multiple sprint performance of a group of games players. Using a non-motorized treadmill, games players completed 30 maximal sprints of 6 s duration separated by a recovery period of 114 s. Between each sprint, the subjects walked and jogged on the treadmill. Mean power output decreased by 8.8 per cent over the 30 sprints. In this study, the compositions of the diets were modified without changing the daily energy intakes of the subjects. Their carbohydrate intakes were 322 g (4.6 g/kg BW), 80 g (1.1 g/kg BW), and 644 g (8.7 g/kg BW) for those consuming the normal mixed diet, the low carbohydrate diet, and the high carbo-hydrate diet respectively. There were no differences in power outputs between the three groups during the first hour of exercise, nor 24 h later when the subjects attempted to improve their per-formance. All performances were lower during the multiple sprint tests following the 24-h recovery period. Although the carbo-hydrate intakes of the subjects were increased during the recovery period, their daily energy intake did not match their energy expen-diture and so their overall performances may have suffered as a consequence of this energy deficit.

In contrast, endurance running capacity can be restored after a day's recovery when the daily carbohydrate and energy intake is increased. For example, when runners ate a high carbohydrate diet (9 g/kg BW) after completing 90 min of treadmill running at speeds equivalent to 70 per cent V_{O_2}max, they were able to repeat this training run the following day, whereas runners who ate their normal mixed diets, with additional energy in the form of extra fat and protein, could only complete 78 per cent of the 90 min run.

Even when the recovery period is as short as 4 h, drinking a carbohydrate solution in the amount recommended to increase muscle glycogen resynthesis helps, but does not entirely restore, endurance capacity. Runners who completed 90 min of treadmill running were able to run for a further 62 min after a 4 h rest during which they drank the equivalent of 1 g/kg BW of carbohydrate from a sports drink immediately after exercise and again after 2 h of recovery. In contrast, a control group drank the same volume of a sweet placebo during the recovery period after which they could only run for 40 min.[99]

There are clear benefits to be gained from drinking a carbo-hydrate solution immediately after exercise, even when the recov-ery period is very short. Coggan and Coyle[100] showed that when their subjects were given glucose solution during exercise, 20 min after cycling to exhaustion at 70 per cent V_{O_2}max they were able to continue for longer than when they consumed only water. During three recovery trials their subjects (a) consumed a water placebo, (b) consumed a glucose polymer solution (3 g/kg BW), or (c) received a glucose infusion throughout the subsequent exercise. The glucose infusion was included as an effective way of main-taining normal blood glucose concentrations during prolonged exercise. Endurance capacity was significantly greater after carbohydrate ingestion (26 min) and after glucose infusion (43 min) than after ingesting a water placebo (10 min). However, even though glucose infusion maintained the blood glucose concentrations of the cyclists, it did not restore their exercise capacity to its former value. Thus the unanswered question is: why does fatigue occur even when the provision of exogenous carbohydrate is sufficient to maintain blood glucose concentrations and cover the rate of glucose metabolism in working muscles?

SUPPLEMENTS

Athletes have a poor knowledge of nutrition and nutritional strategies.[101, 102] Therefore it is not surprising that they are vulnerable to advertisements for nutritional supplements which claim to have performance-enhancing effects (ergogenics).[103] Although most of the supplements cause more financial than physical harm, they can pose a potential health threat. This occurs when athletes ignore sound nutritional advice and believe that supplements can provide them with an easy route to a balanced diet.

Nevertheless, from time to time there are reports about performance-enhancing supplements which appear to have some basis for their claims. This is particularly likely if explanations and evidence offered for the effectiveness of the supplement are consistent with well-established principles of nutritional science. For example, nutritional supplements that purport to improve fat metabolism in skeletal muscle should either raise plasma fatty acid concentrations or increase their metabolism by skeletal muscles. Supplements which appear to achieve either or both these aims are of interest because the ensuing increase in energy production would spare the limited glycogen stores and so increase endurance capacity. There have been several products which claim to improve fat metabolism, but only two have attracted sustained attention by nutritional scientists.

Caffeine

The first consideration is the potential fatty acid mobilizing effect of caffeine. Caffeine is found mainly in tea, coffee, and some soft drinks. The amount of caffeine in tea and coffee depends on their origins and how the drinks have been prepared. For example, tea which is allowed to stand in hot water for several minutes will have a higher caffeine content than tea which has not been brewed in this way. Caffeine is a methylxanthine which inhibits the fat cell enzyme phosphodiesterase. This enzyme normally controls the hydrolysis of triglycerides into long-chain fatty acids and glycerol by inhibiting the activation of the hydrolytic enzyme lipase. Some, but not all, studies have shown an increase in plasma fatty acids and an increase in their metabolism during exercise after the athlete has drunk several cups of coffee. Those studies which have shown an improvement in endurance performance have not necessarily shown an increase in plasma fatty acid concentrations or indeed glycogen sparing.[104] Caffeine also has stimulatory influences on the central nervous system, and so some of its reported performance-enhancing effects may not be linked to changes in energy metabolism per se. Therefore it is difficult to conclude in favour of the fat-metabolism-enhancing effect of caffeine ingestion because of the conflicting evidence. Furthermore, when athletes prepare for competition by carbohydrate loading, the caffeine-induced increase in the concentrations in plasma fatty acids is abolished.[105] As a nutritional strategy, increasing muscle glycogen stores before competition is a more effective practice than attempting to elevate plasma fatty acids.

The usual dosage used in these studies is 5 to 6 g/kg BW, whereas the average daily intake is about 200 mg. This dosage of caffeine is a diuretic, which causes its own problems for athletes. Caffeine per se is not a banned substance, but the International Olympic Committee does put an upper limit on its use. The British Olympic Association advises athletes not to drink more than three cups of coffee before competition and to avoid the consumption of caffeine-containing products which are often sold by health food stores under a variety of trade names.

Carnitine

Carnitine supplementation is, in theory, another method of increasing fat metabolism. Most of the body's carnitine is in skeletal muscle. Carnitine contributes to energy metabolism as a carrier of long-chain fatty acids across the inner membrane of mitochondria. The proposal that carnitine supplementation helps improve fatty acid oxidation presupposes that the amount of carnitine in muscle available for this purpose is limited. Although there are some studies which suggest that carnitine supplementation has potential performance-enhancing benefits,[106] most are, at best, speculative. There is at present no evidence to show that human skeletal muscle has an insufficient store of L-carnitine and that it is significantly reduced during prolonged heavy exercise.[107] Therefore, the weight of available evidence does not support the proposal that supplementation of athletes' diets with L-carnitine improves endurance capacity. The influence of carnitine supplementation on performance during high intensity exercise has not been studied extensively and so it is too early to offer a judgement on its possible effectiveness.

Acid–base balance

Fatigue during maximal exercise of short duration is not a consequence of muscle glycogen depletion. However, this type of exercise produces high concentrations of lactate and hydrogen ions in working skeletal muscles. Fatigue is associated with, though may not be caused by, increased concentrations of hydrogen ions. One of the adaptations to sprint training is an increase in intramuscular buffering capacity.[39] As a result, muscle pH during maximal exercise after training does not fall to such low values as during exercise before training. Rapid removal of hydrogen ions from working muscle would, in theory, allow muscle to continue to work unhindered by the inhibitory influence of this product of glycolysis. The rate of removal of hydrogen ions from working muscles is improved when blood perfusing muscle is alkalotic; however, when muscle is perfused by acidotic blood then both the rate of removal of hydrogen ions and endurance capacity decrease.[108]

Bicarbonate

Increasing the buffering capacity of plasma by ingesting a bicarbonate solution before exercise improves endurance during prolonged heavy exercise. In addition, bicarbonate loading appears to give athletes a distinct advantage during races over distances of 400 and 800 m. In a study of 400 m racing performance, the runners reduced their time for this distance by 1.6 s (58.5 s versus 56.9 s) after bicarbonate loading.[109] The 800 m times of college athletes decreased by 3 s (2 min 2.9 s versus 2 min 5.9 s) after bicarbonate loading compared with their performance over the same distance after consuming a placebo solution.[110] However, the performances of these runners is not of a particularly high standard. Improvements of the magnitude recorded following bicarbonate ingestion could also be achieved by training. Improvements in performance have not been shown during sprinting for less than 30 s.[111] However, it appears that the induced alkalosis is effective during exercise which lasts between 1 and 7.5 min.[108] The effective dose appears to be about 0.3 g/kg BW taken 2 to 3 h before exercise; however, this amount of bicarbonate is not tolerated by everyone without some unpleasant side-effects. Gastrointestinal discomfort is the most common reaction to this treatment, but it is not unusual for people to experience diarrhoea and vomiting. In general, it is not a dietary practice which should be given serious consideration by athletes.

Creatine

Creatine, which is found in meat and fish, contributes to the formation of a high energy compound called phosphocreatine.[112] This compound plays a central role in energy production. It is the most rapid route by which ATP is regenerated from ADP. During brief periods of high intensity exercise, when the rate of ATP utilization is high, the concentration of phosphocreatine is gradually reduced to very low levels. Therefore repeated sprints are tolerable only when skeletal muscles have an adequate concentration of phosphocreatine. Recovery periods of at least 30 s are needed to regenerate sufficient phosphocreatine for skeletal muscles to work maximally, albeit for brief periods. Training does not appear to increase phosphocreatine concentrations in resting skeletal muscle to values significantly above those found in moderately active people.[39] In contrast, training increases resting muscle glycogen concentrations above those found in moderately active people. This response suggests that training and changes to the habitual diets of formerly untrained individuals improves their nutritional status in a direction which prepares them to meet the challenge of exercise successfully. The apparent absence of a natural increase in phosphocreatine in response to training may reflect an optimum level in trained individuals. Alternatively, there may have been too few studies in which untrained rather than active people have undertaken sprint training, and so we do not have the information about the extent of training-induced changes in the phosphocreatine stores of these people. However, creatine supplementation does increase the resting concentration of phosphocreatine in human skeletal muscle.[113] Dosages of about 24 to 30 g of creatine per day, for 2 days, increased phosphocreatine concentrations by about 50 per cent, but much of the ingested creatine is lost in the urine. This amount is equivalent to eating approximately 6 kg (13 pounds) of beef over 2 days.[113]

Nevertheless, an increase in pre-exercise phosphocreatine concentrations would, in theory, give athletes an advantage when participating in multiple-sprint sports. At present there is only a limited amount of information on the effectiveness of creatine supplementation on exercise performance. In one of the few studies on this topic, Greenhaff et al.[114] reported that creatine supplementation improved the capacity of their subjects to perform repeated bouts of maximal exercise. Twelve subjects completed five periods of 30 maximal isokinetic contractions with a 1-min rest between each of the five bouts of voluntary exercise before and after 5 days of supplementation with either creatine (5 g of creatine plus 1 g of glucose four times daily) or a glucose placebo (6 g glucose four times daily). The peak torque values generated during these repeated bouts of exercise decrease with successive contractions as fatigue occurs. After 5 days of supplementation there was no differences in the gradual decrease in peak

torque values generated by the placebo group over the five bouts of isokinetic contractions. In contrast, the creatine group were able to delay the onset of fatigue longer than before supplementation. This was reflected by their ability to maintain higher peak torque value during successive bouts of isokinetic contractions than before supplementation with creatine. As yet there is insufficient information on the dose response to creatine supplementation to judge whether or not it could, or indeed should, be used as a nutritional supplement for those athletes who participate in multiple-sprint sports.

Amino acids

There is considerable interest in the potential ergogenic benefits of supplementing athletes' diets with amino acids. Although there are claims for the performance-enhancing benefits of amino acids there is not, as yet, a consensus view on which, if any amino acids are effective in this role. One line of reasoning, however, suggests that there is a role for amino acid supplementation in delaying the onset of centrally, rather than peripherally, mediated fatigue during prolonged exercise.[115]

Serotonin (5-hydroxytryptamine) is a neurotransmitter in the brain that, when levels are elevated, is associated with drowsiness and lethargy. The amino acid tryptophan contributes to the synthesis of serotonin, and so when the plasma tryptophan concentration increases there is a concomitant increase in the concentration of this neurotransmitter. Tryptophan is transported across the blood–brain barrier by a carrier mechanism which it shares with the branched-chain amino acids (isoleucine, leucine, and valine) and other large neutral amino acids. Therefore the amount of tryptophan crossing the blood–brain barrier is related to the ratio of free tryptophan concentration to the concentration of the large neutral amino acids. Tryptophan is normally loosely bound to plasma albumin; however, it competes for these albumin binding sites with fatty acids. During prolonged exercise plasma fatty acids are released from fat cells and take up ever greater numbers of the binding sites on plasma albumin, and so there are less binding sites available for tryptophan. Therefore, as plasma fatty acids increase, there is also an increase in the concentration of free tryptophan. Furthermore, as exercise progresses, muscle glycogen stores decrease and there is an increase in the oxidation of branched-chain amino acids. Thus, the ratio of free tryptophan to branched-chain amino acids increases, accompanied by greater transport of tryptophan into the brain. The rise in the concentration of serotonin which follows may lead to a decreased central drive to continue exercising. Conversely, decreasing the amount of free tryptophan in plasma during exercise should delay this central contribution to fatigue.

There are a number of indirect ways in which the proposed link between increases in plasma free tryptophan, brain serotonin levels, and fatigue can be explored. One is to alter the levels of serotonin and to measure the changes, if any, in exercise capacity. Wilson and Maughan[116] attempted such an experiment by providing active healthy subjects with a serotonin uptake inhibitor, paroxetine, which is effectively a serotonin agonist. Cycling time to exhaustion was reduced by 19 per cent (94 min versus 116 min) when the subjects exercised 6 h after a standard meal and 20 mg of paroxetine.

Another approach is to depress plasma free tryptophan and so decrease its transport across the blood–brain barrier. Consuming carbohydrate either before or during exercise generally decreases plasma fatty acid concentrations. It is well established that when runners or cyclists consume carbohydrate solutions during exercise there is an improvement in performance.[35, 71, 117] The accepted explanation for the improved exercise tolerance is the contribution of the exogenous glucose to energy metabolism in working muscle. An alternative explanation is, of course, one which includes the role of tryptophan and serotonin synthesis. Plasma fatty acid concentrations are, in general, lower during prolonged submaximal exercise when a carbohydrate solution is consumed rather than water. Lower plasma fatty acid concentrations would free more albumin binding sites for tryptophan and so decrease the concentration of free tryptophan. Reduced concentrations of free tryptophan and increase exercise tolerance have been shown to occur when cyclists consumed a 6 per cent or a 12 per cent carbohydrate solution.[118] The ratio of plasma free tryptophan to branched-chain amino acids decreased during exercise which included carbohydrate consumption, thus providing circumstantial evidence for the hypothesis that carbohydrate ingestion during exercise may have a role in decreasing the centrally mediated contribution to fatigue.

An alternative strategy is to raise the plasma concentrations of branched-chain amino acids and so decrease the transport of free tryptophan into the brain. Blomstrand *et al.*[119] pursued this approach and found that feeding 16 g of branched-chain amino acids (50 per cent valine, 30 per cent leucine, 20 per cent isoleucine) in a 5 per cent carbohydrate solution improved the performance times of marathon runners by 3 per cent (3 h 18 min versus 3 h 27 min). Providing branched-chain amino acids in a carbohydrate solution was not an ideal choice because of the performance-enhancing influence of carbohydrate solutions.[35, 71] Therefore these studies alone do not support claims for the performance-enhancing benefits of supplementing a well-balanced diet with branched-chain amino acids before endurance races.

Rather than enhance endurance performance, branched-chain amino acids may decrease it. During prolonged submaximal exercise when muscle glycogen concentrations are low, the decrease in carbohydrate metabolism will lead to a decrease in the concentrations of intermediates in the citric acid cycle. Oxidation of branched-chain amino acids increases when muscle glycogen stores are low, particularly towards the end of prolonged exercise. This process makes demands on the citric acid cycle intermediate 2-oxoglutarate and as such produces glutamate. Glutamate is converted to glutamine as a consequence of a combination with ammonia, and is released from muscles as the main non-toxic amino group carrier. Therefore Wagenmakers *et al.*[120] propose that, rather than improving endurance performance, branched-chain amino acid supplementation has the potential to decrease it.

The circumstantial evidence supports the hypothesis that there is a link between an increase in plasma tryptophan and the early onset of fatigue, reflecting a central contribution to fatigue. However, the relative contribution of the central drive to the fatigue process is difficult to judge. It is probably less than the effects of a significant reduction in muscle glycogen and may simply increase the 'sensitivity' to the signals emanating from fatiguing muscle.

SUMMARY

The clear message from over a half a century of research on the links between food, nutrition, and exercise capacity is that next to natural talent and appropriate training, a high carbohydrate diet and adequate fluid intake to avoid dehydration are the two most important elements in the formula for successful participation in sport. Of course, there is an underlying assumption that athletes normally eat a well-balanced diet made up of a wide variety of foods and containing sufficient energy to cover their needs. Nutritional strategies to support training and help recovery from exercise will only be fully effective when they are used by athletes who regularly eat a well-balanced diet.

REFERENCES

1. Schobel H. *The Ancient Olympic Games*. New York: Van Nostrand, 1965: 325.
2. Gisolfi CV, Duchman SM. Guidelines for optimal replacement beverages for different athletic events. *Medicine and Science in Sports and Exercise* 1992; **24**: 679–87.
3. Maughan RJ, Noakes TD. Fluid replacement and exercise stress. A brief review of studies on fluid replacement and some guidelines for the athlete. *Sports Medicine* 1991; **12**: 16–31.
4. Noakes TD, Rehrer NJ, Maughan RJ. The importance of volume in regulating gastric emptying. *Medicine and Science in Sports and Exercise* 1991; **23**: 307–13.
5. COMA. *Diet and cardiovascular disease, Report on Health and Social Subjects No. 28*, Committee on Medical Aspects of Food Policy Report of the Panel on Diet, in relation to cardiovascular disease. London: DHSS, 1984.
6. Devlin JT, Williams C. Foods, nutrition and sports performance; a final consensus statement. *Journal of Sports Science* 1991; **9** (Suppl. 1): iii.
7. Jenkins DJA *et al.* Glycemic index of foods: a physiological basis for carbohydrate exchange. *American Journal of Clinical Nutrition* 1981; **34**: 362–6.
8. Coyle EF. Timing and method of increased carbohydrate intake to cope with heavy training, competition and recovery. *Journal of Sports Science* 1991; **9** (Suppl.): 29–52.
9. FAO/WHO/UNU. *Energy and protein requirements* Geneva: World Health Organization, 1985.
10. Lemon PWR. Effect of exercise on protein requirements. *Journal of Sports Science* 1991; **9** (Suppl.): 53–70.
11. Kiens B. Translation nutrition into diet: diet for training and competition. In: Macleod DAD, Maughan RJ, Williams C, Madeley CR, Sharp JCM, Nutton RW, eds. *Intermittent high intensity exercise: preparation, stresses, and damage limitation*. London: Spon, 1993: 175–82.
12. Saris WHM. Exercise, nutrition and weight control. In: Brouns F, ed. *Advances in nutrition and top sport*. Basel: Karger, 1991: 200–16.
13. Inge K, Brukner P. *Good for sport*. London: Kingswood Press, 1988: 242.
14. Clark N. *Nancy Clark's sports nutrition guidebook*. Champaign, Ill: Leisure Press, 1990: 322.
15. van der Beek EJ. Vitamin supplementation and physical exercise performance. *Journal of Sports Science* 1991; **9** (Suppl.): 77–90.
16. Clarkson PM. Minerals: exercise performance and supplementation in athletes. *Journal of Sports Science* 1991; **9** (Suppl); 91–116.
17. Christensen EH, Hansen O. Arbeitsfahigkeit und Ehrnahrung. *Skandinavisches Archiv für Physiologie* 1939; **81**: 160–75.
18. Nilsson LH, Furst P, Hultman E. Carbohydrate metabolism of the liver in normal man under varying dietary conditions. *Scandinavian Journal of Clinical and Laboratory Investigation* 1973; **32**: 331–7.
19. Maughan RJ, Williams C. Differential effects of fasting on skeletal muscle glycogen content in man and on skeletal muscle in the rat. *Proceedings of the Nutrition Society* 1981; **40**: 45A.
20. Bergstrom J. Hultman E. Muscle glycogen synthesis after exercise: an enhancing factor localized to the muscle cell in man. *Nature, London* 1966: **20**: 309–10.
21. Ahlborg B, Bergstrom J, Brohult J, Ekelund L-G, Hultman, E, Maschio G. Human muscle glycogen content and capacity for prolonged exercise after different diets. *Forsavarsmedicin* 1967; **3**: 85–99.
22. Hermansen L, Hultman E, Saltin B. Muscle glycogen during prolonged severe exercise. *Acta Physiologica Scandinavica* 1967; **71**: 129–39.
23. Costill DL, Gollnick PD, Jansson ED, Saltin B, Stein EM. Glycogen depletion pattern in human muscle fibres during distance running. *Acta Physiologica Scandinavica* 1973; **89**: 374–83.
24. Gollnick PD, Armstrong RB, Saubert IV CW, Sembrowich WL, Shep-

herd RE, Saltin B. Glycogen depletion patterns in human skeletal muscle fibers during prolonged work. *Plügers Archiv* 1973; **344**: 1–12.

25. Karlsson J, Saltin B. Diet, muscle glycogen and endurance performance. *Journal of Applied Physiology* 1971; **31**: 203–6.

26. Bergstrom J, Hermansen L, Hultman E, Saltin B. Diet, muscle glycogen and physical performance. *Acta Physiologica Scandinavica* 1967; **71**: 140–50.

27. Åstrand P-O. Diet and athletic performance. *Federation Proceedings* 1967; **26**: 1772–7.

28. Sherman WM, Costill DL, Fink WJ, Miller JM. Effect of exercise-diet manipulation on muscle glycogen and its subsequent utilization during performance. *International Journal of Sports Medicine* 1981; **114**: 114–18.

29. Goforth HW, Hodgdon JA, Hilderbrand RL. A double blind study of the effects of carbohydrate loading upon endurance performance. *Medicine and Science in Sports and Exercise* 1980; **12**: 108A.

30. Bergstrom J, Hermansen L, Hultman E, Saltin B. Diet, muscle glycogen, and physical performance. *Act Physiologica Scandinavica* 1967; **71**: 140–50.

31. Brewer J, Williams C, Patton A. The influence of high carbohydrate diets on endurance running performance. *European Journal of Applied Physiology* 1988; **57**: 698–706.

32. Madsen K, Pedersen PK, Rose P, Richter EA. Carbohydrate supercompensation and muscle glycogen utilization during exhaustive running in highly trained athletes. *European Journal of Applied Physiology* 1990; **61**: 467–72.

33. Costill DL, Jansson, E, Gollnick PD, Saltin B. Glycogen utilization in leg muscles of men during level and uphill running. *Acta Physiologica Scandinavica* 1974; **91**: 475–81.

34. Williams C, Brewer J, Walker M. The effect of a high carbohydrate diet on running performance during a 30-km treadmill time trial. *European Journal of Applied Physiology* 1992; **65**: 18–24.

35. Williams C, Nute MG, Broadbank L, Vinall S. Influence of fluid intake on endurance running performance. *European Journal of Applied Physiology* 1990; **60**: 112–19.

36. Lakomy HKA. The use of a non-motorized treadmill for analysing sprint performance. *Ergonomics* 1987; **30**: 627–37.

37. Lakomy HKA. Measurement of work and power output using friction loaded cycle ergometers. *Ergonomics* 1987; **29**: 509.

38. Boobis LH, Williams C, Wooton, SA. Influence of sprint training on muscle metabolism during brief maximal exercise in man. *Journal of Physiology* 1983; **342**: 36–7.

39. Nevill ME, Boobis LH, Brooks S, Williams C. Effect of training on muscle metabolism during treadmill sprinting. *Journal of Applied Physiology* 1989; **67**: 2376–82.

40. Greenhaff PL, Nevill ME, Soderlund K, Boobis L, Williams C, Hultman E. Energy metabolism in single muscle fibres during maximal sprint exercise in man. *Journal of Physiology* 1992; **446**: 528P.

41. Wootton SA, Williams C. Influence of carbohydrate-status on performance during maximal exercise. *International Journal of Sports Medicine* 1984; **5**: 126–7.

42. Maughan RJ, Poole DC. The effects of a glycogen-loading regimen on the capacity to perform anaerobic exercise. *European Journal of Applied Physiology* 1981; **46**: 211–19.

43. Maughan RJ, Greenhaff PL. High intensity exercise performance and acid-base balance: the influence of diet and metabolic alkalosis. *Medicine and Sport Science* 1991; **32**: 147–65.

44. Boobis LH. Metabolic aspects of fatigue during sprinting. In: Macleod D, Maughan R, Nimmo M, Reilly T, Williams C, eds. *Exercise, benefits, limitations and adaptations*. London: E & FN Spon, 1987: 116–140.

45. Gaitanos GC, Williams C, Boobis LH, Brooks S. Human muscle metabolism during intermittent maximal exercise. *Journal of Physiology* 1993; in press.

46. Essen B, Kaijser L. Regulation of glycolysis in intermittent exercise in man. *Journal of Physiology* 1978; **281**: 499–511.

47. Saltin B. Metabolic fundamentals of exercise. *Medicine and Science in Sports and Exercise* 1973; **15**: 366–9.

48. Jacobs I, Westlin N, Karlsson J, Rasmusson M, Houghton B. Muscle glycogen and diet in elite soccer players. *European Journal of Applied Physiology* 1982; **48**: 297–302.

49. Bangsbo J, Norregaard L, Thorsoe F. The effect of carbohydrate diet on intermittent exercise performance. *International Journal of Sports Medicine* 1992; **13**: 152–7.

50. Costill DL, Miller JM. Nutrition for endurance sport: carbohydrate and fluid balance. *International Journal of Sports Medicine* 1980; **1**: 2–14.

51. Neufer PD, Costill DL, Flynn MG, Kirwan JP, Mitchell JB, Houmard J. Improvements in exercise performance: effects of carbohydrate feedings and diet. *Journal of Applied Physiology* 1987; **62**: 983–8.

52. Sherman WM, Brodowicz G, Wright DA, Allen WK, Simonsen J, Dernbach A. Effects of pre-exercise carbohydrate feedings on cycling performance. *Medicine and Science in Sports and Exercise* 1989; **21**: 598–604.

53. Piearce L. Dietary habits of football players. In: Macleod DAD, Maughan RJ, Williams C, Madeley CR, Sharp JC, Nutton RW, eds. *Intermittent high intensity exercise*. London: E & FN Spon, 1993: 159–73.

54. Sherman WM, Peden MC, Wright DA. Carbohydrate feedings 1h before exercise improves cycling performance. *American Journal of Clinical Nutrition* 1991; **54**: 866–70.

55. Costill DL, Coyle E, Dalsky G, Evans W, Fink W, Hoopes D. Effects of elevated plasma FFA and insulin on muscle glycogen usage during exercise. *Journal of Applied Physiology* 1977; **43**: 695–9.

56. Foster C, Costill DL, Fink WJ. Effects of pre-exercise feedings on endurance performance. *Medicine and Science in Sports and Exercise* 1979; **11**: 1–5.

57. Devlin JT, Calles-Escandon J, Horton ES. Effects of pre-exercise snack feeding on endurance cycle exercise. *Journal of Applied Physiology* 1986; **60**: 980–5.

58. Gleeson M, Maughan RJ, Greenhaff PL. Comparison of the effects of pre-exercise feeding of glucose, glycerol and placebo on endurance and fuel homeostasis in man. *European Journal of Applied Physiology* 1986; **55**: 645–53.

59. McMurray RG, Wilson JR, Kitchell BS. The effects of fructose and glucose on high intensity endurance performance. *Research Quarterly on Exercise and Sport* 1983; **54**: 156–62.

60. Calles-Escandon J, Devlin JT, Whitcomb W, Horton ES. Pre-exercise feeding does not affect endurance cycle exercise but attenuates post-exercise starvation-like response. *Medicine and Science in Sports and Exercise* 1991; **23**: 818–24.

61. Porte D, Williamson RH. Inhibition of insulin release by norepinephrine in man. *Science* 1966; **152**: 1248–50.

62. Wright DA, Sherman WM, Dernbach AR. Carbohydrate feedings before, during, or in combination improve cycling endurance performance. *Journal of Applied Physiology* 1991; **71**: 1082–8.

63. Muckle DS. Glucose syrup ingestion and team performance in soccer. *British Journal of Sports Medicine* 1973; **7**: 340–3.

64. Kirkendall DT, Foster C, Dean JA, Grogan J, Thompson NN. Effect of glucose polymer supplementation on performance of soccer players. In: Reilly T, Lees A, David K, Murphy WJ, eds. *Science and football*. London: E & FN Spon, 1988: 33–41.

65. Ivy JL *et al*. Endurance improved by ingestion of a glucose polymer supplement. *Medicine and Science in Sports and Exercise* 1983; **15**: 466–71.

66. Bjorkman O, Sahlin K, Hagenfeldt L, Wahren J. Influence of glucose and fructose ingestion on the capacity for long-term exercise in well-trained men. *Clinical Physiology* 1984; **4**: 483–94.

67. Coyle EF, Coggan AR, Hemmert MK, Ivy JL. Muscle glycogen utilization during prolonged strenuous exercise when fed carbohydrate. *Journal of Applied Physiology* 1986; **61**: 165–72.

68. Okano G, Takeda H, Morita I, Katoh M, Mu Z, Miyake S. Effect of pre-exercise fructose ingestion on endurance performance in fed men. *Medicine and Science in Sports and Exercise* 1987; **20**: 105–9.

69. Coggan AR, Coyle EF. Metabolism and performance following carbohydrate ingestion late in exercise. *Medicine and Science in Sports and Exercise* 1989; **21**: 59–65.

70. Maughan RJ, Fenn CE, Leiper JB. Effects of fluid, electrolyte and substrate ingestion on endurance capacity. *European Journal of Applied Physiology* 1989; **58**: 481–6.

71. Wilber RL, Moffatt JR. Influence of carbohydrate ingestion on blood

glucose and performance in runners. *International Journal of Sports Nutrition* 1992; **2**: 317–27.

72. Kuipers H, Keizer HA, Brouns F, Saris WHM. Carbohydrate feeding and glycogen synthesis during exercise in man. *European Journal of Applied Physiology* 1987; **410**: 652–6.

73. Coggan AR, Coyle EF. Carbohydrate ingestion during prolonged exercise: effects of metabolism and performance. In: Holloszy JO, ed. *Exercise and sports sciences reviews*. Baltimore: Williams & Wilkins, 1991: 1–40.

74. Tsintzas OK, Williams C, Wilson W. Influence of carbohydrate ingestion on glycogen utilization during prolonged running in man. *Journal of Physiology* 1993; in press.

75. Wahren J. Substrate utilization by exercising muscle in man. *Progress in Cardiology* 1973; **2**: 255–80.

76. Bonen A, Malcolm SA, Kilgour RD, MacIntyre KP, Belcastro AN. Glucose ingestion before and during intense exercise. *Journal of Applied Physiology* 1981; **50**: 766–71.

77. Pascoe DD, Costill DL, Robergs RA, Davis JA, Fink W, Pearson DR. Effects of exercise mode on muscle glycogen restorage during repeated days of exercise. *Medicine and Science in Sports and Exercise* 1990; **22**: 593–8.

78. Kirwan JP *et al.* Carbohydrate balance in competitive runners during successive days of intense training. *Journal of Applied Physiology* 1988; **65**: 2601–6.

79. Wallberg-Henriksson H, Constable SH, Young DA, Holloszy JO. Glucose transport into rat skeletal muscle: interaction between exercise and insulin. *Journal of Applied Physiology* 1988; **65**: 909–13.

80. Adolfsson S, Ahren K. Control mechanisms for the synthesis of glycogen in striated muscle. In: Pernow B, Saltin B, ed. *Muscle metabolism during exercise*. New York: Plenum Press, 1971: 257–72.

81. Piehl K, Adolfsson S, Nazar K. Glycogen storage and glycogen synthetase activity in trained and untrained muscle of man. *Acta Physiologica Scandinavica* 1974; **90**: 779–88.

82. Vollestad NK, Vaage O, Hermansen L. Muscle glycogen depletion patterns in type I and subgroups of type II fibres during prolonged severe exercise in man. *Acta Physiologica Scandinavica* 1984; **122**: 433–41.

83. Yan Z, Spencer MK, Katz A. Effect of low glycogen on glycogen synthase in human muscle during and after exercise. *Acta Physiologica Scandinavica* 1992; **145**: 345–52.

84. Zachwieja JJ, Costill DL, Pascoe DD, Robergs RA, Fink WJ. Influence of muscle glycogen depletion on the rate of resynthesis. *Medical and Science in Sports and Exercise* 1991; **23**: 44–8.

85. Robergs RA. Nutrition and exercise determinants of post-exercise glycogen synthesis. *International Journal of Sports Nutrition* 1991; **1**: 307–37.

86. Ivy JL. Muscle glycogen synthesis before and after exercise. *Sports Medicine* 1991; **11**: 6–19.

87. Blom P, Hostmark AT, Vaage O, Kardel KR, Maehlum S. Effect of different post-exercise sugar diets on the rate of muscle glycogen synthesis. *Medicine and Science in Sports and Exercise* 1987; **19**: 491–6.

88. Roberts KM, Noble EG, Hayden DB, Taylor AW. Simple and complex carbohydrate-rich diets and muscle glycogen content of marathon runners. *European Journal of Applied Physiology* 1988; **57**: 70–4.

89. Zawadzki KM, Yaspelkis BB III, Ivy JL. Carbohydrate–protein complex increases the rate of muscle glycogen storage after exercise. *Journal of Applied Physiology* 1992; **72**: 1854–9.

90. Coyle EF, Montain SJ. Benefits of fluid replacement with carbohydrate during exercise. *Medicine and Science in Sports and Exercise* 1992; **24**(Suppl.): S324–30.

91. Gonzalez-Alonso J, Heaps CL, Coyle EF. Rehydration after exercise with common beverages and water. *International Journal of Sports Medicine* 1992; **13**: 399–406.

92. Neufer PD, Sawka MN, Young AJ, Quigley MD, Latzka WA, Levine L. Hypohydration does not impair skeletal muscle glycogen resynthesis after exercise. *Journal of Applied Physiology* 1991; **70**: 1490–4.

93. Johansson C, Tsai L, Hultman E, Tegelman R, Pousette A. Restoration of anabolic deficit and muscle glycogen consumption in competitive orienteering. *International Journal of Sports Medicine* 1990; **11**: 204–7.

94. O'Reilly KP, Warhol MJ, Fielding RA, Frontera WA, Meredith CN,

Evans WWJ. Eccentric exercise-induced muscle damage impairs muscle glycogen repletion. *Journal of Applied Physiology* 1987; **63**: 252–6.

95. Piehl K. Time course of refilling of glycogen stores in human muscle fibres following exercise-induced glycogen repletion. *Acta Physiologica Scandinavica* 1974; **90**: 297–302.

96. Keizer HA, Kuipers H, van Kranenburg G. Influence of liquid and solid meals on muscle glycogen resynthesis, plasma fuel hormone response, and maximal physical working capacity. *International Journal of Sports Medicine* 1987; **8**: 99–104.

97. Nevill ME, Williams C, Roper D, Slater C, Nevill AM. Effect of dietary manipulation on recovery from maximal intermittent exercise. *Journal of Sports Science* 1993; **11**: 119–26.

98. Fallowfield J, Williams C. Carbohydrate intake and recovery from prolonged exercise. *International Journal of Sports Nutrition* 1993; **3**: 150–64.

99. Fallowfield J, Williams C, Singh R. Unpublished data.

100. Coggan A, Coyle EF. Reversal of fatigue during prolonged exercise by carbohydrate infusion or ingestion. *Journal of Applied Physiology* 1987; **63**: 2388–95.

101. Wootton SA. Nutritional beliefs and eating habits of British athletes and coaches. In: Shrimpton DH, Berry Ottaway P, eds. *Nutrition in sport*. Loughborough: Echo Press, 1986: 64–75.

102. Shoaf LR, McClellan PD, Birkskovich KA. Nutrition knowledge, interests, and information sources of male athletes. *Journal of Nutritional Education* 1986; **18**: 243–5.

103. Williams MH. *Beyond training: How athletes enhance performance legally and illegally*. Champain, IL: Leisure Press, 1989: 214.

104. Conlee RK. Amphetamine, caffeine, and cocaine. In: Lamb DR, Williams MH, eds. *Ergogenics-enhancement of performance in exercise and sport*. Indianapolis: Wm. C. Brown, 1991: 285–319.

105. Weir J, Noakes TD, Myburgh K, Adams B. A high carbohydrate diet negates the metabolic effects of caffeine during exercise. *Medicine and Science in Sports and Exercise* 1986; **19**: 100–5.

106. Cerretelli P, Marconi C. L-carnitine supplementation in humans. The effects on physical performance. *International Journal of Sports Medicine* 1990; **11**: 1–14.

107. Wagenmakers AJM. L-carnitine supplementation and performance in man. In: Brouns F, ed. *Advances in nutrition and top sport*. Basel: Karger, 1991: 110–27.

108. Heigenhauser GJF, Jones NL. Bicarbonate loading. In: Lamb DR, Williams MH, eds. *Ergogenics—enhancement of performance in exercise and sport*. Indianapolis: Wm. C. Brown, 1991: 183–203.

109. Goldfinch J, McNaughton L, Davies P. Induced metabolic alkalosis and its effects of 400-m racing time. *European Journal of Applied Physiology* 1988; **57**: 45–8.

110. Wilkes D, Gledhill N, Smyth R. Effect of acute alkalosis on 800-m racing time. *Medicine and Science in Sports and Exercise* 1983; **15**: 277–80.

111. Gaitanos GC, Nevill ME, Brooks S, Williams C. Repeated bouts of sprint running after induced alkalosis. *Journal of Sports Science* 1991; **9**: 355–70.

112. Sahlin Ch. 2.

113. Harris RC, Soderlund K, Hultman E. Elevation of creatine in resting and exercised muscle of normal subjects by creatine supplementation. *Clinical Science* 1992; **83**: 367–74.

114. Greenhaff PL, Casey A, Short AH, Harris R, Soderlund K, Hultman E. The influence of oral creatine supplementation on muscle torque during repeated bouts of maximal voluntary exercise in man. *Clinical Science* 1993; in press.

115. Newsholme EA, Parry-Billings, M, McAndrews, N, Budgett R. A biochemical mechanism to explain some characteristics of overtraining. In: Brouns F, ed. *Advances in nutrition and top sport*. Basel: Karger, 1991: 79–93.

116. Wilson W, Maughan RJ. Evidence for a possible role of 5-hydroxytryptamine in the genesis of fatigue in man: administration of paroxetine, a 5-HT re-uptake inhibitor, reduces the capacity to perform prolonged exercise. *Experimental Physiology* 1992; **77**: 921–4.

117. Gleeson M, Maughan RJ, Greenhaff P. Comparison of the effects of

pre-exercise feeding of glucose, glycerol, and placebo on performance and fuel homeostasis in man. *European Journal of Applied Physiology* 1987; **55**: 645–53.

118. Davis JM, Baily SP, Woods JA, Galiano FJ, Hamilton MT, Bartoli WP. Effects of carbohydrate feedings on plasma free tryptophan and branched-chain amino acids during prolonged cycling. *European Journal of Applied Physiology* 1992; **65**: 513–19.

119. Blomstrand E, Hassmen P, Ekblom B, Newsholme EA. Administration of branched-chain amino acids during sustained exercise-effects on performance and on plasma concentration of some amino acids. *European Journal of Applied Physiology* 1991; **63**: 83–8.

120. Wagenmakers AJM *et al.* Carbohydrate supplementation, glycogen depletion and amino acid metabolism during exercise. *American Journal of Physiology* 1991; **26**: 883–90.

1.2.3 Fluid and electrolyte loss and replacement in exercise

R. J. MAUGHAN

INTRODUCTION

Fatigue is an inevitable accompaniment of prolonged strenuous exercise, but the nature of the fatigue process will be influenced by many factors. The most important of these is undoubtedly the intensity of the exercise in relation to the capacity of the individual, and the most effective way to delay the onset of fatigue and improve performance is by training. The primary cause of fatigue in exercise lasting more than 1 h but not more than 4 to 5 h is usually the depletion of the body's carbohydrate reserves. This time-scale covers most ball games such as football, hockey, and tennis, and also individual events such as marathon running. Systematic training results in many adaptations to the cardiovascular system and to the muscles, allowing them to increase the extent to which they can use the relatively unlimited fat stores as a fuel and thus spare the rather small amounts of carbohydrate which are stored in the liver and the muscles. Where the availability of carbohydrate fuel limits exercise, this will result in an improved performance.

Many other factors will influence performance, however, and among these are the environmental conditions under which the exercise is performed. When the ambient temperature and humidity are high, the capacity to perform prolonged exercise is reduced: in this situation, dehydration and thermoregulatory problems rather than substrate depletion may be the cause of fatigue. At rest the rate of heat production by the body is low, but at high work rates metabolic heat production can exceed 80 kJ/min (20 kcal/min), and highly trained athletes can sustain these work rates for more than 2 h. The rate of sweating necessary to lose this heat load will result in a rapid loss of body water with an associated loss of electrolytes.

FLUID LOSS AND TEMPERATURE REGULATION

Fluid loss during exercise is linked to the need to maintain body temperature within narrow limits. The resting oxygen consumption is about 250 ml/min, corresponding to a rate of heat production of about 70 W. Thermoregulation is primarily achieved by behavioural mechanisms: the amount of clothing worn is adjusted or the ambient temperature is changed so that the rate of heat production is balanced by the rate of heat loss. During exercise, the rate of heat production can be increased to many times the resting level. Although this does not pose a problem in events of short duration, it represents a major threat to the endurance athlete. In an event such as running on the level, the rate of heat production is determined primarily by running speed and body mass, with individual variations in mechanical and metabolic efficiency being of secondary importance. Running a marathon in 2 h 30 min requires an oxygen consumption of about 4 l/min to be sustained throughout the race for the average runner with a body mass of 70 kg. When the ambient temperature is higher than skin temperature, heat will also be gained from the environment by physical transfer. Despite this, marathon runners normally maintain body temperature within 2 to 3°C of the resting level, indicating that heat is being lost from the body almost as fast as it is being produced.

At high ambient temperatures, the only mechanism by which heat can be lost from the body is evaporation, and even at low ambient temperatures high sweat rates are sometimes observed.[1] Evaporation of 1 l of water from the skin will remove 2.4 MJ (580 kcal) of heat from the body. For the 2-h 30-min marathon runner with a body mass of 70 kg to balance the rate of metabolic heat production by evaporative loss alone would therefore require sweat to be evaporated from the skin at a rate of about 1.6 l/h; at such high sweat rates, an appreciable fraction drips from the skin without evaporating, and a sweat secretion rate of about 2 l/h is likely to be necessary to achieve this rate of evaporative heat loss. This is possible, but it would result in the loss of 5 l of body water, corresponding to a loss of more than 7 per cent of body weight for a 70-kg runner. Water will also be lost by evaporation from the respiratory tract. During hard exercise in a hot dry environment, this can amount to a significant water loss, although it is not generally considered to be a major heat loss mechanism in man. The rise of 2 to 3°C in body temperature which normally occurs during marathon running means that some of the heat produced is stored, but the effect on heat balance is minimal; a rise in mean body temperature of 3°C for a 70-kg runner would reduce the total requirement for evaporation of sweat by less than 300 ml.

It is often reported that exercise performance is impaired when an individual is dehydrated by as little as 2 per cent of body weight, and that losses in excess of 5 per cent of body weight can decrease the capacity for work by about 30 per cent.[2] Prior dehydration will impair the capacity to perform high intensity exercise as well as endurance activities.[3,4] Nielsen *et al.*[3] showed that prolonged exercise, which resulted in a loss of fluid corresponding to 2.5 per cent of body weight, resulted in a 45 per cent fall in the capacity to perform high intensity exercise.

Fluid losses are distributed in varying proportions among the plasma, extracellular water, and intracellular water. The decrease in plasma volume which accompanies dehydration may be of particular importance in influencing work capacity; blood flow to the muscles must be maintained at a high level to supply oxygen and substrates, but a high blood flow to the skin is also necessary to remove heat by convection to the body surface where it can be dissipated.[5] When the ambient temperature is high and blood volume has been decreased by sweat loss during prolonged exercise, there may be difficulty in meeting the requirement for a high blood flow to both these tissues. In this situation, skin blood flow is likely to be compromised, allowing central venous pressure and muscle blood flow to be maintained but reducing heat loss and causing body temperature to rise. [6]

ELECTROLYTE LOSS IN SWEAT AND THE EFFECTS ON BODY FLUIDS

The sweat which is secreted on to the skin contains a wide variety of organic and inorganic solutes, and significant losses of some of these components will occur where large volumes of sweat are produced. The electrolyte composition of sweat is variable, and the concentration of individual electrolytes as well as the total sweat volume will influence the extent of losses. The normal concentration ranges for the main ionic components of sweat are shown in Table 1, along with their plasma and intracellular concentrations for comparison. A number of factors contribute to the variability in the composition of sweat: methodological problems in the collection procedure, including evaporative loss, incomplete collection, and contamination with skin cells, account for at least part of the variability, but there is also a large biological factor.

Table 1 Concentration of the major electrolytes in sweat, plasma, and intracellular water

	Concentration (mmol/l)		
	Sweat	Plasma	Intracellular
Sodium	20–80	130–155	10
Potassium	4–8	3.2–5.5	150
Calcium	0–1	2.1–2.9	0
Magnesium	<0.2	0.7–1.5	15
Chloride	20–60	96–110	8
Bicarbonate	0–35	23–28	10
Phosphate	0.1–0.2	0.7–1.6	65
Sulphate	0.1–2.0	0.3–0.9	10

These values are taken from a variety of sources, but are based primarily on those reported in refs. 7–9.

The sweat composition undoubtedly varies between individuals, but can also vary within the same individual depending on the rate of secretion, the state of training, and the state of heat acclimation.[10] In response to a standard heat stress, the sweat rate increases with training and acclimation and the electrolyte content decreases. These adaptations allow improved thermoregulation while conserving electrolytes.

The major electrolytes in sweat, as in the extracellular fluid, are

sodium and chloride (Table 1), although the sweat concentrations of these ions are invariably lower than those in plasma. Contrary to what might be expected, Costill[11] reported an increased sodium and chloride sweat content with increased flow, but Verde et al.[12] found that the sweat concentration of these ions was unrelated to the sweat flow rate. Acclimation studies have shown that elevated sweating rates are accompanied by a decrease in the concentration of sodium and chloride in sweat.[13,14] The potassium content of sweat appears to be relatively unaffected by the sweat rate, and the magnesium content is also unchanged or perhaps decreases slightly. These apparently conflicting results demonstrate some of the difficulties in interpreting the literature in this area. Differences between studies may be due to differences in the training status and degree of acclimation of the subjects used as well as differences in methodology; some studies have used whole-body washdown techniques to collect sweat, whereas others have examined local sweating responses using ventilated capsules or collection bags.

Because sweat is hypotonic with respect to body fluids, the effect of prolonged sweating is to increase the plasma osmolality, which may have a significant effect on the ability to maintain body temperature. A direct relationship between plasma osmolality and body temperature has been demonstrated during exercise.[15,16] Hyperosmolality of plasma, induced prior to exercise, has been shown to result in a decreased thermoregulatory effector response; the threshold for sweating is elevated and the cutaneous vasodilator response is reduced.[17] In short-term exercise, (30 min) however, the cardiovascular and thermoregulatory response appears to be independent of changes in osmolality induced during the exercise period.[18] The changes in the concentration of individual electrolytes are more variable, but an increase in the plasma sodium and chloride concentrations is generally observed in response to both running and cycling exercise. Exceptions to this are rare and occur only when excessively large volumes of drinks low in electrolytes are consumed over long time periods; these situations are discussed further below.

The plasma potassium concentration has been reported to remain constant after marathon running,[19,20] although others have reported small increases irrespective of whether drinks containing large amounts of potassium[21] or no electrolytes[22,23] were given. Much of the inconsistency in the literature relating to changes in the circulating potassium concentration can be explained by the variable time taken to obtain blood samples after exercise under field conditions; the plasma potassium concentration rapidly returns to normal in the post-exercise period.[24] Laboratory studies, where an indwelling catheter can be used to obtain blood samples during exercise, commonly show an increase in the circulating potassium concentration in the later stages of prolonged exercise. The potassium concentration of extracellular fluid (4–5 mmol/l) is small relative to the intracellular concentration (150–160 mmol/l), and release of potassium from liver, muscle, and red blood cells will tend to elevate plasma potassium levels during exercise despite the losses in sweat.

The plasma magnesium concentration is unchanged after 60 min of moderate intensity cycling exercise,[25] but Rose et al.[26] observed a 20 per cent fall in the serum magnesium concentration after a marathon race and attributed this to a loss in sweat; a fall of similar magnitude was reported by Cohen and Zimmerman.[23] A larger fall in the serum magnesium concentration has been observed during exercise in the heat than at neutral temperatures,[27] supporting the idea that losses in sweat are responsible. However, there are reports that the fall in plasma magnesium concentration that occurs during prolonged exercise is a consequence of redistribution, with uptake of magnesium by red blood cells,[28] active muscle,[11] or adipose tissue.[29] Although the concentration of potassium and magnesium in sweat is high relative to that in the plasma, the plasma content of these ions represents only a small fraction of the whole-body stores; Costill and Miller[30] estimated that only about 1 per cent of the body stores of these electrolytes was lost when individuals were dehydrated by 5.8 per cent of body weight.

CONTROL OF WATER AND ELECTROLYTE BALANCE

The excretion of some of the waste products of metabolism and the regulation of the body's water and electrolyte balance are the primary functions of the kidneys. Excess water or solute is excreted, and where there is a deficiency of water or electrolytes, these are conserved until the balance is restored. Under normal conditions, the osmolality of the extracellular fluid is maintained within narrow limits; since this is strongly influenced by the sodium concentration, sodium and water balance are closely linked.

At rest, approximately 15 to 20 per cent of the renal plasma flow is continuously filtered out by the glomeruli, resulting in the production of about 170 l of filtrate per day. Most (99 per cent or more) of this is reabsorbed in the tubular system, leaving about 1–1.5 l to appear as urine. The volume of urine produced is determined primarily by the action of antidiuretic hormone (ADH) which regulates water reabsorption by increasing the permeability of the distal tubule of the nephron and the collecting duct to water. ADH is released from the posterior lobe of the pituitary in response to signals from the supraoptic nucleus of the hypothalamus: the main stimuli for release of ADH, which is normally present only in low concentrations, are an increased signal from the osmoreceptors located within the hypothalamus, and a decrease in blood volume, which is detected by low pressure receptors in the atria and by high pressure baroreceptors in the aortic arch and carotid sinus. An increased plasma angiotensin concentration will also stimulate ADH output.

The sodium concentration of the plasma is regulated by the renal reabsorption of sodium from the glomerular filtrate. Most of the reabsorption occurs in the proximal tubule, but active absorption also occurs in the distal tubules and collecting ducts. A number of factors influence the extent to which reabsorption occurs, and among these is the action of aldosterone, which promotes sodium reabsorption in the distal tubules and enhances the excretion of potassium and hydrogen ions. Aldosterone is released from the kidney in response to a fall in the circulating sodium concentration or a rise in plasma potassium; aldosterone release is also stimulated by angiotensin which is produced by the renin–angiotensin system in response to a decrease in the plasma sodium concentration. Angiotensin thus acts on the release of both aldosterone and ADH. Atrial natriuretic factor is a peptide synthesized in and released from the atria of the heart in response to atrial distension. It increases the glomerular filtration rate and decreases sodium and water reabsorption leading to an increased loss; this may be important in the regulation of extracellular volume, but it seems unlikely that atrial natriuretic factor plays a significant role during exercise. Regulation of the body's sodium balance has profound implications for fluid balance, as sodium salts account for more than 90 per cent of the osmotic pressure of the extracellular fluid.

Loss of hypotonic fluid as sweat during prolonged exercise usually results in a fall in blood volume and an increased plasma osmolality; both these changes act as stimuli for the release of ADH.[31] The plasma ADH concentration during exercise has been reported to increase as a function of the exercise intensity.[32] Renal blood flow is also reduced in proportion to the exercise intensity and may be as low as 25 per cent of the resting level during strenuous exercise.[33] These factors combine to result in a decreased urine flow during exercise and usually for some time afterwards.[33] However it has been pointed out that the volume of water conserved by this decreased urine flow during exercise is small, probably amounting to no more than 12 to 45 ml/h.[34]

The effect of exercise is normally to decrease the renal excretion of sodium and to increase the excretion of potassium, although the effect on potassium excretion is rather variable.[34] These effects appear to be due largely to an increased rate of aldosterone production during exercise.[33] Although the concentrations of sodium, and particularly potassium, in the urine are generally high relative to the concentrations in extracellular fluid, the extent of total urinary loss in most exercise situations is small.

FLUID REPLACEMENT DURING EXERCISE

The ability to sustain a high rate of work output requires that an adequate supply of carbohydrate substrate be available to the working muscles, and fluid ingestion during exercise has the twin aims of providing a source of carbohydrate fuel to supplement the body's limited stores and supplying water to replace the losses incurred by sweating. Increasing the carbohydrate content of drinks will increase the amount of fuel which can be supplied, but will tend to decrease the rate at which water can be made available; where provision of water is the first priority, the carbohydrate content of drinks will be low, thus restricting the rate at which substrate is provided. The composition of drinks to be taken will thus be influenced by the relative importance of the need to supply fuel and water; this in turn depends on the intensity and duration of the exercise task, the ambient temperature and humidity, and the physiological and biochemical characteristics of the individual athlete. Carbohydrate depletion will result in fatigue and a reduction in the exercise intensity which can be sustained, but is not normally a life-threatening condition. Disturbances in fluid balance and temperature regulation have potentially more serious consequences, and therefore it is possible that, for the majority of participants in endurance events, the emphasis should be on proper maintenance of fluid and electrolyte balance.

AVAILABILITY OF INGESTED FLUIDS

The first barrier to the availablity of ingested fluids is the rate of gastric emptying which controls the rate at which fluids are delivered to the small intenstine and the extent to which they are influenced by the gastric secretions. The rate of emptying is determined by the volume and composition of fluid consumed. The volume of the stomach contents is a major factor in regulating the rate of emptying, and the rate of emptying of any solution can be increased by increasing the volume present in the stomach; emptying follows an exponential time course, and falls rapidly as the volume remaining in the stomach decreases.[35–37] Where a high rate of emptying is desirable, this can be promoted by keeping the volume high by repeated drinking.[38] Dilute solutions of glucose will leave the stomach almost, but not quite, as fast as plain water; the rate of emptying is slowed in proportion to the glucose content, and concentrated sugar solutions will remain in the stomach for long periods. There has been some debate as to the carbohydrate concentration at which an inhibitory effect on gastric emptying is first observed; the conflicting results reported in the literature are caused at least in part by deficiencies in the methodology employed in some studies. It appears that glucose concentrations of greater than 40 g/l will have some slowing effect on the rate of gastric emptying.[39]

It has been proposed that the rate of emptying of nutrient solutions is regulated so as to provide a constant rate of energy delivery to the intestine,[40] but it is clear from Fig. 1 that the rate of energy delivery from glucose solutions of different concentrations is not constant but is proportional to the glucose concentration of the drinks. Even though the volume emptied is decreased, the amount of glucose emptied is increased with more concentrated solutions.

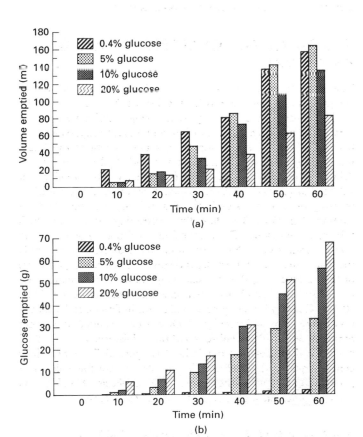

Fig. 1 Gastric emptying of flavoured water and glucose solutions of different concentrations after ingestion of a 200 ml drink: (a) the volume emptied; (b) the amount of glucose emptied. Measurements were made by scintigraphic imaging of technetium added to the drinks.

An increasing osmolality of the gastric contents will tend to delay emptying, and there is some evidence that substitution of glucose polymers for free glucose, which will result in a decreased osmolality for the same carbohydrate content, may be effective in increasing the volume of fluid and the amount of substrate delivered to the intestine. The differences are generally small, with the exception of a report by Foster et al.[41] who found that emptying of a 5 per cent glucose polymer solution was about one-third faster than that of a 5 per cent solution of free glucose; however, the results of this study may be misleading as no account was taken of the volume of fluid secreted into the stomach, and this is now known to be greater for solutions of free glucose than for polymers.[38,42] Sole and Noakes[42] found no significant difference in emptying rates between 5 per cent polymer and free glucose solutions, and Naveri et al.[43] made the same observation on 3 per cent solutions. With more concentrated solutions, Foster et al.[41] found no differences between polymer and free glucose solutions in the concentration range 10 to 40 per cent, but Sole and Noakes[42] found that 15 per cent polymer solutions emptied faster than the corresponding free glucose solution. It thus appears that the results are rather variable, but it is worth noting that there are no reports of polymer solutions being emptied more slowly than free glucose solutions with the same energy density; even when the difference is not significant, there is a tendency for faster emptying of polymer solutions.

The temperature of ingested drinks has been reported to have an influence on the rate of emptying, and it has been recommended that drinks should be chilled to promote gastric emptying.[44] This recommendation is based on a study by Costill and Saltin,[45] who gave subjects 400 ml of a dilute glucose solution at temperatures ranging from 5 to 35°C; the volume emptied in the first 15 min after ingestion was approximately twice as great for the solution at 5°C as for the solution at 35°C. However, more recent reports have cast some doubt on the importance of temperature in affecting emptying of liquids. Sun et al.[46] gave isosmotic orange juice at different temperatures, and found that the initial emptying rate for cold (4°C) drinks was slower than for drinks given at body temperature (37°C); the emptying rate for warm (50°C) drinks was not significantly different from that for the other two drinks. McArthur and Feldman[47] have also shown recently that the emptying rate remained unchanged for coffee drinks given at 4, 37, or 58°C. Lambert and Maughan[48] have recently used a tracer technique to show that fluids ingested at high temperature (50°C) appear in the circulation slightly faster than if they are chilled (4°C) before ingestion. Other factors, such as pH, may play a minor role. Although there is some evidence that emptying is hastened if drinks are carbonated,[49] more recent results suggest that carbonation has no effect;[50] it is probable that light carbonation as used in most sports drinks does not influence the gastric emptying rate, but a greater degree of carbonation, as used in many soft drinks, may be effective by raising the intragastric pressure.

Absorption of glucose occurs in the small intenstine, and is an active energy-consuming process linked to the transport of sodium. There is no active transport mechanism for water, which will cross the intestinal mucosa in either direction depending on the local osmotic gradients. The rate of glucose uptake is dependent on the luminal concentrations of glucose and sodium, and dilute glucose electrolyte solutions with an osmolality which is slightly hypotonic with respect to plasma will maximize the rate of water uptake.[51] Solutions with a very high glucose concentration will not necessarily promote an increased glucose uptake relative to more dilute solutions but, because of their high osmolality, will cause a net movement of fluid into the intestinal lumen. This will result in an effective loss of body water and will exacerbate any pre-existing dehydration. Other sugars, such as sucrose[52] or glucose polymers,[53,54] can be substituted for glucose without impairing glucose or water uptake. In contrast, the absorption of fructose is not an active process in man; it is absorbed less rapidly than glucose and promotes less water uptake.[55]

Several studies have shown that exercise at intensities of less than about 70 per cent of Vo_2max has little or no effect on intestinal function, although both gastric emptying and intestinal absorption may be reduced when the exercise intensity exceeds this level.[45,56] These studies have been reviewed and summarized by Brouns et al.[57]

Some more recent results obtained using an isotopic tracer technique to follow ingested fluids have suggested that there may be a decreased availability of ingested fluids even during low intensity exercise; a decreased rate of appearance in the blood of a tracer for water added to the ingested drinks indicated a decreased rate of appearance of the tracer at an exercise intensity of 40 per cent of Vo_2max.[58]

METABOLIC EFFECTS OF CARBOHYDRATE INGESTION DURING EXERCISE

Once emptied from the stomach and absorbed in the small intestine, carbohydrates ingested during exercise will enter the blood glucose pool, either directly or after metabolism in the liver. A fall in the circulating glucose concentration is commonly observed in the later stages of prolonged exercise and ingestion of glucose during exercise will maintain or raise the blood glucose concentration compared with the situation where no glucose is given.[59-61]. Sugars other than glucose are commonly used in the formulation of sports drinks, and there is some justification for their inclusion as similar effects are seen with short-chain (3–10 glucosyl units) glucose polymers,[62-67] sucrose,[68] or mixtures of sugars.[69-71]

One of the aims of ingesting carbohydrate during exercise is to spare the limited muscle glycogen stores as there is a good relationship between the availability of muscle glycogen and endurance capacity.[72] It is not clear how effectively this aim can be achieved, nor indeed whether sparing of muscle glycogen is necessary for ingested carbohydrate to be effective in increasing endurance capacity. It has been reported that glucose solutions providing 1 g glucose per kilogram body weight can reduce the rate of muscle glycogen utilization by about 30 per cent during 90 min of bicycle exercise at 65 to 70 per cent of Vo_2max.[61] In more prolonged exercise (4 h) consisting of low intensity cycle exercise interspersed with high intensity sprints, Hargreaves et al.[73] fed subjects hourly with either a flavoured placebo or drinks containing 43 g of sucrose together with small amounts of fat and protein; the rate of muscle glycogen utilization was not different between the two trials in the first hour, but over the following 3 h was about 37 per cent lower on the fed trial. Some other studies, which have employed a variety of exercise models and have fed different types and amounts of carbohydrate during exercise, have shown no effect of carbohydrate feeding during exercise on the rate of muscle glycogen utilization.[64,67,74,75] The reason for these different results is not clear, and may be partly explained by differences in the type and amount of carbohydrate given, by the different exercise models used, and by differences in the training status of the subjects. Nutritional status may also be important; Flynn et al.[76] found that carbohydrate feeding during exercise had no effect on the rate of muscle glycogen breakdown when this was elevated prior to exercise by a carbohydrate-loading procedure.

The ready availability of ingested carbohydrate as a fuel for the working muscles is demonstrated by the numerous studies which have followed the appearance in expired air of isotopes of carbon added as tracers. Oxidation of ingested glucose can account for about half the total carbohydrate oxidation after 1 to 2 h of walking at 50 per cent of Vo_2max;[77] after 3 to 4 h ingested glucose can supply as much as 90 per cent of the total carbohydrate oxidation.[78] In this situation, there is clearly some sparing of endogenous carbohydrate, but it is not clear whether the results can be applied to exercise at higher intensities where it appears that total carbohydrate turnover is increased when exogenous carbohydrates are given.

EFFECTS OF FLUID INGESTION ON PERFORMANCE

The effects of feeding different types and amounts of beverages during exercise have been extensively investigated using a wide variety of experimental models. Not all of these studies have shown a positive effect of fluid ingestion on performance but, with the exception of a few investigations where the composition of the drinks administered was such as to result in gastrointestinal disturbances, there are no studies showing that fluid ingestion will have an adverse effect on performance.

Laboratory studies—cycling

Laboratory investigations into the ergogenic effects of the administration of carbohydrate–electrolyte drinks during exercise have usually relied upon changes in physiological function during submaximal exercise or on the exercise time to exhaustion at a fixed work rate as a measure of performance. While this is a perfectly valid approach in itself, it must be appreciated that there are difficulties in extrapolating results obtained in this way to a race situation in which the work load is likely to fluctuate as the pace, the weather conditions, and the topography vary, and where tactical considerations and motivational factors are involved. It is possible to demonstrate large differences in the time for which a fixed work load can be sustained in laboratory tests when carbohydrate solutions are given during exercise; in one study, for example, a 30 per cent increase (from 3 to 4 h) was seen.[64] In a simulated race situation, where a fixed distance had to be covered as fast as possible, the advantage would translate to no more than a few per cent, and in a real competition it would probably be even less. However, a few per cent is often the difference between a world class performance and a mediocre one. To take account of some of these factors, some recent investigations have used exercise tests involving intermittent exercise, simulated races, or prolonged exercise followed by a sprint finish. Because different exercise tests and different solutions and rates of administration have been used in these various studies, comparisons between them are difficult. Some have included a trial where no fluids were given, whereas others have compared the effects of test solutions with trials where plain water or a flavoured placebo drink was administered. These studies have been the subject of a number of extensive reviews which have concentrated on the effects of administration of carbohydrate, electrolytes, and water on exercise peformance, and the results of the individual studies will not be considered in detail here.[79-82] These studies have generally reported an improvement in performance with the ingestion of carbohydrate-containing drinks; although it does not always reach statistical significance, there are no reports of adverse effects on performance.

Laboratory studies—treadmill exercise

As with studies during cycling exercise many different exercise models have been used to investigate the effcts of the administration of carbohydrate–electrolyte solutions during walking and running. Again, conflicting results have been obtained in that a significant effect of glucose ingestion has not always been observed. Sasaki et al.[68] found that running time at 80 per cent of V_{O_2}max was increased, compared with a placebo trial, by ingestion of 90 g of sucrose in a volume of 500 ml. Macaraeg[83] reported an increased endurance time during running at 85 per cent of maximum heart rate when a 7 per cent carbohydrate–electrolyte solution providing 84 g of carbohydrate in a volume of 1.2 l was given compared with either a no-drink or water-only trial. In contrast to these results, however, Fruth and Gisolfi[84] gave subjects a placebo or 150 g of glucose or fructose as a 10 per cent solution during treadmill running at 70 per cent of V_{O_2}max; running time on the fructose trial was less than on the other two runs, but there was no difference in running time between the glucose and placebo trials. Riley et al.[85] also found no difference in running time when a 7 per cent carbohydrate–electrolyte solution was given compared with a placebo trial; in this study subjects fasted for 21 h before exercise tests, and the first drink was given 20 min before exercise. In a study of very prolonged walking, Ivy et al.[86] reported an increased walking time (299 min) when 120 g of glucose polymer was given in a volume of 1.5 l compared with a placebo trial (268 min).

Williams and coworkers[87,88] have used an experimental model in which the subject is able to adjust the treadmill speed while running; the subject can then be encouraged either to cover the maximum distance possible in a fixed time or to complete a fixed distance in the fastest time possible. They showed that ingestion of 1 l of a glucose polymer–sucrose (50 g/l) solution did not increase the total distance covered in a 2-h run, but that the running speed was greater over the last 30 min of exercise when carbohydrate was given compared with a placebo trial.[87] They observed a similar effect when a carbohydrate solution (50 g of glucose–glucose polymer or 50 g of fructose–glucose polymer) or water was given in a 30-km treadmill time trial.[88] The running speed decreased over the last 10 km of the water trial, but was maintained in the other two runs; there was no significant difference between the three trials in the time taken to cover the total distance. As with cycling exercise, the conclusion must be that ingestion of carbohydrate-containing drinks is generally effective in improving performance.

Field studies

There are many practical difficulties associated with the conduct of field trials to assess the efficacy of ergogenic aids, which accounts for the fact that few well-controlled studies of the effects of administration of glucose–electrolyte solutions have been carried out in this way. The main problem is with the design of an adequately controlled trial: if a cross-over design is used, this is likely to be confounded by changes in the environmental conditions between trials, and the use of parallel control and test groups raises the difficulty of matching the groups. Many of the early studies purporting to show beneficial effects of ingestion of carbohydrate-containing solutions on performance in events such as cycling, canoeing, and soccer were so poorly designed that the results are of no value.

Cade et al.[89] gave subjects no fluids, or approximately 1 l of hypotonic saline, or a glucose–electrolyte solution during a 7-mile course consisting of walking and running at an ambient temperature of 32 to 34°C. None of the subjects completed the course when no fluid was given, and the mean distance covered was 4.7 miles; when saline was given, they covered 5.5 miles and all subjects completed the 7-mile course when given the glucose–electrolyte solution.

Studies where matched groups of competitors consumed 1.4 l of either water or a glucose–electrolyte solution during a marathon race[90] or 1.4 l of different carbohydrate-containing drinks during marathon and ultramarathon races[75] have shown no differences between the groups in finishing time. In the study by Maughan and Whiting,[90] subjects were matched on the basis of their anticipated finishing times. Many of these individuals had not previously completed a marathon and so these times must be considered unreliable. Nonetheless, the mean finishing time for the runners ($n=43$) drinking the carbohydrate-electrolyte solution was 220 ± 40 min compared with a predicted finishing time of 220 ± 35 min; for the group drinking water ($n=47$) the actual finishing time was 217 ± 32 min and the predicted time was 212 ± 32 min. Twenty-four runners (60 per cent) in the carbohydrate–electrolyte group ran faster than expected, compared with 19 (40 per cent) in the water group.

Leatt[91] gave 1 l of a 7 per cent glucose polymer solution or a flavoured placebo to soccer players during a practice game. During the match, the group who had been given carbohydrate utilized 31 per cent less glycogen than the placebo group. No measure of performance of the two groups was made, but it was suggested that the players taking the glucose polymer would experience a beneficial effect in the later stages of the game.

PRACTICAL ISSUES IN FLUID REPLACEMENT DURING EXERCISE

Many factors affect the need for fluid replacement during exercise. The composition of the fluid, as well as the volume and frequency of drinks, which will confer the greatest benefit during exercise will depend very much on individual circumstances. As with most physiological variables, there is a large interindividual variability in the rates of fluid loss during exercise under standardized conditions and also in the rates of gastric emptying and intestinal absorption of any ingested beverage. Marathon runners competing under the same conditions and finishing in the same time may lose as little as 1 per cent or as much as 5 per cent of body weight, even though their fluid intake during the race is the same.[1] Under more controlled conditions, the sweat rate during 1 h of exercise at a work load of 70 per cent of V_{O_2}max and an ambient temperature of 23°C ranged from 426 to 1665 g/h.[92] It would seem logical that the need for fluid (water) replacement is greater in the individual who sweats profusely, and any guidelines as to the rate of fluid ingestion and the composition of fluids to be taken must be viewed with caution when applied to the individual athlete. Sweat rate in activities such as running can be predicted from estimates of the energy cost of running, as used by Barr and Costill,[93] but these do not explain the variation which is observed between individuals. A more reliable method might be for the individual to measure body weight before and after training or simulated competition and to estimate sweat loss from the change in body weight.

Many individuals and organizations have issued recommendations as to the most appropriate fluid-replacement regimens.[44,94,95] Olsson and Saltin[94] recommended 100 to 300 ml of a 5 to 10 per cent sugar solution every 10 to 15 min during exercise; they also suggested that the temperature of ingested fluids should be 25°C. At the extreme ends of this range, this would give an intake each hour of 400 to 1800 ml of liquid and 20 to 180 g of sugar. In 1975 the American College of Sports Medicine publishd a Position Statement on the prevention of heat injuries during distance running, in which an intake of 400 to 500 ml of fluid 10 to 15 min before exercise was recommended; although no figures were given, it was also suggested that runners ingest fluids frequently during competition and that the sugar and electrolyte content of drinks should be low (2.5 per cent and 10 mmol/l sodium respectively) so as not to delay gastric emptying.[95] A revised version of these guidelines[44] continued to recommend hyperhydration prior to exercise by the ingestion of 400 to 600 ml of cold water 15 to 20 min before the event. The recommendations as to intake during a race were more specific than previously: cool water was stated to be the optimum fluid, and an intake of 100 to 200 ml every 2 to 3 km was suggested, giving a total intake of 1400 to 4200 ml at the extremes. Again, taking these extreme values, it is unlikely that the elite runners could tolerate a rate of intake of about 2 l/h, and equally unlikely that an intake of 300 ml/h would be adequate for the slowest competitors except when the ambient temperature was low.

Exercise intensity and duration

The rate of metabolic heat production during exercise is dependent on the exercise intensity and the body mass; in activities such as running or cycling this is a direct function of speed. The rate of rise of body temperature in the early stages of exercise and the steady state level which is eventually reached are both proportional to the metabolic rate. Therefore the rate of sweat production is also closely related to the absolute work load. In many sports, including most ball games, short bursts of high intensity activity are separated by variable periods of rest or low intensity exercise.

The time for which high intensity exercise can be sustained is necessarily rather short; the factors limiting exercise performance where the duration is in the range of about 10 to 60 min are not clear, but it does seem that substrate availability is not normally a limiting factor and that performance will not be improved by the ingestion of carbohydrate-containing beverages during exercise. Also, even though the sweat rate may be high, the total amount of water lost by sweating is likely to be rather small. Accordingly, there is generally no need for fluid replacement during very high intensity exercise, although it is difficult to define a precise cut-off point. In a recent study the effects of an intravenous infusion of saline during cycle ergometer exercise to exhaustion at an exercise intensity equivalent to 84 per cent of V_{O_2}max were investigated.[96] In the control trial a negligible amount of saline was infused, whereas an infusion rate of about 70 ml/min was used in the other trial. The saline infusion was effecive in reducing the decrease in plasma volume which occurred in the initial stages of exercise, although it did not completely abolish this response, and the core temperature and heart rate at the point of exhaustion were both lower in the infusion trial. There was no effect on endurance time which was the same in both trials. However, the endurance times were short (20.8 and 22.0 min for the infusion and control trials respectively), although the range was large (from about 9 to 43 min), and these results support the idea that fluid provision will not benefit exercise performance when the exercise duration is short.

There are also likely to be real problems associated with any attempt to replace fluids orally during very intense exercise. The rate of gastric emptying, which is probably the most important factor in determining the fate of ingested fluid, is impaired when the exercise intensity is high, as described above. Even at rest, the maximum rates of gastric emptying which have been reported are only about half the saline infusion rate (70 ml/min) used in the study by Deschamps *et al*.[96] and are commonly much less than this. To achieve a high rate of fluid delivery from the stomach, it is necessary to ingest large volumes, and any attempt to do so when the exercise intensity exceeds about 80 per cent of V_{O_2}max would almost certainly result in nausea and vomiting.

At lower intensities of exercise, the duration of exercise is inversely related to the intensity. In an activity such as running, this holds true for populations as much as for individuals. As the distance of a race increases, so the pace that an individual can sustain decreases;[97] equally, in an event such as a marathon race where all runners complete the same distance, the slower runners are generally exercising at a lower relative (as a percentage of V_{O_2}max and absolute work intensity.[98] Because the faster runners are exercising at a higher work load, in absolute as well as relative terms, their sweat rate is higher, although this effect is offset to some extent by the fact that they generally have a lower body weight; however, because the faster runners are active for a shorter period of time, the total sweat loss during a marathon race is unrelated to the finishing time.[1] Therefore the need for fluid replacement is much the same in terms of the total volume required, irrespective of running speed, but there is a need for a higher rate of replacement in the faster runners. Among the fastest marathon runners, sweat rates of about 30 to 35 ml/min can be sustained for a period of about 2 h 15 min by some runners. The highest sustained rates of gastric emptying reported in the literature are greater than this, at about 40 ml/min.[45,99] These gastric emptying measurements were made on resting subjects, and it is possible that there may be some inhibition of gastric emptying at the exercise intensity (about 75 per cent of V_{O_2}max) at which these elite athletes are running.[45] In the slower runners, the exercise intensity does not exceed 60 per cent of V_{O_2}max, and gastrointestinal function is unlikely to be impaired relative to rest.[36,70] In these runners, sweat rates will also be relatively low.[1]

Therefore, although in theory it should be possible to meet the fluid loss by oral intake, gastric emptying rates of fluids are commonly much lower than the maximum figures quoted above and it is inevitable that most individuals exercising hard, particularly in the heat, will incur a fluid deficit.

Composition of drinks

Despite the definitive statement by the American College of Sports Medicine in their 1984 Position Statement on the prevention of thermal injuries in distance running that cool water is the optimum fluid for ingestion during endurance exercise,[44] some of the evidence presented above indicates that there may be good reasons for taking drinks containing added substrate and electrolytes. In prolonged exercise, performance is improved by the addition of an energy source in the form of carbohydrate; the type of carbohydrate does not appear to be critical, and glucose, sucrose, and oligosaccharides have all been shown to be effective in improving endurance capacity. Some recent studies have suggested that long-chain glucose polymer solutions are more readily used by the muscles during exercise than are glucose or fructose solutions,[100] but others have found no difference in the oxidation rates of ingested glucose or glucose polymer.[38,101] Massicote *et al*.[101] also found that ingested fructose was less readily oxidized than glucose or glucose polymers. Fructose in high concentrations is best avoided because of the risk of gastrointestinal upset. The argument advanced in favour of the ingestion of fructose during exercise, namely that it provides a readily available energy source but does not stimulate insulin release and consequent inhibition of fatty acid mobilization, is in any case not well founded: insulin secretion is suppressed during exercise.

The optimum concentration of sugars to be added to drinks will depend on individual circumstances. High carbohydrate concentrations will delay gastric emptying, thus reducing the amount of fluid that is available for absorption; very high concentrations will result in secretion of water into the intestine and thus actually increase the danger of dehydration. High sugar concentrations (>10 per cent) may also result in gastrointestinal disturbances. However, where there is a need to supply an energy source during exercise, increasing the sugar content of drinks will increase the delivery of carbohydrate to the site of absorption in the small intestine. As the carbohydrate concentration increases, the volume emptied from the stomach is reduced but the amount of carbohydrate emptied is increased.

The available evidence indicates that the only electrolyte that should be added to drinks consumed during exercise is sodium, which is usually added in the form of sodium chloride. Sodium will stimulate sugar and water uptake in the small intestine and will help to maintain extracellular fluid volume. Most soft drinks of the cola or lemonade variety contain virtually no sodium (1–2 mmol/l); sports drinks commonly contain 10 to 25 mmol/l; oral rehydration solutions intended for use in the treatment of diarrhoea-induced dehydration, which may be fatal, have higher sodium concentrations in the range 30 to 90 mmol/l. A high sodium content, although it may stimulate jejunal absorption of glucose and water, tends to make drinks unpalatable, and it is important that drinks intended for ingestion during or after exercise should have a pleasant taste in order to stimulate consumption. Specialist sports drinks are generally formulated to strike a balance between the twin aims of efficacy and palatability, although it must be admitted that not all achieve either of these aims.

When the exercise duration is likely to exceed 3 to 4 h, there may be advantages in adding sodium to drinks to avoid the danger of hyponatraemia which has been reported to occur when excessively large volumes of low sodium drinks are taken. It has often been reported that the fluid intakes of participants in endurance events are low, and it is recognized that this may lead to dehydration and heat illness in prolonged exercise when the ambient temperature is high. Accordingly, the advice given to participants in endurance events is that they should ensure a high fluid intake to minimize the effects of dehydration and that drinks should contain low levels of glucose and electrolytes so as not to delay gastric emptying.[44] In accordance with these recommendations, most carbohydrate–electrolyte drinks intended for consumption during prolonged exercise also have a low electrolyte content, with sodium and chloride concentrations typically in the range 10 to 25 mmol/l. While this might represent a reasonable strategy for providing substrates and water (although it can be argued that a higher sodium concentration would enhance water uptake and that a higher carbohydrate content would increase substrate provision), these recommendations may not be appropriate in all circumstances.

Physicians dealing with individuals in distress at the end of long-distance races have become accustomed to dealing with hyperthermia associated with dehydration and hypernatraemia,

but it has become clear that a small number of individuals at the end of very prolonged events may be suffering from hyponatraemia in conjunction with either hyperhydration[2,102–104] or dehydration.[105]

All the reported cases have been associated with ultramarathon or prolonged triathlon events; most have occurred in events lasting in excess of 8 h, and there are few reports of cases where the exercise duration is less than 4 h. Noakes et al.[102] reported four cases of exercise-induced hyponatraemia; race times were between 7 and 10 h, and post-race serum sodium concentrations were between 115 and 125 mmol/l. Estimated fluid intakes were between 6 and 12 l, and consisted of water or drinks containing low levels of electrolytes; estimated total sodium chloride intake during the race was 20 to 40 mmol. Frizell et al.[104] reported even more astonishing fluid intakes of 20 to 24 l (an intake of almost 2.5 l/h sustained for a period of many hours, which is in excess of the maximum gastric emptying rate that has been reported) with a mean sodium content of only 5 to 10 mmol/l in two runners who collapsed after an ultramarathon run and were found to be hyponatraemic (serum sodium concentration 118 and 123 mmol/l). Hyponatraemia as a consequence of ingestion of large volumes of fluids with a low sodium content has also been recognized in resting individuals. Flear et al.[106] reported the case of a man who drank 9 l of beer, with a sodium content of 1.5 mmol/l, in the space of 20 min; his plasma sodium fell from 143 mmol/l before drinking to 127 mmol/l afterwards, but the man appeared unaffected. In these cases, there is clearly a replacement of water in excess of losses with inadequate electrolyte replacement. However, hyponatraemia associated with dehydration has also been reported to be present in competitors in the Hawaii Ironman Triathlon.[105] Fellmann et al.[107] reported a small but statistically significant fall in serum sodium concentration, from 141 to 137 mmol/l, in runners who completed a 24-h run, but food and fluid intakes were neither controlled nor measured.

These reports are interesting and indicate that some supplementation with sodium chloride may be required in extremely prolonged events where large sweat losses can be expected and where it is possible to consume large volumes of fluid. However, this should not divert attention away from the fact that electrolyte replacement during exercise is not a priority for most participants in most sporting events.

Sodium is also necessary for post-event rehydration, which may be particularly important when the exercise has to be repeated within a few hours; if drinks containing little or no sodium are taken, plasma osmolality will fall, urine production will be stimulated, and most of the fluid will not be retained. When a longer time interval between exercise sessions is possible, replacement of sodium and other electrolytes will normally be achieved as a result of intake from the diet without additional supplementation.

It is often stated that there is an advantage to taking chilled (4°C) drinks as this accelerates gastric emptying and thus improves the availability of ingested fluids. However, the most recent evidence suggests that the gastric emptying rate of hot and cold beverages is not different. Despite this, there may be advantages in taking cold drinks as the palatability of most carbohydrate–electrolyte drinks is improved at low temperatures.

Environmental conditions

The ambient temperature and wind speed will have a major influence on the physical exchange of heat between the body and the environment. When ambient temperature exceeds skin temperature, heat is gained from the environment by physical transfer, leaving evaporative loss as the only mechanism available to prevent or limit a rise in body temperature. The increased sweating rate in the heat will result in an increased requirement for fluid replacement. Other precautions such as limiting the extent of the warm-up prior to competition and reducing the amount of clothing worn will help to reduce the sweat loss and hence reduce the need for replacement. For endurance events at high ambient temperatures, there may also be a need to reduce the exercise intensity if the event is to be successfully completed.

When the humidity is high, and particularly in the absence of wind, evaporative heat loss will also be severely limited. In this situation, exercise tolerance is likely to be limited by dehydration and hyperthermia rather than by the limited availability of metabolic fuel. Suzuki[108] reported that exercise time at a work load of 66 per cent V_{O_2}max was reduced from 91 min when the ambient temperature was 0°C to 19 minutes when the same exercise was performed in the heat (40°C). In an unpublished study in which six subjects exercised to exhaustion at 70 per cent V_{O_2}max on a cycle ergometer, we found that exercise time was reduced from 73 min at an ambient temperature of 2°C to 35 min at a temperature of 33°C. Exercise time in the cold was increased by ingestion of a dilute glucose–electrolyte solution, but in the heat the exercise duration was too short for fluid intake to have any effect on performance. Although fatigue at these work intensities is generally considered to result from depletion of the muscle glycogen stores, this is clearly not the case when the ambient temperature is high; the rate of carbohydrate oxidation was not different during exercise at different ambient temperatures.

From these studies, we can conclude that the supply of water should take precedence over the provision of substrate during exercise in the heat. Therefore, there may be some advantage in reducing the sugar content of drinks to perhaps 2 to 5 per cent, and in increasing the sodium content to something in the range 30 to 50 mmol/l. Conversely, when exercise is undertaken in the cold, fluid loss is less of a problem and the energy content of drinks might usefully be increased.

State of training and acclimation

It is well recognized that both training and acclimation will confer some protection against the development of heat illness during exercise in the heat. Although this adaptation is most marked in response to training carried out in the heat,[109] endurance training at moderate environmental conditions will also confer some benefit. Among the benefits of training is an expansion of the plasma volume.[110] Although this condition is recognized as a chronic state in the endurance-trained individual, an acute expansion of plasma volume occurs in response to a single bout of strenuous exercise; this effect is apparent within a few hours of completion of exercise and may persist for several days.[111,112] This post-exercise hypervolaemia should be regarded as an acute response rather than an adaptation, although it may appear to be one of the first responses to occur when an individual embarks on a training regimen. Circulating electrolyte and total protein concentrations are normal in the endurance-trained individual despite the enlarged vascular and extracellular spaces, indicating an increased total circulating content.[113]

The increased resting plasma volume in the trained state allows the endurance-trained individual to maintain a higher total blood volume during exercise,[114] allowing for better maintenance of cardiac output albeit at the cost of a lower circulating haemoglobin

concentration. In addition, the increased plasma volume is associated with an increased sweating rate which limits the rise in body temperature.[115] These adaptive responses appear to occur within a few days of exposure to exercise in the heat although, as pointed out above, this may not necessarily be a true adaptation. In a series of papers reporting the same study,[115–117] the time course of changes in men exposed to exercise (40–50 per cent V_{O_2}max for 4 h) in the heat (45°C) for 10 days was followed. Although there were marked differences between individuals in their responses, resting plasma volume increased progressively over the first 6 days, reaching a value about 23 per cent greater than the control, with little change thereafter. The main adaptation in terms of an increased sweating rate and an improved thermoregulatory response (with body temperature lower by 1°C and heart rate lower by 30 beats/min in the later stages of exercise) occurred slightly later than the cardiovascular adaptations, with little change in the first 4 days.

Although there is clear evidence that acclimation by exercise in the heat over a period of several days will improve the thermoregulatory response during exercise, this does not affect the need to replace fluids during the exercise period. Better maintenance of body temperature is achieved at the expense of an increased sweat loss. Although this allows for a greater evaporative heat loss, the proportion of the sweat which is not evaporated and which therefore drips wastefully from the skin is also increased.[115] A high sweat rate may be necessary to ensure adequate evaporative heat loss, but it does seem that many individuals have an inefficient sweating mechanism; even in the unacclimated state their rate of sweat secretion appears to be greatly in excess of the maximum evaporative capacity. However, the athlete who trains in a moderate climate for a competition to be held in the heat will be at a disadvantage because of his inability to sustain a high sweat rate.

POST-EXERCISE REHYDRATION

Replacement of water and electrolyte losses in the post-exercise period may be of crucial importance when repeated bouts of exercise have to be performed. The need for replacement will obviously depend on the extent of losses incurred during exercise, but will also be influenced by the time and nature of subsequent exercise bouts. Rapid rehydration may also be important in events such as wrestling, boxing, and weight-lifting where competition is by weight category. Competitors in these events frequently undergo acute thermal and exercise-induced dehydration to make weight; the time interval between the weigh-in and competition is normally about 3 h, although it may be longer. The practice of acute dehydration to make weight should be discouraged, as it reduces exercise performance and increases the risk of heat illness, but it will persist and there is a need to maximize rehydration in the time available.

Ingestion of plain water in the post-exercise period results in a rapid fall in the plasma sodium concentration and in plasma osmolality.[118] These changes have the effect of reducing the stimulus to drink (thirst) and of stimulating urine output, both of which will delay the rehydration process. In the study by Nose et al.,[118] subjects exercised at low intensity in the heat for 90 to 110 min, inducing a mean dehydration of 2.3 per cent of body weight, and then rested for 1 h before beginning to drink. Plasma volume was not restored until after 60 min when plain water was ingested together with placebo (sucrose) capsules. In contrast, when sodium chloride capsules were ingested with water to give a saline solution with an effective concentration of 0.45 per cent

(77 mmol/l), plasma volume was restored within 20 min. In the sodium chloride trial, voluntary fluid intake was higher and urine output was less; 71 per cent of the water loss was retained within 3 h compared with 51 per cent in the plain water trial. The delayed rehydration in the water trial appeared to be a result of a loss of sodium, accompanied by water, in the urine caused by enhanced plasma renin activity and aldosterone levels.[119]

In an earlier study where a more severe (4 per cent of body weight) dehydration was induced in resting subjects by heat exposure, consumption over a 3-h period of a volume of fluid equal to that lost did not restore plasma volume or serum osmolality within 4 h.[120] However, ingestion of a glucose–electrolyte solution did result in greater restoration of plasma volume than did plain water; this was accompanied by a greater urine production in the water trial. Where the electrolyte content of drinks is the same it appears that addition of carbohydrate (100 g/l) or carbonation has no effect on the restoration of plasma volume over a 4-h period after sweat loss corresponding to approximately 4 per cent of body weight.[50] Gonzalez-Alonso et al.[121] have recently shown that a dilute carbohydrate–electrolyte solution (60 g/l carbohydrate, 20 mmol/l Na^+, 3 mmol/l K^+) is more effective in promoting post-exercise rehydration than either plain water or a low electrolyte diet cola; the difference between the drinks was primarily a result of differences in the volume of urine produced.

It is clear from the results of these studies that rehydration after exercise can only be achieved if the sodium lost in sweat is replaced as well as the water, and it might be suggested that rehydration drinks should have a sodium concentration similar to that of sweat. Since the sodium content of sweat varies widely, no single formulation will meet this requirement for all individuals in all situations. However, the upper end of the normal range for sodium concentration (80 mmol/l) is similar to the sodium concentration of many commercially produced oral rehydration solutions intended for use in the treatment of diarrhoea-induced dehydration, and some of these are not unpalatable. The oral rehydration solution recommended by the World Health Organization for rehydration in cases of severe diarrhoea has a sodium content of 90 mmol/l. In contrast, the sodium content of most sports drinks is in the range 10 to 25 mmol/l and is even lower in some cases; most commonly consumed soft drinks contain virtually no sodium.

The requirement for sodium replacement stems from its role as the major ion in the extracellular fluid. It has been speculated that inclusion of potassium, the major cation in the intracellular space, would enhance the replacement of intracellular water after exercise and thus promote rehydration.[122] It appears that inclusion of potassium is as effective as sodium in retaining water ingested after exercise-induced dehydration: addition of either ion will significantly increase the fraction of the ingested fluid which is retained, but when the volume of fluid ingested is equal to that lost during the exercise period there is no additive effect of including both ions as would be expected if they acted independently on different body fluid compartments.[123]

REFERENCES

1. Maughan RJ. Thermoregulation and fluid balance in marathon competition at low ambient temperature. *International Journal of Sports Medicine* 1985; **6**: 15–19.
2. Saltin B, Costill DL. Fluid and electrolyte balance during prolonged exercise. In Horton ES, Terjung RL, eds. *Exercise, nutrition and metabolism*. New York: Macmillan, 1988: 150–8.
3. Nielsen B, Kubica R, Bonnesen A, Rasmussen IB, Stoklosa J, Wilk B. Physical work capacity after dehydration and hyperthermia. *Scandinavian Journal of Sports Sciences* 1982; **3**: 2–10.
4. Armstrong LE, Costill DL, Fink WJ. Influence of diuretic-induced dehydration on competitive running performance. *Medicine and Science in Sports and Exercise* 1985; **17**: 456–61.
5. Nadel ER. Circulatory and thermal regulation during exercise. *Federation Proceedings* 1980; **39**: 1491–7.
6. Rowell LB. *Human circulation*. Oxford University Press, 1986.
7. Pitts RF. *The physiological basis of diuretic therapy*. Springfield, IL: CC Thomas, 1959.
8. Lentner, C, ed. *Geigy scientific tables*. 8th edn. Basle: Ciba-Geigy, 1981.
9. Schmidt RF, Thews, G, eds. *Human physiology*. 2nd edn. Berlin: Springer-Verlag, 1989.
10. Leithead CS, Lind AR. *Heat stress and heat disorders*. Cassell, London, 1964.
11. Costill DL. Sweating: its composition and effects on body fluids. *Annals of the New York Academy of Sciences* 1977; **301**: 160–74.
12. Verde T, Shephard RJ, Corey P, Moore R. Sweat composition in exercise and in heat. *Journal of Applied Physiology* 1982; **53**: 1540–5.
13. Allan JR, Wilson CG. Influence of acclimatization on sweat sodium secretion. *Journal of Applied Physiology* 1971; **30**: 708–12.
14. Kobayashi Y, Ando Y, Takeuchi S, Takemura K, Okuda N. Effects of heat acclimation of distance runners in a moderately hot environment. *European Journal of Applied Physiology* 1980; **45**: 189–98.
15. Greenleaf JE, Castle BL, Card DH. Blood electrolytes and temperature regulation during exercise in man. *Acta Physiologica Polonica* 1974; **25**: 397–410.
16. Harrison MH, Edwards RJ, Fennessy PA. Intravascular volume and tonicity as factors in the regulation of body temperature. *Journal of Applied Physiology* 1978; **44**: 69–75.
17. Fortney SM, Wenger, CB, Bove, JR, Nadel ER. Effect of hyperosmolality on control of blood flow and sweating. *Journal of Applied Physiology* 1984; **57**: 1688–95.
18. Fortney SM, Vroman NB, Beckett WS, Permutt S, LaFrance ND. Effect of exercise hemoconcentration and hyperosmolality on exercise responses. *Journal of Applied Physiology* 1988; **65**: 519–24.
19. Meytes I, Shapira Y, Magazanik A, Meytes D, Seligsohn U. Physiological and biochemical changes during a marathon race. *International Journal of Biometeorology* 1969; **13**: 317.
20. Whiting PH, Maughan RJ, Miller JDB. Dehydration and serum biochemical changes in runners. *European Journal of Applied Physiology* 1984; **52**: 183–7.
21. Kavanagh T, Shephard RJ. Maintenance of hydration in 'post-coronary' marathon runners. *British Journal of Sports Medicine* 1975; **9**: 130–5.
22. Costill DL, Branam G, Fink W, Nelson R. Exercise induced sodium conservation: changes in plasma renin and aldosterone. *Medicine and Science in Sports and Exercise* 1976; **8**: 209–13.
23. Cohen I, Zimmerman AL. Changes in serum electrolyte levels during marathon running. *South African Medical Journal* 1978; **53**: 449–53.
24. Stansbie D, Tomlinson K, Potman JM, Walters EG. Hypothermia, hypokalaemia and marathon running. *Lancet* 1982; **ii**: 1336.
25. Joborn H, Åkerström G, Ljunghall S. Effects of exogenous catecholamines and exercise on plasma magnesium concentrations. *Clinical Endocrinology* 1985; **23**: 219–26.
26. Rose LI, Carroll DR, Lowe SL, Peterson EW, Cooper KH. Serum electrolyte changes after marathon running. *Journal of Applied Physiology* 1970; **29**: 449–51.
27. Beller GA, Maher JT, Hartley LH, Bass DE, Wacker WEC. Serum Mg and K concentrations during exercise in thermoneutral and hot conditions. *Physiologist* 1972; **15**: 94.
28. Refsum HE, Meen HD, Stromme SB. Whole blood, serum and erythrocyte magnesium concentrations after repeated heavy exercise of long duration. *Scandinavian Journal of Clinical and Laboratory Investigation* 1973; **32**: 123–7.
29. Lijnen P, Hespel P, Fagard R, Lysens R, Vanden Eynde E, Amery A. Erythrocyte, plasma and urinary magnesium in men before and after a marathon. *European Journal of Applied Physiology* 1988; **58**: 252–6.

30. Costill DL, Miller JM. Nutrition for endurance sport. *International Journal of Sports Medicine* 1980; **1**: 2–14.

31. Castenfors J. Renal function during prolonged exercise. *Annals of the New York Academy of Sciences* 1977; **301**: 151–9.

32. Wade CE, Claybaugh JR. Plasma renin activity, vasopressin concentration and urinary excretory responses to exercise in men. *Journal of Applied Physiology* 1980; **49**: 930–6.

33. Poortmans J. Exercise and renal function. *Sports Medicine* 1984; **1**: 125–53.

34. Zambraski EJ. Renal regulation of fluid homeostasis during exercise. In: Gisolfi CV, Lamb DR, eds. *Perspectives in exercise science and sports medicine*. Vol. 3. *Fluid homeostasis during exercise*. Carmel: Benchmark Press, 1990: 247–80.

35. Leiper JB, Maughan RJ. Experimental models for the investigation of water and solute transport in man: implications for oral rehydration solutions. *Drugs* 1988; **36**(Suppl 4): 65–79.

36. Rehrer NJ, Beckers E, Brouns F, Ten Hoor F, Saris WHM. Exercise and training effects on gastric emptying of carbohydrate beverages. *Medicine and Science in Sports and Exercise* 1989; **21**: 540–9.

37. Rehrer NJ, Janssen GME, Brouns F, Saris WHM. Fluid intake and gastrointestinal problems in runners competing in a 25-km race and a marathon. *International Journal of Sports Medicine* 1989; **10**(Suppl. 1): S22–5.

38. Rehrer NJ. *Limits to fluid availability during exercise*. De Vrieseborsch: Haarlem, 1990.

39. Maughan RJ, Vist GE. Gastric emptying of dilute glucose solutions in man. *Medicine and Science in Sports and Exercise* 1992: **24**; S70.

40. Brener W, Hendrix TR, McHugh PR. Regulation of the gastric emptying of glucose. *Gastroenterology* 1983; **85**: 76–82.

41. Foster C, Costill DL, Fink WJ. Gastric emptying characteristics of glucose and glucose polymers. *Research Quarterly* 1980; **51**: 299–305.

42. Sole CC, Noakes TD. Faster gastric emptying for glucose-polymer and fructose solutions than for glucose in humans. *European Journal of Applied Physiology* 1989; **58**: 605–12.

43. Naveri H, Tikkanen H, Kairento A-L, Harkonen M. Gastric emptying and serum insulin levels after intake of glucose-polymer solutions. *European Journal of Applied Physiology* 1989; **58**: 661–5.

44. American College of Sports Medicine. Position stand on prevention of thermal injuries during distance running. *Medicine and Science in Sports and Exercise* 1984; **16**: ix–xiv.

45. Costill DL, Saltin B. Factors limiting gastric emptying during rest and exercise. *Journal of Applied Physiology* 1974; **37**: 679–83.

46. Sun WM, Houghton LA, Read NW, Grundy DG, Johnson AG. Effect of meal temperature on gastric emptying of liquids in man. *Gut* 1988; **29**: 302–5.

47. McArthur KE, Feldman M. Gastric acid secretion, gastrin release, and gastric temperature in humans as affected by liquid meal temperature. *American Journal of Clinical Nutrition* 1989; **49**: 51–4.

48. Lambert CP, Maughan RJ. Effect of temperature of ingested beverages on the rate of accumulation in the blood of an added tracer for water uptake. *Scandinavian Journal of Medicine and Science in Sports* 1992; **2**: 76–8.

49. Lolli G, Greenberg LA, Lester D. The influence of carbonated water on gastric emptying. *New England Journal of Medicine* 1952; **246**: 490–2.

50. Lambert CP *et al*. Fluid replacement after dehydration: influence of beverage carbonation and carbohydrate content. *International Journal of Sports Medicine* 1992; **13**: 285–92.

51. Wapnir RA, Lifshitz F. Osmolality and solute concentration—their relationship with oral rehydration solution effectiveness: an experimental assessment. *Pediatric Research* 1985; **19**: 894–8.

52. Spiller RC, Jones BJM, Brown BE, Silk DBA. Enhancement of carbohydrate absorption by the addition of sucrose to enteric diets. *Journal of Parenteral and Enteral Nutrition* 1982; **6**: 321.

53. Jones BJM, Brown BE, Loran JS, Edgerton D, Kennedy JF. Glucose absorption from starch hydrolysates in the human jejunum. *Gut* 1983; **24**: 1152–60.

54. Jones BJM, Higgins BE, Silk DBA. Glucose absorption from maltotriose and glucose oligomers in the human jejunum. *Clinical Science* 1987; **72**: 409–14.

55. Fordtran JS. Stimulation of active and passive sodium absorption by sugars in the human jejunum. *Journal of Clinical Investigation* 1975; **55**: 728–37.

56. Fordtran JS, Saltin B. Gastric emptying and intestinal absorption during prolonged severe exercise. *Journal of Applied Physiology* 1967; **23**: 331–5.

57. Brouns F, Saris WHM, Rehrer NJ. Abdominal complaints and gastrointestinal function during long-lasting exercise. *International Journal of Sports Medicine* 1987; **8**: 175–89.

58. Maughan RJ, Leiper JB, McGaw BA. Effects of exercise intensity on absorption of ingested fluids in man. *Experimental Physiology* 1990; **75**: 419–21.

59. Costill DL, Bennett A, Branam G, Eddy D. Glucose ingestion at rest and during prolonged exercise. *Journal of Applied Physiology* 1973; **34**: 764–9.

60. Pirnay F *et al*. Fate of exogenous glucose during exercise of different intensities in humans. *Journal of Applied Physiology* 1982; **53**: 1620–4.

61. Erickson MA, Schwartzkopf RJ, McKenzie RD. Effects of caffeine, fructose, and glucose ingestion on muscle glycogen utilisation during exercise. *Medicine and Science in Sports and Exercise* 1987; **19**: 579–83.

62. Ivy J, Costill DL, Fink WJ, Lower RW. Influence of caffeine and carbohydrate feedings on endurance performance. *Medicine and Science in Sports* 1979; **11**: 6–11.

63. Coyle EF, Hagberg JM, Hurley BF, Martin WH, Ehsani AH, Holloszy JO. Carbohydrate feeding during prolonged strenuous exercise can delay fatigue. *Journal of Applied Physiology* 1983; **55**: 230–5.

64. Coyle EF, Coggan AR, Hemmert MK, Ivy JL. Muscle glycogen utilisation during prolonged strenuous exercise when fed carbohydrate. *Journal of Applied Physiology* 1986; **61**: 165–72.

65. Maughan RJ, Fenn CE, Gleeson M, Leiper JB. Metabolic and circulatory responses to the ingestion of glucose polymer and glucose/electrolyte solutions during exercise in man. *European Journal of Applied Physiology* 1987; **56**: 356–62.

66. Coggan AR, Coyle EF. Effect of carbohydrate feedings during high-intensity exercise. *Journal of Applied Physiology* 1988; **65**: 1703–9.

67. Hargreaves M, Briggs CA. Effect of carbohydrate ingestion on exercise metabolism. *Journal of Applied Physiology* 1988; **65**: 1553–5.

68. Sasaki H, Maeda J, Usui S, Ishiko T. Effect of sucrose and caffeine ingestion on performance of prolonged strenuous running. *International Journal of Sports Medicine* 1987; **8**: 261–5.

69. Murray R, Eddy DE, Murray TW, Seifert JG, Paul GL, Halaby GA. The effect of fluid and carbohydrate feedings during intermittent cycling exercise. *Medicine and Science in Sports and Exercise* 1987; **19**: 597–604.

70. Mitchell JB, Costill DL, Houmard JA, Flynn MG, Fink WJ, Beltz JD. Effects of carbohydrate ingestion on gastric emptying and exercise performance. *Medicine and Science in Sports and Exercise* 1988; **20**: 110–15.

71. Carter JE, Gisolfi CV. Fluid replacement during and after exercise in the heat. *Medicine and Science in Sports and Exercise* 1989; **21**: 532–9.

72. Ahlborg B, Bergstrom J, Brohult J, Ekelund L-G, Hultman E, Maschio E. Human muscle glycogen content and capacity for prolonged exercise after different diets. *Forsvarsmedicin* 1967; **3**: 85–99.

73. Hargreaves M, Costill DL, Coggan A, Fink WJ, Nishibata I. Effect of carbohydrate feedings on muscle glycogen utilisation and exercise performance. *Medicine and Science in Sports and Exercise* 1984; **16**: 219–22.

74. Fielding RA, Costill DL, Fink WJ, King DS, Hargreaves M, Kovaleski ME. Effect of carbohydrate feeding frequencies on muscle glycogen use during exercise. *Medicine and Science in Sports and Exercise* 1985; **17**: 472–6.

75. Noakes TD, Adams BA, Myburgh KH, Greeff C, Lotz T, Nathan M. The danger of an inadequate water intake during prolonged exercise. A novel concept re-visited. *European Journal of Applied Physiology* 1988; **57**: 210–19.

76. Flynn MG, Costill DL, Hawley JA, Fink WJ, Neufer PD, Fielding RA, Sleeper MD. Influence of selected carbohydrate drinks on cycling performance and glycogen use. *Medicine and Science in Sports and Exercise* 1987; **191**: 37–40.

77. Pirnay F, Lacroix M, Mosora F, Luyckx A, Lefebvre P. Glucose oxidation during prolonged exercise evaluated with naturally labeled [^{13}C]glucose. *Journal of Applied Physiology* 1977; **43**: 258–61.

78. Pallikarakis N, Jandrain B, Pirnay F, Mosora F, Lacroix M, Luyckx AS, Lefebvre PJ. Remarkable metabolic availability of oral glucose during long-duration exercise in humans. *Journal of Applied Physiology* 1988; **60**: 1035–42.

79. Coyle EF, Coggan AR. Effectiveness of carbohydrate feeding in delaying fatigue during prolonged exercise. *Sports Medicine* 1984; **1**: 446–58.

80. Lamb DR, Brodowicz GR. Optimal use of fluids of varying formulations to minimize exercise-induced disturbances in homeostasis. *Sports Medicine* 1986; **3**: 247–74.

81. Murray R. The effects of consuming carbohydrate-electrolyte beverages on gastric emptying and fluid absorption during and following exercise. *Sports Medicine* 1987; **4**: 322–51.

82. Maughan RJ. Effects of CHO-electrolyte solution on prolonged exercise. In: Lamb DR, MH Williams, eds. *Perspectives in Exercise Science and Sports Medicine*. Carmel: Benchmark Press, 1991: 35–85.

83. Macaraeg PVJ. Influence of carbohydrate electrolyte ingestion on running endurance. In: Fox EL, ed. *Nutrient utilisation during exercise*. Columbus, OH: Ross Laboratories, 1983: 91–6.

84. Fruth JM, Gisolfi CV. Effects of carbohydrate consumption on endurance performance: fructose versus glucose. In: Fox EL, ed. *Nutrient utilisation during exercise*. Columbus, OH: Ross Laboratories, 1983: 68–75.

85. Riley ML, Israel RG, Holbert D, Tapscott EB, Dohn GL. Effect of carbohydrate ingestion on exercise endurance and metabolism after a 1-day fast. *International Journal of Sports Medicine* 1988; **9**: 320–4.

86. Ivy JL *et al.* Endurance improved by ingestion of a glucose polymer supplement. *Medicine and Science in Sports and Exercise* 1983; **15**: 466–71.

87. Williams C. Diet and endurance fitness. *American Journal of Clinical Nutrition* 1989; **49**: 1077–83.

88. Williams C, Nute MG, Broadbank L, Vinall S. Influence of fluid intake on endurance running performance: a comparison between water, glucose and fructose solutions. *European Journal of Applied Physiology* 1990; **60**: 112–119.

89. Cade R, Spooner G, Schlein E, Pickering M, Dean R. Effect of fluid, electrolyte, and glucose replacement on performance, body temperature, rate of sweat loss and compositional changes of extracellular fluid. *Journal of Sports Medicine and Physical Fitness* 1972; **12**: 150–6.

90. Maughan RJ, Whiting PH. Factors influencing plasma glucose concentration during marathon running. In: Dotson CO, Humphrey JH, eds. *Exercise physiology*. Vol. 1. New York: AMS Press, 1985: 87–98.

91. Leatt P. *The effect of glucose polymer ingestion on skeletal muscle glycogen depletion during soccer match-play and its resynthesis following a match*. MSc Thesis, University of Toronto, 1986.

92. Greenhaff PL, Clough PJ. Predictors of sweat loss in man during prolonged exercise. *European Journal of Applied Physiology* 1989; **58**: 348–52.

93. Barr SI, Costill DL. Water: can the endurance athlete get too much of a good thing. *Journal of the American Dietetic Association* 1989; **89**: 1629–32.

94. Olsson KE, Saltin B. Diet and fluids in training and competition. *Scandinavian Journal of Rehabilitation Medicine* 1971; **3**: 31–8.

95. American College of Sports Medicine. Position statement on prevention of heat injuries during distance running. *Medicine and Science in Sports* 1975; **7**:

96. Deschamps A, Levy RD, Cosio MG, Marliss EB, Magder S. Effect of saline infusion on body temperature and endurance during heavy exercise. *Journal of Applied Physiology* 1989; **66**: 2799–804.

97. Davies CTM, Thompson MW. Aerobic performance of female marathon and male ultramarathon athletes. *European Journal of Applied Physiology* 1979; **41**: 233–45.

98. Maughan RJ, Leiper JB. Aerobic capacity and fractional utilisation of aerobic capacity in elite and non-elite male and female marathon runners. *European Journal of Applied Physiology* 1983; **52**: 80–7.

99. Duchman SM, Blieler TL, Schedl HP, Summers RW, Gisolfi CV. Effects of gastric function on intestinal composition of oral rehydration solutions. *Medicine and Science in Sports and Exercise* 1990; **22** (Suppl): S89.

100. Noakes TD. The dehydration myth and carbohydrate replacement during prolonged exercise. *Cycling Science* 1990; 23–9.

101. Massicote D, Peronnet F, Brisson G, Bakkouch K, Hillaire-Marcel C. Oxidation of a glucose polymer during exercise: comparison with glucose and fructose. *Journal of Applied Physiology* 1989; **66**: 179–83.

102. Noakes TD, Goodwin N, Rayner BL, Branken T, Taylor RKN. Water intoxication: a possible complication during endurance exercise. *Medicine and Science in Sports and Exercise* 1985; **17**: 370–5.

103. Noakes TD, Norman RJ, Buck RH, Godlonton J, Stevenson K, Pittaway D. The incidence of hyponatraemia during prolonged ultraendurance exercise. *Medicine and Science in Sports and Exercise* 1990; **22**: 165–70.

104. Frizell RT, Lang GH, Lowance DC, Lathan SR. Hyponatraemia and ultramarathon running. *Journal of the American Medical Association* 1986; **255**: 772–4.

105. Hiller WDB. Dehydration and hyponatremia during triathlons. *Medicine and Science in Sports and Exercise* 1989; **21**: S219–S221.

106. Flear CTG, Gill CV, Burn J. Beer drinking and hyponatraemia. *Lancet* 1981; ii: 477.

107. Fellmann *et al.* Enzymatic and hormonal responses following a 24 h endurance run and a 10 h triathlon race. *European Journal of Applied Physiology* 1988; **57**: 545–53.

108. Suzuki Y. Human physical performance and cardiocirculatory responses to hot environments during sub-maximal upright cycling. *Ergonomics* 1980; **23**: 527–42.

109. Senay LC. Effects of exercise in the heat on body fluid distribution. *Medicine and Science in Sports* 1979; **11**: 42–8.

110. Hallberg L, Magnusson B. The aetiology of sports anaemia. *Acta Medica Scandinavica* 1984; **216**: 145–8.

111. Davidson RJL, Robertson JD, Galea G, Maughan RJ. Haematological changes associated with marathon running. *International Journal of Sports Medicine* 1987; **8**: 19–25.

112. Robertson JD, Maughan RJ, Davidson RJL. Changes in red cell density and related parameters in response to long distance running. *European Journal of Applied Physiology* 1988; **57**: 264–9.

113. Convertino VA, Brock PJ, Keil LC, Bernauer EM, Greenleaf JE. Exercise training-induced hypervolemia: role of plasma albumin, renin and vasopressin. *Journal of Applied Physiology* 1980; **48**: 665–9.

114. Convertino VA, Keil LC, Greenleaf JE. Plasma volume, renin, and vasopressin responses to graded exercise after training. *Journal of Applied Physiology* 1983; **54**: 508–14.

115. Mitchell D, Senay LC, Wyndham CH, van Rensburg AJ, Rogers GG, Strydom NB. Acclimatization in a hot, humid environment: energy exchange, body temperature, and sweating. *Journal of Applied Physiology* 1976; **40**: 768–78.

116. Wyndham CH, Rogers GG, Senay LC, Mitchell D. Acclimatization in a hot, humid environment: cardiovascular adjustments. *Journal of Applied Physiology* 1976; **40**: 779–85.

117. Senay LC, Mitchell D, Wyndham CH. Acclimatization in a hot humid environment: body fluid adjustments. *Journal of Applied Physiology* 1976; **40**: 786–96.

118. Nose H, Mack GW, Shi X, Nadel ER. Role of osmolality and plasma volume during rehydration in humans. *Journal of Applied Physiology* 1988; **65**: 325–31.

119. Nose H, Mack GW, Shi X, Nadel ER. Involvement of sodium retention hormones during rehydration in humans. *Journal of Applied Physiology* 1988; **65**: 332–6.

120. Costill DL, Sparks KE. Rapid fluid replacement following thermal dehydration. *Journal of Applied Physiology* 1973; **34**: 299–303.

121. Gonzalez-Alonso J, Heaps CL, Coyle EF. Rehydration after exercise with common beverages and water. *International Journal of Sports Medicine* 1992; **13**: 399–406.

122. Nadel ER, Mack GW, Nose H. Influence of fluid replacement beverages on body fluid homeostasis during exercise and recovery. In: Gisolfi CV, Lamb DR, eds. *Perspectives in exercise science and sports medicine*. Vol. 3. *Fluid homeostasis during exercise*. Carmel: Benchmark, 1990: 181–205.

123. Leiper JB, Owen JH, Maughan RJ. Effects of ingesting electrolyte solutions on hydration status following exercise dehydration in man. *Journal of Physiology* 1993; **459**: 28P.

1.3.1 Biomechanics as applied to sports

BENNO M. NIGG

INTRODUCTION

Biomechanics is the science that studies external and internal forces acting on a biological system and the effects produced by these forces.

The discipline of biomechanics can be subdivided into particular directions including (a) anatomical biomechanics, studying movements of joints and muscular contribution to these movements, (b) neurophysiological biomechanics, studying muscular control and reflex and feedback mechanisms, and (c) mechanical biomechanics, studying the mechanical aspect of human movement by analysing motion and by assessing internal and/or external forces.

This chapter concentrates on mechanical biomechanics, which is concerned with forces, moments, accelerations, velocities, energy, momentum, and other mechanical expressions associated with Newton's three laws of motion. The aim is to help the reader learn to analyse and assess human movement from a biomechanical perspective, and to look at human movement from a biomechanical viewpoint.

Biomechanical questions can be subdivided into two groups: those dealing with performance, and those dealing with load, overload, and associated injuries. Both types of question may require basic or applied research to answer them.

Selected examples of biomechanical topics and questions related to performance are as follows.

1. How is it possible for an athlete to jump over 8 m? Or, in a more general sense, what are the factors that determine the performance of a long jumper? Can these factors be modified through training to improve performance?
2. Gymnasts, divers, and trampolinists perform somersaults with twists. Often they make specific movements with their arms when starting the twisting movement in the air. How can an athlete initiate a twisting movement during a somersault and can movement during the flight phase be modified to improve performance? (It should be noted that there are athletes with no arms who perform twisting somersaults.)
3. Sport shoes vary in their construction depending on the sport activity. What are the factors determining performance in a particular sport activity and can performance be enhanced by sport shoe design? Is it possible to construct shoes which return energy during running?
4. Amputees walking with a prosthesis may desire a walking style which is similar to that of normal people. What are the possibilities for differentiating between normal and abnormal human gait? What are the main factors characterizing normal gait and can they be learned by the amputee?
5. Performance in sports depends often on forces produced by the muscles. What are the rules governing muscular force production? How can muscular forces be transferred into power output to enhance performance?

Examples of biomechanical topics and questions related to load and overload on the human locomotor system are as follows.

1. Forces acting on the anatomical structures of a skier are assumed to be quite high. How high are such forces in specific structures of the human body? How do forces acting on articular cartilage in the knee joint, for instance, compare with the critical limits of articular cartilage? Can strategies be developed to reduce these forces?
2. Tennis is played on clay, concrete, asphalt, artificial turf, sand-filled artificial turf, natural grass, and other surfaces. How does the surface influence the forces acting on specific anatomical structures of a tennis player? Are certain surfaces less likely to produce particular injuries than others?
3. Shin splints (pain at the anterior aspect of the tibia) are common in runners. Long-distance runners, sprinters, and even basketball and volleyball players are often affected with this injury. There are different theories about the aetiology of this condition. Is the type of movement related to the occurrence of shin splints? Are anthropometric factors associated with the development of shin splints?
4. In many sport activities, bones are broken, ligaments are torn, and cartilages are destroyed. Injuries occur in acute situations and as a result of chronic overloading. Arthritis as a result of repeated microdamage to cartilage, for instance, is assumed to be such a chronic 'injury'. What mechanisms are responsible for these injuries? What are the critical limits for human tissue, such as bone, cartilage, ligament, tendon, etc., for acute and chronic injuries? Does age affect these limits?

In the following sections we attempt to provide insight into possibilities for answering such questions. The purpose of this chapter is to illustrate how biomechanics can be used to study human movement or, in more popular terms, to outline how one can look at human movement through biomechanical spectacles.

Selected historical highlights

Aristotle (384 BC–322 BC), a Greek philosopher and 'natural scientist' introduced the term 'mechanics' and is considered to be the initiator of mechanics as we understand it at present. He can be considered the first biomechanist since he wrote about the movement of living beings. It is obvious that his view of the science of nature and mechanical principles is different from the current understanding. However, in his time his ideas aimed at understanding nature, and particularly human movement, were unique and are considered an important step in the development of the understanding of human movement and of the factors influencing it.

Ambroise Paré (1510–90) developed prostheses for legs, hands, and arms. They were often used for soldiers who had lost parts of their upper or lower extremities. Paré was among the first to apply biomechanics to improve the life of the handicapped. He can be considered as one of the first biomechanists or bioengineers.

Alphonso Borelli (1608–79), an Italian scientist, is considered the 'father of biomechanics'. He was a student of Galileo Galilei and had degrees in medicine and mathematics. This enabled him to understand the human body anatomically and to develop the necessary mathematical models to describe human movement

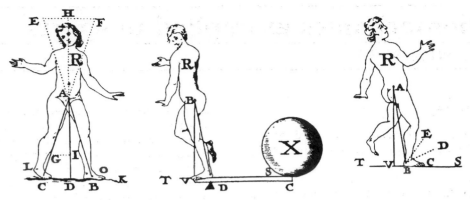

Fig. 1 Forces acting on the human body, an illustration from Borelli's book *De motu animalium*. (Reproduced from ref. 1, with permission.)

mechanically. His research in biomechanics was published in the book *De motu animalium*.[1] He was probably the first to perform gait analysis and to determine the centre of mass of the human body. An illustration from *De motu animalium* is shown in Fig. 1. Borelli discussed many different aspects of human movement as related to the externally visible motion of animals or humans. He described the function of muscles, formulated mechanical lemmas to explain the movements produced by muscles, explained the function of muscles for the motion of the knee joint, and discussed the influence of the direction of the muscle fibres on force production by this muscle. Furthermore, he discussed standing, walking, jumping, flying, and swimming. He formulated his findings or hypotheses in the form of propositions.

Isaac Newton (1642–1727) formulated the mechanical principles which are still used for classical mechanics. Newtonian mechanics is the foundation of biomechanics related to human movement. Newton was not a full time physicist or natural scientist. His main interest was in philosophy, and it is said that he considered his involvement in physics more as a burden than as an interesting task.

Etienne Jules Marey (1838–1904) developed various techniques such as film analysis and pressure-measuring devices which could be applied to biomechanical research. His contribution to the field of biomechanics was not to propose solutions to basic questions as his predecessors had done. Rather it consisted of the development of methodologies and techniques to quantify human movement. His pressure-measuring device (Fig. 2) was the first attempt to quantify forces acting on the foot of a subject during movement. He also developed chronophotography.

Elftmann developed a method of quantifying the centre of pressure under the foot during gait (published in 1938, see ref. 24).

Paul estimated internal forces in the human hip joint in 1965.[2] One of the difficulties in the estimation of internal forces is that the human locomotor system has more force-carrying structures crossing a joint than are needed to perform a specific movement, which results in a mathematical system with more unknowns than equations. Paul solved this problem by reducing the actual number of muscles crossing a joint to equal the number required to perform the movement in question (reduction method).

Hatze[4] expanded the mainly mechanical approach to modelling the human body during movement by developing a model which included a comprehensive mechanical, physiological, and neurological approach.

Currently, several thousand researchers are active in the field of

Fig. 2 Test subject with a pneumatic device to quantify pressure underneath the foot as developed by Marey. (Reproduced from ref. 3, with permission.)

biomechanics, covering fields such as cardiac biomechanics, fluid biomechanics, injury biomechanics, muscle biomechanics, orthopaedic biomechanics, rehabilitation biomechanics, sport biomechanics, and tissue biomechanics.

MECHANICS

Classical mechanics is based on Newton's three laws: the law of inertia (first law), the law of action (second law), and the law of reaction (third law).

Mechanics can be subdivided into statics and dynamics. Dynamics is usually subdivided into kinematics and kinetics. Kinematics describes position, displacement, velocity, and acceleration without an indication of the cause of the movement. Kinetics relates the forces acting on a body to its mass and motion. Conventionally, kinetic analysis predicts the motion of a mass for known forces (direct dynamics) or determines the forces required to produce a given motion (inverse dynamics). The study of statics goes back to the Greek philosophers. The first significant contribution to the study of dynamics was made by Galileo Galilei (1564–1642). His experimental findings on the topic of uniformly accelerated bodies formed the basis for Newton's fundamental laws of motion.

STATICS

Force

Most of our ideas about force are associated with exertion of muscular forces on objects. The meaning of force seems clear because force is experienced every day, whether it be sitting in a chair, pushing against a resistance, jumping, or walking. Force is considered as a 'push' or 'pull' exerted by one body on another. Effects produced by forces can be listed and are descriptions (not definitions). In a static sense, force can be described as follows:

If 'something' is able to keep a spring stretched, this 'something' is called a force.

In biomechanics the expressions external and internal forces are used. To distinguish between external and internal forces the system of interest must first be defined.

Forces acting upon the system of interest are called external forces.
Forces acting within the system of interest are called internal forces.

If, for instance, the human body is the system of interest, all the forces acting upon the human body are external forces (e.g. gravity, ground reaction force, air resistance), while all the forces acting within the body are internal forces (e.g. joint forces, muscle forces, tendon and ligament forces). If, however, the forearm and hand comprise the system of interest, the joint forces in the elbow joint would be external forces.

If the human body is considered to be the system of interest, typical external forces would be gravitational forces, frictional forces, ground reaction forces, and contact forces. Typical internal forces would include bone-to-bone forces in joints, tendon insertion forces, muscular forces, and ligament forces.

Forces acting upon or within the human body while at rest can be called 'static forces'. When a person is standing on both legs the force body weight is acting. Body weight and body mass are not the same. Body weight is analytically determined from body mass by using Newton's second law. To a first approximation the acceleration is 10 m/s^2. This value will be used in all the following examples. Thus the body weight for a human with the mass of 50 kg is 500 N, and the body weight for a human with the mass of 80 kg is 800 N (where N is the symbol for 1 newton).

Often the forces are expressed in units of body weight (BW). Examples of static forces expressed in body weight are as follows:

force in hip joint (standing on both legs) = 0.5 BW
force in hip joint (standing on one leg) = 3–5 BW
force in the Achilles tendon (one leg/toe) = 3–6 BW.

Forces (like all vectors) are often resolved into components in the direction of the axes of the chosen co-ordinate system. Sometimes these axes are related to body posture. Typical conventions used in this text are listed above. Forces may also be resolved into components relative to surfaces. In joints one often distinguishes between forces perpendicular to the joint surface (normal forces) and forces tangential to the surface. Forces in muscles may be resolved with respect to the line of action of muscle fibres (Fig. 3).

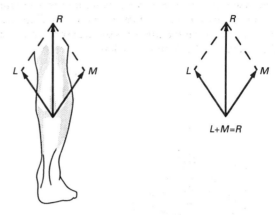

Fig. 3 Illustration of a resultant vector **R** resolved into its components **L** and **M** (adapted from ref. 5).

Moment of force

If the system of interest is not a particle but a rigid body with length, width, and height (dimensions), forces can produce two types of movements: translation and/or rotation. In order to describe the effects of forces which can produce a rotation, the quantity 'moment of force' or 'moment' is introduced. The symbol for moment is M, where M is a vector:

$$M = \text{moment of force} = \text{moment} = \text{torque}.$$

The magnitude of the moment of F about an axis O is

$$M_O = d_O F$$

where

M_O is the magnitude of the moment with respect to point O, d_O is the perpendicular distance from 0 to the line of action of F, and F is the magnitude of force acting on the body. The units of F are newtons (N), the units of d_O are metres (m), and the units of M_O are newton metres (N.m).

Examples of moments

The following numbers are given for an average body mass of 80 kg. The moment produced by the Achilles tendon with respect to the ankle joint while standing on one leg on the toes is about 7–14 N.m. The moment produced by the (idealized) body weight about the sagittal hip joint axis while standing on one leg is about 5–10 N.m. The moment of the (idealized) biceps and brachialis tendon for an assumed mass of 1 kg and an assumed forearm mass of 2 kg is about 0.56 N.m.

Equilibrium of rigid bodies

If a body is at rest and remains at rest it is said to be in equilibrium. In this case the vector sum of all the forces and moments acting on the body is equal to zero:

$$F_1 + F_2 + \ldots + F_n = \Sigma F_i = O \quad i = 1, \ldots, n$$
$$M_1 + M_2 + \ldots + M_n = \Sigma M_i = O \quad i = 1, \ldots, n$$

These equations will be used in the following to estimate internal forces in static situations. The examples will be restricted to two-dimensional questions.

Example of internal forces

To illustrate how the concept of equilibrium can be applied to biomechanical problems the equilibrium equations will be used to calculate muscle and joint forces for a two-dimensional example. A subject is standing on the forefoot of one leg as illustrated in Fig. 4. We wish to determine the bone-to-bone force in the ankle joint and the force in the Achilles tendon.

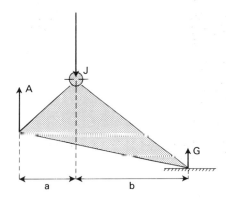

Fig. 4 Free-body diagram of a human foot while standing on the forefoot.

Assumptions

1. The problem can be solved two dimensionally.
2. There is one joint to the neighbouring segment—the ankle joint which is idealized as a hinge joint. The human foot is considered as one rigid body.
3. The forces acting are as follows (in scalar form): ground reaction force $G = 800$ N; force in the Achilles tendon $A =$ unknown; force in the ankle joint $J =$ unknown.
4. All forces act in vertical direction.
5. The distance of the forces from the ankle joint are $a = 0.04$ m and $b = 0.12$ m.
6. Friction in the ankle joint and on the ground can be neglected.
7. The weight of the foot can be neglected.

Solution

The following equilibrium equation can be formulated for the moments with respect to an axis through the ankle joint (note that the force J does not produce a moment about the ankle joint since its line of action is through the axis of the ankle joint):

$$-aA + bG = 0.$$

The only unknown is A; therefore

$$A = (b/a)G$$
$$A = 2400 \text{ N}.$$

The following equilibrium equation can be formulated for the forces:

$$A + G - J = 0.$$

The only unknown is J; therefore

$$J = A + G$$
$$J = 3200 \text{ N}.$$

It should be noted that in this example the joint and muscle forces are larger than the ground reaction force. This result is typical of the relationship between internal and external forces.

DYNAMICS

Kinematics

Movement, in general, is subdivided into translational and rotational movement. A motion is said to be a translation if any straight line within a body keeps the same orientation for any time t_i throughout the motion. The mechanical quantities which are used to describe translational movement are position, displacement, velocity, and acceleration.

The acceleration which is frequently used is the earth acceleration or the acceleration due to gravity. The value for the acceleration due to gravity is 9.81 m/s^2 and is referred to as $1/g$. In many practical applications and for simplicity, the acceleration due to gravity g is often approximated as 10 m/s^2. This approximation is used consistently throughout this chapter.

Examples of speeds and accelerations for various movements are summarized in Tables 1 and 2.

Table 1

Body part	Velocity (m/s)	Movement
CM	30	Take-off in ski-jumping
CM	10	Average speed in sprinting
CM	4.5	Free fall from $H = 1$ m
CM	7.7	Free fall from $H = 3$ m
Heel	2–3	Landing speed in running
Toe	1–2	Landing speed in running
Hand	30	Javelin (maximal speed)
Foot	20	Football kick

CM, centre of body mass.

Table 2

Body part	Acceleration (g)	Movement
Hand	5–10	Boxing (active movement)
Hand	30–60	Boxing (passive impact)
Tibia	10–30	Landing on room surface from 0.5 m
Tibia	30–50	Skiing (30 km/h, squatting)
Hip	2–3	Skiing (30 km/h, squatting)
Head	1–2	Skiing (30 km/h, squatting)
Tibia	5–10	Skiing (30 km/h, standing)
Tibia	200	Skiing (100 km/h, squatting)
CM	0.4	Sprint start

CM, centre of body mass.

Translational considerations can be appropriate for the analysis of running. The average speed in running can be written as:

$$v_a = \text{distance/time} = (\text{stride length}) \times (\text{stride frequency}).$$

For each step the average speed depends on the stride length and the stride frequency. Therefore an increase in stride length and/or stride frequency can be used to increase average running speed. This has been supported experimentally[6] for changes in running speed between 3 and 4 m/s (Fig. 5). However, stride lengths and stride frequencies cannot be increased indefinitely. The increase in stride length may be limited by the fact that the centre of mass is increasingly lowered with an increase in the stride length. Thus additional work must be done for the increased vertical movement of the centre of mass. The increase in frequency is limited by the muscular strength and/or the inertia of the masses involved. While the masses involved usually only change slightly for a given athlete, the muscular strength may increase substantially.

The following set of equations can be used to calculate displacements, speeds, accelerations, and time intervals for movements with constant acceleration:

$$x = x_0 + v_0 + \tfrac{1}{2}\, at^2 \tag{1}$$
$$v = v_0 + at \tag{2}$$

The following three equations can be derived from equations (1) and (2):

$$x - x_0 = \tfrac{1}{2}\, t\, (v + v_0) \tag{3}$$
$$x - x_0 = 1\ 2a\ (v^2 - v_0^2) \tag{4}$$
$$a = v^2 - v_0^2\ 2(x - x_0) \tag{5}$$

where

x_0 is the initial displacement, x is the final displacement, v_0 is the initial speed, v is the final speed, a is the constant acceleration, and t is the time interval between initial and final time.

Examples for movement with (assumed) constant acceleration

1. Determine the average acceleration of a javelin in a world-class throw if the approach speed of the athlete's hand is 5 m/s, the release speed of the javelin from the hand is 30 m/s and the distance over which the javelin is accelerated is 2 m.

$$\begin{aligned} \text{Known: } v_0 &= 5 \text{ m/s} \\ v &= 30 \text{ m/s} \\ x_0 &= 0 \text{ m} \\ x &= 2 \text{ m} \\ \text{Unknown: } a_a &= ? \end{aligned}$$

Solution
Use equation (5):

$$\begin{aligned} a_a &= 900 - 25\ 2(2\text{---}0\text{---})\ \text{m/s}^2 \\ &= 218.75\ \text{m/s}^2. \end{aligned}$$

The average acceleration of the javelin is about 220 m/s² or about 22 g. The actual maximum acceleration in javelin throwing is higher than the average acceleration estimated in this example. A realistic estimate for the maximum acceleration is about twice the calculated average acceleration. This can be illustrated by assuming a triangle distribution of the acceleration–time curve.

2. Determine the take-off speed of a world-class high jumper assuming that the jumper's centre of mass at take-off is at 1.05 m, that it clears the bar by 10 cm, and that the cleared bar is at 2.45 m.

$$\begin{aligned} \text{Known: } x_0 &= 1.05 \text{ m} \\ x &= 2.55 \text{ m} \\ v &= 0 \text{ m/s} \\ a &= -10 \text{ m/s}^2 \\ \text{Unknown } v_0 &= ? \end{aligned}$$

Solution
Use equation (5) and solve it for v_0:

$$2a(x - x_0) = v^2 - v_0^2.$$

Now

$$v = 0.$$

Therefore

$$\begin{aligned} v_0 &= [-2a(x - x_0)]^{1/2} \\ &= (20 \times 1.5)^{1/2}\ \text{m/s} \\ &= 5.5\ \text{m/s}. \end{aligned}$$

The take-off speed of the centre of mass for a world-class high jump is about 5.5 m/s.

3. Determine the average acceleration for a sprinter, assuming that he/she accelerates from no movement to a speed of 10 m/s in the first 20 m.

$$\begin{aligned} \text{Known: } v_0 &= 0 \text{ m/s} \\ v &= 10 \text{ m/s} \\ x_0 &= 20 \text{ m} \\ \text{Unknown: } a &= ? \end{aligned}$$

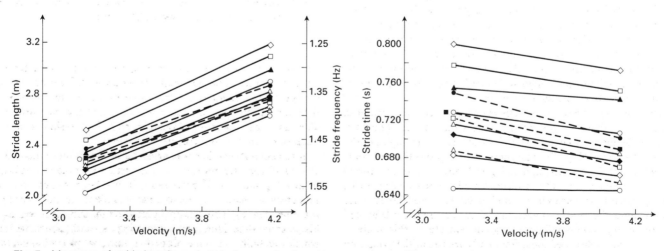

Fig. 5 Measures of stride lengths and stride frequencies for athletes running at different speeds on a treadmill. (Adapted from ref. 6.)

Solution
Use equation (5):

$$a = 100/40 \text{ m/s}^2 = 2.5 \text{ m/s}^2.$$

The average acceleration of a world-class sprinter for the first 20 m is about 2.5 m/s² or about 0.25 g. The maximum acceleration for a sprint start is about twice the average acceleration which corresponds to about 5 m/s² or 0.5g.

A motion is said to be a rotation if any straight line within a rigid body changed its direction during the motion. A normal movement is usually a combination of translation and rotation. In what follows rotations are studied independent of translations. The magnitudes of the mechanical quantities used to describe rotational movements are angular position, angular displacement, angular velocity, and angular acceleration. Rotational movements of the foot or parts of the foot as measured during running are used to illustrate the application of rotational analysis to human movement.

The foot is a complex anatomical structure consisting of 26 bones. It possesses unique qualities. It can be rigid or flexible depending on the task performed. The talus plays a critical role in the mechanics of the link between the lower leg and the foot. On the proximal side it connects to the lower leg. On the distal side it connects to the foot. The joint between the tibia and the talus is called ankle joint, talocrural joint, or talotibial joint. To a first approximation, it is considered to be a hinge joint and to lie approximately in the frontal plane. Measurements[7] have shown that the ankle joint is slightly oblique. The ankle joint axis is rotated laterally in the transverse plane, and inclined downwards and laterally since the fibular malleolus (on the outside) lies more posteriorly (towards the back side) and extends more distally (towards the extremities) than the tibial malleolus. Consequently, rotation around the ankle joint produces dominantly plantarflexion/dorsiflexion, but additionally some minor components or inversion/eversion and adduction/abduction.

The joint between the talus and the calcaneus is known as the subtalar joint. (It should be noted that the talus has additional joints with the navicular bone (the talonavicular joint) and with the cuboid bone, (the talocuboid joint). These joints are not discussed in the following.) It appears to be well established that the motion between the talus and the calcaneus is a rotation about a single oblique axis, the subtalar joint axis.[7] The 'average' subtalar joint axis is inclined upwards anteriorly by approximately 42° and inclined medially approximately by 16° (Fig. 6). A rotation about the subtalar joint axis includes components for inversion/eversion, adduction/abduction, and dorsiflexion/plantarflexion. The rotation about the subtalar joint axis is known as pronation or supination. It is difficult to distinguish *in vivo* between rotations about the ankle and about the subtalar joint.

It has been speculated that many running injuries are related to excessive eversion (or pronation). For this reason a methodology has been developed to quantify angles and changes in angles in the lower extremities (Fig. 7). The method uses projections of markers which are fixed on the human leg and the shoe into the frontal and the sagittal plane. The projected angle which proved to be most frequently implicated in running injuries was the Achilles tendon angle, the angle between the lower leg and the calcaneus.[8] The following examples for running concentrate on this angle.

The two variables used most frequently are the initial pronation and the maximal pronation. The initial pronation is defined as the

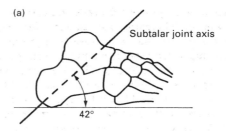

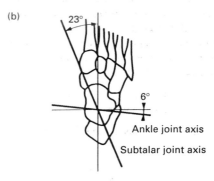

Fig. 6 Illustration of the ankle and subtalar joint axes. (Adapted from ref. 7.) (a) Lateral view. (b) Superior view.

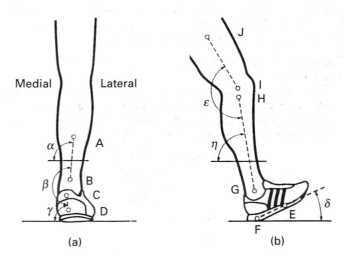

Fig. 7 Illustration of the angles used in two-dimensional analysis of running shoes: (a) posterior view; (b) lateral view.

change in the Achilles tendon angle during the first tenth of ground contact. The maximal pronation is defined as the total change in the Achilles tendon angle between first contact and maximum pronation. It should be noted that the measured variable does not quantify pronation and/or eversion accurately because of the two-dimensional limitation.

It has been shown in a prospective study[9] that more than 50 per cent of all running injuries are associated with excessive values in maximal pronation. Furthermore, it has been shown[8] that the pronatory movement of the foot can be influenced significantly by the shoe (Fig. 8). The maximal pronation for subject A was 21° in his personal shoe. However, when using a special (laboratory) shoe which was built to reduce pronation, maximal pronation could be reduced to only 12° in this subject. The effect is even more

Subject A   Subject B

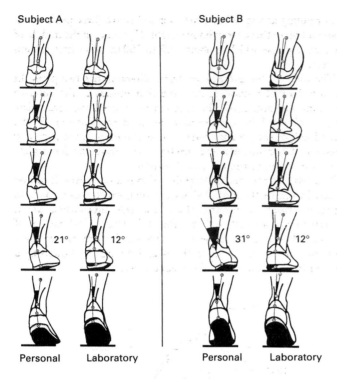

Personal Laboratory   Personal Laboratory

Fig. 8 Illustration of the influence of the running shoe on pronation during ground contact in running at a speed of 4 m/s (heel–toe running). The picture sequences should be read from the top to the bottom.

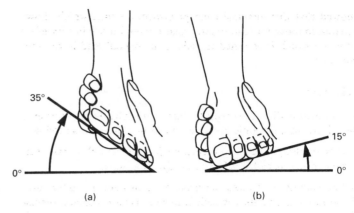

Fig. 9 Illustration of inversion (a) and eversion (b) of the forefoot for a fixed position of the rearfoot. (Adapted from ref. 11.)

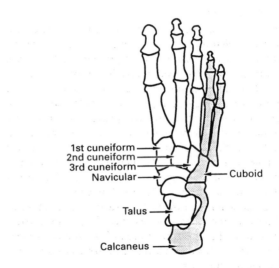

Fig. 10 Illustration of the subdivision of the human foot into two functional units.

pronounced for subject B who shows an excessive maximal pronation of 31° in his personal shoe.

One method of influencing the pronatory movement is to change the construction of the shape of the heel on the lateral side of the running shoe. A second approach is to use different material densities in the construction of the heel of the shoe. The effect of these two strategies on pronation has been studied using nine shoes which were identical except in their heel shape and their material composition in the heel.[10] The shapes (on the lateral side) used were flared, neutral, and rounded. The material compositions used were a combination of 'soft' on the outside and 'medium hardness' on the inside (shore A 25/35, dual density EVA), a 'medium hardness' shoe sole (shore A 35, single density EVA), and a 'hard' shoe sole (shore A 70/45, dual density polyurethane).

The results of the study show that the 'initial pronation' can be reduced by changing (a) the geometry of the lateral side of the heel (reduce the lever) or (b) the material hardness of the heel of the shoe (soft on the outside and hard on the inside).

Initial biomechanical research dealing with sport shoe development concentrated primarily on the ankle and subtalar joint. Limited additional research has been performed for the metatarsal–phalangeal joints. However, studies of the foot have revealed that there is an additional possibility of rotation about the longitudinal axis of the foot. If the heel is fixed the forefoot can be rotated through about 35° into inversion and about 15° into eversion (Fig. 9). This movement is possible because the first three rays and the talus form one 'connected structure' and the fourth and fifth ray together form a second connected structure (Fig. 10). The relative movement of these connected structures allows torsion of the forefoot with respect to the rearfoot.

This aspect is important since each joint of the lower ex-

tremities (hip, knee, ankle, subtalar, etc.) provides an additional possibility for the absorption of forces during locomotion. However, the conventional shoe constructions are rather stiff in the midfoot area and do not allow for torsional movement of the foot. Consequently, rotation of the forefoot with respect to the rearfoot is restricted, and the ankle and subtalar joint complex must primarily compensate to provide the necessary additional rotation. This may result in excessive loading situations. If an athlete lands in an everted position on his/her forefoot (e.g. landing on the foot of a team mate in volleyball or landing on uneven ground in cross-country running), the rearfoot must follow the forefoot eversion and excessive strain may occur on the medial side of the ankle joint complex. This would not occur in the barefoot situation since the forefoot would rotate torsionally with respect to the rearfoot.

The barefoot movement of the foot in the example above is assumed to be less stressful than the same movement in a shoe. Consequently, a solution for the sport shoe has been proposed[12,13] by cutting the shoe sole in the midfoot area (torsion groove), allowing torsional movement of the forefoot with respect to the rearfoot to occur. Additionally, a stiff element has been added (torsion bar) to prevent the shoe from flexing in the midfoot area since the midfoot does not allow for this movement. Research showed that this torsional concept provides a shoe which allows the foot to move more in the way that it would move in a barefoot situation which is assumed to reduce the overall load in the foot complex.

Kinetics

As outlined earlier, the quantity 'fore' cannot be defined but only described. In a dynamic sense force can be described as follows:

If 'something' is able to accelerate (or decelerate) a mass, this 'something' is called a force in a dynamic sense.

A force acting on a mass can produce translational and/or rotational motion. The mass which is moved has a resistance against changes in its state of motion. For translation this resistance is proportional to the mass. For rotation this resistance is proportional to the moment of inertia.

Ground reaction forces

Force platforms are commonly used in biomechanics. They are measuring devices which quantify the forces exerted by subjects onto them. Often, they are installed in the ground and the forces measured with them are called ground reaction forces. The force vector is resolved into three components: a vertical component F_z, an anterior–posterior component F_y, and a mediolateral component F_x. Figure 11 illustrates the three components of the ground reaction force during one ground contact for heel–toe running at a speed of 3.5 m/s. The time axis is normalized. The initial value 0 is the time of first ground contact and the final value 1 is the time of last ground contact. Different aspects of ground reaction forces may be of interest and are discussed in the following.

The force components are associated with the acceleration of the centre of mass in the corresponding directions. The vertical force component describes the acceleration of the centre of mass in the vertical direction. Analogous statements are appropriate for the two horizontal components.

The vertical force component for heel–toe running usually has two force peaks. The first vertical peak occurs about 5 to 30 ms after first ground contact. It is called the vertical impact force peak and is referred to here as F_{zi}. Similar peaks may occur in the two horizontal force-time curves. Impact forces can be defined as follows:

Impact forces are forces due to a collision of two objects with a maximum earlier than 50 ms after first contact of the two objects.

In running, impact forces are connected to the landing of the heel of the foot on the ground. The time occurrence and the magnitude of the impact force peak depends on various factors such as the running speed, the style of running, the geometrical shoe construction, and the material properties of the shoe sole. For running barefoot on steel the impact peak occurs about 5 to 10 ms after first contact, and for running with a soft-soled running shoe on asphalt the impact peak occurs about 20 to 30 ms after first contact. Running on a soft sandy beach would show no impact peak at all.

The second peak in the vertical force–time curve for heel–toe running is called the vertical active force peak and is referred to in this text as F_{za}. Similar peaks may occur in the two horizontal force-time curves. Active forces can be defined as follows:

Active forces are forces due to movement which is entirely controlled by muscular activity.

In running at a speed of 4 m/s vertical active force peaks occur at about two to three times body weight. F_{za} occurs in the middle of the stance phase which is about 100 to 200 ms after first ground contact.

The force in the anterior–posterior direction has two parts. In the first half of ground contact the foot pushes in the anterior direction. Consequently, the reaction force from the force platform is directed in the posterior direction (backwards). In the second half of the ground contact the foot pushes in the posterior direction. Consequently, the reaction force from the force platform is directed in the anterior direction.

The force in the mediolateral direction has two parts. It can be directed towards the medial side (inwards), which corresponds to a positive force, or towards the lateral side (outwards), which corresponds to a negative force (Fig. 11). The typical pattern of the mediolateral force in heel–toe running initially shows for a short time a reaction force in the lateral direction followed by a force of longer duration in the medial direction.

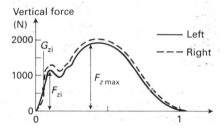

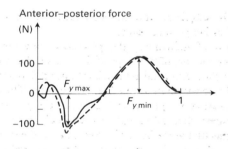

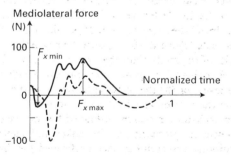

Fig. 11 Example for the components of the ground reaction force during running at a speed of 4 m/s for one subject and one trial (left and right foot).

Forces in the vertical and anterior–posterior direction show a small variability while force components in the mediolateral direction show a large variability. Additionally, ground reaction forces for left and right are often different, which suggests that the movement may not be symmetric. It has been shown that world-class runners show significant differences in their ground reaction forces for left and right feet.

Values of externally measured impact forces have been reported in the literature over the last 15 years. A summary of the reported values is shown in Fig. 12. The external impact forces were assessed using force platforms. The internal impact forces were

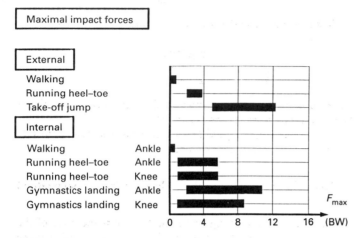

Maximal impact forces

Fig. 12 Summary of the magnitude of the measured external and estimated internal impact force peaks.

estimated using mathematical models of the human body together with kinematic and kinetic data from experiments. As shown, external vertical impact force peaks can exceed 10 BW. Internal impact peaks have been estimated to be as high as 6 BW in running and 10 BW for landings in gymnastics. It is speculated that such forces may be close to or exceed limits beyond which damage to anatomical structures may occur.

Values of active forces which have been reported by various authors are shown in Fig. 13. Maximal active force peaks acting on

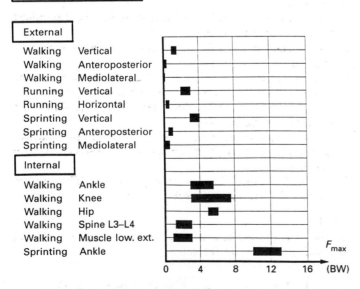

Maximal active forces

Fig. 13 Summary of the magnitudes of the measured external and estimated internal active force peaks as reported in the literature by various authors.

the feet of athletes during various sports activities have been measured with force platforms and do not appear to exceed 4 BW. Internal active forces, estimated from mathematical models, were reported to be larger than the external forces. Maximal forces in the ankle joint during running and sprinting have been estimated to be 10 to 13 BW.[15,16]

A two-dimensional model of the human body in sprinting may explain the difference in magnitude of the internal and external active force peaks. The external force during push-off in sprinting acts with a moment arm (lever) of about 20 cm with respect to the ankle joint axis. The main reaction to this moment on the posterior side of the leg is provided by the Achilles tendon. The moment arm (lever) of the Achilles tendon to the ankle joint axis is about 5 cm. Consequently, the force in the Achilles tendon must be about four times the ground reaction force.

Impact force peaks are affected by the geometric construction and the material properties of the heel of the shoe. Results of the running shoe project which were discussed earlier with respect to angular kinematics are discussed below for kinetics. Results of this experiment show that geometry, as well as material, influences the magnitude of the external vertical impact force peaks. The effect of the material is more pronounced for flared heels but is minimal for rounded heels. The maximal differences in external vertical impact force peaks is between 20 and 25 per cent.

Kinetics and kinematics

Kinetics and kinematics must be combined to allow estimations of effects of external forces. Internal forces acting on structures of the human locomotor system are associated with external forces. If a force acts with a small moment arm, the moments produced about that joint axis are small. If a force acts with a large moment arm, the forces acting in structures of the human body may be a multiple of the external force. An example of the combination of kinematics and kinetics for heel–toe running is illustrated in Fig. 14.

The (idealized) joint axes are indicated by points in the figure, which is an oversimplification of the actual situation. However, the example illustrates that the ground reaction force produces positive and negative (left- and right-turning) moments about the ankle and the knee joint. Furthermore, the sign of the moment about one joint changes several times, throughout foot contact, thus producing changing loads for specific structures of the athlete's leg. Consequently, one can expect changes of compression and tension in the tibia, for example, and changes between high and low forces in the gastrocnemius, soleus, and tibialis anterior muscles and the corresponding tendons during one ground contact.

A graphical representation of the posterior view (Fig. 14) illustrates the influence of the midsole hardness on the relative position of the external ground reaction force with respect to the foot. The general finding as derived from this figure can be summarized as follows. The differences are significant during the first half of ground contact but not during the last half. The ground reaction force is close to the longitudinal foot axis for a soft shoe sole material (shore 25) while a harder material (shore 45) has a ground reaction force towards the lateral side of the shoe. Consequently, one expects larger internal forces in the tendon–muscle structures and in the ligaments around the subtalar joint for the harder shoe compared with the softer shoe.

It has been speculated that the loading of the lower extremities with respect to possible acute and/or chronic injuries is reduced

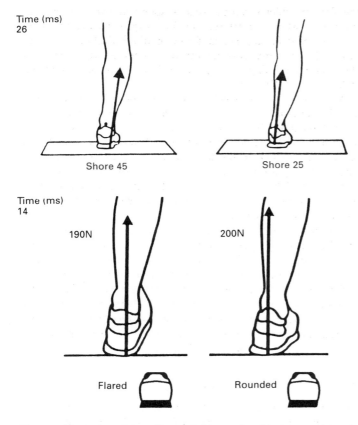

Fig. 14 Illustration of the effects of changes in midsole material and heel geometry on the location of the line of action of the resultant ground reaction force.

when the external force acts with a small moment arm compared with an external force acting with a large moment arm. On the basis of the results mentioned above, there are at least two possible methods of influencing the external forces during landing in running with the shoe. Both possibilities influence the relative position of the line of action of the force with respect to the foot. One approach uses a soft material for the sole of the heel on the lateral side or for the whole heel (Fig. 14(a)). Another possibility uses a rounded or neutral geometrical shape on the lateral heel of the running shoe (Fig. 14(b)). The two solutions have advantages and disadvantages. The soft heel material may be compressed totally under the applied load (bottoming out), which would result in excessive impact forces. Possible increases in impact force peaks may be as high as 30 to 50 per cent. The rounded heel increases the impact peaks by about 10 to 20 per cent. Both changes in shoe construction seem to have the effect of increasing the external ground reaction forces. The increases are in a range from zero to about 60 per cent. In order to understand the effect of such changes in shoe construction, models can be developed which are able to estimate the internal forces in anatomical structures of interest. On the basis of literature data it seems that increases in the moment arm of the acting force with respect to the subtalar joint axis may increase the internal forces by several hundred per cent. Consequently, it seems to be appropriate to work with the rounded (or neutral) heel geometry to reduce the initial pronatory effects of ground reaction forces.

Linear momentum

Consider a particle of mass m acted upon by a force F. Replacing the acceleration by the time derivative of the velocity, one can write Newton's second law as

$$\Sigma F = m \frac{dv}{dt}$$

or, if we assume that the mass is constant,

$$\Sigma F = \frac{d(mv)}{dt}$$
$$mv = L$$

where L is the linear momentum of the system and has units of kilogram metres per second (kg/m/s).

The force acting on a particle with the mass m is equal to the rate of change of its linear momentum:

$$\int_{t_1}^{t_2} F(t)\, dt\ (\text{impulse}) = mv_2 - mv_1 = (\text{change in (linear) momentum})$$

where F is the force acting on the particle, m is the mass of the particle, v_2 is the velocity of the particle at the time t_2, and v_1 is the velocity of the particle at the time t_1.

Linear momentum is of mechanical importance as it can be written in the form of the principle of conservation of linear momentum which can be recognized as another form of Newton's first law.

If the resultant force acting on a particle is zero, the linear momentum of the particle remains constant, in both magnitude and direction, and the change in momentum is zero.

Example

A subject is standing on a frictionless cart which stands on a horizontal plane. The subject throws a brick (mass m_1) in the horizontal direction from the cart. What happens to the person (mass m_2) and the cart (mass m_3)?

Assumptions
1. The cart can move without friction along the line determined by the throw of the brick.
2. Subject and cart do not have any relative movement and can be considered as one mass m_4
3. The masses m_4 and m_1 can be considered as particles and the problem can be dealt with as a two-dimensional problem.
4. No external forces act on the system cart–subject–brick. Air resistance can be neglected.
5. The following values are assumed for the numerical calculations: $m_4 = 100$ kg; $m_1 = 10$ kg; $v_1 = 10$ m/s (in the positive y axis direction).
6. The brick is thrown in the positive y direction. The problem can be solved one-dimensionally.

Solution
Let the times just before and just after the brick has been thrown be t_1 and t_2 respectively. Then

$$\int_{t_1}^{t_2} F_y(t)\, dt = 0 = (m_4 v_4 + m_1 v_1)_2 - (m_4 v_4 + m_1 v_1)_1$$

momentum of the system after the throw   momentum of the system before the throw = 0

$$m_4 v_4 + m_1 v_1 = 0$$
$$v_4 = -(m_1/m_4)v_1$$
$$v_4 = -1\,\text{m/s}.$$

The cart and the subject will move with a velocity of 1 m/s in the direction opposite to the brick (negative y axis).

Work and energy

Work

Work done on a body by the action of an external force producing a motion is an important concept that has been developed in physics. The term 'work' as used in physics is a mechanical term. It is defined as follows.

> The work done by a force that acts on a body is equal to the product of the magnitude of the force and the distance that the body moves in the direction of the acting force, while the force is being applied to the body.

This is expressed as

$$W = \int_{d_1}^{d_2} F.\mathrm{d}s.$$

Work for a constant force in rectilinear motion is

$$W = F d_F$$

where W is the work, F is the force, and d_F is the distance travelled in the direction of the force while the force is being applied.

Work in a mechanical sense, work in a physiological sense, and work in daily life may have a different meaning. A person standing in an airport and holding two suitcases in his/her hands does no mechanical work. However, physiologically he/she does work as can be measured by his/her oxygen consumption. A person sitting in a chair in an office developing new marketing strategies 'works' in the sense of the daily use of the term (at least he/she is paid for it). His/her muscles are active at a low level. Consequently, the physiological work is not zero. However, the mechanical work is zero as long as he/she does not move.

Different forms of mechanical work are distinguished:

work against gravity $\quad W_{gr} = -mg\Delta H$
work to deform a spring $\quad W_{sp} = \frac{1}{2}kx^2$
work to accelerate a mass $\quad W_{acc} = \frac{1}{2}mv^2$.

The unit of work is kg m^2/s^2 = joule (J).

The term 'calorie' or 'kilocalorie' is often used in discussions of losing weight or the mechanical work done during competition. The connection between joules and kilocalories (kcal) is

$$1\ \text{J} = 2.4 \times 10^{-4}\ \text{kcal} = 0.00024\ \text{kcal}.$$

Energy

Energy is the capacity (ability) to perform work.

A body has energy if it has the ability to alter the state or the condition of another body. For example, a flying football has energy because it is able to move or deform another object. A diver standing on the diving board at a height above the ground has energy because of his/her position. When diving, the body will accelerate and attain a certain speed when hitting the water. The body is able to move with water and consequently it has energy.

Different forms of mechanical energy are distinguished:

kinetic translation $\quad E_{kint} = \frac{1}{2}mv^2$
kinetic rotation $\quad E_{kinro} = \frac{1}{2}I\omega^2$
potential gravity $\quad E_{potgr} = mgH$
potential spring $\quad E_{potsp} = \frac{1}{2}kx_2$

where

m is the mass, v is the speed, I is the moment of inertia, g is the earth acceleration, H is the height, k is the spring constant, and x is the deformation of spring. The unit of energy is the same as the unit for work. There are other forms of energy such as chemical, electrical, thermal, nuclear, etc.

One of the main laws of physics is the conservation of energy. In general a group of bodies is selected arbitrarily and called a 'system'. The law of conservation of energy applies to such an arbitrarily selected system.

> The sum of all the energies in a 'system' remains constant.

An example of a mechanical system is a trampolinist. The system is the athlete and the trampoline. An athlete jumping down into a trampoline will rebound. At the start, the athlete has a potential energy due to the position. Next, this potential energy will decrease and the kinetic energy will increase. When contacting the trampoline the potential and the kinetic energy will decrease to zero and the energy is used to deform/stretch the trampoline (spring). This potential energy is then given back and the athlete is accelerated upwards. However, the final height will be less than the initial height since part of the energy is lost in heat due to friction in the trampoline.

In a scientific sense the law of conservation of energy holds for all forms of energy. This is obvious in the subsequent example of a marathon runner. Without energy from outside (food and beverages), an athlete is able to do work of about 1500 kcal (about 6 250 000 J). Consequently, a marathon runner must eat and drink since the total energy needed for a marathon is about 2500 kcal (about 10 400 000 J).

Example

The work done by a runner during a marathon is of the order of about 2000 to 2500 kcal. One may be interested in the additional work an athlete must perform due to the weight of his/her running shoes. The example in this section estimates the additional work for an additional shoe mass of 200 g per shoe.

Assumptions
1. The two forms of additional work of interest are the work against gravity and the work in order to accelerate the shoe.
2. Each foot is lifted 0.2 m in each step.
3. The maximal speed of the swinging leg is 7 m/s.
4. The step length (left toe to right toe) is 2 m which corresponds in a marathon to about 21 000 steps.

Solution
For additional work against gravity

$$W = nmgH$$

where $n = 21\,000$, $m = 0.2$ kg, and $H = 0.2$ m. Therefore

$$W = 21\,000 \times 0.2\,\text{kg} \times 10\,\text{m/s}_2 \times 0.2\,\text{m}$$
$$= 8400\,\text{J}\ 2\,\text{kcal}.$$

For additional work for acceleration of the foot

$$W = n\,(\tfrac{1}{2}mv^2)$$
$$= 21\,000 \times 0.5 \times 0.2\,\text{kg} \times 49\,\text{m}^2/\text{s}^2$$
$$= 102\,900\,\text{J} = 25\,\text{kcal}.$$

The mechanical work to accelerate the foot is about 10 times greater than the work against gravity. This suggests that the movement of the extremities during running is critical with respect to the mechanical work done in running. The additional work to accelerate an additional mass of 0.2 kg on each foot is about 100 000 J or about 25 kcal. This corresponds to about 1 per cent of the total physiological work done during a marathon. If work and time are related linearly this could correspond to an additional 1 to 1.5 min in the total marathon time.

The relative additional work due to the acceleration of the additional shoe mass depends on the actual running speed and on the actual added mass (Fig. 15). An increase in mass corresponds to a linear increase in work required. An increase in speed for a given mass corresponds to a quadratic increase of the required work. Mass considerations for shoes for high performance athletes become increasingly important for sports with high speeds. However, considerations of the effect of shoe mass on performance is less important for recreational marathon runners.

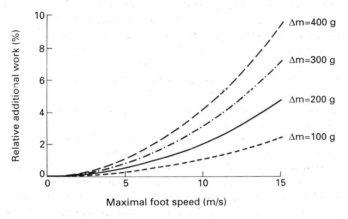

Fig. 15 Relative additional work due to acceleration of additional shoe mass as a function of the maximal speed of the foot.

The mechanical estimation of the additional work due to the acceleration of the additional mass of the shoe assumes that everything is kept constant when changing speed and/or mass. This is obviously a simplification since kinematics, kinetics, and internal muscular activities probably change. Also, the acceleration phase is followed by a deceleration phase, which in itself needs additional work which may well be of the same order of magnitude as the acceleration work. Therefore the estimated additional work is not accurate. However, the estimate seems to be on the conservative side.

BIOMATERIALS (BONE AND CARTILAGE)

General comments

The human body is constructed of different materials such as bone, cartilage, ligament, tendon, muscle, and other soft tissues. Based on their function, the different anatomical components of the human body can be divided into two groups: the first group comprises the passive structures (bone, cartilage, ligament, tendon) and the second group comprises the only active elements, the muscles. The passive structures do not produce force, while the active structures do.

The importance of the musculoskeletal system is documented by its various functions. The musculoskeletal system is one important tool which can be used to influence the environment. The effects of actions due to the musculoskeletal system can be mechanical (e.g. walking, jumping) but also psychological (e.g. expression of joy or pain by facial expression).

In this section we provide some insight into mechanical aspects of construction and function of some of these anatomical structures. Space considerations limit the section to the description of bone and cartilage.

Bone

Selected historical highlights

The understanding of the construction and functioning of bone and bone growth made significant progress in the second half of the nineteenth century. Selected important contributions are listed in Table 3.

Table 3

1855	Breithaupt	Described stress fractures in military recruits of a Prussian military unit.
1856	Fick	Bone is a passive structure; the surrounding muscles determine the form of bone
	Virchow	Bone plays an active role in developing its form and structure
1862	Volkmann ⎫	Pressure inhibits bone growth and
1863	Hüter ⎭	release of pressure promotes it
1867	van Meyer	Relationship between architecture and function of bone
1867	Culman	Similarity of the trabecular arrangement in bone to that of a crane; in both cases the principle of highest efficiency and economy is used
1870	Wolff	Interdependence between form and function of bone; physical laws have strict control over bone growth
1883	Roux	Orientation of the trabecular system corresponds to the direction of tension and compression stresses and is developed using the principle of maximum economy of use of material (as Wolff); architecture of bone follows good engineering principles
1897	Stechow	First radiographic verification of stress fractures

Various illustrations of bone have been published, showing the similarity of theoretical constructions of three-dimensional trajectorial systems and the actual arrangement of the trabeculae in the human bone (Fig. 16).

General comments

The entire human skeleton of an adult consists of 206 distinct bones (Table 4). The structural functions of bone are as follows:

(1) to provide support for the body against gravity;
(2) to act as a lever system to transfer muscular forces;
(3) to act as protection for vital internal organs.

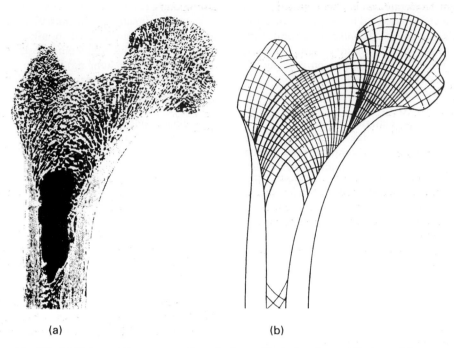

(a) (b)

Fig. 16 (a) Picture of the cross-section of a femur illustrating the architecture of the upper end of a femur of a 31-year-old male and (b) schematic representation of the same picture. (Reproduced from ref. 17, with permission.)

Table 4

Part	Number of bones
Vertebral column, sacrum, and coccyx	26
Cranium	8
Face	14
Auditory ossicles	6
Hyoid bone, sternum, and ribs	26
Upper extremities	64
Lower extremities	62
Total	206

In addition to the structural functions, bone has metabolic functions (e.g. as a repository for calcium).

Bone cells produce two types of tissue, poorly organized woven bone and highly organized lamellar bone. Woven bone has a lower mineral content than lamellar bone and forms rapidly in periods of intensive growth such as adolescence, fracture healing, or periods of rapid bone remodelling. Lamellar bone forms more slowly than woven bone and characteristically comprises thin layers of bone with collagen arranged in a perpendicular matrix. Cortical or compact bone is hard and predominates in the long bones. Trabecular, cancellous, or spongy bone is softer than cortical bone and predominates in bones of the axial skeleton (e.g. vertebrae, ribs).

The bones are divisible into four classes: long, short, flat and irregular. The long bones consist of a hollow cylindrical shaft and two extremities. Examples are the humerus, tibia, fibula, and femur. Long bones are found in the extremities and form a system of levers to transfer forces. The short bones are roughly cubical and are spongy except on their surfaces, where they are compact. Examples of short bones are the carpal and tarsal bones. It is assumed that their function is mainly to provide strength. The flat bones are, as the term indicates, bones where the osseous structure is expanded into broad flat plates. These bones are composed of two thin layers of compact bone with spongy bone enclosed between them. Examples of flat bones include the sternum, ribs, skull bones, ilium, and scapula. The main function of flat bone is to provide protection or a large area for the attachment of tendons or ligaments. The irregular bones are of irregular shape. They consist of a compact outer layer and a spongy inner layer of bone. Their shape seems to be adapted to a special function. Examples of irregular bones include the ischium, pubis, maxilla, and vertebrae.

Wolff's law of functional adaptation

In his classic publication in 1892, Wolff wrote:[17]

> The shape of bone is determined only by the static stressing . . . Only static usefulness and necessity or static superfluity determine the existence and location of every bony element and, consequently, of the overall shape of the bone.

Two comments are appropriate in the context of these statements.

1. Stress may have different effects on bone. It can effect growth as seen in the healing process of a fracture or it can have an inhibitory effect causing absorption of bone. Such atrophy can occur when stress is absent, constant, excessive or when periods of pressure exceed periods of release.
2. Growth of bone is influenced by heredity, among other factors. If a person has an inherited bone deformity, stress will not change the inherited form.

Considerations of this kind suggest that Wolff's law of the functional adaptation of bone should be restated in a more general way. Stress is not the only factor which determines bone growth, but it plays a major role. Thus Wolff's law can be modified as follows.

Physical laws are a major factor influencing bone growth.

Physical properties of bone

The terms used in this section are defined as follows.[18] The density of a material is its mass per unit volume. The mineral content is the ratio of the unit weight of the mineral phase of bone to the unit weight of dry bone. The water content is the ratio of extracted water divided by the volume of the specimen (bone). The elastic modulus E is the ratio of stress divided by strain:

$$\sigma = E\varepsilon$$

where the units of E are N/m^2 or pascals (Pa). The tensile strength or ultimate tensile strength is the maximal force in tension that a material can sustain before failure. The compressive strength or ultimate compressive strength is the maximal force in compression a material can sustain before failure.

Selected physical properties of bone (and other selected materials for comparison) are summarized in Table 5.

Table 5

Variable	Comment	Magnitude	Unit
Density	Cortical bone	1700–2000	kg/m³
	Cortical bone	1.7–2.0	g/cm³
	Lumbar vertebra	600–1000	kg/m³
	Water	1000	kg/m³
Mineral content	Bone	60–70	%
Water content	Bone	150–200	kg/m³
Elastic modulus E	Femur	11–28	GPa
Cortical bone, tensile strength	Femur	80–150	MPa
	Tibia	95–140	MPa
	Fibula	93	MPa
Cortical bone, compressive strength	Femur	131–224	MPa
	Tibia	106–200	MPa
Compressive strength	Wood (oak)	40–80	MPa
	Limestone	80–180	MPa
	Granite	160–300	MPa
	Steel	370	MPa

The elastic modulus E represents the stress needed to double the length of an object. The elastic modulus is a constant which describes the material characteristic. It does not suggest that the particular material can in fact be stretched to its double length. However, this constant can be used to determine the ultimate strength of a material. For bone the typical values for E are as follows: trabecular (spongy) bone, 10^9 Pa = 1 GPa; cortical (compact) bone, 2×10^{10} Pa = 20 GPa; metals, 10^{11} Pa = 10 GPa.

For bone, the elongation to fracture requires only a small fraction of this doubling length. As a guide one can write

$$F_{fracture} = (1/200)\, F_{double}.$$

Example

Using the information given above one can estimate the ultimate tensile force required to break a trabecular bone. Calculate the ultimate tensile force for trabecular bone in general and for the femur in particular.

Assumptions:

$$E = 10^9 Pa$$
$$A = 1\,mm^2 = 10^{-6}\,m^2 \text{ for general calculation}$$
$$\Delta L/L_o = 1/200$$
$$A_{tib} = 800\,mm^2 = 8 \times 10^{-4}\,m^2.$$

Solution

What force is needed to break a bone sample with a diameter of $1\,mm^2$?

$$\varepsilon = (1/E)\sigma = (1/E)\,(F/A)$$
$$= \Delta L/L_o$$
$$F = (1/L_o)\Delta L\, E A$$
$$F = (1/200) \times 10^9 \times 10^{-6}$$
$$F = 5\,N.$$

A force of 5 N is needed to break a bone of $1\,mm^2$ cross-section in tension. For a bone of $800\,mm^2$ cross-section (femur in the part where it is not hollow) a force of about 4000 N is needed to break it in tension.

The ultimate forces for tension and compression are different. The ultimate force for compression $F_u(compr)$ is about 30 per cent higher than the ultimate force for tension $F_u(tension)$:

$$F_u(compr) = 1.3 F_u(tension).$$

There is only a small region of elongation where bone follows a linear law between force and elongation (Hooke's law). Beyond this limit, deformation and force are no longer arithmetically proportional and when force is removed, bone does not return to its original form but some deformity remains. This is demonstrated in conditions such as osteomalacia or rickets.

Estimation of stress

Maximal stress for compression, tension, or torsion can be estimated for bone of circular cross-section by using the formulae in Table 6.

Table 6

Cross-section	Maximal stress for tension or compression σ_{be} (MPa)	Maximal stress for torsion σ_{tor} (MPa)
Solid circle	$32M_{be}/\pi D^3$	$16M_{tor}/\pi D^3$
Hollow circle	$32M_{be}D/\pi(D^4 - d^4)$	$16M_{tor}D/\pi(D^4 - d^4)$

M_{be}, bending moment responsible for compression and tension.
M_{tor}, moment responsible for torsion.
$D = 2R$, outer diameter of bone.
$d = 2r$, inner diameter of the hollow bone.

Example

An idealized bone has the shape illustrated on the left of Fig. 17. The cross-section is assumed to be circular. The radius of the bone column is $R = 1$ cm and a constant force of 100 N is acting as illustrated at a distance $2R$ from the axis of the bone. Estimate the maximal tensile and compressive stress at the cross-section S.

If the force F acts as illustrated in Fig. 17, the bone will be bent to the right. Consequently, the right bone surface will show compression and the left will show tension.

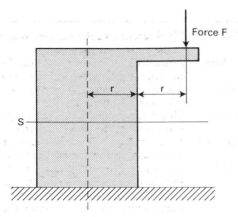

Fig. 17 Schematic illustration of the idealized bone.

Assumptions
1. The weight of the bone can be neglected.
2. Bone is isotropic (same material properties over the whole bone).
3. The problem can be treated two-dimensionally.
4. Bending moments are defined to be positive.

Solution

$$\sigma_{comp} = \sigma_{ax} + \sigma_{be}$$
$$s_{tens} = \sigma_{ax} - \sigma_{be}$$
$$\sigma_{ax} = F/\pi R^2$$
$$\sigma_{be} = 32 M_{be}/\pi D^3 = (32 F \times 2R)/(\pi \times 8R^3) = 8F/\pi R^2.$$

Consequently:

$$\sigma_{comp} = 9F/\pi R^2$$
$$\sigma_{tens} = -7F/\pi R^2.$$

Hence:

$$\sigma_{comp} = 2.87\,\text{MPa} = 2.87\,\text{N/mm}^2$$
$$\sigma_{tens} = 2.23\,\text{MPa} = 2.23\,\text{N/mm}^2.$$

The stress distribution in a cross-section of the bone structure discussed is illustrated in Fig. 18.

The actual maximal compression stress is greater than the actual maximal tension stress. The result is in the same direction as the

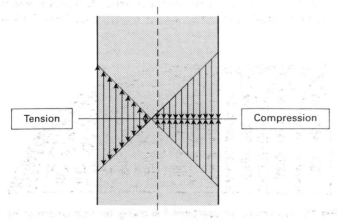

Fig. 18 Stress distribution due to a bending force *F* at the cross-section S.

previously mentioned result, which indicates that the ultimate compression stress is about 30 per cent higher than the ultimate tension stress. It is possible that the higher ultimate stress for compression is a result of an adaptation.

The stresses estimated due to bending in the above example are about an order of magnitude larger than the axial stresses. This illustrates that the geometry of the acting forces is extremely important. If the resultant force is not acting along the axis of a bone, the total stress on the surface of the bone increases and can easily reach multiples of the stresses produced by the axial forces.

This has at least four practical implications.

1. Joint forces which do not act along a bone axis are often compensated by muscular forces reducing the maximal stresses on the surface of the bone.
2. Misalignment of the skeleton may require an increased muscular compensation to reduce the maximal stresses on the bone surfaces. In cases of muscular atrophy as a consequence of injury or ageing, this balance may be disturbed and excessive stresses on the bone surfaces may develop which may lead to fractures.
3. Movements where external forces do not act along the bone axes on the human body (e.g. forces on the foot in different shoes) may produce high internal stresses on the bone surfaces.
4. Commonly, bone is loaded in different modes (tension, compression, torsion, and shear).

Bone and load

Bone can be loaded in different modes—compression, tension, shear, and torsion. Usually, a combination of all these forms of loads are present in bone during human locomotion. The different forms of loading of one finite bone element may change continuously during one ground contact.

Fracture of bone may occur for a variety of reasons:

(1) excessive forces;
(2) weak material;
(3) small bone diameter;
(4) excessive frequency of repetition of load application;
(5) reduced time of recovery between load application.

Bone needs a stimulus. However, the stimuli in the form of forces acting on bone must be in an optimal range.

Integrity of bone

As a living tissue, bone responds to physical, environmental, and biological events. Bone modelling occurring during childhood and adolescence determines the shape and length of bones. In the adult, bone remodelling predominates and provides a continual turnover of bone matrix and bone mineral. Some of the most important factors affecting bone integrity include hormonal, physical, nutritional, ageing, and pathological factors.

Hormonal factors

The hormones primarily responsible for calcium metabolisms and thus skeletal metabolisms are collectively referred to as the calciotropic hormones. These hormones respond to changes in serum calcium balance and will sacrifice calcium from the skeleton in order to restore calcium homeostasis. A second group of hormones, the reproductive hormones include the female sex hormone oestrogen (E2), the male sex hormone testosterone, and progesterone. They act as critical hormones for the balance of the skeletal metabolism. Finally, bone integrity is influenced by a number of other hormones (e.g. thyroid hormones).

Physical factors

Bone integrity is influenced by mechanical loading which affects local stress. Studies with animals and humans have demonstrated that bone responds to low and high mechanical stress and associated strain. The response of bone to stress and strain (modelling, remodelling) depends on the age of the organism and the magnitude, duration, and repetition of the imposed mechanical stress and strain.

Nutritional factors

Bone integrity is influenced by nutritional quality and quantity. Calcium is one of the most widely reported nutritional requirements for optimal bone growth, development, and maintenance. In addition, bones require protein, vitamin C, magnesium, boron, zinc, and in fact most of the nutritional components of food.

Age-related factors

The ageing process indirectly influences skeletal integrity through its impact on hormonal balance, digestion of dietary products, and the degree of physical activity a person engages in. Reproductive function, most obviously in women but also in men, decreases with age. Additionally, calcitonin levels and the production of vitamin D metabolites decrease with increasing age, and thus may affect the adequate absorption of ingested calcium.

Pathological factors

Bone integrity is influenced directly or indirectly by many pathological conditions. Chronic asthma, for instance, results in severe reduction of bone mass in children who are treated with glucocorticosteroids. Diseases such as osteomalacia and osteoporosis have a direct impact on bone remodelling and thus skeletal integrity.

Cartilage

Selected historical highlights

Many outstanding contributions to the understanding of the construction and the functioning of cartilage were made in the latter part of the nineteenth century and the first half of the twentieth century. Selected examples are given in Table 7.

General comments

Cartilage consists of cells (chondrocytes) which are embedded in an extracellular matrix. As described in *Gray's Anatomy*, cartilage is a non-vascular structure which is found in various parts of the human body. The articulating surfaces of bones are covered with articular cartilage (the main focus of this section). Additionally, cartilage is found in various tubes (nostrils, ears) which are to be kept permanently open. Cartilage is classified according to its structure as white fibrocartilage, yellow fibrocartilage, and hyaline cartilage.

White fibrocartilage consists of white fibrous tissue and cartilaginous tissue. The fibrous tissue is mainly responsible for flexibility and toughness, the cartilaginous tissue for elasticity. White fibrocartilage is found in intervertebral discs and articular discs.

Yellow fibrocartilage consists of cartilage cells and a matrix which is pervaded in every direction. It can be found in the external ear, the eustachian tubes, and other areas.

Hyalin cartilage is the main focus of this section. The word hyalin derives from the Greek word *hyalos*, which means glass.

Table 7

1851	Weber	Thickness of articular cartilage is proportional to the pressure it must sustain
1860	Virchow	Intercellular matter in hyaline cartilage is perfectly homogeneous and as clear as water
1876	Rauber	Modulus of elasticity of cartilage is $0.9\,kg/mm^2 = 0.9\,MPa$
1887	Morner	Discovered a substance with a high sulphur content which he called condroit
1898	Hultzkranz	Joint cartilage is under less tension in the transverse than in the longitudinal direction
1907	Fischer	Cartilage can be compressed: results from specimens are from 5 to 2.5 mm and from 2.5 to 1 mm
		Cartilage deformation increases the contact area in a joint and increases the possible range of motion
		Within the limits of its normal function, cartilage can fully regain its normal shape after release of pressure
1911	Fick	Greater elasticity always develops in the direction of joint motion; where the joint pressure is concentrated, the tension lines always run radially from the point of greatest pressure
1918	Fairbank	Moniscectomy results in overloading the articular surface locally with increasing compression of the cartilage and an increase of friction of about 20%

Hyalin cartilage consists of a gristly mass of firm consistency, bluish colour, and considerable elasticity. Hyalin cartilage is externally covered by a fibrous membrane, the perichondrium. It contains no nerves. Hyalin articular cartilage is specialized connective tissue and consists of three elements: cells, intercellular matrix (hyalin substance), and a fibre system.

The schematic representation (Fig. 19) shows the main course of the collagenous fibrils in articular cartilage. The lines of the collagenous fibrils are perpendicular to the underlying bony surface of the deep layers. They bend sharply in the transitional zone and run parallel to the cartilage surface in the superficial zone where they act as a protecting net. The thickness of articular cartilage increases with increasing local stress and is inversely proportional to the congruency of the joint surfaces. The patellar cartilage, for instance, is thickest at the sagittal crest (up to 6 mm).

Articular cartilage is a viscoelastic material. It is rather porous and the interstitium (space between the collagenous fibres) is filled

Fig. 19 Schematic representation[19] showing the main course of the collagenous fibrils in articular cartilage and arcual arrangement of fibrils with the chondrium drawn as black ovals.[20]

with fluid (about 60–80 per cent of the wet weight) which moves in and out under stress. Fluid flows in when the tissue is not under pressure and dilates. Fluid flows out when the tissue is under compression. The mechanical properties of articular cartilage alter with changes in its fluid content. The movement of the fluid in and out of the cartilage seems to be the principal way that cartilage obtains its nutrients.

Physical properties of cartilage

Articular cartilage exhibits a viscoelastic response to loads. Creep and stress relaxation are typical mechanical reactions of cartilage to loads. The viscoelastic response depends on the intrinsic viscoelastic properties of the fibre matrix and the frictional drag arising from the interstitial fluid. Articular cartilage has mechanical material properties which change as a function of load. Selected physical properties of cartilage are summarized in Table 8.

Table 8

Variable	Magnitude	Unit
Density	1300	kg/m^3
	1.3	g/cm^3
Water content (wet)	75	%
Organic content (wet)	20	%
Organic content (dry)	90	%
Coefficient of friction (dry)	0.0025	
Ultimate stress	5	MPa
	500	N/cm^2

The friction coefficient of articular cartilage is the lowest friction coefficient measured for any solid material. No technology has yet been developed which would permit construction of materials with a lower friction coefficient.

The numbers for ultimate stress indicate an order of magnitude. They are based on data from Yamada[21] and indicate the stress beyond which irreversible structural damage will occur to the cartilage tissue.

Cartilage deforms under load. Figure 20 shows a typical deformation–time curve based on an experiment performed by Hirsch.[20] A constant force of 10 N is applied to patellar cartilage for a time of about 5 min. Fluid is initially squeezed out of the cartilage and a deformation close to the maximal deformation in these 5 min is reached quickly. After unloading at about 5 min, the cartilage does not return immediately to its original thickness but needs time to recover. A second experiment under similar conditions (Fig. 21) underlines the importance of the duration of loading. Loading with a constant force for about 1 min allows cartilage to recover its original form in about another minute. However, when the cartilage was loaded with the same force for 2 min it had not recovered after 6 min. In general the longer the loading of cartilage, the slower is the recovery. An example of this compression effect is the change of body height as a function of the time of the day. Humans are 1 to 2 cm taller in the morning than in the evening.

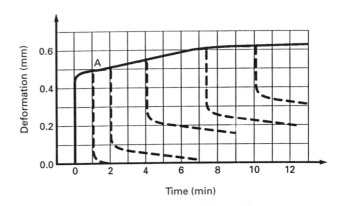

Fig. 21 Schematic illustration of the influence of the duration of loading on the deformation of cartilage. The applied force was 5 N and the area of contact was 10.6 mm[6]. (Adapted from ref. 20.)

Several mechanical functions are often suggested for articular cartilage, including mobility in the joints, stress distribution in the joints, and shock absorption of impact forces. The appropriateness of such suggestions is discussed below.

The suggested function of mobility in the joints seems to be well supported by the fact that articular cartilage has the lowest friction coefficient measured for two solid materials and the observation that joint range of motion is increased when cartilage is compressed.

The suggested function of stress distribution in the joints seems to be well supported by the fact that the contact area increases substantially due to the compression of articular cartilage. Consequently, the local stress is reduced.

The suggested function of shock absorption of impact forces seems to have less support. The human body during locomotion has a soft tissue pad of about 1.5 cm thickness at the heel which can be compressed up to about 1 cm. The thickness of articular cartilage is only a few millimetres. Additionally, articular cartilage can only be compressed by about 1 to 2 mm. Furthermore, articular cartilage in the load-bearing joints of the lower extremities is compressed after a few minutes of standing, walking, or running, and the possible compression due to impact loading is reduced. Consequently, it is suggested that the suggested function of shock absorption is not appropriate and that the following holds.

The two main functions of articular cartilage are (1) providing mobility in the joints and (2) distributing stress in the joints and thus minimizing peak stresses on subchondral bone.

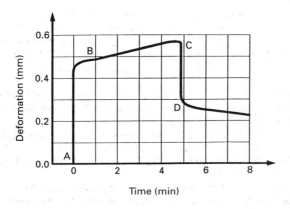

Fig. 20 Deformation–time curve for healthy patellar cartilage. (Adapted from ref. 20.)

The average stress in the hip joint has been estimated for two groups of subjects, subjects who needed a hip replacement because of arthritis and normal subjects of the same age and gender.[22] The estimation found an average stress during walking of 5 MPa for the subjects with arthritis and 1 MPa for the normal subjects with no sign of arthritis. This finding suggests that excessive load may be associated with arthritis and that daily stress situations in joints may be close to the ultimate stresses for cartilage. The following example illustrates this point.

Example
Calculate the maximal average stress in the tibiofemoral joint for running.

Assumptions

1. Maximal bone-to-bone force for running $F_{knee} = 4000$ N.
2. Contact area of the tibiofemoral joint $A = 10$ cm^2
3. Ultimate compression stress for cartilage $\sigma_u = 5$ MPa.
4. The stress is equally distributed over the total contact area.

Solution
The actual maximal average stress in the tibiofemoral joint is

$$\sigma = F_{knee}/A$$
$$\sigma = 4000 \text{ N}/10 \text{ cm}^2 = 400 \text{ N/cm}^2 = 4 \text{ MPa}.$$

Comment
The result supports the previous statement that loading of articular cartilage may often be close to the ultimate stress limits during sport activities.

Articular cartilage is an example of functional efficiency. The lubrication efficiency is of a superior magnitude to the best lubrication mechanisms know in modern engineering. Additionally, the functional efficiency is maintained throughout life in most humans. This is remarkable considering that cartilage is only a few millimetres thick and has limited repair capabilities. If one compares the functional efficiency of articular cartilage with many modern mechanical devices, one rarely finds something which is at the same efficiency level as cartilage.

Integrity of articular cartilage
The integrity of any living organism depends largely on the nutrients it obtains. In normal adult articular cartilage, the nutrients required by the chondrocytes are derived from the synovial fluid. The nutrients must travel a significant distance to reach the cells. The transport occurs by diffusion and/or convection. Cartilage deformation in response to joint loading plays an important role in this process.

The integrity of articular cartilage is significantly influenced by joint motion. It has been shown that the biochemical composition, the tensile properties, and the swelling properties of articular cartilage depend strongly on whether or not the tissue is normally subjected to load. Ingelmark and Ekholm[23] showed an increase of 12 to 13 per cent in the thickness of knee cartilage in animals after running for 10 min compared with the thickness after 60 min of immobilization. However, stress on articular cartilage can be excessive. It is speculated that 'fatigue wear' due to an accumulation of microscopic damages within the cartilage material may occur. A further mechanism responsible for creating excessive stress is speculated to be associated with joint impact loading.[25] One theory suggests that the time between impact forces (e.g. in running) is too short for cartilage to recover and that the mechanical properties of cartilage are reduced and make it prone to injuries. Another theory suggests that impact loading may be particularly dangerous when cartilage is not loaded (e.g. in the morning) and cannot adjust properly by increasing the contact area and consequently reducing the local stresses. The exact mechanisms are not well understood. Nevertheless, impact forces are assumed to be a major factor in the aetiology of damage to articular cartilage.

A number of studies have shown a gradual decrease in material characteristics with age. This change is assumed to be associated with changes in the relative content of collagen and proteoglycan. In the case of osteoarthritis the change is not only in the relative content but is also associated with chaotic changes in the collagen ultrastructure and the proteoglycan organization.

Articular cartilage can rarely be repaired. Research is under way to study possibilities of repairing and/or regrowing articular cartilage. However, current knowledge does not make it possible to repair cartilage in most applications.

REFERENCES

1. Borelli G A. *On the movement of animals* (*De motu animalium*). Macquet P, transl. Berlin: Springer Verlag, 1989.
2. Paul J P. Bioengineering studies of the forces transmitted by joints. In: Kennedy R M, *Engineering analysis, biomechanics and related bioengineering topics*. Oxford: Pergamon, 1965: 369–80.
3. Marey E J *La photographie du movement*. Paris: Centre Georges Pompidou, Musee National d'Art Moderne, 1977.
4. Hatze H. *Myocybernetic control models of skeletal muscle*. Pretoria: University of South Africa, 1981.
5. Williams M., Lissner H R. *Biomechanics of human movement*. Philadelphia W B Saunders, 1977.
6. Cavanagh P, Kram R. Stride length in distance running: velocity, body dimensions, and added mass effects. *Medical Science in Sports and Exercise* 1989; **21**, 467–79.
7. Inman V T. *The joints of the ankle*. Baltimore: Williams & Wilkins, 1976.
8. Nigg B M. *Biomechanics of running shoes*. Champaign, IL: Human Kinetics, 1986.
9. Bahlsen, A H. The etiology of running injuries:a longitudinal, prospective study. PhD Thesis, University of Calgary, 1988.
10. Nigg B M, Bahlsen A H. Influence of heel flare and midsole construction on pronation, supination and impact forces for heel–toe running. *International Journal of Sport Biomechanics* 1988; **4**: 205–19.
11. Debrunner H U. *Oethopaedisches Diagnostikum* Stuttgart: Thieme, 1978.
12. Segesser B, Stuessi E, Stacoff A, Kaelin X, Ackermann R. Torsion—ein neues Konzept im Sportschuhbau *Sportverletzung–Sportschaden* 1989; **3**: 167–82.
13. Stacoff A, Denoth J, Kaelin X, Segesser B. The torsion of the foot in running. *International Journal of Sport Biomechanics* 1989; **5**: 375–89.
14. Nigg B M, Denoth J, Neukomm P A. Quantifying load on the human body: Problems and some possible solutions. In Morecki A., Fidelus K, Kedzior K, Wit A, *Biomechanics VII*. Baltimore: University Park Press, 1981: 88–99.
15. Baumann W, Stucke H. Sportspezifische Belastungen aus der Sicht der Biomechanik. In Cotta H, Krahl H, Steinbrück K, *Die Belastungstoleranz des Bewegungsapparates*. Stuttgart: Thieme, 1980: 55–64.
16. Scott S H, Winter D A. Internal forces at chronic running injury sites. *Medical Science in Sports and Exercise* 1990; **22**: 357–69.
17. Wolff J. *The law of bone remodelling*. Maquet P, Furlong R., transl. Berlin: Springer Verlag, 1986.
18. Cowin S C, ed. *Bone mechanics*. Boca Raton: CRC Press, 1989.
19. Benninghoff, A *Lehrbuch der Anatomie des Menschen*. München: Lehmann, 1939.
20. Hirsch C. A contribution to the pathogenesis of chondromalacia of the

patella. *Acta Chirurgia Scandinavica.* (from Steindler A. *Kinesiology of the human body.* Springfield IL: Charles C. Thomas, 1977.)

21. Yamada H. *Strength of biological materials.* Baltimore: Williams & Wilkins, 1970.
22. Legal H, Reinecke M, Ruder H. Zur biostatischen Analyse des Hüftgelenkes. *Zeitschrift für Orthopaedie and Ihre Grenzgebiete* 1980; **118**: 804–15.
23. Ingelmark BE, Ekholm R. A study on variations in the thickness of articular cartilage in association with rest and periodical load. *Uppsala Läkareförenings Förhandlingar*, 1948; **53**: 61. (From Åstrand P, Rodahl K. *Physiological bases of exercise. Textbook of work physiology.* New York: McGraw-Hill, 1977).
24. Elftman H. The force exerted by the ground in walking. *Arbeitsphysiologie* 1938; **125**: 357–66.
25. Mow WC, Rosenwasser M. Articular cartilage: Biomechanics. In: Woo S L-Y, Buckwalter JA. *Injury and repair of the musculoskeletal soft tissues.* Park Ridge, IL: American Academy of Orthopedic Surgeons, 1988.

1.3.2 Strength

HENRYK K. A. LAKOMY

INTRODUCTION

Strength is defined as the maximum force or torque that a muscle, or a group of muscles can exert on the associated skeletal structure at a specific speed of movement. This chapter will examine the structure of skeletal muscle and a number of factors which affect the maximum ability of muscle to generate force.

Traditionally the term 'contraction' has been used to describe the action of a muscle generating force. This term is confusing as the muscle does not always shorten whilst stimulated. In this chapter, therefore, the term muscle 'action' will be used to describe muscle in its active state.[1]

ARCHITECTURE OF SKELETAL MUSCLE

Skeletal, or voluntary, muscle is composed of individual cells called muscle fibres. Groups of fibres are bundled together to form fascicles. Each muscle fibre is cylindrical with a diameter of 10 to 100 μm. The average fibre length is 0.03 m, ranging from less than 0.001 m up to as long as 0.3 m. The fibre is enclosed within a thin plasma membrane called the sarcolemma. Each muscle cell contains up to several thousand rod shaped structures called myofibrils. These are arranged in parallel with each other and are as long as the cell itself. The myofibrils in turn are made up of a series of repeating units called sarcomeres laid end to end.[2] It is the sarcomere that is the basic contractile unit of the cell (Fig. 1).[3]

Within each sarcomere there are two types of protein filaments. The thin filaments are made up of two chains of G actin which are polymerized into strands of F actin. The two chains are twisted to form a helix. Two rod-like protein structures, called tropomyosin, are twisted around these helical chains and are bound to the chains at intervals by a troponin complex. In the absence of calcium ions the troponin complex holds the tropomosin in a position which covers the active sites on the actin filaments. At the extremes of each sarcomere one end of the actin filaments is attached to form the Z line. The other ends of the filaments extend into the heart of the sarcomere forming a symmetrical pattern and are unattached.

The thick filaments are made up of approximately 200 myosin molecules. Each molecule of myosin is made up of a 'tail or shaft with two heads at one end'. The molecules are arranged in bundles so that all the tails point to the centre of the fibre and all the heads protrude from the filament towards its end. The centre of the filament where there are no heads present is called the bare zone. The arrangement of the filament allows the heads of each myosin filament to interact with the binding sites on six actin filaments.

Calcium is stored in an elaborate structure called the sarcoplasmic reticulum. The sarcoplasmic reticulum envelopes the myofibrils. At the location of each Z line the surface of the fibre is invaginated creating a tube-like structure called the T tubule. This tubule is in contact with the sarcoplasmic reticulum and an action potential passing down it will cause the sarcoplasmic reticulum to release the calcium ions stored in it. The concentration of calcium ions in the sarcoplasmic reticulum is about 10^5 times higher than in the remainder of the cell. When released these ions will diffuse throughout the sarcomere, with some attaching to the binding sites on the troponin complex. If no further impulse arrives the calcium ions are pumped rapidly back to the sarcoplasmic reticulum. Resting concentrations of calcium are restored within approximately 30 ms.

EXCITATION–CONTRACTION COUPLING

The sliding filament theory which has become accepted as the mechanism of muscle action was described by Huxley in 1957.[4] A stimulus of sufficient magnitude by the motoneurone results in a transmitter substance called acetylcholine to be released at the end of the axon. Acetylcholine diffuses across the synaptic cleft between the axon and the surface of the fibre and binds on to the receptor sites on the muscle fibre. This causes an action potential to be created on the surface of the muscle. This electrical signal passes down the T tubules to the sarcoplasmic reticulum, possibly through protein complexes called 'junctional feet', which in turn releases its stored calcium. The free calcium ions bind to the receptor sites on the troponin complex causing translocation of the tropomysium and exposing the previously covered binding sites on the actin filaments. Heads of the adjacent myosin filaments, which are in a high energy state, bind on to these exposed sites creating cross-bridges. Immediately upon attachment these heads undergo a conformational change resulting in relative movement of the two filaments. It is this change which creates the tension in the muscle. In the presence of ATP the myosin heads which have changed to a low energy stage as a result of the 'power stroke', detach from the binding sites on the actin filament. The ATP then hydrolyses to provide energy to return the myosin heads

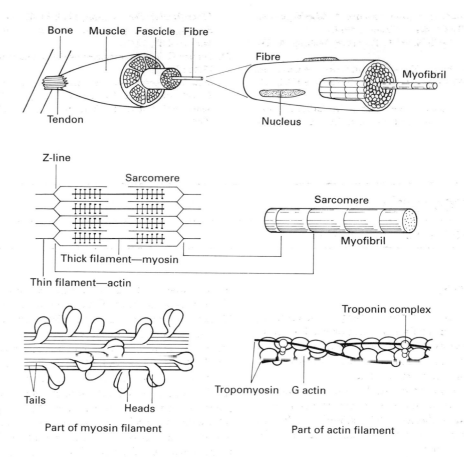

Fig. 1 The structure of skeletal muscle.

to their high energy state. If calcium is still present the binding sites remain exposed and the power stroke cycle can continue. If, however, reuptake of the calcium into the sarcoplasmic reticulum has occurred then the binding sites become covered again preventing further cross bridge formation. In an active muscle tension is produced by the concerted action of more than 10^9 myosin heads interacting with actin asynchronously. During excitation approximately 50 per cent of possible cross-bridges are attached at any time (Fig. 2).

To allow the force created by the cross-bridge formation within a single sarcomere to be transmitted to the tendons and skeletal system a similar number of cross-bridges must be created in each sarcomere in series with it. If this were not the case the shortening occurring in that sarcomere would simply result in lengthening of another in which insufficient cross-bridges had been created. The force transmitted to the tendon is, therefore, that generated by the weakest sarcomere in the myofibril chain. It can be assumed, however, that a neural stimulus activates the whole myofibril equally and therefore the tension created by each myofibril can be regarded as being equal to that produced by a single constituent sarcomere. The overall tension developed by the muscle is the sum of all the myofibrils activated in parallel with each other. The more myofibrils activated the greater will be the number of cross-bridges being created and the higher will be the overall tension developed by the muscle. The number of cross-bridges being created depends on a number of factors.

1. Level and type of muscle activation

A muscle is subdivided into motor units.[5] Each motor unit is made up of a number of fibres innervated by a single motoneurone. The number of fibres comprising a motor unit can range from less than 10 to more than 1000. The excitation of a single motoneurone will result in action of all the muscle fibres belonging to its motor unit. The level of activation and the tension of the muscle reflect the number of motor units being recruited at any time. With increasing recruitment the number of active fibres increases. The more fibres that are excited the more cross-bridges that are created resulting in an increased muscle tension. Each active fibre will make a different contribution to the overall muscle tension depending on its size and type. The larger the cross-sectional area of the fibre the greater will be the number of parallel myofibrils and sarcomeres. When stimulated the larger fibre will, therefore, be able to develop more tension than one with a smaller cross-sectional area. Fibres can also be classified according to the way in which they respond to a stimulus. Some develop high levels of tension quickly and the tension also decays rapidly following activation. These types of fibres are termed fast twitch fibres (Type II) and have properties making them most suited to short bursts of activity in which high power outputs or speeds of movement are required. In contrast slow twitch fibres (Type I) develop tension more slowly and have a longer decay time. These fibres are specialized for activity of extended duration. When a standard

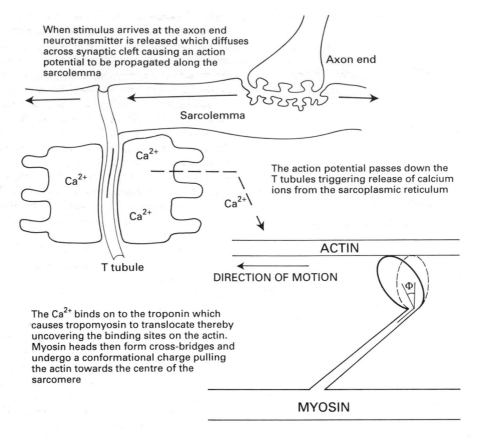

When stimulus arrives at the axon end neurotransmitter is released which diffuses across synaptic cleft causing an action potential to be propagated along the sarcolemma

Axon end

Sarcolemma

Ca^{2+}

Ca^{2+}

Ca^{2+}

Ca^{2+}

Ca^{2+}

The action potential passes down the T tubules triggering release of calcium ions from the sarcoplasmic reticulum

T tubule

ACTIN

DIRECTION OF MOTION

The Ca^{2+} binds on to the troponin which causes tropomyosin to translocate thereby uncovering the binding sites on the actin. Myosin heads then form cross-bridges and undergo a conformational charge pulling the actin towards the centre of the sarcomere

MYOSIN

Fig. 2 Excitation–contraction coupling—a schematic diagram.

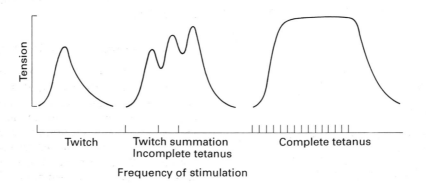

Twitch

Twitch summation Incomplete tetanus

Complete tetanus

Frequency of stimulation

Fig. 3 Effect of stimulus rate on tension development.

number of sarcomeres of each fibre type are compared (cat muscle) an approximate 2.5-fold difference in maximum rate of shortening is found.[6] As the different fibre types show clear differences in maximum speed of shortening their ability to produce tension is affected by the speed of limb movement. However, in a maximal static muscle action the force developed is independent of the fibre type and is only dependent on the cross-sectional area.[7,8]

An average value for the tension produced by muscle is 15 to 42 N/cm^2 with 22.5 N/cm^2 most often used as the mean value for homogenous muscle.[9,10] One major factor influencing maximum muscle tension is, therefore, the total cross sectional area of the muscle, perpendicular to the direction of the fibres.

The way in which a muscle is stimulated affects the magnitude of the tension it develops. A one-off stimulation of the motor units results in a muscle twitch (Fig. 3), resulting in a low force production lasting only a fraction of a second. If, however, the stimulus is repeated sufficiently quickly (20 to 30 Hz) twitches overlap resulting in optimum tension development called tetanus. Tetanic muscle actions are those which are most frequently required during sporting activity and it is this type of action which will be considered in this chapter.

The level to which a muscle has been activated can be examined by monitoring the electrical activity within the muscle. This can be achieved by placing electrodes on prepared sites on the skin and measuring the potential difference between these electrodes. The potential difference reflects the electrical activity of the muscles directly beneath the electrodes (EMG). Figure 4 shows the rela-

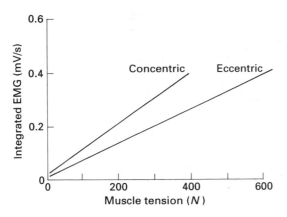

Fig. 4 The relationship between muscle tension and integrated EMG.

tionship between muscle tension and level of activation (expressed as integrated EMG). The illustration shows that as activation increases so does the tension developed by the muscle. This relationship remains consistent for all types of muscle action being performed.[11]

2. Angle of pennation

The fibres of many muscles in the body lie parallel to the long axis of the muscle. These are called longitudinal or fusiform muscles in which the line of pull of the fibres is the same as that of the whole muscle. In other muscles the fibres are arranged so as to be at an angle relative to the line of pull of the muscle. These are called pennated muscles and the angle between the line of pull of the whole muscle and that of the muscle fibres is called the angle of pennation.

Figure 5 shows a very simple model of how an increase in the

Angle of pennation	0	30	60
No. of fibres (=n)	12	16	27
Relative no. of fibres	100	133	225
Efficiency of angle of pull (=e) (%)	100	87	50
Relative tension ($n \times e$) (%)	100	116	113
Relative average length of fibres (=l) (%)	100	75	44
Effective length of fibres ($l \times e$) (%)	100	65	22

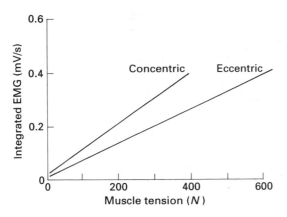

0°   30°   60°

Angle of pennation

Fig. 5 The effect of pennation on the ability of muscle to generate tension.

angle of pennation results in more fibres per unit volume of muscle. The active muscle will, therefore, be able to create a greater number of cross-bridges resulting in an increased force production. The force produced, however, is along the line of the muscle fibres and, therefore, only a proportion of the force produced is transmitted along the long axis of the muscle, the desired direction. If F is the force produced by the muscle fibres then the tension developed in the tendon is $F\cos\Phi$ where Φ is the angle of pennation. The efficiency of the pull of the fibres in the desired direction decreases with increasing angle of pennation. Figure 5 shows the way in which the proportion of the force F being transmitted to the tendon (relative tension) is affected by increasing angle of pennation. The combined effect is that the tension transmitted to the tendons increases to a maximum at angles of pennation of around 30 to 45° and decreases thereafter.

There is however a 'trade-off' for having the fibres pennated. Although the maximum tension improves with pennation the range that a muscle can shorten decreases as the average length of the muscle fibres becomes shorter. This decrease is further magnified when the way in which the fibres are angled to the desired line of shortening is considered (the effective length). If the maximum rate of muscle shortening (V_o) is proportional to the effective length then an increase in pennation angle will result in a reduction of V_o. Those muscles which require a large range of movement without the need for high force outputs, such as the Sartorius muscle, have fibres which are not pennated. If, however, force, not range, is important then the muscles have evolved with significantly pennated fibres. Examples of pennated muscles are the quadriceps femoris group and the deltoids. It should be noted that the actual relationship between angle of pennation and force development is very much more complex than the simple model described above.[12]

3. Force–length relationship

A factor which not only affects the number of cross-bridges that can be created but also the contribution that the elastic structures of the muscle can make to the overall tension development is the length of the muscle at the time of its action. The number of cross-bridges that can be formed is dependent on the extent of the overlap between the actin and myosin filaments. At the natural resting length of the muscle (I_s), a sarcomere length of approximately 2.85 μm, the overlap of the filaments is near optimal for cross-bridge formation. If the muscle is stretched then the overlap decreases until at approximately 4.2 μm (150 per cent of I_s) there can be no cross-bridges formed at all).[13] If, in contrast, the muscle is permitted to shorten then the actin filaments start to overlap each other. This happens at a sarcomere length of less than 2.6 μm. This actin filament overlap reduces the number of sites available to the heads of the myosin filaments, thereby reducing the overall magnitude of the possible tension development. Muscle tension, however, is not governed exclusively by the active cross-bridges. As the muscle is physically stretched beyond its resting length several muscle tissues are themselves stretched resulting in tension being created by these elastic structures. This is regarded as the passive component of muscular force production and the contribution it makes to the overall tension increases with further stretching. The resultant muscle tension is, therefore, the sum of the active and passive components as shown in Fig. 6. It should be noted that *in vivo* the range of muscle sarcomere lengths (functional range) is approximately 80 per cent to 120 per cent of I_s.

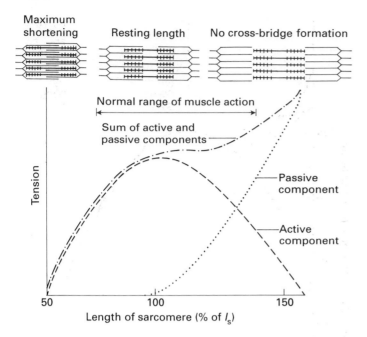

Fig. 6 The relationship between sarcomere length and active, passive, and total tension generation.

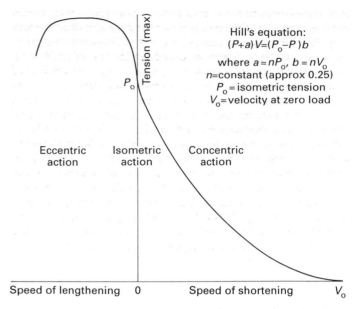

Fig. 7 The relationship between maximum muscle tension and speed of shortening/lengthening during concentric, isometric, and eccentric muscle actions.

4. Type and speed of muscle action

There are three possible results of recruiting a muscle.

1. Concentric action. In this type of action the distance between the origin and insertion of the muscle becomes shorter. The tension developed by the muscle produces a torque which is greater than that resisting movement of the limb resulting in movement of that limb.
2. Isometric action. Although the muscle is active the distance between the origin and insertion of the muscle does not change. The muscle torque matches the resistance applied to the limb which results in no movement occurring. It should be noted that although there is no overall shortening of the entire muscle there will be some shortening of the sarcomeres and some stretching of the elastic structures in series with those sarcomeres, e.g. tendons, connective tissue.
3. Eccentric action. In this type of action the tension developed by the active muscle produces less torque than that caused by the resistance. The result is that the muscle is forced to lengthen, increasing the distance between the origin and insertion of the muscle.

This section of the chapter will examine the effects of both type and velocity of muscle action on the maximum tension that the muscle can generate.

It has been well established that as the velocity of muscle shortening increases during a concentric action the maximum tension that can be exerted decreases. The relationship between maximum tension and velocity of shortening was first described by Fenn and Marsh.[14] This relationship was further explained by Hill[15] and is shown in Fig. 7. Although muscle actions *in vivo* do not appear to follow the curves predicted by Hill's equation exactly the inverse relationships between force and velocity found have similar characteristics.[16] The maximum speed of shortening (V_o) occurs when the load is zero. In contrast maximum force is produced (P_o) when in the isometric state, when no muscle shortening

takes place. Hill's equation does not consider eccentric muscle actions. If the external torque applied to the limb increases further the limb is forced to move in the direction opposite to that during the concentric action. The maximum tension developed in the muscle is found to increase above the level generated during the isometric action. A further increase in external torque will result in an increased velocity of muscle lengthening and an increase in muscle tension until a plateau is reached. Any further increase in load will increase the velocity of lengthening without an increase in maximum tension until the speed of lengthening becomes too great when 'failure' to maintain tension occurs.

Some of the changes in maximum muscle tension with differing velocities of shortening and lengthening can be accounted for by the effects of internal resistance or viscosity of the muscle. When the muscle is shortening during a concentric muscle action some of the tension generated by the cross-bridges is required to overcome the internal viscosity of the muscle and this tends to resist the movement of the internal structures. This results in only a proportion of the total tension produced being transmitted by the tendons to the skeletal structure. The greater the velocity of shortening the greater the effect of the viscosity resulting in further reduction in overall external tension. In contrast, however, when the muscle is being stretched during an eccentric muscle action the act of forcing the muscle to lengthen results in useful tension being created to overcome the muscle viscosity. When this is added to the force being produced by the cross-bridges the effect is an increased tension development which becomes greater as the velocity of lengthening increases.

The effect of internal muscle viscosity does not, however, account for all of the magnitude of the changes in maximum muscle tension with velocity of shortening/lengthening measured. The changes in the duration of cross-bridge attachment may provide a reason for the remainder of the changes seen. When the muscle is shortening the time taken for the conformational change in the myosin head during the power stroke diminishes with increasing velocity of shortening. The contribution made by each cross-bridge to the overall muscle tension is a function both of the magnitude of the force generated and the duration for which it applies. The effect of increasing speed of shortening will be to decrease both the peak force produced and the time of attachment. As the overall muscle tension is the sum of the contributions of each active cross-bridge then peak tension will decrease with increased muscle shortening speed. During eccentric muscle actions the myosin heads are also attempting to undergo conformational changes. During these muscle actions the myosin heads are being forced to move in the opposite direction to that which would result from the conformational change. It is possible that significant deformation of the filaments must then take place to achieve this change which would result not only in a significant increase in the maximum force produced during attachment but also in the duration of the bridging. This may well not only account for the increased tension seen in muscle during eccentric muscle actions but also explain why muscle damage occurs more often during this type of muscle action.

5. Angle of insertion

The strength of a muscle group is often determined by measuring the external torque being produced. This torque is a function of the tension created by the muscle and transmitted to the skeleton by the muscle tendon (T), the distance of the point of insertion from the axis of rotation of the associated joint (d), and the angle of insertion of the tendon on the bone (Φ). It is calculated from the equation:

$$\text{Torque} = Td\sin\Phi$$

The angle of insertion of the muscle tendon on the bone changes as the limb moves through its range of movement. As a consequence a muscle in which a constant tension is being generated will produce an external torque that is varying as the limb moves, as shown in Fig. 8. The optimum position for torque production is when the angle of insertion is 90°.[17]

It should be noted that even if the muscle was constrained to shorten or lengthen at a constant velocity the external torque would not only vary with the change in the angle of insertion but would also be affected by the force–length relationship of muscle discussed earlier. As the limb moves so the length of the muscle changes affecting its ability to produce tension due to the force–length relationship discussed earlier. The net external effect is therefore a combination of the maximum tension that the muscle can generate, the angle of insertion on the bone, and any changes in the distance from the joint axis of the point of insertion.

6. Stretch–shortening cycle

In dynamic activities muscles seldom perform a single isolated muscle action. More often there is a combination of actions resulting from the large variation in applied external forces. By combining actions it is possible to create a sequence of movements resulting in optimum force production. Cavagna, Saibene, and Margaria clearly describe how greater tension can be produced in the final concentric muscle action if it is preceded by active stretching.[18] This sequence of events is called the stretch–shortening cycle.[19]

The enhanced force production seen when the stretch–shortening cycle is performed is due to several factors.

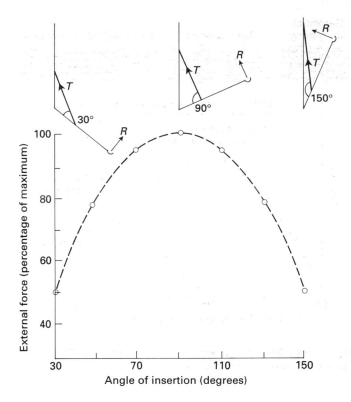

Fig. 8 The effect of angle of insertion on the external torque produced by a muscle generating constant maximum tension (after ref. 22).

1. Maximum muscle tension in an eccentric muscle action is greater than in a concentric muscle action.
2. Energy stored during stretching of the elastic components of the muscle can be subsequently recovered during the concentric phase of the action.[20]
3. The time course of tension build up is optimized.
4. Uninhibited activation of the muscle occurs because of invocation of the myotatic stretch reflex.

In order to achieve the maximum increase in muscle tension during the main phase of the activity the pattern of prior muscle action must be performed in the correct sequence. Before initiating motion in the desired direction the limbs must perform a countermovement in the opposite direction. Muscle activity occurs to overcome the momentum of the countermovement resulting in an eccentric action whilst being forcibly stretched. This eccentric action must be immediately followed, without pause, by a forceful concentric action of the same muscles. Without the active stretch preceding the concentric action performance will rely almost exclusively on the contractile structures of the muscle with no contribution from its elastic properties, thereby greatly attenuating the maximum tension that could be produced.[21]

CONCLUSION

Many factors have been shown to influence the maximum tension that a muscle, or muscle group, can generate. These variables affect the number of cross-bridges being formed, the peak force created by each myosin head and the duration of its attachment to the actin filament. Determinations of strength must be made under carefully controlled conditions so that meaningful and therefore valid results are obtained.

REFERENCES

1. Cavanagh PR. On 'muscle action' vs. 'muscle contraction'. *Journal of Biomechanics* 1988; **22**: 69.
2. Partridge LD, Benton LA. Muscle, the motor. *In*: Brookhart JM, Mountcastle VB, Brooks VB. Geiger SR, eds. *Handbook of Physiology: Section 1. The Nervous System*. Bethesda, MD: American Physiological Society, 1981: 43–106.
3. Barany M. ATPase activity of myosin correlated with speed of muscle shortening. *Journal of General Physiology*, 1967; **50**: 197–216.
4. Huxley AF. Muscle structure and theories of contraction. *Progress in Biophysics and Biophysical Chemistry*, 1957; 7: 255–318.
5. Burke RE. Motor units: anatomy, physiology and functional organisation. *In*: Brooks VB, ed. *Handbook of Physiology. Section I, The nervous system II*. Washington: American Physiological Society, 1981: 345–422.
6. Spector SA, Gardiner PF, Zernicke RF, Roy RR, Edgerton VR. Muscle architecture and force–velocity characteristics of the cat soleus and medial gastrocnemius: Implications for motor control. *Journal of Neurophysiology* 1980; **44**: 951–60.
7. Close RI. Dynamic properties of mamalian skeletal muscle. *Physiological Reviews*, 1972, **52**: 128–97.
8. Saltin B, Gollnick PD. Skeletal muscle adaptability: significance of metabolism and performance. *In*: Peachet L, ed. *Handbook of Physiology*. MD: American Physiological Society, 1983: 555–631.
9. Edgerton VR, Roy RR, Apor P. Specific tension of human elbow flexor muscles. In: Saltin B, ed. *Biochemistry of exercise*. Champaign, IL: Human Kinetics 1986: 487–500.
10. Wickiewicz TL, Roy RR, Powell PL, Perrine JJ, Edgerton VR. Muscle 1984; architecture and force-velocity relationships in humans. *Journal of Applied Physiology* 56: 435–43.
11. Bigland B, Lippold OCJ. The relation between force, velocity and integrated electrical activity in human muscles. *Journal of Physiology* 1954; **123**: 214–24.
12. Otten E. Concepts and models of functional architecture in skeletal muscle. *Exercise and Sport Sciences Reviews*, 1988; **16**: 89–137.
13. Gordon AM, Huxley AF, Julian FJ. The variation in isometric tension with sarcomere length in vertebrate muscle fibres. *Journal of Physiology*, 1966; **184**: 170–92.
14. Fenn WO, Marsh BS. Muscular force at different speed of shortening. *Journal of Physiology* 1935; **885**: 277–97.
15. Hill AV. The heat of shortening and the dynamic constants of muscle. *Proceedings of the Royal Society B* 1938; **126**: 136–95.
16. Wilkie DR. The relationship between force and velocity in human muscle. *Journal of Physiology* 1950; **110**: 248–80.
17. Singh M, Korpovich PV. Strength of forearm fexors and extensors. *Journal of Applied Physiology*, 1968; **25**: 177–80.
18. Cavagna GA, Saibene FP, Margaria R. Effect of negative work on the amount of positive work performed by an isolated muscle. *Journal of Applied Physiology*, 1965; **20**: 157.
19. Norman RW, Komi PV. Electromechanical delay in skeletal muscle under normal movement conditions. *Acta Physiologica Scandinavica*, 1970; **106**: 241–8.
20. Shorten MR. Muscle elasticity and human performance. *In*: van Gheluwe B, Atha J, eds. *Current Research in Sports Sciences. Medicine and Sports Science*. Basel: Karger, 1987; **25**: 1–18.
21. Cavagna GA, Dusman B, Margaria R. Positive work done by the previously stretched muscle. *Journal of Applied Physiology*, 1968; **20**: 157–8.
22. Komi PV. Relationship between muscle tension, EMG and velocity of contraction under concentric and eccentric work. In: Desmedt JE, ed. *New developments in Electromyography and Clinical Neurophysiology*. Basel: Karger, 1973; **1**: 596–600.

1.4.1 Sport psychology

LEW HARDY AND GRAHAM JONES

INTRODUCTION

The beginnings of sport psychology in the Western world can be traced back at least as far as Triplett's studies of the effects upon performance of the presence of other people in 1897.[1] However, despite the prodigious work of Coleman Griffith in the 1920s (see, for example, ref. 2), sport psychology did not really become established as a scientific discipline until the mid-1960s, when the International Society of Sport Psychology (1965), the North American Society for the Psychology of Sport and Physical Activity (1967), and the European Federation of Sport Psychology (1968) were formed. Early work in sport psychology had a very strong research emphasis. However, during the last 15 years sport psychologists have also become heavily involved in applied work with athletes on the maximization of performance. In most countries, this has led to inevitable territorial disputes over who is qualified to do what. Essentially, sport psychologists, who have first degrees in sports science and higher degrees in (sport) psychology, claim that psychologists, who have first (and sometimes higher) degrees in psychology but no background in sports science, do not have the necessary knowledge of sports science.

Conversely, of course, psychologists claim that sport psychologists do not have the necessary breadth of knowledge in psychology (see below).

A rather more constructive development has been the shift in research emphasis away from behavioural approaches towards more process-oriented cognitive approaches. This shift in emphasis follows the lead of mainstream psychology, and has been accompanied by the adaptation and adoption of many cognitive behavioural methods from mainstream psychology for use in applied work with performers.[3–6] Other positive developments include the establishment of a subdiscipline of exercise psychology, and sustained improvements in the quality of both research output and applied work with performers.

Current areas of interest in sport psychology are diverse, but include the identification of factors which influence participation and performance: the development of psychological strategies to enhance performance, the influence of involvement in sport and exercise upon psychological well-being, the psychology of children involved in sport, and the development of ecologically valid methodologies to answer applied questions in the field. However, in view of the space constraints that operate upon a review chapter such as this, it is perhaps inevitable that some aspects of sport psychology have been omitted from it. Indeed, the present authors would readily admit that they have somewhat arbitrarily divided the chapter into seven general sections, which must at least partially reflect their own areas of interest. These seven sections focus upon motivation, stress and performance, individual differences, interpersonal relationships, exercise psychology, mental training, and the practice of sport psychology. Each section provides a review of the current state of knowledge in that area, and then attempts to identify some of the important issues that are likely to be addressed by sport psychologists over the next few years.

MOTIVATION

In behavioural terms, motivation is concerned with the strength of intensity dimension of behaviour. Furthermore, according to motivation theories, this energy is determined by the biological and social needs of the organism.[7] It is often suggested that motivation theories are concerned with the reasons why people behave in certain ways, and to some extent this is true. However, there are many explanations of people's behaviour that do not depend upon needs for their explanatory power; for example social learning theory.[8] Strictly speaking, these are not motivation theories, although the precise distinction between such (social) cognitive theories and motivation theories is not always very clear.

Motives for participation

Early research on motivation in sport examined reasons for participation, and indicated that the majority of young (North American) performers had multiple reasons for their involvement. The most commonly identified were a need for affiliation, to develop skill, for excitement, to succeed, and for fitness/health.[9] More contemporary researchers[10] have reduced these needs to three more fundamental reasons: the activities have intrinsic value (enjoyment), they are claimed to have positive health benefits, and they provide opportunities for participants to gain feelings of competence. It is perhaps worth noting that the last explanation is as much cognitive in nature as it is motivational.

Other research examining the reasons why young people 'drop out' of sport has identified four motives for achievement in sport: to demonstrate competence, to master the task, to gain social approval, and to experience enjoyment and excitement.[11,12] This research suggests that young children are motivated by task mastery and social approval rather than a need to feel competent, but as they grow older so they acquire a greater need to demonstrate competence.[7] Furthermore, Roberts[11] has argued that the existing evidence suggests that children are likely to 'drop out' of sport if their prime motive for participation is to demonstrate their competence by beating others. The extent to which this finding can be generalized to other cultures is as yet unclear.[12] However, if the findings do generalize to all levels of performance in all cultures, then the implications for health promotion and mental training could be very far-reaching indeed.

An interesting question in the area of motives for participation is why people pursue dangerous sports such as climbing, caving, skin diving, parachuting, and motor racing. One possible explanation is that such performers are high sensation seekers[13] who indulge in dangerous activities for 'arousal jags'.[14] An alternative explanation has been put forward by Solomon.[15] He proposed that, although performers habituate to the fear that precedes the performance of dangerous activities, their secondary reaction of elation and satisfaction upon completion of the activity is increased with successive presentations of the feared stimulus (i.e. the dangerous activity). This secondary effect then becomes the motive for continued involvement. Whilst Bakker et al.[10] suggested that this theory has considerable potential for explaining participation in dangerous sports, from a cognitive viewpoint it could be argued that it simply replaces one 'black box' with another. More precisely, it does not explain why the secondary response to negative stimuli does not habituate with repeated presentations of the stimulus; nor does it really explain why people choose to take part in sports where the negative stimulus is physical danger, rather than in competitive sports where the negative stimulus is ego threat.

A rather different sort of explanation can be deduced from the work of Piet (cited by Bakker et al.)[10] and of Lester.[16] Following a qualitative study of elite Himalayan mountaineers, Lester concluded that they find the normal interpersonal situations of everyday life stressful. Coupled with Piet's findings that one of the major reasons given for mountaineering was the great feeling of competence generated by controlling stress responses, this suggests that such mountaineers may have a very strong need to demonstrate their competence at dealing with stressful situations precisely because they find many situations in everyday life stressful. Having said all this, direct empirical tests of both this and Solomon's theory[15] have yet to be performed.

Intrinsic and extrinsic motivation

As indicated above, behaviour can be initiated and maintained for a variety of reasons. Similarly, the rewards that people receive as a result of participating in sport can also be very varied; for example, praise, prizes, financial rewards, satisfaction, enjoyment, and a sense of achievement are all valued rewards that could be obtained through participation in sport. Some of these rewards are external to the task, whilst others are apparently inherent in performing the task itself.[9] Behaviour that is maintained by rewards that are intrinsic to the task is said to be intrinsically motivated, whilst behaviour that is maintained by rewards extrinsic to the task is said to be extrinsically motivated. However, the situation is much more complex than this, since external rewards can sometimes enhance subsequent intrinsic motivation and sometimes reduce it. This paradox has been superbly illustrated by Casady:

[An] old man lived alone on a street where boys played noisily every afternoon. One day the din became too much, and he called the boys into his house. He told them he liked to listen to them play, but his hearing was failing and he could no longer hear their games. He asked them to come around each day and play noisily in front of his house. If they did he would give them each a quarter. The youngsters raced back next day and made a tremendous racket in front of the house. The old man paid them, and asked them to return the next day. Again they made noise, and again the old man paid them for it. But this time he gave each boy only 20 cents, explaining that he was running out of money. On the following day, they got only 15 cents each. Furthermore, the old man told them, he would reduce the fee to five cents each on the fourth day. The boys became angry and told the old man that they would not be back. It was not worth the effort, they said, to make noise for only five cents a day. (ref. 17, p. 52)

De Charms[18] and Deci[19] attempted to explain this phenomenon by arguing that humans have an innate need to feel competent and self-determining (in control of their own lives). Based upon this assumption, the cognitive evaluation theory[7] proposes that external rewards will enhance intrinsic motivation for a particular activity whenever the information that is contained in them infers an increase in self-competence and self-determination. Conversely, extrinsic rewards which convey information indicating a decrease in self-competence, or which attempt to control the performer and thereby decrease self-determination, will lead to a reduction in intrinsic motivation. These basic predictions of cognitive evaluation theory have now received considerable empirical support in the literature.[7]

Cognitive evaluation theory has attracted considerable attention in the context of sport because the success and failure that are implicit in competitive activities can be regarded as rewards that are essentially external to the task of performing.[7,20] Unfortunately, a detailed review of this research is beyond the scope of this chapter. However, the implications of cognitive evaluation theory for sport include the following.

1. Performers should be actively involved in the decision-making processes that determine their training and competition programme.
2. Performers should be encouraged to set goals that are not easily influenced by others. These goals should be difficult, but must be realistic so that performers have the maximum chance of achieving objectives that they perceive to be worthwhile (see section on mental training).
3. Rewards should be used to provide information and not control. In particular, rewarding success and punishing mistakes is likely to lead to a decrease in the intrinsic motivation of performers.

Attributions

Attributions are the reasons that people perceive to be the causes of success or failure. Weiner's attribution theory[21] proposed that attributions could be classified along two dimensions, stability and locus of causality. These two dimensions therefore yielded four stereotypic attributions for success and failure: ability, effort, task difficulty, and luck. Essentially, Weiner's theory predicted that the locus of causality of attributions influences the intensity of the affective response to the outcome, whilst the stability of attributions determines the performer's expectations for future success or failure. Subsequent theorizing [20,23] led to the identification of two further attributional dimensions. The first of these, controllability, was non-orthogonal to Weiner's earlier dimensions and was hypothesized to be concerned with the moral judgements that people make in relation to success and failure. The second dimension, globality, referred to the extent of generalizability of the cause to other areas of behaviour. Considerable research has now been conducted in sport settings using the two basic attributional dimensions, although rather less has been performed on controllability and globality.[24,25] In summary, this research suggests that attributions may be important determinants of affective responses to success and failure,[24,26,27] expectations regarding future performance,[28,29] and intrinsic motivation.[7]

Other research has shown that people's needs and expectations influence the attributions that they make. In particular, the need to preserve self-esteem leads performers to employ a self-serving bias when making attributions for success and failure, such that they generally give more internal attributions for success but more external attributions for failure.[30] Similarly, team cohesion leads to a team-serving bias in attributions for success and failure.[31] Finally, social factors may also influence attributions, since publicly declared attributions have been found to show less self-serving bias than private attributions.[25]

In 1983, Rejeski and Brawley commented that attribution research in sport had to that date been quite narrow in focus.[25] In particular, they criticized its preoccupation with applying Weiner's theory[21] to the study of self-attributions for achievement outcomes. In 1992, this criticism still largely stands, the only real difference being that the literature now applies Weiner's theory[22] to the study of self-attributions for achievement outcomes. This is a shame, because attributions seem to form an important part of most cognitive and motivational theories of emotion and behaviour, both personal and interpersonal. Furthermore, personal construct theorists[32] would no doubt argue that current trends towards subjects mapping their own attributions on to attributional dimensions provided by the experimenter [25,33] should be extended to subjects defining their own attributional dimensions (personal construct system). Sport psychologists appear to be largely unaware of the potential of personal construct theory to both researchers and practitioners alike (for an exception see Butler[34]).

Withdrawal

Apart from the research discussed earlier into the reasons why young people 'drop out' of sport, withdrawal is a much under-researched area. Examination of the effects of enforced withdrawal from sports participation in the health-related exercise domain has led some researchers to propose that performers can become 'addicted' to exercise or sport.[35] Various reasons for this dependency have been proposed, including the release of β-endorphins during and after exercise[36] (see also section on exercise psychology). However, the magnitude of the problem is frequently sensationalized, and several researchers have emphasized the need to distinguish between commitment and addiction.[37,38]

Research has also be conducted on the psychological symptoms of staleness in performers following 'overtraining'.[39] This research has largely relied on measuring disturbances from the so-called 'iceberg' mood profile that elite performers have been argued to possess. The underlying idea behind this research was that, once identified, training loads could be adjusted according to psychological moods so as to minimize the risk of placing too great a training load on athletes. However, the 'iceberg' profile is not without its critics,[40] whilst sports scientists and medical practitioners working in the field generally acknowledge that staleness results from the total sum of stressors that the athlete has to deal with, not just the training load.

There are some clear similarities between staleness, or overstressing, in athletes and psychological burn-out in coaches and other service professions. However, whilst there is an established literature on burn-out in the service professions, and even some literature on burn-out in coaches, there is as yet no empirical literature on burn-out in sports performers.[41] A similar comment can also be made regarding research into strategies for maintaining motivation during prolonged injury, and adjusting motivation towards new directions following retirement.

Current situation

Despite the fact that motivation was one of the earliest areas to be investigated in sport psychology, many interesting and far-reaching questions still await an acceptable answer. In the area of motives for participation, there is a need to perform prospective studies to explore the generalizability of Roberts' finding that children who compete in sport to beat other people are more likely to 'drop out' than other children.[11] Another interesting question which has not yet been satisfactorily answered relates to the reasons why people pursue dangerous sports.

Although the area of intrinsic motivation is one of the few aspects of motivation to receive serious attention from researchers, there is still an urgent need for longitudinal studies which explore the cyclical nature of the relationships that exist between rewards, attributions, self-competence, and intrinsic motivation.[42] Furthermore, such studies may throw some light on the different roles that may be assumed by extrinsic rewards at different stages of the development of intrinsic motivation. Other research on attributions in sport might focus upon actor–observer differences in attributions, attributions for interpersonal as well as achievement outcomes, and the mapping of attributions on to subjects' own personal construct systems.

The whole area of withdrawal from sport is in need of development. In particular, there is an urgent need for research into the role that active regeneration, and other preventative and curative strategies, might have in helping athletes to avoid and recover from overstressing.[43] Related to this is a need for empirical research addressing the question of how athletes might best cope with the stressors of injury[44] and retirement. There is also a need to explore further the differences between, and commonalities of, commitment and addiction.[38] Finally, at a theoretical level, future research will no doubt make a much clearer distinction than previous research between motivational explanations of athletes' behaviour and explanations based on social cognitions.[25,31]

STRESS AND PERFORMANCE

The recent publication of texts on the specialized topic of stress[45] and anxiety[46,47] reflects the considerable amount of research attention devoted to this area in recent years. This endeavour to gain a greater understanding of stress and sports performance represents an extension of the early interest in social facilitation.

Social facilitation

Social facilitation is one of the oldest and most studied topics in social psychology. However, whilst the study of audience effects upon motor performance was a popular area of research for many years, it was also plagued by a general inability to explain the inconsistent findings which were produced. It was not until 1965 that Zajonc went some way towards resolving the situation by explaining the previously inconsistent findings in terms of the emission of dominant responses (a construct of drive theory) as a function of the increased drive or generalized arousal elicited by the mere presence of an audience, i.e. the increased arousal caused by the presence of an audience enhanced performance on well-learned tasks but impaired performance on poorly learned tasks.[48] Zajonc's classic paper rekindled interest in this area, and the subsequent upsurge in social facilitation research led to a restatement of Zajonc's proposals by Cottrell et al.[49] These authors argued that the mere presence of others was not necessarily arousing and that the audience had to be perceived by the performer to have the potential to evaluate performance in order for arousal to be increased and the hypothesized effects to occur. These developments in the latter half of the 1960s encouraged sport psychologists to gain a greater knowledge of the interaction between performers' social environments, their skill levels, and performance. By the 1970s, three different aspects of social facilitation had been identified and studied using different paradigms: audience effects, coaction effects, and competition effects.[50] Much of this research was laboratory based and involved non-interactive audiences. However, researchers experienced considerable difficulty in achieving experimental control over the many variables operating: for example, personality, nature and difficulty of the task, perception of the 'threat' or the 'challenge' of the situation, etc. Other research examined interactive audience effects in more ecologically valid field settings, and studied factors such as home advantage,[51] audience size,[52] and spectator mood.[53] However, largely because of methodological problems, social facilitation research has become unfashionable and more recent work has turned towards examining some of the specific variables which are thought to underlie social facilitation effects.

Competitive state anxiety and performance

Research into competitive state anxiety can be divided into three general areas: the nature and temporal patterning of the competitive state anxiety response, the antecedents or precursors of anxiety, and the relationship between anxiety and performance.

Nature and temporal patterning of the state anxiety response

The competitive state anxiety concept itself has formed the focus of much recent research effort, and opinion has generally fallen into line with the clinical and test anxiety literature in which the anxiety response has been separated into cognitive and somatic components.[54,55] The current multidimensional conceptualization of competitive state anxiety stems largely from the work of Martens et al.[56] and their development of the Competitive State Anxiety Inventory–2 (CSAI–2), a competition-specific questionnaire which measures cognitive and somatic anxiety components together with self-confidence. The inclusion of anxiety and self-confidence as independent scales is interesting since they were largely viewed as representing opposite ends of the same continuum in earlier literature. For example Borkovec[57] described self-confidence as simply an epiphenomenon resulting from a lack of anxiety. Conversely, Bandura[58] proposed that anxiety was an epiphenomenon resulting from a lack of self-confidence, or self-efficacy as he termed it. However, factor analysis of CSAI–2 revealed separate anxiety and self-confidence factors which suggests that they should be regarded as being independent rather than bipolar in nature.

The CSAI–2 has been employed to examine the temporal patterning of its three components during the time leading up to competition. The findings have been fairly consistent in showing that cognitive anxiety remains relatively stable prior to competition; somatic anxiety, however, tends to increase rapidly close to the start of the event. Self-confidence is generally less consistent during the precompetition period, but the theoretical predictions are that neither cognitive anxiety nor self-confidence is likely to change unless expectations of success change during this period. However, recent research findings have shown that the patterning of the multidimensional competitive state anxiety components may differ as a function of individual difference variables, including skill level,[47] type of sport,[47,59] and gender.[60]

Whilst the CSAI–2 has facilitated the development of knowledge about the competitive state anxiety response, it is, like many other state anxiety measures, based on a somewhat limited conceptualization of the anxiety response in that it measures only the intensity of the symptoms and not how they are perceived along a 'facilitative-debilitative to performance' continuum.[3,61,62]. Anxiety has been largely viewed as negative and detrimental to performance, particularly in the North American sport psychology literature. However, as findings from gymnastics have indicated,[63] anxiety can also be seen as a stimulant to enhanced performance. The ways in which sports performers perceive their competitive state anxiety responses is clearly an area that warrants research attention.

Antecedents of competitive state anxiety

It has been proposed that identification of the precursors of competitive state anxiety may provide a valuable means of determining effective methods for achieving optimal performance states.[64] Whilst some research has examined the predictors of unidimensional anxiety,[65] relatively little research has been carried out in the context of multidimensional anxiety. The antecedents of cognitive anxiety and self-confidence are hypothesized to be those factors in the environment which are related to the athlete's expectations of success, including perception of his/her own and his/her opponent's ability. Cues which elicit elevated somatic anxiety, however, are thought to be non-evaluative, of shorter duration, and to consist mainly of conditioned responses to stimuli, such as changing room preparation and precompetition warm-up routines.[47,64] Findings from a sample of good student middle-distance runners[66] suggested that cognitive anxiety and self-confidence do share some common antecedents which contribute to performance expectations, but that there are also factors which may be unique to each. Furthermore, findings reported by Jones et al.[60] suggest that the antecedents of cognitive anxiety and self-confidence may differ between male and female sports performers.

Anxiety–performance relationship

The general lack of precise definitions of, and consequently distinctions between, key concepts such as arousal and anxiety has meant that precise identification of the relationship between anxiety and performance has proved elusive. Many researchers have debated the merits of drive theory and the inverted-U hypothesis, but both of these approaches have been based on the assumption that the anxiety–performance relationship can be explained as a function of changes in a very general arousal system. Despite equivocal research findings and problems associated with using stress, arousal, and anxiety almost interchangeably in this context, the inverted-U hypothesis continues to form the focal point of discussions on anxiety and performance in very nearly every sport psychology textbook. However, numerous objections to the validity of the assumptions underlying the inverted-U hypothesis, particularly when viewed in the context of sports performance, have been raised recently.[67,68]

The adoption of a multidimensional approach to anxiety has been an encouraging and important step forward in the empirical examination of the anxiety–performance relationship in sport psychology as it has also meant the adoption of much more precise definitions and terminology.[47,61] This has, in turn, led to an increasing number of studies which have attempted to examine the relationship between the specific components of the competitive state anxiety response and sports performance.[64,69,70] For example, Burton[69] carried out a study of swimmers and found the following results: cognitive anxiety was related to performance in the form of a negative linear trend, with performance deteriorating as

cognitive anxiety increased; somatic anxiety was related to performance by a quadratic trend, thus supporting earlier findings[70] and suggesting the existence of an inverted-U-shaped curve, although not in terms of a general undifferentiated arousal system; self-confidence was related to performance by a positive linear trend.

Some more recent work has moved away from examining global sports performance and has adopted the 'broad-band' approach[71] to examine the effects of anxiety on different subcomponents of performance, such as working memory and anaerobic output. What is becoming increasingly clear from this line of research is that competitive state anxiety does not necessarily impair performance and can, in some circumstances, enhance it.

The application to sport of models developed within the 'mainstream' psychological literature, such as Apter's reversal theory,[73,74] Sanders's model of stress and performance,[75] and Humphreys and Revelle's model of personality, motivation, and performance,[76,77] has also proved encouraging for advancing knowledge of the competitive anxiety–performance relationship. Furthermore, the catastrophe model[78] of the relationship between competitive state anxiety and sports performance represents a behavioural application of Zeeman's mathematically based catastrophe theory.[79] Hardy and Fazey's proposals[78] stem from a dissatisfaction with the attempts of multidimensional anxiety theory to explain the anxiety–performance relationship by means of two main effects (i.e. separate effects of cognitive and somatic anxiety). They argue, instead, that performance effects are determined by a complex interaction between cognitive anxiety and physiological arousal. Their model has gained some empirical support.[80]

Finally, Gould and Krane[81] have commented on the almost over-reliance on quantitative information in the pursuit of an understanding of the anxiety–performance relationship. They advocated the use of in-depth interviews with athletes in conjunction with questionnaires to gain a broader perspective of their experiences. Certainly, this approach has been lacking and is worthy of greater research attention.

Current situation

Recent research into stress and performance within the realms of sport psychology has resulted in some encouraging conceptual and methodological advances. Current issues include measuring both the 'intensity' and the 'perception' of competitive state anxiety symptoms and examining their relationships with performance, identifying the antecedents of anxiety as a function of individual difference variables such as sex and sport, investigating the complex interaction between cognitive anxiety and somatic anxiety, and in particular testing the predictions of the catastrophe model,[78] adopting both quantitative and qualitative research methodologies in order to gain a clearer picture of the relationship between stress and performance in sport.

Individual differences

The study of individual differences has been popular in sport psychology research. In this section four individual difference variables which have formed the focus of much of this research attention are considered: personality, gender, cognitive style, and skill level.

Personality

The area of personality in sport has been a subject of considerable interest since the late 1960s when it represented the first area of sport psychology to receive systematic research attention.[82] Personality research has centred largely around the person–situation debate, focusing on whether traits or environmental factors are the primary determinants of behaviour. Much of the early work[83] adopted the trait approach, in which traits were viewed as being stable across both time and situations. Mischel, however, later argued that environmental factors were better predictors of behaviour.[84] Current theory favours an approach in which person and situation factors interact.[85] The person–situation debate has continued to be very evident in the sport psychology literature,[86] although as early as 1975 Martens had concluded that the interactional approach was the one that should be favoured in future research.[87]

Unfortunately, research in this area has been plagued by a number of problems, leading Bakker et al. to observe that '. . .in spite of the fact that no other topic in the field of sport psychology has received quite so much attention as the relationship between sport and personality, the findings are not particularly impressive' (ref. 10, p. 54). The reasons for this have been discussed at length by Martens[87] and Morgan,[88] and include a generally atheoretical approach, poor sampling procedures, inappropriate instrumentation, and interpretive errors. Martens[89] reported that only 10 per cent of the empirical studies which he had reviewed involved any experimental manipulation, whilst 89 per cent were correlated in nature and lacking any inferences of causality.

However, in an examination of trends and issues that have emerged since 1975, Vealey has portrayed a somewhat brighter picture of the situation: '. . .sport personality research in the last 14 years has become more theoretical and more balanced methodologically' (ref. 82, p. 226). In particular, the proportion of studies using correlational methods had decreased to 68 per cent of those examined, whilst the proportion using experimental methods had increased to 28 per cent. In addition, Vealy emphasized that significant advances had been made in the form of establishing theoretical bases in several areas, including anxiety, self-confidence, and motivation.

In the case of anxiety, for example, advances in the conceptualization of competitive trait anxiety and, in particular, the development of the Sport Competition Anxiety Test[89] formed the basis of a considerable amount of research. Furthermore, Smith et al. have recently developed a multidimensional measure of competitive trait anxiety, the Sport Anxiety Scale, thus keeping abreast of current developments in competitive state anxiety research[90] (see the section on stress and performance). Other measures of personality which have recently been developed as a result of similar theoretical advances include measures of self-confidence (e.g. the Trait Sport Confidence Inventory)[91] and motivation (e.g. the Sport Orientation Questionnaire)[92] which have proved to be very useful.

Gender

Following the convention used by Deaux,[93] in this section we distinguish between sex, based upon the biological distinction, and gender, which refers to the psychological characteristics associated with males and females. This appears to be an important distinction within the context of both mainstream and sport psychology.

The influential work of Maccoby and Jacklin[94] on sex differences showed males to be more aggressive and to have greater mathematical and spatial ability than females, whereas females have greater verbal ability. However, a meta-analysis by Hyde[95] showed that sex differences accounted for less than 5 per cent of the variance in mathematical and spatial ability. In a similar vein, Gill[96] has argued that sex differences are unlikely to be a powerful predictor of behaviour in sport, and that social psychological aspects of gender in sport are more valid predictors.

Gender differences in sport have generally been found in the areas of achievement orientations, achievement cognitions, and competitive anxiety. Gill[97] concluded that females seem to focus more on personal goals and standards which reflect a non-competitive achievement orientation, whereas males focus more on interpersonal comparisons and winning. Research findings have also demonstrated that females tend to score higher than males on competitive trait anxiety[89,97] and multidimensional competitive state anxiety.[60] Finally, Gill et al.[98] have proposed that competitive sport situations exaggerate gender differences in achievement cognitions, with females generally reporting less confidence and lower expectations of success than males. This is in line with Lenney's assertions[99] that these differences vary according to the task and situation, with gender differences being particularly evident in tasks perceived as masculine; sport has generally been considered a male domain,[100] a gendered cultural form that has been dominated by males and masculinity.[101]

The reference to 'masculinity' introduces an extension to gender research which has been relatively popular in recent years—gender role orientation. Gender role concerns the psychological traits of masculinity and femininity which do not depend at all on sex; males and females can possess both masculine and feminine traits. The plethora of gender role research which emerged in the 1970s was in part due to the development of the Bem Sex Role Inventory[102] and the Personality Attributes Questionnaire.[103] Although these measures have been used widely in sport psychology research, several criticisms have been levelled at them, limiting the usefulness of the research that they have generated.[96]

Cognitive style

Individual differences in sports performers' cognitive styles have attracted a notable amount of research interest. One of the most prominent cognitive style typologies was developed by Nideffer[104] in the context of attentional style.

He proposed that attentional style exists along two dimensions, width (i.e. broad versus narrow) and direction (i.e. internal versus external), based upon which attentional focus was classified into four types: broad external, broad internal, narrow external, and narrow internal. Nideffer's Test of Attentional and Interpersonal Style (TAIS)[104] was developed to measure individual differences in ability to use the various attentional styles and has proved popular in sport psychology research. However, Nideffer's typology makes no distinction between relevant and irrelevant information so that the validity of the TAIS is questionable. Furthermore, the findings of Van Schoyk and Grasha[105] and of Albrecht and Feltz[106] showed that sport-specific measures of attentional style were more valid and reliable than the TAIS. Both these studies also raised questions about the factor structure of Nideffer's model.

Landers[107] was quite damning in his assessment of the worth of the TAIS in sport psychology research, advocating less reliance on questionnaire measures and greater emphasis on physiological and behavioural measures of attention in sport. Several studies have been carried out to examine physiological concomitants of attention in sport.[108,109] Behavioural measures of perception have also been used,[110] and have similarly proved encouraging.

Associative versus dissociative attentional strategies were

identified as a cognitive style dimension which distinguished elite from non-elite marathon runners.[111] Specifically, elite runners were generally found to use associative strategies whereby their direction of attention was internalized towards the body's feedback signals, whilst non-elite runners tended to use dissociative strategies in which attention was externalized. However, it is likely that runners, whether elite or non-elite, do not use solely one or the other strategy but a combination of the two. In any case, subsequent research has shown that associative strategies are not necessarily superior.[112]

Skill level

In an area of study where skill level is clearly a crucial factor, it is not surprising that it has been the focus of considerable research interest. This research has mainly concentrated on differences between elite and non-elite sports performers which are evident across a broad range of areas. In the case of anxiety, for example, differences have been found in both the patterning and perceptions of anxiety. Elite performers have generally been found to experience their highest levels of anxiety prior to the event, whilst non-elite performers experience their highest levels during performance.[63,113] In addition to this apparent ability of elite performers to control anxiety during the crucial moments of competition, Mahoney and Avener[63] have also reported that elite gymnasts used their anxiety as a stimulant to better performance, whereas the less successful gymnasts aroused themselves into near-panic states.

Other areas in which skill level differences are evident include imagery, attention, and perceptual–motor factors. Orlick and Partington[114] have reported that successful Olympians make greater use of imagery than their less successful counterparts. Mahoney and associates have also suggested that elite performers may make greater use than non-elite performers of internal and kinesthetic imagery as opposed to external imagery.[63,115] Boutcher and Zinsser[108] reported that elite performers were more efficient than non-elite performers in terms of attentional control, a finding supported by Mahoney et al.[115] and Orlick and Partington.[114] Elite performers have also been found to be superior on the following perceptual–motor factors: the ability to chunk and recall elements of structured ball games,[116] the use of advanced visual cues,[117] and decision time and accuracy.[116]

Current situation

Because of the large number of individual difference variables operating in sport, it has not been possible to consider all of them in this section. Research in those areas selected demonstrates that the situation is very complex and that there is still much to be learned. One of the major problems with the research which has been carried out is that the vast majority is descriptive. There is a clear need for the use of causal analyses in future research in this area.

INTERPERSONAL RELATIONS IN SPORT

At least four different types of relationship can be identified which are of interest to sport psychologists. These are between coaches and performers, between different performers in the same team, between performers and their parents, and between coaches and the parents of performers. Unfortunately, despite the obvious importance of the last two relationships, very little empirical research has been published in these areas.[118] Consequently, we shall focus strongly on the other two areas.

The coach as a leader

Following the 'Great Man' theory of leaders, early research attempted to identify personality differences between successful and less successful coaches. This research suffered from many of the same deficiencies as early trait personality research on sports performers,[89] not least of which was that it ignored the potential influence of situational factors and the performers being 'led' upon the leadership process.[119] More recent research has been based on a person-by-situation model of behaviour, and has been dominated by the work of Chelladurai. Chelladurai[120] presented a multidimensional model of leadership in sport that emphasized three factors: the personal characteristics of the leader, the group members' needs and desires, and the demands of the situation. The major prediction of the model was that performance and satisfaction are a positive function of the degree of congruence between three different aspects of leadership: the leader's actual behaviour, the leader behaviour preferred by the group, and the leader behaviour required by the situation. Furthermore, according to the model these different types of leader behaviour are influenced by the antecedents of leader characteristics, the group members' characteristics, and the situational characteristics.

In order to measure the coach's leadership behaviour, Chelladurai and Saleh[121] developed the Leadership Scale for Sports. This consist of five subscales which assess training and instruction behaviours, democratic behaviours, autocratic behaviours, social support behaviours, and rewarding behaviours (positive feedback). It is worth noting that three of these types of behaviour focus upon an end-product (training and instruction, social support and reward), whilst the other two focus upon a process (the means by which decisions will be made). The Leadership Scale for Sports has now been used to examine the influence of several antecedents on preferred leader behaviour. These have included gender differences,[122] experience,[123] task characteristics,[124] organizational variables,[119] and cultural differences.[125] The findings of these studies have been generally supportive of Chelladurai's multidimensional conceptualization of leadership behaviour in sport,[120] and have also shown that performers generally prefer coaching that emphasizes training and instruction, together with positive feedback. These findings replicate earlier results obtained with different paradigms.[126,127] Furthermore, studies by Chelladurai and Saleh[122] and by Chelladurai and Carron[123] have also shown that experienced performers prefer more social support than inexperienced performers, and that males prefer more social support than females. However, it is as yet unclear whether this last finding is a true reflection of gender differences, or is due to the fact that female athletes often train with a male coach.

The coach as a decision-maker

Humanistic orientations towards participation in decision-making[19] have led to claims that coaches are generally autocratic and insensitive.[128] However, researchers in organizational psychology[129] have argued that participation is not always the most appropriate decision-making strategy for a leader or manager to use. In line with these arguments, Chelladurai and Haggerty[130] proposed a model of decision-making that matched seven situational attributes of problems to one of three decision-making processes. The seven situational attributes were time pressure, quality requirements, coach's relative information, problem complexity, acceptance requirement, coach's power, and group integration; the three decision-making processes were autocratic, participative, and delegative. Subsequent tests of this model[131] indicated that the delegative style of decision-making was not acceptable to performers of either gender. However, a revised version of the original model, which replaced the delegative style of decision-making with varying degrees of autocratic and democratic decision-making, did receive support when tested.[132] Other results which are of relevance to this model include findings that autocratic decision-making processes are both preferred and more effective under conditions of external stress,[133] male athletes show a greater preference than female athletes for autocratic decision-making,[122] and older more experienced athletes show a greater preference than younger less experienced athletes for autocratic decision-making.[123]

Group dynamics

Like much of the more recent competitive stress and performance literature, interest in group cohesion in sport developed out of research into social facilitation,[50] group motivation,[134] and the Ringelmann effect (later known as 'social loafing').[135] The term 'social loafing' refers to the reductions in effort that have been observed when a group of individuals work together on a task. These effects have been demonstrated across a wide range of tasks in both males and females, but appear to be alleviated by identifiability, the uniqueness of the individual's contribution to the group, group cohesion, and the personal meaning of the task. Nevertheless, social loafing has been shown to be a reliable phenomenon that exists even when tasks are personally meaningful.[136] The converse of social loafing has also been demonstrated, with subjects demonstrating enhanced rather than diminished effort when working in groups.[134]

These findings clearly require some sort of conceptual model in order to be understood. Carron[137] provided such a model which related personal, environmental, and leadership factors to team factors, group cohesion, and personal and group outcomes. Carron's model enables some of the rather equivocal findings regarding team effects upon performance to be tentatively explained.[96] Much of the early literature on team cohesion utilized the Team Cohesiveness Questionnaire[138] which contains items that directly rate the individual's attraction to the group and items that tap interpersonal attraction between individuals. Most of the studies that have demonstrated negative effects for team cohesion upon performance have used an interpersonal measure of cohesion, whilst those producing positive effects have used a direct measure of group attraction.

Other important factors in Carron's model[135] include the interactive nature of the task, the compatibility of the group members, group norms regarding performance, team goals, and the crudeness and conceptual distance of the criterion measure of performance from the independent variables of interest.[96,137] Another area of interest is the circularity of the cohesion–performance relationship. Although the literature is replete with design problems that are implied by Carron's model, the findings generally suggest a stronger influence for performance upon cohesion than for cohesion upon performance.[96]

More recent research on team cohesion has viewed it as a multidimensional construct. Yukelson et al.[139] developed the Multidimensional Sport Cohesion Instrument for measuring cohesion in basketball teams. Factor analysis of the instrument yielded four factors: attraction to the group, valued roles, quality of teamwork, and unity of purpose. Similarly, Carron et al.[140] developed the Group Environment Questionnaire which evaluates attraction to group and integrated group activity along both task and social dimensions. The interested reader will be able to identify links between these two factor structures.

Several researchers have now used the Group Environment Questionnaire to examine the differential effects of team cohesion upon adherence to physical activity,[140] group resistance to disruption,[141] and collective efficacy[142] in performers of different ability levels. Widmeyer et al.[143] have also examined the effects of group size (both in terms of action unit size and total number of players in the team) upon cohesion, enjoyment, and performance. Their results indicated that cohesion, enjoyment, and performance were all decreased in large groups, and suggested the possibility that these effects may be mediated by a lack of influence/responsibility felt by members of large teams, together with poor co-ordination of the team effort.

Aggression

Aggression has not proved to be an easy construct to define. We shall follow Silva's lead[144] and define it narrowly as overt behaviour which is intended to injure psychologically or physically another person or oneself. In using this definition, it is important to distinguish between hostile aggression, where the primary goal of the action is to injure, and instrumental aggression, where the behaviour is directed at the target for some means other than the injury itself. Instrumental aggression is often allowed by the rules of sport, whereas hostile aggression is generally prohibited. Finally, it is necessary to distinguish between aggression and assertiveness. Assertive behaviour is purposeful goal-directed behaviour where there is no intention to harm.

At least three different types of aggression theory can be distinguished in the literature:[96] instinct theories, drive theories, and social learning theories. Instinct theories are usually based on territorial defence and propose that aggressive energy builds up as a result of the 'territorial invasions' that naturally occur during social interactions. Drive theories argue that aggressive drives arise from the frustration that occurs as a result of any blocking of goal-directed behaviour. Social learning theories argue that aggression is a learned response that is acquired through direct reinforcement or observational learning. Instinct and drive theories are currently very unfashionable, largely because they predict that sport should have a cathartic effect upon aggression by providing a constructive outlet for naturally occurring aggressive energy or drive. Empirical research that has addressed this question has been distinctly equivocal.[96,145,146] However, it should be noted that Berkowitz, one of the most prominent researchers in the area, has proposed that learned and innate sources of aggression coexist.[147] Evidence in favour of the social learning theory position is much stronger in terms of both direct reinforcement and observational learning of aggression.[96] Bandura[8] has also argued that disinhibition and de-individuation of the target for aggression could be an important feature of the development of aggression. Eventually, the target of the aggression becomes so stripped of all favourable characteristics that any form of aggression appears to be perfectly justifiable.

Research on the antecedents of aggression also favours the influence of situational factors in the triggering of aggressive behaviour. Such factors include the norms and values of the sport, the period of the game, the stage of the season, and the state of the game.[10] Other factors that are thought to influence aggression include physical appearance,[148] experience,[149,150] and gender,[151] although the influence of gender has been disputed.[152] Unfortunately, research on the effects of aggression upon performance is somewhat confounded by the use of either non-experimental field study designs[153] or provocation to manipulate aggression.[154] A considerable quantity of research has been performed by van der Brug and others on the situational factors that elicit aggression in spectators.[10] However, this research tends to be sociological rather than psychological in nature, and will not be reviewed here.

Finally, Gill has presented a reasoned argument that sport is neither a 'good' nor a 'bad' influence upon moral development.[96] It simply provides opportunities for moral judgement ('good' or 'bad') to be developed. This hypothesis has been empirically explored by Bredemeier and associates[149,155] who have shown that appropriately structured sport experience can indeed aid moral development.

Current situation

It should be clear from the research that has just been reviewed that both Chelladurai's model of leadership behaviour in sport[120] and Chelladurai and Haggerty's model of decision-making styles in sport[130] offer considerable promise for future research. Research on the latter model[132] has shown that situational factors account for approximately three times more of the variance in performers' preferred decision-making styles than individual differences in preferred styles. It has also suggested that performers may prefer democratic styles of decision-making in less than 20 per cent of the situations which are commonly encountered in sport. Whether these preferences result from social evolutionary processes or from considerations of performance effectiveness remains unclear. In the light of these findings, it is disappointing that several of the studies reviewed have examined global preferences and trends in coaches' leader behaviour, without really exploring the details of the situation by group member interaction across different situations that occur in sport. Future research should also explore the situational discrepancy hypothesis in terms of both group performance and performer satisfaction, as well as the necessity and sufficiency of Chelladurai and Haggerty's seven situational attributes.[130]

The processes underlying group dynamics are necessarily complex; nevertheless, progress in this area was held up for a long time by the lack of a theoretical framework. The work which has been reviewed on Carron's conceptual model[137] therefore represents a fairly significant step forward. However, much remains to be learned about the group dynamics of sports teams, and Carron's model is not the only conceptual model that has been proposed.[156] Having said all this, it can be at least tentatively concluded that groups which are not too large and have clearly defined team goals, but recognize and value the individual contributions that are made by their members, are most likely to maximize their potential.

Psychological research into aggression proved difficult to review because, although there is a considerable body of research on the topic, much of it seems to sit on the sociology side of the social psychology 'fence'. From psychological research by Berkowitz[147] and Bandura[8] it is known that aggressive behaviour is acquired by observation and direct reinforcement, but little is known about the psychological processes that trigger this behaviour. It is also known that watching sport does not have a cathartic effect by releasing aggressive urges in spectators. However, despite claims to the contrary,[95] the available literature does not indicate that participation in sport has no cathartic effect upon performers. It merely indicates that a cathartic effect has not been reliably demonstrated. Considering that interventions generally exert their greatest effects upon 'extreme' populations, together with the ethical problems that would have to be overcome to design an experiment to test the catharsis hypothesis in (say) men who regularly abuse their wives, this inability of the empirical literature to demonstrate a cathartic effect is perhaps unsurprising. Finally, in a rather different direction, the recent work of Bredemeier and others,[155] exploring the potential of appropriately structured sport environments for moral development, looks most promising.

EXERCISE PSYCHOLOGY

The role that exercise plays in the prevention of physical health problems has long been recognized by both health professionals and scientists. A significant development in this area during the last decade or so has been the recognition that exercise also has a potentially very important role to play in both the prevention and treatment of mental health problems. This development, which has largely been associated with the recent growth in health psychology and broadening of research interests in the sport and exercise sciences, has led to the emergence of 'exercise psychology'.[157] The increasing prominence of exercise psychology during the 1980s is reflected in the publication of texts on such specialized topics as exercise and mental health[158] and exercise adherence.[36] A significant move in sport psychology was that in 1988 the flagship of sport psychology in North America, the *Journal of Sport Psychology*, was renamed the *Journal of Sport and Exercise Psychology*. The two major areas of research which have emerged in exercise psychology are exercise and mental health, and exercise motivation.

Exercise and mental health

Despite general assertions in the literature that exercise and mental health are positively related in both clinical and non-clinical populations, there is a significant lack of evidence to suggest that this relationship is anything more than associative. The dearth of evidence to imply causality is largely due to problems in research methodology, and the lack of longitudinal studies in particular.[37,159] Nonetheless, Biddle and Fox[157] have identified three areas in which the link between exercise and mental health appears to be relatively robust: a reduction in anxiety[160,167] and depression,[162] an enhancement of self-esteem,[163] particularly in the subdomain of physical self-esteem,[157] and improved reactivity to stress (see Crews and Landers[164] for a meta-analytic review). All the benefits discussed are related predominantly to aerobic exercise, with the most popular exercise activities being running[165] and swimming.[166]

In an attempt to draw together research findings and to unravel specific details of the relationship between exercise and mental health, the American National Institute of Mental Health produced a series of consensus statements of the state of current knowledge and future research needs.[158] These were summarized by Biddle and Fox[157] as follows.

Current knowledge

1. Exercise is associated with reduced state anxiety.
2. Exercise has been associated with a decreased level of mild to moderate depression.
3. Long-term exercise is usually associated with reductions in traits such as neuroticism and anxiety.
4. Exercise may be an adjunct to the professional treatment of severe depression.
5. Exercise results in the reduction of various stress indices.
6. Exercise has beneficial emotional effects across all ages and in both sexes.

Research questions

1. What is the role of exercise in primary prevention?
2. What is the effect of exercise on the rehabilitation of people with physical and mental disorders?
3. What are the mechanisms that mediate the effects of exercise on stress?
4. What are the effects of exercise on the stress reactivity of different groups, such as those differing by age, sex, socio-economic status, and personality?
5. What are the optimal exercise doses required to produce effective responses to mental stress?
6. What is the effect of exercise compared with other interventions?
7. What are the effects of exercise on the mental health of children?
8. What are the mechanisms underlying exercise effects?

The identification of the mechanisms underlying the relationship between exercise and mental health is clearly crucial to understanding the specific details of the relationship. Recent proposals in this context include muscle relaxation, thermogenesis effects, and brainwave pattern change,[167,168] release of endogenous opiates,[169] and enhancement of self-esteem[170] and mastery.[157]

A question of primary concern, of course, relates to the comparison of the effect of exercise with other interventions. In particular, there is a need to know whether exercise provides anything more than a 'time-out'.[171] Whilst several researchers[172,173] have compared exercise with other types of intervention, the findings are somewhat equivocal, leading Biddle[174] to state:

> ...despite impressive survey data[175] and theoretically sound speculation on the mechanisms of mental health effects from exercise,[37,176,177] there remains some doubt about the uniqueness of exercise effects and whether other interventions might not be equally effective. The challenge is there ...' (ref. 174, p. 6)

Finally, it should be emphasized that exercise may not always be associated with mental health benefits and can sometimes have negative psychological effects.[159] Steinberg *et al.*[178] cited evidence which they interpreted as suggesting that regular physical exercise which becomes compulsive can be viewed as a form of dependence or addiction that is similar to addiction to opiates and other drugs.[35] For example, Morris *et al.*[179] found that male runners deprived of running for 2 weeks produced a withdrawal syndrome, although in a form milder than one would expect from opiate withdrawal. This is clearly a cause for concern amongst exercise psychologists which requires extensive systematic investigation. However, as has already been stated in the section on motivation, it is important that future research clearly defines addiction in such a way as to distinguish between commitment and addiction.[180] For example, mothers who are separated from their children frequently demonstrate considerable distress, but does this mean that they are addicted to them?

Exercise motivation

Despite the now wide acceptance of the mental health benefits of exercise, surveys have suggested that a very disappointing proportion of the population (probably no more than 50 per cent) are involved in any significant amount of exercise.[181] Biddle and Fox[157] further noted that an even smaller proportion are likely to undertake an exercise programme or schedule which will significantly enhance health (it should be noted here that the intensity, frequency, and duration of exercise required to produce health benefits remains an area for debate). This situation is exacerbated still further by the fact that relatively few of those who do initiate an exercise programme will persist for very long. In fact, the rapidly expanding amount of research examining adherence to exercise seems to have firmly established that 50 per cent of those who enrol in a supervised exercise programme can be expected to drop out in the first 3 to 6 months.[182,183] Interestingly, those people who would benefit most from exercise (such as the obese) tend to be the very people who drop out.[184]

Early research which attempted to identify factors important in decisions to initiate and adhere to exercise suffered from the lack of a theoretical underpinning. More recently, however, researchers have become increasingly 'theoretically driven', examining models based upon beliefs, attitudes, and self-perceptions.[157] The health belief model,[185] together with the theory of reasoned action,[186] have formed the basis of research into beliefs and attitudes, whilst self-esteem, attribution theory, locus of control, and self-efficacy theory have been amongst the self-perception theories that have been examined.[187] Bandura's social cognitive theory[188] is also becoming increasingly prominent, with the most studied mechanism within the theory being self-efficacy. In fact, recent studies by Dzewaltowski and coworkers[189,190] showed social cognitive theory (based on self-perception) to be a better predictor of exercise behaviour than the theory of reasoned action (based on beliefs and attitudes).

The different approaches to exercise motivation reflect the general lack of consensus among researchers over the definition of motivation.[191] Recently, emphasis has been placed on defining the concept of motivation itself, with Duda[192] proposing a broad-based definition involving three components: direction, intensity, and persistence. Clearly, exercise motivation will continue to provide a fertile area for research in the foreseeable future. However, one issue which will have to be addressed in this area is the definition of adherers and drop-outs, as this has been a source of inconsistency in studies to date.

Current situation

Exercise psychology is clearly an area of massive growth and development. It is beyond the scope of this chapter to discuss the many applications of exercise psychology, other than to note that it has proved to be of great interest and benefit in rehabilitation,[193] corporate,[194] therapeutic,[162] and educational[195] settings. However, many issues remain to be addressed, including the examination of causal relationships between exercise and mental health, the examination of factors affecting exercise adherence, antecedent factors influencing exercise commitment and dependence, the efficacy of exercise versus other forms of therapeutic intervention, and the intensity, frequency, and duration of exercise required for mental health benefits. Clearly, well-designed longitudinal studies are required to address such issues. Finally, the vast majority of the literature and research findings in this area has emanated from North America; the generalization of this work to other populations and cultures has yet to be determined.

MENTAL TRAINING

The last 10 years has seen increasing acceptance of mental training as an integral part of preparation for peak performance, so that sport psychologists working alongside national squads and teams have now become the norm rather than the exception. Research on peak performance and ideal performance states is still in its infancy; however, retrospective studies of peak performance[196] suggest that a number of factors may contribute to such performances. These include intentionality on the part of the performer, clear focus upon the task, total absorption in the task, effortless concentration, feelings of the body performing on its own without fatigue, loss of fear, and certainty of success. Similarly, research into the differences between elite and non-elite performers[63,114,115] has identified that elite performers have clear daily goals, less anxiety, greater confidence, higher levels of motivation, and better attention control strategies than their less successful colleagues.

Following Vealey,[197] we shall make a distinction between psychological skills, such as anxiety control, and basic techniques or methods, such as goal-setting, imagery, physical relaxation, and self-talk. Consequently, after a discussion of these basic techniques, the rest of this section will focus upon their use and application in the development of anxiety control, activation control, self-confidence, motivation, and attention control.

Basic techniques

Goal-setting

Much of the empirical literature on goal-setting has been provided by organizational psychology.[198] It has also been claimed that many of these findings should generalize to sport settings.[199] However, Beggs[200] has quite rightly pointed out that there are a number of important differences between occupational organizations and sport which may well impede generalizations. Nevertheless, following an extensive review of the goal-setting literature, Beggs concluded that goal-setting was potentially a valuable tool for athletes and coaches. The available literature suggests that a number of factors need to be controlled for goal-setting to exert such a beneficial influence upon performance.[68,199–201] Essentially, these are that goals must identify specific targets that lie within the performer's own control, that the performer is committed to, and that the performer perceives to be realistic and worthwhile. Furthermore, long-term goals should be broken down into short-term goals, and feedback regarding goal achievement for the goal-setting process to work.

At least three different types of goal can be identified in the literature.[69,202] Outcome goals focus upon the outcomes of particular events and usually (but not always) involve social comparisons of some sort, for example finishing a race in seventh place. Performance goals specify an end-product of performance that will be achieved by the performer relatively independently of other performers, for example running a 5-min mile. Finally, process-oriented goals specify the processes in which the performer will engage in order to perform at an optimal level, for example

maintaining perceived exertion at a rating of 14 for 1 mile. Whilst outcome goals may possess great motivational value for performers in the short term, Roberts[11] and others have argued that, ultimately, such goals are likely to lead performers to drop out of sport. Burton[69] has also shown that outcome goals are associated with higher levels of competitive state anxiety than are performance goals. Consequently, sport performers are usually encouraged to set performance, rather than outcome, goals. There has been little research so far on process-oriented goals, but Hardy and Nelson[201] have suggested that they may exert their influence upon performance by the allocation of attentional resources. However, this speculation has yet to be empirically tested.

Imagery

Imagery can be defined as symbolic sensory experience that may occur in any sensory mode. However, the visual, kinesthetic, and auditory modes seem likely to be those most relevant to sport. The majority of research that has examined the use of imagery in sports settings has utilized a learning paradigm. The consensus of this research is that mental practice of a motor task is better than no practice at all.[203] A smaller number of studies have investigated the immediate effects of (multimodal) imagery upon performance, either in conjunction with relaxation[204] or without it.[205] Again, the consensus of these studies is that imagery can exert a beneficial effect upon performance. It has also been suggested that imagery can be used to initiate relaxation[54] and enhance self-confidence.[58,206] However, to date, direct empirical support for these suggestions has been limited (see later sections on physical relaxation and self-confidence).

Traditionally, several theories have been called upon to explain the effects of mental rehearsal upon motor performance, including symbolic learning theory, psychoneuromuscular theory, and the activation set hypothesis.[203] Nevertheless, all these theories have been extensively criticized on both conceptual and empirical grounds.[10,203,205,207,208] Furthermore, following a detailed critique of imagery research in sport psychology, Murphy[207] concluded that earlier theorizing on imagery effects had been greatly hampered by its preoccupation with the effects of mental rehearsal upon motor performance. He argued that theories were required which acknowledged the importance of psychophysiology in the imagery process and the meaning of the image for the individual.[209,210]

Physical relaxation

Physical relaxation has long been used as a means of anxiety control, often in the form of progressive muscular relaxation[211] or one of its derivatives.[212,213] Furthermore, physical relaxation is also included as part of most stress management programmes[214–216] and has been readily adapted to mental training in sports settings.[202,217] Whilst there is relatively little literature that shows a direct effect of relaxation upon sports performance, there is a considerable literature that demonstrates the anxiety-reducing properties of relaxation[54,218] and the efficacy of cognitive-behavioural strategies involving relaxation, imagery, and positive self-talk.[4]

A number of researchers have examined the relationship between relaxation and imagery.[204,219–221] Although the results of these investigations are equivocal, it is widely accepted that relaxation enhances imagery and imagery can be used to relax. However, Weinberg et al.[221] interpreted their data as suggesting that the ideal activation state for mentally rehearsing a task depends upon the nature of the task. Whatever the case, relaxation can be regarded as one of the fundamental techniques involved in more complex skills such as anxiety and attention control.

Self-talk

Several researchers have shown that thought content and self-statements are important predictors of sports success.[63,114,222,223] The precise reasons for this relationship are not known, although it seems likely that self-confidence[58] and anxiety control[215,224] are at least partially involved. Furthermore, despite the strong emphasis that mental training programmes often place upon the development of 'appropriate' self-talk,[225] relatively few controlled studies have been performed that would enable any sort of empirically based operationalization of the word 'appropriate' to be attempted. However, it is known that self-defeating self-statements have a negative effect upon performance,[223] that thought-stopping techniques can be used to modify such self-statements,[226] that self-statements can be used to trigger desired actions more effectively,[227] and that self-statements can be used to provide self-reward.[7] It is also thought that self-statements can be used to increase effort,[225] modify mood,[202] and control attention[228] (see also the earlier discussion of process-oriented goals).

Anxiety control

As indicated earlier, the state anxiety response is currently conceived to have at least two components: cognitive anxiety (worry) and somatic anxiety (perceptions of physiological arousal). Traditionally, these different modes of anxiety were thought to be optimally controlled by matching relaxation strategies to the response system in which the anxiety response principally occurred.[54] For example, meditation and imagery-based strategies might be used to control cognitive anxiety, whilst progressive muscular relaxation might be used to control somatic anxiety.[229] However, more recent research has suggested that in practice the different types of anxiety rarely occur in total isolation,[46] and that some sort of multimodal stress management strategy is likely to be necessary.[3]

Most multimodal stress management programmes currently employed in sports settings are derivatives of Meichenbaum's Stress Inoculation Training,[215] or Smith's Stress Management Training.[216] Programmes are usually characterized by helping performers to learn how to use positive self-statements and physical relaxation when they are confronting a problem, imagery to rehearse their coping strategy, and process-oriented goals to enable them to reward themselves for successfully engaging in the coping process regardless of the outcome of their coping efforts.[4] The empirical literature suggests that such multimodal stress management strategies are generally effective at controlling anxiety, but are perhaps not quite as effective as enhancing performance under stressful conditions.[3] There are a number of theoretical reasons why this might be, most of which revolve around the question of whether or not high states of physiological arousal are beneficial to performance.[3,80]

Activation control

In order to consider more fully the issue of optimal physiological arousal, it is necessary to distinguish between activation and physiological arousal. The work of Pribram and McGuinness[230] suggests that physiological arousal is best viewed as a phasic response to any input to the organism. Its primary purpose is to energize the perceptual processes. Conversely, activation is best viewed as a tonic readiness to respond, which includes the preparation of a motor response. Pribram and McGuinness[230] also present evidence that the arousal and activation functions are handled by different neural networks of the brain. The implications of this line of research for mental training are important, since it suggests that increases in general physiological arousal are likely to be beneficial to performance only to the extent that such increases assist the performer to obtain the desired activation state. Therefore 'psyching up' strategies will be effective for some tasks but not for others.[221,231] Furthermore, the likelihood of such activation strategies being successful will almost certainly depend upon their precise content, a fact that does not always seem to have been appreciated by sport psychologists or performers. Individual differences in cognitive style may also be an important variable to consider in determining an appropriate activation strategy. Consequently, mental training programmes usually include strategies based on both self-talk and imagery for performers to adapt to their needs.[3]

Self-confidence

Self-confidence has been thought to be an important influence upon performance for a number of years,[58,232] and more recent research has confirmed this importance.[68,115] Research suggests that this relationship is cyclical. Confident performers set more difficult goals towards which they persist until they have achieved them, thereby gaining feelings of competence and increased self-confidence.[233] Bandura's theory of self-efficacy (situationally specific self-confidence)[58] has now received considerable support in sport and other settings.[95,188] Essentially, the theory predicts that the strongest influence upon self-efficacy is previous experience of success, followed by vicarious experience, verbal persuasion, and physiological arousal in that order. Consequently, most mental training which is focused on enhancing self-confidence teaches performers the following:

(1) to break down long-term into short-term goals which are within their control so as to maximize their experience of success;
(2) regularly to rehearse mentally succeeding at their goals so as to maximize the vicarious experience of success;
(3) to use positive self-talk to encourage themselves and reinforce their feelings of self-confidence.

Motivation

Most serious athletes seem to be naturally highly motivated towards their sport. However, extremely high levels of motivation may be necessary to produce repeatedly the sort of high quality training sessions that are necessary for success in some sports.[115] Furthermore, maintaining motivation throughout the duration of a long season, during periods of injury, or following setbacks to the training programme may severely tax the motivational skills of even the most elite performers. One of the major reasons for this is probably best explained by cognitive evaluation theory.[7]

According to Deci and Ryan[7], humans have an innate need to feel self-competent and self-determining. In accordance with these needs, athletes commit themselves to difficult and demanding goals, the achievement of which will lead to enhanced feelings of competence and greater intrinsic motivation towards their sport. Consequently, any disruption to these goals is highly likely to lead to reductions in motivation by a reversal of this cycle. It is also worth noting the involvement of self-confidence in this motivational cycle. Greater self-confidence means that athletes are likely to set and achieve more difficult goals,[188] which then increase feelings of self-competence and intrinsic motivation. In view of these findings, mental training of motivational skills often focuses upon encouraging performers to set goals that are within their control in order to enhance feelings of self-competence, and upon attributional retraining to encourage performers to attribute their failures and setbacks to external factors or unstable internal factors.[22,29,234] Clearly, positive self-talk and self-reward are also important techniques to develop here.

Finally, a cautionary word is in order regarding the role of goals in the motivational process. As Beggs[220] has pointed out, goals are something of a 'double-edged sword' in sport, since they can themselves become a source of additional stress and anxiety if performers do not adjust them when circumstances change. Anecdotal evidence suggests that this presents a serious problem for committed athletes. One implication of this seems to be that athletes should be taught how to set process-oriented goals that focus on emotional control as well as performance goals that focus upon the end-product of successful performance. They should also be taught how to evaluate their goals regularly and be flexible enough to adjust them when appropriate.[202,235] Having said all this, there is a dearth of empirical research that has directly addressed this question.

Attention control

The need for totally focused concentration is frequently cited by athletes as one of the primary requirements of performance.[45,114] However, the means by which such control of attention might be achieved are far from well understood. Attention has been conceptualized by empirical researchers in a number of different ways: in terms of single versus multiple resources,[236] in terms of scanning, focus, and selectivity,[237] and in terms of breadth and direction.[104] However, the research that has been generated by these conceptualizations is no more than suggestive as regards the control of attention.[197] In practice, desensitization to the competition environment by simulation training and heavy overlearning are thought to be fundamental requirements.[45] Both association and dissociation strategies have been shown to be of potential benefit to performance,[238,239] and some sort of dissociation from stressful stimuli is a common part of many stress management training programmes.[217] Elite athletes also report using process-oriented goals and detailed well-rehearsed competition plans as part of their attention control strategies.[45,109,114,240]

Current situation

Mental training has become established as an important aspect of sport psychology. Furthermore, there is evidence that mental training can be an effective means of enhancing competitive performance.[210] However, empirical research into the optimal content of mental training programmes is still in its infancy. In particular, there is an urgent need for research on the means by which attention can be controlled and concentration enhanced, the precise role of physiological arousal and activation in the performance of sports tasks, the means by which activation can be increased in a controlled fashion if this is appropriate, the use of goal-setting with complex tasks in sports settings, and the role of process-oriented goals in stress management and attention control. Furthermore, in recent years a number of distance-learning mental training packages have been published which purport to enhance athletes' mental skills. Most of these packages consist of either a book or a book plus audiocassette tapes. There is an urgent need to evaluate the effectiveness of these packages when used both with and without tutorial support. Finally, recent research has also indicated a need for investigation of the factors that influence adherence to mental training programmes.[241]

THE PRACTISING SPORT PSYCHOLOGIST

As noted in the previous section, there has been a rapid growth in applied sport psychology through work with sports performers in recent years. This trend has been reflected by the recent formation of the Association for the Advancement of Applied Sport Psychology (AAASP) which focuses upon three aspects of sport psychology: health psychology, social psychology, and intervention/performance enhancement. Furthermore, two journals, *The Sport Psychologist* and the *Journal of Applied Sport Psychology*, have recently been established for the purpose of communicating research and methods pertaining to applied sport psychology. Most applied sport psychology activities have thus far been directed at the enhancement of performance in elite athletes.[95] However, a major problem which has confronted applied sport psychologists in such roles concerns debates over three important issues: (1) precise functional roles; (2) accompanying ethical considerations; (3) the criteria for assuming the title 'sport psychologist'. These debates have raged for more than a decade in North America and have recently come to the forefront in Great Britain.

The debate over sport psychology as a profession was particularly rife in North America in the late 1970s and the first half of the 1980s.[242] The crux of the situation around this time was the fact that many sport psychologists had emerged from a sport and physical education background, without the prescribed qualifications and credentials in psychology which were deemed necessary in some quarters. Nideffer proposed that the title 'sport psychologist' should be 'limited to those individuals who currently qualify as psychologists under existing state and provincial laws' (ref. 243, p. 9). In 1982 the North American Society for the Psychology of Sport and Physical Activity (NASPSPA) and the Canadian Society for Psychomotor Learning and Sport Psychology (CSPLSP) accepted a modified version of the American Psychological Association's Ethical Standards as those that should guide the behaviour of members. A year earlier, however, Danish and Hale had called for the adoption of a more educationally oriented approach emphasizing personal development as opposed to the prevailing 'general, remedial services model' (ref. 244, p. 93). Danish and Hale[244] and Brown[245] developed a rationale for practitioners with clinical training and certification working as 'clinical sport psychologists' and practitioners without clinical training and certification working as 'educational sport psychologists'.

In 1982, the United States Olympic Committee (USOC) became interested in the debate and established a psychology advisory committee which subdivided sport psychology into clinical, educational, and research functions. The clinical sport psychologist's role was deemed to include assessing psychopathology, providing individual and group psychotherapy, providing crisis intervention, treating personality disorders, dealing with drug dependence, eating disorders, and so on. The educational sport psychologist was assigned the role of dealing with 'normal' athletes in terms of providing non-clinical information and training in stress management, concentration skills, etc. in order to facilitate the growth and development of healthy athletes. The research sport psychologist's role was designated as helping to develop and evaluate services, and providing scholarly research contributions to the field.[246] Furthermore, precise minimum standards for the credentials of these activities were established, but the crucial factor was that each individual seeking qualification was required to qualify for full membership of the American Psychological Association (APA).

Heyman's reaction to the USOC's proposals for a Sport Psychology Registry was that it excluded many of the individuals with physical education backgrounds.[247] Further consideration and discussion led the Director of the Sports Medicine Division of the USOC to explain: 'it is clear that the USOC will lean heavily on the standards and ethics of APA but not to the exclusion of those who meet their equivalent, while confirming that neither an advanced degree in physical education or clinical psychology by itself is sufficient qualification for sport psychology' (ref. 248, p. 366).

Recently, Vealey has supported the distinction between personal development and more clinically orientated models, arguing that:

> guidelines for service provision by educational versus clinical psychologists should be based upon the distinction between normal and abnormal behaviour…, not the distinction between performance enhancement and personal development. Individuals adhering to the clinical model should be trained and credentialed to provide legal clinical service based on that model, and educational sport psychologists subscribing to the personal development model should be trained and credentialed with regard to the services derived from their particular model. It seems likely that the adoption of a model to serve as the basis for the practice of educational sport psychology may facilitate our attempts to define our profession. (ref. 193, p. 330).

In Britain, sport psychologists have attempted to establish their professional status without becoming quite so embroiled in acrimonious disputes. Nevertheless, when the British Psychological Society established its rules for chartering in such a way that the vast majority of practising sport psychologists were debarred from qualifying because they did not have a first degree in psychology, the British Association of Sports Sciences was more or less obliged to respond by establishing its own list of Accredited Sport Psychologists. To become accredited, psychologists were normally expected to possess a first degree in a sport-related subject and a higher degree in a psychology-related subject, or vice versa. This accreditation process has now been in operation since 1988, and is currently being reviewed and revised with a view to providing better safeguards for naive customers and greater access to the Register for sport psychologists from a rather broader background than is at present the case.

Other recent developments include Sports Council funding of a Sports Science Support Programme to which national governing bodies can apply for the funding of psychological projects, and the formation by the British Olympic Association (BOA) of a Psychology Advisory Group. The former programme has now funded a substantial number of sport psychology support projects for national squads and teams, thereby establishing the possibility of a genuine career structure for practising sport psychologists. The BOA's Psychology Advisory Group was given the remit of enhancing the provision of psychological support for Olympic sports. It is currently preparing an 'Olympic benchmark' document which will specify the requirements of psychological support services delivered to Olympic sports.

Whilst many issues in the development of applied sport psychology as a profession remain unresolved, the demands for the services of sport psychologists by teams and individuals has risen substantially in recent years. Must has been learned from the experiences of those sport psychologists providing these services. *The Sport Psychologist*, for example, has published special issues on the delivery of psychological services to Olympic teams and athletes at the 1988 Winter and Summer Olympic Games (volume 3, issue 4, 1989) and to professional teams (volume 4, number 4, 1990). The various authors deal with issues relating to the sport psychologists' own philosophies of service delivery, the range and type of services provided, the organization of the services, their effectiveness, the problems they encountered, and how they dealt with them. Amongst these articles was an interesting account of an alternative 'organizational empowerment' consultation model successfully implemented with a professional baseball team.[249] The model adopted differed from the traditional approach, where the sport psychologist works directly with athletes, in that the sport psychologist trained one of the organization's management team (who had formal training in counselling) to implement the mental skills training programme. At a time when demand appears to be outstripping the supply of 'good' sport psychologists, this approach may become more popular in the future.

If the services provided by sport psychologists are to be developed and improved, then the evaluation of their activities is clearly important.[250,251] To this end, Partington and Orlick[251] has developed the sport psychology consultation evaluation form. Following its use with a sample of Canadian Olympic athletes, Partington and Orlick were able to identify several characteristics of the 'effective' sport psychologist: someone who provides clear, practical, and concrete strategies for the athlete to try out in an attempt either to solve problems or to improve the level and consistency of his or her performance, someone who is easy for the athlete to relate to, someone who fits in with everybody connected with the team, someone who provides a minimum of several hours of individual sessions for each athlete during the year, and someone who attends at least two or three competitions with the team or athlete.

Current situation

It is now time for practising sport psychologists to solve their accreditation issues and focus upon the development of a 'multicultural' society which can utilize its diversity of backgrounds to provide an enhanced professional service. Issues which are likely to be addressed in the next few years include the development of new methods for evaluating performers' needs and the effectiveness of psychological programmes, identification of the special needs of performers at different levels, exploration of possible areas for collaboration with other professionals (for example, coaches, team medical officers, and physiotherapists), and the development of services on a broader base than simply working with performers (for example, working with team managers, coaches, and sports officials).

CONCLUSION

In this chapter we have reviewed our current state of knowledge in seven areas of sport psychology: motivation, stress and performance, individual differences, interpersonal relationships, exercise psychology, mental training, and the practising sport psychologist. As indicated in the introduction, it is perhaps inevitable that some important topics have been omitted or only very briefly considered; for example, some would no doubt argue that the psychology of children in sport and the female athlete should have received more prominent positions. Nevertheless, we feel that this remains a fairly broad, if not all-embracing, review of current activity in sport psychology.

At the end of each section, a number of important issues and research questions have been identified. Many of these issues and questions are quite fundamental, and the consequences of their investigation are likely to be far-reaching. It should be clear from all this that the current rate of development in sport psychology, at both a professional and an academic level, is very rapid indeed. Furthermore, this progress looks set to continue until at least the end of this century.

REFERENCES

1. Triplett N. Dynamogenic factors in pacemaking and competition. *American Journal of Psychology* 1897; **9**: 507–33.
2. Griffith CR. *Psychology of athletics*. New York: Scribners, 1928.
3. Burton D. Multimodal stress management in sport: current status and future directions. In: Jones JG, Hardy L, eds. *Stress and performance in sport*. Chichester: Wiley, 1990: 171–201.
4. Mace R. Cognitive behavioural interventions in sport. In: Jones JG, Hardy L, eds. *Stress and performance in sport*. Chichester: John Wiley, 1990: 203–30.
5. Straub WF, Williams JM. Cognitive sport pychology: historical, contemporary and future issues. In: Straub WF, Williams JM, eds. *Cognitive sport psychology*. Lansing, NY: Sport Science Associates, 1984: 3–10.
6. Straub WF, Williams JM. *Cognitive sport psychology*. Lansing, NY: Sport Science Associates, 1984.
7. Deci EL, Ryan RM. *Intrinsic motivation and self-determination in human behavior*. New York: Plenum Press, 1985.
8. Bandura A. *Aggression, a social learning analysis*. Englewood Cliffs, NJ: Prentice-Hall, 1973.
9. Carron AV. *Motivation: implications for coaching and teaching*. London, Ontario: Sport Dynamics, 1984.
10. Bakker FC, Whiting HTA, van der Brug H. *Sport psychology: concepts and applications*. Chichester: Wiley, 1990.
11. Roberts GC. The growing child and the perception of competitive stress in sport. In Gleeson G, ed. *The Growing Child in competitive Sport*. London: Hodder and Stoughton, 1986: 130–44.
12. Whitehead J. Achievement goals and drop-out in youth sports. In Gleeson G, ed. *The Growing Child in competitive Sport*. London: Hodder and Stoughton, 1986: 240–47.
13. Robinson DW. Stress seeking: selected behavioral characteristics of elite rock climbers. *Journal of Sport Psychology* 1985; **7**: 400–4.
14. Carruthers M. *The Western Way of Death*. London: Davis-Poynter, 1974.
15. Solomon RL. The opponent-process theory of acquired motivation. *American Psychologist* 1980; **35**: 691–712.
16. Lester JT. Wrestling with the self on Mount Everest. *Journal of Humanistic Psychology* 1983; **23**: 31–41.
17. Casady M. The tricky business of giving rewards. *Psychology Today* 1974; **8**: 52.
18. De Charms R. *Personal Causation: The Internal Affective Determinants of Behavior*. New York: Academic Press, 1968.
19. Deci EL. *Intrinsic Motivation*. New York: Plenum Press, 1975.
20. McAuley E, Duncan T, Tammen VV. Psychometric properties of the Intrinsic Motivation Inventory in a competitive sport setting: a confirmatory factor analysis. *Research Quarterly for Exercise and Sport* 1989; **60**: 48–58.
21. Weiner B. *Theories of motivation: from mechanism to cognition*. Chicago: Rand-McNally, 1972.
22. Weiner B. A theory of motivation for some classroom experiences. *Journal of Educational Psychology* 1979; **71**: 3–25.
23. Abramson LY, Seligman MEP, Teasdale JD. Learned helplessness in humans: Critique and reformulation. *Journal of Abnormal Psychology* 1978; **87**:49–74.
24. Biddle SJH. Attributions. In: Singer RN, Murphy M, Tennant LK, eds. *Handbook on research in sport psychology*. New York: Macmillan, in press.
25. Rejeski WJ, Brawley LR. Attribution theory in sport: current status and new perspectives. *Journal of Sport Psychology* 1983; **5**: 77–99.
26. McAuley E, Duncan TE. Causal attributions and affective reactions to disconfirming outcomes in motor performance. *Journal of Sport and Exercise Psychology* 1989; **11**: 187–200.
27. McAuley E, Duncan TE. Cognitive appraisal and affective reactions following physical achievement outcomes. *Journal of Sport and Exercise Psychology* 1990; **12**: 415–26.
28. Carver CS, Scheier MF. Outcome expectancy, locus of attribution for expectancy,and self-directed attention as determinants of evaluations and performance. *Journal of Experimental Social Psychology* 1982; **18**: 184–200.
29. Dweck CS. Achievement. In Lamb LE, ed. *Social and personality development*. New York: Holt, Rinehart & Winston, 1978: 114–30.
30. Scanlan TK, Passer MW. Self-serving biases in the competitive sport setting: an attributional dilemma. *Journal of Sport Psychology* 1980; **2**: 124–36.
31. Brawley LR. Attributions as social cognitions: contemporary perspectives in sport. In: Straub WF, Williams JF, eds. *Cognitive Sport Psychology*. Lansing, NY: Sport Science Associates, 1984: 212–30.
32. Bannister D, Fransella F. *Inquiring man: the theory of personal constructs*. London: Croom Helm, 1986.
33. Benson MJ. Attributional measurement techniques: classification and comparison of approach for measuring causal dimensions. *Journal of Social Psychology* 1989; **129**: 307–23.
34. Butler RJ. Psychological preparation of Olympic boxers. In: Kremer J, Crawford W, eds. *The psychology of sport: theory and practice*. Occasional Paper, Leicester: The British Psychological Society, 1989: 74–84.
35. Veale MW. Exercise dependence. *British Journal of Addiction* 1987; **82**: 735–40.
36. Dishman RK, ed. *Exercise adherence: its impact on public health*. Champaign, Ill: Human Kinetics, 1988.
37. Morgan WP, O'Connor PJ. Exercise and mental health. In: Dishman, RK, ed. *Exercise adherence: its impact on public health*. Champaign, Ill: Human Kinetics, 1988: 91–121.
38. Summers JJ, Hinton ER. Development of scales to measure participation in running. In: Unestahl LE, ed. *Contemporary sport psychology*. Orebro, Sweden: VEJE, 1986.
39. Morgan WP, Costill DL, Flynn MG, Raglin JS, O'Connor PJ. Mood disturbance following increased training in swimmers. *Medicine and Science in Sports and Exercise* 1988; **20**: 408–14.
40. Cockerill IM, Nevill AM, Lyons N. Modelling mood states in athletic performance. *Journal of Sports Sciences* 1991; **9**: 205–12.
41. Fender LK. Athlete burnout: Potential for research and intervention strategies. *Sport Psychologist* 1989: 63–71.
42. Kelly HH, Michela JL. Attribution theory and research. *Annual Review of Psychology* 1980; **31**: 459–501.
43. Paikov VB. Means of restoration in the training of speed skaters. *Soviet Sports Review* 1985; **20**: 9–12.
44. Ievleva L, Orlick T. Mental links to enhanced healing. *The Sport Psychologist*, 1991; **5**: 25–40.
45. Jones JG, Hardy L, eds. *Stress and performance in sport*. Chichester: Wiley, 1990.
46. Hackfort D, Spielberger CD, eds. *Anxiety in sports: an international perspective*. Washington, DC: Hemisphere, 1989.
47. Martens R, Vealey RS, Burton D, eds. *Competitive anxiety in sport*. Champaign, Ill: Human Kinetics, 1990.
48. Zajonc RB. Social facilitation. *Science* 1965; **149**: 269–74.
49. Cottrell NB, Wack NB, Sekerak GJ, Rittle RH. Social facilitation of dominant responses by the presence of an audience and the mere presence of others. *Journal of Personality and Social Psychology* 1968; **9**: 245–50.

50. Landers DM, McCullagh PD. Social facilitation of motor performance. *Exercise and Sport Sciences Reviews* 1976; **4**: 125–62.

51. Baumeister RF. Choking under pressure: self-consciousness and paradoxical effects of incentives on skilful performance. *Journal of Personality and Social Psychology* 1984; **46**: 610–20.

52. Schwartz B, Barsky SF. The home advantage. *Social Forces* 1977; **55**: 641–61.

53. Thirer J, Rampey MS. Effects of abusive spectators' behavior on performance of home and visiting inter-collegiate basketball teams. *Perceptual and Motor Skills* 1979; **48**: 1047–54.

54. Davidson RJ, Schwartz GE. The psychobiology of relaxation and related states: a multiprocess theory. In: Mostofsky D., ed. *Behavioural control and modification of physiological activity*. Englewood Cliffs, NJ: Prentice-Hall, 1976: 399–442.

55. Liebert RM, Morris LW. Cognitive and emotional components of test anxiety: a distinction and some initial data. *Psychological Reports* 1967; **20**: 975–8.

56. Martens R, Burton D, Vealey RS, Bump LA, Smith DE. Development and validation of the Competitive State Anxiety Inventory–2. In: Martens R, Vealey RS, Burton D, eds. *Competitive anxiety in sport*. Champaign, Ill: Human Kinetics, 1990: 1117–90.

57. Borkovec TD. Physiological and cognitive processes in the regulation of anxiety. In: Schwartz G, Shapiro D, eds. *Consciousness and self-regulation: advances in research*. Vol. 1. New York: Phelem, 1976: 261–312.

58. Bandura A. Self-efficacy: toward a unifying theory of behavioural change. *Psychological Review* 1977; **84**: 1475–82.

59. Krane V, Williams JM. Performance and somatic anxiety, cognitive anxiety and confidence changes prior to competition. *Journal of Sport Behavior* 1987; **10**: 47–56.

60. Jones JG, Swain A, Cale A. Gender differences in precompetition temporal patterning and antecedents of anxiety and self-confidence. *Journal of Sport and Exercise Psychology* 1991; **13**: 1–15.

61. Parfitt CG, Jones JG, Hardy L. Multidimensional anxiety and performance. In: Jones JG, Hardy L, eds. *Stress and performance in sport*. Chichester: Wiley, 1990: 43–80.

62. Swain A, Jones JG. Intensity, frequency and direction dimensions of competitive state anxiety and self-confidence. *Journal of Sports Sciences* 1990; **8**: 302–3 (abstract).

63. Mahoney MJ, Avener M. Psychology of the elite athlete: an exploratory study. *Cognitive Therapy and Research* 1977; **1**: 135–41.

64. Gould D, Petlichkoff L, Weinberg RS. Antecedents of, temporal changes in, and relationships between CSAI–2 subcomponents. *Journal of Sport Psychology* 1984; **6**: 289–304

65. Scanlan TK. Competitive stress and the child athlete. In: Silva JM, Weinberg RS, eds. *Psychological foundations of sport*. Champaign, Ill: Human Kinetics, 1984:

66. Jones JG, Swain A, Cale A. Antecedents of multidimensional competitive state anxiety and self-confidence in elite intercollegiate middle-distance runners. *The Sport Psychologist* 1990; **4**: 107–18.

67. Jones JG, Hardy L. Stress and cognitive functioning in sport. *Journal of Sports Sciences* 1989; **7**: 41–63.

68. Neiss R. Reconceptualizing arousal; psychobiological states in motor performance. *Psychological Bulletin* 1988; **103**: 345–66.

69. Burton D. Do anxious swimmers swim slower? Reexamining the elusive anxiety–performance relationship. *Journal of Sport and Exercise Psychology* 1988; **10**: 45–61.

70. Gould D, Petlichkoff L, Simons J, Vevera M. Relationship between Competitive State Anxiety Inventory–2 subscale scores and pistol shooting performance. *Journal of Sport Psychology* 1987; **9**: 33–42.

71. Hockey GRJ, Hamilton P. The cognitive patterning of stress states. In: Hockey GRJ, ed. *Stress and fatigue in human performance*. Chichester: Wiley, 1983: 331–62.

72. Parfitt CG, Hardy L. Further evidence for the differential effects of competitive anxiety upon a number of cognitive and motor sub-systems. *Journal of Sports Sciences* 1987; **5**:62–3 (abstract).

73. Apter MJ. *The experience of motivation: the theory of psychological reversals*. London: Academic Press, 1982.

74. Kerr JH. Stress in sport: reversal theory. In: Jones JG, Hardy L, eds. *Stress and performance in sport*. Chichester: Wiley, 1990: 107–31.

75. Sanders A. F. Towards a model of stress and human performance. *Acta Psychologica* 1983; **53**: 64–97.

76. Humphreys MS, Revelle W. Personality, motivation, and performance: a theory of the relationship between individual differences and information processing. *Psychological Review* 1984; **91**: 153–84.

77. Jones JG. A cognitive perspective on the processes underlying the relationship between stress and performance in sport. In: Jones JG, Hardy L, eds. *Stress and performance in sport*. Chichester: Wiley, 1990: 17–42.

78. Hardy L, Fazey JA. The inverted-U hypothesis—a catastrophe for sport psychology and a statement of new hypothesis. Paper presented at the Annual Conference of the North American Society for the Psychology of Sport and Physical Activity, Vancouver, Canada, 1987.

79. Zeeman EC. Catastrophe theory. *Scientific American* 1976; **234**: 65–83.

80. Hardy L, Parfitt CG. A catastrophe model of anxiety and performance. *British Journal of Psychology* 1991; **82**: 163–78.

81. Gould D, Krane V. The arousal–athletic performance relationship: current status and future directions. In: Horn T., ed. *Advances in sport psychology*. Champaign, Ill: Human Kinetics, 1992: 119–141.

82. Vealey RS. Sport personology: a paradigmatic and methodological analysis. *Journal of Sport and Exercise Psychology* 1989; **11**: 216–35.

83. Cattell RB. *Description and measurement of personality*. Yonkers-on-Hudson, NY: World, 1946.

84. Mischel W. *Personality and assessment*. New York: Wiley, 1968.

85. Magnusson D, Endler NS. Interactional psychology: present status and future prospects. In: Magnusson D, Endler NS, eds. *Personality at the cross-roads: current issues in interactional psychology*. Hillsdale, NJ: Erlbaum, 1977: 3–31.

86. Fisher AC. New directions in sport personality research. In: Silva JM, Weinberg RS, eds. *Psychological foundations of sport*. Campaign, Ill: Human Kinetics, 1984: 70–80.

87. Martens R. The paradigmatic crisis in American sport personology. *Sportwissenschaft* 1975; **1**: 9–24.

88. Morgan WP. The trait psychology controversy. *Research Quarterly for Exercise and Sport* 1980; **51**: 50–76.

89. Martens R. *Sport competition anxiety test*. Champaign, Ill: Human Kinetics, 1977.

90. Smith RE, Smoll FL, Schutz RW. Measurement and correlates of sport-specific cognitive and somatic trait anxiety: the Sport Anxiety Scale. *Anxiety Research* 1990; **2**: 225–36.

91. Vealy RS. Sport-confidence and competitive orientations: preliminary investigation and instrument development. *Journal of Sport Psychology* 1986; **8**: 221–46.

92. Gill DL, Deeter TE. Development of the Sport Orientation Questionnaire. *Research Quarterly for Exercise and Sport* 1988; **59**: 191–202.

93. Deaux K. Sex and gender. *Annual Review of Psychology* 1985; **36**: 49–81.

94. Maccoby E, Jacklin C. *The psychology of sex differences*. Dubuque, IA: William C Brown, 1974.

95. Hyde JS. How large are cognitive gender differences? A meta-analysis using w^2 and d. *American Psychologist* 1981; **36**: 892–901.

96. Gill DL. *Psychological dynamics of sport*. Champaign, Ill: Human Kinetics , 1986.

97. Gill DL. Gender differences in competitive orientation and sport participation. *International Journal of Sport Psychology* 1988; **19**: 145–59.

98. Gill DL, Gross JB, Huddleston S, Shifflett B. Sex differences in achievement cognitions and performance in competition. *Research Quarterly for Exercise and Sport* 1984; **55**: 340–6.

99. Lenney E. Women's self-confidence in achievement settings. *Psychological Bulletin* 1977; **84**: 1–13.

100. Harris DV. Femininity and athleticism: conflict or consonance. In Sabo, Runfola, Jock, eds. *Sports and Male Identity*. Englewood Cliffs, New Jersey: Prentice-Hall 1980; 222–39.

101. Theberge N. (1987). Sport and women's empowerment. *Women's Studies International Forum*, 1987; **10**: 387–93.

102. Bem SL. The measurement of psychological androgyny. *Journal of Consulting and Clinical Psychology* 1974; **42**: 155–62.

103. Spence JT, Helmreich RL, Stapp J. The Personality Attributes Questionnaire: a measure of sex role stereotypes and masculinity-feminity. *JSAJ Catalogue of Selected documents in Psychology* 1974; **4**: 127.

104. Nideffer RM. Test of attentional and interpersonal style. *Journal of Personality and Society Psychology* 1976; **34**: 394–404.

105. Van Schoyk SR, Grasha AF. Attentional style variations and athletic ability; the advantages of a sports-specific test. *Journal of Sport Psychology* 1981, 3: 149–65.

106. Albrecht RR, Feltz DL. Generality and specificity of attention related to competitive anxiety and sport performance. *Journal of Sport Psychology* 1987; **9**: 231–48.

107. Landers DM. Beyond the TAIS: alternative behavioral and psychophysiological measures for determining an internal vs external focus of attention. Paper presented at the NASPSPA Conference, Gulfpark, MS, 1975.

108. Boutcher SH, Zinsser NW. Cardiac deceleration of elite and beginning golfers during putting. *Journal of Sport and Exercise Psychology* 1990; **12**: 37–47.

109. Boutcher SH. The role of performance routines in sport. In: Jones J. G, Hardy L, eds. *Stress and performance in sport*. Chichester: Wiley, 1990: 231–45.

110. Allard F, Graham S, Paarsalu MT. Perception in sport: basketball. *Journal of Sport Psychology* 1980; **2**: 14–21.

111. Morgan WP, Pollock ML. Psychologic characterization of the elite distance runner. *Annals of the New York Academy of Sciences* 1977; **301**: 382–403.

112. Gill DL, Strom EH. The effect of attentional focus on performance of an endurance task. *International Journal of Sport Psychology* 1985; **16**: 217–23.

113. Fenz W. Coping mechanisms and performance stress. In: Landes D. M, ed. *Psychology of sport and motor behavior*. II. University Park, PA: Pennsylvania State University, 1975: 3–24.

114. Orlick T, Partington J. Mental links to excellence. *Sport Psychologist* 1988; **2**: 105–30.

115. Mahoney MJ, Gabriel TJ, Perkins TS. Psychological skills and exceptional athletic performance. *Sport Psychologist* 1987; **1**: 181–99.

116. Starkes JL. Skill in field hockey: the nature of the cognitive advantage. *Journal of Sport Psychology* 1987; **9**:146–60.

117. Abernethy B, Russell DG. Expert–novice differences in an applied selective attention task. *Journal of Sport Psychology* 1987; **9**: 326–45.

118. Smoll FL, Magill RA, Ash MJ, eds. *Children in sport*. 3rd edn. Champaign, Ill: Human Kinetics, 1988.

119. Weiss MR, Friedrichs WD. The influence of leader behaviours, coach attributes, and institutional variables on performance and satisfaction of collegiate basketball teams. *Journal of Sport Psychology* 1986; **8**: 332–46.

120. Chelladurai P. Leadership in sports. In: Silva JM, Weinberg RS, eds. *Psychological foundations of sport*. Champaign, Ill: Human Kinetics, 1984: 329–39.

121. Chelladurai P, Saleh SD. Dimensions of leader behavior in sports: development of a leadership scale. *Journal of Sport Psychology* 1980; **2**: 34–45.

122. Chelladurai P, Saleh SD. Preferred leadership in sports. *Canadian Journal of Applied Sport Science* 1978; **3**: 85–92.

123. Chelladurai P, Carron AV. Athletic maturity and preferred leadership. *Journal of Sport Psychology* 1983; **5**: 371–80.

124. Chelladurai P. Discrepancy between preferences and perceptions of leadership behavior and satisfaction of athletes in varying sports. *Journal of Sport Psychology* 1984; **6**: 27–41.

125. Chelladurai P, Imamura H, Yamaguchi Y, Oinuma Y, Miyauchi T. Sport leadership in a cross-national setting: The case of Japanese and Canadian university athletes. *Journal of Sport and Exercise Psychology* 1988; **10**: 374–89.

126. Tharp RG, Gallimore R. What a coach can teach a teacher. *Psychology Today* 1976; **9**: 74–8.

127. Smoll FL, Smith RE. Leadership research in youth sports. In: Silva J M, Weinberg RS, eds. *Psychological foundations of sport*. Champaign, Ill: Human Kinetics, 1984: 371–86.

128. Hendry LB. Human factors in sport systems: Suggested models for analysing athlete-coach interaction. *Human Factors* 1974; **16**: 528–44.

129. Vroom VH, Yetton RN. *Leadership and decision-making*. Pittsburgh: University of Pittsburgh Press, 1973.

130. Chelladurai P, Haggerty TR. A normative model of decision styles in coaching. *Athletic Administrator* 1978; **13**: 6–9.

131. Chelladurai P, Arnott M. Decision styles in coaching: Preferences of basketball players. *Research Quarterly for Exercise and Sport* 1985; **56**: 15–24.

132. Chelladurai P, Haggerty TR, Baxter PR. Decision style choices of university basketball coaches and players. *Journal of Sport and Exercise Psychology* 1989; **11**: 201–15.

133. Rosenbaum LL, Rosenbaum WB. Morale and productivity consequences of group leadership style, stress, and type of task. *Journal of Applied Psychology* 1971; **55**, 343–88.

134. Zander A. *Motives and goals in groups*. New York: Academic Press, 1971.

135. Latane B, Williams KD, Harkins SG. Many hands make light work: the causes and consequences of social loafing. *Journal of Personality and Social Psychology* 1979; **37**: 823–32.

136. Hardy CJ, Latane B. Social loafing in cheerleaders: effects of team membership and competition. *Journal of Sport and Exercise Psychology* 1988; **10**: 109–14.

137. Carron AV. Cohesiveness in sport groups: interpretations and considerations. *Journal of Sport Psychology* 1982; **4**: 123–38.

138. Martens R, Landers DM, Loy JW. *Sport cohesiveness questionnaire*. Unpublished Report, University of Illinois at Urbana-Champaign.

139. Yukelson D, Weinberg R, Jackson A. A multidimensional sport cohesion instrument for intercollegiate basketball players. *Journal of Sport Psychology* 1984; **6**: 103–17.

140. Carron AV, Widmeyer WN, Brawley LR. Group cohesion and individual adherence to physical activity. *Journal of Sport and Exercise Psychology* 1988; **10**: 127–38.

141. Brawley LR, Carron AV, Widmeyer WN. Exploring the relationship between cohesion and group resistance to disruption. *Journal of Sport and Exercise Psychology* 1988; **10**: 199–123.

142. Spink KS. Group cohesion and collective efficacy of volleyball teams. *Journal of Sport and Exercise Psychology* 1990; **12**: 301–11.

143. Widmeyer WN, Brawley LR, Carron AV. The effects of group size in sport. *Journal of Sport and Exercise Psychology* 1990; **212**:177–90

144. Silva JM. Understanding aggressive behavior and its effects upon athlete performance. In: Straub WF, ed. *Sport psychology: an analysis of athlete behavior*. Ithaca, NY: Mouvement,1980: 177–86.

145. Leith LM. The effect of various physical activities, outcome, and emotional arousal on subject aggression scores. *International Journal of Sport Psychology* 1989; **20**: 57–66.

146. Nosanchuk TA. The way of the warrior: the effects of traditional martial arts training on aggressiveness. *Human Relations* 1981; **34**: 435–44.

147. Berkowitz L. Some determinants of impulsive aggression; role of mediated association with reinforcements for aggression. *Psychological Review* 1974; **81**: 165–76.

148. Frank MG, Gilovich T. The dark side of self- and social perception: black uniforms and aggression in professional sports. *Journal of Personality and Social Psychology* 1988; **54**: 74–85.

149. Bredemeier BJ. Moral reasoning and the perceived legitimacy of intentionally injurious acts. *Journal of Sport Psychology* 1985; **7**: 110–24.

150. Ryan MK, Williams JM, Wimer B. Athletic aggression: perceived legitimacy and behavioral intentions in girls' high school basketball. *Journal of Sport and Exercise Psychology* 1990; **12**: 48–55.

151. Silva JM. The perceived legitimacy of rule violating behavior in sport. *Journal of Sport Psychology* 1983; **5**: 438–8.

152. Frodi A, Macauley J, Thome PR. Are women always less aggressive than men? A review of the experimental literature. *Psychological Bulletin* 1977; **84**: 638–60.

153. McCarthy JF, Kelly BR. Aggression, performance variables, and anger self-report in ice-hockey players. *Journal of Psychology* 1978; **99**: 97–101.

154. Silva JM. Behavioral and situational factors affecting concentration and skill performance. *Journal of Sport Psychology* 1979; **1**: 221–27.

155. Bredemeier BJ, Weiss MR, Shields DL, Shewchuk RM. Promoting

amoral growth in a summer sport camp: the implications of theoretically grounded instructional strategies. *Journal of Moral Education* 1986; **15**: 212–20.

156. McGrath JE. *Groups: interaction and performance.* Englewood Cliffs, NJ: Prentice-Hall, 1984.

157. Biddle SJH, Fox KR. Exercise and health psychology: emerging relationships. *British Journal of Medical Psychology* 1989; **62**: 205–16.

158. Morgan WP, Goldston SE (ed.) *Exercise and mental health.* Washington, DC: Hemisphere, 1987.

159. Taylor CB, Sallis JF, Needle R. The relation of physical activity and exercise to mental health. *Public Health Reports* 1985; **100**: 195–202.

160. Berger B, Owen D. Stress reduction and mood enhancement in four exercise modes: swimming, body conditioning, hatha yoga, and fencing. *Research Quarterly for Exercise and Sport* 1988; **59**: 148–59.

161. Raglin JS, Morgan WP. Influence of exercise and quiet rest on state anxiety and blood pressure. *Medicine and Science in Sports and Exercise* 1987; **19**: 456–63.

162. Greist JH. Exercise intervention with depressed patients. In: Morgan WP, Goldston SE, eds. *Exercise and mental health.* Washington, DC: Hemishphere, 1987: 117–21.

163. Gruber JJ. Physical activity and self-esteem development in children: a meta-analysis. Instull G, Eckert H, eds. *Effects of physical activity on children.* Champaign, Ill: Human Kinetics, 1986.

164. Crews DJ, Landers DM. A meta-analytic review of aerobic fitness and reactivity to psychosocial stressors. *Medicine and Science in Sports and Exercise* 1987; **19** (5, supplement): S114–20.

165. Harris DV. Comparative effectiveness of running therapy and psychotherapy. In: Morgan WP, Goldston SE, eds. *Exercise and mental health.* Washington, DC: Hemisphere, 1987: 123–30.

166. Berger B. Stress levels in swimmers. In: Morgan WP, Goldston SE, eds. *Exercise and mental health.* Washington, DC: Hemisphere, 1987: 139–43.

167. de Vries HA. Tension reduction with exercise. In: Morgan WP, Goldston SE, eds. *Exercise and mental health.* Washington, DC: Hemisphere, 1987: 99–104.

168. Hatfield BD, Landers DM. Psychophysiology in exercise and sport research: an overview. *Exercise and Sport Sciences Reviews* 1987; **15**: 351–87.

169. Harber VJ, Sutton JR. Endorphins and exercise. *Sports Medicine,* 1984; **1**: 154–71.

170. Sonstroem RJ. Exercise and self-esteem. *Exercise and Sport Sciences Reviews* 1984; **12**: 123–55.

171. Bahrke MS, Morgan WP. Anxiety reduction following exercise and meditation. *Cognitive Therapy and Research* 1978; **2**: 323–33.

172. Berger B, Friedman E, Eaton M. Comparison of jogging, the relaxation response, and group interaction for stress reduction. *Journal of Sport and Exercise Psychology* 1988; **10**: 431–47.

173. Long BC, Haney CJ. Long-term follow-up of stressed working men: a comparison of aerobic exercise and progressive relaxation. *Journal of Sport and Exercise Psychology* 1988; **4**: 461–70.

174. Biddle SJH. Introduction. In: *Sport, Health, Psychology and Exercise Symposium.* London: Sports Council/Health Education Authority, 1990: 5–7.

175. Stephens T. Physical activity and mental health in the United States and Canada: evidence from four population surveys. *Preventive Medicine* 1988; **17**: 35–47.

176. Dishman RK. Medical psychology in exercise and sport. *Medical Clinics of North America* 1985; **69**: 123–43.

177. Dishman RK. Mental health. In: Seefeldt V, ed. *Physical activity and well-being.* Reston, VA: AAHPERD, 1986.

178. Steinberg H, Sykes EA, Morris M. Exercise addiction: the opiate connection. In: *Sport, Health, Pyschology and Exercise Symposium.* London: Sports Council/ Health Education Authority, 1990: 161–6.

179. Morris M, Steinberg H, Sykes EA, Salmon P. Temporary deprivation from running produces 'withdrawal' syndrome. In: *Sport, Health, Psychology and Exercise Symposium.* London: Sports Council/Health Education Authority, 1990: 161–71.

180. Sheehan G. The best therapy. *Physician and Sports Medicine* 1983; **11**: 43.

181. Sports Council. *Sport in the community: the next ten years.* London: Sports Council, 1982.

182. Dishman RK. Compliance/adherence in health-related exercise. *Health Psychology* 1982; **1**: 237–67.

183. Morgan WP. Involvement in vigorous physical activity with special reference to adherence. *Proceedings of the NCPEAM/NAPECW National Conference.* 1977: 235–46.

184. Dishman RK, Gettman LR. Psychobiologic influences on exercise adherence. *Journal of Sport Psychology* 1980; **2**: 295–310.

185. Janz NK, Becker MH. The Health Belief Model: a decade later. *Health Education Quarterly* 1984; **11**: 1–47.

186. Fishbein M, Ajzen I. *Belief, attitude, intention and behavior: an introduction to theory and research.* Reading, Ma: Addison-Wesley, 1975.

187. Sonstroem RJ. Psychological models. In: Dishman RK, eds. *Exercise adherence: its impact on public health.* Champaign, Ill: Human Kinetics, 1988: 125–153.

188. Bandura A. *Social foundations of thought and action.* Englewood Cliffs, NJ: Prentice-Hall, 1986.

189. Dzewaltowski DA. Toward a model of exercise motivation. *Journal of Sport and Exercise Psychology* 1989; **11**: 251–69.

190. Dzewaltowski DA, Noble JM, Shaw JM. Physical activity participation: social cognitive theory versus the theories of reasoned action and planned behaviour. *Journal of Sport and Exercise Psychology* 1990; **12**: 388–405.

191. Kleinginna PR, Kleinginna AM. A categorized list of motivation definitions with a suggestion for a consensual definition. *Motivation and Emotion* 1981; **5**: 263–91.

192. Duda JL. Goal perspectives and behavior in sport and exercise settings. In: Ames C, Maehr M, eds. *Advances in motivation and achievement.* Vol. 6. Greenwich, CT: JAI Press, in press.

193. Oldridge NB. Compliance with exercise in cardiac rehabilitation. In: Dishman RK, ed. *Exercise adherence: its impact on public health.* Champaign, Ill: Human Kinetics, 1988: 283–304.

194. Shephard RJ. Exercise adherence in corporate settings: personal traits and program barriers. In: Dishman RK, eds. *Exercise adherence: its impact on public health.* Champaign, Ill: Human Kinetic, 1988: 305–19.

195. Corbin CB. Youth fitness, exercise and health: there is much to be done. *Research Quarterly for Exercise and Sport* 1987; **58**: 308–14.

196. Ravizza K. Peak experience in sport. *Journal of Humanistic Psychology* 1977; **17**: 35–40.

197. Vealey RS. Future directions in psychological skills training. *Sport psychologist* 1988; **2**: 318–36.

198. Locke EA, Shaw KN, Saari LM, Latham GP. Goal setting and task performance: 1969–1980. *Psychological Bulletin* 1981; **90**: 125–52.

199. Locke EA, Latham GP. The application of goal setting to sports. *Journal of Sport Psychology* 1985; **7**: 205–22.

200. Beggs WDA. Goal setting in sport. In: Jones JG, Hardy L, eds. *Stress and performance in sport.* Chichester: Wiley, 1990: 135–70.

201. Hardy L, Nelson D. Self-regulation training in sport and work. *Ergonomics* 1988; **31**: 1673–83.

202. Hardy L, Fazey JA. *Mental training.* Leeds: National Coaching Foundation, 1990.

203. Feltz DL, Landers DM. The effects of mental practice on motor skill learning and performance: a meta-analysis. *Journal of Sport Psychology* 1983; **5**: 25–57.

204. Suinn RM. Imagery and sports. In: Straub WF, Williams JM, eds. *Cognitive sport psychology.* Lansing, NY: Sports Science Associates, 1984: 235–72.

205. Ainscoe M, Hardy L. Cognitive warm-up in a cyclical gymnastics skill. *International Journal of Sport Psychology,* 1987; **18**: 269–75.

206. Weinberg RS, Gould D, Jackson A. Expectations and performance: an empirical test of Bandura's self-efficacy theory. *Journal of Sport Psychology* 1979; **1**: 320–31.

207. Murphy SM. Models of imagery in sport psychology. Unpublished manuscript. USOC Olympic Training Center, Colorado, 1987.

208. Pylyshyn Z. The imagery debate: analog media versus tacit knowledge. In: Block N, ed. *Imagery*. Cambridge, Ma: MIT Press; 1981: 151–205.

209. Lang PJ. A bio-informational theory of emotional imagery. *Psychophysiology* 1979; **17**: 179–92.

210. Ahsen A. ISM: the triple code model for imagery and psychophysiology. *Journal of Mental Imagery* 1984; **8**: 15–42.

211. Jacobson E. *Progressive relaxation*. University of Chicago press, 1930.

212. Bernstein DA, Borkovec TD. *Progressive relaxation: a manual for the helping professions*. Champaign, Ill Research Press, 1973.

213. Ost LG. Applied relaxation: description of an effective coping technique. *Scandinavian Journal of Behaviour Therapy* 1988; **17**: 83–96.

214. Wolpe J. *Psychotherapy by reciprocal inhibition*. Stanford, CA: Standford University Press, 1958.

215. Meichenbaum DH. *Cognitive-behaviour modification*. New York: Plenum Press, 1977.

216. Smith RE. A cognitive-affective approach to stress management training for athletes. In: Nadeau CH, Halliwell WR, Newell KM, Roberts GC, eds. *Psychology of motor behavior and sport—1979*. Champaign, Ill: Human Kinetics, 1980: 54–72.

217. Unestahl LE. *Inner mental training*. Orebro, Sweden: Veje 1983.

218. Cooke LE, Alderson GJK. *Stress and anxiety in sport*. Sheffield: Pavic, 1986.

219. Hamberger K, Lohr J. Relationship of relaxation training to the controllability of imagery. *Perceptual and Motor Skills* 1980; **51**: 103–10.

220. Singer JL. *Imagery and daydream methods in psychotherapy and behavior modification*. New York: Academic Press, 1974.

221. Weinberg R, Seabourne T, Jackson. Arousal and relaxation instructions prior to the use of imagery. *International Journal of Sport Psychology* 1987; **18**: 205–14.

222. Klinger E, Bart SJ, Glas RA. Thought content and gap time in basketball. *Cognitive Therapy and Research* 1981; **5**: 109–14.

223. Rotella RJ, Gansneder B, Ojala D, Billing J. Cognitions and coping strategies of elite skiers: an exploratory study of young developing athletes. *Journal of Sport Psychology* 1980; **4**: 350–4.

224. Ellis A. Self-direction in sport and life. *Rational Living* 1982; **17**: 27–33.

225. Rushall BS. The content of competition thinking. In: Straub WF, Williams JM, eds. *Cognitive sport psychology*. Lansing, NY: Sports Science Associates, 1984: 51–62.

226. Meyers AW, Schleser RA. A cognitive behavioral intervention for improving basketball performance. *Journal of Sport Psychology* 1980; **2**: 69–73.

227. Silva JM. Performance enhancement in sport environments through cognitive intervention. *Behaviour Modification* 1982; **6**:433–63.

228. Schmid A, Peper E. Techniques for training concentration. In: Williams JM, ed. *Applied sport psychology: personal growth to peak performance*. Palo Alto, Ca: Mayfield, 1986: 271–84.

229. Schwarz EE, Davidson RJ, Goleman DJ. Patterning of cognitive and somatic processes in the self-regulation of anxiety: effects of medication versus exercise. *Psychosomatic Medicine* 1978; **40**: 321–8.

230. Pribram KH, McGuinness D. Arousal, activation and effort in the control of attention. *Psychological Review* 1975; **82**: 116–49.

231. Shelton TO, Mahoney MJ. The content and effect of 'psyching-up' strategies in weight lifters. *Cognitive Therapy and Research* 1978; **2**: 275–84.

232. Mahoney MJ. Cognitive skills and athletic performance. In: Kendall PC, Hollon SD, eds. *Cognitive behavioral interventions*. New York: Academic Press, 1979: 423–43.

233. Locke EA, Frederick E, Bobko P, Lee C. Effect of self-efficacy, goals, and strategies on task performance. *Journal of Applied Psychology* 1984; **69**: 241–51.

234. Forsterling F. Attributional retraining: a review. *Psychological Bulletin* 1985; **98**: 495–512.

235. Harris DV, Harris BL. *The athlete's guide to sports psychology: mental skills for physical people*. New York: Leisure Press, 1984.

236. Eysenck MW. *Attention and arousal: cognition and performance*. Berlin: Springer-Verlag, 1982.

237. Wachtel PL. Conceptions of broad and narrow attention. *Psychological Bulletin* 1967; **68**: 417–29.

238. Morgan WP, Horstman DH, Cymerman A, Stokes J. Facilitation of physical performance by a cognitive strategy. *Cognitive Therapy and Research* 1983; 7: 251–64.

239. Schomer HH. Mental strategy training programme for marathon runners. *International Journal of Sport Psychology* 1987; **18**: 133–51.

240. Orlick T. *Psyching for sport*. Champaign, Ill: Human Kinetics, 1986.

241. Bull SJ. Personal and situational influences on adherence to mental skills training. *Journal of Sport and Exercise Psychology* 1991; **13**: 121–32.

242. Zeigler EF. Rationale and suggested dimensions for a code of ethics for sport psychologists. *Sport Psychologist* 1987; **1**: 138–50.

243. Nideffer RM. *The ethics and practice of applied sport psychology*. Ithaca, NY: Mouvement, 1981.

244. Danish SJ, Hale BD. Toward an understanding of the practice of sport psychology. *Journal of Sport Psychology* 1981; 3: 90–9.

245. Brown JM. Are sport psychologists really psychologists? *Journal of Sport Psychology* 1982; **4**: 13–18.

246. Nideffer RM. Current concerns in sport psychology. In: Silva J. M., Weinberg RS, eds. *Psychological foundations of sport*. Champaign, Ill: Human Kinetics, 1984: 35–44.

247. Heyman SR. The development of models for sport psychology: examining the USOC guidelines. *Journal of Sport Psychology* 1984; **6**: 125–32.

248. Clarke KS. The USOC sport psychology registry: a clarification. *Journal of Sport Psychology* 1984; **6**: 365–6.

249. Smith RE, Johnson J. An organizational empowerment approach to consultation in professional baseball. *Sport Psychologist* 1990; **4**: 347–57.

250. Boutcher SH, Rotella RJ. A psychological skills educational program for closed-skill performance enhancement. *Sport Psychologist* 1987; **1**: 127–37.

251. Partington J, Orlick T. The sport psychology consultant evaluation form. *Sport Psychologist* 1987; **1**: 309–17.

1.4.2 The acquisition of motor skills

JOHN ANNETT

INTRODUCTION

Skill (or a skill) is behaviour which is (a) purposeful or goal directed, (b) well organized and economical of effort, and (c) acquired through training and practice. The term 'skill' is normally applied to physical activities which are carried out easily and efficiently, but the meaning has been extended to refer to intellectual or 'cognitive' skills[1] and even social skills.[2] Particularly when referring to motor activity, skill may suggest rigid or automatic behaviour which has been practised until perfect as opposed to behaviour which is thoughtful or considered. As we shall see, skills can be both fluent and flexible, and there is a useful sense in which a skill can be seen as a solution to a problem. At first, the novice may wish to achieve a goal but has no appropriate behaviour in the repertoire with which to attain this goal. Riding a bicycle without falling off is a problem if you have never done it before! The idea of skill as a solution to a problem then directs our attention to the analysis of the problem. What is the goal? What kind of obstacles must be overcome? How this can be achieved? We should not, therefore, think of skill acquisition as simply a matter of honing existing behaviour to a higher level of efficiency; we should enquire about the nature of the processes underlying performance and what happens to them in the course of learning.

Let us take the case of bicycle riding as an example. The 'problem' has been described in terms of the physical dynamics of the rider–machine–terrain system.[3] The sensory information available to the rider and a description of the outputs necessary to keep the system within certain goal tolerances, for example upright, moving forward, avoiding obstacles, and so on, define the nature of the problem. Control is dependent on three nested feedback loops which provide information on lateral displacement, heading change, and roll rate. The key problem for the novice cyclist is that the physical dynamics of the machine lead to instability (a high roll rate) when the handlebars are turned to change direction. This is also strongly affected by the forward velocity, but novices typically travel slowly, when instability is higher than at faster speeds, and are anxious about avoiding obstacles. Applying the wrong correction to the handlebar can introduce instability, especially at low speeds, and so the initial stages of learning can be difficult.

The skilled cyclist has learned just how much pressure to apply and how long to apply it to the handlebar to produce an appropriate turn without falling. Perhaps this is best learned by practising in an area free of obstacles and by inducing a sufficiently high forward velocity to increase the natural stability of the machine. When this basic skill of controlling the roll rate has been acquired, the learner can devote more attention to the skills of controlling direction, some of which are probably already in the repertoire having been learned, at least in part, through the normal processes of navigating through foot traffic.

ANALYSING SPORTS SKILLS

Sports skills present a particularly wide range of problems to be solved, with or without the help of a coach. Some, like gymnastics, are problems of generating unusual patterns of body movement, others, like archery, demand fine control, and others, like football, require motor control, perceptual anticipation, and an understanding of strategy. Analysing the nature of each kind of sport skill is the first step in planning an optimal route to skill acquisition, and it is here that the knowledge and experience of the expert coach can be combined with the scientific results from the laboratory even though these may derive from 'artificial' and often quite simple tasks. Bridging the gap between science and practice therefore requires a taxonomy of skills which will help in identifying which scientific theories and results may be applied to which sports.

Knapp[4] distinguished between 'open' and 'closed' skills in sport. The distinction is based on the role of feedback information in performance. 'Open' skills are open-loop, i.e. ballistic. The key movements are preplanned and run off with little or no control from environmental stimuli. 'Closed' skills, however, involve movements which are adjusted to ongoing feedback from the environment and therefore represent 'closed-loop' control. A tennis serve exemplifies an 'open' skill, whilst positioning the racket to return service is very much a 'closed skill'. The significance of this theoretical distinction is that open skills require the kind of

training and practice which will create a reliably accurate movement pattern or 'motor programme'. The movement pattern of a closed skill, however, may be almost infinitely varied but linked to the reception of sensory cues which, if correctly predicted, enable the player to anticipate events, thus overcoming the temporal lag inherent in all feedback-controlled systems.

Holding[5] refers to open skills as 'perceptual' and 'controlled' and to closed skills as 'automatic' and 'motor', and adds another dimension—simplicity–complexity or gross and discrete versus fine and continuous. Morris and Annett[6] attempted an empirical classification of sports skills based on the answers of 104 physical education staff and students to some 35 questions about the skill components of 14 representative sports. In addition to rating the principal skills in a sport on the open–closed dimension using a five-point scale, questions referred to the state of the body and any relevant object (such as a ball) prior to movement, the type of movement principally involved, the main sources of feedback, the principal cognitive processes, the main objectives or criteria for success, and the knowledge required. A hierarchical cluster analysis of the data produced the results shown in Fig. 1. The numbers in the diagram are χ^2 values indicating cluster 'distance'. Given the limited size of the sample of respondents and of sports, it is interesting to note that the major division seems to be between open and closed skills. Other interesting distinctions also emerge, such as the separation of swimming from other whole-body sports and the grouping of hockey with 'implement controlling' games rather than with team games.

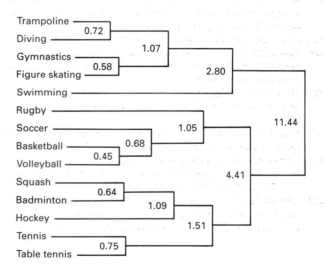

Fig. 1 Cluster analysis of 14 common sports. The numbers are values of χ^2 taken as indicating 'cluster distance' based on clustering of responses of 104 physical education experts to a 35-item questionnaire.

The relationships between items were examined by factor analysis and six factors were extracted. Two factors which together accounted for 38.5 per cent of the variance loads on items which involve processing sensory information and knowing about the behaviour of objects; two more factors (13.1 per cent of the variance) are concerned with gross motor control and interpretation of the behaviour of other players, and two further factors represent manipulative skills and the physical attributes of the player. The seventh factor, which accounts for less than 3 per cent of the variance, represents an assortment of minor features.

The detailed description and classification of sport skills is still in its infancy and is something that the coach in each particular sport probably does using a combination of experience and intuition. Nevertheless, the process of analysing the skill to be acquired is the key to successful training. Analysis enables the coach to specify effectively what aspects of the skill problem the learner still has to solve and to provide various kinds of assistance. The kinds of problems faced may be quite varied. In some cases, the learner who is having difficulty in performing a task simply needs information about the outcome of a particular response pattern or knowledge of a rule, such as what to do under particular circumstances. In other cases it may be helpful to draw attention to particular sensory cues whose significance is not obvious to the novice. In still other cases, when speed or precision is needed, the training solution may be more practice of one particular task component. It is inherent in task analysis that, in practice, the identification of learning problems is the shortest route to identifying the type of training required.

THEORIES OF SKILL ACQUISITION

Research into the processes of skill acquisition recognizes two fundamental paradigms, practice and instruction. In practice experiments the learner is active and makes repeated attempts to perform the task. Under instruction, in contrast, the experimenter (or coach) provides verbal instruction, text, illustrations, models and simulations, advice, and correction, while the learner remains essentially passive, at least whilst receiving instruction. Practice is readily quantified in terms of number of discrete trials or amount of time spent, at least if the central features of the task remain constant. Tracking tasks (in which the learner attempts to align a cursor with a moving target or keep a moving indicator at a constant reading) and linear positioning tasks (in which the subject learns to make a discrete movement of a specific extent) have been the most popular experimental situations for studying the effects of practice. However, almost any task in which uniform responses are required can serve to show how a single feature of performance, such as speed or accuracy, changes as a function of the number of trials. Instruction, however, does not provide such a simple experimental paradigm. Information presented in text material, verbal instruction, or a demonstration is not easily quantified and is at best treated as a binary (all or none) variable. For example, in an early experiment, Judd[7] investigated the effect of instruction in the principles of refraction on the transfer of a dart-throwing task between targets seen through different depths of water. One group of subjects receiving instruction did not learn the initial adjustment any faster than those who received no instruction; however, these subjects were better able to learn subsequently to hit a target at a different depth. The problem with this sort of experiment is that there is no independent measure of the quality of the instruction and hence it is not easy to generalize from a single set of results. Perhaps some other instructor could explain the principles of refraction more simply or perhaps another might not do it so well. Thus we cannot conclude simply from Judd's experiment that instruction will always improve transfer but not acquisition.

The distinction between practice and instruction also reflects the assumption that quite different processes may be responsible for learning. Practice provides the opportunity for a uniform, slow, incremental, and essentially automatic learning process,

whilst instruction achieves its effects through cognitive processes which may include rapid changes in knowledge of relevant information, perceptual organization, or response strategy. One of the fundamental research issues is whether separate theories of acquisition are needed to account for the culturally transmitted and vicariously acquired aspects of skill on the one hand and the personal effects of individual practice on the other. Fitts[8] characterized skill acquisition as a progressive shift of the control of performance from cognitive to non-cognitive processes. In his three-phase theory of skill acquisition the initial cognitive phase is dominated by learning rules and procedures, and other items of factual knowledge are acquired by means of instruction or trial and error. The second and third stages are dominated by practice, during which stimuli become connected with responses (the associative phase) and performance becomes increasingly independent of cognitive control (the autonomous phase). The Fitts sequence, which has been echoed by many writers,[9–12] implies that not only are different training techniques appropriate at different stages in learning, but also that different processes underlie the learning which does occur. Fitts did not claim that sharp distinctions could always be made between these three phases, but for the purposes of this review the main topics will be discussed in the order suggested by the Fitts sequence.

COGNITIVE PROCESSES

The first or 'cognitive' phase, which is dominated by verbal instruction and demonstration, is seen 'as a first step in the development of an executive program' (ref. 13, p. 12). Behavioural elements which are already in the learner's repertoire are selected and rearranged, and other changes may occur, for example changes in attention, particularly focusing on relevant cues. Items of factual information relevant to the task may also be learned. The two principal classes of cognitive methods are verbal instruction and demonstration. The central theoretical problem is how information received 'passively' by these two methods is translated into the capability for action. It is primarily this issue which divides 'cognitive' from 'behaviourist' theories of learning; however, as Adams[14] pointed out, theory in this aspect of motor learning is somewhat underdeveloped. Figure 2 represents an attempt

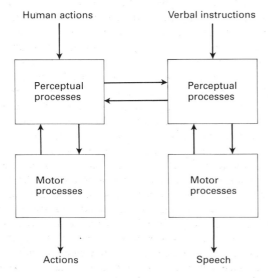

Fig. 2 Hypothetical relationships between motor and verbal systems. See text for explanation.

to formulate the problem of the relationships between cognitive and non-cognitive process in a way that suggests lines of empirical research. The top of the diagram represents two classes of inputs—words and actions. (Other classes of inputs such as stimuli arising from non-human sources are not shown.) The central part of the diagram represents internal processes and the bottom represents the output, either words or actions. The left-hand side of the diagram represents the non-verbal domain of actions, whilst the right-hand side represents words or the verbal domain.

The central part of Fig. 2 is divided into four areas, the top pair representing receptive and interpretive processes and the bottom pair representing productive processes. They are represented separately in recognition of the fact that production can be inhibited; however, as we shall see, a fairly intimate relationship between receptive and productive processes can be assumed.

A number of different experimental paradigms are represented as routes through the system indicated by arrows. Most, but not all, of these are from the top down in the conventional direction of perception → action. Some routes proceed from the top straight down, with the output mode matching the input mode (i.e. actions are imitated and words are repeated), whilst some cross over between the action and verbal systems. In particular, verbal instructions can be translated into actions, a familiar instructional paradigm, but perceived actions can also be translated into words. The latter route corresponds to the less familiar task (at least as far as experimental research is concerned) of giving a verbal account of actions. The most obvious case is the radio sports commentator, but generally we are talking about the production of eyewitness accounts.

Verbal instruction

The model makes two important assumptions: (a) action and language systems can be regarded as separate; (b) representational and productive systems, whilst separable, are closely interconnected. The first assumption is justified in terms of neurological evidence that the ability to learn and perform skilled actions is to a large extent independent of verbal learning and memory. The evidence has been reviewed in detail elsewhere,[15–17] but the most striking examples are found in cases of ideomotor apraxia such as the patient described by Geschwind and Kaplan.[18] This patient was able to follow verbal instructions, such as 'show me how you use a hammer', with the right hand but not with the left, although the latter was not paralysed. Furthermore, when a hammer was placed in his left hand the subject was able to demonstrate its use. On postmortem examination this patient was found to have extensive damage to the corpus collosum such that verbal instructions, which are processed in the left hemisphere, could not be passed to the right hemisphere, which controls the left hand. Translation between verbal and non-verbal codes is undoubtedly more complex than making a simple anatomical connection, but, as this example shows, traffic between the two systems across what I have called the action–language bridge is worthy of investigation. A number of studies[12,16–22] have confirmed that it is often difficult to explain in words how a skilled action is performed. In the case of a familiar skill, such as tying a bow, it is clear that the translation from actions to words is normally mediated by a process of self-observation. Subjects typically rely on 'going through the motions' either overtly by making gestures or by generating images which they are then able to describe. There are also clear limitations on what the subjects are able to describe. Bow tying, for example, is typically described in terms of actions or outcomes, such as making a loop, rather than as the pattern of finger and hand move-

ments which are actually employed. In other words, the representations stored in memory generally do not provide kinematic detail but are expressed in terms of objectives to be achieved, for instance the knot is pulled tight. Looking at the action–language bridge from the opposite view, namely that of turning verbal instructions into actions, it appears that again kinematic detail is not the most successful form of communication about actions. Coaches typically use metaphor and imagery to convey complex movement information. For example, a squash coach of the author's acquaintance describes the stance to be adopted to receive service as 'like a Red Indian on the warpath'. This produces a comprehensive image of an alert stance with feet apart and knees bent, and the right hand raised holding the racket at about shoulder height.

In learning a complex skill a great deal of factual information is also acquired. Sport skills, for example, require knowledge of the rules, and also knowledge of techniques and strategies and about the behaviour of bodies and motion and much more besides. Skilled games players typically exhibit a rich factual database of 'declarative' information;[23] also, French and Thomas[24] have demonstrated that as children gain proficiency in a sport skill so their factual knowledge of the game increases. Of course it does not necessarily follow that simply because theoretical instruction often precedes practice that learning is the conversion of declarative knowledge into procedural knowledge (as proposed by Anderson[11]). The ability to perform a skill and the verbal knowledge relating to it may develop in parallel. It may even be the case that some kinds of verbally accessible knowledge only emerge after the skilled performer has the leisure to reflect on his own or other people's performance. As noted earlier, some aspects of motor skill may never have been and may never be translated into declarative knowledge.

The model in Fig. 2 also distinguishes between representational and production processes in both verbal and non-verbal modalities. The representational processes are involved in perceiving and imagining. Thus a pair of mechanisms (one verbal and none non-verbal) form representations of incoming data, speech, and actions respectively. The production processes are involved in the execution of speech and motor acts, but they can be inhibited when engaging in inner speech or imaginary action. This close association between representation and production is a feature of some recent theories of speech perception and production[25] but has not previously been suggested in relation to action perception and production. The unique aspect of my proposal[15] is that there is a specialized action perception system which serves the dual purpose of interpreting both the actions of others and also of organizing our own. The work of Johansson[25] and others such as Cutting[27] illustrate the way in which movement and intention can be efficiently extracted from minimal visual data. A few point light sources attached to various parts of the body are sufficient to allow an observer with no other cues to identify human movement and even to deduce unseen features, such as the nature of the load carried and the sex of the actor. These findings are consistent with the existence of a finely tuned system such as one might expect of a social species where the behaviour of others is one of the crucial sets of environmental information needed for survival. This is precisely the kind of mechanism which is needed as a basis for understanding imitation and observational learning.

Demonstration and observational learning

The dominance of behaviourist theories of learning has inhibited the development of theories of the cognitive processes underlying imitation. Bandura's theory[28,29] covers the broad span of observational learning, including circumstances in which the learner will adopt another individual as a model. However, a complete theory of observational learning must account for the mechanism by which perceived action is coded in such a way as to be capable of generating action. The model outlined in the previous section suggests that encoding of action information is a specialized process closely linked to action production. An action is 'perceived' when a 'description' has been achieved. A 'description' not only identifies an action but is at the same time a recipe for producing that action. Since the work of Bernstein[30] has become more widely known through 'ecological' writers,[31,32] it has become increasingly apparent that the central representation of actions rarely if ever involves detailed patterns of instruction to individual muscles. The 'description' of an action is more like a program which, when given appropriate data concerning current conditions, can produce a particular result such as moving an object from one place to another.

If successful imitation of an action requires the learner to acquire a description of the action, then it is worth looking at factors which might prevent or distort this process. Conjurers who practise sleight of hand operate largely by misleading the audience into misinterpreting what they see. Even without the intention to deceive, however, the demonstration of a skill may fail to provide the observer with an adequate description. In a systematic review of teaching by demonstration, Sheffield[33] drew attention to the importance of breaking down the demonstration of complex skills into 'natural' units. Most individuals seem to be able to do this intuitively. For example, video recordings of subjects first tying a bow and later demonstrating how to tie a bow show how they actually make characteristically different kinds of movement.[17] In demonstrating the task, the 'natural units' (picking up the loose ends, twisting them together, etc.) tend to be separated out into discrete sections with pauses between them, and the movements are not simply slowed but are often amplified in scale. For instance, in demonstrating the final step in which the knot is pulled tight, the pulling action is made in an exaggerated form which can be two or three times the amplitude of the normal action.

There is a growing body of evidence that action perception typically involves identifying and encoding specific features. Newtson[34] suggested that skilled observers monitor movement features, particularly encoding 'breakpoints' where a particular feature undergoes a significant transformation. In these experiments, subjects were shown filmed movement sequences and were required to detect whether or not short sections of the record lasting up to 0.5 s had been deleted. Over half the deletions that occurred at breakpoints were detected compared with less than a third of those occurring between breakpoints. Whiting et al.[35] studied the use made by subjects of a model in the acquisition of a complex dynamic skill resembling slalom skiing. The subject stood on a 'ski simulator' consisting of a platform mounted on a pair of bowed rails. The platform was attached to springs which tended to hold it in a central position, and the subject's task was to move the platform rhythmically to the left and right against the springs using the legs and trunk with an action pattern similar to skiing. Two groups of subjects practised over five daily sessions: whilst one group was allowed to discover an efficient technique for performing the task, an experimental group was shown a 1.5-min video recording of an expert working on the apparatus during training. The opportunity to observe the skilled model enabled subjects in the experimental group to achieve a more fluent performance than control subjects, although they did not differ in either

the amplitude or frequency of movement. Whiting et al. argued that fluency is the best measure of skill since it represents efficient use of effort. They also pointed out that this result was obtained without subjects necessarily imitating all aspects of the model's performance. It appears that they had been able to extract some higher-order feature of the movement pattern and apply it to their own productions.

Many sport skills are not easily broken down into distinct elements or performed at significantly slower speed, and so it is difficult for the unskilled observer to produce an adequate description. It is here that video recordings or even diagrams accompanied by verbal explanation may enable the observer to 'see' how the skill works and to form an adequate description. There is some evidence that skilled performers do have more detailed perceptions of actions. Imwold and Hoffman[36] found that experienced instructors recognized more components in recordings of handsprings than novices. Vickers[37] showed that when visual fixations were recorded, expert gymnasts were more likely to attend to the relevant body parts as well as being able to make more accurate judgements about filmed performances. These findings in motor learning compared with those of de Groot[38] and Chase and Simon[39] in relation to chess, indicating that skilled players perceive and retain more information about the distribution of pieces on a chessboard than do novices and non-players. Bandura's theory has stressed the importance of internal representations of actions. Carroll and Bandura[40] used recognition tests to assess the strength of a representation of a series of modelled actions showing that both recognition and recall of the series is enhanced by repeated exposure to the model.

Detailed differences between the performances of the model and the imitator can also show how the action has been 'described'. A demonstration which contains enough elementary descriptions to exceed short-term memory will result in an unsuccessful attempt, typically because one or more units are omitted. Depending on the nature of the task this may well result in overall failure. Smyth and Pendelton[41] have recently shown that short-term memory for discrete meaningless movements is only about four items. A typical error made by young children is to produce a mirror reversal of the demonstrated movement, and this too may be interpreted as a failure in producing the appropriate description. Imitation of tongue protrusion and hand gestures has been recorded in neonates[42] and initiation is well documented in other species (chimpanzees[43] and in Japanese macaques[44]). Despite this early onset of ability to imitate, children do improve in their ability to perceive and reproduce action patterns. Thomas et al.,[45] testing girls aged 7 and 9 on a stabilometer skill, found that the younger children gained less benefit from seeing a model than older children, who were also better able than the younger ones to use the model as a source for correcting the partially established skill.

Video recordings have been used both to provide demonstrations by skilled models and also to provide feedback to learners from their own performances. Burwitz,[46] reviewing the use of demonstrations and videotape recordings in teaching gymnastic skills, found that the results were sometimes disappointing and offered a number of possible reasons. Sometimes the time delay between performance and viewing the recording may be too long, but sometimes critical features of the model's performance may not be easy to see. Scully and Newell[47] confirmed that video demonstration was more effective with the Bachman ladder tasks, in which the successful technique was clearly discernible, than with a ball-rolling tasks, in which the difference between successful and unsuccessful trials was not apparent from the gross kinematic pattern visible in the recording.

To summarize, observational learning of skills has been a somewhat neglected research field, no doubt because it did not fit any of the popular models of skill acquisition. However, a theoretical framework is beginning to emerge to which the encoding of motor information is the key. The perception of action appears to be selective in a way which makes sense. As social organisms, we are naturally interested in interpreting the actions of others and so have developed a sensitivity to a variety of features of body movement. The attractive hypothesis is that the ability to perceive an action pattern is closely coupled to the ability to reproduce it, but before this can be adequately tested we need to learn more about the perception and encoding of movement information and how it varies with age and experience.

PRACTICE

Practice is the *sine qua non* of skill acquisition, but the mechanism by which repetition is effective is still a matter of speculation. The negatively accelerated learning curves noted by the earliest workers or the log–log linear function relating performance to the amount of practice identified by later workers[48-51] suggest an underlying process which is both homogenous and slow, such as laying down a memory trace or engram. Theories are traditionally divided into two camps, namely those that suggest that exercise or repetition *per se* is effective and those that emphasize the selective possibilities offered by repeated trials. Exercise theorists propose that each learning trial offers an opportunity to acquire some new information or to strengthen associations between stimuli or between stimuli and responses, whilst selection theorists propose that trials offer the opportunity to strengthen some aspect of behaviour and/or weaken others. It is, of course, possible to propose both kinds of process. For instance, Rumelhart and Norman[9] suggested that new knowledge could be accumulated or 'accreted', and that processes dealing with new information might also be 'tuned', i.e. selectively adjusted to take account of new information, or even 'restructured' (see also Cheng[52]), which is a more drastic kind of reorganization than tuning.

Annett and Kay[53,54] proposed a theory which began with the proposition that there is a fixed capacity to process stimulus information. Since information is proportional to stimulus uncertainty, the apparent improvement in the rate of processing information which comes with increasing skill might be accounted for by a progressive reduction in stimulus uncertainty. This comes about as the learner builds up an internal model of the environment and particularly of the non-random relationships between events. These also include events brought about as a consequence of previous actions. For example, after practice the flight of a dart becomes progressively more predictable from feedback received during the course of preparing for and executing a throw. Two specific predictions follow from the theory, one concerning part-task training and one concerning the withdrawal of knowledge of results. The first prediction is paradoxical in that it suggests that tasks which have high sequential interdependence are best learned initially in parts. The reason for this is simply that when a novice practises such a task his own error will feed forward to generate a more unpredictable environment than would otherwise be the case. The second prediction is that knowledge of results can only be safely withdrawn without performance loss when it has become effectively redundant. When performing a task the operator receives a stream of feedback signals and, as a result of repeated

trials, will build up a probabilistic model of sequential dependencies between them. On the basis of the model, events relating to the final outcome are predictable from events occurring earlier in the sequence. Hence an experienced golfer knows before the swing is complete if the shot is likely to be good.

Amongst other 'elementary' learning principles which have been proposed to account for the log–log linear relationship are 'chunking'[50] and discrimination.[51] The idea of grouping emerged from studies of morse telegraphy by Bryan and Harter,[55,56] in which trainees progressed from transcribing each letter as a separate item to identifying whole words as units. The modern term 'chunking', originating from Miller's[57] notion of chunks of information, suggests that processing resources, such as memory, are limited in the number of separate chunks which can be held in store or actively processed at any one time. A chunk is any set of mental entities (or 'expressions'), perceptual or motor, which can be dealt with (e.g. stored in memory) as a single unit. Chunking is an automatic process and learning progresses as elementary chunks become grouped together as larger chunks. Newell and Rosenbloom[50] argue that such a process offers a good fit to the empirical power law of learning in a variety of perceptual–motor and cognitive tasks.

A different kind of theory based on the effects of pure 'exercise' was proposed by MacKay.[25] His theory is particularly relevant to serial skills, of which speech production is an example. The production of a coherent speech string is seen as being controlled in a hierarchical fashion. At the top of the hierarchy, a sequence of ideas is generated taking account of both semantic and syntactic rules. This feeds down into a phonological system that organizes the ideas into sound patterns. Finally, activation feeds down into a muscle movement system which directly controls the vocal apparatus. This hierarchy is activated from the top down through a network of connections that determine which items are activated and in what order. The learning principle is that when a node is activated by receiving stimulation from other nodes it is 'primed', i.e. its potential for firing is raised. A particular node will fire when its priming exceeds that of all the other nodes in its domain. The theory predicts a number of phenomena found in serial production skills. For instance, sequence errors are seldom random but, like Spoonerisms, appear to result from failures in the sequencing mechanism. Most substitution errors occur within the same category, i.e. noun for noun, verb for verb, and so on. The model also makes some interesting predictions relating to improvements in performance resulting from rehearsing a skill in imagination rather than overtly.

Mental practice

Mental practice deserves a brief digression since learning effects cannot be attributed to any external consequences, such as rewards and punishments, nor can they be attributed to the effects of repeated external stimulation. Change can only take place through the medium of some internal trace or representation of the skill. Mental practice has a long history; William James observed that we learn to skate in the summer and swim in the winter.[58] A substantial number of studies[59,60] have shown that rehearsing a skill in imagination can result in improvements in performance which, although usually less marked than those achieved by physical practice, are nonetheless greater than those found after no practice or rest.

A number of theories advanced to account for mental practice effects were compared by means of a meta-analysis by Feltz and

Landers.[61] A classic theory, illustrated by Jacobson's studies,[62] is that mental practice evokes activity in the motor output system and, although this is largely suppressed, it is detectable in EMG records. According to the theory, this activity is enough to generate minimal kinaesthetic feedback through which some learning is mediated. Whilst EMG activity has been reported in mental practice, for example the mental rehearsal of skiing[63] there is no firm evidence that this activity is related to the specific response pattern's being learned as opposed to generalized activation. A more plausible theory, supported by meta-analysis,[61] is that mental practice permits the rehearsal of cognitive processes associated with task performance. Tasks which involved learning mazes and other sequential skills were found to be much more likely to produce significant improvements with mental practice than others, such as balancing tasks, which were more purely motoric in character. However, a revised analysis[64] failed to confirm this conclusion.

Some results obtained by Johnson[65] (see also Annett[16]) illustrate the specifically cognitive nature of mental rehearsal in one kind of motor task. Johnson used a linear-positioning task to demonstrate the well-established phenomenon of interference in short-term motor memory. If, between learning to make a linear movement of a particular extent and having to recall it, the subject is required to make a movement of a very different extent (say twice as long), then the recalled movement is overestimated. Johnson first showed that instructions to imagine making a movement twice as long produced the same bias in recall as an interpolated overt movement. Then, by adding a variety of secondary tasks to the instruction to imagine making the movement, he showed that only tasks which involved spatial imagery disrupted the effect. Most interestingly, subjects required to tap on the table with the hand that they were simultaneously imagining moving laterally retained the imagery-induced bias. Thus the effect of imaginary movement was shown to be completely isolated from any muscular activity.

MacKay's theory, as well as accounting for features of skill acquisition referred to above, also offers an account of mental practice. MacKay[66] showed that the subvocal repetition of novel sentences gave practice effects which were, if anything, larger than those obtained by overt practice. The argument is that uttering a novel sentence will involve the activation of an unfamiliar pattern of nodes representing the semantic and syntactic structure of the sentence. Since activation leads to priming, these nodes will be rendered more likely to fire in this new pattern, and even relatively few trials will have an effect on the speed with which the whole sequence is run off. The lower-level nodes controlling the muscles to produce familiar morphemes are not much affected even by overt practice since they are already well rehearsed and hence optimally primed. MacKay's results certainly fit the predictions and may be taken as supporting the 'cognitive' explanation of mental practice, but attempts by Beladaci (see Annett[67]) to extend these predictions to typing have met with less success. According to the theory, skilled typists who, by definition, have had a great deal of practice at the perceptual–motor level should show relatively greater benefit from mentally practising unfamiliar sequences of words. This prediction was not confirmed, nor was the prediction that mental practice would bring about more improvement with nonsense material than meaningful sentences. The distinction between open and closed tasks may be helpful in resolving some of the contradictions in research on mental practice. Learning a closed skill requires the learner to adapt responses to varying environmental demands, whilst acquiring an open skill

requires the learner to rehearse a regular motor pattern. McBride and Rothstein[68] confirmed a prediction that the open skill of hitting a stationary ball with a paddle showed relatively greater benefit from mental practice than the closed skill of hitting a mechanically delivered moving ball. It is also important to consider the retention and transfer effects of mental practice. McBride and Rothstein also confirmed that a combination of mental and physical practice gave better retention of the batting skill after one day.

Feedback and knowledge of results

The paradigms of instruction and practice come together in one of the central research issues in skill acquisition. Practising with knowledge of results provided by an instructor, either directly or through some automatic scoring device, is one of the most effective ways of acquiring a skill.[69–71] The central theoretical question about knowledge of results is to establish the nature of the underlying learning process. Is it, as Thorndike[72] and other behaviourists such as Skinner[73] would claim, an automatic process (reinforcement) by which stimuli are linked to responses; or is it, as most later theorists maintain,[69,74,75] a cognitive process in which feedback information is used to modify responses or to store up useful information?

Before attempting to answer this question let us briefly review the basic experimental paradigm and typical results. Thorndike[76] developed the most widely used experimental technique. The subject is required to attempt to draw a line or make a simple linear movement of some specified extent, usually without the aid of vision or other intrinsic cue. After each attempt, the subject is given knowledge of results, which may simply be 'right' or 'wrong', or may be more detailed such as 'N units of distance too long—or too short'. Sometimes more complex tasks, such as tracking, are used. Knowledge of results may be in the form of some continuous signal, such as a light or sound indicating 'on target', or in the form of a time-on-target or an error score provided at intervals between trials. In even more complex tasks knowledge of results might come in the form of scores relating to more than one aspect of performance such as the kinematic pattern of the response.

Thorndike's theory specified that the reinforcing effects of knowledge of results in strengthening the stimulus–response bond were best served when it was provided immediately after the relevant response and on as many occasions as possible. Whilst results generally fit this pattern, there are complications. Strict temporal contiguity can be violated without detrimental effects on learning provided that the interval between response and feedback is not filled with other activities.[77] Some experiments have confounded delay of knowledge of results with intertrial interval, but Bilodeau and Bilodeau,[78] in a comprehensive study independently varying the interval between response and knowledge of results and the delay after knowledge of results, demonstrated decisively that delay of knowledge of results as such was of no consequence.

According to reinforcement theory the strength of a stimulus–response bond is directly proportional to the number of reinforcements, but in animal studies[79] partial reinforcement schedules (i.e. giving a reward on some trials but not others) makes for slower acquisition but also promotes greater resistance to extinction. The principle was applied by analogy to tracking training by Houston[80] and by Morin and Gagne.[81] These experimenters used a gunnery simulator in which the trainee tracks targets projected on to a screen and receives artificial feedback in the form of a filter that makes the target change to red whenever a hit is scored. The results confirmed the prediction that on removal of knowledge of results, by analogy with experimental extinction, the hit score declined less rapidly for those subjects on a 50 per cent schedule than for those receiving the red filter with every hit. However, the filter treatment seems to have acted as a 'crutch' to performance rather than as an aid to learning, since performance tended to decline rapidly once it was removed. Hence, its value as a training aid was seriously in question. Again, Bilodeau and Bilodeau[82] carried out the definitive study using a version of the line-drawing or linear-positioning task with knowledge of results given after every trial or after every two, three, four, five, or ten trials. They found that the rate of learning was directly proportional to the absolute number of trials on which knowledge of results was provided, but unfortunately they did not report on performance after withdrawal of knowledge of results. Annett,[83] in a similar task, found that error on withdrawal of knowledge of results was less, but not significantly so, if subjects had received it on alternate trials only.

Salmoni et al.[71] rightly complained that many investigators have paid more attention to acquisition than retention following the withdrawal of knowledge of results or transfer to the condiiton without knowledge of results. Annett and Kay[54] pointed out that what really counts is what happens when the learner transfers from practising on the training device, or from under the watchful eye of an instructor, to the actual task. The provision of temporary knowledge of results is only of value if the trainee can subsequently obtain all the information needed from cues which are intrinsic to the task. Figure 3 shows some of the principal feedback loops involved in performing a task such as tracking or discrete linear positioning. The upper level represents the human operator, in this context the learner, and the second level down represents the machine or experimental apparatus. The arrows represent feedback loops which are active during or following a response. At the top level, numbers 1, 2, and 3 are internal feedback loops concerned with the central control of attention (1), proprioceptive control (3), and exteroceptive control (2). In a positioning task, for example, loop 2 would represent the situation in which the learner can see whether his response is correct as he makes it. All these feedback loops are intrinsic to the task, hence the term intrinsic feedback.

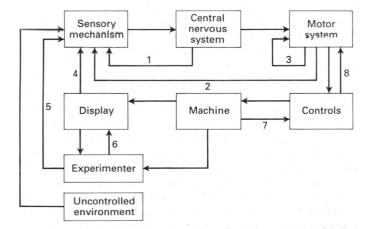

Fig. 3 The principal feedback loops involved in motor control. for 'machine' read 'any object or situation the performer wishes to control' and for the 'experimenter' read 'coach'. See text for detailed explanation of the various loops.

If the learner is operating a machine, then feedback typically comes via loop 4, i.e. through a display such as a moving pointer or some other artificial indicator. Loop 2 may not be available. For example, in driving a car there are two sources of feedback concerning speed: the changing visual field through the windscreen (loop 2) and the speedometer (loop 4). Both normally provide intrinsic feedback, but we tend to rely on the speedometer when precise control of speed is important; it is not required that we learn to judge speed without the help of the instrument. At the bottom of the diagram, the experimenter (or instructor) can form an additional feedback loop, either substituting for the display by giving verbal feedback or embellishing it with additional comments on standards or hints on corrective strategies etc. This feedback loop (5) is extrinsic if it is only used as a temporary measure during training, and so it is important that the trainee not only learns from it but also learns to do without it. Dependence on extrinsic feedback may be reduced by having to do without it on some trials, by being 'weaned' from it, or by the instructor's drawing attention to feedback which is intrinsic to the task—whether it be proprioceptive or some exteroceptive source of feedback. Further experimental confirmation that these techniques give better retention than providing extrinsic feedback on every trial comes from studies by Ho and Shea[84] and Schmidt et al.[85]

The rejection of the reinforcement interpretation of knowledge of results depends not just on the failure of a number of predictions to do with its frequency and timing (see Annett[69] for a detailed review) but on how well an information processing account fits the data. The main evidence comes from findings that acquisition is enhanced by the information content of knowledge of results, where the amount of information is a function of the precision or amount of detail in knowledge of results. Trowbridge and Cason[86] showed that telling subjects not only whether their responses were right or wrong but also the direction and extent of error enhanced both acquisition and retention of a discrete line-drawing task. Although this result supports the information processing viewpoint, a number of subsequent studies[83,87-89] failed to confirm that learning and retention bore any simple relationship to the degree of precision in knowledge of results. In linear-positioning tasks, giving directional knowledge of results is beneficial to a point but further increases in precision typically fail to yield benefits. Annett,[83] using a lever-positioning task, found no difference in the rate of acquisition between knowledge of results given on a three-point scale, a seven-point scale, and a 60-point scale, nor was retention significantly improved by giving knowledge of results to an accuracy greater than on a three-point scale.

Although these results seems to pose a problem for the information processing view, in fact they give an important clue to the learning mechanism. Annett[69] argued that knowledge of results in positioning tasks is used in much the same way as an artilleryman uses ranging shots to locate a target. If the first attempt is an overshoot, the second attempt is shortened by an arbitrary amount; if this turns out to be an undershoot, a third shot halving the differences between the two preceding shots will be very close. Figure 4 shows this strategy in the form of a simple algorithm.

The interesting point here is that it is perfectly possible to learn an accurate response using a short-term memory which contains only the preceding item. The efficiency of learning depends primarily on the ability to distinguish differences in intrinsic feedback between the current response and the immediately preceding response. In tasks of this kind three or four trials provide enough feedback information to enable the subject to produce responses which are as accurate as this discrimination permits. The results

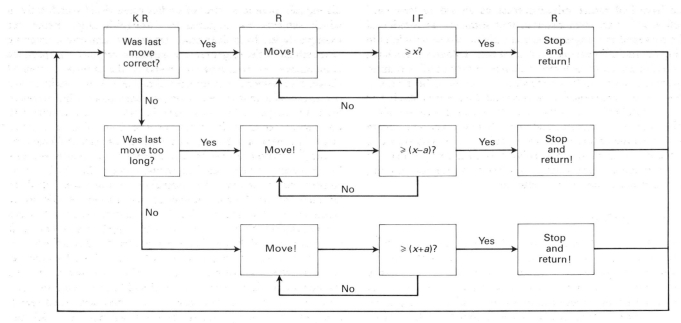

Fig. 4 A simple mechanism capable of learning a response of a prescribed magnitude on the basis of short-term memory for the sensory feedback of the immediately preceding response and knowledge of results following each response: KR, knowledge of results; x, sensory feedback from preceding response; a, an arbitrary constant greater than the difference threshold for the relevant sensory dimension; R, response; IF, intrinsic feedback.

obtained by Annett[83] suggest that there is no further improvement in accuracy after four trials. Although subjects given the least accurate results (to the nearest 40 mm for a target of 60 mm) managed to achieve an accuracy of around 10 mm, those given knowledge of results to the nearest 17 mm and 2 mm did not differ. Retention is slightly but not significantly better for the group given least precise knowledge of results for up to 70 trials after knowledge of results has been withdrawn. This may reflect the fact that providing more information than the learner can handle induces 'hunting' behaviour which could itself interfere with long-term storage. This kind of overcorrection resulting from detailed and immediate performance feedback was found in a tapping task.[90] Providing immediate and detailed time error information reduced constant error (bias), but increased variable error during acquisition and also resulted in poorer overall retention in the absence of knowledge of results.

This view of knowledge of results as providing corrective information, or guidance, requires only that the learner retain a trace of the intrinsic feedback from a response for as long as it takes to compare it with feedback from the next attempt. Two later, and better known theories give knowledge of results a role in establishing long-term memories. Adams closed-loop theory[74] proposed two long-term 'traces', one known as the perceptual trace which is a store of response-produced intrinsic feedback which is laid down and added to on every trial, and a second called (rather confusingly) the memory trace which is a brief motor program required to initiate a response. Responses later come under the control of the perceptual trace as concurrent feedback is compared with the stores information from previous responses. In two studies, Adams and his colleagues demonstrated that (a) learning was primarily due to the strengthening of the perceptual trace, since providing enhanced feedback cues leads to better learning,[91] and (b) the greater the number of practice trials, the better subjects were able to estimate the correctness of their responses.[92]

Schmidt[75] produced a variant of the information processing account of knowledge of results which is known as schema theory. A schema, as understood by Schmidt, is a kind of generalized memory used in the generation of new responses of a given class. The schema notion has the advantage in that it allows for the fact that skilled responses are by no means uniform, but instead are often matched to the varying needs of the occasion. It reflects the flexibility of many motor skills that enables the performer to meet new environmental demands by producing novel responses. As in Adams' theory, two kinds of memory are hypothesized, one motor—the recall schema—and one perceptual—the recognition schema. The recall schema is a record of the relationships between previously executed response instructions (or motor output) under different initial conditions and their outcomes. The recognition schema stores relationships between past sensory consequences and actual outcomes or results. Both kinds of schema are built up by experience, and the greater the variety of experience (within a given class of responses such as linear displacements of the hand), the easier it will be to abstract a general rule from specific cases. Schema theory therefore makes the specific prediction that variability of instances in learning will enhance transfer to new responses of the same class, and this turns out to be generally the case.[93]

It is important to place these theories and experiments dealing with very simple responses in perspective. The provision of knowledge of results for multidimensional tasks, such as gymnastic or flying skills, is rarely a matter of giving precise quantitative information concerning a single response parameter. Whilst overall achievement may be reduced to a single score, this may not be useful if it does not enable the learner to identify specific aspects of performance which should be modified. For example, studies of feedback that gives information about spatial and temporal aspects of performance do not always give better learning than simpler forms of knowledge of results.[94] Moreover, the use of video recordings which provide detailed feedback on complex performance have had rather mixed success as an aid to training.[95] Newell *et al.*[96] have demonstrated that summary knowledge of results packaged as a single numerical variable can be unhelpful in learning the multidimensional task of moving a doubly articulated level mechanism so as to describe a circle of given radius centred on a particular location. Precise feedback concerning the degree of overlap between the subject's response and the criterion circle was not very helpful unless information on the size and location of the criterion was also supplied, and the difference was even more marked when the criterion was an irregular shape. Knowledge of results is useful as an aid to learning only to the extent that the learner can identify the relationship between response output, intrinsic sensory feedback, and outcome.

Automatization

The development of skill is also characterized in the Fitts sequence as a progressive change in the way in which task information is processed, or more precisely in the nature of the control processes involved, such that early in skill acquisition responses are produced under direct conscious control whilst after a great deal of practice performance becomes automatic, taking place with little conscious attention or mental effort. Again, it is tempting to adopt as a general hypothesis that most, if not all, of what we mean by skill acquisition is the process by which controlled processing becomes automatized. Logan[97] has drawn attention to a number of important differences between skilled performance and automatic behaviour. Highly skilled performance can still be very flexible; thus, skilled typists may make errors but typically correct them very quickly[98] Whilst a skill may include automatic procedures, it often also includes a high level of cognitive activity and even metacognitive processes. Much has been made of the difficulty that some skilled performers find in explaining just how they achieve their results,[12,16,19] but it would be wrong to assume that this ability was necessarily present at some earlier stage of practice and then has somehow been lost along the way towards high levels of skill. Neither novices nor skilled swimmers are very good at answering certain kinds of factual questions about swimming technique,[16] nor is it necessarily true that early attempts at a skill are dominated by conscious controlled processes, with every move being thought out in detail. On the contrary, novice swimmers and cyclists may have problems learning to control their automatic, but inappropriate, responses to the novel situation in which they find themselves.

The supposed process of automatization has had to carry a heavy theoretical burden, but the nature of the process is still poorly understood. In the first place, the criteria for automatization are debatable but are often said to include speed (i.e. being faster than controlled processes), relative uniformity of kinematic pattern, being involuntary, being relatively unavailable to introspective analysis, being free from interference by other concurrent tasks, and being independent of load as measured by stimulus or response information. Debates such as that between Neisser *et al.*[99] and Lucas and Bub[100] (1981), or between Cheng[52] and Schneider and Shiffrin,[101] have typically hinged on which criteria are taken as indicating true automaticity.

The nature of automatization has been formulated in a number of different ways. Annett and Kay's theory of stimulus redundancy, Newell and Rosenbloom's chunking theory, and MacKay's node hierarchy theory have all been outlined earlier in this chapter. Schneider and Shiffrin[102] proposed that there are two kinds of process, controlled and automatic, and have sought to distinguish them in a series of studies using visual search tasks in which subjects are required to distinguish target items, for example digits, from distractors, say letters. Consistent mapping of members of the target set to a particular response can produce automaticity, as measured by several of the criteria mentioned above, with quite modest amounts of practice. Automaticity in this context implies that there is a simple computational link between input and output, a kind of private line which is always open and not subject to crosstalk or interference from other concurrent tasks. The view of Schneider and Shiffrin has been challenged by Logan,[97,103] who proposes that, with practice, previously successful responses are retrieved from memory rather than having to be constructed on the basis of an algorithm or set of rules. Called the 'instance' theory because automatic performance is said to be based on the retrieval of previous instances from memory, Logan's theory predicts the log–log linear relationship between mean unit response time and the number of practice trials and also the reduction in variability of response times brought about by practice. However, the two-process theory, the chunking theory, and the instance theory have been developed and largely tested on tasks which are cognitive rather than motor. These are visual search tasks, text-processing, and problem-solving tasks in which the motor component is minimal. Whilst practice curves with mirror drawing,[104] cigar rolling,[48] and choice reaction time[49] show the same general shape, these theories of automatization do not appeal to the mechanisms of motor control.

A different and more traditional account of automaticity appeals to the distinction between closed-loop and open-loop control. In closed skills, such as catching a ball or intercepting a pass, the motor output is linked to and driven by an error feedback signal, whilst in an open-loop task, such as striking a golfball or taking a free kick, the motor output is driven by a once and for all pattern of signals, or motor program, which determines the form and magnitude of the response. Closed-loop tasks usually take longer to perform than open-loop tasks because feedback information is typically subject to a temporal lag and also requires processing capacity. Speed can be traded for accuracy by paying more attention to feedback information and vice versa. The effect of practice may be to make feedback information redundant, as proposed by Annett and Kay,[54] or to create an accurate motor program[105] capable of generating responses without the need for feedback.

The concept of a motor program as a precise set of output instructions does not reflect the considerable flexibility shown by highly skilled performers in adapting responses to detailed variations in task requirements. Yet there is evidence for various degrees of motor preparation in well-practised tasks of short duration. Keele and Posner,[106] for example, found that probe reaction times were predictably slower when probe signals were in competition with program preparation, and Rosenbaum[107] similarly found that time to initiate a response increases proportionately to the complexity of the program required to generate the response. The motor program concept became something of a straw man for those who, like Turvey,[31] argued that motor control is a highly distributed rather than a centralized process and hence requires less 'central' storage of information than the motor program theory seems to demand. This leads to an alternative conceptualization of automaticity in terms of levels of control. Complex tasks, like driving, cannot be adequately described as either a collection of motor programs or simple feedback loops, but are better characterized as hierarchically organized control structures. At the highest level of control, strategic decisions are made about which route to follow and whether to minimize journey time or the risk or accident. At a lower level of control, specific decisions are made about whether to turn off at the next junction, whether to overtake, and so on. At a still lower level decisions (largely unconscious) are made about how far and when to turn the steering wheel, and how hard and when to step on the brake pedal. The development of automaticity in this view refers to the gradual changes in the focus of attention and control away from the lower functions towards higher level goals.

Some clear examples of changes in level of control can be found in keyboard skills. Shaffer,[108] in an elegant analysis of the performances of highly skilled pianists using a specially equipped piano, has shown how the relative timing of notes is subject to high level control. Variations in timing, or rubato, are important in the emotional expression of music. A detailed analysis over different performances showed that the relative timing of individual keystrokes was consistent with varying the rate of an internal 'clock' rather than piecemeal adjustments to individual interkeystroke intervals. The motor programs representing sequences of movements were themselves subject to a timebase that the virtuoso varied to expresses his or her musical intentions.

A computer simulation of typing skill by Rumelhart and Norman[109] also illustrates this principle of different levels of control. The model envisages at least two distinct levels of control: a higher level concerned with the perceptual processes of reading the text to be typed, and a lower level concerned with moving the fingers around to strike individual keys in the correct order. In a skilled typist the interval between successive keypresses can be as little as 60 ms, less than half the time taken in individual reactions to visual signals. The simulation includes two main processes, a perceptual analysis process which identifies words and a motor process which controls hand movement. Each key is represented by a unique schema—a servo unit which, when activated, attempts to drive the finger towards the key making use of an internalized spatial map of the keyboard built up by experience. An activated keypress schema automatically inhibits others until it is fired. This two-level model reproduces a number of features of skilled typing including short interkeypress intervals and some typical errors. The common error of doubling the wrong letter, for instance typing 'bokk' instead of 'book', can be interpreted as implying an intermediate level of control representing double striking—a doubling schema—which, from time to time, is applied to the wrong letter. The theory that automatization refers to the lowest levels of control is consistent with current theories of motor control that strongly suggest, on both behavioural and neurological grounds, that voluntary action involves the integration of a number of semi-autonomous systems, rather like an army in which subordinates have quite a lot of freedom to interpret, or sometimes even reject, orders from above on the basis of their special knowledge of local conditions.

SUMMARY AND CONCLUSIONS

Skill acquisition is best understood if the nature of the physical problem to be solved by performance of the skill is first analysed. Analysis is particularly important in designing training since it enables the instructor/trainer to design training procedures to help the learner solve the specific skill problem. Information processing concepts provide the best available framework within which to analyse specific skills. Sport skills are very varied in character, but a simple classification system suggests that one of the most important ways in which they differ is between open- and

closed-loop control with other factors such as motor co-ordination and the interpretation of motion cues being of considerable importance.

Theories of skill acquisition imply two distinctly different sets of underlying processes, cognitive processes mediated by instruction, and perceptual–motor control processes mediated by practice. Skill is acquired through the automatic effects of repetitive practice and also through various methods of instruction which involve high level cognitive processes. The relationship between cognitive and non-cognitive processes in skill is still not fully understood, but must depend on links between stored representations of action (procedural knowledge) and linguistic units (declarative knowledge). Observational learning is mediated by action–perception processes, and hence effective demonstrations must take into account the way in which the trainee perceives and interprets complex action patterns. Verbal communication about actions is often effective only if it can make use of established action imagery. Learners find it easier to follow instructions which summon up clear movement images.

Practice offers the opportunity to acquire information relevant to the performance of a skill. This may include information about patterns of stimuli presented by the task and about the consequences of different responses. Some theorists have emphasized the perceptual aspects of skill learning, whilst classical learning theory has emphasized the selective effects of consequences, especially in the form of knowledge of results. Knowledge of results has to be placed in the context of all the information which is available to the skilled performer, especially other forms of feedback information which are intrinsic to the task. The principal role of knowledge of results seems to be informative rather than reinforcing in the sense used by behaviourists. In the early stages of learning knowledge of results may be used to identify the essential parameters of a response pattern, especially if they are difficult to ascertain by other means. However, the fact that mental practice has been shown to have some effect on skill acquisition does suggest that not all the effects of practice can be attributed to information feedback, and there is clearly a role for sheer repetition as such in acquiring skill.

Repetitive practice generally leads to automatization, but this does not always mean that performance is inaccessible to cognitive influence. A number of theories have been proposed, but the most compelling suggests that practice allows the development of autonomous perceptual–motor control mechanisms which are subject to supervisory control by the cognitive system which is able to monitor more abstract features of performance, such as the spatial pattern of actions or their timing and rhythm.

REFERENCES

1. Bartlett FC. *Thinking: an experimental and social study*. London: Allen and Unwin, 1958.
2. Argyle N. *Social interaction*. London: Methuen, 1969.
3. Doyle AJR. The essential human contribution to bicycle riding. In Patrick J, Duncan KD, eds. *Training, human decision making and control*. Amsterdam: North-Holland, 351–70.
4. Knapp B. *Skill in sport*. London: Routledge, 1963.
5. Holding D. *Human skills* 2nd. edn. Chichester: Wiley, 1989.
6. Morris A, Annett J. A questionnaire for classifying sports skills. Unpublished report, Department of Psychology, Warwick University, 1982.
7. Judd CH. The relation of special training and intelligence. *Educational Review* 1908; **36**: 28–42.

8. Fitts PM. Perceptual–motor skill learning. In: Melton AW, ed. *Categories of human learning*. New York: Academic Press, 1964.

9. Rumelhart DE, Norman DA. Accretion, tuning and restructuring: three modes of learning. In: Cotton JW, Klatzky RL, eds. *Semantic factors in cognition*. 1978: 37–53.

10. Anderson JR. The acquisition of cognitive skill. *Psychological Review* 1982; 89: 369–406.

11. Anderson JR. Skill acquisition: compilation of weak-method problem solutions. *Psychological Review* 1987; 94: 192–210.

12. Annett J. On knowing how to do things. In Heuer H, Fromm C eds. *Generation of modulation of action patterns*. Berlin: Springer, 1986: 187–200.

13. Fitts PM, Posner MI. *Human performance*. Belmont CA: Brooks/Cole, 1967.

14. Adams JA. Historical review and appraisal of research on the learning, retention and transfer of human motor skills. *Psychological Bulletin* 1987; 101: 41–74.

15. Annett J. Action, language and imagination. In Wankel L, Wilberg RB, eds. *Psychology of sport and motor behavior: research and practice*. Edmonton: University of Alberta, 1982: 271–82.

16. Annett J. Motor learning: a review. In Heuer H, Kleinbeck U, Schmidt K-H, eds. *Motor behavior: programming, control and acquisition*. Berlin: Springer, 1985: 189–212.

17. Annett J. Relations between verbal and gestural explanations. In: Hammond GR, ed. *Cerebral control of speech and limb movements*. Amsterdam: North-Holland, 1990: 327–46.

18. Geschwind N, Kaplan E. A human disconnection syndrome. *Neurology* 1962; 12: 675–85.

19. Bainbridge L. Verbal reports as evidence of the process operator's knowledge. *International Journal of Man-Machine Studies* 1979; 11: 311–436.

20. Berry DC, Broadbent DE. On the relationship between task performance and associated verbal knowledge. *Quarterly Journal of Experimental Psychology* 1984; 36A: 209–231.

21. Berry DC, Broadbent DE. Explanation and verbalisation in a computer-assisted search task. *Quarterly Journal of Experimental Psychology* 1987; 39A: 585–609.

22. Berry DC, Broadbent DE. Interactive tasks and the implicit–explicit distinction. *British Journal of Psychology* 1988; 79: 251–272.

23. Starkes JL, Deakin JM, Lindley S, Crisp F. Skill in field hockey: the nature of the cognitive advantage. *Journal of Sport Psychology* 1987; 9: 222–30.

24. French KE, Thomas JR. The relation of knowledge development to children's basketball performance. *Journal of Sport Psychology* 1987; 9: 15–32.

25. Mackay DG. The problems of flexibility, fluency and speed-accuracy trade-off in skilled behavior. *Psychological Review* 1982; 89: 483–506.

26. Johansson G. Visual perception of biological motion and a model for its analysis. *Perception and Psychophysics*, 1977; 14: 201–211.

27. Cutting JE. Generation of male and female synthetic walkers. *Perception* 1978; 1: 395–405.

28. Bandura A. *Social learning theory*. Englewood Cliffs, NJ: Prentice-Hall, 1977.

29. Bandura A. *Social foundations of thought and action: a social cognitive theory*. Englewood Cliffs, NJ: Prentice-Hall, 1986.

30. Bernstein N. *The co-ordination and regulation of movements*. London: Pergamon Press, 1967.

31. Turvey MT. Preliminaries to a theory of action with reference to vision. In Shaw R, Bransford J, eds. *Perceiving acting and knowing: towards an ecological psychology*. Hillsdale, NJ: Lawrence Erlbaum, 1977.

32. Turvey MT, Kugler PN. An ecological approach to perception and action. In: Whiting HTA, ed. *Human motor actions: Bernstein reassessed*. Amsterdam: North-Holland, 1984: 373–412.

33. Sheffield FD. Theoretical considerations in the learning of complex sequential tasks from demonstration and practice. In Lumsdaine AA, ed. *Student response in programmed instruction*. Washington, DC: NAS-NRC, 1961.

34. Newtson D. Foundations of attribution: the perception of ongoing behavior. In: Harvey J, Ickes W, Kidd R, eds. *New directions in attribution research*. Hillsdale, NJ: Lawrence Erlbaum, 1980.

35. Whiting HTA, Bijlard MJ, den Brinker BPLM. The effect of the availibility of a dynamic model on the acquisition of a complex cyclical action. *Quarterly Journal of Experimental Psychology* 1987; 39A: 43–59.

36. Imwold CH, Hoffman SJ. Visual recognition of a gymnastic skill by experienced and inexperienced instructors. *Research Quarterly for Exercise and Sport Sciences* 1983; 54: 149–55.

37. Vickers JN. Knowledge structures of expert–novice gymnasts. *Journal of Human Movement Science* 1988; 7: 47–72.

38. de Groot AD. *Thought and choice in chess*. The Hague: Mouton, 1965.

39. Chase WG, Simon HA. Perception in chess. *Cognitive Psychology* 1973; 4:55–81.

40. Carroll WR, Bandura A. Representational guidance of action production in observational learning: a causal analysis. *Journal of Motor Behavior* 1990; 22: 85–97.

41. Smyth MM, Pendelton LR. Working memory for movements. *Quarterly Journal of Experimental Psychology* 1989; 41A, 235–50.

42. Meltzoff AN, Moore MK. Imitation of facial and manual gestures. *Science* 1977; 198: 75–80.

43. van Lawick-Goodall J. *In the shadow of man*. London: Collins, 1971.

44. Kawai M. Newly acquired precultural behavior in the natural troupe of Japanese monkeys on Koshima islet. *Primates* 1965; 6: 1–30.

45. Thomas JR, Pierce C, Ridsale S. Age differences in children's ability to model motor behaviour. *Research Quarterly* 1977; 48: 592–7.

46. Burwitz L. The use of demonstrations and video-tape recorders in sport and physical education. In: Cockerill IM, MacGillivary WW, eds. *Vision and sport*. Cheltenham: Stanley Thornes, 1981.

47. Scully DM, Newell KM. Observational learning and the acquisition of motor skills: towards a visual perception perspective. *Journal of Human Movement Studies* 1985; 11: 169–86.

48. Crossman ERFW. A theory of the acquisition of speed skill. *Ergonomics* 1959; 2: 153–66.

49. Seibel R. Discrimination reaction time for a 1023–alternative task. *Journal of Experimental Psychology* 196; 66: 215–26.

50. Newell A, Rosenbloom PS. Mechanisms of skill acquisition and the law of practice. In Anderson JR, ed. *Cognitive skills and their acquisition*. Hillsdale, NJ: Lawrence erlbaum, 1981.

51. Welford AT. On rates of improvement with practice. *Journal of Motor Behavior* 1987; 19: 401–15.

52. Cheng PW. Restructuring vs automaticity: alternative accounts of skill acquisition. *Psychological Review* 1985; 92: 414–23.

53. Annett J, Kay H. Skilled performance. *Occupational Psychology* 1956; 30: 112–17.

54. Annett J, Kay H. Knowledge of results and skilled performance. *Occupational Psychology* 1957; 31: 69–79.

55. Bryan WL, Harter N. Studies in the physiology and psychology of the telegraphic language. *Psychological Review* 1897; 4, 27–53.

56. Bryan WL, Harter N. Studies on the telegraphic language: the acquisition of a hierarchy of habits. *Psychological Review* 1899; 6: 345–75.

57. Miller GA. The magical number seven, plus or minus two: some limits on our capacity for processing information. *Psychological Review* 1950; 63: 81–97.

58. James W. *Principles of psychology*. New York: Henry Holt, 1890.

59. Richardson A. Mental practice: A review and discussion. *Research Quarterly for Exercise and Sport Sciences* 1967; 38: 95–107, 263–73.

60. Corbin CB. Mental practice. In: Morgan WD, ed. *Ergogenic aid and muscular performance*. New York: Academic Press, 1972.

61. Feltz DL, Landers DM. The effects of mental practice on motor skill learning and performance: a meta-analysis. *Journal of Sport Psychology* 1983; 5: 25–57.

62. Jacobson E. Electrophysiology of mental activities. *American Journal of Psychology* 1932; 44: 677–94.

63. Suinn R. Behavior rehearsal training for ski races. *Behavior Therapy* 1972; 3: 210–12.

64. Feltz DL, Landers DM, Becker BJ. A revised meta-analysis of the mental practice literature on motor skill learning. Background paper prepared for the Committee on Techniques for the Enhancement of

Human Performance, National Academy of Sciences, 1988. (Available through the Publication on Request Program.)

65. Johnson P. The functional equivalence of imagery and movement. *Quarterly Journal of Experimental Psychology* 1982; **34A**: 349–65.

66. Mackay DG. The problem of rehearsal or mental practice. *Journal of Motor Behavior* 1981; **13**: 274–85.

67. Annett J. Motor learning and retention. In: Gruneberg MM, Morris PE, Sykes RN, eds. *Practical aspects of memory: current research and issues.* Chichester: Wiley, 1988; **2**: 434–46.

68. McBride E, Rothstein A. Mental and physical practice and the learning and retention of open and closed skills. *Perceptual and Motor Skills* 1979; **49**: 359–65.

69. Annett J. *Feedback and human behaviour.* Harmondsworth: Penguin, 1969.

70. Bilodeau IMcD. Information feedback. In: Bilodeau EA, Bilodeau IMcD, eds. *Principles of skill acquisition.* New York: Academic Press, 1969.

71. Salmoni AW, Schmidt RA, Walter CB. Knowledge of results and motor learning: a review and critical appraisal. *Psychological Bulletin* 1984; **95**: 355–86.

72. Thorndike EL. A theory of the effects of the action of the after effects of a connection upon it. *Psychological Review* 1933; **40**: 434–89.

73. Skinner BF. *Science and human behavior.* New York: Macmillan, 1953.

74. Adams JA. A closed loop theory of motor learning. *Journal of Motor Behavior* 1971; **3**: 111–50.

75. Schmidt RA. A schema theory of discrete motor skill learning. *Psychological Review* 1975; **82**: 225–60.

76. Thorndike EL. *The fundamentals of learning.* New York: Teacher's College, Columbia University, 1932.

77. Lorge I, Thorndike EL. The influence of delay in the after-effect of a connection. *Journal of Experimental Psychology* 1935; **18**: 186–94.

78. Bilodeau EA, Bilodeau IMcD. Variation in temporal intervals among critical events in five studies of knowledge of results. *Journal of Experimental Psychology* 1958; **55**: 603–12.

79. Ferster CB, Skinner BF. *Schedules of reinforcement.* New York: Appleton-Century-Crofts, 1957.

80. Houston RC. The function of knowledge of results in learning a complex motor skill. Unpublished MA Thesis, Northwestern University, Evanston, Ill, 1947.

81. Morin RE, Gagne RM. Pedestal sight manipulation test performance as influenced by variations in type and amount of psychological feedback. *USAF HRRC Research Note P & MS* 51-7, 1951.

82. Bilodeau EA, Bilodeau IMcD. Variable frequency of knowledge of results and the learning of a simple skill. *Journal of Experimental Psychology* 1958; **55**: 379–83.

83. Annett J. Aspects of the acquisition of skill. Unpublished D. Phil. Thesis, University of Oxford, 1959.

84. Ho L, Shea J. B. Effects of relative frequency of knowledge of results on retention of a motor skill. *Perceptual and Motor Skills* 1978; **46**: 859–66.

85. Schmidt R. A., Young D. E., Swinnen S., Shapiro D. C. Summary knowledge of results for skill acquisition: Support for the guidance hypothesis. *Journal of Experimental Psychology: Learning Memory and Cognition* 1989; **15**: 352–9.

86. Trowbridge M. H., Cason H. An experimental study of Thorndike's theory of learning. *Journal of General Psychology* 1932; **7**: 245–58.

87. Bilodeau E. A. Speed of acquiring a simple motor response as a function of the systematic transformation of KR. *American Journal of Psychology* 1953; **60**: 409–420.

88. Bilodeau E. A., Rosenback J. H. Acquisition of response proficiency as a function of rounding error in informative feedback. *USAF Human Resources Research Center Research Bulletin* 1953; **53**–21.

89. Green R. F., Zimilies H. L., Spragg S. D. S. The effects of varying degrees of knowledge of results on knob-setting performance. *SPECDEVCEN Technical Report* 241-76-20, 1955.

90. Lee TD, White MA, Carnahan H. On the role of knowledge of results in motor learning: Exploring the guidance hypothesis. *Journal of Motor Behavior* 1990; **22**: 191–208.

91. Adams JA, Goetz ET, Marshall PH. Response feedback and motor learning. *Journal of Experimental Psychology* 1972; **92**: 391–97.

92. Adams JA, Gopher D, Lintern G. Effects of visual and proprioceptive feedback on motor learning. *Journal of Motor Behavior* 1977; **9**: 11–22.

93. Shapiro DC, Schmidt RA. The schema theory: recent evidence and developmental implications. In: Kelso JAS, Clark JE, eds. *The development of movement control and coordination.* New York: Wiley, 1982; 113–73.

94. Newell KM, Walter CB. Kinematic and kinetic parameters as information feedback in motor skill acquisition. *Journal of Human Movement Studies* 1981; **7**: 235–54.

95. Rothstein AL, Arnold RK. Bridging the gap: application of research on videotape feedback and bowling. *Theory into Practice*, 1976; **1**: 35–62.

96. Newell KM, Carton MJ, Antoniou A. The interaction of criterion and feedback information on learning a drawing tasks. *Journal of Motor Behavior* 1990 **22**: 536–52.

97. Logan GD. Skill and automaticity: relations, implications and future directions. *Canadian Journal of Psychology* 1985; **39**: 367–86.

98. Rabbitt PMA. Detection of error by skilled typists. *Ergonomics* 1978; **21**: 945–58.

99. Neisser U, Hirst W, Spelke ES. Limited capacity theories and the notion of automaticity: Reply to Lucas and Bub. *Journal, of Experimental Psychology: General*, 1981; **110**: 499–500.

100. Lucas M, Bub D. Can practice result in the ability to divide attention between two complex language tasks? Comment on Hirst *et al. Journal of Experimental Psychology: General* 1981; **110**: 495–8.

101. Schneider W, Shiffrin R. Categorisation (restructuring) and automatisation: Two separable factors. *Psychological Review* 1985; **92**: 424–8.

102. Schneider W, Shiffrin RM. Controlled and automatic human information processing 1: Detection, search and attention. *Psychological Review* 1977; **84**: 1–66.

103. Logan GD. Toward an instance theory of automatisation. *Psychological Review* 1988; **95**: 492–527.

104. Snoddy GS. Learning and stability. *Journal of Applied Psychology* 1926; **10**: 1–36.

105. Keele SW. Movement control in skilled motor performance. *Psychological Bulletin* 1968; **70**: 387–403.

106. Keele SW, Posner MI. Processing of visual feedback in rapid Movements. *Journal of Experimental Psychology*, 1968; **767**: 155–8.

107. Rosenbaum DA. Motor programming: a review and scheduling theory. In: Heuer H, Kleinbeck U, Schmidt K-H, eds. *Motor behavior: programming control and acquisition.* Berlin: Springer, 1985: 1–33.

108. Shaffer LH. Performance of Chopin, Bach and Bartok: studies in motor programming. *Cognitive Psychology*, 1981; **13**: 326–76.

109. Rumelhart DE, Norman DA. Simulating a skilled typist: A study of skilled cognitive motor performance. *Cognitive Science* 1982; **6**: 1–36.

1.5.1 Anthropometry and the assessment of body composition

PETER R.M. JONES AND N.G. NORGAN

INTRODUCTION

Anthropometric and body composition techniques allow the body's size and shape and the masses and proportions of its constituents to be described. This information is helpful in acquiring a greater understanding of those attributes which contribute to performance in sport. Physical performance, for example speed, strength, and endurance, depends strongly on the amount of force-producing tissue, the application of forces through the levers of the limbs, and the adaptations to training.

The significance of information about the size, shape, and body composition of athletes is that it correlates with performance and may also indicate 'condition' and 'potential'. In athletics in particular, achievement and success require a particular body configuration; elite sprinters, middle-distance runners, and long-distance runners tend to have characteristic sizes and shapes, although inevitably there are exceptions. Anthropometric and body composition information can therefore contribute to decisions concerning the sport or event in which an individual is most likely to succeed and help in the development of appropriate training schedules and in the management and rehabilitation of those with sports-related injuries. Variation in the intensity of training and the periodic attainment of peaks may be accompanied by changes in size and/or composition which reflect relatively subtle changes in energy balance not readily apparent from measuring body mass alone. Indices of body fatness can contribute to the assessment of the state of training of a sportsman or sportswoman.

Anthropometry and body composition have long been indispensable measurements in exercise and sports science, as they have in human biology, and the proximity and inter-relations of the two disciplines have been mutually beneficial. The study of body composition has benefited from investigations of sports participants. An early stimulus to the development of reproducible measurements of body fatness was the now familiar observation that American professional footballers could be classified as overweight and unfit for military service not because of excess fatness but because of muscularity and skeletal frame size. This led Behnke and colleagues to introduce a method for the measurement of body specific gravity which provided estimates of fat content of the body.[1]

Numerically, the majority of sports participants are not the elite or competitive sports people but modern day urban dwellers with regular working hours who want an enjoyable activity that brings a sense of well-being and may help in regaining or maintaining fitness and appropriate body weight. Anthropometric and body composition techniques must therefore be applicable to the general population and relevant to the monitoring of health status in general. However, the extreme physiques of elite sportspeople test the methodology, leading to improvements which benefit both sport and science. For example, body composition studies in athletes have emphasized the needs and the benefits of multicomponent models of body composition (see later in this chapter).

This chapter considers the techniques available to describe body size, shape, and composition in sports medicine and recommends the use of a minimum of six basic anthropometric parameters. In addition, the validity of the underlying assumptions when dealing with sportsmen and sportswomen is considered.

RATIONALE FOR TAKING SIZE, SHAPE, AND BODY COMPOSITION MEASUREMENTS

Physicians have been prominent in the history of constitutional investigation, especially in studies of the relationships between morphology and susceptibility to disease. Damon[2] reviewed the history of human classification and defined constitution as the sum total of the morphological, physiological, and psychological characters of an individual, in large part determined by heredity but influenced in varying degrees by environmental factors.

The characteristics of body physique associated with success in sports and other types of physical activity have always greatly interested scientists, artists, sports writers, and others. There have been numerous studies which have used measurements describing the morphology of sportsmen to focus on the descriptions and comparisons within and between sporting groups and athletic events. A review by Tittel and Wutscherk[3] cited over a hundred studies and to date we must add at least another hundred.

We would expect to find that the most successful sportsmen and sportswomen have physical characteristics best suited to their particular sport, and that differences in morphology and physique will highlight the importance of measuring aspects of physique such as a somatotype. This general hypothesis has been stated by Tanner[4] who reported that an appropriate body build and composition was important to success in athletic competition.

Early anatomists noted four different body types: (i) the fat abdominal type, (ii) the strong muscular type, (iii) the tall slender-chested thoracic form, and (iv) the more rounded and larger headed cephalic type. Later writers contrasted the well-rounded and compact form of the pyknic contestant with the muscular athletic build and the long thin asthenic leptosome. In the 1940s Sheldon, a pioneer in this area, moved away from the strict typology by introducing the concept of three discrete and continuous variables—endomorphy, mesomorphy, and ectomorphy—to describe the varieties of human physique.[5] Unfortunately, this promising somatotype method came to be virtually abandoned because of the inflexibility in the technique and Sheldon's rigid adherence to his concept of the unchanging genetically determined somatotype. Although important and sound concepts underly his original method, modifications to it by Heath and Carter have produced an extremely useful and widely used research technique for the study of variation in human physique.

The Heath–Carter method uses the somatotype as a phenotypic rating. This allows not only for changes over time, but also recognizes that each of the rating scales for the three components are open ended and apply to physiques of both sexes. The selected anthropometric measurements bring objectivity to the rating scores. Technical issues of this hybrid methodology have been published. The photoscopic and the anthropometric somatotype methods have been described recently.[6]

Measurements of body composition are important in many of the disciplines which contribute to sports medicine. They can, for example, provide sensitive indices of body fatness and therefore a means of describing relatively subtle changes in energy balance which are not readily apparent from measuring body mass alone but which have the potential to influence performance or, in the longer term, health. During an intensive training programme, the body density can increase which is indicative of a reduction in body fat content that, at the same time, would be supported by a reduction in the skinfold thicknesses observed.

The amount of fat in the body is often expressed as a percentage of fat, or the proportion of the mass of fat to the total body mass. Fat distribution may vary considerably even between people with exactly the same percentage of body mass as fat, and some may have a larger proportion of fat subcutaneously whilst others may have a larger proportion internally. Similarly, some people may have a larger proportion of subcutaneous fat on the limbs as opposed to the thorax and abdomen.

CUSTOMIZATION

Anthropometry—its contribution to sports medicine management

A considerable number of physiotherapists in sports medicine clinics use anthropometry in the everyday treatment and rehabilitation of the injured athlete. It is important to understand what kind of anthropometric and body composition techniques are available and useful as well as how they can be applied to aid the clinical practitioner and answer some of the injured athlete's problems.

Muscle has an important protective role in the function and stabilization of joints. If muscle strength and function are reduced because of pain following injury, then new damage can occur more easily. Changes in muscle circumference with strength-related exercises pre- and postoperatively, as well as during and after rehabilitation, can be monitored by measuring circumferences of the quadriceps femoris and hamstring muscle groups at levels of 50, 100, and 150 mm proximal to the superior patella border with the leg straight and the muscles relaxed. These types of muscle bulk measurements would also be invaluable in the assessment of postoperative treatment following meniscectomy by arthroscopy for anterior cruciate, medial, and lateral collateral ligament injuries. It is usual to compare the measurements of the injured limb with those of the contralateral limb.

Where knee pain occurs in tibialis anterior disease, chondromalacia patellae, which has several synonyms including runner's knee, jogger's knee, and cyclist's knee to name but a few, it is usual to see a small difference in the total leg length. This in turn often reflects abnormalities associated with discrepant leg lengths leading to pelvic tilting and hence spinal deformities such as scoliosis or compensatory spinal curvature. It also gives rise to unequal stride length causing pelvic rotation. To identify this condition the subjects are measured in the standing position using a long spirit level placed on skin marks over the anterior superior iliac spines. Two measurements are then made on each leg from the anterior

superior iliac spines to the medial malleolus or datum using a flexible steel tape. Packing boards, of thickness 5 or 10 mm, can be placed under the feet in order to achieve a horizontal pelvic level and hence a correct leg length. These measurements can also be obtained from radiological techniques, provided that the subject's posture is carefully standardized during the examination. However, this method of measurement is undesirable owing to additional radiation hazards, especially where the gonads are unprotected during the procedure.

For severe injuries to the shoulder requiring surgical replacement, it is helpful to measure total arm length. This is measured, using marked points, from the inferior border of the acromial process (acromiale) to the distal point of the ulna bone.

Most joint angles (shoulder, spine, elbow, wrists, hips, knees, and ankles) can be measured using a goniometer. This is particularly useful in cases of suspected hypermobility very often seen in gymnasts. In runners, excessive pronation of the ankle joint is associated with high injury risk.

Anthropometry for biomechanics

The segmental lengths of the body's limbs are known to have a significant effect on the performance of sportsmen. For example, elite endurance runners differ from non-runners not only in factors such as muscle fibre types but also in the mechanics of the limb levers, in particular the ratios of the sitting height to subischial length (leg length) and its divisions (upper and lower leg). Body mechanics are of prime importance in throwing, and the optimal body physique of throwers can be readily interpreted from the study of the mechanics of athletic activities.[7] For example, the length of the humerus, radius, and ulna contribute to the ultimate mechanical advantage which is gained by these skeletal levers and their muscles, thereby affecting the force and speed of movement and the speed of objects delivered by these lever systems.

Many muscles and joints in the body act as third-class levers, i.e. levers in which the fulcrum is farther away from the object to be raised than from the point where the force is being applied. For example, the biceps brachii muscle is inserted into the bicipital aponeurosis and radial tuberosity in a very mechanically disadvantageous way close to the elbow joint. Therefore it follows that a sportsman with a biceps brachii inserted 2 or 3 cm distal to the elbow joint would have considerable mechanical advantages over other throwers. This analogy also applies to the lower appendicular system and has implications for running.

The role of anthropometry in measuring the length of these lever systems seems to be important. What is perhaps more important is the role of new technology such as ultrasound which could be applied to demonstrate the actual points where tendons insert into the bones in relation to the joint space. This may reveal individual differences that exist in leverage arrangement and therefore help us to understand more clearly the differences in performance.

Anthropometry and body composition in selection, estimation, and explanation of performance

There is no doubt that certain shapes, configurations, and compositions are more appropriate and conducive to success in particular sports, events, or team positions. However, determinism should be avoided. The exceptions are almost as well known as the generalizations. The practice of selection for teams or squads based on certain anthropometric or compositional criteria may reflect more a coach's intuition of the degree of application in diet control or training effort than the importance or advantage of characteristics as an indicator of training or for a team role.

In theory, the strength, speed, and possibly endurance are expected to be related to the amount of skeletal muscle. However, a simple linear single-stage model is far too simplistic to account for much of the variation in elite performance. Other biological, anatomical, and psychological factors are also involved, requiring a multicomponent multistage non-linear model. Therefore many variables intervene between anthropometry, body compositions, and performances, for example the level of training, the degree of motivation, etc. It should come as no surprise that the relationships are less strong than might be expected on first sight. There are other reasons for the unexpectedly weak relationships between size and performance. First, the common index of size and composition is the fat-free mass. This is a heterogeneous entity, largely consisting of water. Indices such as skeletal muscle mass or lean leg volume are more appropriate. Second, individuals of different sizes are compared as per kilogram body weight or per kilogram fat-free mass but this is inappropriate as the relationship is not directly proportional. There is a significant intercept term. This means that, for example, the $\dot{V}o_2$ max of small lean individuals would be higher (per kilogram fat-free mass) than those of larger individuals without any necessary difference in other performance variables.

MEASUREMENTS AND CALCULATIONS

The measurements on anthropometry and body composition that may be used in sports medicine are discussed in this section.

Weight and height

Weight and height are the two cardinal measurements of size. They are also widely used as indices of overweight and obesity. Although weight alone or weight and height in combination cannot distinguish between overweight due to excess fat and the muscular athlete or between the underweight of the ill or undernourished and the lean thin endurance runner, they remain the primary descriptive characteristics and should be recorded in all investigations or treatments.

Body mass is widely used as an index of body size and composition, but without other anthropometric measurements its application in the assessment of fatness is limited because it may be fat or lean. It is impossible to discriminate between fatness or leanness on the basis of height and body mass alone. Changes in body mass over a period of time in an adult are more useful as the height and frame size will be constant. However, without further measurements it is still impossible to tell whether the change in body mass is due to a change in fat or muscle. This is particularly true in an exercise or training programme where there is likely to be a change in muscle mass. During inactivity or habitual training, the body mass may remain relatively constant although inactivity often produces a gradual body mass gain. Daily changes in body mass, which could be plus or minus 1 kg, are most likely to be due to variations in the amount of body water or gastrointestinal contents. Tables of excess body mass developed by insurance companies have been used for many years as a simple index of obesity.[8]

In terms of assessing overweight and obesity, the current popular reference values are those of the Metropolitan Life Insurance Company. They are derived from the data on mortality and longevity from more than 25 North American insurance companies for over 4 million men and women insured between 1954 and 1972. Ninety per cent of the weights were obtained by actual weighing and those with major diseases were excluded. A range of weights is given for three frame sizes at each height for the two sexes. The frame sizes are determined by measurement of elbow breadth. The data for the references were not derived from insurance company data but were collected separately in the National Health and Nutrition Examination Surveys. The interquartile range, the range of the 50 per cent of the population in the middle of the range, is used for the medium frame category and the lower and upper quartiles are used for small and large frames. The weight ranges reflect the lowest mortality at each height.

Although the objective assessment of an index of frame size is an improvement, it should be remembered that it was not measured in the insured. The wide ranges for each frame size and height should highlight the fact that this is an imprecise approach. Many factors affect weight. The data on the insured population may not be representative of the general or sports populations. Interestingly, the difference between weights for height associated with the lowest mortality and average weights decreased markedly between 1959 and 1979. Weights for height were some 6 to 7 kg higher but remained below average weights. These weight increases were 10 per cent for short men and women, 5 per cent for medium men and women, and 1 per cent for tall men and women. Furthermore, the weight for a given height associated with the lowest mortality increases with age by some 4 to 5 kg per decade for individuals of average height. There is no systematic sex difference in the weight ranges. Thus weight standards ought to be adjusted for age.

Frame size

The rationale for categories of frame size is that the greater the frame size, the greater the lean body mass will be for a given height and therefore the greater the total weight can be without increasing the risk from adiposity.

The elbow breadth (biepicondylar humerus) is most commonly used. It is easily accessible, it is unaffected by adiposity, and normative data are available. The measurement technique is described below and standard values are given by the Metropolitan Life Insurance Company, New York.

Indices of fatness and leanness

Relative weight

Relative weight (ratio of body weight to midpoint of the weight range for the appropriate sex, height, and frame size) has been used to categorize overweight and obesity in different individuals. Obesity has been defined as more than 120 per cent of standard weight. However, the move is away from reference-based comparators with all their problems to independent indices such as weight-to-height ratios. However, the cut-offs for these usually require recourse to the insurance company data.

Weight-to-height ratios

The simplicity of measurement and availability of normative data have contributed to the widespread use of these indices in population studies and in the clinics. The body mass index (W/H^2,

kg/m^2), or Quetelet's index, is widely used as a simple assessment of fatness and leanness in Western populations. It correlates with fatness more highly ($r = 0.7$) than indices such as W/H or W/H^3 and is poorly correlated with height. This index is useful because it is relatively independent of height, although the requirement for height independence has recently been challenged. The addition of age to the body mass index improves the estimations of the percentage of the body mass as fat.

Weight is more fat-free mass than fat mass, and, not surprisingly, the body mass index is also well correlated with fat-free mass. Also, it does not discriminate between fat-free mass and fat mass so that it cannot separate the muscular from the overweight or obese. For an individual gaining or losing weight, or for setting targets, body mass index may offer no real advantage over weight as height remains constant and changing body mass index is determined by changing weight. Some endurance athletes have very low body mass indexes within the range that nutritionists would interpret as chronic energy deficiency, which may well be the case. This raises the question of form versus function and the importance of long-term outcomes (see below).

Waist-to-hip circumference ratio

The site of deposition of fat is an independent risk factor for health, as important as the amount of fat. The ratio of the waist circumference to the hip circumference is a measure of intra-abdominal fatness. The waist-to-hip circumference ratio is a risk factor in cardiovascular disease.

Skinfold thickness

This measurement is the thickness of a so-called double fold of skin and underlying subcutaneous adipose tissue under a calliper jaw pressure of 10 g/mm^2. Skinfold thickness, when measured properly, correlates well ($r = 0.8–0.9$) with hydrostatic weighing, with a low standard error of estimate, i.e. 3 to 4 per cent of the body weight as fat. Their value in estimating percentage body weight as fat in young adults is well documented. Measurements of limb, trunk, and other skinfolds can be used to detect changes in fat distribution.

Proportionality

Size, shape, and composition cannot be described by single variables. Each has several components. The portrayal of multidimensional data is usually diagrammatic, e.g. somatograms, '0'-scales deviation.[9] The reference data may be obtained from national studies, data collected in single laboratories, or data for single groups of sportspeople. These methods have considerable merit theoretically, but have not proved to be popular with users as opposed to researchers.

Body density

One of the most accurate estimations of body composition is obtained from body density determined by hydrostatic weighing. Here, body density is calculated from body mass divided by body volume. An inexpensive system to measure body volume by underwater weighing has been described by Jones and Norgan.[10] The apparatus is transportable and does not require dedicated space as it is easily stored. It consists of a 3.5 m^3 cylindrical tank containing water heated to 35 to 36 °C. A lightweight plastic chair is suspended from a calibrated force transducer mechanically protected against excess weights. The filtered output allows underwater

weight to be measured to 0.02 kg in habituated subjects and to 0.05 kg in others.

Subjects are equipped with a snorkel and trained to submerge themselves gently to avoid water movement and undue weight oscillation. They learn to expire maximally underwater and carry out the rebreathing sequence needed to measure residual volume after the underwater weight has been taken. Residual volume is measured by the three-breath nitrogen dilution method at the same time as the weight is taken. After the necessary procedures have been learned by the subject, the measurements are made in duplicate or until two results agree to within 1 per cent fat. In our experience in a large series of middle-aged men and women, satisfactory results were obtained in two trials in 40 per cent of subjects and in three trials in 55 per cent of subjects. If more than three trials are required, the reasons for the variations should be established (e.g. fluctuating weight, poor rebreathing) before accepting duplicates. The learning and measurement of body density together with the associated anthropometry takes about an hour. Subjects who are unfamiliar with immersion in water, such as non-swimmers, may require a greater degree of habituation in order to give consistent readings. In two separate studies on women aged 20 to 70 years and 35 to 55 years only four out of the total of 100 were unable to participate in underwater weighing.

The body density D (kg/m^3) is given by

$$D = \frac{\text{mass in air}}{[(\text{mass in air} - \text{mass in water})/\text{density of water}] - \text{RV}} \times 10^3$$

The residual volume RV (litres) at body pressure, ambient temperature, saturated with water vapour (BTPS) is given by

$$\text{RV} = 3.000 \times \frac{N - 0.5}{80 - N} \times \frac{\text{Bp}}{\text{Bp} - 47} \times \frac{310}{273 + t} - \text{DS}$$

when using 3 l of dry 99.5 per cent oxygen (0.5 per cent N_2) at room temperature t (°C), Bp is the barometric pressure, and the saturated water vapour pressure at 37 °C has a value of 47. DS is the dead space of the snorkel and tap, and the nitrogen content of alveolar air is assumed to be 80 per cent. N is the nitrogen content obtained by subtraction after measuring the oxygen and carbon dioxide in the expired and rebreathed air. Siri's equation, the most widely accepted equation for converting body density to body composition, is based on the assumption that fat has a known density of 900 kg/m^3 and that the fat-free mass density is, on average, 1100 kg/m^3.

The proportion (ffm) of the body weight that is fat free is given by Siri's equation:[11]

$$\text{ffm} = 1 - \left(\frac{4950}{D} - 4.5 \right)$$

and the fat-free mass is given by

$$\text{FFM} = \text{ffm} \times \text{body weight}$$

Estimation procedures
Estimation of body density

The technique of hydrostatic weighing imposes severe restrictions of both practicality and convenience, so that the alternative procedure of estimating density from skinfold thicknesses is used for most purposes.

A skinfold measurement includes the double thickness of the dermis and the underlying subcutaneous adipose tissue. The sum of four skinfold thicknesses, invariably biceps, triceps, subscapular, and suprailiac, can be used to estimate body density and hence total body fat or fat-free mass. Many equations are available. For the general population, we recommend the equations of Durnin and Womersley[12] and of Jackson et al.[13] These equations are given in the Appendix. The authors provide similar equations for individual skinfolds and all combinations of the four used in the Appendix, where the standard errors of estimation (SEEs) were 7.3 to 12.5 kg/m^3. Jackson and Pollock's equation for men aged 18 to 60 years is

$$D = 1109.4 - 0.827x + 0.0016x^2 - 0.257 \text{ (age)}$$

where x is the sum of chest, abdomen, and thigh skinfolds.[14] The SEE is 7.7 kg/m^3. The authors provide validated equations for the sum of seven skinfolds (the three above plus axilla, triceps, thigh, and suprailiac) transformed to natural logarithms and waist and forearm circumference.

Their equation for women aged 18–55 years is

$$D = 1099.5 - 0.993x + 0.0023x^2 - 0.139 \text{ (age)}$$

where x is the sum of triceps, thigh, and suprailiac skinfolds.[13] The SEE is 8.6 kg/m^3. The authors provide validated equations for the sum of four skinfolds (those listed above plus abdomen), seven skinfolds (the preceding plus chest, axilla, and subscapular), the natural logarithm of the sum of skinfolds, and the gluteal circumference.

All these equations have been cross-validated and have proved to be widely applicable. The suprailiac site of Durnin and Womersley differs from the usual site by being just above the iliac crest in the midline. The abdominal site described by Jackson and Pollock is adjacent to the umbilicus as opposed to the usual site of 50 mm to the left.[14] For the best estimate, it may be necessary to follow the site description given by the original authors. Whichever technique is used, this should be recorded in the reports of the study.

However, there may be inherent dangers in the use of equations which are derived from populations whose physical characteristics differ markedly from those of athletes. Evidence about the applicability of the commonly used equations to this special group is available from cross-validation surveys. Thorland et al.[15] compared 17 equations for estimating body density from skinfolds, circumferences, and diameters in adolescent male athletes and 15 equations for adolescent female athletes. Linear and quadratic forms gave acceptable accuracy for males whilst only quadratic equations did so for females. The male adolescent athletes seemed more similar in terms of body density to non-athletic populations than did the females. This could explain why the estimations from equations were more applicable to males than females.

Sinning and coworkers[16,17] conducted similar studies using 265 male athletes and 79 female athletes. Of a total of 21 equations, only those of Jackson and Pollock gave estimates of percentage body fat which did not differ significantly from values obtained by hydrostatic weighing.[14] Overall, the equations tended to overestimate the percentage of body mass as fat in men. For women, the equation of Jackson et al.[13] which relied on four skinfolds gave a mean percentage of fat identical to that derived from hydrostatic weighing. No other equation was sufficiently accurate and, as in the men, there was a tendency towards overestimation.

The available evidence therefore suggests that, in athletes, the quadratic equations are more generalizable than the logarithmic or linear equations. If subjects with a greater range of age and density had been studied, this advantage might have been even more pronounced. Nevertheless, estimations derived from some

linear equations were sound, for example Sloan's equation in men, whilst some based on quadratic equations were poor.

Estimation of limb cross-sectional areas

Estimates of limb muscle plus bone areas correlate well with strength and can be calculated from

$$A = (C - \pi S/10)^2/4\pi$$

where A is muscle plus bone area in square centimetres, C is limb circumference in centimetres, and S is the appropriate limb skinfold or mean skinfold in millimetres. This formula can be applied to the mid-upper arm, the thigh, and the calf at the levels described in the next section. For the upper arm, the upper arm circumference, and the biceps and triceps skinfolds are required. For the mid thigh, the thigh circumference and the anterior thigh and posterior thigh skinfolds are measured. For the calf, the calf circumference and the medial and lateral calf skinfolds are required. It is recommended that two skinfolds should be taken at each level.

Subcutaneous adipose tissue (SCAT) areas can be calculated by subtracting A from the total limb area:

$$\text{SCAT} = C^2/4\pi - A \text{ cm}^2$$

ANTHROPOMETRIC TECHNIQUES AND INSTRUMENTATION

The techniques and methods of measuring anthropometry and body composition variables that are useful in sports medicine or related studies are described in this section. The descriptions include the anatomical landmarks, sites, and conventions adopted. Precision and accuracy of measurement are crucial in anthropometry to evaluate measurements, particularly if serial measurements are made to categorize changes. More commonly, measurements are compared with norms or reference values of some type. The deviation from the reference can be calculated or a series of deviations for a series of measurements can be determined, leading to a scale or profile.

Basic anthropometric techniques

The basic anthropometric techniques are as follows:

(1) body mass (weight);
(2) stature (height);
(3) biepicondylar humerus;
(4) upper arm circumference;
(5) triceps skinfold;
(6) subscapular skinfold.

By convention, measurements are taken on the left side of the body where appropriate. Some measurements require landmarks to be identified and marked. These are given at the beginning of the measurement. In many cases an observer can greatly assist the measurer and check the technique. The apparatus required and a list of suppliers is included at the end of this section.

Body mass (kg)   The apparatus required is a weighing machine. The subject should be lightly clothed. The weight of the clothes or representative garments should be recorded and subtracted from the body mass.

Stature (height) (mm)   The apparatus required is a stadiometer (wall-mounted stadiometer or portable anthropometer). The landmark is the Frankfort plane—the position of the head when the imaginary line from the lower border of the left orbit to the upper meatus is horizontal.

The subject should stand on a horizontal platform with the heels together, stretching upwards to the full extent. Gentle upward pressure is exerted on the mastoid processes by the measurer who encourages the subject to 'stand tall, take a deep breath, and relax'. The subject's back should be as straight as possible, which can be achieved by rounding or relaxing the shoulders and manipulating the posture. The head is held in the Frankfort plane. Either the horizontal arm of the anthropometer or a counterweighted board is brought down on the subject's head. If an anthropometer is used, one measurer should hold the instrument vertical with the horizontal arm in contact with the subject's head, while another applies gentle upward pressure. The subject's heels must be watched to ensure that they do not leave the ground.

Biepicondylar humerus (cm)   The apparatus required is an anthropometer or sliding calliper. The subject's arm is bent at a right angle and the width across the outermost parts of the lower end of the humerus is taken. The measurement is usually oblique since the medial epicondyle of the humerus is lower than the lateral.

Upper arm circumference (cm)   The apparatus required is a tape. The landmark is a horizontal mark on the mid left circumference of upper arm, measured halfway between the inferior border of the acromial process and the tip of the olecranon process, with the arm flexed at a right angle.

The measurement is performed with the arm relaxed and hanging beside the body, and the hand supinated. When measuring a circumference the tape should be at right angles to the long axis and contact with the skin should be continuous along the tape, but the skin should not be compressed.

Skinfold thicknesses (mm)   The skinfold is picked up vertically between thumb and forefinger and the calliper jaws are applied at exactly the level marked. The measurement is read 2 s after full pressure of the calliper jaws is applied to the skinfold; if a longer interval is allowed, the reading may 'creep'. Skinfold callipers are used for the measurements.

The landmark for the triceps skinfold measurement is the same as that for the upper arm thickness. The skinfold is picked up at the back of the arm about 1 cm above the level described for the upper arm thickness, directly in line with the olecranon process.

For measurement of the subscapular skinfold, the subject stands with shoulders relaxed and arms by his or her sides. The inferior angle of the scapula is located and the skinfold is lifted slightly inclined downwards and laterally in the natural cleavage lines.

Additional anthropometric techniques

The choice of additional measurements will depend on how comprehensive a description of size and an estimation of composition is required. This is likely to be influenced by the time, resources, and personnel available, and whether the measurements are made in the laboratory or the field. The following additional anthropometric measurements could be useful where the aims and purposes of the study have been carefully considered in relation to either the size or shape of sportsmen.

Chest circumference (cm)   The apparatus required is a tape. The circumference is measured horizontally at the junction of the third and fourth sternebrae. The arms are raised to allow the tape to be passed round the trunk and lowered once it is in position. The measurement is taken during normal light breathing in mid inspiration.

Waist circumference (cm)   The apparatus required is a tape. The landmark is in the mid-axillary line at the midpoint between the costal margin and iliac crest. This is often the natural waist.

The tape is lowered from the position in which the chest circumference is measured to the level of the waist marking.

Hip circumference (cm)   The apparatus required is a tape. The subject stands with feet together and at an angle of approximately 15°. The tape is lowered to the level of the trochanters, which is used as a guide for location of the maximum hip circumference. The observer ensures that the tape remains horizontal, while the tape is raised and lowered until the maximum circumference is found.

Mid-thigh circumference (cm)   The apparatus required is a tape. The landmark is the mid-thigh. With the subject sitting, the mark is at the point between the centre of the inguinal crease and the proximal border of the patella.

The subject stands with the weight on the right leg and the left leg relaxed and slightly bent. The measurement is made round the thigh level with the landmark.

Calf circumference (cm)   The apparatus required is a tape. The subject sits with his lower limbs relaxed. The maximal horizontal circumference is measured, using the belly of the gastrocnemius as a guide.

Sitting height (m)   The apparatus required is an anthropometer or sitting height table. The measurement is made with the subject's back stretched up straight, sitting on a table top with the feet hanging down unsupported over the edge; the backs of the knees should be directly above the edge of the table. Extension of the spine should be encouraged in the same way as for stature. Gentle upward pressure is applied under the chin; the muscles of the thighs and buttocks should be relaxed. The head is held as in the measurement of stature and the anthropometer is stood vertically in contact with the back at the sacral and interscapular regions. If a sitting height table is used, the movable backboard is brought into contact with the subject's back and the headboard is lowered to rest firmly against the head.

Biacromial diameter (cm)   The apparatus required is an anthropometer. The landmark is the inferior edge of the most lateral border of the acromial process.

To give maximum shoulder width the subject stands with shoulders relaxed by not slumping forward. Standing behind the subject, the measurer feels for the outside edge of the acromial process of the shoulder blade which can be detected as a ridge just above the shoulder joint. The measurer then places the edge of one arm of the anthropometer along the lateral border of one acromial process and brings the other arm of the anthropometer inwards until its edge rests on the lateral border of the opposite acromial process.

Bi-iliac diameter (cm)   The apparatus required is an anthropometer. The subjects stands with heels together and the anthropometer arms are brought into contact with the iliac crests at the place which gives the maximum diameter. The measurement should always be taken with the measurer standing behind the subject. As in measurements of the biepicondylar humerus and the biacromial diameter, sufficient pressure should be applied so that the calliper blades compress the soft tissues.

Biceps skinfold (mm) The apparatus required is a skinfold calliper. The landmark is as for the measurement of the upper arm circumference.

The skinfold is measured with the subject's arm relaxed by the side with the palm facing forward. It is measured on the anterior of the arm, over the belly of the biceps muscle, at the level of the mid-upper arm marking described for the upper arm circumference measurement.

Suprailiac skinfold (mm) The apparatus required is a skinfold calliper. The landmark is the iliac crest. The most prominent superior border of the iliac crest is in the mid-axillary line.

The skinfold is picked up vertically approximately 1 cm above the landmark.

Mid-anterior thigh skinfold (mm) The apparatus required is a skinfold calliper. The landmark is the mid-thigh. With the subject sitting, a mark is made midway between the inguinal crease and the proximal border of the patella as for the measurement of the mid-thigh circumference.

The subject stands with the weight on the right leg so that the left leg is relaxed. The skinfold is taken vertically, in the midline, on the anterior aspect of the thigh at the level of the mid-thigh marking.

Some of these measurement descriptions are based on the guide to field methods prepared for the Human Adaptability Section of the International Biological Programme and subsequently updated.[18] These sources contain descriptions of other anthropometric measurements used mainly in growth and physical anthropology and detailed descriptions of instruments. A further source of measurement descriptions is the recent manual edited by Lohman *et al.*[19] A feature of this well-illustrated publication is that the technique, purpose, literature, reliability, and sources of reference data are given for each measurement, and the recommendations are likely to be adopted widely, at least in North America.

Instruments

A list of instruments, together with the names and addresses of the supplier, that are suitable for undertaking anthropometric measurements such as stature (height), lengths, breadths, girths, and skinfold thicknesses, either in the laboratory or for field studies, is given below.

Harpenden stadiometer	Holtain Ltd, Crosswell,
Harpenden anthropometer	Crymmych, Dyfed SA41 3UF,
Harpenden sitting height table	Wales
Harpenden bicondylar callipers	
Flexible 2 m measuring tapes	
Holtain skinfold callipers	
GPM (Martin type)	Pfister Import-Export Inc.,
Anthropometer	450 Barell Avenue, Carlstadt,
	NJ 07072, USA
	Owl Industries Ltd,
	177 Idema Road,
	Markham, Ontario L3R 1A9,
	Canada
Harpenden skinfold callipers	British Indicators Ltd, Quality House, 46–56 Dumfries Street, Luton, Beds LU1 5BP, UK
	H. E. Morse Co., 455 Douglas Avenue, Holland, MI 49423, USA
Harpenden electronic read-out incorporating computer system (HEROICS) skinfold callipers	HUMAG Research Group, Department of Human Sciences, University of Loughborough, Loughborough, Leics LE11 3TU, UK
CMS weighing scales	CMS Weighing Equipment Ltd, 18 Camden High Street, London NW1 0JH, UK
Salter weighing scales	Salter International, Measurement Ltd, George Street, West Bromwich, Staffs, UK
Toledo electronic scales	Toledo Scale, 431 Ohio Pike, Suite 302, Way Cross Office Park, Toledo, OH, USA
Linen measuring tape	Pfister Import-Export Inc., 450 Barell Avenue, Carlstadt, NJ 07072, USA
Anthropometric tape measure	County Technology Inc., PO Box 87, Gays Mill, WI 54631, USA

OTHER BODY COMPOSITION TECHNIQUES

Other body composition techniques, including two relatively new pieces of equipment that can be used either in the field or laboratory, are described in this section. They are useful as quick and socially acceptable body composition measurements where a change in value is of more interest than an answer in absolute terms.

Bioelectrical impedance analysis

The measurement of human body composition by bioelectrical impedance analysis assumes that the body consists of two compartments: the fat-free mass and the fat mass. The fat-free mass contains virtually all the electrolytes and body fluids which are involved in conduction, so that the impedance of the body can be used as an estimate of fat-free mass. The impedance of a conductor is related to its length and cross-sectional area:

$$Z = r\,L/A$$

where Z is the impedance in ohms, r is the volume resistivity in ohm cm, L is the length of the conductor in centimetres, and A is its area in square centimetres. Multiplying by L/L gives

$$Z = r L^2 / AL$$

where AL will be equal to the volume V:

$$Z = r L^2 / V^4 \text{ or } V^4 = r L^2 / Z$$

Bioelectrical impedance was not used in the evaluation of human body composition until 1981 when Nyboer developed a tetrapolar electrode method.[20] Four electrodes, placed on the dorsal surfaces of the hands and feet, are used to minimize the contact impedance. A current of 800 mA at 50 kHz is applied at the distal electrodes, and the voltage drop at the proximal electrodes is measured. Operators should be aware that displacement of the electrodes by 1 cm at the wrist and the ankle will produce a combined change in impedance of 4.1 per cent, equivalent to a 3 per cent change in body mass as fat. As the technique is dependent upon the water content of the body, factors which may alter body water content, such as alcohol consumption and exercise before measurement, are avoided. Regression equations have been developed relating total body water (measured by D_2O dilution) to impedance measurements (usually as Ht^2/Z). As the total body water forms a relatively constant proportion of the fat-free mass, the latter can be calculated from total body water and fat mass calculated from the difference between body mass and fat-free mass.

A number of instruments to measure impedance are available, and a number of equations have been developed to estimate body composition from impedance values. When compared with densitometry, the SEE of percentage body fat calculated using various instruments and equations has been reported to vary widely, between 2.7 and 6.1 per cent. Some workers have found bioelectrical impedance analysis to be a better predictor of hydrostatically determined body fat than anthropometry (SEE 3.5 per cent) and a better predictor of changes in body water and fat-free mass than anthropometry.

The cost effectiveness of bioelectrical impedance analysis equipment, relative to its accuracy, must be seriously questioned. It is unlikely that improvements in a 'high tech' design will overcome the limitations caused by the shape of the human body.

Near-infrared interactance

Near-infrared interactance involves irradiating a sample in the near-infrared spectrum and determining the proportion of energy transmitted which returns to the detector. The interactance depends on the distance through which the radiation travels, as well as the physical properties of the substance irradiated, according to the Beer–Lambert law:

$$\text{interactance} = I/I_0 = 10^{-kcL} \text{ or } \log_{10}(I/I_0) = kcL$$

where I is the intensity of radiation emerging, I_0 is the intensity transmitted, k is the molar absorption coefficient, c is the molar concentration, and L is the path length through which the radiation travels.

Each substance has its own absorption spectrum with characteristic peaks. For example, there is a peak for fat and a trough for water at 930 nm, and a peak for water and a trough for fat at 970 nm. These absorption spectral differences can be exploited for measurement of human body composition because adipose tissue is high in fat and muscle is high in water content. The relative absorptions at two wavelengths, at one of which there is a greater absorption for fat and at the other a greater absorption for muscle, can be used to obtain information about relative proportions of fat

and muscle. The radiation used is non-ionizing and of low intensity, and so the measurement is harmless.

A commercial instrument, the Futrex 5000 (Futrex Inc.), has been developed. It consists of a light wand with four infrared-emitting diodes, two emitting radiation at a wavelength of 940 nm and two at 950 nm. These illuminate a circular diffusing ring which allows radiation to be emitted evenly. In the centre is a silicon detector which measures the intensity of re-emitted light. The light wand is attached to a Hitachi microprocessor from which the optical density readings ($\log 1/I$) can be read. The manufacturers have calibrated each machine against their own hydrostatic weighing measurements to allow estimation of percentage body fat from optical density readings, body weight, height, sex, and activity level. This method had a SEE of 3 per cent body fat in a cross-validation group of 17 men and women when a sophisticated computerized spectrophotometer was used. The criterion method was based on isotope dilution assessment of total body water. Modelling of the performance of a low cost portable apparatus suggests a SEE of 3 to 4 per cent body fat. Comparisons of estimates of percentage fat measurements by Futrex 5000 with hydrostatic weighing in 39 males and females gave a SEE of 3.1 per cent fat for males and 4.3 per cent for females.

In our laboratory we have now had over 3 years' experience with Futrex 5000 and, like other workers, have found that this instrument consistently underestimates percentage fat, to an increasing amount with increasing fatness, when compared with estimates of body fatness from our 'gold standard'—hydrostatic weighing. Despite these limitations, Futrex 5000 is in widespread use by health care professionals whose advice athletes and non-athletes seek to gain information about body fatness and the changes required to maintain the levels recommended and for participation in sport.

Isotope dilution

Measurement of body water alone by single labelled water has provided information on body composition. The analytical standard error is about 1 litre of water equivalent to 1.4 kg of fat-free mass. Factor analysis of data in patients suggests that values for fat-free mass obtained from body water have more random errors than those from skinfolds.[21] Equilibration periods of 3 to 5 h may be a consideration for some investigators.

Naturally occurring and induced radiation

These techniques require apparatus beyond the scope of most laboratories. Detection of the naturally occurring gamma radiation from potassium allows a measurement of body potassium content with a standard error of 4 per cent, equivalent to more than 2 kg of fat-free mass. The calculation of fat-free mass from potassium content has evoked more controversy than other body composition technique.

In-vivo neutron activation analysis provides a means of estimating the body content of up to seven elements (potassium, carbon, calcium, nitrogen, phosphorus, sodium, and chlorine) with one whole-body irradiation. The precision (standard deviation of repeated measures) of fat-free mass and fat mass measurements of 0.7 kg is similar to that for body density, but more confidence may be justified in these more direct measurements of multiconstituents. The use of ionizing radiation and the expense are obstacles to widespread adoption, and its main value may be in the validation of other methods.

Total body electrical conductivity

This is another method with bulky, expensive apparatus. SEE values of 1.4 kg fat-free mass have been reported. Direct carcass analysis of pigs suggest a SEE value of 2.8 kg fat-free mass. Recent cross-validation trials have shown this technique to underestimate percentage body fat by 3.6 in 117 males aged 20 to 90 years and 0.8 per cent in 178 females aged 19 to 89, when compared to underwater weighing.[22]

Computerized tomography and magnetic resonance imaging

CT and MRI are exciting new methods with great potential. However, as with other approaches, the high cost restricts the availability of these methods and CT also uses ionizing radiation. The methods can be used to measure fat mass and fat-free mass and to establish regional fat distribution, although the use of MRI is in the earliest stages. Their immediate value to the field has been validating existing and new simpler techniques as in the case of CT and limb cross-sectional areas.

Soft tissue radiography

These provides an accurate assessment of regional, particularly appendicular, body composition provided that care is taken with tissue magnification, subject positioning, and subject protection from X-radiation. The method should not be used with children and has become less popular because of the radiation exposure, the limited number of sites that can be measured, and the development of other non-hazardous techniques.

Ultrasound

Amplitude-scan ultrasound is not a new technique but several portable machines have appeared recently. Ultrasound should provide measurements of single uncompressed subcutaneous adipose tissue avoiding the interference of varying compressibility from skinfold callipers. To achieve this, uniform and constant transducer pressure at the skin surface is essential. The identification of the adipose tissue–muscle interface can be difficult at some sites and in some individuals, but improved instrumentation, particularly displays, help overcome this problem. For the estimation of body density, the SEE using subcutaneous adipose tissue from ultrasound is only 0.5 kg/m^3 lower than that from skinfolds. When the expense, measurement time, and absence of extensive validation trials and reference data are taken into account, ultrasound does not prove to be superior to callipers.

Photon absorptiometry, broadband ultrasonic attenuation, and contact ultrasonic broadband attenuation

Measurement of bone mineral by the absorption of monoenergetic photon energy from radionuclides is useful for investigating the composition of its fat-free mass. Single-photon absorptiometry of limb bones such as the radius can be made with portable apparatus in the field. Correct positioning of the limb is crucial for high precision and accuracy. As an estimate of skeletal mass, it has an error of 10 per cent. Dual-photon absorptiometry has a high precision (2–3 per cent) in humans and accuracy of 1 per cent for skeletons. The error of estimate compared with in-vivo neutron activation analysis is 113 g of total body mineral, i.e. about 2 per cent. However, the apparatus is expensive and difficult to transport.

Bone density measurements by ultrasound are usually made at the calcaneus or patella since they have a large percentage of cancellous bone and the medial and lateral sides are approximately parallel. These new approaches of broadband ultrasonic attenuation and contact ultrasonic broadband attenuation use non-ionizing radiation and portable equipment. Coefficients of variation of 3 per cent and 5 per cent respectively have been reported.

Dual energy X-ray absorptiometry

This technique was introduced in 1989 and has become a valuable research tool for assessing axial and appendicular bone density in vivo. It was a natural development of dual-photon absorptiometry and uses two photon energies to allow for different absorber thickness and inhomogeneities overlying the sites of measurement in the lumbar spine, neck of the femur, and distal radius. The X-ray source permits greater photon flux and so measurements are quicker and more precise (about 1 per cent) than with dual-photon absorptiometry although the results are very highly correlated.

APPLICATIONS

Applicability of methods

Anthropometry involves simple objective measurement without inherent assumptions, although there are exceptions to this in the calculations of areas and volumes where perfect circles, cones, or pyramids are assumed to represent parts of the body. However, the estimation of body composition in vivo requires assumptions about the relationships of the proportions of constituents of the body. These relationships differ in the sexes, change with age, differ between and within performers and during training, etc. The significance of these variations to the accuracy of the estimates of body composition needs to be considered. Fortunately, in sports the individual and changes in the individual, rather than the group, are usually the units of interest to coaches, doctors, or selectors.

The transformation of physical values of density, resistivity, interactance, and ultrasound attenuation, and the chemical values of water, potassium, or calcium content, to whole or regional body composition estimates requires the relationships between them to be established. Many have been defined for young adult men and women but not for elite athletes or the other stages of the life-span.

The water and mineral contents of fat-free mass are lower in children than in adults. In athletes with high proportions of muscle, fat-free mass density is lower. Both of these will overestimate fatness. Body densities of more than 1100 kg/m^3 have been described, i.e. negative body fat contents. These indicate the inapplicability of the methods to every situation.

To overcome problems such as these, multicomponent models and methods of body composition have been introduced. Commonly, two or more of body density, water content, and mineral content will be measured and fatness and leanness calculated. This improves the accuracy of the estimate. The fundamental equations expressing the relationships between body composition and constituents are reformulated and applied to new groups. This is exemplified by the new equations available for calculating the body composition of children.[23] Suggested revisions to Siri's equation for age, level of nutrition, and activity have appeared over many years, but the evidence for modifications was weak as the topic is characterized by widespread circularity of reasoning. Methods that require constancy of proportions of constituents to be valid are used to show that the proportions change or differ.

Effects on body composition

Of the variable fat-free mass components, for example total body water, protein, and osseous minerals, the one most likely to have the greatest effect on body composition determinations is the skeleton. The substantial body of evidence that bone mineral density is markedly higher in sportsmen and sportswomen than in their less active contemporaries calls into question the validity and applicability of Siri's equation to estimate the percentage of body mass as fat.

Other factors, such as ethnic group, age, and sex, influence the density of the fat-free mass. For example, a decrease in the degree of osseous mineralization of about 1 per cent per year occurs after the age of 50. Durnin and Womersley[12] have calculated that a 15 per cent decrease in the body mineral content results in a fall in the density of the fat-free mass of about 6 kg/m³, approximately a 15 per cent error on percentage body fat. The potential extent of the distortion of the estimates of fat-free mass in athletes is of a similar order if one assumed, quite reasonably, that athletes possessed 18 per cent higher bone mineral density than the population from which Siri's equation was based. Such changes play an important role in influencing the validity of body fat estimations from body density measurements. This has been clearly shown when spuriously low values for the percentage of body mass as fat have been reported for athletes using conventional methods of estimation.

Thus it follows that if sportsmen and women have denser bones and muscles, this will result in an underestimation of their body fat. Conversely, if osteoporosis is present in the skeleton, this reduces the bone density and leads to overestimation of the body fat.

Applicability to changing states

Acute

Many sports competitions are based on weight classes. Individuals attempt to perform at the top end of a weight category as weight is usually an advantage and determinant of performance. Frequently, participants will have to 'make weight'—lose a kilogram or more to meet the requirement.

The composition of weight loss depends on the deviation and degree of water and energy balance. The initial weight loss and high weight loss in a short time will predominantly consist of water or lean tissue. Neither of these are consistent with optimum functioning. Proper and improper weight loss programmes have been the topic of a position statement of the American College of Sports Medicine.[24]

In this context, these weight loss procedures, although accurately reflected by weight changes, compromise the body composition measurements. Composition of fat-free mass changes, affecting the interpretations of density and body water measurements, and probably changes in skinfolds, bioelectrical impedance analysis, near-infrared interactance, etc. as well. Thus, in acute weight loss, weight changes do not reflect losses of lean or fat and

the accuracy of body composition is also affected. The way forward would be to regard these losses as water, often associated with losses of lean tissue or diminution of the glycogen stores. The arithmetic of the energy balance equation should be checked before any claims for sizeable fat losses are claimed.

Chronic

Longer-term efforts by sports participants to modify body composition involve reducing the fat content by strict control of food intake and, to a greater or lesser degree depending on the sport, to increase leanness or muscularity by conditioning, particularly weight training. Both are widespread in recreational participants and elite athletes.

There is more information in the literature on weight loss than on any other aspect of human energetics. Unfortunately, it concentrates on attempts by the overweight and obese to move towards the midpoint of the range. Sports participants are attempting to move beyond this to the end of the range. Whereas the arithmetic and efficiency of energy utilization remain unchanged, the composition of tissue lost, and hence the extent to which weight loss compromises estimates of body composition and changes in composition, depends on the initial body composition. In underfeeding studies of at least 4 weeks' duration, the proportion of lean in weight loss was negatively and exponentially related to initial fatness. Dieting alone is an inappropriate weight loss strategy for most athletes. Some exercise to maintain or increase lean is required. The common experience of weight loss and resistance training regimes is for modest changes that reverse easily. The dedicated participant is more successful, as evinced by the well-known changes in body form and muscle development and definition unaided by artificial means. However, the exact role of agents such as anabolic steroids in modifying body composition remains controversial.

Changes in body composition need to be greater than 1 kg to be detected by current methods. Gains of lean tissue of more than 3 to 5 kg occur slowly and might best be reflected by serial weight measurements. Measurements of total body water would be less affected by changing compositions and could be used to determine whether the changes could be ascribed to lean tissue.

CONTEMPORARY ISSUES

Hazards of thinness

Much is known of the dangers of raised mortality and mortality in the overweight and obese but, in contrast, little is known about the dangers of thinness, natural or induced. The relationships between mortality and body mass index are J-shaped. The body mass index associated with the lowest mortality increases with age. It is commonly assumed that the raised mortality at low body mass index arises from individuals with existing but unrecognized disease or from smokers with their lower body weights. There is a higher than expected mortality from pneumonia and influenza, malignant neoplasms, suicides, and hypertensive heart disease. However, observations such as the increase in life expectancy of populations over the last 200 years coinciding with an increase in body weight, and the rise in the body mass index with lowest mortality in the 1959 and 1979 insurance company data suggest that thinness may have other ill-defined risks. Although induced extreme leanness is regarded by coaches and athletes as increasing performance in, for example, middle distance running, a fitness–health dichotomy follows the hazards of thinness. The groups at risk may be jockeys, runners, and ballet dancers, who participate for many years, and gymnasts.

The association of disturbance of menstruation and other aspects of reproductive function in women athletes is well known and has been reviewed repeatedly. Concerns about the effects of prolonged hard training and dieting on bone mineral content have also arisen.

Body composition and menstrual status in athletes: the hazards of estimations

Many women athletes with high training loads develop menstrual irregularities such as secondary oligomenorrhoea or amenorrhoea. The condition is also common (up to 50 per cent) in ballet dancers, but less so in swimmers and cyclists. Frisch and McArthur[24] proposed that a minimum level of fat is required to initiate menarche (17 per cent) and to maintain regular menstrual function (22 per cent).[25] It is suggested that adequate energy stores as fat are required for successful pregnancy and lactation and this trigger has evolved. The mechanism may be a degree of conversion of androgens to oestrogens in the adipocyte affecting steroid feedback to the hypothalamus or pituitary gland. The evidence advanced for the hypothesis included the observations that early maturers usually weigh more than their peers, that body weight at menarche is independent of menarcheal age, that both ballet dancers and anorectics have a high incidence of amenorrhoea, and that very obese men have raised serum oestradiol, decreased serum testosterone, and an increased conversion of androstanedione to oestrone.

The hypothesis of Frisch and colleagues that body fat is a determinant of mature sexual function in women was anchored on the lower variability in fatness than in body weight at menarche, yet this is also true at other ages during growth. Other evidence against the hypothesis are the considerable ranges in weight and fatness at menarche and in groups of eumenorrhoeic and amenorrhoeic athletes. Also, changing training status without alteration of body weight or fatness and the development of changing menstrual function indicates an absence of a link between body composition and menstrual function. Thus increasing the training distance of runners or introducing training to the untrained resulted in the majority developing abnormal menses. Conversely, ballet dancers report resuming menstruation on vacation or during injury. Thus the stress of the exercise appears a more likely determinant of menstrual function than critical levels of weight or fatness. The mechanism may be neural rather than hormonal, although other hormone responses such as prolactin may be involved. A predisposition to abnormalities may remain. A greater proportion of amenorrhoeic athletes report a prior history of irregular menses than is the case in eumenorrhoeic athletes.

Frisch and coworkers estimated body fatness from measurements of height and weight using a regression equation to estimate body water and from that percentage fat. This equation was based on data from 27 girls and 18 young women that had not been cross-validated and, like all estimation equations, it performs less well at the ends of the range. In particular, it is not accurate for women with less than 22 per cent fat. The work of Frisch and her colleagues has been criticized on statistical grounds of improper data analysis and interpretation of results. When measurements are made with more reliable body composition estimates and controlling for age, size, training load, and fitness, percentage fat values in amenorrhoeic and eumenorrhoeic runners are identical.

This is not the end of the question, however, as athletes may differ from non-athletes in the composition of fat-free mass, affecting estimates of body composition by the traditional techniques usually regarded as reliable. This is not a problem if amenorrhoeic and eumenorrhoeic athletes are equally affected, but low bone mass (osteopenia) is more common in amenorrhoeic athletes than in those with normal functions. Percentage fat will be overestimated if the bone mineral content is less than that assumed.

Osteoporosis

Osteoporosis can be defined as a complex multicausal chronic disease characterized by a reduction in bone mass, leading to bone fragility and debilitating fractures. It is a major public health problem in Europe and North America, being most common in post-menopausal women and elderly persons. The condition is chronic with serious consequences for at least one-third of older women and 15 to 20 per cent of older males.

It has been suggested that physical activities and sports may be effective in the prevention of bone loss due to osteoporosis. For example, Dalen and Olsson[25] showed that the mineral content of the calcaneus, humerus, and distal radius and ulna in a group of cross-country runners with a mean age of 56 years was higher by 20 per cent than that in control subjects matched for age and body size.[26] Whilst these differences may be due in part to mechanical stresses imposed by exercise, the fact that some of these skeletal sites were not stressed during running confirms the importance of constitutional differences between athletes and others. Constitutional differences alone, however, cannot account for the higher bone density observed in athletes. Bone density studies in tennis players, for example, have shown that bone mineralization in the radius is greater in the playing arm than in the non-playing arm.

CONCLUDING COMMENTS

Anthropometric and body composition techniques are becoming commonplace in the sports medicine clinic and the human performance laboratory. The size, shape, and composition of athletes and recreational sports participants correlate with performance and indicate condition and potential. To yield precise, reliable, and, most importantly, useful information, standardized measurement procedures are essential. The validity of the information is strongly dependent on the applicability of these techniques to the sports population and particularly the elite athlete. This has been considered but is an area that will change rapidly as more information becomes available. Future trends are likely to be towards multidimensional scaling systems and multicomponent descriptions of body composition based on population-specific approaches. The coexistence of anthropometry and body composition with exercise and sports science has been mutually beneficial and synergistic, and this is also likely to be the case in the future.

REFERENCES

1. Behnke AR, Feen BG, Welham WC. The specific gravity of healthy men: body weight/volume as an index of obesity. *Journal of the American Medical Association* 1942; **118**: 495–501.
2. Damon A. Constitutional medicine. In: Von Meering O, Kasdan L, eds. *Anthropology and the behavioural and health sciences*. Pittsburgh: University of Pittsburgh Press, 1970: 179–95.
3. Tittel K, Wutscherk H. *Sportanthropometrie*. Leipzig: Barth, 1972.

4. Tanner JM. *The physique of the olympic athlete*. London: Allen and Unwin, 1964.
5. Sheldon WH (with the collaboration of SS Stevens and WB Tucker). *The varieties of human physique*. New York: Harper and Brothers, 1940.
6. Carter JL, Heath BH. *Somatotyping—development and applications*. Cambridge University Press, 1990.
7. Dyson GHG. *The mechanics of athletics*. 2nd edn. University of London Press, 1963.
8. Metropolitan Life Insurance Company. 1983 Metropolitan height and weight tables. *Statistical Bulletin of the Metropolitan Life Insurance Company* 1983; **64**: 1–9.
9. Ross WD, Ward R. Proportionality of Olympic athletes. In: Carter JEL, ed. *Physical structure of Olympic athletes*. Part II. *Kinanthropometry of Olympic athletes*. Basel: Karger, 1984: 110–43.
10. Jones PRM, Norgan NG. A simple system for the determination of human body density by underwater weighing. *Journal of Physiology* 1974; **239**: 71–3P.
11. Siri WE. The gross composition of the body. *Advances in Biological and Medical Physics* 1956; **4**: 239–80.
12. Durnin JVGA, Womersley J. Body fat assessed from total body density and its estimation from skinfold thickness: measurements on 481 men and women aged from 16 to 72 years. *British Journal of Nutrition* 1974; **32**: 77–97.
13. Jackson AS, Pollock ML, Ward A. Generalized equations for predicting body density of women. *Medicine and Science in Sports and Exercise* 1980; **12**: 175–82.
14. Jackson AS, Pollock ML. Generalized equations for predicting body density of men. *British Journal of Nutrition* 1978; **40**: 497–504.
15. Thorland WG, Johnson GO, Tharp GD, Housh TJ, Cisar CJ. Estimation of body density in adolescent athletes. *Human Biology* 1984; **56**: 339–48.
16. Sinning WE, Dolny DG, Little KD, Cunningham LN, Racaiello A, Siconolfi SF, Sholes JL. Validity of generalized equations for body composition analysis in male athletes. *Medicine and Science in Sports and Exercise* 1985; **17**: 124–30.
17. Sinning WE, Wilson JR. Validity of generalized equations for body composition in women athletes. *Research Quarterly* 1984; **55**: 153–60.
18. Weiner JS, Lourie JA. *Practical human biology*. London: Academic Press, 1981.
19. Lohman TG, Roche AF, Martorell R, eds. *Anthropometric standardisation reference manual*. Champaign, Il: Human Kinetics.
20. Nyboer J. Percent body fat measured by four terminal bioelectrical impedance and body density methods in college freshmen. Proceedings of the 5th International Conference in Bioelectrical Impedance, August 1981, Tokyo, Japan.
21. Burkinshaw L. Measurement of body composition *in vivo*. In: Orton CG, ed. *Progress in Medical Radiation Physics* 2. New York: Plenum Press, 1985: 113–37.
22. Pierson RN, *et al.* Measuring body fat: calibrating the rulers. Intermethod comparisons in 389 normal Caucasian subjects. *American Journal of Physiology*, 1991; **261** (*Endocrinology and Metabolism 24*): E103–E108.
23. Lohmann TG. Applicability of body composition techniques and constants for children and youths. *Exercise and Sports Sciences Review*, 1986; **2**: 29–57.
24. American College of Sports Medicine. Position statement on proper and improper weight loss programs. *Medicine and Science in Sports and Exercise* 1983; **15**: ix–xiii.
25. Frisch RE, McArthur JW. Menstrual cycles: fatness as a determinant of minimum weight for height necessary for their maintenance of onset. *Science* 1974; **185**: 949–51.
26. Dalen N, Olsson KE. Bone mineral content and physical activity. *Acta Orthopaedica Scandinavica* 1974; **45**: 170–4.

APPENDIX. Recommended equations for the estimation of the body density of men and women

$D = a + b(x)$ where $x = \log_{10}$ sum (biceps, triceps, subscapular and suprailiac skinfolds) and D is the density; a and b are given in Table 1.

Table 1

	Men		Women	
	a	*b*	*a*	*b*
17–19 years	1162.0	63.0	1154.9	67.8
20–29 years	1163.1	63.2	1159.9	71.7
30–39 years	1142.2	54.4	1142.3	63.2
40–49 years	1162.0	70.0	1133.3	61.2
50 + years	1171.5	77.9	1133.9	64.5
17–70 years	1176.5	74.4	1156.7	71.7

Reproduced from ref. 12 with permission from Cambridge University Press.

The authors provide similar equations for individual skinfolds and all combinations of the four used here. The SEE values were 7.3–12.5 kg/m^3. SEE values for other groups will be higher (see text).

For men aged 18–60 years

$$D = 1109.4 - 0.827\,x + 0.0016x^2 - 0.257\,(\text{age, } y)$$

where x is the sum of chest, abdomen, and thigh skinfolds and SEE = 7.7 kg/m^3.[14] The authors provide validated equations for the sum of seven skinfolds (those above plus axilla, triceps, thigh, and suprailiac) transformed to natural logarithms and waist and forearm circumference.

For women aged 18–55 years

$$D = 1099.5 - 0.993\,x + 0.0023\,x^2 - 0.139\,(\text{age, } y)$$

where x is the sum of triceps, thigh, and suprailiac skinfold and SEE = 8.6 kg/m^3.[13] These authors provide validated equations for the sum of four skinfolds (those above plus abdomen), seven skinfolds (the preceding plus chest, axilla, subscapular), natural logarithm of the sum of skinfolds, and gluteal circumference.

1.5.2 Cardiovascular responses to exercise and training

PETER B. RAVEN AND R. DONALD HAGAN

INTRODUCTION

Physical exercise is associated with increases in the rate and depth of breathing to increase ventilation of the lungs, and increases in heart rate and stroke volume to increase cardiac output. These cardiorespiratory adjustments insure adequate delivery of oxygen to active muscles. Thus physical exercise involves an interaction among the respiratory, cardiovascular, and cellular energy systems of working muscles.

The gas transport mechanisms coupling cellular and pulmonary respiration are schematically illustrated (Fig. 1) using interlocking gears representing the functional interdependence of the major systems involved in oxygen delivery.

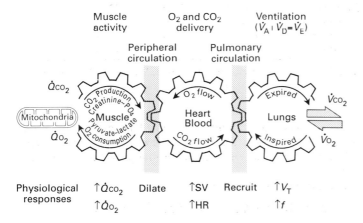

Fig. 1 A scheme illustrating the gas transport mechanisms for cellular (internal) to pulmonary (external) respiration. The gears represent the functional interdependence of the physiological components of the system. The large increase in oxygen utilization by the muscles (Qo_2) is achieved by increased extraction of oxygen from the blood perfusing the muscles, the dilatation of selected peripheral vascular beds, an increase in cardiac output (stroke volume and heart rate), an increase in pulmonary blood flow by recruitment and vasodilatation of pulmonary blood vessels, and finally an increase in ventilation. Oxygen is taken up ($\dot{V}o_2$) from the alveoli in proportion to the pulmonary blood flow and degree of oxygen desaturation of haemoglobin in the pulmonary blood. In the steady state, $\dot{V}o_2 = Qo_2$. Ventilation (tidal volume V_t and breathing frequency f increase in relation to newly produced carbon dioxide (Qco_2) arriving at the lungs and the drive to achieve arterial carbon dioxide and hydrogen ion homeostasis. (Reproduced from K. Wasserman, *et al. Principles of Exercise Testing and Interpretation.* Philadelphia: Lea & Febiger, 1987. Used with permission.)

The increase in oxygen use by active muscle occurs as a result of an increase in oxygen extraction from blood perfusing active muscle, vasodilation of peripheral vascular beds within active muscle, an increase in cardiac output and vasoconstriction in non-active vascular beds to maintain blood pressure, and an increase in pulmonary blood flow and ventilation[1]. In this review, we advance

the hypothesis that regulation of respiratory, cardiovascular, and metabolic function during exercise is based on the detection of error signals related to skeletal muscle oxygen demand. Monitoring these error signals leads to appropriate corrections in systemic blood flow and pressure facilitating the transport of oxygen to active skeletal muscle. Furthermore, we postulate that exercise training alters the respiratory, cardiovascular, and metabolic responses to exercise by altering the sensitivity of those mechanisms regulating systemic blood flow and pressure.

During exercise, working muscle utilizes oxygen to generate free energy to be stored in ATP and produces carbon dioxide as a byproduct. Therefore transfer of oxygen and carbon dioxide between the atmosphere and the mitochondria requires a highly co-ordinated interaction between cardiovascular and respiratory mechanisms linked to muscle metabolic activity to ensure adequate arterial oxygen content and the appropriate perfusion pressure to couple oxygen delivery and blood flow to the metabolic needs of the working muscle (see Fig. 1).

PHYSIOLOGICAL RESPONSES OF STATIC AND DYNAMIC EXERCISE

The acute metabolic and haemodynamic responses associated with exercise and training are dependent upon the type of muscle contraction. In general, the type of muscle contraction that occurs during exercise is either static or dynamic, or some proportional combination of the two. However, the acute cardiovascular responses to exercise and training are quite different to the responses to sustained and habitual use of these types of contractions.

Static muscle contractions

Blood flow is unaffected by isometric contractions which are less than 30 per cent of maximal voluntary contraction. However, isometric contractions above 30 per cent of maximal voluntary contraction increase intramuscular tissue pressure to the point of reducing the effective muscle perfusion pressure, thereby progressively reducing muscle blood flow and oxygen delivery in a graded fashion up to 100 per cent of maximal voluntary contraction.

Static isometric muscle contractions produce a sustained increase in heart rate and in systolic, diastolic, and mean arterial pressures in direct relation to the force or relative intensity of the contraction. The increase in heart rate is due to vagal withdrawal, while muscle ischaemia due to the static contraction potentiates the blood pressure response.[2,3] Static contractions are also associated with a small increase in left-ventricular end-diastolic pressure and an enhanced contractile state.[4] However, there is relatively little change in cardiac output or total peripheral resistance. Thus static exercise results primarily in an afterload pressure on the heart.

161

Cardiovascular adaptations to strength training

Weight lifting and resistance training utilizing sustained isometric contractions and slow muscle contractions of high force output increase muscular strength. The gain in strength is due to both neural factors and muscle hypertrophy. Neural involvement is indicated by an increase in maximal integrated electromyographic activation without a change in force per fibre output or the number of innervated motor units[5]. The increase in muscle hypertrophy is due primarily to an increase in muscle fibre cross-sectional size as a result of increases in the number of myofibrils per fibre, the sarcoplasmic volume, the total protein, and the hypertrophy of connective, tendinous, and ligamentous tissues.[5] The increase in muscle myofibril size actually decreases mitochondrial density. In some cases, muscle hypertrophy can result from an increase in fibre number (hyperplasia) as a result of stimulation of dormant stem cells.[6] Strength training increases intramuscular stores of ATP, creatine phosphate, and glycogen, but produces little change in anaerobic glycolytic and aerobic oxidative enzyme activities. Muscle hypertrophy is generally not as large in women as in men because of lower levels of circulating testosterone.

Cross-sectional studies[7] suggest that strength training increases left-ventricular mass without affecting left-ventricular volume. The increase in left-ventricular mass occurs without normalization to lean body mass. This indicates that chronic strength training alters the LaPlace constant (increased wall stress) of the heart, which in some cases may lead to an increase in myocardial oxygen demand similar to that seen in hypertensive individuals with concomitant cardiac hypertrophy. However, strength training has little or no effect on maximal cardiac output, stroke volume, arteriovenous oxygen difference, and hence maximal oxygen uptake.[7–10]

Dynamic muscle contractions

Dynamic contractions and cellular metabolism

During the onset of dynamic exercise, the heart rate increases due to withdrawal of vagal tone and an increase in sympathetic adrenergic activity.[11,12] The increase in sympathetic activity increases plasma noradrenaline concentration, cardiac contractility, and muscle sympathetic nerve activity.[12] The increase in plasma noradrenaline concentration is due to an increase in spillover from the vascular α-adrenergic receptors which mediate vasoconstriction. The amount of spillover is directly related to the exercise intensity.[13]

The increase in heart rate increases cardiac output, which in turn increases the mean arterial pressure. Vasoconstriction in splanchnic and renal vascular beds redistributes blood to the large central veins, increasing venous return and the central blood volume and thereby providing support for increases in stroke volume and cardiac output.[14] Thus dynamic exercise causes a volume load to the heart.

The accumulation of metabolites as a result of muscle contraction produces local muscle vasodilation.[15] However, the local muscle vasodilation response is not great enough to over-ride the tonic vasoconstrictor tone or totally abolish vasoconstrictor activity.[16] Much speculation exists as to the agent facilitating active muscle vasodilation. Carbon dioxide, adenosine, and potassium have been suggested as possible vasodilator regulators, but no one agent can explain all features of the vasodilator response.[2,3,15,17] It

has also been postulated that arterial baroreceptors and muscle chemoreflexes function to regulate sympathetic outflow, and thus muscle perfusion pressure and blood flow.[2,3] Whatever the mechanism, optimal perfusion of all active tissues is created by vasoconstriction in the non-working tissues which redistributes systemic blood flow to active muscle in accordance with the metabolic demands required by active muscle.

Effect of dynamic contractions on central circulatory responses

Oxygen uptake

In the laboratory, oxygen uptake $\dot{V}O_2$ is determined by indirect calorimetry.[18] Under these circumstances, $\dot{V}O_2$ is equal to pulmonary minute ventilation V_E multiplied by the difference between the inspired and expired oxygen percentages at the mouth as expressed in the following equation:

$$\dot{V}O_2 \text{ (l/min)} = V_E \times \text{(inspired fraction of } O_2 - \text{expired fraction of } O_2). \quad (1)$$

Oxygen uptake at rest is approximately 0.25 l/min. The maximum oxygen uptake $\dot{V}O_2max$ of a sedentary 70-kg man averages 3.0 l/min (43 ml/min.kg). $\dot{V}O_2max$ is determined during graded increases in exercise workload during cycle ergometer or motorized treadmill tests (Fig. 2).[19]

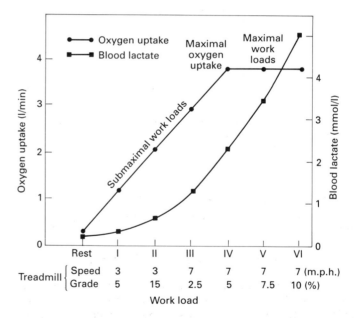

Fig. 2 The objective determination of maximal oxygen uptake $\dot{V}O_2max$ on a treadmill. The plateau of oxygen uptake $\dot{V}O_2max$ occurs while the subject continues to perform work at increasing workloads. The exponential increase in plasma lactate concentrations as workload increases beyond 50 per cent $\dot{V}O_2max$ should be noted. (Reprinted by permission of the *New England Journal of Medicine*, 1971; **284**: 1018–22.)

$\dot{V}O_2max$ is confirmed when additional increases in workload produce no further increase in $\dot{V}O_2$.[8,19] Therefore $\dot{V}O_2max$ is a rate measurement and indicates the maximal level of aerobic power output. Oxygen uptake can also be expressed as cardiac output Q_c multiplied by the systemic oxygen extraction, or the arteriovenous oxygen difference across active tissues, as

$$\dot{V}_{O_2} \text{ (l/min)} = Q_c \text{ (l/min)} \times \text{(arterial } O_2 \text{ content} - \text{venous } O_2 \text{ content)}. \qquad (2)$$

Therefore the maximal oxygen uptake can also be determined by measuring the maximal cardiac output and the arteriovenous oxygen difference. The maximal cardiac output is the product of maximal heart rate and maximal stroke volume, while the maximal arteriovenous oxygen difference depends upon the maximal oxygen content of the arterial blood and minimal oxygen content of the venous blood. The major factors affecting cardiac output are heart rate and stroke volume determinants, which are summarized in Fig. 3.

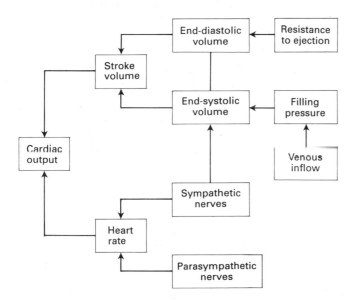

Fig. 3 A schematic block diagram summarizing the major factors determining cardiac output.

Cardiac output

Cardiac output Q_c is equal to heart rate HR times stroke volume SV:

$$Q_c = \text{HR (beats/minute)} \times \text{SV (ml/beat)}. \qquad (3)$$

At rest, cardiac output approximates 5–6 l/min. However, it increases to 20–25 l/min during maximal exercise in a sedentary individual. Cardiac output increases to its maximum as a result of increases in heart rate and stroke volume and the attainment of their respective maxima.[8,14,20] During maximal exercise, the arterial haemoglobin saturation and oxygen content remain relatively constant at 97 per cent and 20 ml/dl of blood respectively. However, the venous oxygen content decreases substantially owing to an increase in muscle oxygen extraction. During maximal exercise, the arteriovenous oxygen difference increases from 6 ml O_2/dl of blood at rest to 16 ml O_2/dl of blood at maximal exercise.

From equation (3), an example calculation for an individual with a maximal heart rate of 180 beats/min and a maximal stroke volume of 140 ml/beat is

$$\dot{V}_{O_2}\text{max} = 180 \text{ beats/min} \times 140 \text{ ml/beat} \times 16 \text{ ml/dl}$$
$$= 4.03 \text{ l/min}$$

Heart rate

Heart rate is the number of ventricular contractions occurring during a 1-min time frame. At rest, heart rate is approximately 70 beats/min in a sedentary individual, and is determined by the relative contributions of the parasympathetic and sympathetic nervous systems.[12] Increases in heart rate from a resting value up to an exercise rate of 100 beats/min are primarily a result of the withdrawal of vagal tone and minor increases in sympathetic activity. However, once the heart rate reaches 100 to 110 beats/min further increases in heart rate occur as a result of an increased sympathetic activity and continual vagal withdrawal up to a heart rate of 150 to 160 beats/min. Above 150 to 160 beats/min only slight adjustments in vagal tone up to a maximal heart rate occur (Fig. 4).

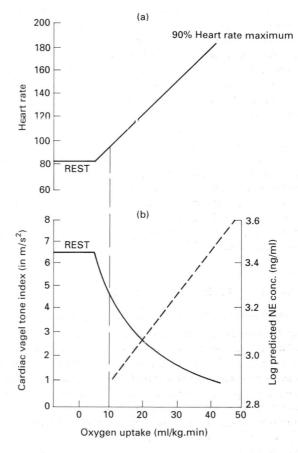

Fig. 4 (a) Changes in heart rate during a progressive increase in workload exercise test to 90 per cent of the maximal heart rate. It should be noted that the initial increase in heart rate from rest to 100 beats/min involves only vagal withdrawal. (b) Between heart rates of 100 beats/min and the maximum heart rate the increase is a result of further vagal withdrawal and increasing sympathetic drive (log NE conc) with the sympathetic drive predominating at heart rates in excess of 150 beats/min.

During exercise, heart rate increases in proportion to the workload and oxygen uptake, and reaches 190 to 200 beats/min at $\dot{V}_{O_2}$max.[11,12] Heart rate during submaximal exercise is elevated in individuals in whom stroke volume is reduced owing to a decrease in myocardial contractility as a result of deconditioning or cardiac disease.[8]

Stroke volume

Stroke volume is the amount of blood ejected from the heart with each heart beat and is equal to the difference between the amount of blood in the heart after completion of filling (the end-diastolic volume) and the amount remaining after ejection (the end-systolic volume).[21] At rest, the stroke volume is 80 to 90 ml/beat, while during maximal exercise stroke volume increases to 110–115 ml/beat (Fig. 5).

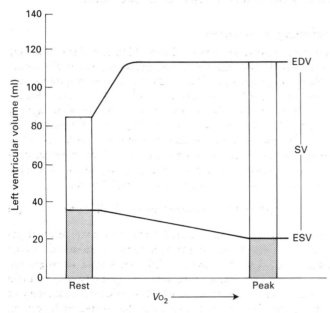

Fig. 5 Changes in end-diastolic volume (EDV) and end-systolic volume (ESV) of the left ventricle during upright bicycle exercise from rest to $\dot{V}O_2$peak. (Modification of the data of Poliner et al.[21] and reproduced with permission from Circulation 1980; **62**: 528–34. Copyright 1980 American Heart Association.)

The rhythmic muscle contractions of dynamic exercise increase the venous return and the intrathoracic and pulmonary capillary blood volumes.[22,23] The increase in these central volumes increases the reserve volume, thus providing for an enhanced preload for the heart. Therefore, increases in stroke volume are achieved by an increase in pulmonary vascular pressure which increases left ventricular filling pressure and end-diastolic volume and by an increase in left-ventricular contractility, as indicated by increases in ejection fraction and decreases in end-systolic volume[21] (Figs 5 and 6).

Myocardial oxygen consumption

During dynamic exercise, increases in heart rate, myocardial contractility, stroke volume, and blood pressure are associated with an increase in myocardial oxygen consumption. The arteriovenous oxygen difference across the heart during exercise is only slightly greater than that at rest.[8] This indicates that the increase in myocardial oxygen consumption is almost totally dependent upon an increase in coronary blood flow. The large increase in coronary blood flow with exercise suggests that a large vasodilator error signal over-rides the existing tonic vasoconstrictor tone.[24] While myocardial oxygen consumption during exercise is difficult to measure, the rate–pressure product is highly correlated to myocardial oxygen consumption and cardiac work and is used in clinical settings to assess the results of pharmacological, surgical, or exercise therapy.

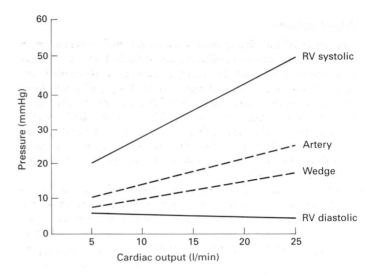

Fig. 6 Changes in right-ventricular (RV) and pulmonary (P) pressures from rest to maximal cardiac outputs during supine exercise. (Adapted from Ekelund and Holmgren,[14] and reproduced with permission from Chapman CB, ed. *Physiology of Muscular Exercise*. American Heart Association Monograph 15. Copyright American Heart Association 1967.)

Systemic arteriovenous oxygen difference

At rest, haemoglobin leaving the lungs for the peripheral vasculature is approximately 97 per cent saturated with oxygen producing an arterial oxygen content of 20 ml/100 ml of whole blood. As blood is transported through metabolically active tissues, large pressure gradients occur between the capillary and mitochondria.[25] These pressure gradients for oxygen establish the release of oxygen from haemoglobin and diffusion into the cell. Not all the oxygen is released from the haemoglobin as it passes through the peripheral tissue. Varying levels of venous oxyhaemoglobin saturation reflect the amount of oxygen extracted at the tissue level. The difference between systemic arterial and venous oxygen contents is called the arteriovenous oxygen difference and is an indicator of oxygen extraction. During exercise, the increase in the arteriovenous oxygen difference is also solely due to an increase in oxygen extraction by working skeletal muscle (Fig. 7).

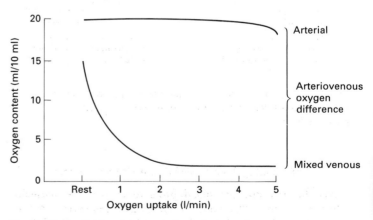

Fig. 7 Change in arteriovenous oxygen content from rest to $\dot{V}O_2$max of 5 l/min. Individuals with high $\dot{V}O_2$max values tend to have some arterial desaturation at maximal capacity; however, they appear to be able to extract more oxygen at the working muscle resulting in mixed venous oxygen contents of 2–3 ml O_2/100 ml of blood at $\dot{V}O_2$max. (Adapted from ref. 28, and reproduced with permission.)

Blood volume

The onset of exercise is usually associated with a reduction in plasma volume due to a redistribution of fluids from the vascular to the interstitial space. However, in most cases this shift is the result of movement to an upright exercise body position or to an increase in mean arterial pressure related to exercise intensity.[26] After the initial fluid shift, an equilibrium is reached between vascular fluid influx and efflux resulting in a constant plasma volume. A 'true' decrease in plasma volume usually only occurs as a result of total body dehydration due to prolonged exercise, high sweat rates, and heat stress.[27]

Either the onset of exercise or the decrease in plasma volume stimulates the release of neuroendocrine hormones associated with fluid and electrolyte balance. The release of these hormones, i.e. arginine vasopressin, atrial natriuretic peptide, and aldosterone, is related to the intensity of the exercise and during prolonged exercise is related to a decrease in cardiac filling pressure. The increase in the plasma concentration of these hormones and the constancy of the plasma volume during continuous exercise suggests an active regulation of the vascular fluid volume for the purpose of maintaining cardiac output and mean arterial pressure.[27]

Vascular resistance and conductance

Mean arterial blood pressure provides the driving force for systemic blood flow, and involves the interplay of cardiac output and peripheral vascular resistance. The major factors determining the mean systemic arterial pressure include the cardiac output and peripheral vascular resistance (Fig. 8). In the laboratory, cardiac output is determined by the product of heart rate and stroke volume, while total peripheral resistance (TPR) is equal to the ratio of mean arterial pressure (MAP) to cardiac output Q_c:

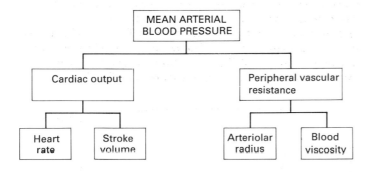

Fig. 8 Physiological factors that affect flow and resistance to produce the resultant mean arterial pressure, i.e. resistance = pressure/flow.

$$TPR \ (mmHg/l.min) = MAP \ (mmHg) \times Q_c \ (l/min) \qquad (4)$$

Systolic and diastolic blood pressure at rest are normally 120 mmHg and 80 mmHg respectively. However, during maximal exercise systolic blood pressure increases to 200–240 mmHg while diastolic pressure changes little or decreases to 60 mmHg, resulting in a widening of the pulse pressure and a moderate increases in mean arterial pressure.

During dynamic exercise, local muscle vasodilation produces a decrease in total peripheral resistance. In order to maintain mean arterial pressure and an adequate muscle perfusion pressure, the cardiac output increases. During maximal exercise, the increase in cardiac output in relation to the change in mean arterial pressure indicates that the decrease in total peripheral resistance is associated with an increase in vascular conductance. Since skeletal muscle vascular conductance can exceed cardiac pumping capacity, reductions in splanchnic and renal blood flow and vasoconstriction in non-active and exercising muscle occur in order to maintain cardiac output at a level high enough to ensure adequate muscle perfusion pressure and muscle blood flow.[2,17,28]

Pulmonary ventilation

Pulmonary ventilation at rest is 10 to 14 l/min and the breathing rate is 10 to 14 breaths/min. However, during maximal exercise ventilation will increase to 100 to 120 l/min, while respiration rate will increase to 40 to 50 breaths/min.[29] Tidal volume also increases to approximately two-thirds of the forced vital capacity with most of the increase occurring by 50 per cent $\dot{V}O_2$max. At the onset of dynamic exercise, pulmonary ventilation increases due to neurogenic and neurohumoral mechanisms which increases both breathing frequency and tidal volume. Thus ventilation is equal to breath rate multiplied by the volume of each breath:

$$\dot{V}_e (l/min) = f_b(breaths/min) \times V_t (l/breath) \qquad (5)$$

During maximal exercise, arterial lactic acid concentration increases in a curvilinear pattern as a result of increased energy production by anaerobic glycolysis (Fig. 2). Arterial hydrogen ion concentration is maintained constant up to approximately 50 to 60 per cent $\dot{V}O_2$max. At this point, which is called the ventilatory or anaerobic threshold,[1,23] it is proposed that increases in carbon dioxide flow stimulate carotid chemoreceptors which further increase pulmonary ventilation (hyperventilation) as a compensatory mechanism to regulate the plasma carbon dioxide partial pressure p_{CO_2}, the hydrogen ion concentration [H$^+$], and the arterial pH. Increases in [H$^+$] produced with lactate during inadequate oxygen delivery to active muscle increases capillary oxygen partial pressure p_{O_2}, and facilitates oxygen diffusion to mitochondria.[23] Muscle chemoreceptors sensitive to the concentration of lactic acid (or other metabolic byproducts) provide another and probably more dominant mechanism contributing to exercise hyperpnoea.[17]

Dynamic exercise produces an increase in venous return as a result of the pumping action of the exercising limbs and diaphragm. The increase in venous return in conjunction with vasodilation of the pulmonary vasculature produces an increase in pulmonary capillary blood volume. The increase in pulmonary capillary blood volume increases the lung oxygen diffusion capacity which serves to maintain arterial p_{O_2} and oxyhaemoglobin at nearly 97 per cent saturation. As long as arterial oxyhaemoglobin saturation is maintained, pulmonary ventilation is not a limiting factor to $\dot{V}O_2$max.

Factors limiting maximal oxygen uptake capacity

There is substantial evidence that in normal man $\dot{V}O_2$max is the objective measure of the supply of oxygen to the exercising muscle, i.e. it is the maximum capacity to deliver oxygen. However, despite the integrative aspects of the oxygen transport system, it is generally held that cardiac output is the major limiting factor in the achievement of $\dot{V}O_2$max.

Cardiac function

During maximal exercise, arterial pO_2 is maintained constant while skeletal muscle vascular conductance is expanded. This indicates that oxygen delivery is greatly increased. However, skeletal muscle vascular conductance capacity is larger than the capacity of the heart to expand cardiac output, and whole-body $\dot{V}O_2$max is produced through engagement of only one-third of the total skeletal muscle mass.[16,28] Thus the limited ability to expand cardiac output in relation to the vascular conductance capacity strongly suggests that cardiac output is the limiting factor to oxygen delivery and $\dot{V}O_2$max. If a larger muscle mass is engaged during exercise, vasoconstriction is thought to occur in the arterioles of the exercising limbs to avoid a reduction in blood pressure (Fig. 9).

If vasoconstriction did not occur, exercise requiring a majority of the muscle mass of the body would require a cardiac output of up to 60 l/min to prevent a drop in blood pressure.[16] However, a major question remains as to what the signal is for the sympathetically mediated vasoconstriction.

Pulmonary function

The lack of dyspnoeic sensations during exhaustive exercise suggests that pulmonary blood gas transport is most efficient. However, during graded exercise there is a progressive widening of the alveolar to arterial pO_2 difference owing to intraregional variations in the relationship between alveolar ventilation and lung blood flow distribution.[22] The presence of arterial oxygen desaturation during maximal exercise suggests that the respiratory function can be a limiting factor to $\dot{V}O_2$max. However, as long as pulmonary blood flow and pulmonary capillary blood volume increase during maximal exercise, the mean pulmonary capillary transit time appears sufficient to permit alveolar–arterial oxygen equilibration (Fig. 10).[22]

Skeletal muscle oxygen utilization

While many researchers contend that central circulatory factors limit $\dot{V}O_2$max, others[18,25] argue that the major limitation to $\dot{V}O_2$max is the rate at which oxygen is moved by diffusion from haemoglobin in the red cell to the muscle mitochondria. This concept is supported by evidence showing $\dot{V}O_2$max reduced by decreases in red cell oxygen-carrying capacity and muscle ischaemia due to peripheral vascular diseases, and by evidence showing little increase in $\dot{V}O_2$max during the breathing of hyperoxic gases.[30]

Factors influencing maximal oxygen uptake capacity

Much of the physiological variation in $\dot{V}O_2$max can be attributed to genetics, gender, age, body composition, and habitual level of physical activity.[8]

Genetics and $\dot{V}O_2$max

Evidence supports the concept that genetic factors are associated with individual differences in $\dot{V}O_2$max capacity and changes in $\dot{V}O_2$max with training.[31] It has also been shown that there is a maternal effect on maximal exercise heart rate, blood lactate, and $\dot{V}O_2$max. Recent evidence[33] suggests that sequence variations in mitochondrial DNA may contribute to individual differences in $\dot{V}O_2$max and its response to training. In addition, the greater cardiorespiratory functional capacity of endurance athletes is usually manifested at an early age prior to extensive training.[34] The selection of endurance sports by some athletes may be related

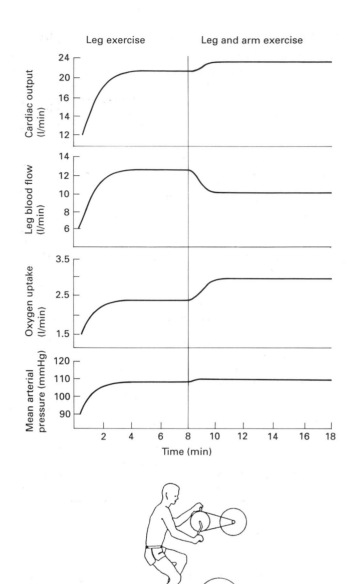

Fig. 9 A schematic representation of the decrease in flow that occurs in exercising legs when additional arm exercises is added. The cardiac output increases to increase flow to exercising arms and the distribution of cardiac output to the legs is reduced as a means of maintaining mean arterial pressure. However, the locus of the error signal necessary to produce and maintain active vasoconstriction in the legs has not been determined. (Modified from the data presented in ref. 28 and reprinted with permission.)

to early success in endurance events due to an already elevated $\dot{V}O_2$max capacity. However, the $\dot{V}O_2$max capacity of endurance athletes is also a result of adaptations in cardiovascular, respiratory, and muscle metabolic structure and function as a result of endurance exercise training.[28] Thus attainment of world-class status in endurance sports is probably due to a combination of genetic endowment and participation in a programme of long-term training.

Body composition and $\dot{V}O_2$max

Maximal oxygen uptake expressed relative to body weight (ml/min.kg) serves to normalize individuals of high and low body

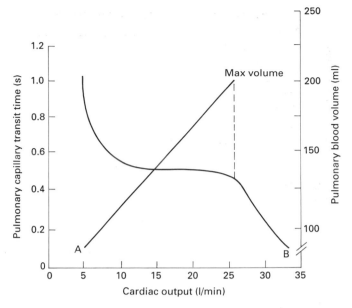

Fig. 10 A schematic outline of progressive increases in (a) pulmonary capillary blood volume in relation to (b) decreases in pulmonary capillary transit time with increasing cardiac output. As can be seen, early on in exercise capillary recruitment increases the measured capillary blood volume and decreases the mean transit time without arterial haemoglobin desaturation. From cardiac outputs of 15 l/min to 25–28 l/min arterial haemoglobin saturation is maintained because of the increasing capillary recruitment, increasing pulmonary capillary volume and maintaining transit time constant. When maximum capillary volume is achieved at cardiac outputs of 25–28 l/min, the transit time decreases to a time which is insufficient to provide complete arterial haemoglobin saturation. (Redrawn from data presented in ref. 28 and reprinted with permission.)

weight and is the most useful indicator of cardiorespiratory functional capacity. Endurance athletes are usually lower in body weight compared with sedentary individuals. However, while the sedentary individual possesses a greater amount of body fat, the lean body weights of sedentary individual and endurance athletes are remarkably similar.[35,36] Maximal oxygen uptake capacity can be altered by changes in body composition. Decreases in body fat due to dieting alone, accompanied by little or no change in lean body weight, will increase $\dot{V}O_2max$ when $\dot{V}O_2max$ is expressed relative to body weight.[37]

Gender and $\dot{V}O_2max$

Maximal oxygen uptake is also influenced by gender. $\dot{V}O_2max$ expressed in absolute units, i.e. litres per minute, increases linearly with increases in height, weight, and age in both boys and girls. The physical stature and body weight of boys and girls is similar up to puberty, making their $\dot{V}O_2max$ capacities almost equivalent.[20] During and after puberty, males experience an increase in lean muscle mass while females increase their percentage of body fat. As a result males experience a greater increase in $\dot{V}O_2max$. The magnitude of these changes in certain individuals may contribute to the type of sport that they choose to perform. Endurance athletes tend to be individuals of low body weight and body fat, while strength and anaerobic athletes tend to be large and muscular. The $\dot{V}O_2max$ of adult females is about 20 per cent lower than that of adult males.[19,38] However, this difference is reduced to about 5 per cent when $\dot{V}O_2max$ is normalized to fat-free lean body mass. This small difference may be due to the lower haemoglobin concentration of females.[39]

Age and $\dot{V}O_2max$

The maximal oxygen uptake of children and adults expressed in litres per minute is proportional to height and body weight, and increases in a linear pattern with body growth.[20] Cross-sectional studies show that after the age of 20 years there is a decline in $\dot{V}O_2max$ ranging from 0.3 to 0.5 ml/min.kg per year.[39,40] The decrease in $\dot{V}O_2max$ is also associated with a decrease in maximal exercise heart rate of 0.5 to 1.0 beats/min.year.[40] The age-related decline in $\dot{V}O_2max$ appears to be due to a decrease in maximum heart rate, cardiac output, and systemic arteriovenous oxygen difference. However, much of the age-related decline in $\dot{V}O_2max$ can be explained by a more sedentary life-style and an increase in body fat. The $\dot{V}O_2max$ of athletes engaged in endurance training is higher than that of sedentary individuals of similar age, and individuals who maintain a physically active life-style have a smaller amount of body fat and a slower rate of decline in aerobic endurance capacity with advancing age.[40]

Physical activity and $\dot{V}O_2max$

Cardiorespiratory functional capacity is most profoundly influenced by the level of habitual physical activity.[42] Prolonged exposure to bed rest and detraining leads to a decrease in $\dot{V}O_2max$ (Fig. 11), while endurance exercise training produces an increase in $\dot{V}O_2max$.[8,19,40,41]

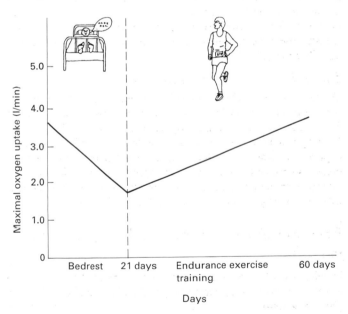

Fig. 11 A representative description of the loss of maximal oxygen uptake $\dot{V}O_2max$ with complete bed rest and the prolonged exercise training necessary to achieve the same level of $\dot{V}O_2max$ that occurred prior to the bed rest. (Adapted from ref. 41 and reproduced with permission from *Circulation* 1969; **38** (Suppl.): 1–78. Copyright 1969. American Heart Association.)

The rate and magnitude of increase in $\dot{V}O_2max$ is related to the age of the individual, the level of $\dot{V}O_2max$ prior to the commencement of training,[16,40,41] and the frequency, intensity, and duration of the exercise training programme.[42]

Cardiovascular adaptations to dynamic exercise training

The adaptive responses to habitual endurance exercise occur throughout training.[43] The adaptations to endurance exercise occur in both central and peripheral circulatory function, and lead to increases in maximal oxygen uptake and muscle oxygen extraction, and an improved ability to sustain submaximal steady state exercise. The increase in $\dot{V}O_2$max is related primarily to an increase in maximal cardiac output and secondarily to an increase in oxygen extraction across the active muscles (Fig. 12).

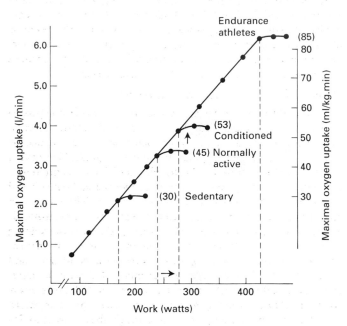

Fig. 12 The range of $\dot{V}O_2$max that exists within a normal healthy adult population. It should be noted that the range includes sedentary individuals with $\dot{V}O_2$max = 30 ml/kg.min, normally active individuals with $\dot{V}O_2$max = 45 ml/kg.min, active people who carry out a 2–3 month training programme with $\dot{V}O_2$max = 53 ml/kg.min, and the world-class endurance performer with $\dot{V}O_2$max = 80 ml/kg.min. (Reprinted from ref. 28, with permission.)

$\dot{V}O_2$max and training

Cross-sectional studies show that the $\dot{V}O_2$max of endurance trained athletes is higher than that of sedentary individuals.[44] The $\dot{V}O_2$max of sedentary individuals is usually less than 40 ml/min.kg. The $\dot{V}O_2$max of world-class male endurance athletes can exceed 80 ml/min/kg, while the $\dot{V}O_2$max of local area champions may vary between 60 and 70 ml/min.kg. There is a strong relationship between training volume, as determined from the product of exercise frequency, duration, and intensity, and $\dot{V}O_2$max capacity.[40–42] In the sedentary individual, endurance training can increase $\dot{V}O_2$max by 15 to 30 per cent during the first 2 to 3 months of training. Further increases up to 40 to 50 per cent can occur over the next 9 to 24 months of training. However, there is little change in $\dot{V}O_2$max after this point. Despite this $\dot{V}O_2$max plateau and because of continued intracellular metabolic adaptations, performance in endurance events may continue to improve with continued endurance exercise training. The termination of training results in a slow and gradual decline in $\dot{V}O_2$max to pretraining

levels over a 6-month period with a more rapid decline in the metabolic energy channelling systems of the skeletal muscle.

Submaximal exercise responses

The cardiorespiratory response to submaximal exercise is improved by endurance exercise training. Heart rate for any given absolute level of submaximal power output is reduced, while stroke volume is increased.[19,41] The reduced heart rate is only partially related to a decrease in sympathetic stimulation because the decline in plasma catecholamines is completed by the third week of training, while the decline in resting and submaximal heart rates continues over a longer period of training. Resting heart rates of world-class endurance runners vary from 30 to 40 beats/min and remain in sinusoidal rhythm.[1,25] However, pre- and post-training concentrations of plasma catecholamines are similar at any given percentage of $\dot{V}O_2$max. These differences imply that the reduction in heart rate during submaximal exercise is due to other factors such as a decrease in the sensitivity of β_1 receptors on the heart,[45] the increased vagal control of the heart,[46] or an increase in filling volume at any given heart rate as a result of training-induced increases in blood volume.[27]

The cardiac output, arteriovenous oxygen difference, and oxygen uptake for any absolute level of exercise power output are unchanged by endurance exercise training. As a result, muscle blood flow and blood flow to inactive tissues during submaximal steady state exercise are not affected by endurance training.

The rate of attainment of the steady state responses for oxygen uptake, cardiac output, and heart rate occur more rapidly after endurance training.[47,48] The more rapid attainment of the cardiorespiratory steady state during submaximal exercise after training is possibly a reflection of the increased muscle mitochondrion volume, which increases ATP replenishment and reduces the magnitude of cellular oxygen deficit, and also to a more rapid increase in blood flow to active muscle.[23]

Haemodynamic adjustments

Exercise training increases $\dot{V}O_2$max by increasing maximal cardiac output and maximal arteriovenous oxygen difference. Endurance training does not change the maximal exercise heart rate. However, the maximal stroke volume is increased as a result of an increase in cardiac filling pressure (Fig. 13). The greater preload and cardiac filling pressure of the endurance athlete during exercise is probably due to a higher pulmonary capillary wedge pressure[49] and an increase in diastolic reserve as a result of an increase in peripheral venous compliance. Thus the greater cardiac filling pressure produces a greater maximum stroke volume which allows the development of a greater maximum exercise cardiac output.[28,49]

Endurance exercise training does not affect maximal systolic blood pressure. However, maximal diastolic blood pressure is decreased by endurance training, probably as a result of an increased capacity for muscular vasodilation.[50] The increase in vasodilator capacity lowers the total peripheral resistance and reduces ventricular after-load. This enables the endurance athlete to generate a higher maximal exercise cardiac output while maintaining a normal mean arterial pressure. The greater vascular conductance of the endurance-trained athlete leads to an increased capacity for muscle oxygen extraction. The increased capacity for muscle oxygen extraction may also be related to an increase in oxygen diffusion capacity owing to an increase in muscle vascularity and capillary density. This increase in oxygen diffusion capacity from vessel to respiratory chain enzymes also increases $\dot{V}O_2$max.[25]

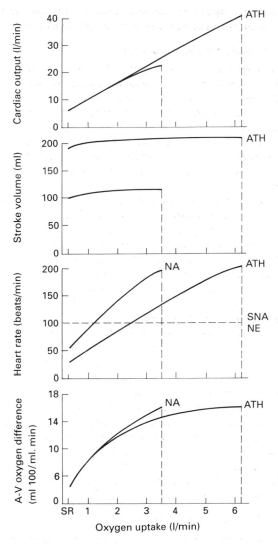

Fig. 13 A description of the circulatory responses to graded exercise in normal active individuals and world-class performers. The large differences between the cardiac output range and resting and exercise stroke volumes of the athletes compared with those of the normally active should be noted. In fact, the primary difference in the greater $\dot{V}o_2$max of the athletes is clearly the difference in resting stroke volumes. This difference is maintained throughout exercise to $\dot{V}o_2$max and is evidence of the training-induced increase in pumping capacity of the heart. NA = normally active, ATH = athletes, NE = norepinephrine (noradrenaline), SNA = sympathetic nerve activity. (Adapted from data presented in ref. 28 and reprinted with permission.)

Respiration function

The endurance-trained athlete can produce a higher maximal pulmonary ventilation and breathing rate.[29] However, the maximal tidal volume is unchanged by training. The higher pulmonary ventilation of endurance athletes can produce arterial hypoxaemia, thereby limiting $\dot{V}o_2$max.[22] In addition, the endurance athlete can generate higher levels of lactic acid and tolerate a lower pH level, i.e. greater metabolic acidosis. Endurance training shifts the oxygen–haemoglobin dissociation relation such that, at any given level of po_2, oxygen is released more readily from haemoglobin. The improved oxygen–haemoglobin dissociation may be due to a greater erythrocyte concentration of 2,3-diphosphoglycerate.[19]

Cardiac structure

Endurance exercise training increases cardiac mass and volume.[7] Left-atrial and right-ventricular dimensions and intraventricular septal and posterior wall thicknesses are consistently larger in endurance athletes. Endurance athletes have a larger absolute left-ventricular mass and left-ventricular end-diastolic volume even when normalized to lean body mass. However, the ratio of left-ventricular mass to left ventricular end-diastolic volume, and the ratio of left-ventricular wall thickness to left-ventricular chamber radius are unchanged. Parallel increases in left-ventricular mass and volume occur to maintain a constant relation between systolic blood pressure and the ratio of left-ventricular wall thickness to ventricle radius.[7,41] Thus wall tension, or stress, is held constant in accordance with LaPlace's law. While the results from humans are equivocal, studies of experimental animals show that chronic physical exercise increases myocardial capillary growth and enlarges extramural vessels.[51] It has been postulated that these changes lead to an increase in myocardial perfusion capacity.[52]

Blood volume and thermoregulation

Chronic endurance exercise training leads to an increase in plasma volume, red cell volume, and hence total blood volume. Initially, the increase in blood volume is confined to an increase in plasma volume,[53,54] but as training progresses there is a gradual increase in the total number of red blood cells. Haemoglobin concentration is unaffected by training, but because of the increase in the number of red cells, total haemoglobin content is increased. The increase in plasma volume with training is related to the level of sympathetic activity and total body dehydration during the exercise training sessions,[53,54] the release and concentrations levels of neuro-endocrine hormones affecting the renal reabsorption of water and electrolytes,[56] and to the intravascular influx of albumin to maintain plasma colloid oncotic and osmotic pressure.[54,55] These alterations in volume-regulating hormones and plasma proteins persist after the training session, and it is through these sustained increases that the blood volume changes occur during recovery from the exercise training session. In addition, endurance training increases the renal sensitivity of these hormone responses producing more efficient reabsorption of water and sodium.[55]

Endurance exercise training increases the capacity to dissipate body heat during exercise.[56] This is accomplished through a reduction in the threshold for sweating at any given level of central thermoregulatory drive, and an increase in the sensitivity of the central mechanisms. Heat acclimatization alone reduces the threshold for sweating of the sweat gland, but does not appear to change the sensitivity of the central mechanism for heat dissipation. However, endurance training and heat acclimatization have little effect on cutaneous blood flow. Accordingly, for any given level of heat stress, an equivalent workload is accomplished with a lower cardiovascular strain. In addition, endurance exercise training is associated with an increased sensitivity of the neuro-endocrine hormones responsible for renal reabsorption of water and electrolyte, and an enhanced perception of the need to drink. These changes serve as the basis for the finding that endurance training increases plasma volume and diastolic reserve.[56]

MECHANISMS OF CARDIOVASCULAR REGULATION DURING STATIC AND DYNAMIC EXERCISE

The primary function of the cardiovascular system is to supply oxygen and energy substrates to tissues of the body and remove metabolic waste products. The matching of cardiovascular and metabolic function during exercise is based on open-loop and closed-loop feedback circuits. The basic pattern of effector activity at the onset of exercise is set by centrally generated somatomotor ('central command') and cardiovascular motor signals (Fig. 14).[2,3,17]

In this model, 'central command' initiates the cardiovascular response and appears to establish the level of sympathetic and parasympathetic autonomic efferent activity to the heart and vascular vessels in parallel with recruitment of the requisite motor units to perform muscular work.[17] In addition, this activity is then

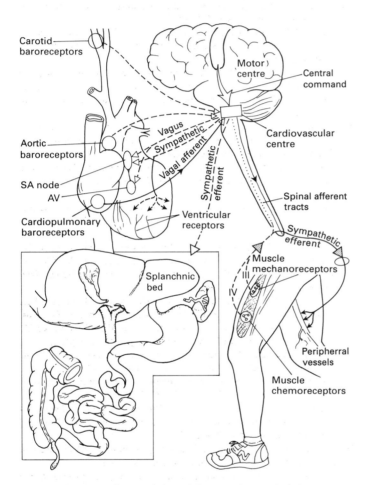

Fig. 14 A schematic description of the neural control of the circulation during exercise. Central command from the cortex establishes the required cardiovascular reference activity at the cardiovascular centre in parallel with the muscle fibre activity (EMG) of the working skeletal muscle in direct proportion to the required force output of the muscle. Muscle mechanoreceptors and muscle metaboreceptors via type 3 and type 4 afferent neural fibres provide the cardiovascular centre with information regarding the adequacy of the perfusion of the muscle. The error signal generated at the cardiovascular centre results in increased sympathetic stimulation of cardiac output (HR and SV) and regional vasoconstriction. Cardiopulmonary baroreceptors regulate peripheral vasoconstriction reflexly and are thought to monitor and maintain cardiac filling pressure. The arterial baroreflexes appear to act as a brake to increases in blood pressure by counteracting heart rate increases and vasoconstriction mediated via the muscle reflexes. Local venovasospinal reflexes and/or encephalinergic modification of the sympathetic ganglia may play a role in regulating flow through the working muscle.

modulated by mechanosensitive and chemosensitive afferent nerve fibres (unmyelinated type 3 and type 4 nerves) in active muscle which are sensing changes in muscle contractile force, metabolism, and perfusion pressure or flow.[2,3,15,17] Furthermore, modulation by mechanosensitive afferents within the carotid sinuses and aortic arch is thought to occur in response to the development of specific blood pressure error signals.[2,3]

Adjustment of error signal between oxygen demand and supply

Exercise increases oxygen utilization and hence oxygen demand. The increase in oxygen demand is met by increases in respiratory and cardiovascular functions to provide and maintain oxygen delivery to the active skeletal muscle for the purpose of maintaining energy production. At the onset of exercise, recruitment of muscle fibres and the increase in oxygen utilization by the increased active muscle causes a large error signal between oxygen demand and oxygen supply of the active muscle for a short period of time. Since the respiratory, cardiovascular, and muscle metabolic systems are partially regulated by negative feedback mechanisms, this error signal (an imbalance between oxygen delivery, fibre activation, and energy metabolism) is detected by peripheral mechano- and chemoreceptor sensors.[2,3] The mechano- and chemoreceptor sensors transmit afferent impulses to the cardiovascular centres in the brain which regulate the dissemination and distribution of efferent signals to target arterioles for the purpose of correcting the error signal. In muscle, the metabolic error signal appears to involve the monitoring of the 'phosphorylation state,' which may be linked to muscle blood flow, while in the respiratory system this involves monitoring the partial pressures of oxygen and carbon dioxide and pH by chemoreceptors which may be located within the muscle as well as the cardiovascular system. In the cardiovascular system, the error signal is linked to alterations in central and peripheral blood pressure caused by variations in cardiac output and total peripheral resistance. The interplay between cardiac output and peripheral resistance forms the basis of feedback from peripheral sensors which monitor blood flow (muscle chemoreceptors) and blood pressure (baroreceptors) in conjunction with 'central command'.

During dynamic exercise, cardiovascular adjustments include a reduction in working muscle vascular resistance, increases in heart rate and cardiac contractility, and an increase in pulmonary ventilation. However, it is not known whether the primary signal is a 'flow error' based on the accumulation of muscle metabolites that activate group 3 and group 4 chemosensitive muscle afferents as a result of a mismatch between blood flow and metabolism, or as a 'blood pressure error' that activates arterial baroreflexes and raises blood pressure as a result of a mismatch between cardiac output and vascular conductance.[2,3,17]

The coupling of oxygen demand and oxygen supply during dynamic exercise

The oxygen uptake of an individual with an arterial oxygen content of 20 ml/dl of blood and a total blood volume of 5 l is 1 l/min. This indicates that 5 l/min of cardiac output is necessary for effective transport of 1 l/min of oxygen. This ratio of 5 l/min of cardiac output per 1 l/min oxygen uptake is constant from rest to maximal oxygen uptake. In normal individuals, the coupling of cardiac output Q_c and oxygen uptake $\dot{V}O_2$max is unaffected by sex, age, body size, exercise position, and aerobic fitness capacity.

The $\dot{V}O_2$–Q_c relationship is also maintained during exposure to high altitude and during anaemia,[8,19] but is disrupted in pathological conditions which hinders the transfer of oxygen from ambient air to the mitochondria.[8,57] Pulmonary disease reduces alveolar diffusion capacity and the oxygen content of the arterial blood. The oxygen content of the blood is also reduced with anaemia, with carbon monoxide poisoning, and in genetic conditions altering the oxygen binding of haemoglobin. Oxygen delivery is reduced in any disease or condition that reduces cardiac output and cardiac function, or reduces venous return.[57] Thus the remarkable aspect of cardiovascular regulation during exercise is the exquisitely sensitive coupling that exists between the metabolic demand for oxygen and the delivery of oxygen to the working muscles by the cardiovascular system.

Oxygen uptake is also limited by defects in cellular metabolism regulating ATP regeneration.[13] Patients with deficiencies in phosphorylase or phosphofructokinase (McCardle's patients) have an impaired ability to utilize intramuscular glycogen as an energy substrate. In these disorders, the coupling of cardiac output with oxygen uptake is disrupted, as evidenced by a slope of increase in cardiac output in relation to oxygen uptake which is two- to threefold greater than normal and by the reduced arteriovenous oxygen difference at maximal exercise. However, the extramuscular delivery of oxidizable substrates (free fatty acids, glucose, lactate, etc.) via the blood serves to increase maximal oxygen uptake during graded exercise and normalizes (although not completely) the relationship between cardiac output and oxygen uptake.[13] Thus activations of metabolically sensitive muscle afferents by changes in muscle oxidative phosphorylation are linked to the regulation of cardiac output.

SUMMARY

The experimental evidence suggests that cardiovascular function plays an integral role in the linking of oxygen demand with oxygen supply. During exercise, the cardiovascular system functions to deliver oxygen and energy substrates to metabolically active skeletal muscle. Increases and decreases in blood flow to active and non-active tissues respectively occur in response to blood flow error and blood pressure error signals. Flow error signals sensed in active muscle and pressure error signals sensed by arterial baroreceptors appear to play a major role in the regulation of cardiac output and distribution of peripheral blood flow. Active muscle ATP concentrations appear to play an important role in the setting of oxygen demand. The level of oxygen demand in turn establishes the appropriate error signals which produce and maintain oxygen delivery. Endurance exercise training increases the capacity to perform exercise. A large portion of the increase in maximal exercise capacity occurs as a result of an increase in oxygen delivery due to an increase in exercise stroke volume, cardiac output, and vasodilatory capacity, while a smaller portion is due to an increased capacity for muscle oxygen extraction due to an increase in muscle mitochondrial volume. Endurance exercise training appears to alter the sensitivity of certain effector tissues to neurohumeral agents which subsequently alter physiological responses at rest and during exercise.

REFERENCES

1. Wasserman K, Hansen JE, Sue DY, Whipp BJ. *Principles of exercise testing and interpretation.* Philadelphia: Lea & Febiger, 1987.

2. Rowell LB, O'Leary DS. Reflex control of the circulation during exercise: chemoreflexes and mechanoreflexes. *Journal of Applied Physiology* 1990; **69**: 407–18.

3. Rowell LB, Sheriff DD. Are muscle 'chemoreflexes' functionally important? *News in Physiological Sciences* 1988; **3**: 250–3.

4. Mullins CB, Blomqvist CG. Isometric exercise and the cardiac patient. *Texas Medicine* 1973; **69**: 53–8.

5. Fox EL. *Sports physiology.* Philadelphia: W. B. Saunders, 1979: 121–58.

6. Mikesky AE, Giddings CJ, Matthews SW, Gonyea WJ. Changes in muscle fiber, size and composition in response to heavy-resistance exercise. *Medicine and Science in Sports and Exercise* 1991; **23**: 1042–9.

7. Longhurst JC, Kelly AR, Gonyea WJ, Mitchell JH. Left ventricular mass in athletes. *Journal of Applied Physiology* 1980; **48**: 154–62.

8. Mitchell JH, Blomqvist G. Maximal oxygen uptake. *New England Journal of Medicine* 1971; **284**: 1018–22.

9. Smith ML, Graitzer HM, Hudson DL, Raven PB. Baroreflex function in endurance- and static exercise-trained men. *Journal of Applied Physiology* 1988; **64**: 585–91.

10. Smith ML, Raven PB. Cardiorespiratory adaptations to training. In: Blair SN, Painter P, Pate RR, Smith LK, Taylor CB, eds. *Resource manual for guidelines for exercise testing and prescription.* Philadelphia: Lea & Febiger, 1988: 62–5.

11. Clausen JP. Circulatory adjustments to dynamic exercise and effect of physical training in normal subjects and in patients with coronary artery disease. *Progress in Cardiovascular Diseases* 1976; **18**: 459–95.

12. Ekblom B, Kilbom A, Soltysiak J. Physical training, bradycardia, and autonomic nervous system. *Scandinavian Journal of Clinical and Laboratory Investigation* 1973; **32**: 251–6.

13. Lewis SF, Haller RG. Skeletal muscle disorders and associated factors that limit exercise performance. *Exercise and Sport Sciences Reviews* 1989; **17**: 67–113.

14. Ekelund LG, Holmgren A. Central hemodynamics during exercise. In: Chapman CB, ed. *Physiology of muscular exercise.* American Heart Association Monograph 15. New York: American Heart Association, 1967: 133–43.

15. Mitchell J, Schmilt RF. Cardiovascular reflex control of afferent fibers from skeletal muscle receptors. In: Shepher JT, Abbound FM, eds. *Handbook of physiology: the cardiovascular system*, Part III, Chapter 17. Bethesda, MD: American Physiological Society, 1983: 623 58.

16. Saltin B. Physiological adaptation to physical conditioning. Old problems revisited. *Acta Medica Scandinavica* 1986; Suppl. 711: 11–24.

17. Mitchell JH. Neural control of the circulation during exercise. *Medicine and Science in Sports and Exercise* 1990; **22**: 141–54.

18. Connett RJ, Honig CR, Gayeski TEJ, Brooks GA. Defining hypoxia: a systems view of $\dot{V}O_2$, glycosis, energetics, and intracellular Po_2. *Journal of Applied Physiology* 1990; **68**: 833–42.

19. Snell PG, Mitchell JH. The role of maximal oxygen uptake in exercise performance. *Clinics in Chest Medicine* 1984; **5**: 51–62.

20. Åstrand P.-O. *Experimental studies of physical working capacity in relation to sex and age.* Copenhagen: Munksgaard, 1952.

21. Poliner LR, Dehmer GJ, Lewis SE, Parkey RW, Blomqvist CG, Willerson JT. Left ventricular performance in normal subjects: a comparison of the responses to exercise in the upright and supine positions. *Circulation* 1980; **62**: 528–34.

22. Dempsey JA, Gledhill N, Reddan WG, Forster HV, Hanson PG, Claremont AD. Pulmonary adaptation to exercise: effects of exercise type and duration, chronic hypoxia and physical training. *Annals of the New York Academy of Sciences* 1977; **301**: 243–61.

23. Wasserman K, Hansen JE, Sue DY. Facilitation of oxygen consumption by lactic acidosis during exercise. *News in Physiological Sciences* 1991; **6**: 29–34.

24. Stone HL. Control of the coronary circulation during exercise. *Annual Review of Physiology* 1983; **45**: 213–27.

25. Wagner PD. The determinants of $\dot{V}O_2$max. *Annals of Sports Medicine* 1988; **4**: 196–212.

26. Hagan RD, Diaz FJ, McMurray R, Horvath SM. Plasma volume changes related to posture and exercise. *Proceedings of the Society of Experimental Biology and Medicine* 1980; **165**: 155–60.

27. Convertino, VA. Blood volume: Its adaptation to endurance training. *Medicine and Science in Sports Exercise* 1991; **23**: 1338–48.

28. Rowell LB. *Human circulation: regulation during physical stress.* New York: Oxford University Press, 1986: 1–416.

29. Folinsbee LJ, Wallace Es, Bedi JF, Horvath SM. Exercise respiratory pattern in elite cyclists and sedentary subjects. *Medicine and Science in Sports and Exercise* 1983; **15**: 503–9.

30. Kaijser L. Limiting factors for aerobic muscle performance: the influence of varying oxygen pressure and temperature. *Acta Physiologica Scandinavica* 1970; Suppl. 346: 1–96.

31. Klissouras V. Keritability of adaptive variation. *Journal of Applied Physiology* 1971; **31**: 338–44.

32. Lesage R, Simoneau JA, Jobin J, Leblanc J, Bouchard C. Familial resemblance in maximal heart rate, blood lactate, and aerobic power. *Human Heredity* 1985; **35**: 182–9.

33. Dionne FT, Turcotte L, Thibault MC, Boulay MR, Skinner JS, Bouchard C. Mitochondrial DNA sequence polymorphism, V_{O_2}max, and response to endurance training. *Medicine and Science in Sports and Exercise* 1991; **23**: 177–85.

34. Murase Y, Kobayshi K, Kamei S, Matsue H. Longitudinal study of aerobic power in superior athletes. *Medicine and Science in Sports and Exercise* 1981; **13**: 180–4.

35. Hagan RD, Gettman LR, Maximal aerobic power and serum lipoproteins in male distance runners matched to sedentary controls by physical characteristics and body composition. *Journal of Cardiac Rehabilitation* 1983; **3**: 331–7.

36. Upton SJ, Hagan RD, Rosentswieg J, Gettman LR. Comparison of the physiological profiles of middle-aged women distance runners and sedentary women. *Research Quarterly of Exercise and Sport* 1983; **54**: 83–7.

37. Hagan RD, Upton SJ, Whittman J, Wong L. The effects of aerobic conditioning and/or caloric restriction on body composition, maximal aerobic power, and serum lipoprotein fractions in overweight adult men and women. *Medicine and Science in Sports and Exercise* 1986; **18**: 87–94.

38. Drinkwater BL. Physiological responses of women to exercise. *Exercise and sports science reviews* 1973; **1**: 126–54.

39. Åstrand I. Aerobic work capacity in men and women with special reference to age. *Acta Physiologica Scandinavica* 1960; **49** (Suppl. 169): 1.

40. Raven PB, Mitchell JH. Effect of aging on the cardiovascular response to dynamic and static exercise In: Weisfeldt ML, ed. *The aging heart: its function and response to stress,* Vol. 12. New York: Raven Press, 1980: 269–96.

41. Saltin B, Blomqvist CG, Mitchell JH, Johnson RL, Wildenthal K, Chapman CB. Response to exercise after bed rest and after training. *Circulation* 1969; **38** (Suppl. 7): 1–78.

42. Adams WC, McHenry MM, Bernauer EM. Long-term physiologic adaptations to exercise with special reference to performance and cardiorespiratory function in health and disease. *American Journal of Cardiology* 1974; **33**: 765–75.

43. Saltin B, Henriksson J, Nygaard E, Andersen P. Fiber types and metabolic potentials of skeletal muscles in sedentary man and endurance runner. *Annals of the New York Academy of Sciences* 1977; **301**: 3–29.

44. Pollock ML. Submaximal and maximal working capacity of elite distance runners. Part I: Cardiorespiratory aspects. *Annals of the New York Academy of Sciences* 1977; **301**: 310–22.

45. Brundin T, Cernigliaro C. The effect of physical training on the sympathoadrenal response to exercise. *Scandinavian Journal of Clinical and Laboratory Investigations* 1975; **35**: 525–30.

46. Smith ML, Hudson DL, Graitzer HM, Raven PB. Exercise training bradycardia: the role of autonomic balance. *Medicine and Science in Sports and Exercise* 1989; **21**: 40–4.

47. Hagberg JM, Coyle EF, Carroll JE, Miller JM, Martin WH, Holloszy JO. Faster adjustment to and from recovery from submaximal exercise in the trained state. *Journal of Applied Physiology* 1980; **48**: 218–24.

48. Holloszy JO. Biochemical adaptations to exercise: aerobic metabolism. *Exercise and Sport Science Reviews* 1973; **1**: 45–71.

49. Blomqvist CG, Saltin B. Cardiovascular adaptations to physical training. *Annual Reviw of Physiology* 1983; **45**: 169–89.

50. Snell PG, Martin WH, Buckey JC, Blomqvist CG. Maximal vascular leg conductance in trained and untrained men. *Journal of Applied Physiology* 1987; **62**: 606–10.

51. Scheuer J. Effects of physical training on myocardial vascularity and perfusion. *Circulation* 1982; **66**: 491–5.

52. Scheuer J, Tipton CM. Cardiovascular adaptations to physical training. *Annual Review of Physiology* 1977; **39**: 221–51.

53. Convertino VA, Brock PJ, Keil LC, Bernauer EM, Greenleaf JE. Exercise training-induced hypervolemia: role of plasma albumin, renin, and vasopressin. *Journal of Applied Physiology: Respiratory, Environmental and Exercise Physiology* 1980; **48**: 665–9.

54. Greenleaf JE, Sciaraffa D, Shvartz E, Keil LC, Brock PJ. Exercise training hypotension: implications for plasma volume, renin, and vasopressin. *Journal of Applied Physiology: Respiratory, Environmnetal and Exercise Physiology* 1981; **51**: 298–305.

56. Wade CE, Freund BJ. Hormonal control of blood volume during and following exercise. In: Gisolfi CV, Lamb DR, eds. *Perspectives in exercise science and sports medicine.* Vol. 3, *Fluid homeostasis during exercise.* Indianapolis: Benchmark Press, 1990: 207–245.

57. Stolwijk JA, Roberts MF, Wenger CB, Nadel ER. Changes in thermoregulatory and cardiovascular function with heat acclimation. In: Nadel ER, ed. *Probelms with temperature regulation during exercise.* New York: Academic Press, 1977: 77–90.

58. Mitchell J. Exercise training in the treatment of coronary heart disease. *Advances in Internal Medicine* 1975; **20**: 249–71.

1.5.3 Assessment of endurance capacity

BERTIL SJÖDIN AND JAN SVEDENHAG

INTRODUCTION

Limitation of performance in endurance sports is naturally primarily related to onset of fatigue. Fatigue, however, is a very complex concept related to several psychological and physiological factors. During prolonged exercise there is certainly an impairment of the active muscles themselves which is likely to affect the sense of fatigue, but central factors may also be independently involved.

Experienced athletes in endurance sports have often an 'inbuilt' feeling for the pace which can be maintained comfortably for a longer period of time. Athletes practising endurance sports have also trained their ability to maintain the feeling of fatigue at an acceptable level during prolonged exercise. In this type of athlete the rate of perceived exertion has been found to be closely related to the accumulation of certain breakdown products in specified energy systems. It seems likely that accumulation of such metabolites, at least during specific conditions, builds up the feeling of fatigue. They may also directly inhibit the function of the muscles. Of course, fatigue may also be related to many other factors such as depletion of local energy stores, loss of liquid, overload of heat, or disturbance of electrolyte balance.

In this chapter we shall evaluate those physiological and biochemical factors which may limit or contribute to high endurance capacity. A short introduction is given describing the energy sources and energy systems available during muscular exercise. The fibre type specific characteristics of the human skeletal muscle and adaptation to endurance training are summarized. A more detailed description of training induced adaptation of skeletal muscle is presented elsewhere (Chapter 1.1.2). Finally, more basic physiological factors which are closely related to endurance performance are described.

ENERGY SYSTEMS FOR MUSCLE EXERCISE

Energy transformation in the actomyosin complex

Muscle tissue has the unique ability to convert chemically bound energy to kinetic energy which is utilized in the muscle contraction process. The primary source of energy exists in the form of adenosine triphosphate (ATP). The energy is released when ATP is hydrolysed to adenosine diphosphate (ADP) and inorganic phosphate at the cross-bridges of the myofibrillar proteins known as the actomyosin complex.[1] The liberated energy permits the cross-bridges from myosin to slide over the thinner actin threads. The consumption of ATP in this contraction mechanism is proportional to the activity of the cross-bridges. The concentration of ATP in the muscle tissue is relatively limited and has been estimated to be around 7 to 10 mmol/kg muscle tissue wet weight. It has been calculated that this ATP store in the muscle tissue is sufficient for only a couple of seconds of relatively moderate

muscle exercise.[2] This means that ATP has to be continuously regenerated during exercise lasting for a longer period of time. ATP can be resynthesized from ADP and inorganic phosphate through coupling reactions in different metabolic systems (Fig. 1).

The creatine kinase system

One of the immediate sources of high energy phosphate for the regeneration of ATP is creatine phosphate. Creatine kinase catalyses the reaction in which the phosphate group from creatine phosphate is transferred to ADP for the formation of ATP. The reaction can also be reversed. Although the amount of creatine phosphate in the muscle cells only permits an extra exercise time of 5 to 10 s, the creatine phosphate makes a valuable buffer for the ATP regeneration that occurs in the region close to the active sites in the actomyosin complex.

The enzyme creatine kinase exists in at least three different isoforms in the muscle tissue. The major isoform, creatine kinase MM, is a cytosolic enzyme but has also been found to be closely associated with the myofibrillar proteins. An even higher binding affinity to the myofibrillar proteins has been found for another isoform, creatine kinase MB. Another form of creatine kinase isoenzyme has been located in the intermembrane space of the mitochondria. Evidence of the existence of a creatine–creatine phosphate shuttle between the myofibrillar and mitochondrial systems in which the three isozymes are involved has been obtained.[3] The mitochondrial creatine kinase mediates the uptake of high energy phosphate from intramitochondrial ATP which is transmitted to creatine. The creatine phosphate formed diffuses to the myofibrillar creatine kinase which catalyses the reversed reaction enabling the formation of ATP from ADP very close to the cross-bridges of the contractile protein complex. Liberated creatine

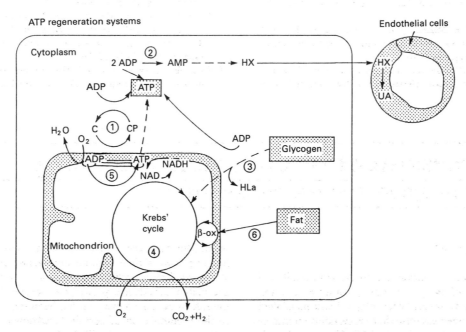

Fig. 1 ATP regenerating pathways: (1) the creatine kinase system; (2) the adenylate kinase system; (3) anaerobic glycolysis; (4) total oxidation of glucose intermediates in the Krebs' cycle; (5) the electron transport chain; (6) oxidation of fatty acids. HX, hypoxanthine; UA uric acid; C, creatine; CP, creatine phosphate; Hla, lactic acid; β-OX, β-oxidation.

moves back to the mitochondria and the procedure is repeated. However, the efficiency of this shuttle system during muscle exercise is difficult to evaluate and thus is still uncertain.

The adenylate kinase system

The adenylate kinase system is another enzyme system able to provide the contractile system with ATP rapidly during intense exercise. In the presence of adenylate kinase 2 mol of ADP are converted to 1 mol of ATP and 1 mol of adenosine monophosphate (AMP). The equilibrium constant for the reaction is very close to unity, which means that the reaction can proceed in both directions. The formation of ATP is favoured when the ADP concentration is high, provided that the secondary product AMP is continuously removed. These conditions are satisfied during intense exercise. A number of enzymes are also available in the cytosol of the muscle cells for the degradation of AMP to hypoxanthine. Hypoxanthine can diffuse through the cell membranes directly to the interstitial space and to the capillary endothelial cells or further to the bloodstream. In the capillary endothelial cells hypoxanthine may be oxidized to the end-product uric acid. Whereas hypoxanthine can be salvaged back to the adenine nucleotide pool in the muscle tissue, formation of uric acid is an irreversible reaction. Because of the limited concentration of free ADP available in the muscle cells and the high rate of ADP consumed by the adenylate kinase system, it can be regarded as a reserve system. The adenylate kinase system can be utilized during extreme conditions, for example during very intense exercise[4] or when the muscle glycogen is exhausted.[5]

Anaerobic glycolysis

The most prominent secondary source of energy in the musculature is available as glycogen. Glycogen is also the most efficient state for glucose storage in the cells. The glycogen concentration in leg musculature of sedentary humans has been estimated to be 10 to 15 g/kg wet weight, but can be extended to as much as 20g/kg wet weight in well-trained athletes.

The process which leads to the formation of 2 mol of lactate and 4 mol of ATP per unit of glucose in the cytosol is called anaerobic glycolysis (Fig. 2). It proceeds in 11 steps and is initiated by a phosphorylytic cleavage in which glucose units phosphorylated by inorganic phosphate are split off from glycogen.[6] This first reaction is called glycogenolysis and is catalysed by the enzyme phosphorylase. The activity of phosphorylase is strictly controlled by metabolites from other energy systems, hormones, and inorganic ions (e.g. Ca^{2+} and inorganic phosphate in the cytosol). A second and rate-regulating step in the glycolysis, where fructose 6-phosphate is phosphorylated by ATP to fructose 1,6-diphosphate (step 4), is catalysed by phosphofructokinase. Phosphofructokinase is also controlled by metabolites from other energy systems and by the hydrogen ion concentration in the cytosol.

Another factor which may inhibit the rate of the glycolysis is the availability of the electron carrier nicotine adenine nucleotide (NAD). NAD is essential for the oxidation of glyceraldehyde 3-phosphate in step 6 of the glycolysis. The concentration of free NAD in the cytosol is limited. Under aerobic conditions (rest and low to medium intensity exercise) the electrons and hydrogen ions from NADH are transferred across the mitochondrial membranes to the oxidative enzyme system inside the mitochondria by specific enzyme shuttle systems.[7] During anaerobic conditions (intensive exercise), however, the rate of reduction of NAD may exceed the transport capacity of the mitochondrial shuttle system. To regenerate NAD at an appropriate rate, the reducing equivalents from NADH have to be accepted by pyruvate which leads to the formation of lactate. Accumulation of lactate and other acidic products of the glycolysis leads to an increased concentration of hydrogen ions in the cytosol. The hydrogen ions inhibit the phosphofructokinase activity which decreases the rate of the glycolysis. Provided that the hydrogen ions can be efficiently removed from the cells, the rate of ATP regeneration is almost twice as high during anaerobic glycolysis as during aerobic glycolysis. However, the glycogen depots would be depleted within 10 min owing to the inefficient energy utilization of the glucosyl units.

Aerobic glycolysis

During aerobic conditions the glycolysis is continued by oxidation in the mitochondrial enzyme systems including the Krebs' cycle and the respiratory chain (Figs. 1 and 2). The combination of these energy systems is one of the most efficient systems for ATP regeneration with respect to the amount of substrate consumed. Thus the net production of ATP is either 38 or 39 mol per mole of glucose or glucosyl units depending on whether the substrate is glucose or glycogen.[6] One of the limitations of this aerobic glycolysis is the availability of molecular oxygen in the muscle tissue. The rate at which oxygen can be transported to the muscle cells is closely related to the capacity of the central circulatory system and the local circulation in the tissue. The maximal capacity to transport and utilize oxygen determines the maximal oxygen uptake (V_{O_2}max) of an individual.

As in the case of anaerobic glycolysis, the magnitude of the glycogen depots in the musculature will also be a limiting factor during aerobic conditions. It has been calculated that glycogen depots of average size (15–20 g/kg wet weight) will be sufficient for at least 90 min of exercise at an intensity of 70 per cent of V_{O_2}max. During such prolonged exercise circulating blood glucose may progressively increase in importance as an energy source. After 90 min of exercise the relative utilization of circulating glucose may increase to as much as 40 per cent of the total oxidative metabolism in the active musculature.[8] These conditions activate the mechanisms for degradation of liver glycogen and release of glucose from the liver to the blood. The amount of glucose available in liver and blood has been estimated to be around 20 g. If this 'extra' glucose were the sole energy source, it would be utilized within 6 min during submaximal exercise. However, during long-term exercise, glycogen in inactive musculature can also be utilized via adrenaline-mediated release of lactate to the blood stream. Released lactate can then be reabsorbed and utilized in the active muscle tissue.

Fat oxidation

By utilizing intracellular fat depots and free fatty acids circulating in blood, muscular exercise of low intensity can be extended to an almost unlimited period of time. In comparison with carbohydrates, the oxidation of fatty acids needs much more oxygen per unit of weight. This is obvious from the respiratory quotient $[CO_2]/[O_2]$ which is 0.74 for 100 per cent fat oxidation compared with 1.00 for carbohydrate oxidation. The high requirement of oxygen for the utilization of fat and the need for intermediates in the Krebs' cycle limits its use primarily to exercises of lower intensities.

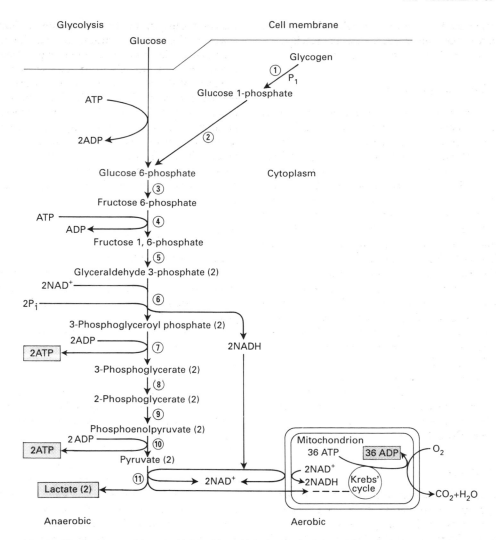

Fig. 2 The pathway of 'anaerobic' and 'aerobic' glycolysis. Anaerobic glycolysis proceeds in 11 steps and leads to the formation of 2 mol of lactate from one glycosyl unit. This sequence of reaction gives rise to the net formation of 3 or 2 mol of ATP depending on whether the substrate is glycogen or glucose. During aerobic glycolysis pyruvate is further oxidized to CO_2 and H_2O in the mitochondria. This leads to the formation of a further 36 mol of ATP from 2 mol of pyruvate.

ENDURANCE TRAINING ADAPTATIONS

Characteristics of the muscle

In human skeletal musculature there is a mixture of two main fibre types: type 1 (slow twitch) muscle fibre and type 2 (fast twitch) muscle fibre.[9] Type 1 fibres are characterized by a relatively higher oxidative capacity as indicated by high respiratory capacity[10] and high activities of oxidative enzymes.[11] Type 2 fibres have a relatively higher glycolytic potential and are more fatiguable than type 1 fibres.[12,13]

During submaximal steady state exercise in the presence of high lactate levels in the musculature, the type 1 muscle fibres will preferentially oxidize lactate. The type 2 muscle fibres, however, will utilize lactate for glycogen resynthesis.[14] The differences in lactate metabolism in type 1 and type 2 muscle fibres may be due to differences in the relative activities between the glycolytic system

in the cytosol and oxidative enzyme system in the mitochondria[15] and also in the activities in enzymes involved in glycogen resynthesis. The lactate metabolism in the different fibre types also seems very closely related to the isozyme pattern of lactate dehydrogenase (LDH). Thus the relative activity of heart-specific LDH (H-LDH), which is preferentially involved in lactate oxidation to pyruvate, predominates in type 1 fibres. Muscle-specific LDH (M-LDH), which preferentially catalyses the reduction of pyruvate to lactate, predominates in type 2 muscle fibres.[16]

The relative activity of H-LDH has been found to increase specifically in type 1 fibres with endurance training (Fig. 3). This will increase the difference between the fibre types in the musculature of endurance-trained individuals and will also increase the efficiency of oxidizing lactate in type 1 muscle fibres.

With endurance training the oxidative capacity of both types of fibres increases. The difference in activities between the two fibre types with regard to the pattern of Krebs' cycle enzyme activities has been reported to be similar[17] or smaller[18,19] in the trained

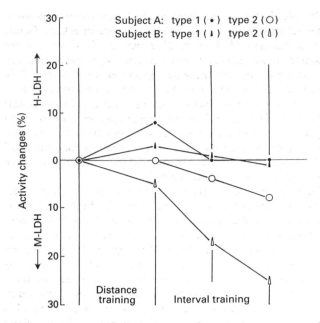

Subject A: type 1 (•) type 2 (O)
Subject B: type 1 (ı) type 2 (◊)

Fig. 3 Relative changes in M- and H-LDH activity in type 1 and 2 muscle fibres in subjects A and B after periods of endurance distance training and more intense interval training.

compared with the untrained state. However, with regard to the enzymes of the fat oxidative pathway, the difference between fibre types may be similar whether or not the individual is endurance trained.[17,18]

The intramuscular triglyceride concentration is greater in type 1 than in type 2 muscle fibres and has been shown to increase with endurance training. The intramuscular lipid stores are located close to the mitochondria and appear to be utilized prior to the utilization of plasma free fatty acids during exercise.

Whereas the enzyme activities of different energy systems in the active musculature change rather quickly in response to training, the fibre type distribution seems to be almost unaffected. Type 1 muscle fibres predominate in the active musculature of endurance athletes, but this may be primarily due to selection factors. A comparison of elite long-distance runners, elite middle-distance runners, and untrained men showed a significantly higher portion of type 1 muscle fibres in the gastrocnemius muscle of the long-distance runners (79 per cent) compared with corresponding values for the other two groups (middle-distance runners, 62 per cent; untrained men, 58 per cent; $p < 0.05$). The oxidative potential, as measured by the activity of the mitochondrial enzyme succinate dehydrogenase, was also higher in the long-distance group.[20] Furthermore, the relative distribution of type 1 muscle fibres and capillary density in the active leg musculature (vastus lateralis) has been found to be positively related to performance capacity in marathon running.[21] Thus a predominance of type 1 fibres (with more efficient microcirculation), together with training-induced increases in oxidative and microcirculatory capacity in the active musculature, is truly beneficial for long-distance running performance. This advantage in endurance may be related to a higher capacity to oxidize fat and lactate, with a sparing effect on carbohydrate stores.

The cardiovascular system

An efficient oxygen transport system seems to be vital for success in endurance sports. Enlarged heart volumes and well-developed vascular systems have been noted in successful endurance athletes.[22] Calculations based on echocardiographic measurements have also indicated larger diastolic volumes and a somewhat thicker left-ventricular wall in marathon runners compared with controls.[23,24] These enlarged dimensions of the heart have been suggested to be important for the cardiac pump performance in order to achieve a large stroke volume and cardiac output which is thought to be the main determinant of the maximal oxygen uptake.[25] Of course, high capacity of the central circulation is important during such prolonged exertion as marathon running, and elite marathon runners have been estimated to use as much as 92 per cent of their maximal cardiac output for 2.1 to 2.5 h.[26] However, cardiovascular regulation is not static during the effort since it has been noted that the stroke volume declines gradually with time while the heart rate slightly increases.[26,27]

PHYSIOLOGICAL FACTORS RELATED TO ENDURANCE PERFORMANCE

Maximal oxygen uptake

Since the 1930s it has been known that the maximal oxygen uptake V_{O_2}max is exceptionally high in elite endurance event athletes. These high values are thought to be due to a combination of training effort and natural endowment. Early studies of elite runners[28,29] measured values of up to 81.5 ml/kg.min in champion athletes. This is comparable with V_{O_2}max in elite runners of today (highest value obtained in our laboratory for a runner, 87.1 ml/kg.min; mean for a 5000–10 000 metre group, 81.5 ml/kg.min). Thus improvements in competitive results in middle and long distances seen in the past 50 years cannot be ascribed to a higher V_{O_2}max of present elite runners. Although evidently important, the maximal oxygen uptake is only one of the factors that determines success in long-distance events. This is illustrated by the large variation in performance between marathon runners with equal V_{O_2}max values.[30]

Oxygen cost of movement

During the last two decades, there has been a growing interest in how best to utilize the maximal aerobic capacity in endurance events. During running, the submaximal oxygen uptake is directly related to the running velocity. However, at a given running speed, the submaximal oxygen requirement (in ml/kg.min) may vary considerably between subjects.[31,32] In contrast, differences may be small or non-existent when groups of elite runners from different distances are compared.[32]

In elite distance runners with a relatively narrow range in V_{O_2}max, the running economy at different speeds has been found to be significantly correlated ($r = 0.79$–0.83) with performance in a 10-km race.[33] There is also a surprisingly wide (about 20 per cent) variation in oxygen cost of running at a given speed between marathon runners of similar performance capacities.[30] A low oxygen cost of running (i.e. good running economy) is thus truly beneficial. As for V_{O_2}max, however, there is a relatively poor correlation between oxygen cost of running and performance (e.g. $r = 0.55$) in a heterogeneous marathon sample.[30]

Because the oxygen cost of running has been thoroughly studied for only two decades, any long-term trend for elite runners cannot be assessed. However, except for obvious improvements in road running shoes, tracks, and spikes, the improved results of the past 50 years (especially in long-distance running) are probably explained by improvements in running economy and/or in the ability to exercise at a high percentage of V_{O_2}max during long periods of time (lactate threshold).

Total aerobic capacity

A given performance in an endurance event such as running can be attained in different ways. Two kinds of elite runners with different physiological characteristics can be distinguished. One category is characterized by a high V_{O_2}max but a relatively poor running economy. The second category involves runners with excellent running economy but a relatively low V_{O_2}max. In many cases the overall result of these differences is a fairly even performance level. Both the training accomplished and various natural abilities ought to explain the differences in the oxygen cost of running and in V_{O_2}max. Only the outstanding runner may have excellent values in both respects (thus forming a third category).

To account better for these individual differences in relation to performance, the fractional utilization of V_{O_2}max when running at a specific speed (e.g. 15 or 20 km/h) can be calculated. The % V_{O_2}max value calculated in this way has been found to be significantly correlated with performance at different long distances. For a runner, this value can be regarded as the total aerobic running capacity of the subject. For instance, in a heterogeneous marathon sample,[30] the relation between fractional utilization of V_{O_2}max at a submaximal speed of 15 km/h and performance was as good as $r = 0.94$ ($n = 35$). Such a good correlation between % V_{O_2}max at a specific submaximal speed and performance is thus attributable to the fact that the % V_{O_2}max value expresses the effects of both V_{O_2}max and running economy, which may be separately related to performance. In recent years, another way of expressing the combined effect of running economy and V_{O_2}max has gained popularity.[34] In this approach the relationship between running speed and oxygen utilization of an individual is extrapolated up to his or her $V_{O_2 \, max}$ value and the running velocity at which this occurs is used. If the running economy extrapolation can be accepted, this may be the preferred mode of expression, particularly for middle-distance runners with high racing velocities.

Anaerobic thresholds

The metabolic response to exercise intensity has been extensively studied by exercise physiologists. Different metabolic markers have been used to mirror the energy metabolism in active musculature.

Both blood lactate concentration and respiratory parameters may reflect the relation between aerobic and anaerobic glycolysis in the muscle tissue. An exponential increase in both the blood lactate concentration and ventilation is seen from exercise intensities corresponding to 70 to 75 % V_{O_2}max, indicating an increased activity in the glycolytic enzyme system. The blood lactate concentration is 0.5 to 1.0 mmol/l at rest but can increase to as much as 4 to 5 mmol/l with a remaining steady state of blood lactate during submaximal exercise.

Preliminary results indicate that the maximal steady state level for blood lactate is lower than 4 mmol/l in endurance-trained subjects and may be higher than 4 mmol/l in intensely trained and untrained subjects. However, in endurance tests performed as graded exercise, 4 mmol/l has been used as an average value for maximal steady state of the blood lactate concentration.

With physical endurance training, the lactate curve is known to be shifted to the right relative to V_{O_2}max, resulting in lower lactate levels at the same % V_{O_2}max. However, the % V_{O_2}max at the 4 mmol/l lactate threshold does not differ significantly between elite marathon runners and elite runners in the 400- and 800-metre events (88 per cent, 84 per cent, and 83 per cent respectively).[32] This and other findings may indicate that the rightward shift of the lactate threshold relative to V_{O_2}max may occur largely as an early response to training. The excessive amounts of training done by some endurance athletes thus appear to have a relatively small effect, or no effect at all, on further shifting of the lactate curve (expressed as % V_{O_2}max) to the right.

The shape of the blood lactate curve related to exercise intensity is also very different between endurance-trained subjects and intensely trained and untrained subjects. In endurance-trained subjects there is a negligible accumulation of blood lactate up to an intensity corresponding to about 75% V_{O_2}max. With increasing intensities above this inflexion point there is a steep increase in the blood lactate concentration. For intensely trained and untrained subjects the blood lactate concentration increases continuously with increasing intensity. The inflexion point around 70 to 75% V_{O_2}max is sometimes indistinct. The differences in the blood lactate kinetics in response to physical training may be due to differences in microcirculation and activity of glycolytic and oxidative mitochondrial enzymes systems in these three categories of subjects.

The high activity of oxidative enzymes and mitochondrial shuttle enzymes as well as the corresponding low activities of glycolytic enzymes in all fibre types of endurance-trained subjects may lead to a low capacity to produce lactate and a high capacity for consuming lactate in the musculature. The diffusion of excess muscle lactate to the blood is facilitated owing to the high capillary density and relatively small cross-sectional area of the muscle fibres in these individuals.

For some years now metabolic parameters measured at submaximal exercise have been used as better indicators of endurance exercise capacity than maximal oxygen uptake. Terms such as anaerobic threshold, aerobic threshold, and onset of blood or plasma lactate accumulation have been introduced by various authors for the corresponding submaximal measurements of threshold velocity. These have been determined from ventilatory parameters and lactate values for blood or plasma (2.0, 2.2, and 4.0 mmol/l and delta lactate), as well as from heart rate measurements. As indicated above, we have used the 4 mmol/l blood lactate concentration as the highest steady state level of lactate that can be obtained during running (expressed as the corresponding running velocity $v_{La \, 4}$ of each subject).

These terms are commonly used even though all the mechanisms underlying the different threshold concepts have not been clarified. Nevertheless, there appears to be a rather close relationship between these thresholds and performance. Thus, in long-distance running, high correlations of the order of $r = 0.88$ to 0.99 have been reported for distances of 3.2 km to the half marathon. In the marathon, correlations of the order of $r = 0.94$ to 0.98 have been found (Fig. 4). Thus, it appears from this that the threshold is the best single predictor of performance in long-distance running, including the marathon.

High correlations are found between the lactate threshold and long-distance running speed because the lactate threshold is dependent on several variables which are all related to performance. The 4 mmol/l lactate threshold (expressed as a velocity) is thus a function of V_{O_2}max, oxygen cost of running, and % V_{O_2}max at v_{La4}

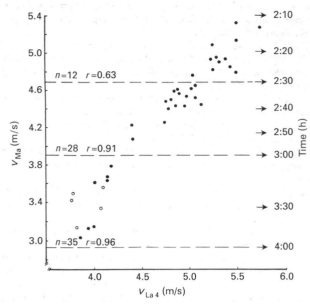

Fig. 4 The relationship between the individual mean marathon velocity v_{Ma} and the running velocity corresponding to the lactate threshold $v_{La\,4}$ in elite runners ($n = 12$), in elite and good runners ($n = 28$), and in elite, good, and recreational runners ($n = 35$).[30]

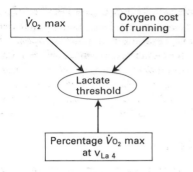

Fig. 5 Inter-relations between different physiological variables of importance for performance in middle-distance and particularly long-distance running.

(Fig. 5). Therefore, an improvement in one of these factors (e.g. oxygen cost of running, others unchanged) will result in an improvement in the lactate threshold to an analogous degree. However, an unbalanced training programme over-emphasizing one or a few training elements, may lead to opposite changes in these factors with, at best, an unchanged lactate threshold as the overall result.

Training at the speed corresponding to the 'anaerobic threshold' is widely used as a means of improving performance in long-distance running. As for the early rightward shift of the lactate curve (see above), such a training effort would appear to be rather fruitless for a previously trained individual. However, an improvement in the anaerobic threshold velocity after threshold training in previously well-trained runners may be mostly due to effects on the running economy and Vo_2max.[35] Evidently, there may be an intricate interplay between training and the different physiological variables determining running performance.

Endurance factor

Another aspect of performance is the question of the closeness to Vo_2max that can be attained during competition. Obviously, this value is dependent on the distance as well as on environmental factors. For the marathon, the % Vo_2max at race pace (% Vo_2Ma/Vo_2max determined on a treadmill, where Ma is the marathon speed) has been reported to range between 60 per cent in slow marathon runners and 86 per cent in elite marathon runners. In our marathon sample[30] the % Vo_2Ma/Vo_2max was the same (on average 80 per cent) in elite runners (mean time 2 hours 21 minutes) and good runners (mean time, 2 hours 37 minutes), but was significantly lower (71 per cent) in the slow runners (mean time, 3 hours, 24 minutes). Likewise, the percentage of lactate threshold velocity (% $v_{La\,4}$) performed during the marathon was significantly lower in slow runners (85 per cent) than in good or elite runners (92–93 per cent). These findings suggest that the

% Vo_2max that can be maintained during a marathon race differs in runners with large differences in performance capacity. These differences would be even greater if the effect of wind resistance (proportional to the third power of velocity) was accounted for. In experienced runners prepared for the event with adequate endurance training (such as our sub–3-hour marathon runners) the % Vo_2Ma/Vo_2max may be similar. The lower % Vo_2max in the slowest runners may be due not only to a lack of adequate training (with a lower capacity to metabolize fat and save glycogen) but is probably also related to the longer period of exertion and/or inadequate running and racing experience.

Expressing oxygen uptake

From a strictly dimensional point of view, Vo_2max (l/min) should be proportional to body mass raised to the 2/3 power.[36] Based on calculated limitations imposed by elastic components of biological material, aerobic power has been suggested to be proportional to the 3/4 power of body mass. This is in conformity with the relationship between resting as well as maximal oxygen uptake and body mass raised to the power 0.73–0.79 as reported for adult mammals from the mouse or dwarf mongoose to the elephant or eland. From calculations for a large series of adult humans, it was recently suggested that submaximal and maximal Vo_2 attained during running is better related to $kg^{2/3}$ or $kg^{3/4}$ than to kg^{1}.[37] Furthermore, quite recent results in adolescent boys suggest that changes in oxygen cost of running and Vo_2max (ml/kg.min) during growth may largely be due to an overestimation of the body weight dependence of Vo_2 during running.[38] Calculations favoured oxygen uptake expressed as $kg^{0.75}$.[38] Thus several lines of evidence suggest that oxygen uptake determined during submaximal or maximal treadmill running should not be related to kg^{1} but rather to $kg^{0.75}$. After calculating new reference values, such a mode of expressing Vo_2 may increase our understanding of differences and changes in Vo_2max and running economy, particularly on the individual level.

CONCLUSION

It can be concluded that the performance capacity of successful athletes in endurance sports is to a great extent related to genetic factors. However, those talents have to be developed through years of systematic training. The response to this training will lead to an increased capacity of the cardiovascular system and specific adaptations of the metabolism to more efficient energy turnover in the active musculature. The main physiological effect will be improved oxygen transport and facilitated oxygen delivery to the active musculature, resulting in a higher Vo_2max. Training will also lead to a greater technical skill which will be reflected in lower oxygen cost of movement at a given intensity. All these physiological changes will result in a higher exercise capacity with a higher turnover rate of ATP in the aerobic energy system and a lower accumulation of lactate and inhibiting hydrogen ions.

REFERENCES

1. Geeves MA. The dynamics of actin and myosin association and the crossbridge model of muscle contraction. *Biochemical Journal* 1991; **274**, 1–14.
2. Sahlin K. Metabolic changes limiting muscle performance. In Saltin B, ed. *Biochemistry of exercise*, Vol VI. International Series on Sport Science 16. Champaign, Il: Human Kinetics Publishers, 1986: 323–45.
3. Bessman SP, Carpenter CL. The creatine–creatine phosphate energy shuttle. *Annual Review of Biochemistry* 1985; **54**: 831–62.
4. Sjödin B, Hellsten-Westing Y. Changes in plasma concentration of hypoxanthine and uric acid in man with short distance running of various intensities. *International Journal of Sports Medicine* 1990; **11**: 493–5.
5. Norman B, Sollevi A, Kaijser L, Jansson E. ATP breakdown products in human skeletal muscle during prolonged exercise to exhaustion. *Clinical Physiology* 1987; **7**: 503–9.
6. Lehninger AL. *Biochemistry*. 2nd edn. New York: Worth, 1975.
7. Schantz PG, Sjöberg B, Svedenhag J. Malate-aspartate and alphaglycerophosphate shuttle enzyme levels in human skeletal muscle: methodological considerations and effect of endurance training. *Acta Physiologica Scandinavica* 1986; **128**: 397–407.
8. Ahlborg G, Felig P, Hagenfeldt L, Hendler R, Wahren J. Substrate turnover during prolonged exercise in man. Splanchnic and leg metabolism of glucose, free fatty acids and amino acids. *Journal of Clinical Investigation* 1974; **53**: 1080–90.
9. Padykula HA, Herman E. The specificity of the histochemical method for adenosine triphosphatase. *Journal of Histochemistry and Cytochemistry* 1955; **3**: 170–83.
10. Ivy JL, Withers RT, van Handel PJ, Elger DH, Costill DL. Muscle respiratory capacity and fiber type as determinants of the lactate threshold. *Journal of Applied Physiology* 1980; **48**: 523–7.
11. Essén B, Jansson E, Henriksson J, Taylor AW, Saltin B. Metabolic characteristics of fiber types in human skeletal muscle. *Acta Physiologica Scandinavica* 1975; **95**: 153–65.
12. Tesch P. Muscle fatigue in man with special reference to lactate accumulation during short term intense exercise. *Acta Physiologica Scandinavica* 1980; Suppl. 480.
13. Thorstensson A. Muscle strength, fibre types and enzyme activities in man. *Physiologica Scandinavica* 1976; Suppl. 443.
14. Nordheim K, Vøllestad NK. Glycogen and lactate metabolism during low intensity exercise in man. *Acta Physiologica Scandinavica* 1990; **139**: 475–84.
15. Sjödin B, Jacobs I, Karlsson J. Onset of blood lactate accumulation and enzyme activities in m. vastus lateralis in man. *International Journal of Sports Medicine* 1981; **2**: 166–70.
16. Sjödin B. Lactate dehydrogenase in human skeletal muscle. *Acta Physiologica Scandinavica* 1976; Suppl. 436.
17. Essén-Gustavsson B, Henriksson J. Enzyme levels in pools of microdissected human muscle fibres of identified type. *Acta Physiologica Scandinavica* 1984; **120**, 505–15.
18. Chi M M-Y, Hintz CS, Coyle EF, Martin WH, Ivy JL, Nemeth PM, Holloszy JO, Lowry OH. Effects of detraining on enzymes to energy metabolism in individual human muscle fibers. *American Journal of Physiology* 1983; **244**: C276–87.
19. Jansson E, Kaijser L. Muscle adaptation to extreme endurance training in man. *Acta Physiologica Scandinavica* 1977; **100**: 315–24.
20. Costill DL. Physiology of marathon running. *Journal of the American Medical Association* 1972; **221**: 1024–9.
21. Sjödin B, Jacobs I. Onset of blood lactate accumulation and marathon running performance. *International Journal of Sports Medicine* 1981; **2**: 23–6.
22. Keul J, Dickhuth H-H, Lehman M, Staiger J. The athletes hearthaemodynamics and structure. *International Journal of Sports Medicine* 1982; **3**: 33–43.
23. Parker BM, Londeree BR, Cupp GV, Dubiel JP. The noninvasive cardiac evaluation of long distance runners. *Chest* 1978; **73**: 376–81.
24. Paulsen W, Boughner DR, Ko P, Cunningham DA, Persaud, JA. Left ventricular function in marathon runners. Echocardiographic assessment. *Journal of Applied Physiology* 1981; **51**: 881–6.
25. Blomqvist G, Saltin B. Cardiovascular adaptations to physical training. *Annual Review of Physiology* 1983; **45**: 169–89.
26. Costill DL, Fink WJ, Pollock, ML. Muscle fiber composition and enzyme activities of elite distance runners. *Medicine and Science in Sports and Exercise* 1976; **8**: 96–100.
27. Saltin B. Aerobic work capacity and circulation at exercise in man. *Acta Physiologica Scandinavica* 1964; Suppl. 230.
28. Robinson S, Edwards HT, Dill DB. New records in human power. *Science* 1937; **85**: 409–410.
29. Åstrand P-O. New records in human power. *Nature*, 1955; **176**: 922–3.
30. Sjödin B, Svedenhag J. Applied physiology of marathon running. *Sports Medicine* 1985; **2**: 83–99.
31. Costill DL, Thomason H, Roberts E. Fractional utilization of the aerobic capacity during distance running. *Medicine and Science in Sports and Exercise* 1973; **5**: 248–52.
32. Svedenhag J, Sjödin B. Maximal and submaximal oxygen uptakes and blood lactate levels in elite male middle- and long-distance runners. *International Journal of Sports Medicine* 1984; **5**: 255–61.
33. Conley DL, Krahenbuhl GS. Running economy and distance running performance of highly trained athletes. *Medicine and Science in Sports and Exercise* 1980; **12**: 357–60.
34. Morgan DW, Baldini FD, Martin PE, Kohrt WM. Ten kilometer performance and predicted velocity at Vo_2max among well-trained male runners. *Medicine and Science in Sports and Exercise* 1989; **21**: 78–83.
35. Sjödin B, Jacobs I, Svedenhag J. Changes in onset of blood lactate accumulation (OBLA) and muscle enzymes after training at OBLA. *European Journal of Applied Physiology* 1982; **49**: 45–57.
36. Åstrand, P-O, Rodahl K. *Textbook of physiology.* New York: McGraw-Hill, 1986.
37. Bergh U, Sjödin B, Forsberg A, Svedenhag J. The relationship between body mass and oxygen uptake during running in humans. *Medicine and Science in Sports and Exercise* 1991; **23**: 205–11.
38. Sjödin B, Svedenhag J. Oxygen uptake during running as related to body mass in circumpubital boys: a longitudinal study. *European Journal of Applied Physiology* 1992; **65**: 150–7.

1.5.4 Assessment of anaerobic power

HENRYK K. A. LAKOMY

INTRODUCTION

Many sports are characterized by their need for short bursts of maximal power output. A high proportion of the required energy is provided by anaerobic metabolism, and it is the intention of this chapter to examine some of the tests which can be used to measure the maximum external power output that can be produced during such activities. It is important to realize that 'it is impossible to measure the rates at which muscles produce and transfer energy internally. Only the external manifestations of the energy are capable of measurement'.[1] In this chapter only tests which measure either the external power output of the body or the rate of energy transfer to external devices are discussed.

The two components of anaerobic performance of most interest are peak and mean external power output. Peak power output is the maximum rate at which energy is being transferred to the external system. The mean power output is the total work done during the performance test divided by the time taken. The total work done in tests lasting for a minimum of 30 s is sometimes termed the 'anaerobic capacity'.[2,3] The term 'capacity' implies that there is a finite exhaustible store of energy that can be used for anaerobic energy metabolism and that tests can determine the size of these stores.[4] This author believes that neither of these statements can be supported and therefore, other than when referring to published texts, values of mean power output and total work done will be discussed rather than anaerobic capacity.

In order to calculate power output two components must be measured—force and velocity. Power is defined as the product of the applied force and the velocity at which that point of application is moving.[5] In dynamic muscular activities in which limb acceleration is permitted neither force nor velocity remain constant. The two quantities are not independent of each other, and in tests which measure maximum anaerobic performance the external power output will be constantly varying due to the rapidly changing limb acceleration and type of muscle action.

Before examining specific tests of anaerobic power several factors that affect power output must be considered.

FACTORS AFFECTING POWER OUTPUT

Exercise duration

The power output that can be sustained during an activity is markedly influenced by the duration of the activity. Figure 1 shows the average power output produced by a series of activities of different durations. The figure shows that the average power output declines with increasing duration of activity. It should be noted that Fig. 1 shows only the power output averaged over entire movement cycles. The actual instantaneous power output during most activities will differ significantly from this average.

Figure 2 shows a typical example of the total amount of work

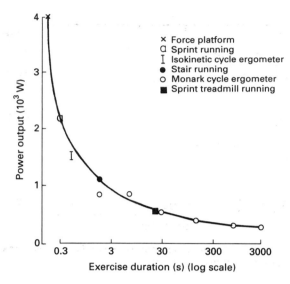

Fig. 1 Relationship between maximum power output and exercise duration for various activities.

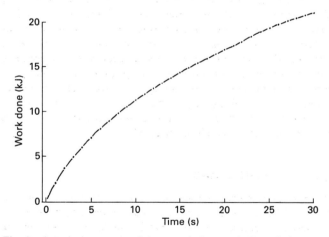

Fig. 2 A typical example of the cumulative work done during a 30-s sprint on a cycle ergometer.

achieved during a test. The total work done increases with increasing duration of activity; however, the rate of increase is not constant.

Speed of contraction

The power generated by the contracting muscle is a function of the product of the force that the muscle is generating and the velocity at which it is contracting. These variables are not independent of each other but have been shown to have a relationship, for concentric contractions, which is either hyperbolic[6] or exponential.[7] However, this relationship does not explain how the muscle behaves when the muscle action is eccentric. It is during maximum eccentric contractions that the highest values of peak tension are generated. Figure 3 shows the force–velocity and power–velocity relationships of muscle contracting concentrically.

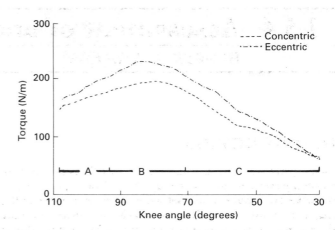

Fig. 4 Relationship between maximum torque and knee angle for concentric and eccentric contractions performed at 30°/s on an isokinetic dynamometer.

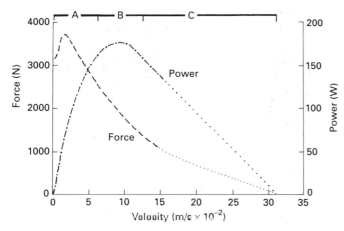

Fig. 3 The force–velocity and power–velocity relationships of the knee extensors performing concentric contractions.

It can be seen in Fig. 3 that the speed at which the muscle is contracting markedly influences the maximum power that can be produced. Peak power has been found to occur at approximately $0.3V_0$, where V_0 is the maximum speed of muscle contraction. Exercise with muscle contraction velocities falling in the bands marked A and C in Fig. 3 will result in a power output below optimum. Maximum power output will occur when the speed of contraction falls within band B; for example, for cycling the speed represents a pedalling speed of approximately 110 rev/min.[8] Therefore to maximize power output the resistance against which the subject will work during the test which results in optimum speed of contraction should be selected.

Type of contraction

The maximum power generated is dependent on the type of contraction. The maximum tension produced during an eccentric contraction is greater than the maximum force which can be attained during isometric or concentric contractions. The mechanical efficiency of the exercise will be strongly influenced by the type of contraction being performed. For the same rate of metabolic energy provision very different external power outputs will be generated for different types of contraction. For the isometric contraction no external power output will be measured as contraction velocity is zero. For the eccentric contractions the muscle will appear to be absorbing power rather than producing it (Fig. 4).

Muscle mass

The amount of power that can be generated in an activity is dependent on the muscle mass involved: the greater the muscle mass, the greater the number of fibres recruited and therefore the

greater the power that can be produced. However, it is important that the muscle recruitment is optimally co-ordinated.

Muscle temperature

Inbar and Bar-Or[9] showed the beneficial effects of warm-up on the ability to produce power in a short-duration explosive activity. Therefore it is clear that sufficient warm-up should precede such tests and that the warm-up procedure should be standardized.

Range of movement

Figure 3 shows a typical relationship between the tension that can be generated by the muscle (muscle group) throughout its range of movement. For a given speed of contraction, movement in the range represented by the area B will result in a greater power output than an equivalent range of movement in areas A and C. Therefore when testing for anaerobic power it is important to consider range of movement, for example when setting the seat height on a cycle ergometer.

Type of activity

In general in repetitive exercise greater average power output can be maintained in those exercises that are cyclic (e.g. cycling), where energy can be stored during the propulsive phase and recovered during the recovery phase, than in those in which the power generated is immediately dissipated (e.g. biceps curls). The storage–recovery of energy (e.g. in the flywheel of the cycle and the rowing ergometer) results in a smaller speed fluctuation than would otherwise be possible, resulting in the subject's maintaining a speed of contraction close to the optimum value required for greatest power output.

TESTS OF INCREASING DURATION

Margaria et al.[10] reported that supramaximal exercise of 10 to 15-s duration could be performed without a significant elevation in blood lactate. This suggested that maximal exercise lasting up to 10 s would use energy derived solely from alactic anaerobic

181

sources and therefore that tests of up to this duration would measure these alactic energy sources. Boobis *et al.*[11] and Jacobs *et al.*[12] showed, using muscle biopsy techniques, that maximum exercise of 6 and 10 s duration respectively resulted in significantly elevated concentrations of muscle lactate, suggesting that glycolysis did in fact occur within these time frames. Therefore it is evident that it is not possible to test the 'alactic' and 'lactic' acid components of energy production independently as both systems appear to be active throughout such tests. No attempt will be made to apportion energy provision from these sources.

Vertical jump test

Tests lasting only a fraction of a second have been shown to result in the highest values of power output. The most common test is the vertical jump test described in 1921 by D. A. Sargeant. In this test the maximum height that the centre of mass of the body is raised during a maximal standing vertical jump (with countermovement) is used as an indicator of the power generated in the knee and hip extensors.[13] The instantaneous values of force and velocity required to calculate the power generated are determined using force platforms. It is assumed that the force accelerating the centre of mass is equal to the reaction force applied to the force plate minus body weight. By integrating the resultant acceleration with respect to time the instantaneous value of the velocity of the centre of mass is obtained. Instantaneous power output is calculated from the product of the accelerating force and velocity of the centre of mass.[14–16] Davis and Young,[17] using this technique, showed a high correlation ($r = 0.92$) between the height achieved in the jump and the peak power output generated.

However, the vertical jump test has been strongly criticized as not being a valid method for measuring power output.[18] The main criticism is that jumping is an impulsive activity with the height jumped being a function of the product of force and time and not the product of force and velocity.

Single dynamic movement of a limb

Isokinetic dynamometers are able to give instantaneous information on torque and limb position (angle) at predetermined angular velocities of the limb. Power is calculated from the product of torque T and the preset angular velocity ω. This can be shown to be the same as the product of force F and velocity V

$$F = T/r$$

and

$$V = \omega r$$

where r is the length of the lever arm. Therefore
$$FV = T/r\ \omega r = T\omega.$$

Perrine[1] showed that the maximum peak power of the knee extensors occurs at angular velocities around 240deg/s (range 192–288deg/s). This is approximately 34 per cent of the projected maximum unloaded velocity of the limbs of 832deg/s. This value of 34 per cent is close to that which would be predicted from the power–velocity relationship of muscle discussed earlier. Perrine[1] also reported unpublished observations of volleyball players showing a correlation coefficient of 0.87 between the peak total leg power per unit bodyweight and the vertical jump height. A high correlation was also found between peak leg power and 100-yard sprint time for women sprinters.

Margaria step test

Margaria *et al.*[19] proposed a test in which anaerobic power is measured over an activity lasting approximately 3 s. The test requires the subjects to sprint up a flight of ordinary stairs, two steps at a time, as quickly as possible after a run-up of 2 m (Fig. 5). It is

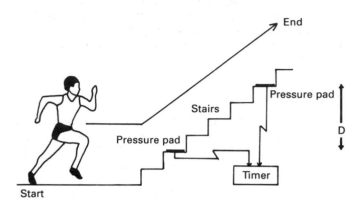

Fig. 5 Margaria Step Test.

assumed that the subject attains maximum speed prior to reaching the first step and that the running speed remains constant throughout the stair climb. The time taken to run up an even number of stairs is accurately measured, usually using pressure mats resting on the steps. The subjects are instructed to maintain their normal sprinting posture throughout the test. It is assumed that all the external work is done in raising the centre of mass of the body and that this rise is the same as the level difference D between the steps. The power output of the subject is calculated from the formula

$$\text{power} = FV = \frac{m_{b}g\,D}{t}$$

where m_{b} is the mass of the subject (kg), g is the acceleration due to gravity (9.81 m/s^2), D is the level difference, and t is the time taken. The test–retest coefficient has been found to be high ($r = 0.85$–0.9) with a variability over a 5-week period of less than 2 per cent.[19–21]

It has been shown that the power output during this type of test can be increased by having the subjects run with weight packs on their backs. This increase in power output has been shown to range from 10 per cent with a loading of 33 per cent of body weight[22] to 16 per cent with a loading of 40 per cent of body weight.[23]

Modifications to the original protocol have been proposed. For example, Kalamen[24] had the subjects take three steps at a time after a run-up of 6 m. In this test, however, the number of 'measured steps' is the same as in the Margaria protocol.

Bosco jump test

This test, which examines the power generated whilst jumping, was suggested by Bosco *et al.*[25] The test duration T_{s} is from 15 to 60 s and requires the exact determination of the time spent in the air during a series of repeated maximum jumps. The performer attempts to jump as high and as many times as possible during the test period. An instrumented pad is required to determine the number n of jumps performed and the total 'flight' or non-contact time F_t. The average power W generated during the test is calculated from

$$W = \frac{F_t T_s g^2}{4n(T_s - F_t)}$$

It is important that the amount of hip and knee flexion is standardized and that the jumps are performed without interruption. Bosco et al.[24] showed high correlations between the power output for a 15-s jump test and the results of a 60-m sprint ($r = 0.84$) and a 15-s cycle ergometer test ($r = 0.87$). A power index can be obtained by dividing the average power obtained by the mass of the performer (kg).

Cycle ergometry

In 1954 von Dobeln[26] described a cheap but accurate cycle ergometer which used the principle of the sinus balance to 'weigh' the resistive torques being applied to the flywheel of the ergometer. A steel band was attached to the back wheel of a stationary cycle to give a cylindrical surface to which a strap brake was fitted, the pull of which could be adjusted. The frictional force acting on the strap brake to retard the flywheel is the difference between the forces at the two ends of the belt (Fig. 6).

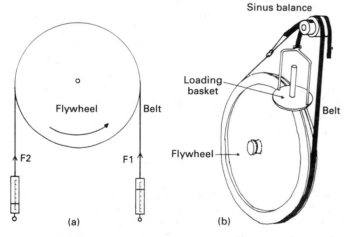

Fig. 6 (a) The frictional forces applied to a strap brake. (b) The modern system adopted by Monark which uses the sinus balance to apply a known frictional loading to the flywheel.

The work done by a mechanical system is given by the product of the applied force and the distance travelled by the point of application of the force. Once the sinus balance has been calibrated using known forces, the work done can be calculated from the formula

$$W \text{ (J)} = N \times \text{flywheel circumference (m)} \times \text{NR}$$

where N is the balance reading and NR is the number of flywheel revolutions. The power output is then calculated by dividing the work done by the time taken to do the work. Monark-Crescent AB have developed a range of friction-loaded cycle ergometers based on the von Dobeln ergometer in which the sinus balance is replaced by a suspended weight. When the sinus wheel is free to rotate and the applied load is $L(N)$, the von Dobeln equation becomes

$$W \text{ (J)} = L(N) \times \text{flywheel circumference (m)} \times \text{NR}$$

This formula enables us to calculate power output

$$\text{power output (W)} = L\omega t$$

where ω is the angular velocity of the flywheel and r is the radius of the flywheel.

When deciding on the test protocol to be used the duration of the exercise must be considered. Gollnick and Hermansen[27] showed that an exercise duration of 10 s was too short to tax the anaerobic processes fully. Although increasing the duration of the exercise will place greater demands on the anaerobic processes, it will also proportionally increase contributions to energy provision from the anaerobic metabolism. Gollnick and Hermansen calculated the relative contributions of the anaerobic metabolism to the total energy output during maximal exercise. They found that for the first 10 s of exercise the anaerobic contribution was approximately 83 per cent. This value dropped to approximately 60 per cent, 40 per cent, and 20 per cent for exercise durations of 1 min, 2 min, and 5 min respectively. Katch[28] reported that 90 per cent of V_{O_2}max was reached in approximately 60 s during supramaximal exercise. As the aerobic system clearly does not achieve steady state in activities lasting only a few seconds, the contribution made to the total energy production by the aerobic metabolism in this type of exercise is difficult to measure precisely, requiring the determination of the oxygen debt which occurred during the test. Although there is general agreement that the tests of 'anaerobic capacity' must be of maximal intensity throughout, there has been no standardization of the duration of the exercise, with test durations lasting from 20 to 240 s having been proposed. The work by Margaria et al.[10,19] has greatly influenced the decisions that researchers have made regarding the duration of such tests. They estimated that the maximum lactic acid production was reached after 40 to 60 s of maximal exercise. Katch et al.[29] examined test durations of 40 and 120 s on the Monark cycle ergometer and found that the total cumulative work at 40 s had a high correlation ($r = 0.95$) with the total cumulative work in 120 s. They also stated that 'it can be calculated that by 40 s the work rate drops to a level that is within the 'aerobic range' of oxygen requirements'.

Using the findings described above, coupled with the developmental work by Cumming[30] and Ayalon et al.,[20] Bar-Or[2] described a test protocol which has become known as the 'Wingate test' or the anaerobic work test. This test requires the subject to cycle at maximum speed against a predetermined resistance for 30 s. The resistance is set at 75 N per 1000 N body weight. During the test the number of flywheel revolutions for each 5-s period is monitored. Using the von Dobeln equation described earlier the power output for each 5-s period of the test could be calculated, as could the total work done during the test.

In the protocol described by Bar-Or[2] the subjects start to pedal as fast as possible against a very low resistance which is increased to the required level during the first 2 to 3-s of the test. Many researchers have subsequently replaced this 'flying start' with either a stationary start or a 'rolling start' at a predetermined submaximal speed, with maximum effort commencing once the required load has been introduced. Irrespective of the method used to start the test, the subject works as hard as possible throughout the test with no attempt made at 'pacing' or energy conservation. Toe clips are used to hold the feet on the pedals. The subjects are required to remain seated throughout the test.

Three indices of anaerobic performance are calculated from the test.

1 Maximal anaerobic power: the highest 5-s power output. This index reflects the peak power generated by the active muscles.
2 Anaerobic capacity: the total work done in the 30-s test.
3 Fatigue index: the difference between the highest and lowest 5-s power outputs divided by the elapsed time.

The test–retest reliability of the protocol was checked for various age, sex, and fitness groups both on the same day and over a period of up to 2 weeks. For tests repeated on the same day correlation coefficients of 0.95 to 0.98 were obtained. Over the 2-week period correlation coefficients of 0.90 to 0.93 were calculated even when environmental conditions were modified. It was concluded that the test was highly reliable.

The method of calculation of power output in the 'Wingate test' has been criticized.[31] This method does not include the work being done in transferring kinetic energy to and from the flywheel of the ergometer. In order to calculate correctly the total power output being generated the work done against the frictional load, as described by von Dobeln, must be added to the work done in accelerating the flywheel

$$\text{total power output} = PO_{TOT} = PO_{FL} + PO_{AF}$$

where PO_{FL} is the power output against the frictional load, PO_{AF} is the power output required to accelerate the flywheel, and

$$PO_{TOT} = \omega \, (Lr + I\delta\omega/\delta t)$$

where I is the moment of inertia of flywheel which is obtained by performing a mechanical run-down test.

It is clear from the equation for total power output that the angular acceleration $\delta\omega/\delta t$ of the flywheel must be known. To achieve this, the speed of the flywheel must be constantly monitored by a computer and the instantaneous values of flywheel acceleration calculated. A simple but effective computerized system for monitoring flywheel speed is described by Lakomy.[31] It comprises a small d.c. generator which is driven by the flywheel. The voltage output from the generator, which is proportional to the speed of revolution ($r^2 = 0.998$), is logged by the computer. The computer is able to monitor the flywheel speed and acceleration and therefore to calculate the correct power output.

When the corrected method of calculation is applied to the standard Wingate test, i.e. where the subject is pedalling at maximum speed before the load is applied, the power output at any time during the test is found to be less by a small (up to 6.2 per cent) but significant amount than that calculated by the conventional method[32] because of the recovery of kinetic energy from the flywheel as it decelerates throughout the test. When a stationary or low speed rolling start is used, large discrepancies in power output between the two methods of calculations are found. Using a loading recommended by Bar-Or[2] of 75 N per 1000 N body weight and a rolling start of 60 pedal rev/min, it was found that not only were the corrected 1-s averaged peak power values approximately 32 per cent higher than the uncorrected values but that peak power output occurred on average 2.1 s earlier.[33] Examination of the Wingate indices revealed that, when using the low speed start, the uncorrected values of the maximal anaerobic power and fatigue index were greatly underestimated when compared with the correct values. However, the total work done in the 30 s, defined as anaerobic capacity, was found to be independent of the method of calculation ($p < 0.005$).

Many researchers have attempted to optimize the power output

generated during cycle ergometer tests by varying the frictional loads being applied to the flywheel whilst using the uncorrected method of calculation.[29,34–36] All found that peak power output was sensitive to changes in applied load and that it increased with increasing loads up to approximately 100 N per 1000 N body weight. In contrast with these findings, Lakomy[37] has shown that when the corrected method of calculation is used the peak power output is independent of the load used. Irrespective of the method of calculation, however, it appears that a loading of approximately 105 N per 1000 N of body weight will result in the greatest amount of work being done during the test.

In addition to the single continuous exercise tests on the cycle ergometer, multiple sprint tests have been developed. In many sports, such as rugby, soccer, hockey, basketball, and volleyball, the performer is required to exercise at maximum or near-maximum intensities for a few seconds with these bouts of activity being interspersed with periods of recovery. The ergometer-based multiple sprint tests are designed to investigate the response of the body to such activities. These tests attempt to mimic the sporting situation. The duration of the sprints and the recovery time can be linked to a temporal analysis of the sport. An example of a test protocol was that adopted by Wootton and Williams[38] to examine the influence of recovery duration on repeated 6-s sprints. Five 6-s maximal sprints were performed with rest intervals of either 30 or 60-s against loadings of 75 N per 1000 N body weight. They found that the capacity to perform repeated bouts of maximal exercise was markedly influenced by the preceding number of sprint bouts and the duration of recovery. In such tests the decline in peak and mean power output per sprint is used to describe the fatigue taking place.

An alternative multiple sprint protocol requires the subject to perform a fixed amount of work whilst performing maximally (Loughborough sprint test).[33] The time taken for each sprint is recorded, with the fatigue taking place being described by the increase in this time. Figure 7 shows a typical example of the time taken to perform set amounts of work, with 100 per cent being the total amount of work that could be performed in a single 6-s sprint.

Dynamic power has been investigated using isokinetic cycle ergometry.[39] On such ergometers the pedalling speed is controlled and the forces exerted on the pedal cranks are detected by foil strain gauges and transmitted to a computer. Torque, work, and power are calculated for each leg in every pedal stroke to obtain data on maximal power and decline in power over the test period, expressed as a fatigue index. All the tests described for the friction-loaded cycle ergometer can also be performed with isokinetic ergometers.

Critical power

A contrasting system for measuring anaerobic work capacity and critical power has been developed by Moritani et al.[40] based on the test proposed by Monod and Scherre.[41] The test involves a series of exhaustive exercises on the cycle ergometer at a range of intensities during which the total amount of work and the time to exhaustion are measured. The tests exercise intensities ranged from 170 to 360 W 'depending on body weight and level of fitness' so that the durations of the exercise (to exhaustion) lasted approximately 1 to 10 min. The pedalling rate was set at 70 rev/min. The test was terminated immediately when the subject was unable to maintain cadence (i.e. when the pedalling speed dropped to 65 rev/min). A regression equation was generated such that the x and y values were the duration of the exercise and the product of

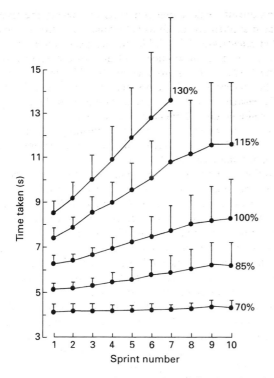

Fig. 7 The time taken to perform a series of fixed work sprints with 30 s of recovery between sprints ($n = 8$; mean ± SD).

exercise intensity W and exercise duration respectively. The slope of the relationship obtained was termed the critical power and the intercept was termed the anaerobic work capacity. The authors suggested that critical power is the exercise intensity which can just be maintained without exhaustion while anaerobic work capacity is the total amount of work that can be performed utilizing only stored energy sources within the muscle. They claim further that critical power is significantly correlated ($p < 0.01$) with the ventilatory threshold ($r = 0.927$) and that the physical work capacity at the fatigue threshold ($r = 0.869$), which has been defined as the highest constant power output that requires no increase in electromyographical activity of the major muscles involved for its maintenance.[40] Although four exercise intensities over a period of 2 days were used to determine the regression equation, Moritani et al.[40] concluded that critical power and anaerobic work capacity could be estimated with reasonable accuracy using only two exercise bouts. They qualified this statement by stipulating that the two bouts should be chosen with care, so that they produce exhaustion in the range 1 to 10 min and differ by at least 5 min. They felt that 'the adoption of this protocol would reduce demands on the subject and investigator thereby improving the practicality of the critical power test in the laboratory and field settings'.

Nebelsick-Gullett et al.[42] examined the relationship between anaerobic work capacity and anaerobic capacity (AC) from the critical power and Wingate tests respectively. Anaerobic work capacity and AC were found to be significantly ($p < 0.05$) related ($r = 0.74$). There was no significant difference ($p < 0.05$) between test–retest means for critical power or anaerobic work capacity. Test–retest correlations for critical power ($r = 0.94$) and anaerobic work capacity ($r = 0.87$) led Nebelsick-Gullett et al.[42] to conclude that the critical power test was a reliable technique for measuring anaerobic capacity as well as the maximum rate of fatigueless work.

Sprint running

Sprint running, in contrast with sprint cycling, is a weight-bearing activity. Therefore the cycle ergometer may be of limited value to those interested in the evaluation of sprint running. Thomson and Garvie[43] described a laboratory test in which subjects sprinted for 15, 30, 45, and 60 s on a motor-driven treadmill up a 5 per cent grade at a speed which would produce exhaustion at around 60 to 70 s. The energy expenditure was determined by indirect calorimetry combining measurement of oxygen uptake with those of peak lactate concentrations.[44] In the sprinters the mean total energy expended in the sprint to exhaustion was 63.9 kcal, of which 72.3 per cent was calculated to be derived from anaerobic metabolism. In contrast, the corresponding values for the marathon runners were 52.8 kcal and 62.6 per cent. The anaerobic work done was then subdivided into alactic (mean 48 per cent) and lactacid (mean 52 per cent) components. These total energy expenditure results were highly correlated with the subject's sprinting performance over 329 m ($r = 0.82$, $p < 0.01$, $n = 14$). The conclusion drawn was that the protocol provided a direct quantitative measurement of anaerobic capacity and partitioned the alactic and lactacid components. It was also sensitive enough to distinguish between groups of subjects (trained versus untrained, sprinters versus endurance) and between subjects within a specific group.

In contrast with the indirect calorimetry test of Thomson and Garvie,[43] an ergometer for the direct measurement of power output during sprint running has been described.[37,45] The ergometer was based on a non-motorized treadmill which (a) allowed the subjects to sprint at speeds similar to those achieved in sprint running, (b) allowed the same variability in instantaneous work rate during sprinting that cycle ergometers permit during cycling, and (c) enabled instantaneous values of power output to be determined throughout the sprint.

The external forces generated by a sprinter can be resolved into vertical and horizontal components. The vertical component raises the centre of gravity during each stride so that leg recovery can take place. Fukunaga et al.[46] found that for a given runner the vertical component of the work being done was independent of the running speed. The horizontal or 'propulsive' component moves the runner along the ground; the greater this propulsive component the faster the sprinter will cover the ground. Therefore the sprinter attempts to maximize his work rate in this direction. Ideally, the total work done by the sprinter, i.e. the sum of the vertical and horizontal components, should be monitored when evaluating the physiological demands of sprinting. It is extremely difficult to measure the vertical component of the work done during sprinting and so the instrumented ergometer measured only the horizontal component of the work rate, i.e. the component that actually maintains treadmill motion.

A commercially available non-motorized treadmill (Woodway model AB), which normally slopes backwards, was levelled by placing its rear feet on supports (Fig. 8). All four feet were anchored securely to a baseboard to prevent excessive lateral movement during use. Measurement of the propulsive force was achieved by applying Newton's third law of motion which states that for every action there is an equal and opposite reaction. If the sprinter does not move relative to the ground, then the force that the sprinter applies to the treadmill belt must be equal to the

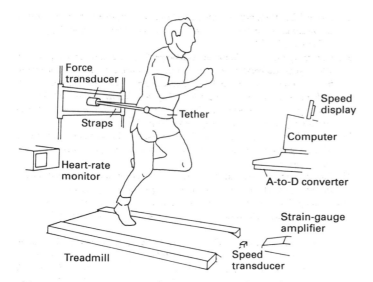

Fig. 8 Non-motorized treadmill for the analysis of the power output generated during sprint running.

horizontal component of the restraining force in the harness. The horizontal restraining tether was connected to a force transducer. The output from the force transducer amplifier and the treadmill belt speed was continuously monitored by a computer. The rate at which work was being done to move the treadmill, i.e. the horizontal power output, was calculated from the product of the restraining force and the belt speed. This calculation assumes that the error resulting from the points of force application and measurement not being the same is small and that little of the subject's weight is detected as a horizontal force due to forward lean.

At the conclusion of a test the following information was calculated and displayed: (a) the mean propulsive power for each second of the sprint; (b) the mean propulsive power for the sprint duration; (c) the total propulsive work done; (d) a fatigue index defined as the difference between the peak and the lowest propulsive power output values expressed as a percentage of the peak value. The same test indices that were described for the cycle ergometer could also be measured on the sprint ergometer.

Peak propulsive power outputs were found to occur during the acceleration phase of the sprint. Mean 1-s values in excess of 1 kW were commonly obtained, with within-stride instantaneous values (0.025 s average) often in excess of 3 kW. Fukunaga *et al.*[46] have shown that this propulsive power represents approximately 80 per cent of the total power output, with the remaining 20 per cent required for the work against gravity.

According to Cheetham *et al.*[47] 'Thus this laboratory-based method for studying sprint running offers an additional way of investigating human response to brief periods of high intensity exercise.'

CONCLUSION

It is clear that there are many different test protocols for measuring anaerobic power. All the tests require the applied force and the velocity of movement to be monitored or indirectly calculated. No test will allow accurate determination of the contribution to total anaerobic energy metabolism by the alactic acid systems. In order

to decide which test to use the experimenter must answer a number of questions including the following.

1 What type of activity is to be measured?
2 What is the total muscle mass which is required to be activated?
3 What is the duration of the activity?
4 Do instantaneous or average values of power output require to be measured?
5 Will the system allow acceleration or will it be limited to a constant speed?
6 Will single or multiple sprints be performed? If multiple sprints, how many are to be performed and what recovery duration is required?
7 Do the force–velocity and power–velocity relationships of the active muscle groups need to be determined?

No single test can encompass all the possible test permutations. However, the greatest versatility is possible using the cycle ergometer. Within a single test instantaneous values of power can be obtained within, as well as over, a pedal revolution. The average power output over the test period can be measured with relative ease. Single or multiple sprints can be performed and monitored. The time required to reset the equipment is very short, making it possible to have recovery periods between sprints as short as 2 s. Usually the tests of anaerobic power require a large muscle mass to be activated. Although other types of ergometers can involve larger muscles masses, pedalling on the cycle ergometer uses a sufficiently large muscle mass for most purposes. The muscle groups most investigated for the force–velocity relationship of muscle are the knee extensors. The ideal tools for examining these relationships are isokinetic dynamometers. However, a great deal of useful information on the force–velocity relationship of the knee extensors can be obtained using the cycle ergometer.[8,35,36]

It is important that the method of calculation of power on the cycle ergometer used incorporates the work being performed against the inertia of the flywheel and that appropriate loads are used to optimize performance.

Finally, it is important to appreciate that the errors of measurement in the tests described are at best ± 5 per cent, and are often very much greater. This error is large when compared with the magnitude of the changes in performance which could result from training etc. Therefore it is possible that such changes may well be masked, bringing into question the validity of using these tests when more direct performance scores may be more accurate and valid.

REFERENCES

1. Perrine JJ. The biophysics of maximal muscle power outputs: Methods and problems of measurement. In: Jones, McCartney, McComas, eds. *Human muscle power*. Champaign, IL: Human Kinetics, 1986: 15–25.
2. Bar-Or O. A new anaerobic capacity test—characteristics and applications. *Proceedings of the 21st World Congress of Sports Medicine, Brasilia, 1978*: 1–27.
3. Katch VL, Weltman A. Interrelation between anaerobic power output, anaerobic capacity and anaerobic power. *Ergonomics* 1979; **22**: 325–32.
4. Simoneau J, Lortie G, Boulay M, Bouchard C. Test of anaerobic alactacid and lactacid capacities: description and reliability. *Canadian Journal of Applied Sport Sciences* 1983; 8(4): 266–70.
5. Knuttgen HG. (1978) Force, work, power, and exercise. *Medicine and Science in Sports* 1978; 227–8.
6. Hill AV. The heat of shortening and the dynamic constant of muscle. *Proceedings of the Royal Society of London* 1938; **126**: 136–95.

7. Fenn WO, Marsh BS. Muscular force at different speeds of shortening. *Journal of Physiology (London)* 1935; **85**: 277–97.

8. McCartney N, Heigenhauser GJ, Jones NL. Power output and fatigue of human muscle in maximal cycling exercise. *Journal of Applied Physiology* 1983; **55**: 218–24.

9. Inbar O, Bar-Or O. The effects of intermittent warm-up on 7–9 year old boys. *European Journal of Applied Physiology* 1975; **34**: 81–9.

10. Margaria R, Cerretelli RP, Mangili F. Balance and kinetics of anaerobic energy release during strenuous exercise in man. *Journal of Applied Physiology* 1964; **21**: 1662–4.

11. Boobis L, Williams C. Wootton S. Human muscle metabolism during brief maximal exercise. *Journal of Physiology (London)* 1982; **338**: 21–2P.

12. Jacobs I, Tesch Per, Bar-Or O, Karlsson J, Dotan R. Lactate in human skeletal muscle after 10 and 30 s of supramaximal exercise. *Journal of Applied Physiology* 1983; **55**(2): 365–67.

13. Cavagna GA. Force platforms as ergometers. *Journal of Applied Physiology* 1975; **39**: 174–9.

14. Davies CTM, Rennie R. Human power output. *Nature London* 1968; **217**: 770–1.

15. Offenbacher EL. Physics and the vertical jump. *American Journal of Physics* 1970; **38**: 7.

16. Davies CTM. The aerobic and anaerobic components of work during submaximal exercise on a bicycle ergometer. *Ergonomics* 1971; **14**(2): 257–63.

17. Davies CTM, Young K. Effects of external loading on short term power output in children and young male adults. *European Journal of Applied Physiology* 1984; **52**: 351–4.

18. Adamson GT, Whitney RJ. Critical appraisal of jumping as a measure of human power. *Medicine and Sport 6: Biomechanics II*. Basel, Karger, 1971: 208–11.

19. Margaria R, Aghemo P, Rovelli E. Measurement of muscular power (anaerobic) in man. *Journal of Applied Phsyiology* 1966; **21**: 1662–4.

20. Ayalon A, Inbar O, Bar-Or O. Relationships among measurements of explosive strength and anaerobic power. *Biomechanics IV*. 1975: 572–77.

21. Sawka MN, Tahamount MV, Fitzgerald PI, Miles DS, Knowlton RG. Alactic capacity and power: reliability and interpretation. *European Journal of Applied Physiology* 1980; **41**: 93–9.

22. Kitiwaga K, Suzuki M, Miyashita M. Anaerobic power output of young obese men: comparison with non obese men and the role of excess fat. *European Journal of Applied Physiology* 1980; **43**: 229–34.

23. Caiozzo VJ, Kyle CR. the effect of external loading upon power output in stair climbing. *European Journal of Applied Physiology* 1980; **44**: 217–22.

24. Kalamen J. Measurement of maximum muscular power in man. Ph.D. Thesis, Ohio State University, 1968.

25. Bosco C, Luhtanen P, Komi P. A simple method for measurement of mechanical power in jumping. *European Journal of Applied Physiology* 1983; **50**: 273–82.

26. von Dobeln W. A simple bicycle ergometer. *Journal of Applied Physiology* 1954; **7**: 222–4.

27. Gollnick PD, Hermansen L. Biochemical adaptations to exercise: anaerobic metabolism. In: Wilmore JH, ed. *Exercise and sports science reviews*. New York: Academic Press, 1973: 1–43.

28. Katch VL. Kinetics of oxygen uptake and recovery for supra maximal work of short duration. *International Z Angew Physiologica* 1973; **31**: 197–201.

29. Katch VL., Weltman A, Martin R, Gray, L. Optimal test characteristics for maximal anaerobic work on the cycle ergometer. *Research Quarterly* 1977; **48**: 319–27.

30. Cumming GR. Correlation of athletic performance and aerobic power in 12–17 year old children with bone age, calf muscle, total body potassium, heart volume and two indices of anaerobic power. In: Bar-Or O, ed. *Pediatric work physiology*. Natanya: Wingate Institute, 1974: 109–37.

31. Lakomy HK. An ergometer for measuring the power generated during sprinting. *Journal of Physiology* 1984; **354**: 33P.

32. Bassett DR Jr. Correcting the Wingate test for changes in kinetic energy of the ergometer flywheel. *International Journal of Sports Medicine* 1989; **10**(6): 446–9.

33. Lakomy HK. Measurement of external power output during high intensity exercise. Ph.D. Thesis, Loughborough University, 1988.

34. Evans J, Quinney H. Determinations of resistance settings for anaerobic power testing. *Canadian Journal of Applied Sports Sciences* 1981; **6**: 43–56.

35. Nadeau M, Brassard A, Cuerrier JP. the bycycle ergometer for muscle power testing. *Canadian Journal of Applied Physiology* 1983; **8**: 41–6.

36. Nakamura Y, Yoshiteru M, Miyashita M. Determination of the peak power output during maximal brief pedalling bouts. *Journal of Sports Science* 1986; **3**: 181–7.

37. Lakomy HK. Effect of load on corrected peak power output generated on friction loaded cycle ergometers. *Journal of Sports Sciences* 1985; **3**(3): 240.

38. Wootton SA, Williams C. The influence of recovery duration on repeated maximal sprints. In: Knuttgen HG, Vogel JA, Poortmans JR, eds. *Biochemistry of exercise. International Series on Sport Sciences* Champaign, IL: Human Kinetics, 1983: 269–73.

39. McCartney N, Heigenhauser GJ, Sargeant A, Jones NL. A constant-velocity ergometer for the study of dynamic muscle function. *Journal of Applied Physiology* 1983; **55**: 212–17.

40. Moritani T, Nagata A, DeVries H, Muro M. Critical power as a measure of physical work capacity and anaerobic threshold. *Ergonomics* 1981; **24**: 338–50.

41. Monod H, Scherre J. The work capacity of a synergic muscular group. *Ergonomics* 1965; **8**: 329–38.

42. Nebelsick-Gullett L, Housh T, Johnson G, Bauge S. A comparison between methods of measuring anaerobic work capacity. *Ergonomics* 1988; **31**(10): 1413–19.

43. Thomson JM, Garvie KJ. A laboratory method for the determination of anaerobic energy expenditure during sprinting. *Canadian Journal of Applied Sport Sciences* 1981; **6**(1): 21–6.

44. Margaria, R, Cerretelli P, di Prampero PE, Massari C, Torelli G. Kinetics and mechanisms of oxygen debt contraction in man. *Journal of Applied Physiology* 1963; **18**: 371–7.

45. Lakomy HK. The use of a non-motorised treadmill for analysing sprint performance. *Ergonomics* 1987; **30**(4): 627–37.

46. Fukunaga T, Matsuo A, Yuas K, Fujimatsu H, Asahina K. Mechanical power output in running. In: Asmussen E, Jorgensen K, eds. *Biomechanics VI-B. International Series on Biomechanics*. Baltimore: University Park Press, 1978: 17–22.

47. Cheetham ME, Williams C, Lakomy HK. A laboratory running test: metabolc responses of sprint and endurance trained athletes. *British Journal of Sports Medicine* 1985; **19**: 81–4.

Sport and the environment 2

2.1　　The underwater environment

DAVID H. ELLIOTT

INTRODUCTION

There are some 50 000 sports divers in the United Kingdom and, according to the American 'Diver Alert Network', some 3 million in the United States. There are more than 100 diving fatalities per annum in the United States alone, and no reliable estimate is available on the prevalence of diving illnesses.

The problems and challenges imposed by the underwater environment are many and complex. The sport of diving is universal, and an understanding of the effects of this environment is essential, for example, when deciding upon the fitness of otherwise seemingly healthy individuals to participate in underwater sports.

THE SPORT OF DIVING

There are two quite distinct underwater activities to be considered. Each shares the challenge of raised environmental pressure, but the physiological response and sequelae can be very different. The two activities are separated by the simple distinction of whether the individual is relying upon his breath-hold duration for his excursion underwater or whether, when submerged, he is able to breath from some form of breathing apparatus.

Breath-hold diving includes snorkel diving in which the use of a J-shaped tube enables a person to breathe while face down on the surface of the water. This equipment together with a mask and fins, is cheap and readily available to the young, the casual, and the untrained. Snorkelling is associated with many hazards but, though it is not safe, the risks are fewer than those of scuba diving.

The well-trained and experienced breath-hold diver can achieve remarkable underwater excursions and it is in this area in particular that the competitive element has arisen. Depth alone is a challenge, and the achievement of more than 105 m (350 ft) on a single breath is associated with a risk of death by drowning that only the foolish would accept. Speed through the water is important in competitive spear-fishing, a breath-hold activity which is still practised in parts of the Mediterranean, whereas competitive underwater photography demands breath-hold control.

The use of self-contained underwater breathing apparatus (scuba) requires more than a bottle of compressed air. The air cylinder is pressurized to around 150 bar (15 MPa) and passes through a first-stage regulator on the cylinder and a pressure hose to the second-stage regulator on the mouthpiece. This regulator provides compressed air to the diver on inspiratory demand at an adequate flow and at the same pressure as the water surrounding him. The scuba diver will also have a snorkel, for swimming on the surface, a mask, fins, and a weight belt with extra weight to compensate for the buoyancy of a thermally protective wet suit or dry suit. A knife, a watch, and a depth gauge are carried, and a buoyancy vest is worn. This does not simply act as a life-jacket at the surface but also, by adjusting the volume of compressed air within it, serves as a buoyancy compensator while at depth.

There is a minority of sports divers who use breathing gases other than compressed air. Closed circuit breathing apparatus using pure oxygen provides a lesser bulk which can facilitate the negotiation of narrow passages when diving in caves. This type of apparatus, which is also favoured by video and film cameramen because of its lack of bubbles, is notoriously dangerous. It exposes the subject to a number of additional hazards not encountered with scuba. The increased partial pressure of oxygen can cause oxygen intolerance in the form of oxygen neurotoxicity or shallow-water black-out. These are described later. The nature of the apparatus itself can lead to the accidental inhalation of a caustic mixture of soda lime and seawater. The soda lime normally scrubs all the carbon dioxide and thus also removes carbon dioxide accumulation which would act as a warning of gas supply failure. When this occurs the oxygen content of the recirculating gas can diminish unnoticed by the diver who passes quietly through hypoxia to an anoxic death. This type of apparatus is rightly banned by many diving organizations.

Oxygen can also be used to enrich the air of self-contained equipment in order to provide a small benefit for the prolongation of bottom-time or shortening of decompression. Such techniques require special gas-mixing techniques or monitoring and the use of specific decompression procedures, and thus are not for the untrained. Oxygen can also be used during decompression stops to facilitate the elimination of dissolved nitrogen. The risk of acute oxygen toxicity is enhanced if there is not meticulous separation of the supply of oxygen to the diver from that to the air.

The relative expense of scuba-diving equipment limits its availability, but it can be hired, particularly from 'diving shops' around the world. Training is essential, but a 1-day 'dive-resort course' is too often provided without a proper pre-dive health check and, though better than no introductory supervision, this is regarded as inadequate for this complex and independent sport.

Those who pursue scuba diving as a recreational sport not only require to have knowledge of the equipment used, its failures, and its maintenance, but also need to understand the environmental hazards which they will encounter. They should pass a recognized training course of several days' duration and, after that, will be able to enjoy practical experience under a degree of supervision while progressing towards more advanced training courses. For all this they need to be mentally, physically, and medically fit. The professional sports diver is primarily concerned with instruction and requires to be in virtually perfect heath.

Underwater sports are widely distributed; divers certainly do not confine their activities to coastal waters. For instance, in the United Kingdom nearly every major town has a branch of a sports diving club which holds weekly training sessions in a local swimming pool. These are usually well conducted and contribute to the safety performance of the sport. The geographical distribution of recreational diving is increased by diving training in lakes and reservoirs, wherever these are available. On one occasion an archaeological interest in a Roman well gave rise to a diving incident on a motorway building site. Practitioners may also encounter problems when returning by air from a vacation overseas. Altitude itself is also hazardous, and thus a diving sports injury can be susceptible to aggravation by bubble growth at a cabin altitude which is normally around 8000 feet. Similar activities and associated incidents occur throughout much of the Western world.

Though a popular image of sports diving may be associated with tropical seas, the practice of diving medicine is in fact universal.

THE ENVIRONMENT

Water has a specific heat some 1000 times greater and a conductivity some 32 times greater than air, and so the exposed skin of the diver can be considered to be at the same temperature as the water around him. Thus those diving without any thermal protection will become cold with time, even in relatively warm waters of the Caribbean (30 °C). The subjective threshold for the onset of feeling cold depends on many intrinsic factors, and each diver must determine the most suitable protection for him in particular diving conditions. One effect of cold water on the face and body is to cause a bradycardia, a feature of the so-called 'diving reflexes'. These phenomena, which are a topic of intense physiological research, may contribute to the response of an individual in an underwater incident, but they have no clinical significance and will not be described further.

Water isolates the individual. Though sound travels well through the water, verbal communication between divers is not normally possible and hand-signals, though limited in their vocabulary, are needed.

In water there is some magnifying distortion at the air–water interface of the diver's mask, but the water is not always clear and indeed when diving in certain locations, visibility can be zero.

Water contains unexpected currents and dangers, such as nylon fishing lines, wrecks, and caves. The surge of waves overhead will affect safe movement and constant vigilance is essential.

However, the feature of the underwater environment that provides most of the problems is that the weight of a 10-m column of water exerts an additional pressure equivalent to the earth's atmosphere above it. As a result of the pressure exerted at depth by the column of water above the diver, his body experiences effects which are in accordance with some fundamental physical laws (Fig. 1). Boyle's law states that the volume of a given mass of gas is inversely proportional to the pressure. Thus 1 l of gas at sea level decreases to 0.25 l at 30 m depth where the pressure is four times atmospheric. Of importance to the diver are the effects of this upon his gas-containing spaces: barotrauma of decent and ascent, when those spaces are unable to compensate for the changes of pressure and volume, and changes of buoyancy as a result of gas volume changes. The application of Dalton's law, that the partial pressure of a gas in a mixture is equal to the product of its fractional concentration and the absolute pressure, shows that the partial pressure of oxygen and nitrogen in compressed air are doubled at 10-m depth, and these partial pressures increase proportionately at deeper depths.

Also, the weight of water is sufficient to create a pressure gradient of physiological significance to the diver, particularly when in a vertical orientation. Yet this pressure is not noticed subjectively. Whether salty or fresh, the density of water is approximately equivalent to that of the body, such that the diver feels neutrally buoyant. The redistribution of venous blood consequent upon this pressure gradient is one of the factors causing a physiological diuresis which, through dehydration, may contribute to subsequent decompression sickness.

For the scuba diver at any environmental pressure, buoyancy

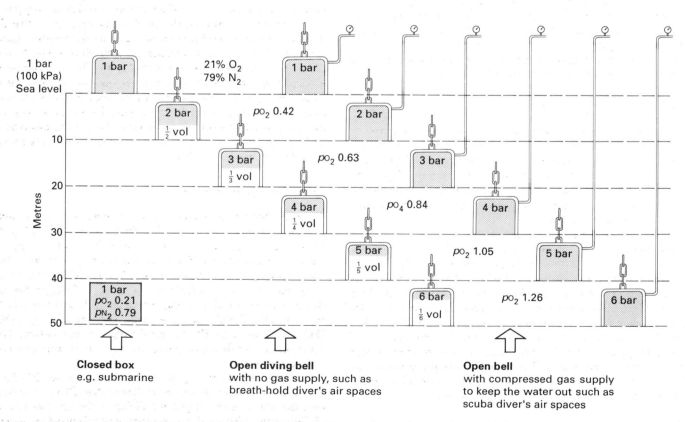

Fig. 1 The volume of gas within an open and inverted jar diminishes on descent in accordance with Boyle's law. If additional compressed gas is admitted, the volume within the jar can be kept constant. The particular pressure of the contained gases increases in proportion to the absolute pressure. If a closed pressure-proof box is submersed, the pressure and volume are maintained at the atmospheric level (1 bar = 100 kPa).

can be finely controlled by the volume of gas held in the chest. However, the support given to the body by being in the water almost eliminates proprioception; for example, if an episode of dizziness or vertigo occurs while underwater, there may be no proprioceptive clues and also no visual clues to orientation. To what might be a transient episode on land, the undersea environment contributes additional factors which may influence the outcome adversely.

The physical laws which determine the nature of the effect of the underwater environment upon a diver need to be fully understood by the diver and his physician: Pascal's law which describes the transmission of pressure throughout a fluid, Dalton's law which describes the partial pressures of gases, and Henry's law which accounts for the uptake and distribution of gases, particularly nitrogen, from the lungs of a diver. Equally fundamental is Boyle's law which affects the air-containing spaces of the body. These are outlined where relevant in the following text.

THE EFFECTS OF SUBMERSION

Cellular effects

The body behaves as is it were a fluid and is regarded as incompressible. Within the shallow depths encountered by sports divers this can be considered true for practical purposes. Nevertheless the direct effects of pressure upon the cell and its constituents give rise to the serious 'high pressure nervous syndrome'[1] in professional divers at depths greater than some 200 m. At this depth and beyond there are also changes, for instance, in red-cell membranes and blood platelet aggregation[2,3] which show that, after a period at raised environmental pressure, man is not the same physiologically as he was previously at the surface. While these deep diving phenomena may seem to have little direct bearing on the clinical aspects of sports diving medicine, they do illustrate that mere exposure to the environmental pressure of sports diving is not without possible effects after a subsequent safe return to the surface.

Air-containing spaces

The effect of submersion upon the gas-containing spaces of the body is more obvious. The gas within them must follow Boyle's law and thus either reduce in volume with increased depth or, to maintain volume, be able to admit additional compressed gas. The adverse consequences of pressure–volume inequality are termed 'barotrauma' and can occur in both the compression and decompression phases of the dive.

The increased density of the gas also has an effect upon respiration: indeed, physical effort underwater tends to be ventilation limited. This limitation affects both expiration and inspiration. Maximum expiratory flow rate is diminished and at any depth is limited due to increased density. Thus the limitation is effort independent. Inspiratory flow is also diminished but this is dependent on effort. A diminished ventilation is associated with a raised arterial carbon dioxide tension. Another consequence of breathing gas at increased density is an increased respiratory heat loss. This loss is not so great that the inspiratory gas requires to be heated as it is in deep oxyhelium diving, but is a significant drain on the body's thermal reserve during an air dive.

Buoyancy

The self-contained diver has equipment which, when properly used, can overcome changes in his buoyancy. A set of weights, which can easily be ditched in an emergency, counterbalance the natural buoyancy of the diver and his suit. A buoyancy compensator (BC) vest can be used for trimming buoyancy as the tank of compressed gas becomes lighter as the gas is used. Misuse of the air inflation of a dry suit, inadequate control of the buoyancy compensator, or unintentional release of the weights can lead to an unexpected, rapid, and hazardous return to the surface.

The breath-hold diver may carry a few weights, especially if he is wearing a wet suit, but equipped normally with only mask, fins, and snorkel he is not capable of making buoyancy adjustments. His buoyancy is controlled by lung volume which, in order to prolong breath-hold duration, will be close to vital capacity when leaving the surface. On plunging down from the surface he quickly reaches a depth where Boyle's law has compressed the full chest and he has become negatively buoyant. It is not sufficiently widely known that, to return to the surface, the breath-hold diver may have to push off from the bottom or even swim his way back up. The importance of such instruction to novices seems obvious.

Latent hypoxia

Another cause of fatalities among breath-hold divers is considered to be excessive hyperventilation before descent. The advantages for breath-hold duration of eliminating carbon dioxide as far as possible seem obvious, but the unique circumstance of breath-hold diving make it hazardous. This is because hyperventilation cannot provide a significant increase in oxygen at the same time.

During the breath-hold dive, which is likely to be continued until the diver nears his carbon dioxide breakpoint, oxygen is being consumed. However, the depth of the dive means that the diminished oxygen level is at an increased partial pressure. Compression of the chest in accordance with Boyle's law further diminishes the desire for the next breath. In some instances, perhaps when the diver has over-ridden his carbon dioxide drive while chasing a fish, the oxygen is consumed further so that on ascent, with the concurrent fall in partial pressure of oxygen, hypoxia and unconsciousness supervene. To prevent this type of accident, no more than two or three deep respirations should precede any breath-hold dive.

Oxygen intolerance

The increased partial pressure of the respiratory oxygen content can lead to a risk of oxygen toxicity which can be manifest beyond 150 kPa in a neurological form as an epileptiform fit or, for longer duration beyond 50 kPa, as a pulmonary form with dyspnoea leading to pulmonary oedema. However, if divers use only compressed air, the threshold for either type of oxygen toxicity should not be reached. In 1990, one diver on air achieved a depth of 137 m (452 feet),[4] where the partial pressure of oxygen (309 kPa) is equivalent to '309 per cent' oxygen at the surface. This demonstrates the foolhardiness of sports record seekers who take unnecessary risks.

Only the use of pure oxygen, or of nitrox (oxygen-enriched air) can pose such threats at depths shallower than 50 m. Associated with the use of pure oxygen for diving is the phenomenon of 'shallow-water black-out', an otherwise unexplained loss of consciousness underwater thought to be due to a lowering of the syncope threshold and perhaps related to the vasoconstrictive effect of oxygen on cerebral blood flow.

Carbon dioxide effects

Carbon dioxide from extrinsic sources should never cause a problem for those using an open-circuit breathing apparatus, but nevertheless a number of experienced divers have been demonstrated to have become 'carbon dioxide retainers'. Such persons do not respond to an increased carbon dioxide tension, whether this arises intrinsically or extrinsically, and, while the increased carbon dioxide threshold might be thought of as an adaptive response, it acts synergistically with other factors to be one cause of apparently unexplained loss of consciousness underwater.

Nitrogen narcosis

Nitrogen is not an inert gas, and at increased partial pressure it behaves like an anaesthetic agent. Decreased performance and behaviour can be quantified at depths as shallow as 30 m, the limit recommended by some for inexperienced amateurs, and are the reason why even a professional diver should not use compressed air at depths greater than 50 m. Beyond such depths euphoria can make the individual irresponsible and less likely to be able to recover from a sudden emergency, thus converting a simple incident in the water into a potential fatality. Even when utterly foolish behaviour has been recorded on video, the diver may be quite unable to recall this after surfacing. However, nitrogen narcosis is directly related to its partial pressure and diminishes equally quickly on ascent. Though undoubtedly another contributory factor for underwater accidents, nitrogen narcosis is self limiting and does not require any specific treatment.

Gas uptake and elimination

The other great problem in sports diving is the safe elimination of the respiratory gases dissolved from the alveoli into the body during the time spent at depth. The amount of gas which dissolves in a liquid is, in accordance with Henry's law, proportional to the partial pressure of that gas. The solubility of gases in the watery and fatty tissues is different and uptake depends on the dynamics of local circulation. Because of bubble formation, the dynamics of gas elimination are more complex than those of uptake. There are many mathematical models relating to the uptake and distribution of dissolved gas in the body but, whether presented to the diver as a printed table or as an on-line computer display, the theoretically safe decompression derived from such predictions cannot be guaranteed. The causative factors and consequences of an inadequate decompression have a clinical importance and will be discussed later.

The phases of submersion

The physiological and pathological effects of pressure, presented only in outline here, are reviewed elsewhere.[5] The clinical consequences of these effects can be considered separately in relation to the compression phase of a dive, the submerged phase, and the decompression phase which extends into a postdive phase for some hours. The subsequent descriptions will follow this sequence. Only a few of these clinical conditions affect the breath-hold diver, particularly those of compression barotrauma to the air-containing spaces of the body.

COMPRESSION ILLNESSES

The air-containing spaces of the head (Fig. 2) are vulnerable to the effects of Boyle's law if the appropriate pressure and volume of

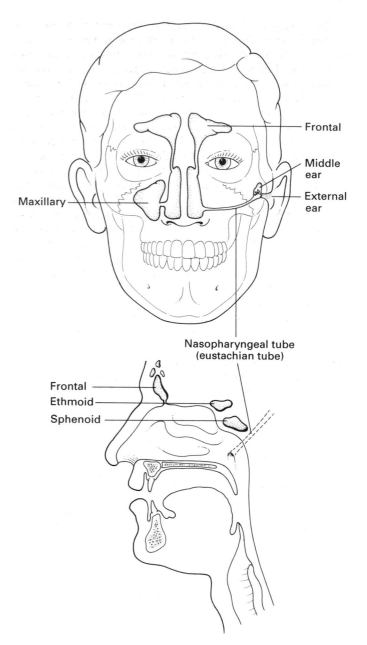

Fig. 2 The air-containing spaces of the head which are vulnerable to the effects of Boyle's law.

the gas within them cannot be maintained during compression. Lack of patency of an orifice into an otherwise closed space denies access to it for additional compressed gas as required. The consequent reduction of relative pressure within that space leads to either a compensating ingress of transudate or rupture of one of its walls.

Middle ear barotrauma

Equalizing the pressure within the middle ear during descent is a procedure known to divers as 'clearing the ears'. Difficulty in doing this is very common. Boyle's law demands that sufficient gas enters the middle ear via the eustachian tube to prevent a relative underpressure there and the consequences of middle ear barotrauma or 'ear squeeze'.[6]

The unsupported medial third of the eustachian tube can become flattened during descent by relative overpressure before any symptoms of pain arise, and the normal tube becomes locked at a depth as shallow as 1 m. Experienced divers know this well and use an equalization manoeuvre immediately on leaving the surface in order to 'keep ahead of the pressure'. If this is unsuccessful and discomfort is felt, it is necessary to ascend to a shallow depth where the ears can be felt subjectively to clear. Another attempt to descend can then be made. It follows that any inflammation of the nasopharynx, whether due to infection or allergy, can reduce the patency of the tube and thereby inhibit pressure equalization.

Equilibration is commonly achieved using the Toynbee manoeuvre, i.e. blowing against closed lips and nose with an open glottis in order to raise the pressure sufficiently within the nasopharynx to encourage air to pass through the eustachian tube. The use of oral decongestants before diving is common and may also help, but the rebound phenomenon from nasal sprays can aggravate the problem. Experienced divers learn the Frenzel manoeuvre which is performed by a swallowing movement with a closed glottis, again with the lips and nose closed.

Failure to equilibrate middle ear pressure leads to damage ranging from a simple mild injection to rupture of the tympanic membrane (Fig. 3). A serosanguinous transudate may have occurred to compensate for the potential reduction in middle ear volume and may be seen through the drum as a fluid level, possibly full of bubbles. If this is the case, the person should not dive until the condition has resolved and a prophylactic antibiotic against secondary infection should be administered. Rupture of the eardrum is more serious, and when it occurs underwater it carries the risk of immediate and debilitating vertigo if relatively cold water enters the middle ear. The individual should be given a prophylactic antibiotic and should not dive for some 6 weeks until the drum is well healed and its mobility easily demonstrated subjectively or by observation.

Inner ear barotrauma

Persistent attempts to clear the ears on descent may cause a greater problem: round or oval window rupture. The pressure from a forced Toynbee or Valsava manoeuvre passes through the cerebrospinal fluid to the perilymph. Rupture of the window results in vertigo and possible total deafness. Its possible association with the forceful Toynbee or Valsava manoeuvres may help to distinguish it from the other causes of vertigo such as acute decompression illness. First aid is by bedrest with the head elevated. Some ear, nose, and throat surgeons may wish to operate immediately; others may wish to postpone the repair for a few days to allow for the possibility of spontaneous improvement. Persistent vertigo postdive or the onset, perhaps delayed by 12 h, of total sensineural hearing loss requires immediate surgery to close the fistula.

Alternobaric vertigo

A more benign phenomenon is a sudden transient disorientation during the dive caused by an inequality of pressure between the right and left ears consequent upon the relative difficulty of air passing through one of the eustachian tubes. There may be a history of some difficulty in clearing the ears during the descent. However, in contrast with the possible association of round or oval window rupture with a forced Valsava or Toynbee manoeuvre, this form of barotrauma commonly manifests itself as transient vertigo after a relatively trouble-free period at depth, just at the start of the ascent. No treatment is required.

'Reversed' ear

The external auditory meatus may occasionally suffer from compression injury. This occurs if the meatus is closed by wax, an ear-plug, or some foreign body, or if it is covered by the impermeable hood of a dry suit. The consequent pressure change may rupture the tympanic membrane outwards.

Sinus and dental compression barotrauma

The free access of compressed air to the paranasal sinuses may be compromised by inflammatory conditions such as infection, allergy, or polyps. Pain may be referred from a sinus and can be intense. The affected sinus may fill with serosanguinous transudate, for which prophylactic antibiotics may be given, but usually there are no other effects.

Dental pain, associated with compression of gas entrapped behind a filling or within an unhealthy tooth, can also be painful.

Thoracic compression

Even with the lungs filled to vital capacity before a breath-hold dive, it is obvious from Boyle's law that the lung volume must have

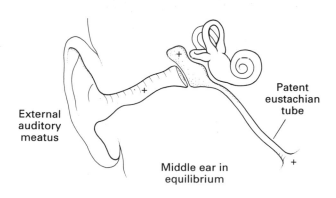

External auditory meatus

Patent eustachian tube

Middle ear in equilibrium

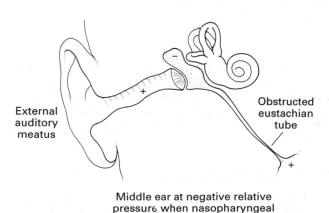

External auditory meatus

Obstructed eustachian tube

Middle ear at negative relative pressure when nasopharyngeal tube fails to admit more gas during the descent

Fig. 3 The effect of an obstruction of the nasopharyngeal tube upon the pressure of the middle ear during descent.

diminished to residual volume long before the maximum recorded depth of 90 m was reached. However, it has been shown that, to compensate for the reduction of breath-hold pulmonary gas volume, the pulmonary blood volume can increase by as much as 1 l at depths of 30 m.[8]

COINCIDENTAL ILLNESS AND INJURY

There is always the small chance of the onset of an acute illness during a dive, although this is much reduced if the diver has been previously screened for fitness to dive and also if he has been taught not to dive if feeling less than '100 per cent'. Conditions such as diabetes, which are absolute or relative contraindications for diving, are reviewed later. Many such conditions would, in their acute phase, be more hazardous if the patient were underwater at the time. Others, such as epilepsy and diabetes, would also be difficult to diagnose differentially from specific diving illnesses.

In practice, myocardial infarction among male sports divers over 50 years old is the most common coincidental condition and in 1990 represented some 10 per cent of all recreational diving fatalities.[9] Among professional divers there have been a number of cerebrovascular accidents at pressure, but there is no evidence that these occur other than by coincidence.

The final endpoint of many non-fatal diving accidents is near-drowning, perhaps complicated by hypothermia. These conditions are described in other chapters and all that needs to be emphasized here is that, if such a condition arises in a patient who has been diving, there may be an urgent need also to combat the effects of omitted decompression or of the embolic consequences of pulmonary barotrauma. Concurrent recompression of the ill diver is compatible with the management of most medical and surgical conditions. This is possible only if a recompression chamber is available and, if so, the appropriate procedures should be well known to the chamber staff and their medical adviser.

Physical injury underwater is uncommon in sports diving and needs to be managed in accordance with basic principles. Even if injured tissues bubble more than uninjured tissues on decompression, there are no special precautions required other than to complete any decompression profile to which the individual is obligated by the nature of the preceding dive.

MARINE ANIMAL INJURY

Another hazard of sports diving in some locations is from animals which injure divers by envenomation or direct physical attack. The majority of these animals are not aggressive and attack only in self-defence. Thus the prevention of injury can be achieved by good predive instruction. The diagnosis and treatment of such injuries has been dealt with fully by Edmonds.[10]

Coelenterates

Coelenterates are very diverse in their variety ranging from static corals to the mobile jellyfish. The common feature of many thousands of species is the nematocyst. This is a stinging cell which shoots out a coiled thread with a poisoned tip in order to immobilize its prey. The structure, action, and consequences to man of nematocysts varies from species to species.

The hazard to sports divers is largely defined by the geographical location of the diving, and the risk within these regions can be affected by the individual's own vigilance and behaviour.

Fire coral and stinging hydroids

These animals are static and thus sting only if touched by the diver. Recognition of their particular plant-like form and the avoidance of contact is thus an effective preventive measure. An immediate slight itch which is persistent for several days is the mildest manifestation. An extensive contact can result in a severe stinging with visible local inflammatory response. Any rubbing of the part can cause the discharge of more of the attached nematocysts and aggravate the condition. Fanning seawater with the hand over the affected part is the only recommended action to be taken to remove the nematocysts while the diver is still underwater. On return to the surface a liberal dowsing with vinegar should denature the toxic proteins and relieve symptoms. The application of alcohol on the part has been recommended, but the consensus view is that alcohol can aggravate the condition.

Coral is sharp and lacerations are common. These wounds may be painful and persistent because of the initial effects of the nematocysts and the likelihood of infection with marine pathogens. The wound can progress over several days to chronic ulceration. A culture of any organisms present should be made prior to the use of a topical antibiotic such as neomycin.

Portuguese man-of-war

Below its floating air sac, sometimes purple in colour, trail tentacles as long as 10 m. A sharp stinging sensation on contact is soon followed by intense pain which tracks centrally and may be associated with swelling of the axillary or inguinal lymph modes. The red weal of the original line of contact with the tentacle may be complicated by the bead-like string of very small blisters. Systemic manifestations are unusual, but hypotension, nausea and vomiting, abdominal cramps, irritability, confusion, and respiratory depression have been reported.[11]

Vinegar and the local application of lidocaine (lignocaine) ointment (5 per cent) are recommended as first-aid measures. Systemic support may be required as described below for the sea wasp.

Sea wasp or box jellyfish

The sea wasp, which is confined to the Indo-Pacific, is a venomous animal. Contact with as little as 7 cm of tentacle has led to death. Some 15 tentacles, up to 3 m long, trail below its box-shaped body which has sides up to 20 cm long. The pain is immediate and so intense that a number of victims have drowned at this early stage. Confusion may proceed to coma and death within 10 min. The prospect of survival increases after the first hour.

The clinical picture is well described elsewhere,[10] and, besides the severe pain, can include hypotension and cardiac shock with transient episodes of hypertension and rapid respiration. Respiratory distress, cyanosis, and pulmonary congestion also occur. The skin lesions and general irritability may persist for weeks.

First-aid measures include vinegar and the removal of tentacles, using gloves to do so. Local anaesthetic ointment and systemic analgesics should be provided and hydrocortisone given (100 mg, 2 hourly). Cardiopulmonary resuscitation may be needed, using oxygen and intermittent positive pressure respiration with intubation and general anaesthesia. Sea wasp antivenom is available in Australia, and it is obvious that the treatment of this condition, which is unknown in northern latitudes, is best left in experienced hands.

Prevention is vital. Consultation with knowledgeable local residents should help with the assessment of the risk which varies according to season and weather. In the same way as for the prevention of other milder stings, significant protection against unforeseen contact can be provided by wearing a wet-suit or other coverall at all times in hazardous waters.

Echinoderms

The sea-urchin is characterized by its many long spines which are brittle and break off on puncturing the diver's skin. Some of the spines are venomous, with systemic effects such as nausea and vomiting, and all can cause an uncomfortable wound. The calcareous material left under the skin may be absorbed in days or may persist for months. Since the spines are very difficult to remove intact it has been suggested that instead they are treated by the application of blunt trauma so that they are broken up and more readily absorbed.[11] It is also said that, for this particular inflammation, the conventional use of rest or immobilization is not as effective as activity.

Stingrays

Because of their habit of settling down and concealing themselves in the sand, stingrays provide an unexpected hazard to the inattentive diver. Though not aggressive to man, the stingray will defend itself by a sudden up-and-over flick of its tail which has a spine that is both sharp and venomous. The spine may penetrate the lung, the abdomen, or a major artery, with possibly fatal consequences. The venom from the spine enters the wound and can cause bradycardia, hypotension, a degree of cardiac ischaemia, and some respiratory depression. This too can be fatal but more usually regresses within 24 h.

The pain of the wound can be intense, but the venom is heat labile and much of the pain, and possibly some of the systemic consequences, can be relieved if the affected part is immersed in hot water (up to 50 °C) for an hour or so. Debridement, with radiography to check for bone injury if this seems a possibility, and antibiotics are indicated.

Later systemic deterioration should be managed symptomatically.

Other stinging fish

The stonefish, which can be some 30 cm long, is well camouflaged and does not move away when approached. Its venomous spines are visible only when erect. The venom is vasoconstrictive, and thus remains fairly localized, but nevertheless can lead to muscular paralysis, respiratory depression, and cardiac arrest. The pain can be sufficiently excruciating to lead to death by drowning.

First-aid measures include immersion of the wound into hot water, because the toxin is heat labile, and cardiopulmonary support as necessary. An antivenom for stonefish is available in Australia but is not necessary for the milder venom of the related European weever fish. A local anaesthetic should be used without adrenaline.

The scorpion fish and other species differ from the stonefish in that they are not camouflaged but warn potential predators of their venomous capability by the use of bright colours. The manifestations are, in general terms, similar and the intensity depends upon the species and degree of envenomation. Even those fish with little or no venom, such as the surgeon fish, can inflict a wound, which,

contaminated by marine pathogens, can become necrotic and take months to heal.

Cone shell

The geographical distribution of venomous molluscs such as the cone shell is such that in Europe there is no such hazard and in warmer waters there is a hazard the consequences of which are known to locally resident medical practitioners. It is therefore advisable for the diver visiting a strange area to make himself aware of the risks and of any necessary first-aid measures.

The cone shell, which is visually attractive, up to 10 cm long, and found in shallow water, has a proboscis which can reach fingers on any part of its outer shell, and if the animal is in a pocket, can penetrate clothing. The mortality rate is some 25 per cent and children are particularly vulnerable. Neurological manifestations, particularly of vision, precede total paralysis of skeletal muscle, including respiration, which can occur within 30 min. Symptomatic support may lead to recovery in some 24 h.

Blue-ringed octopus

This octopus, which is confined to parts of the Indo-Pacific, has a tentacle spread of not more than about 20 cm, and usually less. Its blue striations may be noticed when the animal is excited, but its bite is said to be painless. Death by total paralysis can occur within minutes. Symptomatic cardiopulmonary support may be needed for many hours.

Electric fish

Electric eels and electric rays can discharge as much as 200 V, enough to disable a diver. Other than support, no particular treatment is required. Again, the diver should rely upon local knowledge to assess the risk and take preventive action.

Moray eels

These fish bite. They have a powerful body which they can use to anchor themselves in a hole in a rock and their teeth are angled inwards. They are usually docile and often fed by divers, but they can attack when provoked. Removal of a bitten hand from the mouth of an eel may be difficult and the injuries can be extensive.

Sharks

The vast majority of sharks are harmless. The same is true of the much maligned barracuda. However, a hazard can be present and this depends not only on species but also the particular location within the wider distribution of that species, i.e. some may be harmless in one area, e.g. the Caribbean, but dangerous elsewhere, for example California.

There is no substitute for local knowledge in assessing the potential risk to divers. If a shark attack occurs, the wound can be lethal. If the victim is recovered from the water and survives, first aid should be applied to the wound and, if the equipment is available, to the restoration of blood volume.

Other marine animals

The venomous sea-snake is another hazard of diving, but only in the Indo-Pacific. If envenomation occurs by a bite from some species, death by general ascending paralysis may occur within minutes in some cases, but the onset may be 24 hours later in others. Immobilization and pressure bandages may defer the onset of symptoms. An antivenom is available and knowledgeable medical support is essential. Fortunately, the mouths of some snakes are too small to bite through clothing and a number of those bitten never develop signs of envenomation.

Crocodiles and related reptiles have a much wider geographical distribution. Divers have been killed by them and all should be regarded as dangerous, both on land and underwater.

This overview of marine animals which are dangerous to divers is far from complete but does indicate to the specialist in sports medicine the breadth of knowledge that may be needed by a physician who provides medical support to any sports diving expedition.

IMPAIRED CONSCIOUSNESS UNDER WATER

A number of factors which may lead to loss of consciousness in the water have been mentioned already, for instance latent hypoxia in breath-hold divers. There are many other possible causes of impairment or loss of consciousness under water.

It is rare for the content of the breathing gas to be a contributory factor for those using open-circuit compressed-air breathing apparatus. However, some carbon monoxide occasionally enters the air inlet of the compressor from the exhaust of its engine and this possibility needs to be considered in the differential diagnosis of diving incidents.

Divers should also be aware that if moist compressed air is retained for some months within a steel tank, much of the oxygen may be used up in the formation of rust.

A risk of providing the wrong gas occurs at the time of any change of breathing gas, for instance to pure oxygen for in-water decompression stops. If open-circuit apparatus is being supplied with a premixed gas, the mixture itself may be incorrect and cases of hyperoxia and anoxia have occurred.

Semiclosed-circuit or closed-circuit 'rebreather' breathing apparatus can lead to similar consequences. The presence of a carbon dioxide scrubber eliminates any drive to respiration due to carbon dioxide build-up. Thus failure of the oxygen supply can lead to an insidious loss of consciousness.

'Deep-water black-out' is a phenomenon among divers who go below about 70 m on air. The cause is unknown, but is probably due to a combination of the many factors present as such depths. These include increased oxygen partial pressure (more than 160 kPa at 70 m), nitrogen narcosis, and a build-up of carbon dioxide associated with the increased density of respiratory gases. Some of these divers are found to have an inadequate ventilatory response to carbon dioxide, and one common feature of these incidents is that the sudden loss of consciousness may occur during a period of hard physical work. If the victim retains his air supply he may recover quickly and there will be no serious consequences. However, underwater loss of consciousness is a hazardous event.

DECOMPRESSION BAROTRAUMA OF PARANASAL AND OTHER SPACES

The expansion of gas occurs within the air-containing spaces of the body during decompression in accordance with Boyle's law. If the gas cannot be naturally vented, its expansion may cause tissue damage or barotrauma. Expansion of the gases of the intestines does not usually cause any problem in sports divers but is sufficient to justify a recommendation that an unreduced inguinal hernia is a contraindication to diving.

The expansion of gas from the sinuses (Fig. 2) may occasionally be restricted by a polyp or by swelling. The result is facial pain and the expanded gas may escape into the subcutaneous tissue of the face or into the venous system. Though painful, there are usually no serious sequelae. The expansion of gas from the middle ear is not restricted by the flutter valve of the eustachian tube as it was during compression and, other than alternobaric vertigo which has been described already, no serious sequelae result. Gas retained within a tooth cavity has been know to blow out a dental filling explosively which may be disconcerting but is not necessarily dangerous.

PULMONARY DECOMPRESSION BAROTRAUMA

Pulmonary barotrauma is a very serious event. Damage to the lungs by the expansion of the gases retained within them may cause severe local damage and may also cause arterial gas embolism because of disruption of the pulmonary vasculature (Fig. 4). These two principal consequences of lung overpressure are rarely seen together in one individual and, because the clinical features of gas embolism may be similar to the neurological features of a different decompression illness, know traditionally as 'decompression sickness', the neurological sequelae of both will be described together, later.

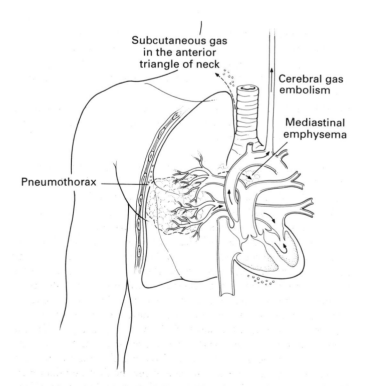

Fig. 4 The effects of localized pulmonary overpressure during ascent (pulmonary barotrauma): mediastinal emphysema and other extrapulmonary gas and arterial gas embolism.

Pathogenesis of pulmonary barotrauma

Pulmonary barotrauma can occur in any type of diving but the subsequent course of events may be complicated by the additional presence of gas which became dissolved in the tissues during the diver's stay at raised environmental pressure. In contrast, cases arising from submarine escape training, a rapid procedure in which no significant gas load is acquired by the tissues, provide the classic clinical presentations of pulmonary barotrauma alone. In 200 000 buoyant ascents from submarine escape training in the Royal Navy there were 88 incidents, including five fatalities.[12]

The obvious cause of pulmonary barotrauma is failure to exhale respiratory gases during ascent. At a steady rate of ascent, the rate of volume expansion increases exponentially and it follows that, from whatever depth the diver ascends, the greatest hazard is close to the surface. Not much overpressure is required and cases of the associated complication of arterial gas embolism have been reported from an exposure as shallow as 1.5 m. Only a single breath of compressed gas is needed to place the lungs in hazard. Another reason for gas retention in the lungs is pulmonary pathology. The most obvious cause, such as asthma and chronic bronchitis, can be eliminated by a full and careful medical history and examination. A recent upper respiratory tract infection is a good temporary reason not to dive. Nevertheless, despite good training, correct procedures, and apparently good pulmonary fitness, cases are still all too frequent.

Another mechanism for localized pulmonary overpressure has been termed 'dynamic airway collapse'. When a rapid flow of gas is driven through the airways by a relatively large pressure differential, the unsupported smaller airways may collapse. The air retained distally will continue to expand and may press upon nearby airways, compounding the condition.

Clinical manifestation

The expanded gas may escape (Fig. 4) by rupturing the pleura, possibly leading to bilateral tension pneumothoraces, or it may pass along the perivascular sheaths to the hilum of the lung. Such gas accounts for the finding of pneumomediastinum or pneumopericardium, each of which may be symptom free. Further tracking of the air can lead to its detection in the retroperitoneum by radiography or in the subcutaneous tissue of the anterior triangle of the neck by palpation. These findings, though not often present, are diagnostic.

The passage of expanded alveolar gas into the pulmonary circulation leads to arterial gas embolism, a condition which will be described with other neurological decompression illnesses.

A history of sudden respiratory embarassment on arrival at surface is usually sufficient to make the diagnosis. Less severe cases may escape notice and, particularly if an associated neurological condition is treated by recompression to depth in a compression chamber, some may reveal themselves only later during the subsequent reduction of environmental pressure. Pulmonary manifestations may be relieved by recompression but the underlying condition will recur when decompression is resumed and experience shows that the onset will be at a deeper depth. Clinical examination of the chest within a compression chamber is made more difficult because raised environmental pressure distorts sound and because the performance of the medical examiner may be impaired by the effect of raised environmental pressure. Diagnostic aids such as radiography and ultrasound, suitably modified, are not always available for use in a compression chamber.

In 64 cases of barotrauma reviewed by the Royal Australian Navy[13] 42 had clinical evidence of pulmonary damage, 21 had arterial gas embolism, and only one had both concurrently. Nevertheless, the urgency of treatment for neurological complications takes precedence and demands immediate recompression. Mediastinal emphysema has been observed radiologically in 1 per cent of asymptomatic healthy individuals following submarine escape ascents.

Treatment of pulmonary barotrauma

The management of pulmonary barotrauma is entirely symptomatic. The most urgent matter may be the relief of respiratory distress due to tension pneumothorax by use of a trochar and cannula with a Heimlich valve. This can, if necessary, be done while the patient is in a recompression chamber, but recompression is not required for the treatment of extra-alveolar gas alone. However, failure to recognize and relieve a unilateral pneumothorax in a compression chamber has led to at least one death. The treatment of the associated neurological manifestations will be described later.

Prevention

In addition to medical fitness screening of all divers and proper instruction , the recent introduction by some sports agencies of an arbitrary 'stop' for scuba divers at a depth of around 5 m is intended to reduce the risk of pulmonary barotrauma and arterial gas embolism due to ascent made too rapidly, often compounded by loss of buoyancy control. The need for a diver to stop in midwater at this depth encourages proper buoyancy control throughout the ascent.

THE DECOMPRESSION ILLNESSES

Nomenclature

There are at least two quite distinct causes of the acute illnesses that arise as a result of the reduction of environmental pressure which is experienced by the diver on returning to the surface. It has been customary to describe these illnesses as distinct entities, but it is becoming increasingly realized that the two conditions may merely represent the extremes of what now appears to be a continuous spectrum of manifestations which are end-results of many pathogenetic pathways.

In order to present a logical description of these decompression illnesses it is necessary first to review the nomenclature of the principal terms which have been in current use.

'Dysbarism' refers to any adverse effect of pressure and, though often used for the decompression illnesses, it can be applied to many other conditions such as middle ear squeeze on descent.

'Pulmonary barotrauma' is the damage to the lung, already described, which is due to the expansion of retained alveolar gas on ascent. The term can also be correctly applied to compression barotrauma, i.e. 'squeeze' of the lungs associated with an uncorrected descent through the water of an old-fashioned helmet diver. Pulmonary compression barotrauma is not seen in sports scuba diving. The term 'pulmonary barotrauma' implies decompression and is also used to refer to the underlying lung pathology causing arterial gas embolism even when there is no clinical evidence of pulmonary damage in that individual.

'Arterial gas embolism' is usually used to refer to the carotid or vertebrobasilar embolism associated with pulmonary barotrauma. However, this use is not exclusive and arterial gas embolism can arise from other sources.

'Decompression sickness' is a term which was first used by Matthew[14] and has since become identified with the clinical conditions associated with the release of gas as bubbles from the tissues in which it became dissolved during the dive.

'Decompression illness' has been adopted as a clinical term to include both the major decompression disorders, together with other manifestations, such as those associated with a foramen ovale which is patent to otherwise innocuous venous gas emboli. To resolve any potential confusion, while a descriptive approach has been adopted for clinical use, descriptions of pathogenesis will continue to use the traditional mechanistic names for the decompression illnesses.

Arterial gas embolism

This traditional category is now better identified as a 'rapid progressive neurological decompression illness'. This is a more accurate term clinically which can be modified descriptively from case to case, for instance to 'spontaneously regressing' or 'following rapid ascent' or 'following negligible gas uptake', as appropriate. The advantage of the seemingly more cumbersome title is that it is accurate. It does not presume to make diagnoses based on a possible pathogenic mechanism. It is becoming increasingly realized that such assumptions can be wrong and so may mislead subsequent management and inhibit progress towards a better understanding of clinical decompression syndromes.

From accidents arising in the course of submarine escape training it is possible to define some characteristics of the neurological illness of decompression which occurs when there is no significant gas uptake in the tissues due to exposure to pressure. Features of this 'pure' form of gas embolism may be found in sports divers but there are differences, as noted by Kidd and Elliott.[15]

The causative exposure in all these naval cases was an excursion 1 or 2 min, often less, to a depth of not more than 30 m. Ascent rate through the water was at approximately 2 m/s, a rate at which the venting of the expanding alveolar gas also eliminates carbon dioxide build-up. The accidents occurred on arrival at the surface where often the casualty's eyes are seen to roll upwards. Onset might be delayed by a few minutes, sufficient for the individual to emerge from the water. The occurrence of one exceptional case after a surface interval of over 2 h can be explained by the speculation that a trapped volume of the lung expanded on ascent but, though under tension, did not burst until some time later.

The presenting manifestation in 65 cases was immediate unconsciousness in 30 and disorientation, giddiness, or observed unsteadiness in 20. Fifteen presented with a paresis: six of one arm, four of one leg, and five with a hemiparesis. Paraplegia, which is characteristic of what may be called 'spinal decompression sickness', was not seen in any of these neurological cases which had had no significant exposure to depth. Blindness developed later in a few victims. Thus the clinical picture may be similar to stroke, suggesting carotid embolism, but vertebrobasilar embolism is also invoked to explain some of the physical signs.[12]

It is important to appreciate that in this and similar series there were no cases of limb pain, no cutaneous manifestations, and no cases characteristic of multiple small venous emboli to the lungs (known as the 'chokes'), all of which conditions are considered to be components exclusively of 'decompression sickness' due to bubbles arising from dissolved gas. These will be described later. Similarly, non-specific constitutional disturbances are not reported in this rapidly progressive illness, but possibly because there is no time for them to be noticed during the rapid onset of neurological deficits.

A similar condition occurs in divers and retention of alveolar gas is a common cause of death in sports diving. While the associated features may include inexperience, inadequate training, equipment failure, and running out of air, the fundamental association seems to be with an ascent which is more rapid than usual. As will be discussed, the clinical picture and subsequent time-course may be more complex if the diver has acquired a significant gas load, but there are often sufficient similarities for the condition to have been called arterial gas embolism due to pulmonary overpressure in the past even though there may be a component in the final clinical picture due to dissolved gas.

After its rapid onset, the normal progression of gas embolism, as it occurs when there is no significant gas load, is either death or spontaneous resolution. Every case must be regarded as a medical emergency. The specific treatment will be described later.

Decompression sickness

In its classic form, many of the features of this condition are quite characteristic but there is a wide variety of clinical presentations.[16] On some occasions, there is also the possible contribution of alveolar gas emboli to the pathogenesis of what primarily is due to the evolution of dissolved gas from the tissues as bubbles. Another factor which may also modify the pathogenesis of decompression sickness is the presence of a small right-to-left shunt such as a patent foramen ovale through which relatively benign venous bubbles can pass to the arterial distribution with a potential for growth in size when they arrive in the peripheral tissues.

Some of the features of decompression illness are seen only when there has been the opportunity for an individual to acquire a dissolved gas load. These features, such as cutaneous manifestations, limb pain, and so-called 'chokes', are not found in those such as submarine escape trainees who have no significant gas-load. These manifestations of decompression illness will be described first while those of a vestibular or neurological nature, which could arise from either of the two classic pathological processes, will be described later.

Cutaneous manifestations

A number of skin conditions may be associated with decompression illness. None is serious, but they can be uncomfortable and should be a warning of the possible onset of a more serious condition.

A simple itching of the skin has been linked to direct absorption of gas into the skin when it is exposed to compressed air during the time at pressure. This can occur in pressure-chamber dives and in those who wear a dry suit during wet dives. This cutaneous condition does not arise in those areas of the skin that were wet throughout the dive. It is totally benign.

Cutis marmorata is the term given to a superficial red and/or purple blotchy rash, usually of the upper trunk. Itching may also be present. The 'marbling' appearance is thought to be due to vasodilation and then stasis. The rash whitens under local pressure. An urticarial response has also been described, but is not common.

Peau d'orange and possibly gross oedema occur and are considered to be due to the obstruction of the lymphatics and lymph glands by bubbles. This has been sufficiently gross to have been diagnosed as mumps, but usually is peripheral in distribution.

Constitutional symptoms

A general malaise may be indicated by a diver losing his appetite for his postdive meal, a warning of possibly more serious manifestations to come. In many persons it is not until they have been successfully treated by recompression that they realize how 'off colour' they have been feeling just previously. Fatigue that is disproportionate for the preceding physical exercise is said to be another warning of further manifestations.

Musculoskeletal manifestations ('bends')

Pain in the region of one or more large joints—bends—is the best known manifestation of decompression illness but is not always present. Indeed, though reported as the presenting symptom in up to 70 per cent of cases,[15] more recent analysis of sports diving at some treatment centres suggest limb pain in less than 25 per cent.[17]

Although the shoulders and the knees are the most commonly affected joints, pain can occur in the region of any synovial joint. In some the pain is mild and transient, while in others it may be intense. The pain may seem to flit from joint to joint and several joints may be affected in one individual. Pain is not easy to classify, but it has been considered that there may be two basic varieties—a superficial pain, fairly precisely located and close to the joint, and a deeper and more diffuse throbbing pain in the region of a joint. A 'niggle' may be defined arbitrarily as a pain which is already beginning to disappear within 10 min of onset. The course of established 'bends', if not treated, is that they slowly resolve over a period of several days though in some cases they may persist painfully for weeks.

The presence of some limitation of movement or of associated paraesthesiae must raise the possibility that there is also a neurological involvement, whereas the presence of redness or oedema in the vicinity of the joint suggests that the pathology is local.

Cardiopulmonary decompression illness—'chokes'

This condition, though not common, may become serious. It is believed to be due to the pulmonary embolism of many intravascular bubbles, and the onset usually occurs very soon after surfacing and may rapidly progress towards total cardiovascular collapse. The classic presentation is a sudden onset of a sharp retrosternal pain that limits deep inspiration and progressive dyspnoea. However, as in other conditions, the classic presentation is exceptional and early diagnosis can be difficult.

Neurological decompression illness

Almost any neurological manifestation can follow decompression and, classically, hemiplegia is associated with gas embolism due to pulmonary barotrauma, whereas paraplegia is associated with decompression sickness of the spinal cord. However, many cases are not so clearly distinguished and each of these classic presentations could be due to the other pathogenic mechanism.

A precise aetiological diagnosis is not possible. Since the emergency treatment for both conditions is essentially the same, all the neurological manifestations must be considered together.

Cerebral manifestations

The anatomical distribution of bubble-induced lesions is widespread and almost any presentation is possible. Psychotic disturbances, disorders of speech and of affect, generalized 'spaciness', and headaches are each presentations that, if they resolve rapidly on therapeutic recompression to raised environmental pressure, can be considered to be due to presence of bubbles. Such a response is diagnostic, but the failure of treatment does not eliminate the decompression bubble as the primary cause since the bubbles can quickly induce many secondary haematological and localized deficits.

Lesions of the individual cranial nerves, particularly V and VII, are not common but have been reported. Visual blurring is likely to be centrally located, and total blindness has been reported by several submarine escape trainees. As mentioned previously, the most common presentation in this particular category is sudden loss of consciousness on surfacing.

The vestibular presentation, commonly referred to as 'staggers', is characteristic of acute decompression illness. Often associated with ipsilateral deafness or tinnitus, and possibly with vomiting, the causative lesions are considered to occur most commonly in the cochlear and vestibular end-organs, though in some cases they may be more centrally located.

The most common peripheral presentation of neurological decompression illness is that usually ascribed to lesions of the spinal cord. While this might not be the site of the lesion in all such cases, there is a large amount of evidence to show that there are multi-level and multifocal discrete lesions which combine to produce paraplegia. This may be identified as being at a specific cord level but, more commonly, may be due to a more diffuse collection of deficits. The presence of a monoparesis or hemiparesis may be a consequence of cerebral gas embolism and, though less common than unconsciousness or disorientation, is a characteristic presentation in those who have just completed a rapid ascent through the water. In those who have also acquired a dissolved gas load during their dive, the clinical picture can be more complex. It is also known that a rapid ascent, perhaps because it initiates a number of intra-arterial bubbles of alveolar gas, may be followed by 'classic spinal cord' manifestations, even after a dive that did not infringe accepted no-stop decompression durations. There is also the evidence that venous bubbles may bypass the pulmonary filter, where their gases are normally excreted, and enter the arterial circulation through a shunt.

The insidious onset of 'pins and needles' in the feet, of a sensation of woolliness', or of some slight weakness is a characteristic mode of onset for what will develop in minutes or hours into a serious paraplegia, possibly ascending to quadriplegia with respiratory difficulties. Retention of urine is common in such cases, and impotence and rectal incontinence may follow.

A sudden 'girdle' pain on one or both sides of the trunk is an uncommon but well-documented onset of acute neurological decompression illness, and is usually followed rapidly by complete loss of function below that level.

It must be concluded that any type of neurological lesion can occur after any dive profile. One single breath of compressed gas at depth is sufficient to expose the individuals to the hazard of 'burst lung' on ascent, but there is no report in the literature of a neurological deficit occurring from depths shallower than 1.5 m (5 feet) of seawater.

Hypovolaemia

Although haemoconcentration due to bubble-induced increased capillary permeability is a feature of serious decompression sickness, postural hypotension is not considered to be a presenting manifestation but develops later, as a complicating feature.

Latency and duration

While the onset of decompression illness may be immediate upon surfacing, or in some even while still making the ascent, a delay of some 36 h may occur before the first onset of symptoms. A failure to recognize or report the condition may account for latencies apparently longer than this. In one review, 85 per cent of neurological presentations occurred within 1 h.[18]

The time-course of the illness is unpredictable. After the onset of the presenting manifestation, there may be a spontaneous recovery or the development of further symptoms or signs. Deterioration may be very rapid or take several hours. Recovery can be very slow and some neurological lesions may become permanent even when treated vigorously and expertly.

Treatment of the decompression illnesses

Early suspicion of the possible diagnosis and a contingency plan for urgent recompression are the essentials of good case management. The unpredictable progression of the illness and the diminution of its responsiveness to treatment if there is delay makes the condition a medical emergency.[19]

The generally accepted view is that anything untoward which occurs after a dive should be considered as decompression illness unless it can be shown otherwise. There is a natural tendency for persons to look for some explanation for their symptoms other than 'bends'—a recent injury, for example, or possibly seasickness.

A recompression chamber is not always immediately available but, if it is, a return of the diver to raised environmental pressure can provide apparent complete relief almost instantaneously. Delay, even for further history and examination, is not justifiable in these circumstances. The history can be taken in detail later and the examination can be made at pressure to check for any residua before the necessary decompression commences. In some, the placement of an intravenous line and urinary catheter may be performed, usually by the doctor associated with the recompression unit.

The remarkable reversal of unconsciousness or paraplegia is one reason why the bubble is still regarded as the primary pathological event; no other explanation of such a response seems feasible. However, the response may be incomplete or the patient may subsequently deteriorate, which are two reasons why there should be access to the patient at raised environmental pressure. Though used in places where nothing else is readily available, a one-compartment chamber has no facility to allow a doctor to 'lock in' to his patient. Thus the management of a tension pneumothorax during the subsequent return to the surface would become very difficult. One-man chambers which are transportable and which can then be locked on to a large pressure chamber elsewhere are available in some parts of the world.

A temptation for the patient's colleagues may be to return the stricken diver to the water, perhaps 'for extra stops'. This, for a number of reasons, may make the patient worse. Back in the water the ill diver may require physical support, there may be a limited duration of gas supply, and all concerned are likely to become very cold. Thus the treatment is likely to be too shallow and too brief to be of value. In-water recompression has been successful, but only when a predive decision has been made to provide additional equipment for use in the event of this emergency. This includes dry suits for thermal protection, a large volume of oxygen, and full facemasks.[20] Even so, the recompression is limited in depth and duration and there is no access for ancillary treatment.

A two compartment recompression chamber capable of taking the patient to an equivalent pressure of 50 m (165 feet) of seawater and providing oxygen-enriched mixtures or pure oxygen by means of a built-in breathing system is preferred and, indeed, required by regulations in some commercial circumstances.

Many decompression accidents occur at sea, miles from any recompression chamber. In these circumstances administration of 100 per cent oxygen by close-fitting mask is essential. Some advocate quickly tipping the subject head down 'to dislodge cerebral bubbles' but, if maintained, this procedure may lead to cerebral oedema. Plenty of fluids, preferably by mouth but intravenously if the subject is neurologically ill, are equally important. These should be continued until the urine is copious and colourless. A fluid balance sheet should also be kept. Catheterization and pleurocentesis may be required.

The use of drugs such as the corticosteroids to reduce cerebral oedema is less clear cut. There is as yet no consensus view on the appropriate dosage, and the efficacy of steroids is relatively unproven and depends upon reports of their use in analogous conditions.

There are, however, a number of contraindications and warnings. These include the use of analgesic nitrous oxide–oxygen mixtures because the nitrous oxide is known to diffuse rapidly into bubbles, thus enlarging them. The use of drugs such as acetylsalicyclic acid and heparin might enhance a haemorrhage within the inner ear. Valium might provide such good symptomatic relief that it would be tempting to omit subsequent recompression which is necessary for the underlying pathology.

A spontaneous recovery may be only transient. The evacuation of a 'recovered' casualty to a treatment centre is important because of the real possibility that a relapse and further deterioration will occur.

Evacuation must take into account the need to avoid exposing the patient to the compounding effects of diminished environmental pressure caused by either driving over a mountain pass or transport by aircraft. Whatever possible the patient should remain at an altitude below 150 m (500 feet).

As for all emergencies, preparedness is the key. When the emergency occurs, it is too late to begin planning. Not only should the correct first-aid equipment be available at the dive site and the potential methods of emergency transport arranged, but the recompression treatment centre must also be checked in advance in case it may be unavailable for casualties for some period of time. Once the subject is in a recompression chamber and being treated by a competent chamber crew, the details of the recompression should be left to the chamber's own doctor who should be well trained and experienced in such treatment. In cases of difficulty, whether at the contingency planning phase or in an emergency, there are a number of on-call physicians in several countries who can be reached for advice by telephone. These sources are known to those in the diver training organizations and, in most countries, to the coastguard and the navy also.

The emergency procedures must be planned in advance and each case treated as an emergency with aggressive optimism.

FITNESS CRITERIA

The unique effects of changes of environmental pressure upon the human body create a wide range of specific pathology that may be found among divers as well as being associated with life-threatening underwater incidents. In order to minimize such potential effects a number of predisposing medical conditions need to be recognized and some individuals advised to avoid this sport. For professional divers, particularly the sports diving instructor upon whom the life of a student in an underwater emergency may depend, the highest standards of physical, medical, and mental fitness are to be expected. Indeed, the criteria on entry to the profession must be very strict since cardiopulmonary and other important functions are likely to deteriorate with age. For the amateur, who can choose when, where, and how to dive, there is a lesser demand for such perfection. Nevertheless, it is not only for the potential diver's own sake that some candidates should be medically disbarred from diving, but also for the sake of a diving partner ('buddy') upon whom the responsibility for a rescue would fall.

Those disorders which compromise the equalization of pressure or volume of the air-containing spaces during descent or ascent through the water provide the most obvious examples of medical conditions incompatible with this sport. Also, because of the need for continuous alertness in this unforgiving environment, any disorder that may lead to altered levels of consciousness or to erratic or irresponsible behaviour are likewise incompatible. Therefore these are regarded as absolute disqualifications (Table 1).

More difficult to evaluate are those disorders which have less obvious effects in diving. There are no clearly defined boundaries and, for conditions such as asthma or diabetes, there are both liberal and conservative interpretations. These relative contraindications (Table 1) to diving should be interpreted by a physician familiar with the environmental hazards and the nature of the diving to be undertaken. With recognition of the need for greater supervision and support while in the water, diving can be a suitable recreation for blind, paraplegic, and other handicapped sportspersons.

Thus the following examples, which are more fully discussed elsewhere,[22] apply to recreational divers. These examples, presented below by organ system, cannot be comprehensive but illustrate the reasoning to be considered.

Neurological system

The absence of any neurological illness or deficit is the obvious ideal for a diver candidate but, in sports diving in particular, there are conditions in which some allowances can be made without compromising safety. The safety of the individual, and thus of any diving companions who might have to effect a rescue, is the prime consideration. Any possibility of a seizure underwater, whether epileptic in origin or as a consequence of previous head injury, is a good example. There is a secondary consideration, given that the condition is not a threat to immediate safety, which is the possible risk to subsequent health. For example, in multiple sclerosis the risk of associated decompression sickness from diving must be seen as potentially reducing still further the diminished function of the central nervous system, and there is also the potential problem that the sudden onset of neurological symptoms could provide difficulties in planning the management of the case. Wise counselling should be used to persuade such persons to seek another sport.

Table 1 Some contraindications to sports scuba diving

Ophthalmological
Inadequate visual acuity (unless diving in a category for the blind)
Ocular surgery within previous 12 months

Ear, nose, and throat
Inability to autoinflate the middle ear
Tympanic membrane perforation or aeration tubes
Obstruction to external ear equilibration
Menière's disease or other vertiginous conditions
Middle ear prostheses and stapedectomy
Inner ear surgery
Chronic mastoiditis
Deformity interfering with retention of mouthpiece
Laryngectomy or laryngocele; tracheostomy
Chronic sinusitis

Respiratory
History of spontaneous pneumothorax
Bronchial asthma including history of childhood asthma unless fully
    investigated
Exercise or cold-induced asthma
Chronic obstructive pulmonary disease
Radiological blebs, bullae, or cysts
Sarcoidosis

Cardiovascular
History of myocardial infarction, angina, or other evidence of coronary artery
    disease
History of cerebrovascular accident
Unrepaired gross cardiac septal defects
Aortic or mitral stenosis; aortic coarctation
Complete heart block; fixed second-degree heart block
Exercise-induced tachyarrhythmias
Wolf–Parkinson–White syndrome with syncope or paroxysmal atrial
    tachycardia
Fixed-rate pacemakers; any drugs which inhibit cardiovascular response to
    exercise
Hypertension requiring drug treatment or with related retinal, cardiac, renal,
    or vascular findings
Inadequate exercise tolerance to cope with any physical emergency in water

Peripheral circulation
Any vascular disease that limits exercise tolerance

Gastrointestinal
Paraoesophageal or incarcerated sliding hiatal hernia
Any abdominal wall hernia with potential for gas trapping until surgically
    repaired
Hepatitis
Bleeding

Genitourinary
Pregnancy (or currently intended pregnancy)
Renal failure or transplant

Endocrine
Insulin-dependent diabetes mellitus
History of hypoglycaemic episodes even if not requiring insulin

Haematological
Sickle-cell disease
Haemophilia
Polycythaemia
Leukaemia
Unexplained anaemia

Central nervous system and psychiatric
Demyelinating diseases
History of head injury with prolonged unconsciousness or with post-traumatic
    amnesia for more than 1 h
Neurological deficits due to illness or injury unless fully investigated
History of epilepsy including childhood fits unless fully investigated
Neurological migraine
Myasthenia gravis
Transient ischaemic attacks
Psychosis, significant anxiety states, manic states
Severe depression, suicidal ideation
Alcoholism and use of mood-altering drugs

Musculoskeletal
Impediments of mobility or dexterity

Gastrointestinal system

An unreduced hernia may contain bowel and has the potential for barotrauma during ascent. No diving should be performed until it is fully repaired. Diverticulitis is similarly disqualifying. Ileostomies and colostomies carry no such risk and are compatible with diving.

Haematological system

Sickle-cell trait, with greater than 45 per cent HbS, is considered to be a contraindication to diving in which inadvertent episodes of hypoxia are a risk.

Musculoskeletal system

Traumatic paraplegics can dive under close supervision provided that they have sufficient control over pulmonary function.

Bone necrosis, though an occupational hazard of professional diving, is a rare complication of sports diving. The lesions may be in the shaft of the long bones or juxta-articular, and the latter can lead to osteoarthritis of the shoulder or hip. Radiological evidence of asymptomatic dysbaric osteonecrosis is not, *per se*, a contraindication to further diving since the necrosis does not affect in-water safety. However, the presence of juxta-articular necrosis is of sufficient long-term concern to advise the individual to discontinue this activity in case further diving leads to deterioration of the condition.

Endocrine system

Insulin-dependent diabetes is not compatible with safe diving even when there is no end-organ disease. Some gentle diving by dietary-controlled diabetics is safer, but the onset of unconsciousness during a dive or after surfacing is likely to make immediate diagnosis and management in the field rather difficult.

Pulmonary system

A history of spontaneous pneumothorax is disqualifying. Repetition is likely to result in a potential fatal expanding pneumothorax during the return to the surface. Traumatic pneumothorax, especially with lung injury, is also a contraindication even when all pulmonary function tests are normal.

Any respiratory disease is disqualifying; the possibility of gas retention behind a mucus plug or constricted airway is ever present. A history of childhood asthma, now outgrown, together with normal pulmonary function tests and normal response to a methacholine challenge test can be considered more favourably after discussion of the risks with the individual.

Vision

Those with good eyesight will find this important for tasks associated with diving such as reading gauges and finding the boat again after returning to the surface. Corrective lenses, contact lenses, and, when healed, lens implants are compatible with diving. Radical keratotomy may be susceptible to barotrauma within the facemask and is a relative contraindication.

Ear, nose, and throat

The ability to equalize middle ear pressures during descent is essential in diving. Middle ear disease, history of stapes surgery,

or radical mastoidectomy are disqualifying. Provided that middle ear equalization is easy, a healed repair of the tympanic membrane is acceptable. Chronic sinusitis may well prevent pressure excursions.

Menière's disease is a contraindication; an attack during the dive could totally disorientate the diver in midwater, and if it occurred after the dive, management of the case would need to focus on the possibility of acute decompression sickness.

Dental health is advantageous because of the need to grip the mouthpiece of the underwater breathing apparatus even though this can also be done adequately after removal of partial dentures. Air pockets in carious teeth and behind fillings can cause local pain.

Cardiovascular system

Controlled hypertensives can dive, but those taking drugs like β-blockers that limit exercise should demonstrate their ability to reach 13 mets (metabolic equivalents) of exercise (around 45 ml of oxygen consumption per kilogram body weight per minute), a level necessary in order to surmount in-water difficulties.

A history of coronary insufficiency contraindicates recreational diving in which a maximal effort may be required unexpectedly.

Aortic and mitral stenosis are disqualifying. Mitral regurgitation and aortic insufficiency, as long as there is no left-ventricular dysfunction, are acceptable. Mitral valve prolapse with symptoms is also acceptable. Septal defects are not necessarily considered to be disqualifying even though there is a possibility of venous bubbles crossing to the left side. This is justified because the one-third of sports divers who have septal defects of some kind far exceeds those few (less than 0.1 per cent) who will ever suffer neurological decompression illness.

Arrhythmias are a particular cause of concern but the ultimate criterion is again the ability to sustain hard exercise.

REFERENCES

1. Bennett PB, Rostain JC. The high pressure nervous syndrome. In: Bennett PB, Elliott DH, eds. *The physiology and medicine of diving*. 4th edn. London: WB Saunders, 1993: 194–237.
2. Thorsen T, Klansen H, Lie RT, Holmsen H. Compression, hyperbaric pressure and decompression: effect on platelet aggregation induced by gas bubbles, adenosine 5' -diphosphate and epinephrine in vivo. *Undersea Biomedical Research*, in press.
3. Hallenbeck J, Andersen JC. Pathogenesis of the decompression disorders. In: Bennett PB, Elliott DH, eds. *The physiology and medicine of diving*. 3rd edn. London: Baillière Tindall, 1982: 435–60.
4. Anonymous. Editorial. *Aquacorps* 1991: **2**, 1–3.
5. Bennett PB, Elliott DH, eds. *The physiology and medicine of diving*. 4th edn. London: WB Saunders, 1993.
6. Farmer JC. Otologic and paranasal sinus problems in diving. In: Bennett PB, Elliott DH, eds. *The physiology and medicine of diving*. 4th edn. London: WB Saunders, 1993: 267–300.
7. Pullen WF. Round window membrane rupture: a cause of sudden deafness. *Transactions of the American Academy of ophthalmology and otology* 1972; **76**: 1444–50.
8. Schaefer KE *et al*. Pulmonary and circulatory adjustment determining the limits of depths in breath-hold diving. *Science* 1968; **162**: 1020–3.
9. Diver Alert Network. Personal communication, 1991.
10. Edmonds C. *Marine animal injuries to man*. Melbourne: Wedneil, 1984.
11. Edmonds C. Marine animal injuries. In: Bove AA, Davis JC, eds. *Diving medicine*. Philadelphia: WB Saunders, 1990: 115–37.
12. Elliott DH, Harrison JAB, Barnard EEP. Clinical and radiological features of 88 cases of decompression barotrauma. In: Shilling CW,

Beckett MW, eds. *Underwater physiology VI, Proceedings of the 6th International Symposium on Underwater Physiology*. Bethesda: FASEB, 1978.

13. Gorman DF. Arterial gas embolism as a consequence of pulmonary barotrauma. In: Desola J, ed. *Diving and hyperbaric medicine. Proceedings of the 9th Congress of the EUBS*. Barcelona: CRIS, 1984: 347–68.

14. Matthews BHC. *Interim report on research on oxygen problems*. London: Air Ministry, 1939.

15. Kidd DJ, Elliott DH. Clinical manifestations and treatment of decompression sickness in divers. In: Bennett PB, Elliott DH, eds. *The physiology and medicine of diving and compressed air work*. 1st edn. London: Baillière, Tindall, and Cassell, 1969: 464–90.

16. Elliott DH, Moon RE. Manifestations of the decompression disorders. In: Bennett PB, Elliott DH, eds. *The physiology and medicine of diving*. 4th edn. London: WB Saunders, 1993: 481–505.

17. Bennett PB, Dovenbarger J, Corson K. Etiology and treatment of air diving accidents. In: Bennett PB, Moon RE, eds. *Diving accident management*. Bethesda: Undersea and Hyperbaric Medical Society, 1990: 12–22.

18. Francis TJR, Pearson RR, Robertson AG, Hodgson M, Dutka AJ, Flynn ET. Central nervous system decompression sickness: latency of 1070 human cases. *Undersea Biomedical Research* 1988; **15**: 403–17.

19. Davis JC, Elliott DH. Treatment of the decompression disorders. In: Bennett PB, Elliott DH, eds. *The physiology and medicine of diving*. 3rd edn. London: Baillière Tindall, 1983: 473–87.

20. Edmonds C, Lowry C, Pennefather J. Underwater oxygen treatment of decompression sickness. In: *Diving and subaquatic medicine*. 2nd edn. Mosman, NSW: Diving Medical Centre, 1981: 171–80.

21. Davis JC. *Medical examination of sport scuba divers*. 2nd edn. San Antonio: Medical Seminars, 1986.

22. Bove AA, Davis JC, eds. *Diving medicine*. 2nd edn. Philadelphia: WB Saunders, 1990.

2.2 Immersion in cold water: effects on performance and safety

MICHAEL J. TIPTON AND FRANK GOLDEN

INTRODUCTION

Up to 7 million people participate in water-based leisure activities on a regular basis in the United Kingdom alone. For individuals involved in sporting activities which require deliberate immersion in cold water, such as swimming, triathlons, or diving, a knowledge of how such an immersion can influence performance, and how performance can be maintained in this environment is clearly advantageous. For others for whom immersion is not planned, such as sailors, canoeists, and fishermen, a knowledge of the hazards associated with immersion in cold water may be life-saving.

It is estimated that 140 000 people drown each year throughout the world;[1] in the United Kingdom the figure is approximately 700.[2,3] An understanding of the responses associated with cold water immersion may be critical and should include knowledge of the methods of protecting against them. It is also important that anyone who may have to treat an immersion casualty has an understanding of the problems that may be encountered during rescue and subsequent management.

This chapter includes information on the responses associated with immersion and how they may influence performance or threaten life. Methods of protecting against the responses and treating those who become immersion casualties are also considered.

THERMOREGULATION: AIR VERSUS WATER

Humans possess an intricate and integrated thermoregulatory system (Fig. 1) which, by varying heat production and heat loss, attempts to regulate the deep body temperature of man at a relatively constant temperature of about 36.9°C (98.4°F).

As man is constantly producing heat as a byproduct of metabolism, this must be lost to the environment for the total heat content of the body to remain constant. For a man in heat balance the following must apply:

$$M \pm R \pm C \pm K - E \pm W \pm S = 0$$

where M is the metabolic rate, R is radiation, C is convection, K is conduction, E is evaporation, W is the work done, and S is body heat storage. Detailed descriptions of both the routes of heat exchange and the thermoregulatory system are outside the scope of this chapter. Interested readers are referred to several excellent publications which deal with these areas.[4-9]

Heat loss from the surface of the naked body follows the same basic principles in air and water: heat passes to the microenvironment adjacent to the skin by conduction and is carried away by convection. In still air, radiation and evaporation are the major pathways for heat loss; the contribution of convective heat loss is increased with air movement ('forced' convection). In water, evaporation of sweat cannot be used and heat loss due to radiation is negligible. Therefore the route of heat transfer in water is principally convective and partly conductive.

Although only two primary pathways for heat loss are available in water, humans cool two to five times more quickly in cold water compared with air at the same temperature.[10] Water can produce an equivalent physiological response to air at a temperature 11°C lower.[11] Furthermore, thermoneutral temperature in water averages about 35°C with a very narrow range;[12,13] this compares with about 26°C and a broader range for still air.[13] (A thermoneutral temperature is one in which the deep body temperature remains stable and thermoregulation is achieved by variations in vasomotor tone alone.) Therefore a naked resting man will be unable to maintain his deep body temperature when immersed in water at a temperature regarded as thermoneutral in air.

The reason for these differences lies in the physical properties of air and water; the thermal conductivity of water is 25 times that of air, and water has a specific heat per unit volume that is approximately 4000 times that of air. Thus water, unlike air, provides practically no insulation at the skin–water interface. Therefore cooling is extremely effective in water and results in rapid

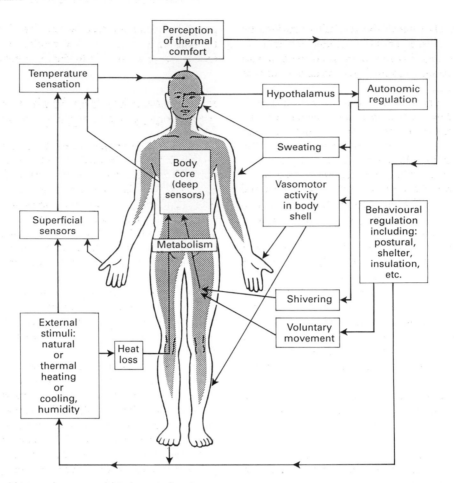

Fig. 1 Schematic diagram of temperature regulation in man. The body 'shell' (the vasomotor labile region) is represented by the non-shaded area.

dissipation of the heat which is delivered to the skin from the deeper tissues. As a consequence, skin temperature quickly approximates water temperature following cold water immersion. In contrast, the lower cooling capacity of air often allows skin temperature to be adjusted independently of air temperature; therefore it can assist in the maintenance of heat balance.

The cooling capacity of water is reflected in the values which have been reported for the combined heat transfer coefficient for convection and conduction. In still water the value is between 44 and 230 $W/m^2.°C$,[14-16] rising to 460 $W/m^2.°C$ at rest in moving water and 580 $W/m^2.°C$ when swimming at any speed.[16] These figures are difficult to obtain and vary greatly between studies largely because of methodological differences. However, they are at least two orders of magnitude greater than those reported for air.

As a result of the capacity of water to remove heat from the surface of a naked body, heat loss must be largely limited by internal insulation between the deep and superficial tissues. Heat produced in the deep tissues of the body is transported to the skin surface by mass flow in blood and conduction through intervening body tissues to the skin. Therefore heat flux is determined by tissue conduction and the temperature gradient between the deep tissues and the skin.

Bullard and Rapp[17] have produced a theoretical model of heat conductance to the skin in which the flow of heat from the deep tissues to the skin is effected by two parallel resistors. One of these

is 'variable' and is represented by the peripheral circulation, and the other is 'fixed' and is determined by the subcutaneous layer of fat.

As peripheral blood flow represents a 'variable' resistance, when this is at its lowest, tissue conductance will also be at its lowest. Forearm blood flows in water at 13°C can be as low as 0.5 ml/min.100 ml tissue.[18,19] This compares with values of around 17.6 ml/min.100 ml tissue in water at 45°C.[18]

There is general agreement that in lean individuals peripheral blood flow—and as a consequence tissue conductance—is lowest and maximum insulation is achieved in water at temperatures below 33 to 30°C.[20-22] The comparative air temperature is 10°C.[23] There is further evidence[21,24,25] to suggest that fatter individuals do not achieve maximum tissue insulation until they are immersed in much lower water temperatures, as low as 12°C in some cases.[21]

As the 'variable' resistance is dependent on vasomotor activity, it can be influenced by factors such as exercise or shivering, both of which can increase peripheral blood flow and thereby increase tissue conductance. There is indirect evidence[19,26,27] to suggest that 70 to 90 per cent of the total body insulation of individuals at rest in cold water is provided by poorly perfused skeletal muscle. This is derived from the finding that the maximum body insulation of a resting individual is approximately four times higher than would be predicted from fat insulation alone.

This has practical implications because skeletal muscle blood flow will increase significantly with exercise, including shivering.

This increase will effectively remove the insulation provided by the muscle when poorly perfused. Thus the 'variable' resistance to heat flow depends on both muscular and cutaneous blood flow with the former altering conductance and the latter the amount of heat delivered to the skin by mass flow.

The effectiveness of peripheral vasoconstriction in heat conservation is further improved by a reduction in the temperature of the arterial blood reaching the extremities through counter-current heat exchange between venae committes and arterial blood in the proximal portion of the limbs.[28] In water at 22°C, heat loss from the hands and feet is negligible;[21] heat loss from the torso is transferred by conduction down the high thermal gradient over the short tissue pathway from the visceral organs to the body surface. Therefore most of the heat loss in relatively cold water takes place from the trunk rather than the limbs.

It has been suggested[29] that even fat individuals are at risk during immersion in very cold water (below 12°C) as a result of a sudden increase in peripheral heat loss brought on by cold-induced vasodilatation. Lewis[30] first described cyclical fluctuations in the skin temperature of fingers immersed in cold water (10–12°C). This response is caused by cold paralysis of vascular smooth muscle.[31] The resulting increase in perfusion by warm blood raises the vasoconstrictor muscle temperature, restoring its contractility and thus vasoconstrictor tone until it becomes subject to cold paralysis once more.

The classic lobster pink coloration of the skin seen following prolonged immersion in water below 12°C is generally attributed to such cold-induced vasodilation. However, the time delay evidenced in capillary filling of a blanched area of skin following digital pressure suggests that haemodynamic stasis and temperature impairment of oxygen dissociation, rather than high blood flow, are responsible for this red skin coloration. Presumably, proximal larger vessels are still vasoconstricted, thereby minimizing blood flow and thus the delivery of heat to the more superficial dilated vascular bed.

The evidence for increased heat loss due to cold-induced vasodilatation has generally been obtained from individuals who are warm but whose hands are immersed in cold water,[32] i.e. from those in whom proximal vasoconstriction is likely to be less intense. Such circumstances rarely apply during whole-body cold water immersion, and these authors do not consider cold-induced vasodilatation to be a significant route of heat loss during such immersions.

With regard to the 'fixed' resistance to heat flow, it is well known that individuals with thicker layers of subcutaneous fat cool more slowly in cold water than those with less fat.[33–35] The insulation provided by fat remains fairly constant with exercise because blood flow to fat is relatively low in all conditions.[36]

Peripheral vasoconstriction will slow the rate of heat loss in cold water but will not prevent an overall reduction in body heat stores and deep body temperature. In response to falling skin and deeper body temperatures, involuntary metabolic heat production through shivering is evoked in an attempt to maintain normal deep body temperature.[4] Both the magnitude and rate of change of skin and deep body temperature combine to stimulate shivering, the severity of which is reflected in oxygen consumption.[17,37] When skin temperature is low, oxygen consumption from shivering increases linearly as deep body temperature falls; the metabolic rate may increase and stabilize at a maximum value of approximately 1.5 1/min, i.e. approximately five times the resting level.[13,33,38]

Although shivering increases heat production, by increasing

peripheral blood flow and the combined heat transfer coefficient at the skin–water interface it may result in an overall reduction in the heat stores of a naked individual immersed in still water.

RESPONSES TO IMMERSION IN THERMONEUTRAL WATER

Head-out seated immersion in thermoneutral water can produce profound changes in cardiovascular, renal, and endocrine functions. These effects are a direct result of the high density of the water and the differential hydrostatic pressure over the immersed body. A negative transthoracic pressure of about 14.7 mmHg is established[39] which will result in negative pressure breathing. There is a cephalad redistribution of blood which, within six heart beats of immersion, can increase central blood volume by up to 700 ml.[40–42] This is associated with enhanced diastolic filling, a raised right atrial pressure and a 32 to 66 per cent increase in cardiac output due entirely to an increase in stroke volume which itself is due to enhanced filling of the heart rather than alterations in afterload and contractility.[41,42] Heart rate has been reported to remain either unchanged or to decrease slightly following immersion.[41,43–45]

The increase in intrathoracic blood volume engorges the pulmonary capillaries and competes with air for space in the lung. The engorgement of the pulmonary capillaries results in a 30 to 50 per cent reduction in static and dynamic lung compliance while pulmonary gas flow resistance is increased by 30 to 58 per cent and impedance by 90 per cent.[46,47] In conjunction with the increase in hydrostatic pressure on the chest, these alterations result in a 65 per cent increase in the work of breathing.[40]

Vital capacity is reduced by an average of 6 per cent, maximum voluntary ventilation by 15 per cent,[48] and expiratory reserve volume by an average of 66 per cent which results in a reduction of functional residual capacity.[46,49] The decrease in functional residual capacity and the increase in intrathoracic pooling of blood produces a small increase in pulmonary shunting and a small but consistent fall in the arterial partial pressure of oxygen.[50]

Opposing some of these reductions in lung function are an improved ventilation–perfusion ratio and improved diffusion capacity.[41,44] From the practical viewpoint there is little evidence that these changes in lung function threaten respiration in fit individuals.[51]

Most of the renal responses seen following immersion in thermoneutral water are due to the shift in blood volume which the body senses as hypervolaemia. These responses include diuresis, natriuresis, and kaliuresis.[52–54] Diuresis is usually manifest by the first or second hour of immersion, and natriuresis peaks by the fourth or fifth hour of immersion.[55]

In fully hydrated sodium-replete individuals head-out upright immersion can result in 200 to 300 per cent increases in sodium excretion and free water clearance, with urine output reaching 350 ml/h and leading to dehydration.[55]

It is concluded that immersion in water at thermoneutral temperatures can result in profound alterations in the physiological function of the body. However, many of these responses are of long duration and vary with factors such as posture, activity, and level of hydration before and during immersion. It is this variability which makes it impossible to draw any firm conclusions about the effect which immersion *per se* may have on in-water performance, much of which is of relatively short duration.

RESPONSES TO IMMERSION IN COLD WATER

Most of the alterations resulting from immersion in water at thermoneutral temperatures also occur in cold water. The changes in lung function remain and may be potentiated because of the effect of cooling on respiratory muscle function.[56]

Cardiac output increases by a similar amount (30–40 per cent) during prolonged immersion in cold water compared with immersion at thermoneutral temperatures; on the basis of this it has been suggested that no further translocation of blood occurs with the vasoconstriction and venoconstriction seen during cold water immersion.[57,58] However, cold and water pressure are thought to act additively to raise urinary output during cold water immersion. Cold-induced diuresis accounts for one-third of the total response and is a result of cold-induced vasoconstriction.[59]

Despite a paucity of information, prolonged exposure to cold is thought to be a powerful stimulus for sympathetic nervous system activity and hormonal secretion in man.[60–62] Long-distance swimming in cold water may produce higher plasma concentrations of catecholamines, cortisol, and thyroxine, and lower glucose, insulin, and growth hormone concentrations than those observed during comparable activities in thermoneutral environments.[61,63]

The response to short-term immersion is less clear. Little evidence of the release of catecholamines was found in early work.[64,65] However, in more recent work employing modern analytical techniques significant increases have been found in plasma levels of noradrenaline and adrenaline after immersion in cold water for 1 to 2 min.[60,66,67] The release of noradrenaline from the sympathetic nervous system is closely correlated with metabolic rate during immersion; it can be quickly activated or suppressed on immersion or removal from cold water and is therefore thought to be evoked by changes in skin rather than deep body temperature.[60]

Hazardous responses

Short term

Sudden immersion in cold water evokes a group of cardiorespiratory responses which are collectively known as the 'cold shock' response. These responses, which have recently been reviewed,[68] are summarized in Fig. 2. They are initiated by rapid falls in skin temperature and are potentially extremely hazardous; they are probably responsible for the majority of deaths resulting from immersion in cold open water.

Respiratory drive is enhanced on immersion in water cooler than 25°C.[69] It is inversely related to water temperature, reaching a maximum level in water at a temperature of about 10°C.[70] The respiratory responses include an inspiratory 'gasp' response of

between 2 and 3 l,[70,71] and uncontrollable hyperventilation which can result in a 10-fold increase of minute ventilation,[72] and significantly reduce the arterial tension of carbon dioxide.[65]

The respiratory drive evoked by cold water immersion can reduce the maximum breath hold times of normally clothed individuals to less than 10 s[73] and overrides the diving response in most people,[74,75] even when clothed.[75] This significantly increases the chance of aspirating water and drowning during the first few minutes of immersion in choppy water. The hypocapnia caused by hyperventilation probably accounts for the tetany, disorientation, and clouding of consciousness observed in individuals, including swimmers and canoeists, on cold water immersion.[76–78]

There is an inspiratory shift in end-expiratory lung volume following cold water immersion which can result in the occurrence of tidal breathing within 1 litre of total lung capacity.[69,70] This response makes breathing very difficult and probably contributes to the sensation of dyspnoea experienced on initial immersion.[69]

The initial cardiovascular responses to immersion include intense vasoconstriction, a 42 to 49 per cent increase in heart rate, and 59 to 100 per cent increase in cardiac output.[64,79] As a result of these responses, arterial and venous pressures are increased.

The early cardiovascular responses to immersion in cold water place a significant and sudden strain on the system. The work of the heart, particularly the left ventricle, is increased; this can result in greater ventricular irritability and cardiac irregularities, and, on rare occasions, can precipitate ventricular fibrillation.[80] Both ventricular and atrial ectopic beats have been reported on immersion in cold water.[80] They have also been reported in young healthy individuals on initial immersion in thermoneutral water,[41,51] when they are probably due to distension and acute straining of the right heart. This response, together with the increase in catecholamine secretion, probably both contribute to the arrhythmias seen on cold water immersion.

The initial cardiovascular responses to cold water immersion are a particular threat to people with coronary heart disease, in whom myocardial ischaemia is more likely to develop with sudden increases in cardiac workload. Hypertensive or aneurysmal individuals are also at risk from the sudden elevations in blood pressure on immersion.

Long term

Prolonged immersion in water at temperatures below thermoneutrality will inevitably result in hypothermia, i.e. a deep body temperature below 35°C. Hypothermia is unlikely to be a problem within 30 min of immersion in water even as low as 5°C, and times to onset of hypothermia will vary between individuals for the reasons given in a later section. As a general rule, however, the deep body temperature of those wearing ordinary clothing will have fallen to 35°C after 1 h in water at 5°C, after about 2 h in water at 10°C, and after 3 to 6 h in water at 15°C.[81]

Before general hypothermia becomes established, locomotor impairment will produce a deterioration in swimming performance through delays in muscle cell membrane repolarization as a consequence of the cooling of intracellular enzymes. The tone of both protagonist and antagonist muscle groups gradually increases to such a level that swimming becomes virtually impossible; shivering becomes intense and eventually the body tends to assume a semirigid, fetal attitude. Unless extraneous aids to flotation are available, the individual will be unable to maintain the airway clear of the water and drowning will result before the deep body temperature falls to a level where cardiac arrest from hypothermia would normally be expected to occur, i.e. a myocar-

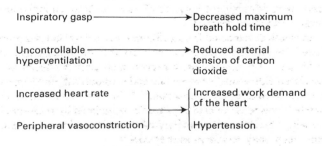

Fig. 2 Summary of the cold shock response.

dial temperature below 28°C but more usually 24 to 26°C.[82]

At cardiac temperatures below 28°C the conduction velocity of the Purkinje tissue is slowed to approximate that of myocardial fibres. This predisposes the heart to ventricular fibrillation. If a flotation aid such as a life-jacket is available, then the airway should remain protected even when unconsciousness from hypothermia occurs (at a deep body temperature of about 30°C).

Normally an individual supported by a life-jacket will keep his back to the waves through a paddling action. However, once cold-induced locomotor incapacitation becomes established, there is a tendency for the relaxed body supported by a life-jacket to be turned to face the oncoming waves. Depending on the frequency and the steepness of the front of the waves, they may break or splash over the face and compromise the airway. Aspiration could result in coughing and hence uncontrollable inspiration, leading to drowning.

An exception to the general sequence of events outlined above can be found in long-distance outdoor swimmers who are habituated to cold water. The ability of these swimmers to withstand prolonged immersion in cold water without apparent ill effect was attributed to their unique combination of physical fitness and substantial thickness of subcutaneous fat.[33] This enabled them to maintain a steady work rate for several hours and to retain much of the heat produced within the body. However, many contemporary outdoor distance swimmers are not as fat as those described by Pugh and Edholm.[33] Before the Windermere International Race in 1974, Golden[83] measured the skinfold thickness of eight swimmers. They were not excessively fat yet all completed the 16.5 mile swim in water at 15°C.

The winner of the race, a male aged 16 (15.9 per cent fat, height 195 cm, and weight 78 kg), completed the distance in 6 h 42 min; the remainder staggered in over the next 6 h (mean swim time, 9 h). None of the swimmers showed subjective signs of hypothermia—a remarkable feat for near-naked people, given that the estimated 50 per cent survival time for a fully clothed 70-kg man in water at 15°C is approximately 6 h.[81]

In a subsequent experiment on three long-distance swimmers the value of cold habituation was clearly demonstrated.[84] Conversely, the apparent absence of cold habituation in other published accounts[63,85,86] may explain why some swimmers failed to complete the required distance because of hypothermia, despite relatively high water temperatures (18–19°C) and short swim times (less than 3 h).[63,85]

Successful outdoor long-distance swimmers would thus appear to be those who are physically fit with a good swimming technique, moderately fat, and habituated to the water temperature at which they are to compete. Lean fast swimmers will be successful in warmer water, but in colder water fatter swimmers are likely to have an advantage.

A negative aspect of the habituation of distance swimmers may be that the incipient onset of hypothermia may go unnoticed in the absence of subjective discomfort and other signs such as shivering.[84] More research is required to identify the deep body temperature at which such individuals become distressed. The potential is there for them to become hypothermic before swimming is seriously impaired.[86] Then, with the cessation of swimming and therefore heat production, heat reserves from the deep body store will continue to be lost to the substantial heat sink in their subcutaneous fat, thereby reducing the interval of useful consciousness from the initial prodromal symptoms to collapse.

Hardwick[86] describes a semiconscious long-distance swimmer reluctantly being removed from the sea after almost 12 h and collapsing pulseless into unconsciousness on arrival in the boat. Three years later the same swimmer attempted an identical swim in water around 13°C; again, she had to be removed from the sea after 7 h 33 min. Throughout the swim her normal stroke rate of 60 strokes/min progressively declined to 45 strokes/min.

Post-immersion

In the absence of flotation assistance, loss of consciousness through hypothermia, uncomplicated by near-drowning, or pure hypothermic cardiac arrest is unlikely in immersion victims. Live hypothermic casualties, or near-drowned cold casualties, may suffer collapse and cardiac arrest during or shortly after the rescue process. This phenomenon, which has been termed 'circum-rescue collapse' is not yet fully understood and has recently been reviewed.[87] Excluding those who may die shortly after rescue from hypoxia caused by drowning, there is evidence to suggest that circulatory collapse and cardiac arrest may occur during rescue as a result of a number of factors, of which the most important are as follows: loss of hydrostatic assistance to venous return and the reimposition of the full effects of gravity; hypovolaemia caused by diuresis and intercompartmental fluid shifts; increased blood viscosity as a result of cooling; diminished work capacity of the hypothermic heart and reduced time for coronary filling; dulled baroreceptor reflexes; unattainable demands to perfuse skeletal muscle; psychological stress and pre-existing coronary disease.

It follows from the mechanisms suggested above that, during rescue, the greatest problems are likely to be encountered by those who are lifted vertically after prolonged immersion in a vertical position. Any problems are likely to be potentiated by a requirement for activity by the casualty during rescue.

Lifting casualties in a horizontal position is likely to be less traumatic and, with the inevitable proviso 'circumstances permitting', immersion victims should be handled with the utmost gentleness and as the potentially critically ill patients that they are.

INDIVIDUAL DIFFERENCES

One of the difficulties in describing the responses associated with immersion in cold water is the wide variation encountered between individuals. This is due in part to methodological differences between investigations which are reflected in both the way and the degree to which subjects have been cooled. In addition, differences in body morphology, sex, age, fitness, and previous exposure to cold are also major sources of variation. As might be expected, many of these factors are interrelated.

As mentioned previously, the fall in deep body temperature during cold water immersion is inversely related to subcutaneous fat thickness. It has been estimated[38] that each additional per cent of body fat equates approximately with a 0.1°C rise in deep body temperature. Additionally, as poorly perfused muscle makes a significant contribution to insulation at rest, differences in body mass are another potential source of variation between individuals during resting immersions.

As would be expected from a response evoked by skin receptors, there is no significant difference between the cold shock responses of fat and lean individuals; however, increased fat thickness is associated with a smaller metabolic and cardiovascular response to prolonged cold water immersion.[21,38] In fatter individuals the metabolic response to cold is primarily stimulated by receptors in the skin, whereas in leaner individuals there is usually also an input from deep body receptors as they cool.[21]

The ratio A_D/wt of surface area to weight may also be a cause of variation between the responses of individuals during cold water immersion. A small ratio has been reported to reduce heat losses in both young[88] and adult groups.[89] In contrast, Toner et al.[90] conclude that, within a given population and sex, differences in A_D/wt have no effect on the responses observed during exercising or resting immersions.

The greater A_D/wt and lower thermoregulatory sensitivity of women may explain why, during cold water immersion, the deep body temperatures of resting females fall by a greater amount than in males with similar subcutaneous fat thicknesses. The lower thermoregulatory thermosensitivity of women is evidenced by the smaller increase in their metabolism, compared to that of men, in response to reductions of deep body temperature by more than 1°C.[38] Thus, although women have greater amounts of subcutaneous fat than men, as a consequence of their greater A_D/wt and lower thermoregulatory sensitivity it has been calculated than lean women require twice the amount of fat of lean men to show similar changes in deep body temperatures in cold water.[38]

Despite these differences, women with similar fat thicknesses to men demonstrate roughly equivalent responses when exercising in cold water.[91,92] This is probably because the exercise enables the women to produce similar levels of heat to the men which, because of a greater thickness of fat over the active musculature[93] and a greater peripheral vasoconstriction, they are better able to retain.[94] These morphological and physiological differences may be of limited benefit at rest, but an advantage when exercise increases peripheral blood flow. During exercise immersions they compensate for the higher A_D/wt of women; during resting immersions they are counteracted by the less sensitive metabolic response.

The initial respiratory response of females to cold water immersion may be smaller than that of males,[95] although this has not been confirmed.[74,79]

Largely because of differences in the methodologies employed, a rather confused picture has emerged regarding the effect of age on the responses of individuals to cold. In general, it appears that ageing is accompanied by a diminished responsiveness to cold. This includes progressive weakening of the vasoconstrictor response,[96,97] possibly because of morphological changes in blood vessels[98] and a reduced shivering response[99]. Older men appear to have lower deep body temperatures on exposure to thermoneutral conditions[96,100] and are more susceptible to cold than younger people; on exposure to cold they do not prevent further falls in their initially low deep body temperatures.[100]

Although fitness does not appear to be related to the hormonal changes induced by cold water immersion,[101] individuals with high levels of aerobic fitness do have attenuated initial respiratory and cardiac responses to cold water immersion.[102]

A cross-adaptation between fitness and long-term thermoregulatory responses to immersion has been reported by several authors.[23,103] The major differences between very fit and less fit individuals appear to be that fitter individuals have a slightly reduced normal body temperature, they experience cold sensation and thermal discomfort at lower mean body temperatures, and they demonstrate a downward shift in their shivering thresholds as a consequence of a downward shift in their thermoregulatory setpoint.[103,104]

Despite the downward shift in the shivering thresholds of fitter individuals, there is evidence[105] to suggest that the maximum intensity of shivering achievable by individuals is related to their maximum oxygen consumption $\dot{V}o_2$max. Furthermore, on exposure to cold air, rather than water, individuals may show a direct relationship between fitness, metabolic heat production, and skin

temperature.[23,106] This may be due to the differences in the physical properties of air and water, or the type of training programme used to increase levels of fitness.

As suggested by the above discussion, many of the responses exhibited by fitter individuals on exposure to cold are similar to those occurring as a result of cold habituation. In the wide range of studies and methodologies which typify the field of human acclimatization to cold, a 'hypothermic' adaptation to cold is the most frequently reported response. This adaptation is characterized by a smaller metabolic response to cold, a greater fall in deep body temperature on exposure to cold, and, despite these alterations, a greater level of thermal comfort during exposure.[107,108] However, this response may not occur if individuals perform exercise which helps maintain deep body temperatures during exposure to cold.[109]

Both the initial respiratory and cardiac responses to cold water immersion can be significantly reduced by as few as six repeated immersions in cold water.[80,109]

EXERCISE IN COLD WATER

Effect of exercise on responses to cold water

The performance of gentle exercise during the first few minutes of cold water immersion does not prevent the cold shock-induced falls in the end-tidal partial pressure of carbon dioxide. At moderate and maximal levels of exercise, however, the fall can be reduced or reversed and therefore hyperpnoea rather than hyperventilation occurs.[77]

At an exercise intensity of 150 W or more the 70 to 90 per cent of total body insulation provided by poorly perfused muscle is almost completely eliminated by exercise hyperaemia.[110,111] This leaves the physical barriers of subcutaneous fat and skin as the principal sources of insulation. On the basis of this, body mass should give the best indication of the insulation available to an individual resting in cold water and skinfold thickness should give the best indication of the insulation available when exercising.

Although exercise will increase heat loss, several factors will determine whether an individual will cool more or less quickly when exercising compared with resting during cold water immersion. These factors include water temperature, water agitation, exercise intensity, type of exercise performed, subcutaneous fat thickness, fitness, and clothing worn.

In general, at water temperatures above 25°C, exercise requiring an oxygen consumption of about 1 l/min will slow the rate of fall of the deep body temperature of an average individual compared with resting immersions. Below 25°C this intensity of exercise will accelerate the fall in deep body temperature.[34,112,113] In fat individuals, however, physical exertion has a smaller effect on falls in deep body temperature in water at temperatures as low as 5°C,[112] although when either fat or thin men rest in water at a temperature at which they are just unable to stabilize deep body temperature, exercise increases their rates of fall of deep body temperature.[21]

Maximal or high intensity exercise in cold water may result in slower rates of fall of deep body temperature than those seen with either lower intensity exercise[112] or during resting immersions.[13,38] Costill et al.[12] observed that when their well-conditioned non-obese subjects performed high intensity exercise (oxygen consumption of 3 l/min) in water at 17°C, they showed a slight increase in deep body temperature after 20 min.

It might be expected that high intensity exercise should be

beneficial in cold water; convective heat losses are maximal when swimming at any speed and the insulation provided by poorly perfused muscle is lost at quite low levels of exercise. Thereafter, any further increases in heat production should be associated with relatively small increases in heat loss and result in a net heat gain.

In most of the studies in which the effect of physical exertion on deep body temperature during cold water immersion has been investigated, whole-body exercise has been employed. In the studies in which either leg-only or arm-only exercise has been examined,[102,114] the results suggest that leg-only exercise results in slower rates of fall in deep body temperature than resting immersions, and that the arms are a greater source of heat loss than the legs during high intensity exercise. This might be expected as the arms have a surface area to mass ratio of approximately twice that of the legs[115] and a shorter conductive pathway from the centre to the surface of the limb. Furthermore, when they are not being exercised, the arms can be placed against the torso, thereby providing some insulation as well as reducing the overall surface area for convective heat loss.[102]

The number of factors determining the influence of exercise on deep body temperature during cold water immersion, and their possible interactions, make it extremely difficult to give general advice concerning whether or not individuals should exercise following accidental immersion. It is fair to say that in most cases, and in particular those involving lean people wearing few clothes in relatively still water, body movements should be avoided.

Effect of cold immersion on exercise performance

On immersion, exercise performance may be significantly impaired at first, as a result of the initial responses to cold. The respiratory responses can cause swimming failure within the first few minutes of immersion. High respiratory frequencies can make the synchronization of breathing and swim stroke impossible and result in the inhalation of water.[116–118]

With the adaptation of the initial responses, after about 3 min of immersion, the next influence on performance is likely to be peripheral cooling. The hands are particularly susceptible to cooling largely because of their high ratio of surface area to mass and, to a lesser extent, their low level of local heat production and variable blood supply. Grip strength can be reduced very quickly following cold water immersion,[119] and both tactile sensitivity and manual dexterity are also quickly affected.[120,121]

The effects of peripheral cooling are primarily due to alterations in muscle and nerve function. Low muscle temperature can affect several chemical and physical processes at the cellular level including metabolic rate, enzyme activity, calcium and acetylcholine release, diffusion rate, and series elastic components.[119] Muscle fibre repolarization becomes adversely affected, resulting in an increased muscle tone of both protagonist and antagonist muscle groups.

The rate of conduction of nervous impulses is slowed, and the amplitude of action potentials is reduced with cooling.[122,123] This occurs at nerve temperatures below about 20°C; in the ulnar nerve a reduction in conduction velocity of 15 m/s per 10°C fall in local temperature can occur. Nerve block may occur at different temperatures with different types of fibres;[124] in general, a local temperature of between 5 and 15°C is required for 1 to 15 min. Such cooling of peripheral motor and sensory nerves leads to a dysfunction equivalent to peripheral paralysis.[123,125]

The result of these alterations is that maximal performance is reduced; maximum power output is reduced by 3 per cent per degree fall in muscle temperature[126,127] and mechanical efficiency is also reduced.[110]

In contrast with the reduction in maximum performance, muscle temperatures down to 27°C increase the duration for which sustained contractions can be maintained. This may be due to a slower production and accumulation of the metabolites causing fatigue.[123] At muscle temperatures below 25°C muscle fatigue occurs earlier as cooling begins to impair neuromuscular function in peripheral muscle fibres, leaving a smaller number of fibres to produce the same amount of force.[123]

The changes in neuromuscular function resulting from cooling explain, in part, the reduction in work capacity in cold water. Other reasons for this decline are the alterations which occur to the central circulation and the reductions which occur in deep body temperature.

The shivering which is evoked by cold water immersion raises oxygen consumption. This increases as more and more muscle groups are recruited as skin and deep body temperatures continue to fall. There is progressive involvement of the muscles of the neck, torso, and finally the extremities. It is possible for shivering to occur during exercise,[13,128] but it is progressively centrally inhibited with increasing exercise intensity.[128] Shivering has been reported to be 80 per cent suppressed by exercise at an intensity of 50 per cent $\dot{V}o_2max$[128] and totally suppressed at workloads requiring oxygen consumptions of between 1.2 and 1.4 l/min.[129,130] In contrast, higher oxygen consumptions have been noted in cold compared with neutral environments during exercise requiring oxygen consumptions of up to 2.0 l/min,[131,132] but not at 3.0 l/min in cold compared with thermoneutral water.[12]

Swimming and ergometry in cool and cold water result in an increased metabolic rate at a given work intensity and a decreased heart rate at a given oxygen consumption.[16,133,134] In water at 25°C and 18°C average oxygen consumption during arm and leg ergometry is increased by 9 per cent and 25.3 per cent respectively when compared with that seen in water at 33°C. The increase in oxygen consumption is greater in leaner individuals.[58]

The twofold increase in the viscosity of water at 0°C compared with 25°C may contribute to the greater energy expenditure seen during swimming in cold water. However, this effect is compensated to some extent by the increased pulling power per stroke which results from the increase in viscosity. Although the efficiency of each stroke may improve, the shivering and increased muscle tone during exercise in cold water may further reduce mechanical efficiency by increasing the activity of antagonist muscles.[128]

It can be concluded that during cold water immersion shivering can coexist with exercise up to moderate intensities of exercise. This results in an increase in the energy cost of submaximal exercise in water cooler than 26 to 28°C.[12,16] As shivering involves both protagonist and antagonist muscles simultaneously, it will itself impair performance.

Shivering may not only impair skilled performance but the metabolic cost it adds to exercise can result in a more rapid depletion of carbohydrate and lipid energy sources and, as a consequence, an earlier onset of fatigue.[135–137]

$\dot{V}o_2max$, during ergometry or swimming, and maximum performance are both reduced during cold water immersion.[16, 58, 134, 138] The reduction in $\dot{V}o_2max$ can occur in water with a temperature as high as 25°C[110] and is approximately linearly related to deep body temperature,[134] with a 10 to 30 per cent reduction following a 0.5 to 2.0°C fall in deep body temperature.[139] Maximal swimming performance can be reduced in water with a temperature as high as 26°C.[16] In water at 18°C the subjective sensations associated with exhaustive swimming are related to muscle function rather than cardiorespiratory distress.[16]

Associated with the decrease in $\dot{V}O_2$max in cold water, lactic acid appears in the blood at lower work-loads and accumulates at a more rapid rate than in thermoneutral conditions.[134,138] With more profound cooling, however, this is also reduced due to muscle cooling and direct impairment of the processes responsible for the anaerobic production of energy.[132] These changes will all contribute to the degradation of performance in the cold.

A decrease in deep body temperature of 0.5 to 1.5°C results in a reduction of 10 to 40 per cent in the capacity to supply oxygen to meet the increased requirements of activity.[139] The mechanisms proposed for this reduction and that observed in $\dot{V}O_2$max in cold water include temperature-dependent reductions in enzyme activity within the muscle, reduced oxygen transport due to decreased respiratory and/or cardiac function, and reduced muscle blood flow.[139]

As with immersion in thermoneutral water, there is little evidence to suggest that the alterations associated with cold water immersion impair respiratory function to the extent that it interferes with oxygen uptake during exercise.[134,140]

With regard to cardiac function, once the cold shock response has subsided, cold water immersion reduces resting, submaximal, and maximal heart rates when compared with those seen in warm water.[134,138,141] Despite the reduction in heart rate, cardiac output appears to be maintained during submaximal exercise in cold water by an elevated stroke volume.[58,134,138] Both $\dot{V}O_2$max and maximum cardiac output[110] are reduced by 10–30 per cent in cold water, but the relationship between them remains linear and similar to that seen in warm water. There is little evidence to suggest that cardiac output limits $\dot{V}O_2$max in cold water.[139] With profound cooling, however, β-receptor activity may be reduced by the direct effect of cold on cardiac muscle and the reduction in $\dot{V}O_2$max may then be related to decreased heart rate, decreased cardiac output, and decreased oxygen delivery due to the reduced capacity of the heart to develop force.[132,142]

The early appearance of lactic acid in the blood and the elevation in diastolic blood pressure seen during exercise in the cold suggest that oxygen delivery to the working muscle may be reduced in cold water. Marked reductions in skeletal muscle blood flow have been measured in subjects in cold water.[139] Therefore it is probable that the exercise hyperaemia seen in normal conditions is attenuated in cooled individuals by a sympathetically mediated vasoconstriction of muscle resistance vessels.[110,134,139] This will not only reduce oxygen delivery, but may also reduce the rate of removal of the end-products of metabolism; these, in turn, may then contribute to the impairment of performance.

It was noted above that immersion in thermoneutral water resulted in a reduction in plasma volume. In cold water this reduction may be as great as 24 per cent.[137] Such changes may further affect muscle perfusion and, as a result of dehydration, reduce work times and compromise thermoregulation.[12]

Therefore a major limitation on performance in the cold is thought to be a progressively lower muscle blood flow with cooling. If the oxygen requirements of the muscle remain the same, then local hypoxia will develop and impair work capacity. This situation may be accentuated by a shift to the left in the oxygen dissociation curve as a result of cooling.[143]

It can be concluded that performance of both short- and long-term exercise may be seriously impaired in cold water. In some cases this impairment may reach the point where survival is threatened.

PROTECTION, AUGMENTATION OF PERFORMANCE, AND TREATMENT

Protection

The use of specialist protective clothing, which either keeps the body dry and/or insulates it, will help to maintain both skin and deep body temperatures during cold water immersion and thereby help to maintain performance and extend the time of useful consciousness and the survival time (Fig. 3).[73,144–146] With such garments a good fit is essential to prevent leakage or 'flushing' of water between the skin and garment; both these will significantly increase heat loss. Protection of the head is particularly important as the poor vasoconstrictor response in this region makes it a major area of heat loss.[147] Finally, shivering is more likely to make a positive rather than a negative contribution to the maintenance of body heat stores in individuals who are wearing specialist protective clothing designed to keep the surface of the skin dry and therefore reduce the heat transfer coefficient at the skin.

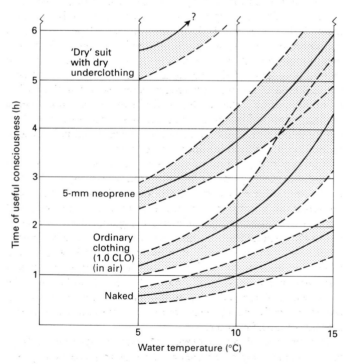

Fig. 3 Times of useful consciousness (deep body temperature 35°C) of individuals immersed in cold water in different clothing assemblies under laboratory conditions. (1 BLO = 0.155 °C/m². w; or the insulation provided by a business suit and standard undergarments).

If no 'external' insulation is available, the most important source of 'internal' insulation for a swimming man is subcutaneous fat. Pugh *et al.*[148] estimate that an extra millimetre of subcutaneous fat may be equivalent to increasing water temperature by 1.5°C. A thick layer of subcutaneous fat is particularly useful if, as in many long-distance outdoor swimmers, it is coupled with a high level of aerobic fitness.[33,148] Such a combination allows heat production to be maintained at a high level and to be retained due to low tissue conductance. Pugh et al.[148] argued that a person who is not fat should not take up swimming because he or she will be unable to endure the cold; however, there is evidence to the contrary in a more recent study of long-distance outdoor swimmers.[84]

The advantage of an insulating layer of fat forms the rationale for the use of grease in outdoor cold-water swimming. In such situations as swimming in cold water, where heat flows are high, it has been calculated that a 1-mm thick layer of lanolin or Vaseline, which has approximately the same thermal conductivity as fat, has the same effect as raising water temperature by about 1.5°C.[17,148]

With the exception of increasing buoyancy, subcutaneous fat, as noted earlier, provides no protection against the initial cold shock response because the receptors which initiate this response are located in the superficial subepidermal tissues above the subcutaneous fat. Apart from specialist protective clothing, the best protection against the initial responses to immersion is to habituate the responses by repeated immersions in cold water.[109]

However, cold habituation may be disadvantageous with regard to long-term immersions if a 'hypothermic' adaptation is developed; although skilled performance may improve due to increased thermal comfort and the attenuation of shivering, the consequent reduction in heat production could result in deep body temperature falling at a faster rate.[109,149] Because of the increased comfort associated with the 'hypothermic' type of adaptation there is a danger of the development of incipient hypothermia. Furthermore, it is not clear whether with such an adaptation normal thermoregulatory responses are eventually actuated (decreased thermoregulatory threshold) or whether they remain attenuated (decrease in 'gain' of thermoregulatory responses). Cold adaptation could be of use during long-term immersion if an 'insulative' adaptation were developed; this results in an increase in insulation on immersion[84,150,151] and a reduction in the rate of fall of deep body temperature. Unfortunately the factors determining which type of adaptation is developed are not clear.

As mentioned previously, increasing fitness levels may be advantageous with regard to protecting against the long- and short-term responses to cold water immersion. If all else fails, the initial responses can be attenuated by slowing the rate of entry into cold water, while swimming performance is significantly improved if the initial ventilatory responses are allowed to subside before swimming is commenced.[152] A fitter individual will also have a greater capacity to cope with the increased work requirements associated with exercise in cold water. With less fit individuals, the combination of the reduction in $\dot{V}o_2$max and increased oxygen requirements for submaximal work in the cold may seriously impair physical performance.

Maintenance of normal levels of muscle glycogen in large muscle groups may be an important factor in protection against the cold; it has been reported that low skeletal muscle glycogen levels are associated with more rapid body cooling during cold water immersion in humans.[153] However, the availability and compensatory use of other metabolic substrates have resulted in failure to confirm this finding.[154,155] In addition, muscle glycogen levels that are higher than normal do not increase tolerance to cold.[153]

Augmentation of performance

Special diets and pharmacological agents have both been suggested as means of improving cold tolerance in humans.[156–158] It is believed that fat-rich diets are superior to carbohydrate or protein-rich diets with regard to the maintenance of body temperature in the cold.[156]

Drinking a moderate amount of alcohol (28 ml) before cold water immersion reduces the rise in heart rate and number of extrasystoles seen on initial cold water immersion; it also improves morale. In addition, because the powerful cold-induced stimulus to vasoconstrict overcomes the vasodilatory effect of the alcohol, a faster rate of fall of deep body temperature is not seen during cold water immersion after a moderate amount of alcohol.[159] Despite these potential 'advantages', the use of alcohol prior to cold water immersion is not recommended because, when combined with exercise in a cold environment, alcohol induces hypoglycaemia which in turn leads to failure of the metabolic processes and, as a result, an accelerated onset of hypothermia and earlier impairment of performance.[160]

Finally, with regard to thermal comfort, heat flux, and maximal swimming performance, the optimal pool temperature for teaching and competitive short-distance swimming is 28 to 30° C.[16,161] For longer distances, water temperature should be lowered to about 25°C; this will enable thermal balance to be maintained. Outdoor long-distance swimmers are advised to habituate to water at temperatures that they expect to encounter in competitive events.

Treatment

Following rescue the casualty should be quickly placed in a horizontal position while a general assessment of his condition is carried out and essential first aid or appropriate resuscitation given. The possibility of other injuries should always be considered, particularly in those cases not showing the expected response to early treatment. It is not uncommon to sustain a traumatic injury when falling into the water or during a capsize. The signs and symptoms of head or spinal injury, or ruptured viscera, may be concealed by confusion or unconsciousness from a subsequent near-drowning.

Management of the hypoxia, hypercapnia, and acidosis associated with near-drowning casualties is outside the scope of this chapter and has been well reviewed elsewhere.[162] Controversy still surrounds the rewarming of hypothermic casualties; the authors' views, giving some simple guidelines, follow.

On rescue prevent further heat loss by providing adequate insulation both from the ground and from the ambient conditions.

If evaluation is not immediately possible and facilities are available, a casualty who is conscious, shivering, and uninjured, may be immersed to the neck in a bath of hot water. The temperature of the bath should approximate but not exceed 40°C (comfortable to one's own elbow and to the casualty) and should be maintained at this temperature by constant stirring and adding hot water as necessary. This treatment will require a good reservoir of hot water.

Cessation of shivering will occur on, or shortly after, immersion, but this should not be interpreted as an indication that rewarming is complete. When subjectively warm, and just before warm flushing or sweating occurs, the casualty should be helped from the bath, dried, and placed in a warmed bed. When fully rested and recovered the casualty may be discharged.

An unconscious or semiconscious cold casualty should be regarded as being in a critical condition. The major objective of management is the prevention of further heat loss to facilitate passive metabolic rewarming, without stressing the cardiovascular system which may precipitate ventricular fibrillation. Therefore all movement of the casualty should be conducted slowly and carefully.

Normal cardiopulmonary resuscitation should be commenced on the way to hospital if indicated. Subsequent hospital management should be similar to that given to normothermic patients with the following provisos: death should not be diagnosed until the casualty is 'warm and dead'; cardioversion is unlikely to be successful until the myocardial temperature exceeds 28°C; in fact, repeated attempts at defibrillation below this temperature are likely to damage the heart. Attempts at surface warming are contraindicated as they may result in 'rewarming collapse',[163] i.e. a fall in systemic blood pressure caused by peripheral local heat inducing vasodilatation before central reflex cardiovascular responses, dulled by hypothermia, are restored.

In patients in cardiac arrest, metabolic rewarming will be virtually non-existent, while surface rewarming will be too slow and largely ineffective. In such patients extracorporeal circulation, when available, has proved successful, as has the less sophisticated peritoneal dialysis. The dialysate should be warmed to 37°C and external cardiac compression should be continued until the deep body temperature has risen above 28 to 30°C when defibrillation can be attempted.

SUMMARY

Immersion in cold water represents a profound change in the environment surrounding the body. The responses which are evoked by this change can significantly influence performance and, when extreme, can threaten survival. An understanding of these responses should help reduce the risk to, as well as maintain the performance of, those who participate in water-related activities.

REFERENCES

1. Pleuckhahn VD. The aetiology of 134 deaths due to 'drowning' in Geelong during the years 1957–1971. *Medical Journal of Australia* 1972; **2**: 1183–7.
2. Home Office. *Report of the working party on water safety.* London: HMSO, 1977.
3. Royal Society for the Prevention of Accidents. *Drownings in the UK 1987.* Birmingham: Royal Society for the Prevention of Accidents, 1988.
4. Burton AC, Edholm OG. *Man in a cold environment.* London: Edward Arnold, 1955.
5. Stlowijk JAJ, Hardy JD. Temperature regulation in man. *Pflügers Archiv* 1966; **291**: 129–62.
6. Bligh J. Neuronal models of temperature regulation. In: Bligh J, Moore RE, eds. *Temperature regulation.* Amsterdam: North-Holland, 1972: 105–20.
7. Carlson LD, Hsieh CL. Temperature and humidity. In: Slonim NB, ed. *Environmental physiology.* St Louis, Missouri: CV Mosby, 1974: 61–83.
8. Fox RH. Temperature regulation with special reference to man. In: Linden RJ, ed. *Recent advances in physiology.* London: Churchill Livingstone, 1974: 340–405.
9. Hensel H. Thermoreception and temperature regulation. *Monographs of the Physiological Society*, No. 38. New York: Academic Press, 1981.
10. Hong SK. Thermal considerations. In: Shilling CW, Carlston CB, Mathias RA, eds. *The physician's guide to diving medicine.* New York: Plenum Press, 1984: 153–78.
11. Gagge AP. Standard operative temperature. A generalized temperature scale applicable to direct and partitional calorimetry. *American Journal of Physiology* 1940; **131**: 93–103.
12. Costill DL, Cahill PJ, Eddy D. Metabolic responses to submaximal exercise in three water temperatures. *Journal of Applied Physiology* 1967; **22**: 628–32.
13. Craig AB, Dvorak M. Thermal regulation during water immersion. *Journal of Applied Physiology* 1966; **21**: 1577–85.
14. Colin J, Timbal J, Guieu JD, Boutelier C, Houdas Y. Combined effects of radiation and convection. In: Hardy JD, Gagge AP, Stolwijk JAJ, eds. *Physiological and behavioural temperature regulation.* Springfield, Il: CC Thomas, 1970: 81–96.
15. Witherspoon JM, Goldman RF, Breckenridge JR. Heat transfer coefficients of humans in cold water. *Journal de Physiologie (Paris)* 1971; **63**: 459–62.
16. Nadel ER, Holmer I, Bergh U, Astrand PO, Stolwijk JA. Energy exchanges of swimming man. *Journal of Applied Physiology* 1974; **36**: 465–71.
17. Bullard RW, Rapp GM. Problems of heat loss in water immersion. *Aerospace Medicine* 1970; **41**: 1269–77.
18. Barcroft H, Edholm OG. The effect of temperature on blood flow and deep temperature in human forearm. *Journal of Physiology (London)* 1943; **102**: 5–12.
19. Rennie DW. Thermal insulation of Korean diving women and non-divers in water. In: Rahn H, Yokoyoma T, eds. *Physiology of breath-hold diving and the Ama of Japan.* Washington, DC: US Government Printing Office 1965: 315–24 (NAS-NRC Publication 1341).
20. Burton AC, Bazett HC. A study of the average temperature of the tissues, of the exchange of heat and vasomotor responses in man by means of a bath calorimeter. *American Journal of Physiology* 1936; **117**: 36–54.
21. Cannon P, Keatinge WR. The metabolic rate and heat loss of fat and thin men in heat balance in cold and warm water. *Journal of Physiology (London)* 1960; **154**: 329–44.
22. Rennie DW, Covino BG, Howell GJ, Hong SH, Kang BS, Hong SK. Physical insulation of Korean diving women. *Journal of Applied Physiology* 1962; **17**: 961–6.
23. Bittel JHM, Nonotte-Varly C, Livecchi-Gonnot GH, Savourey GLMJ, Hanniquet AM. Physical fitness and thermoregulatory reactions in a cold environment in men. *Journal of Applied Physiology* 1988; **65**: 1984–9.
24. Smith RM, Hanna JM. Skinfolds and resting heat loss in cold air and water: temperature equivalence. *Journal of Applied Physiology* 1975; **39**: 93–102.
25. Toner MM, Holden WL, Foley ME, Bogart JE, Pandolf KB. Influence of clothing and body-fat insulation on thermal adjustments to cold water stress. *Aviation, Space and Environmental Medicine* 1989; **60**: 957–63.
26. Veicsteinas A, Ferretti GT, Rennie DW. Superficial shell insulation in resting and exercising man in cold water. *Journal of Applied Physiology* 1982; **52**: 1557–64.
27. Park VS, Pendergast DR, Rennie DW. Decreases in body insulation with exercise in cold water. *Undersea Biomedical Research* 1984; **11**: 159–68.
28. Bazett HL, Love L, Newton M, Eisenberg L, Day R, Forster R. Temperature changes in blood flowing in arteries and veins in man. *Journal of Applied Physiology* 1948; **1**: 3–19.
29. Keatinge WR. *Survival in cold water.* Oxford: Blackwell, 1969.
30. Lewis T. Observations upon the reactions of the vessels of the human skin to cold. *Heart* 1930; **15**: 177–208.
31. Keatinge WR. Mechanism of adrenergic stimulation of mammalian arteries and its failure at low temperatures. *Journal of Physiology (London)* 1964; **174**: 184–205.
32. Nelms JD, Soper JG. Cold vasodilatation and cold acclimatisation in the hands of British fish filleters. *Journal of Applied Physiology* 1962; **17**: 444–8.
33. Pugh LGC, Edholm OG. The physiology of Channel swimmers. *Lancet* 1955; ii: 761–8.
34. Carlson LD, Hsieh ACL, Fullington F, Elsner RW. Immersion in cold water and total body insulation. *Journal of Aviation Medicine* 1958; **29**: 145–52.
35. Keatinge WR. The effect of subcutaneous fat and of previous exposure to cold on the body temperature, peripheral blood flow and metabolic rate of men in cold water. *Journal of Physiology (London)* 1960; **153**: 166–78.
36. Rowell LB. *Human Circulation. Regulation during physical stress.* New York: Oxford University Press, 1986.

37. Nielsen B. Metabolic reactions to changes in core and skin temperature in man. *Acta Physiologica Scandinavica* 1976; **97**: 125–38.

38. McArdle WD, Magel JR, Gergley TJ, Spina RJ, Toner MM. Thermal adjustment to cold water exposure in resting men and women. *Journal of Applied Physiology* 1984; **56**: 1565–71.

39. Lin YC. Hong SK. Physiology of water immersion. *Undersea Biomedical Research* 1984; **11**: 109–11.

40. Hong SK, Cerretelli P, Cruz JC, Rahn H. Mechanics of respiration during submersion in water. *Journal of Applied Physiology* 1969; **27**: 535–8.

41. Arborelius MJr, Balldin UI, Lilja B, Lundgren CEG. Hemodynamic changes in man during immersion with the head above the water. *Aerospace Medicine* 1972; **43**: 592–8.

42. Risch WD, Koubenec HJ, Beckmann U, Lange S, Gauer OH. The effect of graded immersion on heart volume, central venous pressure, pulmonary blood distribution, and heart rate in man. *Pflügers Archiv* 1978; **374**: 115–18.

43. Gauer OH, Henry JP. Neurohumerol control of plasma volume. In: Guyton AC, Crowley AW, eds. *Cardiovascular Physiology II*. Baltimore, MD: University Park Press, 1976: 145–90.

44. Begin R, Epstein M, Sackner MA, Levinson R, Dougherty R, Duncan D. Effects of water immersion to the neck on pulmonary circulation and tissue volume in man. *Journal of Applied Physiology* 1976; **40**: 293–9.

45. Lin YC. Circulatory functions during immersion and breath-hold dives in humans. *Undersea Biomedical Research* 1984; **11**: 123–38.

46. Dahlback GO, Jonsson E, Liner MH. Influence of hydrostatic compression of the chest and intrathoracic blood pooling on static lung mechanics during head-out immersion. *Undersea Biomedical Research* 1978; **5**: 71–85.

47. Reid MB, Banzett RB, Feldman HA, Mead J. Reflex compensation of spontaneous breathing when immersion changes diaphragm length. *Journal of Applied Physiology* 1985; **58**: 1135–42.

48. Flynn ET, Camporesi EM, Nunnely SA. Cardiopulmonary responses to pressure breathing during immersion in water. In: Lanphier EH, Rahn H, eds. *Man water, pressure*. Buffalo: University of New York, 1975:79–94.

49. Jarrett AS. Effect of immersion on intrapulmonary pressure. *Journal of Applied Physiology* 1965; **20**: 1261–6.

50. Craig AB, Dvorak M. Expiratory reserve volume and vital capacity of the lungs during immersion in water. *Journal of Applied Physiology* 1975; **38**: 5–9.

51. Epstein M. Johnson, G. DeNunzio AG. Effects of water immersion on plasma catecholamines in normal humans. *Journal of Applied Physiology* 1983; **54**: 244–8.

52. Behn C, Gauer OH, Kirsch K, Eckert P. Effects of sustained intra-thoracic vascular distension on body fluid distribution and renal excretion in man. *Pflügers Archiv* 1969; **313**: 123–35.

53. Gauer OH, Henry JP, Behn C. The regulation of extracellular fluid volume. *Annual Review of Physiology* 1970; **32**: 547–95.

54. Epstein M, Duncan DC, Fishman LM. Characterization of the natri-uresis caused in normal man by immersion in water. *Clinical Science* 1972; **43**: 275–87.

55. Epstein M. Renal effects of head-out water immersion in man: Implications for an understanding of volume homeostasis. *Physiological Reviews* 1978; **58**: 529–81.

56. Choukroun M-L, Kays C, Varene P. Effects of water temperature on pulmonary volumes in immersed human subjects. *Respiration Physiology* 1989; **75**: 255–66.

57. Rennie DW, DiPrampero P, Cerretelli PC. Effects of water immersion on cardiac output, heart rate and stroke volume of man at rest and during exercise. *Medicine dello Sport* 1971; **24**: 223–8.

58. McArdle WD, Magel JR, Lesmes GR, Pechar GS. Metabolic and cardiovascular adjustments to work in air and water at 18, 25 and 33°C. *Journal of Applied Physiology* 1976; **40**: 85–90.

59. Knight DR, Horvath SM. Urinary responses to cold temperature during water immersion. *American Journal of Physiology* 1985; **248**: R560–6.

60. Johnson DG, Hayward JS, Jacobs TP, Collis ML, Eckeson JD, Williams RH. Plasma norepinephrine responses of man in cold water. *Journal of Applied Physiology* 1977; **43**: 216–20.

61. Galbo H. *et al.* The effect of water temperature on the hormonal response to prolonged swimming. *Acta Physiologica Scandinavica* 1979; **105**: 326–37.

62. Arnett EL, Watts DT. Catecholamine excretion in men exposed to cold. *Journal of Applied Physiology* 1960; **15**: 449–500.

63. Dulac S et al. Metabolic and hormonal responses to long-distance swimming in cold water. *International Journal of Sports Medicine* 1987; **8**: 352–6.

64. Manager WM, Wakim, KG, Bollman JL. *Chemical quantitation of epinephrine and norepinephrine in plasma*. Springfield, Il: CC Thomas, 1959.

65. Keatinge WR, McIlroy MB, Goldfien A. Cardiovascular responses to ice-cold showers. *Journal of Applied Physiology* 1964; **19**: 1145–50.

66. Le Blanc J, Cote J, Jobin M, Labrie A. Plasma catecholamines and cardiovascular responses to cold and mental activity. *Journal of Applied Physiology* 1979; **47**: 1207–11.

67. Buhring M, Spies HF. *Pathogenesis of sudden death following water immersion (immersion syndrome)*. Washington, DC: US Government Printing Office, 1981 (NASA Technical Memorandum 76542).

68. Tipton MJ. The initial responses to cold water immersion in man. *Clinical Science* 1989; **77**: 581–8.

69. Keatinge WR, Nadel JA. Immediate respiratory response to sudden cooling of the skin. *Journal of Applied Physiology* 1965; **20**: 65–9.

70. Tipton MJ, Stubbs DA, Elliott DS. Human initial responses to immersion in cold water at 3 temperatures and following hyperventilation. *Journal of Applied Physiology* 1991; **70**: 317–22.

71. Goode RC, Duffin J, Miller R, Romet TT, Chant W, Ackles A. Sudden cold water immersion. *Respiration Physiology* 1975, **23**. 301 10.

72. Tipton MJ, Golden, FStC. The influence of regional insulation on the initial responses to cold immersion. *Aviation, Space and Environmental Physiology* 1987; **58**: 1192–6.

73. Tipton MJ, Vincent, MJ. Protection provided against the initial responses to cold immersion by a partial coverage wet suit. *Aviation, Space and Environmental Physiology* 1989; **60**: 769–73.

74. Hayward JS, Hay C, Matthews BR, Overweel CH, Radford DD. Temperature effect on the human dive response in relation to cold water near-drowning. *Journal of Applied Physiology* 1984; **56**: 202–6.

75. Tipton MJ. The effect of clothing on 'diving bradycardia' in man during submersion in cold water. *European Journal of Applied Physiology* 1989; **59**: 360–4.

76. Golden FStC, Hervey GR. A class experiment on immersion hypothermia. *Journal of Physiology (London)* 1972; **227**: 35P–6P.

77. Cooper KE. Martin S, Riben P. Respiratory and other responses in subjects immersed in cold water. *Journal of Applied Physiology* 1976; **40**: 903–10.

78. Baker S, Atha J. Canoeist's disorientation following cold immersion. *British Journal of Sports Medicine* 1981; **15**: 111–15.

79. Hayward JS, Eckerson JD. Physiological responses and survival time predictions for humans in ice-water. *Aviation, Space and Environmental Physiology* 1984; **55**: 206–12.

80. Keatinge WR, Evans M. The respiratory and cardiovascular response to immersion in cold and warm water. *Quarterly Journal of Experimental Physiology* 1961; **46**: 83–94.

81. Molnar GW. Survival of hypothermia by men immersed in the ocean. *Journal of the American Medical Association* 1946; **131**: 1046–50.

82. Alexander L. *The treatment of shock from prolonged exposure to cold, especially water*. London: HMSO, 1945 (Combined Intelligence Objectives Sub-Committee APO 413 C105, Item No. 24).

83. Golden FStC. *Physiological changes in immersion hypothermia, with special reference to factors which may be responsible for death in the early rewarming phase*. Ph.D. Thesis, University of Leeds 1979.

84. Golden FStC, Hampton IFG, Smith DJ. Lean long distance swimmers. *Journal of the Royal Naval Medical Service* 1980; **66**: 26–30.

85. Bergh U, Ekblom B, Holmer I, Gullstrand L. Body temperature response to a long distance swimming race. In: *International series on sport sciences*, Vol. 6. *Swimming medicine* IV. Baltimore: University Park Press, 1978: 342–4.

86. Hardwick RG. Two cases of accidental hypothermia. *British Medical Journal* 1962; **1**: 147–9.

87. Golden FStC, Hervey GR, Tipton MJ. Circum-rescue collapse: collapse, sometimes fatal, associated with rescue of immersion victims. *Journal of the Royal Naval Medical Service* 1991; **77**: 139–49.

88. Sloan REG, Keating WR. Cooling rates of young people swimming in cold water. *Journal of Applied Physiology* 1973; **35**: 371–5.

89. Buskirk ER, Kollias J. Total body metabolism in the cold. *New Jersey Academy of Science Special Symposium Issue* 1969: 17–25.

90. Toner MM, Sawka MN, Foley ME, Pandolf KB. Effects of body mass and morphology on thermal responses in water. *Journal of Applied Physiology* 1986; **60**: 521–5.

91. Hayward JS, Eckerson, JD, Collins ML. Thermal balance and survival time prediction of man in cold water. *Canadian Journal of Physiology and Pharmacology* 1975; **53**: 21–32.

92. McArdle WD, Magel JR, Spina RJ, Gergley TJ, Toner MM. Thermal adjustments to cold-water exposure in exercising men and women. *Journal of Applied Physiology* 1984; **56**: 1572–7.

93. Edwards DAW. Differences in the distribution of subcutaneous fat with sex and maturity. *Clinical Science* 1951; **10**: 305–15.

94. Bollinger A, Schlumpf M. Finger blood flow in healthy patients of different age and sex and in patients with primary Raynauld's disease. *Acta Chirurgica Scandinavica (Supplement)* 1976; **465**: 53–7.

95. Malkinson TJ, Martin S, Simper P, Cooper KE. Expired air volumes of males and females during cold water immersion. *Canadian Journal of Physiology and Pharmacology* 1981; **59**: 843–6.

96. Wagner JA, Robinson S, Marino RP. Age and temperature regulation of humans in neutral and cold environments. *Journal of Applied Physiology* 1974; **37**: 562–5.

97. Budd GM, Brotherhood JR, Hendrie AL, Jeffery SE. Effects of fitness, fatness, and age on men's responses to whole body cooling in air. *Journal of Applied Physiology* 1991; **71**: 2387–93.

98. Wagner JA, Matsushita K, Horvath SM. Effects of carbon dioxide inhalation on physiological responses to cold. *Aviation, Space and Environmental Medicine* 1983; **54**: 1074–9.

99. Horvath SM, Radcliffe CE, Hutt BK, Spurr GB. Metabolic responses of old people to a cold environment. *Journal of Applied Physiology* 1958; **8**: 145–8.

100. Wagner JA, Horvath SM. Cardiovascular reactions to cold exposures differ with age and gender. *Journal of Applied Physiology* 1985; **58**: 187–92.

101. Jacobs I, Romet T, Frim J, Hynes A. Effects of endurance fitness on responses to cold water immersion. *Aviation, Space and Environmental Medicine* 1984; **55**: 715–20.

102. Golden FStC, Tipton MJ. Human thermal responses during leg-only exercise in cold water. *Journal of Physiology (London)* 1987; **391**: 399–405.

103. Baum E, Bruck K, Schwennicke HP. Adaptive modifications in the thermoregulatory system of long-distance runners. *Journal of Applied Physiology* 1976; **40**: 404–10.

104. Dressendorfer RM, Smith RM, Baker DG, Hong SK. Cold tolerance of long-distance runners and swimmers in Hawaii. *International Journal of Biometeorology* 1977; **21**: 51–8.

105. Golden FStC, Hampton IFG, Hervey GR, Knibbs AV. Shivering intensity in humans during immersion in cold water. *Journal of Physiology (London)* 1979; **290**: 48P.

106. Adams T, Herberling EJ. Human physiological responses to a standard cold stress as modified by physical fitness. *Journal of Applied Physiology* 1958; **132**: 226–30.

107. Stanton-Hicks C, O'Connor WR. Skin temperatures of Australian aborigines under varying atmospheric conditions. *Australian Journal of Experimental Biology and Medical Science* 1938; **16**: 1–18.

108. LeBlanc J. Evidence and meaning of acclimatization to cold in man. *Journal of Applied Physiology* 1956; **9**: 395–8.

109. Golden FStC, Tipton MJ. Human adaptation to repeated cold immersions. *Journal of Physiology (London)* 1988; **396**: 349–63.

110. Rennie DW, Park Y, Veicsteinas A, Pendergast D. Metabolic and circulatory adaptation to cold water stress. In: Cerretelli P, Whipp B, eds. *Exercise bioenergetics and gas exchange*. Amsterdam: North-Holland, 1980: 315–21.

111. Strong LH, Gee GK, Goldman RF. Metabolic and vasomotor insulative responses occurring on immersion in cold water. *Journal of Applied Physiology* 1985; **58**: 964–77.

112. Keatinge WR. The effect of work and clothing on the maintenance of the body temperature in water. *Quarterly Journal of Experimental Physiology* 1961; **46**: 69–82.

113. Beckman EL. Thermal protection during immersion in cold water. In: *Proceedings of the Second Symposium on Underwater Physiology*. Washington, DC: US Government Printing Office, 1963 (NAS-NRC Publication 1181).

114. Toner MM, Sawka MN, Pandolf KB. Thermal responses during arm and leg and combined arm-leg exercise in water. *Journal of Applied Physiology* 1984; **56**: 1355–60.

115. Burton AC. Human calorimetry II. The average temperature of the tissues of the body. *Journal of Nutrition* 1935; **9**: 261–80.

116. Glaser EM, Hervey GR. Swimming in very cold water. *Journal of Physiology* 1951; **115**: 14P.

117. Keatinge WR, Prys-Roberts C, Cooper KE, Honour AJ, Haight J. Sudden failure of swimming in cold water. *British Medical Journal* 1969; **1**: 480–3.

118. Golden FStC, Hardcastle PT. Swimming failure in cold water. *Journal of Physiology (London)* 1982; **330**: 60P–1P.

119. Vincent MJ, Tipton MJ. The effects of cold immersion and hand protection on grip strength. *Aviation, Space and Environmental Medicine* 1988; **59**: 738–41.

120. Bowen HM. Diver performance and the effects of cold. *Human Factors* 1968; **10**: 445–63.

121. Stang PR, Weiner EL. Diver performance in cold water. *Human Factors* 1970; **12**: 391–9.

122. Douglas WW, Malcolm JL. The effect of localized cooling on conduction in cat nerves. *Journal of Physiology (London)* 1955; **130**: 53–71.

123. Clarke SJ, Hellon RF, Lind AR. The duration of sustained contractions of the human forearm at different muscle temperatures. *Journal of Physiology (London)* 1958; **143**: 454–73.

124. Basbaum CB. Induced hypothermia in peripheral nerve: electron microscopic and electrophysiological observations. *Journal of Neurocytology* 1973; **2**: 171–87.

125. Vanggaard L. Physiological reactions to wet-cold. *Aviation, Space and Environmental Medicine* 1975; **46**: 33–6.

126. Bergh U, Ekblom B. Influence of muscle temperature on maximal muscle strength and power output in human skeletal muscles. *Acta Physiologica Scandinavica* 1979; **107**: 33–7.

127. Sargeant AJ. Effect of muscle temperature on leg extension force and short-term power output in humans. *European Journal of Applied Physiology* 1987; **56**: 693–8.

128. Hong SK, Nadel ER. Thermogenic control during exercise in a cold environment. *Journal of Applied Physiology* 1979; **47**: 1084–9.

129. Stromme S, Andersen KL, Elsner RW. Metabolic and thermal responses to muscular exertion in the cold. *Journal of Applied Physiology* 1963; **18**: 756–63.

130. Andersen KL, Hart LS, Hammel HT, Sabean HB. Metabolic and thermal response of Eskimos during muscular exertion in the cold. *Journal of Applied Physiology* 1963; **18**: 613–8.

131. Hanna JN, Hill PMcN, Sinclair JD. Human cardiorespiratory responses to acute cold exposure. *Clinical and Experimental Pharmacology and Physiology* 1975; **2**: 229–38.

132. Bergh U. Human power at subnormal body temperatures. *Acta Physiologica Scandinavica* 1980; **478**: 1–39.

133. Nielsen B. Metabolic reactions to cold during swimming at different speeds. *Archives des Sciences Physiologiques* 1973; **27**: A207–11.

134. Holmer I, Bergh U. Metabolic and thermal response to swimming in water at varying temperatures. *Journal of Applied Physiology* 1974; **37**: 702–5.

135. Jacobs I, Romet TT, Kerrigan-Brown D. Muscle glycogen depletion during exercise at 9°C and 21°C. *European Journal of Applied Physiology* 1985; **54**: 35–9.

136. Shephard RJ. Adaptation to exercise in the cold. *Sports Medicine* 1985; **2**: 59–71.

137. Martineau L, Jacobs I. Muscle glycogen utilization during shivering thermogenesis in humans. *Journal of Applied Physiology* 1988; **65**: 2046–50.

138. Dressendorfer RH, Morlock JF, Baker DG, Hong SK. Effect of head-out water immersion on cardiorespiratory responses to maximal cycling exercise. *Undersea Biomedical Research* 1976; **3**: 177–87.

139. Pendergast DR. The effect of body cooling on oxygen transport during exercise. *Medicine and Science in Sports and Exercise* 1988; **20**: S171–6.

140. McMurray R, Horvath S. Thermal regulation in swimmers and runners. *Journal of Applied Physiology* 1979; **46**: 1086–92.

141. Denison DM, Wagner PD, Kingaby GL, West JB. Cardiorespiratory responses to exercise in air and under water. *Journal of Applied Physiology* 1972; **33**: 426–30.

142. Davies M, Ekblom B, Bergh U, Kanstrup-Jensen I-L. The effects of hypothermia on submaximal and maximal work performance. *Acta Physiologica Scandinavica* 1975; **95**: 201–2.

143. Gutierrec G, Warley AR, Dantzker DR. Oxygen delivery and utilization in hypothermic dogs. *Journal of Vascular Research* 1986; **60**: 751–7.

144. Goldman RF, Breckenridge BS, Reeves E, Beckman EL. 'Wet' versus 'dry' suit approaches to water immersion protective clothing. *Aerospace Medicine* 1966; **47**: 485–7.

145. Hayward JS. Thermal protection performance of survival suits in ice-water. *Aviation, Space and Environmental Medicine* 1984; **55**: 212–15.

146. Tipton MJ, Stubbs DA, Elliott DS. The effect of clothing on the initial responses to cold water immersion in man. *Journal of the Royal Naval Medical Service* 1990; **76**: 89–95.

147. Froese G, Burton AC. Heat losses from the human head. *Journal of Applied Physiology* 1957; **10**. 235–41.

148. Pugh LGCE *et al.* A physiological study of channel swimming. *Clinical Science* 1960; **19**: 257–73.

149. Muza SR, Young AJ, Sawka MN, Bogart JE, Pandolf KB. Respiratory and cardiovascular responses to cold stress following repeated cold water immersion. *Undersea Biomedical Research* 1988; **15**: 165–78.

150. Skreslet S, Aarefjord F. Acclimatization to cold in man induced by frequent scuba diving in cold water. *Journal of Applied Physiology* 1968; **24**: 177–81.

151. Park YS *et al.* Time course of deacclimatization to cold water immersion in Korean women divers. *Journal of Applied Physiology* 1983; **54**: 1708–16.

152. Golden FStC, Hardcastle PT, Pollard CE, Tipton MJ. Hyperventilation and swim failure in man in cold water. *Journal of Physiology (London)* 1986; **378**: 94P.

153. Martineau L, Jacobs I. Muscle glycogen availability and temperature regulation in humans. *Journal of Applied Physiology* 1989; **66**: 72–8.

154. Young AJ, Sawka MN, Neufer PD, Muza SR, Askew EW, Pandolf KB. Thermoregulation during cold water immersion is unimpaired by low muscle glycogen levels. *Journal of Applied Physiology* 1989; **66**: 1809–16.

155. Martineau L, Jacobs I. Effects of muscle glycogen and plasma FFA availability on human metabolic responses in cold water. *Journal of Applied Physiology* 1991; **71**: 1331–9.

156. Kreider MB. Effect of diet on body temperature during sleep in the cold. *Journal of Applied Physiology* 1961; **16**: 239–42.

157. Wang LCH, Man SFP, Belcastro AN. Metabolic and hormonal responses in theophylline-increased cold resistance in males. *Journal of Applied Physiology* 1987; **63**: 589–96.

158. Vallerand AL, Jacobs I, Kavanagh MF. Mechanism of enhanced cold tolerance by an ephedrine–caffeine mixture in humans. *Journal of Applied Physiology* 1989; **67**: 438–44.

159. Keatinge WR, Evans M. Effect of food, alcohol, and hyoscine on body-temperature and reflex responses of men immersed in cold water. *Lancet* 1960; **ii**: 176–8.

160. Haight JSJ, Keatinge WR. Failure of thermoregulation in the cold during hypoglycaemia induced by exercise and ethanol. *Journal of Physiology (London)* 1973; **229**: 87–97.

161. Robinson S, Somers A. Temperature regulation in swimming. *Journal de Physiologie (Paris)* 1971; **63**: 406–9.

162. Modell JH. *Pathophysiology and treatment of drowning and near-drowning.* Springfield, Il: CC Thomas, 1971.

163. Golden FStC. Rewarming. In: Pozos RS, Wittmers LE, eds. *The nature and treatment of hypothermia.* University of Minnesota Press, 1983: 194–208.

2.3   High altitude

JAMES S. MILLEDGE

INTRODUCTION

The study of high altitude physiology and medicine is of interest to the student of sports medicine for a number of reasons. First as part of a general interest in human physiology, the response of the body to a reduction in oxygen availability is a fascinating example of adaptation. Oxygen being such a vital substrate for animal life, its lack affects every system of the body though some are more sensitive to it than others. The central nervous system may be the most sensitive in terms of irreversible damage but the musculo-cardio-respiratory system suffers in terms of function under even the mild hypoxia of modest altitude such as that experienced in Mexico City (2300 m above sea level).

Secondly altitude training for athletics is of great topical interest to all involved in coaching athletes and the questions of its efficacy, timing, duration, and optimum altitude are all of importance. Finally, with more and more people going to high altitude on trekking and climbing holidays the problems of mountain sickness are of interest to all involved including those doctors asked to advise on such activities.

The high altitude environment of course may include elements other than hypoxia which are deleterious. At extreme altitude cold is almost always a problem whilst at lower altitudes in many mountain regions gastrointestinal problems common to travellers worldwide may well be greater than those of hypoxia. These aspects of the mountain environment are dealt with in other chapters. This chapter therefore, is concerned only with the effects of hypoxia, particularly with chronic hypoxia of some days or longer. The hypoxia of some minutes is the realm of aviation medicine.

THE ATMOSPHERE

The composition of the atmosphere is the same all over the world (apart from local pollutants) and at all altitudes. Air contains 21 per cent of oxygen and 79 per cent of nitrogen in round figures with small quantities of argon, carbon dioxide, and other trace gases. The problem humans face at altitude is not strictly the shortage of oxygen but its low partial pressure which decreases exactly in line with the decreasing barometric pressure.

We live at the bottom of a sea of air and the barometric pressure is due to the weight of air above us. As we go up in altitude there is less air above and the pressure falls. If air were incompressible like water the fall in pressure would be linear with rise in altitude but because it is compressible the relationship is curvilinear. The relationship of pressure to altitude is further complicated by the fall in temperature with altitude. Finally there are local variations of pressure and temperature which complicate matters even more.

Figure 1 shows the relation of pressure to altitude according to the International Civil Aviation Organization (ICAO) Standard Atmosphere. This is a simplified model atmosphere in which certain assumptions are made regarding temperature, humidity, and pressure and their changes with altitude in order to arrive at a standard for the calibration of aircraft altimeters. Altimeters are barometers calibrated in height. Aviation authorities are more interested in making sure that all aircraft use the same calibration than in the actual pressure at any given height. There are other formulae for describing this relationship, for instance that of Zuntz which happen to fit the data from measurements of barometric pressure made in the Himalayas and Andes rather better than the ICAO model.[1] Up to about 5500 m there is little difference but at heights above this the difference becomes greater until at the height of the summit of Everest it becomes very important. The ICAO line would predict a pressure of only 236 mmHg. In using the Zuntz formula one must guess the temperature. Pugh calculated a pressure of 250 mmHg assuming various temperatures at various heights based on measurements made in the

mountains.[2] When it was actually measured on a day of good weather in October 1981 the pressure on the summit of Everest was found to be 253 mmHg.[3]

This difference of about 17 mmHg may seem small but at that altitude it is equivalent to about 470 m of height. Climbers not using supplementary oxygen reach the summit with a rate of climb of less than 50 m (vertical) per hour, so it is unlikely that Everest would ever have been climbed without oxygen had the pressure been only 236 mmHg. The difference between the Standard Atmosphere and the pressures measured in the mountains is greatest near the equator and least near the poles. For a full discussion of this phenomenon see ref. 3.

Pressure is highest in summer and lowest in winter. Figure 2 shows this seasonal variation based on measurements made above Delhi with weather balloons. The January figure of 243 mmHg means that Everest is, physiologically speaking, about 330 m higher in the winter than in the summer which, together with the cold and wind, makes Sherpa Ang Rita's winter ascent without oxygen in December 1987 even more remarkable.

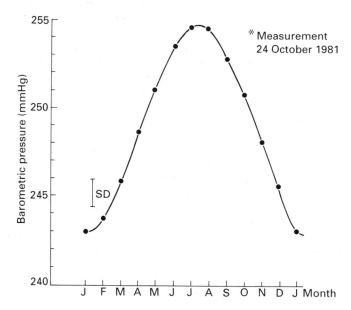

Fig. 2 Mean monthly pressures for 8848 m (the height of Mount Everest) from weather balloons released from New Delhi. The pressure measured on the summit of Everest (*) also shown. (Reproduced from ref. 57 with permission.)

THE PHYSIOLOGICAL RESPONSE TO HYPOXIA

The response of the body to hypoxia depends upon the rate and degree of the hypoxic stress. For instance the effect on the pilot of sudden loss of cabin pressure in an aircraft at 9000 m is quite different from the effect of a similar altitude on a climber who has spent some weeks at altitude. The pilot would lose consciousness in a few minutes (Fig. 3) whereas the climber not only remains conscious but is able to work out his route and climb upward if rather slowly. Perhaps the surprising thing about hypoxia is how little it is felt at rest. Indeed this is why it is so dangerous for an aviator. If at high altitude his oxygen line becomes disconnected the symptoms are so subtle that without training he is likely to lose consciousness before realizing his danger. The time to uncon-

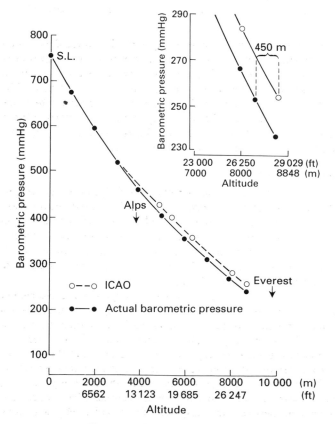

Fig. 1 The relationship of barometric pressure and altitude. The upper curve is calculated from the Zuntz formula assuming a mean air column of 15°C.[1] The lower curve is from the IOAC standard atmosphere. The plotted points are from observations made on the ground in the Himalayas and Andes (after Pugh[2]).

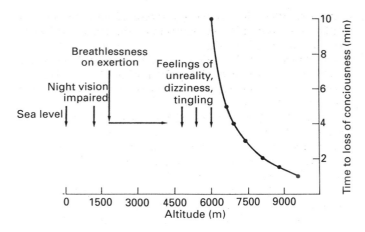

Fig. 3 Effect of sudden exposure to various altitudes on an unacclimatized subject.

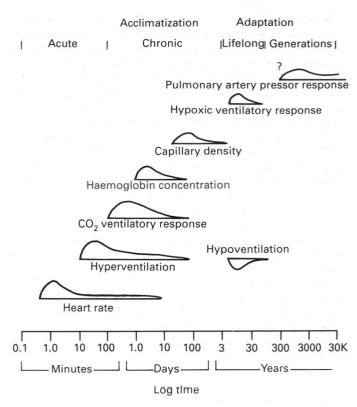

Fig. 4 Time courses of a number of acclimatization and adaptive changes plotted on a logarithmic time-scale, the curve for each response denoting the rate of change, fast at first then tailing off.

sciousness decreases with increasing altitude as shown in Fig. 3. At the height of the summit of Everest this is only about two minutes.

By contrast the effect of hypoxia on an acclimatized person is much less. The main symptom is of breathlessness on exertion. The work rate used by climbers walking up hill at their preferred speed (at sea level) is about half maximum work rate. At sea level a ventilation of about 50 l/min is required whereas at 6300 m the rate is 160 l/min, close to the maximum voluntary ventilation. Below this altitude the climber adopts a discontinuous pattern of climbing with pauses for breath. The difference between the pilot and the climber is due to a series of adaptive changes in the body known as acclimatization. These changes occur in various systems and with varying time courses. Figure 4 illustrates the futility of the question 'How long is required for acclimatization?'. However, the most important changes are in the blood and cardiorespiratory systems with a time course of days or weeks. These processes are described, as far as they are understood, in the next few sections.

Respiratory acclimatization

One of the most important mechanisms of acclimatization is the increase in ventilation which results in as increase in Pa_{O_2} and a decrease in Pa_{CO_2}. Thus the decrease in inspired P_{O_2} due to lower barometric pressure is partially countered. This unconscious increase in breathing is brought about by changes in the chemical control of ventilation.

Changes in the chemical control of ventilation

Hypoxia

Breathing is stimulated by both carbon dioxide and hypoxia. Normally at sea level the hypoxic stimulus to breathing is small. If the carotid bodies are removed the ventilation falls by about 15 per cent so that Pa_{CO_2} rises from the normal of 40 to 46 mmHg (5.3–6.1 kPa), only just outside the normal range. However, if acute hypoxia is imposed by breathing a hypoxic mixture there is normally an increase in ventilation especially if CO_2 is added to the inspired mixture to prevent the Pa_{CO_2} from falling. This is called the hypoxic ventilatory response and is linear with respect to arterial oxygen saturation (Sa_{O_2}) and hyperbolic with respect to Pa_{O_2}.

People born and bred at high altitude are found to have blunted hypoxic ventilatory response compared with lowlanders. Their performance at altitude does not seem to be in any way impaired by this; on the contrary they seem less prone to acute mountain sickness and to function very well at extreme altitude. Presumably, having developed more fundamental changes perhaps at the tissue level, they do not need the emergency response of hyperventilation that is useful to the lowlander. It is this late adaptation in hypoxic ventilatory response which is shown in Fig.4.

Carbon dioxide

If a subject is given a small percentage of CO_2 in the inspired gas, the alveolar and arterial P_{CO_2} increases, the central chemoreceptors in the brain-stem are stimulated, and the minute ventilation increases. Further increases in the inspired P_{CO_2} cause further increases in ventilation. This CO_2-stimulated increase in ventilation is called the CO_2 ventilatory response. After an initial, more flat portion the response is linear.

Figure 5 shows the results of such an experiment conducted at sea level and repeated at altitude. On going to altitude the response line shifts to the left. About half the eventual shift takes place in the first day at altitude and it takes 2 to 3 weeks to complete the adaptive response. For a discussion of the mechanisms underlying these changes in control of breathing see ref. 4.

The net results of these changes is that the breathing is set at a higher level at both rest and during exercise. This reduces the difference between the inspired and alveolar P_{O_2} and thus mitigates the effects of reduced barometric pressure.

Lung diffusion

The next step in the oxygen transport system after the alveolus is transfer of the oxygen from air into blood. This is by diffusion. The

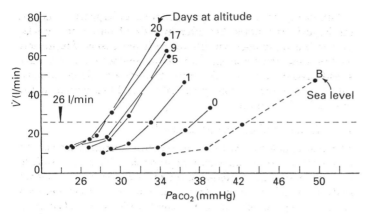

Fig. 5 Effects of acclimatization at 4340 m on the CO_2 ventilatory response. The numbers against each response line indicate the days at altitude. (After Kellogg.[58])

diffusing capacity of the lung does not increase as part of the acclimatization process, apart form a small increase due to the increased haematocrit.[5] This is unfortunate because the diffusing capacity is probably an important limiting factor to exercise at altitude. Highlanders born and reared at altitude appear to have higher lung diffusing capacities.[6]

The cardiovascular response

Heart rate

On exposure to hypobaric oxygen the heart rate is increased both at rest and during exercise. With acclimatization the resting heart rate falls. At altitudes below about 4500 m it falls to within the sea level range, and above this altitude resting heart rates remain modestly elevated. At submaximal work rates a similar pattern is seen. Figure 6 shows the heart rate at a fixed work rate in an individual at sea level and at intervals after arrival at altitude. In the fully acclimatized subject at altitudes up to 4300 m the heart rate for the same absolute work rate is the same as at sea level (or lower as in Fig. 6 if the study induces a training effect).

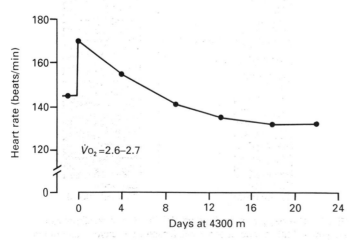

Fig. 6 The heart rate at a fixed work rate at sea level and at various days after arrival at altitude in one subject. (After Astrand and Astrand.[59])

At higher altitudes the heart rates are higher for low work rates but equal to or even lower for high work rates. Figure 7 shows this effect. It should also be noted that since maximum work rate is reduced and since heart rate is related to absolute work rate, the maximum exercise heart rate is progressively reduced with increasing altitude. The maximum heart rate at an altitude equivalent of the summit of Everest was found to be only 118 beats/min compared with 180 to 200 beats/min at sea level.[7] This limitation of maximum heart rate is surprising since an increase in rate (and cardiac output) would increase oxygen supply to the working muscles. Presumably it protects the myocardium from possible hypoxic damage.

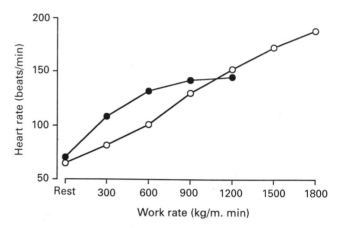

Fig. 7 Effect of increasing exercise on heart rate at sea level (○) and at an altitude of 5800 m (●) in one subject. (Redrawn from ref. 60.)

Cardiac output, stroke volume, and contractility

The early changes in cardiac output on exposure to hypobaric oxygen mirror the changes in heart rate and presumably reflect sympathetic activity. After acclimatization the cardiac output is the same as at sea level for any given work rate in absolute terms.[7,8] It follows that after acclimatization the stroke volume at moderate altitudes is unchanged from that at sea level but is reduced at altitudes where the heart rates are elevated.

Indices of contractility, left and right filling pressures, and left ventricular ejection fraction were found to be well maintained even on exercise up to an equivalent altitude of 8000 m on Operation Everest II.[7]

Systemic blood pressure

In contrast to some animals, altitude hypoxia results in no increase in systemic blood pressure. Indeed some studies report a reduction in blood pressure whilst others found no change.[9] The rise in blood pressure with exercise is more pronounced than at sea level (Fig. 8) but the change is much less than for the pulmonary circulation.

Pulmonary hypertension

One of the most striking cardiovascular changes found at high altitude is pulmonary hypertension. This hypoxic pulmonary pressor response has been much studied since it was first reported in man by Motley et al.[10] It can be demonstrated in isolated lung preparations so is thought to be due to the release of mediators in the lung rather than any neurological mechanism. Despite so

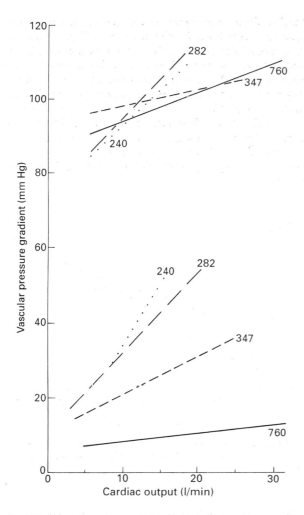

Fig. 8 Effect of increasing cardiac output during exercise on the mean pulmonary artery minus wedge pressure (lower) and on systemic artery minus right atrial pressure (upper) at sea level (760 mmHg) and at two higher equivalent altitudes. Hypoxia has a marked effect on the pulmonary but not on the systemic circulation. (Reproduced from ref. 61 with permission.)

much research the actual mediator of this response is still unknown. Probably the main role of the response is in the fetus. Here the hypoxia is similar to that in a man on the summit of Mount Everest and the response constricts the pulmonary circulation, diverting blood via the ductus to the systemic circulation. After birth the response may also be important in diverting blood away from small areas of lung consolidation or collapse, thus maintaining better matching of blood flow and ventilation.

At altitude, with global hypoxia, the pressor response results in a raised pulmonary artery pressure with very little benefit to the individual. Indeed it merely puts strain on the right ventricle and may well be important in the genesis of acute pulmonary oedema of high altitude. It is of interest that the yak, an animal well adapted to high altitude, has little or no hypoxic pressor response unlike lowland cattle which have brisk responses.[11]

Most studies have looked at animals or man at rest, but probably the importance of this response is that on exercise the pressure goes up very significantly. Figure 8 shows the rise in resistance with increasing cardiac output during progressive exercise. At sea level there is very little rise in resistance, whereas at 8000 m there is a far greater rise with a smaller increase in cardiac output.

Pulmonary hypertension reverses in acute hypoxia as soon as the hypoxia is relieved but after hypoxia of only a few weeks, oxygen breathing does not relieve the hypertension. Presumably by this time structural changes in the pulmonary artery have already taken place. People native to high altitude show muscularization of the pulmonary arterial tree and hypertrophy of the right side of the heart.[12]

The electrocardiogram at altitude

The main changes in the ECG are due to the increased load on the right ventricle from pulmonary hypertension. There is a shift to the right and to the posterior in the QRS vector.[13] The standard leads show this right axis deviation, and the T wave becomes inverted progressively across the chest leads. These changes in general become more pronounced the higher the altitude and are seen in both low- and highlanders. They do not indicate cardiac dysfunction, nor do the inverted T waves have the sinister implication of ischaemia that they might have at sea level.

Haemoglobin and haematocrit

Probably the best known effect of altitude is the increase in haemoglobin concentration in the blood of people and animals at high altitude. This is illustrated in Fig. 9 where it can be seen that, despite a reduced oxygen saturation, the oxygen content of the blood is maintained by increase in the haemoglobin concentration and therefore in the oxygen carrying capacity. The mechanism for this increase is due to a rise in the concentration of erythropoietin due to hypoxia which stimulates the bone marrow to produce more red blood cells.

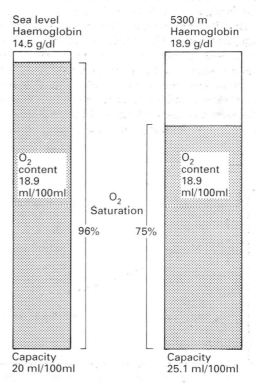

Fig. 9 The oxygen content of arterial blood in an acclimatized subject at 5300 m and at sea level. Despite the oxygen saturation falling to 75 per cent the oxygen content is the same as at sea level because of the increase in haemoglobin and therefore oxygen capacity.

Figure 10 shows the time course of erythropoietin levels in a group of climbers going quite rapidly to altitude, together with the ascent profile and the haematocrit. However it is clear that the initial rise in haematocrit is mainly due to a reduction in plasma volume and that the erythropoietin-stimulated increase in red cell mass takes much longer.

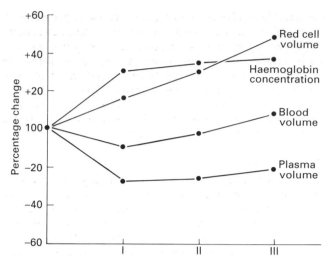

Fig. 11 Changes in haemoglobin concentration, red cell volume and plasma volume in four subjects; I after 18 weeks at between 4000 and 5800 m; II after a further 3 to 6 weeks at 5800 m; III after a further 9 to 14 weeks at or above 5800 m. (After Pugh[14].)

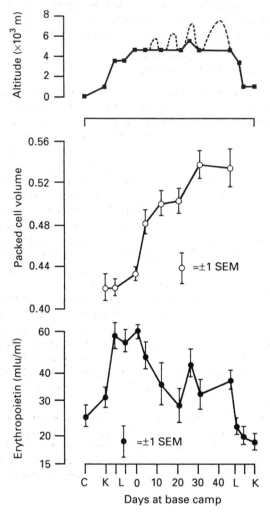

Fig. 10 The effect of ascent to altitude on the serum erythropoietin concentration (lower panel). Upper panel shows the altitude/time profile of ascent; the dotted line indicating ascents above base camp between sampling. The sample at 30 days was taken at 5500 m. The centre panel shows the packed cell volume (PCV). (C) control, sea level, (K) Kashgar, 1200 m, (L) Karakol lakes, 3500 m. (Reproduced from ref. 62 with permission.)

Figure 11 shows the time course of the change in plasma volume, red cell mass, and blood volume. The rise in haemoglobin concentration levels out after about 6 weeks. At altitude the red cell mass is still increasing after 36 weeks and was 67 per cent higher than sea level values. By this time the blood volume was 23 per cent above sea level values when corrected for the reduction in body weight.[14] Similar or greater red cell masses are found in people resident at high altitude.

There is some debate as to the importance of this mechanism in acclimatization. It has been argued that the evolutionary importance of this system is to deal with the problem of blood loss. The fact that altitude hypoxia also activates the mechanism is fortuitous but adds little benefit for the individual. It is pointed out that increase in haemoglobin concentration increases the viscosity of blood and may reduce cardiac output. Further it is argued that some high altitude populations have very little increase in haemoglobin concentration, yet seem to perform well, and that within any group of climbers on an expedition there is no correlation between climbing performance and haemoglobin concentration. These considerations led Winslow and Monge to state that 'Excessive polycythaemia serves no useful purpose. Indeed, it is doubtful whether there is any physiological value in "normal" polycythaemia.'[15]

However, it is becoming clear that an increase in haemoglobin concentration is an advantage in some athletic events even at sea level so there can be little doubt that some increase is beneficial at altitude.

The endocrine response

The renin–aldosterone system

Aldosterone is probably the most important hormone in the regulation of sodium. It acts on the kidney causing retention of sodium. Exercise is a stimulus to renin and aldosterone release. If exercise of some hours duration is taken, significant quantities of sodium are retained, especially if continued for a few days.[16]

Altitude hypoxia has a variable effect on renin levels but in all studies of resting subjects aldosterone levels have been found to be reduced. With exercise the response of aldosterone to the rise in renin is blunted. Subjects who develop acute mountain sickness have higher aldosterone levels on arrival at altitude (and more sodium retention) than subjects who are resistant to acute mountain sickness.

Atrial natriuretic peptide

Atrial natriuretic peptide (ANP) is secreted by the atria of the heart in response to stretching. Its effect on the kidneys is to promote a sodium diuresis. It is thought to play an important part in maintaining the constancy of the plasma volume in the face of a salt or water load. Hypoxia results in an increase in plasma levels of ANP probably by both a direct effect on the heart and secondarily by causing an increase in pulmonary artery pressure and hence in right atrial pressure. Exercise also causes an increase in ANP levels so that it would be expected that subjects climbing to altitude would have greatly increased levels. However, those reported are of only a modest rise.[17,18] The relationship with acute mountain sickness is an open one with some studies showing a positive and some an inverse correlation.

Antidiuretic hormone

Altitude hypoxia in itself seems to cause no change in the level of the antidiuretic hormone in those free of acute mountain sickness. In those who become ill levels are occasionally found to be elevated especially if subjects actually vomit or have developed pulmonary oedema.[19] In this case the rise is thought to be the effect of sickness rather than the cause. After acclimatization at 6300 m the antidiuretic hormone level was low in the face of high osmolality, suggesting a reduced response of this system.[20]

Corticosteroids

On ascent to higher altitude there is stimulation of the adrenal cortex by ACTH with secretion of cortisol as part of a generalized stress reaction. This response wanes over 5 to 7 days. After some weeks at altitudes up to 6300 m, cortisol levels and the response to ACTH is unchanged from response at sea level.

Insulin and glucose control

Acute hypoxia causes a rise in glucose of about 1.7 mmol/1 followed by a fall towards control levels over about a week. In acclimatized subjects fasting glucose tends to be lower than at sea level and insulin sensitivity is increased (as it is by athletic training).

The central nervous system and hypoxia

Psychomotor performance at altitude

Acclimatized subjects seem to perform well even at altitudes up to 6300 m. On the 1953 Everest expedition Bourdillon completed *The Times* crossword puzzle in the Western Cwm (6300 m). However climbers find that any task, mental or physical, requires a much greater effort of will to start and complete. Sensitive tests of psychomotor function will detect diminution in function. At an altitude as low as 1524 m a test that depends on learning a novel task was found to show lower scores in subjects breathing air compared with subjects breathing oxygen.[21] A test of hand–eye co-ordination, moving a stylus along a groove as fast and accurately as possible, showed deterioration at 4000 m even after 10 months though there was some improvement after 25 months.[22] At higher altitudes although simple well-learnt tasks can be carried out quite well, tests will bring out deficiencies in concentration, speed, and dexterity.

Psychomotor function after altitude exposure

Of equal interest as performance at altitude is the question of residual impairment after return to sea level. Anecdotally it has been pointed out that many of those pioneer climbers of prewar Everest expeditions who climbed above 8000 m without oxygen went on to achieve distinguished careers and lived to a ripe old age with mental faculties better than average, Professor Odel and Dr T. H. Somervell being outstanding examples. However, studies which have addressed this question rigorously have found evidence of decrease in some aspects of psychomotor performance after return to sea level. Even 1 year after return from altitude, Townes *et al.* found that a finger tapping test and one out of four tests of memory was significantly reduced compared with scores before the expedition (two other aspects of memory were impaired immediately on return to low altitude but recovered within a year).[23] These findings was confirmed in Operation Everest II, a chamber experiment, thus excluding the possibility that results could have been due to such factors as cold or dehydration.[24]

Weight loss and anorexia

Anorexia and weight loss are features of life at high altitude. In the first few days at altitude anorexia, nausea, and vomiting are likely to be part of the syndrome of acute mountain sickness. After a few days acute mountain sickness passes and below about 4500 m appetites are regained. However, above about 5500 m most people complain of anorexia which can worsen the longer they spend at these altitudes. Above 6500 m anorexia is almost universal and weight loss common. No doubt reduced calorific intake plays a part in the cause of this weight loss. However, energy expenditure must be reduced at these altitudes since V_{O_2}max is reduced, climbing rates are slowed, and even simple tasks of daily living are carried out slowly. When climbers have kept diaries of food eaten, intakes of 2200 to 3000 kcal have been estimated which should have been adequate to maintain body weight, yet weight was lost at about 500 g per week. The weight chart of one well-acclimatized subject is shown in Fig. 12. Most of the time was spent at 5800 m where weight was lost at just under 400 g/week. Descent to 4500 m on two occasions for only a few days resulted in weight gain. Arterial oxygen saturation would have been below 70 per cent at the higher altitude and above 70 per cent at the lower. This seems to be the crucial degree of hypoxaemia needed to produce continued weight loss.

There are probably a number of causes for weight loss above about 5500 m. There is evidence that this degree of hypoxia causes some small bowel malabsorption. Climbers report that their stools tend to be greasy at very high altitude and patients with this degree of arterial desaturation due to congenital heart disease or severe

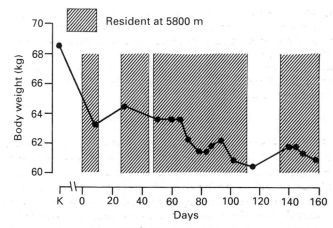

Fig. 12 Weight chart of one subject on the Silver Hut Expedition 1960–61. After the march out from Katmandu (K) and the initial period of preparation he was in residence at 5800 m (hatched areas) or at base camp, 4500 m. Note the loss of weight at 5800 m but weight gain during two breaks at 4500 m.

lung disease had reduced xylose absorption which was improved when their hypoxia was corrected by surgery or oxygen breathing.[25] Boyer and Blume found in climbers at 6300 m that their xylose absorption was less than at sea level and their faecal fat was increased.[26] There is also some evidence of a reduction in muscle protein synthesis at altitude.[27]

Diet at altitude

Views on diet (not only at altitude) are strongly held, and often the strength of opinion is inversely related to the weight of scientific evidence. Climbers at altitude often develop a preference for high carbohydrate/low-fat diets. There is a good physiological reason for this since such a diet results in an increase in the respiratory quotient. This in turn means that for a given P_{CO_2} (or ventilation) the P_{O_2} will be higher. There is also the possibility that fat is less well absorbed than carbohydrate at altitude. In a study at 4300 m it was found that subjects on a high carbohydrate diet had greater exercise endurance than subjects on a normal diet.[28]

There is no evidence that extra vitamins or minerals are beneficial if subjects are on a good mixed diet.

The major dietary problem at high altitude is anorexia. The sense of taste seems to be dulled and a craving for strong flavours, spicy foods, and more sugar develops. Fresh food, after a period on preserved food, is also much desired. Every effort should be made to provide for these wishes on expeditions in order to keep up energy intake and lessen weight loss.

Peripheral tissues

Capillary density

It would clearly be an advantage to increase the number of capillaries per unit of tissue volume since the intercapillary distance would be decreased and the pathway for oxygen shortened. Increased vascularization has been reported in the brain, muscles, and liver of animals at high altitude. However, at least in the case of the muscles, the increased capillary density is probably the result of reduction in muscle fibre diameter. This in turn may be due to a combination of disuse, negative calorie balance, and reduced protein synthesis. The result is a loss of muscle bulk but with the same number of capillaries and therefore more capillaries per unit area.[29]

Muscles

Although there is reduction of muscle fibre size at altitude which must reduce muscle power, myoglobin concentration is increased in the muscle cell in the same way that haemoglobin concentration is increased in the blood. Men native to high altitude had a myoglobin concentration 16 per cent higher than lowlanders.[30] Mitochondria have been reported to be increased in some animals but not in others or in man.[31] These changes, together with an increase in capillary density, will all help to mitigate the effect of arterial hypoxaemia on oxygen transport in muscles.

Intracellular enzymes

A number of the enzymes essential to energy production from glucose and oxygen have been studied in muscle biopsies in men at high altitude. It seems that at intermediate altitudes (up to about 4500 m) levels of enzymes in the Krebs' cycle are increased, but at more extreme altitudes (above 6000 m) they are reduced.[32] The changes at intermediate altitude are similar to those resulting from endurance training and lend support to the idea that in training the changes are caused by hypoxia in the exercising muscles.

ATHLETIC PERFORMANCE AT ALTITUDE

The effect of altitude on athletic performance depends upon the type of sport. The reduction in oxygen partial pressure with altitude increases fatigue and reduces $V_{O_2}max$. This will reduce performance in sports that depend upon aerobic capacity. Thus times for middle and long distance running events will be increased but sports that use mainly anaerobic capacity such as weight lifting will be unaffected below extreme altitudes. Sports such as sprinting and throwing events will show increased performance because of the decrease in air density, which parallels the reduction in barometric pressure. These expected effects were born out by the results in the 1968 Olympics in Mexico City (2300 m) shown in Table 1 and Fig. 13.

Altitude and $V_{O_2}max$

Figure 14 shows the reduction of $V_{O_2}max$ with altitude. At 2300 m (the height of Mexico City) $V_{O_2}max$ is reduced to 84 per cent of the sea level value. Thereafter $V_{O_2}max$ decreases with increasing altitude so that at the summit of Everest it is 25 per cent of the value at sea level. With acclimatization there is an increase again in $V_{O_2}max$ which in one study was shown to be about 10 per cent from the first to the fourteenth day at 4300 m.[33]

Altitude and fatigue

There is a strong impression that in exercise that involves work at rates below maximum, endurance is reduced at altitude. Endurance of this sort is much more difficult to study than $V_{O_2}max$ since

Table 1 Olympic Games—Mexico City 1968

Event	Winner	Time/distance	Comment
Long jump	Beamon	8 m 90 cm	World record
Pole vault	Seagren	5 m 40 cm	World record
100 m	Hines	9.95 s	World record
200 m	Smith	19.83 s	World record
400 m	Evans	43.86 s	World record
800 m	Doubell	1 min 44.3 s	World record
1500 m	Keino	3 min 39.9 s	Olympic record
3000-m steeplechase	Biwott	8 min 51 s	4% slower than world record
10 000 m	Temu	29 min 27 s	7% slower than world record
Marathon	Wolde	2 h 20 min 26 s	8.5% slower than world record

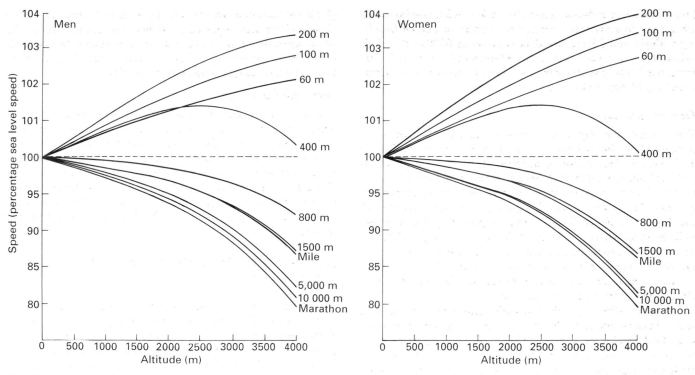

Fig. 13 Theoretical changes in running speeds over various distances, with changes in altitude. Note that the scales are different above and below 100 per cent sea level speed. (Reproduced from ref. 63 with permission.)

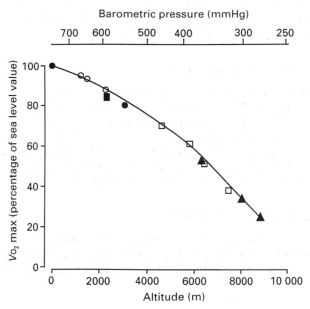

Fig. 14 The reduction in V_{O_2}max at altitude as a percentage of the sea level value. The first three points are from a chamber study (unacclimatized subjects);[64] other points are from acclimatized lowland subjects. The 2300 m point is the mean of five studies,[41,65–68] 3100 m point;[68] open triangles;[60] closed triangles.[57]

roughly the same rate in absolute terms, that is at progressively higher percentage of the prevailing V_{O_2}max, as altitude increases. The endurance time under these conditions of course is much reduced. Thus, an athlete attempting his normal training programme will find he becomes fatigued much more easily at altitude.

Altitude training

The idea that training at high altitude for competitions to be held at low altitude may be beneficial stems from two sources. First it has been observed that a number of very successful middle and long distance runners came from areas of East Africa at altitudes of 1500 to 2000 m and that their life and training there might be part of the reason for their success. Secondly it is presumed that there is benefit in developing an increased haemoglobin concentration or haematocrit as a result of altitude hypoxia. The advantage of this increase achieved by blood transfusion (blood doping) is now well established (see Chapter 3.3.1) but is illegal.

There may well be some advantage in being born and bred at altitude in respect of these aerobic events although it is not clear exactly which physiological parameters are important. It has been suggested that such people have a greater density of capillaries in their muscles. On the one hand this can also be achieved by training. On the other hand altitude hypoxia, in lowlanders at least, causes a loss in muscle mass and the increase in capillary density is merely due to reduction in diameter of muscle fibres. In any case lowland athletes cannot achieve the effect of a lifetime of altitude exposure in a few weeks or months of altitude training. Highlanders also probably have higher pulmonary diffusing capacities and possibly greater ventilatory capacity.

The second reason, to increase haematocrit, is probably uppermost in the minds of those advocating altitude training. The

it is so dependant upon motivation. For work at the same percentage of V_{O_2}max there was a 10 per cent reduction in endurance time on the first day after arrival at 4300 m but this was not statistically significant.[33] With acclimatization there was significant improvement over the first day. In real-life tasks are often attempted at

Table 2 The possible advantages and disadvantages of training at altitude for events held at low altitude

Advantages	Disadvantages
Increased haemoglobin concentration (Hct)	Risk of acute mountain sickness
Increased red cell mass	But only after months at altitude
Increased ventilatory capacity	Reduced V_{O_2}max
Increased capillary density	Reduced training intensity
	Reduced plasma volume

possible benefits of altitude training of a few weeks' duration are shown in Table 2 which also shows the disadvantages of training at altitude.

Haematocrit increase at altitude

Hypoxia stimulates the release of erythropoietin, which in turn stimulates the bone marrow to produce more red cells. However this process takes place over a number of months rather than days. The initial rise in haematocrit is almost entirely due to a reduction in plasma volume, i.e. haemoconcentration. This is an advantage to the acclimatizing subject in that it achieves a rapid increase in the oxygen-carrying capacity of the blood. However, the reduction in blood volume as a consequence may be part of the reason for the reduction in maximum cardiac output found at this stage of altitude exposure. More important for the athlete who is contemplating altitude training for low altitude competition, is that just as plasma volume can be reduced rapidly on going to altitude, it can be increased as rapidly on coming down and the advantage in terms of haematocrit is lost in a few days.

In a person resident at altitude for months or years the red cell mass is increased by as much as 50 per cent of its normal sea level value (Fig. 11) When such a person comes down to low altitude he retains this increased red cell mass for some weeks and this can be of advantage in middle and long distance running events.

Other advantages of altitude training

The other advantages of altitude training listed in Table 2 are rather speculative. The fact that exercise requires much greater ventilation at altitude will mean that the muscles of respiration will be more stressed and therefore possibly the respiratory capacity may be increased, although there is no evidence for this. Training under hypoxic conditions may increase the local effects of training in the working muscles, i.e. their capillary density or metabolism, in some way. This possibility was explored by Davis and Sargeant by having their subjects train one leg under normobaric oxygen and the other under hypobaric oxygen conditions.[34] There was no difference in the effect of training under these two conditions. A recent study of well-trained athletes who trained further at altitude for 2 weeks, found no increase in V_{O_2}max but did find some improvement in short-term running time.[35] Unfortunately this study had no control group so the significance of the last finding is uncertain.

Disadvantages of altitude training

The risk of acute mountain sickness is a possibility (see below). At best this will mean an interruption in training and at worst may mean a life-threatening illness and the need for evacuation to low altitude.

The reduction in maximum work capacity (V_{O_2}max) and the earlier onset of fatigue at altitude mean that training has to be less intensive than is possible at sea level.

Is altitude training worth it?

The short answer to this question is that at present it is not known. There have been few good controlled trials. This is not surprising in view of the difficulty of conducting such trials and the large number of variables of altitude, time, and training schedules. Those trials which have been carried out have given conflicting results. Earlier studies which mostly claimed to show that training at altitude was advantageous were not controlled.[36,37] The improved performance could have been simply due to training at any altitude. Roskamm et al. did use controls and found that subjects trained at 2250 m improved their V_{O_2}max by 17.5 per cent compared with 6.4 per cent in subjects trained at sea level; for subjects at 3450 m the improvement was 10 per cent.[38] The advantage for the altitude group was significant. Hanson et al. also had sea level controls and starting with unfit subjects (V_{O_2}max < 40 ml/kg.min) found no advantage in training at an altitude of 4300 m.

In well-trained subjects the picture is also unclear. Dill and Adams found an increase in V_{O_2}max in high school champion runners after 17 days at 3090 m but used no control sea level group.[40] A well-controlled study by Adams et al. using a crossover design in experienced trained athletes (V_{O_2}max 73ml/kg.min average) showed no significant difference in performance (tested at sea level) between altitude (2300 m) and sea level training.[41] Each leg of the study was 3 weeks' duration of intensive training.

Conclusion from trials

Generally the better controlled trials show no effect of altitude in enhancing the effect of training. If there is a small effect it is probably best to train at modest altitude, i.e. 2300 m rather than higher, although it is possible that residing at about 2000 m and going higher for training might be more beneficial.

Other aspects of altitude training

Optimum duration of stay at altitude is also unclear and there are no trials to guide us. It is possible that repeated trips to altitude may be more beneficial than a single period but there is no firm evidence for this.

It is also unclear how long before an important event the athlete should come down to low altitude. Most subjects after a prolonged period at high altitude feel rather 'slack' for a few days after coming down. Advocates of altitude training seem to agree on a minimum of 2 to 3 days but some suggest 14 to 21 days.[42]

Acute mountain sickness

Acute mountain sickness can be defined as a condition affecting previously healthy individuals who ascend rapidly to high altitude. There is a delay of a few hours to 2 days before symptoms develop. It is characterized by headache (usually frontal), nausea, vomiting, irritability, malaise, insomnia, and poor climbing performance. In the simple or benign form the condition is self limiting, lasting 3 to

5 days. After this time it does not recur at that given altitude though it may do so if the subject goes higher. In a small proportion of individuals there may be progression to the malignant forms of acute mountain sickness, i.e. high altitude pulmonary oedema or cerebral oedema or a mixed form of these two. These conditions if not treated are frequently fatal in a matter of hours.

Chronic mountain sickness

There is also a condition known as chronic mountain sickness which is quite different. It affects residents of high altitude and consists of extreme polycythaemia, with haemoglobin concentrations up to 23 g/dl. The victims become slow mentally and physically and complain of such symptoms as headaches, dizziness, somnolence, fatigue, and difficulty in concentration. On going down to sea level symptoms clear and the polycythaemia disappears only to recur on return to high altitude. Venesection (at altitude) helps to reduce symptoms.

Incidence of acute mountain sickness

The incidence of acute mountain sickness depends upon altitude and the rate of ascent. With the increased accessibility of high altitude resorts and the possibility of getting into high mountains in a very few days the incidence of acute mountain sickness is probably greater than in the more leisurely days of old. A recent survey in alpine huts showed an incidence of 9 per cent at 2850 m, 13 per cent at 3050 m, 34 per cent at 3650 m, and 53 per cent at 4559 m.[43] Amongst trekkers on the way to Everest base camp an incidence of 43 per cent was found at 4300 m and was higher in those who had flown to an airstrip at 2800 m than in those who had walked all the way (49 versus 31 per cent).[44]

Risk factors for acute mountain sickness

Clearly hypoxia of more than a few hours duration is required for the development of acute mountain sickness and speed of ascent is important but there is great variation in susceptibility. Amongst a group of people travelling together to high altitude there will be some unaffected, some mildly, and some severely affected. At present there is no way to predict who is susceptible except that past performance at altitude is a guide.[45] As Ravenhill observed in 1913 'There is in my experience no type of man of whom one can say he will or will not suffer from puna (the South American term for acute mountain sickness). Most cases I have instanced were young men to all appearances perfectly sound. Young, strong and healthy men may be completely overcome. Stout, plethoric individuals ... may not even have a headache'.[46] People of all ages, both male and female, seem to be equally affected. Fitness is no protection; indeed, because the fit are likely to ascend faster, they may be at greater risk. Any respiratory infection is probably a risk factor and may account for the subject who has previously acclimatized well having trouble on one occasion. A brisk hypoxic ventilatory response would be expected to be protective and some studies seem to support this but others do not.[47] Certainly high altitude residents who have a blunted hypoxic ventilatory response are less prone to acute mountain sickness than lowlanders. A brisk pulmonary artery pressor response to hypoxia may well be a risk factor for high altitude pulmonary oedema.

Mechanisms of acute mountain sickness

Although hypoxia is obviously the starting point in the genesis of acute mountain sickness it is not the immediate cause of the symptoms since these are delayed by several hours after arrival, whereas hypoxia is most severe in the first few minutes. It seems that hypoxia sets in train a mechanism which in turn produces symptoms after a few hours. The symptoms are the same as those associated with raised intracranial pressure as seen on neurosurgical wards and at least in cases of high altitude cerebral oedema there is good evidence of increased intracranial pressure. The most popular view is that even in simple acute mountain sickness there is a degree of cerebral oedema (and often subclinical pulmonary oedema) which causes the symptoms of acute mountain sickness. There is frequently dependent or periorbital oedema also. All this points to some disturbance of fluid balance or capillary permeability throughout the body.

Other factors which should be accommodated in an overall scheme include ventilation, since it has been shown that those with the highest P_{CO_2} on arrival at altitude (who have failed to lower their P_{CO_2} by hyperventilation) are likely to go on to develop acute mountain sickness. Exercise is thought to be a risk factor although this is yet to be rigorously tested. The effect of prolonged exercise of the type taken by mountaineers is to cause retention of sodium and water through activation of the renin–aldosterone system.[48] This may place subjects at risk of acute mountain sickness.

Prevention of acute mountain sickness

A slow rate of ascent is the best way to prevent acute mountain sickness. A suggested rule is that above 3000 m ascent should be not more than 300 m a day with a 'rest' day, when no height gain is made, every 3 days. But even this rate will be too fast for some and unnecessarily slow for others. An added rule must be 'If symptoms of acute mountain sickness develop, go no higher. If they become severe, go down'.

Acetazolamide and dexamethasone

In real life it is often impossible or impracticable to plan for this rate of ascent. There may be no camp site between the valley floor and a pass 500 m higher. Some who know themselves to be slow acclimatizers do not wish to delay their companions. In these situations it is justified to recommend the use of acetazolamide. This drug, which is a carbonic acid anhydrase inhibitor probably acts as a respiratory stimulant. It has been shown to decrease $P_{a_{CO_2}}$ thus giving a sort of artificial respiratory acclimatization. There have been a number of good double-blind controlled trials which have shown that acute mountain sickness is reduced in those taking the drug.[49] Of course acute mountain sickness and even high altitude pulmonary oedema or high altitude cerebral oedema is still possible whilst taking the drug and common sense is needed. The dose used in trials has usually been 250 mg 8-hourly but 250 mg twice daily or 500 mg slow release tablets daily is usually recommended. The drug should be started not less than 24 h before a major gain in altitude. The side-effects of the drug include a mild diuresis which tends to diminish if the drug is continued; paraesthesia of the fingers and toes is almost universal. A few subjects find this distressing. Flushing, thirst, headache, rash, and blood dyscrasias are mentioned but are rare; finally beer and all fizzy drinks taste flat! The drug has been very widely used in the treatment of glaucoma for prolonged periods often at higher doses than are recommended for acute mountain sickness prophylaxis so its relative safety is assured.

Dexamethasone has been shown to be an effective prophylactic drug but most would consider using it for this purpose to be unjustified.

Treatment

Simple or benign acute mountain sickness is self limiting and usually lasts about 3 days so treatment is not essential; aspirin or paracetamol can be used to relieve headache but are not very effective. If the condition progresses to high altitude pulmonary or cerebral oedema of course treatment is urgent as indicated in the section on these conditions.

High altitude pulmonary oedema

In the great majority of cases acute mountain sickness is a minor affliction resolving in a few days. However, in a small proportion of people going to high altitude there develops the potentially lethal condition of acute pulmonary oedema or cerebral oedema or a mixture of these. The incidence will depend on the rate of ascent and the population involved. Figures of 0.5 to 2.0 per cent of people going to altitude have been quoted. Individuals with a previous history of high altitude pulmonary oedema are at greater risk of subsequent problems.

Both lowlanders and people resident at high altitude on return to high altitude are susceptible. Men and women of all ages can fall victim although there is an impression that the young male is more at risk than other groups. Athletic fitness affords no protection.

Clinical picture

The patient is typically a previously fit young man who has climbed rapidly to altitude and been very energetic on arrival. He suffers at least a moderate degree of acute mountain sickness. He becomes more breathless than his companions. A cough develops, dry at first, then productive of frothy white sputum later becoming blood tinged. He may complain of chest discomfort. Crackles will be heard at the lung bases and there will be increase in heart and respiratory rates. There may be peripheral oedema and raised jugular venous pressure. A right ventricular heave and accentuated pulmonary second heart sound may be detected. Over a few hours the condition deteriorates; heart and respiration rates rise, breathing becomes 'bubbly' and cyanosis develops. Coma leads to death if no action is taken.

Investigations

The chest radiograph (Fig. 15) typically shows asymmetric blotchy opacities. These clear in a few days if the patient recovers. There is often a mild pyrexia. Blood count usually shows neutrophil leucocytosis. Blood gases show a reduced Po_2 and arterial oxygen saturation compared with fit individuals at the same altitude. Pco_2 is variable but not significantly different from controls. The ECG shows tachycardia, peaked P waves, right axis deviation, and in some cases elevation of the S-T segment, all changes suggestive of pulmonary hypertension. Cardiac catheter findings confirm the high pulmonary artery pressure (81/49 mmHg in one study[50]) but normal wedge pressure. The cardiac output is normal. The oedema fluid is found to have a high protein content with concentrations approaching that of plasma.[51]

Autopsy findings show the lungs to be oedematous but the oedema is very patchy with areas of normal lung adjacent to areas of simple oedema and other areas of haemorrhagic oedema, the pattern corresponding to the radiographic appearance. There are many thrombi and fibrin clots in the small arteries and veins. The alveoli contain fluid, red cells, polymorphs, and macrophages, and there may be hyaline membrane formation.

Mechanism of high altitude pulmonary oedema

The mechanism of this condition is not clear. It is not that of acute left ventricular failure despite the clinical similarity. Catheter studies all agree in their finding of a normal wedge pressure. The most popular hypothesis, originally proposed by Hultgren et al.,[52]

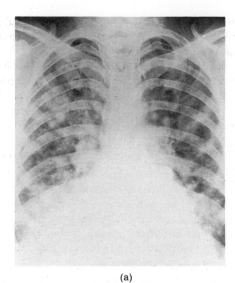

(a)

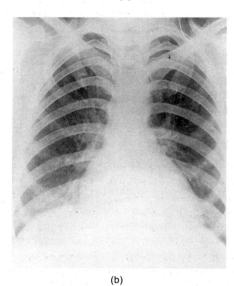

(b)

Fig. 15 Chest radiograph in a case of acute high altitude pulmonary oedema showing typical blotchy asymmetric opacities (by courtesy of Dr T. Norbu of Leh, Jammu, and Kashmir, India).

is that in susceptible subjects who have been shown to have brisk hypoxic pulmonary artery pressor responses,[53] the vasoconstriction is uneven. Those areas with greater vasoconstriction have reduced blood flow and are protected. Those areas with less constriction therefore have greatly increased blood flow. This torrential blood flow causes capillary damage, perhaps by sheer stress on the walls, perhaps by increased capillary pressure. Oedema then results. This accounts for the finding of pulmonary hypertension and the patchy nature of the oedema.

Treatment

The most important measure in a case of high altitude pulmonary oedema is to evacuate the patient to lower altitude. A reduction in altitude of as little as 300 m may make all the difference. If this is not possible, or while awaiting evacuation, oxygen, if available, will help. There is now available a portable, lightweight, rubberized canvas hyperbaric chamber (the Gamow bag) into which the patient can be placed and the pressure increased by 2 p.s.i. using a foot pump. This has the effect, if the patient is at a typical base

HIGH ALTITUDE

camp altitude (4000–4500 m), of reducing his equivalent altitude by almost 2000 m. Use of this bag has given good results although no controlled trials have been reported.

Since pulmonary vasoconstriction is thought to be important in the genesis of the condition, vasodilators have been tried; specifically the calcium channel blocker nifedipine has been shown to be beneficial.[54] A dose of 10 mg sublingually plus 20 mg slow release orally was used.

Diuretics have been advocated but seem to be losing favour amongst those who see many cases. Similarly morphia and digoxin have been suggested but again there is no evidence from controlled trials to support their use.

Outcome

In fully established cases at altitude where evacuation is impossible or is not carried out the result is usually death within a few hours. In cases evacuated promptly signs and symptoms are usually relieved within minutes or hours though the radiograph may take a few days to clear. Patients should be warned to be cautious in reascent to altitude but many have been able to go back to altitude with no recurrence.

Summary

High altitude pulmonary oedema is a serious life-threatening condition. It may be avoided by following the rules for avoiding acute mountain sickness, i.e. a slow ascent, going no higher if symptoms occur, and descending if symptoms persist or worsen. If diagnosed, descent as soon as possible is the first priority. The use of oxygen, nifedipine, and the Gamow bag is likely to be beneficial.

High altitude cerebral oedema

The other malignant form of acute mountain sickness is high altitude cerebral oedema. In the early stages this is indistinguishable from simple acute mountain sickness with headache, nausea, and vomiting being the prominent symptoms. When ataxia (unsteadiness on walking) is added to this symptom cluster the line between benign and malignant acute mountain sickness has probably been crossed. There may follow truncal ataxia (unsteadiness even when sitting), hallucinations, and clouding of consciousness with various neurological signs including extensor plantar reflexes and papilloedema. There are often signs of pulmonary oedema as well. Finally the patient becomes unconscious and dies if not treated.

The incidence is rather lower than acute pulmonary oedema. Hacket and Rennie give an incidence of 1.8 per cent (2.5 per cent for high altitude pulmonary oedema) in 278 trekkers passing through Pheriche (4243 m) on their way to Everest Base Camp.[44]

The mechanism is presumably the same as for simple acute mountain sickness but it is not clear why a few individuals progress to this lethal complication whereas the majority suffer only a self-limiting reversible condition. As with high altitude pulmonary oedema, lowlanders and highlanders, and men and women of any age can become victims. In the few cases where an autopsy has been carried out evidence of antemortem cerebral oedema, raised intracranial pressure, and petechial haemorrhages have been revealed. Venous thrombi have also been found.[55]

Treatment

This is similar to that for acute pulmonary oedema. The most important measure is to get the patient down. While awaiting evacuation oxygen breathing or increasing the ambient pressure in a Gamow bag helps but often not very quickly in more severe cases. Dexamethasone 4 mg intramuscularly in severe cases and orally in less severe cases will help reduce the cerebral oedema.

Outcome

As in acute pulmonary oedema, descent often leads to rapid improvement. However, in some cases recovery is delayed for a number of days and there may even be permanent or at least long-lasting neurological defects.[56]

REFERENCES

1. Zuntz N, Loewy A, Muller F, Caspari W. Atmospheric pressure at high altitudes. In: *Hokenklima und Bergwanderungen in ihrer Wirkung auf den Menschen*, Berlin, Bong and Co., 1906; 37–9. (Translation in High Altitude Physiology. West JB ed.) Stroudsburg: Hutchinson Ross, 1981; 78–80.
2. Pugh LGCE. Resting ventilation and alveolar air on Mount Everest: with remarks on the relation of barometric pressure to altitude in mountains. *Journal of Physiology*, 1957; **135**: 590–610.
3. West JB, Lahiri S, Maret KH, Peters RM, Jr, Pizzo CJ. Barometric pressures at extreme altitudes on Mt. Everest: physiological significance. *Journal of Applied Physiology*, 1983; **54**: 1188–94.
4. Ward MP, Milledge JS, West JB. *High Altitude Medicine and Physiology*. London: Chapman and Hall Medical, 1989; 99–110.
5. West JB. Diffusing capacity of the lung for carbon monoxide at high altitude. *Journal of Applied Physiology*, 1962; **17**: 421–6.
6. Dempsey JA, *et al*. Effects of acute through life-long hypoxic exposure on exercise pulmonary gas exchange. *Respiratory Physiology*, 1971; **13**: 62–89.
7. Reeves JT, *et al*. Operation Everest II: preservation of cardiac function at extreme altitude. *Journal of Applied Physiology*, 1987; **63**: 531–9.
8. Pugh LGCE. Cardiac output in muscular exercise at 5800 m (19000 ft). *Journal of Applied Physiology*, 1964; **19**: 441–7.
9. Ward MP, Milledge JS, West JB. *High Altitude Medicine and Physiology*. London: Chapman and Hall Medical, 1989; 148.
10. Motley HL, Cournand A, Werko L. Himmelstein A, Dresdale D. Influence of short periods of induced acute anoxia upon pulmonary artery pressure in man. *American Journal of Physiology*, 1947; **150**: 315–20.
11. Harris P. Evolution, hypoxia and high altitude. In: Heath D, ed. *Aspects of Hypoxia*. Liverpool University Press, 1986; 207–16.
12. Heath D, Williams DR. *High Altitude Medicine and Pathology*. London: Butterworth, 1989; 102–14.
13. Milledge JS. Electrocardiographic changes at high altitude. *British Heart Journal*, 1963; **25**: 291–8.
14. Pugh LGCE. Blood volume and haemoglobin concentration at altitudes above 18000 ft (5800 m). *Journal of Physiology*, 1964; **170**: 344–54.
15. Winslow RM, Monge C. Hypoxia, polycythemia and chronic mountain sickness. Baltimore: The John Hopkins University Press, 1987; 203.
16. Williams ES, Ward MP, Milledge JS, Withey WR, Older MWJ, Forsling ML. Effect of exercise of seven consecutive days hill-walking on fluid homeostasis. *Clinical Science*, 1979; **56**: 305–16.
17. Milledge JS, Beeley JM, McArthur S, Morice AH. Atrial natriuretic peptide, altitude and acute mountain sickness. *Clinical Science*, 1989; **77**: 509–14.
18. Bartsch P, Shaw S, Franciolli M, Gnadinger MP, Weidmann P. Atrial natriuretic peptide in acute mountain sickness. *Journal of Applied Physiology*, 1988; **65**: 1929–37.
19. Ward MP, Milledge JS, West JB. *High Altitude Medicine and Physiology*. London: Chapman and Hall Medical, 1989; 293–5.
20. Blume FD, Boyer SJ, Braverman LE, Cohen A, Dirkse J, Mordes JP. Impaired osmoregulation at high altitude: studies on Mt. Everest. *Journal of the American Medical Association*, 1984; **252**: 524–26.
21. Denison DM, Ledwith F, Poulton EC. Complex reaction times at simulated cabin altitudes of 5000 feet and 8000 feet. *Aerospace Medicine*, 1966; **57**: 1010–13.
22. Sharma VM, Malhotra MS, Baskaran AS. Variations in psychomotor efficiency during prolonged stay at high altitude. *Ergonomics*, 1975: **18**: 511–16.
23. Townes BD, Hornbein TF, Schoene RB, Sarnquist FH, Grant I. Human cerebral function at extreme altitude. In: West JB, Lahri S, eds. *High*

Altitude and Man. Bethesda: American Physiological Society, 1984: 31–6.

24. Hornbein TF, Townes BD, Schoene RB, Sutton JR, Houston CS. The cost to the central nervous system of climbing to extremely high altitude. *New England Journal of Medicine,* 1989; **321**: 1714–9.

25. Milledge JS. Arterial oxygen desaturation and intestinal absorption of xylose. *British Medical Journal,* 1972; **704**: 557–8.

26. Boyer SJ, Blume FD. Weight loss and changes in body composition at high altitude. *Journal of Applied Physiology,* 1984; **57**: 1580–5.

27. Rennie MJ, *et al.* Effects of acute hypoxia on forearm leucine metabolism. In Sutton JR, Houston CS, Jones NL, eds. *Hypoxia Exercise and Altitude.* New York: Liss, 1983; 317–23.

28. Consolazio CF, Matoush LO, Johnson HL, Krzywicki NJ, Daws TA, Isaac GJ. Effects of high-carbohydrate diets on performance and clinical symptomatology after rapid ascent to high altitude. *American Journal of Clinical Nutrition,* 1972; **25**: 23–9.

29. Ward MP, Milledge JS, West JB. *High Altitude Medicine and Physiology.* London: Chapman and Hall Medical, 1989; 208.

30. Reynafarje B. Myoglobin content and enzymatic activity of muscle and altitude adaptation. *Journal of Applied Physiology,* 1962; **17**: 301–5.

31. Ward MP, Milledge JS, West JB. *High Altitude Medicine and Physiology.* London: Chapman and Hall Medical, 1989; 211–2.

32. Ward MP, Milledge JS, West JB. *High Altitude Medicine and Physiology.* London: Chapman and Hall Medical, 1989; 213–6.

33. Horstman D, Weiskopf R, Jackson RE. Work capacity during a 3-week sojourn at 4,300 m: effects of relative polycythemia. *Journal of Applied Physiology,* 1980; **49**: 311–18.

34. Davis CTM, Sargeant AJ. Effects of hypoxic training on normoxic maximal aerobic power output. *European Journal of Applied Physiology,* 1974; **33**: 227–36.

35. Mizuno M, *et al.* Limb skeletal muscle adaptation in athletes after training at altitude. *Journal of Applied Physiology,* 1990; **68**: 496–502.

36. Balke B, Nagle J, Daniels J. Altitude and maximum performance in work and sports activity. *Journal of the American Medical Association,* 1965; **194**: 646–9.

37. Faulkner JA, Kollias J, Favour CB, Buskirk ER, Balke B. Maximum aerobic capacity and running performance at altitude. *Journal of Applied Physiology,* 1968; **5**: 685–91.

38. Roskamm F, Londry F, Samek L, Schlager M, Weidermann H, Reindell H. Effects of a standardised ergometer training program at three different altitudes. *Journal of Applied Physiology,* 1969; **27**: 840–7.

39. Hanson JE, Vogel JA, Stelter GP, Consoazio F, Oxygen uptake in man during exhaustive work at sea level and high altitude *Journal of Applied Physiology,* 1967: **23**: 511–22.

40. Dill DB, Adams WC. Maximal oxygen uptake at sea level and at 3,090 metres altitude in high school champion runners. *Journal of Applied Physiology,* 1971: **6**: 854–9.

41. Adams WC, Bernauer EM, Dill DB, Bowman JB. Effects of equivalent sea level and altitude training on Vo_2max and running performance. *Journal of Applied Science,* 1975; **39**: 262–6.

42. Dick F. Relevance of altitude training. *Athletics Coach,* 1979: **4**: 11–14.

43. Maggiorini M, Buhler B, Walter M, Oelz O. Prevalence of acute mountain sickness in the Swiss Alps. *British Medical Journal,* 1990; **301**: 853–5.

44. Hacket PH, Rennie D. The incidence, importance and prophylaxis of acute mountain sickness. *Lancet,* 1979; **ii**: 1449–54.

45. Forster P. Reproducibility of individual response to exposure to high altitude. *British Medical Journal,* 1984; **289**: 1269.

46. Ravenhill TH. Some experiences of mountain sickness in the Andes. *Journal of Tropical Medicine and Hygiene,* 1913; **20**: 313–22.

47. Milledge JS, Thomas PS, Beeley JM, English JSC. Hypoxic ventilatory response and acute mountain sickness. *European Respiratory Journal,* 1988; **1**: 948–51.

48. Milledge JS, *et al.* Sodium balance, fluid homeostasis and the renin-aldosterone system during the prolonged exercise of hill walking. *Clinical Science,* 1982; **62**: 595–604.

49. Ward MP, Milledge JS, West JB. *High Altitude Medicine and Physiology.* London: Chapman and Hall Medical, 1989; 377.

50. Antezanan G, Leguia G, Guzman AM, Coudert J, Spielvogel H. Haemodynamic study of high altitude pulmonary edema (12200 ft). In: Brendel W, Zink RA, eds. *High altitude physiology and medicine.* New York: Springer-Verlag, 1982; 232–41.

51. Hacket PH, Bertman J, Rodrigeux G. Pulmonary edema fluid protein in high-altitude pulmonary edema. *Journal of the American Medical Association,* 1986; **256**: 36.

52. Hultgren HN, Robison MC, Wuerflein RD. Over perfusion pulmonary edema. *Circulation,* 1966; **34**: 132–3.

53. Yagi H, Yamada H, Kobayashi T, Sekiguchi M. Doppler assessment of pulmonary hypertension induced by hypoxic breathing in subjects suscepible to high altitude pulmonary edema. *American Review of Respiratory Diseases,* 1990; **142**; 796–801.

54. Oelz O, Maggiorini M, Ritter M, Waber U, Jenni R, Vock P, Bartsch P. Nifedipine for high altitude pulmonary oedema. *Lancet,* 1989; **ii**: 1241–4.

55. Dickinson J, Heath D, Gosney J, Williams D. Altitude related deaths in seven trekkers in the Hiimalayas, 1983. *Thorax,* 1983; **38**: 646–56.

56. Houston CS, Dickenson J, Cerebral form of high-altitude illness. *Lancet,* 1975; **ii**: 758–61.

57. West JB, *et al.* Maximal exercise at extreme altitudes on Mt. Everest: physiological significance. *Journal of Applied Physiology,* 1983; **55**: 678–94.

58. Kellogg RH. The role of CO_2 in altitude acclimatization. In: Cunningham DJC, Lloyd BB, eds. *The Regulation of Human Regulation.* Oxford: Blackwell, 1963; 379–96.

59. Astrand P-O, Astrand I. Heart rate during muscular work in man exposed to prolonged hypoxia. *Journal of Applied Physiology,* 1958; **13**: 75–80.

60. Pugh LGCE, Gill MB, Milledge JS, Ward MP, West JB. Muscular exercise at great altitudes. *Journal of Applied Physiology,* 1964; **19**: 431–40.

61. Groves BM, *et al.* Operation Everest II: elevated high-altitude pulmonary resistance unresponsive to oxygen. *Journal of Applied Physiology,* 1987; **63**: 521–30.

62. Milledge JS, Coates PM. Serum erythropoietin in humans at high altitude and its relation to plasma renin. *Journal of Applied Physiology,* 1985; **59**: 360–4.

63. Peronnet F, Thibault G, Cousineau D-L. A theoretical analysis of the effect of altitude on running performance. *Journal of Applied Physiology,* 1991; **70**: 399–404.

64. Squires RW, Buskirk ER. Aerobic capacity during acute exposure to simulated altitude, 914 to 2286 meters. *Medicine and Science in Sports and Exercise,* 1982; **14**: 36–40.

65. Saltine B. Aerobic and anaerobic work capacity at an altitude of 2,250 metres. In: Goddard RF, ed. *The International Symposium on the Effects of Altitude on Physical Performance.* The Athletic Institute, Chicago. 1967; 97–102.

66. Pugh LGCE. Athletes at altitude. *Journal of Physiology,* 1967; **192**: 619–46.

67. Faulkner JA, Daniels J, Balke B. Effects of training at moderate altitude on physical performance capacity. *Journal of Applied Physiology,* 1967; **23**: 85–9.

68. Daniels J, Oldridge N. The effects of alternate exposure to altitude and sea level on world-class middle-distance runners. *Medicine and Science in Sport,* 1970; **3**: 107–12.

2.4 Physiological and clinical consequences of exercise in heat and humidity

JOHN R. SUTTON

THERMAL EFFECTS AND THERMAL REGULATION

The maintenance of internal temperature is a unique feature of birds and mammals which was acquired only during the last 70 million years of evolution, but it is an adaption which makes these species largely independent of the external environment. The price that we pay for this independence is a high metabolic rate to be able to maintain our body temperature within a fairly narrow range, although the range for survival is much greater. What is interesting is that at 37 °C we are fairly close to the ceiling of thermal viability compared with the lower end of the temperature scale. Therefore it is not surprising that more elaborate temperature-regulating mechanisms are available to prevent over heating than to prevent overcooling.

Figure 1 illustrates the normal responses of resting humans over a wide environmental range. In the centre, CD is the range that requires least thermoregulatory effort. As the environmental temperatures rise in the resting state, heat production remains constant and there is a major increase in evaporative heat loss by sweating. As the environmental temperature is lowered, evaporative heat loss virtually stops and there is a major increase in heat production, particularly by shivering. Exercise changes this balance dramatically by increasing the heat load, sometimes by as much as twentyfold.

There are three bidirectional avenues of thermal exchange between the body and the environment—conduction, convection, and radiation—and one unidirectional change, evaporation. This relationship is expressed by Winslow's heat storage equation[1]

$$S = M + R + C_v + C_d - E$$

where S is heat storage, M is metabolic heat production, R is heat gained or lost through radiation, C_v is heat gained or lost by convection, C_d is heat gained or lost by conduction, and E is heat lost by evaporation. These are important physiological mechanisms, but it should be stated at the outset that, in humans, behavioural factors far outweigh their physical capabilities of withstanding environmental extremes, i.e. putting on or removing clothing, seeking shelter for environmental extremes.

Body core temperature response to exercise

In 1938 Nielsen[2] put forward the idea that the magnitude of the body core temperature rise in steady state exercise was environmentally independent. He came to this conclusion after studying three subjects performing exercise at a variety of intensities under environmental conditions ranging from 5 to 36 °C and low humidity. These findings were extended and substantially supported by the work of Robinson *et al.*[3] and Lind.[4] In essence, the findings revealed the following:

1. With continuous work, rectal temperature increased to a new equilibrium within 60 min;
2. The body temperature depended on the work rate and was higher the higher the work rate;
3. The time taken to reach an equilibrium temperature was longer the higher the exercise intensity;
4. Over a wide range of environmental temperatures the body core temperature was primarily related to work rate but independent of the environment.

Most of these studies concerned work of low intensities for prolonged periods of time and were often designed to examine the work environment rather than sport. However, it was appreciated that under extreme climatic conditions body temperature would continue to rise and the subjects would collapse. A further extension of this work was carried out by Lind,[4] who tried to examine the relationship between environmental conditions and work rate so that safe limits for work environments and work rates (as opposed to recreation) might be constructed. In many work environments, particularly mining, there would be additional problems related to clothing. Again, examining a wide range of ambient conditions and work rates, Lind[4] proposed the so-called prescriptive zone (Fig. 2). This prescriptive zone was determined statistically and naturally there were marked individual variations. What might account for these individual differences? Astrand[5] was the first to report the importance of relative exercise intensity rather than absolute metabolic rate on the rise in body core temperature during exercise, and in 1966 these findings were extended by Saltin and Hermansen[6] (Fig. 3).

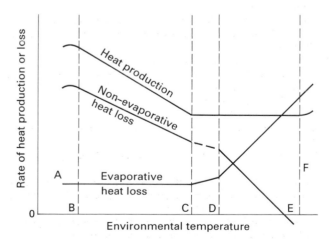

Fig. 1 Relationship between heat production, evaporative and non-evaporative heat loss, and deep body temperature in a homeothermic animal. A = zone hypothermia; B = temperature of summit metabolism and incipient hypothermia; C = critical temperature; D = temperature of marked increase in evaporative loss; E = temperature of incipient hyperthermal rise; F = zone of hyperthermia; CD = zone of minimum metabolism; and BE = thermoregulatory range. These zones were defined by Mount.

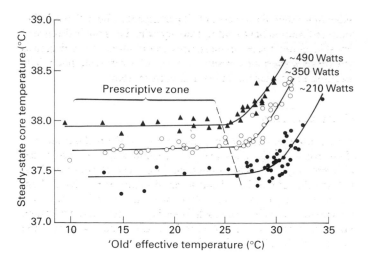

Fig. 2 Relationship of steady-state core temperature responses during exercise at three metabolic rates to the environmental conditions (reproduced from ref. 4 with permission).

Further clarification of the relationship between exercise intensity and ambient conditions was provided in the studies by Davies[7] and Davies and Thompson[8]. Over a wide range of environmental conditions, with dry-bulb temperatures from 5 to 25 °C and relatively low humidity, they examined subjects exercising at between 20 and 90 per cent of V_{O_2}max and demonstrated a curvilinear relationship between steady state core temperature and relative intensity. Even at exercise intensities up to 65 per cent V_{O_2}max, the core temperature was largely independent of dry-bulb temperature within the range 5 to 20°C. These observations were consistent with those of Lind[4], but at intensities of 85 per cent V_{O_2}max the prescriptive zone becomes smaller.

So far we have concentrated on leg cycle ergometry, but work using arm cranking suggests that the body core temperature response may be more closely related to maximal uptake than relative work rate. As early as 1947, Asmussen and Nielsen[9] showed that for the same absolute work intensity performed by arm versus leg cycle ergometry, a lower rectal temperature occurred in the former and the difference became exaggerated as the metabolic rate increased. These observations brought out two points:

1. The effect of different styles of exercise on the rate of body core temperature increase;

2. More importantly is body core temperature and could a rectal temperature be safely assumed to be a good reflection of the body core under all working conditions?

Nowadays, oesophageal temperature is probably considered a more appropriate indicator of body core temperature. In fact, when Nielsen[10] repeated these experiments using oesophageal temperature, the steady state exercise temperatures seemed comparable under two exercise regimes.

So far the emphasis has been on the rates of increase in body core temperature under a variety of ambient conditions. Clearly, while the importance of metabolic rate has dominated our considerations, as ambient conditions increase, particularly as humidity increases, our major route of heat loss, and therefore our ability to thermoregulate in the heat, becomes minimized. Evaporation accounts for the majority of our heat-losing abilities, and as the relative humidity approaches 100 per cent, this avenue of heat loss becomes increasingly reduced, eventually falling to zero. This is because, although we sweat, the sweat cannot evaporate but simply falls to the ground and is lost to the cooling process. In addition, as we become further dehydrated with prolonged exercise in a hot environment, the sweat response to increasing core temperature becomes less effective.

A particularly elegant study contrasting the effects of different body core temperatures on fatigue was that by MacDougall et al.[11] In this study six subjects exercised on a treadmill at 70 per cent of their V_{O_2} max and controlled the thermal environment by use of a lightweight water-perfused suit described by Rowell et al.[12] The subjects ran to exhaustion, and this occurred with similar elevations in rectal temperature which failed to reach a plateau. However, the exercise time to achieve exhaustion was considerably different under the three conditions, with a mean of 75 min under normal conditions and 90.75 min when it was cool, but only 48.25 min under the hot conditions. Additional points of interest in this study were the increase in blood lactate, which were maximum under the hot conditions, and also the observation that immediately prior to exhaustion there was a fall in cardiac output predominantly related to a decrease in stroke volume.

This last problem of circulatory regulation during exercise has received attention in recent years. The circulation has two increasing demands imposed upon it when exercise is performed in a warm environment. First there is the need to deliver appropriate nutrients and oxygen to working muscle, and second there is a need to perfuse the skin for thermoregulation. These dual demands compete. Both these processes require an adequate central

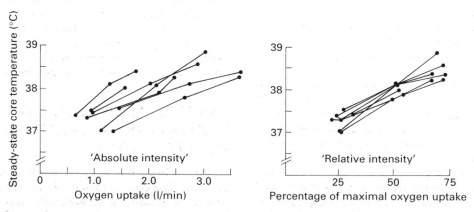

Fig. 3 Relationship of steady-state core temperature responses during exercise to the absolute and relative metabolic rates (reproduced from ref. 6 with permission).

blood volume for the circulation to remain efficient. As the temperature increases there will be a greater displacement of the central circulation to the periphery resulting in blood pooling. In addition, with continued fluid loss by sweating the total circulatory volume may decrease, thus compromising the circulation.[13,14] Under these compromised circulatory conditions, maximum stroke volume and cardiac output will decrease and V_{O_2}max and performance will also be reduced. This results in a decrease in central venous pressure and a redistribution of blood flow from the periphery[15,16] and under these conditions the body is placed in a very compromised situation. Rowell *et al.*[13] concluded that there is a hierarchy of homeostatic mechanisms which favour the maintenance of arterial blood pressure and circulation to the vital organs at the expense of skin vasodilation and thermoregulation. This certainly seems to be the case, and in a recent publication Hales *et al.*[16] demonstrated a marked reduction in skin blood flow in collapsed runners when compared with their control counterparts. They postulated that the skin blood flow was reduced when right heart filling pressure fell and that this was the fundamental regulatory mechanism.

Females appear to sweat at a lower rate than males[17-19] when the greater surface area to weight ratio A_D/WG of females is taken into account. Per unit volume of sweat produced, women reach a lower body core temperature than men per unit volume of sweat produced, and thus have been considered more efficient temperature regulators.[20] In a recent acute heat stress study of males and females with identical cardiorespiratory fitness, surface area, and equal surface area to mass ratios, Avellini *et al.*[21] found that females in the follicular phase of the menstrual cycle had a lower sweat rate, a lower rectal temperature, and a lower heart rate than men. However, following 10 days of acclimatization the differences between the sexes were eliminated.

CLINICAL CONSEQUENCES OF HYPERTHERMIA

Although an increase in body core temperature is a normal concomitant of exercise, the dividing line between normality and abnormality in absolute levels of body core temperature varies enormously. One of the most important issues is the site at which body core temperature is recorded. Clearly, rectal, oesophageal, and tympanic membrane temperatures can all differ considerably, although each of these is a more accurate reflection of the internal body temperature than is the axillary or oral temperature. The last two can be particularly misleading. The use of oral or axillary temperatures in the past has often been so erroneous as to lead clinicians quite incorrectly to a diagnosis of hypothermia when in fact patients were hyperthermic (Fig. 4). In one patient, the oral temperature was 35.5 °C (96 °F) but the rectal temperature was 42 °C (108 °F).

The terms 'heat exhaustion' and heat-stroke' are used by the Medical Research Council of Great Britain and the World Health Organization, although the differentiation between the two is often tenuous as they are part of a spectrum. Serious central nervous system dysfunction and multiple organ failure are probably the most important distinctive features of heat-stroke.

In relation to exercise it is important to realize that the clinical picture differs very much from the classic original description of heat-stroke which included anhydrosis and a rectal temperature in excess of 41 °C. In hyperthermic states associated with 'fun runs' and any exercise of short duration, dehydration is not usually an issue; more than 85 per cent of all participants who collapse in fun runs are found to have cold and sweaty skin. Thus the insistence of dry skin under these circumstances would obscure the diagnosis and more importantly result in delay of treatment, potentially resulting in further morbidity and even mortality.

Heat cramps

Heat cramps are classically intermittent, of short duration, and often excruciating. They have most commonly been associated with prolonged exercise, sometimes over several hours or days, and are thought to be related to an absolute or relative sodium deficiency where there has often been an excessive water replacement in the absence of salt following a period of dehydration. This was observed in the steel mills in Ohio and also during the building of the Hoover dam.[22]

Heat syncope

Heat syncope is a common problem and classically relates to the peripheral pooling of blood when a vasodilated person stops exercising after a run and eliminates the muscle pump. Venous return is therefore reduced and cardiac output and hence cerebral blood flow falls, resulting in syncope.

Heat exhaustion and heat-stroke

Under extreme environmental conditions potential heat-stroke is an ever-present threat to many in the population, but particularly to the aged and the infirm, even at rest.

One of the most graphic descriptions of heat-stroke was reported when a British frigate docked at Liverpool in 1841. The weather gradually became warmer, and the decks were constantly wetted and every precaution used to prevent heat exposure. However, in one day 30 men were lost and it was stated that the decks resembled a slaughterhouse so numerous were the bleeding patients. This was probably one of the first descriptions of disseminated intravascular coagulation associated with heat-stroke.

Nowadays, exercise-related heat-stroke is more common in fun runs and short-duration events rather than in marathons and ultramarathons, and in its classic form is rapid in onset, has multiple organ pathology, and if not treated has a high morbidity and mortality. An additional crucial point to remember is that heat-stroke can occur when environmental conditions are not extreme. In the experience from the Sydney City to Surf Race[23] it is the young fit amateur runner who seems to be at most risk. This is because the relative rate of exertion, not the environmental conditions, is a crucial factor in producing heat-stroke in these individuals. When a person runs at more than 90 per cent of their own V_{O_2}max, their ability to thermoregulate is impaired.

PATHOPHYSIOLOGY OF HEAT STROKE—SIMILARITIES TO SEPTIC SHOCK

In a recent series of cases of heat-stroke in fun runs and football games the clinical picture and the degree of organ damage are also variable. In general, for a comparable rectal temperature, tissue damage to various organs is greater in exertional heat-stroke than in classic heat-stroke.[24] Absolute core temperature is important,

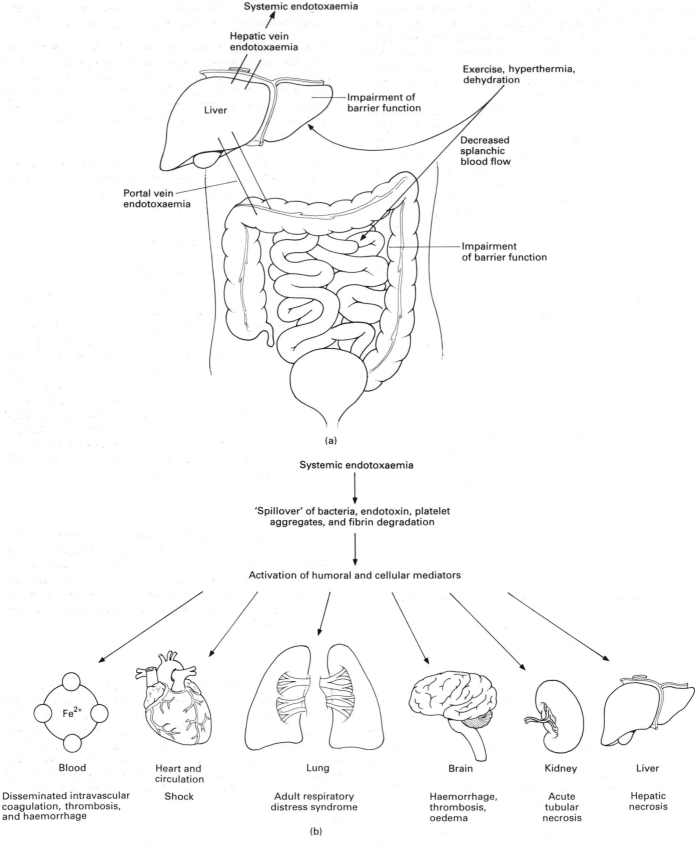

Fig. 4 (a) Mechanisms whereby exercise hyperthermia and dehydration lead to a decreased splanchnic blood flow with impairment of barrier function of the gut and liver—with development of systemic endotoxaemia. (b) The saga of biochemical events following systemic endotoxaemia and resulting cytokine cascade with production of multiorgan failure. Based on an article by A. Ryan, 'Heat stroke and endotoxins sensitization in tolerance to endotoxins'. In: Gisolfi CV, Lamb DL, eds. *Perspectives in Exercise Science and Sports Medicine*. Indiana: Benchmark Press, 1993.

but its duration and the rate of change may be even more important and the integrity of the circulation may well determine organ damage. As with hypotension, the additional ischaemic injury appears to potentiate the problem of heat-stroke. Previously held ideals that heat-stroke was due to the cellular toxicity from the heat *per se* and/or other reflex cardiovascular changes that might be derived from poor cardiac filling pressure in dehydrated subjects must now be modified. Recent evidence implicates endotoxin as an initial trigger in the sequence of events which lead to the clinical scenario of heat-stroke.[25] Endotoxins are lipopolysaccharide-protein complexes (LPS) derived from cell walls of Gram-negative bacteria (occasionally Gram-positive) and are extremely potent stimuli to tumour necrosis factor (TNF) and interleukin-1 (IL-1a)[26] These are endogenous pyrogens which result in fever and are responsible for the shock and tissue injury associated with endotoxaemia[27] During exercise mild elevations of some of these trigger mediators occur,[28] but in those who subsequently go on to develop heat-stroke the increase is several-fold.

In patients with classic (not exertional) heat-stroke, LPS, tumour necrosis factor (TNF), and interleukin-1 (IL-1a) were markedly increased.[29] Although no control measurements were made there was an interesting relationship with increasing temperature in these heat-stroke victims and the plasma concentration of these mediators. Furthermore the plasma concentration fell dramatically upon cooling. Although these were not exercising patients, the lessons almost certainly apply, perhaps even more so, to exertional heat-stroke.

The suggested sequence of events is as follows. During exercise especially if associated with hyperthermia the relatively ischaemic bowel mucosa loses its barrier function and bacterial translocation in the gut will occur, resulting in portal vein endotoxaemia. If the liver also loses its barrier function the systemic endotoxaemia will occur, resulting in increased plasma concentrations of LPS which stimulates release of interleukin and tumour necrosis factor. These important mediators then stimulate a cytokine cascade, viz. complement pathways and intrinsic and extrinsic coagulation pathways, colony stimulating factor, and interferon and numerous other cytokines. However, most evidence suggests that tumour necrosis factor and interleukin-1 are the primary mediators of endotoxic shock. Nevertheless, the situation may be ameliorated when an additional family of proteins (the so-called heat-shock proteins) are released; these may help protect the organism from the destruction mediators. Heat-shock proteins 28 and 70 seem important in the acquired tolerance of mammalian cells to hyperthermia.

Thus more and more evidence is accumulating which supports the view that heat-stroke and the accompanying multiorgan damage is pathogenically very similar to septic shock.

Of all the pathological features of heat-stroke, it is the development of disseminated intrasvascular coagulation, together with rhabdomyolysis, which is probably responsible for the multiorgan pathology resulting in significant morbidity and mortality. In virtually all fatal cases of heat-stroke, disseminated intravascular coagulation is a concomitant. Although it is uncertain precisely how disseminated intravasular coagulation is initiated the following potential mechanisms are suggested.

1. The release of tissue thromboplastins as a result of extensive tissue injury stimulating the coagulation cascade by the extrinsic pathway and resulting in intravascular fibrin deposition.
2. A second possible mechanism is the induction of endothelial damage which activates platelets and the coagulation cascade via the intrinsic pathway.[30,31]

3. Endotoxins released from bowel and stimulating intrinsic and extrinsic pathways of coagulation and also fibrinolysis.[32,29]

In a recent autopsy study 10 patients who died following heat-stroke, disseminated intravasular coagulation was thought to be causally related.[33] Rhabdomyolysis itself may be productive of disseminated intravasular coagulation and certainly with the release of myoglobin and uric acid was thought to have been causally important in the development of progressive renal impairment and acute tubular necrosis.

Few organs have escaped involvement in heat-stroke but the effects on the brain usually determine the outcome, and prolonged unconsciousness in a person with disseminated intravascular coagulation has a particularly bad prognosis.

PATHOPHYSIOLOGY

The central nervous system

Effects on the central nervous system are common, and without some central nervous system impairment the diagnosis of heat-stroke cannot be made. Most commonly there is a transient loss of consciousness, but this may be preceded by impaired judgment, irritability, and hallucinations. Other associated features include status epilepticus, oculogyric crisis and cerebellar symptoms, and hemiplegia. There may even be decerebrate posturing. The duration of unconsciousness is some indication of the prognosis, although even after 10 days to 3 weeks patients have regained consciousness and gone on to full neurological recovery. The pathological changes found in the central nervous system consist of oedema, congestion, and haemorrhages usually associated with disseminated intravascular coagulation.

The kidney

Oliguria, anuria, and acute renal failure with acute tubular necrosis are particularly common.[34,35] The role of rhabdomyolysis in acute renal failure has already been discussed, but disseminated intravascular coagulation may also play a causative role and certainly renal impairment is well documented in patients in whom rhabdomyolysis is not particularly severe.[36] Some patients go on to develop a chronic interstitial nephritis.

The liver

Liver impairment often occurs more than 24 h after the original failure is evident, and the patients will present with jaundice and biochemical or hepatocellular damage with increases in the liver enzymes AST, ALT and GGT as well as elevations of alkaline phosphatase, indicative of liver excretory impairment. Pathological findings include perisinusoidal oedema and centrilobular necrosis; there can also be desquamation of the sinusoidal and lining cells and a ballooning and flattening of the microvilli.[37]

The lungs

Tachypnoea and alveolar hyperventilation are common in heat-stroke, and patients who develop fulminant pulmonary features usually have pulmonary oedema similar to that associated with the adult respiratory distress syndrome. Classically the left atrial pulmonary artery wedge pressure is low distinguishing it from cardiogenic pulmonary oedema. Such patients have a low P_{CO_2} and are hypoxaemic with a widened $(A-a)P_{O_2}$. Mechanical ventilation may be difficult as the lungs are stiff. As there can be cardiovascular problems as well, cardiogenic pulmonary oedema can occasionally be associated with heat-stroke.

Cardiovascular system

Most patients will have a sinus tachycardia and perhaps transient hypertension, although in the early phase after admission to hospital the circulatory state is quite precarious as the patients may have marked dehydration and fluid compartment shifts. Therefore a careful monitoring of central blood volume and cardiac filling pressures with the Swan–Ganz catheter is particularly valuable. Non-specific electrocardiographic changes have been reported including T-wave changes and prolonged or borderline QT interval. Pathologically subendocardial, subpericardial, and myocardial haemorrhages are usually associated with disseminated intravascular coagulation; there can also be fragmentation and rupture of the muscle fibres in the presence of interstitial oedema.[32,33]

The blood

Disseminated intravascular coagulation, a consumption coagulopathy, is very common. Also known as defibrination syndrome, this is a haemorrhagic disorder where there is diffuse intravascular clotting which results in a haemostatic defect because there is excessive utilization of the coagulation factors and the platelets that are essential for clotting—hence the name consumption coagulopathy.[38] As mentioned earlier, Chao et al.[27,32] found disseminated intravascular coagulation inextricably associated in 10 consecutive fatal cases.

The two main clinical features are bleeding and organ damage due to ischaemia. The kidney failure previously reported may well be due to occlusion of small vessels with fibrin deposition. There can also be a haemolytic anaemia associated with the defibrination. Very rarely, there can also be early thrombosis of large veins and arteries, although this is not particularly associated with heat-stroke.

PREVENTION AND MANAGEMENT OF EXERTIONAL HEAT-STROKE

'I am of the opinion that in healthy subjects the only serious potential risk to life from violent exercise is heat-stroke'. Sir Adolphe Abrahams, *British Encyclopaedia of Medical Practice*, 1950.

Forty years later this comment is just as relevant. Both participants and organizers of athletic events must be ever minded of the possibility that bizarre behaviour by a competitor could be the result of heat-stroke. The failure to appreciate that heat-stroke could be the cause of erratic behaviour in athletes or the reason for changing consciousness in someone who has collapsed will delay the diagnosis and therefore the treatment. It cannot be emphasized too much that the speed with which the core temperature is returned to normal and the various other abnormalities are corrected will determine the outcome.

In most cases the measurement of rectal temperature together with full clinical assessment of the cardiovascular, respiratory, and neurological state is vital. Treatment must begin on site. All

too often one hears race organizers alert a neighbourhood hospital where patients are to be transported in the event of some disaster. This results in great delay and often emergency room physicians may not be aware of the likelihood of heat-stroke, thus further delaying the institution of life-saving treatment.

Following the first City to Surf Race in Sydney in 1971 the following suggestions regarding race organizations, medical support, and competitor education, were made.[23]

Race organization

Organized races and sporting events should avoid the hottest summer months and the hottest part of the day. In the northern hemisphere early spring can be particularly hazardous as there are frequently unseasonably hot days. These comments apply to any athletic event.

Athlete education

The following points should be noted.

1. Training and fitness will improve thermoregulation during exercise.
2. Heat acclimatization—athletes are advised to train at the time of day when the competitions will be held. All too often training will be in the morning or evening although the competition is held in the middle of the day. A minimum of 7 to 10 days of heat acclimatization is advisable.
3. Begin the competition hydrated—often football players and athletes will already be one and two litres behind when the competition begins.
4. Only compete when well—hazards include any febrile or dehydrating illness such as upper respiratory tract infections and especially gastroenteritis.
5. During the race or fun run be aware of headache, nausea, dizziness, and lack of co-ordination. These are often the first symptoms of heat injury and the time taken from their onset to the loss of consciousness can be very brief.
6. Run within one's capabilities—in fun runs the victims tend to be younger over-enthusiastic amateur runners striving for a personal best time.

Medical organization and medical facilities

Ideally medical facilities should be available on-site at competitions and the medical personnel should be capable of full-scale resuscitation, and in particular should be able to cool and rehydrate with intravenous therapy as has been described in detail elsewhere.[23]

SUMMARY

Heat-stroke is a serious and potentially fatal event. Organizers of various athletic competitions must be aware of the possibility that heat-stroke could occur in the events they organize—even on apparently cool days. Medical facilities must be available on site and staff must have the ability to make the diagnosis and begin definitive treatment. The full-blown picture of heat-stroke with multiorgan damage may be caused by endotoxaemia and be pathogenically like septic shock. Such an aetiology opens new horizens in the mechanisms of thermal tolerance and intolerance and possible new approaches to the treatment of heat-stroke.

The principal concern in the management of runners who collapse following a race is the establishment of an accurate diagnosis. Anyone who collapses, i.e. falls to the ground and has a temporary impairment of consciousness, is a potential candidate for heat-stroke, and experience has taught us that an accurate measurement of core temperature is essential. Measurements of axillary or oral temperatures can be very misleading, and therefore oesophageal, tympanic, or most practically rectal temperature is essential. Our practice has been to measure rectal temperature, and a wide range of values has been obtained for those who have collapsed. Clearly, not everyone who collapses suffers from heat exhaustion, but unless the rectal temperature is taken it is impossible to establish a logical course of action. As with any seriously ill patient, begin with the ABC (airway breathing and circulation). In most instances airway breathing and circulation are functioning, but in the 1990 City to Surf Race in Sydney three individuals collapsed, were asystolic, and could not be resuscitated.[39]

Having established the diagnosis, the specific treatment is rapid rehydration and cooling. Administration of 1 to 2 litres of intravenous dextrose–saline will restore the circulation and thus can be the most important step in the cooling process. In addition there is some advantage in placing cool packs over the large vessels of the neck, groin, and axilla.

The most effective means of rapid cooling is that established by Weiner and know as the Mecca Cooling Unit in which an atomized spray of warm water is directed on to the individual who is then fanned with warm air. This enables the vasodilatation of the skin to remain and rapid cooling to occur. In contrast, the application of cold sheets or direct cold usually results in vasoconstriction with impairment of heat loss, ensuring that further temperature elevation will result. Some advocate immersion in an ice-cold bath. This will certainly lower body core temperature, but it is usually logistically difficult for large numbers of subjects and where there is the possible need to use defibrillation may expose patients and health care workers to the risk of electrocution.

When the above approach is used most people who collapse in fun runs can be observed for an hour or more and safely discharged. If their recovery is in question they should be transferred to a medical facility for further monitoring. Nevertheless, it should be emphasized that the emergency care of collapsed athletes must be on site as delay in establishing the diagnosis and beginning treatment may increase morbidity and even mortality.

Where there is evidence of multiorgan damage and the developing saga of fulminant heat-stroke, I would recommend bowel sterilization and the intravenous antibiotic regime appropriate for treatment of Gram-negative sepsis.

REFERENCES

1. Winslow CEA, Herrington LP, Gagge AP. Physiological reactions of the human body to varying environmental temperatures. *American Journal of Physiology* 1937; **120**: 1–22.
2. Nielsen, M. Die regulation der korpetemperatuur bei muskelarbeit. *Skandinavisches Archiv für Physiologie* 1938; **79**: 193–230.
3. Robinson S, Dill DB, Wilson JW, Nielsen M. Adaption of white men and Negroes to prolonged work in humid heat. *American Journal of Tropical Medicine* 1941; **21**: 261–87.
4. Lind AR. A physiological criterion for setting thermal environmental limits for everyday work. *Journal of Applied Physiology* 1963; **18**: 51–6.
5. Astrand I. Aerobic work capacity in men and women. *Acta Physiologica Scandinavica* 1960; **49** (Suppl. B): 64–73.
6. Saltin B, Hermansen L. Esophageal, rectal and muscle temperature during exercise. *Journal of Applied Physiology* 1966; **21**: 1757–62.
7. Davies CTM. Thermoregulation during exercise in relation to sex and age. *European Journal of Applied Physiology* 1979; **42**: 71–9.
8. Davies CTM, Thompson M W. Aerobic performance of female marathon and male ultramarathon athletes. *Journal of Applied Physiology* 1979; **41**: 233–48.
9. Asmussen E, Nielsen M. The regulation of the body-temperature during work performed with the arms and with the legs. *Acta Physiologica Scandinavica* 1947; **14**: 373–82.
10. Nielsen B. Thermoregulation during work in carbon monoxide poisoning. *Acta Physiologica Scandinavica* 1971; **82**: 98–106.
11. MacDougall JD, Reddan WG, Layton CR, Dempsey JA. Effects of metabolic hyperthermia on performance during heavy prolonged exercise. *Journal of Applied Physiology* 1974; **36**: 538–44.
12. Rowell LB, Brengelmann GL, Murray JA, Kraning KK, Kusumi F. Human metabolic responses of hyperthermia during mild to maximal exercise. *Journal of Applied Physiology* 1969; **26**: 395–402.
13. Rowell LB, Marx J, Bruce RA, Conn RD, Kusumi F. Reductions in cardiac output, central blood volume and stroke volume with thermal stress in normal men during exercise. *Journal of Clinical Investigation* 1966; **45**: 1801–16.
14. Nadel ER, Cafarelli E, Roberts MF, Wenger CB. Circulatory regulation during exercise in different ambient temperature. *Journal of Applied Physiology* 1979; **46**: 430–7.
15. Johnson JM. Responses to forearm blood flow to graded leg exercise in man. *Journal of Applied Physiology* 1979; **46**: 457–62.
16. Hales JRS *et al.* Lowered skin blood flow and erythrocyte sphering in collapsed fun-runners. *Lancet* 1986; i: 1494–5.
17. Fox RH, Lofstedt BE, Woodward PM, Erikkson E, Werkstrom B. Comparison of thermoregulatory function in men and women. *Journal of Applied Physiology* 1969; **26**: 444–53.
18. Hertig BA, Sargent F. Acclimatization of women during work in hot environments. *Federation Proceedings* 1963; **22**: 810–13.
19. Wyndham CH, Morrison JF, Williams CG. Heat reactions of male and female caucasians. *Journal of Applied Physiology* 1965; **20**: 357–64.
20. Drinkwater BL, Denton JE, Kupprat IC, Talag TS, Horvath SM. Aerobic power as a factor in women's response to work in hot environments. *Journal of Applied Physiology* 1976; **41**: 815–21.
21. Avellini BA, Kamon E, Krajewski JT. Physiological responses of physically fit men and women to acclimatization to humid heat. *Journal of Applied Physiology* 1980; **49**: 254–61.
22. Talbot JH. Heat cramps *Medicine* 1935: 323–76.
23. Richards D, Richards R, Schofield PJ, Ross V, Sutton JR. Management of heat exhaustion in Sydney's *The Sun* City-to-Surf fun runner. *Medical Journal of Australia* 1979; **2**: 457–61.
24. Knochel JP, Reed G. Disorders of heat regulation. In: Kleeman CR, Maxwell MH, Narin RG, eds. *Clinical disorders, fluid and electrolyte metabolism.* New York: McGraw-Hill, 1987: 1197–1232.
25. Gathiram P, Wells MT, Raidoo D, Brock-Utne JG, Gaffin SL. Portal and systemic plasma lipopolysaccharide concentrations in heat-stressed primates *Circulatory Shock* 1988; **25**: 223–30.
26. Old LJ. Tumour necrosis factor. Another chapter in the long history of endotoxin. *Nature,* 1987; **330**: 602–3.
27. Tracey KJ *et al.* Shock and tissue injury induced by recombinant human cachetin. Science 1986; **234** 470–4.
28. Brock-Utne JG *et al.* Endotoxemia in exhausted runners after a long distance race. *South African Medical Journal.* 1988; **73**: 533–6.
29. Bouchama A, Parhar RS, El-Yazigi A, Sheth K, and Al-Sedairy S. Endotoxemia and release of tumour necrosis factor and interleukin 1 in acute heatstroke. *Journal of Applied Physiology,* 1991; **70**: 2640–4.
30. Mustafa KY *et al.* Blood coagulation and fibrinolysis in heat stroke. *British Journal of Haematology* 1985; **61**: 517–23.
31. Sohal RS *et al.* Heat stroke: an electron microscopic study of endothelial cell damage and disseminated intravascular coagulation. *Archives of Internal Medicine* 1968; **122**: 43–7.
32. Choa TC, Sinniah R, Pakiam JE. Acute heat stroke deaths. *Pathology* 1981; **13**: 145–56.
33. Clowes GHA, O'Donnell TF. Heat stroke. *New England Journal of Medicine* 1974; **291**: 564–7.

34. Hart LE, Egier BP, Shimizu AG, Tandan PJ, Sutton JR. Exertional heat stroke: The runner's nemesis. *Canadian Medical Association Journal* 1980; **122**: 1144–50.

35. Savdie E *et al*. Heatstroke following rugby league football. *Medical Journal of Australia*, 1991; **155**: 636–9.

36. Kew MC, Abrahams C, Seftel HC. Chronic interstitial nephritis as a consequence of heatstroke. *Quarterly Journal of Medicine* 1970; **39**: 189–99.

37. Malamud N, Haymaker W, Custer RP. Heatstroke: a clinico-pathologic study of 125 fatal cases. *Military Surgeon* 1946; **99**: 397–449.

38. Sutton JR, Coleman MK, Millar AP, Lazarus L, Russo P. The medical problems of mass participation in athletic competition: the City-to-Surf race. *Medical Journal of Australia* 1972; **2**: 127–33.

39. Richards R, Richards D, Sutton JR. Exertion-induced heat exhaustion: an often overlooked diagnosis. *Australian Family Physician* 1992; **21**: 18–24.

2.5 Circadian rhythms

THOMAS REILLY

INTRODUCTION

Physiological determinants of exercise performance are affected by circadian rhythms. Performance rhythms conform closely in phase with body temperature and also with the level of arousal. These rhythms have implications for elite athletes, for people performing exercise for health purposes, and for the optimal timing of training. They also have repercussions for team managers with responsibilities for the travel plans of athletes competing abroad. There are also consequences for sports medicine personnel who must consider, for example, effects of time of day on joint stiffness and pain perception. Additionally, some drug doses that are safe in the evening may have exaggerated effects if administered early in the morning. Circadian rhythms may be influenced by environmental factors such as ambient temperature, light and darkness, and so on. Endogenous rhythms persist during sleep deprivation, when they are superimposed on an underlying trend towards fatigue. Separate rhythms are desynchronized in time zone transitions and during nocturnal shiftwork, and exercise capability is then temporarily impaired. Coping with desynchronization is helped by adopting different behavioural, dietary, or pharmacological strategies. Attempts to identify individuals with poor tolerance to sleep loss, shift work, or jet lag have largely been unsuccessful.

Biological rhythms should not be confused with 'biorhythms', a theory which has no scientific foundation. According to this theory there are three independent cycles which start for each individual at birth. A physical cycle determines vigour and has a cycle length of 23 days. A cycle of emotion has a period of 28 days, and there is supposedly also a cycle of intellectual ability with a period of 33 days. Half-way through each cycle there is a swing across the baseline from positive to negative. The theory predicts that sports performance will be benefited when cycles are positive and adversely affected when they are negative, but interpreting a mixture of the three rhythms is less straightforward. Advocates of the theory may point to outstanding sports successes (such as the Olympic Games victories of Mark Spitz and the first world record runs of Sebastian Coe) to justify their case, but the theory is easily discredited by retrospective analysis of athletic records. There is no basis for justifying its application to strenuous exercise and the scientific preparation of athletes. In contrast, the scientific study of biological rhythms, which is known as chronobiology, is now a respected field in its own right.

CHRONOBIOLOGY

Biological rhythms

Rhythms are an essential feature of nature and many aspects of human behaviour. Cyclical changes that recur regularly over a given length of time and are related to underlying physiological processes are referred to as biological rhythms. In humans the length of the cycles, known as the period, can range from small fractions of a second, such as in neural firing rates, to slower changes in close harmony with the lunar cycle (circamensal) or with the changes of the season (circannual). Rhythms associated with the solar day are called circadian. Rhythms with periods longer than a day are known as infradian and those recurring repeatedly within a day are known as ultradian.

In addition to the period or length, a rhythm is also characterized by its amplitude, which is equal to half of the variation from peak to trough, and its acrophase, which refers to the time that the peak occurs. The oscillation is deemed to occur around a midpoint which is called the mesor. Where observations are made at unequal intervals throughout 24 h, the mean value may differ from the mesor.

Circadian rhythms

Circadian rhythms influence biological function. They can have a profound impact on performance of physical activity and athletic skills. The major determinant of the rhythms is the spin of the Earth about its vertical axis. Humans have adapted to this over the ages by timing the alternation of sleep and wakefulness to coincide with the periods of darkness and light respectively. In turn, this pattern of rest and activity affects many physiological functions which slow down at night and accelerate with daylight. Human circadian rhythms have been established at levels ranging from cellular and tissue operations to whole-body functions. The myriad of rhythms in the body interact with each other and with the environment. The concept of homeostasis, accepted from the mid-nineteenth century and implying that the internal environment within the body is relatively constant, now acknowledges that the internal environment is constantly changing with a regular oscillatory behaviour. Indeed, the capacity for rhythmic change is accepted as an inherent characteristic of living organisms. Thus when the physiological responses to exercise assume a so-called steady state, this level may depend upon the time of day.

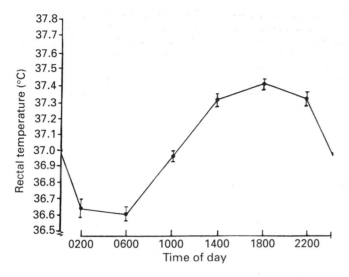

Fig. 1 The circadian rhythm in rectal temperature. Values shown are mean ± SD. (Reproduced from ref. 1, with permission.)

Rhythms in motor performance are closely linked to the circadian curve in body temperature (Fig. 1). Circadian rhythms may interact with habitual daily activities, and the persistence of rhythms in responses to submaximal and maximal exercise should be of interest to sports practitioners. Similarly, circadian variations in joint stiffness and flexibility are relevant to physical therapists. Factors limiting exercise performance, such as thermoregulation, motivation and pain, may operate differently according to time of day. It is important to consider individual preferences for morning or evening effort and the extent to which some performances are vulnerable to ultradian influences. Of course, the existence of circadian rhythms is most easily acknowledged when they are disrupted through transmeridian flight, nocturnal shiftwork, or sleep loss.

PHYSIOLOGY OF CIRCADIAN RHYTHMS

Body temperature and sleep–wake cycles

Those human circadian rhythms that result from changes in the environment are referred to as exogenous. The most prominent environmental changes are alternating day and night. The solar day dictates habits of sleep, rest, and activity, as well as work and leisure-time social activity. Ambient temperature varies with the time of day, potentially accentuating rhythms in physiological processes that are temperature dependent. Light is also an important factor in defining rhythm characteristics and it can be manipulated to alter certain circadian rhythms. These external factors serve to fine tune rhythms into a 24-h period and are referred to as *Zeitgebers* or time-givers.

Certain rhythms persist when environmental conditions are kept constant or in circumstances devoid of time cues. The main ones are body temperature and the sleep–wake cycle. They are recognized as being self-sustaining in character and are referred to as endogenous rhythms. Their identification implicates a biological mechanism which times the duration and sequences of processes involved, thereby operating as a clock. The characteristics of endogenous rhythms are not easily altered by changes in the environment; for example, the rhythm in body temperature persists under conditions of sleep deprivation.

When time cues are removed, which is possible within the Arctic Circle in summer, in isolation chambers, or in underground caves, endogenous rhythms drift to a period between 25 and 27 h. There appear to be two master clocks, one controlling the body temperature cycle and the other the sleep–wake cycles. In conditions of continual isolation the temperature rhythm maintains its period of 25 to 26 h but the sleep–wake cycle drifts to a period of 33 h. This dissociation occurs after only 15 days of isolation.[2]

Many human circadian rhythms result from the combined influences of endogenous and exogenous factors. They are not present at birth but develop in the first year of life. The rhythms are entrained to an exact 24-h period by the *Zeitgebers*. There is probably a hierarchy of clocks, with those governing the body temperature and sleep–wake cycles being the major ones influencing many other rhythmic functions. The major clocks may synchronize the activities of other circadian functions in the manner of non-linear oscillators.[3] Many human performance measures tend to follow closely the circadian rhythm in body temperature,[4] but are also affected by the sleep–wake cycle and local physiological conditions within the active tissues.

Neural and hormonal influences (effects of light and dark)

The suprachiasmatic nucleus cells of the hypothalamus have been cited as the probable source of the master clock. In some animals there is a direct link between the suprachiasmatic nucleus and the retina.[3] There is also evidence of a centre in the lateral hypothalamic area with functions related to circadian rhythm in core temperature and cross-linked to the suprachiasmatic region.[5] The pineal gland is sensitive to changes in light intensity during the day, and important time-keeping functions are attributed to it. In this gland, which in simple organisms is thought to function as a third eye, light exerts time-giving properties via a special visual pathway that synapses with the suprachiasmatic nucleus.[6] The hormone melatonin is synthesized from serotonin in the pineal gland, although there is a secondary source of melatonin within the retina. Human melatonin levels are increased at night, with plasma concentrations rising from about 2 pg/ml diurnally to a nocturnal peak approaching 60 pg/ml.[7] The activity of *N*-acetyltransferase, an enzyme associated with the synthesis of melatonin from serotonin, also shows a pronounced rhythmic fluctuation, with nocturnal activity exceeding daytime levels by a factor of over 1000.

Although environmental and body temperature changes affect the circadian rhythm in arousal, it is influenced mainly by the sleep–wake cycle and by neural traffic through the reticular activation formation in the brain. Circadian phase systems are not completely isolated from other time structures with different periodicities. Nevertheless, it is via environmental signals that rhythms are adjusted to an exact 24-h period.

In northern latitudes near and above the Arctic Circle the incidence of a malaise known as seasonal affective disorder increases when daylight hours are short. The condition is accompanied by decrements in psychomotor performance. Exposure to bright or ultraviolet light may also ameliorate the condition. Elite Scandinavian athletes tend to spend some time during their winter training in southern climates which are warmer and have longer hours of daylight. The shortened hours of daylight in winter affect the training of some athletes; for example, professional soccer players tend to train in the morning, whereas their matches are timed for the afternoon or at night under floodlights. Increasing the duration and intensity of exposure to daylight does seem to affect the phase of the circadian rhythm: in the northern hemisphere the peak occurs 55 min later in June than in December.[9] Changes in ambient temperature do not appear to alter the rhythm.

RHYTHMS IN PERFORMANCE

Sports performance

Most athletic records are set in the late afternoon or evening. For example, world record performances by British athletes in track distance races between 800 and 5000 m from 1979 to date took place between 19:00 and 23:00 hours. This partly reflects the fact that record attempts are usually scheduled for evening meetings when the environmental temperature is more favourable for performance than at mid-day or in the early afternoon. Nevertheless, athletes tend to prefer evening contests and consistently achieve their top performances at this time of day.

This preference has also been manifested in the work rate of soccer players during indoor five-a-side games sustained for four days.[10] Observations on players showed that the pace of play reached a peak at about 18:00 hours and a trough at 05:00 to 06:00 hours. Feelings of fatigue were correlated negatively with levels of activity. The self-paced level of activity conformed closely to the curves in body temperature and in heart rate (Fig. 2); this relation persisted throughout successive days. The fact that the freely chosen level of exercise is highest at about the time the body temperature rhythm reaches a peak has important implications for training as well as for certain competitive sports.

Soccer players are accustomed to playing in the evening and tend to feel at their best at this time. Competitive matches last at least 90 min, and the window of daytime during which professional players can function at their best probably stretches over a 4- to 6-h period from mid-afternoon to evening. The usual routine of professionals is to train in the morning. Soccer coaches might

seriously consider altering this convention by timing the training so that players are taxed maximally in the afternoon or evening.

The same advice may not necessarily always apply to short-term activity, particularly the 'explosive' events. Over the last 50 years the only two track and field world records to have been set in morning meetings were in the men's shot put and the women's javelin. French international sabre fencers had their best scores, as far as they related to speed and skill, around noon.[11] Explosive actions with substantial neuromotor components may be linked more closely to the arousal rhythm than to that of body temperature and perhaps reach a peak earlier in the day.

Sports contests are not amenable to the types of manipulation demanded by experimental designs. Consequently, research workers have tended to concentrate on the effects of time of day on performance in time trials or simulated contests. In the first systematic investigation of diurnal variation in performance, six runners, three weight-throwers, and three oarsmen performed better in the evening than in the morning.[12] Swimmers have produced faster times over 100 m at 17:00 hours compared with 07:00 hours in three out of four strokes studied.[13] The speed of running in a 5-min test varied in close correspondence to the circadian curve in body temperature.[14]

The better performances of swimmers in the evening also applies to multiple efforts. Performances in front crawl were found to be 3.6 per cent and 1.9 per cent faster for 400-m and repeated 50-m swim trials respectively at 17:30 hours than at 06:30 hours.[15] This time-of-day effect was apparent through 3 days of partial sleep deprivation. Even after disrupted sleep swimmers can produce maximal efforts, at least if they are required to do so in the evening.

The diurnal variation in swimming performance is not entirely coincident with the circadian rhythm in core temperature. When front crawl times over 100 m and 400 m were examined at five different times of day from 06:00 hours onwards, it was noted that performances improved steadily throughout the day.[16] There was no turning point evident before the final measurement at 22:00 hours, even though body temperature had peaked some hours earlier (Fig. 3). This reinforces the observations that evening is best for sprint swimmers, particularly if time trial results are attributed importance, such as in achieving championship qualifying standards.

There is a time window close to the acrophase of body temperature in which optimal performance in sports involving gross motor tasks can be attained. This can extend for 4 to 6 h provided that meals and rests are suitably fitted in during the daily routine. Sports requiring fast explosive efforts tend to peak earlier and may be related to the sleep–wake clock rather than to body temperature. Consequently, practices where skills have to be acquired should be conducted early in the day or around mid-day, but more severe training drills and 'pressure training' practices are best timed for later in the day. It is acknowledged that sports performance is determined by many variables and there may be multiple performance rhythms. This can be further examined by looking at the existence of rhythms in components of sports performance.

Components of sports performance

Rhythms in motor performance tasks closely approximate the body temperature curve. Performance rhythms are also related to the state of arousal: a low level of arousal predisposes towards errors and injury risk, whereas heightened arousal promotes readiness for intense physical efforts. Thus, mistakes in motor co-ordination that lead to accidents and injuries are more likely

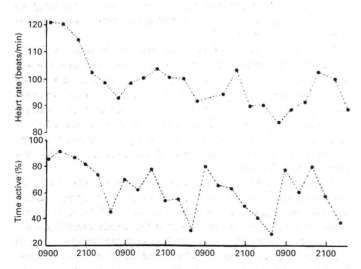

Fig. 2 Mean heart rate and percentage time active in four outfield indoor soccer players every 4 h. Monitoring of heart rate lagged the activity measurement by 1 h. Progression on the horizontal axis indicates successive days of play. (Reproduced from ref. 10, with permission.)

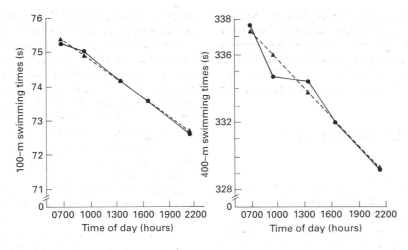

Fig. 3 Performance of swimmers in 100-m and 400-m front crawl at different times of day. The broken lines represent values fitted by a linear trend. (Reproduced from ref. 16, with permission.)

early in the day. Such occurrences later in the day could be due to fatigue resulting from time spent on the task, as in sailing or rally driving, for example.

Isometric muscle force, dynamic muscle activity, neuromotor performance, and gross motor performance are all subject to circadian rhythms.[14] The timing of the peaks tends to follow the phase of the body temperature curve, although this does not imply that they are caused by the changes in temperature. Rhythms in grip strength and back strength have been replicated in dynamic muscular activity such as the vertical jump[14,17] and the standing broad jump. The variation in such tests attributable to time of day ranges from about 3 per cent of the mean value for jumping to 6 to 10 per cent for isometric strength to about 15 per cent for power output on a swim-bench. The amplitude of the rhythm increases with increasing complexity of the task. Although seemingly small in magnitude, an improvement of the order of 3 per cent can have a profound effect on competitive performance. As tests of this type are easy to administer, they can be employed as markers of circadian rhythm in monitoring the effects of desynchronization on performance.

The rhythm in muscular strength is robust and persists under conditions where subjects are deprived of sleep for four consecutive nights. The circadian variation in grip strength is greater than the effects of sleep loss on this function.[15,18] Although the normal rhythm in muscular strength is in phase with that of body temperature, the extent to which alterations in muscle temperature or in motivation of subjects contribute to the rhythm is not clear. Muscle performance is optimal at a muscle temperature of around 39°C and a core temperature of 38.3°C.[19] Competitive athletes may elevate muscle temperature to this level as a result of warming up and so might over-ride an inherent rhythm in muscle performance. Consequently, attention is directed towards a proper warm-up regimen in cases where competitors have to perform in the morning or at night (delayed start) when muscle temperature would otherwise be suboptimal. There is a strong motivational component to muscular strength, and experimental work comparing muscle performance under conditions of maximal voluntary contraction and in response to electrical stimulation is needed in order to establish whether the rhythm in strength is due to central or peripheral factors.

Muscle strength is usually measured by cable tensiometry,

strain gauge assemblies, or isokinetic machines. Although a circadian rhythm in strength is evident in human performance tests, the magnitude of the circadian variation in muscle performance may be beyond the sensitivity of contemporary computer-linked dynamometers. Isokinetic dynamometry has failed to show evidence of a significant rhythm in peak torque for fast and slow actions during maximum concentric and eccentric efforts of the knee extensors.[20] For circadian rhythms to be identified in such actions the measurement error must be small. Similarly, a vigorous warm-up can partly overcome the night-time troughs in performance. In leg exercise the anaerobic capacity (as measured by the Wingate test) is reduced by 8 per cent at 06:00 hours compared with 14:00 hours whereas the peak power value is maintained well;[20] a decline in performance during the 30-s test suggests a motivational component in the circadian variation in anaerobic capacity. When all-out arm exercise for 30 s is preceded by a vigorous warm-up, there is no distinguishable rhythm in mean power or peak power.[22] Since such tests call for total subject compliance in producing maximal efforts, it is difficult to distinguish the motivational basis from a true biological rhythm as an explanation of the time of day effect.

Some psychomotor and muscular strength fluctuations show a bimodal pattern in their circadian rhythms. For example, tapping speed, indicated by moving a stylus as rapidly as possible between two plates, shows a slight dip in performance in the early afternoon (Fig. 4). This is referred to in occupational contexts as the 'post-lunch' dip, although it persists in some tasks even when no lunch is taken.[24] In the study of tapping performance, measurements were performed repeatedly throughout the working day over a 3-month span.

A similar fall has been noted in the isometric strength of the knee extensors.[25] In this case measurements were repeated at 2-h intervals throughout the normal period of wakefulness. It is only when performance tests are repeated frequently throughout the day that this afternoon decline, analogous to the post-lunch dip in ergonomic tasks, is observed. This transient decrement in performance might be due to an underlying 90-min cycle that prevails during sleep and during the day comes closest to the surface of detection in the early afternoon.[26] It may also be linked with the turning point of circulating adrenaline and noradrenaline levels whose peak is attained at about this time.[27] Sports specialists who

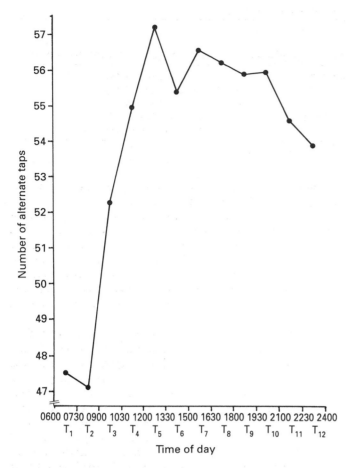

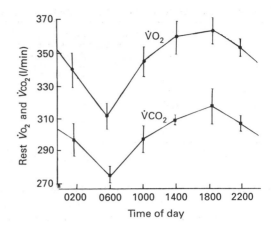

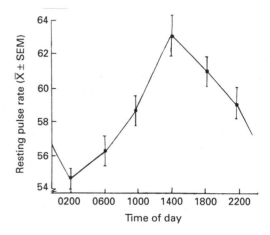

Fig. 4 Speed of tapping performance during the day showing a post-lunch dip in speed. (Reproduced from ref. 23, with permission.)

Fig. 5 Mean values (± SE) for $\dot{V}o_2$ and $\dot{V}co_2$ at rest (top) and pulse rate pre-exercise (bottom) at six different times of the day. (Reproduced from ref. 31, with permission.)

spread their training over, say, three or more separate sessions with long training times overall are the most likely to benefit from an afternoon nap.

Rhythms have been identified in sensory motor (reaction time), psychomotor (hand–eye co-ordination), sensory perceptual, cognitive, and psychological functions.[14, 28–30] It is incorrect to generalize to a single performance rhythm since different types of task can show different circadian rhythms, depending largely on whether they are influenced by the body temperature or the sleep–wake clock.

To determine the links between performance rhythms and physiological processes it is necessary to examine (i) the scale of circadian rhythms in resting conditions and (ii) the persistence of these rhythms during exercise.

CIRCADIAN RHYTHMS IN PHYSIOLOGICAL FUNCTIONS

Physiological rhythms at rest

Many physiological functions are known to show circadian rhythmicity. They include metabolic, cardiovascular, and endocrine functions. The metabolic functions showing cyclical changes include oxygen consumption $\dot{V}o_2$, carbon dioxide production $\dot{V}co_2$, and minute ventilation $\dot{V}_E$. The rhythms in $\dot{V}o_2$ and $\dot{V}co_2$ have an amplitude of about 7 per cent of their mean resting value, whereas that in $\dot{V}_E$ is about 11 per cent (Fig. 5). Only about one-

third of the variations in metabolism can be explained by the circadian rhythm in body temperature, despite the fact that the peaks occur close together in time.[14,13]

Cardiovascular functions also display a rhythm similar in shape to the body temperature curve.[32] The peak in heart rate occurs earlier in the day than that of body temperature but is close to the phase of the rhythm in circulating catecholamines. The peak-to-trough variation at rest is about 8 beats/min; this should be taken into account by athletes and their mentors when resting pulse rate is used as an index of either training state or overtraining. The catecholamine rhythms are probably closely related to changes in arousal; they have been shown to influence rifle shooting performance and are negatively related to fatigue during the day.[33]

Output from endocrine glands, particularly the trophic hormones of the anterior pituitary, exhibit circadian rhythms. Some endocrine secretions, for example growth hormone, prolactin, and testosterone, peak during the night. There is no overall pattern to the rhythms that would help to explain the circadian curves in exercise performance. The rhythms in adrenaline and nonadrenaline are the most closely related to the performance curve.[33] The excretion of electrolytes is also closely related in phase to the body temperature and performance rhythms. The diurnal changes in renal function have only marginal impact during exercise because of the relative shutdown of blood flow to the kidneys and the inactive muscles. Angiotensin and aldosterone are

Table 1 Cosinor parameters for heart rate including rest, exercise, and recovery values ($n=15$)

| | Heart rate (beats/min) | | | |
	Mean	Amplitude	p zero amplitude	Acrophase (95% confidence limits)
Rest	70	4	$p<0.001$	13:50 (12:05–15:35)
Light exercise	105	4	$p<0.001$	13:10 (11:15–15:05)
Medium exercise	134	4	$p<0.001$	13:48 (11:44–15:52)
Maximum	181	2	$p<0.005$	14:10 (12:02–16:18)
3rd minimum	126	5	$p<0.001$	14:04 (12:14–16:54)

Statistically significant rhythm was demonstrated in all cases ($p< 0.005$). Reproduced from ref. 34, with permission.

hormones linked not only with renal function but also with the circadian rhythm in blood pressure, although this function merits most consideration in contexts of exercise for health rather than elite sport.

Physiological responses to exercise

Submaximal exercise

Physiological rhythms detected at rest may be obliterated or attenuated, or maintained or amplified, under exercise conditions. Research reports can be marshalled to support all these possibilities. The conflict is partly due to a failure to control the environment adequately and to masking factors such as diet and previous activity of subjects. When these variables are carefully considered, a consistent picture begins to emerge. For example, the heart rate rhythm that is noted at rest is still evident during light and moderate exercise (Table 1). The consistency of its phase and amplitude applies to both leg[31,34] and arm[35] exercise.

The heart rate rhythm is not paralleled by $\dot{V}O_2$ and $\dot{V}_E$ throughout the range of submaximal exercise intensities. This applies to arm exercise as well as leg exercise. In a longitudinal study of one subject the rhythm in $\dot{V}O_2$ gradually faded away as the exercise intensity increased.[31] The rhythm at a moderate exercise level was accounted for by variations in body weight. No circadian rhythm was found for $\dot{V}CO_2$ or the respiratory exchange ratio during moderate exercise. This indicates that choice of substrate as fuel for exercising muscle is not determined by time of day, once diet, environmental temperature, and activity are controlled. The rhy-

thms in $\dot{V}O_2$ and $\dot{V}CO_2$ may no longer be detectable once the exercise intensity reaches 40 to 50 per cent of $\dot{V}O_2$max.[34]

The most robust rhythm seems to be that of $\dot{V}_E$ which is amplified at light and moderate exercise intensities.[31] Even when $\dot{V}_E$ is expressed as the ventilation equivalent of oxygen (Fig. 6), the rhythm is clearly evident.[36] This may partly explain the mild dyspnoea sometimes associated with exercising in the early morning. At vigorous exercise intensities, the point at which $\dot{V}_E$ begins to increase disproportionately to $\dot{V}O_2$, does not vary with time of day.[31]

The temperature in soft tissues around the joints can affect their resistence to motion and this could alter the energy cost of exercise. The muscular efficiency, which represents the mechanical work done as a percentage of its energy cost, can be computed as gross or net efficiency. Where more than one steady rate of exercise is performed, the delta efficiency can be calculated. For a given exercise mode the muscular efficiency is influenced more by the work rate and the mode of computation then by time of day.[31] The greater joint viscosity at night does not significantly affect the time required to attain 'steady state' metabolic conditions during submaximal exercise.[36]

Responses of catecholamines, aldosterone, cortisol, and plasma renin activity have been examined during 25 min of exercise at 60 per cent of $\dot{V}O_2$max.[37] Time of day generally did not affect the hormonal or haemodynamic responses to exercise, with the exception that plasma renin activity was markedly higher during exercise at 16:00 hours compared with 04:00 hours. This was thought to reflect a greater vasoconstrictor activity in the cutaneous blood vessels.

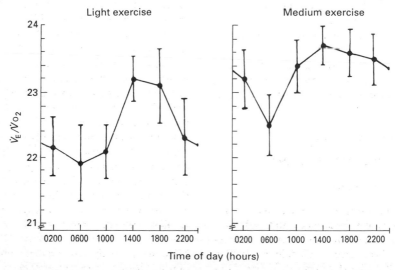

Fig. 6 The rhythm in the ventilation equivalent of oxygen at light and medium exercise intensities. (Reproduced from ref. 36, with permission.)

Maximal physiological responses

When maximal values are being attributed to measurements, the question arises as to whether the ceiling of physiological function was reached during the exercise test. Consequently, recognized criteria are applied when assessing $\dot{V}O_2$max. Otherwise, data collected during graded exercise to volitional exhaustion may merely reflect the reluctance of subjects to work at $\dot{V}O_2$max at night. For a well-trained individual with a body mass of 70 kg the normal amplitude of the $\dot{V}O_2$ rhythm at rest would be less than 0.5 per cent of the mean maximal value. It is difficult to detect this against the background of biological variation and measurement error associated with assessing $\dot{V}O_2$max. Therefore, it should not be surprising that the most carefully conducted studies fail to show circadian variation in $\dot{V}O_2$max. This applied when 12 different times of day and duplicate measurements at each time point were used.[38] It also applied to a longitudinal design used to eliminate variability between subjects and to results of a cross-sectional approach:[31,34] the coefficient of variation of $\dot{V}O_2$max was found to be 2.9 per cent. In these studies it was concluded that $\dot{V}O_2$max is a stable function, independent of time of day. This is in sharp contrast to the value predicted from submaximal heart rate (Fig. 7), which shows an error in estimating $\dot{V}O_2$max that is not acknowledged when the maximal function is predicted from a submaximal test.

The circadian rhythm in $\dot{V}_E$, which is apparent during light and moderate exercise, similarly disappears under maximal aerobic conditions, at least for leg exercise. In arm exercise the highest metabolic measurements generally do not demonstrate a plateau and are lower than observed at $\dot{V}O_2$max because of the decreased muscle mass involved. Consequently, the highest measurements during arm exercise are referred to as peak rather than maximal values. During performance of arm ergometry the $\dot{V}O_2$ peak and highest heart rate demonstrate a circadian rhythm, the highest values being observed close to the crest time of rectal temperature.[35] The results reflect a rhythm in the total work performed.

Studies of the maximal heart rate during exercise have consistently shown an influence of the time of day.[32] The circadian rhythm is similar in phase to that noted at rest and submaximal exercise but its amplitude is reduced. A circadian rhythm in recovery heart rate is evident soon after maximal exercise ceases.[32,34] Therefore fitness indices such as the Harvard test score could

contain an error as large as 5 per cent due to the time of day that the test is performed. This also means that self-monitoring of post-exercise pulse rates by athletes and coaches is subject to at least this degree of error.

In view of the stability of $\dot{V}O_2$max throughout the solar day, a rhythm in maximal aerobic power cannot explain time-of-day effects in all-out performance such as 400-m swim trials and 5-min shuttle runs. It might be accounted for by an ability to sustain a fixed exercise intensity for longer in the evening. Exercise to voluntary exhaustion at an intensity close to $\dot{V}O_2$max does exhibit circadian variation.[39] Subjects tested on a bicycle ergometer cycled for longer in the evening (22:00 hours) than in the morning (06:30 hours), with the mean values being 436 and 260 s respectively. The subjects also tolerated higher blood lactate levels in the evening as a result of the increase in total work performed at that time.

It is possible that rhythms in vigorous exercise performance may be due to a combination of motivation and psychological drive that is reflected in anaerobic power output. There is a circadian rhythm in power output in a stair run test, although its amplitude is only 2.5 per cent.[21] This is compatible with the circadian variation in body temperature, as maximal anaerobic exercise is impaired by about 5 per cent for each 1°C fall in core temperature.[40] Some studies of anaerobic efforts using the Wingate test have failed to show a significant rhythm in anaerobic power and capacity for leg[21] and arm[22] exercise, although a fall in anaerobic capacity (mean power output over 30 s) at night without a corresponding decline in peak power output (highest power output in the first 5 s) was noted in the study of leg exercise. A comprehensive warm-up procedure employed prior to experimental tests can swamp underlying rhythms. The sensitivity of such tests may be insufficient to detect performance rhythms that are small in magnitude. As there is no objective physiological criterion that maximal anaerobic capacity is being employed in tests such as the Wingate model, results may be due mainly to motivation. This would account for the large diurnal variation in performance in those studies reporting positive findings. Power outputs 5 to 8 per cent higher in the day compared with the night have been reported. Even when this difference in performance between 14:00 and 06:00 hours is observed, data throughout the whole day may not conform to a significant rhythm.[22] Power output over 30 s on a

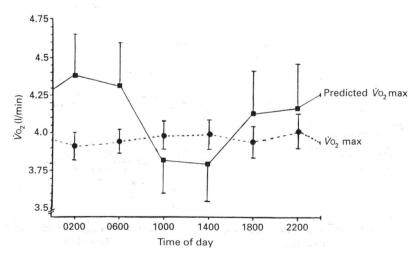

Fig. 7 The contrast between results measured for $\dot{V}O_2$max throughout the day and values predicted from heart rate response to submaximal exercise. (Reproduced from ref. 28, with permission.)

swim-bench does exhibit a circadian rhythm. The acrophase for peak power and mean power (anaerobic power and anaerobic capacity) was found to occur at 16:20 hours with amplitudes of the rhythms being 14 per cent and 11 per cent respectively.[41] These rhythms were linked to both the body temperature and subjective alertness curves, and so the relative effects of these two influences could not be determined. They supplement the observations of circadian variation in short-term efforts such as are required in isometric muscle strength and dynamic jump tests.[14,17]

THERMOREGULATION

Core temperature (reflected in oral, oesophageal, tympanic, or rectal temperature) possesses an endogenous rhythm and so is used in chronobiological studies as a physiological marker of rhythmicity. This rhythm persists during exercise. A 5 per cent impairment in maximal anaerobic power would be predicted at night due solely to the fall in core temperature. The optimal time of day would be in the evening when muscle and core temperatures are at their peaks.

This rationale may not apply to sustained high intensity exercise, such as marathon running, in hot conditions. In such instances thermoregulatory requirements may limit endurance performance. This supports the practice of starting marathon races in the morning rather than in the afternoon in hot climates, although the argument has been based solely on the lower environmental heat stress in the morning. The ideal ambient temperature for marathon running is around 13°C. In cold and wet conditions a morning start would place the slower performers at increased risk of hypothermia. The advantage of starting sustained exercise at a core temperature which would normally be suboptimal is that the onset of heat stress is delayed and the overall strain on thermoregulatory mechanisms is reduced. Experimental support for this view is available.[42] Pre-cooling oesophageal temperature by 0.4°C and mean skin temperature by 4.5°C before 60 min of submaximal exercise causes a 6.8 per cent overall increase in work rate compared with control conditions.

The circadian rhythm in core temperature during exercise represents a fixed thermal load superimposed on the resting baseline temperature.[43] This occurs despite a circadian variation in the mean skin temperature for onset of sweating and in blood flow to the skin.[44] The rhythm in rectal temperature apparent at rest persists in phase and amplitude at different levels of exercise. There are closely related cyclical changes in rectal and skin temperatures.[43] The time course of skin temperature changes varies with the body surface location, and the rhythm in the exercising limb tends to disappear during exercise because of the convective air flow created by the leg movements (Fig. 8).

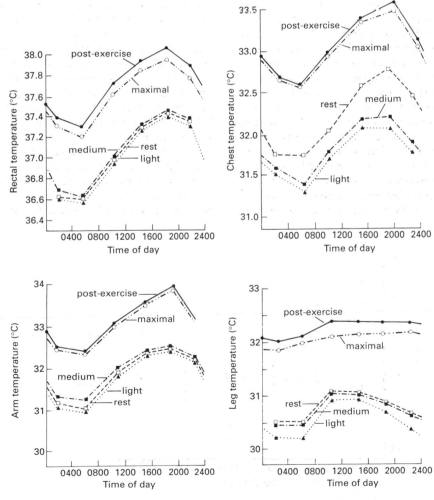

Fig. 8 Mean values for rectal and skin temperatures according to time of day and level of exercise: light (82 W), medium (147 W), and maximal. Rest and post-exercise values are included. (Reproduced from ref. 43, with permission.)

PHYSICAL FACTORS IN CIRCADIAN RHYTHMS AND EXERCISE

Spinal shrinkage

There are physical as well as physiological factors that vary with time of day and have implications for exercise. The day–night cycle of activity and rest provides an alternation of weight bearing associated with the upright posture on the one hand and recovery while sleeping on the other. Weight bearing imposes compressive loading on the spine and leads to a loss of disc height primarily because of extrusion of water through the disc wall.[45] This loss of height is known as shrinkage and is reflected in measurements of changes in stature using appropriate apparatus. Spinal shrinkage during the day is about 1 per cent of stature.[46] The circadian rhythm in shrinkage is of the same order of magnitude in females[47] but is more erratic in subjects with back pain syndrome.[48]

The data in Fig. 9 conform to a cosine curve but a power function best fits the shrinkage during the day, with the rate of loss declining as the discs stiffen.[47] Shrinkage is reversed at night whilst recumbent, with the rate of regain in height being greater in the first part of the night's sleep. An endogenous rhythm in stature could be reinforced or masked by the changes in posture and spinal loading between day and night. The rate of reversal of shrinkage suggests that circadian variation in stature is largely attributable to spinal loading and unloading rather than to an endogenous rhythm.[46]

The loss of disc height could render the spine more vulnerable as its stiffness increases,[49] thus making weight-lifting and similar activities more hazardous as stature recedes throughout the day. Loss in height alters the dynamic response characteristics of the disc and so there is a time-of-day effect on the shrinkage resulting from a fixed exercise regimen. Losses of height as a result of a 20-min circuit weight-training regimen are greater at 07:30 hours than at 23:00 hours. The disc was found to be a more effective shock absorber in the morning when shrinkage was 5.4 mm compared with a mean value of 4.3 mm in the evening.[47] This greater

stiffness in the evening is partly compensated by greater back muscle strength. A high negative correlation was found between back strength and height lost: the greater the muscle strength, the less height was lost.

The habitual activity level and the postures engaged in can influence the amount of shrinkage during the day. Intervention procedures for unloading the spine prior to heavy physical training in the evening have been advocated. An example is the Fowler position at rest with the trunk supine and the legs raised to rest on a bench. Gravity inversion systems have also proved effective,[50,51] although the spinal distension induced before exercise is quickly lost once a strenuous exercise regimen is undertaken.[52]

Joint stiffness and flexibility

Other physical factors that are known to vary with time of day to affect muscular function are joint stiffness and flexibility. The former refers to resistance to motion whilst the latter indicates range of movement about a joint. There is circadian variation in stiffness of the knee joint, with increased stiffness observed late in the evening and early in the morning.[53] Stiffness increases as body temperature decreases, and so the curve may be related to the fall in body temperature towards the end of the day. It may also be affected by activity patterns during the course of the day.

There is a circadian variation in trunk flexibility[16] and hip flexibility.[28] Trunk flexibility measurements were made five times during the day between 07:00 hours and 22:00 hours. Trough values were observed first thing in the morning and peak values in the middle of the day (13:00 hours).[16] Circadian variation has also been found in lumbar flexion and extension, passive straight leg raising, glenohumeral lateral rotation, and the distance from finger tip to floor in forward flexion. The mean acrophase was 18:10 hours, although individual peaks ranged from mid-day to near midnight.[54]

Diseased joints may exert an influence on the circadian rhythm in musculoarticular function. The peak and trough times of the

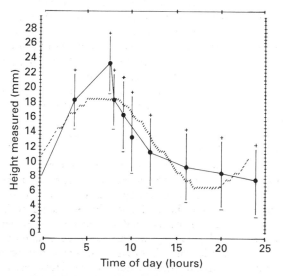
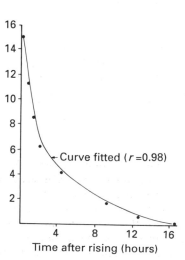

Fig. 9 Changes in stature during a 24-h period from a baseline set at 7.5 mm (left). The full line indicates mean observations with 95 per cent confidence limits. The power function relating time since rising from sleep to the same data for loss of height is shown on the right. This curve overlaps the observed changes in height almost perfectly. (Reproduced from ref. 47, with permission.)

rhythm in the grip strength of patients with rheumatoid arthritis have been found to be similar to those in healthy subjects.[55] Despite this normality in phase, the peak-to-trough variation may be affected: the peak-to-trough difference in rheumatoid arthritic patients is approximately three times the normal range.[56] The magnitude of this variation is reduced with corticosteroid administration, but the timing of the rises and falls in muscular strength is unaffected.[457] The rhythms in muscular strength help to explain the variations in propensity towards physical activity in these patients during the day.

SUBJECTIVE STRAIN

Perceived exertion

A circadian variation in the subjective reaction to exercise might explain why performance is generally better in the evening than in the morning. The perception of effort does not seem to vary at light ergometric loads up to 150 W, although the slope of the relationship between heart rate and perceived exertion changes with time of day.[32] Ratings of exertion have been studied at treadmill running speeds eliciting heart rates of 130, 150, 170 beats/min.[38] As the heart rate response to a fixed submaximal exercise intensity is lowest at night, it follows that more exercise can be performed at a given heart rate at that time. The higher subjective ratings reported at night may be due to the work rate and not to an inherent variation in effort perception.

The circadian variation in the rating of exertion is evident once a high steady rate of exercise is reached. In one investigation of circadian variation in perceived exertion a significant result was found only at one submaximal load (245 W), although several lower exercise levels were rated.[58] This is supported by findings of others whose subjects first cycled at 40 per cent of $\dot{V}O_2$max for 5 min before cycling to exhaustion at 95 per cent of $\dot{V}O_2$max at two different times of day, 06.30 hours and 22:00 hours.[20] No difference in perceived exertion with time of day was noted at either work rate, although the higher work rate was sustained for longer in the evening to the point where the same end-point (exhaustion) was reached. The findings offered support for the concept of a motivational component in the circadian variation in exercise performance.

The predisposition of athletes towards exercise can be examined in another way by allowing them to choose an exercise intensity that they are prepared to tolerate for a predetermined period. This self-selected work rate exhibits a circadian rhythm that is in phase with the body temperature rhythms.[59] In fact it may be more strongly influenced by the arousal cycle. It reinforces the recommendation that strenuous training regimens are best carried out in the late afternoon or evening to obtain optimal compliance from athletes.

Submaximal exercise is perceived to be harder in the morning than in the evening under conditions of partial sleep loss. The time-of-day effect is greater in magnitude than the effect of sleep loss.[15,18] The results suggest that changes in the subjective reactions to exercise with time of day may be tied more closely to the rhythm of arousal than to that of body temperature.

A standard circuit of weight training was rated as harder when conducted at 07:30 hours compared with 22:00 hours. In this instance the perceived exertion was related to back strength ($r = -0.59$), which was higher in the evening than in the morning.[47]

An alternative paradigm which employs light work bouts every 4 h for 24 h may include a cumulative fatigue effect. With this experimental protocol exercise was rated harder at night and perceived exertion was marginally elevated in the afternoon.[14,28] This may represent a subharmonic in the circadian rhythm of biological arousal, which is sometimes reflected in a post-lunch dip in performance[24,26] and in a siesta in Hispanic cultures.

The fact that a set exercise regimen is perceived as harder in the morning than in the evening[47,58] has implications for timing of training. It would be logical to suggest that greater training loads would be tolerated in the evening compared with earlier in the day. Caution should be expressed before arriving at any generalizations, since effects of warm-up, occupational activity during the day, individual characteristics, and the nature of the training stimulus need to be considered. One longitudinal study[60] attempted to establish whether the training effect due to a fixed exercise bout differed with time of day. Although there was some evidence of the specificity of adaptation due to time of day, there were no substantial circadian variations in the training effect that accrued.

Pain perception

There is also a circadian variation in pain perception that might be relevant in the context of sport injury. Self-ratings of pain intensity in patients with painful conditions show definite patterns throughout the waking day. Minimum levels of pain are noted in the morning, with a more or less steady increase during the day to a peak during the evening.[61] Laboratory investigations have shown similar findings, with subjects becoming more sensitive and the pain perception threshold falling as the day proceeds. The highest threshold for epicritic pain is noted at about 03:00–06:00 hours, and that for more diffuse pain at 11:00 hours.[62] These observations have implications for the prescription of analgesics for a variety of sports injuries and other painful conditions.

Time estimation

Estimation of how quickly time passes is also influenced by time of day.[63] The usual method of time estimation is to ask subjects to count to themselves at a 1-s rate for 60 s. The rhythm in subjective time perception is related to a chemical clock hypothesis: the higher the body temperature, the quicker the chemical reaction, the faster the internal clock, and the faster the speed of counting. The chemical clock theory is supported by observations that lowering body temperature in divers leads to overestimation of time lapsed.[64] Whether the circadian variation in time estimation affects co-ordination tasks entailing fine timing of actions or prolonged efforts where motivation may decline has not been investigated.

INDIVIDUAL DIFFERENCES

There is some evidence that the phasing of circadian rhythms is affected by personality. Introverts tend to be better performers in the morning, whereas extroverts are more sluggish at this time. They make up for this by reaching a peak level of performance later in the day and staying alert for longer in the evening. The body temperature curves of these different personalities show a similar form, but that of the introvert peaks earlier in concordance with the performance curves. However, the difference in acrophase between these extremes of personality is only about 1 h.[65]

Individual variability has also been described according to how people phase their habitual activities throughout the day. Circadian phase types have been referred to as 'larks' (morning types) and 'owls' (evening types), indicating a preference for morning or evening work.[66] This preference is likely to be a more influential factor than personality in minor variations in rhythms between subjects, since the majority of people do not exhibit extreme personalities. Being a morning type is more important than extroversion in determining individual differences in the phase of circadian rhythms.[67] The existence of morning and evening types among athletes has not been systematically examined. Most responses to exercise on a cycle ergometer do not differ between the two circadian phase types.[68] This finding also applies to swimmers whose circadian rhythms in power output on a swim-bench are unrelated to circadian phase types.[37] Apparently healthy subjects may commonly display internal desynchronization of a set of circadian rhythms. Competitive athletes with internally synchronized rhythms generally have a better chance of being victorious than individuals with internal desynchronization. This was confirmed in a study of French fencers participating in the 1984 Olympic Games.[11] External desynchronization of rhythms is effected by nocturnal shift work and travelling across time zones. These, together with disruptions due to sleep loss, are discussed next.

DESYNCHRONIZATION

Jet lag

Desynchronization refers to the disruption of circadian rhythms which become out of phase with each other. Rhythms are externally desynchronized if the individual switches to a nocturnal work schedule or rapidly crosses a series of time zones. The physiological rhythms affected include body temperature, the sleep–wake cycle, pulse rate, ventilation, arterial pressure, diuresis, and excretion of electrolytes.[69] The disorientation that results is known as jet lag, and symptoms include fatigue and general tiredness, inability to sleep at night, loss of appetite, loss of concentration, decreased drive, headache, and general malaise. Jet lag can impair the performances of athletes on busy competitive schedules until the major rhythms have all adjusted to the new time zone.[70]

Westward travel, when time is extended, tends to be easier to adjust to than travelling in an easterly direction. This reflects the fact that in conditions where the major environmental signals are absent, the period is lengthened to 25 to 27 h.[3] Performance of psychomotor and athletic skills tends to suffer more after eastward than after westward flights. The difference may be negligible if the time zone shift is 9 to 12 h or near maximal (Fig. 10).

Data on rugby players after travelling to Australia showed a diminution in muscular strength until jet-lag symptoms abated.[71] Recommendations were that training should best be performed in the morning until the body's rhythms are resynchronized.[72] In crossing fewer time zones (phase shifts of 3–5 h) it is advisable to train within the window of time when rhythms in the home and in the local country are above the mesor. For these medium-range phase shifts this time would be mid-day after a westward flight and in the evening after an eastward flight.

Coping with jet lag

A variety of strategies has been suggested for offsetting the effects of jet lag. The first is to tune in mentally to the new local time as soon as possible. This can be done during the flight by altering one's watch, missing meals if inappropriately timed, and avoiding alcoholic and other diuretic drinks such as strong coffee. It is also advisable to drink plenty of fruit juices to avoid the dehydrating effects of inspiring dry cabin air.[69]

Altering bedtime for a few days before departure in the direction of intended travel and for a few days after arrival is sometimes suggested as a means of lessening the disruptive effects of travelling across time zones. Bedtime is delayed for a few nights prior to travelling westwards, with the opposite applying before flying east. Since only the behaviour and sleep–wake cycles, and not environmental variables, are adjusted, the alterations in phase of the new biological rhythms occur at a slower rate than in time zone transition. The entrainment rate for alertness is faster than for core temperature, but even after 8 days on a 3- to 5-h shift of the sleep–wake cycle the adjustment of the alertness rhythm is incom-

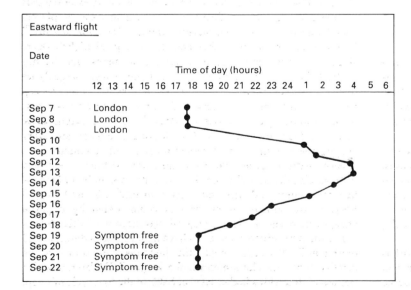

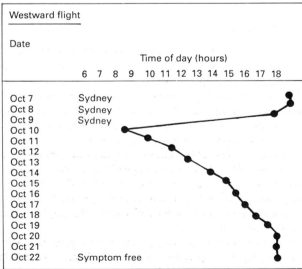

Fig. 10   Time of day at which oral temperature attained a peak plotted according to local time on outward (eastward) and return (westward) journeys between England and Australia. A stopover of one night in Singapore occurred on both journeys. The points were determined using cosinor analysis of data collected throughout each day on one subject. (Reproduced from ref. 71, with permission.)

plete. Entrainment of physiological, performance, and psychological variables in response to phase advance and phase delay of the sleep–wake cycle is easier with phase delay of the cycle, simulating preparation for westward travel, than with phase advance. The conclusions are as follows: shifting the sleep–wake cycle alters rhythms in accord with the direction of the shift; however, motor performance is compromised during the course of such adaptive changes, with both mean and amplitude being depressed.

Sleep is a stronger synchronizer of circadian rhythms than are meal times or social activities. Consequently, prolonged napping at the new location should be avoided as this operates against adaptation by anchoring the rhythm in its previous phase.[73] This tactic of anchoring or maintaining circadian rhythms at the time phase of the country of departure is used by non-sports personnel who regularly cross time zones for short spells abroad. Its use by sports practitioners would be feasible on flying in for single contests, but might be counteracted by the effects of 'travel fatigue'.

Minor tranquillizers (the benzodiazepines) are effective in inducing sleep but not necessarily good at keeping the individual asleep. Also, hangover effects cannot be discounted.[70]

It is important to use natural signals to help resynchronize rhythms. These include both natural daylight and bright artificial light. These work to link the arousal and sleep–wake cycle to the new environment. Exercise stimulates catecholamine production and increases alertness, and so is an effective resynchronizer. Moderate exercise is recommended, even on the day of arrival, unless it is very late in the evening local time. It switches the traveller quickly into local time cues, and is also a good antidote to travel fatigue.

The best plan is to ensure that the athletes arrive in the country of destination in good time for physiological and performance rhythms to resynchronize. Empirical data on runners crossing the Atlantic support the wisdom of allowing 1 day for each time zone shift to allow adaptation to occur. The coping strategies already outlined should help to shorten this time. Rhythms in potassium excretion may take longer than this to readjust,[4] as might rhythms in complex skills such as pistol shooting.[74] Where possible, friendly matches should be arranged in the early stages of team tours and the important competitions later when complete resynchronization should have occurred.

Shift work

About 20 per cent of workers in technologically advanced countries are on some form of shift-work system. Inevitably this number includes sports participants. Although some British and Irish distance runners have gained Olympic representation whilst employed on nocturnal shift-work systems, these have tended to be fast rotating shifts rather than predominantly nightwork.

Training programmes and performances of athletes will be disturbed by the desynchronization of rhythms that shift work causes. The extent of the disturbances depends on the type of shift system employed and whether the individual is allowed to sleep during the shift when 'on call' but not actively engaged. Performances adversely affected include muscular strength, simple motor tasks, and memory-loaded tasks.[75]

Difficulty in ensuring good quality sleep during time off work is a major problem. Apart from sleep disturbances, gastrointestinal problems linked with unusual meal times constitute a major source of discomfort. Long-term issues may be more related to health than to fitness of nocturnal shift workers.

About one-third of workers are intolerant to shift work, and this intolerance is linked to the degree to which rhythms are internally synchronized. Individuals with large amplitudes in their rhythms, and who therefore adjust slowly to altered schedules, are at an advantage in fast rotating shifts.[75] Physical activity contributes to strengthen the internal synchronization,[76] and so athletes are more tolerant of nocturnal shift work than are non-athletes.

The rate at which circadian rhythms revert to their normal cycles seems to be much faster than their adaptation to nocturnal shift work.[77] Even so, it would be advisable to leave a substantial period of time between termination of nightwork and participation in a major sporting event. Individuals with serious sporting aspirations will experience difficulty in realizing their ambitions if their occupations demand working at night. The fast-rotating form of shift-work systems leaves them with some opportunities to organize training and competition schedules. Further, a consistently high level of performance requires a change to daytime work or to full-time engagement in sport. These moves allow the individual to establish the kind of daily routine into which training and competitive programmes, social activity, and leisure can be slotted to dovetail with the body's endogenous rhythms.

THE SLEEP–WAKE CYCLE

Sleep and exercise

Sleep can be described as incorporating a cycle of stages that recurs about every 90 min. Use of electroencephalography (EEG) and electro-oculography (EOG) has provided insights into underlying events. The two major types of sleep are rapid eye movement (REM) sleep, which comprises roughly 20 per cent of total sleep and is discernible by EOG, and non-REM sleep which is subdivided into four stages. Stage 2 is longer than any other stage and usually precedes or follows REM, the phase when dreaming occurs. Stages 3 and 4 together are known as slow wave sleep (SWS) because of their high frequency, low amplitude EEG waves. Early in sleep SWS predominates; REM sleep dominates later.

Although the mean length of human sleep is about 8 h, there is a large variation between individuals in the amount taken, with the coefficient of variation being about 30 per cent. Athletes assume that sound sleep on a regular and habitual basis is an essential part of preparing for top performances. Professional soccer players spend a lot of time resting, meriting the title 'Homo recumbans'.[78] Duration is not the only characteristic of a good night's sleep, since restfulness (indicated by relative movements) and latency (indicated by the time between lights out and onset of stage 2) are important aspects. Aerobically fit athletes display shorter sleep latencies and longer sleep periods than normal, as well as a tendency towards greater levels of SWS.[79] To what extent these features reflect contemporary life-styles of elite athletes and other characteristics rather than aerobic fitness is undetermined.

Sleep patterns of athletes were monitored at times when they were aerobically fit and when they were deemed to be unfit: profiles were compared with sedentary controls.[80] Elevated stage 3 SWS in the athletes when fit was compensated by opposite changes in Stage 4. The biological significance of this shift is unknown. The athletes tended to have a longer sleep duration and more non-REM sleep, although the time in bed was similar to the inactive controls. The athletes had a longer REM latency which was associated with a higher level of SWS in the first cycle of sleep. As the differences between the athletes and non-athletes could not be ascribed to aerobic fitness, it appears that the sleep profiles reflected a trait in athletes rather than an effect of training.

Residual fatigue from daytime exercise is believed to alleviate many problems in sleeping. Subjective reactions after exercise indicate a greater degree of tiredness and sleepiness and, if the exercise is not too vigorous, subjects feel that they sleep better. The intensity and duration of exercise may affect subsequent sleep, as does the interval for recovery before retiring to bed. If vigorous exercise is conducted late in the evening, particularly by unfit individuals, the effect may be a delay in sleep onset rather than an induction of sleep. Physiological recovery processes begin once exercise is ended, and the elevations in metabolism and body temperature induced by exercise may delay onset of sleep. Also, soreness after running on hard surfaces or as a result of physical contact activities may cause discomfort to a level that prevents restful sleep.

Sleep is an enigma in the sense that there is not a consensus among sleep researchers about its essential function. One school of thought relates sleep to the restitution of the body's tissues.[81] Tissue restitution theories have been linked with heightened mitosis during sleep and elevated growth hormone secretion during SWS stages. An alternative view is that the need for sleep is specific to nerve cells.[82] The argument is based on the observations that tissue restitution proceeds during wakefulness, even after exhausting exercise, and convincing evidence such as increased protein turnover in sleep is lacking. Nevertheless, it is obvious that sleep is essential, and this need is apparent when we consider how human operations deteriorate with severe loss of sleep.

Sleep deprivation

There is an interaction between sleep loss and circadian rhythms in that impairments in human performance during sleep deprivation are most pronounced at night. In self-paced activity, consisting of four-a-side indoor soccer sustained for 3 to 4 days, the activity level peaks at about 18:00 hours, coinciding with the daily high point of body temperature. Other variables that follow this curve include grip strength[11] and choice reaction time.[83] The circadian rhythm persists for as many days as the individual can be kept awake. Over 3 to 5 days of total sleep deprivation, there is a trend towards deteriorating performance (Fig. 11) upon which the circadian rhythm is superimposed.[71] For the purposes of statistical analysis this trend has to be removed prior to establishing circadian rhythms by, for example, cosinor or Fourier analysis.[84] The trend is not evident in all functions: gross muscular performances, such as isometric strength, are highly resistant to effects of sleep loss, despite the decrease in muscle enzyme activity noted after the first night.[85] In contrast, cognitive functions are easily affected. Complex and challenging tasks are less affected than monotonous repetitive ones, and strong motivation may overcome the effects of sleep loss, at least for short periods. Nevertheless, after 4 days of sustained activity in which only 2 h sleep were allowed, military subjects were deemed to be ineffective as soldiers.[86] This conclusion was based on performance in a 1-km assault course, a shooting tests, and a 3-km run; provision of a high energy diet was unsuccessful in offsetting these impairments. The decline is likely to have been a fatigue effect rather than attributable to sleep loss; soldiers who are sleep deprived for 2 to 3 nights but not physically fatigued can perform highly demanding tasks at the same work rate as fresh troops.[87]

In one study subjects were kept awake for 64 h under conditions of isolation from external time cues. Activity was classed as sedentary and was kept as constant as possible, as was intake of food and liquids.[88] Adrenaline secretion showed a pronounced

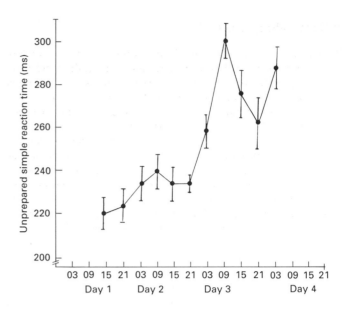

Fig. 11 Unprepared simple reaction time recorded (mean ± SD) on 10 subjects every 6 h for 4 days of indoor soccer play.

circadian rhythm, with the rhythm in noradrenaline being relatively weaker (based on amplitude as a percentage of the mean value). The circadian rhythms in cardiovascular variables (blood pressure, heart rate, contractility, T-wave amplitude, and QRS, PQ, and QT intervals) observed under normal conditions were effectively obliterated, suggesting that alternating between sleeping and waking is their main determinant. The experimental design could not rule out the existence of self-sustained rhythms in cardiovascular variables since the continuous wakefulness may have concealed low amplitude endogenous rhythms. Other studies have shown an apparent haemodilution with sleep deprivation, which may be associated with a slight decline in $\dot{V}O_2max$.[89]

There is also an interaction between sleep loss and environmental stressors, although this effect is not linear. For example, heat compounds the effects of sleep deprivation whereas noise can offset them.[90] Urinary excretion of melatonin follows a circadian pattern in humans during sleep deprivation when subjects are exposed to light, but excretion levels increase with increased sleep loss.[91]

Subjects deprived of sleep for 1 to 3 nights begin to exhibit psychotic-like symptoms and bizarre behaviour. They also experience temporary visual illusions. In such circumstances meaningful physical activity becomes difficult to sustain without error. Sailors will be unreliable on watch duty when suffering from such severe sleep loss. It has been suggested that naturally occurring brain amines may play a role in the cycles of behaviour and mood associated with prolonged sleep deprivation. A circadian rhythm was found in phenylethylamine levels in the urine of sleep-deprived footballers playing indoors: by the third successive night without sleep, the concentrations of this substance being excreted were found to approach the values typically observed in psychiatric patients.[92] Fortunately, such prolonged periods of sleeplessness are experienced only rarely. They may be met by sports medical personnel on hospital duty. In military recruits where such regimens are imposed during training, similar trends superimposed on a circadian rhythm in catecholamine excretion have been noted.[27] The curve in catecholamines coincided with the rises and falls in the accuracy of shooting performance.

Partial sleep loss

Partial sleep loss or disrupted sleep is a more common problem than complete sleep deprivation. It can affect athletes who are restless through anxiety, sailors and yachtsmen during prolonged competitions, and athletes with children who themselves have unsettled sleeping patterns. In view of the variability between individuals in the usual amount of sleep taken, tolerance of sleep loss, sensitivity of laboratory measures of performance to effects of sleep deprivation, and so on, inferences from experimental investigations must be made with caution. Effects of partial sleep loss also depend on motivation, task complexity, stage of sleep most affected, and other factors.[93]

Faculties associated with SWS may be unimpaired unless the duration of sleep is 3 h or less.[93] Consequently the effects of a nightly ration of 2.5 h of sleep were examined on a battery of psychomotor, work capacity, and mental stage tests over 3 nights of sleep loss and after 1 night of subsequent recovery.[83] A 3-day control period was used in a counterbalanced design to eliminate an order effect. Functions that required fast reactions were found to deteriorate significantly. This applied to anaerobic power output in a stair run and choice reaction time at rest and during exercise on a bicycle ergometer. Physical exercise attenuated the effects of sleep loss on reaction time, suggesting the benefits of manipulating arousal level by means of a warm-up. This beneficial effect is likely to be short lived and exercise will be less effective in offsetting sleep loss when the disruptions continue over days. Limb speed, as measured by a reciprocal tapping task, also becomes steadily worse over successive days of partial sleep loss.[83]

Gross motor tasks such as grip strength, lung function, and treadmill run time were unaffected by sleep restriction. As the restricted sleep regimen was found to affect the more complex motor co-ordination tasks whilst leaving gross motor functions relatively intact, the data support the 'nerve restitution' theory of sleep.[90]

These effects of partial sleep loss found in males are replicated in female subjects.[21] These subjects were also limited to 2.5 h of sleep for three successive nights in a counterbalanced experimental design: performances were measured each morning (07:00–09:00 hours) and each evening (19:00–21:00 hours). A circadian rhythm was noted in the majority of measures; for gross motor function this effect was greater than that of sleep loss. The perceived exertion during cycling at 60 per cent of $\dot{V}O_2max$ showed both a diurnal variation and an underlying trend towards increased subjective strain. This coincided with the observations on the sleepiness of subjects, self-rated before exercise. The reduced subjective ratings on the final experimental morning may be explained by anticipation of the end of the experiment.

That sleep is needed more for 'brain restitution' rather than for 'tissue restitution' is further supported by observations on the effects of partial sleep deprivation on swimmers.[18] The sleep ration was restricted to 2.5 h a night for three consecutive nights. Performance over 400 m and over four successive 50-m swims was maintained throughout the experimental period. Swimming times were faster in the evening (17:30 hours) compared with the morning (06:30 hours), replicating the findings that the time-of-day effect of gross motor functions exceeds that of sleep loss.[18] The most pronounced effects of the restricted sleep regimen were deteriorations in mood over the period of the investigation.

Despite the fact that muscular strength may be retained during consecutive days of partial sleep loss, the quality of training may be adversely affected. This applies to training sessions with repeated or multiple maximal efforts, as occurs in weight-training programmes. Thus maximal performances can be reproduced in weight-training exercises executed in the early parts of the session but the quality of performance declines towards the end.[94] The reasons for this are attributed to deteriorations in mood with successive nights of deprived sleep.

A paradoxical result of partial sleep loss is that some tasks show an improvement.[89] Hand steadiness, for example, is generally better after loss of sleep. This is attributable to a decrease in spontaneous contraction of involved muscles as a result of reduced muscle tone. Similarly, tasks with high loadings on short-term memory appear to improve with sleep deprivation owing to a tendency to code information acoustically for mental storage and recall in laboratory tests.[83] Thus care is needed in designing and interpreting sleep deprivation studies and in making assumptions about the effects of disrupted sleep on athletic performance. Individuals forced to reduce their normal sleep ration may adapt to their shortened sleep length without any consequences for exercise performance, provided that the reduction does not exceed about 2 h. Otherwise they may need to reorganize their daily routine to accommodate an afternoon nap. How practical this is depends on personal and occupational circumstances.

Napping

Ultradian cycles with 90 min periods are identifiable during sleep and may be latent during wakefulness. This may explain subharmonics within the circadian phase system, as evidenced in the 'post-lunch dip' in performance.[4,26] To what extent this drop can be offset by napping or reorganization of the work–rest schedule of activity has not been adequately investigated. Individuals on short sleep durations for some time derive considerable refreshment from short naps.[14,95]

It is known that prolonged napping at an inappropriate time can delay resynchronization of rhythms after abrupt phase shifts.[73] A study of nocturnal shift workers showed that a 1-h nap at 02:00 hours was less effective than caffeine in maintaining performance overnight in a range of tasks.[96]

Often we hear people say that they were unable to sleep during the night. Such accounts need independent corroboration. Short periods of sleep snatched unwittingly during the night do serve a restorative function. Individuals deprived of sleep for some time derive considerable benefit from such naps, and those deprived of sleep for 2 to 4 days usually recover from their ordeal after uninterrupted sleep for a complete night. A nap could counteract the fall in arousal underlying any subharmonic in the performance curve linked to ultradian rhythms. There has been no substantive research on the refreshing effects of napping on subsequent exercise performance. A nap taken early in the day should be better than a late afternoon nap (provided that the individual can fall asleep), unless performance is late in the evening. A late afternoon nap would contain more SWS and less REM sleep, and would take longer to rouse from than a nap taken earlier. In preparing mentally for competition athletes would generate the drive to overcome this de-arousal. Advocation of a nap depends on factors such as timing of the contest, pre-competition feeding, and individual preferences.

Sleeplessness

Although true insomnia is rare, a large number of people—perhaps 10 to 15 per cent of the population—do have difficulty in sleeping. Causes may include anxiety, depression, bereavement, stress, overwork, or environmental noise such as motor vehicle traffic. Some of these problems are transient and self-limiting; others may persist and become chronic.

It is generally thought that exercise promotes sleep and so it is recommended as therapy for individuals having difficulty in sleeping. The effect of exercise is likely to be indirect, promoting sleep by alleviating the anxieties that prevented it. Athletes do exhibit different sleep patterns to sedentary individuals according to observations of their EEG. However, EEG characteristics are only marginally altered with a regimen of physical training. Strenuous exercise shortly before retiring to sleep is likely to raise arousal rather than induce drowsiness because of the increased levels of circulating catecholamines. Thus exercise as therapy for sleeping problems should not be strenuous and should be performed early rather than later in the evening.

The usual prescription for insomnia is sleeping pills. People taking sedatives or hypnotics for a prolonged period develop a dependence on them and the drugs lose their effectiveness. Minor tranquillizers (the benzodiazepines) are probably now being over-prescribed in Europe and North America. Habitual users become dependent on benzodiazepines and suffer severe symptoms if treatment is suddenly stopped. Normal doses also impair reaction time and mental concentration the morning after they have been taken.[97] They also affect muscular performance, notably in movements at high velocities. Consequently, their prescription for athletes should be considered only in cases of dire necessity.

Non-pharmacological methods of treating sleeplessness include hypnotherapy. Biofeedback of skin resistance and EEG may also be employed to train the individual to overcome the emotional tension that prevents sleep. Psychological techniques such as visualization of tranquil scenes, concentration on relaxing muscle activity, and deep breathing provide alternative treatments. Stimulus control therapy refers to a mental strategy whereby bed and sleeplessness are dissociated: the individual goes to bed only when sleepy, avoids eating, reading, or watching television in the bedroom, and does not 'sleep in' in the morning. Sensible eating and drinking habits (such as avoiding large meals, heavy alcoholic beverages, or caffeine late at night) should also promote sleep and so help 'knit up the ravell'd sleave of care'.

REFERENCES

1. Reilly TJ. Circadian rhythms in muscular activity. In: Marconnet P, Komi PV, Saltin B, Sejeested OM, eds. *Muscle fatigue mechanisms in exercise and training.* Basel: Karger, 1992: 218–22.
2. Wever R. Zur Zeitgeber-Stärke eines Licht-Dunkel-Wechsels für dier circadiase periodic des Menschen. *Pflügers Archiv* 1970; **321**: 133–42.
3. Minors DS, Waterhouse JM. *Circadian rhythms and the human.* Bristol: John Wright, 1981.
4. Colquhoun WP. *Biological rhythms and human performance.* New York: Academic Press, 1971.
5. Folkard S, Minors DS, Waterhouse JM. Is there more than one internal clock in man? *Journal of Physiology* 1983; **341**, 50P.
6. Moore-Ede MC. Jet lag, shift work and maladaptation. *News in Physiological Sciences* 1986; **1**: 156–60.
7. Arendt J, Minors DS, Waterhouse JM. *Biological rhythms in clinical practice.* London: John Wright, 1989.
8. Sherer MA, Weingartner H, James SP, Rosenthal NE. Effects of melatonin on performance testing in patients with seasonal affective disorder. *Neuroscience Letters* 1985; **58**: 277–82.
9. Horne JA, Coyne I. Seasonal changes in the circadian variation of oral temperature during wakefulness. *Experientia* 1975; **31**: 1296–8.
10. Reilly T., Walsh TJ. Physiological, psychological and performance measures during an endurance record for 5-a-side soccer play. *British Journal of Sports Medicine* 1981; **15**: 122–8.
11. Reinberg A, Proux S, Bartal JP, Levi F, Bicakova-Rocher A. Circadian rhythms in competitive sabre fencers: internal desynchronisation and performance. *Chronobiology International* 1985; **2**: 195–201.
12. Conroy RTWL, O'Brien M. Diurnal variation in athletic performance. *Journal of Physiology* 1974; **236**, 51P.
13. Rodahl A, O'Brien M, Firth PGR. Diurnal variation in performance of competitive swimmers. *Journal of Sports Medicine and Physical Fitness* 1976; **16**: 72–6.
14. Reilly, T. Human circadian rhythms and exercise. *Critical Review in Biomedical Engineering* 1990; **18**: 165–80.
15. Sinnerton S, Reilly T. Effects of sleep loss and time of day in swimming. In: MacLaren D, Reilly T, Lees A, eds. *Biomechanics and medicine in swimming: swimming science (VI).* London: Spon, 1992: 399–405.
16. Baxter C, Reilly T. Influence of time of day on all-out swimming. *British Journal of Sports Medicine* 1983; **17**: 122–7.
17. Reilly T, Down A. Circadian variation in the standing board jump. *Perceptual and Motor Skills* 1986; **62**: 830.
18. Reilly T, Hales AJ. Effects of partial sleep deprivation on performance measures in females. In: Megaw ED, ed. *Contemporary ergonomics.* London: Taylor and Francis, 1988: 509–14.
19. Åstrand PO, Rodahl K. *Textbook of work physiology.* New York: McGraw-Hill, 1986.
20. Cabri J, Clarys JP, De Witte B, Reilly T, Strass D. Circadian variation in blood pressure responses to muscular exercise. *Ergonomics* 1988; **31**: 1559–66.
21. Reilly T, Down A. Investigation of circadian rhythms in anaerobic power and capacity of the legs. *Journal of Sports Medicine and Physical Fitness* 1992; **32**: 342–7.
22. Reilly T, Down A. Time of day and performance on all-out arm ergometry. In: Reilly T, Watkins J, Borms J, eds. *Kinanthropometry III.* London: Spon, 1986: 296–300.
23. Stockton ID, Reilly T, Sanderson FH, Walsh TJ. Investigation of circadian rhythms in selected components of sports performance. *Bulletin of the Society of Sports Sciences* 1980; **1**: 14–15.
24. Rutenfranz J, Colquhoun WF. Circadian rhythms in human performance. *Scandinavian Journal of Work and Environmental Health* 1979; **5**: 167–77.
25. Wit A. *Zayanienia regulacji w procesie rozwoju sily miesnionej na przykladzie zawodnikow uprawiajacych podnoszenie ciezarow.* Warsaw: Institute of Sport, 1979.
26. Horne JA, Gibbons H. Effect on vigilance and sleepiness of alcohol given in the early afternoon ('post lunch') vs early morning. *Ergonomics* 1991; **34**: 67–77.
27. Åkerstedt, T. Altered sleep/wake patterns and circadian rhythms. *Acta Physiologica Scandinavica* 1979; Suppl. 469.
28. Reilly T. Circadian rhythms and exercise. In: Macleod D, Maughan RJ, Nimmo M, Reilly T, Williams C, eds. *Exercise: benefits, limits and adaptations.* London: Spon, 1987: 346–66.
29. Shephard RJ. Sleep, biorhythms and human performance. *Sports Medicine* 1984; **1**: 11–37.
30. Winget CM, De Roshia CW, Holley DC. Circadian rhythms and athletic performance. *Medicine and Science in Sports and Exercise* 1985; **17**: 498–516.
31. Reilly T, Brooks GA. Investigation of circadian rhythms in metabolic responses to exercise. *Ergonomics* 1982; **25**: 1093–1197.
32. Reilly T, Robinson G, Minors DS. Some circulatory responses to exercise at different times of day. *Medicine and Science in Sports and Exercise* 1984; **16**: 477–82.
33. Froberg J *et al.* Circadian variation in performance, psychological ratings, catecholamine excretion and diuresis during prolonged sleep deprivation. *International Journal of Psychobiology* 1972; **2**: 23–36.
34. Reilly T, Brooks GA. Selective persistence of circadian rhythms in physiological responses to exercise. *Chronobiology International* 1990; **7**: 59–67.

35. Cable T, Reilly T. Influence of circadian rhythms on arm exercise. *Journal of Human Movement Studies* 1987; **13**: 13–27.

36. Reilly T. Circadian variation in ventilatory and metabolic adaptation responses to submaximal exercise. *British Journal of Sports Medicine* 1982; **16**: 115–16.

37. Stephenson LA, Kokla MA, Francesconi R, Gonzalez RR. Circadian variations in plasma renin activity, catecholamines and aldosterone during exercise in women. *European Journal of Applied Physiology* 1989; **58**: 756–64.

38. Faria IE, Drummond BJ. Circadian changes in resting heart rate and body temperature, maximal oxygen consumption and perceived excertion. *Ergonomics* 1982; **25**: 381–6.

39. Reilly T, Baxter C. Influence of time of day on reactions to cycling at a fixed high intensity. *British Journal of Sports Medicine* 1983; **17**: 128–30.

40. Bergh U, Ekblom B. Influence of muscle temperature on maximal muscle strength and power in human skeletal muscles. *Acta Physiologica Scandinavica* 1979; **107**: 33–7.

41. Reilly T, Marshall S. Circadian rhythms in power output of swimmers. *Journal of Swimming Research* 1991; **7(2)**: 11–13.

42. Hessemer V, Langusch D, Bruck K, Bodeker RK, Breidenback T. Effects of slightly lowered body temperature on endurance performance in humans. *Journal of Applied Physiology: Respiratory, Environmental and Exercise Physiology* 1984; **57**: 1731–7.

43. Reilly T, Brooks GA. Exercise and the circadian variation in body temperature measures. *International Journal of Sports Medicine* 1986; **7**: 358–62.

44. Stephenson LA, Winger CB, O'Donovan BH, Nadel ER. Circadian rhythm in sweating and cutaneous blood flow. *American Journal of Physiology* 1984: **246**: R321–4.

45. Tyrrell AR, Reilly T, Troup JDG. Circadian variation in stature and the effects of circuit weight-training. *Spine* 1985; **10**: 161–4.

46. Reilly T, Tyrrell A, Troup JDG. Circadian variation in human stature. *Chronobiology International* 1984; **1**: 121–6.

47. Wilby J, Linge K, Reilly T, Troup JDG. Spinal shrinkage in females: Circadian variation and the effects of circuit weight-training. *Ergonomics* 1987; **30**: 47–54.

48. Garbutt G, Boocock MG, Reilly T, Troup JDG. Application of spinal shrinkage to subjects with low back pain. In: Duquet W, Day JAP, eds. *Kinanthropometry IV* London: Spon, 1992.

49. Adams MA, Dolan P, Hutton WP, Porter RW. Diurnal changes in spinal mechanics and their clinical significance. *Journal of Bone and Joint Surgery* 1990; **72B**: 266–70.

50. Troup JDG, Reilly T, Eklund JAE, Leatt P. Changes in stature with spinal loading and their relation to the perception of exertion or discomfort. *Stress Medicine* 1985; **1**: 303–7.

51. Leatt P, Reilly T, Troup JDG. Unloading the spine. In: Oborne D, ed. *Contemporary Ergonomics*. London: Taylor and Francis, 1985: 227–32.

52. Boocock MG, Garbutt G, Reilly T, Linge K, Troup JDG. Effect of gravity inversion on exercise induced spinal loading. *Ergonomics* 1988; **31**: 1631–8.

53. Wright V, Dawson D, Longfield MD. Joint stiffness—its characterisation and significance. *Biology and Medicine Engineering* 1969; **4**: 8–14.

54. Gifford LS. Circadian variation in human flexibility and grip strength. *Australian Journal of Physiotherapy* 1987; **33**: 3–9.

55. Harkness JA *et al.* Circadian variation in disease activity in rheumatoid arthritis. *British Medical Journal* 1982; **284**: 551–4.

56. Job-Deslondre C, Reinberg A, Delbarre F. Chrono–effectiveness of indomethacin in four patients suffering from an evolutive osteoarthritis of hip or knee. *Chronobiologia* 1983; **10**: 245–54.

57. Reinberg A. *et al.* Clinical chronopharmacology of ACTH 1–17 III. Effects of fatigue, oral temperature, heart rate, grip strength and bronchial potency. *Chronobiologia* 1981; **8**: 101–15.

58. Ilmarinen J, Ilmarinen R, Korhonen O, Nurminen M. Circadian variation of physiological functions related to physical work capacity. *Scandinavian Journal of Work and Environmental Health* 1980; **6**: 112–22.

59. Coldwells A, Atkinson G, Reilly T, Waterhouse J. Self-chosen work-rate determines day-night differences in work capacity. *Ergonomics* 1993, **36**: 313.

60. Hill DW, Cureton KJ, Collins MA. Circadian specificity in exercise training. *Ergonomics* 1989; **32**: 79–82.

61. Procacci P, Corte MD, Zoppi M, Marascu M. Rhythmic changes of the cutaneous pain threshold in man. A general view. *Chronobiologia* 1974; **1**: 77–96.

62. Strempel H. Circadian cycles of epicritic and protopathic pain threshold. *Journal of Interdisciplinary Cycle Research* 1977; **8**: 276–80.

63. Ptaff D. Effects of temperature and time of day on time judgements. *Journal of Experimental Psychology* 1968; **76**: 419–22.

64. Baddeley AD. Time estimation at reduced body temperature. *American Journal of Physiology* 1966; **79**: 475–9.

65. Blake MJF. Relations between circadian variation of body temperature and introversion-extroversion. *Nature (Lond)* 1967; **215**: 896–7.

66. Horne JA, Östberg O. Individual differences in human circadian rhythms. *Biological Psychology* 1977; **5**: 179–90.

67. Vidacek S, Kaliterna L, Tadosevic-Vidacek B, Folkard S. Personality differences in the phase of circadian rhythms: a comparison of morningness and extroversion. *Ergonomics* 1988; **31**: 873–8.

68. Hill DW, Cureton KJ, Collins MA, Grisham SC. Diurnal variations in responses to exercise of 'morning types' and 'evening types'. *Journal of Sports Medicine and Physical Fitness* 1988; **28**: 213–19.

69. Winget CM, De Roshia CW, Markley CL, Holley DC. A review of human physiological performance changes associated with desynchronosis of biological rhythms. *Aviation, Space and Environmental Medicine* 1984; **54**: 132–7.

70. De Looy A, Minors D, Waterhouse J, Reilly T, Tunstall Pedoe D. *The coach's guide to competing abroad*. Leeds: National Coaching Foundation, 1988

71. Reilly, T. Time zone shift and sleep deprivation problems. In: Torg JS, Welsh RP, Shephard RJ, eds. *Current theory in sports medicine—2*. Toronto: BC Decker 1990: 135–9.

72. Reilly T, Mellor S. Jet lag in student Rugby League players following a near-maximal time-zone shift. In: Reilly T, Lees A, Davids K, Murphy WJ, eds. *Science and football*. London: Spon, 1988: 249–56.

73. Minors DS, Waterhouse JM. Anchor sleep as a synchroniser of abnormal routines. *International Journal of Chronobiology* 1981; **7**: 165–88.

74. Antal LC. The effects of the changes of the circadian body rhythm on the sports shooter. *British Journal of Sports Medicine* 1975; **9**: 9–12.

75. Reinberg A, Vieux N, Andlauer P. *Night and shift work: biological and social aspects*. Oxford: Pergamon, 1980.

76. Atkinson G, Coldwells A, Reilly T, Waterhouse J. A comparison of circadian rhythms in work performance between physically active and inactive subjects. *Ergonomics* 1993; **36**: 273–81.

77. Van Loon JH. Diurnal body temperature curves in shift workers. *Ergonomics* 1963; **6**: 267–73.

78. Reilly T, Thomas V. Estimated daily energy expenditures of professional association footballers. *Ergonomics* 1979; **22**: 541–8.

79. Griffin SJ, Trinder J. Physical fitness, exercise and human sleep. *Psychophysiology* 1978; **15**: 447–50.

80. Paxton SJ, Turner J, Montgomery I. Does aerobic fitness affect sleep. *Psychophysiology*; 1983; **20**: 320–24.

81. Oswald I. Sleep as a restorative process: human clues. *Progress in Brain Research* 1980; **53**: 279–88.

82. Horne JA. *Why we sleep: the function of sleep in humans and other mammals*. Oxford University Press, 1988.

83. Reilly T, Deykin T. Effects of partial sleep loss on subjective states, psychomotor and physical performance tests. *Journal of Human Movement Studies* 1983; **9**: 157–70.

84. Thomas V, Reilly T. Circulatory, psychological and performance variables during 100 hours of paced continuous exercise under conditions of controlled energy intake and work output. *Journal of Human Movement Studies* 1975; **1**: 149–55.

85. Vondra K. *et al.* Effects of sleep deprivation on the activity of selected metabolic enzymes in skeletal muscle. *European Journal of Applied Physiology* 1981; **47**: 41–6.

86. Rognum TO *et al.* Physical and mental performance of soldiers during prolonged heavy exercise combined with sleep deprivation. *Ergonomics* 1986; **29**: 859–67.

87. Myles WS, Romet TT. Self-paced work in sleep deprived subjects. *Ergonomics* 1987; **30**: 1175–84.

88. Ahnve S, Theorell T, Åkerstedt T, Fröberg JE, Halberg F. Circadian variations in cardiovascular parameters during sleep deprivation. *European Journal of Applied Physiology* 1981; **46**: 9–19.

89. Plyley MJ, Shephard RJ, Davis GM, Goode RC. Sleep deprivation and cardiorespiratory function. *European Journal of Applied Physiology and Occupational Physiology* 1987; **56**: 338–44.

90. Wilkinson RT. Some factors influencing the effect of environmental stressors upon performance. *Biological Bulletin* 1969; **72**: 260–72.

91. Åkerstedt T, Fröberg JE, Friberg Y, Wetterberg L. Melatonin excretion, body temperature and subjective arousal during 64 hours of sleep deprivation. *Psychoendocrinology* 1979; **4**: 219–25.

92. Reilly T, George A. Urinary phenylethylamine levels during three days of indoor soccer play. *Journal of Sports Sciences* 1983; **1**: 70.

93. Reilly T. Exercise and sleep: an overview. In: Watkins J, Reilly T, Burwitz L, eds. *Sports science*. London: Spon, 1986: 414–19.

94. Reilly, T, Piercy, M. The effect of partial sleep deprivation on weight-lifting performance. *Ergonomics* 1993.

95. Dinges DF, Orne MT, Whitehouse WG, Orne EC. Temporal placement of a nap for alertness. Contributions of circadian phase and prior wakefulness. *Sleep* 1987; **10**: 313–29.

96. Rogers AS, Spencer MB, Stone BM, Nicholson AN. The influence of a 1 h nap on performance overnight. *Ergonomics* 1989; **32**; 1193–1205.

97. Reilly T. Alcohol, anti-anxiety drugs and exercise. In: Mottram DR, ed. *Drugs in sport*. London: Spon, 1988: 127–56.

Medical aspects of sport

3

3.1.1 Cardiac adaptations

LEONARD M. SHAPIRO

The ability of an individual to perform vigorous exercise depends on the function of many bodily systems. The cardiovascular system is central to this, since continuous muscle work depends upon the transfer of oxygen to the muscles and removal of waste products. What signals control the cardiovascular response to exercise are unknown, but metabolic and circulatory events are so closely related that blood flow matches the metabolic needs of tissue. The study of a cardiovascular response to exercise is of considerable importance, as more is learned from a system when it is active rather than idle. During maximum exercise, total body oxygen consumption may rise 10- or 12-fold. Cardiac output rises in a linear fashion with oxygen uptake from approximately 6 l at rest to values as high as 42 l/min in endurance athletes. The arteriovenous mixed oxygen difference may double. In addition, there may be redistribution of blood flow from non-exercising areas, such as the splanchnic bed, and an enormous increase in blood flow to exercising muscle. The cardiac response to exercise is complex and ill-understood, but involves the interaction of a number of changes, including heart rate, myocardial contractivity, and the pre- and after-load condition of the heart. The elevation of cardiac output that may be seen during mild exercise is largely made up of increases in heart rate. Stroke volume and left-ventricular volumes show little change. As exercise reaches a maximum, stroke volume increases to approximately twice the resting level, and left-ventricular end diastolic volume and cardiac output rise with the reduction in end-systolic volume (Fig. 1). All forms of exercise increase cardiac output, and in particular endurance sports, such as running, swimming, and cycling, demand prolonged elevation of cardiac output.

A number of terms will be used in this chapter and they are defined as follows.

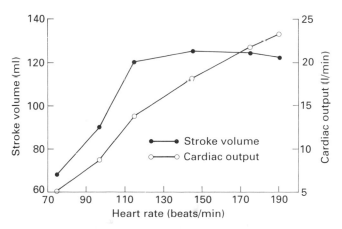

Fig. 1 Graph showing the theoretical plot of stroke volume heart rate and cardiac output during exercise. During exercise, the heart rate rises progressively from approximately 70 beats/min to a maximum of 190 beats/min. During this time, cardiac output rises from 5 to 23 1/min. However, while there is a progressive rise in cardiac output, the stroke volume rapidly rises initially but reaches a plateau at approximately 125 ml, even through cardiac output continues to rise.

V_{O_2}max is the maximum oxygen uptake a person can use whilst performing dynamic exercise. This functional limit is a reproducible measurement in an individual; it is related to age, gender, inherited characteristics, and exercise habits, and shows insignificant day-to-day variations. V_{O_2}max equals the product of maximum values of heart rate/volume and arteriovenous oxygen difference by the Fick principle. It is usually expressed in multiples of sitting–resting requirements. This is the metabolic equivalent (1 MET = 3.5 ml O_2/kg body wt/min).

The cardiac output is the volume of blood ejected by the left or right ventricle into the great arteries per minute. This is usually normalized for body surface area and defined as litres per minute.

Heart rate response: as dynamic exercise is started, heart rate rises due to a decrease in vagal parasympathetic outflow and a later rise in sympathetic outflow. Heart rate rises proportionally to work load and V_{O_2}. There is a decline in maximum heart rate with age, and the relation 220−age (in years) is useful for predicting peak heart rate, but the scatter around the regression line is large as one standard deviation is equal to 12 beats/min.

Stroke volume is the volume of blood ejected from the left ventricle for each heart beat. This is dependent upon the end diastolic volume and the ejection fraction. The ejection fraction is simply a percentage measurement of the difference between the end-systolic and end-diastolic left-ventricular volume.

There are many controversies in the study of 'athlete's heart', including the relationship of training to the development of left-ventricular enlargement and hypertrophy, the nature and functional consequences of ventricular hypertrophy, and whether such hypertrophy may be deleterious in the long term. Technical developments in cardiac imaging have allowed the answer to many of these early questions to be obtained.

The enlarged heart of the athlete was first noted in the last century by Henschen from Uppsala.[1] Using only cardiac percussion, he observed that, in athletes undertaking long-distance skiing events all parts of the heart were enlarged, and he attributed this to physiological hypertrophy. While it may now seem quite obvious that such cardiac hypertrophy is the result of training, as the increased muscle power of a weight-lifter is based on skeletal muscle hypertrophy, during the early parts of the century much of the cardiac enlargement due to training was attributed to underlying heart disease. The term 'athlete's heart' is used in a general sense by both physicians and laymen to describe the cardiovascular effect of training. This is a loosely defined term, but can be regarded as a physiological response of the heart to this training. Athlete's heart is the effect of cardiovascular conditioning and training during repetitive exercise that leads to an increased left-ventricular volume, a reduction of heart rate, and ability to maintain a high level of exercise. Such changes require persistent exercise at a high work-load. This needs to be differentiated from the beneficial effects of moderate recreational or occupational exercise of almost any type that leads to an increase in heart rate and reduces cardiovascular risk factors. This is best shown in Harvard Alumni[2] and Whitehall[3] studies, where the amount of daily vigorous exercise was recorded and related to the later development of coronary artery disease. Data from these studies show that vigorous recreational activity 'protects' against or reduces the risk of coronary artery disease even in smokers (Figs. 2 and 3). The mechanisms by which this occurs are not clear, but exercise tends to reduce other cardiovascular risk factors such as hypertension, hyperlipidaemia, and obesity. Also important is the fact that habitual exercisers tend not to smoke.

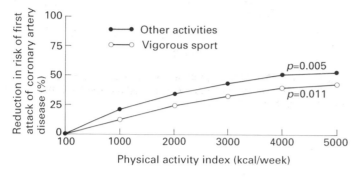

Fig. 2 Graph adapted from the Harvard Alumni Study showing the reduction in the risk of first attacks of coronary artery disease that results from physical activity measured as an index in kilocalories per week.[2]

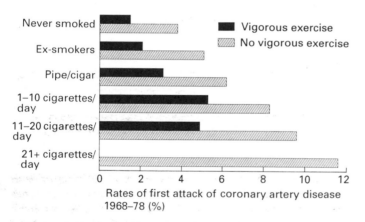

Fig. 3 The percentage rate of first heart attacks. There is a reduction in the rate of a first attack of coronary artery disease in those who exercise vigorously on a habitual basis compared with those who do not (study in 17 944 civil servants). In addition, this demonstrates that exercise is protective in the development of coronary disease, even in smokers.

GENERAL CONSIDERATIONS OF THE EFFECT OF TRAINING ON THE HEART

The effect of training on the heart depends on whether the training is predominantly aerobic (dynamic) or isometric (static) in nature. The majority of sportsmen undertake predominantly aerobic conditioning, which leads to a situation similar to chronic volume overload with reduction in resting heart rate, reduced heart rate at any given exercising cardiac output, a raised left-ventricular end-diastolic volume, and a proportional thickening of the posterior wall and ventricular septum to normalize peak left-ventricular wall stress. In strength sportsmen such as weight-lifters and shot-putters, brief increases in cardiac output against greatly elevated aortic pressure induce a morphological left-ventricular response to pressure overload. This leads to an increased thickness of the posterior wall and ventricular septum without a corresponding elevation of left-ventricular end-diastolic volume.[1–5]

The cardiac adaption in athletes allows a high cardiac output at peak levels of exertion. These adaptations probably act via a number of different and poorly understood mechanisms. They include the fact that a large initial end-diastolic volume allows a given stroke volume to be ejected with less myocardial shortening. Myocardial muscle fibres that are stretched can provide higher tension than unstretched fibres within certain limits. Frictional and tensional energy losses are reduced with an elevated end-diastolic volume and at low rates. These factors are beneficial, but at a given end-diastolic volume myocardial wall tension will rise to maintain a particular left-ventricular pressure, which leads to decreased efficiency at maximum work-loads and will result in wall-thickening and increased left-ventricular mass. Accompanying the effects of training on the heart is an increased vagal tone which leads to a resting and relative exertional bradycardia. It remains debatable whether the resting bradycardia leads to, or results from, the increased ventricular volume. However, at all given levels of cardiac work-load, there is a reduction in heart rate.

METHODS OF ASSESSING CARDIOVASCULAR FUNCTION IN ATHLETES

Physical examination

It is usual to find some abnormal features when examining an athlete. There may be a resting bradycardia and irregularities of the pulse due to a sinus arrhythmia. The nature of these abnormalities will be dealt with later. The pulse is usually normal in character and normal or increased in volume. Blood pressure is usually normal, though a rather wide blood pressure, probably due to the bradycardia, is not unusual. The venous pressure is not elevated. The heart is not enlarged to palpation, but in some long-standing athletes there may be clinical cardiac enlargement. The cardiac impulse may be left-ventricular or heaving in nature, and this is often accentuated in athletes because of their relative leanness. Auscultation may reveal no abnormal features, but third and fourth sounds are common: in particular third heart sounds may be heard in young athletes, and fourth heart sounds in the older athlete. A soft systolic murmur related to pulmonary blood flow and mitral regurgitation may occasionally be heard.

The chest radiograph in athletes is usually normal, but slight cardiac enlargement may be demonstrated.

The electrocardiogram (ECG) can be very useful in examination of athletes at rest. There may be a bradycardia, a prolonged PR interval, and possibly higher degrees of atrioventricular block, which will be dealt with later. Left-ventricular hypertrophy by voltage criteria is quite common and is accentuated by the leanness of athletes. Repolarization changes and ST segment and T-wave flattening and inversion are usual in young athletes but are more common in veterans. T-wave inversion in leads V^1 and V^2 is common, such ECG alterations may be more extensive in black athletes.

The ECG can be recorded over a 24-h period using a miniaturized tape-recorder. Such recordings usually reveal the marked bradycardia that almost invariably occurs in athletes, and other bradyarrhythmias and tachyarrhythmias.

Exercise testing

Exercising is a common physiological stress and may elicit cardiovascular abnormalities not seen at rest. During exercise, respiratory gas exchange, intracardiac pressures, cardiac output, and the ECG can be recorded. However, the last is the simplest and most frequently used. There are a number of methods of performing the exercise required for testing, from the simplest such as walking or running before taking a static ECG, to step-testing, bicycle-ergometry, or treadmill testing with continuous recording of the ECGs. The variable step-test is the simplest and cheapest device available for obtaining maximum exercise in a relatively stationary position. This may be suitable for athletic individuals, but if cardiovascular fitness is high, it may take a long time to reach maximum exertion. There is great distortion of the ECG because of the body movement, and the system is difficult to calibrate. A bicycle ergometer uses little space compared with a treadmill and produces an undistorted ECG because of the lack of body movement. Exercise is again predominantly lower-body based. Mechanical bicycle ergometers are relatively inexpensive and tend to be quiet compared with treadmills. The bicycle ergometer requires external calibration using electronic equipment. Most untrained individuals find it difficult to attain maximum exercise on a bicycle, and once the subject becomes tired adherence to a standard work-rate becomes progressively more difficult. However, in a cyclist or trained athlete, a bicycle ergometer may produce satisfactory ECG recordings. The motor-driven treadmill permits the greatest oxygen consumption rate of the commonly used devices as it involves the legs, torso, and arms. Calibration does not require any special equipment. Exercise carried out on a motor-driven treadmill follows a number of test protocols which utilize the steady state of oxygen uptake that occurs after 2 min at a particular intensity of exercise. The most commonly used for clinical purposes is the Bruce protocol. This goes through a series of stages, commencing at 1.7 miles/h at a 10 per cent grade and rising at stage seven to 6 miles/h at a 22 per cent grade. Each of these stages lasts for 3 min, and these are shown diagrammatically in Fig. 4. Generally, only the ECG is monitored, but it is possible to approximate the oxygen consumption or make direct measurements with expired air collection. Exercise testing continues until the subject is exhausted or reaches predetermined heart rate, such as 75 per cent (in patients with heart disease), 90 per cent, or 100 per cent of the age-predicted maximum heart rate estimated as 220-age (Fig. 5). The development of electrocardiographic morphological abnormalities, arrhythmias, or cardiac symptoms will also terminate a test. While the treadmill exercise test is best used to diagnose and quantify myocardial ischaemia in symptomatic patients, it may be useful for the detecting cardiovascular disease in asymptomatic individuals, both trained and untrained; however the rate of 'false positive' results tend to limit its usefulness. This was readily demonstrated in a study of 75 young isometrically trained athletes: 9 per cent had an abnormal ST–T segment response to exercise compared with 6 per cent of normal control subjects. However, these ECG changes occurred in the absence of other evidence of cardiovascular disease.[6] In athletes, the Bruce protocol produces a rather unusual form of exercise because of the quite marked elevation of the treadmill to produce a 22 per cent grade. Also, the large increments in work make estimation of V_{O_2} less accurate. Use of a longer belt on the treadmill and a faster belt speed on the level rather than elevating the treadmill may be more useful in testing runners. To reproduce results on a treadmill, the athlete may require several trial runs so as to be trained to exercise

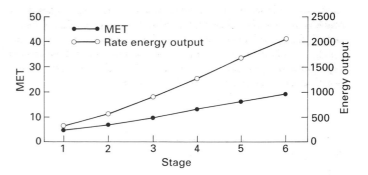

Fig. 4 Theoretical plot showing the first six stages of the Bruce treadmill exercise test with metabolic equivalent (MET) and the energy output for a 70-kg man.

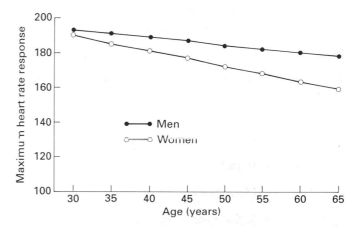

Fig. 5 Maximum heart rate responses during dynamic testing in men and women at various ages. This is calculated as 220–age.

to a high level. While exercise may produce useful information on cardiovascular function, it may also carry risks. Even in apparently healthy individuals or athletes in whom the risks are extremely low, there is a potential for sudden cardiac death. It is estimated that this occurs once in every 375 000 h of activity in healthy individuals, falling as low as once in 6000 h in patients with cardiac disease. In the United States it is widely recommended that individuals wishing to pursue a vigorous exercise programme should undertake a treadmill exercise test. Those over the age of 40 years who show abnormal features on examination, or on an ECG, or who have two or more major coronary risk factors, such as hypertension, hypolipidaemia, smoking, or diabetes, should be considered for testing, but the high rates of false positive and false negative results may again make interpretation difficult.

Non-invasive cardiac imaging

The study of cardiac morphology and function has been widely facilitated by the development of non-invasive techniques such as echocardiography and magnetic resonance imaging (MRI). Echocardiography has been widely used in this application. This involves the placing of a transducer which emits ultrasound waves on the chest. The depth of cardiac structures within the chest are displayed against time. This allows the generation of a time-dependent image (M-mode echocardiogram) and a depth-related image (cross-sectional echocardiogram). Both these allow the measurement of the cavity size and wall thickness of the left and right ventricle, and the size of both atria. In addition, structural abnormalities of the ventricles, the heart valves, the proximal coronary arteries, and the atria may be detected. Transverse measurements are made of posterior wall (PW) and septal thickness (normal, 0.7–1.1 cm), left-ventricular end-diastolic dimension (EDD) (normal, 4.8–6.0 cm), and end-systolic dimension (ESD) (normal, 3 ± 0.4 cm). The ejection fraction is derived as $(EDD^3 - ESD^3)/EDD^3$ with a normal range of 0.5–0.7. Fractional shortening (EDD-ESD/EDD) is often the preferred echocardiographic measurement because it does not require cubing of dimensions. Left-ventricular mass is calculated as $[(2PW + EDD)^3 - EDD^3] \times 1.055$ with a normal value of approximately 180 ± 50 g. Nuclear cardiology involves the injection of a radioisotope to label red blood cells so that imaging over the heart, using a nuclear camera, will give an indication of the function of the left and right ventricles. With the availability of current non-invasive technology it is very unusual to require invasive investigations of the heart of an athlete. Such techniques require the passage of arterial and venous catheters (retrograde or antegrade) into the left and right ventricles and pulmonary circulation to measure pressures and make direct injections of contrast medium.

THE NATURE OF THE ATHLETE'S HEART

The term 'athlete's heart' is ill-defined but reflects the normal physiological response to repetitive exercise. This presents as a constellation of clinical and investigative findings. On examination, a trained athlete will be noted to have reduced body fat and often increased muscle mass. The pulse will be slow, and will possibly have an increase in amplitude due to the elevated stroke volume and bradycardia. The cardiac impulse may be left-ventricular heaving in nature and displaced, representing cardiac enlargement. Mid-systolic murmurs, usually no louder than grade 1/4, may be noted in up to 50 per cent of dynamic athletes.[4] Third heart sounds are common in younger athletes, and fourth heart sounds may also be heard in the older age groups. The ECG may appear very abnormal in athletes, particularly in those undertaking prolonged training. The abnormalities most commonly seen are shown in Figs. 6 and 7.[7,8] The most common are of bradycardia with increased QRS voltages (Fig. 8). In very long-standing athletes, particularly runners in their fifties and sixties very considerable abnormalities may be noted in the ECG with repolarization changes as well as left-ventricular hypertrophy. To be certain that these are physiological and not pathological in origin, exercise testing should be undertaken.

Echocardiography to determine the morphological component of athlete's heart

Left-ventricular mass

Echocardiographic studies have shown that highly trained athletes in a variety of sports have an increased calculated left-ventricular mass (Fig. 9) which is due to increased ventricular wall thickness and cavity size. The average increase is 40 to 50 per cent greater than that in control subjects.[9] There are a variety of methods for estimating left-ventricular mass, and the basis of them is an incorporation of end-diastolic left-ventricular dimension and posterior wall thickness (and septal thickness). In normal individuals or

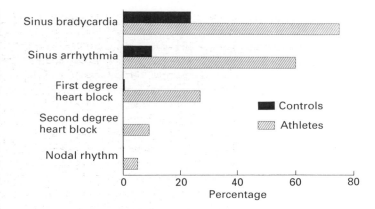

Fig. 6 Graphs comparing ECGs in athletes and controls. ECGs in athletes may reveal a number of disturbances in the cardiac rhythm. Most commonly, a sinus bradycardia may be seen and sinus arrhythmia, which is unusual in adults, is very frequently seen in athletes. First- and second-degree heart block is a frequent finding in athletes and is unusual in normal controls.

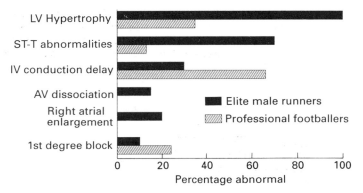

Fig. 7 Morphological abnormalities of the resting ECG are frequently observed in athletes but may differ in various sports. This graph compares the ECGs in runners and footballers. The two studies reported show that left-ventricular hypertrophy is much more common in elite male runners compared with professional footballers, whereas intraventricular conduction delays are more common in professional footballers. One would not expect normal subjects to show any of these abnormalities. (Data from Gibbons et al.[7] and Baladay et al.[8])

athletes these are relatively reliable, but they become progressively less so when there are regional abnormalties in left-ventricular function and hypertrophy.[4, 9–11]

Left-ventricular cavity

There are a large number of studies which compare echocardiographic left-ventricular dimensions in athletes with those in normal subjects. A consistent feature in endurance-trained athletes is an increase in end-diastolic dimension (Fig. 10). Although athletes are larger and often leaner individuals, the end-diastolic dimension remains increased even after correction of the body surface area. The increase in left-ventricular end-diastolic dimension is usually within the normal limits (5.7 cm in our laboratory), although occasionally athletes (particularly professional cyclists) have much larger values. Rost[5] demonstrates an echocardiogram in a cyclist with an end-diastolic dimension of 7 cm. In general, the values for endurance-trained athletes are

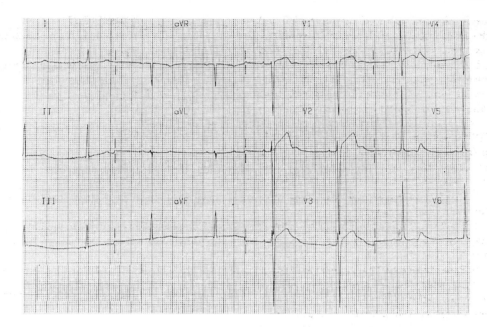

Fig. 8 Twelve-lead ECG in an endurance athlete. This shows the voltage criteria of left-ventricular hypertrophy with very deep S-waves in leads V^1 and V^2, and tall R-waves in leads V^5 and V^6. There is sinus bradycardia with repolarization abnormalities (biphasic T-waves) across the precordial leads (V^4, V^5, V^6).

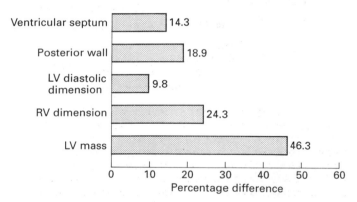

Fig. 9 A graphic summary of the echocardiographic abnormalities of left-ventricular dimensions observed in various published studies (percentage difference compared with normal). Left-ventricular mass is very considerably elevated in athletes, with small increase in left-ventricular diastolic dimensions and thicknesses. (After Maron.[9])

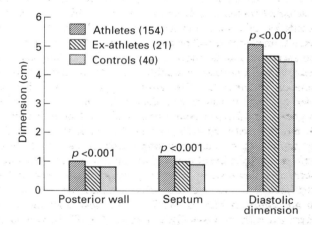

Fig. 10 Chart comparing posterior wall, septal, and left-ventricular diastolic dimension in athletes, ex-athletes, and control subjects. Current athletes have increased thickness of the posterior wall and septum and an increase in left-ventricular cavity size. In comparison, ex-athletes have similar values to normal controls.

approximately 10 per cent above matched sedentary controls. This is equivalent to a 33 per cent increase in left-ventricular volume because of the cube nature of volume measurements. While there are no longitudinal studies to demonstrate progressive increase in left-ventricular end diastolic dimension, it is probable that the more highly trained an individual becomes, the greater will be the increase in left-ventricular volume (Figs. 9–11).

Left-ventricular wall thickness

With the changes in left-ventricular cavity size, there is also an increase in the thickness of the left-ventricular posterior wall and the ventricular septum, which will lead to an elevated left-ventricular mass. In absolute terms, the increase in left-ventricular wall thickness may be small, in the range of 10 to 20 per cent, but is usually within the setting of an increased left-ventricular volume. As with left-ventricular cavity size, the measurements are often within the normal range, and the septum and posterior wall thicknesses seldom exceed 1.2 cm and rarely 1.4 cm. Occasionally, there may be thickening predominantly of the ventricular septum, rather than the posterior wall. The development of septal hypertrophy (asymmetrical septal hypertrophy) is a characteristic feature of hypertrophic cardiomyopathy. However, some athletes develop septal hypertrophy as part of their physiological development. This may occur in endurance athletes and may increase as

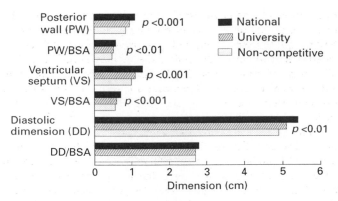

Fig. 11 Graph of echocardiographic left-ventricular wall thicknesses and dimensions in athletes to show the effect of the standard of competition on left-ventricular hypertrophy. The data appear to demonstrate that national standard athletes have more left-ventricular hypertrophy and the posterior wall and septal thickness is greater, even when normalized for body surface area (BSA), compared with athletes of a lesser standard.

training progresses. Values above the normal laboratory range have been recorded in 60 per cent of basketball players and a high proportion of young swimmers.[12,13] The increase in septal thickness results in a raised ratio of septum to posterior wall. This usually remains in the normal range of 1.3:1, but may reach in excess of 1.5:1. While athletes may develop septal hypertrophy disproportionate to the posterior wall, the actual measurements are often within or near the normal range and the characteristic feature of hypertrophic cardiomyopathy is the presence of myocardial hypertrophy which is often severe.[14] There is no evidence in the literature that training in athletes may cause hypertrophic cardiomyopathy; however, some cases of sudden death have been attributed to hypertrophy cardiomyopathy which is, presumably, unrelated to their training.

Right-ventricular cavity

The right ventricle enlarges in athletes, with an approximate 25 per cent increase in diastolic dimensions.

Left atrium

Many athletes would appear to have enlargement of the left atrial cavity according to measurement of transverse dimension. In patients with heart disease, this would be a reflection of abnormal left-ventricular diastolic properties or valvular disease, but this does not appear to be the case in athletes.

The mechanisms involved in the alterations in cardiac dimensions in athletes remain unclear. It has been suggested that this is predominantly due to the training-induced resting bradycardia, with concomitant prolongation of the diastolic filling period. However, the volume changes seen with bradycardia in athletes are much larger than those induced in normal subjects by pharmacologically-induced changes in heart rate, and therefore a reduced heart rate may be an important but not the sole determinant of increasing left-ventricular cavity size in trained athletes.[15]

THE HEART IN VARIOUS SPORTS

While it is clear that endurance training leads to an increase in left-ventricular cavity size and a mild increase in left-ventricular wall thickness, the findings of several investigators suggest that precise alterations in cardiac structures may differ depending on the type of training activity undertaken. There still remains disagreement over whether athletes have an increase in left-ventricular wall thickness and mass. Athletes participating in aerobic exercise are primarily exposed to conditions producing volume loading without large increases in left-ventricular wall thickness. Weight-lifters and shot-putters, in particular, are exposed primarily to a pressure load, and systolic blood pressures in excess of 300 mmHg have been recorded during lifting. The initial work by Morganroth et al.[16] suggested that weight-lifters, shot-putters and judo players had symmetrical left-ventricular hypertrophy without cavity dilatation, whereas endurance athletes have cavity dilation with only moderate increase in wall thickness. There have been a number of publications in the ensuing years which have supported or opposed this view.[5] Proponents have suggested that the enormous increases in blood pressure that occur during strength events lead to symmetrical left-ventricular hypertrophy, as in hypertension, and therefore the degree of hypertrophy is inappropriately increased with respect to the resting blood pressure and left-ventricular dimensions. An alternative hypothesis is that when corrections are made for the enormous body mass of power athletes, little myocardial hypertrophy can be detected. It is rare to find pure power athletes, and these tend to belong to a small select group of individuals with a large body mass. However, a study by Fagard et al.[17] has helped to clarify this point by comparing matched groups of cyclists and long-distance runners (Fig. 12). This confirmed that both runners and cyclists had increased left-ventricular dimensions compared with normal subjects; however, only cyclists have a greater increase in wall thickness, yielding an increased ratio of wall thickness to ventricular cavity dimension representing 'inappropriate' left-ventricular hypertrophy. The explanation for these findings is that cyclists have a higher exercising blood pressure than runners because of the isometric work carried out by the body holding the handle bars.

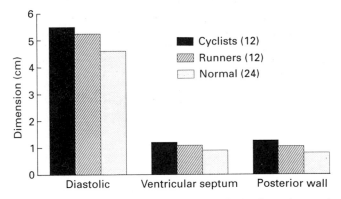

Fig. 12 Graph of echocardiographic left-ventricular dimensions and wall thicknesses in matched groups of cyclists and runners. Cyclists have a greater left-ventricular wall thickness than runners. This is thought to be due to the high exercising blood pressure because of isometric work carried out by the arms while cycling. This leads to further left-ventricular hypertrophy. (After Fagard et al.[17])

TRAINING AND DETRAINING OF THE HEART

A number of studies have shown that changes in left ventricular dimension and hypertrophy can be demonstrated during training and detraining.[18-20] Short-term studies of runners (over a 6-week period) and swimmers (over a 9-week period) showed quite rapid changes in heart rate and left-ventricular dimension (Figs. 13–15).

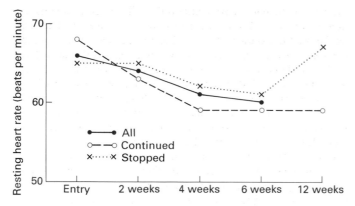

Fig. 13 Graph of resting heart rate changes during a 6-week course of training followed by a further 6-week period of continued training or return to normal sedentary activities. Throughout the period of training, there is a fall in resting heart rates which persists in those who continue to exercise. In the group that stops training, the heart rate promptly return to the pre-exercise levels.

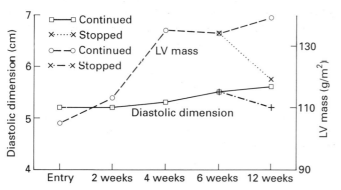

Fig. 14 During a 6-week period of training in a group of initially untrained university students, left-ventricular diastolic dimensions rise by a small amount. This persists on continuation of training and returns towards normal in those who stop. However, there is a significant rise in left-ventricular mass in the same period because of the small increase in septal and posterior wall thickness which occurs commensurate with these increases in left-ventricular dimensions.

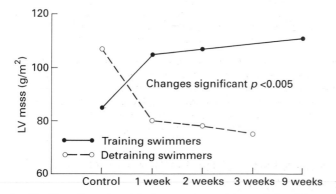

Fig. 15 The effect of ventricular mass on training and detraining swimmers. A group of trained swimmers were detrained for a period of 3 weeks and showed a significant fall in left-ventricular mass. An equivalent group of recently detrained swimmers were trained up for 9 weeks and showed a significant rise in left-ventricular mass.

In swimmers,[18] the end-diastolic dimension increased from 4.87 to 5.2 cm by the ninth week of training ($p > 0.005$) and the posterior wall thickness increased progressively from 0.94 cm at baseline to 1.01 cm by the end of the training period (Fig. 15). This was, of course, associated with a significant increase in left-ventricular mass. Similarly, over a longer period of study of oarsmen, there was a slight and gradual increase in left-ventricular end-diastolic dimension and ventricular wall thickness.[19] Detraining of runners and swimmers shows progressive reduction of the end-diastolic dimension increase and left-ventricular mass.[18–20] Whether the more profound changes seen in veteran athletes rapidly regress is unclear, but it is probable that after a period of prolonged adaption the changes reverse much more slowly. However, only small changes in left-ventricular dimensions, often within the normal range, are observed in non-athletic subjects over a short training programme, usually running or cycling. Only small changes in left-ventricular diastolic dimension occur, the magnitude of which is considerably less than that seen in competitive athletes, but because of the method of calculation there is often a significant increase in left-ventricular mass. It should be noted that, even after the training period, values obtained from these studies were still usually within the normal range. Dedicated and elite athletes who have reached a high level of training are a selected group of individuals who have outstanding development and adaption of skeletal and cardiac muscle. Although the Vo_2max may limit an athlete's ability to achieve competitive status, the limiting factor in whole-body exercise may be the capability of the heart to deliver oxygenated blood to the exercising muscles.[21] It is intriguing to speculate that a predetermined genetic predisposition may affect the potential or achieved level of performance in individual competitive athletes. There are few or no data defining these points, and studies are rather conflicting in the evidence that is available. Morganroth *et al.*[16] found no difference in cardiac size between world-class athletes and their non-elite counterparts. A study comparing national standard athletes with collegiate and recreational sportsmen suggested that the national athletes had a higher level of cardiac adaption.[10] However, even for the elite athletes studied, the values obtained fall largely within the normally accepted ranges; it is therefore likely that, in a potentially successful competitive athlete, performance is more related to genetically determined body structure and composition, which is advantageous, and superior pyschological motivation, which we loosely define as talent, rather than any predetermined, pre-existing, or training-induced adaption.

THE LONG-STANDING ATHLETE

Long-term participation in competitive sports is no longer uncommon. There is some evidence to suggest there is progressive alteration in cardiac dimensions with very prolonged exposure to training, especially into middle age. These findings are, of course, important, as individuals of advancing age are increasingly likely to suffer from organic heart disease. There have been only a few studies of this age group. That by Nishimura *et al.*[22] provided an insight into cardiac morphology in a group of chronically active professional cyclists of ages ranging from 20 to 49 years. The oldest cyclists had been training consistently for up to 27 years. The veteran cyclists showed greater left-ventricular size and mass than the younger athletes. There appeared to be greater left-atrial size and decreased left-ventricular function in the older athletes (Fig. 16).

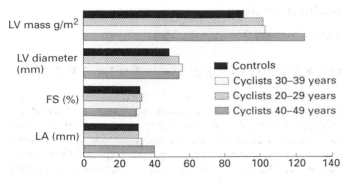

Fig. 16 Comparison of left-ventricular hypertrophy and function in cyclists of various ages. Cyclists aged 40 to 49 years, who had a very long history of professional cycling, appeared to have an increase in left-ventricular mass and left-atrial size with a fall in fractional shortening (FS) compared with younger athletes and controls.

VALVULAR REGURGITATION

Auscultation frequently reveals cardiac murmurs. The most common is a quiet pan-systolic murmur representing mild mitral regurgitation. Doppler echocardiography is very sensitive in the detection of regurgitation and shows regurgitant flow patterns across all valves in some normal individuals. The cause of these regurgitant jets is not known, but it is quite apparent that they are more common in athletes. Studies using colour-flow Doppler mapping[23] showed that in the most highly trained long-distance athletes, 93 per cent had triscuspid regurgitation, 87 per cent has pulmonary regurgitation, and 20 per cent had mitral regurgitation. Lesser trained athletes had a lower prevalence of valvular regurgitation, although 24 per cent of normal controls had triscuspid regurgitation, with 18 per cent having pulmonary regurgitation. Such findings by Doppler echocardiography, although clearly demonstrating valvular regurgitation, are of sufficiently small magnitude not to be of clinical or physiological significance.

LEFT-VENTRICULAR FUNCTION

Systolic function

Left-ventricular ejection fraction in athletes, measured either by echocardiography or by other non-invasive or invasive methods, is usually within the normal range. Stroke volume tends to be increased, which is probably explained by the bradycardia and cavity enlargement. Most measurements have been carried out by M-mode echocardiography which, although giving a high level of precision, is a segmental measurement and does not assess global function. Only in the study by Nishimura *et al.*[22] was a decrease in left-ventricular contraction observed in the older age group of competitive cyclists. Colan *et al.*[24] studied the effects of left-ventricular systolic mechanics in various groups of athletes, all of whom had increased left-ventricular mass. Runners tended to have a dilated left-ventricle with normal wall thicknesses, whereas swimmers has a mildly dilated left-ventricular cavity with increased wall thickness. In contrast, power lifters had a normal systolic cavity size with markedly increased wall thickness. Peak systolic wall stress was normal in runners and swimmers, and reduced in power-lifters. Measurements of myocardial stress (minute stress time integral) were used to show that myocardial

oxygen consumption was normal in runners and swimmers but significantly reduced in power lifters. In runners, fractional shortening was significantly reduced, whereas swimmers and power-lifters had significant augmentation for both the velocity and magnitude of fractional shortening (myocardial fibre shortening). They concluded from this study that physiological hypertrophy resulted in marked alterations in left-ventricular loading conditions with secondary changes in the systolic performance. The left-ventricular contractile state was normal in young athletes despite increased left-ventricular mass. However, different forms of exercise resulted in different patterns of left-ventricular hypertrophy and dilatation.[24] These studies of left-ventricular function were performed with the subjects at rest, but two studies of the effect of exercise on the athlete's heart have recently been published.[25,26] Both showed differences in left-ventricular dynamics during exercise in athletes and sedentary individuals. Long-distance runners at rest had a dilated ventricle, but they utilized the same mechanisms as sedentary adults for increasing cardiac output during exercise. At low levels of dynamic exercise, the Frank–Starling mechanism (ventricular volume increases as a function of pressure) is predominant. However, it has been suggested that at peak exercise, in the presence of preserved fractional shortening, faster relaxation and reduced diastolic left-ventricular filling may contribute to the development of a higher stroke volume in an exercising athlete's heart.

Diastolic function

It is, of course, the property of the left ventricle to fill as well as to empty. At peak exercise, the rate of transmitral blood flow as measured by pulsed Doppler echocardiography exceeds that across the aortic valve. Examination of diastolic function in the physiological hypertrophied heart may be carried out by mitral in-flow velocities using Doppler echocardiography or by digitized M-mode echocardiography.[27–29] These studies show that the peak rate of left-ventricular filling and wall-thinning were within or above the normal range. These diastolic function measurements were associated with a greater ventricular size and systolic performance. By comparison, in patients with pathological hypertrophy due to systolic pressure overload or primary myocardial disease, diastolic function is characteristically abnormal. In particular, isovolumic relaxation (the interval from aortic valve closure to mitral valve opening) may be grossly prolonged (normal 50–80 ms) in pathological hypertrophy (80–120 ms) and the rate of left-ventricular filling reduced, but both are always within the normal range of physiological hypertrophy.

DIFFERENTIATION OF PHYSIOLOGICAL AND PATHOLOGICAL HYPERTROPHY

Added heart sound, systolic murmurs, left-ventricular hypertrophy, morphological electrocardiographic abnormalities, and bradyarrhythmias may occur in athletes. These may also be features of organic heart disease. Sinoatrial disease, hypertrophic cardiomyopathy, and valvular disease may all be simulated, but echocardiography is usually able to differentiate most of these without difficulty. In some life-long athletes, it may be extremely difficult to differentiate whether left-ventricular hypertrophy is due to athletic training or is related to an underlying primary myocardial defect. Certain general rules can be applied to define normality. In particular, athletes have left-ventricular end-diastolic dimensions of less than 6 cm, left-ventricular hypertrophy is symmetrical with the posterior wall or septum not exceeding 1.4 cm, and left-ventricular hypertrophy occurs in the presence of left-ventricular cavity enlargement. If the left-ventricular hypertrophy is observed without cavity enlargement or, in particular, if the cavity is small, the presence of organic disease should be considered. Although asymmetrical septal hypertrophy (septum more greatly hypertrophied than the posterior wall) does occur in some athletes, if the ratio of these measurements exceeds 1.5:1, one may consider the presence of hypertrophic cardiomyopathy although this is usually associated with very marked left-ventricular hypertrophy, a small left-ventricular cavity which is hyperdynamic, and other abnormalities such as systolic anterior motion of the mitral valve. Analysis of left-ventricular diastolic function may be helpful, as in pathological hypertrophy isovolumic relaxation is prolonged and left-ventricular filling is reduced in peak rate and prolonged in duration. In comparison, in athletes these measurements are normal or augmented.[27,28] Myocardial texture as determined by echocardiography is normal in athletes, whereas in pathological states fibrosis occurs which leads to an increase in echo amplitude, which may be detected by a number of commercially available systems.[29] Occasionally, in a patient with marked electrocardiographic abnormalities and chest pain, it may be necessary to proceed to invasive investigations, primarily coronary angiography, to determine the state of the coronary arterial tree.

THE SYMPTOMATIC ATHLETE

Chest pain

Occasionally a young athlete will complain of chest pain on exercise due to myocardial ischaemia from premature coronary atherosclerosis or congenital coronary or valvular abnormalities. However, the majority will have musculoskeletal pain, and the unwary doctor examining a routine ECG and chest radiography will discover abnormalities in both which may lead to an erroneous diagnosis of heart disease.

Collapse

Collapse during exercise requires complete investigation, as a sudden drop in cardiac output is usually due to a defect in cardiac morphology or rhythm. Collapse following strenuous exercise may be physiological, particularly due to hypovolaemia following marathon races and the loss of the venous pump due to muscular movements of the legs in the athlete who stops at the end of an event.

Breathlessness

Breathlessness may be a feature of cardiac disease, but is more commonly due to exercise-induced asthma which may also occur after an upper respiratory tract infection (see Chapter 3.2.2).

Palpitations

Awareness of the heart beat is common in athletes at rest, particularly because of their slow irregular pulse with extrasystoles and marked sinus arrhythmia. Rapid palpitations at rest or of sudden onset during exercise require further investigation.

Sudden death

This is discussed in Chapter 3.2.1.

HEART DISEASE IN ATHLETES

The competitive or elite athlete is identified in society as one of the most healthy individuals. Occasionally, young or, in particular, older athletes may die suddenly during or following athletic activity. The most common cause of sudden death in the younger athlete is hypertrophic cardiomyopathy, but, in addition, aortic rupture of the type seen in Marfan's syndrome, congenital coronary anomalies, and atherosclerotic coronary disease, as well as valvular disease of the heart, especially aortic stenosis, may be present.[30] In addition, particularly those in the older age groups may develop heart disease during many years of training. The most common would be the development of coronary heart disease. Such patients may present with chest pain or other exertional symptoms during or immediately following exertion. Because of the high exercise tolerance that individuals may have developed, symptoms may only present after a considerable amount of exercise. A typical pattern would be that an athlete who usually ran 5 to 10 km at a time would complain of chest pain on running the first 1 to 200 m and then would be able to run through the angina. The athlete would only develop further chest discomfort on running up a hill, or when the weather was particularly cold or windy.

The presence of mild mitral regurgitation, particularly with mitral prolapse, does not preclude competitive athletic training. Other valve diseases, such as pulmonary stenosis or regurgitation, or tricuspid regurgitation, as well as mitral stenosis, are relatively uncommon in training athletes and unless severe would not limit exercise tolerance to any degree. Difficulties arise in the presence of aortic stenosis. Degrees of aortic stenosis which are more than moderate in severity (30–40 mmHg gradient) increase the risk of syncope and sudden death during exercise. With the development of Doppler ultrasound, non-invasive measurement of the aortic valve gradient can be made repeatedly, and if the gradient reaches a significant level, especially with the development of left-ventricular hypertrophy, individuals should be prevented from severe exertion. In aortic regurgitation disease of moderate severity does not preclude exercise, and several well-know runners have been able to exercise to a high level without difficulty.

There remains some concern that severe bradycardia, which may be present in athletes, particularly those who have had lifelong training habits, may develop symptoms related to their slow heart rate. Athlete's bradycardia is presumably due to an increase in vagal tone, though pharmacological denervation does not produce such marked bradycardia independent of vagal tone. Other factors responsible for a training-induced bradycardia include reduced sensitivity to normal sympathetic chronotrophic stimulation and the intrinsic property of large hearts to beat more slowly. Veteran athletes may have quite high degrees of atrioventricular block and prolonged ventricular pauses. Whether athletic training leads to a form of sick sinus syndrome is unclear, but there is some evidence to suggest that atrial fibrillation, which may lead to embolization, may occur more frequently in athletes than in sedentary controls, and pacemaking may occasionally be required.[31, 32]

The presence of hypertension in athletes needs to be considered as some forms of exercise, particularly that involving power sports, can lead to considerable rises in blood pressure. While no cerebral or cardiovascular problems have been reported to be directly due to this, in a hypertensive patient caution would suggest that blood pressure should be reduced before active sport could continue.

Patients with pacemakers *in situ* can continue to exercise without difficulty. Some pacemaker modalities will limit exercise tolerance, and implantation of a pacemaker in an athlete may require more complicated forms of pacing (such as dual–chamber or rate-responsive units) to allow a high level of exercise tolerance.

SCREENING OF POTENTIAL ATHLETES

Most sportsmen undertaking active exercise will be healthy and free from cardiovascular disease. However, a small proportion will have coexisting heart disease or develop it during their training programme. Whether preparticipation screening studies should be carried out remains a matter of conjecture (see above). In an athlete who has been training for many years, and who remains able to exercise to a high level, there is clearly no point in doing this. Whether one should screen new potential athletes, particularly if they are older, is probably a more important subject. Various studies have addressed this problem, particularly that by Maron *et al.*,[33] studied 501 athletes in an intercollegiate athletics programme (Fig. 17). The majority of athletes had no evidence of cardiovascular disease; in those who showed minor abnormalities on physical examination or by ECG, echocardiography revealed no significant abnormalities in the majority. It is apparent that there will be a low frequency of most cardiovascular disease in young athletes, or even in those who are older, as they are self-selected into being healthy. The difficulty is that there will be a large number of false positive observations in young athletes.

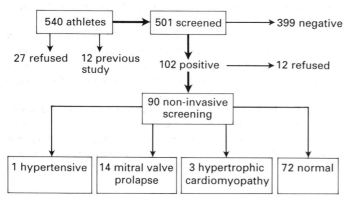

Fig. 17 Diagrammatic representation of the progress of screening previously healthy athletes. Most athletes, though screened, had no abnormal findings, and even the majority of the 102 positive athletes were normal. One was hypertensive, 14 had mitral valve prolapse (MVP), which is usually an insignificant abnormality, and three possibly had hypertrophic cardiomyopathy, but the remainder were normal.

CONCLUSION

Athletes are by definition healthy and it is very unusual to detect significant cardiovascular disease. Training, particularly if prolonged, may lead to cardiovascular alterations which may mimic heart disease.

REFERENCES

1. Henschen S. Skilauf und Skiwettlauf. Eine Medizinische Sportstude. *Mitte Med Klinik, Upsala* 1899; **2**: 15.
2. Paffenbarger RS, Hyde RT, Wing AL, Hsieh C-C. Physical activity, all-causes mortality and longevity of college alumni. *New England Journal of Medicine* 1986; **314**: 605.
3. Morris JN, Pollard R, Everitt MG, Chave SPW, Semmence AM. Vigorous exercise in leisure time: protection against coronary heart disease. *Lancet* 1980; **ii**: 1207–10.
4. Huston TP, Puffer C, Rodney WM. The athlete's heart syndrome. *New England Journal of Medicine* 1985; **313**: 24–9.
5. Rost R. The athlete's heart. *European Heart Journal* 1982; **3** (Suppl. A): 193–8.
6. Spirito P, Maron BJ, Bonmow RO, Epstein SE. Prevalence and significance of an abnormal S–T segment response to exercise in a young athletic population. *American Journal of Cardiology* 1983; **51**: 1663–6.
7. Gibbons LW, Cooper KH, Martin RP. Medical examination and electrocardiographic analysis of elite distance runners. *Annals of the New York Academy of Sciences* 1977; **301**: 283–96.
8. Baladay GJ, Cadigan JB, Ryan TJ. Electrocardiogram of the athlete: an analysis of 289 professional football players. *American Journal of Cardiology* 1985; **53**: 1339–43.
9. Maron BJ. Structural features of the athlete's heart as defined by echocardiography. *Journal of the American College of Cardiology* 1986; **7**: 190–6.
10. Shapiro LM. Physiological left ventricular hypertrophy. *British Heart Journal* 1984; **52**: 130–5.
11. Longhurst JC, Kelly AR, Gonyea WJ, Mitchell JH. Echocardiographic left ventricular masses in distance runners and weight lifters. *Journal of Applied Physiology* 1980; **48**: 154–62.
12. Allen HD, Goldberg SJ, Sahn DJ, Schy N, Wojcik R. A quantitative echocardiographic study of champion childhood swimmers. *Circulation* 1977; **55**: 142–5.
13. Roeske WR, O'Rourke RA, Klein A, Leopold G, Karliner JS. Non-invasive evaluation of ventricular hypertrophy in professional athletes. *Circulation* 1976; **53**: 286–92.
14. Shapiro LM, Kleinebenne A, McKenna WJ. The distribution of left ventricular hypertrophy in hypertrophic cardiomyopathy: comparison to athletes and hypertensive. *European Heart Journal* 1985; **6**: 967–74.
15. Hirshleifer J, Crawford M, O'Rourke RA, Karliner JS. Influence of acute alterations of heart rate and systemic arterial pressure of echocardiographic measures of left ventricular performance in normal subjects. *Circulation* 1975; **52**: 835–41.
16. Morganroth J, Maron BH, Henry WL, Epstein SE. Comparative left ventricular dimensions in trained athletes. *Annals of Internal Medicine* 1975; **82**: 521–4.
17. Fagard R, Aubert A, Staessen J, Eynde EV. Comparative echocardiographic study of cardiac structure and function in cyclists and runners. *British Heart Journal* 1984; **52**: 124–9.
18. Ehsani AA, Hagberg JM, Hickson RC. Rapid changes in left ventricular dimensions and mass in response to physical conditioning and deconditioning. *American Journal of Cardiology* 1978; **42**: 52–6.
19. Weiling W, Borghols EAM, Hollander AP, Danner SA, Dunning AJ. Echocardiographic dimensions and maximal oxygen uptake in oarsmen during training. *British Heart Journal* 1981; **46**: 190–5.
20. Smith RG, Shapiro LM. Effect of training on left ventricular structure and function. *British Heart Journal* 1983; **50**: 534–9.
21. Saltin B. Haemodynamic adaption to exercise. *American Journal of Cardiology* 1985; **55**: 42–6.
22. Nishimura T, Yamada Y, Kawai C. Echocardiographic evaluation of the long-term effects of exercise *Circulation* 1980; **61**: 832–40.
23. Pollack SJ *et al.* Cardiac evaluation of women distance runners by echocardiographic colour Doppler flow mapping. *Journal of American Medical Association* 1988; **11**: 89–93.
24. Colan SD, Sanders SP, Burow KM. Physical hypertrophy: effects on left ventricular systolic mechanics in athletes. *Journal of the American College of Cardiology* 1987; **9**: 776–83.

25. Fagard R, Broeke CVD, Amery A. Left ventricular dynamics during exercise in elite marathon runners. *Journal of the American College of Cardiology* 1989; **14**: 112–18.
26. Ginzton LE, Conant R, Brizenine M, Laks M M. Effects of long-term high-intensity aerobic training on left ventricular volume. *Journal of the American College of Cardiology* 1989; **13**: 364–71.
27. Matsuda *et al.* Effect of exercise on left ventricular diastolic filling in athletes and non-athletes. *Journal of Applied Physiology* 1983; **55**(2): 323–8.
28. Shapiro LM, McKenna WJ. Left ventricular hypertrophy: relation of structure to diastolic function in hypertension. *British Heart Journal* 1984; **51**: 637–42.
29. Shapiro LM, Moore RB, Logan-Sinclair RB, Gibson DG. Relation of regional echo amplitude to left ventricular function and the electrocardiogram in left ventricular hypertrophy. *British Heart Journal* 1984; **52**: 99–105.
30. Maron BJ, Epstein SE, Roberts WC. Cause of sudden death in competitive athletes. *Journal of the American College of Cardiology* 1986; **7**: 204–14.
31. Abdon N-J, Kerstin L, Joahnsson B W. Athlete's bradycardia as an embolising disorder? Symptomatic arrhythmias in patients aged less than 50 years. *British Heart Journal* 1984; **52**: 660–6.
32. Northcote RJ, Rankin AC, Scullion R, Logan W. Is severe bradycardia in veteran athletes an indication for a permanent pacemaker? *British Medical Journal* 1989; **298**: 231–2.

3.1.2 Exercise and the skeleton

ROGER L. WOLMAN AND JONATHAN REEVE

INTRODUCTION

Exercise affects bones in several important ways. The type and intensity of exercise and the anatomical site where the stress is applied are all influential. Intense exercise in premenopausal women can also modify the function of the hypothalamic–pituitary–ovarian axis. This can lead to impaired production of oestrogen from the ovary, which will have a detrimental effect on measured bone density.

With recent advances in bone densitometry it is now possible to make precise and accurate measurements of bone mass which has increased understanding of how the skeleton responds to exercise. In this chapter we shall discuss these factors and the role of sport and exercise in the development and prevention of osteoporosis.

MEASUREMENT OF BONE MASS

The skeleton consists of two types of bone, cortical and cancellous (trabecular). Cancellous bone is more deformable than cortical bone, forming a sort of mechanical buffer between the even more deformable joint cartilage and the comparatively rigid diaphysis of long bones. The proportion of each type of bone varies in different skeletal sites. The highest percentage of cancellous bone is found in the vertebral bodies (about 40 per cent),[1] but there are also significant amounts in the proximal femur and the distal radius. The turnover rate of cancellous bone (i.e. the rate at which old bone is removed and new bone formed) is much greater than that of cortical bone (by a factor of about 5–7). In situations where there is net bone loss, such as amenorrhoea (see below), changes in bone density will appear first and are most marked at sites where cancellous bone predominates (i.e. the spine, the hip, and the wrist).

Plain radiography is an inadequate means of quantifying bone mass except in long bones which are almost entirely cortical. Early attempts were made to improve the estimation of density by incorporating step wedges and other phantoms within the X-ray field for comparison with the bone of interest. Virtually the only techniques to have survived based on plain radiography technology are those that measure cortical width at various sites, for example the measurement of metacarpal cortical width on three or four metacarpals on each hand,[2] or the midshaft of the femur.[3]

Modern measurement of bone mass has been revolutionized by photon absorptiometry. This is based upon the principle that photons (either X-rays or gamma rays) are absorbed more readily by dense than by less dense materials. Thus bone is a better absorber of photons than fat or other soft tissues. The photons emerge to be counted by a detector and are attenuated in inverse proportion to their energy.

Two approaches to measurement are used currently. The first is based on quantitative computed tomography (QCT) in which true bone density can be measured by calculating the total contained mass adjusted for the volume measured. The other approach includes single- and dual-photon absorptiometry and dual-energy X-ray absorptiometry, but these techniques make no allowance for depth, only operating on a single plane. The virtue of dual-photon absorptiometry (and dual-energy X-ray absorptiometry) is that it can make a correction for absorption in soft tissues and therefore can be applied to the axial skeleton at reduced cost and radiation dosage and with higher precision than QCT. Single-photon absorptiometry is suitable for making measurements on the forearm and the os calcis, and has been widely used in that context. Several recent reviews provide further discussion of the techniques currently used in bone densitometry.[4,5]

QCT measures a true bone density (expressed in milligrams of bone mineral per cubic centimetre). However, dual-energy X-ray absorptiometry measurements are expressed in grams per square centimetre. To avoid confusion dual-energy X-ray absorptiometry measurements are increasingly referred to as 'area density' measurements.

PATHOPHYSIOLOGY OF BONE MODELLING AND REMODELLING

During childhood and adolescence, the skeleton has to grow with the rest of the body. It has been shown that a child's skeleton is replaced every 6 months by processes that have been termed modelling and remodelling of the skeleton.[6] Modelling is defined as that process which leads to creation of new bone where soft tissue (e.g. cartilage) was found previously; it controls outside bone diameter, marrow cavity diameter, and cortical thickness, all adjusted to the dictates of longitudinal bone growth and the growth of muscle and body mass. In contrast, remodelling is the process by which a volume of bone is removed and replaced by a new volume of bone.

Where linear growth is required which may be too rapid for the capacity of the modelling process, cartilaginous epiphyseal plates are to be found between the metaphysis and the diaphysis such as in long bones or the vertebral bodies. In the context of athletic activity, these cartilaginous plates are a source of structural weakness through which fracture or shearing injuries (e.g. slipped upper femoral epiphysis) may occur in young people.

The cells responsible for bone remodelling are principally the osteoblast and the osteoclast. However, the osteoblast forms part of a complex lineage of bone cells which include the osteocyte and the lining cell. These cells are generally believed to represent the end-stage of osteoblastic development which, after some 4 to 6 months of synthetic activity have ceased to form new bone. They then become buried in the bone matrix or end up as resting cells on the quiescent surfaces of bone adjacent to blood vessels. It is now believed that these cells are important in directing the activities of osteoclasts towards sites of new bone remodelling[7] and they may be sensitive to mechanical forces through various biochemical mechanisms.[8,9] Once the lining cells allow the osteoclasts access to a mineralized bone surface, these mobile multinucleated cells erode bone to a certain depth (on trabecular or endosteal surfaces) or, in the case of the cortex, drive a tunnel deep into the substance of bone. These tunnels are called Haversian systems when they are again filled by the osteoblasts which appear in the wake of the cutting cone of osteoclasts. At the endosteal and trabecular surfaces, after the osteoclasts have passed on, the scalloped surface becomes a so-called reversal surface, implying that the removal of bone goes into reverse and bone formation commences. Bone formation almost always commences after bone resorption, but the bone that has been removed is not always replaced completely. In disuse osteoporosis, such as occurs on prolonged bed rest for example, there is evidence of impaired osteoblastic function with incomplete refilling of the remodelling space.

Activation of bone remodelling is a subject of considerable current interest.[10] A wide spectrum of cytokines and other locally produced agents (including interleukin 1, tumour necrosis factor, etc.) might be involved in the process of initiating bone remodelling. Certain hormones, such as parathyroid hormone and 1,25-dihydroxyvitamin D (calcitriol), also encourage the remodelling process. In normal circumstances disuse encourages bone remodelling and physical activity discourages it.[11] Thus physical activity may influence both the rate at which remodelling is initiated *per se* as well as the events that take place in the remodelling process once it has begun.

Frost[11] has discussed the relationship between bone strain data and the phenomenon of fatigue fracture. The fracture strain of cortical bone is about 25000 microstrain (μE) in longitudinal tension or compression, but vigorous voluntary activity usually results in strains of only about 3000 μE. The fatigue resistance of bone declines sharply as repetitive cyclic strains rise in amplitude from 2000 uE ($> 10^7$ cycles of fatigue resistance) to 4000 μE (about 20000 cycles). Accumulation of microdamage, as demonstrated by Schaffler et al.,[12] is thought to underlie clinical stress fractures which can only be averted by timely repair of microdamage. This process generally takes 4 months, which is the usual minimum time-span of a cycle of bone remodelling in adult man.

FACTORS THAT INCREASE BONE DENSITY

Exercise and bone density

Wolff[13] recognized 100 years ago that bone tissue adapts to the functional forces acting upon it; however, understanding of the relationship between exercise and bone density has only advanced further in the last 40 years.

Changes in athletes

Nilsson and Westlin[14] showed that athletes have significantly higher bone density in the distal femur than age-matched controls. Amongst the athletes, weight-lifters had the highest and swimmers the lowest bone density, with runners lying in between. In our own series of female athletes, spinal bone density was 10 to 20 per cent higher in normal menstruating athletes than in sedentary controls. Several groups have now investigated the effect of different types of sport on bone mineral density.

Studies of runners have shown that the bone density of the os calcis[15] and the lumbar spine[16] is higher than in matched controls. The bone density of the femoral midshaft is higher in elite runners than in athletes from other sports and in age-matched controls.[17] Furthermore, the usual fall in total bone mass with age seemed to be less rapid in runners.[18]

Tennis players show major differences between the playing and non-playing arms. In active male tennis players bone mineral density is 13 per cent higher in the radius of the playing arm.[19] However, in top class tennis players humeral bone mineral density is up to 40 per cent higher[20,21] and radial bone mineral density is 34 per cent higher in the playing arm.[22] In comparison, radial bone mineral density in non-athletes is similar in the dominant and non-dominant arms.[22]

Swimming is a non-weight-bearing activity and is reported to produce only minor increases in bone density in young adult athletes compared with other sports such as weight-lifting, running, soccer,[14] and tennis.[23] In an older age group swimming has been shown to produce, at most, minor benefit on measured bone density.[23,24]

Female rowers have increased bone density in the spine compared with non-rowing athletes (Fig. 1). This applies to both normally menstruating rowers and those with amenorrhoea when they are compared with non-rowing athletes of the same menstrual status.[25]

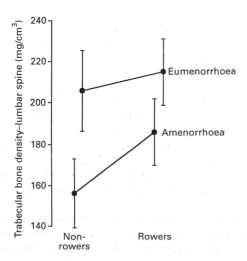

Fig. 1 Trabecular bone density of the lumbar spine of international rowers compared with international endurance runners and elite ballerinas (non-rowers). Comparison of the bone density of the amenorrhoeic and eumenorrhoeic athletes is also shown (taken from work done at the British Olympic Medical Centre, 1988–89).

Relationship with activity level

Various methods of assessing activity level have been used in cross-sectional studies, and these have tended to show a direct relationship between exercise and bone density. Aloia et al.[26] used a motion sensor attached to the trunk of 24 healthy white premenopausal women and showed a significant correlation between activity level and bone density.

Other workers have investigated the relationship between fitness and bone mineral density in healthy peri- and postmenopausal women. A weak positive correlation with bone mineral density was found by those researchers who estimated Vo_2max from the heart rate.[27,28] However, workers who measured Vo_2max directly[29,30] failed to show any relationship in postmenopausal women. Therefore aerobic fitness is likely to be a relatively weak determinant of bone mineral density.

Significant correlations between bone mineral density and muscle strength have also been demonstrated. Cross-sectional studies have shown relationships between spinal bone density and isometric back strength[31] and between spinal and forearm bone density and isometric grip strength,[30] but longitudinal studies have not yet determined the type and frequency of muscle strengthening exercises necessary to increase bone mineral density.

Effect of exercise intervention

It might be argued that the higher bone density seen in athletes allows them to undertake intensive training, i.e. there is an element of selection. The most effective way of showing that bone mineral density does indeed increase in response to exercise is from prospective data.[15] A number of workers have investigated the effect of exercise intervention on age-related bone loss. These studies are discussed below in the section on exercise and osteoporosis after the menopause. The effect of ceasing athletic activity has not yet been investigated in young adults. However, in postmenopausal women, most of the benefits on bone density appear to be lost once exercise intervention stops.[29]

Animal studies

Animal studies have shed further light on the effect of load bearing on bone. Lanyon and Rubin[32] showed that bone remodelling is sensitive to dynamic loads, but not to static loads, and is directly related to the peak strain magnitude.[33] At very low peak strain levels bone loss occurs, but with increasing strain levels increased bone mass results. Work on roosters[34] shows that the frequency at which the load is applied to the bone is also important, with higher frequencies exerting a greater effect. However, in roosters the osteogenic stimulus seems to saturate at loading cycles of as low as 36 per day with no additional benefit at higher frequencies. In sheep[35] the osteogenic stimulus is also affected by the rate at which the strain changes, with higher rates producing the greatest effects.

Research on immobilization and weightlessness (see below) indicates the importance of stress loading in maintaining skeletal integrity. Weight-bearing activity is particularly important in providing the appropriate stress. Work on tennis players has demonstrated that exercise produces a local effect on the bones of the forearm which goes beyond any general benefit to the whole skeleton. Weight-bearing activities such as walking and running are likely to be more effective in maintaining the integrity of the

neck of the femur and the spine than non-weight-bearing activities such as swimming and cycling. However, the specific relationships between the mineral densities of bones of particular interest and the type, intensity, and frequency of exercise has not been delineated in man.

FACTORS THAT PREDISPOSE TO REDUCED BONE DENSITY

Hormonal factors are closely linked to the development of osteoporosis, particularly disorders which lead to impaired sex hormone production. This occurs in postmenopausal women and female endurance athletes in whom low oestrogen status leads to reduced bone density. Other endocrinological disorders, such as thyrotoxicosis and Cushing's syndrome, can also lead to osteoporosis but are far less common. Iatrogenic osteoporosis is seen in association with corticosteroid use, and with immobilization (e.g. in conditions that give rise to reduced levels of physical activity). Reduced bone density also occurs with prolonged weightlessness.

Menstrual abnormalities and reduced bone density

Amenorrhoea and oligomenorrhoea are usually associated with low concentrations of circulating oestrogen. The importance of ovarian function, in particular oestrogen release, in maintaining skeletal integrity was appreciated by Albright et al.[36] when the link between the menopause and osteoporosis was first suggested. Aitken et al.[37] provided further evidence when investigating the effects of oophorectomy for non-malignant disease in 258 premenopausal women showing a significant increase in the prevalence of bone loss within 3 to 6 years of the operation.

Cancellous (trabecular) bone turns over more rapidly than cortical bone (see above). Thus when resorption outweighs formation, as during oestrogen deficiency (at least in its early stages), the losses are relatively greater in cancellous bone. Vertebral body bone density, which includes about 40 per cent cancellous bone, falls rapidly in response to low oestrogen status. The proximal femur and distal radius also contain significant amounts of cancellous bone, and similar, but less severe, losses can be demonstrated at these sites. With the advent of the technology for measuring bone mineral density accurately, it has been possible to show that there is a fall in vertebral bone density of about 2 to 5 per cent per year at the time of the menopause compared with about 0.4 per cent per year in premenopausal women. This may persist for up to 8 years.[38]

Amenorrhoea occurs in patients with anorexia nervosa and with hyperprolactinaemia. In both these conditions bone density decreases as amenorrhoea develops.[39,40] Indeed, osteoporotic spinal crush fractures have been reported amongst anorexia nervosa patients in their twenties.[39]

Luteinizing hormone releasing hormone (LHRH) agonists are used in the treatment of endometriosis by causing hypo-oestrogenaemia and amenorrhoea. Treatment usually continues for a maximum of 6 months during which time there is a significant fall in bone density, particularly of the lumbar spine. Once treatment is discontinued the bone density usually returns towards pretreatment levels within 6 months.[41]

Amenorrhoea also occurs in the context of intense aerobic training and can lead to reductions in bone density (see below).

Changes in bone during immobilization and weightlessness

Deitrick et al.[42] found that during a period of bed rest healthy men showed increased excretion of calcium but no radiologically visible osteoporosis. Young subjects who had been confined to bed for periods of up to 36 weeks showed a negative calcium balance of about 200 to 300 mg per day (equivalent to a loss of 10 per cent of total body bone mineral annually) together with increased urinary hydroxyproline excretion.[43] During immobilization bone mineral density, measured by dual photon absorptiometry, falls by 1 to 2 per cent per week at sites of trabecular bone and by 1 per cent per month at cortical bone sites.[44] Anderson and Nilsson[45] reported an 18 per cent drop in bone density in the proximal tibia in patients operated on for knee ligament injuries which had not fully recovered 1 year later even though full mobility had been regained.

Similar effects have been observed in subjects exposed to a gravity-free environment. Demineralization of bone was demonstrated by Mack and coworkers during the Gemini–Titan orbital flights[46] and the Apollo project.[47] Those astronauts who participated in a programme of isometric and isotonic exercises had smaller losses of bone mineral. Investigations of the Skylab astronauts[48] showed falls in calcium balance similar to the levels in bed-rested subjects.

During immobilization bone losses persist for at least 6 to 9 months. Very few of the studies have exceeded 9 months, but there is some evidence that the losses decline after this time.[44] Bone loss seems to be at least partly reversible but the period of recovery is several times longer than the period of loss.[44] Furthermore, no data are available to permit estimation of a magnitude of loss that would represent a 'point of no return'.

It can be seen from the above studies that weight-bearing activity is important in maintaining the skeleton.

EFFECT OF EXERCISE AND OTHER FACTORS ON THE MENSTRUAL CYCLE

Amenorrhoea is a well-recognized consequence of endurance training.[49] Prior to the late 1970s there were few endurance events for women and intense aerobic training and its associated hormonal complications were unusual. Since then the incidence of amenorrhoea has increased as rigorous training routines have been employed by more female athletes and may be as high as 50 to 60 per cent in some sports. It varies between different sports and is a reflection of the training requirements of each sport (Fig. 2).

Training intensity

Amenorrhoea typically occurs in the endurance athlete and is rarely seen with power training. The classical example of an endurance athlete is the marathon runner, and amongst this group of athletes the incidence of amenorrhoea is related to the intensity of training.[50,51] Between 50 and 60 per cent of athletes running 60 to 80 miles per week develop menstrual abnormalities, but this drops to 20 per cent for those running 20 miles per week.[50] However, endurance exercise forms an important part of training in many other activities such as ballet, gymnastics, and squash, and thus may be partly responsible for the associated incidence of amenorrhoea.

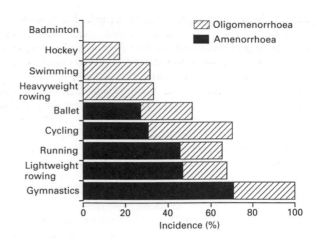

Fig. 2 Incidence of menstrual irregularity amongst elite competitors in different sports (taken from work done at the British Olympic Medical Centre, 1988).[85]

Nutritional status

Calorie deprivation is the single most important factor in the amenorrhoea of anorexia nervosa.[52] However, calorie restriction is important in certain activities that require a high power-to-weight ratio like ballet, cycling, gymnastics, and long-distance running. It is also important in sports where there are weight restrictions. The incidence of amenorrhoea amongst light-weight rowers, who must weigh 59 kg or less to compete, is higher than in their heavyweight counterparts (Fig. 2) despite similar levels of training. The disparity is due to differences in calorie intake.

The calorie intake of amenorrhoeic athletes tends to be lower than that of their eumenorrhoeic counterparts even when matched for training intensity.[53,54] There may also be other nutritional differences between amenorrhoeic and eumenorrhoeic athletes,[55,56] but it is uncertain whether these represent cause or effect.

Body composition

Increased training intensity and calorie restriction lead to weight loss and reduction in body fat which may be important in the pathophysiology of amenorrhoea. There is an association between amenorrhoea and low body weight[57] and low levels of body fat.[58] Frisch and McArthur[59] proposed that a critical percentage of body fat is required for both the onset of menarche (17 per cent) and maintenance of eumenorrhoea (22 per cent). In other studies no differences in body fat between the menstrual groups have been found.[60] An explanation for this discrepancy could lie in the differing selection criteria used in each of the studies or in the well-recognized lack of agreement between the different methods of measuring body fat.[61]

Age

Menstrual dysfunction is more common in young athletes in their late teens and early twenties than in those in their thirties.[57] Intense training in childhood may delay the onset of puberty and lead to primary amenorrhoea.[62] In some cases menarche may be delayed beyond the age of 20. This may have adverse consequences not only on bone mineral density but also on skeletal maturation.[63]

Training in childhood may also increase the risk of later menstrual irregularity.[62] In the perimenarchal period major changes take place in the hypothalamic–pituitary–ovarian axis. Extreme exercise around this time may affect these changes and might help to account for the features of menstrual dysfunction. The very high incidence of amenorrhoea seen amongst gymnasts (Fig. 2) is related to the young age at which they start training (below 10 years) and at which they compete (early to middle teens). Ballerinas start training at a similar age but tend to perform at an older age, and this may explain their slightly lower incidence of amenorrhoea.

Stress

Emotional stress can be associated with menstrual dysfunction, and this is thought by some to be important in the pathogenesis of amenorrhoea, particularly in anorexia nervosa. It has been suggested that top-class female athletes are under increased physical and psychological stress and that this may contribute to the risk of amenorrhoea. There are certain psychological similarities between amenorrhoeic athletes and patients with anorexia nervosa with regard to eating disorders, major affective disorders, and compulsive behaviour.[64] These features apply less to eumenorrhoeic athletes.

Hormonal factors

There is evidence of loss of the pulsatile release of gonadotrophin-releasing hormone (GnRH) from the hypothalamus during exercise.[65] This, in turn, leads to a reduction in the pulsatile release of luteinizing hormone (LH) from the pituitary which will then affect ovarian function. Dopamine, noradrenaline, and endorphin are all thought to play a role in the release of GnRH and changes in these hormones during exercise may be responsible for the development of amenorrhoea.[66] Changes in levels of melatonin and serotonin also occur during exercise and may also affect the menstrual cycle.[67]

BONE DENSITY IN AMENORRHOEIC ATHLETES

Amenorrhoea in endurance athletes was first recognized in the late 1970s. It was initially thought that the high level of exercise would protect such women from developing osteoporosis. However, by 1984 there was clear evidence of reduced bone density in these women.[53,68,69]

Initial reports

Cann et al.[68] investigated 10 amenorrhoeic athletes and compared them with 25 women with amenorrhoea for other reasons and with 50 eumenorrhoeic sedentary controls. All the groups studied were of similar age. Bone mineral density in the lumbar spine (measured by CT scanning) was significantly lower in the amenorrhoeic groups than in the controls, but there was no difference in bone mineral density at the radius (measured by single-photon absorptiometry). There was no difference in bone mineral density between the two amenorrhoeic groups. However, as this study was not designed to investigate bone mineral density changes in athletic amenorrhoea, there were no data regarding training intensity and there was even some doubt as to whether exercise was in fact the cause of amenorrhoea.

Drinkwater et al.[53] measured bone mineral density in the lumbar spine (using dual-photon absorptiometry) and the radius (using single-photon absorptiometry) in 14 amenorrhoeic and 14 eumenorrhoeic runners matched for age, height, and weight. The amenorrhoeics were running 42 miles per week and the eumenorrhoeics 25 miles per week. The nutritional intake was similar in the two groups. Bone mineral density in the lumbar spine was significantly lower in the amenorrhoeic group but there was no difference in radial measurements between the two groups. Lindberg et al.[69] investigated 11 amenorrhoeic runners and reported decreased levels of bone mineral density in both the lumbar spine and the radius when compared with eumenorrhoeic runners and with eumenorrhoeic sedentary controls.

Marcus et al.[70] studied 17 elite female distance runners of whom 11 had amenorrhoea and the remainder were eumenorrhoeic. The amenorrhoeic athletes had significantly lower bone mineral density in the lumbar spine than the eumenorrhoeic athletes but higher levels than a group of amenorrhoeic non-athletes, suggesting that extreme exercise that is also weight bearing partially overcomes the adverse skeletal effects of oestrogen deprivation.

In summary, spinal bone density in particular is reduced in amenorrhoeic athletes who predominantly exercise the lower limbs. However, amongst amenorrhoeic rowers, whose principal activity is targeted at the trunk, spinal bone density is less severely reduced than in runners and dancers.[25]

Nutritional and metabolic studies

Drinkwater et al.[53] and Nelson et al.[54] have shown no difference in calcium intake between amenorrhoeic and eumenorrhoeic athletes. They have also shown that amenorrhoeic athletes have a lower daily energy intake. Lloyd et al.[56] showed that dietary intake of fibre was significantly higher in a group of oligomenorrhoeic athletes compared with the eumenorrhoeic group.

Marcus et al.[70] investigated metabolic features in a group of amenorrhoeic and eumenorrhoeic runners. Serum free tri-iodo-thyronine was lower amongst the amenorrhoeic athletes, suggesting a lower metabolic rate which was possibly a reflection of the lower calorie intake seen in this group. Serum levels of calcium and phosphate were similar in the two groups. Urinary calcium excretion and serum alkaline phosphatase levels were higher in the amenorrhoeic group, but neither difference reached the level of statistical significance.

Longitudinal studies

Parker Jones et al.[71] investigated 39 women with amenorrhoea, including eight athletes, and showed a significant negative correlation between bone mineral density in the radius and the duration of amenorrhoea.

Drinkwater et al.[72] re-evaluated the subjects from their original study of 1984 after a period of 15 months. Nine of the original 14 amenorrhoeic athletes were reassessed together with seven of the 14 eumenorrhoeics (see above). Seven of the nine had regained their menstrual cycles with an associated rise in lumbar spine bone mineral density of 6.3 per cent. The bone mineral density in the two who had remained amenorrhoeic had fallen by a further 3.4 per cent. The bone mineral density in the eumenorrhoeic group remained unchanged.

Lindberg et al.[73] re-evaluated seven of the original 11 amenorrhoeic athletes from 1984 (see above), again after a period of 15 months. In the four who had regained menstruation bone mineral density in the lumbar spine rose by 6.7 per cent whereas it remained unchanged in the three athletes who continued to be amenorrhoeic. In the group who had regained menstruation there had been a reduction in weekly running distance and an increase in body weight.

MANAGEMENT OF THE AMENORRHOEIC ATHLETE

When to investigate amenorrhoea

There are several important causes of amenorrhoea in young adult women in addition to heavy training. For this reason a clear relationship between menstrual irregularity and training should be established. Amenorrhoea usually occurs as training is increased with normal menstruation recommencing within 6 months of stopping training (e.g. during a prolonged injury). However, some athletes never cut down their training sufficiently to allow menstruation to return. It then becomes difficult to exclude other possible causes of amenorrhoea and referral to a specialist should be considered (Table 1).

Table 1 Circumstances when the opinion of a specialist is recommended

1. Athletes with primary amenorrhoea (i.e. 17 years or older with no previous menstrual cycle)
2. Athletes in whom there is no clear-cut association between amenorrhoea and training
3. Athletes in whom amenorrhoea persists for more than a year after intensive training has stopped
4. Athletes with prolonged amenorrhoea (in excess of 3 years)

Treatment of athletic amenorrhoea

Bone mineral density, especially in the spine, is almost invariably reduced in athletes with prolonged amenorrhoea (in excess of 1 year). Bone densitometry is the only effective way of quantifying the bone loss and should be included in the assessment of such athletes.

Nutritional assessment is important, particularly with regard to calorie and calcium intake. Low calorie intake and low body fat are linked with amenorrhoea. In some circumstances increased calorie intake alone may re-establish the menstrual cycle. Dietary calcium might also be important[74] and probably should be maintained at a daily intake of around 1000 mg.

The most effective way of preventing bone mineral loss is with oestrogen replacement. The oestrogen-containing oral contraceptive is probably the best way of providing this. A low dose oestrogen preparation, containing around 20 µg of ethinyloestradiol, is adequate to protect the skeleton. Few side-effects are encountered at this low dosage, although fluid retention, with associated weight gain, occasionally occurs. If contraceptive action is not a consideration, an alternative treatment is hormone replacement therapy (as given to postmenopausal women) with a combined oestrogen–progestagen preparation containing 1.25 µg/day conjugated equine oestrogen or the equivalent. This will fully protect spine and hip bone density.

Bone density and fracture

The risk of fracture increases markedly as bone density decreases below a certain level, often referred to as the fracture threshold. Most women reach this threshold some time after the menopause (Fig. 3). The age at which bone density reaches the fracture threshold is determined by two factors, the peak bone mass and the age-related bone loss. The latter begins shortly before or at the menopause, depending on the site of measurement. The former is achieved in the late twenties or early thirties and is influenced by several factors including diet, oestrogen status, and physical activity (see above). Peak bone mass is likely to be increased by high levels of exercise in the teens and twenties (provided that it does not induce amenorrhoea) (Fig. 3).

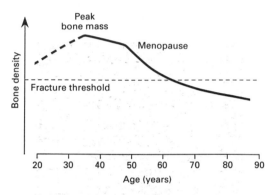

Fig. 3 The change in bone density with age in normal women. Note that peak bone mass is achieved around the age of 35 years; bone mass declines rapidly in the early menopausal years; bone density crosses the fracture threshold from the mid-sixties.

Low oestrogen status in early adult life will reduce the peak bone mass and thus lead to a lower bone density at the time of the menopause with an increased risk of fracture in subsequent decades. The spinal bone density of amenorrhoeic athletes tends to be around 10 to 20 per cent below the levels expected according to age, but is still well above the fracture threshold. However, very occasionally bone density is substantially reduced and may even be as low as that seen in women after the menopause. Amongst the 14 amenorrhoeic athletes in Drinkwater's study,[53] two had vertebral bone densities below the fracture threshold as defined by Riggs et al.[75] In our own series of 25 amenorrhoeic athletes,[74] we also identified two subjects with spinal trabecular bone densities below the fracture threshold as defined by Cann and Genant.[76] Such individuals may well be at increased risk of vertebral compression fracture. However, the only osteoporotic fractures that have so far been reported have been seen in patients with anorexia nervosa.[39]

Stress fracture

Stress fracture is an overuse injury of the skeleton frequently seen in athletes (Fig. 4) (see Chapter 5.2). It is associated with repetitive load-bearing activity; for example, stress fractures in runners are most commonly seen in the metatarsals, tibia, and fibula, and seem to occur more frequently in individuals with malalignment problems such as leg length discrepancy and pes cavus (Table 2).

Although bone tissue adapts to the forces placed upon it, with very high training loads it may not be able to adapt sufficiently to withstand the increased stress (see above). Stress fractures are

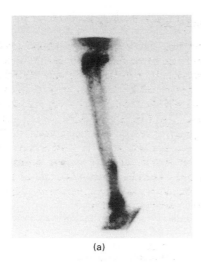

(a)

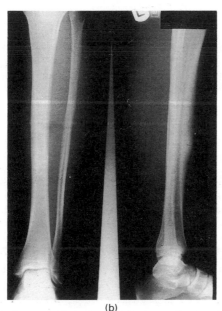

(b)

Fig. 4 (a) Isotope bone scan showing a stress fracture of the tibia in an amenorrhoeic runner. (b) Radiograph in an amenorrhoeic athlete showing a stress fracture of the shaft of the tibia.

Table 2 Factors associated with the development of stress fractures

High training loads
Increasing training intensity too quickly
Training on hard surfaces
Malalignment
    Leg length discrepancy
    Pes cavus

seen more frequently in runners training excessively on hard surfaces. Furthermore, they often occur in athletes who increase their training too quickly, generally after an injury. In this situation the bone does not have sufficient time to adapt to the increasing stress placed upon it.

Stress fractures in women are seen more frequently in amenorrhoeic athletes than in their eumenorrhoeic counter-

parts.[69,70] This may be due in part to their low oestrogen status. However, amenorrhoea is also a reflection of high training load (see above) which could equally be responsible for the higher incidence of stress fracture (Fig. 5). This is supported by the fact that stress fractures occur in cortical bone, whereas most studies on athletic amenorrhoea have shown the greatest bone loss in trabecular bone. Furthermore, bone density levels at sites of both cortical and trabecular bone tend to be similar in athletes with and without stress fractures.[77]

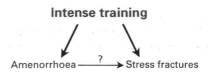

Fig. 5 Relation between intense training, amenorrhoea, and stress fracture.

EXERCISE AND OSTEOPOROSIS AFTER THE MENOPAUSE

Epidemiology of osteoporosis

After the menopause there is an exponential rise in the incidence of hip and vertebral fracture with age (Fig. 6). The rise in the incidence of hip fracture in men occurs some 5 years later than in women. This rise in incidence in both sexes with age is probably attributable to several factors, including sex hormone and nutritional deficiencies, and a decrease in physical activity.

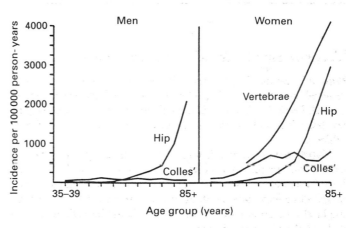

Fig. 6 Incidence rates for the three common osteoporotic fractures (Colles, hip, and vertebrae) in men and women plotted as a function of age at time of fracture. Data are from the community population of Rochester, Minnesota (reproduced from ref. 38, with permission).

The true incidence of hip fractures in the United Kingdom has been rising for three decades, and similar findings have been reported in Scandinavia and Hong Kong.[78] During this period there have probably been decreases in physical activity, and this has been cited as one of the factors responsible for the increasing incidence of osteoporosis.

Exercise and bone density

Exercise may increase bone density in older people. Chow et al.[28] showed a significant correlation between aerobic capacity, measured as Vo_2max, and total body calcium, measured by neutron activation analysis, in postmenopausal women, whereas Pocock et al.[27] showed a significant correlation between estimated Vo_2max and bone density of the femoral neck and the lumbar spine. Bone density in postmenopausal women is also related statistically to the isometric strength of adjacent muscle; for example, wrist bone density relates to grip strength[30] and spinal bone density is correlated with back extensor strength.[79]

Exercise intervention studies in older women have provided encouraging results. Several studies have now demonstrated an increase in bone density with aerobic training compared with non-exercising controls. Spinal bone density increased with as little as 45 min of light aerobic training twice weekly for 8 months in women aged 50 to 73 years.[80] Moreover, total body calcium has been shown to increase in a group performing aerobic exercise three times per week over a period of 1 year.[18] The improvement in bone density that occurs with aerobic exercise can also be demonstrated in women aged 69 to 91 years.[81]

Chow et al.[82] investigated the effect of aerobic training and aerobic plus strength training on total body calcium, measured by neutron activation analysis, in a group of postmenopausal women. Over a 1-year period total body calcium increased in both exercise groups compared with controls. There was no significant difference between the exercising groups, although it was marginally greater in the aerobic plus strength training group. However, Dalsky et al.[29] demonstrated that in women aged 55 to 70 years the improvement in spinal bone density that occurs during almost 2 years of weight-bearing aerobic exercise (walking, jogging, and stair climbing) was not sustained once the exercise was discontinued.

Weight-bearing aerobic exercise on a regular basis (e.g. for 30 min, three times per week) seems to protect bone density in postmenopausal women, and even walking on a daily basis may be sufficient to enhance both spinal and femoral neck bone densities in this age group.[83] These benefits are probably erased once a relatively short-term exercise programme is discontinued, although there are no data relating to exercise which is continued for several years.

Hormone replacement therapy is, at present, the most effective way of protecting the skeleton after the menopause, although recently other medications, such as bisphosphonates, have been proposed as effective alternatives. Exercise intervention in the postmenopausal period may enhance bone density but is likely to be most effective when combined with hormone replacement therapy.[84]

REFERENCES

1. Nottestad SY, Baumel JJ, Kimmel DB, Recker RR, Heaney RP. The proportion of trabecular bone in human vertebrae. *Journal of Bone Mineral Research* 1987; **2**: 221–9.
2. Dequeker J. Precision of the radiogrammetric evaluation of bone mass at the metacarpal bone. In: Dequeker J, Johnston CC, eds. *Non-invasive bone measurements: methodological problems.* Oxford: IRL Press, 1982: 27.
3. Horsman A. Bone mass. In: Nordin BEC, ed. *Calcium, phosphate and magnesium metabolism.* Edinburgh: Churchill Livingstone, 1976: 357–404.
4. Cann CE. Quantitative computed tomography for bone mineral analysis: technical considerations. In: Genant HK, ed. *Osteoporosis update 1987.* Berkeley, Ca: University of California, 1987: 131–44.
5. Eastell R, Wahner HW. Methods for measuring bone mineral density. *Bone: Clinical and Biochemical News and Reviews* 1990; **7**: 81–3.
6. Frost HM. The pathomechanics of osteoporosis. *Clinical Orthopaedics* 1985; **200**: 198–225.
7. Rodan GA, Martin TJ. Role of osteoblasts in hormonal control of bone resorption—an hypothesis. *Calcified Tissue International* 1981; **33**: 349–51.
8. Raisz LG, Pilbeam CC, Klein-Nulend J, Harrison JR (1990). Prostaglandins and bone metabolism: possible role in osteoporosis. In: Christiansen C, Overgaard K, eds. *Osteoporosis 1990*, Vol 1. Copenhagen: Osteopress, 1990: 253–8.
9. Murray DW, Rushton N. The effect of strain on bone cell prostaglandin E_2 release: a new experimental method. *Calcified Tissue International* 1990; **47**: 35–9.
10. Martin TJ. Intercellular communication in bone. In: Christiansen C, Overgaard K, eds. *Osteoporosis 1990*, Vol. 1. Copenhagen: Osteopress, 1990: 221–6.
11. Frost HM. Vital biomechanics: proposed general concepts for skeletal adaptations to mechanical usage (Editorial). *Calcified Tissue International* 1988; **42**: 145–56.
12. Schaffler MB, Radin EL, Burr DB. Mechanical and morphological effects of strain rate on fatigue of compact bone. *Bone* 1989; **10**: 207–14.
13. Wolff JD. *Das Geretz der Transformation der Knochen.* Berlin: Hirchwald, 1892.
14. Nilsson BE, Westlin NE. Bone density in athletes. *Clinical Orthopaedics* 1971; **77**: 179–82.
15. Williams JA, Wagner J, Wasnich R, Heilbrun L. The effect of long-distance running upon appendicular bone mineral content. *Medicine and Science in Sport and Exercise* 1984; **16**: 223–7.
16. Lane NE, Bloch DA, Jones HH, Marshall WH, Wood PD, Fries JF. Long distance running, bone density, and osteoarthritis. *Journal of the American Medical Association* 1986; **255**: 1147–51.
17. Wolman RL, Faulmann L, Clark P, Hesp R, Harries MG. Different training patterns and bone mineral density of the femoral midshaft in elite, female athletes. *Annals of the Rheumatic Diseases* 1991; **50**: 487–9.
18. Aloia JF, Cohn SH, Babu T, Abesamis C, Kalici N, Ellis K. Skeletal mass and body composition in marathon runners. *Metabolism* 1978; **27**: 1793–6.
19. Huddleston AL, Rochwell D, Kulund DN, Harrison RB. Bone mass in lifetime tennis players. *Journal of the American Medical Association* 1980; **244**: 1107–9.
20. Jones HH, Priest JD, Hayes WC, Chin Tichenor C, Nagel DA. Humeral hypertrophy in response to exercise. *Journal of Bone and Joint Surgery A* 1977; **59**: 204–8.
21. Dalen N, Laftman P, Ohlsen H, Stromberg L. The effect of athletic activity on bone mass in human diaphysial bone. *Orthopaedics* 1985; **8**: 1139–41.
22. Pirnay F, Bodeux M, Crielaard JM, Franchimont P. Bone mineral content and physical activity. *International Journal of Sports Medicine* 1987; **8**: 331–5.
23. Jacobson PC, Beaver W, Grubb SA, Taft TN, Talmage RV. Bone density in women: college athletes and older athletic women. *Journal of Orthopaedic Research* 1984; **2**: 328–32.
24. Orwoll ES, Ferar J, Oviatt SK, Huntington K, McClung MR. Swimming exercise and bone mass. Christiansen C, Johansen JS, Rüs BJ, eds. *Osteoporosis* 1987; **1**: 494–8.
25. Wolman RL, Clark P, McNally E, Harries M, Reeve J. Menstrual state and exercise as determinants of spinal trabecular bone density in female athletes. *British Medical Journal* 1990; **301**: 516–18.
26. Aloia JF, Vaswani AN, Yeh JK, Cohn SH. Premenopausal bone mass is related to physical activity. *Archives of Internal Medicine* 1988; **148**: 121–3.
27. Pocock NA, Eisman JA, Yeates MG, Sambrook PN, Eberl S. Physical fitness is a major determinant of femoral neck and lumbar spine bone mineral density. *Journal of Clinical Investigation* 1986; **78**: 618–21.
28. Chow RK, Harrison JE, Brown CF, Hajek V. Physical fitness effect on bone mass in postmenopausal women. *Archives of Physical Medicine and Rehabilitation* 1986; **67**: 231–4.
29. Dalsky GP, Stocke KS, Ehsani AA, Slatopolsky E, Waldon CL, Birge SJ.

Weight-bearing exercise training and lumbar bone mineral content in postmenopausal women. *Annals of Internal Medicine* 1988; **108**: 824–8.

30. Bevier WC, Wiswell RA, Pyka G, Kozak KC, Newhall KM, Marcus R. Relationship of body composition, muscle strength, and aerobic capacity to bone mineral density in older men and women. *Journal of Bone Mineral Research* 1989; **4**: 421–32.

31. Sinaki M, McPhee MC, Hodgson SF, Merritt JM, Offord KP. Relationship between bone mineral density of spine and strength of back extensors in healthy postmenopausal women. *Mayo Clinic Proceedings* 1986; **61**: 116–22.

32. Lanyon LE, Rubin CT. Static versus dynamic loads as an influence on bone remodelling. *Journal of Biomechanics* 1984; **17**: 897–907.

33. Rubin CT, Lanyon LE. Regulation of bone mass by mechanical strain magnitude. *Calcified Tissue International* 1985; **37**: 411–17.

34. Rubin CT, Lanyon LE. Regulation of bone formation by applied dynamic loads. *Journal of Bone and Joint Surgery A* 1984; **66**: 397–402.

35. O'Connor JA, Lanyon LE, MacFie H. The influence of strain rate on adaptive bone remodelling. *Journal of Biomechanics* 1982; **15**: 767–81.

36. Albright F, Smith PH, Richardson AM. Postmenopausal osteoporosis—its clinical features. *Journal of the American Medical Association* 1941; **116**: 2465–74.

37. Aitken JM, Hart DM, Anderson JB, Lindsay R, Smith DA, Speirs CF. Osteoporosis after oophorectomy for non-malignant disease in premenopausal women. *British Medical Journal* 1973; **2**: 325–8.

38. Riggs BL, Melton LJ. Involutional osteoporosis. *New England Journal of Medicine* 1986; **314**: 1676–86.

39. Rigotti NA, Nussbaum SR, Herzog DB, Neer RM. Osteoporosis in women with anorexia nervosa. *New England Journal of Medicine* 1984; **311**: 1601–6.

40. Klibanski A, Neer RM, Beitins IZ, Chester Ridgway E, Zervas NT, McArthur JW. Decreased bone density in hyperprolactinemic women. *New England Journal of Medicine* 1980; **303**: 1511–14.

41. Matta WM, Shaw RW, Hesp R, Katz D. Hypogonadism induced by luteinising hormone releasing hormone agonist analogues: effects on bone density in premenopausal women. *British Medical Journal* 1987; **294**: 1523–4.

42. Deitrick JE, Whedon GD, Shorr E. Effects of immobilization upon various metabolic and physiological functions of normal men. *American Journal of Medicine* 1948; **4**: 3–36.

43. Lockwood DR, Lammert JE, Vogel JM, Hulley SB. Bone mineral loss during bedrest. *Excerpta Medica International Congress Series* 1973; **270**: 261–5.

44. Mazess RB, Whedon GD. Immobilization and bone. *Calcified Tissue International* 1983; **35**: 265–7.

45. Andersson SM, Nilsson BE. Changes in bone mineral content following ligamentous knee injuries. *Medicine and Science in Sports and Exercise* 1979; **11**: 351–3.

46. Mack PB, Lachance PA, Vose GP, Vogt FB. Bone demineralization of foot and hand of Gemini-Titan IV, V & VII astronauts during orbital flight. *American Journal of Roentgenology* 1967; **100**: 503–11.

47. Mack PB, Vogt FB. Roentgenographic bone density changes in astronauts during representative Apollo space flight. *American Journal of Roentgenology* 1971; **113**: 621–33.

48. Whedon GD *et al.* Effect of weightlessness on mineral metabolism; metabolic studies on skylab orbital space flights. *Calcified Tissue International* 1976; **21**: S423–30.

49. Dale E, Gerlach DH, Wilhite AL. Menstrual dysfunction in distance runners. *Obstetrics and Gynecology* 1979; **54**: 47–53.

50. Feicht CB, Johnson TS, Martin BJ, Sparkes KE, Wagner WW. Secondary amenorrhoea in athletes. *Lancet* 1978; **ii**: 1145–6.

51. Feicht CB, Sanborn CF, Martin BJ, Wagner WW. Is athletic amenorrhea specific to runners? *American Journal of Obstetrics and Gynecology* 1982; **143**: 859–61.

52. Walsh BT. The endocrinology of anorexia nervosa. *Advances in Psychoneuroendocrinology* 1980; **3**: 299–312.

53. Drinkwater BL, Nilson K, Chestnut CH, Bremner WJ, Shainholtz S, Southworth MB. Bone mineral content of amenorrheic and eumenorrheic athletes. *New England Journal of Medicine* 1984; **311**: 277–81.

54. Nelson ME, Fisher EC, Catsos PD, Meredith CN, Turksoy RN, Evans WJ. Diet and bone status in amenorrheic runners. *American Journal of Clinical Nutrition* 1986; **43**: 910–16.

55. Deuster PA, Kyle SB, Moser PB, Vigersky RA, Singh A, Schoomaker EB. Nutritional intakes and status of highly trained amenorrheic and eumenorrheic women runners. *Fertility and Sterility* 1986; **46**: 636–43.

56. Lloyd T, Buchanan JR, Bitzer S, Waldman CJ, Myers C, Ford BG. Interrelationships of diet, athletic activity, menstrual status, and bone density in collegiate women. *American Journal of Clinical Nutrition* 1987; **46**: 681–4.

57. Speroff L, Redwine DB. Exercise and menstrual dysfunction. *Physician and Sportsmedicine* 1980; **8**: 42–52.

58. Schwartz B, Cumming DC, Riordan E, Selye M, Yen SS, Rebar RW. Exercise-associated amenorrhea: a distinct entity? *American Journal of Obstetrics and Gynecology* 1981; **141**: 662–70.

59. Frisch RE, McArthur JW. Menstrual cycles: fatness as a determinant of minimum weight for height necessary for their maintenance or onset. *Science* 1974; **185**: 949–51.

60. Sanborn CF, Albrecht BH, Wagner WW. Athletic amenorrhea: lack of association with body fat. *Medicine and Science in Sports and Exercise* 1987; **19**: 207–12.

61. Cumming DC, Rebar RW. Lack of consistency in the indirect methods of estimating percent body fat. *Fertility and Sterility* 1984; **41**: 739–42.

62. Frisch RE *et al.* Delayed menarche and amenorrhoea of college athletes in relation to age of onset of training. *Journal of the American Medical Association* 1981; **246**: 1559–63.

63. Wolman RL, Harries MG, Fyfe I. Slipped upper femoral epiphysis in an amenorrheic athlete. *British Medical Journal* 1989; **299**: 720–1.

64. Gadpaille WJ, Sanborn CF, Wagner WW. Athletic amenorrhea, major affective disorders, and eating disorders. *American Journal of Psychiatry* 1987; **144**: 939–42.

65. Cumming DC, Vickivic MM, Wall SR, Fluker MR. Defects in pulsatile LH release in normally menstruating runners. *Journal of Clinical Endocrinology and Metabolism* 1985; **60**: 810–12.

66. Noakes TD, Van Gend M. Menstrual dysfunction in female athletes. A review for clinicians. *South African Medical Journal* 1988; **73**: 350–5.

67. McCann SM, Snyder GD, Ojeda SR, Lumpkin MD, Ottlecz A. Role of peptides in the control of gonadotrophin secretion. In: McKerns KW, Naor Z, eds. *Hormonal control of hypothalamic–pituitary–gonadal axis.* New York: Plenum, 1984. 3–25.

68. Cann CE, Martin MC, Genant HK, Jaffe RB. Decreased spinal mineral content in amenorrheic women. *Journal of the American Medical Association* 1984; **251**: 626–9.

69. Lindberg JS, Fears WB, Hunt MM, Powell MR, Boll D, Wade CE. Exercise-induced amenorrhea and bone density. *Annals of Internal Medicine* 1984; **101**: 647–8.

70. Marcus R *et al.* Menstrual function and bone mass in elite women distance runners. *Annals of Internal Medicine* 1985; **102**: 158–63.

71. Parker Jones K, Ravnikar VA, Tulchinsky D, Schiff I. Comparison of bone density in amenorrheic women due to athletics, weight loss, and premature menopause. *Obstetrics and Gynecology* 1985; **66**: 5–8.

72. Drinkwater BL, Nilson K, Ott S, Chestnut CH. Bone mineral density after resumption of menses in amenorrheic athletes. *Journal of the American Medical Association* 1986; **256**: 380–82.

73. Lindberg JS, Powell MR, Hunt MM, Ducey DE, Wade CE. Increased vertebral bone mineral in response to reduced exercise in amenorrheic runners. *Western Journal of Medicine* 1987; **146**: 39–42.

74. Wolman RL, Clark P, McNally E, Harries M, Reeve J. Dietary calcium as a statistical determinant of spinal trabecular bone density in amenorrhoeic and oestrogen-replete athletes. *Bone and Mineral* 1992: **17**: 415–3.

75. Riggs BL, Wahner HW, Dunn WL, Mazess RB, Offord KP, Melton LJ. Differential changes in bone mineral density of the appendicular and axial skeleton with aging: relationship to spinal osteoporosis. *Journal of Clinical Investigation* 1981; **67**: 328–35.

76. Cann CE, Genant HK, Kolb FO, Ettinger B. Quantitative computed tomography for prediction of vertebral fracture risk. *Bone* 1985; **6**: 1–7.

77. Carbon R *et al.* Bone density of elite female athletes with stress fracture. *Medical Journal of Australia* 1990; **153**: 373–6.

78. Reeve J. Osteoporosis: epidermiology of risk factors. In Douglas RG, ed. *Assessment and management of risks associated with hyperlipidaemia, osteoporosis and hepatitis-B: effectiveness of intervention*. Philadelphia: Hanley & Belfus, 1991. 13–31.

79. Sinaki M, Offord KP. Physical activity in postmenopausal women: effect on back muscle strength and bone mineral density of the spine. *Archives of Physical Medicine and Rehabilitation* 1988; **69**: 277–80.

80. Krolner B, Toft B, Nielsen SP, Tondevold E. Physical exercise as prophylaxis against involutional vertebral bone loss: a controlled trial. *Clinical Science* 1983; **64**: 541–6.

81. Smith EL, Reddan W, Smith PE. Physical activity and calcium modalities for bone mineral increase in aged women. *Medicine and Science in Sports and Exercise* 1981; **13**: 60–4.

82. Chow R, Harrison JE, Notarius C. Effect of two randomised exercise programmes on bone mass of healthy postmenopausal women. *British Medical Journal* 1987; **295**: 1441–4.

83. Zylstra S, Hopkins A, Erk M, Hreshchyshyn MM, Anbar M. Effect of physical activity on lumbar spine and femoral neck bone densities. *International Journal of Sports Medicine* 1989; **10**: 181–6.

84. Notelovitz M *et al.* Estrogen therapy and variable-resistance weight training increase bone mineral in surgically menopausal women. *Journal of Bone Mineral Research* 1991; **6**: 583–90.

85. Wolman RL, Harries MG. Menstrual abnormalities in elite athletes. *Clinical in Sports Medicine* 1989; **1**: 95–100.

3.1.3 The endocrinology of exercise

TREVOR A. HOWLETT and ASHLEY GROSSMAN

INTRODUCTION

Acute exercise is a powerful modulator of the release of a large number of hormones, and has therefore been widely studied as a stimulus to hormonal secretion.[1] It is a very varied and complex physiological stimulus, however, and as a result the hormonal changes may vary considerably depending on the precise experimental circumstances. Chronic exercise training may modulate the pattern of basal hormone secretion, but may also modify the normal acute hormonal responses to exercise. It is also clear that many normal subjects who exercise regularly and excessively may develop changes in endocrine regulation, in particular gonadal dysfunction in women, which may have harmful long-term sequelae. With the increasing popularity of exercise training for the maintenance of good health and even the treatment of existing diseases, these hormonal changes have achieved increasing relevance in endocrinological practice.

ACUTE HORMONAL RESPONSES TO EXERCISE BEFORE AND AFTER TRAINING

Exercise type, absolute intensity, and relative intensity

Recreational exercise comes in a wide variety of forms, as do the experimental exercise protocols and the types of subjects used by investigators. Hormonal responses have thus been described in untrained subjects, joggers, amateur or professional competitive runners, cyclists, swimmers, and dancers (male and female), using a variety of acute stimuli; these include, in the laboratory, step-climbing, bicycle ergometer, and exercise treadmill, or, in the field, competitive and non-competitive events, short runs and marathons, swimming or other sports, and even military endurance training programmes. There is, of course, no guarantee that these types of experimental or recreational exercise are in any way equivalent, and indeed the large number of factors known to modify the acute response to exercise would suggest the very opposite. Furthermore, there is also evidence that the neuro-endocrinology of recreational exercise is quite different to that seen in elite professional athletes.

Comparison of the intensity of exercise used in different studies can also be difficult. Exercise intensity may be poorly controlled ('20 minutes vigorous exercise', 'exercise to maximum tolerated capacity', etc.), expressed in terms of absolute work-load (in a variety of units), or expressed as a percentage (measured directly or calculated) of the individual subject's maximum work capacity. Work capacity, in turn, may be expressed as either maximum tolerated work-load in the particular experimental system or as the measured maximum aerobic capacity (V_{O_2}max). Since after training a subject may typically increase their V_{O_2}max by 10 to 15 per cent or more, the choice between expressing the exercise intensity in absolute or relative terms is particularly important when studying changes in acute hormonal responses during exercise training. The same absolute work may represent a smaller relative load after training, so that the hormonal response to the same load may alter during training without representing any change in normal endocrine regulation mechanisms. Most authorities therefore consider that reference should be made to the relative intensity of the exercise stimulus, expressed as V_{O_2}max, rather than in terms of absolute work-load. It should be noted, however, that change in the hormonal response to a given absolute stimulus *per se* represents a hormonal response to training, and might be an important part of the body's mechanism of adaptation to exercise. Even when the relative exercise intensity is controlled, however, it seems clear that the duration of exercise must represent an important factor; unfortunately, this has also varied widely in the literature from periods as short as 5 minutes to several hours, or simply 'until exhaustion'.

These inconsistencies in the definition of the exercise stimulus, in addition to the other modifying factors discussed below, may account for many of the apparently contradictory descriptions of responses of individual hormones to acute exercise and chronic exercise training.

Hormonal responses to exercise

Growth hormone

The release of growth hormone (GH) in response to exercise was described in the earliest reports of GH physiology following the introduction of the radioimmunoassay of GH in plasma in the 1960s, and has subsequently been amply confirmed in multiple studies of response to cycling, running, swimming, and rowing. Studies using graded exercise show that GH release can be observed at exercise intensities as low as 30 to 40 per cent V_{O_2}max and increases in proportion to the work-load, although the response may be diminished at very high work loads or at 'exhaustion'. The relationship of the GH response to obesity, diabetes, diet, body temperature, and degree of hypoxia is considered below.

Training may modify the GH response to acute exercise. Thus, while basal levels are unaltered, the GH response to the same absolute exercise intensity is lower in trained than in untrained subjects.[2] A lower GH response to the same relative exercise intensity has also been described in trained male subjects compared with controls,[3] although no change was reported in another study where females entered an intensive training programme and acted as their own controls.[4]

Since exercise is a simple and safe stimulus of GH release, a number of workers have studied its use as a clinical screening test for GH deficiency. Intensive well-controlled exercise may exclude GH deficiency (by a response of > 20 mU/1) in 68 per cent of children with short stature, but less well-controlled exercise is unreliable.[5] Close supervision is therefore necessary, which rather negates the advantage of simplicity.

Cortisol and ACTH

The response of plasma cortisol to acute exercise depends critically on the intensity of the stimulus. During low intensity exercise circulating cortisol falls, reflecting the underlying circadian rhythm and perhaps increased removal from the circulation. More intense exercise, both in the laboratory and during competitive running and swimming, results in an elevation of plasma cortisol proportional to exercise intensity. Most studies are consistent with the view that the critical level of exercise intensity necessary to induce release of cortisol is approximately 60 per cent of V_{O_2}max, with further increases with more severe exercise. Cortisol levels may then fall as the point of collapse from exhaustion is approached. The increase in plasma cortisol occurs despite an increased rate of removal from the circulation and, as expected, follows a rise in plasma ACTH, although this has only been measured in relatively few studies. Many factors, such as psychological stress, circadian rhythm, feeding, fasting, and body temperature, alter the cortisol response to acute exercise and are discussed below.

Changes in the circulating cortisol response to exercise after physical training remain controversial. The cortisol response to the same absolute work-load may fall, but some well-controlled studies have shown a clear potentiation of the plasma cortisol response to the same relative intensity after training,[3,4] although this has been disputed. The cortisol response during marathon running was shown in one study to be directly proportional to the degree of fitness of the runners,[6] but other studies of trained athletes have failed to show that cortisol responds at all to intensive (although often uncontrolled) exercise.[7]

Prolactin

Several groups have reported a small rise in serum prolactin with acute exercise in normal controls, including during pregnancy, and in recreational and competitive runners. The effect of training is again uncertain; some investigations find that the prolactin response occurs only in trained individuals, while others have reported no rise in prolactin in trained runners after a regular daily run, or even during intensive exercise in amenorrhoeic athletes.[7] It

has been suggested that prolactin release only occurs, at least in normal individuals, when the anaerobic threshold is reached. Investigations showing a prolactin response in untrained subjects have largely employed male volunteers, whereas those showing no response have used females; however, the possibility of a sex difference does not seem to have been formally tested.

Thyroid hormones and thyroid-stimulating hormone

Most studies agree that intensive acute exercise causes a rise in plasma thyroxine, free thyroxine index, and free T_4 and T_3, although in one study no change was reported during a marathon run.[6] In addition, isolated studies have reported a fall in either T_4 or T_3. Most workers have found no change in plasma thyroid-stimulating hormone (TSH) during acute exercise, although a slight fall was noted in one study.[8]

Gonadotrophins and gonadal steroids

Female

The acute responses of luteinizing hormone (LH) and follicle-stimulating hormone (FSH) to exercise are controversial: studies have reported no change, a rise, or a fall. The known pulsatile nature of the secretion of these hormones must account, at least in part, for these discrepancies. One study which assessed LH pulsatility in considerable detail during and after acute exercise reported a significant reduction (by nearly 50 per cent) in LH pulse frequency (but not in pulse amplitude) in the six hours following exercise.[9] However, this has not been confirmed by other workers, and a recent study has not shown any major change in gonadotrophin pulse parameters during acute intense exercise.[10]

Plasma oestradiol does not appear to change in response to moderate exercise in the follicular phase, although oestradiol may rise at exhaustion in the midfollicular phase; oestradiol and progesterone may both rise at all exercise intensities in the midluteal phase. Testosterone, androstenedione, and the adrenal androgen dehydroepiandrosterone-sulphate (DHEA-S) all rise during acute exercise, although some or all of this may represent an adreno-cortical response.

We consider the response of the hypothalamopituitary–gonadal axis to exercise training in detail below.

Male

The gonadotrophin response to acute exercise in the male is also controversial. Testosterone rises during short-term exercise, but falls with more prolonged exertion such as marathon running or military training. The acute rise in testosterone seems to be synchronous with the rise in LH, rather than following it, and may therefore not be mediated by gonadotrophin secretion.[11]

Insulin

The changes in insulin secretion and glucose metabolism during acute exercise are complex. Plasma insulin levels fall in normal lean individuals, in obese subjects, and in diabetics. This fall in insulin level is in part a response to the increased utilization of glucose, and also a result of increased sympathoadrenomedullary activity at α_2-adrenoceptors. However, there is also an increase in insulin sensitivity due to an increase in insulin receptor number and affinity. After training, the basal levels of plasma insulin are lower and the percentage fall during acute exercise may be less marked.

Other hormones

Plasma adrenaline and noradrenaline both rise during acute exercise. After training, the exercised-induced release of catecholamines for the same absolute work-load is reduced, but for the same relative work-load the response is probably unchanged.

Acute exercise causes a rise in plasma renin, aldosterone, and vasopressin, associated with a fall in plasma volume and rise in plasma osmolality. After training these responses are less for the same absolute work-load but unchanged for the same relative work-load.

Finally, prolonged (5 h) low intensity exercise has been reported to cause a slight rise in serum parathyroid hormone associated with a small fall in plasma ionized calcium concentrations.

Factors modifying the hormonal response to acute exercise

Changes in many components of the 'exercise stimulus', other than the exercise intensity and degree of training, may modify the hormonal response. Many of these factors remain poorly characterized, and the list below may well be incomplete.

Competition and psychological factors

The adrenocortical response to exercise is clearly enhanced by the psychological 'stress' of competition. Thus, urinary 17-hydroxycorticosteroids, plasma cortisol, and salivary cortisol all show greater increases during a race than during training. Anticipatory rises of cortisol, TSH, and luteinizing hormone (LH) have also been reported before intensive exercise, again presumably related to psychological factors. Such 'stress' is extremely difficult to quantify and thus control. It is therefore unclear to what extent 'stress' may vary before and after training or in response to different exercise-testing protocols, and thus what effect it may have on the hormonal responses under these circumstances.

Circadian rhythm

The relationship between the circadian rhythm of plasma cortisol and its response to exercise has been studied in detail by Brandenberger et al.[12] The plasma cortisol incremental rise during exercise was similar at all times of day when performed during 'quiescent periods' (i.e. without any meal-related or other secretory peaks); however, in view of the lower basal values in the evening, the peak value achieved was lower.

Feeding and fasting

The cortisol response to acute exercise also interacts with the meal-related secretory peaks of plasma cortisol.[12] Thus, exercise performed shortly after the midday meal may result in a diminished cortisol response to exercise; conversely, exercise shortly before lunch may attenuate the meal-related cortisol rise. Prolonged fasting potentiates the exercise-induced secretion in plasma cortisol, GH, and prolactin, and a high fat diet may also increase the GH response.

Menstrual cycle

As mentioned above, the oestradiol and progesterone responses to acute exercise vary during the menstrual cycle, being greater in the luteal phase. Other hormonal responses may also vary: thus, the GH response to bicycle ergometer exercise is much greater in mid-cycle around the time of ovulation. Generally speaking, most workers tend to study exercise in the follicular phase.

Body temperature

Body temperature rises during intense exercise, reaching temperatures as high as 40 to 41 °C during marathon running. If exercise is performed such that body temperature does not rise, for example, by swimming in cool water or running in a cold room, then the catecholamine response is diminished and cortisol and GH responses are abolished or even converted to a fall. Conversely, the cortisol response to marathon running has been noted to be greater in hot than in cool weather. The GH and prolactin responses to passive external body heating are greater than those to exercise producing an equivalent rise in body temperature. This suggests that not only do changes in body temperature represent a major component of the 'exercise stimulus', but also, at least in the case of GH and prolactin, that other components may be exerting the opposite effect. It is unclear to what extent different exercise protocols and exercise training may modify the temperature response to exercise and therefore modulate any hormonal changes.

Other factors

Hypoxia greatly potentiates the GH, cortisol, and insulin responses to acute exercise. The GH response is greater in lean than in obese subjects, and conversely greater in type 1 and type 2 diabetes mellitus where it is normalized by strict control. In view of the complicated nature of the exercise stimulus and the difficulty in controlling for all these factors during training, it is perhaps not surprising that the reported changes in the hormonal responses to acute exercise after training remain controversial.

HORMONAL CHANGES INDUCED BY EXERCISE TRAINING

Hypothalamopituitary–gonadal axis

Female

The gonadal hormone responses to exercise training are certainly the best studied and probably also the most important clinically. Excessive exercise in childhood may delay menarche,[13-15] while menstrual disturbances such as anovulation, oligomenorrhoea, or amenorrhoea are very common in trained female athletes. Up to 50 per cent of women in regular training may be affected,[16] and small prospective studies using intensive training have reported an 80 to 90 per cent incidence of menstrual disorders. Nevertheless, it is becoming increasingly clear that amenorrhoea is exceptional in women undertaking exercise training as a leisure pursuit, as in most such subjects the menstrual abnormalities are relatively subtle; indeed, Prior[17] has suggested that the anovulatory state induced by training can reduce the symptoms of the premenstrual syndrome. However, an inquiry regarding exercise habits has become an essential part of the assessment of any patient with unexplained amenorrhoea.

The incidence of menstrual disturbances is clearly related to the amount of exercise performed.[18] Body weight and composition also contribute;[19] studies have suggested that menstrual disturbances occur when total body weight falls either below 54 kg or by 9 kg, when body fat falls below 22 per cent of total body weight, or when total body fat falls by 30 per cent (for a review, see Baker[16]). However, while athletes as a group are leaner than sedentary women, several studies have shown little or no difference in this parameter between eumenorrhoeic and amenorrhoeic runners.

Furthermore, menstrual disturbances may disappear (or menarche occur) when exercise ceases, without change in body weight or composition, and recur when exercise is resumed.[15] Thus, while changes in body composition may contribute to menstrual dysfunction, they seem unlikely to be the only factor. Previous menstrual history may also contribute, amenorrhoea being associated with prior menstrual disturbance, nulliparity, and youth. Amenorrhoeic runners may also perceive their exercise as more 'stressful' than eumenorrhoeic runners during equivalent training.

Hormonal changes in trained female athletes have been extensively studied. As with acute exercise, a variety of changes in isolated and/or random levels of gonadotrophins have been described: some workers report a fall, some a rise, and some no change. Nevertheless, more detailed studies have clearly documented a decrease in the frequency, and possibly also the amplitude, of LH pulsatility in both amenorrhoeic and eumenorrhoeic runners (Fig. 1),[20,21] and a reduction or abolition of cyclical changes such as the midcycle LH surge.[22] In addition to these changes in LH pulsatility, presumably mediated at the level of the hypothalamus, there may also be changes in pituitary responsiveness to hypothalamic luteinizing-hormone-releasing hormone (LHRH) release. One study clearly demonstrated a progressive fall in the gonadotrophin response to exogenously administered LHRH (100 μg IV) with increasing training mileage in female runners. In contrast, another group reported increased pituitary sensitivity to submaximal boluses of LHRH (2.5–10 μg) in amenorrhoeic runners compared with controls. It seems likely that these changes are secondary to a decrease in the pulsatile

release of LHRH; early on, there may be an increase in the readily releasable pool of LH and FSH, while eventually there is a fall in biosynthesis and thus secretion.

Plasma oestradiol is low in amenorrhoeic athletes and the ratio of oestrone to oestradiol may be increased, indicating increased formation of oestrogens from peripheral conversion compared with follicular synthesis. A number of studies have recently reported decreased vertebral bone density in amenorrhoeic runners compared with eumenorrhoeic runners or sedentary controls, thus demonstrating important biological effects of prolonged oestrogen deficiency in such women. The luteal progesterone rise is abolished in amenorrhoeic subjects, but is often also diminished in athletes who continue to menstruate. This 'inadequate luteal phase' is the first abnormality to appear when women are subjected to an intensive exercise programme[22] and may represent an important clinical problem as a relatively occult cause of infertility.[23] As mentioned above, this may also be therapeutically useful in the treatment of premenstrual tension.

The precise mechanism of these changes in the LH pulse generator remains uncertain; the role of opioid peptides has been the subject of considerable interest and is considered in detail below. The release of prolactin during exercise has also been suggested as a possible mechanism, but studies showing normal 24-h profiles of prolactin in amenorrhoeic runners argue against this possibility. The release of melatonin during exercise might also suppress LHRH release. We have recently investigated the possible hormonal factors involved in exercise-induced menstrual disturbances in a cross-sectional study of the endocrine profile of a group of 36 Olympic squad athletes.[24] In an attempt to correlate menstrual state with hormone level, we carried out a stepwise discriminant factor analysis on 17 neuroendocrine parameters. Each hormone or metabolite assessed (including anterior pituitary hormones, adrenal androgens, glucose, insulin, and catecholamines) was related to a measure of menstrual regularity. The most important finding was that only two major orthogonal factors were revealed: serum cortisol and the insulin-like growth factor–1 (IGF–1) binding protein (BP–1) (Fig. 2). As previously demonstrated, the level of serum cortisol was directly related to the presence of amenorrhoea, suggesting that central activation of the

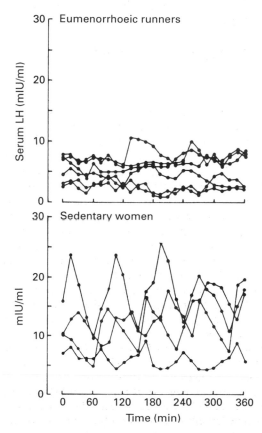

Fig. 1 Reduced LH pulsatility in eumenorrhoeic female runners compared with sedentary controls (reproduced from ref. 20 with permission).

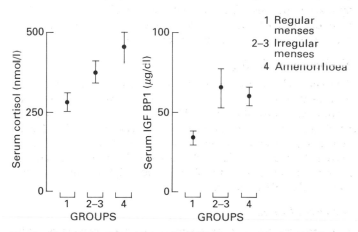

Fig. 2 The relationship of mean (± SEM) serum cortisol and IGF BP–1 to menstrual disturbances induced by exercise. The group with regular menses have lower mean cortisol and IGF BP–1 levels than those with menstrual irregularity, and these two factors appear to be independently related according to multiple factor analysis (reproduced from ref. 24 with permission).

pituitary–adrenal axis (probably by the hypothalamic peptide CRH–41) may be the 'stress' factor responsible for abnormal menses. However, BP–1 is inversely related to insulin, and may therefore reflect the metabolic state or 'fuel supply' of the individual. It is thus possible that BP–1 is a major peripheral signal regulating LHRH release, although this requires formal experimental testing.

Male

Despite anecdotal reports of decreased libido during training, the long-term effects of exercise on the male hypothalamopituitary–gonadal axis are poorly studied. The best study, comparing highly trained athletes with sedentary controls, found small but significant decreases in mean levels of testosterone, free testosterone, and prolactin in athletes, but no change in gonadotrophins or sex-hormone-binding globulin.[25]

Other effects on hypothalamopituitary function

In addition to changes in the exercise-induced hormonal responses discussed elsewhere in this chapter, there are indications that severe training may affect basal hormonal levels and the hypothalamopituitary response to stimuli other than exercise. Basal prolactin levels are reported to be reduced in trained male athletes, and also in female swimmers and runners; however, detailed studies of 24-h prolactin profiles in female runners have failed to show a difference from sedentary controls.

Exercise training does not alter the basal levels of T_4 or TSH, although the TSH response to thyrotrophin-releasing hormone (TRH) may be increased; the rate of degradation of T_4 is 75 per cent higher in trained subjects so that its production rate may be increased.

As previously noted, women runners have higher 09.00-h serum cortisol and 24-h urinary cortisol levels than sedentary controls, while retaining normal dexamethasone suppressibility. Furthermore, a profound alteration in the hypothalamic response to 'stress' may occur after severe training. Barron et al.[26] studied four athletes suffering from the 'overtraining syndrome' compared with five asymptomatic marathon runners: in the overtrained athletes there was a markedly diminished response of plasma ACTH, cortisol, prolactin, and GH to insulin-induced hypoglycaemia, despite normal prolactin, TSH, and LH responses to TRH and LHRH. Similarly, amenorrhoeic runners, unlike eumenorrhoeic runner controls, may show no cortisol or prolactin response to intense exercise, suggesting that the disturbance of hypothalamic function in these women is not confined to LHRH secretion.[7]

Insulin and diabetes

The acute effects of exercise on glucose metabolism are maintained by exercise training, with decreased plasma insulin basally and in response to glucose, but increased insulin sensitivity and therefore unchanged, or even improved, glucose tolerance. These beneficial effects of exercise training may also be seen in pathological conditions, and indeed advice to increase exercise has long been part of the management of diabetic patients. Intensive exercise training may achieve modest improvements in fasting hyperglycaemic and glucose tolerance in type 2 diabetics, and possibly also in type 1 diabetics.

OPIOID PEPTIDES AND EXERCISE

There is a certain amount of evidence which suggests that the endogenous opioid peptides are involved in the body's response to exercise training: these are summarized below.

Opioid mediation of hormonal changes

Gonadotrophins

Opioidergic pathways are known to modulate the hypothalamo-pituitary–gonadal axis, causing inhibition of the release of hypothalamic LHRH and a reduction in LH pulse frequency and possibly amplitude.[27,28] Since exercise training also appears to inhibit LH pulsatility,[9,20,21] the possibility arises that these effects might be mediated via an opioidergic mechanism. The only direct evidence in favour of this possibility was provided by McArthur et al.,[29] who administered the opioid receptor blocking drug naloxone to amenorrhoeic female athletes and noted a prompt restoration of LH and FSH pulsatility. However, this effect has not been confirmed in several other studies. This may relate to changes in body composition, as we have suggested that opioid-related amenorrhoea will not occur in subjects of low percentage body fat.[30]

Growth hormone and prolactin

Opioid peptides stimulate the release of both prolactin and GH;[27,28] the release of both hormones during exercise in untrained individuals is unaltered by administration of naloxone, suggesting that opioid pathways are not involved in this response. In highly trained male athletes, however, the release of prolactin and GH during exercise is practically abolished by simultaneous infusion of high doses of naloxone (Fig. 3).[31] This suggests that a fundamental change in hypothalamic opioid control has occurred during the training process, and is another indication that elite athletes may differ in their hormonal responses to exercise compared with recreational exercisers. In a study of normal subjects trained intensively for 8 weeks, we were unable to demonstrate alterations in naloxone reversibility (P. Bouloux and A. Grossman, unpublished observations).

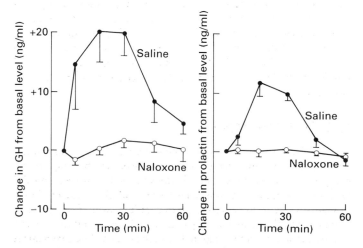

Fig. 3 The effect of naloxone infusion on the mean (± SEM) prolactin and GH responses to acute exercise in eight highly trained male athletes (reproduced from ref. 31 with permission).

Release of opioid peptides during exercise

The opioid peptide β-endorphin is synthesized from the ACTH precursor, pro-opiomelanocortin, and released from the anterior pituitary in parallel with ACTH in response to a wide variety of stimuli, including exercise. Not surprisingly, a number of groups have reported a rise in plasma β-endorphin levels during exercise. Metenkephalin, another opioid peptide, has also been reported to be released in some, but not all, untrained females during treadmill exercise[32] although this was not confirmed in another study which used bicycle ergometer exercise in male subjects.[8] It is unclear whether this difference is related to gender, training, or intensity or type of exercise. There is no consistent change in plasma levels of the potent opioid peptide, dynorphin, during exercise. Furthermore, whether these changes in plasma opioids are directly related to the hormonal changes observed, particularly gonadotrophin secretion, is also uncertain. Opioid modulation of hypothalamic releasing-hormone secretion may occur, at least in part, at sites which are outside the blood–brain barrier and therefore accessible to the effects of circulating opioids. However, in most cases the absolute levels of plasma opioids achieved are extremely low, and it is unclear whether such levels would be sufficient to stimulate known opioid receptors. The release of metenkephalin may simply reflect activation of the adrenal medulla, a major source of opioid peptides, while β-endorphin is coreleased with ACTH. It is perhaps more likely that changes in the activity of opioid pathways within the hypothalamus, and/or elsewhere in the brain, are responsible.

Opioid modulation of the experience of exercise

Several workers have pointed to a relationship between opioid peptides and the subjective experience of exercise. The release of opioids into plasma during exercise has been suggested to be the physiological basis of what runners recognize as the 'runner's high',[33] or what others might consider 'addiction to running'; however, there is little direct experimental evidence to confirm this concept. Changes in plasma β-endorphin levels showed a statistical correlation with changes in feelings of 'pleasantness' in one study; another noted analgesia to a proportion of stimuli after a 6-mile run which was partially blocked by naloxone, as well as increases in psychometric parameters corresponding to 'joy' and 'euphoria' (which were blocked by naloxone) and 'co-operation' and 'conscientiousness' (which were not). We, and others, have also reported that infusion of naloxone during exercise results in an increase in the perceived effort of exercise.[8] Changes in central opioids may therefore modulate the perception of the exercise stimulus.

CONCLUSIONS

Exercise has been shown to alter the secretion of many hormones, both acutely and in response to chronic exercise training. However, there remain many contradictions in the literature about the responses of individual hormones. Many of these uncertainties probably relate to the complex and variable nature of the exercise stimulus, which remains poorly understood and therefore difficult to measure and thus to compare between studies. The mechanism of the hormonal adaptation to exercise is also poorly understood; the suggestion of a role for endogenous opioids is of great interest, but confirmation of their importance awaits further studies and in

particular the advent of long-acting opioid antagonists. One possible explanation for certain of the discrepancies in the literature is that the profound neuroendocrine disturbance of exercise-induced amenorrhoea actually reflects two processes: one mechanism is central activation of CRH–41, a 'stress' pathway, which in turn inhibits LHRH release via an opioidergic link (Fig. 4). This pathway may be particularly activated in elite athletes. A second mechanism relates the amenorrhoea to changes in body composition, specifically a decrease in percentage body fat.[30] In this situation, akin to prepuberty and anorexia nervosa, a peripheral metabolic signal (possibly BP–1) inhibits LHRH release. Whatever the cause, exercise-induced oestrogen deficiency has serious consequences and requires treatment.

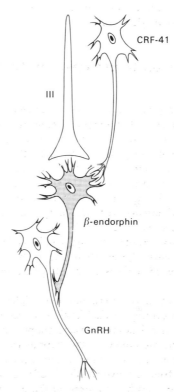

Fig. 4 Schematic representation of the activation of central CRH (CRF–41) neurones, leading to inhibition of pulsatile gonadotrophin releasing hormone (GnRH) release via an opioid interneurone in the region of the third ventricle (III). Thus stress may cause amenorrhoea by activating a central opioid pathway.

Finally, the physiological importance of these hormonal changes under normal circumstances remains unclear, although the subject of much speculation. The adverse effects of the hypogonadism of many female athletes have now become more obvious, but is the muscular development produced by exercise in any way related to hormonal changes such as release of androgens or growth hormone? Certainly, competitors often think it worthwhile to take large doses of exogenous androgenic 'anabolic' steroids or even growth hormone. Similarly, are some female competitors in field events naturally androgynous, or do they become so because of the release of androgens during training, or again is it simply due to the 'anabolic' steroids? Does the activation of opioid pathways during exercise allow the athlete to 'break the pain barrier' to achieve a more spectacular performance? Many questions remain unanswered, but the basic description of exercise-induced hormonal changes is now broadly known, at least in outline.

REFERENCES

1. Howlett TA. Hormonal responses to exercise and training: a short review. *Clinical Endocrinology* 1987; **26**: 723–42.
2. Sutton JR, Lazarus L. Growth hormone in exercise: comparison of physiological and pharmacological stimuli. *Journal of Applied Physiology* 1976; **41**: 523–7.
3. Bloom SR, Johnson RH, Park DM, Rennie MJ, Sulaiman WR. Differences in the metabolic and hormonal response to exercise between racing cyclists and untrained individuals. *Journal of Physiology* 1976; **258**: 1–18.
4. Bullen BA, Skrinar GS, Beitins IZ, Carr DB, Reppert SM, Dotson CO, Fencl MM, Gervino EV, McArthur JW. Endurance training effects on plasma hormonal responsiveness and sex hormone excretion. *Journal of Applied Physiology* 1984; **56**: 1453–63.
5. Lin T, Tucci JR. Provocative tests of growth hormone release: a comparison of results with seven stimuli. *Annals of Internal Medicine* 1974; **80**: 464–9.
6. Dessypris AG, Wager G, Fyhrquist F, Mäkinen T, Welin MG, Lamberg BA. Marathon run: effects on blood cortisol-ACTH, iodothyronines-TSH and vasopressin. *Acta Endocrinologica* 1980; **95**: 151–7.
7. Loucks AB, Horvath SM. Exercise-induced stress responses of amenorrheic and eumenorrheic runners. *Journal of Clinical Endocrinology and Metabolism* 1984; **59**: 1109–20.
8. Grossman A *et al.* The role of opioid peptides in the hormonal responses to acute exercise in man. *Clinical Science* 1984; **67**: 483–91.
9. Cumming DC, Vickovic MM, Wall SR, Fluker MR, Belcastro AN. The effect of acute exercise on pulsatile release of luteinizing hormone in women runners. *American Journal of Obstetrics and Gynecology* 1985; **153**: 482–5.
10. McArthur JW *et al.* The effects of submaximal endurance exercise upon LH pulsatility. *Clinical Endocrinology* 1990; **32**: 115–26.
11. Cumming DC, Brunsting LA, Strich G, Ries AL, Rebar RW. Reproductive hormone increases in response to acute exercise in men. *Medicine and Science in Sports and Exercise* 1986; **18**: 369–73.
12. Brandenberger G, Follenius M, Hietter B, Reinhardt B, Simoni M. Feedback from meal-related peaks determines diurnal changes in cortisol response to exercise. *Journal of Clinical Endocrinology and Metabolism* 1982; **54**: 592–6.
13. Frisch RE *et al.* Delayed menarche and amenorrhea of college athletes in relation to age of onset of training. *Journal of the American Medical Association* 1981; **246**: 1559–63.
14. Frisch RE, Wyshak G, Vincent L. Delayed menarche and amenorrhoea in ballet dancers. *New England Journal of Medicine* 1980; **303**: 17–19.
15. Warren MP. The effects of exercise on pubertal progression and reproductive function in girls. *Journal of Clinical Endocrinology and Metabolism* 1980; **51**: 1150–7.
16. Baker ER. Menstrual dysfunction and hormonal status in athletic women: a review. *Fertility and Sterility* 1981; **36**: 691–6.
17. Prior JC. Physical exercise and the neuroendocrine control of reproduction. *Ballière's Clinical Endocrinology and Metabolism* 1987; **1**: 299–317.
18. Feicht CB, Johnson TS, Martin BJ, Sparkes KE, Wagner WW. Secondary amenorrhoea in athletes. *Lancet* 1978; **ii**: 1145–6.
19. Frisch RE, McArthur JW. Menstrual cycles: fatness as a determinant of minimum weight for height necessary for their maintenance or onset. *Science* 1974; **185**: 949–51.
20. Cumming DC, Vickovic MM, Wall SR, Fluker MR. Defects of pulsatile LH release in normally menstruating runners. *Journal of Clinical Endocrinology and Metabolism* 1985; **60**: 810–12.
21. Veldhuis JD, Evans WS, Demers LM, Thorner MO, Wakat D, Rogol AD. Altered neuroendocrine regulation of gonadotropin secretion in women distance runners. *Journal of Clinical Endocrinology and Metabolism* 1985; **61**: 557–63.
22. Bullen BA, Skrinar GS, Beitins IZ, von Mering G, Turnbull BA, McArthur JW. Induction of menstrual disorders by strenuous exercise in untrained women. *New England Journal of Medicine* 1985; **312**: 1349–53.
23. Prior JC, Ho Yeun B, Clement P, Bowie L, Thomas J. Reversible luteal phase changes and infertility associated with marathon running. *Lancet* 1982; **ii**: 269–70.
24. Jenkins J, *et al.* IGFBP-1: a metabolic signal associated with exercise-induced amenorrhoea. *Neuroendocrinology* 1993; in press.
25. Wheeler GD, Wall SR, Belcastro AN, Cumming DC. Reduced serum testosterone and prolactin levels in male distance runners. *Journal of the American Medical Association* 1984; **252**: 514–16.
26. Barron JL, Noakes TD, Levy W, Smith C, Millar RP. Hypothalamic dysfunction in overtrained athletes. *Journal of Clinical Endocrinology and Metabolism* 1985; **60**: 803–6.
27. Grossman A. Brain opiates and neuroendocrine function. *Clinics in Endocrinology and Metabolism* 1983; **12**: 725–46.
28. Howlett TA, Rees LH. Endogenous opioid peptides and hypothalamo-pituitary function. *Annual Review of Physiology* 1986; **48**: 527–36.
29. McArthur JW, Bullen BA, Bettins IZ, Pagane M, Badger TM, Klibanski A. Hypothalamic amenorrhoea in runners of normal body distribution. *Endocrine Research Communications* 1980; **7**: 13–25.
30. Grossman A. Exercise and the gonadotrophin pulse generator. In: Imura H, Shizume K, Yoshida S, eds. *Progress in endocrinology 1988*, Amsterdam: Excerpta Medica, 1988: 441–8.
31. Moretti C *et al.* Naloxone inhibits exercise-induced release of PRL and GH in athletes. *Clinical Endocrinology* 1983; **18**: 135–8.
32. Howlett TA *et al.* Release of β-endorphin and met-enkephalin during exercise in normal women: response to training. *British Medical Journal* 1984; **288**: 1950–2.
33. Appenzeller O. What makes us run? *New England Journal of Medicine* 1981; **305**: 578–9.

3.2.1 Heart and exercise: clinical aspects

ROBIN J. NORTHCOTE

INTRODUCTION

Rarely are views on one topic so divergent. The role of exercise in the prevention and modification of cardiovascular disease has provided material for debate over the last 40 years. Only recently have the short- and long-term effects of exercise been elucidated. Because the results of enquiry have not always shown overwhelming and conclusive benefit, the role of exercise in the pre- vention and management of cardiovascular disease has been less palpable than measures such as drug treatment or discontinuation of smoking. The dangers of exercise have been aired in the media from time to time, culminating in the death of James Fixx, the author of *The Complete Book of Running*, while jogging in 1984. Such isolated events have a more profound effect on the public perception of the effects of exercise than a dozen scientific papers in medical journals. This unfortunate death followed a few years after a Californian pathologist declared that 'fatal coronary artery disease had never been found in a marathon runner'.[1] Such exaggerated claims have since been disproved[2,3] and do little to foster the confidence of the public and medical profession in what is a very potent preventive and therapeutic tool.

This chapter details the cardiovascular effects of exercise, including metabolic, structural and functional changes, and attempts to place in perspective the risk–benefit equation in relation to coronary atherosclerosis and other cardiovascular pathology.

CARDIOVASCULAR EFFECTS OF EXERCISE

Metabolic effects

Vigorous habitual exercise is known to have a number of beneficial effects, including reduction of body weight, percentage body fat, skinfold thickness, and girth. Prolonged exercise training will also result in changes in the lipoprotein profile. No marked changes occur in total plasma cholesterol or low density lipoprotein (LDL) cholesterol, but profound effects are seen in total triglyceride levels, which fall, and in high density lipoprotein (HDL) levels, which rise. In one study of veteran runners,[4] HDL2, which is the lipoprotein subfraction with the most powerful inverse relationship with the presence and severity of coronary artery disease, was grossly elevated (Fig. 1). If the Framingham predictive data are applied to this group, the risk of future coronary events is virtually nil! Other acute metabolic responses to exercise include a rise in serum catecholamines, which can result in myocardial ischaemia and arrhythmias in the presence of coronary disease.[5] Plasma potassium also increases during exercise,[6] and in the immediate post-exercise period free fatty acids rise[7] and may interact with the high catecholamine and potassium levels pertaining at that time, resulting in a further increased risk of cardiac

dysrhythmias. This period after acute vigorous exercise has been labelled the 'post-exercise vulnerable period', and those who smoke immediately after exercise, thereby increasing free fatty acids and catecholamine release, place themselves at even greater risk.

Recently it has been suggested that opioid peptides may be instrumental in bringing about the sense of well-being that exercise produces and may even cause diminished perception of exercise-induced fatigue and discomfort. The negative effect is the 'withdrawal phenomenon' that many habitual exercisers experience when they stop training. Consequently, exercise has even been used successfully in the treatment of patients with depressive illness.

Electrocardiographic effects

The resting, exercise, and continuous electrocardiogram of the athlete exhibits many changes, most of which have been regarded as 'benign'. Studies have shown that bradydysrhythmias, heart block patterns, and ventricular ectopy are common in athletic or exercising populations.[8]

Athlete's bradycardia is a well-documented phenomenon, and most studies indicate that, despite heart rates that are often very low, circadian variation is maintained (Fig. 2). The cause of bradycardia in athletic subjects is speculative. Most evidence suggests that the sinoatrial node and the atrioventricular nodes are suppressed by an increase in vagal tone. Of course, this can be abolished by pharmacological denervation (atropine plus propranolol) in humans and surgical denervation in animals. This work shows that the bradycardia is independent of vagal tone. Thus the intrinsic heart rate is lower. Although athlete's bradycardia is currently recognized as a benign adaptation, it has been associated with unexplained cerebrovascular accidents and syncope. Certainly, when the bradycardia is associated with atrioventricular block, symptoms occasionally develop, necessitating permanent pacemaker implantation. Some authorities have suggested that continuous habitual exercise throughout life may result in a bradycardia which, rather than being a benign

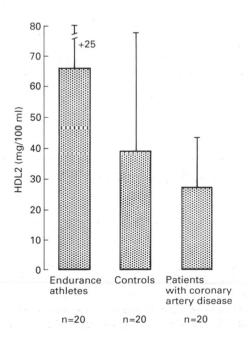

Fig. 1 Lipoprotein effects of vigorous exercise: a comparison of veteran runners and controls with and without evidence of coronary artery disease.

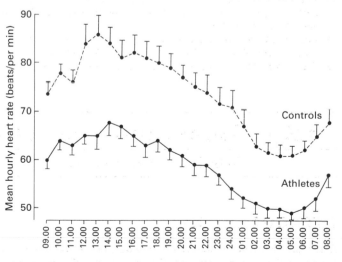

Fig. 2 Circadian variation of heart rate in athletes versus controls (mean hourly rate ± standard error). (Reproduced from Northcote RJ, Canning GP, Ballantyne D. Electrocardiographic findings in male veteran endurance athletes. *British Heart Journal*, 1989; **61**: 155–60, with permission.)

adaptation has been modified into a condition resembling the sick sinus syndrome. However, until follow-up studies of patients with exercise-induced bradycardia and atrioventricular block have been published, cardiologists will have no firm guidelines regarding the correct management of this condition. If an athlete has no desire to stop exercising, then pacemaker implantation needs to be considered in severe cases, even in the absence of symptoms.

Severe bradycardia with asymptomatic asystolic pauses has been reported in a number of trained athletes. Typically, these pauses last from 1.6 to 2.8 s, but several cases have been documented of pauses lasting 15 to 30 s. Most of the pauses occur nocturnally at very low intrinsic heart rates. It could be argued that discontinuation of exercise would cause resolution in the bradycardia and asystolic periods. This is not always found in practice, and one series reported that despite abandoning exercise almost 50 per cent of young sportsmen presenting with syncope ultimately required implantation of a permanent pacemaker.[8]

Although first-degree and, to a lesser extent, second-degree atrioventricular block has been found in younger athletes, complete heart block in a symptomatic individual is exceedingly rare. A characteristic of block patterns occurring in athletes is the reversion to sinus rhythm during exercise, although as has already been stated this is not always the case. The frequency of first-degree heart block varies from 3 to 55 per cent in published series, but is typically around 20 per cent. Second-degree atrioventricular block is less common and is found in 13 to 22 per cent of athletes. The occurrence of block patterns and bradycardia is probably related to increased vagal tone, but this is unlikely to be the sole explanation. In some lifelong exercisers it may coexist with the sick sinus syndrome and thus lead to a profound bradycardia. Almost without exception the resting electrocardiogram is unhelpful in identifying those with atrioventricular block—the athletic heart is probably under the influence of fluctuating vagal tone and thus it is important to perform Holter monitoring in the symptomatic individual to detect block patterns and asystolic pauses.

Exercise electrocardiography

Many physicians remain sceptical of the value of exercise electrocardiograms for the screening of asymptomatic healthy individuals because of the high frequency of false-positive tests (5–35 per cent). Such scepticism is increased when this test is applied to athletes. As many as 15 per cent of athletes, whether power trained or endurance trained, have significant ST segment depression on resting 12 lead electrocardiography and an even greater number exhibit an ischaemic response to treadmill exercise. These responses occur in the absence of symptoms and in patients with normal radioisotope myocardial perfusion scans or coronary arteriograms. The reason for the high frequency of false-positive response is not clear. It has been suggested that athletic ventricles repolarize in a non-homogeneous fashion, with repolarization of the epicardium occurring first. This asymmetry of repolarization may be related to reduced sympathetic tone. Nevertheless, from a clinical standpoint, exercise electrocardiography in athletes is probably not a totally reliable indicator of the presence or future development of coronary artery disease. This does not negate the unequivocal role of this test for individuals who are planning to participate in vigorous and intense physical activities. The exercise ECG will provide useful information regarding the diagnosis of coronary artery disease and identification of the high risk individual, and also a guideline to that individual's functional capacity. For example, any sport or exercise that requires an exercise workload, expressed in metabolic equivalents (METs) beyond the individual's physical capacity, as determined by the exercise ECG, is unquestionably hazardous. When serious ventricular arrhythmias are provoked by the exercise test, particularly with low workloads, the individual should be advised against vigorous exercise until the arrhythmia is fully evaluated and all aspects of cardiac status are assessed.

Arrhythmias

At rest, habitual exercisers or athletes have a similar frequency of both supraventricular ectopic beats and ventricular premature beats. During exercise the frequency of both forms increase in some individuals. Intense exercise, capable of raising the heart rate to above 150 beats/min, may be more likely to provoke ectopy than more gentle exercise. Of 21 young asymptomatic healthy male squash players, three had a burst of ventricular tachycardia during a squash match.[9] Similar events have not yet been reproducibly documented in less intense exercise such as running. Ventricular premature beats may be more prevalent in the immediate post-exercise period, which is referred to as the post-exercise vulnerable period perhaps because a sizeable proportion of sudden exercise-related deaths occur at this time.

The cause of these arrhythmias must be a matter for conjecture, as there are several possible explanations. There is good evidence that exercise can provoke arrhythmias in the coronary-prone population, and thus it is possible that asymptomatic coronary atherosclerosis and resultant myocardial ischaemia is a likely precipitant.

Cardiac arrhythmias have been shown to occur in response to thermal stress (e.g. sauna baths). Thermal stress may interact with other factors, mostly biochemical, to cause an arrhythmia. Plasma catecholamines rise during exercise and on their own may precipitate both myocardial ischaemia and arrhythmias in the presence of coronary artery constriction. Other metabolic changes, such as hyperkalaemia during exercise, may precipitate arrhythmias. If there is a sudden cessation of activity at the end of the exercise session, venous pooling occurs in the tissues resulting in a diminished cardiac return. Continued sympathetic nervous system stimulation this may precipitate coronary artery insufficiency and therefore an arrhythmia. This may be exacerbated by the thermal stresses of a hot bath or shower and in the presence of increased catecholamine levels, which remain elevated in the post-exercise period, and raised free fatty acids. The likelihood of an arrhythmia is greater.

Cardiac structure and function

The structural and functional cardiac adaptations to exercise have been described using echocardiography. Longitudinal studies of the hearts of exercising adults are few, and the remaining cross-sectional studies support the view that, in general, dynamic or aerobic exercise, such as running, leads to left ventricular dilatation and power exercise, such as weight training, results in muscle hypertrophy with little change in the internal ventricular size (Fig. 3) (see Chapter 3.1.1).

In many forms of exercise (e.g. cycling) there is an overlap in power and dynamic components. Competitive athletes usually supplement their training programme with a power component such as resistance running or weight training. However, popular mass exercise such as running consists solely of the dynamic component.

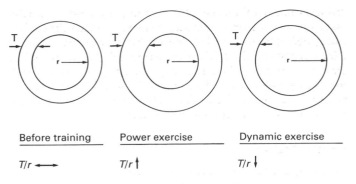

Before training   Power exercise   Dynamic exercise

$T/r \longleftrightarrow$ $T/r \uparrow$ $T/r \downarrow\uparrow$

Fig. 3 The relative effects of power exercise and endurance exercise on left-ventricular internal diameter and wall thickness: *T*, thickness of left-ventricular posterior wall; *R*, internal radius of left ventricle.

Increases in left-ventricular wall thickness and volume have been shown to occur after only a few weeks of exercise training, but probably take longer to resolve if training is ceased.

In a large group of predominantly young athletes from a variety of sports, Shapiro[10] found increased left-ventricular cavity size and increased thickness of the posterior left-ventricular wall and interventricular septum, and consequently an increased left-ventricular mass.

Triathletes (swimming, cycling, and running) also showed increased left-ventricular wall thickness and mass, but presumably because of the predominant power component of their training, exhibited no change in the internal ventricular dimension compared with controls.[11] Fagard *et al*.[12] have confirmed that the changes are seen in women as well as men.

One study of exclusively 'power-trained' athletes found that bodybuilders had thicker left-ventricular walls when compared with controls and that the use of anabolic steroids increases the difference, independent of their effect on body mass.[13]

It is possible that if exercise is continued throughout life, the difference between power and dynamic exercise is less apparent. Older cyclists seem to have a greater tendency to left-ventricular hypertrophy, and veteran runners have been found to have cardiac structural changes more akin to those found in young power-trained athletes;[14] thus the cumulative effects of exercise on cardiac structure may be quite different from the short-term effects.

Ventricular hypertrophy is probably due to a combination of intermittent pressure and volume overload on the ventricle. A volume overload will result in an increase in the left-ventricular internal dimension with a proportional increase in septal and free-wall thickness to normalize wall stress. The intermittent pressure overload occurring during training sessions possibly results in muscle hypertrophy (but not at the expense of reducing left-ventricular size). This also results in a normalization of myocardial wall stress (La Place's law), Shapiro[10] suggests that left-ventricular hypertrophy is related to the amount rather than the type of exercise. It is possible that the response to exercise training is highly individual and variable, rather than being due to specific exercise techniques or, indeed, volume of training.

In terms of function, most exercising adults develop a bradycardia with compensatory increase in the stroke volume, thus maintaining the cardiac output. However, there is no demonstrable change in the left-ventricular ejection fraction. Although resting cardiac output may not differ from that of non-exercising in-

dividuals, the trained individual is capable of increasing the cardiac output to between 20 and 30 l/min, and highly trained athletes may even be able to sustain an output of 40 to 50 l/min for short periods.

The structural changes induced by exercise need not be detrimental. As far as we are aware, they represent a physiological variant. Theoretically, in patients with coronary artery disease there may be a risk of jeopardizing myocardial perfusion because of an increased left-ventricular wall thickness, but the benefits of exercise probably outweigh these risks and it is even possible that exercise encourages coronary collateralization under these circumstances.

CARDIOVASCULAR DANGER OF EXERCISE

Sudden death

Exercise-related sudden death captures media attention despite the relative infrequency of this unfortunate event. Although exercise provides some protection against the development or progression of coronary artery disease in such individuals, it will not provide immunity and causes a transient increase in risk during periods of activity. Studies from Seattle[15] have shown that regular vigorous exercise protects against primary cardiac arrest, but if arrest occurs it is more likely to occur during exercise training. In other words, persons who exercise have a reduced likelihood of developing coronary atherosclerosis and a lower risk of coronary death, but some individuals have an 'atherogenic force' so strong that severe coronary disease develops despite regular exercise and are at even greater risk during exercise sessions.

Frequency of sudden death

Published findings on the relation between sudden death and degree of activity are inconsistent. In many cases of death considered to have been precipitated by exercise, the nature of the activity was not sufficiently vigorous to constitute 'sport'. The determination of a true incidence of sudden death in sport is hindered by inadequate registration and investigation of each case. The level of activity at the time of death is not recorded in the official figures. Most of these deaths are the subject of forensic investigation, but in the United Kingdom the place of death is normally recorded as the hospital at which death was certified. It is also true that such inquiries are concerned with the pathological cause of death rather than the activities surrounding the death.

Despite these difficulties, attempts have been made to evaluate the statistical risk of sudden death during sport. Opie[16] estimated that in rugby football a player will die every 50 000 rugby hours and a referee will die every 3000 referee hours. Even acknowledging the greater age of rugby referees, the latter figure seems unrealistically high. Koplan[17] calculated that 0.77/100 000 joggers die each year by chance alone without there being any relation to exercise. Using these estimates, a study of the incidence of death during jogging in Rhode Island revealed a death rate of 17 times Koplan's estimate.[18] Nevertheless, only one out of 7620 joggers died each year, emphasizing the rarity of such events in this group.

The statistical risk of death during sport in the United Kingdom is not available. The present author has collated a personal series of sudden deaths (Table 1) by using a blanket enquiry of all registered sports clubs, centres, and sports governing bodies. What

Table 1 Exercise-related sudden death in the mainland United Kingdom 1978–1987

Sport/activity	Number		Mean age (±SD) and age range (years)
	Male	Female	
Squash	124	2	44 ± 8 (22–70)
Soccer	53	—	32 ± 11 (17–61)
Swimming	50	6	53 ± 17 (14–84)
Running	38	1	37 ± 17 (11–71)
Badminton	26	—	49 ± 13.6 (27–82)
Rugby	14	—	30 ± 15 (15–52)
Other sports	33	4	NA
Total	338	13	
		351	

NA, not available.

becomes apparent is that there is a low statistical risk of sudden death, but a significant number die unnecessarily. These figures do not imply that squash, football, or swimming are more dangerous, as greater numbers may participate in these activities.

In addition to these statistical studies, a large number of cases are reported in the medical literature.[19] Of 145 cases associated with sport (Table 2), 30 occurred during or immediately after squash, 51 were associated with running, 32 with rugby football, soccer, and American football, nine with basketball, four with tennis, and 13 with various other pursuits. It is noteworthy that sudden death in women is exceedingly rare.

Causes

Coronary artery disease

The features attending sudden death during exercise have been analysed in a series of case reports. Amongst these, Opie[16] described 21 deaths, predominated in rugby players and referees, occurring within 1 h of onset of symptoms, of which 18 were attributed to coronary disease. In a study of 18 deaths during or after jogging, Thompson et al.[20] concluded that 13 had been caused by coronary disease.

Results of numerous series confirm that coronary disease is responsible for the majority of sudden deaths in exercisers. Those dying from coronary artery disease are significantly older than those dying from structural cardiovascular abnormalities (mean age 60 ± 9 years versus 19 ± 6 years). In addition, of the group as a whole, 34 subjects (31 per cent) suffered from prodromal symp-

Table 3 Prodromal symptoms

Symptom	Number
Chest pain/angina	15*
Increasing fatigue	12
Indigestion/heart burn/gastrointestinal symptoms	10
Excessive breathlessness	6[†]
Ear or neck pain	5
Vague malaise	5
Upper respiratory tract infection	4
Dizziness/palpitation	3
Severe headache	2[‡]
None	5

* One subject, an international squash player, had recently consulted a physician because of chest pain after squash play.
[†] One subject, who ran 1 mile every morning, was extremely breathless on three successive mornings before he died, but was not deterred from further strenuous exertion.
[‡] Cause of death in one subject was intracranial haemorrhage.

toms such as angina or dyspnoea (Table 3) and a large proportion exhibited one or more risk factors for coronary artery disease.

Structural anomalies

Structural cardiovascular abnormalities are the most common cause of sudden death in young sportsmen. Maron et al.[21] reported 29 cases of sudden death in young (13–30 years) highly conditioned competitive athletes. In 22 of them death occurred during

Table 2 Sudden death occurring during or within 1 h of sporting activity (n = 145)

	Cause of death*						
	Coronary artery disease[†]	Other					
		Hypertrophic obstructive cardiomyopathy	Coronary artery anomaly	Myocarditis	Conduction disorders	Ruptured aorta	VHD
No. of subjects	110	14	8	3	4	1	3
Male	109	14	8	3	4	1	3
Female	1	2	0	0	0	0	0
Mean age (SD) (years)	40 (9.1)	← ————————————— 19 (6.2) ————————————— →					

VHD, valvular heart disease.
* In 56 of all deaths subjects were reported to have prodromal symptoms.
[†] 57 subjects had positive risk factors for coronary atherosclerosis.

or in the first hour after exercise. Twenty-eight of them were found at autopsy to have structural cardiovascular disease, such as hypertrophic cardiomyopathy which was present in 14 (50 per cent).

Other recorded abnormalities include congenital anomalies of the coronary arteries, the most frequent of which is anomalous origin of the left main stem from the right coronary cusp. The risk of sudden death during exercise or otherwise with this condition is 27 per cent.

Few causes are attributable to valvular heart disease. This probably reflects the easier detection of such abnormalities at routine medical examinations.

Other conditions

In the past, myocarditis was thought to be an important factor in the cause of sudden exercise-related deaths, but this view is not supported by the evidence. Clearly, most cases are related to coronary disease or other overt structural cardiac conditions and it is unlikely that many cases are explained by myocarditis alone, although isolated reports describe myopericarditis and chronic myocarditis in young sportsmen. It is probable that a subclinical myocarditis can accompany or result from a respiratory tract infection, which may result in significant morbidity without sudden death. It is not possible to establish whether an individual with a febrile illness is at increased risk if he or she takes part in exercise. Such foolhardy behaviour must take place often, particularly in professional sport, and fatal sequelae are exceedingly rare.

Dysfunction of the cardiac conduction pathways may lead to fatal arrhythmias. In some young athletes the arteries supplying the sinus and atrioventricular nodes have exhibited bizarre medial hyperplasia and intimal proliferation, thus producing ischaemia. It is important that if no obvious cause of death is discovered at autopsy, a detailed histopathological examination of the conduction system should be made. In older athletes the sick sinus syndrome may become overt in the face of increased vagal tone. This has been reported as the cause of death in some veteran cyclists who died during sleep with no obvious pathological abnormalities.

Personality factors

Competitive sport tends to attract aggressive competitive type A individuals. Likewise, some series of sudden death in sport have noted the high frequency of such personality features in those who died. It is possible that competitive sports such as squash and football may be more likely to precipitate sudden death because of the greater release of catecholamines and free fatty acids associated with the additional psychological stress when compared with exercise alone. Although such a hypothesis appears attractive, there is no definitive evidence that this occurs.

Some sports, particularly running, are thought to be addictive. This may result from recurrent release of cerebral opiates which occurs during running and may be responsible for the training withdrawal effects which include irritability, depression, anxiety, and generalized fatigue. Reintroduction of training causes re-establishment of cerebral opioid stimulation and usually rapid resolution of these unpleasant effects. Addictive runners have been compared with anorexic patients and found to have similar characteristics. For example, they have had dominant mothering and they have extraordinarily high self-expectation, tolerance of physical discomfort, and denial of potentially serious debility.

This denial of prodromal symptoms have been noted previously and may be a phenomenon to which aggressive sportsmen are more prone (see Chapter 1.4.1).

Preventable or inevitable

As all deaths categorized as exercise-related are unnecessary, the possibility that some, at least, may be preventable, must be examined. The totally nihilistic will argue that some exercise-related deaths are inevitable, and any attempt to identify those at risk is unjustifiable and unworkable and may discourage those who would benefit most from exercise.

In youth or childhood, sudden death is rare and frequently related to structural disease. These abnormalities should generally be apparent at routine school medical examination. It is doubtful if any serious preventive measures would reduce exercise-related deaths in this age group.

Most deaths occur in men over the age of 35 years and are due to coronary artery disease. There is a high prevalence of prodromal symptoms, coronary artery disease risk factors, and medical conditions such as hypertension. It may make sense to target our efforts on this group. One argument against this is cost. Would not all general practitioners be overworked by the demands placed on them? Let us take the example of squash. This sport is probably associated with 25 to 40 deaths per year in the United Kingdom. There may be as many as 1.5 to 2.5 million squash players in the United Kingdom (figures from the Squash Rackets Association). This represents a large proportion of the adult population. The Squash Rackets Association have estimated that, of these, only 21 per cent are over 35 years old and 5 per cent over 45 years old. Thus the sudden deaths are occurring in a small portion of the squash-playing public (probably around 150 000). This should be the target population.

What form should prevention/screening take? Probably the cheapest, most practicable, and most successful scheme originated in Canada and takes the form of a three-tiered system of evaluation. The first stage is a simple self-administered questionnaire designed to detect medical conditions and coronary artery disease risk factors which should stimulate further assessment. This questionnaire is called the PAR-Q test or Physical Activity Readiness Questionnaire (Table 4). Trials of this test in Canada have shown that it detects at least half those in whom an increase in physical activity would cause an increase in risk of a cardiac event.

If the response to this test is positive then the second stage is assessment by either a paramedic or general practitioner. This should involve symptom evaluation and a brief physical examination including cardiac auscultation and recording of blood

Table 4 Modified Physical Activity Readiness Questionnaire (PAR-Q test)

1 Are you over 60 years of age?	YES/NO
2 Do you suffer from chest pains?	YES/NO
3 Do you suffer from breathlessness or dizziness during or after exercise?	YES/NO
4 Are you diabetic?	YES/NO
5 Do you have heart condition or heart murmur or have you had heart surgery?	YES/NO
6 Have you ever been told you have high blood pressure?	YES/NO
7 Do you smoke?	YES/NO
8 Have you ever suffered from black-outs?	YES/NO

pressure. If an abnormality is detected, only then should a specialist-supervised medical examination with access to sophisticated tests such as exercise electrocardiography or echocardiography be performed. Such a scheme is relatively inexpensive and would probably have detected most of the individuals who died during exercise enumerated earlier in this chapter.

Having identified those at increased risk, this system would allow counselling of the individual on the desirability of exercise and choice of sport. It is not intended that a subject who is found to have coronary artery disease should stop exercising. In fact, the contrary is true, but that individual would be best advised to avoid totally physically exhausting exercise with excessive cardiovascular demands. For example, a 66-year-old man with triple coronary artery bypass grafting, a permanent pacemaker, moderate aortic stenosis, and a positive exercise test might have been encouraged to take regular light to moderate rhythmic exercise, rather than playing competitive squash, during which he ultimately died in the final of a veterans' tournament. Clearly this is an extreme case, but more than a third of individuals who die on the squash court have cardiac conditions (e.g. valvular heart disease, coronary artery disease, and hypertension) known to themselves and their doctors.

Although exercise should be encouraged in most people, the form and intensity should be tailored to suit the individual's general fitness, health, and age.

Education

Many sports centres and sports governing bodies would do well to follow the example of the Squash Rackets Association. Led by the adage 'Get fit to play squash. Don't play squash to get fit', they have circulated all affiliated clubs with information pamphlets covering such things as 'warning signs' and 'when not to play squash'. They are fulfilling the requirements for more adequate education and supervision of the embryo sportsman. All sportsmen should be aware of the small risks attached to vigorous exercise and should be able to recognize warning symptoms.

As mentioned earlier, competitive sportsmen are known to deny prodromal symptoms. Individuals should be encouraged to refrain from strenuous exercise if such symptoms occur. It has been suggested that individuals should not exercise when suffering from an upper respiratory tract infection or other pyrexial illness, as this can be accompanied by potentially lethal subclinical myocarditis. Although viral myocarditis is not documented in the literature as an important cause of death in sportsmen, it is probably reasonable to discourage vigorous exercise at these times.

Useful precautions

For the healthy and those with cardiac disease some routine habits may help minimize risk. Adequate warming up and warming down should be encouraged to reduce the number of dysrhythmic deaths. Likewise, the discouragement of smoking and avoidance of hot baths, saunas or showers immediately after exercise may reduce the harmful effects of increased levels of catecholamines, free fatty acids, and potassium. Because of the risk of hyperpyrexia and heat stroke, vigorous activity and extreme heat should be avoided.

DOES EXERCISE PREVENT CORONARY ARTERY DISEASE?

Having examined the potentially hazardous cardiac effects of acute exercise, the probable beneficial role of exercise in prevent-

ing coronary disease must be addressed to place the risk–benefit equation in perspective.

The earliest scientific data that suggested that exercise may significantly alter tendency to atheroma were published in 1952, in a study showing that coronary deaths were most common in professional men and least common in unskilled manual workers. Thus, sedentary occupations were thought to predispose to coronary death. The landmark paper followed a short time afterwards. Morris et al.[22] published data from 31 000 London transport workers comparing the drivers with the more active conductors of double-deck buses. The drivers had a threefold greater frequency of sudden death.

Both these early studies presented a hopeful hypothesis—that exercise prevented coronary disease and sudden cardiac death—but it became clear that other factors were also having an influence and so weakened the conclusions of these papers. For example, it became clear that, even on recruitment, the bus drivers were more obese and had higher plasma cholesterol and blood pressure. Thus these were self-selected groups.

Twenty years after these early studies, Morris et al.[23] published further supportive data from a group of 18 000 male civil servants. In this study, leisure time activity was compared within a group with desk jobs whose activity did not differ significantly while at work. Men who exercised vigorously in their leisure time had half as many coronary events; light exercise (walking, bowls, cricket, etc.) provided no advantage. Again, the groups were not matched perfectly for coronary risk factors and took no account of the inherent fitness of the individuals at the start of the study. Nevertheless, these data strengthened the argument that exercise protects against coronary disease.

Other studies published in the 1970s from Framingham, Israel, and the United States also support an inverse relationship between the level of physical activity and coronary disease.

All these studies suffered from the same fault: they did not adequately take account of other factors such as smoking, lipids, body weight, blood pressure, and general health. However, carefully controlled longitudinal studies with multivariate analysis of the data have been performed by Paffenbarger and coworkers.[24,25] In a study of Californian longshoremen[24] they showed that men with more sedentary jobs expended 925 fewer calories each day and had coronary death rates one-third higher than manual workers. They also showed that sudden cardiac death was three times more frequent in light workers than in heavy workers.

Paffenbarger et al.[25] studied Harvard alumni in even greater detail and showed conclusively that lifelong exercise, but not college exercise, protects against a coronary event. However, if post-college exercise is commenced in a previously non-exercising individual, then he will acquire low risk in time.

Not only does exercise probably protect against coronary artery disease, it probably also protects against fatal cardiac arrest if a myocardial infarction is to occur. A recent report from the Seattle Heart Watch Programme[15] has shown that persons engaging in vigorous exercise have a reduced risk of primary cardiac arrest; low and moderately intense leisure time exercise was not clearly related to reduced risk of cardiac arrest but high intensity exercise (jogging, swimming, tennis) was. The likelihood of cardiac arrest in persons in the two upper quartiles of high intensity exercise was only 40 per cent of that of non-exercisers.

Only one study, from Finland, has suggested that regular vigorous exercise may be harmful. Pansar and Karvonen[26] showed that lumberjacks had a higher coronary mortality than farmers.

However, the lumberjacks smoked more, had higher blood cholesterols, and were in a lower socioeconomic class.

In terms of coronary disease, although healthier individuals probably become exercisers, there appears to be justification for assuming that exercisers also become healthier.

Therefore, although vigorous exercise causes a transient increase in risk of a cardiac event, long-term regular exercise confers some protection against atherosclerosis which, in population terms, certainly outweighs the acute risk.

However, it is entirely fallacious to assume that vigorous exhausting exercise performed daily (e.g. marathon training) will confer immunity to coronary atherosclerosis. Many lifelong marathon runners and other sportsmen have been found to have severe coronary disease at autopsy, although it is likely that exercise retarded development of atherosclerosis and may even have led to its early detection.

Exercise as a tool in the management of heart disease

It has now become clear that exercise prescription is a valuable measure in treatment of cardiac disease. There are a wide range of applications: angina pectoris, after myocardial infarction, congestive cardiac failure, and after interventions such as coronary angioplasty, bypass grafting, and cardiac transplant. Exercise has become synonymous with rehabilitation in respect of cardiac disease, but probably goes beyond rehabilitation to the area of therapy. Most individuals are not being rehabilitated; rather, they are being introduced to a completely new life-style and therapeutic strategy.

There is little doubt that exercise training results in an increase in well-being, prolonged treadmill exercise time, and an improvement in the plasma lipid profile and other risk factors. What is less certain is the effect that this therapy has on cardiac function and myocardial perfusion. Early animal studies suggested that vigorous exercise improved collateralization around a stenosed coronary artery, leading to the belief that an ischaemic stimulus was useful. However, published studies of humans have failed to provide positive results in this regard, although recent work from Glasgow[27] using TI 201 myocardial perfusion scintigraphy in patients with stable angina does show increased perfusion in previously underperfused myocardium in response to a 1 year supervised training programme.

In a similar way, patients with congestive cardiac failure have benefited from training, but it has not been shown that cardiac function improves with training although exercise capacity and symptoms improve.

Another area of uncertainty concerns the exercise protocol which is most beneficial to these patient groups. Basically, exercise can be divided into three categories.

1. Dynamic (sometimes called aerobic exercise);
2. Power (muscle contraction against a movable resistance, but exercising the muscle through its full range);
3. Isometric (muscle contraction against a fixed resistance).

It is important to distinguish 'power' from 'isometric' exercise, which has until now been contraindicated in cardiac patients. Although there is a reluctance to utilize isometric exercise because of the profound haemodynamic effects, this should not necessarily also preclude power exercise. Both power exercise and dynamic exercise probably have different effects on myocardial muscle, with power exercise including a thicker left-ventricular wall with no change in the internal volume, and dynamic exercise resulting in an increase in left-ventricular volume without the change in left-ventricular posterior wall or septal thickness. It is hoped that ongoing research in this area will establish the cardiac effects of these exercise protocols and lead to more informed and appropriate exercise prescription for these patient groups.

Contraindications

After a myocardial infarction, certain patients may be at risk from exercise. For example, those with unstable angina, exercise-induced arrhythmias, and severe congestive heart failure should not be exercised. Usually a pre-exercise assessment including a treadmill exercise test (postmyocardial-infarction protocol such as the Balke or Newcastle) will identify those in whom exercise is undesirable. Treadmill testing also allows assessment of functional capacity.

Exercise protocols

The aim of exercise training should be to increase cardiovascular fitness. To do this the individual should be able to attain 70 to 80 per cent of his or her predicted maximum heart rate. This training can be expected to result in an average increase of 20 to 30 per cent in Vo_2max. Exercise capable of inducing heart rate increments beyond this may expose the patients to increased risk without providing a commensurate benefit.

The exercise protocols vary from centre to centre, but should generally last 30 to 40 min and be performed at least three times per week. In Glasgow both power exercise (overhead pulldown, trunk curls, quadriceps extension, bench press, etc.) and aerobic exercise (Canadian Airforce 5BX training programme) are used. A graduated programme of activity designed on an individual basis is carried on for 6 months, although this period is arbitrary, in a gymnasium supervised by physiotherapists and a doctor. Thereafter the patient is encouraged to continue his new life-style in local sports centres.

In Canada, one centre's goal for post-infarct patients is to complete the Hawaii marathon, where there are separate registration booths for cardiac patients. Such extremes of exertion will provide no additional benefit and probably exposes the patient to unnecessary risk. Nevertheless, such exploits emphasize the formidable achievements that are possible.

OVERVIEW

Exercise has profound effects on the cardiovascular system, most of which are beneficial. There is a transiently increased risk of sudden death during exercise in normal individuals, which is increased further in those with previous cardiovascular problems. However, the long-term benefits of exercise far outweigh the short-term risk. If sufficient effort is made by the medical profession, sports governing bodies, and sportsmen themselves any risk will be minimized.

Exercise will probably become a potent therapeutic tool in the rehabilitation of patients with congestive heart failure and after myocardial infarction, coronary angioplasty, and cardiac surgery, including cardiac transplantation. There is good evidence that such a prescription is both more effective and safer than many pharmaceutical alternatives.

REFERENCES

1. Bassler TJ. Marathon running and immunity to atherosclerosis. *Annals of the New York Academy of Sciences* 1977, **301**: 579–92.

2. Opie LH. Long distance running and sudden death. *New England Journal of Medicine* 1975, 293: 941–2.

3. Waller BF, Roberts WC. Sudden death while running in conditioned runners aged 40 years or over. *American Journal of Cardiology* 1980; 45: 1292–1300.

4. Northcote RJ, Canning GC, Todd IC, Ballantyne D. Lipoprotein profiles of elite veteran endurance athletes. *American Journal of Cardiology* 1988; 81: 934–6.

5. Raab W, Van Lith P, Lepeschin E, Herrlich HC. Catecholamine induced myocardial hypoxia in the presence of impaired coronary dilatability independent of external cardiac work. *American Journal of Cardiology* 1962; 9: 455–70.

6. Lim M, Linton RAF, Wolff CB, Band DM. Propranolol, exercise and aterial plasma potassium. *Lancet* 1979; ii: 591.

7. Johnson RH, Walton JL, Krebs HA, Williamson DM. Metabolic fuels during and after severe exercise in athletes and non-athletes. *Lancet* 1969; ii: 452–5.

8. Ector H *et al*. Bradycardia, ventricular pauses, syncope and sports. *Lancet* 1984; ii: 591–4.

9. Northcote RJ, MacFarlane P, Ballantyne D. Ambulatory electrocardiography in squash players. *British Heart Journal* 1983; 50: 372–7.

10. Shapiro LM. Physiological left ventricular hypertrophy. *British Heart Journal* 1984; 52: 130–5.

11. Douglas PS, O'Toole ML, Miller EDB, Reichel N. Left ventricular structure and function by echocardiography in ultra endurance athletes. *American Journal of Cardiology* 1986; 58: 805–9.

12. Fagard R, Brooke C Van Den, Vanhees L, Staessen J, Amery H. Non-invasive assessment of systolic and diastolic left ventricular function in female runners. *European Heart Journal* 1987; 8: 1305–11.

13. McKillop G, Todd IC, Ballantyne D. The effects of bodybuilding and anabolic steroids on LV structure and function. *International Journal of Cardiovascular Ultrasound* 1989; 8(1): 23–9.

14. Northcote RJ, McKillop G, Canning GP, Todd IC. The effect of habitual sustained endurance exercise on cardiac structure and function: A study of male veteran runners. *European Heart Journal* 1990; 11: 17–22.

15. Siscovick DS, Weiss NS, Helstrom AP, Inui TS, Peterson DR. Physical activity and primary cardiac arrest. *Journal of the American Medical Association* 1982; 248: 3113–17.

16. Opie LH. Sudden cardiac death and sport. *Lancet* 1975; i: 263–6.

17. Koplan JP. Cardiovascular deaths while running. *Journal of the American Medical Association* 1979; 242: 2578–97.

18. Thompson PD, Funk EJ, Carleton RA, Styurner WQ. Incidence of death during jogging in Rhode Island from January 1975 through 1980. *Journal of the American Medical Association* 1982; 247: 2535–8.

19. Northcote RJ, Ballantyne D. Cardiovascular implications of strenuous exercise. *International Journal of Cardiology* 1985; 8: 3–12.

20. Thompson PD, Stern MP, Williams P, Duncan K, Haskell W, Wood P. Death during jogging or running: a study of 18 cases. Journal of the American Medical Association 1979; 242: 2265–7.

21. Maron BJ, Roberts WC, McAllister HA, Rosing DR, Epstein SE. Sudden death in young athletes. Circulation 1980; 62: 218–29.

22. Morris JN *et al*. Coronary heart disease and physical activity at work. *Lancet* 1953; ii: 1053–7.

23. Morris JN *et al*. Vigorous exercise in leisure time and the incidence of coronary heart disease. *Lancet* 1973; ii: 333–9.

24. Paffenbarger RS, Laughlin ME, Gima AS, Black RA. Work activity of longshoremen as related to death from coronary heart disease and stroke. *New England Journal of Medicine* 1970; 282: 1112–3.

25. Paffenbarger RS, Wing AL, Hyde RT. Physical activity as an index of heart attack risk in college alumni. *American Journal of Epidemiology* 1978; 108: 161–75.

26. Pansar S, Karvonen MJ. Physical activity and coronary heart disease in populations from east and west Finland. *Advances in Cardiology* 1976; 18: 196–207.

27. Todd IC, Bradnam MS, Cooke MBD, Ballantyne D. Effects of daily high intensity exercise on myocardial perfusion in angina pectoris. *American Journal of Cardiology* 1991; 68: 1593–9.

3.2.2 Asthma

MARK HARRIES

INTRODUCTION

Asthma refers to the condition of subjects with widespread narrowing of the bronchial airways, which changes in severity over short periods of time either spontaneously or under treatment, and is not due to cardio-vascular disease.

This statement about the clinical syndrome asthma was made in 1957.[1] The statement defined no measurements with which a diagnosis could be established, nor did it contain any reference to possible mechanisms. Now that so much more is known about the pathophysiology of asthma, it is clear that any brief definition will necessarily lack precision.

PATHOPHYSIOLOGY

Essential in the pathology of asthma is inflammation of the bronchial mucosa and submucosa which results in narrowing of the lumen. If asthma persists, these changes become chronic with bronchial wall thickening and progressive obstruction of the bronchi by accretions of inflammatory tissue. Inflammation also leads to the other fundamental abnormality found in asthma, namely hyper-reactivity of bronchial smooth muscle.[2] Infection of the respiratory tract is a common antecedent, but asthma also results from inhaling a variety of allergens.[3,4] Some of the early events in the development of bronchial inflammation are now understood. However, the complex interplay between antigen presentation by the dendritic cells, the macrophages, and the peripheral blood leucocytes remain a subject of intensive study.

The role of allergy

Allergy is a state of altered reactivity in the host that results from interaction between antigen and antibody. An antigen is an agent that stimulates the production of antibody, while an allergen is defined as an agent which has been shown to initiate an allergic response. The antibody classically associated with the allergy is IgE immunoglobulin, the production of which may be controlled by a gene or genes at a single locus.[5] It is present unbound in peripheral blood only in very small amounts compared with other immunoglobulins. Most is attached either to leucocytes or to mast cells lying just beneath the epithelial surface.

After inhalation, allergen penetrates the respiratory epithelium to combine with pairs of IgE molecules attached to underlying mast cells, thus providing the signal to release a number of mediators. It is these that initiate an inflammatory cascade. Individuals who readily make IgE antibody are prone to allergic reactions of the respiratory tract, such as rhinitis (see later) and asthma. They constitute around 20 per cent of the population. Those who make IgE to great excess may also suffer eczema and tend to be more sensitive to a wider variety of environmental antigens.

Subjects who develop sensitivity to common environmental antigens are described as atopic and the tendency to react in this way is termed atopy. The atopic status declines with advancing age. IgE antibody to house mite excrement is detectable in the first few months of life; however, it may be years before symptoms develop and then predominantly in atopic subjects. Continuing exposure to allergen gives rise first to bronchial hyper-reactivity and ultimately to asthma. Once the bronchial smooth muscle is rendered hyper-reactive, bronchoconstriction may result after inhaling a variety of unrelated agents such as dusts or pollutants or, in the case of exercise-induced asthma, after inhaling cold air alone.

Allergy to environmental antigens can be tested by pricking a soluble extract of each into the skin. Sensitive individuals react within 10 min by developing a wheal and flare at the site of inoculation. This is due to local release of histamine from underlying mast cells. The response is exactly analogous to that which takes place in the nose where inoculation is unnecessary because the mucous membranes are not covered by skin (see section on rhinitis). In the northern hemisphere allergy to grass pollen is manifest only around June and July, while exposure to airborne allergens in the home, such as house mite excrement or dander from domestic pets, occurs all the year round.

Respiratory infections

Damage to the respiratory epithelium is probably necessary before airborne allergens can penetrate. This may result initially from a viral infection of the respiratory tract, although irritant gases such as nitric oxide and sulphur dioxide have also been implicated. Infection of the respiratory tract also provokes bronchial hyper-reactivity that may persist for several weeks or even months.[6] Transient wheeze is common during the course of chest infections from which recovery is expected. However, for some, a single respiratory infection can mark the start of asthmatic symptoms that may never completely clear.

The effects of cold air and exercise

All asthmatics wheeze on exercise, i.e. exercise-induced asthma is no more than an index of underlying bronchial hyper-reactivity. The asthmatic response is exaggerated by either cooling or drying the inspired air and by increasing the respiratory rate. Non-asthmatic subjects may also develop transient bronchoconstriction if they exercise in air made sufficiently cold (Fig. 1(a)), but the effect on the asthmatic is always more pronounced and may be induced by inhaling cold air alone, without the added element of an exercise test (Fig. 1(b)).

Air pollutants

Continued bronchial irritation gives rise to hyper-reactivity of the smooth muscle. Ozone and the oxides of sulphur and nitrogen are potent bronchial irritants. These gases are ubiquitous, forming as a result of the photochemical action of sunlight on the products of

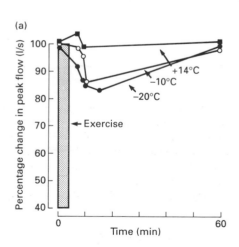
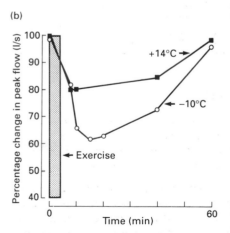

Fig. 1 Bronchoconstriction caused by cold air.
(a) Peak flow was measured in a non-asthmatic before and at intervals after running for 3 min on a treadmill in an environmental chamber. The tests were conducted at ambient temperature (14°C), at −10°C, and finally at −20°C. The percentage change in peak flow was plotted against time. After exercising at ambient temperature, peak flow rose transiently before returning to normal. At −10°C there was a 15 per cent fall in peak flow, while at −20°C the fall was 20 per cent. (b) The same test was conducted on an asthmatic subject. In this case significant bronchoconstriction occurred at ambient temperature with a 20 per cent fall in peak flow. The fall was even greater after exercising in air at −10°C. However, when the temperature was lowered to −20°C bronchoconstriction resulted simply from entering the cold chamber. No exercise was necessary.

combustion of fossil fuels. They are present in the highest concentrations in city locations such as Los Angeles, Mexico City, and Tokyo, where surrounding hills or mountains trap the products of combusted fuels and the local climatic conditions ensure many hours of strong sunlight. Long-term exposure to sulphur dioxide or ozone at levels as low as 50 parts per billion (p.p.b.) has been shown to increase bronchial reactivity. Again, subjects with pre-existing asthma or bronchial hyper-reactivity are more sensitive to the effects of atmospheric pollution than those with normal lungs.

CLINICAL ASPECTS OF ASTHMA

Severe asthma

Symptoms and signs

Cough, tightness in the chest, wheeze, and shortness of breath are all symptoms of asthma. These are usually worse at night or in the early morning, and a history of sleep disturbance indicates moderately severe disease. Symptoms often appear for the first time following a viral infection of the respiratory tract. The cough is unproductive; despite this the patient may have received several courses of antibiotics without benefit. On auscultation, the crackles so typical of a lung infection are absent. Wheeze is elicited by asking the patient to breath out forcibly from full inspiration to full expiration with the mouth open and is best heard with the bell of the stethoscope placed directly over the trachea rather than on the chest wall.

The patient experiences difficulty with breathing out rather than breathing in. Breathlessness is a feature of the severe attack such that even holding a conversation may prove difficult. The breathlessness results from hypoxia which is associated with a sinus tachycardia. A respiratory rate in excess of 25 breaths/min with a heart rate above 110 beats/min, indicates severe asthma.[7] A rise in intrathoracic pressure during expiration impedes venous return to the chest, causing a concomitant rise in jugular venous pressure. On inspiration systolic blood pressure falls, sometimes by as much as 15 to 20 mmHg.[8] This so-called paradoxical pulse, far from being a paradox, is an exaggeration of the small fall in pulse pressure that occurs normally during inspiration. It is not thought to be a useful indicator of clinical severity.

Investigations

Blood gases

Arterial blood gas measurement is essential in severe asthma. Hypoxia develops early because of a mismatch of ventilation with perfusion. Arterial oxygen tension Pa_{O_2} measuring below 8 kPa is a medical emergency. An increase in respiratory rate stimulated by hypoxia results in a fall in arterial carbon dioxide in the early stages of acute severe asthma. If treatment is delayed, the respiratory bronchioles become filled with inflammatory exudate. The alveoli beyond ventilate poorly, causing Pa_{CO_2} to return towards normal or, much more serious, to rise. Pa_{CO_2} also rises as the patient begins to tire and underventilate. Hypercapnia (Pa_{CO_2} > 6 kPa) together with hypoxia (Pa_{O_2} < 8 kPa), despite receiving 60 per cent oxygen, is a sign of impending respiratory failure and may signal the need for assisted ventilation.[6]

Lung function tests

The patient may be in such respiratory distress that formal assessment of lung function becomes impossible; however, peak expiratory flow (PEF in l/min) can usually be measured. Values below 40

per cent of the predicted normal indicate moderately severe asthma. Both the forced expiratory volume in 1 s (FEV_1) and the forced vital capacity (FVC) are reduced by roughly the same margin. Peak flow is the index of lung function usually chosen to monitor response to treatment because measurement can easily be made at the bedside.

Diurnal variations in PEF increase as asthma develops, with the lowest readings in the early hours of the morning (known as the morning dip) and highest around mid-afternoon. These cyclical changes are lost during severe asthma when bronchoconstriction is maximal but re-emerge during the recovery phase. The mid-afternoon measurement is the first to return towards normal, but it may be many days before the morning dip follows the same trend. Peak flow measurements made at 6 a.m. may be grossly abnormal at a time when the midday reading is within the normal range. Recovery should be judged thereafter on early morning peak flow reading before a bronchodilator is given (Fig. 2). Many physicians regard large morning dips as an early sign of an impending severe attack.

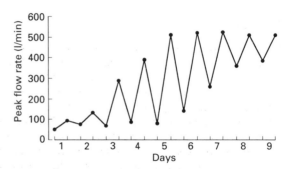

Fig. 2 Recovery in PEF following severe asthma. Diurnal variation in PEF is seldom more than 8.5 per cent in normal individuals. Differences of more than 20 per cent between values obtained at any time during the day is strongly suggestive of asthma. This record of a young patient admitted to hospital with severe asthma shows how peak flow can be used to monitor the response to treatment. Measurements are made before bronchodilator therapy at 6.00 a.m. and 12.00 noon. Variation in peak flow does not develop until the third day after admission and paradoxically marks the beginning of the recovery phase. By the fourth day the 12.00 noon measurement has returned almost to normal, but the 6.00 a.m. reading does not start to recover until the ninth day in hospital. Had clinical response been judged on the 12.00 noon measurement alone, the patient could have been sent home on the fourth day while still as much at risk as when she was first admitted.

Measurements of lung volume only become possible as the patient improves. Initially, residual volume (RV) increases due to air trapping. Hyperinflation causes the total lung capacity to increase, but gas transfer usually remains normal.

Chest radiography

The chest radiograph in acute asthma is usually either normal or shows hyperinflation. Justification for ordering a chest film appears to lie with excluding a tension pneumothorax which may present with similar symptoms. However, there is no evidence to suggest that pneumothorax is any more common in acute asthma than it is during the course of the other respiratory emergencies. It can be argued that a radiograph may assist in revealing a possible cause such as pneumonia, but beyond that a plane film is probably not an essential early investigation.

Mild asthma

Diagnosis

Diagnostic difficulties may arise when the patient is seen at a time when he or she is relatively free of symptoms or when symptoms appear only under special circumstances, for example after exercise or the ingestion of certain food.

Home monitoring of peak expiratory flow

The diurnal variation of peak flow is seldom more than 8 per cent in normal individuals. Differences of more than 20 per cent in values obtained at any time during the day is strongly suggestive of asthma.[9] Measurement is easy, requiring very little patient skill and the most simple of mechanical devices. Frequent measurement of peak flow at home is useful not only in making a diagnosis of asthma, but also in monitoring the response to treatment.

Exercise testing

All asthmatics develop bronchospasm following exercise. An exercise test can therefore be particularly valuable where the diagnosis of asthma lies in doubt. The test is simple to perform and inherently safe. What is more, it is relevant to the athlete because it provides clues to performance impairment.

Slow release bronchodilators such as aminophylline should be stopped 24 h beforehand. No bronchodilator should be inhaled on the morning of the test. It is important to ensure that those taking inhaled steroids are allowed to continue their medication. Failure to do this may result in a severe attack. A change into light clothing and running shoes is a sensible arrangement. Baseline lung function tests are measured immediately before the run and should include either FEV_1 or PEF.

The subject should run outdoors as hard as possible either over a set distance or for a set time. Exercise for 3 to 4 min is usually sufficient. Bronchoconstriction is maximal between 5 and 10 min after stopping running. Measurements of FEV_1 or PEF need only be made at 5, 10, and 15-min intervals after exercise. Normal subjects show a transient rise in PEF due to bronchodilatation and then a return to pretest levels. A significant fall at any time is abnormal, and a fall of 15 per cent or more indicates a diagnosis of exercise-induced asthma. Recovery within an hour is usual, but can be hastened by inhaling a bronchodilator.

Diagnostic value of flow loop spirometry

Two important clues to the diagnosis of exercise-induced asthma are available from the expiratory flow loops. The first is an abrupt decay in flow during the forced expiratory manoeuvre which results in 'peaking' of the expiratory loop. This may be apparent even before exercising. The second is sagging of the downward portion of the loop during expiration which is attributed to airways collapse induced by the rise in intrathoracic pressure with bronchospasm. These changes are amplified by increasing bronchoconstriction so that fall in the mid-expiratory flow F_{50} and F_{25} is proportionately greater than the fall in FEV_1. For example, in Fig. 3 FEV_1 falls by 40 per cent in response to exercise, while F_{50} and F_{25} fall by 60 per cent and 65 per cent respectively.

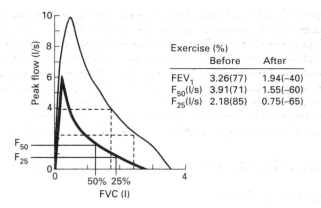

	Exercise (%)	
	Before	After
FEV_1	3.26(77)	1.94(−40)
F_{50}(l/s)	3.91(71)	1.55(−60)
F_{25}(l/s)	2.18(85)	0.75(−65)

Fig. 3 Diagnostic value of flow loop spirometry. The flow loop is performed with the subject breathing out forcibly from full inspiration to full expiration. This volume measures the FVC (litres). The flow (l/s) is zero at the commencement of the expiratory effort and again at the end, but reaches a maximum, i.e. PEF (l/s) almost immediately. This example of exercise-induced asthma in a track athlete shows the flow loops measured before and 10 min after vigorous running. It illustrates the typical changes with fall in both PEF and FVC of greater than 15 per cent. However, in addition there are two important clues to the diagnosis which are only apparent from the flow loops. These are peaking and sagging of the downward limb. The sagging results in a larger percentage fall in flow than that in FEV_1 when both 50 per cent and 25 per cent of FVC remain to be expired (F_{50} and F_{25} respectively). In this instance FEV_1 fell by 40 per cent while F_{50} fell by 60 per cent and F_{25} by 65 per cent.

EFFECT OF ASTHMA ON LIMITING AEROBIC POWER

High aerobic power is desirable in endurance events. By convention, one measure of aerobic power is the rate of oxygen consumption V_{O_2}. Any increase in V_{O_2} must be matched by an increase in ventilation VE (Fig. 4). However, at maximum oxygen consump-

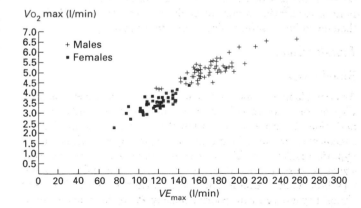

Fig. 4 Relationship between minute ventilation and oxygen consumption. Expired ventilation per minute (VE l/min) was measured at the point when oxygen consumption was maximal (V_{O_2}max l/min) in 99 Olympic class athletes. These group data show that, for individuals with normal lung function, a high V_{O_2} can only be achieved if the minute volume is also large ($R = 0.94$). Any impairment in ventilation such as occurs in asthma will cause a fall in the minute volume. This may also result in a fall in maximal oxygen consumption. (Data from the British Olympic Medical Centre.)

$$V_{O_2}max = (VE\ max \times 0.0263) + 0.4374\ (R = 0.94)$$

tion V_{O_2}max ventilation can be increased still more by voluntary effort. For this reason it is assumed that the lungs do not present any limitation on aerobic performance. In asthma, however, ventilation is impaired and so this assumption no longer holds true.

Ventilation is expressed as the product of the breathing rate and breath volume measured over 1 min, i.e. the minute volume. The maximum breathing frequency reached by young adults during intense exercise is relatively constant, lying between 50 and 60 breaths/min. The relationship between FEV_1 and minute volume is linear. Therefore maximum exercise ventilation MVV can be estimated from the formula

$$MVV = FEV_1 \times 35$$

This formula has been shown to underestimate maximum achievable ventilation in patients with severe obstructive lung disease, say with FEV_1 below 1 l.[10] An analysis of 57 Olympic athletes suggests that for those with normal lungs, the estimate is, if anything, a little high (Fig. 5): $MVV = (FEV_1 \times 30) + 23$. Any fall in FEV_1 will necessarily result in a fall in MVV and hence in aerobic power.

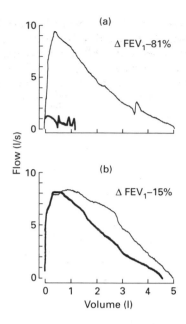

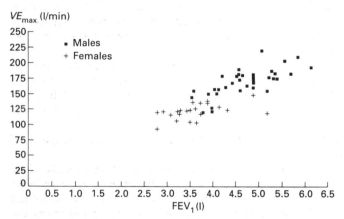

Fig. 5 Relationship between minute ventilation and FEV_1. Expired ventilation per minute (VE l/min) was measured in 57 Olympic athletes during a maximal exercise test and plotted against FEV_1. From these data (British Olympic Medical Centre)

VE (l/min) = $FEV_1 \times 30 + 23$ ($R = 0.82$)

Fig. 6 An example of severe bronchoconstriction induced by exercise in an athlete with a clear history of sleep disturbance. Flow is plotted against volume in a forced expiratory manoeuvre from full inspiration to full expiration. Two expiratory efforts are shown; one measured before exercise (thin line) and the second around 5 min after a 3 min run (thick line) at a time when bronchoconstriction was maximal. Before exercise, peak flow measured 9 l/s. After exercise, peak flow measured only 1 l/s, a fall of more than 80 per cent (pretreatment, (a)). Oscillations in flow during expiration are caused by coughing. The athlete was given prednisolone 30 mg daily for 2 weeks. Improvement was maintained with beclomethasone dry powder 400 mcg twice daily. A month later the exercise test was repeated under identical conditions (post-treatment (b)). There was no change in peak flow but, more important, pretest FEV_1 had risen 20 per cent from 4.031 to 5.31. The exercise test was still marginally abnormal with a small fall in FEV_1 post-exercise and a larger fall in mid-expiratory flow. However, the athlete was sufficiently improved to win a medal at a major international competition just 2 months later.

Therefore performance may be impaired in the asthmatic in three ways: first if FEV_1 falls during the course of the event, second if FEV_1 is below the predicted normal at the commencement of competition, and third if bad asthma interferes with training. These points are well illustrated in the following examples.

Case 1

A middle-distance runner complained of difficulties whilst training. His coach asked him to sprint 100 m, rest for 10 s, and then sprint again, and to do this repeatedly. After the third and fourth repeat the athlete began to experience difficulty in recovering his breath during the 10-s rest. Other members of the squad were unaffected. Direct questioning revealed that for months he had been waking in the early hours of the morning with tightness in the chest. He obtained relief by getting out of bed and passing the rest of the night sitting in an upright chair.

An exercise test during which the athlete ran for 3 min resulted in a fall in both FEV_1 and FVC of more than 80 per cent (Fig. 6).

He was prescribed prednisolone 30 mg daily for 2 weeks and then beclomethasone dry powder 400 μg twice daily as maintenance therapy.

A month later he was sleeping through the night without difficulty and had experienced no further problems with his training sessions. The exercise test was repeated under similar conditions, this time with a fall in FEV_1 of only 12 per cent. Both his improved sleep pattern and his ability to train better resulted in a marked improvement in self-confidence. Eight weeks after the commencement of effective treatment, he won a medal at a major international competition.

Case 2

An elite cyclist complained of a persistent cough that troubled him during his races. Several course of antibiotics had not helped. He gave no history of waking at night, nor of wheeze or tightness in the chest. However, questioning about the timing of his cough was more revealing. He was an endurance competitor, racing over distances in excess of 100 miles. During such races, tactics are all important and consist of a series of sprint attacks. It was following these sprints that coughing was most troublesome.

An exercise test proved that the competitor had exercise-induced bronchoconstriction with a fall in FEV_1 of 31 per cent (Fig. 7). He was treated with inhaled corticosteroids only, using beclomethasone dry powder 400 mcg, twice daily. Two weeks later the exercise test was repeated and showed no bronchoconstriction after exercise. The cough cleared within 4 days of commencing treatment.

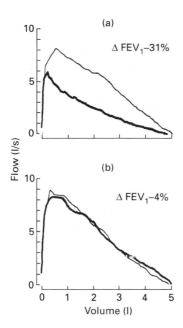

Fig. 7 Exercise-induced asthma. A pursuit cyclist complained of coughing during the sprints but gave no other history to suggest exercise-induced asthma. However, the exercise test is clearly positive with a fall in peak flow of 29 per cent from 8.5 to 6.0 l/s after a 3-min run. FEV_1 also fell by 30 per cent from 4.4 to 3.0 l. It should also be noted that the 'peaking' of the expiratory loop and pressure-dependent airways collapse as exercise asthma develops (a). In this example, although the fall in peak flow was 29 per cent, the fall in F_{50} was 54 per cent. (b) Following treatment with beclomethasone dry powder 400 mcg twice daily his symptoms cleared.
The same test performed 3 weeks later showed no fall in FEV_1 after exercise.

TREATMENT OF ASTHMA

General principles

Medication is directed towards treatment of the two essential pathophysiological elements of asthma: inflammation of the bronchi and bronchial constriction due to hyper-reactivity of bronchial smooth muscle. Corticosteroids are the most effective anti-inflammatory agents available, while β-receptor agonists remain the most potent bronchodilators. Other drugs such as the methyl xanthenes and atropine also bronchodilate, but without stimulating the β receptor. β-Agonists provide only short-term symptomatic relief, but the key to successful treatment is suppression of the inflammatory response with inhaled steroids.

Remission from severe asthma is induced with systemic steroids given either by mouth or by intravenous injection. Prednisolone 30 to 60 mg daily for 10 to 14 days is usually adequate. Nebulized β-agonists form a useful adjunct but the steroids are essential. Remission is then maintained using inhaled steroids with inhaled β-agonists providing symptomatic relief only if required.

Oxygen

All patients with severe asthma become hypoxic; indeed, hypoxia is probably the only cause of death. Oxygen saturation cannot be estimated by the appearance of the mucous membranes, and arterial gas measurement is essential. High concentration oxygen (up to 60 per cent) is safe for all below the age of 50 years. Only patients with asthma complicating chronic air flow obstruction may need any reduction in oxygen concentration.

Corticosteroids

Corticosteroids form the basis of the management of asthma and are considered to be first-line therapy. Remission of an acute severe attack is induced with systemic medication in doses around 40 to 60 mg by mouth daily for 2 to 3 weeks. Oral medication is banned. The athlete may not engage in competition while taking steroids in tablet form. Treatment can be stopped abruptly without fear of adrenal insufficiency unless the patient has been taking systemic steroids long term. Remission is then maintained with inhaled therapy. Most patients can be controlled with a twice or three times daily regimen. For those with persisting severe asthma in whom treatment with inhaled steroids has failed, regular oral medication may be the only recourse despite the risk of side-effects.[11]

In contrast with oral medication, inhaled steroids cause few systemic side-effects. Adrenal suppression is detectable in children with a dose exceeding $400\,\mu g/m^2.day$[12] and in adults with dosages above $1600\,\mu g$ daily.[13] Bone loss from the lumbar spine is also recorded.[14] However, these effects are rare with the modest doses which are adequate for the great majority of patients.

Local side-effects occur on doses above $1600\,\mu g$ daily. They include soreness of the tongue and throat, retrosternal pain on exertion due to tracheitis, dysphonia, and auropharyngeal candidiasis (thrush). Soreness is eased by gargling and rinsing the mouth with saline for 1 min immediately after medication. Candida can be cleared by sucking antifungal lozenges such as nystatin or amphotericin four times daily for 7 to 10 days. Persistent or relapsing infection with candida may require longer periods of treatment. The dysphonia improves when the inhaler is stopped.

Adrenoceptor agonists

β-Agonists should not be used to treat severe asthma without the addition of corticosteroids. Oral medication offers few advantages over inhaled therapy and causes considerably more side-effects. It is banned by the International Olympic Committee (IOC) Medical Commission: an athlete may not engage in competition while taking β-agonists in tablet form. The intravenous route may be preferred in an acute attack, particularly when inhalation is difficult due to respiratory distress (see IOC Doping Rules). Drug administration by nebulizer delivers a larger dose than administration by metered aerosol. Particle size is important. A diameter of between 2 and 5 μm is needed to ensure deposition beyond the central airways.

Inhaled β-agonists act very rapidly, usually reaching maximal effectiveness within 15 min. In contrast, corticosteroids require 2 to 4 h to take effect. One of the few advantages of β-agonists is their speed of action, but they have two major disadvantages. First, unlike corticosteroids, they play no part in alleviating the underlying cause of asthma, namely inflammation of the airways. Second, they have a short duration of action, seldom more than 6 h. A patient needing medication more than four times a day requires the addition of inhaled steroids. Salmeterol is the first of a new generation of longer-acting compounds which are likely to be effective when taken twice or even only once daily.

Side-effects are common but are readily tolerated. They include tremor, headache, and sinus tachycardia. Tremor is dose related and almost invariably occurs with oral medication. The newer β-agonists are highly selective and ventricular arrhythmias are no longer a problem.

Anticholinergic agents

Atropine is a cholinergic receptor antagonist which inhibits the constrictor actions of the vagus nerve on bronchial smooth muscle. The vagus nerve also supplied the goblet cells which secrets mucus. Therefore vagal inhibition with atropine dries secretions in the lung and this can sometimes be an advantage. However, the systemic side-effects of atropine are such that its use in the treatment of asthma is unpractical.

Synthetic atropine-like drugs such as ipratropium are relatively free of atropinic side-effects and are well tolerated. Dry powder, liquid, and metered dose formulations are available. Ipratropium is a less effective bronchodilator than β-agonists in younger people, but may be useful when excessive mucus production is a problem or as adjuvant therapy with β-agonists. The drug is slower in action than β-agonists, with full effects developing over an hour or more. The action is short lasting and the drug must be taken three to four times daily.

Sodium cromoglycate (cromolyn)

Sodium cromoglycate inhibits asthma induced by allergens, exercise, and irritants such as air pollution.[15] The drug is thought to act, at least in part, by inhibiting the release of inflammatory mediators from mast cells and eosinophils, but this does not provide a full explanation of its effects. It is only active when taken prophylactically. Therefore for exercise-induced asthma, cromolyn should be inhaled 15 min before the event. The duration of action is less than 6 h and medication at least three times daily is required for full protection. Sodium cromoglycate is most useful in children with a clear history of allergic symptoms. Dry powder, liquid, and metered dose formulations are available. The drug is almost devoid of side-effects, bronchospasm having been reported only rarely. Cromolyn plays no part in the treatment of severe asthma.

Drug delivery systems for inhaled therapy

There is a trend towards exclusive use of inhaled medicines because effective treatment can be achieved with around a hundredth of the equivalent oral dose. The delivery system and formulation are all important since they decide drug distribution and particle size. Only around 10 per cent of the drug delivered reaches the lungs; the rest is swallowed and ingested. Corticosteroids, cromolyn, β_2-agonists, and atrovent may all be inhaled as either a dry powder or an aerosol. Aerosol delivery is from either a metered dose inhaler or by means of a nebulizer.

Technical difficulties with timing the inspiratory effort are encountered frequently with metered dose inhalers owing to the high velocity of the gases leaving the container. This can largely be overcome by introducing a dead space between the patient and the inhaler by using a plastic device which can vary in size, with the largest (Volumatic or Nebuhaler) serving as a reservoir from which the aerosol is inhaled. Even more elegant systems have been developed to deliver the dry powders. These include the Spinhaler (for cromolyn), the Turbohaler (for budesonide), and the Diskhaler (for beclomethasone).

Other medications

Methylxanthines

Intravenous aminophylline has a place in severe asthma. Initially 250 mg can be given by very slow intravenous injection taking up to 10 minutes to complete the dose. Continuous intravenous infusion (0.5–0.9 mg/kg.h) may be useful in the first 24 h. Death due to ventricular fibrillation may result if the drug is given too quickly, particularly if the patient is hypoxic.[16] Aminophylline must not be given to those already taking a slow-release preparation because of the risk of toxicity.

Slow release oral preparations and suppositories are used mainly for relief of nocturnal breathlessness,[17] but offer no advantage over slow release β-agonists. There are some serious disadvantages. The plasma concentration needed for maximum therapeutic effect is narrow, lying between 10 and 20 mg/l. Monitoring of blood level is essential; anything much above 20 mg/l is toxic. Indigestion and nausea are common and may be intolerable.

Nedocromil

Nedocromil is a cromolyn-like medicine with antiasthma properties. The drug is seldom used alone and is not as effective as cromolyn in preventing exercise-induced symptoms.

Antihistamines

Antihistamines are useful in treating nasal stuffiness, particularly when there is a clear history of allergy (allergic rhinitis), but they have no action in asthma.

Immunotherapy

Immunotherapy (desensitization) is not effective in the treatment of asthma, possibly because the antigens used lack specificity. Advances in treatment may favour the development of vaccines directed towards regulating IgE antibody production[18] (see section on allergic rhinitis).

ANTI-ASTHMA DRUGS AND THE IOC DOPING RULES

Introduction

Treatment of diseases of the respiratory tract provides the greatest possible opportunity to violate unwittingly the IOC Doping Rules. This is due to the fact that a large number of medicines active in the respiratory tract contain banned substances and that in many instances they can be obtained without a physicians prescription. During the 1984 Los Angeles Olympic games 65 per cent of all visits to the doctor by members of the Great Britain team concerned complaints about the nose, throat, or respiratory tract.

The drugs banned in competition are discussed fully in Chapter 3.3.1. Corticosteroids and adrenoceptor agonists must not be given by mouth or by injection. These drugs are only permitted by inhalation. It goes without saying that when asthma is poorly controlled the health of the athlete is at risk. Systemic medication may then be the only recourse, but will result in exclusion from competition.

Corticosteroids

High blood levels of corticosteroids cause euphoria and relieve muscle stiffness, and may therefore enhance performance. Therefore administration of corticosteroids by the oral, intramuscular, or intravenous routes is banned in competition. Inhaled steroids may be used provided that a declaration similar to that made for β_2-agonists is given. Corticosteroids may also be injected locally into a joint space or tendon provided that a similar written declaration is made.

At times when asthma is bad, systemic steroids must be used even though traces may be detectable in the urine for 6 weeks or more after the last oral dose. Effective treatment should never be withheld for fear of making an athlete ineligible for competition. However, there may be instances, subject to special appeal, in which such an individual might be allowed to compete.

A trend in treatment is towards exclusive use of inhaled corticosteroids, none of which appear on the banned list. The inhaled powders are highly effective. Both beclomethasone and budesonide are permitted and effective control is usually achieved with around $400\,\mu g$ given twice daily. As with the use of β-agonists, the athlete must be provided with a letter of authority stating the type of drug in use and the dosage. He must also make this declaration at the commencement of competition (Table 1).

Adrenoceptor agonists

Selective agonists

Selective β_2-agonists are highly effective bronchodilators. They are permitted in competition by the inhaled route only and specifically for the treatment of asthma. A written declaration must be made to the Medical Commission at the commencement of the competition, stating the type of medication and the route by which it is to be given, the dose and frequency of administration, and finally the diagnosis. For example: 'Joe Bloggs is a patient under my care with exercise-induced asthma. His current medication is beclomethasone diproprionate via Diskhaler, $400\,\mu g$ twice daily and salbutamol by Diskhaler $400\,\mu g$ to be inhaled 15 min before the event'. Any competitor using a β_2-agonist without the appropriate authorization risk disqualification.

There is little to choose between the inhaled formulations, but an important exception is fenoterol. This is banned because it is metabolized to parahydroxyamphetamine, a substance thought to have stimulant properties. β-Agonists are effective in attenuating exercise-induced asthma when taken around 15 to 20 min before an event. Salbutamol (albuterol), terbutaline, pirbuterol, and rimiterol are all permitted (Table 1).

Non-selective agonists

Non-selective agonists have both α- and β-stimulant effects. It is the α-stimulant effects that are believed (without proof) to enhance performance by speeding reaction time and stimulating competitiveness through aggression. Adrenaline (epinephrine), noradrenaline (norepinephrine), isoprenaline (isopral), and related compounds with both α- and β-stimulant properties are all banned. Synthetic 'look-alikes' such as ephedrine, pseudoephedrine, norpseudoephedrine, and phenylpropanolamine are also banned.

The α-stimulant activity causes constriction of blood vessels in the upper respiratory tract and nose. Therefore drugs in this group are amongst the more common ingredients to be found in a host of medications for colds and influenza in which blockage of the nose and sinuses is a prominent symptom. Many can be obtained without a physician's prescription; all should be avoided.

Other drugs with bronchodilator properties

There is no restriction on the use of aminophylline and related compounds or on atropine and atropine analogues such as iprat-

Table 1 IOC ruling on drugs used in the treatment of asthma

Drug group	Permitted (other names)	Banned (other names)
Corticosteroids	Beclomethasone* Budesonide*	All oral or parenteral preparations
Adrenoceptor agonists	Salbutomol (albuterol)* Terbutaline* Pirbuterol* Rimiterol*	All oral or parenteral preparations Adrenaline (epinephrine) Noradrenaline (norepinephrine) Fenoterol Phenylpropanolamine Isoprenaline Ephedrine Pseudoepherine Phenylephrine
Anticholinergic agents	Ipratropium	
Sodium cromoglycate	Sodium cromoglycate (cromolyn)	
Methyl xanthines	Theophylline Choline theophyllinate Aminophylline	Caffeine (see Chapter 3.3.1)

*These drugs may only be taken by the inhaled route.

ropium. Neither aminophylline or ipratropium are quite as effective against exercise-induced asthma as β_2-agonists.

Cough suppressants

Cough is a symptom of disease of the respiratory tract. It is important to realize that cough suppressants treat only the symptoms and not the underlying disease. Codeine is used to suppress cough, but like all opiates it is banned in competition. Both codeine and ephedrine-like compounds are commonly found in over-the-counter medicines. In view of the lack of therapeutic benefit and the high risk of giving a banned substance, all cough mixtures should be avoided.

Antihistamines

Certain antihistamines have sedative side-effects which may prove useful in events that require a steady hand such as shooting. For this reason sedative antihistamines are banned in the modern pentathlon (although this is currently the subject of appeal). Antihistamine preparations have no place in the treatment of asthma, but they may be useful in relieving the symptoms of rhinitis which is why they are used in cold cures, cough medicines, and tablets to relieve nasal stuffiness. It is wise to use only preparations with a low incidence of sedation such as terfenadine and astemizole.

Drug combinations

One formulation of sodium cromoglycate (cromolyn) is compounded with isoprenaline which gives it bronchodilator activity. Isoprenaline is a banned substance and this mixture must be avoided. It is distinguished from plain cromolyn by the colour of the lower half of the capsule which is red instead of yellow. Other aerosol inhaler mixtures containing isoprenaline or adrenaline include Bronchilator, Brovon, Duo-Autohaler, Medihaler-duo, and Rybarvin. Duovent is a mixture of ipratropium and the banned drug fenoterol. Unless the constituents are known with confidence, drug combinations should be avoided.

RHINITIS

Introduction

Inflammation of the nasal mucosa leads very quickly to obstruction of the upper airway. Aerobic performance is not impaired because ventilation switches from nose to mouth breathing when the minute volume rises much above 25 l. However, blockage of the nose, nasal discharge, and sneezing is a nuisance, especially in skilled events such as shooting. This common condition has assumed a place out of all proportion to its clinical importance, if only because many of the medicines taken to alleviate the symptoms contain banned substances. Over the years careless use of these drugs has resulted in numerous disqualifications for athletes and the expulsion of at least one team doctor.

Pathophysiology

Allergic rhinitis (hay fever)

Release of inflammatory mediators occurs in precisely the same way as it does in the lung. The immediate result is vasodilatation on the submucosal capillaries of the nose and sinuses, probably due to the action of histamine. The dilated blood vessels occupy more space and the air passages become narrowed. The capillaries then leak into the submucosal space, resulting in oedema and thus worsening the nasal blockage. Finally, the inflammatory component of the allergic response develops, giving rise to chronic symptoms. Chronic obstruction to drainage of the sinuses is a cause of recurrent sinus infection.

Local release of inflammatory mediators, including histamine, causes irritation and itching in the nose, throat, or ears, followed by bouts of sneezing. Vasodilatation of the submucosal vessels of the nose and the maxillary and antral sinuses is followed by a profuse watery discharge. Drainage of the facial sinuses is to the nose. Obstruction of drainage occurs readily, causing a rise in pressure with intense facial pain made worse by bending, coughing, or straining at stool.

Perennial and seasonal causes

Antigens from house mite and household pets cause perennial symptoms, but pollens and fungal spores are abundant only at certain times of the year. Some antigens are specific to certain parts of the world. Ragweed, for example, is an antigen common in the United States but is seldom encountered in Europe. In the northern hemisphere, rhinitis developing early in spring is likely to be due to tree pollens. Allergy to the pollen of Timothy grass peaks around June and July. Fungal spores are shed towards the end of autumn.

Associated conditions

Around 20 per cent of patients with allergic rhinitis also have asthma. However, a rather higher proportion of asthmatics have allergic rhinitis. Chronic inflammation in the nose causes mucosal thickening, leading to prolapse and polyp formation. Polyps are often multiple and may block the nose completely. They often reform after surgical removal. The sinuses drain poorly and may become infected repeatedly. *Haemophilus influenzae* or *Streptococcus pneumoniae* are common pathogens. Infected material drips from the posterior pharynx and may be inhaled, particularly at night. Postnasal drip is thought to be a source of recurrent chest infections or bronchiectasis.

Investigations and diagnosis

Skin testing and measurement of specific IgE antibody

IgE antibody is present in the plasma in minute amounts by comparison with the other antibody classes. Measurement of total IgE is not helpful in diagnosis for several reasons. Amounts vary with age, being immeasurable at birth, reaching a peak of about 20 kU/ml at around 10 years, and then showing a steady decline as age advances. Total serum IgE may also increase up to a hundredfold in eczema or atopy. Finally, high antibody levels do not necessarily correlate with disease.

Measurement of antibody specific to a particular antigen may be useful and is detected by radioimmuno techniques (radioallergosorbent test, RAST). Since there is good correlation between the radio-allergosorbent test and skin tests, measurement of specific IgE antibody by the former is used more often in research than in clinical practice. The results of skin prick tests may be of practical value in antigen identification, particularly if exposure occurs only occasionally or can be avoided.

Treatment

There is a trend towards a greater use of corticosteroids administered antranasally. This type of medication is relatively free from side-effects. The principal action is anti-inflammatory. The results of treatment are not immediate, taking up to 2 weeks to reach full therapeutic benefit. In contrast, decongestants act within minutes by constriction of the nasal blood vessels. Rebound vasodilatation occurs after regular use, ultimately making the symptoms worse. Medicines acting in this way include adrenoceptor agonists banned by the IOC.

Treatment of rhinitis and the IOC doping rules

Corticosteroids

The anti-inflammatory properties of corticosteroids makes them a highly effective treatment. Beclomethasone, betamethasone (flubenisolone), budesonide, and flunisolide are all permitted. Systemic steroids should not be used. A depot injection may render the athlete ineligible to compete for some months (Table 2).

Sodium cromoglycate (cromolyn)

Cromolyn is thought to act in the nose in a similar way to its action in the lung. It is most effective in children with a history strongly suggesting an allergic cause. The drug may be administered as nasal drops, spray, or insufflation, and should be given at least three times a day. The drug is also compounded with a decongestant (oxymetazoline). This mixture is permitted under the IOC rules but should not be confused with its sister preparation for asthma which is a compound of cromolyn and the banned substance isoprenaline.

Antihistamines

Antihistamines can be helpful in relief of symptoms of allergic rhinitis, but they are not effective in the treatment of asthma. Sedation is a major problem, and newer preparations such as terfenadine (Triludan) are relatively free of such side-effects. Currently, drugs with sedative actions, including antihistamines, are banned in the modern pentathlon, although this ruling is under review (see IOC doping rules).

Nasal decongestants

Dilation of the nasal blood vessels contributes to stuffiness by increasing the thickness of the submucosa. Adrenoceptor agonists with α-effects cause vasoconstriction within seconds and afford rapid relief. All are banned in competition. 'Look-alikes' such as oxymetazoline or xylometazoline are permitted.

Immunotherapy (hyposensitization)

Immunotherapy is a device used by some in the management of allergic rhinitis. The principle of treatment is to induce tolerance (desensitization) to a specific antigen by repeated subcutaneous inoculation of ever-increasing concentrations of allergen. There is no agreement on the way in which immunotherapy may be working.

Eleven deaths due to anaphylaxis linked directly with hyposensitization injections were recorded in the United Kingdom over a 10-year period from 1980 to 1990. The Committee on Safety on Medicines has recommended that such treatment should only be carried out where 'facilities for full cardiopulmonary resuscitation are immediately available, and patients should be kept under medical observation for at least 2 hours after treatment'.[19]

With the advent of more effective local therapy, the risks posed by desensitization injections in the treatment of what is a completely benign disease are now unjustifiable. However, the use of more specific antigens such as bee or wasp venom may be acceptable by reducing the incidence of adverse reactions. Vaccination therapy may yet be shown to have a role in treating allergic diseases by down-regulating the production of IgE antibody.

REFERENCES

1. Fletcher CM, Gilson JG, Hugh-Jones P, Scadding JG. Terminology, definitions and classification of chronic pulmonary emphysema and related conditions: a report of the conclusions of a CIBA guest symposium. *Thorax* 1959; **14**: 286.
2. Barnes PJ. A new approach to the treatment of asthma. *New England Journal of Medicine* 1989; **321**: 1517–27.
3. Halperin SA *et al.* Exacerbations of asthma in adults during experimental rhinovirus infection. *American Review of Respiratory Disease* 1985; **132**: 976–80.
4. Boushey HA, Holtzman MJ, Sheller JR, Nadel JA. Bronchial hyperreactivity. *American Review of Respiratory Disease* 1980; **121**: 389–913.
5. Sandford AJ, *et al.* Localisation of atopy and beta subunit of high-affinity IgE receptor (FaeRI) on chromosome llq. *Lancet* 1993; **341**: 332–4.
6. Empey DW, Laitinen LA, Jacobs L, Gold WM, Nadel JA. Mechanisms of bronchial hyperreactivity in normal subjects after upper respiratory tract infections. *American Review of Respiratory Disease* 1976; **113**: 131–9.
7. British Thoracic Society. Acute severe asthma in adults and children: guidelines on the management of asthma. *Thorax* 1993; **48** (supplement): S12–S17.
8. Knowles GK, Clark TJH. Pulsus paradoxus as a valuable sign indicating severity of asthma. *Lancet* 1973; **ii**: 1356–9.

Table 2   IOC ruling on drugs used in the treatment of allergic rhinitis

Drug group	Permitted (other names)	Banned (other names)
Corticosteroids	Beclomethasone* Betamethasone* (flubenisolone)* Budesonide* Flunisolide*	Any oral or parenteral preparation
Sodium cromoglycate	Sodium cromoglycate (cromolyn)	
Antihistamines	Terfenadine	All except terfenadine currently banned in modern penthalon
Decongestants	Xymetazoline Xylometazoline	All adrenoceptor agonists banned for the use of asthma (see Table 1)

*These drugs may only be taken by the intranasal route.

9. Hetzel MR, Clark TJH. Comparison of normal and asthmatic circadian rhythms in peak expiratory flow rate. *Thorax* 1980; **35**: 732–8.

10. Spiro SG, Hahn HL, Edwards RHT, Pride NB. An analysis of the physiological strain of submaximal exercise in patients with chronic bronchitis. *Thorax* 1975; **30**: 415–25.

11. British Thoracic Society. Chronic asthma in adults and children; guidelines on the management of asthma. *Thorax* 1993; **48** (supplement): S3–S11.

12. Priftis K, Milner AD, Conway E, Honour JW. Adrenal function in asthma. *Archives of Diseases of Childhood* 1990; **65**: 838–40.

13. Toogood JH. Complications of topical steroid therapy for asthma. *American Review of Respiratory Disease* 1990; **141**: S89–96.

14. Reid DM, Nicoll JJ, Smith MA, Higgins B, Tothill P, Nuki G. Corticosteroids and bone mass in asthma: comparisons with rheumatoid arthritis and polymyalgia rheumatica. *British Medical Journal* 1986; **293**: 1463–6.

15. Harries MG, Parkes PEG, Lessof MH, Orr TSC. Role of bronchial irritant receptors in asthma. *Lancet* 1981; **i**: 5–7.

16. Sessler CN, Cohen MD. Cardiac arrhythmias during theophylline toxicity: a prospective continuous electrocardiographic study. *Chest* 1990; **98**: 672–8.

17. Rhind GB, Connaughton JJ, McFie J, Douglas NJ, Flenley DC. Sustained release choline theophyllinate in nocturnal asthma. *British Medical Journal* 1985; **291**: 1605–7.

18. Stanworth DR, Jones VM, Lewin IV, Nayyer S. Allergy treatment with a peptide vaccine. *Lancet* 1990; **336**: 1279–81.

19. Committee on Safety of Drugs. Desensitising vaccines. *British Medical Journal* 1986; **293**: 948.

3.2.3 Skin disease and sport

SALEEM K. GOOLAMALI

INTRODUCTION

The skin prevents the penetration of foreign material and radiation, resists mechanical shocks, regulates temperature, and mediates sensation. In sport all these functions are stressed to their limit, and it is not surprising that a variety of cutaneous disorders arise because of infection, inappropriate climatic conditions, injury, increased perspiration, or indeed insect bites. This chapter deals primarily with those entities which are commonly encountered (Table 1).

FUNGAL INFECTION

Athlete's foot (tinea pedis)

Athlete's foot (Plate 1) is a euphemism so commonly used that it tends to be applied to virtually any eruption on the feet. The condition has little relationship to athletic prowess, but is more common amongst those who bathe communally,[1] sportsmen,[2] students, and members of the armed forces. Migration of labour, wartime troop movements, and increased intercontinental travel are thought to have influenced the spread of the most commonly encountered dermatophyte responsible for most superficial ringworm infection, the anthropophilic fungus *Trichophyton rubrum*.

The most common form of tinea pedis, the interdigital variety (trichophyton interdigitale) (Plate 2), causes perennial peeling with maceration and fissuring, usually between the third and fifth toes. In warmer weather multiloculated vesicular lesions (podopompholyx) may form between the toes and extend to the arch of the foot. Tinea pedis is most common in the shod populations of the world, indicating the effect of shoes or 'trainers' in producing the incubatory environment that the fungus (*T. rubrum*, *T. mentagrophytes*, or *Epidermophyton floccosum*) thrives upon. All these organisms may invade the toenails, which become discoloured, brittle, thickened, or onycholytic. The most frequent complication of tinea pedis is secondary infection with pyogenic organisms. This may start within the fissures between the toes, or in vesicles, and result in cellulitis, lymphangitis, or inguinal lymphadenopathy. The cause of many a recurrent cellulitis has been missed because the feet have not been examined closely enough.

The differential diagnosis would include a foot dermatitis, which may be allergic in origin and result from dye from shoes or socks, chemicals in shoe leather, or rubber compounds within the shoes. Unlike tinea pedis, the eruption is symmetrical, generally spares the interdigital spaces, and affects mainly the dorsum and sides of the feet. Localized or recurrent pustules over the soles

Table 1 Common skin disorders in sport

1. **Skin infections**
   Fungal infection
   Bacterial infection
   Viral infection
2. **Disorders as a result of injury**
   Acute
       Blisters
       Calcaneal petechiae (black heel, talon noir)
       Jogger's nipples
   Chronic
       Callosities and corns
3. **Climate related skin disease:**
   Sunlight
   Acute
       Sunburn
       Photosensitivity
   Chronic
       Keratoses, premature skin ageing, melanoma
   Heat
       Intertrigo
       Miliaria
   Cold
       Perniosis
       Raynaud's phenomenon
4. **Exacerbation of pre-existing skin disease**
   Acne
   Psoriasis
   Atopic dermatitis
5. **Insect bites and stings**

may be due to pustular psoriasis, and careful examination of the rest of the skin is then warranted.

Treatment

As a general rule acute inflamed 'wet' forms of superficial fungal infection should be treated with a combination of soothing soaks and topical antifungals. The soaks should be tepid and contain potassium permanganate (1: 10 000 dilution), and the feet should be immersed in the solution for 15 min three times daily. After soaking, the affected areas are treated with any of the available broad spectrum antifungals, such as clotrimazole, miconazole, econazole (all imidazoles), or ciclopirox olamine. In countries where treatment cost may influence therapy, a modified Whitfield's ointment (6 per cent benzoic acid and 3 per cent salicylic acid in a vanishing cream base)[3] may be a suitable alternative. The latter preparation, though effective against tinea, has a less pronounced action against Candida or Gram-positive bacteria.

Nail infections (tinea unguium)

The toenails are infected some four times more commonly than the fingernails (Plate 3). The curious phenomenon of the sparing of some nails and the asymmetrical involvement of nails that are diseased remains unexplained. The diagnosis should be confirmed by examination of a deep scraping of the nail plate or nail clipping under 20 per cent potassium hydroxide solution and fungal culture (nail specimen placed on Sabouraud's agar for 2 weeks). If the organism is susceptible, the treatment of choice has been oral griseofulvin. However, with this drug fingernail infections require approximately 6 months of therapy whilst toenails require a year or longer. Some 20 to 40 per cent of affected nails fail to respond, and relapse rates of 40 to 70 per cent are recorded for toenails.[4]

Ketoconazole can cause transient elevation of serum liver enzymes and symptomatic hepatic abnormalities.[5] Its incidence has been reported as 1 in 10 000 patients, which limits the usefulness of this drug in onychomycosis although it is as effective as griseofulvin.[6]

Results with a new allylamine drug, terbinafine, in the treatment of dermatophyte nail infections are more encouraging.[7] The drug appears to achieve mycological cure more readily than griseofulvin, and of 20 patients followed up for a year, all remained in mycological remission. In the treatment of chronic tinea pedis, 6 weeks of treatment with terbinafine produced 100 per cent recovery compared with 42 per cent with griseofulvin.[8]

Tinea versicolor

Tinea versicolor (Plate 4) is caused by the saprophytic yeast *Pityrosporon orbiculare*, which develops into its parasitic fungal form *Malassezia furfur*.[9] This transformation may be provoked by humidity or ambient heat, as in the tropics, as a result of exercise or malnutrition, or by systemic disease such as uncontrolled diabetes mellitus or undiagnosed Cushing's syndrome. As the organism grows in the keratin layer of the skin, Scotch tape stripping of the affected stratum corneum provides a quick and painless specimen for microscopy examination. The main disadvantage of tinea versicolor is cosmetic. The skin affected develops light tan scaly patches, which become hypopigmented on exposure to sunlight.

Treatment

All the effective topical antiversicolor agents, ranging from the well-tried Whitfield's ointment to the more recent selenium sulphide shampoo[10] to the very modern broad spectrum antifungals suffer from the disadvantage that relapse after treatment is common.[11] Treatment is usually prolonged and the preparation is applied to skin beyond the apparent infection. Wood's light examination of the skin helps to delineate the extent of infection. Griseofulvin is ineffective in tinea versicolor, but in a recent study the triazole antifungal itraconazole (100 mg twice daily for 7 days) was found to be extremely effective in eradicating the organism.[12]

BACTERIAL INFECTIONS

Impetigo and folliculitis may result from minor trauma to the skin, such as abrasions or insect bites. High ambient temperature, humidity, low altitude, and poor hygiene favour development and transmission of these infections. Streptococcal impetigo has spread amongst footballers[13] and those playing North American football.[14] *Staphylococcus aureus* has been claimed as the more common aetiological agent in impetigo,[15] though advocates of the streptococcal cause consider the staphlococcus to be a secondary invader and draw support from the fact that the staphylococcus is often isolated from the nose and unaffected skin long before it is isolated from lesions of impetigo. Once the lesions of impetigo are established, they spread with remarkable speed on both the patient and those within close contact.

Treatment

The patient will need to be temporarily isolated and topical and systemic antibiotic therapy instituted. The crusts are gently soaked off four times daily with compresses of potassium permanganate solution (1: 10 000 dilution), and a thin film of antibiotic cream (neomycin, tetracycline in adults, or mupirocin) is applied after each removal. The entire skin, scalp included, should be washed with an antiseptic soap, and a systemic antibiotic (erythromycin 250 mg four times a day for 10 days) instituted, unless antibiotic sensitivity dictates otherwise.

Pitted keratolysis

Pitted keratolysis, which is believed to be caused by a species of Corynebacterium, presents with numerous superficial erosions on the soles of the feet in association with hyperhidrosis. Patients involved in sport are particularly susceptible, and maceration of the skin and malodour are frequent associated findings. Control of sweat production with antiperspirants containing aluminium chlorhexahydrate and control of bacteria with antibiotic ointments are effective.

Otitis externa

Otitis externa can become recalcitrant and presents with scaly erythema of the concha and meatus. Trauma to the skin—scratching and picking—leads to secondary infection with bacteria, often Gram-negative enterobacteriaceae. Patients with seborrhoeic dermatitis are particularly prone to develop otitis externa. It is common in swimmers and ear plugs may exacerbate the condition. Treatment is topical and should be directed at the cause as well as the secondary infection. Gentle debridement of the meatus is useful, but overzealous use of cotton-wool sticks can perpetuate the problem. Application of antibiotic–corticosteroid cream for a week is usually sufficient. Resistant cases should be referred for specialist advice.

VIRAL INFECTIONS

Warts

Warts are epidermal tumours caused by infection by the papova group of viruses. The virus may be found in the surface layer of the wart, and so spread of infection can occur if infected debris comes into contact with abraded skin (indirect contact). Infection may also occur as a result of direct contact or autoinoculation as might occur with nail biting or shaving. Warts may be more common in callouses which develop in sport.[16]

It is argued that as common warts (verruca vulgaris) resolve spontaneously sooner or later, treatment should be deferred and the wart simply covered with collodion, plaster, or a clear nail varnish to prevent spread. However, in most cases warts are socially unacceptable or painful and require treatment.

Treatment

1. Patients should be advised that no certain or quick cure exists for warts, and no matter which treatment is chosen, the time for cure can extend to 3 months or longer.
2. Many wart paints, most containing salicylic acid, are available. One formula consisting of salicylic acid and lactic acid in flexible collodion (Salactol, Duofilm) is useful. It is important that the paint is applied to the lesion itself, that the surrounding skin is protected with soft paraffin, and that the surface of the wart is abraded with a pumice stone or manicure emery board prior to paint application. In one study,[17] 67 per cent of hand warts had cleared after 12 weeks. Plantar warts (Plate 5) and those in the anogenital region may be treated with compound Podophyllin paint containing 15 per cent Podophyllum resin. With genital warts, the paint is left on for a maximum of 6 h and then washed off with ordinary soap and water. Very occasionally a malignant melanoma (Plate 6) may be confused with a plantar verruca, and it is vital that any lesion out of the ordinary (pigmented, bleeds with gentle trauma, or rapid increase in size without hyperkeratosis) should be referred to a specialist.
3. Liquid nitrogen ($-196\,°C$) has replaced carbon dioxide snow as the cryotherapy treatment of choice for warts apparently resistant to paint therapy and for lesions on the face. The warts are frozen until a halo appears around the base of the lesion, 5 to 20 s depending on the size of the lesion. One series[18] showed that 69 per cent responded to frequent light applications of liquid nitrogen.
4. In selected patients, for example those with large warts on the face or groin, curettage and cauterization under local anaesthetic may be the treatment of choice.

Molluscum contagiosum

Molluscum contagiosum (Plate 7) is a pox virus infection which may be spread by person-to-person contact. Occasional epidemics in institutions have been recorded. The incubation time from inoculation to clinical appearance of lesions is from 2 to 8 weeks.[19] Techniques for eradication of molluscum include the following:

1. light freezing of lesions with liquid nitrogen;
2. touching the lesions with 25 per cent podophyllin in 95 per cent ethyl alcohol twice weekly;
3. puncturing the surface of lesions with a sharpened orange stick to release the cheesy material within the centre of the lesion.

Herpes simplex

Herpes simplex (Plate 8), which is caused by Herpes virus hominis, is acquired by close personal contact. In rugby players it may spread by body contact,[20] and the close skin-to-skin contact of wrestlers occasionally results in what has poetically been described as 'herpes gladiatorum'.[21] In the vesicular phase and until the crusts have separated, the sufferer should avoid sports which could involve physical contact.

Treatment

Topical acyclovir treatment (5 per cent acyclovir cream five times a day) has a limited role in the treatment of herpes simplex infection. Once ulcers or vesicles have formed, treatment is unlikely to be helpful.

Patient-initiated treatment, for example drying agents such as alcohol applied during the prodrome, is as likely to help as not. Herpes simplex is a latent infection of nerve root ganglia. When viral replication occurs the virus moves from the ganglia to the peripheral nerve and finally to the skin or mucous membrane. Even if topical therapy is started as soon as symptoms appear, it is conjectural if the disease course is shortened though the duration of viral shedding may be reduced.

DISORDERS AS A RESULT OF INJURY

Blisters

Heat and humidity allied with unaccustomed localized friction result in foot blisters. They are best treated by sterile aspiration, leaving the overlying skin intact. The area can then be treated with an antibiotic cream or an antiseptic such as povidone–iodine.

Calcaneal petechiae (black heel, talon noir) are specks of blood in the skin which form as a result of rupture of dermal capillaries. They are common in sports such as basketball, squash, or football, where frequent 'stop–go' movements are the norm, and these cause a shearing effect on the skin.

No treatment is required apart from reassurance, though it is paramount to distinguish the condition from malignant melanoma. Chinese weightlifters are reported to have developed 'black palms' caused by haematoma from friction and pressure.[22]

Jogger's nipple

Jogger's nipple is a painful and often fissured dermatitic eruption over the nipples produced by friction from unyielding vests or T-shirts and may also occur in women who do not wear brassieres whilst running or jogging.[23] Treatment is with an emollient, such as petroleum jelly, or a mild topical steroid, such as 1 per cent hydrocortisone cream, applied two or three times daily for a few days.

Callosities and corns

Callosities (Plate 9) are well-defined plaques of hyperkeratosis which result from repeated friction of the skin. Corns are localized callosities over bony prominences. Both may be caused by unsuitable footwear or abnormal weight bearing or both.

First the abnormal mechanical stress needs to be corrected—rear foot varus may be present—and paring of the corn or the callosity with a blade, salicylic acid plasters, or curettage of the central core of a corn may help. Occasionally, a bony spur or exostosis may be underlying, and the advice of an orthopaedic surgeon should then be sought.

CLIMATE-RELATED SKIN DISEASE

Effects of sunlight

Sunburn occurs all too frequently in holiday travellers, though modern athletes recognize the need for high sun protection factor sunscreens to prevent acute sunburn and delay the damaging effects of chronic sun exposure. Individuals with red hair and freckles are by far the most susceptible to burning[24] and need to be particularly vigilant.

Photosensitivity (Plate 10) reactions may occur from systemic or topical drugs, perfumes, or cosmetics. Even brief exposure to sunlight in warm or cold weather may cause intense reactions in patients who have used photosensitizing agents, and this sensitivity can persist long after cessation of the offending agent. The most frequent reaction is phototoxicity, resulting in a 'sunburn' reaction with erythema, blisters, desquamation, and peeling. Less often a photoallergic response occurs, resembling contact allergy with an immediate wheal and flare reaction or a delayed eczematous eruption. Some drugs may cause both phototoxic and photoallergic reactions. Drug photosensitizers include antimicrobials (demeclocyline, nalidixic acid, sulphonamides), antiparasitic drugs (bithionol), antipsychotic drugs (phenothiazines), diuretics (thiazides), oral hypoglycaemics, and non-steroidal antiinflammatory drugs. Topical agents include bergamot oil, oils of citron, lavender, lime, and sandalwood (used in many perfumes and cosmetics), musk ambrette (used in perfumes), and 6-methylcoumarin (used in perfumes, aftershave lotions, and sunscreens).

Polymorphic light eruption

Occasionally a non-drug-, non-chemical-related photosensitivity eruption may form, and the most common of these is known as polymorphic light eruption. This can present with a combination of skin lesions ranging from macules to vesicles, which are very pruritic. It is delayed (hours or days after exposure) photodermatosis which is confined to exposed areas. The history points to the diagnosis, but it is important to view the rash and exclude lupus erythematosus. Deliberate exposure to sunlight is the most obvious way of reproducing the eruption. Polymorphic light eruption resolves within a few days and at most in 1 to 2 weeks. The differential diagnosis also includes solar urticaria which forms during or shortly after exposure to sunlight and lasts for a few hours and rarely more than a day.

Chronic sun exposure induces premature skin ageing, abnormal pigmentation, and skin cancers, including melanoma. Both UVB (wavelength 290–320 nm) and UVA (wavelength 320–400 nm) are harmful. UVA radiation passes through window glass and is least harmful on a dose-for-dose basis. Nonetheless, it produces redness, skin ageing, and skin cancer, but at doses approximately a factor of 1000 greater than that for UVB. On a summer's day UVA comprises 95 per cent of terrestrial UV radiation and UVB the remaining 5 per cent.

Cyclist's melanoma

Cycling is a familiar pastime, and the sport has been cited as an aid to healthy living[25] whilst others have considered it as a 'risk' factor for health.[26] Cycling has been associated with melanoma on an exposed part of the lower limb.[27] Melanoma may be caused by intermittent intense exposure to UV light in youth,[28] and it is proposed that such exposure may occur during cycling when the skin over the calf and thigh receive maximum irradiation as the leg adopts the flexed cycling position. Cyclists should be aware of this potential risk and wear protective clothing and high protective factor sunscreen on areas exposed to sunlight.

Treatment

Acute sunburn is treated with cold-water compresses and analgesics. Local anaesthetics are best avoided as they may have undesirable side-effects. In acute photosensitivity (chemical, drugs, or polymorphic light eruption) topical corticosteroids, antihistamines, and occasionally oral steroids may be needed in the short term. Polymorphic light eruption may also respond prophylactically to psoralen plus UVA (PUVA) if sunscreens are not helpful.[29] Clearly, avoiding sunlight is the best prophylactic. Certainly, curtailing outdoor activities between 11.00 a.m. and 3.00 p.m. when the sun's UV rays are at their strongest is wise. Most clothing acts as a sunscreen, and material which casts a deep shadow when held up to light is best as an antisunburn measure. If this is impractical, then sunscreens with high sun protection factors (> 15) must be used and those that absorb both UVA and UVB are best.

Heat

Intertrigo

Intertrigo (Plate 11) is defined as an inflammatory dermatosis affecting the body folds, in particular the submammary and genitocrural regions. Obesity, sweat, and friction predispose to its development, and the clinical appearances can range from mild erythema to frank dermatitis with secondary bacterial infection with *Staphyloccus pyogenes*, *E. coli*, or Candida. In refractory cases diabetes should be excluded. Avoidance of tight clothing and undue exercise, carefully applied wet dressings in the acute phase, and the application of one of the imidazole preparations alone, or with a topical corticosteroid, controls the condition.

Miliaria

Miliaria (prickly heat) is induced by profuse sweating and results from obstruction of the sweat ducts. As might be expected, its incidence is highest in hot humid conditions and measures to reduce sweat production (cool room, less clothing), discomfort (calamine lotion), and in some cases vitamin C (1 g daily) are helpful.

Cold

Perniosis (chilblains)

Chilblains (Plate 12) arise as an abnormal reaction to cold, which causes constriction of arterioles and venules. They usually affect the fingers and toes, and are self-limiting. Prophylaxis includes warm clothing and regular exercise. UV radiation has been recommended, and some patients derive benefit from UVB therapy sufficient to produce a marked erythema.

Raynaud's phenomenon

In 1862 Maurice Raynaud described the syndrome now named after him as an 'asphyxia of the fingers'. It is now recognized as digital vasospasm with the familiar triad of signs—pallor, cyanosis, and rubor—provoked by cold or emotional stress.

303

Women are affected five to eight times more frequently than men, and Raynaud's phenomenon may occur in 10 per cent of otherwise healthy females.[30]

Nifedipine—a calcium-channel blocking agent—in doses of 10 to 20 mg four times a day has been found to be useful in Raynaud's phenomenon.[31,32]

EXACERBATION OF PRE-EXISTING SKIN DISEASE

Acne

Acne may be exacerbated by a hot humid environment. Acne-prone troops in the Second World War suffered an exacerbation of the condition—so called 'tropical acne'—when posted to the Far East. It is thought that hydration of the pilosebaceous pores occurs, which correspondingly increases duct blockage and this triggers the development of inflammatory lesions.

Acne 'mechanica' occurs at sites of physical trauma. Friction from head bands and brassiere straps produces acne lesions at areas of contact.

Psoriasis

In sport, injury to the skin is particularly common. In psoriasis, epidermal injury may precede the development of a local psoriatic lesion by 2 weeks, the so-called 'isomorphic or Koebner' phenomenon.[33]

Atopic dermatitis

Many patients find that sweating aggravates their condition. Seasonal exacerbations in spring and autumn are also frequent. Exposure of the skin to irritant chemicals or physical trauma should be avoided as far as possible.

INSECT BITES AND STINGS

Insect bites and stings are rarely life-threatening, but the discomfort they cause can be considerable. Sport is for the most part an outdoor pursuit and to venture out is to 'invite a bite'. Biting insects include blackflies, sandflies, midges, mosquitoes, deer- and horse-flies, ants, bees, and wasps. Blackflies are blood-sucking insects, and a species of blackfly is responsible for the transmission of the nematode causing human onchocerciasis. Sandflies transmit verruga peruana and leishmaniasis. Only a minority of midges are blood feeders. In those sensitized to midge bites, irritation starts immediately and can result in blisters and regional lymphadenopathy. Mosquitoes transmit malaria, yellow fever, dengue fever, and filariasis. Deer-flies transmit loa-loa and tularaemia.

The immediate treatment of an insect bite requires frequent applications of calamine lotion or cold-water compresses. These alleviate itch, and corticosteroid creams can be substituted later. Oral antihistamines are also helpful. The stings from ants, bees, and wasps are not treated similarly. Bee venom contains formic acid, and the bee also leaves part of its sting in the wound so that care is necessary when handling the sting site. A 'home' remedy is to apply sodium bicarbonate to the injured area. For wasp stings, which are alkaline, an appropriate antidote is vinegar or lemon juice. Prophylaxis rests with insect repellants, and those con-

taining dimethylphthalate or diethyltoluamide are the most effective. The latter is more acceptable cosmetically and comes closest to the ideal.

REFERENCES

1. Gentles J, Evans E, Jones G. Control of *Tinea pedis* in a swimming bath. *British Medical Journal* 1974; i: 577–80.
2. Gentles J, Jones G, Roberts D. Efficacy of miconazole in the topical treatment of *Tinea pedis* in sportsmen. *British Journal of Dermatology* 1975; **93**: 79–84.
3. Logan R, Hay R, Whitefield M. Antifungal efficacy of a combination of benzoic acid and salicyclic acids in a novel aqueous vanishing cream formulation. *Journal of the American Academy of Dermatology* 1987; **16**: 136–7.
4. Davies RR, Everall JD, Hamilton E. Mycological and clinical evaluation of griseofulvin for chronic onychomycosis. *British Medical Journal* 1967; ii: 464–8.
5. Jacobs PH, Nall L. Action and safety of ketoconazole: brief literature review. *Cutis* 1988; **42**: 276–82.
6. Zaias N, Drachman P. A method for the determination of drug effectiveness in onychomycosis. *Journal of the American Academy of Dermatology* 1983; **9**: 912–19.
7. Goodfield MJD, Rowell NR, Forster RA, Evans EGV, Raven A. Treatment of dermatophyte infections of the fingers or toe nails with terbinafine (SGF 86–327, Lamisil) an orally active fungicidal agent. *British Journal of Dermatology* 1989; **12**: 753–8.
8. Savin R. Successful treatment of chronic *Tinea pedis* (mocassin type) with terbinafine (Lamisil). *Clinical and Experimental Dermatology* 1989; **14**: 116–19.
9. Roberts SOB. *Pityriasis versicolor*. A clinical and mycological investigation. *British Journal of Dermatology* 1969; **81**: 315–26.
10. Albright SD, Hitch JM. Rapid treatment of *Tinea versicolor* with selenium sulphide. *Archives of Dermatology* 1966; **93**: 460–2.
11. Faergemann J, Fredriksson T. *Tinea versicolor*: some new aspects of etiology, pathogenesis and treatment. *International Journal of Dermatology* 1982; **21**: 8–22.
12. Cauwenbergh G, DeDoncker P. 'The clinical use of itraconazole in superficial and deep mycoses. In: Fromtling RA, ed. *Recent trends in the discovery, development and evolution of antifungal agents*. South Africa: J.R. Prous Science Publishers, 1987.
13. Dorman J. 'Scrum strep'. *New England Journal of Medicine* 1981; **305**: 467.
14. Bartlett P, Martin R, Cahill B. Furunculosis in a high school football team. *American Journal of Sports Medicine* 1982; **10**: 371–4.
15. Coskey R, Coskey L. Diagnosis and treatment of impetigo. *Journal of the American Academy of Dermatology* 1987; **17**: 62–3.
16. Kantor G, Bergfeld W. Common and uncommon dermatologic diseases related to sports activities. *Exercise and Sports Science Reviews* 1988; **16**: 215–53.
17. Bunney MH. A rational approach to the management of warts. *Prescribers Journal* 1974; **14** (6): 118–25.
18. Barr A. *et al*. Transactions of the St. John's Hospital Dermatological Society 1969; **55**: 69.
19. Goldschmidt H, Kligman AM. Experimental inoculation of humans with ectodermotropic viruses. *Journal of Investigative Dermatology* 1958; **31**: 175.
20. White W, Grant-Kels J. Transmission of herpes simplex virus Type 1 infection in rugby players. *Journal of the American Medical Associations* 1984; **252**: 533–5.
21. Becker T *et al*. Grappling with herpes: herpes gladiatorum. *American Journal of Sports Medicine* 1988; 16: 665–9.
22. Izumi A. Pigmented purpuric petechiae. *Archives of Dermatology* 1974; **109**: 261.
23. Levit F. Jogger's nipples. *New England Journal of Medicine* 1977; **297**: 1127.
24. Azizi E, Lusky A, Kushelevsky A, Schewach-Millet M. Skin type, hair

color and freckles are predictors of decreased minimal erythema ultra-violet dose. *Journal of the American Academy of Dermatology* 1988; **19**: 32–8.

25. Anonymous. Cycling and sanitation. *British Medical Journal* 1906: 1: 1429.
26. Clarke C. The bicycle as a risk factor. *Journal of the Royal College of Physicians* 1988; **22**: 92.
27. Williams H, Brett J, DuVivier A. Cyclist's Melanoma. *Journal of the Royal College of Physicians* 1989; **23** (2): 114.
28. Mackie RM. Links between exposure to ultraviolet radiation and skin cancer. *Journal of the Royal College of Physicians* 1987; **21**: 91.
29. Murphy G, Logan RA, Lovell RA, Morris RW, Hawk JLM, Magnus IA.

Prophylactic PUVA and UVB therapy in polymorphic light eruption. A controlled trial. *British Journal of Dermatology* 1987: 531–8.
30. Hines EA, Christensen NA. Raynaud's disease among men. *Journal of the American Medical Association* 1945; **129**: 1.
31. Smith DC, McKendry RJR. Controlled trial of nifedipine in the treatment of Raynaud's phenomenon. *Lancet* 1982; **ii**: 1299.
32. Rodeheffer RJ *et al.* Controlled double-blind trial of nifedipine in the treatment of Raynaud's phenomenon. *New England Journal of Medicine* 1983; **308**: 880.
33. Pedace FJ, Muller JA, Winkelmann RK. The biology of psoriasis. An experimental study of the Koebner phenomenon. *Acta Dermato-venereologica* 1969; **49**: 390–400.

3.2.4 Infections in sports medicine

GEOFFREY PASVOL

Competitive sport requires peak performance, and even minor physical or psychological injury can blunt achievement. Although physical injury remains one of the most important hazard to good athletic performance, infections of many varieties can lead to deterioration in competitive ability for a number of reasons. Whilst some infections may be of a major acute nature, it is often minor or chronic low grade infections that are implicated in the unexplained failure of top athletes. Unfortunately, it is the latter that on a numerical basis are of relatively great importance, and yet our knowledge of them is sadly deficient.

Whilst infections are commonly associated with early childhood and old age, a number of factors predispose the predominantly young adult sportsperson to infection. These include reduced immunity as a result of stress and overtraining (see below), close contact with other sportspeople facilitating person-to-person spread, trauma, especially of the skin, foreign travel, and sexual activity. Most of the infections encountered in the physically young and fit are of a minor nature, but their effect on performance may be considerable.[1] One particular fascinating area of recent interest has been the effect of both exercise and psychological stress on the immune system and in turn their effect on the resistance or susceptibility to infection.[2]

EXERCISE STRESS AND IMMUNITY

Moderate exercise

It is believed by many that moderate exercise protects against infection, apart from its many other beneficial effects such as those on the cardiovascular system and the psychological well-being of the individual. Certainly studies of the effect of moderate exercise in mice would indicate enhanced humoral immunity when compared to controls. However, most studies when undertaken in man, vary in the selection and the level of fitness of the subjects under scrutiny, making direct comparison difficult.[3] Most agree that moderate exercise increases the granulocyte count which could result from haemoconcentration, mobilization of the marginal granulocyte pool, or release of these cells from marrow. These effects may in turn be the result of increased catecholamine and cortisol levels observed during exercise. However, whether increased neutrophil counts result in functional improvement is not clear, and in one study the granulocytes obtained from well trained athletes were indistinguishable from those of sedentary individuals. Exercise also appears to increase the circulating lymphocyte count especially T-cells. As with granulocyte function however, there is no conclusive evidence that these raised counts alter host defence to any important extent. Moreover, it has been shown that exercise-induced leucocytosis is short-lived. Plasma levels of α-interferon, interleukin-1, endorphin, and met-enkephalin all increase during exercise. Preliminary work would indicate that moderate exercise has no major effect on immunoglobulin or complement levels. Thus, although a number of changes within the immunological system have been noted in moderate exercise, the basic question of whether these changes contribute to resistance to infection remains unanswered.

Overtraining

Whilst moderate exercise stimulates the immune system, overtraining appears to have a deleterious effect on host immune response and may ultimately be a limiting factor in athletic performance. Overtraining is often blamed by top athletes for frequent persistent colds, sore throats, and influenza-like illnesses which can lead to states resembling the postviral fatigue syndrome. An athlete might miss an entire sports session or give up sport altogether because of such infections.

A number of changes in the immune system have been documented in intense exercise. The salivary immunoglobulins IgA and IgM appear to be suppressed for up to 24 h, and this could perhaps account for the anecdotal statements by athletes that severe exercise increases their susceptibility to upper respiratory tract infections. Similar changes can also occur in students during an examination period. Low resting levels of IgG have also been found in elite ultradistance runners at the end of the season. Natural killer cell activity appears to be decreased for up to 24 h after severe exercise, largely due to a decrease in the percentage of these cells, although on a per cell basis natural killer activity is increased. Whilst T lymphocytes are increased by exercise, there is some evidence the ratio of helper (T_4) to suppressor (T_8) may be reduced, and that in fact these T cells may not be capable of responding normally to mitogens. Low lymphocyte counts

(< 1500 per μl) have been found in marathon runners at rest. Maximal physical exercise was found to reduce bactericidal and adherence capacity of neutrophils and monocytes when compared with controls.[4] Overtrained Olympic athletes were found to have significantly low plasma glutamine levels when compared with controls, and glutamine is regarded as an important substrate for lymphocyte metabolism.[5] Whilst all these observations could account for a temporary susceptibility to infection during overtraining, they provide no information on functional outcome.

There are limited data supporting the argument that strenuous exercise leads to increased susceptibility to infection. In mice infected with coxsackievirus B, for example, enforced exercise increased mortality from 5 per cent in the non-exercised to 50 per cent in the exercised group. Monkeys exercised during the incubation of poliomyelitis had a higher incidence of paralysis than controls. In an outbreak of poliomyelitis in the United States, all nine boys who became ill were participating in strenuous sports. In a study of 150 ultramarathon runners, symptoms of respiratory tract infections were most common in those who had achieved the fastest times and who as a group had the highest weekly training mileages.

Psychological stress

A complicating factor in immunity to infection in sport is psychological stress, particularly when competitive sport is involved, as it is thought to result in immunosuppression.[6] Anecdotes abound of individuals who become ill following stressful situations. Many studies over the last 20 years have indicated that psychological stress and psychiatric illness can compromise immunological function, and it appears that the final common pathway is via stimulation of adrenocortical secretion by adrenocorticotrophin (ACTH) and the sympathetic nervous system with the subsequent release of catecholamines.

Overall, the complex relationship between moderate or excessive exercise, psychological stress, and disease makes it extremely difficult to predict the final outcome of these interacting factors.

SPECIFIC INFECTIONS

The young sportsperson is most susceptible to those infections which are spread mainly by droplets (e.g. upper respiratory infections), via the orofaecal route (e.g. hepatitis A and E), or occasionally by sexual activity (e.g. hepatitis B or HIV). Whilst common bacterial infections generally have few sequelae, viral infections, though initially milder, may produce longer-lasting effects which often compromise athletic performance.

Upper respiratory tract infections

Upper respiratory tract infections can be extremely worrisome to the athlete, particularly when they become recurrent. They are most commonly due to viruses such as enteroviruses (e.g. echovirus and coxsackievirus), adenoviruses, and influenza viruses, and therefore frequent consumption of antibiotics for these infections is often unnecessary. Unfortunately, there are very few cost-effective and rapid methods available to distinguish these conditions from one another, other than by throat culture for β-haemolytic streptococci or a heterophile antibody (Paul-Bunnell) test for infectious mononucleosis. Viral culture and serology in the diagnosis of other causes of an acute upper respiratory tract infection is seldom helpful. In any event, infectious mononucleosis and bacterial pharyngitis can coexist.

Management

Symptomatic relief using analgesics, antihistamines, or decongestants may help in some cases. A throat swab should be taken in all athletes with an upper respiratory tract infection and a positive result can usually be obtained within 24 h. Group A β-haemolytic streptococcal sore throats are amenable to treatment with penicillin, or erythromycin in the case of penicillin allergy. Treatment should continue for at least 10 and possibly 14 days to avoid recurrence.

The major issue which arises with regard to sport is whether an individual with an upper respiratory tract infection should refrain from exercise. The conventional guidance is that in the presence of fever, tachycardia at rest, or severe myalgia or lethargy athletes should not participate in sport. There is still controversy as to whether premature resumption of exercise leads to delayed recovery or a postviral fatigue condition. The occurrence of complications, such as a sudden arrhythmia following exercise during a viral illness, has been overemphasized, and sudden cardiac death in the presence of a viral myocarditis is a rare event. Sudden cardiac deaths during exercise are most commonly due to hypertrophic cardiomyopathy or coronary artery disease.

The glandular fevers

The glandular fevers include a number of conditions grouped together because of similarities in clinical presentation which include sustained fever and glandular enlargement. Amongst the more important are infectious mononucleosis (due to the Epstein–Barr virus), toxoplasmosis, and cytomegalovirus infection. To this list should perhaps be added the primary seroconversion illness of HIV infection, which may be similar in presentation, but is more often accompanied by a rash.

The glandular fevers are important in sportspeople for a number of reasons. Firstly the condition is common in this younger age group in developed countries. Second, the glandular fever may predispose to certain complications which are of particular relevance to sportspeople, e.g. splenic rupture or myocarditis. Finally the glandular fevers have achieved some notoriety as being one group of infections which may lead to a post 'viral' fatigue syndrome.

Infectious mononucleosis

Infectious mononucleosis due to the Epstein–Barr virus is an important and common infection in athletes, affecting mainly adolescents and young adults in developed countries, and young children, often asymptomatically, in developing countries. Spread of the infection is mainly by intimate close contact, and for this reason isolation of proven or suspected cases is unnecessary when it occurs in the context of a large sports gathering.

The most common clinical presentation is a sore throat with fever and generalized lymphadenopathy with or without an enlarged spleen. In a minority of cases a diffuse maculopapular rash may be present. Patients may present with jaundice, a haematological disorder (e.g. thrombocytopenia or haemolytic anaemia), or any neurological disorder ranging from an encephalitis to a peripheral neuropathy.

Complications
There are two complications of this disease which are of particular relevance to athletes.

Splenic rupture
Up to 40 per cent of cases of traumatic splenic rupture have occurred in athletes who have or have subsequently been found to

Plates for Chapter 3.2.3   Skin disease and sport

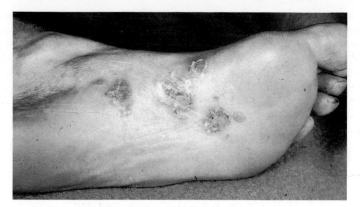

Plate 1   Tinea pedis spreading to the sole of the foot.

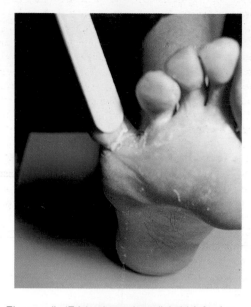

Plate 2   Tinea pedis (*Trichophyton interdigitale*) infection characterized by peeling and maceration of the toe cleft.

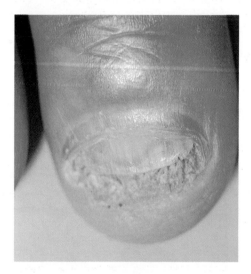

Plate 3   Onychomycosis (tinea unguium) due to *Trichophyton rubrum*.

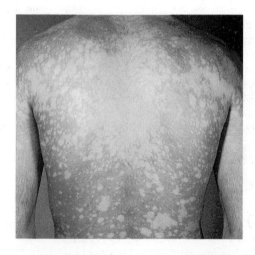

Plate 4   Widespread pityriasis ('tinea') versicolor producing variegated skin pigmentation.

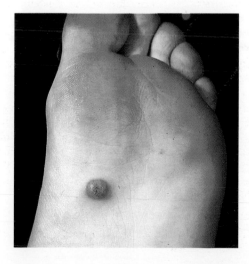

Plate 5   Plantar verruca— well-defined keratotic mass with underlying capillaries.

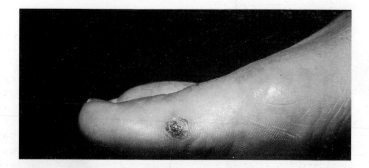

Plate 6   Nodular malignant melanoma exhibiting an irregular border and colour variation.

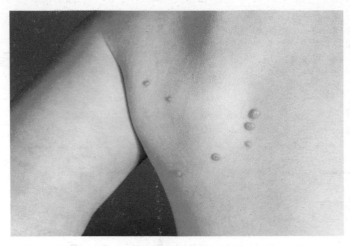

Plate 7   Pink pearly umbilicated papules of *Molluscum contagiosum*.

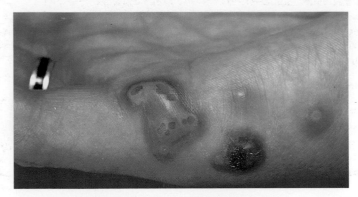

Plate 8   Herpes simplex in a physiotherapist assisting in a sports injury clinic.

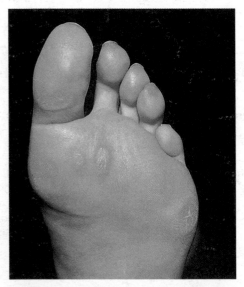

Plate 9   Callosities predominantly over weight-bearing areas.

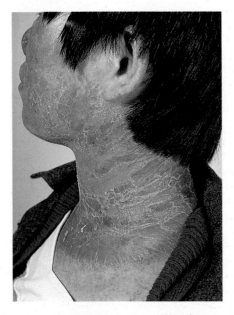

Plate 10   Photosensitive dermatitis. Note sparing of non-light exposed skin.

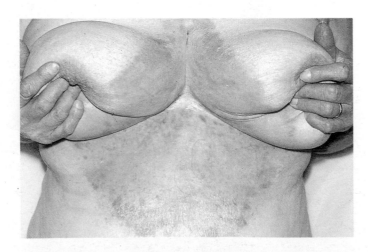

Plate 11   Acute intertrigo commonly Candida-induced with bacterial copathogens.

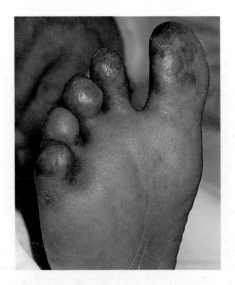

Plate 12   Perniosis ('chilblains').

© S.K. Goolamali

have infectious mononucleosis. Whilst rupture occurs mostly in patients with an enlarged spleen, it may occur in the absence of splenomegaly. Moreover there appears to be no correlation between the severity of infectious mononucleosis and the chance of rupture.

Persistent fatigue

Whilst the majority of episodes of infectious mononucleosis are of relatively short duration (about 2–6 weeks), there appear to be a few cases in whom symptoms, predominantly those of fatigue and lethargy, persist for an indefinite period. These cases may comprise a proportion of individuals with the chronic fatigue syndrome (see below).

Diagnosis

The diagnosis of infectious mononucleosis is made on the basis of (a) the clinical findings, (b) the presence of a significant percentage (at least 15 per cent) of atypical lymphocytes on peripheral blood film, and (c) a positive heterophile antibody test for infectious mononucleosis such as the Monospot® .

Such a test identifies the ability of the patient's serum to agglutinate horse red cells after it has been absorbed with guinea-pig kidney. The latter procedure greatly reduces the number of false positive results. It must be remembered that the Monospot® test may be negative in the first week or two of illness, and that in a proportion of cases (10–15 per cent) the test for such heterophile antibodies might remain negative. The serum may also be tested for antibodies against the viral capsid antigen. A positive viral capsid antigen IgM would indicate a current infection and remains positive for about 2 months. A positive IgG test (whatever the titre) only indicates past infection unless a fourfold rise or fall can be demonstrated which would indicate more recent infection.

Some patients may have a coexistent group A streptococcal throat infection, and for this reason throat swabs for bacterial culture should be taken. Most cases of infectious mononucleosis will have abnormal liver function tests, especially the transaminases and lactate dehydrogenase. The serology of chronic fatigue due to Epstein–Barr virus has not been consistent, but has included increased antibodies to both Epstein–Barr virus early antigen and Epstein–Barr viral capsid antigen IgM.

Management

In the main infectious mononucleosis lasts for 2 to 6 weeks and is self-limiting. Treatment is mainly symptomatic. If the throat swab turns out to be positive for group A streptococci, this should be treated with penicillin or, if the patient is penicillin hypersensitive, with erythromycin. Treatment should continue for at least 10 days to avoid recurrence. Ampicillin or amoxycillin should be avoided as over 90 per cent of infectious mononucleosis patients develop a florid, often troublesome, rash.

Good evidence for the usefulness of corticosteroids in infectious mononucleosis has only been established in the case of obstructive pharyngitis where the patient has difficulty in speaking, swallowing, or breathing, whilst their usefulness in cases with hepatitis, neurological involvement or haematological involvement is unproven. In particular, steroids have not been shown to reduce the risk of splenic rupture or to shorten the course of those with fatigue thought to be due to the Epstein–Barr virus.

Return to sporting activity after infectious mononucleosis should always be graded and limited to the exercise tolerance of the patient.[7] Total bedrest is unnecessary and may even delay recovery, as was demonstrated in a controlled trial amongst university students.[8] Since the risk of splenic rupture is greatest in the first months following infectious mononucleosis, strenuous exercise and alcohol consumption should be avoided during this period. A more cautious return to activity is advisable in contact sports such as rugby and wrestling, and they should not be resumed before resolution of splenic enlargment.

Toxoplasmosis

Toxoplasma gondii is another organism which may produce a glandular fever syndrome with fever, hepatosplenomegaly, and generalized or localized lymphadenopathy. This intracellular protozoan infection is most commonly acquired from animals, particularly cats (via ingestion of the cysts in their faeces), and undercooked or raw meat. Most infections in man are asymptomatic. Up to 60 per cent of healthy adults in certain countries (e.g. France) may be antibody positive without any past history of illness. Clinical presentation as a glandular fever syndrome is the most common observation in young adults and is generally self-limiting. However, it has recently gained notoriety by being implicated in the prolonged fatigue and poor performance of certain outstanding athletes and as an opportunistic infection in patients infected with the HIV virus.

Diagnosis

The clinical diagnosis can be confirmed by serology, and a fourfold rise or fall in the toxoplasma latex test or a positive toxoplasma IgM regardless of titre is indicative of recent infection, although a few cases of 'chronic' toxoplasmosis where the toxoplasma IgM remains elevated have been documented. A positive latex or IgM titre may occasionally be very high, but without a fourfold rise or fall current or recent infection cannot be confirmed.

Management

Most episodes of toxoplasmosis are self limiting and do not require specific treatment. However, when systemic signs are severe and prolonged the patient may be treated with a combination of pyrimethamine (25 mg daily by mouth) and sulphadiazine (500 mg–1 g four times a day by mouth) which is given for up to 4 weeks. However, there is no good evidence that such treatment is beneficial in the immunocompetent host or when the disease is believed to be chronic . A full blood count, looking out for leucopenia and thrombocytopenia, should be carried out weekly during treatment.

Cytomegalovirus

Glandular fever syndrome due to cytomegalovirus is even less frequent in the immunocompetent individual than toxoplasma infection, but may present with a syndrome that is indistinguishable from that caused by Epstein–Barr virus or toxoplasma. The illness is usually self-limiting and the treatment in such cases is symptomatic only. Occasionally the presentation is that of jaundice.

Hepatitis

Acute viral hepatitis may be due to a number of viral causes (Table 1) and is not an uncommon infection in the young adult sportsperson.

Hepatitis A

Hepatitis A virus is a relatively common cause of hepatitis in the young adult and in many cases subclinical infection may occur. Spread is via the orofaecal route. Nausea, loss of appetite,

Table 1 Commoner causes of hepatitis in man

Organism	Transmission route
Hepatitis A virus	Orofaecal
Hepatitis B virus	Blood or sexual contact
Hepatitis C virus	Blood products
Hepatitis D virus (delta agent)	Only in the presence of hepatitis B virus
Hepatitis E virus	Orofaecal
Non-A, non-B hepatitis	Presumed orofaecal

Less common
Epstein–Barr virus
Cytomegalovirus
Toxoplasma gondii
Leptospira spp.

vomiting, and abdominal pain often precede the appearance of jaundice by a number of days, and it is during this period that the individual is most infectious. The urine becomes dark because of the presence of bilirubin and the stools light because of intra-hepatic cholestasis. Jaundice may then appear and remain for a few days or weeks followed by a variable period to complete recovery. Hepatitis A is contagious and spreads rapidly. However, by the time jaundice appears the need to isolate the patient in order to prevent spread is usually unnecessary as viral excretion is decreasing and may no longer be detectable.

Diagnosis

The diagnosis of hepatitis A is made clinically with confirmation in the laboratory of a positive hepatitis A virus IgM antibody test. The illness can be monitored by the measurement of liver function tests such as the transaminases and lactate dehydrogenase, but these do not necessarily correlate with the severity of disease or outcome. A positive hepatitis A virus IgG on its own indicates past infection only.

Management

The treatment of hepatitis A virus infection is symptomatic. Close contacts of those who have not had the infection may be given passive protection with gammaglobulin. The usual dose is 250 mg given intramuscularly. Gammaglobulin has been shown to reduce the severity and duration of illness in those exposed, especially when given early. A vaccine against the hepatitis A virus has recently become available and can be used in seronegative subjects at increased risk; in the case of sportspeople this amounts to travel to countries of medium to high endemicity or those who stay for more than 3 months (see below).[9]

Patients with hepatitis A virus infection should rest and only participate in limited activities until the symptoms have subsided, but need not wait until liver function tests have returned to normal. Moderate activity, even with abnormal tests, has been shown to have no deleterious effect on the rate of recovery or relapse in hepatitis A virus infection. Whilst alcohol has been traditionally prohibited during most acute forms of hepatitis and into convalescence, there are no data to suggest that moderate alcohol intake leads to a worsening of acute hepatitis or predisposes to chronic hepatitis.

Hepatitis B

Hepatitis B virus (HBV) is a relatively rare infection amongst athletes and the major risk occurs by sexual contact. There is also the minimal risk of spread in contact sports. The prodromal symptoms are similar to hepatitis A virus infections but occasionally there may be a preceding skin rash and/or arthralgia. In the context of hepatitis B and sport, the precautions taken by first aid workers and accompanying sports staff should be similar to those outlined for HIV infection as laid out below. If exposed to the blood of an individual known to be infected then the use of hepatitis B vaccine (usually 500 International Units for adults) to confer active immunity as well as specific hepatitis B immunoglobulin (HBIG) should be used. These should be given at different sites. Immunization with HBIG does not suppress an active immune response to the vaccine.

Treatment of acute cases is symptomatic and the majority recover spontaneously. Return to sporting activity must be graded and full activity should not be resumed until the symptoms have subsided. Careful follow-up of patients with hepatitis B virus infection is important to ensure that they eliminate the virus, and do not develop any of the important sequelae of infection such as chronic carriage (occurring in 5 to 10 per cent of cases), chronic active hepatitis, cirrhosis, and hepatoma. Patients with any of these sequelae require further specialist follow-up since, at the present time, treatment with interferon may be indicated according to infection status; this can eradicate infection in up to 40 per cent of those treated. There is at present no indication that chronic carriers of hepatitis B virus should be prevented from participation in sport, except in close contact sports such as boxing, wrestling, and rugby.

Hepatitis C

The hepatitis C virus has recently gained much publicity. It is transmitted primarily by blood and blood products and many haemophiliacs are infected but sexual contact does not appear to be a major route of spread. Clinically there is little to distinguish it from the other causes of viral hepatitis. The diagnosis is made serologically in a commercially available antibody test (C100) but this may take up to 6 months after infection to become positive. 'Second generation' recombinant blot assay (RIBA) tests are now available which are more specific than the C100 test.[10] Hepatitis C virus is an important cause of hepatitis as infection can lead to prolonged liver dysfunction and cirrhosis. However, it is uncommon amongst sportspeople,

Sporadic and enteral non-A, non-B hepatitis

About 40 per cent of cases of non-A, non-B hepatitis have no known source but a 'hepatitis E virus' thought to be responsible for some cases of enterically transmitted non-A, non-B hepatitis has been recently identified and cloned.[11] Hepatitis E appears to be particularly common in the Indian subcontinent but probably occurs worldwide.

Chronic fatigue syndrome

The chronic fatigue syndrome has now become a major controversial area especially because of the debate regarding whether it is of organic or functional origin. Whatever its cause it is of major importance to sportspeople.[12] Undoubtedly some, if not most individuals suffer from a variable period of fatigue, malaise, and depression after viral illness such as influenza, infectious mononucleosis, and hepatitis. However, the extent to which more prolonged symptoms may be attributed to a viral infection remains open to debate. In 1991 an operational definition for chronic fatigue syndrome was published which emphasizes a primary complaint of fatigue for at least 6 months in the absence of neurological signs with myalgia, psychiatric symptoms, and previous viral infections as common associated features, although none are exclusions or necessary for the diagnosis.[13] Our recent studies have shown that the duration of symptoms did not relate to outcome, and that a number of patients with symptoms for less than 6 months subsequently went on to have the syndrome.[14]

Attributing a cause to chronic fatigue has been difficult and it is almost certain that it is a heterogeneous group of conditions. However, evidence in favour of an organic cause includes a number of objective laboratory measures, some of which are highly specialized and have only been used in the context of research. These include the finding of abnormal muscle biopsies in some of these patients together with the findings of virus particles in muscle. Abnormal intracellular acidosis in the muscles of the forearm have been detected using magnetic resonance imaging. Antibodies to the coxsackievirus and a component viral antigen (VP1) have been found in increased frequency in patients with chronic fatigue syndrome. Markers of persistent Epstein–Barr virus infection such as an IgM response to the viral capsid antigen, absence of a response to the nuclear antigen (EBNA antibody negativity), and increased response to the early antigen (EA antibody) have been presented as evidence for prolonged fatigue following infectious mononucleosis. Toxoplasmosis has also achieved recent publicity with regard to its potential of producing prolonged symptoms, and a few individuals may continue to maintain levels of IgM in their serum which may indicate continued replication of the organism. Brucellosis is a good example of a case where a persistent infection is commonly thought to produce prolonged symptoms.

Equally, there are abundant data which argue against an organic cause for chronic fatigue syndrome. In one study of 100 adults with chronic fatigue, nearly 70 per cent were found to have a psychiatric disorder that was thought to be the cause of their fatigue; 5 per cent had medical conditions (e.g. a seizure disorder, sleep apnoea, polymyalgia, and asthma) while in 30 per cent the cause of chronic fatigue was unexplained.[15] Many other published studies have shown clear association between chronic fatigue syndrome and self-reported depression and anxiety.

In many cases patients are able to date their illness back to an upper respiratory tract infection or a diarrhoeal illness. This is followed by a variable period of symptoms such as fatigue, muscle weakness and pain, poor concentration, sleep abnormalities, irritability, joint pain, headaches, forgetfulness, painful lymph glands, photophobia, sore throat, etc. These can often be severe enough to produce gross impairment of athletic performance.

Management

None of the interventions used in controlled studies such as acyclovir or steroids for Epstein–Barr virus infection, definitive treatment outside the acute phase for toxoplasmosis, or more generalized treatment such as gammaglobulins have been shown to be of benefit in chronic fatigue syndrome. Treatment with magnesium has been used, as have interventions to eradicate yeast infection by diet or the use of antifungals. However these treatments have not as yet been supported by adequate studies. Treatment should be directed towards the relief of symptoms. In all cases a positive and optimistic attitude towards outcome, coupled with continued support and reassurance is of far greater importance in the gradual return to full activity. Sportspeople should exercise within their effort tolerance and increase their exercise in a graded fashion. There is no evidence to support the concept of total bedrest.

HIV infection

Infection by HIV and the disease it causes (AIDS) is now a global health problem. Sportspeople generally have the same risk of infection as the general population, but travel abroad and increased sexual activity may increase this risk. The risks would be proportional to the prevalence of HIV positive individuals in a population which in turn would differ from place to place. In certain countries AIDS will become the major killer of individuals aged between 15 and 45. The number of AIDS cases and HIV positive individuals is growing rapidly and is often inadequately documented so that published figures are often misleading. Whilst initially the at-risk groups included homosexual and bisexual men, intravenous drug abusers, haemophiliacs, transfusion recipients, prostitutes, and the sexual partners of all these, infection has now spread into the heterosexual population so that it is no longer possible to predict reliably who may be at risk.

Having said this, however, the risk of contracting HIV infection in sport-related activities must be exceedingly small and to date no cases have been reported. The risk to sportspeople must lie mainly in sexual intercourse, and in this respect the regular use of condoms and restriction of the number of sexual partners must be advocated. The sharing of razors and tooth-brushes has the theoretical possibility of transmitting the virus and should be discouraged. At the same time it should be emphasized that normal social contact, the sharing of changing facilities, and swimming pools constitute no risk of infection.

There are no documented cases of spread of HIV infection by contact sports. However, cases have been recorded where seroconversion occurred following contact of infected blood with open skin lesions. Thus this route could theoretically pose a risk.

In injuries resulting in bleeding from wounds, participants should be aware of the risk and in all cases such wounds should be covered or the player excluded from further participation. All participants, first-aid workers, and accompanying sports staff should realize that, apart from sexual contact and other high risk practices, the risks of acquiring HIV infection are very small but the general recommendation in Table 2 may apply.

Table 2 General recommendations for the handling of injuries on the sports field involving sports persons who may be HIV positive [21]

1. Assume that all casualties are HIV antibody positive
2. Wear gloves for all procedures involving contact with blood or other body secretions
3. Cover all cuts and abrasions where possible
4. Wear protective glasses where blood may be splashed into the face
5. Wash skin immediately after contamination with blood or secretions
6. Dispose of sharp objects safely; never attempt to resheathe needles
7. dispose of waste materials by burning
8. Contaminated cloths should be presoaked in hot ($> 70\,°C$) soapy water for 30 min and then washed in a hot cycle washing machine; alternatively they may be soaked in household bleach (1 in 10 dilution) or Milton® solution for 30 min
9. All contaminated equipment or surfaces may be treated with bleach as above
10. Communal items in the first-aid kit no longer have a place in the care of injured sportspeople (e.g. bucket and sponge)
11. No cases of HIV infection by mouth-to-mouth resuscitation have been recorded; however, simple devices which prevent direct contact between operator and patient are now available to assist in ventilation

TRAVEL ABROAD

Wherever and whenever travel is undertaken for sport, business, or leisure, there is increased mortality and morbidity. Not surprisingly, excess mortality abroad is mainly due to traffic accidents and drowning rather than the scourges of exotic infectious diseases. In one study of 2500 deaths of American travellers of all ages whilst abroad, 50 per cent were due to cardiovascular events, 25 per cent due to injury of one kind or another, and only 1 per cent were due to infection, although it should be noted that these data referred to travellers whilst abroad and not to illnesses which manifested after return. However, this does emphasize the need for anyone travelling abroad to have adequate medical insurance.

Travellers to developing countries suffer a high morbidity which is mainly due to diarrhoeal disease. The relative risk of some other infections is shown in Table 3. Malaria outstrips the others by far. The most frequent occurring disease preventable by immunization is hepatitis A followed by hepatitis B. The risk of acquiring cholera abroad is exceedingly small. Moveover, some of the illnesses may only be minor (e.g. a short episode of traveller's diarrhoea), some will only manifest themselves some time after return from abroad (e.g. hepatitis), while others can be extremely severe and even life-threatening (e.g. cerebral malaria).

Table 3 Morbidity and mortality of certain infections in 1 000 000 non-immune travellers visiting developing countries[22]

Infection	Incidence rate/month	Mortality rate/month
Malaria (without chemo-prophylaxis)		
West Africa	24 000	480
East Africa	15 000	300
Hepatitis A	3000	3
Hepatitis B	800	16
Typhoid		
Overall	30	0.3
India	300	3
Poliomyelitis	1	0.2
Cholera	3	0.06

Many of the measures (immunization or prophylaxis) taken against these diseases are not without side-effects, often minor but also sufficient to interfere with sports performance. Thus if they are to be administered they should be given as early as possible before the time of competition. In the final analysis it is judicious when instituting preventive measures to weigh the benefits against the side-effects, cost, and inconvenience caused. Not all the measures implemented are completely effective. No antimalarial measure can provide absolute protection: gammaglobulin is said to prevent only 70 to 90 per cent of clinical attacks and cholera vaccine is only 60 to 70 per cent effective. It is also often difficult to decide whether or not to give a preventive measure because of the insufficient information available as to the risks. Thus it is often the case that where the risks and costs are thought to be small, the particular preventive measure is administered. Certainly, pre-travel advice with regard to food hygiene, exposure to insect bites, etc. is as important as measures that are ultimately instituted such as vaccination or drug prophylaxis.

Travellers' diarrhoea

Travellers' diarrhoea is by far the most common illness afflicting travellers and varies with destination: below 8 per cent in the United States, Canada, Northern and Central Europe, Australia, and New Zealand; 8 to 20 per cent in the Caribbean, Southern Europe, Israel, Japan, and South Africa; 20 to 55 per cent in developing countries. Enterotoxigenic *Escherichia coli* are most commonly responsible. A wide range of other bacteria, viruses, and protozoa make up the remainder (Table 4).

Table 4 Principal cause of travellers' diarrhoea

Bacteria
  Enterotoxigenic *E. coli*
  Shigella spp.
  Salmonella spp.
  Campylobacter jejuni
  Vibrio cholera
  Non-cholera vibrios, e.g. *Vibrio parahaemalyticus*
Viruses
  Rotavirus
  Small round viruses, e.g. Norwalk agent
Protozoa
  Giardia lamblia
  Entamoeba histolytica
  Cryptosporidium

Symptoms most frequently start on the third day abroad—20 per cent will have a second bout in the second week. The symptoms, apart from watery diarrhoea of varying degree, include cramps, nausea, vomiting, fever in a few cases, and sometimes frank dysentery. Passage of blood and/or mucus implies bowel inflammation or ulceration and raises the likelihood of an invasive organism such as *Shigella* species or *Entamoeba histolytica*, although Salmonella and Campylobacter can produce such a picture. Giardiasis has a longer and often more variable incubation period (frequently measured in weeks rather than days), and often produces persistent diarrhoea, flatulence, abdominal distension, and lactose intolerance.

Management

Prevention

Dietary precautions such as care in selecting well-cooked food and only fresh fruit and vegetables which require peeling are particularly important. Salads and uncooked shellfish are considered high risk. Only sterilized water should be consumed, including toothbrushing and ice in drinks. Because of their low pH, bottled carbonated drinks are safe. However, studies have shown that travellers very soon relinquish these restrictions and eat salads, use ice cubes in drinks, and even indulge in raw uncooked foods.

Prophylaxis

Many drugs have been proposed for the prophylaxis of travellers' diarrhoea, but only antimicrobials such as cotrimoxazole, trimethoprim, and ciprofloxacin have been shown to have proven efficacy. However, the medical profession has been loath to advocate their widespread use mainly because of fear of the development of resistant organisms. Prophylactic antimicrobials may be indicated in athletes who are staying abroad for less than 2 weeks and in whom it is vital that peak performance is assured. In this case the drug choice would be ciprofloxacin (500 mg twice daily) since it covers the majority of gut pathogens, including Campylobacter. Both cotrimoxazole (960 mg twice daily) or trimethoprim (200 mg twice daily) for 5 days have been shown to be effective in the treatment of travellers' diarrhoea.

Treatment

In most cases the management of travellers' diarrhoea is symptomatic. The patient should rest and drink plenty of clear fluids, particularly water with sugar and electrolytes (e.g. Dioralyte®). A simple alternative can be made by adding a pinch of salt and a teaspoon of sugar to 250 ml of water. Potassium can be provided by fruit juice. Milk should be avoided when symptoms are severe. Painful spasms may be treated with diphenoxylate (Lomotil® (four tablets at once then two every 4 h until the diarrhoea has stopped) or loperamide (Imodium® (two capsules at once and then one with each diarrhoeal motion)). These drugs should be used with caution and only for short periods of time, since nausea, vomiting, and sedation are associated with increasing doses, especially in the presence of renal impairment. The use of codeine and morphine has been banned by the International Olympic Committee. If used, the dose is codeine phosphate (two 30 mg tablets every 4 h until diarrhoea has stopped). All these anti-diarrhoeal agents may prolong the course of Shigella and Salmonella infections, and should not be used when invasive disease is suspected (e.g. in the presence of blood or mucus in the stool).

Malarial chemoprophylaxis

The spread of drug-resistant *Plasmodium falciparum* malaria has complicated malarial chemoprophylaxis, as well as the awareness that some of the more effective combination drugs, such as Fansidar® , Maloprim® , and amodiaquine, may have severe and sometimes fatal side-effects. Thus the risk of contracting malaria in any given country or situation needs to be constantly weighed against the risk involved with chemoprophylaxis. In the absence of adequate data this becomes difficult.[16] However, it appears that compliance is of extreme importance, since those who comply poorly have a similar rate of attack to those unprotected individuals and an increased relative risk of death.

It is most important to emphasize to travellers that antimosquito measures are probably at least as important as antimalarial chemoprophylaxis. Thus in endemic areas travellers should take the following precautions:

1. sleep in properly screened rooms;
2. use mosquito nets without holes which are tucked in under the mattress well before nightfall;
3. wear long-sleeved clothing and long trousers when out of doors after sunset;
4. consider using other adjuncts such as insect spray (usually containing pyrethrum), mosquito coils, or repellents such as diethyltoluamide (DEET).

A brief guide to antimalarial chemoprophylaxis is shown in Table 5. If there is any doubt, specialist advice should be sought. Chemoprophylaxis should start a week before departure (to ensure adequate blood levels and to evaluate any potential side-effects), whilst away, and for 4 weeks after return (except for mefloquine). The simplest and safest regimen to use at present (1992) for most malarial endemic parts of the world is chloroquine, 2 tablets (150 mg base each) once a week, together with proguanil (Palu-

drine®) 2 tablets (200 mg) daily. These drugs have only minor side-effects with the most common being difficulty in visual accommodation in the case of chloroquine and mouth ulcers in the use of proguanil. In Papua New Guinea and the Solomon Islands paludrine is replaced by pyrimethamine–dapsone (Maloprim®) 1 tablet weekly and is taken in addition to chloroquine. Short-term travellers to East Africa (less than 3 months) may use mefloquine 250 mg (1 tablet) weekly. More details and specialist advice should be sought in other circumstances as follows:

1. long-term visitors;
2. children under 12 years of age;
3. individuals with drug allergies;
4. pregnancy.

The possibility of malaria should be considered in any person with a fever who is or has been in a malarious area whether or not they have been taking antimalarial chemoprophylaxis. At present, no antimalarial chemoprophylaxis can guarantee absolute protection.

Vaccination of travellers

Vaccination of travellers has become routine and is often undertaken without any consideration of the risks or benefits involved. Such an analysis is often not possible for a given individual, even when the destination is known. Furthermore, there is often not sufficient time before departure to complete a vaccination schedule (e.g. hepatitis B). In this circumstance it should be assumed that, where indicated, some vaccination is better than none. Smallpox is believed to have been eradicated thus making vaccination against it unnecessary. A brief outline of the use of vaccination for travellers is given in Table 6.

Vaccination against poliomyelitis and tetanus should be kept up to date since these illnesses occur worldwide. It is recommended that poliomyelitis vaccination be updated every 5 years where travel to developing countries is involved[17] and tetanus vaccination every 10 years even though the additional protection provided if a full basic course has been given remains unproven. Booster doses at less than 10-year intervals are not recommended since they have not been shown to be necessary and can lead to unpleasant local reactions which could certainly interfere with sporting performance. Even when given, vaccination should be administered well ahead of a sporting event.

Typhoid vaccine is no longer as unpleasant as it was previously.[18] It is now a monovalent vaccine against *Salmonella typhi* only (i.e. not against Paratyphi A and B as before) and has been shown to be as effective when 0.1 to 0.2 ml is given intradermally as compared with 0.5 ml given intramuscularly. This regimen also appears to reduce the side-effects of redness and swelling at the site of injection and the more generalized symptoms of an influenza-like illness. The risk of contracting the disease is variable depending on the destination, nutrition characteristics, duration of stay, and gastric acidity (assumed to be normal in healthy athletes). The vaccine is only regarded as 70 to 80 per cent effective and needs to be given on two occasions a month apart. A booster is recommended every 3 years, but this will depend on circumstances. Areas of 'high risk' appear to be Middle East, Africa, and Asia, especially the Indian subcontinent. Amongst American citizens, for example, travel to Mexico accounted for the majority of cases. Only 7 per cent of those American citizens contracting typhoid had been vaccinated.[18]

Recently two further vaccines for typhoid have become available. The first, an oral typhoid vaccine, has the advantage of being

Table 5 Brief guidelines for malarial chemoprophylaxis[16]

Chemoprophylaxis	Area to be visited	Dose/Comment
None	North Africa (Morocco, Algeria, Tunisia, Libya, tourist areas of Egypt) Tourist areas of south-east Asia (Thailand, Philippines, Hong Kong, Singapore, Bali, China)	
Chloroquine or	Middle East (Including summer months in rural Egypt and Turkey) Central America Rural Mauritius	300 mg base (two tablets) once per week
Proguanil (Paludrine®)		200 mg once per day
Chloroquine and proguanil	Subsaharan Africa Indian subcontinent Afghanistan and Iran Rural areas of south-east Asia South America	Doses as above
Chloroquine and Malaprim® (pyrimethamine 12.5 mg and dapsone 100 mg)	Papua New Guinea, Solomon Islands, and Vanuatu	Chloroquine dose as above Malaprim® one tablet once a week
Mefloquine (Lariam®)	Africa-Cameroon, Kenya, Malawi, Tanzania, Zaire, Zambia, Uganda; all rural areas of south-east Asia	250 mg (one tablet) an alternative to chloroquine and proguanil in areas of high risk and multiple resistance where visit is of short period (<3 months) only; contraindicated in people with epilepsy or psychiatric disorder

more acceptable and the second, a subunit vaccine is said to have fewer side-effects and needs to be given only once rather than twice for the primary course. However, both are expensive.

The oral vaccine is a live attenuated strain of the *Salmonella typhi* (strain Ty 21A) and can be used for active immunization of adults and children older than 6 years. A capsule is swallowed before a meal with a cold or lukewarm drink on alternate days for three doses and the manufacturers advise that the capsules should not be chewed and should be swallowed immediately. The reported side-effects are few but mild nausea, vomiting, abdominal cramps, diarrhoea, and urticaria may occur. The same contraindications as for other live vaccines apply. As with other typhoid vaccines, this vaccine is not 100 per cent effective and travellers should still take precautions against contact or ingestion of potentially contaminated food or water. A limiting factor in its use remains the cost.

The second vaccine is the subunit vaccine prepared from the Vi capsular polysaccharide of *Salmonella typhi* and has the advantage of a single dose schedule which protects for 3 years. It is also said to have a lower incidence of local and systematic reaction compared to the standard whole-cell inactivated typhoid vaccine. However, the vaccine is not licensed for children under 18 months of age and is more expensive than the standard vaccine.

The use of cholera vaccine remains controversial.[20] Cholera vaccines are only about 60 to 70 per cent effective and last only 3 to 6 months. The risk of travellers contracting cholera is extremely low (Table 2). Certain countries sporadically demand a certificate of vaccination against cholera, and it has been argued that possession of a certificate is more relevant than the vaccine itself. Athletes are unlikely to work or live in highly endemic areas which may demand vaccination from time to time. In 1973 the World Health Organization (WHO) waived the requirement for a cholera vaccination certificate. Vaccination, if given, should be by deep subcutaneous or intramuscular injection using two doses a month apart. The first dose is given by intramuscular injection whereas all subsequent doses are given intradermally. Intradermal vaccination results in fewer side-effects but may result in less protection. On balance, cholera vaccine is not necessary for ordinary tourists or athletes visiting most countries.

Active vaccination against hepatitis A has recently become available. It consists of a formaldehyde inactivated hepatitis A virus (HM 175 strain) grown on human diploid cells and absorbed on aluminium hydroxide as adjuvant. The vaccine is an alternative to human normal immunoglobulin for frequent travellers to areas of moderate or high endemicity or for those who stay longer than 3 months. The immunization regimen consists of two doses of 1 ml of vaccine given intramuscularly and spaced 2 to 4 weeks apart, and provides antihepatitis A virus antibodies for at least 1 year. To

Table 6 Some more commonly used vaccines for travel abroad

Vaccine	Dose	Comments
Polio*	3 drops on a sugar lump	Primary course: 3 doses 1 month apart. Boost every 5 years
Tetanus (adsorbed tetanus toxin)	0.5 ml subcutaneously	Primary course: 3 doses 1 month apart. Boost every 10 years unless at special risk
Typhoid (a) Monovalent typhoid vaccine	0.5 ml intramuscularly for first dose then 0.2 ml intradermally	Primary course: 2 doses preferably a month and not less than 10 days apart. Boost every 5 years unless at special risk
(b) Live oral typhoid* vaccine strain Ty21A	3 enteric coated capsules on alternate days	Boost every 3 years unless at special risk
(c) Vi capsular polysaccharide typhoid vaccine	0.5 ml subcutaneously or intramuscularly once only	Single boost every 3 years
Cholera	0.5 ml intramuscularly (first dose), then 0.1 ml intradermally	Primary course: 2 doses preferably a month and not less than 10 days apart. Booster every 6 months
Yellow fever*	0.5 ml subcutaneously	Single injection—valid 10 days after vaccination for 10 years. Must be from a recognized Yellow Fever Vaccination Centre
Hepatitis A (human diploid cell)	1 ml intramuscularly	Primary course: 2 doses 2 to 4 weeks apart. For immunity up to 10 years, a booster at 6 to 12 months
Hepatitis B	1 ml intramuscularly	Primary course: 3 doses at 0, 1 and 6 months. Booster every ?2 years
Rabies (human diploid cell vaccine)	1 ml intramuscularly or 0.1 ml intradermally †	Primary course: 3 doses 1 month apart. Booster every ?2 years

* Indicates live vaccine.
† Not standard recommendation but probably as effective and cheaper.

obtain more persistent immunity of up to 10 years, a 1-ml booster is recommended at 6 to 12 months after the initial dose. The reported side-effects so far have been mild and amount to mainly local symptoms. Because of the expense, travellers should first be screened for antibodies to hepatitis A virus before receiving the vaccine. Those who possess hepatitis A virus IgG do not need the vaccine.

Vaccination against hepatitis B is not considered routine as yet for travellers. However, with increasing knowledge of the safety of the vaccine, falling costs, and an indication as to definite risk in the travelling population (Table 2), it may become more routinely administered. Vaccination (1 ml intramuscularly) should be given at time zero, 6 weeks, and 6 months and an adequate antibody response is confirmed by taking serum at 2 to 4 weeks after the last dose.

Yellow fever vaccine is the only vaccine regarded by the World Health Organization as requiring an International Certificate of Immunization of travellers to Subsaharan Africa and certain parts of South America. A single dose of 0.5 ml is given subcutaneously. It is a safe effective vaccine and is valid for 10 years beginning 10 days after vaccination.

Other vaccines

From time to time indications to give other vaccines may arise in sportspeople. These will include diphtheria, influenza, Japanese B encephalitis, measles, meningococcal vaccine, plague, typhus, tick-borne encephalitis, and rabies. All of these would require further specialist advice according to the circumstances. At the present time smallpox is regarded as being eradicated.

Gammaglobulin

Other than malaria, hepatitis A emerges as the disease in which intervention can be most effective in travellers. Since the incubation period may be prolonged (up to 2 months), symptoms may only begin well after return from abroad but the effects on high level sporting performance may be marked. Administration of pooled gammaglobulin is usually in a dose of 250 mg intramuscularly to those abroad for 2 months or less, and 500 mg to those abroad for longer than 2 months. Gammaglobulin is given as late as possible before departure so as to maintain antibody levels as high as possible whilst at risk of infection. It is important to emphasize that from the point of view of the risk of HIV infection, injection of gammaglobulin in entirely safe and that no cases of transmission of HIV have been reported using these preparations.

REFERENCES

1. Roberts J, Wilson J, Clements G. Virus infections and sports performance—a prospective study. *British Journal of Sports Medicine* 1988; **22**: 161–2.
2. Simon H. The immunology of exercise. A brief review. *Journal of the American Medical Association* 1984; **252**: 2735–8.
3. Keast D. Exercise and the immune response. *Sports Medicine* 1988; **5**: 248–67.
4. Lewicki R, Tchorzewski H, Majewska E, Nowak Z, Baj Z. Effect of maximal physical exercise on T-lymphocyte subpopulations and on interleukin (IL1) and interleukin 2 (IL2) production *in vitro*. *International Journal of Sports Medicine* 1988; **9**: 114–17.
5. Parry-Billings M, Blomstrand E, McAndrew N, Newsholme E. A communicational link between skeletal muscle, brain, and cells of the immune system. *International Journal of Sports Medicine* 1990; **11**: S122–8.
6. Khansari D, Murgo A, Faith R. Effects of stress on the immune system. *Immunology Today* 1990; **11**: 170–4.
7. Haines J. When to resume sports after infectious mononucleosis. How soon is safe? *Postgraduate Medicine* 1987; **81**: 331–3.
8. Dalrymple W. Infectious mononucleosis—II. Relationship of bedrest and activity to prognosis. *Post graduate Medical Journal* 1964; **35**: 345–9.
9. Tilsey A, Banatvala J. Hepatitis A. Changing prevalence and possible vaccines. *British Medical Journal* 1991; **302**: 1552–3.
10. Anonymous. Hepatitis C upstanding. *Lancet* 1990; **335**: 1431–2.
11. Reyes G *et al.* Isolation of a cDNA clone from the virus responsible for enterically transmitted non-A, non-B hepatitis. *Science* 1990; **247**: 1335–9.
12. Budgett R. The post-viral fatigue syndrome in athletes. IN: Jenkins R, Mowbray J, eds. *Post-viral fatigue syndrome.* London Wiley, 1991: 345–62.
13. Sharpe M, Archard L, Banatvala J. Chronic fatigue syndrome: guidelines for research. *Journal of the Royal Society of Medicine* 1991; **84**: 118–22.
14. Sharpe M, Hawton K, Seagroatt V, Pasvol G. Chronic fatigue: A follow-up study to an infectious disease clinic. *British Medical Journal* 1992; **305**: 147–52.
15. Manu P, Lane T, Matthews D. The frequency of chronic fatigue syndrome in patients with chronic fatigue. *Annals of Internal Medicine* 1988; **109**: 554–6.
16. Bradley D, Phillips-Howard P. Prophylaxis against malaria for travellers from the United Kingdom. *British Medical Journal* 1989; **229**: 1087–9.
17. Kubli D, Steffen S, Schar M. Importation of poliomyelitis to industrialised nations between 1975 and 1984: evaluation and conclusion for vaccination recommendations. *British Medical Journal* 1987; **295**: 169–71.
18. Steffen R. Typhoid vaccine, for whom? *Lancet* 1982; i: 615–16.
19. Taylor DN, Pollard RA, Blake PA. Typhoid in the United States and risk to the international traveller. *Journal of Infectious Diseases* 1983; **148**: 599–602.
20. Morger H, Steffen R, Schar M. Epidemiology of cholera in travellers and conclusion for vaccination recommendations. *British Medical Journal* 1983; **286**: 184–6.
21. Payne SDW. *Medicine, sport and the law.* Oxford: Blackwell, 1990.
22. Steffen R. Travel medicine—prevention based epidemiological data. *Transactions of the Royal Society of Tropical Medicine and Hygiene* 1991; **85**: 156–62.

3.3.1 Drug abuse

DAVID A. COWAN

INTRODUCTION

There are so many powerful medications available today that man has almost come to expect that there is a 'pill for every ill'. People often think that many of these substances can act as ergogenic aids to put one into a supranormal position, the position required by the sportsman if he is to succeed.

It may sometimes be difficult to distinguish ethically why some tablets, e.g. vitamins, amino acids, and minerals, are considered by sport as permissable whereas the use of other tablets, e.g. amphetamine, may be restricted. Indeed, the fact that they are frequently difficult to distinguish in appearance, often being white in colour, does nothing to simplify the matter. When one considers that the sportsman may require supplementation of, for example, the water-soluble vitamins to meet his daily requirements, then it is not surprising that sport does not wish to restrict the use of such substances.

The improvement in performance that an athlete may require to win an event is often extremely small and less than the level of significance, which may be demonstrated by normal scientific methods. For example, the 4-minute mile was broken by Sir Roger Bannister in 1954 with a time of 3 minutes 59.4 seconds (before electronic time-keeping). In 1981 Sebastian Coe reduced the record to 3 minutes 47.33 seconds and in 1985 Steve Cram beat this with a time of 3 minutes 46.32 seconds. Thus Cram's performance was 5.5 per cent better than Bannister's and 0.4 per cent better than Coe's, averaging less than 0.2 per cent per year. Therefore, it is hardly surprising that, when a scientific test fails to show a difference between a placebo or a comparator drug in an athletic performance, the sports community fails to consider this as relevant. Lest this comparison be misconstrued, it should not be considered to imply that any of these record-breakers have ever misused any drug. Coyle[1] has pointed out that a good training programme is the most effective means of improving physical

performance and estimates that, in previously sedentary individuals, improvements of 50 per cent in muscle strength or
speed are achievable in long-distance running events.

INTERNATIONAL OLYMPIC COMMITTEE RULES

To determine what is and what is not acceptable in sport one must
look at the doping regulations; the word 'doping' is the accepted term
for drug abuse in sport. The International Olympic Committee
(IOC) provides a set of doping regulations which have been adopted
by all Olympic sports and by most international federations.

The current IOC list of doping classes and methods is given in
the Appendix. It is divided into three groups:

 I Doping Classes, which comprise the main groups of substances whose administration is banned by most international
   federations;
 II Doping Methods, which include blood doping and pharmacological, chemical, and physical manipulation;
 III Classes of Drugs Subject to Certain Restrictions which cover
   additional classes controlled by certain sports, e.g. alcohol in
   shooting, or whose use may be permitted only under specially
   controlled conditions, e.g. corticosteroids.

ACCEPTABILITY OF DRUG CONTROL PROGRAMMES

Drug control programmes implemented by most sports governing
bodies include the collection and analysis of urine samples from
their competitors. Although the great majority of athletes accept
the need for controls and indeed wish drug control to succeed,
those who are taking banned substances do not wish this use to be
detected. In addition, too many people are prepared to sell these
substances on the black market and to provide information about
their use. The *Underground Steroid Handbook* circulated in California is just one such document. Unfortunately, the information
contained therein is not always entirely accurate. For example, it
states that methyltestosterone is undetectable in the steroid tests.
This is somewhat surprising since the very first tests for the detection of synthetic anabolic steroid use were based on radioim-
munoassays using antibodies raised by Professor R. V. Brooks at
St Thomas's Hospital, London. These antibodies were raised to
methyltestosterone.

At present, although there is much discussion about collecting
blood from athletes, urine is collected in drug controls to determine whether a substance from any of the IOC banned classes may
have been taken. Although, in order to ensure that a valid sample
has been collected, most protocols require that the athlete be
observed providing the urine sample, the collection is otherwise
non-invasive and can be performed by sampling officers who have
no medical training.

EFFECTIVENESS OF ANALYTICAL METHODS

Some indication of what may be detected is apparent by inspection
of the findings of IOC accredited laboratories (Table 1). These
figures should not be considered as representative of the scale of
misuse. Most of the samples have been collected at competitions
and may be further biased by the sampling protocol, which may,
for example, be to sample the first, second, and third for one event.

Neither should these figures be used to determine the relative
abuse of the different substances. The detectability of the substance depends not only on the analytical technique employed and
the limit of detection of the substance but also on its elimination
profile, which may be represented by the plasma half-life, and on
its formulation. For example, cocaine has a plasma half-life of
about 1 h and caffeine has a plasma half-life of about 3.5 h, whereas
an oily injection of a nandrolone ester has a half-life of about 22
days.

The IOC Medical Commission has stated that it wishes to control those drugs which may be harmful when misused and to do
this with the minimum interference to their normal therapeutic
use. Far more drugs are permitted than are banned; this is very
different from the sport of horse racing where nearly everything
that is not a normal nutrient is banned.

The number of samples collected worldwide has increased over
the last few years, but even in 1990 was only about 70 000, representing only a very small proportion of athletes in top-level
sport. Approximately 1 to 2.5 per cent of the samples analysed
have been found to contain one or more substance from the banned
classes. The most common group of substances detected has been
the anabolic steroids, with nandrolone being the commonest substance in that group and testosterone the second most common in
most years. The ephedrines are the most frequently found substances in the stimulant category. It should be noted that
β-blockers are controlled only in certain sports and hence the
figures relating to their finding depends on whether participants in
those sports have been tested.

PERFORMANCE-ENHANCING EFFECTS OF THE SUBSTANCES CONTROLLED IN SPORTS

For convenience, the performance-enhancing effects and the side-
effects of the substances controlled in sport will be discussed
grouped under the various IOC categories. In addition, a brief
description of the methods of detection currently used by IOC
accredited laboratories will be given. Although the categories are
intended to represent pharmacological classifications, they are
very broad descriptions.

Analytical methods

The analytical methods used to detect and to confirm the presence
of substances banned in sport rely very much on chromatography
to separate the various components extracted from the urine
sample. Immunoassays are sometimes used for certain substances,
such as benzoyl ecgonine (the metabolite of cocaine) and
cannabinoids, and at present are the only suitable methods for the
detection of the protein hormones. The IOC requires the use of
mass spectrometry to confirm the presence of the substance before
issuing a formal analytical report. Unless the IOC accepts other
confirmatory methods, or advances in mass spectrometry makes it
possible to use the technique for protein hormones, the presence
of any substance from this class cannot be confirmed.

DOPING CLASSES

Stimulants

The first of the IOC categories of doping classes is called 'stimulants'. Examples of substances which are included in this category
are cocaine, amphetamine, the ephedrines, and caffeine.

Table 1 IOC accredited laboratories: summary of samples analysed 1986–1990

Year	No. of samples	No. of negative samples	No. of analytically positive A-samples	Percentage	No. of laboratories
1986	32 982	32 359	623	1.89	18
1987	37 882	37 028	854	2.25	21
1988	47 069	45 916	1153	2.45	20
1989	52 371	51 165	1206	2.30	20
1990	71 341	70 409	932	1.31	21
1991	84 088	83 283	805	0.96	21

Summary of identified substances

IOC Category		1991	%	1990	%	1989	%	1988	%	1987	%	1986	%
A	Stimulants	221	23.9	340	32.0	508	40.4	420	31.0	300	31.9	177	26.3
B	Narcotics	72	7.8	62	5.8	76	6.1	58	4.3	55	5.8	23	3.4
C	Anabolic steroids	552	59.6	579	54.4	611	48.6	791	58.5	521	55.4	439	65.3
D	β-blockers	10	1.1	8	0.8	6	0.5	8	0.6	32	3.4	31	4.6
E	Diuretics	47	5.1	37	3.5	45	3.6	57	4.2	9	1.0	2	0.3
	Masking agents	1	0.1	6	0.6	10	0.8	19	1.4	24	2.6	—	—
	Peptide hormones	1	0.1	1	0.1								
	Ethanol	7	0.8	11	1.0								
	THC	14	1.5	20	1.9								
	Phenobarbital	1	0.1										
Total		926		1064		1256		1353		941		672	

NB Some samples contain more than one substance from banned classes.

IOC Category	Commonest finding	Year				
		1991	1990	1989	1988	1987
Stimulants	Ephedrines	104	230	374	297	215
Narcotics	Codeine	36	32	34	35	26
Anabolic steroids	Nandrolone	165	192	224	304	262
β-blockers	Propranolol	187	4	3	7	19
Diuretics	Frusemide	7	15	15	35	8
Masking agents	Probenecid	1	6	10	19	—

hCG was reported under peptide hormones
Ethanol is classified under substances subject to certain restrictions.
THC (cannabis metabolite) and phenobarbital are classified as 'not on IOC list'.

Cocaine

Although cocaine is one of the most frequently used recreational drugs, apart from anecdotal reports of Peruvians in the Andes chewing leaves of the coca plant, there appear to be no reported studies of the effects of cocaine on athletic performance. The drug acts on catecholaminergic neurones to prevent reuptake of both noradrenaline and dopamine.[2] The physiological effects are similar to those of amphetamine, but the mood effects are more profound.

Immunoassays for benzoyl ecgonine, the main metabolite found in urine, are the most convenient to screen for cocaine since they can be applied directly to a small volume of urine (0.1 ml). Any sample which fails this screen will be submitted to a confirmatory analysis using gas chromatography coupled to mass spectrometry (GC–MS). For this confirmation, a somewhat complex extraction and chemical derivatization procedure has to be used to make the sample suitable for GC–MS analysis.

Amphetamine

The term amphetamine is often used to describe amphetamine itself and the homologous substances methylamphetamine, dimethylamphetamine, benzylamphetamine, and other substances known as 'masked amphetamines' which are metabolized to amphetamine in the body. The term is sometimes used incorrectly to describe ephedrine and its homologues, which are discussed separately below, probably because some immunoassays for amphetamine cross-react with the ephedrines, thus giving rise to possible misinterpretation of the results.

Amphetamines have been shown to enhance certain sporting performances. For example, amphetamines administered 2 to 3 h before a swimming competition produced an increase in time to exhaustion and a small but consistent increase in speed.[3] Other studies have shown that sprinting speed and muscular strength are not affected by amphetamines.[4,5] Large doses in rats[6] (10–20 mg/kg) have been effective in increasing time to exhaustion in swimming whereas doses of 1.25 to 5 mg/kg have not produced a significant effect. Treadmill endurance tests have produced similar results.[7]

Amphetamine can be readily detected by extracting the urine with diethyl ether at pH 13 and by gas chromatography of the extract using a nitrogen-selective detector.

Ephedrines

Although most people accept the need to control drugs such as amphetamine and cocaine, the necessity of restricting the use of the 'ephedrines', i.e. ephedrine, pseudoephedrine, and phenylpropanolamine, is frequently questioned. These substances are readily available in most countries in over the counter 'cold remedies' because the national regulatory authorities consider their use by the public to be relatively safe.

Consideration of the structure–activity relationship of the ephedrines aids the understanding of their effects as sympathomimetics relative to the endogenous neurotransmitter noradrenaline.[8] Substitution at the amino terminal group (Fig. 1) increases β-receptor activity, and the smaller the substituent the greater is the selectivity for α-receptor activity. However, N-methylation increases the potency compared with the corresponding primary amine. The presence of hydroxyl groups in the 3 and 4 positions of the aromatic ring results in maximal α- and β-receptor activity. The absence of the polar phenolic hydroxyl groups permits the compound to cross the blood–brain barrier more readily, producing greater central activity. Thus amphetamine, ephedrine, and phenylpropanolamine exhibit considerable central nervous system activity.

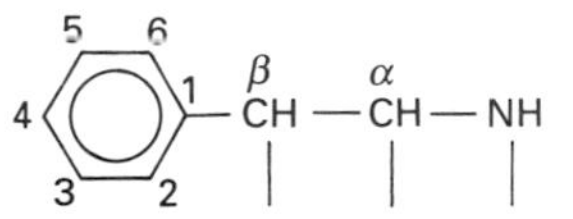

		β	α	
Metaraminol	3–OH	OH	CH₃	CH₃
Ephedrine		OH	CH₃	CH₃
Amphetamine		H	CH₃	H
Phenylpropanolamine		OH	CH₃	H

Fig. 1 Structure–activity relationship of ephedrines.

Substitution on the α-carbon atom blocks metabolism of these compounds by monoamine oxidase, allowing them to persist in nerve terminals and hence prolonging the stimulation of the release of noradrenaline from storage sites. A hydroxyl group on the β-carbon atom tends to decrease central nervous system activity because of reduced lipid solubility. Thus phenylpropanolamine, which is probably the weakest sympathomimetic of the ephedrines, has more marked α-activity than β-activity because it is a primary amine and because of the hydroxyl group on the β-carbon atom. It has considerable, but often underestimated, central nervous system activity because of the absence of aromatic substituents. However, this activity is less than that of amphetamine because of the β-substituent which reduces the lipid solubility. Phenylpropanolamine induces hypertension characterized by an increase in cardiac output, peripheral vascular resistance, stroke volume, ejection fraction, and a decrease in heart rate.[9] Even in normal individuals a typical dose of 75 mg can have a significant effect, and severe hypertension is common after excessive doses and may result in hypertensive encephalopathy, intracerebral haemorrhage, and death.

The ephedrines can be readily detected like amphetamine by extracting the urine with diethyl ether at pH 13 and by gas chromatography of the extract using a nitrogen-selective detector. IOC accredited laboratories are advised to ignore small concentrations of ephedrines, provided that the pH and specific gravity of the urine is within normal limits.

Caffeine

There is a lack of good dose–response data for caffeine, and evidence for its stimulant effect on the central nervous system is conflicting. However, caffeine has been shown to have direct effects on muscle contraction during exercise in vivo. At low stimulation frequencies, increased muscle tension was observed 1 h after 50 mg caffeine was administered orally, but there was no apparent change in endurance time.[10] Most studies have used a single dose of caffeine ranging from less than 1 mg/kg to more than 14 mg/kg. Caffeine tolerance may be an explanation of the conflicting results. However, considering the published data, doses greater than 400 mg are probably capable of increasing endurance and physical performance.

The mechanism for its action is also not certain. Caffeine has been shown to increase calcium permeability which is essential for muscle contraction.[11] Its action on the central nervous system may mask fatigue and increase the capacity for sustained intellectual effort.[12] Although it may decrease reaction time, fine motor coordination and the ability to judge distance may be impaired. It stimulates the medullary respiratory centres but may produce emesis via the central nervous system.

Caffeine significantly increases the availability of free fatty acids from fat by lipolysis.[13,14] When available, fatty acids are the primary substrate for aerobic metabolism, thus sparing glycogen. Unlike fatty acids, glucose (produced from glycogen) can be metabolized either aerobically or anaerobically, and thus the glycogen spared by the alternative metabolism of fatty acids can be made available at a time when the oxygen supply to the tissues is insufficient for aerobic metabolism.[15]

Caffeine has been shown in vitro to produce a translocation of intracellular calcium to inhibit phosphodiesterase, thus producing an accumulation of cyclic nucleotides, and also to block the actions of adenosine at adenosine receptors. Among other actions, adenosine strongly inhibits hormone-induced lipolysis, reduces the release of noradrenaline from nerve endings, and may inhibit the release of excitatory neurotransmitters in the central nervous system. The concentrations of caffeine required for the translocation of intracellular calcium and the inhibition of phosphodiesterase are greater than are thought to be achieved from a therapeutic dose.[16] Thus this leaves the blocking of the adenosine receptor as at least one of the most likely routes for caffeine's actions.

The IOC makes it an offence to have a urinary caffeine concentration greater than 12 mg/l (60 μmol/l) and this may be exceeded with a 400-mg dose. Only about 1 per cent of administered caffeine appears in the urine unchanged; most is metabolized in the liver. A significant proportion of caffeine is N-demethylated to form dimethylxanthines, and the concurrent administration of a dimethylxanthine theophylline or theobromine (found in chocolate) may reduce the metabolism of caffeine.[17] The concentration of caffeine in urine is readily measured by high pressure liquid chromatography (HPLC) with UV detection at 280 nm.

β₂-Agonists

The IOC Medical Commission recognizes that the choice of medication in the treatment of asthma and respiratory ailments has posed many problems. Some years ago, ephedrine and related substances were administered quite frequently. However, these substances are prohibited because they are classed in the category of 'sympathomimetic amines' and therefore are considered as stimulants. However, the IOC permits the use of certain specified β_2-agonists in aerosol form, i.e. bitolterol, orciprenaline, rimiterol, salbutamol, and terbutaline. These are substances which act relatively selectively at the β_2-adrenergic receptor sites in human bronchial muscle and have relatively little activity at cardiac β_1-adrenoreceptor sites. However, the β_2-agonist fenoterol is not permitted because it is metabolized to 4-hydroxyamphetamine which is banned. Newer β_2-agonists need to be considered by the IOC before they will be listed as permitted.

Narcotic analgesics

The second category of doping classes is the narcotic analgesics. Narcotic analgesics are not perceived by most people as ergogenic drugs. The chronic use of narcotics might seem to lead to an impairment of athletic skills. However, the available evidence does not support this assumption. No significant difference has been demonstrated with age-matched controls and addicts in tests of motor strength, rapid alternating movements, eye–hand co-ordination, visual perception, and cognitive skills.[18] Encephalins and endorphins are endogenous peptides with potent analgesic activity which bind to the sites in the brain to which morphine and other potent analgesics bind avidly. A correlation between exercise and endorphin activity has been demonstrated.[19] Much of the euphoria experienced by athletes, sometimes known as a 'runner's high' can be explained by a release of some of these endogenous opiates. Individuals who participate in running may even have a type of addiction to these endogenous opiates.[20] On stopping exercising many athletes experience symptoms including anxiety, restlessness, irritability, nervousness, guilt, muscle twitching, and sleep disturbances which may be an 'abstinence syndrome'.

According to the IOC Medical Commission there is evidence indicating that narcotic analgesics have been and are being abused in sports and hence should be banned. The IOC also justifies their ban on the use of these substances because of the international restrictions affecting movement of these compounds, and comments that their action is in line with the regulations and recommendations of the WHO regarding narcotics.

Many of the narcotic analgesics appear in the urine as glucuronic acid conjugates and some (e.g. morphine) are amphoteric aminophenols and require hydrolysis of the conjugate before extraction with a moderately polar solvent such as ether plus propan-2-ol (9:1 v/v) at pH 9.2–9.5 which is close to the isoelectric point of the aminophenols. Chemical derivatization is usually necessary to facilitate good chromatography, and the derivatized extract can be screened by gas chromatography with a nitrogen-selective detector or by GC–MS with selected ion monitoring (SIM).

β-Blockers

In normal individuals β-blockers adversely affect both anaerobic endurance and aerobic power as measured by V_{O_2}max, endurance, and time for a 2-km run.[21] Athletes in events such as archery and shooting may gain an advantage from the anxiolytic, bradycardic, and antitremor effects of the β-blockers.

The IOC Medical Commission has reviewed the therapeutic indications for the use of β-blocking drugs and considers that there is a wide range of effective alternative preparations available in order to control hypertension, cardiac arrhythmias, angina pectoris, and migraine. Because of continued misuse of β-blockers in some sports where physical activity is of no or little importance, it

reserves the right to test those sports which it deems appropriate and states that these are unlikely to include endurance events. The IOC lists as examples the lipid-soluble β-blockers such as propranolol, which may act via the central nervous system and have anxiolytic activity, as well as those which are water soluble, such as atenolol, nadolol, and sotalol, and hence are not centrally active. Some β-blockers such as atenolol and metroprolol have less effect on the β_2 (bronchial) receptors and are relatively cardioselective but not cardiospecific. These agents are not exempted from control, and hence presumably the IOC wish to control β-blockers because of their effect of slowing the heart rate. Table 2 indicates those sports which in September 1987 the IOC Medical Commission decided would be tested for β-blockers at the Olympic Games. Recently (February 1992) the International Amateur Athletic Federation, because of concern about potential misuse, have asked by way of a survey for tests for β-blockers to be performed on samples from competitors taking part in field events.

Table 2 IOC Medical Commission: sports to be tested for β-blockers

Winter games	Summer games
Biathlon	Archery
Bobsled	Diving and synchronous swimming
Figure skating—compulsory event	Equestrian
	Fencing
Luge	Gymnastics
Ski jumping	Modern pentathlon—shooting only
	Sailing
	Shooting

Most β-blockers are excreted as glucuronic acid conjugates of the parent substance and its metabolites, and urine samples can be screened for β-blockers following enzyme hydrolysis of the conjugates and then by solvent extraction at alkaline pH and analysis by HPLC, or the extract may be chemically derivatized and analysed by GC–MS. Alternatively, a solid phase extraction of the urine can be used followed by GC–MS of the derivatized extract.

Diuretics

The most obvious reason why athletes may misuse diuretics is to lose body fluid and hence body weight rapidly in order to be able to compete in a lower weight category in those sports where weight is controlled, e.g. judo, boxing, and rowing. The second and somewhat more obscure reason for misusing diuretics is to reduce the urinary concentration of drugs through rapid diuresis to decrease the likelihood of detection of those drugs in a urine test. Diuretics listed as examples by the IOC include the potent loop diuretics frusemide and bumetanide as well as thiazide diuretics such as hydrochlorothiazide and benzthiazide. These substances cause a pronounced diuresis resulting in a rapid loss of body water. The carbonic anhydrase inhibitor diuretic acetazolamine also increases the urine pH and will thereby reduce the excretion rate of basic drugs. For example, the peak excretion rate of the stimulant drug phentermine is delayed from the normal 2 to 4 h after administration to 30 to 36 h.[22,23] There appears to be no study which indicates any enhancement in performance as a result of diuretic use.

Despite their effect on reducing the concentration of other substances in urine, the diuretics producing this effect are readily detectable. The urine can be extracted with an organic solvent at acid and also at alkaline pH, and the extracts combined and analysed by HPLC, or the extract can be chemically derivatized and analysed by GC–MS. Alternatively, a solid phase extraction of the urine can be used followed by GC–MS of the derivatized extract.

Anabolic steroids

The anabolic steroids, or perhaps more correctly the anabolic androgenic steroids, comprise probably the most notorious of the banned categories of drugs in sport. Although the misuse of anabolic steroids was originally thought to be a problem unique to competitive heavy sports, there is increasing evidence that they are being misused in endurance events and by individuals who do not participate in sports. A survey in the United States[24] revealed that more than 6 per cent of high school males had taken anabolic steroids to make them appear more masculine to their girlfriends.

Testosterone is the principal biologically active androgen circulating in the blood of both men and women, although women produce about one-twentieth of the amount produced by men. The 5α-reduced form of testosterone is a more potent androgen than testosterone. The reduction predominates in androgen-dependent tissues, for example the prostate, seminal vesicles, and epididymus. Gonadotrophin-releasing hormone (GnRH) stimulates the release of luteinizing hormone (LH) which in turn stimulates the production of testosterone. Circulating androgens exert a negative feedback on the adenohypophysis and hypothalamus, suppressing LH and GnRH release and thus controlling the production of the androgens.

Many chemical modifications of testosterone (Fig. 2) have been attempted to alter the anabolic to androgenic ratio (or more muscle per whisker) in order to reduce the virilizing androgenic

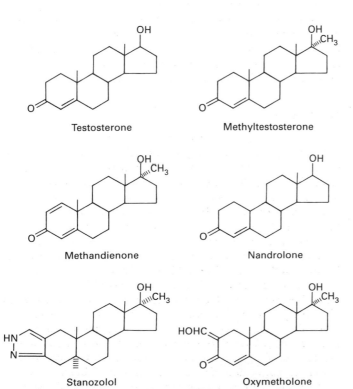

Fig. 2 Structure of testosterone and some synthetic anabolic steroids.

effects or to make the substance orally active. Alkylation at the C17 position reduces hepatic oxidative metabolism and makes the compound orally active, e.g. 17α-methyltestosterone and methandienone. Removal of the methyl group at C10 produces nandrolone (19-nortestosterone) which exhibits more anabolic than androgenic activity. Other modifications to increase the anabolic-to-androgenic ratio are the inclusion of a second double bond in the A ring, e.g. methandienone, attachment of a pyrazole ring to the A ring (stanozolol), or a hydroxymethylene group at C2 (oxymetholone). Despite these modifications, it has not proved possible to remove the androgenic (virilizing) effects from these synthetic steroids which is why this group is called the anabolic androgenic steroids.

Many studies attempting to demonstrate a positive effect of anabolic steroids on sports performance have been conducted over the last 40 years and the reader should consult one of the many reviews of the subject for more details.[25-27] Perhaps the firmest conclusion that can be drawn from these studies is that if anabolic steroids do have any effect, then their effect is small. Although fluid retention and hence increase in body weight may be one of the first effects observed with anabolic steroid use, some studies have shown a significant increase in body size and weight which is not due to fluid retention. These studies have observed experienced weightlifters who continued training while taking anabolic steroids and whose strength was also shown to increase more than when they were not taking anabolic steroids. Generally, the actions of anabolic steroids which would benefit the athlete are due to anabolic, anticatabolic, and motivational or behavioural effects. Athletes can develop a negative nitrogen balance during excessive training, and anabolic steroids may block the effects of the glucocorticoids released from the adrenal cortex in response to the stress of training. Anabolic steroids stimulate erythropoiesis, but it is questionable whether the effect is sufficiently significant to enhance performance. However, now that erythropoietin produced by recombinant DNA technology is readily available, despite being more expensive, this is more likely to be misused to increase haemoglobin.

Adverse effects of anabolic steroids include testicular atrophy, sterility, virilization, and gynaecomastia, and in adolescents premature closure of the epiphyses resulting in permanent short stature. Dermatological disorders include oily skin and acne, hirsutism, and alopecia.[28] Behavioural effects,[29] including major depression or psychosis,[30] may occur quite frequently. Choi et al.[31] have reviewed the subject. In addition, common adverse effects in females include amenorrhoea, hirsutism, enlargement of the larynx with consequent deepening of the voice, clitoral hypertrophy, and breast atrophy.[32] Anabolic steroids decrease serum high density lipoprotein cholesterol (HDL-C) and either do not change or increase low density lipoprotein cholesterol (LDL-C). Thus the ratio of LDL-C to HDL-C will usually increase, and this is a cardiovascular risk factor.[33] Cases of stroke and infarction may reflect an altered thrombogenesis caused by use of anabolic steroids.[34] These and other adverse effects of anabolic steroids are discussed in the review by Hickson et al.[35]

Testosterone

At present, only an untimed urine sample is available from an athlete to determine whether doping with a prohibited substance has occurred. All that is required for detecting the use of synthetic anabolic steroids is the unequivocal identification of the parent compound or its diagnostic metabolite in the urine. In the case of natural hormones, where they or their metabolites are normally present in urine, detection depends upon the effect of administration upon the system controlling their endogenous secretion and the consequent alterations in the pattern of steroid excretion. The administration of supraphysiological doses of testosterone to men results in the inhibition of LH release and, as a consequence, inhibition of the secretion of testosterone and other testicular steroids. The resulting decrease in the urinary excretion rates of LH and its 17-epimer epitestosterone (17α-hydroxyandrost-4-en-3-one), together with the increase in that of testosterone (from the injection), is reflected by elevated ratios of testosterone to epitestosterone[36] and testosterone to LH[37] in urine. A testosterone to epitestosterone ratio greater than 6 is not allowed under the regulations of the IOC (see Appendix) and may result in disqualification of the athlete and disciplinary action. It is possible to evade detection by the administration of combined testosterone–epitestosterone preparations. In normal men, 30 times more testosterone compared with epitestosterone is produced endogenously, but only 1 per cent of testosterone is excreted unchanged compared with 30 per cent of epitestosterone, resulting in a urinary testosterone to epitestosterone ratio of unity. It follows that athletes could take supraphysiological doses of testosterone with epitestosterone in a ratio of about 30:1 and still maintain a normal urinary ratio. In comparison with the testosterone to epitestosterone ratios, the testosterone to LH ratios would be expected to be abnormally high, and in the United Kingdom the ratio of testosterone to LH is measured in addition to that of testosterone to epitestosterone. Pharmaceutical preparations of epitestosterone are not available, but non-proprietary preparations would be easy to prepare and in the age of designer drugs it would be complacent to assume that they are not available to athletes. In the future LH measurement may also be important in the detection of the administration of dihydrotestosterone, which is a particularly active metabolite of testosterone.[38]

Some male athletes who have denied using testosterone have produced testosterone to epitestosterone ratios between 6 and 9.[39] The possibility that these individuals excrete smaller quantities of epitestosterone, perhaps due to an enzymatic deficiency in their biosynthesis of epitestosterone, must not be excluded. These athletes would be expected to have normal urinary testosterone to LH ratios because the homeostatic mechanism tends to keep the concentration of testosterone in the blood constant.

Dihydrotestosterone

Administration of dihydrotestosterone, an active metabolite of testosterone, also inhibits LH secretion,[40] but, because of the reduced excretion of testosterone as well as of LH and epitestosterone, it is unlikely to alter these ratios significantly. Thus dihydrotestosterone is a likely candidate for use by athletes. Furthermore, dihydrotestosterone can be regarded as more potent than testosterone since it is known to bind more strongly than testosterone to the androgen receptor.[41]

Under normal circumstances the concentration of circulating testosterone in men is as much as 10 times that of dihydrotestosterone, so that in skeletal muscle testosterone is the predominant ligand bound to the androgen receptor. In contrast with muscle, sexual tissue contains much greater 5α-reductase activity which converts testosterone to dihydrotestosterone. In these tissues dihydrotestosterone is bound by the receptor, thus amplifying the effect of circulating testosterone. Therefore large plasma concentrations of dihydrotestosterone (e.g. following dihydrotestosterone administration) would have a greater effect in muscle compared with sexual tissue, thereby giving dihydrotes-

tosterone a greater myotrophic to androgenic ratio than testosterone.

No internationally agreed method currently exists to prove dihydrotestosterone administration. However, a test may be possible based on the relative increase in concentration of dihydrotestosterone, 5α-androstane-3α,17β-diol (3α-diol), and 5α-androstane-3β,17β-diol (3β-diol) compared with that of testosterone, 5β-androstane-3α,17β-diol (5β-diol), epitestosterone, LH, and follicle-stimulating hormone (FSH) following dihydrotestosterone administration.[38]

Peptide hormones and analogues

In 1989 the IOC Medical Commission introduced the new doping class of 'peptide hormones and analogues' which include human chorionic gonadotrophin and related compounds, adrenocorticotrophic hormone (ACTH), human growth hormone (hGH), all the releasing factors of these listed hormones, and erythropoietin. Most of these hormones are produced endogenously, and several are now produced synthetically using recombinant DNA technology. In each case, the administered substance is identical to that produced naturally. Currently there are no IOC approved definitive tests for these hormones, but highly specific immunoassays combined with suitable purification techniques may be sufficient to warrant IOC approval. Kicman and Cowan have reviewed the misuse and detection of peptide hormones in sport.[42]

Human chorionic gonadotrophin

Human chorionic gonadotrophin is a glycopolypeptide secreted in large amounts by the developing placenta (chorion) and by certain types of tumour. Although hCG and LH are immunologically distinct, these two gonadotrophins have essentially the same biological action at the cellular level. The approximate molecular weight of hCG is 37 000 and that of LH is 28 000 with very similar α-subunits in primary structure. The hCG α-subunit consists of 92 amino acids and has two carbohydrate moieties which are branch-chained to asparagine residues, giving an average total molecular weight of 14 500. The β-subunit of hCG consists of 145 amino acid residues with 30 additional residues at the C-terminal region compared with the β-subunit of LH. It is within this region that a specific antigenic determinant is located which allows for immunological distinction as opposed to biological distinction between hCG and LH.

Six carbohydrate moieties are attached to the β-hCG subunit, giving an average total molecular weight of 22 000. Two branch-chain moieties are linked to asparagines and four are linked to serines within the unique C-terminal peptide region.

The complete molecule has eight carbohydrate moieties in total, making it approximately 30 per cent by weight in carbohydrate content. Each moiety contains two attached N-acetylneuramic acid groups, which is a sialic acid, giving rise to 16 sialic acids per hCG molecule. However, there is microheterogeneity in the carbohydrate content of the gonadotrophins which has not yet been fully characterized, and hence the total number of sialic acid groups per hCG molecule will only represent an average amount and this number will vary between reported studies. The greater number of sialic acid groups on hCG leads to a greater ionization constant and hence a smaller pI value when compared with the other gonadotrophins. The pI of hCG is about 3.5 and that of LH is about 5.5.

Small concentrations of hCG or hCG-like material have been measured in pituitary and urine extracts, and have also been found in the serum of normal men (less than 60 years, < 1.3 IU/l; more than 60 years, < 2.3 IU/l) and non-pregnant women (pre-menopausal, < 1.0 IU/l; postmenopausal, < 4.8 IU/l).[43] The biochemical and endocrinological features of hCG, together with its detection and misuse in sport have been reviewed by Kicman et al.[44]

In the male, LH and hCG interact with specific target receptors on the surface of Leydig cells in the testes. The hormone-bound receptors activate cyclic AMP and calcium ion secondary messenger systems which in turn activate various protein kinases, finally stimulating steroidogenesis and protein synthesis. Phospholipid metabolites, in particular the leukotrienes, may also act as secondary messengers in LH/hCG-induced steroidogenesis. hCG stimulation of testicular steroidogenesis in healthy adult men is very rapid. A 50 per cent increase in plasma testosterone concentration has been measured 2 h after an intramuscular injection of 6000 IU of hCG.[45] However, there is no direct correlation between plasma concentrations of hCG and testosterone; the rise in plasma testosterone is biphasic. Pharmaceutical preparations of hCG may be used by some male athletes to stimulate testosterone production and also to prevent testicular atrophy during and after prolonged courses of androgen administration. The use of pharmaceutical preparations of LH are not as suitable for these purposes because of the much shorter plasma half-life of LH and the small amount supplied per ampoule (e.g. 75 IU) compared with hCG (e.g. 5000 IU) (1 IU of hCG has approximately the same biological activity as 1 IU of LH). The administration of hCG has been banned by the IOC since December 1987.

Various immunometric assays are available to measure the presence of hCG in urine. The use of an ultrafiltration method may be suitable as part of a confirmatory procedure.[46] However, at present, it is not possible to determine by an assay of hCG in urine whether a concentration greater than normal has arisen from hCG administration or from pregnancy or an hCG-secreting tumour.

Growth hormone

Human growth hormone is secreted episodically by the somatotrope cells of the anterior pituitary. The major physiological form is a single-chain peptide with a molecular weight of 22 000 (22 kDa) consisting of 191 amino acid residues and two intrachain disulphide bridges. Several variants of hGH also exist, possibly due to artefacts generated by proteolytic digestion in the pituitary, enzymatic processing, and the formation of dimers and oligomers. Also a 20-kDa form exists which may result from variation in the processing of the mRNA precursor.

No studies to date have shown an increase in strength or endurance in association with growth hormone (GH) administration.[47,48] GH has a short half-life of 20 to 30 min. The production of growth hormone currently utilizes recombinant-DNA technology and although initially it was not an exact duplication, with the end amino acid being methionine, it now has an identical sequence. Thus it would be extremely difficult to devise a test based on the detection of GH itself to prove that it had been administered in contravention of the rules of sport.

Prior to puberty, excessive GH gives rise to gigantism. The sufferer tends to die in early adult life from infection, debility, or hypopituitarism. After puberty and closure of the long bones, excessive GH can cause acromegaly with characteristic features such as spade-like hands, lengthened and thickened jaw, coarse and leathery skin, increase in coarse body hair, and enlargement of the heart.[49]

GH release is stimulated by a variety of substances including GH-releasing hormone, noradrenaline, adrenaline (and hence stress), levodopa, arginine, and insulin (hypoglycaemia), and by sleep. Blood concentrations are reduced by somatostatin and by GF-1.

The mechanism of action of GH is far from fully understood. The GH receptor is widely distributed throughout the body and has been identified on the cell surface membranes of hepatocytes, adipocytes, fibroblasts, lymphocytes, and chondrocytes.[50] GH has been shown, using rat hepatocytes, to cause phosphorylation of certain proteins. This has been demonstrated to be important for the proliferation, differentiation, and growth processes, and can alter enzymatic activity and cytoskeletal mobility and may modulate events in the nucleus such as gene expression.

Erythropoietin

Erythropoietin is an acidic glycoprotein hormone mainly secreted by the kidney consisting of a 166 amino-acid chain. It has an estimated molecular weight of 34 000 kDa and exists in two distinguishable forms, α and β. The α and β forms have a carbohydrate content of approximately 31 per cent and 24 per cent by weight respectively. These two forms of erythropoietin have similar biological and antigenic properties.

Erythropoietin exerts a specific-receptor-mediated effect on committed erythroid stem cells, inducing these target cells (colony-forming unit or erythroid) to proliferate approximately 30 times into mature erythrocytes. The mechanism by which erythropoietin causes its target cells to undergo division and maturation is not fully understood, although high and low affinity receptors have been identified and signal transduction may involve Ca^{2+} and arachidonic acid. Erythropoietin production is regulated by the relative amount of oxygen in arterial blood; serum erythropoietin concentration increases in hypoxia and decreases in hyperoxia. Nevertheless, erythropoiesis is also stimulated by blood loss despite normal arterial oxygen pressure, which has led to the hypothesis that it is tissue oxygen tension as determined by the supply of and demand for oxygen by cells rather than arterial oxygen per se which regulates erythrocyte production (Fig. 3).

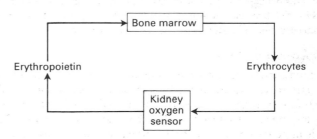

Fig. 3 Diagrammatic representation of erythropoietin production.

The mean serum erythropoietin concentration in normal individuals is 35 IU/l, whereas individuals with aplastic anaemia may have erythropoietin concentrations up to approximately 3000 IU/l and individuals with anaemia caused by renal disease have erythropoietin concentrations which are abnormally small. However, individuals with polycythaemia vera may have normal or small serum erythropoietin concentrations. Hypopituitarism and hypothyroidism reduce metabolic rate and erythropoiesis.

With the recent success in producing recombinant human erythropoietin, sufficient quantities are now available for patient treatment (e.g. anaemia due to renal failure) and it is obtainable as a licensed product from pharmaceutical manufacturers. Recombinant human erythropoietin is of therapeutic value when endogenous erythropoietin production is less than expected for the degree of haemoglobin and/or the volume of erythrocytes relative to whole blood (haematocrit) and provided that erythroid progenitors are available. The average dose necessary to maintain haematocrit between 30 and 36 per cent in anaemic dialysis patients is 50 IU/kg body weight administered three times weekly. Recombinant human erythropoietin may also be administered to patients wanting to deposit blood for transfusion during elective surgery.

Erythrocyte production increases dramatically in response to the administration of supraphysiological doses of recombinant human erythropoietin. Maximal oxygen uptake has been shown to increase in healthy individuals who have undergone erythrocyte infusion to induce a state of erythrocythaemia. Therefore athletes may choose to use recombinant human erythropoietin or erythrocyte infusion (blood doping) to increase their oxygen uptake in the hope of improving their endurance capacity and recovery during training and competition. In contrast with patients who constantly require recombinant human erythropoietin replacement therapy, an athlete would only need a short course of recombinant human erythropoietin to gain a possible advantage against fellow competitors. Thus, relative to the expense of recombinant human erythropoietin replacement therapy, the cost of recombinant human erythropoietin abuse to gain a short-term sporting advantage would be small.

Erythropoietin has minimal toxicity and has been used in clinical trials up to doses of 500 IU/kg.day. Recombinant human erythropoietin appears not to be immunogenic. Some patients report a sensation of cold and ache in the long bones 1 to 2 h after recombinant human erythropoietin administration. In patients receiving recombinant human erythropoietin there may be an increase in peripheral vascular resistance which may lead to hypertension. In athletes who induce erythrocythaemia by recombinant human erythropoietin administration or blood doping, there is a danger of thrombosis and haemorrhage.

Although erythropoietin administration is included in an IOC banned doping class and blood doping in a doping method, currently there is no IOC approved confirmatory procedure to detect these practices. To detect erythropoietin doping simply by determining abnormally large serum erythropoietin concentrations using specific immunoassay techniques may not be effective due to the short half-life of recombinant erythropoietin of approximately 4 to 6 h when administered intravenously. Only about 10 per cent of erythropoietin is excreted into the urine. However, as erythropoietin administration or blood doping would be expected to raise erythrocyte count, haemoglobin concentration, and haematocrit, the measurement of these parameters could be used as criteria for the detection of both types of practice. Additionally, a serum erythropoietin concentration below or above the normal range could be used as an additional criterion in the detection of erythropoietin administration or blood doping. Nevertheless, this profile may be indistinguishable from that of athletes who have undergone high altitude training. In any case, the informed athlete would know that the effects of erythropoietin administration would result in a sustained increase in the erythrocyte population long after administration had ceased, since the average life of erythrocytes is about 107 days. The detection of erythropoietin administration by determining abnormal erythropoietin concentrations, whether in blood or urine, may therefore be difficult because the athlete could take steps to avoid giving a sample while abnormal concentrations are present in his or her body, for example by going on 'holiday'.

Adrenocorticotrophic hormone

ACTH is a polypeptide released from the corticotrope cells of the anterior pituitary, resulting in stimulation of the secretion of adrenal androgens as well as the glucocorticoids. It is composed of 39 amino acids and has a molecular weight of 4500 kDa. However, only the first 24 N-terminal amino acids are necessary for full biological activity, and it is this synthesized form, called tetracosactrin (e.g. Synacthen®) which is used for diagnostic purposes. Pituitary-derived ACTH (e.g. Acthar®) is now only supplied in the United Kingdom as an unlicensed product.

The non-therapeutic administration of corticosteroids prompted the IOC Medical Commission to ban their use except for topical treatment, inhalation therapy, and local or intra-articular injections. Therapeutic use of corticosteroids may be particularly beneficial in the treatment of local inflammation and pain associated with sports injuries, but such treatment may cause some entry of these drugs into the bloodstream. However, any systemic effects are likely to be insignificant except with excessive topical applications where treatment can occasionally result in suppression of the hypothalamic–pituitary–adrenal axis and Cushing's syndrome.[51] As ACTH stimulates the secretion of cortisol and other glucocorticoids into the systemic circulation, its administration is considered to be equivalent to an oral, intramuscular, or intravenous application of a corticosteroid and hence is banned.

The predominent therapeutic use of tetracosactrin is for the differential diagnosis of adrenocortical failure. However, ACTH replacement therapy is not suitable in the treatment of secondary adrenocortical insufficiency, not least because tetracosactrin is not orally active, and in such cases treatment is with cortisol or a synthetic analogue (e.g prednisolone). Tetracosactrin is also used for short-term therapy to induce excessive secretion of glucocorticoids rapidly, for example, in severe asthma attacks. The depot form of tetracosactrin is used in cases where the results of short-term stimulation with tetracosactrin for the diagnosis of adrenocortical insufficiency has been inconclusive or in patients who are unable to tolerate glucocorticoid therapy.

Currently, there is no IOC approved method to detect ACTH administration which represents a complex problem as it is particularly liable to rapid degradation by endopeptidases present in blood, resulting in the excretion of inactive oligopeptides. Hence the development of a confirmatory test using urine samples is unlikely. However, with the future possibility of blood as well as urine samples being collected from athletes, detection could be based on chromatographic methods followed by immunoassays to distinguish between serum ACTH and tetracosactrin. Clinically, special procedures are adopted for analytes such as ACTH and tetracosactrin which are susceptible to enzyme degradation in serum or plasma. Plasma is immediately separated in a refrigerated centrifuge and then stored at −20 °C or lower throughout storage and transport and only thawed prior to analysis. Clearly, this approach is not practical for the collection of samples at competitions and training centres, and alternative procedures for maintaining stability need to be investigated. A more feasible approach could include the presence of enzyme inhibitors in the sample collection tube, together with reducing agents such as dithiothreitol, as ACTH is also particularly susceptible to oxidation at methionine residues.[52]

Releasing factors

The releasing factors are neuropeptide hormones secreted episodically by the hypothalamus, which in turn regulate the secretion of LH, hGH, and ACTH by the anterior pituitary. Single administration of any of these releasing hormones or their analogues stimulates a prompt dose-related response in the secretion of the corresponding pituitary hormone. The response is of short duration, and, to maintain the effect, dose and timing of repeated administration is critical as sustained high concentrations of GnRH or corticotrophin-releasing factor (CRF) suppress the release of the associated pituitary hormones due to the down regulation of receptors. Thus the chronic use of GnRH and CRF are disadvantageous to the athlete as the pituitary requires the pulsatile release of these hormones to respond normally. Nonetheless, long-term pulsatile administration of GnRH or one of its analogues could be used by male athletes in an attempt to stimulate and maintain testicular function during and after prolonged courses of androgen administration. In contrast with GnRH and CRF, continued exposure to GH releasing hormone causes only partial down regulation of the corresponding receptors, probably because of the continuing pulsatile secretion of somatostatin. With the future availability of GH releasing hormone for therapeutic use, athletes may administer supraphysiological doses of this releasing hormone or its more potent analogues to increase their secretion of GH.

DOPING METHODS

Blood doping

Following the revelation after the Olympic Games in Los Angeles in 1984 that the United States cycling team had had blood transfusions, the practice of blood doping in sport was banned by the IOC. Although older studies failed to show clear improvements in performance, this may be because refrigerated citrated blood deteriorates whereas modern procedures enable the freezing of red blood cells to preserve them indefinitely with minimum deterioration. Following infusion of 900 ml of autologous red blood cells collected some 7 weeks previously, highly trained elite athletes experienced increases in run time to exhaustion (35 per cent), V_{O_2}max (5 per cent), and haemoglobin (7 per cent).[53]

Any form of induced erythrocythaemia increases the risk, like the clinical condition of polycythaemia, of hypertension, congestive heart failure, and stroke.

The detection of heterologous blood doping requires a blood sample and proof that the detected phenotypes differ from those of the individual sampled. In this case, all that may be required is proof of the presence of heterologous erythrocytes. This is the principle employed in the Finnish approach.[54] The detection of autologous blood doping is far more complex. The most likely approach to detect blood doping and also erythropoietin administration at present would seem to be to determine a haematocrit threshold value which may not be exceeded. This would need to be set at different values for males and females, and further differentials would probably be necessary since haemoglobin is known to vary widely in athletes.[55] Of course, this approach would control high altitude training and it would be difficult to deal with individuals who live at high altitudes.

Pharmacological, chemical, and physical manipulation

The IOC ban pharmacological, chemical, and physical methods which alter the integrity and validity of the urine sample collected in a doping control. Specifically, as a pharmacological method, they ban probenecid and related compounds.

Probenecid

Probenecid inhibits the transport of organic acids across epithelial barriers, and this is of great importance in the renal tubule in which tubular secretion of many drugs and their metabolites is inhibited. However, the excretion of uric acid is increased since its reabsorption in the renal tubule is inhibited by probenecid. Many anabolic steroids are excreted as glucuronic acid conjugates, and probenecid can reduce their renal clearance and hence urinary concentration. To be effective, about 2 g of probenecid needs to be taken and it is not difficult for the laboratory to detect its presence in a urine sample.

CLASSES OF DRUGS SUBJECT TO CERTAIN RESTRICTIONS

Alcohol

The IOC do not prohibit the use of alcohol. However, because the governing body of some sports such as pistol shooting and motor racing limit alcohol use, the IOC will determine breath or blood alcohol levels at the request of an International Federation.

Marijuana

Although the IOC does not ban marijuana, at the Olympic Games held in Seoul in 1988 tests for the presence of its metabolites in urine were carried out, and this will be continued at subsequent Games if so requested by an International Federation.

Local anaesthetics

The IOC permit the use of certain injectable local anaesthetics such as procaine, xylocaine, and carbocaine, but not cocaine, by local or intra-articular injections when medically justified. The IOC Medical Commission requires the details, including diagnosis, dose, and route of administration, to be submitted immediately in writing to them. There is no ban on the topical use of local anaesthetics.

Corticosteroids

One of the important effects of glucocorticoids is on the central nervous system processes which underlie behaviour.[56] In patients with Addison's disease and Cushing's syndrome psychiatric disturbances are often manifested. In Cushing's syndrome depression is more commonly manifested than mania, but curiously the converse is often the case in patients treated with corticosteroid therapy.[57] Corticosteroid treatment can result in mood-elevating energizing effects, and upon treatment withdrawal the symptoms of tiredness, malaise, and depression.[58] These effects have some analogies with those observed with amphetamine administration and withdrawal. Consequently, corticosteroids may be misused in sport for their immediate psychoendocrine effects in overcoming tiredness and lethargy and to give a 'pepped-up' feeling during training and competition. Short-term use of tetracosactrin, by stimulating excessive corticosteroid secretion, would be expected to give similar effects. Of course, long-term administration of either tetracosactrin or corticosteroids would diminish performance because of inducement of protein catabolism resulting in skeletal muscle wasting and weakness.

The IOC permit the use of corticosteroids by inhalational therapy to treat asthma and allergic rhinitis, and also allow the use of aural, ophthalmological, and dermatological preparations and intra-articular injections. However, written notification must be given to the IOC Medical Commission to justify such use medically.

By analogy with synthetic androgens, the approach to detecting the presence of synthetic glucocorticoids, such as dexamethasone or prednisolone, in athletes' urine is straightforward. Likewise, athletes may switch from using synthetic glucocorticoids to administering tetracosactrin and cortisol to evade detection. As with testosterone, detection of tetracosactrin and cortisol administration must be based on more than elevated concentrations of these hormones alone. The problem is further compounded in that the secretion of ACTH and cortisol increase in response to stress during strenuous exercise.[59,60]

A detection method for cortisol administration using the urinary ratio of the isomer 11β-hydroxyaetiocholanolone (3α,11β-dihydroxy-5β-androstan-17-one) to the isomer 11β-hydroxyandrosterone (3α,11β-dihydroxy-5α-androstan-17-one), defined as the 5β:5α ratio, has been investigated.[61] Cortisol and the androgen 11β-hydroxyandrostenedione (11β-hydroxyandrost-4-en-3,17-dione) are exclusively secreted by the adrenal, but cortisol is metabolized in greater amounts to 11β-hydroxyaetiocholanolone (5β) compared with 11β-hydroxyandrosterone (5α) whereas the reverse is the case for 11β-hydroxyandrostenedione. The study showed that the 5β:5α ratio was a useful criterion of cortisol administration and that tetracosactrin stimulation, to mimic the physiological response to stress, only caused a small increase in this ratio. The minor effect of tetracosactrin administration on the 5β:5α ratio was encouraging, but it meant that athletes could evade detection by switching from cortisol to tetracosactrin administration.

MISCELLANEOUS SUBSTANCES

Bicarbonate

People have used bicarbonate with the aim of increasing the buffering capacity of blood and hence counteracting the build-up of lactic acid resulting from anaerobic glycolysis which would inhibit glycolysis and cause fatigue. Provided that the exercise is not primarily aerobic and the dose is adequate, bicarbonate has been shown to enhance exercise time to exhaustion.[62] Large amounts are required (more than 300 mg/kg) and this will cause diarrhoea in most subjects. Chronic administration can cause hypercalcaemia and alkalosis which can cause changes in electrolyte balance and respiration, potentially with serious consequences.

Bicarbonate has previously been used to alkalinize the urine in order to increase the reabsorption of basic drugs in the renal tubules, and hence to reduce the urinary excretion and hence concentration of these substances. However, modern analytical methods are sufficiently sensitive to make this technique ineffective in evading detection. At the present time, the use of bicar-

bonate is not normally controlled in sport, although urinary pH is sometimes checked at the time of sample collection.

Clenbuterol

Clenbuterol is a β-selective adrenergic agonist which has been shown to stimulate the deposition of body protein and inhibit that of body fat in animals.[63] It has a very small therapeutic dose in man (10–40 μg/day) and a relatively long half-life of about 30 h. Its anabolic and antilipogenic actions are mechanistically distinct, and there is growing concern that it is being misused in sport for its anabolic properties. Since clenbuterol is an adrenergic agonist, it is banned by the IOC under the stimulant and also under the androgenic anabolic steroids categories as a related compound.

CONCLUDING REMARKS

The athlete does not appear to be deterred by any lack of scientific evidence that the substance is efficacious nor by the documented risk of potentially hazardous side-effects. The whole area is dynamic, advancing as science advances. There is a definite need to improve our understanding of whether and how the substances work and, although many are detectable by current analytical methods, new techniques are needed to meet and to anticipate the new challenges. Ultimately, however, what is accepted use and what represents misuse is a moral question to be answered not by science but by all of us in society and especially by those in sport.

APPENDIX

IOC Medical Commission list of doping classes and methods (May 1992)

I Doping Classes
   A Stimulants
   B Narcotics
   C Androgenic anabolic steroids
   D β-blockers
   E Diuretics
   F Peptide hormones and analogues
II Doping methods
   A Blood doping
   B Pharmacological, chemical, and physical manipulation
III Classes of drugs subject to certain restrictions
   A Alcohol
   B Marijuana
   C Local anaesthetics
   D Corticosteroids

NOTE:
The doping definition of the IOC Medical Commission is based on the banning of pharmacological classes of agents.

The definition has the advantage that new drugs, some of which may be especially designed for doping purposes, are also banned.

The following list represents examples of the different dope classes to illustrate the doping definition. Unless indicated, all substances belonging to the banned classes may not be used for medical treatment, even if they are not listed as examples. If substances of the banned classes are detected in the laboratory the IOC Medical Commission will act. It should be noted that the presence of the drug in the urine constitutes an offence, irrespective of the route of administration.

Examples and explanations

The international non-proprietary name (INN), proposed INN, or recommended INN is listed and, where different, the British approved name is given in parentheses.

I Doping classes

A Stimulants

Amfepramone (diethylpropion)
Amfetamine
Amfetaminil
Amineptine
Amiphenazole
Benzfetamine (benzphetamine)
Caffeine*
Cathine
Chlorphentermine
Clobenzorex
Clorprenaline
Cocaine
Cropropamide
Crotetamide (crotethamide)
Dimetamfetamine
Ephedrine
Etafedrine
Etamivan (ethamivan)
Etilamfetamine
Fencamfamin
Fenetylline (fenethylline)
Fenproporex
Furfenorex
Mefenorex
Mesocarb
Metamphetamine
Methylephedrine
Methylphenidate
Morazone
Nikethamide
Pemoline
Pentetrazol
Phendimetrazine
Phenmetrazine
Phentermine
Phenylpropanolamine
Pipradrol
Prolintane
Propylhexedrine
Pyrovalerone
Strychnine
and related compounds

Stimulants comprise various types of drugs which increase alertness, reduce fatigue, and may increase competitiveness and hostility. Their use can also produce loss of judgement, which may lead to accidents to others in some sports. Amphetamine and related compounds have the most notorious reputation in producing problems in sport. Some deaths of sportsmen have resulted even when normal doses have been used under conditions of maxi-

*For caffeine the definition of a positive depends upon the following: if the concentration in urine exceeds 12 µg/ml

mum physical activity. There is no medical justification for the use of 'amphetamines' in sport.

One group of stimulants is the sympathomimetic amines of which ephedrine is an example. In high doses, this type of compound produces mental stimulation and increased blood flow. Adverse effects include elevated blood pressure and headache, increased and irregular heart beat, anxiety, and tremor. In lower doses, some (e.g. ephedrine, pseudoephedrine, phenylpropanolamine, norpseudoephedrine) are often present in cold and hay fever preparations which can be purchased in pharmacies and sometimes from other retail outlets without the need of a medical prescription.

THUS NO PRODUCT FOR USE IN COLDS, FLU, OR HAY FEVER PURCHASED BY A COMPETITOR OR GIVEN TO HIM/HER SHOULD BE USED WITHOUT FIRST CHECKING WITH A DOCTOR OR PHARMACIST THAT THE PRODUCT DOES NOT CONTAIN A DRUG OF THE BANNED STIMULANTS CLASS.

β_2-Agonists

The choice of medication in the treatment of asthma and respiratory ailments has posed many problems. Some years ago, ephedrine and related substances were administered quite frequently. However, these substances are prohibited because they are classed in the category of 'sympathomimetic amines' and therefore considered as stimulants.

The use of only the following β_2-agonists is permitted in the aerosol form:

Bitolterol
Orciprenaline
Rimiterol
Salbutamol
Terbutaline

B Narcotic analgesics

Alphaprodine
Anileridine
Buprenorphine
Codeine
Dextromoramide
Dextropropoxyphene
Diamorphine
Dihydrocodeine
Dipipanone
Ethoheptazine
Ethylmorphine
Levorphanol
Methadone
Morphine
Nalbuphine
Pentazocine
Pethidine
Phenazocine
Trimeperidine (trimeperidinum)
and related compounds.

The drugs belonging to this class, which are represented by morphine and its chemical and pharmacological analogues, act fairly specifically as analgesics for the management of moderate to severe pain. However, this description by no means implies that their clinical effect is limited to the relief of trivial disabilities.

Most of these drugs have major side-effects, including dose-related respiratory depression, and carry a high risk of physical and psychological dependence. There exists evidence indicating that narcotic analgesics have been and are abused in sports, and therefore the IOC Medical Commission has issued and maintained a ban on their use during the Olympic Games. The ban is also justified by international restrictions affecting the movement of these compounds and is in line with the regulations and recommendations of the WHO regarding narcotics.

Furthermore, it is felt that the treatment of slight to moderate pain can be effective using drugs—other than the narcotics—which have analgesic, anti-inflammatory, and antipyretic actions. Such alternatives, which have been successfully used for the treatment of sports injuries, include anthranilic acid derivatives (such as mefenamic acid, floctafenine, glafenine, etc.), phenylalkanoic acid derivatives (such as diclofenac, ibuprofen, ketoprofen, naproxen, etc.), and compounds such as indomethacin and sulindac. The Medical Commission also reminds athletes and team doctors that aspirin and its newer derivatives (such as diflunisal) are not banned but cautions against some pharmaceutical preparations where aspirin is often associated with a banned drug such as codeine. The same precautions hold for cough and cold preparations which often contain drugs of the banned classes.

NOTE: DEXTROMETHORPHAN AND PHOLCODINE ARE NOT BANNED AND MAY BE USED AS ANTITUSSIVES. DIPHENOXYLATE IS ALSO PERMITTED.

C Androgenic anabolic steroids

Bolasterone
Boldenone
Chlordehydromethyltestosterone
Clostebol
Fluoxymesterone
Mesterolone
Metandienone (methandienone)
Metenolone (methenolone)
Methyltestosterone
Nandrolone
Norethandrolone
Oxandrolone
Oxymesterone
Oxymetholone
Stanozolol
Testosterone*
and related substances

The anabolic androgenic steroid (AAS) class includes testosterone and substances that are related to it in structure and activity. They have been misused by the sports world both to increase muscle strength and bulk, and to promote aggressiveness. The use of AAS is associated with adverse effects on the liver, skin, cardiovascular, and endocrine systems. They can promote the growth of tumours and induce psychiatric syndromes. In males AASs decrease the size of the testes and diminish sperm production. Females experience masculinization, loss of breast tissue, and diminished menstruation. The use of AASs by teenagers can stunt growth.

The IOC Medical Commission, while pleased that the testing

programme is decreasing the use of anabolic steroids, is nevertheless concerned that some athletes are attempting to cheat by administering testosterone, testosterone precursors, and epitestosterone. Accordingly, the IOC Medical Commission recommends giving consideration to a medical examination together with endocrine tests and longitudinal studies to evaluate the possibility that testosterone or any other endogenous steroid has been administered.

In order to assist in this evaluation the IOC accredited laboratories shall report every case to the proper authorities in accordance with the following criteria:

A negative, if the ratio is less than 6, or
B T/E greater than 6 and not greater than 10, or
C T/E greater than 10.

In the case of B the IOC Medical Commission recommends that further tests be conducted before considering the result as positive or negative. Such investigations may include:

- review of previous tests,
- endocrinological investigations,
- unannounced testing over several months.

D β-Blockers

Acebutolol
Alprenolol
Atenolol
Labetalol
Metoprolol
Nadolol
Oxprenolol
Propranolol
Sotalol
and related compounds

The IOC Medical Commission has reviewed the therapeutic indications for the use of β-blocking drugs and noted that there is now a wide range of effective alternative preparations available in order to control hypertension, cardiac arrhythmias, angina pectoris, and migraine. Owing to the continued misuse of β-blockers in some sports where physical activity is of no or little importance, the IOC Medical Commission reserves the right to test those sports which it deems appropriate. These are unlikely to include endurance events which necessitate prolonged periods of high cardiac output and large stores of metabolic substrates in which β-blockers would severely decrease performance capacity.

E Diuretics

Acetazolamide
Amiloride
Bendroflumethiazide (bendrofluazide)
Benzthiazide
Bumetanide
Canrenone
Chlormerodrin
Chlortalidone (chlorthalidone)
Diclofenamide
Etacrynic acid (ethacrynic acid)
Furosemide (frusemide)
Hydrochlorothiazide
Mersalyl
Spironolactone
Triamterene
and related compounds

---

* The presence of a testosterone (T) to epitestosterone (E) ratio greater than six (6) to one (1) in the urine of a competitor constitutes an offence unless there is evidence that this ratio is due to a physiological or pathological condition.

Diuretics have important therapeutic indications for the elimination of fluids from the tissues in certain pathological conditions. However, strict medical control is required.

Diuretics are sometimes misused by competitors for two main reasons, namely to reduce weight quickly in sports where weight categories are involved and to reduce the concentration of drugs in urine by producing a more rapid excretion of urine to attempt to minimize detection of drug misuse. Rapid reduction of weight in sport cannot be justified medically. Health risks are involved in such misuse because of serious side-effects which might occur.

Furthermore, deliberate attempts to reduce weight artificially in order to compete in lower weight classes or to dilute urine constitute clear manipulations which are unacceptable on ethical grounds. Therefore the IOC Medical Commission has decided to include diuretics on its list of banned classes of drugs.

NB For sports involving weight classes, the IOC Medical Commission reserves the right to obtain urine samples from the competitors at the time of the weigh-in.

F Peptide hormones and analogues

Chorionic gonadotrophin (HCG—human chorionic gonadotrophin)

It is well known that the administration to males of human chorionic gonadotrophin (HCG) and other compounds with related activity leads to an increased rate of production of endogenous androgenic steroids and is considered equivalent to the exogenous administration of testosterone.

Corticotrophin (ACTH)

Corticotrophin has been misused to increase the blood levels of endogenous corticosteroids notably to obtain the euphoric effect of corticosteroids. The application of corticotrophin is considered to be equivalent to the oral, intramuscular, or intravenous application of corticosteroids. (See section III, D.)

Growth hormone (HGH, somatotrophin)

The misuse of growth hormone in sport is deemed to be unethical and dangerous because of various adverse effects, for example, allergic reactions, diabetogenic effects, and acromegaly when applied in high doses.

All the respective releasing factors of the above-mentioned substances are also banned.

Erythropoietin (EPO)

Erythropoietin is the glucoprotein hormone produced in the human kidney which regulates, apparently by a feedback mechanism, the rate of synthesis of erythrocytes.

II Methods

A Blood doping

Blood transfusion is the intravenous administration of red blood cells or related blood products that contain red blood cells. Such products can be obtained from blood drawn from the same (autologous) or a different (non-autologous) individual. The most common indications for red blood transfusion in conventional medical practice are acute blood loss and severe anaemia.

Blood doping is the administration of blood or related red blood products to an athlete other than for legitimate medical treatment. This procedure may be preceded by withdrawal of blood from the athlete who continues to train in this blood-depleted state.

These procedures contravene the ethics of medicine and of sport. There are also risks involved in the transfusion of blood and related blood products. These include the development of allergic reactions (rash, fever, etc.) and acute haemolytic reaction with kidney damage if incorrectly typed blood is used, as well as delayed transfusion reaction resulting in fever and jaundice, transmission of infectious diseases (viral hepatitis and AIDS), overload of the circulation, and metabolic shock.

Therefore the practice of blood doping in sport is banned by the IOC Medical Commission.

The IOC Medical Commission bans erythropoietin as a method of doping (see section I, F).

B Pharmacological, chemical, and physical manipulation

The IOC Medical Commission bans the use of substances and of methods which alter the integrity and validity of urine samples used in doping controls. Examples of banned methods are catheterization, urine substitution and/or tampering, inhibition of renal excretion, e.g. by probenecid and related compounds, and epitestosterone application.*

III Classes of drugs subject to certain restrictions

A Alcohol

Alcohol is not prohibited. However, breath or blood alcohol levels may be determined at the request of an International Federation.

B Marijuana

Marijuana is not prohibited. However, tests may be carried out at the request of an International Federation.

C Local anaesthetics

Injectable local anaesthetics are permitted under the following conditions:

(a) that procaine, xylocaine, carbocaine, etc. are used but not cocaine;
(b) only local or intra-articular injections may be administered;
(c) only when medically justified (i.e. the details including diagnosis, dose, and route of administration must be submitted immediately in writing to the IOC Medical Commission).

D Corticosteroids

The naturally occurring and synthetic corticosteroids are mainly used as anti-inflammatory drugs which also relieve pain. They influence circulating concentrations of natural corticosteroids in the body. They produce euphoria and side-effects such that their medical use, except when used topically, require medical control.

Since 1975, the IOC Medical Commission has attempted to restrict their use during competitions by requiring a declaration by the doctors, because it was known that corticosteroids were being used non-therapeutically by the oral, rectal, intramuscular, and even the intravenous route in some sports. However, the problem was not solved by these restrictions and therefore stronger measures designed not to interfere with the appropriate medical use of these compounds became necessary.

The use of corticosteroids is banned except for topical use (aural, ophthalmological, and dermatological), inhalational therapy (asthma, allergic rhinitis) and local or intra-articular injections.

---

* If the epitestosterone concentration is greater than 150 ng/ml, the laboratories should notify the appropriate authorities. The IOC Medical Commission recommends that further investigations be conducted.

ANY TEAM DOCTOR WISHING TO ADMINISTER CORTICOSTEROIDS INTRA-ARTICULARLY OR LOCALLY TO A COMPETITOR MUST GIVE WRITTEN NOTIFICATION TO THE IOC MEDICAL COMMISSION.

REFERENCES

1. Coyle EF. Ergogenic aids. *Clinical Sports Medicine* 1984; **3**: 731–42.
2. Kuchenski R. Biochemical actions of amphetamine and other stimulants. In: Creese I, ed. *Stimulants: neurochemical, behavioral, and clinical perspectives.* New York: Raven Press, 1983: 31–61.
3. Smith GM, Beecher HK. Amphetamine sulfate and athletic performance. I. Objective effects. *Journal of the American Medical Association* 1959; **170**: 542–57.
4. Williams MH. *Drugs and athletic performance.* Springfield, IL: Charles C Thomas, 1974.
5. Chandler, JV, Blair SN. The effects of amphetamines on selected physiological components related to athletic success. *Medicine and Science in Sports and Exercise* 1980; **12**: 65–9.
6. Bhagat B, Wheeler N. Effect of amphetamine on the swimming endurance of rats. *Neuropharmacology* 1973; **12**: 711–13.
7. Gerald MC. Effects of (+)-amphetamine on the treadmill endurance performance of rates. *Neuropharmacology* 1978; **17**: 703–4.
8. Bravo EL. Phenylpropanolamine and other over-the-counter vasoactive compounds. *Hypertension* 1988; **11**(3)(Suppl. II): 7–10.
9. Pentel PR, Asinger RW, Benowitz NL. Propranolol antagonism of phenylpropanolamine-induced hypertension. *Clinical Pharmacology and Therapeutics* 1985; **37**: 488–94.
10. Lopes JM, Aubier M, Jardim J, Aranda JV, Macklem PT. Effect of caffeine on skeletal muscle function before and after fatigue. *Journal of Applied Physiology: Respiratory, Environmental and Exercise Physiology* 1983; **54**:(5): 1303–5.
11. Wood DS. Human skeletal muscle: Analysis of Ca^{2+} regulation in skinned fibers using caffeine. *Experimental Neurology* 1978; **58**: 218–30.
12. Goldstein A, Warren R, Kaizer S. Psychotropic effects of caffeine in man. I. Individual differences in sensitivity to caffeine-induced wakefulness. *Journal of Pharmacology and Experimental Therapeutics* 1965; **149**: 156–9.
13. Costill DK, Dalsky GP, Fink WJ. Effects of caffeine ingestion on metabolism and exercise performance. *Medicine and Science in Sports* 1978; **10**: 155–8.
14. Ivy JL *et al.* Influence of caffeine and carbohydrate feedings on endurance performance. *Medicine and Science in Sports* 1979; **11**: 6–11.
15. O'Neill FT, Hynak-Hankinson MT, Gorman T. Research and application of current topics in sports nutrition. *Journal of the American Dietetic Association* 1986; **86**: 1007–15.
16. Rall TW. Evolution of the mechanism of action of methylxanthines: from calcium mobilizers to antagonists of adenosine receptors. *Pharmacologist* 1982; **24**: 277–87.
17. Trang JM, Blanchard J, Conrad KA, Harrison GG. Relationship between total body clearance of caffeine and urine flow rate in elderly men. *Biopharmaceutics and Drug Disposition* 1985; **6**: 51–6.
18. Brown R, Partington J. A psychometric comparison of narcotic addicts with hospital attendants. *Journal of General Psychology* 1942; **27**: 71.
19. Appenzeller O *et al.* Neurology of endurance training. v. Endorphins. *Neurology(NY)* 1980; **30**: 418–19.
20. Glasser W. *Positive addiction.* New York: Harper & Row, 1976.
21. Kaiser P. Physical performance and muscle metabolism during β-adrenergic blockade in man. *Acta Physiologica Scandinavica* 1984; **536**: (Suppl): 1.
22. Delbeke FT, Debackere M. The influence of diuretics on the excretion and metabolism of doping agents. II. Phentermine. *Arzneimittelforschung Drug Research* 1986; **36**: 134–37.
23. Delbeke FT, Debackere M. The influence of diuretics on the excretion and metabolism of doping agents. III. Etilamfetamine. *Arzneimittelforschung Drug Research* 1986; **36**: 1413–16.
24. Terney R, McLain LG. The use of anabolic steroids in high school students. *American Journal of Diseases in Children* 1990; **144**: 99–103.
25. Haupt HA, Rovere GD. Anabolic steroids: a review of the literature. *American Journal of Sports Medicine* 1984; **12**(6): 469–84.
26. Wilson JD. Androgen abuse by athletes. *Endocrine Reviews* 1988; **9**: 181–99.
27. Ryan AJ. Athletics. *Handbook of Experimental Pharmacology* 1976; **43**: 525–34, 723–25.
28. Scott MJ Jr, Scott MJ III. Dermatologists and anabolic–androgenic drug abuse. *Cutis* 1989; **44**: 30–5.
29. Choi PYL, Parrott AC, Cowan D. High-dose anabolic steroids in strength athletes: effects upon hostility and aggression. *Human Psychopharmacology* 1990; **5**: 349–56.
30. Pope HG Jr, Katz DL. Bodybuilder's psychosis. *Lancet* 1987; i: 863.
31. Choi P, Parrott AC, Cowan DA. Adverse behavioural effects of anabolic steroids in athletes: a brief review. *International Journal of Sports Medicine* 1989; **1**: 183–7.
32. Strauss RH, Liggett MT, Lanese RR. Anabolic steroid use and perceived effects in ten weight-trained women athletes. *Journal of the American Medical Association* 1985; **253**: 2871–5.
33. Hurley BF *et al.* High-density-lipoprotein cholesterol in bodybuilders and powerlifters. Negative effects of androgen use. *Journal of the American Medical Association* 1984; **252**: 507–13.
34. Ferenchick GS. Are anabolic steroids thrombogenic? *New England Journal of Medicine* 1990; **322**: 476.
35. Hickson RC, Ball KL, Falduto MT. Adverse effects of anabolic steroids. *Medical Toxicology and Adverse Drug Experiments* 1989; **4**: 254–71.
36. Donike M. Bärwald KR, Klosterman K, Schänzer W, Zimmermann J. The detection of exogenous testosterone. In: Heck H, Hollmann W, Liesen H, eds. *Sport: Leistung und Gesundheit.* Köln: Deutscher Artze-Verlag, 1983: 293–300.
37. Kicman AT *et al.* Criteria to indicate testosterone administration. *British Journal of Sports Medicine* 1990; **24**: 253–64.
38. Southan GJ, Brooks RV, Cowan DA, Kicman AT, Unnadkat N, Walker CJ. Possible indices for the detection of the administration of dihydrotestosterone to athletes. *Journal of Steroid Biochemistry and Molecular Biology* 1992; **42**: 87–94.
39. Catlin DH, Hatton CK. Use and abuse of anabolic and other drugs for athletic enhancement. *Advances in Internal Medicine* 1991; **36**: 399–424.
40. Stewart-Bentley M, Odell W, Horton R. The feedback control of luteinizing hormone in normal adult men. *Journal of Clinical Endocrinology and Metabolism* 1974; **38**: 545–53.
41. Toth M, Zakar T. Relative binding affinities of testosterone, 19-nortestosterone and their 5α-reduced derivatives to the androgen receptor and to other binding proteins: a suggested role of 5α reductive steroid metabolism in the dissociation of 'myotrophic' and 'androgenic' activities of 19-nortestosterone. *Journal of Steroid Biochemistry* 1982; **17**: 653–60.
42. Kicman AT, Cowan DA. Peptide hormones and sport: misuse and detection. *British Medical Bulletin* 1992; **48**: 496–517.
43. Stenman U, Alfthan A, Ranta T, Vartianen E, Jalkanen J, Seppala M. Serum levels of human chorionic gonadotropin in nonpregnant women and men are modulated by gonadotropin-releasing hormone and sex steroids. *Journal of Clinical Endocrinology and Metabolism* 1987; **64**: 730–6.
44. Kicman AT, Brooks RV, Cowan DA. Human chorionic gonadotrophin and sport. *British Journal of Sports Medicine* 1991; **25**: 73–80.
45. Saez JM, Forest MG. Kinetics of human chorionic gonadotropin-induced steroidogenic response of the human testis. I. Plasma testosterone: implications for HCG stimulation test. *Journal of Clinical Endocrinology and Metabolism* 1979; **49**: 278–83.
46. Cowan DA, Kicman AT, Walker CJ, Wheeler MJ. Effect of administration of human chorionic gonadotrophin on criteria used to assess testosterone administration in athletes. *Journal of Endocrinology* 1991; **131**: 147–54.
47. Vanhelder WP, Radomski MW, Goode RC. Growth hormone responses during intermittent weight lifting exercise in men. *European Journal of Applied Physiology* 1984; **53**: 31–4.
48. Vanhelder WP *et al.* Growth hormone regulation in two types of aerobic exercise of equal oxygen uptake. *European Journal of Applied Physiology* 1986; **55**: 236–9.
49. Daughaday WH. The anterior pituitary. In: Wilson JD, Foster DW, eds.

Williams textbook of endocrinology, 7th edn. Philadelphia: WB Saunders, 1985: 568–613.

50. Hughes JP, Friesen HG. The nature and regulation of the receptors for pituitary growth hormone. *Annual Review of Physiology* 1985; **47**: 483–99.

51. Haynes RC. Adrenocorticotropic hormone; adrenocortical steroids and their synthetic analogs; inhibitors of the synthesis and actions of adrenocortical hormones. In: *Goodman and Gilman's pharmacological basis of therapeutics.* New York: Pergamon, 1990: 1431–62.

52. Chard T. Requirements for binding assays–extraction of ligand from biological fluids, and collection and storage of samples. In: Burdon RH, van Knippenberg PH, eds. *An introduction to radioimmunoassay and related techniques.* Amsterdam: Elsevier, 1987: 137–52.

53. Buick FJ *et al.* Effect of induced erythrocytaemia on aerobic work capacity. *Journal of Applied Physiology* 1980; **48**: 636–42.

54. Videman T, Sistonen P, Stray-Gundersen J, Lereim I. Experiences in blood dope testing at the 1989 world cross-country ski championships in Lahti, Finland. In: Bellotti P, Benzi G, Ljungqvist A, eds. *Official Proceedings of 2nd International Athletic Foundation World Symposium on Doping in Sport—1989.* London: International Athletic Foundation, 1990: 5–12.

55. Clement DB, Asmundson RG, Medhurst CW. Hemoglobin values: comparative survey of the 1976 Canadian Olympic team. *Canadian Medical Association Journal* 1977; **117**: 614–16.

56. McEwen BS. Influences of adrenocortical hormones on pituitary and brain function. In: Baxter JD, Rousseau GG, eds. *Glucocorticoid hormone action.* Berlin: Springer-Verlag, 1979: 467–92.

57. Rose RM. Psychoendocrinology. In: Wilson JD, Foster DW, eds. *Williams textbook of endocrinology.* Philadelphia: WB Saunders, 1985: 653–81.

58. Myles AB, Daly JR. Corticosteroid withdrawal. In: *Corticosteroid and ACTH treatment.* London, Edward Arnold, 1974: 89–97.

59. Few JD. Effect of exercise on the secretion and metabolism of cortisol in man. *Journal of Endocrinology* 1974; **62**: 341–53.

60. Kuoppasalmi K, Näveri H, Härkönen M, Adlercreutz H. Plasma cortisol, androstenedione, testosterone and luteinizing hormone in running exercise of different intensities. *Scandinavian Journal of Clinical and Laboratory Investigation* 1980; **40**: 403–9.

61. Southan GJ, Brooks RV. The detection of synthetic and natural corticosteroids. In: Shipe JR, Savory J, eds. *Drugs in competitive athletes.* Oxford: Blackwell Scientific, 1991: 33–7.

62. Wilkes D, Gledhill N, Smyth R. Effect of acute induced metabolic alkalosis on 800 m racing time. *Medicine and Science in Sports and Exercise* 1983; **15**: 277–80.

63. Reeds PJ *et al.* The effect of β-agonists and antagonists on muscle growth and body composition of young rats (*Rattus* sp.). *Comparative Biochemistry and Physiology* 1988; **89C**: 337–41.

3.3.2 Gender verification

MALCOLM A. FERGUSON-SMITH

INTRODUCTION

In the Olympic Games of ancient Greece which date from the 8th century BC, the athletes had to abide by strict rules. They had to be freemen, not slaves, without criminal conviction, and had to have trained for 10 months. Strict penalties for bribery, cheating, and for transgressing the rules were exacted and each athlete affirmed by solemn oath at the start of the Games that they were eligible to compete. Only male athletes took part and the custom was that they, and later also their trainers, entered the stadium naked. Women could not take part or even watch the Games on penalty of death. However, women could enter teams into the equestrian events. Women also took part in their own Games, the Heraia, in honour of the goddess Hera. The Heraia were short foot races for which only virgins were eligible, and the event was held separately from the Olympic Games. It was a great honour to take part in the Olympic Games and to win the crown of wild olive branches. This was the only prize, although all champions were glorified throughout Greece.

When the modern Olympic Games were revived in 1896 Baron Pierre de Coubertin followed the ancient tradition and only male athletes competed. Perhaps he foresaw problems with eligibility for women. None the less, in 1912 women swimmers were accepted at the Stockholm Olympics and by 1928 in Amsterdam women were competing separately in most athletic events. It has been accepted, and is well documented that there are substantial differences in performance between men and women.[1,2] These differences tend to be less in highly trained athletes but they are sufficient to justify the continued segregation of men and women in all but a few sports. Equestrian events and shooting remain the only exceptions.

Success in sport is not only the result of dedication and intense training. Athletes of both sexes tend to have inherent physical advantages over the average person in terms of stature, body build, and neurological and motor attributes. These inherent factors are genetically determined and often direct the athlete to a particular sport. Tall athletes have obvious advantages in netball and long jump whereas short and squat athletes tend to excel in weightlifting and in some field events. Some of these attributes, especially height, are influenced by the sex chromosomes but the main differences in performance between male and female athletes can be attributed to the increased muscle mass in males due to the action of androgenic steroids (especially testosterone) secreted by the testes. This hormonal difference is the main justification for separating the sexes in sport. In a few sports such as boxing and weightlifting, weight is used to classify athletes so that they can compete with one another on a fair basis. A similar argument might be advanced for classifying high and long jumpers by height. However, there would appear to be a limit to the quest for absolute fairness in sport, and genetically-determined height is not at present regarded as a reasonable criterion.

Notwithstanding the high moral standards of the Olympic athlete, it is natural for a competitor to wish to maximize the chances of success by whatever permissible means are available. Intensive training and improved diet are obviously acceptable, but measures which involve taking steroids to increase the muscle mass or erythropoietin to increase the red blood cell count amount to cheating, may be damaging to health, and are rightly proscribed. However, pressures on individual athletes are great. It is not only a matter of individual prestige as in ancient Greece, but there are also other factors including national and political pressure, the prospect of financial gain, self-aggrandizement, and other selfish motives, which sometimes lead the athlete to cheat even in the most recent times when sophisticated methods for screening athletes for anabolic steroid use, for example, are routinely employed at international events.

THE INTRODUCTION OF GENDER VERIFICATION

The anecdotal evidence that on several international occasions individuals competed as females who later changed their sexual identity to live as men, and even father children, has undoubtedly led to rumour and innuendo about those successful female athletes considered to have increased muscle mass, a male habitus, and hirsutism. It was frequently suggested that some form of verification was necessary to ensure that males were not masquerading as females. As early as 1948, the British Women's Amateur Athletic Association required a doctor's letter verifying the sex of women competitors. This was later dropped as being too ineffective against a determined imposter.

There are at least six well documented examples of individuals with testes who competed successfully in international women's events.[3]

1. The earliest recorded case was the winner of the women's 100-m sprint in the 1932 Olympic Games in Los Angeles. Forty-eight years later she was killed in a shooting incident and at autopsy was found to have testes.[4] It is not recorded whether she was virilized and it is probable that she had the androgen insensitivity syndrome.
2. The winner of the women's 800-m race and world record holder in 1934 was found to be a male pseudohermaphrodite who had a sex-change operation at the age of 30 years. From the account of the operation it is clear that there was considerable virilization of the external genitalia, and thus the diagnosis could not have been androgen insensitivity.[5]
3. A high jumper who came fourth in the Berlin Olympics in 1936 was barred from competition by the German Athletic Federation after the 1938 European Championships in Vienna, when she was found to have both male and female organs. It is reported that the Nazi youth movement had forced the athlete to pose as a women for 3 years.[6] He returned to male gender thereafter.
4. Two members of the women's relay team that came second in the European Championships in 1946 subsequently had sex-change operations and lived afterwards as men.[6] One fathered several children.
5. In 1964 another runner broke world track records in races for women at 400- and 800-m distances which were never ratified. He was later recognized by his father as the son he had lost in the war.[7] This may have been an example of a male simply masquerading as a female.
6. The winner of the women's world downhill ski title in 1966 was identified as being male as a result of a medical examination in 1967.[7] This revealed undescended testes and, after surgical correction, the skier married and became a father.

In several of these cases there is documentation of ambiguity of the external genitalia and testicular maldescent. The diagnosis of male intersex seems likely and their androgen levels may have contributed to their success. They were certainly regarded as having an unfair physical advantage over normal women athletes. and there was considerable pressure on the various international organizations of sport to ensure that this form of cheating did not occur in the future. The anecdotes also alerted competitors to the possibility of males masquerading as females and, as indicated above, anybody with an apparently masculine body habitus tended to be regarded with suspicion.

Rumours of this sort were rife at the Rome Olympics in 1960 and this prompted the International Amateur Athletic Federation (IAAF) and later the International Olympic Committee (IOC) to consider establishing rules of eligibility for women athletes in order to ensure that the athletes were competing on an equal basis, considering their physical status.[8] At the same time there was growing concern about the use of hormones and stimulant drugs by athletes, and this led to the introduction of measures designed to stop all forms of cheating.

The drug problem and gender verification of women athletes were both considered by the IOC at the 1964 Olympic Games in Tokyo. A decision was taken to form a Medical Commission to take charge of medical aspects, including the taking of drugs. Competitors were required in future to sign statements accepting the arrangements for drug and 'femininity' controls before they were admitted to the Games. This was to protect the athletes' health, to ensure fair play and equal conditions for all and, in the case of gender verification, to eliminate cases in which a genetic alteration leading to masculinization in a competitor registered as a female gave that competitor an unfair advantage. It was specifically intended that cases of intersexuality or hermaphroditism should be barred from competition. The signed statement also served to absolve the organizers from responsibility should an athlete be debarred in error.

The early attempts at gender verification were clumsy and insensitive. The first was made at the European Athletics Championships in Budapest in 1966. Women athletes were required to undergo a parade in the nude in front of three gynaecologists.[9] All 234 competitors passed the test but it was noted that five world record holders failed to appear at the Championships. Their absence was unexpected, and it was widely assumed that this was because they believed that they might have failed the test. They included a double Olympic champion in the shot put and discus, a double Olympic champion in the 80-m hurdles and pentathlon, a European long jump champion, a European 400-m champion, and an Olympic high jump champion.

In the Commonwealth Games in Kingston in 1966 the feminity control took the form of a manual examination of the external genitalia by a gynaecologist.[10] This treatment created great indignation among the athletes. Further resentment was expressed at the Pan American Games in Winnipeg in 1967 when another 'on sight' inspection was carried out by three doctors sitting at a desk in front of which each athlete was asked to pull up her shirt and push down her pants.[9] It is not known if any of these athletes were barred. These degrading forms of examination have made a lasting impression on the athletic community and have led to resistance to any form of clinical examination in the context of eligibility control.

THE SEX CHROMATIN TEST

The experience with these forms of physical examination prompted the IOC to follow the advice of Bunge and adopt the buccal smear test as an alternative method of femininity control.[11] The test involves scraping a sample of epithelial cells from inside the cheek with a wooden spatula, and transferring the cells to a microscope slide which is then fixed and stained to reveal the presence or absence of the Barr body in the epithelial cell nuclei. The Barr body (or X-chromatin) is a small dense mass of chromatin which can be observed under the oil immersion lens of a light microscope in approximately 30 per cent of nuclei in female somatic cells (Fig. 1). It represents one of the two X-chromosomes which is genetically inactive and becomes condensed within the nucleus. Male cells do not show the Barr body as they have only one active X-chromosome. The test therefore indicates the number of X-chromosomes in the cell nucleus and thus the sex chromosome constitution of the individual, normally XX in females and XY in males. As the short arm of the Y-chromosome carries the primary genetic determinants for the development of testes, normally only individuals with a Y-chromosome develop as males. Male differentiation occurs because the testes secrete the male hormones necessary for the development of male internal ducts including prostate and seminal vesicles, and male external genitalia consisting of penis and scrotum. In the absence of a Y-chromosome and testes, the individual develops along female lines. The sex chromatin test might therefore be considered a satisfactory means of distinguishing XY males from XX females and thus of detecting males who are attempting to masquerade as females. Unfortunately, there are two problems. First, it is possible to make a technical mistake, particularly if the sample has not been properly collected and fixed. XX females may be wrongly classified as male, and XY males may be missed. Both errors have occurred during the course of feminity control, the former with disastrous results for the unfortunate athlete. Secondly, there are several genetic disorders which interfere with the normal process of sexual development and lead to individuals of normal appearance but with paradoxical sex chromatin (and sex chromosome) findings. These cases illustrate the principle that a person's anatomical sex (and thus social and legal sex) is the most important factor in establishing gender, which need not necessarily conform to the chromosomal sex as determined by the sex chromatin test.

The most important genetic conditions that cause problems in femininity control occur in individuals who have an apparently normal male chromosomal constitution but who develop to adulthood as perfectly normal-looking girls. These individuals are sometimes referred to in the medical literature as XY females. They are quite easily recognized after puberty because they fail to menstruate and remain infertile; in some the breasts fail to form and the vulva and vagina do not mature. Many, however, grow into normal women with well-developed breasts and, as they are healthy (apart from being amenorrhoeic) and sometimes taller than average, they tend to do well in sport. In fact, tests for feminity control suggest that at least one in 500 women athletes have one or other of these conditions and are, unknowingly, XY

females.[3] The two most common types of XY female are those with gonadal dysgenesis, in which only vestiges of the gonads remain and no male hormones are produced, and those with the androgen insensitivity syndrome (or testicular feminization syndrome) in which the uterus is not formed and the tissues fail to be masculinized by normal amounts of male hormone produced by intra-abdominal testes.

GONADAL DYSGENESIS

The individual with XY gonadal dysgenesis has normal female internal and external genitalia but fails to mature at the time of puberty as the ovaries are only represented by thin streaks of ovarian tissue which cannot secrete sufficient oestrogenic steroids. Estimation of serum hormones therefore reveals low levels of oestrogens and high levels of gonadotrophins. There is usually little or no breast development and no menstruation. As the streak gonads are liable to malignant change (gonadoblastoma, dysgerminoma) arising from nests of abnormal XY germ cells, it is usual to remove them in late adolescence or early adulthood. Secondary sexual characteristics can be induced by regular replacement therapy with oestrogens to the point of inducing withdrawal bleeding from the uterus; as induced menstruation serves no useful purpose it is usually avoided. Oocyte donation and *in-vitro* fertilization have allowed a number of affected women to have children. Because of the genetic determinants for stature carried by the Y chromosome[12] women with XY gonadal dysgenesis are taller than the average and may do well in sport. The author knows of a successful hockey player with the condition, but it is not known how often femininity controls have debarred individuals with gonadal dysgenesis from international sport. It should be emphasized that the streak gonads do not secrete male hormone and that those affected show no evidence of masculinization. They have no advantage in sport over other women of the same height.

There are several genetic causes of XY gonadal dysgenesis. Some cases lack male determinants due to mutations within the testis-determining region of the Y chromosome.[13] Others may be due to X-linked and autosomal mutations. Similar conditions in animals are known to be due to mutations of the autosomes.

ANDROGEN INSENSITIVITY SYNDROME

Those with the androgen insensitivity syndrome are also XY females who tend to be taller than average women. The condition may be recognized occasionally in childhood because of the presence of testes in the labia or because the child develops an inguinal hernia. Most cases, however, are not recognized until adolescence when the presenting feature is failure to menstruate despite the development of breasts and other secondary sex characteristics. Gynaecological examination reveals that the uterus is absent and that the vagina is only one-third its normal length, ending blindly in the absence of the cervix uteri. Normal-sized testes are usually found in the pelvis or at the internal inguinal ring. As these cryptorchid testes are liable to malignancy (usually dysgerminoma) they should be removed once secondary sex characteristics are fully developed. Scanty pubic and axillary hair is a characteristic sign which should alert the clinician to the correct diagnosis in the adult. It is estimated that the frequency of the condition is approximately 1 in 62 000 males.

The androgen insensitivity syndrome is due to a mutation of the androgen receptor gene carried by the X-chromosome. Normal

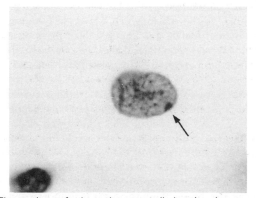

Fig. 1 The nucleus of a buccal smear cell showing the sex chromatin or Barr body (arrowed) which represents the condensed, inactivated X chromosome characteristic of cells with two X chromosomes. Approximately 30 per cent of cells in a buccal smear from a normal female will show such a sex chromatin body.

androgen receptors are not present in the tissue cells which therefore cannot respond to the normal amounts of testosterone produced by the testes. Consequently, the external genitalia, secondary sex characteristics, and musculature are female. As the development of sex hair is also dependent on the presence of normal androgen receptors, pubic and axillary hair is sparse. The tissues are unresponsive to treatment with androgenic steroids, and this is one condition in which anabolic steroids have no effect in building up muscle mass. Serum hormone estimations will reveal levels of testosterone within the normal male range.

The above description applies to the complete androgen insensitivity syndrome but it has been apparent for many years that there are variants due to different mutations at the same X-linked locus. These appear to be 'leaky' mutations in that the abnormal androgen receptors formed are capable of handling small amounts of male hormone, sufficient to cause virilization of the external genitalia and a normal distribution of pubic and axillary hair. These variants are referred to as the incomplete androgen insensitivity syndrome. It is doubtful if the amount of masculinization produced has a significant effect on athletic performance.

As androgen insensitivity is inherited as an X-linked recessive trait, it is usually transmitted through normally fertile women (who may have a patchy distribution of sex hair) to half their XY offspring. Enquiry should therefore be made for the condition in sisters and maternal aunts of affected individuals.

The androgen insensitivity syndrome is the most common abnormality detected by sex chromatin tests for gender verification in sport. Although those detected are aware of their amenorrhoea, they have had no doubts about their femininity and the pronouncement that they are genetic males comes as a complete shock. The response is invariably to withdraw from competition in great distress and to reject any further testing or examination. The Medical Commission of the IOC recommend that those affected with the complete syndrome are eligible to compete, but that before doing so they must submit to chromosome analysis, gynaecological examination, and hormone tests to ensure that there is no masculinization which might indicate that they have an unfair advantage over normal women. Despite the worthy intentions of the IOC Medical Commission, it is clear that athletes who fail the sex chromatin test are too distressed to follow their advice and as a result are disqualified unjustly. The IOC Medical Commission must accept responsibility for this as they have failed to provide any clear guidelines to medical committees at international events as to the eligibility of athletes with androgen insensitivity or one of its variants.

While almost all such cases retire permanently from their sport, one recent case has aroused special interest in that the athlete concerned has fought successfully against the decision and has been granted an eligibility certificate. This has not been without considerable personal distress and embarrassment. The case concerns an athlete in the 60-m hurdles who 'passed' the sex chromatin test at the 1983 World Championships in Helsinki. However, another test was taken at the World Student Games in Kobe (Japan) in 1985 and the athlete was found to have androgen insensitivity and debarred.[14] This led to the withdrawal of her University Scholarship and unwelcome publicity in the press. After 4 months waiting for the results of endocrine tests, the athlete was confirmed to be a woman, although infertile; she felt humiliated by the whole investigative procedure but was encouraged to fight for her eligibility in order to assist other women whom she knew to be in the same predicament. The case illustrates the fallibility of the sex chromatin test, the level of ignorance among sports physicians about the eligibility rules, and the trauma experienced by individual athletes caught by an unjust and ineffective system.

SEX CHROMOSOMAL ABNORMALITIES

The sex chromatin test may give a paradoxical result in individuals who have gross numerical or structural abnormalities of the sex chromosomes. One of the best known of these conditions is Turner's syndrome in which those affected have primary amenorrhoea, short stature, and developmental malformations including webbing of the neck. Gonadal dysgenesis is invariably present and is the cause of the sexual infantilism. Chromosome analysis reveals a 45,X karyotype without a Y-chromosome. As there is only one X-chromosome, the Barr body is absent from somatic cell nuclei and so affected individuals fail the sex chromatin test. The short stature and developmental disabilities are such as to preclude success in most sports and there is no record of anyone with Turner's syndrome being debarred. However, variants of Turner's syndrome are not uncommon where chromosome analysis reveals a more complex sex chromosome constitution and those affected may be normal in height and have few disabilities. Some are chromosomal mosaics, that is their somatic cells are made up of two or more cell lines with different karyotypes. The most common are 45,X/46,XX and 45,X/46,XY mosaics. The latter may have rudimentary testes and ambiguous genitalia associated with significant levels of androgens. Other cases have structural rearrangements of the X- or Y-chromosomes including deletions, isochromosomes, and duplications; these are often accompanied by 45,X mosaicism. It is clear that those with only X rearrangements would have no advantage over normal women in sports, but those whose Y chromosome rearrangements lead to testis differentiation and masculinization may well have increased muscle mass and might be considered to be at an advantage.

Of more relevance to the present discussion are a group of individuals with Klinefelter's syndrome. These are infertile males with small testes (and commonly gynaecomastia) due to a 47,XXY karyotype. Most of those affected are rather taller than average males and tend to show clinical evidence of androgen deficiency such as diminished body hair and reduced frequency of shaving. However, many cases are fully masculinized and the muscle mass would be regarded as normal for an average male. As in Turner's syndrome sex chromosome mosaicism is not uncommon, usually either 46,XY/47,XXY, or 46,XX/XXY. In the last case, the presence of a normal female cell line may lead to ambiguity of the internal and external genitalia and a few cases have been described that are true hermaphrodites with both ovarian and testicular tissue.

One of the first women to be excluded from international events because she failed the current femininity test would appear to have had 46,XX/47,XXY mosaicism. This athlete won gold and bronze medals at the Tokyo Olympics in 1964 and held the world record for the women's 100 metres. In 1966 she passed the inspection at the European Games at Budapest but in the following year in Kiev was ruled ineligible for the European Cup women's track and field competition. On this occasion the test involved a close inspection of the external genitalia. A six-man medical commission investigated her case and karyotype analysis revealed cells with a 47,XXY sex chromosome complement.[9] It is understood that gonads had been removed at an earlier stage and that she was being treated with oestrogens. She was disqualified from competing in women's athletics and in 1970 the IAAF removed her name from the record books. It is of interest to consider what would have happened under current gender verification conditions. She would have passed the sex chromatin test and would not have required to submit to a gynaecological examination or chromosome analysis. This means that she would have been eligible to compete and her name would be honoured in world athletics. Instead she has suffered public humiliation.

Another group of infertile patients, known as XX males, have a rather similar condition to Klinefelter's syndrome. Chromosome analysis reveals an apparently normal female karyotype but molecular studies show that the testis-determining part of the Y-chromosome has been transferred to one of the two X-chromosomes. The Barr body test is positive, but the individual usually develops as a normal-appearing male. Height is within the normal female range, gynaecomastia is frequent and, in a few cases, the genitalia are ambiguous.[15] It is possible that in those cases with ambiguity of the genitalia, the abnormality at birth is sufficient to raise enough doubt about the gender and the sex for the individual to be reared as female. In the closely related condition of XX true hermaphroditism, about half the cases are raised as females. Once again it is doubtful if the degree of masculinization due to the testicular development can confer an advantage in terms of muscle mass over normal women, although the ambiguous phenotype may be a cause for genuine concern among competitors. Of course, all individuals in this group would pass the sex chromatin test and would therefore not be detected under present rules unless the polymerase chain reaction test for the sex determining region (SRY) of the Y chromosome is used (see below).

MALE INTERSEX

The term male intersex (or male pseudohermaphrodite) is used to describe individuals with testes who have ambiguity of the external genitalia and who are assigned female gender at birth usually because of inadequate development of the penis. Although they are reared as girls, problems frequently arise because of masculinization at puberty. In some cases orchidectomy is performed and others undergo sex change operations in adolescence. The IOC believe that there have been cases where such individuals have been selected for training as women athletes and have been deprived of proper treatment. There appears to be no documentation of this in the medical literature. However, it is likely that in some of the early reports of males competing in women's events, the individuals concerned have had a variety of untreated male intersexuality.

Male intersexes have varying degrees of testicular development and usually respond at least partially to androgens secreted by the testes. In addition to the cases of incomplete androgen insensitivity and the cases of sex chromosome mosaicism mentioned above, there are several rare conditions which must be considered in the differential diagnosis. In 5α-reductase deficiency there is defective virilization of the external genitalia at birth because the testes are unable to convert testosterone to dihydrotestosterone. This leads to perineal hypospadias and a blind-ending perineal opening resembling a small vagina. In most families the ambiguity of the external genitalia is sufficient to lead to assignment of female gender at birth. However, at puberty masculinization of the external genitalia occurs due to the action of testosterone and there is no breast development. The testes enlarge and semen can be produced although beard growth may be less than normal. Most

patients undergo reassignment of sex and live as males. The condition is inherited as an autosomal recessive disorder limited to males, i.e. male sibs have a one in four risk of being affected.

The testicular regression syndrome is also likely to be a sex-limited autosomal recessive disorder. The condition occurs in XY individuals and is due to regression of the embryonic testes. If this occurs early, sex differentiation is female with hypoplasia of the Mullerian ducts and normal female external genitalia. If testicular regression occurs later, there may be absence of the uterus and tubes and mild masculinization of the external genitalia. Occasionally, the phenotype is entirely male with absence of testes (anorchia).

CONGENITAL ADRENAL HYPERPLASIA

There are a number of metabolic defects which lead to virilization in women in the form of hirsutism and increased muscle mass. The most common form is 21-hydroxylase deficiency, an autosomal recessive disorder found in at least one in 20 000 births. The most severe form results in virilization of the female fetus before birth and precocious puberty in boys. The metabolic defect is due to a failure to convert 17-hydroxyprogesterone to 11-deoxycortisol in the adrenal cortex. Lack of cortisol production induces the anterior pituitary to secrete adrenocorticotrophic hormone (ACTH) which in turn leads to hyperplasia of the adrenal cortex and over-production of adrenal androgens. The affected newborn female has normal ovaries and internal genital ducts but may appear as a male with well-developed penis (and penile urethra) but empty scrotum. Affected males are normal at birth but develop penile enlargement and pubic hair in early childhood; rapid growth is followed by early closure of the epiphyses and short stature. These changes can be prevented by cortisone therapy which suppresses ACTH production. Milder forms of 21-hydroxylase deficiency appear to be quite common in infertile, but otherwise healthy, adult women.[16] They tend to be muscular, have a male body habitus, and are often hirsute. The diagnosis is made by the estimation of urinary hormone levels following ACTH stimulation.

Female athletes who are well built, muscular, and masculine in bodily habitus are often suspected of having a mild form of congenital adrenal hyperplasia. Those affected with 21-hydroxylase deficiency are, of course, genetic females and, as they would pass the sex chromatin test, are unlikely to be recognized. There is certainly no record of an affected individual excelling in any sport and thus it is not known whether the suspicions are unfounded. Affected males might also benefit in sport from the additional adrenal androgens, but information on this is also not available.

MALE TRANSSEXUALS

Males who have severe problems of gender identity may seek a sex change operation with removal of testes and genitalia and often with the construction of an artificial vagina. Replacement therapy with oestrogens may allow such male transsexuals to appear and to live as women. At least two male transsexuals have taken part in women's national events and both have aroused considerable interest in the sporting media. In 1976, a player who had previously undergone a sex change operation was allowed to enter the United States Open Tennis Tournaments in 1976 after a Supreme Court Justice ruled that she was now a female.[17] That this was a controversial decision was evident from the subsequent action of the United States Tennis Association which required those entering their tournaments to pass the sex chromatin test. (This requirement has since been abandoned.) The United States Golf Association discussed the same issue in 1988 when a male transsexual was accepted for entry to the qualifying rounds of the Mid-Amateur Championship and United States Women's Open Golf Tournament. The question at issue was whether or not a person who had bilateral orchidectomy and 7 years of treatment with oestrogens could be regarded as having any advantage over normal women competitors in a golf tournament. The outcome of these discussions is not known.

While many people would regard the male transsexual as being eligible on the grounds that she was now registered and living as a female without the advantage of male androgens, the case of the male transvestite would appear to be quite different. No physical sex reassignment has been made and it would not be realistic to allow anyone who chooses to cross-dress to masquerade as a female athlete.

In India there is a large sect of individuals (numbering half a million) known as the Hijra or eunuchs who live together in small communities throughout the continent and practise their own special religious rites. Their origin can be traced back several centuries to the custom of abandoning newborn infants with sexual ambiguity to the care of this sect who come to collect their 'new member' whenever they hear about such a birth. In more recent times the membership has been augmented by stealing and forcibly castrating male children and by admitting transvestites. The Hijra earn their living by telling fortunes, giving advice on curing infertility, by entertaining at family weddings and births, and by prostitution. It is not impossible that the more athletic of the castrated eunuchs could be recruited into sport, or that castration and reassignment of sex might be used to obtain the financial benefits associated with success in sport. It would seem important to exclude even the remotest possibility that a young male athlete might be persuaded to undergo a sex change operation, however unrealistic this might seem to those fortunate enough to live in a developed country.

THE INTRODUCTION AND USE OF SEX CHROMATIN TESTS

The buccal smear test was first tried out in a small proportion of athletes chosen at random at the Winter Olympic Games at Grenoble in 1968. The apparent success of the procedure led to its use among almost 1000 female competitors in the Olympic Games in Mexico City the same year. There is no official information about the results of testing on either of these occasions as the IOC Medical Commission has taken a decision to maintain strict secrecy about sex chromatin findings in order to avoid innuendo and rumour. However, it is understood that while one athlete failed the test at Grenoble there were no ineligible athletes at Mexico City. Technical and organizational problems at Mexico City may have led to incomplete testing[18] and thus the failure to prevent 'ineligible' persons from competing. The situation was reviewed by the IOC before the Olympic Games in Munich and stricter guidelines were introduced. In addition to the Barr body test, the buccal smears were also tested for fluorescent Y-chromatin revealed by quinacrine staining. The Y-body represents the heterochromatic and fluorescent distal segment of the long arm of the Y-chromosome which is visible in about 80 per cent of male nuclei. As it varies in size between individuals, it is sometimes difficult to detect and can be missed altogether. In addition, other non-sex chromosomes may occasionally carry large variant fluorescent segments which stain with quinacrine and can be confused with the Y-chromatin. Thus false-positive and false-negative results are common, and it is recorded that a swimmer at the World Student Games in 1985 failed the feminity control because of such a false-positive result on the Y-chromatin test.[19] She was told that she might not be able to have children and suffered considerable anxiety and distress before detailed chromosome analysis at home some months later revealed that the fluorescent spot was due to a variant autosome which had no clinical significance.

It is known that three of the 1280 female athletes at Munich failed the X and Y chromatin tests but no details are available.[18] The tests apparently did not prevent all attempts at fraud for it is alleged that one of the Asian women's volleyball teams included a man.[9] Eighteen hundred tests were performed at the 1976 Olympics in Montreal; some reports suggest that all results were normal while another indicates that four athletes were found to be ineligible. No information is available about the numbers tested at the Moscow Olympics in 1980 although two athletes are thought to have failed the feminity control. In 1984, six athletes out of approximately 2500 failed the test at Los Angeles and several others also failed at the Winter Olympics in Calgary. There is no information as yet on the results of testing at the 1988 Olympic Games in Seoul.

The above information on the results of sex chromatin testing at the Olympic Games has been gleaned largely from unofficial sources and is therefore unreliable. The most reliable figures suggest that at least one in 500 athletes fail the sex chromatin test.[3] As indicated earlier in this chapter, the available information suggests that the diagnosis in the majority of cases is the androgen insensitivity syndrome in XY females. Although athletes with this condition are eligible to compete, they cannot do so without submitting themselves to further investigation and, almost without exception it seems that they choose not to submit to additional testing and examination (see above). Consequently these athletes are debarred unjustly from competition and suffer needless distress and anxiety. On the other hand, there is no evidence that the tests have identified males masquerading as females, or intersex individuals who have a substantial advantage over normal females due, for example, to the production of male hormones.

The author believes that athletes who have to submit to gender verification by sex chromatin tests have a right to the complete information about results of such testing in the past, so that they may understand the limitations of the test and the problems that may occur. Correct information should be made available about the number of athletes who have failed the sex chromatin test, the number who have returned for further investigation after failing the test, the diagnoses obtained on complete investigation, the number of athletes who have retired from competition without returning for further investigation, and the number of tests which gave incorrect results. This information can be supplied without relating it to individual competitions and thus there need be no concern about breaching confidentiality or initiating rumours about individual athletes. The IOC Medical Commission are believed to have the necessary records to provide this data but so far have not responded to requests to make summaries available. Until this information is available, most women athletes will remain ignorant of the injustices associated with current procedures for gender verification and the hazards of erroneous results and misinterpretations to which every athlete is subject.

ALTERNATIVE APPROACHES TO GENDER VERIFICATION

The buccal smear test was introduced as a scientific and practical means of feminity control which avoided the need for the demeaning types of genital inspection which were so resented by athletes in 1966–67. However, the use of sex chromatin was criticized from the outset by one of the pioneers of nuclear sex determination,[20] as being an inappropriate measure of the sex of an individual. In a letter declining an invitation to undertake sex chromatin tests for the Commonwealth Games in Edinburgh in 1970, the author pointed out that there was a greater possibility of detecting an unsuspecting female with androgen insensitivity than a normal male masquerading as a female. It was further suggested that if the athletes found a physical inspection by a physician unacceptable, it would be wiser to drop the establishment of sex as a criterion than to use a wholly inappropriate test. The same view has been taken by a number of medical geneticists and others[21,22] and experience over the past 20 years has fully substantiated all the fears that were expressed. The current position of the IOC Medical Commission is that femininity control has been successful in that it has stopped the rumour and innuendo that was rife among sportswomen in the 1960s and that it has deterred individuals with an unfair advantage from competing in women's events. Athletes have in general been satisfied with the system as they believe that individuals have been deterred from cheating by a test which does not invade their privacy. However, owing to the secrecy imposed about those unjustly debarred from competition, athletes are unaware of the serious deficiencies of the present system.

In the face of the determination of the IOC to continue with the sex chromatin test, the author and others have subsequently agreed to undertake gender verification by sex chromatin testing on the grounds that they can help ensure that those who fail the test unjustly can be advised promptly and correctly. In this way a number of athletes with androgen insensitivity have been granted eligibility certificates in good time to compete without anyone other than the athlete and the scientists involved in the testing being aware of the diagnosis. Not all those involved in testing are adequately informed about these matters and so it cannot be guaranteed that mistakes and tragedies can always be avoided. Indeed, competitors continue to be debarred unjustly despite such efforts. Clearly a more rational approach to gender verification is urgently needed.

THE IAF MONTE CARLO WORKSHOP ON GENDER VERIFICATION

In November 1990 the International Athletic Foundation (IAF) convened an international group of individuals concerned with gender verification in sport at a Workshop in Monte Carlo. The membership included sports physicians, former world-class female athletes, a medical geneticist, paediatrician, obstetrician, pathologist, psychiatrist, and several representatives of international governing bodies of sport including a representative of the IOC Medical Commission. The Workshop was chaired by Professor Arne Ljungqvist, Chairman of the IAAF Medical Committee, and its aim was to review the system of gender verification, discuss the need for it in sport and, if necessary, propose more suitable alternative methods. The group made a number of important recommendations for submission to the IAAF and IOC.[23] These are summarized below.

1. That general information concerning the results of sex chromatin testing since 1968 be made available in a form which protected the confidentiality of individual athletes. It was felt that lack of such data prevented the athletic community from a proper consideration of the need for gender verification and the problems associated with it.

2. That the purpose of gender verification was to prevent normal men from masquerading as women in women's competition. XY females who were raised as girls and live with female gender should not be excluded. Individuals undergoing male-to-female sex reassignment before puberty should not be excluded. Decisions about male transsexuals who have been reassigned to female gender after puberty should be made by the relevant medical body within the sports organization concerned.

3. In view of the unfair exclusion of women athletes in the past due to erroneous sex chromatin testing and medical ignorance, gender verification by this means should be abandoned. So long as concerns exist about males masquerading as females, a revised and medically sound form of eligibility certification is required.

4. All men and women selected to participate in international competition should have a medical examination, performed under the auspices of the respective national organization and according to international guidelines, to ensure that there is no form of ill health or physical reason why it would not be in their interest to compete. The examination would include a simple inspection of the external genitalia and the certificate issued after the examination should be entitled the 'Health and Gender Certificate'.

5. Quality control must be conducted at international competition to combat possible abuse. This should be done randomly and by repeating the examination conducted at national level. The number of examinations should be strictly limited and, once undertaken, should not be required of the same athlete again.

The important conclusion of the Workshop was that female competitors are eligible irrespective of their genetic, chromosomal, gonadal, or hormonal sex. Those who have some form of masculinization due to any cause other than taking drugs are not to be barred from women's events. This includes for example 21-hydroxylase deficiency, 5α-reductase deficiency, androgen insensitivity, and all forms of sex chromosomal mosaicism provided that the individual has been reared as a female. Only those individuals (of whatever genetic sex) who have been reared as male are ineligible. The opinion of the Workshop was that any form of genetic testing, whether by sex chromatin or the more modern polymerase chain reaction technique using Y-specific DNA probes, was inappropriate.

The IAF Council considered and approved the recommendations of the Monte Carlo Workshop in January 1991 and implemented the new procedures at the IAAF World Championships in Athletics in Tokyo in August 1991. Thus, 16 of the 20 teams who had at least 20 participants (male and female) had conducted a full health check on all athletes before they went to Tokyo. Women athletes who had not been subjected to a health check at home and who did not already carry an eligibility certificate took part in the health check. Although male participants were not examined in Tokyo, at least 556 of the 925 male athletes had had a health check at home. Most of the major countries participating in Tokyo had understood the new procedures very well. However, some athletes saw the health check as an unwelcome return to the unacceptable genital examinations used for gender verification over 25 years ago, and there was uncertainty among team physicians about what constituted a health check.

The IOC Medical Commission considered the Monte Carlo Report in February 1991 and decided to recommend to the IOC Executive Board that they accept the proposal to allow National Olympic Committees to issue health certificates. The Medical Commission could not agree to accept the proposals in full and recommended that a sex test should continue to be carried out on all female athletes at the Olympic Games. A decision was taken to use the polymerase chain reaction test at the 1992 Winter Olympics at Albertville and so check the validity of the health and gender certificates issued by the National Olympic Committees.

At Albertville 557 women athletes were tested by polymerase chain reaction amplification of the SRY (sex determining region of the Y chromosome) gene using an appropriate control.[24] Two samples from the inside of the left and right cheeks were taken from each athlete by women assistants and the laboratory tests were run under rigorous conditions. It was stated that the test was 99 per cent reliable and polymerase chain reaction amplification is known to have failed in only a comparatively small number of cases. This implies that in 5 to 6 cases an erroneous result was obtained and that an unstated number of tests had to be repeated. Nonetheless, in 98 per cent of cases the results were available within 24 h. The Medical Commission of the IOC required those involved to undertake not to disclose the results; the number of those who failed the eligibility test and were disqualified is therefore unknown.

More complete details are known about the arrangements for gender verification at the 1992 Summer Olympics at Barcelona. Two samples of buccal mucosal cells were taken from both sides of the mouth of each female competitor. One sample was tested using the polymerase chain reaction for a repetitive DNA sequence (DYZ1) carried by the Y chromosome as well as a control sequence for mitochondrial DNA (Mit). It was expected that the DYZ1 sequence would be present exclusively in XY individuals and that the Mit sequence would give a positive reaction in all individuals irrespective of their sex chromosome constitution. However, 12 of the samples from the 2406 female atheletes gave a positive result with the DYZ1 sequence. When these 12 samples were retested on duplicate samples with the DYZ1 sequence the positive results were confirmed in all except one (false-positive) case. When the 11 true-positive samples were tested for the SRY-specific sequence, only five gave a positive result. The five athletes from whom these samples were taken were asked to attend for physical examination. Four athletes had clinical findings consistent with the diagnosis of an XY female. All were allowed to compete and so it is presumed that the likely diagnosis was the androgen insensitivity syndrome. One athlete refused clinical examination and withdrew from the competition. She is reputed to be the mother of three children born after normal pregnancies, but this is not confirmed.

These results indicate that 1 in 481 (5 of 2406) of the female athletes gave a test result positive for the SRY sex determining gene. This is not significantly different from the 1 in 504 estimated from anecdotal information obtained from previous international events.[3] The six athletes in whom the buccal cell samples were positive for the DYZ1 sequence carried by the Y chromosome but negative for the SRY sequence raise interesting questions. It is possible that their genotypes contain fragments of the Y chromosome, present either as a chromosomal rearrangement or as an additional abnormal Y chromosome carrying Y-linked genes influencing height.

It is reported that 711 samples failed to give results using the polymerase chain reaction test with either the Y-specific sequence or the Mit control sequence. Retesting using purified DNA from the samples gave good results except in three cases which had to be resampled. A recall rate of 1 in 800 would seem reasonably satisfactory in such circumstances.

In May 1992, the IAAF convened a second Working Group in London to consider the views of team physicians and others on the working of the health check proposal. All agreed that in view of the arrangements in place for doping control, which required athletes to void a urine sample in the presence of an observer, there was no longer any need for an additional inspection for gender verification to exclude males masquerading as females. The Group recommended unanimously to the IAAF Council that femininity certification was no longer required. This recommendation was formally accepted at the annual meeting of the IAAF Council in Toronto on 29–31 May, 1992. IAAF members are advised to monitor the health of their athletes before participation in international events, but this is not compulsory and certificates will not be issued in future. In June 1993 the organizers of the 1993 World University Games announced that, following representation from the IAAF and others, they had decided to abandon any form of gender verification at the forthcoming Games in Buffalo.

Despite the stand taken by the IAAF, the IOC remain hesitant to abandon genetic tests as members of the Medical Commission seem unconvinced that women with genetic disorders associated with an XY sex chromosome constitution have no advantage in competition. No evidence has been produced to support this assertion and, until there is such evidence, the Medical Commission should accept medical opinion and abandon testing. The sooner athletes realize that genetic tests are highly discriminating and subject a substantial number of women unnecessarily to emotional and social injury, the sooner will it be possible to achieve international competition among women according to the best principles of the Olympic tradition.

REFERENCES

1. Wilmore JH. The application of science to sport: physiological profiles of male and female athletes. *Canadian Journal of Applied Physiology*, 1979; **27**: 25–31.
2. Dyer KF. The trend of the male–female performance differential in athletics, swimming, and cycling, 1948–1976. *Journal of Biosocial Sciences*, 1977; **9**: 325–8.
3. Ferguson-Smith MA, Ferris EA. Gender verification in sport: the need for change? *British Journal of Sports Medicine*, 1991; **25**: 17–21.
4. Anonymous, Athlete's sex secret. *The Guardian*, 1981; **26 January**.
5. Tachezy R. Pseudohermaphroditism and physical efficiency. *Journal of Sports Medicine and Physical Fitness*, 1969; **9**: 119–22.
6. Donohoe T, Johnson N. Drugs and the female athlete In: *Foul Play*. Oxford: Basil Blackwell, 1986; 66–79.
7. Ryan AJ. Sex and the singles player. *Physician and Sports Medicine*, 1976; **4**: 39–41.
8. Hay E. Sex determination in putative female athletes. *Journal of the American Medical Association*, 1972; **221**: 998–9.
9. Larned D. The femininity test: a woman's first Olympic hurdle. *Womensports*, 1976; **3**: 8–11, and 41.
10. Turnbull A. Woman enough for the Games? *New Scientist*, 1988; **15 September**: 61–4.
11. Bunge RG. Sex and the Olympic Games. *Journal of the American Medical Association*, 1960; **173**: 196.
12. Ferguson-Smith MA. Genotype-phenotype correlations in individuals with disorders of sex determination and development including Turner's syndrome. In: Goodfellow PN, Lovell-Badge R, eds. *Sex determination and the mammalian Y chromosome. Seminars in Development Biology*, 1991; 2, 265–76.

13. Berta P, Hawkins JR, Sinclair AH, Taylor A, Griffiths BL, Goodfellow PN, Fellous M. Genetic evidence equating SRY and the testis determining factor. *Nature*, 1990; **348**: 448–50.

14 Sakamoto H, Nakanoin K, Komatsu H, Michimoto T, Takashima E, Furuyama J. Femininity control of the XXth Universiade in Kobe, Japan. *International Journal of Sports Medicine*, 1988; **9**: 193–5.

15. Ferguson-Smith MA, Cooke A, Affara NA, Boyde E, Tolmie JL. Genotype–phenotype correlations in XX males and their bearing on current theories of sex determination. *Human Genetics*, 1990; **84**: 198–202.

16. New M, Levine LS. Recent advances in 21-hydroxylase deficiency. *Annual Review of Medicine*, 1984; **35**: 649–3.

17. Caldwell F. Victorious in court, she's defeated on court. *Physician and Sportsmedicine*, 1977; **5**: 26–8.

18. Schwinger E. Problems of sex differentiation in athletics. *Atleticastudi*, 1981; **5**: 72–80.

19. Carlson A. Chromosome count. *Ms* (New York), 1988; **October**: 40–4.

20. Moore KL. The sexual identity of athletes. *Journal of the American Medical Association*, 1968; **205**: 787–8.

21. de la Chapelle A. The use and misuse of sex chromatin screening for 'gender identification' of female athletes. *Journal of the American Medical Association*, 1986, **256**: 1920–3.

22. Simpson JL. Gender testing in the Olympics. *Journal of the American Medical Association*, 1986; **256**: 1938.

23. Ljungqvist A, Simpson JL, Medical examination for health of all athletes replacing the need for gender verification in international sports. *Journal of the American Medical Association*, 1992; **267**, 850–2.

24. Dingeon B, Hamon P, Robert M, Schamasch P, Pugeat M. Sex testing at the Olympics. *Nature*, 1992; **358**, 447.

Acute sports injuries

4

4.1   An introduction and brief overview

EJNAR ERIKSSON

The increased interest in sports worldwide has led to an increase in acute sports injuries.[1] The number of patients seen at emergency departments who have sustained injuries from sport has increased from 1.4 per cent of all injuries in 1955[2] to 10 per cent in 1988.[1] However, the number of people participating in sports has also undergone an enormous increase; it has been suggested that some 25 per cent of the population in cities in the Scandinavian countries participate in some types of sports activity at least twice per week. The corresponding figure for the rural population is somewhat lower.[3] It is believed that over 100 million people in North America participate in some type of sports.[4] Therefore it is not surprising that a number of sports injuries occur. Many acute sports injuries, such as contusions, haematomas, fractures, are the same as ordinary orthopaedic injuries. However, some are specific for sports as such and are often specific for a special sports type.

In Sweden about 1.6 million people, out of a total population of 8 million, are insured by the co-operative insurance company Folksam. Whenever a Swedish person wants to become a member of a sports association or to participate in a particular sports event, such as the Stockholm Marathon or the 85-km cross-country Wasa ski race, he or she must pay a fee. Part of that fee goes to a group insurance in the Folksam Insurance Company. Thus since all sports injuries are covered by Folksam, they have been able to determine the injury frequency in different sports. The most recent report from Folksam was published in 1985 and summarizes almost 27 000 sports injuries reported to this insurance company between 1976 and 1983.[5] They report the occurrence of sports injuries as the number of injuries per thousand insurances. As can be seen from Table 1, ice hockey is responsible for the largest number of injuries and claims have also tended to increase over the years. However, Börje Eriksson, the research co-ordinator of Folksam maintains that this increase cannot be taken to reflect a true increased risk of sustaining an injury in the various sports. The general economic standard of Sweden has deteriorated over the period 1976–83, and Eriksson believes that some of the increase in the number of claims may be due to greed, i.e. more minor injuries are being reported in order to recover as much money as possible from the insurance company. However, bandy has not shown this increase in injuries. Bandy, which is played on an ice rink as large as a soccer field with 11 players in each team is very similar to field hockey but is played on skates. It is a non-contact sport which has tightened up its rules so that no tackling whatsoever is permitted. It is interesting that bandy does not show the same increase in sports injuries that can be seen in the other reported sports (Table 1).

A major interest for the sports world is whether and to what extent acute injuries lead to permanent disability. During the period 1976–83 Folksam received a total of 10 659 claims for injuries sustained in soccer, which is the most popular sport in Sweden. Permanent disability occurred as a result of 270 (2.5 per cent) of these injuries. Eight of these 270 injuries were due to road accidents during transport to or from a match, 63 were sustained during training, and 187 occurred during matches. No more than 12 of the 262 disablement cases that occurred in sport had a degree

Table 1   Sports injuries in the Folksam insurance survey (all types of claim)

Sport	Injury frequency/1000 insurances							
	1976	1977	1978	1979	1980	1981	1982	1983
Soccer	3.2	6.9	8.0	8.3	9.2	9.5	10.1	12.3
Ice hockey	21.8	23.2	25.8	25.0	28.7	31.1	33.5	40.5
Bandy	5.1	5.5	7.8	10.7	8.3	6.6	6.9	7.8
European team handball	3.2	3.4	6.5	10.2	10.2	13.3	14.8	18.8
Basketball	2.3	2.5	2.5	2.8	4.8	10.1	9.6	9.9

of disablement greater than 15 per cent; 130 had 5 per cent or less. Fifty-six per cent (150) of all the disablement cases in soccer were due to knee injuries which cost the insurance company more than injuries to any other part of the body. The author, who was the orthopaedic consultant for the Folksam study, contacted the hospitals and obtained the records of 144 of those cases; 80 per cent of them were anterior cruciate ligament injuries in combination with injuries to structures such as menisci and other ligaments. Thus the most severe soccer injuries are those to the knee. They lead to the highest degree of disablement and also require the longest period for recovery. In their summary Folksam concluded that, in view of this, they would support research into the prevention and improved treatment of knee injuries. Together with the Swedish Soccer Pool, they also donated funding for the first full-time professorship of sports orthopaedic surgery in Sweden.

One of the drawbacks of a retrospective study of the type just quoted is that it does not record the total number of injuries. Some minor injuries are probably not reported to this insurance company. However, our group at the Karolinska Hospital has performed a prospective study of injuries to players in a third-division amateur ice hockey team in the Stockholm area during the 1987–8 season.[6] All the players were carefully examined before the season. Their isokinetic concentric and eccentric muscle torques, flexibility, physical conditioning, and records of earlier injuries were obtained. One physician attended every single game during the 1987–8 season, and treated and recorded all acute injuries that occurred during training or matches. During this particular year six real injuries occurred during 594 h of playing, which corresponds to 10 injuries per 1000 h of playing. The total number of injuries that occurred during the season was 68, but 62 (92 per cent) of those were very minor and did not lead to any absence from sport. If all those 68 injuries are counted this would correspond to an injury frequency of 116 injuries per 1000 h of playing. These figures are for playing matches. During the 891 h of training only two injuries occurred which corresponds to an injury frequency of 2.2 injuries per 1000 h of training. These figures are in good agreement with other reports on ice hockey injuries. Therefore, although ice hockey has a relatively high number of injuries, most of them are minor. Ekstrand[7] has reported that many soccer injuries were due to rough play by the player himself and that the injuries were correlated with the standard of refereeing. Our 68 injuries could not be related to any of these factors; they were not correlated to playing against a better or poorer team, nor were they correlated to good or poor refereeing, or to rough play. We recorded a drop of isokinetic concentric and eccentric muscle torque during the 1-year study period, but this drop in muscular torque was not related to the injury frequency.

HOW MUCH DO SPORT INJURIES COST SOCIETY?

Sandelin[1] attempted to estimate the costs of acute sports injuries for the City of Helsinki and reported a yearly cost in 1988 of FIM 5 590 544 which corresponds to US $1 352 551. Helsinki has a population of 500 000; hence, by extrapolation, the cost for a city the size of Stockholm (1 million inhabitants) would be US $2 705 102 and that for a city with a population of 10 million would be $27 million. Sanderlin also calculated that the average cost of a sports injury that could be treated in the casualty department was FIM 1364 (US $330) and the average cost of an injury requiring hospital care was FIM 17 268 (US $4178). Since 8 to 12 per cent of all acute injuries seen in hospital emergency wards are caused by sport, it is not surprising that these costs have frightened some groups in society. In Sweden a discussion has recently been held about the benefits as opposed to disadvantages of sport in which opponents of sport claimed that the costs of sports injuries were too high. However, supporters of sport pointed out that the cost to society of osteoporosis is 100 times that of sports injuries. Sports and an active life have such advantages that the Swedish National Board of Health and Welfare has recommended people to walk for 1 hour every day, and either to walk rapidly or jog two or three times a week in order to maintain health. Therefore if we accept sport as 'medicine', we must also accept that, like pharmaceutical drugs, it has some side-effects. Sports injuries are one of the side-effects of sports but, as noted above, their frequency is relatively low. Doctors see a relatively large number of sports injuries because there is an enormous number of people participating in some form of sports activity.

PREVENTION OF ACUTE SPORTS INJURIES

Some acute sports injuries can be prevented. Studies of downhill skiing carried out in Sweden[8,9] have shown that it is possible to reduce the rate of ski injuries by providing information on testing release bindings and skiing with caution. In a randomized prospective study on soccer injuries Ekstrand[7] showed that he could reduce the different types of injuries in soccer by some 75 per cent by applying relatively simple measures such as having a physiotherapist teach the soccer players how to warm up and stretch before and after matches, by using shin-guards, and by taping the players' ankles.

REFERENCES

1. Sandelin J. *Acute sports injuries—a clinical epidemiological study.* Thesis, Medical Faculty, University of Helsinki, 1988.
2. Peräsalo O, Vapaavuori M, Louhimo I. Über die Sportverletzungen. *Annales Chirurgiae et Gynaecologiae Fenniae* 1955; **44**: 256–62.
3. Seppänen P. *Urheilujärjestöistä ja liikuntatoiminnasta tämän päivän Suomessa.* Helsinki: SVUL, 1977.
4. Nicholas JA, Reilly JP. Orthopaedic problems in athletes. *Comprehensive Therapy* 1985; **11**: 48–56.
5. Folksam. *Sports injuries 1976–1983—a report from Folksam.* Stockholm: Folksam, 1985.
6. Posch E, Haglund Y, Eriksson E. Prospective study of concentric and eccentric leg muscle torques, flexibility, physical conditioning and variation of injury rates during one season of amateur ice-hockey. *International Journal of Sports Medicine* 1989; **10**: 113–17.
7. Ekstrand J. *Soccer injuries and their prevention.* Medical Dissertation 130, Linköping University, 1982.
8. Eriksson E, Danielsson K. A national ski injury survey. In: Figueras J, ed. *Skiing safety II.* Baltimore: University Park Press, 1978: 47–55.
9. Danielsson K, Eriksson, E, Johnson E, Lind E, Lundqvist S. Attempts to reduce the incidence of ski injuries in Sweden. *ASTM Special Technical Publication* 1985; **860**: 326–37.

4.2.1 Acute knee injuries—introduction

ROBERT J. JOHNSON

INTRODUCTION

The most common career-ending injury of athletes probably involves destruction of one or more of the major ligaments of the knee. Injuries to the knee result in more problems for athletic individuals than injuries to any other joint.[1] Thus the subject of the acute knee injury is all too frequently the focus of attention of the sports medicine community. Fortunately, the majority of acute knee injuries are relatively minor in nature or can be resolved with simple non-surgical therapy and thus result in relatively short periods of disability or reduced activity. However, even relatively minor injuries can present chronic problems. While they may not interfere with the activities of daily living, they can be very frustrating for those athletes who require stressful performance of their knees. Thus, a full understanding of the mechanism of the injury, the biomechanics of the normal knee, the pathophysiology of the injury and repair process as it applies to the knee, diagnostic procedures, surgical and non-surgical treatment regimens, and rehabilitation principles and methods is vitally important for those who care for the injured athlete's knee. It is our purpose in this section of the text to discuss the material necessary to diagnose and treat appropriately not only the major knee ligament injuries but also those lesser injuries which, as far as the athlete is concerned, may be either career threatening or only a nuisance for a relatively short period of time.

ACUTE INJURIES OF THE MENISCI

It has been stated that the most common injuries sustained by athletes involves damage to the menisci.[1,2] Meniscectomy is the most frequently performed orthopaedic operative procedure, accounting for 10 to 20 per cent of all surgery.[1] Before the turn of the century the meniscus was considered to be a functionless remnant of muscle origin.[3] Many surgeons in the early to middle portion of the present century suggested that the menisci were functionless and that when they were damaged they should be totally removed.[1,4–9] In fact, Smillie[8] advised that even if damage to the meniscal structure was only suspected, the meniscus should be totally removed. It was believed that unless the meniscus was totally extirpated, a fibrocartilaginous replacement would not form or the retained portion of a partially resected meniscus would result in degenerative arthritis. In 1948, Fairbank wrote his classic paper on the meniscus in which he implied that the structure did, in fact, serve a function by sharing in the transmission of load across the joint.[10] It was not until the 1960s and 1970s that several papers were published showing that removal of the meniscus was not necessarily a benign procedure.[11–16] Innumerable recent studies have confirmed that the meniscus does serve a valuable biomechanical function for the knee.[17–25] Thus, the concept of partial meniscectomy (removal of only the damaged portion of the structure), made possibly by the development of practical arthroscopic techniques, became well substantiated as superior to resection of the entire structure.[26–28] Arthroscopic procedures have also allowed the next logical step in the treatment of damaged menisci, which is the repair of certain lesions rather than their resection (Figs. 1, 2).[29–33] There is little doubt that preservation of all components of the torn meniscus is superior to removal if the healed structure retains the ability to function biomechanically. Recently, several investigations have sought to improve the surgeon's ability to effectively retain fragments of torn menisci.[34–35] However, preservation of crushed and deformed fragments of non-functioning menisci certainly do not serve a useful purpose. At the present time there is little evidence that the healed meniscus still serves its original biomechanical purpose, although it appears logical to

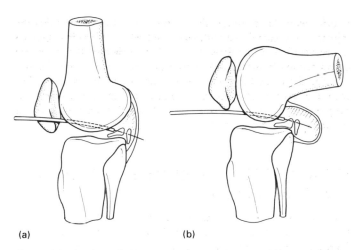

Fig. 2 Meniscal repair. The needle traverses a tear of the medial meniscus in extension (a) to avoid compromise of the meniscus (b). (Redrawn from ref. 115, with permission.)

assume that it does retain its ability to function in the appropriate manner.[22,32,36,37] Efforts have recently been made to devise techniques to replace menisci which require excision.[38–45] Allografting procedures are being performed and the implantation scaffold devices which encourage the ingrowth of the patient's own cells to produce a viable replacement meniscus are being studied. Only time will tell if these measures can result in the re-establishment of biomechanical functions of these 'new' menisci.

LIGAMENT INJURIES

Acute knee ligament injuries have long been recognized as a major problem for the athlete, but probably no subject in sports medicine has resulted in more controversy and confusion. Like other topics discussed in this section of the text, there has been a dramatic change in the way that these injuries have been treated through the years. Almost everyone, including athletes and sports fans, realize that a major injury to the knee ligaments could be the death knell to an athletic career. It is hoped that recent advances in the recognition and treatment of some, if not the majority, of these severe injuries can lead to the continuation of athletic careers.

Ligament injuries about the knee were recognized before the turn of the century. Most physicians believed that major ligament injuries should be treated non-operatively, often with prolonged periods of immobilization in casts or splints. Campbell, in the first edition of his text (1939) stated that: 'operative measures for acute rupture of the lateral ligaments of the knee are seldom, if ever required, as the most extensive ruptures usually undergo repair if properly treated by conservative measures. Repair is indicated in the presence of symptoms of derangement from complete or partial rupture or elongation of long duration'.[4] In 1935 Milch observed that in his opinion 'the anterior crucial ligament is not an essential structure as its loss is compatible with relatively normal function of the joint'.[46] Thus the majority of early operative procedures described in the literature discussed reconstruction of ligament injuries which had either been overlooked at the time of injury or deliberately treated by non-operative means. Those individuals perhaps most responsible for the development of opinion tending to treat acute ligament disruptions with surgical repair were O'Donoghue in North America and Palmer in Europe.[47,48]

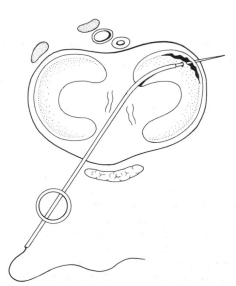

Fig. 1 Meniscal repair. A curved cannula is inserted through the contralateral portal and used for suturing to avoid popliteal vessels and nerve. (Redrawn from ref. 115, with permission.)

After these individuals and many other investigators demonstrated that labourers and athletes were probably better served by early surgical intervention, the standard treatment for most complete knee ligament ruptures, especially if more than one was destroyed, was early operative intervention. Various criteria were developed as to who was a candidate for surgery and who should be non-surgically treated, and certainly no unanimity was established as to which ligament should be repaired and which could be ignored.

Although not universally true, until recently the majority of knee surgeons believed that only the collateral ligaments and the posterior cruciate ligament were of significant structural and biomechanical importance. With regard to this problem many of the great educators taught that the anterior cruciate ligament was unimportant.[46,49] This, coupled with the fact that the vast majority of physicians, including orthopaedic surgeons, could not accurately recognize the presence of a complete tear of the anterior cruciate ligament, led to the feeling that this structure could be ignored or simply cut away if it was found to be torn. Over the past 10 to 15 years no subject of sports orthopaedics has received as much attention as the injured anterior cruciate ligament. A Medline search performed on anterior cruciate ligament injury and treatment from 1984 to the present revealed that over 500 articles on these subjects have been published in the English literature. For the last 10 years approximately one-third of the papers submitted for presentation at the conferences of the American Orthopaedic Society for Sports Medicine have dealt with subjects relating to the anterior cruciate ligament.[50] Clearly, during the same period of time it has become almost universally accepted that the anterior cruciate ligament does contribute to the functional stability of the knee, and that its disruption often leads to devastating problems for the injured athlete. However, there can be no doubt that many patients function quite satisfactorily without their anterior cruciate ligament.

The natural history of the anterior cruciate ligament deficient knee, although intensively studied, is still unknown.[51] No-one can predict with certainty whether a patient who has sustained an acute anterior cruciate ligament disruption will function almost normally or develop significant symptoms if the ligament is not successfully surgically repaired or reconstructed. One group of authors has suggested that the majority of patients can recover with non-operative treatment and return successfully to sport and occupational activity with little or no restriction.[52-56] Others have implied that return to vigorous sport and stressful work leads to a high incidence of re-injury events (pivot shift episodes) which require patients to alter their life-styles substantially in order to avoid significant degeneration of their knee joint.[57-60] Obviously, various authors have assigned differing degrees of significance to the re-injury events and the pain, swelling, and disability that they cause. Some investigators believe that restriction of activity which requires an athlete to alter his or her level of competition or switch to a less vigorous sport is unimportant, while others feel that such change is of great concern. Some have stated that episodes of re-injury are not likely to lead to further degenerative changes of the knee, while others express concern about the almost inevitable destruction of menisci and eventually the articular cartilage if pivot shift events continue to occur. It has been suggested that even unrecognizable dysfunction associated with knees without anterior cruciate ligaments may lead to joint degeneration.[61,62] Depending on the interpretation of these factors and others, various physicians have formed widely divergent opinions as to what constitutes an indication for surgical intervention in the case of the patient with a torn anterior cruciate ligament.

Another matter which adds to the confusion in deciding whether or not to operate, is that outcome studies reveal that a significant number of the surgical interventions performed for anterior cruciate ligament injuries do not succeed, and that even if they do appear to restore subjectively satisfactory function, they are not capable of preventing degenerative changes later.[58,63] Even if patients eventually return to sport following repairs or reconstructions, no one can be certain that they will continue to do well indefinitely. Many methods of treatment are ardently praised by enthusiastic investigators who find that early postoperative results appear to be excellent, only to find that at later follow-up the results have deteriorated. The literature is replete with short-term follow-up studies, but very few long-term results have been published.[58] Results followed for 10 years or more are probably necessary before the orthopaedic community will have a truly accurate assessment of the natural history of anterior cruciate ligament repairs or reconstructions. Many procedures, including innovate use of prosthetics, allografts, and augmentations, have not yet stood the 'test of time'.

The determination of the treatment of choice in the case of the destroyed anterior cruciate ligament is still evolving. At the time of writing, the 'gold standard' for treatment of these individuals with a 'high risk life-style' following injury to the anterior cruciate ligament is probably the complete replacement of the ligament with a bone-patellar tendon–bone autograft (Fig. 3).[64,65] Some believe that the incidence of postoperative parapatellar pain is unacceptably high following this procedure and have advocated alternative procedures.[66] At the present time, the repair of a ruptured anterior cruciate ligament without augmentation has been demonstrated to be an inferior treatment approach.[67-70] Although many papers report satisfactory results following repair with autograft augmentation, this approach seems to be becoming less popular.[67,71-73] It appears that the use of extra-articular reconstruction in replacing the function of the anterior cruciate ligament has also fallen out of favour.[74] Some investigators still advocate its use in the treatment of young patients with open epiphyses, as an adjunct to intra-articular reconstruction, and for

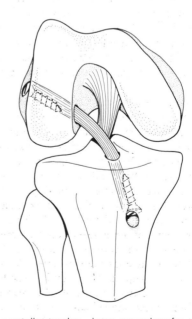

Fig. 3 Bone—patellar tendon—bone procedure for anterior cruciate ligament insufficiency. (Redrawn from ref. 116, with permission.)

older sedentary patients with chronic anterior cruciate ligament dysfunction. In theory, an allograft reduces the morbidity of the replacement of the anterior cruciate ligament, but concerns revolving around the transmission of viral disease and healing problems have not yet been totally overcome. Prosthetic ligament augmentation devices and pure prosthetic replacement are as yet unproven as viable alternatives and have been associated with a high risk of failure. Much more research is necessary before the place of allografts and prosthetic ligaments in the treatment of knee ligament injuries can be established.

There can be little doubt that the occurrence of a truly 'isolated' tear of any ligament is essentially an impossibility, but certainly most surgeons have observed a number of cases where the clinical and even operative evidence indicates that only one structure has sustained severe damage. Perhaps the most dramatic change in approach to the treatment of such injuries is that rendered to those who sustain tears of the medial collateral ligament. Complete tears of the medial collateral ligament are suspected when valgus stress applied to the fully extended knee reveals a mild increase in laxity and a soft end-point, and a dramatic increase in laxity on application of valgus stress at 20 to 30° of flexion. In the 1960s and 1970s most surgeons would probably have treated such injuries with surgical repair in the acute setting. Now many investigators are advocating non-operative treatment and quite rapid rehabilitation.[75] Until quite recently 'isolated' second-degree tears of the medial collateral ligament were treated with prolonged casting, but now it has been well documented that these injuries can be treated very satisfactorily with aggressive conservative rehabilitation and return to vigorous activity in 4 to 6 weeks.[76]

The posterior cruciate ligament can be disrupted in essentially an 'isolated' fashion. The functional results of such injuries treated without repair or reconstruction appear to be much more satisfactory than those of the isolated anterior cruciate ligament injury.[77,78] This finding, together with the almost universal observation that posterior cruciate ligament reconstructions and repairs are often unsuccessful, has led to a much more conservative approach in the management of these lesions than for those of the anterior cruciate ligament.[79] Although there are few advocates of acute repair or reconstruction of 'isolated' posterior cruciate ligament injuries, the majority of opinion probably now favours treating them non-surgically.[80,81] However, it appears that the management of posterior cruciate ligament injuries, when coupled with the destruction of other major stabilizers or the posterior lateral components of the knee, requires an aggressive surgical approach to avoid disability.[78]

When injured in a relatively isolated fashion the lateral collateral ligament can be managed conservatively, but care must be taken to avoid overlooking significant damage to the arcuate ligament and the popliteus tendon, for if they are also damaged non-surgical management may be very disappointing.[82]

The lack of well-controlled randomized prospective long-term (10 years or more) outcome studies comparing one surgical technique with another has made it almost impossible for anyone seriously using the literature to determine which approach, if any, is superior.[57] Only when such studies become available will information exist which allows us to establish the method of choice for treatment of these knee ligament injuries. Also, the lack of standardization of follow-up methodology has resulted in an inability to compare the relative merits of one treatment regimen with another. Thus, those individuals who treat knee ligament injuries remain frustrated when they assess their own approach to their patients' problems. The surprising lack of knowledge of the detailed biomechanics of the knee has also led to much confusion concerning the treatment methods. Proliferation of information concerning the complexities of knee function will undoubtedly assist in advancing our approaches concerning the proper treatment of knee ligament injuries, but this demands that all those concerned with these matters must keep critically abreast of the explosion of information now occurring and apply these changes to their own practices.

FRACTURES AND DISLOCATIONS

Fortunately for the athlete, the incidence of fractures about the knee is relatively low in sports activities.[83] However, those mechanisms of injury which result in ligament disruptions in young adults may produce fractures of the epiphyseal plate of the distal femur or proximal tibia in children with open epiphyses or in tibia plateau fractures in older age groups. Ligament injuries may result in avulsions of relatively small fragments of bone from their attachment sites in patients of all age groups. Growing children relatively often avulse the inferior attachment of their anterior cruciate ligaments with a fragment of bone, which on occasions is quite large (Fig. 4). Such fractures are quite rare in adults. Children also more frequently avulse small fragments of bone with the medical collateral ligament sprains, most frequently from the proximal end. Tibial tubercle avulsion occurs almost exclusively in children. Avulsions of the fibular head are not uncommon in severe lateral knee injuries in adults. The Segond fracture or lateral capsular avulsion, arising from the inferior attachment of the mid-lateral capsule on to the tibia just anterior to the fibular head, is a relatively common occurrence and is almost always associated with the complete disruption of the anterior cruciate ligament (Fig. 5).[84] The 'lateral notch sign' described by Warren et al.[85] probably represents an infraction of the lateral femoral condyle (an exaggeration of the normally present linea terminalis) which is occasionally observed in patients with chronic anterior cruciate insufficiency. Fractures of the patella are most frequently the results of falls and vehicular accidents than sports activities, but the patella is the bone most commonly fractured in sports injuries

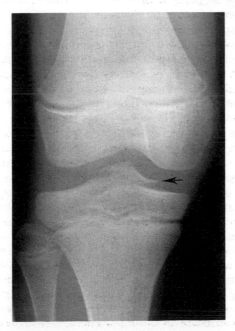

Fig. 4 Avulsion of the tibial spine (arrow).

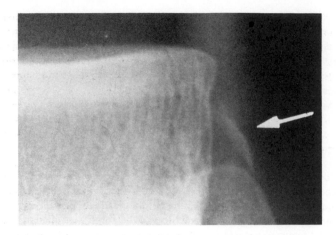

Fig. 5 Segond's fracture (arrow).

involving the knee.[83] Osteochondral fractures of the distal femur occasionally occur in association with dislocations of the patella. More severe intra-articular fractures of the femur are very rare. Stress fractures involving the patella occasionally occur, but stress fractures of the femur and tibia involving the joint surfaces are uncommon. Stress fractures in distance runners are not infrequent, and most often involve the anteromedial tibia 2.5 to 5.00 cm (1–2 inches) below the joint when they occur within the vicinity of the knee. During the past few years several investigators have observed occult fractures of the tibia and femur during investigations of ligamentous injuries of the knee using MRI techniques.[86,87] At the present time the clinical significance of these radiographically indeterminable injuries remains to be defined. Osteochondritis dissecans and osteonecrosis are beyond the scope of this section of the text.

The treatment of fractures involving any joint surface demands accurate reduction and maintenance of near perfect position while healing occurs. Thus open reduction and internal fixation of fractures of the patella, distal femur, and tibia plateau are frequently required. Bone grafts are very often necessary in tibia plateau fracture management. The principles of strong fixation and early protected motion, often with prolonged avoidance of weight bearing (tibia plateau fractures), are advocated. The treatment of epiphyseal injuries also demands special care to avoid damaging the already vulnerable epiphyseal plate. The treating physician must be aware of the potential of total or partial growth arrest and prepare the parents as well as the patient for these eventual complications, should they occur.

Avulsion fractures produced by ligament disruptions can often be ignored if they are small, thus requiring no alteration in the treatment which is necessary for the ligament itself. Avulsions of the anterior cruciate ligament, which free large fragments from the tibia plateau, have advocates of open re-attachment, but the older literature suggested that minimal to moderately displaced fragments could be treated closed by simple extension of the knee.[88] Present knowledge suggests that this approach is probably incorrect for two reasons. Firstly, in full extension the intact anterior cruciate ligament is under relatively high strain and thus actually lifts the bony fragment from its bed. Secondly, fragments of bone and interspersed soft tissue often make the reduction impossible unless the debris is removed surgically. Arthroscopic or open debridement of the bed and fixation of the bone fragments probably provide the best means of treatment in all but the most

minimally displaced of this type of fracture.[83]

Knee dislocations are also quite rare in sports events, but these devastating injuries are certainly the epitome of the knee ligament injury. Results from closed treatment with prolonged casting, as advocated in the early part of the century, were often surprisingly good. Although most authors in recent years have advocated repair, repair with augmentation, or immediate reconstruction for as many of the totally disrupted ligaments as possible in the acute phase of treatment, the presence of contaminated open wounds and other life-threatening injuries frequently makes acute surgery impractical.[89,90] In such cases relatively stable knees often result from prolonged immobilization of the joint, but residual abnormal laxity of at least some of the major ligaments frequently necessitates late reconstructions. Ideally, once the vascularity of the knee has been assured, open repair or primary reconstruction of all the destroyed ligaments should be accomplished. Because of the dire consequences (amputation) of missing significant arterial injury at the level of the knee, arteriography or, at the very least, very close monitoring of the vascular status of the leg should be performed before extensive surgery is performed on these knees.[91] Delay in surgery until the acute reaction is diminished and the range of motion has returned to nearly normal as is now advocated for the treatment of 'isolated' or double-ligament injuries, is probably impractical in these severely traumatized knees. Thus surgery is best performed after a few days, but within 2 weeks of the original injury if possible. Depending on the success in establishing strong fixation of the various injured structures, the postoperative management of such extensive surgery often requires careful and prolonged immobilization if healing is to be successful.

Dislocation of the proximal tibiofibular joint is rare in sports trauma.[92] When it is recognized early, non-surgical management is usually sufficient. Even in cases of recurrent subluxation or dislocation, surgical reconstruction of the joint is rarely necessary.[93]

Ligament injuries of the knee are not infrequently associated with fractures of the tibia plateau, femur, and tibial shaft.[95] Thus these injuries must be suspected and carefully ruled out whenever there is a major fracture in the vicinity of the knee.

REHABILITATION

Rehabilitation following knee ligament surgery or serious injury is universally accepted as being very important to the final outcome of the treatment. However, the performance of such rehabilitation programmes is tempered by one of the most frustrating dilemmas faced by any knee surgeon. On the one hand it is necessary to protect the injured structures (usually ligament repairs, reconstructions, or partial or even complete disruptions) so that they can heal without further injury. On the other hand there are numerous problems resulting from immobilizing the joint while healing occurs. The detrimental effects of even short-term immobilization, including periatricular muscle atrophy, articular cartilage damage, capsular arthrofibrosis, weakening of the non-injured bone and ligamentous tissue, and loss of range of motion, have been documented in a voluminous literature.[96] The ongoing problem for the knee surgeon is how to compromise between these seemingly diametrically opposed circumstances. As documented in other chapters in this section of the text, there have been dramatic changes with time in how various surgeons have managed their patients' rehabilitation. In the early part of this century, prolonged casting (6–10 weeks) following major ligament injury, whether operated or treated conservatively, was often the standard.

Although it was appreciated that muscle rehabilitation and re-establishment of functional range of motion was necessary, very little detail was provided on how this was to be achieved. By the 1950s and 1960s, emphasis was placed on quadriceps rehabilitation, but again little detail was offered. By the late 1970s a number of surgeons were beginning to allow motion as early as 2 to 4 weeks following the surgery or injury, but not until the late 1980s were the majority of physicians allowing immediate motion, which now appears to be the standard of care. How such motion is performed (passively administered by a therapist or patient, continuous passive motion, or with active muscle control) is much more variable. Burks et al. [97] sounded a note of caution in the use of continuous passive motion machines when they reported a number of early failures of anterior cruciate ligament reconstructions performed in vitro.[97] In contrast, Reinecke et al.[98] demonstrated that, when properly positioned, the continuous passive motion machines do not appear to have harmful effects on nearly isometrically positioned anterior cruciate ligament grafts. The advocates of continuous passive motion machines have stated that swelling and pain are minimized in the early postoperative phase, but those who question the cost effectiveness of this treatment point out that there is no difference between groups of patients treated with or without continuous passive motion machines within a few weeks of the surgery.[99]

Quadriceps muscle activity in an unlimited fashion as soon as practical after ligament surgery was often advocated, until several investigators revealed that the quadriceps acts antagonistically to the anterior cruciate ligament between approximately 50° of flexion and full extension.[100–103] In recent years many surgeons have suggested that isotonic quadriceps loading into full extension be avoided for several months after anterior cruciate ligament reconstruction.[104] The use of antishear devices in quadriceps and hamstring cocontracture to reduce the loading on a freshly injured, repaired, or reconstructed anterior cruciate ligament has been advocated, but the effect of these techniques is yet to be proved.[102,105–107] The vast number of different muscle rehabilitation programmes advocated following knee ligament surgery implies that no one method can readily be established as the most appropriate. 'Buzzwords' such as isokinetic, eccentric, closed kinetic chain, and sports specific training all have their enthusiasts, but the timing and safety of these programmes must be carefully considered if harm to the patient is to be avoided. Advocates of aggressive rehabilitation do not report detrimental effects from early quadriceps activity, but there can be no doubt that repaired, reconstructed, or even injured ligaments are vulnerable to overloading if rehabilitation programmes are too aggressive.

Casting was the most common means of protecting complex knee ligament repairs and reconstructions until the 1980s when various knee rehabilitation braces became the standard. Cast bracing did have some advocates in the 1970s.[108] In recent years, with stronger graft material and fixation techniques, some surgeons have avoided casts and braces completely in their post-operative management of their patients. In the past, prolonged avoidance of weight bearing for up to 4 months following ligament injury or surgery was often suggested. As recently as 1981, the majority of surgeons questioned by Paulos et al.[103] did not allow weight bearing until 8 weeks or more after their anterior cruciate ligament repairs or reconstructions. Shelbourne and Nitz[109] now allow their patients to begin weight bearing as tolerated immediately after bone–patellar tendon–bone grafts. The trend towards better fixation techniques and reconstructing rather than repaired injured ligaments has made such practices at least technically possible.

As patients increase their activity following knee ligament surgery or injury, the question of whether or not to prescribe a functional knee brace is often considered. Although much anecdotal information concerning their effectiveness exists, very little hard data proving that these devices can stress shield the injured, repaired, or reconstructed ligaments adequately have been published. Much more investigation using strict scientific principles is necessary before answers concerning the place of functional knee braces in our armamentarium during rehabilitation can be established.

The question of when to return an athlete to sport has remained controversial. During the 1950s and 1960s the main criteria for success following ligament surgery appeared to be simply how quickly the athlete could return to sport. In the 1970s and early 1980s a much more conservative approach was advocated as more was learned about the healing properties of repaired and reconstructed ligaments and the potential dangerous effects of early muscle activity. During that time many investigators were not advocating return to full activities for a year or more.[103] In recent years a definite trend towards more rapid rehabilitation and return to full activities has been advocated as several investigators have reported no deleterious effects in their patient populations when their athletes return to full activities 4 to 6 months after bone–patellar tendon–bone grafts of their anterior cruciate ligaments.[109,110] These authors have stressed that a bone–patellar tendon–bone graft placed isometrically and firmly fixed can probably tolerate vigorous early rehabilitation and early return to sports, but this same rapid rehabilitation may not be successful with other operative procedures.

There can be no standard recipe for rehabilitation programmes following any knee ligament injury, repair, or reconstruction. The trend towards aggressive advances in weight bearing, range of motion, muscles exercises, and return to functional activities has potential dangers. This is particularly true in the case of the cruciate ligaments. It has been well documented in many studies of healing in animal models that normal tensile strength of repaired or reconstructed anterior cruciate ligaments returns slowly and never completely.[111,112] It has also been shown that vigorous quadriceps contractions from 50° of flexion to full extension places significant loads on the injured cruciate ligament.[100–104] It is frequently stated that some tensile loading of healing tissue is probably beneficial (Wolff's law), but no one knows how much is detrimental and how much is safe. It is not known when full muscle rehabilitation and return to activities is safe. Those advocating rapid rehabilitation believe that present fixation methods, accurate placement of the graft at the time of surgery, and the use of stronger graft material makes it possible for aggressive rehabilitation to succeed. Initially, the weak link in any repair or reconstruction is the fixation of the repair or graft.[113,114] Modern techniques (interference fit, screw and washer, and other forms of fixation) allows at least some early motion and muscle function, but no one knows how much. Within a few weeks of surgery, fixation of cruciate grafts is no longer the primary concern because of the deterioration of the strength of that portion of the graft which courses from tunnel to tunnel. This portion of the reconstruction is at its weakest 6 to 8 weeks after surgery. It is at this time that those advocating a rapid rehabilitation are markedly advancing the activities of their patients.[112] Care must also be taken in applying rapid rehabilitation methods to cases where weaker grafts, repaired ligament tissues, and less than ideal surgical techniques have been employed. At the present time our limited knowledge of the healing process, the detailed biomechanics of the knee, and many technical variables which are difficult to control during surgery (accurate placement and fixation of the graft, tensioning of the graft, and the effects of active and passive motion on healing) make it impossible to develop an ideal universally appropriate rehabilitation programme. Care must be taken to base decisions on the development of one's own rehabilitation programme on scientific fact rather than on anecdotal statements of how rapidly someone else's patients were able to return to sports. Marketing pressures rather than sound scientific principles may have more to do with the present trend towards the rapid return of athletes to sports activities following major ligament surgery. Each surgeon should carefully weigh all the data presently available and be prepared to improve his or her own programmes as the results of long-term randomized and controlled studies clearly show that one programme is superior to another. Unfortunately, such outcome studies are not yet available.

REFERENCES

1. P. Renström, Johnson RJ. Anatomy and biomechanics of the menisci. *Clinics in Sports Medicine* 1990; **9**: 523–38.
2. Sonne-Holm S, Fledelius I, Ahn N. Results after meniscectomy in 147 athletes. *Acta Orthopaedica Scandinavica* 1980; **51**: 303–9.
3. King D. The function of semilunar cartilages. *Journal of Bone and Joint Surgery* 1936; **18A**: 1068.
4. Campbell WC. *Operative orthopaedics.* St. Louis: CV Mosby, 1939: 406, 415.
5. Dandy DJ, Jackson RW. Meniscectomy and chondromalacia of the femoral condyle. *Journal of Bone and Joint Surgery* 1975; **57A**: 1116–19.
6. Dandy DJ, Jackson RW. The diagnosis of problems after meniscectomy. *Journal of Bone and Joint Surgery* 1975; **57B**: 349–52.
7. Quigley TB. Knee injuries incurred in sports. *Journal of American Medical Association* 1959; **171**: 1666.
8. Smillie JS. *Injuries of the knee joint.* 4th edn. Edinburgh, Churchill Livingstone, 1971: 68.
9. Watson-Jones OR. *Fractures and joint injuries*, Vol. 2. Edinburgh, Livingstone, 1955; 769–73.
10. Fairbank TJ. Knee joint changes after meniscectomy. *Journal of Bone and Joint Surgery* 1948; **30B**: 664–70.
11. Appel H. Late results after meniscectomy in the knee joint. A clinical and roentgenological follow-up. *Acta Orthopaedica Scandinavica* 1970; Suppl.133: 1–111.
12. GEAR MWL. The late result of meniscectomy. *British Journal of Surgery* 1967; **54**: 270–2.
13. Huckel J. Is meniscectomy a benign procedure? *Canadian Journal of Surgery* 1965; **8**: 254–60.
14. Jackson JP. Degenerative changes in the knee after meniscectomy. *British Medical Journal* 1968; **2**: 525.
15. Johnson RJ, Kettelkamp DB, Clark W, Leaverton P. Factors affecting late results after meniscectomy. *Journal of Bone and Joint Surgery* 1974; **56A**: 719–29.
16. Tapper EM, Hoover NW. Late results after meniscectomy. *Journal of Bone and Joint Surgery* 1969; **51A**: 517–26.
17. Ahmed AM, Burke DL. *In vitro* measurement of static pressure distribution in synovial joints: I. Tibial surface of the knee. *Journal of Biomechanical Engineering* 1983; **105**: 216–25.
18. Baratz ME, Rehak DC, Fu FH, Rudert MJ. Peripheral tears of the meniscus. The effect of open versus arthroscopic repair on intra-articular contact stresses in the human knee. *American Journal of Sports Medicine* 1988; **16**: 1–6.
19. Bargar WL, Moreland JR, Markoff KL, Shoemaker SC, Amstutz HC, Grant TT. *In-vivo* stability testing of post-meniscectomy knees. *Clinical Orthopaedics* 1980; **150**: 247–52.
20. Krause WE, Pope MD, Johnson RJ, Wilder DG. Mechanical changes in the knee after meniscectomy. *Journal of Bone and Joint Surgery* 1976; **58A**: 599–604.
21. Kurosawa H, Fukuboyashi T, Hakajima H. Load-bearing mode of the knee: physical behavior of the knee joint with and without menisci. *Clinical Orthopaedics* 1980; **149**: 283–90.
22. Newman AP, Anderson DR, Daniels AU. Mechanics of the healed meniscus in a canine model. *American Journal of Sports Medicine* 1989; **17**: 164–75.
23. Newman AP, Anderson DJ, Daniels AU, Jee KW. The effect of medical meniscectomy and coronal plane angulation on in vitro load transmission in the canine stifle joint. *Journal of Orthopaedic Research* 1989; **7**(2): 281–91.
24. Seedholm BB, Hargreaves DJ. Transmission of the load in the knee joint with special reference to the role of the menisci: II. Experimental results, discussion and conclusions. *English Medicine* 1979; **8**: 220–8.
25. Walker PS, Erkman MH. The role of the menisci in force transmission across the knee. *Clinical Orthopaedics* 1975; **109**: 184–92.
26. Jackson RW. The virtues of partial meniscectomy. *Second International Seminar on Operative Arthroscopy, Maui,* 1980; 89–95.
27. Jackson RW, Dandy DJ. Partial meniscectomy. *Journal of Bone and Joint Surgery* 1976; **58B**: 142.
28. McGinty JB, Lawrence FD, Marwin R. Partial or total meniscectomy. *Journal of Bone and Joint Surgery* 1977; **59A**: 763–6.
29. Cassidy RE, Shauffer AJ. Repair of peripheral meniscus tears. A preliminary report. *American Journal of Sports Medicine* 1981; **9**: 209–14.
30. DeHaven KE. Peripheral meniscus repair. An alternative to meniscectomy. *Orthopaedic Transactions* 1981; **5**: 399–400.
31. Henning CE. Arthroscopic repair of meniscal tears. *Orthopaedics* 1983; **6**: 1130.
32. DeHaven KE, Black KP, Griffiths HJ. Open meniscus repair. Technique and two to nine year results. *American Journal of Sports Medicine* 1989; **17**: 788–95.
33. Wirth CR. Meniscus repair. *Clinical Orthopaedics* 1981; **157**: 153–60.
34. Arnoczky SP, Warren RF, Spivak JM. Meniscal repair using an exogenous fibrin clot. An experimental study in dogs. *Journal of Bone and Joint Surgery* 1988; **70A**: 1209–17.
35. Henning CE, Lynch MA, Yearout KM, Vequist SW, Stallbaumer RJ, Decker KA. Arthroscopic meniscal repair using an exogenous fibrin clot. *Clinical Orthopaedics* 1990; **252**: 64–72.
36. Ferro TD, Gershuni DH, Danzig LA, Hargens AR, Oyama BK, O'Hara R. The mechanical strength of healed tears in canine menisci. *Proceedings of 34th Annual Meeting, Orthopaedic Research Society, Atlanta, GA, 1–4 February 1988.* Park Ridge, IL: The Orthopedic Research Society: 146.
37. Krause WR, Burdette WA, Loughran TP. Properties of the normal and repaired canine meniscus. *Proceedings of the 35th Annual Meeting, Orthopaedic Research Society, Las Vegas, NV, 6–9 February 1989.* Park Ridge, IL: The Orthopedic Research Society: 207.
38. Arnoczky SP, McDevitt CA, Schmidt MB, Mow VC, Warren RF. The effect of cryopreservation on canine menisci: a biochemical, morphologic, and biomechanical evaluation. *Journal of Orthopaedic Research* 1988; **6**(1): 1–12.
39. Jackson DW, McDevitt CA, Atwell EA, Arnoczky SP, Simon TM. Meniscal transplantation using fresh and crypopreserved allografts—an experimental study in goats. *Proceedings of 36th Annual Meeting, Orthopaedic Research Society, New Orleans LA, 5–8 February 1990.* Park Ridge, IL: The Orthopedic Research Society: 221.
40. Keating EM, Malinin TI, Belchic G. meniscal transplantation in goats: an experimental study. *Proceedings of 34th Annual Meeting, Orthopaedic Research Society, Atlanta, GA, 1–4 February 1988.* Park Ridge, IL: The Orthopedic Research Society: 147.
41. Milachowski KA, Weismeier K, Wirth CJ. Homologous meniscus transplantation. Experimental and clinical results. *International Orthopaedics* 1989; **13**(1): 1–11.
42. Milton J *et al.* Transplantation of viable, cryopreserved menisci. *Proceedings of 36th Annual Meeting, Orthopaedic Research Society, New Orleans, LA, 5–8 February 1990.* Park Ridge, IL: The Orthopedic Research Society: 220.
43. Minns RJ. The Minns mensical knee prosthesis: biomechanical aspects of the surgical procedure and a review of the first 165 cases. *Archives of Orthopaedic and Traumatic Surgery* 1989; **108**(4): 44–8.

44. Stone KR, Rodkey WG, Steadman JR, Webber RJ, McKinney L. Meniscal regeneration using copolymeric collagen scaffolds. *American Journal of Sports Medicine* 1992; **20**: 104–11.

45. Zukor DJ *et al.* Allotransplantation of frozen irradiated menisci in rabbits. *Proceedings of 36th Annual Meeting, Orthopaedic Research Society, New Orleans, LA, 5–8 February 1990*. Park Ridge, IL: The Orthopedic Research Society: 219.

46. Milch H. Injuries to the crucial ligaments. *Archives of Surgery* 1935; **30**: 805.

47. O'Donoghue DH. Surgical treatment of fresh injuries to the major ligaments of the knee. *Journal of Bone and Joint Surgery* 1950; **32A**: 721–38.

48. Palmer I. On the injuries to the ligaments of the knee joint. *Acta Chirurgica Scandinavica* 1938; **53** (Suppl. 8).

49. Hughston JC, Barrett GR. Acute anteromedial instability: long-term results of surgical repair. *Journal of Bone and Joint Surgery* 1983; **65A**: 145–53.

50. Brown D. Personal Communication, 1991.

51. Johnson RJ. The anterior cruciate: a dilemma in sports medicine. *International Journal of Sports Medicine* 1983; **3**: 71–9.

52. Chick RR, Jackson DE. Tears of the anterior cruciate ligament in young athletes. *Journal of Bone and Joint Surgery* 1978; **60A**: 970–3.

53. Giove TP, Miller SJ, Kent BE, Sanford TL, Garrick JG. Nonoperative treatment of the torn anterior cruciate ligament. *Journal of Bone and Joint Surgery* 1983; **65A**: 184–92.

54. Hughston JC. Acute knee injuries in athletes. *Clinical Orthopaedics* 1962; **23**: 114.

55. Kennedy JC, Weinbert HW, Wilson AS. The anatomy and function of the anterior cruciate ligament. *Journal of Bone and Joint Surgery* 1974; **56A**: 223–35.

56. McDaniel W, Dameron TB. Untreated ruptures of the anterior cruciate ligament. *Journal of Bone and Joint Surgery* 1980; **62A**: 696–705.

57. Kannus P, Jarvinen M. Conservatively treated tears of the anterior cruciate ligament. Long-term results. *Journal of Bone and Joint Surgery* 1987; **69A**: 1007–12.

58. Johnson RJ, Eriksson E, Häggmark T, Pope MH. Five to ten year follow-up after reconstruction of the anterior cruciate ligament. *Clinical Orthopaedics* 1984; **183**: 122–40.

59. Hawkins RJ, Misamore GW, Merritt TR. Follow-up of the acute nonoperative isolated anterior cruciate ligament tear. *American Journal of Sports Medicine* 1986; **14**: 205–10.

60. Noyes FR, Mooar PA, Matthews DS, Butler DL. The symptomatic anterior cruciate deficient knee Part I: The long-term functional disability in athletically active individuals. *Journal of Bone and Joint Surgery* 1983; **65A**: 154–62.

61. Gerber C, Matter P. Biomechanical analysis of the knee after rupture of the anterior cruciate ligament and its primary repair. *Journal of Bone and Joint Surgery* 1983; **65B**: 391–9.

62. Marans JH, Jackson RW, Glossop ND, Young C. Anterior cruciate ligament insufficiency: a dynamic three-dimensional motion analysis. *American Journal of Sports Medicine* 1989; **17**: 325–32.

63. Fried JA, Bergfeld JA, Weiker G, and Andrish JT. Anterior cruciate reconstruction using the Jones–Ellison procedure. *Journal of Bone and Joint Surgery* 1985; **67A**: 1029–33.

64. Clancy WG, Jr. Nelson DA, Reider B, Narechania RG. Anterior cruciate ligament reconstruction using one-third of the patellar ligament, augmented by extra-articular tendon transfers. *Journal of Bone and Joint Surgery* 1982; **64A**: 352–9.

65. O'Brien SJ, Warren RF, Paulos H, Panariello R, Wickiewicz, TL. Reconstruction of the chronically insufficient anterior cruciate ligament with the central third of the patellar tendon. *Journal of Bone and Joint Surgery* 1991; **75A**: 278–86.

66. Sachs RA, Daniel DM, Stone ML, Gorfein RF. Patellofemoral problems after anterior cruciate ligament reconstruction. *American Journal of Sports Medicine* 1989; **17**: 760–5.

67. Andersson C, Odensten M, Good L, Gillquist J. Surgical or non-surgical treatment of acute rupture of the anterior cruciate ligament. *Journal of Bone and Joint Surgery* 1989; **71A**: 965–74.

68. Engerbretsen L, Renum P, Sundalswall S. Primary suture of the anterior cruciate ligament. A 6 year follow-up of 74 cases. *Acta Orthopaedica Scandinavica* 1989; **60**: 561–4.

69. Feagin JA Jr, Curl WW. Isolated tear of the anterior cruciate ligament: 5 year follow up study. *American Journal of Sports Medicine* 1976; **4**: 95–100.

70. Sandberg R, Balkfors B, Nilsson B, Westlin N. Operative versus non-operative treatment of recent injuries to the knee ligaments. *Journal of Bone and Joint Surgery* 1987; **69A**: 1120–6.

71. Jonsson T, Peterson L, Renstrom R, Althoff B, Myrhage R. Augmentation with longitudinal patellar retinaculum in the repair of an anterior cruciate ligament rupture. *American Journal of Sports Medicine* 1989: **17**: 401–8.

72. Sgaglione NA, Warren RF, Wickiewicz TL, Gold DA, Panariello RA. Primary repair with semitendinous tendon augmentation of acute anterior cruciate ligament injuries. *American Journal of Sports Medicine* 1990; **18**: 64–73.

73. Straub, T, Hunter RE. Acute anterior cruciate ligament repair. *Clinical Orthopaedics* 1987; **227**: 238–50.

74. Pearl AJ, Bergfeld JA, eds. *Extra-articular reconstruction in the anterior cruciate deficient knee*. Champaign, IL: Human Kinetics Publishers, 1992: 1–55.

75. Mok WW, Good C. Nonoperative management of acute grade III medial collateral ligament injury of the knee. *Injury* 1989; **20**: 277–80.

76. Clancy WG, Bergfeld J, O'Connor GA, Cox JS. Functional rehabilitation of isolated medial collateral ligament sprains. *American Journal of Sports Medicine* 1979; **7**: 206–13.

77. Parolie JM, Bergfeld JA. Long term results of nonoperative treatment of isolated posterior cruciate ligament injuries in the athlete. *American Journal of Sports Medicine* 1986; **14**: 35–8.

78. Torg JS, Barton TM, Pavlov H, Stine R. Natural history of the posterior cruciate deficient knee. *Clinical Orthopaedics* 1989; **246**: 208–16.

79. Kennedy JC, Galpin RD. The use of the medial head of the gastrocnemius muscle in the posterior cruciate deficient knee. *American Journal of Sports Medicine* 1982; **10**: 63–74.

80. Clancy WG, Shelbourne KD, Zoellner GB, Keene JS, Reider B, Rosenberg TD. Treatment of knee joint instability secondary to rupture of the posterior cruciate ligament. *Journal of Bone and Joint Surgery* 1983; **65A**: 310–22.

81. Hughston JC, Bowden JA, Andrews JR, Norwood LA. Acute tears of the posterior cruciate ligament. Results of operative treatment. *Journal of Bone and Joint Surgery* 1980; **62A**: 438–50.

82. Kannus P. Nonoperative treatment of grade II and III sprains of the lateral compartment of the knee. *American Journal of Sports Medicine* 1989; **17**: 83–9.

83. Cohn SL, Sotta RP, Bergfeld JA. Fractures about the knee in sports. *Clinics in Sports Medicine* 1990; **9**: 121–39.

84. Woods GW, Stanley RF, Tillos HS. Lateral capsular sign: X-ray clue to a significant knee instability. *American Journal of Sports Medicine* 1979; **7**: 27–39.

85. Warren RF, Kaplan N, Bach BR Jr. The lateral notch sign of anterior cruciate insufficiency. *American Journal of Knee Surgery* 1988; **1**(2): 119–24.

86. Jackson DW, Jennings LD, Maywood RM, Berger PE. Magnetic resonance imaging of the knee. *American Journal of Sports Medicine* 1988; **16**: 29–38.

87. Lee JK, Yao L. Occult intraosseous fracture: magnetic resonance appearance versus age of injury. *American Journal of Sports Medicine* 1989; **17**: 620–3.

88. Meyers MH, McKeever FM. Fracture of the intercondylar eminence of the tibia. *Journal of Bone and Joint Surgery* 1959; **41A**: 209–22.

89. Kennedy JC. Complete dislocation of the knee joint. *Journal of Bone and Joint Surgery* 1963; **45A**: 889–904.

90. Meyers MH, Moore TM, Harvey JP. Traumatic dislocation of the knee. *Journal of Bone and Joint Surgery* 1975; **57A**: 430–3.

91. McCutchan JD, Gillham NR. Injury to the popliteal artery associated with dislocation of the knee. palpable distal pulses do not negate the requirement for arteriography. *Injury* 1989; **20**: 307–10.

92. Tureo VJ, Spinella AJ. Anterolateral dislocation of the head of the fibula in sports. *American Journal of Sports Medicine* 1985; **13**: 209–15.

93. Weinert CR, Raczka R. Recurrent dislocation of the superior tibiofibular joint. *Journal of Bone and Joint Surgery* 1986; **68A**: 126–8.

94. Delamarter RB, Hohl M, Hopp E. Ligament injuries associated with tibia plateau fractures. *Clinical Orthopaedics* 1990; **250**: 226–33.

95. Templeman DC, Marder RA. Injuries of the knee associated with fractures of the tibial shaft. Detection by examination under anesthesia: a prospective study. *Journal of Bone and Joint Surgery* 1990; **72A**: 1392–5.

96. Johnson RJ. The effect of immobilization on ligaments and ligamentous healing. *Contemporary Orthopaedics* 1980; **2**: 237–41.

97. Burks R, Daniel D, Losse G. The effect of continuous passive motion on anterior cruciate ligament reconstruction stability. *American Journal of Sports Medicine* 1984; **12**: 323–7.

98. Reinecke S, Arms S, Renström P, Johnson RJ, Pope MH. Continuous passive motion and its effect on knee ligament strain. In: Johnson B, eds. *International series of biomechanics*, Vol. 6A. Champaign, IL: Human Kinetics, 1987: 111–18.

99. Noyes FR, Mangine RE, Barber S. Early knee motion after open and arthroscopic anterior cruciate ligament reconstruction. *American Journal of Sports Medicine* 1987; **15**: 149–60.

100. Arms SW, Pope MH, Johnson RJ, Fischer RA, Arvidsson I, Ericksson I. The biomechanics of anterior cruciate ligament rehabilitation and reconstruction. *American Journal of Sports Medicine* 1984; **12**: 8–18.

101. Henning CE, Lynch MA, Glick KR. An *in vivo* strain gauge study of elongation of the anterior cruciate ligament. *American Journal of Sports Medicine* 1985; **13**: 22–6.

102. Maltry JA, Noble PC, Woods GW, Alexander JW, Feldman GW, Tullos HS. External stabilization of the anterior cruciate deficient knee during rehabilitation. *American Journal of Sports Medicine* 1989; **17**: 550–4.

103. Paulos LE, Noyes FR, Grood ES, Butler DL. Knee rehabilitation after anterior cruciate ligament reconstruction and repair. *American Journal of Sports Medicine* 1981; **9**: 149.

104. Huegel M, Indelicato PA. Trends in rehabilitation following anterior cruciate ligament reconstruction. *Clinics in Sports Medicine* 1988; **7**: 801–11.

105. Malone T. Clinical use of the Johnson anti-shear device: how and why to use it. *Journal of Orthopaedic Sports Physical Therapy* 1986; **7**: 304–9.

106. Nisell R, Ericson MO, Nemeth G, Elsholm J. Tibiofemoral joint forces during isokinetic knee extension. *American Journal of Sports Medicine* 1989; **17**: 49–54.

107. Renstrom P, Arms SW, Stanwyck TS, Johnson RJ, Pope MH. Strain within the anterior cruciate ligament during hamstring and quadriceps activity. *American Journal of Sports Medicine* 1986; **14**: 83–7.

108. Häggmark T, Eriksson E. Cylinder or mobile cast brace after knee ligament surgery: a clinical analysis and morphologic and enzymatic studies of changes in the quadriceps muscle. *American Journal of Sports Medicine* 1979; **7**: 48–56.

109. Shelbourne KO, Nitz P. Accelerated rehabilitation after anterior cruciate ligament reconstruction. *American Journal of Sports Medicine* 1990; **18**: 292–9.

110. Clancy WG. Personal Communication, 1991.

111. Paulos LE, Payne, FC, Rosenberg TD. Rehabilitation after anterior cruciate ligament surgery. In: Jackson DW, Drez D, eds. *The anterior cruciate deficient knee*. St. Louis: CV Mosby, 1987: 291–314.

112. Butler DL *et al*. Mechanical properties of primate vascularized vs. non-vascularized patellar tendon grafts: changes over time. *Journal of Orthopaedic Research* 1989; **7**: 68–79.

113. Daniel DM. Principles of knee ligament surgery: graft fixation. In: Danel D, Akeson W, O'Connor, J. eds. *Knee ligament structure, function, injury and repair*. New York, Raven Press; 1990: 20–26.

114. Kurosaka M, Yoshiya S, Andrish JT. A biomechanical comparison of different surgical techniques of graft fixation in anterior cruciate ligament reconstruction. *American Journal of Sports Medicine* 1987; **15**: 225–9.

115. Rosenberg TP, Paulos LE, Parker RD, Abbott PJ. Arthroscopic knee surgery of the knee. In: Chapman MW, Madison M, eds. *Operative Orthopaedics*. Philadelphia: J. B. Lippincott Co. 1988: 1585–1604.

116. Feagi JA, Lambert KL, Cunningham RR. Repair reconstruction of the anterior cruciate ligament. In: Chapman MW, Madison M, eds. *Operative Orthopaedics*. Philadelphia: J. B. Lippincott Co., 1988: 1641–50.

4.2.2 Acute injuries of the meniscus

OWEN H. BRADY AND BRIAN J. HURSON

INTRODUCTION

The advent of knee arthroscopy has provided orthopaedic surgeons with a greater understanding of knee pathology. This, combined with the increasing awareness, of the functional significance of the meniscus[1–7] and the knowledge of the long-term sequelae of meniscectomy[4,8–11] has led to major changes in the management of meniscal injuries. As a result, the removal of this once little thought of 'vestigial fibrocartilage' is avoided at all costs. The meniscus is indeed 'A fibrocartilage of some distinction'.[12]

EMBRYOLOGY

The menisci develop from the proximal tibia. Embryologically, the lower limb bud appears at 4 weeks' gestation.[13] Chondrifica-tion of the femur, tibia, and fibula begins at 6 weeks, at which stage the blastema of the knee joint is present. This consists of a continuous structure which has no joint space. However, there are two identifiable chondrogenic layers which sandwich a less dense layer between them. It is this less dense layer which develops into the menisci and cruciate ligaments. Menisci can be identified at 7 weeks and are well defined by 8 weeks' gestation.[13]

The growth of the menisci proceeds simultaneously with that of the tibia and femur. Growth configuration changes occur in response to alteration in the intra-articular condylar surfaces.[13] Histologically, the collagen fibre alignment alters in response to change in biomechanical function.[1] The longitudinal fibres develop in response to the circumferential stress and the radial fibres develop in response to tension stresses.

Prenatally, the menisci or semilunar cartilages are very cellular and have an abundant blood supply.[13] With growth, there is a

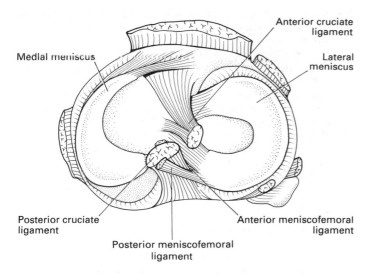

Fig. 1 The tibial plateaux showing the menisci and their relationship to intracapsular ligaments.

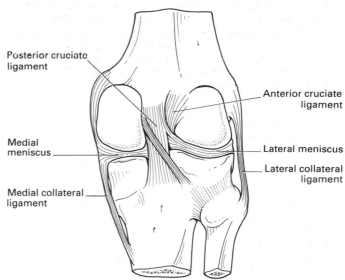

Fig. 2 Posterior view of the knee joint showing the menisci and their relationship to cruciate and collateral ligaments.

gradual decrease in the vascularity of the central area.[6] This avascular area spreads radially.

ANATOMY

The menisci are concentric in shape and have a triangular cross-section (Fig.1). They have a thick peripherally attached base which tapers centrally towards a thin free apex. Their superior concave surfaces articulate with the femoral condyles throughout all ranges of movement, while their flat inferior surfaces articulate with the tibial plateaux. The menisci cover two-thirds of the articulating surface of the tibia and transmit about 50 per cent of axial load across the joint.[7] They are attached to the intercondylar area of the tibia by tough fibrous anterior and posterior horns. Peripherally, they are attached to the synovial membrane and capsule. The coronary ligament also attaches the circumference of the menisci to the tibia, preventing excessive peripheral displacement. The anterior horns are attached to each other by the transverse ligament (when present).

Medial meniscus

The medial meniscus is semicircular, and is broader posteriorly than anteriorly (Fig. 1). Its anterior horn is attached in front of the origin of the anterior cruciate ligament. The posterior horn is attached between the posterior horn of the lateral meniscus and the origin of the posterior cruciate ligament. At its periphery, the medial meniscus is attached to the deep fibres of the medial collateral ligament (Fig. 2).

Lateral meniscus

The lateral meniscus is more circular in shape (Fig. 1). It makes up four-fifths of a circle and, unlike the medial meniscus, it has the same breadth throughout. It is attached to the tibia anteriorly by its anterior horn, which is situated immediately posterolateral to the origin of the anterior cruciate ligament. The posterior horn is inserted into the tibia in front of the posterior horn of the medial meniscus. It is also attached to the medial femoral condyle by the anterior and posterior meniscofemoral ligaments—the ligaments

of Humphrey and Wrisberg respectively[14] (Fig. 1). These run anteromedially in a parallel fashion and are separated only by the posterior cruciate ligament. The lateral meniscus is grooved in its periphery by the tendon of popliteus, whose most medial fibres are inserted to its posterior horn. Here, the coronary ligament is deficient. In contrast with the medial meniscus, the fibular collateral ligament is not attached to its underlying cartilage (Fig. 2).

Blood supply

The blood supply of the menisci is derived from the terminal branches of the superior and inferior medial and lateral geniculate arteries. These vessels supply the connective tissue adjacent to the periphery of the meniscus.[6] Arnoczky and Warren,[15,16] as well as Danzig et al.[17] have shown that the collateral branches perforate the outer 20 to 30 per cent of the meniscus[13] (Figs 3 and 4). This

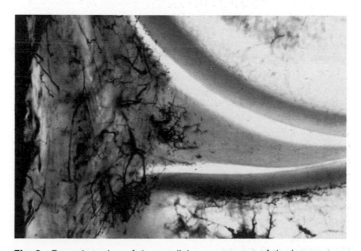

Fig. 3 Frontal section of the medial compartment of the knee demonstrating perimeniscal capillary plexus penetrating the peripheral border of the medial meniscus. (Reproduced from ref. 15 with permission.)

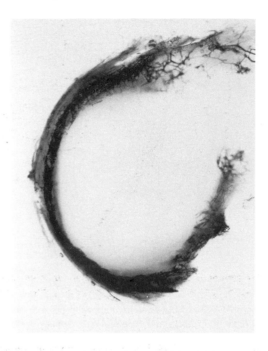

Fig. 4 Superior aspect of the medial meniscus, demonstrating the peripheral vasculature as well as the highly vascular synovial tissues that cover the anterior and posterior horns. (Reproduced from ref. 15 with permission.)

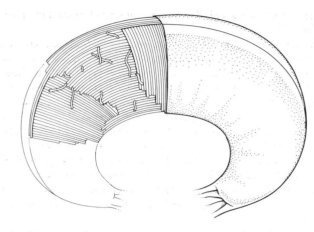

Fig. 5 Ultrastructure of the meniscus. The collagen fibre alignment pattern is seen from the superficial to deep layers of the meniscus. (Adapted from ref. 3, with permission.)

leaves an inner 70 to 80 per cent, which is the 'avascular segment'. The central meniscus is more vascular than the superior or inferior surfaces. The anterior and posterior horns are enveloped by synovial tissue, and are therefore more highly vascularized than the middle segment.

ULTRASTRUCTURE OF THE MENISCI

The meniscus is composed of cells which are surrounded by an extracellular matrix. The basic cell of the meniscus is the fibrochondrocyte. Two distinct types of fibrochondrocyte have been observed,[18] and are identified by their round or oval shape. The meniscal surface fibrochondrocytes are usually oval, while those of the deeper layers tend to be more rounded. Both contain few mitochondria, suggesting that anaerobic glycolysis is their main respiratory pathway.[6]

The extracellular matrix is composed of collagen, proteoglycans, matrix glycoproteins, and elastin. Bullough et al.,[3] using polarized light microscopy, showed that the extracellular matrix is composed mainly of circumferentially arranged collagen fibres. These provide tensile strength and, to a lesser extent, radially arranged cross-links which provide shear strength[1] (Fig. 5). The radially orientated fibres are more densely packed in the meniscal surface layer, which is 30 to 120 mm thick. Collagen makes up about 60 to 70 per cent of the dry weight of the meniscus. Type I collagen makes up about 98 per cent of the total collagen concentration. The other 2 per cent is made up of types II, III, and V. The collagen fibres are arranged in bundles which are 50 to 150 μm in diameter.[6]

Proteoglycans are hydrophilic negatively charged micromolecules held together by collagen fibrils. They provide the meniscus with a high capacity to resist large compressive loads.

However, they do not contribute significantly to its tensile strength. Little is known, as yet, about the types and functions of the matrix glycoproteins. They are thought to be active in the process of repair and regeneration of the torn meniscus.[6] Elastin acts as a cross-link between collagen fibres. This connective tissue component enables the meniscus to recoil towards its normal anatomical shape when the meniscal displacing circumferential stress has subsided.

MOBILITY OF THE MENISCI

Flexion and extension movements of the knee joint take place above the menisci, i.e. between menisci and femoral condyles. Rotation of the joint takes place below the menisci, i.e. between the menisci and tibial plateaux. When the knee joint is extended, the menisci are separated, while in flexion their natural elastic recoil tends to approximate them.

The lateral meniscus is more mobile than the medial meniscus and can move by up to 1 cm. This mobility is due to a combination of factors which include the close proximity of its anterior and posterior horns (which allow a greater degree of pivot), the fact that the posterior horn is not attached to the capsule or coronary ligament, and finally the fact that it is not stabilized by its adjacent collateral ligament, as is the case with the medial meniscus. During flexion, the menisci are displaced posteriorly and are simultaneously internally rotated. During knee extension the menisci are displaced anteriorly and are externally rotated.

Functions of the menisci

Once regarded as useless vestigial organs, the menisci are now known to play an important role in transmission of forces,[3,5,19-21] joint stabilization,[7] shock absorption,[22] nutrition of articular cartilage, and lubrication of the knee joint.[23,24] Considerable clinical evidence indicates that removal of, or damage to, the menisci can have detrimental effects on the knee joint. Fairbank[9] has demonstrated radiography signs of joint degeneration following meniscectomy. The severity of these changes is related to the amount of meniscal tissue removed.[25,26]

The menisci act as spacers which fill the dead space between the incongruous femoral condyles and tibial plateaux. They resist excessive movements of the tibia on the femur. In the cruciate-deficient knee the menisci serve as important secondary knee stabilizers.[27]

The arrangement of the collagen bundles within the meniscus appears to be ideal for absorption and translation of vertical compression loads into circumferential stresses. They distribute these forces across the joint by increasing the contact area between the femoral condyles and tibial plateaux. Distribution of the load result in a decrease in the magnitude of pressure, i.e. force per unit area, between the opposing hyaline cartilage surfaces. Thus the irreplaceable hyaline cartilage is protected from excessive damaging pressures. Meniscectomy dramatically alters the pattern of static loading across the knee joint.[28] Several studies have demonstrated higher peak stresses,[29] greater stress concentration,[5,30] and decreased shock-absorbing capabilities [20,31] after total meniscectomy.

Walker and Erkman[7] have shown that the menisci take between 50 and 70 per cent of the total knee load. More specifically, 90 per cent of the load across the knee joint passes through the medial compartment and 50 per cent of this is taken by the medial meniscus. Conversely, 10 per cent of the total load across the knee joint passes through the lateral compartment, of which 70 per cent is taken by the lateral meniscus.

Synovial fluid functions as a joint lubricant as well as a medium through which substances can diffuse. The fibrocartilage, as well as hyaline cartilage, can thus absorb nutrients and excrete endproducts of metabolism.

Meniscal injury mechanisms

When the shear stress within the meniscus exceeds the tissue strength, a tearing injury results. A tear occurs in one of two ways. Either an excessive abnormal force takes place within a normal meniscus, or a normal force takes place within an abnormal or degenerate meniscus. The former causes a vertical type tear, while the latter causes a horizontal cleavage tear.[32] Vertical tears are more commonly seen in sporting injuries, whereas horizontal cleavage tears occur in patients with degenerative menisci.

As previously stated, the menisci are displaced posteriorly in flexion and anteriorly in extension. With hyperflexion the posterior halves of both menisci are compressed between the posterior aspects of the femoral condyles and tibial plateaux. The medial meniscus is prone to injury when the tibia is externally rotated relative to the femur. The lateral meniscus is more prone to injury when the tibia is internally rotated.

When the femur is internally rotated while the knee is flexed, the posterior horn of the medial meniscus is forced radially towards the intercondylar area by the medial femoral condyle. If the knee is then suddenly extended, the posterior horn may be torn. When the femur is externally rotated while the knee is flexed, the posterior horn of the lateral meniscus is forced radially towards the intercondylar area. If the knee is then suddenly extended, the meniscus is straightened and may be torn.

Medial meniscal injuries are seen up to five times more commonly than lateral meniscal injuries in most sports.[33] This is related to the higher incidence of 'side-on' tackles which apply a valgus strain to the knee. It is also due to its firm attachment to the deep fibres of the medial collateral ligament, which decreases its mobility. Because of the valgus nature of these injuries, there is often a concomitant medial collateral ligament injury (Fig. 6).

Baker et al.[33] have shown the predominance of medial meniscal injuries in a variety of sports: football, 75 per cent; basketball, 75 per cent; skiing, 78 per cent; baseball, 90 per cent. The only

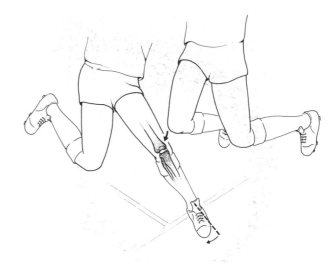

Fig. 6 The side-on tackle applies a valgus knee strain which may cause meniscal injury as well as medial collateral ligament injury.

exception is wrestling which has an almost equal incidence at 55 per cent.

PATHOLOGY

Ageing of the meniscus

In youth, the meniscus is white, translucent, and has an abundant blood supply. It is mobile, elastic, and supple. With time, the smoothly contoured parallel collagen fibres which run concentrically around the meniscus degenerate. The meniscus becomes brittle and yellowish in colour. Its vascularity decreases, and as it loses its translucency it becomes less elastic and supple. With hardening, it becomes more prone to injury. This loss of shear strength is a direct result of an increase in cross-linkage between parallel collagen fibres.[34] Dystrophic calcification, hyaline acellular degeneration, and myxoid degeneration contribute to the ageing process.[35] Progressive fissuring and cleavage tearing ensue. In general, the meniscus of the active child can withstand repeated insult without injury. The meniscus of the elderly is seldom acutely injured because of the relative inactivity that accompanies old age. However, the meniscus of the sportsperson is 'at risk' as this is the stage when it is undergoing degeneration from translucent white to opaque yellow.

CLASSIFICATION OF MENISCAL TEARS

There are many classifications of meniscal tears. However, all are based on two distinct tear patterns, i.e. vertical and horizontal (Fig. 7). Vertical tears may be longitudinal or transverse (radial). Horizontal tears were originally described by Smillie[32] as horizontal cleavage tears. Flap or oblique tears result from a combination of vertical and horizontal tearing. Tears may occur in association with discoid menisci or cysts of the meniscus.

The majority of sports-related tears are of the vertical type. Tears of the horizontal type are much less common. Flap or oblique tears are sometimes referred to as 'parrot beak tears'.

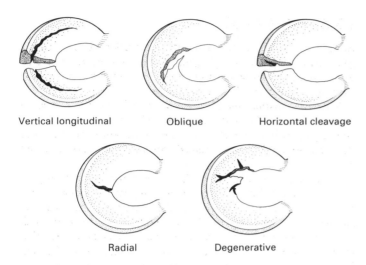

Vertical longitudinal   Oblique   Horizontal cleavage

Radial   Degenerative

Fig. 7 Classification of meniscal tears.

Vertical longitudinal tears

A vertical longitudinal tear occurs as a result of an excessive force acting upon a normal meniscus. These tears usually occur in young patients, often as a result of significant trauma, and are frequently associated with anterior cruciate ligament injuries. The thin concave edge of the meniscus is most commonly involved. The classical vertical longitudinal tear is the 'bucket handle tear' so frequently responsible for the 'locked knee'. It is usually caused by a combination of compression with simultaneous rotation of the femur on a fixed tibia. The longitudinal tear is more common than the transverse tear.

Transverse tears are usually caused by a compression force which attempts to distract and displace the anterior and posterior horns of a meniscus. In so doing, an excessive traction force is applied to the thin concave edge of the meniscus, resulting in a tear of the middle third of the lateral meniscus. The transverse tear originates centrally and runs radially towards its vascular periphery.

Horizontal tears

The horizontal tear most commonly occurs in a degenerate meniscus, and was first termed, by Smillie in 1962, a horizontal cleavage tear.[32] It occurs as a result of a normal force acting on an abnormal or degenerate meniscus and is most commonly seen in those over 40 years of age. Smillie[36] has shown that the posterior half of the medial meniscus or the middle segment of the lateral meniscus[35] are most commonly involved. Two out of every three horizontal tears involve the lateral meniscus. In the sportsperson, this type of injury occurs in an already degenerate meniscus. There is a close association with osteoarthritis.[38] Other features associated with meniscal degeneration are chondrocalcinosis[39,40] and cyst formation.[27]

Histological changes seen in the degenerative meniscus are similar to those seen in degenerative joint hyaline cartilage injury,[41] i.e. fibrochondrocyte necrosis, fibrochondrocyte proliferation, fibrillation, and loss of matrix protein polysaccharide.

It is very difficult to distinguish microscopically between traumatic and degenerative tears. Hough and Webber[41] suggest that gross examination is the easiest and probably the most accurate method of deciding on the actual type of tear.

Ferrer *et al.*[35] have demonstrated that the extension of the horizontal cleavage type tear into the parameniscal region would appear to be the cause of cyst formation.

REGENERATION

King in 1936 was the first to document that tears limited to the avascular segment of the semilunar cartilage probably never healed.[25] Tears involving the vascularized area of the meniscus cause haemorrhage and clot formation within the torn area. This is followed by migration of fibroblasts and ingrowth of new capillaries. Fibrous healing results which resembles mature fibrocartilage.

Arnoczky and Warren[16] have shown that regeneration may occur following meniscectomy but this is inconsistent. The new meniscus is composed of fibrous tissue. It is narrower, thinner, and has a smaller surface area than the original. The concave edge is less well defined and no obvious line of cleavage exists between capsule and meniscus. Regeneration may also occur following partial excision of a meniscus, provided that the vascular periphery is involved.

DIAGNOSIS OF MENISCAL INJURIES

A careful history and thorough physical examination are the most important prerequisites to making a diagnosis of a meniscal injury or, indeed, any knee complaint. Most patients with meniscal tears will present some time following the initial injury and after the acute symptoms have subsided. In obtaining the history of a knee disorder, careful documentation should be made of the nature of the initial injury as well as the time and duration of symptoms. The actual mechanism of injury should be ascertained, i.e. whether the injury was a result of a contact or non-contact activity. It is also helpful to determine the position of the extremity at the time of the original injury.

As pain is the predominant complaint, it is important to determine what activities precipitate this pain. The severity of the pain and the functional level that the patient reaches before pain prevents any further activity are noted. Most mechanical knee pain is intermittent in nature. Pain that is constant and persistent, particularly at night may be caused by the presence of arthritis or, more rarely, a bone tumour.

PRESENTING COMPLAINTS

Pain

It is helpful to ask the patient to point to the specific area of knee pain. This is the least confusing method of determining its anatomical location. Pain is the most common presenting complaint in patients with meniscal pathology. It can usually be localized to the medial or lateral side of the knee. It is commonly precipitated by twisting and turning movements, or by forced flexion or extension. In some patients the symptoms are only present during sporting activities, while others experience symptoms during activities of daily living.

Swelling

Knee swelling associated with a meniscal tear is usually mild and increases with activity. The swelling gives rise to stiffness, and a complaint of decreased flexibility. This stiffness must not be confused with 'locking'.

Locking

Locking is a symptom indicating inability to extend the knee fully. It implies displacement of a torn meniscus into the intercondylar notch—usually a bucket handle tear. Forced extension or flexion is associated with further severe pain, either in the medial or lateral joint line. Locking may last from seconds to minutes to hours or days. It may be relieved spontaneously. Commonly, patients are able to 'wriggle' the knee into extension with perceptible relief of pain.

The presence of a loose body within the knee may cause true locking episodes. Difficulty in extending the knee may also be seen in the immediate post-injury phase in patients with acute injuries, secondary to the presence of pain and spasm. It may take a number of days to regain full knee extension in patients who have sustained an anterior cruciate ligament injury.

Instability

A patient with a meniscal tear may complain that his or her knee 'gives' way. This 'giving way' is always preceded by pain, i.e. the pain causes the 'giving'. This is in contrast with true instability, which is usually associated with anterior cruciate ligament deficiency. In anterior cruciate deficient knees, episodes of instability occur without warning and are followed, rather than preceded, by pain.

History of presenting complaints

An accurate history of the presenting knee complaint should be carefully elicited and documented. Almost all younger patients will be able to recall accurately a history of some type of twisting or overloading episode (e.g. during football, running, racquet sport, basketball, etc.). Some patients may continue their activity, while others may find it too painful to bear weight on the affected knee. Most patients will complain of pain and difficulty moving the knee following the initial injury. The difficulty with knee movement is usually pain related, although in a small number of patients this difficulty will be due to true 'locking' secondary to a displaced torn meniscal fragment.

Patients with an acute meniscal tear usually develop a gradual mild knee effusion over a period of approximately 12 h. This is in contrast with patients who sustain anterior cruciate ligament injuries, who will almost always develop a large knee effusion immediately following the original injury (haemarthrosis).

A number of patients, usually older people, will present with symptoms suggestive of meniscal pathology, but will not recall a specific twisting or overloading event. Many of these will have developed their symptoms of pain and swelling over weeks or months. Some may recall experiencing their first episode of joint pain while squatting or while ascending from a squatting or sitting position. These patients are commonly between 40 and 60 years old and may demonstrate at least some evidence of degenerative joint disease on radiographs. Patients with overt degenerative arthritis may develop degenerative meniscal tears and suffer symptoms which are indistinguishable from those associated with the underlying arthritis.

EXAMINATION

Examination of the knee starts with observation of the patient's gait. Any evidence of antalgic gait, a flexion contracture, or hip or back disease should be carefully noted, remembering that hip or back disease may manifest with knee pain.

Knee examination should be performed with both legs exposed entirely, so that the findings for the affected knee can be adequately compared with those for the normal knee. Alignment of the lower extremity is best judged with the patient standing and walking.

In addition to looking for specific signs of meniscal pathology, a careful search is also made for ligament injuries as well as patella problems. Evidence of bruising, skin abrasions, and localized swelling should be observed. General palpation for swellings, such as a bursa over the tibial tubercle, popliteal cysts, or palpable masses which may represent tumours, should be performed. Examination of the proximal tibiofibular joint is particularly important in patients presenting with lateral knee pain.

Early presentation

Early examination, i.e. soon after the initial injury has occurred, usually reveals generalized knee tenderness, mild effusion, and limitation of knee movements. Palpation of the medial or lateral joint line may be associated with marked tenderness. However, the majority of patients will also have tenderness above or below the joint line, indicating some element of either a medial or lateral collateral ligament sprain, thus making it difficult to differentiate between pure ligament sprain and ligament sprain associated with meniscal injury. It is reasonable to direct initial treatment to the sprained ligaments in these patients. Most patients with isolated ligament sprains should be able to return to their sporting activities within 2 to 4 weeks. Those patients who still complain of persistent joint-line pain after this time should be suspected of having a meniscal tear.

Patients presenting with a locked knee following a twisting injury should be suspected of having a bucket handle tear of one of the menisci. The side of the meniscal tear is usually clarified by eliciting joint-line tenderness or by eliciting pain on passive extension. Inability to extend the knee may also be seen in patients who have an acute anterior cruciate ligament tear. However, these patients, usually present with a large swollen knee (haemarthrosis).

After the acute symptoms have settled, a patient presenting with a meniscal tear will usually have some persistent knee swelling, complain of joint-line pain, and may experience episodes of intermittent locking. More rarely, a patient may have been unable to straighten the knee since the initial twisting injury, i.e. have persistent locking.

Clinical signs

Patients with meniscal tears may exhibit one or a number of clinical signs.

Joint-line tenderness

Joint-line tenderness is usually a reliable indicator of meniscal injury. This is more specific if the tenderness is confined to the joint line, i.e. the periphery of the meniscus. Tenderness above or below the joint line may indicate the presence of a concomitant collateral ligament sprain, or the presence of degenerative joint disease.

Swelling

Effusions associated with meniscal tears are usually mild. Large chronic effusions are not commonly seen with isolated meniscal tears, and usually indicate the presence of arthritis. The patellar tap sign may not be present in patients with small or mild knee swellings. In these knees, a visible ripple of fluid in the parapatella fossa can be demonstrated by first expressing fluid from one side of the joint and then gently stroking the opposite side with the palm of the hand.

Tibial rotation tests

Tibial rotation tests have been described by McMurray[19] and Apley.[42] They were devised in attempts to differentiate a torn meniscus from other knee pathology. Both aim at trapping or catching the torn meniscus between the femoral condyle and the tibial plateau. A positive rotation test is associated with either a palpable or an audible click which may or may not reproduce the patient's painful symptoms. These tests are helpful when positive, but not when negative, and may be difficult to perform on an acutely painful knee. It is important to perform the rotation test on the opposite knee, as quite frequently similar clicking may also be present. The McMurray manoeuvre[19] is performed by externally rotating the leg with the knee fully flexed (Fig. 8). A varus stress is then applied as the knee is gradually extended. A palpable, audible or painful click may indicate a tear of the posterior horn of the medial meniscus. The manoeuvre is then repeated with forcible internal rotation, while applying a valgus stress, to elicit posterolateral meniscal clicks.

The Apley grinding test[42] is performed with the patient in the prone position. The examiner rotates the flexed tibia on the femur while exerting downward pressure along the long axis of the tibia. With pressure exerted during rotation, pain may be felt in the region of a torn meniscus.

In addition to the tibial rotation tests, Insall[34] has shown that squatting and duck-walking may reproduce joint-line pain. On occasion, a loose meniscal fragment can be palpated in the area of the coronary ligaments.

Quadricep wasting

Quadricep wasting is a non-specific finding which may reflect knee pathology in general rather than a meniscal tear in particular.

In addition to looking for the foregoing signs, it is important that a careful overall assessment of the knee is performed. The integrity of the medial and lateral collateral, as well as the anterior and posterior cruciate ligaments should be assessed. Weight-

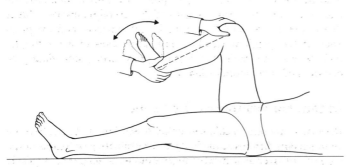

Fig. 8 McMurray's test: the hip is flexed to 90°, the knee is flexed greater than 90°, and the tibia is internally and externally rotated. The joint line is palpated for clicks.

bearing radiographs should exclude the presence of fractures, loose bodies, osteochondral injuries, or arthritis.

Meniscal cysts and associated tears

Meniscal tears may occur in association with cysts of the menisci. These are more commonly seen on the lateral side and are invariably associated with a radial or horizontal cleavage tear of the lateral meniscus. Patients usually present with pain in the lateral joint-line region which persists for a number of hours after activity. Most patients are conscious of a small swelling on the lateral joint line which fluctuates in size. Examination will reveal a tender palpable mass on the lateral joint line. This 'lump' is maximally distended at approximately 30° of flexion. The differential diagnosis is that of a ganglion, an osteophyte in an arthritic knee, or a small soft tissue tumour such as synovial cell sarcoma.

Meniscal cysts have traditionally been treated by total meniscectomy and open removal of the cyst. Currently, these cysts can now be treated effectively by arthroscopic methods. The torn meniscus is excised, allowing decompression of the cyst. The resection will usually require removal of a large segment of the meniscus because of the fragmented multiplane nature of the associated tear. It is not necessary to remove the cyst through a separate skin incision.

Discoid meniscal tears

A discoid meniscus is characterized by a round rather than a crescent shape. About 90 per cent occur on the lateral side of the knee. Symptoms may occur following a tear, causing pain and discomfort, clicking, swelling, or a mechanical block to extension. The majority of patients presenting are under the age of 15 years. Arthroscopic management is recommended for symptomatic tears. Resection of the central portion of the discoid is performed, leaving a rim about the width of a normal meniscus behind. The remaining rim will retriangulate and look much like a normal meniscus in time.[43]

HAEMARTHROSIS

Immediate knee swelling following a twisting injury implies bleeding into the joint—haemarthrosis. In the absence of a fracture, the patient has a 75 to 80 per cent chance of sustaining a tear of the anterior cruciate ligament and a 50 per cent chance of having a peripheral tear of one of the menisci.[44,45] There is a smaller chance of having sustained a patella dislocation, with tearing of the medial retinaculum. Such patients usually give a very clear history of being aware that the patella was dislocated to the lateral aspect of the joint. The knee in those circumstances would have been in a flexed position. Most patella dislocations reduce spontaneously as the knee is extended.

Examination is difficult in the presence of a large painful swollen knee. It is important that the knee be re-examined following aspiration. Aspiration should be performed under sterile conditions using a large bore needle—16-gauge or preferably a 14-gauge. It may be impossible to aspirate blood and clots through small bore needles. A local anaesthetic, which may be combined with adrenaline, may be injected prior to aspiration. Special attention to injecting the synovium and capsule is important, as the most painful aspect of the procedure is the stretching of these regions. In the presence of a large effusion, the procedure can be carried out with equal facility from either medial or lateral

Table 1 Diagnostic accuracy of meniscal tears

Investigators	Clinical examination (%)	Arthrography (%)	Arthroscopy (%)	Combined clinical examination (arthrography and arthroscopy) (%)	CT scan (%)	MR imaging (%)
Nicholas et al. 1970[51]	80	97				
Casscells 1971[65]			80			
Jackson and Abe 1972[50]	68.5	68.2	95			
De Haven and Collins 1975[47]	72	78	94			
McGinty and Metza 1978[53]		Lateral 72 Medial 84	91 under general anaesthetic 95 under local anaesthetic			
Tegtmeyer et al. 1979[54]		Single Contrast 96.5 Double Contrast 95				
Gillies and Seligson 1979[48]	85	83	68			
Ireland et al 1980[49]	64	86	92	98.3		
Daniel et al. 1982[46]	72	Medial 89 Lateral 85				
Selesnick et al. 1982[52]	82.6	73.2	94.4	96.8		
Passariello et al. 1985[62]					Medial 89.2 Lateral 96.1	
Manco et al. 1986[57]					92.2	87.5
Silva and Silver 1988[60]			92			65
Polly et al. 1988[58]						Medial 98 Lateral 90
Fischer et al. 1991[55]						Medial 89 Lateral 88
Raunest et al. 1991[59]						78
Kelly et al. 1991[56]						88

aspects. Technically, there is no difficulty in aspirating large knee effusions.

The knee should be re-examined following aspiration. Particular attention should be paid to the medial and lateral collateral ligaments, as well as to the anterior and posterior cruciates. A radiograph will rule out any bony injury.

The question of performing an urgent arthroscopy on patients with a haemarthrosis is controversial. We feel that urgent arthroscopy is not usually warranted. The treatment of a painful haemarthrosis involves initial aspiration affording pain relief, followed by further clinical examination. Attention should be directed to regaining range of movement, muscle strength, and resolution of swelling. Any major injury to the medial or lateral collateral ligaments should be treated early. Many patients will rehabilitate and go on to uneventful recovery without any further residual symptoms. Clinical evidence of meniscal tears will become obvious with time. It is appreciated that, at times, this 'wait and see' policy may delay treatment of meniscal tears. This may be of particular importance in some groups of patients such as high level or professional sports persons. In these circumstances an arthroscopy may be indicated early during rehabilitation.

In the past, it was felt that urgent arthroscopy was warranted in order to treat anterior cruciate ligament ruptures in the acute phase. However, it is now well documented that reconstruction of acute anterior cruciate ruptures is associated with considerable morbidity, particularly in terms of difficulty in restoring range of movement. Therefore it is important to allow return of full range of movement, muscle strength, and resolution of swelling prior to performing an anterior cruciate ligament reconstruction.

DIAGNOSTIC ADJUVANTS

The reliability of clinical examination for the diagnosis of meniscal disorders of the knee has ranged from 64 to 85 per cent in previous reports (Table 1).[46–52] In an attempt to improve diagnostic accuracy, some authors recommend the use of diagnostic adjuvants, such as arthrography, computer axial tomography (CT), and magnetic resonance imaging (MRI).

Arthrography

The diagnostic accuracy of knee arthrography (Table 1) is dependent on the expertise of the radiologist, and ranges from 68 to 97 per cent.[46–54] It is more useful in the diagnosis of medial meniscal pathology. In a minority of patients it fails to demonstrate tears involving the posterior horn of the lateral meniscus. This is due to the greater mobility of the lateral meniscus and the presence of the popliteus tendon. The advantages of arthrography include ready availability to practising physicians, outpatient procedure, and low complication rate.

Magnetic resonance imaging

Because of its remarkable soft tissue resolution, MRI has become an important diagnostic modality in the evaluation of lesions of the menisci. Its overall accuracy ranges from 65 to 98 per cent (Table 1).[55–60] A negative predictive value for the diagnosis of a torn meniscus is 83 to 97 per cent,[55,59] which means that 13 to 17 per cent of the time a negative finding on the MRI study may hide the presence of a torn meniscus. MRI has the advantage of being non-invasive, but has the disadvantages of being expensive and time consuming (up to 60 min for a single study). Furthermore, it requires that the knee can fully extend. The authors do not advocate the routine use of MRI for the evaluation of meniscal tears. If the clinical history and physical signs suggest meniscal pathology, arthroscopy is indicated. MRI is only indicated if a diagnosis is unclear and when it is felt that the treatment of the patient will be affected by the result. It is important that this modality is used wisely and that unnecessary expense is avoided.[59,61]

Computer axial tomography

CT scanning has achieved a diagnostic accuracy of 90 per cent (Table 1).[57,62] However, it has the disadvantage of using ionizing radiation. The quality of the image is diminished in the presence of a haemarthrosis.[63]

ARTHROSCOPY

The advent of arthroscopy of the knee over the past 10 to 15 years has revolutionized the diagnosis and treatment of meniscal tears. It allows visualization and direct palpation of the various intra-articular structures. A probe can be used to assess the integrity of the medial and lateral menisci, the anterior cruciate ligament, and the status of the articular surfaces of the femoral condyles, tibial plateaux, and patella, as well as the condition of the synovial lining of the joint. The widespread use of arthroscopy has eliminated the necessity for exploratory arthrotomies and enables meniscal surgery to be performed as an outpatient procedure. This surgery can be performed under general, spinal, or local anaesthesia. It is usually associated with minimal pain, allowing patients to walk immediately following surgery without the necessity for crutches. Rehabilitation is rapid, allowing patients return to work within days and to sport in 2 to 3 weeks. In addition to the clinical benefits, there are significant economic advantages in terms of containment of hospital costs, absence from work, and early return to sport.

Indications for arthroscopy

Diagnosis and treatment of meniscal tears are the main indications for arthroscopy of the knee. However, not every painful knee requires a diagnostic arthroscopy. Before considering arthroscopy, a careful history should be obtained, a thorough clinical examination should be made, and weight-bearing radiography of the knee should be performed.

Patients presenting with typical symptoms of joint-line pain and swelling, with or without episodes of locking, may be considered for arthroscopy. Patients with positive clinical signs, including joint-line tenderness, effusion, and meniscal catching signs, such as a positive McMurray sign,[19] are the most obvious candidates. Arthroscopy is also considered for those patients who have persistent symptoms suggestive of a meniscal tear, despite the absence of clinical signs. Prudent clinical judgement is paramount. If symptoms are vague and diffuse, the success rate from arthroscopy is similarly reduced. Arthroscopy should not be considered unless there is a reasonable chance of providing benefit to the patient in terms of either providing more information than was known clinically or a chance of providing treatment. Performing arthroscopy on patients who have little chance of gaining from the pro-

cedure only serves to increase their expectations, which in turn will lead to further disappointment.

HISTORICAL TREATMENT OF MENISCAL TEARS

For many years total meniscectomy was the gold standard for any meniscal pathology. Smillie[32] recommended total meniscectomy in preference to partial meniscectomy, irrespective of the type of lesion encountered. He maintained that, following meniscectomy, a new meniscus was regenerated, creating a perfect replica of the excised meniscus. This has subsequently been contested and disproved.[28]

Knowledge of the consequences of total meniscectomy[4,8,9,11] and the functional role of the meniscus in force transmission and shock absorption has led to the principle of preserving as much functional meniscal tissue as possible. In partial meniscectomy, meniscal resection is confined to the loose unstable fragment, such as a displaceable inner edge of a bucket handle tear or the flap in an oblique tear. Following partial meniscectomy, a stable or balanced rim of healthy meniscal tissue is preserved. This intact balanced peripheral rim provides joint stability and protects the articular surfaces by its load-bearing functions. Preservation of this peripheral cartilage rim is particularly important in patients who have already had major ligamentous injuries.[64]

Many studies have shown that there is an increased incidence of degenerative change following either partial or total meniscectomy.[4,8,9,23,65–70] This has led to the perceived need to consider repairing torn menisci. Several studies have documented that the peripheral one-third of the meniscus has a vascular supply[15,16] and that tears in that zone have a potential to heal.[25,71,72] The first report of a meniscal repair was by Annandale in 1885.[73] In 1936 King showed that peripheral tears of a meniscus could heal and that tears confined to the avascular meniscus substance probably never healed.[25] More recent studies have shown that the process of healing is initiated by the formation of a blood clot at the site of the tear.[71] This clot contains fibrin, serum factors, and blood cells. The fibrin forms a scaffolding which anchors the clot to the torn meniscal edges. Meniscal cells at the edge of the tear proliferate and then migrate into the fibrin scaffolding under the influences of growth factor and chemotactic factors, thus resulting in repair. King's findings have been confirmed by Heatley[74] in the rabbit and by Cabaud et al.[75] in the dog and monkey. The healing process is not initiated if the tear is limited to the avascular segment of the meniscus.

Vascular injection studies have provided documented evidence of the vascularity of the outer 10 to 30 per cent of the adult human meniscus.[15,76] There is even greater penetration of the vasculature in the skeletally immature individual.[6,13] This documented vascularity of the outer zone of the meniscus has provided the fundamental basis for meniscal suturing. Healing of meniscal tears following suture has been shown to be further enhanced by the application of an exogenous blood clot to the area of meniscal repair.[77] Arnoczky et al.[71] demonstrated the role of fibrin clot formation in the healing of meniscal tissue. The recent development of meniscal repair techniques has taken advantage of this healing potential and it is hoped that it will prevent degenerative knee changes which are seen following either total or partial meniscectomy. The rationale behind meniscal repair is to return the knee joint to its original integrity without removal of any meniscal substance. The aim of repair is to reduce opposing torn meniscal edges anatomically and to maintain this reduction with sutures until healing is complete.[78]

CURRENT MANAGEMENT OF MENISCAL INJURIES

Although diagnostic adjuncts such as arthrography[49,51,52,54] and MRI [55,57,58,60,63,79] can be helpful in diagnosing meniscal tears, the ultimate treatment decision is always based on the exact type, location, and extent of meniscal tears, which are best determined by direct visualization and palpation at arthroscopy.[47,48,50,80]

Arthroscopic management of the torn meniscus is now standard, and had virtually replaced arthrotomy and open meniscectomy. Arthroscopy affords much better exposure than can be gained by arthrotomy. Tear patterns can be readily appreciated and explored, following which appropriate management strategies can be planned.

Current management strategies for meniscal tears include partial meniscectomy, meniscal repair, and the most conservative treatment of all, which is to leave the tear alone. Nowadays, it is very rare to resort to total meniscectomy. Decisions as to the best option will depend on clinical evaluation, associated lesions, and the type, location, and extent of the tear, which can best be evaluated at arthroscopy.

The first decision to be made following the diagnosis of a meniscal tear is whether the tear should be treated surgically or left alone. Partial thickness split tears and full thickness (10 mm or less) or short (5 mm or less) radial tears which are stable on probing can be left alone. Instability of a meniscal tear is demonstrated if the inner portion of the torn meniscus can be displaced into the joint. Most commonly, such tears are found incidentally, in knees with more extensive pathology, such as tears of the other meniscus, or in association with anterior cruciate ligament tears. Weiss et al.[81] followed 52 patients who had partial thickness or short radial tears. Their average follow-up was over 4 years. They found that only six patients (12 per cent) required subsequent surgical treatment.

Once a torn meniscus is found to require surgical treatment, it is necessary to determine whether or not repair of the tear is feasible. The tear is evaluated for its length, width, and depth. A careful assessment of the quality of the torn meniscus is made, i.e. whether or not it has significant alteration of shape. Frequently, tears which have been displaced into the intercondylar notch for some time have shortened and cannot be properly repositioned to the original rim of the torn meniscus. Meniscal tears which have undergone alteration in shape and size are not suitable for repair and should be resected.

Partial meniscectomy

Meniscal tears which are not suitable for repair because they are located within the avascular zone or because of the extent of meniscal substance damage are treated by partial meniscectomy.[82] When partial meniscectomy is performed, only the unstable offending segment is removed, leaving behind an intact peripheral rim which maintains stability and protection to opposing articular surfaces (Fig. 9).

Once the decision is made to resect the torn mobile part of the meniscus, it is removed 'en bloc' or by morcellization (piecemeal) with a basket forceps or punch (Fig. 10). Following removal of the obviously mobile torn fragment, a further check of the remaining meniscus is made to ensure that the rim is balanced and stable.

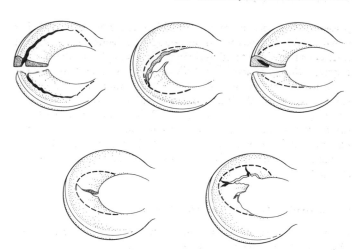

Fig. 9 Partial meniscectomy resection lines.

Metcalf and coworkers[83,84] have elegantly outlined the classical principles of partial meniscectomy. Only the torn portion of the meniscus that can protrude into the centre of the joint need be excised. They describe an imaginary line that represents the inner margin of a normal meniscus (Fig. 11). Any meniscal fragment that can be pulled past this boundary is likely to become caught between the joint surfaces during weight bearing and cause symptoms. As much meniscal rim as possible is preserved, even though this rim may be of poor quality at times and may become unstable later, and give rise to further symptoms. Patients are advised, in general, of the benefits of keeping this rim, if at all possible, and are made aware of the possibility of developing symptoms some time in the future. In reality, this is more possible then probable.

Partial meniscectomy is now performed almost exclusively through the arthroscope and rarely through a formal arthrotomy. Arthroscopic surgery requires considerable training, in both the identification of the various intra-articular structures and the different surgical techniques. Surgeons who are at the learning phase of arthroscopic surgery are better advised to perform an open partial meniscectomy if they are encountering difficulties with closed techniques.

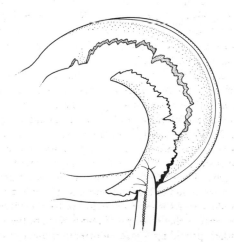

Fig. 10 Partial meniscectomy: removal of the mobile torn fragment *en bloc.*

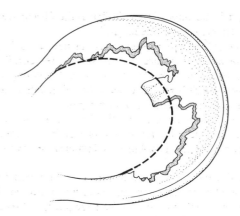

Fig. 11 The imaginary line represents the normal inner margin of the medial meniscus. A torn fragment protruding beyond this line requires resection as it will cause symptoms as it catches in the joint interface.

Meniscal repair

Tears suitable for repair are those within the vascular zone[85] (the outer third of the meniscus) that are unstable on probing, are longer than 7 mm, and have not sustained major surgical damage.[82] Retears of a previously repaired meniscus may also be considered for further repair if the meniscus has not undergone alteration in shape and size.[86]

Repair can be performed by open direct suture of tears which are at the very periphery of the meniscus, or by using variously described arthroscopic techniques. Both open and closed techniques have been reported with equally promising results.[77,87] Traumatic tears within the vascular zone have healing rates of up to 90 per cent whether the repairs are performed by open or by arthroscopic methods.[64,85,87–92] De Haven's study[82] has indicated that the time from injury to surgery does not alter the rate of healing in isolated meniscal tears. Isolated tears of up to 8 years' duration have been successfully repaired. However, acute tears, i.e. of less than 2 months' duration, associated with anterior cruciate ruptures had significantly greater rates of healing following repair than chronic tears seen after many months.

Henning et al.[77] have recently reported their indications for meniscal repair. They would consider all medial and lateral tears for repair, except where the repair would replace less than 25 per cent of the missing area. When a tear suitable for repair is seen in an anterior cruciate ligament deficient knee, it is recommended that the meniscus be repaired at the time of the anterior cruciate ligament reconstruction.[77,87] If reconstruction is not performed, there is a strong possibility that the meniscal repair will break down.

REHABILITATION FOLLOWING MENISCAL SURGERY

At the completion of the operative procedure, we routinely inject bupivacaine 0.25 per cent with adrenaline into the knee joint. This provides early postoperative analgesia, thereby decreasing the necessity for oral analgesia. The skin punctures are closed with a single suture. Adhesive strips can be used instead of sutures. A light compression bandage is applied around the knee.

Isometric quadriceps exercises are started as soon as the patient is awake in the recovery room. Early active range of movement is started on the day of surgery. Full weight bearing is allowed as soon after surgery as recovery from the anaesthetic allows. Crutches are not used. All meniscal surgery is performed as an outpatient procedure, i.e. the patient is discharged home on the day of surgery. Patients are given a prescription for oral analgesia as well as anti-inflammatory medication on discharge. They are advised to remove the light compression dressing 24 h after surgery. Flexion and extension exercises are continued. A full range of knee movement should be restored by 5 to 7 days. We have found that patients regain full range of movement and full quadriceps strength more rapidly if their postoperative exercise programme is monitored by a physical therapist.

Early return to work is encouraged, for example after 2 to 3 days for patients with sedentary occupations and after 7 to 8 days for patients engaged in heavy manual labour. Sporting activity is allowed when there is a full range of knee motion, 75 per cent quadriceps strength, no pain, and minimal or no residual effusion. This usually takes 2 to 4 weeks. Exceptionally, patients who have had a medial bucket handle tear excised may return to sport within 7 to 10 days.

Postoperative effusions usually resolve over a period of 2 to 3 weeks. Our personal observations has been that effusions in patients who have had lateral meniscal surgery seem to last longer than in those who have had surgery on the medial meniscus.

Rehabilitation programmes following meniscal repair vary. We allow patients to start weight bearing, with crutches, immediately following surgery. Early full range of movement is encouraged. Crutches are discarded after a few days. Running and contact sports are allowed after 6 months. Cannon, who uses the modified Henning technique of arthroscopic repair, recommends no weight-bearing for 4 weeks and partial weight bearing for a further 4 weeks.[93] Crutches are discarded at 8 weeks, return to straight running is allowed at 5 months, and return to contact sports at 9 months.

RESULTS

A more rational conservative approach to meniscal surgery has evolved over the last few decades as the long-term effects of total meniscectomy have been demonstrated by such workers as Fairbank, Jackson, Dandy, and Johnson. In 1948, Fairbank[9] published a study of 107 cases which were reviewed over a 14-year period after total meniscectomy. He noted radiographic changes in 67 per cent of patients after medial meniscectomy and in 50 per cent of patients after lateral meniscectomy. These changes included formation of an anteroposterior ridge projecting downwards from the margin of the femoral condyle over the old meniscus site, flattening of the marginal half of the femoral articular surface, and narrowing of the joint space. In a 17-year follow-up study, Johnson et al.[68] showed that only 42 per cent of 99 patients had satisfactory results after total meniscectomy. He also noted that meniscectomy may lead to knee instability. These typical long-term results, together with a fuller understanding of the functional significance of the meniscus, led to the concept of partial meniscectomy.

In the mid-1970s Dandy and Jackson[8] compared the results of partial meniscectomy with those of total meniscectomy: 91 per cent of the partial meniscectomy group were asymptomatic after 5 years, while 65 per cent of the total meniscectomy group were asymptomatic. McGinty et al.,[28] in a 5.5 year follow-up study of 128 patients, found that there was no subjective or objective evidence to favour total meniscectomy over partial excision in the management of bucket handle or anterior horn tears. At follow-up, Fairbank radiography changes occurred in 60 per cent of the total meniscectomy group and in only 30 per cent of the partial meniscectomy group. Their results showed that there was a lower postoperative morbidity, a shorter hospital stay, and less time spent on crutches in those patients who had a partial meniscectomy. The major complication rate was 2.6 per cent compared with 14.6 per cent in favour of partial meniscectomy.

Several authors have reported consistently good results following meniscal repair. In a long-term follow-up of 33 repaired menisci, De Haven[82] reported an overall retear rate of 21 per cent. He noted that 14 per cent of acute repairs and 33 per cent of chronic repairs sustained retears. There was a retear rate of 11 per cent in patients with stable knees and 42 per cent in patients with unstable knees. He emphasizes the need for early repair. In a previous study,[64] he had shown that the retear rate in menisci that were repaired with simultaneous anterior cruciate ligament reconstruction was less than 10 per cent.

Hamberg et al.,[90] in an 18-month follow-up study of 50 cases of meniscal repair, showed that there was an 11 per cent incidence of retear in chronically torn menisci and a 9 per cent incidence of new rupture at a different site. There was a 6 per cent incidence of new rupture and no cases of retear in menisci repaired within 2 weeks of being torn.

In a 2- to 10-year follow-up of 71 meniscal repairs, Hanks et al.[87] found that the failure rate of open repair versus arthroscopic repair were very similar, 11 per cent and 8.8 per cent respectively. The failure in anterior cruciate deficient knees was 13 per cent compared with 8 per cent in anterior cruciate stable knees. Cannon and Vittari[94] found a 93 per cent meniscus healing rate in patients who had had a simultaneous anterior cruciate reconstruction. This compared with a 50 per cent success rate in those patients whose anterior cruciate was not reconstructed.

Henning et al.,[77] in a series of 153 arthroscopic meniscal repairs using an exogenous fibrin clot, showed an overall healing rate of 64 per cent; 24 per cent were incompletely healed and 12 per cent had failed. Follow-up was 4 months. In anterior cruciate ligament deficient knees, the failure rate was 1.5 per cent if the repair was performed within 2 months of injury. This rose to 20 per cent if the repair was performed more than 2 months after the initial injury. They found that using an exogenous fibrin clot improved their healing rate from 59 to 92 per cent.

FUTURE DIRECTIONS

The use of laser technology in meniscus surgery is still at the development stage. The carbon dioxide laser is being superseded by the neodymium—YAG laser. The latter is more compact and can be used in an aqueous medium. It has precise thermal tissue effects which causes only minimal thermal injury to surrounding healthy meniscal tissue.

Cisa et al.[95] have demonstrated that vascularized synovial flaps sutured into the meniscal defect can induce a reparative response. Similarly, carbon fibre implants have been inserted into meniscal defects. However, they induce a rather exaggerated tissue response resulting in excessive fibrous tissue generation.[72]

Arnoczky and Warren[71] have used an exogenous fibrin clot to effect healing of the meniscal defect that is limited to the avascular segment. They state that the fibrin clot appears to act as a chemotactic and mitogenic stimulus for reparative cells and provides a scaffold for the reparative process. Although the reparative tissue is grossly and histologically different from the normal men-iscal tissue, it is morphologically similar to the fibrous tissue seen in the vascular area of the meniscus.[16]

Webber et al.[95] are performing exciting new work in the development of an ideal culture system, which theoretically should allow complete healing of midsubstance tears. Their organ culture model makes it possible rapidly to identify and quantify the factors necessary in effecting meniscal wound repair.

Stone et al.[97] are presently investigating the use of a collagen-based meniscal prothesis. These act as resorbable regeneration templates and offer the possibility of inducing regrowth of new menisci.

Probably the most adventurous current experimental work is that of meniscal transplantation. Meniscal transplantation has two main objectives: firstly to stabilize the knee joint, and secondly to prevent the progression of established degeneration of hyaline articular cartilage. The early independent results reported by Garrett and Stevenson[98] and by Milahowshi and coworkers[99,100] are encouraging. They have noted an arrest in the progress of degenerative changes in articular cartilage. Meniscal transplantation may be of particular benefit in the anterior cruciate deficient knee,[100] as meniscectomy in these patients potentially renders these knees even more unstable.[68,70,101,102] Although no specific complications have been observed, there is shrinkage of transplanted material.[100] Canham and Stanish[103] have shown that a minimum of shrinkage occurs in transplanted tissue-culture-stored allogenic menisci

Garrett and Stevensen[98] recently reported on a 2.5-year follow-up study of six fresh meniscal transplants in the human knee. Four patients were reviewed arthroscopically and found to have healed transplant–rim junctions. No patient required further surgery.

The meniscus appears to be 'immunologically privileged', as no short-term rejection problems have been observed.

REFERENCES

1. Aspden RM, Yarker YE, Hukins DWL. Collagen orientations in the meniscus of the knee joint. *Journal of Anatomy*, 1985; **140**: 371.
2. Aspden RM. A model for the function and failure of the meniscus. *Engineering in Medicine* 1985; **14**: 119.
3. Bullough PG, Munueral L, Murphy J, Weintein AM. The strength of the menisci of the knee as it relates to their fine structure. *Journal of Bone and Joint Surgery* 1970; **52B**: 564–70.
4. Krause WR, Pope MH, Johnson RJ, Wilder DG. Mechanical changes in the knee after meniscectomy. *Journal of Bone and Joint Surgery* 1976; **58A**: 599–604.
5. Kurosawa H, Fukubayashi T, Nakajima H. Load-bearing mode of the knee joint: physical behaviour of the knee joint with or without menisci. *Clinical Orthopaedics* 1980; **149**: 283.
6. McDevitt CA, Webber RJ. Ultrastructure and biochemistry of meniscal cartilage. *Clinical orthopaedics* 1991; **252**: 8–18.
7. Walker PS, Erkman MJ. The role of the menisci in force transmission across the knee. *Clinical Orthopaedics* 1975; **109**: 184–92.
8. Dandy DJ, Jackson RW. The diagnosis of problems after meniscectomy. *Journal of Bone and Joint Surgery* 1975; **57B**: 349.
9. Fairbank TJ. Knee joint changes after meniscectomy. *Journal of Bone and Joint Surgery* 1948; **30B**: 664–70.
10. Goodfellow J. Editorials and annotations: he who hesitates is saved. *Journal of Bone and Joint Surgery* 1980; **62B**: 1–2
11. Northmore-Ball MD, Dandy DJ. Longterm results of arthroscopic partial meniscectomy. *Clinical Orthopaedics* 1982; **167**: 34.
12. Ghosh P, Taylor TKF. The knee joint meniscus: a fibrocartilage of some distinction. *Clinical Orthopaedics* 1987; **224**: 52–63.
13. Clark CR, Ogden JA. Development of the menisci of the human knee joint: morphological changes and their potential role in childhood meniscal injury. *Journal of Bone and Joint Surgery* 1983; **65A**: 538–47.

14. Heller L, Langman J. The menisco-femoral ligaments of the human knee. *Journal of Bone and Joint Surgery* 1964; **46B**: 307–13.

15. Arnoczky SP, Warren RF. Microvasculature of the human meniscus. *American Journal of Sports Medicine* 1982; **10**: 90–5.

16. Arnoczky SP, Warren RF. The microvasculature of the meniscus: its response to injury: experimental study in the dog. *American Journal of Sports Medicine* 1983; **11**: 131.

17. Danzig L, Resnick D, Gonsalves M, Akeson WH. Blood supply to the normal and abnormal menisci of the human knee. *Clinical Orthopaedics* 1983; **172**: 271–6.

18. Ghadially FN, Thomas I, Young N, Lalonde JMA. Ultrastructure of rabbit semilunar cartilage. *Journal of Anatomy* 1978; **125**: 499.

19. McMurray TP. The semilunar cartilages. *British Journal of Surgery* 1941; **29**(116): 407–14.

20. Seedhom BB, Hargreaves DJ. Transmission of the load in the knee joint with special reference to the role of the menisci. *Engineering in Medicine* 1979; **8**: 220.

21. Shrine N. The weight-bearing role of the menisci of the knee. *Journal of Bone and Joint Surgery* 1974; **56B**: 381.

22. King D. The function of semilunar cartilages. *Journal of Bone and Joint Surgery* 1936, **18**: 1069–76.

23. Appel H. Later results after meniscectomy in the knee joint. A clinical and roentgenologic follow-up investigation. *Acta Orthopaedica Scandinavica* 1970; **133** (Suppl.): 1–111.

24. Huckell JR. Is meniscectomy a benign procedure? A long term follow-up study. *Canadian Journal of Surgery* 1965; **8**: 254–60.

25. King D. The healing of semilunar cartilages. *Journal of Bone and Joint Surgery* 1936; **18**: 333–42.

26. Cox JS, Nye CE, Schaefer WW, Woodstein IJ. The degeneration effects of partial and total resection of the medial meniscus in dogs' knees. *Clinical Orthopaedics* 1975; **109**: 178.

27. Fithian DC, Kelly MA, Mow VC. Material properties and structure—function relationships in the menisci. *Clinical Orthopaedics* 1990; **252**: 19–31.

28. McGinty JB, Geuss LF, Marvin RA. Partial or total meniscectomy. A comparative analysis. *Journal of Bone and Joint Surgery* 1977; **59A**: 763–6.

29. Ahmad AM, Burke DL. *In vitro* measurement of static pressure distribution in synovial joints. Part 1: Tibial surface of the knee. *Journal of Biomechanical Engineering* 1983; **105**: 216.

30. Fukubbayashi T, Kurosawa H. The contact area and pressure distribution pattern of the knee. *Acta Orthopaedica Scandinavica* 1980; **51**: 871.

31. Voloshin AS, Wosk J. Shock absorption of the meniscectomized and painful knees. A comparative *in vivo* study. *Journal of Biomedical Engineering* 1983; **5**: 157.

32. Smillie IS. *Injuries of the knee joint*. 3rd edn. London: Churchill Livingstone, 1962.

33. Baker BE, Peckham AC, Pupparo F, Sanborn JC. Review of meniscal injury and associated sports. *American Journal of Sports Medicine* 1985; **13**: 1.

34. Insall JN. *Surgery of the knee*. New York: Churchill Livingstone, 1984: 230

35. Ferrer-Roca O, Vilalta C. Lesions of the meniscus. Part 1: Macroscopic and histologic findings. *Clinical Orthopaedics* 1980; **146**: 289–300.

36. Smillie LS. *Injuries of the knee joint*. 4th edn. London: Churchill Livingstone, 1970.

37. Ferrer-Roca O, Vilalta C. Lesions of the meniscus. Part II: Horizontal cleavages and lateral cysts. *Clinical Orthopaedics* 1980; **146**: 301–7.

38. Noble J, Hamblen DL. The pathology of the degenerate meniscus lesion. *Journal of Bone and Joint Surgery* 1975; **57B**: 180–6.

39. Pritzker KPH, Renlund RC, Cheng PT. Which came first: crystals or osteoarthritis? A study of surgically removed femoral heads. *Journal of Rheumatology* 1983; **10**: 38.

40. Sokoloff L, Varma AA. Chondrocalcinosis in surgically resected joints. *Arthritis and Rheumatism* 1988; **31**: 750.

41. Hough AJ, Webber RJ. Pathology of the meniscus. *Clinical Orthopaedics* 1990; **252**: 32–40.

42. Apley AG. The diagnosis of meniscal injuries: some new clinical methods. *Journal of Bone and Joint Surgery* 1947, **29**: 78–84.

43. Vandermeer RD, Cunningham FK. Arthroscopic treatment of the discoid lateral meniscus: results of long term follow up. *Arthroscopy* 1989; **5**: 101–9.

44. Hurson JB, Kessopersadh E. Natural history of 380 anterior cruciate deficient knees. In preparation.

45. Noyes FR, Bassett RW, Grood ES, Butler DL. Arthroscopy in acute traumatic haemarthrosis of the knee: incidence of anterior cruciate tears and other injuries. *Journal of Bone and Joint Surgery* 1980; **62A**: 687–95.

46. Daniel D, Daniels E, Aronson D. The diagnosis of meniscus pathology. *Clinical Orthopaedics* 1982; **163**: 218–24.

47. De Haven KE, Collins HR. Diagnosis of internal derangements of the knee: the role of arthroscopy. *Journal of Bone and Joint Surgery* 1975; **57A**: 802–10.

48. Gillies H, Seligson D. Precision in the diagnosis of meniscal lesions: a comparison of clinical evaluation, arthrography, and arthroscopy. *Journal of Bone and Joint Surgery* 1979; **61A**: 343–6.

49. Ireland J, Trickey EL, Stoker DJ. Athroscopy and arthrography of the knee: a critical review. *Journal of Bone and Joint Surgery* 1980; **62B**: 3–6.

50. Jackson RW, Abe I. The role of arthroscopy in the management of disorders of the knee: an analysis of 200 consecutive examinations. *Journal of Bone and Joint Surgery* 1972; **54B**: 310–22.

51. Nicholas JA, Freiberger RH, Killoran PJ. Double-contrast arthrography of the knee: its value in the management of two hundred and twenty-five derangements. *Journal of Bone and Joint Surgery* 1970; **52A**: 203–20.

52. Selesnick FH, Noble HB, Bachman DC, Steinberg FL. Internal derangement of the knee: diagnosis by arthrography, arthroscopy, and arthrotomy. *Clinical Orthopaedics* 1985; **198**: 26–30.

53. McGinty JB, Metza RA. Arthroscopy of the knee. Evaluation of an out-patient procedure under local anaesthesia. *Journal of Bone and Joint Surgery* 1978; **60A**: 787.

54. Tegtmeyer CJ, McCue FC, Higgins SM, Ball DW. Arthrography of the knee: a comparative study of the accuracy of single and double contrast techniques. *Radiology* 1979; **132**: 37–41.

55. Fischer SP, Fox JM, Del Pizzo W, Friedman MJ, Snyder SJ, Ferkel RD. Accuracy of diagnosis from magnetic resonance imaging of the knee. *Journal of Bone and Joint Surgery* 1991; **73A**: 2–10.

56. Kelly MA *et al.* MR imaging of the knee: clarification of its role. *Arthroscopy* 1991; **7**: 78–85.

57. Manco LG, Lozman J, Coleman ND, Kavanaugh JH, Bilfield BS, Dougherty J. Noninvasive evaluation of knee meniscal tears: preliminary comparison of MR imaging and CT. *Radiology* 1987; **163**: 727–30.

58. Polly DW, Callaghan JJ, Sikes A, McCabe JM, McMahon K, Savory CG. The accuracy of selective magnetic resonance imaging compared with the findings of arthroscopy of the knee. *Journal of Bone and Joint Surgery* 1988; **70A**: 192–8.

59. Raunest J, Oberle K, Loehnert J, Hoetzinger H. The clinical value of magnetic resonance imaging in the evaluation of meniscal disorders. *Journal of Bone and Joint Surgery* 1991; **73A**: 11–16.

60. Silva I, Silver DM. Tears of the meniscus as revealed by magnetic resonance imaging. *Journal of Bone and Joint Surgery* 1988; **70A**: 199–202.

61. Senghas RE. Editorial: Indications for magnetic resonance imaging. *Journal of Bone and Joint Surgery* 1991; **73A**: 1.

62. Passariello R, Trecco F, Paulis P, Masciocchi C, Bonanni G, Zobel BB. Meniscal lesions of the knee joint: CT diagnosis. *Radiology* 1985; **157**: 29–34.

63. Reichier MA, Hartzman S, Duckwiler GR, Bassett LW, Anderson LJ, Gold RH. Meniscal injuries: detection using MR imaging. *Radiology* 1986; **159**: 753–7.

64. De Haven KE. Meniscus repair in the athlete. *Clinical Orthopaedics* 1985; **198**: 31–5.

65. Casscells SW. The torn or degenerated meniscus and its relationship to degeneration of the weight-bearing areas of the femur and tibia. *Clinical Orthopaedics* 1978; **132**: 196–200.

66. Cox JS, Cordell LD. The degenerative effects of medial meniscus tears in dogs' knees. *Clinical Orthopaedics* 1977; **125**: 236–42.

67. Jackson JP. Degenerative changes in the knee after meniscectomy. *British Medical Journal* 1968; **2**: 525–7.

68. Johnson RJ, Kettlekamp DB, Clark W, Leaverton P. Factors affecting late results after meniscectomy. *Journal of Bone and Joint Surgery* 1974; **56A**: 719.

69. Seedhom BB, Dawson D, Wright V. Function of the menisci, preliminary study. *Journal of Bone and Joint Surgery* 1974; **56B**: 381–2.

70. Tapper EM, Hoover NW. Late results after meniscectomy. *Journal of Bone and Joint Surgery* 1969; **51A**: 517–26.

71. Arnoczky SP, Warren RF, Spivak JM. Meniscal repair using an exogenous fibrin clot: an experimental study in dogs. *Journal of Bone and Joint Surgery* 1988; **70A**: 1209–16.

72. Veth RPH, Den Heeten GJ, Jansen HWB, Nielsen HKL. An experimental study of reconstructive procedures in lesions of the meniscus: use of synovial flaps and carbon fibre implants for artificially made lesions in the meniscus of the rabbit. *Clinical Orthopaedics* 1983; **181**: 250–4.

73. Annandale T. An operation for displaced semilunar cartilage. *British Medical Journal* 1885; **1**: 779.

74. Heatley FW. The meniscus—can it be repaired? An experimental investigation in rabbits. *Journal of Bone and Joint Surgery* 1980; **62B**: 397–402.

75. Cabaud HE, Rodkey WG, Fitzwater JE. Medial meniscus repairs. An experimental and morphologic study. *American Journal of Sports Medicine* **9**: 129–34.

76. Scapinell R. Studies of the vasculature of the human knee joint. *Acta Anatomica (Basel)* 1968; **70**: 305.

77. Henning CE, Lynch MA, Yearout KM, Vequist SW, Stallbaumer RJ, Decker KA. Arthroscopic meniscal repair using an exogenous fibrin clot. *Clinical Orthopaedics* 1990; **252**: 64–72.

78. Kawai Y, Fukubayashi T, Nishino J. Meniscal suture: an experimental study in the dog. *Clinical Orthopaedics* 1989; **243**: 286–93.

79. Stoller DW, Martin C, Crues JV, Kaplan L, Mink JH. Meniscal tears: pathologic correlation with MR imaging. *Radiology* 1987; **163**: 731–5.

80. Dandy DJ, Jackson RW. The impact of arthroscopy on the management of disorders of knee. *Journal of Bone and Joint Surgery* 1975; **57B**: 346–8.

81. Weiss CB, Lunberg M, Hamberg, P, De Haven KE, Gillquist J. Nonoperative treatment of meniscal tears. *Journal of Bone and Joint Surgery* 1989; **71A**: 811.

82. De Haven KE. Decision-making factors in the treatment of meniscus lesions. *Clinical Orthopaedics* 1990; **252**: 49–54.

83. Metcalf RW. Arthroscopic meniscal surgery. In: McGinty JB *et al.*, eds. *Operative arthroscopy*. New York: Raven Press, 1991: 203–36.

84. Metcalf RW, Coward DB, Rosenberg PD. Arthroscopic partial meniscectomy: a five year follow up study. *Orthopaedic Transactions* 1983; **7**: 504.

85. De Haven KE. Peripheral meniscal repair: an alternative to meniscectomy. *Journal of Bone and Joint Surgery* 1981; **63B**: 463.

86. De Haven KE, Lohrer WA, Lovelock JE. *Long term results of meniscal repair*. Presented at the Combined Congress of the International Arthroscopic Association and the International Society of the Knee, Toronto, May 1991.

87. Hanks GA, Gause TM, Sebastianelli WJ, O'Donnell CS, Kalenak A. Repair of peripheral meniscal tears: open versus arthroscopic technique. *Arthroscopy* 1991; **7**: 72–7.

88. Cassidy RE, Shaffer AJ. Repair of peripheral meniscus tears: a preliminary report. *American Journal of Sports Medicine* 1981; **9**: 209–14.

89. Dolan WA, Bhaskar, G. Peripheral meniscus repair: a clinical and pathologic study of 75 cases. *Orthopaedic Translations* 1983; **7**: 503.

90. Hamberg P, Gillquist J, Lysholm J. Suture of new and old peripheral meniscus tears. *Journal of Bone and Joint Surgery* 1983; **65A**: 193–7.

91. Rosenberg TD, *et al.* Arthroscopic meniscal repair evaluation with repeat arthroscopy. *Arthroscopy* 1986; **2**: 14.

92. Scott GA, Jolly BJ, Henning CE. Combined posterior incision nd arthroscopic intra-articular repair of the meniscus. *Journal of Bone and Joint Surgery* 1986; **68A**: 847.

93. Cannon WD. Arthroscopic meniscal repair. In: McGinty JB *et al.*, eds. *Operative arthroscopy*. New York: Raven Press, 1991; 237–51.

94. Cannon WD, Vittari JM. *Arthroscopic meniscal repair results in anterior cruciate ligament reconstructed knee versus isolated repairs in stable knees*. Presented at the Combined Congress of the International Arthroscopic Association and the International Society of the Knee, Toronto, May 1991.

95. Cisa, J, Basora J, Madarnas P, Ghibely A. *Meniscus healing, experimental methods in the rabbit model with a synovial flap*. Presented at the Combined Congress of the International Arthroscopic Association and the International Society of the Knee. Toronto, May 1991.

96. Webber RJ, York JL, Van Der Schilden JL, Hough AJ. An organ culture model for assaying wound repair of the fibrocartilaginous knee joint meniscus. *American Journal of Sports Medicine* 1989; **17**: 393–400.

97. Stone KR, Rodkey WG, Webber RJ, McKinney L, Steadman JR. Future directions: collagen based prostheses for meniscal regeneration. *Clinical Orthopaedics* 1990; **252**: 129–35.

98. Garrett JC, Stevenson RN. Meniscal transplantation in the human knee: a preliminary report. *Arthroscopy* 1991; **7**: 57–61.

99. Milahowski KA, Weismeier K, Erhard TW, Remberger K. Transplantation of meniscus: an experimental study in sheep. *Sport Verlet Zung Sportschaden* 1987; **1**: 20.

100. Wirth CJ, Kohn D, Milachowski KA. Meniscus transplantation in knee joint instability. Presented at the Combined Congress of the International Arthroscopy Association and the International Society of the Knee, Toronto, May 1991.

101. Levy IM, Torzilli PA, Warren RF. The effect of medial meniscectomy on anterior–posterior motion of the knee. *Journal of Bone and Joint Surgery* 1982: **64A**; 883–8.

102. Hsieth HH, Walker PS. Stabilising mechanisms of the loaded and unloaded knee joint. *Journal of Bone and Joint Surgery* 1976; **58A**: 87–93.

103. Canham W, Stanish W. A study of the biological behaviour of the meniscus as a transplant in the medial compartment of a dog's knee. *American Journal of Sports Medicine* **14**: 376–9.

104. Bickerstoff DR, Wyman A, Laing RW, Smith TWD. Partial meniscectomy using the neodymium: YAG laser; an *in vitro* study. *Arthroscopy* 1991; **7**: 63–7.

105. Bourne RB, Finlay JB, Papadopoulos P, Andreae P. The effects of medial meniscectomy on strain distribution in the proximal part of the tibia. *Journal of Biomedical Engineering* 1983; **5**: 157.

106. Cargill AOR, Jackson JP. Bucket-handle tear of the medial meniscus: a case for conservative surgery. *Journal of Bone and Joint Surgery* 1976; **58A**: 248–51.

107. Casscells SW. Arthroscopy of the knee joint: review of 150 cases. *Journal of Bone and Joint Surgery* 1971; **53A**: 287.

108. Gershuni DH, Skyhar MJ, Danzig LA, Camp J, Hargens AR, Akeson WH. Experimental models to promote healing of tears in the avascular segment of canine knee menisci. *Journal of Bone and Joint Surgery* 1989; **71A**: 1363–9.

109. MacConaill MA. The function of intra-articular fibrocartilages, with special references to the knee and inferior radioulnar joints. *Journal of Anatomy* 1931; **66**: 210.

110. Miller DB. Arthroscopic meniscus repair. *American Journal of Sports Medicine* 1988; **16**: 315–20.

111. Tregonning RJA. Closed partial meniscectomy: early results for simple tears with mechanical symptoms. *Journal of Bone and Joint Surgery* 1983; **65B**: 378–82.

4.2.3 Knee ligament sprains—acute and chronic

WILLIAM D. STANISH, PATRICK O'GRADY, AND JADE E. DILLON

INTRODUCTION

Knee ligament injuries are common in sport. These injuries may occur as a consequence of athletic activity, or secondary to motor vehicle collisions. However, knee ligament damage caused by a violent insult during sport has attracted the most attention.

Since the classic works of O'Donoghue et al.[1] Hughston,[2] and others, much controversy has been created as to the best technique for the medical and surgical management of knee ligament sprains. Specifically, treatment of the torn anterior cruciate ligament of the knee has attracted most attention. Unfortunately, most studies of the natural history of knee ligament instabilities are limited by subjective bias, tend to be retrospective in nature, and usually have a very brief follow-up. Despite this, valuable information can be obtained from the majority of these investigations.

This chapter is designed to place maximum emphasis on the clinical aspects of the problem of knee ligament sprain. Each type of knee ligament tear will be profiled to facilitate an understanding of the mechanism of injury, the clinical signs and symptoms, and the options for treatment. Distinctions are drawn between acute ligament sprain and the chronically unstable knee. The basic science of ligament injury and repair has been included in the text to provide a platform for enhanced clinical interpretation of the ligament injury.

LIGAMENT HEALING

Healing is a matter of time, but is also sometimes also a matter of opportunity. (Hippocrates)

The following discussion will deal specifically with the healing process of ligaments and the factors which influence the rate and degree of success of that process. It is intended that a review of recent experimental and clinical studies will establish a time–strength relationship for healing ligaments. This process will afford a better appreciation of treatment regimens which provide the most opportune conditions for repair. Furthermore, it is expected that, by first presenting the fundamental concepts of soft tissue healing and then discussing those factors pertaining particularly to ligament healing, newer treatment concepts will be better understood.

The healing process in ligaments is generally divided into three phases: the substrate phase, the proliferative phase, and the remodelling phase.

Phase I

The substrate phase occurs during the first 4 days following injury. This phase involves the vascular response to the insult, haemostasis, and cellular response. Cells, including polymorphonuclear leucocytes, lymphocytes, macrophages, and mast cells, as well as fibroblasts, aggregate at the wound site during this inflammatory process. By the end of this phase, neovascularization has begun at the edges of the wound.

Phase II

The proliferative phase commences approximately 4 days postinjury and continues for 2 to 3 weeks. This period is also termed the fibroblastic phase to reflect the proliferation of fibroblasts at the site to the extent that they become the dominant cell type. The result of this cellular proliferation is the production of collagen, which in turn relates directly to the rapid gain in wound strength which occurs during this time.

As stated by previous authors, including Peacock,[3] this phase has features which must be recognized when a specific tissue such as ligament is being considered. Peacock[3] states:

It has been observed that the rate at which all wounds gain strength is the same during the first fourteen to twenty-one days after wounding. However, the percentage of normal, unwounded strength that is gained by the wound varies markedly with the tissues. In general, there is an inverse ratio between normal breaking strength of the tissue and the percentage of that strength that the wound regains in fourteen to twenty-one days. This suggests that the absolute gain in strength in the early phase of wound healing is related to the chemical events occurring in all tissues. The gain in strength is apparently limited by the rapidity with which these universal events occur.

Figure 1 illustrates the significance of this phenomenon. Although ligament tissue was not specifically represented in this study, its healing rate would be expected to approximate most closely that of tendon. It is apparent that factors which affect the biochemical events of the proliferative phase will be an important consideration in the determination of optimum healing conditions for ligament tissue.

Phase III

The remodelling of wound repair differs from the two earlier phases in that its duration is indistinct and probably never ends. Early in this phase the number of fibroblasts and the collagen production peak and then begin to decrease gradually. The result is the formation of a collagen scar which undergoes remodelling as the fibres become aligned parallel to the direction of tension on the scar. There is continued turnover of collagen during this phase and the wound strength continues to increase as the collagen realigns, matures, and shifts from the type III collagen of healing wounds to the type I collagen of normal tissue. The persistence of this phase is considered to be due primarily to the stimulus of mechanical stresses on the wound. It is universally conceded that, although this phase is of long duration, the repaired tissue never regains the architecture or strength of the tissue that was present prior to injury. Figure 2 gives a graphic summary of the ligament healing process.

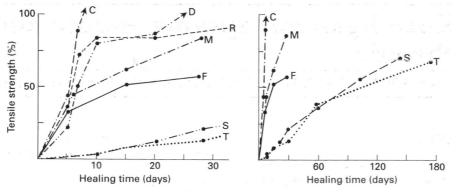

Fig. 1 Relative healing rates for linear incisional wounds in different tissues, the tensile strength being calculated in percentage of that of the respective intact tissues: C: corpus ventriculi (rat), D: duodenum (rat), R: rumen ventriculi (rat), M: lateral abdominal wall muscle (rabbit), F: fascia (linea alba, rabbit), S: skin (dorsal thoracic, rat), and T: tendon (peroneus brevis, rabbit). (Redrawn from ref. 83, with permission.)

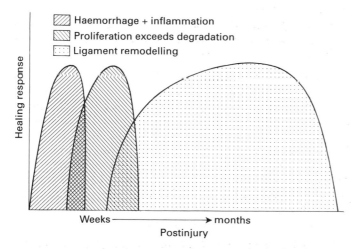

Fig. 2 Theoretical representation of phase of ligament healing showing relative durations and overlap of stages. Treatment is aimed at optimizing proliferation and remodelling in as short a time as possible. (Redrawn from ref. 84, with permission.)

FACTORS INFLUENCING LIGAMENT HEALING

The phases of wound healing are influenced by many factors, the control of which provides the potential for establishing ideal repair conditions. It should be realized that the induction of the so-called 'superhealing' is probably not attainable; however, factors discouraging proper healing can be controlled.

An injury of the ligament is the initial stimulus to repair. Dunphy[4] suggests that although there can be an optimum stimulus to repair in the clinical setting, the situation following a clean sweep of the knife affords the environment for ideal repair. Any additional trauma is inhibitory. Johnson[5] makes the valid point that the use of surgically incised animal ligaments as experimental models cannot reproduce the traumatically torn ligament in the athlete. Such injuries are known to demonstrate various degrees of disruption throughout the entire ligament unit; thus it is obvious that in the majority of such clinical cases, the healing process of injured ligaments begins under less than ideal conditions.

The degree of damage incurred is also related to the mode of injury. Incomplete ligament tears or small wounds, though healing by the same means as complete disruptions, will not produce as profound a deficit and consequently will probably yield a more functional tissue.

Adequate nutrition and blood supply are essential throughout the healing process.[6,7] Vitamin C is particularly important, and its deficiency can result in a wound with decreased tensile strength, secondary to a prolonged substrate phase, and decreased collagen synthesis. Availability of protein is also important, with protein deficiency resulting in slow healing although, with time, the constituents will be eventually the same. The production of type III collagen is particularly dependent upon an adequate supply of cystine during scar formation.

The detrimental effect produced by poor blood supply is due to a number of factors. Apart from carrying the original inflammatory cells that initiate wound healing, blood supply and oxygen availability to the wound are inherently related. Either poor blood supply or low oxygen concentration will slow the rate of healing and can increase the risk of infection. If infection does ensue, wound healing can be significantly deterred. Since all surgical procedures result in some degree of bacterial contamination, the presence of other factors which predispose to poor wound healing must be controlled to reduce the chance of sepsis.

The effect of electrical stimulation on healing rates has been studied for many soft tissues. Stanish et al.[8] suggest that the use of electrical or electromagnetic stimulation may prove to be a valuable adjunct to the healing of surgically repaired ligaments. The use of electrical muscle stimulation during the early periods of healing results in increased cross-sectional area of the repaired tissue which consequently provides a greater tensile strength.

Undoubtedly, the factor affecting the healing rate of ligaments which received the most attention is the issue of wound stressing.[9–14] The stressing of ligaments during healing has been considered in many studies dealing with immobilization, angle of immobilization, and scheduling of remobilization. In essence, it has been established that the primary concern is to protect the wound site during the first two phases of healing so that a certain degree of strength can be regained. It has also been demonstrated that mobilization, or application of stress upon the wound, in the latter phase of healing is essential if maximum repair with respect to biochemical and biomechanical parameters is to be achieved.

LIGAMENT STRUCTURE AND BEHAVIOUR

In order to appreciate fully the healing process of ligaments, it is fundamental to establish their normal biochemical and biomechanical properties.

Ligaments are composed of approximately 60 to 80 per cent by weight of water. Collagen accounts for 70 to 80 per cent of the dry mass; approximately 90 per cent of this is Type I collagen, with the remainder being type III. Type III collagen is an immature form of collagen and reflects the dynamic behaviour in ligamentous tissue. There is also an elastin component and small amounts of proteoglycans. There is only a small cellular component. The collagen content is the most crucial constituent and is responsible for the mechanical properties of ligaments. The strength of collagen is directly related to its structure which consists of both intermolecular and intramolecular cross-links. Failure will occur when the resistance of the fibres or their cross-links is exceeded. Also of particular importance when considering the ability of a ligament to resist deformation is the crimped pattern in which it is arranged. It is suggested that this microscopic phenomenon provides a buffer which will allow for slight elongation of the ligament with full return to its natural state when the force is removed. When the limits of this buffering structure are exceeded, irreversible damage to the ligaments occurs.

The biomechanical behaviour of ligaments is most interesting. Ligaments are considered to be viscoelastic materials and, according to Fung,[15] viscoelasticity is a term used to describe the combined properties of hysteresis, relaxation, and creep. Hysteresis describes the phenomenon by which the stress–strain relationship of a cyclically loaded substance differs during loading and unloading. Relaxation is the characteristic by which the stress in a material decreases in time after the material is suddenly strained and the strain is maintained. Creep is defined as the continued deformation of a body when it is suddenly stressed and the stress is maintained.

According to Frank and coworkers,[12,13,16] the resultant of these properties allows for 'some joint displacement with relatively little effort, but provides increasing resistance as deformation increases'.[12] The resultant of the biochemical structure and the consequential biomechanical properties yields the typical load–deformation curve (Fig. 3).

As soon as injury occurs, the healing process begins. The assessment of the healing properties of ligaments has been the focus of many studies which have analysed the biochemical and biomechanical parameters in animal models. Animal studies have addressed many aspects of healing under controlled conditions and provide the advantage of postmortem study. By comparison, human studies are rather limited in yielding objective data, but do provide valuable information with respect to the degree of function which can be regained subject to the various treatment regimens.

Danielsen[11] has studied the *in-vitro* maturation of collagen and provided baseline data concerning the rate of collagen production. This study concluded that the strength gains plateau at approximately 3 months, as illustrated in Fig. 4. Furthermore, it provides a correlation between the number of reducible cross-links and the mechanical structure of the fibres. It is conceded that, although *in-vitro* maturation occurs in the same pattern as the *in-vivo* process, the rate of maturation in living tissue would be expected to be relatively slower. In fact, this has been verified in a number of *in-vivo* experiments.

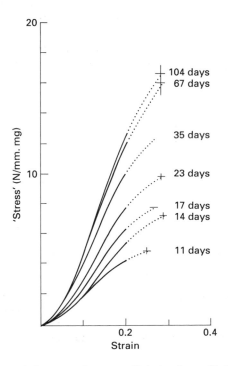

Fig. 4 Stress-strain curves for reconstituted collagen fibrils matured *in vitro* at 37°C for different time periods after aggregation. (Redrawn from ref. 11, with permission.)

The study of healing ligaments under controlled conditions has been divided into two categories; non-repaired and repaired ligaments. Non-repaired ligaments are allowed to heal without surgical intervention, and as a result scar tissue forms between the opposing ends of the severed ligaments. Macroscopically the repair site appears not to change beyond 6 weeks post-injury. However, changes in histochemical and biomechanical parameters may continue for months or years. Surgically repaired ligaments tend to heal by primary intention due to the close apposition of the severed ends. In reality, however, it has been shown that there is indeed scar tissue formation between the ligament ends in these repairs. The results of the surgical repair hastens healing during the early phase of the repair process, and during this time the biomechanical properties will exceed those of the non-repaired

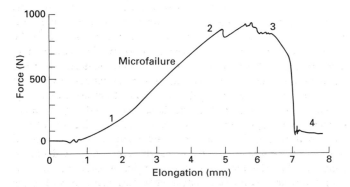

Fig. 3 Nearly 8 mm of joint displacement occurred before the ligament reached complete failure. (Redrawn from ref. 84, with permission.)

ligaments. Following these two groups over the long term demonstrates that there is no difference between the repaired and the non-repaired ligaments. However, it must be recognized that surgically repaired ligament injuries will provide superior results under certain circumstances, such as those injuries involving the anterior cruciate ligaments of the knees.

Although it is apparent that in the final analysis the mode of healing is relatively unimportant, it is convenient to keep the two groups (repaired and non-repaired) separate when evaluating the completeness of the healing process.

In studies detailing untreated ligament injuries, Frank and coworkers discussed many of the measurable parameters for mid-substance medial collateral ligament injuries in rabbits. Histologically, even after 40 weeks of healing ligaments show a disorganized matrix, increased fibroblasts, and abnormal crimp. Biochemically, the significant findings are higher levels of type III collagen, increased collagen turnover, and total collagen approaching only 70 per cent of normal tissue. Peak failure stresses were found to be only 60 per cent of normal, and this value reached a plateau 14 weeks after injury. The typical load–deformation curve compares normal and healed tissue.

Frank *et al.*[12] made the following definitive statement subject to their findings regarding untreated injuries of the medial collateral ligament (Fig. 5).

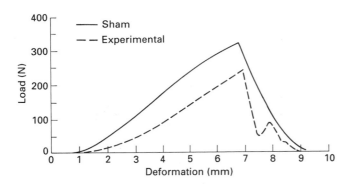

Fig. 5 Biomechanical (structural) results. Typical load deformation diagrams of experimental and sham ligaments from the same 14-week animal. Note that the first millimetre of deformation for these samples represents their 'laxity'. (Redrawn from ref. 13, with permission.)

'Based on the vast difference between the properties of the medial collateral ligament scar and normal ligament tissue, even at forty weeks (without treatment), it is probably safe to assume that the scar will never be mechanically normal and its components possibly never the same as normal medial collateral ligament.'

In situations where surgical repair is indicated, it has been determined that a combination of repair, short-term immobilization, and early remobilization provides for optimum healing. Vailas *et al.*[14] evaluated the healing of surgically repaired medial collateral ligaments in four groups of rats. It was concluded that surgical repair, followed by immobilization for 2 weeks and progressive exercise for 6 weeks, yielded the maximum values for all parameters evaluated. Although this study ended at 8 weeks post-operatively and no long-term follow-up was provided, Vailas *et al.* state that the regimen described 'enhanced the healing process by inducing a more rapid return of tissue DNA, collagen synthesis, and separation force to within normal limits'.

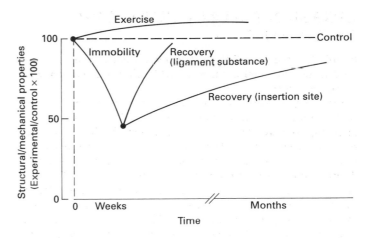

Fig. 6 The relative effects of exercise, immobilization, and remobilization on ligament complexes. (Redrawn from ref. 20, with permission.)

The use of immobilization is an important factor in the healing process regardless of whether surgical repair or non-repair is used. Amiel and coworkers[9,10] have studied the effect of immobilization on collagen turnover. They found that the immobilization of intact collateral ligaments of rabbits resulted in increased collagen turnover, causing decreased stiffness and inferior mechanical properties compared with controlled ligaments. Earlier work by Noyes[17] on primates indicated that, after immobilization for 8 weeks, a decrease of 39 per cent in load to failure could be documented (Fig. 6). Furthermore, Noyes suggested that a 12-month interval after this immobilization was necessary before ligament strength returned to normal. The theoretical curves in Fig. 6 illustrate the relative effects of exercise, immobilization, and remobilization on ligament complexes.[14]

Although these studies were performed on perfectly intact ligaments, it is obvious that the dilatory effects caused by prolonged immobilization must be taken into account when attempting to predict healing rates.

Cabaud *et al.*[18] studied the effects of different treatment regimens on acutely injured anterior cruciate ligaments in dogs and monkeys. It was found that monkeys that had suffered a transected anterior cruciate ligament, which was sutured and then immobilized for 6 weeks, were able to achieve a return of approximately 60 per cent of the pre-injury strength, even with 4 months of rigorous exercise.

In a separate study[19] it has been shown that, independent of the type of treatment, all healing ligaments have the same histological appearance at 6 weeks. By 12 weeks there is a trend towards the cell pattern and collagen fibre orientation resembling normal ligamentous tissue. An associated return of biomechanical properties has been documented. Exercise has the most beneficial effect on mechanical properties during the period 6 to 12 weeks post-injury.

Woo *et al.*[20] have studied the long-term results of different treatments in dogs with isolated medial collateral ligament tears. Treatment ranged from non-repair and normal activity to surgical repair and immobilization for 6 weeks. This study, and others, conclude that the optimum treatment for isolated medial collateral ligament tears was conservative management with early mobilization. More important, however, were the results which demonstrated that even with this ideal management, i.e. early mobilization, the medial collateral ligament regained only 68 per

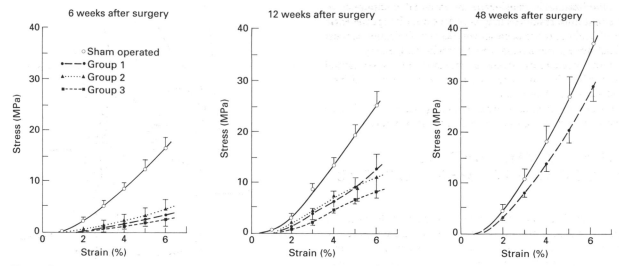

Fig. 7 The mechanical properties of the medial collateral ligament substance from sham-operated controls and experimental knees at 6, 12, and 48 weeks after surgery. (Redrawn from ref. 85, with permission.)

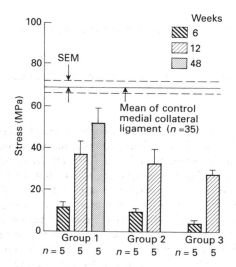

Fig. 8 Tensile strength of medial collateral ligament for all animal groups. The mean ± SEM of 35 sham-operated control medial collateral ligaments are shown. At 13 weeks, significant improvement in tensile strength was seen over that for 6 weeks ($p < 0.01$). All of the experimentals had tensile strengths significantly less than that for the intact controls ($p < 0.01$). At 48 weeks, a continuing increase was noticed for the Group I experimentals, but values were still well below that for the controls. (Redrawn from ref. 85, with permission.)

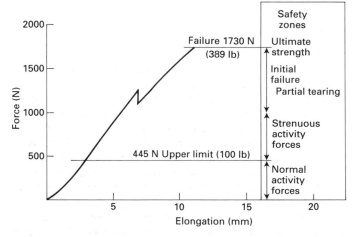

Fig. 9 Hypothetical load-elongation curve for the anterior cruciate ligament/bone unit is shown with the idealized safety zones to be considered in evaluating a ligament replacement. (Redrawn from ref. 86, with permission.)

cent of the mechanical properties of the control at 48 weeks post-injury (Figs. 7 and 8).

Clinical studies differ from controlled studies in that it is difficult to define accurately the degree of injury and completeness of healing that is achieved. As such, clinical studies cannot provide the definitive data offered by experimental investigations. It should also be recognized that a lower quality of healing may be acceptable to the non-athlete, as opposed to very active patients who expose their ligaments to greater stress. The theoretical curve in Fig. 9 was obtained by Noyes.[17] This curve demonstrates that during normal activity the force exerted on the anterior cruciate ligament of the knee is about 30 per cent of the force required to

rupture the intact ligament. It can be assumed that during the process of repair the ligament can withstand functional activities very early in healing. However, ligament stiffness and joint laxity are also important aspects of overall joint function.

SUMMARY OF THE LIGAMENT HEALING PROCESS

A framework has been provided to understand ligament healing. Clinical and experimental data regarding ligament healing can be summarized as follows.

1. Ligament healing follows the basic phases of soft tissue healing, with the strength of the healed ligament being proportional to the collagen content. An important stimulus to healing is progressive stress loading.
2. In long-term studies, both non-repaired and repaired ligaments are capable of achieving equally successful repair.

3. Early mobilization is considered an essential component to achieve optimum results as long as it does not disrupt the early phases of healing. It has been shown that remobilization is most beneficial during the period 6 to 12 weeks post-injury.
4. Maximum mechanical strength regained in experimental healing models does not exceed 70 per cent of a normal ligament at 48 weeks follow-up.
5. Human ligament healing follows the same pattern of healing as in most animal models, but proceeds at a slower rate.

SPRAINS OF THE MEDIAL COLLATERAL LIGAMENT OF THE KNEE

Sprains of the medial collateral ligament complex of the knee are very common. In most surveys, injuries to the ligament are the most frequent of all knee ligament tears. Historically, most medial collateral ligament sprains were treated with forced immobilization, combined with a prolonged programme of non-weight-bearing. More recently, a functional approach has been adapted, leading generally to reduced morbidity and earlier enhanced function.

Basic anatomy

Static

The architecture of the medial collateral ligament of the knee is very interesting (Fig. 10). The ligament is basically constructed of two distinct sheets—the superficial and deep components of the medial collateral ligament complex. These two layers originate from the femur and tibia respectively. The deep layers intimately attach to the medial meniscus along the periphery. Specifically, the medial collateral ligament complex merges into the posterior medial conjoint called the arcuate complex. The more anterior reflection of the medial collateral ligament is invaginated by the retinaculum and the upper portion of the pes anserinus.

The circulation to the medial collateral ligament is provided by a cascade of vessels[21] essentially originating from the superior medial geniculate artery and the inferior medial geniculate. The vessels serving the medial collateral ligament provide the arterial

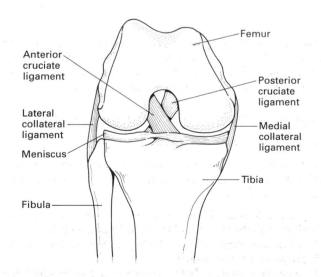

Fig. 10 Medial collateral ligament.

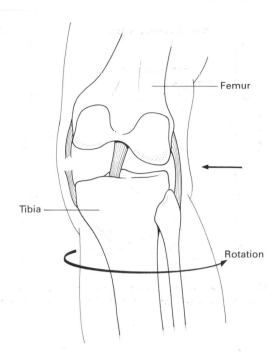

Fig. 11 Complete disruption of medial collateral ligament with valgus force.

supply to the peripheral rim of the medial meniscus, which is fundamental to meniscus repair.

Dynamic

The medial collateral ligament of the knee is vital to normal knee biomechanics, as it is the primary restraint to valgus forces about the knee. Furthermore, when the tibia is externally rotated on the femur, the medial collateral ligament absorbs this force and thus controls a major degree of external rotation. Of course, the medial collateral ligament functions in harmony with the posteromedial complex and the cruciate ligaments in most of these movements.

The medial collateral ligament of the knee is very strong and is able to withstand a force of 100 N before failing in continuity or disrupting grossly. The medial collateral ligament complex becomes strained whenever the knee is forced into a valgus position (Fig. 11).

If the knee is at 0° (straight), the medial collateral ligament shares this valgus stress with the posterior medial structures and the cruciate ligaments. However, once the knee is flexed to 30°, and further to 60°,[22] the medial collateral ligament becomes more responsible for absorbing the valgus stress.

If, or when, the medial collateral ligament fails completely under valgus load, the anterior cruciate ligament will be the next structure to absorb the further challenge of the continuing stress. It is vital to understand that, as the knee fails in valgus, the tibia rotates externally on the femur which is also controlled by the medial collateral ligament and the posterior medial complex. Failure of the medial collateral ligament will allow excessive, and thus pathological, external rotation between the tibial and femoral surfaces.

Mechanism of injury to the medial collateral ligament

The most frequent mechanism of injury of the medial collateral ligament complex is a forced valgus injury to the knee.[23] Typically, this trauma occurs when the athlete is hit from the lateral side upon a fixed foot and the knee is driven medially. Obviously, the injury may occur in any sporting activity; however, contact sports such as soccer, rugby, ice hockey, and football are notorious.

Sprains of the medial collateral ligament follow the traditional classification of ligament injuries and are graded as follows.

Grade I—a microscopic structural injury to the ligament without gross disruption and manifesting a firm end-point on stressing;

Grade II—a partial macroscopic disruption of the ligament demonstrating obvious injury on inspection, and although an end-point is apparent on stressing, it is softer and less distinct than in grade I;

Grade III—a complete gross disruption of the ligament with loss of integrity on stressing, manifesting a lack of end-point.

The medial collateral ligament may be the sole victim in the valgus/external rotation type of injury. However, it is important to understand that the force may continue after the medial collateral ligament fails completely (grade III injury). If the force is not totally dissipated while tearing the medial collateral ligament, then it may continue and disrupt the anterior cruciate ligament (Fig. 12).

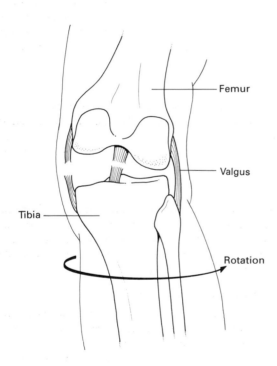

Fig. 12 Complete disruption of medial collateral ligament and anterior cruciate ligament with valgus/external rotation force.

Clinically, this type of injury occurs commonly and results in marked knee instability with a completely disrupted medial collateral ligament and anterior cruciate ligament, and peripheral detachment of the medial meniscus. Far less frequently, complete knee dislocation may occur if the force is so great that it disrupts the medial collateral ligament together with the anterior and posterior cruciate ligaments. Although such a situation is rare, it is very serious.

Signs and symptoms of sprains of the medial collateral ligament

As mentioned previously, sprains of the medial collateral ligament are the most commonly seen knee ligament injuries. These medial collateral disruptions, or sprains, generally are grade I or grade II. As with all ligament sprains which are not complete (grade III insults), the knee joint remains stable but has distinct clinical manifestations.

Grade I sprains

Typically, the athlete will present with a history of having been struck on the lateral side of the knee with the knee straight (or bent slightly) and fixed. Frequently, the injury occurs so quickly that the patient/athlete has only a vague recollection of the specifics of the incident. Of course, pain is immediate and, depending on the degree of ligament disruption, may be modest or quite marked. Swelling is usually mild and without major haemarthrosis. Usually the athlete will immediately cease sporting activities, but there are exceptions depending on the constitution of the athlete, whose pain threshold may be very high. Clinical examination to determine whether the injury is acute, subacute, or chronic is very helpful.

In the acute phase (day 1 to day 10) the knee is usually held in a position of mild flexion. The bony prominences of the knee are somewhat obscured by the swelling which is evident medially. There may be a mild sympathetic effusion within the knee joint but, as stated previously, a dramatic haemarthrosis is rare. The knee is warm to touch medially; however, the remainder of the knee can be unusually normal to palpation.

On physical stressing of the knee, the examiner must isolate the medial collateral ligament and then apply gentle force. This manoeuvre is designed to mimic the injury mechanism. The physical examination is best conducted with the knee in 45° of flexion, with the heel resting on the examining table. With the grade I injury, the medial collateral ligament demonstrates a firm and distinct end-point to a valgus stress applies at this angle (Fig. 13).

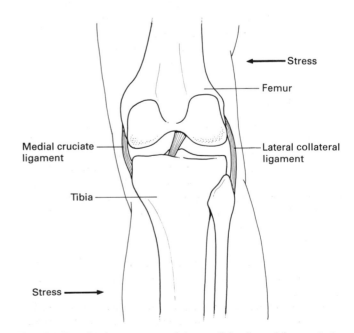

Fig. 13 Examination to stress of the medial collateral ligament at 45° flexion.

Grade II sprains

A moderate sprain of the medial collateral ligament of the knee is termed a grade II sprain of the ligament, resulting in increased laxity of the joint. However, when the valgus stress test is applied to the knee at 30 to 45° of flexion, the end-point is softer but is still present. A word of caution is necessary. It is essential to examine the opposite normal knee initially to gain valuable information as to the character and quality of the normal ligaments. Each individual is surprisingly different in the quality of their tissues and their inherent laxity.

With the grade II sprain, there is usually a more severe degree of ecchymosis and swelling than with the grade I insult, and the knee joint may demonstrate significant effusion. Pain is obviously greater than with the more minor grade I sprain.

Grade III sprains

A severe and complete sprain of the medial collateral ligament of the knee is a serious injury. At the point of complete disruption, the applied force has overwhelmed both the tensile strength of the medial collateral ligament and associated restraints such as the anterior cruciate ligaments.

The patient/athlete usually has very severe pain, gross swelling of the soft tissues around the knee, and immediate profound disability. Indeed, these athletes are unable to finish the game! Remarkably, the knee joint itself may not have a haemarthrosis. With disruption of the capsule which is intimate with the medial collateral ligament, the blood commonly evacuates itself from the joint, reducing the patient's discomfort. The physical examination demonstrates that the medial collateral ligament is completely torn, with widening of the joint on valgus stressing and a very indistinct end-point. In order to conduct this examination in the acute phase, it is critical to have the patient relaxed or sedated. Muscle spasm and apprehension may thwart an effective and satisfactory evaluation.

In more chronic circumstances, with a complete medial collateral ligament tear, the patient will usually demonstrate clinical signs of medial collateral ligament and anterior cruciate ligament insufficiency which will be discussed in the next section. Pure grade III instability of the medial collateral ligament without an associated anterior cruciate ligament tear is very unusual.[24] Invariably, the patient will show medial opening on valgus testing, but will have very little, if any, functional disability.

Clinical mimes

Frequently, injuries to the medial collateral ligament of the knee may be confused with other forms of knee pathology. Patients with dislocations or subluxations of the patella commonly present with a history of receiving a valgus force to the knee. Subsequently the knee is swollen on the medial side and can be irritable to valgus stressing; furthermore, the injury usually occurs in the active population. Palpation of the joint, performed carefully, will provide helpful clues to differentiating the medial collateral sprain from patella instability.

1. Tenderness will be over the medial retinaculum adjacent to the patella rather than over the anatomical course of the medial collateral ligament.
2. The apprehension test applied to the patella will be positive. This test is performed with lateral pressure on the patella, directed laterally, with the knee held passively at 45° of flexion.
3. There is no joint gapping with patella dislocation, in contrast

with grade II and grade III sprains to the medial collateral ligament.

Tears of the medial meniscus may occur in isolation or in combination with sprains of the medial collateral ligament. Isolated medial meniscus injuries are normally a consequence of a low energy injury such as rotation in a squatting position, in contrast with medial collateral ligament disruption which demands significant force.

Once again palpation and stressing of the joint will provide the most valuable information. Joint line tenderness is usual with tears of the medial meniscus but not with medial collateral ligament tears. With isolated injuries to the medial meniscus there is some, albeit mild, apprehension to valgus stressing of the knee.

Treatment

Grade I tear of the medial collateral ligament

Because of the relative stability of the knee joint, extensive clinical experience has dictated that the grade I injury should be treated conservatively with emphasis on a functional approach.[25]

The natural history of this problem is quite predictable. The patient usually suffers a modest degree of discomfort for 1 to 2 weeks, gradually resolving over 4 to 6 weeks. The athlete is able to return to full activities when the knee is painless. However, the athlete may experience mild discomfort with acute unexpected valgus. This is not unusual and may linger on for several months.

During the acute phase of injury the following measures will prove helpful.

Acute phase: day 1–14

1. Ice to the knee four times daily for 20 to 30 min.
2. A firm elastic wrap to the knee, allowing range of motion within the limits of pain.
3. A cane or crutches to allow partial weight bearing within the limits of pain.

Subacute phase: day 14–28

1. A gradual increase in the range of motion and weight bearing.
2. Progressive strengthening with organized physiotherapy to enhance the strength of the quadriceps and hamstring mechanism. Electrical muscle stimulation may prove helpful during this phase.

Chronic phase: after 28 days

1. Emphasis is placed on strength and power training specifically for the quadriceps and hamstring mechanisms, combined with the complementary muscle groups.
2. Sports-specific training, as well as more vigorous controlled challenges such as skipping, running, slalom, and eccentric loading.
3. The intermittent use of a knee brace to control valgus may be helpful in this phase to facilitate return to sports.

Grade II sprains of the medial collateral ligament

The treatment plan for medial collateral ligament tears follows the general format described for grade I disruptions. However, the recovery period will usually be longer—upwards of 3 months before a full return to sporting activities. Supportive knee bracing can be used in the latter stages of rehabilitation to support the healing medial collateral ligament. Orthotic bracing may also enhance proprioception and allow the athlete enhanced security

Fig. 14 A knee orthosis to provide valgus stability.

(Fig. 14). The athlete must continue his or her exercises for many months as muscle function can be extremely retarded, particularly in chronic injury which is assessed late by the sports medicine practitioner, the older athlete, and the more severe grade II tears.

Grade III medial collateral ligament sprains

Historically, complete tears of the medial collateral ligament were managed with plaster immobilization for 6 weeks, followed by rigorous and lengthy physiotherapy. The disaster of prolonged immobilization was quite predictable and led to severe muscle atrophy, which was sometimes irreversible. Furthermore, the medial collateral ligament would heal in a lengthened state, not dissimilar from the pretreatment situation.

To make matters more complex, the grade III tear is frequently associated with a complete disruption of the anterior cruciate ligament. The history of this combined instability is very perplexing, commonly leaving the athlete with a chronic knee instability which is most disabling.

Recent advances in our understanding of grade III tears of the medial collateral ligament can be summarized as follows.

1. If the grade III tear of the medial collateral ligament occurs in isolation, which is rare, it should be treated with pain control combined with functional bracing and finally aggressive rehabilitation.
2. The recovery period to full athletic activities can be upwards of 20 weeks after an isolated grade III sprain.
3. If a medial collateral grade III injury occurs with a complete disruption of the anterior cruciate ligament, surgical repair with reconstruction of the anterior cruciate ligament is

commonly necessary in order to reduce morbidity and produce a more complete restoration of function.

Summary

The medial collateral ligament complex of the knee is commonly injured in athletics as a consequence of a valgus/external rotation force. Usually the ligament injury is incomplete and can be treated with early motion and strengthening. Prolonged immobilization can lead to impaired healing and increased morbidity.

The complete grade III tear of the medial collateral ligament is invariably associated with a partial or complete tear of the anterior cruciate ligament. The combination of complete tears of the anterior cruciate ligament and medial collateral ligament usually demand surgical repair and/or reconstruction in order to restore knee joint stability.

SPRAINS OF THE ANTERIOR CRUCIATE LIGAMENT

Introduction

Over the past decade injuries to the anterior cruciate ligament have received more attention than any other athletic injury.[26–30] This overwhelming attention has been generated by the high prevalence of anterior cruciate ligament tears in athletes, coupled with the associated morbidity. An athlete with anterior cruciate ligament instability is frequently unable to re-establish the pre-injury performance and may never be able to return to sport.

Much has been written about the best technique for treatment of anterior cruciate ligament tears.[26,29,31–34] However, the results have frequently proved disappointing. Many frustrated surgeons have returned to a less surgical approach, emphasizing exercise and bracing as the cornerstones of definitive treatment.

The anterior cruciate ligament is the fulcrum of knee stability. Complete disruption of this ligament leaves the athlete with considerable functional instability. Rotary instability of the knee ensues, exposing the adjacent supporting ligaments and menisci to further degeneration.

Anatomy of the anterior cruciate ligament

The anterior cruciate ligament is located in the centre of the knee joint, intertwined with its partner the posterior cruciate ligament (Fig. 15). These structures are covered with a synovial sheath, except for the posterior aspect of the posterior cruciate ligament which is essentially extrasynovial. The anterior cruciate ligament is approximately 3 cm long and takes its origin from the superior lateral bony extreme of the intercondylar notch. Some fibres at the origin of the anterior cruciate ligament actually roll around the lateral femoral condyle, in the position referred to as 'over the top'.[11,17,18,36,37]

From the bony origin, the anterior cruciate ligament of the knee traverses distally and medially to attach along the interspinous region of the tibia. The insertion of the anterior cruciate ligament of the tibia may cover an area as long as 3 to 5 cm between the tibial spines.

According to Girgis and other workers,[35,38] the anterior cruciate ligament is constructed of two distinct bands, the anteromedial band and the posterolateral band. These portions of the anterior cruciate ligament are intimately attached and are not always readily distinguished on gross inspection.

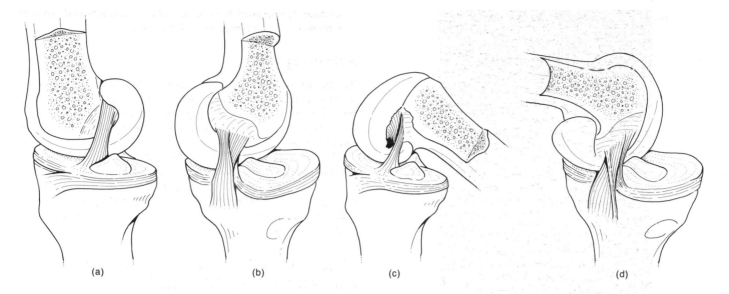

Fig. 15 (a) The anterior cruciate ligament with the knee in full extension; (b), the posterior cruciate ligament with the knee in full extension; (c) the anterior cruciate ligament with the knee in flexion (the arrow indicates the anteromedial band); (d) the posterior cruciate ligament with the knee in flexion (most of the ligament is taut, as a small posterior portion loosens). (Redrawn from ref. 35, with permission.)

The nerve supply and circulation of the anterior cruciate ligament have been discussed in the literature.[39] However, from a clinical standpoint it is vital to understand that the network of vessels permeating the anterior cruciate ligament essentially originates from the synovium and the bone–ligament interface. The intraligament blood circulation[40–42] of the anterior cruciate ligament is rather scanty, which parallels the very poor ability of the anterior cruciate ligament to heal itself.

The medial and lateral menisci,[43,44] as well as the posterior cruciate ligament, must be viewed as operating in concert with the anterior cruciate ligament. The mechanics of this intimate arrangement will be discussed in the next section.

The anatomy of the anterior cruciate ligament can be summarized as follows:

1. It originates from a broad base on the medial side of the lateral femoral condyle and traverses to insert in the intraspinus area of the tibia;
2. It is very strong (withstands a force of 1700 N) and is approximately 3 cm long;
3. Its blood supply is very meagre, and hence its ability to heal, if injured, is rather poor.

The dynamic anterior cruciate ligament

Major reviews of the biomechanics of the anterior cruciate ligament of the knee have been written. From a clinical standpoint, the basics of its behaviour must be understood.

The anterior cruciate ligament is the major restraint controlling anterior translation of the tibia on the femur. This translation calls upon secondary restraints to assist the anterior cruciate ligament in this role; specifically the posterior cruciate ligament and the collateral ligaments constrain the anterior movement of the tibial surface as it slides forward on the femoral condyles.[45]

Furthermore, the anterior cruciate ligament is able to twist upon itself as the knee moves from complete flexion to full extension (Fig. 16). With the two bands intertwined (anteromedial and

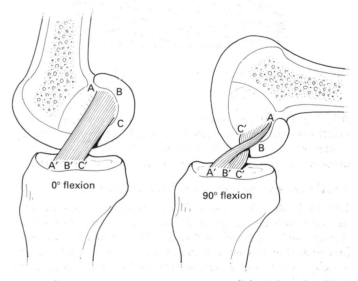

Fig. 16 In flexion there is lengthening of the anteromedial band of the anterior cruciate ligament A′ and shortening of the posterior lateral aspect of the anterior cruciate ligament C (C′). B(B′) represents the intermediate band, which in fact is the transition between the anteromedial and the posterior lateral portion. (Redrawn from ref. 35, with permission.)

posterolateral), a portion of the anterior cruciate ligament remains taut throughout this complete range of motion. The rotation of the femur and tibia is controlled by the anterior cruciate ligament, working in synchrony with the posterior cruciate ligament, collateral ligaments, and menisci.

The anterior cruciate ligament possesses very little inherent elasticity. Application of a force straining it by more than 5 per cent of its resting length will result in rupture. This rupture may be complete and obvious on gross inspection, or it may be partial, demonstrating failure in continuities. The anterior cruciate ligament is able to resist a force of 1700 N before failure.[46–49]

Mechanisms of injury of the anterior cruciate ligament

The anterior cruciate ligament is disrupted most frequently as a consequence of a forced valgus stress to the knee. This force, with a blow to the outer aspect of the knee, initially disrupts the medial collateral ligament and then tears the anterior cruciate ligament as the second component of the injury. The third component of the injury complex is detachment of the medial meniscus. Obviously, this injury is quite common in soccer, American football, and rugby, as tackles from the side are intrinsic to these activities. Isolated tears of the anterior cruciate ligament are believed to be relatively rare, and some authors have debated as to whether this injury could occur without associated soft tissue disruption.

Isolated tears of the anterior cruciate ligament can occur in sports or as a result of an injury in the work-place. The knee is usually fully extended when the insult occurs. With the foot firmly planted, the athlete rotates or changes direction about the fixed foot (Fig. 17). If this unusual stress continues, the femur continues to rotate externally over the tibia at which time the anterior cruciate ligament will disrupt. Eventually, the posterolateral capsule and lateral collateral ligament will also be stretched and injured. In athletics, this is the most common mechanism of injury to the anterior cruciate ligament. Furthermore, the medial and/or lateral meniscus may be damaged during the same manoeuvre.

Fig. 17 An isolated tear of the anterior cruciate ligament occurs with the knee in full extension, coupled with forced internal rotation of the tibia on femur.

In combination with the posterior cruciate ligament, the anterior cruciate ligament may be completely disrupted with an insult of hyperextension to the knee. This may occur when the athlete is clipped from behind, which is an illegal technique in most sports.

In summary, the anterior cruciate ligament is most frequently injured in sport as a consequence of a forced external rotation of the femur on the fixed tibia with the knee in full extension. A complete tear of the anterior cruciate ligament necessitates an insult of high energy (1700 N), which commonly occurs in high velocity sporting activities.

The pathology of the anterior cruciate ligament may be a complete disruption (grade III sprain) or a more moderate sprain (grade I or grade II).

Signs and symptoms of sprains of the anterior cruciate ligament

The athlete suffering an injury to the anterior cruciate ligament has invariably absorbed a blow of significant violence, e.g. tackled from the side or twisted while skiing. The patient generally has significant, if not severe, pain and usually falls to the ground immediately. When the anterior cruciate ligament is completely disrupted, a snap or pop is frequently audible and can sometimes be heard by the spectators. The athlete is rarely able to continue and retreats to the sidelines with assistance. The knee becomes swollen very rapidly, which suggests active bleeding, in contrast with a simple effusion which usually takes hours to accumulate. As the knee continues to distend with blood, the injured athlete complains of increasing pain compatible with the degree of swelling. Even with a chronic injury, the patient/athlete is frequently able to recollect the specifics of the initial injury because the severity of the pain makes the experience quite indelible.

Occasionally, the athlete will describe a lesser insult with a minor feeling of instability, followed by a further twist to the knee, triggering the snap, followed by acute pain and swelling. This scenario suggests that a partial tear of the anterior cruciate ligament occurred with the first insult which became a complete disruption with the subsequent trauma.

Consistently, an athlete with an injured anterior cruciate ligament will describe a high energy twist to the knee, provoking a rather sickening feeling of the knee 'going one way and my body going the other'. A direct blow to the knee very rarely, if ever, injures the anterior cruciate ligament . It is usually a trauma secondary to rotation, with or without body contact.

The physical examination of the knee must be performed very carefully. If a haemarthrosis is evident, the examination will be difficult and frequently inconclusive. Aspiration of the knee to remove the haemarthrosis is appropriate. Furthermore, it may be necessary to sedate the patient to examine the extremity adequately. The patient/athlete with muscle spasm or anxiety renders the situation very tense for all concerned. This predicament can be rectified by mild sedation or general anaesthesia. The leg/knee should be unencumbered with bandages or a tourniquet, which may disguise subtle findings. In more chronic cases, inspection for muscle wasting, knee sag, or bony abnormalities is essential.

Palpation is the most important part of the physical examination of the injured knee and may provide the most useful clues to the diagnosis. It must be remembered that the knee has been injured and that movement will provoke pain and anxiety. The examiner can comfort the patient and quell further apprehension by initially examining the uninjured extremity. Palpation of the injured knee should follow a very specific routine.

1. After a complete history has been taken, the patient should point out the area which is most tender.
2. The examiner should avoid areas of acute tenderness until the final stages of examination. (Areas of distinct tenderness generally determine the underlying tissue damage.)

3. Palpation of the medial collateral ligament, anteromedial corner, lateral collateral ligament, posterolateral corner, and adjacent retinaculum offers clues as to which structures or combination of structures have been injured.

4. The presence or absence of an effusion or haemarthrosis should be noted. The absence of an effusion does not necessarily mean that a major ligament sprain has not occurred. If there is a tear in the capsule, the haemarthrosis may have escaped to the surrounding tissues.

5. Determining the range of motion of the knee is not helpful at this point, as restricted range of motion is predictable in view of the trauma.

6. The integrity of the medial collateral ligament and the lateral collateral ligament should be examined carefully with the knee at 30° flexion (Fig. 18). Gently springing the knee into valgus at 30° flexion will assess the integrity of the medial collateral ligament, and a varus manoeuvre can help determine the status of the lateral collateral ligament.

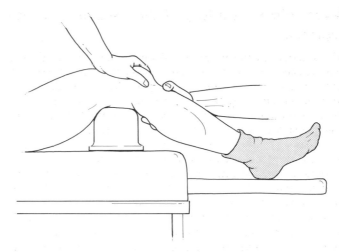

Fig. 19. The Lachman test: the knee is flexed at 30° and, with the femur stabilized, the tibia is drawn forward. The degree of instability is based on the abnormal movement compared with the contralateral extremity.

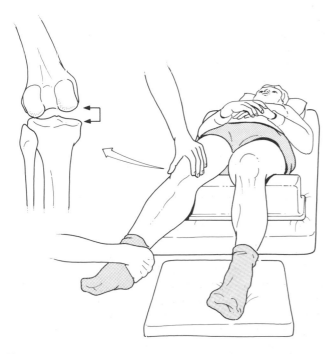

Fig. 18 The valgus stress test: the knee is held at 30° of flexion and the tibia is pulled laterally as the femur is stabilized with the opposite hand. This assesses the integrity of the medial collateral ligament.

7. The integrity of the anterior cruciate ligament is analysed using the anterior drawer manoeuvre, the Lachman test, or the pivot shift manoeuvre as described below. The anterior drawer manoeuvre is performed at 90° knee flexion with a gentle pressure behind the tibial plateau, drawing the tibia forward on the femur. This technique must be performed with a progressive full rather than aggressive tugging because reflex hamstring spasm will limit its value. (NB When the anterior cruciate ligament has been disrupted completely but the collateral ligaments and the menisci are intact, the anterior drawer manoeuvre will frequently be negative.)

The Lachman test is conducted at 30° of flexion with a stress identical with that used in the anterior drawer manoeuvre (Fig. 19). This stress, directed from behind the upper tibia, is designed to slide the tibia forward on the femur. The opposite hand of the examiner stabilizes the femur firmly as the stress is applied to the tibia. A positive test, compared with the opposite knee, is determined by an abnormal excursion of the tibia upon the distal femur. A minimal increase in the anteroposterior movement (less than 10 mm) is considered a mild grade I instability. A moderate grade II anterior cruciate ligament instability exists when the excursion is approximately 10 to 15 mm, and a major anteroposterior instability (grade III) offers abnormal movement of greater than 15 mm. With this degree of instability, a very soft end-point occurs on testing. The specificity of the Lachman manoeuvre[50] in determining the integrity of the anterior cruciate ligament is very high. At least 98 per cent of all complete anterior cruciate ligament disruptions will be obvious on utilizing the Lachman test. However, tight hamstrings, an anxious patient, or a joint effusion may thwart the accuracy of this examination.

The pivot shift manoeuvre[51] is difficult to perform properly; however, it is a precise test for determining the quality of the anterior cruciate ligament (Fig. 20). It demands very precise handling of the patient, particularly after the acute injury. The patient must be relaxed and without hamstring spasm to comply with the test.

The pivot shift stress test is designed to mimic the mechanism of the injury which produces an isolated anterior cruciate tear. The examiner attempts to demonstrate abnormal movement of the lateral tibial plateau on the fixed femur. With the knee in full extension, the examiner stabilizes the femur and then forcibly internally rotates the tibia on the femur. While maintaining pressure, the examiner firmly places a valgus stress on the knee while gradually flexing to 30°. Abnormal displacement of the tibia on the femur occurs when the knee is in full extension under valgus load with internal rotation. The tibia then returns to its normal position on the femur as the knee is flexed to 30°. This abnormal excursion on the tibia is seen as an abnormal slide, depending on the degree of instability and the chronicity of the anterior cruciate ligament tear.

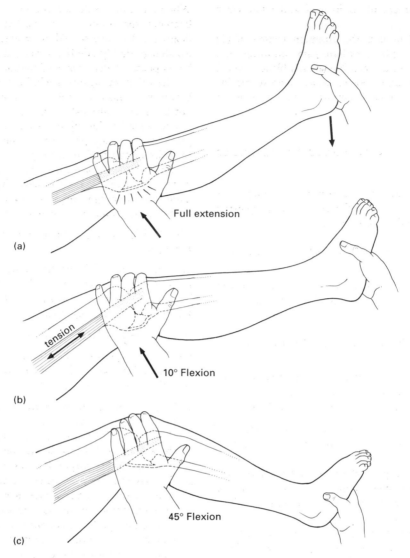

Fig. 20 The pivot shift phenomenon: the abnormal subluxation of the lateral tibial plateau occurs in full extension with further flexion, the iliotibial band loosens, and the lateral tibial plateau reduces to its normal anatomical position.

This test is the most accurate in determining the integrity of the anterior cruciate ligament; however, considerable skill and experience is required when testing knees with instability.

Differential diagnosis

Confusion can occur when the patient/athlete presents with a history of having incurred a high energy injury to the knee, resulting in immediate pain, swelling, and profound disability. The patient will often state that the knee 'gave way' at the time of the trauma. Other causes of acute knee swelling and pain should be considered.

Patella dislocation/subluxation Patella dislocation may provoke immediate swelling or haemarthrosis of the knee. During dislocation, the patella may injure the surface of the lateral femoral condyle, resulting in an osteochondral fracture. The resulting haemarthrosis will invariably be fat laden and will frequently recur, even after aspiration. Fat globules are rare in anterior cruciate disruption, except if an avulsion of the tibial spine has occur-

red. Furthermore, the haemarthrosis does not generally recur once aspirated from the knee joint.

Injuries to the menisci Detachment of the menisci from their peripheral rim can provoke an acute haemorrhage into the knee. This injury usually occurs in conjunction with a major insult to the collateral ligaments, but can exist in isolation. If the physical examination reveals stability of the ligaments, then the meniscal tear or detachment, as confirmed by clinical deduction, arthroscopy, or magnetic resonance imaging (MRI), can be treated expectantly.

Osteochondral fractures Acute osteochondral fractures, or osteochondritis dissecans, can occur in isolation or in conjunction with an acute ligament sprain. An acute osteochondral fracture, usually of the femur, occurs as a consequence of major trauma and can provoke a major knee haemorrhage. The diagnosis may be difficult to make from strictly a clinical standpoint and commonly necessitates arthroscopic examination, CT, or MRI assessment. If the fragment is large, the prudent approach necessitates surgery to secure the osteochondral fragment to its bed and allow early movement thereafter.

These clinical mimes all possess the common factors of (1) provocation by major trauma, (2) immediate pain and incapacity, (3) rapid accumulation of joint fluid which is usually blood, and (4) requirement of examination under anaesthesia, arthroscopic inspection, or specialized radiological examination to delineate the pathology accurately.

Treatment of injuries to the anterior cruciate ligament

The treatment for the disruption of the anterior cruciate ligament is very controversial. However, several features of this perplexing problem have gained international consensus.

Grade I and grade II sprains of the anterior cruciate ligament

Mild and moderate sprains of the anterior cruciate ligament may occur alone or in combination with sprains of the collateral ligaments.

Isolated grade I and grade II sprains of the anterior cruciate ligament

The history of these injuries is benign. Both these sprains are partial disruptions of the anterior cruciate ligament, leaving the knee intrinsically stable. The treatment protocol is as follows.

1. Establish the diagnosis clinically and be prepared to confirm it with examination under anaesthesia combined with arthroscopy.
2. When the patient has acute pain, the following measures will prove helpful. Aspirate the haemarthrosis and then apply ice packs directly to the knee; follow this with partial weight bearing as well as oral analgesia.
3. After 48 h on the above programme, or when the acute episode starts to resolve, support the knee with a wrap or a brace. These measures enhance the confidence of the athlete and facilitate the rehabilitation process.
4. After a week, and as the swelling and pain resolve, start the process of increasing the stress to the knee. This part of the rehabilitation, which emphasizes strength and power training, should be closely monitored by a trained therapist.
5. As the strength and power of the quadriceps muscle improves, the next phase of the rehabilitation is directed towards skill acquisition—running, changing direction, and jumping. Less strenuous skills such as swimming and cycling should be accomplished before a more aggressive programme is adopted.
6. Unrestricted return to sports can occur when the athlete is able to perform all sports specific skills in a satisfactory fashion, as judged by the therapist, the coach, and in some circumstances the parent.

Even with a partial tear of the anterior cruciate ligament (grade I or grade II) the recovery period may be very lengthy, sometimes approaching 20 weeks. Prolonged immobilization after the acute injury, an anxious patient or parent, and an inaccurate diagnosis should always be considered if the rehabilitation process is slow.

Grade I and grade II sprains in the anterior cruciate ligament combined with a grade III sprain of a collateral ligament

When an athlete suffers a complete tear of the medial collateral ligament, for instance, and a partial tear of the anterior cruciate ligament, the knee should be treated in the same fashion as the isolated grade I/grade II anterior cruciate ligament sprain. It must be emphasized that this combination of sprains is rare.

The collateral ligament will heal, albeit in a lengthened state; however, the history of this injury is benign in as much as it should not be treated surgically. Likewise, prolonged immobilization is unnecessary for this injury, as mandatory casting will predictable produce tissue atrophy. This atrophy may be irreversible and recalcitrant to physical therapy.

A complete tear of the lateral collateral ligament with an incomplete tear of the anterior cruciate ligament is most uncommon. Thus, a thorough physical examination, with or without anaesthesia, should be conducted to confirm the diagnosis.

Grade III complete sprains of the anterior cruciate ligament

A complete tear of the anterior cruciate ligament is a very serious injury and there is international debate regarding ideal treatment. The following is a personal protocol for the treatment of a grade III anterior cruciate ligament tear.

Isolated grade III anterior cruciate ligament sprains

Once the diagnosis is established by physical examination and confirmed by examination under anaesthesia and/or arthroscopy, the athlete is categorized as follows.

1. The immature athlete with open epiphysis is treated nonsurgically, according to the treatment protocol advocated for the grade I/II anterior cruciate ligament sprain. However, if there is an avulsion of the ligament, one would be tempted to treat this injury surgically using a transarthroscopic approach.
2. The athlete who, regardless of age, wishes to continue with sports participation is deemed a surgical candidate. Surgery for reconstructing the anterior cruciate ligament generally utilizes a biological graft. This graft can be harvested from the semitendinosis/gracillis, the iliotibial band, or the patellar tendon. The postoperative programme is identical with the physiotherapy designed for the grade I/grade II anterior cruciate ligament sprain.
3. The purely recreational athlete with modest sport-related expectations is treated with knee bracing, early weight bearing, and progressive strengthening. This patient/athlete can return to the sport of choice when her or she personally feels able, on the basis of objective input from the physiotherapist. If such a patient feels that his or her knee disability is profound, surgical reconstruction of the anterior cruciate ligament is the treatment of choice.

There are a number of controversies surrounding the treatment of anterior cruciate sprains.[28,30,58-60]

1. Direct surgical repair of the torn anterior cruciate ligament should not be the sole method of restoring anterior cruciate ligament integrity. If the anterior cruciate ligament is avulsed with a fragment of the tibia or femur, then securing the bony fragment is accepted as a legitimate surgical measure.
2. Knee bracing provides a valuable aid to the patient taking part in a sport with an anterior cruciate ligament tear.[61] Most knee braces will control approximately 40 per cent of the anterior cruciate instability (Fig. 21).
3. Physiotherapy following reconstruction of the anterior cruciate ligament is a vital part of the rehabilitation process. Early

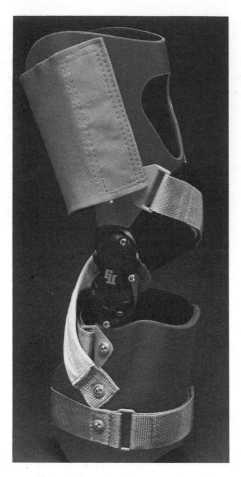

Fig. 21 Custom-made derotation brace designed to control anterior cruciate instability.

mobilization with a progressive increase in the stress to the anterior cruciate ligament reconstruction is fundamental. Current difficulties following anterior cruciate reconstructive surgery involve the knee's becoming stiff and contracted, rather than recurrence of the instability.

Grade III sprains of the anterior cruciate ligament in combination with sprains of the collateral ligaments

Although the anterior cruciate ligament commonly disrupts in isolation, this injury may be accompanied by sprains of the medial collateral ligament, the lateral collateral ligament, and/or the posterior cruciate ligament. When an athlete presents with an acute or chronic complete disruption of the medial collateral and anterior cruciate ligaments, reconstruction of the latter with repair or reconstruction of the former is most appropriate. The medial collateral ligament acts as a major stress shield for the anterior cruciate ligament, which is thus valuable to reconstruction of the anterior cruciate ligament.

In the sedentary individual, functional cast bracing after the acute injury may be sufficient to facilitate rehabilitation.[54] A knee brace may be required to provide functional stability during more challenging activities.

Complete sprains of the lateral collateral ligament of the knee usually occur with disruption of the anterior cruciate ligament and frequently the posterior cruciate ligament. This circumstance requires surgical repair and reconstruction of all structures when possible. The posterolateral capsule and support structures must be reconstructed to thwart posterolateral instability. However, it must be remembered that the anterior cruciate ligament remains the cornerstone of the knee, and thus usually commands surgical attention if completely disrupted in combination with its neighbours the medial collateral ligament, lateral collateral ligament, and/or posterior cruciate ligament.[27–29,30,32,62–66]

POSTERIOR CRUCIATE LIGAMENT

As stated previously, the knee is one of the most complex joints within the human body. Considerable demands are placed upon the knee during everyday activities, particularly in view of our bipedal gait. By virtue of its bony construction, the knee is unstable, but it is supported by a number of muscles, ligaments, and cartilages that hold the osseous structures in proper alignment throughout all phases of movement. To reiterate, these structures act in synergy as a unit, but their functions can be separated out (to a certain degree) during clinical examination.

We shall discuss the posterior cruciate ligament in isolation, but its dependence on the other soft tissue structures must be remembered.

Anatomy

The posterior cruciate ligament arises from the tibial plateau in a rather narrow attachment in the most posterior aspect of the intraarticular region, essentially in a groove in the posterior aspect of the tibia well below the joint line. From there, it travels upwards, forwards, and medially, and fans out to a wide insertion on the lateral surface of the medial femoral condyle, i.e. the medial aspect of the intercondylar notch (Fig. 22).

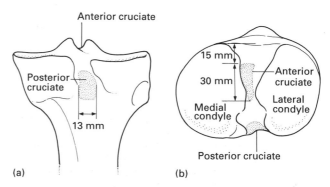

Fig. 22 Drawing of (a) the posterior upper portion of the tibia and (b) the superior surface of the tibia to demonstrate the average measurements and relationships of the tibial attachments of the anterior and posterior cruciate ligaments. (Redrawn from ref. 35, with permission.)

The average length of the posterior cruciate ligament has been determined to be 38 mm, and its width is 13 mm although it fans out proximally to 32 mm[35] at its insertion on the medial femoral condyle. Most of its blood is from the medial geniculate artery, branches of which reach the posterior cruciate ligament via the synovial membrane which surrounds it.[67]

The femoral attachment of the posterior cruciate ligament is in the form of a semicircle; the proximal edge is straight and perpendicular to the long axis of the femur, while the distal edge is convex and lies in the transverse plane. By virtue of the position of its attachments, the posterior cruciate ligament can be separated into two functional units which blend into one another and are not anatomically separable. These are simply called the anterior and posterior portions of the posterior cruciate. Both sections arise from the rather narrow tibial attachment, and as the ligament fans out the anterior portion inserts into the distal convex part of the femoral attachment while the much smaller posterior portion extends to the upper limit of the femoral insertion (Fig. 23).

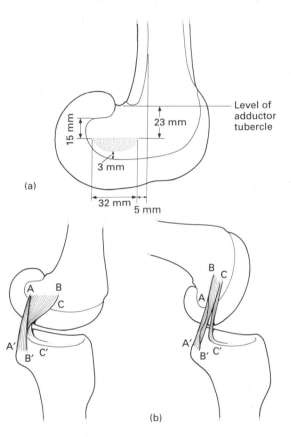

Fig. 23 Structure and attachments of posterior cruciate ligament. (a) Lateral surface of medial condyle of femur showing average measurements and relationships of femoral attachment of posterior cruciate ligament (shaded area). (b) Change in shape and tension of posterior cruciate ligament components in extension and flexion. With flexion there is tightening of bulk of ligament, B–B', but less tension on small band, A–A'. C–C' is ligament of Humphry attached to lateral meniscus. (Redrawn from ref. 35, with permission.)

In 70 per cent of knees a band of fibres extends from the posterior cruciate ligament to the posterior horn of the lateral meniscus; this structure is known as the anterior or posterior meniscofemoral ligament, depending on its relationship to the posterior cruciate ligament as it passes to the meniscus. It is also known as the ligament of Humphrey or the ligament of Wrisberg. Only rarely will both structures be present in the same knee, but occasionally one may be quite robust and well developed.[68] In these cases, they may actually be strong enough to mask classic findings on clinical examination.

Biomechanics

The posterior cruciate ligament works in concert with the anterior cruciate ligament (and other soft tissue structures) to stabilize the knee. The bulk of the fibres of the posterior cruciate ligament (chiefly its anterior portion) are taut when the knee is fully flexed and relaxed as the knee is extended, while the majority of the fibres of the anterior cruciate ligament are loose in flexion and taut in full extension[35] (Fig. 24).

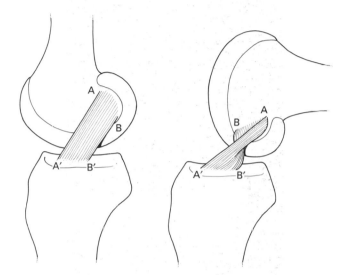

Fig. 24 Drawing presenting changes in shape and relative tension of the anterior (A–A') and posterior portions (B–B') of the anterior cruciate ligament in extension and 90° of flexion. (Redrawn from ref. 35, with permission.)

The main function of the posterior cruciate ligament is to resist posterior displacement of the tibia on the femur in the flexed knee—it provides 95 per cent of the restraining force to this posterior drawer movement.[69] The posterior cruciate ligament also helps to resist varus and valgus forces, and to a lesser extent helps to prevent internal rotation of the tibia on the femur in the flexed knee.[35] It also appears to quiet the 'screw home' mechanism of internal rotation of the femur on the tibia which locks the knee in full extension, preventing any significant rotation in this position.

Overall, the posterior cruciate ligament is one of the most important, if sometimes underrated, stabilizers of the knee.[69,70] Its tensile strength at the point of failure is normally twice that of the anterior cruciate ligament[71] (average of 3400 N).

Injury to the posterior cruciate ligament

Because of the interdependence of all the knee-supporting structures, isolated posterior cruciate ligament rupture is unusual. Associated injuries found during surgery for posterior cruciate ligament repair include anterior cruciate ligament tears, medial collateral ligament tears, posterior oblique capsular ligamentous damage, lateral compartment damage, and meniscal tears (usually medial meniscal).[72]

Injury to the posterior cruciate ligament occurs in the flexed knee, as most of the fibres are taut, when the tibia is driven posteriorly on to the femur with substantial force. The most common scenario is a motor vehicle accident, when the tibia strikes the

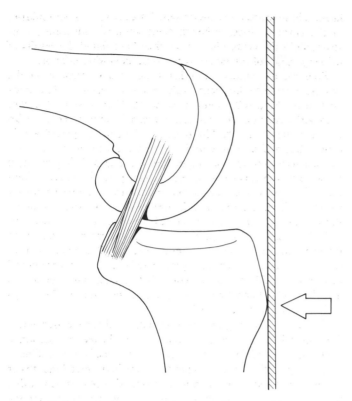

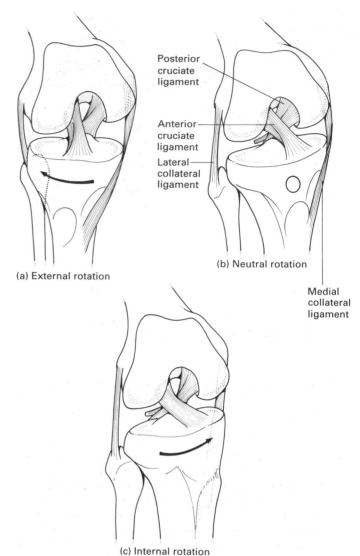

Fig. 25 Injury of the posterior cruciate ligament when the athlete falls on to a flexed knee with the foot plantar flexed.

dashboard and is suddenly stopped while the femur continues forward with the inertia of the body, still travelling at speed, behind it. In sports injuries, the posterior cruciate ligament is damaged when the athlete falls on to a flexed knee with the foot plantar flexed. In this case force is delivered up through the tibial tubercle, causing a shear force on the posterior cruciate ligament. If the foot is dorsiflexed most of the force is absorbed by the patella, which may be fractured while transferring the energy to the femur, and the posterior cruciate ligament is spared (Fig. 25).

A similar injury may be caused by hyperextension and rotation on a planted foot, where the posterior cruciate ligament ruptures before the anterior cruciate ligament [68] (Fig. 26).

Treatment

Before any course of treatment is embarked upon, the exact nature of the injury must be established. A physiological classification of acute versus chronic can be determined from the patient's history and physical examination, but more invasive methods are required to define the tear anatomically; the defect may be in the substance of the posterior cruciate ligament, or a bone fragment may be avulsed from either the femoral or (more commonly) the tibial attachment.[73] In most cases, diagnostic arthroscopy can define the nature of the ligamentous damage and also discover the extent of the associated injuries which commonly occur. This must be performed cautiously as some injuries may escape detection, and during arthrotomy one must be prepared to modify techniques as the local anatomy demands.

Whatever the mechanism of injury, the force required to disrupt the posterior cruciate ligament is considerable. Before the weakest portion of the ligament fails, stretching occurs, resulting in diffuse

Fig. 26 In addition to their synergistic functions, cruciate and collateral ligaments exercise basic antagonistic functions during rotation. (a) In external rotation it is collateral ligaments that tighten and inhibit excessive rotation by becoming crossed in space. (b) In neutral rotation none of the four ligaments is under unusual tension. (c) In internal rotation collateral ligaments become more vertical and are more lax, while cruciate ligaments become coiled around each other and come under strong tension (Redrawn from Müller, W. *The knee: form, function, and ligamentous reconstruction*. New York: Springer-Verlag, 1983, with permission).

interstitial damage and overall lengthening of the posterior cruciate ligament. This has been documented for injuries involving bone avulsions,[58] and is probably equally true for rupture through the substance of the ligament. Therefore, any repair should shorten the posterior cruciate ligament to stabilize the knee effectively. Theoretically, this can be done by overlapping the torn ends of the ligament or advancing the avulsed fragment of bone as the repair is performed.

If the patient presents in the acute phase, within 2 or 3 days of the injury, immediate primary repair is warranted,[73–75] as this seems to give the best functional results in the long term. Interstitial tears should be repaired first, incorporating the posterior capsule for added support. This will also tend to increase the blood

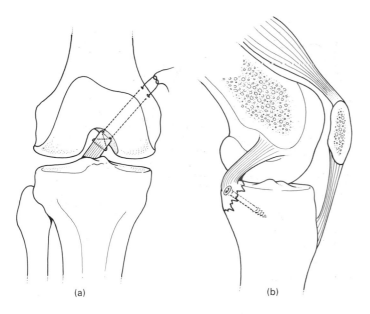

Fig. 27 (a) Reattachment of posterior cruciate ligament to medial femoral condyle. (b) Screw reattachment of bone fragment avulsed with posterior cruciate ligament from posterior tibia.

supply to the damaged ligament. If a piece of bone has been avulsed, it should be placed in position (slightly advanced beyond its anatomical position, as mentioned above), and stabilized with a screw or staple (Fig. 27).

The treatment of more chronic injuries is the subject of much controversy. Some authors[74,76-78] have reported good results with conservative treatment in a limited number of patients. This therapy is based on an intensive physiotherapy programme, with the aim of strengthening the quadriceps femoris, gastrocnemius, and hamstring muscles. These act as dynamic stabilizers of the knee, and when strengthened can give good symptomatic relief in patients who place only moderate demands on their knees. The knee does not return to normal with this treatment and will show evidence of instability on examination and a subjective feeling of unsteadiness, particularly on bearing weight on a semiflexed knee, but 50 to 75 per cent of patients will judge the functional result as at least adequate for the activities of daily living, not requiring any further intervention. Some of these patients have been followed for 2 years,[78] and have demonstrated good maintenance of symptomatic relief up to that point. These results, as applicable to a limited number of patients, are encouraging, but until long-term follow-up is available recommendations arising from these reports should be viewed with caution.

However, long-term difficulties arise in these patients. The abnormal tracking of the knee joint in the absence of the posterior cruciate ligament accelerates degenerative changes in the articular surfaces with resultant early osteoarthritis.[68] The fact that these changes are not seen within the first 1 to 2 years post-injury, but are quite commonly found after that, supports the hypothesis that this degeneration is produced by abnormal wear rather than by any damage to the articular surface that was produced at the time of injury. Most recently, MRI imaging of the knee has suggested that both factors may be implicated in producing joint damage.

Competitive athletes and others who place heavy demands on their knee joints cannot accept the disability associated with conservative therapy, and many different operative procedures have been developed in an attempt to recreate a normal or near-normal knee. These procedures are designed mainly to deal with injuries in the chronic phase, although some authors[68] advocate reconstruction in the acute phase since they have found the results of primary repair of interstitial tears to be somewhat inferior.

Free patellar tendon graft has been used to reconstruct the posterior cruciate ligament with excellent results.[68] The midportion of the patellar tendon (some 10–12 mm wide), together with both osseous attachments, is harvested and then secured with the knee flexed at 90°. The bone attachments are placed within drill holes, and animal experiments have shown that these will revascularize readily. The alignment of the graft is important; it must lie in the exact line of the original posterior cruciate ligament. Just prior to closure of the arthrotomy, a posterior drawer stress is placed on the knee in order to ensure static stability. Postoperatively, the patient should have his or her knee immobilized for 7 to 10 days, after which he or she is started on an intensive course of specialized physiotherapy. The total recovery time until the knee can be safely subjected to moderate stresses is 5 to 6 months. When this procedure is followed, static stability is usually excellent and most patients eventually return to their original level of athletic activity.[58]

The free tendon graft, as inserted, is obviously ischaemic. Histological studies have documented a degree of ischaemic necrosis, followed by revascularization and finally remodelling.[79] The strength of the graft actually decreases in the first 3 months or so, during the ischaemic and revascularization phases. Clinically, this may lead to a transient increase in posterior–anterior instability. During the remodelling phase, the graft has the potential to increase its strength in response to its environment. Andrish and Woods[80] augmented their grafts with Dacron in an attempt to protect the reconstruction during the initial stages when the graft is weakest. However, they observed significant resorption of the biological graft as well as failure of the synthetic material. Only four of ten experimental knees maintained the ligament and Dacron graft intact, and in only two of those four was the tensile strength more than that of the contralateral control knee. This atrophy of the ligament was attributed to stress shielding, and demonstrates the graft's dependence on external stresses during the remodelling period.

Another popular procedure involves the use of the gastrocnemius muscle for the reconstruction.[81,82] The medial head of the gastrocnemius is removed from its femoral attachment, with or without an attached bone block, and directed anteriorly through a hole in the posterior capsule. A drill hole is made in the medial femoral condyle from just anterior to the origin of the medial collateral ligament to the old site of insertion of the posterior cruciate ligament in the intercondylar notch. The gastrocnemius is pulled through the drill hole, with the bone block (if one is used) lying in a slot cut adjacent to the hole. A heavy non-absorbable suture or cancellous screw is used for fixation. In this position, the gastrocnemius acts as a dynamic stabilizer, pulling the femur posteriorly on the tibia as it contracts (Fig. 28). Weight bearing can begin as soon as the patient is relatively comfortable, and most activities can be resumed within 4 to 6 weeks. Approximately 80 per cent of patients will have good to excellent results,[81,82] with substantially decreased pain and instability.

Iliotibial band, semitendinosis tendon, popliteus tendon, and meniscus have all been used for posterior cruciate ligament reconstruction, although not as successfully as the procedures mentioned above. Various synthetic replacements have been tried, but 50 per cent or more are unsuccessful because of breakdown of the implant.[82]

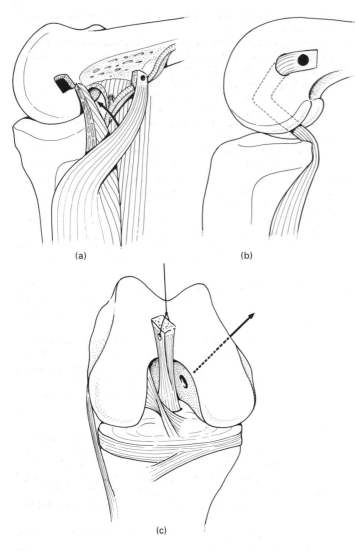

(a)   (b)

(c)

Fig. 28 The transfer of the medial head of the gastrocnemius for posterior cruciate reconstruction employs the posterior entry of the bone plug to sit in the original site of the posterior cruciate origin on the inner aspect of the medial femoral condyle. (Redrawn from ref. 81, with permission.)

Summary

In a joint as complex as the knee, it is difficult to separate out an individual structure such as the posterior cruciate ligament for study and treatment. Any method of repair or reconstruction must take into account the interdependence of all the supporting structures of the knee. Many different operative procedures have been developed over the years of reconstruction of the posterior cruciate ligament. Data for the long-term analysis of these procedures are still pending, but as these become available we shall have a clear idea of the best therapy for posterior cruciate ligament injuries with and without damage to associated structures.

REFERENCES

1. O'Donoghue DH *et al*. Repair of the anterior cruciate ligament in dogs. *Journal of Bone and Joint Surgery* 1966; **48A**: 503–19.
2. Hughston JC. The anterior cruciate deficient knee. *American Journal of Sports Medicine* 1983; **11**: 1.
3. Peacock EE. *Wound repair*. 3rd edn. Philadelphia: WB Saunders, 1984.
4. Dunphy JE. *Wound healing*. New York: Medcom Press, 1974.
5. Johnson RJ. The anterior cruciate ligament problem. *Clinical Orthopaedics and Related Research* 1983; **172**: 14.
6. Bucknall TE, Ellis H. *Wound healing for surgeons*. London: Bailliere Tindall, 1984.
7. O'Donoghue DH. A method for replacement of the anterior cruciate ligament of the knee. *Journal of Bone and Joint Surgery* 1963; **45A**: 905–24.
8. Stanish WD, Rubinovich M, Kozey J. McGillivary G. The use of electricity in ligament and tendon repair. *Physician and Sportsmedicine* 1985; **13**(8): 109–16.
9. Amiel D, Akeson WH, Harwood FL, Frank CB. Stress deprivation effect on metabolic turnover of the medial collateral ligament collagen. *Clinical Orthopaedics and Related Research* 1983; **172**: 265–70.
10. Amiel D, Wood SLY, Harwood FL, Akeson WH. The effect of immobilization of collagen turnover in connective tissue: a biochemical–biomechanical correlation. *Acta Orthopaedica Scandinavica*, 1982; **53**: 325–32.
11. Danielsen CC. Mechanical properties of reconstituted collagen fibrils. *Connective Tissue Research* 1983; **9**: 219–25.
12. Frank C, Amiel D, Woo SLY, Akeson WH. Normal ligament properties and ligament healing. *Clinical Orthopaedics and Related Research* 1985; **196**: 15–25.
13. Frank C *et al*. Medial collateral ligament healing. *American Journal of Sports Medicine* 1983; **11**(6): 379–89.
14. Vailas AC *et al*. Physical activity and its influence on the repair process of medial collateral ligaments. *Connective Tissue Research* 1981; **9**: 25–31.
15. Fung YC. *Biomechanics—mechanical properties of living tissues*. New York: 1981. Springer-Verlag.
16. Frank C, Amiel D, Akeson WH. Healing of the medial collateral ligament of the knee. *Acta Orthopaedica Scandinavica* 1983; **54**: 917–23.
17. Noyes FR. Functional properties of knee ligaments and alterations induced by immobilization. *Clinical Orthopaedics and Related Research* 1977; **123**: 210–42.
18. Cabaud HE, Rodkey WG, Feagin JA. Experimental studies of acute anterior cruciate ligament injury and repair. *American Journal of Sports Medicine* 1979; **7**: 18–22.
19. Gomez MA *et al*. Medial collateral ligament healing subsequent to different treatment regimens. *Journal of Applied Physiology* 1989; **66**(1): 245–52.
20. Woo SLY, *et al*. The biomechanical and morphological changes in the medial collateral ligament of the rabbit after immobilization and remobilization. *Journal of Bone and Joint Surgery* 1987; **69A**: 1200–11.
21. Arnoczky SP, Warren RF. Microvasculature of the human meniscus. *American Journal of Sports Medicine* 1982; **10**: 90–5.
22. Shaperio MS, Markolf KL, Finerman GAM, Mitchell PW. The effect of section of the medial collateral ligament on force generated on the anterior cruciate ligament. *Journal of Bone and Joint Surgery* 1991; **73A**(2): 248–56.
23. Fetto JF, Marshall JL. Medial collateral ligament injuries of the knee: a rationale for treatment. *Clinical Orthopaedics and Related Research* 1978; **132**: 206–18.
24. Seering WP, Piziali RL, Nagel DA, Schurman DJ. The function of the primary ligaments in the knee in varus–valgus and axiorotation. *Journal of Biomechanics* 1980; **13**: 785–94.
25. Indelicato PA. Non-operative treatment of the complete tears of the medial collateral ligament of the knee. *Journal of Bone and Joint Surgery* 1983; **65A**: 323–9.
26. Andrews JR, Carson WG (eds). Symposium on the anterior cruciate ligament, Part II. *Orthopaedic Clinics of North America* 1985; **16**: 2.
27. Arnold JA, Coker TP, Heaton LM, Park JP, Harris WD. Natural history of anterior cruciate tears. *American Journal of Sports Medicine* 1979; **7**: 305–13.
28. Balkfors B. The course of knee ligament injuries. *Acta Orthopaedica Scandinavica (Suppl.)* 1982; **198**: 1–99.
29. Clancy WG, Ray JM, Zoltan DJ. Acute third degree anterior cruciate ligament injury. A prospective study of conservative non-operative treat-

ment and operative treatment with repair and patellar tendon augmentation. *American Journal of Sports Medicine* 1985; **13**: 435–6.

30. Fetto JF, Marshall JL. The natural history and diagnosis of anterior cruciate ligament insufficiency. *Clinical Orthopaedics and Related Research* 1980; **147**: 29–38.

31. Barrett GR, Jiminez WP, Thomas JM. Aggressive rehabilitation protocol following anterior cruciate ligament reconstruction (bone–patella–bone). *Journal of the Mississippi State Medical Association* 1991; **32**(2): 45–8.

32. Eriksson E. Sports injuries of the knee ligaments. Their diagnosis, treatment, rehabilitation, and prevention. *Medicine and Science in Sports* 1976; **8**: 133–44.

33. Noyes FR, Bassett RW, Grood ES, Butler DL. Arthroscopy in acute traumatic hemarthroses of the knee: incidence of anterior cruciate tears and other injuries. *Journal of Bone and Joint Surgery* 1980; **62A**: 687–95.

34. Odensten M, Lysholm J, Gillquist J. Sutures of fresh ruptures of the anterior cruciate ligament—a five year follow up. *Acta Orthopaedica Scandinavica* 1984; **55**: 270–2.

35. Girgis FG, Marshall JL, Al Monajen ARS. The cruciate ligaments of the knee: anatomical functional and experimental analysis. *Clinical Orthopaedics and Related Research* 1975; **106**: 216–31.

36. Daniel D, Akeson W, O'Connor J. *Knee ligaments: structure, function, injury, and repair.* New York: Raven Press, 1990.

37. Slocum DB, Larson RL. Rotatory instability of the knee. Its pathogenesis and clinical test to demonstrate its presence. *Journal of Bone and Joint Surgery* 1968; **50A**: 211–25.

38. Kennedy JC, Weinberg HW, Wilson AS. The anatomy and function of the anterior cruciate ligament as determined by clinical and morphological studies. *Journal of Bone and Joint Surgery* 1974; **56A**: 223–35.

39. Barrack RL, Skinner HB. The sensory function of knee ligaments. In: Daniel D, Akeson W, O'Connor J, eds. *Knee ligaments: structure, function, injury and repair.* New York: Raven Press, 1990: 95–114.

40. Arnoczky SP. Anatomy of the anterior cruciate ligament. *Clinical Orthopaedics and Related Research* 1983; **172**: 19–25.

41. Arnoczky SP. Blood supply to the anterior cruciate ligament and supporting structures. *Orthopaedic Clinics of North America* 1985; **16**(1): 15–28.

42. Arnoczky SP, Rubin RM, Marshall JL. Microvasculature of the cruciate ligament and its response to injury. *Journal of Bone and Joint Surgery* 1979; **61A**: 1221–9.

43. Brantigan OC, Voshell AF. The mechanics of ligaments and menisci of the knee joint. *Journal of Bone and Joint Surgery* 1941; **23**: 44–66.

44. Brantigan OC, Voshell AF. The tibial collateral ligament: its function, its bursae and its relation to the medial meniscus. *Journal of Bone and Joint Surgery* 1943; **25**: 121–31.

45. Woo SLY, Young EP, Kwan MK. Fundamental studies in knee ligament mechanics: In: Daniel D, Akeson W, O'Connor J, eds. *Knee ligaments: structure, function, injury and repair.* New York: Raven Press, 1990: 115–34.

46. Noyes FR, DeLucas JL, Torvik PJ Biomechanics of anterior cruciate ligament failure: an analysis of strain rate sensitivity and mechanisms of failure in primates. *Journal of Bone and Joint Surgery* 1974; **56A**: 236–53.

47. Noyes FR, Grood ES. The strength of the anterior cruciate ligament in humans and Rees monkeys: age related and species related changes. *Journal of Bone and Joint Surgery* 1976; **58A**: 1074–82.

48. Smith JW. The elastic properties of anterior cruciate ligaments of the rabbit. *Journal of Anatomy* 1954; **88**: 369–80.

49. Trent PS, Walker PS, Wolf B. Ligament length patterns, strength and rotational axes of the knee joint. *Clinical Orthopaedics and Related Research* 1976; **117**: 263–70.

50. Torj JS, Conrad W, Kalen V. Clinical diagnosis of anterior cruciate ligament instability in the athlete. *American Journal of Sports Medicine* 1976; **4**: 84–93.

51. Galway RD, Beaupre A, MacIntosh DL. Pivot shift. A clinical sign of symptomatic anterior cruciate insufficiency. Proceedings of the Canadian Orthopaedic Association. *Journal of Bone and Joint Surgery* 1972; **54B**(4): 763–4.

52. Jarvinen M, Kannus P. The clinical and radiological long term results

after primary knee ligament surgery. *Archives of Orthopaedic and Traumatic Surgery* 1985; **104**: 1–6.

53. Jokl P, Kaplan N, Stovell P, Keggi K. Non-operative treatment of severe injuries to the medial and anterior cruciate ligaments of the knee. *Journal of Bone and Joint Surgery* 1984; **66A**: 741–4.

54. Kannus P, Jarvinen M. Conservatively treated tears of the anterior cruciate ligament. *Journal of Bone and Joint Surgery* 1987; **69A**: 1007–12.

55. Lysholm J, Gillquist J. Evaluation of knee ligament surgery results with special emphasis on use of a scoring scale. *American Journal of Sports Medicine* 1982; **10**: 150–4.

56. McDaniel WJ, Dameron TB. The untreated anterior cruciate ligament rupture. *Clinical Orthopaedics and Related Research* 1983; **172**: 158–63.

57. O'Donoghue DH. An analysis of the end results of surgical treatment of major injuries to the ligaments of the knee. *Journal of Bone and Joint Surgery* 1955; **37A**: 1–13.

58. Clancy WG. Knee ligament injury in sports: the past, present and future. *Medicine and Science in Sports* 1983; **15**: 9–14.

59. Feagin JA, Curl WW. Isolated tear of the anterior cruciate ligament: five year follow-up study. *American Journal of Sports Medicine* 1976; **4**: 95–100.

60. Hirshman HP, Daniel DM, Miyasaka K. The fate of unoperated knee ligament injuries. In: Daniel D, Akeson W, O'Connor J, eds. *Knee ligaments: structure, function, injury and repair.* New York: Raven Press, 1990; 481–503.

61. Marans HJ, *et al.* Functional testing of braces for anterior cruciate ligament-deficient knees. *Canadian Journal of Surgery* 1991; **34**(2): 167–72.

62. Nisonson B. Anterior cruciate ligament injuries. Conservative versus surgical treatment. *Physician and Sports Medicine* 1991; **19**(5): 82–9.

63. Noyes FR, McGinniss GH, Grood ES. The variable functional disability of the anterior cruciate ligament deficient knee. *Orthopaedic Clinics of North America* 1985; **16**: 47–67.

64. O'Brien SJ, Warren RF, Pavlov H. Reconstruction of the chronically insufficient anterior cruciate ligament with the central third of the patellar tendon. *Journal of Bone and Joint Surgery* 1991; **73A**(2): 278–85.

65. Palmer I. On the injuries to the ligaments of the knee joint. A clinical study. *Acta Chirurgica Scandinavica (Suppl. 53)* 1938; **81**: 3–282.

66. Smillie IS. *Injuries of the knee joint.* 5th edn. Edinburgh: Churchill Livingstone, 1978: 189–253.

67. Scapinelli R. Studies on the vasculature of the human knee joint. *Acta Anatomica* 1968; **70**: 305–31.

68. Clancy WG, Shelbourne KD, Zoellner GB, Keen JS, Reider B, Rosenberg TD. Treatment of knee joint instability secondary to rupture of the posterior cruciate ligament. *Journal of Bone and Joint Surgery* 1983; **65A**: 310.

69. Butler DL, Noyes ER, Grood E. Ligamentous restraints to anterior-posterior drawer in the human knee. *Journal of Bone and Joint Surgery* 1980; **62A**: 259–70.

70. Hughson JC, Andrews JR, Cross MJ, Moschi A. Classification of knee ligament instabilities, Part 1. *Journal of Bone and Joint Surgery* 1976; **58A**: 159–72.

71. Kennedy JC, Hawkins RJ, Willis RB, Danylchuk KD. Tension studies on human knee ligaments. Yield point, ultimate failure and disruption of the cruciate and tibial collateral ligament. *Journal of Bone and Joint Surgery* 1976; **58A**: 350–5.

72. Loos WC, Fox JM, Blazina ME, Del Pizzo W, Friedman MJ. Acute posterior cruciate ligament injuries. *American Journal of Sports Medicine* 1981; **10**: 150–4.

73. Bianchi M. Acute tears of the posterior cruciate ligament: clinical study and results of operative treatment in 27 cases. *American Journal of Sports Medicine* 1983; **11**(5): 308–14.

74. Cross MJ, Powell JF. Long term follow-up of posterior cruciate ligament rupture: a study of 116 cases. *American Journal of Sports Medicine* 1984; **12**(4): 292–7.

75. Strand T, Molster AO, Engesaeter LB, Raugstad TS, Alho A. Primary repair in posterior cruciate ligament injuries. *Acta Orthopaedica Scandinavica* 1984; **55**: 545–7.

76. Dandy DJ, Pusey RS. The long term results of unrepaired tears of the

posterior cruciate ligament. *Journal of Bone and Joint Surgery* 1982; **64B**: 92–5.

77. Satku K, Chew CN, Seow H. Posterior cruciate ligament injuries. *Orthopaedica Scandinavica* 1984; **55**: 26–9.

78. Tegner Y, Lysholm J, Gillquist J, Oberg B. Two year follow-up of conservative treatment of knee ligament injuries. *Acta Orthopaedica Scandinavica* 1984; **55**: 176–80.

79. Chiroff RT. Experimental replacement of the anterior cruciate ligament. A histological and microradiographic study. *Journal of Bone and Joint Surgery* 1975; **57A**: 1124.

80. Andrish JT, Woods LD. Dacron augmentation in anterior cruciate reconstruction in dogs. *Clinical Orthopaedics and Related Research* 1984; **183**: 298–302.

81. Insall JN, Hood RW. Bone block transfer of the medial head of the gastrocnemius for posterior cruciate insufficiency. *Journal of Bone and Joint Surgery* 1982; **64A**: 691–9.

82. Kennedy JC, Galpin RD. The use of the medial head of the gastrocnemius muscle in the posterior cruciate deficient knee. Indications, techniques, results. *American Journal of Sports Medicine* 1982; **10**: 63–74.

83. Schmid-Schonbein GW, Woo SLY, Zweifock BW, eds. *Frontiers in biomechanics.* New York: Springer-Verlag, 1986.

84. Hunter LY, Funk JF, eds. *Rehabilitation of the injured knee.* St. Louis: CV Mosby, 1984.

85. Woo SLY, *et al.* Treatment of the medial collateral ligament injury. *American Journal of Sports Medicine* 1987; **15**(1).

86. Noyes FR, *et al.* Advances in the understanding of knee ligament injury, repair, and rehabilitation. *Medicine and Science in Sports and Exercise* 1984; **16**(5).

4.2.4 Fractures and dislocations

JEAN-LOUIS BRIARD AND J. ROBERT GIFFIN

INTRODUCTION

As a result of increasing participation and level of competitiveness in both organized and unsupervised sporting activities, the number of individuals presenting with acute knee injuries continues to rise. Compared with the incidence of other soft tissue knee trauma, fractures and dislocations about the knee occur infrequently in the sports population; however, they can be devastating to the athlete in terms of rehabilitation and eventual return to sport. Therefore physicians who deal with athletes must be able to suspect, diagnose, and properly treat these rare injuries. The key to sound management is early recognition, and most of the skeletal injuries encountered about the knee will be briefly discussed in this chapter.

DISLOCATIONS OF THE KNEE

Traumatic injury to the lower extremities of athletes frequently occurs in sports. Athletic injuries are the second most common cause of knee dislocation,[1] and, although this type of knee trauma is rare, disaster may occur with a disruption of the popliteal artery. Vascular compromise could result in an amputation, as the worst possible case scenario, if blood flow is not restored within 6 to 8 h.

The primary care physician must conduct a neurovascular examination during initial assessment of an athlete who has sustained significant trauma to the knee. Alert recognition of this potential limb-threatening injury by the attending physician will ensure thorough investigation and surgical correction of any associated vascular injuries, thereby reducing the likelihood of a disastrous situation.

Emergency transport of the athlete to a hospital with appropriate radiography facilities and both orthopaedic and vascular consultants available is vital to avoid the development of severe residual sequelae. The potential seriousness of the knee dislocation is not restricted to the vascular system; other crucial anatomical structures may also be injured. Peripheral nerve damage could lead to grave functional sequelae, and ligamentous injuries may result in residual laxity or severe instability.

Anatomy

The popliteal artery is at risk of injury during knee trauma because it is relatively fixed within the popliteal fossa (Fig. 1). Proximally it is tethered at the adductor magnus hiatus and distally it is tethered at the fibrous arch of the soleus prior to its division into the anterior and posterior tibial arteries. Despite its collateral

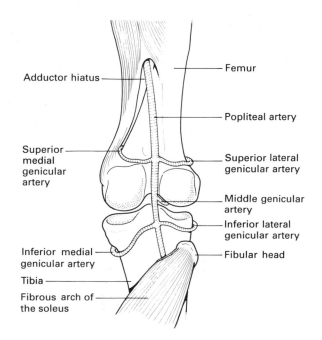

Fig. 1 The popliteal artery is tethered at the adductor hiatus and the fibrous arch of the soleus.

The authors thank the Radiology Department of the Victoria General Hospital for providing the radiographs presented in this chapter.

arteries, perfusion of tissues could be insufficient to maintain the viability of the deep leg musculature in the event that the popliteal artery is stretched and torn. For this reason the popliteal artery is considered to be an end artery.

Other significant neurovascular structures that accompany the artery through the fossa are not as firmly fixed and therefore are less likely to be injured. When neurological damage occurs, it is usually caused by tractional forces stretching the peroneal nerve around the posterior femoral condyle.

Classification

Knee dislocations are classified by the direction in which the tibia is displaced in relation to the femur. The five major types of dislocation (Fig. 2) are described as anterior, posterior, lateral, medial, and rotatory. Rotatory dislocations are further classified as anteromedial, anterolateral, posteromedial, and posterolateral.

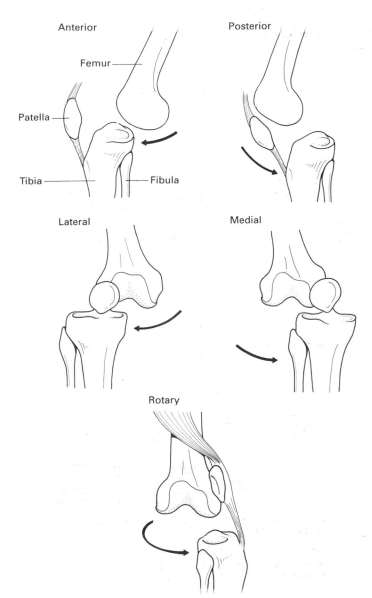

Fig. 2 Classification of knee dislocations. Arrows indicate direction in which the tibia is displaced on the femur.

Mechanism of injury

High speed vehicular trauma is the most common mechanism of injury involved in knee dislocations. However, as mentioned previously, low velocity dislocations of the knee can also occur on the athletic field.

Hyperextension of the knee by more than 30° from neutral will usually result in rupture of the anterior cruciate ligament and the posterior capsule. The tibia then translates forward on the femur causing subsequent disruption of the posterior cruciate ligament. At 50° of hyperextension the popliteal artery will also be torn and the injury becomes a surgical emergency.[2]

Posterior dislocations usually occur when a force is applied directly to the anterior tibia. The force that is required to produce this translocation of the tibia posteriorly is much larger than the force necessary to cause an anterior dislocation.[1] The posterior displacement of the bone stretches the popliteal artery which may become disrupted just proximal to its trifurcation.

A lateral dislocation occurs with a severe valgus force that disrupts the medial supporting structures and the cruciate ligaments and displaces the tibia laterally on the femur. With a medial dislocation, a varus force damages the lateral ligamentous structures as well as both cruciate ligaments.

The most common rotatory dislocation occurs in the posterolateral direction. The mechanism of injury is commonly an anteromedial and varus force applied to the proximal tibia of an extended knee, as in football tackling. Rotatory dislocations are more infrequent and may cause a 'buttonhole' of the femoral condyle through the medial capsule and tibial collateral ligament. If these structures become trapped within the joint, an open reduction must be performed.[3]

Pathology

Knee dislocations result in extensive soft tissue injuries with rupturing of the primary ligaments and stripping of the capsule from the bone. Sisto and Warren[4] reported on the operative findings in 16 surgically reduced knee dislocations. Both the anterior and posterior cruciate ligaments were torn in conjunction with one or both collateral ligaments in each knee examined under open reduction. The anterior cruciate ligament was avulsed from the femur in five cases, avulsed from the tibia in five cases, and torn in the mid-portion of the ligament in the remaining six cases. However, the posterior cruciate ligament was avulsed from the femur in 12 knees, avulsed from the tibia in two knees, and torn in its mid-substance in the other two cases.

It was noted that pure anteroposterior dislocations usually spared the collateral ligaments from injury. Tendon injuries included rupturing of the popliteus in three knees, the lateral head of the gastrocnemius in one knee, and the biceps tendon in another. Injuries to the medial meniscus were noted in five knees and tears of the lateral meniscus in four other cases.

Dislocations that occur in the anterior or posterior plane are more likely to result in the disruption of popliteal artery flow, and the overall calculated risk of vascular injury in all knee dislocations is 33 per cent.[5,6] Arterial injury during anterior dislocations is caused by tractional forces resulting in multiple transverse lacerations or internal damage along a segment of the artery.[1] Experiments[7] have shown that traction applied to an artery will rupture the intima and media prior to tearing of the outer elastic adventitial layer.

This mechanism of arterial injury may result in following:[8] a circumferential intimal tear with resultant thrombus occluding the vessel; an intimal dissection producing a flap of intima and thrombus; a separation of the layers of the arterial wall with resultant intramural hematoma (Fig. 3). Such injuries may result in thrombosis of the vessel prior to the initial examination resulting in absence of distal pulses. However, transmission of a pulse through a thrombotic area may give the false impression of an intact popliteal pulse. Furthermore, arterial compromise could manifest itself in a delayed fashion since a thrombosis may not occur for hours or even days following trauma. Physicians should also be aware that, despite the presence of an intact pedal pulse, an arterial injury may exist if a non-occlusive intimal tear is present.[1]

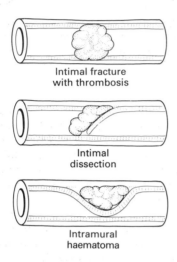

Intimal fracture
with thrombosis

Intimal
dissection

Intramural
haematoma

Fig. 3 Potential mechanisms of vascular occlusion following dislocations.

The consequence of thrombosis is a resultant ischaemia of the tissue supplied by the occluded artery. Animal studies have revealed that irreversible damage occurs after 6 to 8 h of ischaemia. DeBakey and Simeone,[9] in a review of arterial injuries during the Second World War, found an amputation rate of 72.5 per cent when the popliteal flow was disrupted. Green and Allen[6] demonstrated that in knee dislocations, amputation resulted in 86 per cent of the limbs in which the flow of the popliteal artery was not restored within 8 h. More recently, Bloom[5] reported on 11 patients with popliteal artery occlusion in which flow was restored within 8 h; 10 of the 11 had a successful revascularization but the remaining patient's leg had to be amputated.

As Cone[8] states:

> The diagnosis of vasospasm deserves mention here for the sole purpose of condemnation. While vasospasm does occur, it must be a diagnosis of exclusion. A diagnosis of vasospasm should never be accepted as the etiology of diminished perfusion without angiographic confirmation. Even with compatible angiogram, if the spasm does not respond promptly to appropriate pharmacologic agents or if the limb is not well perfused, the vessel should be explored surgically.

Signs and symptoms

The importance of an accurate patient history, diligent observation, and a thorough examination when attempting to diagnose a problem correctly must be emphasized. The position of the athlete's extremity following the traumatic insult will dictate the order of the assessment process.

If the extremity is in a position of obvious deformity, the circulatory status of the injured limb is evaluated first. In most instances, the dislocation can easily be reduced at the scene by gently applying traction to the limb and pressure on the femur in the opposite direction to the deformity. The circulatory status must be assessed again following the reduction of the dislocation and care must be taken never to hyperextend the knee. Finally, the extremity should be immobilized in slight flexion for transport.

If a deformity of the injured leg is not obvious or reduced, careful assessment of the ligamentous stability is indicated with gentle manipulation. Vascular and neurological status should also be evaluated, remembering that sensory disturbances and paralysis could be caused by ischaemia only and not infarction. Severe trauma to the knee must be evaluated quickly and the patient referred to the most appropriate centre because of the high likelihood of major ligamentous disruption or dislocation with associated potential vascular injury.

Radiographs—arteriogram

Standard radiographs should be obtained in order to assess the position of the tibia in relation to the femoral condyles and to check for the presence of any associated fractures. The decision as to whether to perform arteriography must then be made.

Some vascular surgeons believe that in the case of obvious arterial disruption performing an arteriogram merely delays the surgery. Others believe that if a loss of popliteal artery flow is suspected, the patient should have an arteriogram not only to confirm the diagnosis of occlusion or arterial tear, but also to confirm the level of arterial injury.

Since one-third of patients with knee dislocation also have associated popliteal artery damage, every patient with a proven or suspected knee dislocation should have an arteriogram on an emergency basis. Even if the vascular status of the limb does not seem compromised initially, an intimal tear may have occurred, predisposing the patient to a delayed occlusion. High quality biplane sequential arteriography remains the gold standard even when investigating a patient with a seemingly well perfused but injured extremity.

Arterial repair

Although the details of vascular repair are beyond the scope of this chapter, fundamental surgical principles will be discussed. Vascular repair or the use of an intravascular shunt should be of first priority in order to decrease the duration of ischaemia. A properly performed vascular anastomosis is resistant to traction or manipulation and will tolerate the remainder of the operation.

Following distal and proximal balloon embolectomy, the damage segment of the vessel should be resected. The resultant defect is usually too large for primary repair; thus the interpositionary graft of choice is a reversed saphenous vein harvested from the contralateral limb. Intraoperative angiography should be employed routinely upon completion of the repair to check the integrity of the vascular reconstruction.

Ligament repair

Following a vascular repair, all the ligament, tendon, and meniscal injuries are corrected. In cases where the dislocation has been reduced and it has been determined that the popliteal artery has escaped injury, the knee should be kept immobilized in a posterior splint for a few days before ligamentous reconstruction surgery is undertaken.

In athletes in particular, the general practice has been to repair or reattach any injuries of the ligaments, tendons, or menisci. The cruciates, when torn in mid-substance or failing to produce sufficient tension following reattachment, must be reinforced with a portion of adjacent tissue. The knee is then immobilized in full extension for healing and subsequently started on an active range of motion in a cast brace.

Prognosis and complications

Vascular complications can be very serious, causing ischaemia which may lead to amputation. Complications may include acute renal tubular necrosis from myoglobinaemia or possible compartment syndrome if a fasciotomy is not performed at the time of the vascular repair. Late complications of ischaemic contracture and ulceration may develop unexpectedly; thus the vascular status of the limb should be monitored regularly.[1]

Ligament reconstruction greatly improves the stability about the knee. Sisto and Warren[4] reported good postoperative stability following ligamentous repair. They reported that 77 per cent of their patients had a negative posterior drawer sign, 38 per cent had a positive Lachman sign, and 8 per cent had a positive pivot shift. Sisto and Warren[4] also reported that 77 per cent of their patients managed to return to sporting activities including basketball, and skiing.

DISLOCATION OF THE PROXIMAL TIBIOFIBULAR JOINT

Although dislocation of the proximal tibiofibular joint is rare, the sports physician must be attuned to this potential source of dysfunction in the athletic population. It is encountered in a wide variety of sports including soccer, baseball, football, handball, rugby, racquetball, track, gymnastics, wrestling, judo, ski, dance, and parachuting.[10–14]

Anatomy

The proximal tibiofibular articulation is a diarthrodial joint composed of two oval surfaces.[13] The articular surfaces vary in size, shape, and inclination (Fig. 4). The articular facet for the fibula is on the posteroinferior aspect of the lateral tibial condyle and is bordered anteriorly by the lateral tibial sulcus. The fibular facet is usually elliptical and almost flat. The articular surfaces are covered with hyaline cartilage, surrounded by a synovial membrane and a joint capsule that is usually separate from that of the knee. The capsule is reinforced by both anterior and posterior tibiofibular ligaments, with the former being the much stronger of the two. Superior support is provided by the lateral or fibular collateral ligament. The joint's motion consists of proximal–distal translation as well as axial rotation and is related to movement occurring in the ankle joint.[14]

Classification of the mechanism

There are three types of acute traumatic dislocation of the proximal tibiofibular articulation (Fig. 5). The most frequent (77 per cent of dislocations) occurs in the anterolateral direction secondary to a fall on a flexed and adducted leg with the foot in an

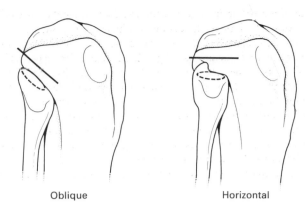

Fig. 4 Ogden identified two basic types of proximal tibiofibular joints, oblique and horizontal. In the oblique joint, the articular surfaces are inclined at an angle greater than 20° and the greater the angle, the smaller the surface area of the joint.

inverted position. Loss of articulation in the posteromedial direction (7 per cent of dislocations) is the result of either a direct blow (horseback rider's knee) or a twisting injury. Peroneal nerve injuries are frequently associated with this type of dislocation. Finally, a disarticulation in the superior direction does not occur as an isolated lesion and is always associated with a fracture of the tibia or a severe ankle injury.[10]

Lyle[15] described the subluxation as a fourth type of disruption of the proximal tibiofibular joint. These subluxations, which occur in 14 per cent of the patients, are chronic in nature and are the result of neglected anterolateral or posteromedial dislocations. They are often associated with peroneal nerve disturbances and must be differentiated from the idiopathic recurrent dislocations.

Signs and symptoms

Following the trauma, the patient can usually describe the mechanism of injury and will state that the joint feels unstable.

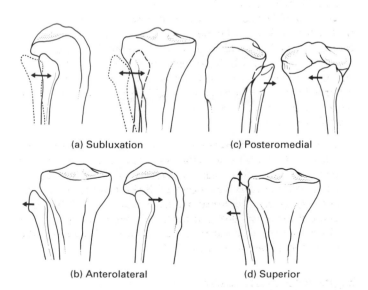

Fig. 5 Ogden classified disruptions of the proximal tibiofibular joint into subluxations (14 per cent) and three types of dislocations—anterolateral (77 per cent), posteromedial (70 per cent), and superior (2 per cent).

Clinical presentation can range from the complete disability and non-weight-bearing to slight discomfort over the lateral aspect of their knee or ankle.

On examination, the dislocated fibular head is usually prominent and tender to palpation. Soft tissue swelling is mild during the acute stages and mobilization of the ankle may produce knee pain. The fibular head may be hypermobile compared with the contralateral side and peroneal nerve function must be evaluated.

Radiographic findings are often subtle, and comparison of the injured side with the normal knee is helpful in determining the diagnosis. Films of an anterior dislocation show that the fibula is laterally displaced and the proximal interosseous span is widened on the anteroposterior view. The lateral view reveals that the fibular head is displaced anteriorly and overlapped by the tibia (Fig. 6).

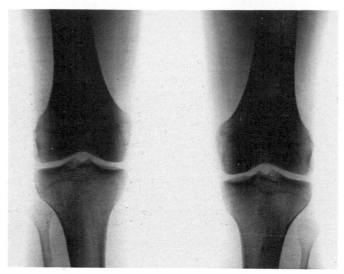

(a)

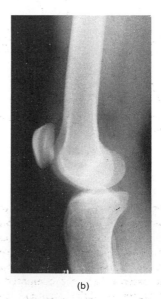

(b)

Fig. 6 (a) Anteroposterior and (b) lateral radiographs of a proximal tibiofibular dislocation. The radiographic findings are often subtle and comparison of the injured side with the normal knee is helpful in determining the diagnosis. The lateral view reveals that the fibular head is displaced anteriorly and overlapped by the tibia.

Chronic tibiofibular dislocation

The patient typically presents with a history of derangement in maximum flexion and has symptoms of clicking and pain over the lateral compartment of the knee. The pain, which can radiate distally to the foot, is aggravated by walking and jumping.[16] Clinically it may mimic a lateral meniscal injury which would be included in the differential diagnosis. However, the diagnosis can be confirmed by the presence of tenderness and swelling directly over the tibiofibular joint. When assessing the mobility of the fibular head, the physician wants to determine the quality of the joint end feel and to determine if an anteroposterior draw sign is present. Radiographs and CT scans may also be employed to demonstrate the dislocation.

Treatment

Acute anterior derangements are nearly always reducible by direct manipulation; however, it may be necessary to perform the reduction under general anaesthesia on occasion.[14] The knee is flexed to 90°, the foot is inverted, and direct manipulation reduces the fibular head with an audible snap. The knee is strapped and the patient begins progressive weight bearing on crutches over a period of 3 weeks. The presence of an associated ankle injury may require the use of a cast immobilization. If the closed manipulation fails or if the dislocation recurs, open reduction is warranted. Ligament reconstruction and fixation with K wires may be required. A short leg cast is then applied for a period of 6 weeks.

In chronic cases, the presence of disability or peroneal palsy warrants surgical intervention. Some surgeons prefer to resect the head and neck of the fibula coupled with a careful reconstruction of perosteal tissue sleeve. Others prefer to perform an arthrodesis of the superior tibiofibular joint and 2 cm resection of the fibular shaft.

In conclusion, the diagnosis of superior tibiofibular joint dislocation should not be overlooked as a cause of post-traumatic lateral knee pain. Furthermore, chronic derangement of the proximal tibiofibular articulation may be mistaken for cartilagenous or meniscal pathology, lateral ligament tears, or popliteal tendinitis.

OSTEOCHONDRITIS DISSECANS

Osteochondritis dissecans is a lesion in which a fragment of articular cartilage and subchondral bone separate partially or completely from the underlying bony matrix. These lesions tend to occur on the convex articular surface, thereby affecting the smooth integration of motion and force transmission within the joint.

The aetiology and treatment of osteochondritis dissecans are still a matter of controversy. There is both a juvenile and an adult form. A very favourable prognosis can be expected when treating the young patient prior to the closure of his or her physis. Intervention in adults is much more aggressive and results of treatment are less satisfactory.

Aetiology

There is considerable speculation regarding the aetiology of osteochondritis dissecans[17] and historically the most accepted theories have included ischaemia, constitutional factors, and endogenous and exogenous trauma as potential causes.

Ischaemia

The obstruction of end arteries to the femoral condyle has been suggested as a possible cause of osteochondritis dissecans. The work of Rogers and Gladstone[18] demonstrated that the distal femur has a network of anastomosing vessels in the subchondral bone. They concluded that ischaemia of the femoral condyle was an unlikely cause of osteochondritis dissecans. However, intraosseous circulatory disorders can occur in the elderly (more than 60 years old).[17]

The pathology of the avascular necrosis differs from that of osteochondritis dissecans. In avascular necrosis, fragmentation occurs between necrotic layers of bone, whereas in osteochondritis dissecans the fragment separates from a completely normal bony matrix.

Constitutional

While there have been multiple reports of familial occurrence suggesting a hereditary factor, the usual presentation of osteochondritis dissecans is not familial in nature.[19,20] However, multiple epiphyseal dysplasia, with its autosomal dominant or recessive pattern of inheritance, must always be considered in the differential diagnosis of patients who have osteochondritis dissecans.[21]

In 1937, Ribbing[21] studied the abnormalities of endochondral ossification of the epiphysis. Localized delay of the ossification process commonly occurs in childhood[22] and usually resolves without any long-term sequelae. In the interim, the delayed ossification process may give the radiological appearance of a crater or an accessory epiphysis. The crater is filled with radiolucent hyaline cartilage which may become partially calcified along its deep surface. Compressive forces may shear the edges of the crater and progressively detach the deeper partially calcified portion. This accessory nucleus of bone may separate with subsequent partial reattachment. Further trauma during childhood could result in a complete separation.

Endogenous and exogenous trauma

Forty per cent of patients relate the onset of their problems to a single traumatic event. The proximity of the lateral aspect of the medial femoral condyle to the tibial spine, the insertion of the posterior cruciate ligament, and the medial aspect of the patella in full flexion has fostered theories of endogenous trauma as the aetiology of osteochondritis dissecans.[19]

Acute unstable osteochondral fractures may proceed to nonunion with radiographic findings similar to osteochondritis dissecans. However, it is debatable whether or not the force required to produce an osteochondritis dissecans lesion actually occurs at the clinical level. In the laboratory it has been impossible to reproduce the 'classic site' of osteochondritis dissecans with experimental forces.[24]

Osteochondritis dissecans is a chronic pathological process quite different from an acute osteochondral fracture. The osteochondral fracture is usually caused by a single traumatic event and is often accompanied by a haemarthrosis. In contrast, osteochondritis dissecans proceeds gradually, resulting in the formation of a loose body over time. Furthermore, if there is an associated loose body in an old osteochondral fracture, it should have a convex cartilaginous surface with a flat base and a corresponding flattened area on the epiphysis.[24]

Cahill[25] has insisted on the role of minimal or insignificant trauma in the production of osteochondritis dissecans. He reported that the investigation of the mechanical properties of bone offers substantiated evidence that the initial lesion of osteochondritis dissecans could be a fatigue failure of previously normal subchondral bone. Joint scintigraphy of patients with juvenile osteochondritis dissecans demonstrated a healing pattern typical of fractures, and thus Cahill *et al.*[26] concluded that juvenile osteochondritis dissecans is a failure of the subchondral bone as a result of cumulative stresses. In summary, the current trend of thought is to recognize a multifactorial cause or several predisposing factors when considering the aetiology of osteochondritis dissecans.

Pathology

Osteochondritis dissecans can be staged anatomically (Fig. 7) according to the progression of the lesion.[27–29] A stage I lesion is a well-formed defect in the subchondral bone that is readily

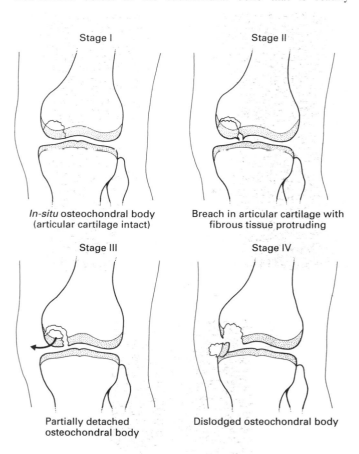

Stage I
In-situ osteochondral body (articular cartilage intact)

Stage II
Breach in articular cartilage with fibrous tissue protruding

Stage III
Partially detached osteochondral body

Stage IV
Dislodged osteochondral body

Fig. 7 Stages of osteochondritis dissecans.

apparent on radiographs. These lesions are kept in place by the integrity of the overlying articular cartilage. Stage II is characterized by a breach in one aspect of the articular cartilage from which fibrous tissue can protrude. The overlying cartilage is mostly intact, and repair and resorption are apparent in the osseous nucleus. The osseous beds are composed of mesenchymal tissue in varying stages of differentiation, ranging from very primitive spindling fibrous tissue to histologically normal fibrocartilage, as can be found in a healing fracture site.

A stage III defect is a partially detached lesion usually held by only a hinge of articular cartilage. Stage IV is a completely detached lesion resulting in the formation of a loose body. Once separated from its bed, the loose body continues to grow by accretion and the crater slowly becomes filled with fibrous tissue.

The incidence of osteochondritis dissecans is estimated at 15 to 21 cases per 100 000 knees.[30] The greatest incidence of the disease occurs in the first half of the second decade of the physiologically active adolescent, though osteochondritis dissecans may affect anyone from 5 to 50 years old. The distribution among sexes is not equal, with males being mainly affected (70 per cent) and females to a lesser extent (30 per cent). Distributions and locations have been extensively studied (Fig. 8). It affects the medial condyle in 62 per cent of cases, the lateral condyle in 22 per cent of cases, and the patella in 16 per cent of cases,[31,32] and a few cases involving the trochlea[33] have been reported. Furthermore, osteochondritis dissecans occurs bilaterally in 30 per cent of cases.[17]

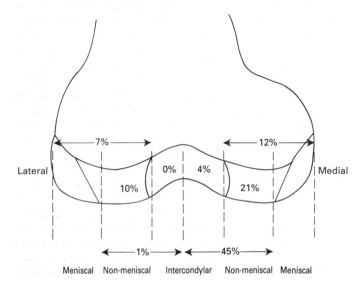

Fig. 8 Hughston's distribution of osteochondritis dissecans according to location on the anteroposterior radiography. The meniscal area represents the outer aspects of the femoral condyle that is directly superior to the meniscus. The intercondylar area is part of the articular surface that does not directly bear weight and extends into the intercondylar notch. The non-meniscal area represents part of the weightbearing area of the condyle between meniscal and intercondylar areas.

Hughston *et al.*[30] distinguished between lesions on the most distal weight-bearing aspect of the femoral condyle and those that occurred more posteriorly. They found that a great majority of defects on the lateral condyle were equally distributed between posterior and distal location. Outerbridge[34] reviewed 14 cases of osteochondritis dissecans affecting the posterior area of both the medial and lateral condyles. It has been his experience that the presence of a defect in this posterior location results in a poor prognosis for the patient.

Clinical features

Patients usually present with early symptoms of knee pain which develop insidiously and are aggravated by physical activity. Onset may be acute in nature, particularly in older patients, and all age groups may complain of joint effusion. Symptoms of giving way, painful buckling, catching, or locking may be reported when the lesion becomes unstable and forms a loose body.

Physical examination may reveal quadriceps atrophy, effusion, limitation of full passive range of motion, and localized tenderness to palpation directly over the site of the lesion with the knee in a flexed position. Wilson[35] has described a sign in which the pain can be reproduced by first flexing and internally rotating the knee and then extending the leg slowly. The test will be positive if a lesion is located in the common posterolateral site of the medial femoral condyle. A positive sign can be confirmed if the pain is subsequently relieved by externally rotating the tibia.

Radiographic findings

The lesion usually appears as a well-defined area of subchondral bone separated from the remaining femoral condyle by a crescent-shaped radiolucent line. The classic location on the posterolateral medial condyle may not be apparent on a standard anteroposterior view and is best visualized using the tunnel view (Fig. 9). Although the lateral view is sometimes difficult to read, it allows an estimation of the size of the defect. Because of the potential variations in the location of an osteochondritis dissecans lesion, if the disorder is suspected the radiographic examination should include anteroposterior, lateral, oblique, notch, and skyline patellar views. CT scans may be useful.

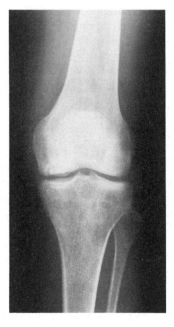

Fig. 9 Osteochondritis dissecans involving the lateral aspect of the medial femoral condyle in a skeletally mature patient.

Prognosis

Once a diagnosis has been established, the prognosis of the patient depends on several variables. The patient's age (physis open or closed), integrity of the articular cartilage, and stability of the lesion are very reliable indicators of clinical outcome. The severity of the athlete's subjective complaints (i.e. catching, buckling, locking), in addition to radiographic findings, is a good parameter for predicting loosening.

Scintigraphic evaluation

The role of joint scintigraphy in the diagnosis and treatment of osteochondritis dissecans has been stressed by Cahill *et al.*[26] and

Mesgarzadeh *et al.*[36] It allows differentiation between osteochondritis dissecans and an accessory ossification centre. Technetium pyrophosphate uptake is a measure of both osteoblastic activity and regional blood flow. The level of scintigraphic activity in a symptomatic knee indicates the remaining potential for healing of the osteochondritis dissecans fragment. Cahill *et al.*[26] insist on the value of repeated bone scintigraphy to assess the progression of the healing and to determine if and when more radical procedures must be adopted.

Magnetic resonance imaging evaluation

More recently, MRI has been employed to analyse the mechanical stability of the lesion in osteochondritis dissecans. It provides direct visualization of any loosening or displacement of the fragment.[36,37]

Arthroscopic evaluation

Arthroscopy provides direct visualization and a technique for manual testing of the integrity of the cartilage. Although arthroscopy is rarely indicated for diagnosis only, lesions can be classified in four stages: intact; early separation; partially detached; craters with loose bodies that are either salvageable or non-salvageable.[28]

Natural history

The juvenile form of osteochondritis dissecans was previously considered to have an excellent prognosis for healing without complication. However, recent reports have demonstrated that healing does not always occur spontaneously and that loosening or complete detachment of the fragment can occur.[25,30]

Osteochondritis dissecans of the skeletally mature patient may be complicated by mechanical detachment of the lesion. The formation of loose bodies and tibial plateau alteration could possibly diminish the performance of the athlete. Long-term follow-up has shown the development of osteoarthritis in the majority of cases even though the athletes have remained asymptomatic for an extended interval of time. The osteoarthritis secondary to an osteochondral defect tends to begin a decade earlier than does primary gonarthrosis.[38]

Results following surgical procedures are not consistent because of ensuing secondary degeneration, particularly in patients who have been diagnosed and treated after skeletal maturity. As recognized by Hughston *et al.*[30] 'We are currently unable to treat the condition with consistently satisfactory results.'

Treatment

The object of treatment is to promote healing of the lesion and prevent detachment in an attempt to preserve a smooth articular surface. The treatment of osteochondritis dissecans can range from observation to surgical intervention which depends on the patient's age and the degree of involvement.

Conservative measures may include a trial of orthotics or the use of splints and crutches as symptoms require. The prudent use of physiotherapy may help to maintain the patient's strength and retard atrophy.

The aim of surgical treatment is to improve the stability of the fragment by enhancing its blood supply. Numerous surgical techniques have been described, including excising the fragment, drilling, crater debridement, grafting,[39,40] and fixation with pins,

screws,[41] or bone pegs.[42] The goal is to restore the surface congruity which may require grafting, debridement, or the repositioning of loose bodies.[43,44]

The child with open epiphyses is usually treated with conservative measures and good results can be expected. Cahill *et al.*[26] suggest following the healing process with a bone scan. When the scan shows decreased activity, healing is assumed to have taken place. Surgery is only indicated if the lesion remains active and symptomatic. Arthroscopic drilling is the treatment of choice for intact lesions.[17,28]

In older adolescents and adults, the treatment must be more aggressive and the prognosis remains guarded. Indications for operative treatment in the skeletally mature patient include a lesion larger than 1 cm and involvement in a weight-bearing surface.[28]

Lesions with intact articular cartilage are simply drilled. Large intact lesions may require grafting, particularly when the defects are posterior.[34] Those with early separation should be debrided at their border, drilled, and pinned in place. Partially detached lesions require debridement of the crater, reduction, and fixation. Salvageable loose bodies should be prepared and replaced when possible, particularly if they involve a portion of a weight-bearing surface. Cancellous allografts are used in selective cases to restore surface congruity. Although most of the procedures are performed arthroscopically, open surgery may be required to deal with large lesions and difficult locations.

Rehabilitation consists of a programme of range of motion exercises as well as strengthening of the musculature about the knee. The length of time during which weight bearing is restricted can range from 2 or 3 weeks to several months depending on the procedure performed. The fibrocartilage forming at the base of a crater on a weight-bearing surface must be given sufficient time to mature before being subjected to body force. Grafting requires cast immobilization for 4 to 6 weeks and weight bearing should be restricted until there is radiographic evidence of small trabecular bone formation.

In Guhl's[28] study of 58 knees, 56 healed and only five required additional surgery. Twenty patients reported their surgical result as excellent, 26 rated their result as good, eight reported their result as fair, and only two were dissatisfied. At the time of the surgery the technetium-99 bone scans remained active. Removal of the lesion was formerly the treatment of choice in adults. Recently it has been shown that knees in which the lesion healed have done significantly better than knees from which the fragment was removed.[30] Osteochondritis dissecans is best diagnosed and treated before skeletal maturity and restoration of the articular surface should be attempted whenever possible.

FRACTURES ABOUT THE KNEE

Major fractures in adults

Fractures about the knee are relatively rare in athletics and are usually limited to contact and high velocity sports. Associated vascular disruption is again rare, but the potential does exist. Arterial injury is a surgical emergency; therefore immediate recognition and investigation is essential.

Fractures about the knee can be very serious and may terminate an athlete's career. Proper treatment is critical if the athlete is to be able to return to sport and compete at his or her previous level. The patella is the most frequently fractured bone in knee injuries.[45]

The evaluation of a fracture should include examination of all the soft tissues and neurovascular structures around the knee. A precise description of the fracture, including its displacement and stability, must be documented and potential complications addressed. Radiographs and occasionally CT scans must be employed to allow proper description and categorization of the fracture into one of the accepted classification schemes.

The most widely accepted classification scheme of supracondylar fractures was devised by Müller and colleagues.[46] This system is relatively simple to use and identifies three general types of fractures; each with three subtypes (Fig. 10). The revised classification (Fig. 11) for tibial plateau fracture dislocations proposed by Hohl *et al.*[46] is mainly used in North America, whereas Europeans tend to use Muller's or Duparc's[78] classification schemes.

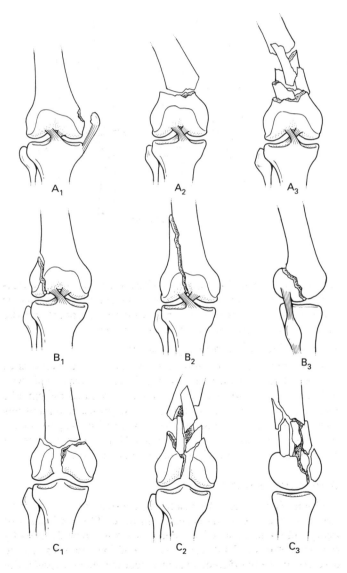

Fig. 10 Müller's comprehensive classification of supracondylar fractures.

Fractures of the proximal end of the fibula occur mainly in association with proximal tibial fractures, particularly split compression, bicondylar, and subcondylar types. The integrity of the peroneal nerve, the anterior tibial artery, the biceps tendon, and the lateral collateral ligament must be assessed following a proximal fibular fracture to rule out any associated injury to these anatomical structures.

The indications for non-operative and operative treatment depend on many factors, such as fracture displacement or depression, the patient's health and skin condition, associated soft tissue injuries, additional fractures, and the character of the fracture (open or closed). Internal fixation decreases the rehabilitation time of the athlete by facilitating the early return of knee function.

Operative methods for displaced articular fractures emphasize accurate restoration of the articular surface through open reduction and internal fixation. Displaced extra-articular fractures are treated by internal fixation in order to maintain overall limb alignment and to enable early motion. In some cases articular fractures can be treated with arthroscopy and cannulated screws to lessen the trauma caused during open reduction.[47,48]

Approximately 22 per cent of plateau fractures have associated ligament damage which must always be addressed in any treatment plan.[49,50] Neurovascular status must always be a concern with knee trauma and restoring circulation takes precedence over fracture treatment.

Major fractures in children

The paediatric musculoskeletal system differs from that of adults in a number of significant ways. The bones of a child differ in both architecture and physiology when compared with those of an adult. A child has a cartilaginous epiphyseal plate that is weaker than bone and ligament.[51] The distal femur is responsible for 70 per cent of the longitudinal growth of the femur and the proximal tibia accounts for 60 per cent of the longitudinal growth of the tibia.[52] The high proportion of cartilage in bone during the growing years leads to a unique vulnerability to both trauma and infection of the growth plate. Knee trauma that is overlooked in the presence of a fracture of the femur or tibia could lead to growth arrest. Adolescents who have sustained fractures in their lower limbs that do not initially appear to have physeal plate involvement should nevertheless be evaluated and followed for possible physeal injury about the knee that can only be detected after additional growth has taken place.[53]

Ligament injuries are often associated with physeal fracture about the knee.[54] Physeal injury may cause a rapid arrest of growth[55] and the consequent development of serious bony deformity without proper intervention. The Salter–Harris[56] classification of epiphyseal plate injuries (Fig. 12) is based on the following: the mechanism of injury; the relationship of the fracture line to the growing physis; the method of treatment; the prognosis of the injury with regard to growth disturbances. There is a poor prognosis for growth following a type IV epiphyseal injury unless a perfect surgical reduction is obtained, and a type V injury causes great concern because of the crushing of the plate.

Distal femoral epiphysis fractures

The distal femoral epiphysis is the most common fracture site in young athletes. Abduction and adduction physeal injuries occur more frequently but the hyperextension injury is more dangerous. The popliteal artery, the medial and lateral popliteal nerves, and other soft tissues are easily injured when the distal end of the femoral shaft is driven posteriorly into the popliteal fossa during a hyperextension physeal fracture. Stress radiographs may be

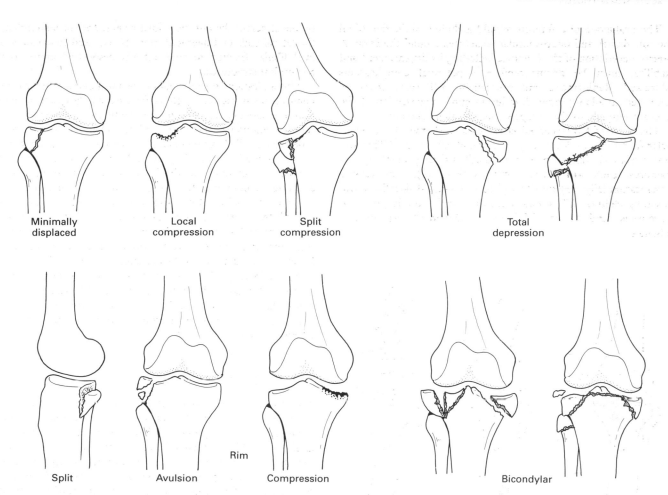

Fig. 11 Hohl's revised classification of tibial articular surface fractures.

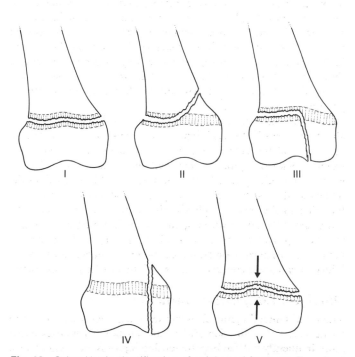

Fig. 12 Salter-Harris classification of epiphyseal fractures. Type I, separation of epiphysis. Type II, fracture-separation of epiphysis. Type III, fracture of part of epiphysis. Type IV, fracture of epiphysis and epiphyseal plate. Type V, crushing of epiphyseal plate.

required to confirm the diagnosis and treatment must be meticulous to avoid complication. Proper anatomical reduction is achieved through open reduction and internal fixation[57] and the ligament stability of the knee must also be checked.

Proximal tibial epiphysis fractures

Proximal tibial physis fractures are rare since the insertion sites for the collateral ligament are distal to the growth plate. This type of injury only occurs after severe trauma and stress radiographs may be required for confirmation of the diagnosis. Complications include popliteal artery disruption and peroneal palsy, particularly in a hyperextension mechanism of injury. Growth disturbances are frequent following a proximal tibial physeal fracture.

Avulsion fractures

Avulsion fractures are the result of excessive stress on specific structures that are firmly anchored to bone. The mechanism of injury and the direction of the traumatic stress can be deduced by examining the resultant joint laxity. Avulsion fractures can be observed in many different sites including the lateral capsule, the tibial eminence, the tibial tubercle, the patella, the popliteus tendon, collateral ligaments, the biceps tendon, and the fascia lata.

Lateral capsular ligament avulsion fracture (Segond's fracture)

As described by Woods et al.,[58] the fleck of bone chipped off the tibia is posterosuperior to Gerdy's tubercle and represents an avulsion of the meniscotibial portion of the middle third of the lateral capsular ligament. Recognition of this fracture in a traumatized knee provides substantial evidence of a significant injury to the lateral capsule (Fig. 13). This fracture has a very strong association with a rupture of the anterior cruciate ligament and usually presents with a haemarthrosis. Meniscal injuries, both medial and lateral, and medial collateral ligament tears are also common.[59]

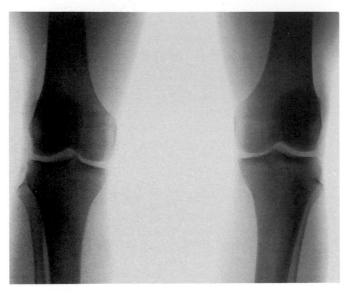

Fig. 13 Anteroposterior radiograph of a Segond fracture showing avulsion of the meniscotibial portion of the middle third of the lateral capsular ligament. This indicates a severe lateral capsular injury and should alert the examiner to the high probability of injury to the anterior cruciate and medial ligaments.

Anterior tibial eminence fracture

Fracture of the tibial eminence in children results from a forceful avulsion of the anterior cruciate ligament. The anterior horn of the medial meniscus lies in front of the cruciate attachment and may block the reduction of a tilted or elevated fragment (Fig. 14).

Avulsion of the anterior tibial eminence frequently occurs as an isolated lesion, but ligaments, menisci, and other soft tissues can also be damaged. These injuries are seen in both adults and adolescents; however the lesions are most frequently observed in the 8 to

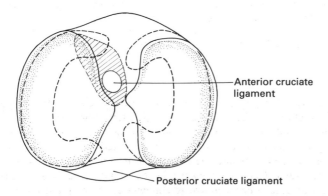

Fig. 14 Tibial plateau attachments of the cruciates and menisci.

15-year-old athlete involved in bicycling or skiing. The usual mechanism of injury involves application of excessive tensile force to be anterior cruciate ligament, leading to an avulsion of the tibial eminence. Because of the relative joint laxity in children, the femoral condyle may 'knock off' the tibial eminence in some injuries.[60] Meyers and McKeever[61] described a radiographic classification of anterior tibial eminence fractures according to their displacement (Fig. 15).

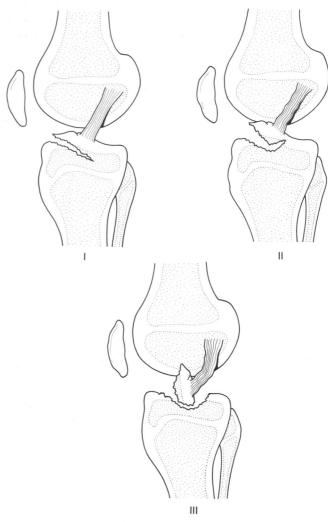

Fig. 15 Meyers and McKever classification of fractures of the intercondylar eminence of the tibia. Type I, minimal displacement of the avulsed fragment. Type II, displacement of the anterior portion of the avulsed fragment, producing a beaklike appearance. Type III, complete displacement of the fragment.

Treatment requires aspiration, reduction when necessary, and immobilization of the knee in 10° of flexion for 6 weeks. A type II lesion can be reduced with full extension followed by cast immobilization. The reduction of a type III fracture may be prevented by the anterior horn of the medial meniscus or floating of the avulsed fragment. Reduction of these lesions can be accomplished through either open or arthroscopic surgery.

Fracture of the posterior intercondylar eminence

The posterior intercondylar eminence declines steeply from the tibial spines to the posterior surface of the proximal tibial metaphysis and provides attachment for the posterior cruciate ligament. The mechanism of injury involved in a fracture of the posterior intercondylar eminence is forced hyperextension (with a concomitant rupture of the anterior cruciate ligament) or a blow to the anterior proximal end of the tibia. This injury is observed in older adolescents or adults. The patient typically presents with a swollen knee that is painful in flexion and is tender to palpation in the posterior fossa.

The fracture is best visualized on the lateral radiography projection when displacement has taken place. It is best reduced through a posteromedial approach, thereby preserving the medial gastrocnemius head, and fixing the avulsed fragment with suture, screw, or staple.

The extensor mechanism

The extensor mechanism can fail at multiple locations. Avulsion fractures may occur at the superior pole of the patella, along its medial border, or at the insertion of the patellar ligament on the tibial tuberosity.

Fractures of the tibial tuberosity have been observed in adults but are usually encountered in late adolescence in gymnasts, basketball players, or high jumpers. The injury occurs during the period in which the proximal tibial epiphysis and the secondary ossification centre of the tubercle are undergoing modification. Columnated bone cells are replacing most of the fibrocartilaginous elements making them more susceptible to tensile stress. An avulsion fracture of the tuberosity leaves the germinal cartilage cells exposed on the tubercle.

This situation is completely different from Osgood–Schlatter's disease[62] which has no physeal involvement. The lesion in Osgood–Schlatter's disease (Fig. 16) is caused by multiple tears at the insertion of the patellar tendon on the most anterior aspect of the tuberosity. On occasion a fragment of bone may be avulsed and produced calcification in the midst of the tendon insertion.

It has been suggested that Osgood–Schlatter's disease predisposes the athlete to further injury of the extensor mechanism. A few cases have been reported but no firm association has been demonstrated; therefore patients with Osgood–Schlatter's disease should not be withheld from athletics.[45,63]

The mechanism of injury can be forced flexion of the knee against an eccentric quadriceps contraction. Patients present with pain, swelling, and tenderness about the knee. The integrity of the extensor mechanism must be checked. A careful examination of the neurovascular structures must be performed to avoid a superior compartment syndrome. Radiographic examination including oblique films will allow for proper description of the fracture according to Ogden's classification (Fig. 17).[64]

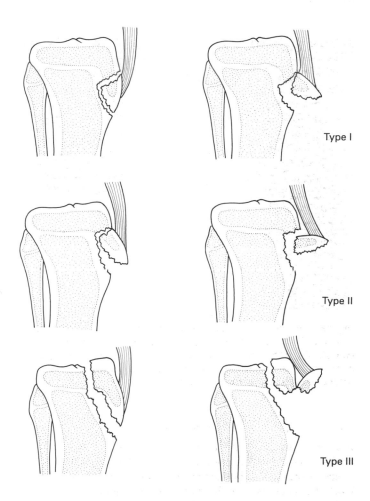

Fig. 17 Ogden–Murphy classification of tibial tuberosity fractures. Three types of fractures have been described depending on the distance of the fracture from the distal tip of the tubercle. Each type is then divided into two subtypes depending on the severity of displacement and comminution. Type I, the fracture is distal to the normal junction of the ossification centres of the proximal tibia and tuberosity. Type II, the separation occurs anteriorly through the area bridging the ossification centres of the tibial tubercle and the proximal tibial epiphysis. Type III, the separation extends across the proximal tibial epiphysis into the knee joint.

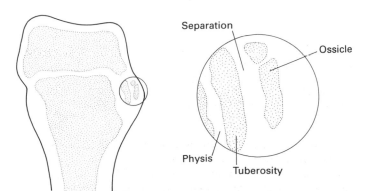

Fig. 16 Ogden–Southwick aetiology of Osgood-Schlatter's lesion. The primary lesion appears to be an avulsion of ossicles at the insertion of the patellar tendon on the anterior aspect of the developing ossification centre of the tibial tuberosity. This results in callus formation in the intervening area, thus enlarging the anterior portion of the tuberosity. The physis of the tibial tuberosity remains intact.

A type III injury in Ogden's classification is a Salter–Harris type III fracture[65] of the proximal tibial epiphysis. Non-displaced fractures should be cast in extension, whereas a displaced fracture requires open reduction and internal fixation (pin, screw, or tension band). Despite the concern of potential growth disturbances, none has been reported in the literature to date. These injuries usually occur in late adolescence as the epiphyseal plates are closing.

Avulsion of the popliteus tendon

This isolated lesion may be a rare cause of haemarthrosis. The mechanism of injury usually involves a twisting motion that is never well described by the patient. Radiographs may reveal a small fragment on the lateral condyle and arthroscopic examination clearly demonstrate the lesion.[66] The potential benefit of surgical reinsertion of the displaced fragment has yet to be determined.

Avulsion of the collateral ligaments

Although isolated collateral ligament avulsion can occur, this injury is more frequently associated with significant soft tissue trauma. The medial collateral ligament usually avulses from its proximal attachment. The lateral collateral ligament is pulled off distally, generally producing a type I physeal fracture of the proximal fibula. A type III avulsion fracture, in which the fibular styloid is pulled upwards, has also been described. The integrity of the peroneal nerve and anterior tibial artery must be assessed.

Avulsion of the fascia lata

Avulsion of the fascia lata can occur at the level of Gerdy's tubercle. This lesion is usually associated with additional rupturing of other lateral structures. On radiographs the avulsion fragment from Gerdy's tubercle is usually located more anteroinferior than the Segond fracture (Fig. 18).

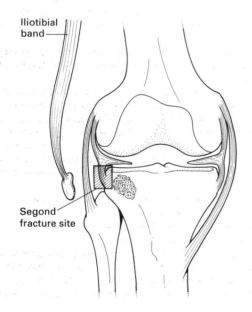

Fig. 18 Avulsion fracture of the iliotibial band from Gerdy's tubercle. This lesion is usually located more inferiorly and anteriorly than the Segond fracture.

Osteochondral and chondral fractures

Landells[67] noted that subchondral bone does not develop its structural integrity until skeletal maturity. The adolescent has little calcified cartilage; therefore tangential forces are directed to the subchondral region, which explains the high incidence of osteochondral fractures at this age group. In adults the articular cartilage tends to tear along the junction of calcified and uncalcified cartilage, which is referred to as the tidemark.[67,68] For

this reason, adults are more likely to sustain chondral fractures than osteochondral lesions when exposed to similar tangential forces (Fig. 19).

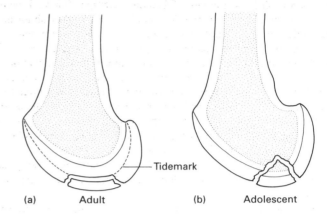

Fig. 19 (a) Adults tend sustain chondral fractures since the articular cartilage tears along the tide mark where the shear forces are dissipated. (b) Adolescents have little calcified cartilage; therefore, tangential forces are directed to the subchondral region explaining the high incidence of osteochondral fractures in this age group.

Osteochondral fractures

Osteochondral fractures have been defined by Cohn et al.[45] as 'fracture of the articular cartilage and underlying bone within area of patella, femur and tibia devoid of load transmitting soft tissue attachments'.

The mechanism of injury usually involves a tangentially directed force that combines both shear and compression. However, the force can be purely compressive in nature as would occur during impaction.[68] The majority of osteochondral fractures arise in adolescents. The most frequent osteochondral fractures are related to patellar dislocations.[45] In addition to the patella, most other lesions are located in the lateral[69] and medial condyles.

Clinical presentation is that of an acute injury resulting from a direct blow or twisting motion with the knee in a more or less extended position. Some patients can feel a crack and pain may preclude weight bearing. Swelling occurs within the hour and aspiration of the knee reveals a blood-stained synovial fluid containing fat globules. A displaced fragment may cause locking of the joint.

Radiographs confirm the diagnosis by demonstrating the osteochondral fragment but several views may be required. The defect may be more difficult to visualize in the early phase. Surgical treatment is required. Arthroscopy will allow confirmation of the diagnosis, assess the location, size, and stability of the lesion, and enable a course of treatment to be decided.

If both cartilagenous and bony portions are intact, reduction and internal fixation should be performed. This could prove to be difficult more than 10 days after the injury because of the fibrous tissue formed in the crater. If the bony portion is too small or too fragmented to accommodate an internal fixation, the entire fragment will be excised and the bed debrided.

Early rehabilitation should allow a good early result, but long-term prognosis remains unknown. Recently, MRI has allowed a better detection of osteochondral injuries;[70,71] the overlying cartilage is either continuous or fractured.

Chondral fractures

The articular cartilage of the knee is exposed to acute and repetitive trauma from both endogenous and exogenous sources. Chondral fractures represent a failure of the tensile and shock-absorbing capabilities of hyaline cartilage. These lesions can occur in isolation or may be associated with other abnormal mechanics such as a rupture of the anterior cruciate ligament.

Complete chondral fractures can present as a stellate crater or as a flap with a vertical margin.[72,73] Incomplete chondral fractures can form an incomplete flap or a defect not reaching the subchondral bone.[74] Some lesions have an intact surface in which the disruption is of the deep cartilage structure and is caused by large shear forces. These lesions are difficult both to evaluate and to treat. The defects are usually located on the femoral condyles both medially and laterally. They can be found in the non-meniscal weight-bearing area in extension, but also more posteriorly in flexion.[75] Several cases on the lateral tibial plateau have been reported.[75]

Most patients are able to recall a specific mechanism of injury contact, impaction or torsion. However, some are unable to recall any specific information regarding their injury. Symptoms are usually rather non-specific. Pain can be either localized or diffuse. Intermittent pain exacerbated by physical activity is the most common complaint.

Physical examination reveals a picture of mechanical internal derangement, suggestive of a meniscal lesion.[76,77] Signs and symptoms include joint line tenderness, clicking, catching, and sometimes locking. Knee effusion may occur acutely following trauma or be delayed in presentation. Radiographs show no evidence of osseous injury and arthrography, when performed, is normal.

The diagnosis is frequently delayed and not made prior to surgery. Because of persistent symptoms or the possibility of a meniscal lesion, the patient finally undergoes an arthroscopic examination. One of the benefits of arthroscopy has been the delineation of chondral injuries.

Terry[77] insists on examining the posterior condyles in deep flexion. In addition to locating and describing the lesion, the rest of the internal structures of the knee should be examined for associated pathology. The following rationale for the treatment of chondral lesions must be kept in mind. Hyaline cartilage has no capacity for healing. Healing can occur after subchondral bone has been exposed. The response to injury of the subchondral bone will be to set up an inflammatory response leading to production of fibrous tissues or fibrocartilage which will never have the same quality as the original hyaline cartilage (endurance, shock-absorbing, and gliding properties).

Incomplete lesions will be debrided to avoid abrasion. Complete lesions will be debrided until the margins are stable and vertical, adjacent subchondral bone will be slightly abraded or perforated to promote the healing response.

The postoperative course is usually longer than after an arthroscopic partial meniscectomy. The prognosis depends on the extent and location of the lesion and any associated pathology.

REFERENCES

1. Cohn SL, Taylor WC. Vascular problems of the lower extremity in athletes. *Clinics in Sports Medicine* 1990; **9**: 449–70.
2. Kennedy JC. Complete dislocation of the knee joint. *Journal of Bone and Joint Surgery* 1963; **45A**: 889–904.
3. Quinlan AG. Irreducible posterolateral dislocation of the knee with buttonholing of the medial femoral condyle. *Journal of Bone and Joint Surgery* 1966; **48**: 1619–21.
4. Sisto DJ, Warren RF. Complete knee dislocation. A follow-up study of operative treatment. *Clinical Orthopaedics and Related Research* 1985; **198**: 94–101.
5. Bloom MH. Traumatic knee dislocation and popliteal artery occlusion. *Physician and Sports Medicine* 1987; **15(10)**: 143–55.
6. Green NE, Allen BL. Vascular injuries associated with dislocation of the knee. *Journal of Bone and Joint Surgery* 1977; **59A**: 236–9.
7. Rich NM, Spencer FC, eds. Concomitant fractures and nerve trauma. In: *Vascular trauma*. Philadelphia: WB Saunders, 1978: 549.
8. Cone JB. Vascular injury associated with fracture-dislocations of the lower extremity. *Clinical Orthopaedics and Related Research* 1989; **243**: 30–5.
9. DeBakey ME, Simeone FA. Battle injuries of arteries in World War II, an analysis of 2471 cases. *Annals of Surgery* 1946; **123**: 534–79.
10. Crothers, OD, Johnson JTH. Isolated acute dislocation of the proximal tibiofibular joint. *Journal of Bone and Joint Surgery* 1973; **55A**: 181–3.
11. Lord CD, Coutts JW. A study of typical parachute injuries occurring in two hundred and fifty thousand jumps at the parachute school. *Journal of Bone and Joint Surgery* 1944; **26**: 547–57.
12. Ogden JA. Subluxation and dislocation of the proximal tibiofibular joint. *Journal of Bone and Joint Surgery* 1974; **56A**: 145–54.
13. Parkes JC, Selko RR. Isolated acute dislocation of the proximal tibiofibular joint. *Journal of Bone and Joint Surgery* 1973; **55A**: 177–80.
14. Turco VJ, Spinella AJ. Anterolateral dislocation of the head of the fibular in sports. *American Journal of Sports Medicine* 1985; **13**: 209–15.
15. Lyle HHM. Traumatic luxation of the head of the fibular. *Annals of Surgery* 1925; **82**: 635–9.
16. Benazet JP, Saillant G, Cazeneuve JF, Lazennec JY, Roy Camille R. Le traitement chirurgical de l'instabilité chronique de l'articulation péronéotibiale supérieure. *Journal de Traumatologie du Sport* 1989; **6**: 97–102.
17. Clanton TO, DeLee JC. Osteochondritis dissecans. History, pathophysiology and current treatment concepts. *Clinical Orthopaedics and Related Research* 1982; **167**: 50–64.
18. Rogers WM. Gladstone H. Vascular foramina and arterial supply of the distal end of the femur. *Journal of Bone and Joint Surgery* 1950; **32A**: 867–74.
19. Mubarak SJ, Carroll NC. Juvenile osteochondritis dissecans of the knee. Etiology. *Clinical Orthopaedics and Related Research* 1981; **157**: 200–11.
20. Roy Petrie PW. Aetiology of osteochondritis dissecans. Failure to establish a familial background. *Journal of Bone and Joint Surgery* 1977; **59B**: 366–7.
21. Ribbing S. The hereditary multiple epiphyseal disturbance and its consequences for the aetiogenesis of local malacias—particularly the osteochondrosis dissecans. *Acta Orthopaedica Scandinavia* 1955; **24**: 286–99.
22. Barrie HJ. Osteochondritis dissecans 1887–1987. *Journal of Bone and Joint Surgery* 1987; **69B**: 693–5.
23. Steiner ME, Grana WA. The young athlete's knee: recent advances. *Clinics in Sports Medicine* 1988; **7**: 527–46.
24. Bradley J, Dandy DJ. Osteochondritis dissecans and other lesions of the femoral condyles. *Journal of Bone and Joint Surgery* 1989; **71B**: 518–22.
25. Cahill BR. Treatment of juvenile osteochondritis dissecans and osteochondritis dissecans of the knee. *Clinics in Sports Medicine* 1985; **4**: 367–84.
26. Cahill BR, Phillips MR, Navarro R. The results of conservative management of juvenile osteochondritis dissecans using joint scintigraphy. *American Journal of Sports Medicine* 1989; **17**: 601–6.
27. Chiroff RT, Cook CP. Osteochondritis dissecans. A histologic and microradiographic analysis of surgically excised lesions. *Journal of Trauma* 1975; **15**: 689–96.
28. Guhl JF. Arthroscopic management of osteochondritis dissecans in arthroscopic surgery. In: McGinty JB, ed. *Arthroscopic surgery update*. Rockville: Aspen Systems Corporation, 1985: 63–84.
29. Vince KG. Osteochondritis dissecans of the knee. In: Scott WN, ed. *Arthroscopy of the knee: diagnosis and treatment*. Philadelphia: WB Saunders, 1990: 175–91.
30. Hughston JC, Hergenroeder PT, Courtenay BG. Osteochondritis dissecans of the femoral condyles. *Journal of Bone and Joint Surgery* 1984; **66A**: 1340–8.

31. Edwards DH, Bentley G. Osteochondritis dissecans patellae. *Journal of Bone and Joint Surgery* 1977; **59B**: 58–63.

32. Schwarz C, Blazina ME, Sisto DJ, Hirsch LC. The results of operative treatment of osteochondritis dissecans of the patella. *American Journal of Sports Medicine* 1988; **16**: 522–9.

33. Kurzweil PR, Zambetti GJ, Hamilton WG. Osteochondritis dissecans in the lateral patellofemoral groove. *American Journal of Sports Medicine* 1988; **16**: 308–10.

34. Outerbridge RE. Osteochondritis dissecans of the posterior femoral condyle. *Clinical Orthopaedics and Related Research* 1983; **175**: 121–9.

35. Wilson JN. A diagnostic sign in osteochondritis dissecans of the knee. *Journal of Bone and Joint Surgery* 1967; **49A**: 477–80.

36. Mesgarzadeh M. *et al.* Osteochondritis dissecans; analysis of mechanical stability with radiography, scintigraphy, and MR imaging. *Radiology* 1987; **165**: 775–80.

37. Dipaola JD, Nelson DW, Colville MR. Characterizing osteochondral lesions by magnetic resonance imaging. *Arthroscopy* 1991; **7**: 101–4.

38. Linden B. Osteochondritis dissecans of the femoral condyles. A long-term follow-up study. *Journal of Bone and Joint Surgery* 1977; **59A**: 769–76.

39. Lee CK, Mercurio C. Operative treatment of osteochondritis dissecans in situ by retrograde drilling and cancellous bone graft: A preliminary report. *Clinical Orthopaedics and Related Research* 1981; **158**: 129–36.

40. Yamashita F, Sakakida K, Suzu F, Takai S. The transplantation of an autogeneic osteochondral fragment for osteochondritis dissecans of the knee. *Clinical Orthopaedics and Related Research* 1985; **201**: 43–50.

41. Johnson LL, Uitvlugt G., Austin MD, Detrisac DA, Johnson C. Osteochondritis dissecans of the knee.Arthroscopic compression screw fixation. *Arthroscopy* 1990; **6**: 179–89.

42. Gillespie HS, Day B. Bone peg fixation in the treatment of osteochondritis dissecans of the knee joint. *Clinical Orthopaedics and Related Research* 1979; **143**: 125–30.

43. Anderson AF, Lipscomb AB, Coulam C. Antegrade curettement, bone grafting and pinning of osteochondritis dissecans in the skeletally mature knee. *American Journal of Sports Medicine* 1990; **18**: 254–61.

44. Lipscomb PR Jr, Lipscomb PR Sr, Bryan RS. Osteochondritis dissecans of the knee with loose fragments. *Journal of Bone and Joint Surgery* 1978; **60A**: 235–40.

45. Cohn SL, Sotta RP, Bergfeld JA. Fractures about the knee in sports. *Clinics in Sports Medicine* 1990; **9**: 121–39.

46. Hohl M, Johnson EE, Wiss DA. Fractures of the knee. In: Rockwood CA Jr, Green DP, Bucholz RW, eds. *Fractures in adults.* 3rd edn. Philadelphia: JB Lippincott, 1991: 1725–97.

47. Mast J, Jakob R, Ganz R. *Planning and reduction technique in fracture surgery.* Heidelberg: Spring-Verlag, 1989

48. Siliski JM, Mahring M, Hofer HP. Supracondylar–intercondylar fractures of the femur. Treatment by internal fixation. *Journal of Bone and Joint Surgery* 1989; **71A**: 95–104.

49. Dejour H, Chambat P, Caton J, Melere G. Les fractures des plateaux tibiaux avec lésion ligamentaire. *Revue de Chirurgie Orthopédique et Réparatrice de l'appareil Moteur* 1981; **67**: 593–8.

50. Delamarter RB, Hohn M, Hopp E Jr. Ligament injuries associated with tibial plateau fractures. *Clinical Orthopaedics and Related Research* 1990; **250**: 226–33.

51. Speer DP, Braun JK. The biomechanical basis of growth plate injuries. *Physician and Sports Medicine* 1985; **13(7)**: 72–8.

52. Mayer PJ. Lower limb injuries in childhood and adolescence. In: Micheli LJ, ed. *Pediatric and adolescent sports medicine.* Boston: Little Brown, 1984: 80–106.

53. Hresko MT, Kasser JR. Physeal arrest about the knee associated with non-physeal fractures in the lower extremity. *Journal of Bone and Joint Surgery* 1989; **71A**: 698–703.

54. Bertin KC, Goble EM. Liagment injuries associated with physeal fractures about the knee. *Clinical Orthopaedics and Related Research* 1983; **177**: 188–95.

55. Stephens DC, Louis E, Louis DS. Traumatic separation of the distal femoral epiphyseal cartilage plate. *Journal of Bone and Joint Surgery* 1974; **56A**: 1383–90.

56. Salter RB, Harris WR. Injuries involving the epiphyseal plate. *Journal of Bone and Joint Surgery* 1963; **45A**: 587–622.

57. Muller ME, Allgower M, Schneider R, Willenegger H. *Manual of internal fixation.* 2nd edn. New York: Springer-Verlag, 1979.

58. Woods GW, Stanley RF, Tullos HS. Lateral capsular sign: X-ray clue to a significant knee instability. *American Journal of Sports Medicine* 1979; **7**: 27–33.

59. Goldman AB, Pavlov H, Rubinstein D. The Segond fracture of the proximal tibia: a small avulsion that reflects major ligamentous damage. *American Journal of Roentgenology* 1988; **151**: 1163–7.

60. Strizak AM, Stoberg AJ. Knee injuries in the skeletally immature athlete. In: Nicholas JA, Hershman EB, eds. *The lower extremity and spine in sports medicine.* St Louis: CV Mosby, 1986, 1262–91.

61. Meyers MH, McKeever FM. Fracture of the intercondylar eminence of the tibia. *Journal of Bone and Joint Surgery* 1970; **52A**: 1677–84.

62. Ogden JA, Southwick WO. Osgood–Schlatter's disease and tibial tuberosity development. *Clinical Orthopaedics and Related Research* 1976; **116**: 180–9.

63. Singer KM, Henry J. Knee problems in children and adolescents. *Clinics in Sports Medicine* 1985; **4**: 385–97

64. Ogden JA, Tross RB, Murphy MJ. Fractures of the tibial tuberosity in adolescents. *Journal of Bone and Joint Surgery* 1980; **62A**: 205–14.

65. Abrams J, Bennett E, Kumar SJ, Pizzutillo PD. Salter–Harris type III fracture of the proximal fibula. A case report. *American Journal of Sports Medicine* 1986; **14**: 514–16.

66. Gruel JB. Isolated avulsion of the popliteus tendon. *Arthroscopy* 1990; **6**: 94–5.

67. Landells JW. The reactions of injured human articular cartilage. *Journal of Bone and Joint Surgery* 1957; **39B**: 548–62.

68. Kennedy JC, Grainger RW, McGraw RW. Osteochondral fractures of the femoral condyles. *Journal of Bone and Joint Surgery* 1966; **48B**: 436–40.

69. Matthewson MH, Dandy DJ. Osteochondral fractures of the lateral femoral condyle. *Journal of Bone and Joint Surgery* 1978; **60B**: 199–202.

70. Lee JK, Yao L. Occult intraosseous fracture: magnetic resonance appearance versus age of injury. *American Journal of Sports Medicine* 1989; **17**: 620–3.

71. Maywood RM, Jackson DW, Berger P. Athletic injuries to the knee. Evaluation using magnetic resonance imaging. *Physician and Sports Medicine* 1988; **16(5)**: 81–95.

72. Dzioba RB. The classification and treatment of acute articular cartilage lesions. *Arthroscopy* 1988; **4**: 72–80.

73. Hubbard MJS. Arthroscopic surgery for chondral flaps in the knee. *Journal of Bone and Joint Surgery* 1987; **69B**: 794–6.

74. Johnson-Nurse C, Dandy DJ. Fracture-separation of articular cartilage in the adult knee. *Journal of Bone and Joint Surgery* 1985; **67B**: 42–3.

75. Terry GC, Flandry F, Van Manen JW, Norwood LA. Isolated chondral fractures of the knee. *Clinical Orthopaedics and Related Research* 1988; **234**: 170–7.

76. Gerard Y, Segal P, Henry C. Lésions traumatiques cartilagineuses pures du condyle interne du genou en pratique sportive. *Revue de Chirurgie Orthopedique et Reparatrice de l'appareil Moteur* 1976; **62**: 245–52.

77. Hopkinson WJ, Mitchell WA, Curl WW. Chondral fractures of the knee. Cause for confusion. *American Journal of Sports Medicine* 1985; **13**: 309–12.

78. Huten D, Duparc J, Cavagna R. Fractures des plateaux tibiaux de l'adulte. *Encyclopédie Médico-Chirurgicale.* Paris, 1990; **14082**: A10, 1–12.

4.2.5 The patella: its afflictions in relation to athletics

JEFFREY MINKOFF AND BARRY G. SIMONSON

INTRODUCTION

The patellofemoral articulation is among the most complex and least understood articulations in the body. Its disorders, to which have been ascribed the frequently used, though nebulously defined, diagnoses of chondromalacia, maltracking, malalignment, and instability, are among the more commonly seen in sports medicine practices comprising young and middle-aged recreational and subelite competitive athletes. Conversely, these disorders more rarely present for treatment in elite amateur and professional populations. This observation does not necessarily imply an absence of developmental and structural disorders of the patellofemoral joint in these athletes. However, it follows as a natural corollary of the doctrine of natural selection that the more severely afflicted athletes are less likely to achieve an elite status. In some instances of more moderate structural handicaps, athletes of elite destiny may override or compensate for their physical disability by virtue of exceptional motor strength and/or technical skills.

Of course, there are the usual exceptions to customary trends. Gymnasts and some dancers are characteristic of athletes whose excellence is predicated upon a measure of joint laxity and flexibility. These performers often achieve an elite status despite their possession of frank patellofemoral instability or even recurrent patellar dislocation.

Parapatellar disorders, such as patellar tendinitis and retinacular inflammation, are quite common among elite athletes. They are particularly prevalent among athletes whose sports require repetitive percussion and/or deceleration in an eccentric format. Basketball and high jumping are prime examples of such sports, and it is from the latter that the name 'high jumper's knee' is derived. 'Combined physicals' have been initiated by the National Basketball Association in very recent years. The purpose of these physicals is to pre-evaluate, for all National Basketball Association professional teams, those college players most likely to be drafted. Eighty-six players were examined in the 1990 physicals. Injuries recorded were not for a single playing year but for the span of the player's basketball life to date. Patella-related problems were the most numerous among all recorded injuries and diagnosis. These problems included chondromalacia with maltracking or alta, status after patellectomy (for the latter), patellar tendinitis, patellar fracture, and ruptured patellar tendon. Despite the range and apparent severity of some of the entities listed, no player manifested a degree of disability which precluded competitive play.[1]

Despite the common presence of patellar laxity and alta among elite basketball players, patellar tendinitis is frequently the only symptomatic manifestation (versus instability or retropatellar pain) in this group.

A number of patellofemoral disorders are predicated upon chronology. Certain congenital and/or development disorders (e.g. dysplasias) are evidenced in the preadolescent period. Osgood–Schlatter's disease and the Sinding-Larsen–Johansson syndrome are manifested in the periadolescent period. Patellar tendinitides are most often seen in younger adults. Distal quadriceps tendinitis and impingements of patella magna are most characteristic of middle-aged athletes. Quadriceps rupture is most commonly observed in the seventh decade. Chondromalacia and manifestations related to maltracking and instability are seen at all ages.

Fractures of the patella are uncommonly observed in the athletic population. Symptoms related to multipart patellae are a consequence of trauma and are seen less rarely. Iatrogenic fractures are created in the process of autogenous bone–patellar tendon–bone harvesting for anterior cruciate ligament reconstructions.

What follows below is a detailed discussion of patellofemoral disorders, their presentation, diagnosis, and treatment, and their relationship to athletic participation.

PATELLA

The patella is a sesamoid bone which develops within the quadriceps muscle–tendon unit. Its development begins during the ninth embryonic week. The shape of its cartilaginous anlage is well defined by birth, but ossification is not initiated until between the ages of 3 and 6 years. The ossification may occur in as many as 6 separate centres which gradually coalesce. Ossification is usually complete by the start of the second decade.

PATELLAR MORPHOLOGY

Anthropological studies by Vriese[2] disclosed no apparent racial differences in the morphological dimensions of the patella. Its vertical length varies between 47 and 58 mm, and its width between 51 and 57 mm. Anteroposterior thickness is more variable, with a range of 2 to 3 cm when measured from the subchondral area of the median ridge to the superficial cortex. The chondral surface is also of variable thickness; its central portion is thickest, and in fact is the thickest in the body, being rather more than a quarter of an inch.[2]

The anterior or extensor surface of the patella is slightly convex and is divisible into three parts. The superior third of this surface is rough. It receives the insertion of the quadriceps tendon, a portion of which continues in the distal direction over the anterior patellar surface to form the deep fascia overlying and adherent to the latter. The middle third is replete with vascular foramina and the distal third is incarcerated within the patellar tendon.[2]

The posterior surface of the patella is divisible into superior and inferior portions, comprising roughly 75 per cent and 25 per cent of the surface respectively. The superior portion is exclusively articular. The inferior portion is dotted with vascular orifices related to the normally adherent infrapatellar fat pad.[2]

The ovoid articular surface is comprised of two major facets divided by a vertically oriented promontory, the 'median ridge'. The lateral facet is usually wider and the medial facet shows the greatest variability. The latter is subdivided into the medial facet proper and a smaller odd facet segregated by a small vertical ridge which corresponds to the curve of the lateral border of the medial femoral condyle of the fully flexed knee. The medial ridge conforms to the straight medial border of the lateral femoral condyle. Interesting discussions of the patellar facets and their relations are available in the literature.[2-6]

The patella serves two major functions. One is to protect the femoral condyles and the other is to enhance the effectiveness of extensor apparatus of the knee joint. The latter is accomplished in two ways. The first is by increasing the movement arm of the extensor mechanism and keeping centralized the force being generated. The second is by providing a low friction articular surface, thereby increasing the efficiency of glide of the extensor apparatus.[7]

CONGENITAL AND DEVELOPMENTAL ABNORMALITIES

Several developmental abnormalities of the patellofemoral mechanism either impact upon the efficiency of its functions or may produce symptoms.

Hypoplastic or absent patella

The most dramatic abnormality is absence of the patella. This is an extremely rare condition which is not encountered in the elite athlete by virtue of the performance disability with which it is associated. Absence of the patella is most commonly one component of arthro-onychodysplasia or the nail-patella syndrome. The mechanical disability engendered by a congenitally absent patella is less severe than might be imagined because the associated enlargements of the femoral condyles and tibial tubercle enhance the leverage of a centralized extensor mechanism. However, when absence or hypoplasia of the patella occurs as an isolated entity, there may be a decentralization of the extensor mechanism.[8,9]

Severe lateral dislocation of the extensor mechanism (and patella, if present) is not so rare. The authors have seen at least one such case in which hypoplastic femoral condyles and laterally dislocated hypoplastic patella and extensor mechanism failed to preclude participation as a first-term player in a high school basketball squad (Fig. 1).

Multipartite patellae

Multipartite patellae, particularly bipartite patellae, are very common. The incidence of bipartite patellae in adolescents has been reported to be between 0.2 and 6 per cent.[10] It is usually evident by 12 years of age. Secondary ossification centres responsible for this entity are most commonly seen superolaterally, but can be seen at the distal pole, at the lateral margin, anterocentrally, or anterodistally. The incidence is predominant in males by a 2:1 ratio.[10]

While the bipartite patella has been said to be bilateral in most instances, Green[11] reports a unilateral occurrence of 57 per cent. The variance in demographic reports may in part be ascribable to a clinical differential between the developmental and post-traumatic varieties. Indeed, the greatest import of the developmental bipartite patella, typically an asymptomatic entity, is its differentiation from an acute fracture of the patella. In the skewed circumstances in which a patient presents with patellar pain, the

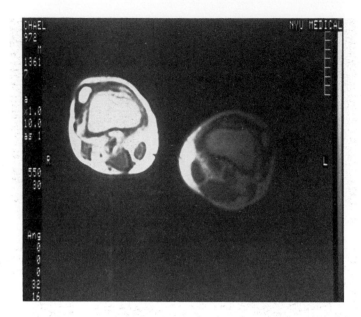

Fig. 1 MRI axial view showing hypoplastic femoral condyles and a congenital laterally dislocated patella.

distinction between entities is often discernible by plain radiography. The fractured patella demonstrates irregular margins of cancellous bone, while the developmental entity is hallmarked by smooth margins of cortical bone (Fig. 2). Bilateral mirror-image lucencies may establish the diagnosis of the developmental form. MRI, CT, or bone scans can be used for differentiation if necessary.

Osgood–Schlatter's disease

The most frequently encountered symptomatic condition of the patellofemoral mechanism during development is that known as Osgood–Schlatter's disease. Osgood–Schlatter's disease is more frequent in males than in females (3:2 ratio), more predominant on the left side, and bilateral in up to 25 per cent of cases.[12] Characteristically, a periadolescent male between the ages of 11 and 15 years will present with pain and swelling about the tibial tubercle of one or both knees.[13] Pain is frequently exacerbated by impact and decelerating activities such as running, jumping, and cutting. The most commonly described physical findings are exquisite tenderness about an obviously enlarged tibial tubercle with overlying soft tissue swelling (Fig. 3).

What is of importance, though not commonly reported or appreciated, is that the syndrome (not a disease) is often associated with patella alta and/or instability, tibiofemoral rotational laxity, and tenderness about the medial patellar retinaculum, inferior pole, and/or the patellar tendon. These additional findings may implicate instability of the patella and maltracking as causative agents in an entity which hedges the issue of acquired versus developmental afflictions.

Pursuant to the foregoing paragraph is the commonly provided explanation that Osgood–Schlatter's disease is a 'traction apophysitis'.[14] What this means in terms of deducing a mechanism by which this syndrome develops is not altogether clear. In 1903, when Osgood and Schlatter independently described the syndrome which bears their names, they suggested that trauma was the inciting cause.[12]

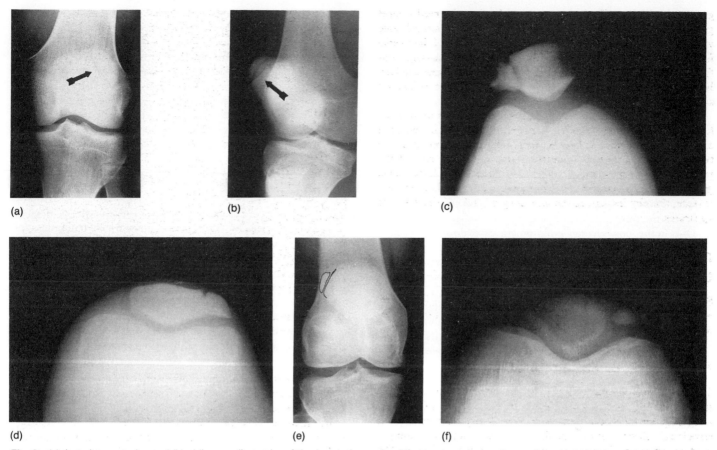

Fig. 2 (a) Anterior-posterior and (b) oblique radiographs of the knee of a professional hockey player after a direct impact injury to the knee. The superolateral location of the lucent line (arrow) and its relatively smooth margins leave doubt as to whether this is a fracture or a bipartate patella. Pain severity, effusion, and a bone scan confirmed the suspicion of acute patellar fracture. (c) Axial radiograph demonstrating a patellar fracture. The irregularity of fragment margins should be noted. (d) Axial radiograph in which a bipartate fragment of the lateral facet was incidentally noted. Despite slight irregularity of the fragment margins, the patient manifested no patellar symptoms and denied a history of direct trauma to the patella. (e) Anterior–posterior radiograph demonstrating a more traditional bipartate fragment of relatively small size and smoothly marginated (see marker lines). (f) Axial radiograph of an obviously fractured patella. The fragment is situated medially and has been sheared from the patella as a consequence of an acute patellar dislocation.

Fig. 3 Lateral radiograph of an adolescent (open epiphyses) male with pain over the tibial tubercle accompanied by severe tenderness and swelling. The irregularity of the tubercular apophysis and the small opacification anterior to it should be noted.

With respect to the pathogenesis of Osgood–Schlatter's disease, Uhry[15] determined that it is traumatically induced. He indicated that the trauma resulted in a laceration of the interface between the patellar tendon and the tubercle, and in some distances at the cartilaginous plate between the tuberosity and the metaphysis.

During the apophyseal stage (1b in the classification of Ogden and Southwick (1976)),[12] the tuberosity is susceptible to injury. Microavulsions can take place throughout the bone and/or cartilage of the secondary ossification centre because they are weaker than the distal fibrous tissue and adjoining bone.[12] La Zerte and Rapp (1958) reported that histological sections from the tuberosity defect show the presence of granulation tissue and osteoid, suggesting the presence of a healing fracture.[12] Apparently, forceful contractions of the quadriceps by active adolescents can produce these 'fractures' in susceptible tubercles.

According to this explanation, it would not be difficult to assume that dysfunctional patellofemoral mechanisms (e.g. instability and maltracking) played an aetiological role, propagating eccentric tensile forces to the tubercle and thereby producing an indirect traumatic force to the tubercle. Additional credence for these suppositions is provided by the frequent reduction in symptoms produced by bracing, strengthening, and quadriceps stretching programmes. These hypotheses also explain the cessation of symptoms with apophyseal fusion. Obviously, the exclusive occurrence of this syndrome during early adolescence rules out its interference in the performance of elite athletes other than gymnasts and perhaps the youngest of female elite athletes.

Plain radiography often reveals swelling anterior to the tibial tubercle, thickening of the patellar tendon, and, not infrequently, alta or malposition of the patella on special lateral and axial views respectively. In more chronic cases or in older adolescents enlargement of the tubercle and opaque densities near the distal insertion of the tendon may be observed. The most obvious radiographic finding, when present, is fragmentation of the tuberosity. It is also the most variable finding.[12]

Only in the most severely symptomatic cases do afflicted adolescents refrain from activities voluntarily. More typically, parental concern is the stimulus for a visit to the paediatrician or orthopaedic surgeon. Treatment is first targeted at the parents by allaying any fears that Osgood–Schlatter's disease is a foreboding condition, despite the pain and the ominous-sounding name. Typically the parents are assured that it is a self-limiting syndrome, rarely lasting more than 1 to 2 years from its inception. This is a half truth. The truthful half relates to the almost inviolate trend of a reduction in symptoms confined to the tubercle *per se*. However, it should be appreciated that the patellar and tendon manifestations often associated with this syndrome may regress, persist, or become more predominant with continued growth and development of the adolescent.

Treatment of the child is conducted from among several limited options which include activity moderation, soothing physical therapeutic modalities, brace or compression bandage facsimiles, and strengthening and stretching exercises for the muscles con-

trolling the knee. The application of extended knee cylinder casts is no longer an option because of its atrophic effects. In the past, recalcitrant cases, or those in which expedient functional return was demanded, were often treated by tubercular excision with a risk of affecting the growth of the subjacent proximal tibial physis. Occasionally there is a rationale for excising loose tubercular fragments which are producing symptoms. Such surgical exercises should be discouraged until after growth has been completed to avoid growth plate arrest and the development of recurvatum and/or valgus of the knee(s).

Patellar tendinitis and Sinding-Larsen–Johansson syndrome

Sinding-Larsen (1921) and Johansson (1922) each described an entity in adolescents in which there is inferior pole tenderness and radiographic fragmentation without any consistent history of trauma.[12] This entity is another in the continuum of chronologically related traction lesions within the patellofemoral apparatus. Its symptomatic focus is at the distal pole of the patella, a predilected site in adolescents and young adults as opposed to the tubercular locus (Osgood–Schlatter's disease) of younger adolescents, the quadriceps tendinitides of older adults, or the degenerative quadriceps ruptures of septuagenarians.

Typically, the afflicted patient is between 10 and 13 years of age. The stages of radiographic progression of the disorder have been characterized by Medlar and Lyne (1978)[12] (Fig. 4):

Stage I   Normal findings

Stage II   Irregular inferior pole calcification

Stage III   Coalescence of calcification

Stage IVa   Incorporation of calcification by the inferior pole

Stage IVb   A calcified mass, separate from the patella

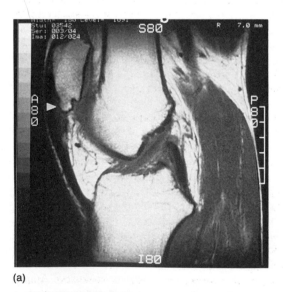
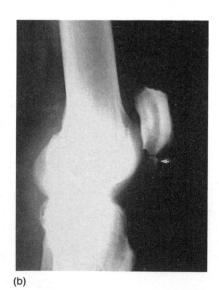

(a)   (b)   (c)

Fig. 4 (a) Sagittal magnetic resonance (MR) image of the knee of a professional basketball player with severe inferior pole pain and tenderness. This appearance is consistent with stage IVa findings of the Medlar and Lyne classification, in which there is incorporation of calcification by the inferior pole (arrow). (b) Lateral radiograph of the knee of a professional basketball player with minimal inferior pole symptoms. This radiograph demonstrates stage IVb in the Medlar and Lyne classification, in which the calcific mass is separate from the patella (arrow). (c) Lateral radiograph of the knee of a professional basketball player with no opacities about the inferior pole. Instead, there is a lucent zone (arrow). This player's symptoms were so severe and recalcitrant that surgery, with debridement of the inferior pole, was necessary.

The onset of symptoms (focal pain and swelling) are often attributed to the patient's having participated in repetitive jumping or other activities responsible for eccentric and impulsion loading of the extensor mechanism. This is consistent with the theory of Curwin and Stanish[16] who suggested that overload occurs in the eccentric phase of quadriceps usage, as when landing from a jump. It is also for this reason that Blazina coined the name 'jumper's knee'.[17]

Sinding-Larsen–Johansson syndrome and patellar tendinitis are similarly derived from overuse traction formats. The high school, college, and adult athletes afflicted with these entities present with pain and swelling at the distal pole of the patella and the subjacent portion of the patellar tendon. On physical examination, tenderness is most easily elicited by placing the knee in extension to relax the extensor mechanism and then palpating the deepest portion of the distal tip of the patella which has been made to stand proud by applying a distally and posteriorly directed force to the superior pole with the opposite hand.

Pathologically, the entity represents a partial tear (or tears) of the patellar tendon just distal to the inferior pole.[17] The deep central fibres of the tendon are those most typically involved. Histologically, the involved areas demonstrate mucoid degeneration and fibrinoid necrosis.[12] Plain radiography is usually unremarkable in cases of less than several months duration.[17] In more chronic cases, fibrocartilaginous metaplasia within the tendon may result in a zone of opacification contiguous with the inferior pole (Fig. 3).

Blazina et al.[18] classified the clinical progression in three phases.

Phase 1 Pain after activity only

Phase 2 Pain during and after activity

Phase 3 Same as in phase 2, but performance is diminishing.

Roels et al. (1978) suggested a similar progression, but with tendon rupture as the ultimate phase.[12]

Ultrasonography may have some worth in the evaluation of patellar tendinitis. It is non-invasive and may document the presence of oedema, haemorrhage, partial rupture, or calcification which may not be evident on plain radiographs.[19] The usual tendon thickness at the inferior pole (usually less than 5 mm) may be increased to more than 10 mm by oedema or haemorrhage and is calculable by ultrasonography.[19]

Bone scans can be used to demonstrate the uptake of tracer at the inferior or superior poles of the patella, with activity at the bone–tendon interface implying a chronic microfailure–reparative process.[19] Abnormal sonograms and positive bone scans usually indicate a chronic and refractory disease process.[19] Jackson et al. (1988) revealed the value of MRI in documenting the components of patellar tendinitis.[19]

The obvious dilemma for the elite athlete afflicted with an inferior pole syndrome who must practise or compete virtually daily throughout a well-defined season, and whose performance is his or her future, is how to keep symptoms minimized while averting a progression in pathology. This is sometimes impossible without sacrificing a portion of the season.

Options for treatment may include each of the four major categories (pharmacological, physiotherapeutic, immobilizing, and surgical) depending upon the severity, chronicity, and expediency of the case in point. Activity modification (curtailment of impulsion activities) is often necessary for the abatement of any inflammatory stimulus. A more rapid reduction of inflammatory signs and symptoms may be accomplished by physiotherapeutic modalities (ice, ultrasound, phonophoresis, iontophoresis, electrotherapy) and non-steroidal anti-inflammatory drugs (NSAIDs). These treatments should not be administered for the purpose of allowing participation (i.e. merely to mask symptoms while allowing damage to accrue), except under the most limited circumstances when participation in a vital event (e.g. a championship or a career-determining event) must not reasonably be abandoned. Steroidal injections are to be avoided for the same reason, as well as for fear of creating iatrogenic degeneration within the tendon. Physical therapy is a standard of treatment. Apart from the application of modalities, the strengthening of all muscle groups of the extremity must be undertaken, titrating the arcs and loads to levels which do not provoke symptoms during or after the sessions. Eccentric strengthening of the quadriceps must be included in the regimen to help make the extensor mechanism more resistant to (i.e. absorb more of the energy of) those impulsion forces which perpetrate the onset of symptoms. Traction effects upon the painful insertion site of the inflamed tendon into the inferior pole may be reduced by maintaining a vigil of quadriceps stretching. In some cases the implementation of patellar restraining braces may impede symptoms by reducing patellar mobility and recoil, and by diffusing the forces at the inferior pole (Fig. 4).

Recalcitrant cases (and some with a special need for expedient resolution) may solicit surgical intervention, though such invasive treatment is rarely necessary. When surgery is performed, it is directed at the excision of degenerated or necrotic tendon and any calcified metaplastic tissue. The nature and magnitude of surgical pathology is a function of the duration of the condition. A representative progression may be as follows: degenerated tissue at 3 to 4 months, haemorrhage and cyst formation at 6 to 12 months, and gritty metaplastic cartilage after 1 year.[17] Surgery for this entity may be performed using local anaesthesia. When the local anaesthetic is administered layer by layer, the focus of tenderness can often be identified intraoperatively.

Postoperatively, motion of the knee is initiated immediately. As the acute symptoms subside, quadriceps stretching to prevent 'hitches' in extensor mechanism glide and hamstring stretching to inhibit a block to terminal extension are initiated. Once full painless motion is achieved, a slow progression of resistive exercises (see above) is begun. Ultimately, agility, proprioceptive, and pleiometric activities are progressed towards a return to normal athletic function. The foregoing sequence is normally accomplished in a period of 3 to 4 months.

It must be understood that many individuals whose activities induce an inferior pole–patellar tendon pain syndrome have predisposing structural assemblages of their extensor mechanisms. Patellae which are alta, too lax, or too tight, which have excessive static or dynamic Q angles, and which have various malalignments are among the structural formats which may assist activities in the production of tensile injuries to the extensor apparatus. The severity of underlying aberrations in patellofemoral function may determine the need to consider treatments above and beyond those necessary for the tendinitis alone, even to the extent of considering derotational bracing or patellar mechanism realignment.

Elite basketball players are among those athletes who commonly manifest an inferior pole syndrome. They are stereotypically tall with high-riding patellae and elongate tendons subjected to a whipping action by virtue of their jumping and cutting manoeuvres. One of us (JM) has attended only two players who required surgical treatment for severe recalcitrant symptoms in his experience with two professional basketball teams spanning a 10-year period.

Iatrogenic considerations

A greater awareness of the rising incidence of iatrogenically induced inferior pole syndromes developed in the 1980s. It was during this decade that the popularity of bone–patellar tendon–bone substitutes for the anterior cruciate ligament reached an apex. After harvests of the central third of the tendon (Fig. 5), inferior pole syndromes arose with some frequency and often persisted for periods of up to 12 to 18 months postoperatively. It was

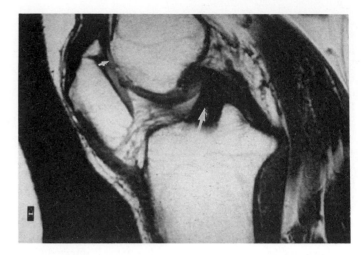

Fig. 6 Sagittal MR image demonstrating a dramatic patella baja. A fluid-containing tissue occupies the interval between the patella and anterior condylar surface of the femur which demonstrates articular irregularity (small arrow). Incidentally noted is the posterior position of the anterior cruciate ligament (large arrow).

Fig. 5 Patellar tendon autograft with bone from the patella and the tibial tubercle at its respective ends.

not clear whether these symptoms were primarily the result of the patellar fracture, harvesting, the reaction to the tendon defect created, the magnitude of the overall surgery, atrophic effects, the nature of the initiated physical therapy programmes, or other factors. As the decade progressed, the magnitude of surgery was reduced by the advent of arthroscopically assisted reconstruction; motion was started immediately and many of the atrophic effects of the surgery were eliminated. Nevertheless, a reasonable incidence of the syndrome persisted, leaving patellar fracture and tendon harvest site responses, therapy formats, and generalized inflammatory responses with tardy restoration of patellar glide as probable culprits. The use of allografts, iliotibial band, or pes tendon substitutes resulted in a reduced incidence and severity of inferior pole syndromes. However, some incidence was observed, leaving failure of patellar glide restoration (by virtue of cicatrix about the tendon and fat pad, or captured knee motion by virtue of anisometrically tensioned grafts) and therapy formats as probable sources. In view of these suppositions, it is interesting that autogenous bone–patellar tendon–bone substitute procedures yield an estimated 5 to 10 per cent incidence of arthrofibrotic responses with curtailments of terminal extension and flexion of the knees operated upon. A characteristic finding in these cases is pain with activity in terminal extension at the inferior pole of the patella, which must often be alleviated by additional surgery to free patellar glide. Frequent findings include a relative patella baja, patellar tendon fibrosis and adherence to the anterior capsule and fat pad, adhesions in and about the suprapatellar pouch, and degenerative changes of the anterior portions of the femoral condyles (Fig. 6).

That physical therapy can result in iatrogenically induced inferior pole syndromes has become well known in the wake of the

rise of strengthening machinery throughout the 1980s. The forces generated by the use of some machine therapy forms can result in micro- or macrotears of the variety described above. Machines with ballistic high speed formats, notably the Cybex (Lumex Corporation, Ronkonkama, NY) have been particularly implicated.

PATELLAR TENDON AND QUADRICEPS DISRUPTION

Disruptive injuries to the extensor mechanism were reported by such ancients as Galen.[20] Nevertheless, quadriceps and patella tendon ruptures are relatively uncommon injuries, relatively more common in males, and relatively specific with respect to the age bracket above and below which each is likely to occur. Siwek and Rao[20] retrospectively studied 117 patients with patellar tendon or quadriceps ruptures. The results of their study revealed that 80 per cent of patients with patellar tendon ruptures were aged 40 or less, while 88 per cent of those with quadriceps ruptures were more than 40 years of age.

There is a general consensus that extensor mechanism ruptures are more likely to occur as a sequel to pre-existing pathology at or near the site of rupture.[8] Predisposing pathology includes chronic tendinitis, repetitive traumatic microtears, and degenerated tissue (from wear and tear or ageing). Each of these pathologies is producible by 'overuse' activities and/or by poor structural mechanics. McLaughlin (1949) reported that biopsies were routinely performed on freshly ruptured tendons at the New York Presbyterian Hospital and that microhistological sections repeatedly revealed a decrease in the collagen content of the tendon fibres, fibrotic degeneration, and a marked loss of nuclei, and Speed (1950) added that fatty degeneration was often seen as rupture sites in obese patients.[21]

When ruptures of the extensor mechanism occur without a prodrome or history of predisposing pathology or when spontaneous bilateral ruptures occur, an underlying systemic disorder must be considered (Fig. 7).[22,23] Some of the disorders which have been associated with such ruptures are rheumatoid arthritis, systemic

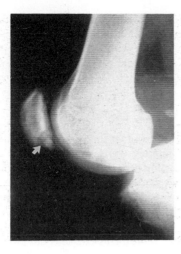

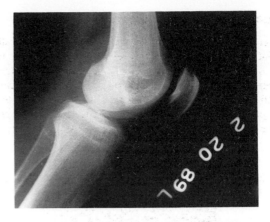

Fig. 7 Lateral radiograph 10 years subsequent to rupture of the patellar tendon from the distal pole in a teenage male under active treatment for leukaemia. It reveals patella alta and irregularity of the inferior pole to which the avulsed tendon was reattached by sutures through drill holes.

Fig. 9 Lateral radiographs of the knee of a high school 'All American' hockey player. The defect (arrow) in the superior segment of the patella sustained as a result of an impact to the anterior aspect of the extremity with the knee in flexion should be noted.

lupus erythematosis, gout, chronic renal failure, secondary hyperparathyroidism, diabetes mellitus, and peripheral vascular disease.

The use of local and systemic steroids has also been implicated in the production of extensor mechanism ruptures. Nevertheless, Kelly et al.[24] found no correlation with the use of steroid injections in a series ($n = 14$) of ruptures.

Most commonly, tears of the quadriceps occur through the rectus tendon at its insertion into the proximal pole of the patella. Tears may begin several centimetres more proximally and not infrequently they may extend to the aponeurotic expansions of the vasti (Fig. 8).[21] Sometimes, the rectus tendon will avulse a small piece of bone from the patella (Fig. 9).

Patellar tendon disruptions are most common at the inferior pole of the patella where associated avulsions of bone fragments may be seen. Mid-substance rupture tears of the quadriceps or patellar tendons are the least common varieties.[25] Disruptions at

the tibial insertion are rare in adults, but do occur in adolescents. In 1990, Chow et al.[26] reported that 150 avulsion fractures of the tibial tubercle had been reported up to 1986. The injury is of concern because it involves the growth plate. Chow et al. reviewed 16 patients after an average follow-up of 3.7 years. They classified the injuries in accordance with the classification of Ogden and Southwick (1976) (Fig. 10). Most cases can be treated conservatively. Twelve patients with a combination of type I and type II avulsions could actively extend their knees and required no surgery. Review of their series revealed a male and a left-sided predominance. Only one patient had an appreciated pre-existing condition, which was Osgood–Schlatter's disease. The present authors have experienced two such avulsions over a 15-year period, one in a patient with leukaemia (Fig. 7) and the other with

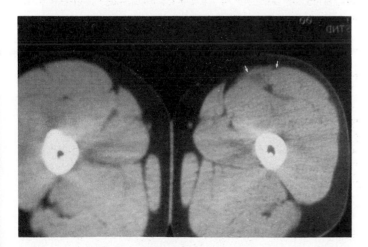

Fig. 8 Axial MR image of the mid-thigh of a college soccer player who sustained a rupture of the rectus muscle (between arrows) several inches proximal to the patella

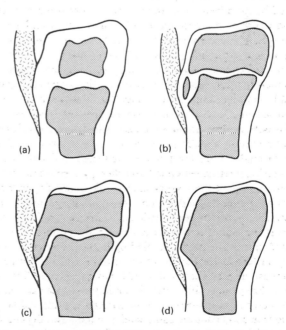

Fig. 10 Stages of proximal tibial development. (a) Cartilaginous stage. (b) Apophyseal stage. (c) Epiphyseal stage. (d) Bony stage. (Redrawn from ref. 26, with permission.)

a pre-existing 'jumping' patella, which is a flagrant manifestation of J-sign tracking.

Type II and higher grades of avulsion require internal fixation. Chow et al.[26] recommend a cancellous screw and tension band wiring. They reported no late occurrences of genu recurvatum in any of their cases.

Most disruptions are due to indirect trauma. Anzel et al. (1959) corroborated this contention, stipulating that the majority of ruptures in their large series were due to indirect trauma.[22] A strong eccentric contraction of the quadriceps muscle with the knee in flexion is the usual indirect traumatic cause. In a classic report, McMaster (1933) contended that without prior degenerative attenuation, the enormous tensile strength of the tendon will preclude rupture.[23] Jobe et al. calculated that the force needed to rupture the infrapatellar tendon is about 17.5 times body weight. However, it is clear that repetitive microtrauma and systemic conditions may substantially weaken these tendons, lessening the force needed to rupture them.[27]

Patients who have sustained a quadriceps tendon rupture traditionally present with a history of having fallen or stumbled. Pain is often acute and immediate, as is an unsettling sensation of instability within the extremity. The pain may preclude an effective evaluation of knee extension ability. An injection of local anaesthetic may facilitate the physical examination when diagnosis of a rupture is not corroborated by finding a gap in the tendon or by radiographic identification of a patella baja (particularly with a fragment of bone avulsed from its proximal pole). Lesser tears may not preclude an ability to extend the knee completely, making the decision to explore the injury site surgically more difficult. Tardy presentations (after several weeks) are usually more for dysfunction than pain, and any defect in the tendon which may have existed has often become partially or wholly obliterated by collagenous ingrowth (though tenderness usually persists). Apart from the most innocuous of tears, surgical repair is a virtual necessity for the restoration of adequate knee function, no less athletic function. Primary suture techniques, passing the sutures through drill holes in the patella, are the most frequently practised. Ruptures through pathological or degenerated tissue may require reinforcement of the suture repair with such augmentations as Dacron tapes, fascia lata, pes tendons, or reflected flaps from the quadriceps apparatus itself. If the repair and the tissue appear sound, early passive motion should be initiated, as should submaximal quadriceps setting exercises. Active motion without loading may be initiated within a few weeks in many instances, but weight bearing should be curtailed for at least 6 to 8 weeks to avoid an unwitting transmission of force across the suture line before some tensile strength has developed at the repair site. Naturally, the size of the defect, the quality of the tissues, the expediency of recovery, and the personality of the patient will help determine the rates of progressing motion and loading. Athletes will generally not defer treatment for a major injury disability and, for this reason, late reconstruction of quadriceps tendons will not be discussed here.

Patients with patellar tendon ruptures also present after having sustained injury as a result of violent eccentric quadriceps mechanism stress, as in basketball rebounding. As with the quadriceps rupture, there is immediate pain, instability, inability to extent the knee, and a defect at the rupture site. There may be a fragment avulsed from the inferior pole and/or alta appearance of the tendon on a plain radiograph. The principles of repair (with or without augmentation) and postoperative management are similar to those described for ruptures of the quadriceps tendon. Undue

loading of the patellar mechanism must be avoided for several months at least. Early restoration of range of motion is necessary to avoid contractures within the extensor apparatus and subsequent loss of motion and patellar tendon inflammation.

CHONDROMALACIA, INSTABILITY, AND MALALIGNMENT

The related entities of chondromalacia, patellar instability, and patellar malalignment are among the banes of the practising orthopaedic surgeon. For the most part, the orthopaedic literature has dealt with each of these entities as if they were primary or isolated pathologies. Fortunately, patellar pathology alone has rarely been a cause for permanent abandonment of participation once an elite status has been achieved. One of us (JM) has witnessed no forced retirements due solely to patellar pathology in a combined professional team experience totalling more than 40 years (including soccer, American football, basketball, and ice hockey). Nevertheless, Hayes et al.[28] cite Schneider (1962) in stating that articular cartilage degeneration is a leading cause of restricted sports activity.

Chondromalacia patella, in particular, has been treated as an isolated patellar pathology by many authors (idiopathic chondromalacia).[29] Chondromalacia, which literally means softened cartilage, has inappropriately become a receptacle for all manner of anterior knee pain. Hayes et al.[28] state that patellofemoral contact pressures have been implicated in the pathogenesis of articular cartilage degeneration, citing Pfeil (1966) and Ohno (1988) among others. In many instances it is a known consequence of direct trauma to the patella (Fig. 11), infection, post-injury and postoperative atrophic dystrophies, and patellar instability or maltracking syndromes. It is not unreasonable to surmise that the

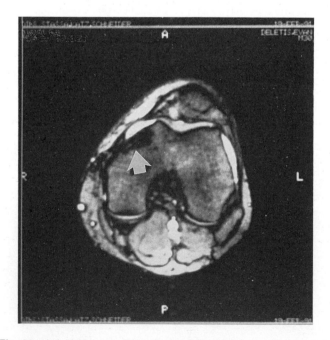

Fig. 11 Axial MR image of the knee of an individual whose flexed knee impacted an immovable object with substantial force. Patellar pain and crepitation followed shortly thereafter, consistent with the development of chondromalacia. A bony defect or 'bone bruise' is noted just medial to the trochlear groove (arrow), attesting to the magnitude of impact.

seemingly idiopathic forms are consequent to less evident forms of the above listed entities.

Chondromalacia has been staged in accordance with the degree of damage which is present, but without positive correlations to a spectrum of symptom severity and functional disability. The articular surface of the patella has no sensory innervation.[2] Nevertheless, patients with chondromalacia are typically reported to complain of retropatellar pain. Whether this pain derives from the underlying spongiosa or from the adjacent retinaculae is a matter of speculation. Many patients are more disturbed by the gritty crepitation experienced while extending the knee than by any associated pain. The flaking and fibrillated articular surface frequently incites the synovium to produce effusions. Fulkerson and Shea[30] rightly indicate that correlations between articular cartilage degeneration (with crepitus) and patellar pain are inconsistent.

Supportive treatments are similar to those described above for other patellar afflictions. NSAIDs may reduce inflammatory sequelae and have superseded the aspirin regimens thought to prevent a progression of chondromalacia.[31] Physical therapy for the institution of modalities, stretching, and strengthening should be initiated empirically. The response to physical therapy is less predictably successful than it is for some patellar entities with a more obvious mechanical basis. Bracing is sometimes a detriment, with the compression increasing the sensation of painful crepitation. In other instances, as when chondromalacia is an accompaniment of patellar instability, bracing may provide a sense of increased stability as well as a reduction in pain or swelling.

When instability or maltracking of the patella is not identifiable as a cause of symptomatic chondromalacia, the role of surgery is limited and its success is not quite predictable.

According to Buckwalter et al.[32] loss of large areas of articular cartilage or the presence of full thickness 'holes' will compromise joint function. However, the issue is how to deal with these entities. Buckwalter et al. point out that articular cartilage injury response has been studied for a quarter of a millennium since Hunter (1743) observed that articular ulcerations are not repaired. They indicate that current research reaffirms Hunter's contention that while cartilage is repaired under certain conditions, for the most part repair attempts fail to restore its normal molecular composition and durability.

Several methods of chondrogenesis, including cartilage shaving, abrasion chondroplasty, articular surface load alteration, gels, and electrostimulation, have been proposed.[32] O'Donoghue[33] and Johnson[34] reported that patellar articular shaving and/or abrasion may relieve symptoms. However, Buckwalter et al.[32] caution that the efficacy of these frequently performed procedures has not been established for parameters apart from the symptomatic relief that may occur for variable periods. Historically, most surgery has been preoccupied with smoothing of the chondral surface. In the era prior to the therapeutic arthroscopy era (i.e. prior to the mid-1970s) an arthrotomy was performed: the patella was released and everted, and a scalpel blade was used to pare damaged facets down to shiny cartilage. Whether the apparent improvements noted in some cases were due to the shaving or to inadvertent realignment from having released the patella or to other factors is again not clear. The postoperative course was often protracted, restoration of motion was often difficult, and results were often not gratifying. In some instances in which the most severe damage was confined to a limited area of the articular surface, such as the odd facet, facetectomy was performed with some success.[33] In the most flagrant and recalcitrant

cases patellectomy was not uncommon. Symptoms were often alleviated but sometimes were superseded by knee instability due to a reduction in leverage of the extensor mechanism.

The advent of therapeutic arthroscopy, and of power shaving instrumentation in particular, at the end of the 1970s created a more palatable option to arthrotomy. Using chondral abrasion of varying depth and surface area, as well as selective synovectomy, debridement and chondroplasty could be effected by closed methods as an outpatient. Results were better than those with its open counterpart, morbidity was substantially lower, and the initiation of early motion was dramatically expedited. However, whether arthroscopic chondral abrasion or skilled scalpel planing of the softened or fibrillated articular surface were used, controversy pertinent to the principles of these procedures persists. Though the removal of damaged cartilage may temporarily reduce symptoms and eliminate the likelihood of synovial inflammation by flaking chondral debris, the articular surface is being reduced in thickness, accelerating potential exposure of the underlying bone.

In their preoccupation with achieving a smooth articular surface, the Chinese have reported success (reduced crepitation and pain) in using silicone plugs to fill articular divets.

Since the advent of arthroscopy, facetectomy has essentially become a procedure of historical interest and the salvation of the patella has become a universal doctrine. For many surgeons who may have considered patellectomy in the past, the Macquet procedure has become its replacement.[35–38] Elevation of the tibial tubercle affords greater leverage to the extensor mechanism, and reports of great satisfaction in relieving the signs and symptoms of chondromalacia pervade the literature.[35–39] The good results may be fortuitous, since it has yet to be proved that the manifestations of chondromalacia derive from inadequate extensor leverage. Furthermore, the Macquet operation produces a protuberant and often painful tubercular region as well as other complications (Fig. 12).

'Maltracking' of the patella is a particularly difficult subject to comprehend and discuss. Its complexities are profound, and the interface of its dynamics has yet to be reasonably elucidated. Maltracking may be represented by abnormal glide patterns attributable to patellar instability. The instability may result in maltracking due to the extremes of patellar laxity, medial and/or lateral, or due to weakness or dysplasia of the extension mechanism. The instability may be manifested in a limited portion of the arc of knee motion (most commonly near the extended position) or through larger portions of the arc. Small or alta patellae, hypoplastic or dysplastic femoral condyles, and rotational laxities of the tibiofemoral articulation are among the additional factors which have been found to contribute to some patellar instabilities (Fig. 13).

Unfortunately, there are no absolute parameters which define the limits of normal versus abnormal. Attempts to define these limits by radiographic indices have been numerous and helpful, but leave large voids in intelligence towards therapeutic planning. For a comprehensive review of the values and limitations of radiography in the evaluation of chondromalacia, instability, and the maltracking syndromes, the reader is referred to ref. 40 (Fig. 13).

Maltracking may also be a consequence of fixed malpositioning of the patella at one or many points in the arc of motion. An extreme example of such malpositioning was reported in the section on aplastic and hypoplastic patellae, i.e. a patella and extensor mechanism which resides in a position of permanent lateral dislocation. Individuals with excessive Q angles are virtually assured of some quantity of maltracking (malalignment), the

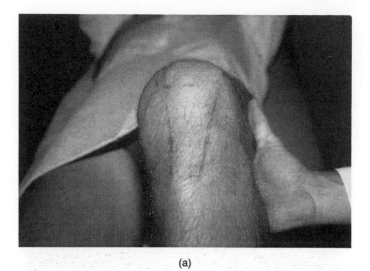

(a)

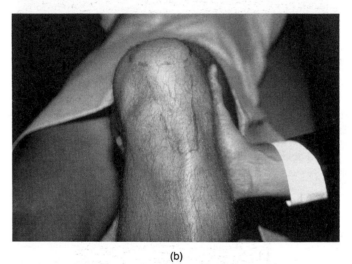

(b)

Fig. 12   Bone scan of the knee of a patient with proximal anterior tibial pain 2 years after a Macquet operation. A dramatic uptake is noted over the area of tibial cortical elevation where the patient manifested severe pain and incomplete union. Excision of the elevated section of cortex led to complete resolution of symptoms.

Fig. 13   Young adult male with patellar symptoms and patellar laxity. Maximally internally and externally rotated photographs of the tibia with the knee flexed are shown. The wide rotational excursion adds to the potential for patellar displacement by altering the angles of pull of the patellar tendon.

responsibility for which can be shared between malpositioning and instability.

The indiscriminate use and interchange of verbiage implying the presence of a patellar disorder has added measurably to the confusion on the subject (Table 1). 'Maltracking' is not 'instability' but may be due to the latter. 'Malalignment' may produce 'maltracking'; either may produce 'chondromalacia' or be affiliated with 'instability', but need not be. 'Patellar tendinitis' may be produced by repetitive overuse of the extensor mechanism, but it may also be an accompaniment of and partially induced by 'malalignment' and/or 'instability'. Similarly, the term 'subluxation' is frequently misused. It is used by most to imply that the patella has an ability to dislocate partially from the trochlea, but is used by others in referring to a fixed partial dislocation of the patella. In the former instance it should be said that the patella is 'subluxable', while in the latter it should be stated that the patella is permanently 'subluxed' (i.e. manifests a fixed malalignment) (Fig. 14).

With respect to patellar instability (conditions of 'subluxability' and 'dislocatability'), the most flagrant forms may present at a very young age, often in pre- or periadolescent females, and often as part of a syndrome of ubiquitous ligamentous laxity (Fig. 15).[41] Other findings may be associated with this syndrome. Included among them is an alta patella which adds to instability because the patella rests precariously on the anterior femoral surface proximal to the stabilizing trochlear groove. Patella alta is easily identified on clinical examination and by the use of one of several radiographic indices (Fig. 13).

When youngsters of lax habitus exhibit a trilogy of femoral anteversion, external tibial torsion (productive of a large Q angle), and pronated feet, an ungainly malalignment is present about the knee (Fig. 16). It is termed the 'malicious malalignment syndrome' by virtue of its pernicious symptoms and its reluctance to

respond to non-operative (or even operative) treatment. In some cases operative treatments have been so bold as to include proximal femoral external derotational osteotomies, sometimes in combination with internal derotation osteotomies of the proximal end of the tibia. Individuals with severe anatomical disfigurations about the knee are generally incapable of achieving a competitive level of athletic accomplishment.

Traumatic dislocations of the patella are less frequent in patients with seemingly normal patellofemoral and knee joint anatomy than in individuals with obvious anatomic predispositions. Generally, a violent force is required. Occasionally, dislocations are medial. However, most commonly they are lateral, having been produced either by a laterally directed blow to the medial side of the patella or by the dynamics of a strong quadriceps contraction acting upon an extending knee with the tibia (and tubercle) in external rotation relative to the femur. Once dislocated, the patella may 'hang up' on the lateral aspect of the lateral wall of the trochlea with the knee partially flexed or spontaneous reduction may occur. The greater the angle of knee flexion at

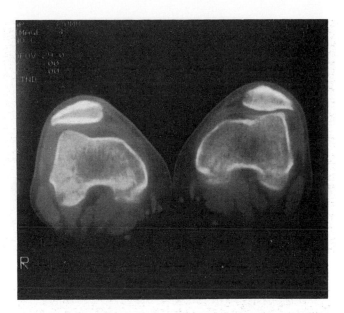

Fig. 14 CT scan showing fixed displacements (malalignments) of the patellae bilaterally.

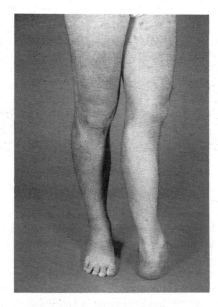

Fig. 15 Loose-jointedness in a male with symptoms of (and prior surgery for) patellar subluxation.

which the traumatizing force is applied, the greater is the likelihood that an osteochondral fracture of the patella or femoral condyles will occur because of the more stable seating of the patella within the trochlea at progressive angles of flexion. If the patella is dislocated upon presentation, reduction is accomplished by extending the knee while coaxing the patella medially over the lateral condylar eminence. Analgesia and/or sedation may be necessary adjuncts to the reduction. When the patella has reduced spontaneously prior to presentation for treatment, the diagnosis is based upon the historical description of the event, the presence of retinacular tenderness, patient apprehension about any manipulation of the patella, and any evidence from the radiographs which must be taken when the diagnosis is known or suspected. Fractures

Table 1 Confusion of definitions by pathological, radiological, and clinical criteria

Bentley and Dowd	Chondromalacia is a syndrome comprising irregularity of the patellar undersurface, retropatellar and 'movie sign' pains, and crepitation.
Stougard	Chondromalacia is poorly correlated with anterior knee pain
Bentley and Dowd	Abnormal patellar tracking is a rare cause of chondromalacia
Insall et al.	Malalignment is a frequent finding (excessive Q angles) in patients with chondromalacia
Brattstrom	Recurrent patellar dislocation is often associated with femeropatellar dysplasia seen as a foreshortened lateral femoral condyle on axial radiography
Insall and Salvati	Patella alta exists with a 98% confidence level if the LT:LP ratio is 1.2 or less
Lancourt and Cristini	Greater incidence of patella alta with both instability and chondromalacia
Marks and Bentley	Alta correlated with sex (female) rather than with chondromalacia
Bentley and Dowd	LT:LP ratio mean of 1.25 with subluxation, and normal with chondromalacia
Merchant et al.	Patellar subluxation is 95% assured if the congruence angle is greater than +16°
Laurin et al.	93% of patients with chondromalacia have a patellofemoral index of more than 1.6 (at 20° knee flexion) and 100% of patients with subluxation have an index greater than 1.6
Moller et al.	For both unilateral chondromalacia and unilateral subluxation, side-to-side comparisons of the congruence angle are more important than the raw numbers
Bentley and Dowd	The mean congruence angle of subluxation cases is normal in contrast with the findings of Merchant and of Aglietti and Cerulli who found mean angles of +16° or more
Dowd and Bentley	Patients with instability had increased PT:P ratios and sulcus angles but often normal congruence angles
Sikorski et al.	Abnormal patellar tilts and rotations are observed in chondromalacia at given intervals of the flexion arc
Imai et al.	Malalignment is present when the patella is not centred by 60° of flexion using axial arthrography
Schutzer et al.	The patella is subluxated if the congruence angle exceeds 0° or the patella tilt angle is less than 8° with knee evaluated by CT at 10° of flexion.

Reproduced from ref. 40, p. 205 with permission.

resulting in a splitting of the chondral articular surface may only be appreciated by arthroscopy (Fig. 17). Those representing compressions of the subchondral bone might only be appreciated on an MR image as an area of altered signal which may be expected to resolve over a period of several weeks (Fig. 18). Thereafter, only technetium bone scanning may reveal evidence of its having been present. Plain radiographs may reveal the fragments most often

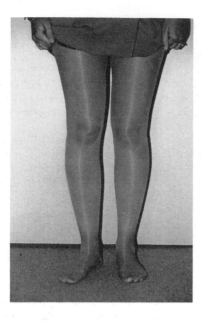

Fig. 16 A relatively benign form of malicious malalignment.

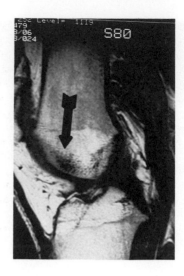

Fig. 18 Sagittal MR image of the knee of a professional basketball player who suffered a hyperextension injury with subsequent anterior knee pain. The 'bone bruise' of the anterior portion of the femoral condyle (arrow), where patellofemoral compression occurred, should be noted.

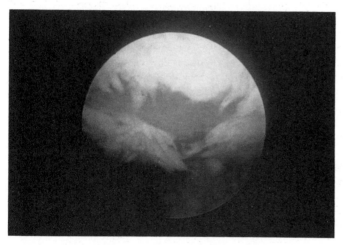

Fig. 17 Chondral fracture of undersurface of patella viewed by arthroscopy. (Reproduced from ref. 42, with permission.)

deriving from the inferior portion of the medial facet of the patella or those emanating from fractures of the femoral condyles.

There is some controversy surrounding the treatment of an acute patellar dislocation. Following an avulsion of bone from the medial aspect of the patella, some surgeons will intervene with excision of the fragment and repair of the torn medial retinaculum (with a lateral release when necessary to the maintenance of the reduction).[43] In the past decade it has been discovered that arthroscopy after acute dislocations may reveal a rent in the medial retinacular tissue; this may then be repaired by arthroscopic suturing techniques, with the aim of preventing potential recurrences attributable to progressive stretching of a retinaculum which had already been torn.[44]

A common procedure among other orthopaedists has been to immobilize the knee in extension for up to 6 weeks before initiating physical therapy toward a restoration of motion.[43] The rationale for this approach is the hope that scarring of the torn tissues will prevent recurrence. This is probably as unlikely to inhibit future

patellar dislocation as postdislocation immobilization of the shoulder has been in preventing further dislocation of that joint.

Unfortunately, there are no randomized prospective studies comparing the results of the various treatment options to guide the treating physician. For the elite competitive athlete, temporal considerations may dictate the treatment option selected. Such an individual afflicted in the midst of the season or approaching a critical event can most often be restored to a competitive level by the immediate initiation of physical therapy to reconstitute motion, strength, and agility (as long as no loose fracture fragment is present). Gymnasts, who often fit the description of the hyperlax 'predisposed' individual mentioned above, may dislocate the patella with virtually no symptomatic consequence and require no treatment.

Treatments for malalignments and instability have been varied. Whether thought to be intractable or not, almost all cases deserve trials of non-operative treatment inclusive of anti-inflammatory agents, physical therapy, and bracing. Surgical treatments have been numerous in variety. They have included realignments of proximal patellar mechanisms intended to improve the influence of the pull of the quadriceps muscle and retinaculae, distal realignments designed to alter the direction of pull of the patellar tendon, simple retinacular (lateral) releases, and adjunctive procedures such as derotational osteotomies or tendon transfers to the patellar tendon to inhibit the patella, its tendon, or the tibial tubercle from lateral or external displacement. There are no published reports reflecting the relative successes of any of these procedures in an athletic population.

It must be recognized that surgery cannot enlarge a small patella, deepen a hypoplastic trochlea, effectively lower a patella alta, or correct other developmental anatomical factors predisposing to fixed or unstable malalignments. Surgery can release, tighten, or redirect the pulls of tissues influencing patellar glide, but with little hope of achieving normality rather than merely a beneficial modification. There are many factors to consider. For example, a patella may be too lax, producing pain when it is displaced medially and reproducing a familiar sense of instability upon being displaced laterally (while the knee remains passively extended). Ignoring for the moment the existence of any other influencing factors, does the surgeon release the lateral retinaculum and tighten the medial or tighten both retinaculae to different degrees? Are the intended tightenings or releases performed in extension or at the angle of knee flexion at which symptoms are most commonly manifested? What length of release or tightening is optimum to control the symptoms? How can the effects of these procedural titrations be evaluated intraoperatively when the patient is not functioning? Vision may assess passive alignment in a very gross fashion, but affords little ability to assess the functional dynamics of a 'corrected' extensor mechanism (Fig. 19). Furthermore, these surgical procedures are not excisional; they alter the way that the mechanism works, and the alteration may progressively change the dynamics for better or worse for many months following the operation.

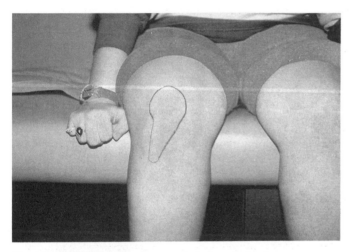

Fig. 19 The knee shown has an alta but centralized appearance. There is an obviously large Q angle. Examination may reveal patellar laxity either medially or laterally. Intraoperative radiography to determine patellar indices is not practised. Therefore the only parameter that can be used to assess dynamic alterations during surgery is gross visual inspection which is unlikely to appreciate medial–lateral shifts, tibial rotatory excursions, and patellar tilts occurring at various angles in the passive mode created by general or spinal anaesthesia. Local anaesthesia and/or sensory epidural anaesthesia which retains muscle control and patient perception, to a degree, may be helpful.

Little objective scientific insight regarding the anticipated influences of commonly practised procedures is available; one study of particular interest should be described. Hayes et al.[28] used pressure-sensitive film to study patellofemoral contact patterns in cadaveric knees subjected to artificially induced quadriceps loading. The effects of several surgical procedures were studied in this way. Results implied that lateral and medial capsular plication produced increased pressures on the medial and lateral facets respectively, and that the changes were dependent upon the angle of flexion of the knee. However, lateral release did not produce a consistent reduction of peak pressure or alteration in contact area. Alterations of the Q angle may alter contact patterns and create areas of high peak stresses; however, a prediction as to what angle changes might be beneficial or detrimental is not forthcoming. Tibial tubercle elevations produced modest alterations of contact

patterns but no substantial reductions in average or peak contact pressures.

Semidynamic intraoperative monitoring of surgical procedures is possible with the use of 'sensory epidural anaesthesia'. The patient is awake and motor ability is retained but attenuated. Active knee motion assists visual assessments and on occasion the patient may provide feedback relative to the sense of function and glide.

In view of the above results, it requires little imagination to realize that a single procedure is unlikely to produce gratifying results for the great variety of dysfunctional patellar entities which present for surgery. Unfortunately, the generally benevolent advent of arthroscopy has had a somewhat questionable influence upon the treatment of these entities. Its simplicity and low morbidity have led to an almost excessive performance of arthroscopic lateral retinacular releases. Early reports of success were high,[45] but reported success rates have diminished substantially. Hayes et al.[28] cite Osborne and Fulford (1982) who reported a time-related failure pattern of lateral release for symptomatic patients with chondromalacia. From an early result of pain relief in almost 90 per cent of cases, success dropped to under 40 per cent after 3 years. There are several obvious reasons for this diminution. One is the inability of a simple release to address a complexity of considerations such as those discussed above. Hayes et al.[28] assert that, despite the recent popularity of lateral releases for patellar alignment problems and chondral damage, these procedures have been used 'with no particular regard for differences in etiologic mechanisms or locations of the lesions'. A second reason for diminishing success is the not infrequent consequence of making a lax patella even looser by a release of restraining tissue. A third relates to the Q angle, both static and dynamic. When the Q angle is large (e.g. greater than 15–20°), the resultant forces acting upon the patella will favour lateral displacement even if a lateral release has been performed (as an isolated procedure). If, with flexion of the knee, external rotatory laxity of the tibia permits the Q angle to enlarge even more, then a procedure to reduce this thrust must be considered. A modification of the Trillat procedure,[46] by which the tubercle is osteotomized at its proximal end (to which the patellar tendon is attached) and is rotated medially and transfixed, is often successful for this purpose (Fig. 20).[47]

The conclusion is that the design and anticipated results of

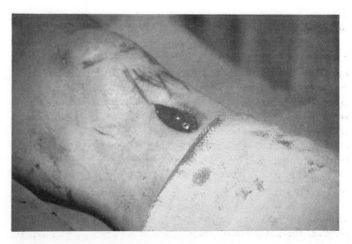

Fig. 20 Examples of axial radiographic techniques used to evaluate the tilt or displacement of the patella.

surgery to alter patellar glide should be based upon the specific dysfunctions involved in the individual case. As already indicated, examination provides limited clues. The most commonly used guides to the design of a procedure are derived from the radiographic indices. Most are based upon plain axial radiography with or without arthrography (Fig. 21), and some upon CT axial scans which permit evaluations in an extended position of the knee (less than 20° of flexion) (Fig. 22). In fact, Fulkerson and Shea[30] imply that malalignment and instability syndromes are classifiable and distinguishable by CT imaging techniques alone, and further that the categories discernible by imaging allow the clinician to anticipate a quantity of associated chondral damage and a definitive treatment plan. They describe five categories of patellar imaging findings (three of which have subtypes). The concept is

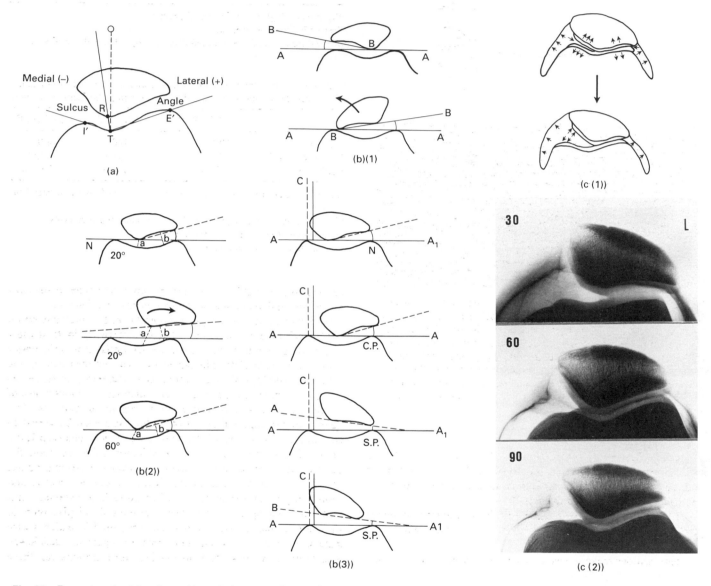

Fig. 21 Examples of axial radiographic techniques used to evaluate the tilt or displacement of the patella.

(a) Merchant technique[48]: technique with patient supine and the knee flexed to 45°. Using reference line TO, the congruence angle (OTR) is negative if R is medial to TO and positive if it is lateral to TO. An angle of +16° was determined to be abnormal at the 95th percentile (meaning subluxed) (see Tables 2 and 3).

(b) Laurin method[49]: supine technique with the knee in 20° of flexion (a more extended angle than that used for all other plain radiographic axial techniques). Three indices are determinable (see Table 4). (b(1)) The lateral patellofemoral angle (LPS): the angle is formed by ABB and is open laterally in most normal knees. (b(2)) The patellofemoral index (PFI): a ratio of medial and lateral patellofemoral interspaces. The ratio is more than 1.6 in most cases of chondromalacia. (b(3)) The lateral patellar displacement (LPD): disorders are based on the relative displacements of the patella from perpendicular line C which rises from the promontory of the lateral wall of the groove. Lateral displacements are seen in a minor quantity of knees with chondromalacia and a greater quantity of knees with subluxation (versus merely tilt).

(c) Axial views in conjunction with arthrography. The best known of these methods is that of Imai et al.[50] (c(1)) Imai et al. make the point that arthrographic enhancement of axial views helps delineate the chondral surface outline, permitting a distinction between apparent tilt and true tilt. They also make the point that abnormal tilts will disappear at increasing angles of knee flexion, whereas they may persist in knees with subluxation. (c(2)) Axial arthrographic views performed at three angles of knee flexion demonstrate significant tilt at 30° and apparent patellar reduction at 60° and 90°. (From ref. 50.)

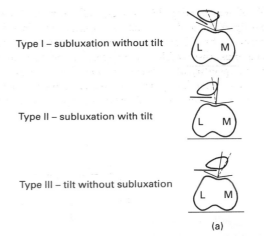

Type I – subluxation without tilt

Type II – subluxation with tilt

Type III – tilt without subluxation

(a)

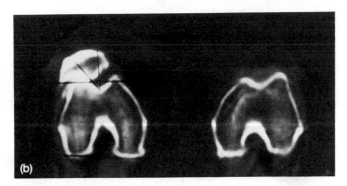

(b)

Fig. 22 Schutzer *et al.*[51] ascertained normal and abnormal indices for axial CT imaging with regard to patellar subluxation. In type I, the angle is greater than 0°. In type II, the trochlia is very shallow and the congruence angle is greater than 0°. In type III, the congruence angle is less than 0°. The type of treatment (e.g. lateral release or anteriorization of the tubercle) is based on the severity of the type. (Redrawn from ref. 51, with permission.) (b) Type II (subluxation and tilt) subluxation with osteophyte formation. (By courtesy of John P. Fulkerson.)

interesting, but naive, in assuming that formulaic approaches to patellofemoral conditions are likely to succeed consistently. Most of these studies present the patella in an adynamic state at only one or two angles of flexion (Tables 2–5).

Ciné CT or MRI have been used[52] in an attempt to reveal patellar displacements at progressing angles of knee flexion, but still in an adynamic fashion.

These realities make it wise to delay surgery in elective instan-

Table 3 Distinguishing normal patellae from chondromalacia and patellar instability using a variety of plain radiographic indices

Study	Normal patella (mean value)	Chondromalacia	Instability
Congruence angle	–8.1°	–6.48°	–1.8°
Sulcus angle	139°	138°	147°[a]
Pt:P ratio	1.03[b]	1.09	1.25[a]

[a]Significantly higher than in normal patellae.
[b]Generally higher in women.
Adapted from ref. 25. Reproduced from ref. 40, p. 224 with permission.

Table 4 Percentages of normal and abnormal indices in normal, chondromalacic, and subluxatable patellae using the Laurin method

	LPD (%)[a]	LPA (%)[b]	PFI (%)[c]
Normal	100/0	97/3/0	100/0
Chondromalacia }	70/30	90/10/0	3/97
Subluxation	47/53	0/60/40	0/100

[a]Lateral patellar displacement: per cent medial to line C/per cent lateral to line cc.
[b]Lateral patellofemoral angle: open laterally/parallel/open medially.
[c]Patellofemoral index: 1.6 or less/more than 1.6.
Adapted from ref. 49 with permission.

ces until the patient has been seen and studied sufficiently to gain a maximum understanding of the complaints and dynamics.

The above doctrine is equally applicable to the implementation of physical therapy and bracing. There are virtually limitless alterations to the strengthening programmes initiated to gain better control of patellar glide. The quantity of load and the nature of the load (free weight, isokinetics, cam accommodations, elastic resistance, water resistance, arcs through which load is applied, use of concentric versus eccentric loading, repetition patterns, frequency of loading, and speed of loading) can each be varied to accommodate symptoms. The situation in which rotational laxity of the tibiofemoral joint exerts a whipping influence upon the patella and its tendon can be taken as an example. Attempted use of ballistic machines such as the Cybex is often provocative and results in peripatellar pain. It is often possible to overcome this problem by using an antishear bar to curtail tibial displacement in the same fashion in which it is customarily used in patients with instability due to an anterior cruciate ligament deficiency. The same principles apply to bracing considerations for these

Table 2 Comparative congruence angles reported by authors studying normal, chondromalacic, and subluxating patellae

Authors	Mean normal CA	Abnormal 95% limits	Mean CA in chondromalacia	Mean CA in subluxation
Merchant *et al.*	–6°	+16°	—	+16°[a]
Aglietti and Cerulli	–9°	—	0°	+17°
Bentley and Dowd	—	—	–8.1°	Normal[b]

CA = congruence angle.
[a]Within 95% confidence limits.
[b]Of 33 cases, four were outside the 95th percentile of normal subjects.
Reproduced from ref. 40, p. 223 with permission.

Table 5 Use of arthroscopy and radiographic indices to distinguish normal chondromalacic, and subluxatable patellae

Group	Presenting symptoms	Q Angle in extension (±2.SD)*		Sulcus angles (±2 SD)	Congruence angles (±2 SD)	Insall-Salvati index (±2 SD)	Median angle of alignment on arthroscopy		
		Women	Men				Lateral facet	Median ridge	Medial facet
1. Control (n = 17)	None in the patellofemoral joint	12–16°	7–11°	134–143°	−13° to −6°	0.94–1.10	20°	35°	50°
							All patients with medial facet contact by 45° could centre patella with quadricep contraction		
2. Idiopathic (n = 20)	Retropatellar pain	13–20°	10–15°	136–145°	−8° to −2°	1.07–1.27	20°	35°	50°
							All could centre patellar by isometric quad at 45°		
3. Instability (n = 13)	Sense of instability + apprehension sign	16–24°	12–20°	140–153° (significantly higher)	−11° to +8° (median +4)	1.13–1.44 (significantly higher)	30°	55°	85°
							All with medial facet contact by 85° could centre with quadriceps contraction†		

*Flexion to 30° increased Q angle. External rotation increased Q angle.
†External rotation of tibia had an adverse effect on centring for group 3. Chondromalacia was present in a large percentage of groups 2 and 3.
(Adapted from Sojbjerg JO, *et al*. Arthroscopic determination of patellofemoral malalignment. *Clinical Orthopedics*, 1987; **215**: 243–47, with permission.)

problems. Whereas standard patellar braces may be unsuccessful in controlling symptoms in patellar instability due to a rotational laxity of the tibia upon the femur, derotational braces used for tibiofemoral instabilities are often successful (Fig. 23).

The foregoing information was intended to review highlights and principles of patellofemoral glide and tracking disorders, their treatment, and their relative roles in athletics. Indeed, complete books have been written on the subject of the patella. It is hoped that the reader will at least have been made familiar with the gamut of existing disorders and provided with an insight into the complexities governing their management.

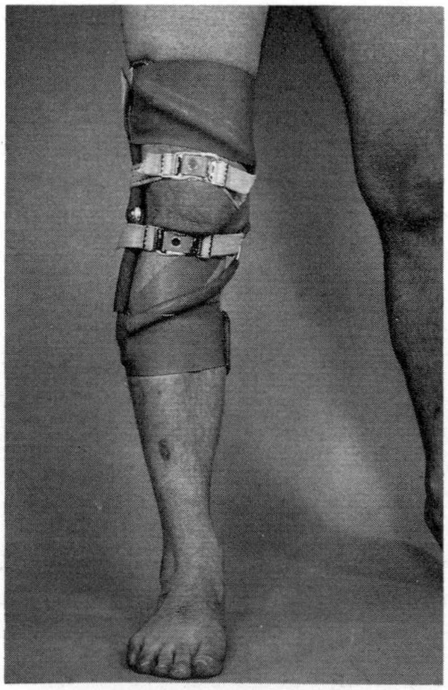

PATELLOFEMORAL ARTHROSIS

This condition, with symptoms of pain, stiffness, and swelling in varying proportions, is infrequently encountered among competitive athletes (recreational or otherwise). Osteophytes widen the margins of the patella, creating a patella magna (Fig. 24). Broadening and incongruity of the patella compromise patellar excursion and accentuate pressure at contact points. The result is pain and reduced extensor leverage. The knee lacks full extension and the gait is rather stiff. Stair climbing and descent are often particularly difficult as a result of pain, weakness, and crepitation.

Fig. 24 Axial radiographs of a retired professional hockey defenceman with a unilateral patella magna and no history of patellar fracture.

The arthrosis is most commonly manifested about the lateral portion of the patellofemoral joint consistent with the lateral vector of force deriving from the physiological valgus common to the knee joint. The lateral facet shows a greater incidence of wear both clinically and at necropsy.[2] Isolated medial patellofemoral arthritis is much less commonly observed.

The diagnosis of patellofemoral arthrosis is usually easily deduced by physical examination by inspecting and/or palpating the

Fig. 23 Derotational braces reduce tibiofemoral rotational laxity and hence reduce displacing pulls upon the patella.

irregularly enlarged patella, the crepitation with motion, and the lack of passive full extension. Plain radiographs demonstrate the enlarged irregular perimeter of the patellar on anterior–posterior views, and often show projectile promontories extending proximally from the superior margin and/or distally from the inferior margin of the patella. However, the incongruency and enlargements are best observed on axial views, either plain films or CT images. When the arthrosis is incipient, technetium bone scanning may be of value in deducing the presence of arthritic disease (Fig. 24).

Non-operative treatment consists of strengthening and stretching of the muscles about the knee at levels of loading and in arcs which are non-provocative. NSAIDs may allay symptoms, but their use for the purpose of participation in sports should be construed as being potentially deleterious. Bracing is often counterproductive; the pressure from the brace frequently magnifies the pre-existing perception of stiffness.

For the most part (and depending upon the magnitude and focus of the arthrosis) surgical treatments hold little promise of substantial or protracted relief of symptoms. Arthroscopic abrasion chondroplasty can occasionally restore the eroding articular surface with inconsistent short-lived benefits. Focal painful impingements can sometimes be modified by open and/or arthroscopic cheilotomy or partial facetectomy.

Associated retinacular releases may facilitate glide of the patella. The Macquet operation, the purpose of which is to obtain offending patellofemoral loading forces and to enhance the leverage of the extensor mechanism, offers a faint hope of persistently improved mechanics. However, the procedure is accompanied by a protracted recuperative period and a multifaceted potential morbidity.[35,36]

REFLEX SYMPATHETIC DYSTROPHY

Reflex sympathetic dystrophy is an entity shrouded in confusion by virtue of its broad array of presenting manifestations, and its frequent resilience to multiform treatment approaches. Its relevance to this chapter is the reported finding of Katz and Hungerford (1987) that 65 per cent of reflex sympathetic dystrophies about the knee originate at the patellofemoral joint.[3]

Reflex sympathetic dystrophy appears to be a relative of the causalgia classically described by Mitchell, Morehouse, and Keen (1964) during the American Civil War.[2] It is commonly known that porotic bone changes are a typical consequence of reflex sympathetic dystrophy. At the beginning of the twentieth century, Sudeck recognized an osteoporosis which could occur in 'younger' individuals and yet not be consequent to disuse (Sudeck's atrophy). It was soon concluded that the mechanism of production was related to the automatic nervous system.[3] De Takats (1937) applied the term 'reflex dystrophy' to describe a wide clinical spectrum of reflex sympathetic dystrophy.[3] The classical presentation of reflex sympathetic dystrophy includes a swollen painfully immobile limb, which is hypersensitive to touch.[53] However, not all presentations are so clearly marked. Homans (1940) discussed less or minor causalgic forms,[3] and Casten and Betcher (1955) recognized three grades of severity of the syndrome.[53] It is contended that reflex sympathetic dystrophy of the tibiofemoral and patellofemoral articulations is more frequent than commonly thought, but that lack of suspicion, particularly in cases at the lower end of the spectrum of overt manifestations, has been a deterrent to its detection.[2] The patellofemoral articulation is predisposed to reflex sympathetic dystrophy on the basis of its subcutaneous location and its consequent vulnerability to concussive trauma. The rich and anastamotic circulation of the patella, subject to the influence of autonomic control, is an additional predisposing cause.[2]

The autonomic nervous system is involved to some degree with any injury, with its response generally being proportional to the magnitude of injury.[2] The autonomic disturbance usually abates with progression of the injury repair process. However, for reasons which have not been clearly elucidated, the usual abatement sometimes fails to occur, and the result is reflex sympathetic dystrophy. Onset without trauma has been recognized, most typically in the upper extremities. However, the traumatically induced onset is much more typical. Direct trauma to the patella (as in a direct blow to or fall upon the knee) is the most frequent inciting cause of reflex sympathetic dystrophy of the patella. Reflex sympathetic dystrophy of the patella and its environs may also result from more insidious traumatic affectations (e.g. overloading with weights in quadriceps building, or recurrent subluxation).[3]

Theories of the pathogenesis of reflex sympathetic dystrophy have been numerous, but none has been accepted with universal accord. Hungerford and Fulkerson[3] suggest that the most favoured theory is that of Lorente de No (1938) as applied by Livingstone (1941). According to this theory the painful stimulus enters the cord through afferent fibres which stimulate the internuncial pool, leading to the spread of the stimulus in multiple directions within the cord; the resulting efferent consequence at the periphery serves to restimulate the afferents, thereby creating a vicious circle. Among the efferent consequences are local circulatory disturbances. The circulatory changes were recognized by Leriche (1923) nearly two decades earlier. He postulated that an initial vasoconstriction was produced by the initiating trauma and that this was superseded by a secondary phase of vasodilatation. The result, in effect, was a simulated inflammatory response (heat and rubor) due to heightened regional vascularization.[3] Increased blood flow was confirmed by later investigators (e.g. de Takats (1937)) using objective laboratory apparatus.[3]

The implication of the above discussion is that the vascular alterations of reflex sympathetic dystrophy are associated with a disturbance of the sympathetic nervous system. Furthermore, recent investigations by Roberts provide evidence that the pain patterns of reflex sympathetic dystrophy may also have a derivation in the sympathetic nervous system disturbances which characterize the syndrome.[53]

Contemporary theories of pathogenesis are reviewed by Omohundro and Payne[53] and are listed chronologically below.

1. Sunderland (1978): nerve injury results in abnormal sympathetic afferent discharges.
2. Devor (1983): spontaneous ectopic discharges may be germinated at the nerve injury site where neuroma formation or focal demyelination may occur.
3. Barasi and Lynn (1983) and Roberts and Elardo (1985): sympathetic discharges result in hypersensitization of the peripheral mechanoreceptors and/or nociceptors.
4. Janig (1985): artificial synapses (ephapses) are created secondary to injury which lead to a shunting of sympathetic efferent to sympathetic efferent stimulation near the site of injury.

Omohundro and Payne caution that none of the theories proposed account for the variety of clinical manifestations expressed in reflex sympathetic dystrophy. They claim that Roberts' theory

(1986) is best capable of explaining the clinical features of reflex sympathetic dystrophy. This theory suggests that sympathetic efferent activity may be normal and that the symptoms expressed in reflex sympathetic dystrophy are a function of a reduced threshold in the mechanoreceptors near the injury site. Trauma, activating the C-nociceptive sensory units (e.g. mechanoreceptors), sensitizes the wide dynamic range neurones in the cord, which in turn produce spontaneous activity and a lower threshold for firing. Such a mechanism could result in the experience of hypersensitivity (hyperpathia) with a seemingly normal touch or stimulus.[53]

Regardless of the pathogenic theorem accepted, the clinical onset of reflex sympathetic dystrophy is divisible into three types.[3]

1. Pain begins immediately and is out of proportion to the inciting trauma.
2. A typical post-traumatic course and pain level is noted for several days, and then persists or worsens to preclude movement of the knee due to the severity of pain.
3. A typical post-traumatic course is observed with a progressive resolution of symptoms, or even their disappearance. Subsequently, there is a resurgence of pain and an expression of reflex sympathetic dystrophy features.

When reflex sympathetic dystrophy has established itself, there may be evidence of variably expressed symptoms and signs. The *sine qua non* symptom is severe pain. Night pain is quite characteristic. Stiffness, in conjunction with painful motion, is also quite typical. Hyperpathia is a common cardinal feature, as is sensitivity to cold.[53] The often disproportionate severity of symptoms in relation to the initiating trauma, and the known exacerbation of symptoms as a result of emotional stress,[53] have subjected reflex sympathetic dystrophy victims to speculation that their affliction is of a functional rather than organic aetiology. Mayfield and Devine (1945) evaluated reflex sympathetic dystrophy patients and determined that, for the most part, they were psychiatrically stable.[3]

Stiffness of the knee, in varying proportions, is the most characteristic physical finding. The parapatellar soft tissues are indurated as well as tender, adding to the painful restriction of motion and stiffness;[3] at different stages, rubor and vascularity of the involved area may predominate, or coldness and a cyanotic appearance may prevail. The various complexes of signs and symptoms associated with reflex sympathetic dystrophy may wane over a period of months or persist as long as a few years.[3] These symptoms and signs are characteristic and alerting, but not pathognomonic. Therefore diagnostic tests are often performed to gain corroboration.

Plain radiographs typically demonstrate a honeycombed microcystic, and osteopenic appearance of the patella (and/or other bones) on axial views. However, this almost (but not quite) pathognomonic appearance does not become apparent until fairly late in the disease.

Bone scans are positive in 49 to 92 per cent of cases in reports reviewed by Omohundro and Payne[53] and they are characterized by an increased uptake.[3] Hungerford and Fulkerson[3] indicate that thermography is reliable and practical in establishing local vascular temperature alterations and in ruling out non-organic disorders. Temperature differentials of 2°C or more are suggestive of reflex sympathetic dystrophy.[53]

Citing Ficat *et al.* (1973), Hungerford and Fulkerson[3] indicate that intraosseous pressure readings and venography and core biopsies demonstrate stasis and ischaemia in reflex sympathetic dystrophy. Neither these tests nor bone scanning or thermography are unequivocally diagnostic; they are just corroborative.

A potentially therapeutic endeavour—a sympathetic blockade—is the best diagnostic test for establishing the presence of reflex sympathetic dystrophy.[53] Sympathetic blocks can be performed in numerous ways (which will not be described here). If a block (or blocks) is effective in relieving the manifestations of reflex sympathetic dystrophy, the diagnosis is corroborated and a potentially long siege may be averted by a relatively simple and innocuous exercise. Toumey[54] stated that failure of a single block is insufficient evidence of that blocks will not work. With the use of long-acting agents and repetitive blocks, the cycle of pain may ultimately be broken.[3] While there is a positive correlation between the results of a sympathetic block and those of sympathectomy (permanent variety), Toumey demonstrated an inverse relationship between the latter and the duration of the reflex sympathetic dystrophy syndrome.[2,3,54] Furthermore, sympathectomy may be successful in relieving the sensitivity to cold and any regional cyanosis, but without necessarily relieving the pain.[2]

Relief of pain is the critical aim of treatment for reflex sympathetic dystrophy, for without relieving the pain, initiation of successful physical therapy to restore motion and to reverse atrophic bone and soft tissue changes would be difficult and delayed at the very least. In addition to the use of blocks to accomplish this end, numerous drug therapy programmes have been proposed. Omohundro and Payne[53] list the following groups of drugs as being potentially useful:

1. Oral sympatholytic drugs (e.g. prazosin, phenoxybenzamine, propanalol);
2. Anti-inflammatory drugs (prednisolone and non-steroids);
3. Tricyclic drugs (e.g. amitriptyline, desipramine, doxepin, and imipramine);
4. Anticonvulsants (e.g. phenytoin and carbamazepine);
5. Calcium—channel blockers (e.g. nifedipine);
6. Narcotic analgesics—to break the pain cycle, but with care to avoid addiction by virtue of the length of treatment which may be required.

Various combinations of drugs may be helpful, although individually the drugs used may be valueless. Bircher (1967, 1971) reported success in animal models using a combination of Tandearil (anti-inflammatory), Hydergine (vasodilator), and Valium, whereas none of these drugs was effective when used alone.[3]

The principal liability of this condition to athletes is the protracted recovery time. Significant functional residua are essentially nil in most cases. Physical therapy is the mainstay of treatment. Intermittently continuous passive motion alternated with frequent intervals of active motion within a tolerable arc is suggested. Strength development as an instigator of regional blood flow and as an aid to overcoming stiffness is also important. Hydrotherapy with the performance of resistive exercises in water is gentle and often painlessly effective. In an attempt to break cycles of vascular spasms and to stimulate local circulation, contrast treatments, using alternating applications of heat and cold, have been reported to be of value.[2] Facilitation of range of motion, when severely curtailed, by manipulation under general or regional anaesthesia with or without arthroscopy has been discouraged as being provocative.[2,3] When physical therapy, pharmacological treatments, and other supportive measures fail to bring about a progression of improvement after several months, a continuous epidural block should be performed.[53]

The cardinal concepts are those of early suspicion of reflex sympathetic dystrophy and an aggressive approach to the reduction of pain so that physical therapy can be undertaken to restore function.

PATELLA MAGNA

Patella magna' is traditionally a sequel to prior infection or trauma. Trauma may be represented by fracture or an open patellar operation (rarely performed today) for malalignment or general exploration of the knee (Fig. 25). It is rarely seen prior to middle age, although it may be seen in young adults, and it is a frequent accompaniment of a patellofemoral or more ubiquitous

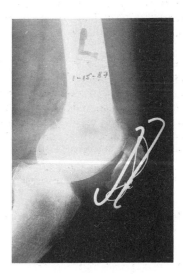

Fig. 25 Splaying of fragments, callus formation, and altered patellar dynamics consequent to patellar fractures frequently result in a secondary patella magna.

degenerative arthritis of the knee joint. Congenital forms of patella magna have been described,[2] but the overwhelming majority are a consequence of osteophytic broadening of the medial and/or lateral facets. Liabilities of the condition include pain with athletic activities and a reduced leverage of the extensor mechanism by virtue of entrapment and reduced excursion of the oversized patella and its relative incongruence with respect to the trochlear sulcus. Since patella magna is rarely seen in highly competitive athletes, treatments other than physical therapy are rarely considered. Occasionally, and if pain is significant, a miniaturization may be considered, or if symptoms are very focal, a facetectomy at the site of pain may be appropriate.

PLICA SYNDROME

Three arthroscopically discernible synovial folds are identifiable within the knee: a suprapatellar fold, a medial fold, and an infrapatellar fold (ligamentum mucosum).[55] Asymptomatic plicae have been reported to be presented in between 18.5 and 55 per cent of arthroscopically inspected knees.[55] A direct trauma to the flexed knee can convert an asymptomatic plica (usually the medial and less often the superior) into an enlarged symptomatic plica.[55,56] The trauma, whether an athletically related contusion or a dashboard injury, produces haemorrhage and synovitis. Eventually, hyalinization and fibrotic thickening can produce a hard

bowstrung plica, which can erode the medial facet of the patella and the medial femoral condyle.[55] The resulting plica syndrome consists of anteromedial knee pain, pseudolocking retropatellar crepitation, and often the presence of a tender palpable cord (the plica) in the interval between the medial aspects of the patella and the medial femoral condyle (Fig. 26). Symptomatic plica are not encountered with great frequency. Broom and Fulkerson[56] reported their presence in 29 of 730 knee arthroscopies performed

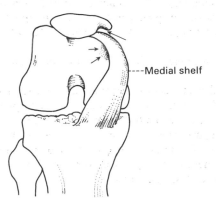

Fig. 26 Medial shelf. (Redrawn from ref. 55 with permission.)

for anterior knee pain. Little more than half the patients in their series reported a history of the trauma, and about three-quarters of the patients were consistently involved in athletics. As with the arthroscopic lateral release for patellar maltracking, plical excision is performed to excess. An MRI or CT arthrogram may demonstrate the presence of a large plical cord in advance of any surgery considered (Fig. 27). In instances in which plical symptoms are severe, arthroscopic resection can produce dramatic relief.

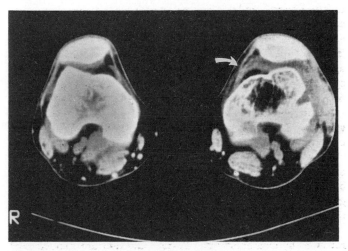

Fig. 27 MRI axial view showing a large medial shelf (arrow) which was confirmed and excised by arthroscopy.

REFERENCES

1. Hefferon J. Personal communication, 1991.
2. Ficat RP, Hungerford DS. *Disorders of the patellofemoral joint*. Baltimore: Williams & Wilkins, 1977.

3. Hungerford DS, Fulkerson JP. *Disorders of the patellofemoral joint*. Baltimore: Williams & Wilkins, 1990: 247–64.

4. Outerbridge RE. Further studies on the etiology of chondromalacia patellae. *Journal of Bone and Joint Surgery* 1964; **46B**: 179–190.

5. Outerbridge RE. The etiology of chondromalacia patella, *Journal of Bone and Joint Surgery* 1961; **43B**: 752.

6. Wiberg G. Roentgenographic and anatomic studies on the femoropatellar joint with special reference to chondromalacia patellae. *Acta Orthopaedica Scandinavica* 1941; **12**: 319.

7. McCarrol JR, O'Donoghue DH, Giana WR. The surgical treatment of chondromalacia of the patella. *Clinical Orthopaedics* 1983; **175**: 130.

8. Crenshaw AH, ed. *Campbell's operative orthopaedics*. 7th edn. St Louis: CV Mosby, 1987: 2233–6.

9. Stanislavjevic S, Zemenick G, Miller D. Congenital, irreducible, permanent lateral dislocation of the patella. *Clinical Orthopaedics* 1976; **116**: 190–9.

10. Rockwood CA, Wilkins KA, King RE. *Fractures in children*. Philadelphia: JB Lippincott, 1984: 248.

11. Rockwood CA, Green DP. *Fractures in adults*. 2nd edn. Philadelphia: JB Lippincott, 1984: 948.

12. Graf BK, Fujisaki CK, Reider B. Disorders of the patellar tendon. In: Reider B, ed. *Sports medicine: The school-age athlete*. Philadelphia: WB Saunders, 1991: 355–64.

13. Osgood RB. Tibial tubercle occurring during adolescence. *Boston Medical Science Journal* 1903; **148**: 114,

14. Jacob RP, von Gumppenberg S, Engelhorst P. Does Osgood Schlatter's disease influence the position of the patella? *Journal of Bone and Joint Surgery* 1981; **63B**: 579.

15. Uhry E. Jr. Osgood–Schlatter disease. *Archives of Surgery* 1944; **48**: 406.

16. Curwin S, Stanish WD. Jumper's knee. In *Tendinitis: its etiology and treatment*. Lexington: DC Heath, 1984.

17. Jackson RW. Etiology of chondromalacia patella. *AAOS Instructional Course Lectures* 1976; **25**: 36.

18. Blazina ME, Kerlan RK, Jobe FW, Canter VS, Carison GJ. Jumper's knee. *Orthopedic Clinics of North America* 1973; **4**: 665m.

19. Ray JM, McCombs W, Sternes RA. Basketball and volleyball. In Reider B, ed. *Sports medicine: the school-age athlete*. Philadelphia: WB Saunders, 1991; 601–31.

20. Siwek CW, Rao JP. Ruptures of the extensor mechanism of the knee joint. *Journal of Bone and Joint Surgery* 1981; **63A**(6): 1932.

21. Ramsey RN, Muller GE. Quadriceps tendon rupture: a diagnostic trap. *Clinical Orthopaedics* 1970; **70**: 161.

22. Kamali M. Bilateral traumatic rupture of the infrapatellar tendon. *Clinical Orthopaedics* 1979; **142**: 131–4.

23. Sherlock DA, Hughes A. Bilateral spontaneous concurrent rupture of the patellar tendon in the absence of associated local or systemic disease. *Clinical Orthopaedics* 1988; **237**: 179–83.

24. Kelly DW, Carter VS, Jobe FW, Kerlan RK. Patella and quadriceps tendon ruptures—jumper's knee. *American Journal of Sports Medicine* 1984; **12**(5): 375–80.

25. Dowd GSE, Bentley G. Radiographic assessment in patellar instability and chondromalacia patellae. *Journal of Bone and Joint Surgery* 1986; **68B**: 297–300.

26. Chow SP, Lam JJ, Leong JCY. Fracture of the tibial tubercle in the adolescent. *Journal of Bone and Joint Surgery* 1990; **72B**(2): 231–4.

27. Goodrich MS, DiFiore RJ, Tippens JK. Bilateral simultaneous rupture of the infrapatellar tendon: a case report and literature review. *Orthopedics* 1983; **6**(11): 1472–4.

28. Hayes WG *et al*. Patellofemoral contact pressures and the effects of surgical reconstructive procedures: articular cartilage and knee joint function. In: Ewing JJ, ed. *Basic Science and Arthroscopy*. New York: Raven Press, 1990: 57–77.

29. Insall J, Falvo KA, Wise DW. Chondromalacia patella. *Journal of Bone and Joint Surgery* 1976; **58A**: 1–8.

30. Fulkerson JP, Shea KP. Mechanical basis for patellofemoral pain and cartilage breakdown; articular cartilage and knee joint function. In: Ewing JW, ed. *Basic science and arthroscopy*. New York: Raven Press, 1990: 93–102.

31. Chrisman OD, Snook GA. Studies on the protective effect of aspirin against degeneration of human articular cartilage. *Clinical Orthopaedics* 1968; **56**: 77.

32. Buckwalter JA, Rosenberg LC, Hunziken BB. Articular cartilage: composition, structure, response to injury and methods of facilitating repair, articular cartilage and knee joint function. In: Ewing JW, ed. *Basic science and arthroscopy*. New York: Raven Press, 1990: 19–56.

33. O'Donoghue RB. Treatment of chondral damage to the patella. *American Journal of Sports Medicine* 1981; **9**: 1.

34. Johnson LL. Arthroscopic abrasion arthroplasty: historical and pathologic perspective: present status. *Arthroscopy* 1986; **2**: 54–69.

35. Macquet P. Advancement of the tibial tuberosity. *Clinical Orthopaedics* 1986; **115**: 225.

36. Macquet P. Mechanics and osteoarthritis of the patellofemoral joint. *Clinical Orthopaedics* 1979; **144**: 70–3.

37. Radin EL. A rational approach to the treatment of patellofemoral pain. *Clinical Orthopaedics* 1979; **144**: 107.

38. Radin EL. Anterior tibial tubercle elevation in the young adult. *Orthopedic Clinics of North America* 1986; **17**: 297–302.

39. Ferguson AB, Brown TD, Fu FH, Rutkowski R. Relief of patellofemoral contact stress by anterior displacement of the tibial tubercle. *Journal of Bone and Joint Surgery* 1979; **61A**: 159.

40. Minkoff JM, Fein L. The role of radiography in the evaluation and treatment of common anarthrotic disorders of the patellofemoral joint. *Clinics in Sports Medicine* 1989; 8(2): 203–60.

41. Carter D, Sweetnam R. Familial joint laxity and recurrent dislocation of the patella. *Journal of Bone and Joint Surgery* 1958; **40B**(4): 664.

42. Dandy D. *Arthroscopy of the knee*. Vol. 4. New York: Gower Medical, 1984.

43. Bassett FN. Surgery of the patellofemoral joint: acute dislocation of the patella, osteochondral fractures and injuries to the extensor mechanism of the knee. *AAOS Instructional Course Lectures* 1976; **25**: 46.

44. Cash JD, Nughston JC. Treatment of acute patellar dislocation. *American Journal of Sports Medicine* 1988; **16**(3): 244.

45. McGinty JB, McCarthy JC. Endoscopic lateral retinacular release: a preliminary report. *Clinical Orthopaedics* 1981; 158: 120.

46. Brown DE, Alexander AH, Lichtman DA. The Elmslie–Trillat procedure: evaluation in patellar dislocation and subluxation. *American Journal of Sports Medicine*, 1984; **12**(2): 104–9.

47. Cox JS. Evaluation of the Elmslie–Trillat procedure for knee extension realignment. *American Journal of Sports Medicine* 1982; **5**: 303.

48. Merchant AC *et al*. Roentgenographic analysis of patellofemoral congruence. *Journal of Bone and Joint Surgery* 1974; **56A**: 1391–6.

49. Laurin CA, Dussault R, Levesque HP. The tangential X-ray investigation of the patellofemoral joint: X-ray technique, diagnostic criteria and their interpretation. *Clinical Orthopaedics* 1979; **144**: 16–26.

50. Imai N *et al*. Clinical and roentgenologic studies on malalignment disorders of the patellofemoral joint: K. Classification of patello-femoral alignments using dynamic sky-line view arthrography with special consideration of the mechanism of the malalignment disorders. *Journal of the Orthopaedic Association* 1987; **61**: 1–15.

51. Schutzer SF, Ramsby GR, Fulkerson JP. Computed tomographic classification of patellofemoral pain patients. *Orthopedic Clinics of North America* 1986; **17**: 235–47.

52. Shellock FG, Mink JN, Fox JM. Patellofemoral joint: kinematic MR imaging to assess tracking abnormalities. *Radiology* 1988; **168**: 551–3.

53. Omohundro PH, Payne R. Reflex sympathetic dystrophy. Cincinnati, OH: Fellowship Project, unpublished.

54. Toumey JW. Occurrence and management of reflex sympathetic dystrophy. *Journal of Bone and Joint Surgery* 1948; **30A**: 883.

55. Patel D. Plica as a cause of anterior knee pain. *Orthopedic Clinics of North America* 1986; **17**(2): 273–7.

56. Broom MJ, Fulkerson JP. The plica syndrome: a new perspective. *Orthopedic Clinics of North America* 1986; **17**(2): 279–81.

4.3.1 Introduction

GARRY GREENFIELD and WILLIAM D. STANISH

Historically, disorders of the shoulder joint were extremely difficult to diagnosis and manage. Commonly, the diagnoses for various shoulder complaints were lumped into broad categories. It was difficult to differentiate, for example, a rotator cuff lesion from other sources of shoulder pain such as glenohumeral instability. Contemporary radiographic techniques have expanded our understanding of these aetiologies. The judicious use of shoulder arthroscopy, coupled with CT scanning and MRI, have defined shoulder diseases in a more precise fashion. With current technology the physician is able to achieve a more thorough understanding of the pathomechanics related to disorders of the shoulder. Faced with this improved 'vision' the practitioner is able to improve upon the success of a surgical or non-surgical treatment.

Before these contemporary radiographic techniques were available, a surgeon relied on clinical skills, combined with an understanding of anatomy. A complete history and physical examination remain the cornerstones for obtaining a proper diagnosis, although non-invasive imaging has provided a very important advancement.

Treatment of instability of the shoulder joint, including frank dislocations and mild subluxations, is one example that has undergone rapid improvement in technique.

In the remote past the ancient scholars had no difficulty in diagnosing an overt shoulder dislocation. In the Greek literature, between 450 BC and 100 AD[1] physicians recognized that shoulder instability, i.e. dislocations, were related to joint laxity. The treatment regimen would include cauterizing the axillary tissues with a hot iron, thus fibrosing the adjacent fat and fibrous tissue. Hippocrates stated, in his explanation of the surgical technique, that 'careful attention to detail must be taken in order avoid major vessels'. Obviously inherent scarring with this technique would produce a contracture, decreasing the intracapsular volume and thus thwarting the possibility of recurrent dislocations. This form of treatment has been modified through time.

The basic premise of these 'primitive techniques' was to decrease the available capsular volume, coupled with a programme of postoperative immobilization. The contemporary treatment for shoulder instability is directed towards obtaining the same goals dictated by the ancient Greeks. The objective was, and is to decrease the intracapsular volume through a precise inferior capsular shift to provide a soft tissue buttress, thus stabilizing the joint. In the future laser cautery of the inferior capsule after transarthroscopic plication of the capsule to eliminate shoulder instability can be envisaged.

Dislocations of the shoulder joint, which are very commonly seen as a product of athletic injuries, can also be treated by non-surgical means. Some methods have relied on mechanical devices to provide traction and abduction to the upper extremity to reduce the dislocated joint.[2] Other methods were strictly manipulative. Hippocrates described a method not dissimilar from a current technique. The classical Hippocratic method would have the patient standing on tiptoes, with the axilla resting on a horizontal beam. Traction was applied to the arm by the practitioner, while countertraction was supplied by the patient's body or by an assistant. These techniques for closed reduction of the shoulder have been illustrated on the wall paintings of the tomb of Ramses II (1200 BC). Ancient Egyptians actually reduced shoulders which were dislocated using the Kocher manoeuvre.[3]

Open surgical techniques, as mentioned previously, have advanced since the Hippocratic technique of cautery. In 1894 Ricord described the open surgical method for capsular repair.[4] A multitude of surgical techniques were described in the early part of this century, all purporting to have a better solution for the problem. These techniques have included the Clairmont ligamentous sling for suspension,[5] the Bankart glenoid labrum repair,[6] the Nicola tenodesis of the head of the biceps,[7] the Magnuson and Stack subscapular transfer,[8] the Putti-Platt subscapularis suturing,[9] DuToit staple capsulorrhaphy,[10] the coracoid transfer described by Bristow,[11] and more current procedures described by Charles Neer[12] and others.

The work of Codman in 1911 provided major advancement in our understanding of shoulder disease. Codman described a rupture of the rotator cuff which was at that point an undiscovered clinical entity.[13] Smith in 1934 observed defects in the supraspinatus tendon, although the frequency and significance of this problem was not appreciated.[14] It was the work of Codman and others that helped clarify the diagnosis, pathology, and treatment of injuries to the rotator cuff and to this day there are but few changes to his original observations, although there are refinements in diagnoses and operative techniques.

Since the original reports of Codman, various methods of rotator cuff repair have been devised. Mayer recommended a fascia repair for large rotator cuff disruptions;[15] Bosworth recommended transplantation of the infraspinatus and the supraspinatus into a more proximal defect in the greater tuberosity;[16] Mclauglin excised the non-viable rotator cuff edges and inserted the healthy edge into a raw bed of bone.[17] Prior to the work of Neer, complete acromioplasty to decrease the rotator cuff was conducted but this has been superseded by the more contemporary procedure, which includes a partial resection of the acromion to relieve classical impingement.[12,18]

The advancement in our understanding of the shoulder joint in athletics have derived from a greater understanding of biomechanics and more precise imaging techniques. Injuries of overuse and pathology from high velocity forces are usually elucidated with a precise physical examination, augmented with CT scanning and/or MRI. The chapters that follow detail the understanding and treatment of shoulder disorders, whether a product of overuse or a single episode of trauma.

BIBLIOGRAPHY

1. Corpus Hippocrates. In Bick EM. *History and Source Book of Orthopaedic Surgery*. New York: Hospital for Joint Diseases, 1933: 13.
2. Brockbank W, Griffiths D. Orthopaedic surgery in the sixteenth and seventeenth centuries. I. Luxations of the shoulder. *Journal of Bone and Joint Surgery*, 1948; **30**: 365.
3. Hussein MK. Kocher's method is 3000 years old. *Journal of Bone and Joint Surgery*, 1968; **50-B**: 669.

4. Ricord: traitement des luxations recidivantes de l'epaule par la suture de la capsule articulaire ou arthrorraphie. *Gazette des Hôpitaux* 1894: 49. In: Bick EM. *History and Source Book of Orthopaedic Surgery*. New York: Hospital for Joint Diseases, 1933: 179.

5. Clairmont P, Ehrlich H. Ein neues operations-verfahren zur behandlung der habituellen schuterluxation mittels muskelphastik., *Archiv für Klinische Chirurgie*, 1909; **89**: 798.

6. Bankart AB. The pathology and treatment of recurrent dislocation of the shoulder joint. *British Journal of Surgery*, 1938; **26**: 23.

7. Nicola J. Recurrent anterior dislocation of the shoulder. A new operation. *Journal of Bone and Joint Surgery*, 1929; **11**: 128.

8. Magnuson PB, Stack JK. Recurrent dislocation of the shoulder. *Journal of the American Medical Association*, 1943; **123**: 889.

9. Osmond-Clarke H. Habitual dislocation of the shoulder. The Putti-Platt operation. *Journal of Bone and Joint Surgery*, 1948; **30-B**: 19.

10. DuToit GT, Roux D. Recurrent dislocation of the shoulder. *Journal of Bone and Joint Surgery*, 1956; **38-A**: 1.

11. Helfet J. Coracoid transplantation for recurring dislocation of the shoulder. The W. Rowley Bristow operation. *Journal of Bone and Joint Surgery*, 1958; **40-B**: 198.

12. Neer CS. Anterior acromioplasty for the chronic impingement syndrome in the shoulder. A preliminary report. *Journal of Bone and Joint Surgery*, 1972; **54A**: 41–50.

13. Codman EA. Complete rupture of the supraspinatus tendon. Operative treatment with report of two successful cases. *Boston Medical and Surgical Journal*, 1911; **164**: 7087–710.

14. Smith JG. Pathological appearances of seven cases of injury of the shoulder joint with remarks. *London Medical Gazette*, 1834; **14**: 280; (reported in *Amiercan Journal of Medical Science*, 1834; **16**: 219–24.

15. Mayer L. Rupture of the supraspinatus tendon. *Journal of Bone and Joint Surgery*, 1937; **19**: 640–2.

16. Bosworth DM. The supraspinatus syndrome. Symptomatology pathology and repair. *Journal of the American Medical Association*, 1941; **117**: 422–38.

17. Mclaughlin HL. Rupture of the rotator cuff. *Journal of Bone and Joint Surgery*, 1962; **44A**: 979–83.

18. Armstrong JR. Excision of the acromion in the treatment of the supraspinatus syndrome. *Journal of Bone and Joint Surgery*, 1949; **31B**: 436–42.

4.3.2 Glenohumeral instability

R. MITCHELL RUBINOVICH

INTRODUCTION

The glenohumeral joint is a mixed blessing. On the one hand it has a greater range of movement than any other joint in the human body. On the other, it is more susceptible to dislocation and recurrent instability than any other articulation. This chapter will deal with this problem of instability in its acute and subsequent forms. Other common problems of the shoulder such as rotator cuff tears or lesions of the acromioclavicular joint, except in those situations where the lesions coexist with shoulder instability, will not be considered.

The population dealt with in this chapter encompasses all ages. Some may suggest that this does not fit into the perspective of a text book on sports medicine, but experience shows that athletes are becoming older. It is not unusual in any sports medicine clinic to see patients in their sixties and seventies. Furthermore, as the interest in physical fitness continues to grow, such individuals will account for a greater percentage of sports injuries.

HISTORICAL PERSPECTIVE

Interest in dislocation about the shoulder dates back to antiquity. The Edwin Smith Papyrus describes the treatment of a dislocation in an epileptic patient.[1] A drawing in the tomb of Upuy (1200 BC) depicts what would seem to be the earliest recorded use of the Kocher method of reduction.[2] Hippocratic texts from 400 BC describe not only the diagnosis and treatment of acute dislocations but also the problem of recurrent dislocation and its treatment. Indeed, only in the past few hundred years has the Hippocratic technique of scarring the inferior joint capsule with a hot poker been supplanted by less dramatic methods. Textbooks on medicine from the sixteenth and seventeenth century detail many elaborate contraptions used to reduce a dislocated shoulder, some of which bear an uncanny resemblance to devices used during the Spanish Inquisition (Fig. 1).[3] All this rich history has prompted

Fig. 1 Using the rack to reduce a dislocated shoulder. (Reproduced from ref. 3, with permission.)

Rockwood to say that 'nothing is new' regarding this problem.[1] Nevertheless, an attempt will be made to describe some of the newer material.

CLASSIFICATION

There are almost as many published classifications of shoulder instability as there are methods of treating the problem. A good classification should possess a rationale additional to being an encyclopaedic listing of the problem. Current classifications are based on initial presentation pattern, degree of initial trauma, or suspected underlying pathoanatomy. The classification presented here is based on direction of instability (Table 1). It has the advantage of being easy to remember so that, when faced with a clinical situation, the clinician can mentally review all the possibilities and proceed appropriately. Although it may appear to leave out certain situations (e.g. posterior voluntary dislocation), these omissions will be explained in subsequent sections.

Table 1   Classification of glenohumeral instability

I ANTERIOR
  Acute
    Dislocation
    Subluxation
    Labral instability
  Recurrent
    Subluxation
    Dislocation
  Chronic
  Voluntary or habitual

II POSTERIOR
  Acute
    Subluxation
    Dislocation
  Recurrent
    Subluxation
    Dislocation
  Chronic

III INFERIOR
  Luxatio erecta
  Subluxation
    Traumatic
    Atraumatic

IV MULTIDIRECTIONAL

EPIDEMIOLOGY

The shoulder is the most commonly dislocated joint in the human body. Two prospective epidemiological studies have been conducted, one in Sweden[4] and one in the United States.[5] Their findings were remarkably similar. The overall incidence of dislocations is between 12 and 17 per cent per 100 000 per year. When only primary dislocations were included, the rates were almost equal at 11.3 and 12.3 per 100 000 per year. Rowe[6] has pointed out that there are equal numbers of dislocations before and after the age of 45, but there is a distinct bimodal pattern to the distribution with a first peak in the third decade and a second peak between the ages of 60 and 80. The early peak is associated with sports-related injuries. The second peak is in the geriatric population and is related to falls in the home. Anterior dislocations are by far the most common, accounting for 97.2 per cent of all dislocations. Posterior dislocations accounted for 2.8 per cent.[5] It is presumed that the rarer forms of dislocation, such as luxatio erecta, were not seen because of the small sample size. Neurological lesions were seen in 7.4 per cent of patients, with 2.8 per cent involving the axillary nerve. While vascular damage is always mentioned as a complication of shoulder dislocation, it appears to be exceedingly rare. In a 5-year period accounting for 216 dislocations, only one patient, an 85-year-old female with multiple injuries, suffered an injury to the axillary artery.[5] Injuries to the rotator cuff may occur, particularly in the older population. One study quotes an incidence of 57 per cent, documented by arthrography, in middle-aged and elderly patients.[7]

ANATOMY AND PHYSIOLOGY

The shoulder joint, and its cousin the hip, are ball-and-socket joints. The socket of the hip is the acetabulum. Its configuration allows almost complete coverage of the head of the femur, giving the hip enormous stability. The cost of this stability is a very limited range of movement. However, the shoulder has a very shallow socket. The glenoid covers only 25 per cent of the humeral articular surface at any one time. This allows a wide range of motion, but sacrifices stability.

The shoulder's stability derives for the most part from soft tissue extensions and reinforcements. These can be roughly divided into static and dynamic stabilizers. The static group includes the labrum, the glenohumeral ligaments, and the joint capsule. Dynamic stabilizers include a superficial layer of muscles, the deltoid and teres major, and the more important deep layer, consisting of the subscapularis, supraspinatus, infraspinatus, and teres minor muscles.

In the normal scapula the glenoid has a forward inclination of between 2° and 12°. The scapula itself is rotated forward through 45°, thereby giving an overall anterior inclination to the glenohumeral joint.[8] This gives the shoulder significantly greater stability posteriorly then anteriorly. The glenoid labrum is attached circumferentially to the glenoid. It deepens the glenoid cavity and allows attachment of the glenohumeral ligaments and the long head of the biceps. It is more developed anteriorly and thus contributes more to anterior than to posterior stability.[9]

There are three glenohumeral ligaments. The superior and middle glenohumeral ligaments are most involved in stability with the arm at 0° to 90° of abduction.[10] The inferior glenohumeral ligament originates from the anteroinferior, inferior, and posteroinferior rim of the glenoid labrum. This ligament is the most important static stabilizer of the shoulder.[11] In selective cutting experiments in cadavers, the shoulder could not be dislocated until the entire inferior glenohumeral ligament was incised including its posterior attachment.[8]

The deltoid and supraspinatus work in concert to inhibit inferior movement of the humeral head. The subscapularis acts to rotate the humeral head inwards and thereby resist anterior subluxation. It is of note that with the shoulder abducted and externally rotated (a position of inherent instability), the tendon of the subscapulis rotates superiorly so that it can no longer function (actively or passively) to resist minor subluxation.[8]

Orthopaedic literature is filled with speculation as to the 'essential' pathological lesion in shoulder instability. The first mention of this concept was by Bankart in 1938, when he described the lesion that still bears his name.[12] This lesion is essentially an avulsion of the anterior glenoid labrum and its attached glenohumeral ligaments. The capsule and ligaments can now strip off the anterior neck of the glenoid, forming a pouch into which the humeral head can recurrently dislocate. The labrum develops as a separate structure embryologically, which would explain why it is possible to tear the anterior capsule through a plane between the glenoid and labrum.[13] Alternatively, the head can dislocate anterior to the labrum through the capsule and glenohumeral ligaments, which would result in stretching of the anterior capsule. This redundant tissue would then allow for recurrent dislocations on the basis of incompetent ligamentous restraints. This is splitting hairs. Whether the head is going anterior or posterior to the labrum is not really that important. Recognizing the incompetence of the anterior static stabilizers of the shoulder is much more to the point.

Once the anterior capsule has failed, several bony changes may take place secondarily. These are important to recognize for the following reasons: first, in difficult diagnostic situations they may be important clues as to the exsistence of instability and its predominant direction; second, the erosion of bone on the humerus or glenoid may compound the problem of an already unstable joint.

There are two important radiological signs of anterior instability. The Hill–Sachs lesion was described in 1940.[14] This is a posterior impaction fracture of the humeral head caused by impingement on the anterior glenoid rim. It can be found in almost all cases of recurrent anterior dislocation if it is sought with appropriate radiological views (see diagnostic section).[15] When large, a Hill–Sachs lesion may articulate with the glenoid in the abducted externally rotated position and contribute to recurrent episodes of instability. Injury to the anterior glenoid can occur by avulsion or erosion. In an anterior dislocation with a Bankart lesion, bone may be pulled off with the labrum. This may involve only small flakes of bone, or alternatively significant portions of the articular surface. Recurrent instability may cause the formation of bony deposits anterior to the glenoid, marking areas of injury and bleeding.

The glenoid may also be eroded by recurrent episodes of in-stability. Each time the head slides forward over the rim the bone is progressively scuffed. This may lead to alarming amounts of bone loss in long-standing recurrent cases, occasionally requiring bone grafting to restore sufficient bony stability.

Posterior pathoanatomy is much less well defined. Reverse Bankart lesions are not common. The static ligamentous stabilizers are less important and the muscular support is less developed than anteriorly. The common thread in most cases of posterior recurrent instability is a redundant posterior capsule.[9]

One bony lesion should be mentioned here. McLaughlin[16] has described a second form of impaction fracture seen anteriorly. This usually occurs in posterior dislocations either acutely or in cases of missed diagnosis. Like the Hill–Sachs lesion it may also contribute to instability by articulating with the glenoid.

DIAGNOSTIC TESTS

Diagnostic tests are divided into radiological and surgical procedures, with the latter including examination under anaesthesia and diagnostic arthroscopy.

Radiographic tests

The standard trauma series for the shoulder should include a true anteroposterior view, a trans-scapular lateral view (Neer's), and an axillary view of the shoulder.[17] When all three views are obtained and interpreted correctly, it is unlikely that any significant pathology will be missed.

The anteroposterior view of the shoulder should be made at 90° to the surface of the glenoid (Fig. 2(a)–(c)). Because the scapula is rotated forward 35 to 40° it is necessary to rotate the patient's

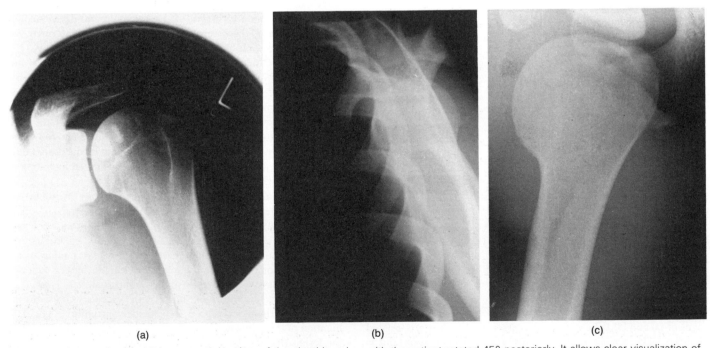

(a)   (b)   (c)

Fig. 2 The trauma series. (a) Anteroposterior view of the shoulder taken with the patient rotated 45° posteriorly. It allows clear visualization of the joint space and the relationship between the humeral head and the glenoid. (b) The trans-scapular lateral (Neer's) view. The acromion, coracoid process, and body of the scapula are clearly seen. The glenoid fossa, while not actually visualized, is at the junction of the Y formed by the three bony elements. The humeral head lies directly over the junction. (c) The axillary view. Again, the acromion and coracoid process are clearly seen. The glenoid articular surface can also be seen, as can its relationship to the humeral head. The three views together constitute a full trauma series and all should be obtained routinely.

affected side posteriorly to match this inclination. A small amount of inferior angulation will also help clear the clavicle and allow better visualization. Attention should be paid to the greater tuberosity as 10 per cent of all anterior dislocations will have an associated fracture here.[13]

The trans-scapular view was described by Neer (Fig. 3).[18] This can be obtained with the patient sitting or supine and does not

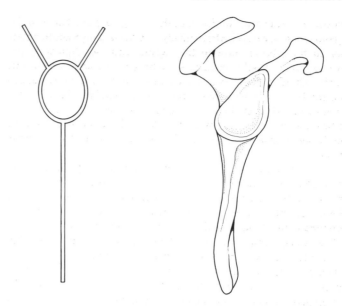

Fig. 4 The scapula as seen on the transcapular (Neer's) view. The coracoid process, acromion, and body of the scapula intersect at the glenoid fossa forming a Y junction. While the glenoid cannot be seen on this view, its position will be known from the other three bony landmarks.

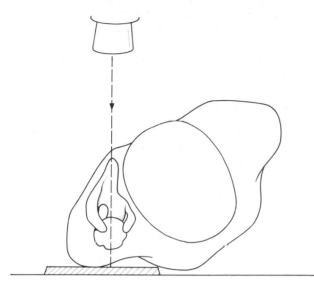

Fig. 3 The transcapular lateral view. The X-ray beam passes along the axis of the body of the scapula.

require the arm to be moved or removed from a sling. The patient is rotated 45° forward on the affected side. The cassette is placed anterior to the shoulder and the X-ray is projected along the body from posterior to anterior. The view obtained will show the coracoid and acromion processes, marking anterior and posterior aspects of the shoulder respectively. The body of the scapula is projected as a vertical line inferior to the junction of the coracoid and the acromion. Although not actually seen in this view, the glenoid fossa is situated at the junction of this Y (Fig. 4). The humeral head should be superimposed over the Y junction in the glenoid fossa. While clearly showing instances of anterior dislocation, this view may be quite misleading if the humeral head is posteriorly dislocated. This demonstrates the necessity of the third component of the trauma series, the axillary view.

The axillary view is often omitted because it is believed that it is too painful to perform in the trauma situation. This is not true. The patient is placed supine and the shoulder is abducted only far enough to position the X-ray tube between the hip and the arm. This is often only 20 to 30° (Fig. 5). The arm can be supported on a table or by an intravenous pole placed beside the patient. The cassette is placed above the acromion and the film exposed. This view is essential to verify the true position of the humeral head and must not be deleted.

Further views of the shoulder are useful in patients suspected of having recurrent instability.

The anteroposterior view of the shoulder in full internal rotation may reveal the presence of a Hill–Sachs lesion evident as a depression or linear density at the superolateral margin of the humeral head.

The Stryker notch view is an even more sensitive technique for

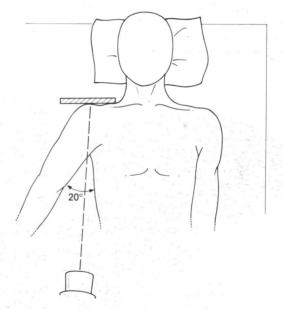

Fig. 5 The axillary view. The arm only needs to be abducted 20° to obtain an adequate view.

picking up Hill–Sachs defects. The patient lies supine with the arm forward flexed 100°. The elbow may be flexed and the hand placed beside the patient's head to help position the arm. The X-ray tube is then angled 45° cephalad (Fig. 6(b)). The lesion will be demonstrated as a circular bite laterally on the humeral head adjacent to the articular surface. In cases of recurrent anterior dislocation this view will show a Hill–Sachs lesion in more than 90 per cent of patients.[15] It is therefore very useful where the diagnosis is unclear. A similar view, the craniocaudal, can be taken with the arm at the patient's side. The patient is supine in the anatomical position and the tube is angled 45° caudal (Fig. 6(c)). While this view has the advantage of not having to move the patient's arm, it is slightly less sensitive to the presence of Hill–Sachs lesions.[15]

The West Point prone axillary view gives the best detail of the anterior glenoid. The patient is placed prone with the arm abducted and the elbow flexed over the edge of a supporting table. The X-ray beam is directed from inferior to superior through the axilla and angled 25° anteriorly from the horizontal (Fig. 6(a)). This gives excellent detail of the anterior glenoid rim. Lesions in this area will be present in 10 per cent of patients with anterior instability.[13] The lesion may vary from small flake fractures and bony deposits (the so-called bony Bankart lesion) to true fractures of the anterior glenoid.

If a diagnosis is still in doubt, various stress films can be taken in the hope of catching the humeral head in a subluxated position. Posterior instability can often be difficult to identify. The cephalo-scapular projection can be useful in this situation. The patient flexes forward 45° at the waist and rests the elbow on a table in front of him or her (Fig. 6(d)). Leaning on the elbow exerts posterior pressure on the humeral head, subluxating it over the posterior glenoid rim. The X-ray beam is projected horizontal to the floor. This view allows clear visualization of the humeral head while displacing the shadows of the acromion and coracoid pro-

cesses. The clavicle still overlaps the joint, but is usually quite radiolucent and only partially visible.[19]

Another stress view is the Stripp axial.[20] The patient sits on a stool, with his or her back against a solid object, allowing him or her to lean posteriorly. The arm is forward flexed about 40° and the X-ray cassette is placed superior to the shoulder joint (Fig. 6(e)). An angled foam sponge may be used here to help keep the cassette level to the ground. The X-ray beam is then projected upwards parallel to the vertical. Because the patient is leaning backwards, the beam is able to clear the hip and buttocks. This angle generates a clear view of the relationship between the humeral head and the glenoid. This view can be employed as a stress film for patients who feel that they can reproduce their instability by actively contracting their muscle or by passively pushing on the humeral head.

Arthrography of the shoulder is of some academic interest in unstable shoulders, but its use as a diagnostic tool is limited. The normal volume of the intact glenohumeral joint is 15 to 18 cm^3.[13] Low volume shoulders are associated with contracture of the capsule, as seen in conditions such as adhesive capsulitis. Volume in excess of 20 cm^3 is associated with paralytic conditions of the shoulder (as in syringomyelia) and recurrent dislocation. This fact in itself would not make arthrography meaningful. However, up to 30 per cent of acute dislocations can be shown to have ruptures of

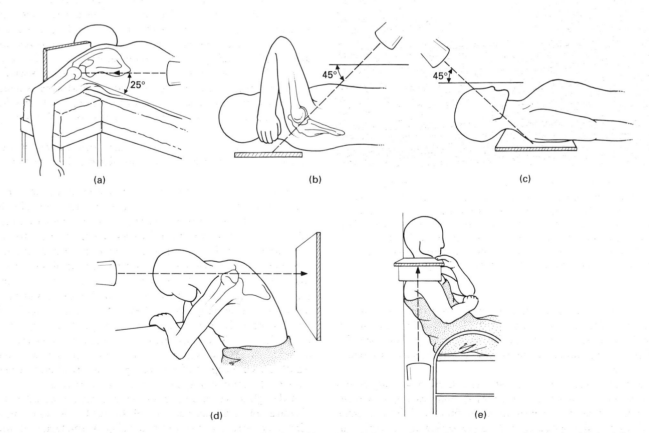

(a)   (b)   (c)

(d)   (e)

Fig. 6 Instability views. In cases where the presence or direction of instability is in question, various views may be obtained to look for stimata of recurrent dislocation, or to catch the humeral head in a subluxated position. (a) The West Point axillary view: this view gives excellent detail of the anterior glenoid rim. (b) The Stryker notch view: this is the most sensitive view for demonstration of a Hill–Sachs lesion. (c) The craniocaudal view (also used for demonstration of Hill–Sachs lesions): while not as sensitive as the Stryker view, the patient does not have to position his or her arm overhead. (d) The cephaloscapular view: the patient leans forward on his or her elbow, applying a posterior stress to the glenohumeral joint. In cases of recurrent posterior subluxation, the humeral head may slide over the posterior glenoid lip, establishing the diagnosis. (e) The Stripp axial view: this view clearly demonstrates the direction of the instability.

the rotator cuff.[13] This percentage climbs significantly when only patients over 50 years old are considered. While most of these ruptures heal spontaneously (or at least are not clinically significant), suggestions are that any patient over the age of 40 who does not progress after reduction should have an early arthrogram (preferably within 4 weeks) to identify lesions of the rotator cuff.[21]

While plain arthrography is of limited use in shoulder instability, arthrotomography can be very helpful. Here, a small amount of dye is injected into the shoulder, followed by 10 cm³ of room air. Tomographic cuts of the glenoid are then carried out in a modified axillary projection.[22] The resulting radiograph shows the profiled glenoid in an inferosuperior projection, clearly revealing the anterior and posterior labrum (Fig. 7). When this technique is used, lesions of the labrum can be delineated very accurately. In one study comparing preoperative arthrotomography with intraoperative findings, a positive correlation was found in 16 out of 17 patients.

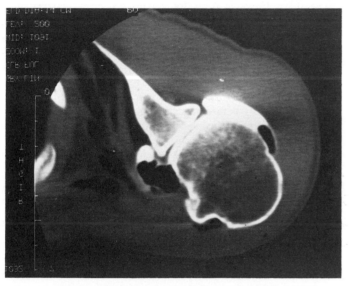

Fig. 7 CT arthrogram of the shoulder. This is an axillary cut showing both the humeral head and the scapula. The glenohumeral articular surface is clearly seen. The labrum is seen at the anterior and posterior margins of the glenoid as dark triangular wedges. They are outlined on their deep surface by the bony glenoid and superficially by dye injected into the joint. Both are normal. Dye passing into the substance of the labrum would signify a tear.

Arthrotomography may also be useful in picking up patients with lesions of the glenoid labrum, but who do not have anatomical instabilities. These patients, described as having 'functional instability',[22] would otherwise be completely undetected (see below under Labral Disorders).

The latest addition to diagnostic visualization techniques has been magnetic resonance imaging (MRI). This exciting new technique offers superior visualization and definition of the soft tissue in and around the shoulder without requiring the injection of radio-opaque material into the joint. MRI can easily visualize Hill–Sachs lesions and erosions of bony glenoid.[23] This alone would not make the technique attractive. However, it is possible to use MRI to visualize lesions of the labrum, including Bankart lesions, stripping of the anterior glenoid neck,[24] tears of the glenohumeral ligaments, and attenuation or atrophy of the sub-

scapularis tendon.[25] The latter lesion cannot be seen by any other means. In addition, MRI is superior to arthrography in identifying tears of the rotator cuff.[26] However, current cost and availability make universal access to MRI impossible. The clinically relevant information available through MRI is also available through conventional radiographic techniques.

Surgical tests

If, after thorough radiological assessment (as well as a carefully conducted history and physical examination), the diagnosis is still in doubt, an examination under anaesthesia can be carried out. A great deal of experience is required to be able to distinguish true pathological instability from variations of the normal shoulder. Palpation of an anaesthetized shoulder (with complete muscle relaxation) is quite different from palpation of a shoulder in the standard clinical situation. Anyone planning to use examination under anaesthetic as a diagnostic test should take every opportunity to examine normal shoulders of patients anaesthetized for other procedures in order to build up an accurate perspective of normal.

With the patient anaesthetized in the supine position the shoulder is abducted between 70° and 100°. The examiner holds the arm tucked between his hip and his elbow, allowing both hands to be free to manipulate the shoulder. The humerus is grasped proximally near its neck. The humeral head is then gently levered anteriorly, posteriorly, or inferiorly, feeling carefully for a clunk as the head slips over the glenoid rim. While any amount of anterior slide may be considered pathological, up to 50 per cent posterior slide is within normal limits.[27] Even more credence can be attributed to findings on the symptomatic shoulder when they are absent from the contralateral normal shoulder.[28] Norris[11] has added a further dimension to examination under anaesthetic. He advocates the use of a neuro-rest for the head and a C-arm fluoroscopy unit to allow direct visual confirmation of any palpated instability.

The final diagnostic tool is the arthroscope. Interest in and use of this instrument regarding the shoulder has flourished in recent years.

Arthroscopy will certainly give a clear view of the interior of the shoulder joint. It can recognize bony lesions. The Hill–Sachs lesion is easily seen from a posterior portal, although care must be taken not to confuse the normally present bare area on the back of the humeral head with a pathological impaction fracture. The articular surface of the humerus and glenoid can also be evaluated for chondromalacia. The glenoid labrum can be seen circumferentially, and its attachment verified both visually and by probing with an arthroscopic hook. Tears of the labrum not involving attachment of the glenohumeral ligaments, but causing functional instability, can also be identified. The glenohumeral ligaments can be seen, as can the superior edge of the subscapularis tendon. Another useful aspect of arthroscopy is its capability to identify tears of the rotator cuff, either complete or partial.

Although it may seem that all this information would make the diagnosis of instability very straightforward, this is not the case. First, the arthroscopist must develop a catalogue of normal anatomy. Despite our growing expertise, such a catalogue is not readily available, and the arthroscopist must often develop it through extensive experience. Seeing the lesion is easy. Deciding whether it represents true pathology is not.

Secondly, while clues to the direction of one instability may be present, other instabilities may have left no evidence. For example, while an isolated erosion of the anterior glenoid labrum would lead one to suspect anterior instability, the associated inferior dislocation of a multidirectional instability may not have left any intra-articular stigmata. Overdependence on arthroscopic findings could therefore lead the surgeon to overlook a very important component of the patient's problem.

Finally, with the shoulder in position for an arthroscopic examination and using skin traction to distract the joint, it is often difficult to assess the degree or even direction of instability. A recent experimental study showed that the introduction of room air into an otherwise stable cadaver shoulder caused the shoulder to subluxate without any division of muscle or ligament.[29] The author surmises that the shoulder was partially stabilized by negative intra-articular pressure. Thus information obtained from the shoulder after introduction of irrigating cannula and arthroscopic equipment should be treated with caution.

Summary

In summary, the acute shoulder must be evaluated using a three-view trauma series. Additional information can be added in suspected cases of recurrent instability using selected instability views. While arthrography can be helpful in certain situations (particularly in older patients with dislocations), the arthrotomogram is generally more useful. MRI (when available) gives the most information without the use of invasive techniques. Examination under anaesthesia (when used by experienced examiners) is helpful in difficult cases, particularly cases of multidirectional instability. Arthroscopy, while potentially a powerful tool, needs to be interpreted cautiously and by an experienced eye.

ACUTE ANTERIOR DISLOCATION

As the shoulder is the most commonly dislocated joint, and anterior dislocation is the most common, the patient with an anteriorly dislocated shoulder is seen frequently in most emergency rooms. Although diagnosis and treatment appear to be straightforward several pitfalls await the unwary. This section will help familiarize readers with this common clinical situation and point out some of the traps.

Presentation

There are three common mechanisms of injury: a fall on the outstretched hand, a blow against the anterior arm forcing it into extension while it is abducted and externally rotated, or a blow from behind directly on to the posterior aspect of the humeral head. The first usually occurs as an attempt to cushion a fall. The last two are more common in sports-related injuries as an athlete tries to block an object in front of him (e.g. basketball) or tries to tackle another athlete who is running by (e.g. football or rugby). The injury is often quite painful. The patient presents to the emergency room clutching the elbow and forearm in an attempt to support and stabilize the limb. The history must include not only current injury, but also any antecedent injury or surgery that the patient has had in the past.

Physical examination

Once the patient is undressed the deformity is visibly obvious even in muscular individuals. From the front, there is a squaring off of the shoulder. This is caused by medial displacement of the

humeral head. The deltoid muscle, which usually drapes over this round prominence, now falls directly downwards over the tip of the acromion. There is often a visible fullness in the vicinity of the coracoid process. In addition, the arm will appear longer, owing to inferior displacement of the humerus. This will be most easily recognized by comparing the relative level of the two elbows. From behind, the shoulder will also appear square, with the posterior glenoid rim tenting outward against the supraspinatus and infraspinatus.

The arm itself will be held slightly abducted and internally rotated. Further internal rotation is painful, and the patient will not be able to touch the opposite shoulder.

Palpation of the shoulder will help to define more closely the relationship of the various bony prominences. A thorough examination of all bones and joints around the shoulder must also be undertaken to look for associated fractures and injuries.

The next step is a thorough neurovascular examination. The lower portions of the brachial plexus can be fully assessed by a motor and sensory examination of the hand and forearm. The upper plexus becomes more difficult to assess as motor function about a dislocated shoulder cannot be accurately tested. Therefore sensory examination is extremely important.

Two areas of special interest are the autonomous sensory area of the axillary nerve, over the lateral deltoid, and the lateral aspect of the proximal forearm innervated by the terminal sensory branch of the musculocutaneous nerve. Documentation of sensory examination is mandatory before any attempts at reduction are carried out. This has both medical and legal implications. Medically, if a nerve lesion is evident only post-reduction, more aggressive investigation and treatment will be warranted than when the patient presents with an established deficit. Legally, a patient who has a neurological deficit post-reduction, and who did not have documentation of this injury pre-reduction, may well assume that the neurological lesion was caused by the physician and not the injury. While most authors emphasize this point, Goss[30] has stated that 'sensory exam for axillary nerve lesions is completely unreliable'. He recommends EMG evaluation for suspected lesions. Unfortunately this is not possible until 3 weeks after the injury. Other investigators have shown denervation potentials in patients with normal sensory examinations.[31] Despite this spectre of unreliability, a thorough documented examination is still highly desirable.

The vascular status of the limb can be assessed by pulses, colour, and capillary refill. Although vascular injury is rare, any patient who demonstrates signs of vascular compromise warrants a full investigation including arteriography.[8]

Radiography

A complete trauma series should be carried out before an attempt at reduction is made. The exception to this rule is the patient who presents to the physician shortly after suffering the dislocation. This most commonly occurs at a sporting event when the on-field physician examines an athlete. In this situation a gentle reduction may be carried out before spasm and swelling make it more difficult. However, the clinician should be experienced and feel relatively comfortable with the diagnosis. There have been instances where an athlete was referred for care after multiple attempts to reduce the shoulder in the dressing room had failed. The athlete did not have a shoulder dislocation, but rather a third-degree acromioclavicular dislocation. Prereduction radiographs are particularly useful in older individuals where fractures of the humeral shaft, often undisplaced and undiagnosed, may complicate management.

The anteroposterior radiograph will show the humeral head displaced medially, overlapping the glenoid neck and lying either in the subcoracoid or subglenoid position (Fig. 8(a)). The dislocation itself is often distracting, and therefore a careful search for associated pathology is sometimes neglected. Special attention must be paid to the greater tuberosity, the surgical neck of the humerus, and the glenoid. It should be remembered that these fractures may be undisplaced and subtle.

The trans-scapular lateral will confirm the anterior nature of the dislocation (Fig. 8(b)). While the anteroposterior often shows a typical configuration, a definite diagnosis of anterior displacement cannot be made unless two views 90° to each other are compared.

The axillary view, while not essential in the diagnosis of anterior dislocation, should be made routinely. As will be discussed later in the section on posterior instability, the most common cause of a missed posterior dislocation is the omission of the axillary view. This view should be obtained in all trauma series.

Once the shoulder has been reduced, post-reduction films are mandatory. Here an anteroposterior and Neer's view will be sufficient. It is essential to prove that the head is actually in the glenoid fossa before the patient leaves the emergency room. If the patient returns for a follow-up visit with the shoulder dislocated, it will be difficult to prove that the dislocation occurred while the arm was immobilized unless all the documentation is available.

Reduction

While there are numerous ways of reducing a shoulder, several important principles apply in all situations.

1. Spasm of the muscle around the joint is the opposing force to reduction. Once the muscle is relaxed, reduction will be easy (except in unusual circumstances).

2. Spasm is fuelled by pain and augmented by the stretch reflex. Any time that striated muscle is elongated suddenly the involuntary reaction is contraction of the muscle and increased spasm.

3. Spasm can be broken by fatiguing the muscle and by medication.

If reduction is attempted by using brute force, regardless of how much stronger than the patient the doctor is, the attempt is doomed to failure. Essential to a successful reduction is that the anatomical relationships are restored without causing further damage to bone or soft tissue. While the audible and palpable clunk of the shoulder popping back into place may seem satisfying, the ideal reduction occurs so gently that only re-examination of the patient and subsequent radiographs reveal that a successful reduction has indeed taken place.

The easiest way to fatigue skeletal muscle is with traction. This can be accomplished either by attaching weight to the patient's arm and letting it dangle, or by attaching one's own body weight and leaning away. The present author calls this 'water skiing'. Grabbing and pulling simply means that the doctor's muscle fatigues as quickly as the patient's, resulting in deadlock. However, leaning away from the patient allows gravity to do the work.

In terms of pharmacology, narcotic analgesics or muscle relaxants can be used either separately or in combination. Both act centrally to decrease pain, relax the patient, and reduce muscle spasm. The drugs can be given either intramuscularly or intravenously. Intravenous administration allows a more rapid onset of action and therefore easier titration of dosage. When using these medications the physician must be completely familiar with the drug dosages and potential complications. As both groups cause respiratory depression and potential apnoea, they must be administered in a setting where assisted ventilation can be easily administered. An Ambu bag, oral airway, and, preferably, equipment for endotracheal intubation should be present and available. Because of this possibility of respiratory depression the combined use of

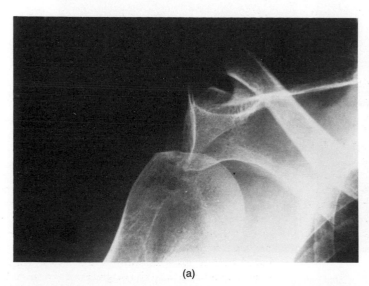
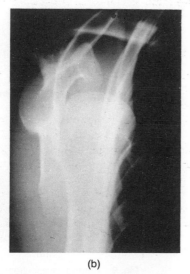

(a)   (b)

Fig. 8 Anterior dislocation. (a) Anteroposterior view showing typical subglenoid position of the humeral head. The glenoid fossa is vacant. (b) Trans-scapular (Neer's) view of an anterior dislocation in the subcoracoid position. (This is not the same shoulder as in (a). With the head dislocated, the glenoid becomes visible. The position of the head under the hooked silhouette of the coracoid process identifies the anterior nature of the dislocation.

narcotics with muscle relaxants should be avoided. Narcotics have the advantage of being reversible by using intravenous naloxone. However, these drugs are less effective as muscle relaxants and therefore the benzodiazapine diazepam is the drug of choice. An intravenous line is set up with the patient positioned for the reduction manoeuvre. Increments of 1 mg of diazepam are slowly administered. The patient is observed for several minutes between doses, and dosage is increased until the patient becomes relaxed and slightly drowsy. The reduction is then carried out. Post-reduction it is important to observe the patient closely as the half-life of many of these medications may be quite long.

All reduction techniques are based on traction, leverage, and rotation, either alone or in combination. All can be used effectively, but each has inherent risks. These risks must be understood when deciding on a technique.

As mentioned earlier, traction reduces spasm. It also helps distract the humeral head from the glenoid. If the head is wedged under the glenoid or if a Hill–Sachs fracture has occurred, traction may avoid further damage to bony structures. It also allows reduction of the joint while keeping the delicate articular cartilage away from the edge of the glenoid. However, indiscriminate traction (and counter-traction applied over a small surface area) may injure vascular and neurological structures in the axilla.

Leverage over a fulcrum is a powerful reduction tool. However, it can be quite dangerous if the bones are not disengaged first. For example, if the glenoid edge hooks the posterior aspect of the humeral head, leverage may cause a fracture of the anatomical neck, reducing the shaft into the glenoid while leaving the articular surface dislocated (Fig. 9).

Rotation shares the same danger as leverage. It also exposes the humeral shaft to rotational torque, a situation that can easily result in a spiral fracture of the shaft. Osteoporotic individuals are particularly susceptible to this latter complication. A further problem with rotation is the possibility of increased stripping of the capsular structures causing further soft tissue damage. At least one author has suggested that rotational techniques may increase the risk of recurrent dislocation.[32] With all this in mind, let us review some of the reduction methods in common use.

The Hippocratic technique is one of the oldest and most widely known methods (Fig. 10). The patient lies supine and the physi-

Fig. 10 A seventeenth century woodcut demonstrating the Hippocratic technique of shoulder reduction. (Reprinted from ref. 3, with permission.)

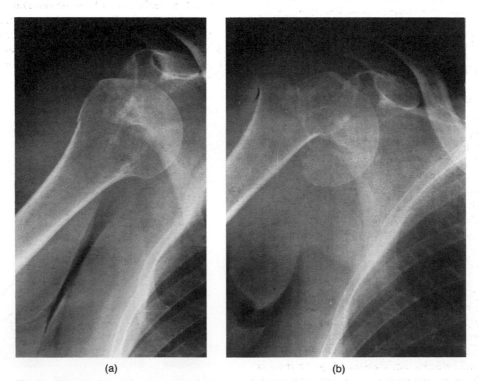

(a) (b)

Fig. 9 Anterior dislocation associated with a fracture of the surgical neck. (a) While the dislocation is obvious, the surgical neck fracture is not. It is best seen as a break in the cortex laterally. It was missed at presentation in the emergency room. (b) The same patient after an attempted closed reduction. The fracture has displaced leaving the humeral head in a subcoracoid position, while the humeral shaft was reduced into the glenoid fossa. The patient required open reduction and internal fixation.

cian places his unshoed heel in the patient's axilla. (It is unclear whether Hippocrates meant to place the heel against the chest wall or directly into the axilla.) The arm is then pulled inferiorly and abducted about 45°. Once distracted, the arm is adducted and the heel is used to help lever the head around the inferior glenoid and back into the joint. Several problems exist with this technique. Control of the amount of traction is difficult. Because of the operator's position it is hard to maintain traction for the 5 to 10 min often necessary to fatigue the muscles fully. The counter-traction provided by the heel is dangerously close to the neurovascular structures of the axilla, and is applied over too small a surface area. Therefore this technique is not recommended.

Milch has described a technique using pure leverage.[33] While palpating the humeral head in the axilla, the arm is slowly abducted. A firm upward and lateral pressure is constantly applied to the head. When fully abducted the arm is externally rotated and traction applied. The head can now be pushed into the glenoid by manual pressure. The danger in this technique is that the humerus and glenoid are not distracted before they are moved, allowing for the possibility of articular surface damage.

The Kocher manoeuvre is a similar leverage technique.[33] With traction applied to the elbow, the arm is externally rotated to about 60°. The elbow is then adducted across the chest. Finally, the arm is internally rotated allowing the head to reduce. The acronym TEAM has been applied to this manoeuvre (traction, external rotation, adduction, and medial rotation). As with Milch's technique, leverage on the undistracted humerus can be dangerous. It has the further disadvantage of using rotation in two different directions.

Hanging methods use traction as the sole mechanical tool. Counter-traction is either by the patient's weight or through a large flat surface applied to the patient. Therefore they are relatively safe.

The classic hanging technique was described by Stimson (Fig. 11). The patient lies prone on a stretcher with a pad anterior to the affected shoulder and pectoral region. The arm dangles downward in a forward flexed position. A weight (4.5 kg or 10 pounds) is attached to the wrist and allowed to hang freely. (Having the patient grasp the weight would cause muscle contraction and oppose complete relaxation.) The operator checks the patient's axilla from time to time, and as the head descends attempts to push it gently into the glenoid fossa.[31] Alternatively, a supraclavicular block can be used and the patient positioned without weights. The patient's relaxed arm will act as the distracting force.[34] The advantage of the Stimson technique is that it allows a controlled gentle reduction comfortably, and with little risk. However, it is time consuming and requires that the patient be prone, thus making it difficult to monitor ventilation. Nevertheless, it is probably the technique of choice for the inexperienced operator.

One further traction method deserves mention. The so-called Eskimo technique was shown to a physician in Greenland by native hunters. It was subsequently taught to non-medical staff in outlying emergency rooms and resulted in a success rate of 74 per cent when used by inexperienced operators.[35] The patient lies laterally on the floor, with the affected side up. Two people grasp the wrist and bring the arm slowly to 90° of abduction, applying enough upward force to suspend the other shoulder several centimetres from the ground. If the shoulder does not reduce in a few minutes, manual pressure is applied to the axilla as in Stimson's method. Although inelegant and requiring two people to apply traction, this technique does have a certain appeal, not least of which is its

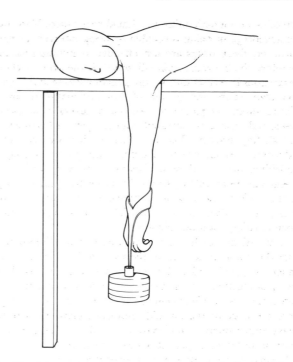

Fig. 11 The Stimson manoeuvre. The patient lies prone with a weight attached to the wrist. The technique is easy to perform and comfortable for the patient. However, it is time consuming.

application in a setting where an experienced surgeon is unavailable. However, it is impractical for older patients and others who might have difficulty in lying on the floor.

The author's preference is for the traction–counter-traction technique (Fig. 12). The patient lies supine on a stretcher. A sheet is wrapped around the chest and axilla on the affected side. It is then attached to the stretcher on the other side of the patient, or wrapped around the waist of an assistant. The elbow is flexed,

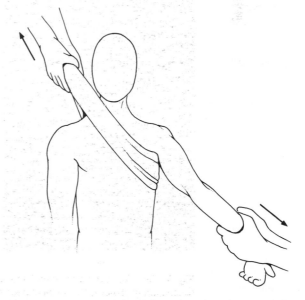

Fig. 12 The traction–counter-traction reduction. This is the safest and most effective method of reducing an anterior dislocation. If an assistant is unavailable to apply counter-traction, the sheet around the patient's body can be attached to the stretcher.

allowing the operator to grip the patient's forearm. This gives a handle to pull on and helps control rotation of the humerus. The arm is abducted 45°. The operator now slowly leans back, maintaining steady traction on the arm. Once fully distracted, gentle pressure in the axilla will aid its reduction. The technique uses pure traction with a broad area of counter-traction. Furthermore, if properly placed, the counter-traction will be against the chest wall and not the delicate axillary contents. The operator maintains rotational control and can speak to the patient, thus adding a little 'verbal anaesthesia'. The patient is supine and therefore both comfortable and accessible.

If several attempts using at least two reduction methods have failed with the patient fully relaxed, a reduction under general anaesthesia will be required. Factors which may lead to a reduction under general anaesthesia include an impacted fracture of the humeral head (locking it into place), an obese or muscular individual (making manipulation difficult), or an exceptionally nervous patient who is unable to relax. In addition, soft tissues may be trapped in the glenohumeral joint, blocking concentric reduction. A further indication for reduction under general anaesthesia is a fracture of the humeral shaft. In this situation it is often possible to manipulate the shoulder without further displacement of the fracture, and the complete muscle relaxation afforded by general anaesthesia is recommended. In any event, the principle of a gentle atraumatic reduction still applies and must be adhered to in order to avoid iatrogenic injury to the shoulder.

Open reduction

Open reduction of the shoulder is indicated in the following circumstances:

1. Failure of closed reduction under general anaesthesia. The head will often be found button-holed through the capsule. Entrapment of the long head of the biceps may also block reduction.
2. Fractures of the greater tuberosity which do not reduce anatomically along with the head. Any displacement of over 1 cm should be considered non-anatomical. Not only will the bony fragment impinge on abduction of the shoulder, but there will certainly be an associated tear of the rotator cuff that will need repair.
3. Fractures of the glenoid comprising more than 30 per cent of the articular surface. This will require internal fixation at the time of surgery.

Post-reduction

Once reduced, the shoulder should be immobilized, adducted to the body, and internally rotated. It can be held in position by a commercially available shoulder immobilizer which is easy to apply, comfortable, and relatively inexpensive. Several models are available. Alternatively a Velpeau sling can be made from materials found in most emergency rooms. The arm is placed in a triangular bandage in the reduced position. The sling is then strapped to the patient's body by wrapping several 15-cm (6 inch) elastic bandages around the chest and affected arm. This keeps the arm at the side and impedes external rotation of the humerus. The thorny question of how long the patient should be immobilized will be dealt with in the section on recurrent dislocation. As mentioned earlier, a post-reduction anteroposterior and trans-scapular lateral radiograph are taken to verify the reduction.

Complications

The major complications to watch for in acute anterior dislocations are redislocation, associated fractures, tears of the rotator cuff, neurological injury, and vascular damage. Recurrent dislocation is more common in the young and decreases with age. The inverse is true for all other complications. Recurrence is sufficiently important that it merits its own section and will be dealt with later in the chapter.

The Hill–Sachs lesion, a posterior impaction fracture of the humeral head, has been reported in 10 to 55 per cent of anterior dislocations (Fig. 13).[13] This rises to nearly 100 per cent of

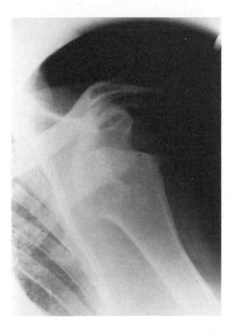

Fig. 13 Anteroposterior view demonstrating an anterior dislocation and a large Hill–Sachs lesion. The latter is a posterior impaction fracture of the humeral head caused by impingement of the anterior glenoid lip against the dislocated humeral head. It is seen as a crescentic lucency at the superolateral margin of the humeral head.

patients with recurrent episodes.[15] It does not require specific treatment but does give important clues as to diagnosis and prognosis. As special views may be required to see the lesion, it may be overlooked in the acute situation. Fractures of the greater tuberosity occur in approximately 10 per cent of acute dislocations.[13] They are best seen on the anteroposterior view (Fig. 14). With reduction of the glenohumeral joint the majority reduce spontaneously. Any displacement of more than 1 cm merits open reduction and internal fixation. If left displaced, the fragment will impinge on the underface of the acromion, blocking full abduction. Late repair is more difficult and the results may be poor. These fractures are usually associated with tears of the rotator cuff and neglect may lead to rotator cuff insufficiency. If the tuberosity reduces anatomically, the shoulder can be treated as an uncomplicated dislocation without fear of redisplacement. Fractures of the humeral shaft and surgical neck are not uncommon, but are beyond the scope of this chapter.

Fractures of the glenoid rim will be present in 10 to 20 per cent of acute dislocations.[8] For the most part these are small flakes of bone that accompany an avulsed glenoid labrum—the so-called

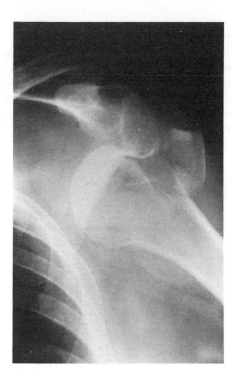

Fig. 14 Anteroposterior view showing an anterior dislocation associated with a displaced fracture of the greater tuberosity. Fracture of the tuberosity occur in approximately 10 per cent of anterior dislocations and should be looked for carefully.

bony Bankart lesion. If small, they are insignificant and require no specific treatment. Fractures of 30 per cent or more of the articular surface will lead to inherent bony instability, and therefore indicate acute open reduction and internal fixation. If left unattended, they will lead to difficult reconstructive situations requiring bone grafting.

Tears of the rotator cuff occur in between 30 and 50 per cent of patients with anterior dislocations.[7,13] The clinician must be vigilant for this injury as the dislocation may make diagnosis difficult. While it is possible to demonstrate the lesion arthrographically in a very high percentage of patients, the incidence of long-term insufficiency drops to only 5 per cent.[13] This would imply that with reduction of the shoulder, the cuff has the ability to heal. Rowe *et al.*[36] have pointed out the potential cleavage plane between the subscapularis and the supraspinatus tendons. It may be here that the rotator cuff tears, allowing it to heal on its own, but this is only conjecture. While acute repair of the rotator cuff is not necessary, any patient with demonstrated insufficiency who has associated pain and functional disability should be considered for delayed surgery. Such repair is definitely easier before atrophy, and retraction of the edges makes anatomical suturing impossible. Therefore arthrography is indicated after 4 to 6 weeks in patients who are not progressing in their post-reduction course.

Brachial plexus injuries, if carefully looked for, will be found in approximately 5 per cent of patients. The majority are infraclavicular and incomplete, with full recovery expected. Isolated lesions of the axillary nerve will be found in 5 to 10 per cent of patients.[7] This usually occurs as a traction neuropraxia with the humeral head trapping the axillary nerve in the quadrilateral space. As Goss[30] has pointed out, sensory examination is not totally reliable and in suspected cases EMG examination may be needed. This examination will look for denervation potentials in the deltoid muscle, but will not be positive until at least 3 weeks post-injury. The majority of these lesions are transient and recovery should be expected without treatment.[8] Axillary nerve lesions may occur in conjunction with tears of the rotator cuff making the latter diagnosis difficult. The presence of an axillary nerve lesion (particularly in an older individual) should make one suspect concurrent tear of the cuff.

Vascular injury with anterior dislocation occurs but is very uncommon. It may be associated with multiple injuries or high velocity trauma. Despite the low incidence, the clinician must always be vigilant for vascular compromise, as any delay in recognition and treatment will lead to disastrous end results. Any patient displaying signs of vascular injury—including decreased pulses, expanding haematomas, or pallor of the arm—should be evaluated aggressively, with inclusion of an arteriogram if deemed necessary.

RECURRENT ANTERIOR DISLOCATION

Recurrent subluxation

Before discussing recurrent anterior dislocation, it is necessary to introduce one further concept: recurrent subluxation. This was first described by Blazina and Satzman in 1969, and has since been defined further by Rowe. The patient is usually a young active athlete who recalls an episode of forced extension and rotation with the arm abducted. The patient's chief complaint is recurrent episodes of weakness or giving way in the shoulder with overhead activities like throwing or serving in tennis. Often the patient describes a complete lack of strength—hence the name dead arm syndrome.[37] The symptoms may have been initiated by an episode of frank dislocation, but usually the patient denies any such episodes. Approximately 50 per cent of these patients are unaware that the shoulder is unstable.[38] The remaining 50 per cent will describe a sensation of the shoulder slipping out of place followed by spontaneous reduction. Such patients are never able to demonstrate the instability voluntarily, but are usually aware of the position causing the sensation (i.e. abduction and external rotation).

On physical examination the patient will have a full range of motion, no tenderness, and no signs of rotator cuff insufficiency. The apprehension sign will be universally positive. To do this test, abduct the patient's arm between 90° and 120°. While standing behind the patient, externally rotate the arm and apply gentle pressure to the back of the humeral head. A reproduction of the patient's symptomatology or a feeling that the shoulder is about to pop out is taken as a positive test. The patient may resist external rotation or flex the shoulder by leaning the body forward, also indicating a positive test. Examination of the opposite shoulder is helpful as a comparison. Another useful test can be carried out with the patient supine on an examining table. This manoeuvre is described in the section on examination under anaesthesia. While actual subluxation may be difficult to detect in the awake subject, the patient may experience the symptomatology signifying a positive test.

The instability series may provide clues to diagnosis. The West Point view may show changes of the anterior glenoid—either fuzziness of the rim or actual soft tissue calcification. The Stryker notch view may show a Hill–Sachs defect even in patients who have not had a frank dislocation.[28] The most important factor for accurate diagnosis is awareness of the syndrome and an index of suspicion. Once the syndrome is recognized, treatment proceeds as with recurrent dislocation.

Recurrence rate

Many factors appear to affect the rate of recurrence in acute anterior dislocation. One, which is universally accepted, is age at first dislocation. All authors agree that the younger the patient, the greater the likelihood of later recurrence. Rowe's classic paper of 1956 stated that for patients under the age of 20 the recurrence rate was 94 per cent; between the ages of 20 and 40 it fell to 74 per cent, and after the age of 40 only 14 per cent recurred.[6] This trend has been subsequently corroborated by many authors, although most find rates lower than those reported by Rowe. The reason for this age distribution remains a subject of debate. Some authors attempt to explain it on the basis of associated pathology. When tears of the rotator cuff or posterior capsule occur, the shoulder may dislocate in a way which spares the anterior capsular structures. If the superior or posterior lesions heal, there will be no residual Bankart lesion or anterior capsular redundancy. Such tears may be more likely in older individuals who are known to have incipient weakening of the cuff on an age-related basis. Other authors suggest that as younger patients are more active in sports, they are more likely to be exposed to traumatic events predisposing them to recurrent episodes.

The mechanism of injury has also been implicated in recurrence rate. Bankart[12] believed that abduction external rotation injuries caused a dislocation through a tear in the anterior capsule. He believed that this tear would heal and therefore have a lower recurrent rate. A direct blow posteriorly would tear the glenoid labrum off the glenoid (i.e. a Bankart lesion), which would not heal, thereby leading to recurrent dislocation. However, Rowe's study does not show any correlation between mechanism and recurrence.[39]

I believe that patients who have dislocated at an early age have a particular anatomical predisposition to instability. This could be due to insufficient strength of capsular material, or perhaps improper attachment of the glenoid labrum. Most people are exposed to a certain number of traumatic episodes during their youth. Some with anatomical predisposition will dislocate, thereby creating a cohort prone to subsequent instability. Those of 'heartier stock' who do not dislocate in their youth may dislocate later in life when normal anatomy is eventually exposed to an overpowering traumatic event. In this situation, the shoulder will heal with little propensity for redislocation. The theory would allow for a spectrum of predisposition with those at one extreme dislocating younger, intermediate types dislocating later, and a stable type which would dislocate only much later in life, perhaps after the normal attrition of aging leads to weakening of surrounding tissues.

Many authors have attempted to reduce redislocation rates by immobilization immediately after primary injury. Earlier authors suggested an immobilization period of 6 weeks, particularly in the young, basing this on the projected healing time of soft tissues. Subsequent studies by several authors have failed to show this to be effective.[31,37,40] Hovelius[41] has shown no change in redislocation rates with up to 4 weeks of immobilization.

Despite a lack of statistical evidence, most authors still suggest a 3-week period of immobilization for patients under the age of 30, and a 1-week period of immobilization for older patients. The patient must be informed from the outset of the high redislocation rate, regardless of immediate therapy.

Certain bony lesions appear to influence recurrence rate. It patients with fractures of the greater tuberosity the rate fell to between 3 and 7 per cent,[30,37] while fracture of the surgical neck was not associated with any recurrence.[30] The Hill–Sachs lesion was not found to influence the redislocation rate in younger patients, but in patients over the age of 23 the recurrence rate was significantly higher.[41]

Treatment options

When the patient presents with episodes of recurrent dislocations, several options are available. First, the patient may wish to accept the instability. While most patients who have two or more redislocations can expect to continue dislocating without treatment, one prospective study showed that 20 per cent of patients with two or more redislocations during the first 2 years post-primary dislocation had no further dislocations in the subsequent 3 years.[41]

No study has conclusively shown that stabilizing a recurrently unstable shoulder would change the incidence of late degeneration. Indeed, two studies have shown an equal incidence of degeneration in untreated and surgically stabilized patients.[42,43] Furthermore, the severity of the arthritis was not related to the number of recurrences.[42]

Conservative measures can be instituted. Some authors believe that strengthening exercises would be of benefit for recurrent dislocation.[8] They suggest exercises for both internal and external rotators in order to improve dynamic stability of the shoulder. Other authors disagree.[27] As stated previously in the section on anatomy, the subscapularis muscle, a prime dynamic stabilizer, is superior to the shoulder with the arm abducted and externally rotated. Consequently, in the shoulder's most vulnerable position, the subscapulis would be useless in preventing dislocation. Furthermore, when the arm is abducted and externally rotated, those muscles that would be most useful in preventing dislocation are neurogenically inhibited to allow full range of motion at the shoulder. Nevertheless, for patients unwilling to accept their instability, but not yet ready for surgery, a temporizing course of physiotherapy and strengthening exercises seems to be reasonable.

Bracing is also available. These braces, which are used almost exclusively during sporting activities, attempt to block abduction and external rotation by means of a harness. The brace straps around the patient's chest, with a second strap encircling the arm just distal to the axilla. While they may be effective in preventing dislocation, they place a dangerous fulcrum over the humerus against which a fracture can occur. They should be used with caution, particularly in contact sports.

The third and final option is surgical correction of the instability. When should surgery be advised? There are no absolute indications for stabilization. Although recurrent dislocation is associated with an increased risk of late degenerative arthritis, this risk is not related to the number of recurrences, nor is it reduced by surgical correction. Surgery does not guarantee a return to normal stability as the overall rate of redislocation post-operatively is approximately 10 per cent.[13] The prime indication for surgery is the patient's dissatisfaction with the *status quo*, coupled with his or her acceptance of the surgical risks involved.

Some authors, including Jobe,[2] have concluded that in the young competitive athlete—particularly when a throwing shoulder is involved—primary repair is indicated with the first dislocation. Although this may seem radical, statistics lead to this conclusion. Early repair is also recommended for patients whose occupation or sport may expose them to severe bodily harm if their shoulder should dislocate at a potentially dangerous moment. This applies to people working on roofs and ladders, or particularly in sports like long-distance open-water swimming.

Surgery

Once a surgical course is selected, the physician is faced with choosing from literally hundreds of procedures and their variations. Although the choice may seem endless, there are certain principles involved which can be evaluated to help choose the most appropriate procedure. Generally speaking, all operations for recurrent instability approach the problem via one of three mechanisms. They may block external rotation, thereby avoiding the position of vulnerability. They may try to maintain stability by buttressing the front of the joint with bone or soft tissue. They may attempt to identify the specific pathology and reconstruct the normal anatomy. In addition, any of these procedures may involve the use of metal fixation implants.

Current thinking is that restriction of motion is not only unnecessary but undesirable, and is often listed as a postoperative complication. Buttressing operations are often performed extra-articularly, thereby not allowing full intra-articular evaluation (a factor associated with increased recurrence rates). Metal staples and screws have the disadvantage of breaking and migrating, and can impinge on the articular surface. This may result in severe degenerative changes postoperatively.[42] Consequently, the ideal operation would identify and correct pathological anatomy without metal implants or designed restriction of movement. It would have a low recurrence rate. It would not be associated with major complications. It would permit visualization of the joint and be relatively easy to perform. While many procedures do not achieve all the prerequisites of the perfect operation, most have been performed over the decades with surprisingly similar success rates. Perhaps the original Hippocratic technique of hot poker scarification merits a second look!

The actual surgical techniques are not discussed in this chapter. However, some of the currently popular procedures will be outlined to make it easier for physicians to relate to patients who have undergone previous surgical repairs.

The Magnusen–Stack operation is a transfer of the subscapularis tendon distally on the humerus lateral to the biceps tendon. It is designed to cause an internal rotation contracture and to provide an anterior sling to prevent inferior movement of the head. Because of its goal and because it utilizes a large metallic staple, this procedure is not recommended. Once popular, it is now rarely used. A radiograph showing a staple in the humerus distal to the joint usually signifies a previous Magnusen–Stack repair.

The Putti–Platt procedure is an operation that first detaches the subscapularis tendon near its insertion on the humerus. The joint is then opened and the stump of the tendon on the lesser tuberosity is sutured to the glenoid labrum. This provides an intra-articular reinforcement to the anterior capsule. It can be combined with reattachment of the glenoid labrum. This procedure addresses most requirements. Unfortunately, it may also lead to an internal rotation contracture of the shoulder by shortening the excursion of the subscapularis. This operation has frequently been performed in the past with excellent results, but is no longer common (particularly in throwing athletes) because of the potential restriction in range of movement. Nevertheless it remains an excellent choice for non-throwing individuals.

The Eden–Hybinette procedure is a buttressing operation that attaches an anterior bone block to the glenoid. It is fraught with many problems, not least of which is the screw fixation necessary to hold the graft. The bone may be resorbed and may migrate or impinge on the humeral head. For these reasons it has been supplanted by other techniques.

Another buttress type operation is the Bristow repair. Here the tip of the coracoid is detached with the short head of the biceps and the coracobrachialis. The bone is then transferred to the glenoid neck and screwed on via a split in the subscapularis tendon. Not only is the bone block present as a buttress, but the attached muscles are intended to work as an anterior sling, stabilizing the front of the joint as the arm is brought up into abduction and external rotation. While the operation is still popular, it has fallen into disrepute in recent years for several reasons. The musculocutaneous nerve pierces the coracobrachialis just distal to its attachment to the coracoid. The nerve is at risk during transfer. The muscles moved are important flexors of the shoulder and may not function as effectively from their new location. There is no way of addressing any associated intra-articular pathology with this procedure. Finally, there is the problem of metallic screws.

The Dutoit procedure uses a large metal staple placed through a longitudinal split in the subscapularis to reattach the capsule to the anterior glenoid neck. Although quick, this procedure offers no visualization of the intra-articular structures. Malposition and migration of Dutoit staples have made this procedure the prototype for iatrogenic shoulder arthrosis. Thus it has been widely condemned.

The Bankart repair as popularized by Rowe is the gold standard by which most other techniques are measured. The actual procedure may be of interest.[37] Basically the subscapularis is detached and the shoulder opened via a T-shaped incision. The glenoid and labrum are visualized, and if a detachment is noted it is reattached with interrupted sutures via drill holes in the glenoid rim. The procedure fulfils most of the prerequisites but is very difficult to perform. Rowe himself claims an average operative time of 2.75 h. This is well beyond the average of 1 to 1.5 h, even in the hands of a very experienced surgeon. However, it is designed to allow early range of movement and little if any restriction in external rotation.

The author's procedure of choice is a technique recently described by Neer: the capsular shift (Fig. 15).[18] Here, the subscapularis is detached and the capsule is opened in a T-shaped fashion. The

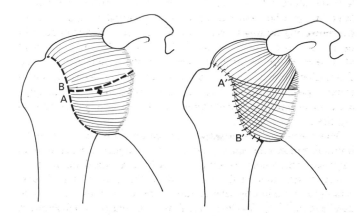

Fig. 15 Neer's capsular shift procedure. The capsule is opened in a T-shaped fashion. The capsule is then closed by shifting the inferior half upwards from A to A' and the superior half downwards from B to B'.

shoulder is inspected and if necessary intra-articular work is performed. The capsule is closed by shifting the lower portion of the T upwards and laterally. The upper flap is then moved laterally and inferiorly, double vesting the repair. Placement of the flaps is made with the shoulder in neutral rotation so as to avoid an internal rotation contracture. The main advantage is that it not only complies with all the prerequisites, but also allows obliteration of inferior capsule redundancy. It can be performed in a similar fashion through a posterior capsulotomy, making it adaptable to cases of posterior instability. It is also the procedure of choice for multidirectional instability (see later). Finally, the operation is easier to perform than a classic Bankart procedure.

Postoperatively the patient is immobilized in a sling identical with that used in the acute situation. Duration of immobilization must be tailored to the patient and the operation. Procedures that rely on an internal rotation contracture often keep the patient in a sling for up to 6 weeks. Most other techniques suggest beginning movement at or near the 3-week mark. In the throwing athlete, where any loss of external rotation may compromise function, earlier mobilization is the rule.

Complications

Postoperative complications fall into five major categories. The first is general postoperative complication such as haematoma and infection. As in all surgical procedures, the incidence of these complications can be reduced by meticulous technique. The second is recurrence. According to various authors, the recurrence rate may range from 0 to 10 per cent. The actual rate probably lies closer to the higher figure. The third complication is neurovascular injury. The musculocutaneous nerve is particularly at risk where it passes through the coracobrachialis. The nerve can be injured in procedures which release the tendon, or by overzealous retraction. The axillary artery is inferior to the humeral neck and is surprisingly close to the surgical field. Fortunately, injury to this large vessel and the accompanying brachial plexus is exceedingly rare. The fourth complication is restriction of movement. This is generally related to the specifics of the technique and the length of postoperative immobilization. Most patients who complain of limitation of movement can regain significant amounts of range by means of aggressive physiotherapy if the problem is recognized early. The final complication is late arthrosis or degenerative change. While some degree of degeneration may be inherent in the recurrently unstable shoulder, much is related to surgical errors causing the shoulder to be overly tight, or to misplacement or migration of metallic implants.

Arthroscopy

One final group of surgical procedures should be mentioned: arthroscopic stabilization techniques. This relatively new and rapidly expanding field holds great promise for the future. The surgery involves the reattachment of the glenoid labrum and/or the glenohumeral ligaments to the glenoid. This can be accomplished by the use of metallic staples, cannulated screws, arthroscopic sutures, and, most recently, absorbable fixation devices. Any sports medicine surgeon finds any technique which allows performance of surgery without major skin incisions seductively attractive. However, experience has shown that caution must be exercised before committing to arthroscopic stabilization.

First the surgeon must have mastered diagnostic arthroscopy of the shoulder. Before he can repair the shoulder, he must be able to see and recognize the pathology. In addition to the usual operative complications already described, he must add those peculiar to arthroscopic surgery, including puncture of vital structures by errant portals, injuries to the brachial plexus by improper positioning of the patient, and soft tissue dissection by irrigating fluids. Furthermore, it seems that, even in the best hands, arthroscopic stabilization has a higher rate of recurrence than open procedures. Hawkins[44] has reported a recurrence rate of 16 per cent. However, in a series of 100 consecutive cases, the recurrence rate was close to 30 per cent.[45] Why it should be so high may be explained by the theory of congenitally abnormal anatomy in patients with recurrent dislocation described earlier. Arthroscopic surgery is so relatively atraumatic that it restores the predislocation anatomy without modifying it in any way. Even a Bankart repair requires that the anterior shoulder be dissected and then reapproximated, causing an obligate amount of anterior scarring. If an individual is predisposed to dislocation by his nascent insufficient anatomy, restoration of this anatomy will not be enough to prevent recurrence. However, if a higher recurrence rate can be accepted by the patient and the surgeon, there are certain other advantages to arthroscopic surgery. It is easier and less time consuming than open procedures. It can be done on an outpatient basis (no mean consideration in today's economically pressing times). Assistants are not required. Finally, cosmesis is much improved over the standard deltopectoral approach.

Summary

The personal approach of the author to the patient with recurrent instability is as follows: First, establish a diagnosis by history, physical examination, and radiography. If the patient is functionally impaired and fully understands the risks of surgery, surgical stabilization is undertaken. Patients unwilling to undergo surgery are sent for rehabilitative exercises. Braces are rarely used and are avoided in contact sports. The author's surgical procedure of choice is the capsular shift. On occasion, arthroscopic stapling is done, but this procedure is generally restricted to low demand individuals who are concerned about cosmesis and who understands the increased risk of recurrence with this form of stabilization.

ACUTE POSTERIOR DISLOCATION

Posterior dislocation of the shoulder is extremely uncommon. The incidence in reported series varies from 1.5 to 3.7 per cent. Because of its rarity and because of the difficulty in interpreting standard radiographic films, many acute dislocations are missed at initial presentation only to be picked up later as chronic dislocations. This is unfortunate, as late treatment is more difficult and generally associated with poorer outcome. It is also easy to avoid if the clinician follows a few simple rules.

Presentation

As with anterior dislocation, the patient will present with pain and decreased movement about the shoulder. The mechanism of injury is often a fall on the outstretched hand with the shoulder flexed, adducted, and internally rotated. A direct blow anteriorly may dislocate the shoulder, but this is less common since it is resisted by both the anterior tilt of the glenoid and the surrounding prominences of the coracoid and acromion. Other situations which should alert the physician to a possible posterior dislocation are seizures and electroshock therapy. These events cause forcible contraction of the shoulder musculature with the stronger anterior muscles forcing the shoulder out posteriorly. Multiple trauma should also alert the physician, as should alcohol-related injuries which always seem to result in unusual phenomena.

Physical examination

Physical examination will show a patient holding the arm internally rotated and slightly flexed. The anterior shoulder may lack its usual fullness, and the acromion and coracoid processes will seem overly prominent. Attempts at external rotation will be painful. The most striking feature will be the inability to rotate the humerus externally. Another feature is the patient's inability to supinate the hand with the shoulder forward flexed.[38]

Radiography

The anteroposterior view of the shoulder may be misleading. The humeral head does not fully dislocate from the glenoid, but usually lies in full internal rotation with a varying amount of articular surface remaining in the glenoid (Fig. 16(a)). The demarcation between that portion of the humeral head in the glenoid and that portion dislocated is often marked by an impaction fracture of the humeral head (Fig. 17). Again, this fracture may not be evident on the anteroposterior view. Marked internal rotation of the humerus and a slight overlap of the head in the glenoid may be the only signs noted.

The trans-scapular lateral may also show only subtle findings. The head is posteriorly displaced, but only marginally (Fig. 16(b)). As the glenoid position is only inferred on this projection, the dislocation may be missed entirely.

The key diagnostic clue is the axillary view. This point cannot be overemphasized. The common thread linking all missed posterior dislocation cases is the omission of this one radiograph. Not only does this projection clearly show the dislocation, but it also delineates the presence and size of any impaction fracture (Fig. 18).

Reduction

Reduction of the posterior dislocation is made difficult by the impaction fracture and is compounded by the unfamiliarity of most physicians with this entity. Unless the humerus can be gently disengaged from the posterior lip of the glenoid, reduction will be either impossible or disastrous, resulting in a shearing off of the articular surface through the impaction fracture.

Once the patient is suitably relaxed, as with an anterior reduction, a manoeuvre combining elements of flexion, adduction, and lateral traction is carried out. The latter is necessary to disengage the head from the glenoid and is effected by pushing against the inside of the humerus high on the shaft near the axilla. Once disengaged, the head can be reduced by external rotation. Should closed reduction be unsuccessful under intravenous sedation, an attempt under general anaesthesia will be needed. As with anterior

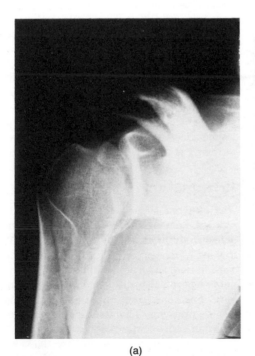
(a)

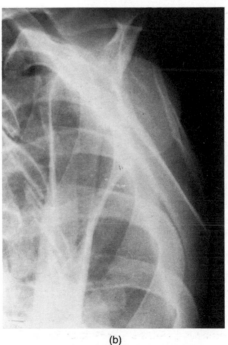
(b)

Fig. 16 Posterior dislocation. (a) The anteroposterior view showing very subtle signs of a posterior dislocation. The joint space is obliterated and there is slight overlap of the glenoid and the humeral head. This gives the articular surface of the glenoid a sclerotic appearance. In addition, the humeral head is internally rotated, foreshortening the neck and causing the lateral aspect of the head to appear radiolucent. (b) The trans-scapular (Neer's) view of the same shoulder. The articular surface is directed posteriorly, demonstrating the dislocation and its direction, but the head remains superimposed over the area of the glenoid fossa. While detectable, the dislocation could easily be overlooked. The subtle nature of the posterior dislocation the radiograph underlines the need for the axillary view in a trauma series.

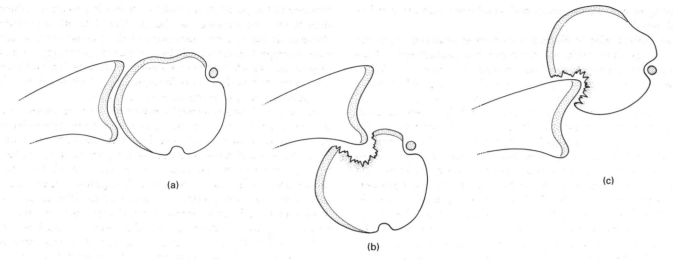

Fig. 17 Impaction fractures of the humeral head: (a) reduced position of the glenohumeral joint; (b) posterior dislocation with an anterior impaction fracture or McLaughlin lesion; (c) anterior dislocation with a posterior impaction fracture or Hill–Sachs lesion.

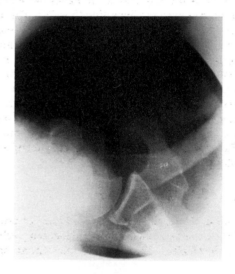

Fig. 18 McLaughlin lesion. An axillary view of the shoulder in the reduced position. A large defect is noted anteriorly in the humeral head, adjacent to its articular surface. This is an impaction fracture caused by the posterior lip of the glenoid while the shoulder is posteriorly dislocated.

dislocations, a gentle hand will be rewarded with fewer complications and better long-term results.

Post-reduction

After reduction the shoulder is assessed for stability. If stable, it is immobilized in moderate external rotation for a period of 4 to 6 weeks. This may be shortened for older individuals. If there is no impaction fracture, a situation said to be rare,[47] recurrent instability will be low. In the usual instance, i.e. cases with impaction fractures, the prognosis is related to the size of fracture. Since posterior dislocation is so rare, there are few studies of actual recurrence rates. However, one such study found recurrences in nine out of 24 patients.[17] As with anterior instability, age plays an important role with younger patients more likely to suffer recur-

rences. Operative repair with specific reference to the treatment of the impaction fracture will be dealt with in the section on chronic dislocation.

POSTERIOR SUBLUXATION

Recurrent posterior subluxation is much more common than posterior dislocation of the shoulder.[47,48] The humeral head slips posteriorly over the glenoid rim and spontaneously reduces, a situation analogous to recurrent anterior subluxation. Most commonly the patient presents complaining primarily of pain rather than instability.[48] The disability is usually mild, rarely interfering with activities of daily living (although occasionally it is severe enough to interfere with sports, particularly in the dominant arm of a throwing athlete). There may be a history of an antecedent event, but this is rare. Most patients present with insidious onset of symptoms perhaps related to various overuse phenomena.

Presentation

The posterior subluxators fall into several categories based on physical examination. The first are those patients who can actively demonstrate their instability by contraction of muscles or positioning of the arm. This usually includes a combination of forward flexion, adduction, and internal rotation. These patients are not to be confused with patients who can actively dislocate anteriorly. The latter group often have underlying emotional or psychiatric problems. This issue will be taken up later in the section on wilful and voluntary dislocators.

The next group of patients are those who cannot reproduce their symptoms at will, but who can be subluxated by the examiner. Again, flexion, adduction, and internal rotation accompanied by axial load may reproduce the subluxation. Most patients subluxate between 70° and 110° of forward flexion. Beyond this point the shoulder reduces spontaneously, perhaps owing to the winding up of the capsule and tightening of posterior structures. The final and most difficult group of patients are those in whom the subluxation cannot be demonstrated on clinical examination. This group may well be confused with anterior subluxators, as symptoms occur in both conditions with the arm forward flexed.

Any patient who is diagnosed as having posterior instability must be carefully evaluated for associated multidirectional instability, since correction of one component alone may fail to reduce symptoms and may even exacerbate the problem. Once again, this will be dealt with in a subsequent section.

Radiographs of the shoulder should include anteroposterior, lateral, and axillary projections. These are used mainly to rule out other pathology as radiographic findings are usually absent. Recurrent subluxation rarely progresses to frank posterior dislocation and so will not lead to impaction fractures of the head.[17] Up to 20 per cent of such patients may have calcification of the posterior capsule, or even small flake fractures of the posterior glenoid rim,[17] but these are generally small and often missed. CT scanning is much more sensitive to their presence. A CT scan is also helpful in assessing the orientation of the glenoid relative to the body. As discussed in the section on diagnostic techniques, stress films are very useful particularly in patients who can actively or passively subluxate their shoulder (Fig. 19(a),(b)).

When all other tests prove negative, examination under anaesthesia can be helpful. It should be re-emphasized that up to 50 per cent posterior glide of the humeral head is within the normal range.[11]

Initial treatment

Initial treatment should always be conservative, as most patients will respond to an aggressive programme of strengthening exercises.[17] These include strengthening of the rotator cuff musculature along with the deltoid and stabilizers of the scapula. Only when an extended trial of conservative management fails should the patient be considered for surgical repair. Remember that demonstrable instability *per se* is not an indication for correction. The patient must also have significant functional disability to warrant intervention.

The pathoanatomy of posterior instability is less well defined than its anterior counterpart. Reverse Bankart lesions are rare. The capsule is not reinforced by specific named ligaments, and the function of the posterior musculature is less important to stability than at the anterior aspect of the shoulder. The prime posterior stabilizer is the anterior inclination of the glenoid articular surface. Perhaps because of this lack of specific pathology, surgical repairs are less reliable than for anterior instability. This difficulty is compounded by the relative infrequently of the problem and the inexperence of most surgeons in correcting it.

Surgical techniques generally consist of bony operations, soft tissue operations, or a combination of both. The posterior glenoid osteotomy is carried out by first detaching the external rotators from the humeral head. The neck of the glenoid is cut partially through with an osteotome and then levered forward, increasing the anterior inclination of its surface. It is propped open using a bone graft from the posterior acromion or the iliac crest. This procedure is technically demanding and care must be taken to avoid completely cutting across the neck or cracking the osteotomy into the glenoid articular surface. It is recommended for those patients in whom the primary deficit is felt to be relative retroversion of the glenoid.

A second bony correction can be carried out by taking a block of bone from the posterior iliac crest and attaching it to the posterior glenoid neck in such a way that it overhangs the humeral head providing a buttress against subluxation (Fig. 20). This technique has been used with good, though certainly not universal, success. One patient who incorporated the graft radiologically presented 3 years postoperatively with recurrent symptoms. On examination the humeral head could be seen sliding laterally over the graft before it subluxated out posteriorly.

Soft tissue procedures have also been described. These are generally procedures which either overlap or shift redundant posterior capsule. They also have a high rate of recurrence, in part related to the poor quality of the tissue normally found at the back of the shoulder.

Summary

Posterior subluxation is the most common form of posterior instability of the shoulder. While many patients can easily demonstrate their instability, difficult cases may need specialized

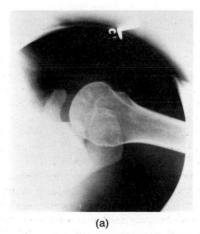 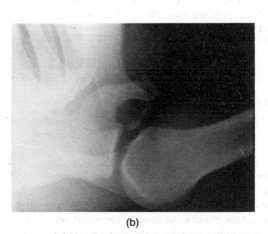

(a)   (b)

Fig. 19 Posterior subluxation. Modified axillary views of a patient able to subluxate his shoulder voluntarily by positioning his arm and contracting his muscles. (a) Reduced position: note the anterior attitude of the articular surface, and the close proximity of the head to the coracoid process. (b) Subluxated: the articular surface is now posterior to the glenoid and the greater tuberosity appears to lie in the glenoid fossa. The space between the coracoid process and humeral head is increased owing to the posterior translation of the head.

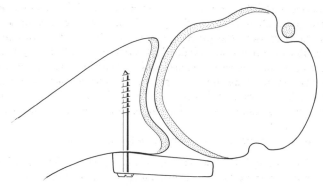

Fig. 20 Posterior bone block procedure for recurrent posterior instability. The rectangular graft is taken from the iliac crest or the posterior acromion and attached to the posterior glenoid. By slightly overlapping the head of the humerus it acts as a buttress against posterior translation.

radiography or even examination under anaesthesia before diagnosis can be made. Strengthening exercises are often curative and should be attempted before any surgical procedure is undertaken. Repair is only for patients with functional disability. Failure of operative techniques with resultant recurrence of subluxation is much more common in posterior than anterior repairs.

CHRONIC DISLOCATION

For the purposes of the following discussion a shoulder will be considered chronically dislocated after a period of 3 weeks. This arbitrary cut-off is used as there is no general consensus available in the literature. The common presenting complaint is lack of range of motion with pain only on attempted movement.[49] The most common cause of a missed posterior dislocation is the omission of an axillary radiograph. The most common cause of a missed anterior dislocation is the lack of any radiographs. Many of these patients are referred late after being treated for varying amounts of time as frozen shoulders. Treatment of the dislocation must be individualized and considered in relation to the patient's symptoms, age, and disability.

Presentation

The physician must be alert to the diagnosis in any patient with chronic limitation in range of shoulder movement. A history of seizure, electroshock therapy, multiple trauma, or alcoholism should raise suspicion. In addition, the elderly always seem to be at risk of neglect of acute problems.

Rowe and Zarins[50] have suggested simple tests designed to identify the unreduced shoulder.

1. Flex both elbows to 90°: this manoeuvre will show the rotational limitation characteristic of chronic dislocation. Patients dislocated posteriorly will have marked internal rotation deformities. This deformity is further demonstrated by flexing the shoulder forward and asking the patient to turn the palm upwards. While the forearms will supinate, the internal rotation deformity of the humerus keeps the palm facing medially. The patient dislocated anteriorly will be locked in external rotation. This is even more disabling than an internal rotation contracture as the patient will be unable to bring the hand to their head for grooming or eating, or to their back for personal hygiene.

2. Observe the patient from above. Comparison of the contralateral shoulder will demonstrate a fullness anteriorly or posteriorly depending on the direction of the dislocation. This may be made more dramatic by the disuse atrophy of the shoulder girdle which often accompanies the chronic deformity.

3. Palpate the angle of the scapula as the shoulder is manipulated. As the humeral head is usually locked against the glenoid, the scapula will be felt to move in unison with the humerus.

4. Radiography: a full trauma series is indicated with particular emphasis on the axillary view. The most probable cause of a missed dislocation is an incomplete or absent radiological investigation at the time of the original injury. Though stated at the outset, this point merits repetition.

Examination of the patient must also include a complete neurological assessment. The axillary nerve may have been damaged at the time of injury or by continued pressure placed on it by the dislocated head. Injuries to the median and ulnar nerve have also been reported.

Treatment

Treatment is dependent on the patient's needs. The age of the dislocation must be established. In shoulders dislocated for more than 6 months the viability of the articular cartilage is questionable and total joint arthroplasty has to be considered.

Posterior dislocations are usually complicated by the presence of a posterior impaction fracture. The presence and size of such fractures will determine the therapeutic course and can be assessed by axillary views or CT scanning.

The following recommendations can be made with regard to treatment.

1. For defects less than 20 per cent of the articular surface, an attempt at gentle closed reduction should be made, followed by an open reduction if unsuccessful. Undue force is absolutely contraindicated in the closed reduction. Damage to vascular structures can lead to disastrous complications. In addition, the bone will be osteoporotic and easily fractured. Post-reduction, some authors suggest transarticular fixation with screws and pins.[49] However, Rowe and Zarins[50] suggest that the arm be immobilized at the side with the humerus in slight extension for 4 to 6 weeks. This modification simplifies care and does not seem to affect the outcome adversely.

2. For defects between 20 and 40 per cent, open reduction and transfer of the subscapularis tendon into the defect is advised. This technique was described by McLaughlin in 1952.[16]

3. For defects greater than 40 per cent of the articular surface, or for any dislocation older than 6 months, an arthroplasty is indicated. If the glenoid is intact a hemiarthroplasty will suffice. Otherwise total shoulder arthroplasty will be required.

Patients with untreated anterior dislocation can be managed similarly. In the anterior dislocation, however, the main complicating factor is erosion of the anterior glenoid rim. This may require a reconstruction by a total shoulder arthroplasty, with or without bone grafting.[51] A concomitant tear of the rotator cuff may also be present, requiring repair at the time of reduction. While recurrence of posterior instability is rare, reconstruction of anterior capsule structures may be required to combat recurrent anterior instability.[51]

Those who refuse treatment may still obtain reasonable results.

Rowe and Zarins[50] reported eight shoulders treated in this way with one good, four fair, and three poor results. However, the same study demonstrated superior results in patients receiving active care.

Summary

An ounce of prevention is worth a pound of cure. Almost all chronic dislocations can be avoided through proper radiographic evaluation. Once established, the problem will tax even the most experienced surgeon. Proper treatment demands careful assessment of the patient's needs and a skilled pair of hands.

MULTIDIRECTIONAL INSTABILITY

While awareness of shoulder instability dates back to the time of the Pharaohs,[1] the first recognition of multidirectional instability was by Neer in 1980.[18] Although not as common as anterior instability alone, it is by no means rare. Missing a patient with multidirectional instability and treating them for a unidirectional instability is a frequent cause of surgical failure.[18,48] Inferior subluxation is the hallmark of multidirectional instability.[11]

Presentation

The patient with multidirectional instability may present primarily with an anterior or posterior pattern of recurrent instability. There may be an antecedent history of trauma or repetitive overuse injuries. The latter is common in sports requiring extreme flexibility such as gymnastics or swimming. In addition, patients often manifest signs of inherent ligamentous laxity in joints other than the shoulder. Genetic disorders that cause systemic ligamentous laxity, such as Ehlers–Danlos syndrome, should alert the physician to the possibility of multidirectional instability. As well as the usual manifestations of instability, the patient may complain of a dull ache in the arm when it hangs at the side or with overhead use. Carrying heavy objects may be uncomfortable. Patients with multidirectional instability sometimes describe vague intermittent paraesthesias, leading the observer to suspect emotional or psychiatric disorders.[18]

Physical examination

The cardinal feature of multidirectional instability is inferior subluxability. This can be documented by producing a positive sulcus sign. The patient sits with the arm comfortable at the side and the shoulder completely relaxed. The examiner grasps the elbow and pulls downwards using steady firm pressure. A dimple or gap will form at the edge of the acromion process in affected individuals. This dimple is caused by the inferior subluxation of the humeral head and resultant negative pressure in the subacromial space. Multidirectional instability can occur as anteroinferior, posteroinferior, or a combined anterior, posterior, and inferior instability.

Treatment

The initial management of multidirectional instability should be conservative. Neer[18] recommends repeated examinations over a 1-year period combined with an aggressive strengthening programme before any consideration of surgery. The patient's emotional stability, motivation, and actual disability should be carefully evaluated during this period. Once a surgical approach has been chosen, the procedure must be able to deal with the combined elements on an individual patient basis. Neer's capsular shift operation has been designed with this aim in mind.[52] It can be performed from an anterior approach to correct anteroinferior instability, or from behind for posteroinferior instability. For tridirectional instability a combined anterior and posterior approach can be used.

Surgical repairs may fail for the following reasons.

1. The surgery is instituted for the wrong pathology. The identification of multidirectional instability does not imply that it is causing the patient's symptoms. The complaints may be vague and non-specific. Associated skeletal pathology, including rotator cuff tendinitis or acromioclavicular arthritis, may be the primary cause of symptoms. Correction of multidirectional instability without addressing the primary pathology will result in continuing symptoms.
2. The surgery corrects one element of the instability while failing to correct instability in another direction.
3. The surgery corrects one element of the instability and either increases another element of the instability or forces the shoulder to remain subluxated in a secondary direction. For example, a patient with combined anterior, posterior, and inferior multidirectional instability who undergoes a Putti–Platt repair may develop an internal rotation contracture forcing the head to remain posteriorly subluxated.

Summary

The patient with multidirectional instability presents a diagnostic and therapeutic challenge. The physician must identify all the components of the instability, decide which patients require surgery, and then address all demonstrable pathology. An error at any one stage may result in failure.

INFERIOR INSTABILITY

Luxatio erecta

Direct inferior dislocation of the shoulder, luxatio erecta, is exceedingly rare and must be differentiated from inferior subluxation. The latter is commonly seen after shoulder injury. In luxatio erecta the humeral head dislocates directly inferiorly, lodging in a completely inverted position relative to its normal anatomy. The head is directed downwards and the humeral shaft upwards. The incidence of such a dislocation is so small that it has not been calculated in literature. It can occur at any age and has a marked predisposition for men, with 92 per cent of all recorded cases occurring in males.[53] The mechanism of injury is hyperabduction of the arm. The humerus comes into contact with the acromion as the arm is fully elevated, and with continued abduction this bony prominence acts as a fulcrum, levering the humeral head out of the glenoid. The patient presents with the arm fully abducted, the elbow flexed, and the hand and forearm resting on the patient's head (Fig. 21).

Pre- and post-reduction radiographs are important since luxatio erecta has a very high incidence of associated injury including fractures about the shoulder.[54] Reduction is usually easy using a traction–counter-traction technique. With the patient supine a sheet is passed over the top of the affected side to keep the patient

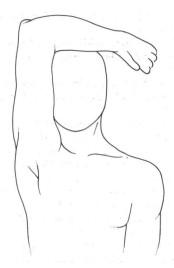

Fig. 21 Luxatio erecta. The classic hyperabducted position of the arm in an inferior dislocation of the shoulder.

from sliding upwards. The shoulder is hyperabducted gently and pulled upwards. Reduction with an audible click signifies that the head has returned to the glenoid and the arm can be brought to the side. Mallon et al.[53] have suggested the use of fluoroscopy to ensure that the head has been reduced prior to lowering the arm. This may be useful but is not mandatory. Fractures of the tuberosity are common but are usually reduced anatomically. Other fractures reported include the acromion, the glenoid, the body of the scapula, and osteochondral fractures of the humeral head. Neurological injury is very common. A literature search of 80 patients revealed an overall incidence of 59 per cent,[53] with the axillary nerve being most commonly involved. Spontaneous resolution over a period of months is to be expected. Injury to the rotator cuff is also possible, and post-reduction evaluation and follow-up is necessary. Vascular injury, while not common, is more likely to occur in luxatio erecta than in any other dislocation. As stated in previous sections, any suspicion of vascular compromise should be evaluated immediately by clinical examination, Doppler studies, and, if necessary, an arteriogram. Post-reduction immobilization is suggested for 4 weeks, although the literature does not report a single case of recurrence.

Inferior subluxation

Inferior subluxation is a clinical entity totally unrelated to luxatio erecta. The patient has usually sustained trauma to the shoulder, either bony or soft tissue, and presents over a period of days to weeks with gradual inferior drooping of the shoulder. On occasion this may be seen in the emergency room at the time the initial trauma. A sulcus sign may develop and radiography will reveal the humeral head partially subluxated from the glenoid fossa in an inferior direction. There is never a component of anterior or posterior displacement. This situation is distinct from the patient with multidirectional instability who may show inferior subluxation, but only with stress downwards on the arm. The incidence of inferior subluxation post-fracture may be as high as 10 to 20 per cent.[55] It may also occur in patients with neurological lesions (axillary and others) about the shoulder. It has been suggested that the aetiology is an inhibition of shoulder musculature by pain or paralysis, although solid documentation is lacking. The natural

history of the condition is spontaneous reduction and recovery over a period of weeks to months, depending on the underlying associated pathology. However, it should be noted that in the non-traumatic situation, inferior subluxation has been associated with more ominous pathology including Pancoast's tumour of the lung.[56] A careful history should easily exclude the latter variant.

LABRAL DISORDERS

With advances in arthroscopy and arthrotomography, a new category of instability is developing, that of lesions of the glenoid labrum. These tears can be associated with recurrent dislocation or subluxation of the shoulder (so-called anatomical instability) or occur as isolated lesions causing functional instability.[22] This last category has only recently been identified. Patients present with painful episodes of clicking and locking in the shoulder, often during overhead activities. They have no history of dislocation and may have no history of a traumatic event. Physical examination may show tenderness on deep palpation, although this is not a prerequisite to diagnosis. A clunk may be appreciated by the manipulating shoulder in various degrees of abduction and rotation. The whole picture is quite analogous to meniscal tears in the knee.

Diagnosis may be confirmed by arthrotomography. This examination is quite sensitive to lesions of the glenoid labrum. CT scan or MRI (when available) may also be helpful in elucidating the pathology. Confirmation and treatment is by arthroscopy. Those patients with functional instability (i.e. tears of the labrum, but no demonstrable instability of the glenohumeral joint) may be rendered completely asymptomatic by resecting loose portions of labrum. This is easily accomplished by an arthroscopic route. The number of such patients recognized in the literature so far is small; however, with increased awareness and better diagnostic techniques this may become a more commonly identified cause of instability in the future.

HABITUAL AND VOLUNTARY DISLOCATION

An undefined percentage of patients presenting with glenohumeral instability will be able to demonstrate their instability voluntarily during clinical examination. Subsequent care of these patients can be simplified by dividing them into several subgroups.

The habitual or wilful dislocator

This group of patients can usually dislocate or subluxate their shoulder anteriorly. The dislocation and subsequent reduction are painless and cause the patient little consternation. Approximately 50 per cent will be shown to have psychiatric or emotional problems.[30] This subset has been shown to have a much lower incidence of the usual pathological lesions associated with recurrent anterior instability.[57] Careful assessment of the patient's emotional stability and motivation are important. Surgical results in this group are universally poor. A determined patient will be able to undo almost anything that can be done to stabilize the joint. Also, narcotic addicts sometimes develop the ability to dislocate their shoulder in the hopes of receiving intravenous medication. Patients who present repeatedly to the emergency room with dislocations should be evaluated carefully for stigmata of drug dependency.

Voluntary anterior dislocators

This subgroup may be difficult to differentiate from the wilful dislocator without full psychiatric evaluation. However, once defined this group will respond well to physiotherapy. Those who fail in conservative treatment should be considered surgical candidates and can be expected to have results similar to the general population.[30]

Voluntary posterior subluxation

This subgroup is distinct from the anterior voluntary stability group. Patients with recurrent posterior subluxation are able to demonstrate their instability by muscular contraction or by positioning of the arm. They should not be considered emotionally unstable, and treatment can proceed as discussed in the section on recurrent posterior subluxation.

REFERENCES

1. Rockwood CA, Green DP. *Fractures*. Philadelphia: JB Lippincot, 1975.
2. Jobe FW. Unstable shoulders in the athlete. *American Academy of Orthopedic Surgeons Instructional Course Lectures* 1985; **34**: 228–31.
3. Brockbank W, Griffiths DL. Orthopedic surgery in the sixteenth and seventeenth centuries. *Journal of Bone and Joint Surgery* 1948; **30B**: 365–75.
4. Kroner K, Lind T, Jensen J. The epidemiology of shoulder dislocations. *Archives of Orthopedic and Trauma Surgery* 1989; **108**: 288–90.
5. Simonet WT, Melton LJ III, Cofield RH, Ilstrup DM. Incidence of anterior dislocation in Olmstead County, Minnesota. *Clinical Orthopedics and Related Research* 1984; **186**: 186–91.
6. Rowe CR. Prognosis in dislocation of the shoulder. *Journal of Bone and Joint Surgery* 1956; **38A**: 957–77.
7. Johnson JR, Bayley JLL. Early complications of acute anterior dislocation of the shoulder in the middle-aged and elderly patient. *Injury* **13**: 431–4.
8. Daulton SE, Snyder SJ. Glenohumeral instability. *Baillière's Clinical Rheumatology* 1989; **3**: 511–34.
9. Hawkins RJ, Belle RM. Posterior instability of the shoulder. *American Academy of Orthopedic Surgeons Instructional Course Lectures* 1989; **38**: 211–15.
10. O'Brien SJ, Warren RF, Schwartz E. Anterior shoulder instability. *Orthopedic Clinics of North America* 1987; **18**: 395–408.
11. Norris, TR. Diagnostic techniques for shoulder instability. *American Academy of Orthopedic Surgeons Instructional Course Lectures* 1985; **34**: 239–57.
12. Bankart ASB. The pathology and treatment of recurrent dislocation of the shoulder joint. *British Journal of Surgery* 1938; **26**: 23–9.
13. Tullos HS, Bennett JB, Brayly WG. Acute shoulder dislocations; factors influencing diagnosis and treatment. *American Academy of Orthopedics Surgery Instructional Course Lectures* 1984; **33**: 364–85.
14. Hill HA, Sachs MD. The grooved defect of the humeral head. A frequently seen unrecognized complication of dislocation of the shoulder joint. *Radiology* 1940; **35**: 690–700.
15. Rozing PM, De Bakker HM, Obermann WR. Radiographic views in recurrent anterior shoulder dislocation. *Acta Orthopedica Scandinavia* 1986; **57**: 228–330.
16. McLaughlin HL. Posterior dislocation of the shoulder. *Journal of Bone and Joint Surgery* 1952; **34A**: 584–90.
17. Schwartz E, Warren RF, O'Brien SJ, Fronek J. Posterior shoulder instability. *Orthopedic Clinics of North American* 1987; **18**: 409–19.
18. Neer CS II. Involuntary inferior and multi-directional instability of the shoulder; etiology, recognition and treatment. *American Academy of Orthopedic Surgeons Instructional Course Lectures* 1985; **34**: 232–8.
19. Oppenheim WL, Dawson EG, Quinlan C, Graham SA. The cephaloscopular projection. *Clinical Orthopedics and Related Research* 1985; **195**: 191–3.
20. Horsfield D, Stutley J. The unstable shoulder—a problem solved. *Radiography* 1988; **54**: 74–6.
21. Hawkins RJ, Bell RH, Hawkins RH, Koppert GJ. Anterior dislocation of the shoulder in the older patient. *Clinical Orthopedics and Related Research* 1986; **206**: 192–5.
22. Pappas AM, Goss TP, Kleinman PK. Symptomatic shoulder instability due to lesions of the glenoid labrum. *American Journal of Sports Medicine* 1983; **11**: 279–88.
23. Seeger LL. Magnetic resonance imaging of the shoulder. *Clinical Orthopedics and Related Research* 1989; **244**: 48–59.
24. Tsai JC, Zlatkin MB. Magnetic resonance imaging of the shoulder. *Radiologic Clinics of North America* 1990; **28**: 279–91.
25. Holt RG, Helms CA, Steinbach L, Neumann C, Munk PL, Genant HK. Magnetic resonance imaging of the shoulder; Rational and current applications. *Skeletal Radiology* 1990; **19**: 5–14.
26. Habibian A *et al.* Comparison of conventional and computed arthrotomography with MR imaging in the evaluation of the shoulder. *Journal of Computer Assisted Tomography* 1989; **13**: 968–75.
27. Johnson PH. Recurrent subluxation of the shoulder. *Journal of the Arkansas Medical Society* 1988; **84**: 335–7.
28. Hastings DE, Coughlin LP. Recurrent subluxation of the glenohumeral joint. *American Journal of Sports Medicine* 1981; **9**: 352–5.
29. Kumar VP, Balasubramanian P. The role of atmospheric pressure in stabilizing the shoulder. *Journal of Bone and Joint Surgery* 1985; **67B**: 720–1.
30. Goss TP. Anterior glenohumeral instability. *Orthopedics* 1988; **11**: 87–95.
31. Cofield RH, Kavanagh BF, Frassica FJ. Anterior shoulder instability. *American Academy of Orthopedic Surgeons Instructional Course Lectures* 1985, 34: 210 27.
32. Plummer D, Clinton J. The external rotation method for reduction of acute shoulder dislocation. *Emergency Medicine Clinics of North America* 1989; **7**: 165–75.
33. Beattie TF, Steedman DJ, McGowan A, Robertson CE. A comparison of the Kocher and Milch techniques for acute anterior dislocations of the shoulder. *Injury* 1986; **17**: 349–52.
34. Rollinson PD. Reduction of shoulder dislocations by the hanging method. *South African Medical Journal* 1988; **73**: 106–7.
35. Poulsen SR. Reduction of acute shoulder dislocations using the Eskimo technique: a study of 23 consecutive cases. *Journal of Trauma* 1988; **28**: 1382–3.
36. Rowe CR, Southmayd WW, Patel D. The Bankart procedure—a long term and result study. *Journal of Bone and Joint Surgery* 1978; **60A**: 1–16.
37. Rowe CR. Acute and recurrent anterior dislocations of the shoulder. *Orthopedic Clinics of North America* 1980; **11**: 253–70.
38. Zarins B, Rowe C. Current concepts in the diagnosis and treatment of shoulder instability in athletes. *Medicine and Science in Sports and Exercise* 1984; **16**: 444–8.
39. Rowe CR. Recurrent anterior transient subluxation of the shoulder. *Orthopedic Clinics of North America* 1988; **19**: 767–72.
40. Kiviluoto O, Pasila M, Jaroma H, Sundholm A. Immobilization after primary dislocation of the shoulder. *Acta Orthopedica Scandinavia* 1980; **51**: 915–19.
41. Hovelius L. Anterior dislocation of the shoulder in teenagers and young adults. *Journal of Bone and Joint Surgery* 1987; **69A**: 393–9.
42. Samilson RL, Prieto V. Dislocation arthropathy of shoulder. *Journal of Bone and Joint Surgery* 1983; **65A**: 456–60.
43. Sutro CJ, Sutro W. Delayed complications of treated reduced recurrent anterior dislocations of the humeral head in young adults. *Bulletins of the Hospital for Joint Diseases* 1982; **42**: 187–216.
44. Hawkins RJ. Arthroscopic stapling repair of shoulder instability: A retrospective study of 50 cases. *Journal of Arthroscopic and Related Surgery* 1989; **5**: 122–8.
45. Rubinovich M, Coughlin LP. Unpublished data.
46. May VR. Posterior dislocation of the shoulder: Habitual, traumatic and obstetrical. *Orthopedic Clinics of North America* 1980; **11**: 271–85.
47. Hawkins RJ, McCormack RG. Posterior shoulder instability. *Orthopedics* 1988; **11**: 101–7.

48. Fronek J, Warren RF, Bowen M. Posterior subluxation of the glenohumeral joint. *Journal of Bone and Joint Surgery* 1989; **71A**: 205–16.
49. Neviaser TJ. Old unreduced dislocations of the shoulder. *Orthopedic Clinics of North America* 1980: **11**: 287–94.
50. Rowe CR, Zarins B. Chronic unreduced dislocations of the shoulder. *Journal of Bone and Joint Surgery* 1982; **64A**: 494–505.
51. Hawkins RJ. Unrecognized dislocation of the shoulder. *American Academy of Orthopedic Surgeons Instructional Course Lectures* 1985; **34**: 258–63.
52. Neer CS II, Foster CR. Inferior capsular shift for involuntary inferior and multi-directional instability of the shoulder. *Journal of Bone and Joint Surgery* 1980; **62A**: 897–908.

53. Mallon WJ, Bassett FH III, Goldrev RD. Luxatio erecta: the inferior glenohumeral dislocation. *Journal of Orthopedic Trauma* 1990; **4**: 19–24.
54. Saxema K, Stavas J. Inferior glenohumeral dislocation. *Annals of Emergency Medicine* 1983; **12**: 718–20.
55. Yosipovitch Z, Goldberg I. Inferior subluxation of the humeral head after injury to the joint. *Journal of Bone and Joint Surgery* 1989; **71A**: 751–3.
56. Lev-Toaff AS, Karasick D, Rao VM. Drooping shoulder—non-traumatic cause of glenohumeral subluxation. *Skeletal Radiology* 1984; **12**: 34–6.
57. Rosaaen BJ, De Lisa JA. Voluntary shoulder dislocation: case study. *Archives of Physical Medicine and Rehabilitation* 1983; **64**: 326–8.

4.3.3 Injuries of the rotator cuff

MARK K. BOWEN AND RUSSELL F. WARREN

INTRODUCTION

The shoulder plays an important role in the performance of nearly all athletic activities. There are many popular sports in which repetitive overhead motion is critical. This activity is well tolerated by most participants; however, in some the stresses are extreme and not physiological. Pain and dysfunction may result and limit effective participation. Shoulder disability in the athlete is frequently related to injury to or alterations in the function of the rotator cuff.

Rotator cuff injury in the general population has been attributed to attritional degeneration that usually occurs over several years. The impingement syndrome has been implicated and the continuum of the pathological process has been well described.[1,2] Rotator cuff tears frequently represent the end result of this process and most commonly occur after the fifth and sixth decades. The cuff often tears following minimal trauma in an area weakened by the degenerative process.

Rotator cuff injuries in the athlete appear to result from several different pathological processes. They are commonly seen in sports requiring repetitive overhead motions such as swimming, baseball, and tennis. The repetitive forceful overhead motions that take place injure the rotator cuff by creating high loads within the tendon. Cuff injury may be related to overuse, collagen failure, and progressive attrition of the tendon(s). The aetiological role of glenohumeral joint instability in rotator cuff injury has received increased attention recently.[3,4] When the static restraints of the bony and capsuloligamentous anatomy of the glenohumeral joint are exceeded, the rotator cuff may be subjected to high eccentric loads. Less commonly, an acute macrotraumatic injury may be the mechanism of cuff disruption. Impingement, or mechanical compression, in the athlete is now felt to occur more commonly as a secondary condition related to tendon fatigue, failure due to high loads, or increased glenohumeral translations that occur with instability. In some patients coracoacromial arch anatomy probably plays some pathological role.

ANATOMY

The rotator cuff is comprised of four muscles (supraspinatus, infraspinatus, teres minor, and subscapularis), whose musculotendinous insertions envelope the humeral head. The insertion of the supraspinatus contains an area of relative risk for injury owing to its blood supply (Fig. 1). In 1939, Lindblom and Palmer[6] demonstrated that there are areas of decreased vascularity near the insertion of the supraspinatus tendon[5]. Moseley and Goldie[7] described this as a 'critical zone' or watershed area where vessels in the muscle belly anastamosed with those from the insertion. Rothman and Parke[8] showed that this area is hypovascular relative to the remainder of the cuff and suggested that it was prone to degeneration. In 1970, Rathburn and MacNab[9] used microangiographic

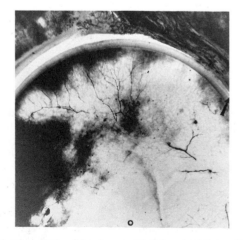

Fig. 1 Injection study of the humeral head and rotator cuff demonstrates the area of decreased vascularity in the tendinous insertion. (Reproduced from ref. 5 with permission.)

techniques and observed that this avascular zone is constant and precedes the development of degenerative changes. They concluded that arm position affects vascularity of the cuff, with adduction, 'wringing out' the blood vessels by compression. The actual importance of this vascular anatomy appears to relate to tendon healing following injury, rather than representing an area susceptible to injury.

ROTATOR CUFF FUNCTION

Motion of the shoulder is the result of complex bony relationships and muscle interactions. The shoulder is comprised of three joints (the sternoclavicular, acromioclavicular, and glenohumeral) and one articulation (the scapulothoracic). While each muscle of the rotator cuff has a specific function attributed to it as a mover of the shoulder, it appears that the primary function of the entire cuff is as a stabilizer of the glenohumeral joint. The rotator cuff performs this role by generating a joint compressive load with muscle contraction. Poppen and Walker[10] studied the forces at the glenohumeral joint and estimated that the joint reaction force of the 90° abducted unweighted arm approached body weight. They further analysed rotator cuff function, and observed that the subscapularis and infraspinatus have small lever arms and act at nearly 90° to the glenoid face. The supraspinatus has a slightly larger lever arm, and acts at approximately 80° to the glenoid. They concluded that since these muscles act nearly perpendicularly to the glenoid with small lever arms, they function primarily as joint compressors, with the supraspinatus playing a small additional role as an arm abductor.[10]

The cuff appears to function by creating a stable fulcrum through which the deltoid can act to achieve arm elevation. If rotator cuff function becomes impaired for whatever reason, upward displacement of the humeral head may occur and the normal fulcrum of the head on the glenoid is lost. With superior translation impingement of the humeral head and the rotator cuff occurs under the acromion. Altchek et al.[11] and Paletta et al.[12] used radiographic techniques to document the superior migration of the humeral head that occurs in patients with stage II and stage III impingement.

The most critical function of the shoulder is to elevate the position of the hand. This occurs by way of movement of the scapula on the thorax as well as rotations at the glenohumeral joint. This interaction has been studied extensively by several investigators.[13–16] Poppen and Walker[16] conducted a radiographic analysis of shoulder abduction and found an overall ratio of glenohumeral joint to scapulothoracic motion of 2:1. In the first 20° of elevation the ration is approximately 4:1; above this the contributions of glenohumeral and scapulothoracic motions are almost equal (5:4).

ROTATOR CUFF BIOMECHANICS—THROWING

Overhead motion is an integral part of many sports and is best exemplified by the act of throwing which has been described as occurring in five phases: wind-up, cocking, acceleration, deceleration, and follow-through. Throwing involves a transfer of energy from the body to the arm and the object being propelled. Part of this kinetic energy is absorbed by the rotator cuff. This is seen to the greatest extent during the follow-through phase of throwing where the cuff muscles act eccentrically to decelerate the arm and

contain the humeral head on the glenoid surface. At this point the rotator cuff tendons are subjected to high tensile and compressive forces. The distraction force during follow-through is in the range of 80 per cent of body weight.[17] Dillman et al.[17] have shown that during the throwing motion, the angular velocity of the arm is 7150°/s as it moves from maximal external rotation to internal rotation. The force of anterior glenohumeral translation at the fully cocked position is calculated at approximately 40 per cent of body weight.

The wind-up is mostly attributed to activity of the deltoid, while the cocking phase is concluded by activation of the subscapularis and pectoralis major.[18,19] During the cocking phase, a torque of 17 000 kg/cm is generated. In a skilled athlete the cuff remains relatively quiet during arm acceleration. However, it may be quite active when technique and training are suboptimal. During follow-through the supraspinatus, infraspinatus, and teres minor become active to aid in decelerating the humeral head.

ROTATOR CUFF IN ATHLETES

The rotator cuff in the overhead athlete is subjected to high repetitive loads. Cuff injury may range from fatigue to primary collagen failure. Aetiologies that have been implicated in rotator cuff injury include overuse, glenohumeral joint instability, acute trauma, and mechanical compression from 'impingement' (Fig. 2). In addition to throwing sports, other sports with a high

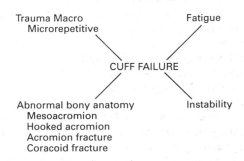

Fig. 2 Diagram demonstrating the many different possible aetiologies of rotator cuff failure.

incidence of shoulder pain and dysfunction include tennis, volleyball, swimming, and water polo. Overuse and fatigue due to eccentric loading of the cuff commonly occur in athletes participating in these sports. Of additional interest is the high incidence of suprascapular nerve palsy recently observed in volleyball players.[20]

Competitive swimmers maintain rigorous training schedules, swimming 7000 to 20 000 m/day. Swimmers complete a stroke every 0.6 s and travel through the water at up to 1.9 m/s.[21] Rotator cuff overload and 'tendinitis' are frequently seen, as is associated glenohumeral joint instability and labral damage. Whether instability occurs primarily or becomes more prominent with cuff fatigue and failure is not clear. It is apparent that alterations in glenohumeral joint function lead to mechanical impingement of the cuff and consequent pain. Despite the high incidence of shoulder pain in swimmers, progression to complete rotator cuff tear is infrequent.

Similar mechanisms of cuff injury occur in tennis players. In contrast with throwing sports and swimming, tennis is frequently played by an age group with a higher risk of cuff degeneration. It is not uncommon to see tennis players in their forties and fifties with a history of shoulder pain who have progressed to a rotator cuff tear.

Although cuff injury in athletes is most frequently the result of microtraumatic overuse, an acute macrotraumatic episode may also be seen as a cause of rotator cuff disruption. An acute traumatic injury to the cuff may occur in the setting of an anterior shoulder dislocation. This has been referred to as the 'posterior mechanism' of anterior shoulder instability.[22,23] It is explained by the relative weakness of the rotator cuff, caused by attrition, compared with the anterior capsular structures. While this has been seen most frequently in the older athlete, we have treated younger athletes who have sustained a cuff tear with a traumatic anterior dislocation. We have seen this occur in American football, where a player dislocated his shoulder making a tackle (Fig. 3). We have also treated a number of skiers with a range of ages who presented with a cuff tear following a fall. In some a dislocation did not occur.

Fig. 3 Young American football player who presented with marked weakness of shoulder abduction 2 weeks after an anterior dislocation. Arthrography revealed a large rotator cuff tear.

It is important to appreciate that an anterior dislocation which is associated with an avulsion fracture of the greater tuberosity will generally heal without weakness or recurrent instability if the tuberosity is well reduced. Impingement, like pain, may persist in patients with a small undisplaced fracture. However, an unrecognized or untreated cuff disruption following a traumatic episode has a poor prognosis for return of function. We recommend that individuals over 40 who have persistent pain and weakness after an anterior dislocation be evaluated aggressively.

Electromyography and an arthrogram or magnetic resonance imaging (MRI) may be used to rule out a nerve or rotator cuff injury. In the rare case where a rotator cuff tear occurs with a nerve injury, we recommend early surgery to repair the rotator cuff and observation of the nerve injury.

ROTATOR CUFF IMPINGEMENT

Rotator cuff impingement in an overhead athlete may occur primarily, but more commonly develops secondary to tendon overload or glenohumeral instability. The classic impingement syndrome, which is well described by Neer,[1,2] is implicated much less frequently in this population. Primary impingement is most likely to occur during the cocking phase of the throwing motion as the cuff passes beneath the coracoacromial arch. Bony anatomy may play some role by predisposing to the impingement process; however, irregularities of the coracoacromial arch generally develop as a result of abnormal forces placed on the bone. Variations in acromial anatomy such as a hooked acromion or os acromiale may limit the available subacromial space. Bigliani observed three different acromial shapes and correlated the presence of a hooked acromion with rotator cuff tears.[24] However, these findings are felt to be developmental and not aetiological. Nevertheless, the study correctly emphasizes that the space beneath the coracoacromial arch may be compromised by acromial anatomy, acromial joint or coracoacromial ligament abnormalities, or by prominence of the greater tuberosity. Rotator cuff tears that occur with primary impingement are associated with mechanical compression, and may be observed on either side of the cuff.

While impingement may precede the development of a tear in an overhead athlete, it is more likely that repetitive high loading of the cuff tendons during throwing causes gradual failure of tendon fibres. This may lead to a cycle of inflammation, fatigue, and cuff weakness. While tendinitis is a commonly used diagnostic term, inflammation probably plays only a small role in the later stages of cuff injury where collagen degeneration appears to be the primary pathological process.

Cuff dysfunction may result in diminished effective control of the humeral head on the glenoid. This has the potential to stress the cuff further as continued activity is associated with greater translations and strain of the cuff tendons. Weakness of the cuff results in superior translation, reducing the available space under the coracoacromial arch and encroaching on that necessary for smooth passage of the rotator cuff[11,12] Thus, in addition to tendon failure, there is also mechanical impingement. Although the impingement process does not generally progress to a complete tear in the young athlete, partial tears are not unusual and a college or professional athlete who continues forceful throwing may complete a tear. Rotator cuff tears due to tendon overload are most frequently seen to develop on the intra-articular side of the cuff.

INSTABILITY AND CUFF INJURY

The shoulder maintains a uniquely delicate balance between mobility and stability. Static and dynamic factors interact to limit translation of the humeral head on the glenoid. The critical dynamic role of the rotator cuff in providing joint compression has been discussed. The most critical components of the static stabilizers are the glenohumeral ligaments. The various ligaments function in a load-sharing manner, and their relative roles vary with arm position, rotation, and direction of applied stress. Excessive translation may occur with injury or failure of the dynamic containment mechanism, or with capsuloligamentous injury. It remains unclear which factor is most critical to glenohumeral stability. It is likely that the static and dynamic restraints function synergistically with varying roles depending on the demands placed on the shoulder. Excessive translation due to failure of static constraints may cause cuff injury by direct mechanical compression or through the increased loads on the cuff that occur during distraction or shear of the humeral head versus the glenoid. Tendon failure related to underlying instability is referred to as secondary tensile overload. In simple terms, increased strain is produced in the rotator cuff as it works to limit excessive translation and maintain stability.

Anterior subluxation is the most frequent instability observed in these athletes; however, posterior and multidirectional instability are also seen and may represent a group with an underlying predisposition. Throwers are commonly noted to have increased external rotation with a corresponding loss of internal rotation. Increased tension in the posterior capsule may play some role. Harryman et al.[25] have shown that tightening of the posterior capsule causes increased anterior translation with arm abduction and flexion. Thus the development of tight posterior structures may increase anterior shear forces in a thrower.

An athlete with recurrent anterior subluxation may present with the classic description of 'dead arm' symptoms such as sudden sharp pain and weakness when the arm is abducted and externally rotated. These patients frequently have a past history of macrotrauma to the shoulder. Patients with instability as the underlying cause of rotator cuff pathology most commonly present with pain related to a specific overhead activity. A sense of movement or looseness of the shoulder may also be described. Characteristic findings on physical examination include a positive impingement sign, anterior apprehension, and relocation test that eliminates the apprehension. At times it may be difficult to distinguish between pain with forward elevation (impingement) and apprehension (subluxation).

DIAGNOSIS OF ROTATOR CUFF INJURIES

History

In the evaluation of athletes with shoulder problems, pain is the most frequent presenting complaint. They may also have complaints as vague as a loss of pitching speed, power, or endurance. Instability other than recurrent dislocation is not usually appreciated by patients. Pain may occur superiorly, but more frequently it is poorly localized. The pain may radiate down the arm to the area of the deltoid insertion of the biceps tendon. The pain is generally increased with overhead motion, particularly the inciting activity. It is helpful to determine at what time in the cycle of overhead motion pain occurs. Pain that occurs during the cocking phase suggests that anterior subluxation may be taking place. Pain on follow-through with the arm flexed, adducted, and internally rotated raises concern over the possibility of posterior instability. Pain at night is quite common in patients with rotator cuff tears. Pain frequently diminishes with a period of rest; however, this is often not well accepted by an athlete.

A history of trauma to the shoulder can be divided into recurrent microtrauma or macrotrauma depending on an analysis of the insult to the shoulder. When onset of symptoms is related to a clear traumatic event it is useful to ascertain the mechanism and the position of the arm at the time of injury. A microtraumatic aetiology is further investigated for details of training, overuse, and technique.

Differential diagnosis

Several pathological processes may present with similar symptoms to those seen with rotator cuff injuries. These include neurological conditions such as cervical radiculopathy and brachial plexus or suprascapular nerve injuries. A careful physical examination and an arthrogram or electrodiagnostic studies will often provide sufficient information to clarify the diagnosis. Recurrent glenohumeral subluxation (anterior, posterior, or multidirectional) may cause symptoms that mimic those seen with injury to the rotator cuff. Similarly, it may be difficult to distinguish injuries to the labrum, superior labral anterior posterior lesions, or abnormalities of the biceps tendon.

Examination

Examination of the shoulder should begin with an inspection for symmetry, atrophy, and deformities. Infraspinatus atrophy may occur with suprascapular nerve palsy or a large cuff tear. The supraspinatus is more difficult to evaluate for atrophy as it is covered by the trapezius muscle. A rupture of the biceps tendon produces a characteristic bulge in the upper arm with contraction. It is useful to observe patients from behind while they elevate their arms as this may bring out subtle cuff dysfunction.

Palpation should begin medially at the sternoclavicular joint and proceed laterally including the clavicle, the acromioclavicular joint, and the bicipital groove. The greater tuberosity and supraspinatus are palpable just distal to the anterolateral aspect of the acromion with the arm extended. Tenderness of the acromioclavicular joint is the most reliable indicator of active pathology in the joint. Tenderness in the area of the biceps is not unusual and may be related to mechanical impingement or rarely to tensile injury to the biceps tendon. Interpretation of tenderness in this area may be difficult as it may occur in the absence of pathology if palpation is too vigorous.

Range of motion is recorded and compared with the contralateral shoulder. For any deficits in active motion, passive motion should be evaluated as well. Motions tested include forward elevation, internal and external rotation at 0° and 90° of abduction. Many athletes, particularly throwers, demonstrate an asymmetric range of motion, and typically have increased external rotation and loose internal rotation. This may predispose the throwing athlete to rotator cuff injury. It is also not unusual to observe what appears to be a symmetrical increased range of motion in some athletes, particularly swimmers. While joint laxity may contribute to success in their sport, it may also increase their susceptibility to injury.

Muscle strength testing is performed in an attempt to assess the function of individual muscles as well as the cervical nerve roots and brachial plexus. Of particular interest to the evaluation of the rotator cuff is testing forward elevation, external and internal rotation, and biceps strength.

Stability testing is performed in both the sitting and supine position. The humeral head is manually translated in the anterior, posterior, and inferior directions on the glenoid. We grade anterior and posterior translation as follows: 1+, increased translation; 2+, humeral head jumps over the glenoid edge with a grind; 3+, humeral head locks out over the glenoid. Inferior translation or the 'sulcus sign' is graded by the number of millimetres of inferior displacement with respect to the lateral edge of the acromion (1+, <1 cm; 2+ 1–2 cm; 3+ >2 cm). Apprehension in abduction and external rotation is a sensitive test for anterior subluxation. The relocation test may also be positive. In this test a posteriorly directed force applied while the arm is abducted and externally rotated eliminates previously demonstrated apprehension. In addition pain and/or crepitation produced on stressing the shoulder should be noted as it may indicate the presence of a labral injury. Apprehension related to posterior instability with the arm adducted and flexed is less frequent, and is usually present only after an acute instability episode.

There are several 'special tests' that are useful in evaluating the rotator cuff. The impingement sign is positive when pain is produced near terminal forward elevation. Impingement pain may also be elicited with the abduction test by abducting to 90° and internally rotating the arm. The impingement test is performed by injecting approximately 10 ml of local anaesthetic into the subacromial space. The test is considered positive if the impingement sign is eliminated. While this test can be very useful in localizing the source of pain, it may contribute to confusion over the diagnosis as it may be positive in a patient with impingement secondary to anterior instability of the shoulder. The Speed and Yerguson tests may be helpful in localizing pain to the biceps tendon. Injection in the area of the bicipital groove can at times provide useful diagnostic information.

Diagnostic tests

In the majority of young athletes plain radiographs will be normal. We use an instability series which includes an anterior–posterior view in internal and external rotation, a West Point axillary view, and a Stryker notch view.[26] Possible radiographic findings include a Bankart lesion, glenoid erosions, a Hill–Sachs lesion, or ossification abnormalities of the acromion. A coracoacromial arch or outlet view is performed as a trans-scapular lateral of the scapula with the beam tilted caudally approximately 10°. Classic findings of advanced rotator cuff disease include subacromial spurs, acromial hooking, decreased acromiohumeral distance, and sclerosis or cystic change in the greater tuberosity. These radiographic findings are uncommon in a young athlete.

Ultrasonography has been used effectively in many centres as a non-invasive method of confirming the presence of a rotator cuff tear.[27–29] However, it has proved to be highly examiner-dependent in terms of sensitivity and specificity, particularly in the diagnosis of incomplete tears.

Arthrography has been the traditional gold standard with a high sensitivity for detecting full-thickness rotator cuff tears.[30] It has the disadvantage of an invasive procedure, but may provide other potentially useful information regarding capsule volume and, when combined with a CT scan, an excellent view of the glenoid labrum. Visualization of incomplete rotator cuff tears is unreliable but can be improved with double-contrast techniques.[31,32]

MRI is being used more frequently but, although several studies have assigned fairly high sensitivity and specificity to its use,[33–37] our experience has been that it tends to overestimate the degree of rotator cuff pathology. Large tears of the supraspinatus with retraction can be readily visualized; however, scans are frequently misinterpreted for partial thickness tears.[38] Newer technology and surface coils are improving the diagnostic accuracy. The advantages of MRI are that it is non-invasive, provides an assessment of location and size of tear, and may delineate other areas of shoulder pathology.

The role of arthroscopy in the diagnosis and treatment of rotator cuff injuries has increased with experience with shoulder arthroscopy. Partial thickness tears are easily identified. Labral and ligament pathology is also more clearly defined on direct inspection during arthroscopy. Evaluation of the subacromial space and cuff on the bursal side may also provide diagnostic information. Diagnostic arthroscopy is indicated in patients who have not responded to 3 to 6 months of intensive physical therapy or in a thrower with undiagnosed pain.

Determining cuff tear size

Determination of the size of a rotator cuff tear may be difficult despite a careful physical examination and modern radiographic techniques. A helpful clinical finding is the presence of a significant deficit in external rotation strength. In the absence of limiting pain, such weakness is usually consistent with a large tear involving the infraspinatus in addition to the supraspinatus. While glenohumeral joint arthrography is excellent in confirming the presence of a rotator cuff tear, assessment of tear size is frequently unreliable. MRI provides a clearer representation of rotator cuff anatomy; however, its ability to quantify tear size predictably is also limited.

TREATMENT CONSIDERATIONS

Treatment of rotator cuff injuries and symptoms referrable to the rotator cuff is preventative, non-operative, or operative, and may encompass one or all of these modalities. In addition to making an accurate diagnosis it is critical to assess the sport(s), level of participation, and demands placed on the athlete's shoulder. Incorporation of this knowledge will contribute to the development of the most effective treatment programme.

Clearly, the most successful treatment for the rotator cuff is prophylaxis, best achieved with an intensive upper and lower extremity conditioning programme and sound technique. Prevention of injury by carefully developed off-season and preseason programmes is essential. Several elements are critical: overall body strength and conditioning, upper and lower extremity flexibility, shoulder and rotator cuff strengthening, and an analysis of technique. Fatigue and overuse are probably involved in almost every rotator cuff injury. A properly designed programme will produce benefits in endurance and prevention of injury, while inappropriate exercises may actually predispose to injury. Rotation contractures are common in throwing athletes and may cause subtle muscular compensations. Stretching is performed to achieve a full and symmetrical range of motion. The rotator cuff may be strengthened using surgical tubing, isokinetic machines, or free weights. Because the rotator cuff in an overhead athlete functions critically in an eccentric manner, we emphasize exercises that work the rotator cuff eccentrically. The periscapular muscles, including the latissimus dorsi, the rhomboids, the serratus anterior, and the trapezius, create a stable platform for the shoulder joint and are also emphasized in the strengthening programme.

Once a rotator cuff injury is established, the majority will respond to rest, avoidance of painful positions and activities, and a rehabilitation programme that emphasizes stretching and eccentric and concentric strengthening for control of the humeral head and the scapula. The initial phase of therapy is directed at treating the 'tendinitis' with rest, ice, non-steroidal anti-inflammatory medications (NSAIDs), and infrequently injection of corticosteroids. Stretching designed to address contractures or limitations of motion is also started early. It is particularly helpful to address the limited internal rotation seen in throwers with the aim of restoring a painless full range of motion to the extremity. We are careful not to push stretching in positions of impingement. The early judicious use of isometric exercises also plays a role in the recovery process.

As with our preventative therapies, the primary treatment goal is to restore health and strength to the rotator cuff. This may benefit patients who have had primary tensile failure of the rotator cuff as well as those with subtle instabilities. Once pain is well controlled and range of motion is restored, strengthening and more sport-specific exercises can progress. The emphasis of the strengthening phase is to improve power, endurance, timing, and muscular control of the arm. Rotator cuff strengthening with internal and external rotation exercises using theraband is begun first with the arm at the side and later advanced to 90° of abduction, provided that there is no pain. Scapulothoracic stabilizers, including the seratus anterior, rhomboids, latissimus dorsi, and trapezius, are also included in the strengthening programme. Trunk and lower extremity exercises are not neglected in the total rehabilitation programme. As symptoms are eliminated and strength improves, activities that develop co-ordination and proprioception are added. Finally, return to overhead sport activities is begun.

SURGICAL MANAGEMENT—ROTATOR CUFF TEARS

As discussed earlier, the most predictable and gratifying results in the management of rotator cuff injuries occur with non-operative treatment. While surgical interventions may give excellent results, depending on the underlying pathology, return to overhead sports following any operative procedure is unpredictable and variable.

Primary tensile failure of the rotator cuff will most probably respond to a non-operative programme such as that outlined earlier. Surgical intervention in the form of examination under anaesthesia, glenohumeral joint, and subacromial arthroscopy may be useful diagnostically. Debridement of partial cuff tearing may have a therapeutic effect; however, response is more probably related to postoperative rehabilitation than to debridement of failed rotator cuff tissue.

Andrews reported that 85 per cent of young throwing athletes had satisfactory results and were able to return to competitive activities following rotator cuff debridement alone.[39] However, Ogilvie-Harris and Wiley[40] found that satisfactory results were obtained with only half of 57 patients treated with cuff debridement for partial thickness tearing. Ellman[32] treated 20 patients with partial thickness rotator cuff tears with debridement and subacromial decompression , and obtained 75 per cent satisfactory short-term results. One possible benefit of cuff debridement may be derived from a reduction in pain that will then allow athletes to participate in a rehabilitation and strengthening programme.

If rotator cuff failure presents as a complete tear, operative repair should be considered for best results in an active person. An older athlete might manage by modifying his or her activities; however, pain is frequently a persisting complaint. In the athletic population with an overuse injury the tear is frequently small. When there is a history of significant trauma to the shoulder, a large avulsion type tear from the insertion site may be encountered. In the older athlete the impingement process may have led to tendon degeneration and a large tear. Despite excellent surgical reconstructive techniques, predicting the success of return to sports participation following rotator cuff repair is difficult.[32,41–46] Results are dependent on the quality of the remaining

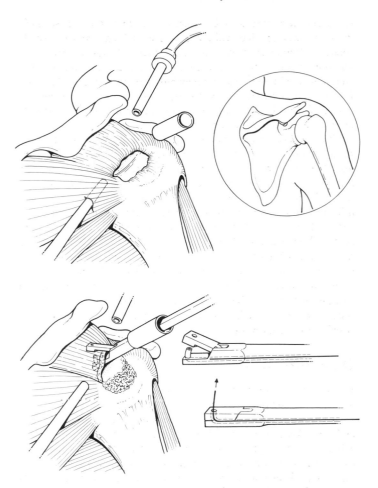

Fig. 4 Arthroscopic-assisted repair of the rotator cuff requires excellent fluid inflow and visualization of the subacromial space. Special suture-passing instruments are used to place sutures in the torn rotator cuff.

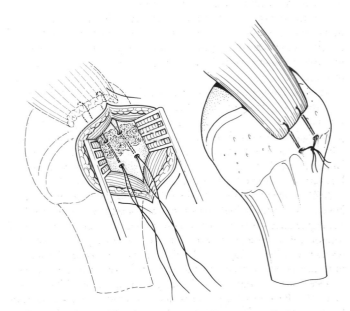

Fig. 5 A deltoid splitting approach is used, incorporating the lateral arthroscopy portal. A bleeding bony bed has been prepared at the greater tuberosity. Sutures placed arthroscopically using a suture punch are secured through drill holes and are tied over a bony bridge.

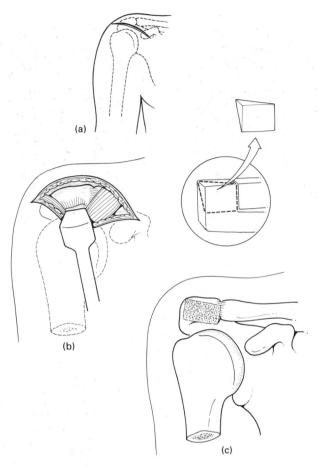

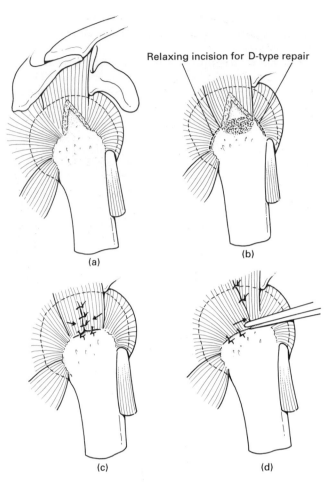

Fig. 6   We explore the rotator cuff through a superior approach.
(a) The deltoid is detached along the anterior aspect of the acromion
only, from the anterolateral corner to the acromioclavicular joint.
(b) A 3-cm longitudinal split is extended along the deltoid fibres from
the corner. (c) An anterior acromioplasty is then performed.
(Redrawn from ref. 46 with permission.)

Fig. 8   (a) Smaller tears less than 2 cm often form a V-shape as the
supraspinatus retracts; (b) preparation of a broad bleeding bony bed
for reattachment of the tendon; (c) a V-shaped tear may be
converted to a Y-shape and advanced laterally, with repair to bone;
(d) if a V-shaped tear is difficult to close, one edge may be advanced
to fill the defect. (Redrawn from ref. 46 with permission.)

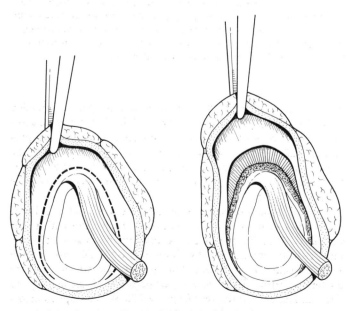

Fig. 7   It is often necessary to perform an intra-articular release of
the capsule and rotator cuff, particularly when the rotator cuff is
large or retracted. (Redrawn from ref. 46 with permission.)

cuff, tissue retraction, and the adequacy of repair. Care must be
taken not to overtighten the cuff as this may also predispose to
failure.

Our approach to rotator cuff tears is based on an assessment of
size of the tear. If preoperative physical examination and radiographic evaluation (MRI versus arthrogram) suggest that the tear
is small or medium, we proceed with an arthroscopic evaluation.
At arthroscopy we attempt to assess the size of the tear, the quality
of the tissue, and its retraction and mobility. If the tear appears to
be partial, involving the articular surface, a suture can be passed
via a spinal needle placed through the tear from laterally. This
suture can then easily be located in the subacromial space and
facilitates assessment of the cuff on this side. If the tear is less than
4 cm long and the tissue is satisfactory, we attempt to repair it
using arthroscopic techniques. Successful surgical reconstruction
of rotator cuff tears requires adherence to several principles including decompression, tissue mobilization, and repair to a
bleeding bony bed.

The first step of an open or arthroscopic reconstruction is an
adequate acromioplasty in order to remove any present or
potential source of impingement. If the acromioclavicular joint is
prominent inferiorly, the spurs are removed with a burr. We pre-

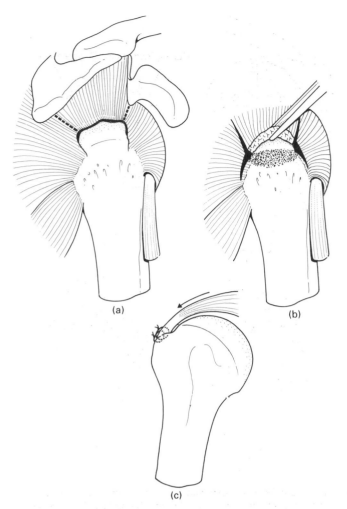

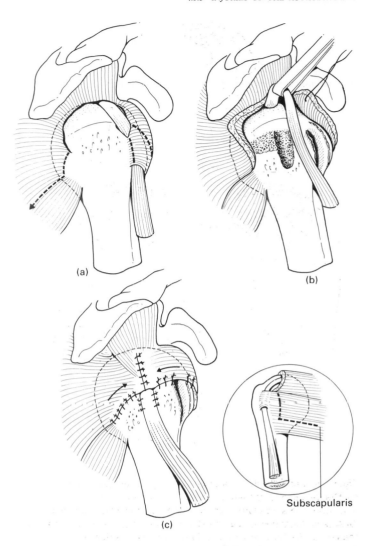

Fig. 9 (a) Moderately sized tears (2–5 cm) frequently require extensive dissection on both sides of the rotator cuff to achieve sufficient length of the cuff tissue; (b) relaxing incisions parallel to the cuff fibres towards the spine of the scapula and the coracoid may be required to advance retracted tissue; (c) the tendon is secured to a bony surface with non-absorbable sutures. (Redrawn from ref. 46 with permission.)

Fig. 10 (a) Large retracted tears of the rotator cuff may require more extensive mobilization of the cuff tissues, including a superior transfer of the upper subscapularis tendon and/or mobilization of the infraspinatus and teres minor; (b) repositioning the biceps tendon into a more posteriorly positioned groove to aid in closure of the defect; (c) superior advancement of infraspinatus and subscapularis to the biceps tendon. (Redrawn from ref. 46 with permission.)

serve the joint unless there are symptoms which can be referred to the acromioclavicular joint and there is direct tenderness pre-operatively. Next, a broad bleeding bony surface is created at the greater tuberosity using an arthroscopic shaver and burr. Sutures are then placed in the edge of the torn cuff tendon using a suture passer (Fig. 4). A small split is made in the deltoid laterally, to allow drill holes to be placed in the greater tuberosity and repair of the tendon to bone (Fig. 5). The technique offers the advantage of minimal soft tissue dissection and preservation of the attachment of the deltoid to the acromion. Our short-term follow-up with overhead athletes returning to sports has been encouraging. Other arthroscopically assisted fixation techniques using an absorbable tack and TAG suture anchors to secure sutures from the cuff are also being developed.

When preoperative evaluation suggests that the rotator cuff tear is large and probably fixed and retracted, we proceed with an open reconstruction. We explore the rotator cuff through a superior approach (Fig. 6). The deltoid is detached along the anterior aspect of the acromion only, from the anterolateral corner to the

acromioclavicular joint. A 3 cm longitudinal split is extended along the deltoid fibres from the corner. The first step is an acromioplasty to remove the source of impingement and also to provide for sufficient visualization of the rotator cuff tissue. Once the rotator cuff tear is adequately identified, it is graded by size. A tear less than 2 cm in length is small, 2 to 5 cm is moderate, and more than 5 cm is large.

A broad elevator or scalpel is used to release tissues both intra- and extra-articularly in order to mobilize the rotator cuff tissues sufficiently (Fig. 7). This is critical to allow the rotator cuff to be repaired without tension and with the arm at the side. We prepare a broad bleeding bony surface at the greater tuberosity as a bed for the tendon repair. The type of repair performed is dictated by the size and anatomy of the tear (Figs. 8–10). We attempt to incorporate some repair to bone in our rotator cuff reconstructions. Non-absorbable sutures are placed in the leading edge of the tendon, passed through drill holes in the prepared bony surface, and tied over sturdy bone bridges. If the quality of the bone is good, we have at times used TAG suture anchors in the prepared bony surface to co-apt the cuff to bone. The biceps is not tenodesed unless it is grossly disrupted. Repair of the anterior deltoid is crucial and should be through drill holes in the acromion.

Large tears, although unusual in the younger athlete, are more common in the older athlete and often present a more challenging and technical problem. This type of patient is represented by the 60-year-old tennis player who plays several days a week and has noticed a loss of power. Pain radiographs may document proximal migration of the humeral head consistent with a massive rotator cuff tear. In such situations, when pain is reasonably controlled, activity modification and rehabilitation may be the most prudent approach. If surgery is attempted for massive cuff tears, aggressive mobilization of the tissues is critical. Pain relief is quite predictable; however, return of strength and function is largely dependent on the quality of the tissue and the performance of the repair. In these difficult reconstructions, cuff debridement may be considered in an older athlete; however, functional results are frequently poor and continued participation in an overhead sport is unlikely.

Our postoperative rehabilitation for rotator cuff repairs is guided by our findings at the time of surgery. For moderate to large rotator cuff tears we use a splint that holds the arm in approximately 30° of abduction. This allows the tendons to heal without tension and in a favourable position with regard to the vascularity of the supraspinatus tendon.[9] In general, tendon attachment requires 6 weeks, following which further healing continues to take place. Passive range of motion, including pendulum exercises and forward elevation, are begun early postoperatively. The pace of therapy is guided by the size of the tear, the quality of tissue, and a judgement of the security of the repair.

In patients with small to moderate tears we begin use of a pulley for overhead elevation at approximately 1 week. Passive motion is thus continued for 6 weeks postoperatively, following which active assisted motion is permitted. Rotator cuff strengthening exercises with theraband rubber tubing are begun at 6 weeks. Isotonic exercises at the side are begun at about 8 weeks, and are progressed to 90° of abduction at 12 weeks. Patients with large tears are splinted for 6 weeks, following which passive range of motion, including pendulum exercises and use of a pulley, is begun. The major cause of failure following cuff repair is the use of excessive weight during the first 3 to 4 months. It is better to work on motion and use light strengthening exercises with theraband while tendon healing occurs. Gentle swimming may be started in the third month. In general, throwing and other overhead activities such as serving should be avoided until 7 to 9 months. Return to throwing involves a gradual progression, controlling the variables of distance, speed, and number of throws. These are increased over an approximately 3-month period, making only one change at a time.

SURGICAL MANAGEMENT— IMPINGEMENT

Primary impingement in the young athlete is unusual and is more likely to occur secondary to instability or cuff failure with resulting loss of downward displacement of the humeral head. Similarly, a primary anatomical abnormality of the coracoacromial arch is rare. Eccentric overload of the rotator cuff is probably the most common inciting cause in the overhead athlete. The mainstay of management is a well-designed rehabilitation programme. Occasionally, conservative treatment fails and a decompressive

procedure is considered. A decompression is indicated when symptoms continue despite therapy and work-up reveals a rotator cuff tear or clear impingement persists in the absence of detectable instability. We combine all subacromial decompressions with a careful examination under anaesthesia and arthroscopy to confirm that a subtle instability is not present.

Open acromioplasty has a well-documented history of good results in a non-athletic population. High demand overhead athletes represent a greater challenge to our diagnostic and surgical skills. Tibone reported on young patients who underwent anterior acromioplasty for stage II impingement; only 43 per cent returned to sports, with fewer returning to overhead throwing.[47] These results may reflect untreated subtle instabilities in some of their patients, many of whom were athletes engaged in sports involving competitive throwing.

Arthroscopic decompression is a more recently developed technique with the potential advantage of being less traumatic and allowing earlier rehabilitation. It also allows for an intra-articular examination of the glenohumeral joint, and assessment for signs of occult instability. The results of arthroscopic decompression are rewarding in terms of pain relief; however, its impact on athletes has not been as favourable. Ellman[48] reported good to excellent results in 88 per cent of his patients without comment on their activity level. Altchek et al.[49] reported an 83 per cent good or excellent rate in patients with stage II impingement, and 76 per cent of athletic patients returned to full participation in sports. However, 55 per cent of throwers and swimmers had significant residual symptoms, and 83 per cent of these had inferior labral tears at arthroscopy. Fly et al.[50] noted that 77 per cent of athletes less than 40 years old returned to overhead sports following arthroscopic subacromial decompression; however, many were unable to return to their previous level of participation. They also noted a high failure rate associated with labral pathology. Failures in each of these studies can probably be attributed to errors in diagnosis or inconsistent removal of acromial bone.

Following arthroscopic subacromial decompression, rehabilitation is begun on the initial postoperative day. Rehabilitation goals included elimination of pain and restoration of range of motion, normal strength, and function in desired sports. We begin with passive stretching and strengthening exercises and progress to a resisted programme as tolerated.

SURGICAL MANAGEMENT— INSTABILITY

The diagnosis of shoulder instability in an athlete with shoulder pain and loss of function is frequently difficult. The player frequently gives a history characteristic of a pattern of overuse and findings are consistent with rotator cuff pathology. It is unusual for an overhead athlete to describe a sensation of instability. A physical examination noting apprehension and a positive relocation test is suggestive of the diagnosis of anterior subluxation. Manual stress testing may also demonstrate the subluxation. The mainstay of management of the athlete with glenohumeral joint instability and secondary rotator cuff injury (either tensile failure or impingement) is an aggressive rehabilitation programme. Emphasis is on stretching and strengthening of the internal and external rotators and periscapular muscles.

If non-operative treatments fail and surgery is required, the underlying pathology of the instability should be addressed directly. As discussed earlier, subacromial decompression, both open and arthroscopic, has not produced satisfactory results in the majority of athletes. It is unclear what role, if any, arthroscopic decompression has in this population of patients. It would be a significant advantage if we could define a population of athletes with underlying subluxation who might still benefit from decompression rather than undergoing more extensive shoulder-stabilization surgery. If instability is not clearly present, the capsule and labrum are competent, and cuff degeneration is observed, a limited decompression and cuff debridement may provide symptomatic relief.

At the time of surgery a careful examination under anaesthesia is performed to confirm the preoperative diagnosis. In a patient with primarily anterior instability we also perform an arthroscopic evaluation. If a Bankart lesion is observed, repair is directed towards repairing this pathological lesion. This can be achieved either by arthroscopic techniques using one of several different fixations methods (Fig. 11) or by using motion-sparing open stabilization techniques. Regardless of repair method, in the high

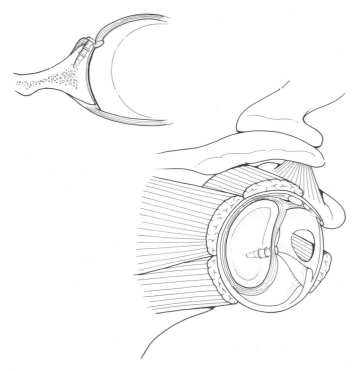

Fig. 11 Arthroscopic Bankart repair with an absorbable cannulated fixation device. Following preparation of the anterior glenoid, the capsulolabral detachment is secured to the glenoid using one or more Suretacs®.

demand overhead athlete it is important not to overtighten the capsule or advance the subscapularis tendon. In open reconstructive surgery exposure of the capsule is achieved either by splitting the subscapularis muscle or partially incising the upper half of the tendon just medial to its insertion. If the subscapularis is released, considerable care is taken in its repair in order to avoid shortening the tendon. A transverse incision in the capsule allows exposure of the Bankart lesion, if present. The capsulolabral repair can then be performed from the inside out, and the capsule can also be tensioned in the superior direction. Tension is set with the arm abducted to 40° with 30–40° to external rotation. This approach

minimizes shortening of the capsule in the medial–lateral direction and maintains the critical range of motion.

When anterior instability is due to capsular insufficiency and not a Bankart lesion, or when multidirectional instability is detected by the presence of a 'sulcus sign', a capsular shift procedure is indicated. Arthroscopic techniques for releasing the capsule and shifting it medially and superiorly exist; however, they are more technically demanding and may provide less predictable results. An open capsular shift is performed either laterally or medially at the glenoid (Fig. 12).

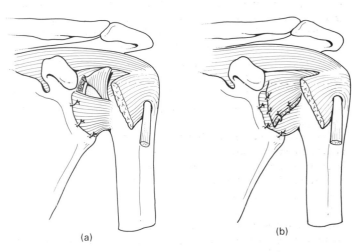

(a)   (b)

Fig. 12 Illustration depicting T-plasty capsular shift performed at the glenoid side involving a superior shift of the inferior capsular flap (a) and an inferior shift of the superior flap (b). (Redrawn from ref. 51 with permission.)

Postoperative rehabilitation in an overhead athlete is begun immediately. The pace of progression of the rehabilitation depends on the technique used and the quality of the repair achieved. Passive elevation, internal rotation, and external rotation to the limits of the repair is allowed. At 4 weeks external rotation is increased and internal and external rotator strengthening is commenced. In a thrower stretching is progressed with the aim of achieving a full range of motion 6 weeks after surgery.

CONCLUSIONS

Many athletic activities require shoulder motions that place a high degree of stress on the rotator cuff. Shoulder complaints in the athlete are frequently referred to rotator cuff dysfunction or injury. Rotator cuff failure may occur as the result of one or a combination of different potential aetiological processes. In addition to tensile failure of the rotator cuff from overload, glenohumeral instability is frequently recognized as a contributing factor to cuff injury. An accurate diagnosis is critical to successful treatment of these athletes. A well-designed shoulder and total body rehabilitation programme remains the mainstay of treatment for the majority of pathologies encountered. Surgical interventions should correct specific pathologies, be well designed and minimally traumatic, and allow early restoration of shoulder motion.

REFERENCES

1. Neer CSI. Anterior acromioplasty for the chronic impingement syndrome in the shoulder. A preliminary report. *Journal of Bone and Joint Surgery* 1972; **54A**: 41.

2. Neer CSI. Impingement lesions. *Clinical Orthopaedics and Related Research* 1983; **173**: 70–7.

3. Warren RF. Subluxation of the shoulder in athletes. *Clinics in Sports Medicine* 1983; **2**: 339–54.

4. Jobe FW, Kvitne RS. Shoulder pain in the overhand or throwing athlete. The relationship of anterior instability and rotator cuff impingement. *Orthopaedic Review* 1989; **18**: 963–75.

5. Arnoczky SP, Altchek DW, O'Brien SJ. Anatomy of the shoulder. In: McGinty JB, ed. *Operative arthroscopy*. New York: Raven Press, 1991: 425.

6. Lindblom K, Palmer F. Rupture of the tendon aponeurosis of the shoulder joint—the so-called supraspinatus rupture. *Acta Chirurgica Scandinavica* 1939; **82**: 133.

7. Moseley HF, Goldie I. The arterial patterns of the rotator cuff of the shoulder. *Journal of Bone and Joint Surgery* 1963; **45B**: 780.

8. Rothman RH, Parke WW. The vascular anatomy of the rotator cuff. *Clinical Orthopaedics and Related Research* 1965; **41**: 176.

9. Rathburn JB, MacNab I. The microvascular pattern of the rotator cuff. *Journal of Bone and Joint Surgery* 1970; **52B**: 540.

10. Poppen NK, Walker PS. Forces at the glenohumeral joint in abduction. *Journal of Bone and Joint Surgery* 1978; **58A**: 165.

11. Altchek DW, Schwartz E, Warren RF. Radiologic measurement of superior migration of the humeral head in impingement syndrome. *Annual Meeting of the American Shoulder and Elbow Surgeons, New Orleans, La, 1990.*

12. Paletta GA, Warner JP, Warren RF. Biplanar X-ray evaluation of the shoulder in patients with instability and rotator cuff tear. *Annual Meeting of the American Shoulder and Elbow Surgeons, Anaheim, Ca, 1991.*

13. Doddy SG, Waterland JC, Freedman L. Scapulohumeral goniometer. *Archives of Physical Medicine and Rehabilitation* 1970; **51**: 711.

14. Freedman L, Munro RH. Abduction of the arm in scapular plane: scapular and glenohumeral movements. *Journal of Bone and Joint Surgery* 1966; **18A**: 1503.

15. Inman VT, Saunders JR, Abbott LC. Observations on the function of the shoulder joint. *Journal of Bone and Joint Surgery* 1944; **26A**: 1.

16. Poppen NK, Walker PS. Normal and abnormal motion of the shoulder. *Journal of Bone and Joint Surgery* 1976; **58A**: 195.

17. Dillman C, Andrews JR. Sports medicine focus: the throwing athlete's shoulder. *Orthopaedic News* 1991; **13**: 7.

18. Jobe FW, Moynes DR, Tibone JE, Perry J. An EMG analysis of the shoulder in pitching: a secondary report. *American Journal of Sports Medicine* 1984; **12**: 218–20.

19. Jobe FW, Tibone JW, Perry J, Hoynes D. An EMG analysis of the shoulder in throwing and pitching: a preliminary report. *American Journal of Sports Medicine* 1983; **11**: 3–5.

20. Ferretti A, Cerullo G, Russo G. Suprascapular neuropathy in volleyball players. *Journal of Bone and Joint Surgery* 1987; **69A**: 260–3.

21. Counsilman JE. Forces in swimming: two types of crawl stroke. *Research Quarterly* 1955; **26**: 127.

22. McLaughlin HL. Dislocation of the shoulder with tuberosity fracture. *Journal of Bone and Joint Surgery* 1963; **43A**: 1615.

23. Craig EV. The posterior mechanism of acute anterior shoulder dislocations. *Clinical Orthopaedics and Related Research* 1984; **190**: 212.

24. Bigliani LV, Morrison SD, April EW. The morphology of the acromion and its relationship of rotator cuff tears. *Orthopaedic Transactions* 1986; **10**: 228.

25. Harryman DT, Sidles JA, Clark JM, McQuade KJ, Gibb TD, Matsen FA. Translation of the humeral head on the glenoid with passive glenohumeral motion. *Journal of Bone and Joint Surgery* 1990; **72**: 1334–43.

26. Pavlov H, Warren RF, Weiss CB Jr, Dines DM. The roentgenographic evaluation of anterior shoulder instability. *Clinical Orthopaedics and Related Research* 1985; **184**: 153–8.

27. Harcke HT, Grissom LE, Finkelstein MS. Evaluation of the musculoskeletal system with sonography. *American Journal of Roentgenology* 1988; **150** 1253–61.

28. Mack LA, Kilcoyne RS, Matsen FA III. Sonographic evaluation of rotator cuff. *Radiology* 1984; **153**: 23.

29. Middleton WD, Reinus WR, Tatty WU, Melson CL, Murphy WA. Ultrasonic evaluation of the rotator cuff and biceps tendon. *Journal of Bone and Joint Surgery* 1986; **68A**: 440.

30. Goldman AB, Ghelman B. The double-contrast shoulder arthrogram. A review of 158 studies. *Radiology* 1978; **127**: 655–63.

31. Mink JH, Harris E, Rappaport M. Rotator cuff tears: Evaluation using double-contast shoulder arthrography. *Radiology* 1985; **157**: 621–3.

32. Ellman H. Diagnosis and treatment of incomplete rotator cuff tears. *Clinical Orthopaedics and Related Research* 1990; **254**: 64–74.

33. Evancho AM et al. MR imaging diagnosis of rotator cuff tears. *American Journal of Roentgenology* 1988; **151**: 751–4.

34. Iannotti JP, Zlatkin MB, Esterhai JL, Kressel HY, Dalinka NK, Spindler KP. Magnetic resonance imaging of the shoulder. *Journal of Bone and Joint Surgery* 1991; **73A**: 17–29.

35. Kieft GJ, Bloem JL, Rozing PM, Obermann WR. Rotator cuff impingement syndrome: MR imaging. *Radiology* 1988; **166**: 211–14.

36. Kneeland JB et al. MR imaging of the shoulder: diagnosis of rotator cuff tears. *American Journal of Roentgenology* 1987; **149**: 333–7.

37. Zlatkin MB et al. Rotator cuff disease: diagnostic performance of MR imaging. *Radiology* 1989; **172**: 223–9.

38. Seeger LL, Gold RH, Bassett LW, Ellman H. Shoulder impingement syndrome: MR findings in 53 shoulders. *American Journal of Roentgenology* 1988; **150**: 343.

39. Andrews JR, Broussard TS, Carson WG, Arthroscopy of the shoulder in the management of partial tears of the rotator cuff: preliminary report. *Journal of Arthroscopy* 1985; **1**: 117.

40. Ogilvie-Harris DJ, Wiley AM. Arthroscopic surgery of the shoulder. *Journal of Bone and Joint Surgery* 1986; **68B**: 201.

41. Ellman H, Hanker G, Bayer M. Repair of the rotator cuff. End-result study of factors influencing reconstruction. *Journal of Bone and Joint Surgery* 1986; **68A**: 1136.

42. Gore DR, Murray MP, Sepic SB, Gardiner GM. Shoulder-muscle strength and range of motion following surgical repair of full thickness rotator cuff tears. *Journal of Bone and Joint Surgery* 1986; **68A**: 266.

43. Hawkins RH, Misamore GW, Hobeika PE. Surgery of full-thickness rotator-cuff tears. *Journal of Bone and Joint Surgery* 1985; **67A**: 1349.

44. Kimmel J, Bigliani LB, McCann PD, Wolfe I. Repair of rotator cuff tears in tennis players. *Annual Meeting of the American Academy of Orthopaedic Surgeons, New Orleans, La, 1990.*

45. Tibone JE et al. Surgical treatment of tears of the rotator cuff in athletes. *Journal of Bone and Joint Surgery* 1986; **68A**: 887.

46. Warren RF. Surgical considerations for rotator cuff tears in athletes. In: Jackson DW, ed. *Shoulder surgery in the athlete*. Rockville, Md: Aspen, 1985: 73.

47. Tibone JE et al. Shoulder impingement syndrome in athletes treated by an anterior acromioplasty. *Clinical Orthopaedics and Related Research* 1985; **198**: 134.

48. Ellman H. Arthroscopic subacromial decompression: analysis of 1–3 year results. *Arthroscopy* 1987; **3**: 173.

49. Altchek DW, Warren RF, Wickiewicz TL, Skyhar MJ, Ortiz G, Schwartz E. Arthroscopic acromioplasty. *Journal of Bone and Joint Surgery* 1990; **72A**: 1198–1207.

50. Fly WR, Tibone JE, Glousman RE, Jobe FW, Yocum LA. Arthroscopic subacromial decompression in athletes less than 40 years old. *Orthopaedic Transactions* 1990; **14**: 250.

51. Altchek DW, Warren RF, Skyhar MJ, Ortiz G. T-plasty modification of the Bankhart procedure for multidirectional instability of the anterior and inferior types. *Journal of Bone and Joint Surgery* 1991; **73A**: 105–12.

4.3.4 Injuries of the acromioclavicular joint

JAY S. COX AND PAUL S. COOPER

INTRODUCTION

Injuries to the acromioclavicular joint are common in young athletic individuals. These injuries have long been recognized as a potentially difficult diagnostic and therapeutic problem: in 400 BC Hippocrates reported on the diagnosis and conservative management.[1]

The most common mechanism of injury is a force directed to the point of the involved shoulder, causing a displacement of the distal clavicle. This force may result from a fall on the shoulder and may be directed anteriorly, posteriorly, or inferiorly. The direction of force determines the direction of displacement of the distal clavicle. If the fall on the point of the shoulder occurs with the arm relatively close to the side, the scapula and the clavicle attached by the coracoclavicular ligament are driven downward and medially. If the force continues, the clavicle impacts against the first rib, establishing counter-resistance to the direction of major stress and concentrating the injuring force at the acromioclavicular and coracoclavicular ligaments. Continuation of this downward and medial excursion of the scapula in opposition to the clavicle tears these two major ligaments.[2]

The amount of displacement of the distal clavicle following this injury depends upon the extent of rupture of the acromioclavicular ligament, the acromioclavicular capsule, the coracoclavicular ligaments, and the trapezius and deltoid muscles. Horizontal stability of the clavicle depends upon the acromioclavicular ligament and vertical stability depends upon the coracoclavicular ligament. The coracoclavicular ligament is indeed the primary suspensory ligament of the upper extremity. When it is ruptured, the shoulder is actually displaced downward. The clavicle may be slightly elevated because of the pull of the trapezius muscle, but the main characteristic is a downward displacement of the shoulder and arm.

Hoyt[3] described a mechanism for an indirect injury of the acromioclavicular joint which is uncommon. The injury is indirect when the force is transmitted from its point of application through an outstretched hand or the point of the elbow causing an impact to the acromioclavicular joint by the humerus. The force traverses the glenohumeral joint which is stable in the position of moderate flexion and abduction, and therefore is concentrated on the acromion process. The scapula is forced superior and medially and the capsular investment of the acromioclavicular joint receives the brunt of the force, creating capsular stretch and tear. The coracoclavicular ligaments are seldom damaged by this particular mechanism because they are in a relaxed position. A clinical picture is that of a sprain or a type I acromioclavicular injury.

CLASSIFICATION (OR MECHANISM OF INJURY)

One of the original classifications was that of Cooper[4] which was later modified by Allman.[5] In their classification, a type I injury is a partial tear of the acromioclavicular ligament. The distal clavicle is not displaced and the coracoclavicular ligament is intact. In a type II injury there is a rupture of the acromioclavicular ligament

with a partial tear of the coracoclavicular ligament. Displacement in relation to the acromion is less than the full width of the clavicle. In a type III injury the acromioclavicular and coracoclavicular ligaments are completely torn, allowing complete displacement of the distal clavicle in relation to the acromion, greater than the width of the clavicle. There may be some detachment of the deltoid and trapezius musculature over the distal end of the clavicle, but the muscle aponeurosis is still grossly intact. For many years the classification of this type of injury was all encompassing, describing injuries of the acromioclavicular joint that displaced the distal clavicle posteriorly through the trapezius musculature and also superiorly completely through the deltoid and trapezius aponeuroses.

To clarify the more severe injuries, Rockwood and Green have described three additional types of acromioclavicular dislocations (Fig. 1). Their classification of type I and type II injuries remains the same as previous descriptions. A type III injury is a complete rupture of the acromioclavicular ligament and the coracoclavicular ligament, but in this new classification the deltoid and trapezius musculature aponeurosis over the distal clavicle is essentially intact.

A type IV injury is an injury in which the clavicle is grossly displaced posteriorly into or through the trapezius musculature. The clavicle may not be able to be reduced because it is trapped in the muscle fibres in a 'buttonhole' mechanism.

A type V injury is a previously described type III injury with complete rupture of the deltoid and trapezius musculature over the distal half or two-thirds of the clavicle. This allows gross disparity between the clavicle and the acromion. The distal clavicle is covered only by skin and subcutaneous tissue.

The type VI injury is very rare and is a disruption of the sternoclavicular joint and the acromioclavicular joint with a downward displacement of the distal clavicle. This occurs by a very severe direct force to the superior surface of the distal clavicle along with abduction of the arm and retraction of the scapula.[6] The distal clavicle becomes trapped underneath the muscle aponeurosis arising from the tip of the coracoid. In 1987 Gerber and Rockwood[7] reported three cases of inferior subcoracoid dislocations of the distal clavicle: one in a child and two in adults

PHYSICAL DIAGNOSIS

Type I injury

Because there is no disruption of the acromioclavicular joint, there is no prominence of the distal clavicle. There may be swelling over the acromioclavicular joint and there is certainly tenderness over this area. There is some pain on motion of the shoulder because of the movement at this joint.

Type II injury

There is disruption of the acromion and distal clavicle, and so some prominence of the distal clavicle may be noted. Again, there is swelling and tenderness directly over the distal clavicle and the

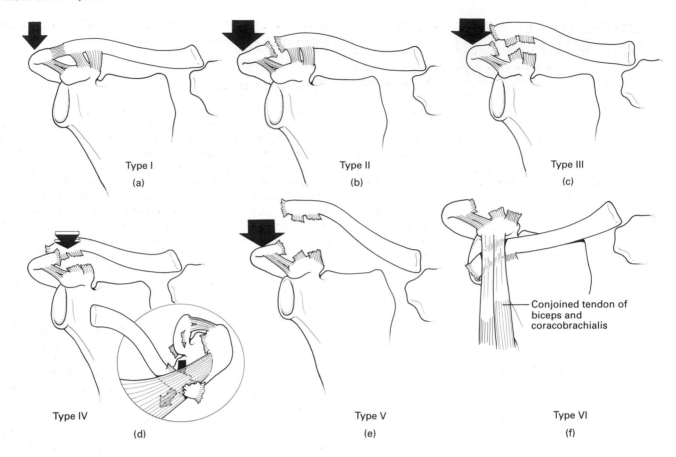

Fig. 1 The types of acromioclavicular joint injuries. (a) In the type I injury a mild force applied to the point of the shoulder does not disrupt either the acromioclavicular or the coracoclavicular ligaments. (b) A moderate to heavy force applied to the point of the shoulder will disrupt the acromioclavicular ligaments, but the coracoclavicular ligaments remain intact. (c) When a severe force is applied to the point of the shoulder both the acromioclavicular and the coracoclavicular ligaments are disrupted. (d) In a type IV injury not only are the ligaments disrupted but the distal end of the clavicle is also displaced posteriorly into or through the trapezius muscle. (e) A violent force applied to the point of the shoulder not only ruptures the acromioclavicular and coracoclavicular ligaments, but also disrupts the muscle attachments and creates a major separation between the clavicle and the acromion. (f) This is an inferior dislocation of the distal clavicle in which the clavicle is inferior to the coracoid process and posterior to the biceps and coracobrachialis tendons. The acromioclavicular and coracoclavicular ligaments are also disrupted. (Redrawn from ref. 12, with permission.)

acromioclavicular joint. Motion of the shoulder in this injury causes more severe pain because of the incongruence of the acromioclavicular joint.

Type III injury

The complete dislocation of the distal clavicle in relation to the acromion causes prominence of the distal clavicle. It is easily palpable and visible, and is accompanied by some tenderness over this area. The upper extremity is notably displaced downwards in relation to the opposite uninjured shoulder. There may also be tenderness along the coracoclavicular space at the site of rupture of the coracoclavicular ligament. There is moderate to severe pain on abduction and other shoulder motions.

Type IV injury

This clinical diagnosis is made by viewing the patient either from above or posteriorly. The distal clavicle is posterior in relation to the acromion and can be visualized and even palpated in this

position. There is acute tenderness over the clavicle and severe pain on shoulder motion. The distal clavicle is often unable to be reduced because it is trapped in the trapezius musculature. There may be some tenting of the skin over the posterior aspect of the shoulder.

Type V injury

This injury is usually quite easily visualized as the distal clavicle is covered only by skin and subcutaneous tissue. There is no overlying musculature to give the shoulder any contour. Again, the entire upper extremity is displaced downwards. There is pain on shoulder motion, particularly over the distal clavicle, since this is the area that is stripped of the deltoid–trapezius aponeuroses.

Type VI injury

This injury usually diagnosed by the inferior displacement of the distal clavicle. The acromion is very readily visualized and there appears to be a step-off at the acromioclavicular joint.

RADIOGRAPHIC EXAMINATION

Radiographic examination of the acromioclavicular joint does not require the normal X-ray penetration necessary for the shoulder joint. Examination of this joint usually requires one-third to one-half of the normal X-ray penetration. This prevents a dark or overpenetrated radiograph, and gives a better examination of the acromioclavicular joint.

The routine view is an anterior–posterior view, and this should preferably be obtained with the patient in the standing position. Care must be taken not to allow the individual to shrug the shoulder or support the injured extremity with the opposite hand. This would allow some upward reduction of the involved acromioclavicular joint. Routinely, these views are obtained without stress to the acromioclavicular joint. Stress is applied by weights strapped to the wrists. If the individual is allowed to hold the weight, he or she may be able to shrug the shoulder and thus cause some reduction of the dislocation of the acromioclavicular joint. In addition, these views should be taken with the X-ray beam at an upward tilt of 10° to 15°, as described by Zanca,[8] a true anterior–posterior view of the acromioclavicular joint. In a normal anterior–posterior view of the acromioclavicular joint the distal clavicle and the acromion may be superimposed on the spine of the scapula. This would preclude visualizing the entire distal clavicle and small fractures would be overlooked. Because a type IV injury may be missed in these routine views, Alexander[9] described a lateral stress view of the acromioclavicular joint to help identify the posterior dislocation. This is a shoulder forward view, and will show the displacement posteriorly of the distal clavicle in relation to the acromion. An axillary view may also be useful in discerning any posterior displacement of the distal clavicle.

On reading the regular anterior–posterior views with 10° to 15° of cephalic tilt, the position of the distal clavicle in relation to the acromion is assessed. In a type I injury there will be no change in relation to the opposite uninjured extremity. In a type II injury the distal end of the clavicle is displaced superiorly in relation to the acromion, but less than the full width of the clavicle. In a type III injury there is complete acromioclavicular dislocation with the distal end of the clavicle displaced completely above the superior edge of the acromion. This coracoclavicular interspace is significantly greater than in the normal shoulder. Bearden et al.[10] found a range of normal values for the coracoclavicular interspace of 1.1 to 1.3 cm. Bosworth[11] reported the average distance between the clavicle and the coracoid process as 1.3 cm. Even though Bearden et al.[10] stated that an increase in the coracoclavicular distance of 50 per cent over the normal side signified complete acromioclavicular dislocation, Rockwood et al.[12] have documented complete dislocation with an increase of as little as 25 per cent in the coracoclavicular distance.

TREATMENT OF ACUTE ACROMIOCLAVICULAR INJURIES

Type I injuries

The treatment of type I injuries is certainly not controversial and involves symptomatic treatment only. This includes temporary immobilization with a sling for comfort, ice application for control of pain and swelling, oral anti-inflammatory medications, and an early rehabilitation programme. The patient is allowed active range of motion exercises immediately and usually achieves full painless range of motion within 7 days. At this time rehabilitation of the shoulder girdle musculature is started and is continued until full strength is obtained.

Type II injuries

There are two options for the treatment of type II injuries. The first is symptomatic treatment, similar to that for type I injuries. The second is manual reduction and immobilization in some type of a shoulder harness for 4 to 6 weeks. The reduction must be performed manually and the immobilizer must be applied to retain the reduction. If the clavicle cannot be reduced manually, there is probably soft tissue intervention and the immobilizer should not be used. Care must be taken to prevent skin maceration and ensure compliance if this treatment option is to be successful. Complications with the acromioclavicular immobilizer are loss of position, skin problems, such as maceration, irritation, or ulceration, poor patient compliance, and stiffness of the shoulder, particularly in older patients. Recently, compression of the anterior interosseous nerve after the use of an immobilizer has been reported.[13] This immobilizer must be worn and the position must be maintained for a minimum of 4 weeks to ensure adequate healing of the soft tissues around the acromioclavicular joint. The aim of rehabilitation is to regain first range of motion and then strength of the musculature. Several authors have reported that residual symptoms in type II injuries are much higher than had been previously recognized.[14-17] Cox[15] found that treatment of type II injuries with an acromioclavicular immobilizer reduced the incidence of residual symptoms when compared with symptomatic treatment. We believe that residual symptoms in the type II injuries are caused by the chronic subluxation of the joint and incongruence of the joint allowing abnormal stress forces. This may result in a late degenerative condition in the joint. Therefore the best type of treatment for this type II injury, if compliance can be obtained from the patient, is to wear this sling to maintain normal position of the joint surfaces. In the Cox series in 1981,[15] 52 type II injuries were treated, 32 symptomatically and 20 in an acromioclavicular immobilizer. Sixty-three per cent of those treated symptomatically had residual symptoms but only 25 per cent of those treated with the acromioclavicular immobilizer had late symptoms. If the symptomatic treatment is chosen by the patient, then he or she should be told that there is a high incidence of late symptoms. Of course, these late symptoms can be resolved with a surgical excision of the distal clavicle by either open or arthroscopic techniques. Because of poor compliance by patients and their unwillingness to accept this form of treatment, we still treat the majority of type II injuries with symptomatic treatment followed by a late excision of the distal clavicle if warranted.[18]

TREATMENT OF COMPLETE DISLOCATIONS OF THE ACROMIOCLAVICULAR JOINT

In general, most orthopaedic surgeons agree that injuries of types IV, V, and VI are best managed operatively. However, a significant amount of controversy centres on the optimal treatment of the type III injury.

Following Cooper's description of an operative repair for acromioclavicular dislocation in 1861,[4] surgical treatment of dislocations remained the method of choice for over 100 years. In 1974 Powers and Bach[19] published a survey of the preferred method of treatment of acromioclavicular injuries in all orthopaedic residency training programmes in the United States. At that time, 91.5 per cent of the programmes were advocating surgical treatment for the complete dislocation and non-operative treatment was rarely recommended. The orthopaedic chairmen were also asked about their choice of surgical treatment and, of the 126 responses received, 60 per cent were advocating fixation across the acromioclavicular joint.

Powers and Bach also reported on 42 of their own cases and found that non-operative treatment in 28 patients had produced better overall results than in those treated operatively. Since that time, there have been numerous other papers reporting successful management of these complete dislocations by non-surgical techniques. In 1974 Rosenorn and Pederson[20] evaluated 24 patients: 11 were treated using a Bosworth screw and 13 were treated non-operatively. No significant functional or radiographic differences between the two groups could be detected. Those treated non-operatively returned to full activity more rapidly. In 1975 Imatani et al.[21] compared a series of patients treated prospectively, either operatively or non-operatively, and recommended a combination of minimal immobilization and early rehabilitation of the shoulder. Glick et al.[22] reported good results in a follow-up study of 35 unreduced type III acromioclavicular injuries in athletes treated non-operatively. In 1983 Bjerneld et al.[23] reported on a 5-year follow-up study of 33 patients with complete dislocation treated conservatively by minimal immobilization and no attempt at reduction. Satisfactory results were obtained in 30 of these patients. In a comparison of operative versus non-operative treatment of complete acromioclavicular dislocations, Galpin et al.[24] concluded that non-operative treatment provided an equal if not superior result, with an earlier return to activities, when compared with a coracoclavicular screw fixation. Taft et al.[25] compared 52 patients with type III injuries treated operatively with 75 patients treated non-operatively. Based on subjective, objective, and radiographic criteria, they concluded that surgical reduction of the acromioclavicular joint was not required in order to obtain good results and that surgery was associated with a higher complication rate. In 1987 Dias et al.[26] reviewed a series of 44 patients who had been treated non-operatively for acromioclavicular dislocation. Subjective and objective results were satisfactory in all cases. Recently, having reviewed the current literature, Dias and Gregg[27] concluded that conservative management is the treatment of choice for most acromioclavicular joint injuries because of the complication rate associated with various surgical procedures. In 1989 Bannister et al.[28] conducted a prospective randomized study of 60 patients with acute type III injuries, allocating patients to non-operative treatment of surgical open reduction and internal fixation with a coracoclavicular screw. With conservative treatment patients regained motion faster and more fully, thus allowing them to return to work and sport earlier.

A more recent survey by Cox[29] has shown that a shift towards non-operative treatment of acromioclavicular dislocations has occurred since the survey by Powers and Bach.[20] A questionnaire was circulated to all chairmen of orthopaedic residency training programmes in North America. Of the 187 chairmen who responded, 135 (72 per cent) advocated non-operative treatment. In a subgroup of 59 orthopaedists who specialized in the care of athletes, 86 preferred non-operative treatment of a type III injury.

Walsh et al.[16] were the first to examine endurance and strength after complete acromioclavicular dislocation injuries. Using isokinetic testing, they showed no significant difference in residual shoulder weakness in non-operative versus operative patients.

They noted a strength deficit in vertical abduction at high speeds in the postoperative shoulder when compared with the uninjured side. MacDonald et al.[30] studied the recovery of shoulder strength and function in 20 male patients with complete acromioclavicular dislocations. Ten of these patients had been treated non-surgically and 10 had been treated surgically. The majority of strength and flexibility tests showed no significant difference between the non-surgical and surgical groups, but some strength tests indicated that the non-surgical treatment of the complete acromioclavicular separation might be superior in restoring normal shoulder function following injury. Recently, Wojtys and Nelson[31] examined 22 patients treated non-operatively with isokinetic testing. At a mean follow-up of 2.6 years, they demonstrated that labourers and athletes recover strength and endurance with non-operative treatment. However, they noted that the strength and endurance advantage of the dominant shoulder may be lessened or lost if an acromioclavicular dislocation occurs on the dominant side.

NON-OPERATIVE TREATMENT OF TYPE III INJURIES

Many methods of non-operative treatment for complete acromioclavicular dislocations have been reported.[2] They include variations on the use of straps, slings, harnesses, traction, or casts as described for type II injuries. Two methods have gained popularity in recent decades: manual reduction and maintenance with an immobilizer, or symptomatic treatment with early motion.

With the sling-and-harness method, continuous use is required for a minimum of 6 weeks, and the complications are the same as described for the treatment of type II injuries. A failure rate of up to 20 per cent due to poor patient compliance has been reported.[32]

'Skilful neglect', or the symptomatic treatment of type III injuries, offers the advantages of convenience and shorter rehabilitation while obtaining a good functional outcome. The growing popularity of this form of treatment among orthopaedists was reflected in Cox's recent survey: 66 per cent of team physicians and 72 per cent of orthopaedic chairmen preferred the use of symptomatic treatment if non-operative treatment was chosen.[29]

Success rates ranging from 90 to 100 per cent with up to 7 years of follow-up has been reported in type III injuries treated non-operatively.[21,23,26,33] Several authors[15,27,34] have advocated symptomatic treatment for athletes, allowing them a rapid return to competition. Emphasis is placed on early and aggressive strengthening of the shoulder.

OPERATIVE TREATMENT OF ACROMIOCLAVICULAR INJURIES

Three main categories of procedures are currently recommended for acute acromioclavicular dislocations. These include acromioclavicular repairs, coracoclavicular repairs, and primary distal clavicle excision with joint debridement. (Fig. 2). Rockwood et al.[2] have cited over 200 surgical modifications and combinations of the three categories in their most recent edition.

Any operative technique for surgical repair of acromioclavicular injuries must address the following: reduction of the distal clavicle and maintenance of that reduction by fixation between the coracoid and the clavicle or across the acromioclavicular joint; repair of the coracoclavicular ligament if possible; repair of the acromioclavicular ligaments if possible; debridement of the acromioclavicular joint; excision of the distal clavicle if

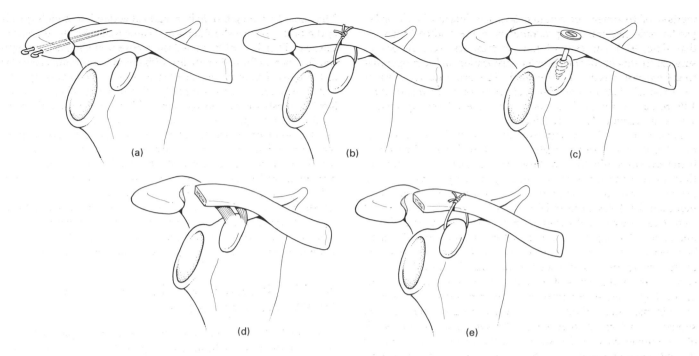

Fig. 2 Various operative procedures for injuries to the acromioclavicular joint. (a) Steinmann pins across the acromioclavicular joint. (b) Suture between the clavicle and the coracoid process. (c) A lag screw between the clavicle and the coracoid process. (d) Resection of the distal clavicle when the coracoclavicular ligaments are intact. (e) Resection of the distal clavicle with suture, fascia, or ligament between the clavicle and the coracoid process when the coracoclavicular ligaments are missing. (Redrawn from ref. 12, with permission.)

there are existing articular cartilage changes in the acromioclavicular joint or if the distal clavicle is damaged; repair of the deltoid and trapezius muscle aponeuroses.

Acromioclavicular joint repair

These procedures generally involve maintaining reduction of the acromioclavicular joint using K wires, Steinman pins, screws, suture wires, plates, or a combination of these. The coracoclavicular and/or acromioclavicular ligament may also be repaired primarily or reconstructed. K wires have been reported both to break and to migrate, and most authors advocate the use of threaded Steinman pins for fixation across the acromioclavicular joint.[35-38]

In 1987 Eskola et al.[39] compared various fixation techniques in a prospective randomized study of 100 patients. Comparison of smooth pins, threaded pins, and cortical screws showed that a significantly higher portion of symptomatic osteolysis developed using screw fixation. Most authors report superior results with the use of adjunct fixation to acromioclavicular fixation alone.[40-44] Pins are removed after 6 to 8 weeks, after which range of motion and strengthening exercises are allowed.

Extra-articular coracoclavicular repairs

In 1941 Bosworth[45] first described the placement of a coracoclavicular screw to maintain reduction. There have been many modifications of Bosworth's original technique[46-48] involving debridement of the acromioclavicular joint and/or primary repair of the coracoclavicular ligament in addition to coracoclavicular fixation. Variations in fixation techniques of the coracoid to the

clavicle have been described. These include the use of wire, a Dacron graft, fascia lata, or part of the conjoined tendon.[49-51]

Complications pertaining to coracoclavicular fixation include anterior translation of the clavicle with reduction and fracture of the coracoid process. Clavicular erosion with wire and graft loop fixation has been described by several authors.[52-54]

Rehabilitation is begun on the first postoperative day to avoid stiffness at the glenohumeral joint. Generally, heavy lifting and contact athletics are not to be resumed until 8 weeks postoperatively.

Comparisons of acromioclavicular and coracoclavicular repairs have yielded mixed results. Lancaster et al.[55] found a higher failure rate with coracoclavicular fixation, while others have demonstrated improved results with a Dacron coracoclavicular loop.[53,56,57] Kiefer et al.[58] showed biomechanically that the Bosworth screw provided the most rigid fixation, yet they advocated acromioclavicular fixation.

In the original survey by Powers and Bach,[19] 60 per cent of the physicians were using fixation across the acromioclavicular joint. Cox's study 15 years later showed that of 28 team physicians who indicated their surgical choice, 27 (96 per cent) preferred fixation between the coracoid and clavicle.[29] The majority (19) favoured fixation using either tape or suture, with the others using wire (two) or a screw (six). Of the chairmen of the orthopaedic residency training programmes, 101 indicated their choice. In this group 28 (27.7 per cent) were still using fixation across the acromioclavicular joint and 73 (72.3 per cent) were using fixation between the coracoid and the clavicle.

Our technique involves a modification of the method described by Park et al.[59] The acromioclavicular joint is exposed by a straight incision from the acromion to the coracoid process. The clavicular origin of the deltoid is elevated with the periosteum and the acromioclavicular joint is exposed. The acromioclavicular joint is debrided if necessary. Most surgeons recommend that, if there is evidence of an intra-articular fracture or pre-existing arthritis, the distal 1.5 cm of the clavicle be excised and bevelled in such a way as to ensure that there is no impingement with 90° or greater abduction of the humerus. However, we routinely perform a distal clavicle resection at the time of surgery to avoid any joint incongruity. An attempt is made to repair the coracoclavicular ligaments, and this is usually possible. If there is an avulsion from the periosteum, Bunnell-type sutures are placed in the ligaments and these sutures are subsequently fixed to the periosteum or through drill holes in the clavicle or coracoid. Four strands of no. 2 absorbable sutures are braided and used to secure the fixation. The braided suture is pulled through a drill hole in the clavicle and around the base of the coracoid and then secured with the clavicle fully reduced. It is important to place the knot anteriorly and/or inferiorly so that it will not be prominent subcutaneously over the superior portion of the clavicle. The braided absorbable suture affords the advantages of decreased erosion through either the clavicle or the coracoid and no hardware to remove afterwards.

The acromioclavicular ligament and capsule are repaired, followed by a meticulous repair of the deltoid and trapezius muscle aponeurosis. The skin is then closed with a subcuticular suture. Drains are not routinely used.

Postoperative immobilization consists only of a sling for comfort. Active range of motion exercises are instituted as soon as pain permits, usually in 3 to 5 days. The patient continues with these exercises until full painless range of motion is achieved. Strengthening exercises are then instituted and continued until the strength equals that of the opposite non-injured shoulder musculature.

When considering any type of coracoclavicular fixation, it is imperative to rule out any coracoid process fracture. This may best be accomplished using an axillary radiograph preoperatively. Surgical intervention in this situation has varied from reduction and acromioclavicular K-wire fixation to internal fixation of the coracoid using a lag screw technique.[60–62] Surgical results in this type of injury indicate a comparable prognosis to midsubstance coracoclavicular ligament injuries.[2,60]

Primary resection of the distal clavicle in the acute type III injury is usually indicated for fracture or arthritis involving the acromioclavicular joint. Powers and Bach[19] showed less than 1 per cent of orthopaedic surgeons favoured advocated immediate excision. In 1977 Browne et al.[63] reported that excision of the distal clavicle in surgical treatment of acromioclavicular dislocations did not offer significant improvement over coracoclavicular fixation alone. In 1988 Cooke and Tibone[64] reported on an objective analysis of function following excision of the distal clavicle and found little effect on the strength decrease or performance following the procedure. However, Cox's study revealed that only 34 per cent of team physicians and 22 per cent of orthopaedic chairmen advocated primary excision of the distal clavicle at the time of surgery.[29]

There is little controversy regarding the treatment of the rarer type IV, V, and VI injuries. In general, surgical repair or reconstruction as described for type III injuries is indicated. These injuries typically have extensive soft tissue dissection and it is imperative that the deltotrapezial fascia is closed. Prognosis with open surgical procedures for these complicated dislocations is favourable.[2] In type IV injuries, the clavicle may occasionally be manipulated from its entrapment in the trapezius muscle and then be treated as a type III acromioclavicular dislocation by closed techniques.

Chronic acromioclavicular dislocations

The choices for operation in symptomatic chronic acromioclavicular dislocations include excision of the outer clavicle, transfer of the coracoid end of the coracoacromial ligament to the lateral clavicle to reconstruct the acromioclavicular ligament, and transfer of the acromial end of the coracoacromial ligament to reconstruct the coracoclavicular ligaments.

Distal clavicle excision as described by Mumford[65] and Gurd[66] is indicated for symptomatic type II and III injuries. However, it is not advocated as a sole procedure for more extensive injuries (types III–IV). This could lead to further instability of the remaining portion of the clavicle and, by irritating soft tissues around the neck and shoulder, increase the patient's symptoms. Therefore we recommend a combined coracoclavicular ligament reconstruction for these more severe injuries. Recently, an arthroscopic technique for distal clavicle excision has been described which avoids excessive soft tissue stripping of the joint capsule.[18]

Since 1965 some authors have advocated use of a dynamic muscle transfer to hold the clavicle down for chronic type III injuries.[67,68] The technique usually involves transfer of the acromial end of the coracoclavicular ligament to the lateral edge of the resected clavicle. Dewar and Barrington[67] claimed good results in five patients with chronic acromioclavicular joint dislocation. Berson et al.[69] reported their experience with this procedure in 29 patients, and claimed 28 good or excellent results, only six of which had chronic dislocations. With nearly 50 per cent of their patients complaining of residual symptoms and a number of minor complications, Ferris et al.[70] recently advised caution in using this procedure.

Distal clavicle resection and transfer of the acromial insertion of the coracoacromial ligament into the clavicle was first described by Weaver and Dunn in 1972[71] They estimated a 10 per cent complication rate involving the pull-out of the transferred coracoacromial ligament from the resected distal end of the clavicle. Shoji et al.[72] modified the technique by harvesting a bone block attachment to the coracoacromial ligament and transferring it into the medullary canal of the resected distal end of the clavicle. Using this modified technique, one failed reconstruction was reported in 15 consecutive cases. All but one of the patients regained painless full range of motion. However, Larsen and Peterson[73] noted that 50 per cent of their patients were symptomatic over the bone block. Rockwood et al.[2] reported on 25 patients treated with coracoacromial ligament transfer and coracoclavicular lag screw fixation with favourable results.

CONCLUSION

Acromioclavicular joint dislocations can present both diagnostic and therapeutic challenges. Injuries of types I and II are best treated symptomatically with an awareness of delayed development of degenerative or post-traumatic changes in the joint. The current preferred treatment of uncomplicated type III injuries is by non-operative methods. Most orthopaedic surgeons prefer symptomatic treatment rather than reduction and immobilization for this injury. Complicated injuries of types III, IV, V, and VI frequently require surgical treatment, and temporary fixation between the coracoid and clavicle is currently favoured over fixation across the acromioclavicular joint. Most surgeons recommended against primary excision of the clavicle.

Stable, chronic, and painful acromioclavicular injuries can be treated by distal clavicle excision with good results. In chronically dislocated acromioclavicular joints, a coracoclavicular reconstruction using either a dynamic muscle transfer or a transfer of the acromioclavicular ligament is advocated. The latter procedure is considered superior since dynamic transfers may allow pistoning of the clavicle during stressful activity. When these methods fail, a synthetic graft may be necessary.

REFERENCES

1. Dameron TB, Rockwood CA. Fractures and dislocations of the shoulder. In: Rockwood CA, Wilkins KE, King RE, eds. *Fractures in children*. Philadelphia: JB Lippincott, 1984: 624–53.

2. Rockwood CA, Green DP, Bucholtz RW eds. *Fractures in adults*. 3rd edn. Philadelphia: JB Lippincott, 1991: Ch. 14.

3. Hoyt WA Jr. Etiology of shoulder injuries in athletes. *Journal of Bone and Joint Surgery* 1967; **49A**: 755–66.

4. Cooper ES. New method of treating long standing dislocations of the scapuloclavicular articulation. *American Journal of Medical Science* 1861; **41**: 389.

5. Allman FL. Fractures and ligamentous injuries of the clavicle and its articulation. *Journal of Bone and Joint Surgery* 1967; **49A**: 774–84.

6. McPhee IB. Inferior dislocation of the outer end of the clavicle. *Journal of Trauma* 1980; **20**: 709–10.

7. Gerber C, Rockwood CA. Subcoracoid dislocation of the lateral end of the clavicle: a report of three cases. *Journal of Bone and Joint Surgery* 1987; **69A**: 924–7.

8. Zanca P. Shoulder pain: involvement of the acromioclavicular joint: analysis of 1000 cases. *American Journal of Radiology* 1971; **112**: 493–506.

9. Alexander OM. Dislocation of the acromio-clavicular joint. *Radiography* 1949; **15**: 260.

10. Bearden JM, Hughston JC, Whatley GS. Acromioclavicular dislocation: method of treatment. *Journal of Sports Medicine* 1973; **1**(4): 5–17.

11. Bosworth BM. Complete acromioclavicular dislocation. *New England Journal of Medicine* 1949; **241**: 221–25.

12. Rockwood CA Jr, Williams GR, Young DC. Injuries to the acromioclavicular joint. In: Rockwood CA Jr, Green DP, Bucholz RW, eds. *Fracture in adults*. 3rd edn. Philadelphia: JB Lippincott, 1991.

13. O'Neill DB, Zairins B, Gelberman RH, Keating TM, Louis D. Compression of the anterior interosseous nerve after the use of a sling for dislocation of the acromioclavicular joint. *Journal of Bone and Joint Surgery* 1990; **72A**(7): 1100–2.

14. Bergfeld JA, Andrish JT, Clancy WG. Evaluation of the acromioclavicular joint following first and second degree sprains. *American Journal of Sports Medicine* 1978; **6**: 153–9.

15. Cox JS. The fate of the acromioclavicular joint in athletic injuries. *American Journal of Sports Medicine* 1981; **9**: 50–3.

16. Walsh WM, Peterson DA, Shelton G, and Newmann RD. Shoulder strength following acromioclavicular injuries. *American Journal of Sports Medicine* 1985; **13**: 153–8.

17. Babe JG, Valle M, Couceiro, J. Treatment of acromioclavicular disruptions: trial of a simple surgical approach. *Injury* 1988; **19**: 159–61.

18. Gartsman GM, Combs AM, Davis PF, Tullos MS. Arthroscopic acromioclavicular joint resection. *American Journal of Sports Medicine* 1991; **19**(1): 2–5.

19. Powers JA, Bach PJ. Acromioclavicular separation—closed or open treatment. *Clinical Orthopaedics and Related Research* 1974; **104**: 213–23.

20. Rosenorn M, Pedersen EB. A comparison between conservative and operative treatment of acute acromioclavicular dislocation. *Acta Orthopaedica Scandinavica* 1974; **45**: 50–9.

21. Imatani RJ, Hanlon JJ, Cady GW. Acute complete acromioclavicular separations. *Journal of Bone and Joint Surgery* 1975; **57A**: 328–31.

22. Glick JM, Milburn LJ, Haggerty JF, Nishimoto D. Dislocated acromioclavicular joint: follow-up study of 35 unreduced acromioclavicular dislocations. *American Journal of Sports Medicine* 1977; **5**: 264–70.

23. Bjerneld H, Hovelius L, Thorling J. Acromio-clavicular separations treated conservatively: a 5-year follow-up study. *Acta Orthopaedica Scandinavica* 1983; **54**: 743–5.

24. Galpin RD, Hawkins RJ, Grainger RW. A comparative analysis of operative vs. nonoperative treatment of grade III acromioclavicular separations. *Clinical Orthopaedics and Related Research* 1985; **193**: 150–5.

25. Taft TN, Wilson FC, Oglesby JW. Dislocation of the acromioclavicular joint: an end-result study. *Journal of Bone and Joint Surgery* 1987; **69A**: 1045–51.

26. Dias JJ, Steingold RA, Richardson RF, Tesfayohannes B, Gregg PJ. The conservative treatment of acromioclavicular dislocation: review after five years. *Journal of Bone and Joint Surgery* 1987; **69B**: 719–22.

27. Dias JJ, Gregg PJ. Acromioclavicular joint injuries in sport: recommendations for treatment. *Sports Medicine* 1991; **11**(2): 125–32.

28. Bannister GC *et al*. A prospective study of the treatment of acromioclavicular dislocations. In: Bateman JE, Welsh RP, eds. *Surgery of the shoulder*. St. Louis: CV Mosby, 1984.

29. Cox JS. Acromioclavicular joint injuries and their management principles. *Annales Chirurgiae et Gynaecologiae* 1991; **80**: 155–9.

30. McDonald PB, Alexander JJ, Frejuk J, Johnson GE. Comprehensive functional analysis of shoulders following complete acromioclavicular separation. *American Journal of Sports Medicine* 1988; **16**: 475.

31. Wojtys EM, Nelson G. Conservative treatment of grade III acromioclavicular dislocations. *Clinical Orthopaedics and Related Research* 1991; **268**: 112.

32. Urist MR. Complete dislocations of the acromioclavicular joint. *Journal of Bone and Joint Surgery* 1946; **28**: 813–37.

33. Hawkins RJ. The acromioclavicular joint. Presented at: *AAOS Summer Institute*, Chicago, July 1980.

34. Bannister GC, Wallace WA, Stableforth PG, Hutson MA. The management of acute acromioclavicular dislocation. *Journal of Bone and Joint Surgery* 1989; **71B**: 848–50.

35. Mazet RJ. Migration of a Kirschner wire from the shoulder region into the lung: report of two cases. *Journal of Bone and Joint Surgery* 1943; **25A**: 477–83.

36. Norrell H, Llewellyn RC. Migration of a threaded Steinmann pin from an acromioclavicular joint into the spinal canal: a case report. *Journal of Bone and Joint Surgery* 1965; **47A**: 1024–6.

37. Lindsey RW, Gutowski WT. The migration of a broken pin following fixation of the acromioclavicular joint: a case report and review of the literature. *Orthopaedics* 1986; **9**: 413–16.

38. Eaton R, Serletti J. Computerized axial tomography—a method of localizing Steinmann pin migration: a case report. *Orthopaedics* 1981; **4**: 1357–60.

39. Eskola A, Vainionpää S, Korkala O, Rokkanen P. Acute complete acromioclavicular dislocation: a prospective randomized trial of fixation with smooth or thread Kirschner wires or cortical screw. *Annales Chirurgiae et Gynaecologiae* 1987; **76**: 323–6.

40. Neviaser JS. Acromioclavicular dislocation treated by transference of the coracoacromial ligament: a long term follow up in a series of 112 cases. *Clinical Orthopaedics and Related Research* 1968; **58**: 57–68.

41. Ho WP, Chen JY, Shih CH. The surgical treatment of complete acromioclavicular joint dislocation. *Orthopaedic Reviews* 1988; **17**: 1116–20.

42. Inman VT, McLaughlin HD, Neviaser J, Rowe C. Treatment of complete acromioclavicular dislocation. *Journal of Bone and Joint Surgery* 1962; **44A**: 1008–11.

43. Amstrom JP Jr. Surgical repair of complete acromioclavicular separation. *Journal of the American Medical Association* 1971; **217**: 785–9.

44. Augereau B, Robert H, Apoil A. Treatment of severe acromioclavicular dislocation: a coracoclavicular ligamentoplasty technique derived from Cadenat's Procedure. *Annales de Chirurgie* 1981; **35**: 720–2.

45. Bosworth BM. Acromioclavicular separation: new method of repair. *Surgery, Gynecology and Obstetrics* 1941; **73**: 866–71.

46. Kennedy JC, Cameron H. Complete dislocation of the acromioclavicular joint. *Journal of Bone and Joint Surgery* 1954; **36B**: 202–8.

47. Lowe GP, Fogarty MJP. Acute acromioclavicular joint dislocation: results of operative treatment with the Bosworth screw. *Australian and New Zealand Journal of Surgery* 1977; **47**: 664–7.

48. Tsou PM. Percutaneous cannulated screw coracoclavicular fixation for acromioclavicular dislocations. *Clinical Orthopaedics and Related Research* 1989; **243**: 112–21.

49. Aldredge RH. Surgical treatment of acromioclavicular dislocations (abstract). *Journal of Bone and Joint Surgery* 1965; **47A**: 1278.

50. Fleming RE, Tomberg DN, Kiernan HA. An operative repair of acromioclavicular separation. *Journal of Trauma* 1978; **18**: 709–12.

51. Bunnell S. Fascial graft for dislocation of the acromioclavicular joint. *Surgery, Gynecology and Obstetrics* 1928; **46**: 563–4.

52. Goldberg JA *et al.* Review of coracoclavicular ligament reconstruction using Dacron graft material. *Australian and New Zealand Journal of Surgery* 1987; **57**: 441–5.

53. Dahl E. Follow up after coracoclavicular ligament prosthesis for acromioclavicular joint dislocation. *Acta Chirurgica Scandinavica* 1981; **506**: 96.

54. Dust WN, Lenczner EM. Stress fracture of clavicle leading to non-union. *American Journal of Sports Medicine* 1989; **17**(1): 128–9.

55. Lancaster S, Horowitz M, Alonso J. Complete acromioclavicular separations: a comparison of operative methods. *Clinical Orthopaedics and Related Research* 1987; **216**: 80–8.

56. Kappakas GS, McMaster JH. Repair of acromioclavicular separation using a Dacron prosthesis graft. *Clinical Orthopaedics and Related Research* 1978; **131**: 247–51.

57. Nelson CL. Repair of acromioclavicular separations with knitted Dacron graft. *Clinical Orthopaedics and Related Research* 1979; **143**: 289.

58. Kiefer H, Claes L, Burri C, Holzworth J. The stabilizing effect of various implants on the torn acromioclavicular joint: a biomechanical study. *ARCM Orthopedic Trauma Surgery* 1986; **106**: 42–6.

59. Park *et al.* Treatment of acromioclavicular separations. *American Journal of Sports Medicine* 1980; **8**: 251–6.

60. Bernard TN, Burnet ME, Haddad RJ. Fractured coracoid process in acromioclavicular dislocations. *Clinical Orthopaedics and Related Research* 1983; **125**: 227–32.

61. Wilson KM, Colwin JC. Combined acromioclavicular dislocation and coracoid process fracture. *American Journal of Sports Medicine* 1989; **17**: 697–8.

62. Lancourt JE. Acromioclavicular dislocation with adjacent clavicle fracture in a horseback rider. *American Journal of Sports Medicine* 1990; **3**(18): 321–2.

63. Browne JE, Stanley RF, Tullos MS. Acromioclavicular joint dislocations: comparative results following operative treatment with and without primary distal clavisectomy. *American Journal of Sports Medicine* 1977; **5**: 258–63.

64. Cooke FF, Tibone JE. The Mumford procedure in athletes: an objective analysis of function. *American Journal of Sports Medicine* 1988; **16**: 97–100.

65. Mumford EB. Acromioclavicular dislocation. *Journal of Bone and Joint Surgery* 1941; **23**: 799–802.

66. Gurd FB. The treatment of complete dislocation of the outer end of the clavicle: a hitherto undescribed operation. *Annals of Surgery* 1941; **113**: 1094–8.

67. Dewar FP, Barrington TW. The treatment of chronic acromioclavicular dislocation. *Journal of Bone and Joint Surgery* 1965; **47B**: 32–5.

68. Bailey RW. A dynamic repair for complete acromioclavicular joint dislocation (abstract). *Journal of Bone and Joint Surgery* 1965; **47A**: 858.

69. Berson BL, Gilbert MS, Green S. Acromioclavicular dislocations: treatment by transfer of the conjoined tendon and distal end of coracoid process to the clavicle. *Clinical Orthopaedics and Related Research* 1978; **135**: 157–64.

70. Ferris BD, Bhamra M, Paton DF. Coracoid process transfer for acromioclavicular dislocations: a report of twenty cases. *Clinical Orthopaedics and Related Research* 1989; **242**: 184–7.

71. Weaver JK, Dunn HK. Treatment of acromioclavicular injuries; especially complete acromioclavicular separations. *Journal of Bone and Joint Surgery* 1972; **54A**: 1187–97.

72. Shoji H, Roth C, Chuinard R. Bone block transfer of coracoacromial ligament in acromioclavicular injury. *Clinical Orthopaedics and Related Research* 1986; **208**: 272–7.

73. Larsen E, Peterson V. Operative treatment of chronic acromioclavicular dislocation. *Injury* 1987; **18**: 55–6.

4.3.5 Shoulder rehabilitation: principles and clinical specifics

TERRY R. MALONE

INTRODUCTION

The shoulder is the most mobile, and thus the least constrained, major joint in the human body. Stability is provided through a combination of active and passive structures. No other joint is so dependent upon musculature for stability and control. The inherent dichotomy of stability and mobility requires the rehabilitative specialist to approach patients with shoulder pathologies with care and precision. In this chapter we shall attempt to provide general principles of rehabilitation based upon anatomy and pathology in relationship to the functional demands of the shoulder. Other chapters in this textbook address specific pathologies (rotator cuff, shoulder instability, and injuries of the acromioclavicular joint), and thus we shall not attempt to be all-encompassing but rather eclectic in providing clinical insight to the rehabilitation challenges presented by the shoulder complex.

ANATOMY

The shoulder complex must be considered as a whole, as single structures cannot be recognized without recognizing the impact provided by each component of the 'complex'. Additionally, many compensatory patterns/positions are seen in athletes/workers which may have evolved to allow better performance of specific athletic or work activities (e.g. tennis serve, throwing, typing, etc.). The importance in interpreting these alterations has been described by numerous authors.[1,2] Clinical recognition of these alterations requires awareness of the inter-relatedness and linkage of the shoulder in activity pursuits.

SKELETAL ANATOMY

Three bony structures (clavicle, scapula, and humerus) comprise the shoulder skeleton. Unfortunately, many individuals do not recognize that these inter-related structures function as a single entity through the three synovial joints (sternoclavicular, acromioclavicular, and glenohumeral) and one physiological joint (scapulothoracic) to allow the multiple synchronous actions seen in the upper extremity. Many movements seen during shoulder function are provided through the contributions of each individual joint rather than action at the glenohumeral joint alone. Failure to realize these distinct contributions allows the patient to fall into the 'robbing Peter to pay Paul' sequence. Clinician awareness is critical, as the pattern of the musculofascial structures which 'bind' the shoulder to the trunk may be altered and thus result in structural or functional alteration in upper-extremity activities.

The clavicle is anchored to the trunk at the sternoclavicular joint which has two synovial cavities separated by a fibrocartilaginous disc.[3,4] This joint functions primarily as a 'saddle', allowing rotation but also multiple other displacements. The acromioclavicular joint provides a gliding action and does not overly constrain the scapula.[3] The clavicle acts as a strut (yardarm) or crankshaft to enable greater function of the shoulder complex.[5]

The scapula is a complex triangular structure which provides sites for muscular/ligamentous attachment and bony articulation. Interestingly, its axial/appendicular orientation is completely controlled by musculature or fascial constraint. Clinicians frequently forget the enormous neuromuscular demands of shoulder function, and thus do not recognize the need for a stable base (proximal control) for normal distal activity.[6] The glenoid fossa is deepened by the fibrocartilaginous labrum which increases the limited congruency presented by the glenohumeral joint. The increase is quite substantial and extremely important in the high demands of athletic participation.[7,8]

The proximal humerus includes the humeral head with the greater and lesser tubercles, as well as the proximal body or shaft. Examination of the glenohumeral joint demonstrates the inherent lack of stability and the resultant demands placed on neuromuscular control (Fig. 1).

STABILITY

The stability of the shoulder complex can be divided into passive and active components. Inherent passive stabilizers include numerous fibrous structures, but their contributions vary with composition and joint position. Individual variation in composition is obvious and dictates the level of laxity seen with direct measures. Joint position is critical, as the ability of a structure to control movement may be correlated to fibre alignment/orientation. An example of this is the need for patients to be urged to work on increasing rotation at multiple areas or points of flexion rather than working only in the extremes of the range of motion.

The arthroscope has allowed a much greater appreciation of the shoulder complex to emerge during the 1980s and 1990s.[10,11] Shoulder ligamentous restraints are frequently described by location and orientation. The glenohumeral ligaments are divided into superior, middle, and inferior (Fig. 2), with each portion

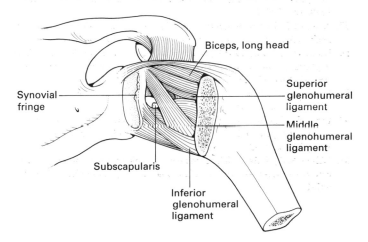

Fig. 2 The anterior glenohumeral ligamentous/capsular structures as seen from a posterior exposure looking forward. (Redrawn from ref. 9, with permission.)

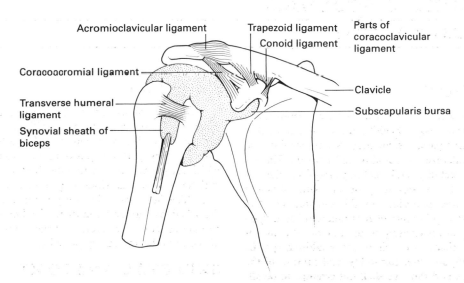

Fig. 1 Orientation of the glenohumeral joint and ligamentous/bony orientation. (Redrawn from ref. 9, with permission.)

exhibiting different levels of tension throughout the range of motion. O'Brien *et al.*[10] have demonstrated the importance of viewing the capsule in total rather than addressing only specific portions. The function of these structures has been stated beautifully by Terry *et al.*: 'The static restraints of the scapulohumeral joint provides stability for the humeral head in the glenoid cavity, limit extremes of motion of the glenohumeral joint, and guide positioning of the humerus during normal shoulder movement'.[11] It should also be noted that these structures work in a complementary fashion with the glenoid labrum. Many of the problems related to throwing can be ascribed not only to the tremendous demands but also to the concept of mechanical links. We frequently talk of lower-extremity patterns being dominated by closed kinetic chain activities while the upper extremity functions very much in an open pattern. This places extreme demands on musculature, as the greatest muscular activity is present during deceleration (Fig. 3).[12,13]

Fig. 3 The open kinetic chain pattern of function is seen in the upper extremity. (Reproduced from ref. 2, with permission.)

Active stabilizers

Hughston nicely describes the functionality of the complex as being almost completely dependent on the 'synergism of musculotendinous units'. Six muscles support and anchor the scapula to the thorax, while nine muscles provide the free movements seen at the glenohumeral joint. The complex inter-relationships of the musculature have been described quite eloquently by Hollinshead,[3] Kent,[15] and Perry.[16] Hollinshead describes the glenohumeral actions as being best delineated as the short muscles acting to retain the humerus in its proper orientation, with the longer muscle responsible for the freedom of movement of the humerus upon the glenoid. Complex force couples are required for normal patterns of movement with minimal changes requiring major adaptive responses which may predispose the individual to

future injury or degeneration. Because of the complex synergistic relationships, it is often difficult to determine whether a muscle is acting as a prime mover or a stabilizer, thus preventing undesired proximal movement which would alter length-to-tension ratios. We believe that the supraspinatus is the key to shoulder rehabilitation, and is also the muscle most frequently unable to perform the 'fine tuning' required for normal athletic activities. Additionally, the external rotators are quite synergistic with the supraspinatus and require attention during most rehabilitative programmes. The absolute importance of the musculature is displayed vividly by the early inferior subluation which occurs in a flaccid shoulder or a shoulder inhibited by joint effusion. Not only is musculature frequently inhibited by an effusion, there is a loss of proprioceptive feedback following injury to capsular structures, and a normalization of this neural mechanism should not be overlooked.[17] Neural activity (drive, recruitment, synchronization) has a direct relationship with musculature, and thus they are interwoven and demand attention during the rehabilitation process.

SPECIAL EVALUATION OF THE ATHLETIC SHOULDER

General evaluation of the shoulder complex has been presented by numerous authors.[18–21] The throwing shoulder requires additional evaluation because of the enormous stresses presented and the adaptations seen in the musculoskeletal system. Throwing athletes typically demonstrate a large increase in external rotation with a corresponding loss of internal rotation when the throwing shoulder is compared with the normal shoulder. This relative increase is even further enhanced after the athlete has warmed up, and this action appears to be primarily an adaptation to allow greater distance for acceleration, thus allowing greater velocity to be achieved by the throwing athlete. A second compensatory change is the hypertrophy of the upper-extremity musculature of the throwing shoulder. Possibly associated with this finding is the characteristic drooping or depression of the dominant shoulder.[22] Subtle postural and mechanical anomalies are thus the rule rather than the exception when evaluating the throwing shoulder. Clinicians would do well to realize the 'normalcy' of these changes and their implications, including alteration of length-to-tension ratios, and the need for these adaptive responses if an athlete is to be successful following rehabilitation.

A functional evaluation of the shoulder is probably one of the most important facets of the total evaluation scheme which is frequently not utilized by many clinicians. It is imperative to determine when the chief complaint appears during a specific action, as well as determining its severity and reproducibility. As stated previously, it is sometimes difficult to determine whether a muscle is acting as a prime mover or as a stabilizer, but adding information as to where in the range of motion and what action is being performed may allow the clinician to determine whether the muscle is acting as a stabilizer (isometric or eccentric muscle activation) or as a prime mover (concentric contraction). This information is imperative for a proper structuring of the rehabilitation sequence. It is not enough for the clinician to determine that the shoulder pops or catches; rather, specific information as to when the pop or catch occurs, whether it occurs consistently or inconsistently, and whether there is pain associated with this action gives much information that is relevant and important.

PRINCIPLES OF SHOULDER REHABILITATION

The following principles are presented to allow development of a well-structured and sequenced rehabilitation scheme.

Principle 1: let pain be the guide

Pain must be recognized as the controlling factor in our rehabilitation sequence, as pain which is range of motion specific or action specific requires the modification or elimination of activity. Pain is frequently a sign of inflammation to a specific structure, or impingement on or catching of a specific structure through a specific movement. The use of a functional classification scheme of pain for patient progression (Table 1) is recommended.

Table 1 Functional sequence of pain

Level 1	Pain after specific activity
Level 2	Pain during and after specific activities, but not affecting performance
Level 3	Pain during and after specific activity, and affecting performance of such activity
Level 4	Pain with activities of daily living
Level 5	Pain at rest

A functional classification pattern allows the patients to communicate what is happening in their rehabilitation in relationship to activities, and also allows them to become aware of their progression. It is imperative that the patients become active participants within the rehabilitation sequence and take 'ownership' of their specific problem. As previously stated, inflammation may play a predominant role and thus must be controlled by minimizing continued or secondary insult (elimination of offending exercise), pharmacological intervention (steroidal or non-steroidal anti-inflammatory drugs), or cryotherapy modulation. Severe pain may also necessitate the use of pain modulation techniques, including manual therapy techniques (oscillations), transcutaneous neuromuscular stimulation (TENS), high voltage, microamperage stimulated (MENS), and/or other physical therapy pain modulation techniques/modalities. A guiding/overriding pattern is that the rehabilitation programme must reach the cause of the problem rather than treating the symptoms presented. Pain control must be followed by an appropriate intervention/rehabilitation programme.

Principle 2: exercise must be performed in a pain-free/available range of motion

Muscles are absolutely critical to the control of the upper extremity. Since the shoulder is dependent upon musculature for functional stability, rehabilitation protocols must address early strengthening within the available range of motion provided by specific problems. An example is the use of isometrics at multiple angles in patients whose pathology dictates the restriction of the range of motion. As soon as isometric exercise is pain free, partial range of motion isotonic activities can be performed, as well as what may be called isodynamic actions involving the use of surgical tubing (Fig. 4).

It is important to note that exercises of this type (using surgical tubing) can be performed with an emphasis on concentric or eccentric muscle activation. We attempt to determine the type of lesion (cause of problem) and address it by matching the muscular demand to functional activity. An example of this process is the use of early activities below 60° of abduction and forward flexion with minimal external rotation performed as soon as tolerated by patients with anterior inferior shoulder dislocations. We have found the use of surgical tubing to be very helpful as well as being a very portable means of exercise. We also attempt to ensure that the patient is provided with a means of performing the exercises requested, and have found the BREG shoulder therapy kit to be extremely helpful. This comprises a collapsible bar, a rope-and-pulley set, and multiple 'surgical tubing' sets with different resistances (Fig. 5).

Principle 3: increase range of motion at multiple planes

Clinicians frequently attempt to increase rotation only at the extremes of the range of motion. In order to recognize the rotational components and their effect on the multiple structures of the shoulder capsule, it is very helpful to work on internal and external rotation at different portions of forward flexion or abduction. Also, with athletes it is imperative to work on rotation in functional positions. We frequently view throwing as involving overhead activity when, in fact, the glenohumeral actions occur from 70° to 100° of abduction.[23]

Specificity of range of motion would thus dictate that internal/external rotation work is performed with the arm in a 70° to 100° abduction position, thus duplicating the throwing position (Fig. 6).

A combination of Principles 1 and 3 is involved in the treatment of the patient with adhesive capsulitis. Clinicians frequently use thermotherapy modalities applied over the shoulder when the area of maximal involvement is in the inferior fold of the shoulder capsule. Thus clinicians should apply their modality to the axilla, but may also wish to apply a similar modality over the shoulder for generalized relaxation. This is particularly true if these patients have been treated previously, and they may question the treatment if they are not receiving similar patterns or at least a consistency of application!

Principle 4: strengthen proximal musculature

Proximal stability is required for normal upper-extremity function. Clinicians must address weakness of scapular stabilizers, as well as recognizing the important role that upward rotation plays in allowing enhanced functional athletic activities (Fig. 7).

Clinicians must also address the endurance needs of the patient involved in activity that requires repetitive or continued positioning (swimming, typing, overhead work, etc.), as well as addressing the absolute maximum effort demanded of athletes in events such as pitching or tackling in power-dominated sports.

Principle 5: strengthen individual muscles

The ability of the practitioner to isolate and thus provide appropriate strengthening of an individual muscle is paramount. Teaching the patient to avoid compensatory muscular action or abnormal substitutions is absolutely vital to long-term success.

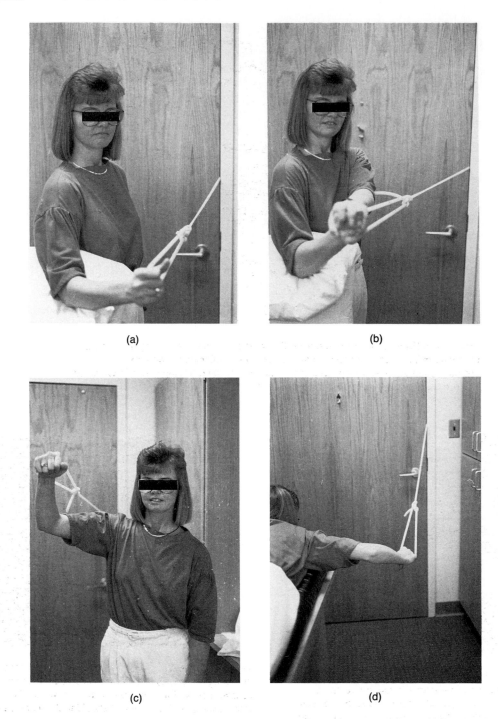

(a)   (b)

(c)   (d)

Fig. 4 The use of surgical tubing is extremely flexible: (a) external rotation in a standing position with the humerus slightly forward flexed and abducted, a position found more comfortable for many patients with impingement syndrome when strengthening the external rotators; (b) Subjects can emphasize the eccentric component by stretching the tubing with the contralateral extremity and then slowly allowing a return, thus eccentrically controlling the movement; (c) specific positioning and speed can be included in the work-out by working on a throwing pattern, as well as incorporating a plyometric (stretch–contraction cycle) through surgical tubing; (d) strengthening in a sport-specific position is extremely critical for rotational strength and endurance as demonstrated in the freestyle swimming position (the athletic can work on both internal and external rotation in this position). A similar example of this process is for patients to work from 20° to 30° of abduction in combination with slight forward flexion into what is referred to as the 'scapular plane position' and for them move only to approximately 60°, thus staying in a very comfortable part of the range of motion. Any impingement symptoms are minimized and functional strengthening is allowed, and this is a plane of motion that is frequently used.

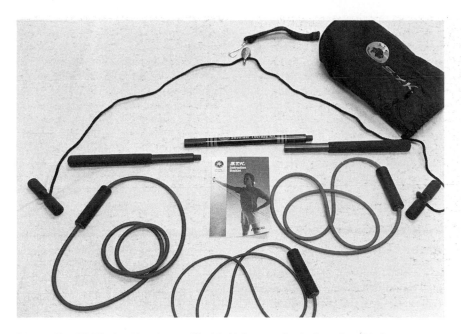

Fig. 5 The BREG shoulder therapy kit: this kit is comprised of a collapsible bar, an overhead pulley with mounting apparatus for door positioning, three different thicknesses of surgical tubing, an instructional booklet, and a canvas bag in which all these materials can be stored and transported.

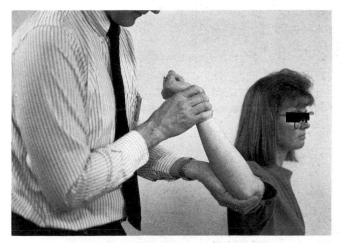

Fig. 6 Stretching should be performed in a specific range of motion which is demanded by the activity-external rotation stretching performed in a thrower in a 90° glenohumeral abduction position. (The increased external rotation seen in most throwers should be noted.)

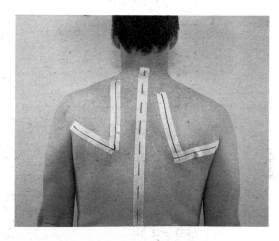

Fig. 7 The slightly upwardly rotated position of the dominant (throwing) extremity which may be an adaptation to allow greater acceleration and thus more efective throwing. The depression (lower rest position) of the dominant shoulder should also be noted.

Also, individual muscles must be strengthened at a variety of length-to-tension ratios since their activity may demand different levels of performance/activation, as well as requiring different functions (eccentric versus concentric or stabilizing) throughout the range of motion. It is also important for the clinician to apply appropriate resistance, but mixing and matching the effort required is frequently useful. An example of this is to use a variety of sets and repetition combinations to 'bombard' the neuromuscular system with differing demands, thus emphasizing motor learning throughout the rehabilitation sequence. An example is to use not only isotonic (concentric and eccentric) activities, but also

to supplement these with isodynamic (surgical tubing) and isokinetic movements during the rehabilitation progression. Another interesting factor is the use of eccentric activity, which is available with newer active isokinetic dynamometers. These activities can be performed in a pattern against the motorized equipment which may actually give a different demand and message to the neural system than those seen with the eccentric activity performed isotonically. This is an assumption, but it may be quite helpful in allowing a variety of demands to be placed on the neuromuscular system. One of the advantages of the use of isokinetic equipment is that it allows the patient to work in isolated rotational patterns but does not allow them to substitute inappropriately or demand that they attempt to perform at a level greater than that at which they are comfortable. Isokinetic exercise that is performed concentrically is extremely safe and is completely dependent upon the forces applied by the individual. We have much less information and experience with eccentric dynamometry (controlled patterns), but the application of these devices is increasing with excellent clinical results. Some of these devices allow both an active eccentric mode (i.e. the machine pushes as long as the minimum force required of the patient is present) and resistance against the passive movement mode in a 'resist as one is able' pattern. The passive movement mode is also very comfortable and allows additional exercise applications.

Principle 6: strengthen muscles in groups—within functional patterns

Once the clinician has ensured that proximal stability and adequate strength are established in individual muscles, the patient must be progressed into functional patterns. Strengthening in patterns has been espoused very strongly by the proponents of proprioceptive neuromuscular facilitation.[24] An interesting expression is the functionalization of individual muscular strength. This is a very apt description of this phase of the rehabilitation, as the patient works in diagonal patterns which are much more appropriate and require neuromuscular coordination of rotation and fine movements distally with proximal stability.

Principle 7: functional rehabilitation— functional progression

Once the patient has proceeded through the rehabilitation sequence and has moved through the various levels of the functional classification of pain, the culmination of these activities should be a functional progression. Functional progressions are designed to allow the individual to return to their activities of choice through specific adaptation to imposed demands.[25] During this phase of rehabilitation, attention is directed to specific activities such that rehabilitation addresses velocity, level of resistance, type of contraction, and endurance required for specific performance. An example is the progression of the tennis player from controlled ground strokes to controlled overheads, with a slow progression on to a flat serve prior to the initiation of either a slice or twist serve. The progression allowed must proceed according to the reaction of the patient as he or she attempts to perform the activity properly, as well as the response of the tissue on the following day (i.e. not only the immediate response but also the residual seen the next day should be monitored). The rehabilitation/functional progression must be designed with intermediate as well as with long-term aims in mind.

We conclude this chapter with a variety of case studies demonstrating the application of these principles and examples of how modification of the exercise sequence is required to fit the individual needs of the specific patient.

CASE STUDY 1

A 17-year-old white male interscholastic swimmer presented with shoulder pain of 6 months' duration. The pain had worsened dramatically over the last 3 weeks, with a decrease in the ability to perform. Pain was now present during normal daily activities but

the symptoms were greatly exacerbated during swimming. He was most symptomatic with freestyle, which unfortunately was his primary competitive stroke. Upon questioning, this patient related that his routine had recently changed to include the use of hand paddles for 2500 to 3000 m daily. Interestingly, this coincided with the increase in his symptoms. This individual had sought medical attention approximately 3 months earlier, and had been placed on a strengthening routine as well as an attempt to decrease his swimming through the use of a kickboard over a 3- to 4-week time-frame. Unfortunately, the strengthening routine implemented included primarily activities which increased internal rotation strength and were performed in non-swimming-oriented positions. He had also used the kickboard in an extended position (arms in front of the body), which will not improve but rather maintain symptoms associated with this condition. Examination of his present routine also revealed two daily work-outs of approximately 5000 to 6000 m each, with approximately 1200 to 1300 m being accomplished with the use of hand paddles. It is our experience that the use of hand paddles will greatly increase swimming shoulder problems, as rotation is the least well controlled of the movements and thus the most demanding.

Initial plan

This athlete was exhibiting level 5 to level 4 pain, as he was experiencing pain with activities of daily living and minimal pain at rest. Isokinetic evaluation of his internal/external rotators revealed that the external rotators were only capable of generating approximately 40 per cent of the internal rotators in a concentric pattern (Fig. 8). Falkel and Murphy[21] have reported similar findings in the evaluation of swimmers with shoulder pain. Our initial management thus centred on the following modalities.

1. Controlling inflammation (oral anti-inflammatory drugs.
2. Maintenance of aerobic base through the use of a kickboard under the abdomen, and discontinuation of the use of hand paddles.

Fig. 8 Isokinetic evaluation-test position of internal/external rotators.

3. Modification of strength training to emphasize external rotation in multiple angles of abduction and forward flexion and swimming position patterns (Fig. 9).

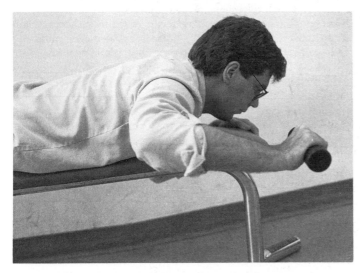

Fig. 9 Prone weight-training positions for strengthening of the external rotators (see also Fig. 4(d) on the use of surgical tubing). This position is ideal for the swimmer but would not be as applicable to the thrower, i.e. the thrower would exercise in an upright standing position.

4. Use of modalities to decrease impingement types of problems at the suprahumeral space (moist heat and ultrasound)—these were applied with the arm in an abducted and slightly forward flexed position to minimize vascular problems (Fig. 10).[26]
5. Alteration in swimming practice to emphasize quality of stroke rather than quantity, thus allowing only a few hundred metres to be performed using the shoulder but emphasizing quality

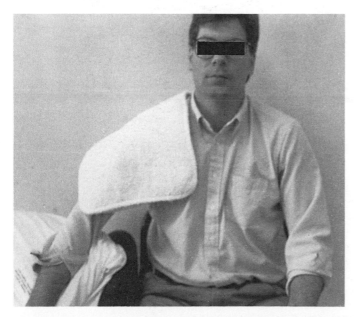

Fig. 10 Positioning of the upper extremity during the application of modalities to minimize vascular compromise.

during that time-frame. It is our opinion that coaches have too often emphasized distance, and thus quantity rather than quality of stroke. This has led to Olympic swimmers actually being survivors rather than the best trained. The only swimmers able to survive these exhausting work-outs were those who were fortunate in having a very large suprahumeral space and excellent external rotator/supraspinatus musculature.

6. After work-outs, ice massage to the suprahumeral area was performed for 10 to 12 min.

After 10 days of this sequence, increasing endurance was emphasized in the subject's external/internal rotator strengthening sequence. Twenty to fifty repetitions were being performed, and he was able to increase from two to three sets up to four to five sets utilizing the modified base position at a velocity of approximately 240°/s (Fig. 11). This position helps to minimize the vascular

Fig. 11 Modified base position for internal/external rotation to minimize vascular compromise.

compromise which may be present, as well as providing greater space to the suprahumeral area. This athlete used this routine for approximately 3 weeks on a thrice weekly basis, and was also using pull cords and surgical tubing in a swim-stimulated stroke position to work again on endurance as well as the specific strengthening needed for the external/internal rotation component. He was allowed to increase his swimming as his symptoms permitted, and was now swimming approximately 3000 m daily with relatively minimal symptoms. Over the next 3 weeks, he increased to where he was able to swim 5000 m daily and remained at that level with no return of symptoms. After 6 weeks of modified work-outs, his isokinetic strength assessment revealed that his external rotators were now 55 per cent of his internal rotators as measured concentrically.

This case presentation highlights the integration of sports-specific rehabilitation with appropriate medical management in the presentation of a young swimmer with shoulder pain. The actual aetiology of this problem is frequently thought to be an impingement or supraspinatus tendonitis. The actual entity is not as important as the appropriate management of symptoms and attention to the causative factors. Too often, medical practitioners will either tell the swimmer to stop swimming or only address the pain. The pain that is present is a signal to the clinician to find and address the cause of the patient's pain. Previous intervention had been unsuccessful, as the causative factors had not been addressed (external rotational weakness, supraspinatus weakness, use of hand paddles, and use of a kickboard with the shoulders in an extended position). One of the areas in sports medicine which demands further attention is the integration of medical and coaching expertise in the formulation of training programmes. Swimming is one of the areas where this is easily seen but not easily achieved.

CASE STUDY 2

A 62-year-old white man was self-referred to the clinic for evaluation of right shoulder pain. Ten days previously, while attempting to start his lawnmower, he experienced an acute onset of pain when the starter 'caught' as he was attempting to pull on the starter cord. He continued to have pain both during the day and at night (Level 5—pain at rest). Examination revealed inability to abduct the arm actively, and the drop arm test was positive. An arthrogram was performed which demonstrated a complete thickness tear, and the patient chose to have surgical intervention. A separation of approximately 1 cm in the superior rotator cuff was repaired. Postoperative management began with sling immobilization for 4 weeks. During this period, he was allowed to forward flex and abduct assistively, utilizing the opposite hand to support the weight of the arm (designed to minimize distraction forces). Additional recommendations during the early part of postoperative management included supporting the arm when writing or seated with pillows, thus propping and supporting the weight of the limb. This is particularly important if the person is going to be in an automobile for extended periods of time.

We frequently instruct the patient in self-oscillation techniques, but again they must be performed without distraction forces in these initial weeks. Typically, between the third and fourth week of their postoperative course, patients are allowed to begin more active/assistive patterns through the use of a pool. This can also be accomplished utilizing a stick (dowel rod, cane, tennis racket, golf club, etc.) or assistive pulley patterns (Fig. 12). The use of water, when available, is preferable as buoyancy allows fairly rapid redevelopment of neuromuscular co-ordination and minimizes the development of dissociation seen with scapular and glenohumeral actions. The normal two-to-one ratio of glenohumeral-to-scapulothoracic contributions to abduction are frequently not present following rotator cuff repairs. Patients will have great difficulty moving in normal patterns, and will attempt to maintain the glenohumeral position and move primarily through the scapulothoracic joint. Water appears to be a very good medium to minimize a continuation of this adaptive or compensatory movement pattern. At approximately 6 weeks after the operation, the patient is allowed to begin a more aggressive active programme, and weights are often attached to the wrist rather than held in the hand (Fig. 13). This helps to minimize the formation of abnormal movement patterns or compressive activities frequently seen with these patients.

The primary function of the supraspinatus may be to position the head of the humerus properly during shoulder movements. These actions may be extremely demanding, and thus any lack of neural control leads to abnormal force couple relationships and abnormal shoulder function. A very interesting finding in these patients is the difficulty in moving from a concentric contraction into an eccentric or lowering muscle activation pattern. These patients frequently say that 'It hurts the most when I try to lower the arm after I have raised it.' We believe that this is related to the transfer of increased stress to the tendon as the contribution of the contractile element of the muscle is decreased. Thus it is very important to work on both concentric and eccentric activities, but these must be done lower in the range of motion rather than in the more demanding side-arm positions. Another technique which is

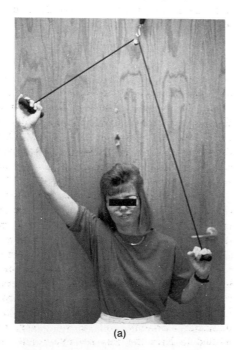
(a)

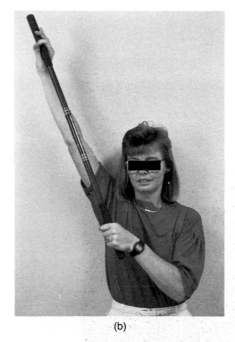
(b)

Fig. 12 (a) Assistive pulleys which allow the person to perform active movement when possible, assisted if necessary; (b) the use of a BREG stick to assist in abduction. It may often be helpful to supinate rather than use the pronated position.

can be performed and little equipment is required. One additional active assistive technique that is very helpful in regaining full abduction patterns is for the patient to forward flex with the hands interlocked and, rather than returning in a forward flexion pattern, to rotate the extremity into an abduction pattern and controllably lower through that range with the assistance of the opposite extremity (Fig. 15). This seems to be quite useful and helpful for many patients.

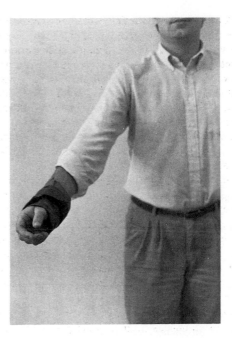

Fig. 13 It is frequently helpful to suspend the weight rather than allowing the patient to grasp the weight, thus minimizing abnormal muscular patterns. The scapular plane positioning (functional pattern) should be noted.

frequently used is to teach the patient self-mobilization/distraction patterns at 8 to 10 weeks after the operation (Fig. 14). This may be quite helpful in those patients who have had chronic rather than traumatic development of shoulder dysfunction, but it is most important in those patients tending to have impingement problems associated with rotator cuff lesions. Surgical tubing is used in these patients, as both concentric and eccentric activities

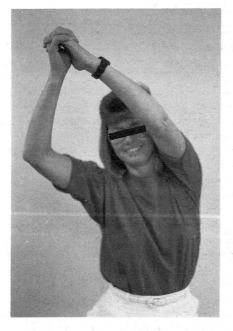

Fig. 15 Teaching the individual to 'roll' assistively from a forward flexed position into external rotation and abduction as he or she returns from the forward flexed position. The term to describe this to the patient may be to 'roll out and down' or 'roll out and over' as the arm is lowered.

We attempt to educate our rotator cuff repair patients into a 3-month, 6-month, 1-year sequence. We tell them that it will be 3 months before they begin to have functional ranges and control of those ranges, while at 6 months they will know how much range and control they can achieve. The most sobering information for them is that between 6 months and 1 year they will gain an additional 10 to 15 per cent of strength, and thus they will not know how well they will be able to do for 1 year.

This case study presents many common-sense activities that can be provided following rotator cuff injury and repair. One of the important factors is to warn patients (and the general public) of the danger of starting motors with a pull cord, as we find this to be a fairly common mechanism of injury. In this individual's case he did not wish to wait to see if he would be able to tolerate the limitations presented by his injury in a conservative approach, but many individuals should be allowed an opportunity to determine what would be their own natural course. In this case, it was his dominant extremity, and he was a very active individual and wished to have the best possible result. At 6 months he had essentially a full range of motion, and was swimming three times weekly. At 1 year assessment, he was asymptomatic and doing extremely well.

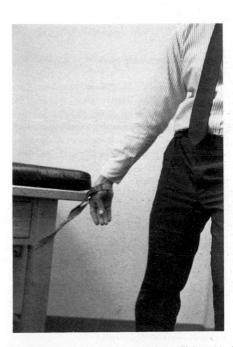

Fig. 14 Self-mobilization/distraction patterns utilizing a belt and a stabilizing object.

CASE STUDY 3

A 42-year-old university professor came to the clinic complaining of pain and stiffness in the left shoulder. His examination revealed the classic pattern seen with adhesive capsulitis: marked loss of abduction, external rotation, and forward flexion. He was also diabetic, which we have frequently seen to be related to this condition. He could not relate any acute episode of onset, but rather complained of a slow progression over 3 to 4 months.

As he preferred not to have an injection, he was provided with oral anti-inflammatory medication and started on an aggressive physical therapy programme which included a supervised sequence three times weekly to supplement his thrice daily independent programme. He was instructed in self-mobilization techniques (Fig. 14), moist heat was applied over and to the axilla in an attempt to reach the inferior capsular recess (Fig. 16), and an assistive–passive–active programme was utilized for forward flexion, abduction, and external rotation. These activities included a wall stretch (facing and away from the side) and utilizing a door frame for external rotation patterns. He also worked at multiple levels of rotation at different levels of forward flexion and abduction.

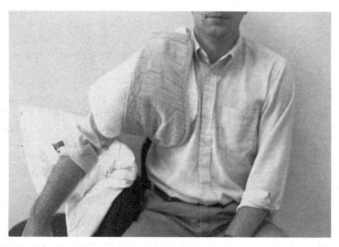

Fig. 16 In adhesive capsulitis the application of moist heat should be superior and inferior to the glenohumeral joint, thus reaching the axillary fold.

His supervised therapy sequence included ultrasound to the inferior fold/axilla, moist heat (as previously described), mobilization (inferior glide, glides in abduction, and rotational patterns), and stretching routines. After 1 month, the patient had achieved a very marked increase in his abduction (120° combined glenohumeral and scapulothoracic actions) and was able to forward flex actively approximately 140°. He was also using the forward flexion roll into abduction sequence quite effectively (Fig. 15). His external rotation had improved from 5–10° to approximately 35–40°. He was placed on a maintenance/supervised physical training sequence once a week for the next month, and continued using his thrice daily programme independently. His clinical progress reduced to a plateau but his functional level continued a very gradual improvement over the next few months.

This case study demonstrates the usual onset of adhesive capsulitis of the shoulder. It has an insidious onset, women present earlier than men as their activities of daily living demand greater activities of the hand behind the back, and it has an apparent link to other disease processes (diabetes etc.). It is imperative to educate these patients as to the control of their condition, and to make them aware that they may have this problem in the future. There is debate as to how aggressively this entity should be treated, but we have found conservative measures to be helpful in the majority of cases.

CONCLUSIONS

The great mobility and lack of inherent stability at the shoulder make it a very challenging complex for the rehabilitation specialist. We have recommended that the clinician becomes aware of the functional sequence of pain and uses this as a guide to the rehabilitation process. A series of principles interlinks the total process, with the final principle being that of functional rehabilitation leading return to the desired activity. Clinicians must learn to examine the specific actions required and the type of contraction imposed on the shoulder complex during shoulder activities. During rehabilitation concentric activity alone is frequently utilized when, in reality, eccentric activity is symptomatic and required (see Chapter 5 on tendon injuries). Finally the aim of the rehabilitation process must be function, and patients must become responsible for their ultimate return to the functional state.

REFERENCES

1. Kibler B. Scapular position in throwing athletes. Presented to American Orthopaedic Society for Sports Medicine, Traverse City, MI, 1989; and Kibler WB. Role of the scapula in the overhead throwing motion. *Contemporary Orthopaedics*, 1991; **22**: 525–32.
2. Kegerreis S, Jenkins L. In: Malone TR, ed. Sports injury management. Vol. 2, *Throwing injuries*. Baltimore: Williams & Wilkins, 1989.
3. Hollinshead WH. *Functional anatomy of the limbs and back*, 3rd edn. Philadelphia: WB Saunders, 1969.
4. Gould JA. *Orthopaedic and sports physical therapy*, 2nd edn. St. Louis: CV Mosby, 1990.
5. Hughston JC. Functional anatomy of the shoulder. In: Zarins B, Andrews, JR, Carson WG, eds. *Injuries to the throwing arm*. Philadephia: WB Saunders, 1985.
6. Soderberg GL. *Kinesiology: application to pathological motion*. Baltimore: Williams & Wilkins, 1986.
7. Bost FC. The pathological changes in recurrent dislocation of the shoulder. *Journal of Bone and Joint Surgery* 1942; **26**: 595–613.
8. Saha AK. Mechanics of elevation of the glenohumeral joint. Its application in rehabilitation. *Acta Orthopaedica Scandinavia* 1973; **44**: 668–78.
9. *Grant's Atlas of Anatomy*. 7th edn. Baltimore: Williams & Wilkins, 1978.
10. O'Brien SJ, Neves MC, Arnoczky SP, Rozbruch SR, DeCarlo EF, Warren RS, Schwartz R, Wiekiewicz TL. The anatomy and histology of the inferior glenohumeral complex of the shoulder. *American Journal of Sports Medicine* 1990; **18(5)**: 449–56.
11. Terry GC, Hammon D, France P, Norwood LA. The stabilizing function of passive shoulder restraints. *American Journal of Sports Medicine* 1991; **19(1)**: 26–34.
12. Jobe FW, Tibone JE, Perry J, Moynes D. An EMG analysis of the throwing shoulder in throwing and pitching. *American Journal of Sports Medicine* 1983; **11(1)**: 3–5.
13. Jobe FW, Tibone, JE, Perry J, Moynes D. An EMG analysis of the throwing shoulder in throwing and pitching: a second report. *American Journal of Sports Medicine*, 1984; **12(3)**: 218–20.
14. Malone T., ed. Sports injury management. Vol. 2. Throwing injuries, no. 4. Baltimore: Williams & Wilkins, 1989.
15. Kent BE. Functional anatomy of the shoulder complex. *Physical Therapy* 1971; **51**: 867–87.

16. Perry J. Shoulder anatomy in biomechanics. In: *Clinics in sports medicine—injuries to the shoulder in the athlete*. Philadelphia: WB Saunders, July 1983.

17. Smith RL, Brunolli J. Shoulder kinesthesia after anterior glenohumeral joint dislocation. *Physical Therapy* 1989; **69**: 106–12.

18. Davies, GJ, Gould JA, Larson RL. Functional examination of the shoulder girdle. *Physician in Sports Medicine* 1981; **9**: 82–102.

19. Hoppenfeld S. *Physical examination of the spine and extremities*. New York: Appleton-Century-Crofts, 1976.

20. Magee DJ. *Orthopaedic physical assessment*. Philadelphia: WB Saunders, 1987.

21. Falkel JE, Murphy TC. In: Malone TR, ed. *Sports injury management—shoulder injuries*. Baltimore: Williams & Wilkins, 1988.

22. Priest JD, Nagel DA. Tennis shoulder. *American Journal of Sports Medicine* 1976; **4**: 213–42.

23. Atwater AE. Biomechanics of overarm throwing movements and of throwing injuries. *Exercise and Sports Science Review* 1979; **7**: 43.

24. Voss DA, Ionta MK, Meyers VJ. *Proprioceptive neuromuscular facilitation: patterns and techniques*, 3rd edn. Philadelphia: Harper & Row, 1985.

25. Keggereis ST. The construction and implementation of functional progressions as a component of athletic rehabilitation. *Journal of Orthopaedic and Sports Physical Therapy* 1983; **4**: 14.

26. Rathbun JB, MacNab I. The microvascular pattern of the rotator cuff. *Journal of Bone and Joint Surgery* 1970; **52B(3)**: 540–53.

4.4.1 Ankle injuries—acute and overuse

JACK TAUNTON AND PETER A. FRICKER

Ankle injuries are extremely common: they occur in acute and chronic overuse situations. In this chapter we review the epidemiology and anatomy of ankle injuries. Overuse and acute injuries will be discussed, together with their mimics. Details of diagnosis and management will be included and the role of shoes, orthotics, taping, and braces in the prevention and management of ankle injuries will also be discussed.

EPIDEMIOLOGY

Ankle injuries are the single most common sporting injury, resulting in the greatest time lost from practice and games. Garrick[1] has performed some of the most extensive studies of the epidemiology of ankle injuries. He evaluated the injuries from four United States high schools during the 1973–4 and 1974–5 school years. He identified 3049 students participating in 19 sports with a total of 1197 injuries. The three most common sites were the thigh, knee, and ankle, representing respectively 14.6 per cent, 14.5 per cent, and 14.1 per cent of all injuries. The most common injury was the ankle ligament sprain which represented 11.8 per cent of all injuries. A total of 161 ankle injuries were observed, of which 83.9 per cent were ligament sprains, 6.2 per cent were muscle strains, 2.5 per cent were lacerations, and 2.5 per cent were fractures. The only identifiable difference between the males and females was in the strain injury. This was believed to be related to American football where 90 per cent of the musculotendinous strains occurred.

The sports most commonly resulting in ankle injuries were basketball, American football, soccer, and cross-country running for males, and cross-country running, basketball, badminton, and gymnastics for females. Basketball resulted in an injury rate of 13.1 per cent ankle injuries per season for the males, and 11.4 per cent for the females. The rates for football were 10.9 per cent, and those for cross-country running were 7.8 per cent for males and 11.5 per cent for females.

Garrick[1] also evaluated 11 141 sports injuries sustained from 1980 to 1984 inclusive. Figure skating, basketball, and soccer produced the greatest number of ankle injuries. Ankle sprains were the most common injuries seen in basketball and soccer, whereas other overuse injuries (often associated with excessive boot pressure) were the most common in figure skating. A significant number of ankle injuries were also reported in gymnastics and ballet. In gymnastics the most common cause of injury was repetitive forced dorsiflexion of the ankle from 'landing short' in practice, producing anterior impingement injuries. In ballet, dancing in a hyperplantar flexed position resulted in posterior impingement injuries including os trigonum stress.

Overuse ankle injuries are commonly seen in our clinic at the University of British Columbia. Clement *et al.*[2] reported a retrospective survey of the clinical records of 1650 runners seen over a 3-year period. The lower leg and foot accounted for 46.7 per cent of the total injuries. The most common injuries in the ankle region were distal tibial stress fractures, tibialis posterior tendinitis, and peroneal tendinitis, which were the eighth, ninth, and tenth most common individual injuries seen. These injuries accounted for 2.6 per cent, 2.5 per cent, and 1.9 per cent respectively of all the injuries documented. Aetiological factors fell into five categories: training errors, lack of strength or flexibility, biomechanical and anatomical factors, poor running shoes, and poor training surfaces.

We have recently updated this study with a clinical study of 4175 runners seen over a 4-year period (1985–88).[3] These athletes were grouped as recreational, marathon, and middle-distance competitive runners. The pattern of injuries had changed in the time between the two studies with a higher proportion of knee injuries and a relatively lower frequency of lower leg, ankle, and foot injuries. Much of this change appears to be attributable to footwear improvements. Bony stress injuries were more common among the middle distance runners, possibly because of more intense training. The highest proportion of ankle overuse injuries were reported from cycling, ice skating, ballet, and running with added speed and interval sessions. In addition, this group carried out more of their training using spikes and racing flats with lower shock-attenuation and stabilization capacity.

Walter *et al.*[4] reported that 22 per cent of all running related injuries affected the foot and ankle. Marti *et al.*[5] reported a higher incidence of foot and ankle injuries in their studies of runners. They found that 40 per cent of the total injuries were to the foot and ankle, with some 15 per cent being lateral ankle ligament sprains. Macintyre *et al.*[3] found that only 2.1 per cent of the total injuries were ankle sprains. Garrick and Requa,[6] in their investigation of a huge series of ankle injuries over 7 years, found that 27.6 per cent were due to overuse, with the highest proportion being due to cycling, ice skating, ballet, and running.

Stress fractures are now commonly seen in sports medicine clinics. We analysed 320 athletes with bone scan positive stress fractures seen over a period of 3.5 years.[7] The most common site of stress fracture was the tibia (49.1 per cent), followed by the tarsals (25.3 per cent), metatarsals (8.8 per cent), femur (7.2 per cent), and fibula (6.6 per cent). Tarsal stress fractures were seen in older athletes, and a history of trauma was significantly more common in injuries to the tarsal bones. Tarsal stress fractures also took the longest time to diagnosis (16.2 weeks) and to recovery (17.3 weeks). The average time to diagnosis for all stress fractures was 13.4 weeks and the average time to full recovery was 12.8 weeks. Pronated feet were most commonly found in tibial stress fractures and tarsal bone fractures, and cavus feet were most frequently seen with metatarsal and femoral stress fractures. Running was by far the most common activity involved, providing 221 of the 320 cases, with fitness classes the next most commonly involved, accounting for 25 cases.

ANATOMICAL CONSIDERATIONS

The ankle joint is a ginglymus or hinge joint. Dorsiflexion and plantar flexion occur at the ankle joint and the range of motion varies from 50° to 90°. The ankle joint comprises the distal tibia and fibula which form the mortise for the articulation with the talus. The transverse axis of the ankle joint is slightly oblique.[8] The mortise itself is formed by the articular facet of the distal tibia and its extension the medial malleolus plus the articular facet of the fibula with its extension, the lateral malleolus, which is longer than the medial malleolus. The wedge-shaped talus is wider anteriorly, which results in maximum stability in dorsiflexion but vulnerability to injury in plantar flexion. Weight-bearing transfer of force occurs from the talus to the foot. The talus is positioned on the anterior two-thirds of the calcaneus. The anterior head of the talus is rounded and articulates with the tarsal navicular. The navicular also articulates with the medial column of the foot comprising the three cuneiforms and the first to third metatarsals. The calcaneus articulates with the cuboid and the fourth and fifth metatarsals to comprise the lateral column of the foot.

The ankle joint has a capsule which Akesson described as being thin both anteriorly and posteriorly.[9] The capsule is reinforced by both medial and lateral collateral ligaments. The lateral ligament complex comprises the anterior talofibular ligament, the calcaneofibular ligament, and the posterior talofibular ligament. The strong medial ligament is also known as the deltoid ligament and can be divided into the anterior tibiotalar, tibionavicular, tibiocalcaneal, and posterior tibiotalar ligaments.

The ankle mortise is stabilized by very strong anterior and posterior tibiofibular ligaments. These are supported by the inferior transverse and interosseus ligaments forming the distal tibiofibular syndesmosis.

The subtalar joint is the site of the bulk of the inversion and eversion movements of the foot. This joint is supported by lateral, medial, and posterior talocalcaneal ligaments. There is also an interosseus talocalcaneal ligament within the sinus tarsi.

The functional anatomy of the ankle ligaments is summarized in Gray.[8] Briefly, the deltoid ligament is so strong that a bony avulsion fracture usually occurs with ankle eversion before significant ligament damage is done. The middle portion of the deltoid liga-

ment, in combination with the calcaneal fibular ligament, control ankle eversion and inversion. The anterior and posterior aspects of the deltoid ligament limit plantar flexion and dorsiflexion of the foot. In addition, ankle abduction is limited by the anterior fibres. The posterior talofibular ligament of the lateral complex assists the calcaneal fibular ligaments in controlling posterior displacement of the foot. Anterior displacement is controlled primarily by the anterior talofibular ligaments. The anterior talofibular ligament also limits plantar flexion.

Ankle movement is a result of muscle actions that have origins from the distal end of the femur, the tibia, and the fibula, and which are applied to the foot. Dorsiflexion of the ankle is accomplished by the muscles of the anterior tibial compartment. These muscles are the tibialis anterior, extensor digitorum longus, and extensor hallucis longus. The last two muscles also extend (dorsiflex) the toes. This muscle compartment is innervated by the deep peroneal nerve. Plantar flexion of the foot and ankle is effected by the muscles of the superficial calf compartment: the gastrocnemius, soleus, and plantaris. Their insertion is by a common tendon, the Achilles tendon, attached to the posterior calcaneus. There is a retrocalcaneal bursa which separates this tendon from the calcaneus proximal to its insertion. These muscles are supplied by the tibial nerve. The deep posterior calf compartment has three muscles which act upon the foot and the ankle. The tibialis posterior acts as an inverter and plantar flexor of the foot. Its tendon travels in a tunnel over the posterior aspect of the medial malleolus and is held in place with the tendons of flexor digitorum longus and flexor hallucis longus by the flexor retinaculum, thus forming the tarsal tunnel. Also travelling in this tunnel is the tibial nerve which innervates the deep posterior compartment. This can be a site of nerve entrapment. The nerve then divides into the medial and lateral plantar nerves after it passes deep into the abductor hallucis muscle. The popliteus is the fourth muscle of the deep posterior compartment and is proximal within this compartment.

The lateral compartment of the lower leg everts the foot. This compartment is innervated by the superficial peroneal nerve and comprises the peroneus longus and peroneus brevis muscles. Their tendons travel in the posterior groove of the lateral malleolus and are held in place by the superior peroneal retinaculum. The peroneus brevis inserts into the tubercle on the base of the fifth metatarsal bone and the peroneus longus passes through a groove on the plantar aspect of the cuboid bone to insert into the base of the first metatarsal and on to the medial cuneiform.

The sensory nerve supply to the anterior aspect of the lower leg, ankle, and foot is shown in Fig. 1. The vascular supply to the ankle is derived from the malleolar branches of the anterior tibial and peroneal arteries.

BIOMECHANICS OF THE ANKLE

The action at the ankle is primarily that of dorsiflexion and plantar flexion, with inversion and eversion being primarily a function of the subtalar joint. A discussion of the biomechanics of the ankle must include those of the foot, particularly when considering the gait cycle. A knowledge of normal and abnormal gait is important in understanding overuse injuries in particular. A full discussion of the biomechanics is beyond the scope of this chapter, however, and the reader is referred to the literature.[10–12]

Nuber[13] pointed out that the obliquity of the ankle's motion permits dorsiflexion and plantar flexion together with other motions of the foot. Lateral deviation and pronation are associated with dorsiflexion of the ankle, while medial deviation and supi-

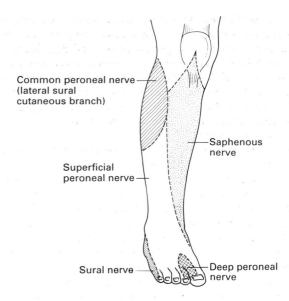

Fig. 1 Sensory supply to anterior aspect of lower leg. (Redrawn from ref. 50 with permission.)

nation accompany plantar flexion. In reality, pronation is a complex motion of the ankle, subtalar joint, and midtarsal joints producing dorsiflexion, eversion, and abduction of the foot.

During running, the gait cycle is that of a stance (or support) phase alternating with a swing phase. Slocum and James[14] were among the first in the orthopaedic literature to describe the biomechanics of running and point out the differences from walking. They divided the stance phase into foot contact, mid-support (or full weight bearing until plantar flexion starts), and finally toe-off. Mann[15] has studied in detail the function of the ankle during running. He has shown that the foot is dorsiflexed some 10° at heel strike, and then continues to dorsiflex through half the stance phase to a total of 20°. During toe-off there is a rapid plantar flexion to approximately 25°. This cycle is repeated through some 1000 footstrikes per mile (675 per kilometre).

Mann *et al.*[16] performed an electromyographic study of the actions of the muscles during running. They showed that the tibialis anterior muscle is active at heel strike and continues to be active through the first half of the stance phase, primarily contracting concentrically. The anterior group is active during 75 per cent of the swing phase associated with dorsiflexion. The gastrocnemius and deep calf muscles function during the first 60 per cent of the stance phase and the last 25 per cent of the swing phase. The gastrocnemius is functioning eccentrically in the early part of the stance phase contributing to ankle stability. Mann believes there is minimal contribution of the deep calf muscles at toe-off, with the gastrocnemius terminating their activity with the beginning of ankle plantar flexion. The peroneal muscles contract during the mid-support phase of stance and stop firing just prior to toe-off.

As the foot descends at the end of the swing phase, it is in a supinated position. It then pronates through mid-stance to supinate prior to toe-off. Pronation of the foot after heel strike, associated with ankle dorsiflexion, is an important component of shock absorption. This action occurs in combination with knee flexion and hip extension and absorbs the ground reactive force of some 2 to 2.5 times body weight with each heel strike. The complex action of pronation also enables the foot to adapt to the surface upon which it is walking or running. Resupination of the

foot prior to toe-off is vital to ensure a rigid lever for propulsion. Associated with pronation of the foot is internal tibial rotation. This early portion of the stance phase is also the beginning of knee flexion and hip extension. During supination in the last half of the stance phase knee extension and further hip extension occurs. The supinated or pes cavus foot which does not pronate sufficiently results in poor shock absorption. The restricted subtalar motion in the cavus foot transfers more stress of rotation to the ankle joint. In contrast, excessive pronation results in greater ankle dorsiflexion, subtalar eversion, and forefoot abduction. In addition, there is excessive internal tibial rotation which can result in tibial stress syndrome and patellofemoral pain syndrome. Toe-off is also less effective in the pronated foot as the foot is not locked and this can produce excessive strain on the plantar fascia.

INJURY PREVENTION

As stated, ankle inversion injuries are extremely common and result in considerable loss of time from training and competition. Preventive measures include exercise programmes, adhesive taping, bracing, and shoe modifications. Vast amounts of money are allotted each year to university, national, and professional teams for materials and medical staff in an attempt to minimize ankle injuries.

Exercise programmes

An ankle injury prevention programme has been developed by Dale Ohman, Head Coach of the Men's Varsity Volleyball Team at the University of British Columbia.[17] In response to a large number of ankle inversion injuries in one season he developed a very practical programme which requires 10 min of explanation and demonstration plus practice during the first week of training. Thereafter, 5 min per practice session is required to implement and maintain the programme. The essential elements are flexibility and balance drills for proprioception training, with some strengthening components.

There are four specific exercises in his programme. The first requires the athlete to balance on one leg while bending forward with the head up and the back horizontal. The arms are spread out to the side, parallel to the floor. The other leg is held extended at the hip and parallel to the floor. The athlete balances for 20 s initially and progresses to 30 s on each leg; the standing (support) leg is alternated in each training session. These features are similar for the remaining three exercises. The second exercise requires the athlete to balance on one foot, and then to side-bend at the hip with the free leg parallel to the floor and the arms perpendicular to the floor. In the third exercise the athlete balances on one leg, leaning backwards on a slightly bent knee with the free leg raised in front parallel to the floor. Both arms are held out sideways and parallel to the floor. In the fourth exercise the athlete balances on the same hand and foot. The free arm is raised perpendicular to the floor and the free leg is abducted as high as possible. Since this programme was introduced, there has been a significant reduction in the number and severity of ankle sprains.

Gastrocnemius and soleus muscle stretching and specific ankle inversion, eversion, and dorsiflexion exercises can be added to this (essentially balance) programme to provide a comprehensive preventive programme.

Proprioceptive training on a wobble board, used as a balance board, is also useful. The exercise programme can employ simple resistance devices such as surgical tubing or a bicycle inner tube. A programme of exercises progressing to three sets of 20 repetitions is effective and develops strength and endurance.

Taping and bracing

Garrick and Requa[18] and Emerick[19] have documented the effectiveness of ankle taping in the prevention of ankle sprains. Fumich et al.[20] took these studies a step further by evaluating the role of taping the ankle in neutral inversion and in inversion with plantar flexion—the most vulnerable position. Their taping system yielded good control in both positions. A major problem with taping is the known loss of effectiveness after approximately 20 min of exercise.[20,21] This may be a result of loss of skin adhesion or mechanical failure of the tape or both.

Bunch et al.[22] utilized the ankle support made of a polyurethane foot form to which an inversion torque was applied to study the effective control of taping, cloth wrapping, and five different lace-on braces. In addition, they studied the loss of control after 350 inversion cycles. They found that initially adhesive tape offered the best support, with a 25 per cent advantage over the two best lace-on braces and a 75 per cent improvement over cloth wraps and two lace-on braces. There was a significant difference between the control offered by the lace-on braces studied, with the Mikros 9 inch and Swedo-O types being the best. After 350 inversion cycles over a 20-min period, there was no significant difference in the control offered by adhesive taping and the two best braces. They utilized a low-cut Oxford style shoe in this study to reduce the potential additional control of a high cut shoe. The tape lost 21 per cent effectiveness over the 20-min inversion stress period. The taping technique used 3.8 cm (1.5 inch) tape with the Gibney basket weave and heel lock method. It was interesting to note that in the preliminary laboratory studies over one-third of the control was lost as a result of inexperienced ankle taping.

More recently, Rovere et al.[23] reported on a 6-year retrospective comparison of ankle taping and a laced ankle stabilizer in the prevention of ankle injuries in collegiate football. During the study period the athletes used high-cut or low-cut boots. Ankle taping was found to be much less effective in preventing ankle injuries and reinjuries. The cost of a pair of the ankle-stabilizing lace-on braces was $32, and they used two pairs per player for the season. The cost of the tape was $400. Interestingly, the best combination to prevent injuries was a low-cut shoe with the ankle stabilizers. No significant difference was noted between the high-cut shoe with a stabilizer and the low-cut shoe with tape.

The use of an Airstirrup in the prevention and treatment of ankle inversion sprains has recently come into favour. Kimura et al.[24] studied subtalar ankle inversion on a specifically designed inversion platform, with and without an Airstirrup ankle-training brace, using high speed cinematographic techniques. Significant control of ankle inversion was seen in the 18 subjects. A mean reduction of ankle inversion of 9.8° was seen. Gross et al.[25] compared the Airstirrup and ankle taping in providing ankle support before and after exercise. Twenty-two ankles were randomly treated with either tape or the Airstirrup, and then the subjects exercised for 10 min on a 5 m × 10 m figure-of-eight course followed by 20 toe raises on a 15.25 cm step. The ankles were then tested on the Cybex II goniometer measuring three trials of maximum inversion and eversion. The results indicated that both treatment conditions significantly reduced ankle motion, but the Airstirrup produced a significantly greater restriction of range of motion, particularly eversion, than taping in pre-exercise testing.

Post-exercise analysis again showed a significant loss of control with taping but not with the Airstirrup.

An ankle brace (Don Joy ankle ligament protector), described as a semirigid orthosis, has been compared with adhesive ankle taping before, during, and after a 3-h volleyball practice.[26] Fourteen ankles were treated with both methods of support. Maximal losses in taping restriction were noted for both inversion and eversion of the ankle at 20 min into exercise. The orthosis (ankle brace) lost restriction on eversion after 3 h but showed no loss of restriction of inversion over this period. Neither support system had a significant effect on jumping ability.

The authors are not aware of any long-term prospective and retrospective comparison of Airstirrups, ankle taping, or lace-on braces.

Orthotics

A study of the control features of a lateral ankle brace system attached by Velcro straps to the ankle and to a semirigid foot orthotic designed to keep the ankle and subtalar joint in neutral is now under way in Canada. This brace has been utilized for both the men's and women's national field hockey teams with good clinical results in reducing ankle inversion sprains. This system has been particularly useful for athletes with generalized ligament laxity, excessive foot pronation, and a history of recurrent ankle inversion sprains. The semirigid orthotic can be modified with a deeper heel seat to offer more control. A number of reports have documented the effectiveness of soft and semirigid orthotics in reducing foot and ankle injuries.[2,27,28]

Taunton et al.,[29] using a triplanar electrogoniometer, investigated the effectiveness of semirigid orthotics in reducing pronation in a group of endurance runners. They demonstrated a decrease in calcaneal eversion during the support phase of running. They also noted a significant difference in the response of the right and left legs in many subjects, suggesting the need for 'custom' orthotics for each foot.

Bates et al.[30] had previously reported a reduction in the period of pronation and the maximum extent of pronation with the use of orthotics. Smith et al.[31] utilized high speed cinematography to evaluate pronation control of soft and semirigid orthotics. The maximum velocity of calcaneal eversion was significantly reduced by both orthotics, but the maximum amount of calcaneal eversion was only significantly reduced with the semirigid orthotics. McKenzie[32] stated that as soft orthotics are clinically effective, controlling the velocity of eversion may be more important than controlling the degree or amplitude of eversion.

Orthotics are designed to achieve biomechanical control, and with the pronated foot this means neutralizing the subtalar point and maintenance of the longitudinal arch. In mild degrees of pronation a motion control shoe or soft orthotic may be sufficient. For more severe pronation, semirigid devices are made from varying grades and thicknesses of plastics. With sports involving rapid lateral changes of direction, we limit the rear-foot posting to 4° to prevent inversion sprains. We measure the degree of subtalar varus,[27,33] subtract 3°, which we have found can be well tolerated, and then post half this value. In addition, we measure the additional forefoot varus and then subtract the tolerable level of 2 to 3° and post an additional half of this value to the forefoot. If there is a significant degree of tibial varum (over 5°), we take this into consideration in the rearfoot posting. As mentioned, caution is used to prevent overcorrection which can result in inversion ankle sprains and lateral knee and leg pains (including iliotibial band friction syndrome and peroneal tendinitis). While waiting for an orthotic to be constructed, foot pronation can often be controlled with low dye arch taping.

As previously described, the cavus foot suffers from decreased motion of the subtalar joint and hence insufficient pronation and poor shock absorption. Orthotics for the cavus foot are most effective if made of a soft material with good shock absorptive properties and flexibility. Heel lifts to unload the tight gastrocnemius and soleus muscles are important. The cavus foot often has a subtalar varus and forefoot valgus. The forefoot is treated with a lateral or valgus post and the rearfoot varus with a modest rearfoot varus medial post. The cavus foot is often associated with a fixed or partially fixed plantar flexed first ray. This can be similarly treated with a posting under the second to fifth metatarsal heads. The thickness of the posting or padding should be equal to the degree of plantar flexion of the first ray. Lutter[34] used a flexible longitudinal orthotic with a lateral heel lift.

Shoes

Garrick and Requa[6] completed a prospective study of ankle inversion sprains in intramural basketball players. As previously discussed, they reported a lower incidence of ankle injuries among those players with prophylactic adhesive taping compared with those without any taping. They also found that the risk of ankle injury was lowered in patients wearing high-cut shoes alone. The best combination for reducing ankle injuries was high-cut shoes and ankle taping. This was not seen in a retrospective study of football ankle injuries[23] where it was found that the combination allowing the fewest ankle injuries was low-cut shoes with a lace-up ankle stabilizer. The low-cut shoe with an ankle stabilizer was better in reducing ankle injuries than the high-cut shoe with a stabilizer. The poorest combination in their study was the high-cut shoe with tape. They felt that it was easier to tighten the lace-up brace when wearing a low-cut shoe, and this may explain their surprising data on shoe type and injury. A review of the newer supposedly more stable high-cut shoes being produced would be useful. The relationship of this footwear to prevention of ankle injury needs to be determined together with the possible benefits of simultaneous ankle taping or bracing.

Overuse injuries have been attributed, at least in part, to foot type, particularly of either excessive pronation or supination. We now see shoes specifically designed for these two extremes. McKenzie et al.[28] and, more recently, Moore and Taunton[35] have detailed new design features in athletic shoes.

Excessive pronation is partially controlled by the 'motion control' shoes produced by the reputable shoe manufacturers. Such shoes have a straighter last with board-lasting rather than slip-lasting. Straight-lasted shoes reduce pronation whereas curve-lasted shoes promote pronation and hence are more suitable for the rigid pes cavus (supinated) foot. With board-lasting offering more torsional rigidity, the upper material of the shoe is attached to the firm fibreboard inner sole. In contrast, with slip-lasting the upper is stitched in a 'moccasin' fashion down the middle of the sole, with no firm inner sole, and is then attached to the premoulded middle sole portion. The 'motion control' shoe also possesses a high density thermoplastic heel counter which may be reinforced externally by additional material. The stable heel counter is the key to control of calcaneal eversion. Additional support can be added by medial strapping in the form of leather cradles in the midfoot. The midsole is constructed over shock-absorptive ethylene vinyl acetate (EVA) of various densities (or hardnesses, as

measured by a durometer); more recently shoe manufacturers have returned to the more durable polyurethane for the midsole, sacrificing some shock absorption but gaining more durability and stability. Shock absorption is maintained by the inclusion of rearfoot, and often forefoot, air pockets, pillars, or gels. In some shoes more medial pronation control has been attempted by additional higher density EVA medially in the dual-density concept of midsole construction. We have seen more iliotibial band friction injuries as a result of these shoes where there was an extreme difference in the medial–lateral densities of the EVA, with the lateral side collapsing.

The width and shape of the sole and rearfoot can also modify motion control. In the early 'motion control' shoes very wide heel flares were used, which controlled ankle motion but permitted rapid calcaneal eversion upon heel strike with more internal tibial torsion and reportedly more patellofemoral pain. Nigg[12] has shown that initial pronation can be reduced by a medial support applied more to the posterior side of the medial arch with a round heel shape ('negative heel flare') on the lateral and posterior sides of the heel. Frederick et al.[36] have shown that the best combination of rearfoot control and cushioning occurred in a shoe with a thicker midsole, with a durometer reading of 35, and a rearfoot flare with a reading of 15. With motion control shoes the challenge is to achieve a balance between torsional (pronation and supination) control and shock absorption. The latter is often lost in the attempt to gain more pronation control.

The cavus foot needs more shock absorption and features to accentuate pronation. Hence a curved last with slip-lasting construction is best. The midsole should have a softer EVA with durometer readings in the range of 25 to 30 and a narrower heel flare. A higher heel lift and additional shock absorption insole made of Neoprene, or possibly the newer viscoelastic materials Akton or Zekon, are useful. The air soles in the rearfoot and particularly the forefoot, as seen in Nike shoes, are also attractive features for the cavus foot. Nigg[12] has shown that initial pronation can be significantly increased by using more flare on the lateral and/or posterior side of the heel. In addition, pronation is increased by relatively hard sole material laterally and softer material medially. He has also shown that the geometry of the shoe influenced the ankle and foot alignment on take-off. Oversupination during take-off was reduced by a lateral wedge in the forefoot.

More recently, biomechanical research has been applied to achieve more ankle stability for tennis, basketball, volleyball, and aerobic shoes.[12] Aerobic shoes require the combination of forefoot shock absorption with lateral ankle support. Soccer and the various other football boots offer some ankle support but have poor pronation control features. The recent provision of removable insoles, a feature in some hockey skates and hiking boots, makes it much easier to employ a soft or semirigid orthotic to suit the wearer.

PAEDIATRIC ANKLE PROBLEMS

Trott[37] reported that 12.1 per cent of all adolescent injuries seen were in the foot and ankle. Micheli[38] and Beauchamp[39] have published excellent reports on the common paediatric foot and ankle injuries. They relate these injuries to congenital and growth factors and to specific acquired traumatic and overuse injuries. Ankle injuries and dysfunction related to congenital factors are seen with club foot (talipes), excessive internal or excessive external tibial torsion (equinus), contracture, and subtalar coalition.

The club foot (or talipes equinovarus) is unstable laterally and is vulnerable to inversion sprains of the ankle and recurrent injury to the anterior talofibular ligament. Protective ankle taping is recommended for the club foot with minor degrees of varus positioning.

Excessive external tibial torsion can result in a valgus heel and excessive foot pronation. As a result of hereditary equinus contracture, the walker also develops an unstable ankle and subtalar joint with an increased risk of ankle inversion sprains. Gastrocnemius–soleus stretching and a more stable shoe are helpful. Beauchamp[39] also points out that tight Achilles tendons are seen with pronated feet and standing heel valgus in children, and these respond well to calf stretching and a foot orthotic. Night splints, occasional casting in maximum ankle dorsiflexion, and, rarely, Achilles tendon lengthening may be needed.

Subtalar coalition by either a bony or fibrous bridge between two tarsal bones can produce ankle and foot pain. Clinically, subtalar motion is markedly reduced and the affected child often has a spastic pronated pes planus foot. Peroneal spasm can be seen with subtalar motion testing. The diagnosis is made with oblique radiography of the foot, but a CT scan may be required to identify the coalition fully. A flattened talar dome is an associated finding. Treatment ranges from semirigid orthotics, to cast immobilization from 4 to 6 weeks to reduce the pain and spasm, to surgical division of the coalition and other procedures as necessary.

Foot and ankle complaints can be seen with variance of lower extremity alignment. One example is that of persistent femoral neck anteversion with internal femoral rotation, which is associated with excessive external tibial torsion and excessive foot pronation. This complex has been termed the miserable malalignment syndrome by James et al.[27] Similarly, persistent genu valgum can produce excessive foot pronation with resultant medial ankle pain. In the developing child, physiological genu valgum occurs at 3 to 4 years, with a more neutral knee alignment developing by age 6 to 7 years.

The ankle is similar to the metatarsals, knees, hips, and elbows, and, during growth, can develop osteochondritis dissecans which is seen in the talus; although its aetiology is uncertain, repetitive microtrauma may be a factor. Management is initially conservative provided that there is an adequate remaining growth period. The use of a functional cast for 6 to 8 weeks with maintenance of fitness by cycling or swimming is often enough to provide for resolution of the problem. If pain persists, especially with episodes of catching or locking, a CT scan is indicated to look for a loose fragment within the ankle joint. An unstable fragment is best managed by transarthroscopic fixation followed by continuous passive motion therapy over 6 to 12 weeks.

Epiphyseal ankle injury is common with an ankle inversion injury in the child. Salter and Harris[40] have classified epiphyseal injuries and have outlined their management with emphasis on proper reduction. Rotational injuries about the ankle can injure epiphyses and the distal tibia. Salter I and II fractures from supination and ankle plantar flexion are examples. Initially, local swelling may be minimal, but marked tenderness with rotational injuries must lead to a high degree of suspicion, even in the face of 'normal' radiographs. A below-knee cast and a repeat radiograph in 2 weeks is appropriate in such cases. The common Salter II fracture of the distal tibia is managed by closed manipulation (often under anaesthesia) and 6 weeks of immobilization (initially in a non-walking cast for 2–3 weeks). Growth abnormalities are unusual.

Salter III and IV fractures, if not managed properly, can lead to premature epiphyseal closure or malalignment. Good reduction is

essential, and open reduction and internal fixation may be necessary. A unique fracture seen in the adolescent with closing epiphyses is the Salter III or Tillaux fracture. The medial aspect of the distal tibial epiphysis is closed and, with an external rotation force, the inferior tibiofibular ligament avulses the anterolateral quadrant of the ankle complex. Reduction is achieved by internally rotating the foot, but open reduction and internal fixation avoiding the epiphyseal plate, are mandatory for any displacement over 2 mm.

The most common fractures in children are the Salter I and II fractures of the distal fibula, seen with supination–inversion ankle sprains. These can be adequately treated with closed reduction and 6-weeks' cast immobilization, initially non-weight-bearing for 3 weeks, with a total healing time of 6 weeks. In the adolescent it is often difficult to distinguish the inversion sprain with a tender anterior talofibular ligament from a Salter I fibular fracture. In the case of fracture tenderness is usually directly over the tip of the fibula, rather than more anterior or inferior over the sinus tarsi as in the case of the ligament sprain. Ankle sprains can be managed with the use of ice, elevation, and an Air-Cast ankle brace with rapid early mobilization. Protected return to sports with the Air-Cast yields good results.

ACUTE INJURIES

Ankle fractures

Chandler[41] presents a clear outline of the management of ankle fractures based on the Danis–Weber classification. This classification system is related to the fracture morphology (with emphasis on the fibula) and divides the fractures into types A, B, and C according to the three zones of the fibula (Fig. 2) A type A fracture is below the level of the tibiotalar joint, and often disrupts the talofibular articulation and is associated in many cases with medial ligament disruption. Simple non-displaced type A fibular fractures can be managed with 6 weeks of cast immobilization or functional bracing, with weight bearing permitted in the last 3 weeks. After this time, aggressive rehabilitation is recommended with emphasis on ankle strength, flexibility, and balance–proprioceptive exercises.

Displaced fractures in the athlete are best managed with open reduction and internal fixation allowing early rehabilitation. Chandler[41] states that a low transverse fracture of the type A variety can be managed with a tensor band technique or an intramedullary fixation of the fibula. Medial soft tissue damage may also need to be repaired. A type B fracture of the ankle is at the tibiotalar joint and involves partial disruption of the tibiofibular syndesmosis. In addition to initial radiography with anteroposterior, lateral, and oblique views of the mortise, these more complex fractures are more fully evaluated by CT scans. Type B fractures require fixation with lag screws, or lag screws and neutralization plating, compression plating, or neutralization plating and bone grafting. With these fractures Chandler[41] advised a medial incision for joint debridement of chondral or osteochondral talar loose bodies and deep medial ligament repair before fracture fixation. The type C fracture is defined as being proximal to the tibiofibular syndesmosis with disruption of the syndesmosis and the interosseous ligament to the level of the fibular fracture. There is often extensive ligamentous damage. These, together with pilon distal tibial fractures, are amongst the most severe ankle fractures and have the poorest prognosis. Internal fixation techniques are similar to those for the type B fracture. The mortise is

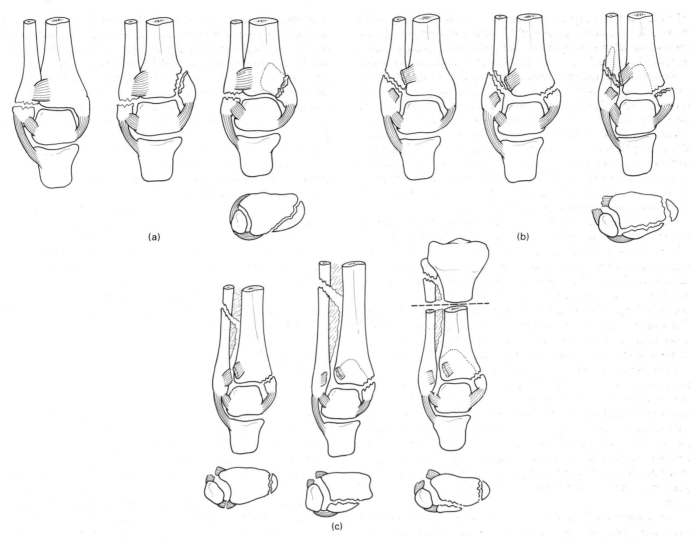

(a)

(b)

(c)

Fig. 2 The Danis–Weber classification of ankle fractures. (a) Type A. (b) Type B. (c) Type C. (Redrawn from ref. 41 with permission.)

initially stabilized by medial exploration and lateral plate fixation. The medial ligament repair and medial bony stabilization of the fracture are completed with a single cortical screw. Stabilization of the distal tibiofibular syndesmosis is achieved if the screw is placed above the level of the fibular sulcus of the tibia. Chandler[41] advised removal of this screw at 6 to 8 weeks, before weight bearing takes place.

Following internal fixation, these fractures are supported in a posterior splint and a U-splint. Elevation of the leg is needed for 2 weeks initially, and non-weight-bearing mobility with the aid of crutches is started on the second postoperative day. Early ankle and toe dorsiflexion exercises are commenced on the first post-operative day. At the sixth week weight bearing is permitted with a return to sports at about the end of the third or fourth month after surgery. Pool running and stationary cycling to maintain cardiova-scular fitness can be resumed after satisfactory wound closure at 3 weeks. As previously mentioned, full strength, flexibility, and balance must be regained prior to return to sport. In addition, sport-specific skills must be performed before practices are allowed. For a basketball or soccer player this should involve backward running and 'crossover' and 'cutting drills' with repeated changes of direction. If there has been extensive soft tissue damage

and repair, an external brace or splint is advisable for the first year until full ligamentous healing has taken place.

Ankle sprains

As discussed, ankle inversion sprains are the most common sports injuries. The majority of ankle sprains are of the inversion variety, occurring when one ankle is plantar flexed and the foot is supinated, which is the position of least bony stability for the ankle with support achieved by ligament and musculotendinous func-tion. In this position the peroneal muscles and anterior talofibular ligament are the prime stabilizers of the joint, and McConkey[42] emphasizes that in this position the axis of the anterior talofibular ligament is much closer to the long axis of the fibula. In addition, in the plantar fixed position the tension in the anterior talofibular ligament maximally increases as the attachment sites of the liga-ment from the fibula and talus separate. This results from the location of the centre of rotation of the ankle at a point anterior and inferior to the lateral malleolar origin of the ligament.

As well as these functional anatomical considerations, other aetiological factors have been outlined for inversion ankle sprains. Walsh and Blackburn[43] discussed the roles of uneven playing surfaces, poor shoes with inadequate heel counters, and conditions of increased shoe–surface reactive forces such as newly polished floors and artificial turf. Other factors involved are poor peroneal muscle strength, inadequate gastrocnemius–soleus flexibility, and poor balance and proprioception. The pes cavus foot and ankle, with increased forefoot valgus, is at greater risk of injury as the forefoot makes early contact medially (at the first metatarsal head) and not laterally. Athletes with generalized ligament laxity (subtalar, talar, and midtarsal laxity) are also more at risk of inversion sprains.

Classification and diagnosis

Ankle inversion sprains are usually classified from grade I to grade III. Grade I sprains are mild in nature with little swelling, mild tenderness, slight pain on inversion stress, and only mild loss of function. The ankle is stable to an anterior drawer test or radiographic examination for 'talar tilt' with only minor ligament damage. From grade I to grade III ankle sprains there is a progression of injury, first to the anterior talofibular ligament (grade I), then involving the calcaneus fibular ligament (Grade II), and with more severe inversion stress (grade III) including the posterior talofibular ligament.

Although grade I sprains are primarily to the anterior talofibular ligament, with greater degrees of force and more rotational stress the injury may involve the distal tibiofibular syndesmosis and medial (deltoid) or subtalar ligaments. Grade II sprains are of moderate degree with more functional loss and diffuse swelling. Walking and hopping are difficult. Grade III sprains are severe with complete lateral ligament complex rupture, consequent instability, and marked functional impairment. Walking is impossible, and of course there is marked swelling and tenderness. The active range of motion of the ankle is markedly reduced and there is gross instability.

Ankle stability is assessed clinically with the anterior drawer test. This is a sensitive indicator of the stability of the anterior talofibular ligament in particular. The anterior drawer test is performed with the ankle in the neutral position and is compared with the non-injured ankle. Laurin and Mathieu[44] have evaluated the anterior drawer test of up to 4 mm with the ankle in the neutral position, and this is reduced with the ankle in the plantar flexed position. A visible and palpable drawer, which is not reduced by plantar flexion and is greater than the non-injured ankle, indicates complete anterior talofibular ligament rupture.

Clinical diagnosis is aided by radiography to rule out fracture, avulsion injury to bone, osteochondral injury, or syndesmosis separation. Controversy still exists regarding the value of 'talar tilt'; an inversion stress to the ankle, 5 to 10° greater than the non-injured side is considered abnormal. A positive 'talar tilt' is indicative of significant, if not complete, rupture of the anterior talofibular ligament, which is the major restraint on inversion.

Treatment

Plaster casting is almost a thing of the past in our clinic with the use of the Airstirrup. Rehabilitation follows a three-phase plan: phase I, over the first 48 to 72 h, is instituted to limit the extent of the injury, phase II is designed to regain strength and full range of motion of the ankle, and phase III is for the restoration of proprioception, agility, speed, and endurance. Post-exercise ice applications continue throughout the programme and return to sport is not permitted until the athlete has regained full range of ankle motion, full strength, and balance, with minimal tenderness, pain, or swelling. In addition, the athlete must be able to perform agility drills, sport-specific jumping, and change of direction drills. Post-injury drills of strength, flexibility, and balance, together with protective taping or bracing, are advised for the remainder of the competitive season.

Treatment of grade I and grade II ankle sprains has changed considerably in the past decade with the development of the Airstirrup plus the concept of early mobilization including relative rest, ice, compression, and elevation (RICE). Many grade III sprains also do well managed non-operatively, but the resprain rate may be increased compared with those undergoing early repair.

McConkey[42] reviewed the literature on surgical and nonsurgical management of severe ankle sprains. He summarized the indications for surgical management of severe grade III ankle sprains as follows.

1. Young athletes and workers requiring near normal ankle function for safety and athletic success;
2. Patients with an associated lesion such as an osteochondral fracture, peroneal tendon dislocation or distal syndesmosis disruption;
3. Patients who have suffered an ankle joint dislocation.

Physiotherapy plays a very strong role in the management of grade I and grade II injuries; in particular, electrotherapy with ultrasound, short wave diathermy, and interferential therapy (and perhaps laser therapy for those with experience in this technique) combine well with strength, power, and proprioceptive drills to promote early activity and functional recovery.

Lane[45] has recently described a very successful ankle treatment programme for 26 patients who had suffered radiographically confirmed grade III ankle sprains. The programme consisted of a 3-week period using an ankle–foot orthosis at all times, except during physical therapy, with immediate weight bearing as tolerated. Patients received physiotherapy, consisting of isometric exercises, whirlpool, electric stimulation, ultrasound, and Jobst compression, daily for a week and then three times per week. After 3 to 5 weeks the ankle–foot orthosis was replaced with a lace-up brace and range of motion exercises, strengthening exercises, and stationary cycling were introduced. After 5 to 7 weeks agility drills using the lace-up brace were added along with proprioceptive exercises with a balance board. After 7 weeks the athlete was introduced to sport-specific drills. The ankle was taped or the lace-up brace was used for 6 to 12 months. With this programme 96 per cent of patients returned to activity without restriction after 10 weeks.

Re-injury following a significant ankle sprain is common, with reports of as many as 40 per cent who had been treated by early protected mobilization, plaster immobilization, or surgical repair.[46] Jackson et al.[47] saw an 8 per cent resprain rate at 6 months for all grades of ankle sprains managed non-operatively. The causes of resprains seem primarily to result from inadequate proprioceptive retraining, lack of peroneal strength, and residual lateral ligament laxity. Freeman[46] believes that the feeling of instability is largely due to proprioception loss. McConkey[42] adds other potential sources of functional ankle instability including adhesion formation and ossicle impingement at the talofibular junction, and anteroposterior or rotational instability, rather than varus instability of the talus in the mortise with subtalar joint laxity. The majority of athletes with recurrent sprains are controlled well with strength and proprioception drills together with one of the better ankle braces (either lace-on or stirrup type). A small percentage of athletes with recurrent instability require late ankle reconstruction.

The control of instability by conservative or operative means is very important in the prevention of secondary osteoarthritis and impingement osteophytes, with the possible development of loose fragments.

Associated injuries

Associated injuries must always be sought with acute ankle sprains. Such injuries include osteochondral lesions to the talus, peroneal tendon dislocation, tibialis posterior tendon injury, disruption of the distal tibiofibular syndesmosis, and nerve traction injury (particularly of the peroneal nerve). McConkey[42] adds occult and/or unusual fractures which result from excessive inversion stress. These can be mimics of ankle sprains and result from excessive lateral tension forces. Such injuries often do not produce any ligament sprain or medial compressive forces and include avulsion fracture by peroneus brevis at the base of the fifth metatarsal, fracture separation of the distal fibular epiphysis in children, capsular avulsion fracture from the lateral calcaneocuboid joint, navicular compression fracture, and fracture of the neck or body of the talus. Forced eversion with dorsiflexion and compression can produce a lateral process fracture of the talus which is often not seen on initial radiography examination. Bone scans with follow-up CT scans are often required in the presence of persistent disability and tenderness inferior to the lateral malleolus. Fracture fragments may require internal fixation or removal. Other osteochondral fractures are seen with a forced inversion injury, and the majority of these are to the posterior medial corner of the talus. Less frequently, outer lateral talar surface lesions are seen with talar friction and shear against the fibula on inversion and subluxation of the ankle. Talar dome fractures are seen as a result of a combination of ankle inversion, plantar flexion, and internal rotation of the foot. Avascular necrosis can be a consequence of medial talar fractures, and it is often only at this stage that the fracture is visible on a radiograph. Early diagnosis requires a bone scan or CT arthrogram. The stability of an osteochondral talar fracture dictates its management. Thompson and Loomer[48] have shown that lateral dome fractures are usually unstable and should be treated with arthroscopic excision of the fragment. Many fractures of the medial corner and dome of the talus are stable and can be treated non-operatively in an orthotic ankle brace. Unstable fractures, particularly if they have undergone avascular necrosis, require arthroscopic excision and perhaps drilling.

Peroneal tendon subluxations can be seen with forced ankle inversion and even more frequently with forced dorsiflexion, as in a forward fall while skiing. With such an injury the peroneal retinaculum is torn from its fibular attachment and is best managed by acute surgical repair. Radiography may show a small avulsion fragment some 2 to 3 cm above the tip of the fibula. Clinically, the tenderness and maximal swelling is located posterior to the lateral malleolus, with additional tenderness along the peroneal muscles. With resisted eversion, particularly with the foot in plantar flexion, tendon subluxation can be identified.

Stanish et al.[49] have described injuries to the tibialis posterior tendon associated with ankle injuries. They discussed subluxation of the tibialis posterior tendon as a result of severe ankle sprain, particularly in an athlete with a shallow tibial bony trough for the tendon. This dislocation can lead to a sensation of ankle instability on forceful toe-off, because at toe-off the posterior tibial muscle tendon unit acts as the prime dynamic stabilizer of the ankle and the prime inverter and supporter of the arch. In addition, medial ankle instability, secondary to a severe eversion injury, may lead to

chronic tibialis posterior tendon dislocation with tendinitis and eventual tendon rupture. This requires tendon repair, tendon anastomosis, or even free tendon transplant followed by orthotic foot support. If the tendon repair fails, extensive rehabilitation and eventual ankle fusion is required.

Distal syndesmosis injuries can be seen with grade III ankle inversion sprains and also with forced eversion with external rotation or abduction. The eversion injury progressively ruptures the syndesmosis ligaments and interosseous membranes, then fractures the fibula as a spiral or transverse fracture, and eventually, as the talus shifts further, ruptures the deltoid ligament. Diffuse swelling is seen, with maximal local tenderness over the distal syndesmosis (rather than over the anterior talofibular ligament). Mid-leg and distal compression of the tibia and fibula should be performed by the examiner—and is painful. A talar shift and/or fibular fracture should be sought on radiographs, but may not be present if only a partial syndesmosis sprain has occurred. Complete disruptions of the syndesmosis require surgical reduction and internal fixation to prevent recurrent instability and chronic pain.

Finally, upon initial clinical examination of the acutely injured ankle, peripheral nerve involvement should be excluded. Grade III ankle sprains have a high incidence of traction injuries to both the peroneal and posterior tibial nerves. Naturally, this can result in prolonged inversion, eversion, and post-injury plantar flexion weakness. Less frequently, peroneal nerve injuries are seen with milder degrees of inversion ankle sprains.

Schamberger[50] has comprehensively reviewed nerve injuries about the foot and ankle. In brief, inversion–supination forces stretch the sural and common peroneal nerves, often with compression of the posterior tibial nerve in the tarsal tunnel. Eversion–pronation forces stretch the tibialis posterior nerve with lateral compression forces to the sural nerve. In addition, a stretching (traction) force can be applied to the superficial branch of the common peroneal nerve with excessive plantar flexion of the ankle. On physical examination, a positive Tinel sign can often be elicited, and diagnosis can be confirmed by local anaesthetic nerve blocks and by electromyography and nerve conduction studies.

OVERUSE INJURIES

Stress fractures

Stress fracture commonly affects the ankle, and in a series of 320 athletes with this injury the tibia was involved in 49.1 per cent, the tarsals in 25.3 per cent, and the fibula in 6.6 per cent.[7] Many of these injuries were sited at the ankle joint. Runners are by far the most affected group of patients, and biomechanical factors play an important part in the aetiology of such lesions, emphasizing the fact that stress fractures are overuse in origin. Matheson et al.[7] found that pronated feet were particularly associated with tibial stress fractures and tarsal bone fractures.

The presentation of a stress fracture is typically that of a gradual onset of pain accompanied by tenderness, which is quite easily localized to a particular site such as the lateral malleolus. Pain is associated with activity and is relieved by rest. With progression of the stress fracture, pain develops earlier with activity and takes longer to settle with rest and it is painful to hop. Eventually constant pain develops and activity is prevented. There is normally little swelling (if any) to be found and no evidence of bruising or trophic changes of the skin. Vascular and neurological signs are not associated with isolated stress fractures.

The examiner should take note of biomechanical factors such as excessive pronation of the foot and inquire as to recent increases in activity, particularly on hard surfaces such as asphalt or Tarmac court surfaces or non-spring floors used at aerobic classes. Attention to footwear may reveal inappropriate shoes which lack 'motion control' (if required) and shock absorption properties, as discussed earlier in this chapter.

The diagnosis of stress fracture is supported by the observation of an area of increased isotope uptake on a technetium-99 bone scan at the site of bone pain and tenderness. Radiographs are typically negative for up to 3 weeks after development of the lesion, but a CT scan may be useful in defining the architecture of the bone and the lesion itself when a positive bone scan has been returned.

Management includes active rest (modifying exercise to limit painful activities and substituting swimming, cycling, and weight training to maintain cardiovascular fitness and strength) with a programme of rehabilitation which grades a return to full sporting activity over a period of 6 to 8 weeks. Typically, non-weight-bearing exercise is encouraged for 2 weeks, and then small increments in weight-bearing activity at weekly intervals are permitted. The diminution in local bone pain, tenderness, and pain when hopping is used as a guide to these gradual increases in the programme. Physiotherapy modalities which may be of use include pulsed magnetic field therapy and electrical stimulation via superficial (skin-mounted) electrodes. An Air-Cast ankle stirrup is particularly useful in stress fractures about the ankle, particularly where early mobilization and weight bearing is being encouraged. The Air-Cast should be worn at least until initial recovery, and for a period thereafter to facilitate return to sport. Biomechanical factors such as excessive pronation and length discrepancy should be attended to, and the fitting of correctly prescribed orthotic devices is recommended together with the provision of appropriate footwear.

Although stress fractures typically present as overuse injuries and therefore are gradual in onset, stress fracture may occasionally be manifested via acute injuries such as inversion sprains or trauma to the ankle or foot. Navicular and talar fractures may present in this way, and the physician must remember to consider these diagnoses as appropriate and investigate accordingly.

Stress fracture of the talus is notorious for presenting in an insidious fashion and is often not accompanied by impressive local tenderness. There may be some effusion of the ankle and some limitation of range of movement, but many patients continue to weight bear, albeit in some discomfort for some time after the onset of symptoms. Typically, hopping is painful and this should alert the clinician to the diagnosis. Bone scan is positive and a CT scan is recommended, remembering that a CT scan may be negative in talar stress fractures of traumatic origin.[7]

Management depends upon the severity of the lesion and may range from reduced weight-bearing activities in an Air-Cast stirrup to ankle protection in a non-weight-bearing cast for up to 8 weeks. Pain when hopping used as a guide to the reintroduction of weight-bearing activities after an initial rest period where cycling, swimming, or 'pool running' has been used to maintain fitness. Any lesion affecting the articular cartilage must be reviewed by an orthopaedic surgeon.

Similarly, stress fractures of the tarsal navicular can present as ankle problems. They may present in an acute fashion associated with an ankle sprain or with a more typical history of gradually increasing pain, tenderness, and disability associated with activity. Technetium-99 bone scanning and CT scanning are recommended, and appropriate aggressive management should be instituted immediately.

Tendinitis

Overuse of the ankle, in running, jumping, and dance for example, often produces tendinitis of the main functional stabilizers and strong plantar flexors of the ankle. The posterior tibial tendon and the tendons of the peroneal muscles are particularly affected in this respect.

As discussed earlier in this chapter, the posterior tibial tendon is a strong supporter of the medial arch of the foot and is active as a plantar flexor and a controller of pronation during gait. It also works eccentrically with its muscle in landing to absorb shock. The tendon extends along the distal third of the medial tibial margin and inserts on to the medial tarsal navicular surface, with variable attachments to adjacent tarsal bones. It passes behind the medial malleolus and is enclosed in a synovial sheath in this part of its course, where it lies deep to the flexor retinaculum.

Tendinitis is typically of gradual onset and is precipitated by increased or high levels of activity, with limited periods of rest or recovery. The pain is localized to the medial tibial margin or to the tendon adjacent to the medial malleolus. Often there is swelling of the distal tendon and sheath, and pain may be reproduced by resisted plantar flexion and inversion of the foot and ankle.

Biomechanical factors play a significant role in the aetiology of this condition and relief is often rapid once appropriate orthotics have been fitted. Overpronation is perhaps the most important fault in this respect. Attention to footwear is recommended, and sport shoes which control pronation and provide for shock absorption, particularly in basketball and netball, should be considered essential.

Treatment consists of modification of painful activity until symptoms and signs subside, regular applications of ice, physiotherapy which employs electrotherapeutic modalities to reduce inflammation, and an exercise programme to stretch and strengthen the posterior tibial muscle and tendon, particularly utilizing an eccentric phase.[49] Non-steroidal anti-inflammatory medication can be very useful, and in recalcitrant cases the judicious administration of a local corticosteroid injection may be helpful, taking extreme care not to inject the tendon. A rest from jumping and running for 10 days is advised after a corticosteroid injection to minimize the risk of a corticosteroid-associated tendon rupture.

Posterior tibial tendinitis should be distinguished from the medial tibial stress syndrome (shin-splints), the deep calf compartment pressure syndrome, and stress fracture (of the tibial shaft, the medial malleolus, and the tarsal navicular bone). Technetium-99 bone scanning and/or compartment pressure studies may be indicated here.

The tendons of the peroneus longus and brevis provide functional lateral stability of the ankle as everters and plantar flexors of the foot. Although they are more commonly injured in acute inversion sprains, overuse through jumping, running, and dancing may occasionally induce tendinitis or tenosynovitis, typically at the lateral malleolus where the tendons pass deep to the superior and inferior retinacula and curve around the malleolus to insert at their respective sites on the medial cuneiform and metatarsals. Repetitive friction through their common tendon sheath and around the 'pulley' of the malleolus, produces pain of gradual onset with associated local tenderness and swelling of the sheath. Forced plantar flexion and eversion may reproduce the pain.

The treatment of tendinitis in this situation is the same as for posterior tendinitis with local applications of ice, physiotherapy, and exercises for the tendon(s) and occasional injection of corticosteroid along the tendon(s) or within the sheath. Occasionally a heel raise provides relief, and a gradual return to sport in appropriate footwear should be possible over 6 to 8 weeks.

The differential diagnosis of peroneal tendinitis or tenosynovitis includes injury to the ligaments of the ankle, bony injury to the fibular shaft and lateral malleolus, and disruption of structures such as the peroneal retinacula.

Impingement syndrome

Footballer's ankle

Footballer's ankle is a condition whereby repeated microtrauma to the margins of the anterior ankle joint, through repetitive forceful plantar flexion and dorsiflexion, produces traction osteophytes at the capsular ligament attachments and subsequent formation of loose bodies within the joint. The condition is not a true osteoarthritis as the articular cartilage is spared. Any sport which involves this repeated ankle movement can produce such a condition, and soccer, basketball, triple jumping, long jumping, and dance are all well-known causes of footballer's ankle.

Symptoms are usually of gradual onset, with diffuse anterior ankle joint pain; occasionally swelling is noted after activity. There is often local tenderness, and pain can be reproduced by forced ankle plantar flexion or dorsiflexion which stretches the ligament and impinges the ligament attachments respectively. Plain radiography demonstrates the osteophystes at the joint margins and loose bodies may be seen.

Management depends upon symptoms, as the only definitive treatment is surgical excision of osteophytes with or without removal of loose bodies. If possible, limitation of painful activity should be recommended, and soccer players should be encouraged not to slide on the plantarflexed foot or kick with the dorsum of the foot when moving the ball forwards. Kicking with the medial surface of the foot is preferred. Anti-inflammatory medication, physiotherapy with electrotherapeutic modalities, ice applications, and occasionally local injection of corticosteroid can settle the inflammation but should not be seen as curative.

Os trigonum and talar spur (posterior process)

In the normal development of the talus, a posterior process may appear or may present as a separate ossicle, the os trigonum. This occurs in approximately 10 per cent of the population.

Impingement of the ossicle or posterior process can result from forced plantar flexion of the ankle and produce a deep pain posterior to the ankle joint, localized deep to the Achilles tendon. This is particularly common amongst ballet dancers, jumpers, and fast bowlers in cricket. Tenderness is usual and forceful plantar flexion reproduces the pain. There may be a little swelling, but usually no bruising and the Achilles tendon is typically unaffected. Plain radiographs demonstrate the ossicle or talar process and a bone scan may show increased isotope uptake, particularly if there has been disruption of the fibrous union between the os trigonum and the talus, or a fracture of the posterior talar process.

Management consists of initial rest from painful activity, ice applications, anti-inflammatory medication, electrotherapy, and a gradual return to activity as symptoms permit. Occasionally a local injection of corticosteroid may be indicated and provides relief in most cases. If symptoms persist, surgical excision of the offending ossicle or spur yields good results.[51]

Bursitis at the malleoli

A problem common amongst ice skaters and ice hockey players is bursitis over the lateral or medial malleolus. Wearing tight boots for long periods produces a combination of compression and friction about the malleolus, which results in the development of a painful, tender, and swollen bursa. The diagnosis is simple and no investigations are necessary.

Malleolar bursitis responds quickly to adjustment of footwear to minimize pressure and friction, perhaps with the addition of an orthotic device to correct an underlying biomechanical fault, such as excessive foot pronation, supported by regular applications of ice, non-steroidal anti-inflammatory medication, and physiotherapy. If necessary, a local injection of corticosteroid may be helpful. Persistent cases may require surgical excision of the offending bursa.

REFERENCES

1. Garrick JG. Epidemiology of foot and ankle injuries. In: Shepard RJ, Taunton JE, eds. *Medicine and sport science* Vol. 23. *Foot and ankle in sport and exercise.* Basel: Karger, 1987: 1–7.
2. Clement DB, Taunton JE, Smart GU, McNicol KL. A survey of overuse running injuries. *Physician and Sportsmedicine* 1981; 9: 47–58.
3. Macintyre JG *et al.* Running injuries: a clinical study of 4173 cases. *Canadian Journal of Sports Medicine* 1989; 1: 7–11.
4. Walter SD *et al.* Training habits and injury experience in distance runners: age and sex related factors. *Physician and Sports Medicine* 1988; 16: 101–13.
5. Marti B *et al.* On the epidemiology of running injuries. *American Journal of Sports Medicine* 1988; 16: 285–94.
6. Garrick JG, Requa RK. Role of external support in the prevention of ankle sprains. *Medicine and Science in Sports* 1973; 5: 200–3.
7. Matheson GO, Clement DB, McKenzie DC, Taunton JE, Lloyd-Smith DR, Macintyre JG. Stress fractures in athletes. A study of 320 cases. *American Journal of Sports Medicine* 1987; 15: 46–58.
8. Gray H. *Anatomy of the human body.* 28th edn. Philadelphia: Lea & Febiger, 1966.
9. Akesson EJ. Anatomy of foot and ankle. In: Shepard RJ, Taunton JE, eds. *Medicine and sport science* Vol. 23. *Foot and ankle in sport and exercise.* Basel: Karger, 1987: 8–21.
10. Cavanagh PR, Lafortune MA. Ground reaction forces in distance running. *Journal of Biomechanics* 1980; 13: 397–406.
11. Procter P, Berne N, Paul JP. Ankle joint biomechanics. In: Morechi, Fidelus, Kedzier, Wit, eds. *Biomechanics VII-A.* Baltimore: University Park Press, 1981: 52–6.
12. Nigg BM. Biomechanical analysis of ankle and foot movement. In: Shephard RJ, Taunton JE, eds. *Medicine and Sport Science* Vol. 23. *Foot and ankle in sport and exercise.* Basel: Karger, 1987: 22–9.
13. Nuber GO. Biomechanics of the foot and ankle during gait. *Clinics in Sports Medicine* 1988; 7: 127–42.
14. Slocum DB, James SL. Biomechanics of running. *Journal of the American Medical Association* 1958; 205: 97–104.
15. Mann RA. Biomechanics of running. In: AAOS Symposium on the Foot and Leg in Running Sports. St. Louis: CV Mosby, 1982, 1982: 30–44.
16. Mann RA, Moran GJ, Dougherty SC. Comparative electromyelography of the lower extremity in jogging, running and sprinting. *American Journal of Sport Medicine* 1986: 14: 501.
17. Ohman D. Injury prevention program for the ankle. *SportsAider* 1989; 6: 1–5.
18. Garrick JG, Requa RK. The epidemiology of foot and ankle injuries in sports. *Clinics in Sports Medicine* 1988; 7: 29–36.

19. Emerick CE. Ankle taping: prevention of injury or waste of time? *Athletic Training* 1979; 149–150, 188.
20. Fumich RM *et al.* The measured effect of taping on combined foot and ankle motion before and after exercise. *American Journal of Sports Medicine* 1981; **9**: 165–70.
21. Rarick GL, Bigley GK, Ralph MR. The measurable support of the ankle joint by conventional methods of taping. *Journal of Bone and Joint Surgery*, 1962; **44A**: 1183–90.
22. Bunch RP, Bednarski K, Holland D, Macinanti BA. Ankle joint support: a comparison of reusable lace-on braces with taping and wrapping. *Physician and Sportsmedicine* 1985; **13**: 59–62.
23. Rovere GD, Clarke TJ, Yates CS, Burley K. Retrospective comparison of taping and ankle stabilizers in preventing ankle injuries. *American Journal of Sports Medicine* 1988; **16**: 228–33.
24. Kimura IF, Nawoczenski DA, Epler M, Owen MG. Effect of the air stirrup in controlling ankle inversion stress. *Journal of Orthopaedic Sports Physical Therapy* 1987; **9**: 190–3.
25. Gross MT, Bradshaw MK, Ventry LC, Weller KH. Comparison of support by ankle taping and semi-rigid orthosis. *Journal of Orthopaedic Sports Physical Therapy* 1987; **9**: 33–9.
26. Greene TA, Hillman SK. Comparison of support provided by a semi-rigid orthosis and adhesive ankle taping before, during, and after exercise. *American Journal of Sports Medicine* 1990; **18**: 498–506.
27. James SL, Bates BT, Ostering LR. Injuries to runners. *American Journal of Sports Medicine.* 1978; **6**: 40–50.
28. McKenzie DC, Clement DB, Taunton JE. Running shoes, orthotics and injuries. *Sports Medicine* 1985; **2**: 334–7.
29. Taunton JE, Clement DB, Smart GW, Wiley JP, McNicol KL. A triplanar electrogoniometer investigation of running mechanics in runners with compensatory overpronation. *Canadian Journal of Applied Sport Science* 1985; **10**: 104–15.
30. Bates BT, Osternig LR, Mason B, James SL. Foot orthotic devices to modify selected aspects of lower extremity mechanics. *American Journal of Sports Medicine* 1979; **7**: 338–42.
31. Smith L, Clarke T, Hamill C, Santopietro F. The effects of soft and semi-rigid orthoses upon rearfoot movement in running. *Medicine and Science in Sport and Exercise* 1983; **15**: 171.
32. McKenzie DC. The role of the shoe and orthotic. In: Shephard RJ, Taunton, JE, eds. *Medicine and Sport Science* Vol. 23. *Foot and ankle in sport and exercise.* Basel: Karger, 1987: 30–8.
33. Brody D. Running injuries. *Ciba Foundation Symposium* 1980; **32**: 1–36.
34. Lutter LD. Cavus foot in runners. *Foot and Ankle* 1981; **1**: 225–8.
35. Moore PH, Taunton JE. Medically based athletic footwear. Design and selection. *New Zealand Journal of Sports Medicine* 1991; **19**: 22–5.
36. Frederick CC, Clarke TC, Hamill CL. The effect of running shoe design on shock absorption. In: Frederick EC, ed. *Sport shoes and playing surfaces.* Champaign, IL: Human Kinetics, 1984: 190–8.
37. Trott AW. Foot and ankle problems in adolescents: sport aspects. *American Academy of Orthopaedic Surgeons Symposium on Foot and Ankle.* St Louis: CV Mosby, 1979: 47.
38. Micheli LJ. Overuse in children's sports: the growth factor. *Orthopaedic Clinics of North America* 1983; **14**: 337–60.
39. Beauchamp R. Pediatric foot and ankle problems. In: Shepard RJ, Taunton JE, eds. *Medicine and sport science* Vol 23. *Foot and ankle in sport and exercise.* Basel: Karger, 1987: 128–44.
40. Salter RB, Harris WR. Injuries involving the epiphyseal plate. *Journal of Bone and Joint Surgery* 1963; **45**: 587–622.
41. Chandler RW. Management of complex ankle fractures in athletes. *Clinics in Sports Medicine* 1988; **7**: 127–42.
42. McConkey JP. Ankle sprains, consequences and mimics. In: Shephard RJ, Taunton JE, eds. *Medicine and sport science* Vol. 23. *Foot and ankle in sport and exercise.* Basel: Karger, 1987: 39–55.
43. Walsh M, Blackburn T. Prevention of ankle sprains. *American Journal of Sports Medicine* 1977; **5**: 243–5.
44. Laurin C, Mathieu J. Sagital mobility of the normal ankle. *Clinical Orthopaedics* 1975; **108**: 99–104.
45. Lane SC. Severe ankle sprains. Treatment with an ankle foot orthosis. *Physician and Sportsmedicine* 1990; **18**: 43–51.
46. Freeman MAR. Treatment of ruptures of the lateral ligaments of the ankle. *Journal of Bone and Joint Surgery* 1965; **47B**: 661–684.
47. Jackson D, Ashley RL, Powell J. Ankle sprains in young athletes. *Clinical Orthopaedics* 1974; **101**: 201–14.
48. Thompson JP, Loomer RL. Osteochondral lesions of the talus in a sports medicine-clinic. *American Journal of Sports Medicine* 1984; **12**: 460–3.
49. Stanish WD, Ratson G, Curwin S. Tendinopathies about the foot and ankle. In: Shephard RJ, Taunton JE, eds. *Medicine and sport science* Vol. 23. *Foot and ankle in sport and exercise.* Basel: Karger, 1987: 80–98.
50. Schamberger W. Nerve injuries around the foot and ankle. In: Shephard RJ, Taunton JE, eds. *Medicine and sport science* Vol. 23. *Foot and ankle in sport and exercise.* Basel: Karger, 1987: 105–20.
51. Fricker PA, Williams JGP. Surgery to the os trigonum and talar spur in sportsmen. *British Journal of Sports Medicine* 1978; **13**: 55–7.

4.4.2   The acute ankle sprain

ANGUS M. MCBRYDE

INTRODUCTION

In competitive or recreational activity, e.g. basketball, acute ankle sprains are the single most frequent specific injury.[1–4] The lateral ligament complex of the ankle is the most frequently injured single musculoskeletal structure in the body (Fig. 1).[5]

Eighty-five per cent of ankle injuries are sprains. These sprains involve athletes in virtually all sports at all levels of competition and affect both males and females.[6,7] Twenty-five per cent of all time loss is attributed to ankle sprain.[8]

An estimated 30 per cent of milder (grade I and grade II) sprains and perhaps 15 per cent of more severe sprains never enter the health care system. Most of these acute ankle sprains not seen by a practitioner are recurrent acute sprains. Acute sprains occur primarily in persons between the ages of 10 and 35.

Ankle sprains (not strains)[9] are secondary only to back sprains and strains as a cause of disability. In the United States alone 20.7 million restricted activity days occur per year; 5.8 million work and school days and 500 000 bed disability days are lost to ankle sprains.[10]

Foot and ankle injuries comprise up to 25 per cent of sports injuries.[8,11,12] Of these injuries, 9.7 per cent involve the ankle alone. Youth soccer (12–17 years old), for instance, produced an overall 23.1 per cent incidence of ankle injuries, with the percentage rising with the increasing age of the players.[13]

Acute ankle sprains are treated by the entire range of medical-school trained and non-medical-school trained practitioners. Primary care physicians, such as family physicians, paediatricians and emergency room physicians, plus physician assistants, nurse practitioners, physical therapists, and athletic trainers—all are

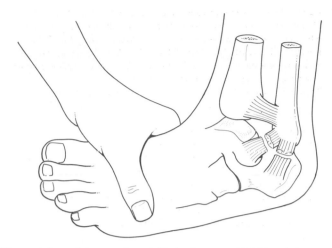

Fig. 1 Injury of the anterior talofibular ligament (ATFL) and the calcaneofibular ligament (CFL) is a most common injury.

called on to see and treat the acute ankle sprain with all its severities.[14] Orthopaedic surgeons see, at some point in treatment, an estimated 60 per cent of ankle sprains that are formally seen by any practitioner.

Ninety per cent of ankle sprains can be handled by the primary care practitioner. The problem arises with (a) the need to recognize and select the more severe 10 per cent and (b) the universal need to diagnose adequately and treat the remaining 90 per cent. A detailed history and physical examination is clearly necessary for the more severe 10 per cent of ankle sprains. This is supplemented by stress films or other studies which are frequently necessary to determine whether more aggressive treatment with implied referral is indicated. At this point in the decision-making process a suboptimal diagnosis is frequently made, with a resultant insufficient acute treatment and rehabilitation programme. Even when the need for referral is recognized, many are 'late referrals', i.e. after 3 weeks. In essence, the frequency of ankle sprain has helped breed casualness in treatment.[9]

All these factors, i.e. self-treatment, numerous care-givers, injury frequency, late referrals, etc., cause major difficulties in any study of ankle sprain. Therefore it becomes difficult to select, digest, study, and then quantitate specifically or standardize the injury patterns, diagnosis, management, and rehabilitation.

The history is indispensable to subsequent diagnosis, treatment, and rehabilitation of an acute ankle sprain. It points up the typical musculoskeletal and usually sports-related injury. Ankle sprain is both a sports-specific injury, e.g. soccer, and a sports-generic injury, e.g. any leg-based sport. Both documented and undocumented histories indicate this fact. Most ankle sprains occur with an agility move while loaded (with weight-bearing). Other acute sprains occur when the foot is planted on a non-level surface which throws the foot and ankle into supination. There are pre-existing tendencies due to previous injury, increased height and weight,[6] and inherent joint laxity. Other performance factors also bear on the predisposition to injury, i.e. conditioning, selective muscle group strength (posterior tibial) or weakness (peroneals), and shoe and/or orthosis ankle protection or modification of foot plant.

More importance should be attached to the history, the physical examination, the quality of diagnosis, the treatment, and the rehabilitation of acute ankle sprains.[14]

ANATOMY AND MECHANICS OF THE ANKLE JOINT

The ankle (talocrural joint) is a complex hinge joint which is also a mortise joint. There is an inherent stability in its interlocking configuration. The joint and the talus are wider anteriorly with the talus congruently fitting into the tibial fibula mortise (Fig. 2). Dorsiflexion bony stability is maximum as the talus locks into this

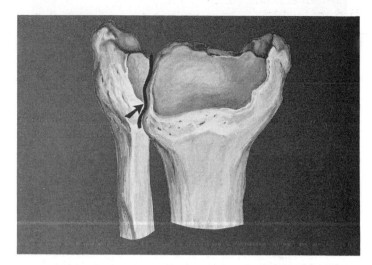

Fig. 2 With weight bearing the talus locks into its tibiofibular bed and there is 40 per cent more stability. Integrity of the opposing articular cartilage at the tibia–fibula junction (arrow) must be maintained. Grade IV sprain can compromise this integrity.

mortise. Plantar flexion stability therefore relies more on ligamentous integrity. This decreased bony stability in plantar flexion is particularly evident on the medial side since the longer lateral malleolus offers better buttress for the ankle and subtalar joint in any position. Thus the bony anatomy partially determines plantar flexion and inversion to be the most vulnerable direction for injury.

The lateral collateral ligament complex binds the talus and calcaneus to the fibula (Fig. 3). The anterior talofibular ligament is the weakest ligament and has a high strain rate prerupture.[15] This correlates with its physiological function of allowing increased ankle plantar flexion and internal rotation.[16] The anterior talofibular ligament checks the forward subluxation of the talus in a sagittal plane. It becomes parallel to the tibia and fibula with full plantar flexion as its lateral talar neck distal attachment moves beneath the lateral malleolus.

The calcaneofibular ligament is extracapsular (Fig. 3). It runs obliquely to insert on the posterior lateral aspect of the os calcis. It is perpendicular to the horizontal axis of the posterior subtalar joint. Thus the calcaneofibular ligament acts as a stabilizer of the subtalar joint as well as the ankle joint as it bridges both joints. It is a primary ankle stabilizer with dorsiflexion and gives major increased support. It prevents talar tilt, primarily with straight inversion. The calcaneofibular ligament is four times stronger than the anterior talofibular ligament. Its tear often involves a tear of the peroneal tendon sheath which is always connected and in proximity.

The posterior talofibular ligament (Fig. 2) is the strongest ligament and runs from the distal fibular to the lateral talar tubercle or

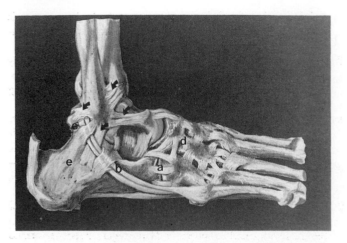

Fig. 3 The lateral ankle hindfoot and midfoot ligamentous complex. Grade I–III ankle sprain primarily involves the calcaneofibular, anterior and posterior talofibular, and anterior tibiofibular ligaments (arrows). Its differential diagnosis can include (a) calcaneocuboid sprain, (b) peroneal tendon partial tear, (c) subtalar sprain, (d) midfoot ligament sprain, and (e) stress fracture of the os calcis.

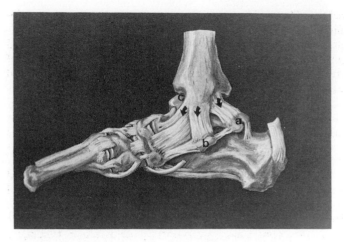

Fig. 4 The medial ankle, hindfoot, and midfoot ligamentous complex. Grade IV ankle sprains involve the deltoid ligament (arrows) and the syndesmosis. Grades I–IV ankle sprain can involve other medial structures: (a) posterior medial and posterior lateral talar processes; (b) sustentaculum tali; (c) the talar dome.

posterior to the lateral articular surface of the posterior subtalar joint. Its isolated tear promotes only slightly increased hyper-extension, but when coupled with anterior talofibular and calcaneofibular ligament tears promotes major instability[17] including effects on the subtalar joint.[18]

The lateral talocalcaneal ligament protects the subtalar joint and merges fibres with the anterior talofibular ligament and the calcaneofibular ligament.

A range of 20 to 30° of dorsiflexion is necessary, depending on sport-specific needs, with the mean value for loaded dorsiflexion being 32.5°.[19] The average full range of passive extension is 21° and the average full range of flexion is 23°.[20] When the ankle moves from plantar flexion towards dorsiflexion, initial internal rotation of the talus changes during the last 10° to external rotation towards a neutral position.[21]

The ankle is not a uniaxial joint. There are small but definite adaptive rotations and limited pronation in the ankle joint in the plantar flexion phase. This helps allow accommodation for non-level ground without losing push-off.[21] Rotation is normally only 12° in the ankle joint and increases with anterior talofibular ligament injury.[22,23] Midfoot joints contribute substantially to transverse axis rotation in 30° of plantar flexion.[21] The articular surface becomes extremely important in rotation when loaded, and decreases that rotation. It affects 30 per cent of stability in rotation and 100 per cent in inversion.[24]

There is a large angle between the calcaneofibular ligament and the anterior talofibular ligament. This may result in reduced resistance to inversion since neither the anterior talofibular ligament nor the calcaneofibular ligament offers maximum protection at neutral position.

The deltoid ligament stabilizes the medial aspect of the ankle (Fig. 4). The tibionavicular, tibiocalcaneal, and anterior and posterior tibiotalar ligaments constitute the superficial portion. The deep portion consists of the transverse tibiotalar ligament which runs anteriorly and posteriorly from the medial malleolus to the talus. The deltoid ligament limits inversion and external rotation. It fans out to a wide insertion on the medial aspect of the talus and calcaneus.

The anterior and posterior tibiofibular ligaments (Fig. 3), together with the intraosseous membrane, anchor the fibula to the tibia. The posterior tibiofibular ligament melds with the inferior transverse (tibiofibular) ligament. Both the posterior and anterior tibiofibular ligaments permit ankle rotation during running and walking gait.

Little is known about the information process—the kinesthetic information about joint position, function, and speed.[25] There is no consistent relationship between lack of co-ordination of functional muscle and mechanical instability. Delayed peroneal proprioceptive response to quick inversion stress reduces protection and predisposes to injury.

In essence, the ankle joint has no single ligament that is dominant in stability. The direction of applied forces, the position of the ankle, and the load on the ankle all help determine which stabilizers are most important.[26,27]

MECHANISM OF INJURY

Eighty-five per cent of acute ankle sprains follow excess inversion and plantar flexion. Fifteen per cent follow excess dorsiflexion and eversion. Slight internal rotation accompanies the usual inversion and plantar flexion injury. Often a plantar fixed shoe (cleated or studded) allows the body to rotate over that fixed point and the classic injury occurs. When the foot is plantar fixed, and particularly when the triceps surae does not eccentrically decelerate the tibia, there is even more predisposition to lateral ankle ligament disruption.

The inversion, plantar flexion, and internal rotation injury can also be described as supination–inversion with external rotation of the tibia on the fixed foot. The anterior capsule ruptures first, followed by the anterior talofibular ligament,[28] the anterior tibiofibular ligament,[29] and the calcaneofibular ligament. If total lateral disruption occurs, the posterior talofibular ligament also tears. The deep anterior deltoid ligament fibres can tear at the extreme of internal rotation and plantar flexion.[28]

Forty per cent of anterior talofibular ligament injuries will be accompanied by calcaneofibular ligament injury. Except for the rare pure inversion injury with the ankle in neutral position, the calcaneofibular ligament remains intact until the anterior talofibular ligament is torn.[28] The 'blunt ridge'[15] situated at the mid-portion of the anterior talofibular ligament tents the ligament and predisposes to rupture at that point. The lateral articular facet of the talus and its sunken neck make this blunt ridge an 'anatomical exostosis'.

In inversion injuries in the immature skeleton the distal fibular epiphysis may be injured without or occasionally with ligament injury. This is usually a Salter–Harris type II injury.[30] The mechanism of injury also puts stress on a possibly present accessory fibular epiphysis and can cause an avulsion.

Lateral ligamentous injury with an associated fibular or lateral malleolar fracture can occur in the same way as medial collateral ligament injuries associated with tibial plateau fractures at the knee.[31] Although concomitant rupture with fracture is uncommon, the more aggressive and early mobilization techniques currently used make it important to recognize these associated ligamentous injuries. Immobilization for the fracture is usually more than enough protection.

HISTORY

Initial evaluation of the acute ankle sprain requires a precise history. Appropriate questions must be asked.

1. When did it happen? Was it during recreation or competition? How long before presentation? What were the surface characteristics, e.g. non-level, cement, sand, etc.? Sports-specific injury tendencies are often present. For instance, running in European cross-country events with non-level surfaces promotes inversion injuries.

2. How did it occur? Was the foot planted? Was it full weight bearing? Was it at foot strike or at push-off with off-loading? Was another person involved? Did someone fall on it? Was an object involved? Associated injuries with ankle sprain are more likely to occur with contact sports, e.g. American football or rugby ruck or soccer collision.

3. How did the foot move? Was the stress twisting or angular? Did the body twist to the right or the left? Did the foot stay planted at the moment of injury? What were the shoe characteristics? Orthosis? Taping? If the body rotated to the left at the time of a right ankle sprain a grade IV sprain with possible syndesmosis/interosseous injury should be suspected.

4. How rapidly did the ankle swell? Was it iced and elevated immediately? Was compression applied? If so, what type? The basic principles, when enacted immediately, minimize both the post injury appearance and the residual functional disability. When an equinus posture, dependence, and heat are used, the sprain at 72 h may seem deceptively severe. Less localized and more diffuse soft tissue swelling may indicate a more severe injury with the global spread of haemorrhage.

5. What was felt or heard? Was there a 'pop', 'crack', or 'snap'? Was there immediate acute pain? Weakness? Tingling? The subjective sensation at the time of injury can suggest medial, lateral, or interosseous ligament rupture, talar dome fracture, or an acute peroneal dislocation, all in the absence of tibial and/or fibular bony injury.

6. Was it possible to bear weight? Was it possible to continue to play? Did weight bearing cause pain? Was there a sensation of instability? Was the ankle 'wobbly' and did it 'give way'? Could weight be borne with a stick or cane? It is unlikely that a more severe grade II or grade III sprain or an additional injury could permit effective weight bearing. Grade I injuries do allow weight bearing. Ankle haemarthrosis or subtalar haemarthrosis effectively prevent weight bearing and can indicate intra-articular chondral or osteochondral injury.

7. Could you walk, run, or jump? Could you perform agility moves, i.e. lateral moves, cutting, deceleration? A grade I injury often permits agility moves with sturdy taping or bracing and supportive footwear.

8. Where was the initial tenderness? Medial? Lateral? Posterior? Leg? Lateral tenderness alone suggests a single-ligament grade I or mild grade II injury. Global tenderness suggests a severe grade II or a grade III injury. Tenderness over the distal fibula epiphysis or anterior tibial fibula interval or midfoot suggests a misdiagnosis or an underdiagnosed associated injury.

9. Was there previous injury? How was it treated? How long ago? How many times previously has the ankle been sprained? Had the ankle returned to 'normal'? Had surgery been performed previously? Recurrent acute sprain suggests inadequate rehabilitation, an unstable ankle, an associated undiagnosed injury, or faulty mechanics or footwear. This type of patient should be followed with full sports-specific return and with a monitored programme. Ankle instability detected on examination could be chronic and not acute in the presence of recurrent acute sprain.

The history, when properly requested, gives factual information which can be used for (a) grading the injury, (b) treatment (c) rehabilitation, and (d) prevention of subsequent injury.

EXAMINATION AND DIAGNOSTIC STUDIES

Physical examination includes inspection, palpation, range of motion examination, stability examination, and specific testing, when indicated, for possible associated or concurrent injury. Swelling, pain, and disability correlate with the degree of injury only within the first few minutes unless vigorously and immediately treated with rest, ice, compression, elevation (RICE), and protection. As noted earlier, the immediate history in the initial moments after sprain remains most helpful in further diagnosis and stability determination.

Local tenderness laterally at the 4, 6 and 8 o'clock positions (looking at the lateral malleolus) gives immediate information about the three lateral ligaments. A positive drawer test in plantar flexion implies anterior talofibular ligament disruption. In a positive test the talus will move forward at least 4 mm more than the uninjured ankle. There may be simple absence of a film end point. A positive anterior drawer test in the neutral position suggests that the calcaneofibular ligament and the posterior talofibular ligament have been injured. Otherwise, the test would be negative.

The talar tilt should be attempted during examination by forcefully and smoothly inverting and everting the hindfoot (by holding the os calcis) and comparing with the contralateral ankle. Radiography may be needed to confirm the positive tilt test by bilateral anterior–posterior talar tilt stress films (Fig. 5).

Midfoot and other ligamentous instabilities, i.e. subtalar, midfoot, and even tarsal–metatarsal, can confuse the 'ankle sprain' picture. Specific tenderness found with careful palpation along with motion pain in the hindfoot, midfoot and forefoot suggests something other than talocrural joint injury. Heel impaction pain suggests ankle or subtalar haemarthrosis, and the need for further investigation and clinical correlation, e.g. possible radiographic talar tilt stress or dye studies.

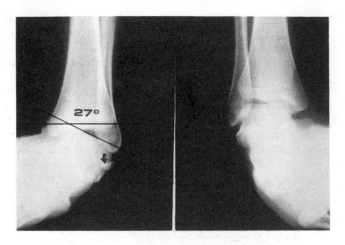

Fig. 5 Talar tilt stress radiography is reliable and reproducible in trained hands. A measurement of 27° is clearly abnormal (more than 15° above normal) and indicates tear of the calcaneofibular ligament and significant lateral instability. The avulsed fibular attachment of the calcaneofibular ligament (arrow) should be noted.

A major injury (severe pain out of proportion to clinical findings) coupled with a mildly positive drawer test suggests lateral talocalcaneal ligament and subtalar joint injury. The 'squeeze test' produces distal pain with proximal compression when the syndesmosis is interrupted and interosseous ligament damage has occurred.[32] Ankle injection with local anaesthetic and steroid is often performed, but is not recommended. Local block with anaesthetic alone can be used for differential diagnosis such as trauma to the interosseous membrane, subtalar joint, or calcaneocuboid joint, and to perform a better stress test. Specific examiner consistency and repeated use of the differential block optimizes interpretation.

Although the primary examination is physical,[33,34] anterior–posterior, lateral, and mortise views are mandatory when there is any significant injury. In the immature skeleton with suspected distal epiphyseal injury, contralateral comparison views should be taken. In addition, lateral stress views and talar tilt views can be indicated if the ankle injury is clearly grade II or questionably grade III with major instability.[35,36] A talar tilt of 15° or 5° more than the opposite ankle on anterior–posterior stress projection indicates probable anterior talofibular ligament and calcaneofibular ligament injury. Anterior–posterior talar tilt stress views for varus and anterior instability[37,38] are more helpful than the sagittal stress views. In addition, small flake fractures of the talar dome (osteochondral fractures) may show only on anterior talar tilt stress films.[39] A forward movement of the talus (beneath the tibia) by 4 mm, as seen on the lateral sagittal stress film with the anterior drawer test is significant. Although not widely used, quantitative stress methods such as graded stress radiography, where practical, can increase accuracy[37,40] and can be done regardless of the time of injury. Tenography and arthrography can be used effectively separately.[41] When used together there is a 100 per cent positive predictive value for calcaneofibular ligament tear.[42]

Arthrography after the first 24 h[15] is usually unsuccessful since clotting occurs and possible abnormal dye distribution is thwarted. Arthrography has both its proponents[15,31,43] and opponents.[44,45] Likewise, tenography or even stress tenography has proponents[42,46] and opponents. Stress tenography may pick up more double-ligament ruptures but may also give more false posi-

tive results.[47] Often, neither tenography nor arthrography is carried out.[14]

Bone scan can help localization of injury as early as 72 h. Distal tibiofibular and interosseous injury can be picked up on bone scans for up to 10 weeks.

Magnetic resonance imaging (MRI) is beneficial for in-depth differential diagnoses. MRI can delineate injuries which may co-exist with ankle sprain as well as document the ligamentous damage itself. However, its current use is primarily indicated in selective non-resolved ankle sprains and in those with failure to respond to a full functional rehabilitation programme.[48] As MRI becomes more widely available and more reliable, it will replace many of the imaging techniques now being used.

TYPES OF INJURY

Clinical stability testing and radiographic stability patterns of the ankle are helpful with diagnosis and grading.[9] There are numerous classification systems.[49]

A grade I injury is a partial tear of the anterior talofibular ligament.[50] Only local tenderness is present. This is a 'single-ligament' injury. Heel and toe walking is satisfactory. There is no significant subjective or objective instability. A full range of motion is permitted and there is definite but minimal weight-bearing pain. Thirty per cent of all ankle sprains meet these criteria.

A grade II injury implies complete anterior talofibular ligament disruption. The drawer test is positive. Moderate decreased range of motion, pain, swelling, and minimally positive talar tilt of less than 15° can be present. The calcaneofibular ligament is horizontal and out of harm's way in plantar flexion. However, it may be injured with a grade II injury. This would be a 'double-ligament' injury. A double-ligament injury always involves a talocalcaneal ligament insult.[38] Forty per cent of all ankle sprains meet these criteria. This is the ankle sprain most common in dancers.[51]

A grade III injury implies damage to three ligaments (anterior and posterior talofibular and calcaneofibular ligaments) plus varus instability. Both functional treatment and operative repair allow stabilization of this unstable joint. The patient presents in a non-weight-bearing status with significant pain and swelling. Anterior and lateral instability is present on examination, and the range of voluntary and involuntary motion is reduced. Talar tilt is greater than 15°. Fifteen per cent of all ankle sprains meet these criteria.

A grade IV injury is an eversion, external rotation, and inconsistent dorsiflexion injury. In successive order, the anterior and posterior tibiofibular ligaments, the deltoid ligament, and the interosseous membrane can all be involved. This can be a more significant injury, frequently with a delayed (2–4 months) return to sport compared with the more common lateral sprain.[52] There is minimal swelling and pain with isolated anterior tibiofibular ligament sprain. In contrast, posterior tibiofibular ligament injury is accompanied by severe pain frequently accompanied by a partial or near complete interosseous tear.[53] All ankle sprains with medial findings are grade IV type. Indoor soccer players incur ligamentous diastasis without fracture as a grade IV injury. The tibiofibular diastasis occurring with grade IV sprain warrants confirmation by stress radiography under anaesthesia. These stress films can show widening and indicate the need for surgery.[32] There may be a radiographic and even a clinical plastic deformation or bend in the tibia or fibula. Fifteen per cent of all ankle sprains meet these grade IV injury criteria.

Open sprains are quite rare and need immediate surgical repair and closure.[54]

DIFFERENTIAL DIAGNOSIS AND ASSOCIATED INJURIES

There are numerous acute injuries which may be overlooked or disguised at the time that an injured ankle presents.[55] Many of these diagnoses are established late and with difficulty.[56]

1. *Calcaneocuboid ligament sprain*   This is a foot sprain. Appropriate anterior–posterior and lateral films of the foot must be ordered. A bony fleck is often seen on the anterior–posterior view laterally adjacent to the cuboid.

2. *Subtalar joint sprain*[57]

   Type I   Forced supination of the hindfoot with plantar flexion can tear the anterior talofibular ligament with calcaneofibular ligament and lateral capsule tear.

   Type II   Forced supination of the hind-foot with plantar flexion resulting in an anterior talofibular ligament tear with talocalcaneal ligament tear.

   Type III   Forced supination with dorsiflexion of the ankle can rupture the calcaneofibular ligament, cervical ligament and the interosseous talocalcaneal ligament.

   Global ligament damage with severe ankle and subtalar sprain occurs with maximum inversion in a neutral ankle position.

3. *Sinus tarsi syndrome*   Lateral fat pad necrosis causes anterior lateral ankle pain post-injury. This scarring can usually be controlled by icing, ultrasound, and non-steroidal anti-inflammatory drugs.

4. *Synovial pinch*   The synovium can hypertrophy with 'pinching' or impingement at the anterior lateral ankle joint line, primarily with dorsiflexion.

5. *Meniscoid lesion*   Soft tissue entrapment occurs as a sequelae to acute ankle sprain. Either a synovial fold and or a piece of the torn anterior talofibular ligament becomes fixed in position and impinges between the lateral malleolus and the lateral talus.[58,59]

6. *Anterior capsular impingement*   Capsular ligamentous tissue can become shaggy and be caught or impinged at the anterior ankle joint.

Items (3)–(6) are entities overlapping in aetiology and regional anatomy. The arthroscope has enabled and helped categorization of these lesions. More aggressive and universal functional treatment of acute ankle sprains, particularly recurrent acute ankle sprains, may have increased their incidence.

7. *Os calcis anterior process fracture with avulsion of the bifurcate ligament ('beak fracture')*   Inversion and internal rotation can cause this injury. Radiography confirms the fracture. Rocking the hindfoot in the medial lateral plane and local tenderness when painful should call attention to the possibility of this diagnosis.

8. *Compression fracture of the anterior calcaneal articular facet*   Local tenderness and CT or polytomography may be necessary to detail this injury

9. *Common peroneal nerve injury*[37,60] *superficial peroneal nerve or crural nerve injury*   Stretch nerve fracture occurs prior to rupture of the epineurium since connective tissue is strange and will stretch. However, myelin and axoplasm have only a 6 per cent elongation before neuropraxia/axonotomesis occur.[60] Sensory hypaesthesia or hyperaesthesia in the involved sensory distribution is usually present. Late reflex dystrophy and slow proprioceptive return can accompany this injury.

10. *Talar dome fractures (transchondral or osteochondral)*[61]   Initial films may appear negative. Repeat radiographs are often necessary for diagnosis. Osteochondritic lesions can create diagnostic confusion.

11. *Lateral talar process fracture with or without lateral talocalcaneal ligament injury*[62,63]   This is a common injury seen on the anterior–posterior ankle films. The size of the avulsed fragment is variable. The fracture line is vertical. Closed treatment is used.

12. *Posterior impingement, posterior lateral process fracture (tubercle fracture) or os trigonum fracture/separation*[63,64]   This area has the most variable anatomy in the hindfoot. Close radiograph examination and occasional CT or linear tomography scan plus the clinical picture is necessary for diagnosis (Fig. 6).

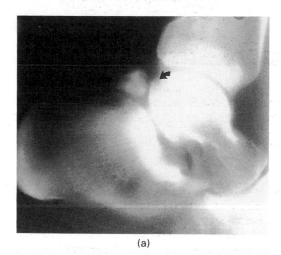

(a)

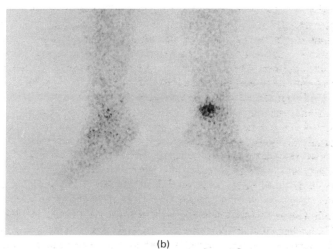

(b)

Fig. 6   This 20-year-old patient continued to have posterior medial ankle pain and was unable to play agility sports 4 months after a grade II sprain. Both radiography and bone scan suggested a posterior impingement syndrome. Surgical excision completely relieved his symptoms. (a) The opposing lucency (arrow) on tomography should be noted. (b) The bone scan shows increased uptake at the area of impingement (separation of the synchondrosis of the os trignonum).

13. *Tibiotalar ligament posterior medial process avulsion*   This talar fracture lesion is much less common than posterior lateral process injury. Bilateral CT scanning may be necessary for accurate diagnosis.

14. *Peroneal tendon subluxation*   Radiographs will usually show a small bone chip. Eversion and dorsiflexion allow the tendon attachment to the fibula to separate, often with a vertical fleck

or periosteum and bone. Immobilization for 3 weeks can obviate recurrent subluxation and the need for early or late lateral repair and/or bony buttress.

15. *Fracture of the base of the fifth metatarsal* (Fig. 7(b)) Avulsion fractures or more distal fractures at the proximal diaphysis can easily be seen on both anterior–posterior lateral and oblique views of the foot. However, clinical tenderness without verification by radiography often occurs with acute or stress fracture. This clinical picture warrants radiographic re-examination in 10 to 14 days. Even with minimal late symptoms there can be delayed or non-union of the unrecognized fracture. A high index of suspicion is necessary.

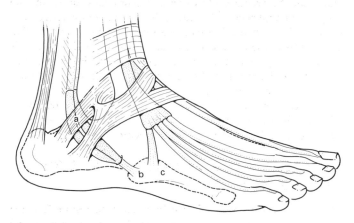

Fig. 7 Tears of the lateral retinaculum and tears of the peroneal tenosynovium occur with tears of the calcaneofibular ligament. This allows tenography and arthrography to help with differential diagnosis (a). The fifth metatarsal is often fractured with supination inversion injuries. The avulsion fractures of the base of the fifth metatarsal (b) or fractures of the proximal diaphysis (c) can occur with or without ankle sprain.

16. *Inferior tibiofibular ligament injury*[53,65] This occurs with or without associated bony injury.
17. *Interosseous ligament injury (syndesmotic sprain)*[66–69] The anterior tibiofibular ligament is often torn with grade II ankle sprain. When a grade IV sprain occurs with forced dorsiflexion and external rotation, the posterior tibiofibular ligament injury together with interosseous injury can occur. High fibula tenderness suggests an associated 'Maissoneuve' fracture. Mortise views (20° internal rotation oblique) can confirm widening. Surgery with a transmortise syndesmotic screw may be necessary. These injuries are only seen in 1 to 2 per cent of all ankle sprains.
18. *Stress fracture of the ankle or hindfoot* (Fig. 7(c)) There will have been symptoms prior to the injury. A good history and a high index of suspicion is necessary to recognize this repetitive stress problem at presentation.
19. *Distal fibular epiphyseal injury in children*[31] Local anterior and posterior physeal area tenderness make this a potential diagnosis even when radiographs are negative. Many adolescents with major ankle 'sprains' will present with a type I or type II Salter–Harris epiphyseal plate injury rather than the ligamentous injury characteristic of the mature skeleton.[90]
20. *Stenosing peroneal tenosynovitis* (Fig. 7(a))[71] Tenosynovial post-traumatic fracture scarring and inflammation occasionally requires tenosynovectomy.

21. *Avulsion fracture of the lateral malleolus*[72] An anterior talofibular ligament tear can avulse an anterior lateral chip of fibula bone. A substantial 4- to 6-mm fragment can be avulsed at the fibula attachment of the calcaneofibular ligament.
22. *Fracture of the lateral malleolus*[43].

TREATMENT

General principles

Ideal treatment for acute ankle sprain has evolved and is evolving.[38,73] The pneumatic compression brace and other orthoses (Figs 8(a) and 8(b))[74–77] are successful with early protected

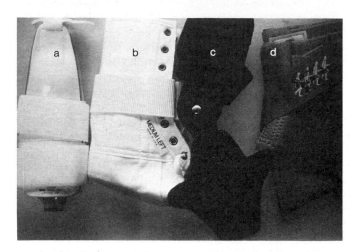

Fig. 8 Ankle braces and supports of many types can be helpful. The pneumatic compression brace (a) is a cornerstone of the functional treatment of ankle sprain. The variably flexible (with or without medial and lateral firm buttresses) lace brace (b) competes with taping for late post-injury protection and prevention. The hinge brace (c) is less often used but appropriate for protection even though inversion/eversion mobility is more rigidly controlled than with the other braces. Elastic braces (d) of several types provide less rigid and more dynamic support. (From ref. 55; reprinted by permission of WB Saunders Co.)

mobilization[78] and promote less post-traumatic atrophy and stiffness.[79,80] Thus a quicker, more efficient, and competent return to sports is achieved. It is apparent that an early intense post-injury programme can avoid a longer, and usually more expensive, rehabilitation programme.[44,81] This ideal treatment cannot be achieved in all patients, particularly for those with recreational injuries in a setting of socioeconomic obligations, poor availability of care, and/or other factors. For this reason immobilization, i.e. a short-leg cast, permits the patient to remain in the work-place and can be considered appropriate treatment in many cases. Thus treatment for the employed person with a routine ankle sprain can differ considerably from that for an elite or lower-level competitive athlete participating on a high school, college, professional, or even 'mid-career' level. The earlier more aggressive programme with the athlete requires more initial resources, i.e. additional time, a facilitating environment, trainer/therapy personnel, etc. This early, aggressive, and time-intensive treatment/rehabilitation results in prompt weight bearing, quicker strength regain, earlier return to competition, and reduced treatment requirements.[29,78,81,82] Early mobilization with this functional treatment must be accompanied by protective ankle support (Fig. 8(c)).[55,81,83,84] Full rehabilitation may take as long as 20 weeks.[7,85]

Stemming from Broström's experience in 1976, when all but 3 per cent of acutely operated ankle sprains became normal, there has been a trend towards primary repair of sprains with proven anterior talofibular ligament and calcaneofibular ligament tears. These observations and the fact that surgical treatment of acute ankle sprain is successful with minimally increased temporary morbidity is not debated.[33,37,45,86,87] However, there seems to be no inherent advantage in primary operative repair.[14,46,72,78,88–91]

The clear trend and sense of the 1990s is towards non-operative treatment of the sprained ankle. Although there is continuing debate, the aggressive non-operative approach is less expensive and makes surgery unnecessary.[37,40,83,92,93] Even in the series with a more radical approach, only 12 per cent were surgically repaired.[29]

Certain authors have felt and still feel that a grade III sprain with a double-ligament tear needs surgical repair[8,18,49,64,72,80,86,94–96] in selected athletes, i.e. highly competitive athletes including performing artists.[64]

It has been shown that ligamentous tissue is largely retained even with recurrent sprains.[97,98] This fact implies the equally successful results with later reconstruction which have now been verified.[36,37,43] This chronic reconstruction, when coupled with better protective devices, reduces the necessity for primary acute surgery and argues for functional treatment only.[4,33,39,77,92] Surgical repair in children has been reported and can be successful, but in keeping with the current philosophy is generally not necessary.[99]

To summarize, current thinking and consensus in the 1990s is that functional treatment or management is indicated for essentially all grade I, II, and III ankle sprains with surgery reserved only for isolated highly competitive athletes with proven gross instability/significant subluxation or if there is documented bone or cartilage injury.

A severe eversion grade IV sprain involving the interosseous membrane requires 3 to 6 weeks of cast immobilization. Tibiofibular diastasis, when present and proved radiographically, warrants transtibiofibular fixation in the form of a syndesmotic screw.[39] Heterotopic ossification can occur in up to 50 per cent of interosseous injuries but usually does not require surgical excision of the ossification unless frank synostosis is present.[100]

Specific treatment principles

Many algorithms, and specific protocols have been suggested.[14,48,101] All encourage a systematic approach and are generically appropriate. By consensus, there are three phases in the treatment and rehabilitation of an acute ankle sprain.

Phase I is cryotherapy with protected range of motion and weight bearing to tolerance. Compression and elevation enhance venous and lymphatic return and help control oedema. A pneumatic type splint is integral to this early functional treatment (Fig. 8(a)).[75]

Phase II is continuing cryotherapy, isometric strengthening, gradually increasing range of motion, proprioceptive work, and proceeding from full linear stress to early agility work.

Phase III is continued strengthening, flexibility, and proprioceptive work, followed by progressive sports-specific return to competitive or full recreational function. Phases I and II (disability phases) are estimated to take approximately 8 days for grade I injuries and 15 days for grade II injuries.[102]

Many modalities are used for the treatment of acute ankle sprains. Ultrasound, ice, and heat delivered in appropriate ways have legitimate therapeutic benefits.[29,103,104] High voltage pulse stimulation does not add to ankle sprain treatment.[105] Cryotherapy with ice or an ice substitute[106] is essential for acceptable initial care and later rehabilitation. Ice is effective for analgesic purposes.[29] Ice increases the pain-free range of motion, decreases muscle fatigue, increases blood flow, and forestalls oedema and inflammation. If ice is left on for too long a reflex increase in blood flow to the skin occurs. A significant cooling of the muscle in thin athletes with less than 1 cm of fat occurs within 10 min to a maximum depth of approximately 2 cm into the muscle (or presumably ligament in the case of ankle sprain). Athletes with more than 2 cm of fat require 20 to 30 min. Cryotherapy (direct and not simple cold application)[106] tends to deliver cold in the range of 3° to 10°C. Cryokinetics is a combination of ice and exercise. This combination, when coupled with early range of motion and early weight bearing, constitutes functional treatment. Heat should be in the form of contrast baths (alternating with ice) or used alone following ice and elevation.

Non-steroidal anti-inflammatory drugs have a minor treatment role with their mild anti-inflammatory and analgesic properties.[107] Treatment for 1 to 4 days during the acute phase is sufficient.

REHABILITATION OF AN ACUTE ANKLE SPRAIN

The aim of post-sprain ankle rehabilitation is to restore function and agility and to regain maximum strength in both concentric and eccentric modes.[108,109] The muscles involved are the triceps surae, posterior tibial, peroneal, and anterior tibial. There also needs to be a maximum pain-free range of motion and at least 10° of dorsiflexion. This implies that isometry of rehabilitation in the ankle is important, just as it is in the knee. The heel cord needs to be supple to permit adequate dorsiflexion, a normal running gait, and prevention of secondary injury.

Proprioception allows the athlete to have a sense of movement and a sense of the position of the ankle (or any part) in space. The importance of proprioception is not understood or used in principle or in fact by most practitioners handling ankle sprains. Only 2 to 3 per cent of emergency room physicians and only 25 per cent of orthopaedists have a grasp of, and utilize, the appropriate proprioceptive principles.[14] The necessity for proprioceptive recovery is often ignored in rehabilitation.[63] Figure-eights, one-legged stands, tilt boards, and other agility efforts accelerate, promote, and allow testing to define proprioceptive return.[52,85] Agility and a full return to the sports arena can be achieved with such proprioceptive normalizers.[110]

Occasionally a traditional brace with a single or even a double upright and a lateral T-strap can be used for ambulatory function in working individuals. These selected individuals might need protection in a postoperative situation or with a major grade III sprain with the need for a prolonged protective range of motion and weight bearing. They are most likely to be recreational athletes without access to aggressive functional treatment and the need for major ankle stability during work. The most appropriate functional orthoses are generally also cheaper and easier to use.

A final note on the athlete's rehabilitation effort is that he or she should not be allowed to exceed the prescribed protocol. Overmotivation breeds overcompliance, and this can be self-defeating and injurious to the athlete and even delay return to recreation or competition.

Rehabilitation by any protocol must be systematic and grad-uated. A functional optimal progression of a baseball player with grade II–III ankle sprain and a properly supervised and aggressive functional treatment programme might include the following:

partial weight bearing at 3–5 days;
full weight bearing at 4–7 days;
walking limp-free slowly at 7–10 days;
linear slow atraumatic running at 10–14 days;
linear speed work at 14–17 days;
careful cutting, figure-eight, and lateral moves at half-speed at 15–21 days;
agility drills at full speed at 20–25 days;
sport-specific protected practice with sport-specific return near the pre-injury level of function at no earlier than 3 weeks.

PREVENTION OF ACUTE ANKLE SPRAINS

Training of the ankle is tantamount to injury prevention and is extremely important. Maximum strength is the cornerstone of prevention. This strength can be developed by numerous isomet-ric, concentric, and eccentric methods. Numerous pieces of equipment are available and range from low cost to expensive 'high tech' varieties. All can be used. Benefit will occur with appropriate supervision and programme prescription.

Appropriate footwear enhances stability of the ankle position and causes/facilitates decreased inversion/ankle motion through muscle action. There is some question as to whether the more sophisticated modern athletic shoe increases or in some way masks or absorbs some of the required proprioceptive sensations. High top shoes (Fig. 9) help reduce inversion ankle injuries.[7] There

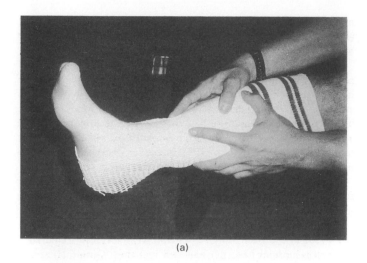

(a)

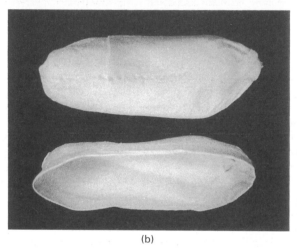

(b)

Fig. 10 (a) Individually customized 'in-house', but easily fabricated and cheap, orthoplast semi-rigid supports provide excellent and early protection and support; (b) customized semi-rigid foot orthoses made from plaster moulds are often indicated post-rehabilitation, i.e. with forefoot valgus and medial post, to help prevent recurrent inversion sprain of the ankle.

Fig. 9 High top shoes of an 'off the shelf' or specified type can provide additional support. The higher hind uppers coupled when necessary with strong heel counters and an in-shoe orthosis provide significant ongoing stability.

seems to be no difference in protection between standard high top shoes such as modified basketball shoes and standard lightweight infantry boots.[6] Semirigid custom-made orthoses can integrate into an aggressive rehabilitation and prevention programme (Fig. 10). Rigid orthoses are rarely indicated.

Since higher values of $W + H^2$, where W is weight and H is

height, causes higher lateral ankle sprain morbidity, a normal weight is both desirable and preventive.[6]

Ankle taping is generally positive. Elastic ankle guards are of no significant help.[89] The disadvantages of taping are that it decreases performance in certain high level sports participants, encourages disuse and decreased strength of the ankle-supporting muscula-ture, and decreases subtalar joint motion. Taping, although better than certain 'ankle guards', quickly becomes ineffective[20,111] and is also expensive.[112] In addition, it theoretically preloads other joints, i.e. the forefoot, and predisposes to turf toe and meta-tarsalgia.[113] There are a number of studies comparing laced ankle stabilizers with taping etc.[111] Low-top shoes and laced ankle stabilizers were found best in one study.[50] Others[114] found taping better with a 30 to 50 per cent decreased range of motion but a loss of some effectiveness after 1 h.[88] Neither causes significant inter-ference with performance, although vertical jump and sprinting are measurably altered.[112]

Functional stability and dynamic postural control are critical (ankle disc co-ordination training) and must not be subordinated to mechanical support such as orthoses.[115] The latter is supplemental and mainly temporary. Both have a place in the overall treatment programme.

Ankles with previous sprains are twice as likely to be sprained as an ankle with no prior ankle sprain history.[6] Preventive taping or laced stabilizers are recommended, particularly in high risk athletes, e.g. soccer players. Recurrent sprains induce an inversion lever. Since relatively weak peroneals allow the heel to plant in mild varus,[71] a protective orthosis as well as a full strength and agility programme is necessary to prevent an acute sprain in an ankle with a previous sprain.

Friden et al.[116] use stabilometry techniques which suggest significant post-sprain agility improvement with functional treatment and a pneumatic brace. Clinical impressions and studies leave similar suggestions.

Fatiguing activities, as well as simple 'lack of sleep' fatigue, can cause local predisposition to sprain. Logical sleep habits adequately meeting individual sleep requirements are desirable.

CONCLUSION

The acute ankle sprain is treated by the whole range of medical personnel who deal with recreational and competitive athletic injuries. Many associated and secondary injuries are often initially unrecognized. Much of the post-sprain problem of pain, instability, and impaired performance are due to untreated or inadequately treated acute and recurrent ankle sprains.

All practitioners involved in treating sports injuries should have or should develop a reasonable protocol for the severe ankle sprain. This protocol includes early diagnosis, aggressive protected and ongoing weight bearing, and functional treatment. This should be done by the patient at home or in a formal physical therapy setting. Although functional treatment is generally indicated, surgery may be necessary for selected elite athletes and performers.

There should be a clear and conclusive termination point for rehabilitation, at which time sports can be safely resumed.

With the above points in mind, the ankle is a reasonably forgiving joint (i) if the index of suspicion for injury such as in the inferior tibial–fibular joint is high, (ii) if the appropriate diagnosis is made,[56,90] (iii) if treatment is not necessarily shortened or compromised, (iv) if adequate rehabilitation is required before return to competition or full use, and (v) if ongoing protection is used if indicated. Acute ankle sprain maintains its importance in sports medicine. Ongoing and increased cognizance of the importance of acute ankle sprain is central to proper prevention, and treatment.

REFERENCES

1. Henry JH, Lareau B, Neigut D. The injury rate in professional football. *American Journal of Sports Medicine* 1982; **10**: 16–18.
2. Colliander E, Eriksson E, Herkel M, Sköld P. Injuries in Swedish elite basketball. *Orthopedics* 1986; **9**: 225–7.
3. Zelisko AJ, Noble HB, Porter M. A comparison of men's and women's professional basketball injuries. *American Journal for Sports Medicine* 1982; **10**: 297–9.
4. Kannus P, Renström P. Treatment for acute tears of the lateral ligaments of the ankle. *Journal of Bone and Joint Surgery* 1991; **73A**: 305–23.
5. Garrick JG. The frequency of injury, mechanism of injury, and epidemiology of ankle sprains. *American Journal of Sports Medicine* 1977; **5**: 241–2.
6. Milgrom C, *et al.* Risk factors for lateral ankle sprain: a prospective study among military recruits. *Foot and Ankle* 1991; **12**: 26–30.
7. DeHaven KE, Allman FL, Cos JS, Fowler PJ, Henning CE. Symposium: ankle sprains in athletes. *Contemporary Orthopaedics* February 1979: 56–78.
8. Boruta PM, Bishop JO, Braly WG, Tullos HS. Acute lateral ankle ligament injuries: A literature review. *Foot and Ankle* 1990; **11**: 107–13.
9. Calliet R. Injuries to the ankle. *Foot and ankle pain*. Philadelphia: FA. Davis, 1968: 117–25.
10. Holbrook TL. *et al.* The frequency of occurrence, impact and cost of selected musculoskeletal conditions in the United States. *American Academy of Orthopaedic Surgeons Annual Meeting* 1984: 101.
11. Garrick JG, Requa RU. Epidemiology of foot and ankle injuries in sports. *Clinical Sports Medicine* 1988; **7**: 29–36.
12. Ekstrand J, Tropp H. The incidence of ankle sprains in soccer. *Foot and Ankle* 1990; **11**: 41–4.
13. Schmidt-Olsen S, Jøøorgensen U, Kaalund S, Søørensen J. Injuries among young soccer players. *American Journal of Sports Medicine* 1991; **19**: 273–5.
14. Kay DB. The sprained ankle: current therapy. *Foot and Ankle* 1985; **6**: 22–8.
15. Kelikian H, Kelikian AS. Disruption of the fibular collateral ligament. *Disorders of the ankle*. Philadelphia: W. B. Saunders, 1988: 437–95.
16. Attarian DE, DeVito DP, Garrett, Jr. W. A biomechanical study of human ankle ligaments and autogenous reconstructive grafts. *Surgical Rounds for Orthopaedics* April 1987: 24–7.
17. Leonard MH. Injuries of the lateral ligaments of the ankle. A clinical and experimental study. *Journal of Bone and Joint Surgery* 1949; **31A**: 373–7.
18. Anderson ME. Reconstruction of the lateral ligaments of the ankle using the plantaris tendon. *Journal of Bone and Joint Surgery* 1985; **67A**: 930–4.
19. Lindsjo U, Danckwardt-Lilliestrom G, Sahlstedt B. Measurement of the motion range in the loaded ankle. *Clinical Orthopaedics and Related Research* October 1985; **199**: 68–71.
20. Sammarco J. Biomechanics of the ankle. I. Surface velocity and instant center of rotation in the sagittal plane. *American Journal of Sports Medicine* 1977; **5**: 231–4.
21. Lundberg A, Goldie I, Kalin B, Selvik G. Kinematics of the ankle/foot complex: plantarflexion and dorsiflexion. *Foot and Ankle* 1989; **9**: 194–200.
22. McCullough CJ, Burge PD. Rotatory stability of the load-bearing ankle. *Journal of Bone and Joint Surgery* 1980; **62B**: 460–4.
23. Shojhi H, Ambrosia RD, Parlaska R. Biomechanics of the ankle. II. Horizontal rotation and ligamentous injury states. *American Journal of Sports Medicine* 1977; **5**: 235–7.
24. Stormont DM, Morrey BF, An KN, Cass JR. Stability of the loaded ankle. Relation between articular restraint and primary and secondary static restraints. *American Journal of Sports Medicine* 1985; **85**: 295–300.
25. Burgess PR, Wei JY, Clark FJ, Simon J. Signalling of kinesthetic information by peripheral sensory receptors. *Annual Review of Neuroscience*. 1982; **5**: 171–87.
26. Bulucu C, Thomas KA, Halvorson TL, Cook SD. Biomechanical evaluation of the anterior drawer test: the contribution of the lateral ankle ligaments. *Foot and Ankle* 1991; **11**: 389–93.
27. Nigg BM, Skarvan G, French CB, Yeadon MR. Elongation and forces of ankle ligaments in a physiological range of motion. *Foot and Ankle* 1990; **11**: 30–40.
28. Dias LS. The lateral ankle sprain: an experimental study. *Journal of Trauma* 1979; **19**: 266–9.
29. Cox JS. Surgical and nonsurgical treatment of acute ankle sprains. *Clinical Orthopaedics* 1985; **188**: 88–96.
30. Ogden JA. *Skeletal injury in the child*. Philadelphia: WB Saunders, 1990: 853–6.
31. Whitelaw GP, Sawka MW, Wetzler M, Segal D, Miller J. Unrecognized injuries of the lateral ligaments associated with lateral malleolar fractures of the ankle. *Journal of Bone and Joint Surgery* 1989; **71A**: 1396–9.
32. Hopkinson WJ, St Pierre P, Ryan JB. Syndesmosis sprains of the ankle. *Foot and Ankle* 1990; **10**: 325–30.
33. Henry JH. Lateral ligament tears of ankle. One to six years follow-up study of 202 ankles. *Orthopaedic Review* 1983; **12**: 31–9.
34. Auletta A. G., Conway W. F., Hayes C. W., Guisto D. F., Gervin AS.

Indications for radiography in patients with acute ankle injuries: role of the physical examination. *American Journal of Radiology* 1991; **157**: 789–91.

35. Anonymous. X-rays under pressure reveal hidden injury to ankle. *Medical World News* 1967 Oct: 50–1.

36. Brand RL, Collins MF, Templeton T. Surgical repair of ruptured lateral ankle ligaments. *American Journal of sports Medicine* 1981; **9**: 40–4.

37. Rijke AM, Jones B, Vierhout PA. Injury to the lateral ankle ligaments of athletes. *American Journal of Sports Medicine* 1988; **16**: 256–9.

38. Staples OS. Ruptures of the fibular collateral ligaments of the ankle. *Journal of Bone and Joint Surgery* 1975; **67A**: 101–7

39. Bergfeld JA, Cox JS, Drez Jr D, Raemy H, Weiker GG. Symposium: management of acute ankle sprains. *Contemporary Orthopaedics* 1986;**13**: 83–116.

40. Rijke AM. Lateral ankle sprains. Graded stress radiography for accurate diagnosis. *Physician and Sports Medicine* 1991; **19**: 107–18.

41. Ehrensperger J. Arthrography of the ankle joint in injuries of the fibular ligament system in children and adolescents. *Zeitschrift fur Kinderchirurgie* 1985; **40**: 71–5.

42. Bleichrodt RP, Kingma LM, Binnendijk LM, Klein JP. Injuries of the lateral ankle ligaments: classification with tenography and arthrography. *Radiology* 1989; **173**: 347–9.

43. Lassiter TE, Malone TR, Garrett Jr WE. Injury to the lateral ligaments of the ankle. *Orthopedic Clinics of North America* 1980; **20**: 629–40.

44. Drez D, Young JC, Waldman D, Shackleton R, Parker W. Nonoperative treatment of double lateral ligament tears of the ankle. *American Journal of Sports Medicine* 1982; **10**: 197–200.

45. Ruth CJ. The surgical treatment of injuries of the fibular collateral ligaments of the ankle. *Journal of Bone and Joint Surgery* 1961; **43A**: 229–39.

46. Evans GA, Hardcastle P, Frenyo AD. Acute rupture of the lateral ligament of the ankle. *Journal of Bone and Joint Surgery* 1984; **66B**: 209–12.

47. Evans GA, Frenyo SD. The stress-tenogram in the diagnosis of ruptures of the lateral ligament of the ankle. *Journal of Bone and Joint Surgery* 1979; **61B**: 347–51.

48. Edelman B. MRI plus clinical exam most sensitive to chronic pain after ankle sprain. *Orthopedics Today* 1991; **11**: 14–15.

49. Jahss MH. *Disorders of the foot and ankle*. Vol. III, 2nd edn. Philadelphia: WB Saunders, 1991: 2408–10.

50. Rovere GD, Clarke TJ, Yates CS, Burley K. Retrospective comparison of taping and ankle stabilizers in preventing ankle injuries. *American Journal of Sports Medicine* 1988; **16**: 228–33.

51. Hamilton WG. Foot and ankle injuries in dancers. *Clinics in Sports Medicine* 1988; **7**: 143–73.

52. Ryan AJ, Cox JS, Inniss B, Rice LE, Woodward EP. Round table discussion: ankle sprains. *Physician and Sports Medicine* 1986; **14**: 101–18.

53. Colville MR, Marder RA, Boyle JJ, Zarins B. Isolated injury of the distal tibiofibular ligaments: a clinical and experimental report. *American Journal of Sports Medicine*, 1990; **118**: 196–200.

54. White J. Rare open ankle sprain sidelines Weiss. *Physician and Sports Medicine* 1991; **19**: 47–8.

55. McBryde AM. Disorders of the ankle and foot. In: Grana WA, Kalenak A, eds. *Clinical sports medicine*. Philadelphia: WB Saunders, 1991: 466–89.

56. Grana WA. Chronic pain persisting after ankle sprain. *Journal of Musculoskeletal Medicine* June 1990: 35–49.

57. Meyer JM, Garcia J, Hoffmeyer P, Fritzchys D. The subtalar sprain. A roentgenographic study. *Clinical Orthopaedics and Related Research* 1988; **226**: 169–73.

58. McCarroll JR, Schrader JW, Shelbourne KD, Rettig AC, Bisesi MA. Meniscoid lesions of the ankle in soccer players. *American Journal of Sports Medicine* 1987; **15**: 255–7.

59. Bassett FH, Gates HS, Billys JB, Morris HB, Nikolaou PK. Talar impingement by the anterior-inferior tibiofibular ligament. *Journal of Bone and Joint Surgery* 1990; **72A**: 55–9.

60. Nitz AJ, Dobner JJ, Kersey D. Nerve injury and grades II and III ankle sprains. *American Journal of Sports Medicine* 1985; **13**: 177–82.

61. Alexander AH, Barrack RL. Arthroscopic technique in talar dome fractures. *Surgical Rounds for Orthopaedics* January 1990: 27–35.

62. Amis JA, Gangl PM, Graham CE. Inversion ankle injuries: an accurate diagnosis requires a high level of suspicion. Poster presentation at *American Association of Orthopaedic Surgeons Annual Meeting, Cincinnati*, 1991.

63. Arendt E. Inversion injuries to the ankle. *Surgical Rounds for Orthopaedics* June 1989: 15–22.

64. Bergfeld JA, Fu F, Garrick J, Hamilton W, Weiker G. Musculoskeletal problems of dance and gymnastics. *American Orthopaedic Society for Sports Medicine Annual Meeting, Palm Springs*, 1988.

65. McMaster JH. Tibiofibular synostosis: a cause of ankle disability. *Journal of Bone and Joint Surgery* 1975; **57A**: 1035.

66. Manderson EL. The uncommon sprain. Ligamentous diastasis of the ankle without fracture or bony deformity/ *Orthopaedic Review* 1986; **15**: 77–81.

67. Hopkinson WJ, St Pierre P, Ryan JF, Wheeler JH. Syndesmosis sprains of the ankle. *Foot and Ankle* 1990; **10**: 325–30.

68. Marymount JV, Lynch MA, Henning CE. Acute ligamentous diastasis of the ankle without fracture. Evaluation of radionuclide imaging. *American Journal of Sports Medicine* 1986; **14**: 407–9.

69. Liever L, Gross M. Frank ligamentous diastasis of the ankle. *Orthopaedic Grand Rounds* 1985; **2**: 10–15.

70. Letts RM. The hidden adolescent ankle fracture. *Journal of Pediatric Orthopedics* 1982; **2**: 161–4.

71. Andersen E. Stenosing peroneal tenosynovitis symptomatically simulating ankle instability. *American Journal of Sports Medicine* 1987; **15**: 258–9.

72. Brand RL. Operative management of ligamentous injuries. In: Torg, JS, Welsh RP, Shephard RJ, eds. *Current therapy in sports medicine 2*. Toronto: BC Decker, 1990: 239–41.

73. Moseley HF. Traumatic disorders of the ankle and foot. *Ciba Foundation Symposium* 1965; **17**: 3–30.

74. Carne P. Nonsurgical treatment of ankle sprains using the modified Sarmiento brace. *American Journal of Sports Medicine* 1989; **17**: 256–7.

75. Kimura IF, Nawoczenski DA, Epler M, Owen MG. Effect of the Air Stirrup in controlling ankle inversion stress. *Journal of Orthopaedic and Sports Physical Therapy* 1987; **8**: 190–3.

76. Lane SE. Severe ankle sprains. Treatment with ankle-foot orthosis. *Physician and Sports Medicine* 1990; **18**: 43–51.

77. Fritschy D, Junet C, Bonvin JC. Functional treatment of severe ankle sprain. *Journal de Traumatologie du Sport* 1987; **4**: 131–6.

78. Konradsen L, Høolmer P, Søondergaard L. Early mobilizing treatment for grade III ankle ligament injuries. *Foot and Ankle* 1991; **12**: 69–73.

79. Neumann H, O'Shea P, Nielson J, Climstein M. A physiological comparison of the short-leg walking cast and an ankle–foot orthosis walker following six weeks of immobilization. *Orthopedics* 1989; **12**: 1429–34.

80. Baxter DE. Traumatic injuries to the soft tissues of the foot and ankle. In: Mann R, ed. *Surgery of the foot* 5th edn. St Louis: CV Mosby 1986: 456–501.

81. van den Hoogenband CR, van Moppes FI, Coumans PF, Stapert JWJL, Geep JM. Study on clinical diagnosis and treatment of lateral ligament lesions of the ankle joint: a prospective clinical randomized trial. *International Journal of Sports Medicine* 1984; **5**(suppl.): 159–61.

82. Rettig AC, Kraft DE. Treat ankle sprains fast—it pays. *Your Patient and Fitness* 1991; **4**: 6–9.

83. Linde F, Hvass I, Jürgensen U, Madsen F. Early mobilizing treatment in lateral ankle sprains: course and risk factors for chronic painful or function-limiting ankle. *Scandinavian Journal of Rehabilitation Medicine* 1986; **18**: 17–21.

84. Raemy H, Jakob RP. Functional treatment of fresh tibular ligament lesions using the aircast splint. *Swiss Journal of Sports Medicine* 1983; **31**: 1–5.

85. Vegso JJ. Non-operative management of ankle injuries. In: Torg JS, Welsh RW, Shephard RJ. *Current therapies in sports medicine 2*. Toronto: BC Decker, 1990: 234–9.

86. Korkala O, Rusanen M, Jokipii P, Kytömaa J, Avikainen V. A prospective study of the treatment of severe tears of the lateral ligament of the ankle. *International Orthopaedics* 1987; **11**: 13–17.

87. Jaskulka R, Fischer G, Schedl R. Injuries of the lateral ligaments of the ankle joint. Operative treatment and long-term results. *Archives of Orthopaedic and Traumatic Surgery* 1988; **107**: 217–21.

88. Myburg KH, Vaughan CL, Isaacs SK. The effects of ankle guards and taping on joint motion before, during and after a squash match. *American Journal of Sports Medicine* 1984; **12**: 441–6.

89. Sommer HM, Arza D. Functional treatment of recent ruptures of the fibular ligament of the ankle. *International Orthopaedics* 1988; **13**: 69–73.

90. Bassett FH. The treatment of ankle and foot sports injuries. In: Schneider RC, Kennedy JC, Plant ML, eds. *Sports injuries: mechanisms, prevention and treatment.* Baltimore: Williams & Wilkins, 1985: 788–96.

91. Ciullo PV, Jackson DW. Track and field. In: Schneider RC, Kennedy JC, Plant ML, eds. *Sports injuries: mechanisms, prevention and treatment.* Baltimore: Williams & Wilkins, 1985: 212–46.

92. Renström P. Sports traumatology today. A review of common current sports injury problems. *Annales Chirurgiae et Gynaecologiae* 1991; **80**: 81–93.

93. Schaap GR, deKeizer G, Marti K. Inversion trauma of the ankle. *Archives of Orthopaedic and Trauma Surgery* 1989; **108**: 273–5.

94. Peterson L. Ankle ligament injuries and operative treatment principles. *Annales Chirurgiae et Gynaecologiae* 1991; **80**: 168–76.

95. Leach RE, Schepsi A. Ligamentous injuries. In: Yablon IG, Segal D, Leach RE, eds. *Ankle injuries.* New York: Churchill Livingstone, 1983: 193–230.

96. Cass JR, Morrey BF, Katoh Y, Chao EYS. Ankle instability: comparison of primary repair and delayed reconstruction after long-term follow-up study. *Clinical Orthopaedics* 1985; **198**: 110–17.

97. Broström L, Sundelin P. Sprained ankles. IV. Histologic changes in recent and 'chronic' ligament ruptures. *Acta Chirurgica Scandinavica* 1966; **132**: 248–53.

98. Gould N, Seligson D, Gassman J. Early and late repair of lateral ligament of the ankle. *Foot and Ankle* 1980; **1**: 84–9.

99. Vahvanen V, Westerlund M, Kajanti M. Sprained ankle in children. A clinical follow-up study of 90 children treated conservatively and by surgery. *Annales Chirurgiae et Gynaecologiae* 1983; **72**: 71–5.

100. Speir KP, Warren RF, Wall DJ. Update on football injuries. *Mediguide to Orthopaedics* 1991; **10**: 1–6.

101. Larkin, J, Brage M. Ankle, hindfoot and midfoot injuries. In: Reider B, ed. *Sports medicine. The School-age athlete.* Philadelphia: WB Saunders 1991; 365–79.

102. Jackson DW, Ashley RL, Powell JW. Ankle sprains in young athletes. Relation of severity and disability. *Clinical Orthopaedics* 1974; **101**: 201–15.

103. Lehman LE. *Therapeutic heat and cold rehabilitation.* 3rd edn. Baltimore: Williams & Wilkins, 1982.

104. Halvorson GA. Therapeutic heat and cold for athletic injuries. *Physician and Sports Medicine* 1990; **18**: 87–94.

105. Michlovitz S, Smith W, Watkins M. Ice and high voltage pulsed stimulation in treatment of acute lateral ankle sprains. *Journal of Orthopaedic and Sports Physical Therapy* 1988; **9**: 301–14.

106. Hocutt JR, Jaffe R, Rylander CR, Beebe JK. Cryotherapy in ankle sprains. *American Journal of Sports Medicine* 1982; **10**: 316–19.

107. Fredberg U, Hansen PA, Skinhøoj A. Ibuprofen in the treatment of acute ankle joint injuries. *American Journal of Sports Medicine* 1989; **17**: 564–6.

108. Rettig AC, Kraft DE. Treat ankle sprains fast—it pays. *Your Patient and Fitness* 1991; **15**: 6–9.

109. Hunter S. Rehabilitation of ankle injuries. In: Prentice WE, ed. *Rehabilitation techniques in sports medicine* St. Louis: Times Mirror/Mosby, 1990: 331–8.

110. Garrick JG. 'When can I . . . ?' A practical approach to rehabilitation illustrated by treatment of ankle injury. *American Journal of Sports Medicine* 1987; **15**: 258–9.

111. Bunch RP, Bednarski K, Holland D, Macinati R. Ankle joint support: a comparison of reusable lace-on braces with taping and wrapping. *Physician and Sports Medicine* 1985; **13**: 59–62.

112. Burks RT, Bean BG, Marcus R, Barker H. Analysis of athletic performance with prophylactic ankle devices. *American Journal of Sports Medicine* 1991; **19**: 104–6.

113. Carmines DV, Nunley JA, McElhancy JH. Effects of ankle taping on the motion and loading pattern of the foot for walking subjects. *Journal of Orthopaedic Research* 1988; **6**: 223–9.

114. Garrick JG, Requa R. Role of external support in the prevention of ankle sprains. *Medicine and Science in Sports* 1973; **5**: 200–3.

115. Tropp H, Askling C, Gillquist J. Prevention of ankle sprains. *American Journal of Sports Medicine* 1985; **13**: 259–62.

116. Fridén T, Zätterström R, Lindstrand A, Moritz UA stabilometric technique for evaluation of lower limb instabilities. *American Journal of Sports Medicine* 1989; **17**: 118–22.

4.4.3 Tendon–ligament basic science

BARRY W. OAKES

INTRODUCTION

In musculoskeletal practice it is important to develop clinical skills and to establish an accurate diagnosis with anatomical precision to manage a patient optimally. Management regimens are based on the previous practical experience of both the clinician and others (as reported in the world literature) and on the results of animal and human experiments. Fundamental data relevant to the structure and biomechanics of tendons and ligaments are reviewed in this chapter, and mechanisms of injury to ligaments and tendons and their repair response are discussed. These data are related to practical clinical patient management. The remodelling of human anterior cruciate ligament autografts and allografts and goat anterior cruciate ligament patellar tendon autografts is also discussed.

STRUCTURE AND BIOMECHANICS OF LIGAMENTS AND TENDONS

There are subtle differences between ligament and tendon morphology,[1] but in this discussion they will be treated as very similar tissues.

Mature adult ligaments and tendons are composed of large diameter type I collagen fibrils (>150 nm in diameter) tightly packed together with a small amount of type III collagen dispersed in an aqueous gel containing small amounts of proteoglycan and elastic fibres (Figs 1 and 2). The outstanding feature of these unique load-bearing tissues is the collagen 'crimp' which is a planar wave pattern found extending in phase across the width of all tendons and ligaments (Fig. 3). The crimp appears to be built into the tertiary structure of the collagen molecule and is probably

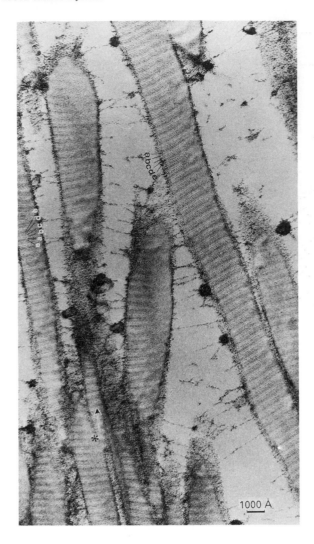

Fig. 1 Adult rat anterior cruciate ligament collagen fixed with ruthenium red after treatment with elastase at pH 8.8 for 12 h Typical regular banding periodicity (labelled) and filaments probably hyaluronate, which link the fibres between the c and d bands can be seen. The fibre marked with an asterisk appears to be dividing into a smaller fibre. Proteoglycan granules are attached to the fibres in the region of the c and d bands. Magnification, 98 000×.

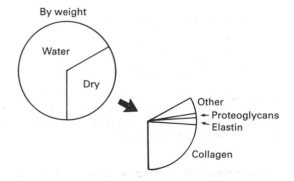

Fig. 2 Approximate amounts of main tendon and ligament components. Collagen forms the main protein component of the dry weight with smaller amounts of elastin and proteoglycans.

maintained *in vivo* by inter- and intramolecular collagen cross-links as well as by a strategically placed elastic fibre network. The crimp may help to attenuate muscle loading forces at the tendoperiosteal and musculotendinous junctions.

Ligament and tendon injury can be closely correlated with the load–deformation curve.[2,3] The load–strain curve can be divided into three regions (Fig. 4).

1. The 'toe' region or initial concave region represents the normal physiological range of ligament/tendon strain up to about 3 to 4 per cent of initial length and is due to the flattening of the collagen crimp. Repeated cycling within the toe region or physiological strain range of 3 to 4 per cent (which may approach 10 per cent in cruciate ligaments owing to intrinsic macrospiral of collagen cruciate fibre bundles) can normally occur without irreversible macroscopic or molecular damage to the tissue.
2. The second part of the load–deformation curve is the linear region where pathological irreversible ligament/tendon elongation occurs due to partial rupture of intermolecular cross-links. As the load is increased further intra- and intermolecular cross-links are disrupted until macroscopic failure is evident clinically. The early part of the linear region corresponds to mild or grade 1 ligament tears (0–50 per cent fibre disruption) and the latter part corresponds to grade 2 tears (50–80 per cent fibre disruption) where there is obvious clinical laxity on stress testing. There is always some pain associated with grade 1 and 2 injuries after the initial trauma, and an athlete with a grade 2 injury usually cannot continue. The severity of the pain provides a rough guide to the clinical severity of the injury.
3. In the third region the curve flattens and the yield or failure point is reached at 10 to 20 per cent strain depending on the macro-organization of the ligament/tendon fibre bundle. In this region complete ligament/tendon rupture occurs at the maximum breaking load; this is the dangerous grade 3 ligament rupture (Fig. 4) on clinical testing.

An enormous amount of work has been done on the biomechanical properties of the human knee ligaments with the anterior cruciate ligament dominating research because of its key role in the stability of the knee. Rotational injury to the knee with the foot fixed appears to be one of the key mechanisms involved in anterior cruciate ligament disruption. Forces of the order of 2000 N are required to disrupt the anterior cruciate ligament. Direct falls on to the tibia or collisions with opponents such that there is a posterior displacement force of the tibia on the femur appears to be a common mechanism for posterior cruciate ligament injury. Collateral ligament injury involves excessive varus, valgus, or rotational forces. The ligament–bone junction with its special fibrocartilage transition zone is a common region of clinical failure. Recent animal experiments indicate that ligament midsubstance strain is much lower than at the insertion sites, and this appears to be due to a difference in collagen fibre crimp amplitude and angle. This higher strain at the insertion sites together with ligament insertion geometry could explain the preferential failure of some ligament insertion sites, particularly the femoral attachment of the medial collateral ligament of the knee which has an almost 90° insertion into the region of the medial femoral epicondyle compared with its tibial periosteal insertion.

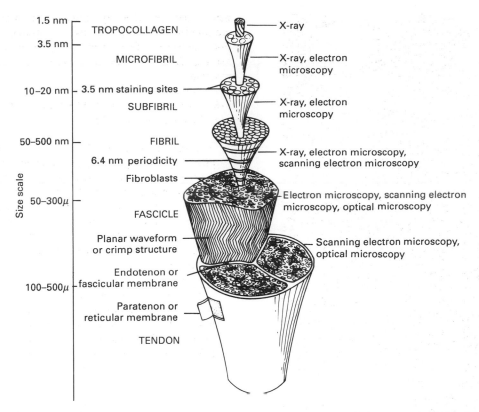

Fig. 3 Structural organization of tendons and ligaments. The planar crimp seen at the light microscopic level should be noted. (Modified from ref. 11.)

CORRELATION OF COLLAGEN FIBRIL SIZE WITH THE MECHANICAL PROPERTIES OF THE TISSUES

Parry *et al.*[4] have completed detailed quantitative morphometric ultrastructural analyses of collagen fibrils from a large number of collagen-containing tissues in various species. They came to a number of conclusions that can be summarized as follows.

1. Type I orientated tissues such as ligaments and tendons have a bimodal distribution of collagen fibril diameters at maturity.
2. The ultimate tensile strength and mechanical properties of connective tissues are positively correlated with the mass average diameter of collagen fibrils. In the context of response of ligaments to exercise, the collagen fibril diameter distribution is closely correlated with the magnitude and duration of tissue loading (Fig. 4).

Variation of collagen fibril diameter with age and correlation with the tensile strength of the anterior cruciate ligament

Oakes[5] measured collagen fibrils in the rat at various ages from 14 days fetal to 2-year-old senile adults. The mean fibril diameter and the range of diameters are plotted against age in Fig. 5. The mean fibril diameter reaches a plateau at about 10 weeks after birth. The age dependence of the separation force required to rupture the anterior cruciate ligament in the rat is also plotted in Fig. 5. Special grips were used to obviate epiphyseal separation and 70

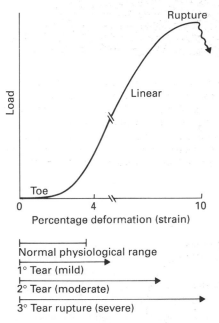

Fig. 4 Load–deformation (strain) curve for ligament/tendon and the clinical correlation with the grading of the injury. The toe region of the curve is entirely within the normal physiological range but strains greater than 4 per cent cause tissue damage.

per cent of the failures occurred within the anterior cruciate ligament. The two curves almost coincide, indicating a close correlation between the size of the collagen fibrils and the ultimate tensile strength of the anterior cruciate ligament, as has been

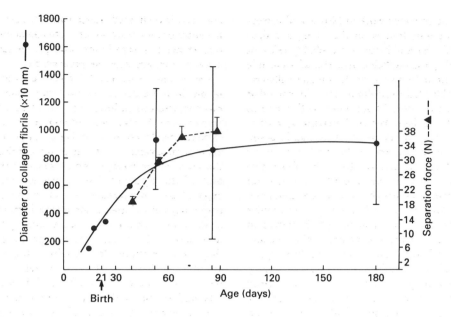

Fig. 5 Increase in diameter of collagen fibrils (full curve through mean diameters with bar representing the largest and smallest fibrils) and tensile strength (broken curve with triangles) with age in the growing rat anterior cruciate ligament. Both curves reach a plateau at about 90 days post-conception or about 70 days after birth.

suggested by Parry *et al.*[4] This rapid increase in collagen fibril size over 6 weeks in the growing rat is not seen during normal ligament/tissue repair or remodelling.[6,7]

Recent work by Shadwick[8,9] has also elegantly demonstrated a clear correlation between collagen fibril diameter and the tensile strength of tendons. He found that the tensile strength of pig flexor tendons was greater than that of extensor tendons and that this greater flexor tendon tensile strength was correlated with a population of larger-diameter collagen fibrils not present in the weaker extensor tendons.

THE EFFECTS OF IMMOBILIZATION ON THE CAPSULE AND SYNOVIAL JOINTS

Akeson *et al.*[10] have recently reviewed investigations of the effects of immobilization on joints. Articular cartilage atrophy occurs with proteoglycan loss and the development of an associated fibrofatty connective tissue that is synovially derived which adheres to the articular cartilage. Ligament insertion sites are weakened, and the ligament itself has increased compliance with reduced load to failure due to loss of collagen mass which may occur after only 8 to 12 weeks of immobilization but may take up to a year to recover after mobilization.[11,12,15] Capsular changes include loss of water due to loss of glycosaminoglycans and hyaluronic acid which leads to joint stiffness. Clearly, joint immobilization should be avoided if possible to prevent occurrence of the above changes which take many months to reverse (Fig. 6).

EFFECT OF IMMOBILIZATION ON LIGAMENT TENSILE STRENGTH

The most important study in this context is that of Noyes *et al.*[12] who used the anterior cruciate ligament in monkeys as a part of a

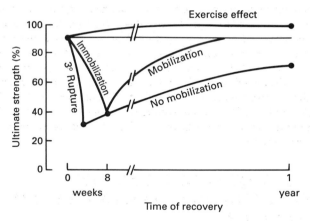

Fig. 6 Effects of immobilization, mobilization, and exercise on recovery of ligament ultimate tensile strength. The loss in ligament tensile strength after a relatively short period of immobilization (about 8 weeks) requires many months to recover—with no mobilization it may take up to 1 year. The effect of exercise on ligament ultimate tensile strength is small. (Modified from ref. 15.)

NASA investigation of the effects of prolonged unloading of ligaments in astronauts. They demonstrated clearly that lower limb cast immobilization for 8 weeks led to a substantial loss of ligament tensile strength which took at least 9 months to recover even with a reconditioning programme. The predominant mode of failure in these experiments was ligament failure rather than avulsion fracture failure. The latter mode of failure was more predominant in the immobilized group because of the resorption of Haversian bone at the ligament attachment, but after 5 months of reconditioning femoral avulsion fractures did not occur. It should be noted that there was no surgery, just simple immobilization.

Another important study with direct implications for clinical practice is that of Amiel *et al.*[1] Although this work was done in rabbits, in the context of the previous work of Noyes and coworkers it is very relevant. The time frames for the changes in ligament tensile strength are similar and hence are probably relevant to clinical orthopaedics and sports medicine. Amiel *et al.* were able to show that 12 weeks of immobilization of the medial collateral ligament in the growing rabbit led to profound atrophy such that there was a decrease of approximately 30 per cent in collagen mass as a result of increased collagen degradation. Remarkably, most of this atrophy occurred during weeks 9–12 of immobilization. Again, there was no trauma or surgery to the medial collateral ligament in this study. Hence, it appears from the above two studies that prolonged immobilization, i.e. 6 to 12 weeks, without trauma can itself lead to profound atrophy of both collateral and cruciate ligaments of the knee joint, and recovery may require many months or even a year. This time frame must be kept in mind when managing patients after prolonged knee immobilization and advising them when they can best return to full competitive sport.

Amiel *et al.*[13] have also shown that there is a close relationship between joint stiffness induced by immobilization and a decrease of total glycosaminoglycan in the periarticular connective tissues. They demonstrated alleviation of this joint stiffness in a rabbit model using intra-articular hyaluronate.

Apart from the original classic work of Noyes *et al.*,[12] in which the effects of immobilization on the anterior cruciate ligament in monkeys were examined, there have been investigations of the cause of the decreased strength and elastic stiffness of the anterior cruciate ligament in response to immobilization. Tipton *et al.*,[14] in an investigation using optical microscopy, reported that collagen fibre bundles in dogs were decreased in number and size; this may also be the cause of the decreased cross-sectional area seen in immobilized medial collateral rabbit ligaments.[15] The explanation for the decreased strength and elastic stiffness in these immobilized ligaments may be found at the collagen fibril level. Binkley and Peat[16] reported a decrease in the number of small diameter fibrils after 6 weeks of immobilization of the rat medial collateral ligament.

EFFECT OF MOBILIZATION (EXERCISE) ON LIGAMENT TENSILE STRENGTH

There have been a large number of investigations of the effects of mobilization, and there are several literature reviews available.[2,17–19]

Normal ligaments

The results in experimental animals generally indicate an increase in bone–ligament–bone preparation strength as a response to endurance type exercise. However, some workers have reported no change in ligament or tendon strength, and this may reflect different exercise regimens, methods of testing, or species differences.

The observation[19–21] that ligament strength depends on physical activity prompted an ultrastructural investigation of mechanism of this increase in tensile strength within ligaments. Increased collagen content was found in the ligaments of exercised dogs, and this correlated with increased cross-sectional area and larger fibre bundles. This accounts for the increased ligament tensile strength, but whether this increased collagen content was due to the deposition of collagen on existing fibres or the synthesis of new fibres has

not been investigated. Larsen and Parker[22] have shown that both the anterior and posterior cruciate ligaments of young male Wistar rats showed a significant strength increase ($p < 0.05$) after a 4-week intensive exercise programme.

Oakes and coworkers[5,23] measured the collagen fibril populations in the anterior and posterior cruciate ligaments of young rats subjected to an intensive 1-month exercise programme. This study was performed in an attempt to explain the increased tensile strength produced in these ligaments by the intensive endurance exercise programme and to determine if this could be explained at the level of the collagen fibril, which is the fundamental tensile unit of the ligament. Five pubescent rats (30 days old) were exposed to a progressive 4-week exercise programme of alternating days of swimming and treadmill running. At the conclusion of the programme the rats were running for 60 to 80 min at 26 m/min on a 10 per cent treadmill gradient, and on alternate days they were swimming for 60 min with a 3 per cent body weight attached to their tails. Five caged rats of similar age and initial body weights were controls. After 30 days the exercise and control rats underwent total body perfusion fixation, and the anterior cruciate ligaments and posterior cruciate ligaments were removed and prepared for electron microscopy. Analysis of ultrathin transverse sections cut through collagen fibrils of the exercised anterior cruciate ligaments revealed: (1) a larger number of fibrils per unit area examined (20 per cent increase, $p < 0.05$) compared with the non-exercised caged controls; (2) a fall in mean fibril diameter from 9.66 ± 3.0 nm in the control anterior cruciate ligaments to 8.30 ± 3.0 nm in the exercised anterior cruciate ligaments ($p < 0.05$); (3) as a consequence of (1) and (2) the major cross-sectional area of collagen fibrils was found in the 112.5-nm diameter group in the exercised anterior cruciate ligaments and in the 150.0-nm diameter group in the controls. However, the total collagen fibril cross-section per unit area examined was approximately the same in both the exercised and the non-exercised control anterior cruciate ligaments. Similar changes occurred in the exercised and control posterior cruciate ligaments. These results are shown in Figs 7 to 9. In the exercised posterior cruciate ligaments, the DNA content of collagen was almost double that of the control, suggesting that the posterior cruciate ligament was loaded more than the anterior cruciate ligament in this exercise regime. The conclusion of this study is that anterior cruciate ligament and posterior cruciate ligaments 'fibroblasts' deposit tropocollagen as smaller-diameter fibrils when subjected to an intense programme of intermittent loading (exercise) for 1 month rather than the expected accretion and increase in size of the pre-existing larger-diameter collagen fibrils. Very similar ultrastructural observations have been made for collagen fibrils from exercised mouse flexor tendons.[24]

The mechanism of the change to a smaller-diameter collagen fibril population is of interest and may be related to a change in the type of glycosaminoglycans and hence proteoglycans synthesized by ligament fibroblasts in response to the intermittent loading of exercise. It is well recognized since the original work of Toole and Lowther[25] that glycosaminoglycans have an effect on fibril size *in vitro*, and this has recently been confirmed *in vivo* by Parry *et al.*[26] Merrilees and Flint[27] demonstrated a change in collagen fibril diameters between the compression and tension regions of the flexor digitorum profundus tendon as it turns though 90° around the talus. Amiel *et al.*[28] have shown that rabbit cruciate ligaments contain more glycosaminoglycans than the patellar tendon and hence it is likely that they also play an important role in determining collagen fibril populations in cruciate ligaments.

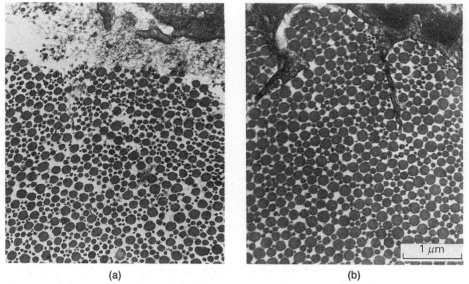

Fig. 7 Transverse sections through (a) exercised and (b) non-exercised control anterior cruciate ligaments. Magnification, 21 600×. The inset shows histogram profiles of mean data for the number of fibres (left) and the percentage area occupied for each diameter group (right).

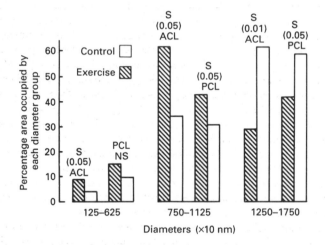

Fig. 8 Comparison of percentage area occupied by three diameter groupings used for statistical analysis for exercised and control anterior and posterior cruciate ligaments (ACL and PCL).

Surgically repaired ligaments

Tipton *et al.*[14] demonstrated a significant increase in the strength of surgically repaired medial collateral ligaments of dogs which were exercised for 6 weeks after 6 weeks cast immobilization. However, they emphasized that at 12 weeks post-surgery (6 weeks of immobilization and 6 weeks of exercise training) the repair was only approximately 60 per cent of that for normal dogs, and then results suggested that at least 15 to 18 weeks of exercise training may be required before return to 'normal' tensile strength is achieved. Similar observations have been made by Piper and Whiteside[29] using the medial collateral ligament of dogs. They observed that mobilized medial collateral ligament repairs were stronger and stretched out less, i.e. there was less valgus laxity than in medial collateral ligament repairs managed by casting and delayed mobilization. This conclusion is supported by the recent work of Woo *et al.*[15]

Some insight into the biological mechanisms involved in the repair response to exercise has come from the elegant work of

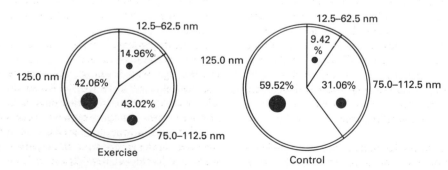

Fig. 9 Comparison of percentage area occupied by the three diameter groupings used for statistical analysis in the exercised and control posterior cruciate ligaments.

Vailas *et al.*[30] By using ^{3}H-proline pulse labelling to measure collagen synthesis in rat medial collateral ligament surgical repairs and combining this with DNA analyses and tensile testing of repaired ligaments subjected to exercise and non-exercise regimens, they were able to show that exercise commencing 2 weeks after surgical repair enhanced the repair and remodelling phase by inducing a more rapid return of cellularity, collagen synthesis, and ligament tensile strength to within normal limits. Further, Woo *et al.*[15] have recently elegantly demonstrated almost complete return (98 per cent) of structural properties of the transected canine femoral–medial collateral ligament–tibial complex at 12 weeks post-transection without immobilization. Canines immobilized for 6 weeks and the femoral–medial collateral ligament–tibial complex tested at 12 weeks had mean loads to failure that were 54 per cent of the control. However, the tensile strength of the medial collateral ligament was only 62 per cent of the control at 48 weeks. This apparent paradox in the non-immobilized dogs was explained by the doubling in cross-sectional area of the healing medial collateral ligament (and hence increased collagen deposition) during the early phases of healing. This repair collagen probably consisted of small diameter fibrils and would account for the poor strain performance of the medial collateral ligament scar collagen which would be similar to the lower curve on Fig. 3.

Woo *et al.*[15] examined the effects of prolonged immobilization and then mobilization on rabbit medial collateral ligament. Both the structural properties of the femoral–medial collateral ligament–tibial complex and the material properties of the medial collateral ligament were examined. After immobilization, there were significant reductions in the ultimate load and energy-absorbing capabilities of the bone–ligament–bone complex. The medial collateral ligament became less stiff with immobilization, and the femoral and tibial insertion sites showed increased osteoclastic activity, bone resorption, and disruption of the normal bone attachment to the medial collateral ligament. With mobilization, the ultimate load and energy-absorbing capabilities improved but did not return to normal. The stress–strain characteristics of the medial collateral ligament returned to normal, indicating that the material properties of the collagen of the medial collateral ligament in the rabbit return relatively quickly after remobilization but that the return of the ligament–bone junction strength to normal took many months as shown by the Woo–Akeson–Amiel curves (see Fig. 6).[11]

The detailed biological cellular mechanisms involved in this enhancement and remodelling of the repair are not understood but may involve prostaglandin and cyclic-AMP synthesis by fibroblasts subjected to repeated mechanical deformation by exercise. Amiel *et al.*[31] have recently shown that maximal collagen deposition and turnover occurs during the first 3 to 6 weeks post-injury in the rabbit. Chaudhuri *et al.*[32] used a Fourier domain directional filtering technique to determine collagen fibril orientation in repairing ligaments. The results indicated that ligament collagen fibril reorientation does occur in the longitudinal axis of the ligament during remodelling. It has also been shown that collagen remodelling of the repairing rabbit medial collateral ligament appears to be encouraged by early immobilization, but after 3 weeks collagen alignment and remodelling appears to be favoured by mobilization.[33,34]

With this basic biological knowledge there is now a rationale for the use of 'early controlled mobilization' of patients with ligament trauma. The use of a limited motion cast with an adjustable double-action hinge for the knee joint is now accepted and enhances more rapid repair and remodelling as well as preserving quadriceps muscle bulk.

COLLAGEN FIBRIL POPULATIONS IN HUMAN KNEE LIGAMENTS AND GRAFTS

In order to gain some biological insight into collagen repair mechanisms within human cruciate ligament grafts, biopsies were obtained from autogenous anterior cruciate ligament grafts from patients subsequently requiring arthroscopic intervention because of stiffness, meniscal and/or articular cartilage problems, or removal of prominent staples used for fixation. Most of the anterior cruciate ligament grafts were from the central third of the patellar tendon as a free graft ($N=33$); in some they were left attached distally ($N=8$) and in others the hamstring or the iliotibial tract was used ($N=7$). These biopsies represented approximately 20 per cent of the total free grafts performed over the 3 years of this study. The clinical anterior cruciate ligament stability of the biopsy group differed little from the remainder. All had a grade 2 to 3 pivot shift (jerk) preoperatively which was eliminated postoperatively in 87 per cent of patients. Subsequent clinical review at 3 years showed an increase in anterior instability with a return of a grade 1 pivot shift in 20 per cent.

A total of 48 biopsies have been quantitatively analysed for collagen fibril diameter populations in patients aged 19 to 42 years. These data were compared with collagen fibril populations obtained from biopsies of cadaver anterior cruciate ligaments ($N=5$) and also from biopsies of anterior cruciate ligaments from young (<30 years, $N=10$) and old patients (>30 years, $N=6$) who had sustained a recent tear. Biopsies were also obtained from normal patellar tendons at operation ($N=7$) and from cadavers ($N=3$) (Fig. 10).

The results (Figs 11 and 12) from the morphometric analysis of the collagen fibril diameter in all the anterior cruciate ligament grafts clearly indicated a predominance of small diameter collagen fibrils. Absence of regular crimping of collagen fibrils was observed by both optical and electron microscopy, as was a less ordered parallel arrangement of fibrils. In most biopsies capillaries were present and most fibroblasts appeared viable.

Recent biochemical analysis of a human patellar tendon autograft *in situ* for 2 years indicated a large amount of type III as well as type V collagen. This confirmed our suspicion that a large amount of collagen in remodelled grafts of this age may be type III and not type I as is normally found in the patellar tendon and adult anterior cruciate ligament.[35] Recent studies of biopsies obtained from anterior cruciate ligament human allografts utilizing fresh frozen Achilles tendons or tibialis anterior tendons ranging in age from 3 to 54 months indicated a similar predominance of small diameter collagen fibrils.[36,37]

Before discussing the biopsy data it is of interest to compare the collagen profiles for the patellar tendon with the normal anterior cruciate ligament. It can be seen that the profiles are different in that the distribution in the patellar tendon is skewed to the right with a small number of large fibrils which are not present in the normal anterior cruciate ligament. Recent elegant work by Butler *et al.*[38] has shown that the patellar tendon is significantly stronger than the human anterior cruciate ligament, posterior cruciate ligament, and lateral collateral ligament from the same knee in terms of maximum stress, linear modulus, and energy density to maximum strength. The larger fibrils observed in the patellar tendon

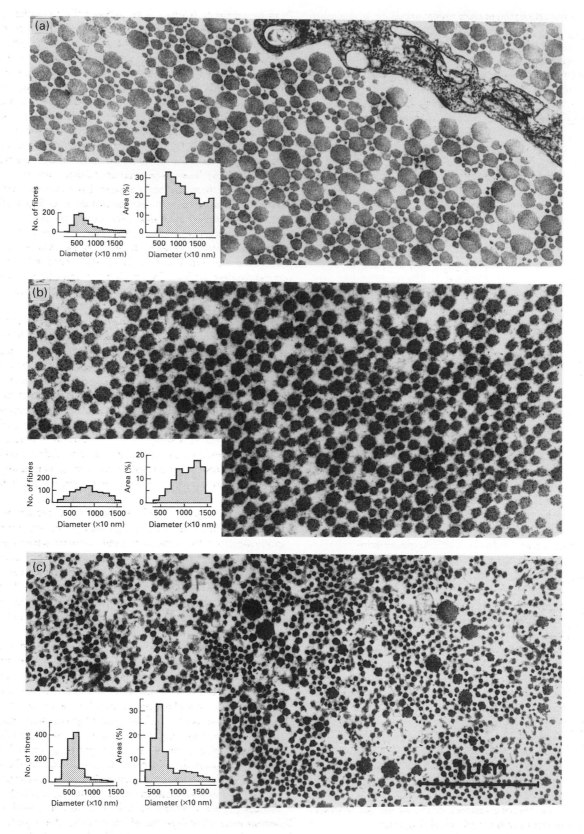

Fig. 10 Transverse sections through collagen fibres of (a) normal young adult patellar tendon (mean of six biopsies), (b) normal young adult anterior cruciate ligament (mean of six biopsies), and (c) Jones free graft (mean of nine biopsies). Magnification, 34 100×. The insets show the number of fibres versus diameter (left) and the percentage area occupied per diameter group (right). The preponderance of small diameter fibres in the graft (c) and large fibres in the patellar tendon (a) which are not seen in the 'normal' anterior cruciate ligament should be noted.

Human anterior cruciate ligament replacement/grafts

Young normal human anterior cruciate ligament (28-year-old male)

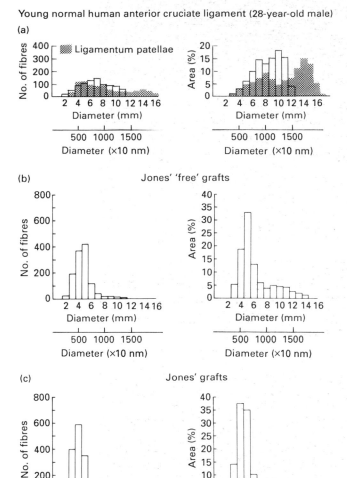

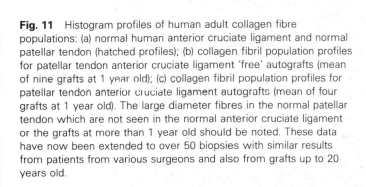

Fig. 11 Histogram profiles of human adult collagen fibre populations: (a) normal human anterior cruciate ligament and normal patellar tendon (hatched profiles); (b) collagen fibril population profiles for patellar tendon anterior cruciate ligament 'free' autografts (mean of nine grafts at 1 year old); (c) collagen fibril population profiles for patellar tendon anterior cruciate ligament autografts (mean of four grafts at 1 year old). The large diameter fibres in the normal patellar tendon which are not seen in the normal anterior cruciate ligament or the grafts at more than 1 year old should be noted. These data have now been extended to over 50 biopsies with similar results from patients from various surgeons and also from grafts up to 20 years old.

and not found in the anterior cruciate ligament are an obvious explanation for the stronger biomechanical properties.

The biopsies from the grafts were obtained from patients with a good to fair rating in terms of a moderate anterior drawer (0–5 mm) and correction of the pivot shift, but both these tests of anterior cruciate ligament integrity showed an increasing laxity of the anterior cruciate ligament at the 3-year clinical review. The length of time that the grafts were *in vivo* prior to biopsy varied

from 6 months to 6 years. The collagen fibril population did not alter very much for the older grafts (i.e. more than 3 years), which is not what was hoped for or expected but is in keeping with the clinical observation that the anterior cruciate ligament grafts 'stretched out' postoperatively. The most striking feature of all the biopsies from the grafts, irrespective of whether they were free grafts, Jones grafts, fascia lata, or hamstring grafts, was the invariable prevalence of small diameter fibrils among a few larger fibrils which were probably the original large diameter patellar fibrils. The packing of the small fibrils in the grafts was not as tight as is usually observed in the normal patellar tendon (compare Figs 10(a) and 10(c)). The Jones free grafts had more large diameter fibrils than the Jones grafts. It appears from direct observations of electron micrographs and the quantitative collagen fibril data using a non-isometric surgical procedure that the large diameter fibrils of the original patellar tendon graft are removed and almost entirely replaced by smaller diameter fibrils which are less well packed and orientated than the larger diameter fibrils found in the normal patellar tendon. The smaller-diameter fibrils are probably recently synthesized because they are thinner than those found in the original patellar tendon.

Is gentle mechanical loading of the anterior cruciate ligament grafts an important stimulus to fibroblast proliferation and collagen deposition? Inadequate mechanical stimulus may occur, particularly if grafts are non-isometric and are 'stretched out' by the patient before they have adequate tensile strength. A lax anterior cruciate ligament graft may not induce sufficient mechanical loading on graft fibroblasts to alter the glycosaminoglycans–collagen biosynthesis ratios to favour large diameter fibril formation. Certainly in this study there was anterior cruciate ligament graft laxity which increased postoperatively. This would lend credence to the above notion. However, use of continuous passive motion in grafted primates does not increase the strength of grafts.

Another more likely possibility is that the 'replacement fibroblasts' in the anterior cruciate ligament grafts are derived from stem cells from the synovium (and synovial perivascular cells) which are known to synthesize hyaluronate which in turn favours the formation of small diameter fibrils.[26] The strong correlation of small diameter fibrils with a lower tensile strength has been observed by Parry *et al.*[4] and Shadwick,[9] and the observations in this study confirm this and correlate with the observations of Clancy *et al.*[39] and Arnoczky *et al.*[40]

The recent observations by Amiel *et al.*[41] indicates that collagenase may play a role in the remodelling of anterior cruciate ligament tears/grafts. The conclusion from this study is that the predominance of the small diameter collagen fibrils (< 75 nm) and their poor packing and alignment in all the anterior cruciate ligament grafts, irrespective of the type of graft, their age, and the surgeon may explain the clinical and experimental evidence of a decreased tensile strength in such grafts compared with the normal anterior cruciate ligament. It appears that, in the adult, the 'replacement fibroblasts' in the remodelled anterior cruciate ligament graft cannot re-form the large diameter, regularly crimped, and tightly packed fibrils seen in the normal anterior cruciate ligament even after 6 years, which was the oldest graft analysed. The origin of the 'replacement' fibroblasts which remodel the anterior cruciate ligament grafts is not known at present. It is likely that they will not come from the actual graft itself, although some of these cells may survive owing to diffusion. However, most of the stem cells involved in the remodelling process are probably derived from the surrounding synovium and its vasculature.

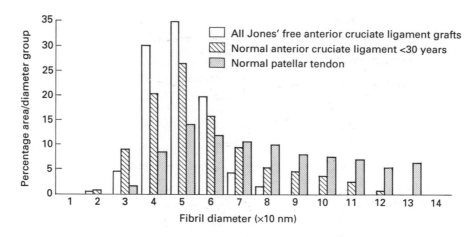

Fig. 12 Summary histograms of normal human anterior cruciate ligament, patellar tendon, and anterior cruciate ligament patellar tendon autografts expressed as percentage area occupied for the collagen fibrils versus fibril diameter.

GOAT ANTERIOR CRUCIATE LIGAMENT PATELLAR TENDON AUTOGRAFT COLLAGEN REMODELLING—QUANTITATIVE COLLAGEN FIBRIL ANALYSES OVER 1 YEAR

The aim of this study was to quantify in detail the collagen fibril remodelling process in adult goat patellar tendon anterior cruciate ligament autografts over a period of 12 months.

Ten mature female adult goats were used in the study. They were anaesthetized and the middle third of the right patellar tendon was harvested via a central arthrotomy. A V-shaped patellar bone component removed by hand saw was left attached at the femoral end. The tibial end was also left attached. The anterior cruciate ligament was removed and fixed for time zero quantitative collagen fibril analyses. The lateral femoral condyle was exposed and the posterolateral capsule was opened above the lateral femoral condyle from within the intercondylar notch. A Gigli saw was introduced into the notch and the posteromedial corner of the lateral femoral condyle was grooved to allow for isometricity of the anterior cruciate ligament graft and also to create a suitable vascular bone bed for anterior cruciate ligament graft attachment. The patellar tendon anterior cruciate ligament graft was then routed via a drill hole or slot in the tibia under the anterior horn of the medial meniscus through the intercondylar notch and posterolateral capsule 'over the top' and lodged within the groove created by the Gigli saw. The patellar bone was stapled under a periosteal flap to the lateral femoral condyle.

At times $T = 0, 3, 6, 12, 24$, and 52 weeks two goats were culled and anterior cruciate ligament grafts were obtained and prepared for quantitative ultrastructural collagen fibril analyses. Tissue sampling was performed as follows. The normal anterior cruciate ligament grafts ($T=0$, $n=5$) and anterior cruciate ligament patellar tendon grafts ($n=7$) were divided into thirds and 1 mm thick sections were cut from the femoral, middle, and tibial thirds. Each section was then cut into strips such that four sections were obtained: two were deemed 'superficial' and contained a synovial surface, and two were deemed 'deep'. Two ultrastructural thin

sections from each section were analysed systematically on copper 400 mesh grids. Systematic random sampling was carried out such that about 10 per cent of the grid spaces that the sections covered were photographed for later quantitative ultrastructural collagen fibril analyses. The patellar tendon was sampled in the central region and two blocks per tendon were analysed as above ($n=3$). The collagen fibril profiles were directly measured from electron micrograph negatives taken at a magnification of 20 000× using a specifically designed software program for automated computerized image analysis within a constant-size inclusion grid. A calibration grid was included at each sitting to determine magnifications accurately. The ASCII data files were imported into a Framework 3 database and the frequency of fibrils within 20 diameter size classes and the percentage area occupied for each diameter group of fibrils were automatically calculated as a mean (Fig. 13). The results of the study were as follows.

Normal adult goal anterior cruciate ligament (T=0). The distribution was clearly bimodal with a large number of small fibrils less than 100 nm in diameter and a group of larger fibrils greater than 100 nm in diameter. A small number of the larger fibrils contributed about 45 per cent of the total collagen fibril area; these large fibrils seen in adult goat anterior cruciate ligament are not seen in normal adult human anterior cruciate ligament.

Normal adult goat patellar tendon (T=0). The collagen fibril distribution was quite different to that in the anterior cruciate ligament with a more unimodal distribution. There were less small diameter fibrils and more larger diameter fibrils (>100 nm) than in the anterior cruciate ligament, and these larger fibrils contributed 65 per cent of the total collagen fibril area (Fig. 14).

Patellar tendon anterior cruciate ligament graft: tibial region versus femoral region (Fig. 15). At 6 weeks there was a large increase in the number of small diameter collagen fibrils (<100 nm) which was greatest at the tibial end. The number of small diameter fibrils was increased at both ends of the graft at 52 weeks. A loss of the large diameter fibrils (>100 nm) could be seen at 6 weeks, and at 52 weeks only a few large fibrils remained in the femoral region. This is shown by the ratio of large diameter (>100 nm) to small diameter (<100 nm) fibrils (Fig. 16).

Patellar tendon anterior cruciate ligament graft: superficial region versus deep region. The major observation was the complete loss of

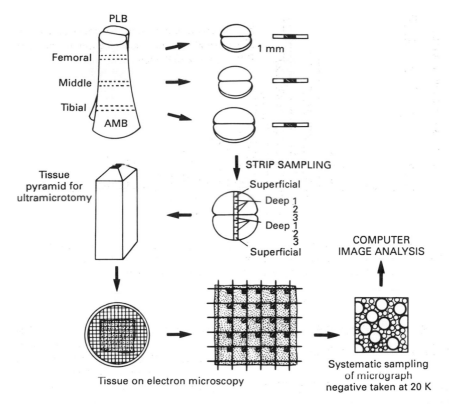

Fig. 13 Methodology of sampling the goat anterior cruciate ligament and the tissue blocking and electron microscope grid random sampling technique.

the large diameter fibrils from the superficial regions of the graft after 52 weeks. There was also a concomitant increase in the number of small diameter fibrils in the deep region at 52 weeks which was not seen in the superficial regions. If the ratio of large diameter (>100 nm) to small diameter (<100 nm) fibrils is plotted, the major fall in the ratio is seen in the superficial region, reflecting the complete removal of the large diameter fibrils in this region at 52 weeks (Figs 15–17).

This study has enabled the anatomical regions of the anterior cruciate ligament graft which undergo remodelling to be determined and has shown that this process continues for up to 52 weeks after grafting. The remodelling process changes the collagen fibril profile of the original patellar tendon to one containing a greater proportion of small diameter fibrils. The remodelling process occurs from the outside inwards and is more vigorous in the tibial region of the graft. This is consistent with synovial revascularization. The rapid depletion of the large fibrils in the grafts at 12 weeks is also consistent with the dramatic decrease in the mechanical and material properties of these grafts (Fig. 18). This collagen fibril study in the goat parallels that in the human anterior cruciate ligament patellar tendon grafts and appears to be a useful model for following collagen remodelling in anterior cruciate ligament grafts.

CLINICAL AND ULTRASTRUCTURAL OBSERVATIONS ON ACHILLES TENDON INJURIES

In this section we attempt to relate the three phases of healing of soft tissues to observations in Achilles tendon injuries. These in-juries can be classified as described above for ligament injury, i.e. grades 1–3 with the last being complete rupture. Several patient histories are used to illustrate these three phases of healing and attempted repair.

Grade 3 complete rupture

Patient profile An Australian Rules Football rover, aged 26, accelerating to avoid an opponent. Biopsy was taken during surgical repair. Clinical signs were a palpable defect in the Achilles tendon and a positive Thompson's sign.

With this injury there is both collagen bundle failure and vascular disruption, and hence bleeding is a feature when seen at an early surgical repair. This trauma to the tendon initiates the acute inflammatory phase and hence microscopically there is massive red cell extravasation, fibrin clot formation, and collagen fibril disruption. The damaged tendon becomes oedematous, and poly morphonuclear monocytes migrate into this area and release their lysosomal contents as well as actively phagocytosing cellular and other debris. Macrophages also move into the rupture site and commence phagocytosis of damaged cells and tissue. This phase lasts for 72 h or more and is followed by the repair phase (Fig. 19).

Grade 2 tear

Patient profile An Australian Rules football ruckman with a painful thickened Achilles tendon for 6 months.

Operation Excision of paratenon and tendon incision to remove damaged, haemorrhagic, and necrotic regions. Biopsy taken for optical (Fig. 20) and electron microscopy.

Optical microscopy demonstrated a thickened paratenon and an oedematous thickened tendon. Ultrastructurally many fibroblasts have a dilated rough endoplasmic reticulum and prominent

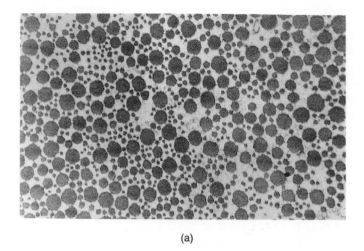

(a)

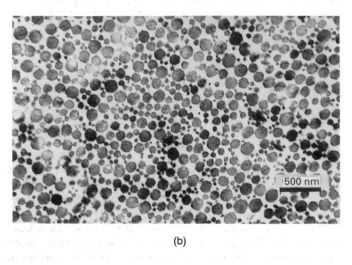

(b)

Fig. 14 Electron micrographs of (a) normal adult goat patellar tendon and (b) normal adult goat anterior cruciate ligament.

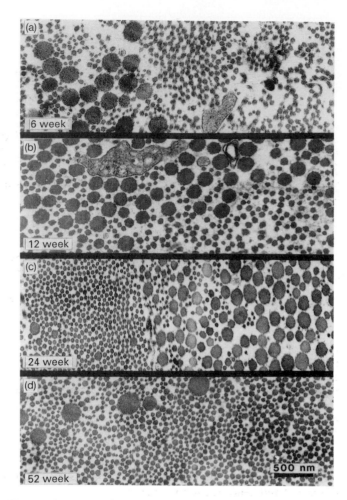

Fig. 15 Representative electron micrographs of anterior cruciate ligament patellar tendon autografts at $T=0$, 3, 6, 12, 24, and 52 weeks after grafting. There is progressive loss of the large diameter fibrils present at $T=0$ and accumulation of the small diameter fibrils with increasing graft age.

nucleoli indicative of increased collagen synthesis. Apart from the many free red cells the other feature was the prevalence of many small diameter (18–20 nm) collagen fibrils not aligned or closely packed among the older larger pre-existing fibrils with diameters ranging from 80 to 150 nm. Polymorphonuclear monocytes and macrophages were not common at this stage, and this may reflect the slowness of repair in this unique tissue.

This biopsy demonstrated the features of the repair phase which follows the acute inflammatory phase and lasts from 72 h to 4 to 6 weeks. In this patient the repair phase had been perpetuated because of continued activity, a common problem with these athletes because the Achilles tendon pain usually subsides when they run and then returns on 'cool-down', often with a vengeance. In other patients, where there is less florid tendinitis and the tendon was clinically painful but not obviously enlarged, discrete areas of increased cellularity were seen in the tendon in the form of free red cells, cell debris, and viable fibroblasts which were surrounded by large diameter and many small diameter collagen fibres. These areas were almost isolated from the densely packed collagen of the rest of the tendon by a fibrin precipitate. These discrete areas are probably due to collagen fibre ruptures or 'microtears' corresponding to the early part of region 2 of the

load–deformation curve. As yet there is no evidence that the use of massage or 'deep friction' enhances this repair phase.[42]

Grade 1 injury

Patient profile A runner, aged 25, with a painful tender lump in the Achilles tendon for 18 months. The biopsy was obtained at open operation.

The lump at operation was firmer than the rest of the tendon and was slightly darker in colour. On light microscopy of the biopsy taken from the nodule the changes in the collagen bundles were very subtle. There was less regular crimping of the collagen bundles and they were not as tightly packed. However, the cause of this less regular collagen crimp was obvious ultrastructurally in that between the large diameter fibrils there were many small diameter fibrils which were less well orientated longitudinally. These tendons were interpreted as being in the remodelling phase as there was no increase in fibroblast numbers in the nodules and no inflammatory cells were observed (Fig. 21).

The mechanism of acute complete rupture in young athletes accelerating during sprinting indicates that the gastrocnemius–soleus complex can generate sufficient force to rupture the tendon.

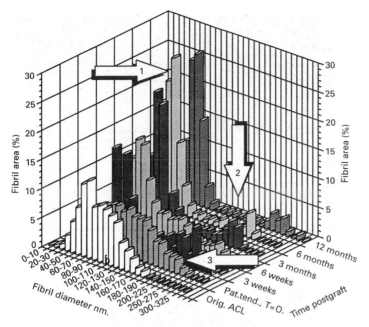

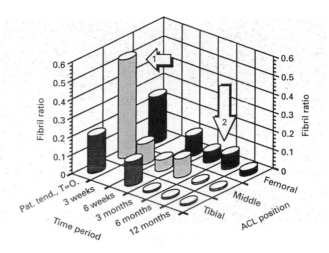

Fig. 16 Three-dimensional histograms of normal goat patellar tendon, anterior cruciate ligament, and anterior cruciate ligament autografts expressed as percentage area covered by the collagen fibrils versus age and fibril diameter. Again, there is progressive loss of the large diameter fibrils and rapid replacement of them as early as 6 weeks after grafting (arrows 2 and 3) with a predominance of small fibrils (arrows 1).

Fig. 17 Three-dimensional histograms of normal goat patellar tendon and anterior cruciate ligament patellar tendon autografts at T=0, 3, 6, 12, 24, and 48 weeks in relation to the femoral middle and tibial regions of the anterior cruciate ligament autografts. The vertical axis is the ratio of large diameter (>100 nm) to small diameter (<100 nm) fibrils. Large fibrils are lost from the tibial end of the graft as early as 6 weeks post-grafting but some large fibrils still remain at the femoral end of the graft at 48 weeks (~ 12 months— arrow 2). Note predominance of large fibrils in the original patellar tendon used as the donor graft (arrow 1).

However, tendon strength usually exceeds that of its muscle by a factor of 2 and hence rupture is unusual.

Viidik[43] has shown that rat tendons *in vitro* undergo increasing deformation or 'plasticity' if cycled to loads less than one-tenth of the failure load, and the strain or deformation is well before the beginning of region two or the linear part of the load–strain curve (Fig. 22). Similar observations have been made both *in vivo* and *in vitro* for rat knee joint ligaments.[44] It is possible that in distance runners a similar fatigue plasticity and elongation occurs in the tendon, and this causes the microruptures and repair nodules already described.

The notion that some running athletes may not have an Achilles tendon of sufficient cross-sectional area to sustain the repetitive tendon loading of distance without injury has been investigated by Engstrom *et al.*[45] In an elegant study they used ultrasound to measure the cross-sectional area of the human Achilles tendon *in vivo* and, using cadaver Achilles tendons, validated this technique

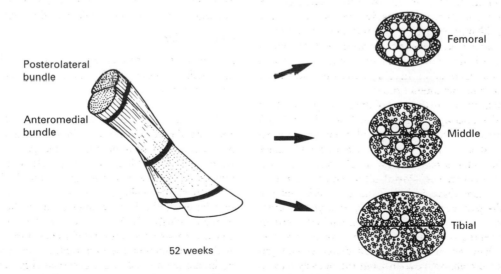

Fig. 18 Summary diagram of goat anterior cruciate ligament graft fibril profiles at 12 months post-grafting. The preponderance of small fibrils at the tibial end of the graft and some large fibrils remaining in the femoral end of the graft at 12 months post-grafting should be noted.

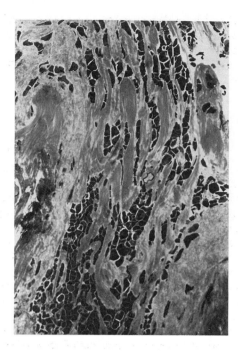

Fig. 19 Optical micrograph of acute Achilles tendon rupture at 1 week post-injury showing red cell extravascular extravasation together with many polymorphs and macrophages. Epon-Araldite, Azure A–methylene blue.

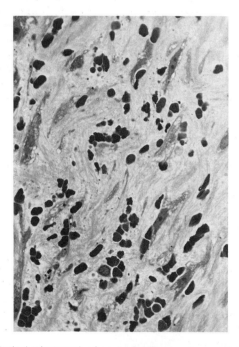

Fig. 20 Optical micrograph of acute Achilles tendon grade 2 partial tear after 6 months of pain and swelling. The tendon is oedematous and very cellular, and the fibroblasts are dilated with an enlarged rough endoplasmic reticulum indicative of active collagen synthesis. Epon-Araldite, Azure A–methylene blue.

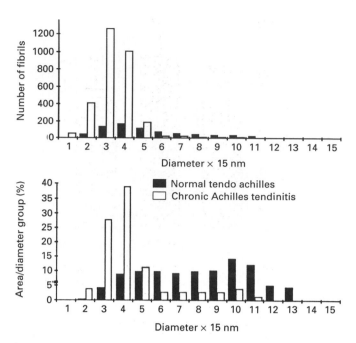

Fig. 21 Number of fibrils versus fibril diameter in patients with chronic Achilles tendinitis and percentage area occupied for each diameter group versus diameter. The preponderance of small diameter fibrils in 'repairing' chronic Achilles tendinitis should be noted. The large normal fibrils are not replaced.

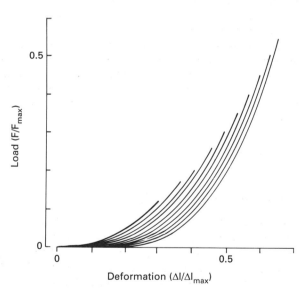

Fig. 22 The effect of fatigue failure or plasticity of load–strain curves with repeated loadings of a tendon to successively higher loads (the curve of the next loading is shifted to the right) within the toe region of the curve and before the linear part of the curve. This effect may be operating in chronic human Achilles tendinitis, particularly in those athletes with tendons of small cross-sectional area. (Adapted from ref. 43.)

as a reliable method. Two groups of distance athletes with and without grade 1 Achilles tendinitis, who were matched for age, weight, and distance, had their Achilles tendon cross-sectional area measured using the previously validated ultrasound tech-

nique. It was found that athletes with grade 1 type Achilles tendinitis had about a 30 per cent decrease in the cross-sectional areas of their Achilles tendons ($p < 0.05$) (Fig. 23). This indicates that a major mechanism in this type of common injury may simply be fatigue creep failure of Achilles tendon collagen as shown in

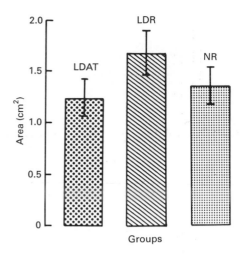

Fig. 23 Cross-sectional area (cm²) of human Achilles tendons measured by ultrasound for two running groups, matched for age, weight, and distance, compared with a sedentary control group (NR): LDAT, runners with grade 1 Achilles tendinitis; LDR, runners without Achilles tendinitis. The smaller cross-sectional area of the Achilles tendons in the LDAT group ($p<0.05$) should be noted. (Reproduced with the permission of Professor A. W. Parker.)

Fig. 20. Komi *et al.*,[46] have recently developed an *in-vivo* buckle transducer which they located around the Achilles tendon in a number of subjects. Direct force measurements were made on several subjects who were involved in slow walking, sprinting, jumping, and hopping after calibration of the transducer. During running and jumping forces close to the previous estimated ultimate tensile strength of the tendon were recorded, indicating that fatigue creep in a small cross-sectional tendon is a possible mechanism of injury without the need to invoke other lower limb biomechanical pathology as has been suggested by Clement *et al.*[47] and Williams.[48]

BASIC BIOMECHANICS OF TISSUE INJURY

Muscle–tendon–bone injury

The basic causes of intrinsic muscle injury are still not entirely clear, but have been attributed to inadequate muscle length and strength, e.g. 'tight' hamstrings, particularly in adolescent boys, muscle fatigue, and inadequate muscle skills. It is also clear that most muscle injuries occur in the lower limb and most involve the 'two-joint muscles' (the hamstrings and the rectus femoris), probably because of the complex reflexes involved in simultaneous co-contraction and co-relaxation involved with these two muscle groups.[49]

Both concentric and eccentric muscle–tendon unit loading can cause muscle–tendon–bone junction injury. The use of eccentric muscle loading to cause increased muscle hypertrophy rather than the more conventional concentric loading has led to the phenomenon of eccentric muscle soreness which is now known to be due in part to muscle sarcomere disruption at the Z lines.[50] Eccentric muscle–tendon–bone load can generate more force than concentric contractions and may be the mechanism by which the patellar tendon and its attachments lead to tendoperiosteal partial disruptions at both the superior and inferior poles of the patellar.

Recent studies[51] have demonstrated that when the inferomedial collagen fibre bundles of the human patellar tendon are subjected to mechanical analysis they fail at loads which are much less than the lateral fibre bundles. The biological reasons for this are not clear at the moment.

Muscle–tendon junction

Failure at the muscle–tendon junction is common clinically. There is an increased folding of the terminal end of the last muscle sarcomere which has important mechanical implications for reducing the stress at this critical junctional region. It has been found that a typical vertebrate fast-twitch cell can generate about 0.33 MPa of stress across the cell. The maximum stress placed on the cell's junctional complex at the muscle–tendon junction by the complex folding of the terminal sarcomeres experiences a maximal stress of 0.015 MPa which is much less than 0.33 MPa, and this difference may determine whether mechanical failure occurs at this junctional region. With muscle injury at this junctional site it is probable that this complex sarcomere muscle membrane infolding to increase the surface area and hence substantially decrease the stress is probably not reproduced following repair and may be an explanation for the occurrence of 're-tears' at this junction in athletes with previous injury and repair to this region.

Tendon injury

Tendon injury in sport is not unusual because of the large loads applied. Komi[52] has recently measured the high forces generated in the human Achilles tendon using a calibrated buckle transducer introduced surgically for short periods of time. Forces of up to 4000 N were recorded in the Achilles tendon with toe running and hence it is not surprising that with repetitive loading of this magnitude microfatigue failure could occur with long distance running, particularly in a tendon of small cross-sectional area.[45]

Spontaneous tendon rupture

Spontaneous tendon rupture is uncommon in the young athlete. It generally occurs in the older sportsman and is usually associated with degenerative pathology of the collagen fibrils.

LIGAMENT/TENDON REPAIR

Repair to ligaments other than the anterior cruciate ligament are discussed below (Figs 24–26). Anterior cruciate ligament injuries appear to be unique in that the chondrocyte-like cells in this special ligament apparently have a limited capacity to proliferate and synthesize a new collagen matrix and hence repair appears to be limited. Collagenase release may also affect the effectiveness of the repair process.[41]

Acute inflammatory phase

The gap in the ligament/tendon is immediately filled with erythrocytes and inflammatory cells, particularly polymorphonuclear leucocytes. Within 24 h monocytes and macrophages are the predominant cells and actively engage in phagocytosis of debris and necrotic cells. These are gradually replaced by fibroblasts from either intrinsic or extrinsic sources and the initial deposition of the type III collagen scar commences. At this stage

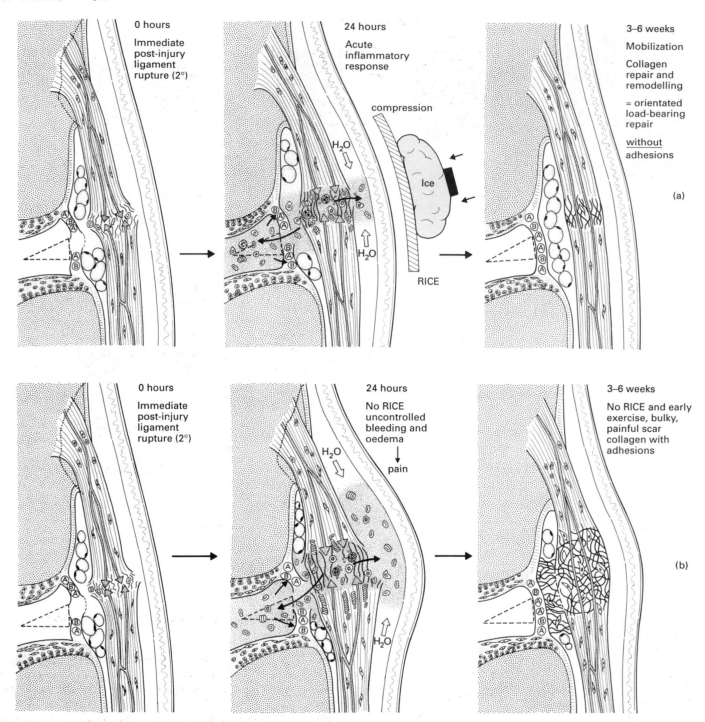

Fig. 24 Ligament repair (a) with and (b) without RICE (rest, ice, crepe, and elevation of the injured limb) management and early mobilization.

collagen concentration may be normal or slightly decreased but the total mass of ligament collagen scar is increased. Glycosaminoglycan content, water, fibronectin, and DNA content are increased (Fig. 27).

Proliferation

Fibroblasts predominate, the water content remains high, and the collagen content increases and peaks during this phase (3–6 weeks). Type I collagen now begins to predominate and glycosaminoglycan concentration remains high. The increasing amount of scar collagen and reducible cross-link profile has been correlated with the increasing tensile strength of the ligament matrix. Recent quantitative collagen fibril orientation studies indicate that early mobilization of a ligament at this stage (within the first 3 weeks) may be detrimental to collagen orientation. After this time frame there is experimental evidence that mobilization increases the tensile strength of the repair and probably enhances this phase and the next phase of remodelling and maturation.[15,30,54] With this basic biological knowledge there is now a rationale for the use of 'early controlled mobilization' of patients with ligament trauma.

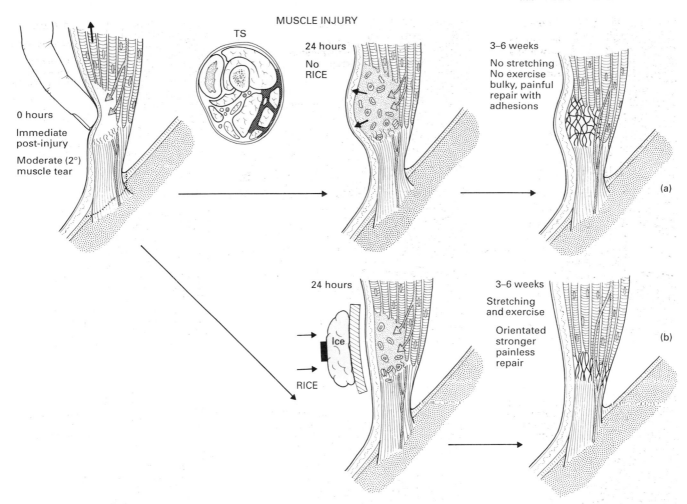

Fig. 25 Muscle repair at the muscle–tendon junction (a) without and (b) with RICE (rest, ice, crepe, and elevation of the injured limb) and mobilization.

Remodelling and maturation (6 weeks–12 months)

There is a decreasing cell number and hence decreased collagen and glycosaminoglycan synthesis. Water content returns to normal and collagen concentration returns to slightly below normal, but the total collagen content remains slightly increased. With further remodelling there is a trend for scar parameters to return to normal but the matrix in the ligament scar region continues to mature slowly over months or even years. Scar collagen matrix and adjacent 'normal' collagen matrix may actually shorten the repair region perhaps by an interaction of ligament/tendon fibroblasts with their surrounding collagen matrix.[56] Collagen fibril alignment in the longitudinal axis of the ligament occurs, even though small diameter collagen fibrils are involved (Fig. 28).

Achilles tendon injuries, particularly partial tears, present a dilemma for the clinician in that they are often intractable to management, although Stanish *et al.*[57] claim good clinical results from graded eccentric loading regimes. We have examined Achilles tendon biopsies of patients with chronic localized tears and more generalized thickened tender chronic Achilles tendons. The feature which characterized the pathology ultrastructurally

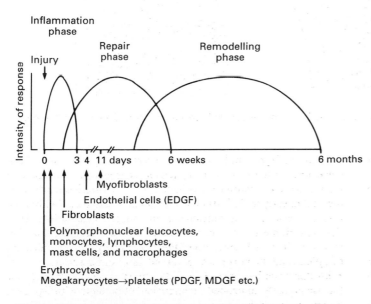

Fig. 26 The three phases of healing and the cells involved.

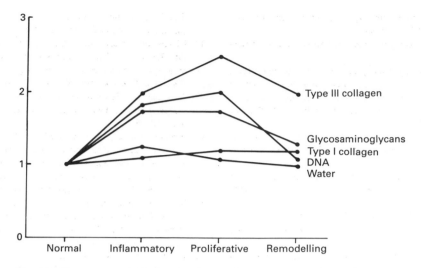

Fig. 27 Ligament repair during the phases of healing and the normalized content of type I and III collagen, water, DNA, and glycosaminoglycans. (Reproduced with permission from ref. 53.)

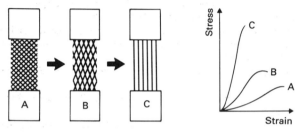

Fig. 28 The effect of collagen repair with identical fibrils but different geometry and the corresponding load–strain response to tensile testing. (Modified from ref. 55.)

was the persistence of small diameter collagen fibrils. The large fibrils of the original tendon do not appear to be replaced in either a repairing tendon or a ligament (Fig. 21).

REFERENCES

1. Amiel D, Akeson WH, Harwood FL, Frank CB. Stress deprivation effect on the metabolic turnover of the medial collateral ligament collagen: a comparison between nine and 12-week immobilization. *Clinical Orthopaedics and Related Research* 1983; **172**: 265–70.
2. Butler DL, Grood ES, Noyes FR, Zernicke RF. *Exercise and Sport Sciences Reviews* 1979; **6**: 125–81.
3. Oakes BW. Acute soft tissue injuries—nature and management. *Australian Family Physician* 1981; **10** (Suppl.): 1–16.
4. Parry DAD, Barnes GRG, Craig AS. A comparison of the size distribution of collagen fibrils in connective tissues as a function of age and a possible relation between fibril size and distribution and mechanical properties. *Proceedings of the Royal Society of London, Ser. B* 1978; **203**: 305–21.
5. Oakes BW. Ultrastructural studies on knee joint ligaments: quantitation of collagen fibre populations in exercised and control rat cruciate ligaments and in human anterior cruciate ligament grafts. In: Buckwalter J, Woo SL-Y eds. *Injury and repair of the musculoskeletal tissues*. Section 2. Park Ridge, IL: American Academy of Orthopedic Surgeons: 66–82.
6. Oakes BW, Knight M, Mclean ID, Deacon OW. Goat ACL autograft collagen remodelling–quantitative collagen fibril analyses over 1 year. Paper 65, *Transactions of the Combined Meeting of the Orthopedic Research Societies of USA, Japan and Canada*. Banff, Alberta, Canada, October 1991: 60.

7. Matthew C, Moore MJ, Campbell L. A quantitative ultrastructural study of collagen fibril formation in the healing extensor digitorum longus tendon of the rat. *Journal of Hand Surgery* 1987; **12B**: 313–20.
8. Shadwick RE. The role of collagen crosslinks in the age related changes in mechanical properties of digital tendons. *Proceedings of the North American Congress on Biomechanics* 1986; **1**: 137–8.
9. Shadwick RE. Elastic energy storage in tendons: mechanical differences related to function and age. *Journal of Applied Physiology* 1990; **68**: 1033–40.
10. Akeson WH *et al.* Effects of immobilization on joints. *Clinical Orthopaedics and Related Research* 1987; **219**: 28–37.
11. Kastelic J, Galeski A, Baer E. The multicomposite structure of tendon. *Connective Tissue Research* 1978; **6**: 11–23.
12. Noyes FR, Torvic PJ, Hyde WB, De Lucas JL. Biomechanics of ligament failure. 2. An analysis of immobilization, exercise and reconditioning effects in primates. *Journal of Bone and Joint Surgery* 1974; **56A**: 1406–18.
13. Amiel D, Frey C, Woo SL-Y, Harwood F, Akeson W. Value of hyaluronic acid in the prevention of contracture formation. *Clinical Orthopaedics and Related Research* 1985; **196**: 306–11.
14. Tipton CM, James SL, Mergner W, Tcheng TK. Influence of exercise on the strength of the medial collateral knee ligament of dogs. *American Journal of Physiology* 1970; **218**: 894–902.
15. Woo SL-Y, Gomez MA, Sites TJ, Newton PO, Orlando CA, Akeson WH. The biomechanical and morphological changes in the medial collateral ligament of the rabbit after immobilization and remobilization. *Journal of Bone and Joint Surgery* 1987; **69A**: 1200–11.
16. Binkley JM, and Peat M. The effects of immobilization on the ultrastructure and mechanical properties of the medial collateral ligament of rats. *Clinical Orthopaedics and Related Research* 1986; **203**: 310–18.
17. Tipton CM, Vailas AC. Bone and connective tissue adaptions to physical activity. In: Bouchard C, ed. *Proceedings of the International Conference on Exercise, Fitness and Health, Toronto, 1988*. Champaign, IL: Human Kinetics 1989; 331–44.
18. Tipton CM, Vailas AC, Matthes RD. Experimental studies on the influences of physical activity on ligaments, tendons and joints: a brief review. *Acta medica Scandinavica* 1986; **711** (Suppl.): 157–68.
19. Parker AW, Larsen N. Changes in the strength of bone and ligament in response to training. In: Russo P, Gass G, eds. *Human adaptation*. Sydney, Australia: Department of Biological Sciences, Cumberland College of Health Sciences, 1981: 209–21.
20. Tipton CM, Matthes RD, Maynard JA, Carey RA. The influence of physical activity on ligaments and tendons. *Medicine and Science in Sports* 1975; **7**: 165–75.
21. Cabaud HE, Feagin JF, Rodkey WG. (1980). Acute anterior cruciate

ligament injury and augmented repair: experimental studies. *American Journal of Sports Medicine* 1980; **8**: 79–86.

22. Larsen N, Parker AW. *Proceedings of the Australian Sports Medicine Federation* 1982; **8**: 63–73.

23. Oakes BW, Parker AW, Norman J. Changes in collagen fibre populations in young rat cruciate ligaments in response to an intensive one month's exercise program. In: Russo P, Gass G, eds. *Human adaptation.* Sydney, Australia: Department of Biological Sciences, Cumberland College of Health Sciences, 1981: 223–30.

24. Michna M. Morphometric analysis of loading-induced changes in collagen-fibril populations in young tendons. *Cell Tissue Research* 1984; **236**: 465–70.

25. Toole BP, Lowther DA. The effect of chondroitin sulphate-protein on the formation of collagen fibrils *in vitro. Biochemical Journal* 1968; **109**: 857–66.

26. Parry DAD, Flint MH, Gillard GC, Craig AS. A role for glycosaminoglycans in the development of collagen fibrils. *FEBS Letters* 1982; **149**: 1–7.

27. Merrilees MJ, Flint MH. Ultrastructural study of the tension and pressure zones in a rabbit flexor tendon. *American Journal of Anatomy* 1980; **157**: 87–106.

28. Amiel D, Frank CB, Harwood F, Fronek J, Akeson WH. Tendon and ligaments: a morphological and biochemical comparison. *Journal of Orthopaedic Research* 1984; **1**: 257–65.

29. Piper TL, Whiteside LA. Early mobilization after knee ligament repair in dogs: an experimental study. *Clinical Orthopaedics and Related Research* 1980; **150**: 277–82.

30. Vailas AC, Tipton CM, Matthes RD, Gart M. Physical activity and its influence on the repair process of medial collateral ligaments. *Connective Tissue Research* 1981; **9**: 25–31.

31. Amiel D, Frank CB, Harwood FL, Akeson WH, Kleiner JB. (1987). Collagen alteration in medial collateral ligament healing in a rabbit model. *Connective Tissue Research* 1987; **16**: 357–66.

32. Chaudhuri S, Nguyen H, Rangayyan RM, Walsh S, Frank CB. A Fourier domain directional filtering method for analysis of collagen alignment in ligaments. *IEEE Transactions on Biomedical Engineering* 1987; **34**: 509–18.

33. MacFarlane BJ, Edwards P, Frank CB, Rangayyan RM, Liu Z-Q. Quantification of collagen remodelling in healing nonimmobilized and immobilized ligaments. *Transactions of the Orthopaedic Research Society* 1989; **14**: 300.

34. Frank C *et al.* A quantitative analysis of matrix alignment in ligament scars: a comparison of movement versus immobilization in an immature rabbit model. *Journal of Orthopaedic Research* 1991; **9**: 219–27.

35. Deacon OW, McLean ID, Oakes BW, Cole WG, Chan D, Knight M. Ultrastructural and collagen typing analyses of autogenous ACL grafts—an update. *Proceedings of the International Society of the Knee, Toronto, May 1991.*

36. Shino K, Oakes BW, Inoue M, Horibe S, Nakata K, Ono K. Human ACL allograft: collagen fibril populations studied as a function of age of the graft. *Transactions of the 36th Annual Meeting of the Orthopaedic Research Society* 1990; **15**: 520.

37. Shino K, Oakes BW, Inoue M, Horibe S, Nakata K. Human ACL allografts. An electronmicroscopic analysis of collagen fibril populations. *Proceedings of the International Society of the Knee, Toronto, May 1991.*

38. Butler DL, Kay MD, Stouffer DC. Comparison of material properties in

fascicle–bone units from human patellar tendon and knee ligaments. *Journal of Biomechanics* 1985; **18**: 1–8.

39. Clancy WG, Narechania RG, Rosenberg TD, Gmeiner JG, Wisnefske DD, Lange TA. Anterior and posterior cruciate reconstruction in rhesus monkeys: a histological microangiographic and biochemical analysis. *Journal of Bone and Joint Surgery* 1981; **63A**: 1270–84.

40. Arnoczky SP, Warren RF, Ashlock MA. Replacement of the anterior cruciate ligament by an allograft. *Journal of Bone and Joint Surgery* 1988; **63A**: 376–85.

41. Amiel D, Ishizue KK, Harwood FL, Kitayashi L, Akeson W. Injury of the anterior cruciate ligament: the role of collagenase in ligament degeneration. *Journal of Orthopaedic Research* 1989; **7**: 486–93.

42. Walker J. Deep transverse frictions in ligament healing. *Journal of Orthopaedic and Sports Physical Therapy* 1984; **6**: 89–94.

43. Viidik A. Functional properties of connective tissues. *International Review of Connective Tissue Research* 1973; **6**: 127–215.

44. Weisman G, Pope MH, Johnson RJ. Cyclical loading in knee ligament injuries. *American Journal of Sports Medicine* 1980; **8**: 24–30.

45. Engstrom CM, Hampson BA, Williams J, Parker AW. Muscle–tendon relations in runners. In: Oakes BW, ed. *Abstracts of the Proceedings of the National Conference of the Australian Sports Medicine Federation, Ballarat, 1985.* Melbourne, Australia: Australian Sports Medicine Federation, p. 56.

46. Komi PV, Salonen M, Jarvinen M, Kokko O. *In vivo* registration of Achilles tendon forces in man. Methodological development. *International Journal of Sports Medicine* 1987; **8**: 3–8.

47. Clement DB, Taunton JE, Smart GW. Achilles tendinitis and peritendinitis: aetiology and treatment. *American Journal of Sports Medicine* 1984; **12**: 179–84.

48. Williams JGP. Achilles tendon lesions in sport. *Sports Medicine* 1986; **3**: 114–35.

49. Oakes BW. Hamstring injuries. *Australian Family Physician* 1984; **13**: 587–91.

50. Friden J, Sjostrom M, Ekblom B. Myofibrillar damage following intense eccentric exercise in man. *International Journal of Sports Medicine* 1983; **4**: 170–6.

51. Chun KJ *et al.* Spatial variation in material properties in fascicle–bone units from human patellar tendon. *Transactions of the Orthopaedic Research Society* 1989; **14**: 214.

52. Komi PV. Neuromuscular factors related to physical performance. In: Russo P, Balnave R, eds. Muscle and nerve, factors affecting performance. *Proceedings of the 6th Biennial Conference, Cumberland College of Health Sciences, 1987.*

53. Woo SL-Y, Buckwalter JA, eds. *Injury and repair of the musculoskeletal soft tissues.* Chicago: American Academy of Orthopedic Surgeons, 1988.

54. Hart DP, Danhers LE. Healing of the medial collateral ligament in rats. *Journal of Bone and Joint Surgery* 1987; **69A**: 1194–9.

55. Viidik A. Interdependence between structure and function. In: Viidik A, Vuust J, eds. *Biology of Collagen.* London: Academic Press, 1980: 257–80.

56. Danhers LE, Barnes AJ, Burridge KW. The relationship of actin to ligament contraction. *Clinical Orthopaedics and Related Research* 1986; **210**: 246–51.

57. Stanish W, Rubinovich RM, Curwin S. Eccentric exercise in chronic Achilles tendinitis. *Clinical Orthopaedics and Related Research* 1986; **208**: 65–8.

4.4.4   The aetiology and treatment of tendinitis

SANDRA L. CURWIN

INTRODUCTION

Soft-tissue overuse injuries are common in sport and are a challenge for health professionals because of the difficulty in balancing patient demands for maintained mobility with the injured tissue's requirements for ideal healing. This balance is not always achieved, and conditions like chronic tendinitis can develop. Despite much recent progress in understanding and treating soft tissue injuries, tendinitis remains a difficult clinical problem for both the people affected and those involved in their treatment. Controversy and disagreement still abound. A wide range of descriptions of tendinitis exist and many treatment protocols have been described. Some authors question whether tendinitis is an appropriate term to use, particularly in the cases of chronic degenerative tendinitis that actually pose the most difficult clinical problem. It is easy for the inexperienced clinician, and even easier for the patient, to become confused when confronted with seemingly contradictory descriptions of the cause of tendinitis and its treatment. One person may suggest reduced training, another may propose resisted exercise, and yet another may suggest modalities such as ultrasound or laser or electrical stimulation, or may advocate the use of corticosteroid injections. Surgery may even be proposed in very chronic cases.

The purpose of this chapter is to describe briefly the physiology and mechanics of tendon, and to explain how this information can be used to design a successful treatment programme for chronic tendinitis. Rather than providing a description of the signs and symptoms of various examples of tendinitis and outlining protocols for treatment, a procedure is developed whereby any clinical example of tendinitis can be treated once it has been identified. Faced with a case of tendinitis in the clinic, the physician should focus on the mechanics of that particular case, decide why the patient has tendinitis, and use his or her knowledge of normal and abnormal tendon behaviour to tailor treatment for the individual.

Prescribing or following a treatment regime for a particular type of tendinitis may provide short-term relief; it is far better to design a strategy that will both treat the injured tendon and protect it from re-injury. This requires knowledge and understanding of tendon structure and function, how they are altered in pathological conditions such as tendinitis, and how exercise and tissue physiology can be used in the treatment of tendon injuries. Most cases of tendinitis can be successfully treated in 6 to 8 weeks, and most patients should never experience the same tendinitis more than once. The aim of this chapter is to provide the reader with the necessary information to make this possible.

STRUCTURE AND FUNCTION

Structure of the tendon and its components

While it is likely that the fine structure of tendons varies between species and, indeed, even between different tendons in the same animal, many features appear to be common to most.[1-3] Tendons are rope-like structures that attach muscle to bone at each end of the muscle, although typically the distal tendon is much larger and better developed. They are complexes composed primarily of collagen and other proteins, glycosaminoglycans, elastin, glycolipids, cells, and water.[4-6] There are many types of collagens, but that in tendons is mainly type I. It comprises 70 to 80 per cent of the dry weight of the tissue after all water is removed. Collagen is the major load-bearing component in dense connective tissues such as ligament and tendon.[7] Like all proteins, collagen is synthesized inside the cell, where it is subject to a variety of influences.[8] However, unlike many other proteins that remain largely unchanged after leaving the cell, collagen undergoes a number of modifications outside the cell that contribute to its function in tissues. These modifications include cross-linking between molecules and organization into the fibrillar structure that gives collagen its unique load-bearing properties.[9,10] These processes are summarized in Fig. 1.

Cross-linking between and within collagen molecules contributes strongly to the tensile strength of the collagen, prevents enzymatic, mechanical, or chemical breakdown, and helps direct the organization of collagen molecules into fibrillar structures.[9,10] Both intramolecular cross-links, formed within the cell, and intermolecular cross-links, formed later after the collagen is extruded from the fibroblast, are present as shown in Fig. 2. The intermolecular, or mature, cross-links, are believed to be an important indicator of the tendon's ability to withstand high levels of tensile force.[9] Any defect in the number or quality of cross-links in the collagen leads to flaws in the mechanical behaviour of connective tissue. There are numerous clinical examples illustrating the importance of collagen cross-linking, most based on genetic defects or nutritional inadequacies. Decreased cross-linking results in tissues that are low in tensile strength and elongate readily when tensile force is applied, as in some forms of Ehler–Danlos syndrome.[11] Excessive cross-linking, however, results in much stiffer tissues that require more force to stretch. This occurs in diabetes and also in joint capsules after immobilization, leading to decreased joint range of motion and increased resistance to movement.[12-14]

Also found within the tendon is the so-called ground substance, a mixture of water and glycosaminoglycan compounds. The ground substance provides friction that helps the collagen fibrils to adhere to one another, yet at the same time provides lubrication and spacing that allows them to slide past one another and not to become excessively cross-linked.[4,15] The glycosaminoglycans, while forming only about 1 per cent of the tissue dry weight, are important because of their ability to bind water which forms 65 to 75 per cent of the total weight of the tendon. The glycosaminoglycans are amino sugar chains of varying length which may combine with other glycosaminoglycans, or with a protein core, to form proteoglycans.[16] These proteoglycans are similar in basic organization to those found in cartilage, but are much smaller. Interestingly, different types and amounts of proteoglycans are found in areas of tendon that are subject to compressive, rather than tensile, forces, and these differences are accompanied by variations in the amount, type, and organization of the collagen in those areas.[5,6,16-18] Generally, increased amounts of glycosaminoglycans or proteoglycans accompany small diameter collagen fibrils and a lower collagen concentration.[16,19] Thus there is an

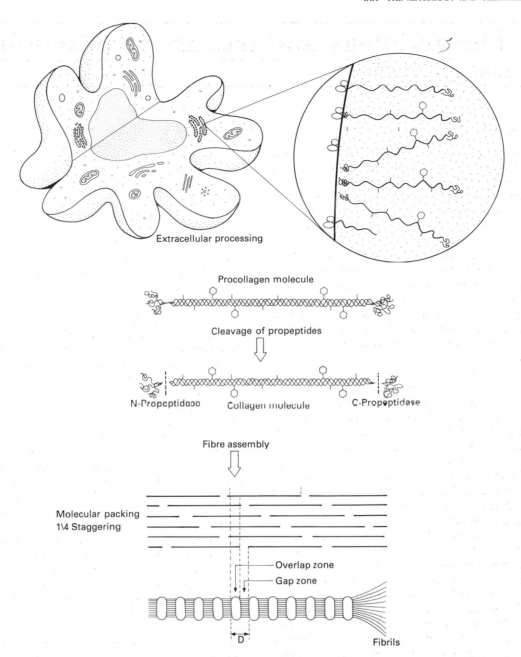

Procollagen molecule

Cleavage of propeptides

N-Propeptidase   Collagen molecule   C-Propeptidase

Fibre assembly

Molecular packing
1\4 Staggering

Overlap zone
Gap zone

D

Fibrils

Fig. 1 Collagen α chains of amino acids are synthesized in ribosomes inside the cell and then enter the endoplasmic reticulum, where several modifications occur and the chains associate in groups of three to form procollagen molecules. Propeptidase enzymes in cellular infoldings of the extracellular space remove part of the procollagen to form the tropocollagen molecule. The molecules can then assemble into collagen fibrils and fibres. (Modified from ref. 8, with permission.)

observable relationship between the composition of the tendon and its mechanical environment.

Outside the cell, collagen molecules, alone or in small groups called microfibrils, continue the aggregation that began inside the cell.[3,20] Molecules overlap in a head-to-tail fashion dictated by the ability of certain amino acids to cross-link with each other in the presence of the necessary enzymes and cofactors, and by the type and amount of glycosaminoglycans already present.[4] As the molecules assemble into fibrillar structures, they become capable of resisting load. While tissue strength is closely correlated with collagen content, this collagen must be organized before it can function as a load-bearing unit. Masses of non-organized collagen molecules are incapable of resisting tensile force application.[6] The new microfibrils may be added to existing nearby fibrils, increasing their size, or they may associate to form new fibrils, which are typically much smaller in diameter than the older longer-forming fibrils. In a manner as yet unexplained, the ground substance modifies the organization of collagen in the tissue, both directing and somehow limiting fibril organization.[6,15] As the collagen fibrils enlarge, the fibroblasts are squeezed flat between them and end up

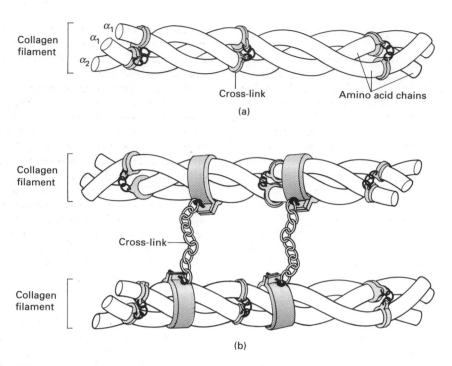

Fig. 2 (a) Intramolecular and (b) intermolecular cross-links are largely responsible for connective tissue tensile strength. (Redrawn from ref. 10, with permission.)

as columns of spindle-shaped cells with thin cytoplasmic processes extending around the fibrils. The strongest tendons are those which contain large numbers of large diameter fibrils. These are generally the oldest fibrils, which therefore contain the largest numbers of mature cross-links.[6,21]

Collagen fibrils group together to form successively larger primary bundles, which are also known as fibres. Groups of primary bundles, surrounded by a loose connective tissue sheath called the endotenon which also encloses the nerves, lymphatic system, and blood vessels supplying the tendon, form a fascicle (secondary bundle). While collagen fibrils are the smallest tendon structures capable of bearing a load, the fascicles appear to be the smallest load-bearing units in a functional setting. Individual fascicles are associated with discrete groups of muscle fibres or motor units and thus may be stressed independently of other fascicles if only those muscle fibres are activated. This might result in uneven stress distribution within the tendon when only certain motor units are activated during a particular movement or activity, which would be particularly true for the fast motor units that are activated only during very rapid movements or situations requiring a large force output from the muscle. Several fascicles may form a larger group (tertiary bundle), also surrounded by endotenon, while surrounding all the secondary bundles and their endotenon coverings is another sheath, the epitenon. These sheaths may be differentiated by their location and also by their collagen type content, with the endotenon containing much more type III collagen than the epitenon. An additional double-layered sheath of areolar tissue, the peritenon or parateneon, is loosely attached to the outer surface of the epitenon. The peritenon may become a sheath filled with synovial fluid, the tenosynovium, in tendons that are subjected to friction. The hierarchial organization of the tendon is illustrated in Fig. 3.[3]

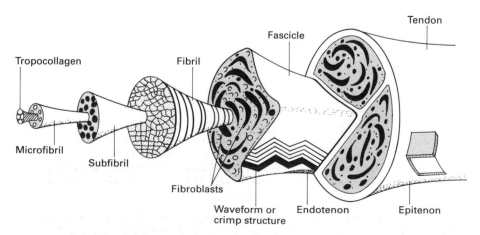

Fig. 3 The hierarchial organization of the tendon, showing the various stages from molecule to tendon. (Redrawn from ref. 3, with permission.)

Mechanical behaviour of the tendon

The organization of the tendon is responsible for its mechanical behaviour. The presence of the ground substance causes the tendon at rest to have a slightly crimped or wave-like appearance.[1,2] The tendon is very elastic at low loads, which means that little tensile force is required to elongate the tendon, and removal of these low loads allows immediate recovery of the tendon to its undeformed state. This elastic behaviour creates the 'toe' region of the force–elongation curve (Fig. 4) where small magnitudes of force result in relatively large length increases of the tendon. It is the straightening of the crimp that is responsible for mechanical behaviour in this region of the curve, and little or no physical deformation of the fibrillar structure takes place.[1]

Further application of force beyond the toe region (2–4 per cent elongation) results in progressive deformation of the tendon's components and structure. There is a linear relationship between applied force and resulting deformation after the toe region, and stress is applied directly to the now straightened collagen fibrils. Short-time X-ray diffraction techniques have shown that intrafibrillar slippage takes place initially, followed by interfibrillar slippage and finally gross disruption of the collagen fibrils or fibres themselves.[22–24] The linear region of the curve ends when some fibrils begin to rupture, and an irregular plateau region is rapidly followed by complete rupture of the tendon as the remaining fibres are subjected to even larger loads, usually after the tendon has elongated about 8 to 10 per cent further than its original length.[25]

Most of the tendon injuries classified as tendinitis probably involve loading in the linear region. The increase in severity of tendon injuries appears to be a progressive collapse of lateral cohesion between components, beginning at the fibrillar level. This collapse can even result in a reduction of tensile strength that is greater than that suggested by the number of torn fibrils or fibres observed on tissue examination.[23] Such damage can occur not only

during tissue loading but also during rapid unloading, perhaps as a result of shearing within the tendon.[22] This may explain why both the application and release of sudden or unexpected force is often associated with tendinitis.

Historically, the breaking strength of the tendon, i.e. the magnitude of the force being applied at the point of tendon rupture, has been used to assess the tensile strength of the tendon.[1] However, there are considerable technical difficulties with ensuring proper gripping of many tendon specimens for tensile testing, which can lead to large errors in measurement of deformation and force at the end-points of mechanical testing. This fact, together with the knowledge that the tendon probably functions in the toe and linear regions under physiological loading conditions, has led some authors to suggest that the slope $\Delta F/\Delta L$ of the linear portion of the curve, i.e. the stiffness, may be a better indicator of the mechanical behaviour of the tendon *in vivo*.[25] Since tendons rarely rupture under normal loading conditions, unless previously injured or diseased,[26] we can probably assume that they are seldom loaded to the breaking point, and that most tendons *in vivo* are rarely subjected to the maximum loads that are employed in laboratory testing.[2] Thus an appreciation of the physical changes that occur during the linear region of the stress–strain curve is probably more important in understanding the tissue damage that occurs during tendinitis than knowledge of the maximum load that the tendon can tolerate before breaking.[25] The latter situation is more representative of tendon rupture, which fortunately is much less common clinically.

The force–elongation curve of the tendon can be affected by a number of factors. The actual amounts of force or length change will depend on the size (cross-sectional area) and length of the tendon, as shown in Fig. 5. Larger tendons will be able to tolerate larger forces, and longer tendons will undergo greater absolute (but not percentage) changes in length before fibril disruption than will shorter tendons. These size-dependent features are referred to as the structural properties of the tendon.[1]

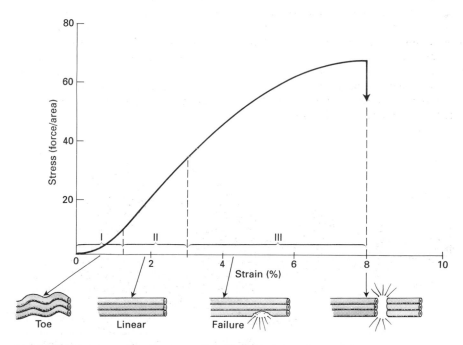

Fig. 4 The stress–strain curve for tendon. The crimped tendon straightens in the toe region (0–2 per cent) and then force is applied to the fibrils in the linear region (2–8 per cent). Some fibrils at the end of the linear region, and the remaining fibrils rapidly follow if force continues to be applied.

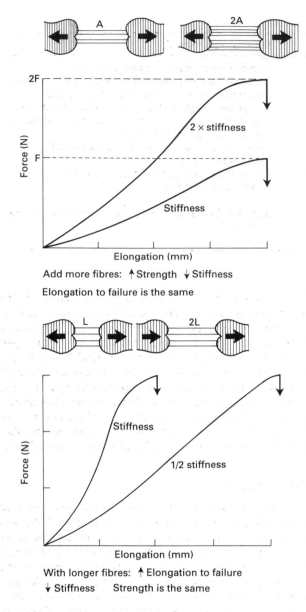

Add more fibres: ↑Strength ↓Stiffness
Elongation to failure is the same

With longer fibres: ↑ Elongation to failure
↓ Stiffness   Strength is the same

Fig. 5 The structural properties of tendon, which depend on its area and length.

Variability in tendon composition, such as differences in collagen concentration or cross-link number, makes comparison of the mechanical behaviour of different-sized tendons difficult, unless the influence of these factors is somehow removed. This is done by calculating the force relative to the cross-sectional area of the tendon, i.e. stress, and by expressing length changes as a percentage of the original length, i.e. strain. The force–elongation curve becomes the stress–strain curve, a representation which is independent of area or length. Differences in the stress–strain curves of different tendons are said to reflect the material properties of the tendons.[1,27]

Thus functional mechanical behaviour of a tendon, i.e. its function in the normal anatomical setting, can be influenced by a number of factors, including size, length, and composition. For example, an increase in cross-linking would be expected to alter the material properties of a tendon and make it stronger than a tendon of the same size containing fewer cross-links. Similarly, an

overall increase in area, but no change in cross-linking or collagen concentration, could also result in a tendon that could withstand larger loads. Both these adaptations take place in tendons exposed to altered physiological conditions.

Effects of exercise and disuse

The effects of disuse and immobilization on tissues such as muscle, ligament, joint capsule, and tendon are well established.[12–14, 28–34] All musculoskeletal tissues atrophy under conditions of decreased load. Collagen turnover increases, but degradation exceeds synthesis such that there is a net decrease in collagen.[14] Both collagen and cross-link concentration decline, and the tissue becomes weaker[20] both structurally and materially. These findings have led to the concept of using early motion and gradual stress application in treating many hard and soft tissue injuries.[35–39]

Both healing and normal tendons can also adapt to increased loads, either by becoming larger and hypertrophying as muscle does, or by changing their material properties to become stronger per unit area.[1,27,40–42] In fact, almost all musculoskeletal tissues respond to increased load by increasing their tensile strength.[1,43,44] Muscle rapidly hypertrophies under increased load conditions, and may also increase its connective tissue content.[29,45] Ligament and tendon have been shown to behave in a similar fashion.[27,34,37,42,43,46,47]

Increased loading is usually associated with exercise such as running and jumping. It is now well documented that many sports-related activities stress tendons to large percentages of the theoretical maximum for mammalian tendon (Table 1).[48,49] The association of tendinitis or tendon rupture with particular sports has confirmed the high demands placed on these tissues during many movements in sports and dance.[49,50–55] The incidence of

Table 1 Forces on tendon during activities

Activity	Tendon	Force (N)
Running (slow)	Achilles	4000–5000
Running (fast)	Achilles	8000–9000
Running (fast)	Patellar	7500–9000
Walking	Achilles	1000–3000
Walking	Patellar	500
Push-off	Achilles	4000
Kicking	Patellar	5000
Jumping (take-off)	Patellar	2500
Jumping (landing)	Patellar	8000

tendon injuries after sudden increases in the amount of training, or when training is resumed unchanged after a period of inactivity, suggests that the tendon is being subjected to loads which exceed its tensile strength and thus damage it. Sometimes, however, it is difficult to recognize any change in training pattern which may explain the onset of tendinitis. Athletes may develop tendinitis for no apparent reason while training at the same level of intensity, and it is difficult to understand why these loads, previously well tolerated by the tendon, should now produce an injury.

Such occurrences suggest that the relationship between training intensity, load, and tendon physiology is more complex than previously thought. The issue of whether all types of exercise have the same influence on tendon requires further investigation.[56] There is evidence to suggest that the cross-linking in the Achilles tendon

Table 2 Factors predisposing tendon to injury

Factor	Change in tendon	Clinical relationship
Decreased collagen synthesis	Decreased collagen concentration	Immobilization, disuse, systemic factors, nutritional deprivation (vitamin C, copper, protein)
Increased collagen degradation	Decreased collagen concentration	Immobilization, disuse, stressful exercise, surgery, systemic factors such as renal disease, rheumatoid arthritis, hyperparathyroidism, hormonal influence (e.g. oestrogen, catecholamines)
Decreased cross-linking	Decreased tensile strength	Decreased loading, increased synthesis of collagen, nutritional factors
Decreased size—the 'weak' tendon	Increased physical deformation under same loading conditions—fibril rupture	Chronic low loading, or factors as described for decreased synthesis
Decreased length—the 'tight' tendon	Increased physical deformation under same loading conditions—fibril rupture	Adaptive shortening due to immobilization or pattern of use

may be increased by chronically increased loads but decreased by an intermittent strenuous running programme, even though the latter would be expected to load the Achilles tendon and indeed is often used as a model of increased tendon and ligament loading.[43,46,47,57-59] Are there other factors that may induce changes in the tendon such that previously safe levels of loading are now capable of damaging the tendon?

TENDON INJURY AND HEALING

The injured tendon

Armed with a thorough understanding of normal tendon structure and function, we can predict changes that would make the tendon susceptible to injury. Some of these changes and their clinical correlates or causes are shown in Table 2. The most basic principle in the aetiology of tendinitis is that the tendon is exposed to forces which can damage it.[35,50,60,61] The key lies in determining whether the cause is external to an otherwise healthy tendon, or whether the tendon itself is 'sick'. The nature of the actual damage probably varies with the type of force (compressive versus tensile) as well as its magnitude and pattern of application.[35,49]

Extrinsic versus intrinsic tendinitis

Tendinitis resulting from forces outside the tendon is often called 'extrinsic' tendinitis.[35,62,63] Some examples are shown in Fig. 6. The cause is usually excessive compressive force on the tendon. This compression may come from an article worn by the patient, for example, tight laces in a high-top training shoe or skate may cause tenosynovitis of the extensor tendons at the ankle joint. Pressure from the patient's own anatomical structures may also be responsible.[64] A large acromion process may cause pressure on the supraspinatus tendon, leading to shoulder impingement syndrome. Similarly, retinacula at the wrist can cause various forms of tenosynovitis, usually in combination with repeated use during occupational or recreational activities.[51] These cases of tendinitis are best treated by early removal of the compressive forces.

Tendinitis may also result from changes or inadequacies within the tendon, the so-called 'intrinsic' forms of tendinitis. In such cases, there are no readily identifiable external causes and tendinitis is attributed to a change in tendon structure.[60] However, it should be recognized that factors outside the tendon are almost always inducing the change, except in rare cases involving genetic abnor-

malities or diseases affecting connective tissue structure.[8,11] The tendon is simply not strong enough to tolerate the tensile loads to which it is subjected. There are two possible solutions: (1) remove or reduce the tensile forces; (2) cause the tendon to become stronger. Frequently the loading pattern is within normal limits for this patient, and force reduction is not a viable long-term solution.[35] Many athletes and dancers suffer from chronic tendon pain because they are unable or unwilling to reduce the forces applied to their injured tendons.[50,54] This creates a situation where the tendon must be modified to match its environment. Many forms of intrinsic tendinitis fall in this category. Most are referred to as overuse injuries.

Overuse injury

Overuse injury is probably the most familiar description of chronic tendinitis. An otherwise normal tendon is chronically subjected to relatively large loads, perhaps extending into the linear region of the stress–strain curve.[65] The overuse theory holds that this causes partial rupture (microscopic failure) of some of the fibrils within the tendon, or slippage between fibrils, and leads to tendon injury.[66] Such injuries have been likened to the stress fractures which occur in bones subjected to chronic loads.[60] The injury is believed to be the result of fatigue of the loaded structure, just as metal beams will fatigue with repeated loading. Individual loads may be within the physiological range, but are repeated so often that recovery cannot occur and the structure fatigues. Since tendon structure recovers with rest, even after loading in the linear region, time is probably an important element in producing these injuries. This is the type of tendinitis which seems to develop very gradually and is often related to high training levels, such as distance running.[35,54,60] It may be appropriate to consider this another form of overtraining, and it can be helpful to use this analogy when explaining this disorder to athletes since most are familiar with situations where high intensities of training result in no improvement, or even a decline, in performance.[67-69]

It is not only high level athletes who are afflicted with this type of tendinitis, although this group does account for most of the eponyms, such as tennis elbow and jumper's knee, that are used for many cases of chronic tendinitis. Many older individuals involved in recreational sports also suffer, as do non-athletes.[70] One of the most common examples of tennis elbow is the lateral epicondylitis that develops in the handyperson who has been remodelling his or her house.[71] This is another form of overtraining or overuse.

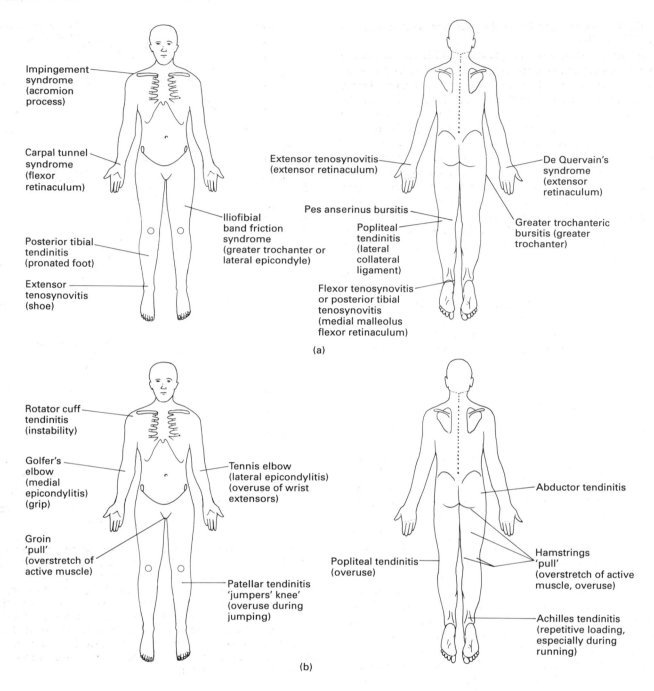

Fig. 6 Examples of (a) extrinsic and (b) intrinsic tendon disorders. Both are related to external force, but extrinsic disorders can only be successfully treated by removing the outside cause, while intrinsic tendinitis can be treated by removing or reducing the force, or by inducing adaptation in the tendon to meet the increased mechanical demand. The cause of the disorder is enclosed in parentheses.

Sudden loading or excessive force

The tendon may also be damaged by loading patterns other than chronic. Sudden force application, particularly involving lengthening (eccentric) muscle contractions, is another way that tendons and muscles are commonly injured.[35,53,72,73] A sudden maximum muscle activation results in the rapid application of a larger than normal force to the tendon, causing partial or complete rupture. Frequently used examples are the jogger who misjudges a curb and lands too far forward on his or her foot whereupon the heel drops rapidly, stretching the Achilles tendon and invoking a reflex contraction of the calf muscles, or the Achilles tendon rupture in badminton as the athlete suddenly changes from backward to forward motion.[55] One competitive weight-lifter was coincidentally filmed in a lift during which his patellar tendon ruptured; later analysis demonstrated that the disruption occurred as the lifter changed from downward to upward motion, i.e. the end of the eccentric phase.[74] Lifting the largest possible weight certainly demands maximum force from the muscle, and would therefore apply a very large force to the tendon. In the case of the weight-lifter, the force on the patellar tendon was estimated at about 17 times body weight!

As explained earlier, sudden application of force can cause more damage than a gradual increase in force to the same level of loading. The sudden removal of a given force level is also more likely to cause disruption than is a gradual reduction of the same force. The reasons for this are not entirely known, although it appears to involve the disruption of the relationship between the collagen fibrils and their surrounding matrix.[22,23] Also, the distribution of stress within the tendon is largely unknown and is generally assumed to be symmetrical across the tendon cross-sectional area. However, not every motor unit within a muscle fires during each activation of the muscle. Slow movements or maintenance of posture requires the use of small motor units composed of slow muscle fibres. Therefore the tendon fascicles associated with these motor units are regularly loaded during daily activities. Very rapid loading may cause the firing of fast motor units which are never used at lower force levels. Presumably, the tendon fascicles associated with these seldom used motor units have not been exposed to the same loading history as other fascicles within the tendon. It is possible that these tendon fascicles may actually be weaker as a result of little or no loading during daily activities, and that they are thus more easily injured when a sudden demand on the muscle for rapid force production is made.

Role of eccentric muscle activation

Muscle physiology experiments on both isolated and *in-vivo* human muscle have shown that force increases as the velocity of active muscle lengthening increases, while the opposite is true during concentric (shortening) muscle activations.[75,76] This may explain the frequent connection between eccentric loading and tendon injury.[35,73] The tendon is exposed to larger loads during eccentric loading, especially if the movement occurs rapidly.[35,49,75,76] This is exactly the situation that occurs in landing from or preparing for a jump (patellar tendinitis),[50,53] midstance during running or during a demiplié in ballet[54] (Achilles tendinitis), and when hitting a backhand in tennis (lateral epicondylitis).[71] In fact almost all shortening activations of muscle–tendon units are preceded by lengthening while the muscle is active. This activation pattern stretches the muscle–tendon unit, creating a passive force in the muscle owing to the elongation of its elastic elements, and providing elastic energy if the muscle is allowed to shorten immediately after being lengthened.[75–77] This storage of elastic energy allows the muscle to produce more force at less metabolic cost. Unfortunately, it also results in simultaneous application of maximum force and maximum elongation to the tendon, which may cause damage. Figure 7 shows the loading patterns on the Achilles tendon associated with some different activities and exercises.[49] It is well known that muscle damage is closely associated with eccentric muscle contraction, presumably because of higher load levels, and this appears to be true for tendon as well.[73,78]

Other factors

Endocrine

It is often difficult to explain to patients why they have developed tendinitis when no apparent change in loading or training has occurred, and there have been no examples of sudden unexpected application of force (although this may have happened, but gone unnoticed by the patient). This suggests that there may sometimes

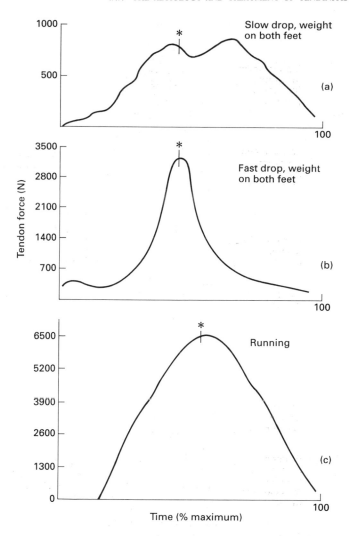

Fig. 7 Achilles tendon forces during an exercise programme used to treat Achilles tendinitis, and during running at a moderate speed. The peak in force that occurs both during the rapid exercise movement (dropping heels over the edge of step (b)) and running should be noted whereas force is more gradually applied in the slow exercise movement (a) even after time differences are considered. All movements are expressed relative to the total time required to complete the movement. Forces are much larger in running (c). These steps in the exercise programme would be regarded as beginning (a) and intermediate (b).

be other reasons why tendons, previously pain free, become symptomatic. The emphasis is on the words apparent change, since only after all external influences have been eliminated as potential causes should other factors be considered as a possible primary cause. Endocrine responses to stress, such as increased glucocorticoid and catecholamine release, may have negative effects on connective tissue,[79,80] increasing turnover and thus resulting in decreased cross-linking.[81,82] While we know that these hormones can influence connective tissues such as tendon, no relationship has yet been demonstrated in cases of tendinitis. However, it is interesting to speculate that an endocrine response to chronic levels of overtraining[67–69,79,80,83] may be at least a partial explanation for those 'mysterious' cases of tendinitis that develop for no other apparent reason. Similar influences have been suggested for lateral epicondylitis (lack of oestrogen),[71] and the relationship between chronic corticosteroid administration and tendon rupture is well known.[84,85]

Compressive loading

The tensile strength of the tendon may gradually decline over time if it is subjected to chronic compressive loads, or if turnover is increased markedly, resulting in less mature cross-linking. Less collagen, fewer collagen cross-links, and more ground substance all result in decreased tissue tensile strength.[4,5,17,19] This could be a factor in rotator cuff tendinitis, and may suggest a role for loading the supraspinatus tendon with tensile forces during rehabilitation.[35] Such loading would be expected to induce a change in tendon composition if applied gradually, such that collagen content and cross-linking were actually increased. Compressive loading may also result in a decrease in blood flow through the tendon, as has been demonstrated for supraspinatus tendinitis.[86]

Nutritional

The amounts of amino aids supplied by a normal healthy diet are adequate for all protein synthesis, and this of course includes collagen. Cofactors such as vitamin A, vitamin C, and copper are known to be important in collagen synthesis and cross-linking,[8] and iron deficiency can have a negative influence on healing.[87] There is also some evidence to suggest that collagen synthesis is more severely affected than that of some other proteins during fasting,[82] which may have some implications for athletes involved in activities that emphasize a slender build, such as gymnastics and ballet. However, the nutritional influences on chronic tendinitis remain largely unexplored.

Referred pain

This is a largely unrecognized cause of tendinitis-like pain, and is due to peripherally produced (somatic referred) pain from irritation of spinal structures such as joint capsule and ligament. It is probably most common in cases of tennis elbow, where degenerative changes in the cervical spine cause symptoms which exactly mimic those of tennis elbow.[71] One case seen by the author involved a 35-year-old male patient complaining of right calf pain which, from the history and examination, appeared to be a strain of the medial gastrocnemius. Failure to respond to treatment led to a more detailed examination of the lumbar spine which revealed that the symptoms were reproduced by specific stretching of a lumbar facet joint. Restoration of pain-free lumbar movement eradicated symptoms. These cases are probably not common, but the clinician should be very suspicious if a 'tendinitis' fails to respond to treatment, and a spine examination should always be conducted in parallel with peripheral examination, particularly if the patient falls in the appropriate age category for degenerative spinal changes.

TENDON HEALING

Much is known about the healing of severed tendons. This work comes from models where the tendon has been divided, surgically or accidentally, and the severed ends reopposed and held in place by immobilization or suture.[20,26,88–91] These models have told us a great deal about tendon healing under these conditions. For example, we know the timing of the biochemical and mechanical changes that take place in the healing tendon. The initial inflammatory stage triggers an increase in glycosaminoglycan synthesis within days that is rapidly followed by collagen synthesis, such that the healing wound can be subjected to low levels of force

within a matter of days.[36,92] We know that the application of some force (but not too much!) is ideal in encouraging the new collagen fibrils to align with the direction of force application, and that healing tissues subjected to loading are almost always stronger than unloaded tissues, whether this be skin, ligament, or tendon.[34,36–38,43,59,91,93] The application of these principles in plastic surgery has led to the design of early motion programmes after tendon repair, and a rehabilitation programme which is truly based on scientific principles.

Unfortunately, a great deal of confusion remains when it comes to applying this information to tendinitis. Just what is tendinitis? Does tendinitis follow the same healing pattern described for complete lacerations? Little is known about the inflammatory response to the type of mechanical trauma that produces chronic tendinitis. Any injury that involves the tendon or its associated sheaths may be referred to as tendinitis, but a plethora of other terms also exists: tendinosis (degeneration of the tendon without inflammation), tenosynovitis (inflammation of the sheath surrounding the tendon), paratenonitis, and peritendinitis. Clancy[60] has recommended the classification system shown in Table 3.

Table 3 Classification of tendon injury

I Tenosynovitis and tenovaginitis—an inflammation of only the paratenon, either listed by synovium or not

II Tendinitis—an injury or symptomatic degeneration of the tendon with a resultant inflammatory reaction of the surrounding paratenon
  (1) Acute—symptoms present less than 2 weeks
  (2) Subacute—symptoms present longer than 2 weeks but less than 6 weeks
  (3) Chronic—symptoms present 6 weeks or longer
    (a) Interstitial microscopic failure
    (b) Central necrosis
    (c) Frank partial rupture
    (d) Acute complete rupture

III Tendinosis—assymptomatic tendon degeneration due to either ageing, accumulated microtrauma, or both
  (1) Interstitial
  (2) Partial rupture
  (3) Acute rupture

Some authors argue that, with cases of chronic tendinitis that develop very gradually and never appear to have an acute stage, there is no associated inflammation of the tendon or the tendon sheath and so the condition should be called tendinosis.[60,61,94] This conclusion is based primarily on observations of human tissue obtained during surgical procedures aimed at relieving chronic tendinitis. These observations probably represent the late fibrotic stage of tendinitis, and do not eliminate the possibility of the earlier presence of inflammation.

One of the major difficulties in evaluating chronic tendinitis has been the lack of a suitable animal model. Racehorses frequently develop chronic tendinitis, and, indeed, are responsible for much of what we now know about the condition.[66,95,96] However, they cannot be used by most researchers because of the expense, and because the amount (time and/or magnitude) of loading responsible for the tendinitis is unknown. Recently, the first animal model to simulate the type of overloading which often leads to human tendon dysfunction was designed.[97,98] The ankle joints of anaesthetized rabbits were repeatedly flexed and extended for 2 h

Table 4 Classification of tendon disorders

Intensity	Level	Pain	Performance
Mild	1	None	Does not affect performance
	2	With extreme exertion only	
		Not intense	Does not affect performance
		Disappears immediately when activity stops	
Moderate	3	Starts with activity	
		Lasts 1–2 h after activity	Performance may be affected
	4	With any athletic activity	Performance level significantly decreased
		Increases during activity	
		Lasts 4–6 h afterwards	
Severe	5	Immediately upon any activity involving tendon	Performance markedly curtailed or prevented
		Sudden increase in pain if activity is continued	
		Lasts 12–24 h afterwards	
	6	During daily activities	Unable to participate

daily while the triceps surae muscle was electrically stimulated. After 4 weeks of exercise, palpation of the exercised tendons showed that all had irregular thickening, with palpable nodules a short distance from the tendon insertion into the calcaneus. Light microscopy of the tendons showed degenerative changes in the tendon and thickening of the tendon sheath (paratenon).[97] Blood flow to both the paratenon and the tendon was increased (about two-fold).[98] Most cellular changes suggestive of an inflammatory response were found in the tendon sheath, with degenerative changes in the central portion of the tendon. This model is the first to produce tendon dysfunction through reproducible loading. Other animal models have used partial laceration or chemical injection to induce a tendon injury and thus may not be as closely representative of naturally occurring injuries.

The findings from human surgical and postmortem specimens, and those of Backman et al.,[97] suggest that most of the inflammatory changes take place in the paratenon and are accompanied (preceded? followed?) by areas of focal degeneration within the tendon.[94,97] However, these findings also suggest that injury to the tendon structure does not occur in isolation, and that it would be uncommon for tendon degeneration to occur without accompanying inflammatory changes. In other words, it should be possible to have paratenonitis without tendinosis (as can occur with extrinsic tendinitis), but not tendinosis without some inflammation of the tendon sheath.

It is probably unlikely that tendon damage takes place without some symptoms being present. While patients may not complain of pain serious enough to prevent them from participating in activities they consider enjoyable (or unavoidable), careful questioning will often reveal that symptoms were/are present during activities that load the tendon, or that a more acute episode did take place at some earlier point in time. The clinical signs of tenderness with palpation and pain on loading can be interpreted to reflect the presence of an inflammatory response, since it is difficult to see why such signs would be present otherwise.[99] The response may be mostly in the tendon sheath, but it reflects and accompanies the structural damage to the tendon, although the exact relationship is not yet known.

In clinical cases, it is usually impossible to determine the exact portion of the tendon that is the source of pain. Palpation involves both the tendon and its sheath, and muscle contraction will apply force to an inflamed sheath as well as a damaged tendon. We can perhaps define tendinitis as a syndrome of pain and tenderness localized over an area of tendon, aggravated by activities that require activation of the particular muscle–tendon unit and thereby apply tensile force to the tendon. The syndrome can include inflammation of the tendon sheath, as in tenosynovitis and tenovaginitis, as well as actual inflammation of the tendon substance itself, and is either caused or followed by degenerative changes in tendon structure. Since it is unlikely that pathological changes can be determined for most patients, it may be helpful to use a classification system which is based on pain and function, which can be determined clinically, even though the exact relationship between symptoms and pathology remains unknown. Such a system is presented in Table 4.[35]

TREATING THE INJURED TENDON

Acute tendon injuries

The guide to treating the severed tendon has been well-established and is summarized in Table 5. The main aims are to avoid disruption of the repairing tendon early in healing, and then, once the danger of re-rupture has passed, to prevent the atrophy of other joints and muscles that would accompany limb immobilization prolonged beyond the minimum amount necessary to ensure tissue integrity.[35,92] Later, as these aims are achieved, the emphasis shifts to the healing tissue itself. This guide provides a framework on which to base all tendon injury rehabilitation, and is particularly useful in treating acute tendon injuries where the exact time of injury is known.[35,91,92]

Chronic tendinitis

When dealing with chronic tendinitis, the date of the initial injury may be uncertain and the difficulty then becomes deciding where in the healing process the injured tendon lies, for this often determines how to treat the tendinitis. Can we assume, based solely on the length of time since injury, that the injury is in the remodelling phase, in which case exercise would be indicated? Or should we assume that the injury is really in an earlier stage of healing, if the patient has continued to subject the tendon to the same loads that caused the original injury, and so treat the injury with rest, ice, and/or application of modalities? These questions are impossible to answer at this time; we simply have no accurate means of assessing human soft tissue injuries such as tendinitis. There are several promising techniques, such as magnetic resonance imaging, ultrasound, and thermography, which may prove useful in assessing

Table 5 Overview of principles governing clinical intervention during tendon healing

	Stage of healing		
	Inflammatory	Fibroblastic/proliferation	Remodelling/maturation
Time (days)	0–6	5–21	20 days and onwards Progressive stress on tissue
Suggested therapy	Rest, ice Anti-inflammatory modalities Decreased tension	Gradual introduction of stress Modalities to increase collagen synthesis	
Physiological rationale	Prevent prolonged inflammation	Increase collagen	Increase cross-linking (tendons and ligaments)
	Prevent disruption of new blood vessels and collagen fibrils	Increase collagen cross-linking	Decrease cross-linking (joint capsule)
	Promote ground substance synthesis	Increase fibril size and alignment	Increase fibril size
Main aims	Avoid new tissue disruption	Prevent excessive muscle and joint atrophy	Optimize tissue healing

these injuries, but they are not yet cost effective or widely available clinically.[100–107]

Some assumptions about chronic tendinitis

Since our understanding of chronic tendinitis remains inadequate, the development of a rational treatment strategy requires making some assumptions that we hope will be proved to be true. First, there is probably no harm in assuming that all cases of tendinitis are in an acute stage if treatment is progressed accordingly. Appropriate progression means that treatment may be prolonged by 2 or 3 weeks for some chronic injuries, but also ensures that injuries that are really in the inflammatory or proliferative phases are treated correctly. This leads to the second assumption, that the chronically injured tendon will heal in much the same manner as the severed (acute) tendon. Third, let us assume that inflammation reflects the degree of damage to the tendon, and that the signs of inflammation (pain, tenderness, function) can be monitored to assess the progress of tendon damage or recovery. This assumption is very important, because the successful treatment of chronic tendinitis is based on the theory that pain is a reflection of the inflammatory response occurring in the tendon, monitoring pain and function to determine whether improvement is occurring, and modifying treatment as the pain and inflammation changes. While there are limitations to this approach, in that pain is a very subjective perception, it appears to be the most clinically useful way to monitor patient progress. The fourth assumption is that the effects of exercise and disuse will be the same for chronic tendinitis as for normal and severed tendons, and for ligaments, and that exercise will therefore be beneficial in the treatment of chronic tendinitis.

General principles for treating tendinitis

There are many principles, physiological and mechanical, that should be considered when treating chronic tendinitis—too many to discuss each thoroughly, although most have been touched upon earlier in this chapter. Some of the most important guidelines for treating all forms of tendinitis are discussed below.

Identify and remove all negative external forces/factors

In the case of extrinsic tendinitis, the outside force is usually pressure on the tendon. Identification and removal of the source of pressure is the fundamental treatment for this form of tendinitis,

and is imperative if further tendon damage is to be avoided.[62] A simple example is tenosynovitis of the extensor tendons at the ankle joint caused by pressure from tight laces in a skate or shoe. The pressure can be removed by loosening the laces or designing a device to redistribute the force. A more complicated example is the classic shoulder impingement syndrome. The appropriate treatment is removal of the cause of impingement. If shoulder motion is restricted, this may involve physiotherapy to restore normal elasticity to the tight joint capsule of a hypomobile shoulder, a case where impingement occurs when the head of the humerus fails to glide downwards on the glenoid as the humerus rolls during abduction. In cases where degenerative rotator cuff changes have occurred that require repair, or where the configuration of the acromion is at fault, surgery may be necessary. In the case of extrinsic tendinitis, all other forms of treatment, however helpful they may be in relieving symptoms or restoring range of motion, can only be considered temporary or adjunctive. Eliminating the cause is the fundamental treatment.

It is not only extrinsic forms of tendinitis that can be affected by outside forces; there are also cases where external factors may be contributing to, or even causing, intrinsic tendinitis. Excessive foot pronation can cause the medial side of the Achilles tendon, or the tibialis posterior muscle–tendon unit, to be overstretched. The anatomical configuration of the Achilles tendon, whereby it twists during descent, may play a role in Achilles tendinitis.[108] Similarly, excessive lateral shoe wear may lead to iliotibial band syndrome as the lateral aspect of the lower limb is subjected to increased stress. Many of these problems are easily corrected with a shoe orthotic, or simply by buying new shoes.[35,52,60]

Probably the single most common contributing factor in almost all cases of chronic tendinitis is a lack of flexibility of the muscle–tendon unit involved. For this reason, a thorough and specific stretching programme is nearly always an essential part of the rehabilitation strategy.

Estimate the phase of healing (stage of tendinitis)

This is a very imprecise process, and requires clinical judgement based on the examiner's clinical experience to be truly successful. Generally, the more severe the patient's symptoms, the more closely the choice and timing of treatment should resemble that used for an acute tendon injury. Treatment should also progress as it would for an acute injury.

Determine the appropriate focus for initial treatment

This involves matching treatment with the stage of healing. Most cases of chronic tendinitis should be in the remodelling phase of healing, where force application is the most effective treatment. However, as noted above, more severe cases may have to be treated with judicious rest, ice, and modalities for a short period of time, followed by gradual stress increase, as would be the case for an acute tendon injury or repair.

Institute an appropriate tensile loading programme

The healing tendon must be loaded if collagen synthesis, alignment, and maturation via cross-linking are to be ideal.[97] The more acute or severe the injury, the lower is the force applied.[20,36,65,92,93] Passive movement produces very little tensile force, is safe immediately after injury, and is known to have beneficial mechanical effects on tendon.[38] The next step is gentle stretching, followed by increased stretching force and then active exercise.[63]

Control pain and inflammation

While the appropriate use of loading during healing should ensure that mechanical disruption and re-injury, provoking inflammation, do not occur, there are cases where additional help is needed to reduce a prolonged inflammatory response. In such cases, drugs, ice, and modalities can be used as adjuncts to treatment.[99,109–111]

Specific forms of treatment

Modalities

Physiotherapists employ a wide variety of modalities in treating soft tissue disorders, including ultrasound, laser, ice, heat, pulsed electromagnetic current, electromagnetic field therapy, high voltage galvanic stimulation, acupuncture, interferential current, etc. Most are proposed to 'decrease inflammation and promote healing'. Unfortunately there is only limited evidence as yet to support many of these claims.

Ultrasound is one of the most commonly used modalities. Generally, pulsed ultrasound is recommended for acute injuries to avoid a thermal effect, and continuous ultrasound is used for more long-standing injuries.[112,113] While there have been reports of no influence on healing tendon,[88] most studies have shown that ultrasound increases collagen synthesis by fibroblasts,[114] speeds wound healing,[109,112] and results in increased tensile strength in healing tendons.[115,116] Ultrasound has little or no effect on inflammation.[110,117] None of these models simulates the clinical situation of chronic tendinitis, so that we are again forced to use our second assumption that the chronically injured tendon will heal in a manner similar to the severed tendon. Given that one of the explanations for chronic tendinitis is 'failed healing response',[60] it is uncertain whether this assumption will always be true. It would appear that ultrasound has its most important effect when the synthetic activity of the fibroblasts is maximum, i.e. the proliferative stage of healing. However, because of the nature of chronic injuries, it is probably most widely used clinically during the remodelling stage. It would be interesting to see whether ultrasound increases the synthesis of collagen during all stages of healing, particularly during remodelling when the synthesis rate has declined. This appears to be its most widespread clinical use in 'promoting healing', and would, in effect, prolong the proliferative phase of healing. Given the timing of the normal healing response, there would seem to be little indication for using ultrasound for prolonged periods of time.

Another modality widely used in Europe and Canada is the laser.[118] Like ultrasound, lasers have been shown to increase fibroblast synthesis of glycosaminoglycans and collagen, and to speed superficial wound healing, but, unlike ultrasound, they have also been shown to decrease inflammation.[110] The use of lasers in non-superficial cases like chronic tendinitis remains speculative, however, and there is as yet no clinical or scientific evidence to support its use for treating deeper tissues like tendons or to suggest its superiority to ultrasound. Given the wide variety of laser types and dosages, much more research is needed on the effects of this modality.

Electrical stimulation is a modality that has also been demonstrated to have a positive influence on tendon healing.[119–121] Again, results were obtained from an acute tendon healing model and may not necessarily represent chronic tendon injuries. Both direct electrical stimulation[120,121] and indirect current via electromagnetic field induction[118] seem to augment tendon healing. Pulsed electromagnetic fields can treat both deep and superficial tissues, and cover larger areas then ultrasound or lasers.[122] Questions about timing and dosage remain unclear, but it seems that this treatment needs to be prolonged for several hours daily to have an influence on tissue since clinical use for shorter periods of time has not shown the same positive effects.[122]

One of the most widely used modalities for all soft tissue injuries is ice. Its use is recommended immediately after injury to prevent excessive soft tissue swelling, and it is thought to act mainly by decreasing the activity of inflammatory mediators and the overall metabolic rate of the injured tissue.[111] Another important effect is analgesia, which allows the use of appropriate forms of exercise, such as passive motion, that otherwise might be uncomfortable for the patient. The use of ice with chronic injuries is less clear, although it can be used for its analgesic effect and may help offset inflammatory changes induced by mechanical injury to the tendon during exercise.[35]

The use of modalities, although widespread, remains largely speculative in the treatment of chronic soft tissue injuries. Most scientific studies suggest that increased synthetic activity by fibroblasts is the major effect of most modalities, except ice and perhaps laser. Clinicians should always remember that this synthetic activity is mainly part of the proliferative phase of healing, and that mechanical forces are also required during remodelling if the newly synthesized collagen is to assemble and cross-link into a structure capable of withstanding tensile loading. The effects of modalities on chronic injuries remains largely unexplored, and well-designed clinical trials are needed to determine efficacy in the clinical setting.

Drugs

The most potent anti-inflammatory drugs are the corticosteroids, which are sometimes used, via local injection, to treat chronic tendinitis. The negative effects of systemic corticosteroid use are well-known,[81,84,85] but the effects of local injection are less clear. Both negative and zero effects have been reported;[123] however, it is generally agreed that injection into the tendon substances should be avoided.[35,60,71] This is due to the effects of the drug (which decreases collagen synthesis), the mechanical disruption caused by the needle, and the irritant effect of the solvent in which the drug is dissolved.[124] Given the potent anti-inflammatory effect, and the fact that most inflammatory changes occur in the paratenon, injection into the tendon sheath paratenon may be indicated if inflammation is marked or prolonged.[124,125] If the tendon has been injected, tensile force should be reduced for 10 to 14 days afterwards and the tendon treated as if it has suffered an acute injury, i.e. ice, rest, and modalities, followed by progressive loading starting at about 2 weeks.[35,71] Repeated steroid injections into a tendon are almost certain to result in substantial mechanical disruption and should be avoided.[124]

Non-steroidal anti-inflammatory agents (NSAIDs) are also widely used in the treatment of acute soft tissue injuries, and less so for chronic injuries such as tendinitis.[126] They are thought to limit inflammation by inhibiting prostaglandin synthesis, although other mechanisms are also involved. Unlike corticosteroids, they do not inhibit fibroblast or macrophage activity. While there is some evidence to show that preventive use of indomethacin may reduce subsequent muscle injury, most clinical studies have been poorly designed and it is impossible to conclude whether the use of NSAIDs has a beneficial effect on post-injury recovery.[126-128] Since paracetamol (acetaminophen) has only an analgesic effect, and patients may not distinguish between aspirin and paracetamol, it is important to distinguish exactly what drugs the patient is taking, since many of these agents may be self-administered, and anti-inflammatory action may be assumed when none is actually present.

Some patients may also be self-administering anabolic steroids, another class of drugs known to affect connective tissues. While the exact effects of these agents are unknown owing to difficulties in determining use and dosage, there is both scientific and anecdotal evidence to suggest that anabolic steroid users are more likely to develop a tendon injury.[129,130] The effect of anabolic steroids on the healing of these injuries is unknown. Clinicians should be alert to the possibility of anabolic steroids since their use by young males has become widespread.[129]

Surgery

There is little place for surgery in the treatment of chronic tendinitis unless the tendon ruptures and the ends need to be approximated.[71,90] Some authors advocate the removal of scar tissue and repair of the injured tendon,[52,65,95] but it is unlikely that it is the surgery, rather than the postoperative healing response and carefully progressed treatment, which causes the improvement in the patient's condition. Almost all cases of chronic tendinitis can be successfully treated without surgery.[35]

THE ROLE OF EXERCISE IN CHRONIC TENDINITIS

As previously discussed, there are many forms of treatment for chronic tendinitis. A recent 'progress note' from a physiotherapist accompanied a tennis elbow patient to clinic and included ultrasound, interferential, laser, ice, transcutaneous electric nerve stimulation (TENS), acupuncture, deep friction massage, muscle stimulation, pulsed magnetic field, and an exercise programme! After 2 months of this treatment, the patient's condition remained unimproved. To make matters worse, the patient was subjected to a strength test involving maximal eccentric loading, which is almost certain to reproduce the injury if used before full recovery has occurred.

This scenario, which is not uncommon, and a growing appreciation of tendon physiology and mechanics, suggests that the modality-based approach to the treatment of chronic tendinitis is not correct. This does not imply that the use of modalities and drugs should be abandoned altogether, but rather that they should no longer form the basis of a treatment strategy for chronic tendinitis. An understanding of tendon's ability to adapt to increased loads, and the belief that chronic tendon injuries are the result of tensile loads exceeding the tendon's mechanical strength, suggest that exercise should be the cornerstone of treatment.

Basic principles of exercise

For any exercise programme to succeed, whether its aim is to strengthen muscle or tendon, basic exercise principles must be followed. The following are most important for the treatment of chronic tendinitis.

Specificity of training

Training must be specific to the affected muscle–tendon unit, and must also be activity specific in terms of the type of loading (tensile, eccentric) and the magnitude and speed of loading. Specificity is achieved by simulating the movement pattern associated with maximal tendon forces, i.e. a lengthening of the active muscle–tendon unit followed by shortening contraction. The initial magnitude and speed of loading are based on the estimated stage of healing. The more acute the injury, the lower is the magnitude of the force and the lower is the speed of eccentric loading. The affected tendon must be subjected to tensile loading, not just a generalized exercise involving use of the affected muscle–tendon unit. For example, the Achilles tendon is loaded by having the patient stand at the edge of step and drop his or her heels downwards, rather than running. This specific exercise is aimed at countering the potentially negative effects associated with stressful training (e.g. hormonal) with the positive effects of exercise on the muscle–tendon unit (increased strength).

Maximal loading

Maximal loading is essential to induce adaptation in musculoskeletal tissues.[29] Clinically, the maximum load is determined by the tendon's tolerance, which is judged by the patient's pain level during exercise. It has been determined empirically that the patient should experience some pain between the twentieth and thirtieth repetition of the movement.[35] Pain before this seems to indicate overloading and causes worsening of the patient's condition, while no pain by 30 repetitions results in no change in the patient's symptoms, suggesting that the stimulus is inadequate to induce a change in the tendon.

Progression of loading

As the tendon becomes stronger, loading must be progressed so that maximal loads continue to be applied and the tissue will continue to have a stimulus for adaptation. This progress can be made by increasing the speed of movement (eccentric muscle activation) or increasing the magnitude of the tensile force by changing the external resistance (isometric, concentric, and eccentric muscle activation). The progression is determined by the patient's symptoms (see Fig. 8).

Eccentric exercise programme

The principles outlined above have been incorporated into an 'eccentric exercise programme' for treating chronic tendinitis.[35] The principles can be applied to any injured tendon, following the guidelines in Fig. 8. The overall programme has five steps, which are performed in the order listed below.

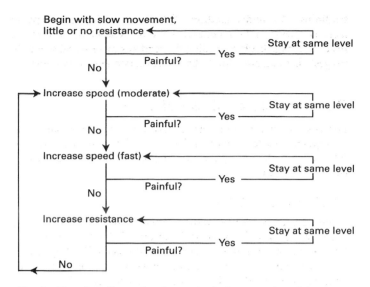

Fig. 8 Flowchart illustrating the method of progression of the eccentric exercise programme.

1. Warm-up: a generalized exercise like cycling or light jogging is used to increase body temperature and increase circulation. This exercise is not intended to load the tendon and should not be uncomfortable.
2. Flexibility: as noted earlier, lack of flexibility is a common finding in chronic tendinitis. It is recommended that the patient perform at least two 30-s static stretches of the muscle–tendon unit involved and its antagonist. More stretching can be done if this is felt to be a major factor in causing the patient's symptoms.
3. Specific exercise: this is done following the guidelines in Fig. 8 based on the principles outlined above. It is suggested that three sets of 10 repetitions be performed, with a brief rest and sometimes a stretch between each set. The patient should feel a reproduction of his or her symptoms after 20 repetitions. If pain is felt earlier, reduce the speed of movement or decrease the load; if no pain is experienced, increase speed or load (not both). If this is the first exercise session and the initial level of loading is being determined, the intensity of exercise may be increased and the 30 repetitions repeated until the appropriate level of intensity is reached.
4. Repeat flexibility exercises.
5. Apply ice for 10 to 15 min to the affected (painful to palpation) area. It is hoped that this will help prevent any inflammatory response provoked by microscopic damage to the tendon that might occur during the exercise.

Since this programme is designed for patients with chronic tendinitis, most people are able to participate in athletic activity but find participation painful (level 2/3) or their performance impaired (level 4/5). These patients do **not** need to cease participating in sports unless they are unable to perform. In fact, it is most desirable that patients change nothing about their activity except adding the exercise programme. A decrease in physical activity usually results in a parallel decrease in symptoms, and this makes effects success of the exercise difficult to assess. The exercises are performed daily, with continuous progression, until symptoms are no longer present during activity. Strength testing should not be performed until after treatment is completed and the

patient is asymptomatic, since the maximum force levels generated may damage the healing tendon. However, there is some suggestion that strength deficits may exist in tendinitis patients after symptoms are reduced, suggesting that continued strength training and eventual testing may be beneficial in some cases.[131]

For patients with severe or constant pain (level 5/6), modification of activity may be necessary. Often, these patients will already have reduced their activity level, since professional attention is seldom deemed necessary until function is affected. This level of pain is interpreted as reflecting a process of acute tendinitis, even though symptoms may be long-standing, and treatment is commenced at a very low level: ice, gentle stretching, passive movement, modalities to stimulate collagen synthesis, etc. This treatment is progressed as healing progresses, so that by 2 weeks, more vigorous exercise can usually be introduced as the patient's symptoms will have subsided.

Like all forms of rehabilitation, the stage between ending a therapeutic programme and resuming full activity is the greatest challenge for the patient and clinician because so little is known about the best way to return to activities. This is not a problem for patients able to continue sports participation throughout treatment. These patients can often be successfully treated with a home programme and periodic re-checks, and should be asymptomatic in 6 to 8 weeks. However, for those more severely affected, the athletic activity must be resumed gradually. There are no rules to guide reintroduction to activity, but certainly the patient should be asymptomatic during non-athletic activities and should be performing the eccentric exercise rapidly. Athletic involvement should probably be started at about 25 per cent of the pre-injury level (duration or intensity, depending on the sport and movement), and should be done on alternate days to avoid muscle soreness and to allow evaluation of the tendon's response to training. Assuming that few or no symptoms are produced, progression can be made in approximately 10 per cent increments until full training has been resumed. This should take about 8 weeks, making the entire treatment period for a patient with severe tendinitis about 10 to 12 weeks.

Success or failure of the programme

We monitored over 200 chronic tendinitis patients treated using the eccentric exercise programme, and found that most had minimal or no symptoms after 6 weeks.[35] Similar improvements have been noted by others using eccentric loading slightly different from that suggested in this programme.[131] Most patients will require at least 6 weeks of treatment, a few will experience complete resolution in 2 to 3 weeks, and a very few others will need to continue for 12 to 16 weeks. Modalities can be employed if desired, particularly if the patient's symptoms are acute or prolonged and the physiotherapist suspects that the synthetic activity of the tendon is decreased. A similar rationale can be used by the physician to decide on the use of NSAIDs if the inflammatory phase is prolonged.

If the eccentric exercise is not successful in treating the patient's tendinitis, a number of explanations are possible. The patient may not have tendinitis, or there may be unrecognized external factors that are causing or perpetuating the problem. For exercise treatment to be successful, symptoms must be related to tensile loading, usually during eccentric muscle activation. The most common reason for lack of success is incorrect programme progression: the patient is either started at too high a level or is not progressed to the next level of intensity. While some patients may experience a slight increase in symptoms during activity in the first 2 to 3 weeks of carrying out the exercise, this should not be a progressive increase and can be minimized by having the patient avoid carrying out the exercise programme immediately before or after athletic activity.

SUMMARY

Chronic tendinitis remains a clinical dilemma for many, and the best treatment is not always possible or recognized. Ideally, treatment should combine resolution and prevention of symptoms, and should be based on good science and common sense. The use of exercise to treat chronic tendinitis relies on both of these—the science of tendon adaptation to increased stress and the common sense of patient participation based on ability to perform. In order to be successful, exercise cannot be used in isolation if other factors are also involved, but a knowledge of the beneficial effects of tensile loading on tendon can be used to treat almost all tendon injuries.

REFERENCES

1. Butler DL, Grood ES, Noyes FR, Zernicke RG. Biomechanics of ligaments and tendons. *Exercise and Sports Science Review* 1978; **6**: 125–82.
2. Elliott DH. Structure and function of mammalian tendon. *Biological Reviews of the Cambridge Philosophical Scoiety* 1965; **40**: 392–421.
3. Kastelic J, Galseki A, Baer. The multicomposite structure of tendon. *Connective Tissue Research* 1978; **6**: 11–23.
4. Flint MH. Interrelationships of mucopolysaccharides and collagen in connective tissue remodelling. *Journal of Embryology and Experimental Morphology* 1972; **27**: 481–95.
5. Gillard GC *et al*. A comparison of the glycosaminoglycans of weight-bearing and nonweight-bearing human dermis. *Journal of Investigative Dermatology* 1977; **69**: 257–61.
6. Scott JE, Hughes EW. Proteoglycan-collagen relationships in developing chick and bovine tendons: influence of the physiological environment. *Connective Tissue Reserach* 1986; **14**: 267–78.
7. Evans JH, Barbenel JC. Structure and mechanical properties of tendon related to function. *Equine Veterinary Journal* 1975; **7**: 1–8.
8. Prockop D, Guzman NA. Collagen diseases and the biosynthesis of collagen. *Hospital Practice* 1977; **12**: 61–8.
9. Bailey AJ, Robins SP, Balian G. Biological significance of the inter-molecular crosslinks of collagen. *Nature, London* 1974; **251**: 105–9.
10. Hardy MA: The biology of scar formation. *Physical Therapy* 1989; **69**: 1014–24.
11. Ihme A. *et al*. Ehler–Danlos syndrome type VI: collagen type specificity of defective lysyl hydroxylation in various tissues. *Journal of Investigative Dermatology* 1984; **83**: 161–5.
12. Akeson WH, Amiel D, Abel MF, Garfin SR, Woo SLY. Effects of immobilization on joints. *Clinical Orthopaedics and Related Research* 1987; **219**: 28–37.
13. Amiel D *et al*. The effect of immobilization on the types of collagen synthesized in periarticular connective tissue. *Connective Tissue Research* 1980; **8**: 27–35.
14. Amiel D, Woo SLY, Harwood FL, Akeson WH. The effect of immobilization on collagen turnover in connective tissue: a biochemical–biomechanical correlation. *Acta Orthopaedica Scandinavica* 1982; **53**: 325.
15. Scott JE. The periphery of the developing collagen fibril: quantitative relationships with dermatan sulphate and other surface-associated species. *Biochemical Journal* 1984; **218**: 229–33.
16. Vogel KG, Trotter JA. The effect of proteoglycans on the morphology of collagen fibrils formed *in vitro*. *Collagen and Related Research* 1987; **7**: 105–14.
17. Gillard GC, Reilly HC, Bell-Booth PG, Flint MH. The influence of mechanical forces on the glycosaminoglycan content of the rabbit flexor digitorum profundus tendon. *Connective Tissue Research* 1979; **7**: 37–47.
18. Koob TJ, Vogel KG. Proteoglycan synthesis in organ cultures from different regions of bovine tendon subjected to different mechanical forces. *Biochemical Journal* 1987; **246**: 589–98.
19. Flint MH *et al*. Collagen fibril diameters and glycosaminoglycan content of skins—indexes of tissue maturity and function. *Connective Tissue Research* 1984; **13**: 69–81.
20. Postacchini F, De Martino C. Regeneration of rabbit calcaneal tendon: maturation of collagen and elastic fibers following partial tenotomy. *Connective Tissue Research* 1980; **8**: 41–7.
21. Doillon CJ *et al*. Collagen fiber formation in repair tissue: development of strength and toughness. *Collagen and Related Research* 1985; **5**: 481–92.
22. Knorzer E. *et al*. New aspects of the etiology of tendon rupture: an analysis of time-resolved dynamic-mechanical measurement using synchotron radiation. *Archives of Orthopaedic and Traumatic Surgery* 1986; **105**: 113–20.
23. Mosler E. *et al*. Stress-induced molecular rearrangement in tendon collagen. *Journal of Molecular Biology* 1985; **182**: 589–96.
24. Stevens FS, Minns RJ, Finlay JB. Evidence for the location denaturation of collagen fibril during the mechanical rupture of human tendons. *Injury* 1975; **6**: 317–19.
25. Butler DL, Grood ES, Noyes FR, Zernicke RG, Barckett K. Effects of structure and strain measurement technique on the material properties of young human tendons and fascia. *Journal of Biomechanics* 1984; **17**: 579–96.
26. Barfred T. Experimental rupture of the Achilles tendon: comparison of various types of experimental rupture in rats. *Acta Orthopaedica Scandinavica* 1971; **42**: 528–43.
27. Woo SLY *et al*. The biomechanical and biochemical properties of swine tendons—long-term effects of exercise on the digital extensors. *Connective Tissue Research* 1980; **7**: 177–83.
28. Booth FW. Effect of limb immobilization on skeletal muscle. *Journal of Applied Physiology* 1982; **52**: 1113–18.
29. Booth FW, Gould EW. Effects of training and disuse on connective tissue. *Exercise and Sports Science Review* 1975; **3**: 83–107.
30. Jozsa L, Kannus P, Thoring J, Reffy A, Jarvinen M, Kvist M. The effect of tenotomy and immobilization on intramuscular connective tissue. A morphometric and microscopic study in rat calf muscles. *Journal of Bone and Joint Surgery* 1990; **72**: B:293–7.
31. Klein L, Sawson MH, Heiple KG. Turnover of collagen in the adult rat after denervation. *Journal of Bone and Joint Surgery* 1977; **59A**: 1065–7.
32. Noyes FR, Torvik PJ, Hyde WB, DeLucas JL. Biomechanics of ligament failure II. An analysis of immobilization, exercise, and reconditioning effects in primates. *Journal of Bone and Joint Surgery* 1974; **56A**: 1406–18.
33. Vailas AC, Deluna DM, Lewis LL, Curwin SL, Roy RR, Alford EK. Adaptation of bone and tendon to prolonged hindlimb suspension in rats. *Journal of Applied Physiology* 1988; **65**: 373–8.
34. Zuckerman J, Stull GA: Ligamentous separation force in rats as influenced by training, detraining and cage restriction. *Medicine and Science in Sports* 1973; **5**: 44–49.
35. Curwin SL, Stanish WD. *Tendinitis: its etiology and treatment*. Lexington, MA: Collamore Press, DC Heath, 1984.
36. Hitchcock TF *et al*. The effect of immediate constrained motion on the strength of flexor tendon repairs in chickens. *Journal of Hand Surgery* 1987; **12A**: 590–5.
37. Karpakka J, Vaananen K, Virtanen P, Savolainen J, Orava S, Takala TES. The effects of remobilization and exercise on collagen biosynthesis in rat tendon. *Acta Physiologica Scandinavica* 1990; **139**: 139–45.
38. Loitz BJ, Zernicke RJ, Vailas AC, Kody MH, Meals RA. Effects of short-term immobilization versus continuous passive motion on the biomechanical and biochemical properties of the rabbit tendon. *Clinical Orthopaedics and Related Research* 1989; **244**: 265–71.
39. Matsuda JJ *et al*. Structural and mechanical adaptation of immature bone to strenuous exercise. *Journal of Applied Physiology* 1986; **60**: 2028–34.
40. Becker H, Diegelman RF. The influence of tension on intrinsic tendon fibroplasia. *Orthopaedic Reviews* 1984; **13**: 65–71.
41. Blanchard O *et al*. Tendon adaptation to different long-term stresses and collagen reticulation in soleus muscle. *Connective Tissue Research* 1985; **13**: 261–7.

42. Woo SLY, Gomez MA, Amiel D, Ritter MA, Gelberman RH, Akeson WH. The effects of exercise on the biomechanical and biochemical properties of swine digital flexor tendons. *Journal of Biomechanical Engineering* 1981; **103**: 51–6.

43. Tipton CM, Matthes RD, Maynard JA, Carey RA. The influence of physical activity on ligaments and tendons. *Medicine and Science in Sports* 1975; **7**: 165–75.

44. Vailas AC *et al.* Adaptation of rat knee meniscus to prolonged exercise. *Journal of Applied Physiology* 1986; **60**: 1031–4.

45. Wong TS, Booth FW. Skeletal muscle enlargement with weight-lifting exercise by rats. *Journal of Applied Physiology* 1988; **65**: 950–4.

46. Heikkinen E, Vuori I. Effect of physical activity on the connective tissue metabolism in mice. *Scandinavian Journal of Clinical and Laboratory Investigation (Suppl.)* 1979; **113**: 36–41.

47. Kiiskinen A. Physical training and connective tissues in young mice: physical properties of Achilles tendons and long bones. *Growth* 1977; **41**: 123–37.

48. Alexander R McN, Vernon A. The dimensions of knee and ankle muscles and the forces they exert. *Journal of Human Movement Studies* 1975; **1**: 115–23.

49. Curwin SL. force and length changes of the gastrocnemius and soleus muscle–tendon units during a therapeutic exercise program and three selected activities. MSc Thesis, Dalhousie University, 1984.

50. Blazina M. Jumper's knee. *Orthopedic Clinics of North America* 1973; **2**: 665–78.

51. Casanova J, Casanova J. 'Ninten-dintis'. *Journal of Hand Surgery* 1991; **16A**: 181.

52. Clancy WG, Neidhart D, Brand RL. Achilles tendinitis in runners: a report of five cases. *American Journal of Sports Medicine* 1976; **4**: 46–56.

53. Colosimo AJ *et al.* Jumper's knee. Diagnosis and treatment. *Orthopaedic Reviews* 1990; **19**: 139–49.

54. Fernandez-Palazzi F, Rivas S, Muica P. Achilles tendinitis in ballet dancers. *Clinical Orthopaedics and Related Research* 1990; **257**: 257–61.

55. Jorgensen U, Winge S. Injuries in badminton. *Sports Medicine* 1990; **10**: 59–64.

56. Vailas AC *et al.* Patellar matrix changes associated with aging and voluntary exercise. *Journal of Applied Physiology* 1985; **58**: 1572–6.

57. Curwin WL, Vailas AC, Wood J. Immature tendon adaptation to strenuous exercise. *Journal of Applied Physiology* 1988; **65**: 2297–301.

58. Michna H. Morphometric analysis of loading-induced changes in collagen-fibril populations in young tendons. *Cell and Tissue Research* 1984; **236**: 465–70.

59. Vailas AC. *et al.* Physical activity and its influence on the repair process of medial collateral ligaments. *Connective Tissue Research* 1981; **9**: 25–31.

60. Clancy WG. Tendon trauma and overuse injuries. In: Leadbetter WB, Buckwalter JA, Gordon SL, eds. *Sports-induced inflammation*. Park Ridge: American Academy of Orthopaedic Surgeons, 1990: 609–18.

61. Puddu G, Ippolito E, Postacchini F. A classification of Achilles tendon disease. *American Journal of Sports Medicine* 1976; **4**: 145–50.

62. Fredenburg MN, Tilley G, Yagoubian E. Spontaneous rupture of posterior tibial tendon secondary to chronic non-specific tenosynovitis. *Foot Surgery* 1983; **22**: 198.

63. Stanish WD, Ratson G, Curwin S. Tendinopathies about the foot and ankle. *Medicine and Science in Sports and Exercises* 1987; **23**: 80–98.

64. Ray JM, Clancy WG Jr, Lemon RA. Semimembranosus tendinitis: an overlooked cause of medial knee pain. *American Journal of Sports Medicine* 1988; **16**: 346–51.

65. Silver IA, Rossdale PD. A clinical and experimental study of tendon injury, healing and treatment in the horse. *Equine Veterinary Journal* (Suppl. 1) 1983; **1**: 1–43.

66. Fackelman BE. The nature of tendon damage and its repair. *Equine Veterinary Journal* 1973; **5**: 141–9.

67. Bonen A, Keizer HA. Pituitary, ovarian and adrenal hormone responses to marathon running. *International Journal of Sports Medicine* 1987; **8** (Suppl. 3): 161–7.

68. Bosenberg AT *et al.* Strenuous exercise causes systemic endotoxemia. *Journal of Applied Physiology* 1988; **65**: 106–8.

69. Vailas AC, Morgan WP, Vailas JC. Physiologic and cellular basis of overtraining. In: Leadbetter WB, Buckwalter JA, Gordon SL, eds. *Sports-induced inflammation*. Park Ridge: American Academy of Orthopaedic Surgeons, 1990: 677–86.

70. Matheson GO, Macintyre JG, Taunton JE, Clement DB, Lloyd-Smith R. Musculoskeletal injuries associated with physical activity in older adults. *Medicine and Science in Sports and Exercise* 1989; **21**: 379–85.

71. Nirschl RP. The etiology and treatment of tennis elbow. *Journal of Sports Medicine* 1974; **2**: 308.

72. Stanish WD, Rubinovich RM, Curwin SL. Eccentric exercise in chronic tendinitis. *Clinical Orthopaedics and Related Research* 1986; **208**: 65–8.

73. Stauber WT. Eccentric action of muscles: physiology, injury and adaptation. *Exercise and Sports Sciences Reviews* 1989; **17**: 157–85.

74. Zernicke RF, Garhammer J, Jobe FW. Human patellar tendon rupture. *Journal of Bone and Joint Surgery* 1977; **59A**: 179–83.

75. Komi PV. Measurement of the force–velocity relationship in human muscle under concentric and eccentric contractions. *Medicine in Sport* 1974; **8**: 224–9.

76. Komi PV. Neuromuscular performance; factors influencing force and speed production. *Scandinavian Journal of Sports Science* 1979; **1**: 2–15.

77. Bosco C, Komi PV. Potentiation of the mechanical behaviour of the human skeletal muscle through pre-stretching. *Acta Physiologica Scandinavica* 1982; **14**: 543–50.

78. Horswill CA *et al.* Excretion of 3-methyl-histidine and hydroxyproline following acute weight-training exercise. *International Journal of Sports Medicine* 1988; **9**: 245–8.

79. Alen *et al.* Responses of serum androgenic-anabolic and catabolic hormones to prolonged strength training. *International Journal of Sports Medicine* 1988; **9**: 229–33.

80. Kjaer M. Epinephrine and some other hormonal responses to exercise in man: with special reference to physical training. *International Journal of Sports Medicine* 1989; **10**: 2–15.

81. Newman RA, Cutroneo KR. Glucocorticoids selectively decrease the synthesis of hydroxylated collagen peptides. *Molecular Pharmacology* 1978; **14**: 185–98.

82. Oxlund H. Manthorpe R. The biomechanical properties of tendon and skin as influenced by long-term glucocorticoid treatment and good restriction. *Biorheology* 1982; **19**: 631–46.

83. Davis JM *et al.* Stress hormone response to exercise in elite female distance runners. *International Journal of Sports Medicine* 1987; **8** (Suppl 2): 132–5.

84. Agarwal S *et al.* Tendinitis and tendon ruptures in successful renal transplant recipients. *Clinical Orthopaedics and Related Research* 1990; **252**: 270–5.

85. Murison MS *et al.* Tendinitis—a common complication after renal transplantation. *Transplantation* 1990; **48**: 587–9.

86. Rathbun JB, MacNab I: The microvascular pattern of the rotator cuff. *Journal of Bone and Joint Surgery* 1970; **52B**: 540–53.

87. Andrews FJ *et al.* Effect of nutritional iron deficiency on acute and chronic inflammation. *Annals of the Rheumatic Diseases* 1987; **46**: 859–65.

88. Abrahamsson S-O, Lundborg G, Lohmander LS. Tendon healing *in vivo*. An experimental model. *Scandinavian Journal of Plastic and Reconstructive Surgery* 1989; **23**: 199–205.

89. Morcos MB, Aswad A. Histological studies of the effects of ultrasonic therapy on surgically split flexor tendons. *Equine Veterinary Journal* 1978; **10**: 267.

90. Nistor L. Surgical and non-surgical treatment of Achilles tendon rupture. *Journal of Bone and Joint Surgery* 1981; **63A**: 394–9.

91. Steiner M. Biomechanics of tendon healing. *Journal of Biomechanics* 1982; **15**: 951–8.

92. Enwemeka CS. Inflammation, cellularity and fibrillogenesis in regenerating tendon: implications for tendon rehabilitation. *Physical Therapy* 1989; **69**: 816–25.

93. Enwemeka CS, Spielholz NI, Nelson AJ. The effect of early functional activities on experimentally tenotomized Achilles tendons in rats. *American Journal of Physical Medicine and Rehabilitation* 1988; **67**: 264–9.

94. Kannus P, Jozsa L. Histopathological changes preceding spontaneous rupture of a tendon. A controlled study of 891 patients. *Journal of Bone and Joint Surgery* 1991; **73A**: 1507–25.

95. Watkins P *et al.* Healing of surgically created defects in the equine superficial digital flexor tendon: collagen-type transformation and tissue morphologic reorganization. *American Journal of Veterinary Research* 1985; **46**: 2091–6.

96. Watkins JP *et al.* Healing of surgically created defects in the equine superficial digital flexor tendon: effects of pulsing electromagnetic filed therapy on collagen-type transformation and tissue morphologic reorganization. *American Journal of Veterinary Research* 1985; **46**: 2097–103.

97. Backman C, Boquist L, Friedn J, Lorentzon R, Toolanen G. Chronic Achilles paratenonitis with tendinosis: an experimental model in the rabbit. *Journal of Orthopaedic Research* 1990; **8**: 541–7.

98. Backman C, Friden J, Widmark A. Blood flow in chronic Achilles tendinosis. Radioactive microsphere study in rabbits. *Acta Orthopaedica Scandinavica* 1991; **62**: 386–7.

99. Hargreaves KM. Mechanisms of pain sensation resulting from inflammation. In: Leadbetter WB, Buckwalter JA, Gordon SL, eds. *Sports-induced inflammation*. Park Ridge: American Academy of Orthopaedic Surgeons, 1990: 383–92.

100. Bodne D. *et al.* Magnetic resonance images of chronic patellar tendinitis. *Skeletal Radiology* 1988; **17**: 24–8.

101. Davies SG, Baudoin CJ, King JB, Perry JD. Ultrasound, computed tomography and magnetic resonance imaging in patellar tendinitis. *Clinical Radiology* 1991; **43**: 52–6.

102. Fornage BD, Rifkin MD. Ultrasound examination of tendons. *Radiological Clinics of North America* 1988; **26**: 87–107.

103. Matin P. Basic principles of nuclear medicine techniques for detection and evaluation of trauma and sports medicine injuries. *Seminars in Nuclear Medicine* 1988; **18**: 90–112.

104. Mourad K, King J, Guggiana P. Computed tomography and ultrasound imaging of jumper's knee-patellar tendinitis. *Clinical Radiology* 1988; **39**: 162–5.

105. Panageas E, Greenberg S, Franklin PD, Carter AP, Bloom D. Magnetic resonance imaging or pathologic conditions of the Achilles tendon. *Orthopaedic Reviews* 1990; **19**: 975–80.

106. Pochaczevsky R. Thermography in posttraumatic pain. *American Journal of Sports Medicine* 1987; **15**: 243–50.

107. Turner TA. Thermography as an aid to the clinical lameness evaluation. *Veterinary Clinics of North America, Equine Practice* 1991; **7**: 311–38.

108. Cummins EJ *et al.* The structure of the calcaneal tendon (of Achilles) in relation to orthopaedic surgery. *Surgery, Gynecology and Obstetrics* 1946; **83**: 107–16.

109. Dyson M, Suckling J. Stimulation of tissue repair by ultrasound: a survey of the mechanisms involved. *Physiotherapy* 1978; **64**: 105–8.

110. Enwemeka CS. Laser biostimulation of healing wounds: specific effects and mechanisms of action. *Journal of Orthopaedic and Sports Physical Therapy* 1988; **9**: 333–8.

111. Knight KL. Cold as a modifier of sports-induced inflammation. In: Leadbetter WB, Buckwalter JA, Gordon SL, eds. *Sports-induced inflammation*, Park Ridge: American Academy of Orthopaedic Surgeons, 1990: 463–7.

112. Dyson M. *et al.* The stimulation of tissue regeneration by means of ultrasound. *Clinical Science* 1968; **35**: 273–85.

113. Gieck JH, Saliba E. Therapeutic ultrasound: influence on inflammation and healing. In: Leadbetter WB, Buckwalter JA, Gordon SL, eds. *Sports-induced inflammation*. Park Ridge: American Academy of Orthopaedic Surgeons, 1990: 479–92.

114. Harvey W, Dyson M, Pond JB, Grahame R. The stimulation of protein synthesis in human fibroblasts by therapeutic ultrasound. *Rheumatology and Rehabilitation* 1975; **14**: 237.

115. Enwemeka CS. The effects of therapeutic ultrasound on tendon healing. A biomechanical study. *American Journal of Physical Medicine and Rehabilitation* 1989; **68**: 283–7.

116. Frieder S *et al.* A pilot study: the therapeutic effect of ultrasound following partial rupture of Achilles tendons in male rats. *Journal of Orthopaedic and Sports Physical Therapy* 1988; **10**: 39–46.

117. Snow CJ, Johnson KA. Effect of therapeutic ultrasound on acute inflammation. *Physiotherapy Canada* 1988; **40**: 162–7.

118. Basford JR. Low-energy laser therapy. In: Leadbetter WB, Buckwalter JA, Gordon SL, eds. *Sports-induced inflammation*. Park Ridge: American Academy of Orthopaedic Surgeons, 1990: 499–508

119. Frank C. *et al.* Electromagnetic stimulation of ligament healing in rabbits. *Clinical Orthopaedics and Related Research* 1983; **175**: 263–72.

120. Nessler JP, Mass DP. Direct-current electrical stimulation of tendon healing *in vitro. Clinical Orthopaedics and Related Research* 1987; **217**: 303–12.

121. Stanish WD *et al.* The use of electricity in ligament and tendon repair. *Physician and Sports Medicine* 1985; **13**: 109–16.

122. Binder A. *et al.* Pulsed electromagnetic field therapy of persistent rotator cuff tendinitis: a double-blind controlled assessment. *Lancet* 1984; i: 695–8.

123. Kennedy JC, Willis RB. The effects of local steroid injections on tendons; a biomechanical and microscopic correlative study. *American Journal of Sports Medicine* 1976; **4**: 11–21.

124. Leadbetter WB. Corticosteroid injection therapy in sports injuries. In: Leadbetter WB, Buckwalter JA, Gordon SL, eds. *Sports-induced inflammation*. Park Ridge: American Academy of Orthopaedic Surgeons, 1990: 527–45.

125. Da Cruz DJ *et al.* Achilles paratendonitis: an evaluation of steroid injection. *British Journal of Sports Medicine* 1988; **22**: 64–5.

126. Abramson SB. Nonsteroidal anti-inflammatory drugs: mechanisms of action and therapeutic considerations. In: Leadbetter WB, Buckwalter JA, Gordon SL, eds. *Sports-induced inflammation*. Park Ridge: American Academy of Orthopaedic Surgeons, 1990: 421–30.

127. Almedkinders LC, Gilbert JA. Healing of experimental muscle strains and the effects of nonsteroidal anti-inflammatory medication. *American Journal of Sports Medicine* 1987; **15**: 357–61.

128. Salminen A, Kihlstrom M. Protective effect of indomethacin against exercise-induced injuries in mouse skeletal muscle fibers. *International Journal of Sports Medicine* 1987; 8: 46–9.

129. Haupt HA. The role of anabolic steroids as modifiers of sports-induced inflammation. In: Leadbetter WB, Buckwalter JA, Gordon SL, eds. *Sports-induced inflammation*. Park Ridge: American Academy of Orthopaedic Surgeons, 1990: 449–54.

130. Michna JH. Tendon injuries induced by exercise and anabolic steroids in experimental mice. *International Orthopaedics* 1987; **11**: 157–62.

131. Jensen K, DiFabio RP. Evaluation of eccentric exercise in treatment of patellar tendinitis. *Physical Therapy* 1989; **69**: 11–216.

Chronic and overuse sport injuries

5

5.1   An introduction to chronic overuse injuries

PER A. F. H. RENSTRÖM

INTRODUCTION

Most people in society are engaged in sports activities and physical exercise of some kind. These activities invoke a variety of stress responses in the human body, of which many, including improved cardiovascular function, altered body composition, improved muscular strength, increased physical stamina, and a sense of well-being or fitness, are beneficial.[1] However, sports also have damaging effects such as overuse injuries.

The term 'Overuse injury', as we know it today, originates from a paper by Slocum and James on running injuries, published in 1968.[2] However, examples of overuse injuries first appeared in the literature in 1855 when Breithaupt reported on stress fractures of the metatarsal bone. Runge described the tennis elbow syndrome in 1873, and Albert described Achilles tendinitis in 1893.

Overuse injuries are increasingly common in relation to sports activities, constituting at least 50 to 60 per cent of all sports injuries, but the true incidence is not known. They are related to extrinsic factors such as training errors, poorly instrumented sporting equipment, unsuitable environmental conditions, and intrinsic factors such as malalignments and muscle imbalance.

Overuse injuries constitute a large diagnostic and therapeutic problem because the symptoms are often diffuse and uncharacteristic. Despite early recognition of these injuries, our understanding of them is still very limited, and little scientific evidence concerning their diagnosis and treatment is available. Therefore, the diagnosis is often based on the examining physician's clinical experience in sports medicine. The treatment should be based on specific diagnosis and scientific background, but is still often performed by trial and error. Thus, overuse conditions provide a challenge to sports medicine physicians.

AETIOLOGY OF OVERUSE SYNDROMES

Overuse injuries in athletes are generally due to overload or repetitive trauma of the muscular skeletal system. Factors implicated in overuse injuries can be classified as extrinsic or intrinsic.

Extrinsic factors

A number of activities are implicated with overuse injuries in sports: one is associated with endurance training, which is part of most sports, and another is associated with repetitive performances requiring skill, technique, power, as in gymnastics, high jump, and weight-lifting. Overuse syndromes appear to occur about three times more frequently in the former group than in the latter group.[3]

Many overuse injuries are associated with repetitive pounding activities such as running and jumping. A runner weighing 75 kg (150 pounds) absorbs a total of 220 000 kg (440 625 pounds) on each foot per mile assuming a shock absorption of 250 per cent of body weight at ground contact.[4] Running injuries are generally associated with the stance phase portion of the running cycle. At a pace of 1 mile per 7 minutes, stance time is approximately 0.2 seconds, which translates into approximately 5100 contacts per hour of running.[5] These huge repetitive forces suggest that even small biomechanical abnormalities and deviations can result in a significant concentration of stress and load on the human tissue and result in an overuse injury.

Extrinsic factors are present in 60 to 80 per cent of reported injuries to runners.[6-8] The most common causes are changes in running activities such as increased mileage, increased intensity, or excessive hill work (Table 1). Examples of this are a novice runner who has started to run or an athlete returning to running after an injury. Harvey[6] found that 80 per cent of the athletes visiting his clinic had only recently taken up the sport, or had markedly increased their training intensity a few days to 2 weeks prior to the onset of symptoms. Occasionally, an athlete had started participating in a second sport or another activity which had similar biomechanics, providing added stress which initiated the injury. Adult athletes can develop overuse injuries after 2 years of regular daily training during which both the amount and intensity progressively increase.[3]

Running downhill can produce problems in the knee joint. During such activity, there is an increased knee flexion, extensor movement, and patellar femoral forces, increased power absorption, and increased eccentric contraction of the knee extensors.[10] The main bodyweight is behind the knee joint, which produces increased loads on the extensor muscles resulting in increased risk of fatigue and hence decreased capacity to protect the knee. Downhill running can produce and increase the symptoms from patellofemoral pain and iliotibial band friction syndromes.

The types and conditions of the sports surface used may be of

Table 1   Extrinsic factors related to injuries in sports

Excessive load on the body
   Type of movement
   Speed of movement
   Number of repetitions
   Footwear
   Surface
Training errors
   Excessive distance
   Fast progression
   High intensity
   Hill work
   Poor technique
   Monotonous or asymmetric training
   Fatigue
Unsuitable environmental conditions
   Dark
   Heat/cold
   Humidity
   Altitude
   Wind
Poor equipment
Ineffective rules

Reproduced from ref. 9, with permission.

importance. There is a much higher force at first contact for running on asphalt compared with running on grass or sand.[11] The incidence of injuries associated with running on hard surfaces is much greater than that associated with running on wood chips, soft dirt, or a composite surface. Running on uneven or artificial surfaces and on slippery roads can also cause overuse injuries. Running on a banked surface or a cambered road in one direction will probably cause abnormal stress on one side of the body, resulting in a short–long syndrome, with increased secondary pronation on the inner leg, and possibly injuries such as iliotibial band friction syndrome or trochanteric bursitis.

Overuse injuries are more common in soccer played on artificial turf than on grass or gravel.[11,12] Both adolescent and adult tennis players may sustain more overuse injuries, such as medial tibial stress syndrome, Achilles tendinitis, and plantar fasciitis, on fast surfaces with high friction than on slower and more yielding clay courts.

Inadequate or worn shoes can cause increased stress and overuse. It is not acceptable to wear tennis, basketball, or soccer shoes during regular running programmes as they do not have the appropriate features to protect a runner from injuries.

Poorly instrumented sporting equipment can cause overuse injuries, particularly if used in combination with poor technique. Oversized tennis rackets absorb vibration better from tennis balls that are hit off-centre along the vertical axis.[13] The material, size, stringing, and grip size of the racket may all be important factors in the development of tennis overuse injuries. Studies suggest that players should use a light, middle-sized, or oversized racket, with a gut stringing of about 22.7 to 25 kg (50–55 pounds), and play on slow courts with light balls in order to prevent lateral epicondylitis.

Unsuitable environment conditions may play a role. Cold and hot environments may influence the metabolic rates and vascular supply of the soft tissues, thus limiting their ability to work efficiently for any length of time.[14]

Rules may also be a factor introducing overuse syndromes, particularly in young athletes. Thus in junior baseball in the United States the number of pitches that a player can make per season is limited to prevent the occurrence of the condition known as Little Leaguer's elbow.

Intrinsic factors

Intrinsic predisposing factors may be present in 40 per cent of athletes with running injuries, but only in 10 per cent are they the only demonstrable factor.[5] The most common intrinsic factors related to overuse injuries are alignment abnormalities, leg length discrepancy, muscle weakness and imbalance, decreased flexibility, joint laxity and instability, female gender age, overweight, and predisposing diseases (Table 2).[16]

Malalignments of various types are the most important and common intrinsic factor. The majority of malalignment problems are minor and subtle, and can be corrected with applied external forces.[17] James et al.[7] found that increased pronation was present in about 60 per cent of a group of injured runners. However, it is not known what percentage of uninjured runners have increased pronation. The constraints of pronation are the shape of the subtalar joints, the ligamentous support, and, to a lesser degree, the muscle support (Fig. 1)[4] Maximum muscle participation of the tibialis anterior, the tibialis posterior and the soleus is not enough to control excessive eversion forces.

During walking, there is approximately 6 to 8° of subtalar joint

Table 2 Intrinsic factors related to injuries in sports

Malalignments
  Foot hyperpronation/hypopronation
  Pes planus/cavus
  Forefoot varus/valgus
  Hindfoot varus/valgus
  Tibia vara
  Genu valgum/varum
  Patella alta/baja
  Femoral neck anteversion
Leg length discrepancy
Muscle weakness/imbalance
Decreased flexibility
Joint laxity and instability
Female gender
Youth/old age
Overweight
Predisposing diseases

Reproduced from ref. 9, with permission.

motion in normal individuals. The pronation that occurs at the time of ground contact is not an active event. It is the result of loading the body weight on to the foot. With increased pronation, there is an increase of 10 to 12° of subtalar joint motion.

During running, approximately 80 per cent of people have heel initial contact at distance running speeds, while the other 20 per cent have midfoot initial contact.[5] Therefore, running should be differentiated from sprinting, which is characterized by such factors as increased velocity, decreased shock absorption in early stance, and initial contact with the toe. Maximum pronation is usually reached after 40 per cent of the stance phase portion. The foot gradually levels over into supination at around 60 per cent to prepare for push-off.

Some increased pronation of the foot is often physiological, but

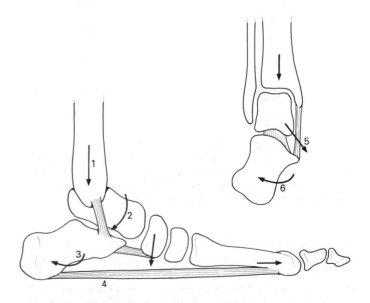

Fig. 1 Pronation of the foot is a complicated mechanism. The load on the foot is from the tibia (1) down towards the talus, which is gliding forward (2) and medially (5). The calcaneus is moving forward–downward (3) and is rotated down into the valgus (6). The plantar aponeurosis (4) is stretched out. There is increased tension in the insertions.

excessive pronation is potentially harmful. Compensatory hyper-pronation may occur for anatomical reasons, such as a tibia vara of 10° or more, forefoot varus, leg length discrepancy, or ligamentous laxity, or because of muscular weakness or tightness in the gastrocnemius and soleus muscles. Excessive pronation will have secondary effects on the lower extremities such as an increased compensatory internal rotation of the tibia resulting in lower leg and knee problems (Fig. 2). The degree of subtalar eversion, and therefore pronation, determines the degree of compensatory internal tibia rotation. This increased rotation of the tibia may also give more proximal effects through the femur and pelvis. In other words the lower extremity should be regarded as a functional unit when carrying out clinical examinations.

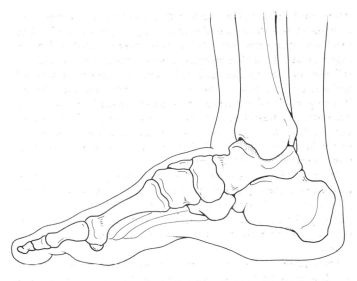

Fig. 3 The cavus foot is a foot which is very inelastic with a very rigid midfoot resulting in increased loads on the anterior and posterior part of the foot. (Redrawn from Peterson L, Renström P. *Injuries in Sports*. London: Martin Dunitz, 1985, with permission.)

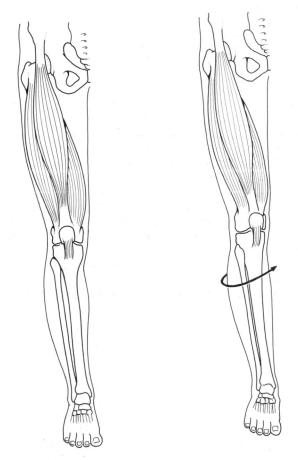

Fig. 2 Excessive pronation of the foot results in compensatory mechanism of the lower extremities such as internal rotation of the tibia. Left, a normal leg right: a compensatory internal rotation of the tibia. (Redrawn from Peterson L, Renström P. *Injuries in Sports*. London: Martin Dunitz, 1985, with permission.)

Excessive pronation will predispose for injuries on the medial aspects of the lower extremities. An increased pronation is associated with injuries such as medial tibial stress syndrome, tibialis posterior tendinitis, Achilles bursitis or tendinitis, plantar fasciitis, patellofemoral disorders, iliotibial band friction syndrome, and lower-extremity stress fractures. It mush be pointed out that specific anatomic abnormalities and abnormal biomechanics of the lower extremity are not correlated with specific injuries on a predictable basis.[18]

Cavus feet (Fig. 3) are also associated with overuse injuries.

James *et al.*[7] found that cavus feet were present in about 20 per cent of their injured group of runners. Athletes with cavus feet have decreased motion at the subtalar and midtarsal joints with resulting decreased flexibility of the foot. At foot strike the heel remains in varus, the longitudinal arch is maintained, and the foot does not unlock.[19] There is decreased internal rotation of the tibia and the lower extremity, which means that the tibia remains in external rotation, and the net result is increased stress since the arch continues to be rigid through the midstance phase of running.[8] With this lack of compensatory internal tibia rotation, stress is passed through the lateral foot and knee resulting in injuries of the lateral side of the lower extremity such as the iliotibial band friction syndrome, trochanteric bursitis, stress fractures, Achilles tendinitis, muscle strain, and metatarsalgia.

Many athletes have a mild genu varum. A lower leg varus alignment of more than 8 to 10° is considered non-physiological and results in compensatory functional foot pronation. There is a low incidence of injury if the total varus is less than 8° but a marked incidence of running-related injuries if the varus alignment is more than 18°.[20]

An athlete can have a combination of malalignments. Athletes with the so-called miserable malalignment syndrome (Fig. 4), including femoral neck anteversion with internal rotation of the hip and genu varum with or without hyperextension, squinting patella, excessive Q-angle, tibial varum, functional equinus, and compensatory foot pronation, have an increased risk for overuse injuries. A miserable malalignment syndrome may cause such problems with running on a regular basis that it is feasible to recommend that some people with this syndrome should not run long distances.

Leg length discrepancy is commonly discussed with respect to overuse injuries. Leg length discrepancy can be secondary to a difference in leg length, but this condition is quite uncommon. This malalignment is usually functional, and is caused by such activities as running on cambered roads. Leg length discrepancy can give secondary effects such as pelvic tilt to the short side, functional lumbar scoliosis, increased abduction of the hip,

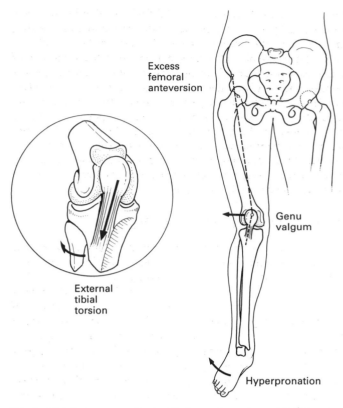

Fig. 4 Malalignment syndrome. (Redrawn from Grana W, Kalenak A. *Clinical Sports Medicine*. Eastbourne: W. B. Saunders, with permission.)

excessive pronation, increased knee valgus, and outward rotation of the hip.[21] Leg length discrepancy has been suggested to be a factor in the development of injuries such as iliotibial band friction syndrome, trochanteric bursitis, low back pain, and stress fractures. From an orthopaedic viewpoint, discrepancies of less than 13 mm are cosmetic in most people. However, in top-level athletes a discrepancy of more than 5 to 10 mm may be symptomatic and should then be treated.

The significance of muscle imbalance as an injury-causing factor is a matter of debate. Muscle imbalance means that there is an asymmetry between the agonist and antagonist muscles in one extremity, asymmetry between the extremities, or a difference from the anticipated normal value.[22] An athlete with over 10 per cent difference in quadriceps or hamstring strength between the right and left sides is believed to be at a greater risk of muscular tendon injury. Also, an athlete who has a ratio of hamstring to quadriceps strength in one leg of 60 per cent or less is believed to have a propensity for sustaining muscle injury.[23] In a study of an American football team, hamstring injuries were found in 6.7 per cent with a recurrence rate of 13 per cent. Muscle imbalance was found, but after correction the occurrence and recurrence of these injuries disappeared.[24]

Muscle weakness due to scarring and fibrosis from previous injury or surgery may predispose to recurrent injury because the scar tissue is not as strong or elastic as the other components of the muscular tendinous unit. Athletes with a previous joint injury have persistent and long-lasting deficits in muscular strength, power, and endurance in the extremity. The affected joints may be at a greater danger of reinjury than uninjured joints. However,

some studies have not found any direct relationship between muscle weakness and injury.[22]

Decreased flexibility and joint laxity may contribute to overuse injuries. Joint instability after anterior cruciate ligament injuries in the knee joint, for example, may result in serious long-term consequences such as post-traumatic osteoarthritis.

Gender may be a factor as there seems to be a higher incidence of overuse injuries among women.[25] The reason for this may be that women have a weaker muscular skeletal system, 25 per cent less muscle mass per bodyweight, less bone, a wider pelvis, and more mobile joints than men. Menstrual irregularities, which are much more common among female athletes than among non-athletes, constitute a risk factor for certain overuse injuries. There is an increased incidence of stress fractures among amenorrhoeic athletes compared with menorrhoeic athletes in the same sport.[26] A prolonged hypo-oestrogenic state may result in a loss of bone mass and therefore may increase the risk of osteoporotic acute fractures.

Age may be a predisposing factor for injuries. During the growth spurt which occurs in most girls at about 12 years of age and in boys at about 14 years, there is a great imbalance between muscle strength, tightness, joint mobility, and co-ordination. In this phase, there is an increased risk of both acute and chronic overuse injuries such as traction apophysitis of the tibial tubercle (Osgood–Schlatter's syndrome) or of the humeral medial epicondyle in young pitchers (Little Leaguer's elbow). In gymnastics, intensive training over a long period of time may produce hypermobility of the spinal column and other joints, and the end result may be earlier osteoarthritis.[27]

Stanitski[28] considered that factors that lead to injuries in adolescent athletes include the athlete's own psychobiology, inappropriate equipment, the sports environment, such as playing surfaces and temperature, training and coaching errors, and parental influences. Most sport-related injuries in children are minor and self-limiting. The appropriate aims of children and most sports must remain enjoyment with acquisition of sports-specific skills and thereby prevention of overuse injuries.

With ageing, various functions of the body gradually deteriorate. However, active elderly people do better than those who are inactive. Elderly runners have less physical disability, maintain a better functional capacity, and have fewer visits to the physician per year.[29] In elderly athletes, sports injuries are more frequently overuse-related than acute, with the most common injuries found in the muscles and tendons of the lower leg.[30]

Overweight may be a factor in producing overuse symptoms. The development of knee and hip symptoms from osteoarthritis is associated with overweight. Physical activity may further accelerate the osteoarthritic process.[31] Weight reduction in severely obese individuals often leads to a significant relief of their muscular skeletal symptoms.[32]

Predisposing diseases may make athletes more prone to injury. A patient with diabetes who has a low blood sugar level may lose his concentration or co-ordination capability. Children with Perthes' disease may lose the rotatory function of their knee and may secondarily develop overuse symptoms.

FREQUENCY AND TYPE OF OVERUSE INJURIES

The actual incidence of overuse injuries is unknown, since frequently they do not require the athlete to visit a physician. The majority of injuries seen in sports medicine clinics are overuse injuries. The most extensive investigation of overuse injuries has been presented by Orava.[3] He found that in Finland top level athletes aged 20 to 29 years and recreational athletes aged 30 to 49 years most frequently visited sports medicine clinics for treatment of overuse injuries, with 15 per cent of the injured athletes being female.

Overuse injuries occur in most sports and often develop when athletes try to advance their training too rapidly. They generally occur in repetitive pounding activities such as long distance running or in overhead activities such as throwing. Injuries in 'gliding' sports such as cross-country skiing and swimming are not as common and the injury patterns are different from those in running. In Orava's study of the injured individuals 86.5 per cent participated in endurance sports.[3] Endurance activities increase the incidence of overuse injuries: 91 per cent of ultra-endurance triathlete competitors sustained at least one soft tissue overuse injury during the previous year of training.[33]

Most overuse injuries involve the lower extremities. The most frequent locations in Orava's study[3] were the knee (28 per cent), ankle, foot, and heel (21 per cent), and lower leg and shin (17 per cent). In a large survey of 4173 runners, the knee was involved in 40 to 42 per cent of the injuries depending on distance and gender.[34] The lower leg (11 per cent in men and 30 per cent in women) and foot (10 per cent and 27 per cent) were the most commonly injured regions, followed by the upper leg (3 per cent and 10 per cent), hip (5 per cent and 11 per cent), and low back (2 per cent and 5 per cent). The area of the body most frequently involved in ultra-endurance athletes was the back, and the most common pattern was to have multiple areas involved.

The most common structures involved are the muscles and fascias (27 per cent), tendon and muscle insertion (22 per cent), joint surfaces (17 per cent), tendons and tendon sheaths (15 per cent), and bursae, bones, and nerves (21 per cent).[3]

Patellofemoral pain syndrome was diagnosed in one patient out of four (24 per cent in men and 30 per cent in women) with a running injury.[34] Iliotibial band friction syndrome and plantar fascitis were the second (7 per cent and 8 per cent) and third (5 per cent and 4 per cent) most common injuries. Achilles tendinitis constituted 5 per cent (men) and 3 per cent (women) of the total number of injuries, which is a reduction in frequency compared with reports from 1981.[35] This reduction in Achilles tendon disorders correlates well with the introduction of appropriate motion control running shoes having heel wedging of more than 12 mm and increased rear foot support.[8] In addition, an appreciation of the importance of gastrocnemius–soleus flexibility and the need to develop a pattern of warm-up and cool-down activities involving stretching has contributed to the reduced incidence of this disorder.

The injury data in the literature have shown an alarming rise in the incidence of knee pain in runners—from 18 to 50 per cent of injuries in 13 years.[8,35] MacIntyre et al.[34] believed that an imbalance or insufficiency in the quadriceps–hamstring muscle groups, or lower extremity malalignment with subsequent abnormal patellofemoral tracking, contributed significantly to knee pain in their patients. According to McKenzie et al.,[8] errors in training judgment with excessive loading, particularly in runners with compromised biomechanical features, represent the primary aetiological factor for this increase in the incidence of knee pain.

The elderly seem to have an increased incidence of overuse injuries. Kannus et al.[36] found that 70 per cent of injuries in a group of injured elderly athletes were overuse injuries compared with 41 per cent in a younger population. In both groups, the knee joint was most frequently affected; 36 per cent in young athletes and 21 per cent in the elderly. Shoulder injuries were present in 18 per cent of the elderly athletes and Achilles tendon injuries were present in 20 per cent, which was significantly more common than among young athletes. Matheson et al.[37] compared young athletes with relatively senior athletes and found that the frequency of tendinitis was similar in both age groups while metatarsal, plantar fasciitis, and meniscal injuries were more common in the older population. Patellofemoral pain syndrome, stress fractures, and periostitis were more common in the younger population. In the older population, the prevalence of osteoarthritis was 2.5 times higher than the frequency of osteoarthritis as the source of activity-related pain. Injuries in elderly athletes occur more frequently in endurance sports, are more frequently overuse related, and are more often degenerative in nature.

DIAGNOSTIC PRINCIPLES

The diagnosis of overuse injuries is based on clinical experience and limited science. Therefore the diagnosis is often a challenge.

A correct history of these injuries is of greater importance than for other injuries because it forms the main basis on which diagnosis is made. The clinician must explore in depth the history of the athlete's complaints and the duration and nature of onset of most problems. The pain-causing situations should be analysed, as well as any swelling, locking, or popping. It is important to penetrate the patient-training programme, eliciting information, for example, on changes in mileage and intensity, number of workouts, type of strengthening or stretching exercises, equipment, location for activities, and competitions. The shoes that the athlete has been wearing and whether there is any specific abnormal shoewear pattern should be discussed. Use of orthotics should be discussed, as well as the type of surface or turf on which training has occurred. Nutritional aspects may be of importance. The physician should be aware of the demands and biomechanics of the sport involved in order to be able to analyse how the injury occurred and to determine whether excessive loads are present to cause this injury. In most cases the doctor should be able to suspect the diagnosis accurately from the history.

Physical examination of a leg injury should be thorough; it should include the entire lower extremity and back and be carried out with the patient in standing, sitting, prone, lying, and supine positions. In the standing position, it is possible to examine the back and the overall alignment of the whole lower extremity. The patient should also be observed while walking. In the sitting position, the patient's knee alignment can be evaluated as well as ankle stability and foot configurations. In the supine position, the range of motion of the hip and knee joints can be evaluated. In the prone position, the Achilles tendon, the heel–leg alignment, the subtalar motion, and the sole can be carefully examined. Examination of the functional anatomy with the suspicion of a biomechanical imbalance is the basis for the diagnosis and management of overuse injuries. With a basic understanding of the biomechanics of the extremities and of the stresses which occur during physical activity, most overuse problems can be managed properly.

Radiological examination can sometimes be valuable, and the indications of conventional radiography include suspicion of degenerative changes in a joint, loose intra-articular bodies, healing stress fractures 2 to 3 weeks after the onset of symptoms, etc. Special views can sometimes be indicated in, for example, patella and shoulder injuries. According to Merchant et al.[38] axial patella views at 45° evaluate the congruence angle which indicates a patella subluxation. Axial views at 30° indicate different malformations of the patella and trochlea. Subacromial space views of the shoulder show the Bigliani types of the acromion which can be helpful in establishing indications for subacromial decompressive surgery. Stress radiographs demonstrating chronic joint instability in the ankle, knee, and elbow can sometimes be of value.

Magnetic resonance imaging (MRI) is a non-invasive imaging technique that provides excellent soft tissue contrasts in multiple planes without exposure to ionizing radiation. The MRI technique has been used in sports medicine mainly in the shoulder and knee joints. It provides an excellent non-invasive method for identifying the type, site, and extent of rotator cuff overuse injuries and impingement problems. MRI is also commonly used in the diagnosis of knee pathology with almost complete accuracy in evaluating meniscus and ligament tears. It is not as accurate in the diagnosis of articular cartilage lesions or the detection of loose bodies. MRI is increasingly used in evaluating tendon pathology, but more research is needed. It allows the tendons to be identified easily because of the large differences in the relaxation properties of hydrogen in the water molecules in various tissues and fat. The disadvantages of MRI include patient positioning, claustrophobia, and contraindications of metal artefacts such as clips or cardiac pacemakers. MRI will continue to develop and will become the tool of choice as it gradually improves its accuracy, when it can be used for functional evaluation, and when it becomes affordable. MRI spectroscopy has potential for the non-invasive measurement of important metabolites in muscle and therefore opens new possibilities.

CT is used in the diagnosis of ankle and elbow joint overuse injuries. With this technique, it is possible to evaluate the position and frequency of loose bodies and osteophytes. CT of the patella, when properly performed, allows precise reproducible imaging of patellofemoral relationships and reveals normal alignment, excessive lateral tilt, and subluxation of the patella. CT arthrograms are accurate in identifying labrum tears in the shoulder joint, as well as synovial plicae in the knee joint.

Ultrasound evaluation has been shown to be valuable in the diagnosis of Baker cysts, jumper's knee, medial meniscus lesions, and meniscal cysts in the knee. However, diagnosis of lateral meniscus lesions with ultrasound appears to have limitations. Ultrasound is also used to evaluate tendon pathology with good accuracy. This technique requires special experience for evaluation.

Bone scans allow accurate evaluation of stress reactions in the bone. A bone scan is positive 2 to 8 days after the onset of stress fracture symptoms.

The development of the arthroscopy during the last 20 years has meant a revolution in the diagnosis of different joint problems, particularly the knee joint. Experience from knee arthroscopy has been carried over to most of the joints in the body, particularly the shoulder, ankle, wrist, and elbow joints.

Overuse injuries constitute a major diagnostic problem because the symptoms are often diffuse and uncharacteristic. The diagnosis of an overuse injury rests with the identification of not only the affected tissue but also the underlying predisposing conditions. An accurate diagnosis is necessary for a successful treatment. Appropriate diagnosis followed by adequate treatment is required in order to improve or eliminate most of these conditions.

PRINCIPLES OF TREATMENT OF OVERUSE INJURIES

When treating overuse injuries, it is important to consider not only the symptoms but also the cause. Therefore it is important to distinguish the primary and secondary problems.

Treatment of the cause of the injury—the primary problem

1. Training routines should be analysed and changed if necessary. The athlete should only participate in the sport within the limits of pain. For runners, this can include a decrease in the training distance or a change in the stride and pace. A change of surface to a soft less sloping surface with minimal curves can be effective for a runner.
2. Malalignments should be corrected if possible. As there are individual combined malalignments, different approaches are available. Sometimes referral to an expert is recommended. Foot orthotics are often used to correct malalignments. Orthotic devices have a significant effect on the amount of maximum pronation, time to maximum pronation, maximum pronation velocity, period of pronation, and movement or rear-foot angle in the first 10° of foot contact.[39] It is important not to overcorrect; it is more reasonable to undercorrect, and the exact rearfoot and fore-foot posting necessary to attain a neutral subtalar joint is rarely provided. One possible approach is to post the subtalar position to a maximum of 4° varus as this provides adequate control and minimizes the possibility of initiating a lateral ligament sprain of the ankle.[8] The use of orthotics is still controversial, and there are failures with this type of treatment. Failures may occur because the orthotic devices have not been adjusted or changed as needed from time to time. Although there are no prospective randomized studies available showing the effect of the use of orthotics, vast experience is available in sports medicine concerning the value of orthotics in compensating and neutralizing different malalignment positions. Sperryn and Rostan[40] followed 50 runners for whom orthotic devices had been prescribed. After 3.5 years, 64 per cent of the runners had relief of symptoms, but only 54 per cent were still using the orthotic device. This indicates that care should be taken to identify athletes with syndromes that will benefit from the use of these devices.
3. Shoes are of great importance in reducing the large impact forces. This shock absorption can be increased by heel confinement.[41] A good shoe is also a stable shoe, giving motion control with a well-fitted heel counter. Modern running shoes are effective, and give a maximum point of pronation during the stance phase which occurs significantly later with bare feet than with the shoe. Shoes should be flexible in the forefoot, or they can cause forefoot overuse problems during push-off. In summary, the shoes should provide cushioning, support, and friction.
4. If the athlete has tight muscles and joints or muscular imbalance, stretching exercises are indicated. If the athlete is excessively flexible, strengthening exercises are indicated.
5. Braces and taping can be used when joint instability is present.
6. Surgical correction of a specific malalignment, such as a patella lateral tilt, can sometimes be indicated.

Treatment of the symptoms

General principles

1. When treating overuse injuries, it is important to have an exact diagnosis before specific treatment is initiated.
2. The treatment of overuse injuries depend on the healing process. The different stages of healing determine the different

modes of treatment. An injured tissue such as a tendon initially heals during the first 48 h with an inflammatory response characterized by increased vessel permeability leading to swelling in the adjacent area. The weeks following the initial phase are known as the proliferative phase; during this phase new collagen and mature cross-links are formed. This is followed by a formative phase involving tissue remodelling which can continue for a year. Tendon strength is a direct function not only of the number and size of collagen fibres but also of their orientation. These fibres respond favourably to tension and motion, and therefore it is important to stimulate protected motion as soon as possible and at the latest during the proliferative phase, i.e. in the third week after the injury.[14]

3. The treatment should be individualized. Patient compliance is the single largest factor determining the success or failure of treatment.

4. The treatment should start as early as possible. When overuse injuries have become chronic, they are very difficult to treat.

5. Treatment of overuse injuries is difficult. Early motion and exercises are usually required to stimulate the healing process. However, long experience is required to balance the need for rest against the need for motion. If the injury becomes resistant to therapy, early consultation with more experienced colleagues is advised.

6. It is important to adopt a multidisciplinary approach to these injuries. Teamwork, including the physician, the physiotherapist, the trainer, the coach, and sometimes the family, is often preferred in developing a suitable treatment and rehabilitation programme.

Treatment of the symptoms—a basic treatment programme

1. Pain and swelling in a joint, for example, will cause muscle inhibition resulting in muscle wasting and weakness. The initial treatment of pain, swelling, and inflammation is of great importance and may include rest, avoidance of weight-bearing, elevation, ice, and compression bandaging and protection. Ice treatment is useful during the first 48 h. Ice relieves pain by the direct effect of cold on pain reception and nerve fibre transmission and by a secondary decrease in swelling and inflammation. Ice reduces the chemical activity and therefore can reduce the inflammatory response. The injured part should be protected and rehabilitated in parallel with the healing process. The injured tissue should be activated but protected from significant stress which may incur further damage. Therefore activities should be carried out below the pain threshold.

2. Heat is effective 48 h after the acute phase and in chronic phases as it increases the extensibility of the collagen in connective tissue, decreases joint stiffness, relieves muscle spasm, and gives pain relief. Local heat can be applied in different ways. Heat retainers or neoprene sleeves are valuable tools in both prevention and rehabilitation of muscle and tendon injuries as they can also be applied during activity. Heat can also be applied by other modalities such as high voltage galvanic stimulation, infrared lamps, diathermy, and paraffin baths. Ultrasound is extensively used in addition to electrical stimulation. Application of moist heat by hydrocolater, whirlpool, or thermophore pads may be effective.

3. Anti-inflammatory medication may be used for the treatment of overuse injuries, but medical modalities do not affect healing. They are often effective against pain and stiffness, but the athlete should be aware that these medications may mask the symptoms.

4. Exercise is the key to a successful treatment programme of overuse injuries. Immobilization should be avoided as it may result in the loss of ground substances found in collagenous connective tissue, which will result in poor reorientation of the collagen fibres and a reduction in the tensile strength. Joint immobilization will also result in muscle wasting and weakness, joint stiffness, and diminished proprioception. Injured tissues should be protected from significant stress, which may incur further damage, but mobilization should start as early as possible. Repeated exercise will increase the mechanical and structural properties of the tissue. It is important to start an exercise programme very carefully with a gradual increase of the load within the limits of pain. Restoration of the mechanical properties in particular often requires prolonged treatment and guidance by experienced physiotherapists or trainers is recommended.

5. Steroid injections should generally be avoided in overuse injuries as they will cause a decrease in metabolic activity and there is a risk of necrosis. Tendon ruptures have been reported secondary to injections of steroids into tendons. However, steroid injections can sometimes be used for specific indications such as chronic conditions at tendon insertion sites or in bursae. A steroid injection should be combined with reduced physical activity for 5 to 10 days.

6. Braces and tape can often be used successfully to reduce the load and weight bearing in the injured area.

7. Surgery should be avoided until conservative therapy has failed. Excision of scar and degenerative tissue secondary to failed healing response in tendons and muscles has often given good results for lateral epicondylitis and in chronic partial tears of tendons such as the Achilles, the adductor longus, or rotator cuff. The rehabilitation period following this type of surgery may be rather long, but the overall results are good. Arthroscopic surgery of the knee joint has proved to be an extremely effective treatment, allowing an early return to sport. The arthroscope can be used in the shoulder to carry out subacromial decompression surgery for impingement problems to repair tears in the labrum and the rotator cuff, and to evaluate instability. Elbow arthroscopy allows removal of loose bodies. Ankle arthroscopy is an effective surgical procedure in patients with synovitis and transchondral defects of the talus. The benefits and long-term results are less predictable in the arthroscopic treatment of ankle osteophytes and arthrosis.

Training programme

Strength training

Strength training should start cautiously as early as possible. The aim of strengthening exercises is recovery of muscle wasting and strength, and the enhancement of recruitment and firing rates of the motor units.[42]

Careful isometric exercises may be used initially and should be carried out without load. Gradually increasing loads can then be applied. In the early stages it is often enough to use the athlete's injured limb as the load. Isometric contractions can be facilitated with electrical muscle stimulation against persistent pain, weakness, and immobilization. Strength improvement by isometric exercises is limited to the joint angle used during the exercise. When isometric exercises can be carried out without pain at multiple joint angles, active dynamic motion exercises of the injured area and gradual increase in strength training may start. Dynamic exercise correlates better with improved functional performance.

Eccentric exercises can be effective in the treatment of chronic overuse injuries. The tendon is subjected to larger loads which maximize the force production and minimizes time delays and energy expenditure.[43] Eccentric exercises should be avoided early when pain, inhibition, and patient confidence are factors of importance. There is a risk of overloading the tissue. Treatment regimens for chronic overuse injuries have been used by Curwin and Stanish[44] with good clinical results. They have used pain as a regulator for the exercise programme. In chronic overuse injuries, it is sometimes possible to break through a therapy-resistant chronic condition by allowing the athlete to exercise above the pain threshold.

When carrying out strength training, the specificity of the training should be kept in mind. Muscle exercises in different positions will affect the neural factors. These must be trained before the full effect of the muscle training programme can be achieved. The initial strength gain is caused by neural activation and not by muscle hypertrophy.

Flexibility training

Strength training alone has a negative effect on joint flexibility which can be counteracted by flexibility training. Therefore it is important to combine strength training with flexibility training. The aim is to restore adequate flexibility on a regional basis together with specific strength and movement.[45] An ideal state of flexibility in the athlete would heighten the sensory feedback mechanism with the advantage of increased proprioceptive accuracy and sensitivity. Studies of the biomechanical effects of stretching shows that it will result in greater flexibility and in increased length of the muscle–tendon unit.[46]

There are different methods of flexibility training including ballistic stretching, slow static stretching and contract–relax stretching. The most widely used method in the treatment of overuse injuries is slow static stretching. Contract–relax stretching is based on the proprioceptive neuromuscular facilitation technique and is widely used by physiotherapists. The placement of the stretching is debated. As heat increases the extensibility of the collagen, stretching should be carried out not only during warm-up but also after the training session when the athlete is warm.

Proprioceptive training

Specific exercises that require balance, weight shift, stimulation of antigravity reflexes, and co-ordination appear to facilitate proprioceptive feedback mechanism. Proprioceptive training of ankle functional instability using ankle discs gave maximum results after 10 weeks of training.[14] This type of exercise is more important than strength training for chronic ankle problems.

Functional training

Functional training such as jogging, running straight and figure of eight, and jumping on two legs should start as early as the injury permits.

Sports-specific training

Sports-simulated exercises of the non-injured body parts should be carried out as early as possible following an injury. It is often possible to perform non-gravity exercises by cycling or swimming if other treatments fail. Rowing, cross-country skiing, or roller skiing are other useful alternatives. Pool running is effective in

maintaining cardiovascular muscular fitness. General conditioning is helpful, not only as stimulation but also for increasing the circulation in the injured areas.

DIFFERENT TYPES OF CHRONIC OVERUSE INJURIES

Overuse injuries occur in most body tissues. The most common sites are in the muscle–tendon units, but problems also occur in other soft tissues such as bursae, fascia, and nerves, as well as in the bones and joints.

Muscle overuse injuries

Injuries may occur at any point in the muscle–tendon unit—within the muscle belly, within the tendon, at the muscle–tendon junction, or at the origin or insertion of the muscle or tendon into the bone (Fig. 5). Failure will occur at the weakest point within the unit. The location and severity of the injury are influenced by the athlete's age.

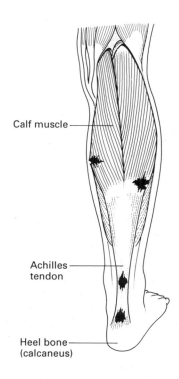

Calf muscle

Achilles tendon

Heel bone (calcaneus)

Fig. 5 Muscle tendon injuries can occur at different locations. (Redrawn from Peterson L, Renström P. *Injuries in Sports*. London: Martin Dunitz, 1985, with permission.)

Avulsion fractures through the apophyseal plate or traction and fragmentation of the bone at the attachment site are most likely to occur in young athletes and adolescents as bone is relatively strong in growing individuals. After 25 to 30 years of age, there is progressive degeneration of the collagen fibres, particularly within the tendon, making the tendon itself a weak area and more susceptible to both traumatic and overuse injuries. The muscle is susceptible to injury in all age groups, although muscle injuries seem to be more common in the elderly.[17,30]

Muscle strain

Garrett[47] has shown that partial and complete overload injuries to the muscle exhibit disruption of muscle fibres near the muscle–tendon junctions. Healing of partial injuries is characterized by an initial inflammatory response followed by a healing phase marked by fibrosis. Treatment of these strains follows the general principles already mentioned, and the prognosis is good. Chronic conditions may occur, but are not common.

Muscle contusion injuries are characterized by intra- or intermuscular haematomas. The intermuscular injuries heal well with rapid recovery, but intramuscular injuries often heal with fibrosis. In the repair process, there is not only a regeneration of muscle fibres but also competitive simultaneous production of granulation tissue. An inelastic fibrotic scar tissue localized in the muscle may thus develop. The healed muscle consequently includes areas of varying elasticity which often result in chronic overuse problems. Pain during or after muscle activity may be experienced by soccer players, for example, in the muscles of their shooting leg. Frequent stretching and strength exercises are necessary as prevention and treatment.

Muscle soreness

Muscle exercise commonly results in injury to fibres in the active muscles, particularly when the exercise is relatively intense, is of long duration, and/or includes eccentric contraction.[48] Friden *et al.*[49] have suggested that myofibrillar lesions are a direct result of mechanical tearing of the z disc, which is the weak link in the myofibrillar contraction chain during high muscle tension. These tears may result in the formation of protein components and a consequent release of protein-bound irons which, by osmosis, results in oedema and soreness. The precise mechanism underlying these injuries is not well known. This soreness caused by overuse seldom becomes chronic.

Chronic compartment syndrome

The lower leg is the location for chronic compartment syndromes usually caused by muscle hypertrophy as a result of repeated exercise. In the presence of appropriate clinical findings one or more of the following intramuscular pressure criteria is diagnostic of chronic compartment syndrome of the leg: (1) a pre-exercise pressure more than 15 mmHg, (2) a 1-min post-exercise pressure of about 30 mmHg, and (3) a 5-min post-exercise pressure of about 20 mmHg.[50] The treatment for this syndrome is specific, with alteration of the training programme, orthosis, medication, and sometimes decompression. Other causes of pain such as stress fractures or reactions or medical tibial stress syndrome should be ruled out.

Tendon overuse injuries

The tensile force applied to the tendon is resisted primarily by the collagen which is characterized by high mechanical strength but poor elasticity. According to Curwin and Stanish,[44] overuse in a tendon means that it has been strained repeatedly until it is unable to withstand further loading, at which point damage occurs. Repeated overload with increasing strain in the tendon will cause the cross-links within the collagen fibres to break. The shearing force will then cause the collagen fibrils to slide, resulting in an injury to the tendon at the microscopic level.

In its resting state, a tendon has a wavy configuration which disappears on 4 per cent stretching. At 4 to 8 per cent strain, the collagen fibres will slide past one another as the cross-links start to break. There is now an inflammatory reaction in the tendon. At 8 to 10 per cent strain, the tendons begin to fail and resist less force as more cross-links are broken. The weakest fibres rupture and the tendon loses its composite structure at the molecular level.[44]

The clinically observed development of sports-induced soft tissue inflammation includes spontaneous resolution, fibroproductive healing, regeneration, or chronic inflammatory response.[51] The chronic inflammatory conditions are associated with persistent structural alterations. They are characterized by either a failure to develop adequate scar tissue restoration or an excessive fibroproductive scar response. It is not clear whether inflammation is truly present in all forms of the pathology in the diagnosis of tendinitis. Historically, tendinitis has been the clinical term used to describe painful conditions in tendons. In the acute injury there is an inflammatory response. An inflammatory cell infiltration seems to be absent in the chronic tendon injury which instead is characterized by degeneration. This degeneration causes a weakened tendon which is more vulnerable to cyclic overloading, resulting in mechanical fatigue and failure. These findings have led to the definition of a new classification system of tendon injury and to the coining of the word 'tendinosus'. It is now possible to describe four pathological conditions (Table 3) based on the anatomy of the tendon and its surrounding tissues.[51]

Chronic Achilles tendon injuries

Tendon injury classification can also be based on the clinical entities observed, i.e. paratenonitis, tendinosis, paratenonitis with tendinosis, tenoperiostitis, partial rupture, and complete rupture.

Paratenonitis

Paratenonitis, which in other locations is called tenosynovitis or tenovaginitis, is an inflammation of the sheath or paratendon surrounding the tendon caused by tissue friction or by external mechanical irritation. This injury is usually associated with tendinosis.

Tendinosis

Tendinosis is a degenerative condition in the tendon without inflammation, characterized by collogen degeneration, fibre disorientation, and sometimes calcification, which is often seen in the elderly.

Chronic Achilles tendon injury

Chronic Achilles tendon problems are very common in athletes and are often secondary to malalignment. The most common malalignment is excessive pronation, which results in a compensatory internal rotation of the tibia. This hyperpronation results in a whipping action resulting in an increased stress within the Achilles tendon.[52] Furthermore the Achilles tendon rotates in its distal 15 cm, resulting in a sawing action during walking and running.

In 1982, Clancy[53] suggested two potential causes for overuse tendon injury: (1) repeated loading of the muscular tendinous unit leads to fatigue of the muscle and shortening and decreased flexibility which may result in passive increased loading of the tendon during the state of relaxation; (2) repetitive active loading of the muscular tendinous unit leads to collagen failure.

The diagnosis of chronic tendinitis is made by a typical history characterized by a gradual onset of pain. There is usually a diffuse tenderness and swelling of the tendon. The so-called pain cycle (Fig. 6) is typical. Pain disappears during warming up, allowing the athlete to go on with his activities. The athlete develops pain after activity but not enough to prevent performance the following

Table 3 Terminology of tendon injury[51]

New	Old	Definition	Histological findings	Clinical signs and symptoms
Paratenonitis	Tenosynovitis Tenovaginitis Peritendinitis	An inflammation of only the paratenon, whether or not lined by synovium	Inflammatory cells in paratenon or peritendinous areolar tissue	Cardinal inflammatory signs: swelling, pain, crepitation, local tenderness, warmth, dysfunction
Paratenonitis with tendinosis	Tendinitis	Paratenon inflammation associated with intratendinosis degeneration	Same as above, with loss of tendon collagen fibre disorientation, scattered vascular ingrowth but no prominent intratendinous inflammation	Same as above, with often palpable tendon nodule, swelling, and inflammatory signs
Tendinosis	Tendinitis	Intratendinous degeneration due to atrophy (ageing, microtrauma, vascular compromise, etc.)	Non-inflammatory intratendinous collagen degeneration with fibre disorientation, hypocellularity, scattered vascular ingrowth, occasional local necrosis, or calcification	Often palpable tendon nodule that can be asymptomatic, but may also be point tender. Swelling of tendon sheath is absent
Tendinitis	Tendon strain or tear A. Acute (less than 2 weeks) B. Subacute (4–6 weeks) C. Chronic (over 6 weeks)	Symptomatic degeneration of the tendon with vascular disruption and inflammatory repair response	Three recognized subgroups: each displays variable histology from purely inflammation with acute haemorrhage and tear, to inflammation superimposed upon pre-existing degeneration, to calcification and tendinosis changes in chronic conditions. In the chronic stage there may be: 1. Interstitial microinjury 2. Central tendon necrosis 3. Frank partial rupture 4. Acute complete rupture	Symptoms are inflammatory and proportional to vascular disruption, haematoma, or atrophy-related cell necrosis. Symptom duration defines each subgroup

Data taken from Clancy WG. Tendon trauma and overuse injuries. *In* Leadbetter WB, Buckwalter JA, Gordon SL, eds. *Sports-Induced Inflammation*. Park Ridge, IL; American Academy of Orthopaedic Surgeons, 1990; and Puddu G, Ippolito E, Postacchini P. A classification of Achilles tendon disease. *American Journal of Sports Medicine* 1976; 4: 145–50.

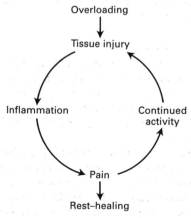

Fig. 6 The pain cycle.

day. When the pain no longer disappears during the warm-up, the injury is chronic. Pain is a sign of injury, and the pain cycle should be interrupted by rest and treatment of the injury.

Partial chronic tear

These injuries are characterized by a history of sudden onset of pain and clinically by a distinct palpable tenderness which in most cases is combined with a localized swelling. The diagnosis should be made by history and clinical examination. Soft tissue radiography can be valuable. An oedema of more than 10 cm decreasing the Kagers triangle is indicative of a partial tear.[54] Ultrasound is helpful, but requires experience to use. MRI will probably be the diagnostic method of choice in the future. If these injuries are not adequately treated initially, delayed healing will occur and they will become chronic and more difficult to treat, often requiring late surgical intervention. It is not known why there is delayed healing or non-healing in these chronic injuries.

Explanations for the delayed healing observed in these chronic tendon lesions include the following:

(1) diminished vascularity with subsequent decreased cellularity due to attritional cell death;

(2) failure of reparative cells to postulate an area of tendon injury, a finding supported by biopsies in younger patients with chronic tendinitis that have failed to demonstrate any cellular activity other than a more metabolically active tenocyte;

(3) accelerated ageing of the tenocyte;

(4) age-dependent changes in the matrix with adverse tendon biomechanical properties[51] (This type of reaction can occur in Achilles and rotator cuff tendons).

Conservative therapy will often give unsatisfactory results. Allenmark *et al.*[55] found in a group of athletes with Achilles tendon partial tears verified by bursography that even after 10 years, 74 per cent still had severe clinical problems which made physical activity impossible. Surgical treatment with excision of the scar tissue give good results in more than 80 per cent of cases.

Tenoperiositis

Tenoperiositis, which is an overuse injury in the tendon insertion into bone, is not very common in adults. In adolescents it is recognized as calcaneal apophysitis (Sever–Haglund's disease) when there is pain at the insertion of the Achilles tendon into the calcaneus, or as Osgood–Schlatter's disease when there is pain at the tendon insertion at the tibial tuberosity.

Chronic patella tendon injuries

Chronic patella tendon injuries follow the same principles as Achilles tendon overuse injuries. The chronic tears are almost always located on the undersurface of the tendon where it inserts into the patella.

Shoulder tendon injuries

Rotator cuff and biceps tendon injuries are common in throwers and in older squash players. The rotator cuff tendons show early degenerative changes and have poor vascularity. Therefore they are difficult to treat and have a long healing and rehabilitation period. These shoulder tendon injuries are often associated with shoulder instability and impingement.

Overuse injuries in joints

The joint most susceptible to overuse injuries is the knee. It is a major power absorber of negative work as twice as much work is done by the knee during running, for example, as is done by the ankle and hip joints.

Knee overuse injuries

Intra-articular injuries

Intra-articular pathology of the knee is often experienced in team sports and individual sports such as skiing, but is less common in running. Collision or twisting trauma is usually involved, but there may also be repetitive minor trauma in which a degenerative meniscus is damaged. Other injuries that can occur are osteochondritis dissecans and unusual disorders such as the plica syndrome. Articular cartilage lesions can cause long-lasting problems in athletes. Arthroscopic surgery is often used to treat intra-articular injuries, generally with good results.

Patellofemoral disorders

The patellofemoral pain syndrome is the most common problem in running. This pain syndrome is often associated with pain in downhill running, and with getting up after squatting for a while. Patellofemoral disorders are usually associated with some kind of malalignment such as the increased Q-angle secondary to genu valgum, formal neck anteversion, and external tibial torsion. Muscle imbalance with tight retinaculum, poor vastus medialis tone, and tight quadriceps muscles can occur. Individuals with patella malalignment should be identified. If this condition is present it is important to separate patella subluxation from patella tilt. Patella subluxation may lead to problems of extensor mechanism instability, increased risk of dislocation, apprehension, and some risk of articular cartilage damage. Abnormal patella tilt creates a pattern of increased lateral loading, adaptive shortening of the lateral retinaculum, and patella articular breakdown and increased retinacular strain.

These injuries can often be well managed conservatively by control of pain and inflammation. The quadriceps, particularly the vastus medialis and the hamstrings, should be subjected to slowly progressive exertion which will stimulate healing. Patella brace or taping, orthotic foot control and motion-controlled shoes, and gradual return to sports are helpful. Surgery may be indicated if conservative treatment fails, but a mechanical disorder must be present.

Iliotibial band friction syndrome

The iliotibial band friction syndrome is quite common and is known as runner's knee; it usually occurs in runners who have been running for less than 4 years and regularly run more than 14 km per week.

At 30° of knee flexion, which is often the case in downhill running, the iliotibial band glides over the lateral knee epicondyle. In downhill running, there is an excessive stride causing an increased compression of the iliotibial band against the lateral epicondyle. Structural abnormalities, such as a prominent lateral epicondyle, a tight iliotibial band, or an excessive genu varum, can also cause this problem, as can foot abnormalities such as excessive pronation, heel varus, cavus feet, and forefoot pronation. Excessive pronation will result in compensatory increased internal rotation of the tibia which draws the insertion site of the iliotibial band anteriomedially, resulting in a tightness. Leg length discrepancy is also associated with this syndrome.

As always, it is important to treat the cause of the injury and therefore the use of a corrective training regime is often effective. Orthotics and good shoes may also be used. If conservative therapy will not help, surgical incision of the posterior iliotibial band, 2 cm proximate of the epicondyle, is effective.

Hip joint arthritis

Degenerative arthritis of the hip joints is common in the elderly. There is no increased incidence in athletes, and overuse is probably not an important aetiological factor. Intensive running activities for more than 30 to 40 years did not result in an increased incidence of hip arthritis.[56]

Shoulder overuse injuries

Athletes who throw overarm are at risk because of the repetitive high velocity chronic mechanical stress placed on their shoulders. According to Jobe *et al.*,[57] there is a progressive continuum of shoulder pathology: overuse leads to microtrauma which leads to instability, subluxation, and impingement, which may result in rotator cuff tears.

The impingement syndrome without instability is common in throwers and swimmers. The treatment of this problem is increasingly aggressive with the development of arthroscopic-assisted subacromial decompression. The results are good and allow more athletes to return to sports more quickly than before.

The most pervasive disorder in the young athlete today is due to the lack of shoulder stability. Primary shoulder instability can be caused by chronic labral microtrauma and can result in secondary impingement problems. Undetected instability should be ruled out in any athlete with impingement findings. Arthroscopy is the most accurate diagnostic tool for shoulder instability. By understanding the delicate balance between mobility and stability in a normal shoulder, the clinician is better able to understand the aetiology and biomechanics of the problem and can design an optimal treatment programme.

Fig. 7 A pitcher and a tennis player have a major valgus overload in their pitch or serve.

Elbow overuse injuries

Overuse injuries in the elbow are quite common in baseball and other throwing sports, gymnastics, power events, and squash (Fig. 7).

During the cocking and acceleration phases the baseball pitcher's elbow is subjected to valgus overload which results in medial tension, lateral compression, and extension overload. The muscles in the vicinity of the elbow do not protect against overloading, particularly valgus loads, and therefore the main focus is on static stability. Medial tension can cause a medial epicondylitis. In young pitchers this tension causes traction and apophysitis, resulting in Little Leaguer's elbow. If valgus instability has occur-

red by attenuation of the anterior oblique ligament, the ulnar nerve may be secondarily injured by repeated mechanical stretching, friction, and compression, resulting in ulnar neuritis. The lateral compression of the radial head against the capitulum may cause osteochondritis dissecans and loose bodies in the elbow. The extension overload often seen in the tennis serve may result in impingement posteriorly and cause osteophytes and loose bodies. The valgus motion can be combined with pronation and will then cause lateral overload with lateral epicondylitis as a result. Medial epicondylitis can be caused by a twisted serve in tennis or by a heavy topspin on forehand. This injury must be carefully treated as the recovery and healing time is long. These injuries should be treated according to the general principles discussed earlier with early motion. Elbow muscle counterforce braces may be effective. Arthroscopy is indicated if there is a suspicion of osteochondritis dissecans, loose bodies, or osteophytes.

Overuse syndromes in other soft tissues

Bursitis

Direct trauma to the elbow, the patella, or the knee may result in acute bursitis within an already extended bursa or lead to development of an adventitious bursa or haemobursa (Fig. 8). If the haematoma is not absorbed or evacuated in the acute phase, chronic bursitis will rapidly result as calcification, free bodies, and adhesions often form and cause irritation. These injuries can be prevented by appropriate protective equipment and padding.

Friction bursitis can occur in the subacromial bursa in the shoulder, the retrocalcaneal bursa or subcutaneous bursa around the Achilles tendon, the iliopsoas bursa, and different bursae around the knee.

Trochanteric bursitis frequently occurs after intensive running on banked roads, resulting in a functional leg length discrepancy. This type of bursitis has a tendency to become chronic and can be very difficult to treat. Therefore early treatment and a correction of training errors and malalignment should be considered. In addition to infection, chemical products from degenerative tendon tissue can cause bursitis.

The treatment of bursitis generally involves rest, local decom-

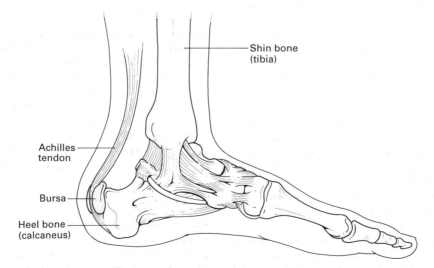

Shin bone (tibia)

Achilles tendon

Bursa

Heel bone (calcaneus)

Fig. 8 Chronic bursitis can be secondary to trauma or friction. The figure shows inflammation of the deep bursa located anterior of the attachment and the distal part of the Achilles tendon to the calcaneus (Redrawn from Peterson L, Renström P. *Injuries in Sports*. London: Martin Dunitz, 1985, with permission.)

pression and sometimes aspiration of fluid, protection, and anti-inflammatory medication. In a long-lasting chronic bursitis surgical excision will give excellent results.

Nerve injuries

Nerve entrapment syndromes

Nerve entrapment syndromes are not very common, but may develop from overuse as a result of swelling in the surrounding soft tissues secondary to trauma or malalignment. The most common injury is the carpal tunnel syndrome in the hand. The ulnar nerve may be entrapped at the level of the wrist and cause problems for the cyclist, and entrapment of the posterior interosseus nerve in the forearm may be experienced by tennis players. Soccer players may experience entrapment of nerves such as the genitofemoralis, the ilioinguinalis, and the cutaneous femoris lateralis within the groin region. Medial and lateral plantar nerves may be entrapped, resulting in the tarsal tunnel syndrome seen in runners.

Neuromas

The Morton's neuroma or interdigital neuroma can produce pain in the toe area, particularly in runners. Typically, pain is experienced during running but is often relieved after running, particularly if the shoes are removed.

Neuritis

The ulna nerve may be injured by repeated mechanical stretching, friction, and compression secondary to the large tensile forces generated during throwing. Ulnar neuritis may develop in an unstable elbow secondary to an attenuated ulnar collateral ligament.

The treatment of these nerve injuries includes rest and anti-inflammatory medication and flexibility training. Orthotic treatment may be indicated in nerve entrapments in the foot. Sometimes releasing surgery is necessary.

Overuse injuries in bones

Stress fractures

Stress fractures can occur in normal bone during normal situations. There is no single aetiology, but muscle fatigue and biomechanical imbalance are believed to be associated with stress fractures. Skeletal asymmetry, leg length discrepancy, poor conditions, variations of gait, running on hard or sloping surfaces, or prior injury can also predispose to this condition. Women have a decreased bone density and a wider pelvis, resulting in different foot plants and running gaits, and thereby predisposing for stress fractures.

Stress fractures can occur in most bones. In runners, 34 per cent occur in the tibia, 24 per cent in the fibula, 18 per cent in the metatarsals, 14 per cent in the femur, 6 per cent in the pelvis, and 4 per cent in other bones. These stress fractures are diagnosed by distinct tenderness on palpation. This is verified by a radionuclide bone scanning, which provides the diagnosis 2 to 8 days after onset of symptoms. Because of continuous remodelling of the bone, the bone scan can be positive for up to a year.

Sports activities can often be continued within the limits of discomfort. If running can be carried out during the healing phase, it should be done on soft surfaces. Some specific fractures need special management. If a femoral neck fracture does not become pain free in initial walking or running with or without crutches, surgery is indicated to avoid a full fracture and vascular necrosis.

Special attention should also be paid to navicular stress fractures, anterior tibial stress fractures, and proximal fifth metatarsal stress fractures as these often show delayed healing if not treated properly.

Occult bone lesions

MRI can confirm bone bruises, stress fractures, tibia fractures, femoral fractures, and osteochondral lesions. These occult bone lesions may be responsible for pain and discomfort, but are not evident on routine radiographs. The sensitivity and specificity of these lesions are not yet recognized or correlated with a specific diagnosis. However, it is known that they are present in 60 to 70 per cent of acute anterior cruciate ligament injuries. In patients with continuous pain after an initial trauma, these lesions should be suspected.

Specific injury site overuse problems

Foot problems

The forefoot can be the site of many overuse problems. They are often localized in the skin and soft tissues, and appear as blisters, corns, etc. Examples of other common overuse injuries are bunions and hammer toes, black toenails, metatarsal phalangeal joint pain, metatarsalgia, or sesamoiditis.

In the midfoot the longitudinal arch can often be high and stiff, and be related to arch pain, peroneal tendinitis, and midtarsal pain.

Plantar fasciitis is a major problem in the hindfoot, and painful heel pad, stress fractures in the calcaneus, and entrapment of nerves are not uncommon. Most of these injuries are treated using orthotics and shoes. Plantar fasciitis is an inflammatory reaction which is the result of a fatigue failure of the plantar fascia (Fig. 9). It can be caused by hyperpronation at midstance and is due to inward rotation of the foot and a pull on the medial fascia.[4] The cavus foot produces problems because of its inflexibility just as it reaches the flat stage. The rigidity of the midfoot diverts into the calcaneal origin and causes plantar fasciitis. Treatment consists of rest, heat, stretching, and sometimes crutches. Carefully made orthotics which unload the injured area are often indicated and helpful. Rigid orthotics should not be used in the cavus foot.

Lower leg pain

Medial tibial stress syndrome is a painful condition with the pain localized in the medial area of the tibia. This may be due to involvement of the tibialis posterior tendon and muscle secondary to increased pronation.[18] There is a greater angular displacement in subtalar passive mobility in patients with medial tibial stress syndromes than in controls.[58] Stress changes may also occur at the attachment of the soleus muscle. Biomechanically, the soleus is eccentrically stressed with heel eversion. These problems are treated with rest, stretching, and orthotics with varying results. Compartment syndromes and stress fractures can also cause lower leg pain.

Groin pain

Hip and groin pain are common in sports. Groin pain constitutes around 4 per cent of all injuries in soccer.[59] It is mostly caused by overuse injuries in the adductor muscles and tendons, particularly the adductor longus. Sometimes muscles and tendons from the rectus abdomini, rectus femoris, and iliopsoas may be causing the pain.

Intra-articular hip problems may occur in sports, although they

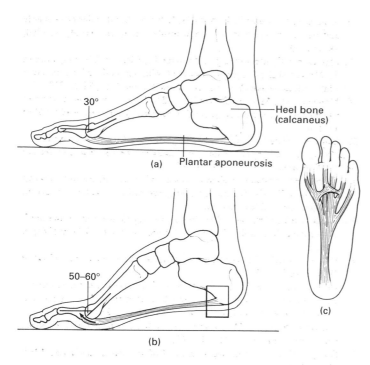

Fig. 9 Plantar fasciitis. (a) The foot and the plantar aponeurosis (fascial) shows the whole foot loaded against the surface. (b) The plantar aponeurosis is stretched during take-off. The marked area indicates the seat of inflammation at the origin of the plantar aponeurosis and the calcaneus. (c) The plantar aponeurosis seen from underneath (Redrawn from Peterson L, Renström P. *Injuries in Sports*. London: Martin Dunitz, 1985, with permission.)

are rare. Dancers can have early phases of arthritic changes. Intra-articular abdominal problems, genital infections, inflammatory conditions of the sacroiliac joint, osteitis pubis, nerve entrapments, pubic stress fractures, and other conditions may give pain syndrome similar to those caused by conventional groin overuse injury syndromes.

Spine overuse injuries

Lower back pain is a common problem in many sports. It may be caused by overuse of the lower back muscles or by imbalance of the back muscles in relation to other muscle groups. Other causes of back problems are sciatica, spondylosis, and stress fractures. Gymnasts, riders, and rowers may develop intervertebral joint changes in the lumbar spine.[27,60] Wrestlers can demonstrate low back pain[61] as well as cervical spine changes. Overuse injuries in the back region can be avoided with a general and well-balanced exercise programme.

Back pain in athletes must be taken seriously and assumed to be a significant problem until proved otherwise. If the athlete is out of competition for more than 10 days, a full and comprehensive evaluation is indicated.[1]

CONCLUSIONS

Overuse injuries are becoming more common in sports because of the increasing number of people in training and recreational activities, and the higher requirements and higher intensity at all levels of competition. Overuse injuries constitute a problem for all kinds of athletes, as they may stop them from training effectively.

The injuries may become chronic if intensive training is continued. If this occurs, the injuries are very difficult to treat and unfortunately have forced many athletes into early retirement from their main sport.

Most overuse injuries are caused by extrinsic factors, but intrinsic factors are also common. Malalignments, including excessive pronation and cavus feet, result in secondary compensatory changes in the lower extremity which, therefore, must be regarded as a functional unit. However, specific anatomic abnormalities and abnormal biomechanics of the lower extremity are not correlated with specific injuries on a predictable basis. Other factors such as leg length discrepancy, poor flexibility, muscle weakness and imbalance, deficit in neuromuscular co-ordination, and ligamentous laxity may be other important aetiological factors. Acquired or secondary factors such as kinetic chain dysfunctions are more common than previously acknowledged. Many overuse injuries in running, for example, are manifestations of dysfunction of the kinetic chain, particularly in athletes with recurrent and previous injuries.

The importance of finding a specific diagnosis and identifying kinetic chain dysfunctions in order to secure a correct and successful treatment cannot be emphasized enough.

Most of these injuries can heal with conservative regimes, but this requires good co-operation between the doctor and a very compliant patient. These injuries do not cause problems in everyday life, but the physician must understand the importance of the athlete's wish to continue his or her activities and to try to understand the injury from the athlete's point of view.

There is a need for more investigation of the basic biomechanics of soft tissues and how the tissues react to trauma and overuse. All musculoskeletal injuries are defined by the reparative response of the cell matrix. The distinction of macrotraumatic (acute) and microtraumatic (clinical overuse) injury patterns is implied by the pathological evidence. Overuse injury is a generalization that is more understood in terms of tendon load and use, for example, and how these influence matrix morphostasis versus degradation and subsequent degeneration. More research is needed to study the aetiology, diagnosis, and treatment of these increasingly common injuries. Studies of changes in collagen as a function of free-radical-mediated events secondary to disuse or ischaemia–reperfusion should be encouraged.[62] Overuse injuries are unnecessary and can usually be prevented, which implies that much more attention should be directed to this area.

REFERENCES

1. Weiker GG. Evaluation and treatment of common spine and trunk problems. *Clinics in Sports Medicine* 1989; 8(3): 399.
2. Slocum DB, James SL. Biomechanics of running. *Journal of the American Medical Association* 1968; **205**: 720–8.
3. Orava S. *Exertion injuries due to sports and physical exercise. A clinical and statistical study of nontraumatic overuse injuries of the musculoskeletal system of athletes and keep-fit athletes*. Thesis, University of Oulu, Finland, 1980.
4. Mann RA, Baxter DE, Lutter LD. Running Symposium. *Foot and Ankle* 1981; **1**: 190–224.
5. Cavanagh PR, Kram R. Stride length in distance running. *Medicine and Science in Sports and Exercise* 1990; **21**(4): 467–79.
6. Harvey JS. Overuse syndromes in young athletes. *Clinics in Sports Medicine* 1983; **2**(3): 595–607.
7. James SL, Bates BT, Osternig LR. Injuries to runners. *American Journal of Sports Medicine* 1978; **6**: 40–50.
8. McKenzie DC, Clement DB, Taunton JE. Running shoes, orthotics and injuries. *Sports Medicine* 1985; **2**: 324–7.

9. Renström P, Kannus P. Prevention of sports injuries. In: Krauss RH, ed. *Sports medicine*. Philadelphia: WB Saunders, 1991.

10. Ounpu S. *The biomechanics of running*. Instructional Course 322. Anaheim, Ca: American Society of Sports Medicine, 1991.

11. Nigg B. Causes of injuries: extrinsic factors. In: Dirix A, Knuttgen HG, Tittel K., eds. *The Olympic book of sports medicine*. Oxford: Blackwell, 1988: 363–375.

12. Renström P, Edberg B, Olofson B, Peterson L, Svennang J. Faktorer med betydelse for skador i football (Factors of importance for injuries in soccer). In: *Fotbollsspel pa Konstgras*. Stockholm: Statens Naturvardsverk (State Board of Nature and Recreation), 1977.

13. Elliott BC, Blaksby BA, Ellis R. Vibration and rebound velocity characteristics of conventional and oversized tennis rackets. *Research Quarterly on Exercise and Sports* 1980; **51**: 608–615.

14. Teitz C. Overuse injuries. In: Teitz C., ed. *Scientific foundation of sports medicine*. Oxford: Blackwell, 1990: 295–328.

15. Lysholm J, Wilander J. Injuries in runners. *American Journal of Sports Medicine* 1987; **15**: 168–71.

16. Renström P, Johnson RJ. Overuse injuries in sports. A review. *Sports Medicine* 1985; **2**: 316–33.

17. Stanish W. Overuse injuries in athletes: a perspective study. *Medicine and Science in Sports and Exercise* 1989; **16**(1): 1–7.

18. James, SL, Jones DC. Biomechanical aspects of distance running injuries. In: Cavanagh P, ed. *Biomechanics of distance running*. Champaign, IL: Human Kinetics, 1990; 249–69.

19. Lutter LD. Cavus foot in runners. *Foot and Ankle* 1981; **1**(2): 225–8.

20. Ross CF, Schuster RO. A preliminary report in predicting injuries in distance runners. *Podiatric Sports Medicine* 1983; **73**: 275–7.

21. Lorentzon R. Causes of injuries: intrinsic factors. In: Dirix A., Knuttgen H. G., Tittel K., eds. *The Olympic book of sports medicine*. Oxford: Blackwell, 1988: 376–90.

22. Grace TG. Muscle imbalance and extremity injury. A perplexing relationship. *Spots Medicine* 1985; **2** 77–82.

23. Safran M, Seaber A, Garnett W. Warm-up and muscular injury prevention. An update. *Sports Medicine* 1989; **8**(4): 239–49.

24. Heiser TM, Weber J, Sullivan G, Clare P, Jacobs RR. Prophylaxis and management of hamstring muscle injuries in intercollegiate football players. *American Journal of Sports Medicine* 1984; **12**: 368–70.

25. Kannus P, Nittymaki S, Jarvinen M. Sports injuries in women: a one-year prospective follow-up study at an outpatient sports clinic. *British Journal of Sports Medicine* 1987; **21**: 37–9

26. Marcus R, Cann C, Madvig P. Menstrual function and bone mass in elite women distance runners. *Annals of Internal Medicine* 1985; **102**: 158–63.

27. Sward L. *The back of the young top athlete; symptoms, muscle strength, mobility, anthropometric and radiological findings*. Thesis, Göteborg, Sweden, 1990.

28. Stanitski CL. Common injuries in preadolescent and adolescent athletes. Recommendations for prevention. *Sports Medicine* 1989; **7**: 32–41.

29. Lane NE, Block DA, Wood PD. Aging, long distance running and the development of musculoskeletal disability. A controlled study. *American Journal of Medicine* 1987; **82**: 772–80.

30. Peterson L, Renström P. Varldsmasterskap tor veteraner—en medicinsk utmaning (Championships for veterans—a medical challenge). *Lakartidningen* 1980; **77**: 3618.

31. Felson DT, Anderson JJ, Naimvk A, Walker AM, Meenan RF. Obesity and knee osteoarthritis. The Framingham Study. *Annals of Internal Medicine* 1988; **109**: 18–24.

32. McGoey BV, Deitel M, Saplys RJF, Likman ME. Effect of weight loss on musculoskeletal pain in the morbidly obese. *Journal of Bone and Joint Surgery* 1990; **72B**: 322–3.

33. O'Toole M, Hiller DB, Smith R, Sisk T. Overuse injuries in ultra-endurance tri-athletes. *American Journal of Sports Medicine* 1989; **17**(4): 514–18.

34. MacIntyre J. G., Taunton J. E., Clement D. B., Lloyd-Smith DR, McKenzie DC, Morrell RW. Running injuries. A clinical study of 4,173 cases. *Clinical Journal of Sports Medicine* 1991; **1**: 81–7.

35. Clement DV, Taunton JE, Smart GE, McNicol KL. A survey of overuse running injuries. *Physician and Sportsmedicine* 1981; **9**: 47–58.

36. Kannus P, Nittymaki S, Jarvinen M, Lehto M. Sports injuries in elderly athletes: A three-year perspective, controlled study. *Age and Ageing* 1990; **18**: 263–70.

37. Matheson GO, MacIntyre JG, Taunton JE, Clement DB, Lloyd-Smith R. Musculoskeletal injuries associated with physical activity in older adults. *Medicine and Science in Sports and Exercise* 1989; **21**(4): 379–85.

38. Merchant AC, Mercer RL, Jacobson RH. Roentgenographic analysis of patellofemoral congruence. *Journal of Bone and Joint Surgery* 1974; **56A**: 1391–6.

39. Bates BT, Osternig LR, Mason BR, James SL. Foot orthotic devices to modify selected aspects of lower extremity mechanics. *American Journal of Sports Medicine* 1979; **7**(6): 338–42.

40. Sperryn PN, Rostan L. Podiatry and sport physician. In: Bachl H, Prokop L, Suchert E, eds. *Current topics in sports medicine*. Baltimore: Urban, Schwarzenberg, 2984: 930–40.

41. Jorgensson U, Ekstrand J. Significance of heel pad finement for the shock absorption at heel strike. *International Journal of Sports Medicine* 1988; **9**: 468–73.

42. Dillingham MF. Strength training. *Physical Medicine and Rehabilitation* 1987; **1**(4): 555–68.

43. Komi PV. Physiological and biomechanical correlates of muscle function: effects of muscle structure and stretch-shortening cycle on force and speed. *Exercise Sports Science Review* 1984; **12**: 81–121.

44. Curwin S, Stanish WD. *Tendinitis: Its etiology and treatment*. Lexington: Collamore Press—D. C. Health, 1984.

45. Saal JS. Flexibility training. *Physical Medicine and Rehabilitation* 1987; **1**(4): 537–54.

46. Taylor DC, Dalton JD, Seaber AV, Garret WE. Viscoelastic properties of muscle-tendon units. *American Journal of Sports Medicine* 1990; **18**(3): 300.

47. Garrett WE. Muscle strain injuries: clinical and basic aspects. *Medicine and Science in Sports Exercise* 1990; **22**(4): 436.

48. Armstrong RB. Initial events in exercise-induced muscular injury. *Medicine and Science in Sports and Exercise* 1990; **22**(4): 429.

49. Friden J, Sjöstrom M, Ekblom B. Myofibrillar damage following intense eccentric exercise in man. International Journal of Sports Medicine 1983; **4**: 170.

50. Pedowitz RA, Hargens AR, Mubarak SJ, Gershuni DH. Modified criteria for the objective diagnosis of chronic compartment syndrome of the leg. *American Orthopaedic Society for Sports* 1990; **18**(1): 35.

51. Leadbetter WB, Buckwalter JA, Gordon SL. Sports-induced inflammation. *American Orthopaedic Society for Sports Medicine Symposium*. Chicago: American Academy of Orthopaedic Surgery, 1990.

52. Clement DB, Taunton JE, Smart GW. Achilles tendinitis and peritendinitis: etiology and treatment. *American Journal of Sports Medicine* 1984; **12**: 179–84.

53. Clancy WG. Tendinitis and plantar fasciitis in runners. In: D'Ambrosia R, Drez D, eds. *Prevention and treatment of running injuries*. Thorofare, NJ: Charles B. Slack, 1982: 77–88.

54. Irstam L. Personal communication, 1990.

55. Allenmark C, Renström P, Peterson L, Irstam L. Ten-year follow-ups of verified partial achilles tendon ruptures. *Proceedings of the AOSSM Meeting, New Orleans*. Chicago: AOSSM, 1990.

56. Konradsen L, Hansen EM, Sondergaard L. Long distance running and osteoarthrosis. *American Journal of Sports Medicine* 1990; **18**(4): 379.

57. Jobe FW, Bradely JP. The diagnosis and nonoperative treatment of shoulder injuries in athletes. *Clinics in Sports Medicine* 1989; **8**(3): 419.

58. Viitasalo JR, Kvist M. Some biomechanical aspects of the foot and ankle in athletes with and without shin splints. *American Journal of Sports Medicine* 1983; **11**: 125–30.

59. Renström P, Peterson L. Groin injuries in athletes. *British Journal of Sports Medicine* 1980; **14**: 30.

60. Jackson DW, Wiltse LL, Dingeman R, Hayes M. Stress reactions involving the pars intaricularis in young athletes. *American Journal of Sports Medicine* 1981; **9**(5): 304–12.

61. Granhed H, Morelli B. Low back pain among retired wrestlers and heavy weight lifters. *American Journal of Sports Medicine* 1988; **16**: 520–3.

62. Gordon GA Stress reactions in connective tissues: a molecular hypothesis. *Medical Hypotheses* 1991; **36**(3): 289–94.

5.2 Stress fractures

GIANCARLO PUDDU, GUGLIELMO CERULLO, ALBERTO SELVANETTI, AND FOSCO DE PAULIS

INTRODUCTION

Stress fractures are an overloaded pathology due to repetitive exogenous or endogenous microtrauma and/or the application of a few heavy loads whose effects exceed the biological capacities of functional adjustment with resulting final partial or complete bone collapse.[1] Two types of stress fracture are classically distinguished:

'fatigue fractures' as defined above;[2,3]
'insufficiency fractures',[2,4] in which normal repetitive loads cause fractures in bones which are less resistant as a result of pathological conditions such as osteoporosis, osteoarthrosis, osteomalacia, Paget's disease, bone tumours.

In this chapter we consider the first condition.

The first report of stress fractures in the literature appeared in 1855 when the Prussian Army Physician Breithaupt described the syndrome of a painful swollen foot associated with marching.[5] In 1897 Stechow published the first radiograph report of this condition in the metatarsal bones (Deutschlander's fracture).[6] However, in the last few years an increasing number of reports describing stress fractures in runners and other sports persons have appeared in the sports medicine and orthopaedic literature, probably because the increased number of players participating in sports and the improvement of diagnostic techniques such as bone scintigraphy, computed tomography (CT) and magnetic resonance imaging (MRI).

Stress fractures comprise 10 per cent of sports injuries,[3,7] and are particularly common in the weight-bearing sports.

BIOMECHANICAL AND HISTOLOGICAL ASPECTS

Stress fractures in athletes are due to the inability of healthy bone to withstand chronic submaximal repetitive mechanical stress or a sudden increase in loads during sports activities.

Bone is a dynamic tissue which, according to Wolf's law, can adapt to load variations (compressive, distractive, rotational and shearing forces) to ensure an equal stress distribution. This adjustment occurs through continuous structural remodelling depending on load characteristics (intensity, volume, time of application) and begins with prevalence of osteoclastic activity with consequent weakening of the bone. Osteoblastic activity increases to balance the reabsorption and to supply the required resistance to the bone. If the stress is not eliminated or reduced during this process, plastic deformation can occur in the bone with the possibility of a stress fracture.[8–11] Jones et al.[12] called this process 'stress reactions' and divided it into five stages.

Grade 0 (normal remodelling) is characterized by a thin new periosteal bone, not visible on radiographs, which is clinically asymptomatic. However, the bone scan reveals a small linear area of increased uptake.

Grade 1 (mild stress reaction) is also present as a cortical rearrangement (tunnelling). The subject complains of local pain exacerbated by activity. Tenderness is absent, radiographs are negative, and the bone is intact. However, the bone scan is positive.

Grade 2 (moderate stress reaction) Cortical reabsorption prevails on periosteal reaction. Pain and tenderness are present, and some indistinct signs may be visible on radiographs. The bone scan is positive and the bone should be intact.

Grade 3 (severe stress reaction) Periosteal reaction and cortical tunnelling are extensive, and pain persists at rest. Radiographs show cortical thickening and the bone scan is positive.

Grade 4 (stress fracture) Bone biopsy reveals necrotic areas, trabecular microfractures and granulation tissue. Weight bearing is sometimes impossible because of the pain. Radiographs show the fracture and early signs of callus formation. Scintigraphy is positive.

AETIOLOGY AND PATHOGENESIS

Two theories have been proposed to explain stress fractures. According to overload theory[13] rhythmic and repetitive contractile activity of the muscles produces stresses at their osseous insertions which reduce the mechanical resistance of the bone. This theory can explain stress fractures of the chest and upper body which are non-weight-bearing areas. According to muscle fatigue theory[14–17] progressive exhaustion during sports activity makes the muscle less effective as a 'shock absorber'. Abnormal load distribution occurs with stress concentration in restricted areas. This mechanism is applicable to stress fracture in weight-bearing activities.

Other factors associated with stress fractures are as follows.[7,12,18]

Genetic factors Large series which reveal predisposing genetic factors have not been reported in the literature. However there is some anecdotal evidence regarding monozygotic twins submitted to the same stresses.[19]

Race Stress fractures are less frequent in black people, possibly because of their higher bone density.[9,20,21]

Somatotype The risk of stress fracture is higher in large subjects and those whose morphology is inadequate for the required performance.[22,23]

Sex Women are more at risk then men (from 3.5:1 to 10:1).[20,24,25] This difference is probably due to their lower percentage of lean mass and lighter skeletal structure, disadvantageous morphotype for running (large pelvis, coxa vara, genu valgum, lower height, shorter step), anorexia nervosa, and menstrual disorders.[26–30] It has been shown that consumption of the oral contraceptive pill reduces the risk of stress fracture in long-distance runners.[31]

Age The risk increases with age, with a peak between 18 and 28 years old when lamellar cortical bone gradually transforms into adult osteonic bone.[32,33] Greenstick fractures are more common in subjects aged less than 16 years.[34]

Physical fitness Stress fractures occur more frequently in sedentary subjects who have just begun sports activity or in athletes who have returned to sport after prolonged inactivity.[35]

Training errors The history of athletes with stress fractures frequently reveals a sudden increase in loads during training, particularly if the subject is not in good physical condition.

Equipment Footwear which is too small or too worn can increase the risk of fracture because the capacity to absorb stresses is reduced.[18] Combat boots are often inadequately designed to absorb stresses from the ground.[36]

Playing surface Running or walking on a hard surface such as cement or asphalt can increase the risk of stress fractures

Anatomical abnormalities (structural and/or postural), such as malalignment or leg length discrepancy, excessive femoral anteversion, flat foot, varus forefoot, and pronation, can result in a non-physiological load distribution.[37]

CLINICAL, DIAGNOSTIC, AND THERAPEUTIC ASPECTS

History

The characteristic history is a dull aching localized pain with insidious onset and progressive worsening with activity. The patient often reports some recent changes in training such as increasing duration or distance, a harder playing surface, or new footwear. In some cases, such as ballet dancers, onset is acute.

Clinical examination

In most cases clinical examination is non-specific or negative, particularly if the area is surrounded by soft tissues. However, tenderness, swelling, and percussion tenderness localized to the fracture site may be noted. If a long time has elapsed since the onset of symptoms, callus may be present.

Plain radiographs

In 70 per cent of cases radiographs are negative in the early stages of the pathological changes, and no more than 50 per cent are positive in the later stages.[34] Although radiology is not very sensitive to stress fractures, its usefulness can be increased by using X-rays in two planes, oblique views, high magnifications and tomograms.

A period from 2 to 12 weeks, depending on the site of the fracture, is required before a stress fracture can be visualized by radiography.[38] Characteristic signs are the local periosteal reaction with addition of new bone (evident after 6 weeks)[29]and endosteal sclerosis. In cancellous bone, early lesions are of the compressive type. Sometimes they are visible on radiographs after 24 h. The classical aspect of a stress fracture is a radiotransparent line which interrupts the cortical continuity (Fig. 1).[39,40]

Bone scanning

Phosphate labelled with technetium–99m is incorporated in osteoblasts and a 'hot spot' appears in areas of bone stress reaction. Scintigraphy is less specific than radiology but is much more sensitive to stress fractures. It can give positive results 6 to 72 h after the onset of pain.[18,41] The most sensitive technique is the triple-phase bone scans which can date the lesion, distinguish partial from complete fractures, and distinguish stress fractures from 'shin splints'.[42] It is useful to perform anteroposterior, lateral, and oblique scans, and to compare them with the contralateral scan.

In the early stages (grade 1) bone scans show a slightly increased uptake over a poorly defined area which becomes progressively more obvious (grade 2) with well-defined margins. Initially only one cortex is involved (grade 3); eventually the other cortex is incorporated (grade 4). Diagnosis of stress fracture is only possible in the last two stages.[43,44] Bone scanning can also be used to

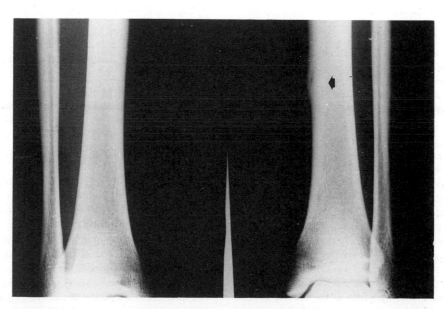

Fig. 1 Classical aspect of a stress fracture on a radiograph: a radiotransparent line interrupts the cortical continuity. The local periosteal reaction, with apposition of new bone, is also evident.

monitor the healing process which is demonstrated by progressive return to normal uptake.[45]

False positive scans are sometimes obtained for muscle strains, bone cysts, osteoid osteoma, sickle-cell disease, and osteomyelitis. In some cases, when cortical reabsorption dominates over osteoblastic activity, scintigraphy is temporarily negative.[46] However, a bone scan may reveal areas of increased uptake in asymptomatic patients.

Computed tomography

Because of its high definition, image clarity, and axial vision, CT reveals stress lesions more effectively than traditional tomography, particularly in specific sites such as the lumbar spine, and provides useful information for differential diagnosis (Fig. 2).

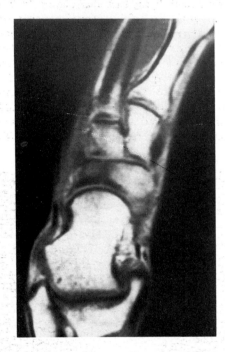

Fig. 2 Stress fracture of the tarsal navicular: CT shows the line of fracture, which interrupts the anterior and posterior cortex, with endosteal sclerosis. The patient had negative radiographs and positive bone scan. When he returned to sports activity (hurdling) after 2 months of treatment with pulsating magnetic fields, he inverted the impact foot while running. Two months later he developed stress fracture of the contralateral scaphoid.

Magnetic resonance imaging

MRI provides a very sensitive method for monitoring variations accompanying stress fractures. It is more precise than scintigraphy and radiography since it shows the exact site of the lesions and can give good chronology of the pathology, as the signal's characteristics change with the passage of time from the onset of symptoms. A high intensity area surrounding a band at low intensity is visualized on T_2-weighted images during the first 3 weeks. Oedema and haemorrhage resolve within 3 weeks but low intensity signals still remain on T_1-weighted images.

Stress fracture is characterized by a line in the cortex, at very low signal intensity, with a surrounding area of bone at reduced intensity on T_1-weighted images. The intensity of the signal on T_2-weighted images is increased to an extent depending on the associated pathology (oedema, haemorrhage, medullar fibrosis, and periosteal and endosteal reaction).

MRI also provides useful information for differential diagnosis (osteoid osteoma, Garre's osteomyelitis, osteogenic sarcoma, ischaemic necrosis, and intraosseous occult fractures), and in acute injuries it may differentiate strains or ligamentous insults associated with or responsible for the symptoms.[47-49]

General principles of treatment

Conservative treatment of stress fractures is generally successful. The athlete must suspend activity in his or her sport, but other exercises (cycling, swimming, upper-body ergometrics) are encouraged to limit deconditioning and demoralization. Avoidance of all activities that involve impact loading is mandatory in stress fractures of the lower limbs. Weight-bearing can be maintained if it is not painful. When pain is present at rest, it may be necessary to immobilize the limb in plaster or splints. Analgesic drugs or non-steroidal anti-inflammatory drugs and ice can be used to relieve pain.

When the athlete has been free of pain for 2 or 3 weeks, percussion tenderness is negative, full weight bearing is normal (in stress fractures of lower limb), and plain radiographs show bone healing, he or she can gradually return to sport, using pain as a criterion for monitoring effective recovery.

It is necessary to identify the presence of the risk factors discussed previously. Surgical treatment is indicated when there is a high risk that the fracture will not consolidate or in the presence of a non-union.

UPPER LIMBS

Coracoid process

Stress fractures of the coracoid process typically occur in trapshooters and are due to the repetitive percussive action of the butt of the rifle and the cyclic contraction of the coracoid muscles.[50] The athlete complains of an ache in the anterior shoulder while shooting and also at rest if the pathology progresses. Physical examination reveals local tenderness of the coracoid process, pain on resisted adduction and flexion of the shoulder with negative signs for rotary cuff pathology or instability.

Plain radiographs are negative but the axillary view reveals the fracture at the base or the middle third of the coracoid process (in rare cases).[51] Treatment is based on avoidance of shooting and painful movements of the shoulder for 2 to 3 months.

Humerus

Stress fractures of the humerus have been described in young throwing athletes and participants in racket sports who have immature bone and muscle.[52,53] These fractures rarely occur in athletes who have been involved in sport for many years and have developed muscle and cortical hypertrophy. The causes of this injury are the high acceleration forces operating on the humeral shaft.

The athlete complains of pain during throwing or at the end of the game or training session. Clinical examination reveals local tenderness on deep palpation of the humeral shaft but movement of the shoulder and elbow is painless. Radiographs may be negative but bone scans are positive.

Treatment is based on curtailment of throwing sports and improvement of technique on return to training. If the athlete does not rest fracture may become complete with dislocation of fragments and the necessity for surgical treatment, compromising the athlete's career.

Olecranon

Stress fractures of the olecranon in throwing athletes or gymnasts are due to the repetitive tension generated by the triceps brachii tendon (avulsion-type fracture).[8,54,55] Pain is present during activity and in maximal flexion or extension. Because of the high risk of delayed union or non-union of this fracture it is best treated by immobilization.[56,57] If the fracture is complete, surgical fixation is preferred.

In some cases, when the fracture occurs at the distal end of the olecranon, pain may persist during activity even after healing. In this case surgical excision is indicated.

Ulna

Stress fractures of the ulnar shaft have been called 'lifting fractures' because they were initially described in farmers involved in digging of lifting heavy objects with a pitchfork. These fractures are occasionally reported in volleyball and tennis players, bodybuilders, and softball pitchers.[58–62] Pain is referred during activity and there is local tenderness in the forearm. Radiographs are usually negative in the early stages but scintigraphy confirms the diagnosis. Rest from athletics for 4 to 6 weeks is generally sufficient to restore normal condition.

Radius

Only one case of stress fracture of the radius, which was bilateral, has been reported in the literature. It was described in a sailor after training for 'field gun running'.[34] Treatment consisted of 6 weeks of rest and had a good outcome.

Carpal scaphoid

Stress fracture in the carpal scaphoid has been reported in gymnasts and shot-putters and is due to the forced dorsal flexion that crushes the navicular against the radial styloid.[63,64] The onset is usually insidious, but the symptoms may worsen such that the athlete cannot continue his or her sport. Palpation over the snuffbox generates pain. The same effect is produced by radial deviation and hyperextension. Oblique radiographs can show the fracture in the middle third of the bone,[65,66] and there is an intense uptake in bone scans. CT is also useful.[67] A thumb spica is recommended for 2 to 4 months.

Metacarpals

Only one case of stress fracture of the metacarpals has been reported.[68] A tennis player who held the racket incorrectly developed a stress fracture at the base of the second metacarpal which healed after rest from sport for a month.

THORACIC CAGE

Sternum

Keating[69] reported stress fracture of the sternum in a wrestler presenting with chest pain approximately a week after performing a hyperextension trunk stretching exercise. Examination revealed tenderness and a slight palpable prominence 2 cm distal to the sternum angle. A lateral radiograph showed a stress fracture, which was confirmed by a hot spot on the bone scan. The wrestler was allowed to resume his sport 2 months after diagnosis.

Ribs

First rib

Stress fractures of the first rib have been reported in baseball, tennis, American football, and basketball players.[70,71] The fracture generally occurs at the level of the subclavian groove, which is the thinnest part of the rib and is exposed to the strong action of the scalenii, serratus anterior, and intercostal muscles. There may be insidious or acute onset of pain at the base of the neck, sometimes extending under the scapula or towards the pectoral area. Pain may increase during deep breathing. Clinical examination shows an area of local tenderness and pain on resisted shoulder lifting. The diagnosis is confirmed by radiography and bone scan. Occasionally this fracture is complicated by non-union (Fig. 3).

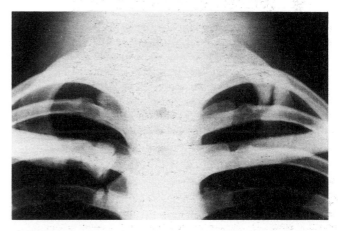

Fig. 3 Non-union of a stress fracture of the left first rib. The injury was occasionally revealed in a 25-year-old basketball player during the routine preseason examination. No treatment was necessary.

Other ribs

Stress fractures of the other ribs have been described in rowers, gymnasts, golfers, and tennis players.[72,32] They usually occur in the posterolateral area, where the serratus anterior muscle exerts a strong bending moment during scapular motion. The athlete complains of a chest pain which is exacerbated by deep breathing, activity and local palpation. Differential diagnosis between muscle strain and fracture is required, but radiographs and bone scans provide confirmation. Rest from athletic activity for 4 to 8 weeks is sufficient for healing to take place.

LUMBAR SPINE

The lumbar spine is subjected to considerable stress during the maintenance of certain postures in ballet and the performance of

various athletic movements. Stress fractures of the lumbar vertebral arch are relatively common, with the site of the injury depending on the direction in which stress is applied. Fractures of
the lamina, the pedicles, and the articular processes have been
described,[74-76] but injuries of the pars interarticularis, are most
frequent.[77]

Spondylolysis is a loss of continuity of the pars interacticularis
(or isthmus) of the vertebra; it most frequently involves L5,
though it can also be found at L4 and, rarely, more proximally.

Spondylolisthesis involves relative anterior slipping of one vertebra over those below it. This may be associated with spondylolysis or it may arise alone.

Of the five forms of spondylolysis and spondylolisthesis
classified by Wiltse et al.[78] we shall deal only with those most
frequently found in young people and athletes, i.e. dysplastic (type
I) and isthmic (type II).

Epidemiology

The incidence of spondylolysis in the general population is 5 per
cent,[79] but this figure may be higher among athletes.[164] According
to one survey,[80] the incidence in young gymnasts is 11 per cent
(four times higher than the 2.3 per cent reported in the general
female population) and 6 per cent also suffer from spondylolisthesis. The incidence in adult linemen (American football)
is 24 per cent compared with 6.4 per cent in white men and 2.8 per
cent in black men.[81]

The incidence of spondylolysis and spondylolisthesis increases
from about the age of 5 to 6 to about the age of 20 and then remains
steady; as regards sex and race, it is lowest among black women
(1.1 per cent) and highest among white men (6.4 per cent). The
incidence reaches 54 per cent in certain Eskimo tribes.[82]

The dislocation starts at about the age of 8 in girls and about the
age of 12 in boys and increases with age, particularly during the
adolescent growth spurt (10–15 years of age). Girls are more often
symptomatic and require surgical stabilization of the lesion more
frequently.

Aetiology

Congenital anomalies of the sacrum and spina bifida are associated
with spondylolysis and/or lumbosacral instability. However, spondylolysis and spondylolisthesis have never been found at birth or in
subjects who have never walked.[83] The possibility of genetic predisposition may explain why cases of spondylolysis are sometimes
found among relatives of patients suffering from this pathology.

Lack of continuity of the pars interacticularis is considered to be
stress fracture. It is unilateral in 20 to 25 per cent of cases, and is
caused, in predisposed subjects, by repeated forced cyclical movements involving flexion–extension of the lumbar spine, implying
functional overloading.[82,84] This initiates a concentration of high
shear forces at the isthmus level which increase in movements
involving both extension and lateral inclination and/or rotation of
the spine.

Athletes most likely to be affected by these pathological changes
are those involved in the following activities: gymnastics, weight-
lifting, wrestling, American football (collisions of linemen produced in the 'three point position; with a lumbosacral
hyperextension), soccer, mountain skiing, handball, judo, swimming, diving, basketball, rugby, parachuting, track and field
athletics (pole-vaulting, javelin throwing, hurdling), tennis,
baseball (pitcher), ballet, lacrosse, karate, and rowing. The follow-

ing factors may predispose to the condition: lumbar hyperlordosis,
high body weight, and a strong paravertebral musculature
opposed by a relative deficit of abdominal muscles.

Clinical description

Symptoms are usually absent in children, particularly if they do
not practise sport. Discomfort arises at the onset of the adolescent
growth spurt. However, the lesion is usually only noticed adven-
titiously following radiography. A typical example is an athlete
practising one of the sports listed above who has had months of
lumbar pain with insidious onset caused and/or exacerbated by
physical activity (particularly repeated flexion and extension
and/or rotation of the lumbar spine), occasionally extending to the
gluteus and the thigh. The symptoms may be temporarily relieved
by resting from sport and taking analgesics, but if the activity is
continued the pain becomes more frequent and intense, sometimes interfering with athletic practice and even with normal daily
activities. There is not usually a history of serious trauma, although sometimes the onset of symptoms is acute and may coincide with a minor injury.[85]

Objective examination may be negative in spondylolysis or first
or second degree spondylolisthesis; sometimes the only evidence
available is tightened hamstrings (present in 80 per cent of symptomatic patients). In more advanced spondylolisthesis, the muscular contracture may prevent full flexion of the thigh on the pelvis
so that the patient walks in small steps, with the pelvis rotating at
each step ('pelvic waddle'). Children run or walk on the tips of
their toes with knees semiflexed.

When the lumbar region is palpated, particularly if the patient is
symptomatic, tenderness and spasms of the paravertebral muscles
maybe present at the level of the vertebral defect and the surrounding segments.

The pain is induced and increased by the following movements
of the trunk: anterior flexion (if the contraction of the hamstrings
is intense, the patient cannot touch the floor with his palms),
contralateral rotation and inclination ipsilateral to the injury, and
extension while bipodal bearing (if the defect is bilateral) or on the
corresponding leg (if unilateral). In gymnasts and dancers the pain
can be induced by making them adopt the scale or arabesque
position respectively. In both cases the patient will feel pain when
resting in a contralateral monopodalis weight-bearing position if
the injury is unilateral.[86] Straight leg raising is reduced when
hamstrings are very contracted.

In the more advanced (third and fourth) degrees of spondylolisthesis examination of the orthostatis posture reveals flattening of the lumbar lordosis, a 'step-off' or depression which can
be felt on the middle line over the spinous process of the lumbar
vertebra and which is the site of the injury, a short trunk with a low
chest, a transverse transumbilical groove, flared ilia, and flattened
and heart-shaped buttocks.

Lumbar scoliosis is more often found accompanying spondylolisthesis (particularly the dysplastic type), in women and in cases of
severe slippage. It is secondary to muscular spasm and disappears in
asymptomatic periods (sciatic scoliosis) or after surgical treatment
of the injury. Idiopathic structural dorsal or dorsolumbar scoliosis
is reported in a third of all cases of spondylolisthesis.[82]

Plain radiographs

Anteroposterior, lateral, and oblique underload projections,
sometimes augmented by tomograms, are used to examine a spon-

dylolysis. In a relatively advanced spondylolysis, the anteroposterior view shows 'décalage' of the alignment of the spinal processes above the lysis resulting from the advancement of the contralateral superior articular process, and the stress hyperosthosis of the isthmus and the pedicle opposite the site of the injury (in this last case, the differential diagnosis must consider osteoid osteoma and it is thus necessary to look for the nidus).[87] The lateral view may occasionally reveal the lysis. However, this is better shown by an oblique projection with the typical 'decapitated dog' or 'Scotty dog' image; if the injury is recent the gap will be narrow with irregular margins, but if it is chronic the borders of the injury will be blunt and smooth (Fig. 4).

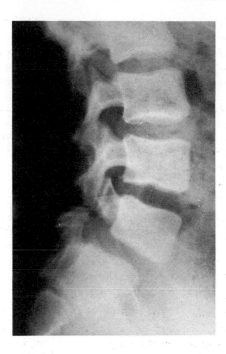

Fig. 5 Second-degree spondylolisthesis in a 24-year-old pole vaulter occasionally suffering from lumbar pain. The pathology did not influence his sports career.

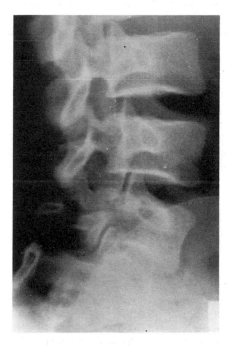

Fig. 4 Spondylolysis: an oblique radiograph shows the typical 'decapitated dog' image. The blunt smooth borders of the injury indicate that it is chronic.

In examining a case of spondylolisthesis, oblique projections are useful only in those cases of first-degree slip.[88] When L5–S1 spondylolisthesis is advanced, the overlapping of the lumbosacral body over the sacrum may produce the image of an 'inverted Napoleon's hat' on the anteriorposterior projection and the isthmic lysis may appear as an interruption of bone continuity just below the pedicle. In the lateral projection under load we see lysis or lengthening of the isthmus, and the degree of subluxation of the vertebra can be measured in terms of tangential slipping (four degrees according to Wiltse's classification[89] (Table 1)) and sagittal rotation of the L5

Table 1 Wiltse's classification of spondylolisthesis (percentage
of slipping of the upper displaced vertebra upon the lower)

First degree	< 25%
Second degree	25%–50%
Third degree	50%–75%
Fourth degree	>75%

vertebra over the sacrum (Fig. 5). As the slip progresses, anterior and posterior rounding of the sacral promontory, a wedge-shaped aspect of the L5 body (lumbar index), a vertical position of the sacrum, an increase in lumbar lordosis, and, in adults suffering from first and second degree spondylolisthesis, a reduction of the L5–S1 intervertebral space and sclerosis of the anterior margin of the sacrum are observed.

Lateral weight-bearing views at the highest degree of flexion and extension are also useful for evaluating the degree of translatory instability of the unstable vertebra. Radiographs also allow investigation of associated anomalies (spina bifida and congenital aplasia of the proximal part of the sacrum or of the superior articular facets). In dysplastic spondylolisthesis the pedicles and pars interarticularis appeared lengthened, and in the lateral projection the whole vertebra slips anteriorly and thus produces lysis of the isthmus or the lamina, which favours further anterior slippage of the vertebral body.

Bone scintigraphy

Bone scintigraphy is useful in symptomatic patients with negative radiographs to reveal a stress reaction of the pars interarticularis which cannot be visualized radiographically, or when the radiographs show a stress fracture of the isthmus to estimate the age of the injury and classify the activity.

In fact, in a asymptomatic patient with negative radiographs and a positive bone scan, preventive treatment may stop the development of the process, whereas, when the radiographs show spondylolysis and the bone scan is normal, the injury is a pseudoarthrosis which has been present for about 6 to 12 months and which cannot 'heal'.[90–92] Hence there is no point in performing scintigraphy on those patients who have been symptomatic for more than a year unless tumours or osseous infections are suspected or who are asymptomatic.[82]

Computed tomography

CT is useful in adults to visualize lysis and sclerosis of the isthmus and to establish the degree of slippage, spinal stenosis, and radicular compression.

Myelography

Myelography is indicated where an osseous neoplasia or discitis is suspected or in those cases with clinical signs of cauda equina syndrome.

Treatment of spondylolysis

If the injury is acute (recent onset of symptoms, normal radiographs, and positive bone scan) recovery may be promoted by immobilizing with plaster (corset or single-leg pantaloon body cast) or thoracolumbosacral orthosis for at least 3 months and avoiding activities which may challenge the lumbosacral spine. The efficacy of the treatment is monitored by disappearance of pain and decreasing scintigraphic activity. Once the plaster is removed, it is necessary to wait for another 3 months before a gradual return to sporting activity is allowed, provided that the patient remains asymptomatic and that scintigraphy is normal.

The athlete follows a programme of stretching exercises for the paravertebral muscles and hamstrings as well as abdominal and hamstring strengthening exercises avoiding hyperextension of the lumbar spine. The functional demands are gradually increased. It may be necessary to forbid training with weights in the upright position. If the bone scan is positive and radiographs show a recent fracture, the symptomatic patient may be treated as described above even if the radiological evidence of recovery of the injury is less clear.

When the injury to the isthmus is chronic (positive radiography and negative scintigraphy), the treatment is symptomatic, including analgesics and temporary rest from athletic activity, avoiding painful posture. The patient may be confined to bed or immobilized in plaster for a short time, and must start a gradual programme of exercises to strengthen the abdominal muscles and to stretching and strengthening the hamstrings and back muscles. It is sometimes necessary to resort to surgical stabilization of spondylolysis in those rare patients who are refractory to conservative treatment. This may allow a return to sport after about a year.

When spondylolysis is diagnosed in a child, it is necessary to carry out clinical and radiological checks at least every 6 months until the skeleton is fully developed to ensure early recognition of the initial signs of a progression of slippage. Sporting activity need not be severely limited in the child or adolescent with asymptomatic lysis, but the patient must be informed of the possible consequences of participating in high risk sports. The risk of the development of spondylolisthesis in a child suffering from spondylolysis increases between the ages of 10 and 15 years, particularly if he or she participates in sport, and if recurring episodes of lumber pain with a dysplastic spondylolisthesis are present. A worsening of the condition is rare after the age of 20 and has never been documented during pregnancy.[82]

Treatment of spondylolisthesis

In first-degree asymptomatic spondylolisthesis the patient must be informed of the possibility of an evolution of the pathology although specific sporting activities need not be curtailed. Clinical and radiographical checks should be performed every 6 to 12 months, and exercises to correct muscular imbalance should be prescribed. When a 'silent' spondylolisthesis exceeds the first degree, sport involving a high risk of physical contact and trauma to the back should be forbidden.

In symptomatic spondylolisthesis, long-term conservative measures are ineffective in over 50 per cent of cases, particularly if the slippage is second degree or above. Patients with slippage of 25 per cent or more and an early degenerative discopathy are more likely to suffer from acute relapsing episodes and to require surgical treatment in adult years.[93] Competitive athletes suffering from spondylolisthesis have severe functional limitations and often have to change discipline or interrupt their athletic careers.[94]

In first- and second-degree symptomatic spondylolisthesis, surgery is indicated when no result is obtained by reducing physical activity, applying a corset, and maintaining muscular tone. Surgery is also indicated in third- and fourth-degree spondylolisthesis when pain is refractory to conservative treatment or when there are neurological manifestations. A bilateral arthrodesis is generally carried out. If there is radiological evidence of a consolidated osseous fusion (80–100 per cent of patients undergoing surgery), a gradual return to sports activity is allowed.[82]

It follows from the above discussion that an early diagnosis of isthmic overload pathology is very important. It might be useful to screen predisposed patients participating in sports involving risks in this respect so that injuries can be identified at an early stage and suitable treatment prescribed.

LOWER LIMBS

Pelvis

Stress fractures of the pelvis are most frequently in joggers, long-distance and marathon runners, and military recruits.[95] They generally occur in the pubic ramus where it joins the ischium near the symphysis and comprise only 1.25 per cent of stress fractures in endurance runners.[96]

It is believed that these fractures are the result of repeated tensions produced in this area during a prolonged race by the rhythmic contraction of the adductors and the obturator muscles. They are more frequent in women, probably because their racing technique is different and/or because of the different geometry of the female pelvis. They can be bilateral and may be complicated by an avulsion fracture at the level of adductor insertion.[95]

Pelvic stress fractures must be considered and suspected when a runner complains of characteristic pain in the buttock or thigh during or after training which increases if training is continued until the runner has to slow down and starts limping.[97] It is important to enquiries whether there has been a sudden change in the running distances, a variation in the racing technique, or a change of training site. There is marked pain when pressure is exerted in the region of the inferior pubic ramus near the symphysis, and upright posture with weight bearing on the limb on the side of the injury is uncomfortable and painful (positive standing sign).[98] Slight pain may limit movements of the hip in abduction and/or external rotation. Differential diagnosis may consider insertional tendinopathy of the adductors, muscular strain of the adductors, pubic osteitis, trochanteric bursitis, degenerative pathology of the hip, pain in the lumbosacral spine, tendinopathy, or muscular strain of the hamstrings.

Radiographs are positive 2 to 3 weeks after the fracture, but if too long a period elapses before they are carried out, the frequently exuberant aspect of the callus may suggest osteosarcoma. For a further evaluation it is better to combine the usual anteroposterior view with two further projections orthogonal to each other with the patient horizontal and the X-ray beam inclined at 45° to the longitudinal axis of the body in the cephalic–caudal and caudal–cephalic directions. Tomograms may also be useful. An early bone scan is recommended and is very useful in the 'preradiographic' stage of the injury which may be complicated by delayed union and recurrence; the sclerosis surrounding the margins of the fracture indicates non-union which recovers on resting.

Treatment requires a period of abstinence from sport, and all activities causing pain are forbidden for 8 to 12 weeks. Crutches are sometimes required for about 4 to 5 weeks, particularly when there is pain under load. Stretching exercises are allowed, if they can be tolerated. A gradual return to sport is allowed only if the patient is completely asymptomatic and radiographs show no signs of delayed union or non-union.

Femoral neck

Fracture of the femoral neck is the typical 'insufficiency' fracture which affects elderly people and which can be found as a stress fracture in long-distance runners, dancers, military recruits, hurdlers, football players, and cross-country skiers. During walking the proximal femur is subject to loads up to six times the body weight with high compressive loads on the concave side of the femoral neck and tensile loads on the convex side. Thus loading increases substantially during running.[99]

This type of fracture must be considered when the athlete, usually a runner who has recently intensified his training programme and/or started to run on a different surface, complains of a pain in the groin extending to the anterior thigh and sometimes to the knee, which starts on weight bearing during or immediately after physical activity and is temporarily relieved by rest and functional unloading of the limb. However, if physical training continues, pain starts earlier and intensifies, eventually becoming constant, preventing training, and sometimes causing limping.[100] In some cases the pain starts at night and, in the absence of symptoms, the athlete may present to the doctor when the fracture is already complete and displaced. Antalgic gait and discomfort when standing on the foot of the fractured limb (positive standing sign) are observed. Palpation of the groin region above the hip joint is painful. Sometimes this area appears swollen, there is pain in the area of the injury because of percussion or compression of the heel or the great trochanter, and a slight antalgic limitation of the highest degrees of articular excursion (particularly in flexion and internal rotation).[101] Differential diagnosis should consider strain of the iliopsoas or the adductors, hernia of the groin, pubic osteitis, bursitis, or synovitis.

Plain radiographs of a femoral neck fracture may be negative during the first 2 to 6 weeks after the onset of symptoms; if the fracture is suspected clinically it may be useful to perform a bone scintigraphy which will allow early discovery of a hot spot in the area of the femoral neck. If this is the case, serialized radiographs can be taken to check the radiological evolution of the injury. Anteroposterior, lateral, and oblique radiographs of the hip are obtained and compared with those of the controlateral hip (fractures are frequently bilateral). Tomograms may be necessary. There are many radiographic classifications of femoral neck fractures.[27,102,103] From a therapeutic viewpoint 'subradiographic'

cases, where the clinical picture and scintigraphy suggest a fracture although plain radiographs are still negative (conservative treatment by avoidance of weight bearing until symptoms disappear and scintigraphy becomes normal), must be distinguished from those in which the radiographs show unmistakable signs of stress fracture. In the latter case we adopt the Devas classification which distinguishes two kinds of femoral neck stress fracture.[102]

Compression fractures

Compression fractures are more frequent in young athletes[104] and are located at the cortex of the lower medial margin of the femoral neck. Initially they appear radiographically as a rather opaque area which, if loading continues, becomes more sclerotic; sometimes a small central 'crack' appears whose margins in time become thicker. In the more advanced stages of recovery, the anteroposterior tomogram shows gradual healing of the injury until the smooth aspect of the concave margin of the neck is re-established. This type of fracture seldom displaces, unless stress continues.

If, after diagnosis, radiographs do not show the fracture, treatment is conservative and the patient must be restricted until he or she becomes asymptomatic and active and passive motion of the hip is completely regained and no longer painful. At this point the patient may be allowed to walk with crutches, first non-weight-bearing for about 6 weeks and then partially loaded. When full weight bearing is asymptomatic the patient may abandon one crutch and start swimming and walking in water. He or she may also use a bicycle, gradually abandon the other crutch, and increase the walking distance on soft ground. At all times the guidelines for treatment must be the absence of pain and/or limping and the gradual recovery of the fracture as shown by serial radiography. Finally, the patient is allowed to run, avoiding hard surfaces, first on alternate days for short distances interrupted by walking, and then for longer distances.

If initial radiographs show a fracture of the cortex, it is advisable to admit the patient to hospital and, if necessary, apply skeletal traction. If radiographs show deepening and/or widening of the rima of fracture, internal fixation must be carried out, allowing unloaded walking with crutches from the first day after the operation. The osteosynthesis is removed after 6 months and a gradual rehabilitation programme is carried out for another 3 months during which participation in contact sports is forbidden.

Distraction-type fracture

Distraction-type fractures occur most frequently in elderly people and military men. Onset is at the superior margin of the femoral neck with interruption of continuity of the cortex (seen radiographically in the appropriate anteroposterior view if it is not hidden by the great trochanter) which, if not recognized early, tends to deepen, crossing the force lines of the femoral neck perpendicularly and evolving towards a complete and displaced fracture. If untreated, complications are likely, including aseptic necrosis of the femoral head, delayed union, non-union, and in varus consolidation. Initial radiographs sometimes show a rather large break with fragments which still fit together but callus is rarely observed at the site of the fracture. A fracture at this site must be considered a surgical emergency. Early diagnosis is essential because the consequences of a delay in starting treatment may have disastrous effects on the athlete's career.[105]

An undisplaced fracture of the femoral neck due to distraction requires the patient to be confined to bed until passive movements of the hip no longer cause pain and radiographs show evidence of the formation of endosteal callus.[106] At this point unloaded walking with crutches is allowed, progressing partial weight bearing when the callus has filled the fracture gap. Walking without crutches can be started when the fracture is entirely consolidated. If the patient refuses bed rest or if radiographic monitoring during conservative treatment indicates widening of the fracture site, internal osteosynthesis will be necessary.

Surgical reduction and internal fixation are mandatory for displaced fractures. Following surgery, it may be useful to evaluate the vascularization of the femoral head by means of scintigraphy.

Femoral shaft

Stress fractures of the femoral shaft are less frequent than those in the femoral neck, but have been reported in runners, hurdlers, skiers, and baseball and basketball players.[107,108] These fractures generally occur in the proximal third of the diaphysis (usually in the subtrochanteric area) because the degree of compressed strain to which the medial side of the loaded femoral diaphysis is subject gradually decreases in a proximal–distal direction and increases in the subtrochanteric region because of the powerful action of the vastus medialis and adductor brevis. Moreover, the distractive strains acting on the lateral cortex of the diaphysis are further reduced by the vastus lateralis and the iliotibial tract.[99,109]

The athlete complains of a deep-seated diffuse pain in the groin and/or the thigh or knee, depending on the site of the fracture. The pain increases during or after physical activity, but is not intense enough to cause limping. There is no acute trauma; in runners the history sometimes reveals a recent increase in the intensity of training and/or a change in the surface used. Examination reveals a widespread pain in the thigh, and palpation sometimes produces pain in the injured region; the range of motion of the hip and knee is normal. Tendinitis and muscular injury are the most common initial diagnoses.

According to Provost and Morris[110] stress fractures of the femoral shafts can be classified radiographically into three groups depending on the site of the injury.

Group I: oblique fracture of the medial cortex of the proximal third of the diaphysis with periosteal reaction and sclerosis. The bone scan shows a medial increased uptake focus in the subtrochanteric region. Its development into a displaced fracture is rare. Treatment consists of bed rest until the patient becomes asymptomatic followed by gradual restoration of the load, swimming, cycling and, if the athlete remains asymptomatic, running after 6 to 8 weeks.

Group II: oblique spiral displaced fracture of the middle third of the diaphysis. Because of its rapid development, the interval between the onset of symptoms and the diagnosis is often very short (a few days to 1–2 weeks). In these cases it is advisable to admit patients to hospital and apply skeletal traction. Unloaded walking with crutches is begun, and partial weight bearing is allowed later depending on the outcome of the radiographs used to monitor rehabilitation. Surgery may be necessary.

Group III: supracondylar transverse fracture of the distal third of the epiphysis which may or may not be displaced. If this fracture is not suspected initially, it may not be diagnosed until it has become completely displaced and osteosynthesis will be necessary. If early diagnosis is made the patient must be immobilized and load bearing initially forbidden to avoid a complete fracture.

Conservative treatment (temporary cessation of specific sporting activities, bed rest or crutches when walking is painful,

and limitation of contact and jumping sports for 8–12 weeks with a gradual return to training after 8–14 weeks) is usually sufficient for undisplaced stress fractures of the femoral shaft.

Patella

Although initially described as a complication of prosthesis of the knee, stress fracture of the patella has been observed in healthy subjects as the result of functional overload.[111] It is an avulsion fracture characterized by anterior knee pain which increases when extending the leg. The patella is painful when palpated or struck, and is sometimes inflamed.

Radiography may show a transverse fracture subject to the risk of displacement due to the traction exercised by the quadriceps (therefore if the fracture is not displaced it is necessary to immobilize the patient, whereas if it is displaced it is necessary to reduce and fix it), or a vertical fracture, generally of the lateral facet (this may require excision of the fragment if it is displaced or there is non-union).

Tibia

Plateau

Stress fracture may occur in the medial plateau as this bears most of the body weight in the stance phase and in the absence of axial alterations.[112,113] Onset is painful and gradual, and pain in the anteromedial tibial area just below the medial compartment above the metaphysis is inclined to increase during activity. This pain is worsened by load and alleviated by rest. The area of lesion is painful under finger pressure and is sometimes oedematous. There are no signs of meniscal and/or capsuloligamentous pathology and the knee is not swollen.

Differential diagnosis should consider injuries of the medial collateral ligament at the tibial insertion level with tendinitis, and/or bursitis of the pes anserinus. In the first case the common valgus stress tests are positive; in the second case pain is felt more posteriorly. However, radiographs and bone scintigraphy are necessary.

Radiographs become positive about 3 weeks after the onset of symptoms and show an area of hyperostosis or sclerosis 2–3 mm thick (related to the endosteal callus) under and parallel to the internal tibial plateau. The periosteal callus is rarely visible.

A period of rest from sport is usually sufficient (an average of 1 month); if walking is painful or the patient limps, crutches may be advisable. Complications are rare.

Shaft

About 50 per cent of stress fractures in athletes (runners, dancers, baseball players, and swimmers) are in the tibial shaft.[18] The predisposing factors seem to be a narrow tibia[114] and pronated foot. In the latter an increase of the tibial torsion is produced when the foot is laid down during running. Sometimes the runner's history reveals too much training on steep or sloping ground (the different ways in which the feet are laid down on inclined surfaces require different leg muscle action to maintain balance). Fractures may occur at any point on the shaft but are more frequent between the middle third and distal third of the posteromedial tibial cortex. They are frequently (15–46 per cent) multiple and/or bilateral.[115–116] However, in young runners occurrence is most common in the proximal third.

A typical case is that of the runner who, after having intensified training and/or changed to hard and/or inclined running surfaces, feels a dull aching pain in the anteromedial or posterolateral region of the tibia when he or she stops training. The pain increases with further training on subsequent days. Rest from athletic activity for a few days temporarily relieves the symptoms, but they reappear when training is started again. Despite repeated temporary breaks from athletic activity, onset of pain occurs earlier, and the pain itself becomes more intense, lasts longer, and eventually running becomes difficult or even impossible. Pain continues for longer periods after training is stopped, and is felt even during normal daily activities under load and finally at night, though it does not prevent sleep.

The site of the fracture is particularly painful when palpated or under finger pressure. The region is seldom oedematous and the callus is palpated only in the chronic phase. Sometimes pain in the region of the injury can only be induced by striking the tibia. The contralateral tibia should always be examined to determine whether there is a bilateral fracture. Radiographs (anteroposterior, lateral, and oblique projections augmented by tomograms if necessary) show fracture of the proximal third, the presence of an endosteal callus,[117] and, in fractures at the junction of the middle third with the distal third, a periosteal callus with a thickening of the cortex along the posteromedial tibial margin (Figs. 1 and 6).[34]

Differential diagnosis should consider entrapment of the popliteal artery, shin splints, posteromedial tibial stress syndrome (osseous reaction to stress which is preradiographic but revealed by scintigraphy), and compartment syndrome (symptoms are independent of the application of loads, the possible presence of muscular hernia, a delayed local paraesthesia and/or dysaesthesia, negative radiographs and scintigraphs, but compartment pressure measurements should be made before, during, and after stress).

Treatment requires a temporary suspension of specific sports and all activities in which heavy loads are applied to the lower limbs. If walking is painful, a period of non-weight-bearing is necessary and a functional brace (Aircast type) may be required.[118] If the patient is asymptomatic when at rest and radiographs in-

dicate satisfactory improvement of the injury, a gradual rehabilitation programme is started, involving low frequency activity gradually increasing in intensity provided that the patient remains asymptomatic. Swimming and cycling are useful for this purpose. At this point it can be useful to wear viscoelastic insoles which improve stress absorption. Provided that the patient is asymptomatic, a gradual return to sport is allowed after 4 to 8 weeks (sometimes 3 months in distal tibial fractures).

Particular mention must be made to fractures of the middle third of the anteromedial tibial cortex which occur in activities involving jumping (basketball, figure-skating, volleyball, football, and ballet) but may also be seen in long-distance runners.[119–122] These fractures are caused by repeated traction tension to which this region is physiologically subject because of its own anatomical curve and which is exaggerated by the movements involved in jumping, particularly because the axis is generally more resistant to compression than to traction, and because this region is not protected by surrounding muscle which reduces the tensile stress (tension-type stress fracture).

In lateral and oblique radiographs and tomograms this fracture generally appears as a radiotransparent horizontal wedge or V-shaped defect open in the front with its peak pointing into the anterior cortex of the middle third of the tibial crest ('sawtooth aspect' or 'dreaded back line'), sometimes surrounded by a sclerotic hypertrophic cortex and/or osteoporotic reabsorption areas. Owing to its subcutaneous position it is sometimes possible to feel a painful swelling. Because of the pathomechanics of the fracture and the poor vascular supply to this area, this injury is often complicated, particularly when it is not protected from further stress, and non-union or complete fracture may result.[123,124] Scintigraphy is useful to define the degree of activity of the injury.

When this fracture is diagnosed, it is necessary to immobilize the patient using a non-weight-bearing or partially unloaded splint or plaster for at least 6 months to encourage consolidation. If radiography shows no evidence of recovery after 4 to 6 months, bone grafting should be considered.[122] Since such a procedure may

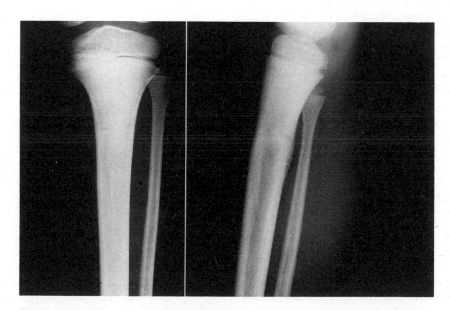

Fig. 6 Anteroposterior and lateral radiographs of the tibia of a 14-year-old soccer player. Stress fractures typically occur at this level (posteromedial aspect of the proximal third of the shaft) at this age and heal after a month of athletic rest.

seem too lengthy, particularly for an athlete, a valid alternative is to consider surgical treatment initially to facilitate an early return to sport.[120]

Medial malleolus

Stress fracture of the medial malleolus has been reported in long-distance runners, footballers, and basketball players. It appears to be due to repeated overloads experienced by the ankle during running when, at the moment when the heel hits the ground (heel strike), the talus rotates internally, bumping against the internal malleolus and thus transmitting the torsion to the tibial diaphysis.[125,126]

In the absence of a specific acute trauma, this condition manifests itself as a medial pain in the ankle with insidious onset and increasing severity. It is worsened by running and jumping, and is temporarily alleviated by rest, but if tension continues athletic activity is restricted and walking may be painful.

Objectively the ankle appears inflamed and the medial malleolus is painful under finger pressure; sometimes there is painful movement with dorsiflexion. Radiographs in two projections augmented by tomograms show an endosteal callus or a vertical fracture at the junction between the medial malleolus and the tibial diaphysis which may propagate supromedially towards the distal metaphysis of the tibia (Fig. 7). If radiographs are negative, bone scans may reveal a subradiographic stage.

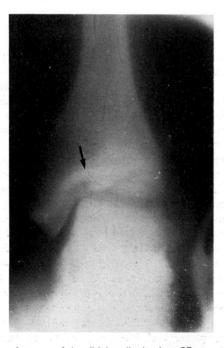

Fig. 7 Stress fracture of the tibial malleolus in a 27-year-old long-distance runner revealed by an anteroposterior tomogram (arrow).

When radiographs show the fracture, surgical osteosynthesis is indicated followed by immobilization in plaster and early start of range of movement exercises. After 2 to 3 weeks a splint is applied and after 6 weeks the patient may start running again. The athlete is usually ready to resume sport after 8 weeks.

If radiographs are negative and scintigraphy is positive, a functional splint is prescribed for 6 to 8 weeks until the patient becomes asymptomatic. Weight bearing is allowed if it is not painful, but running and jumping are forbidden. This is followed by flat ground jogging for gradually increasing distances until the athlete is completely recovered and no longer feels pain.

Fibula

Stress fractures of the fibula are most frequent in the lower part, 4 to 7 cm above the lateral malleolus. They are due to repeated strains caused by eversion of the foot and/or the action of the calf muscles which, pushing the fibula towards the tibia, produce high tension on the distal fibula.[127] These fractures have been observed in runners, gymnasts, ballet dancers, and ice skaters.[128] They are often associated with a pronated foot.

Clinically, there is an insidious onset of pain in the lateral region of the ankle, which may even give rise to discomfort in walking. Locally there is tenderness and swelling. It is necessary to look for possible subluxation of the peroneal tendons. Radiographs may be positive only 2 or 3 weeks after the onset of symptoms (callus of the posterolateral margin).[129] Thus if a fracture is suspected a bone scan would be useful. Rest from athletic activity is sufficient (swimming, cycling, and walking are allowed if they do not cause pain). Running and jumping must be avoided for 3 to 4 weeks, and stretching and strengthening the calf and the peroneal muscles is important to avoid stiffness or instability of the ankle. If walking is painful, it is necessary to resort to plaster or partial weight bearing for 2 to 3 weeks. The athlete may return to his sport in 6 to 8 weeks.

Stress fractures of the proximal third of the fibula are less frequent. They are usually reported in jumpers and parachutists,[130,131] and are probably due to the combined effects of compression loads, traction strains of the biceps muscle, shear forces, and/or the cyclic activity of the flexor hallucis longus, soleus, tibialis posterior, and peroneus longus muscles.[132] Pain is experienced under effort extending to the lateral proximal region of the leg and sometimes even posteriorly. The differential diagnosis considers entrapment of the peroneal nerve, compartment syndrome, and insertional tendinopathy of femoral biceps. Radiographs may show a periosteal callus on the side of the peroneal neck and scintigraphy is positive. Conservative treatment is sufficient and recovery usually occurs in about 6 weeks.

Talus

Stress fracture of the talus is rare and usually occurs in runners, both novice and expert.[133] It is assumed that the injury is the result of a fulcrum effect which occurs at the neck of the talus when, at heel strike, the subtalar joint is strained in valgus with resulting plantar flexion and internal rotation of the talus head. It is often associated with a pronated foot. The athlete suffers from pain in the dorsal aspect of the foot which increases on running. There may be an antalgic limp, pain, and oedema at the site of the injury.

Radiographs show a callus, or a crack in the cortex, at the neck of the talus, parallel to the talonavicular joint, but this is not generally observed until 2 to 3 weeks after the onset of symptoms. This fracture can occur at the posterior apophysis (Fig. 8).

Healing is promoted by immobilization with a weight-bearing cast. Recurrence can be avoided by applying an orthosis inside footwear. In some cases with positive scintigraphy and negative radiography, CT reveals areas of cancellous bone reabsorption which have been described in the knee as 'occult fracture' (Fig. 9).[134]

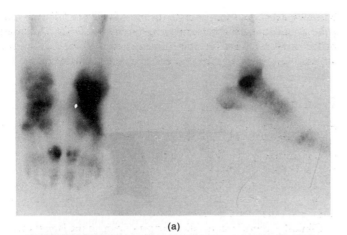

(a)

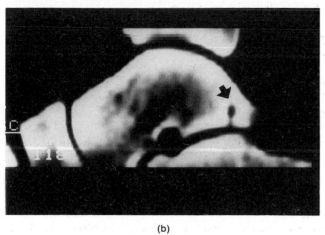

(b)

Fig. 8 Stress fracture of the talus in a 22-year-old triple jumper. Radiographs were negative. The bone scan revealed a hot spot at the anke (a), but only a CT scan in sagittal reconstruction (b) showed the exact location of the injury (arrow).

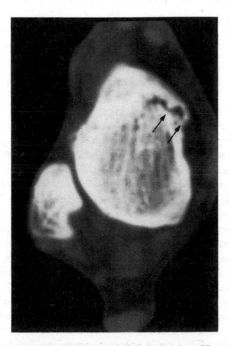

Fig. 9 Intraosseous occult fracture of the talus in a 27-year-old discus thrower. CT shows an area of bone reabsorption with integrity of the cortex.

Calcaneus

Stress fracture of the calcaneus is characteristically reported in military recruits. It is produced by rigid footwear and by the particular style of the parade march, particularly in subjects in poor physical condition.[135,136] Sometimes it is bilateral. It is rarer in athletes and is manifested as heel pain. After an insidious onset, the pain rapidly becomes so severe that it causes limping or compels the patient to use crutches. The pain spreads to the back of the foot and increases under finger pressure on the medial and lateral facets in an area halfway between the malleolus and the posterior tuberosity of the heel. The 'heel squeeze test' is positive.[137,138]

Differential diagnosis must include pathology of the subtalar joint, plantar fasciitis (in this case pain is more severe on the plantar and medial facet of the heel), entrapment of the plantar nerve, retrocalcaneal bursitis (which has a characteristic scintigram),[139] and Achilles tendonitis. The bone scan is positive at an early stage, and radiographs (axial, lateral, and dorsoplantar) taken 7 to 30 days from the onset of symptoms may show an endosteal callus perpendicular to the longitudinal axis of the heel between the posterior facet of the subtalar joint and the tuberosity of the heel.[140]

Symptoms generally regress in 3 to 4 weeks with temporary suspension of sport activity and the use of a heel pad and an absorption insole to protect the site of the fracture from excessive shocks during walking. Provided that the athlete is asymptomatic, he or she is usually able to return to sport within 6 to 8 weeks.

Tarsal navicular

Stress fractures of the tarsal navicular are typically reported in athletes practising activities requiring an explosive force in which sprints and jumps are the main athletic movements (basketball, volleyball, triple jump, high jump, long jump, sprinting, hurdling, middle-distance running, ballet, American football, soccer, and figure-skating), and also in long-distance runners, particularly if they use the forefoot in the footstrike.[138,141–144]

Early recognition of the fracture is important because of the severe complications characterizing its development. However diagnosis is frequently delayed (on average 7.2 months from the onset of symptoms) because of the initially vague symptoms and the frequent lack of fracture signs in routine radiographs.[145] It is a vertical fracture which occurs on the sagittal plane in the middle third of the navicular, at relatively low vascularity, starting on the dorsal surface. Sometimes the foot shows biomechanical alterations that change the normal distribution of loads and/or concentrate them on the navicular (short first metatarsus and/or long second metatarsus, adducted metatarsus, reduced dorsiflexion of the ankle, and/or limited subtalar motion).[145] The shape of the plantar arch does not appear to be a determinant, but an excessive inclination of the foot in the support phase overloads the talonavicular joint.[146] This is also true in the equinuus positions of the foot (figure-skating).[121]

Symptoms occur insidiously without acute trauma. There is vague soreness or cramping in the medial dorsal region of the midfoot or in the area of the internal side of the longitudinal arch which becomes worse on physical activity or weight bearing and is alleviated by rest, but which reappears as soon as sport is taken up again. In the advanced stages training becomes difficult and sometimes even walking is uncomfortable.

Deep palpation of the dorsal aspect of the foot and/or the longitudinal medal arch is painful, but the greatest pain is felt in the area above the navicular; oedema is sometimes present and walking on the toes is uneasy and painful, particularly when resting on the injured foot. Examination of the range of motion often shows reduced dorsiflexion of the ankle and/or lack of mobility in the subtalar joint.

If a fracture of the navicular is suspected, weight-bearing anteroposterior, lateral, and oblique radiographs must be obtained, possibly augmented with a coned-down anteroposterior projection, centred on the bone and, if necessary, with tomograms and magnifications. However, navicular fractures frequently do not show up in routine views because the partial fracture deepens for about 5 mm in the dorsal surface of the bone and, even with projections centred on the navicular, it is not clearly seen because in the complete fracture the lateral fragment is interpreted, in the anteroposterior view, as a normal tarsal (the continuity of the tarsus should be carefully observed).

However, when routine radiographs suggest a fracture of the navicular they show mechanical alterations often associated with this injury: the anteroposterior view, short first metatarsal, abducted metatarsus, hyperosthosis or fracture of the last four metatarsals, sclerosis of the proximal articular navicular surface, narrowing of the medial side of the interarticular talus-navicular space; in the lateral view, plantar slippage of cuneiform and metatarsals with respect to the talus and the navicular, with consequent malalignment of the dorsal surfaces of the talonavicular and cuneiform–navicular joints. Further findings are talar beaks and accessory ossicles.[147] The bone scan of both feet (frontal, medial, and/or lateral plantar projections) shows increased uptake in the navicular at the site of the injury.

Pavlov's tomographic technique[147] is used for radiographic visualization of stress fractures of the navicular. This technique places this bone in its own true anatomical anteroposterior position, with the tomographic plane parallel to the longitudinal axis of the talus and to the dorsal surface of the navicular itself, and reveals even partial sagittal fractures of the middle third of the bone. To do this, the forefoot is kept raised and the whole foot is slightly supinated until, under fluoroscopic control, the navicular is no longer completely visible in the medial-lateral direction (at this point the X-ray beam is tangent to the talonavicular joint). The injury then appears linear and sagittal at the middle third of the bone. If the fracture is partial, it is limited to the first 5 mm of the dorsal surface and the proximal articular facet is involved; extension to the distal articular facet is less frequent and a transverse fracture is the least common injury. A deeper tomographic view allows us to establish whether the fracture is limited to the dorsal surface of the navicular or whether it extends through the bone and also involves the plantar surface (complete fracture). CT may also be useful for diagnostic purposes (Fig. 2).

Because of late diagnosis and the relative avascularity of the central third of the navicular, this fracture may develop into a complete fracture which is sometimes displaced (particularly if the athlete continues to train regardless of pain) or complicated by delayed union, non-union, or refracture.[148]

According to Torg et al.[149] treatment should be differentiated as follows.

1. Non-complicated partial fracture and undisplaced complete fracture: non-weight-bearing plaster for 6 to 8 weeks, allowing subsequent weight bearing and a gradual return to sport only if the patient is continually asymptomatic and there is a radiographic demonstration of union of the fracture. Once the cast is removed, rehabilitation starts with restoration of the motion of the ankle and the application of orthotic devices to footwear for the gradual introduction of weight bearing.
2. Displaced complete fracture: treatment as above or, alternatively, surgical reduction and fixation followed by non-weight-bearing immobilization in plaster for 6 weeks.
3. Fracture complicated by delayed union or non-union: curettage and inlaid bone grafting with internal fixation of unstable fragments (without attempting reduction because in general there is already a fibrous union). Any sclerotic fragments found must not be removed but must be fixed. After the operation, a non-weight-bearing cast must be applied for 6–8 weeks. Recovery is monitored by radiographs (sometimes 3–6 months are necessary).
4. Partial fracture complicated by a small transverse dorsal fracture: the dorsal fragment may have to be removed.
5. Complete fracture complicated by a widespread transverse dorsal fracture: recovery takes place by immobilization.

Dorsal talar beaks must be removed during surgery.

Metatarsals

Stress fractures of the metatarsal bones were first described in military recruits and were caused by marching. They have been reported in long-distance runners and ballet dancers in particular.[138,150] They usually involve the second and third metatarsals. Because of its larger diameter and relatively short length, the first metatarsal is more resistant to bending stress, while the second and third metatarsals are weaker and their heads experience the greatest strain in the propulsive stage of running. Moreover, shearing forces are concentrated at the second metatarsal level.[151] In younger athletes stress fractures are most frequently found in the second metatarsal because of the greater mobility of the first metatarsal associated with excessive inclination of the foot.

The patient feels pain in the dorsal surface of the forefoot which increases with physical activity and is alleviated by rest. Tenderness is felt on palpation of the area of the fractured metatarsal and on passive dorsiflexion of the toe. Swelling is often present and in more advanced cases callus is palpable. Radiographs are positive 10 to 14 days after the onset of symptoms if the first metatarsal is involved but are delayed further for the others.

Fracture of the first metatarsal usually occurs in the proximal metaphysis, and this is seen radiographically as a linear band of sclerosis perpendicular to the direction of application of the load. Fractures of the other metatarsal bones are generally located at the diaphysis or the neck and appear as breaks in the cortex progressively surrounded by a periosteal reaction which reaches its peak with the formation of the callus (Fig 10).[152] Scintigraphy is useful in preradiographic cases and confirms those in which there is some doubt.

Treatment is conservative, involving temporary abstinence from all activities which cause pain. It may be useful to insert an orthosis inside the shoe to reduce strains on the site of the injury (for instance, a 'dome' metatarsal applied just proximally to the head of the metatarsal). Rigid shoes are suggested, and swimming and cycling are allowed. Recovery takes place in 4 to 6 weeks. If pain persists it is necessary to apply a well-moulded weight-bearing cast for 4 to 6 weeks.

Female ballet dancers are predisposed to stress fracture of the proximal part of the base of the second metatarsal, involving the medial and volar surface of the second metatarsal–tarsal joint (Lisfranc's joint), which is the most rigid of the five.[153,154] At this

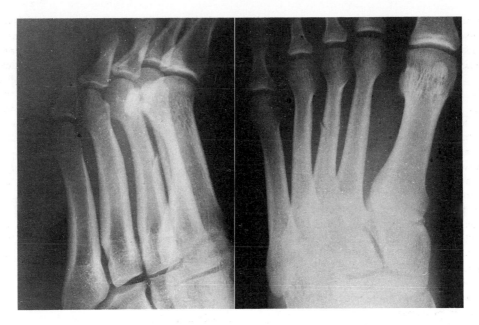

Fig. 10 Stress fracture of the third metatarsal in a long-distance runner. When radiographs show lateral and medial breaks in the cortex, as in this case, it may be necessary to apply a well-moulded weight-bearing cast for 4 to 6 weeks to accelerate the healing process.

level, the three cuneiform bones provide an osseous niche to receive the base of the second metatarsal, which articulates proximally at the bottom of the niche with the second cuneiform, more posteriorly with respect to the other two, and medially and laterally with respect to the first and third cuneiform bones that form the wall of this pocket. Therefore the second metatarsal, anchored to the base, is less mobile; this explains why it tends to fracture just distally, at the base of the second metatarsal bone, when subjected to prolonged and repeated strain.

Factors predisposing to this stress fracture are Greek foot with a short first metatarsal, cavus foot, amenorrhoea, eating disorders (anorexia), and plantar dislocation of the metatarsals due to anterior impingement of the ankle. This fracture must be considered in a female ballet dancer who feels pain on the dorsal aspect of the foot, with an insidious onset which increased on jumping and taking up the *en pointe* position, when the foot is maximally plantar flexed and weight bearing is on the plantar face and on the tip of the distal phalanx of the first two toes.

Tenderness is most obvious on the dorsal aspect of the foot at the first intermetatarsal space around and/or near the proximal part of the second metatarsal. Radiographs (anteroposterior, lateral, and two oblique views, augmented by a dorsoplantar tomogram or magnification) typically show an oblique fracture of the proximal part of the base of the second metatarsal. If the radiographs are negative and clinical suspicion remains, scintigraphy may reveal a hot spot at the site of the injury.

If an early diagnosis is made, immobilization for 6 to 8 weeks in a non-weight-bearing cast is sufficient. If the patient is asymptomatic and radiographs show recovery, he or she must undertake a programme of rehabilitation consisting of a range of motion and flexibility exercises, strengthening of the ankle (initially without weight bearing), swimming, and long-distance walking which is gradually increased with partial weight bearing. When symptoms continue for many months and the fracture develops into non-

union, it is necessary to resort to surgical excision of the necrotic fragment or bone grafting.

Fatigue fractures of the proximal portion of the fifth metatarsal within 1.5 cm distal of the tuberosity where the peroneus brevis muscle is inserted (Jones' fracture) are found in basketball, American football, soccer, and baseball players, and in long-distance runners. The site of the fracture corresponds to the area distally located at the point at which the fifth metatarsal is joined to the fourth metatarsal and the cuboid bone. It is often associated with varus forefoot or varus hindfoot so that when the foot hits the ground, distraction forces which may eventually produce a stress fracture of the more lateral metatarsal are concentrated on its external part along the base of the fifth metatarsal.[149,155]

Clinically, the athlete feels a vague pain over the fifth metatarsal area or lateral forefoot–midfoot area which increases in intensity insidiously or suddenly, sometimes after an acute trauma, and which may inhibit sporting activity. There is tenderness over the proximal third of the fifth metatarsal. In the initial stage of the injury radiographs (anteroposterior, lateral, and oblique views, augmented by tomograms and magnification) may be negative, and thus it is necessary to use scintigraphy which shows a hot spot at the site of the injury. Complications associated with this fracture are delayed union, non-union, and refracture, probably because of sclerotic obliteration of the medullary channel.[156]

Torg *et al.*[156] have outlined the principles of the basic treatment and the clinical and radiographic classification of this injury.

Acute fracture

There is no previous fracture, although the subject may have had prodromic symptoms of pain and/or discomfort, an absence of intramedullar sclerosis, a narrow line of fracture with clear margins, and a minimum cortex hypertrophy or slight periosteal reaction. The injury begins in the lateral cortex and develops into a complete fracture, or it may be a fracture which has taken place in an area already abnormally strained. Treatment requires immobilization in a non-weight-bearing case for 6 to 8 weeks.[157] Sometimes clinical recovery may require an average of 20 weeks. If the fracture is displaced, it must be reduced and fixed with a screw.

Delayed union

There is pre-existing trauma and/or fracture, a wide transcortical line of fracture with areas of bone reabsorption, a periosteal reaction with cortical hypertrophy, and a moderate degree of intramedullar sclerosis. Conservative treatment is efficacious but takes rather a long time (an average of 22 weeks), and is generally not acceptable to highly motivated athletes. Therefore surgical treatment is necessary: curettage of the sclerotic medullary channel to re-establish the permeability of the channel, followed by an inlay tibial bone graft and then immobilization in a non-weight-bearing cast for 6 weeks,[156] or else percutaneous fixation with an intramedullary screw, followed by immobilization with a non-weight-bearing or slipper cast for 2 weeks and then gradual weight bearing with rigid shoes such as post-bunionectomy clogs or tennis shoes with semi-rigid steel soles.[36]

The technique involving osteosynthesis with screw allows return to sport in about 7 to 14 weeks. It is sometimes necessary to wear shoes with an orthotic device protecting the lateral margin of the foot at the level of the base of the fifth metatarsal to reduce pressure on the head of the screw. Complications due to this technique are fracture of the screw or its dislocation outside the medullary channel, too distal positioning, or a screw which is too long.

Non-union

This is characterized by recurrent traumas and symptoms, a large fracture line, a periosteal reaction with cortical hypertrophy, bone reabsorption, and, in particular, complete sclerotic obliteration of the medullary channel. Treatment is exclusively surgical by means of one of the techniques described above. Sport may be taken up again when the patient is asymptomatic and radiographs show a solid union of the fracture and recanalization of the medullary channel without sclerosis.

Great toe

Stress fracture of the proximal phalanx has been described in athletes (fencers, sprinters, and football players) with the hallux valgus deformity.[158] It is probably an avulsion fracture at the level of the insertion area of the medial collateral ligament caused by the numerous stresses on the great toe from the tendons of the adductor muscles. The effect is like 'bowstringing', and is mainly evident during positions (tiptoe) or exercises (sprint on place) that exacerbate pain more than any others. In other cases, stress fracture of the proximal phalanx have been ascribed to the numerous stresses present during dorsal hyperflexion and rotation of the great toe in the propulsive and 'toe-off' phase of running, which are more intense on the medial side of the toe.[159,160]

There may be tenderness and swelling at the level of the dorsomedial aspect of the interphalangeal joint, which manifests as a reduced range of motion. Plain radiographs, which may be negative during the first 2 weeks, typically show an oblique dorsomedial fracture, with more or less sclerotic borders, near the interphalangeal joint or extended to its proximal articular surface. It is sometimes visualized as an intra-articular loose body. When radiographs are negative or ambiguous, a bone scan is useful.

It is sufficient to suspend sports activities temporarily, to tape the great toe, and to use shoes with rigid soles. In the case of osteochondral fractures it may be necessary to remove the sclerotic loose body. When the hallux valgus deformity resists conservative treatment, corrective osteotomy of the first metatarsal is indicated.

Great toe sesamoids

Stress fractures of the great toe sesamoid have been reported in runners (spring and long-distance), dancers, basketball players, tennis players, and figure skaters.[138,153,161,162] They are more frequent in the medial ossiculum, probably because it is located immediately under the head of the first metatarsal. These are traction-type fractures which are due to high repetitive tensile forces at the level of the plantar surface of the first metatarsophalangeal joint, which is often forced into dorsiflexion during body elevation in the propulsive phase of the gait preceding toe-off from the ground. This pathology can occur when the first metatarsal is malaligned.

The onset is insidious, manifesting as medial plantar pain at the level of the first metatarsophalangeal joint which is exacerbated by activity and relieved with rest. The plantar surface of this joint can appear oedematous and painful; there is often a painful limitation of active and passive dorsiflexion which is attributed to synovitis of the first metatarsophalangeal joint.

Differential diagnosis includes sesamoiditis, hallucis tendonitis, metatarsalgia, isolated synovitis or arthrosis of the first metatarsophalangeal joint (in this case radiographs show intra-articular space reduction and bone spurs), sesamoid osteochondritis (more frequent in lateral ossiculum), acute fracture or non-union, and entrapment neuropathies (positive Tinel sign).[163] Intra-articular injection with anaesthetic helps to differentiate the extra-articular pathology.

Radiographs are obtained in anteroposterior, lateral, oblique, and axial views with a dorsiflexed great toe, but it is difficult to distinguish a stress fracture from a bipartite sesamoid. Bone scans are useful because the bipartite sesamoid is 'cold'.

If this fracture is untreated, it can evolve into delayed union or non-union. Athletic rest and foam-padding orthesis are indicated. In the case of severe pain a walking cast may be useful. If symptoms persist, surgical excision of the sesamoid may be indicated, whereas bone grafting is better in non-union.

REFERENCES

1. McBryde AM. Stress fractures in athletes. *American Journal of Sports Medicine* 1975; **3**: 212–17.
2. Devas MB. *Stress fractures.* Edinburgh: Churchill Livingstone, 1975.
3. McBryde AM. Stress fractures in runners. *Clinics in Sports Medicine* 1985; **4**: 737–52.
4. Goergen TG, Venn-Watson EA, Rossman DJ, Resnick D, Gerber K. Tarsal navicular stress fractures in runners. *American Journal of Roentgenology* 1981; **136**: 201–3.
5. Briethaupt MDS. Zur Pathologie des menschlichen Fusses. *Medizinische Zeitung* 1855; **24**: 169–71, 175–7.
6. Stechow AW. Fussödom und Röntgenstrahlen. *Deutsch, Mil. Aerztl. Zeitg.* 1987; **26**: 465.
7. Hulkko A, Orava S. Stress fractures in athletes. *International Journal of Sports Medicine* 1987; **8**: 221–6.
8. Hulkko A., Orava S., Nikula P. Stress fractures of the olecranon in javelin throwers. *International Journal of Sports Medicine* 1986; **7**: 210–13.

9. Burr DB, Milgrom C, Boyd RD, Higgins WL, Robin G, Radin EL. Experimental stress fractures of the tibia. Biological and mechanical aetiology in rabbits. *Journal of Bone and Joint Surgery* 1990; **72B**: 370–5.

10. Johnson LC, Stradford HT, Geis RW, Dineen JR, Kerley E. Histiogenesis of stress fractures. *Journal of Bone and Joint Surgery* 1963; **45A**: 1542.

11. Li G, Zhang S, Chen G, Chen H, Wang A. Radiographic and histologic analyses of stress fracture in rabbit tibias. *American Journal of Sports Medicine* 1985; **13**: 285–94.

12. Jones BH, Harris JM, Vinh TN, Rubin C. Exercise-induced stress fractures and stress reactions of bone: epidemiology, etiology and classification. In: Pandolf KB, ed *Exercise and Sport Sciences Reviews*. *American College of Sports Medicine Series*, vol. 17. Baltimore: Williams & Wilkins, 1989: 379–422.

13. Stanitski CL, McMaster JH, Scranton PE. On the nature of stress fractures. *American Journal of Sports Medicine* 1978; **6**: 391–6.

14. Baker J, Frankel VH, Burstein A. Fatigue fractures: biomechanical considerations. *Journal of Bone and Joint Surgery* 1972; **54A**: 1345–6.

15. Blickenstaff LD, Morris JM. Fatigue fracture of the femoral neck. *Journal of Bone and Joint Surgery* 1966; **48A**: 1031–47.

16. Clement DB. Tibial stress syndrome in athletes. *American Journal of Sports Medicine* 1974; **2**: 81–5.

17. Frankel VH. Editorial comment. *American Journal of Sports Medicine* 1978; **6**: 396.

18. Matheson GO, Clement DB, McKenzie DC, Taunton JE, Lloyd-Smith DR, MacIntyre JG. Stress fractures in athletes. A study of 320 cases. *American Journal of Sports Medicine* 1987; **15**: 46–58.

19. Singer A, Ben-Yehuda O, Ben-Ezra Z, Zaltzman S. Multiple identical stress fractures in monozygotic twins. Case report. *Journal of Bone and Joint Surgery* 1990; **72A**: 444–5.

20. Brudvig TJS, Gudger RD, Obermeyer L. Stress fractures in 195 trainees: a one-year study of incidence as related to age, sex and race. *Military Medicine* 1983; **148**: 666–7.

21. Trutter M., Broman GE, Peterson RR. Densities of white and negro skeletons. *Journal of Bone and Joint Surgery* 1960; **42A**: 50–8.

22. Lysens RJ, Ostyn MS, Auweele YV, Lefevre J, Vuylsteke M, Renson L. The accident-prone and overuse-prone profiles of the young athlete. *American Journal of Sports Medicine* 1989; **17**: 612–19.

23. Taimela S, Kujala UM, Osterman K. Stress injury proneness: a prospective study during a physical training program. *International Journal of Sports Medicine* 1990; **11**: 162–5.

24. Protzman RR. Physiologic performance of women compared to men: observations of cadets at the United States Military Academy. *American Journal of Sports Medicine* 1979; **7**: 191–4.

25. Reinker KA, Ozburne S. A comparison of male and female orthopaedic pathology in basic training. *Military Medicine* 1979; **144**: 532–6.

26. Cook SD, Harding AF, Thomas KA, Morgtan EL, Schnurpfeil KM, Haddad RJ. Trabecular bone density and menstrual function in women runners. *American Journal of Sports Medicine* 1987; **15**: 503–7.

27. Drinkwater BL, Nilson K, Chesnut CH III, Bremner WJ, Shainholtz S, Southworth MB. Bone mineral content of amenorrheic and eumenorrheic athletes. *New England Journal of Medicine* 1984; **311**: 277–81.

28. Highet R. Athletic amenorrhoea. An update on aetiology, complications and management. *Sports Medicine* 1989; **7**: 82–108.

29. Markey KL. Stress fractures. *Clinics in Sports Medicine* 1987; **6**: 405–25.

30. Warren MP, Brooks-Gunn J, Hamilton LH, Warren LF, Hamilton WG. Scoliosis and fractures in young ballet dancers: relation to delayed menarche and secondary amenorrhea. *New England Journal of Medicine* 1986; **316**: 1348–53.

31. Barrow GW, Saha S. Menstrual irregularity and stress fractures in collegiate female distance runners. *American Journal of Sports Medicine* 1988; **16**: 209–16.

32. Evans FG, Riolo ML. Relations between the fatigue life and histology of adult human cortical bone. *Journal of Bone and Joint Surgery* 1970; **52A**: 1579–86.

33. Worthen BM, Yanklowitz BAD. The pathophysiology and treatment of stress fractures in military personnel. *Journal of American Podiatric Association* 1978; **68**: 317–25.

34. Hershman EB, Mailly T. Stress fractures. *Clinics in Sports Medicine* 1990; **9**: 183–241.

35. Belkin SC. Stress fractures in athletes. *Orthopedic Clinics of North America* 1980; **11**: 735–42.

36. De Moya RG. A biomedical comparison of the running shoe and the combat boot. *Military Medicine* 1983; **148**: 666–7.

37. James SL, Bates BR, Osternig LR. Injuries to runners. *American Journal of Sports Medicine* 1978; **6**: 40–50.

38. Wilcox JR, Moniot AL, Green JP. Bone scanning in the evaluation of exercise-related stress injuries. *Radiology* 1977; **123**: 699–703.

39. Keats TD. *Radiology of muscoloskeletal stress injury*. Chicago: Year Book, 1990.

40. Pavlov H., Torg JS. *The running athlete: roentgenographs and remedies*. Chicago: Year Book, 1987.

41. Matin P. The appearance of bone scans following fractures, including immediate and long-term studies. *Journal of Nuclear Medicine* 1979; **20**: 1227–31.

42. Martire JR. The role of nuclear medicine bone scan in evaluating pain in athletic injuries. *Clinics in Sports Medicine* 1987; **6**:13–37.

43. Milgrom C. *et al*. Multiple stress fractures: a longitudinal study of a soldier with 13 lesions. *Clinical Orthopaedics and related research* 1985; **192**: 174–9.

44. Zwas ST, Elkanovitch R, Frank G. Interpretation and classification of bone scintigraphic findings in stress fractures. *Journal of Nuclear Medicine* 1987; **28**: 452–7.

45. Roub LW, Gumermann LW, Hanley EN Jr, Clark MW, Goodman M, Herbert DL. Bone stress: a radionuclide imaging perspective. *Radiology* 1979; **132**: 431–8.

46. Milgrome C *et al*. Negative bone scan in impeding tibial stress fractures. A report of three cases. *American Journal of Sports Medicine* 1984; **12**: 488–91.

47. Lee JK, Yao L. Stress fractures: MR imaging. *Radiology* 1988; **169**: 217–20.

48. Mink JH, Deutsch AL. Occult cartilage and bone injuries of the knee: detection, classification and assessment with MR imaging. *Radiology* 1989; **170**: 823–9.

49. Stafford SA, Rosenthal DI, Gebhardt MC, Brady TJ, Scott JA. MRI in stress fracture. *American Journal of Roentgenology* 1986; **147**: 553–6.

50. Boyer DW Jr. Trapshooter's shoulder: stress fractures of the coracoid process. Case report. *Journal of Bone and Joint Surgery* 1975; **57A**: 562.

51. Sandrock AR. Another sports fatigue fracture: stress fracture of the coracoid process of the scapula. *Radiology* 1975; **117**: 274.

52. Allen ME. Stress fracture of the humerus. A case study. *American Journal of Sports Medicine* 1984; **12**: 244–5.

53. Rettig AC, Beltz HF. Stress fracture in the humerus in an adolescent tennis tournament player. *American Journal of Sports Medicine* 1985; **13**: 55–8.

54. De Haven HE, Evarts CM. Throwing injuries of the elbow in athletes. *Orthopaedic Clinics of North America* 1973; **4**: 801–3.

55. Miller JE. Javelin thrower's elbow. *Journal of Bone and Joint Surgery* 1960; **42B**: 788–92.

56. Torg JS, Moyer RA. Non-union of a stress fracture through the olecranon epiphyseal plate observed in an adolescent baseball pitcher. A case report. *Journal of Bone and Joint Surgery* 1977; **59A**: 264–5.

57. Wilkerson RD, Johns JC. Nonunion of an olecranon stress fracture in an adolescent gymnast. A case report. *American Journal of Sports Medicine* 1990; **18**: 432–4.

58. Hamilton KH. Stress fracture of the diaphysis of the ulna in a body builder. *American Journal of Sports Medicine* 1984; **12**: 405–6.

59. Iwaya T, Takatori I. Lateral longitudinal stress fractures of the patella. Report of three cases. *Journal of pediatric orthopaedics* 1985; **5**: 73–5.

60. Mutoh Y, Mori T, Suzuki Y, Sugiura Y. Stress fractures of the ulna in athletes. *American Journal of Sports Medicine* 1982; **10**: 365–7.

61. Orava S. Stress fractures. *British Journal of Sports Medicine* 1980; **14**: 40–4.

62. Rettig AC. Stress fracture of the ulna in an adolescent tournament tennis player. *American Journal of Sports Medicine* 1983; **11**: 103–6.

63. Hanks GA, Kalenak A, Bowman L, Sebastianelli WJ. Stress fractures of the carpal scaphoid: a report of four cases. *Journal of Bone and Joint Surgery* 1989; **71A**: 938–41.

64. Manzione M, Pizzutillo PD. Stress fracture of the scaphoid waist. A case report. *American Journal of Sports Medicine* 1981; **9**: 268–9.

65. Recht MP, Burk L Jr, Dalinka MK. Radiology of wrist and hand injuries in athletes. *Clinics in Sports Medicine* 1987; **71A**: 938–41.

64. Manzione M, Pizzutillo PD. Stress fracture of the scaphoid waist. A case report. *American Journal of Sports Medicine* 1981; **9**: 268–9.

65. Recht MP, Burk L Jr, Dalinka MK. Radiology of wrist and hand injuries in athletes. *Clinics in Sports Medicine* 1987; **6**: 811–28.

66. Tehranzadeh J, Davenport J, Pais MJ. Schaphoid fracture: evaluation with flexion–extension tomography. *Radiology* 1990; **176**: 167–70.

67. Biondetti PR, Vannier MW, Gilula LA, Knapp P. Wrist coronal and transaxial CT scanning. *Radiology* 1987; **163**: 148–51.

68. Murakami Y. Stress fracture of the metacarpal in an adolescent tennis player. *American Journal of Sports Medicine* 1988; **16**: 419–20.

69. Keating TM. Stress fracture of the sternum in a wrestler. *American Journal of Sports Medicine* 1987; **15**: 92–3.

70. Barrett GR, Shelton WR, Miles JW. First rib fractures in football players. A case report and literature review. *American Journal of Sports Medicine* 1988; **16**: 674–6.

71. Lankenner PA Jr, Micheli LJ. Stress fracture of the first rib. A case report. *Journal of Bone and Joint Surgery* 1985; **67A**: 159–60.

72. Holden DL, Jackson DW. Stress fractures of the ribs in female rowers. *American Journal of Sports Medicine* 1985; **13**: 342–8.

73. McKenzie DC. Stress fracture of the rib in an elite oarsman. *International Journal of Sports Medicine* 1989; **10**: 220–2.

74. Abel MA. Jogger's fractures and other stress fractures on the lumber sacral spine. *Skeletal Radiology* 1985; **13**: 221–7.

75. Ireland ML, Micheli LJ. Bilateral stress fracture of the lumbar pedicles in a ballet dancer. A case report. *Journal of Bone and Joint Surgery* 1987; **69A**: 140–2.

76. Omar MM, Levinsohn EM. An unusual fracture of the vertebral articular process in a skier. *Journal of Trauma* 1979; **19**: 212–13.

77. Lamy C, Bazergui A, Kraus H, Farfan JF. The strength of the neural arch and etiology of spondylolysis. *Orthopedic Clinics of North America* 1975; **6**: 215–31.

78. Wiltse LL, Newman PH, MacNab I. Classification of spondylolysis and spondylolisthesis. *Clinical orthopaedics and Related Research* 1976; **117**: 23–9.

79. McCarroll JR, Miller JM, Ritter MA. Lumbar spondylolysis and spondyloisthesis in college football players. A prospective study. *American Journal of Sports Medicine* 1986; **14**: 404–6.

80. Jackson DW, Wiltse LL, Cirincione RJ. Spondylolysis in the female gymnast. *Clinical Orthopaedics and Related Research* 1976; **117**: 68–73.

81. Roche MB, Rowe GG. The incidence of separate neural arch and coincident bone variations. *Journal of Bone and Joint Surgery* 1952; **34A**: 491–4.

82. Hensinger RN. Spondylolysis and spondylolisthesis in children and adolescent. *Journal of Bone and Joint Surgery* 1989; **71A**: 1098–1107.

83. Rosenberg NJ, Bargar WL, Friedman B. The incidence of spondylolysis and spondylolisthesis in non ambulatory patients. *Spine* 1981; **6**: 35–8.

84. Walsh WM, Huurman WW, Shelton GL. Overuse injuries of the knee and spine in girl gymnastics. *Orthopedic Clinics of North America* 1985; **16**: 329–50.

85. Eyres KS, Salam A. Unilateral traumatic spondylolysis in tennis players. *Clinical Sports Medicine* 1989; **1**: 211–16.

86. Weiker GG. Evaluation and treatment of common spine and trunk problems. *Clinics in Sports Medicine* 1989; **8**: 399–417.

87. Maldague B, Malghem J. Aspect radio-dynamiques de la spondylolyse lombaire. *Acta Orthopaedica Belgica* 1981; **47**: 441–57.

88. Milbauer D, Patel S. Roentgenographic examination of the spine. In: Nicholas JA, Hershman EB, eds. *The Lower Extremity and Spine in Sports Medicine*. St Louis: C.V. Mosby, 1986: 1205–7.

89. Wiltse LL, Winter RB. Terminology and measurement of spondylolisthesis. *Journal of Bone and Joint Surgery* 1983; **65A**: 768–72.

90. Ciullo JV, Jackson DW. Pars interarticularis stress reaction, spondylosysis and spondylolisthesis in gymnasts. *Clinics in Sports Medicine* 1985; **4**: 95–110.

91. Hutson ES, Wastie ML. Bone scintigraphy in the assessment of spondylolysis in patients attending a sports injury clinic. *Clinical Radiology* 1988; **39**: 269–72.

92. Wiltse LL, Widell EH Jr, Jackson DW. Fatigue fracture: the basic lesion in isthmic spondylolisthesis. *Journal of Bone and Joint Surgery* 1975; **57S**: 17–22.

93. Saraste H. Long-term clinical and radiological follow-up of spondylolysis and spondylolisthesis. *Journal of Pediatric Orthopaedics* 1987; **7**: 631–8.

94. Karpakka J, Takala T, Orava S. The long-term consequences of spondylolysis or spondylolisthesis on athletic activities. *Clinical Sports Medicine* 1989; **1**: 89–93.

95. Pavlov H, Nelson TL, Warren RF, Torg JS, Burstein AH. Stress fractures of the pubic ramus. A report of twelve cases. *Journal of Bone and Joint Surgery* 1982; **64A**: 1020–5.

96. Latshaw RF, Kantner TR, Kalenak A, Baum S, Corcoran JJ Jr. A pelvic stress fracture in a female jogger. *American Journal of Sports Medicine* 1981; **9**: 54–6.

97. Selakovich W, Love L. Stress fractures of the pubic ramus. *Journal of Bone and Joint Surgery* 1954; **36A**: 573–6.

98. Noakes TD, Smith JA, Lindenberg G, Willis CE. Pelvic stress fractures in long distance runners. *American Journal of Sports Medicine* 1985; **13**: 120–3.

99. Oh I, Harris WH. Proximal strain distribution in the loaded femur. An *in vitro* comparison of the distributions in the intact femur and after insertion of different hip-replacements femoral components. *Journal of Bone and Joint Surgery* 1978; **60A**: 75–85.

100. Hajek MR, Noble HB. Stress fractures of the femoral neck in joggers: case reports and review of the literature. *American Journal of Sports Medicine* 1982; **10**: 112–16.

101. Lombardo SJ, Benson DW. Stress fractures of the femur in runners. *American Journal of Sports Medicine* 1982; **10**: 219–27.

102. Devas MB. Stress fractures of the femoral neck. *Journal of Bone and Joint Surgery* 1965; **47B**: 728–38.

103. Fullerton LR, Snowdy HA. Femoral neck stress fractures. *American Journal of Sports Medicine* 1988; **16**: 365–77.

104. Kaltsas D. Stress fractures of the femoral neck in young adults. *Journal of Bone and Joint Surgery* 1981; **63B**: 33–7.

105. Johansson C, Ekenman I, Tornkvist H, Eriksson E. Stress fractures of the femoral neck in athletes: the consequence of a delay in diagnosis. *American Journal of Sports Medicine* 1990; **18**: 524–8.

106. Aro H, Dahlstrom HA. Conservative management of distraction-type stress fractures of the femoral neck. *Journal of Bone and Joint Surgery* 1986; **66B**: 65–7.

107. Hershman EB, Lombardo J, Bergfeld JA. Femoral shaft stress fractures in athletes. *Clinics in Sports Medicine* 1990; **9**: 111–19.

108. Sullivan D, Warren RF, Pavlov H, Kelman G. Stress fractures in 51 runners. *Clinical Orthopaedics and Related Research* 1984; **187**: 188–92.

109. Butler JE, Brown SL, McConnell BG. Subtrochanteric stress fractures in runners. *American Journal of Sports Medicine* 1982; **10**: 228–32.

110. Provost RA, Morris JM. Fatigue fractures of the femoral shaft. *Journal of Bone and Joint Surgery* 1969; **51A**: 487–98.

111. Devas MB. Stress fractures of the patella. *Journal of Bone and Joint Surgery* 1960; **42B**: 71–4.

112. Cahil BR. Stress fracture of the proximal tibial epiphysis. A case report. *American Journal of Sports Medicine* 1977; **5**: 186–7.

113. Engber WD. Stress fractures of the medial tibial plateau. *Journal of Bone and Joint Surgery* 1977; **59A**: 767–9.

114. Giladi M *et al.* Stress fractures and tibial bone width. A risk factor. *Journal of Bone and Joint Surgery* 1987; **69B**: 326–9.

115. Blatz DJ. Bilateral femoral and tibial shaft stress fractures in a runner. *American Journal of Sports Medicine* 1981; **9**: 322–5.

116. Donati RB, Echo BS, Powell CE. Bilateral tibial stress fractures in a six-year-old male. A case report. *American Journal of Sports Medicine* 1990; **18**: 323–5.

117. Daffner RH, Martinez S, Gehweiler JA, Harrelson JM. Stress fractures of the proximal tibia in runners. *Radiology* 1982; **142**: 63–5.

118. Dickson TB Jr, Kichline PD. Functional management of stress fractures in female athletes using a pneumatic leg brace. *American Journal of Sports Medicine* 1987; **15**: 86–9.

119. Blank S. Transverse tibial stress fractures. A special problem. *American Journal of Sports Medicine* 1987; **15**: 97–602.

120. Green NE, Rogers RA, Lipscomb AB. Nonunions of stress fractures of the tibia. *American Journal of Sports Medicine* 1985; **13**: 171–6.

121. Pecina M, Bojanic I, Dubravcic S. Stress fractures in figure skaters. *American Journal of Sports Medicine* 1990; **18**: 277–9.

122. Rettig AC, Shelbourne KD, McCarroll JR, Bisesi M, Watts J. The natural history and treatment of delayed union and non-union stress fractures of the anterior cortex of the tibia. *American Journal of Sports Medicine* 1988; **16**: 250–55.

123. Brahms MA, Fumich RM, Ippolita VD. Atypical stress fracture of tibia in a professional athlete. *American Journal of Sports Medicine* 1980; **8**: 131–2.

124. Orava S, Sulkko A. Delayed unions and nonunions of stress fractures in athletes. *American Journal of Sports Medicine* 1988; **16**: 378–82.

125. Rettig AC, Shelbourne KD, Beltz HF, Robertson DW, Arfken P. Radiographic evaluation of foot and ankle injuries in the athlete. *Clinics in Sports Medicine* 1987; **6**: 905–19.

126. Shelbourne KD, Fisher DA, Rettig AC, McCarroll JR. Stress fractures of the medial melleolus. *American Journal of Sports Medicine* 1988; **16**: 60–3.

127. Devas MB, Sweetnam R. Stress fractures of the fibula. A review of 50 cases in athletes. *Journal of Bone and Joint Surgery* 1956; **38B**: 818–29.

128. Ingersoll CF. Ice skater's fracture. *Radiology* 1943; **50**: 469–79.

129. Castillo M, Tehranzadeh J, Morillo G. Atypical healed stress fracture of the fibula masquerading as chronic osteomyelitis. A case report of magnetic resonance distinction. *American Journal of Sports Medicine* 1988; **16**: 185–8.

130. Daffner RH. Stress fractures. Current concepts. *Skeletal Radiology* 1978; **2**: 221–9.

131. Symeonides PP. High stress fractures of the fibula. *Journal of Bone and Joint Surgery* 1980; **62B**: 192–3.

132. Blair WF, Hanley SR. Stress fracture of the proximal fibula. *American Journal of Sports Medicine* 1980; **8**: 212–13.

133. Hontas MJ, Haddad RJ, Schlesinger LC. Conditions of the talus in the runner. *American Journal of Sports Medicine* 1986; **14**: 586–90.

134. Apple JS, Martinez S, Allen NB, Caldwell DS, Rice JR. Occult fractures of the knee: tomographic evaluation. *Radiology* 1983; **148**: 383–7.

135. Hopson CN, Perry DR. Stress fractures of the calcaneus in women Marine recruits. *Clinical Orthopaedics and Related Research* 1977; **28**: 159–62.

136. Leabhart JW. Stress fractures of the calcaneus. *Journal of Bone and Joint Surgery* 1959; **41A**: 1284–90.

137. Dalby RD. Stress fractures of the os calcis. *Journal of American Medical Association* 1967; **200**: 131–2.

138. Davis AW, Alexander IJ. Problematic fractures and dislocations in the foot and ankle of athletes. *Clinics in Sports Medicine* 1990; **9**: 163–81.

139. Rupani MD, Molder LE, Espinola DA. Three phases of radionuclide bone imaging in sports medicine. *Radiology* 1985; **156**: 187–96.

140. Pilgaard S. Stress fracture of the os calcis. *Acta Orthopaedica Scandinavica* 1968; **39**: 270–2.

141. Campbell G, Warnekros W. Tarsal stress fracture in a long-distance runner. A case report. *Journal of the American Podiatric Association* 1983; **72**: 532–5.

142. Fitch KD, Blackwell JB, Gilmour WN. Operation for non-union of stress fracture of the tarsal navicular. *Journal of Bone and Joint Surgery* 1989; **71B**: 105–10.

143. Hulkko A., Orava S., Petokallio P., Tultaoura I, Walden M. Stress fracture of the navicular bone: nine cases in athletes. *Acta Orthopaedica Scandinavica* 1988; **56**: 303–5.

144. Towne LC, Blazina ME, Cozen LN. Fatigue fracture of the tarsal navicular. *Journal of Bone and Joint Surgery* 1970; **52A**: 376–8.

145. Torg JS *et al.* Stress fractures of the tarsal navicular. A retrospective review of twenty-one cases. *Journal of Bone and Joint Surgery* 1982; **64A**: 700–12.

146. Ting A *et al.* Stress fractures of the tarsal navicular in long-distance runners. *Clinics in Sports Medicine* 1988; **7**: 89–101.

147. Pavlov H, Torg JS, Freiberger RH. Tarsal navicular stress fractures: radiographic evaluation. *Radiology* 1983; **148**: 641–5.

148. Coughlin L., Kwok D., Oliver J. Fracture dislocation of the tarsal navicular. A case report. *American Journal of Sports Medicine* 1987; **15**: 614–15.

149. Torg JS, Pavlov H, Torg E. Overuse injuries in sport: the foot. *Clinics in Sports Medicine* 1987; **6**: 291–320.

150. Drez D Jr, Young JC, Johnston RD, Parker WD. Metatarsal stress fractures. *American Journal of Sports Medicine* 1980; **8**: 123–5.

151. Gross TS, Bunch RP. A mechanical model of metatarsal stress fracture during distance running. *American Journal of Sports Medicine* 1989; **17**: 669–74.

152. Levy JM. Stress fractures of the first metatarsal. *American Journal of Roentgenology* 1978; **139**: 679–81.

153. Hamilton WG. Foot and ankle injuries in dancers. *Clinics in Sports Medicine* 1988; **7**: 143–73.

154. Micheli LJ, Sohn RS, Soloman R. Stress fractures of the second metatarsal involving Lisfranc's joint in ballet dancers: a new overuse of the foot. *Journal of Bone and Joint Surgery* 1985; **67A**: 1372–5.

155. Santopietro FJ. Foot and foot-related injuries in the young athlete. *Clinics in Sports Medicine* 1988; **7**: 563–89.

156. Torg JS, Balduini FC, Zelko RR, Pavlov H, Peff TC, Das M. Fractures of the base of fifth metatarsal distal to the tuberosity. Classification and guidelines for non-surgical and surgical management. *Journal of Bone and Joint Surgery* 1984; **66A**: 209–14.

157. Zogby RG, Baker BE. A review of non operative treatment of Jones' fracture. *American Journal of Sports Medicine* 1987; **15**: 304–7.

158. Yokoe K, Taketomo M. Stress fracture of the proximal phalanx of the great toe. A report of three cases. *American Journal of Sports Medicine* 1986; **14**: 240–2.

159. Jones P. Fatigue failure osteochondral fracture of the proximal phalanx of the great toe. *American Journal of Sports Medicine* 1987; **156**: 616–18.

160. Orava S, Weitz H, Hulkko A, Karpakka J, Takata T. Intra-articular stress fracture of the proximal phalanx of the great toe in athletes. *Clinical Sports Medicine* 1989; **1**: 105–7.

161. McBryde AM, Anderson RB. Sesamoid foot problems in the athlete. *Clinics in Sports Medicine* 1988; **7**: 51–60.

162. Van Hall ME, Keene JS, Lange TA, Clancy WG Jr. Stress fractures of the great toe sesamoids. *American Journal of Sports Medicine* 1982; **10**: 122–8.

163. Chillag K, Grana WA. Medial sesamoid stress fracture. *Orthopaedics* 1985; **8**: 819–21.

164. Alexander MJL. Biomechanical aspects of lumbar spine in athletes. *Canadian Journal of Applied Sport Sciences*, 1985; **10**: 1–10.

5.3   Compartment syndromes

NANCY E. VINCENT

DEFINITION

Each muscle, or group of muscles, is enclosed in a 'compartment' by borders which usually consist of one or more bones and a tough unyielding muscular fascia. Thus the compartment is relatively rigidly defined, leaving little room for expansion. If the muscles within the compartment enlarge, either by the normal swelling associated with exercise or if there is bleeding or fluid leakage into the compartment, as occurs from damaged tissues, the borders may not be capable of sufficient expansion to accommodate these changes. If this occurs, the pressure within the compartment rises. If the pressure rises above the level of the end arterial inflow pressure to the compartment, blood flow into the compartment ceases. Lack of oxygenated blood, and lack of removal of waste products of cell metabolism results in pain and/or decreased peripheral sensation in the area involved. The resulting symptom complex is what has become recognized as 'compartment syndrome'.[1–14]

ANATOMY

All muscles in the extremities are arranged in collaborative groups. Each group comprises a compartment bound by relatively rigid walls of bone and/or fascia. The actual contents of the various compartments may vary to some extent; however there is always at least one muscle–tendon unit and its neurovascular supply (Fig. 1).

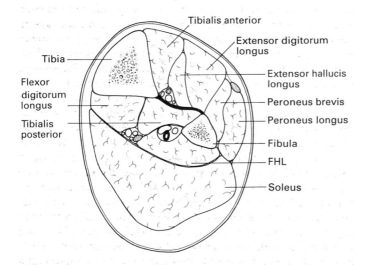

Fig. 1   Cross-section of lower leg.

For some reason certain muscle compartments of the body appear to be more likely to develop compartment syndromes. For example, the compartments of the lower leg and the volar forearm seem particularly prone to this injury pattern. Examination of these two areas and assessment of their anatomical similarities reveals that each compartment has one virtually rigid boundary made up of two bones joined by an inflexible interosseous membrane. In addition, tough retinaculum or fascia is present where the structures of the compartment pass over a joint. The compartments of the thigh, buttock, and upper arm are less restricted. Their investing fascia tends to be thinner and more flexible, and often there are no bony boundaries.

PHYSIOLOGY

The basic principle underlying the development of compartment syndromes is easily explained by the concept of critical closing pressures. Simply, the pressure surrounding the capillary network within the muscle must be low enough to enable the capillary to fill and empty. As the compartment pressure rises it eventually reaches a point where the venous pressure is exceeded and the outflow tract is blocked. Initially, the body is capable of compensating by raising the mean arterial pressure to keep a constant blood flow to the muscles. If the pressure within the compartment continues to rise, a point is reached where no further compromise is possible. With closure of the outflow tract the capillary tends to leak, causing release of fluid and cells into the interstitial space which contribute to the rising intracompartmental pressure. As the removal service for the cellular wastes of metabolism is lost the cells tend to swell, contributing to the increase in compartment pressure. Up to a certain point the inflow of fresh blood continues, allowing continued metabolism. Eventually the end arterial pressure is exceeded, preventing any further inflow of fresh blood carrying essential oxygen. Cellular death occurs, resulting in cell membrane breakdown and thus leakage of intracellular contents, further contributing to the rise in interstitial pressure.

The whole process is reversible if it can be corrected prior to cell necrosis. This appears to occur 6 to 8 h after the onset of oxygen debt for muscle tissue, and perhaps after a longer period for neurovascular structures.[1,2,8] The peripheral neurological structures have some capacity to regenerate: however, stimulation of the fibrous scar replacing the necrotic muscle does not result in muscle twitch.

CLASSIFICATION (AETIOLOGY) OF COMPARTMENT SYNDROMES

Compartment syndromes can be classified into traumatic and exercise-induced. The traumatic group can be further subdivided into acute, subacute, and chronic (Volkmann's ischaemic contracture).

The physiology and anatomy are identical for the various forms of compartment syndrome; only the inciting force, the rate of onset, and the severity of symptoms will vary. The most commonly recognized variety of compartment syndrome is that associated with trauma, either a soft tissue crush injury alone or a fracture which is always associated with some soft tissue damage. Usually the signs and symptoms of an evolving compartment syndrome start to appear in the first 24 to 48 h after injury; thus the history and the physical evidence of trauma are obvious to the

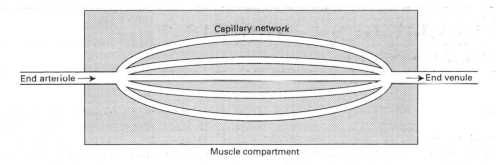

Fig. 2 Diagram of a muscle compartment.

health-care worker who is assessing the patient, and the possible diagnosis of compartment syndrome should always be close to the top of the differential diagnosis list. Unfortunately, the possibility of development of a compartment syndrome after a pure soft tissue injury does not seem to be a well-established item in the knowledge base of all primary care personnel.

There appears to be a subacute version of the traumatic compartment syndrome in which the trauma is relatively minor and the signs and symptoms start to appear after 48 h, and are less dramatic in severity. This scenario has been observed by the author on three separate occasions, each associated with a minor fracture. Two patients had minimally displaced fractures of the radial head, and one had a nightstick fracture of the ulna (transverse fracture of the shaft of the ulna, without involvement of the radius). Whether or not the problem was initiated by a cast applied too tightly, each patient had documented raised intracompartmental pressures. Decompressions of the volar forearm muscle groups were carried out, and in each instance evidence of muscle necrosis was visible. Because of the delayed onset and the less dramatic symptoms and signs, this form of compartment syndrome is more likely to be overlooked. If release of the plaster does not eliminate the patient's pain, investigation is warranted. Close follow-up is mandatory.

The chronic form of trauma-induced compartment syndrome is classic Volkmann's ischaemic contracture. This stage is irreversible, but there may be ways of improving the patient's function.

Like so many other sports related illnesses exercise-induced compartment syndrome is becoming better recognized because of the rise in interest in all aspects of sport and sports medicine within the last 10 years.

SIGNS AND SYMPTOMS

The signs and symptoms are similar whether the patient presents to the emergency ward with a trauma-related injury or the athlete gives a history of inability to perform sustained intense exercise. First and foremost, the most reliable symptom is pain, which is persistent and intense and is often described as deep and aching in nature. If the syndrome is exercise induced, the pain can usually be controlled by decreasing work-out intensity or ceasing all activity. The trauma victim will often experience uncontrollable unremitting pain which will not be responsive to normal doses of narcotics. The level of pain can be exacerbated by passively stretching the muscles of the compartment involved. Paraesthesia, or numbness, is an unreliable symptom. If objective evidence of sensory deficit is present, the condition is quite far advanced. Thus, intervention should be considered before this stage is reached. Similarly, if

there are objective findings of paralysis, muscle necrosis has probably already occurred. The only early reliable symptom is pain, and the only early reliable sign is increased pain on passive stretching of compartment muscles.

The compartment will feel tense on palpation. Skin, colour vascular return, pulses, and even Doppler verification of a pulse are all unreliable findings. The major arteries may be filling by collateral circulation, bypassing the affected compartment completely.

DIFFERENTIAL DIAGNOSIS OF COMPARTMENT SYNDROMES

Trauma-induced compartment syndrome is rarely confused with other conditions. A cast applied too tightly may seem to be the cause of pain, and certainly that in itself is an iatrogenic form of compartment syndrome. However, it is easily eliminated by cutting the cast down to skin to allow for swelling. Sometimes a compartment syndrome will masquerade as a deep vein thrombosis, and if the deep posterior compartment is involved in the suspected compartment syndrome, it may be very difficult to identify without a venogram. Both the entities may be present, one causing the other. A hysterical patient, or one who cannot give a proper history and is difficult to examine, such as an unconscious patient, a child, or a mentally deficient individual, will be a challenge to diagnose. Any question of doubt should be investigated.

The differential diagnosis of exercise-induced compartment syndrome most commonly includes shin splints, stress fractures, deep vein thrombosis, and occasionally tendinitis. Because no actual injury is involved, the athlete with exercise-induced compartment syndrome may not be taken seriously. For example, an elite downhill skier who performs well on shorter courses, but who does not seem to be able to hold the tuck on the longer courses, may be misdiagnosed as not conditioned well enough, when in fact he or she must stand up on the skis because the quadriceps are in oxygen debt. Most physicians are comfortable enough when faced with a patient with overt disease, but when assessing an obviously fit individual who has trouble finishing work-outs or finishing a race or maintaining race speed may react inappropriately by dismissing the individual as unfit, lazy, or psychologically unprepared for competition.

Usually the patient will have been sitting in the waiting room in a relaxed manner and thus will appear completely normal when assessed. No physical signs will be present. Given an appropriate history consistent with exercise-induced compartment syndrome, the patient should be exercised to the point of pain in the office or clinic. Upon re-examination, the compartment will appear tense

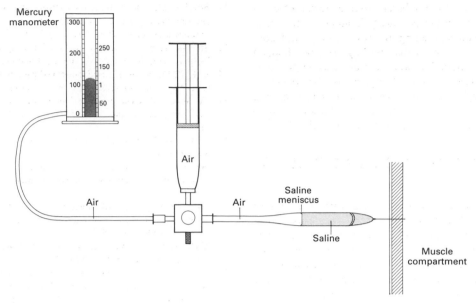

Mercury manometer

300
250
200
150
100
50
0

Air

Air

Air

Saline meniscus

Saline

Muscle compartment

Fig. 3 The Whiteside method for measuring compartment pressures.

and there will be increased pain on passive stretch. The symptoms and signs will disappear with rest. This simple stress test will differentiate exercise-induced compartment syndrome from most other entities. The overuse injuries most likely to be confused with exercise-induced compartment syndrome are shin splints, periostitis, and inflammation of the enthesis (attachment of muscle to bone).

Shin splints will remain painful and tender long after activity ceases. The tenderness is localized to the bony attachment of the muscle(s) involved, and rarely does enough swelling occur to present as tenseness of the compartment. A bone scan will reveal increased uptake along the anterior ridge of the tibia.

Stress fractures demonstrate well-localized pain on a specific area of the affected bone. There will be very little effect with passive stretching of the muscles, and while the associated pain may increase with exercise, it does not disappear with rest. Again, a bone scan will show a classic pattern of uptake.

Deep vein thrombosis is rarely confused with exercised-induced compartment syndrome, as an active individual rarely spontaneously develops a deep vein thrombosis. The pain associated with a deep vein thrombosis is constant and not specifically associated with exercise. A deep vein thrombosis will cause swelling and tightness of the compartments, and passive stretching of the posterior compartment will cause an increase in the associated pain. A deep vein thrombosis almost always occurs in the posterior compartment of the lower leg, or in the thigh, while a compartment syndrome isolated to the posterior compartment of the leg is an unusual entity.

It would be unusual to confuse tendinitis with a compartment syndrome. Usually the pain and tenderness of tendinitis is fairly well localized and is associated with exercise, but does not disappear so completely with rest. Tendinitis is easy to localize by actively testing the strength of each tendon in the area, i.e. by resisted active contraction.

Bone tumours and infections which may affect the limbs are uncommon. While the pain may be similar to that of a compartment syndrome in that it is felt to be deep and intense, it is not so consistently affected by exercise. Most of these conditions will be detected by plain radiography.

INVESTIGATION OF COMPARTMENT SYNDROMES

The investigation of trauma-related compartment syndromes is relatively simple. A history of trauma to the area and the classic signs and symptoms associated with an evolving compartment syndrome is sufficient evidence and the patient needs surgical release. If there is any doubt about the diagnosis, compartment pressures can be measured. The standard level accepted as the upper limit of normal for compartment pressures in the lower leg is less than 30 mmHg.[1,2,5–7,9,14–16] Accepted levels for other areas have not yet been agreed upon. Because of the physiological basis underlying the development of a compartment syndrome, it would seem reasonable to take the patient's mean blood pressure into account. According to a study by Heppenstall et al.,[9] it is the difference between mean arterial blood pressure and the measured compartment pressure which is important: acceptable levels are 30 mmHg in normal muscle and 40 mmHg in traumatized muscle. The patient's signs and symptoms must take precedence.

When the signs and symptoms are not straightforward it is recommended that comparison pressure studies be performed on the unaffected limb. In the case of a patient with suspected exercise-induced compartment syndrome pressure studies should be performed pre-exercise, during exercise, immediately after exercise, and 5 and 10 min after exercise. Typically, the patient will show a resting pressure higher than normal (15 mmHg), an elevated exercise pressure (30 mmHg), and a prolonged elevation in pressure after exercise (at least 20 mmHg after 5 min).[17]

There are various methods for objectively documenting compartment pressures, but all involve the insertion into the compartment of a needle or catheter connected to fluid-filled tubing which is then connected to a manometer or a transducer attached to a recorder (Fig. 3).

Portable hand-held compartment pressure recorders which allow the use of disposable syringes and needles and give a very easily reproducible digital read-out are commercially available.

The investigation of exercise-induced compartment syndrome is rather more complicated because the symptoms are intermittent and associated with exercise. Thus the pressure readings must be obtained during exercise or immediately after cessation of exercise. In addition, the condition is usually bilateral so that it is impossible to obtain a control reading. In these cases an implanted pressure sensor must be used so that readings can be obtained throughout the exercise period as well as afterwards or the commercially one-shot devices can be used to give repetitive readings. As previously discussed, baseline readings are often 10 to 15 mmHg higher then normal, but the key finding is that the level fails to return to baseline within 5 min of finishing the exercise.[10,13,17-24]

MANAGEMENT OF COMPARTMENT SYNDROMES

The patient with traumatic compartment syndrome is at risk of permanent damage to all compartment structures because, compared with the patient with exercise-induced compartment syndrome, he or she cannot control the compartment pressure by simply remaining inactive. Thus surgical release by fasciotomy is indicated. In most instances all compartments in the area should be released. Details of the surgical approach for each region are reported elsewhere.[1,2,15]

The patient with exercise-induced compartment syndrome can attempt a stretching and retraining programme, a change in exercise shoes, or a change in the type of sport practised. If these are unacceptable alternatives or if they are tried and fail, the other option is surgical fasciotomy. Release of only the compartment involved will probably be sufficient, and this can often be performed subcutaneously through a small incision.

SUMMARY

Compartment syndrome is a symptom complex initiated by raised pressure within a muscle compartment resulting in dysfunction of all tissues within the compartment. It is easily recognized and also easily treated. The worst case scenario can mean the difference between a normally functioning limb and an amputation. To the athlete plagued with exercise-induced compartment syndrome, it can make a world of difference in performance.

REFERENCES

1. Crenshaw AH, ed. *Campbell's operative orthopedics*, 7th edn. Toronto: CV Mosby Company, 1987.
2. Evarts CM, ed. *Surgery of the musculoskeletal system*, 2nd edn. New York: Churchill Livingstone, 1990.
3. Anderson JE, ed. *Grant's atlas of anatomy*, 7th edn. Baltimore: Williams & Wilkins, 1978.
4. Almdahl SM, *et al.* Compartment syndrome with muscle necrosis following repair of hernia of tibialis anterior. *Acta Chirurgica Scandinavica* 1987; **153**: 695.
5. Blick SS, *et al.* Compartment syndrome in open tibial fractures. *Journal of Bone and Joint Surgery* 1986; **68A**(9).
6. Bourne RB, Roabeck CH. Compartment syndromes of the lower leg. *Clinical Orthopedics* 1989; **240**: 97–104.
7. Gershuni DH, *et al.* Fracture of the tibia complicated by acute compartment syndrome. *Clinical Orthopedics* 1987; **217**: 221.
8. Heppenstall RB, *et al.* A comparative study of the tolerance of skeletal muscle to ischemia. *Journal of Bone and Joint Surgery* 1986; **68A**(6): 820–8.
9. Heppenstall RB, *et al.* The compartment syndrome. *Clinical Orthopedics* 1988; **226**: 138.
10. Pedowitz RA, *et al.* Chronic exertional compartment syndrome of the forearm flexor muscles. *Journal of Hand Surgery* 1988: **13A**(5): 694–6.
11. Straehley D, Jones WW. Acute compartment syndrome (anterior, lateral, and superficial posterior) following tear of the medial head of the gastrocnemius muscle. *American Journal of Sports Medicine* 1986; **14**(1): 96–9.
12. Styf JR, Korner LM. Chronic anterior-compartment syndrome of the leg. *Journal of Bone and Joint Surgery* 1986; **68A**(9): 1338–47.
13. Styf JR. Diagnosis of exercise-induced pain in the anterior aspect of the lower leg. *American Journal of Sports Medicine* 1988; **16**(2): 165–9.
14. Tischenko GJ, Goodman SB. Compartment syndrome after intramedullary nailing of the tibia. *Journal of Bone and Joint Surgery* 1990; **72A**(1): 41–4.
15. Rockwood CA, Green DP, eds. *Fractures.* 2nd edn. New York: JB Lippincott, 1984.
16. Shakespeare DT, *et al.* The slit catheter: a comparison with the wick catheter in the measurement of compartment pressure. *Injury* 1987; **13**: 404–8.
17. Pedowitz RA, *et al.* Modified criteria for the objective diagnosis of chronic compartment syndrome of the leg. *American Journal of Sports Medicine* 1990; **18**(1): 35–40.
18. Allen MJ, Barnes MR. Exercise pain in the lower leg. *Journal of Bone and Joint Surgery* 1986; **68B**(5): 818–23.
19. Awbrey BJ, *et al.* Chronic exercise-induced compartment pressure elevation measured with a miniaturized fluid pressure monitor. *American Journal of Sports Medicine* 1988; **16**(6): 610–15.
20. Fronek, *et al.* Management of chronic exertional anterior compartment syndrome of the lower extremity. *Clinical Orthopedics* 1987; **220**: 217.
21. Gertsch P, *et al.* New Cross-country skiing technique and compartment syndrome. *American Journal of Sports Medicine* 1987; **15**(6): 612–13.
22. McDermott AGP, *et al.* Monitoring dynamic anterior compartment pressures during exercise. *American Journal of Sports Medicine* 1982; **10**(?) 83–9.
23. Rorabeck CH, *et al.* The role of tissue pressure measurement in diagnosing chronic anterior compartment syndrome. *American Journal of Sports Medicine* 1988; **16**(2): 143–6.
24. Styf JR, *et al.* Intramuscular pressure and muscle blood flow during exercise in chronic compartment syndrome. *Journal of Bone and Joint Surgery* 1987; **69B**(2).

BIBLIOGRAPHY

Abdul-Hamid AK. First dorsal interosseous compartment syndrome. *Journal of Hand Surgery* 1987; **12B**(2): 269–72.

Amendola A. *et al.* The use of magnetic resonance imaging in exertional compartment syndromes. *American Journal of Sports Medicine* 1990; **18**(1): 29–34.

Bell S. Repeat compartment decompression with partial fasciectomy. *Journal of Bone and Joint Surgery* 1986; **68B**(5): 815–17.

Bonutti PM, Bell GR. Compartment syndrome of the foot. *Journal of Bone and Joint Surgery* 1986; **68A**(9): 1449–50.

Brumback RJ. Compartment syndrome complicating avulsion of the origin of the triceps muscle. *Journal of Bone and Joint Surgery* 1987; **69A**(9): 1445–6.

Brumback RJ. Traumatic rupture of the superior gluteal artery without fracture of the pelvis causing compartment syndrome of the buttock. *Journal of Bone and Joint Surgery* 1990; **72A**(1): 134–7.

Due J, Nordstrand K. A simple technique for subcutaneous fasciotomy. *Acta Chirurgica Scandinavica* 1987; **153**: 521–2.

Gershuni DH, *et al.* Ultrasound evaluation of the anterior musculofascial compartment of the leg following exercise. *Clinical Orthopedics* 1982; **167**: 187.

Gibson *et al.* Weakness of foot dorsiflexion and changes in compartment pressures after tibial osteotomy. *Journal of Bone and Joint Surgery* 1986; **68B** (3): 471–5.

Goldie BS, *et al.* Recurrent compartment syndrome and Volkmann contracture associated with chronic osteomyelitis of the ulna. *Journal of Bone and Joint Surgery* 1990; **72A**(1): 131–3.

Hastings H, Misamore G. Compartment syndrome resulting from intravenous regional anaesthesia. *Journal of Hand Surgery* 1987; **12A**(4): 559–62.

Hieb LD, Alexander AH. Bilateral anterior and lateral compartment syndromes in a patient with sickle cell trait. *Clinical Orthopedics* 1988; **226**: 190.

Jenkins NH, Mintowt-Czyz WJ. Compression of the biceps-brachialis compartment after trivial trauma. *Journal of Bone and Joint Surgery* 1986; **68B**(3): 374.

Johnell O, *et al.* Morphological bone changes in shin splints. *Clinical Orthopedics* 1982; **167**: 180.

Jones WG, *et al.* Changes in tibial venous blood flow in the evolving compartment syndrome. *Archives of Surgery* 1989; **124**: 801–4.

Knezevich S, Torch M. Streptococcal toxic shock-like syndrome leading to bilateral lower extremity compartment syndrome and renal failure. *Clinical Orthopedics* 1990; **254**: 247.

Kym MR, *et al.* Compartment syndrome in the foot after an inversion injury to the ankle. *Journal of Bone and Joint Surgery* 1990; **72A**(1): 138–9.

Luk KDK, Pun WK. Unrecognized compartment syndrome in a patient with tourniquet palsy. *Journal of Bone and Joint Surgery* 1987; **69B**(1): 97–9.

Lutz LJ, *et al.* Chronic compartment syndrome in a nonathletic elderly man. *Journal of Family Practice* 1988; **27**(4): 417.

McLaren AC, *et al* Crush syndrome associated with use of the fracture-table. *Journal of Bone and Joint Surgery* 1987; **69A**(9): 1447–9.

Miniaci A, Rorabeck CH. Compartment syndrome as a complication of repair of a hernia of the tibialis anterior. *Journal of Bone and Joint Surgery* 1986; **68A**(9): 1444–73.

Mubarak SJ, *et al.* The medial tibial stress syndrome. *American Journal of Sports Medicine* 1982; **10**(4): 201–5.

Naito M, Ogata K. Acute volar compartment syndrome during skeletal traction in distal radius fracture. *Clinical Orthopedics* 1989; **241**: 234.

Nixon RG, Brindly GW. Hemophilia presenting as compartment syndrome in the arm following venipuncture. *Clinical Orthopedics* 1989; **244**: 176.

Ottolenghi CE. Vascular complications in injuries about the knee joint. *Clinical Orthopedics* 1982; **165**: 148.

Peck D, *et al.* Are there compartment syndromes in some patients with idiopathic back pain? *Spine* 1986; **11**(5): 468–75.

Petrik ME, *et al.* Posttraumatic gluteal compartment syndrome. *Clinical Orthopedics* 1988; **231**: 127.

Raether PM. Lutter LD. Recurrent compartment syndrome in the posterior thigh. *American Journal of Sports Medicine* 1982; **10**(1): 40–3.

Rimoldi RL, *et al.* Pseudoaneurysm of the radial artery as a cause of a late compartment syndrome. *Clinical Orthopedics* 1990; **251**: 263.

Rorabeck CH. Exertional tibialis posterior compartment syndrome. *Clinical Orthopedics* 1986; **208**: 61.

Rorabeck CH, Fowler PJ. The results of fasciotomy in the management of chronic exertional compartment syndrome. *American Journal of Sports Medicine* 1988; **16**(3): 224–7.

Shakespeare DT, Henderson NJ. Compartmental pressure changes during calcaneal traction in tibial fractures. *Journal of Bone and Joint Surgery* 1982; **64B**(4): 498–9.

Shall J, *et al.* Acute compartment syndrome of the forearm in association with fracture of the distal end of the radius. *Journal of Bone and Joint Surgery* 1986; **68A**(9): 1451–3.

Simms M, *et al.* Compartment syndrome associated with staphylococcus endotoxin. *Orthopedic Review* 1986; **15**(6): 383–6.

Skyhar MJ, *et al.* Hyperbaric oxygen reduces edema and necrosis of skeletal muscle in compartment syndromes associated with hemmorrhagic hypotension. *Journal of Bone and Joint Surgery* 1986; **68A**(8): 1218–24.

Stack C. Superficial posterior compartment syndrome of the leg with deep venous compromise, *Clinical Orthopedics* 1987; **220**: 233.

Stockley I, *et al.* Acute volar compartment syndrome of the forearm secondary to fractures of the distal radius. *Injury* 1986; **19**: 101.

Strecker WB, *et al.* Compartment syndrome masked by epidural anaesthesia for postoperative pain. *Journal of Bone and Joint Surgery* 1986; **68A**(9): 1447–8.

Styf JR, Korner LM. Microcapillary infusion technique for measurement of intramuscular pressure during exercise. *Clinical Orthopedics* 1986; **207**: 253.

Styf JR, Lysell E. Chronic compartment syndrome in the erector spinae muscle. *Spine* 1987; **12**(7): 680–2.

Styf JR, *et al.* Chronic compartment syndrome in the first dorsal interosseous muscle. *Journal of Hand Surgery* 1987; **12A**(5): 757–61.

Tarlow SD, *et al.* Acute compartment syndrome in the thigh complicating fracture of the femur. *Journal of Bone and Joint Surgery* 1986; **68A**(9): 1439–42.

Viegas SF, *et al.* Acute compartment syndrome in the thigh. *Clinical Orthopedics* 1988; **234**: 232.

Werbel GB, Shybut GT. Acute compartment syndrome caused by a malfunctioning pneumatic-compression boot. *Journal of Bone and Joint Surgery* 1986; **68A**(9): 1445–6.

Wiley JP, *et al.* Ultrasound catheter placement for deep posterior compartment pressure measurements in chronic compartment syndrome. *American Journal of Sports Medicine* 1990; **18**(1): 74–9.

Willey RF, *et al.* Non-invasive method for the measurement of anterior tibial compartment pressure. *Lancet*; **i**: 595–6.

5.4   Overuse injuries of the knee

WILLIAM D. STANISH AND ROBERT M. WOOD

INTRODUCTION

The knee joint is commonly afflicted with overuse injuries. The knee may be predisposed to injury in circumstances where its architectural design is abnormal, i.e. genu valgum, patella alta, etc. If the knee joint is morphologically normal, it may become injured when challenged beyond its capacity, as may occur when a jogger increases his or her mileage too quickly. In this chapter we review problems of overuse related to the knee joint and then offer some proposals for managing these difficulties.

ANATOMY OF THE KNEE

Despite its single joint capacity, it is convenient to describe the human knee as two condylar joints between the corresponding condyles of the femur and the tibia. Furthermore, a joint exists between the patella and the trochlea of the femur. The condylar joints are separated by two fibrocartilage menisci between the corresponding articular surfaces (Fig. 1). The plateau of the tibia possesses two separate articular facets, each slightly concave. The medial facet lies completely on the superior surface of the condyle, but the lateral facet curves backwards over the posterior aspect of the tibial condyle. This bevelled margin allows withdrawal of the lateral meniscus by the popliteus muscle. The femoral condyles are separated posteriorly by a deep notch but fuse anteriorly into a trochlear groove for articulation with the patella. The lateral ridge of the trochlear is very prominent. Viewing the condyles laterally, it can be seen that they are cam-shaped, i.e. flatter at the end of the femur and more highly curved at the free posterior margin. The articular surface of the medial condyle is narrower, longer, and more curved than that of the lateral condyle. This is important in the passive rotation that occurs in extending the knee.[1,2]

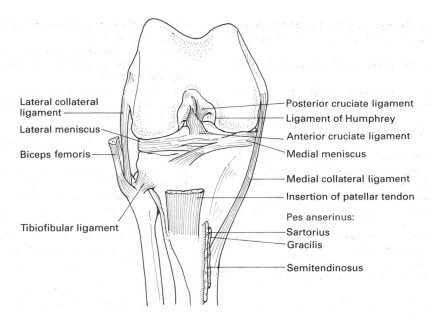

Fig. 1 The anterior view of the knee demonstrating the two menisci separating the femur from tibia.

The articular surface of the patella is divided by a vertical ridge into a lateral and a medial surface; this medial surface is further divided by a second vertical ridge into two smaller areas. The lateral articular surface of the patella is in contact with the lateral condyle of the femur in all degrees of flexion. In extension, the area next to it lies on the trochlea, and the most medial of the three articular surfaces of the patella is not in articulation with the femur. In flexion, this surface glides into articulation with the medial condyle and the middle of the three surfaces lies in the intercondylar notch of the femur.[3]

The capsule which invests the knee joint is usually thin and in some areas is deficient. It is attached to the femur above the intercondylar fossa, to the margins of the femoral condyles, to the margins of the patella, and to the patellar tendon, as well as to the margins of the tibial condyles. The patellar tendon and patella serve as a capsule to the anterior aspect of the joint. As the capsule extends from the femur to the tibia, it is attached to the outer aspects of the menisci. That part of the capsule between the tibia and the menisci is known as the coronary ligaments. Medially, the capsule is fused with the medial collateral ligament. Laterally, a strong thickening of the capsule extends from the lateral epicondyle to the head of the fibula. This lies deep to the lateral ligament and forms part of the origin of the popliteus tendon. On both sides, the capsule is strengthened by aponeurotic expansions of the vasti muscles and overlying fascia. These facial–aponeurotic sheets are known as the medial and lateral retinaculum of the patella.[4] The capsule is reinforced by four main ligaments. These ligaments are the patella retinacula, the medial and lateral collateral ligaments, and the oblique popliteal ligaments.

The patellar retinacula extend from the patella to the lower margins of the condyles of the tibia; these are fibrous expansions from the quadriceps tendon and from the lower margins of the vastus medialis and lateralus.[4] In front of the collateral ligaments, they blend with the capsule; anteriorly they are attached to the margins of the ligaments and patella below the patellar attachment of the capsule.

The medial collateral ligament is attached to the epicondyle of the femur below the adductor tubercle and to the subcutaneous surface of the tibia, roughly a hand's breadth below the knee. The anterior margin of this ligament lies free except at its attached extremities. This part of the ligament is not attached to the medial meniscus and, in fact, is separated from it by a small bursa. The posterior margins of the medial collateral ligament converge to insert the medial meniscus. Over the condyle of the tibia the ligament is separated from bone by the forward extension of the semimembranosus tendon in the intervening bursa. From its tibial attachment, the ligament slopes slightly back as it passes up to be inserted behind the axis of flexion of the medial femoral condyle. The medial collateral ligament is drawn taut by full extension of the knee.[1,2]

The lateral collateral ligament is attached to the lateral epicondyle of the femur in continuity with the short external lateral ligament (Fig. 2). It slopes down and back to the head of the fibula. It lies free from the capsule and lateral meniscus, being separated from the meniscus by the popliteus tendon. Like the medial collateral ligament, it is attached just behind the axis of flexion of the femoral condyle and is drawn taut by full extension of the knee.[1,2]

The oblique popliteal ligament is a thick rounded band of great strength, perforated by the middle genicular vessels. It is an expansion from the insertion of the semimembranosus as it slopes up to the popliteal surface of the femur. The oblique nature of this ligament limits extension–rotation in the locked position of the knee.[1] A structure which is closely related to the capsule but functionally distinct from it is the ligamentum patellae, or patellar tendon. It lies in a smooth area, slightly oblique on the tibial tuberosity. The patellar retinacula, fibrous extensions from the tendon of the quadriceps femoris, are inserted into the edges of the patella and ligaments of the patella, as well as into the inferior borders of the tibial condyles. Several bursae lie in close association with the ligaments and the patella, notably the superficial and deep infrapatellar bursae.[5,6]

Several intra-articular structures are relevant with respect to both the static and dynamic anatomy of the knee. The cruciate ligaments are two very strong ligaments which lie within the

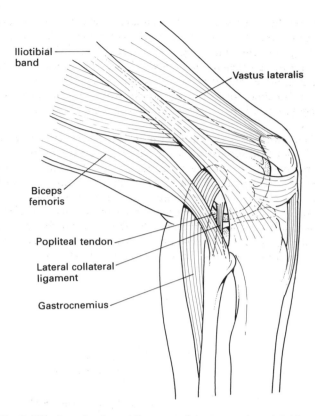

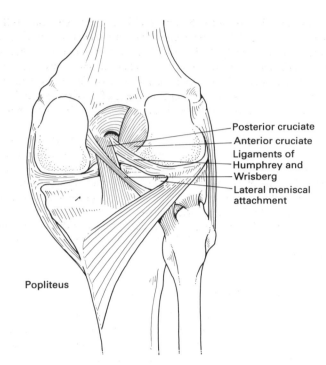

Fig. 2 The lateral collateral ligament of the knee where it bridges the lateral femoral condyle and the proximal fibula in proximity to the popliteal tendon and the biceps femoris.

Fig. 3 The posterior cruciate ligament attaches to the posterior part of the proximal tibia and passes anteriorly and superior to the anterior cruciate ligament.

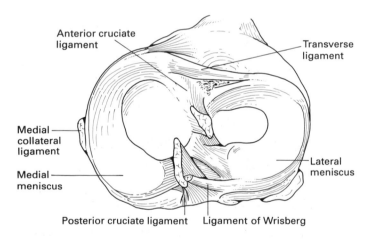

Fig. 4 The medial and lateral menisci are composed of fibral cartilage and are fixed with anterior and posterior horns.

capsule of the knee, but outside the synovial membrane. The anterior cruciate ligament attaches to the anterior part of the tibial plateau, in front of the tibial spine, and extends superiorly and posteriorly to a smooth insertion on the lateral femoral condyle, well back in the intercondylar notch. The posterior cruciate ligament attaches to the posterior part of the head of the tibia between the condyles and passes anteriorly and superiorly medial to the anterior cruciate ligament. It is attached to a smooth impression on the medial femoral condyle, well forward in the intercondylar notch. The cruciate ligaments are essential to the anteroposterior stability of the knee joint. The posterior cruciate ligament is the main stabilizing factor preventing the femur from sliding backwards off the tibial plateau in a weight-bearing flexed knee. The anterior cruciate prevents forward displacement of the femur on the tibial plateau. This ligament also limits extension of the lateral condyle of the femur and causes medial rotation of the femur in the act of fully extending the knee (Fig. 3).[1,7,8]

The menisci are kidney-shaped cartilages composed of dense fibrous tissues. Each meniscus lies on the superior surface of its respective femoral condyle. The are avascular except at their attachments. The medial meniscus is fixed at its anterior and posterior horns by fibrous tissue invested in the tibia. The lateral meniscus is likewise attached to the tibia and both its horns (Fig. 4). The circumference of the meniscus is attached by very lax capsule to the articular margins of the femur and the tibia, except beneath the popliteal tendon. Here there is a gap in the capsule through which the popliteal tendon and bursa migrate.

The popliteal tendon lies between the capsule and the synovial membrane. This tendon does not lie free within the cavity of the knee joint but is firmly adherent to the capsule. The adherence of

this tendon makes a prominent ridge on the internal surface of the capsule. The ridge is invested with the synovial membrane of the joint cavity, both above and below the lateral meniscus. The bare popliteal tendon is in contact with the bare lateral meniscus between the upper and lower synovial reflections. The tendon may even be attached to the lateral meniscus, but this is rare.[9-11]

Since the knee is a condylar joint, its movements are complex. The shapes and curvatures of the articular surfaces are such that hinge movements are combined with gliding and rolling, coupled with rotation about a vertical axis. With the knee extended, the axis of rotation extends from the head of the femur to the medial intercondylar tubercle of the tibia. Hence the lateral condyle swings around this vertical axis through the medial condyle.[12,13]

When the knee is flexed with the leg remaining fixed, the thigh rotates laterally during the first part of flexion and the femur rolls backward on the tibia. Conversely, when the knee is extended the thigh rotates medially during the final part of extension. This is know as the locking or screw-home mechanism and is attributed to the vastus medialis. This action makes the articular surfaces more congruent and puts the joint in a position of maximum stability. The popliteus muscle initiates flexion of the knee by unscrewing the locked knee joint.[2,12,14]

When the thigh is fixed and the leg is free to move, the tibia rotates on the femur. Since medial rotation of the tibia is equivalent to lateral rotation of the femur, it can be deduced that in this case initiation of flexion of the knee is accompanied by medial rotation of the leg and the terminal part of extension of the knee is accompanied by lateral rotation of the leg.[1,12,14]

Before describing changes observed in the articular surface in various overuse injuries of the knee, the histology of normal cartilage must be considered.

Histology of articular cartilage

Articular cartilage is a composite material consisting of a cellular and an extracellular component. The chondrocytes compose the cellular component while the extracellular component consists of an intricate arrangement of mucopolysaccharide ground substance and collagen fibres.[15.] Three zones which differ in relative size and composition have been identified within articular cartilage (Fig. 5)

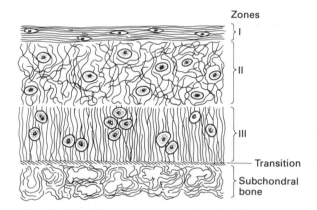

Fig. 5 Normal articular cartilage possesses three zones in transition to subchondral bone.

Zone I, which is the outermost layer, forms a thin tough skin over the surface of the articular cartilage. In this zone the collagen fibres have a definite orientation, lying parallel and presenting a relatively smooth surface. Also, the collagen fibres in this zone are rather small.[16,17] Zone II, which lies immediately deep to zone I, presents a more disorganized arrangement of collagen. In this zone the collagen fibres tend to be coiled, presenting large open spaces in the mesh work which are filled with ground substance.[16,17] Zone III, which is the deepest zone, consists of large cellular fibres which are radially oriented and arranged more densely, or closer together, than zone II collagen fibres.

The different arrangement of the collagen fibres in each of the three zones allows each zone to take a different responsibility. The circumferentially arranged fibres in zone I present a low diffusion effect upon compression, while the more randomly oriented collagen fibres of zone II act as an area of energy storage. The fibres of zone III serve as an attachment between the articular cartilage and the underlying subchondral bone.[16,17] These fibres also present a geometric pattern which provides resistance to the shearing forces of articulation. The three layers also differ with respect to chondrocyte activity—those in zone I tend to be relatively inactive, while chondrocytes in the deeper layers are more active.

Joint lubrication and cartilage nutrition are maintained by the ability of lubricants and nutrients to permeate articular cartilage, a property attributed to the ultrastructure and biochemical composition of the cartilage. With ageing, the mechanism by which articular cartilage obtains its nutrients changes slightly. In the child, nutrition of the cartilage is achieved by diffusion of nutrients from within the joint and from subchondral bone into the articular cartilage (Fig. 6).[17] After maturity, nutrition is dependent solely upon diffusion from superficial to deep layers. Studies have shown

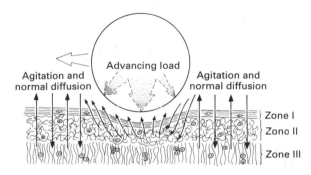

Fig. 6 Joint lubrication and cartilage nutrition are maintained by the ability of lubricants and nutrients to permeate articular cartilage.

that the permeability of the articular cartilage tends to change with the compressive load imposed.[18] During periods of non-compression, fluids are able to penetrate the cartilage by simple diffusion. However, when stresses are placed on the cartilage, several changes take place. First, the collagen fibres of the middle zone (zone II) form a densely oriented structure; in other words, the fibres become more densely packed together. Second, compressing the proteoglycan component of the ground substance tends to increase its density. Both these changes have the effect of decreasing the permeability of the articular cartilage. When a stress is placed on the articular cartilage fluid exudes from it, and when the stress is removed the fluid is resorbed by the cartilage. This phenomenon is important in the maintenance of joint lubrication.[17]

The integrity of articular cartilage is essential to the normal functioning of each and every joint, and any alteration in either the biochemical composition of the ground substance or the collagen architecture will have undesirable effects on articular cartilage function, with respect to permeability, lubrication, and biomechanical function.

CHONDROMALACIA PATELLA

Chondromalacia patella is a condition which affects young healthy individuals. Invariably these patients complain of pain about the patella, and in fact manifest degenerative changes to the articular cartilage of the patella. As a distinct entity it was first described by

Budinger in 1906. Although much has been written about this condition, it continues to pose a major problem to the practitioner in terms of both diagnosis and treatment.

The most notable problem is differentiating chondromalacia patella from other conditions of the knee such as disorders of the meniscus. There appears to be no single sign which is pathognomonic of either condition. Several points must be established in dealing with chondromalacia patella.[17] First, the aetiology of chondromalacia patella is multifactorial and frequently related to malalignment of the lower extremity. This results in abnormal stresses to the patella. The ultimate effect of this abnormal stress is damage to the articular surface. Secondly, other possible causes of anterior knee pain must be eliminated before chondromalacia patella is diagnosed. Third, conservative non-surgical treatment is indicated in most circumstances.

PATHOLOGY

Chondromalacia patella is a pathological finding of cartilage on the posterior aspect of the patella.

The most common lesion is fibrillation of the articular cartilage at the junction of the medial and odd facets. This fibrillation may spread to involve most of the medial facet. Histologically, chondromalacia patella is characterized by maintenance of the superficial zone of articular cartilage with the initial changes occurring primarily in the deeper zones. The main changes observed include specific alterations of the mucopolysaccharide ground substance and loss of the normal energy-storing capacity of the middle zone. These changes result in the application of abnormal forces to the underlying subchondral bone and subsequent fissuring, and fibrillation occurs throughout the layers of articular cartilage. In severe cases, all layers of articular cartilage may erode to the subchondral bone (Fig. 7).[19]

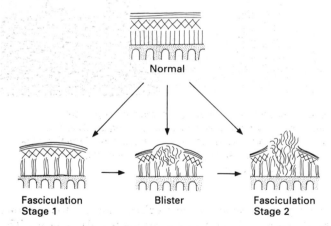

Fig. 7 Progression of basal degeneration of the articular cartilage.[19]

Goodfellow et al.[19] have differentiated between changes in the articular cartilage, which are commonly seen in older individuals, and the alterations seen in the adolescent with chondromalacia patella. Surface degeneration with superficial fibrillation is the characteristic change seen in the articular cartilage of old individuals. It is described as changes initially involving the superficial layer with eventual involvement of all layers, including the subchondral bone. These changes are also seen in youth in areas such as the medial articular margin of the odd facet of the patella.

This is usually a region of non-contact, and since cyclic loading is necessary for normal articular cartilage nutrition, the area will be nutritionally deficient, resulting in degenerative changes. These degenerative changes may lead to osteoarthritis.

Chondromalacia patella in the adolescent involves degeneration of the articular cartilage and allegedly does not progress to osteoarthritis. This allegation is controversial. Stage I of degeneration is known as blistering and involves normal gross appearance of the articular cartilage. However, some softening of the cartilage occurs and the normal architecture of zone II is disrupted. The superficial and deeper layers remain relatively intact. Alterations in the ground substance are seen histologically, with loss of normal staining characteristics and chondrocyte proliferation. The superficial layer of articular cartilage may eventually be involved, with disruption of the tangentially oriented collagen fibres which present a rough, uneven, and deeply fissured surface.[17]

There has been difficulty in explaining the development of pain from pathological conditions such as those already described, because articular cartilage is devoid of the nerve fibres mediating pain and usually at this stage there is no evidence of synovial reaction. Because the overlying articular cartilage has lost some of its ability to absorb energy, the subchondral bone is subjected to abnormal forces. These forces on the subchondral bone can explain the pain observed, since it is richly endowed in pain fibres. It has been suggested by Darracott and Vernon-Roberts[20] that the initial changes with chondromalacia patella occur in the subchondral bone and are characterized by chondrocyte hyperplasia with vascularization and advancing ossification into the deeper zone of the articular cartilage. This process is accompanied by new bone formation and either focal or diffuse osteoporosis. This process is thought to precede changes within the articular cartilage and may be secondary to an alteration of blood supply in the subchondral bone. Involvement of the subchondral bone in such a fashion seems consistent with the theory of pain production and may well have merit in understanding the aetiology of this condition.

FACTORS IN AETIOLOGY

There are numerous aetiological factors of chondromalacia patella. These factors include direct trauma to the knee, incongruency of the patellofemoral joint, internal derangements of the knee, recurrent subluxation of the patella, malrotation of the tibia on the femur, malalignments of the extensor mechanism, or extended periods of immobilization following surgery.

Direct trauma to the knee has been implicated as the major aetiological factor in the pathogenesis of chondromalacia patella. In 1986, Chrisman and O'Donald[21] confirmed the previous hypothesis that mechanical trauma can lead to degradation of articular cartilage. Using a weighted pendulum, a blow was delivered to the normal articular cartilage. Within 2 h a fourfold increase in the concentration of free arachidonic acid in the cartilage was found by gas chromatography. Arachidonic acid is an unsaturated fatty acid which is essential for human nutrition, and is a major component of the bipolar phospholipid cell membranes. When released, it is a precursor for local hormones and chemotactic factors such as prostaglandins. Prostaglandin E_2 is the major prostaglandin released under these circumstances and, by its activation of cAMP, articular cartilage deterioration may ensue due to the release of catheptic proteases. These enzymes are exuded into the cartilage matrix where they split the protein links attaching chondroitin sulphates to major components of the cartilage matrix. The loss of matrix leads to the softening and possibly the fibrillation of articular cartilage seen in chondromalacia patella.

A discussion of the aetiological factors of chondromalacia patella would be incomplete without mention of lower-extremity malalignment. Rotational and angular malalignments of the lower extremity have a significant influence on patellofemoral joint mechanics. Persistent anteversion through the femoral neck frequently remains unrecognized and is often associated with a series of compensatory growth disturbances throughout the limb.

James and coworkers[17,22] have described an alignment abnormality in active young individuals which is known as the 'miserable malalignment'. These patients show significantly increased internal rotation of the hip while it is in the extended position. Examining individuals who are standing with the feet parallel reveals bilateral squinting of the patellae and apparent genu varum, and often an associated recurvadum. More distally there is a real tibial varum and the foot presents a compensatory pronation. Very frequently this anterior knee pain, once initiated, is extremely difficult to manage.

PRESENTATION OF CHONDROMALACIA PATELLA

The hallmark of chondromalacia patella is anterior knee pain. The symptoms occur in two groups of patients[23] The first group involves the somewhat inactive teenage female. In this group the symptoms are frequently triggered by prolonged periods of sitting with knees flexed, for example in an automobile or in the classroom. The patient's symptoms may be alleviated by walking for a short period of time. The second group of patients with symptomatic chondromalacia patella is comprised of the highly active male in the late teenage years or early twenties. In this group symptoms are aggravated by activity, in particular those activities which involve stresses on the knee joint in flexion and twisting. In both groups ascending stairs or hills causes exacerbation of the condition, as do deep knee bending exercises such as squats.

The pain itself is always experienced in the retropatellar area and is described by the patients as aching. Roughly one-third of the cases are bilateral. A sensation of catching, grating, or even false locking is commonly described by patients, and these may be accompanied by the feeling of instability or giving way. Most patients will give a history of a rather insidious onset of pain without a major trauma. The pain may occur with activity, but more commonly it will occur with a period of rest following vigorous activity.

PHYSICAL EXAMINATION

It must be stressed that the physical examination of a patient suspected of having chondromalacia patella must include the entire lower extremity. A comprehensive examination of the lower extremity must be divided into four phases: standing, sitting, supine, and evaluation while prone. Inspection and palpation are the major diagnostic tests, combined with more specific techniques to complete the comprehensive examination of the lower extremity.[24]

Standing examination

In this examination the patient stands upright, facing the examiner, with the feet held slightly apart and aligned directly towards the examiner. The patient should be wearing running shorts or comparable clothing so that the entire lower extremity can be viewed by the examiner. General alignment of the lower extremity is first observed with attention directed to mechanical malalignment such as genu varum or valgum deformities. Next, the angle of the lower tibia to the floor is measured. An angle of 10° or more demands excessive subtalar joint pronation to offer a plantar grade foot. The excessively pronated foot is accompanied by a compensatory internal rotation of the tibia. Commonly, this increased amount of rotation triggers stress through the peripatellar tissues of the knee.[17,22,24] The status of the foot, particularly as it relates to excessive pronation, is then determined by first stabilizing the subtalar joint in the neutral position. This is accomplished by palpating the talar head between the thumb and index finger over the anterior aspect of the ankle, while the patient transfers the weight to the lateral border of the foot and the medial aspect of the heel is lifted from the floor. The position in which the talar head is palpated equally on the medial and lateral aspects of the ankle is considered to be the neutral position of the subtalar joint, and is the optimal position for normal lower extremity weight bearing.[17,22,24] An individual with a foot in excessive pronation will have an increased internal tibial rotation. This will result in prolonged and increased forces absorbed by the soft tissues of the knee (Fig. 8).[24]

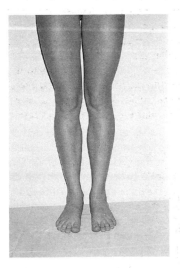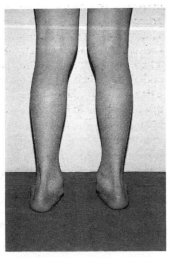

Fig. 8 A patient demonstrating miserable malalignment, characterized by femoral anteversion, squinting patellae, genu varum, patella alta, increased Q-angle, external tibial rotation, tibia varum, and compensatory pronation of the feet. (Reproduced from ref. 17, with permission.)

Noting the position of the patellae, with the feet pointing directly ahead, will help detect rotational malalignments. The so-called 'squinting patellae', where both patellae point medially, is a sign of increased femoral torsion or excessive femoral anteversion. A close observation of the gait pattern of the patient will be valuable in detecting rotational abnormalities.

The standing examination should also include viewing the lower extremity from the lateral aspects. This is the best position for observing flexion contractures or genu recurvatum.

Sitting examination

This part of the physical examination concentrates on the patellofemoral joint and is performed while the patient is seated on a normal examining table with the legs hanging freely at 90° of flexion.

The position of the patella, as it sits over the distal aspect of the femur, is first evaluated. The normal patella sinks deeply into the patellofemoral sulcus at 90° of flexion.[24] A patient with patella alta will demonstrate a protuberant patella which appears to point towards the ceiling. Patella alta and patella infera (baja) have been associated with numerous patellofemoral disorders and these conditions should be sought in patients with patellofemoral complaints.[24,25]

Next, the alignment of the patella and patellar tendon is assessed. As the knee is flexed, the tibia normally derotates and decreases the quadriceps angle as the patellar tendon orients itself in the same longitudinal axis as the anterior crest of the tibia. The 'grasshopper-eye' patella is one sign of abnormal patellar tracking. This is a lateral tilting of the patella and can be easily seen as the knee is flexed to 90°. Patella tracking is further evaluated by palpating the patella as the knee is passively flexed and extended. The tracking should be smooth, longitudinal, and accompanied by only small amounts of rotation. Any abrupt or sudden movements of the patella should be considered abnormal.[24]

Finally, the sitting examination should be concluded by palpation of the patella for crepitus, as the knee is actively flexed and extended. The presence of crepitus does not always correlate with pain or the degree of patellar involvement as there may be minimal pain with marked crepitus and vice versa.

Examination while supine

This part of the examination is important in determining the mobility of the patella. The quadriceps angle can also be measured at this time.

Patella mobility is tested with the knee in full extension and at 30 to 40° of flexion. Most patellae demonstrate a moderate amount of lateral mobility in full extension; however, instability of the patella must be suspected if the patella displaces more than half its width in this position. At 30° of flexion there should be little or no lateral mobility of the patella. If the patient has a grossly unstable patella, it may be possible to sublux it over the lateral femoral condyle. This will often cause apprehension in the patient, as the same sensation is felt with episodes of giving way.[24,26]

All borders of the patella should be palpated for tenderness. Facets can be examined by displacing the patella medially and laterally with the knee extended and palpating the border of the medial and lateral facets. This will elicit varying degrees of tenderness with chondromalacia.

The final component of the examination while supine involves compressing the patella against the femoral condyles while the patient is tensing the quadriceps. Once again, varying degrees of discomfort will be reported and this is often accompanied by crepitus in the presence of chondromalacia patella.

Examination while prone

The examination of the lower extremity and patellofemoral joint is concluded with the patient in the prone position with the feet extended beyond the end of the examining table. Hip motion is measured with the knees in 90° of flexion, while preventing the pelvis from tilting. Most adults will have more external than internal rotation; however, a discrepancy of over 30° may be considered normal.[24]

Next, measurements are taken to determine the leg–heel and heel–forefoot alignment. These measurements must be taken with the subtalar joint in the neutral position. The neutral position is achieved by application of pressure beneath the fourth metatarsal head and dorsiflexion of the ankle until resistance is met. The foot is then inverted and everted to the point at which the foot appears to fall off to one side. This point in the arc of motion is the neutral position of the subtalar joint.[24] With the subtalar joint held in its neutral position, leg–heel alignment can be evaluated as follows. A line is drawn along the longitudinal axis of the distal portion of the posterior leg and a second line is drawn over the longitudinal axis of the posterior portion of the calcaneus. A normal value for leg–heel alignment is 2 to 3° of varus.[24]

The heel–forefoot alignment is then assessed, once again with the foot held in the neutral position. This is determined by noting the relationship of the transverse plane of the forefoot at the metatarsal head in relation to the vertical axis of the heel. Normally the transverse plane of the forefoot should be perpendicular to the vertical axis of the heel. If the plane of the forefoot of the metatarsal head is such that the medial side of the foot rises above a perpendicular plane to the heel, the forefoot is supinated (varus). However, if the medial side of the foot falls below the perpendicular plane of the heel, the foot is in pronation (valgus). A flat or severely pronated foot often has forefoot supination and subtalar varus, while the varus foot often has plantar flexed first ray and subtalar varus.[24]

TREATMENT

It is generally agreed that the patellofemoral pain syndrome, more specifically chondromalacia patella, should be managed initially by conservative measures. Several authors, including DeHaven et al.[27] in a prospective study of chondromalacia patella, have demonstrated that a non-surgical programme was successful in 82 per cent of patients. Furthermore, 66 per cent of conservatively treated patients were able to return to unrestricted athletic activities. Only 8 per cent required surgical intervention. Most authorities considered rest, quadriceps muscle strengthening, and medication important ingredients of any conservative programme.

Before commencing a treatment programme for managing patellofemoral pain, the physician must take the necessary time to educate the patient about the condition. He or she must be reassured that the condition is not serious and reduction of painful activities for a period may be necessary. The nature of the patellofemoral programme and the specific treatments must be explained. These practices can prove to be time-consuming: however, they are essential in maximizing patient co-operation and compliance.

Acute pain must be managed by restriction of activities associated with increased patellofemoral pressure such as weight bearing with a flexed knee. The value of immobilization of the knee is outweighed by its harm in causing muscle atrophy and weakness, and possibly interfering with the nutrition of the joint cartilage.[28]

Mild or moderate pain can generally be managed by relatively minor modifications of activities. In most cases only activities which cause knee pain in each individual should be avoided, whereas all other activities are permitted and in fact encouraged. If the patient must sit for long periods of time, they must be informed that sitting with the knee in relaxed extension will reduce pain afterwards. Swimming can be substituted for jogging as a means of promoting general physical fitness.[28,29]

Strengthening of the quadriceps muscle group is the most important part of the conservative management of chondromalacia patella. Atrophy and weakness of the quadriceps almost always accompanies disorders of the patellofemoral joint, and although the mechanism is not completely clear, quadriceps strengthening helps alleviate patellofemoral pain. It has been suggested that quadriceps strengthening may relate to alteration of the contact surfaces of the patella with the trochlea, thereby relieving pain.[28,30]

Although several exercise programmes have been suggested for quadriceps rehabilitation, isometric and progressive resistive exercises, performed with the knee in full extension, have been found to be the most useful. The programme may be altered according to the severity of the pain in each individual. Patients with severe pain should be prescribed an isometric exercise regime consisting of maximum isometric quadriceps recruitment for 5 s with the leg in the fully extended position. This exercise should be repeated at least 50 to 100 times per day. When this exercise can be performed in a pain-free fashion, progressive resistive exercises may be instituted. Straight leg raising exercises, with weights around the ankle, have proved to be effective in further increasing the strength of the quadriceps group. The maximum weight can be lifted through 10 repetitions to an angle of 45° at the hip, and is used initially for a maximum of 30 to 40 repetitions per day. The weight can be increased to 20 to 30 pounds (9–14 kg) as strength increases. If must be emphasized that these exercises must be performed with the knee in the fully extended position. Even short arc (less than 30°) knee flexion has been shown to be detrimental in some circumstances.[28]

When the exercise programme is followed, a gradual return to full activities is usually possible 4 weeks after starting it. Some patients will find it necessary to be reminded to continue a maintenance programme of periodic quadriceps exercises to prevent recurrent patellofemoral pain.

Aspirin and other non-steroidal anti-inflammatory drugs are currently the only medications used in the management of chondromalacia patella. Although Chrisman and O'Donald[21] demonstrated that aspirin did not promote matrix healing with established cartilage defects, there is some rationale for the use of aspirin in patellofemoral disorders. Aspirin or other anti-inflammatory drugs are chiefly used as a means of reducing the synovitis that follows cartilage destruction in chondromalacia patella, and possible as a means of preventing further breakdown of articular cartilage. A dose of 600 mg, four times daily, is recommended in the patients who do not respond to rest and quadriceps muscle strengthening alone. Intra-articular steroid injections should be reserved for those unique circumstances where synovitis is a major component.

Other conservative measures, such as knee bracing which provides support for the knee or prevent lateral subluxation of the patella, have been utilized with varying degrees of success. Orthotics within shoes have provided useful correction in runners with excessive foot pronation. Gruber suggests that conservative treatment for chondromalacia patella should be abandoned if there is no improvement after 3 months, or if the symptoms worsen after 1 month.[31] However, the physician should be wary of the patient who has a very low pain threshold.

SURGICAL TREATMENT OF CHONDROMALACIA PATELLA

In less than one-third of patients with chondromalacia patella, the pain associated with the condition is so severe and disabling that operative treatment is considered. Operative treatments are generally preceded by arthroscopy and photography of the lesions of the patella. Several operative procedures have been suggested and varying degrees of success have been recorded. The most encouraging procedure is that devised by Insall et al.[32] in which, following a lateral release of the retinaculum, the medial and lateral components of the quadriceps expansion are brought together to medialize the patella. These authors report a very high success rate.

Hughston and Walsh[33] have taken a different approach to the operative treatment of chondromalacia patella. They emphasize the importance of a proximal realignment of the patella by advancement of the vastus medialis. They report a 71 per cent success rate over a 15-year period. An alternative approach to realignment has been to transpose the patellar tendon medially, coupled with a lateral release of the quadriceps retinaculum. However, it is clear that the scientific explanation for the success of some of these operations remains rather vague.

Debridement of the patella, performed in a transarthroscopic fashion, has also resulted in varying degrees of success. This procedure has been performed extensively by Wiles et al.[34] with satisfactory results. However, Bentley[35] reports only a 25 per cent success rate in over 60 patients.

In general, surgery for chondromalacia patella must be approached in a very cautious fashion.

ILIOTIBIAL BAND FRICTION SYNDROME

Iliotibial band friction syndrome is an overuse injury frequently affecting individuals in long-distance athletic activities. This syndrome, which commonly manifests itself as a poorly localized pain over the lateral femoral condyle, has rarely been described in the literature. However, as the number of people involved in marathon type activities increases, there is an increasing patient population presenting with iliotibial band friction syndrome.

In terms of anatomy, the iliotibial band is a thickened strip of fascia lata that receives part of the insertion of the tensor fascia lata and the gluteus maximus (Fig. 9). It passes distally down the lateral aspect of the thigh in continuity with the lateral intermuscular septum and inserts into Gerdy's tubercle on the lateral tibial condyle. When the knee is in full extension the iliotibial band lies anteriorly to the line of flexion of the knee. Since the iliotibial band is free of bony attachments between the lateral femoral epicondyle and Gerdy's tubercle, it is free to move posteriorly to this axis upon flexion of the knee (Fig. 10). Thus, when the knee is repeatedly flexed and extended, the iliotibial band migrates posterior and then anterior to the line of flexion of the knee. It is postulated that this movement against the lateral femoral condyle leads to inflammation of the iliotibial band, resulting in the discomfort associated with this friction syndrome.[36]

Sutker et al.[37] report that the iliotibial band friction syndrome is an overuse injury which most commonly affects males between the ages of 20 and 40 who have been running 20 to 40 miles per week (32–34 km) for at least 3 years. A common physical finding in these men was a varus alignment of the knee. Sutker et al. also found that it was predominantly a unilateral disease.

The pain associated with iliotibial band friction syndrome occurs on the outer aspect of the knee, in close relation to the lateral femoral condyle. In a paper describing iliotibial band friction syndrome in 16 military trainees, Renne[36] noted that the

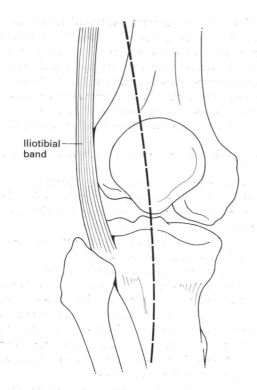

Fig. 9 The iliotibial band is the thickened strip of fascia lata that passes distally over the lateral femoral condyle to insert into Gerdy's tubercle. The alignment of the femur and tibia is denoted with the dotted line.

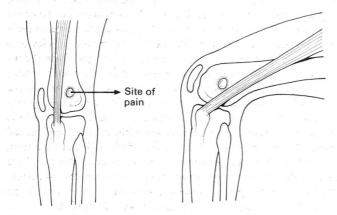

Fig. 10 The iliotibial band moves behind the lateral femoral condyle on flexion of the knee and returns to the anterior position in extension.

patients were first seen because of a limp and pain in the lateral aspect of the knee during walking or running. The symptoms commonly appear after a long run or hike, with the pain worsening with increased distance. The pain usually worsens when walking upstairs and downstairs, and with hills. It is postulated that any activity which calls for an increased length of stride while running can cause excessive compression between the iliotibial band and the lateral femoral condyle, thereby aggravating symptoms. In all studies reviewed these findings were observed in the absence of any report of direct or indirect trauma to the knee.[36-38]

Both Renne[36] and Sutker et al.[37] reported the following physical findings. All patients had focal tenderness at a point over the lateral femoral epicondyle, approximately 3 cm proximal to the knee joint. The pain was reproduced by having the person support their whole bodyweight on the affected leg, with the knee held in 30° of flexion, gradually rocking over top of the weight-bearing extremity. This manoeuvre is known to bring the iliotibial band into prominence. A full range of motion of the hip and knee is present. In a select group of patients, i.e. those with refractory symptoms, a peculiar 'creak' was felt while palpating the epicondyle during flexion and extension of the knee. Renne described this sensation as resembling the rubbing of a finger over a wet balloon.

Tenderness or effusion of the knee, ligament laxity, swelling, and McMurray's test were all absent. Varus stressing of the knee did not predictably trigger discomfort. The functional examination revealed that the patients could jog in place, hop, squat, and rise without significant discomfort.[36-38] As mentioned earlier, the knee alignment of individuals presenting with iliotibial band friction syndrome is either neutral or varus; it is extremely uncommon for a patient with genu valgum to present with this particular syndrome, presumably because less stress is put on the iliotibial band in that particular alignment.

In Renne's study of 16 military trainees who were diagnosed as having the iliotibial band friction syndrome,[36] the radiographs of the symptomatic knees could not be distinguished from those of age-matched controls. No degenerative changes or prominent osteophytes were observed. Sutker et al.[37] report that three of the runners in their study had arthrograms and all were within normal limits.

In the treatment of an athlete with iliotibial band friction syndrome, it must be stressed that this disorder is in fact an overuse injury. Perhaps the most valuable component of the treatment regime for this condition is reduction of the stress to the knee. The patient should also be told to avoid conditions which make the symptoms worse, such as running up or down hills. Shortening of the running stride may also be helpful; however, this has not been proven.[36-38]

In addition to reducing the athletic stress, oral non-steroidal anti-inflammatory medications have proved to be useful in the reduction of inflammation associated with this syndrome. With persistence of the symptoms, steroid injections may be given at 2-week intervals until the pain disappears.[36] Renne found that, with these measures, all the patients were able to complete the remaining 3 months of military training. Lateral shoe orthotics have also been shown to be effective in relieving the symptoms of this condition.[36,38]

If the treatment format described above is not successful, a period of total rest (4–6 weeks) is recommended. Surgery remains as the very last resort of treatment. The most widely accepted surgical technique involves splitting transversely the posterior 2 cm of the iliotibial band at the area of the lateral femoral condyle so that this portion of the band is not taut at 30° of flexion.[38]

PATELLAR TENDINITIS

Patellar tendinitis, or 'jumper's knee', refers to a clinical syndrome seen frequently in athletes who participate in sports which particularly involve jumping, thus excessive stress to the extensor apparatus of the knee. Generally, the pathology associated with jumper's knee is triggered by incessant jumping, sometimes with single-foot landings. Volleyball players, basketball players, and high jumpers will frequently present with symptoms consistent with patellar tendinitis.

The actual pathology involved with jumper's knee is localized at

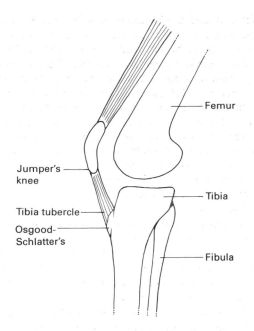

Fig. 11 This schematic demonstrates the site of maximum tenderness in jumper's knee in contrast to Osgood Schlatter's disease.

the level of the bone–tendon junction, and thus suggests that jumper's knee is an overload lesion comparable with other insertional tendinopathies (Fig. 11). The normal junction between bone and patellar tendon shows four distinct zones. These zones are tendon, fibrocartilage, mineralized fibrocartilage, and bone. A well-demarcated borderline, the 'blue line' separates the fibrocartilage from the mineralized fibrocartilage; none of the zones has a thickness exceeding 2 mm. In jumper's knee, or patellar tendinitis, the 'blue line' is absent, the fibrocartilage is much thicker than normal, and pseudocysts can be present at the borderline between mineralized fibrocartilage and bone.[40] Patellar tendinitis is characterized by the insidious onset of aching in the knee centred over the infrapatellar or suprapatellar region, usually localized to the inferior pole of the patella. A study[40] of volleyball players presenting with symptoms of jumper's knee showed that 65 per cent had pain over the inferior pole of the patella, 25 per cent had discomfort over the superior pole, and 10 per cent had pain at the insertion of the patellar tendon into the tibial tuberosity. In mild cases pain was predominantly felt during, and more frequently after, the athletic activity. Day-to-day activities such as climbing and descending stairs can frequently aggravate the pain. The aching type of discomfort usually disappears after a period of rest ranging from a few hours to several days depending on the patient and the severity of the symptoms. Some patients may report swelling of the knee, but upon further questioning it appears that they are describing a feeling of fullness in the area of the lesion. Rarely, the onset of pain may be related to a discrete injury during take-off or landing, followed by the development of persistent aching in the knee. Less frequently, the onset of pain can be attributed to a direct blow to the knee joint such as being kicked or falling on hard ground.[41]

If untreated, the symptoms experienced by the patient may progress to become more frequent and more severe. As it worsens, the condition leads to persistent discomfort which completely disrupts athletic activity. In these cases pain may be present even

when the patient walks, causing a disturbed gait and decrease in the excursion of the knee joint. The discomfort may be relieved when the patient lies in a supine position with the knee in full extension. The pain may also occur after prolonged sitting, so that the individual athlete feels compelled to extend the knee.

If the patient is permitted to continue with intense athletic activity, a catastrophic circumstance may occur with a complete rupture of the tendinous attachment to the involved pole. This is rare.

Jumper's knee can be classified according to symptoms as follows:[41,42]

stage I pain after practice or competitive athletics;

stage II pain at the beginning of the activity which disappears after warm-up and reappears after completion of the activity;

stage III pain remains during and after activity and the patient is unable to participate in the sports;

Stage IV complete rupture of the patellar tendon.

The diagnosis of patellar tendinitis can be established on the basis of physical examination by demonstrating exquisite tenderness upon palpation at the inferior/superior pole of the patellar or the insertion of the patellar tendon to the tibial tuberosity. At times, a cystic fluctuation may also be noted in that area. However, a generalized effusion of the knee is rather unusual. Signs of injuries of the meniscus or ligaments are absent.

Certain anatomical abnormalities, such as patellar hypermobility, patella alta, Osgood-Schlatter's disease, or genu recurvatum, may be present; however, their role in the pathological process of patellar tendinitis is not yet understood.[41]

In evaluating the patient radiographically, views of the knee including tangential views would be appropriate. The radiographs may be reported as negative initially, or lucency at the involved pole of the patellar may be observed. Adolescents and teenagers should show irregular centres of ossification in the involved poles or a periosteal reaction at the anterior surface of the patella in the involved area.[41] Elongation of the involved pole, occasionally with a stress fracture at the junction with the main portion of the patella, has also been observed. Other alterations observed in radiographs of patients with patellar tendinitis include fatigue fractures of the inferior pole of the patella, calcification of the involved tendon, and patella alta. A complete avulsion of the inferior pole has been seen, but exclusively in the end-stage of this condition. Attempts have been made, through xeroradiography, to delineate the soft tissue attachments to the patella; however, the value of this method in the study of patellar tendinitis has not yet been determined.

As mentioned earlier, patellar tendinitis is classified according to the severity, onset, and duration of symptoms. The treatment of patellar tendinitis is dependent upon the state of the disease.[41]

Stage I

The individual manifesting early phases of patellar tendinitis, who has pain only after the activity, can be treated with mild restriction of activity, non-steroidal anti-inflammatory agents, and ice massage. In most instances, the symptoms will disappear and remain quiescent after the medication is terminated. In other instances, the symptoms will disappear while the medication is being taken, only to reappear upon discontinuation of those same medications. Local corticosteroid injections are not recommended for the individual with stage I patellar tendinitis. Some symptoms may be alleviated with the use of a long elastic knee support or a horseshoe type of orthotic inserted around the patella. This can be fabricated from felt, which is readily available.

Stage II

The patients should use the same treatment modalities as the individual with less pain. They should use moist heat. If the aching becomes more intense and the athlete is becoming apprehensive about his or her performance, a local low dose corticosteroid injection may be employed. The athlete should be made aware of the potential hazard of corticosteroid injections. A programme of frequent injections of cortisone should not be condoned.

Stage III

These patients suffer with pain which is persistent before, during, and after athletic performance. These individuals should be advised to curtail their athletic activities, while being placed on an aggressive exercise programme to strengthen the patellar tendon. This has been well described by Curwin and Stanish.[43] It should be noted that, even for those individual with stage I or stage II patellar tendinitis, an aggressive exercise programme is an essential component of rehabilitation after the inflammatory phase of the disorder has been controlled.

Stage IV

A complete rupture of the patellar tendon is treated surgically in some situations. After a brief period of immobilization the patient is gradually started on a progressive exercise programme, finally terminating with eccentric loading to the tendons. It is generally accepted that it will be 6 to 12 months before the athlete can return to aggressive ballistic activities. The surgical procedures have been described by many authors, including Blazina et al.[41] Excision of the degenerative portion of the involved tendon has been described by some workers, while others have suggested that a resection of the involved pole of the patella may prove successful.

MEDIAL COMPARTMENT OVERLOAD

Introduction

Osteoarthritis is a degenerative disease of synovial joints. The initial changes in an osteoarthritic joint are still the subject of speculation; however, several observations have been made. Freeman and Meachian[44] have suggested that the initial lesion is in the cartilage fibre framework of the articular surface, causing an abnormally wide separation of the fibres, a deterioration in the mechanical strength of the matrix, and thus an increased susceptibility to further structural damage during use of the knee joint. A second hypothesis[44,46] considers the importance of proteoglycan in giving cartilage its unique properties and protecting the collagen fibre framework from damage during use of the joint. A change in the proteoglycan synthesis by chondrocytes occurs in the early stages of experimentally induced osteoarthritis in dogs, and is accompanied by an ultrastructural deterioration in the collagen framework of the articular cartilage.[47] Another biomechanical hypothesis suggests that the initial event may be an enzymatic degradation of the collagen in the articular cartilage by collagenase activity, or enzymatic degradation of the proteoglycan resulting in loss of its protective effect against collagen framework damage. The common factor in all theories of the initiation of osteoarthritis is change in the collagen framework of articular cartilage.[48]

When a person with normally aligned lower extremities stands on both legs, the line of weight bearing travels from the centre of the femoral head to the centre of the knee and through the centre of the ankle.[49] In patients with abnormally aligned lower ex-

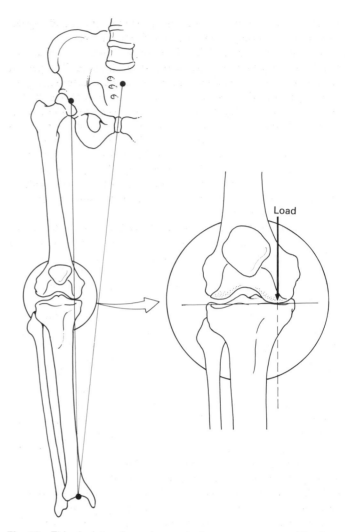

Fig. 12 This depicts a knee demonstrating genu varus resulting in medial compartment overload.

tremities, however, the line of force is very different. In a patient with genu varum the weight-bearing line will pass through the medial tibial plateau (Fig. 12), while in a patient with genu valgum the weight-bearing line will pass through the lateral tibial plateau. During normal walking, a force of approximately three times body weight is transmitted through the knee, with the largest portion of this force being born on the medial side of the knee.[50] Other activities such as running or climbing stairs increases the force transmitted through the knee to approximately four to five times body weight.[50] Therefore it should not be surprising that individuals who are involved in regular vigorous physical activity tend to have knee problems, particularly medial knee problems. This so-called medial compartment overload may result in several disease processes, the most important of which is osteoarthritis.

Of all the joints, the knee best illustrates the biomechanical contribution to osteoarthritis and its progression. The enormous stresses which are put on the knee joint during physical activities evokes a response by the musculoskeletal system, resulting in osteoarthritis. Any pre-existing problems to the knee joint which produce increased articular surface stress result in a predictable osteoarthritis of the knee. These problems include meniscal tears, instability secondary to ligament disruption, irregularity of the articular surface secondary to a tibial plateau fracture, and angular deformities following fractures to the femur or tibia.[51]

Pain, usually with activity, is virtually always a presenting complaint of a patient with osteoarthritis of the knee. The patient may also complain of stiffness in the morning and after sitting for a prolonged period. This stiffness tends to lessen with activity but returns later in the day.[52] The history obtained from the patient is extremely important in determining the extent of the problem and the mode of therapy which should be employed. The most important piece of information, which must be extracted, is whether or not there is pain with activity. The physician must determine, with the help of the patient, if there is pain with all weight bearing and the location of the pain, and the specific limitations are important.

Some patients will actually have osteoarthritis, but will demonstrate extremely mild symptoms. Frequently there is no history of an initiating cause such as trauma. Many patients will describe intermittent difficulty with the knee that has gradually worsened over a period of years. The patient/athlete may also describe an apparent increase in deformity of the lower extremities.

Physical examination of the osteoarthritic knee should be directed towards demonstrating the location and degree of articular cartilage degeneration. Initially, any patient presenting with a lower-extremity problem should be carefully observed standing, walking, and then running on the spot. Males, who are more prone to varus deformity of the knee, tend to have more medial compartment osteoarthritis than the female, because varus alignment tends to overload the medial compartment of the knee. The range of motion of the knee should be noted for further reference. Care should be taken in differentiating mechanical disturbance of the knee seen with loose bodies or meniscal damage. The knee should be checked for injuries to ligaments, particularly to the anterior and posterior cruciate ligaments. Osteoarthritis can occur as a consequence of long-standing knee instability. Finally, examination of the osteoarthritic knee should include an examination of the hip and ankle. The status of the muscle strength and neurological examination of the vascular tree should be routine.

In radiographic analysis to delineate medial compartment osteoarthritis, plain radiographs of the knee should be augmented with weight-bearing radiographs. For a more in-depth analysis, CR scanning may be helpful and confirmatory in determining the presence of ligament and/or meniscal disturbance.

The treatment of osteoarthritis of the knee depends strongly on the severity of the symptoms, the structural abnormalities present, and associated factors such as the age of the patient and concomitant medical difficulties.[51] Most patients should initially be treated with a non-surgical programme consisting of exercise modification. Intra-articular corticosteroid injections should be used cautiously to control the inflammatory component of the osteoarthritis. Emphasis should be placed on isometric exercises designed to increase the strength of the surrounding musculature. Those exercises specific for quadriceps strengthening are similar to those previously described for patients with chondromalacia patella.

One treatment rationale which is frequently overlooked by the practitioner is the importance of weight loss by the athlete. Usually this approach is fundamental to the more mature athlete. As mentioned previously, the knee receives a compressive force during weight bearing which approximates three times body weight, depending on the position of the knee. It follows that if a patient/athlete loses 1 kg of body weight, the force transmitted through the knee joint may be decreased by as much as 3 kg. The

patient must understand that increased stress on the knee joint means increasing pain.

Another non-surgical technique for treatment of medial compartment osteoarthritis of the knee is the use of lateral heel wedges in shoes. Shoe orthotics, although cumbersome, are effective in relieving symptoms of medial compartment osteoarthritis.

Surgical intervention in medial compartment overload, with secondary osteoarthritis, is reserved for those patients who are refractory to non-surgical measures and are in considerable pain. Some examples of surgical procedures which may be performed are debridement, high tibial osteotomy, and, if all else fails, complete knee arthroplasty.

OSTEOCHONDRITIS DISSECANS

Introduction

Osteochondritis dissecans is a condition in which part of the articular surface of a joint separates due to a plane of cleavage through the subchondral bone. This condition is found most commonly in the knee, most specifically in the medial femoral condyle. However, it may occur in other joints such as the hip, ankle, and elbow.

In almost 75 per cent of cases the patient suffering with this condition is a young athletic male. The symptoms are usually present unilaterally. The literature reveals that there are two different clinical subsets which frequently present with osteochondritis dissecans of the knee:

(1) the child, or younger adolescent (5–15 years of age) who demonstrates open growth plates;
(2) the older adolescent or adult (15–50 years of age).

Theories

The most widely accepted theories regarding the aetiology of osteochondritis dissecans are trauma and ischaemia.

Trauma was first suggested as a cause for osteochondritis dissecans by Paget[52] and König.[53] König believed that trauma caused necrosis of part of the underlying bone and that this was followed by a dissecting inflammation which caused the fragment of bone and articular cartilage to separate from the underlying bone. In 1933, Fairbanks[54] published a paper which strongly supported the theory that trauma plays at least a part in the triggering of osteochondritis dissecans. Fairbanks suggested that the tibial spine could impinge against the medial femoral condyle during rotational strains (applied to the tibia or femur) and thus would cause a fracture through the subchondral bone. The articular surface could remain intact for a period of time which could depend upon how extensively the knee was being used. If, in fact, the knee continued to be used in its normal fashion, repeated movement could provoke a non-union and the fracture would eventually extend through the articular surface.[55]

Direct trauma to the knee has always been proposed in the aetiology of osteochondritis dissecans. The medial articular facet of the patella has been shown to contact the classic site of osteochondritis dissecans when the knee is fully flexed. In an experimental study in dogs, Rehhein[56] produced lesions which histologically and radiographically resembled those of osteochondritis dissecans by repeated minor trauma to the anterior aspect of the knee.[55]

Many researchers have suggested ischaemia as a major factor in the development of osteochondritis dissecans. The theory presented is that obstruction of end arteries to the femoral condyle, at the site of involvement, could precipitate a separation between the cartilage and bone. Everything from fat emboli to bacterial infection has been proposed to cause such vascular obstructions.[57]

Enneking[58] compared the blood supply of subchondral bone with that of bowel mesentery with its end-arterial arcade and found that terminal branches of vessels to subchondral bone anastomose poorly with their neighbours. Therefore infarction will result in necrosis of a wedge-shaped piece of bone immediately beneath the articular cartilage. A zone of granulation tissue is formed between the viable bone and the necrotic wedge, as in growth of vascular buds, and mesenchymal cells initiate resorption of the necrotic bone. The intact overlying articular cartilage usually holds the wedge in place; however, additional trauma may cause fracture of the articular cartilage and loosening of the wedge. The articular cartilage will remain intact since its nutrient supply is the synovial fluid, but the subchondral bone will undergo necrosis because of loss of blood supply.[57]

Symptomatology

The symptoms of osteochondritis dissecans are often poorly localized and non-specific. Patients usually report pain, but of varying degrees. Not infrequently stiffness and possible knee swelling is evident. As the condition progresses, a sensation of catching, locking, or giving way may be described. The symptoms more commonly appear with exertion and if a loose body is formed in the joint, symptoms will become more specific.[57]

Physical examination

Physical examination may reveal quadriceps atrophy. Rarely, an effusion may be obvious. The involved femoral condyle, usually medial, is frequently tender to palpation when the knee is flexed. The loose body may be palpated but this is extremely rare. The patient may walk with the tibia externally rotated in order to avoid impinging the tibial eminence with the lateral aspect of the medial femoral condyle. Flexing the knee to 90° while internally rotating the tibia will usually result in pain elicited at 30° of flexion—this is known as a positive Wilson test.[59] This pain is relieved by externally rotating the tibia.

Radiographic evaluation

Radiographic examination of the affected knee will reveal a well-circumscribed area of subchondral bone, sometimes separated from the remaining femoral condyle by a crescent-shaped radial lucent line.[57] An intercondylar notch or a view of the intercondylar notch—a tunnel view—is often the most useful radiography, as the classic location of the lesion is the posterolateral aspect of the medial femoral condyle. Haring[60] reports that the defect of osteochondritis dissecans is commonly located in this subarticular bone of the medial femoral condyle between two radiographic lines. One line extends anteriorly from the density of the roof of the intercondylar notch, and the other extends distally from the posterior cortex of the distal femoral diaphysis. Osteochondritis dissecans may also involve the articular surface of the patella.

Treatment of osteochondritis dissecans

The treatment of osteochondritis dissecans depends on the skeletal maturity of the patient and the stage of the lesion. When the individual is skeletally immature the knee has a greater capacity for healing. Thus the non-displaced osteochondritis dissecans lesion in patients under 15 years of age can be treated with immobilization and observation for up to 16 weeks.[61]

Aggressive surgery, such as drilling or pinning of the osteochondritic defect is reserved for partially or completely separated fragments, or for those lesions which do not completely heal after the prescribed 16 weeks of immobilization. The function of drilling the bone in a child who does not respond quickly to immobilization is merely to speed up the revascularization and to reduce the period of morbidity. Most cases do extremely well regardless of the treatment protocol. Unfortunately, the story is quite different for adults with osteochondritis dissecans.

Linden[62] evaluated 67 joints in 58 patients with an average follow-up of 33 years. None of these patients had loose fragments replaced or internal fixation of any kind. Linden concluded that children did well and generally had no secondary degenerative changes of complications. In contrast, adults often had pain, instability, and decreased range of motion. Osteoarthritis of the affected knee eventually appeared in 100 per cent of patients who developed osteochondritis dissecans after closure of the physis.

Arthroscopic surgery is an invaluable tool in the diagnosis and treatment of osteochondritis dissecans. With the aid of an arthroscope, the lesion may be visualized, drilled, curetted, or pinned. Loose bodies can be removed with minimal morbidity. Guhl[63] classifies lesions by location, percentage of weight-bearing surface, and degree of separation. The treatment chosen depends on these factors.

Operative treatment is indicated in a symptomatic adult knee with a lesion greater than 1 cm and involvement of the weight-bearing surface. Lesions with intact articular cartilage are drilled, while those which involve early separation of a fragment are drilled and may be pinned in place. Postoperative therapy is individualized and depends on the extent and severity of the original lesion. Those patients with intact lesions may bear weight immediately, while those with loose bodies or detached fragments can begin weight bearing when the lesions are stable.

The younger patient with osteochondritis dissecans involving a significant portion of the weight-bearing surface of the femoral condyle poses a serious therapeutic dilemma, particularly if the segment is completely separated and cannot be replaced. These patients are not candidates for knee arthroplasty, and other modes of treatment have also proved quite ineffective. Gross[64] has done some work with fresh osteochondral allografts in this condition, and has reported excellent results in patients who had osteoarthritis, post-traumatic osteoarthritis, and osteonecrosis. Other tools which are potentially quite useful in the treatment of osteochondritis dissecans are tibial or femoral osteotomies. In these cases an osteotomy may have some effect in unloading the overloaded medial compartment, decreasing the pressure on the osteochondritic lesion.

Clanton and DeLee[57] have devised the following summary of the treatment for osteochondritis dissecans. In the first instance it is essential to differentiate between childhood and adult forms. The symptomatic child is initially treated by decreasing activity. Multiple epiphyseal dysplasia and irregular ossification must be ruled out and then arthroscopy is indicated. This permits a direct visualization of the lesion and drilling of the soft, but intact, articular cartilage. Flap fragments are reattached with pins after the base is curetted under arthroscopic control. Loose bodies are treated in the same way as in the adult.

The treatment of the symptomatic adult is more aggressive and is based upon the radiographic and arthrographic appearance of the lesion. Soft but intact cartilage is drilled via the operative arthroscope. The arthroscope also allows drilling, curettage, and peg stabilization of separated but undisplaced fragments. Loose bodies in non-weight-bearing areas are removed during arthroscopy and the crater is debrided to offer a bleeding surface. Any larger loose bodies consisting of articular cartilage and bone must be reattached, particularly if a weight-bearing area of the femoral condyle is involved—this usually requires arthrotomy. Cancellous bone graft is used in the base of the crater when needed to support a loose segment and to elevate it for restoring joint congruity. Osteotomy, or the use of an allograft, is considered when the defect remains after surgical debridement and is followed by none-weight-bearing ambulation coupled with continuous passive motion.

Conclusion

Osteochondritis dissecans can be very perplexing, particularly when the joint surface damage is extensive. Surgery is reserved as a very last resort but may be necessary in the very severe case.

REFERENCES

1. Last RJ. *Anatomy: regional and applied.* 7th edn. New York: Churchill Livingstone, 1984.
2. Kaplan EB. Some aspects of functional anatomy of the human knee joint. *Clinical Orthopaedics* 1962; **23**: 18–29.
3. Hungerford DS, Perry M. Biomechanics of the patellofemoral joint. *Clinical Orthopaedics* 1979; **144**: 9–15.
4. Turek SL. *Orthopedics—principles and their applications.* Philadelphia: JB Lippincott, 1984.
5. Teider B, Marshall JL, Koslin B, Girgis FG. The anterior aspect of the knee joint. *Journal of Bone and Joint Surgery* 1981; **63A**(3): 351–6.
6. Fulkerson JP, Hungerford DS. *Disorders of the patellofemoral joint.* Baltimore: Williams & Wilkins, 1990.
7. Girgis FG, Marshall JL, Al Monajem ARS. The cruciate ligaments of the knee joint: anatomical, functional, and experimental analysis. *Clinical Orthopaedics* 1975; **106**: 216–31.
8. Odensten M, Gilliquist J. Functional anatomy of the anterior cruciate ligament and a rationale for reconstruction. *Journal of Bone and Joint Surgery* 1985; **67A**: 257–61.
9. Smillie IS. *Injuries of the knee joint.* Edinburgh: Livingstone, 1970.
10. Last R. J. The popliteus muscle and the lateral meniscus. *Journal of Bone and Joint Surgery* 1950; **32B**: 93–9.
11. Cohn AK, Mains DB. Popliteal hiatus of the lateral meniscus. *American Journal of Sports Medicine* 1979; **7**(4): 221–6.
12. O'Rahilly R, Gardiner ED, Gray DJ. *Anatomy. A regional study of human structure.* Philadelphia: WB Saunders, 1986.
13. Shaw JA, Eng M, Murray DG. The longitudinal axis of the knee and the role of cruciate ligaments in controlling transverse rotation. *Journal of Bone and Joint Surgery* 1974; **56A**: 1603.
14. Barnett CH. Locking at the knee joint. *Journal of Anatomy* 1953; **87**: 91–5.
15. McCall J. In: Wright V, ed. *Lubrication and wear in joints.* Philadelphia: JB Lippincott, 1969.
16. Weiss C, Rosenberg L, Helfet AJ. An ultrastructural study of normal adult human cartilage. *Journal of Bone and Joint Surgery* 1968; **50A**: 663–74.
17. James SL. Chondromalacia of the patella in the adolescent. In: Kennedy J. C., ed. *The injured adolescent knee.* Baltimore: Williams & Wilkins, 1979.
18. Mansour JM, Mow VC. The permeability of articular cartilage under compressive strain and at high pressures. *Journal of Bone and Joint Surgery* 1976; **58A**: 509–10.
19. Goodfellow J., Hungerford D. S., Woods C. Patellofemoral joint mechanics and pathology. Chondromalacia patella. *Journal of Bone and Joint Surgery* 1976; **58B**: 291–9.
20. Darracott J, Vernon-Roberts B. The bony changes in chondromalacia patellae. *Rheumatology and Physical Medicine* 1971; **11**: 175–9.
21. Chrisman, O'D. The role of articular cartilage in patellofemoral pain. *Orthopedic Clinics of North America* 1986; **17**(2): 231–3.
22. James SL, Bates BT, Ostering LR. Injuries to runners. *American Journal of Sports Medicine* 1978; **6**: 40–50.
23. Bently G, Dowd G. Current concepts of etiology and treatment of chondromalacia pagellae. *Clinical Orthopaedics* 1984; **189**: 209–27.
24. Carson W, James SL, Singer KM, Winternitz WW. Patellofemoral disorders: physical and radiographic evaluation. *Clinical Orthopedics* 1984; **185**: 165–77.
25. Hughston JC. Subluxation of the patella. *Journal of Bone and Joint Surgery* 1968; **50A**: 1003.
26. Dimon JH. Apprehension test for subluxation of the patella. *Clinical Orthopedics* 1974; **103**: 39.
27. Dehaven KE, Dolan WA, Mayer PJ. Chondromalacia patellae in athletes. Clinical presentation and conservative management. *American Journal of Sports Medicine* 1979; **7**(1): 5–11.
28. Fisher RL. Conservative treatment of patellofemoral pain. *Orthopedic Clinics of North America* 1986; **17**(2): 269–72.
29. Smillie JS. *Diseases of the knee joint.* London: Churchill Livingstone, 1974.
30. Radin EL. A rational approach to the treatment of patellofemoral pain. *Clinical Orthopaedics* 1979; **144**: 107–9.
31. Gruber MA. The conservative treatment of chondromalacia patella. *Orthopedic Clinics of North America* 1979; **10**: 105–15.
32. Insall J, Falvo K, Wise D. Chondromalacia patella. A prospective study. *Journal of Bone and Joint Surgery* 1976; **58A**: 1–8.
33. Hughston JC, Walsh WM. Proximal and distal reconstruction of the extensor mechanism for patellar subluxation. *Clinical Orthopaedics* 1979; **144**: 36.
34. Wiles P, Andrews PS, Bremmer RA. chondromalacia of the patellae: a study of the latest results of excision of the articular cartilage. *Journal of Bone and Joint Surgery* 1960; **42B**: 65.
35. Bently G. The surgical treatment of chondromalacia patellae. *Journal of Bone and Joint Surgery* 1978; **60B**: 74.
36. Renne JW. The iliotibial band friction syndrome. *Journal of Bone and Joint Surgery* 1975; **57A**(8): 1110–11.
37. Sutker AN, Jackson DW, Pagliano JW. Iliotibial band syndrome in distance runners. *Physician and Sports Medicine* 1981; **9**(1)).
38. Noble CA. Iliotibial band friction syndrome in runners. *American Journal of Sports Medicine* 1980; **8**(4): 232–4.
39. Ferretti A, Appolito E, Mariani PP, Puddu G. Jumper's Knee. *American Journal of Sports Medicine* 1983; **11**: 58–62.
40. Ferretti A, Papandrea P, Conteduea F. Knee injuries in Volleyball. *Sports Medicine* 1990; **10**(2): 132–8.
41. Blazina ME, Karlan RK, Jobe FW. Jumper's knee. *Orthopaedic Clinics of North America* **4**: 665–73.
42. Roels J, Martens M, Mulier JC, Burssens A. Patellar tendinitis (jumper's knee). *American Journal of Sports Medicine* 1978; **6**: 362–8.
43. Stanish WD, Curwin S. *Tendinitis: its etiology and treatment.* Lexington MA: Collamore Press, 1984.
44. Freeman MAR, Meachian G. Aging and degeneration. In: Freeman MAR, *Adult articular cartilage.* 2nd end. Tunbridge Wells: Pitman Medical, 1979: 487–543.
45. Muir IHM. Biochemistry. In: Freeman MAR. *Adult articular cartilage.* 2nd end. Tunbridge Wells: Pitman Medical, 1979: 145–214.
46. Marondas A. Physicochemical properties of articular cartilage. In: Freeman MAR. *Adult articular cartilage.* 2nd edn. Tunbridge Wells: Pitman Medical, 1979: 215–90.
47. McDevitt CA, Muir H. Biochemical changes in the cartilage of the knees in experimental and natural osteoarthritis in the dog. *Journal of Bone and Joint Surgery* 1976; **58B**: 94–101.
48. Meachian G, Brooke G. The pathology of osteoarthritis. In: Moskowitz RWD, Howell DS, Goldberg VM, Mankin HJ, eds. *Osteoarthritis, diagnosis and management.* Philadelphia: WB Saunders, 1984.

49. Maquet PGJ. *Biomechanics of the knee*. New York: Springer-Verlag, 1976.
50. Morrison JB. *The forces transmitted by the human knee joint*. Thesis, University of Strathclyde, Glasgow, 1967.
51. Kettefkamp DB, Colyer RA. Osteoarthritis of the knee. In: Moskowitz RW, Howell DS, Goldberg VM, Mankin HJ, eds. *Osteoarthritis, diagnosis and management*. Philadelphia: WB Saunders, 1984.
52. Paget J. On the production of some of the loose bodies in joints. *St Bartholomew's Hospital* 1870; **6**: 1.
53. König F. Verber freie Korper in den Gelentren. *Deutsche Zeitschrift für Chirurgie* 1887–8; **27**: 90.
54. Fairbanks HAT. Osteochondritis dissecans. *British Journal of Surgery* 1933; **21**: 67.
55. Green JP. Osteochondritis dissecans of the knee. *Journal of Bone and Joint Surgery* 1966; **48B**(1): 82.
56. Rebhein F. Die Entstehung der Osteochondritis dissecans. *Archiv fur klinische Chirurgie* 1950; **256**: 69.
57. Clanton TO, DeLee JC. Osteochondritis dissecans. History, pathophysiology and current treatment concepts. *Clinical Orthopaedics* 1982; **167**: 50.
58. Enneking WF. *Clinical musculoskeletal pathology*. Gainesville, FL: Shorter, 1977: 147.
59. Wilson JN. A diagnostic sign in osteochondritis dissecans of the knee. *Journal of Bone and Joint Surgery* 1967; **49A**: 477.
60. Harding WG. III. Diagnosis of osteochondritis dissecans of the femoral condyles. *Clinical Orthopaedics* 1977; **123**: 25.
61. Smillie IS. Treatment of osteochondritis dissecans. *Journal of Bone and Joint Surgery* 1957; **39B**: 248.
62. Linden B. Osteochondritis dissecans of the femoral condyles. *Journal of Bone and Joint Surgery* 1977; **59A**: 769.
63. Guhl JF. Arthroscopic treatment of osteochondritis dissecans *Clinical Orthopaedics and Related Research*, 1982; **167**: 65–74.
64. Gross AE. *Course on rehabilitation of articular joints by biological resurfacing*. St Louis, MO, November 1979.

5.5 Overuse injuries of the spine

LYLE J. MICHELI AND ROBERT A. YANCEY

INTRODUCTION

Most reviews of sports-related injuries show a relatively low incidence of injury to the spine.[1] While there are some sports in which the risk of injuries to the back is higher than in others, the overall level is generally low. Exceptions to this, of course, may include gymnastics, dance, American football, rowing, and weight-lifting. What these sports have in common are repetitive flexions, extensions, or rotations of the spine, particularly the lumbar spine, which increase the risk of injury to the structures of the spine.[2–19]

While specific injuries to the spine are relatively uncommon in sports, back pain is a relatively common complaint shared by both athletes and non-athletes. It is imperative that the physician dealing with an athletically active individual complaining of back pain be aware that the pattern of injury and back pain as an athlete is quite different from that in a non-athlete. The diagnoses are quite different from those seen in the general population. Failure to appreciate the difference between the pattern of diagnosis and the structures injured in the athlete in contrast with the general population result in many misdiagnoses and delays in diagnosis which have had a serious impact not only on athletic performance but also on the future potential for healing in a number of these injuries.

As with injuries at any site, two very different mechanisms, or occasionally a combination of the two, may be responsible for complaints of pain in the spinal region. Acute traumatic injuries to the spine are the result of direct blows, twists, or sudden applications of force. Needless to say, many of these may be medical emergencies and may ultimately be diagnosed as acute traumatic fracture or dislocation, as well as sprain of the ligamentous structures of the spine. Overuse injuries—the result of repetitive activity—are much more specific to the training phase of sports participation.

The onset of pain may actually occur during either sports participation or training, but the mechanism of injury is usually a combination of repetitive flexion, extension, or rotation of the spine occurring in actual competition as well as in preparation for the competition. Most commonly the pattern of pain will be that of slow gradual onset. Often the athlete cannot remember exactly when he or she first experienced back pain, but it frequently increases in severity and duration in association with progressive training. Occasionally, an injury which ultimately turns out to be an overuse injury may have an acute onset such as one particular twist or fall. With subsequent diagnostic assessment and evaluation, it is evident that the weakening of the structure which was ultimately injured was probably the result of a progressive subtle structural change with perhaps a single superimposed injury.

SPINE ANATOMY AND BIOMECHANICS

The spinal elements consist of the seven cervical vertebrae, the 12 thoracic vertebrae, and the five lumbar vertebrae, perched upon the sacrum and pelvis. The structure and function of each segment of the spine is specific to demands placed upon it anatomically and physiologically. The vertebrae of the neck have demands for both range of motion and structural integrity. The proximal three levels of the cervical spine and the junction of the cervical spine with the skull account for most of the motion about the spine, particularly rotation. The apparent focus of repetitive mechanical activity in the cervical spine appears to be at the C5–C6 and C6–C7 disc and joint levels. Progressive anatomical changes appear to occur at these levels in particular , and as a result impingement of the nerve roots at the neural foramina may occur. The majority of the 'overuse injuries' of the cervical spine are the result of a combination of discogenic and facet deterioration and arthrosis with secondary bony overgrowth and impingement. Impingement, as noted, is generally that of the exiting nerve roots, but sometimes a frank myelopathy may occur. While rare, this should always be considered in the differential diagnosis of neck pain in the older athlete.

Overuse injuries of the thoracic spine are relatively rare, and this is undoubtedly related to the structure and function of the thoracic spine. There are usually 12 osseous segments in the thoracic region. Each osseous segment progresses in size from cranial to cephalad and forms a protective bony ring around the neural tube. The osseous component of the thoracic spine, the vertebral bodies, are joined anteriorly by the discs posteriorly at the facet joints with associated ligamentous structures. There is a normal posterior angulation, or kyphus, of the thoracic spine which varies normally between 20° and 40° as measured by the Cobb technique. It is noteworthy that, while direct injury to the structures of the thoracic spine from repetitive overuse activities in sports is rare, the structural anatomy of the thoracic spine can indirectly contribute to the occurrence of overuse injuries in both the cervical and lumbar spine.

As an example, a significant decrease in the kyphus of the thoracic spine with a 'flat-back' alignment of the spine may result in a relative hypolordosis of the lumbar spine with an apparent increase in the mechanical forces at the high lumbar and thoracolumbar junction. We have observed that repetitive overuse injuries at the thoracolumbar junction, called by some authors 'atypical Scheuermann's disease', is invariably associated with a hypokyphosis of the thoracic spine and hypolordosis of the lumbar spine.[20,21] In contrast, increased postural thoracic kyphosis is often associated with a relative forward head thrust and hyperlordosis of the cervical spine. This is frequently associated with chronic strain of the posterior cervical and cervicothoracic muscles seen in athletes involved in sports such as tennis or swimming. There also appears to be an association between the thoracic kyphus and the development of long-term degenerative changes at the lower cervical spine.

The five bony elements of the lumbar spine are joined to the pelvis at the sacrum. As opposed to the cervical spine, the more proximal elements of the lumbar spine contribute relatively less to the flexion, extension, and rotation of the lumbar spine. Major components of lumbar motion, as well as the concentration of relative stresses in the lumbar spine, occur near the base from the L3–L4 juncture through L5–S1. This in turn is reflected in the pattern of degenerative changes seen from repetitive activity in the lumbar spine. Failure of both the anterior elements of the spine, in particular the disc and surrounding plates, and the posterior elements of the lumbar spine, and in particular the pars interarticularis, occur near the base of the spine. The most common level is the L5–S1 level followed sequentially by L4–L5 and L3–L4.[22]

There is growing evidence that overuse stress to the posterior elements concentrates primarily at the pars interarticularis. Cadaver studies and computer analogue research have suggested that the incidence of posterior element failure is overwhelmingly at the pars interarticularis.[23] Pedicle overuse injury, while rare, has certainly been reported in the literature.[14]

Several recent studies have suggested that the duration and intensity of training is directly related to lumbar spine failure in young athletes.[8,24] Unfortunately, we are still not in a position to give coaches, athletes, and parents an accurate statement of how much training is enough and how much is 'too much' for a given sport or a child at a given age.

As with overuse injuries in general, a number of risk factors for overuse injury of the lumbar spine can be identified in the occurrence of injury (Table 1). These risk factors include training, in particular its duration and intensity as well as its rate of progression, muscle–tendon imbalances about the spine, and anatomical factors such as pre-existent lumbar lordosis. One study of spine

Table 1 Overuse injury risk factors

Training error
Muscle–tendon imbalance
Anatomic malalignment
Footwear
Playing surface
Associated disease state
Nutritional factors
Cultural deconditioning

injuries in young athletes in Scandinavia suggested that lumbosacral inclination was the one major factor which could be related to the occurrence of back pain. In sports in which there is an element of impact, such as gymnastics, the impact characteristics of surface are definitely a factor in the occurrence of injury, although they have not yet been measured in any sequential fashion. Of course, the shoewear used in certain sports activities, in particular its ability to dissipate force, is also a factor in these injuries. Gender has not yet been identified as an obvious factor in sports-related injuries, but there is no question that gender factors appear to play a role in the occurrence of back pain in industrial populations. Many of the cases of back pain in athletes have been identified in young female gymnasts. Pre-existent cultural conditions which include extended periods of sedentary activity, such as sitting in school, riding in cars or buses, or sitting in front of computer modules, are undoubtedly a factor. A back which has been deconditioned by these activities and is then put into short periods of flexion, extension, and rotation may be at additional risk of overuse injury. Finally, the level of skeletal maturation may play an important role in the occurrence of overuse injury. Studies of young Italian weight-lifters, gymnasts, interior linemen in American football, and children in general have strongly suggested that there is increased risk of overuse injury, particular of the posterior elements of the spine, when repetitive stresses are applied to the growing spine.[2,4,5,25] As is becoming increasingly evident in many sports throughout the world, intense high level training is being applied to younger and younger athletes in such sports as gymnastics, figure skating, tennis, and more traditional team sports such as hockey, soccer, field lacrosse, and field hockey.

Studies of spondylolysis have strongly supported the concept that this is an acquired condition, although congenital predisposition may exist. In a study of 143 non-ambulatory institutionalized patients over 10 years of age Rosenberg et al.[26] found zero occurrence of spondylolysis. Other studies have suggested a genetic predisposition to this condition, such as in the Inuit population of northern Canada.[27,28] However, these studies were performed on adult spines and therefore in no way answer the question as to whether these were truly congenital conditions, i.e. they occurred at birth or were detected within the first year of life, or were acquired during growth and development, given the special demands of the lower lumbar spine in this particular population.

The majority of spine complaints in the athletically active patient are secondary to chronic overuse injuries resulting in repetitive microtrauma to the thoracolumbar spine. Many of these complaints have an insidious onset and are classified as 'chronic back strain'; thus there are delays in diagnosis and definitive treatment. While repetitive microtrauma and discogenic pain make up the majority of the diagnoses in this category, the more serious problems of metabolic, neoplastic, and infectious aetiologies must

Table 2 Differential diagnosis of spine pathology in athletes less than 20 years old

Developmental
Scoliosis
Kyphosis
Spondylolisthesis
Spondylolysis

Acquired
Acute
Musculotendinous sprain/strain
Stress fracture
   Spondylolysis
   Vertebral end-plate fractures
Fractures
   Ring apophysis fracture
   Transverse processes
   Spinous processes
   Compression fracture
Herniated nucleus pulposus
Infection
   Disc space
   Vertebral osteomyelitis

Chronic
Musculotendinous sprain/strain
Stress fracture
   Spondylolysis
   Vertebral end-plate fractures
Spondyloarthropathy
Neoplasm
   Osteoblastoma
   Osteoid osteoma
   Metastatic neoplasm
Infection

Table 3 Differential diagnosis of spine pathology in athletes more than 20 years old

Acute
Musculotendinous sprain/strain
Stress fracture
   Spondylolysis
Fractures
   Transverse processes
   Spinous processes
   Compression fracture of vertebral body
Discogenic pain—acute herniated nucleus pulposus
Infection

Chronic
Musculotendinous sprain/strain
Stress fracture
   Spondylolysis
   Lumbar pedicle fracture
Mechanical
   Facet arthropathy
   Spinal stenosis
   Herniated nucleus pulposus—degenerative disc disease
Infection
Spondyloarthropathy—Reiter's syndrome, ankylosing
   spondylitis
Metastatic or primary neoplasm
Referred pain
   Intra-abdominal
   Retroperitoneal
   Pelvic

be entertained with persistent symptoms over several weeks, despite the fact that the pain began in association with sport (Tables 2 and 3).

STRESS REACTION OF THE PARS INTERARTICULARIS AND SPONDYLOLYSIS IN THE YOUNG ATHLETE

Mechanical injury to the pars interarticularis is probably the most frequently encountered anatomical lesion diagnosed in the young athletic population. It has been frequently described as a stress fracture resulting in a bony defect in the pars interarticularis at one or both sides of a given vertebral level.[28–30] The instability created by this lesion results in pain, particularly when stresses are placed across the posterior elements of the spine.

Epidemiology

Studies have shown that approximately 6 per cent of adults in the general population have evidence of spondylolysis[25,31] and that the mean age of the symptomatic population is between 15 and 16 years of age. Eighty-five per cent of the lesions occur at the L5 vertebral level. Although pars defects have never been identified in the newborn infant, a genetic predisposition to these lesions has

been documented in multiple studies.[11,32–34] There is considerable variation of incidence between races, with an incidence of only 2 per cent in black subjects and as high as 50 per cent in some Inuit communities.[30,33] In the athletic population, it is found much more frequently in athletes who sustain repetitive traumatic stresses on the lumbar spine, as is seen in blocking and sled training in American football, or perform specific repetitive lumbar motions, as seen in ballet dancing, competitive diving, pole vaulting, hurdling, and gymnastics.

Pathogenesis

Defects of the pars interarticularis have generally been classified into dysplastic, isthmic (traumatic), and degenerative types. The pathogenesis of this lesion in athletes is believed to be due to microtrauma and resultant stress fractures of the posterior elements of the spine rather than a congenital condition.[27–30,35] It has been noted in biomechanical studies that shear stresses across the pars interarticularis are increased when the spine is extended and accentuated with lateral flexion manoeuvres from a hyperlordotic posture.[23,,27,35] Similar to other fatigue fractures, the diagnosis of a stress reaction of the pars interarticularis is initially made with a bone scan before development of a plain radiograph.[36] However, in contrast with other fatigue fractures, pars interarticularis defects can develop at an earlier age and have a hereditary predisposition.[11,32–34] They rarely develop an exuberant periosteal reaction and the radiographic evidence of the lesion can persist even in the stable asymptomatic state.

   The history of onset of pain is very important in the diagnosis of spondylolysis. In many instances, the onset of symptoms coincides closely with the adolescent growth spurt. The pain usually has an insidious onset and athletes complain of a dull backache which is exacerbated by activity. It eventually becomes more severe, is associated with daily activities and is relieved by rest. Rarely, patients complain of radicular symptoms. A history of repetitive strenuous activity involving flexion/extension and rotation of the spine is almost always elicited.

On physical examination, 80 per cent of patients are noted to have relatively tight hamstrings.[37] Subtle changes in hamstring flexibility may be noted in hyperflexible athletes such as gymnasts and dancers. Palpation usually elicits tenderness localized at the involved vertebral level. Pain is reproduced by asking the patient to extend the lumbar spine against resistance. Active hyperextension and rotation of the lumbar spine while in a one-legged stance specifically stresses the ipsilateral pars and elicits pain.

The diagnostic work-up consists initially of anteroposterior, lateral, and oblique radiographic views of the lumbosacral spine to assess the integrity of the posterior elements. The defect can be seen as a narrow gap with irregular edges in the pars interarticularis. Reactive sclerosis may be seen opposite the lesion in unilateral cases. Radiographic changes may not be apparent initially, and radioisotope bone scans are very useful in the prompt confirmation of the diagnosis.[36] Single-photon emission CT has also been useful in further localizing spondylolytic defects.[31] When the diagnosis is indeterminate, CT scanning can be used to assess the integrity of the pars interarticularis. The diagnosis of osteoid osteoma, facet arthropathy, and infection must be entertained and ruled out in the initial work-up when the bone scan is noted to be positive.

Spondylolysis in the athletic population is generally a mechanically stable lesion, and the usual problem for the patient is the potential for activity-related pain rather than spine instability. Since the isthmic spondylolysis seen in athletes is thought to be the equivalent of a stress fracture of the posterior elements, the current initial management consists of a restriction of activities and immobilization of the lumbar spine with a rigid polypropylene antilordotic lumbosacral brace (Fig. 1).

A brace with 0 to 15 per cent lumbar flexion acts to flatten the lumbar lordosis while immobilizing the spine, relieves the pain, and promotes healing of the lesion. After a brace has been fitted, the patient wears it for 23 out of 24 h a day for 6 months. Concurrently, abdominal strengthening, pelvic tilts, and antilordotic and lower-extremity flexibility exercises are prescribed. Most patients are able to resume limited activities to maintain their aerobic fitness and muscle strength when they become pain free in the brace in 3 to 4 weeks. The presence of hamstring tightness has been suggested as an indicator of the success of a treatment programme. Physical therapy intervention at this point focuses on abdominal strengthening. Bone scans are a useful adjunct to the clinical examination for following the status of the lesion. Brace treatment can be reduced in those patients with an initially negative bone scan who become asymptomatic after 4 to 6 weeks of brace wear. Usually patients need 6 months to wean themselves from full-time brace wear. About 32 per cent of patients treated in this manner healed their lesion and 88 per cent were able to resume sports activities that were previously painful, even though their lesions had not healed.[38] In patients who are asymptomatic, despite an apparent non-union of the lesion, full return to activities including contact sports is allowed. The clinical status of the patient takes precedence over radiographic examination in the follow-up of these lesions.

Athletes who are unable to be weaned from their brace without a recurrence of their symptoms may require surgery. A posterolateral transverse process fusion is the classic form of treatment, although direct osteosynthesis of the lesion has been described.[22,39-41]

After surgery, the patient can expect to be immobilized in a cast or brace for 6 months and sports activities are not allowed for 12 months following fusion. Contact sports are contraindicated after lumbar fusion.[42]

Cost containment

Serial radiographs are unnecessary in the follow-up of spondylolytic defects. After an initial screening radiographic series of the lumbar spine, a physician may want to confirm the diagnosis with a technetium bone scan. Patients may then be followed clinically over the next several months of brace wear. When patients are being weaned from their brace or brace wear is being terminated, another bone scan may be helpful to assess the status of the lesion. After an initial evaluation, physical therapy may be readily instituted at home with monthly follow-up intervals.

Spondylolisthesis occurs as a vertebral body slips forward on the one below it, subsequent to a pars interarticularis defect (Fig. 2). The dysplastic type, which features spondylolysis with malformed or dystrophic posterior elements, has a much higher incidence of slippage than the isthmic type seen in the athletic population. It is most frequently noted in non-athletic females during their adolescent growth spurt between 10 and 15 years of age. The L5–S1 vertebral level is involved in 85 to 90 per cent of cases. Studies have suggested that there is a genetic predisposition to spondylolisthesis. The incidence of high grade spondylolisthesis (more than 75 per cent slippage) is much higher in females than in males.[33,34]

These patients may give a history similar to that elicited with spondylolysis, although in some instances presentation may be with tight hamstrings rather than pain. Severe grades of spondylolisthesis with a palpable step-off are rare in the athletic population, and associated pain and flexibility loss preclude most athletic participation. The diagnosis is made by reviewing standing lateral radiographs of the lumbar spine and calculating the percentage of slip.

Most authors suggest that patients with less than 30 per cent asymptomatic spondylolisthesis should have no athletic restrictions and may be able to participate in contact sports. Patients with minimal symptoms are treated with a conservative programme of physical therapy emphasizing antilordotic exercises, abdominal strengthening, and modification of their sports activities. Brace management with a rigid antilordotic polypropylene lumbosacral orthosis is carried out for 3 to 6 months in those patients who fail initial treatment with physical therapy. Antilordotic and abdominal strengthening exercises are continued while the patient is still in the brace. Patients may resume their activities in a brace when they become asymptomatic.[22,38] Between 30 and 50 per cent of spondylolisthesis in skeletally immature patients should be closely followed for signs of progression. Surgical management is reserved for those patients whose vertebral slippage increases over 50 per cent and posterior in-situ fusion is the procedure of choice.[22,39,43,44] Contact sports are contraindicated after lumbar fusion.[42]

SCHEUERMANN'S DISEASE

Some adolescents will compensate for tight lumbodorsal fascia and hamstrings by developing a roundback deformity. This postural deformity is usually transient; however, in some cases

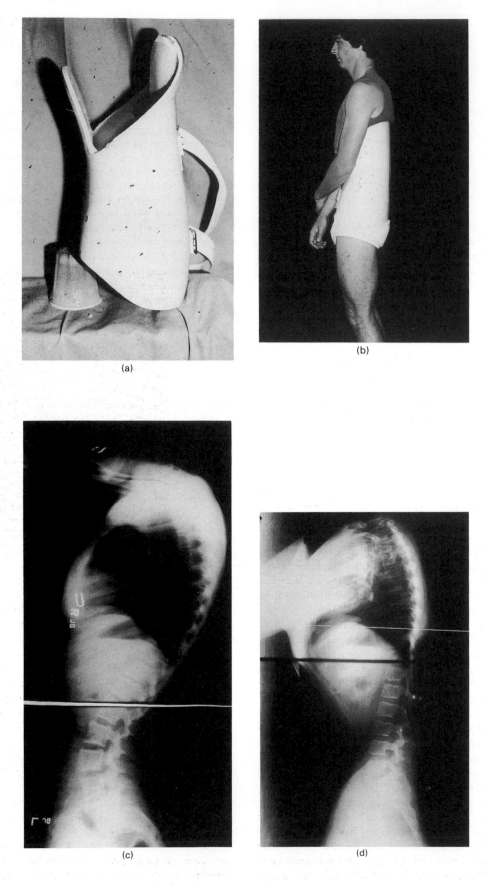

Fig. 1 An antilordotic thermoplastic brace (a) can be used as part of the treatment regimen of young athletes with back pain (b). A lateral radiograph of the lumbar spine (c) before and (d) after application of a 0° Boston brace.

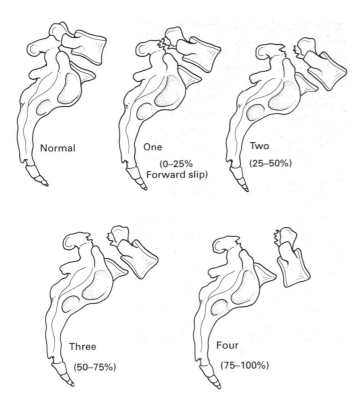

Fig. 2 Spondylolisthesis is graded into four degrees of severity based upon the extent of forward slip.

APOPHYSEAL MICROTRAUMA, ATYPICAL SCHEUERMANN'S DISEASE

In young athletes a radiographic picture which resembles the wedged vertebra and irregular vertebral end-plates seen in thoracic Scheuermann's disease is sometimes obtained in the mid-thoracic to the mid-lumbar spine.[20] This is most commonly seen in adolescent athletes with repetitive flexion/extension of the spine such as rowers and gymnasts in very rigorous training programmes. This has been referred to as 'atypical' or lumbar Scheuermann's disease since it appears to be the result of repetitive microtrauma in athletes and does not meet all the radiographic criteria of true Scheuermann's disease (Fig. 3). The peak age is between 15 and 17 years and there is a 2:1 male predominance.

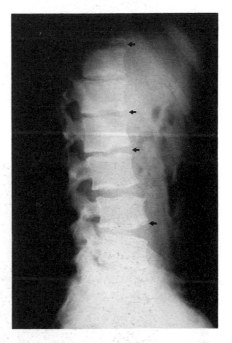

Fig. 3 Atypical Scheuermann's disease with anterior vertebral wedging and irregularity of the vertebral end-plates.

patients will develop anterior wedging of the vertebral bodies.[21] The radiographic criteria for Scheuermann's disease are met when three or more vertebra are wedged more than 5°.[45,46] Classic Scheuermann's disease, or juvenile thoracic kyphosis, is usually painless and rarely seen in its true form in the athletic population.[20,21] In contrast, 'atypical' or lumbar Scheuermann's disease presents with irregular vertebral end-plates at the thoracolumbar junction and is seen more frequently in the athletic population. This form of the disease is painful and is associated with activities that produce repetitive microtrauma to the thoracolumbar spine.

Thoracic Scheuermann's disease usually presents without a distinct traumatic history and with deformity rather than pain. The radiographic picture which is diagnostic includes irregular vertebral end-plates, Schmorl's nodes, narrowed disc spaces, and anterior wedging of three consecutive vertebral bodies.[46]

Patients present with a roundback deformity which they cannot reverse with forced hyperextension. Many of these patients have a relatively flat lumbar spine with tight hip flexors and hamstrings.

Treatment initially addresses the tight lumbodorsal fascia and hamstrings with flexibility exercises. Abdominal strengthening exercises are added to this programme. Progressive thoracic kyphosis above 50° in a skeletally immature athlete is an indication for 18 h/day treatment with a Milwaukee or modified Boston brace which precludes most sports activities. Further progression beyond 70° is an indication for spinal fusion and instrumentation. These patients are then restricted from sports activities, except swimming, for 1 year postoperatively. Contact sports and gymnastics are not allowed after surgery but patients can eventually return to light non-contact activities.[42]

Repetitive flexion and extension of the thoracolumbar junction results in multiple growth-plate fractures and secondary bony deformation of the vertebra. Apophyseal fragments at the anterior margin of the vertebral body may be avulsed, resulting in Schmorl's node formation and irregular vertebral end-plates (Fig. 3).

The history is usually significant for complaints of transient nondescript pain at the thoracolumbar junction followed by complaints of moderately severe pain accentuated by forward flexion and relieved by rest. Radicular complaints are rare. Invariably, these patients are consistently noted to have tight lumbodorsal fascia and hamstrings and relative thoracic hypokyphosis and lumbar hyperlordosis (flatbacks). Neurological examination is unremarkable. Radiographs reveal multiple irregularities of the vertebral end-plates, signs of chronic vertebral end-plate wedging, and subsequent changes in the disc space.

The hallmark of treatment is rest and avoidance of the inciting activity. Persistent symptoms may be managed with non-steroidal anti-inflammatory drugs (NSAIDs). A bracing regimen may be instituted using a semirigid thermoplastic brace with 15° of lumbar lordosis to immobilize the patient until remodelling is seen on plain radiographs.[47] Patients may return to sports while they are braced and are prescribed a flexibility programme. No restrictions are placed on the patients when they are weaned from their brace. Surgical intervention is rarely indicated in these patients.

Preparticipation sports screening is an important avenue for evaluating the young athlete for scoliosis.[48,49] Idiopathic scoliosis does not cause pain and does not functionally impair the athlete. Congenital scoliosis has associated renal and cardiac anomalies that should be thoroughly excluded before the young athlete is allowed to participate in sports. Patients with congenital anomalies of the cervical spine and cervicothoracic junction should be advised not to participate in contact sports.

Diagnosis

In the skeletally immature child, curves are usually noted by the child's parents or coach as an asymmetry in the child's shoulders or thoracic spine accentuated by forward bending or shoulder protraction. These patients should be referred to an orthopaedic surgeon for a full evaluation (Fig. 4).

Management

If a curve is noted to progress rapidly or reaches 25 to 30° in the skeletally immature child, a corrective bracing programme is initiated to control the progression of the curve. Patients generally wear the brace for at least 18 h per day and are allowed to participate in sports activities in or out of the brace without restrictions. An active strengthening and flexibility programme is encouraged during brace wear. When growth is completed and the athlete is weaned out of the brace, there are no restrictions placed on patients with residual idiopathic scoliosis. Patients whose curves progress beyond 50° have a high incidence of progression and require anterior or posterior spinal fusion and corrective instrumentation. Since the fusion involves significant motion segments of the thoracic and lumbar spine with subsequent loss of mobility, contact sports, gymnastics, and diving sports are prohibited following this treatment.[42]

Painful scoliosis may also be the presenting physical finding associated with disc herniations, osteoid osteomas, osteoblastomas, spondylolisthesis, infections, and intraspinal tumours in the child. A painful scoliosis should always raise suspicion in the evaluation of a child with back pain.

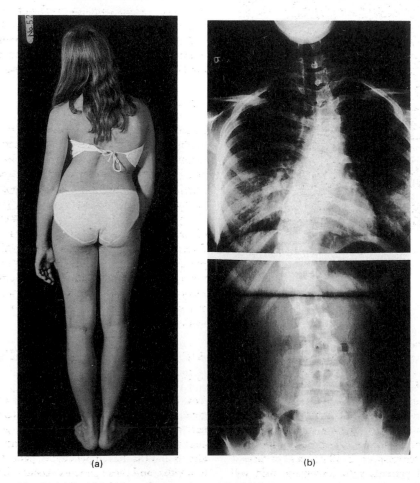

(a)   (b)

Fig. 4 (a) A child with scoliosis and secondary spinal asymmetry; (b) a standing coronal radiograph of the spine demonstrating the spinal curvature.

DISCOGENIC BACK PAIN IN THE ADULT ATHLETE

Compression of a nerve root along its anatomical course causes radicular pain. In the athletic population under 40 years old, the most common cause of radicular pain is a herniated disc. This condition is rare in the prepubescent child and the incidence increases from adolescence to the adult population.[50-52]

Sharp sciatic pain radiating from the buttock and extending below the knee in the distribution of the affected nerve root are the classic symptoms. Initially, symptoms are typically exacerbated by increased activity and the pain may eventually become a dull ache in the buttock or hip region. Subtle signs noted by the athlete or the trainer are slight symmetric decreases in hamstring flexibility, paravertebral spasm, scoliosis, and changes in running patterns. Physical examination typically reveals positive straight-leg raising and occasionally sensory and motor changes in the involved nerve root distribution. The diagnostic evaluation of choice at present is magnetic resonance imaging (MRI) (Fig. 5). Myelography and CT scans are now reserved for those patients whose MRIs are equivocal.

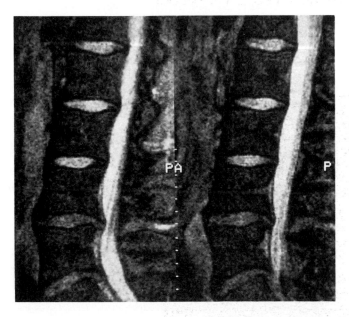

Fig. 5 MRI of the lumbar spine of a young athlete demonstrating disc prolapse.

Management

Conservative therapy remains the primary mode of treatment in these patients. NSAIDs and a relative period of rest for 2 to 7 days usually result in definite improvements in symptoms. Traction, braces, corsets, and manipulations have been advocated in the initial treatment; however, none has proven long-term effects on decreasing the symptoms. In athletes who did not initially respond to conservative management, bracing therapy using a flexible polyethylene brace with 15° of lumbar lordosis has been shown to be effective in returning athletes to sports in 50 per cent of cases.[47]

Only 5 to 10 per cent of patients with discogenic pain will eventually require further intervention. Treatment modalities using epidural steroids, chymopapain chemonucleolysis, per-

cutaneous discectomy, and disc excision have been described.[53-57] It is not known whether participation in vigorous contact sports following discectomy accelerates degenerative changes.

REFERRED BACK PAIN

Adult athletes may complain of chronic low grade back pain that may appear to be unrelated to sports activities. A thorough orthopaedic evaluation may be negative. Additional causes of back pain in the adult include pain that is referred from neoplastic or infectious origin and intra-abdominal, retroperitoneal, or pelvic sources. Further evaluation of these possibilities is warranted in these cases.

MECHANICAL BACK PAIN

Mechanical back pain is generally a diagnosis of exclusion in the athletic population. The symptoms are frequently non-specific and include nondescript low back pain that is exacerbated by activity and relieved by rest. Some authors have suggested that this mechanical pain syndrome is secondary to overuse or stretch injuries to the soft tissues, ligaments, joint capsules, and the facets themselves.[1,58]

Diagnosis

Excessive athletic activity with poor conditioning and training methods, insufficient pre-work-out stretching, and improper technique are commonly noted in the history. Plain radiographs and bone scans are normal. Patients with a predisposition to these injuries are noted to have weak abdominal muscles and tight lumbodorsal fascia, hip flexors, and hamstrings.

Management

Acutely, these patients respond well to a programme of rest and modification of activities. NSAIDs and buffered aspirin may be helpful during the acute phase. Hydrotherapy, ultrasound, and electrical stimulation to break up paraspinal spasm are useful physical therapy adjuncts. Some authors have suggested that injection of steroid and local anaesthetic into the facet joint relieves the symptoms in the majority of patients.[1,59] After the acute phase of pain is over and the diagnosis has been established, an individualized rehabilitation and stretching programme must be prescribed for the athlete. Suggestions must be made to the athlete for modifications in technique and training schedules to prevent recurrence of the injury. Persistent pain despite a careful attempt at conservative therapy necessitates a more complete diagnostic work-up.

ATHLETIC REHABILITATION OF THE SPINE

Low back pain is a very common musculoskeletal ailment in the general population and generates an enormous expense in health care and loss of productivity. In the general population, 50 per cent of patients recover in 2 weeks and 90 per cent are pain free after 3 months. Seven per cent of patients have pain for more than 6 months, and these patients account for 85 to 90 per cent of the compensation for low back pain.

In the athletic population, the rehabilitation of spinal injuries presents the clinician with a complex therapeutic challenge. Few of these patients eventually require surgical intervention. Most of them are not sufficiently debilitated from their injuries to refrain from normal daily activities, yet their athletic performance and enjoyment is significantly restricted. Their motivation to return to sport is high. The ultimate goal for the physician is to return the patient safely to the repetitive demands of athletics in a pain-free state as quickly as possible without exacerbating the problems. In the case of spinal rehabilitation, clear initial goals must be set regarding the time course of immobilization and rehabilitation for each problem. A clear understanding of the natural history of the specific problems and the expected time course of rehabilitation is essential for the patient to avoid the psychological pitfalls of a chronic back injury.

Many clinicians stress the role of exercise as a mainstay in the treatment of lumbar spine problems.[61,62] It has been postulated that exercise decreases pain by increasing endorphin levels. A decreased incidence of back pain and injury recurrence has been clearly associated with increased fitness levels.[63] For this reason, we emphasize the maintenance of aerobic cardiovascular fitness in the prevention and rehabilitation of lumbar spine problems.

Thermoplastic antilordotic thoracolumbar spine braces, such as the Boston brace, have previously been mentioned in a number of therapeutic modalities.[47] These lightweight braces have allowed patients to resume light sports activities while wearing the brace. However, it must be emphasized that all brace prescriptions are accompanied by a physical therapy programme designed to strengthen the abdominal musculature and increase flexibility of the spine and hips. An excellent description of a complete set of spine flexibility and strengthening exercises is outlined by Torg *et al.*[64]

REFERENCES

1. Spencer GW, Jackson DW. Back injuries in the athlete. *Clinics in Sports Medicine* 1983; **2**(1): 191–216.
2. Aggrawal ND, Kaur R, Kumar S, Mathur DN. A study of changes in the spine in weight lifters and other athletes. *British Journal of Sports Medicine* 1979; **13**(2): 58–61.
3. Brady TA, Cahill BR, Bodnar LM. Weight training-related injuries in the high school athlete. *American Journal of Sports Medicine* 1982; **10**(1): 1–5.
4. Ciullo JV, Jackson DW. Pars interarticularis stress reacion, spondylolysis, and spondylolisthesis in gymnasts. *Clinics in Sports Medicine.* 1985; **4**(1): 95–110.
5. Ferguson RH, McMaster JF, Stanitski CL. Low back pain in college football linemen. *American Journal of Sports Medicine* 1974; **2**(2): 63–9.
6. Garrick JG, Requa RK. Epidemiology of women's gymnastics injuries. *American Journal of Sports Medicine* 1980; **8**(4): 261–4.
7. Goldberg MA. Gymnastics injuries. *Orthopaedic Clinics of North America* 1980; **11**: 717–24.
8. Granhed H, Jonson, Hansson T. The loads on the lumbar spine during extreme weight lifting. *Spine* 1987; **12**(2): 146–9.
9. Granhed H, Morelli B. Low back pain among retired wrestlers and heavyweight lifters. *American Journal of Sports Medicine* 1988; **16**(5): 530–5.
10. Hall SJ. Mechanical contribution to lumbar stress injuries in female gymnasts. *Medicine and Science in Sports and Exercise* 1986; **18**(6): 599–602.
11. Hensinger RN. Back pain in children. In: Bradford DS, Hensinger RM, eds. *The pediatric spine.* New York: Thieme, 1985; 41–60.
12. Howell DW. Musculoskeletal profile and incidence of musculoskeletal injuries in lightweight women rowers. *American Journal of Sports Medicine* 1984; **12**(4): 278–81.
13. Hresko MT, Micheli LJ. Sports medicine and the lumbar spine. In: Floman Y, ed. *Disorders of the lumbar spine.* Rockville, MD: Aspen, 1990: 879–94.
14. Ireland ML, Micheli, LJ. Bilateral stress fracture in the lumbar pedicle in a ballet dancer. *Journal of Bone and Joint Surgery* 1987; **69A**(1): 140–2.
15. Jackson DW, Wiltse LL, Cirincione RJ. Spondylolysis in the female gymnast. *Clinical Orthopaedics and Related Research* 1976; **117**: 68–73.
16. McCarroll JR, Miller JM, Ritter MA. Lumbar spondylolysis and spondylolisthesis in college football players. *American Journal of Sports Medicine* 1986; **14**(5): 404–6.
17. Micheli LJ. Back injuries in gymnastics. *Clinics in Sports Medicine* 1985; **4**(1): 85–93.
18. Semon RL, Spengler D. Significance of lumbar spondylolysis in college football players. *Spine* 1981; **6**(2): 172–4.
19. Techakapuch S. Rupture of the lumbar cartilage plate into the spinal canal in an adolescent. A case report. *Journal of Bone and Joint Surgery* 1981; **63A**(3): 481–2.
20. Hensinger RN. Back pain and vertebral changes simulating Scheuermann's disease. *Orthopaedic Transactions* 1982; **6**(1): 1–6.
21. Micheli LJ. Low back pain in the adolescent: differential diagnosis. *American Journal of Sports Medicine* 1979; **7**(6): 362–4.
22. Bradford DS. Spondylolysis and spondylolisthesis in children and adolescents: current concepts in management. In: Bradford DS, Hensinger RM, eds. *The pediatric spine.* New York: Thieme, 1985: 403–23.
23. Cyron BM, Hutton WC. The fatigue strength of the lumbar in spondylolysis. *Journal of Bone and Joint Surgery* 1984; **60B**(2): 234–8.
24. Goldstein JD, Berger PE, Windler GE, Jackson DW. Spine injuries in gymnasts and swimmers. An epidemiologic investigation. *American Journal of Sports Medicine* 1991; **19**(5): 463–8.
25. Baker DR, McHolick W. Spondylolysis and spondylolisthesis in children. *Journal of Bone and Joint Surgery* 1956; **38A**(4): 933–4.
26. Rosenberg NJ, Bargar WL, Friedman B. The incidence of sponsylolysis and spondylolisthesis in non-ambulatory patients. *Spine* 1981; **6**(6): 35–8.
27. Troup JDG. Mechanical factors in spondylolisthesis and spondylolysis. *Clinical Orthopaedics and Related Research* 1976; **147**: 59–67.
28. Wiltse LL. The etiology of spondylolisthesis. *Journal of Bone and Joint Surgery* 1962; **44A**(3): 539–60.
29. Jackson DW. Low back pain in young athletes: evaluation of stress reaction and discogenic problems. *American Journal of Sports Medicine* 1979; **7**(6): 364–6.
30. Wiltse LL, Widell EH, Jackson DW. Fatigue fracture: the basic lesion in isthmic spondylolisthesis. *Journal of Bone and Joint Surgery* 1975; **57A**(1): 17–22.
31. Collier BD *et al.* Painful spondylolysis or spondylolisthesis studies by radiography and single photon emission computed tomography. *Radiology* 1985; **154**(1): 207–11.
32. Frederickson BE, Baker D, McHolick WJ, Yuan HA, Lubicky JP. The natural history of spondylolysis and spondylolisthesis. *Journal of Bone and Joint Surgery* 1984; **66A**(5): 669–707.
33. Pizzutillo PD. Spondylolisthesis: etiology and natural history. In: Bradford DS, Hensinger RM, eds. *The pediatric spine.* New York: Thieme, 1985: 395–402.
34. Winney-Davies R, Scott JHS. Inheritance and spondylisthesis—a radiographic family survey. *Journal of Bone and Joint Surgery* 1979; **61B**(3): 301–5.
35. O'Neill DB, Micheli LJ. Post-operative radiographic evidence for fatigue fracture as the etiology of spondylolysis. *Spine* 1989; **14**(12): 1342–55.
36. Papanicolaou N, Wilkinson RH, Emans JB, Treves S, Micheli LJ. Bone scintigraphy and radiography in young athletes with low back pain. *American Journal of Roentography* 1985; **145**: 1039–44.
37. Phalen GS, Dickson JA. Spondylolysis and tight hamstrings. *Journal of Bone and Joint Surgery* 1961; **43A**(4): 505–12.
38. Micheli LJ, Steiner ME. Treatment of symptomatic spondylolysis and spondylolisthesis with the modified Boston brace. *Spine* 1985; **10**: 937–43.
39. Bradford DS, Iza J. Repair of the defect in spondylolysis or minimal degrees of spondylolisthesis by segmental fixation and bone grafting: *Spine* 1985; **10**(7): 673–9.

40. Buck JE. Direct repair of the defect in spondylolisthesis. *Journal of Bone and Joint Surgery* 1970; **52B**(3): 432–43.

41. Buring K, Fredensborg N. Osteosynthesis of spondylolysis. *Acta Orthopaedica Scandinavica* 1973; **44**(1): 91.

42. Micheli LJ. Sports following spinal surgery in the young athlete. *Clinical Orthopaedics and Related Research* 1985; **198**: 152–7.

43. Bradford DS. Treatment of severe spondylolisthesis: a combined approach for reduction and stabilization. *Spine* 1979; **4**(5): 423–9.

44. Hensinger RN, Lang JR, MacEwen GD. Surgical management of spondylolisthesis in children and adolescents. *Spine* 1976; **1**: 207–16.

45. Bradford DS, Moe J, Montalvo JF, Winter RB. Scheuermann's kyphosis and roundback deformity. *Journal of Bone and Joint Surgery* 1974; **56A**(4): 740–58.

46. Sorenson HK. *Scheuermann's juvenile kyphosis.* Copenhagen: Munksgaard, 1964.

47. Micheli LJ, Hall JE, Miller ME. Use of modified Boston brace for back injuries in athletes. *American Journal of Sports Medicine* 1980; **8**(5): 351–6.

48. Lonstein JE. Natural history and school screening for scoliosis. *Orthopedic Clinics of North America* 1988; **19**(2): 227–37.

49. Micheli LJ. Preparticipation evaluation for sports competition: musculoskeletal assessment of the young athlete. In: Kelley VC, ed. *Practice of pediatrics.* Philadelphia: Harper & Row, 1984; 1–9.

50. DeOrio JK, Bianco AJ. Lumbar disc excision in children and adolescents. *Journal of Bone and Joint Surgery* 1982; **64A**(7): 991–5.

51. Garrido E, Humphreys RP, Hendrick EB, Hoffman JH. Lumbar disc disease in children. *Neurosurgery* 1978; **2**: 22–6.

52. Kurihara A, Kataoka O. Lumbar disc herniation in children and adolescents. A review of 70 operated cases and their minimum 5 year follow-up studies. *Spine* 1980; **5**(5): 443–51.

53. Brown FW. Epidurals—management of discogenic pain using epidural and intrathecal steroids. *Clinical Orthopaedics and Related Research* 1977; **129**: 72–8.

54. Day AL, Friedman WA, Indelicato PA. Observations on the treatment of lumbar disc disease in college football players. *American Journal of Sports Medicine* 1987; **15**(1): 72–5.

55. Green P, Burke A, Weiss C, Langan P. The role of epidural cortison injection in the treatment of discogenic low back pain. *Clinical Orthopaedics and Related Research* 1980; **153**: 121–5.

56. Jackson DW, Rettig A, Wiltse LL. Epidural cortisone injections in the young athletic adult. *American Journal of Sports Medicine* 1980; **8**(4): 239–43.

57. Nordby EJ. Chymopapain in intradiscal therapy. *Journal of Bone and Joint Surgery* 1983; **65A**: 1350–3.

58. Jackson DW, Wiltse LL. Low back pain in young athletes. *Psysican and Sportsmedicine* 1974; **2**(11): 53–60.

59. Fairbank JCT, Park WM, McCall IW, O'Brien JP. Apophyseal injection of local anesthetic as a diagnostic aid in primary low back syndromes. *Spine* 1981; **6**(6): 598–605.

60. Kahanovitz N. Lumbar spine. In: *Orthopaedic knowledge update 3: Home study syllabus.* Park Ridge, IL: American Academy of Orthopedic Surgeons, 1990.

61. Jackson CP, Brown MD. Analysis of current approaches and a practical guide to the prescription of exercise. *Clinical Orthopaedics and Related Research* 1983; **179**: 46–54.

62. Jackson CP, Brown MD. Is there a role for exercise in the treatment of patients with low back pain? *Clinical Orthopaedics and Related Research* 1983; **179**: 39–46.

63. Cady LD Jr, Thomas PC, Karwasky RJ. Program for increasing health and physical fitness of fire fighters. Journal of Occupational Medicine 1985; **27**: 110–4.

64. Torg JS, Vegso JJ, Torg E. The low back. In: *Rehabilitation of athletic injuries: an atlas of therapeutic exercise.* Chicago: Year Book Medical Publishers, 1987.

Sports injuries of special groups

6

6.1 Introduction

LYLE J. MICHELI

Athletically-active individuals or individuals participating in organized exercise activities of all ages and backgrounds can benefit from safe and injury-free exercise. The health benefits of sports and exercise have received traditional support in the medical community and have recently been supported by scientific investigation. The particular benefits to growing children, elderly individuals, whether male or female, and the handicapped cannot be overemphasized. In each of these groups, exercise and organized sports can provide important physical and psychological benefits for participants. It is imperative in these groups, in particular, that careful study of sports activities be done with an eye toward prevention of unnecessary injury, as well as rapid diagnosis, treatment, and rehabilitation when injuries do occur.

Exercise has been demonstrated to be extremely important for normal growth and development of children. Increasingly, with the growing complexities of urbanized societies, the only exercise many children will get will be in the organized sports or exercise setting. In North America, the growth of organized sports for children and adolescents has continued apace. While a relatively recent phenomenon, with the organization of sports beginning with baseball soon after the end of the Second World War, the rapid growth of organized participation in sports as diverse as soccer, ice hockey, gymnastics, figure skating, and baseball, as well as gridiron football, is evident in nearly every community in North America. In Europe, club sports are increasingly being enriched by the addition of youth and children's sports teams. This is certainly the trend in much of Europe and Asia with athletics, rugby, and soccer as well as other sports benefiting from this development.

The organization of sports participation for children and adolescents can provide opportunities for the prevention of injury. The organized sports setting allows for a careful assessment of mechanisms of injury and, in particular, techniques which may contribute to injury. Steps can then be taken to eliminate or modify playing techniques which increase the risk of injury.

In addition to the well-recognized risk factors for both acute and overuse injury seen in the adult, children have additional risk for injury because of their more vulnerable growth cartilage at the physeal plates, joint surfaces, and at the sites of major muscle and tendon insertions. In addition, the child is subject to the growth process. Evidence is accumulating to show that growth, and, in particular, the adolescent growth spurt, presents an increased risk of musculoskeletal injury to the child and, in particular, to the child participating in repetitive training activities in sport. Finally, the inconsistency and variability of coaching and training for this age group may well be an additional risk factor for injury. The development of coaching education and certification for youth sports coaches should receive high priority for injury prevention.

While some critics of organized sports for children have suggested that we should return to informal exercise and sports for this age group, it is doubtful whether this can occur. Admittedly, overuse injuries from repetitive training are only seen in the organized situation; they rarely occur in the free play, physical education, or informal sports setting; however, the growing social and economic constraints on informal sports and exercise, as well as the concerns for the physical safety of children, will most prob-

ably increase the trend toward organized children's sports. It is therefore important for all of us who care about safe and effective sports participation by children and adolescents to ensure that a systematic assessment of risk factors for injury, as well as the institution of preventive techniques, be initiated for this age group.

While there is a general consensus and recognition that physical exercise has many benefits for the growing child or adolescent, it is only very recently that scientific evidence has demonstrated that continuation of regular exercise and sporting activities are extremely beneficial and may even be lifesaving for the geriatric population. It has been known for many years that ageing is associated with progressive decrease in strength, aerobic capacity, and lean muscle mass. It has been recently documented, however, that these changes are not an inevitable result of ageing, but may frequently be due to the physical inactivity of the aged population which is particularly prevalent in urbanized and industrialized societies. As Evans and colleagues have determined, men and women in the seventh, eighth, and even ninth decades can show a significant increase in strength and lean body mass in response to a properly designed strength training programme. This, in turn, results in marked functional improvements. Geriatricians have determined that falling is often the pivotal event which can result in serious injury, loss of ambulation, and institutionalization of the geriatric individual. A geriatric man or woman who was previously independent, living on their own at home or with friends or family, may be rendered bedridden, institutionalized in a nursing home or tertiary care facility, or even subject to early death as a result of a simple fall. While falling can be multifactorial in this population, a number of studies have recently suggested that basic physical weakness may be a major contributor to this event.

The psychological benefits of sports and exercise for the elderly rival the physical and functional improvements achieved. In our own state of Massachusetts, a public health initiative sponsoring organized walking clubs at the community level has reaped immeasurable benefits for this population. These observations suggest that organized sport, walking, and dance activities can be as essential to geriatric care as medication, food, and shelter.

While the benefits of organized sports and sports medicine are being increasingly recognized for children, adolescents, and our geriatric population, the systematic promotion of sports and exercise for the handicapped is a relatively recent phenomenon. The great impetus to the development of therapeutic sports and exercise has come via two very different avenues: paediatric medicine and military medicine. The needs of children with physical disabilities resulting from hereditary or drug-induced defects, cerebral palsy, or certain other childhood acquired diseases, such as poliomyelitis, received early attention. The Crippled Children Services of the United States were formed in the 1930s. However, the active development and promotion of sports for handicapped children is a relatively recent development. Needless to say, this goes well beyond merely assisting the child to obtain the functional level of 'community ambulator'.

A major impetus in the United States to the comprehensive approach to the handicapped child's well-being was Public Law

94–142, the Education for All Handicapped Children's Act, passed in 1975. As a result of this law and its interpretations, the handicapped child has the right to be assessed by an 'individualized education program committee' in order to implement the 'free appropriate education guarantee', which was secured by law.

The implementation of this wide-reaching law popularized the development of a special committee consisting of physical therapists, occupational therapists, speech pathologists, nurses, social workers, and psychologists, as well as physicians, to provide a multifaceted approach to meet the needs of the handicapped child. Not only would the musculoskeletal problems of the handicapped be addressed, but also the physical and cognitive problems, including traditional education, physical education, and sports.

The most recent addition to the care team for the handicapped child has been the physical educator, sports coach, or dance teacher with specific skills and an interest in sports and fitness programmes for the handicapped. These programmes go one step beyond pure physical therapy. While many of the games and dance movements may incorporate therapeutic exercises or patterns, the structure is indeed that of game or sport, and, as such, requires the special skills of the sport specialist. Needless to say, as injuries were systematically incurred in these 'new games' for these special needs children, the need for specialized sports medicine training and understanding has also grown.

Changes in the care of handicapped military veterans has also paralleled the developments in the care of the handicapped child. Dramatic improvements were made in prosthetic and orthotic design and in the development of rehabilitative services and programmes for the patient with spinal cord injury and the amputee. While the initial primary goal of rehabilitation was, once again, to obtain the level of 'community ambulator', it was soon recognized that handicapped servicemen had many other emotional and social needs, including the need for regular and competitive sports. Initially, organized sports activities were barely tolerated, as they all too often resulted in broken prostheses or damaged wheelchairs.

One of the pioneer efforts in the incorporation of sports and systematic exercise into rehabilitation took place at the Veteran's Administration Hospital in Boulder, Colorado, with the development of handicapped skiing and riding programmes for the amputee. This programme, in turn, was actively expanded to the civilian population, including paediatric patients with acquired or congenital amputations.

Technical advances and design of wheelchairs, prosthetics, outrigger skis, and special weight-training machines have now made sports a mechanical possibility for the handicapped. Additionally, special needs coaches, teachers, and the athletes themselves have collaborated to devise appropriate rules and technique modifications.

Therapeutic exercises aimed at improving the range of motion and developing strength or co-ordination of a child or handicapped adult in a more traditional venue, the hospital or outpatient physical therapy unit, have sometimes been perceived as laborious, potentially painful, or 'boring'. The same exercises, when incorporated into a competitive sport, dance programme, or 'workout session' in a fitness centre, may become a challenge to be mastered or a source of active enjoyment to be pursued by the participants.

Much of the impetus for this new development of specific sports programmes for the handicapped has come from the handicapped themselves. The development of special ski equipment, lightweight pylons for canoeing or kayaking, and lightweight, low-friction wheelchairs made from thermoplastics and aluminum are a few examples of equipment designed from client demand.

It is imperative that physicians dealing with the child, geriatric, or handicapped athlete provide a maximally supportive approach to their special needs as well as their incurred injuries. In our general population, the sports medicine discipline has evolved from an environment in which all too many physicians suggested that an athlete should avoid further sports injuries by ceasing sports participation. It can be detrimental to the social, physical, and emotional growth of the child, the health maintenance of the older adult, or the overall 'rehabilitation' of the handicapped if a restrictive approach is used following the occurrence of an overuse or acute traumatic sports injury. The physician must bring an open mind to sports participation in this setting. He or she must be prepared to learn the details of the physical demands of the given sports and their potential for injury and make every effort to co-operate with the participants and the coaches in the prevention or early recognition of injuries sustained in this extremely important sports setting.

In summary, much research remains to be done in the study of sports and exercise activity for the young, elderly, and handicapped. In particular, careful assessment of injury risk must be made in these settings so that the maximal health benefits of these sports may be attained without sustaining unnecessary injury. Initial clinical observations have, in general, given strong impetus to the further exploration of ways in which sports and exercise can be incorporated into the education and care of these athletes. Coaches, physical educators, and sports medicine specialists must co-ordinate their efforts to ensure the maximal benefit of safe sports participation without unnecessary risks.

6.2   The ageing athlete

DARRELL MENARD

Age is not a barrier to performance, only an inconvenience.

INTRODUCTION

For centuries society has supported the paradigm that people should grow old gracefully. Ageing citizens should retire quietly to their rocking chairs and watch the grass grow. Saturday night bingo should be the most strenuous activity that an elderly person undertakes. Running around the block could overtax the heart, and running a marathon is a physical impossibility for anyone over the age of 30. It has taken years of effort by masters athletes, sports organizers, promoters, coaches, scientists, and the media to change the way that our society views older adults and exercise. The ageing athlete is now recognized as a legitimate entity in the world of sports.

It is estimated that there are more than 30 million North Americans over the age of 65 and that this number is increasing on a daily basis.[1] Both the absolute and the relative number of older individuals is increasing so rapidly that by 2030 as many as 20 per cent of the population will be over 65 years of age.[2] In addition, the most rapidly growing segment of our society is the group aged over 85.[3] To confirm that our society is ageing, one only has to realize that over 50 per cent of the people who have ever been older than 65 are alive today.[4] This enormous segment of humanity is no longer content with the prospect of retiring to a rocking chair and living a life of leisure. Instead, many of them are turning to sports in search of fun, fitness, self-fulfilment, and new challenges. Some go so far as to complete the Ironman triathalon, run marathons, climb mountains, win Olympic medals, swim the English Channel, or cycle across North America.

At one time, including a masters age group event at a sports competition was a concession reluctantly made by the event director. Things have evolved to the point where masters categories are now considered a routine feature of most competitions and the number of senior participants frequently rivals that of the young. Globally, we now find events organized exclusively for older athletes. These include Oldtimers events, senior leagues, masters races, tri-masters events, and mature athletes leagues.

The first World Senior Games were held in 1970 with 200 competitors taking part. Nineteen years later, Eugene, Oregon, hosted the 8th World Veterans Games in which 4951 athletes from 58 countries gathered to test their athletic abilities. This was over twice the number of track and field athletes who competed at the 1988 Olympic Games, and it is rated as the largest track and field event in history. Participants ranged in age from 40 to 96, and when the event was over they had established over 124 world age class records. In the articles written about these games, it was remarkable how often the authors referred to the competitors as possessing child-like enthusiasm despite their obvious ages. Many of the athletes came to test themselves and were not preoccupied with the 'win at all costs' philosophy that is corrupting our younger generation of athletes. This does not suggest that ageing athletes do not take their sports participation seriously. It is only necessary to watch them perform to be convinced of their intensity and determination. What they do possess is a realistic perspective on their efforts. The masters athletic movement offers its participants more than the obvious fitness and social advantages. It also offers them one of the elixirs of life—fun and something to look forward to as they age. In addition, they have a chance of interacting with athletes many years their junior, and both age groups benefit from these exchanges. The older competitors are pushed a little harder and stay in touch with the younger generation, while the young athletes are inspired by the efforts of their elders and also benefit from the social interaction.

The masters athlete has become a focus of considerable media interest in the last 15 years. It appears that the general public is genuinely captivated by the life stories and achievements of many older competitors. This is possibly because these individuals serve as role models, reminding all of us of what can be accomplished if we accept no physical limits and perform to our limits. Perhaps some people feel a little younger and more worthwhile when they see a senior competitor outperform opponents 20, 30, or 40 years their junior. When an ageing athlete enjoys success it reminds all of us that anything is possible with a little effort, dedication, and

self-confidence. People may be inspired to question if they really are 'over the hill'.

Consider, for example, the story of Priscilla Welch. Thirty pounds overweight, a heavy smoker, and never before active, she took up jogging at the age of 34. She did not remain a jogger for long and at the age of 39 was selected for the 1984 British Olympic Team. She finished sixth in the marathon at the Los Angeles Olympics, establishing a British national record. Her marathon personal best of 2 h 26 min and 51 s was run at the age of 42. At the age of 46, she continues to be one of the finest distance runners in the world. In her mid-40s she remained one of the finest distance runners in the world and believes that age has been a minor factor in her performances. Individuals like this provide important role models for our older citizens. Publicizing their efforts will serve to convince all of us that much can be accomplished beyond the age of 40.

The scientific community has also taken an interest in the ageing process and how it affects sports performances. This is reflected in the increasing volume of literature that is being written on the subject. Sports magazines, scientific journals, and reference texts frequently include discussions of the ageing athlete, a fact which attests the increasing importance of this large group of competitors. There is a National Institute on Ageing in the United States, and this organization actively encourages research and symposia on the ageing process. Unfortunately, until recently, much of the research and literature in this field was focused on the benefits of exercise in terms of cardiovascular health and longevity. Very little was focused specifically on the ageing athlete. This has left many important questions unanswered, offering the scientific community an almost unlimited opportunity to perform meaningful research. This chapter is dedicated to addressing a number of these questions in the hope of providing a more thorough understanding of this unique entity in the world of sports—the ageing athlete.

AGEING

Considerable research has been directed towards understanding the ageing process and the many changes it produces in the human body. One of the most surprising results of recent research is that almost everything we once believed about the ageing process is highly suspect. Many of the changes that were attributed to the ageing process are now known to be the results of disease processes, environmental influences, and physical inactivity. Earlier researchers often failed to control for these variables, and as a result many of their experimental conclusions were incorrect. As more masters athletes push the 'envelope of ageing' to its limits and more research is performed, it is almost certain that we shall find that many of the changes currently attributed to the ageing process are actually the result of other factors.

Before the structural and functional changes associated with the ageing process can be appreciated adequately, it is necessary to understand the concept of ageing. Contrary to popular belief, ageing is not a metamorphosis that suddenly occurs on reaching some significant milestone in life, such as the age of 40. Rather, it is an inevitable and continuous process that begins at conception and continues through infancy, childhood, adolescence, maturation, and old age. It cannot be avoided, reversed, or even postponed. Research shows that, with certain life-style changes, the best that can be hoped for is a reduction in the rate at which the process occurs.

There are almost as many definitions of ageing as there are

authors who write on the subject. Most of them state something to the effect that ageing involves an impairment of the ability to respond appropriately to environmental stresses. After careful consideration, however, many of these definitions fall considerably short of being complete. In fact, what many of them do is simply state the net result of ageing. Strehler recognized this fact and offers one of the simplest and yet most complete descriptions of the ageing process. He states that the ageing process has four basic properties: 'Ageing is universal, discremental, progressive and intrinsic.'[5] In other words, we all age; structural and functional losses occur and relentlessly progress with the passage of time, and the ageing phenomenon is innate to our genetic make-up and is not a pathological process. Any change that is observed in an older person that fails to meet these criteria cannot be attributed to the ageing process. The net result of ageing is a gradual impairment of all our organ systems such as that their built-in functional reserves are eroded to the point where it is progressively more difficult to compensate for environmental stresses, metabolic disturbances, disease processes, and anything else that can disrupt homeostasis. Several investigators have shown that there is a linear decline in the functional reserves of most body systems after the third decade of life.[6,7] These decrements will remain obscure until the individual is placed under a sufficient volume of physiological stress. At that point, the individual's ability to respond will be less effective than it was when he or she was younger. Ageing is an extremely complex process that involves changes at the molecular and cellular levels. The exact mechanism by which it occurs has been an academic preoccupation for centuries.

Ageing theories

The concept of ageing and why it occurs has probably been a subject of curiosity to mankind since time began. The gifted intellectual Leonardo da Vinci was one of the first people to take a scientific interest in the investigation of ageing. He conducted an experiment where he dissected the bodies of 30 older adults in an effort to discern what anatomical changes accompanies the ageing process. On completion of his investigations, he concluded that ageing was due to thickening of the blood vessels.[8] Since then, many scientists have unsuccessfully attempted to unravel the mystery behind why organisms age. With the recent explosion of knowledge in the fields of genetics, nuclear medicine, biochemistry, immunology, and molecular biology, tools and investigational techniques are being developed that could answer many of the questions related to ageing. Techniques that may prove important include recombinant DNA, cell fusion, and cell hybridization. Until a definitive explanation can be provided, we shall have to content ourselves with a number of hypotheses.

As a general rule, if something is poorly understood there will be numerous theories attempting to provide all the answers, and the ageing process is no exception. Despite the many explanations proposed for the mechanism of ageing, none adequately explains what has been observed experimentally, although each of them contains an element of apparent truth. It would seem reasonable to assume that the mechanisms for a process as complicated as ageing, would be more complex than could ever be imagined. The following is a brief review of some of the more popular theories which attempt to explain the mechanism of ageing.

The free-radical theory

Free radicals are atoms or molecules that contain an unpaired electron and so are highly reactive. Within a cell, free radicals will randomly react with other structures and damage them. Free-radical damage in connective tissues can induce the cross-linkage of macromolecules such as collagen. This can render these tissues increasingly brittle. Free radicals can also inactivate enzymes, break DNA molecules, peroxidize lipids, and injure cellular membranes. In order to combat free-radical damage, the cell is equipped with a number of defence mechanisms, including antioxidants and DNA repair enzymes. Over time, however, the continual bombardment of the cellular components by free radicals will render the cell progressively less able to function under stress. In addition to damaging cellular structures, these reactions are costly in terms of the energy and materials that are required to effect repairs. The net result of these cellular injuries is a reduced ability to function efficiently and replicate genetic material reliably. Since this effect is occurring throughout the individual's body, there is a global reduction in his or her ability to cope with physical stresses.

The ageing programme theory

This theory contends that the ageing process is actually programmed into each organism's genetic make-up. Our chromosomes may contain 'ageing' or 'senescence' genes which function as biological chronometers setting the unique pace at which each of us ages. In support of this, researchers have observed that members of various species exhibit a relatively constant life-span, which suggests some genetic control over longevity. The strongest evidence supporting this theory was provided when Hayflick and Moorhead[9] demonstrated that human embryonic fibroblasts grown in culture lived a finite period of 50 ± 10 replications. In addition, they also found that the cells taken from older donors were capable of fewer replications before dying. Their findings have been reproduced by other investigators.[10] There are also a number of diseases such as diabetes, progeria, and Down's syndrome in which premature ageing occurs at the cellular level. Perhaps understanding the genetics behind these conditions will help us understand the ageing mechanism better. While this theory is appealing, no one has explained what evolutionary advantage ageing offers an organism.

The neuroendocrine theory

The central premise of this theory is that because the neuroendocrine system is essential to nearly all of an organism's functions, any alterations in the controls exerted by this system will have far-reaching effects at the molecular, cellular, and organ levels. This system declines in function with age, which could explain the drop in physiological efficiency noted during the ageing process. While this theory can potentially explain the wide range of changes that occur with ageing, it fails to explain how cells which have been isolated from the influence of the neuroendocrine system can be grown in culture and still undergo typical age-related alterations.[10]

The altered protein theory

Proponents of this theory conjecture that, as we age, altered protein formation occurs. Since proteins play a central role in the structure and function of all organisms, it seems reasonable that alterations in normal protein synthesis will have widespread ramifications in terms of cellular function. Consider, for example, enzymes which are specialized proteins responsible for effecting a multitude of bodily reactions. Alterations in their structure will greatly interfere with the normal physiological operation of the cell. It is possible that intrinsic or extrinsic factors which cause DNA damage lead to errors in the synthesis of various proteins and enzymes. Defective cellular products are not as efficient and so handicap the cell's ability to adapt to the stresses it encounters.

The waste product accumulation theory

This theory contends that with ageing there is a progressive accumulation of ineffective biological materials within each cell and that this ultimately impedes the optimal functioning of the cell. One such compound is lipofuscin, which is a byproduct of free-radical damage. The volume of this material increases in ageing cells and can account for 6 to 7 per cent of the intracellular volume by the age of 90.[10] This proposal breaks down in that there is no definite evidence that lipofuscin has any deleterious effect on cellular function. However, it is possible that in this process the cell may accumulate a number of noxious materials. As these harmful substances accumulate, they could inflict progressively greater amounts of damage on the cell, rendering it less able to function efficiently.

The cross-linkage theory

Intramolecular and intermolecular cross-linkages form to stabilize the macromolecules within cells. In this theory, it is believed that this process is enhanced with ageing, leaving macromolecules increasingly resilient to normal degradation and turnover. Cross-linkages can affect many types of macromolecules including collagen, elastin, and other connective tissue elements. As a result of this, tissues are rendered increasingly brittle and membranes more difficult to cross. Cross-linkages also occur in molecules that are essential for cellular replication, such as DNA and RNA. It is possible that linkage of these molecules interferes with or prevents the transcription of genetically encoded information and so drastically interferes with normal cellular functioning. Further study is required in this area.

The immunological theory

Walford proposed that as we age alterations occur in our immunoregulatory genes such that organisms lose their ability to discriminate self from non-self.[11] The net result of this change is an increase in the number of autoimmune reactions. Consequently, there is a reduction in the efficiency with which the various target organs function and the individual becomes increasingly susceptible to diseases, infections, and neoplasms. Further investigation is required to determine whether this theory can explain the diversity of changes seen in ageing.

The above theories represent the most popular attempts at explaining the mechanisms by which ageing occurs. Other possible explanations include DNA methylation, protein racemization, non-enzymatic glycosylation, dysdifferentiation, codon restriction, loss of ribosomal RNA genes, genetic mutation, and errors in transcriptional and translational processes.

While the above theories do not provide any definitive explanations as to how and why we age, they do serve to illustrate that ageing is an extremely complex event. Ageing is not a pathological process but rather an inevitable event in which numerous factors play a role. Everyone ages in a unique pattern and it is impossible to predict the course that individuals will follow. Regardless of the individual pattern, the structural and physiological changes associated with ageing have important implications for athletic performance. The very nature of the ageing process dictates that the ageing athlete differs biologically from the younger competitor.

INACTIVITY

An inactive life-style is a far greater threat to continued health than the ageing process. This is disconcerting because inactivity is almost inherent in our industrialized world, partly because the natural behaviour pattern of all animals is to become less active as they age. No less important is the fact that in our highly automated society inactivity is almost a cultural objective. Why take the stairs when you can use the escalator? Why push a lawnmower when you can drive it? Why rake leaves when you can blow them away with a machine? Why walk to the corner store when you can drive? There is a time and a place for every labour-saving innovation, and individually they do not account for a significant altering of our daily physical activity. However, collectively these devices greatly reduce the activity required of each individual. Labour-saving devices were originally intended to improve efficiency and free people to pursue more pleasurable pastimes. Instead of dedicating some of this free time to physical exercise, many people elect to pursue largely sedentary interests and assume the accompanying health risks. In fact, data from the Framingham study strongly suggest that physical inactivity may be an independent health risk factor that increases with age.[7] Inactivity and the health problems that it fosters are so endemic to our culture that people have begun to use the term 'hypokinetic diseases'. Conditions such as osteoporosis, atherosclerosis, obesity, hypertension, hyperlipidaemias, depression, inflexibility, and chronic fatigue are often closely associated with a sedentary life-style.

In addition, it is of great concern that the majority of our society accepts inactivity as such a natural consequence of ageing that older adults who remain physically active are often thought of as odd, eccentric, or juvenile. This paradigm is so firmly established that ageing participants are thought to be gambling with their lives whenever they exercise. The ageing athletic enthusiast is frequently given the distinct impression that others believe that what he or she is doing is abnormal. Exercising regularly may well be abnormal, considering that the average older adult lives a largely sedentary life. Why should someone be embarrassed by doing something that is fun and good for their health? Remaining physically active may be viewed as an option for the young, but it is imperative for the maintenance of an ageing individual's health. Failing to remain active permits an excessive rate of structural and physiological loss that will threaten the person's ability to remain independent in later years. Regular exercise is one way in which the elderly can continue to live and enjoy life rather than simply existing.

Many of the changes attributed to the ageing process are actually the product of a sedentary life-style. It is estimated that inactivity accounts for more than 50 per cent of the structural and physiological decrements that can be demonstrated in a sedentary adult.[6,12] This relationship was not recognized until a few years ago, and consequently much of the ageing research performed prior to this recognition is biased by this 'inactivity factor'. Kasch et al.[1] conducted a study of the effect of exercise and inactivity on the aerobic power of older men that provides an excellent illustration of the relative roles of ageing and inactivity in the deterioration of physiological function. The study consisted of two sets of 15 men and took place over a period of 23 years. One group participated in an aerobically oriented programme, while the other group did not take any regular exercise over the study period. At the conclusion of the study the two groups were tested and the results were as follows.[1]

1. The exercisers reduced their body weight by an average of

3.4 kg while the non-exercisers gained an average of 3.2 kg.
2. The exercisers had an average of 15.9 per cent body fat while the non-exercisers averaged 25.7 per cent.
3. The exercisers had an average resting pulse rate 10 beats/minute lower than that of the non-exercisers.
4. Blood pressures were noted to be higher in the non-exercisers, with nine of the 15 being clinically hypertensive at the conclusion of the study.
5. The maximum oxygen uptake of the exercisers decreased by an average of 13 per cent, while the non-exercisers showed an average decline of 41 per cent.
6. The maximum attainable heart rate of the exercisers was an average of 20 beats/min higher than that of the non-exercisers.

Despite the small number of subjects involved in this study, the results strongly suggest that inactivity makes a large contribution to the physical changes that we see in ageing individuals. The maximum oxygen uptake results are particularly striking because they suggest that 34 per cent of the loss seen in the non-exercisers could be attributed to ageing, with the remaining 66 per cent of the loss resulting from disuse. This suggests that the effect of disuse on structural and physiological decrement may be even greater than is currently believed to be the case. Dr Cooper, the originator of aerobics, has conducted extensive physiological evaluations of numerous world class athletes. Based on his findings, he concludes that 'Most of the decline that comes with age isn't inevitable at all. It's caused by disuse'.[13]

One question that needs to be addressed is: 'Why do these disuse changes occur in the first place?' The answer lies in the fact that one of the basic rules governing the economical operation of the body is the 'use it or lose it' principle. Management of the entire body is controlled by this unavoidable biological law. In practical terms, this principle states that if you are not using a particular tissue then why maintain it. It is inefficient to nurture and maintain tissues that are not being utilized regularly. Anyone who has worn a cast can attest to the enthusiasm with which the body adheres to this principle. After only 6 weeks of immobilization the injured limb usually emerges with marked atrophy and a significantly reduced range of motion. The body makes no distinction between inactivity and immobilization. In both cases, a given body part is not being utilized and therefore some of its components could be better used elsewhere or eliminated. In our current climate of financial restraint, many corporations are having to employ the same principle in order to remain competitive.

STRUCTURAL AND PHYSIOLOGICAL CHANGES

A number of well-documented structural and physiological changes occur in the ageing body. These changes are probably initiated at the cellular level and associated with a reduced number of parenchymal cells, qualitative changes in those cells, and alterations in the forces that regulate their function. The net result is a reduction in the ageing individual's functional reserves such that he or she is less able to handle physiological stresses. It should also be noted that changes have been identified in virtually every organ system, and individually these changes can have significant implications. However, since the effective operation of the body depends on a collection of highly complex and integrated activities, no change in one component should be considered in isolation from changes in the others. Undeniably, changes in one system can have far-ranging effects in many others. A good example of this would be changes in the neuroendocrine system which impacts on virtually every cell in the body. Thus, even if researchers identify the specific changes that occur as a result of ageing, interpreting their significance will be much more complicated. When reviewing the changes associated with ageing, it is important to recognize that ageing is not a disease and that there are no specific symptoms or sensations permitting an individual suddenly to identify that he or she is ageing. All the alterations occur so subtly that the individual fails to recognize them. It is only when the person is evaluated with a sophisticated collection of medical equipment or is subjected to the maximal demands of an athletic event that functional losses become undeniable.

Cardiovascular changes

Age-related alterations in the cardiovascular system have been the focus of attention for many years. This is undoubtedly because cardiovascular disease is a major cause of morbidity and mortality. There is much contradictory information in this area because it is often difficult to distinguish between changes due to ageing, inactivity, disease processes, or some combination of these factors. Physiologists consider that an individual's maximum oxygen uptake V_{O_2}max is the single most reliable indicator of fitness. Evidence indicates that beyond the age of 35 there is an inevitable decline in an individual's V_{O_2}max such that by the age of 60 it is often reduced to 80 per cent of that at the age of 20.[14-16] This represents a rate of decline of 0.5 to 1.0 per cent annually and has important implications in events requiring maximum aerobic effort. The V_{O_2}max of deconditioned or sedentary individuals may decrease at a rate in excess of 1 per cent annually. The more active an individual is, the slower will be the decline in V_{O_2}max. For elite masters athletes the annual rate of decline may be as low as 0.1 per cent, a tenth of the rate seen in inactive people.[7] Several investigators contend that this decline in V_{O_2}max is due to the reduction in maximum attainable heart rate that occurs with maturation. In fact, maximum attainable heart rate shows a linear decline with age.[17] As a general rule, maximum attainable heart rate (MAHR) can be estimated by the formula:

MAHR = 220 − years of age

This is only a rough estimate, and individuals can show considerable variation. V_{O_2}max is a complex value that depends on a number of physiological variables including cardiovascular function, pulmonary function, aerobic fitness level, and muscle function. Since ageing is associated with diminished capacity in all these areas, it is reasonable to conjecture that the cause of the age-related decline in V_{O_2}max is multifactorial.

Cardiac output is a measure of how much blood the heart can pump in a given period of time. Mathematically it is calculated as the product of stroke volume and heart rate. Since research suggests that both stroke volume and heart rate are reduced at high workloads, cardiac output must also decline.[17-19] This has major implications in terms of aerobic work capacity because it reduces the amount of oxygen-rich blood that can be delivered to the working tissues. With ageing there is also a decrease in the rate at which the heart rate returns to resting values following maximal and submaximal efforts. This is important to understand, as it has implications for the interval training of ageing competitors. In view of the above, it would seem appropriate to allow masters athletes longer recovery times between intervals than would be allowed for younger athletes of similar fitness status. Therefore to avoid unnecessarily long recovery periods, athletes should be taught to monitor their heart rate and to begin another interval as soon as it returns to a predetermined value. Although there has been controversy over haematological values in ageing, a recent study by Zauber and Zauber[20] strongly suggests that the haematological values of healthy subjects over the age of 85 are the same as those of young adults. This includes values for blood volume, red blood cell count, haemoglobin, haematocrit, serum iron, erythrocyte sedimentation rate, and white blood cell count. Zauber and Zauber[20] also showed no significant reduction in haematological response to stress occurs with ageing. This suggests that the ageing competitor will probably be able to elevate his or her haemoglobin levels in response to regular aerobically demanding exercise.

Excluding disease processes, the cardiac functional changes noted above occur with only minor structural alterations to the heart itself. For example, myocardial weight appears to remain constant even in studies which have included subjects aged 100.[21] However, some compositional changes have been reported. The interstitial volume of collagen, reticulin, and elastin fibres increases, resulting in a condition known as fibrosis. While these increases are generally small, they do render the myocardium less compliant. This may be functionally insignificant at rest, but when the cardiovascular system is operating at the maximum level it could limit ventricular filling and so reduce cardiac output. In support of this, several authors have noted an age-related decline in cardiac ejection fraction at high workloads only.[??] Ageing myocardium accumulates lipofuscin, the significance of which is uncertain. Amyloid is another protein which is found in the hearts of most people over 90 years of age but seldom in those under 60.[21] This change is also of uncertain significance. The cardiac conduction system is also influenced by the ageing process. In general, collagen, elastin, reticulin, and fatty tissues infiltrate the conducting tissues. This fibrotic process occurs in conjunction with a loss of the specialized conducting tissues. For instance, between the ages of 20 and 75, up to 90 per cent of the pacemaker cells located in the sinoatrial node may disappear.[21] While the significance of these changes is unknown, they may contribute to the age-related decline in maximum heart rate. Controversy exists as to whether the contractile properties of the myocardium decrease with age.[17,21,23] In addition to the changes discussed above, the following are also observed.

1. A reduction in myocardial responsiveness to catecholamine stimulation.[17,19,21]
2. The pericardium is essentially a bag of collagen in which the heart resides. With ageing, these collagen fibres straighten, rendering the pericardium thicker and less compliant.[21]
3. The valvular structures experience collagen deposition and degeneration, lipid accumulation, and calcium deposition. These changes serve to stiffen these tissues, but the effect that these changes have on performance is unknown.[21]
4. The circumference of all four cardiac valves increases.[21]
5. The ageing heart may undergo geometric alterations: it may not be as long, ventricular septal thickness increases, the left atrium dilates, and the volume of the left ventricle is diminished.[21]
6. There are reduced levels of noradrenaline, acetylcholine, and adenylcyclase in cardiac tissue.[17]

It is difficult to be certain whether these changes are the direct result of the ageing process, disease, inactivity, or some combination of factors. It is also difficult to determine whether many of these changes impact on performance in any way. However, when considering the issue of maximum performances, even minor alterations may have considerable implications. On the whole, the observed cardiac changes serve to reduce myocardial compliance, reduce maximum attainable heart rate, decrease cardiac output, and lower V_{O_2}max. A direct consequence of this is that the ageing heart must work harder to meet the metabolic demands of the body at any given workload. This loss of cardiovascular efficiency is a major factor in the progressive decline in aerobic performance noted with increasing age and is clearly evident in the world marathon records for various ages (Fig. 1).

The vascular system is by no means exempt from the relentless alterations of the ageing process. In fact, this is one area where

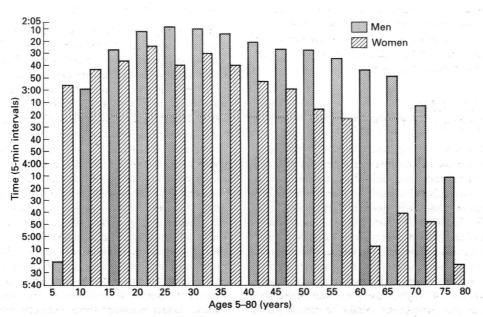

Fig. 1 Marathon records by age—United States records taken at ages 5 to 80 in 5-year intervals, and correct to 1 January 1983.

considerable change is often noted. However, care must be taken to differentiate age-related modifications from those caused by pathological processes such as atherosclerosis. While atherosclerosis is a disease process, its universality has led some authors to question whether it should be considered as a normal part of ageing. Although this distinction may be academic, the reality is that virtually every ageing adult, regardless of his or her lifetime activity pattern, will have some degree of atherosclerotic change in his or her vascular system. The extent to which this disease exists in asymptomatic individuals varies enormously. This insidious disease narrows the lumens of blood vessels, leaving them less able to provide blood to a given area. At the same time, it also reduces vessel-wall extensibility. These changes begin well before the age of 40 and have been noted in autopsies of teenagers. Independent of the disease process, the vasculature experiences enhanced collagen deposition. This collagen also becomes progressively more rigid as its cross-linkages mature and undergo stabilization. In addition, calcium is deposited into elastin fibres, a phenomenon referred to as elastocalcinosis. The end results of these changes are narrower vessel lumens and increased vessel wall resistance. In terms of maintaining adequate circulation, greater kinetic energy will be required to overcome vascular resistance and therefore cardiac workload is noticeably increased without providing an increase in cardiac output.

Injection studies have also shown that ageing is associated with a concomitant decline in the microvascular supply to muscle, major organs, and peripheral tissues.[15] This has important implications in a number of areas. First, it is a major factor in limiting the body's functional organ reserves. This makes it increasingly difficult to maintain homeostasis when under stress. Second, it reduces the volume of blood that can be delivered to the working tissues. Without a continuous supply of oxygen and substrates, these tissues will be unable to function at an intense level for very long. In addition, these changes reduce the volume which can be delivered to an injury site and so may alter the rate of recovery.

Connective tissue changes

Connective tissues are those elements that assist in providing both support and structure to the various components of our body. Since connective tissues are major constituents of most body tissues, they make a major contribution to their mechanical properties. In fact, the different mechanical properties of numerous tissues are largely dictated by their unique combination of connective tissues. This is an economical design because by using the same basic ingredients and arranging them in unique combinations the body is able to meet and adapt to its various structural and support requirements. With this amazing versatility, the connective tissues are able to satisfy the radically different physical requirements of structures as dissimilar as the cornea and the medial meniscus. Connective tissues such as fasciae, tendons, and ligaments were once thought to be entirely inert structures. After synthesis they were believed to reside passively in the body, playing their structural and support role. Research has since shown that these tissues are highly dynamic in nature. They constantly undergo modifications in an effort to adapt to the multitude of stresses to which they are subjected. As a rule, physical activity serves to stimulate connective tissue hypertrophy while inactivity promotes atrophic changes. This is yet another example of the 'use it or lose it' principle that is such an innate part of our physical being.

It is obvious from the above that not all connective tissues are the same. Their compliance and tensile strength are determined by their unique combination of proteins such as elastin, connectin, proteoglycans, and collagen. These basic macromolecules are arranged in a variety of combinations to create the framework for most of the body's tissues. Collagen accounts for approximately 30 per cent of our total body protein, making it by far the most abundant connective tissue element. It is a major component of basement membranes, intervertebral discs, blood vessels, teeth, bone, cartilage, tendon, skin, and ligaments. It is also present in nearly all organs where it serves to hold cells together in discrete units. Five major types of collagen are recognized. Each differs in its amino acid composition and sequencing and the type and extent of cross-linkage bonding.[24,25] Their individual molecular designs permit each type of collagen to possess unique properties of compliance and tensile strength. The amount of either of these properties is determined by the amount of cross-linkage bonding present. The greater the amount of tissue rigidity required, the greater is the amount of cross-linkage. Thus the collagen found in bone is far more heavily cross-linked than that found in tendinous tissues. These cross-linkages occur both inter- and intramolecularly, and their existence provides collagen with a greater tensile strength than steel. It is estimated that a load of 10 to 40 kg is required to rupture a collagen fibre 1 mm in diameter.[26,27]

Many of the age-related alterations that occur in collagen are focused on the cross-linkages. As collagen ages, its molecular stability increases and so the tissue in which it lies becomes less compliant. The physical evidence supporting this change is that ageing collagen is increasingly less soluble and has increased thermal stability.[28] This enhanced stability was initially attributed to an increase in the number of cross-linkages present in the collagen molecule. It has since been shown that shortly after being synthesized a collagen fibre processes all the cross-linkages that it will ever have. It is believed that, during maturation, cross-linkages that were once reducible undergo stabilization, leaving the collagen fibre progressively less compliant.[17,25] It should also be noted that protein cross-linkage also occurs via free-radical damage and non-enzymatic glycosylation processes.[29] The rate at which collagen loses its compliance can be influenced by several factors, including hormonal changes and physical activity. Exercise seems to enhance the rate of collagen turnover. In this way, exercise shortens the life-span of collagen molecules and retards the process of maturational stabilization. This suggests that exercise will help maintain the youthfulness of these tissues. Hormones such as insulin, thyroxine, and corticosteroids have also been implicated in the ageing of collagen. Hamlin et al.[30] have found that collagen isolated from the tissues of a 40-year-old diabetic resembled those found in healthy individuals over 100 years old. Excessive non-enzymatic glycosylation occurs in the diabetic state and may be the mechanism by which the collagen of a diabetic person ages so rapidly. Since hormonal changes accompany the normal ageing process, it is reasonable to postulate that these changes may influence the collagen structure of the senior athlete.

Elastin is found in most connective tissues. It is the major component of the elastic fibres found in large amounts in skin, the walls of blood vessels, and ligaments. These unique fibres possess physical properties permitting them to stretch to several times their length and then return to their original size when the traction forces are withdrawn. Elastin molecules undergo age-related changes similar to those seen in collagen. More specifically, their cross-linkages undergo maturational stabilization. They also experience an increase in their polar amino acid composition.[31]

Microscope examination shows that ageing elastic tissues lose their regularity and their lamellae are visibly ragged and slender. The net result of all these changes is that ageing elastic fibres become increasingly brittle and more easily subject to fracture.[25] This has significant implications for the ageing athlete who depends on his vascular and ligamentous tissues during all athletic performances.

Ageing connective tissues also experience a reduction in their water content. The consequences of this change include a reduction in the shock-absorbing capacity and a loss of tissue compliance. Considered in their totality, all the known alterations in connective tissue that occur with ageing are important because they substantially alter the mechanical properties of many tissues and render the ageing competitor increasingly vulnerable to injury.

Skin changes

Skin is not an organ in the classic sense, but it is often referred to as a functional organ because of the number of critical roles that it plays in the effective operation of the body. This is particularly evident when a person experiences extensive burns and requires extraordinary medical efforts to ensure survival. Skin functions as a protective barrier against trauma, as an energy storage site, as a boundary against infection, as a barrier against environmental insults such as ultraviolet light and rain, and as an important component of our thermoregulatory system. With so many roles to play, any age-related changes in skin may have far-reaching implications. Skin ages as a result of intrinsic structural and functional changes in combination with extrinsic influences such as ultraviolet light, wind, and thermal stresses. While these extrinsic factors are not innate to the ageing process, they are mentioned because of the universality with which people are exposed to them. The very nature of most sporting activities exposes most ageing competitors to more sun, wind, and cold than the average person.

Wrinkling is an obvious superficial feature that we all identify as a sign of ageing. However, many age-related changes occur below the skin surface and have direct applications to the ageing athlete. One of the most evident morphological changes is the thinning of the epidermal, dermal, and subcutaneous layers of the skin.[32] This substantially compromises the skin's insulatory ability, leaving the person more vulnerable to cold injury. Additionally, it reduces the trauma cushion that skin provides for the underlying tissues. Thus a blow to the thigh of insufficient magnitude to cause a contusion in a younger person might cause one in an older competitor. Ageing skin also experiences a loss of elastin fibres and increasing maturational stabilization of its collagen fibres. These changes alter the viscoelastic properties of ageing skin such that it is less able to respond to deforming forces. Thus the ageing process leaves skin increasingly fragile and so more susceptible to damage from even mild trauma. Bumps, scratches, shearing forces, and wear and tear are unavoidable facts of life for every athlete. When exposed to these forms of mechanical trauma, ageing skin will often tear at intensities that would leave younger skin unaffected. This should be kept in mind, particularly when applying adhesive tape to the skin of an ageing athlete. All adhesive tape should be removed very carefully to avoid tearing the underlying skin, as its adhesive properties may exceed the skin's mechanical strength.

The interdigitations, or rete pegs, between the epidermis and the underlying dermis (Fig. 2) ensure that these two skin layers adhere to each other. With ageing, these interdigitations diminish, making it easier to separate the epidermis from the dermis, a phenomenon that occurs when blisters form. This increased propensity to develop blisters will hamper performance in sports where friction forces are a major factor. In the older competitor, these blisters will heal more slowly and are more likely to become infected. With ageing, the capillary density in the peripheral tissues is diminished and the remaining vasculature is thinner and more fragile. These changes increase the ageing competitor's tendency to bruise following trauma. Some loss of sensory acuity may occur in ageing peripheral tissues, which means that ageing athletes can experience greater injury to the skin before becoming aware of it. The number of immunocompetent cells present in the skin is also reduced during ageing. There is diminished T-cell function and Langerhans cells die off, rendering the older individual more vulnerable to skin infections and neoplasms. Fingernails and toenails are extensions of the skin and are also altered with age. Ageing nails are thinner and so more prone to injury following minor trauma. This type of injury is a common occurrence in many sports. Since ageing nails grow more slowly, the recovery from such an injury is often delayed.

Ageing skin also experiences changes that alter its ability to offer protection from the damaging effects of ultraviolet light. After the age of 30, an individual loses approximately 2 per cent of his or her melanocytes annually.[32] Melanocytes are specialized skin cells that produce the pigment melanin in response to exposure to ultraviolet light. The resultant 'sun tan' is actually a protective mechanism designed to shield the body from further ultraviolet injury. The age-related reduction in melanocyte population leaves a person increasingly susceptible to sun damage. To make matters worse, ageing skin displays a reduced inflammatory response to injury. Sunburn is a typical example of such a response. Thus the ageing person will tolerate longer exposures to the sun before becoming sunburned. What is potentially dangerous is that people frequently misinterpret this change as indicating that they have developed greater sun tolerance. In reality, they are still experiencing ultraviolet-induced damage but are no longer receiving one of the body's early warning signals. Thus older skin is at increased risk of injury, not only because it has a reduced defence capacity but also because its early warning systems do not work as well. This is of concern for two reasons. First, athletes tend to spend more time in the sun than non-athletes. Second, the cumulative effect of sun damage is important, and the older people become, the more these changes will accumulate. Ageing athletes can minimize these risks in several ways.

1. Wear a hat or cap when out in the sun, particularly for long events such as 18 holes of golf or a triathlon.
2. Wear light coloured clothing to reflect as much of the sun as possible.
3. Try to exercise during periods of the day when the sun is the least intense, i.e. before 10 a.m. and after 4 p.m.
4. If any sores are slow to heal or moles show changes, obtain a medical opinion from the family physician.

Skeletal changes

Some of the best understood age-related alterations occur in the skeletal system. Bony tissue is basically a complex matrix of organic and inorganic materials. This combination yields a product that is lightweight and yet capable of withstanding a lifetime of mechanical stress. Bones are highly dynamic tissues which undergo structural adaptations in direct response to the physical demands placed upon them. This quality was recognized 100 years

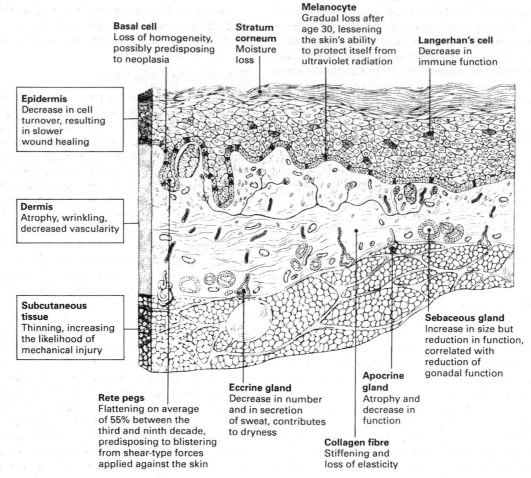

Basal cell
Loss of homogeneity, possibly predisposing to neoplasia

Stratum corneum
Moisture loss

Melanocyte
Gradual loss after age 30, lessening the skin's ability to protect itself from ultraviolet radiation

Langerhan's cell
Decrease in immune function

Epidermis
Decrease in cell turnover, resulting in slower wound healing

Dermis
Atrophy, wrinkling, decreased vascularity

Subcutaneous tissue
Thinning, increasing the likelihood of mechanical injury

Sebaceous gland
Increase in size but reduction in function, correlated with reduction of gonadal function

Rete pegs
Flattening on average of 55% between the third and ninth decade, predisposing to blistering from shear-type forces applied against the skin

Eccrine gland
Decrease in number and in secretion of sweat, contributes to dryness

Apocrine gland
Atrophy and decrease in function

Collagen fibre
Stiffening and loss of elasticity

Fig. 2 Anatomy of skin and the ageing process. (Reproduced from Femtu NA, *et al*. Common problems of aging skin. *Patient Care*, 1990; 33, with permission.)

ago when Wolff's law of anatomy was first proposed. This law basically states that: 'The robustness of bone is in direct proportion to the physical forces applied to it.'[33] Bones become mechanically stronger when regularly stressed and weaker when left unchallenged. This hypertrophic response will occur in the specific area of the skeleton that is experiencing the extra forces. The best example of this is the unilateral bony enlargement that occurs in the dominant arms of baseball pitchers and tennis players. This is undoubtedly a protective response aimed at preparing the dominant arm to be able to cope with the stresses to which it is subjected. Bassett and Becker[34] suggest that the mechanism for the local control of bony growth is electrical. They believe that bone tissue functions like a piezoelectric crystal that converts mechanical stresses into electrical energy. When a bone is mechanically stressed, the segment that undergoes compression produces a negative electric charge while the extended segment becomes positive. It is these charges that control the level of cellular activity in bone. This mechanism permits bones to respond and adapt to the specific stresses that they encounter (Fig. 3). Any condition that compromises a bone's architecture or its adaptability will reduce its resilience to mechanical stresses and leave it vulnerable to structural breakdown. These breakdowns can occur gradually, as seen in overuse-induced stress fractures, or instantly, as occurs in traumatic fractures.

Without a doubt, the greatest threat to the skeletal integrity of

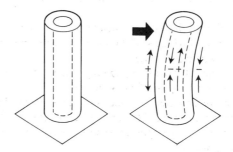

Fig. 3 The response of bone to compressive and tensile forces. (The large arrow represents a force applied to the bone.) The concave surfaces of the bone cylinder carry a negative charge, and the convex surfaces carry a positive charge. These charges affect the loss and gain of bone tissue (Redrawn from ref. 37, with permission.)

the ageing adult is osteoporosis. Osteoporosis can be defined as an ideopathic condition in which the rate of bone resorption exceeds the rate of bone formation. This imbalance leads to a progressive loss of bone mass and is a major concern because bone strength is directly related to the volume of bone mass. It is a condition that affects over 25 million North Americans and often progresses insidiously until a fracture occurs. These initial fractures are often of the hip or vertebral body, and are associated with a high morbidity and mortality rate. Until recently, it was believed that the rate of bone loss in osteoporosis was gender specific, with women experiencing losses two to three times more rapidly than men. It now appears that there are two different types of osteoporosis. Type I osteoporosis affects women only and occurs with the loss of oestrogen production that accompanies menopause. This process involves an annual reduction in bone mass of approximately 0.6 per cent.[35] Type II osteoporosis affects both sexes and occurs as a direct result of the ageing process. It is believed that this loss is due to an age-related decline in osteoblast activity and accounts for an annual reduction in bone mass of approximately 0.3 per cent.[35] It is obvious from the above that the bone mass depletion experienced by women exceeds that of men. The injustice does not end here. Men often do not appear to experience losses until the age of 50 and usually experience no sequelae until the eighth decade of life. However, women frequently begin losing bone mass in their early thirties. As a result of this earlier onset and higher rate of loss, by the time many women have reached their seventies they have lost more than 30 per cent of their total bone mass and are considerably more vulnerable to fracture than they were at age 20.[36-38] The sedentary life-style so commonly associated with advancing age is also a major factor in the loss of bone mass. In studies involving individuals who have been immobilized for various reasons (astronauts, quadriplegics, and people with casts) dramatic bone mass depletion has been observed over relatively short periods of time. Clearly, age-related bone loss is a serious threat to every ageing individual, including ageing athletes.

The aetiology of osteoporosis appears to be multifactorial. Reductions in skeletal blood flow, reduced calcium intake, decreased gastrointestinal calcium absorption, hormonal factors such as reduced oestrogen, levels of physical activity, and genetics have all been implicated in this process. As far as ageing athletes are concerned, several critical concepts should be kept in mind.

1. Age-related loss of bone mass is a universal process and it must be assumed that every ageing athlete is osteoporotic to some degree.
2. Considering all the variables involved in this process, it is clear that no two individuals of the same age will have experienced the same volume of losses.
3. Almost without exception, ageing female athletes will show greater loss than men of the same age.
4. There are no obvious symptoms indicating that an individual is seriously osteoporotic. The first indication of a problem usually occurs when the competitor experiences a fracture following minimal trauma.
5. Loss of bony mechanical strength is the greatest threat to individuals who have lived largely sedentary lives and begin training when they are 50 or older. It is critical that these individuals control their initial enthusiasm and 'start slowly and progress gradually'. In this way, they will allow their skeletal system sufficient time to adapt to the new stresses that it is experiencing and so reduce the incidence of stress fractures.
6. Age does not appear to be a limiting factor in bone's ability to respond to mechanical stresses. Individuals in their eighties have shown bony changes in response to muscular traction and gravitational forces.[39] Some of these changes include increases in thickness, strength, calcium concentration, nitrogen concentration, and hydroxyproline and DNA content. Thus it appears that regular physical activity not only slows down the rate of bony demineralization but to some extent may even reverse it. Thus it appears that exercise offers the ageing individual a greater opportunity to maintain skeletal integrity than any mediation that medical science has to offer.

Respiratory changes

While a functioning respiratory system is a requisite for life, an efficient respiratory system is essential for successful athletic performance in many sports. Unfortunately, with the passage of time, the respiratory system falls prey to the deleterious influence of a number of factors including previous pulmonary illnesses, environmental pollutants, tobacco smoke, a sedentary life-style and the ageing process itself. No one is certain how much each of these factors contributes to the respiratory alterations that accompany ageing. However, studies have controlled for the above variables and concluded that there is a progressive deterioration in respiratory function that is directly attributable to the age process.[40] Many of these changes manifest themselves as an increased sense of respiratory effort during physical exertion. Consequently, ageing athletes tend to experience breathlessness at lighter workloads than when they were younger.

A number of architectural alterations to the pulmonary tissues occur during ageing of the respiratory system. For example, there is a progressive degradation of the collagen and elastin fibre content of the pulmonary parenchyma.[25,36,41] In its most advanced state this deterioration is referred to as senile emphysema. The rate of degradation is considerably increased in individuals whose lungs are repeatedly insulted by cigarette smoke and the multitude of environmental pollutants infesting the atmosphere. While this is not truly an age-dependent phenomenon, it is included here because these noxious agents have a cumulative effect on everyone living in an industrial society. This damage to the pulmonary tissues and to collagen and elastin fibres significantly compromises the structural integrity of the lung and reduces its capacity for elastic recoil. The lung depends on elastic recoil to aid expiration. Since recoil requires no energy, it reduces the active workload associated with breathing. As progressively greater amounts of elastic recoil are lost, the work associated with breathing increases. This loss of structural support for the pulmonary tissues will also permit airway closure at higher lung volumes and so increase the older adult's residual lung volumes. While this is occurring, there is a generalized increase in the rigidity of many of the chest wall structures. The intercostal musculature becomes less compliant, the costovertebral joints stiffen and the costochondral cartilage becomes less elastic. Consequently, greater physical effort is needed to overcome chest wall resistance during inspiration. Thus ageing is clearly associated with an increased workload during both inspiration and expiration. deVries[42] believes that this could account for as much as a 20 per cent increase in respiratory effort. This has major implications for older athletes who participate in aerobically demanding sports.

Ageing respiratory musculature experiences both a loss of strength and a progressive increase in its connective tissue content. These weaker and stiffer muscles are less able to generate the inspiratory pressures needed to ventilate the lower lung lobes where blood profusion is the greatest. This leads to ventilation–perfusion mismatches whereby blood is delivered to lung tissues that are not being inflated with oxygen-rich air. Consequently, respiratory gas exchange is less efficient. With ageing there is a reduction in the small vessel density of the lung parenchyma. It is also possible that collagen deposition occurs at the level of the alveolar capillary basement membrane. If this occurs, it will increase the diffusion barrier over which O_2 and CO_2 must be exchanged during respiration.

The above changes in the respiratory system manifest themselves in a number of specific functional decrements.

1. Reduced inspiratory capacity;
2. Reduced forced expiratory volume in 1 second (FEV_1) (a measure of the volume of air that can be forcibly exhaled in 1 s);
3. Increased residual lung volume;
4. Reduced total lung capacity;
5. Reduced tidal volume (the volume of air exchanged with each normal respiration) which means that the older person must breathe more frequently in order to maintain an adequate volume of oxygen exchange;
6. Reduced vital capacity (vital capacity may decline by as much as 25 ml annually beyond the age of 20[6,41]);
7. Reduced inspiratory air flow.

The importance of many of the specific changes that occur in the ageing respiratory system is uncertain. Collectively, however, these changes produce a gradual reduction in the respiratory system's functional reserve. This reserve is normally large enough to ensure that there are no apparent functional changes while the individual is at rest. However, when ageing athletes begin an aerobically demanding activity such as cross-country skiing they will experience respiratory distress at a lower physiological workload than they would have in their twenties or thirties. Basically, the ageing competitor's respiratory system is less efficient and progressively less able to deliver oxygen to the body's working tissues. This is one of the major reasons for the age-related performance deterioration seen in events demanding maximum aerobic efforts. Comparison of the world records for the marathon at age 40 (2 h 11 min 4 s) and age 80 (4 h 23 min 55 s) illustrates how significant these changes are in terms of performance. It must be clearly understood that the rate at which many of these respiratory changes occur is considerably increased in the sedentary individual. Regular exercise substantially reduces their extent, with the most significant contribution being the strengthening of the respiratory musculature.

Skeletal muscle changes

Skeletal muscle is a vital element in athletic performance regardless of the age of the competitor or the event. There has been considerable research in this area and a great deal is known about the ultrastructural features of muscle fibres and how they function to permit muscular contraction. Unfortunately, our understanding of the age-related changes in muscle tissue has been confused by the failure of researchers to control for the effects of inactivity. This issue is made more confusing by the fact that many of the muscle changes attributed to the ageing process occur in individuals who are subjected to prolonged immobilization. It seems that the ageing process is accompanied by a number of alterations in muscle tissue and that a sedentary life-style enhances the rate at which they occur.

Ageing is accompanied by a loss of muscle tissue. Superficially this loss may not be apparent because many muscle fibres are replaced by fat or connective tissues and muscle girth is maintained. At the cellular level there is a loss of muscle fibres and atrophic changes occur in those that remain.[42] It should be noted that the elements remaining show no apparent alterations in contractile protein.[43] Larsson et al.[44] have shown a selective atrophy of the type II (fast-twitch) fibre population, with the type I (slow-twitch) fibre population remaining fairly stable. As a result of this shift, there is a relative increase in the type I fibre population. This change is unlikely to be related to the ageing process; it is probably a reflection of the differential use made of these fibre types during ageing. Type I fibres are recruited extensively for the postural and low intensity activities that dominate the lives of most elderly people, whereas type II fibres are employed in more explosive activities such as sprinting and are utilized less frequently on ageing. Thus, as ageing progresses, less muscle is available for use and what is available is increasingly dominated by the type I fibre population. Research has shown that regular use of the fast-twitch fibre population will help prevent its relative loss.[43,45]

Muscle remains biochemically stable during ageing. The levels of both the aerobic and anaerobic muscle enzymes are consistent with those of younger individuals in terms of their activity per unit of muscle weight.[43] In other words, while the other athlete has less muscle tissue available, the muscle they do have retains its normal enzyme levels. Thus it seems that the metabolic capacity of muscle tissue is maintained with age, and there is a considerable potential for improvements in endurance and strength with appropriate training. Capillary-to-fibre ratio is a term referring to the number of capillaries that make contact with a given muscle fibre. This is an important concept because it determines the maximum distance over which oxygen and nutrients must diffuse to reach each working muscle fibre. The greater this ratio the more oxygen can be delivered to working tissues and the greater is the aerobic work capacity. Plyley[46] has reported a very small decrease in the capillary-to-fibre ratio in older sedentary men compared with younger sedentary men. However, he did not compare younger and older athletes. Plyley comments that ageing individuals can respond to the stimulus of regular exercise by increasing their capillary density. Based on this, there is no reason to suspect that ageing is responsible for a significant change in capillary-to-fibre ratios.

A motor unit consists of a motor nerve and all the muscle fibres that it innervates. With ageing, there are fewer available muscle fibres in each motor unit and possibly fewer motor units.[23] Because of these changes, less contractile tissue can be utilized when a motor unit is recruited. These changes are associated with a progressive loss of strength, but these losses are not as dramatic as might be anticipated. Strength decreases very slowly until approximately age 50 when it begins to fall off more rapidly. Despite this, the anticipated loss of maximum voluntary strength at age 60 should not normally exceed 10 to 20 per cent of an individual's maximum strength.[41,43] This assumes that the individual has remained active; for sedentary individuals their losses would be greater. Studies show that, with training, older adults can experience significant improvements in their strength. There seems to be a fundamental difference in the way that the ageing athlete achieves this gain. Young individuals acquire strength increases primarily through muscular hypertrophy, whereas older individuals appear to increase their strength mainly through improved motor unit recruitment. Thus young athletes can improve by developing more contractile elements, while ageing athletes rely on using the muscle that they have more efficiently.[23] These observations were made on the basis of studies of relatively short duration. Studies of longer duration are required to see whether ageing athletes are indeed capable of muscular hypertrophy.

The speed of muscular contraction is reduced in ageing muscle, and this may in part result from a delay in nerve impulse transmission at the motor end-plate. This has important implications for athletic performance. Power is the ability to do work over time. In an explosive event such as throwing the javelin, the most successful competitors can apply the greatest amount of force to the implement over the maximum velocity of muscle contraction. Damon[47] has shown that the maximum velocity of muscle contraction possible against any given mass decreases with age. These detriments exist for isometric, concentric, and eccentric muscle contractions. In other words, ageing javelin throwers will be unable to accelerate the javelin as rapidly as they once could and so their performance will suffer. This may serve to explain why power events are the area in which ageing athletes have traditionally fared the poorest in comparison with their younger opponents.

Some of the other muscle structure changes that have been attributed to the ageing process are as follows.

1. Grouping of fibre types has been observed on muscle biopsy. This supports the possibility that neurogenic changes occur as muscle ages.[48]
2. Alnaqeeb et al.[28] have shown there is an increase in the collagen fibre content of ageing muscle. Since collagen is a relatively non-compliant substance, the musculature of ageing athletes becomes progressively stiffer. This will hamper performance and predispose them to injury.
3. Ultrastructural alterations have been reported by Sato et al.[49], but their athletic significance is uncertain. Type I fibres show Z-band streaming and the formation of nemalin-like structures. Type II fibres exhibit fragmentation, increased lipofuscin, and loss of Z-materials.
4. Lipofuscin granules develop, but the significance of this is unknown.[29]
5. Mitochondria are one of the principal intracellular organelles responsible for energy production. Studies have shown no change in their density within ageing muscle, but it is uncertain whether or not their volume decreases.[50]

To reiterate, many of the structural alterations attributed to ageing muscle have also been noted in the musculature of immobilized younger individuals. The net result of the observed age-related changes is a loss of strength, endurance, mass, compliance, and speed of contraction. These losses translate into a significant deterioration in athlete performance. However, the ability of ageing muscle tissue to respond to the stimulus of training does not appear to be lost. Structural and functional improvements can be achieved at any age with appropriate training.[23]

Nervous system changes

The nervous system is an extremely complex collection of several billion cells that function as the body's internal communication network. Even more impressive are the innumerable synaptic connections that permit the nervous system to communicate with every cell in the body. Any process that affects the operation of this system will have far-reaching effects on the functioning of the body. Research shows that structural and functional changes occur in the nervous system with increasing age. The structural changes include the following:

1. A reduced rate of cerebral spinal fluid production and turnover;[51]
2. Increased size of the brain's ventricles;[52]
3. Decrease of brain weight by as much as 20 per cent between the ages of 45 and 85, most of which is attributed to a loss of extracellular fluid rather than neurones;[52]
4. Loss of 50 000 to 100 000 neurones daily from the cerebral cortex, spinal cord, and peripheral nerves.[52]

The significance of these alterations to the operation of the nervous system is uncertain.

The senses are also affected by the ageing process. Age-related hearing loss is referred to as presbycusis and is extremely common. To compound matters, our ears are continually bombarded by numerous sources of sound pollution. These two factors combine to produce a hearing decrement in virtually every older individual. Hearing loss is particularly relevant to athletic performance because many of the cues that athletes rely upon are auditory in nature. The opponent's breathing pattern, the team-mate calling for a pass, the dribble of a basketball, or the scream of the crowd are all important to athletic success.

Normal age-related changes in vision are known as presbyopia and result in decreased visual acuity and accommodation. As we age the crystalline lens of the eye becomes increasingly brittle and so will not change shape as readily to permit accommodation. Good vision is a prerequisite to success in most athletic events because we rely heavily upon visual cues to influence our actions and reactions. Fortunately for ageing competitors with visual or auditory deficits, the technology exists to help them overcome all but the very worst of losses.

One of the most valued of athletic abilities is an individual's reaction time, and this clearly slows with advancing age.[42] This slowing is the result of several changes within the ageing neuromuscular system:

1. A reduce rate of muscular contraction;
2. A decrease in the rate of nerve impulse propagation;
3. A reduction in the rate of perceptual processing;
4. A decrease in the conduction rate of sensory nerves;
5. An increase in the central processes time for sensory stimulation.

In other words, every element in the reflex loop is hindered to some extent by the ageing process. It appears that the central processing component is the major contributor to the overall slowdown. Research also shows that when a choice of possible reaction-timed responses exists, the older person's reaction time slows much more than that of a younger person. This is the result of increasing even further the central processing required for a response to occur. Thus the ability to make split-second decisions and react is significantly impaired with age, particularly if it involves a novel situation. This has major implications for the older athlete's ability to perform in the read-and-react situations so commonly encountered in sports like hockey, soccer, football, squash, basketball, and volleyball. It should be noted that the reaction times of older athletes have been shown to be shorter than those of younger sedentary individuals, but not shorter than those of younger athletes.[53] This strongly suggests that reaction time, like so many other physical abilities, will also deteriorate with disuse.

There is evidence that there are differences in the way that ageing athletes cope with the attentional or focusing stresses demanded of them, particularly in highly stressful competitive situations. Molander and Backman[45] conducted a study in which older and younger athletes were monitored during the performance of a precision sporting activity. Although both groups reported an increase in their state of arousal during the competitive phase, the older athletes differed in two notable ways.

1. During the competitive phase of the activity their heart rates did not decrease as much as for the younger subjects.
2. Their motor performance deteriorated from the training to the competitive situation while that of the young group remained unchanged or improved.

Molander and Backman conjecture that this deterioration in performance may reflect an age-related change in attentional functioning (focusing skills). It seems that older athletes are less efficient at focusing their attention on a specific task during a stressful situation. They also postulate that the deteriorating motor skills of the ageing athlete require the commitment of greater amounts of mental energy towards rehearsing and attempting to retrieve the required neural patterns. This leaves them less mental energy to commit to dealing with external variables such as wind speed and direction, strategy, and the crowd. Regardless of the actual mechanism, these results are important because they suggest that even the most highly skilled ageing athlete will probably experience performance deficits while competing in highly stressful situations. These deficits will probably increase significantly as the skill level of the athletes decreases because they must commit even greater amounts of their mental concentration to performing the skill rather than dealing with the event in its entirety. This can be very frustrating for older athletes, particularly if they find that their training performances are superior to their competitive efforts. They begin to consider themselves mistakenly, as 'chokes' who cannot perform under pressure. If these individuals become frustrated, their performance will deteriorate even further and a vicious circle of decaying performance is established. A reasonable coaching approach to this situation would involve making the athlete aware that this phenomenon is a normal consequence of ageing. This may serve to reduce the frustration that the athlete is experiencing. The coach should also structure training such that the athletes are exposed to stressful situations and have the opportunity to learn to cope better with distractions and pressures.

Crystallized intelligence refers to the knowledge that individuals accumulate through their lifetime experience. This form of knowledge is maintained into old age and may even improve with the passage of time.[54] Fluid intelligence refers to a person's ability to solve new and complex problems, particularly those requiring abstract thought. This form of intelligence declines with age, lending support to the old saying that "you can't teach an old dog new tricks", or at least not easily. This change is very relevant to coaching and competitive situations. On one side, the coach will generally find it more difficult to teach skills and tactics to an older competitor. On the other side, the ageing athlete will have greater difficulty coping with novel situations that arise during competition. Both these changes will handicap the older competitor's performance, particularly in complex sporting activities.

Sleep architecture is altered with age such that total sleep remains relatively unchanged but sleep efficiency decreases.[55] In other words, more time must be spent in bed to attain the same volume of sleep. Older individuals have more trouble remaining asleep than falling asleep, they wake more often, and, once awakened, they remain awake longer than younger individuals. Their sleep pattern becomes more fragmented. This may occur because of a gradual loss of deep delta sleep, a change which provides them with a lighter quality of sleep. The sleep requirement for an individual does not appear to change with increasing age;[55] rapid eye movement (REM) sleep remains constant throughout life. These changes in sleep patterns may be relevant because they can interfere with an ageing athlete's ability to recover from training and competitive efforts.

In summary, the nervous system undergoes many age-related changes. In addition, there appears to be considerable evidence suggesting that the 'use it or lose it' principle governs the functions of neurones as much as it does muscle fibres. Neurophysiologists have demonstrated that regular stimulation delays the involution of neurones, and that hypertrophy may occur with sufficient stimulation.[42] Decreases in fluid intelligence, selective and divided attention, the senses, and reaction time all affect athletic performance. These changes occur very subtly and may not be clinically apparent in a sedentary person until he or she is over 65. However, they will be evident in the athletic arena at a much earlier age because in this environment microadvantages separate winners from losers.

Tendon and ligament changes

Ligaments and tendons are important structures because they play essential roles in the day-to-day operation of the musculoskeletal system. By definition, tendons are bands of connective tissue that anchor muscles to bones. In doing so, they function as energy transmission mechanisms. Ligaments are connective tissue bands that attach one bone to another. As connective tissue structures, both consist primarily of collagen and will experience all collagen's age-related alterations. Ageing tendons and ligaments become progressively less compliant and increasingly more vulnerable to injury. Once injured, ageing tendons and ligaments are relatively unforgiving. Severely traumatized ligaments will never return to their original length, their stress–strain characteristics will be permanently disrupted, and microscope examination will reveal evidence of collagen fibre failure.[56] Ageing also appears to be associated with a reduction in the glycosaminoglycans concentration found in tendons.[57] The significance of this change is uncertain.

Ligaments and tendons are by no means exempt from the deleterious effects of an inactive life-style. Research has identified a number of alterations including the following:

1. A reduced rate of collagen turnover so that cross-linkage stabilization has a greater opportunity to occur, rendering these structures increasingly brittle;
2. Bony resorption at the site of tendon/ligament insertion so that less tension is required to produce an avulsion fracture;
3. A decreased number of cellular elements;
4. Diminished collagen fibre thickness;
5. Reduced tissue capillarization;
6. Decreased glycosaminoglycans concentration;
7. Lower tissue water content.

All these changes considerably hinder the ability of tendons and ligaments to withstand the stresses applied to them during periods of physical exertion and leave them at an increased risk of damage. However, these changes have all been shown to regress when tendons and ligaments are exposed to the stresses and strains of regular physical exercise. In fact, the ageing exerciser will develop stronger, thicker, and more supple tendons and ligaments. Viidik[25] has developed several animal models which illustrate that training increases the tensile stress tolerance of both ligaments and tendons. In this state, these structures are able to tolerate considerably more wear and tear before tissue breakdown occurs. This is important for all athletes, but particularly for ageing athletes involved in sports requiring a high volume of repetitive movement such as swimming.

Flexibility changes

Flexibility can be thought of as the range of motion possible for any joint (i.e. the hip) or set of joints (i.e. the spinal column). In any individual, flexibility varies considerably from one joint to another. The individual's specific flexibility needs are dictated by the athletic event in which he or she participates. For instance, superior hip flexibility may be essential to a successful hurdler, but of little value to an archer. The movement possible about any joint is determined by a number of factors including the following.

1. Bony structures—the olecranon in the elbow is a classic example of a bony structure that restricts joint movement;
2. Muscle and fascia—inflexible muscle tissue will limit joint movement;
3. Ligaments, tendons, and joint capsules—non-compliant connective tissues will restrict the range of movement of the joint.

Any changes in the body which promote the above alterations will limit flexibility. The consequences of reduced flexibility for an athlete are numerous and potentially costly. For example, limited shoulder range of motion will interfere with a tennis player's serve. Athletes who cannot assume certain positions will be less effective. Consider, for example, the ice hockey goal-tender who cannot do the 'splits'. Inflexible people encounter greater resistance to work and so expend more energy to accomplish a given task. The final consideration is that injury rates increase as flexibility decreases. Non-compliant tissues appear to be more easily damaged.

There is a paucity of well-controlled research addressing whether flexibility changes with age and quantifying those changes in specific joints. Controversy exists over whether flexibility is altered by the ageing process at all or whether inactivity is responsible for the losses that occur. While some studies report increased passive resistance to movement, others do not.[2] It seems reasonable that if muscle, fascia, ligaments, tendons, and joint capsules become less compliant with age, flexibility must also diminish. This is consistent with the frequently noted observation that flexibility declines steadily after childhood. While there is undoubtedly an innate component to the flexibility losses noted with age, it appears that inactivity is the dominant factor in the average person. Generally, the less active individuals are, the more rapidly their flexibility will deteriorate. While the age-induced losses are permanent, the inactivity-induced losses are not. Studies in which older individuals were given range of motion exercises noted significant improvements in flexibility in a relatively short period of time.[23] The greatest gains observed were in individuals who had lived the most sedentary lives. It should be noted that studies have shown that increasing the temperature of a joint will enhance tissue compliance and so improve flexibility.[2] This supports the importance of a warm-up, particularly for the ageing athlete whose tissues are usually less compliant. Failing to warm up adequately is a mistake that the younger athlete may get away with, but for the ageing athlete it often proves costly in terms of unnecessary time lost to injury.

Cartilaginous changes

Articular cartilage is an ingenious combination of collagen fibres embedded in a matrix of ground substance rich in chondroitin sulphates and mucoproteins. This blend provides it with the physical properties of a gel that is anchored to the underlying bony tissues and permits joints to function as lubricated bearings during movement. It also gives articular cartilage mechanical properties ideally suited to shock absorption.

Articular cartilage differs from most body tissues in that it lacks a direct blood supply and therefore depends entirely on imbibition and diffusion processes to meet its nutritional needs. Both these processes are facilitated by the mechanical loading and unloading of the joint surface. Given regular mechanical stimulation, articular cartilage will respond by thickening in much the same way as bone does. Hall's research[58] supports the contention that articular cartilage will remain healthy if regularly subjected to compressive and decompressive forces.[58] His work also illustrates that, deprived of such stimulation, cartilage will undergo disuse atrophy and become more vulnerable to injury. Regardless of age, cartilage is fatiguable. It can fail structurally when exposed to excessive mechanical stress. This overuse causes fractures of collagen fibres, reduces the volume of proteoglycans on the cartilaginous surface, and damages chondrocytes. The damage will be even greater if some degree of joint malalignment exists as a result of a congenital problem or previous joint injury. For instance, a torn medial meniscus substantially alters the mechanics within the knee joint such that whenever the knee is extended, the femur is forced into the tibial plateau. This high contact force causes excessive wear and tear of the underlying articular cartilage and the development of early degenerative changes. Following a complete menisectomy, the contact pressure on the menisectomized side of the joint is increased and once again the articular cartilage is subjected to excessive wear and tear. The partial menisectomies that are performed today attempt to preserve as much of the viable meniscus as possible in an effort to minimize the disruption of normal knee joint mechanics.

Research demonstrates that articular cartilage experiences a number of age-related alterations. The most important of these is the loss of tissue compliance that occurs as a result of the cross-linkage stabilization of collagen molecules. This stabilization renders cartilage increasingly brittle and less able to cope with repetitive stresses. Articular cartilage in this state is increasingly vulnerable to the structural failure that can occur with overuse injuries. Ageing cartilage also contains less water and has increased concentrations of keratosulphate and chondroitin sulphate. All these changes reduce the compliance of articular cartilage, rendering it less able to perform its critical role of shock absorption. While the ageing process does produce cartilaginous changes, inactivity makes the greatest contribution to cartilaginous deterioration.

PSYCHOSOCIAL FACTORS

Psychosocial issues play a major role in the performance of all athletes and are particularly relevant to the success and failure of the ageing athlete. It is critical that health care professionals, coaches, and trainers recognize that the ageing sports enthusiast lives, works, trains, and competes in a psychosocial environment very different from that of his or her younger opponents. Compare, for example, the pressures on a 46-year-old mother of three who works full time and is training for a triathalon with those of the average 20-year-old college basketball player. Ageing competitors almost always have more financial, professional, social, and family obligations than their younger rivals. This translates into more distractions, disruptions, preoccupations, concerns, commitments, time compression, and fatigue. None of these factors is compatible with optimal training or competing. Pressured athletes often hurry their warm-ups, ruminate during their work-outs, and then cut their cool-downs short. This puts them at an increased risk of becoming injured. In addition, harried athletes often fail to derive the pleasure and relaxation commonly associated with a good work-out. Despite the potential for this problem, many ageing athletes actually look to their training as a form of stress management which helps them cope with the demands of the rest of their life. This section will review a number of psychosocial factors that are apparent in the masters sporting environment, in the hope that this knowledge will prepare the reader better to understand ageing athletes and their problems.

Sports participation not only provides ageing citizens with many structural and physiological benefits, but also offers them a number of psychosocial rewards. Studies have shown that elite and recreational runners, regardless of their weekly mileage, score lower on measures of tension, fatigue, depression, confusion, and anger, and higher on items that estimate vigour (Fig. 4).[59] These benefits occur regardless of the runner's age. One of the unpleasant realities of ageing is that it leaves many individuals alone in the world. In the United States, one in seven males and one in three females live alone.[60] This is very disconcerting for many individuals and provides significant motivation for participation in team sports. A sense of belonging with other human beings is an essential feeling for most people. Kavanagh and Shephard[61] conducted studies on the participants at the 1985 World Veterans Games and found that the top three reasons for participating in the games were as follows:

1. To belong to a group, 92.8 per cent;
2. To enhance mood, 90 per cent;
3. Fitness, 54 per cent.

These are radically different from the top three responses that would have been given if the athletes interviewed had been young Olympic hopefuls. It appears from the responses that the ageing competitor is generally more concerned with the psychosocial benefits of sports involvement than the physical benefits. Sports participation offers people the opportunity to get out of the house, meet people, have fun, and keep in touch with athletes many years their junior. Regular exercise often leads to improvements in physical appearance and in some instances athletic success. Both of these serve to enhance an individual's feeling of self-confidence and worth. In a culture which extols the virtues of remaining eternally youthful, maintaining a positive self-image can be very difficult as we age. Thus, it appears that participating in a sports programme is one way to add life to your years.

Why is it uncommon for world class athletes to carry on to become world class masters athletes? Why are there not more athletes like Gordie Howe, Nolan Ryan, Precious McKenzie, Archie Moore, and Arnold Palmer competing into their forties and fifties? The physiological and structural changes that occur with ageing are not sufficient to explain the large drop-out of high level competitors before the age of 40. The answer lies in a loss of motivation. Many of these individuals devoted their younger years to the endless amount of fatiguing training it takes to become and remain a world-class athlete, postponing education, marriage, family, business, recreation, and socializing. These people can often justify deferring their obligations by promising themselves that it will be over after the Olympics or whatever other athletic goal that they are striving to achieve. Once these goals are reached there is often considerable pressure and little incentive to resist retiring from sports and living like a normal person. As one athlete put it: 'My motto used to be that I wouldn't let my job interfere with my running. Now I do not let my running interfere with my job. My family also takes precedence over my running'.[62] In this particular case, sport has gone from first to third on this person's priority list. This change in status greatly affects the quality and quantity of training that a person will consistently perform. This attitude differs from that of individuals who begin their athletic careers late in life. In many cases, these people gave up sport during their youth, while they attended to the other distractions of life. Once they are educated and secure in their jobs and their children have grown up, they find themselves looking for a new challenge and sport provides the answer.

Numerous terms have been used to describe the loss of internal drive which often affects the ageing athlete's ability to perform. Burn-out, lack of motivation, losing the competitive edge, lacking the killer instinct, and just not having it any more are just a few of the excuses frequently made for declining physical performance. These individuals often retain their desire to achieve but lose their drive. This phenomenon is not unique to the sporting world; it is a well-documented fate for many young business executives who spend years working day and night and eventually can no longer handle the load. Russell and Branch[63] summarized this process nicely.

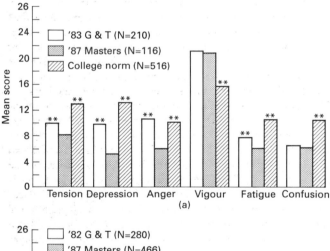

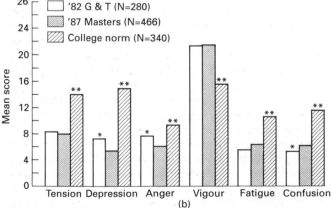

Fig. 4 Comparison of group means on the Profile of Mood States (a) Men athletes **$p<.001$, two-tailed t tests comparing this group's mean scores with masters' mean scores. (b) Female athletes *$p<.05$, **$p<.001$, two-tailed t tests comparing this group's mean scores with the masters' group mean scores. (Reproduced from ref. 77, with permission.)

Rarely will you see an athlete who hasn't put on 10–15 pounds over a full career but even rarer are the ones who don't put on the same amount of mental fat. That's the biggest killer of ageing champions because it works on your concentration and mental toughness, which are the margins of victory; it prevents you from using your mind to compensate for your diminishing physical skills.

'Mental fat' or lack of motivation has forced the retirement of many ageing athletes long before it would have been required by physical deterioration.

Several authors have studied the psychological reactions of people who suffer significant losses.[64] While the terms used may vary, the results consistently point to a typical grief reaction containing three basic phases: (a) denial; (b) anger; (c) acceptance. While these grief reactions were first described in individuals who were told that they were dying, they are no less applicable to an athlete's having to face the reality that he or she is no longer capable of performance improvement. Every athlete will eventually slow down, and some will slow down faster than others. When this point is finally reached, athletic improvement is no longer possible. This can be extremely frustrating because improvement is often the athlete's *raison d'être*, and when this motivator no longer exists many fail to see a reason for continuing with athletic involvement. Such a realization can represent a major psychological crisis in athletes who have always overvalued their athletic performances. It is easy to believe that athletes can undergo grief like reactions on experiencing a loss of this magnitude. These reactions can occur in athletes who face the reality of no future improvement, regardless of their age. Consider, for example, how a National Hockey League draft pick feels when he experiences a career-ending internal derangement injury to one of his knees. People who have competed for decades will often experience this reaction in their late thirties and forties when they realize that they are struggling to maintain their position rather than improve. However, individuals who do not begin competing until they are older often enjoy several years where they physiologically adapt to their sport and their performances continue to improve. These people often describe themselves as feeling almost immortal because while everyone around them is physically deteriorating, they are actually achieving personal bests. When these individuals reach the point beyond which athletic improvement is no longer physically possible, they often undergo an intense grief reaction.

The initial response of a grieving older athlete is to deny that they have reached their physiological peak. They do not accept that their inability to establish personal best performances is the result of biological changes associated with the ageing process. Instead, they attribute their faltering performances to a variety of factors including unsatisfactory training, a bad season, not enough miles run, insufficient weight training, not enough competitions, too little speed work, preoccupation with the family, work-related problems, or illness. They convince themselves that all that is needed to correct the situation is to train harder, longer, faster, further, and more frequently. When athletes adopt this approach, they overtrain, their performances worsen, and their chances of becoming injured increase dramatically. In any of these instances, the athlete usually ends up in the second phase of grieving, which is anger. Unable to accept that they can no longer perform at a given level, these athletes feel that they are the victims of a great injustice. They are angry because the skills that they have worked so hard to develop are slowly slipping through their hands and, despite their best efforts, there is absolutely nothing that they can

do about it. No amount of training, dieting, self-sacrifice, or dedication can restore them to their former ability level. They may also be angry because others around them continue to improve while they cannot. They have lost control of the situation and begin to ask: 'Why is this happening to me? I have done nothing to deserve this punishment.' It is in this state of mind that the grieving athlete is at the greatest risk of abandoning sport entirely. 'Why bother if you can't perform at your best? I don't want to be remembered as a has-been. I want to retire quietly before someone asks me to' are commonly heard expressions from individuals in the anger phase. It is at this critical point that the ageing competitor would benefit most from contact with someone who has survived the grieving process and is still actively enjoying sports. This role model can help the athlete make a successful transition into the acceptance phase of the grieving process. Unfortunately, such people are few and far between.

During the acceptance phase, the ageing athlete essentially realizes that nothing can be done to change his or her situation—the ageing process affects everyone. The maturation that is required to accept this situation often permits the individual to focus on new challenges and accept that the quality of his or her future performances must be considered in the light of physiological limits. Basically, a set of realistic performance expectations are developed. These athletes are often the happiest because they have come to accept that the greatest satisfaction that any athlete can derive from a contest is to give his or her best effort, regardless of winning. While the above sounds theoretical, it is in fact a very normal occurrence that has been repeating itself ever since people began competing. Anyone involved in the coaching, training, management, and treatment of older competitors must be aware of this process so as to ensure that their athletes are effectively managed. With some understanding and guidance many individuals will survive this phase to enjoy fuller sporting lives.

The ageing process can offer to the athlete two major competitive advantages: maturity and experience. The knowledge acquired through a lifetime of experience is a treasure that every coach values and would love to be able to instil automatically in their novices. Unfortunately for coaches, experience cannot be acquired from a pill, injection, ointment, textbook, or a well-delivered 'pep talk'. It can only come from success and failures experienced during years of training and competing. This is where ageing competitors often make major contributions in team sports. They are reservoirs of knowledge to whom the younger players can turn for instruction and inspiration. The younger players often have the raw physical talent but lack the refined physical skills, the maturity, and the gamesmanship required to be successful in a given sport. Skills like bluff, surprise, anticipation, patience, concentration, focusing, taking advantage of an opponent's lapses, and making opponents play your game all take time to develop. In this regard, the older athlete often functions as a teacher and stabilizing influence on the team. Their experience enables them to know what must be done in order to be victorious in a given game situation. Experienced competitors often possess an advanced level of self-understanding. As individuals they understand their body responses, when they need to work hard, when they need to rest, and the price that must be paid to be successful. This intimate knowledge goes a long way towards negating the physical advantages of youth. Sheehan[65] summarized the value of experience when he stated:

In the September of my life I am all that the past has taught me. I know things that were a mystery before. Every year has become an asset, every experience a treasure. I am no longer a rookie, no longer a neophyte and I no longer look to veterans for guidance—I am a veteran, a master at my own game.

With the growing popularity of masters sports has come an increasing volume of media coverage and the greater possibility of monetary rewards. It is now possible for a successful ageing athlete to become both rich and famous. The lure of these desirable elements is providing considerable motivation for many older competitors to continue competing or to come out of retirement. The American Road Running Circuit and the Senior Golf Tour are examples of two athletic arenas where both fame and fortune are available. The Senior Golf Tour currently offers several million dollars in prize money, and some golfers are earning more than they did on the Open PGA Tour. For example, Al Geiberger joined the tour at the age of 50 and in his first year earned more than in any of his previous 28 years on the PGA Tour. The American Road Running Circuit provides us with possibly the best example of the fame that can be attained by an ageing competitor. John Campbell had retired from running on three occasions and before becoming a masters competitor could have been described, at best, as a relatively obscure distance runner from New Zealand. Since turning 40, however, this determined competitor has achieved the status of an international running superstar. He currently holds the masters marathon record of 2 h 11 min 4 s, a world class performance at any age. In 1990 he was undefeated in 20 races and established age class world records at six different distances. The secret of his success appears to be his arduous training schedule in which he runs 140 miles a week. While the availability of money and recognition will help the masters sporting movement grow at the elite level, there is an increasing possibility that the problems which are corrupting the younger athletic world will diffuse into the masters sporting movement. Problems such as excessive training, blood doping, steroids, and other ergogenic aids may not be far off, if they have not already invaded the realm of the ageing athlete.

A review of the literature concerning ageing athletes reveals that an individual's attitude towards ageing is of paramount importance in determining how well they cope. If the person views ageing as an inevitable tragedy in which all the virtues of youth are lost, then they will probably handle the ageing process poorly. We live in a world which cherishes the attributes of youth and billions of dollars a year are spent convincing people that they should use various products to maintain a youthful image. This negative attitude towards growing older has convinced many people that ageing is something to be feared. This attitude is by no means new. Shakespeare sums up our society's attitude towards ageing in his poem *The Passionate Pilgrim*:[66] 'Age, I do abhor thee, youth, I do adore thee'. Berman *et al.*[52] state that, as a result of this pervading attitude, a significant number of older adults suffer from the 'brainwashed elderly syndrome'. In this mental set, the individual believes that ageing is associated with wrinkles, retirement, old clothes, rocking chairs, inactivity, unattractiveness, diminished mental capacity, and lack of worth. These people are also convinced there is nothing that can be done to avoid this fate and they frequently live a self-fulfilling prophecy. Tragically, this conditioned thinking is so powerfully ingrained in our culture that people are frequently convinced that they are decrepit by the age of 30. The biographies of successful older athletes provides a striking contrast to this self-deprecating attitude. These people are remarkably consistent in their positive mental outlook. They are not focused on all the things that ageing is taking away from them but rather strive to enjoy the many things it provides. They frequently think younger, act younger, dress younger, and partake of activities commonly identified with the younger generation. There appears to be more to growing old successfully than maintaining good physiology; it would also appear that maintaining a positive attitude is a necessary prerequisite.

AGEING AND ALTITUDE

Since people are retiring earlier and the opportunity for travel is increasing, larger numbers of ageing individuals will be participating in high altitude recreation and sport. This includes hiking excursions in mountainous regions, mountain climbing, and running, cycling, and triathalon events held at altitude. While mountain climbing is undeniably one of the most physically demanding athletic events, this has not prevented ageing athletes from accomplishing some impressive feats. At the age of 52, Dick Bass conquered Mount Everest, a climb of 8848 m. Hielda Crook did not take up serious exercise until the age of 70. Since then, she has climbed Mount Whitney (4418 m) over 22 times.[67] She also runs in local road races and during the summer months climbs a mountain every month just to keep in shape. Her guide for these adventures is a man over 80 years of age.

Surprisingly little research has been done on the effect of ageing and altitude tolerance. The investigations that have been performed suggest that increasing age offers mixed blessings in terms of coping with the stresses of altitude. On the positive side, researchers have found that ageing is associated with a decreased incidence and severity of acute mountain sickness.[67] This benefit appears to be unrelated to the rate of ascent, disproving the commonly held belief that the young are more susceptible to altitude illness because they climb faster than their older colleagues. Acute mountain sickness is a condition commonly experienced by individuals who ascend to altitudes in excess of 3000 m. The afflicted person experiences symptoms of headache, anorexia, nausea, vomiting, weakness, and insomnia within 2 to 4 h of arriving at altitude. This condition is usually self-limiting; however, some individuals go on to develop high altitude pulmonary oedema, a potentially life-threatening condition. The mechanism by which increasing age protects against altitude-related illnesses is uncertain.

On the negative side, altitude will compound the oxygen delivery problems of the ageing athlete. Ageing is associated with a reduction in an individual's arterial partial pressure of oxygen.[67] Basically, older competitors are unable to deliver the same volume of oxygen to their working muscles as younger people in the same state of health. To complicate matters, as altitude increases the partial pressure of oxygen in the atmosphere decreases and so less oxygen is available for the respiratory system to capture and carry to the tissues that need it. This double disadvantage serves to handicap, or at least greatly challenge, the ageing athlete who is exercising at altitude. However, it should be noted that research indicates that an individual's ability to tolerate the physiological demands of altitude depend more on their general health and fitness than on their age.[67]

THERMAL STRESS

Thermoregulation is an essential body function and can be defined as the ability to maintain a normal body temperature by balancing heat-dissipating and heat-generating mechanisms. This is necessary because the body is designed such that the operation of cellular structures, enzyme functions, and chemical reactions occur optimally within a narrow band of temperatures. In humans, the optimal physiological operating temperature is approximately 37 °C. When thermoregulatory mechanisms fail to hold the body temperature at this level, optimal physiological function is impossible and athletic performance deteriorates. Independent of age, this functional deterioration will be greatly facilitated if the athlete starts the competition underhydrated, ill, hung over, tired, unacclimatized, or in poor physical condition. Thermoregulatory failure leaves the individual vulnerable to thermal injuries such as heat cramps, heat exhaustion, heat stroke, frostbite, or hypothermia. A wide variety of athletic events are held in environments where the ambient temperatures range from −30 °C to +40 °C. At the extremes of these temperatures the risk of thermal-stress-related problems is high. With more ageing athletes participating in these events, it behoves us to determine whether the ability to tolerate thermal stresses is altered with age.

Ageing has traditionally been associated with reduced heat tolerance. It has been observed that during heat waves older individuals are more susceptible than younger people to thermal injuries and related fatalities.[68] However, this trend does not indicate whether the reduction in tolerance is the result of the ageing process or other variables. Relatively little research has been done in this area and, of the studies that have been performed, many were poorly designed in that they failed to control for important variables such as cardiovascular fitness, state of acclimatization, disease processes, and body composition. These variables significantly influence an individual's ability to tolerate heat stress. A review of the well-controlled investigations suggests that the ageing process is responsible for some reduction in heat tolerance but that this loss is not as significant as the losses attributable to poor physical fitness, disease processes, and lack of acclimatization.[68] The exact mechanism by which this age-related decline occurs is uncertain, but it seems reasonable to assume that, since thermoregulation is a complex process, the cause is multifactorial.

Researchers have pointed to changes in the ability of the ageing athlete to dissipate heat through perspiration. Apparently, sweat responses to exercise are both delayed and diminished on ageing.[69] Each of these changes hinders the athlete's ability to dissipate heat to the environment. However, it has been shown that, with physical training, the sweat glands will respond to exercise earlier and show no decline in their capacity for sweat production.[68] Kenny and Anderson[70] have demonstrated that ageing athletes competing in hot and humid environments sweat the same volumes as younger competitors. However, when the environment is hot and dry, ageing competitors produce substantially less sweat than their younger opponents. This difference could represent an inability to handle the volume of sweating required because of the relatively underhydrated state of the average older person. Goldman[15] reported that total body water decreases from 62 per cent of body weight at age 25 to 53 per cent at age 75. Berman et al.[17] put it another way, stating that the average 35 year old weighing 70 kg has as much as 7 to 8 l more body water than a 75-year-old man of the same weight. This has considerable relevance to athletic performance, particularly since ageing is also associated with a reduced sensation of thirst. Thus ageing athletes competing in hot dry environments could become significantly dehydrated before the sensation of thirst motivates them to drink. Experience from the 1989 World Veterans Games supports this, in that the competitors were reported to be less aware of when they were overheating and in need of fluids.[71]

In view of the serious implications of heat injuries and the ageing athlete's unique physiology, a number of important recommendations can be made for training and competing in hot environments.

1. Athletes should hydrate themselves well before their event. During the competition, they should begin taking in replacement fluids long before they feel thirsty. By waiting for thirst to develop, they will already be in trouble and unable to compensate enough to correct their fluid depletion.
2. Athletes should continue drinking fluids after the event in order to replenish their losses.
3. Ideally, competitions and training sessions should be scheduled during the coolest time of the day.
4. Fluid intake should never be restricted during hot weather practices.
5. Athletes should remain in the shade as much as possible and wear light coloured clothing.
6. If permitted, the competitors should be encouraged to wear hats to protect their heads from the sun.
7. Participants should realistically assess their fitness level and understand that the less fit they are the lower their heat tolerance will be.
8. Athletes should be familiar with the early symptoms of heat stress such as muscle cramps, dizziness, cool skin, dry skin, elevated pulse, nausea, excessive thirst, weakness, and fatigue.

Very little research has addressed the issue of age-related alterations in cold tolerance. However, some facts are available for consideration. The ageing competitor is at increased risk of experiencing localized cold injuries such as frostbite for several reasons. First, although it is not specifically an age-related change, atherosclerosis reduces the peripheral blood flow in the extremities. This reduces the volume of warm blood that can be provided to cooling peripheral tissues. It has also been suggested that as individuals age, their perception of cold may be diminished.[17] Finally, there is a gradual reduction in subcutaneous adipose tissue during ageing, so that the layer of fatty insulation is reduced. These changes render the ageing competitor's peripheral tissues more vulnerable to cold injury. It also seems reasonable that, if ageing is associated with a reduction in our ability to handle physiological stresses, it will be associated with diminished physical performance in cold environments. More research is required before many of the questions in this area can be answered.

INJURIES

Ageing not only renders masters athletes physiologically, structurally, and psychosocially different from younger athletes, but also leaves them increasingly vulnerable to injury. Stiffer tendons, muscles, and joint capsules, brittle ligaments, a reduced rate of tissue repair, decreased joint flexibility, and loss of bone mass all contribute to this vulnerability. As athletes age, they can often maintain specific performance levels only by working closer to their physiological maximum. Working at this intensity places any athlete at an increased risk of injury. To make matters worse, injured ageing athletes also appear to be at greater risk of developing complications such as infections. This may be the result of the ageing person's altered physiological ability to respond to the stress of injury. In view of these factors, the injury pattern experienced by masters competitors will probably differ significantly from that of their younger opponents. In this section, we shall not consider specific injuries but rather focus on how the ageing athlete's injuries differ from those seen in younger adults.

613

Masters competitors are potentially the victims of two distinct types of injury: those resulting from their current training and competing, and those that occurred in their youth and return to haunt them. Either type of injury can be sufficient to cause considerable discomfort, hamper proper training, and ultimately force a premature retirement from the athletic arena. Consider, for example, athletes who experienced major joint injuries such as shoulder dislocations, anterior cruciate ligament disruptions, or meniscal tears during high school. Regardless of the treatment received, the affected joints will have their integrity compromised for ever. The problems frequently remain quiescent until middle age when the individual returns to athletics. At this time, the individual notices that his or her knee can no longer withstand the repetitive trauma of an activity such as running. This is an all too frequent frustrating situation for many ageing competitors. The conclusion is that significant injuries incurred in a person's youth often dictate what athletic activities they can tolerate in later years.

Clinical experience suggests that younger athletes suffer from a much higher incidence of traumatic musculoskeletal injury than their older colleagues. This occurs despite the younger athlete's connective tissues being generally more supple. Since the injuries that athletes experience are largely dictated by the sports in which they compete, it is not surprising to find that younger athletes tend to participate in sports with a greater potential for body contact and violence. Additionally, their relative lack of competitive experience and their tendency to participate with more reckless intensity puts them at greater risk of experiencing traumatic injuries. Masters athletes are by no means immune to traumatically induced injuries such as internal joint derangements; however, they are uncommon and when they occur the individual may not present with a classic history. This is one reason why health care professionals often overlook the diagnosis of a significant joint injury and mistakenly attribute the ageing athlete's signs and symptoms to degenerative processes. This can lead to delays in obtaining definitive treatment and cause major disruptions to training schedules.

The greatest threat to ageing athletes comes from degenerative tissue changes induced by excessive wear and tear (overuse). Postural malalignments serve to exacerbate this threat by overloading various components of the musculoskeletal system. For example, runners with genu varus (bowed legs) repeatedly subject the medial compartments of their knees to excessive mechanical stresses. This repetitive overloading of the medial compartment will considerably shorten its life-span. Eventually, osteoarthritic degeneration develops and the medial meniscus will break down as shown in Fig. 5. Research by DeHaven and Littner[72] indicates that the incidence of inflammatory problems increases on ageing until by the age of 70 the top five athletically associated maladies are inflammatory in nature. Anyone working with masters athletes should beware of the 'itises'—arthritis, tenosynovitis, fasciitis, bursitis, capsulitis, and tendinitis. These injuries commonly begin as small nagging problems. If they are identified at this stage and treated effectively, they usually resolve with no sequelae. If they are ignored or inadequately treated, these small problems can progress to become major injuries and the athlete may lose much valuable training time.

The most effective strategy for dealing with injuries in any athletic group is preventing their occurrence in the first place. This is no less true for ageing competitors; in fact, it is a more important consideration for them because their injury recovery rate is longer. Many of the structural changes associated with

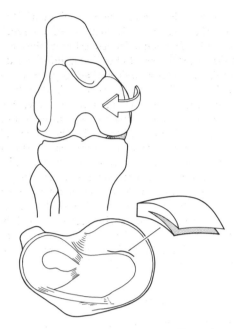

Fig. 5 Degenerative horizontal cleavage tear. (Reproduced from Sutton JR, Brook RM. *Sports Medicine for the Mature Athlete.* Philadelphia: Benchmark Press, 1986; 283, with permission.)

ageing leave an individual's tissues with reduced compliance and a diminished blood supply. In view of these changes, it is important for ageing athletes to perform an adequate warm-up before every work-out or competition. This increases the blood supply and temperature of their working tissues. Both these adjustments prime the tissues for work, enhance their viscoelastic properties, and so prepare them to be used. Compliant muscles and tendons are far more difficult to injure. The value of having compliant tissues reinforces the importance of including some flexibility training in every ageing athlete's programme. The time invested in stretching will be more than reimbursed by the time saved through injury prevention.

Healing and ageing

Research indicates that ageing has a detrimental affect on the healing process. The body's normal response to injury is described as occurring in three phases:

(1) the inflammatory phase;
(2) the proliferative phase;
(3) the remodelling phase.

All three phases are detrimentally affected by the ageing process. With ageing there is dampening of the inflammatory phase. There is also a delay in the cellular migration, proliferation, and maturation that occur during the proliferative phase. In the remodelling phase, collagen is laid down less rapidly, in smaller volumes, and with altered binding patterns.[4] The net result of these changes is that, as we age, the events associated with the healing process begin later, proceed more slowly, and do not achieve the same level. Studies involving rats have shown that following a muscle contusion the proliferation of muscle fibres, fibroblasts, and blood vessels occurs more rapidly in the young.[4] Some researchers have suggested that the age-related delays in the healing process occur because of increases in the time that it takes to prime the system for the regenerative process.[4] Regardless of the aetiology, the conclusion is that, when injured, the older athlete's body still effects a quality repair but it takes longer than it used to. Al Oerter, the great discus champion, noticed this in his training, in that the injuries that he developed in his forties and fifties seemed to require a longer time to heal. While there are large variations in the rate at which individuals age, there are also wide variations in the rate at which they recover from injury. To complicate matters, the rate of healing is also influenced by tissue changes that have resulted from disease processes, disuse changes, environmental insults, and the presence of other body stressors. Any process that reduces tissue capillary density will reduce the rate at which oxygen and repair substrates can be delivered to an injury site. It is interesting to note that healing delays do not appear to occur in bony tissues. The speed at which bony union occurs decreases rapidly until skeletal maturity is achieved in late adolescence. Beyond this point, there appears to be no age-related decline in the rate of fracture healing.[73]

Treatment considerations

The standard treatments for most injuries do not vary from patient to patient, but there are a number of issues related to the treatment of ageing athletes that merit discussion. First, beware of Ageism! Ageism is defined as discrimination against individuals on the basis of their being old. It is ageism that prompts some health care professionals to brush off an older competitor's complaints with the all too common phrase: 'What do you expect for someone your age?' This approach is unacceptable as there is no scientific evidence suggesting that healthy older individuals cannot participate in a wide variety of sporting events without experiencing pain. Pain is not a normal part of the ageing process, and when it occurs it often indicates that something is wrong. When pain occurs in an older athlete it is just as likely to be athletically induced as it would be if the patient were younger. Many older people will not complain of pain because they fear treatment or do not want to 'give in'. This is particularly true of ageing competitors who often view injuries as a threat to their continued participation. Health care professionals are obliged to take every athlete's complaints of injury seriously, regardless of age. They all deserve a comprehensive evaluation, and the decision to refer for specialist consultation should always be based on the individual's symptoms and not his or her age.

It is important to recognize that dedicated masters athletes will react to injury in the same way as most athletes and so will be reluctant simply to discontinue their training and competitive schedules. Their athletic involvement has become a highly valued component of their self-image and this is seriously threatened when an injury restricts their active participation. Before health-care professionals can effectively manage these individuals they must recognize the existence of this 'athletic mind set'. To pronounce dogmatically to an athlete, 'The solution is simple: if it bothers you to throw the discus then stop throwing the discus' is both inappropriate and insensitive. This type of attitude does little to dissuade many athletes from participating in their sport, and in fact challenges some individuals to attack their sport with even greater fervour in an effort to prove the medical world wrong.

As an alternative, health-care professionals should take the time to explain the problem carefully and to indicate that continuing an unrestricted training programme could not only cause more extensive damage but might permanently interfere with the athlete's ability to compete and improve in their particular sport. The objective is to convince athletes that by accepting a short-term loss they will ultimately enjoy a long-term gain. The extent of this short-term loss can be minimized by proposing alternative athletic activities that permit competitors to maintain or even improve their fitness skills without hampering their recovery. The more sport specific these activities are, the greater is their potential benefit. A good example of this would be water running sessions for a marathon runner with a metatarsal stress fracture. Technique modification is another avenue that can be pursued. For example, a tennis player suffering from recurrent supraspinatous tendonitis may find switching from an overhead to a side-arm serve is all that is needed to permit continued participation with minimal discomfort. Although the problem is not directly addressed, what the athlete desires—continued participation in his or her chosen sport—is accomplished. If necessary, bracing can be used to support the body parts where the normal musculo-skeletal structures have been compromised. Only as a last resort should a complete change of activity be recommended.

It should be remembered that injuries of little significance to a sedentary person may cause considerable problems for an elite athlete. This is no less true when the patient happens to be an elite masters athlete. An injury that could normally be satisfactorily treated with a conservative approach may require definitive correction in ageing athletes because it hinders their ability to perform at a desired level of competence. Left untreated, these individuals will often experience the same depressive symptoms as seen in younger competitors under the same circumstances. To dismiss this reaction as immature, exaggerated, or hypochondriacal is irresponsible. One must be aware that ageing athletes often take their sporting involvement very seriously and deserve at least the same level of concern from the health-care profession.

As noted earlier, inactivity poses a considerable threat to the structural integrity of the ageing individual because it clearly accelerates the rate at which many age-related changes occur. This is a critical consideration to bear in mind in the management of the injured ageing athlete. Immobilization is frequently relied upon in the treatment of certain injuries such as fractures and severe ligamentous sprains. While in many instances immobilization cannot be avoided, it should be used only when necessary and for as short a period of time as possible. Not only does immobilization threaten the ageing competitor with all the changes associated with inactivity, it is also a major source of frustration. Knortz[50] has pointed out that immobilized geriatric patients are at a dual disadvantage in terms of muscle fibre alterations. Not only do they continue to experience the selective fast-twitch fibre atrophy that occurs with ageing, but they also experience slow-twitch fibre atrophy in response to their immobility. This indicates the need for early and aggressive physiotherapy to ensure that joint range of motion is maintained, muscular strength is preserved, and bone mass loss is minimized. This is essential to ensuring an ageing athlete's rapid and complete recovery.

A regular exercise programme is one way in which individuals can reduce their risk of developing problems such as coronary artery disease, but it is by no means an absolute guarantee that these problems will never develop. The adage 'we all must die of something' is as true today as it was five thousand years ago. Health-care professionals must keep this in mind when dealing with ageing athletes. These people can present for the first time with the signs and symptoms of a significant underlying illness, and their involvement in sport can confuse the issue. For example, shortness of breath on exertion is a classic feature of advanced coronary artery disease. It is also commonly experienced by athletes participating in aerobically demanding sports. Shortness of breath in elderly athletes should never be superficially dismissed as indicating that they are unfit, particularly if it has never occurred before and is hindering their ability to train. These individuals merit a thorough examination to establish whether there are any pathological explanations for their respiratory problem such as coronary artery disease, anaemia, exercise-induced asthma, etc. This caution extends to a number of symptoms including unexplained weight loss, chronic fatigue, dramatic decline in physical performance, loss of motivation, and persistent pain. All these symptoms can occur in the young or be attributed to stress of training. However, in the ageing individual, they have a higher probability of indicating the presence of disease and so merit more than a superficial review.

Rehabilitating an injured athlete is perhaps the most important aspect of treatment. It often determines whether and how quickly the athlete can return to competition safely and effectively. Its aim is to restore normal function in the shortest possible time. This is critically important to ageing athletes because the longer that they are unable to train properly, the harder they must work to recover their performance ability. Rehabilitative efforts should be initiated as soon as possible after an injury and the entire range of treatment modalities should be considered. As noted earlier, the ageing athlete's physiology results in slower healing and the rehabilitation programme must be adjusted accordingly. As a general guideline, when estimating the rehabilitation time required for an older athlete twice as much time should be allowed for someone aged 60 as for someone aged 20. An athlete aged 75 or more will probably require three times the standard time.[71] These estimates hold for almost all injuries and can be a source of frustration to both the athlete and the therapist if neither is aware of them from the start. It is important to recognize that successful rehabilitation depends more on the person's motivation than on their age.

Medication considerations

The use of medication is a concern for athletes of any age. However, it is a particularly important topic for the ageing athlete. The entire pharmacotherapy process is significantly influenced by age, disuse, disease, and environmental related alterations to a person's structure and physiological function. Generally speaking, in our society the older people become, the greater the amount of medication that they consume on a regular basis. This fact and the altered anatomy and physiology of the older individual combine to increase their incidence of adverse drug reactions. A number of pharmacokinetic changes are associated with the age of process.

1. Reduced absorption—this occurs as a result of changes such as decreased gastric acid production, increased stomach emptying time, reduced intestinal absorptive surface area, and decreased intestinal mucosal blood flow.[22] In theory these changes will reduce the rate at which many drugs are absorbed, but in practice the healthy older person appears to experience only minor reductions.
2. Altered distribution—with ageing there is usually a reduction in total body water, an increase in body fat, and a loss of lean body mass.[22] Thus water-soluble medications will be distributed over a smaller volume and will have a greater effect for a given dose. However, fat-soluble medications will be dispersed over a larger body volume and so have a diminished effect for a given dose.

3. Plasma binding proteins—as their name suggests, these proteins bind to medications and assist in their distribution throughout the body. The concentration of these proteins remains unchanged in many older adults and if reductions are noted, they are usually minor.[22]
4. Reduced renal clearance—ageing is clearly associated with a loss of renal tissue and a reduced glomerular filtration rate.[22] As a result, drugs that require renal clearance are handled more slowly as we age. This is particularly true of many antibiotics.
5. Reduced hepatic clearance—on ageing there is a loss of hepatocytes and a reduction in the volume of blood flow they receive.[22] Thus drugs that depend on hepatic metabolism to be broken down will be handled more slowly.
6. Reduced cell receptors—the number of drug receptors on various cells may be reduced on ageing.[22] This reduces the amount of drug that can enter these cells and so reduces the effectiveness of the medication.

The above changes combine to have a significant influence on how an older individual responds to certain medications. Since drugs are eliminated from the body principally through hepatic metabolism and renal excretion, these changes serve to increase the half-life of many medications in ageing individuals.

Inflammatory conditions are the most common problems affecting ageing athletes and non-steroidal anti-inflammatory drugs (NSAIDs) are commonly used in their treatment. These agents include ibuprofen, acetylsalicylic acid, naproxen, indomethacin, and several others. Although these medications are frequently prescribed, they should be used with caution for several reasons. First, NSAID-induced gastrointestinal complications (e.g. gastric ulceration) occur more frequently in older people, particularly women or individuals with a history of peptic ulcers.[74] This is partially due to the decreased gastric vascularization that occurs with ageing. This compromises gastric mucosal protective mechanisms which are also compromised by the NSAID-induced reduction in prostaglandin synthesis. With gastric mucosal protection diminished, the stomach is at increased risk of injury. Second, ageing is associated with an increased incidence of adverse reactions to NSAIDs. Third, the anti-inflammatory and analgesic effect of NSAIDs may mask the warning pains that an individual normally recognizes as suggesting that it is time to stop. Without these signals, people often train longer and harder and potentially inflict greater degrees of injury upon themselves. This is particularly true when the individual is suffering from a degenerative joint condition. A safer approach in these cases is to prescribe a very light dose or none at all prior to exercise and then a full dose after the work-out to limit pain and inflammation.

The general principles for prescribing medications for ageing athletes include the following:

1. Use as few medications as possible;
2. Use as low a dose as possible;
3. Medication responses may be less dramatic than in the young;
4. Beware of side-effects;
5. If a medication is not accomplishing what is required, discontinue it and try another.

TRAINING PRINCIPLES

Many volumes have been written on the principles of training safely and effectively so as to guarantee maximum athletic performance. While it is beyond the scope of this chapter to discuss all these principles in detail, it is worthwhile highlighting a number of issues of particular importance in the training of an ageing athlete.

Just as ageing athletes experience decrements in their organs' functional reserves, their ability to compensate for training errors is also reduced. Basically, the ageing competitor cannot expect to commit the same training indiscretions that many younger athletes do and survive uninjured. Competing too frequently, training too long and hard, failing to rest and permit recovery, ignoring flexibility work, training inconsistently, and avoiding strength training are all examples of common training errors. These errors often leave younger athletes performing suboptimally or result in minor injuries. However, masters competitors may pay a much heavier price for making the same mistakes. Frequently their performances will deteriorate considerably and their chances of developing a serious injury are much greater. Frank Shorter, the 1972 Olympic marathon champion, discovered this when he entered masters competition. He stated:[75]

You can't train the same way you did when you were 23. It just isn't going to work. All you do when you're 23 is go out the door, turn left, turn right and go as hard as you can till you start to break down. Then you take a day or two of easy running till you heal up. But that's not what happens when you get to be a master.

At the age of 43, Nolan Ryan remains one of baseball's finest pitchers and the all-time leader in strike-outs. He openly credits his competitive longevity to his commitment to a year-round training programme. His programme emphasizes a balance of weight-training, running, stretching, and regularly scheduled rest days. He consistently dedicates himself to this programme regardless of how his pitching game is going. Ageing athletes are undeniably unique and must be mature enough to recognize and accept this if they intend to enjoy the many fruits of competitive effort.

If there were a set of commandments governing the rules for training older athletes, the first commandment would be 'Thou shalt start low and progress slowly'. This is a critical concept that applies to the development of all athletes, regardless of their age, sport, or level of expertise, but it is of particular importance for the older competitor. All body tissues are capable of adapting to the various training stresses to which they are subjected. However, two points should be remembered. First, athletes will adapt to the stresses of training at their own unique rate. Since older athletes have a greater potential for structural and functional diversity, the variation in their rate of adaptation will be greater than that seen in younger competitors. Second, there are considerable differences in the rate at which different body tissues adapt to the stresses of an exercise programme. For instance, the cardiovascular and respiratory systems respond much faster to training stimuli than do muscle and bony tissues. Thus people who have been sedentary for a prolonged period must be careful not to be overly ambitious in the early stages of their programme.

Consider the sedentary individual who wants to become a runner. It would make sense to start off with a walking programme in which he or she progressively introduced longer sessions and faster cadences. Once the individual is comfortable with this, he or she can begin to introduce brief periods of running in a pattern such as walk for 5 min, jog for 2 min, walk for 5 min, etc. Gradually, the individual increases the distances and the relative amount he or she is running until several months later the entire work-out consists of running. While this may sound like a cautious approach, it makes good physiological sense. During this process

the individual's cardiovascular and respiratory systems will respond rapidly and he or she will feel capable of enduring greater volumes of exercise. Unfortunately, the muscle and bony tissues, which absorb much of the mechanical trauma associated with exercise, will not have had sufficient time to undergo the structural reinforcement necessary to handle heavier work-loads. This difference in the rates of adaptive response leaves the ageing athlete vulnerable to injury during this time. By adhering to the principle of starting low and progressing slowly, an athlete can often avoid unnecessary injuries and enjoy a more rewarding sporting life. Athletes who ignore this principle may enjoy a rapid initial improvement but often find their further improvement hindered by injury. As a general rule, the older the individual and the more unfit they are, the lower they should start out and the more slowly they should progress.

One of the questions addressed in the literature is how responsive older athletes are to training stimuli. Do they respond in the same way and to the same extent as their younger opponents? There are a number of studies demonstrating that, regardless of age, older individuals who begin training are capable of making noteworthy physiological improvements.[39,76] These improvements include the following.

(1) reduced resting heart rate;
(2) improved V_{O_2}max;
(3) increased vital capacity;
(4) decreased percentage body fat;
(5) improved work capacity;
(6) reduced blood pressure;
(7) increased strength;
(8) increased endurance;
(9) increased flexibility.

These benefits occur even in individuals with clinically evident cardiovascular disease, although to a lesser extent than in disease-free individuals.[39] It has long been recognized that participating in aerobic activities such as cycling, cross-country skiing, walking, swimming, and running can provide substantial physiological benefits regardless of the individual's age. However, strength training was an area of uncertainty in this regard. Studies of strength training in older subjects have demonstrated that older athletes are capable of strength improvements at a rate equal to or faster than some younger athletes.[44] This is encouraging because the ability to make strength improvements is an important factor in athletic performance. The benefits are obvious for events such as shot-put, discus, and power lifting, but are no less important in curling, swimming, mountain climbing, and soccer. The need for strength training is perhaps more imperative for ageing athletes than for the young. As noted earlier, ageing skeletal muscle undergoes a deterioration of its fast-twitch fibre population, including a loss in fibre number and a decrease in fibre diameter. The rate of this deterioration is considerably enhanced by an inactive lifestyle. This deterioration can be greatly minimized if the individual maintains a training programme that utilizes the fast-twitch muscle fibre population. This is one aspect of training that many athletes neglect, and it often proves to be the deciding factor in winning or losing a given event. If this holds true for younger competitors, it is no less so for the ageing athlete.

One of the most frequently employed training guides is monitoring the response of the athlete's heart rate to exercise. People are taught to take their pulse before, during, and after work-outs. Heart-monitoring devices are available on the market, some of which provide a continuous breakdown of the heart rate response throughout a training session. A person's maximum heart rate can be estimated using the formula MAHR = 220 − age in years. For example, a 40 year old can expect to have a maximum heart rate of approximately 180 beats/min. Using this number, it is possible to estimate the heart rate that a person should maintain during workouts of various intensity. For example, during long swims aimed at improving aerobic capacity a heart rate of 70 to 80 per cent of the estimated maximum attainable heart rate should be maintained. It is important to remember that this formula is only approximate, and that there are many individuals for whom it will not provide a reasonable estimate. In people over 60 years of age, maximum heart rates can range from over 200 to as low as 105 beats/min.[23] This is an enormous variation. Thus target heart rates based on age alone will often underestimate or overestimate an individual's ideal exercise intensity, neither of which is desirable. In addition, individuals who are taking β-blockers have their maximum attainable heart rate pharmacologically suppressed. These people are often frustrated because they cannot exercise hard enough to attain heart rates in the target zone. This is not only discouraging but potentially dangerous. A better operating guide is to advise participants to exercise initially at an intensity that permits them to carry on a reasonable conversation. Once a reasonable level of fitness has been achieved, they can then begin to test the parameters of their cardiac response to exercise and so discover their limits.

'No pain, no gain' is a training philosophy that is encountered in many of the sports centres of the world. While many understand this motto to imply that it takes sacrifice and dedication to improve, some people interpret it literally. The latter approach is dangerous as it encourages people to ignore one of the body's important warning signals that something may be wrong. Encouraging individuals to train to the point of pain is inappropriate when dealing with adults who have just abandoned their sedentary lifestyle and taken up an exercise programme. Many of these people have never been in good physical condition and every exposure they have ever had to exercise has been painful. Many of them believe that everyone who exercises experiences pain and they cannot comprehend why anyone would want to do this on a regular basis. This is one reason why many people who become involved in fitness programmes give up before they are fit enough to realize how enjoyable it can be. Physical education and health-care professionals should encourage older adults to begin training at a level they enjoy and will want to come back to. If this approach is used, more older adults will become enthusiastic about fitness and perhaps the news will spread that exercise does not have to be a 'near death' experience.

Rest is a critical component of every athlete's training programme. The very nature of a training programme involves subjecting the body to stresses sufficient to cause tissue breakdown. The athlete must then ease off to allow the body to replace, repair, reinforce, and recover. The body undergoes its adaptive responses during the recovery phase. Athletes who fail to incorporate an element of rest and recovery into their programme often experience tissue breakdown at a rate faster than tissue repairs can be effected. With this imbalance, it will not be long before injury occurs and the athlete is forced to rest in order to recover. The recovery time that an athlete requires is extremely variable. Some individuals can only work intensely twice a week and require easier sessions on the other 5 days. Others can work hard for 6 days, take a day off, and then start again. Despite this wide variety of rest requirements, one thing is certain: the capacity to recover from hard work-outs is significantly reduced on ageing. Athletes who were training hard on alternate days in their thirties often find that they can only work hard every fourth or fifth day in their forties. The reason for this phenomenon is uncertain, but it could be conjectured that the body's ability to initiate the protein synthesis needed for cellular repair may slow with increasing age. Regardless of the mechanism, anyone involved in coaching ageing athletes must recognize that realistically they cannot be placed on the same training schedule as younger athletes. If they are, it is probable that insufficient recovery time will be allowed and the athletes will soon be injured or overtrained.

The final comment on training focuses on the need for a preprogramme medical evaluation. It is recommended that all older individuals who have never been active or who are returning to an active life-style be given a thorough medical assessment. Many of these individuals will have disease processes and orthopaedic problems that should be addressed before they begin a vigorous training programme. An evaluation of this type should include a number of features.

1. An assessment of risk factors for cardiovascular disease, respiratory problems, and diabetes;
2. A general medical examination with attention focused on the cardiovascular, respiratory, and musculoskeletal systems;
3. A general assessment of strength and flexibility;
4. An electrocardiogram;
5. A stress test if anything in the history or physical examination suggests potential cardiovascular compromise.

This evaluation also affords the assessor the opportunity to reinforce some of the basic training principles with the individual. It is hoped that using this screening process will identify individuals with problems before they injure themselves. It is also hoped that those who are given clean bills of health will feel confident as they embark on their fitness programme.

CONCLUSION

In this chapter we have addressed the phenomenon of ageing and how it potentially influences the athletic performance of the older competitor. After careful consideration, it appears indisputable that the ageing athlete is truly a unique entity in the world of sport. The masters athletic movement is growing rapidly, and it will continue to grow as our population continues to age. A number of individuals who have performed at world class levels and won medals in international competition have come from its ranks. We have seen that ageing is a universal, intrinsic, progressive, and deleterious process. Notwithstanding this, the specific mechanism by which ageing occurs remains uncertain. A well-defined collection of structural and physiological changes occur with ageing, and many of these alterations are associated with decline in athletic performance. In addition, the ageing athlete lives in a psychosocial environment very different from that of his or her younger opponents. In this environment, issues such as loss of motivation can prove to be more effective than all the physical alterations combined in producing a deterioration in athletic performance. The ageing competitor appears to be less able to handle the stresses associated with performing in the heat, the cold, or at altitude, but the age-related changes that are involved here are not nearly as important as the effects of diminished cardiovascular fitness, disease processes, and acclimatization. All these changes influence injury patterns, wound healing, and treatment concerns. The reader is reminded that much of the research in this area has been biased by poorly designed studies that failed to control for confounding variables such as disease processes, environmental influences, medications, life-style habits, and inactivity. Of these, the influence of inactivity is probably the most underestimated and potentially the most easily corrected.

The structural and physiological changes associated with an inactive life-style may be the greatest health threat facing our ageing population. Aristotle recognized this fact several thousand years ago when he wrote that 'life is motion'. In the ensuing years science has produced no evidence to dispute this basic truth. More recently, Shephard, a leader in the field of exercise physiology, has stated that:[61] 'Physical activity has more potential for promoting healthy ageing than anything else science or medicine has to offer today.' There can be no denying that the body was designed so that its structure and function are reinforced by physical activity and that, deprived of this stimulus, it will undergo many dilapidative changes. Exercise is a prescription for life, and it is one of those miracle medicines that does everything we ask of it with very few side-effects. It is truly a fountain of youth that is available to us all. In sharing his secret of successful ageing, Sheehan[65] offers this inspiring advice.

The best way to play the aging game is to concede nothing. Never make it easy for yourself. Should your body suggest it is too old for this effort, say 'Nonsense! Should your mind decide it is too late to learn new tricks, say 'Balderdash!' Should your soul say it needs a respite from duty and obligation, say 'Rubbish'.

If we follow this advice it may be possible to live up to the motto of the American Health Foundation which states: 'The art of living consists of dying young, but as late as possible'.

REFERENCES

1. Kasch FW, Boyer, JL, Van Camp SP, Verity LS, Wallace JP. The effect of physical activity and inactivity on aerobic power in older men (a longitudinal study). *Physician and Sportsmedicine* 1990; 18: 73–83.
2. Chesworth BM, Vandervoort AA. Age and passive ankle stiffness in healthy women. *Physical Therapy* 1989; 69: 217–24.
3. Gorman KM, Posner JD. Benefits of exercise in old age. *Clinics in Geriatric Medicine* 1988; 4: 181–92.
4. Eaglestein WH. Wound healing and ageing. *Clinics in Geriatric Medicine* 1989; 5: 183–8.
5. Strehler BL. *Time, cells and ageing*. New York: Academic Press, 1962.
6. Siegel AJ, Warhol MJ, Lang E. Muscle injury and repair in ultra long distance runners. In: Sutton JR, Brock RM, eds. *Sports medicine for the mature athlete*. Indianapolis: Benchmark Press, 1986: 35–43.
7. Stauss RH. *Sports Medicine*. Philadelphia: WB Saunders, 1984.
8. Belt E. Leonardo da Vinci's studies of the ageing process. *Geriatrics* 1952; 7: 205–10.
9. Hayflick L, Moorhead PS. The serial cultivation of human diploid cell strains. *Experimental Cell Research* 1961; 25: 585–621.
10. Makinodan T. Biology of ageing. In: Meakins JL, McClaran JC, eds. *Surgical care of the elderly*. Chicago: Year Book Medical Publishers, 1988.
11. Walford RL. *The immunologic theory of ageing*. Copenhagen: Munksgaard, 1969.
12. McArdle WD, Katch FI, Katch VL. *Exercise physiology—energy, nutrition and human performance*. 2nd edn. Philadelphia: Lea & Febiger, 1986.
13. Fixx JF. The test of time. *Runner* 1984; May: 58–62.
14. Astrand PO. Exercise physiology of the mature athlete. In: Sutton JR, Brock RM, eds. *Sports medicine for the mature athlete*. Indianapolis: Benchmark Press, 1986: 3–16.
15. Goldman R. Speculations on vascular changes with age. *Journal of the American Geriatric Society* 1970; 18: 765–9.
16. Schuman JE. Some changes of ageing. *Journal of Otolaryngology* 1986; 15: 211–13.
17. Berman R, Haxby JV, Pomerantz RS. Physiology of ageing. Part 1: Normal changes. *Patient Care* 1988; 22: 20–36.
18. Bortz WM II. Disuse and ageing. *Journal of the American Medical Association* 1982; 248: 1203–8.
19. Sheppard JG, Pacelli LC. Why your patients shouldn't take ageing sitting down. *Physician and Sportsmedicine* 1990; 18: 83–91.
20. Zauber NP, Zauber AG. Hematologic data of healthy very old people. *Journal of the American Medical Association* 1987; 257: 2181–4.
21. Kitzman DW, Edward WD. Age-related changes in the anatomy of the normal human heart. *Journal of Gerontology* 1990; 45: M33–9.
22. McClaren JC. Clinical profile of the elderly. In: Meakins JL, McClaran JC, eds. *Surgical care of the elderly*. Chicago: Year Book Medical Publishers, 1988.
23. Brooks GA, Fahey TD. *Exercise physiology, human bioenergetics and its applications*. New York: Macmillan, 1985.
24. Torp S, Baer E, Friedman B. Effects of age and mechanical deformation on the ultra structure of tendon. *Proceedings of 1974 Colston Conference on Structure of Fibrous Biopolymers*. London: Butterworths, 1975: 223–50.
25. Viidik A. Connective tissues—possible implications of the temporal changes of the ageing process. *Mechanisms of Ageing and Development* 1979; 9: 267–85.
26. Stryer L. *Biochemistry*. 2nd edn. San Francisco: WH Freeman, 1981.
27. Turek SL. *Orthopedics—principles and their application*. 4th edn. Philadelphia: JB Lippincott, 1984.
28. Alnaqeet MA, Alzrid NS, Eoldspink E. Connective tissue changes and physical properties of developing and ageing skeletal muscle. *Journal of Anatomy* 1984; 139: 677–89.
29. Balin AK, Allen RG. Molecular basis of biologic ageing. *Clinics in Geriatric Medicine* 1989; 5: 1–18.
30. Hamlin CR, Kohn RR, Luschin JH. Apparent accelerated ageing of human collagen fibers. *Diabetes* 1975; 24: 902.
31. Nejjar I, Pieraggi MT, Thiers JC, Bouissou H. Age-related changes in the elastic tissue of the human thoracic aorta. *Atherosclerosis* 1990; 80:-199–208.
32. Fenske NA, Lober CW. Skin changes of ageing: pathological implications. *Geriatrics* 1990; 45: 27–32.
33. Wolff J. *Das Gesetz der Transformation der Knochen*. Berlin: A Hirschwald, 1892.
34. Bassett CA, Becker RO. Generation of electric potentials by bone in response to mechanical stress. *Science* 1962; 137: 1063–4.
35. Stillman RJ, Lohman TG, Slaughter MH, Massey BH. Physical activity and bone mineral content in women aged 30 to 85 years. *Medicine and Science in Sports and Exercise* 1986; 18: 576–80.
36. Beyer Re, Huang JC, Wilshire EB. The effect of endurance exercise on bone dimensions, collagen and calcium in the ageing male rat. *Experimental Gerontology* 1985; 20: 315–23.
37. Smith EL. Exercise for the prevention of osteoporosis: a review. *Physician and Sportsmedicine* 1982; 10: 72–9.
38. Twomey L, Taylor J. Age changes in lumbar intervertrebal discs. *Acta Orthopaedica Scandinavica* 1985; 56: 496–9.
39. Sager K. Senior fitness—for the health of it. *Physician and Sportsmedicine* 1983; 11: 31–6.
40. Knudson RJ, Clark DF, Kennedy TC, Knudson DE. Effect of ageing alone on mechanical properties of the normal adult human lung. *Journal of Applied Physiology*, 1977; 43: 1054–62.
41. Jones PL, Millman A. Wound healing and the aged patient. *Nursing Clinics of North America* 1990; 25: 263–77.
42. deVries HA. *Physiology of exercise for physical education and athletics*. 4th edn. Dubuque, Iowa: WC Brown, 1986.
43. Wiswell RA, Jaque SV, Hamilton-Wessler M. Exercise and muscle strength. In: Morley JE, Glick Z, Rubenstein LZ, eds. *Geriatric nutrition*. New York: Raven Press, 1990: 447–56.
44. Larsson L, Sjodin B, Karlsson J. Histochemical and biochemical changes in human skeletal muscle with age in sedentary males, age 22–65 years. *Acta Physiologica Scandinavica* 1978; 103: 31–9.
45. Molander B, Backman L. Age differences in heart rate patterns during concentration in a precision sport: implications for attentional functioning. *Journal of Gerontology* 1989; 44: P80–7.

46. Plyley MJ. Fine-tuning muscle capillary supply for maximum exercise performance. *Cardiology* 1990; **6**: 25–34.

47. Damon EL. *An experimental investigation of the relationship of age to various parameters of muscle strength.* Doctoral dissertation, University of Southern California, 1971.

48. Green HJ. Characteristics of ageing human skeletal muscles. In: Sutton JR, Brock RM, eds. *Sports medicine for the mature athlete.* Indianapolis: Benchmark Press, 1986: 17–26.

49. Sato T, Akatsuka H, Kito K, Tokoro Y, Tauchi H, Kato K. Age changes of myofibrils of human minor pectoral muscle. *Mechanisms of Ageing and Development* 1986; **34**: 297–304.

50. Knortz KA. Muscle physiology applied to geriatric rehabilitation. *Topics in Geriatric Rehabilitation* 1987; **2**: 1–12.

51. May C, Kaye JA, Atack JR, Schapiro MB, Friedland RP, Rapoport SI. Cerebrospinal fluid production is reduced in healthy ageing. *Neurology* 1990; **40**: 500–2.

52. Berman R, Haxby JV, Pomerantz RS. Physiology of ageing. Part 2: Clinical implications. *Patient Care* 1988; **22**: 39–66.

53. Spirduso WW, Clifford P. Replication of age and physical activity effects on reaction and movement time. *Journal of Gerontology* 1978; **33**: 26.

54. Hofland B, Willis S, Baltes P. Fluid intelligence performance in the elderly: intra-individual variability and conditions of assessment. *Journal of Educational Psychology* 1981; **73**: 573–86.

55. Dinner DS, Erman MK, Roth T. Help for geriatric sleep problems. *Patient Care* 1990; **2**: 24–31.

56. Booth FW, Gould EW. Effects of training and desire on connective tissue. *Exercise and Sports Science Review* 1975; **3**: 83–112.

57. Vailas AC, Perrini VA, Pedrini-Mille A, Holloszy JO. Patellar tendon matrix changes associated with ageing and voluntary exercise. *Journal of Applied Psychology* 1985; **58**: 1572–6.

58. Hall MD. Cartilage changes after experimental relief of contact in the knee joint of the mature rat. *Clinical Orthopedics* 1969; **64**: 64–76.

59. Ungerleider S, Porter K, Golding J, Foster J. Mental advantages for masters. *Running Times*, 1989; July: 18–20.

60. Edelstein JE. Foot care for the ageing. *Physical Therapy* 1988; **68**: 1882–6.

61. Kavanagh T, Shephard RJ. Can regular sports participation slow the ageing process? Data on masters athletes. *Physician and Sportsmedicine* 1990; **18**: 94–104.

62. Tymn M. Carl Hatfield: never beaten by a McCoy. *National Masters News* 1988; May: 18.

63. Russell B, Branch T. *Second wind.* New York: Ballantine, 1980.

64. Kubler-Ross E. *On death and dying.* New York: Macmillan, 1969.

65. Sheehan G. *Personal best.* Emmaus, PA: Rodale Press, 1989.

66. *The Works of Shakespeare.* Shakespeare Head Press Edition, Oxford University Press, 1938: 1249.

67. Balcomb AC, Sutton JR. Advanced age and altitude illness. In: Sutton JR, Brook RM, eds. *Sports medicine for the mature athlete.* Indianapolis: Benchmark Press, 1986: 213–24.

68. Kenny MJ, Gisolfi CV. Thermal regulation: effects of exercise and age. In: Sutton JR, Brock RM, eds. *Sports medicine for the mature athlete.* Indianapolis: Benchmark Press, 1986: 133–41.

69. Astrand PO, Rodahl K. *Textbook of work physiology.* Singapore: McGraw-Hill, 1986.

70. Kenny WL, Anderson RK. Responses of older and younger women to exercise in dry and humid heat without fluid replacement. *Medicine and Science in Sports and Exercise* 1988; **20**: 155–60.

71. Brown M. Special considerations during rehabilitation of the aged athlete. *Clinics in Sports Medicine* 1989; **8**: 893–901.

72. DeHaven KE, Littner DM. Athletic injuries: comparison by age, sport and gender. *American Journal of Sports Medicine* 1986; **14**: 218–24.

73. McRae R. *Practical fracture treatment.* New York: Churchill Livingstone, 1981.

74. Huang SHK. Preventing NSAID-induced gastropathy before it becomes a pain in the gut. *Canadian Journal of Geriatrics* 1989; **5**: 29–34.

75. Tymn M. Frank Shorter is looking ahead. *National Masters News* 1987; May: 6.

76. Thomas SG, Cunningham DA, Rechnitzer PA, Donner AP, Howard JH. Determinants of the training response in elderly men. *Medicine and Science in Sports and Exercise* 1985; **17**: 667–72.

77. Ungerleider S. Perceptual and motor skills. *Mood Profiles of Masters Track and Field Athletes* 1989; **68**: 607–17.

6.3   The growing athlete

J. C. HYNDMAN

INTRODUCTION

The participation of children in sport is accepted as an important component in the development of healthy well-rounded adults. Traditionally, participation in sport is believed to be important not only in the establishment of a healthy physical life-style but also in the overall development of healthy attitudes with respect to work, competition, preparation, and fair play, all of which are ultimately beneficial to the individual. However, many paradoxes are evident when the problems of the paediatric athlete are addressed, and health care professionals must be well aware of these when attending to the needs of these children.

Even a cursory perusal of both the print and electronic media demonstrates the increasing role that sport plays in our society. The influence of sport permeates virtually all aspects of our culture including fashion, life-styles, and institutional and even national goals. Increasing professionalism and globalization of sport all contribute to the tremendous impact that modern high level participation in sport has upon our young people. Of course,

similar influences are exerted upon coaches, administrators, and parents, and consequently planning in the management of children's sports is all too frequently carried out with the high level elite athlete as the model. As a result of this top-down planning many inappropriate decisions are made with respect to training, administration, and the management of children's injuries. However, young children form the largest group participating in sports, and a bottom-up planning process would probably be more appropriate if a positive result is to be achieved.

Considerable confusion exists with respect to the correct approach that the health care professional should adopt in dealing with children. Fundamental to establishing the most appropriate course is a clear definition of what constitutes paediatric sport. The paediatric athlete is a youngster who is still growing. The definition of a child in terms of the concept of 'still growing' is so simple that it is often forgotten or confused with other ill-defined terms such as adolescence. The phenomenon of growth is not age or size related, and the cessation of growth varies very considerably with sex and genetic background. According to this definition a child differs from an adult by the presence of growth plates which are necessary for longitudinal growth (Fig. 1). Apart from secondary sexual characteristics, the existence of growth plates in

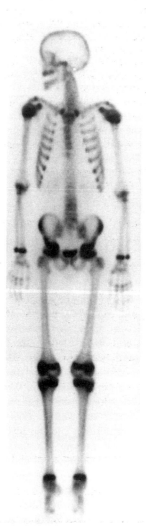

Fig. 1 High resolution technetium bone scans clearly show the different metabolic activity reflective both of the growth plates present in children and of the importance of growth in determining the absolute difference of the child's skeleton.

bones is the only absolute difference between the anatomical structures of children and adults. This absolute difference in structure is clearly implicated in the different injury patterns observed in children compared with adults (Fig. 2).

Principles developed in the treatment of the elite adult athlete are not necessarily transferrable to the problems of the growing athlete. No scientific data reported to be obtained from children should be acceptable unless they are clearly defined as dealing with that population which is still growing and in whom growth plates are present. Failure to make this simple observation leads to further misunderstandings in dealing with the problems of the growing athlete.

The presence of growth plates in bone results in definable biomechanical differences from the mature bone of adults. This

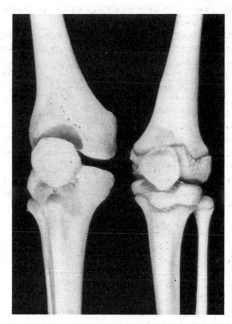

Fig. 2 The knee of the skeleton of an adult and of a child showing clearly the absolute difference between children and adults.

essential difference frequently leads to a different expression of injury when the body is subjected to a similar trauma. Generally, this results in more frequent bone injuries compared with the soft tissue injuries that are so common in adults (Fig. 3). Indeed, even within different age groups of growing children the expression of bone injury may be predicated on the variability of growth velocity. For example, fractures of the distal radial epiphysis occur less frequently in older children than in younger and slower growing children.[1]

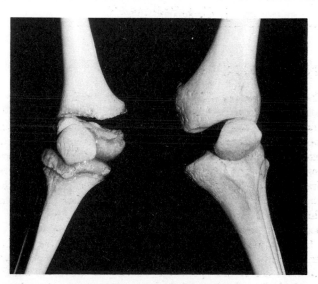

Fig. 3 Similar valgus (type 2 injury) injuries to the knee of an adult (right) and a child (left) may result in an absolute difference in injury produced, with the adult developing massive ligament disruption and the child a fracture separation of the growth plate.

621

INJURY IN CHILDHOOD

Injury can be defined as the response of tissue to kinetic energy applied to the body. The degree of injury which occurs directly parallels this kinetic energy and therefore follows basic physical laws, largely relating to the mass and particularly the velocity involved. Therefore, in the assessment of an injury it is important to be aware of the circumstances surrounding it in order to be able to anticipate and assess the effects of this force. The reaction of tissue to injury is manifested by inflammation. Inflammation is characterized by swelling, redness, increase in temperature, and loss of function of the injured part. The more severe the injury, the more severe is the inflammation. By definition, inflammation in the extremities can easily be observed by those in attendance. Significant tissue damage will always be associated with a proportional degree of observable inflammation (Fig. 4).

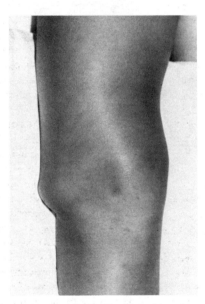

Fig. 4 Fracture dislocation of the elbow showing the easily demonstrated signs of inflammation.

The ways in which injuries can occur can be classified into three groups.

Type 1 injuries

Type 1 injuries occur as a result of a direct blow. This type of injury is quite common in sport, and can result from being struck by a ball, a stick, or a racket, or by striking the knee directly on the playing surface. This type of injury is the easiest to prevent in that vulnerable areas which are sports specific can be identified and protected. Examples of protective equipment are helmets in American football and shin pads in soccer or ice hockey.

Type 2 injuries

Type 2 injuries occur as a result of an indirect blow, for example when a fall on the outstretched hand leads to an injury of the elbow or shoulder. In this case the force is transmitted along the skeleton to a point distant from the original contact. Similarly, twisting of the body around the planted foot frequently leads to injuries of the knee. These injuries can be prevented by altering footwear or playing surfaces. Administrative changes may also help to reduce these injuries by, for example, establishing rules preventing blocking from behind.

Type 3 injuries

Type 3 or chronic repetitive injuries are the common overuse injuries which are so frequently presented to health care professionals.[2] The confusion between adults and children is most prevalent in this type of injury. Many therapeutic modalities which are affective in these types of injuries in adults are inappropriate when used in the growing child. Again, this largely relates to the more frequent bony expression of injury resulting from chronic repetitive injury compared with soft tissue expressions in adults. For example, patellar tendinitis, which is common in adult jumping athletes, would be more likely to express itself as Osgood–Schlatter's disease, a bony injury, in the growing athlete.

EPIDEMIOLOGY

Introduction

Unprecedented interest in high performance sporting activity has developed in the second half of the twentieth century. This is the result of increased leisure time, the pervasive influences of television in our society, and, unfortunately, the political influence of competing cultures and economic units. Many national sports organizations have encouraged high level competition with professional coaching at a very young age, creating a subclass of unidimensional athletes who are streamed into highly specialized sporting activity while still very young. This has resulted in successful international competition and the achievement of national sporting goals focused on gaining international success. These programmes rely on the development of child athletes as the raw material from which to develop these young elite performers. Although these programmes have been superficially successful, there have been significant failures. Intensive training and increased frequency of exposure has led to career-ending injuries.[3,4] Tragically, and perhaps less well known, many athletes have failed to develop into well-rounded and mature adults or have had difficulties in adjustment following the completion of a highly profiled career. Cheating as a result of the credo that 'winning is everything' are further contradictions of the aims of sport in childhood.

Research into injuries in growing athletes is poorly developed. However, some general observations can be made. Clearly, the rate of injuries increases exponentially with age; the frequency of injury is low in very young children and the injuries which are incurred tend to be of a minor degree of severity.[5–8] Therefore it appears that as increasing maturity, size, and competitiveness result in increasing collision forces, the injury levels increase appropriately.[9–12] Evidence exists suggesting that the greater the frequency of participation, the greater is the rate of injury.[13]

The difficulty of assessing the epidemiology of athletic injuries is further compounded by the variability of the maturation rate. In the interests of fairness, sports administration bodies insist upon careful categorization of participation based upon chronological age. This is frequently enforced rigidly with the utilization of birth certificates. However, this practice is confounded by the observation that in contact sports such as ice hockey, a chronological age range of 13 to 15 years may in fact represent a biological variation of 13 to 19 years.[14] Further, high school football players have been shown to be physically more mature than age-matched non-participants. These biological variations certainly influence participation and potential injury rates in growing children. Sporting activities requiring aerobic efficiency are also affected by biological variation. Aerobic efficiency is not size related, and, regardless of age, immature athletes cannot generate the same efficiency as those who are physically mature.[16,17]

Therefore the epidemiology of injuries must take into account not only the paradox of physical differences between growing athletes and mature athletes but also the biological variation in the rate of this maturation process. Recognizing these paradoxes then emphasizes that the whole area of understanding, treating, and preventing injuries in the child athlete is a very complex affair. While maturation screening can be used to match athletic participation to biological maturity, this methodology has not been adequately tested scientifically nor is it likely to be sociologically acceptable.[18]

The development of preventive measures and equipment have been confounded by these variables, and, with notable exceptions, many of the measures recommended have not been properly evaluated despite their well-intentioned institution. Essentially, the factors influencing the occurrence of injury in athletes can be divided into two groups. First there are intrinsic factors, i.e. those factors which are related to the athletes themselves, and second there are extrinsic factors, i.e. those factors relating to the nature of the activity itself and to the competitive environment.

Intrinsic factors

Much research effort has been expended in attempting to define whether or not certain types of body habitus or joint stability are factors in the genesis of injury. Despite these efforts, no conclusive evidence has been obtained to show that there are intrinsic risk factors in the physical make-up of some athletes which makes them susceptible to injury. There is also no conclusive evidence that time-honoured preparation strategies, including pre-performance stretching and warming-up, have a salutory effect on the prevalence of injury.[5,19–22]

Interestingly, one area where there is increasing evidence of a direct relationship with the prevalence of injury is in the area of stressful life events. These psychological factors are clearly implicated in the prevalence of injury, a fact that must be understood by coaches, trainers, and physicians.[23]

Extrinsic factors

The assessment of extrinsic factors in the genesis of sports injuries has been documented. Injuries of type 1 (due to a direct blow) are the easiest to prevent.

Recognition of the sometimes tragic consequences of eye injuries in sports such as squash and ice hockey has led to changes in protective equipment, resulting in a substantial decrease in the frequency of injuries.[24–28] The definition of the problem, the institution of remedial action and the resultant effect is well described in three papers by Pashby.[29]

Extrinsic factors involving injuries of type 2 (indirect force) are, by their nature, more difficult to prevent. Protective equipment may not have the desired effect as the forces are transmitted longitudinally through the skeleton, with their adverse effect taking place somewhere distant from the origin of the force. An example

of this difficulty is the use of prophylactic knee braces in American football. What seemed to be a reasonable idea has in fact been shown to be ineffective; indeed, it may even have adverse effects.[30] However, significant injuries to the knee in American football may well be minimized by rule changes such as making blocking from behind illegal. The interface between the shoe and the playing field is also important in the genesis of injuries due to indirect forces.

Paradoxically, changes in equipment designed to prevent type 1 injuries may enhance the potential for type 2 injuries because the protection provided may lead to a loss of fear of injury. Consequent high velocity collisions can lead to type 2 injuries such as the neck injuries associated with spearing in American football.[31,32]

Type 3 injuries (chronic repetitive injuries) in growing athletes have clearly been shown to have the potential to lead to serious injury and disability such as 'little leaguer's elbow'. The recognition of this has led to administrative control of the frequency of pitching, thereby minimizing the adverse effects. Other examples where chronic repetitive activity can lead to significant complications include the use of inappropriate footwear by elite immature distance runners, epiphyseolysis of the distal radius and potential growth arrest in gymnasts, and spondylolysis and spondylolisthesis which appear to be associated with repetitive hyperextension of the lower spine. Careful examination should be performed to ensure identification and early recognition of these potentially serious injuries in order to try and minimize adverse results of sporting activity.[33–35]

A large study of sports injuries in Ireland showed that only 35 per cent of injuries in children can be related to facilities and equipment.[36] When this result is coupled to the inability to define an injury-prone habitus, the variability of injury at/various developmental stages, the wide biological variation in achievement of these developmental stages, and the increasing evidence of the importance of psychological factors in the genesis of injury, all health care professionals, coaches, trainers, and athletes must recognize the innate complexity of the epidemiology of injury in the growing athlete. It should be clearly emphasized that a holistic approach to the assessment, treatment, and prevention of injury in growing athletes is essential if the desired aims of the athlete, the team, and society are to be balanced with the healthy development of a well-rounded mature adult.

INJURIES TO THE GROWING ATHLETE

Clinical application

Having reviewed the anatomy of the growing athlete and the basic mechanisms and epidemiology of injuries, we must now discuss the practical management of common problems of the child athlete. The management of macroinjury of type 1 (direct force) or type 2 (indirect force) is beyond the scope of this book and the reader is referred to standard paediatric fracture texts. Macroinjuries which have major implications for sport will be described in detail. Particular attention will be paid to type 3 injuries (chronic repetitive stress) as these form by far the largest group presenting for management.

Clavicle

The clavicle is frequently injured in sports activities. In the growing athlete fractures to the clavicle can be treated simply with a triangular bandage sling and/or the addition of a figure-of-eight bandage. There is virtually no risk of non-union, although a palpable mass of clavicle is frequently associated with the early stages of healing (Fig. 5). The most difficult part of the management of fractures of the clavicle in athletes is when to allow them to return to sport. The risk of refracture is real, and before return is allowed the athlete should demonstrate the ability to stress the clavicle repeatedly by, for example, performing 15 to 20 unrestricted

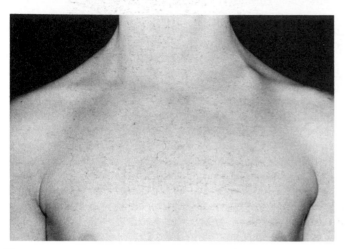

Fig. 5 Typical prominence associated with healing fractured clavicle in a 13-year-old hockey player.

push-ups. Fractures of the distal clavicle do not have the same ominous implications as they do in adults, and can usually be simply treated using the same methods as those applied to the shaft of the clavicle. Rare cases of fracture dislocation involving the sternoclavicular joints may occur, but they are difficult to define other than with specialized techniques such as CT scans. Symptomatic treatment is all that is necessary (Fig. 6).

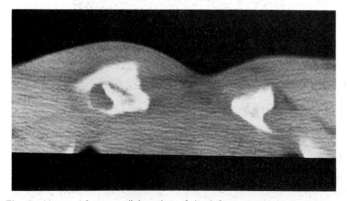

Fig. 6 Unusual fracture dislocation of the left sternoclavicular joint. Note soft tissue swelling anteriorly. The injury healed promptly with non-operative treatment.

Shoulder

Dislocations in the shoulder region, including acromioclavicular joint and glenohumeral joint, are unusual in patients with open growth plates. The different shoulder injuries sustained by these athletes are a good example of the pattern of injuries in children

being distinct from injuries in adults caused by the same mechanism. Fractures of the proximal humerus are easy to treat and have very few long-term sequelae (Fig. 7). The remodelling potential of the proximal humerus is well known, and therefore most frac

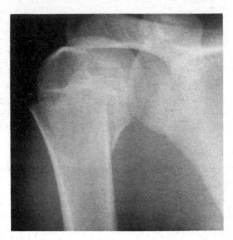

Fig. 7 Typical fracture of the proximal humerus with minimal displacement.

tures can be treated simply with a Velpeau bandage or some similar symptomatic treatment with the expectation of fairly early return to vigorous sporting activities (Fig. 8.).

Pathological fractures of the proximal humerus secondary to a unicameral bone cyst frequently present significant concerns to attending personnel as to the suitability of the patient to participate in sports following healing of the fracture. This is a matter of judgement. There is no good evidence that the pathological fractures which can occur consequent to these injuries are associated with any long-term complications. These fractures usually heal in 2 to 3 weeks and, depending on careful informed consent, it may be appropriate to allow these young athletes to return to their sport. The complications of refracture must be balanced against the legitimate aims and wishes of the athlete. Ancillary treatment such as the injection of hydrocortisone may speed the healing of these benign cysts (Fig. 9).

The elbow

It is well known that sports such as baseball which involve repetitive stressful overhand throwing can produce a non-specific painful and stiff elbow which may have important implications in the long term. This is a good example of a type 3 injury which can be prevented by limiting the activity of the athlete by legislative means in order to prevent or minimize the likelihood of this potentially career-ending injury (Fig. 10).

Wrist

Repetitive injuries to the wrist are analogous to repetitive injuries to the elbow (Fig. 11). These injuries have been well documented in young gymnasts undergoing extreme training who can develop changes in the distal radial growth plate that are related to the severity and intensity of the training programmes.[35] As with the elbow injuries, they need to be identified and can be prevented by careful attention to training regimens. Periods of rapid growth

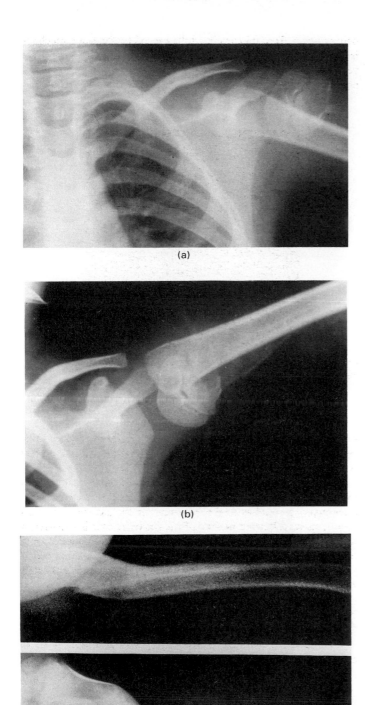

(a)

(b)

(c)

Fig. 8 A series of radiographs indicating a grossly displaced fracture of the proximal humerus treated without reduction, showing prompt healing and ultimate remodelling potential. While not the ideal management this is a good example of remodelling seen in growing children.

may be associated with an increased vulnerability to this type of injury. This is analogous to other chronic stress-related growth-plate disturbances such as Osgood–Schlatter's disease.

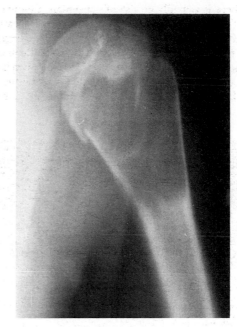

Fig. 9 Simple bone cyst of the proximal humerus showing pathological fracture.

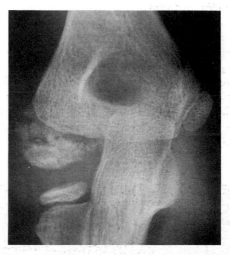

Fig. 10 Irregular ossification of elbow aggravated by overhand throwing.

Hand

Fractures involving the small metacarpal frequently result from both sanctioned and unsanctioned pugilistic endeavours in young athletes (Fig. 12). Many fracture texts recommend aggressive therapy for these punch fractures. However, in children a rapid return to sports with no long-term complications is expected unless the adjacent finger is rotated in which case the rotational deformity must be corrected (Fig. 13).

Spine

The two areas of concern with regard to spinal problems in the growing athlete, in addition to macroinjuries, are Scheurmann's disease and spondylolisthesis.

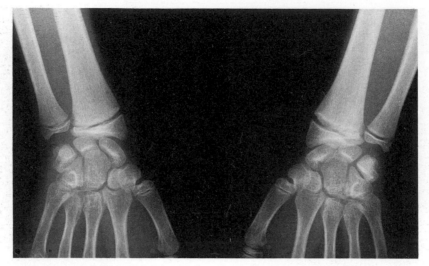

Fig. 11 Note widening of growth plate in this painful wrist of a highly competitive gymnast.

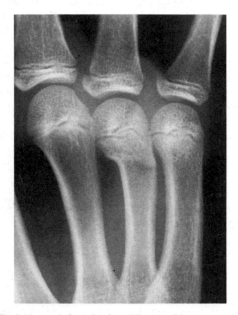

Fig. 12 Typical punch fracture in a 12 year old.

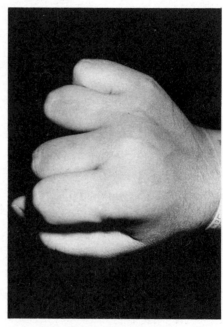

Fig. 13 Rotational malalignment secondary to malunion of a metacarpal fracture.

Scheurmann's disease

Much confusion exists with respect to this so-called disease. Essentially, Scheurmann's disease is yet another expression of the rapid growth of the athlete, specifically the vertebral bodies. In athletes Scheurmann's disease usually effects the lumbar spine and presents with chronic activity-related back pain and irregular ossification of the vertebral bodies (Fig. 14). It is important to recognize that this is a self-limiting condition with no evidence of any long-term complications or predisposition to chronic back syndromes. Treatment is simple and symptomatic. It is not necessary to preclude participation in sport, but it may be advisable to limit it to some extent. In some cases an external support such as a thoracolumbar brace may be used to allow continued participation. With maturation, symptoms and radiographic changes disappear.

Spondylolisthesis

Spondylolisthesis is the other major cause of mechanical back pain in children. Evidence has accumulated that spondylolisthesis is associated with chronic repetitive hyperextension activities such as those performed by gymnasts. As spondylolisthesis may be progressive, it can be associated with very significant long-term spinal problems and its presence may be sufficient to preclude ongoing participation in competitive gymnastics. However, spondylolysis without spondylolisthesis does not necessary require discontinuing gymnastics but rather symptomatic treatment and close follow-up (Fig. 15).

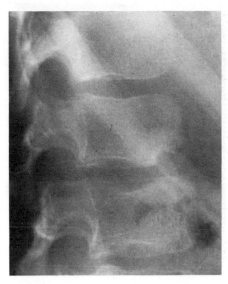

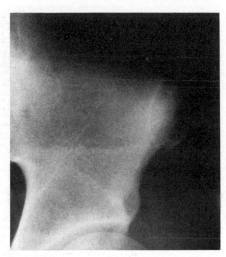

Fig. 16 Avulsion fracture of anterior superior iliac spine. The patient is a 15-year-old soccer player.

Fig. 14 Typical changes of Schuerman's disease in a 12 year old.

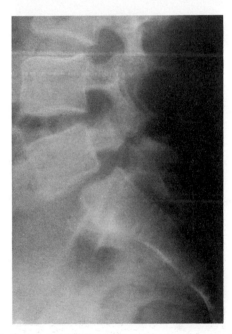

Fig. 15 Grade I spondylolisthesis in an 11-year-old competitive gymnast.

Pelvis and hip

Avulsion fractures of the pelvis

Violent muscle contractions occur in sports such as sprinting and occasionally result in the avulsion of the attachment of major muscle groups from the pelvis. This is associated with acute pain and is a good example of a type 2 injury (indirect force). Symptomatic treatment with rest and a gradual return to activity is all that is required (Fig. 16).

Slipped epiphysis

Slipped epiphysis, or more correctly slipped proximal femoral physis, may present either acutely or chronically in association with athletic participation. Frequently the symptoms may be sub-

tle and associated with a limp and decreased performance. As the process increases the pain experienced by the athlete may be referred to the knee. These athletes tend to be in the rapidly growing adolescent period and are frequently overweight. Slipped epiphysis is a potentially very serious problem which requires acute treatment and securing of the epiphysis with some form of fixation to prevent further slippage. Failure to recognize and treat this relatively common condition can result in serious arthritis and consequent permanent disability. Athletes in whom this condition is suspected should be issued with crutches immediately and referred for radiological confirmation (Fig. 17). Operative treatment with pin fixation is the standard treatment for this condition.

Knee

Knee pain in athletes is perhaps the most common presentation to coaches, trainers, and physicians. The human knee is subject to tremendous mechanical stress. Flexing the knee increases the forces across it exponentially and thereby makes it very vulnerable to chronic repetitive stress. When the mechanical weakness inherent in the growing knee is added to this, the frequency with which problems occur in the knee joint can be readily understood. The symptoms may relate to a variety of well-defined clinical problems such as Osgood–Schlatter's disease, osteochondritis dissecans, jumper's knee, and patella femoral instability. All these conditions can easily be diagnosed either clinically or with simple radiography.

Osgood–Schlatter's disease

Osgood–Schlatter's disease is particularly common in adolescent boys. Much has been written about this condition, and it can best be envisaged as a stress fracture. It results from chronic repetitive stress due to tension developed in the extensor mechanism as it attaches to the growing tibial tubercle.[37] Developmental age is important in that during rapid growth the bone structure of the tibial tubercle is mechanically weak and can fail. Treatment ranges from short-term immobilization to restricted activity. Some athletes may suffer pain which is severe enough to preclude competition. In such patients it has been found that a faster return to competition is obtained with the use of plaster casts than with treatment by other modalities (Fig. 18).[38]

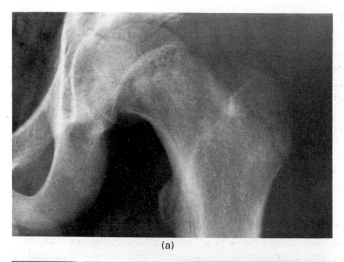

(a)

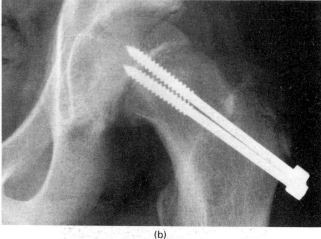

(b)

Fig. 17 (a) Slipped proximal femoral physis. (b) Following fixation with Knowles' pins.

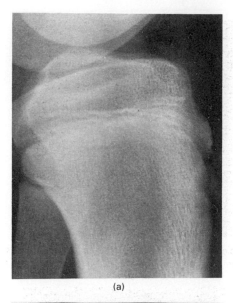

(a)

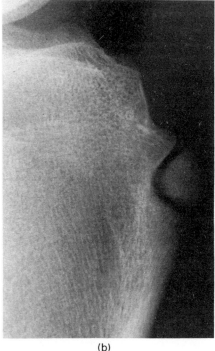

(b)

Fig. 18 (a) Early changes of Osgood—Schlatter's disease in a 13 year old. Note fragmentation of tibial tubercle. (b) Residual Osgood—Schlatter's disease with loose bone formation in an adult.

Osteochondritis dissecans

Osteochondritis dissecans presents with pain, a tendency for the knee to give way, and sometimes a limp. The diagnosis is quickly established with plain radiography of the knee. In children it is possible to correct the condition by immobilization. Treatment by fixing these fragments with various forms of pins has not been adequately tested with clinical trials. In some cases the fragment will become loose and require excision (Fig. 19).

Jumper's knee (Sinding-Larsen–Johansson disease)

Jumper's knee is another example of a chronic repetitive stress (type 3) injury at the patellar attachment to the patellar tendon. The injury is treated in a similar fashion to Osgood–Schlatter's disease. In this age group it is self-limiting.

Patella femoral instability

Patella femoral instability is also relatively common in the growing athlete. It frequently presents with an acute dislocation which requires reduction (Fig. 20). Following reduction, a radiograph should be taken to define whether or not an osteochondral fracture has occurred. Joint aspiration may be necessary for comfort. After a short period of immobilization, rehabilitative exercises are very important to prevent recurrences of this condition. Isometric exercises in extension are very helpful in preventing chronic

symptoms in those patients who present with instability but without dislocation. Persistent instability despite adequate rehabilitation may require soft tissue surgical reconstruction as the growth plates preclude bony procedures.

Tibia femoral instability

Major ligamentous disruptions of the knee can occur in children.[39] However, an injury which would lead to an unstable knee in an adult generally results in a fracture in a child (Fig. 21). Avulsion fractures of the tibial spine are by definition isolated anterior cruciate ligament injuries. Because the ligament itself is detached from its insertion and can be replaced by either operative or non-operative means the long-term implications of this type of injury are less significant than they are in an adult.

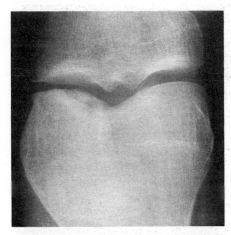

Fig. 19 Typical osteochondritis dissecans on the medial femoral condyle. Note early evidence of arthritis in this untreated patient.

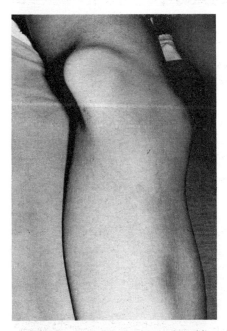

Fig. 20 Acutely dislocated patella. Note lateral position of the patella.

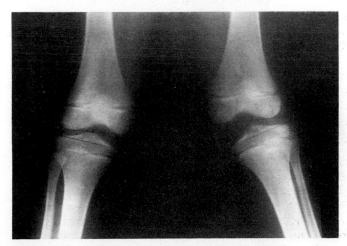

Fig. 21 Stress radiographs showing that, while unusual, ligamentous injuries can occur in growing athletes.

Peripatellar pain syndrome

The most problematic of all the injuries that occur in this age group are those associated with chronic knee pain without evidence of obvious clinical abnormality. Considerable experience is required to define the true pathogenesis of the pain in syndromes of this type. The development of arthroscopic examination has assisted tremendously in defining internal derangements of the knee such as torn menisci, discoid menisci, and other soft tissue abnormalities. It is in these enigmatic syndromes where the athlete is disabled and yet there are no objective findings in any of the usual parameters that clinicians must be aware of the subleties of pain interpretation and the many factors which may modify it. This is discussed further in the section on somatization disorders.

Stress fractures

Stress fractures occur in growing athletes in the same way as in adults. Treatment is symptomatic, with footwear and activity modifications. Persistent symptoms can easily be treated by a short period (2 weeks) of immobilization in plaster and resolution is rapid (Fig. 22). Stress fractures in children may be associated

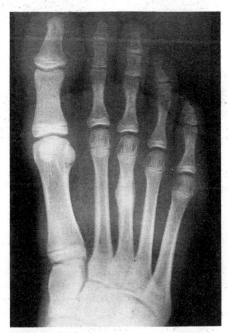

Fig. 22 Stress fracture of third metatarsal in a cross country runner.

with other adjacent pathologies such as benign tumours, tarsal coalition, and presence of internal fixations. The judicious use of plaster casts in growing athletes should not be feared (Figs 23–25).

Myositis ossificans

Myositis ossificans is a relatively common problem and almost always results from a type 1 direct blow to the thigh. The history is very characteristic: consequent to this type of injury there is persistent pain and dysfunction in the extensor apparatus of the thigh; later, radiography shows characteristic formation of new bone outside the periosteum. Thus confusion in diagnosis is unlikely. No long-term effects are expected (Fig. 26), but care should be taken not to attempt aggressive rehabilitation of the muscle as this

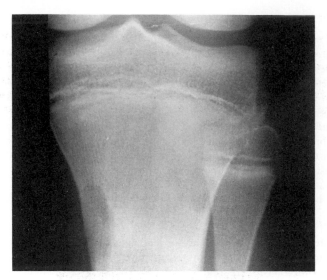

Fig. 23 Stress fracture of the proximal tibia through a non-ossifying fibroma.

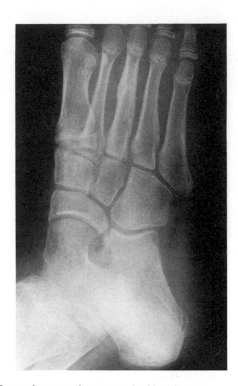

Fig. 25 Stress fracture of metatarsal with adjacent calcaneonavicular coaction. Stiffness in the foot increases stress on adjacent bones.

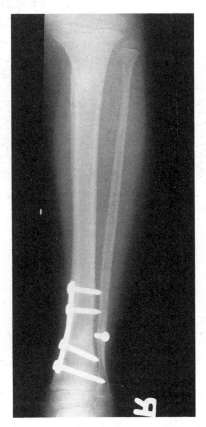

Fig. 24 Unusual open reduction of a fracture of tibia showing stress shielding of plate and stress fracture of the proximal tibia. The presence of the plate affects the normal bone mechanics.

will delay return to sport as a result of recurrent microinjury and pain.

Osteochondroses

Osteochondroses is a regrettable term which is used to identify the similarly appearing radiographs obtained for a wide group of conditions such as Osgood–Schlatter's disease, Legg–Perthes' disease, osteochondritis dissecans, Scheurmann's disease, Frieberg's disease, and Panner's disease. Actual pathology differs, despite those similar radiographic images.

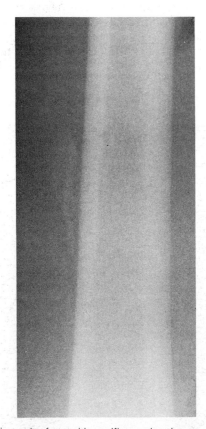

Fig. 26 Radiograph of myositis ossificans showing new bone formation within the muscle well differentiated from the bone by the periosteal space.

Osteochondritis dissecans

Osteochondritis dissecans can occur in the ankle joint as well as the knee. In many instances it can be treated symptomatically with either a plaster cast or, preferably, an ankle orthosis. If the lesion is displaced, surgical removal may be necessary (Fig. 27).

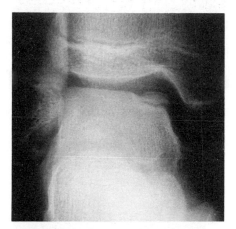

Fig. 27 Separated osteochondritis dissecans fragment. Pain and ankle instability are frequent.

Frieberg's disease

Frieberg's disease is an unusual condition that usually affects the second or third metatarsal head of 13-year-old females. There is collapse of the subchondral bone and swelling and discomfort in this region (Fig. 28). Treatment is symptomatic, and a properly designed orthosis with a metatarsal pad will usually alleviate the symptoms. Surgical treatment is necessary in very rare instances.

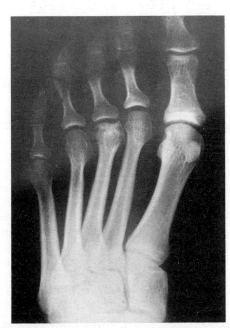

Fig. 28 Typical example of Frieberg's disease of the third metatarsal head.

Severs disease

Severs disease is similar to Osgood–Schlatter's disease but occurs at the insertion of the Achilles tendon and is associated with rapid growth of the calcaneus. It is frequently confused with tendinitis. Normal irregular ossification is often interpreted as pathology (Fig. 29). Treatment is symptomatic and is designed to decrease impact loading of the heel. The condition is self-limiting and there are no long-term sequelae.

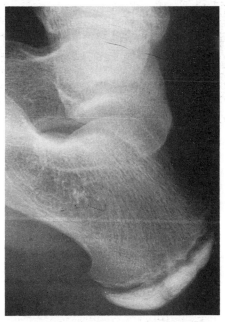

Fig. 29 Normal irregular ossification of calcaneal apophysis. Rapid growth is associated with this appearance.

Drugs and rehabilitation

Prescription drug therapy should be used rarely in the growing athlete. An understanding of the pathogenesis of symptoms and their frequent bony origin should modify the enthusiasm for non-steroidal anti-inflammatory drugs and other drugs designed for the treatment of soft tissue inflammation. There is little evidence that these drugs have any salutary effect on conditions encountered with growing athletes, and not only is their cost–benefit ratio questionable but, more importantly, their use frequently prevents a true understanding of the pathogenesis of the problem from being obtained.

Perhaps the most difficult problem with the rehabilitation of growing athletes after injury is tempering their enthusiasm for an early return to their particular sport. The physician and coaching personnel must understand that objective evidence of sufficient healing must be demonstrated prior to return to training if further injury is to be avoided. Overenthusiastic early return is frequently associated with a sequence of nagging injuries and consequent diminished performance for a period longer than would have been the case with a more cautious return.

The objective evidence of healing is the reversal of those signs of tissue damage which indicated the presence of the injury in the first instance. Full return of range of movement following immobilization, absence of swelling, and atrophy of adjacent muscle groups are prerequisites for return to pre-injury training regimes. Traditional physical therapy methods may be helpful in enhancing range of motion and strength rehabilitation. Many other modalities frequently used have not been scientifically tested and are probably of little value in this age group.

Weight training is a common method for enhancing strength and athletic performance. It is often felt that this training will assist in the prevention of injury, although there is no evidence to support these claims. Injuries in the growing athlete usually occur in the bone component of the musculoskeletal system. Therefore weight training as part of an overall regime is reasonable to the limit defined by the absence of bone pain syndromes. Regimens that stress strength with less resistance are preferred to very heavy power lifting.

Clearly injury rates increase with increasing performance.[11] Higher, faster, stronger does have limits, and the challenge is to recognize that limit for the individual athlete.

Somatization disorders

Psychological factors are clearly important in achieving high level athletic performance and successful coaches are often known for their skill in psychological motivation. The recognition that psychological factors are important in the genesis of injury underlines the importance of recognizing when such factors can become negative. Extreme extrinsic psychological stress placed on the young athlete by either parents or coaches can be a form of child abuse.

The expression of these factors has to be measured against the more obvious tissue psychological injury which has been discussed previously. This area of expression of psychological stress as manifested by somatic complaints is called somatization.

Somatization disorders are syndromes of somatic complaints expressed by patients who are experiencing underlying psychological stress. They must clearly be differentiated from malingering as the patients do not appreciate the dynamics of the pathophysiology of their complaints. Most frequently the patient will complain of pain. Usually there is a definable cause for the pain, but the definable pathology cannot validate the degree of disability which is frequently associated with these conditions. This syndrome is extremely common, particularly in growing athletes whose physical maturity is almost complete. It is important to re-emphasize that these disorders are not voluntary but rather, through mechanisms which are not well understood, express and amplify somatic complaints to the point of disability when in fact there is no definable objective pathological process present which would normally lead to the level of disability observed.

In some cases these disabilities are very dramatic, such as the presentation of a patient with the knee locked in full extension. Evidence clearly shows that any acutely injured knee would never lock in full extension but, in order to achieve comfort, in some degree of flexion. Therefore presentation of a young athlete with the knee acutely locked in full extension is never an indication of serious intra-articular pathology, but rather a significant indication of a stressful life experience being expressed as, or somatized as, an acute musculoskeletal problem. Not all somatization disorders present in such a dramatic fashion. Surprisingly, one of the major advances in our understanding of these disorders has come with the development of the arthroscope. The ability to obtain a complete history and to perform a careful physical examination coupled with a careful review of up-to-date radiographs,

supplemented by a definition of the internal architecture of the knee joint with the arthroscope, has allowed a much greater depth of understanding of the pathogenesis of chronic pain in the knee, a complex symptom frequently seen in the growing athlete. The ability objectively to grade inflammation with widely disparate clinical histories has allowed recognition that many of the painful clinical syndromes such as chondromalacia of the patella are in fact related to no definable pathology either within or adjacent to the knee. In fact, they may reflect a more global problem with respect to the interpretation of pain.

Technetium polyphospate bone scanning has also been helpful in interpreting these syndromes in that it is extremely sensitive to bone pathology. Thus a normal examination almost excludes substantial underlying pathology of the bone (Fig. 30).

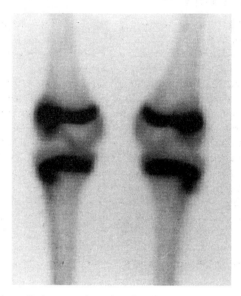

Fig. 30 Normally increased uptake of radioisotopes in growing knees clearly reflecting the concentration of metabolic activity in the growth plates.

Much unscientific and expensive treatment has been prescribed for these ill-defined pain syndromes. Frequently these treatments are successful in that they provide at least transient alleviation of the symptoms. However, this is not always of benefit to the patient in that the underlying problem has not been addressed. Clearly, coaches, trainers, and health care professionals must be aware that somatization disorders exist, and, whilst they may not reflect serious musculoskeletal pathology, they may well indicate a perverted expression of psychological stress which may result from any number of adverse life experiences. In some instances these may be of a minor nature such as those associated with competition anxiety. However, they may reflect serious background psychopathology resulting from family separation, abusive family settings, real or perceived school failure, and conflicts in the development of the unique personality which is characteristic of mature adults.

Thus the growing athlete is unique because he or she is growing physically and hence expresses a specific pattern of injuries unique to his or her age. Additionally, the young athletes are developing concepts of self-worth and of their place in the world around them. They are growing persons as well as growing bodies. Adults who are charged with the responsibility for encouraging the full potential of young athletes and for evaluating and preventing injury must recognize the scope of these responsibilities. The ultimate challenge is to encourage maximum performance without adversely affecting the development of the whole person and to encourage the growth of a healthy well-rounded adult citizen who can apply the positive experiences attained in childhood athletics to the greater challenges of life. Failure to understand the dynamics of growth and development can, and does, lead to an enhanced rate of injury and confusion of the true pathology. Mature clinicians, therapists, and coaches can best be part of the solution by having a full understanding of the potential problems and their complex inter-relationships.

CONCLUSION

Child athletes are different from adult athletes simply because they are in fact growing. The difference in the bone tissue necessitated by the ability to grow creates a completely different reaction to physical stress and injury. The rate of growth in the growing athlete and the attainment of full maturity are widely disparate and the injury patterns are defined by these very disparities. Adult physicians, trainers, and coaches must exercise mature judgement in assessing the genesis of athletic dysfunction. Failure to understand the nature of the presenting complaints may lead at best to failure to achieve ultimate athletic performance, and at worst to a lifetime of disability caused by potentially unnecessary injury. Perhaps even more important than physical injury is the failure to achieve the possibly greater aim of enabling the young athlete to develop into a well-rounded and productive member of society. There is little doubt that many of these undesirable traits originate early on, and that only through a clear understanding of the differences in growing children can these effects be avoided.

REFERENCES

1. Bailey DA *et al.* The relationship of fractures of the distal radius to growth velocity in children. *Canadian Journal of Sports Sciences* 1988; **3**: 40–1.
2. Kannus P, Nittymaki S, Jarvinen M. Athletic overuse injuries in children—a 30 month prospective follow-up study at an outpatient sports clinic. *Clinical Pediatrics* 1988; **27**(7): 333–7.
3. American Academy of Pediatrics Committee on Accident and Poison Prevention. Skateboard injuries. *Pediatrics* 1989; **83**(6): 1070–1.
4. Kvist M *et al.* Sports related injuries in children. *International Journal of Sports Medicine* 1989; **10**: 81–6.
5. Paulson JA. The epidemiology of injuries in adolescents. *Pediatric Annals* 1988; **17**(2): 84–6, 89–96.
6. Sutherland GW. Increased hockey injuries with age. *American Journal of Sports Medicine* 1976; **4**: 264–8.
7. Buncher CR. Statistics in sports injury research. *American Journal of Sports Medicine* 1988; **16**(9): 105–12.
8. Backous DD, Friedl KE, Smith NJ, Parr TJ, Carpine WD Jr. Soccer injuries and their relation to physical maturity. *American Journal of Diseases of Children* 1988; **142**(8): 839–42.
9. deLoes M, Goldie I. Incidence rate of injuries during sport activity and physical exercise in a rural Swedish community; incidence rate in 17 sports. *International Journal of Sports Medicine* 1988; **9**(6): 461–7.
10. Wetterhall SF, Waxweiler RJ. Injury surveillance at the 1985 Boy Scout Jamboree. *American Journal of Sports Medicine* 1988; **16**(5): 534–8.
11. Watson MD, DiMartino PP. Incidence of injuries in high school track and field athletes and its relation to performance ability. *American Journal of Sports Medicine* 1987; **15**(2): 151.
12. Chalmers DJ, Cecchi J, Langley J, Silva PA. Injuries in the 12th and 13th years of life. *Australian Pediatric Journal* 1989; **25**(1): 14–20.
13. McCauley E *et al.* Injuries in women's gymnastics. The state of the art. *American Journal of Sports Medicine* 1988; **16**(9): 124–31.
14. Hyndman JC, Armstrong K. Unpublished data.
15. Malina R *et al.* Athletes tend to be more physically mature than non-athletes. *Pediatric Clinics of North America* 19; **29**(6): 1321.
16. Storey WB. Evaluation in athletics—cardiovascular. *Pediatric Clinics of North America* 19; **29**(6): 132.
17. Rowland TW. Developmental aspects of physiological function relating to aerobic exercise in children. *Sports Medicine* 1990; **10**: 255–66.
18. Kreipe RE, Gewanter ML. Physical maturity screening for participants in sports. *Pediatrics* 1985; **75**(6): 1076–80.
19. Gallagher SS, Finison K, Guyer B, Goodenough S. The incidence of injuries among 87,000 Massachusetts children and adults: results of the 1980–1981 Statewide Childhood Injury Prevention Program Surveillance System. *American Journal of Public Health* 19; **74**(12): 1343.
20. Tames *et al.* Injuries to runners. Stretching good—no good. *American Journal of Sports Medicine* 1978; **6**: 40–50.
21. Snellock FG, Prentice WE. Warming-up, stretching for improved physical performance of sports related injuries. *Sports Medicine* 1985; **2**: 267–78.
22. Taimela S, Kujala UM, Osterman K. Intrinsic risk factor and athletic injuries. No conclusive evidence exists re these factors. *Sports Medicine* 1990; **9**(4): 205–15.
23. Kerr G, Fowler B. *Sports Medicine* 1988; **6**(3): 127–33.
24. Maestrello-deMova MG, Primosch RE. Orofacial trauma and mouth protector wear among high school varsity basketball players. *ASDC Journal of Dentistry for Children* 1989; **56**(1): 36–9.
25. Sane J. Comparison of maxillofacial and dental injuries in four contact team sports: American football, bandy, basketball and handball. *American Journal of Sports Medicine* 1988; **16**(6): 647–51.
26. Sim FH, Simonet WT, Melton LJ III, Lehn TA. Hockey injuries. *American Journal of Sports Medicine* 1988; **16**(9): 86–96.
27. De Respinis PA, Caputo AR, Fiore PM, Wagner RS. A survey of severe eye injuries in children. *American Journal of Diseases of Children* 1988; **143**(6): 711–16.
28. Jones N. Eye injuries in sports—90% preventable. *Sports Medicine* 1988; **7**(3): 163–81.
29. Pashby TJ. Ocular injuries in hockey. *Canadian Medical Association Journal* 1979; **121**: 643–4.
30. Grace TG, Skipper BJ, Newberry JC, Nelson MA, Sweeter ER, Rothman ML. Prophylactic knee braces and injury to the lower extremity. Journal of Bone and Joint Surgery 1988; **70A**(3): 422–7.
31. Goldberg B, Rosenthal PP, Robertson LS, Nicholson JA. Injuries in youth football. Pediatrics 1988; **81**(2): 255–61.
32. Thompson N, Halpern B, Curl WW, Andrews JR, Hunter SC, MacLeod WD. High school football injuries evaluation. American Journal of Sports Medicine 1988; **16**(9). 97–104.
33. Sowinski *et al.* Minimize age levels for young marathon runners. New Studies in Athletics 1986; **4**: 91–100.
34. Cahill BR *et al.* Little League shoulder: lesions of the proximal humeral epiphyseal plates. Journal of Sports Medicine and Physical Fitness 19; **2**: 150–4.
35. Albanese SA *et al.* Wrist pain and distal plate closure of the radius in gymnasts. Journal of Pediatric Orthopedics 1989; **9**: 23–8.
36. Watson A. Sports injuries in 6799 Irish schoolchildren. American Journal of Sports Medicine 1984; **12**: 65–71.
37. Osgood RB. Lesions of the tibial tubercle occurring during adolescence. Boston Medical and Surgical Journal 1903; **148**: 114–19.
38. Ehrenborg G. The Osgood–Schlatter lesion: a clinical and experimental study. Acta Chirurgica Scandinavica 1962; **28**: 1–36.
39. Brown DCS, Hyndman JC. Ligamentous injuries in children. 1978.

6.4   Athletes with disabilities

DARREN W. BOOTH

HISTORICAL PERSPECTIVE

The origin of organized competitive sport for the disabled is directly related to the rehabilitation of Second World War veterans with spinal cord injury. There are earlier examples of outstanding disabled athletes and of sports organizations for the disabled.[1] However, it was the renewed interest in sport as therapy in post-war hospitals in the United Kingdom and the United States that led to the present-day state of sport for the disabled.

The development of this philosophy in the United Kingdom is largely credited to the efforts of Sir Ludwig Guttman. While Director of the National Spinal Injuries Centre at Stoke Mandeville in 1948, he introduced archery as a therapeutic measure for paraplegic war veterans. The first international sporting event for such individuals took place at Stoke Mandeville 4 years later in the form of an archery competition between resident paraplegics and similarly disabled Dutch athletes.

Concurrent development in the United States involved the beginnings of organized wheelchair basketball competition between Veterans Administration Hospitals.[2] A prominent figure in this movement was Professor Timothy J. Nugent of the University of Illinois. He was responsible for organizing the first National Wheelchair Basketball Tournament in Galesburg, Illinois, in 1949.

The Stoke Mandeville Games grew in success to include more participating countries in an increasing number of events. In 1956 the Games were formally recognized by the International Olympic Committee with the presentation of the Fearnley Cup, a symbol of the aim of the modern Olympic movement. Wheelchair sport in the United States was also expanding to include other events, and in 1957 the first National Wheelchair Games in New York saw competition in basketball, track and field, swimming, table tennis, and archery.

Several sport organizations for the disabled were subsequently born of the movements in the United States and the United Kingdom. The establishment of the National Wheelchair Athletic Association (NWAA) in 1958 provided further opportunities for disabled athletes to compete in a number of sports. The International Stoke Mandeville Games Committee (ISMGC) was formed in 1960 to meet the needs of international interest in sport for the disabled. The NWAA initiated a formal affiliation with the ISMGC in the same year, demonstrating the need for co-operative efforts in the development of the international sports movement.

The International Stoke Mandeville Games grew to such a proportion that a change to a larger venue became necessary. In 1960 the Games were moved to Rome and held after the completion of the Olympic Games. Approximately 400 competitors from all over the world lived in the Olympic Village and competed in the Olympic Stadium.

International competition in the subsequent 3 years returned to Stoke Mandeville. In 1964 the Games moved to the Olympic site at Tokyo, Japan, establishing the pattern of holding the Paraplegic Olympic Games (or Paralympics) in the Olympic year in the host country of the able-bodied Games.[3] Approximately 450 athletes from 25 countries participated in an event which promoted international competition amongst the disabled according to the Olympic creed.

In 1968 the organizers of sport for the disabled in Mexico were unable to receive the support necessary from their government to hold the Paralympics in the host Olympic country. The 1968 Paralympic Games were thus held in Tel Aviv after the ISMGC accepted an invitation from Israel to host the competition. A total of 750 athletes from 29 countries participated in an event for which the host country had less than 12 months of preparation.

Regional conferences became organized during this time to facilitate additional international competition for disabled athletes. In 1967 the Pan American Paraplegic Games were established, as were equivalent events in Europe, the Far East, and South Pacific regions.

The 1972 Paralympic Games were hosted by the City and University of Heidelberg, Germany. The Olympic site in Munich could not be used because of a lack of wheelchair-accessible facilities. Approximately 1000 paraplegic competitors from 44 countries participated in a variety of sports. Demonstration sports involving amputee and blind athletes of the German Disabled Sports Movement aptly conveyed their equal need for organized international competition.

The 1976 Paralympics in Toronto, Canada, were the first true Olympic Games for the physically disabled. A total of more than 1500 paralysed, amputee, and blind athletes from 44 countries participated in a truly independent international sporting event.[4] Unfortunately, these were the first international games for the disabled to fall victim to political intrusion. The inclusion of a fully integrated and freely chosen South African team resulted in the withdrawal of athletes by governments of eight countries. Also, the Canadian government withdrew its financial and moral support from the Games. However, they were completed successfully and without a financial deficit thanks to voluntary support from the Canadian public and private sectors. Most important, the Games provided to more than 100 000 spectators an education in and example of what can be accomplished by disabled individuals through the medium of sport.

The inclusion in the 1976 Paralympics of amputee, blind, and cerebral palsy athletes was largely a result of the efforts of the International Sport Organization for the Disabled (ISOD). ISOD was created in 1964 to co-ordinate sports for all disabled athletes and incorporate future groups into its organization.[1] International rules of sport established by ISOD were first implemented in 1974 at the first World Games for the Disabled held in Stoke Mandeville. Amendments and alterations in these rules were subsequently adopted and applied at the 1976 Paralympics.

The 1980 Paralympic Games were not held in the Soviet Union in conjunction with the Olympics for able-bodied athletes. The Soviet Union officially claimed that 'they had no disabled' and thus did not offer to host competition for disabled athletes.[3] Instead, the Olympics for the Disabled were held in Arnhem, Holland, and with 2500 competitors were the second-largest sporting event of that year.

In 1984 there was no single unified international sports event that could be termed the Olympic Games for the Disabled. The

International Stoke Mandeville Games Federation (ISMGF), now the governing body for athletes with spinal-cord injuries, staged the Seventh World Wheelchair Games at Stoke Mandeville. Approximately 1100 individuals from 41 countries competed separately from the blind, amputee, and other paralysed athletes. These groups assembled for the 1984 International Games for the Disabled in Long Island, New York. Over 1700 athletes from 45 countries participated in this event. This year also witnessed the first events representative of sports for the disabled at an able-bodied Olympic Games. Women's 800-m and men's 1500-m wheelchair track races were included as demonstration sports at the Summer Olympics in Los Angeles.

The 1988 Paralympics in Seoul, South Korea, reunited wheelchair, blind, amputee, and other paralysed (primarily cerebral palsied) athletes. These largest-ever Olympic Games for the Disabled were held in the Olympic host city for the first time since 1964. More than 4000 competitors from over 60 countries participated in 16 sports over a 10-day period.

TYPES OF ATHLETES WITH DISABILITIES

Disabled athletes may have impaired physical, sensory, or cognitive function. The discussion which follows is limited in its references to physically and sensory impaired athletes and does not address the mentally disabled. In making this distinction it should be recognized that the majority of athletes with physical and sensory impairment do not have reduced cognitive abilities.[5] They function on a normal intellectual level and generally have little difficulty interacting with others.

Athletes with sensory impairment

The deaf athlete

The deaf athlete's hearing impairment is often a result of sensorineural deficits caused through cochlear damage.[6] Equilibrium deficits with a concomitant loss of balance and co-ordination may compound the athlete's disability if there has been damage to the semicircular canals or vestibular apparatus. However, the greatest limitation which deaf athletes usually confront is their inability to communicate effectively with other individuals. This inability can be overcome by the use of sign language and other methods of visual cueing. Deaf athletes can also compensate for their hearing loss by maximizing their visual abilities through training powers of observation and peripheral vision. Acquisition of these skills enables most deaf persons to participate in almost any athletic or fitness activity.

The blind athlete

Blind athletes have a partial or complete loss of sight. Eligibility for athletic competition is granted only to those individuals who have less than 10 per cent of useful vision, the legal limit for blindness. Fair competition in any single event is ensured by allowing participation of individuals with a similar degree of visual impairment. Classification of blind athletes for competition involves testing the individual's better eye while he or she is wearing required contact or corrective lenses. Blind athletes are assigned to one of three classes established by the International Blind Sports Association (Table 1).

Blind athletes compete in a wide variety of sports including baseball, bowling, cycling, judo, marathon racing, cross-country

Table 1 International Blind Sports Association classification system for blind athletes (less than 10 per cent of useful vision)

Class	Description of visual impairment
B1	No light perception up to light perception but inability to recognize the shape of a hand at any distance or in any direction.
B2	Ability to recognize the shape of a hand up to a visual acuity of 2/60 and/or a field of vision of less than 5 degrees.
B3	Visual acuity greater than 2/60 up to 6/60 and/or a field of vision greater than five degrees but less than 20 degrees.

Reproduced from ref. 7, with permission.

and downhill skiing, swimming, track and field, and wrestling. Individual events for blind athletes are the same as those for more normally sighted individuals with modification of some rules to facilitate participation by blind competitors. Examples of these modifications include the use of sighted guide-runners for blind runners competing in events of more than 100 m. The rules for the 100-m event allow the blind athlete to run freely from the starting position to his coach who calls instructions from beyond the finish line. Recent world record performances by blind athletes attest to their ability to attain levels of performance almost equal to those of more normally sighted individuals. Winning times for the 100-m sprint at the 1988 Seoul Summer Paralympics included 11.00 seconds and 12.43 seconds for men and women respectively.[8]

Blind athletes also participate in sports originated for visually impaired persons. Goalball, for example, is a game between two teams of three competitors (Fig. 1). The centre of one team aggressively bowls a large heavy ball towards the opposing team. The defenders orient themselves to the ball's trajectory by listening to the ringing of a bell placed inside the ball. The object of goalball is to use the body to block the ball from entering the net, thus preventing the opposing team from scoring points against the defenders.

Athletes with physical disability

The athlete with spinal cord injuries

Athletes with spinal cord injury have attracted more scientific and public attention than any other group of disabled athletes. They were the first to participate in international athletic competition for the disabled and today comprise the largest group of physically disabled athletes. The majority of athletes with spinal cord injury are wheelchair dependent, thus giving them the label 'wheelchair athletes'. This designation is commonly used for these individuals; however, it is important to realize that athletes in other disability groups also require wheelchairs for competition.

The traditional ISMGF classification system for athletes with spinal cord injury favoured an anatomical description of disability.[9] Individuals were assigned to one of three classes for quadriplegics and one of five classes for paraplegics. Each class was categorized according to the level and completeness of spinal cord injury and the resulting muscle function (Table 2).

Functional classification systems are now more widely accepted and used for athletes with spinal cord injury. Functional classification is based on the quality and quantity of muscle function and the resulting ability to perform movement patterns required for

Fig. 1 Goalball players blocking an opponent's shot. All competitors wear opaque goggles to eliminate any advantage to partially sighted athletes.

Table 2 ISMGF traditional classification system for athletes with spinal-cord injury

Class	Spinal-cord level	Anatomical/functional characteristics
IA	Cervical 4–6	Triceps grade 0–3 on manual testing; severe weakness of trunk and lower extremities interfering with sitting balance and ability to walk
IB	Cervical 4–7	Triceps grade 4–5 on manual testing; wrist flexion and extension may be present; generalized weakness of trunk and lower extremities interfering with sitting balance and ability to walk
IC	Cervical 4–8	Triceps grade 4–5 on manual testing; finger flexion and extension grade 4–5, with grasping and releasing; generalized weakness of trunk and lower extremities interfering with sitting balance and ability to walk
II	Thoracic 1–5	Abdominal muscles grade 0–2 on manual testing; no functional lower intercostal muscles; no useful sitting balance
III	Thoracic 6–10	Good upper abdominal muscles; no useful lower abdominal or trunk extensor muscles; poor sitting balance.
IV	Thoracic 11–lumbar 3	Good abdominal and spinal extensor muscles; some hip flexors and adductors; weak or non-existent quadriceps strength, limited gluteal control (grade 0–2); points 1–20 traumatic, 1–15 polio*
V	Lumbar 4–sacral 2	Good or fair quadriceps control; points 21–40 traumatic, 1–15 polio
VI	Lumbar 4–sacral 2	Points 41–60 traumatic, 36–50 polio (class VI is a subdivision of class V, applied only for athletes in swimming competitions)

*The scoring systems for classes IV–VI assigns up to 5 points (5 = normal strength) per side for hip flexors, extensors, abductors, adductors, knee flexors and extensors, and ankle plantar and dorsiflexors. This gives a potential score of 40 points per side.

Reproduced from ref. 10, with permission.

any given sport or activity. Functional classification was first used for wheelchair basketball players, but more recently has been applied to shooting, swimming, table tennis, and track and field participants. The functional approach also considers other factors which may influence an athlete's performance. These include muscle spasticity, spinal fusion, and the use of supportive strapping and other functional aids.[11] Functional classification also provides disabled athletes with a common ground and opportunity to compete against others who have similar functional capabilities but differing aetiologies of disability. A wheelchair athlete with spinal cord injury can compete with amputees or cerebral palsied athletes who have a similar level of function. A detailed explanation of functional classification systems for different sports is beyond the scope of this discussion. However, the present ISMGF classification system for track athletes is provided to facilitate comparison with the traditional ISMGF system (Table 3).

Athletes with spinal cord injury compete in many sports, but track and field and swimming are the most popular. Track and field participants compete in wheelchairs, with the exception of athletes who are able to throw field implements from a standing position. Other competitive sports for athletes with spinal cord injury include archery, basketball, fencing, marathon racing, shooting, snooker, table tennis, volleyball, and weight-lifting. The rules of all sports are based on those used for able-bodied competition with some modifications necessary to allow competition by wheelchair athletes. Wheelchair basketball, for example, involves two teams of five players on the basketball court during playing time. Two points are awarded for a goal from the field, except when three points are awarded for a successful shot from the three-point area.[13] The two large rear wheels of the chair must be within the three-point area for an attempt to be recognized. One point is awarded for a shot from the field goal line.

The cerebral palsied athlete

Athletes with cerebral palsy have a centrally mediated non-progressive neurological disorder which results in varying degrees of

Table 3 ISMGF functional classification system for track athletes with spinal-cord injury

Class	Old class	Spinal-cord level	Anatomical/functional characteristics
T1	IA complete	C6	Functional elbow flexors and wrist extensors; no functional elbow extensors or wrist flexors May have shoulder weakness; may use elbow flexors to start (back of wrist behind push rim); hands stay in contact or close to the push rim, with the power coming from the elbow flexors
T2	IB/IC complete	C7–8	Functional elbow and wrist flexors and extensors; may have finger flexors and extensors; functional pectoral muscles Usually use elbow flexors to start, but may use elbow extensors; power for pushing technique from elbow and wrist extensors, upper chest muscles
T3	IC incomplete/2/upper 3	T1–7	Normal or near-normal upper limb function; no abdominal muscles; may have weak upper spinal extensors Trunk may rise with the pushing action; power for pushing technique from hand flick technique (or friction technique); may use shoulder to steer around curves; may interrupt pushing movements to steer, resuming pushing position with difficulty
T4	Lower 3/4/5/6	T8–S2	Usually both upper and lower spinal extensors as well as abdominal muscle function Backwards, and usually rotational trunk movements; may use trunk to steer around curves; pushing technique usually not interrupted around curves; trunk moves to upright position with quick stop; abdominal muscles used when starting and pushing wheelchair.

Reproduced from ref. 12, with permission.

motor dysfunction. The variable nature and extent of the disorder is reflected in the traditional classification system established for cerebral palsied athletes by the Cerebral Palsy—International Sports and Recreation Association. The details of this complex system are outlined elsewhere,[12] but generally individuals are assigned to one of eight disability groups. Functional classification systems are also used for cerebral palsied athletes, allowing them to compete in several sports against amputees and other disabled athletes who have similar functional capabilities.

Track and field and swimming are popular sports for athletes with cerebral palsy. Participants may be ambulatory or compete in a wheelchair depending on the extent of their motor dysfunction. Cerebral palsied athletes who compete in sports such as Boccia (a strategic competition with similarities to lawn bowling and curling) are more severely disabled and wheelchair dependent. Seven-a-side soccer participants are all ambulatory, with some individuals requiring the use of walking aids.

Athletes with cerebral palsy also compete in cycling, power-lifting, and shooting. The rules for these sports vary little from those governing able-bodies competitions. Modifications, including the use of the tricycle in cycling competitions, make participation possible for more severely affected individuals.

Athletes with cerebral palsy have demonstrated their ability to perform on a level almost equal to that of able-bodied competitors. Recent world record performances for cerebral palsied athletes in the 100-m sprint at the 1988 Seoul Summer Paralympics were 11.79 seconds and 15.07 seconds for men and women respectively.[8]

The amputee athlete

Amputee athletes have a partial or complete loss of one or more limbs. The traditional ISOD classification system for amputee athletes describes nine different classes and its use has recently become more limited to groups such as track and field competitors (Table 4). Loss of one or both upper extremities is distinguished as

Table 4 ISOD classification system for amputee athletes

Class	Description of limb loss
A1	Double leg amputation above or through the knee joint
A2	Single leg amputation above or through the knee joint
A3	Double leg amputation below the knee, but through or above the talocrural joint
A4	Single leg amputation below the knee, but through or above the talocrural joint
A5	Double arm amputation above or through the elbow joint
A6	Single arm amputation above or through the elbow joint
A7	Double arm amputation below the elbow, but through or above the wrist joint
A8	Single arm amputation below the elbow, but through or above the wrist joint
A9	Combined lower and upper limb amputation

Reproduced from ref. 14, with permission.

being above or below the elbow. Loss of one or both lower extremities is distinguished as being above or below the knee. General minimum disability requirements for track and field participants stipulate that upper extremity loss must be through or above the wrist joint and lower extremity loss through or above the talocrural joint. Functional classification systems for amputee athletes are now becoming more widely used and have application to competitors in sports such as swimming.[15]

Track and field and swimming have been popular sports for amputee athletes. The rules for individual events are generally the same as rules for able-bodied competition. Amputee athletes may use a prosthesis for some track and field events, but no other assistive device is allowed.[16] Prosthesis use is optional in events including the high jump and is not permitted in others including the long jump. Double above-knee amputees compete in a wheelchair.

Sports participation by amputee athletes has expanded in recent years because of an increased awareness of their athletic ability and improvements in prosthetic design. Amputee athletes now compete in basketball, cycling, shooting, cross-country and downhill skiing, table tennis, volleyball, and weight-lifting. Recent world record performances by amputee athletes at the 1988 Seoul Summer Paralympics included 11.93 seconds and 15.30 seconds for the 100-m sprint by men and women respectively.[8] These performances further attest to the ability of disabled athletes to compete at or near a level equal to that of able-bodied individuals.

Les Autres

The title 'Les Autres' designates disabled athletes who do not fit into any of the disability groups previously described. The functional impairment evident in these athletes is caused by less common disorders including muscular dystrophy, multiple sclerosis, and dwarfism.[17] Until recently individual athletes were only eligible for competition against other Les Autres. The advent of functional classification has provided Les Autres with the opportunity to compete against athletes from the other disability groups. Les Autres now participate with amputees, athletes with spinal cord injuries, and cerebral palsied athletes in sports including archery, cycling, power-lifting, shooting, swimming, table tennis, track and field, and volleyball. The functional classification schemes used for Les Autres are described elsewhere.[12] They generally evaluate these individuals for functional capabilities comparable with other disabled athletes.

COMMON INJURY PATTERNS IN DISABLED ATHLETES

There is not sufficient published information to make definitive statements regarding sports-related injury patterns in disabled athletes. However, general trends of musculoskeletal and other soft tissue problems have been identified for the physically disabled and especially the wheelchair athlete.[18,19] Injuries to these athletes are often of similar aetiology to those of able-bodied individuals and can be managed using similar methods.[5] Disabled athletes are also vulnerable to medical problems not commonly associated with able-bodied individuals. These problems are discussed in more detail later in this chapter.

The incidence of sports-related injuries in wheelchair athletes has been documented in only a few studies. Of 128 wheelchair athletes surveyed in 1981, 72 per cent reported having had at least one injury during their sports history.[18] Soft tissue problems (sprains, strains, tendinitis, bursitis) were the most common, accounting for 33 per cent of the 291 injuries reported (Table 5). Similarly, sprains and strains accounted for 53 per cent of all injuries reported in a retrospective study of 61 wheelchair athletes.[19] Injuries such as fractures and other more serious traumas had a much lower incidence in both groups.

The risk of sports-related injury, accidents, or other complications in disabled athletes is controversial and not well studied. The medical staff at the 1976 Olympiad for the Disabled treated 184 of approximately 1500 competitors for a variety of problems.[4] They concluded that disabled athletes were more vulnerable to stress and fatigue than their able-bodied counterparts. However, Nilsen et al.[19] stated that the risk of sports-related or other medical complications in disabled athletes was low, being 2.51 per 1000 hours of training. Comparative figures given for able-bodied soccer players ranged from 4 to 23 per 1000 playing hours.

High risk sports for wheelchair athletes have been identified despite the disagreement regarding general risk of injury in sports for the disabled. Curtis and Dillon[18] reported that track, basketball, and road racing accounted for almost 75 per cent of all injuries incurred by their 128 respondents. Lower-risk sports included (in descending order) tennis, weight-lifting, field events, swimming, and archery. Slalom track events, billiards, and bowling demonstrated the lowest risk of injury to participants. Factors related to the greater incidence of injury to wheelchair athletes in the high risk sports included a history of participation in more than one sport and a greater number of hours per week spent in training.

THE WHEELCHAIR ATHLETE

Wheelchair athletes include individuals disabled by spinal cord injury, cerebral palsy, lower-extremity amputation, or any of the disorders included with Les Autres. The common denominator for these disabled athletes is their mobility impairment and need of a wheelchair for sports participation.

Wheelchair locomotion is not an efficient means of transportation, even when employed by experienced athletes. The mechanical efficiency of wheelchair locomotion is at best 5 per cent, compared with a minimum of 20 per cent for walking or cycling at similar velocities.[20] Mechanical efficiency depends on numerous intrinsic factors including wheelchair mass and design, the athlete's mass, propulsion technique, and physiological efficiency, and extrinsic factors such as the nature and grade of the ground surface.

The physiological inefficiency of wheelchair locomotion is

Table 5 Injuries sustained by a group of 128 wheelchair athletes

Type of injury	Percentage of total (N=291)
Soft tissue (sprains, strains, tendinitis, bursitis)	33
Blisters	18
Lacerations and abrasions (including complications such as skin infections)	17
Pressure sores	7
Arthritis and other joint disorders	5
Fractures	5
Hand weakness or numbness	5
Temperature regulation disorders	3
Head or dental injury	3

Data from ref. 18.

largely a result of the dependence on the upper extremities to provide the necessary propulsion force. Upper-extremity muscles generally require more energy and tire more quickly for any given task than do lower-extremity muscles. Therefore wheelchair locomotion is more costly in terms of energy requirements and working muscles are more susceptible to injury. A more thorough understanding of the mechanics and physiology of wheelchair locomotion will aid in the development of a more efficient wheelchair stroke and will help recognize, treat, and prevent upper-extremity injuries to wheelchair athletes.

Sports wheelchair design

The design of sports wheelchairs is constantly evolving. Recent design modifications are intended to improve the mechanical efficiency of the wheelchair, facilitate a more effective wheelchair stroke, and minimize the risk of upper-extremity injuries. Current wheelchairs generally combine a low centre of mass with a high degree of manoeuvrability.[9] Arm rests and handles are minimized in size or eliminated completely. The lightweight frames are narrower and have a low forward-placed seat which is angled to the horizontal. Chairbacks are greatly reduced in size and footrests may be rigid or adjustable according to the needs of individual athletes. The wheels are cambered inwards and are fitted with tubular tyres inflated to high pressures.

The push rim diameter of sports wheelchairs varies according to the nature of the competitive event for which the wheelchair is to be used. A small diameter push rim is preferred, particularly by track athletes, because of the potential for greater stroke efficiency and speed of wheelchair locomotion (Fig. 2). The high drive ratio propulsion system maximizes muscle force–velocity relationships and lowers the stroke velocity which needs to be produced by the upper extremities.[21] This results in reduced metabolic and cardiopulmonary demands for wheelchair racing.[20] Conversely, a large diameter push rim is used by most other wheelchair athletes. This provides a low drive ratio system which requires less propulsive force and allows the wheelchair greater manoeuvrability. Greater manoeuvrability is an obvious advantage to athletes competing in sports such as wheelchair basketball (Fig. 3).

The competitive wheelchair stroke

The competitive wheelchair stroke can be described according to drive and recovery phases.[22] The drive phase involves the application of propulsive force to the wheelchair push rims by the synchronous action of the upper extremities. The method of force application is subject to numerous variations of style and technique. Sprint athletes generally use a shuttle wheeling motion with limited hand contact on the push rims.[23] Distance racers employ a circular pattern of arm motion which facilitates longer hand contact and application of propulsive force.

The actions of the upper extremities during the drive and recovery phases of wheelchair locomotion consist primarily of flexion and extension. Synergistic actions of the shoulder girdle provide positioning of the glenohumeral joint to minimize excessive abduction and rotational movements. The forward thrust of the arms during the drive phase causes a reverse trunk momentum which is restrained through stabilization by trunk and shoulder girdle muscles. Further restraint is provided by placing the lower extremities in a seated tuck position.

The drive phase of the competitive wheelchair stroke requires co-ordinated upper-extremity actions to place the hands in con-

Fig. 2 A track athlete demonstrating the use of a competitive wheelchair designed for speed and stroke efficiency.

Fig. 3 Basketball players demonstrating the use of competitive wheelchairs designed for manoeuvrability.

tact with the push rims and apply propulsive force to the wheelchair wheels.[22] Initially, the wrist extensors, biceps, anterior and medial deltoid, and pectoralis major work co-operatively to produce hand contact with a minimal interruption of the forward momentum of travel. The same muscles continue to move the upper extremities until the hands have passed the midline of the body. The wrist flexors, triceps, latissimus dorsi, middle and posterior deltoid, and upper trapezius then become active until the drive phase is completed. The pectoralis major continues to work until the hand breaks contact with the push rim.

The recovery phase of the competitive wheelchair stroke requires co-ordinated upper-extremity actions to break hand contact with the push rims and return the arms to the position required for the next drive phase. The same muscles working at the end of the drive phase co-operatively relax to release the hands from the push rims with minimal interruption of the forward momentum of travel. The wrist extensors, biceps, and medial and anterior deltoid then become active and again bring the hands into contact with the push rims. The upper trapezius may also be active, particularly if larger push rims are being used, necessitating shoulder elevation for positioning of the upper extremities.

Training guidelines and recommendations

Physical disability does not prevent an individual from potentially gaining the benefits of fitness or athletic training. Programmes to improve flexibility, strength, power, endurance, skill, and co-ordination are commonly used with disabled athletes. These programmes generally employ the same methods used with able-bodied athletes. However, there are recommendations which should be considered in the training of wheelchair athletes who have higher levels of spinal cord injury. The following discussion addresses these recommendations as well as the general means by which upper-body flexibility and strength may be developed. Stretching and resistance exercises designed for these purposes are not only valuable in the preparation of the wheelchair athlete for sports participation but can also help to prevent injury to muscles, tendons, and ligaments.

Stretching exercises for the wheelchair athlete are designed to improve and/or maintain muscle flexibility and joint mobility in the trunk and upper extremities. The joints of the shoulder girdle in particular need to be moved through extreme ranges of motion. Routine stretching of the muscles and tendons producing these motions will help to prevent injury to the soft tissues. Stretching programmes for the trunk, shoulders, elbows, and wrists of wheelchair athletes have been developed[24,25] and should be part of the warm-up and cool-down associated with sports participation. Individual exercises are usually performed by athletes while sitting in their wheelchairs. Stretching exercises which are difficult to perform in a sitting position are often made more effective with the assistance of a partner.

Resistance exercises for the wheelchair athlete are designed to improve the strength, power, and endurance of the trunk and upper-extremity muscles. Resistance exercises can also improve technical efficiency (i.e. the wheelchair stroke) and prevent injury to muscles not balanced with stronger opposing muscles. Resistance training programmes for the trunk, shoulders, elbows, and wrists of wheelchair athletes have been developed[26,27] and can be implemented employing the same weight-training facilities as used by able-bodied athletes. Free-weight training will provide a good quality work-out and requires minimal modification of existing equipment. The need for a spotter or partner is essential to ensure safety when using free weights. The use of other equipment including various isotonic and isokinetic exercise systems should produce comparable benefits for the wheelchair athlete. Assistance may be required for transfers; otherwise, safety and stability are usually ensured through the use of securely fitting straps and seat belts.

Special recommendations should be considered when establishing training goals and limitations for wheelchair athletes with high thoracic or cervical level spinal cord injury. Autonomic nervous system dysfunction in these athletes will reduce the ability of their cardiovascular and other systems to respond to exercise demands.[20] The cardiac output, particularly of quadriplegic athletes, will be limited owing to loss of the reflex redistribution of blood to working muscles and diminished sympathetic control of heart rate and myocardial contractility. Exercising muscles thus become more dependent on anaerobic metabolism to meet energy demands. Lactic acid and other waste products accumulate more quickly and fatigue becomes more of a limiting factor to training. Therefore wheelchair athletes with high spinal cord injury will tire more quickly and require longer rest periods, particularly with training methods dependent upon anaerobic metabolism. It should also be noted that measurement of heart rate responses to exercise will not be an accurate indicator of the training progress of these athletes.[27] Training times, distances, and number of repetitions are examples of parameters which should be recorded instead during training sessions.

A number of other factors may limit the wheelchair athlete's capacity for fitness and athletic training. A reduced thermoregulatory response to exercise is evident, particularly in temperature extremes. This factor is discussed in more detail in the next section. Dizziness, ataxia, and depression are potential side-effects of medications used to treat muscle spasticity in persons with spinal cord injury.[20] Other means of controlling muscle spasticity should be employed if wheelchair athletes are to train and perform to their full potential.

UPPER-EXTREMITY INJURIES IN WHEELCHAIR ATHLETES

Upper-extremity injuries in wheelchair athletes are the most widely recognized group of sports-related disorders affecting disabled athletes. However, much of the information regarding these injuries is anecdotal or derived from publications investigating the same problems in able-bodied athletes. Many of the soft tissue injuries are common to both populations and are similar in presentation.[5] The important difference in the consideration of injuries to wheelchair athletes is in determining the specific cause of the injury with respect to the individual's disability and sports activity. This may affect the management and rehabilitation of injuries to these individuals.

Upper-extremity injuries in wheelchair athletes can be of traumatic or non-traumatic origin. Direct injury to the soft tissues may be caused by the impact of the participant with the ground, with an opponent's body or wheelchair, or with other fixed or moving objects. More serious injuries such as fractures are not common.[28] Non-traumatic injuries may be caused by chronic overuse of muscles, tendons, and ligaments, or by forceful exertion upon these tissues without proper warm-up. The overuse phenomenon is attributed to the repetitive or prolonged use of the arms for wheelchair locomotion. Chronic or recurrent injuries in particular will impede athletic performance and occasionally end sports participation.

There is very little information regarding the treatment of injuries to disabled athletes. Upper-extremity injuries in wheelchair athletes can be managed using the same general methods as employed with able-bodied individuals. Special considerations in the rehabilitation of injuries to disabled athletes are discussed in more detail at the end of this section.

The prevention of injuries is important for all athletes regardless of ability. The prevention of upper-extremity injuries in wheelchair athletes can generally be managed by the proper fitting of sports wheelchairs, wearing adequate protective equipment, and taping or splinting previous injuries requiring support. Other preventive measures for wheelchair athletes include adequate training of strength, power, or endurance prior to competition, development of proper sports-specific techniques, and the use of appropriate warm-up and stretching routines in pre-event preparation.

Shoulder injuries

Injuries caused by direct trauma to the shoulder region of wheelchair athletes commonly include contusion of muscle and other soft tissues, tearing of the ligaments, the joint capsule, or the rotator cuff, and occasional fracture of bone (Table 6). Non-traumatic aetiologies secondary to overuse or inadequate warm-up may result in varying degrees of muscle and tendon strain. Inflammatory responses to these excessive forces are pathognomonic of tendinitis of the shoulder region.

Rotator cuff injuries have been reported in non-athletic wheelchair users.[29] The prevalence of similar disorders in wheelchair athletes has been suggested.[5,30] Repetitive use of the upper extremities, particularly in a prolonged or overhead manner, may lead to tears and inflammatory disorders of the rotator cuff muscles and tendons. Wheelchair athletes with inadequate strength, flexibility, or endurance of these tissues are particularly susceptible to these problems. An inefficient technique for pushing the wheelchair rims may also contribute to these overuse phenomena as the shoulders assume both weight-bearing and locomotor functions.

Overuse injuries of the shoulder are less likely to occur in wheelchair athletes who are well prepared for the demands of their sport. Adequate strength and endurance training of the muscles of the shoulder girdle will help to prevent overuse syndromes. An efficient wheelchair stroke minimizing abduction and rotational movements of the shoulder will also reduce the incidence of overuse injury. Shoulder derangements have not been reported in wheelchair marathon competitors despite the repetitive nature of the propulsion technique used in their event.[31]

Stretching and tears of the glenohumeral joint ligaments and capsule may be caused by acute injury or repetitive microtrauma.[32] Acute instability (subluxation or dislocation) of the glenohumeral joint of wheelchair athletes can result after falls from wheelchairs moving at high velocities. The acromioclavicular joint may also be injured in this way. Acquired instability of the glenohumeral joint can occur in athletes who repetitively stress the joint, particularly in an overhead manner. This is evident in sports such as swimming (i.e. butterfly stroke) and field events (i.e. javelin throw (Fig. 4)).

The skin of the axilla and the inner aspect of the upper arm of wheelchair athletes is vulnerable to abrasion injuries.[5] Abrasion injuries are most common in wheelchair racers and can be caused by incidental contact of the skin with the wheelchair tyres or by purposeful use of the arms to apply pressure to the wheels in a braking manoeuvre. Simple means of prevention include protecting the skin of the upper arm with a friction-reducing material or wearing a shirt which covers the chest wall. Wheelchair athletes should also be discouraged from using their upper arms to slow their wheelchairs.

Elbow and wrist injuries

Elbow and wrist injuries in wheelchair athletes generally have the same aetiologies as injuries to the shoulder region. Acute soft tissue trauma and overuse injuries of these joints do occur,[30] but few have been examined in any detail. Carpal tunnel syndrome is common in the non-athletic wheelchair population.[33] Carpal tunnel syndrome may be even more prevalent in wheelchair athletes

Table 6 Common shoulder disorders in wheelchair athletes including causes of injury and preventive measures

Injury	Cause	Prevention
Traumatic		
Muscle contusion (i.e. deltoid contusion)	Fall from wheelchair	Ensure proper wheelchair fitting and maintenance
Ligamentous tear (i.e. acromioclavicular joint sprain)	Physical contact with opponent's body or wheelchair	Wear adequate protective equipment
		Supportive taping and splinting
Capsular/ligamentous tear (i.e. glenohumeral joint dislocation)		
Rotator cuff tear (i.e. supraspinatus tear)		
Non-traumatic		
Capsular/ligamentous strain (i.e. glenohumeral joint subluxation)	Repetitive overuse of muscles and tendons	Adequate strength and endurance training of muscles of propulsion
Rotator cuff strain (i.e. supraspinatus tendinitis)	Over-exertion of muscles and tendons without proper warm-up	Efficient wheelchair stroke
		Appropriate warm-up, cool down, and stretching routines

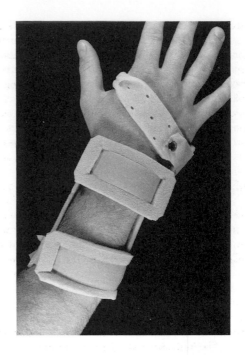

Fig. 4 The overhead use of the upper extremity in the javelin throw by a wheelchair athlete.

who use a stroke dependent on the propulsive force of the heels of the hands on the wheelchair rims.[18] Pain or paraesthesia in the thumb and first two fingers are symptoms which aid in the recognition of this disorder. Rest from the offending activity, and the use of anti-inflammatory medication and various therapeutic modalities are indicated in the treatment of carpal tunnel syndrome in its early stages.[5] Splinting of the wrist and hand (Fig.5) and rest from any wheelchair pushing may be indicated for more advanced cases of the disorder. Strengthening of the wrist flexors and extensors and the use of padded push rims and gloves may aid in the prevention of carpal tunnel syndrome.

Hand and finger injuries

The hands and fingers of wheelchair athletes are frequently injured.[28,30] Acute sprains of any of the small joints may occur when fingers are caught between the rim and spokes of a single wheelchair or between opponents' wheelchairs in sports such as wheelchair basketball. Overuse injury of the metacarpophalangeal joint of the thumb may be caused by the repetitive contact of the hand with the wheelchair push rim. Chronic instability of this joint will develop as a result of the recurrent injury to the ulnar collateral ligament.

The hands of wheelchair athletes are particularly vulnerable to skin injuries. Cuts, abrasions, and blisters may be caused by abrupt contact with other surfaces or as a result of repetitive contact with the wheelchair push rims or tyres. The hypothenar eminence of the hands is particularly subject to abrasions and blisters. Protective calluses will eventually form if blisters are properly managed. However, thick calluses may crack and split, leaving open wounds susceptible to infection.

Skin injuries can be prevented by the use of padded wheelchair push rims and protective gloves (Fig. 2). Areas of chronic irritation may have to be further protected with friction-reducing materials such as moleskin or by using adhesive tape with a layer of petroleum jelly.

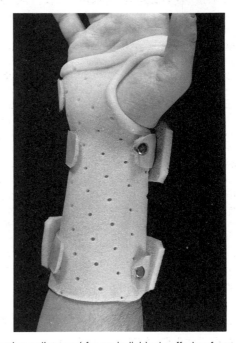

Fig. 5 Resting splint used for an individual suffering from carpal tunnel syndrome.

Special considerations in injury management and rehabilitation

Disabled athletes generally expect and deserve the same consideration and attention as that given to able-bodied athletes. The sports medicine community is now recognizing that the same principles of management and rehabilitation of injuries can and should be employed with athletes regardless of their ability. However, there are some special considerations which need to be recognized when working with wheelchair and other disabled athletes.

Rest, medication, and various therapeutic modalities are often indicated in the care of injuries to disabled athletes. The use of rest in the treatment of upper-extremity injuries in wheelchair athletes is difficult owing to the individual's dependence on the upper extremities for providing normal daily locomotion. The importance of adequate rest in the healing of injured tissues must be explained to the athlete. Severe injury may necessitate that another person push the athlete's wheelchair during the recovery period. Not all individuals will be compliant with instructions for rest, and the time required for rehabilitation of an upper-extremity injury may be longer than initially expected.[5]

Medication must be carefully prescribed in the treatment of sports-related injuries or other medical problems in disabled athletes. The deliberate or inadvertent use by any athlete of substances banned or prohibited by the International Olympic Committee can result in his or her suspension from national or international competition.[6] Many cough and cold remedies, pain killers, hay fever medications, nasal sprays, and decongestants contain banned or restricted substances that would result in a positive doping control test. Physicians, coaches, parents, and the athletes themselves must be aware of the contents of any medication, vitamin, herbal remedy, or nutritional supplement being used. Responsible individuals should refer to current authorized information on the use of banned and safe drugs to ensure that the athlete does not use any substance which will conflict with anti-doping regulations.

The same indications should be considered in the use of various physiotherapeutic modalities for the treatment of sports-related injuries in disabled and able-bodied athletes. The same precautions and contraindications must also be considered when using heat-producing modalities, cryotherapy, and electrotherapy. This is particularly important in the application of these methods of treatment to areas of paraesthetic or anaesthetic skin in disabled athletes.

The general philosophy guiding the management of sports-related injuries in disabled athletes should not differ from that used with their able-bodied counterparts. Sports participants who are disabled must first and foremost be recognized as athletes. Their disability must be secondary in consideration when examining their injuries and determining the most effective treatment. Progressive attitudes and methods of management and rehabilitation are necessary to return these athletes to their previous level of athletic performance.

SPECIAL MEDICAL PROBLEMS IN WHEELCHAIR ATHLETES

Sports participation has a positive effect on the maintenance of good health in individuals with spinal cord injury.[34,35] Wheelchair athletes in particular have demonstrated an ability to incur fewer of the medical complications for which they are at risk. They tend to seek medical attention less frequently and require fewer hospital admissions for medical complications. However, these athletes still have certain vulnerabilities which sports medicine personnel must be aware of and prevent if at all possible.

Urinary tract complications

Neurological control of the urinary tract is usually lost after spinal cord injury. Resulting complications of significant risk in paraplegics and quadriplegics include bladder and kidney infections and stones, bladder distension, and urethral fistulae.[36] Kidney damage secondary to infection alone is the main cause of death in persons with spinal cord injury.

Wheelchair athletes have demonstrated a lower incidence and frequency of urinary tract complications when compared with sedentary wheelchair users.[34] They may use a method of urinary management (i.e. intermittent versus indwelling catheterization) which results in fewer complications, and their lesser reliance upon medical and health care practitioners may indicate a greater sense of responsibility for looking after themselves.[35] However, education of wheelchair athletes regarding the signs and symptoms of urinary tract complications remains an important responsibility of persons entrusted with their health care.

Pressure sores

Pressure sores are one of the most common and costly complications of spinal cord injury.[37] This form of skin breakdown is usually caused by prolonged pressure and compromise of the blood supply to the affected tissues. Additional causes of skin breakdown in wheelchair athletes includes external trauma such as bumps, bruises, and scrapes.

Wheelchair athletes have demonstrated a tendency towards fewer pressure sores when compared with sedentary wheelchair users.[34] In particular, skin breakdown caused by prolonged pressure is less frequent. When wheelchair athletes do suffer from pressure sores, the skin breakdown is often not extensive and individuals do not require hospitalization as frequently as non-athletes.

Normal hydrostatic pressure in subepidural capillaries averages 25 mmHg.[30] Prolonged pressure of more than 25 mmHg causes collapse and thrombosis of these capillaries. The time required for tissue damage under these conditions has been estimated at 2 h.[38] Tissue damage may occur more or less quickly depending on the individual's body weight, build, and nutritional status.

Wheelchair athletes have demonstrated skin pressures of up to 60 mmHg over the sacrum and ischial tuberosities.[30] Wheelchair athletes who sit in their chairs with their knees higher than their buttocks are particularly prone to these high pressures. Wheelchair athletes participating in sports such as basketball or tennis are also prone to pressure sores due to the shear forces imposed upon the skin of the sacrum, buttock, and hips. Further risk of pressure sore development is inherent as the athlete's skin becomes damp with sweat and other moisture.

Prevention of pressure sores is important in all persons with spinal cord injury. Intermittent shifting and lifting of the buttocks from the wheelchair seat relieves pressure and may prevent skin breakdown.[39] Wheelchair users should frequently perform a wheelchair push-up or lift-off while sitting. Wheelchair seats should be well padded or incorporate a cushion customized to the individual athlete. Wheelchair seat cushions should be fitted by a qualified professional.[40,41]

Further preventive measures for pressure sores include wearing moisture-absorbing clothing to reduce skin maceration and friction forces. Frequent skin checks of the trunk, sacrum, buttocks, and legs should be performed to identify problematic pressure points or areas of breakdown. Finally, good nutrition and personal hygiene practices must be followed at all times by wheelchair athletes if they are to minimize the risk of developing pressure sores.

There are many methods of managing pressure sores if they

develop despite preventive measures. These methods are described in detail elsewhere.[42,43] Most physicians now advocate a conservative approach to the management of most pressure sores. Sports medicine personnel working with wheelchair athletes must be capable of recognizing sites of potential skin breakdown. They should also be prepared to assist other medical and health professionals more experienced in the treatment of general medical problems in persons with spinal cord injury.

Autonomic hyper-reflexia

Autonomic hyper-reflexia is a potential complication of spinal cord injury above the fourth to sixth thoracic vertebrae.[30] Although autonomic hyper-reflexia occurs in more than 80 per cent of high level paraplegics and quadriplegics, it is often misinterpreted by medical personnel not familiar with its presentation.

Autonomic hyper-reflexia occurs as a result of the loss of central inhibitory control over the isolated distal spinal cord. A generalized sympathetic hyperactivity may be triggered in response to numerous sensory stimuli including those associated with bowel or bladder distension, catheterization, or urinary tract infection.[44,45] High or low temperatures, sunburn, pressure sores, and thrombophlebitis have also been implicated in autonomic hyper-reflexia.

Autonomic hyper-reflexia in persons with spinal cord injury classically presents with sudden hypertension, bradycardia, headache, anxiety, and profuse sweating through reflex perspiration of insensate skin. Wheelchair athletes suffering from autonomic hyper-reflexia should cease any activity and be closely examined by medical personnel. Most acute episodes of the disorder will respond to conservative non-pharmacological intervention.[45] All wheelchair athletes should be aware of the existence of autonomic hyper-reflexia and be taught how to prevent its occurrence through elimination of precipitating stimuli.

Temperature regulation disorders

Impairment of thermoregulatory function is a significant complication of spinal cord injury. The loss of autonomic nervous system influence in high paraplegics and quadriplegics greatly reduces vasomotor and sudomotor responses of insensate skin.[46] The loss of sensory afferent impulses from the spinal cord and muscles below the level of the lesion may limit hypothalamic responses to exercise and temperature. Finally, the loss of lower-extremity skeletal muscle pump activity reduces venous return to the heart during exercise and further compromises thermoregulatory responses.

Impairment of thermoregulatory ability in wheelchair athletes is dependent upon the level and completeness of spinal cord injury. High level paraplegics and quadriplegics will be the most affected, particularly when exercising under conditions of extremes of temperature. The core temperatures in these individuals tend to be higher than normal in the heat and lower than normal in the cold. Wheelchair athletes are generally at a thermoregulatory disadvantage and certain precautions must be taken to prevent the occurrence of hyperthermia and hypothermia.

Hyperthermia

The regional loss of cutaneous circulatory and sweating responses has been defined in persons with spinal cord injury according to the level of their lesions.[47] Limited sweating may occur over insensate skin under hot conditions, but it is not necessarily synchronous with sweating over normally innervated skin.[48,49]

Therefore body core temperature regulation in exercising wheelchair athletes is dependent upon evaporative heat loss from the head, neck, and arms, and dry heat exchange from the remainder of the body.

Heat loss under hot humid conditions may be assisted in a number of ways. Minimal clothing should be worn to facilitate heat dissipation. Cool damp towels or water spray bottles should be used to keep the athlete's skin moist and to aid heat convection. Cooler shaded locations should be used by athletes when not competing. Finally, wheelchair athletes must force fluids during training and competition.[30] Thirst is not an adequate stimulus for fluid replacement and prevention of dehydration and hyperthermia.

Hypothermia

Excessive heat loss resulting in hypothermia may occur in wheelchair athletes competing in cold, damp, or windy conditions. Hypothermia may also occur after wheelchair athletes have finished competition and sit waiting in their wheelchairs. Their skin remains vasodilated and damp with sweat and other moisture. The protective shivering reaction which generates heat in able-bodied persons under such circumstances will be minimal or absent in wheelchair athletes.

Hypothermia in wheelchair athletes can be prevented if they wear additional layers of clothing under adverse weather conditions. Insulating blankets should be wrapped around athletes who must wait after competition before taking a warm bath or shower and changing into dry clothing. Adequate fluid replacement remains important in maintaining body core temperature. Wheelchair athletes should be encouraged to drink water and other replacement fluids regardless of weather conditions.

THE FUTURE OF SPORT FOR THE DISABLED

Sports for the disabled are not well understood by the general population or even by many of the medical and health professionals who assist the efforts of disabled athletes. This lack of understanding exists for many reasons. Partial blame must be attributed to the large number of sports organizations and the resulting broad scope of athlete disability types and sports events which have defined sports for the disabled. The simple categorization of disabled athletes into groups with cognitive, sensory, or physical impairment can be confusing to many who watch competition by these athletes. The separation of each group into smaller subgroups according to the aetiology of the disability creates even more confusion. This becomes particularly apparent when attempting to appreciate the different classes of athletes within each subgroup as defined by anatomical description and degree of disability.

The separation of disabled athletes into different disability groups and classes has created a multiplicity of sports events and awards of recognition for winning performances. Wheelchair sports, in particular, have been subject to the declaration of different champions for the same athletic event. The 100-m sprint alone declares men's and women's champions for each of the amputee, cerebral palsy, and spinal cord injury disability groups. Individual winning performances for each event may not appear to differ significantly when viewed by spectators or by the athletes themselves. This situation has contributed to the lack of understanding and even credibility of sports for the disabled. The multiplicity of winning standards for athletes with similar functional capabilities has also served to demean elite performances by individual athletes.

The historical tendency to organize sports for the disabled according to aetiology and degree of disability has been valuable in providing a framework by which disabled athletes can participate in equitable competition. Recently, however, there have been attempts to reorganize sports for some groups of disabled athletes, particularly those who are wheelchair ambulatory. The underlying premise for the reorganization of wheelchair sports assumes that differences in locomotor ability are made irrelevant by the use of a wheelchair for sports participants. Equitable competition for athletes with different aetiologies of disability then becomes possible. This premise also relies on the assumption that competitors will not have a functional advantage over their opponents. Fairness of competition is promoted by classification of athletes based on evaluation of similar degrees of athletic potential. In summary, wheelchair sports are being reorganized to reflect levels of functional performance rather than aetiology of disability.

The reorganization of wheelchair sports provides elite athletes with the opportunity to compare their performances directly with those of competitors who have similar functional capabilities. For many events, a single winning standard will be set by one champion. The reduction in numbers of similar events, multiple standards, and champions should facilitate a greater degree of understanding and credibility by the public for wheelchair sports. Victorious athletes should feel more legitimate in that they are the true champions of their event.

The integration of disabled athletes with similar functional capabilities is an evolutionary step towards the integration of some disabled athletes into competition with able-bodied athletes. This process is necessary if society believes that disabled persons should be accorded the same opportunities as the able-bodied. The application of this belief to the realm of sport implies that disabled athletes should participate in normal competitive experiences. This requires that disabled athletes participate in sports which will not require rule changes to accommodate individuals with differences in functional capability. Target archery and shooting are examples of sports which can provide equitable competition between disabled and able-bodied athletes. Wheelchair athletes without upper-extremity impairment may challenge their able-bodied counterparts on an equal basis as differences in functional mobility become irrelevant when considering the demands of the sport.

The true future of sport for the disabled must also provide normalized competitive experiences for other disabled athletes. When disabled athletes cannot compete equitably with able-bodied athletes, they must be given the opportunity to participate with athletes of similar capability in sports with minimal adaptation of the original. Only then can these disabled athletes receive the same psychological, physical, and social benefits afforded to the able-bodied by normal competition. Wheelchair basketball and marathon racing are examples of sports which can provide near-normal athletic experiences for disabled athletes. The growing popularity and success of these sports is evident in many ways. Most ironically, their success is evident from the growing interest of able-bodied individuals in participation in these wheelchair sports. Although this interest is not without opposition from the disabled and able-bodied sport communities, it is a good indicator of the acceptance and credibility which sports for the disabled are gaining worldwide.

REFERENCES

1. Guttman L. The development of sport for the disabled—historical background. In: Guttman, L. ed. *Textbook of sport for the disabled*. Aylesbury, UK: HM & M Publishers, 1976: 14–20.
2. Labanowich, S. The physically disabled in sports: tracing the influence of two tracks of a common movement. *Sports 'n' Spokes* 1987; **12**: 33–42.
3. Jackson RW, Davis GM. The value of sports and recreation for the physically disabled. *Orthopedic Clinics of North America* 1983; **14**: 301–15.
4. Jackson RW, Fredrickson A. Sports for the physically disabled: the 1976 Olympiad (Toronto). *American Journal of Sports Medicine* 1979; **7**: 293–6.
5. Magnus BC. Sports injuries, the disabled athlete, and the athletic trainer. *Athletic Training* 1987; **22**: 305–10.
6. Shapira W. Competing in a silent world of sports. *Physician and Sportsmedicine* 1975; **3**: 99–105.
7. *International Blind Sports Association Handbook*, October 1989.
8. Seoul Paralympic Organizing Committee. *Results summary, 1988 Seoul Paralympics.*
9. General rules. *International Stoke Mandeville Games Federation Handbook 1989.* Section 13.
10. *ISMGF guide for doctors.* Aylesbury: Stoke Mandeville Games Federation, 1982.
11. Shephard RJ. Sports medicine and the wheelchair athlete. *Sports Medicine* 1988; **4**: 226–47.
12. *General and functional classification guide. IX Paralympic Games, Barcelona 1992.*
13. Official Wheelchair basketball rules. *International Wheelchair Basketball Federation Handbook 1990–1994.*
14. *International Sport Organization for the Disabled Handbook*, February 1987.
15. Functional classification for swimming. *International Sport Organization for the Disabled Handbook 1989*, Section III, Ch. 6.
16. Kegel B. Physical fitness, sports and recreation for those with lower limb amputation or impairment. *Journal of Rehabilitation Research and Development* 1985; Clinical Supplement 1: 1–125.
17. Labanowich S, Karmon P, Veal LE, Wiley BD. The principles and foundations for the organization of wheelchair sports. *Sports 'n' Spokes* 1984; **9**: 26–32.
18. Curtis LA, Dillon DA. Survey of wheelchair athletic injuries; common patterns and prevention. *Paraplegia* 1985; **23**: 170–5.
19. Nilsen R, Nygaard P, Bjorholt PG. Complications that may occur in those with spinal cord injuries who participate in sports. *Paraplegia* 1985; **23**: 152–8.
20. Glaser RM. Exercise and locomotion for the spinal cord injured. In: Tergung, RL, ed. *Exercise and sports sciences reviews.* New York: MacMillan, 1985: 263–303.
21. Glaser RM, Sawka MN, Brune RF, Wilde SW. Physiological responses to maximal effort wheelchair and arm crank ergometry. *Journal of Applied Physiology, Respiratory, Exercise and Environmental Physiology* 1980; **48**: 1060–4.
22. Davis R, Ferrara M, Byrnes D. The competitive wheelchair stroke. *National Strength and Conditioning Association Journal;* **10**: 4–10.
23. Higgs C. Propulsion of racing wheelchairs. In: Sherrill G, ed. *Sport and Disabled Athletes.* Champaign, IL: Human Kinetics Publishers, 1986: 165–72.
24. Curtis KA. Wheelchair sports medicine. Part 3: Stretching routines. *Sports 'n' Spokes* 1981; **7**: 16–18.
25. Walsh CM, Hoy DJ, Holland LJ. *Get fit. Flexibility exercises for the wheelchair user.* Edmonton: Research and Training Centre for the Physically Disabled, University of Alberta, 1982.
26. Walsh CM, Steadward RD. *Get fit. Muscular fitness exercises for the wheelchair user.* Edmonton: Research and Training Centre for the Physically Disabled, University of Alberta, 1984.
27. Skuldt A. Exercise limitations for quadriplegics. *Sports 'n' Spokes* 1984; **10**: 19–20.
28. Botvin Madorsky JG, Curtis KA. Wheelchair sports medicine. *American Journal of Sports Medicine* 1984; **12**: 128–32.
29. Bayley JC, Cochran TP, Sledge CB. The weight-bearing shoulder. The

impingement syndrome in paraplegics. *Journal of Bone and Joint Surgery* 1987; **69A**: 676–8.

30. Schaffer RS, Proffer DS. Sports medicine for wheelchair athletes. *American Family Physician* 1989; **39**: 239–45.

31. Corcoran PJ *et al.* Sports medicine and the physiology of wheelchair marathon racing. *Orthopaedic Clinics of North America* 1980; **11**: 697–716.

32. Neer CS. *Shoulder reconstruction*. Philadelphia: WB Saunders, 1990.

33. Aljure J, Eltoria I, Bradley WE, Lin JE, Johnson B. Carpal tunnel syndrome in paraplegic patients. *Paraplegia* 1985; **23**: 182–6.

34. Stotts KM. Health maintenance: paraplegic athletes and nonathletes. *Archives of Physical Medicine and Rehabilitation* 1986; **67**: 109–14.

35. Curtis KA, Hall KM. Health, vocational and functional status in spinal cord injured athletes and nonathletes. *Archives of Physical Medicine and Rehabilitation* 1986; **67**: 862–5.

36. Donovan WH. Spinal cord injury. In: Stolov WC, Clowers MR, eds. *Handbook of severe disability*. Washington, DC: US Government Printing Office, 1981.

37. Young JS, Burns PE. Pressure sores and spinal cord injured: part II. *Model Systems' Spinal Cord Injury Digest* 1981; **3**: 11–26.

38. Cooper DM, Watt RC, Alterescu V. *Guide to wound care*. Chicago: Hollister, 1983: 59–60.

39. Merbitz CT, King RB, Bleiberg J, Grip JC. Wheelchair push-ups: measuring pressure relief frequency. *Archives of Physical Medicine and Rehabilitation* 1985; **66**: 433–8.

40. Garder SL, Krouskop TA. Wheelchair cushion modification and its effect on pressure. *Archives of Physical Medicine and Rehabilitation* 1984; **65**: 579–83.

41. Seymour RJ, Lacefield WE. Wheelchair cushion effect on pressure and skin temperature. *Archives of Physical Medicine and Rehabilitation* 1985; **66**: 103–8.

42. De Lisa JA, Mikulic MA. Pressure ulcers. *Postgraduate Medicine* 1985; **77**: 182–220.

43. Seilor WO, Stahelim HB. Decubitus ulcers: treatment through five therapeutic principles. *Geriatrics* 1985: **40**: 30–44.

44. Johnson B, Thompson R, Pallares V, Sadore MH. Autonomic hyper-reflexia: a review. *Military Medicine* 1975; **140**: 345–9.

45. Erickson RP. Autonomic hyperreflexia: pathophysiology and medical management. *Archives of Physical Medicine and Rehabilitation* 1980; **61**: 431–40.

46. Sauka MN, Latzka WA, Pandolf KB. Temperature regulation during upper body exercise: able-bodied and spinal cord injured. *Medicine and Science in Sports and Exercise* 1989; **21**: 132–40.

47. Normell LA. Distribution of impaired cutaneous vasomotor and sudomotor function in paraplegic man. *Scandinavian Journal of Clinical Laboratory Investigation* 1974: **33** (Suppl. 138): 25–41.

48. Randall WC, Wurster RD, Lewin RJ. Responses of patients with high spinal transection to high ambient temperature. *Journal of Applied Physiology* 1966; **21**: 985–93.

49. Huckaba CE *et al.* Sweating responses of normal, paraplegic, and anhidrotic subjects. *Archives of Physical Medicine and Rehabilitation* 1976; **57**: 268–74.

6.5   The child and adolescent

LYLE J. MICHELI

INTRODUCTION

Physical activity during childhood and adolescence is an accepted component of normal growth and development in all cultures. Traditionally, physical activity has been incorporated into free play and game activities, including sports. In addition, physical labour performed by children, often in the setting of the family home or work-place, is also widely practised in many cultures. However, the participation of the growing child and adolescent in organized sports activities, with set regimens of training, competition, and teams, is a relatively recent phenomenon.[1]

In North America, the first organized sport aimed directly at the child was Little League baseball, which had a short history prior to the Second World War but was then recognized and spread rapidly after its reinstitution in 1947. At the present time, it is estimated that 1.2 million children are participating in organized Little League baseball in the United States. Thus the growth of this sport and many other similar programmes has reached astonishing proportions.

The growing interest in Olympic sports competitions has also fuelled participation by children in organized sports programmes. In many countries, promising prospects for elite level competition are identified early and then are provided with a special programme of training and competition. This may include enrolment in special sports academies or access to seasonal supervised sports programmes at centralized facilities. In nations such as the former German Democratic Republic, the former Soviet Union, and Romania, early recognition of promising children and then careful training and supervision of their progress in certain sports was felt to be essential to promotion of elite athletic programmes in a great variety of sports and competitions. The development of organized sports for children and adolescents has occurred in three different ways.

In some countries, the local, regional, and central development of organized sports has been on a sport by sport basis. This is typical of North America and the European nations where adult sports, such as rugby football, association football, field hockey, netball, athletics, etc., have developed children's programmes as an adjunct to their senior or adult programmes. Thus youth rugby in the United Kingdom is the purview of the regional and national rugby associations.

A second organizational structure in common use is that of an overall national governing body which attempts to supervise and promote development of all sports in a systematic fashion, with similar approaches to changes of rules, field dimensions, equipment, and coaching requirements for each of the youth sports. This model has gained acceptance in Australia, New Zealand, and, to some extent, Canada.

The third organizational structure widely used is that of a centralized system of candidate selection and supervision with the early identification of athletic prospects and their invitation to participate in structure programmes at central facilities. This has been most widely used in countries such as the former German Democratic Republic, the former Soviet Union, Romania, and, to a lesser extent, nations such as the United States and Canada.

Needless to say, the aims of these various structured sports programmes vary. Some have as their obvious purpose the development of safe and enjoyable sports participation by all children, or at least the majority of children, in these societies. Others have a more specific objective: the early identification and development of the skills of elite young athletes with an aim towards international or Olympic competition. In some nations there have been attempts to develop both types of programmes. Confusion often arises as to the relative objectives and priorities of these two very different approaches to children's sports.

While many of these organized sports training programmes for children were initially seen as adjuncts to normal growth and development within the more general aspects of physical activity indulged in by children, there is certainly a growing trend for many children to obtain most of their physical activity through organized sports programmes.[2] This trend appears to be accelerated by certain social and cultural trends that have resulted in a decreased opportunity for free play and unorganized sports participation by children. As a result, many children would undertake little if any regular physical activity except for their participation in organized sports activities. Critics of the growth of organized sports participation by children cite concerns about both the physical and psychological impacts of these programmes. As will be discussed subsequently, a great variety of new injuries are being encountered as a result of these sports activities and training programmes.[3] In addition, there is evidence that some children may sustain unacceptable levels of psychological stress in the organized sports setting.[4] Despite these concerns, the growth of organized sports for children appears to be increasing throughout the world. Attempting to reverse this trend would be neither successful nor realistic. It would appear much more appropriate to study the structure of competition and training in these sports to ensure that they are providing proper, safe, and enjoyable exercise for children.

It is essential when discussing the child and adolescent participating in organized sports to be aware of the unique characteristics of this age group. Growth is the basic physiological process of the child and adolescent. Musculoskeletal growth, in particular, is central to an understanding of the sports-active child.

PHYSIOLOGY OF MUSCULOSKELETAL GROWTH

It is imperative for the physician dealing with sports-active children or adolescents to be very familiar with the time course and sequential development of the musculoskeletal system. In the past, there has been an oversimplified approach to sports medicine and sports traumatology. In this scheme, a division has been made between sports injuries in the child and sports injuries in the adult. In practice the differential patterns of injury and response to injury are much more complex than this. It is much more useful for the clinician dealing with sports-active children to divide patterns of development and associated patterns of injury into four phases in this age group: (1) prepubescence; (2) early pubescence; (3) mid-pubescence; (4) late pubescence.

The primary sites of growth in the musculoskeletal system are in the bones, with growth secondarily occurring in the muscle–tendon units, nerves, and ligaments spanning these bony structures. Growth tissue in the child is present at three anatomical sites: the physis (growth plate), the articular surface of the joints, and the apophyses, which are the sites of major muscle–tendon insertion. Each of these sites undergoes an endochondral process

in which cartilaginous cells are formed from germinal layers, increase in size, and then are invaded by bone-forming tissue. In addition to these sites of endochondral ossification, both the endosteum and periosteum in the child have special metabolic qualities which differ from those of the adult. They are also responsible for sequential growth in the metaphysis and diaphysis of the long bones in particular and have special characteristics which differ from those of the adult.[5]

The growth plate, or physis, is the primary site of longitudinal growth in the human. The physes throughout the skeleton have a similar structure. This consists of longitudinal columns of cells that can be divided functionally into four zones: growth, maturation, transformation, where ossification takes place, and the zone of remodelling, where the structure of the bony layers becomes more complex and more mechanically stable.

It is important to remember that this same pattern occurs on the joints surface of the growing child. It is both biologically and biomechanically different from the articular cartilage of the adult. This accounts for the different patterns of injury which are seen at this site.

A great variety of factors can affect normal growth and development of a given physeal plate or extremity. These include both intrinsic and extrinsic factors. Endogenous hormones, which include thyroxine, growth hormone, sulphation factor, and testosterone, appear capable of stimulating skeletal growth. Oestrogen has also been demonstrated to have an effect on physeal growth, although this may be dosage related and it may in fact be inhibitory. Similarly, excessive dosages of testosterone may have an inhibitory effect upon longitudinal growth.

Mechanical factors, including both intrinsic tension due to muscle force across joints and intrinsic factors, such as repetitively applied force, may also affect growth rate.[6]

In addition, vascular changes, direct traumatic changes to the metaphysis or epiphysis itself, and genetic differences in stimulation may affect growth. Neurological factors may also play a role in the relative rate of growth. These are not fully understood, but leg length discrepancies are frequently seen in certain neurological disorders.

Each individual will demonstrate a characteristic pattern of growth rate which is, to some extent, unique. However, each member of the species goes through the same sequence of growth change. The relative stage of the individual in this growth sequence is referred to as 'skeletal age' and it may differ significantly from chronological age. Once again, this becomes an important factor in the occurrence of injury in so far as skeletal age, which corresponds closely to maturation of a wide variety of tissues including the neurological, musculoskeletal, and osseous tissues, may play a role in the occurrence and patterns of injury.

There is a progressive decline in the rate of growth in each individual following birth. The growth rate is quite high in the first year of life and then progressively decreases until it reaches a relatively low rate of no more than 5 or 6 cm a year through the prepubescent quiescent period of approximately age 6 to 10 in boys and 5 to 9 in girls. This prepubescent period corresponds approximately to stages I and II in the Tanner classification; the first signs of genital development and secondary sexual characteristics which reflect changes in the hormonal pattern in the child are seen in stage II. The stage of early pubescence corresponds to stage III of the Tanner classification and occurs at the time when the growth rate begins to accelerate in the child and secondary sexual characteristics begin to become further delineated. The third developmental stage which is important for the clinician is the stage of midpubescence which corresponds to stages III and IV and corresponds to the time of peak height velocity in the child. The final stage of development, late pubescence, corresponds to the time when the growth rate again begins to decline and ultimately the child goes on to physiological epiphyseodesis, in which the physeal plates close, the joint surface articular cartilage takes on the characteristics of adult articular cartilage, and the apophyses also close.

Needless to say, these developmental changes are a continuum, but they occur sequentially in each individual based upon development of the secondary ossification centres of the long bones and, later, the age of physeal closure of the major long bones. As examples, complete closure of the distal femoral epiphysis or proximal tibial epiphysis may not occur until age 16 to 19 years, but the majority of growth in this physes is already completed by skeletal age 15 in the male and 13.5 in the female.

For a better understanding of injury patterns, it is also useful to know the relative contributions of individual growth regions to overall lengths of individual bones. For example, the proximal humeral epiphysis is responsible for 80 per cent of the growth of the humerus, whereas the distal humeral epiphysis is responsible for the remaining 20 per cent. In addition, the distal radial epiphysis is responsible for the growth of 75 to 80 per cent of the radius, while the proximal radial epiphysis is responsible for 20 to 25 per cent. In the lower extremities, the proximal femoral epiphysis is responsible for the growth of 30 per cent and the distal femoral epiphysis is responsible for 70 per cent. In the tibia and fibula, however, the proximal epiphysis is responsible for 55 per cent of the growth and distal epiphysis for 45 per cent. In the case of selected injury to one of the pair of epiphyses, such as at the wrist where there may be selective injury to the distal radial epiphysis but little, if any, injury to the distal ulnar epiphysis, relative ulnar overgrowth or positive ulnar variance may occur. This is particularly important in the young gymnast's injury pattern.

There is growing evidence that delayed onset of menarche in young female athletes, whether due to nutritional or intrinsic physiological factors, can cause significant delay in skeletal maturation and a significant discrepancy between chronological age and developmental age.[7] Thus, a 16-year-old gymnast may have a skeletal maturation level of 11.5 to 12 years and therefore may be subject to injuries characteristic of the prepubescent or early pubescent individual.

The practical application for the clinician relates to the differing patterns of injury seen at different stages of development. In the growing child, the stage of relative maturation may affect location of tissue disruption and the pattern of injury. For example, a fall on the outstretched hand in the prepubescent child (stage I or II in the Tanner classification) usually results in a fracture at the junction of the diaphysis and metaphysis of the radius and ulna distally. In the midpubescent adolescent, a similar fall would result in a fracture through the distal radial physis and ulnar styloid. During late pubescence, the result might be a fracture of the metaphysis of the radius and ulna or a fracture of the carpal scaphoid. Another important example of the way in which the stage of maturation and development affects patterns of injuries can be seen in knee injuries in children and adolescents. In the prepubescent child, the physis and its attachment site to adjacent hard bone—the zone of Ranvier—may be stronger than the ligamentous structures of the contiguous joints. In this stage of development a macrotrauma injury, such as a lateral blow to the knee, may not result in a growth plate injury, but actually in injuries to the collateral or cruciate ligaments of the knee. This high incidence of unsuspected liga-

mentous injuries of the knee in the prepubescent child is receiving increased attention.[8]

The same mechanism of injury in the first two stages of pubescence may result in a growth plate injury. Alexander,[9] in animal studies, and Bailey et al.,[10] in a recent study of wrist injuries in children and adolescents, have demonstrated a decreased strength of the physeal tissue during the adolescent growth spurt or period of peak height velocity. Finally, in late pubescence ligamentous injuries once again predominate, with the physis being relatively stronger than the contiguous ligaments.

While these examples demonstrate the maturation stage specificity of macrotraumatic injuries in the growing child, there is now a growing body of evidence that repetitive microtrauma may result in similar variable secondary injury. Many of these data are coming from studies of young gymnasts, particularly girls where repetitive impact from landing or dismounts appears to be responsible for partial arrest of the radial epiphysis, with little if any effect upon the growth of the ulnar epiphysis.[11] The result is a relative overgrowth of the ulnar with positive ulnar variance. A number of recent case studies suggest that if this occurs earlier, during the stage of early pubescence, simple arrest will result in a restoration of normal growth.[12] Occurrence in midpubescence may result in permanent loss of growth and inability to correct the deviation of these contiguous structures.

Patterns of musculotendinous injuries may also be related to stages of relative growth in the child or adolescent. Clinical observations have suggested that the incidence of musculotendinous injury appears to increase the time of peak height velocity, particularly in relatively musculoskeletally tight individuals who often appear to be in a subgroup with enhanced motor skills. During this time of early and midpubescence, a relative decrease in the intensity of training and an increase in the proportion of training devoted to stretching would appear to be a significant factor in reducing potential injury.[1]

TYPES OF INJURY

Injury is a major concern to any athlete. In the sports setting, injury can occur from two basic but different mechanisms or sometimes a combination of the two. Injuries which result from macrotrauma are the more familiar acute injuries of sports. These are the torn ligaments, strained muscle–tendon units, or even fractured bones which occur from a single exposure to a major force. Sports-active children and adolescents are certainly subject to such injuries also; however, since the tissue of the child is different from that of the adult, the patterns of injury are actually quite different. The growing bone is more porous and more plastic than adult bone. This bone can actually undergo a significant amount of deformation before frank fracture occurs. If a fracture does occur, the healing callus may actually serve as a stimulus to bone growth and result in problems of overgrowth or angular deformity. However, non-unions of fractures are quite rare.[13]

Injuries to the physis or growth plate of the child have often received much attention in discussions of children's sports injuries (Fig. 1). In practice, physeal injuries are relatively uncommon. In addition, the prepubescent child actually has a very strong physis, and in this age group ligament injuries may occur while the physis remains intact. At the time of the adolescent growth spurt, however, physeal injury and separation may indeed occur, in so far as the ligaments may be stronger than the physis during this period of vulnerability. Very careful radiographs and very careful attention to physical examination are necessary at this time to ensure that a

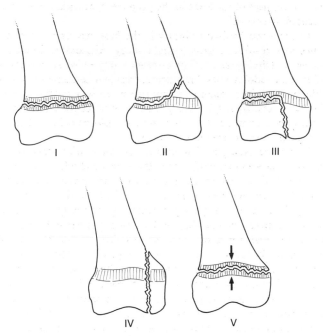

Fig. 1 Salter–Harris classification of epiphyseal injuries.

physeal injury has not occurred. Similarly, macrotraumatic injury to the muscle–tendon units may result in contusions or strains in the prepubescent similar to those in the adult, but a similar level of force from intrinsic contracture of the muscle or extrinsic elongation may result in an avulsion of the apophysis in the adolescent (Fig. 2).

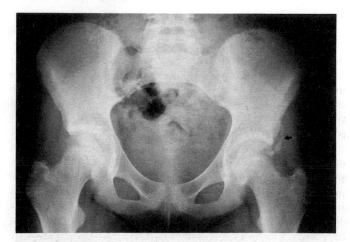

Fig. 2 Avulsion of the rectus femoris apophysis in a young soccer player.

The second mechanism of injury which is particularly pertinent to athletic training and participation is that of repetitive microtrauma. Injuries which result from repetitive microtrauma are often classed as overuse or fatigue injuries. The occurrence of overuse injuries in children is very specific to repetitive sports training.[3] Prior to the advent of organized sports training and competition for children, overuse injuries in this age group were extremely uncommon, as they rarely occurred in the free play situation. These overuse injuries—tendinitis of muscle–tendon units, stress fractures of bone, 'chondromalacia' of articular cartilage, and bursitis or fascitis—all of which are commonly seen in adults involved in repetitive training, are now also seen in children. In addition, there is growing evidence that the special tissue of children—the growth tissue—may also be subject to special patterns of overuse injury. Repetitive microtrauma applied to the physis of the growing child may actually result in injury and decreased rate of growth at the injured physis. If this condition is not recognized and training continues, a permanent loss of length may occur. This has now been widely reported at the wrist in sports such as gymnastics,[11] and certainly raises concern about repetitive impacting of the growth plates of the lower extremity in activities such as distance running or jumping. Similarly, repetitive microtrauma applied to the growing articular cartilage of the child, which is quite different biomechanically from the articular cartilage of the adult, appears once again to have a different pattern of injury. Osteochondritis dissecans appears to be the injury which results from repetitive microtrauma to the softer and more plastic articular cartilage of the child. This growing articular cartilage, which differs biomechanically from the adult articular cartilage, transmits the forces applied to it to the subjacent subchondral bone, which results in fracture and death of this tissue. Chondromalacia, or frank fissuring destruction of the articular cartilage, has not been encountered by the author in the growing child. It may be seen at the later stages of adolescence, but even then it is quite rare and appears to be a specific pattern of tissue injury of the articular cartilage of the adult and not of the growing child.[14]

The third site of growth cartilage in the child, the apophysis, also appears to have a specific pattern of response to repetitive microtrauma. The result is a series of small avulsions at this weaker bone–cartilage junction and repetitive rehealing. The class of injuries seen at these sites are called the apophysitises, and in practice represent microfractures with rehealing and not really an inflammatory process at all but rather a fracture process. Thus Osgood–Schlatter's disease at the knee, Sever's disease at the heel, accessory navicular disease at the foot, and iliac crest apophysitis at the pelvis are examples of this special type of overuse injury seen in the young athlete.[15] Tendinitis, a common overuse injury in the adult, may be encountered in the child, but is relatively rare.[16]

RISK FACTORS FOR INJURY

The clinician dealing with the sports-active child has a primary responsibility not only for the proper diagnosis and initial treatment of sports injuries, but also for the prevention and rehabilitation of these injuries. A clear understanding of the factors which contribute to the occurrence of injury is essential to the proper commissioning of all three of these responsibilities. In particular, the factors which contribute to the occurrence of injury, either macrotrauma injury or overuse injury, are particularly pertinent to its prevention. While this should be a concern in all athletic participation, it would appear to be particularly pertinent to the child athlete, in so far as prevention of injuries in this age range are essential to the continued performance of healthy exercise by the child and the prevention of tissue injury which may be very detrimental to the life-style of the adult with or without subsequent athletic participation.

Factors which appear to contribute to the occurrence of macrotrauma in this age group include coaching expertise, quality of athletic facilities and athletic equipment, including protective equipment, and careful matching of participants by size, weight, and level of maturation. Steps which can be taken to facilitate this level of prevention include proper training and education of coaches. Mandatory education or certification of coaches for children's sports may also be necessary. Careful attention should be paid to the sports equipment used by children, ensuring that they are not using adult-size implements, in particular, or second-hand items. Thirdly, participants should be matched by size and maturity level. For example in American football leagues in the United States, in which children participate, different league levels are based on chronological age but also on body weight. This would appear to be particularly appropriate in contact sports such as ice hockey and rugby football.

OVERUSE INJURY

A number of host or environmental factors have been identified which may contribute to the occurrence of overuse injury in the growing child or adolescent. A detailed assessment of the history of injury occurrence, combined with careful physical examination and, in some instances, imaging techniques, can help to elucidate the factors which have contributed to the occurrence of overuse injury. Needless to say, this determination can be beneficial not only in the treatment of such injuries, but also in their prevention. An overuse injury in a child may often be the result of a combination of factors, no one of which can be determined to be solely responsible, which result in tissue injury. (Table 1) (Fig. 3).

Table 1 Risk factors for overuse injury

Training error
Muscle–tendon imbalance
Anatomical malalignment
Footwear
Playing surface
Associated disease state
Nutritional factors
Cultural deconditioning

Reproduced from ref. 3, with permission.

Training errors

Levels of training and, in particular, increases in its volume or intensity appear to be major contributors to many overuse injuries in children. Determining the level of specific sports training for a given child is essential not only to prevent injury but also to enhance performance.[17] In addition, this is a central question of all children's exercise. How much exercise is enough for normal growth and development in a child? Conversely, how much exercise, particularly of a very sports-specific type, is too much for the child and may result in deterioration of performance or tissue injury? The minimal amount of physical activity and exercise which is necessary for normal growth and development in the child has not yet been determined with certainty. Some observers have suggested 25 min of sustained exercise, three to four times a week, as a minimum for healthy children. This would appear to be a very low value indeed. Observations of spontaneous activity by children have suggested that the average child requires at least an hour of physical activity and motion each day for good health.

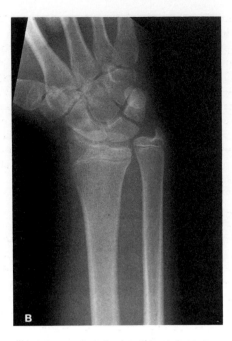

Fig. 3 Repetitive injury to the distal radial epiphysis in a young gymnast, resulting in relative overgrowth of the ulna.

At the other end of the spectrum, one of the great challenges to the sports physiologist and clinician is to determine the upper limit on athletic participation by children. Several recent observations and studies have provided some evidence in this regard. Observation of young female gymnasts suggested that young women training more than 16 h a week in gymnastics had a dramatic increase in occurrence and severity of injury to the lumbar spine.[18] A more conservative approach was suggested by a study at Tokashima University, Japan. A review of low back injuries in over 600 children seen in their sports clinic suggested that children in the growing age should participate in extension–flexion sports for no more than 1.5 h per day, 3 days per week.[19] Similarly, a study of elbow injuries resulting from overhand throwing in baseball reported from the same university recommended that the growing child involved in baseball should not perform overhand pitching more than 50 times daily or 350 times per week.[20]

At present, we know all too little about the dose–response curve of the growing tissues to repetitive sports training. However, these are the essential questions of sports participation by children. How much is enough? How much is too much? Future research into this question will only be possible with careful collection of epidemiological data and injury surveillance techniques, and must be a high priority in any nation where sports participation by children and adolescents is increasing.[21]

Muscle–tendon imbalance

Changes in the balance of strength, flexibility, and endurance between muscle groups spanning joints and within certain intrinsic components of muscle groups appears to be a major contributor to the occurrence of injury in this age group. In particular, and as noted above, changes in the range of motion of the joints associated with the adolescent growth spurt, which is a time when many young athletes are at a high level of intense training, may be a further contributor towards injury or derangement of the muscle–tendon units.[1]

A four-part progression has been suggested for the development of muscle–tendon imbalances in young athletes: growth decreases flexibility (particularly during growth spurts); strength increases with growth in both girls and boys; however, this increased strength may not be uniform, contributing to imbalances about the joints; overstress of particular muscle–tendon units with repetitive training may cause subtle neurovascular injury and further contribute to the imbalance of the muscles about a strained joint.

Anatomical malalignment

Abnormal alignments of the low back and lower extremities, in particular, may increase the risk of overuse injury. Excessive lumbar lordosis appears to be a risk factor for lumbar spine injury. Excessive femoral anteversion, often in combination with external tibial torsion, genu valgum, and pronation of the lower extremities, may result in a lower extremity which is at a disadvantage for repetitive running activity. Since running is common to a wide variety of sports, including not only distance running but also field sports, this may increase the chance of overuse injuries, particularly about the knee.

Some anatomical malalignments may be compensated by specific exercises to reduce their impact by increasing the range of motion or strength about adjacent joints; others may be compensated by orthotic devices in the shoes or at the knees which may alter the force transmission or biomechanics of force transmission.

Footwear

In sports involving impact delivery to the lower extremity, such as running or field sports, properly constructed, designed, and fitted shoes appear to be important in dissipation of force and enhancement of function. In any sport involving lower-extremity activity, footwear characteristics should include attention to impact absorption, mechanical support, and ability to compensate for changes in alignment of the specific athlete.

Playing surface

The relative hardness of the playing surface may be an additional risk factor for injury. Children moving their training from turf or grass to a synthetic or concrete surface often note an increased level of pain. The origin of many stress fractures has as one of its aetiological factors changes in surface compositions. Of course, these factors can be elicited in a carefully recorded history.

Associated disease state

A careful history to determine pre-existent illness or injury will help to determine any additional health factors which have contributed to the occurrence of injury. Musculoskeletal injury may be increased in the child suffering from chronic infections or chronic debilitating states or illnesses and should be carefully assessed in addition to the biomechanics of their extremities.

Nutritional and hormonal factors

The child or adolescent participating in sport must have adequate nutritional intake for the increased demands made by the participation. Sports training and competition should never be allowed to compete with growth for nutritional intake, including very basically overall caloric intake. The caloric demands of sports training should be calculated in order to supplement, if necessary, the calories required for growth and development at the stage the child is in. In addition, recent investigations have revealed deficient intake of calcium and vitamins in amenorrhoeic distance runners and ballet dancers.

A condition of special concern in young female athletes who are training very hard is the development of amenorrhoea. Controlled studies have shown significantly lower levels of bone mineral density in amenorrhoeic versus amenorrhoeic athletes matched by age, weight, sport, and training regimen. Decreases in bone mineral density increase the susceptibility of the amenorrhoeic athlete to overuse injury in general and to stress fractures in particular.

Cultural deconditioning

The child who lacks the opportunity for general exercise and activity may sustain a subtle but definite deconditioning of their overall musculoskeletal structure and composition. Of course, this is a product of the more developed countries where increased leisure and, in particular, excessive amounts of sedentary activity, such as watching television, being driven rather than walking, and the use of video games, have eroded the general physical activity of children. A child who is being systematically deprived of physical activity with the resultant weakening of tissues and who is then exposed to an intense period of athletic training of a very specialized nature may actually have an increased chance of injury. Data from four national surveys in the United States indicate a pronounced increase in the prevalence of paediatric and adolescent obesity.[22] These increased levels of obesity are felt to reflect a declining level of physical activity in these children in so far as nutritional intake has remained relatively constant during the period of observation. Coaches and athletic trainers who are dealing with physically deconditioned children must take this into account when designing their specific sports training programmes. Unfortunately, in addition to teaching children the specifics of a given sport activity, coaches at this level may now have the added responsibility for improving the basic fitness of the children prior to the institution of specific sports fitness training activities.

Growth

As noted above, the growth process can affect the relative function, strength, and balance of the musculoskeletal system at every stage of the growth and development of the child. This can affect the capacity of the child for training levels and performance, as well as having an impact on the potential for injury, particularly during periods of rapid growth such as the adolescent growth spurt. An understanding of this 'growth factor' should contribute not only to enhancement of performance but also to prevention of unnecessary injury of both the macrotrauma and microtrauma type.[1]

PREVENTION

Careful and accurate diagnosis of athletic injuries and appropriate treatment and rehabilitation is extremely important in any well-designed sports programme. In addition, and in particular with respect to the young athlete, prevention of sports injury must have absolutely the first priority in our youth sports programmes. Unfortunately, this is rarely the case. Children are often exposed to sports programmes which are modelled in a general fashion on those of adults. In addition, very little attention is paid to the psychological differences between the child and the adult with regard to attention span, sense of self-worth, and exercise compliance when developing youth sports programmes. Identification of the factors, both host and environmental, that appear to have contributed to the occurrence of sports injury is essential for proper treatment and prevention of occurrence of the same or similar injuries, sometimes of greater severity.

The preparticipation assessment of the young athlete constitutes one of the few occasions when the physician may actually prevent injuries. This examination should include not only a careful history and physical examination and assessment of the child's general health, but a determination of whether the child is fit to participate in a particular sport and at what level of intensity of training. Careful attention should be paid to anatomical and musculoskeletal factors which may increase the risk of injury in sport, and exercise techniques, training alterations, or special sports equipment which may alter these factors and decrease the chance of injury should be instituted.[23]

The relative levels of strength and flexibility of these young athletes, who are quite susceptible to variability in these factors, should be determined. Physicians responsible for young athletes should be prepared to provide very specific training regimens and advice to the athletes and their coaches regarding flexibility and strength training techniques. These must include not only a general recommendation for strengthening and flexibility, but very specific details as to how these exercises are to be performed and how many sets and repetitions are to be instituted in order to maximize the strength and flexibility of the child and decrease injury.

An apparent guideline for training progression which has proved very useful in distance running is the '10 per cent per week rule'. This common-sense rule suggests that the amount of training should be increased by no more than 10 per cent per week. For example, a young runner running for 20 min, four times per week, can probably safely run for 22 min, four times per week, in the following week. Progression of the rate of intensity of training by more than 10 per cent per week invites a chance of tissue overuse and resultant injury.

At the present time, the maximum and minimum levels of safe athletic training and physical activity for children are still being determined. Until we have very clear guidelines for exercise prescription in this age group, the physicians dealing with these young athletes should be very aware of the relative risk factors for sports injury and attentive to prevention of injuries in this vulnerable age group.

REFERENCES

1. Micheli LJ. Overuse injuries in children's sports: the growth factor. *Orthopedic Clinics of North America* 1983; **14**: 337–60.

2. Backx FJG, Hein JMB, Bol E, Erick WBM. Injuries in high-risk persons and high-risk sports. A longitudinal study of 1818 school children. *American Journal of Sports Medicine* 1991; **19**: 124–30.

3. O'Neill DB, Micheli LJ. Overuse injuries in the young athlete. *Clinics in Sports Medicine* 1988; **7**: 591–610.

4. Pillemer FG, Micheli LJ. Psychological considerations in youth sports. *Clinics in Sports Medicine* 1988; **7**: 679–89.

5. Rallison ML. *Growth disorders in infants, children, and adolescents.* Philadelphia: Wiley, 1986.

6. Larson RL, McMahon RO. The epiphysis and the childhood athlete. *Journal of the American Medical Association* 1966; **196**: 607–12.

7. Malina RM, Harper AB, Avent HH, Campbell DE. *Medicine and Science in Sports and Exercise* 1973; **5**: 11–17.

8. DeLee JC. ACL deficiencies in children. In: Feagin JA Jr, ed. *The cruciate ligaments.* New York: Churchill Livingstone, 1988: 439–47.

9. Alexander CJ. Effect of growth rate on the strength of growth plate shaft junction. *Skeletal Radiology* 1976; **1**: 76–80.

10. Bailey DA, Wedge JH, McCulloch RG, Martin AD, Berghardson SC. Epidemiology of fractures of the distal end of the radius in children as associated with growth. *Journal of Bone and Joint Surgery* 1989; **71A**: 1225–31.

11. Roy S, Caine D, Singer KM. Stress changes of the distal radial epiphysis in young gymnasts. A report of twenty-one cases and a review of the literature. *American Journal of Sports Medicine* 1985; **13**: 301–8.

12. Hurley DL. Wrist pain—female elite gymnast. Physician case presentation at the Annual Meeting of the American College of Sports Medicine, Orlando, FL, May 1991.

13. Currey JD, Butler G. The mechanical properties of bone tissue in children. *Journal of Bone and Joint Surgery* 1975; **57A**: 810–19.

14. Bright RW, Burstein AH, Elmore SM. Epiphyseal plate cartilage: a biochemical and histological analysis of failure modes. *Journal of Bone and Joint Surgery* 1974; **56A**: 688.

15. Micheli LJ. The traction apophysitises. *Clinics in Sports Medicine* 1987; **6**: 389–404.

16. Micheli LJ, Fehlandt AF. Overuse injuries to tendons and apophysis in children and adults. *Clinics in Sports Medicine.* 1992, **11**: 713–26.

17. Micheli LJ, Jenkins MD. *Sportswise: An Essential Guide for Young Athletes, Parents, and Coaches.* Boston: Houghton Mifflin, 1990.

18. Goldstein JD, Berger PE, Windler GE, Jackson DW. Spine injuries in gymnasts and swimmers: an epidemiologic study. *American Journal of Sports Medicine.* 1991; **19**: 463–8.

19. Akimoto T, Kohno S. The etiology of spondylolysis with reference to athletic activities during the growing period. Paper presented at the American Orthopaedic Society for Sports Medicine/Japanese Orthopaedic Society for Sports Medicine Trans-Pacific Meeting, Kauai, HI, January 1991.

20. Iwase T, Ikata T, Kashiwaguchi S. Elbow osteochondral injuries of young baseball players. Paper presented at the American Orthopaedic Society for Sports Medicine/Japanese Orthopaedic Society for Sports Medicine Trans-Pacific Meeting, Kauai, HI, January 1991.

21. Micheli LJ, Slater JA, Woods E, Gerbino PG. Patella alta and the adolescent growth spurt. *Clinical Orthopaedics* 1986; **213**: 159–62.

22. US Department of Health and Human Services, Office of Disease Prevention and Health Promotion. *Summary of findings from national children and youth fitness study.* Washington, DC: US Government Printing Office.

23. American Academy of Pediatrics, Committee on Sports Medicine and Fitness. Recommendations for participation in competitive sports. *Pediatrics* 1988; **81**: 737–9.

Special problems in sports injuries $\quad$ 7

7.1 Introduction

WILLIAM D. STANISH

Over the years it has become obvious that there are areas of special concern in sport medicine. These include emergencies of the musculoskeletal system, maxillofacial injuries, pulmonary and abdominal injuries, injuries to the spine and head, and injuries to the wrist and carpus.

Historically fractures about the face were treated with a 'tincture of time' which would frequently result in considerable residuum. A malunion of the mandible, for example, would commonly result in malocclusion of the jaw triggering degeneration of the temporomandibular joint. Diplopia would occur if, and when, the zygomatic arch was fractured and not totally realigned. Current oral medicine/surgery emphasizes the importance of intricate appraisal and definitive management of fascial injuries. Furthermore, considerable attention has been directed towards preventative measures. For example, protective mouth guards have been fundamental in decreasing injuries to the teeth, jaw, and brain. The ability of the semirigid mouth orthotic to decrease the effects of impact during sport has been well documented. However, an orderly physical examination, combined with systematic radiographic analysis, still afford the best means for achieving a successful clinical outcome.

Head and spine trauma, incurred during athletics, can have devastating morbidity. Protective head gear, more stringent and regulations in contact sports such as ice hockey, and improved medical care have been vital to the reduced incidence of sport related head and spine trauma. For example, American football was notorious for coaching techniques which emphasized the use of the head/helmet as a battering ram. Rules have been introduced which outlaw this technique (spearing) have reduced trauma to the head and neck.

However, when an injury does occur, proper first aid care must be initiated. The on-field medical analysis of the injured athlete is very important. An improper technique for transporting the athlete with a neck injury, for instance, may add considerably to the long-term morbidity for that participant. An unrecognized unstable fracture of the cervical spine could result in complete quadriplegia, if not handled in a proper fashion.

Somewhat more subtle, but equally concerning, are athletic injuries to the chest and abdomen. Blunt trauma to the thorax or flanks is commonly underestimated in terms of severity. Particularly troublesome, in terms of diagnosis and management, are visceral injuries in the adolescent athlete. For example, because the younger individual has enormous physiological reserve, a ruptured spleen as a consequence of a blunt trauma in sport, may remain indolent for hours until haemorrhagic shock suddenly occurs.

Historically, injuries to the hand and wrist have been treated with a relative cavalier attitude by most sport medicine personnel. However, an unrecognized scaphoid fracture and/or an unstable wrist will provoke functional difficulties in the long term. Osteoarthrosis of the wrist could be a career ending injury in a baseball player or cricketer. Diminished range of motion in the wrist can cripple the performance of a skilled ice hockey player. A contemporary approach to the analysis and treatment of injuries to the carpus is vital.

These special problems in sports medicine are evaluated in the following Section.

7.2 The Fallen Athlete: emergencies of the musculoskeletal system in sport

ROBERT M. BROCK

INTRODUCTION

Definition and scope

The title 'The Fallen Athlete' refers, with apologies, to Rodin's sculpture of the same name. I attended the 1989 World Figure Skating Championships in Paris, as a sports physician. At the Rodin Museum, I was moved by the amazing sensitivity with which Rodin depicted the pain, strain, and anxiety experienced by the injured athlete. The figure is real. The reality of sport is that it encompasses all the hopes and dreams of mankind. When sudden injury occurs, the sports physician will most certainly have a 'fallen athlete'.

Fortunately, the incidence of true emergency situations is low. Crisis situations are only emergencies if immediate intervention can alter the clinical outcome. The Oxford Dictionary defines an emergency as a '... sudden state of danger, requiring immediate action ...'[1] This is the reason for inclusion of these topics in this chapter. Practical remarks will focus on early recognition, transport, acute 'on-the-field' clinical assessment and 'sideline' treatment. Athletes, coaches, and parents always want to know when it will be possible to resume active participation in sport. Some injured athletes will be able to return to the same sport after appropriate management, but others will only be able to return to other sports.

Approach to the acutely injured athlete

It is easy to be distracted by the gross deformity of a fracture, but it must be remembered that the trauma causing the limb injury may have damaged the airway or caused life-threatening head, chest, or abdominal injuries. Multiple injuries are expected in motorcycle racers. However, these risks are no less prevalent in the downhill skier who is travelling at 100 km/h, constantly on the edge of a sudden catastrophe. In orthopaedics we often become so focused on the problems of the extremities that we forget that there is a human body attached to the injured limb. All trauma courses teach the 'A,B,C,D,E,F' of trauma management. It should be noted that 'F' is listed last for a good reason.

Emergencies of the musculoskeletal system are often seen in conjunction with emergencies of other systems. In dealing with an injured athlete at the scene of the accident, listen to the athlete and identify the primary area of concern while assessing the 'A,B,C'. The history of the mechanism of injury and the degree and direction of forces are of great importance in determining the severity and the nature of the injury. Often, information from team mates or officials will be useful. Advance knowledge of any pre-existing conditions in the athlete is usually helpful.

After taking the history and assessing the A.B.C, a limited examination of the injured limb should be performed. Look for gross deformity and feel for crepitus. Note the gross function of the joint, including the range of motion and neuromuscular performance. Examine the pulses and capillary perfusion. Are there any other injuries to the same limb or to other limbs? Decisions about the form of splinting needed and the mode of transport are made only after adequate assessment.

The athlete's needs should take precedence over the needs of the competition. Amid the confusion of acute emergencies, several groups have their own priorities. Unfortunately, in many sporting activities the media, particularly television, influence the timing of events. They become very upset if there is a pause in the action. There are examples at various major competitions where the needs of the injured athlete may have been neglected by officials in order that the event could proceed. The primary concern of the sports physician should be to ensure that proper medical attention is given to the injured participant.

Preparation

Preparation for a disaster is necessary both before and during the event. Because the incidence of critical injuries is low, there is often general confusion when a crisis does occur. Be prepared! In fact, this may be the most important way of dealing with an emergency.

Before the event

One of the simplest parts of any disaster plan, which is often forgotten or underestimated, is the list of all the emergency telephone numbers. A copy must be provided beside all the telephones that are available at the site. Before the event, local hospitals and physicians should be notified about the chances of serious injuries presenting on their doorsteps. The availability of emergency transport systems differs. All this must be arranged in advance of the competition.

Another important aspect is a written and practised disaster plan flowsheet covering the sudden unexpected emergency. Every written plan has problems when put into action. At one event all the supplies, including oxygen, a fracture board, and a telephone to make the necessary calls, were collected in one room. When an emergency occurred and the supplies were required, no one could find the key to the room. Simple tasks can easily be overlooked if the plan is just written or, worse still, talked about and not actually practised. At a major event there are frequently many volunteers who have good medical skills and great interest because their children are involved in the competition, but no experience in actual sports medicine. Taking a skater off a busy practice rink is different from removing an injured diver from a pool. The injury may be the same, but the practical implementations of the disaster plan are very different.

Similarly, when travelling with a team outside its own health care system (country), it is very important that before the event the sports physician should familiarize himself or herself with local arrangements. We have been at world competitions where attending doctors were not practising clinicians. They were good administrators or parents of participants, but they had not treated a clinical patient for many years. Being forewarned is part of being prepared.

At the event

Being prepared for a competition means having the proper coverage. This includes personnel as well as equipment and communications systems.

Personnel

Sports medicine is a recognized area of medical expertise. In Canada many individuals have gone through an accreditation process[2] and are interested in covering competitions. A person with this or similar training should form the core of the medical staff. Emergency physicians are often physicians who are interested and adroit in the early management of acute injuries. Such people are excellent sources of physician coverage. Sport physiotherapists and athletic trainers are invaluable. Some jurisdictions have personnel who are specialists in the evacuation of injured people and can often contribute to the team. It should be remembered that medical care, particularly sports medicine care, is really a team approach.

Equipment

Each event has its own special needs. The equipment required for aquatic championships is different from that for a skiing event or a marathon. A knowledge of injury types and rates of occurrence will help to determine the personnel and equipment required. Obviously, emergency resuscitation equipment, splints, and emergency transportation must be provided. Minor trauma equipment may be helpful. In environmentally stressful situations such as marathons, hyperthermia and hypothermia are potential problems. Specific therapeutic modalities should be available at the site of competition. Equipment and personnel requirements are dictated by the specific events.

Communications

Communication is very important in any crisis situation and should be in both the written and oral form. Good communications systems are the cornerstone of any disaster plan.

Many events take place on sites that extend over a large area. Examples include road cycling and downhill skiing. It may be important to ensure that an injury sustained on any part of the course can easily be identified and reached. Internal communications by walkie-talkies may be necessary. The location of emergency telephones and emergency telephone numbers should be known. This alone could save a life.

A patient taken from the site of an event to a medical facility must be accompanied by written notes. Information about the general state of the patient and the status of the neurovascular assessment is of vital importance. Vital signs recorded on arrival at the institution have much more meaning when they can be compared with the situation at the time of injury. Good written communication is also necessary in our litigious society. If the athlete has sustained a significant injury which requires treatment, it is important to provide a written report for his or her own doctor.

The essential element of management of emergencies of the musculoskeletal system is to be prepared to deal with crisis when it arises, as it always does, unexpectedly. It is necessary to recognize the inherent risks in each sport. On approaching the fallen athlete, we need to use the clinical skills basic to all medicine. Taking a good history and performing the appropriate physical examination require practice enhanced by experience. Acute treatment and communication are skills to be practised.

In this chapter, acute injuries of the shoulder, elbow, and knee are discussed. Compartment syndromes are an additional cause of tragic outcomes in sports injuries which can be prevented. Frostbite is another preventable emergency condition which is usually seen in weekend athletes who are not as well trained or equipped as national or elite athletes. The best approach is to be as practical as possible with each crisis situation. As in most medicine, the approach to the problems is as important as achieving the actual diagnosis. (This has similarities with the approach of some judges of athletic performance who feel that style is more important than technique!) Recognition, on-the-field clinical assessment, transportation, sideline management, and prognosis is discussed in each section.

Most emergencies start suddenly with a shout 'player down'. Frantic officials, glaring lights, concerned coaches, distraught parents, and scrambling confusion follows. The sports physician needs to analyse the crisis objectively with clinical clarity and to bring order to the management of the musculoskeletal emergency.

SHOULDER

Recognition

The setting

The shoulder is the second most common site of musculoskeletal injury in both emergency departments and sports injury clinics. Acute shoulder injuries are particularly common in contact sports. The patient has had contact with either another player or the playing surface. The actual mechanism may not be witnessed by officials or spectators on the sideline. The injury usually involves a fall on an outstretched arm or collision with another player, or is the result of a 'pile-up' in either American football or ice hockey. Each sport has its own shoulder risk: a volleyball player 'digging' for a volley, a figure skater missing a landing from a triple axel jump, an ice hockey player receiving a check into the boards, or an American football player crunched during a tackle.

On-the-field clinical recognition

In all subsequent discussion, we shall assume that the athlete has only one injury at a time. However, it should never be forgotten that in a real situation multiple areas of injury often occur and that other areas of the patient (airway, bleeding, etc.) take precedence over the musculoskeletal areas.

The medical team usually has to deal with a player lying on the playing surface. Initial inability to move the arm and a great deal of pain are the common symptoms. Glenohumeral dislocation is the most common problem seen in the 15 to 30 age group. In younger patients fractures of the proximal humerus or clavicle are more common. Glenohumeral dislocation is uncommon in children under the age of 15.

The patient usually complains of pain in the shoulder area. At the outset, the patient should be asked about the presence of neck pain or dysfunction in the other extremities. He or she may complain of pain in the sternoclavicular area or down the arm. Questions about the distribution of the pain and the character of the pain are important. The ability to function is equally important. The advantage of initial assessment includes the ability to note significant defects of neurological function, often seen in the deltoid or various cords of the brachial plexus, secondary to an acute shoulder injury. Many of these are transient after initial injury and may only last for seconds or minutes; very few persist for a long period of time. The patient may complain of inability to move the arm, either in external rotation or abduction, or of numbness in terms of a 'dead arm' syndrome.

The details of the mechanism of injury are often difficult to ascertain in the heat of the moment. Distinguishing trauma associated with a fall on the shoulder itself from trauma associated with a fall on the outstretched arm is very helpful in deciding the actual diagnosis. In individual athletic sports such as track and field, the actual mechanism is often seen. However, in multiple-person collisions, as seen in team sports, the actual mechanism is sometimes more difficult to identify. A previous history of injury can be helpful.

At the scene of injury, one looks for deformity. Squaring of the shoulder with the associated anterior bulge or fullness is associated with anterior dislocation of the glenohumeral joint. Sometimes athletic clothing prevents visualization of the shoulder until the competitor is removed to the sidelines. There may be obvious deformity of the clavicle, or the patient may be short of breath with marked pain over the medial aspect of the clavicle. Crepitus is not usually felt in a simple shoulder dislocation, but is obviously present in fracture dislocations of the glenohumeral joint and fractures of the clavicle.

It is reasonable to make a quick assessment of gross function of the neurological status on the field, looking for paraesthesias and gross weakness. In the event of any kind of brachial plexus injury, the neurological deficit may not follow a specific pattern but may be patchy in its presentation.

An open wound is unlikely in an acute shoulder injury, although a clavicular fracture may occasionally either tent the skin or be associated with an abrasion and should be considered as an open wound until proved otherwise.

A decision as to the seriousness of the problem should be made at the scene. In this situation the differential diagnosis includes the following:

(1) glenohumeral dislocation;
(2) acromioclavicular separation;
(3) fracture of the clavicle;
(4) fracture of the proximal humerus;
(5) brachial plexus lesion;
(6) acute posterior glenohumeral dislocation (rare).

Based on history and physical findings it is usually possible to determine whether the injury is a straightforward dislocation or a fracture dislocation of the proximal humerus. Most fractured clavicles are easily discernible. Shoulder tip pain with or without deformity indicates shoulder separation. The brachial plexus injury may not be accompanied by any gross deformity at time of first assessment. Patchy weakness of the shoulder associated with severe pain may be seen. Often the shoulder dislocates but reduces spontaneously.

Transportation

The patient is usually able to walk from the playing field. Given that the shoulder is the only area of injury, it is usually reasonable for the athlete to splint his or her forearm with the opposite hand or for another person to splint the forearm, steadying it to decrease the motion at the shoulder area, while he or she leaves the playing field. If the nature of the sport allows, a sling and swath can be applied. If the athlete is multiply injured, splinting is essential before transport from the playing area.

Treatment

Sideline management

At the sidelines the history and physical examination should be re-evaluated. A more definitive physical examination should be made together with a more complete neurovascular examination carried out with better exposure of the injured area. After this examination, better splinting can be applied.

A diagnosis of anterior dislocation is often made on the basis of the sideline examination. In the case of a glenohumeral dislocation, particularly in the recurrent dislocator, it is often very easy to reduce the shoulder gently. The patient should be removed to a quiet area and made comfortable in a supine position; firm but gentle traction should then be applied. When the shoulder is reduced promptly and gently, less soft tissue damage and subsequent swelling will occur and rehabilitation will be swifter, particularly if there is neurovascular compromise. The earlier the reduction the less damage there will be to the vital structures. However, if the diagnosis is uncertain or fracture is suspected or any significant resistance is met in attempted reduction, the patient should be transported to a facility for more definitive diagnosis and treatment. It is important to note the neurovascular status of the patient at time of injury (even if it is normal) and to follow this until more definitive care can be arranged. Once the shoulder is reduced, a modified Velpeau shoulder dressing and ice pack will keep the athlete comfortable until further treatment is given.

Fracture of the clavicle can be managed by ice pack and sling until radiography is performed. Shoulder separation with or without fracture of the distal clavicle can be treated as a fractured clavicle. If no neurovascular defect is present there is no need to evacuate the patient immediately.

Fortunately, the most critical of all situations is rare. When present it is potentially deadly. Posterior sternoclavicular dislocations occur secondary to direct force at the medial end of the clavicle. The initial symptoms are pain and possibly respiratory distress. Immediate reduction with traction, abduction, and extension of the shoulder is required. Sometimes it is necessary to use a towel clip to reach the medial end of the clavicle and pull it anterior. If the situation is still unstable after reduction, the patient should lie with rolled towels or sand bags between the scapulae and the shoulder should be extended.

A 25 per cent (16/60) rate of serious complications involving the trachea, oesophagus, or great vessels has been reported.[3] Anterior dislocation of the sternoclavicular joint is a more

common injury. This is indicated by pain and prominence of the sternoclavicular joint.

Sometimes no deformity can be detected by examination on the sideline, there is minimal pain, and a full range of motion and function is present. Unless a major deficit was present initially, return to the game may be allowed.

Definitive treatment

More definitive treatment of the dislocated shoulder is indicated if reduction is not accomplished on the sideline. The patient suffering dislocation for the first time is best managed with 6 weeks of adduction in the internal rotation position. The recurrent dislocator needs early rehabilitation with consideration of surgical repair. If the first-time dislocator is more than 30 to 35 years old, there is little to be gained by 6 weeks immobilization as the recurrence rate is relatively low.[4]

It can be argued that radiological evaluation is not necessary for the recurrent dislocator if reduction was obtained very easily and quickly. However, in all other situations plane radiographs should be taken to look for evidence of fracture. The timing of this obviously involves balance between the patient's needs and the availability of facilities.

Fractured clavicles may need a clavicle strap or a simple sling depending on the character of the fracture. Non-unions are rare unless open reductions are performed. Most centres do not generally perform open reductions on acromioclavicular separations.[5] The exceptions are the types that are caught in the trapezius (Rockwood type IV + V) or the extremely rare type VI when the distal clavicle is below the coracoid.[6]

It is necessary to advise the patient and interested parties that there is a significant injury. The player will be unable to return to play, particularly if there is any suspicion of a fracture or neurological deficit. This holds true even in the situation of recurrent dislocation if there is any suspicion of neurological deficit or fracture.

A patient with a recurrent shoulder dislocation will be able to return to sport within 10 to 14 days. Aggressive physiotherapy is used in the first few days for progressing to graduated functional utilization for both activities of daily living and sport. In our clinic, a first-time anterior dislocator is usually immobilized for 4 to 6 weeks in internal rotation and adduction. However, there has been a great deal of discussion about the length of time that the first-time dislocator should be immobilized. In theory, the damage is already done, particularly to the anterior glenoid labrum. Hence many surgeons do not immobilize the shoulder for any length of time. In either case, during that period of time isometric exercises can be done in the sling. Following release from immobilization, range of motion and progressive exercise muscle rehabilitation can be undertaken.

In the skeletally mature athlete fractured clavicles heal in about 6 weeks. The amount of exercise relative to shoulder permitted during this period depends upon the amount of displacement and the pain experienced by the patient. Although non-union of the clavicle is frequently feared, it is not very common in practice. Early range of motion of the shoulder does not inhibit fracture healing. Muscle strengthening exercises that do not cause any pain and hence do not involve motion at the fracture site should be encouraged. In children who are not skeletally mature the clavicle heals very rapidly. Return to sport is allowed on the basis of clinical and radiological evidence of healing. However, most athletes should not resume their sport until the clavicle is clinically solid and the appropriate muscles are rehabilitated. A period as long as 12 weeks may be required to ensure optimum clavicle solidity and rehabilitation of muscle strength.

Much has been written about the separated shoulder. In our practice rehabilitation is performed very quickly, allowing a few days of rest for the initial discomfort to ease followed by progressive exercises, particularly strengthening of the external rotator. The defect in external rotator strength is usually greater than in other muscles and needs the most rehabilitation. We have seen a number of 'failed' conservatively treated shoulder separations. When a regime of infraspinatus strengthening has been undertaken, non-operative care has been successful.

Prognosis

As with all injuries the prognosis depends upon both the actual diagnosis and the requirements of the sport. Overall, when the patient has reached approximately 85 per cent of the capacity of the non-injured shoulder, he or she can return to full activity.

Having an injured shoulder does not preclude a patient from maintaining cardiovascular fitness. If the patient ignores cardiovascular fitness, return to sport will take much longer. Upper-extremity injuries allow lower-extremity workouts to be continued. Bicycle riding and running must continue for the maintenance of cardiovascular fitness. Weight-lifting with the non-injured limbs is also important.

ELBOW

Recognition

The setting

Falls from heights as in gymnastic dismounts, particularly non-intentional dismounts, often cause significant injuries to elbows. Wrestlers, high-jumpers, and pole-vaulters also commonly present with serious elbow injuries.

On-the-field clinical recognition

After an acute elbow injury pain is usually in the local area, but it may not be well localized to one aspect of the elbow. Injuries involving fractures, as in the radial head or olecranon, are often well localized to that anatomical area; however, subluxations, dislocations, and supracondylar fractures may have pain and swelling which is not well localized in the area of injury.

Many elbow injuries appear in the locked position, i.e. there is great reticence for the patient to move the elbow. A history of previous injury is usually non-contributory in terms of elbow injuries as recurrent dislocations are very unusual. The mechanism of injury is very important in that fracture of the olecranon usually occurs with direct trauma and most of the other injuries occur with indirect trauma, i.e. landing on the outstretched hand. Numbness, tingling, or lack of power are common complaints at the time of initial injury to the elbow. These may be due to either specific neurological deficit or more likely the trauma of the injury. Seven per cent of supracondylar injuries are accompanied by some form of neurological defect.[7]

On examination, deformity is often present at the elbow. At 90° of flexion, anatomically the lateral epicondyle, the medial epicondyle, and the tip of the olecranon form an equilateral triangle. In dislocations this relationship is altered. In the common types of transverse supracondylar fracture this relationship is often maintained. Radial head fractures can occur either alone or in combination with a dislocation. When they occur alone, pain is usually localized to the radial head and is worse with supination or pronation of the forearm. In the presence of an olecranon or supracondylar humeral fracture, crepitus can sometimes be felt. Palpable defects are felt in a fractured olecranon or if there is an acute triceps tear (as occurs in older patients). The gross function is usually poor after a significant elbow injury because of spasm alone, let alone anatomical derangements. Open wounds are more common because of the proximity of the olecranon to the subcutaneous tissue in the skin. Even small abrasions over an acute fracture area must be considered an open fracture until proved otherwise.

Differential diagnosis of these elbow injuries includes fracture and/or dislocation. On the basis of history and physical examination, a single bone fracture, i.e. radial head or olecranon, can be diagnosed. More serious emergency dislocations can also be identified. Often an osteochondral fracture is combined with dislocation. This will be confirmed by radiological investigation.

Transportation

When fracture dislocation is suspected, a decision must be made as to the method of transporting the patient. Splinting before removal from the playing field is advised. Care must be taken because many neurovascular complications are possible with elbow injuries. The ulnar nerve often has some transient defect.[7] In supracondylar fractures there must always be concern about vascular supply, even if the vessel is not acutely torn.[8,9] Recording the presence or absence of pulses and the gross function of the nerves during the on-field examination is important for subsequent evaluation.

After the serious injury is splinted, some athletes will be able to move under their own power. However, if this is not possible, a stretcher should be used.

Treatment

Sideline treatment

The elbow may have fairly normal alignment. Localized pain over a specific anatomical area will indicate a specific fracture. Radial head fractures present with local pain and decreased pronation and supination. Fractures of the olecranon have local pain and decreased flexion–extension. The skeletally immature patient may present with or rapidly develop a 'purple football'. This is a painful swollen purple-blue elbow which has a fracture even if it cannot be seen on a radiograph. Hence, if the injured player develops early swelling but has a good range of motion and minor pain, he or she should not be permitted to return to action until a better radiological examination has been performed.

Neurovascular monitoring must continue on the sideline. If there is gross deformity and the anatomical triangle is not intact, posterior dislocation should be suspected. An immediate reduction might be considered. The athlete should be removed to a quiet place and a reduction attempted. Early gentle reductions are often very easy. The pain, swelling, and complications are reduced by this early reduction. Gentle longitudinal traction is applied to the forearm while the elbow is forward flexed. If any significant force is necessary or there is crepitus, it is better to move the patient to a facility where an orthopaedic examination can be performed. Appropriate splinting and monitoring of the neurovascular status needs to be continued. This may be forgotten if a 'sideline reduction' was attained easily. In essentially all significant elbow trauma situations a radiological evaluation must be performed to look for evidence of fracture and adequacy of reduction. Identification of a displaced fracture or a loose body can prevent congruent reduction. Unstable reduction requires open reduction with internal fixation.

Gross deformity may be accompanied by an intact anatomical triangle. Here the deformity usually represents a displaced supracondylar fracture. In the emergency situation no pulse may be felt. If that is the case, gross overall alignment should be gently attained with traction and a little flexion whereupon the pulse should return. If, during attempt to restore gross alignment, the pulse disappears as the elbow is flexed, the reduction should be discontinued until the pulse can be felt again. Needless to say, in clinical evaluation the adequacy of capillary circulation is more important than the presence or absence of pulses. A patient with inadequate capillary circulation, regardless of pulses, is in a true emergency situation and should be treated appropriately with immediate evacuation from the playing field to a facility of definitive care. Fortunately, these vascular emergencies are rare.[10] Constant vigilance will decrease even further the complications of inadequate blood supply to a limb.

Definitive treatment

Displaced fractures and dislocations need the appropriate reduction. Congruency of the elbow joint must be obtained without loose pieces in the joint. Only anatomical reductions can be accepted in supracondylar fractures. Treatment often requires operative intervention either with closed reduction and fluoroscopic pinning or open reduction and internal fixation.

Prognosis

Again, the prognosis for return to sport depends on the injury and the sport. Patients with stable subluxations can be started on early range of motion exercise as the amount of pain dictates. Of course, this should be active range of motion and not passive stretching of already damaged muscles. As pain subsides and the range of motion and strength increase, the patient can return to sport. In the case of gymnastics a large range of painless motion must be achieved before return to sport can be contemplated. This may take 3 to 6 weeks or more if the subluxation is significant.

Patients with elbow dislocations can be started on early protected range of motion exercise without compromising the final result. Patients with displaced fractures need 3 weeks of immobilization after open reduction, followed by active, rather than passive, range of motion exercise. The treatment is completed by progressive strengthening and functional rehabilitation. It may be impossible for the athlete to return to the same sport. However, a number of determined elite gymnasts have returned to their original high level of competition following displaced supracondylar fractures. The time required may be 3 to 6 months.

The patient with an undisplaced fractured radial head needs early mobilization and usually returns to full sports function within 6 weeks. These athletes may not regain full range of motion, but usually lack only a few degrees in extension or flexion. In radial head fractures, pronation and supination are reduced slightly. This may not inhibit athletic performance. However, in throwing sports any decrease in range of motion is significant, and in sports such as artistic gymnastics, figure skating, and synchronized swimming, a lack of 10 to 20° of extension may cost the athlete marks for artistic impression. Similarly, a patient with a

fractured olecranon can start on early active range of motion exercise once the fracture has been stabilized, usually by open reduction, and clinical and radiological healing are often sufficiently advanced after 6 to 12 weeks to allow return to sport.

The whole athlete needs rehabilitation. Cardiovascular training can continue without interruption. Other joints and their muscle groups need functional exercises just as much as the injured parts. Exercise of all non-involved limbs and joints should be continued throughout the post-injury period. Then the athlete will be completely ready to return to competition as soon as the elbow injury is healed.

KNEE

Recognition

Setting

Emergency knee injuries are more common than hip or ankle problems. Injuries in contact sports (football, hockey, and rugby) and high velocity sports (motor sports and downhill skiing) can result in permanent limb dysfunction. This type of injury also occurs in jumping sports (long jump and gymnastics).

On-the-field clinical assessment

A contact sport collision, a gymnast falling from an apparatus such as the high bar, and a long jumper who misses a landing are all specific examples of situations where serious knee injuries should be expected. The foot is planted but the upper thigh keeps moving due to either a direct collision or the body's own momentum. Violent force is necessary to dislocate a knee. Cadaver studies have shown that the force necessary to dislocate knees and rupture cruciate ligaments is sufficient to cause complete rupture of the popliteal artery above the trifurcation. Other life-threatening injuries may also occur. Supracondylar fracture of the femur is not uncommon. In the skeletally immature athlete, the distal femoral or proximal tibial epiphyseal plate is weaker than either the bone in the supracondylar area or the ligamentous structures about the knee.[11,12] In this case separation of the epiphyseal growth plate will occur before diaphyseal fracture or ligamentous disruption. All these scenarios are true emergencies.

The fallen athlete who complains of severe pain diffusely around the knee may have a gross deformity. However, a knee dislocation can spontaneously reduce before medical attention arrives. In fact, many complex knee disruptions are knee dislocations that have spontaneously reduced at the time of injury. A history of previous injury will sometimes be recorded if there has been a previous anterior cruciate injury. However, recurrent tibial–femoral dislocations are not a problem. If the patient claims to 'have had this problem before', recurrent patellar dislocation rather than tibial–femoral disassociation should be considered as the diagnosis.

Deformity is usually present if either the knee dislocation is not reduced or there is a fracture. Sometimes the deformity is relatively subtle to the unpractised eye, particularly if immediate swelling is assumed to be its only cause. Crepitus is felt as a fracture is realigned. Functional assessment at the scene is grossly abnormal and the range of motion is poor. Approximately half the dislocations are anterior and the major vessel area is at greatest risk in this direction.[13] Neurovascular function must be assessed thoroughly; 20 to 35 per cent of patients exhibit vascular impairment.[6] Pulses may or may not be present. Capillary function is very important in assessing vascular adequacy. However, neither intact capillary perfusion nor the presence of pulses guarantees vascular continuity. Neurological function, drop foot, paraesthesia, etc., may be noted in 25 to 35 per cent of patients.[6] Posterolateral dislocations stretch the peroneal nerve over the lateral femoral condyle.[6] Popliteal swelling should be investigated as a sign of popliteal artery injury. A dislocated patella can also cause deformity, but the violence required is much less and careful examination makes patellar dislocation obvious. The knee will be flexed, the medial femoral condyle will be prominent, and it will be possible to observe and palpate the laterally displaced patella.

Puncture wounds from displaced femoral fractures are occasionally present. It should be remembered that even the smallest of puncture wounds represents a compound fracture.

Differential diagnosis

Possible diagnoses are as follows:
(1) knee dislocation;
(2) displaced fracture of distal femoral epiphysis;
(3) displaced fracture of proximal tibial epiphysis;
(4) supracondylar femoral fracture.

Exact diagnosis may be difficult at the scene. On-the-field assessment can indicate the occurrence of a very significant lesion. A dislocation and a displaced fracture can sometimes be differentiated by determining the point of maximum deformity.

Transportation

Many complications can result from the initial injury. The immediate treatment can also result in further problems. Systemic shock, further compounding, or further damage to the neurovascular structures can occur. Traction to realign either a dislocation or a displaced fracture is appropriate, particularly if the dislocation is associated with arterial embarrassment. If a pulse is initially present but disappears after traction, the traction should be loosened or removed and some deformity allowed until the pulse returns. Alignments on field should be carried out gently and easily. If marked resistance is met or great force is needed, then it is better to leave the deformity as it is. Some posterolateral dislocations are not reducible by closed means. Forcing them can result in further damage to the peroneal nerve. Medial capsular infolding can prevent reductions.[14]

Splinting should be carried out using proper long leg splints. Many types are available, but whichever is chosen should be familiar to the physicians applying it. A splint should be light, easily positioned, and non-constricting. Air splints and splints made of cardboard or bristleboard are frequently used. The latter are easily shaped to the appropriate size. If no prefabricated splint is available, splinting to the good leg with cloth ties is an option. If air splints are not inflated to the appropriate pressure severe constriction of the limb can occur. In addition, pressure blisters are frequently caused by splints which are too tight. Immediate transport to a definitive facility using a fracture board and/or a stretcher is mandatory. The neurovascular status should be documented. Appropriate medical support staff should accompany the patient to the hospital with written information regarding the neurovascular status. Competition officials should not be allowed to influence medical decisions. Stabilization of the injury should not be hurried although there will be many pressures to resume competition to satisfy other parties such as the media or sponsors.

Treatment

Sideline management

The application of ice is appropriate on the sideline. The splint should be checked; proper splinting should be well padded and not too rigid to allow for the massive swelling that will develop. The neurovascular status should be continuously monitored. Patient evacuation to the appropriate facility is a matter of urgency. When there is vascular compromise, the complications increase exponentially with time.

As suggested above, many of these serious injuries are disabling enough to require splinting on the field. However, it is possible to be surprised by an athlete who is helped to the sidelines with a knee injury without deformity. This type of injury may be recognized as a multiple ligament disruption. Occasionally knees will be examined which have damage to both cruciates with or without a collateral ligament injury. These knees were probably dislocated at the time of impact. The danger of vascular complications in the case of multiple ligament disruption is as real as in the knee that is found displaced at the scene. Hence if the knee demonstrates a severe degree of laxity or multidirectional instability, a dislocation of the tibial–femoral joint should be assumed, and should be treated in the same way as an emergency.

Definitive management

Motor cycle racers and downhill skiers often sustain major injuries to more than one system. Catastrophic knee injuries are associated with other major injuries of chest and head. Obviously, these other areas take priority in treatment. However, while other injuries are being assessed or treated, fractures and dislocations should be stabilized.[15] Continuous monitoring of the vascular supply and the development of compartment syndromes should not be forgotten.

Not all vascular injuries are initially obvious. Welling et al.[16] described 14 cases of dislocated knee, of which seven had palpable pulses. Constant vigilance must be undertaken and vascular complications should be dealt with promptly. Arterial perfusion must be restored within 6 h. This time window will allow salvage of most functions.[13,17] Fracture stabilization after vascular repair is the current preferred course of action.[6] Compartment decompression by fasciotomy is crucial if vascular injury has occurred.[18] Intraoperative intra-arterial bolus injection of contrast appears to be as good as a formal angiogram in these injuries. The time needed to arrange the angiogram can be excessive. The most common method of intervention is to identify the area of lesion in the operating room with an intraoperative arteriogram. This is followed by a medial approach. Lesions are dealt with by a reverse saphenous graft. Some method of knee stabilization is then used to protect the graft.[9] The patient is at risk of developing a compartment syndrome for at least 72 h. Most vascular surgeons will perform a fasciotomy at the time of initial vascular repair.

Epiphyseal displacements can sometimes be stabilized by closed reduction with or without pin fixation. In patients with closed epiphyseal plates, fractures about the knee are treated by open reduction and internal fixation.[11,12]

Neurological defects may be present. Most lesions are neurapraxias or lesions in continuity. Unfortunately the common peroneal nerve is very unforgiving. Nizt et al.[19] showed with grade 3 ankle sprains that nerve function will be adversely affected by only a 6 per cent change in the length (elongation). Some neurosurgeons will elect to explore these nerves, usually on a delayed basis.

Fractures in and around the knee are treated by open reductions with internal fixation. Different methods of fixation are used depending on the character of the fracture. If the fracture is in the diaphysis, a locked intramedullary rod is used. If it is intercondylar or truly supracondylar, a blade plate is fixed to the lateral cortex of the femur. Occasionally a cast brace may be used if the fracture is not stable. The aim is to move the knee through a range of motion as early as possible. If possible, early partial, or at least feather, weight bearing is better than no weight bearing.

There is controversy over the definitive care of knee dislocations. A difference of opinion exists on primary versus delayed ligament repair.[20] Some orthopaedic surgeons repair the ligaments after 2 weeks when collateral circulation is established.[6] Others prefer to cast the injuries in 30° of flexion for 6 weeks. A delayed repair of functional defects can then be performed as necessary.

Prognosis

Epiphyseal fractures have a better prognosis than those patients with multiple ligament disruption. These injuries usually occur in younger athletes, and long-term complications of growth are possible.[11] Angular problems do not usually occur with a well-reduced femoral epiphysis. The femoral epiphysis usually closes after this injury. However, the tibial side is not as forgiving. Even with a perfect reduction, abnormal growth can occur secondary to the changed blood flow in this area (either increased or decreased). As the patients with tibial epiphyseal injury are younger, closure is farther off and hence there is longer time for abnormal growth to occur. For example, the increased blood supply to the proximal metaphysis will sometimes stimulate increase growth in the epiphyseal plate.

Adult fractures should bear little or no weight for 6 to 12 weeks. If the fracture reduction is stable, range of motion and muscle strengthening can often be commenced immediately post-operatively. In the adult, fractures will need at least 12 weeks to become solid enough to allow full weight bearing. Water exercises and cycling can be started earlier than this. Both are necessary to maintain cardiovascular fitness. Cycling and swimming will help regain hip, knee, and ankle range of motion. Return to sport is possible after most fractures. If an internal fixation device is used, it should be removed before full sport participation. The whole process may take as long as 18 months to 2 years depending on the fracture.

Complex ligamentous injuries have poor prognoses in terms of regaining full function. This seems to be true regardless of treatment. Stiffness of range of motion or instability of residual ligament pathology is common. However, if there are no complications (artery or nerves), return to some sport is a reasonable aim. In our practice we make extensive use of bracing. Some sports, such as rugby, and some patients will not tolerate bracing. However, we recommend bracing for 2 years after injury.

If the knee injury is accompanied by vascular injury, the prognosis is poor. In 1963 the amputation rate with knee dislocations was about 50 per cent.[21] More recently, with early recognition, improved imaging, and more vascular surgeons, this rate should be much lower. Savage[9] reported an 85 per cent salvage rate of knee dislocation with arterial embarrassment.

Although nerve injuries are usually lesions without gross disruption, generally neurapraxia and axonotmesis, peroneal nerve problems are slow to recover. Many need surgical exploration for intraneural neuromas.

Sport is one of the main sources of catastrophic knee injuries. When armed with suspicion, early recognition is usually possible.

A hidden injury may be a multiple ligament disruption that was dislocated at the time of injury but reduced at the time of medical examination. If early aggressive treatment is instituted, return to sport may be possible. However, if any complications are present, full return to the demands of sport is unlikely.

COMPARTMENT SYNDROME

Recognition

The setting

The end results of compartment syndromes were first identified by Volkmann in 1881. He described the paralyses and contracture that occur in the upper extremity with bandages that are too tight: 'The paralysis depends on the fact that the primary muscle groups die when deprived of oxygen for too long'.[22] In 1909 Thomas reviewed the world literature, noting only a few cases of leg compartment syndromes.[23] Since that time ischaemic contracture has been identified with bandages and casts that are too tight. Other causes have been recognized,[6,24] including primary vascular compromise, toxins, burns, etc. However, in sports medicine closed injury by direct blunt trauma is the common source of acute compartment syndromes.

Compartment syndrome is defined as a condition in which the circulation and function of tissues within a closed space are compromised by an increased pressure within that space. It usually occurs in the calf or the forearm. Foot and thigh compartment syndromes have also been described.[25,26] The precipitating event can be relatively minor trauma or trauma without fracture. Contusions sustained in contact sports such as soccer[27] or ruptured muscles in racquet sports[28] are examples of sport situations causing acute compartment syndromes.

On-the-field clinical assessment

The initial injury can be a fractured tibia, forearm, or distal humerus. The large vessel supply can be interrupted directly or by the pressure of immobilization which is too tight. However, the initial injury may not be serious. A kicked shin or contused calf are aetiologies that have presented in our sports clinic. For example, a tennis player was accelerating towards the net and instead of tearing her Achilles tendon, tore the midportion of the gastrocnemius. Massive swelling in the closed superficial posterior compartment resulted in a compartment syndrome.

At the scene of the sporting event complaints may be minimal. However, bleeding into the osteofascial enclosed space continues for several hours after injury, raising the intracompartment tension. The player may wake up in the middle of the night in agony. Perhaps the swelling begins after the game and hot shower. Pain may increase on the bus ride home. The signal of increasing intracompartmental pressure is pain which is out of proportion to the injury. Even a fracture, once splinted, does not usually cause a great deal of pain. If the normal doses of analgesic are not sufficient, compartment syndrome should be assumed until proved otherwise. Up to 72 h can elapse after injury before a compartment syndrome manifests itself. Complaints of numbness, tingling, coldness, and paralysis are all late symptoms.

The pathophysiology has been well described. Compartments are bounded by bone and dense unforgiving fascia. Pressure will increase as the bleeding continues. When the venule pressure is exceeded by the intracompartmental pressure the venules collapse, causing further increase in interstitial and compartment pressure. Next the arterioles close, causing muscle ischaemia and releasing histamine-like substances. These substances increase capillary permeability, further increasing the compartment pressure.

The ischaemia is primarily from the arteriole level. Therefore the major clinical findings of pallor, absence of pulse, paraesthesias, and paralysis are late signs. The first sign is pain associated with passive stretching of the muscles in the affected compartment. This will occur long before the pressure is high enough to cause permanent damage to the enclosed structures. Sometimes this early stretch pain is associated with a cyanotic tinge to the extremity. Normal pulse and nerve function is found in the early salvageable stages. Palpating the tenseness of a compartment is very subjective and is not reliable. Accurate pressure measurements are easily obtained. Manometric evaluation of the pressures should be performed if there is any question of the diagnosis.

Differential diagnosis

Possible diagnoses are as follows:
(1) compartment syndrome;
(2) major arterial injury;
(3) isolated partial nerve injury;
(4) ruptured muscle;
(5) undiagnosed fracture.

Acute compartment syndrome can be an isolated event or occur secondary to a major arterial injury. If blood supply is not restored with an arterial repair within 6 h, the muscles undergo cloudy swelling. This further compromises the compartment blood supply by increasing the pressure. Occasionally athletes will have an isolated contusion or stretched peroneal nerve causing severe pain and neural dysfunction. This may be confused with a primary compartment problem. Isolated partial rupture of muscles (e.g. medial head of the gastrocnemius) are sometimes seen. These usually cause pain, but this is relatively easily managed with rest, ice, compression, and elevation. Undisplaced fractures may not be diagnosed or suspected without radiological evaluation. They can be very painful if they are not splinted because fracture is not suspected. In addition, low velocity trauma, enough to cause a fracture but not enough to cause disruption of the compartment sheath or interosseous membrane, will predispose to a compartment susceptible to increased pressure. This is because in a displaced fracture the compartments are sufficiently disrupted to self-decompress.

Transportation

As stated above, compartment syndrome is more likely to present after the competition. However, it is one of the few true emergencies in sport medicine. When compartment syndrome is seriously suspected, it must be treated as such until proved wrong. If a circumferential cast or dressing for a fracture or wound has been applied, the cast or bandage should be cut so that it is not circumferential, i.e. skin can be seen from one end of the cast to the other, while the patient is being transferred to hospital.

The position of the leg during transportation is also important. If there is vascular impairment, the limb should probably be kept at a neutral level with respect to the heart. There is a relative decrease in arterial flow to an elevated limb. This increases venous congestion which would compound the problems in compartment syndromes.

Treatment

Sideline management

Preventive measures are possible on the sidelines. Elevation of an injured extremity will decrease the likelihood of venous congestion. Early application of ice and compression of a contusion to stop bleeding will also decrease the risk of an expanding compartment.

Definitive management

The diagnosis can usually be made clinically. When compartment syndrome is present, pain out of proportion to the injury and pain with passive stretch of muscles in the tight compartment are the classical findings. Sometimes there is doubt about the clinical findings. There are several methods of measuring compartment pressures directly. Static methods of measuring compartment pressure have been described.[18,29] Occasionally, a patient may be obtunded from a head injury or be in a period of cardiovascular shock. In these situations a dynamic method using an indwelling wick catheter to monitor the pressures is indicated.

Normal resting compartmental pressure is 4 mmHg. In the emergency situation a compartment pressure of over 30 mmHg is considered high enough to demand fascial decompression. If the compartment pressure is measured within 10 to 30 points of the diastolic pressure, when the patient is in shock, a fasciotomy is necessary. With normal exercise the readings can reach 50 to 60 mmHg, but it falls quickly with cessation of activity. In the presence of chronic compartment syndromes the pressures elevate to 75 to 100 mmHg and their resting pressures are 15 to 30 mmHg.[30] Some recent work[31] suggests that there is a place for magnetic resonance imaging in the evaluation of chronic compartment syndromes.

It has been shown experimentally that a unilateral cast split reduces the compartment pressures by 30 per cent.[32] When a cast, a soft roll, and a dressing are bivalved so that there are no circumferential constricting elements from one end of the dressing/cast to the other, a reduction in compartment pressures of up to 85 per cent is obtained.

Compartment syndrome is a surgical emergency as soon as the diagnosis is made either clinically or by objective measurement. The definitive treatment is fasciotomy. In the calf there are four compartments (five if the posterior tibial is assumed to be in a separate compartment).[12,27] If the diagnosis is made objectively and only one or two compartments are involved, it may be reasonable to release only those compartments. However, Rorabeck[18] believes that if the pressure is over 30 mmHg, even if only one compartment is involved, all four compartments should be decompressed. Double-incision fasciotomy is advocated by some authors.[18,29] Another method of approaching the four compartments is to excise the fibula.[33] In our centre we use the two-incision method. Delayed primary closure, with or without skin grafting, is performed 5 to 7 days later.

Prognosis

Delay in making the diagnosis or providing treatment can be catastrophic to the athlete. Research has shown that early muscle changes will be produced within 2 h.[34] Classical Volkmann's ischaemic contracture develops after 12 h of ischaemia. In practice this has been shown to be accurate; Rorabeck[18] has obtained excellent results in patients who were treated less than 12 h after injury.

The prognosis for return to sport depends slightly on the cause of the syndrome. In cases of muscle rupture the prognosis for return to a high level of competition is not good. Irreversible changes to at least some of the muscle occur after 6 h. The percentage of defect affects the ability to return to competition.

The muscle rehabilitation necessary after fasciotomy is prolonged. The limb may show some chronic swelling as a result of chronic venous insufficiency. Muscle often prolapses, unconstrained by fascia. This may influence the ultimate power that can be generated.

Our definition of an emergency is a situation in which immediate intervention will produce a difference in outcome. Compartment syndrome is a good example. Delayed diagnosis and treatment is devastating to the athlete. If the diagnosis is made early and treatment is carried out within 6 h, it is reasonable to expect a full return to function and sport at a competitive level.

FROSTBITE

Recognition

The setting

In some parts of Canada the conditions for frostbite are present in some areas for 12 months of the year. The Canadian national ski team can train all year long on glaciers in the Rocky Mountains. Downhill and cross-country skiing or hiking are possible for 6 months of the year in many populated areas of Canada. Therefore there are many possibilities for cold injuries to occur.

There are several types of cold injuries. Either the whole individual is cold, as in hypothermia, or localized parts are cold, as in frostbite. In hypothermia the core body temperature drops. In frostbite, the core temperature may be normal while the temperature of the peripheral tissues drops. In this case there is local freezing of tissues.

The end result of frostbite is death and destruction of tissue. The body attempts to preserve core temperature by decreasing blood flow to parts exposed to cold. Vasoconstriction, mediated by the sympathetic system, is the mechanism by which this is carried out. Unfortunately some patients are predisposed to problems. Smoking (or any use of tobacco), extreme youth or age, fatigue, poor nutrition, and alcohol intake all make people more likely to suffer from frostbite.

Destruction due to frostbite is the result of a sequence of events. The first serious pathological change is the formation of ice crystals in the cells of tissue exposed to cold. This mechanically distorts the cell and adversely affects its functions. Local vascular stasis occurs, followed by clotting and distal local hypoxia, ischaemia, and acidosis. Rewarming will cause a rebound vasodilation and some capillary breakdown will occur. Oedema and some further tissue damage will follow. However, if the exposure to cold continues, deeper tissues become involved. If a large vessel is frozen, then a mummification process accompanied by gangrene takes place.[35]

Obviously the skier is at risk. Not so obvious are speed skaters who have to train outdoors. The football player who plays outdoors in November in the middle of a snowstorm is also at risk. Even hunters who hike over varied terrains in the autumn and spring are at risk. Wet marshland and fields predispose to cold, wet feet. These conditions are perfect for the development of chilblains (trench foot). These hunters do not want to appear weak by stopping for wet feet, and hence repetitive insult occurs.

Another potential victim is the snowmobile rider out racing with the boys, stopping only for a cigarette or a swig from the beer bottle. Cold injuries sneak up on these individuals.

The weekend athlete who is inexperienced in outdoor winter athletic activities is at risk. An example is the novice skier who, in an attempt to be 'cool' (but not cold), wears fashionable clothing that is thinly insulated and tight. Such inexperienced individuals frequently wear ski boots that are stylish but too tight. Similarly, gloves that match the outfit may not allow enough finger movement. Experienced athletes encounter problems due to neglect. Their favourite pair of cross country ski boots may be comfortable, but the insulation in the toe and heel areas has probably worn through.

The environmental temperature does not need to be excessively low. Repetitive exposure to water around 0°C is enough to cause trench foot. Skin will freeze if the tissue is at −2°C. The wind chill factor at temperatures of 5 to 10°C can produce actual temperature levels capable of freezing tissues. A sophisticated measurement of risk developed for the Glacier Patrol 100-km ski race over the high terrain of Switzerland is available. Known as the Arolla index, this collates time of exposure, wind chill factors, and measured temperature into one index.[36]

On-the-field clinical assessment

The first effects of local cold are not usually noticed by the affected individual. Frostnip, where reversible ice crystals form on the skin surface, is the first step in local cold injury. The frosted appearance and numb feelings are reversible if warming procedures are carried out. If they are not, the process continues and deeper tissues are frozen. Slight initial swelling and redness may precede skin prickling sensations. As the process continues the skin becomes yellow and waxy. When the cold stimulus is unchecked the subcutaneous tissue freezes and the skin feels solid. Serous blisters appear which represent superficial freezing. If capillaries are involved and rupture, the blisters become haemorrhagic. This represents a deeper level of involvement. If, on initial assessment, dry gangrene is present, larger vessels than capillaries have been compromised. In that case some tissue destruction is complete.

Trench or immersion foot occurs on repetitive exposure of bare skin to cold and wet at about 0°C. The skin is red in the early stages and becomes mottled grey-blue or white after further exposure. The skin is sensitive and easily 'burns' with any temperature change. On rewarming, the foot becomes swollen and oedematous, and the skin reddens and reacts to minimal stimuli.

Transportation

A patient with frostbite is a true emergency. If the frozen area is in the foot and any thawing occurs, he or she becomes a stretcher case. A thawed area of tissue that has been frostbitten is defenceless against mechanical forces. The trauma of simple weight bearing will increase local tissue destruction at an alarming rate. Patients with frostnip can be rewarmed at the site and no evacuation is necessary. Sufferers from trench foot need protection from repeated exposure to cold and wet.

Treatment

Sideline management

Prevention is the best treatment of any sports injury. Cold injury is a good example of an injury that can be prevented. Unfortunately, a key element of prevention is common-sense and pharmacies do not carry remedies for lack of common-sense!

Preventative lubrication of chronically dehydrated skin is useful in decreasing the effects of dryness and flaking incurred with frostnip. Adequate insulation in the form of dry properly fitted (not tight) clothing is a major step in prevention. Hands and feet should be kept moving. If they become numb, something should be done to enable them to be moved (remove boots). A participant should always ski or hike with a companion who can help check face, ears, and nose for early signs of blanching or frosting. Shelter from the wind should be sought if cold injury is suspected. This will increase the relative temperature in the tissues. Skiing should be performed in areas that are protected by trees and higher altitudes should be avoided until the cold part is thoroughly rewarmed. Beer or other alcohol should not be drunk in the belief that it will help in the rewarming process. Alcohol intake stimulates heat loss but does not produce significant peripheral dilatation in the presence of cold-induced vasoconstriction; it only compounds the problem. Similarly, smoking is absolutely contraindicated.

Frostnip cases need early rewarming. This is easily achieved by blowing warm exhaled air across the affected area. The axilla or the crotch are two areas which can be used to rewarm a cold hand or foot on site. Continued monitoring by a companion is important.

Trench foot should be kept dry. Great care must be taken in skin management as the skin may be friable. Any blisters or skin breaks should be covered with dry padded dressing as there is a potential for superficial infection.

A worse emergency than a frozen limb is a limb which has been frozen, thawed, and then frozen again. If frostbite occurs in a situation where there is the possibility of refreezing after thawing, the part should be left frozen until definitive treatment is available. The frozen limb can bear some weight. However, as soon as it is thawed meticulous care of tissues is necessary. The thawed foot will swell rapidly and is unlikely to fit into the original boot. The mountain climber, cross-country trekker, or patient with a previously frozen extremity becomes a great liability to the whole group.

Definitive management

The patient with frostbite may well have central hypothermia as well as local frostbite. Cconditions such as confusion, slowness of mental processes, or weakness, that are associated with hypothermia must be treated. The patient should be rewarmed as rapidly as possible. However, ventricular fibrillation may occur during this phase so that one should be prepared to deal with this potential complication. The extremity should be rewarmed by immersion in warm water. Under ideal conditions the water should be sterile, but at a campsite clean water is satisfactory. The water temperature should be maintained at about 40 to 42°C. The water container should be large enough to accommodate the whole frozen part. Warm water should be added as the frozen extremity cools the original water. The part should not be warmed against a fireplace as the affected part will thaw unevenly and the initial flow of blood into the area will make the part 'burn'. The nerves will be sensitive and capillary fragility will increase oedema into the area. Blisters should be left intact, but if they are broken they should be debrided and the wounds dressed with sterile dressing.

Various agents for enhancing revascularization have been investigated. Heparin, intra-arterial reserpine, chemical and operative sympathectomy, and low molecular weight dextran have all been tested but without convincing results. The damage is done by the initial intracellular freezing. The cell wall membranes rupture, releasing the contents. Extracellular fluid becomes hyperosmotic and clotting follows. Upon rewarming there is increased vascularity. Local haemorrhage and oedema causes further anoxia and cellular damage. If the freezing is in the calf or tibial area, the rewarming may produce compartment syndromes.

If gangrene is present initially, it is usually dry gangrene, i.e. it is not infected. The wounds should be dressed and monitored as the rewarming may demarcate the area of gangrene at a different level than initially anticipated. Live tissue may be present under the layers of necrotic skin. Most workers suggest leaving amputation for weeks or months as long as secondary infection does not intervene. If long-term pain is a problem, sympathectomy is sometimes of benefit.

Prognosis

By definition, frostnip is the reversible freezing of very superficial layers of skin. Hence if it is treated early and adequately, a full return to function can be expected. Obviously this is the form of 'frostbite' that recreational skiers have had when they relate to the gathering at après ski that 'they had frostbite out on the hill today'.

Skin flaking and redness often follow tench foot rewarming. Patients who have experienced trench foot or immersion foot have a sensitive extremity for a long time which may also be prone to future cold injury. Increased vascular changes in the skin and chronic hypersensitivity are common sequelae.

Frostbite causes permanent damage, the extent of which depends on the amount and depth of tissue initially involved and the length of time that the area was frozen. The response to rewarming is also extremely important. Large amounts of oedema and the presence of a compartment syndrome are associated with a poor prognosis. If secondary infection occurs, the outcome depends on the organism and the extent of the infection. When frostbite occurs in growing bones it affects the epiphyseal plates. Characteristically, the growth plates fuse, giving rise to irregularly shortened deformities of the digits.

Frostbite is like many sports injuries. Its presentation can be incidental or emergency. If the clinical diagnosis is ignored by the athlete or if there is misdiagnosis or mismanagement by the doctor, the results can be devastating. Fortunately, like many sports injuries, frostbite is uncommon and can usually be prevented by common-sense measures.

SUMMARY

Sport is its own medicine! Fortunately, major emergencies are rare; experience has shown that more than half the battle is to be prepared for disaster.

The patient with a major limb injury may well have airway or cardiovascular lesions and these take precedence over the injured limb. Arterial injury, neurological defects, and compound fractures are the most common emergency limb injuries in sport. Appropriate clinical examination is the second step in their management. (Being prepared is the first step.) The extent and methods of rehabilitation vary with the injury, the athlete, and the sport. Prognosis varies with the specific injury. Management includes rehabilitation of the whole individual, for the sportsman with a significant injury is truly a 'Fallen Athlete'.

REFERENCES

1. Allen RE, ed. *The Oxford Dictionary of Current English*. Oxford University Press, 1985: 239.
2. Pipe A. Canadian Academy of Sport Medicine: Accreditation Committee. Examination, 1990.
3. Worman LW, Leagus C. Intrathoracic injury following retro-sternal dislocation of the clavicle. *Journal of Trauma* 1967; 7: 416–23.
4. Simonet WT, Cofield RH. Prognosis in anterior shoulder dislocation. *American Journal of Sports Medicine* 1984; 12: 19–24.
5. Hawkins RJ. Personal communication, 1991.
6. Rockwood CA, Green DP. *Fractures in adults*. Vol 2. Philadelphia: JB Lippincott, 1984.
7. Pritchard DJ, Linscheid RL, Svien HJ. Intra-articular median nerve entrapment with dislocation of the elbow. *Clinical Orthopaedics and Related Research* 1973; 90: 100–3.
8. Axe MJ. Limb threatening injuries in sport. *Clinics in Sports Medicine* 1989; 8: 101–9.
9. Savage R. Popliteal artery injury associated with knee dislocation: Improved outlook? *American Surgeon* 1980; 46: 627–32.
10. Hurley JA. Complicated elbow fractures in athletes. *Clinics In Sports Medicine* 1990; 9: 39–57.
11. Burkhart SS, Peterson HA. Fractures of the proximal tibial epiphysis. *Journal of Bone and Joint Surgery* 1979; 61A: 996–1002.
12. Shelton WK, Canale CF. Fractures of the proximal tibia epiphyseal cartilage. *Journal of Bone and Joint Surgery* 1979; 61A: 167.
13. Green NE, Allen BL. Vascular injuries associated with dislocation of the knee. *Journal of Bone and Joint Surgery* 1977; 59A: 236–9.
14. Sisto DJ, Warren RF. Complete knee dislocation. *Clinical Orthopaedics and Related Research* 1985; 198: 94–101.
15. Allman FL Jr, Ryan AJ. The immediate management of sports injuries. In: Ryan AJ, Allman FL Jr, eds. *Sports medicine*. 2nd edn. San Diego: Academic Press, 1989: 281–318.
16. Welling RE, Kakkasseril J, Cranley JJ. Complete dislocations of the knee with popliteal vascular injury. *Journal of Trauma* 1981; 21: 450–3.
17. Cohen SL, Taylor WC. Vascular problems of the lower extremity in athletes. *Clinics in Sports Medicine* 1990; 9: 449–70.
18. Rorabeck CH. The treatment of compartment syndromes of the leg. *Journal of Bone and Joint Surgery* 1984; 66B: 93–7.
19. Nitz AJ, Dobner JJ, Kersey D. Nerve injury and grade 2 and 3 ankle sprains. *American Journal of Sports Medicine* 1985; 13: 177–82.
20. Meyers MH, Moore TM, Harvey JP Jr. Traumatic dislocation of the knee joint. *Journal of Bone and Joint Surgery* 1975; 57A: 430–3.
21. Kennedy JC. Complete dislocation of the knee joint. *Journal of Bone and Joint Surgery* 1963; 45A: 889–91.
22. Volkmann R. Die ischaemischem muskellamungen und kontrakturen. *Chirurg* 1881; 8: 801.
23. Thomas JJ. Nerve involvement in the ischemic paralysis and contracture of Volkmann. *Annals of Surgery* 1909; 49: 330–70.
24. Matsen FA. Compartment syndrome: A unified concept. *Clinical Orthopaedics and Related Research* 1975; 113: 8–14.
25. Gardner AMN, Fox RH, Lawrence C, Bunker TD, Ling RSM, MacEachern AG. Reduction of post-traumatic swelling and compartment pressure by impulse compression of the foot. *Journal of Bone and Joint Surgery* 1990; 72B: 810–15.
26. Myerson MS. Experimental decompression of the fascial compartments of the foot: The basis for fasciotomy in acute compartment syndromes. *Foot and Ankle* 1988; 8: 308–14.
27. Leach RE, Corbett M. Anterior tibial compartment syndrome in soccer players. *American Journal of Sports Medicine* 1979; 7: 258–9.
28. Davies JAK. Peroneal compartment syndrome secondary to rupture of peroneus longus: case report. *Journal of Bone and Joint Surgery* 1979; 61A: 783–4.
29. Mubarak SL, Owen CA. Double-incision fasciotomy of the leg for decompression in compartment syndromes. *Journal of Bone and Joint Surgery* 1977; 59A: 184–7.
30. Mubarak SJ, Hargens AR, Owen CA, Garetto LP, Akeson WH. The wick catheter technique for measurement of intramuscular pressure: a new research and clinical tool. *Journal of Bone and Joint Surgery* 1976; 58: 1016–20.
31. Vellet 1991. Personal communication
32. Garfin SR, Mubarak SJ, Evans KL, Hargens AR, Akeson WH. Quantification of intracompartmental pressure and volume under plaster cast. *Journal of Bone and Joint Surgery* 1981; 63: 449–53.
33. Kelly RP, Whitesides RE Jr. Transfibular route for fasciotomy of the leg. *Journal of Bone and Joint Surgery* 1967; 49A: 1022–3.
34. Jepson PN. Ischemic contracture, experimental study. *Annals of Surgery* 1926; 84: 785–95.
35. Fritz RL, Perrin DH. Cold exposure injuries: prevention and treatment. *Clinics in Sports Medicine* 1989; 8: 111–27.
36. Reymond M, Rigo M, The 'Arolla' index: a study of 88 cases of frostbite during a high mountain competition. *Journal de Chirurgie (Paris)* 1988; 125: 239–44.

7.3 Maxillofacial injuries in sport

G. F. GOUBRAN

INTRODUCTION

As contact sports have become more competitive in recent years and as there is strong encouragement to participate in them from an early age, sporting injuries to orofacial and dental structures have become more common. In a recent Australian study, one-third of dental trauma was due to participation in sport.

Maxillofacial injuries due to sporting involvement may include fractures of the facial skeleton, intra- and extraoral lacerations, and dental injuries, of which the last mentioned are the most common. The use of professionally fitted (custom-made) mouthguards and sports helmets (as worn by cricketers, American footballers, etc.) in contact and non-contact sports can effectively prevent many such injuries and make a great contribution to safety in sport (Plate 1).

A resilient plastic mouthguard acts by absorbing some of the force of a blow to the mouth at the impact site and by distributing the remaining energy throughout the mouthguard, i.e. to a much greater surface area than that of the actual impact. Of course, if the contact is severe, this device will be inadequate to deal with the forces in question and injury will occur. The mouthguard can then only help to reduce the damage. In general, however, the custom-made mouthguard protects the upper anterior teeth during frontal impact and helps to avoid their avulsion, guards against intraoral lacerations by separating the upper teeth from the soft tissues of the tongue, lips, and cheeks, protects the upper and lower

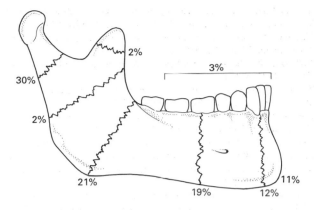

Fig. 1 Percentage of fracture 'mandibular site'.

opposing teeth upon impact when the mandible is involuntarily and forcibly closed, and reduces the likelihood of a fractured mandible following impact from below. Perhaps most importantly, it reduces the force of mandibular impact transmitted through the mandibular joints to the skull and brain, thus lessening the risk of concussion and other more serious head injuries.

A bimaxillary mouthguard obviously increases protection of the intraoral soft tissues and protects the mandibular teeth. This mechanism also stabilizes the mandible to the maxilla, thus reducing the risk of its fracture—a property that has been confirmed experimentally.[1] The bimaxillary mouthguard may be of special value in the boxing ring where many punches may be aimed at the mandible (uppercuts) with the intention of causing a knockout.

As well as benefiting physically from wearing a custom-made mouthguard and a sports helmet, sportsmen may experience a significant improvement in their self-confidence.

FRACTURES OF THE FACIAL SKELETON

The degree of maxillofacial trauma can vary from a simple crack fracture, requiring no treatment and causing minimum inconvenience, to major disintegration of the facial skeleton, with involvement of overlying soft tissues and adjacent structures (e.g. eyes, sinus, tongue, and teeth). Maxillofacial injuries are relatively easy to recognize by the obvious bony deformity seen immediately after the impact before overlying tissue becomes oedematous and swollen. Fractures should always be suspected if teeth fail to occlude normally.

For convenience the facial skeleton is divided into three parts:

(1) the lower third—the lower jaw (mandible);
(2) the middle third—the area between the superior orbital margin above and the occlusal plane below (in the case of an edentulous patient the maxillary alveolus);
(3) the upper third—the area above the superior orbital margin.

Since surgical treatment of these injuries is a highly specialized subject, in this chapter we shall concentrate on recognition and diagnosis of maxillofacial trauma.

Fractures of the lower third (fractured mandible)

The second most common cause of mandibular fractures is sports injury (assault being the most common).

Fractures of the mandible can occur in either the ramus or the body of the mandible (Fig. 1). Fractures of the ramus are usually closed injuries, while those of the body are often compound. Fractures of these two main areas differ in their clinical presentation and thus present different management problems.

Ramus area fractures include condylar region fracture, coronoid process fracture, and ramus fracture. Fractures of the body of the mandible include fracture of the angle, midbody fracture (molar and premolar area), midline fracture, fracture lateral to the midline in the incisor area, and dentoalveolar fractures. These fractures can occur singly or in any combination; the most common is a bilateral fracture involving the condylar neck on one side and the opposite angle.

Fractures of the ramus area

Condylar region fracture

The condylar region is the most common site of mandibular fracture and unfortunately the most frequently undiagnosed. The importance of diagnosis of fractures of the mandibular condyles cannot be overemphasized, particularly in children. If they are undetected and mismanaged, they may lead to gross asymmetry of growth or ankylosis of the joint (Fig. 2).

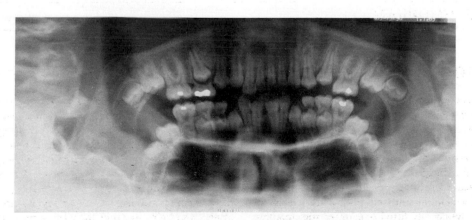

Fig. 2 Orthopantogram showing fracture of left condyle in an 11-year-old boy with mixed dentition.

The fractured condyle can be unilateral or bilateral and intra-capsular or extracapsular. Most are extracapsular fractures of the condylar neck which may or may not be dislocated. There is pain and swelling over the fractured temporomandibular joint (Plate 2). It is tender to palpate and invariably the mandibular mid-line is deviated towards the fractured side. Movements are limited, particularly lateral excursion away from the injured side. Anterior open bite and gagging of the occlusion on the posterior molars are indications of bilateral condylar fractures (Plate 3).

Coronoid process fracture

Fractures of the coronoid process are very rare and cannot be diagnosed on clinical examination alone. Apart from pain and limitation of mandibular movement, there is intraoral tenderness and ecchymosis over the coronoid process region.

Ramus fracture

Ramus fractures are also very rare. Apart from swelling and dis-comfort, there is little else to see as the fractured segment is sandwiched between the masseter and the pterygoid muscles.

Fractures of the body of the mandible

Fracture of the angle

Fracture of the angle is the second most common mandibular fracture, and often occurs through an unerupted lower-third molar. The displacement is caused by the pull of the masseter and/or medial pterygoid muscles, depending on the direction of the fracture line through the bone (Fig. 3).

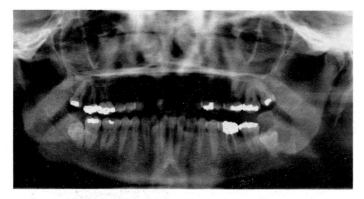

Fig. 3 Orthopantogram showing a fracture of the right mandibular angle through an unerupted right wisdom tooth following a fall.

Midbody fracture (molar and premolar area)

Unilateral fractures present with very little displacement as the muscles on either side of the fractured site tend to counteract each other.

Midline fractures (symphyseal)

Midline fractures can present with little displacement as the frac-ture line passes between the genial tubercles. The pull of the genioglossus and geniohyoid muscles tends to impact the bone ends together. Although it is difficult to demonstrate such a frac-ture radiologically, it should be suspected in all cases where patients sustained trauma to the point of the chin and particularly where bilateral condylar fractures are present.

Fracture lateral to the midline fractures (parasymphyseal)

Unlike the midline fracture, lateral fractures present with con-siderable displacement, as the muscles attached to the genial tubercles pull the larger fragment lingually (Plate 4).

The diagnosis of mandibular fractures is usually obvious as there is pain and discomfort accompanied by swelling and ecchymosis. Fractures of the body of the mandible are often compound in the mouth and in the cases of severe facial lacerations can be com-pound in the face. Invariably there is evidence of haemorrhage at the fracture site in compound fractures. Unlike the middle third, the presence of powerful muscles of mastication inserted into the mandible produces gross displacement of the fractured fragments unrelated to the direction of the traumatic force. In fractures of the body of the mandible abnormal movement across the fracture site can be elicited on gentle pressure. Derangement of occlusion and limitation of mandibular movement caused by pain and trismus are present as well as bloodstained saliva. If the inferior dental branch of the mandibular nerve is involved in the fracture, there will be paraesthesia or anaesthesia of the lip. However, the precise physical signs and symptoms of mandibular fractures vary according to the site of the fracture.

Dentoalveolar fracture

In dentoalveolar fractures the teeth are avulsed, subluxed, or frac-tured with or without an associated fracture of the supportive alveolar bone and without a demonstrable fracture of the body of the mandible. The diagnosis of such a fracture is usually obvious because of the derangement of the occlusion, pain, and discom-fort. The fractured segment is usually loose (Plate 5). It is manda-tory to obtain a radiograph of the chest if teeth are missing following injury in order to exclude inhalation (Fig. 4).

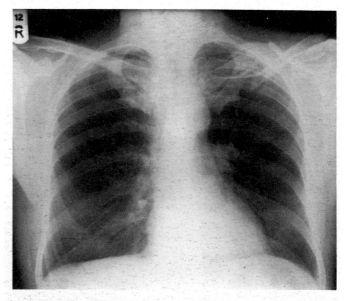

Fig. 4 Chest radiograph showing an inhaled central incisor following a 'hockey' accident.

Examination of the mandible

Both sides of the lower border of the mandible are palpated from behind the patient using the fingers of both hands, starting from the midline and working backwards. Step deformities, particularly at the region of the angle, should be noted.

Swelling of the temporomandibular joint is noted by standing in front of the patient. Gentle palpation of the joints will reveal any tenderness. The absence or presence of movement of the condylar heads is detected by placing the little fingers in both external auditory meati with the pulp of the finger facing forward. The patient is then asked to move the mandible in all directions. Pain, discomfort, and restricted movement are present in the region of the affected condyle. Attempting to move the jaw will worsen the pain. Intraorally, the presence of ecchymosis sublingually following trauma is pathognomonic of mandibular fracture (Plate 6). On suspected fracture sites, particularly those of the midline or lateral to the midline, the thumb and forefinger of each hand are placed on each side of the suspected fracture site and gentle pressure is used to elicit any mobility across the fracture line.

Radiological investigation
Views at right angles to each other are needed to diagnose a fractured mandible.

Lateral oblique (right and left)
The lateral oblique is centred over the angle of the mandible. As well as showing the fractured body of the mandible it may reveal fractures of the condyle (Fig. 5).

Posteroanterior
The posteroanterior view will demonstrate fractures of the body and angle, together with any displacement of the fractured fragments (Fig. 6). Undisplaced fractures of the midline or lateral to

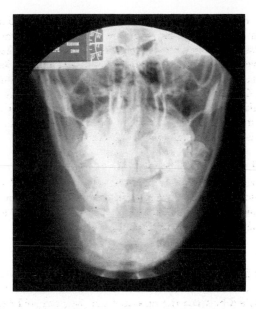

Fig. 6 Posteroanterior film showing a marked displacement of the distal fragment in a fractured mandible.

midline are not easily demonstrated on this view. The condylar head is obscured by the superimposition of the mastoid process.

Intraoral radiographs
Intraoral radiographs are required to demonstrate the condition of the teeth in the fracture line. Occlusal films are invaluable for demonstrating midline or lateral to midline fractures (Fig. 7).

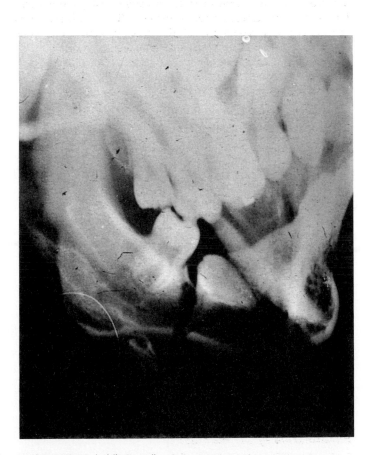

Fig. 5 A lateral oblique radiograph showing a fracture between an unerupted first molar and the second molar.

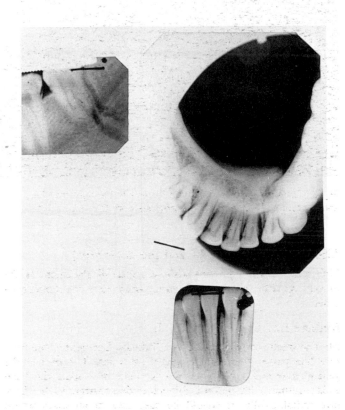

Fig. 7 Intraoral radiographs showing a fractured mandible between the central and lateral incisors as well as the third molar region.

Orthopantomogram

The orthopantomogram is invaluable in detecting fractures anywhere in the mandible. It demonstrates injuries in the condylar region unseen on other radiographic standard views, particularly those of the condyles. Unfortunately this view cannot be obtained on bedridden or unconscious patients (Fig. 8).

Reverse Town's views

Fractures of the condylar neck are best seen on this view.

Temporomandibular joint views

Temporomandibular joint views are used to demonstrate condylar dislocations. If these are taken with the mouth closed and then open, they will demonstrate the functioning of the joint.

Tomography

Tomography is the only helpful radiological view which demonstrates intercapsular fracture of the temporomandibular joint.

Fractures of the middle third

Fractures of the middle third are less common than fractured mandibles. The middle third extends backwards to the frontal bone above and the body of the sphenoid below. It is made up of a number of bones:

(1) two maxillae;
(2) two palatine bones;
(3) two inferior conchae;
(4) the vomer;
(5) the ethmoid and its attached superior and middle conchae;
(6) two nasal bones;
(7) two lacrimal bones;
(8) two zygomatic bones;
(9) two zygomatic processes of the temporal bones;
(10) the pterygoid plates of the sphenoid.

The majority are wafer-thin fragile bones and thus comminute easily. They also articulate with one another in a complex manner that makes it impossible to fracture one bone without disrupting its neighbours (Plate 7).

Fractures of the middle third are usually closed fractures and therefore are difficult to visualize. In severe facial fractures the facial bones can disintegrate into tens of fragments.

Classification of middle-third fractures

The middle third can be further divided into the right and left lateral block (zygomaticomaxillary) and the central block (nasomaxillary) (Fig. 9).

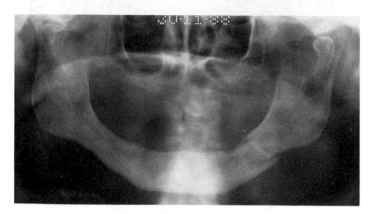

Fig. 8 Orthopantogram in an edentulous mandible showing a bilateral fractured condyle as well as left body of mandible.

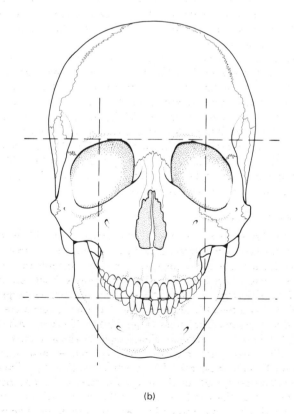

(a)   (b)

Fig. 9 Subdivision of the middle third of the facial skeleton into central and lateral blocks.

Fractures of the lateral middle third

Zygomatic complex fracture

The zygomatic bone (malar bone, cheek bone) usually fractures in the proximity of the zygomaticomaxillary, zygomaticotemporal, and zygomaticofrontal sutures involving the related parts of the maxillary, temporal, and frontal bones. Therefore the term 'zygomatic complex fracture' is more appropriate than zygomatic bone fracture. The bone is unlikely to fracture by itself except in severe trauma.

The zygomatic bone is usually driven inwards into the maxillary sinus. The depression varies in degree according to the fracturing force. Fractures vary from minimally displaced to severe, giving rise to an unsightly flattening of the cheek prominence, best seen by standing behind and above the patient before it is masked by oedema and swelling (Plate 8). There is usually epistaxis on the fractured side as the maxillary sinus fills with blood.

Circumorbital ecchymosis develops within a short time of injury. Unlike a black eye, which is patchy, circumorbital ecchymosis is a generalized swelling around the eye with a uniform intensity limited by the orbicularis oculi (Plate 9).

In the zygomatic complex, the subconjunctival ecchymosis (bloodshot eye) occupies the outer quadrant of the eye with no limit to its extension posteriorly (Plate 10). This is demonstrated by asking the patient to look inwards. Unlike bruising elsewhere in the body, subconjunctival ecchymosis stays bright red in colour until it disappears after 4 to 6 weeks because oxygen can pass from atmospheric air through the thin conjunctiva to oxygenate the haemoglobin.

If the fracture involves the nearby infraorbital nerve, the patient invariably complains of paraesthesia or anaesthesia of the area supplied by this nerve (lower eyelid, lateral side of the nose, the cheek, and the related half of the upper lip). The inward displacement of the zygomatic bone may impinge on the mandibular coronoid process, thus interfering with the lateral movement of the lower jaw.

Temporary diplopia for 1 or 2 days is a common symptom in early stages of zygomatic fractures and is caused by oedema and effusion in and around the eye. Involvement of the extraocular muscles or their nerve supply in the fracture site results in diplopia. If a fracture occurs in the vicinity of the frontozygomatic suture above the Whitnall's tubercle, the globe of the eye is displaced with the fractured zygomatic bone (Plate 11). Severe fracture of the zygomatic complex involving the floor of the orbit can result in herniation of the orbital fat and contents in the maxillary antrum, giving rise to enophthalmos.

'Blow-out' fracture

'Blow-out' fracture is a fracture of the orbital floor without fracture of the orbital rim. It is caused by a sudden rise in intraorbital pressure, such as might occur when a ball or a fist hits the rim of the orbit and forces back the orbital contents without rupturing the globe of the eye. As a result the very thin orbital floor (approximately 0.5 mm thick) is easily disrupted and its contents are displaced into the maxillary sinus (Figs. 10,11). If excess orbital fat herniates into the maxillary sinus, enophthalmos occurs. Occasionally extraocular muscles (inferior rectus and inferior oblique) become incarcerated in the fracture line, thus limiting ocular movement, particularly in upward gaze, and giving rise to diplopia. Lacerations and abrasions of the lids as well as circumorbital and subconjunctival ecchymosis can also occur. The presence of blood in the antrum can give rise to unilateral epistaxis.

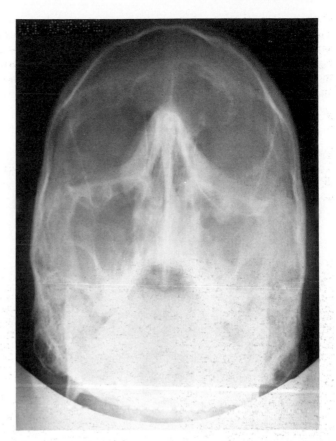

Fig. 10 Occipitomental view showing 'pearl drop' in a blow-out fracture.

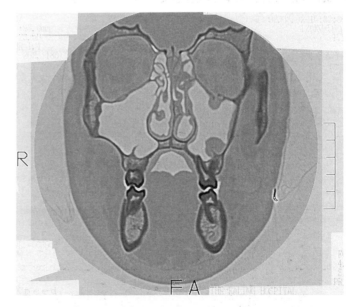

Fig. 11 CT scan of the same patient showing the 'pearl drop'.

Twenty per cent of blow-out fractures occur through the thin orbital plate of the ethmoids. Entrapment of the medial rectus is extremely rare.

Fracture of the zygomatic arch

It is not uncommon for a fracture of the zygomatic arch to occur without any other facial bone fracture. Invariably there is a circular depression (dimple) of depth 1.2 cm overlying the zygomatic arch (Plate 12). The displaced fracture arch impinges on the coronoid process of the mandible, thus limiting the lateral mandibular excursion towards the side of the fracture (Fig. 12). If the fractured zygomatic arch is part of a more extensive zygomatic complex fracture or facial fracture, the above sign is replaced with the more gross physical signs of facial fractures.

Fig. 12 Submentovertex radiograph showing left zygomatic arch.

Fractures of the central block

Dentoalveolar fractures

Dentoalveolar fractures present as a marked derangement of occlusion without a demonstrable fracture of the maxilla. The fractured segment is usually loose (Plate 13).

Le Fort type fractures

Le Fort types of fractures were first described in 1900 by Rene Le Fort in Paris, following experiments on cadavers.[2] Since then, the pattern of these fractures has repeatedly been confirmed both clinically and radiologically (Plate 14). In all the Le Fort type fractures the backward displacement of the tooth-bearing portions leads to positioning of the upper teeth behind the lower incisors.

Le Fort type 1 (Guerin or low transverse fracture) (Plate 14(b)).

This is a horizontal fracture of the tooth-bearing segment of the maxilla. The line of fracture starts from the lower anterolateral edge of the nasal cavity and runs across the canine fossa to the pterygoid maxillary fissure below the zygomatic buttress, involving the lower third of the pterygoid plates. The fracture line crosses medially above the floor of the nose to meet the lateral fracture line behind the maxillary tuberosity. Apart from the mobility and downward displacement of the tooth-bearing alveolus and palate, there are none of the physical signs usually associated with Le Fort type 2 and type 3 fractures (Plate 15). Occasionally the fractured segment is not mobile and only a derangement of the occlusion is observed. This usually takes the form of a deviated maxillary midline.

Le Fort type 2 (pyramidal fracture) (Plate 14(b))

Le Fort type 2 fractures separate the whole of the central block of the middle third of the face from its cranial support. The line of weakness passes through the middle part of the face inside and below the zygomatic bones. It commences in the inferior half of the nasal bone and crosses the frontal process of the maxilla on either side to reach the lacrimal bone above the nasolacrimal canal. The fracture then runs forward along the very thin floor of the orbit, crosses the inferior orbital margin in the region of the zygomaticomaxillary suture, often involving the infraorbital nerve, and traverses the anterior lateral wall of the antrum, crossing the pterygoid maxillary fissure and involving the pterygoid plates halfway.

Le Fort type 3 (high transverse fracture) (Plate 14(b))

This fracture separates the entire bony facial structure from the base of the skull and occurs when a blow of great severity crashes the face backwards, involving the zygomatic bone and the naso-ethmoidal buttress. The line of weakness passes through the naso-frontal suture separating the nasal bones and the frontal processes of the maxilla from the frontal bones. It then extends across the superior half of the lacrimal bone into the orbital plate of the ethmoid and continues backwards to the optic foramen. The strong ring of compact bone around the optic foramen deflects the line downwards to the posterior part of the inferior orbital fissure where the fracture line bifurcates. One limb continues backwards over the upper part of the maxilla in the sphenopalatine fossa to reach the upper limit of the pterygomaxillary fissure to the pterygoid plates, causing them to fracture from the sphenoid bone. The other limb of the bifurcation follows another line of weakness from the anterior part of the inferior orbital fissure to the lateral wall of the orbit, crossing the frontozygomatic suture to meet the outer line of fracture on the infratemporal surface of the greater wing of the sphenoid. The separation from the bone of the skull is completed by fracturing both zygomatic arches and the nasal septum, which is usually comminuted in such fractures.

On superficial examination Le Fort fractures of types 2 and 3 appear very similar. These two fractures can only be differentiated clinically after careful examination and palpation of the zygomatic bone, which is not fractured in Le Fort type 2. This should be confirmed by radiological investigations.

Displacement of the fragments in fractures of the middle third of the face is independent of the muscles of facial expression, but is determined by the degree of violence and the direction of the blow. Within a few hours of injury, patients with severe middle-third fractures assume a very characteristic appearance with three basic clinical signs (Plate 16).

1. Bilateral circumorbital and subconjunctival ecchymosis: this develops rapidly and becomes very pronounced, reaching its peak in 48 h.
2. Balloon face: the enriched blood supply to the face in the absence of deep cervical fascia causes the facial oedema to be gross, giving rise to characteristic ballooning of the face.
3. Lengthening of the face: the frontal bone and the body of the sphenoid bone form an inclined plane which lies at an angle of about 45° to the occlusal plant. In Le Fort type 2 and 3 fractures the facial skeleton is driven down this inclined plane. As a result of the displacement the face is pushed in (dished face), which is more obvious when the oedema has subsided. The posterior teeth of the maxilla push open the mandible, causing a lengthening of the face.

On examination, patients may complain of inability to open their mouths, but in point of fact the mouth is already wide open with bilateral gagging of the molar teeth (Fig. 13). Closure can be achieved by elevating the displaced upper jaw forwards and upwards.

The mobility of the central block is confirmed by grasping the

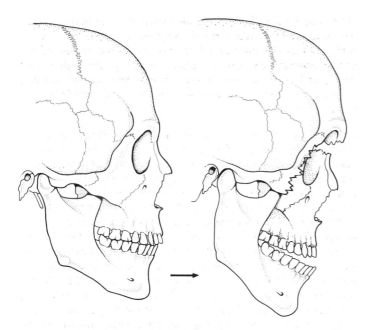

Fig. 13 The downward and backward displacement of the middle third of the facial skeleton forces the lower jaw to gag open.

upper incisor teeth and/or the anterior part of the maxilla between the index finger and the thumb of one hand and moving it gently to and fro against the middle finger and the thumb of the other hand which is located firmly on the frontonasal region.

Damage to the infraorbital nerve may lead to paraesthesia of the cheek, the lateral wall of the nose, the lower eyelid, and the upper lip. Depending on whether there is neuropraxia or neurotmesis, recovery may take up to 18 months. Comminution of the ethmoid bone may lead to dural tear and cerebrospinal fluid rhinorrhoea (Plate 16).

Herniation of the orbital contents through a comminuted orbital floor can result in enophthalmos. Should the inferior rectus and oblique muscles become entrapped in the comminuted orbital floor, mechanical interference will prevent upward and outward rotation of the eye and cause vertical diplopia. Alteration in the level of the globe of the eye will occur if the fracture passes above the origin or the insertion of the Lookwood suspensory ligament which passes from the lacrimal bone medially to Whitnall's tubercle, situated laterally just below the frontozygomatic suture. As the globe of the eye drops the upper lid follows it downwards, giving rise to the physical sign known as 'hooding of the eye'.

Depressed fracture of the zygomatic complex in Le Fort type 3 fractures may impinge on the coronoid process of the mandible and prevent its lateral excursion towards the fractured side.

Nares are invariably blocked with fresh blood and/or dried blood. Occasionally in Le Fort type 2 and 3 fractures severe and complex nasal injuries involve the nasolacrimal duct, resulting in epiphora. A severe cleaving blow directed up the centre of the upper jaw can result in a split palate (Plate 17).

Nasal complex fractures

It is uncommon to fracture the nasal bones alone without fracturing the nasal process of the maxilla with disruption and comminution of the nasal septum—hence the expression 'nasal complex'.

Severe nasal complex fracture can involve the cribriform plates,

resulting in dural tear and cerebrospinal fluid rhinorrhoea. The displacement of the fractured nasal complex will depend on the direction of the fracturing force. If the fracturing force is applied directly to the bridge of the nose, the nasal complex is pushed in and the maxilla is forced out. If the fracturing force is applied laterally, the nasal complex is displaced to one side (Plate 18). Obvious deformity of the nose makes the diagnosis of such a fracture easy. However, oedema and bruising can mask the deformity. The skin on the bridge of the nose often splits and occasionally fragments of the nasal bones are seen through it. Epistaxis occurs from both nostrils. Bilateral circumorbital and subconjunctival ecchymosis are confined to the medial half of the eye.

Rhinorrhoea

If the fracture of the perpendicular and cribriform plate of the ethmoid bone is associated with dural tears, cerebrospinal fluid will escape through the nostrils. This is difficult to recognize in recently injured patients because of the presence of epistaxis, blood clot, and dried blood in the nasal cavities and nares.

The discharge of clear serum following the organization of the blood clot or if the patient has been suffering acute coryza before injury can lead to confusion. However, if there is any doubt about the secretion from the nostrils following a middle-third injury it should be considered as cerebrospinal fluid rhinorrhoea and treated accordingly.

Occasionally, and if the nostrils are blocked, leaking cerebrospinal fluid escapes down the throat to the posterior third of the tongue. In this case a conscious patient may complain of a salty taste. As penicillin and its derivatives do not cross the blood–brain barrier in therapeutically effective concentrations, sulphadiazine should be administered either orally or parenterally to safeguard against meningitis. This therapy should be continued for at least 2 days after the cerebrospinal fluid leakage has ceased. Sulphadiazine is used in conjunction with any other antibiotics employed in treating associated injuries.

Naso-orbital deformity (telecanthus)

Severe displacement of the frontal process of the maxilla, which carries with it the insertion of the medial canthal ligament, produces telecanthus by widening of the nasal bridge.

Radiological investigations

Four radiographs are needed to diagnose all middle-third fractures (Fig. 14).

10 Occipitomental projection

This view gives an indication of the amount of downward displacement of the facial bones. In this view the radiography tube is angled to the feet. Therefore the central ray emerges through the infraorbital margin, displacing the petrous temporal bone downwards and thus clearing the alveolar segment of the maxilla.

30 Occipitomental projection

This view shows the associated backward displacement of the facial bones. In effect, the 10 and 30 occipitomental views act on the same principle applied to radiography of long bones, i.e. they are taken at right angles to one another.

True lateral skull

In general, this is the most useful of the four views. Fortunately, in the unconscious, ill, or unco-operative patient, it is also the most

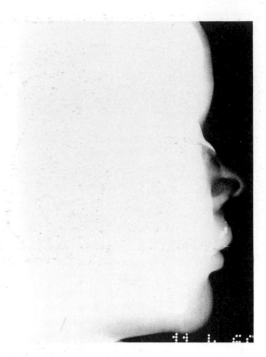

Fig. 15 Soft tissue lateral projection showing a foreign body (a broken tooth) in the upper lip.

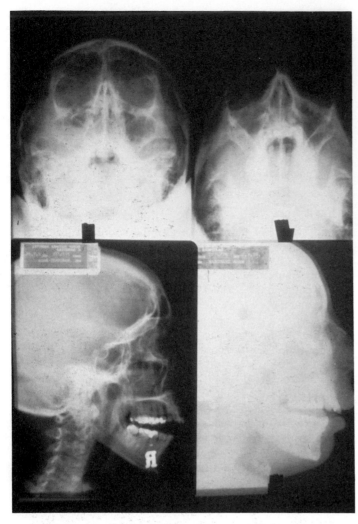

Fig. 14 The four standard radiographs for diagnosis of facial fractures.

accessible view. Nevertheless, complete and careful radiological diagnosis of the facial fractures can only be determined after careful examination of all four projections.

The true lateral projection can show fracture of the inner and/or outer plate of the frontal sinus in conjunction with fracture of the nasal bones. Severe injuries in this region might also show a fracture of orbital roof and cribriform plates and fractures of the anterior cranial fossa. The true lateral view also shows widening of the frontozygomatic suture in high transverse facial fractures (Le Fort type 3); backward and downward displacement of the central block in Le Fort types 1, 2, and 3 can also be seen.

Whenever the central block or the zygomatic bone is fractured, the posterior walls of the antral cavity are disrupted. Disruption of the dense line of the floor of the nose is an indication of fracture of the hard palate.

The presence of the low transverse fracture (Le Fort type 1) can be checked at the anterior nasal aperture. If the lateral view is projected in the supine position (blow-up), fluid levels in the antral cavity can be seen as well as disruption of the pterygoid plates.

Soft tissue lateral projection

From this lateral view, fractures of the nasal bones and spine are clearly visible. Foreign bodies (e.g. broken teeth, road grit, etc.) in the soft tissues of the face, particularly the lips, can also be seen (Fig. 15).

System of interpretation

Except for a fractured zygomatic arch, complete fracture lines are not seen in facial trauma. Radiological diagnosis of facial trauma depends on a knowledge of the fracture patterns that may occur. An overall assessment of the four projections from a distance is important before a closer study is made of the minutiae of the facial bones, since a discrepancy between the size and shape of the orbits, antra, etc., or an open anterior bite from a displacement of either jaw will be immediately noticeable.

The occipitomental projections are viewed first and their interpretation should be based on the five curvilinear lines (Campbell lines) (Fig. 16). The first line curves from one frontozygomatic suture on one side along the supraorbital ridges to the opposite side. The second line follows the curve of one zygomatic arch and the infraorbital margin of one side across the nasal bones and the zygomatic complex of the other side. The third line curves through the condylar neck, the coronoid process, and the lateral and medial walls of the antrum to the opposite side. The fourth line follows the line of occlusion from one side to the other, and the fifth line follows the lower border of the lower jaw from one angle to the opposite.

Facial fractures will show on radiographs as a disturbance of the continuity along one of Campbell's lines. In addition to the standard projections, it is often useful to rely on additional views.

1. Orbital tomography: to confirm blow-out fractures.
2. Submentovertical: an extremely useful view to confirm zygomatic arch fractures.
3. Intraoral radiographs: (a) periapical film to show the condition of teeth in the line of fracture; (b) occlusal film to investigate fractured palate.

Plates for Chapter 7.3 Maxillofacial injuries in sport

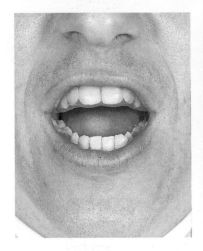

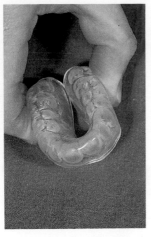

Plate 1 Custom-made, vacuum-formed, soft-bite guard, made of 4-mm thick thermoplastic sheet.

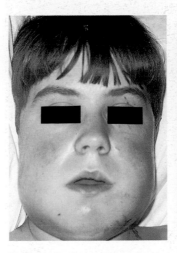

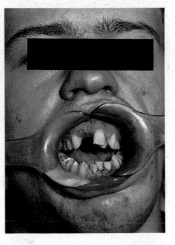

Plate 2 Facial swelling following a bilateral fracture of the condyles.

Plate 3 Anterior open bite in a patient with bilateral condylar fracture.

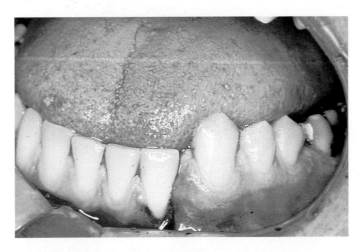

Plate 4 Mandibular parasymphyseal fracture between the lower canine and lateral incisor.

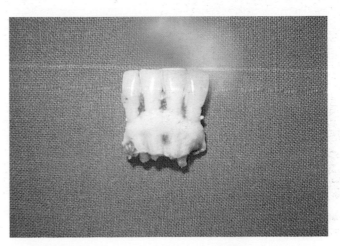

Plate 5 A dissected mandibular dentoalveolar fracture including the four lower incisors.

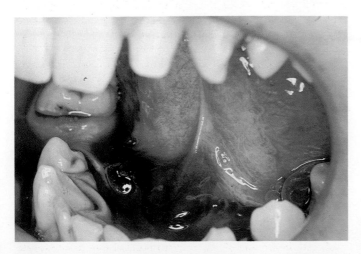

Plate 6 Sublingual ecchymosis is an indicative sign of fractured mandible.

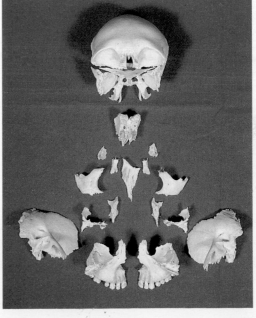

Plate 7 Components of the middle and upper third of the facial skeleton.

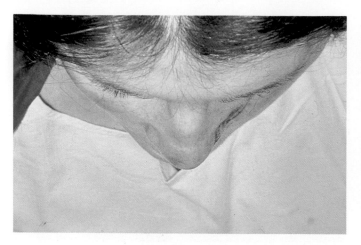

Plate 8 Flattening of the cheek following fracture of left zygomatic complex.

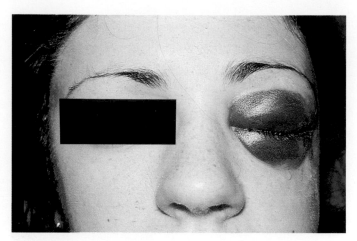

Plate 9 Circumorbital ecchymosis in a fractured left zygomatic complex.

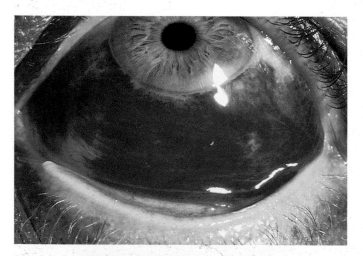

Plate 10 Subconjunctival ecchymosis.

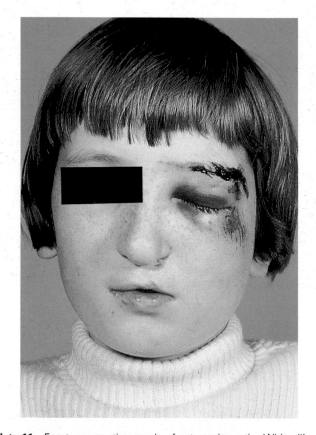

Plate 11 Frontozygomatic complex fracture above the Whitnall's tubercle. Note the drop of the level of the left orbit.

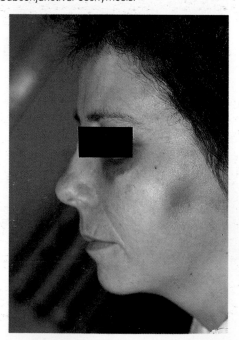

Plate 12 Circular depression 'dimple' overlying a zygomatic arch.

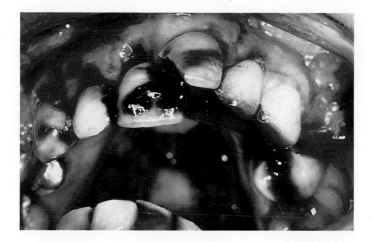

Plate 13 Dentoalveolar fracture of the upper right central and lateral incisors. Note the derangement of occlusion and the dried blood on the palate.

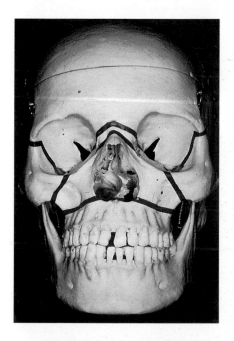
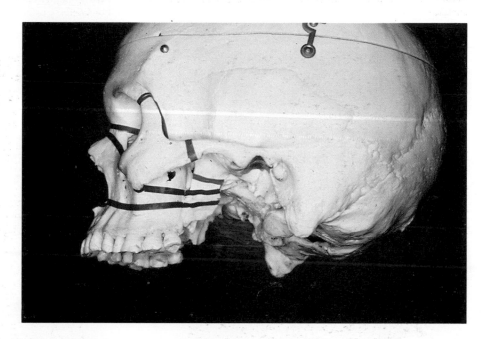

Plate 14 Le Fort lines of fractures of the middle third of the facial skeleton. (a) Le Fort 1 (red). (b) Le Fort 2 (blue). (c) Le Fort 3 (green). (Skull front and lateral views side by side.)

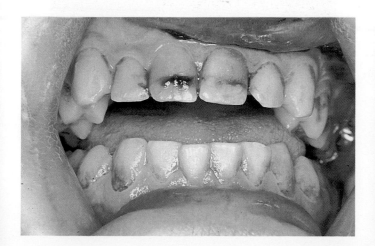
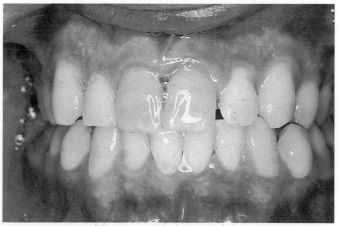

Plate 15 (a) Occlusion in Le Fort fracture before treatment. (b) Same patient after treatment.

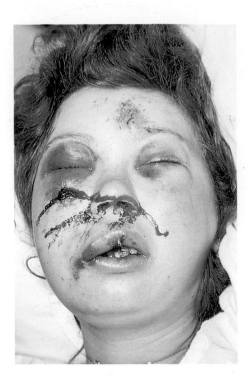

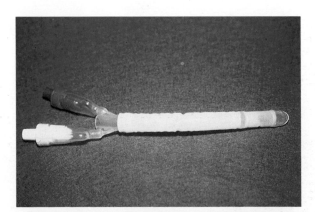

Plate 17 Split palate, missing incisors, and sheared teeth in the Le Fort 2 fracture.

Plate 16 Le Fort 3 fracture. Note the bilateral circumorbital ecchymosis, ballooned face, flattened nasal bridge, and bloodstained CSF and rhinorrhoea from the left nostril.

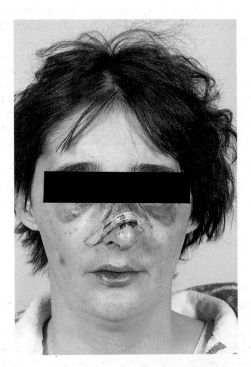

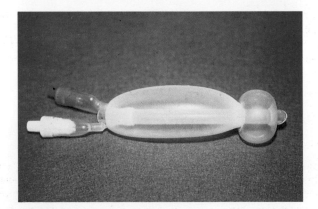

Plate 18 Laterally deviated fractured nasal complex. (Note the bilateral circumorbital ecchymosis.)

Plate 19 Epitek nasal catheters.

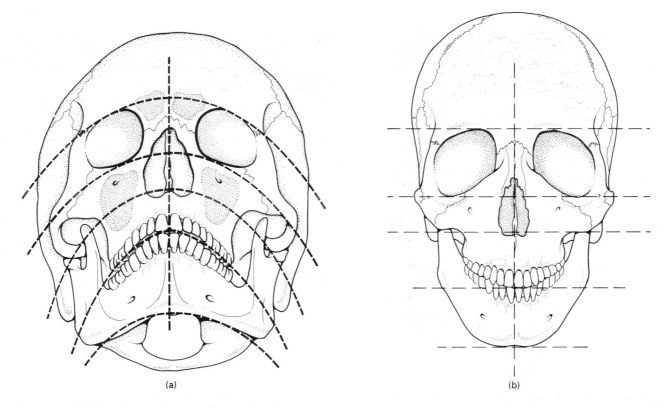

(a) (b)

Fig. 16 Campbell lines system of interpretation of occipitomental radiographs for facial fractures.

Computerized tomography and MRI

CT scans and MRI are able to show considerable detail of soft tissue relating to bone and air spaces in a remarkable anatomical display which is unmatched by any other non-invasive technique. Therefore their use in maxillofacial injury, particularly in extensive fractures in the region of the midface and orbits, is invaluable.

MANAGEMENT OF FACIAL FRACTURES

Facial fractures are never a life-threatening injury unless they interfere with the patency of the airway or are bleeding profusely. The immediate treatment should be directed to the patient's general medical welfare with a thorough physical examination.

Priorities in treatment

Urgent treatment

Airway

Injuries to the middle third of the face can be a source of upper-airway embarrassment, particularly if the middle third is impacted backwards and downwards, and the soft palate is in contact with the posterior third of the tongue and the posterior pharyngeal wall. In such cases two fingers are inserted behind the soft palate into the posterior nasal apertures and the whole of the middle third is forcibly pulled forwards and upwards. The oral cavity should be examined for dentures, broken or whole, together with avulsed loose or broken teeth. It is important to clear the oral cavity of blood and other foreign bodies.

Constant supervision is necessary to confirm that the naso-pharyngeal airway is patent and this is done by frequent aspiration. Occasionally it is necessary to pass an endotracheal tube to secure the patient's airway. The assurance of the airway by intubation is directly related to its patency. Thus regular suction of the lower third of the tube and beyond is of vital importance to secure patency of the lumen against occlusion by blood clot and mucosa. Conscious patients should be nursed in a sitting position with the head forward, provided that there is no medical contraindication to this position, to allow secretion to dribble out of the mouth. Unconscious patients are better nursed on their side so that blood and saliva can drain from the mouth.

The indications for tracheostomy are fewer since the introduction of soft flexible plastic tubes which cause less irritation to the trachea and which can be left *in situ* for days. However, tracheostomy should be considered if the middle third of the facial skeleton is impacted and cannot be brought forward manually or in cases of uncontrollable postnasal haemorrhage or severe oedema of the glottis.

Massive bleeding

Severe bleeding is often internal. Therefore exploration may be urgently required. However, persistence of profuse nasal or oropharyngeal bleeding may be a sign of damage of the large vessels associated with fractures of the base of the skull or to the anterior and posterior ethmoid vessels. The insertion of a post-nasal pack containing a haemostatic agent or Surgitek (Reuter Epitek) is very effective in these cases. The advantage of the latter is that drainage as well as aspiration is possible (Plate 19).

Intracranial bleeding

Close monitoring of the patient using the Glasgow coma scale is important to detect any deterioration. CT or NMR scanning may also confirm intracranial bleeding. Extradural bleeding requires urgent decompression.

Treatment needed promptly

This category includes serious injuries which need prompt attention, perhaps by multidisciplinary specialities. Examples are fractured limbs, pneumothorax, wounds of the bowels, and extensive wounds of the muscles.

Non-urgent treatment

Maxillofacial injuries come under this category. However, it must be emphasized that fractures of the facial skeleton must be diagnosed and treated as soon as the patient's condition permits. If untreated for 3 weeks they are virtually uncorrectable. At best, cosmetic manoeuvres to camouflage the initial injury can be utilized. Furthermore, there may well be a functional disability in biting or vision (diplopia) or difficulty in breathing or lacrimation (epiphora).

FRACTURES OF THE UPPER THIRD

Fractures and clinical manifestations of the upper third are beyond the scope of this chapter. However, depressed and compound fractures of the frontal bone are serious injuries which require special management and treatment.

REFERENCES

1. Chapman PJ. Mouthguards and the role of sporting team dentists. *Australian Dental Journal* 1989; **34**(1); 36–43.
2. Le Fort R. Etude expérimentale sur les fractures de la machôire supérieure. *Revue de Chirurgie* 1900–1; **23**: 208.

FURTHER READING

1. Chapman PJ. The prevalence of orofacial injuries and the use of mouthguards in Rugby Union. *Australian Dental Journal* 1985; **30**(5): 364–7.
2. Jagger RG, Clarke DA. Mouthguards for contact sports. *Dental Technician* 1974; **27**: 88–90.
3. Ranks P, ed. *Killey's fractures of the middle third of the facial skeleton.* 4th edn. Bristol: Wright, 1981.

7.4 Cardiopulmonary and abdominal emergencies in sports medicine

MICHAEL F. MURPHY

INTRODUCTION

The ill or injured athlete is no stranger to emergency departments. Both competitive and recreational athletes constitute a significant proportion of an emergency department's annual census. The vast majority of these athletes have a minor orthopaedic injury that prevents participation for a brief period to allow recovery. Less commonly, the condition poses a permanent threat to long-term athletic participation. It is exceedingly uncommon for such conditions to constitute a true emergency which threatens the life or limb of the athlete. A discussion of cardiopulmonary and abdominal emergencies in sports medicine must concentrate on 'life threats', and therefore are the focus of this chapter.

It is important for the health-care provider involved in the care of the athlete to have an appreciation of the setting and types of life-threatening emergencies that may be encountered. Armed with this index of suspicion, the provider must then develop a strategy to enable a rational and effective approach to the situation. This approach should take into account limitations in personnel and equipment imposed by the particular location of the interaction, i.e. at the scene of the sporting event or in the emergency department. The aim of this chapter is to address these issues.

SCOPE OF THE PROBLEM

The incidence of cardiopulmonary and abdominal emergencies in sports medicine is related to the underlying health of the athlete, the inherent risks of the activity, and the intensity of the exertion.

Jokl, in his exhaustive review 'Sudden death of athletes', presents the historical perspective.[1] The general impression that exercise had the potential to produce primary cardiac pathology and death persisted into this century. What has become apparent is that the vast majority of cardiac-related emergencies, such as sudden death and myocardial infarction, occur in individuals with underlying cardiac disease. Conditions such as hypertrophic cardiomyopathy, rheumatic heart disease, and ischaemic heart disease are examples. Diabetes mellitus, asthma, sickle-cell anaemia, and infectious mononucleosis (glandular fever) are underlying conditions that may predispose the athlete to an acute emergency.

The inherent risks of the activity relate specifically to the potential for high energy impact and the degree of impact.[2] Motorized vehicular events such as automobile, motorcycle, and snowmobile racing are clearly high risk activities for both participants and spectators. The pattern of cardiothoracic and abdominal injury are the same as those with any other vehicular trauma victim. Augmented speed sports include skiing, equestrian events, sledging of all types, cycling, skateboarding, and roller skating. The risk to spectators in these events is negligible, while that to participants remains significant. There is a significant incidence of cardiothoracic and abdominal traumas in contact sports. Fortunately, the majority are trivial. However, the potential for serious injury is real, particularly in sports such as ice hockey, lacrosse, football, and rugby. Even in sports where impact potential is low (soccer, baseball, and basketball) serious and fatal cardiothoracic and abdominal injuries have occurred.[2] Spirited partisan enthusiasm in spectator ranks may result in significant illness or injury, and should be taken into account by the sports medicine health-care provider in planning or providing for care at sporting events. More detailed discussion of this topic is available in the literature, but is beyond the scope of this chapter.

In attempting to delineate the risk of significant illness or injury, it is natural to analyse the intensity of exertion. The competitive athlete is an individual who participates in an organized team or individual sport in which regular competition is a component, a high priority is placed on excellence and achievement, and vigorous systematic training is required.[3] However, the recreational athlete is more focused on personal well-being and enjoyment. In fact, the intensity of exertion fails to stand alone as an independent risk factor and must be related to the presence of underlying disease and the inherent risks of the activity.

OVERVIEW OF EVALUATION AND MANAGEMENT

The level of sophistication of the evaluation and management of the acutely ill or injured athlete is dependent upon three factors: the training and skills of the care provider, the equipment and supplies available, and the degree to which support personnel and services are available. Much of this is dependent on where these activities take place, in the field or in the emergency department, and whether the emergency is traumatic or non-traumatic.

Field

The delivery of medical care in the field is typified as first aid, basic life support, and advanced life support. The distinctions may be blurred as one moves from one echelon of care to the next. The first-aid provider provides just that—first aid. The provider has a basic knowledge of anatomy, physiology, and pathophysiology, and some slight familiarity with medical terminology. The emphasis is on splinting, bandaging, haemorrhage control, the prevention of further injury, and preparation for transport. Some first-aiders are capable of performing cardiopulmonary resuscitation. These individuals constitute the backbone of field medical support units for mass gatherings such as sporting events. Typically, they are committed volunteers who see their role as purely supportive of a more definitive system of medical care delivery.

Basic life support is the minimum level of qualification for ambulance personnel in much of the developed world. These providers typically have been trained in basic and advanced first aid, supplemented with some anatomy, physiology, and pathophysiology. They are minimally literate in medical terminology, although to a utilitarian degree. Principles of safe extrication, splinting, haemorrhage control, and transport are well developed. More advanced basic airway maintenance skills utilizing oral and nasal airway adjuncts and bag-and-mask ventilation, combined with cervical spine stabilization, are expected at this level. Basic life support providers reliably monitor vital signs and perform cardiopulmonary resuscitation if necessary. They should be able to identify an athlete with a potentially life- or limb-threatening condition, which distinguishes them from the first-aider. It is this ability that allows the basic life support provider some latitude in defining how quickly the victim must be moved to more definitive

levels of care. The basic life support provider does not perform tasks normally carried out by physicians, such as endotracheal intubation, defibrillation, and intravenous therapy.

Advanced life support personnel deliver sophisticated medical care and may be physicians. These individuals have a firm grasp of anatomy, physiology, pathophysiology, relevant pharmacology, and medical terminology. What essentially sets them apart is their ability to perform life-saving skills ('medical acts') that have traditionally been the domain of physicians. These skills fall into three categories: advanced airway management (endotracheal intubation), direct current cardioversion and defibrillation, and intravenous fluid and drug therapy. Thus they have the ability to deliver substantial supportive, and in some cases definitive, therapy.

The skills of the provider will generally determine equipment and supply needs. In most cases the providers will carry the necessary equipment with them. However, the sports medicine health-care provider must ensure that the personnel, equipment, and supplies available at the scene match the potential case mix.

Principles of field management

The field management of sports-related injuries generally focuses on the prevention of further injury and the comfort of the athlete. The care provided in this setting is not medically complex, and there is no particular necessity for speed. Cardiopulmonary and abdominal emergencies are quite different because of the potential for a lethal outcome. At the first aid/basic life support level medical care is not complex and the emphasis is on basic care and rapid transport for definitive therapy. However, as advanced life support providers and physicians become involved in field management, the potential exists to deliver very complex care in the field. Consequently, delivery of the athlete to a health-care facility may be delayed. Thus complexity of care and speed become competing priorities. The resolution of this dilemma rests in understanding the distinction between supportive and definitive care. Advanced life support providers deliver definitive care for many 'medical' emergencies such as cardiac arrest, acute respiratory failure, and hypoglycaemia. Time spent at the scene is justified to deliver such care and achieve an optimum outcome. However, the definitive care of the trauma victim occurs in a hospital, and often in an operating theatre. For this reason, only essential supportive care such as cervical spine immobilization and definitive airway management should precede expeditious transport of the athlete with cardiothoracic or abdominal trauma.

Emergency department

The principles of management of the athlete with a cardiopulmonary or abdominal emergency are no different from those of any other emergency patient. Knowledge of the patterns of illness or injury associated with various athletic endeavours are essential in planning a diagnostic and therapeutic course. As in the field, it is important to understand the distinction between supportive and definitive care. This is particularly true of the injured athlete where resuscitation and diagnosis occur simultaneously, while emergency physicians and surgeons determine the most appropriate course of action.

MEDICAL EMERGENCIES

Pulmonary emergencies

Exercise-induced asthma (exercise-induced bronchospasm), pulmonary embolism (thromboembolism and air embolism), and spontaneous pneumothorax constitute the most common life-threatening emergencies in this category.

Exercise-induced asthma is discussed in detail elsewhere in the text. However, it constitutes an important life-threatening entity and should be mentioned here. Asthma is defined as the presence of intermittent wheezing, chest tightness, and cough, together with bronchial hyper-responsiveness.[4] Pathologically, most asthmatics demonstrate infiltrates of immunocompetent cells, particularly eosinophils, neutrophils, lymphocytes, and mast cells, in the bronchial mucosa, and inflammation is a prominent feature of the disease. Certain triggers cause the release of preformed substances from these inflammatory cells leading to bronchospasm. In the case of exercise-induced asthma, it is believed that exercise and hyperventilation with dry air cause mucosal drying (mucosal cooling may have an additional effects), leading to changes in osmolarity and release of mediators. Sodium chromoglycate and β-agonists are useful in the mitigation of exercise-induced asthma.

Despite advances in our understanding of the disease and its management, the death rate from asthma is rising. Fatal or near-fatal asthma has been associated with profound emotional upsets, thermal inversions, the use of β-blockers, and the ingestion of acetylsalicylic acid or non-steroidal anti-inflammatory drugs (NSAID) in sensitive patients. Thus far, exercise alone has not been implicated.[5] The vast majority of deaths are preventable and result from a failure of the physician, the patient, or both to recognize the seriousness of the episode.[6]

Standard management of an athlete with exercise-induced asthma should include inhaled sodium chromoglycate and β-agonists as prophylaxis, as indicated. The mainstay of therapy for an acute attack is inhaled β-agonists. Failure of a moderate to severe attack to respond promptly to such measures should motivate urgent referral to an emergency department.

Pulmonary thromboembolism in otherwise healthy individuals is not common. It is decidedly rare in young and vigorous athletes. However, the syndrome of pulmonary thromboembolism as a complication of 'effort thrombosis' of the upper or lower extremity is well described in athletes.[7] The death rate from untreated pulmonary thromboembolism is in the range 20 to 30 per cent. If diagnosed and treated appropriately, the mortality is less than 10 per cent. The diagnosis of pulmonary thromboembolism is frequently elusive, in large part because of the vagaries of its presentation. Pleuritic chest pain, dyspnoea, cough, haemoptysis, or syncope may be present. Tachypnoea and tachycardia are often, but not invariably, present. Hypoxaemia is usually present on arterial blood gas measurement. Having said all this, the key to the diagnosis of pulmonary thromboembolism remains vigilance and a high index of suspicion, particularly in an athlete with acute or subacute pulmonary symptoms.

Air embolism, in the context of sports medicine, is isolated to recreational scuba diving. The cause of air embolism in scuba divers has been attributed to barotrauma from overdistension of the alveoli and pulmonary venous vasculature, with subsequent air embolization induced by the negative intrathoracic pressures which occur during inspiration after surfacing.[8] Predisposing factors include uncontrolled ascent, breath-holding on ascent, lung cysts or blebs, or obstructive pulmonary disease. Once air has entered the pulmonary venous system, cardiac and distal arterial air embolization becomes the risk. The usual presentations include focal neurological deficits, seizures, coma, and cardiac arrest at or near the time of surfacing. Definitive therapy is recompression in a hyperbaric oxygen chamber. However, in the interim the diver should be given 100 per cent oxygen and maintained head down in the left lateral position unless cardiopulmonary resuscitation is under way. This position is believed to trap the air in the apex of the left ventricle and prevent the buoyant bubbles from travelling to the brain.

A spontaneous pneumothorax, heralded by the relatively sudden onset of pleuritic chest pain and dyspnoea but not associated with trauma, is uncomfortable but usually no cause for alarm. However, if the collection of air comes under tension as additional air is forced out of the parenchymal defect into the pleural space during expiration, mediastinal shift and impedance to venous return produce a life-threatening situation. The picture is one of air hunger, shock, distended neck veins, and decreased air entry on the affected side. Tracheal deviation away from the affected side is an inconsistent sign. The immediate management is by needle thoracostomy in the second intercostal space of the midclavicular line. Small asymptomatic spontaneous pneumothoraces may be followed in the outpatient department, provided that compliance with post-discharge instructions can be assured. Alternatively, small calibre chest tubes attached to one-way flutter valves (Heimlich valve[R]) may be managed in the outpatient department. Patients with large or persistent pneumothoraces, or significant symptoms, require tube thoracostomy and hospital admission.

Cardiovascular emergencies

Sudden collapse and death in the athlete is a rare event, but when it does occur the predominant causes are cardiac in nature. More rarely, it is related to an intracranial catastrophe or a disorder of temperature regulation.[1] The definition of sudden death is not straightforward. The World Health Organization defines sudden death as that occurring within 24 h of the onset of illness or injury.[9] The American Heart Association has adopted a modified definition that accounts for the majority of deaths due to cardiac disease excluding vascular causes.[10] 'Arrhythmic death' is the abrupt loss of consciousness and disappearance of pulse without prior collapse of the circulation. 'Death due to myocardial failure' is the gradual circulatory failure and collapse of the circulation before disappearance of the pulse. Time frames are not defined, nor do they appear to be relevant. For the most part, athletes who die suddenly have underlying cardiovascular disease with myocardial ischaemia as the proximate event manifesting itself as a malignant ventricular arrhythmia with collapse.[11] These cardiac arrest rhythms are ventricular fibrillation, pulseless ventricular tachycardia, ventricular asystole, and electromechanical dissociation. Cardiac arrest occurs during or immediately after exertion, and is often unheralded by premonitory symptoms. Athletes aged less than 35 who die suddenly usually have some form of congenital heart disease. Acquired disorders predominate in those aged over 35.

Sudden death at age less than 35 years

These athletes are predominantly male, 1 to 18 years of age, and involved in interscholastic sports.[3] The underlying cardiac disorders include hypertrophic cardiomyopathy, idiopathic concentric left-ventricular hypertrophy, and congenital anomalies of the coronary arteries. Mitral valve prolapse, myocarditis, aortic valvular stenosis, sarcoidosis, cardiac conduction system abnormalities, coronary artery disease, and aortic rupture due to 'cystic medial necrosis' (e.g. Marfan's syndrome) are rare causes of sudden death in the young athlete and are mentioned only for completeness.

In most cases hypertrophic cardiomyopathy, also known as asymmetric septal hypertrophy or idiopathic hypertrophic subaortic stenosis, is a genetically transmitted disorder following an autosomal dominant pattern with a high degree of penetrance. Pathologically, the left ventricle is markedly hypertrophied, particularly in the ventricular septum. Increased numbers of abnormal intramural coronary arteries with thickened walls and narrowed lumens are present.[3] Functionally, obstruction to left-ventricular ejection may occur and is potentiated by a reduction in left-ventricular end-diastolic volume or increased myocardial contractility (e.g. exercise). Dyspnoea, angina, presyncope, and syncope are the most common presenting symptoms, when they exist. Unfortunately, sudden death is a frequent presentation of this disorder in the athlete. The probable sequence of events is obstruction to ejection, leading to ischaemia and a malignant ventricular arrhythmia. Sudden death in a young athlete should prompt echocardiographic evaluation of first-degree relatives, at a minimum.

Some investigators have found severe concentric left-ventricular hypertrophy in young athletes who die suddenly.[3] It is not characterized by genetic transmission or asymmetric left-ventricular hypertrophy and may be related to severe systemic hypertension in the athlete.

Congenital abnormalities of the coronary arteries have been incriminated as the cause of sudden death in young athletes. Abnormal origin of the vessel or functional obstruction to flow are believed to lead to myocardial ischaemia and cardiac arrest.

Idiopathic mitral valve prolapse is a very common cardiac valvular disorder with a prevalence of about 5 per cent in the general population. Most (66 per cent) are female and the vast majority are asymptomatic. Although exceedingly rare, the most feared complication of mitral valve prolapse is sudden death. Available data would suggest that it is reasonable to limit athletic endeavours in mitral valve prolapse patients with a history of syncope, disabling chest pain, complex ventricular arrhythmias (particularly if induced or worsened by exercise), significant mitral regurgitation, prolonged QT interval on the ECG, Marfan's syndrome, or a family history of sudden death.[12]

The American Heart Association has published recommendations regarding recreational and occupational activity levels for young patients with heart disease to aid physicians in counselling.[13]

Sudden death at age over 35 years

Increasing numbers of competitive recreational athletes are aged over 35. As is the case with younger athletes, these athletes may have underlying cardiovascular disease. However, the condition is more often 'known' and 'acquired' in this population, than 'occult' and 'congenital' as is the case with younger athletes. Up to 50 per cent of sudden deaths in this age group, during or immediately after physical activity, occur in persons with known or symptomatic cardiovascular disease.[14] The vast majority of deaths are due to coronary heart disease with severe atherosclerotic narrowing of one, two, or three epicardial coronary arteries.

Perhaps the most disturbing contribution to the incidence of sudden death among athletes is the growing usage of performance-enhancing medications by athletes. Predictably, the most dangerous medications are those 'stimulants' which appear to improve performance and stamina, with cocaine achieving much notoriety. All stimulant-type medications, including amphetamines, phenylpropanolamines, ephedrine and its derivatives, and cocaine have the potential to cause large increases in heart rate and blood pressure with lethal outcomes. It is clear that a lethal outcome is not necessarily related to dose, route of administration, or the presence of underlying cardiovascular disease, although a correlation exists.

TRAUMATIC EMERGENCIES

Much of the focus of sports medicine is on the care of the injured athlete. Far and away the most common injuries are those to the musculoskeletal system, particularly the extremities. However, the most lethal are those to the central nervous system (including spinal cord), the chest, and the abdomen. Although serious cardiopulmonary and abdominal trauma can occur in all sports, the risk of serious injury is directly proportional to the potential for high energy impact and its amount. Motorized vehicular events, augmented speed sports, and contact sports are most commonly implicated.

Death from trauma has a trimodal distribution.[15] The first peak is within second to minutes and is associated with such severe injury that salvage is only rarely possible even in the most sophisticated emergency medical services systems. The second peak occurs from minutes to hours after the injury and may be significantly blunted by applying the fundamental principles of trauma care. The third peak occurs days to weeks after the initial injury and is ordinarily related to sepsis and multiorgan failure.[15]

Even in a high volume trauma centre emergency department, immediately life-threatening injuries are not common. The keys to detection and optimum outcome are index of suspicion, the recognition of important, but sometimes subtle, symptoms and signs, and the ability to access definitive care in a timely fashion. A prioritized approach that is rapid and comprehensive, emphasizing simultaneous evaluation and management of the injured athlete, is essential. Initially, the life threats, the settings where they are likely to occur, and their presentation will be analysed. Later, an organized and consistent approach to the injured athlete will be presented.

Cardiopulmonary trauma

Although blunt or penetrating trauma to the airway and chest would be expected to be hazards of virtually any athletic endeavour, there is scant literature on the topic. What is available relates only to blunt trauma and, in particular, blunt trauma to the heart owing to its potential lethality. The American College of surgeons in their Advanced Trauma Life Support Course categorized airway and chest injuries into 'immediately' and 'potentially' life-threatening categories.[15]

The immediate life threats are those that produce acute cardiopulmonary failure leading to cellular hypoxia and death, and include airway obstruction, tension pneumothorax, open pneumothorax, massive haemothorax, flail chest, and cardiac tamponade. Potentially life-threatening injuries include myocardial contusion, pulmonary contusion, aortic disruption, traumatic diaphragmatic hernia, tracheobronchial disruption, and oesophageal disruption.

Airway obstruction is the major life-threatening consideration in any patient presenting with neck or maxillofacial trauma. Since gunshot and stab wounds are decidedly uncommon in the athlete, the mechanism of injury will be blunt trauma. Motorized vehicular sports may produce dramatic maxillofacial or neck injuries in the setting of multiple trauma, with the potential to overlook subtle but important laryngeal signs. Projectiles such as hockey pucks, lacrosse balls, and baseballs have produced potentially life-threatening blunt laryngeal trauma. Any contact or racket sport possesses the potential to inflict serious neck injury. Regardless of how benign these injuries may appear, the potential for rapid deterioration to total airway obstruction and death must be appreciated. The threat to airway integrity with oral and maxillofacial injuries is usually evident, and is secondary to the pooling of blood and saliva in the hypopharynx, massive oedema, or foreign bodies such as teeth or dentures. Blunt neck trauma may be much more insidious. The classic presentation of impending upper airway obstruction has the patient seated, leaning forward, drooling saliva, and stridorous. It is important to realize that the athlete who is simply 'a little hoarse' after blunt neck trauma may also rapidly deteriorate. This hoarseness may be a subtle sign of serious laryngeal injury or oedema. These athletes must be quickly transported to a facility where a full assessment of airway integrity, including radiography and laryngoscopy, can be performed. In the unlikely event that the sports medicine care-giver is faced with a penetrating neck injury, it is axiomatic that the wound should not be probed with fingers or instruments, in case a non-life-threatening situation is made into one that is. Formal evaluation in a setting where definitive care can be rendered is imperative in such situations.

Tension pneumothorax is an uncommon clinical condition most often encountered in the setting of a patient who is being manually or mechanically ventilated, particularly if high pressures are required to achieve adequate ventilation. Blunt chest trauma is the aetiology in the context of sports medicine. Athletes with a past history of pneumothorax or those with underlying obstructive pulmonary disorders, such as asthma, may be at increased risk, particularly following a blow to the chest. Sports that create a setting for multiple trauma (motorized vehicular events and augmented speed sports) are also implicated. The condition develops when the visceral pleura is disrupted, allowing air to escape into the pleural space. If this air leak acts as a one-way valve with additional air entering the pleural space with each expiration (but unable to escape from the pleural space), the pressure in the pleural cavity eventually rises and compresses mediastinal structures and the opposite lung. Mediastinal compression leads to inadequate cardiac filling and shock. The loss of lung volume in the ipsilateral lung and the compression of the opposite lung lead to acute respiratory failure with hypoxaemia and hypercarbia. Tension pneumothorax is a clinical, not a radiological, diagnosis and immediate decompression is mandated.[15] The classic clinical findings include marked respiratory distress, decreased breath sounds on the affected side, tracheal deviation to the opposite side, distended neck veins, and shock. The patient who is alert, with a single system injury, is easily identifiable and likely to fit this mould. In the author's experience, the classic presentation is more difficult to appreciate in the setting of multiple trauma when it is necessary to work through the diagnostic possibilities for the patient with moderate to severe respiratory distress, shock, and distended neck veins (or a central venous pressure that is normal

or elevated if such a line is in place). In the setting of blunt chest trauma these possibilities include tension pneumothorax, cardiac tamponade, and myocardial contusion. The diagnosis is made when the chest is vented and produces a gush of air, and the vital signs improve. Definitive therapy is tube thoracostomy with a chest tube 36 French or larger.

Open pneumothorax is due to a penetrating or avulsing type of injury, and has not been mentioned in the sports medicine literature to date. The priorities include consideration of additional injuries, particularly to lung, heart, and vascular structures, and the management of the pleural space. Particular hazards relate to the development of a tension pneumothorax, blood loss, or a 'sucking chest wound'. A sucking chest wound is typified by air preferentially entering the pleural cavity through the chest wall defect rather than entering the lung via the trachea during inspiration. Management involves covering the wound with an occlusive dressing. A one-way valve that allows the escape of pleural space air can be fashioned by leaving one side of the dressing unsecured. A tension pneumothorax can thus be avoided. Definitive management will range from tube thoracostomy and routine wound care to formal surgical closure, depending on the characteristics and behaviour of the wound.

Massive haemothorax is defined as the loss of more than 1500 ml of blood into the pleural space. The disorder is exceedingly uncommon, and in civilian emergency practice is almost always the result of penetrating chest trauma. Its presentation in sports medicine is in the athlete with severe blunt chest trauma. The rate and volume of blood loss are variable and depend upon the size of the injury in the vessel and the pressure within it. Tube thoracostomy and volume resuscitation constitute definitive therapy unless ongoing blood loss exceeds 200 ml/h, at which time thoracotomy may be indicated.

Flail chest should be considered not as just a chest-wall disorder secondary to multiple rib fractures, but as a syndrome of segmental chest wall instability combined with visceral thoracic injury of varying severity. The syndrome should be considered in any athlete sustaining significant blunt trauma. Flail chest is produced by a blunt force sufficient to fracture three or more adjacent ribs, each at two points. The flail segment may be on either side of the thorax or may be central if it involves the sternum. If the disorder is unilateral, the lung is the injured organ, leading to hypoxia of varying degrees depending on the severity of the pulmonary contusion underlying the flail segment. Paradoxical motion of the unstable chest wall segment with inspiration and expiration contributes to this hypoxia. The flail segment is often not visualized acutely because of the time course of compliance changes related to the lung injury. Chest wall palpation, radiographic findings, and hypoxia on arterial blood gas determination point to the correct diagnosis. Supplemental oxygen is the initial step in management, and may culminate in endotracheal intubation and mechanical ventilation if respiratory failure progress. A central flail segment has the potential to produce significant trauma to the underlying heart, most commonly myocardial contusion which will be discussed in more detail below.

Cardiac tamponade most commonly results from penetrating trauma and must be exceedingly rare in recreational or competitive athletics, although the scenario can be imagined. The pericardial sac is a rigid fibrous structure that encases the heart. The introduction of even small amounts of blood, other fluids, or air into this sac compresses the heart and limits its ability to fill with blood. The result is a reduction of cardiac output and, potentially, shock. The classic presentation is known as Beck's triad and consists of hypotension, elevated venous pressure, and muffled heart sounds. The key to diagnosis is index of suspicion. Initial resuscitation consists of intravenous fluids. Pericardiocentesis is indicated if these measures fail to produce haemodynamic stability. The author uses a large (e.g. 16 gauge) intravenous catheter over needle arrangement and leaves the plastic cannula in the pericardial space to provide continuous drainage.

In the setting of trauma, the patient who presents with shock and a normal or elevated venous pressure is suffering from one of three disorders: tension pneumothorax, cardiac tamponade, or myocardial contusion. If deterioration is rapid, in addition to providing supplemental oxygen and intravenous fluids, the clinician should initially needle both sides of the chest to detect and manage a tension pneumothorax. If this manoeuvre fails to produce a diagnosis, the next step is to attempt a periocardiocentesis.

Potentially life-threatening disorders include myocardial contusion, pulmonary contusion, aortic disruption, traumatic diaphragmatic hernia, tracheobronchial disruption, and oesophageal disruption. Presentation is usually insidious rather than dramatic, although if overlooked these disorders produce an increase in mortality. With the exception of myocardial contusion, none are specifically mentioned in the sports medicine literature. However, the potential exists for any of these disorders to occur in the traumatized athlete and thus they deserve mention.

The ability of cardiac trauma to result in dysrhythmias and dysfunction must be appreciated. Sudden death has been reported in young athletes who have received blows to the chest. Some chest injuries, such as those resulting from a ball (thrown or hit) or a punch during a boxing match, are seemingly trivial in degree.[16] Another injury has involved falling with considerable force on a football and delivering a blow to the chest.[17] Blunt myocardial injury can be divided into three entities: cardiac concussion, myocardial contusion, and cardiac rupture. Cardiac concussion, the existence of which is in some dispute, is not associated with morphological evidence of injury, and cardiac enzymes are not elevated on serum testing. However, lethal arrhythmias such as ventricular fibrillation may result. Myocardial contusions, however, while difficult to diagnose, are associated with morphological cell damage. Cardiac enzymes may be elevated, the ECG may be abnormal (sinus tachycardia, non-specific ST segment and T wave changes, and disturbances of conduction and rhythm), and echocardiography typically demonstrates segmental wall motion abnormalities. Lethal cardiac arrhythmias may result. Large contusions with significant cardiac dysfunction may lead to shock and must be differentiated from other causes of non-hypovolaemic shock in the setting of trauma, such as tension pneumothorax and cardiac tamponade. Cardiac rupture in the setting of blunt trauma is a uniformly lethal condition. However, penetrating cardiac injury may be potentially survivable, but only in the most efficient emergency medical services systems.

A pulmonary contusion is characterized by intrapulmonary haemorrhage and oedema. Gas exchange is compromised, and the damaged area of lung becomes stiff and non-compliant. Hypoxaemia and hypercarbia result. It may take some time for these changes to develop fully, necessitating a high index of suspicion and ongoing monitoring of the athlete at risk. In fact, initial arterial blood gases and chest radiography may be normal. The association of pulmonary contusion with flail chest is described above.

Ninety per cent of those who sustain a traumatic aortic injury die at the scene of the accident. The mechanism of injury is rapid deceleration, such as a skier hitting a tree or falls from height as may occur in many sports. Of those who survive to reach hospital,

50 per cent will die if the injury is left unrecognized or unrepaired.[15] Aortic rupture most commonly occurs where the ligamentum arteriosum attaches to the distal arch. Less commonly, it occurs at the aortic root or where the aorta pierces the diaphragm. The diagnosis is the result of a high index of suspicion combined with radiological findings. Symptoms and signs are non-specific and vague. Chest or back pain, dyspnoea, hoarseness, and dysphagia may be present. A difference in blood pressure between arms, or between the arms and the legs, as well as a systolic murmur or bruit may be noted. A supine anteroposterior chest radiograph taken as a matter of routine in all cases of significant trauma may reveal any of the following: superior mediastinal width exceeding 8 cm, distortion or blurring of the aortic outline, presence of a left pleural cap, elevation of the right and depression of the left mainstem bronchi, fractures of the first or second ribs, obliteration of the clear space between the pulmonary artery and the aorta, or deviation of a nasogastric tube to the right.[18] The gold standard for diagnosis remains aortography, and it should be performed if rupture is suspected. Computed tomography, although useful for screening, is not definitive and may lull the physician into a false sense of security. Definitive therapy is surgical.

Trauma to the diaphragm, tracheobronchial tree, and oesophagus must be exceedingly rare in athletes. Even if they occur, their time frames and modes of presentation make their diagnosis and management a matter of concern for emergency physicians and surgeons rather than sports medicine personnel.

Abdominal trauma

Much more has been published in relation to abdominal trauma in the athlete than on cardiopulmonary trauma. It has been reported that 10 per cent of abdominal injuries are sports related.[19] Motorized vehicular events, augmented speed sports, and team contact sports have produced the majority of these injuries. As with cardiopulmonary injuries, penetrating trauma is rare and the vast majority of cases are the result of blunt trauma. Both solid and hollow viscera may be injured.[20]

The evaluation and management of an athlete with abdominal trauma is particularly challenging. Even in the face of life-threatening intra-abdominal injury, symptoms and signs may be absent or subtle. Up to 20 per cent of patients with blood in the peritoneal cavity will have a benign abdominal examination when first examined in the emergency department. Such patients often have a head injury or other painful injury that overshadows or masks the abdominal findings. The importance of a high index of suspicion in patients with abdominal trauma cannot be overemphasized. The aim of the initial assessment of patients with serious abdominal trauma is to determine that an intra-abdominal injury exists and that operative intervention is required, not to diagnose accurately injury to a specific organ.[15] Once it is determined that immediate surgery is not required, more time-consuming and specific diagnostic manoeuvres can be undertaken to pin-point specific organ injuries.

Blunt and penetrating trauma produce much different patterns of injury. Penetrating trauma tends to be rather easier to identify in that there is outward evidence of abdominal injury, i.e. the entry or exit wound. In general terms, the injury tends to be limited to the abdominal cavity, although this is not always the case, and the organs inured tend to lie in the path of the penetrating object. Any gunshot wound violating the periotoneum, and the vast majority do, requires operative exploration because of the frequency of significant intra-abdominal injury. Peritoneal violation is more difficult to document in stab wounds unless the patient exhibits signs of peritoneal irritation or blood loss with haemodynamic instability, and operation is indicated. In most cases this is not present and the wound is explored under local anaesthesia. If the wound track is clearly entirely extraperitoneal, no further evaluation is necessary. However, if the exploration is equivocal or demonstrates peritoneal violation, a diagnostic peritoneal lavage is indicated. Some centres utilize CT scanning instead of peritoneal lavage at this stage, particularly in children. The probability of organ injury can be determined by these procedures and a decision made as to the necessity for operation.

While penetrating trauma does not discriminate between hollow and solid viscera, blunt trauma has a decided predilection to affect solid organs such as the spleen, liver, kidney, and pancreas. Direct blows coupled with shear forces related to sudden deceleration are the probable mechanisms. Hollow viscus injury is believed to occur when a sudden increase in intraluminal pressure leads to rupture. Ruptures of the large and small bowel, the gallbladder and the urinary bladder due to sports-related trauma have all been reported.

The appropriate evaluation and management of abdominal trauma depends on a clear understanding of the anatomy involved and the patterns of injury as discussed above. The abdomen has several distinct regions: the peritoneal cavity, the retroperitoneum, and the pelvis. The peritoneal cavity is divided into intrathoracic and abdominal compartments. At the limit of expiration the intrathoracic abdomen extends from the fourth intercostal space to the inferior costal margin. Thus the rib cage affords an element of protection for the liver, spleen, stomach, and transverse colon if a blunt force is applied. However, it is important to recognize that penetrating chest trauma below the fourth intercostal space (nipple level anteriorly, inferior tip of scapula posteriorly) may violate the diaphragm and produce intra-abdominal injury. It is important to realize that such diaphragmatic injuries heal poorly and produce an opportunity for internal herniation of gut into the chest to occur. Subsequent strangulation is not infrequently fatal, emphasizing the importance of diagnosing and repairing such injuries at the time that they present. CT scanning and diagnostic peritoneal lavage are manoeuvres employed to establish the diagnosis; the latter is more sensitive in the author's experience. The retroperitoneum contains the great vessels—the pancreas, the kidneys and ureters, and portions of the duodenum and colon. The pelvis houses the rectum, the urinary bladder, the major pelvic vessels and, in the female, the internal genitalia.

APPROACH TO AN ATHLETE WITH AN ACTUAL OR POTENTIAL LIFE THREAT

The evaluation and resuscitation of the critically ill or injured athlete is strongly dependent on the ability of the provider to undertake both tasks simultaneously. The ultimate outcome will rely on the ability to undertake a rapid and prioritized assessment, intervene appropriately, and make key decisions as necessary. The knowledge base and skills of the provider, the equipment and supplies available, the presence of qualified help, and the setting of the resuscitation are powerful determinants of the outcome. Preplanning has the effect of favourably influencing some or all of these. The standardized approach developed by the American Heart Association for Advanced Cardiac Life Support (ACLS) and the Committee on Trauma of the American College of Surgeons for Advanced Trauma Life Support (ATLS) provides an excellent preparation for such providers.[10,15] The principles of simultaneous prioritized evaluation and treatment are equally applicable to trauma and non-trauma situations.

The initial phase of the evaluation addresses the immediate priorities of airway, breathing, and circulation (ABCs), is known as the primary survey and is common to both types of emergencies. Trauma and non-trauma strategies tend to diverge at this point. Life-threatening non-perfusing cardiac arrest rhythms make up the majority of the non-trauma emergencies and lend themselves well to early definitive electrical and pharmacological interventions. Successful resuscitation is followed by post-resuscitation management. For the traumatized athlete the primary survey identifies the immediate life threats, intervenes with definitive airway management, and initiates haemodynamic support as needed. The primary survey then blends imperceptibly into the resuscitation phase where the issues of haemorrhage control, shock management, and respiratory support continue to be addressed. The secondary survey or head-to-toe evaluation, is then undertaken to determine potential threats to life or limb and the requirements for definitive care. This final phase addresses the indications for, and urgency of, operative and non-operative care, as well as stabilization of the patient in preparation for transfer to a facility that can provide a higher level of care.

Primary survey

The primary survey is a rapid evaluation designed to detect and attempt to reverse immediate life threats. The ABCs are the focus, and at each step of the evaluation the provider employs the 'look, listen, and feel' format. Life-saving interventions are performed as the need is determined. On approaching the ill or injured athlete the provider immediately surveys the scene for hazards, and if safe proceeds to the victim's head. While ascertaining the circumstances of the event, the provider observes the victim and gently places a hand on either side of the head to maintain cervical spine alignment and provide reassurance. Depending on the setting and expertise of available personnel, two large-bore intravenous infusions should be initiated, 100 per cent oxygen applied, and the patient placed on a monitor and undressed. While listening to available history, verbalization by the patient is noted and if not spontaneous, solicited. If the patient is able to speak, airway and breathing are sufficient to render immediate intervention unnecessary. If the patient does not speak, the chest is observed for respiratory effort and a hand is placed near the mouth and nose to feel for air movement. One listens for a stridorous or obstructive quality to the breathing. A decision is then made as to the adequacy of the airway.

Management of the airway is coupled with a consideration of the likelihood of cervical spine instability because airway manoeuvres may also cause cervical spine movement. In most cases the decision as to whether precautions are necessary is clearly evident. However, when in doubt one should err on the side of caution and maintain cervical spine immobility. The cervical spine moves in six directions: flexion and extension, right and left lateral flexion, and right and left rotation. A variety of immobilizing devices are commercially available but offer little advantage over simply taping the forehead to either side of the backboard or stretcher, and in the case of the non-co-operative patient having an attendant apply manual immobilization.

An inadequate airway requires immediate intervention. Ideally, such intervention would follow an adequate cross-table lateral cervical spine radiography film, although this is not always possible. The airway may just require simple opening manoeuvres such as a held tilt or jaw thrust, coupled with upper airway suctioning. In the patient with a potential cervical spine injury the jaw thrust manoeuvre is modified. With a hand stabilizing each side of the head, the index fingers of both hands elevate the angles of the mandible while cervical spine alignment is maintained. Simple airway manoeuvres may be augmented or replaced by the use of nasopharyngeal or oropharyngeal airways. The former are much better tolerated by patients with active airway protective reflexes such as an intact gag reflex.

Failure to secure an adequate airway or demonstrate effective breathing leads to the institution of manual ventilation and the consideration of endotracheal intubation. Manual ventilation may be in the form of mouth to mouth, mouth to mask, or bag and mask. The route of endotracheal intubation will depend on patient condition and operator skill. When cervical spine stability is not a concern, traditional direct laryngoscopy and orotracheal intubation is the norm. If cervical spine integrity is in question, blind nasal intubation is preferred unless significant midfacial trauma provides the potential for the tube to be passes intracranially. In this case cricothyroidostomy or percutaneous transtracheal jet ventilation should be considered. For those unfamiliar with the last two techniques, in-line cervical spine stabilization applied by an assistant is an acceptable adjunct to direct laryngoscopy and orotracheal intubation in preventing significant spine motion. Orotracheal intubation aided by a light wand is a superior technique that produces little spine movement and is an additional consideration.

No discussion of endotracheal intubation is complete without reference to the principles of 'rapid sequence intubation'. The aims of this technique are rapid airway control, maintenance of oxygenation, airway protection, and attenuation of adverse cardiovascular and intracranial pressure responses to intubation. The complete algorithm can be seen in Table 1. The skill set and

Table 1 Algorithm for rapid sequence intubation

Prepare and check equipment
Adequate suction

↓

Preoxygenate (3–5 min)

↓

Pretreat pancuronium 0.01 mg/kg
Lidocaine 1.5 mg/kg
Alfentanyl 30 µg/kg

↓

β-Blocker esmolol 1.5 mg/kg

↓

Sodium thiopentone 1–3 mg/kg
Succinylcholine 1.5 mg/kg

↓

Cricoid pressure
No ventilation

↓

Intubate
Verify tube position

↓

Release cricoid pressure

knowledge base of the operator and the clinical scenario affect the application of the components of the algorithm. For some patients the entire sequence is appropriate, while for others none of it may be (e.g. cardiopulmonary arrest).

Once airway patency has been established or ensured, adequacy of breathing must be evaluated following the 'look, listen, and feel' approach. If the patient is apnoeic, manual ventilation must be instituted as described above. Three traumatic conditions should be considered: tension pneumothorax, open pneumothorax, and flail chest.

Circulation is evaluated by assessing the level of consciousness, skin colour, the presence and quality of the pulse, and blood pressure. In the absence of vital signs cardiopulmonary resuscitation is initiated unless blunt trauma is the proximate event. In such cases, the success rate of even maximal resuscitation are so dismal that measures short of immediate thoracotomy, open-chest cardiac massage, and thoracic aortic cross-clamping are futile.

External haemorrhage should be controlled and obvious fractures splinted during the primary survey. It is important at this juncture to recall the principles of field management articulated earlier as they apply to the athlete with chest or abdominal trauma. In the field, only essential supportive care such as cervical spine immobilization and definitive airway management should precede expeditious transport to a centre that is capable or providing definitive care.

Advanced cardiac life support

The primary survey in an athlete who suddenly collapses in cardiopulmonary arrest is neither complex nor difficult. It is usually clear that the situation is non-traumatic and primarily neurological or cardiac in nature. Once it is recognized that the athlete is not breathing and has no pulse, cardiopulmonary resuscitation is initiated and help summoned. Ideally, a direct current defibrillator and medications needed for resuscitation will be available at the scene of the event or arrive with first responders (e.g. fire and police) or ambulance personnel.

There are four common cardiac arrest rhythms: ventricular fibrillation, pulseless ventricular tachycardia, asystole, and electromechanical dissociation. The factors that most affect the chances for successful resuscitation are, in order of importance, rapid defibrillation for ventricular fibrillation or pulseless ventricular tachycardia, early onset and continued effective cardiopulmonary resuscitation, early definitive airway management with endotracheal intubation and 100 per cent oxygen, and adrenaline in adequate dosage to maintain coronary and cerebral blood flow.[10]

The aetiology of the arrest in an athlete is usually related to the presence of underlying heart disease, as discussed earlier. However, aggravating or initiating factors such as dehydration or hypovolaemia, electrolyte abnormalities, acidosis, hypoxaemia, tension pneumothorax, or cardiac tamponade may be present and correctable.[10]

Resuscitation phase

This term applies specifically to the period of time following the primary survey of a traumatized athlete. Interventions initiated during the primary survey are continued during this phase. The specific actions that occur at this time relate to the management of shock and obtaining essential radiographs.

Shock is initially managed with warmed crystalloid solutions such as Ringer's lactate or normal saline, ideally delivered under pressure to the unstable patient. Hypotonic solutions are not used for resuscitation. These fluids are continued to a limit of 40 ml/kg or about 3 l in the average adult, at which point ongoing shock management involves the use of blood products.

At this time a severely traumatized athlete may have a nasogastric tube inserted to decompress the stomach, if not contraindicated by a midface fracture. A urinary catheter may also be inserted unless contraindicated. The patient in shock should also be considered for central venous pressure line insertion and measurement, particularly if there has been blunt trauma to the chest and cardiac tamponade and myocardial contusion are diagnostic possibilities.

Three plain film radiographs are indicated at this time in all blunt trauma cases: a cross-table lateral cervical spine, an anteroposterior chest, and an anteroposterior pelvis film. These are obtained in the resuscitation suite using fixed or portable apparatus. Additional films may be of interest in an athlete who has sustained a gunshot wound to the trunk or head in order to determine the trajectory of the missile and potential organs injured.

Secondary survey

The secondary survey does not begin until immediate life threats have been addressed, resuscitation initiated, and the need for immediate operation dispelled. This evaluation is a head-to toe survey employing the 'look listen, and feel' format to determine the extent of injury and the need for definitive care.

Evaluation of the head and neck documents pupil size, reaction, and equality, the presence or absence of blood behind the tympanic membranes, symmetry of neurological function and a Glasgow coma scale score (Table 2). Airway patency is re-evaluated and cervical spine status confirmed radiologically.

Table 2 Glasgow Coma Scale

	Reaction	Score
Eye opening	Spontaneous	4
	To voice	3
	To pain	2
	None	1
Verbal response	Oriented	5
	Confused	4
	Inappropriate words	3
	Incomprehensible sounds	2
	None	1
Motor response	Obeys command	6
	Localizes pain	5
	Withdraw (pain)	4
	Flexion (pain)	3
	Extension (pain)	2
	None	1

The chest is inspected for symmetry of motion, the presence of segmental paradoxical motion, and adequacy of ventilation. Palpation and auscultation confirm bony integrity and symmetry, detect tenderness, and assess equality and adequacy of air entry.

Arterial blood gases are indicated to confirm adequate oxygenation and ventilation. The chest radiograph should be carefully studied.

The spleen and liver are the most frequently injured intraperitoneal organs. The fragile nature of the spleen makes it particularly susceptible to damage from blunt trauma sustained in all types of sporting events, but particularly contact sports and augmented speed sports such as skiing and cycling. The increased susceptibility of the spleen to rupture in athletes contracting infectious mononucleosis is a matter of some debate. However, a review of the literature suggests that the following recommendations are reasonable: in the absence of marked symptomatic splenomegaly mild activity may be resumed after the twenty-first day of illness; non-contact sports may be resumed a month after the onset of symptoms; resumption of strenuous contact sports may also begin after a month provided that the spleen is neither palpable nor enlarged by abdominal ultrasound examination.[21,22]

The abdominal examination may be difficult and misleading. The aim is to determine whether surgical intervention is required. Close observation and frequent re-evaluation are important principles in the ongoing management of abdominal trauma. An athlete with an altered sensorium is a particular dilemma. Inspection, palpation, auscultation, and rectal examination are integral aspects of the examination. Positive findings are useful and may dictate management. Negative findings are suspect and demand re-evaluation. Rectal examination is of particular importance and should address specific issues: the quality of rectal sphincter tone, the presence or absence of blood, and, in the male, prostate position. Diagnostic peritoneal lavage is a useful adjunct in the evaluation of abdominal trauma and in determining the need for laparotomy. However, it is an diagnostic procedure that significantly alters subsequent abdominal examination. Therefore it should be performed by or on the direction of the surgeon who will act on its results.

Physical examination of the pelvis consist of three manoeuvres: compressing the iliac wings, distracting the iliac wings, and applying pressure to the symphysis pubis in an anteroposterior direction. Radiography of the pelvis is essential in ruling out a pelvic fracture.

Injuries to the kidney are relatively common in athletes sustaining blunt trauma to the trunk, particularly the flanks. The hallmark of trauma to the urinary system is haematuria. Dipsticks used to detect blood in the urine actually react to haemoglobin released into the urine by lysed red cells. The reaction is not specific for haemoglobin, as it also gives a positive response to myoglobin. For this reason it is essential to confirm the presence of red cells in a dipstick-positive urine by microscopy. The severity of genitourinary trauma does not correlate with the amount of blood in the urine; thus an intravenous pyelogram and a cystogram must be obtained if haematuria is present in an athlete who presents with trauma. The issue is clouded by the fact that occult haematuria is commonly detected in athletes after many types of sporting events (e.g. boxing, basketball, football, long-distance running). Of course, the difference is that those who present for evaluation have sustained an injury that is perceived as being out of the ordinary and a full examination is thus indicated. Fortunately, the majority are renal contusions or minor parenchymal injuries which produce microscopic haematuria and can be managed in the outpatient department. A most devastating injury which requires quick diagnosis and operative intervention to salvage the kidney is a renal pedicle injury. This injury may reduce or eliminate blood flow to the kidney, leading to organ necrosis if not reversed within 4 to 6 h.

The usual means of obtaining a urine specimen from a trauma victim is via a urethral catheter. Injury to the female urethra is uncommon even in perineal trauma, and the most frequently encountered impediment to bladder catheterization is an inability to gain access to the urethral meatus due to pelvis, hip, or femur fracture. The male urethral meatus is not usually very difficult to find, but careful consideration is necessary before inserting a urethral catheter into a traumatized male. Blood at the meatus or a high-riding prostate on rectal examination mandate a urethrogram prior to the procedure. If radiography shows a pelvic fracture, the prostate is normal, and there is no blood at the meatus, a careful attempt at catheterization can be made. If resistance is encountered, the attempt should be abandoned and a urethrogram obtained.

Traumatic pancreatitis has been associated with bicycle handlebar injuries in children, contact sports such as karate and football, and non-contact sports including soccer and skiing. The diagnosis may be overlooked on initial examination as symptoms and signs may be gradual in onset. An index of suspicion for pancreatic injury in cases of blunt upper abdominal trauma would aid in earlier diagnosis. Typical symptoms and signs of abdominal pain and tenderness combined with elevated serum amylase establish the diagnosis. Associated injuries such as duodenal haematoma and rupture must be considered. Surgical consultation is essential for all patients with traumatic pancreatitis owing to the incidence of surgically correctible pancreatic lesions and the development of serious complications such as haemorrhage, pseudocyst formation, and pancreatic abscesses.

The evaluation of the extremities includes the palpation of all bones and joints, with particular attention being paid to the elbows and knees for occult dislocation and relocation. A particular concern in these cases is the integrity of vascular and neurological structures in proximity to the joint. All peripheral pulses are palpated and major peripheral nerves evaluated where possible.

Definitive care

Following the secondary survey an inventory of injuries sustained by the athlete can be compiled and prioritized for management. Fracture stabilization and operative interventions that are required are undertaken in this phase. In addition, stabilization of the patient in preparation for transfer to a more appropriate facility is completed.

Post-resuscitation care

In the immediate post-resuscitation period attention must be focused on several priorities. Oxygenation must be maintained and the athlete ventilated effectively as long as he or she fails to do so unaided. Ongoing failure of the victim to recover consciousness and spontaneous respiration despite an adequately perfusing rhythm should alert the provider to the possibility of anoxic brain injury. Basic manoeuvres to manage intracranial hypertension should be initiated, including hyperventilation to a $PaCO_2$ of 30 mmHg, 15° of head-up tilt, and the maintenance of oxygenation. Hypotension is not uncommon immediately after a rhythm is re-established, and often recovers with time, intravenous fluids,

or pressors as required. Hypertension may exist, particularly if resuscitation is prompt and adrenaline was used. This usually settles quickly without specific therapy. Antiarrhythmic agents utilized during the resuscitation are continued as infusions. Whether or not they should be instituted as a prophylactic measure is controversial, but is not the habit of the author. Malignant arrhythmias such as ventricular tachycardia with a pulse, multiform premature ventricular contractions, and R-on-T premature ventricular contractions are treated as they arise with the appropriate antiarrhythmic agents.

SUMMARY

Cardiopulmonary and abdominal emergencies of sufficient magnitude to constitute a threat to the life of the athlete are not common. However, the fact that they exist demands that those who care for these athletes have an appreciation of the disorders that may present, the settings in which they occur, and the management required. When the care-giver is armed with this knowledge, his or her ability to serve the athletic community is greatly enhanced as is the safety of the individual athlete.

REFERENCES

1. Jokl E. *Sudden death of athletes*. Springfield: Charles C Thomas, 1985.
2. Mustalish AC, Quash ET. Sports injuries to the chest and abdomen. In: Scott NW, Nisonson B, Nicholas JA, eds. *Principles of Sports Medicine*. Baltimore: Williams & Wilkins, 1984.
3. Maron BJ, Epstein SE, Roberts WC. Causes of sudden death in competitive athletes. *Journal of the American College of Cardiology* 1986; 7: 204–14.
4. Woolcock AJ, Asthma. In: Murray J, Nadel JA, eds. *Textbook of respiratory medicine*. Philadelphia: WB Saunders, 1988.
5. Benatar SR. Fatal asthma. *New England Journal of Medicine* 1991; 314: 423–9.
6. McFadden ER. Fatal and near fatal asthma. Editorial. *New England Journal of Medicine* 1991; 324: 409–11.
7. Johnson PR, Krafcik J, Green JW. Massive pulmonary embolism in a varsity athlete. *Physician and Sports Medicine* 1984; 12: 61–3.
8. Cales RH, Humphreys N, Pilmanis HA, Heilig R. Cardiac arrest from gas embolism in scuba diving. *Annals of Emergency Medicine* 1981; 10: 589–92.
9. World Health Organization. *Manual of the international statistical classification of diseases, injuries and causes of death*: based on the recommendations of the Ninth Revision Congress, 1975, and adopted by the Twenty Ninth World Health Assembly. 1975 Revision. Geneva: World Health organization, 1977: 40.
10. American Heart Association. *Textbook of advanced cardiac life support*. Dallas: American Hearth Association, National Centre, 1987.
11. Cobb LA, Weaver D. Exercise: a risk for sudden death in patients with coronary heart disease. *Journal of the American College of Cardiology* 1986; 7: 215–19.
12. Jeresaty RM. Mitral valve prolapse: definition and implications in athletes. *Journal of the American College of Cardiology* 1986; 7: 231–6.
13. Recreational and occupational recommendations for young patients with heart disease. A statement for physicians by the Committee on Congenital Heart Defects of the Council on Cardiovascular Disease in the Young, American Heart Association. *Circulation* 1986; 74: 1195A–8A.
14. Northcote RJ, Flannigan C, Ballantyne D. Sudden death and vigorous exercise—a study of 60 deaths associated with squash. *British Heart Journal* 1986; 55:198–203.
15. *Advanced trauma life support student manual*. Chicago: American College of Surgeons, 1989.
16. Green ED, Simson LR, Kellerman HH, Horowitz RN, Sturner WQ. Cardiac concussion following a softball blow to the chest. *Annals of Emergency Medicine* 1980; 9: 155–7.

17. Finn WF, Byrum JE. Fatal traumatic heart block as a result of apparently minor trauma. *Annals of Emergency Medicine* 1988; **17**: 59–62.
18. Rosen P, Murphy MF. Thoracic vascular pathologies. *Emergency Medicine Clinics of North America* 1983; **1**: 417–30
19. Bergqvist D, Hedelin H, Karrlson G. Abdominal trauma during 30 years: analysis of a large case series. *Injury* 1981; **13**: 93–9.
20. Diamond DL. Sports related abdominal trauma. *Clinics in Sports Medicine* 1989; **8**: 91–9.
21. Maki DG, Reich RM. Infectious mononucleosis in the athlete: prognosis, complications and management. *American Journal of Sports Medicine* 1982; **10**: 162–73.
22. Haines JD. When to resume sports after infectious mononucleosis. How soon is safe. *Postgraduate Medicine* 1987; **81** 331–3.

7.5 Spine injuries

DAVID C. REID

Spine trauma, even without neurological injury, presents a challenging clinical situation, and when associated with paralysis is one of the more serious medical problems. The paradox of these injuries occurring in association with sports and recreation lends even more tragic overtones. Spine fractures are encompassed within the classification of catastrophic injury.

The term catastrophic injury is defined as any injury incurred during participation in sport in which there is a permanent severe functional neurological disability (non-fatal) or a transient but not permanent functional neurological disability (serious).[1] An example of a serious catastrophic injury would be a fractured vertebra with no paralysis, while a similar injury with associated partial or complete quadriplegia would be a non-fatal catastrophic injury. Fatalities are the final group of catastrophic injuries and are subdivided into those resulting directly from participation in the skills of sport, or indirect fatality which is caused by a systemic failure as a result of exertion while participating in a sport. In a review of 1081 spine fractures, 134 (12 per cent) had sporting or recreational causes[2] (Table 1). Sports and recreation was the fourth most common cause of sustaining a spine fracture and second only to motor vehicle accidents in producing paralysis (Fig. 1). In many areas, these catastrophic injuries are decreasing in numbers and severity, while in some sports there is a considerable concern about the possibility of rising statistics.[1,3–6]

SPINE FRACTURE IN SELECTED SPORT

Diving and swimming

World-wide, diving is the sport with the highest frequency of catastrophic injury, accounting for about 25 per cent of the fractures in our series; an alarming 71 per cent of these were associated with neurological deficiencies.[2,5,7] Thus diving injuries account for about 45 per cent of the spine trauma in sports and recreation or about 9.2 per cent of overall spinal cord injuries from all causes. While to some degree the figure correlates with hours of sunshine, private pools, and prevalence of water sports on lakes and rivers, most series still feature an incidence of diving accidents (Table 2).

The profile of an average diving injury is a cervical fracture, usually between C4 and C6 with an associated complete motor and sensory lesion with a very poor prognosis for functional useful recovery (Fig. 2). It is this last fact that makes it such a disastrous statistic. Diving off a beach or a pier or into a private pool and

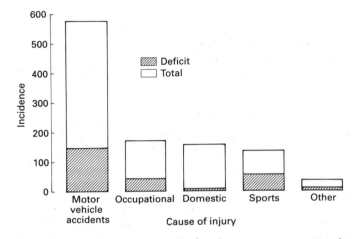

Fig. 1 Sports and recreation are the fourth most common cause of spine fracture, and the second most frequent cause of sustaining associated paralysis.

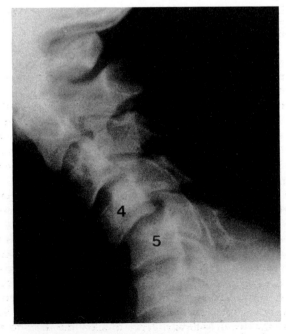

Fig. 2 Typical cervical fracture from diving with associated paralysis and poor prognosis for recovery.

Table 1 Spine fractures and neurological deficits related to sport

Sport/recreation	Absolute frequency	Percentage	Percentage with neurological deficit for particular sport Absolute frequency	Percentage
Diving	34	25	24	71
Snowmobile	16	12	7	44
Equestrian	16	12	4	25
Parachute/skydiving*	14	10	3	21
All-terrain vehicles†	12	9	4	33
Toboggan	11	8	1	10
Bicycle	4	4	1	25
Rugby	4	4	1	25
Ice hockey	3	2	1	33
Downhill skiing	3	2	1	33
Surfing	3	2	1	33
Football	2	1	0	0
Mountaineering	2	1	1	50
Other‡	10	8	4	40
Total	134	100	53	

* Includes hang-gliding and ultralite plane.
† Includes only 1 case of trampoline injury.
‡ Includes motocross racing, dirtbike and all-terrain (3-wheel) vehicles each with 4 cases.

Table 2 Diving as a cause of traumatic spinal cord injury*

Site	Reference	Total cases (all causes)	Percentage due to Sport	Diving
United States	6	318	5.0	2.2
Norway	9	725	8.2	4.4
England	10	619	6.9	5.3
Austria	11	112	10.7	8.0
Australia (Victoria)	12	325	12.6	8.3
Alberta	13	262	21.0	9.2
Ontario	7	358	15.4	10.6
Australia (Brisbane)	14	207	17.9	14.0

* Taken for other sports from Tator and Palm[7] and Kurtzke.[8]

striking the bottom is the most common mechanism, invariably secondary to water which is too shallow.

In the recreational setting, alcohol abuse often contributes to poor judgement. From this perspective, it is obvious that most of these injuries are preventable and public education programmes will play an important role. A group of neurosurgeons have embarked upon a very aggressive education campaign aimed mainly at teenagers using the slogan 'feet first—first time' and they claim to have reduced the incidence of spinal cord injuries by about 40 per cent. Similar slogans have been adopted in various parts of the world in attempts to reduce this unnecessary statistic.

In a report by Mueller and Cantu,[1] swimming accounts for a second significant number of catastrophic injuries which involved swimmers practising racing starts in the shallow end of the pool. With the current use of a deeper dive for the racing start, it is imperative that, in both practice and competition, racing starts are carried out in the deep end of the pool only. Most pools with a diving tank are of sufficient depth to allow the diver to avoid collision with the bottom. Nevertheless, poor technique from a high dive may result in contact even in adequately designed areas.

The water does not slow the diver's speed significantly until at least a depth of 1.5 m (5 feet) has been reached. The force of impact with the water spreads the unskilled diver's arms apart and the head may contact the bottom. Many drownings are due to quadriplegia secondary to the burst on impact. Once again, emphasis on technique is the key to reducing risk.

Rugby football

Rugby spine trauma parallels the popularity of the sport in different countries. Thus in British Columbia, Savio et al.[15] reviewed 390 spine fractures and found only nine due to rugby. However, figures reported in the United Kingdom, New Zealand, and Australia are sufficiently high to raise considerable concern.[16–22] Frequently these injuries are severe enough to cause either death or complete quadriplegia. There is a suggestion that the incidence is increasing. The vulnerable age for a rugby player is between 15 and 21, and the injuries are often related to aggressive and dangerous play, particularly in the scrums. For this reason new rules have been introduced to 'de-power' the scrums, rucks, and mauls. These rules are gradually being adopted for international competition. The rule changes include keeping the head and shoulders above the level of the hips, no charging, and penalties for collapsing, popping, or wheeling the scrum. The importance of these rule changes cannot be overemphasized since rugby is probably a major cause of sport-related spine trauma worldwide.[16,21] In the club system, when a substitute is required for a regular member of the scrum, care should be taken to ensure that the individual has sufficient experience and neck musculature, and is relatively strength matched, if unnecessary risk is to be avoided. Furthermore, there are now safer techniques that can be used in 'locking' the scrum, so that there is less chance of injury during inadvertent collapse.

American football

Football has always been associated with a large number of injuries and this is reflected in the catastrophic injury data.[23–31] In 1931, the American Football Coaches Association (AFCA) initiated the first annual survey of football fatalities, and in 1977 the National Collegiate Athletic Association (NCAA) initiated a national survey of catastrophic football injuries.[1,31] As a result of these projects, important contributions to the safety of the sport have been made. The most noticeable of these were the 1976 rule changes concerning tackling and blocking with the head.[1] They also established safety standards for football helmets and suggested improvements in medical care and coaching techniques.[32] In more current series, the low frequency of football-related catastrophic injuries is encouraging. This is particularly interesting in view of the high frequencies reported in many of the older sources.

The mechanism of injury is usually axial loading with the cervical spine slightly flexed (Fig. 3).[33] This aligns the spine in such a way that a burst type injury is frequently the result of a massive impact. In the past, the fulcrum supplied by a single-bar face-mask (Schneider's hyperflexion injury), as well as the guillotine mechanism of hyperextension with the posterior ridge of the football helmet forming a fulcrum in the upper cervical area have gradually been eliminated by changes in equipment manufacturing and fitting.[26] The number of permanent quadriplegics in our study decreased from 34 in 1976 to five in 1984, supporting Murphy's contention[5] that preventive programmes involving rule and equipment changes are effective.

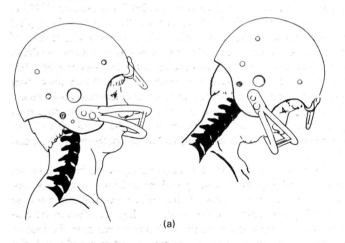

(a)

(b)

Fig. 3 (a) The neutral position of the cervical spine is characterized by lordosis. With slight forward flexion the cervical vertebrae are reduced to a vertical column. (b) In tackles during which the head is the first point of contact (spearing), the cervical spine is at risk.

As well as the potential efficacy of preventive programmes involving rules and equipment changes, the importance of cervical muscle strength has been repeatedly emphasized and incorporated more frequently into preseason training and as ongoing conditioning in the season. These rule changes and careful training in tackling techniques are particularly important in the younger age group. While neck rolls may protect against cervical plexus root and cord traction injuries to some degree, they offer very little protection against cord damage secondary to fracture and fractured dislocation.

A recent report[34] on late progressive instability of the cervical spine in a high jumper illustrates the potential for insidious onset of neurological signs, and we should be aware of this with football.[34] Sports which involve repetitive flexion stresses to the cervical spine can cause this problem. Complaints of intermittent peripheral paraesthesia, neck pain, and central neurological signs, and the presence of slight anterior subluxation, wedging, or subtle kyphotic angulation at one level should alert the physician to this diagnosis.

Gymnastics and trampolining

Most fatal injuries in gymnastics are the result of head trauma, with males sustaining injury on the high bar or parallel bars in particular and females injured performing on the uneven bars.[1,35] Occasionally spinal injury is the result of a mistimed vault. The use of trampolines either in association with gymnastics and diving training or as a recreational pursuit has also resulted in an ever-increasing number of spine injuries, and the high risks with this equipment have been discussed.[24,35,36] A virtual ban on trampolines in the high school system and better awareness of spotting techniques and safety rules have led to a decreased incidence. However, inasmuch as trampolines are often used in practising very difficult moves in gymnastics and diving, it is important that ongoing appreciation of the safety precautions must be present. Furthermore, care to ensure adequate supervision, particularly in the younger age groups, and the presence of experienced or well-trained spotters and catchers during practice sessions will eliminate many injuries.

Skydiving

Skydiving accounted for 10 per cent of the injuries in our series and these were all thoracolumbar fractures. Parachuters usually land feet first with the legs together as they absorb the shock with the parachute landing fall. This involves slight flexion of the hips and spine, sudden rotation to the side, and a roll onto the ground. Thus the injury mechanism theoretically usually involves flexion and rotation producing either a wedge compression or burst type fracture.[37] Many of the injuries involve a combination of minor equipment malfunction and inexperience. Inasmuch as many programmes require very little training before the first jump is allowed, it is surprising that this figure is as low as the data suggest.

Pole vault

Because of the number of injuries in high school pole vaulting, the National Federation of State High School Associations passed a rule that, beginning with the 1987 season, all individual units in the pole vault landing area must include a common cover or pad extending over all sections of the pit.[1] This has had the effect of dramatically reducing the number of catastrophic injuries in this sport and is a special reminder that physicians and therapists in attendance at meetings have a duty to ensure and check the safety of equipment. This in turn requires a detailed knowledge of the sport.

Winter sports

Many sports have their main season during the winter months, but the traditional winter sports are those activities requiring ice or snow as a playing surface.[38] Because of the physical characteristics of this frozen water, most winter sports (with notable exceptions) have evolved into high speed events. Velocity in sports such as skiing, the hard ice surface and the physical contact of hockey, the speed and height of landing in ski-jumping, and the bullet-like but relatively exposed projectile in bob-sleigh would lead one to suspect that there would be an unacceptably high incidence of spine trauma. Nevertheless, considering the large number of participants in these pastimes, the relative incidence of spine fracture is low. However, the increasing popularity of the snowmobile as a recreational vehicle has added an alarming dimension to the previously unremarkable statistic.[38,40]

Ice hockey

There has been concern that the incidence of spine fractures in ice hockey is increasing.[41] Indeed, on a *per capita* basis, ice hockey now causes approximately twice as many cases of quadriplegia annually in Canada as does football in the United States.[42–44] There has been much speculation on the reason for this apparent sudden surge in hockey-related spine trauma including larger stronger players, changes in the attitude of coaches and referees, and improved helmets and face-masks leading to a false sense of security. Although the hockey helmet protects the skull, it does not appreciably alter the dynamics of the neck.[45] Inevitably, the mechanism of injury is an axial loading or flexion injury as the individual collides with the boards. Ice hockey has now implemented a new rule forbidding checking from behind into the boards as well as more enforcement of boarding, cross-checking, and other stick penalties. There also has been a very serious effort to educate people involved in hockey—players, coaches, and as referees—about the seriousness of the problem.[46] As with so many hockey injuries, abiding by and enforcement of existing rules would make a major impact on the statistics.

Alpine skiing

Spine fractures comprise a small but devastating percentage of skiing injuries. The incidence in alpine (downhill) skiing has ranged from zero to 5.2 per cent in the reported series.[47–50] The number appears to be increasing slowly over the 7 years of our study, but this may simply reflect a larger number of skiers.[2] Non-experienced and professional skiers seem to be injured with equal frequency.[50–52] The most common mechanisms include attempting a jump and landing poorly, and losing control and hitting a tree (Table 3). The other contributing factors are (1) collision with stationary or slower skiers owing to their location just over the edge of a slope where they were invisible until the last minute, (2) novice skiers crossing the slopes in the paths of fast skiers on the main run, (3) overcrowding of the slopes, particularly at the junction of two runs, with fast skiers coming into the same flow pattern as the slow skiers, and (4) collision with the lift supports, rocks, or grooming equipment.

These aetiological factors are basically of two varieties. The

Table 3 Factors related to spine injury with skiing

Landing poorly from a jump
Losing control; skiing too fast
Collision with trees, lift-posts, or rocks
Collision with hidden skiers over the crest of a hill
Collision where side-runs join main runs
Collision due to overcrowding of slopes
Novice skiers crossing the slopes in path of fast skiers
Alcohol use
Wearing earphones and being unaware of auditory clues
Poorly groomed slopes; inadequate snow
Inadequately taught or poorly practised ski etiquettes

first is essentially related to poor judgement and this in turn is linked to youth, alcohol, and in many cases poor teaching, lack of awareness of others, poor ski etiquette, or pure selfishness. The second factor is the ski slope design, with too many runs converging and overcrowding of some parts of the slope. Added to this is inadequate policing of the activities of skiers when dangerous mini-jumps have been made or when obvious and inappropriate behaviour has developed on crowded slopes.

The same factors that generate spine injuries also lead to head injury and death. Morrow reported on 22 fatalities among alpine skiers in Vermont from 1979–80 to the 1987–8 ski seasons.[53] However, alpine skiing fatalities do not usually occur in the typical recreational skier but in the highly skilled young adult, capable of skiing at very high speeds. There is an estimated rate of one death per 1.6 million skier days. In addition to the factors cited, Shealy[53] noted that these accidents also resulted from falls on slopes that were rated above the skier's abilities and crashes while racing informally with other skiers, as well as while making practice runs prior to competition.[54] The cause of death is mainly from blunt trauma to the head (82 per cent) and occasionally to the chest and abdomen.

While cervical spine fractures occur frequently, burst fractures of L1, signifying actual loading with or without a flexion component, predominate in most series. When cervical fractures do occur they are usually midcervical.[2,55] They are often associated with head injuries and normally with direct impact to the face or forehead. The skier's position at high speed is usually with the head forward and the shoulders raised, and it is suggested that a helmet with an extended posterior rim design may afford some protection against cervical spine injuries.[56,57]

In our series, the average age was 20.2 years with the youngest being 13 years. This series include several teenagers but no one over the age of 26 years.[36] Ellison[56] and Margreiter et al.[51] reported 1.8 per cent and 4.7 per cent respectively in their series of skiing injuries involving the spine in children and teenagers. Injuries to the spine in young individuals require considerable energy input, but are frequently less severe with a relatively less prolonged morbidity.[58]

Associated injuries are present in about 60 per cent of individuals sustaining a spine fracture, with approximately one-third having a neurological deficit (Table 4). Long bone fractures, thoracic injuries, abdominal injuries, and ligament damage to the knee are frequent concomitant injuries and point to the need for careful evaluation before moving the injured individual as well as skilled management of the transport down off the ski slopes.[57] There does not appear to be a specific relationship to the time of day, weather, or, with the exception of head injuries, helmet and equipment used.

Table 4 Associated injuries with spine fractures

Degree and site of injury	Snowmobiles	Toboggan	Alpine skiing	Ice hockey	All winter sports
None present	10	7	5	5	27(56%)
Associated injuries	10	4	6	1	21(44%)
Another vertebrae	—	2	1	—	3
Thoracic trauma	4	2	2	—	6
Head injury	2	2	—	1	5
Long bone fracture	5	1	2	—	8
Abdominal trauma	3	—	1	—	4
Pelvic trauma	2	—	—	—	2
Facial injury	1	—	1	—	2
Brachial plexus	1	—	1	—	1
Urinary tract	—	—	1	—	1
Knee ligaments	—	—	1	—	1
Total injuries	17	5	10	1	33
Multiple trauma	6	1	3	0	10 (21%)

Adapted from ref. 38.

One of the most important factors for ski patrollers is to understand that, firstly, spine fracture does not have to be associated with paralysis. Hence, careful assessment is always necessary before moving an individual. Secondly, spine fractures are often accompanied by at least one other major injury.[58-61] These associated injuries are frequently more obvious. A rapid but diligent assessment of the entire skier is necessary in these cases. Otherwise, dramatic peripheral injuries may lead to overlooking the spine fracture with its potentially disastrous result. Ski patrollers must be trained carefully in the difficult task of transporting the skier with spine fractures and associated multiple trauma.

Freestyle skiing

The acrobatic nature of freestyle skiing and the temptation for unskilled individuals to mimic the dramatic manoeuvres seen in the professional results in the greatest number of spine injuries in freestyle skiing. As the sport gains in popularity, we can expect the incidence of injuries to rise. Thoracolumbar fractures alone account for 8 per cent of all time-lost injuries in some series. This high figure, cited by Dowling,[62] was obtained without the inclusion of inverted aerials (flips). These aerials have caused a number of serious spine fractures and have been banned by the United States Skiing Association. Inverted aerials are still included in world cup competitions held in Canada and European countries. This event in freestyle skiing is controversial, and careful statistics of injuries, related to the number of exposures, should be monitored. The risk from freestyle skiing may be reduced to some extent by good training programmes that include adequate dry land gymnastic skills and carefully supervised competition with attention to weather conditions, visibility, and snow conditions in the take-off and landing areas.

Ski jumping

Cross-country and alpine skiing are enjoyed by millions of people for recreational purposes. In contrast, ski jumping is almost exclusively a competitive sport with a limited number of performers.[62] These jumpers may perform on average about 400 jumps per year. Special facilities are required for nordic ski jumping (Fig. 4). A skier begins at the top of the in-run, which is a ramp supported by scaffolding or conformed ground. On the in-run, the jumper crouches to minimize wind resistance and hence maximize take-off

speed. At the end of the in-run the skier must time the jump perfectly and, almost simultaneously, press the body forward over the skis to make an airfoil and generate lift. This leaning position is maintained as long as possible (Fig. 4). Towards the end of the flight, the hips are pressed forward, the shoulders are raised, and the trunk is extended into a position perpendicular to the slope of the hill. With a crouched movement, one foot is brought slightly ahead of the other as the ski touches the landing hill. The run is completed by skiing through the transition curve and into the long flat out-run. This allows deceleration. Two jumps are completed, and points are awarded for style and distance.

Fig. 4 The design of a ski jump allows landing with minimal impact by maximizing the transition area. This, together with airfoil techniques, has made ski jumping safer.

To the casual observer the ski jump would appear to be a natural setting for catastrophic injuries. However, because of the skill level of the ski jumper, proper hill maintenance, and good judgement on the part of the officials, spine fracture is rare.[63-68] Indeed, in the 1980 Lake Placid Winter Olympics, over 5000 jumps were made with only two injuries—a mild concussion and a fractured clavicle. At the Intervale ski jump complex, the largest ski jumping complex in North America a 5-year record of ski jumping injuries does not include a cervical spine fracture or dislocation in its statistics. Nevertheless, in sports where take-off speeds of 50 to 56 miles/h are achieved on 70-m hills and 45 to 55 miles/h on the 50 to 60 m hills, followed by up to 70 to 90 m in the air, there is always the potential for catastrophic injuries.[67–73]

In Wester's study[65] of a series of ski jumping injuries in Norway, the risk of being seriously injured was approximately 5 per cent in a 5-year period (1977–81) and was higher in the age group 15 to 17 years. The first jump of the day is particularly dangerous, but most of the serious injuries occur in jumps where the jumper has previously experimented. It is possible that the jumper meets unexpected snow conditions on a jump that is felt to be familiar. Other possibilities include the need for a couple of jumps to 'remember' or 'get the feeling of' a particular jump. When spine fractures do occur, they are usually middle to low cervical and are associated with concussion and paralysis. Only six fatalities occurred in a 50-year period in the United States; four of these were associated with cervical spine injuries and at least three were in the C1–C2 area. The overall fatality rate for nordic jumping is estimated at about 12 per 100 000 participants annually, which is within the range for other 'risky' outdoor sports.[69]

Injuries are not common in the age group 12 and under. These youngsters compete on relatively small jumps, with lengths up to 30 to 35 m, which are naturally associated with correspondingly lower speeds. With flexibility of the young spine structures and light body weights, the resultant kinetic energy is low. Age 15 to 17 seems to present the greatest risk, with poor judgement and attempts at longer jumps than those for which they are physically or technically qualified being the main contributing factor. The need for careful observations and control by coaches, and progression only once sufficient skill has been obtained, are obvious factors in reducing injuries. Despite the fact that many skiers blame personal faults such as rotation of the take-off, too early take-off, and asymmetric ski placement as the main causes of injury, it is frequently poor judgement regarding their own ability that results in problems for the skier. Lack of practice may account for early season problems. Changing snow conditions are more prevalent at both ends of the season. The quality of the in-run, the development of ruts, and the uniformity of packing can all alter the velocity and balance during the critical moments prior to take-off. At the end of the season or after a heavy snowfall, the landing area can develop gnarls or irregularities, or the snow can allow sinking of the skis upon impact with the ground, all of which may disturb balance or cause deceleration. Only careful inspection and grooming of the run will overcome these difficulties.

Increasing confidence and efforts to achieve or surpass personal best distances may influence the rise of injuries towards the end of the season. This factor, along with the fact that over 50 per cent of severe injuries occur on the first jump, indicates the need for jumpers to examine all aspects of the jump thoroughly, including the in-run and landing areas, before committing themselves to attempting the course. Equipment is another key factor and there has been a remarkable improvement in this area. Skis, bindings, clothing, and above all good head protection have added a dimension of protection that unfortunately can sometimes lead to recklessness. Previous experiments with high heel blocks allowing early retainment of the floating position seems to increase the incidence of dangerous falls during take-off, particularly when small decelerations are encountered on the in-run. Apart from the skier's equipment, the construction of rails, shields, and fences in all areas of the jump should make it impossible for skis or body parts to become stuck in the case of a fall.[66]

One of the most significant factors in reducing ski jumping injuries is the improvement in technique. The older less aerodynamic-efficient ski jumping techniques, requiring that the jumper be projected high into the air and then free fall while flying horizontally, has been replaced with a modern jump in which the idea is to mimic an airfoil and generate lift. Therefore it is possible to fly further without needing so much altitude. The vertical component of the flight curve relative to the horizontal component has been decreased. This in turn has allowed the jumping hill to be redesigned. The modern ski jumping hill is not so steep and as a result the transition area is flatter and longer. This has increased the margin of safety since the landing hill more closely resembles the flight curve of a broader range of jumps, including the very long and the very short attempts.

Off-road vehicles

Off-road motor vehicles made their appearance in the late 1960s and early 1970s and have gained rapidly in both popularity and as a cause of spine fracture.[38,39,71] The predominant reason for this increase is the all-terrain vehicle. They form up to 10 per cent of sport and recreational induced spine fractures, and in some publications 35 per cent of these have associated neurological injury. However, there are also over 2 million snowmobiles in use in the United States and a larger *per capita* number in Canada.[64] Typically, the snowmobiler is injured while driving at night (52 per cent), driving under the influence of alcohol (53 per cent), and driving in unfamiliar terrain.[70–72] In 1973 Wenzel and Peters[71] reported that 11 per cent of the injuries occurred in children aged 10 or less (Table 5). The incorrect image of the snowmobile as a safe piece of equipment and easy to drive leads many parents to view it almost as a toy.[71–80] Ultimately the key to reducing the number of snowmobile injuries and fatalities may be setting an age limit and

Table 5 Sport and recreational acquired spine fractures related to age

Sport	Mean	SD	Range
Diving	22.00	5.85	12–39
Equestrian	35.81	11.69	17–52
Snowmobile	26.93	7.71	12–39
Parachute/skydiving	26.00	5.72	19–41
All-terrain vehicle	27.33	9.35	16–46
Toboggan	20.82	9.29	8–35
Bicycle	42.00	26.91	12–64
Rugby	26.75	2.50	24–30
Downhill skiing	18.33	6.81	13–26
Ice hockey	19.33	4.51	15–24
Surfing	40.00	17.06	26–59
Mountaineering	24.00	5.66	20–28
Football	17.00	1.41	16–18
Soccer	24.00*		
Water skiing	28.00*		
Frisbee	27.00*		
Dancing	14.00*		
Badminton	72.00*		
Caber throw	50.00*		
Trampoline	5.00*		
Wrestling	20.00*		
Rope swing	21.00*		

*One subject only.
Adapted from ref. 2.

educating drivers.[72–76, 81] Going off embankments, tipping on steep terrain, colliding with another snowmobiler, hitting an object (frequently a tree), or becoming airborne are the most common mechanisms of sustaining spine fractures. Occasionally the machine may roll on to the driver. Because of the cold weather conditions, compliance with wearing protective headgear is good but frequently the impact is so great that these do not eliminate spine injury, head injury, or even fatalities.

Throughout North America, there is a general paucity of legislation relating to the use of off-road motor vehicles. Legislation, better equipment, and public education has made snowmobiling a safer sport over the last few years.[83–87] The following recommendations may further improve this record:

(1) a review of snowmobile legislation to ensure the requirement for safety and training programmes for all drivers;

(2) stricter enforcement and regulations of snowmobilers on public highways;

(3) the development of designated areas and trails for snowmobiles;

(4) improved safety designs of equipment including helmets as well as education for the snowmobiler.

Tobogganing

As would be anticipated, tobogganing injuries frequently involve very young individuals and approximately one-third of the spine fractures in our series occurred in children under the age of 15 years. Poor judgement features highly in the mechanism of injury which frequently involves collision with a post, fence, or tree.[2] Properly cleared and designated areas are the single most important factor in reducing injury.

TRANSIENT QUADRIPLEGIA

In 1986, Torg et al.[88] reported 32 cases of athletes who had experienced transient quadriplegia. The quadriplegia lasted for as little as 1 min to as long as 48 hs and in all cases resolved completely. Magnetic resonance imaging (MRI) was performed for only one patient and this did not reveal intrinsic cord abnormalities. The mechanism of injury was postulated to be a neuropraxia of the cervical cord. Four of the 32 athletes showed evidence of ligament instability, six had acute or chronic intervertebral disc disease, and five had congenital cervical anomalies. Seventeen of the 32 athletes were identified as having relative cervical spine canal stenosis.

Torg's criterion for relative spinal stenosis was that the ratios of spinal canal to vertebral body in these individuals were less than 0.80 (Fig. 5). This is compared with a ratio of 0.98 or more in a control group of athletes. The athletes did not go on to have further significant problems and Torg's conclusion was that the young patient who had an episode of neuropraxia of the cervical spinal cord, with or without transient quadriplegia, was not necessarily predisposed to permanent neurological injury. However, this does not mean that these individuals may not require a full neurological work-up. Indeed, depending on the severity of the transient symptoms, they may require either a CT or MRI examination.[88,89]

Ladd and Scranton[90,91] have gone so far as to suggest that any individual with identifiable relative spinal stenosis of the cervical spine should be advised to discontinue participation in contact sport. However, this area is still very controversial, particularly in the light of a more recent study by Odor et al. who reviewed the

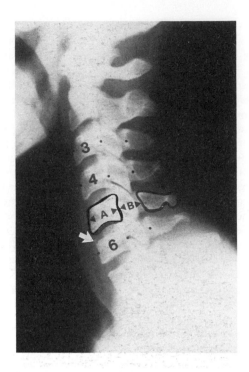

Fig. 5 Relative spinal stenosis as indicated by the ratio of the spinal canal to the width of the vertebral body. A value of less than 0.8 is considered to be abnormal.

Torg ratio in 224 football players and noted that approximately 33 per cent of them had a ratio of less than 0.8 at one or more levels of the cervical spine. It is possible that the large size of the cervical vertebral body of these heavily developed individuals gives apparent narrowing of the canal with the Torg ratio when indeed the canal width is more than adequate. In any event, in view of the fact that a large number of football players apparently have ratio of less than 0.80, it is not immediately clear that this parameter predisposes the athlete to either transient or permanent neurological sequelae. Thus any decision for return to play or advice regarding vulnerability to further injury should include factors other than the simple plain radiograph.

BURNERS AND STINGERS (TRAUMATIC RADICULOPATHY)

In 1983 a survey of over 3000 secondary school and varsity football teams, 25 per cent of the players had a history of neck injury.[93] Most of these reported neck pain, numbness, tingling, burning, or weakness in one upper limb. Usually this was related to tackling using the head as the first point of contact, or in collision with forced side flexion of the neck and depression of the shoulder of the contralateral side (Fig. 6).[94,95] These episodes were clumped together under the heading of burners and stingers.[96] They have often tended to be regarded as benign and are treated with an uncharacteristic lack of respect compared with that given to most neurological traumas.

The symptoms range from a brief 'knife-like' shooting pain confined to the shoulder to transient numbness and paralysis of the whole arm. For the most part the symptoms clear up quickly, although with careful testing it can be seen that there is frequently detectable pathology lasting over 48 h and electrophysiological

Table 6 Duration of symptoms from burners

Symptoms	Duration	Frequency (%)
Initial dysthetic pain	⩽ 1 min	57
	5 min or more	35
	Unknown	18
Continued subjective weakness	None	34
	8 h	12
	24 h	12
	⩾ 2–3 weeks	42

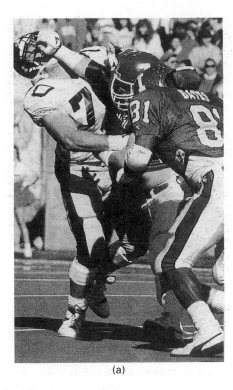

(a)

(b)

Fig. 6 While direct compressive loading of the cervical spine may be a common cause of the 'burner' syndrome, (a) forced extension with shoulder depression and (b) rotary torques may also generate symptoms in the susceptible individual.

changes which may persist for several months (Table 6). At various times the pathophysiology of the lesion has been located in the spinal cord, the nerve root, the brachial plexus, or the peripheral nerve level. It is common for the athlete to admit to sustaining burners as frequently as one per game for an entire season without

ever having received a detailed neurological assessment or any specific therapy.

The management and advice varies from uninterrupted continuation of play to complete cessation of sport. Typically, however, there is a brief assessment on the field followed by return to play. The athlete may receive bolstering of the cervical spine or shoulder and instruction on improving tackling techniques. Rarely is there rigid immobilization until fracture or dislocation has been ruled out. In our series a significant number of the athletes tested had persistent neurological signs and symptoms 3 weeks after their latest burner episode.

The C5–C6 area is most commonly affected with both radiographic abnormalities and electrophysiological changes. The lesions are in the upper trunk in over half the cases, and nerve root lesions and peripheral nerve involvement are equal in frequency.[97–100] Thus the locations and severity of the neurological injury following a forced cervical compression or flexion injury is variable and the diagnosis of burner or stinger does not imply a specific nerve injury or prognosis. Individuals sustaining recurrent burners should have a thorough investigation, the findings should be reviewed, and the athlete should be advised regarding the potential risks of the neural system.

In any event, when such an episode occurs the person responsible for medical care should make sure that the athlete is removed from play and properly assessed and not allowed to return until a definitive diagnosis is made.

Naturally, the major concern is to separate mild transient injuries from those which may lead to permanent paralysis if inadequately managed. The following points are helpful.

1. When in doubt do not make the final decision on the field or sidelines where things are often hectic. The subtle pressures from athlete and coach are significant and there is a tendency to be hurried. The player should be taken to the dressing room and carefully reassessed. This should be done immediately and not at the end of the game.
2. The player should be adequately undressed in order to palpate and inspect the area, and this often requires careful removal of equipment.
3. Where there is concern that the symptoms do not fit the classic pattern or are more than transient, send the player for radiography.
4. If there appear to be any residual symptoms, make a decision which errs on the side of safety.[101–103]

The following points should arouse suspicion of a more severe injury. Significant tenderness over the trapezius and sternomastoid, particularly if associated with a restricted range of functional motion, should be noted. Any obvious midline deformity to palpation, specifically a step or a gap between the spinous processes, should be taken as evidence of spinal instability, and appropriate splinting and precautions taken. Even extreme tenderness in the midline, over the ligament nuchae and the spinous processes, is uncharacteristic of the usual burner and stinger and must be treated with caution. Weakness, paralysis, and paraesthesia in more than one extremity should be assumed to be spinal cord damage until proved otherwise.

During preseason medicals, those individuals with a history of multiple burners and stingers, or those who are in vulnerable positions such as linebackers or fullbacks, must have a very careful neurological examination and appropriate radiography. These individuals require a special neck-strengthening programme and their contact time in practice should be minimized. The emphasis should be on teaching correct techniques of blocking and tackling which minimize the use of the head as a major weapon. Occasionally neck rolls and cervical collars seem to reduce the frequency of episodes.[104] Nevertheless in those injury-prone individuals who continue to have problems, wherever possible a change in position should be contemplated which may have the effect of placing them at less risk.

Torg et al.[33] have estimated that the compressive load limits for the cervical vertebral bodies are between 750 and 1000 pounds ($1.65-2.2 \times 10^3$ kg). If this load has reached a maximal compressive mode, with the head slightly flexed, a compression fracture of the vertebral body may occur. Extending the neck to keep the face up may create uneven forces, and tackling with the head laterally flexed may also leave the cervical spine in a vulnerable position. Thus it is not always possible to protect the neck. The main points of emphasis should be anticipating contraction of the large neck muscles, hunching the shoulders, and using the arms for the initial contact in blocking and tackling.

Where the radiographs show changes in the shape of the vertebral body, which may be related to old or new trauma, flexion and extension films may help to reveal minor instabilities (Fig. 7).

Where the symptom complex suggests a more proximal pathology than the traditional midcervical area, careful views of the C1–C2 area are required. Individuals with degenerative spondylosis need to be considered on an individual basis. There is an increased incidence of facet joint change, joint space narrowing, and osteophyte formation in many individuals who have spent a large number of years playing football, wrestling, or soccer, and, while there is no direct relationship between radiographic changes and symptomatology, significant degenerative changes combined with chronic nerve root irritation can ultimately affect their activities of daily living. Thus athletes should be alerted to this possibility so that they can decide whether or not to continue their sport should there be a definite risk of the development of significant symptomatic degenerative changes. Individuals with degenerative spondylosis need to be counselled on an individual basis. Similarly, any congenital abnormality must be considered on an individual basis (Fig. 8).

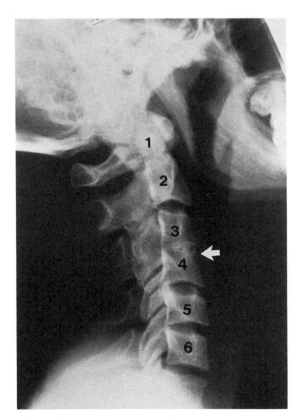

Fig. 8 Congenital block vertebrae at C3–C4. In the presence of multiple episodes of 'burners' this may form a contraindication to continued participation.

Essentially, the common symptom complex known as 'burners or stingers' should never be taken lightly and no player should be returned to the field of contact until careful neurological examination has been carried out.

PRINCIPLES OF MOVING A SPINE INJURY ATHLETE

It is incumbent upon the senior medical person attending any contact sporting event with risk of spinal injury to ensure that a

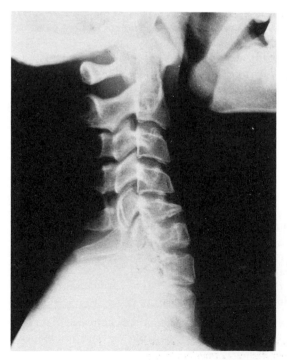

Fig. 7 Old trauma to C5 has resulted in deformity. Careful flexion and extension films will reveal any associated instability.

protocol has been set up for the handling of significant emergencies. The location of the nearest telephone, hospital, and access for stretcher or ambulance must be established. Furthermore, the field of play must be examined for hazards and safety. Should a potentially catastrophic injury to the spine occur, the playing surface will dictate potential difficulties with transferring the athlete safely.[3] Rescuing somebody from the water or a jumping pit presents a very different challenge to that of moving an ice hockey player on an ice surface, moving someone from the ski slopes, or dealing with the cumbersome helmet and shoulder

pads of football players. Only familiarization, training, and rehearsal will equip the medical personnel sufficiently to handle the situations smoothly and efficiently.

The unconscious patient with a potential neck injury presents a further challenge. All unconscious athletes must be assumed to have a cervical spine injury and be handled accordingly (Fig. 9). The initial screen of the individual establishes the adequacy of the airway and then spinal tenderness, the athlete's ability to move all limbs, and the presence or absence of sensation are carefully sought. The most experienced person present should take

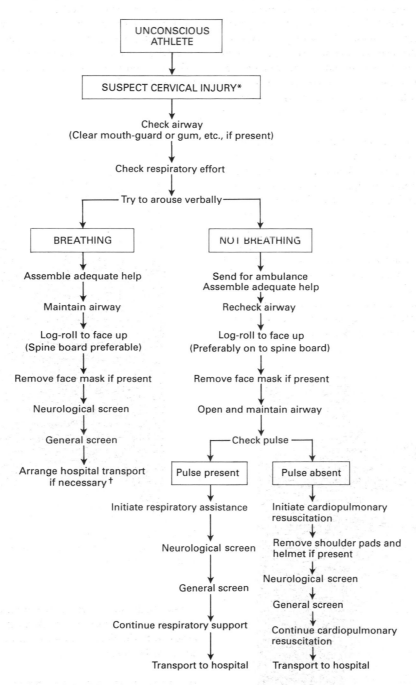

Fig. 9 Decision matrix for on-site management of the unconscious athlete.
*Any significant suspicion of cervical injury should prompt all spinal precaution and arrange for transport to medical facility. †Other than very transient loss of consciousness or grade I to II concussions, full assessment at hospital is advisable.

command, and any evidence of head and neck problems requires immediate stabilization of the cervical spine while a more detailed secondary screen is carried out to assess the extent of the injury. If necessary, turning is organized along with transfer to a spine board or stretcher. It is essential to have enough people available to do this properly. It should be remembered that high speed motor sports and skiing produce combined injuries to the viscera and appendicular skeleton as well as the cervical spine. Thus it is important not to be distracted totally from the assessment of the total athlete upon identifying a cervical spine injury and, conversely, a seriously deformed fractured limb should not divert suspicion away from the cervical spine completely until the injury to this area has been specially ruled out.

Finally it should be pointed out that significant disruption of the vertebral column can occur without associated paralysis and often the clinical signs of local tenderness may be the only clue to an underlying unstable spine.

SUMMARY

In this chapter we have emphasized the high incidence of potentially serious and catastrophic injuries in sport. Several points have emerged, including the fact that spine trauma need not necessarily be associated with paralysis and thus may be overlooked unless a diligent and careful examination is made. Secondly, the events leading up to a spine injury in high velocity sports such as skiing may be associated with many other problems. Thus in the excitement of managing a peripheral injury, there may be a tendency to miss the cervical or lumbar spine problem.

Most individuals do not encounter spine injuries on a frequent basis and thus on the occasion that it occurs when they are in charge of supervising a sport, they are often caught unprepared. Rescue and immobilization procedures on ice, gymnastic mats, ski slopes, and football fields should be practised on a regular basis to ensure familiarity with the problems and nuances of these different environments. It is a sad fact that many of the injuries that occur in sports and recreation are eminently preventable if elementary precautions are taken. Nowhere is this more true than in the frequent injuries associated with diving or incurred on recreational all-terrain vehicles.

Finally this chapter should serve a reminder for anybody in charge of a sports facility to ensure and check that adequate triage procedures are in place before any game or competition commences so that in the event of a serious injury, expeditious and safe transport can be arranged.

REFERENCES

1. Mueller FO, Cantu RD. Catastrophic injuries and fatalities in high school and college sports. Fall 1982–Spring 1988. *Medical Science in Sports and Exercise* 1990; **22**: 737.
2. Reid DC, Saboe L. Spinal trauma in sports and recreation. *Journal of Clinical Sports Medicine* 1991; **1**: 75.
3. Reid DC. *Sports Injury, assessment and rehabilitation.* New York: Churchill Livingstone, 1991.
4. Tator CH, Edmonds VE. Sports and recreation are a rising cause of spinal cord injury. *Physician and Sportsmedicine* 1986; **14**: 157.
5. Murphy P. Still too many neck injuries. *Physician and Sportsmedicine* 1985; **13**: 29.
6. Key AG, Retief PJM. Spinal cord injuries: An analysis of 300 new lesions. *Paraplegia* 1970; **7**: 243.
7. Tator CH, Palm J. Spinal injuries in diving. Incidence high and rising. *Ontario Medical Review* 1981; **48**: 628.
8. Kurtzke JF. Epidemiology of spinal cord injury. *Experimental Neurology* 1975; **88**: 163.
9. Zrubecky 1974
10. Kraus JF, Franti CE, Riggin RS, Richard D, Borhani NO. Incidence of traumatic spinal cord lesions. *Journal of Chronic Diseases* 1975; **28**: 471.
11. Gjone RN, 1974. Personal communication: Sunnaas Sykehus, Nesodden, Norway. Cited in Kurtzke JF. Epidemiology of spinal cord injury. *Experimental Neurology* 1975; **18**: 3, 12.
12. Cheshire DJ. *The complete and centralized treatment of paraplegia: a report on the spinal injuries centre for Victoria, Australia.* Proceedings of the sixteenth Annual Spinal Cord Injury Conference. 1987: 39–49.
13. Reid DC, Saboe LA. Spinal trauma in sports and recreation. *Clinical Journal of Sports Medicine* 1991; **1**: 75.
14. Sutton NG. *Injuries of the Spinal Cord: The Management of Paraplegia and Tetraplegia.* London: Butterworth, 1973, 185.
15. Sovio OM, Van Peteghan PK, Schweigel JF. Cervical spine injuries in rugby players. *Canadian Medical Association* 1984; **130**: 36.
16. Burry HC, Gowland H. Cervical injury in rugby football-A New Zealand survey. *British Journal of Sport Medicine* 1981; **15**: 56.
17. Carvell JE, Fuller DJ, Duthie RB, Cockin J. Rugby football injuries to the cervical spine. *British Medical Journal* 1983; **286**: 49.
18. Kewalramani LS, Drauss JF. Cervical spine injuries resulting from collision sports. *Paraplegia* 1981; **19**: 303.
19. McCoy GF, Piggot J, Macafee AL, Adair IV. Injuries of the cervical spine in schoolboy rugby football. *Journal of Bone and Joint Surgery* 1984; **66B**: 500.
20. O'Carroll PF, Sheehan JM, Gregg TM. Cervical spine injuries in rugby football. *Irish Medical Journal* 1981; **74**: 377.
21. Scher AT. Rugby injuries to the cervical spinal cord. *South African Medical Journal* 1980; **57**: 37.
22. Silver JR. Injuries to the spine sustained in rugby. *British Medical Journal* 1984; **288**: 37.
23. Albright JP, McAuley E, Martin RK. Head and neck injuries in college football: an eight year analysis. *American Journal of Sports Medicine* 1985; **13**: 147.
24. Alley R H, Jr. Head and neck injuries in high school football. *Journal of American Medical Association* 1964; **188**: 418.
25. Clarke KS. An epidemiologic view. In: Torg JS, ed. *Athlete injuries to the head, neck and face.* Philadelphia: Lea & Febriger, 1982: 15–25.
26. Schneider RC (Ed). *Head and neck injuries. Football: mechanisms, treatment and prevention.* Baltimore: Williams & Wilkins, 1973: 77–125.
27. Funk FJ, Wells RE. Injuries of the cervical spine in football. *Clinical Orthopaedics and Related Research* 1975; **109**: 50.
28. Nine KM, Veqsgo JJ, Sennett B, Torg JS. Prevention of cervical spine injuries in football. A model for other sports. *Physician and Sportsmedicine* 1991; **19**: 54–64.
29. Schneider RC. Serious and fatal neurosurgical football injuries. *Clinical Neurosurgery* 1964; **12**: 226.
30. Schneider RC, Reifel E, Crisler HO, Oosterbaan BG. Serious and fatal football injuries involving the head and spinal cord. *Journal of American Medical Association* 1961; **177**: 362.
31. Torg JS, Veqso JJ, Sennett B, Das M. The National Football Head and Neck Injury Registry 1971–1984. *Journal of American Medical Association* 1985; **254**: 3439–3.
32. Clarke KS, Powell JW. Football helmets and neurotrauma, an epidemiological overview of three seasons. *Medicine and Science in Sports and Exercise* 1979; **11**: 138.
33. Torg JS, Vegso JJ, O'Neill MJ, Sennett B. The epidemiologic, pathologic, biomechanical and cinematographic analysis of football induced cervical spine trauma. *American Journal of Spine Medicine* 1990; **18**: 50–7.
34. Paley D, Gillespie R. Chronic repetitive unrecognized flexion injury of the cervical spine (high jumpers' neck). *American Journal of Sports Medicine* 1986; **14**: 92.
35. Hodgson VR. Reducing serious injury in sports. *Interschool Athletic Administration* 1980; **7**: 11.
36. Saboe LA, Reid DC, Davis L, Warren S, Grace MG. Spine trauma and associated injuries. *Journal of Trauma* 1991; **31**: 43–8.

37. Rodrigo J, Boyd R. Lumbar spine injuries in military parachute jumpers. *Physicians and Sportsmedicine* 1979; **7**: 9.

38. Reid DC, Saboe L. Spine fractures in winter sports. *Sports Medicine* 1989; **7**: 393.

39. Reid DC, Saboe LA, Allan DG. Spine trauma associated with off-road vehicles. *Physician and Sportsmedicine* 1987; **16**: 143.

40. Allan DG, Reid DC, Saboe LA. Off-road recreational motor vehicle accidents: Hospitalization and deaths. *Canadian Journal of Surgery* 1988; **31**: 233.

41. Feriencik K. Trends in ice hockey injuries: 1965 to 1977. *Physician Sportsmedicine* 1979; **7**: 81.

42. Tator CH, Ekong CE, Rowed DW, Schwartz ML, Edmonds VE, Cooper PW. Spinal injuries due to hockey. *Canadian Journal of Neurological Science* 1984; **11**: 34–41.

43. Reid DC, Saboe SA. Spine injuries resulting from winter sports. In Torg JS, ed. *Athletic injuries to the head, neck and face.* 2nd edn. Chicago: Mosby Year Book, 1991.

44. Tator CH, Edmonds VE. National survey of spinal injuries in hockey players. *Canadian Medical Association Journal* 1984; **30**: 875.

45. Bishop PJ, Norman RW, Wells R, Raney D, Skleryk B. Changes in the centre of mass and movement of inertia of a headform induced by a hockey helmet and face shield. *Canadian Journal of Applied Sports Science* 1983; **8**: 19.

46. Hayes D. Reducing risks in hockey. Analysis of equipment and injuries. *Physician and Sports Medicine* 1978; **6**: 67.

47. Frymoyer JW, Pope MH, Kristiansen T. Skiing and spinal trauma. *Clinics in Sports Medicine* 1982; **1**: 309.

48. Davis MW, Litman T, Drill FE, Mueller JK. Ski injuries. *Journal of Trauma* 1977; **17**: 802.

49. Gutman J, Weisbuch J, Wolf M. Ski injuries in 1972–73. a report analysis of a major health program. *Journal of American Medical Association* 1974; **230**: 1423.

50. Howarth B. Skiing injuries. *Clinical Orthopaedics and Related Research* 1965; **43**: 171.

51. Margreiter R, Raas E, Luger LJ. The risk of injury in experienced alpine skiers. *Orthopedic Clinics of North America* 1976; **7**: 51.

52. Tapper EM. Ski injuries from 1939 to 1976: The Sun Valley experience. *American Journal of Sports Medicine* 1978; **6**: 114.

53. Morrow PL. Downhill ski fatalities: The Vermont experience. Presented at the National Association of Medical Examiners Annual Meeting, Boston, 5 November 1988.

54. Shealy JE. How dangerous is skiing and who's at risk? *Ski Patrol Magazine* 1985; **2**: 21.

55. Oh S . Cervical injury from skiing. *International Journal of Sports Medicine* 1984; **5**: 268.

56. Ellison AE. Skiing injuries. *Clinical Symposium* 2977; **29**: 1.

57. Clancy WG, McConkey JP. Nordic and alpine skiing. In Schneider RCS, Kennedy JCK, Plant MLP, eds. *Sports injuries—mechanisms, prevention and treatment.* Baltimore: Williams & Wilkins, 1985.

58. Garrick JG, Requa RK. Injury patterns in children and adolescent skiers. *American Journal of Sports Medicine* 1979; **7**: 245.

59. Henderson RL, Reid DC, Saboe LA. Multiple contiguous spine fractures. *Spine* 1991; **16**: 128–31.

60. Keen JS. Thoracolumbar fractures in winter sports. *Clinical Orthopaedics and Related Research* 1987; **216**: 39.

61. Harris JB. Neurological injuries in winter sports. *Physician and Sports Medicine* 1983; **11**: 111.

62. Dowling PA. Prospective study of injuries in United States Ski Association freestyle skiing, 1976–77 to 1979–80. *American Journal of Sports Medicine* 1982; **10**: 268.

63. Reif AE. Risks and gains. In: Vinger PF, Hoerner EF, eds. *Sports injuries: the unthwarted epidemic.* 2nd edn. Littleton, MA: PSG, 1986.

64. Rich P. Canada leads field in sports-related spinal injuries. *Medical Post* 1985 Sep 3: 50.

65. Wester K. Serious ski-jumping injuries in Norway. *American Journal of Sports Medicine* 1985; **13**: 124.

66. Wright JR, Hixson EG, Rand JJ. Injury patterns in nordic ski-jumpers. A retrospective analysis of injuries occurring at the Intervale Ski Jump Complex from 1980–1985. *American Journal of Sports Medicine* 1986; **14**: 393.

67. Wester K. Improved safety in ski-jumping. *American Journal of Sports Medicine* 1988; **16**: 499.

68. Eriksson E. Ski injuries in Sweden: a one year survey. *Orthopedic Clinics of North America* 1976; **7**: 3.

69. Wright JR, Hixson EG, Rand JJ. Injury patterns in nordic ski jumpers. *American Journal of Sports Medicine* 1986; **14**: 292.

70. Write JR. Nordic ski-jumping fatalities in the United States: a 50 year summary. *Journal of Trauma* 1988; **28**: 848.

71. Wenzel FJ, Peters RA. A ten-year survey of snowmobile accidents, injuries and fatalities in Wisconsin. *Physicians and Sportsmedicine* 1986; **14**: 140.

72. Kritter AE, Carnesale PG, Prusinski D. Snowmobile: Fund and/or folly. *Wisconsin Medical Journal* 1972; **71**: 230.

73. Trager GW, Grayman G. Accidents and all-terrain vehicles (C). *Journal of American Medical Association* 1986; **225**: 2160.

74. Hamidy CR, Dhir A, Cameron B, Jones H, Fitzerald GWN. Snowmobile injuries in Northern Newfoundland and Labrador. An 18 year review. *Journal of Trauma* 1988; **28**: 1232.

75. Haynes CD, Stroud SD, Thompson CE. The three wheeler (adult tricycle): An unstable, dangerous machine. *Journal of Trauma* 1986; **26**: 643.

76. Chism SE, Soule AB. Snowmobile injuries: hazards from a popular new winter sport. *Journal of American Medical Association* 1969; **209**: 1672.

77. Wiley JJ. The dangers of off-road vehicles to young drivers. *Canadian Medical Association Journal* 1986; **135**: 1345.

78. Dominici RH, Drake EH. Speed on snow: The motorized sled. *American Journal of Surgery* 1970; **119**: 483.

79. Damschroder AD, Kleinstiver BS. *Homo snomoblilius. American Journal of Sports Medicine* 1976; **4** 249.

80. Percy EC. The snowmobile. Friend or foe? *Journal of Trauma* 1972; **12**: 444.

81. Stevens WS, Rodgers BM, Newman BM. Pediatric trauma associated with all-terraine vehicles. *Journal of Pediatrics* 1986; **109**: 25.

82. Smith SM, Middaugh JP. Injuries associated with three-wheeled all-terrain vehicles. Alaska 1983 and 1984. *Journal of American Medical Association* 1986; **255**: 2454.

83. Speca JM, Cowell HR. Minibike and motorcycle accidents in adolescents. A new epidemic. *Journal of American Medical Association* 1975; **232**: 55.

84. Golladay ES, Slezak JW, Mollitt DL, Siebert RW. The three wheeler—a menace to the preadolescent child. *Journal of Trauma* 1985; **25**: 232.

85. Westman JA, Morrow G. III. Moped injuries in children. *Pediatrics* 1984; **74**: 820.

86. Wiley JJ, McIntyre WM, Mercier P. Injuries associated with off-road vehicles among children. *Canadian Medical Association Journal* 1986; **135**: 136.

87. Withington RL, Hall LN. Snowmobile accidents: a review of injuries sustained in the use of snowmobiles in northern New England during the 1968–69 season. *Journal of Trauma* 1970; **10**: 760.

88. Torg JS, Quedenfeld TC, Burstein A, Spealman A, Nichols C III. National Football Head and Neck Registry: Report on cervical quadriplegia. 1971 to 1975. *American Journal Sports Medicine* 197; **7**: 127.

89. Torg JS et al. Neurapraxia of the cervical spinal cord with transient quadriplegia. *Journal of Bone and Joint Surgery* 1986; **68A**: 1354–70.

90. Ladd AL, Scranton PE. Congenital cervical spinal stenosis presenting as transient quadriplegia in athletes. Report of two cases. *Journal of Bone Joint Surgery* 1986; **68A**: 1371–4.

91. Ladd AL, Scanton PE. Congenital cervical stenosis presenting as transient quadriplegia in athletes. *Journal of Bone Joint Surgery* 1986; **68A**: 1731.

92. Odor JM, Watkins RG, Dillin WH, Dennis S, Saberi M. Incidence of cervical spinal stenosis in professional and rookie football players. *American Journal of Sports Medicine* 1990; **18**: 507–9.

93. Gerberich SG et al. Spinal trauma and symptoms in high school football players. *Physician and Sportsmedicine* 1983; **11**: 122.

94. Bergfeld JA, Hershman EB, Wilbourn AJ. Brachial injuries in athletes. *Orthopaedic Transactions* 1988; **12**: 743–4.

95. Chrisman OD *et al.* Lateral-flexion neck injuries in athletic competition. *Journal of American Medical Association* 1965; **192**: 117–19.

96. Poindexter DP, Johnson EW. Football shoulder and neck injury. A study of the 'Stinger'. *Archives of Physical Medicine and Rehabilitation* 1984; **65**: 601–2.

97. Robertson WC Jr, Eichman PL, Clancy WG. Upper trunk, brachial plexopathy in football players. *Journal of American Medical Association* 1976; **241**: 1480–2.

98. DiBenedetto M, Markey K. Electrodiagnostic localization of traumatic upper trunk brachial plexopathy. *Archives of Physical Medicine and Rehabilitation* 1984; **65**: 15–17.

99. DiBenedetto M, Chardhry U, Markey K. Proximal nerve conduction. Brachial plexus. *Muscle and Nerve* 1982; **5**: 564.

100. Hu R, Burnham R, Reid DC, Grace M, Saboe L. Burners in contact sports. *Clinical Journal of Sports Medicine* 1991; **1**: 236–42.

101. Masoon JC, Kevin T, Rehhopf P. System for preventing acute neck injury. *Physician and Sportsmedicine* 1977; **5**: 76.

102. Feldick HG, Albright JP. Football survey reveals 'missed' neck injuries. *Physician and Sportsmedicine* 1976; **4**: 77.

103. Reid DC, Henderson R, Saboe L, Miller JDR. Etiology and clinical course of missed spine fractures. *Journal of Trauma* 1987; **27**: 980–986.

104. Gibbs R. A protective collar for cervical radiculopathy. *Physician and Sportsmedicine* 1984; **12**: 139.

7.6 Head injuries in athletics

MICHAEL L. SCHWARTZ and CHARLES H. TATOR

INTRODUCTION

In industrial societies today there is a greater democratization of athletic activity than ever before. With a shorter working week, increased leisure time, and the perception by the public that physical activity need not cease when formal education finishes, older people are remaining active or are resuming sporting activities. Furthermore, through television, there is a greater emphasis on high performance competition that influences the behaviour of athletic participants of all ages and predisposes to more serious injuries.

EPIDEMIOLOGY

Athletic activities vary from one part of the world to another and are influenced by climate, geography, and culture. A review of catastrophic sporting injuries in the province of Ontario, Canada in 1986,[1] found that head injuries comprised a significant proportion of those collected (Fig. 1) and resulted from a range of activities illustrated in Fig. 2. Motor sports proved most dangerous, accounting for 30 of the 129 head injuries, with snowmobiling (12), off-road motor-biking (9), and use of off-road all-terrain

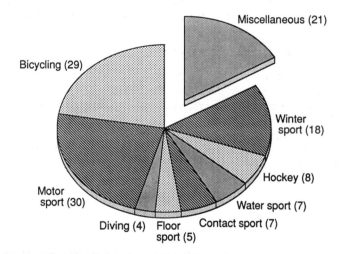

Fig. 2 Head injuries occurred in participants in a variety of activities with motor sport, bicycling, winter sports, and a miscellaneous category accounting for 76 per cent of 129 head injuries.

vehicles (9) making up the 30 injuries. Pedal bicycling accounted for 29 injuries, winter sports including tobogganing (11), skating (4), and skiing (3) accounted for 18 injuries, and a miscellaneous group accounted for 21 additional head injuries (Fig. 3). This report, produced by surveying Ontario neurosurgeons, orthopaedic surgeons, emergency physicians, and other health care professionals who might be expected to see catastrophic injuries, is by no means complete. Nevertheless, the identification of the use of all-terrain vehicles, especially those with three wheels, as a particularly dangerous activity led to withdrawal of the three-wheel all-terrain vehicles from the market. It is desirable that individuals who have an interest or supervisory role in athletics survey the patterns of injury where they live and work so that dangerous activities may be identified and safe practices promoted.

ROLE OF THE PHYSICIAN

Because head injuries, even apparently minor ones, may be so devastating and irrevocable,[2,3] the physician's role must include

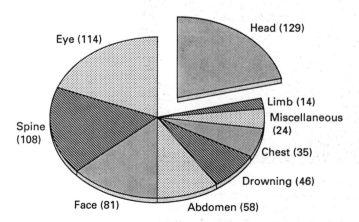

Fig. 1 A review of catastrophic sporting injuries in the Province of Ontario, Canada, in 1986, found that head injuries accounted for 20 per cent of catastrophic sporting injuries.

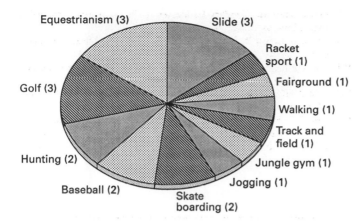

Fig. 3 Miscellaneous activities in which 21 head injuries occurred.

the advocacy of safe practices which enhance prevention and mitigation of brain injuries. Physicians involved in athletics must also understand the pathophysiological mechanisms of brain injury and be familiar with the principles and practice of early treatment.

PATHOPHYSIOLOGY

The vast majority of sporting head injuries are blunt injuries, caused by falls or collisions. In general, if the head is buttressed and acceleration of the cranium and its contents is prevented, there is no significant injury to the brain unless there is mechanical failure of the cranium, that is, a depressed skull fracture. In such a case, there may be a focal injury to the brain underlying the depressed fragment.

In the majority of head injuries that occur in athletics there is violent acceleration (or deceleration) imparted to the head. Even though it is encased within the rigid, bony shell of the cranium and floats in the watery medium of the cerebrospinal fluid, the human brain has evolved by increasing in size and complexity, to the point where it is vulnerable to injury when the head is subjected to violent acceleration or deceleration. Diffuse damage may occur, even without violation of any of the coverings of the brain. The brain parenchyma is composed of a delicate feltwork of interconnecting axons. Because of the high metabolic requirements of neurones, there is a rich, but very delicate, vascular network of fine capillaries that conduct oxygen and glucose to them. There is no fibrous or tough internal structure to the brain. As a result, when a person suffers a blow to the head, or falls and strikes his head on the ground, the brain is literally torn to pieces by the internal shearing forces that are generated on impact.[4]

Figure 4 shows the computed tomographic (CT) scan of a somewhat unusual case where a hockey player, properly equipped with a helmet, fell, striking his head on the ice. The white, irregular haemorrhages that resulted from tearing of blood vessels within the brain parenchyma can be seen. As these lesions are located subjacent to the cerebral cortex subserving leg function, the boy's legs were stiff and spastic but his arms (the relevant cortex is shown in the CT slice on the right in Fig. 4) were unaffected. Because of the disparity between arm and leg function it was thought at first that the athlete had suffered a spinal cord injury.

The damage caused by deceleration injury may be widespread but the shape of the cranial cavity tends in addition to focus the shearing forces in the frontal and temporal lobes as indicated in Fig. 5. Figure 6 shows the CT scan of a person who has suffered an injury that is predicted by the model illustrated in Fig. 5. The left hand CT slice passes through the inferior portion of the frontal

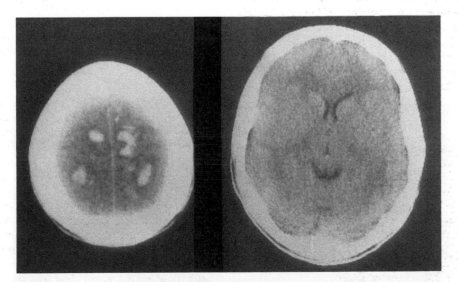

Fig. 4 On the left, the white irregular regions within the brain parenchyma represent blood that has extravasated from vessels within the brain parenchyma torn by shearing forces generated on impact when this hockey player's head struck the ice. These lesions are subjacent to the cerebral cortex subserving leg function. As a result, his legs were stiff and spastic but his arms were unaffected. The CT scan slice on the right represents a cross-section through the portion of the motor strip subserving arm function where there has been no significant injury.

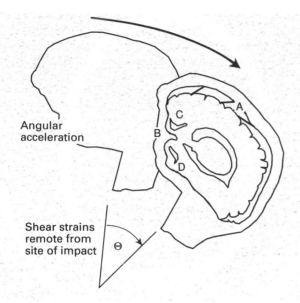

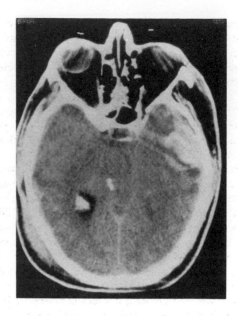

Fig. 5 When angular acceleration or deceleration of the cranium and its contents occur, the brain rotates within the cranial cavity. Bridging (A) veins from the surface of the brain to the major venous sinuses may be stretched and torn with the release of blood into the potential subdural space. The sphenoid wing (B) impedes rotation and focuses shear strains within the frontal lobe (C) and the temporal lobe (D). The traumatic intracerebral haematomas that occur within the substance of the brain, remote from the site of impact, are called 'contre-coup' lesions.

Fig. 7 On the observer's left over the patient's right parieto-occipital region, a subgaleal haematoma can be seen. In the contralateral temporal lobe, a traumatic intracerebral haematoma, a so-called 'contre-coup' injury, is visible.

lobes where fresh haematomas (white) are surrounded by (dark) oedema. The middle slice, passing through the temporal lobes, shows bruising anteriorly at the poles. The right hand slice at the level of the midbrain where the ambient cerebrospinal fluid cisterns can no longer be clearly seen surrounding the brain-stem, indicates raised intracranial pressure.

The focusing effect of skull shape on the shearing forces may also produce damage on the side of the head opposite to where the blow is struck, a so-called 'contre coup' injury. Figure 7 illustrates

a subgaleal haematoma on the observer's left over the patient's right parieto-occipital region. In the contralateral temporal lobe, a traumatic intracerebral haematoma is visible.

Biomechanical studies show that punches and kicks to the head may exceed a force of 100 G.[5] From impact tests on head forms containing accelerometers used in helmet design, it has been determined that the peak force on impact in an unrestrained fall from standing height when the head strikes a hard flat surface may exceed 300 G. In fact, the Canadian Standards Association (CSA) standard for cycling helmets requires only that in a fall from a height of approximately 1.5 m (80 J impact) the force of acceleration be mitigated to 250 G.[6] This corresponds to the force to which the head of an unrestrained passenger in an automobile

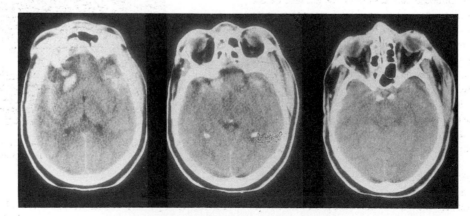

Fig. 6 The left-hand CT slice passes through the inferior portion of the frontal lobes where fresh haematomas (white) are surrounded by (dark) oedema. The middle slice, passing through the temporal lobes, shows bruising anteriorly at the poles. The right-hand slice at the level of the midbrain where the ambient cerebrospinal fluid cisterns can no longer be clearly seen surrounding the brain-stem, indicates raised intracranial pressure.

collision might be subjected in striking the dashboard. In short, very severe forces may be applied to the brain as a result of blows to the head and falls in which the head strikes the ground.

In the evaluation of patients with head injuries there are two aspects to consider. The first is the focal injury, that is, the damage done to a specific part of the brain that subserves a particular function. For example, an injury to the posterior part of the left frontal lobe results in impaired control or even paralysis of the right side of the body, and in most people, produces a deficit in the production of speech. An injury to the occipital lobe may cause hemianopia and so on. If a focal injury is caused without acceleration of the whole brain, there may be no loss of consciousness. The second aspect to consider is the amount of diffuse brain damage caused by acceleration. In the case of focal or diffuse brain damage, there is some loss of function caused by brief mechanical distortion that makes neurones refractory to stimulation. This is, indeed, reversible. More severe shearing forces result in the tearing of axons and the death of neurones. The initial effect is coma and/or amnesia. The lasting effect of diffuse axonal injury, particularly to the frontal lobes, is a lack of resilience in adapting to novel situations with slow and erratic performance.[7] Focal and diffuse injury often coexist and may result in persisting deficits that range from subtle personality changes through to obvious deficits of intellect and memory, or even persistent coma.

A 'second injury'[8,9] may result from hypoxia caused by inadequate ventilation immediately after impact, or from ischaemia resulting from cerebral oedema and raised intracranial pressure. If intracranial blood vessels are torn, then an expanding intracranial haematoma may compress and distort the brain producing focal damage at the site of the haematoma or more diffuse damage by intracranial hypertension.

Temporal skull fractures may tear the middle meningeal artery with release of blood outside the dura, an epidural haematoma. Figure 8 shows a right epidural haematoma (on the observer's left). The blood clot, which appears lentiform or lens shaped on cross-section, has that shape because it is bounded laterally by the inner table of the skull and medially by the dura. Despite the limiting effect of the dura, the brain is compressed and distorted. The lateral ventricles, seen in cross-section, which should straddle the midline, are compressed and displaced towards the patient's left.

Rupture of bridging veins from the surface of the brain to dural sinuses may release blood beneath the dura over the surface of the brain, that is, a subdural haematoma. An immense right acute subdural haematoma is illustrated in Fig. 9. There is a dramatic shift of the ventricles toward the patient's left.

Shearing forces that tear significant blood vessels within the substance of the brain may produce traumatic intracerebral haematomas as illustrated in Figs. 4, 6, and 7. Any of the above processes may produce both focal and diffuse effects. Increasing distortion and rising intracranial pressure generally result in an increasing neurological deficit and a declining level of consciousness, which should prompt a neurosurgical consultation and possibly a CT scan.

Distortion of the oculomotor (cranial III) nerve by transtentorial herniation of the temporal lobe causes pupillary dilatation and is an ominous sign requiring immediate neurosurgical attention.

Once intracranial pressure has been relieved, resolution of cerebral oedema and recovery may begin. Recovery is thought to occur by sprouting of the axons of surviving neurones to form new connections, that is, 'rewiring' of the brain, or by an increase in the role of surviving neurones which subserve new functions formerly

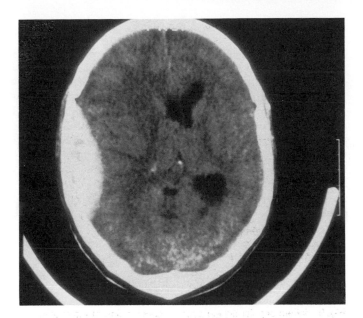

Fig. 8 On the observer's left, a right epidural haematoma can be seen. The blood clot, which appears lentiform or lens-shaped on cross-section, has that shape because it is bounded laterally by the inner table of the skull and medially by the dura. Despite the limiting effect of the dura, the brain is compressed and distorted. The lateral ventricles, seen in cross-section, which should straddle the midline, are compressed and displaced toward the patient's left.

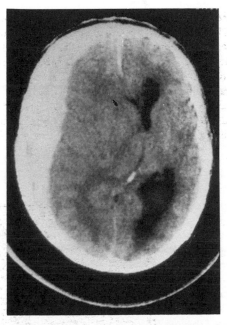

Fig. 9 An immense right acute subdural haematoma is illustrated. There is a dramatic shift of the ventricles toward the patient's left.

carried out by the neurones that have been lost.[10] Recovery from a brain injury continues for many months, although, in general, the greatest improvement occurs within the first weeks after injury. Even in the face of a relatively good recovery, effective return to preinjury activities may be hampered by the psychological effects of diffuse brain injury.[11]

INITIAL EVALUATION

It is the task of the physician who begins the resuscitation of a patient with a head injury to determine the likelihood of a condition requiring treatment, like a traumatic intracranial haematoma. One begins by obtaining a history of the mechanism of injury. A rugby player who has been stunned by a blow to the head and is dazed only momentarily, is less likely to have suffered a severe injury than a cyclist who has come off his bike at 50 km/h. If the patient was initially conscious and is sinking into coma, or if he initially moves all his limbs and has now become hemiplegic, it is likely that there is an expanding intracranial haematoma that is progressively distorting the brain and raising the intracranial pressure. Such a patient must be transferred immediately to a neurosurgical unit for assessment.

The Glasgow Coma Score[12,13] (see Table 2, Chapter 7.4) is useful in deciding whether a patient is improving, staying the same, or deteriorating. In effect, it is an operational definition of consciousness that has the observer repeat at intervals a series of easy, stereotyped observations. By assessing and reporting whether a patient opens his eyes to certain stimuli, answers questions appropriately or moves his limbs in a certain way, the observer can base therapeutic decisions on the patient's improvement or decline. A 15-point scale has been developed.

If the patient's eyes are spontaneously open, a full score of 4 is assigned by the observer. If the eyes are closed during the period of observation but the patient responds to voice (not necessarily an instruction to open the eyes) by opening them, a score of 3 is given. If a painful stimulus is required for eye-opening, then 2 is given. No eye-opening receives a score of 1.

Compliance with a simple verbal instruction, the best motor response (on the good side if the patient is hemiparetic) is assigned a score of 6. Localization of a painful stimulus, such as reaching up to attempt to remove the examiner's hand pressing on the superior margin of the orbit, receives a score of 5. Withdrawal of the hand from nailbed compression is scored as 4 and spastic flexion of the elbow, a stereotyped, reflex response is scored as 3. Spastic extensor posturing receives 2 and no motor response is scored 1.

If in conversation, the patient is oriented to person, place, and time, a score of 5 is given. A score of 4 is assigned to the patient who converses but is disoriented. A score of 3 is given for single words (often expletives) and a score of 2 for incomprehensible sounds. A score of 1 indicates no verbal response. As may be readily appreciated, the verbal portion of the Glasgow Coma Score may be at variance to the eye and motor scores if an apparently conscious patient is very young, lacks facility in the examiner's language, or is aphasic.

The sum of the eye (E), motor (M), and verbal (V) scores may be computed or the three may be evaluated and plotted on a graph separately hour by hour or more frequently as required. Practical advice on the use of the scale is documented in many sources. A deteriorating Glasgow Coma Score is an indication for prompt transfer of an athlete under observation to a medical facility.

FIRST AID

In the management of any head injury, certain priorities govern treatment.[14] Of prime importance is maintenance of the airway, insuring that there is adequate ventilation and making certain that the blood pressure is adequate. On the field, a sphygmomanometer and the means to maintain an intravenous line may not be available. In the vast majority of cases, blood loss will not be an issue and the person rendering first aid will deal only with airway and ventilation. As a blow sufficiently severe to render a person unconscious is also strong enough to cause a cervical spine fracture, one should assume that any unconscious person also has a broken neck. Distortion of the neck should be avoided and removal of helmets and other protective equipment should be done with gentle in-line manual traction so as to maintain cervical alignment. Unconscious competitors should be transferred forthwith to the emergency department of a hospital with a neurosurgical service.

EVALUATION OF SEVERITY

In virtually every significant diffuse brain injury, there is a period of amnesia that surrounds the impact that caused it.[15,16] As we record new memories, they are initially somewhat volatile. With a significant blow to the head, the last memories prior to impact are lost. This backward loss of memory is called retrograde amnesia. The retrograde amnesia may initially be long and tend to shrink with the passage of time. It is generally considered to be less reliable an indicator of the severity of injury than the memory gap that follows the injury, the period of post-traumatic amnesia. Even though a head injury patient may superficially appear to be normal, if the subject is questioned, say 3 months after injury, it may be discovered that a period of several days or even several weeks has elapsed, of which the patient has no recall. In the mildest cases, there may be no loss of consciousness but there may be a period of amnesia. Concussion may be classified as mild, moderate, and severe, as indicated in Table 1.[17] In the mild category, there is only a brief memory gap without loss of consciousness. In severe concussion there is coma and post-traumatic amnesia.

Table 1   Severity of concussion

Grade	Loss of consciousness	Post-traumatic amnesia
1 Mild	No	< 30 min
2 Moderate	< 5 min	> 30 min
3 Severe	> 5 min	> 24 h

Modified from ref. 17.

In the past, it was believed that brief loss of consciousness was the result of a physiological disruption of brain function only and that 'concussion' was a completely reversible condition. We now know that recovery after a concussion is not complete; rather there is death of some neurones and a cumulative effect if concussion is repeated.[18,19]

Since 1975, neuropsychological tests which evaluate the subject's ability to process information rapidly have been available.[20] These tests, applied to athletes who have suffered repeated concussions, indicate that the effect of repeated injury is cumulative and that recovery occurs more slowly and less completely with each successive injury. It has been well-known for a long time that injury severity correlates well with the length of a boxing career and the number of bouts fought.[21] Figure 10 shows the severe cerebral atrophy and cavum septum pellucidum that may result from repeated blows to the head.

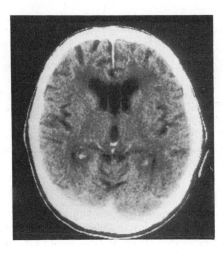

Fig. 10 There is severe cerebral atrophy and a cavum septum pellucidum, findings that may result from repeated blows to the head. Such changes may be correlated with the length of a boxing career and the number of bouts fought.

RETURN TO PLAY GUIDELINES

Any confusion, loss of consciousness, or focal neurological deficit in a player is sufficient cause for withdrawal from the game. A series of guidelines for return to play after concussion is shown in Table 2. The guidelines are predicated on the observations that the effect of repeated concussion is cumulative and that infrequently, a minor blow to the head shortly after a significant head injury may be catastrophic.[22,23] It is known from animal experimentation that following a brain injury, autoregulation (the maintenance of constant cerebral blood flow through a range of systemic arterial pressure) is lost.[24] It is thought that impaired autoregulation may predispose to fulminant cerebral oedema following a concussion. The recovery of autoregulation after a cerebral insult may take several weeks.[25] We propose that 6 weeks is likely to be sufficient time for restitution of normal cerebral vascular reactivity.

After a mild concussion, a player may return to play if he is well for 1 week. Persistent headache, difficulty in concentrating, memory impairment, or neurological deficit after 1 week not only precludes return to play but is indication for a complete neurological examination and possibly a CT scan. A second mild concussion warrants a longer break from competition and a third concussion, longer still. For moderate concussions, a break of 2 weeks is recommended after the first injury, 6 weeks after the second, and suspension for the season after the third. For severe concussions, a suspension of 6 weeks is recommended to allow for the restoration of autoregulation. A second severe concussion, at minimum, requires suspension of play for the duration of the season and should lead to consideration of abandoning the sport altogether. A third severe concussion precludes further play for the individual in all contact sports. It should be recognized that certain individuals are at greater risk than their companions in performance of particular athletic activities. Those who suffer repeated injuries should be directed to other less dangerous sports.

ADVOCACY

Many sports can be made safer by rule changes, improved equipment or the recognition by coaches that certain practices are dangerous. For example, North American football players were at one time directed to tackle by striking the numbers on the opposing player's jersey with their helmets. While often effective in stopping the oncoming runner, the practice sometimes resulted in a broken neck for the tackler. As a result of coaching and rule changes, the incidence of such injuries in North American football has declined.[26] Some activities, by their very nature, are unacceptably hazardous and should be discontinued. With the knowledge that repeated head blows are cumulative in their effect, boxing cannot be condoned as a civilized sporting activity. Proponents and opponents of boxing tend to focus on the infrequent catastrophic ring injuries but in fact the repeated head blows suffered by thousands of participants, in the belief that they are getting healthful exercise and building character, are of much greater concern.[27] Since the object in a boxing match is to knock out one's opponent, that is, to inflict permanent brain damage, the activity must be repugnant to all responsible physicians. In preventing injury, the ringside physician is about as effective as the priest at a judicial hanging.

Bicycling, which has always been a popular sport in Europe, has recently been taken up by an increasing number of people in North America. It is estimated that 85 million people in the United States ride bicycles and that bicycling injuries accounted for an estimated 574 000 emergency room visits and 1300 deaths in the United States in 1985.[28] The most common cause of death and serious disability in those bicycle accidents was head injury. The use of bicycle helmets significantly reduces the severity of head injuries among cyclists[28] but many people, especially children, are reluctant to wear them.

Helmets produce their effect as the liner is crushed on impact. Deceleration of the head is spread over the time and distance that the liner is crushed and hence the peak acceleration of the skull and its contents is reduced. Helmets containing head forms equipped with accelerometers are tested by dropping them on hard surfaces or anvils.[29] In this way, the effect of design and material changes upon deceleration can be evaluated. The crushable material is chosen to be appropriate for the particular impact against which it is intended to protect. Motorcycle helmets, for example, have

Table 2 Return to play after concussion

	First concussion	Second concussion	Third concussion
Grade 1 (mild)	Play if well for 1 week	Play if well for 2 weeks	Off 6 weeks, then play if well for 1 week
Grade 2 (moderate)	Play if well for 2 weeks	Off for 6 weeks, then play if well for 1 week	Off for the season, play next year if well
Grade 3 (severe)	Off for 6 weeks, then play if well for 1 week	Off for the season, play next year if well	Desist from contact sports

Modified from ref. 17.

much stiffer linings than those intended for bicycling because they must cushion against impact at higher speed. At the high speed impacts expected in motor vehicle collisions the wearer's head would completely crush the polystyrene foam (styrofoam) liner and 'bottom out' against the hard outer shell resulting in an abrupt deceleration and worse brain injury. Bicycle helmets are intended to mitigate the impact when a cyclist falls off the bicycle and strikes the pavement. If the cyclist is swept up on the hood of an automobile in a collision and accelerated to the speed of the car, the helmet will be insufficient protection on impact with the pavement. All helmets are a compromise in that sufficient padding to render a blow completely harmless would be too bulky and heavy to be worn.

The most effective helmets currently available have certain features. An external hard plastic shell is necessary to diffuse over a wider area the impact of striking or being struck by a relatively small object. Penetration of a relatively sharp object may also be prevented by a hard but not a 'soft shell' of netting or cloth. As the crushable liner has been designed to be appropriate to the expected deceleration forces that are most likely to occur on impact during a particular activity, a helmet intended for one activity should not be used for another purpose. Once crushed, the liner no longer cushions adequately and the helmet should not be reused. As a rule, one should choose a helmet certified for a particular sport by a responsible agency in one's own country, for example the Canadian Standards Association (CSA), the American National Standards Institute, Inc. (ANSI), or the Snell Memorial Foundation, Inc. These agencies also specify testing procedures for the chin strap and harness. Approval means that the helmet is less likely to slip or fly off at the moment of impact.

After the safety criteria have been met, style and comfort are also factors to consider. Look for adequate ventilation, unobstructed vision, and attractive colour. If the user looks and feels good, the helmet is more likely to be worn. Advocacy and cost subsidy programmes have already been successful in modifying population behaviour. In Seattle, where an educational campaign was undertaken, bicycle helmet use among children increased from 5.5 to 15.7 per cent.[30] It remains to be seen whether increased use will translate into lower morbidity as would be expected from another study by some of the same authors.[28] Responsible physicians should be vocal in advocating preventive measures shown to be effective.

Helmets do not attenuate the forces applied to the cervical spine sufficiently to prevent neck injuries. After the introduction of helmets to ice hockey there was an increase in the number of cervical spine fractures because players felt invulnerable. Physicians should caution against unreasonable expectations for protective equipment.

Recreational athletic activity is fun and healthy for almost all participants. High level competition adds increased enjoyment for those athletes who are sufficiently trained and skilful. Head injuries, even apparently mild ones, are so devastating that they must be avoided at all costs. Responsible physicians must promote only those recreational activities where the risk of head injury does not outweigh the benefits of participation.

REFERENCES

1. Tator CH. Report of the Ontario Sport Medicine Advisory Board, Volume II. Ministry of Tourism and Recreation, Province of Ontario. 1987.
2. Benton AL. Historical notes on the postconcussion syndrome. In: Levin HS, Eisenberg HM, Benton AL, eds. *Mild Head Injury*. New York: Oxford University Press, 1989: 3–7.
3. Dacey RG, Jr. Complications after apparently mild head injury and strategies of neurosurgical management. In: Levin HS, Eisenberg HM, Benton AL, eds. *Mild Head Injury*. New York: Oxford University Press, 1989: 83–101.
4. Strich SJ. Cerebral Trauma. In: Blackwood W, Corsellis JAN eds. Greenfield's Neuropathology. Chicago: Year Book Medical Publishers, Inc. 1976: 327–360.
5. Schwartz ML, Hudson AR, Fernie GR, Hayashi K, Coleclough AA. Biomechanical study of full-contact karate contrasted with boxing. Journal of Neurosurgery 1986; **64**: 248–252.
6. Canadian Standards Association. *Cycling Helmets, Canada September 1989*. Toronto, Ontario: Publication No CAN/CSA-D113.2-M89.
7. Stuss DT, *et al*. Subtle neuropsychological deficits in patients with good recovery after closed head injury. Neurosurgery 1985; **17**: 41–7.
8. Jennett B, Teasdale G. Dynamic Pathology. In: Jennett B, Teasdale G, eds. *Management of Head Injuries*. Philadelphia: F.A. Davis Company, 1981: 45–75.
9. Alexander MP. The role of neurobehavioral syndromes in the rehabilitation and outcome of closed head injury. In: Levin HS, Grafman J, Eisenberg HM, eds. *Neurobehavioral Recovery from Head Injury*. New York: Oxford University Press, 1987: 191–205.
10. Devor M. Plasticity in the adult nervous system. In: Illis LS, Sedgwick EM, Glanville HJ, eds. Rehabilitation of the Neurological Patient. Oxford: Blackwell Scientific Publications, 1982: 44–84.
11. Prigatano GP. Psychiatric aspects of head injury: problem areas and suggested guidelines for research. In: Levin HS, Grafman J, Eisenberg HM, eds. *Neurobehavioral Recovery from Head Injury*. New York: Oxford University Press, 1987: 215–31.
12. Teasdale G, Jennett B. Assessment of coma and impaired consciousness: A practical scale. Lancet 1974; **2**: 81–84.
13. Jennett B, Teasdale G. Assessment of impaired consciousness. In: Jennett B, Teasdale G. *Management of Head Injuries*. Philadelphia: F.A. Davis Company, 1981: 77–93.
14. Committee on Trauma. *Advanced Trauma Life Support Instructor Manual*. Chicago: American College of Surgeons, 1991.
15. Corkin SH, Hurt RW, Twitchell TE, Franklin LC, Yin RK. Consequences of nonpenetrating and penetrating head injury: retrograde amnesia, posttraumatic amnesia, and lasting effects on cognition. In: Levin HS, Grafman J, Eisenberg HM, eds. *Neurobehavioral Recovery from Head Injury*. New York: Oxford University Press, 1987: 318–29.
16. Crovitz HF. Techniques to investigate posttraumatic and retrograde amnesia after head injury. In: Levin HS, Grafman J, Eisenberg HM, eds. *Neurobehavioral Recovery from Head Injury*. New York: Oxford University Press, 1987: 330–40.
17. Cantu RC. Guidelines for return to contact sports after a cerebral concussion. The Physician and Sportsmedicine, October 1986; **14(10)**: 75–83.
18. Gronwall D, Wrightson P. Cumulative effect of concussion. The Lancet November 1975; 995–7.
19. Gronwall D. Cumulative and persisting effects of concussion on attention and cognition. In: Levin HS, Eisenberg HM, Benton AL, eds. *Mild Head Injury*. New York: Oxford University Press, 1989: 153–62.
20. Gronwall D, Wrightson P. Memory and information processing capacity after closed head injury. Journal of Neurology, Neurosurgery and Psychiatry 1981; **44**: 889–95.
21. Corsellis JAN, Bruton CJ, Freeman-Browne D. The aftermath of boxing. Psychological Medicine 1973; **3**: 270–303.
22. Saunders RL, Harbaugh RE. The second impact in catastrophic contact-sports head trauma. Journal of the American Medical Association 1984; **252 (Jul 27)**: 538–9.
23. Kelley JP, Nichols JS, Filley CM, Lillehei KO, Rubinstein D, Kleinschmidt-DeMasters BK. Concussion in sports: guidelines for the prevention of catastrophic outcome. *Journal of the American Medical Association*, 1991; T266: 2867–9.
24. Lewelt W, Jenkins LW, Miller JD. Autoregulation of cerebral blood flow after experimental fluid percussion injury of the brain. Journal of Neurosurgery 1980; **53**: 500–11.

25. Paulson OB, Lassen NA, Skinhøj E. Regional cerebral blood flow in apoplexy without arterial occlusion. *Neurology*, 1970; **20**: 125–38.
26. Schneider RC, Peterson TR, Anderson RE. Football. In: Schneider RC, Kennedy JC, Plant ML, eds. *Sports Injuries*. Baltimore: Williams & Wilkins, 1985: 1–63.
27. Barth JT, Alves WM, Ryan TV, *et al.* Mild head injury in sports: neuropsychological sequelae and recovery of function. In: Levin HS, Eisenberg HM, Benton AL, eds. *Mild Head Injury*. New York: Oxford University Press, 1989: 257–75.
28. Thompson RS, Rivara FP, Thompson DC. A case-control study of the effectiveness of bicycle safety helmets. *New England Journal of Medicine* 1989; **320**(21): 1361–7.
29. Bishop PJ, Briard BD. Impact performance of bicycle helmets. *Canadian Journal of Applied Sports Science*, 1984; **9**(2): 94–101.
30. Rogers LW, Bergman AB, Rivara FP. Promoting bicycle helmets to children: a campaign that worked. *Journal of Musculoskeletal Medicine*, 1991; **8**: 64–77.

7.7   Injuries to the wrist and carpus

GARY R. McGILLIVARY

INTRODUCTION

The wrist is very prone to injury in all walks of life, and athletic endeavours are no exception to this. In collision sports, the wrist is frequently injured because we naturally defend ourselves by raising our arms and hands. In non-collision sports, falls are very common and most of them occur in an outstretched hand in a dorsiflexed position with the wrist bearing the brunt of the force. In sports in which clubs, bats, or sticks are used, both direct and indirect blows to the wrist may be incurred. In indirect blows, the force is transmitted through the club to the wrist. Given the frequency with which this area is injured, it is essential that anyone treating athletes becomes familiar with the large variety of injuries that may occur.[1] Many of the injuries are rather unusual, and exact diagnosis and treatment may only be obtainable from someone seeing large numbers of patients. Nonetheless, it is important for those involved in the primary care of athletes to have an appropriate level of awareness and concern about the types of pathology that may be present.

As entire textbooks have been written about wrist injuries, we shall highlight only the diagnosis and treatment possibilities in this chapter. For an in-depth review of any given injury, the reader is referred to the large number of textbooks and treatises available on the various subjects (see Further Reading and refs. 2,4).

ANATOMY

The bony and ligamentous anatomy of the wrist is very complex and in many ways is still not completely understood. Controversy surrounds the exact description of the various ligamentous structures and the functional anatomy of the write.[2–4] For the most part, however, the details of these controversies will not affect the day-to-day treatment of athletes, at least at this point, and therefore will not be dealt with in detail.

In teaching health professionals, physicians, residents, and other orthopaedic surgeons about the wrist and its anatomy, the single most important anatomical fact to be addressed is that it is not a single joint. Thus we refer to the wrist area and not to the wrist joint.

The wrist is a complicated arrangement of the radius and ulna articulating with one another at the distal radio-ulnar joint and in turn articulating with the scaphoid, lunate, and triquetrum at the radiocarpal joint. The radiocarpal joint itself also has specific joints each with its own problems and pathology, such as the radioscaphoid joint, the radiolunate joint, the ulnolunate joint, and the ulnotriquetral articulation. In addition, the scaphoid, lunate, and triquetrum all articulate with each other. The pisiform resides as a sesamoid in the flexor carpi ulnaris tendon and articulates with the triquetrum at the pisotriquetral joint. Moving distally, one encounters the so-called midcarpal joint, which again is a series of articulations comprised of the scaphotrapeziotrapezoid joint and the scaphocapitate and capitolunate joints, and finally the four-quadrant area of the wrist where the capitate, lunate, triquetrum, and hamate all meet. Distal to this are the carpometacarpal articulations which are best dealt with in a separate chapter.

This network of bones and joints is supported by a complicated series of ligamentous structures. The most vital ligaments are the stout palmar-sided wrist ligaments. The exact course and appropriate nomenclature for each of these ligaments is still controversial, but there is no controversy surrounding the fact that these ligaments are critical in maintaining the normal relationships between carpal bones and therefore maintaining normal wrist kinematics.

In addition to the palmar ligaments, there are significant interosseous ligaments between the scaphoid and the lunate and between the lunate and the triquetrum (Fig. 1). These run in a sheet-like fashion from dorsal to volar between the carpal bones discussed above. The mid-portion of these ligaments is less stout and more of an interosseus membrane than a ligament. The dorsal ligaments are less well defined and of much less clinical importance than the volar and interosseus ligaments.

Arising from the ulnar surface of the radius and extending distal to the distal ulna and blending with the ulnar and volar wrist ligaments is the triangular fibrocartilage complex (Fig. 2). This structure is comprised of a number of different components and serves a number of vital functions in this area, ranging from acting as a shock absorber on the ulnar side of the carpus to aiding in the stabilization of the carpus and the distal radio-ulnar joint.[5]

In summary, the anatomy of the wrist is very complex and

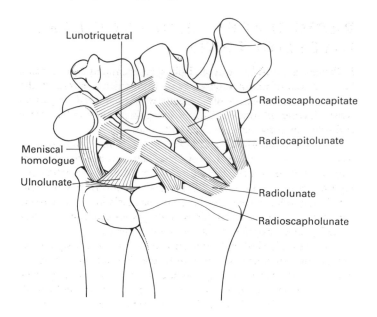

Fig. 1 A schematic of volar wrist ligaments (redrawn from Green DP. *Operative Hand Surgery*, New York: Churchill Livingstone, 1985: 14, with permission).

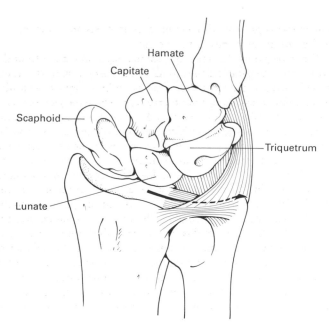

Fig. 2 A diagrammatic representation of the structures comprising the triangular fibrocartilage complex (redrawn from Green DP. *Operative Hand Surgery*, New York: Churchill Livingstone, 1988: 946, with permission).

intricate. Multiple articulations are present, all of which must be considered individually when discussing any significant wrist pathology, and the ligamentous structures in the wrist are significant structures which contribute in an important fashion to normal wrist kinematics and therefore to wrist function. The importance of envisaging the wrist not as a joint but as a series of small joints cannot be overemphasized.

HISTORY AND PHYSICAL EXAMINATION

As in all patient encounters, the initial history and physical examination are the most important part of arriving at a diagnosis and embarking on a plan of treatment in dealing with an athlete with an acutely injured or chronically painful wrist.

The history should include whether there has been a significant injury recently or remotely and the exact nature of this injury, if possible. Details of the amount of force involved and if possible the position of the hand and wrist and upper extremity at the time of the injury should be obtained. In addition to this information, it is critical to enquire as to the exact nature and location of the pain, i.e. it is not enough simply to ascertain that it is the wrist that is sore, but it should also be ascertained whether it is the dorsal radial or the volar radial aspect of the wrist. The patient should simply be asked to indicate the area from which the pain appears to be emanating. In this way information is obtained as to where the patient locates the source of the pain. In addition to this, in many patients who have complaints of chronic pain rather than an acute injury, the magnitude of the pain and the amount of disability it produces are important pieces of information. It is important to decide whether the patient is having pain that can be described as a nuisance, or whether the pain is disabling the individual and has prevented him or her from performing normal activities.

Of course, inquiries must also be made about the presence or absence of neurological symptoms, vascular symptoms, symptoms elsewhere in the extremity, and general functions as would be done in any initial encounter with a patient.

Physical examination of the wrist should be careful, methodical, and very precise, as, unlike most other areas of the body, we are dealing with a very large number of structures confined in a very small space. Precise examination is of the utmost importance.

The first component of the examination is inspection. On initially examining the wrist one should look carefully for any signs of deformity and any areas of localized discoloration or swelling, and note how the patient uses the hand and wrist and entire upper extremity as he or she carries on normal activities throughout the interview. Valuable information can be gained in this fashion. Noting the exact location and the exact nature of any deformities or swellings is of critical importance.

The next step in the physical examination should be careful assessment of the neurological and vascular status of the hand involved. This includes not only checking pulses and capillary refill but examining the extremity to determine whether it is warmer or colder than the opposite extremity, again noting any discoloration present. The neurological examination should check specific muscle function, in particular the thenar mesculature innervated by the median nerve and the ulnar innervated musculature. Two-point discrimination in all digits is quickly and easily assessed and is a reasonably sensitive method for assessing sensation in the hand.

A careful assessment of both an active and a passive range of motion of the wrist should be performed and it should always be compared with the opposite wrist, noting in the history whether there had been any problems with the opposite extremity. Specifically, all planes of motion in the wrist should be checked, including dorsiflexion, palmar flexion, radial deviation, and ulnar deviation pronation and supination. A note should be made of whether any discomfort is present with any of these motions and whether it is limited to the extremes of movement.

When palpating the wrist, one should begin well away from the area where the patient claims to have the most discomfort. All the specific areas of the wrist should be carefully examined with pinpoint thumb pressure and the source from which the pain seems to emanate should be noted as exactly as possible. If this is done quickly or in a cursory fashion, not much information will be gained. If this technique is practised and is applied carefully and precisely, a great deal of information can be gained about the specific and exact location of pathology from this part of the examination. Specific areas that should be carefully examined are the distal radio-ulnar joint, the distal ulna, the region just distal to the distal ulna, where the triangular fibrocartilage complex lies, the distal radius itself, the radioscaphoid area, the scapholunate area, the radiolunate area, and the lunotriquetral area. In addition to this, the pisiform should be grasped and compressed against the triquetrum and moved in a radial and ulnar direction to see whether there is any pain or tenderness in the pisitriquetral joint. This pinpoint examination can be carried out dorsally, radially, ulnarly, and volarly.

There are a number of special tests that can be carried out on physical examination of the wrist which should also be carefully checked, particularly in patients presenting with a history of a recent or remove injury. The first of these is the triquetral ballotment test. This is a specific test for lunotriquetral ligament integrity and whether or not there is lunotriquetral instability. To perform this test the lunate and carpus of the involved wrist are stabilized with the thumb and index finger of one hand. The triquetrum and pisiform can easily be grasped and moved in a dorsal and volar direction with the thumb and index finger of the other hand. Some practice is required to perform this test properly and to learn to assess it, but with experience one can determine whether the amount of motion present is excessive and also whether there is any crepitation associated with this manoeuvre and whether it produces pain. A great deal of information can be obtained in this fashion.

The next specific test is known as the Watson manoeuvre. This test (first described by Dr Kirk Watson) is a specific test for the scapholunate ligamentous disruption, or scaphoid instability. The manoeuvre is performed by having the patient supinate the forearm; the affected wrist is then moved into ulnar deviation and pressure is applied to the volar aspect of the scaphoid, i.e. the scaphoid tubercle, with one thumb. With pressure applied in this manner, the wrist is then forcibly moved from ulnar deviation into radial deviation. In changing from ulnar to radial deviation the normal scaphoid tubercle will move in a volar direction as the scaphoid palmar flexes to avoid impinging upon the radial styloid as the wrist moves into radial deviation. If a scapholunate dissociation is present or if the scaphoid is unstable, pressure on the volar aspect of the scaphoid with the thumb will prevent this normal palmar flexion of the scaphoid and an audible 'clunk' will be produced and/or the patient will experience pain.

The final special test is for midcarpal stability, which is very uncommon. The examiner grasps the patient's distal forearm with one hand, and with the other takes the patient's hand. The patient is asked to relax. Once the patient has relaxed, pressure is applied to the distal hand in a volar direction. If there is any associated midcarpal instability, there will be an obvious 'clunk' with a zigzag deformity produced in the wrist. This is a rather dramatic physical finding when present.

RADIOGRAPHIC AND ANCILLARY INVESTIGATIONS

Following a careful history and physical examination, one has a large amount of information available with which to approach the diagnosis and treatment of a patient with an injured or painful wrist. Further information can now be gained using radiographic and other investigations as seem to be indicated.

Simple anteroposterior and lateral radiographs of the involved wrist are very useful. Review of these radiographs will allow not only the diagnosis of some of the more obvious and dramatic wrist injuries, but also may reveal very subtle abnormalities or signs which may offer clues as to the exact diagnosis in many patients with subtle or rare injuries. These radiographs can only be interpreted after appropriate teaching and experience.

Following physical and radiographic examination a significant number of individuals will require further investigation to determine their diagnosis. A large number of ancillary investigations are possible and not all of these are useful in any one individual. Therefore the investigation should probably proceed on an individual basis.

Technetium pyrophosphate scanning has become increasingly useful to help localize wrist pathology in some patients and has also been used as a screening device in chronic wrist pain. Cineradiography, CT, magnetic resonance imaging (MRI), and cine-MRI are all possible methods which can be used in the investigation of individual patients, but the indications for these modalities are somewhat limited.

Wrist arthrography has been used extensively to investigate wrist pathology, particularly for injuries to various ligamentous structures such as the lunotriquetral ligaments, the scapholunate ligaments, and the triangular fibrocartilage complex (Fig. 3). In recent years techniques have been improved and the information gained has been more beneficial. However, because of the advent of wrist arthroscopy, the present author has not used wrist arthrography to any great extent. Arthroscopy of the wrist has become increasingly popular over the past few years as a means of not only arriving at specific diagnoses but also carrying out arthroscopic surgery to treat certain kinds of pathology.

A number of studies have shown that wrist arthroscopy provides more specific and better information than wrist arthrography. The radiocarpal joint, including the volar wrist ligaments and interosseus ligaments as well as the triangular fibrocartilage complex, can all be directly assessed through a variety of portals. In addition, the midcarpal joint can also be carefully examined directly using the arthroscope. This has led to a better understanding of the various kinds of pathology affecting the wrist, particularly ligamentous injuries and injuries to the triangular fibrocartilage complex.

FRACTURES

Fractures about the wrist are extremely common and athletes are no exception to this. Falls on the outstretched dorsiflexed hand are the most common source of these injuries, but some are produced by indirect trauma. Detailed accounts of the treatment of many of these injuries are given in standard orthopaedic textbooks and no attempt will be made here to supplant these.

Distal radial fractures

Fractures of the distal radius are the most common fracture occurring in the upper extremity. The diagnosis of these injuries is

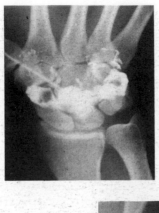
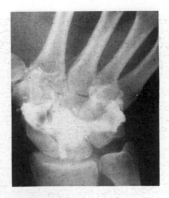

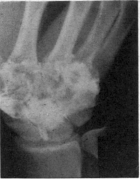

Fig. 3 A typical wrist arthrogram.

generally straightforward, with the patient presenting after a fall on the outstretched hand. The patient will describe an immediate onset of pain, swelling, and frequently deformity.

Physical examination will reveal marked swelling and often a significant deformity through the region of the wrist. The neurological and vascular status of the hand should be very carefully examined as some of these injuries will be associated with significant soft tissue injury, including injuries to the nerve and blood vessels crossing the region of the injury. Radiography confirms the diagnosis and reveals the degree of angulation, displacement, and comminution present at the fracture site. All this information should be carefully and specifically noted. Particular attention should be paid to whether the fracture extends into any of the joints, and specifically whether it is a radioscaphoid joint or, more importantly, the lunate fossa of the radius. In addition, one must remember to assess whether the fracture is intra-articular at the distal radio-ulnar joint. Frequently, the fact that these fractures are intra-articular is not noted.

Traditional treatment has been closed reduction and cast immobilization. More recently, however, clinical reviews of fractures treated in this fashion have shown a high incidence of unsatisfactory results; specifically in one series only approximately 5 per cent of all patients sustaining a fracture of the distal radius has what could be considered an excellent outcome.[6] Studies have generally shown poor results in these fractures, particularly in young active patients. This has led to a more aggressive approach to the treatment of 'simple' fractures, particularly in young people.

When an athlete presents with such an injury the aim is obviously to restore function to as normal a state as possible as quickly as possible. Most studies support the fact that the best way of achieving this aim is to restore anatomy as close as possible to normal, to attain early union, and, if possible, to begin early motion.

A number of classification systems have been developed for fractures of the distal radius and a large number of eponyms are associated with them. The present author prefers to avoid the use of eponyms when dealing with fractures of the distal radius, because they imply a certain amount of familiarity and disdain toward some very complex fractures. More often than not, one will hear phrases such as 'It's just a Colles' fracture' to describe what can be a potential career-ending injury for a young athlete. After the initial history, the physical examination, and a careful review of the radiographs have been completed, a plan should be formulated as to how best to manage the individual fracture. Consideration is given to such factors as fracture stability, comminution, and whether the fracture is intra-articular.

As in all fractures, the initial aim is to obtain an acceptable reduction. In a young adult this would include restoring the length of the radius to normal and correcting to neutral the angulation of its distal articular surface as viewed on a lateral radiograph, i.e. ensuring that the distal articular face is perpendicular to the long axis of the radius. Normally, the distal articular surface of the radius tilts 10 to 14° in a palmar direction (Fig. 4). Correction to neutral is not an arbitrary requirement. Long-term follow-up of fractures of the distal radius with dorsal angulation, i.e. angulation of the distal articular surface beyond neutral, reveals that these patients develop further problems with the carpus. The dorsal tilt of the distal radial surface leads to a dorsal tilt of the lunate and subsequently to palmar flexion of the capitate so that the hand is lined up with the radius. This produces the so-called zigzag collapse deformity of the capitate, lunate, and distal radius. This collapse deformity can subsequently lead to pain, ligamentous instability, and occasionally degenerative change in the carpus. If

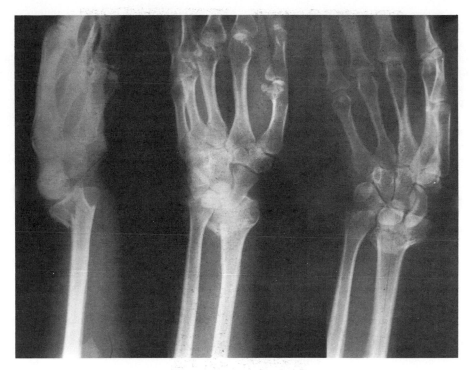

Fig. 4 An extra-articular fracture of the distal radius.

there is an intra-articular component to the fracture it has been shown that a step deformity of greater than 1 to 2 mm is unacceptable as this will lead to accelerated degenerative change.[7] The most critical joints to restore anatomically are the radiolunate joint, or the lunate fossa of the distal radius, and the distal radio-ulnar joint as reconstructive procedures for pain and arthrosis are often based around these two joints.

A closed reduction of the fracture is attainable and this reduction must be maintained. As stated earlier, it is generally believed that the results of cast immobilization in young patients, except for the most minor fractures, are unacceptable as the fractures lose their excellent reduction with simple cast immobilization (Fig. 5). Therefore the majority of these patients should be treated with an external fixation device to maintain the reduction in an acceptable position. If a reduction cannot be attained or maintained in this fashion, then formal open reduction with internal fixation may be required (Fig. 6). This may take the form of limited internal fixation, where an external fixator is used in conjunction with open reduction and internal fixation of one or two small fragments. In more difficult or unstable fractures, formal open reduction and internal fixation with plates and screws is often required.[8]

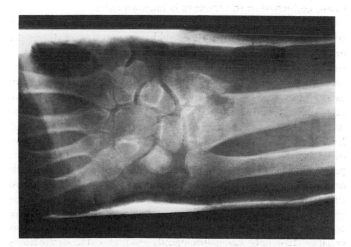

Fig. 5 An intra-articular fracture of the distal radius.

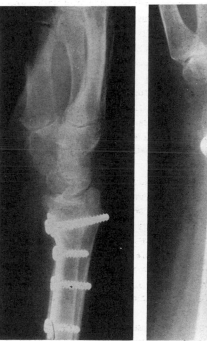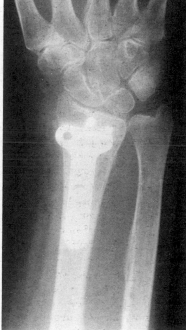

Fig. 6 A distal radius fracture following open reduction internal fixation.

Complications of these fractures are frequent. Neurological injury, particularly median nerve dysfunction, occurs although fortunately, vascular injury and compartment syndrome are both unusual. The more common complications include restriction of motion, arthrosis from non-congruent articular surfaces or as a result of excessive residual dorsal angulation, and ulnar-sided wrist pain and impingement due to lack of restoration of adequate radial length.

In summary, then, cavalier treatment of these fractures may lead to significant hand and wrist dysfunction, particularly in the young athlete. An aggressive approach seems indicated from the outset.

Scaphoid fractures

Scaphoid fractures occur very frequently. Again, these mainly occur as a result of a fall on the outstretched hand.

The initial presentation of a fractured scaphoid may be relatively minor symptoms of slight wrist discomfort on the radial aspect of the wrist with little or no swelling; the only physical finding may be slight discomfort over the volar tubercle of the scaphoid or in the region known as the anatomical snuff-box. If the initial radiograph of the wrist reveals a fracture, then the diagnosis is fairly straightforward. However, if the initial radiographs are normal, things may be more complicated. A patient with the above symptoms and physical findings and a normal radiograph is best treated by immobilization for 2 weeks followed by repeat radiographs in an effort to identify the fracture. If the patient remains symptomatic and the radiographs are still normal, a bone scan may be beneficial.

The aim of treating a scaphoid fracture is to attain anatomical union (Fig. 7). The best treatment depends on a number of factors, such as the location of the fracture within the scaphoid and whether it is displaced or undisplaced. Displacement is far more important than location.

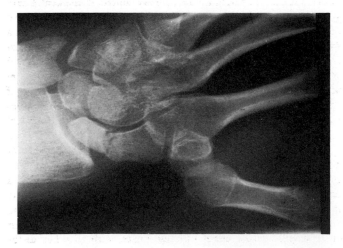

Fig. 7 A fractured scaphoid.

According to standard dogma 99 per cent of undisplaced scaphoid fractures will unite if treated appropriately from the time of injury. Appropriate treatment is said to consist of cast immobilization from the time of initial presentation until fracture union has occurred. The type of immobilization has varied over the years from a below-elbow thumb spica with the interphalangeal joint of the thumb free to above-elbow casting to include immobilization of the interphalangeal joint of the thumb as well as the index and long fingers. More recent data suggest that the use of an above-elbow cast to include immobilization of the interphalangeal joint of the thumb for 3 weeks and subsequently a below-elbow version of the same spica would be adequate treatment.

The total time of immobilization required to gain union is extremely variable. It may be as short as 6 weeks or as long as a year, and the patient should be made aware of this fact from the time of initial presentation. The patient should be informed that the average time to union is 16 weeks and that only 50 per cent of scaphoid fractures are healed in 12 weeks. Of course, this means that a certain number of fractures takes a very long time to unite. Because such prolonged immobilization is unacceptable to many athletes (and workers), there may be some benefit in using early internal fixation of scaphoid fractures.[10] This allows short-term cast immobilization and hence earlier return to wrist motion and activity. This approach is currently under investigation in a number of centres, and may become more popular in subsequent years as methods of fixing the scaphoid internally are improved.

The difficulty in attaining union in such a seemingly minor fracture is directly related to the somewhat precarious blood supply to the scaphoid. Because the majority of the scaphoid consists of articular surface there are only a limited number of points where blood vessels enter this bone. In general, the blood supply is better in the distal portion of the bone and therefore fractures that occur proximally tend to be slower to heal. These fractures also have a higher incidence of non-union, but the exact incidence of this is not known.

The treatment of displaced scaphoid fractures is quite different from that for undisplaced fractures. Even apparently small amounts of displacement (1 mm) are considered to be significant. Such minor amounts of displacement are taken seriously because of the major ligamentous attachments to the scaphoid, as discussed in the section on anatomy. Displacement of fracture fragments by even 1 mm indicates that there is some form of ligamentous injury and therefore instability. This leads to difficulty in attaining union, but can also create long-term problems with carpal instability. Carpal instability is a very complex problem which requires a entire chapter of its own. It is sufficient to state that carpal instability occurs when there have been significant ligamentous injuries in the wrist which may lead to changes in the angular relationships between the carpal bones. Subsequently, this leads to abnormal wrist kinematics and secondary osteoarthritic change.

To avoid these difficulties, displaced fractures in the scaphoid are best treated with early open reduction and internal fixation to restore wrist anatomy to as close as possible to normal (Fig. 8). Obviously, depending on the amount of displacement of the fracture, the associated ligamentous injury is variable. This will be discussed in more detail in the section on ligamentous injuries.

The major specific complication associated with fractures of the scaphoid is that of non-union. This is said to occur in only 1 per cent of cases if undisplaced fractures are treated from the time of injury. However, this number is based on very old data and may be too low. Proving that union has occurred is difficult, and there are many patients with scaphoid fractures said to have been united on the basis of plain radiographs when tomography reveals that the fracture line in the scaphoid is still clearly present and has not healed. Many of these patients are not symptomatic in the early stages.

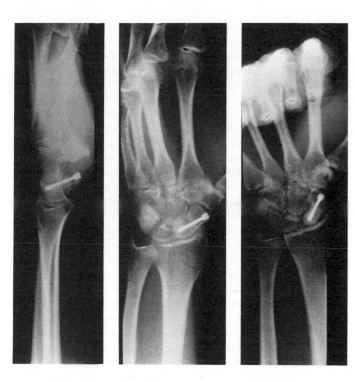

Fig. 8 A scaphoic fracture following open reduction internal fixation and bone grafting

The difficulty with the non-union scaphoid is that, again, it may lead, to carpal instability. The scaphoid is a critical link between the proximal and distal rows of the carpus. With the link disrupted, wrist kinematics change and this leads to a specific pattern of degenerative arthritis in the wrist. Osteoarthritic change begins at the radioscaphoid joint and progresses through the scaphocapitate and capitolunate joints. Normally the radiolunate joint is spared and normal articular cartilage is present. All this degenerative change leads to a painful and stiff wrist. This specific form of degenerative arthritis is known as scapholunate advanced collapse (SLAC) or sometimes as a SLAC wrist. A salvage or reconstructive surgical procedure may be required to give these individuals relief from their pain. The exact frequency with which this complication occurs is not known. It has been suggested that it is the ultimate fate of all non-union scaphoids, but patients followed in all studies on this subject have been highly selected. The incidence of asymptomatic non-union of long standing in the general population is not known. However, it does occur frequently, and therefore aggressive treatment for fresh scaphoid fractures and non-unions is probably justified in the young athlete.

Fractures of other carpal bones

Significant fractures of the other bones in the carpus are generally rare. Minor dorsal chip or capsular avulsion fractures are seen relatively frequently. After ascertaining that they are not associated with major ligamentous disruptions in the carpus (discussed in detail in the section on ligament injuries), they can be treated with short-term immobilization for comfort, with an early return to motion and activity anticipated. Major fractures can be treated with casting alone if they are undisplaced; if they are displaced they will require open reduction and internal fixation to restore joint surface to normal and produce a stable normally functioning carpus.

One specific fracture that should be discussed here is that of the hook of the hamate. This rare injury is usually incurred by striking something solid with a club or a bat, and the force transmitted down the shaft results in fracture of the hook of the hamate. Patients will present with a typical history; frequently they are golfers who have struck a hidden tree root. They have pain and tenderness in the ulnar side of the palm. Symptoms of ulnar nerve irritation or compression may be present as this structure lies in close proximity within Guillon's canal. The diagnosis is confirmed with radiography, but the physician examining the patient must suspect this condition because a special radiographic view known as a carpal tunnel view is required to visualize the hook of the hamate adequately and see the fracture site.

If the diagnosis is made early and the fracture is undisplaced, a period of casting for 6 weeks, followed by repeat examination and radiography, is justified. If the presentation is late, the fracture is displaced, or union has not occurred following casting, surgical treatment is required. Treatment options consist of either excision of the fracture fragment or internal fixation of the fracture. Since no difference in the outcome of these two treatments has been shown, excision is favoured as internal fixation can be fraught with major complications such as ulnar nerve palsy.[11]

The major long-term complication of this fracture is rupture of the flexor digitorum profundus tendon to the little finger. This tendon, which passes directly across the fracture site as it moves through the carpal tunnel, becomes frayed and attenuated, and ultimately ruptures. It is for this reason that surgical treatment is justifiable, even if the patient is asymptomatic.

LIGAMENTOUS INJURIES

As discussed in the section on anatomy, a large number of functionally important ligaments support and maintain intercarpal relationships. It has recently been recognized that these ligaments are frequently injured. Unfortunately, these injuries are usually recognized late, either because the patient has assumed that the injury was minor or because the primary contact in the health care system makes the same assumption. Many of these patients have been told not to be concerned about the injury as it is only a sprained wrist. This information is partly correct in that the wrist is sprained; the problem lies in the fact that most of these so-called sprains are significant ligamentous tears that lead to long-term problems and complications. Consideration of each subtype of ligamentous injury and the various classifications available is beyond the scope of this chapter, but a discussion of the most common injuries is certainly warranted.

It is first necessary to understand that there are degrees of injuries that can occur. At one end of the spectrum is the 'sprained wrist', which follows a benign course, and at the other end is a carpal dislocation or a dislocation of the lunate. Obviously the spectrum of treatment will be as broad as the range of injuries. The major ligamentous injuries are obviously dislocations of the wrist or of specific carpal bones (usually lunate, with all other dislocations being extremely rare).

These injuries are usually readily diagnosable as the patient presents with a swollen, often deformed, and very painful wrist associated with concomitant radiographic abnormalities. These injuries are devastating and are best treated with open reduction of the carpus, temporary pinning of the reduced carpus, repair of any ligaments that are reparable, and immobilization for 6 to 8 weeks. The aims of treatment are not to restore a normal wrist, but to attain a reasonable range of motion (60–80 per cent of normal) and

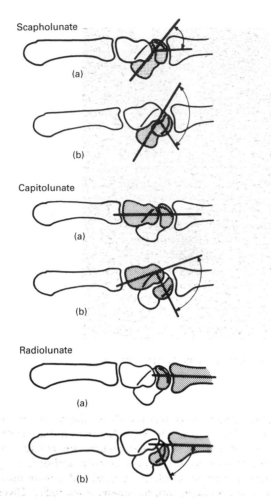

Scapholunate

(a)

(b)

Capitolunate

(a)

(b)

Radiolunate

(a)

(b)

Fig. 9 A diagrammatic representation of various carpal angles as measured on the lateral radiographs (reproduced from Green DP. *Operative Hand Surgery*, New York: Churchill Livingstone, 1988: 891, with permission).

perhaps 75 per cent of normal grip strength (Fig. 9). Attainment of normal carpal relationships on radiographs is very important. Other methods of treatment, such as closed reduction and casting, are outmoded because of the difficulties of maintaining normal carpal relationships. This inability to maintain critical relationships alters wrist kinematics irrevocably and leads to pain, stiffness, weakness, and in some instances degenerative arthritis, as has been discussed previously. This phenomenon is known as carpal collapse and is easily diagnosed radiographically by those familiar with the normal carpal relationships and the measurement of specific carpal angles (e.g. the scapholunate angle, radiolunate, and capitolunate angles). Ultimately, the aim of the treatment of all ligamentous injuries about the wrist is the avoidance of significant carpal collapse.

The most common and more subtle ligamentous injury occurring in the wrist is scapholunate dissociation. This injury involves the ligaments of the scapholunate complex. Disruption of these ligaments allows the scaphoid to dissociate from the lunate. This means then that, in its most extreme form, the scaphoid rotates into an abnormal position of palmar fixation. These patients present with findings similar to those of a scaphoid fracture. Radiographs show no fractures, but radiographic abnormalities are often present. The fact that the scaphoid rotates into palmar flexion results in an abnormal gap between the scaphoid

and the lunate on an anteroposterior plain radiograph of the wrist. This gap should be greater than 2 mm to be considered abnormal. In addition, the scaphoid appears to be foreshortened. On the lateral view of the wrist, specific angle changes occur which can be measured. (For further discussion, the reader is referred to more detailed works.[2,4]) These patients have an unstable scaphoid statically, as clearly the scaphoid is unstable even at rest with these abnormalities appearing on plain radiographs. In many instances, no radiographic abnormalities are identified on static films even though significant ligamentous injury has occurred.

The diagnosis of these injuries is often difficult even in experienced hands. The history is usually quite straightforward, with a fall on the outstretched hand being the mechanism of injury. Cineradiographs, an abnormal bone scan, and wrist arthroscopy may all be required to obtain an exact diagnosis and evaluation of scaphoid stability in these patients. Many of them have dynamic scaphoid instability, i.e. at rest the scaphoid maintains its normal relationships to the rest of the carpus, but when the wrist is loaded the attenuated ligaments do not maintain their position and the scaphoid becomes unstable. Arthroscopy shows obviously attenuated and disrupted ligaments, or disruption of one set of the three sets of ligaments that stabilize and support the scaphoid.

Treatment of scaphoid instability depends upon many factors, including the age and activity level of the patients and whether the diagnosis is made early or late. Early diagnosis is generally defined as within 3 weeks of the original injury. The degree of symptomatology and the wishes of the patient are also critical. The available treatments are varied and controversial and just one approach is presented here. The best treatment for an athlete with an acute scapholunate dissociation is to reduce the scaphoid to its normal position and K-wire it to the lunate and capitate temporarily. The reduction can be performed either closed or open. In addition, the patient is immobilized in a thumb spica case for 6 to 8 weeks. This allows time for some ligamentous healing to occur in the anatomical position. The patient is then generally mobilized and followed carefully with radiographs for signs of carpal collapse. Thumb spica immobilization with careful radiographic follow-up seems to be adequate treatment for those patients who may have an acute ligament injury but whose radiographs are normal. If radiographic abnormalities develop at any time in the early stages more aggressive treatment can be applied.

There are significant differences in the individual presenting late with an unstable scaphoid. Ligamentous healing cannot occur, even if the scaphoid is reduced. Ligament reconstructions have been attempted in a number of centres, but without much success.

The first step in treating a patient with a late-presenting scapholunate dissociation is to have a detailed discussion with him or her about the magnitude of the symptoms. Treatment of the unrecognized unstable scaphoid involves significant surgical procedures, and many patients will want to leave well alone as they find their symptoms to be more of a nuisance than a major disability. However, if the patient's wrist is significantly disabled and he or she wishes something to be done, the next step should be a wrist arthroscopy (Fig. 10). This will allow diagnosis of the scapholunate ligament disruption to be confirmed, and will also allow the radioscaphoid joint to be assessed for degenerative change. Patients with scapholunate dissociation are at risk of the same long-term complications as those patients with a non-union of the scaphoid, i.e. the specific form of degenerative arthritis known as scapholunate advanced collapse. The presence of degenerative change at the radioscaphoid joint will significantly alter the surgical management of these patients.

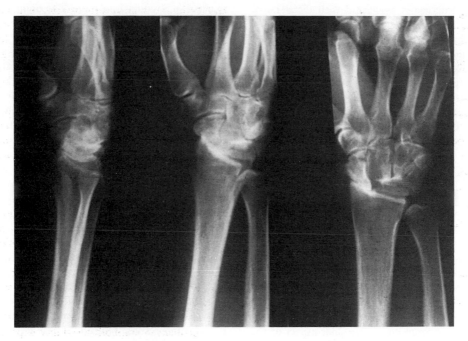

Fig. 10 A typical scapholunate advanced collapse (note advanced degenerative change at radioscaphoid and capitolunate joints).

Given that no degenerative change is present and that the patient wishes some treatment to be applied, two choices are available.

The first, and certainly the most frequent, choice for most hand surgeons in North America is to attempt to stabilize the scaphoid in a good position by a limited carpal arthrodesis. Specifically, the most common procedure is an arthrodesis between the scaphoid, trapezium, and trapezoid, which is known as a triscaphe fusion or an STT arthodesis.[12] This is a significant surgical procedure producing reduction in wrist motion of approximately 35 per cent with a complication rate in published series approaching 52 per cent, including delayed union, non-union, infection, and neurological and vascular injury. At 10 year follow-up acceptable results were attained in approximately 80 per cent of cases.[13,14] It should be noted that these patients will develop excessive motion through other carpal joints and therefore are at high risk for the development of degenerative change elsewhere in the wrist following this procedure.

The other choice is an attempt to stabilize the scaphoid with a flap of the dorsal wrist capsule.[15] This is a much smaller operation in the sense that the reduction in wrist motion is far less, the time of immobilization is far shorter, and there are fewer complications. The problem with this particular procedure is that there are no large series available indicating how effective it is and what the results of long-term follow-up of the patients will be. Therefore patients must be told that the results of this procedure are basically unpredictable but that a certain number of them will benefit from the procedure. They should also be apprised of the fact that the downside risks of this operation are small, with a low complication rate, little restriction in wrist motion, and no risk of degenerative change elsewhere in the carpus as far as we are aware. For these reasons capsuldesis is the preferred first line of treatment in the very young patient or the very young athlete with scaphoid instability. Should this procedure fail, limited carpal arthrodesis can be considered as a second choice.

Other ligamentous injuries about the wrist are much less common than those of the scapholunate complex. In addition, the majority of other ligamentous injuries are not known to have major late sequelae such as degenerative arthritis. They tend to produce symptoms which are generally of a more minor nature than those associated with scapholunate instability. Given all this, the treatment of acutely injured ligaments can be short periods of immobilization for symptomatic relief. In the patient presenting with late ligamentous injuries, such as those with lunotriquetral instability, treatment can be predicted on the basis of the patient's symptoms. Specific instability can always be eliminated by some form of limited carpal arthrodesis, but with varying degrees of effectiveness in terms of symptomatic relief. For this reason these treatments are only embarked upon after careful discussion with the patient regarding the percentage of successful results, complications, and the amount of dysfunction produced by the surgical procedure. These discussions are as important as, or more important than, the specific procedure carried out.

In summary, ligamentous injuries to the wrist represent a wide spectrum of injuries, many of which have been recognized with increasing frequency because of better awareness and improved diagnostic techniques such as wrist arthroscopy. Early recognition may lead to better treatment and the avoidance of long-term complications such as carpal collapse, degenerative arthritis, and the need for extensive surgical reconstruction.

THE TRIANGULAR FIBROCARTILAGE COMPLEX

The triangular fibrocartilage complex is another commonly injured structure in the wrist area. It is probably most frequently injured in association with fractures of the distal end of the radius. This injury may be represented by avulsion fracture of the ulnar styloid, as seen on plain radiographs in distal radius fractures. It can also be injured in isolation.

When injury to this structure is associated with distal radial fractures, it is obviously overshadowed by other symptoms and physical findings. In such cases adequate treatment of the fracture is all that is necessary, and these patients will rarely present with a fracture that has done well and a symptomatic tear of the triangular fibrocartilage complex. This does happen, but it is very uncommon. Patients presenting with an acute injury to the triangular fibrocartilage complex will have dorsal ulnar wrist pain and localized tenderness, and occasionally localized swelling. They may have symptoms of catching or locking with pronation and supination or with other wrist motions. Radiographs are normal, and it is often difficult to distinguish these patients from patients who have lunotriquetral tears and distal radio-ulnar joint pain.

One approach is to treat these individuals symptomatically whereupon most will improve with time. Those that have persistent and aggravating symptoms can be considered as candidates for wrist arthroscopy, which is becoming the gold standard diagnostic tool for pathology in this region. During arthroscopy, the location and nature of the tear are evaluated. Classification systems for tears of the triangular fibrocartilage complex are available, but a significant number of patients will obtain symptomatic improvement from arthroscopic debridement of the tear. Other injuries, specifically avulsions from either the ulnar aspect or the radial aspect of the triangular fibrocartilage complex, will require reattachment to the radius or the ulna using either open or arthroscopic procedure for symptomatic relief to be obtained.

DISTAL RADIO-ULNAR JOINT

The distal radio-ulnar joint is also frequently injured, but the injuries are generally associated with more significant injuries such as distal radial fractures and tears of the triangular fibrocartilage complex. Isolated injuries are less common and frequently unrecognized. The patient will present acutely with pain and swelling, but generally little, if any, associated deformity. Initial physical examination will reveal tenderness well localized at the distal radio-ulnar joint. The pain is often exaggerated with full pronation and supination.

In the most dramatic cases, a frank dislocation of the joint occurs, and on a true lateral radiograph of the radius, the ulna is identified most frequently dislocated dorsally and much less frequently dislocated volarly (Fig. 11). Acute injuries are easily treated with cast immobilization. Dorsally dislocated or unstable ulnas are treated with an above-elbow cast in full supination for 3 weeks, followed by 3 weeks of a below-elbow cast. Placing the arm in full supination reduces the dorsal dislocation and allows healing

to occur in a reduced position. The opposite is true of the volarly unstable distal ulna where pronation is the position of stability for immobilization.

Most frequently, however, the patient will present late with vague symptoms of wrist pain with activity. Again, the late presentation is a result of the patient's lack of appreciation of the original injury or of failure to recognize this injury by the primary health care contact. Generally the only abnormal physical finding is well localized tenderness to the distal radio-ulnar joint. Some patients will have an obvious prominence of the distal ulna. In others, the distal ulna can be excessively mobile in a palmar dorsal plane compared with the opposite wrist. Radiographic examination is generally unhelpful. A very small and highly selected number of patients may have a coronal CT scan performed to demonstrate that the affected wrist is asymmetrical through the distal radio-ulnar joint compared with the unaffected wrist. These patients are generally best treated non-surgically with physiotherapy and taping to try and aid stability and reduce their symptoms during activity.

A large number of surgical procedures which attempt to stabilize and reduce the distal radio-ulnar joint have been described. In general, these procedures do not achieve consistently reproducible results. Therefore these procedures are reserved as a final effort to try and stabilize the distal radio-ulnar joint in those patients with extremely disabling symptoms.

FURTHER READING

Green DP, ed. *Operative Hand Surgery*. 2nd edn. New York: Churchill Livingstone, 1988.

Rockwood CA Jr, Green DP, eds. *Fractures in Adults*. Philadelphia: JB Lippincott.

REFERENCES

1. Linscheid RL, Dobyns JH. Athletic injuries of the wrist. *Clinical Orthopaedics and Related Research* 1985; **198**: 141–51.
2. Lichtman DM, ed. *The wrist and its disorders*. Philadelphia: WB Saunders, 1988.
3. Mayfield JK, Wrist ligamentous anatomy and pathogenesis of carpal instability. *Orthopedic Clinics of North America* 1984; **15**: 209–16.
4. Talesnik J, ed. *The wrist*. New York: Churchill Livingstone, 1985.
5. Palmer AK, Werner FW. The triangular fibrocartilage complex of the wrist—anatomy and function. *Journal of Hand Surgery* 1981; **6**: 153–62.
6. Cooney WP, Dobyns JH, Linscheid RL. Complications of Colles' fractures. *Journal of Bone and Joint Surgery* 1980; **62A**: 613–19.
7. Knirk JL, Jupiter JB. Intra-articular fractures of the distal radius in young adults. *Journal of Bone and Joint Surgery* 1986; **68A**: 657–9.
8. Axelrod TS, McMurtry RY. Open reduction and internal fixation of comminuted, intra-articular fractures of the distal radius. *Journal of Hand Surgery* 1990; **15A**: 1–11.
9. Cooney WP, Dobyns JH, Linscheid RL. Fractures of the scaphoid: a rational approach to management. *Clinical Orthopaedics and Related Research* 1980; **149**: 90–7.
10. Herbert TJ. Management of the fractured scaphoid using a new bone screw. *Orthopaedic Transactions* 1982; **6**: 464–5.
11. Watson HK, Rogers WD. Non-union of the hook of the hamate: an argument for bone grafting the non-union. *Journal of Hand Surgery* 1989; **14A**: 486–90.
12. Watson HK, Hempton RF. Limited wrist arthrodesis I. The triscaphoid joint. *Journal of Hand Surgery* 1980; **5**: 320–7.
13. Kleinman WB. Long term study of chronic scapholunate instability treated by scapho-trapezio-trapezoid arthrodesis. *Journal of Hand Surgery* 1989; **14A**: 425–8.
14. Kleinman WB, Carroll C IV. Scapho-trapezio-trapezoid arthrodesis for

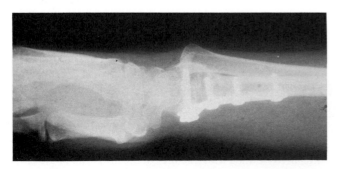

Fig. 11 A persistent dislocation of the distal radio-ulnar joint following open reduction internal fixation of a distal radius fracture (note the dorsal location of the distal ulna on the lateral view).

treatment of chronic, static and dynamic scapho-lunate instability: a 10-year perspective on pitfalls and complications. *Journal of Hand Surgery* 1990; **15A**: 408–14.

15. Green DP. Carpal dislocations and instabilities. In: Green DP, ed. *Operative hand surgery*. 2nd edn. New York: Churchill Livingstone, 1988: 875–938.

16. Bowers WH. The distal radioulnar joint. In: Green DP, ed. *Operative hand surgery*. 2nd edn. New York: Churchill Livingstone, 1988: 939–89.

Index

Since the subject of this book is sports medicine, index entries beginning 'sport' have been kept to a minimum, and readers are advised to seek more specific references.

Page numbers in **bold** refer to principal discussions in the text. Page numbers in *italics* refer to pages on which tables are to be found. In disease-related entries the use of 'vs'. refers to differential diagnosis. This index is in a letter-by-letter alphabetical order.